AF829309

INFORMATION SERVICES DIVISION

Clifford Whitworth Building
Salford M5 4WT

For renewals please go to
www.isd.salford.ac.uk/library
or telephone 0161 295 2444

THIS ITEM MUST BE RENEWED OR RETURNED TO THE LIBRARY ON OR BEFORE THE LAST DATE STAMPED BELOW OR **A FINE WILL BE PAYABLE.**

It is a condition of loan that borrowers may be charged for items lost or damaged. If this is reserved you will be allowed a shorter period; please return on demand.

09988246

HANDBOOK OF STATISTICS
VOLUME 25

Handbook of Statistics

VOLUME 25

General Editor

C.R. Rao

ELSEVIER
NORTH-HOLLAND

AMSTERDAM • BOSTON • HEIDELBERG • LONDON • NEW YORK • OXFORD
PARIS • SAN DIEGO • SAN FRANCISCO • SINGAPORE • SYDNEY • TOKYO

Bayesian Thinking: Modeling and Computation

Edited by

D.K. Dey
Department of Statistics
University of Connecticut
Storrs, CT, USA

C.R. Rao
Center for Multivariate Analysis
Department of Statistics, The Pennsylvania State University
University Park, PA, USA

ELSEVIER
NORTH-HOLLAND

2005

AMSTERDAM • BOSTON • HEIDELBERG • LONDON • NEW YORK • OXFORD
PARIS • SAN DIEGO • SAN FRANCISCO • SINGAPORE • SYDNEY • TOKYO

ELSEVIER B.V.
Sara Burgerhartstraat 25
P.O. Box 211, 1000 AE Amsterdam
The Netherlands

ELSEVIER Inc.
525 B Street, Suite 1900
San Diego, CA 92101-4495
USA

ELSEVIER Ltd
The Boulevard, Langford Lane
Kidlington, Oxford OX5 1GB
UK

ELSEVIER Ltd
84 Theobalds Road
London WC1X 8RR
UK

© 2005 Elsevier B.V. All rights reserved

This work is protected under copyright by Elsevier B.V., and the following terms and conditions apply to its use:

Photocopying
Single photocopies of single chapters may be made for personal use as allowed by national copyright laws. Permission of the Publisher and payment of a fee is required for all other photocopying, including multiple or systematic copying, copying for advertising or promotional purposes, resale, and all forms of document delivery. Special rates are available for educational institutions that wish to make photocopies for non-profit educational classroom use.

Permissions may be sought directly from Elsevier's Rights Department in Oxford, UK: phone (+44) 1865 843830, fax (+44) 1865 853333, e-mail: permissions@elsevier.com. Requests may also be completed on-line via the Elsevier homepage (http://www.elsevier.com/locate/permissions).

In the USA, users may clear permissions and make payments through the Copyright Clearance Center, Inc., 222 Rosewood Drive, Danvers, MA 01923, USA; phone: (+1) (978) 7508400, fax: (+1) (978) 7504744, and in the UK through the Copyright Licensing Agency Rapid Clearance Service (CLARCS), 90 Tottenham Court Road, London W1P 0LP, UK; phone: (+44) 20 7631 5555; fax: (+44) 20 7631 5500. Other countries may have a local reprographic rights agency for payments.

Derivative Works
Tables of contents may be reproduced for internal circulation, but permission of the Publisher is required for external resale or distribution of such material. Permission of the Publisher is required for all other derivative works, including compilations and translations.

Electronic Storage or Usage
Permission of the Publisher is required to store or use electronically any material contained in this work, including any chapter or part of a chapter.

Except as outlined above, no part of this work may be reproduced, stored in a retrieval system or transmitted in any form or by any means, electronic, mechanical, photocopying, recording or otherwise, without prior written permission of the Publisher.
Address permissions requests to: Elsevier's Rights Department, at the fax and e-mail addresses noted above.

Notice
No responsibility is assumed by the Publisher for any injury and/or damage to persons or property as a matter of products liability, negligence or otherwise, or from any use or operation of any methods, products, instructions or ideas contained in the material herein. Because of rapid advances in the medical sciences, in particular, independent verification of diagnoses and drug dosages should be made.

First edition 2005

Library of Congress Cataloging in Publication Data
A catalog record is available from the Library of Congress.

British Library Cataloguing in Publication Data
A catalogue record is available from the British Library.

ISD
SITE CW
CLASS No. 005.74
SUFFIX DAT

ISBN: 0-444-51539-9
ISSN: 0169-7161 (Series)

⊗ The paper used in this publication meets the requirements of ANSI/NISO Z39.48-1992 (Permanence of Paper).
Printed in The Netherlands.

Preface

Fisher and Mahalanobis described Statistics as the key technology of the twentieth century. Since then Statistics has evolved into a field that has many applications in all sciences and areas of technology, as well as in most areas of decision making such as in health care, business, federal statistics and legal proceedings. Applications in statistics such as inference for Causal effects, inferences about the spatio- temporal processes, analysis of categorical and survival data sets and countless other functions play an essential role in the present day world. In the last two to three decades, Bayesian Statistics has emerged as one of the leading paradigms in which all of this can be done in a unified fashion. There has been tremendous development in Bayesian theory, methodology, computation and applications in the past several years.

Bayesian statistics provides a rational theory of personal beliefs compounded with real world data in the context of uncertainty. The central aim of characterizing how an individual should make inferences or act in order to avoid certain kinds of undesirable behavioral inconsistencies and consequent are all successfully accomplished through this process. The primary theory of Bayesian statistics states that utility maximization should be the basis of rational decision-making in conjunction with the Bayes' theorem, which acts as the key to the basis in which the beliefs should fit together with changing evidence scenario. Undoubtedly, it is a major area of statistical endeavor, which has hugely increased its profile, both in context of theories and applications.

The appreciation of the potential for Bayesian methods is growing fast both inside and outside the statistics community. The first encounter with Bayesian ideas by many people simply entails the discovery that a particular Bayesian, method is superior to classical statistical methods on a particular problem or question. Nothing succeeds like success, and this observed superiority often leads to a further pursuit of Bayesian analysis. For scientists with little or no formal statistical background, Bayesian methods are being discovered as the only viable method for approaching their problems. For many of them, statistics has become synonymous with Bayesian statistics.

The Bayesian method as many might think is not new, but rather a method that is older than many of the commonly, known and well formulated statistical techniques. The basis for Bayesian statistics was laid down in a revolutionary paper written by Rev Thomas Bayes, which appeared in print in 1763 but was not acknowledged for its significance. A major resurgence of the method took place in the context of discovery of paradoxes and logical problems in classical statistics. The work done by a number of authors such as Ramsey, DeFinetti, Good, Savage, Jeffreys and Lindley provided a more thorough and philosophical basis for acting under uncertainty. In the developments that went by, the subject took a variety of turns. On the foundational front, the concept of rationality was explored in the context of representing beliefs or choosing actions where uncertainty creeps in. It was noted that the criterion of maximizing expected utility is the only decision criterion that is compatible with the axiom system. The statistical inference problems are simply the particular cases, which can be visualized in general decision theoretic framework. These developments led to a number of other important progresses on Bayesian front. To name a few, it is important to mention the Bayesian robustness criterion, empirical and hierarchical Bayesian analysis and reference analysis etc. that all deepen the roots of Bayesian thoughts. The subject came to be the forefront of practical statistics with the advent of high-speed computers and sophisticated computational techniques especially in the form of Markov chain Monte Carlo methods. Because of that it is evident that a large body of literature in the form of books, research papers, conference proceedings are developed during the last fifteen years. This is the reason we felt that it is indeed the right time to develop a volume in the Handbook of Statistics series to highlight recent thoughts on theory, methodology and related computation on Bayesian analysis. With this specific purpose in mind we invited leading experts on Bayesian methodology to contribute for this volume. This in our opinion has resulted in a volume with a nice mix of articles on theory, methodology, application and computational methods on current trend in Bayesian statistics. For the convenience of readers, we have divided this volume into 10 distinct groups: Foundation of Bayesian statistics including model determination, Nonparametric Bayesian methods, Bayesian computation, Spatio-temporal models, Bayesian robustness and sensitivity analysis, Bioinformatics and Biostatistics, Categorical data analysis, Survival analysis and software reliability, Small area estimation and Teaching Bayesian thought. All chapters in each group are written by leading experts in their own field.

We hope that this broad coverage of the area of Bayesian Thinking will only provide the readers with a general overview of the area, but also describe to them what the current state is in each of the topics listed above.

We express our sincere thanks to all the authors for their fine contributions, and for helping us in bringing out this volume in a timely manner. Our special thanks go to Ms. Edith Bomers and Ms. Andy Deelen of Elsevier, Amsterdam, for taking a keen interest in this project, and also for helping us with the final production of this volume.

Dipak K. Dey
C.R. Rao

Table of contents

Ch. 30. Bayesian Methods and Simulation-Based Computation for Contingency Tables 869

James H. Albert

Ch. 31. Multiple Events Time Data: A Bayesian Recourse 891

Debajyoti Sinha and Sujit K. Ghosh

Ch. 32. Bayesian Survival Analysis for Discrete Data with Left-Truncation and Interval Censoring 907

Chong Z. He and Dongchu Sun

Contributors

Albert, James H., *Department of Mathematics and Statistics, Bowling Green State University, Bowling Green, OH 43403*; *e-mail: albert@bgnet.bgsu.edu* (Ch. 30).

Arellano-Valle, Reinaldo B., *Departamento de Estatística, Facultad de Matemáticas, Pontificia Universidad Católica de Chile, Chile*; *e-mail: reivalle@mat.puc.cl* (Ch. 22).

Baladandayuthapani, Veerabhadran, *Department of Statistics, Texas A&M University, College Station, TX 77843*; *e-mail: veera@stat.tamu.edu* (Ch. 24).

Bernardo, José M., *Departamento de Estadística e I.O., Universitat de València, Spain*; *e-mail: jose.m.bernardo@uv.es* (Ch. 2).

Bolfarine, Heleno, *Departmento de Estatistica, IME, Universidad de Sao Paulo, Brasil*; *e-mail: hbolfar@ime.usp.br* (Ch. 22).

Branco, M.D., *University of São Paulo, Brazil*; *e-mail: mbranco@ime.usp.br* (Ch. 23).

Brandwein, Ann Cohen, *Baruch College, The City University of New York*; *e-mail: ann_brandwein@baruch.cuny.edu* (Ch. 8).

Branscum, Adam J., *Department of Statistics, University of California, Davis, CA 95616*; *e-mail: ajbranscum@ucdavis.edu* (Ch. 9).

Carter, Chris, *CSIRO, Australia*; *e-mail: Chris.Carter@csiro.au* (Ch. 18).

Chen, Ming-Hui, *Department of Statistics, University of Connecticut, Storrs, CT 06269-4120*; *e-mail: mhchen@stat.uconn.edu* (Ch. 15).

Chib, Siddhartha, *John M. Olin School of Business, Washington University in St. Louis, St. Louis, MO 63130*; *e-mail: chib@wustl.edu* (Ch. 29).

Choudhuri, Nidhan, *Department of Statistics, Case Western Reserve University*; *e-mail: nidhan@nidhan.cwru.edu* (Ch. 13).

Cripps, Edward, *Department of Statistics, University of New South Wales, Sydney, NSW 2052, Australia*; *e-mail: ecripps@unsw.edu.au* (Ch. 18).

Damien, Paul, *McCombs School of Business, University of Texas at Austin, Austin, TX 78730*; *e-mail: paul.damien@mccombs.utexas.edu* (Ch. 10).

Datta, Gauri Sankar, *University of Georgia, Athens, GA*; *e-mail: gauri@stat.uga.edu* (Ch. 3).

Dunson, David B., *Biostatistics Branch, MD A3-03, National Institute of Environmental Health Sciences, Research Triangle Park, NC 287709*; *e-mail: dunson1@niehs.nih.gov* (Ch. 25).

Ferreira, Marco A.R., *Instituto de Matemática, Universidade Federal do Rio de Janeiro, Brazil*; *e-mail: marco@im.ufrj.br* (Ch. 19).

Gamerman, Dani, *Instituto de Matemática, Universidade Federal do Rio de Janeiro, Brazil*; *e-mail: dani@im.ufrj.br* (Ch. 19).

Ghosal, Subhashis, *Department of Statistics, North Carolina State University, NC 27695*; *e-mail: sghosal@stat.ncsu.edu* (Ch. 13).

Ghosh, Jayanta, *Indian Statistical Institute, 203 B.T. Road, Kolkata 700 108, India*; *e-mail: jayanta@isical.ac.in and Department of Statistics, Purdue University, West Lafayette, IN 47907*; *e-mail: ghosh@stat.purdue.edu* (Ch. 5).

Ghosh, Malay, *Department of Statistics, University of Florida, Gainesville, FL 32611*; *e-mail: ghoshm@stat.ufl.edu* (Ch. 27).

Ghosh, Sujit K., *Department of Statistics, North Carolina State University*; *e-mail: sghosh@stat.ncsu.edu* (Ch. 31).

Gustafson, Paul, *Department of Statistics, University of British Columbia, Vancouver, BC, Canada, V6T 1Z2*; *e-mail: gustaf@stat.ubc.ca* (Ch. 26).

Hanson, Timothy E., *Department of Mathematics and Statistics, University of New Mexico, Albuquerque, NM 87131*; *e-mail: hanson@math.unm.edu* (Ch. 9).

He, Chong Z., *Department of Statistics, University of Missouri-Columbia, Columbia, MO 65210*; *e-mail: hezh@missouri.edu* (Ch. 32).

Hossain, Shahadut, *Department of Statistics, University of British Columbia, Vancouver, BC, Canada, V6T 1Z2*; *e-mail: shahadut@stat.ubc.ca* (Ch. 26).

Iglesias, P., *Pontificia Universidad Católica de Chile, Chile*; *e-mail: pliz@mat.pic.cl* (Ch. 23).

Johnson, Timothy D., *University of Michigan, School of Public Health*; *e-mail: tdjtdj@umich.edu* (Ch. 28).

Johnson, Valen E., *Institute of Statistics and Decision Sciences, Duke University, Durham, NC 27708-0254*; *e-mail: valen@stat.duke.edu* (Ch. 28).

Johnson, Wesley O., *Department of Statistics, University of California-Irvine, Irvine, CA 92697*; *e-mail: wjohnson@uci.edu* (Ch. 9).

Kohn, Robert, *University of New South Wales, Sydney, NSW 2052, Australia*; *e-mail: r.kohn@unsw.edu.au* (Ch. 18).

Kuo, Lynn, *Department of Statistics, University of Connecticut, Storrs, CT 06269-4120*; *e-mail: lynn@stat.uconn.edu* (Ch. 33).

Liseo, Brunero, *Dip. studi geoeconomici, liguistici, statistici e storici per l'analisi regionale, Università di Roma "La Sapienza", I-00161 Roma, Italia*; *e-mail: brunero.liseo@uniroma1.it* (Ch. 7).

Lopes, Hedibert F., *Graduate School of Business, University of Chicago*; *e-mail: hlopes@gsb.uchicago.edu* (Ch. 19).

Maiti, Tapabrata, *Department of Statistics, Iowa State University, Ames, IA*; *e-mail: taps@iastate.edu* (Ch. 34).

Mallick, Bani, *Department of Statistics, Texas A&M University, College Station, TX 77843*; *e-mail: bmallick@stat.tamu.edu* (Ch. 24).

Marin, Jean-Michel, *Universite Paris Dauphine, France*; *e-mail: marin@ceremade.dauphine.fr* (Ch. 16).

Martín, Jacinto, *Department of Mathematics, U. Extremadura, Spain*; *e-mail: jrmartin@unex.es* (Ch. 21).

McCandless, Lawrence, *Department of Statistics, University of British Columbia, Vancouver, BC, Canada, V6T 1Z2*; *e-mail: lawrence@stat.ubc.ca* (Ch. 26).

Mengersen, Kerrie, *University of Newcastle*; *e-mail: k.mengersen@qut.edu.au* (Ch. 16).

Migon, Helio S., *Instituto de Matemática, Universidade Federal do Rio de Janeiro, Brazil*; *e-mail: migon@im.ufrj.br* (Ch. 19).

Mira, Antonietta, *Department of Economics, University of Insubria, Via Ravasi 2, 21100 Varese, Italy*; *e-mail: antonietta.mira@uninsubria.it* (Ch. 14).

Mukherjee, Bhramar, *Department of Statistics, University of Florida, Gainesville, FL 32611*; *e-mail: mukherjee@stat.ufl.edu* (Ch. 27).

Müller, Peter, *Department of Biostatistics, The University of Texas, M.D. Anderson Cancer Center, Houston, TX*; *e-mail: pm@stat.duke.edu* (Ch. 17).

Pericchi, Luis Raúl, *School of Natural Sciences, University of Puerto Rico, Puerto Rico*; *e-mail: pericchi@goliath.cnnet.clu.edu* (Ch. 4).

Purkayastha, Sumitra, *Theoretical Statistics and Mathematics Unit, Indian Statistical Institute, Kolkata 700 108, India*; *e-mail: sumitra@isical.ac.in* (Ch. 5).

Ray, Shubhankar, *Department of Statistics, Texas A&M University, College Station, TX 77843*; *e-mail: sray@stat.tamu.edu* (Ch. 24).

Ríos Insua, David, *Decision Engineering Lab, U. Rey Juan Carlos, Spain*; *e-mail: david.rios@urjc.es* (Ch. 21).

Robert, Christian P., *Universite Paris Dauphine, France*; *e-mail: xian@ceremade.dauphine.fr* (Ch. 16).

Roy, Anindya, *Department of Mathematics and Statistics, University of Maryland, MD 21250*; *e-mail: anindya@math.umbc.edu* (Ch. 13).

Rubin, Donald B., *Department of Statistics, Harvard University, Cambridge, MA 02138*; *e-mail: rubin@stat.harvard.edu* (Ch. 1).

Ruggeri, Fabrizio, *CNR-IMATI, Milano, Italy*; *e-mail: fabrizio@im.imati.cnr.it* (Chs. 11, 21).

Samanta, Tapas, *Applied Statistics Unit, Indian Statistical Institute, Kolkata 700 108, India*; *e-mail: tapas@isical.ac.in* (Ch. 5).

Sinha, Debajyoti, *Department of Biostatistics, Bioinformatics & Epidemiology, MUSC*; *e-mail: sinhad@musc.edu* (Ch. 31).

Sinha, Samiran, *Department of Statistics, Texas A&M University, College Station, TX*; *e-mail: ssinha@stat.tamu.edu* (Ch. 27).

Sinharay, Sandip, *MS 12-T, Educational Testing Service, Rosedale Road, Princeton, NJ 08541*; *e-mail: ssinharay@ets.org* (Ch. 6).

Stangl, Dalene K., *Institute of Statistics and Decision Sciences, Duke University*; *e-mail: dalene@stat.duke.edu* (Ch. 35).

Stern, Hal S., *Department of Statistics, University of California, Irvine*; *e-mail: sternh@uci.edu* (Ch. 6).

Strawderman, William E., *Department of Statistics, Rutgers University, New Brunswick, NJ 08903*; *e-mail: straw@stat.rutgers.edu* (Ch. 8).

Sun, Dongchu, *Department of Statistics, University of Missouri-Columbia, Columbia, MO 65210*; *e-mail: sund@missouri.edu* (Ch. 32).

Sweeting, Trevor J., *University College London*; *e-mail: trevor@stats.ucl.ac.uk* (Ch. 3).

Vidakovic, Brani, *Department of Industrial and Systems Engineering, Georgia Institute of Technology*; *e-mail: brani@isye.gatech.edu* (Ch. 11).
Vidal, I., *Universidad de Talca, Chile*; *e-mail: ividal@utalca.cl* (Ch. 23).
Walker, Stephen, *Institute of Mathematics, Statistics and Actuarial Science, University of Kent, Canterbury, CT2 7NZ, UK*; *e-mail: S.G.Walker@kent.ac.uk* (Ch. 12).
Waller, Lance A., *Department of Biostatistics, Rollins School of Public Health, Emory University, Atlanta, GA 30322*; *e-mail: lwaller@sph.emory.edu* (Ch. 20).

Handbook of Statistics, Vol. 25
ISSN: 0169-7161
© 2005 Elsevier B.V. All rights reserved.
DOI 10.1016/S0169-7161(05)25001-0

1

Bayesian Inference for Causal Effects

Donald B. Rubin

Abstract
A central problem in statistics is how to draw inferences about the causal effects of treatments (i.e., interventions) from randomized and nonrandomized data. For example, does the new job-training program really improve the quality of jobs for those trained, or does exposure to that chemical in drinking water increase cancer rates? This presentation provides a brief overview of the Bayesian approach to the estimation of such causal effects based on the concept of potential outcomes.

1. Causal inference primitives

Although this chapter concerns Bayesian inference for causal effects, the basic conceptual framework is the same as that for frequentist inference. Therefore, we begin with the description of that framework. This framework with the associated inferential approaches, randomization-based frequentist or Bayesian, and its application to both randomized experiments and observational studies, is now commonly referred to as "Rubin's Causal Model" (RCM, Holland, 1986). Other approaches to Bayesian causal inference, such as graphical ones (e.g., Pearl, 2000), I find conceptually less satisfying, as discussed, for instance, in Rubin (2004b). The presentation here is essentially a simplified and refined version of the perspective presented in Rubin (1978).

1.1. Units, treatments, potential outcomes

For causal inference, there are several primitives – concepts that are basic and on which we must build. A "unit" is a physical object, e.g., a person, at a particular point in time. A "treatment" is an action that can be applied or withheld from that unit. We focus on the case of two treatments, although the extension to more than two treatments is simple in principle although not necessarily so with real data.

Associated with each unit are two "potential outcomes": the value of an outcome variable Y at a point in time when the active treatment is applied and the value of that outcome variable at the same point in time when the active treatment is withheld. The objective is to learn about the causal effect of the application of the active treatment relative to the control (active treatment withheld) on Y.

For example, the unit could be "you now" with your headache, the active treatment could be taking aspirin for your headache, and the control could be not taking aspirin. The outcome Y could be the intensity of your headache pain in two hours, with the potential outcomes being the headache intensity if you take aspirin and if you do not take aspirin.

Notationally, let W indicate which treatment the unit, you, received: $W = 1$ the active treatment, $W = 0$ the control treatment. Also let $Y(1)$ be the value of the potential outcome if the unit received the active version, and $Y(0)$ the value if the unit received the control version. The causal effect of the active treatment relative to its control version is the comparison of $Y(1)$ and $Y(0)$ – typically the difference, $Y(1) - Y(0)$, or perhaps the difference in logs, $\log[Y(1)] - \log[Y(0)]$, or some other comparison, possibly the ratio.

We can observe only one or the other of $Y(1)$ and $Y(0)$ as indicated by W. The key problem for causal inference is that, for any individual unit, we observe the value of the potential outcome under only one of the possible treatments, namely the treatment actually assigned, and the potential outcome under the other treatment is missing. Thus, inference for causal effects is a missing-data problem – the "other" value is missing.

How do we learn about causal effects? The answer is replication, more units. The way we personally learn from our own experience is replication involving the same physical object (ourselves) with more units in time. That is, if I want to learn about the effect of taking aspirin on headaches for me, I learn from replications in time when I do and do not take aspirin to relieve my headache, thereby having some observations of $Y(0)$ and some of $Y(1)$. When we want to generalize to units other than ourselves, we typically use more objects.

1.2. Replication and the Stable Unit Treatment Value Assumption – SUTVA

Suppose instead of only one unit we have two. Now in general we have at least four potential outcomes for each unit: the outcome for unit 1 if unit 1 and unit 2 received control, $Y_1(0, 0)$; the outcome for unit 1 if both units received the active treatment, $Y_1(1, 1)$; the outcome for unit 1 if unit 1 received control and unit 2 received active, $Y_1(0, 1)$, and the outcome for unit 1 if unit 1 received active and unit 2 received control, $Y_1(1, 0)$; and analogously for unit 2 with values $Y_2(0, 0)$, etc. In fact, there are even more potential outcomes because there have to be at least two "doses" of the active treatment available to contemplate all assignments, and it could make a difference which one was taken. For example, in the aspirin case, one tablet may be very effective and the other quite ineffective.

Clearly, replication does not help unless we can restrict the explosion of potential outcomes. As in all theoretical work, simplifying assumptions are crucial. The most straightforward assumption to make is the "stable unit treatment value assumption" (SUTVA – Rubin, 1980, 1990) under which the potential outcomes for the ith unit just depend on the treatment the ith unit received. That is, there is "no interference between units" and there are "no versions of treatments". Then, all potential outcomes for N units with two possible treatments can be represented by an array with N rows and two columns, the ith unit having a row with two potential outcomes, $Y_i(0)$ and $Y_i(1)$.

There is no assumption-free causal inference, and nothing is wrong with this. It is the quality of the assumptions that matters, not their existence or even their absolute correctness. Good researchers attempt to make assumptions plausible by the design of their studies. For example, SUTVA becomes more plausible when units are isolated from each other, as when using, for the units, schools rather than students in the schools when studying an educational intervention.

The stability assumption (SUTVA) is very commonly made, even though it is not always appropriate. For example, consider a study of the effect of vaccination on a contagious disease. The greater the proportion of the population that gets vaccinated, the less any unit's chance of contracting the disease, even if not vaccinated, an example of interference. Throughout this discussion, we assume SUTVA, although there are other assumptions that could be made to restrict the exploding number of potential outcomes with replication.

1.3. Covariates

In addition to (1) the vector indicator of treatments for each unit in the study, $W = \{W_i\}$, (2) the array of potential outcomes when exposed to the treatment, $Y(1) = \{Y_i(1)\}$, and (3) the array of potential outcomes when not exposed, $Y(0) = \{Y_i(0)\}$, we have (4) the array of covariates $X = \{X_i\}$, which are, by definition, unaffected by treatment. Covariates (such as age, race and sex) play a particularly important role in observational studies for causal effects where they are variously known as potential "confounders" or "risk factors". In some studies, the units exposed to the active treatment differ on their distribution of covariates in important ways from the units not exposed. To see how this can arise in a formal framework, we must define the "assignment mechanism", the probabilistic mechanism that determines which units get the active version of the treatment and which units get the control version.

In general, the N units may not all be assigned treatment 1 or treatment 0. For example, some of the units may be in the future, as when we want to generalize to a future population. Then formally W_i must take on a third value, but for the moment, we avoid this complication.

1.4. Assignment mechanisms – unconfounded and strongly ignorable

A model for the assignment mechanism is needed for all forms of statistical inference for causal effects, including Bayesian. The assignment mechanism gives the conditional probability of each vector of assignments given the covariates and potential outcomes:

$$\Pr\bigl(W \mid X, Y(0), Y(1)\bigr). \tag{1}$$

Here W is a N by 1 vector and X, $Y(1)$ and $Y(0)$ are all matrices with N rows. An example of an assignment mechanism is a completely randomized experiment with N units, with $n < N$ assigned to the active treatment.

$$\Pr\bigl(W \mid X, Y(0), Y(1)\bigr) = \begin{cases} 1/C_n^N & \text{if } \sum W_i = n, \\ 0 & \text{otherwise.} \end{cases} \tag{2}$$

An "unconfounded assignment mechanism" is free of dependence on either $Y(0)$ or $Y(1)$:

$$\Pr\big(W|X, Y(0), Y(1)\big) = \Pr(W|X). \tag{3}$$

With an unconfounded assignment mechanism, at each set of values of X_i that has a distinct probability of $W_i = 1$, there is effectively a completely randomized experiment. That is, if X_i indicates sex, with males having probability 0.2 of receiving the active treatment and females probability 0.5, then essentially one randomized experiment is described for males and another for females.

The assignment mechanism is "probabilistic" if each unit has a positive probability of receiving either treatment:

$$0 < \Pr\big(W_i = 1|X, Y(0), Y(1)\big) < 1. \tag{4}$$

A "strongly ignorable" assignment mechanism (Rosenbaum and Rubin, 1983) satisfies both (2) and (3): it is unconfounded and probabilistic. A nonprobabilistic assignment mechanism fails to satisfy (4) for some units.

The assignment mechanism is fundamental to causal inference because it tells us how we got to see what we saw. Because causal inference is basically a missing data problem with at least half of the potential outcomes not observed, without understanding the process that creates missing data, we have no hope of inferring anything about the missing values. Without a model for how treatments are assigned to individuals, formal causal inference, at least using probabilistic statements, is impossible. This does not mean that we need to know the assignment mechanism, but rather that without positing one, we cannot make any statistical claims about causal effects, such as the coverage of Bayesian posterior intervals.

Randomization, as in (2), is an unconfounded probabilistic assignment mechanism that allows particularly straightforward estimation of causal effects, as we see in Section 3. Therefore, randomized experiments form the basis for inference for causal effects in more complicated situations, such as when assignment probabilities depend on covariates or when there is noncompliance with the assigned treatment. Unconfounded assignment mechanisms, which essentially are collections of distinct completely randomized experiments at each distinct value of X_i, form the basis for the analysis of observational nonrandomized studies.

1.5. Confounded and ignorable assignment mechanisms

A confounded assignment mechanism is one that depends on the potential outcomes:

$$\Pr\big(W|X, Y(0), Y(1)\big) \neq \Pr(W|X). \tag{5}$$

A special class of possibly confounded assignment mechanisms are particularly important to Bayesian inference: ignorable assignment mechanisms (Rubin, 1978). Ignorable assignment mechanisms are defined by their freedom from dependence on any missing potential outcomes:

$$\Pr\big(W|X, Y(0), Y(1)\big) = \Pr(W|X, Y_{\text{obs}}), \tag{6}$$

$$\text{where } Y_{\text{obs}} = \{Y_{\text{obs},i}\}$$
$$\text{with } Y_{\text{obs},i} = W_i Y_i(1) + (1 - W_i) Y_i(0).$$

Ignorable assignment mechanisms do arise in practice, especially in sequential experiments. Here, the next unit's probability of being exposed to the active treatment depends on the success rate of those previously exposed to the active treatment versus the success rate of those exposed to the control treatment, as in "play-the-winner" designs (e.g., see Efron, 1971).

All unconfounded assignment mechanisms are ignorable, but not all ignorable assignment mechanisms are unconfounded (e.g., play-the-winner designs). Seeing why ignorable assignment mechanisms play an important role in Bayesian inference requires us to present the full Bayesian approach. Before doing so, we place the framework presented thus far in an historical perspective.

2. A brief history of the potential outcomes framework

2.1. Before 1923

The basic idea that causal effects are the comparisons of potential outcomes seems so direct that it must have ancient roots, and we can find elements of this definition of causal effects among both experimenters and philosophers. For example, Cochran (1978), when discussing Arthur Young, an English agronomist, stated:

> A single comparison or trial was conducted on large plots – an acre or a half acre in a field split into halves – one drilled, one broadcast. Of the two halves, Young (1771) writes: "The soil is exactly the same; the time of culture, and in a word every circumstance equal in both."

It seems clear in this description that Young viewed the ideal pair of plots as being identical, so that the outcome on one plot of drilling would be the same as the outcome on the other of drilling, $Y_1(\text{Drill}) = Y_2(\text{Drill})$, and likewise for broadcasting, $Y_1(\text{Broad}) = Y_2(\text{Broad})$. Now the difference between drilling and broadcasting on each plot are the causal effects: $Y_1(\text{Drill}) - Y_1(\text{Broad})$ for plot 1 and $Y_2(\text{Drill}) - Y_2(\text{Broad})$ for plot 2. As a result of Young's assumptions, these two causal effects are equal to each other and moreover, are equal to the two possible observed differences when one plot is drilled and the other is broadcast: $Y_1(\text{Drill}) - Y_2(\text{Broad})$ and $Y_1(\text{Broad}) - Y_2(\text{Drill})$.

Nearly a century later, Claude Bernard, an experimental scientist and medical researcher wrote (Wallace, 1974, p. 144):

> The experiment is always the termination of a process of reasoning, whose premises are observation. Example: if the face has movement, what is the nerve? I suppose it is the facial; I cut it. I cut others, leaving the facial intact – the control experiment.

In the late nineteenth century, the philosopher John Stuart Mill, when discussing Hume's views offers (Mill, 1973, p. 327):

> If a person eats of a particular dish, and dies in consequence, that is, would not have died if he had not eaten of it, people would be apt to say that eating of that dish was the source of his death.

And Fisher (1918, p. 214) wrote:

> If we say, "This boy has grown tall because he has been well fed," we are not merely tracing out the cause and effect in an individual instance; we are suggesting that he might quite probably have been worse fed, and that in this case he would have been shorter.

Despite the insights evident in these quotations, there was no formal notation for potential outcomes until 1923, and even then, and for half a century thereafter, its application was limited to randomized experiments, apparently until Rubin (1974). Also, before 1923 there was no formal discussion of any assignment mechanism.

2.2. *Neyman's (1923) notation for causal effects in randomized experiments and Fisher's (1925) proposal to actually randomize treatments to units*

Neyman (1923) appears to have been the first to provide a mathematical analysis for a randomized experiment with explicit notation for the potential outcomes, implicitly making the stability assumption. This notation became standard for work in randomized experiments from the randomization-based perspective (e.g., Pitman, 1937; Welch, 1937; McCarthy, 1939; Anscombe, 1948; Kempthorne, 1952; Brillinger et al., 1978; Hodges and Lehmann, 1970, Section 9.4). The subsequent literature often assumed constant treatment effects as in Cox (1958), and sometimes was used quite informally, as in Freedman et al. (1978, pp. 456–458).

Neyman's formalism was a major advance because it allowed explicit frequentistic probabilistic causal inferences to be drawn from data obtained by a randomized experiment, where the probabilities were explicitly defined by the randomized assignment mechanism. Neyman defined unbiased estimates and asymptotic confidence intervals from the frequentist perspective, where all the probabilities were generated by the randomized assignment mechanism.

Independently and nearly simultaneously, Fisher (1925) created a somewhat different method of inference for randomized experiments, also based on the special class of randomized assignment mechanisms. Fisher's resulting "significance levels" (i.e., based on tests of sharp null hypotheses), remained the accepted rigorous standard for the analysis of randomized clinical trials at the end of the twentieth century. The notions of the central role of randomized experiments seems to have been "in the air" in the 1920's, but Fisher was apparently the first to recommend the actual physical randomization of treatments to units and then use this randomization to justify theoretically an analysis of the resultant data.

Despite the almost immediate acceptance of randomized experiments, Fisher's significance levels, and Neyman's notation for potential outcomes in randomized experiments in the late 1920's, this same framework was not used outside randomized experiments for a half century thereafter, and these insights were entirely limited to randomization-based frequency inference.

2.3. *The observed outcome notation*

The approach in nonrandomized settings, during the half century following the introduction of Neyman's seminal notation for randomized experiments, was to build mathematical models relating the observed value of the outcome variable $Y_{\text{obs}} = \{Y_{\text{obs},i}\}$

to covariates and indicators for treatment received, and then to define causal effects as parameters in these models. The same statistician would simultaneously use Neyman's potential outcomes to define causal effects in randomized experiments and the observed outcome setup in observational studies. This led to substantial confusion because the role of randomization cannot even be stated using observed outcome notation. That is, Eq. (3) does not imply that $\Pr(W|X, Y_{\text{obs}})$ is free of Y_{obs}, except under special conditions, i.e., when $Y(0) \equiv Y(1) \equiv Y_{\text{obs}}$, so the formal benefits of randomization could not even be formally stated using the collapsed observed outcome notation.

2.4. *The Rubin causal model*

The framework that we describe here, using potential outcomes to define causal effects and a general assignment mechanism, has been called the "Rubin Causal Model" – RCM by Holland (1986) for work initiated in the 1970's (Rubin, 1974, 1977, 1978). This perspective conceives of all problems of statistical inference for causal effects as missing data problems with a mechanism for creating missing data (Rubin, 1976).

The RCM has the following salient features for causal inference: (1) Causal effects are defined as comparisons of a priori observable potential outcomes without regard to the choice of assignment mechanism that allows the investigator to observe particular values; as a result, interference between units and variability in efficacy of treatments can be incorporated in the notation so that the commonly used "stability" assumption can be formalized, as can deviations from it; (2) Models for the assignment mechanism are viewed as methods for creating missing data, thereby allowing nonrandomized studies to be considered using the same notation as used for randomized experiments, and therefore the role of randomization can be formally stated; (3) The underlying data, that is, the potential outcomes and covariates, can be given a joint distribution, thereby allowing both randomization-based methods, traditionally used for randomized experiments, and model-based Bayesian methods, traditionally used for observational studies, to be applied to both kinds of studies. The Bayesian aspect of this third point is the one we turn to in the next section.

This framework seems to have been basically accepted and adopted by most workers by the end of the twentieth century. Sometimes the move was made explicitly, as with Pratt and Schlaifer (1984) who moved from the "observed outcome" to the potential outcomes framework in Pratt and Schlaifer (1988). Sometimes it was made less explicitly as with those who were still trying to make a version of the observed outcome notation work in the late 1980's (e.g., see Heckman and Hotz, 1989), before fully accepting the RCM in subsequent work (e.g., Heckman, 1989, after discussion by Holland, 1989). But the movement to use potential outcomes to define causal inference problems seems to be the dominant one at the start of the 21st century and is totally compatible with Bayesian inference.

3. Models for the underlying data – Bayesian inference

Bayesian causal inference requires a model for the underlying data, $\Pr(X, Y(0), Y(1))$, and this is where science enters. But a virtue of the framework we are presenting is that it separates science – a model for the underlying data, from what we do to learn about science – the assignment mechanism, $\Pr(W|X_1 Y(0), Y(1))$. Notice that together, these two models specify a joint distribution for all observables.

3.1. *The posterior distribution of causal effects*

Bayesian inference for causal effects directly confronts the explicit missing potential outcomes, $Y_{\mathrm{mis}} = \{Y_{\mathrm{mis},i}\}$ where $Y_{\mathrm{mis},i} = W_i Y_i(0) + (1 - W_i)Y_i(1)$. The perspective simply takes the specifications for the assignment mechanism and the underlying data (= science), and derives the posterior predictive distribution of Y_{mis}, that is, the distribution of Y_{mis} given all observed values,

$$\Pr(Y_{\mathrm{mis}}|X, Y_{\mathrm{obs}}, W). \tag{7}$$

From this distribution and the observed values of the potential outcomes, Y_{obs}, and covariates, the posterior distribution of any causal effect can, in principle, be calculated.

This conclusion is immediate if we view the posterior predictive distribution in (7) as specifying how to take a random draw of Y_{mis}. Once a value of Y_{mis} is drawn, any causal effect can be directly calculated from the drawn values of Y_{mis} and the observed values of X and Y_{obs}, e.g., the median causal effect for males: $\mathrm{med}\{Y_i(1) - Y_i(0)|X_i$ indicate males$\}$. Repeatedly drawing values of Y_{mis} and calculating the causal effect for each draw generates the posterior distribution of the desired causal effect. Thus, we can view causal inference completely as a missing data problem, where we multiply-impute (Rubin, 1987, 2004a) the missing potential outcomes to generate a posterior distribution for the causal effects. We have not yet described how to generate these imputations, however.

3.2. *The posterior predictive distribution of Y_{mis} under ignorable treatment assignment*

First consider how to create the posterior predictive distribution of Y_{mis} when the treatment assignment mechanism is ignorable (i.e., when (6) holds). In general:

$$\Pr(Y_{\mathrm{mis}}|X, Y_{\mathrm{obs}}, W) = \frac{\Pr(X, Y(0), Y(1)) \Pr(W|X, Y(0), Y(1))}{\int \Pr(X, Y(0), Y(1)) \Pr(W|X, Y(0), Y(1))\, \mathrm{d}Y_{\mathrm{mis}}}. \tag{8}$$

With ignorable treatment assignment, Eqs. (3), (6) becomes:

$$\Pr(Y_{\mathrm{mis}}|X, Y_{\mathrm{obs}}, W) = \frac{\Pr(X, Y(0), Y(1))}{\int \Pr(X, Y(0), X(1))\, \mathrm{d}Y_{\mathrm{mis}}}. \tag{9}$$

Eq. (9) reveals that under ignorability, all that needs to be modelled is the science $\Pr(X, Y(0), Y(1))$.

Because all information is in the underlying data, the unit labels are effectively just random numbers, and hence the array $(X, Y(0), Y(1))$ is row exchangeable. With essentially no loss of generality, therefore, by de Finetti's (1963) theorem we have that the distribution of $(X, Y(0), Y(1))$ may be taken to be i.i.d. (independent and identically distributed) given some parameter θ:

$$\Pr\big(X, Y(0), Y(1)\big) = \int \left[\prod_{i=1}^{N} f\big(X_i, Y_i(0), Y_i(1)|\theta\big) \right] p(\theta)\, \mathrm{d}(\theta) \tag{10}$$

for some prior distribution $p(\theta)$. Eq. (10) provides the bridge between fundamental theory and the practice of using i.i.d. models. A simple example illustrates what is required to apply Eq. (10).

3.3. *Simple normal example – analytic solution*

Suppose we have a completely randomized experiment with no covariates, and a scalar outcome variable. Also, assume plots were randomly sampled from a field of N plots and the causal estimand is the mean difference between $Y(1)$ and $Y(0)$ across all N plots, say $\overline{Y}_1 - \overline{Y}_0$. Then

$$\Pr(Y) = \int \prod_{i=1}^{N} f\big(Y_i(0), Y_i(1)|\theta\big) p(\theta)\, \mathrm{d}\theta$$

for some bivariate density $f(\cdot|\theta)$ indexed by parameter θ with prior distribution $p(\theta)$. Suppose $f(\cdot|\theta)$ is normal with means $\mu = (\mu_1, \mu_0)$, variances (σ_1^2, σ_0^2) and correlation ρ. Then conditional on (a) θ, (b) the observed values of Y, Y_{obs}, and (c) the observed value of the treatment assignment, where the number of units with $W_i = K$ is n_K $(K = 0, 1)$, we have that when $n_0 + n_1 = N$ the joint distribution of $(\overline{Y}_1, \overline{Y}_0)$ is normal with means

$$\frac{1}{2}\left[\bar{y}_1 + \mu_1 + \rho\frac{\sigma_1}{\sigma_0}(\bar{y}_0 - \mu_0)\right],$$

$$\frac{1}{2}\left[\bar{y}_0 + \mu_0 + \rho\frac{\sigma_0}{\sigma_1}(\bar{y}_1 - \mu_1)\right],$$

variances $\sigma_1^2(1-\rho^2)/4n_0$, $\sigma_0^2(1-\rho^2)/4n_1$, and zero correlation, where $\bar{y}_1$ and $\bar{y}_0$ are the observed sample means of Y in the two treatment groups. To simplify comparison with standard answers, now assume large N and a relatively diffuse prior distribution for $(\mu_1, \mu_0, \sigma_1^2, \sigma_0^2)$ given ρ. Then the conditional posterior distribution of $\overline{Y}_1 - \overline{Y}_0$ given ρ is normal with mean

$$E\big[\overline{Y}_1 - \overline{Y}_0 | Y_{\mathrm{obs}}, W, \rho\big] = \bar{y}_1 - \bar{y}_0 \tag{11}$$

and variance

$$V\big[\overline{Y}_1 - \overline{Y}_0 | Y_{\mathrm{obs}}, W, \rho\big] = \frac{s_1^2}{n_1} + \frac{s_0^2}{n_0} - \frac{1}{N}\sigma_{(1-0)}^2, \tag{12}$$

where $\sigma_{(1-0)}^2$ is the prior variance of the differences $Y_i(1) - Y_i(0)$, $\sigma_1^2 + \sigma_0^2 - 2\sigma_1\sigma_0\rho$. Section 2.5 in Rubin (1987, 2004a) provides details of this derivation. The answer given by (11) and (12) is remarkably similar to the one derived by Neyman (1923) from the randomization-based perspective, as pointed out in the discussion by Rubin (1990).

There is no information in the observed data about ρ, the correlation between the potential outcomes, because they are never jointly observed. A conservative inference for $\overline{Y}_1 - \overline{Y}_0$ is obtained by taking $\sigma^2_{(1-0)} = 0$.

The analytic solution in (11) and (12) could have been obtained by simulation, as described in general in Section 3.2. Simulation is a much more generally applicable tool than closed-form analysis because it can be applied in much more complicated situations. In fact, the real advantage of Bayesian inference for causal effects is only revealed in situations with complications. In standard situations, the Bayesian answer often looks remarkably similar to the standard frequentist answer, as it does in the simple example of this section:

$$(\bar{y}_1 - \bar{y}_0) \pm 2\left(\frac{s_1^2}{n_1} + \frac{s_0^2}{n_0}\right)^{1/2}$$

is a conservative 95% interval for $\overline{Y}_1 - \overline{Y}_0$, at least in relatively large samples.

3.4. *Simple normal example – simulation approach*

The intuition for simulation is especially direct in this example of Section 3.3 if we assume $\rho = 0$; suppose we do so. The units with $W_i = 1$ have $Y_i(1)$ observed and are missing $Y_i(0)$, and so their $Y_i(0)$ values need to be imputed. To impute $Y_i(0)$ values for them, we need to find units with $Y_i(0)$ observed who are exchangeable with the $W_i = 1$ units, but these units are the units with $W_i = 0$. Therefore, we estimate (in a Bayesian way) the distribution of $Y_i(0)$ from the units with $W_i = 0$, and use this estimated distribution to impute $Y_i(0)$ for the units missing $Y_i(0)$.

Since the n_0 observed values of $Y_i(0)$ are a simple random sample of the N values of $Y(0)$, and are normally distributed with mean μ_0 and variance σ_0^2, with the standard independent noninformative prior distributions on (μ_0, σ_0^2), we have for the posterior of σ_0^2:

$$\sigma_0^2/s_0^2 \sim \text{inverted } X^2_{n_0-1}/(n_0 - 1);$$

and for the posterior distribution of μ_0 given σ_0:

$$\mu_0 \sim N\big(\bar{y}_0, s_0^2/n_0\big);$$

and for the missing $Y_i(0)$ given μ_0 and σ_0:

$$Y_i(0) \ni W_i \neq 0 \overset{\text{i.i.d.}}{\sim} N\big(\mu_0, s_0^2\big).$$

The missing values of $Y_i(1)$ are analogously imputed using the observed values of $Y_i(1)$.

When there are covariates observed, these are used to help predict the missing potential outcomes using one regression model for the observed $Y_i(1)$ given the covariates, and another regression model for the observed $Y_i(0)$ given the covariates.

3.5. *Simple normal example with covariate – numerical example*

For a specific example with a covariate, suppose we have a large population of people with a covariate X_i indicating baseline cholesterol. Suppose the observed X_i is dichotomous, *HI* versus *LO*, split at the median in the population. Suppose that a random sample of 100 with $X_0 = HI$ is taken, and 90 are randomly assigned to the active treatment, a statin, and 10 are randomly assigned to the control treatment, a placebo. Further suppose that a random sample of 100 with $X_i = LO$ is taken, and 10 are randomly assigned to the statin and 90 are assigned to the placebo. The outcome Y is cholesterol a year after baseline, with $Y_{i,\text{obs}}$ and X_i observed for all 200 units; X_i is effectively observed in the population because we know the proportion of X_i that are *HI* and *LO*.

Suppose the hypothetical observed data are as displayed in Table 1.

Table 1
Final cholesterol in artificial example

Baseline	$\bar{y}_1$	n_1	$\bar{y}_0$	n_0	$s_1 = s_0$
HI	200	90	300	10	60
LO	100	10	200	90	60

Then the inferences based on the normal-model are as follows:

Table 2
Inferences for example in Table 1

	HI	*LO*	Population $= \frac{1}{2}HI + \frac{1}{2}LO$
$E(\overline{Y}_1 - \overline{Y}_0 \mid X, Y_{\text{obs}}, W)$	-100	-100	-100
$V(\overline{Y}_1 - \overline{Y}_0 \mid X, Y_{\text{obs}}, W)^{1/2}$	20	20	$10\sqrt{2}$

Here the notation is being slightly abused because the first entry in Table 2 really should be labelled $E(\overline{Y}_1 - \overline{Y}_0 \ni X_i = HI \mid X, Y_{\text{obs}}, W)$ and so forth.

The obvious conclusion in this artificial example is that the statin reduces final cholesterol for both those with *HI* and *LO* baseline cholesterol, and thus for the population which is a 50%/50% mixture of these two subpopulations. In this sort of situation, the final inference is insensitive to the assumed normality of $Y_i(1)$ given X_i and of $Y_i(0)$ given X_i; see Pratt (1965) or Rubin (1987, 2004a, Section 2.5) for the argument.

3.6. *Nonignorable treatment assignment*

With nonignorable treatment assignment, the above simplifications in Sections 3.2–3.5, which follow from ignoring the specification for $\Pr(W \mid X, Y(0), Y(1))$, do not follow in general, and analysis typically becomes far more difficult and uncertain. As a simple illustration, take the example in Section 3.5 and assume that everything is the same except that only Y_{obs} is recorded, so that we do not know whether baseline is *HI* or *LO* for anyone. The actually assignment mechanism is now

$$\Pr\bigl(W \mid Y(0), Y(1)\bigr) = \int \Pr\bigl(W \mid X, Y(0), Y(1)\bigr)\, dP(X)$$

because X itself is missing, and so treatment assignment depends explicitly on the potential outcomes, both observed and missing, which are both correlated with the missing X_i.

Inference for causal effects, assuming the identical model for the science, now depends on the implied normal mixture model for the observed Y data within each treatment arm, because the population Y values are a 50%/50% a mixture of those with *LO* and *HI* baseline cholesterol, and these subpopulations have different probabilities of treatment assignment. Here the inference for causal effects is sensitive to the propriety of the assumed normality and/or the assumption of a 50%/50% mixture, as well as to the prior distributions on μ_1, μ_0, σ_1 and σ_0.

If we mistakenly ignore the nonignorable treatment assignment and simply compare the sample means of all treated with all controls, we have $\bar{y}_1 = .9(200) + .1(100) = 190$ versus $\bar{y}_0 = .1(300) + .9(200) = 210$; doing so, we reach the incorrect conclusion that the statin is bad for final cholesterol in the population. This sort of example is known as "Simpson's Paradox" (Simpson, 1951) and can easily arise with incorrect analyzes of nonignorable treatment assignment mechanisms, and thus indicates why such assignment mechanisms are to be avoided whenever possible.

Randomized experiments are the most direct way of avoiding nonignorable treatment assignments. Other alternatives are ignorable designs with nonprobabilistic features so that all units with some specific value of covariates are assigned the same treatment. With such assignment mechanisms, randomization-based inference is impossible for those units since their treatment does not change over the various possible assignments.

4. Complications

There are many complications that occur in real world studies for causal effects, many of which can be handled much more flexibly with the Bayesian approach than with standard frequency methods. Of course, the models involved, including associated prior distributions, can be very demanding to formulate in a practically reliable manner. Here I simply list some of these complications with some admittedly idiosyncratically personal references to current work from the Bayesian perspective. Gelman et al. (2003), especially starting with Chapter 7, is a good reference for some of these complications and the computational methods for dealing with them.

4.1. Multiple treatments

When there are more than two treatments, the notation becomes more complex but is still straightforward under SUTVA. Without SUTVA, however, both the notation and the analysis can become very involved. The exploding number of potential outcomes can become especially serious in studies where the units are exposed to a sequence of repeated treatments in time, each distinct sequence corresponding to a possibly distinct treatment. Most of the field of classical experiment design is devoted to issues that arise with more than two treatment conditions (e.g., Kempthorne, 1952; Cochran and Cox, 1957, 1992).

4.2. *Unintended missing data*

Missing data, due perhaps to patient dropout or machine failure, can complicate analyzes more than one would expect based on a cursory examination of the problem. Fortunately, Bayesian/likelihood tools for addressing missing data such as multiple imputation (Rubin, 1987, 2004a) or the EM algorithm (Dempster et al., 1977) and its relatives, including data augmentation (Tanner and Wong, 1987) and the Gibbs sampler (Geman and Geman, 1984) are fully compatible with the Bayesian approach to causal inference outlined in Section 3. Gelman et al. (2003), Parts III and IV provide, guidance on many of these issues from the Bayesian perspective.

4.3. *Noncompliance with assigned treatment*

Another complication, common when the units are people, is noncompliance. For example, some of the subjects assigned to take the active treatment take the control treatment instead, and some assigned to take the control manage to take the active treatment. Initial interest focuses on the effect of the treatment for the subset of people who will comply with their treatment assignments. Much progress has been made in recent years on this topic from the Bayesian perspective, e.g., Imbens and Rubin (1997), Hirano et al. (2000). In this case, sensitivity of inference to prior assumptions can be severe, and the Bayesian approach is ideally suited to not only revealing this sensitivity but also to formulating reasonable prior restrictions.

4.4. *Truncation of outcomes due to death*

In other cases, the unit may "die" before the final outcome can be measured. For example, in an experiment with new fertilizers, a plant may die before the crops are harvested and interest may focus on both the effect of the fertilizer on plant survival and the effect of the fertilizer on plant yield when the plant survives. This problem is far more subtle than it may at first appear to be, and valid Bayesian approaches to it have only recently been formulated following the proposal in (Rubin, 2000); see (Zhang and Rubin, 2003) for simple large sample bounds. It is interesting that the models also have applications in economics (Zhang et al., 2004).

4.5. *Direct and indirect causal effects*

Another topic that is far more subtle than it first appears to be is the one involving direct and indirect causal effects. For example, the separation of the "direct" effect of a vaccination on disease from the "indirect" effect of the vaccination that is due solely to its effect on blood antibodies and the "direct" effect of the antibodies on disease. This language turns out to be too imprecise to be useful within our formal causal effect framework. This problem is ripe for Bayesian modelling as briefly outlined in Rubin (2004b).

4.6. *Principal stratification*

All the examples in Sections 4.3–4.5 can be viewed as special cases of "principal stratification" (Frangakis and Rubin, 2002), where the principal strata are defined by partially unobserved intermediate potential outcomes, namely in our examples: compliance behavior under both treatment assignments, survival under both treatment assignments, and antibody level under both treatment assignments. This appears to be an extremely fertile area for research and application of Bayesian methods for causal inference, especially using modern simulation methods such as MCMC (Markov Chain Monte Carlo); see, for example, Gilks et al. (1995).

4.7. Combinations of complications

In the real world, such complications typically do not appear simply one at a time. For example, a randomized experiment in education evaluating "school choice" suffered from missing data in both covariates and longitudinal outcomes; also, the outcome was multicomponent as each point in time; in addition, it suffered from noncompliance that took several levels because of the years of school. Some of these combinations of complications are discussed in Barnard et al. (2003) in the context of the school choice example, and in Mealli and Rubin (2003) in the context of a medical experiment.

Despite the fact that Bayesian analysis is quite difficult when confronted with these combinations of complications, it is still a far more satisfactory attack on the real scientific problems than the vast majority of ad hoc frequentist approaches in common use today.

It is an exciting time for Bayesian inference for causal effects.

References

Anscombe, F.J. (1948). The validity of comparative experiments. *J. Roy. Statist. Soc., Ser. A* **61**, 181–211.

Barnard, J., Hill, J., Frangakis, C., Rubin, D. (2003). School choice in NY city: A Bayesian analysis of an imperfect randomized experiment. In: Gatsonis, C., Carlin, B., Carriquiry, A. (Eds.), *Case Studies in Bayesian Statistics, vol. V*. Springer-Verlag, New York, pp. 3–97. (With discussion and rejoinder.)

Brillinger, D.R., Jones, L.V., Tukey, J.W. (1978). Report of the statistical task force for the weather modification advisory board. In: *The Management of Western Resources, vol. II: The Role of Statistics on Weather Resources Management. Stock No. 003-018-00091-1*. Government Printing Office, Washington, DC.

Cochran, W.G. (1978). Early development of techniques in comparative experimentation. In: Owen, D. (Ed.), *On the History of Statistics and Probability*. Dekker, New York, pp. 2–25.

Cochran, W.G., Cox, G.M. (1957). *Experimental Designs*, second ed. Wiley, New York.

Cochran, W.G., Cox, G.M. (1992). *Experimental Designs*, second ed. Wiley, New York. Reprinted as a "Wiley Classic".

Cox, D.R. (1958). *The Planning of Experiments*. Wiley, New York.

de Finetti, B. (1963). Foresight: Its logical laws, its subjective sources. In: Kyburg, H.E., Smokler, H.E. (Eds.), *Studies in Subjective Probability*. Wiley, New York.

Dempster, A.P., Laird, N., Rubin, D.B. (1977). Maximum likelihood from incomplete data via the EM algorithm. *J. Roy. Statist. Soc., Ser. B* **39**, 1–38. (With discussion and reply.)

Efron, B. (1971). Forcing a sequential experiment to be balanced. *Biometrika* **58**, 403–417.

Fisher, R.A. (1918). The causes of human variability. *Eugenics Review* **10**, 213–220.

Fisher, R.A. (1925). *Statistical Methods for Research Workers*, first ed. Oliver and Boyd, Edinburgh.

Frangakis, C.E., Rubin, D.B. (2002). Principal stratification in causal inference. *Biometrics* **58**, 21–29.

Freedman, D., Pisani, R., Purves, R. (1978). *Statistics*. Norton, New York.

Geman, S., Geman, D. (1984). Stochastic relaxation. Gibbs distributions, and the Bayesian restoration of images. *IEEE Trans. Pattern Anal. Machine Intelligence* **6** (November), 721–741.

Gelman, A., Carlin, J., Stern, H., Rubin, D. (2003). *Bayesian Data Analysis*, second ed. CRC Press, New York.
Gilks, W.R., Richardson, S., Spiegelhalter, D.J. (1995). *Markov Chain Monte Carlo in Practice*. CRC Press, New York.
Heckman, J.J. (1989). Causal inference and nonrandom samples. *J. Educational Statist.* **14**, 159–168.
Heckman, J.J., Hotz, J. (1989). Alternative methods for evaluating the impact of training programs. *J. Amer. Statist. Assoc.* **84**, 862–874. (With discussion.)
Hirano, K., Imbens, G., Rubin, D.B., Zhou, X. (2000). Assessing the effect of an influenza vaccine in an encouragement design. *Biostatistics* **1**, 69–88.
Hodges, J.L., Lehmann, E. (1970). *Basic Concepts of Probability and Statistics*, second ed. Holden-Day, San Francisco.
Holland, P.W. (1986). Statistics and causal inference. *J. Amer. Statist. Assoc.* **81**, 945–970.
Holland, P.W. (1989). It's very clear. Comment on "Choosing among alternative nonexperimental methods for estimating the impact of social programs: The case of manpower training" by J. Heckman, V. Hotz. *J. Amer. Statist. Assoc.* **84**, 875–877.
Imbens, G., Rubin, D.B. (1997). Bayesian inference for causal effects in randomized experiments with noncompliance. *Ann. Statist.* **25**, 305–327.
Kempthorne, O. (1952). *The Design and Analysis of Experiments*. Wiley, New York.
McCarthy, M.D. (1939). On the application of the z-test to randomized blocks. *Ann. Math. Statist.* **10**, 337.
Mealli, F., Rubin, D.B. (2003). Assumptions when analyzing randomized experiments with noncompliance and missing outcomes. *Health Services Outcome Research Methodology*, 2–8.
Mill, J.S. (1973). A system of logic. In: *Collected Works of John Stuart Mill, vol. 7*. University of Toronto Press, Toronto.
Neyman, J. (1923). On the application of probability theory to agricultural experiments: Essay on principles, Section 9. Translated in *Statistical Science* **5** (1990), 465–480.
Pearl, J. (2000). *Causality: Models, Reasoning and Inference*. Cambridge University Press, Cambridge.
Pitman, E.J.G. (1937). Significance tests which can be applied to samples from any population. III. The analysis of variance test. *Biometrika* **29**, 322–335.
Pratt, J.W. (1965). Bayesian interpretation of standard inference statements. *J. Roy. Statist. Soc., Ser. B* **27**, 169–203. (With discussion.)
Pratt, J.W., Schlaifer, R. (1984). On the nature and discovery of structure. *J. Amer. Statist. Assoc.* **79**, 9–33. (With discussion.)
Pratt, J.W., Schlaifer, R. (1988). On the interpretation and observation of laws. *J. Econometrics* **39**, 23–52.
Rosenbaum, P.R., Rubin, D.B. (1983). The central role of the propensity score in observational studies for causal effects. *Biometrika* **70**, 41–55.
Rubin, D.B. (1974). Estimating causal effects of treatments in randomized and nonrandomized studies. *J. Educational Psychology* **66**, 688–701.
Rubin, D.B. (1976). Inference and missing data. *Biometrika* **63**, 581–592.
Rubin, D.B. (1977). Assignment of treatment group on the basis of a covariate. *J. Educational Statistics* **2**, 1–26.
Rubin, D.B. (1978). Bayesian inference for causal effects: The role of randomization. *Ann. Statist.* **7**, 34–58.
Rubin, D.B. (1980). Comment on "Randomization analysis of experimental data: The Fisher randomization test" by D. Basu. *J. Amer. Statist. Assoc.* **75**, 591–593.
Rubin, D.B. (1987). *Multiple Imputation for Nonresponse in Surveys*. Wiley, New York.
Rubin, D.B. (2000). The utility of counterfactuals for causal inference. Comment on A.P. Dawid, 'Causal inference without counterfactuals'. *J. Amer. Statist. Assoc.* **95**, 435–438.
Rubin, D.B. (1990). Comment: Neyman (1923) and causal inference in experiments and observational studies. *Statist. Sci.* **5**, 472–480.
Rubin, D.B. (2004a). *Multiple Imputation for Nonresponse in Surveys*. Wiley, New York. Reprinted with new appendices as a "Wiley Classic."
Rubin, D.B. (2004b). Direct and indirect causal effects via potential outcomes. *Scand. J. Statist.* **31**, 161–170; 195–198, with discussion and reply.
Simpson, E.H. (1951). The interpretation of interaction in contingency tables. *J. Roy. Statist. Soc., Ser. B* **13**, 238–241.

Tanner, M.A., Wong, W.H. (1987). The calculation of posterior distributions by data augmentation. *J. Amer. Statist. Assoc.* **82**, 528–550. (With discussion.)

Wallace, W.A. (1974). *Causality and Scientific Explanation: Classical and Contemporary Science, vol. 2.* University of Michigan Press, Ann Arbor.

Welch, B.L. (1937). On the z test in randomized blocks and Latin squares. *Biometrika* **29**, 21–52.

Zhang, J., Rubin, D.B. (2003). Estimation of causal effects via principal stratification when some outcomes are truncated by 'death'. *J. Educational and Behavioral Statist.* **28**, 353–368.

Zhang, J., Rubin, D., Mealli, F. (2004). Evaluating the effects of training programs with experimental data. Submitted for publication.

Handbook of Statistics, Vol. 25
ISSN: 0169-7161
© 2005 Elsevier B.V. All rights reserved.
DOI 10.1016/S0169-7161(05)25002-2

2

Reference Analysis

José M. Bernardo[1]

Abstract
This chapter describes reference analysis, a method to produce Bayesian inferential statements which only depend on the assumed model and the available data. Statistical information theory is used to define the *reference* prior function as a mathematical description of that situation where data would best dominate prior knowledge about the quantity of interest. Reference priors are not descriptions of personal beliefs; they are proposed as formal *consensus* prior functions to be used as standards for scientific communication. Reference posteriors are obtained by formal use of Bayes theorem with a reference prior. Reference prediction is achieved by integration with a reference posterior. Reference decisions are derived by minimizing a reference posterior expected loss. An information theory based loss function, the *intrinsic discrepancy*, may be used to derive reference procedures for conventional inference problems in scientific investigation, such as point estimation, region estimation and hypothesis testing.

Keywords: amount of information; intrinsic discrepancy; Bayesian asymptotics; Fisher information; objective priors; noninformative priors; Jeffreys priors; reference priors; maximum entropy; consensus priors; intrinsic statistic; point estimation; region estimation; hypothesis testing

1. Introduction and notation

This chapter is mainly concerned with statistical inference problems such as occur in scientific investigation. Those problems are typically solved conditional on the assumption that a particular statistical model is an appropriate description of the probabilistic mechanism which has generated the data, and the choice of that model naturally involves an element of subjectivity. It has become standard practice, however, to describe as "objective" any statistical analysis which only depends on the model assumed and the data observed. In this precise sense (and only in this sense) reference analysis is a method to produce "objective" Bayesian inference.

[1] Supported by grant BMF2001-2889 of the MCyT, Madrid, Spain.

Foundational arguments (Savage, 1954; de Finetti, 1970; Bernardo and Smith, 1994) dictate that scientists should elicit a unique (joint) prior distribution on all unknown elements of the problem on the basis of available information, and use Bayes theorem to combine this with the information provided by the data, encapsulated in the likelihood function. Unfortunately however, this elicitation is a formidable task, specially in realistic models with many nuisance parameters which rarely have a simple interpretation. Weakly informative priors have here a role to play as approximations to genuine proper prior distributions. In this context, the (unfortunately very frequent) naïve use of simple proper "flat" priors (often a limiting form of a conjugate family) as presumed "noninformative" priors often hides important unwarranted assumptions which may easily dominate, or even invalidate, the analysis: see, e.g., Hobert and Casella (1996a, 1996b), Casella (1996), Palmer and Pettit (1996), Hadjicostas and Berry (1999) or Berger (2000). The uncritical (ab)use of such "flat" priors should be strongly discouraged. An appropriate *reference* prior (see below) should instead be used. With numerical simulation techniques, where a proper prior is often needed, a proper approximation to the *reference* prior may be employed.

Prior elicitation would be even harder in the important case of scientific inference, where some sort of *consensus* on the elicited prior would obviously be required. A fairly natural candidate for such a consensus prior would be a "noninformative" prior, where prior knowledge could be argued to be dominated by the information provided by the data. Indeed, scientific investigation is seldom undertaken unless it is likely to substantially increase knowledge and, even if the scientist holds strong prior beliefs, the analysis would be most convincing to the scientific community if done with a consensus prior which is dominated by the data. Notice that the concept of a "noninformative" prior is *relative* to the information provided by the data.

As evidenced by the long list of references which concludes this chapter, there has been a considerable body of conceptual and theoretical literature devoted to identifying appropriate procedures for the formulation of "noninformative" priors. Beginning with the work of Bayes (1763) and Laplace (1825) under the name of inverse probability, the use of "noninformative" priors became central to the early statistical literature, which at that time was mainly objective Bayesian. The obvious limitations of the principle of insufficient reason used to justify the (by then) ubiquitous uniform priors, motivated the developments of Fisher and Neyman, which overshadowed Bayesian statistics during the first half of the 20th century. The work of Jeffreys (1946) prompted a strong revival of objective Bayesian statistics; the seminal books by Jeffreys (1961), Lindley (1965), Zellner (1971), Press (1972) and Box and Tiao (1973), demonstrated that the conventional textbook problems which frequentist statistics were able to handle could better be solved from a unifying objective Bayesian perspective. Gradual realization of the fact that no *single* "noninformative" prior could possibly be always appropriate for all inference problems within a given multiparameter model (Dawid et al., 1973; Efron, 1986) suggested that the long search for a *unique* "noninformative" prior representing "ignorance" within a given model was misguided. Instead, efforts concentrated in identifying, for each particular inference problem, a specific (joint) *reference prior* on all the unknown elements of the problem which would lead to a (marginal) *reference posterior* for the quantity of interest, a posterior which would always be dominated by the information provided by the data (Bernardo, 1979b). As will later be described in detail, statistical information theory was used to provide a precise meaning to this dominance requirement.

Notice that reference priors were *not* proposed as an approximation to the scientist's (unique) personal beliefs, but as a collection of formal *consensus* (not necessarily proper) prior functions which could conveniently be used as standards for scientific communication. As Box and Tiao (1973, p. 23) required, using a reference prior the scientist employs the jury principle; as the jury is carefully screened among people with no connection with the case, so that testimony may be assumed to dominate prior ideas of the members of the jury, the reference prior is carefully chosen to guarantee that the information provided by the data will not be overshadowed by the scientist's prior beliefs.

Reference posteriors are obtained by formal use of Bayes theorem with a reference prior function. If required, they may be used to provide point or region estimates, to test hypothesis, or to predict the value of future observations. This provides a unified set of objective Bayesian solutions to the conventional problems of scientific inference, objective in the precise sense that those solutions only depend on the assumed model and the observed data.

By restricting the class $\mathcal{P}$ of candidate priors, the reference algorithm makes it possible to incorporate into the analysis any genuine prior knowledge (over which scientific consensus will presumably exist). From this point of view, derivation of reference priors may be described as a new, powerful method for *prior elicitation*. Moreover, when subjective prior information is actually specified, the corresponding subjective posterior may be compared with the *reference* posterior – hence its name – to assess the relative importance of the initial opinions in the final inference.

In this chapter, it is assumed that probability distributions may be described through their probability density functions, and no notational distinction is made between a random quantity and the particular values that it may take. Bold italic roman fonts are used for observable random vectors (typically data) and bold italic greek fonts for unobservable random vectors (typically parameters); lower case is used for variables and upper case calligraphic for their dominion sets. Moreover, the standard mathematical convention of referring to functions, say f_x and g_x of $\boldsymbol{x} \in \boldsymbol{\mathcal{X}}$, respectively by $f(\boldsymbol{x})$ and $g(\boldsymbol{x})$ will be used throughout. Thus, the conditional probability density of data $\boldsymbol{x} \in \boldsymbol{\mathcal{X}}$ given $\boldsymbol{\theta}$ will be represented by either $p_{\boldsymbol{x}|\boldsymbol{\theta}}$ or $p(\boldsymbol{x}|\boldsymbol{\theta})$, with $p(\boldsymbol{x}|\boldsymbol{\theta}) \geqslant 0$ and $\int_{\boldsymbol{\mathcal{X}}} p(\boldsymbol{x}|\boldsymbol{\theta})\,\mathrm{d}\boldsymbol{x} = 1$, and the posterior distribution of $\boldsymbol{\theta} \in \boldsymbol{\Theta}$ given $\boldsymbol{x}$ will be represented by either $p_{\boldsymbol{\theta}|\boldsymbol{x}}$ or $p(\boldsymbol{\theta}|\boldsymbol{x})$, with $p(\boldsymbol{\theta}|\boldsymbol{x}) \geqslant 0$ and $\int_{\boldsymbol{\Theta}} p(\boldsymbol{\theta}|\boldsymbol{x})\,\mathrm{d}\boldsymbol{\theta} = 1$. This admittedly imprecise notation will greatly simplify the exposition. If the random vectors are discrete, these functions naturally become probability mass functions, and integrals over their values become sums. Density functions of specific distributions are denoted by appropriate names. Thus, if x is an observable random variable with a normal distribution of mean μ and variance σ^2, its probability density function will be denoted $\mathrm{N}(x|\mu, \sigma)$. If the posterior distribution of μ is Student with location $\bar{x}$, scale s, and $n-1$ degrees of freedom, its probability density function will be denoted $\mathrm{St}(\mu|\bar{x}, s, n-1)$.

The reference analysis argument is always defined in terms of some *parametric model* of the general form $\mathcal{M} \equiv \{p(\boldsymbol{x}|\boldsymbol{\omega}), \boldsymbol{x} \in \boldsymbol{\mathcal{X}}, \boldsymbol{\omega} \in \boldsymbol{\Omega}\}$, which describes the conditions under which data have been generated. Thus, data $\boldsymbol{x}$ are assumed to consist of one observation of the random vector $\boldsymbol{x} \in \boldsymbol{\mathcal{X}}$, with probability density $p(\boldsymbol{x}|\boldsymbol{\omega})$ for some $\boldsymbol{\omega} \in \boldsymbol{\Omega}$. Often, but not necessarily, data will consist of a random sample $\boldsymbol{x} = \{\boldsymbol{y}_1, \ldots, \boldsymbol{y}_n\}$ of fixed size n from some distribution with, say, density $p(\boldsymbol{y}|\boldsymbol{\omega})$, $\boldsymbol{y} \in \boldsymbol{\mathcal{Y}}$, in which case $p(\boldsymbol{x}|\boldsymbol{\omega}) = \prod_{j=1}^{n} p(\boldsymbol{y}_j|\boldsymbol{\omega})$ and $\boldsymbol{\mathcal{X}} = \boldsymbol{\mathcal{Y}}^n$. In this case, reference priors relative to model $\mathcal{M}$ turn out to be the same as those relative to the simpler model $\mathcal{M}_y \equiv \{p(\boldsymbol{y}|\boldsymbol{\omega}), \boldsymbol{y} \in \boldsymbol{\mathcal{Y}}, \boldsymbol{\omega} \in \boldsymbol{\Omega}\}$.

Let $\boldsymbol{\theta} = \boldsymbol{\theta}(\boldsymbol{\omega}) \in \boldsymbol{\Theta}$ be some vector of interest; without loss of generality, the assumed model $\mathcal{M}$ may be reparametrized in the form

$$\mathcal{M} \equiv \big\{p(\boldsymbol{x}|\boldsymbol{\theta}, \boldsymbol{\lambda}), \boldsymbol{x} \in \boldsymbol{\mathcal{X}}, \boldsymbol{\theta} \in \boldsymbol{\Theta}, \boldsymbol{\lambda} \in \boldsymbol{\Lambda}\big\}, \tag{1}$$

where $\boldsymbol{\lambda}$ is some vector of nuisance parameters; this is often simply referred to as "model" $p(\boldsymbol{x}|\boldsymbol{\theta}, \boldsymbol{\lambda})$. Conditional on the assumed model, all valid Bayesian inferential statements about the value of $\boldsymbol{\theta}$ are encapsulated in its posterior distribution $p(\boldsymbol{\theta}|\boldsymbol{x}) \propto \int_{\boldsymbol{\Lambda}} p(\boldsymbol{x}|\boldsymbol{\theta}, \boldsymbol{\lambda}) p(\boldsymbol{\theta}, \boldsymbol{\lambda})\, d\boldsymbol{\lambda}$, which *combines* the information provided by the data $\boldsymbol{x}$ with any other information about $\boldsymbol{\theta}$ contained in the prior density $p(\boldsymbol{\theta}, \boldsymbol{\lambda})$. Intuitively, the *reference prior function* for $\boldsymbol{\theta}$, given model $\mathcal{M}$ and a class of candidate priors $\mathcal{P}$, is that (joint) prior $\pi^{\boldsymbol{\theta}}(\boldsymbol{\theta}, \boldsymbol{\lambda}|\mathcal{M}, \mathcal{P})$ which may be expected to have a minimal effect on the posterior inference about the quantity of interest $\boldsymbol{\theta}$ among the class of priors which belong to $\mathcal{P}$, *relative* to data which could be obtained from $\mathcal{M}$. The reference prior $\pi^{\boldsymbol{\theta}}(\boldsymbol{\omega}|\mathcal{M}, \mathcal{P})$ is specifically designed to be a reasonable *consensus* prior (within the class $\mathcal{P}$ of priors compatible with assumed prior knowledge) for inferences about a *particular quantity of interest* $\boldsymbol{\theta} = \boldsymbol{\theta}(\boldsymbol{\omega})$, and it is always conditional to the *specific experimental design* $\mathcal{M} \equiv \{p(\boldsymbol{x}|\boldsymbol{\omega}), \boldsymbol{x} \in \boldsymbol{\mathcal{X}}, \boldsymbol{\omega} \in \boldsymbol{\Omega}\}$ which is assumed to have generated the data.

By definition, the reference prior $\pi^{\boldsymbol{\theta}}(\boldsymbol{\theta}, \boldsymbol{\lambda}|\mathcal{M}, \mathcal{P})$ is "objective", in the sense that it is a well-defined mathematical function of the vector of interest $\boldsymbol{\theta}$, the assumed model $\mathcal{M}$, and the class $\mathcal{P}$ of candidate priors, with no additional subjective elements. By formal use of Bayes theorem and appropriate integration (provided the integral is finite), the (joint) reference prior produces a (marginal) *reference posterior* for the vector of interest

$$\pi(\boldsymbol{\theta}|\boldsymbol{x}, \mathcal{M}, \mathcal{P}) \propto \int_{\boldsymbol{\Lambda}} p(\boldsymbol{x}|\boldsymbol{\theta}, \boldsymbol{\lambda}) \pi^{\boldsymbol{\theta}}(\boldsymbol{\theta}, \boldsymbol{\lambda}|\mathcal{M}, \mathcal{P})\, d\boldsymbol{\lambda}, \tag{2}$$

which could be described as a mathematical expression of the inferential content of data $\boldsymbol{x}$ with respect to the value of $\boldsymbol{\theta}$, with no additional knowledge beyond that contained in the assumed statistical model $\mathcal{M}$ and the class $\mathcal{P}$ of candidate priors (which may well consist of the class $\mathcal{P}_0$ of *all* suitably regular priors). To simplify the exposition, the dependence of the reference prior on both the model and the class of candidate priors is frequently dropped from the notation, so that $\pi^{\boldsymbol{\theta}}(\boldsymbol{\theta}, \boldsymbol{\lambda})$ and $\pi(\boldsymbol{\theta}|\boldsymbol{x})$ are written instead of $\pi^{\boldsymbol{\theta}}(\boldsymbol{\theta}, \boldsymbol{\lambda}|\mathcal{M}, \mathcal{P})$ and $\pi(\boldsymbol{\theta}|\boldsymbol{x}, \mathcal{M}, \mathcal{P})$.

The reference prior function $\pi^{\boldsymbol{\theta}}(\boldsymbol{\theta}, \boldsymbol{\lambda})$ often turns out to be an *improper* prior, i.e., a positive function such that $\int_{\boldsymbol{\Theta}}\int_{\boldsymbol{\Lambda}} \pi^{\boldsymbol{\theta}}(\boldsymbol{\theta}, \boldsymbol{\lambda})\, d\boldsymbol{\theta}\, d\boldsymbol{\lambda}$ diverges and, hence, cannot be renormalized into a proper density function. Notice that this is not a problem provided the resulting posterior distribution (2) is proper for all suitable data. Indeed the declared objective of reference analysis is to provide appropriate reference *posterior* distributions;

reference prior *functions* are merely useful technical devices for a simple computation (via formal use of Bayes theorem) of reference posterior *distributions*. For discussions on the axiomatic foundations which justify the use of improper prior functions, see Hartigan (1983) and references therein.

In the long quest for objective posterior distributions, several requirements have emerged which may reasonably be requested as *necessary* properties of any proposed solution:

(1) *Generality*. The procedure should be completely general, i.e., applicable to any properly defined inference problem, and should produce no untenable answers which could be used as counterexamples. In particular, an objective posterior $\pi(\boldsymbol{\theta}|\boldsymbol{x})$ must be a *proper* probability distribution for any data set $\boldsymbol{x}$ large enough to identify the unknown parameters.

(2) *Invariance*. Jeffreys (1946), Hartigan (1964), Jaynes (1968), Box and Tiao (1973, Section 1.3), Villegas (1977, 1990), Dawid (1983), Yang (1995), Datta and J.K. Ghosh (1995b), Datta and M. Ghosh (1996). For any one-to-one function $\boldsymbol{\phi} = \boldsymbol{\phi}(\boldsymbol{\theta})$, the posterior $\pi(\boldsymbol{\phi}|\boldsymbol{x})$ obtained from the reparametrized model $p(\boldsymbol{x}|\boldsymbol{\phi}, \boldsymbol{\lambda})$ must be coherent with the posterior $\pi(\boldsymbol{\theta}|\boldsymbol{x})$ obtained from the original model $p(\boldsymbol{x}|\boldsymbol{\theta}, \boldsymbol{\lambda})$ in the sense that, for any data set $\boldsymbol{x} \in \boldsymbol{\mathcal{X}}$, $\pi(\boldsymbol{\phi}|\boldsymbol{x}) = \pi(\boldsymbol{\theta}|\boldsymbol{x})|\mathrm{d}\boldsymbol{\theta}/\mathrm{d}\boldsymbol{\phi}|$. Moreover, if the model has a sufficient statistic $\boldsymbol{t} = \boldsymbol{t}(\boldsymbol{x})$, then the posterior $\pi(\boldsymbol{\theta}|\boldsymbol{x})$ obtained from the full model $p(\boldsymbol{x}|\boldsymbol{\theta}, \boldsymbol{\lambda})$ must be the same as the posterior $\pi(\boldsymbol{\theta}|\boldsymbol{t})$ obtained from the equivalent model $p(\boldsymbol{t}|\boldsymbol{\theta}, \boldsymbol{\lambda})$.

(3) *Consistent marginalization*. Stone and Dawid (1972), Dawid et al. (1973), Dawid (1980). If, for all data $\boldsymbol{x}$, the posterior $\pi_1(\boldsymbol{\theta}|\boldsymbol{x})$ obtained from model $p(\boldsymbol{x}|\boldsymbol{\theta}, \boldsymbol{\lambda})$ is of the form $\pi_1(\boldsymbol{\theta}|\boldsymbol{x}) = \pi_1(\boldsymbol{\theta}|\boldsymbol{t})$ for some statistic $\boldsymbol{t} = \boldsymbol{t}(\boldsymbol{x})$ whose sampling distribution $p(\boldsymbol{t}|\boldsymbol{\theta}, \boldsymbol{\lambda}) = p(\boldsymbol{t}|\boldsymbol{\theta})$ only depends on $\boldsymbol{\theta}$, then the posterior $\pi_2(\boldsymbol{\theta}|\boldsymbol{t})$ obtained from the marginal model $p(\boldsymbol{t}|\boldsymbol{\theta})$ must be the same as the posterior $\pi_1(\boldsymbol{\theta}|\boldsymbol{t})$ obtained from the original full model.

(4) *Consistent sampling properties*. Neyman and Scott (1948), Stein (1959), Dawid and Stone (1972, 1973), Cox and Hinkley (1974, Section 2.4.3), Stone (1976), Lane and Sudderth (1984). The properties under repeated sampling of the posterior distribution must be consistent with the model. In particular, the family of posterior distributions $\{\pi(\boldsymbol{\theta}|\boldsymbol{x}_j), \boldsymbol{x}_j \in \boldsymbol{\mathcal{X}}\}$ which could be obtained by repeated sampling from $p(\boldsymbol{x}_j|\boldsymbol{\theta}, \boldsymbol{\omega})$ should concentrate on a region of $\boldsymbol{\Theta}$ which contains the true value of $\boldsymbol{\theta}$.

Reference analysis, introduced by Bernardo (1979b) and further developed by Berger and Bernardo (1989, 1992a, 1992b, 1992c), appears to be the only available method to derive objective posterior distributions which satisfy all these desiderata. This chapter describes the basic elements of reference analysis, states its main properties, and provides signposts to the huge related literature.

Section 2 summarizes some necessary concepts of discrepancy and convergence, which are based on information theory. Section 3 provides a formal definition of reference distributions, and describes their main properties. Section 4 describes an integrated approach to point estimation, region estimation, and hypothesis testing, which is derived from the joint use of reference analysis and an information-theory based loss function, the *intrinsic discrepancy*. Section 5 provides many additional references for further reading on reference analysis and related topics.

2. Intrinsic discrepancy and expected information

Intuitively, a reference prior for $\boldsymbol{\theta}$ is one which maximizes what it is *not known* about $\boldsymbol{\theta}$, *relative* to what *could* possibly be learnt from repeated observations from a particular model. More formally, a reference prior for $\boldsymbol{\theta}$ is defined to be one which maximizes – within some class of candidate priors – the *missing information* about the quantity of interest $\boldsymbol{\theta}$, defined as a limiting form of the amount of information about its value which repeated data from the assumed model could possibly provide. In this section, the notions of discrepancy, convergence, and expected information – which are required to make these ideas precise – are introduced and illustrated.

Probability theory makes frequent use of *divergence measures* between probability distributions. The total variation distance, Hellinger distance, Kullback–Leibler logarithmic divergence, and Jeffreys logarithmic divergence are frequently cited; see, for example, Kullback (1968, 1983, 1987), Ibragimov and Khasminskii (1973), and Gutiérrez-Peña (1992) for precise definitions and properties. Each of those divergence measures may be used to define a type of convergence. It has been found, however, that the behaviour of many important limiting processes, in both probability theory and statistical inference, is better described in terms of another information-theory related divergence measure, the *intrinsic discrepancy* (Bernardo and Rueda, 2002), which is now defined and illustrated.

DEFINITION 1 (*Intrinsic discrepancy*). The *intrinsic discrepancy* $\delta\{p_1, p_2\}$ between two probability distributions of a random vector $\boldsymbol{x} \in \mathcal{X}$, specified by their density functions $p_1(\boldsymbol{x})$, $\boldsymbol{x} \in \mathcal{X}_1 \subset \mathcal{X}$, and $p_2(\boldsymbol{x})$, $\boldsymbol{x} \in \mathcal{X}_2 \subset \mathcal{X}$, with either identical or nested supports, is

$$\delta\{p_1, p_2\} = \min\left\{\int_{\mathcal{X}_1} p_1(\boldsymbol{x}) \log \frac{p_1(\boldsymbol{x})}{p_2(\boldsymbol{x})}\, \mathrm{d}\boldsymbol{x},\ \int_{\mathcal{X}_2} p_2(\boldsymbol{x}) \log \frac{p_2(\boldsymbol{x})}{p_1(\boldsymbol{x})}\, \mathrm{d}\boldsymbol{x}\right\}, \tag{3}$$

provided one of the integrals (or sums) is finite. The intrinsic discrepancy between two parametric models for $\boldsymbol{x} \in \mathcal{X}$, $\mathcal{M}_1 \equiv \{p_1(\boldsymbol{x}|\boldsymbol{\omega}), \boldsymbol{x} \in \mathcal{X}_1, \boldsymbol{\omega} \in \boldsymbol{\Omega}\}$ and $\mathcal{M}_2 \equiv \{p_2(\boldsymbol{x}|\boldsymbol{\psi}), \boldsymbol{x} \in \mathcal{X}_2, \boldsymbol{\psi} \in \boldsymbol{\Psi}\}$, is the minimum intrinsic discrepancy between their elements,

$$\delta\{\mathcal{M}_1, \mathcal{M}_2\} = \inf_{\boldsymbol{\omega}\in\boldsymbol{\Omega}, \boldsymbol{\psi}\in\boldsymbol{\Psi}} \delta\{p_1(\boldsymbol{x}|\boldsymbol{\omega}), p_2(\boldsymbol{x}|\boldsymbol{\psi})\}. \tag{4}$$

The *intrinsic discrepancy* is a new element of the class of *intrinsic loss functions* defined by Robert (1996); the concept is *not* related to the concepts of "intrinsic Bayes factors" and "intrinsic priors" introduced by Berger and Pericchi (1996), and reviewed in Pericchi (2005).

Notice that, as one would require, the intrinsic discrepancy $\delta\{\mathcal{M}_1, \mathcal{M}_2\}$ between two parametric families of distributions $\mathcal{M}_1$ and $\mathcal{M}_2$ does not depend on the particular parametrizations used to describe them. This will be crucial to guarantee the desired invariance properties of the statistical procedures described later.

It follows from Definition 1 that the intrinsic discrepancy between two probability distributions may be written in terms of their two possible Kullback–Leibler *directed divergences* as

$$\delta\{p_2, p_1\} = \min\bigl\{\kappa\{p_2 | p_1\}, \kappa\{p_1 | p_2\}\bigr\}, \tag{5}$$

where (Kullback and Leibler, 1951) the $\kappa\{p_j | p_i\}$'s are the nonnegative invariant quantities defined by

$$\kappa\{p_j | p_i\} = \int_{\mathcal{X}_i} p_i(\boldsymbol{x}) \log \frac{p_i(\boldsymbol{x})}{p_j(\boldsymbol{x})} \, \mathrm{d}\boldsymbol{x}, \quad \text{with } \mathcal{X}_i \subseteq \mathcal{X}_j. \tag{6}$$

Since $\kappa\{p_j | p_i\}$ is the expected value of the logarithm of the density (or probability) ratio for p_i against p_j, when p_i is true, it also follows from Definition 1 that, if $\mathcal{M}_1$ and $\mathcal{M}_2$ describe two alternative models, one of which is assumed to generate the data, their intrinsic discrepancy $\delta\{\mathcal{M}_1, \mathcal{M}_2\}$ is the *minimum expected log-likelihood ratio in favour of the model which generates the data* (the "true" model). This will be important in the interpretation of many of the results described in this chapter.

The intrinsic discrepancy is obviously *symmetric*. It is nonnegative, vanishes if (and only if) $p_1(\boldsymbol{x}) = p_2(\boldsymbol{x})$ almost everywhere, and it is invariant under one-to-one transformations of $\boldsymbol{x}$. Moreover, if $p_1(\boldsymbol{x})$ and $p_2(\boldsymbol{x})$ have strictly nested supports, one of the two directed divergences will not be finite, but their intrinsic discrepancy is still defined, and reduces to the other directed divergence. Thus, if $\mathcal{X}_i \subset \mathcal{X}_j$, then $\delta\{p_i, p_j\} = \delta\{p_j, p_i\} = \kappa\{p_j | p_i\}$.

The intrinsic discrepancy is *information additive*. Thus, if $\boldsymbol{x}$ consists of n independent observations, so that $\boldsymbol{x} = \{\boldsymbol{y}_1, \ldots, \boldsymbol{y}_n\}$ and $p_i(\boldsymbol{x}) = \prod_{j=1}^{n} q_i(\boldsymbol{y}_j)$, then $\delta\{p_1, p_2\} = n\delta\{q_1, q_2\}$. This statistically important additive property is essentially unique to logarithmic discrepancies; it is basically a consequence of the fact that the joint density of independent random quantities is the product of their marginals, and the logarithm is the only analytic function which transforms products into sums.

EXAMPLE 1 (*Intrinsic discrepancy between binomial distributions*). The intrinsic discrepancy $\delta\{\theta_1, \theta_2 | n\}$ between the two binomial distributions with common value for n, $p_1(r) = \mathrm{Bi}(r | n, \theta_1)$ and $p_2(r) = \mathrm{Bi}(r | n, \theta_2)$, is

$$\begin{aligned}
&\delta\{p_1, p_2\} = \delta\{\theta_1, \theta_2 | n\} = n\delta_1\{\theta_1, \theta_2\}, \\
&\delta_1\{\theta_1, \theta_2\} = \min\bigl[\kappa\{\theta_1 | \theta_2\}, \kappa\{\theta_2 | \theta_1\}\bigr], \\
&\kappa(\theta_i | \theta_j) = \theta_j \log[\theta_j / \theta_i] + (1 - \theta_j) \log\bigl[(1 - \theta_j)/(1 - \theta_i)\bigr],
\end{aligned} \tag{7}$$

where $\delta_1\{\theta_1, \theta_2\}$ (represented in the left panel of Figure 1) is the intrinsic discrepancy $\delta\{q_1, q_2\}$ between the corresponding Bernoulli distributions, $q_i(y) = \theta_i^y (1 - \theta_i)^{1-y}$, $y \in \{0, 1\}$. It may be appreciated that, specially near the extremes, the behaviour of

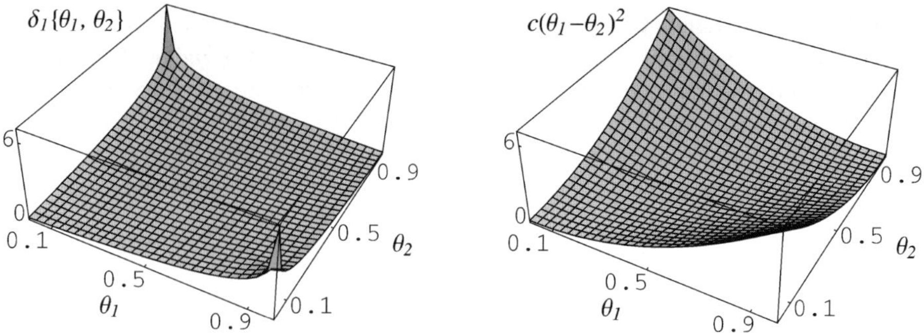

Fig. 1. Intrinsic discrepancy between Bernoulli variables.

the intrinsic discrepancy is rather different from that of the conventional quadratic loss $c(\theta_1 - \theta_2)^2$ (represented in the right panel of Figure 1 with c chosen to preserve the vertical scale).

As a direct consequence of the information-theoretical interpretation of the Kullback–Leibler directed divergences (Kullback, 1968, Chapter 1), the intrinsic discrepancy $\delta\{p_1, p_2\}$ is a measure, in natural information units or *nits* (Boulton and Wallace, 1970), of the *minimum* amount of expected information, in Shannon (1948) sense, required to discriminate between p_1 and p_2. If base 2 logarithms were used instead of natural logarithms, the intrinsic discrepancy would be measured in binary units of information (*bits*).

The quadratic loss $\ell\{\theta_1, \theta_2\} = (\theta_1 - \theta_2)^2$, often (over)used in statistical inference as measure of the discrepancy between two distributions $p(\boldsymbol{x}\,|\,\theta_1)$ and $p(\boldsymbol{x}\,|\,\theta_2)$ of the same parametric family $\{p(\boldsymbol{x}\,|\,\theta), \theta \in \Theta\}$, heavily depends on the parametrization chosen. As a consequence, the corresponding point estimate, the posterior expectation is not coherent under one-to-one transformations of the parameter. For instance, under quadratic loss, the "best" estimate of the logarithm of some positive physical magnitude is *not* the logarithm of the "best" estimate of such magnitude, a situation hardly acceptable by the scientific community. In sharp contrast to conventional loss functions, the intrinsic discrepancy is invariant under one-to-one reparametrizations. Some important consequences of this fact are summarized below.

Let $\mathcal{M} \equiv \{p(\boldsymbol{x}\,|\,\boldsymbol{\theta}), \boldsymbol{x} \in \boldsymbol{\mathcal{X}}, \boldsymbol{\theta} \in \boldsymbol{\Theta}\}$ be a family of probability densities, with no nuisance parameters, and let $\tilde{\boldsymbol{\theta}} \in \boldsymbol{\Theta}$ be a possible point estimate of the quantity of interest $\boldsymbol{\theta}$. The intrinsic discrepancy $\delta\{\tilde{\boldsymbol{\theta}}, \boldsymbol{\theta}\} = \delta\{p_{\boldsymbol{x}|\tilde{\boldsymbol{\theta}}}, p_{\boldsymbol{x}|\boldsymbol{\theta}}\}$ between the estimated model and the true model measures, as a function of $\boldsymbol{\theta}$, the loss which would be suffered if model $p(\boldsymbol{x}\,|\,\tilde{\boldsymbol{\theta}})$ were used as a proxy for model $p(\boldsymbol{x}\,|\,\boldsymbol{\theta})$. Notice that this directly measures how different the two *models* are, as opposed to measuring how different their *labels* are, which is what conventional loss functions – like the quadratic loss – typically do. As a consequence, the resulting discrepancy measure is independent of the particular parametrization used; indeed, $\delta\{\tilde{\boldsymbol{\theta}}, \boldsymbol{\theta}\}$ provides a natural, *invariant* loss function for estimation, the *intrinsic loss*. The *intrinsic estimate* is that value $\boldsymbol{\theta}^*$ which minimizes $d(\tilde{\boldsymbol{\theta}}\,|\,\boldsymbol{x}) = \int_{\boldsymbol{\Theta}} \delta\{\tilde{\boldsymbol{\theta}}, \boldsymbol{\theta}\}\, p(\boldsymbol{\theta}\,|\,\boldsymbol{x})\, d\boldsymbol{\theta}$, the posterior expected intrinsic loss, among all $\tilde{\boldsymbol{\theta}} \in \boldsymbol{\Theta}$.

Since $\delta\{\tilde{\boldsymbol{\theta}}, \boldsymbol{\theta}\}$ is invariant under reparametrization, the intrinsic estimate of any one-to-one transformation of $\boldsymbol{\theta}$, $\boldsymbol{\phi} = \boldsymbol{\phi}(\boldsymbol{\theta})$, is simply $\boldsymbol{\phi}^* = \boldsymbol{\phi}(\boldsymbol{\theta}^*)$ (Bernardo and Juárez, 2003).

The posterior expected loss function $d(\tilde{\boldsymbol{\theta}}|\boldsymbol{x})$ may further be used to define posterior *intrinsic p-credible regions* $R_p = \{\tilde{\boldsymbol{\theta}}; d(\tilde{\boldsymbol{\theta}}|\boldsymbol{x}) < d_p^*\}$, where d_p^* is chosen such that $\Pr[\boldsymbol{\theta} \in R_p|\boldsymbol{x}] = p$. In contrast to conventional highest posterior density (HPD) credible regions, which do *not* remain HPD under one-to-one transformations of $\boldsymbol{\theta}$, these *lowest posterior loss* (LPL) credible regions *remain* LPL under those transformations.

Similarly, if $\boldsymbol{\theta}_0$ is a parameter value of special interest, the intrinsic discrepancy $\delta\{\boldsymbol{\theta}_0, \boldsymbol{\theta}\} = \delta\{p_{\boldsymbol{x}|\boldsymbol{\theta}_0}, p_{\boldsymbol{x}|\boldsymbol{\theta}}\}$ provides, as a function of $\boldsymbol{\theta}$, a measure of how far the particular density $p(\boldsymbol{x}|\boldsymbol{\theta}_0)$ (often referred to as the *null* model) is from the assumed model $p(\boldsymbol{x}|\boldsymbol{\theta})$, suggesting a natural invariant loss function for precise hypothesis testing. The null model $p(\boldsymbol{x}|\boldsymbol{\theta}_0)$ will be rejected if the corresponding posterior expected loss (called the *intrinsic statistic*) $d(\boldsymbol{\theta}_0|\boldsymbol{x}) = \int_{\Theta} \delta\{\boldsymbol{\theta}_0, \boldsymbol{\theta}\} p(\boldsymbol{\theta}|\boldsymbol{x})\, d\boldsymbol{\theta}$, is too large. As one should surely require, for any one-to-one transformation $\boldsymbol{\phi} = \boldsymbol{\phi}(\boldsymbol{\theta})$, testing whether of not data are compatible with $\boldsymbol{\theta} = \boldsymbol{\theta}_0$ yields precisely the same result as testing $\boldsymbol{\phi} = \boldsymbol{\phi}_0 = \boldsymbol{\phi}(\boldsymbol{\theta}_0)$ (Bernardo and Rueda, 2002).

These ideas, extended to include the possible presence of nuisance parameters, will be further analyzed in Section 4.

DEFINITION 2 (*Intrinsic convergence*). A sequence of probability distributions specified by their density functions $\{p_i(\boldsymbol{x})\}_{i=1}^{\infty}$ is said to *converge intrinsically* to a probability distribution with density $p(\boldsymbol{x})$ whenever the sequence of their intrinsic discrepancies $\{\delta(p_i, p)\}_{i=1}^{\infty}$ converges to zero.

EXAMPLE 2 (*Poisson approximation to a Binomial distribution*). The intrinsic discrepancy between a Binomial distribution with probability function $\mathrm{Bi}(r|n, \theta)$ and its Poisson approximation $\mathrm{Po}(r|n\theta)$, is

$$\delta\{\mathrm{Bi}, \mathrm{Po}|n, \theta\} = \sum_{r=0}^{n} \mathrm{Bi}(r|n, \theta) \log \frac{\mathrm{Bi}(r|n, \theta)}{\mathrm{Po}(r|n\theta)},$$

since the second sum in Definition 1 diverges. It may easily be verified that $\lim_{n\to\infty} \delta\{\mathrm{Bi}, \mathrm{Po}|n, \lambda/n\} = 0$ and $\lim_{\theta\to 0} \delta\{\mathrm{Bi}, \mathrm{Po}|\lambda/\theta, \theta\} = 0$; thus, as one would expect from standard probability theory, the sequences of Binomials $\mathrm{Bi}(r|n, \lambda/n)$ and $\mathrm{Bi}(r|\lambda/\theta_i, \theta_i)$ both intrinsically converge to a Poisson $\mathrm{Po}(r|\lambda)$ when $n \to \infty$ and $\theta_i \to 0$, respectively.

However, if one is interest in approximating a Binomial $\mathrm{Bi}(r|n, \theta)$ by a Poisson $\mathrm{Po}(r|n\theta)$ the rôles of n and θ are far from similar: the important condition for the Poisson approximation to the Binomial to work is that the value of θ must be small, while the value of n is largely irrelevant. Indeed (see Figure 2), $\lim_{\theta\to 0} \delta\{\mathrm{Bi}, \mathrm{Po}|n, \theta\} = 0$, for all $n > 0$, but $\lim_{n\to\infty} \delta\{\mathrm{Bi}, \mathrm{Po}|n, \theta\} = \frac{1}{2}[-\theta - \log(1-\theta)]$ for all $\theta > 0$. Thus, arbitrarily good approximations are possible with any n, provided θ is sufficiently small. However, for fixed θ, the quality of the approximation cannot improve over a certain limit, no matter how large n might be. For example, $\delta\{\mathrm{Bi}, \mathrm{Po}|3, 0.05\} = 0.00074$

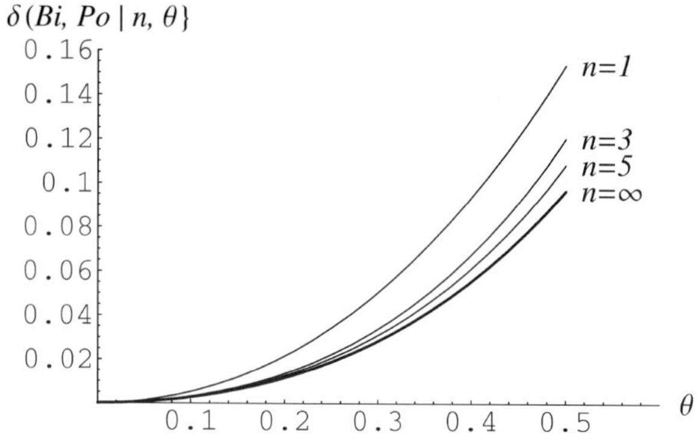

Fig. 2. Intrinsic discrepancy $\delta\{\text{Bi}, \text{Po}|n, \theta\}$ between a Binomial $\text{Bi}(r|n, \theta)$ and a Poisson $\text{Po}(r|n\theta)$ as a function of θ, for $n = 1, 3, 5$ and ∞.

and $\delta\{\text{Bi}, \text{Po}|5000, 0.05\} = 0.00065$, both yielding an expected log-probability ratio of about 0.0007. Thus, for all $n \geqslant 3$ the Binomial distribution $\text{Bi}(r|n, 0.05)$ is quite well approximated by the Poisson distribution $\text{Po}(r|0.05n)$, and the quality of the approximation is very much the same for any value n.

Many standard approximations in probability theory may benefit from an analysis similar to that of Example 2. For instance, the sequence of Student distributions $\{\text{St}(x|\mu, \sigma, \nu)\}_{\nu=1}^{\infty}$ converges intrinsically to the normal distribution $\text{N}(x|\mu, \sigma)$ with the same location and scale parameters, and the discrepancy $\delta(\nu) = \delta\{\text{St}(x|\mu, \sigma, \nu), \text{N}(x|\mu, \sigma)\}$ (which only depends on the degrees of freedom ν) is smaller than 0.001 when $\nu > 40$. Thus approximating a Student with more than 40 degrees of freedom by a normal yields an expected log-density ratio smaller than 0.001, suggesting quite a good approximation.

As mentioned before, a reference prior is often an improper prior function. Justification of its use as a *formal* prior in Bayes theorem to obtain a reference posterior necessitates proving that the reference posterior thus obtained is an appropriate limit of a sequence of posteriors obtained from proper priors.

THEOREM 1. *Consider a model* $\mathcal{M} \equiv \{p(\boldsymbol{x}|\boldsymbol{\omega}), \boldsymbol{x} \in \boldsymbol{\mathcal{X}}, \boldsymbol{\omega} \in \boldsymbol{\Omega}\}$. *If* $\pi(\boldsymbol{\omega})$ *is a strictly positive improper prior,* $\{\boldsymbol{\Omega}_i\}_{i=1}^{\infty}$ *is an increasing sequence of subsets of the parameter space which converges to* $\boldsymbol{\Omega}$ *and such that* $\int_{\boldsymbol{\Omega}_i} \pi(\boldsymbol{\omega})\, d\boldsymbol{\omega} < \infty$, *and* $\pi_i(\boldsymbol{\omega})$ *is the renormalized proper density obtained by restricting* $\pi(\boldsymbol{\omega})$ *to* $\boldsymbol{\Omega}_i$, *then, for any data set* $\boldsymbol{x} \in \boldsymbol{\mathcal{X}}$, *the sequence of the corresponding posteriors* $\{\pi_i(\boldsymbol{\omega}|\boldsymbol{x})\}_{i=1}^{\infty}$ *converges intrinsically to the posterior* $\pi(\boldsymbol{\omega}|\boldsymbol{x}) \propto p(\boldsymbol{x}|\boldsymbol{\omega})\pi(\boldsymbol{\omega})$ *obtained by formal use of Bayes theorem with the improper prior* $\pi(\boldsymbol{\omega})$.

However, to avoid possible pathologies, a stronger form of convergence is needed; for a sequence of proper priors $\{\pi_i\}_{i=1}^{\infty}$ to converge to a (possibly improper) prior function π, it will further be required that the *predicted* intrinsic discrepancy between the corresponding posteriors converges to zero. For a motivating example, see Berger and

Bernardo (1992c, p. 43), where the model

$$\left\{p(x|\theta) = \frac{1}{3},\ x \in \left\{\left[\frac{\theta}{2}\right],\ 2\theta, 2\theta + 1\right\},\ \theta \in \{1, 2, \ldots\}\right\},$$

where $[u]$ denotes the integer part of u (and $[\frac{1}{2}]$ is separately defined as 1), originally proposed by Fraser et al. (1985), is reanalyzed.

DEFINITION 3 (*Permissible prior function*). A positive function $\pi(\boldsymbol{\omega})$ is an *permissible* prior function for model $\mathcal{M} \equiv \{p(\boldsymbol{x}|\boldsymbol{\omega}), \boldsymbol{x} \in \boldsymbol{\mathcal{X}}, \boldsymbol{\omega} \in \boldsymbol{\Omega}\}$ if for all $\boldsymbol{x} \in \boldsymbol{\mathcal{X}}$ one has $\int_{\Omega} p(\boldsymbol{x}|\boldsymbol{\omega})\pi(\boldsymbol{\omega})\,\mathrm{d}\boldsymbol{\omega} < \infty$, and for some increasing sequence $\{\boldsymbol{\Omega}_i\}_{i=1}^{\infty}$ of subsets of $\boldsymbol{\Omega}$, such that $\lim_{i\to\infty} \boldsymbol{\Omega}_i = \boldsymbol{\Omega}$, and $\int_{\Omega_i} \pi(\boldsymbol{\omega})\,\mathrm{d}\boldsymbol{\omega} < \infty$,

$$\lim_{i\to\infty} \int_{\boldsymbol{\mathcal{X}}} p_i(\boldsymbol{x})\delta\{\pi_i(\boldsymbol{\omega}|\boldsymbol{x}), \pi(\boldsymbol{\omega}|\boldsymbol{x})\}\,\mathrm{d}\boldsymbol{x} = 0,$$

where $\pi_i(\boldsymbol{\omega})$ is the renormalized restriction of $\pi(\boldsymbol{\omega})$ to $\boldsymbol{\Omega}_i$, $\pi_i(\boldsymbol{\omega}|\boldsymbol{x})$ is the corresponding posterior, $p_i(\boldsymbol{x}) = \int_{\Omega_i} p(\boldsymbol{x}|\boldsymbol{\omega})\pi_i(\boldsymbol{\omega})\,\mathrm{d}\boldsymbol{\omega}$ is the corresponding predictive, and $\pi(\boldsymbol{\omega}|\boldsymbol{x}) \propto p(\boldsymbol{x}|\boldsymbol{\omega})\pi(\boldsymbol{\omega})$.

In words, $\pi(\boldsymbol{\omega})$ is a permissible prior function for model $\mathcal{M}$ if it always yields proper posteriors, and the sequence of the *predicted* intrinsic discrepancies between the corresponding posterior $\pi(\boldsymbol{\omega}|\boldsymbol{x})$ and its renormalized restrictions to $\boldsymbol{\Omega}_i$ converges to zero for some suitable approximating sequence of the parameter space. All proper priors are permissible in the sense of Definition 3, but improper priors may or may not be permissible, even if they seem to be arbitrarily close to proper priors.

EXAMPLE 3 (*Exponential model*). Let $\boldsymbol{x} = \{x_1, \ldots, x_n\}$ be a random sample from $p(x|\theta) = \theta\,\mathrm{e}^{-\theta x}$, $\theta > 0$, so that $p(\boldsymbol{x}|\theta) = \theta^n\,\mathrm{e}^{-\theta t}$, with sufficient statistic $t = \sum_{j=1}^{n} x_j$. Consider a positive function $\pi(\theta) \propto \theta^{-1}$, so that $\pi(\theta|t) \propto \theta^{n-1}\,\mathrm{e}^{-\theta t}$, a gamma density $\mathrm{Ga}(\theta|n, t)$, which is a proper distribution for all possible data sets. Take now some sequence of pairs of positive real numbers $\{a_i, b_i\}$, with $a_i < b_i$, and let $\Theta_i = (a_i, b_i)$; the intrinsic discrepancy between $\pi(\theta|t)$ and its renormalized restriction to Θ_i, denoted $\pi_i(\theta|t)$, is $\delta_i(n, t) = \kappa\{\pi(\theta|t)|\pi_i(\theta|t)\} = \log[c_i(n, t)]$, where $c_i(n, t) = \Gamma(n)/\{\Gamma(n, a_i t) - \Gamma(n, b_i, t)\}$. The renormalized restriction of $\pi(\theta)$ to Θ_i is $\pi_i(\theta) = \theta^{-1}/\log[b_i/a_i]$, and the corresponding (prior) predictive of t is $p_i(t|n) = c_i^{-1}(n, t)t^{-1}/\log[b_i/a_i]$. It may be verified that, for all $n \geqslant 1$, the expected intrinsic discrepancy $\int_0^{\infty} p_i(t|n)\delta_i(n, t)\,\mathrm{d}t$ converges to zero as $i \to \infty$. Hence, all positive functions of the form $\pi(\theta) \propto \theta^{-1}$ are permissible priors for the parameter of an exponential model.

EXAMPLE 4 (*Mixture model*). Let $\boldsymbol{x} = \{x_1, \ldots, x_n\}$ be a random sample from $\mathcal{M} \equiv \{\frac{1}{2}\mathrm{N}(x|\theta, 1) + \frac{1}{2}\mathrm{N}(x|0, 1), x \in \mathbb{R}, \theta \in \mathbb{R}\}$. It is easily verified that the likelihood function $p(\boldsymbol{x}|\theta) = \prod_{j=1}^{n} p(x_j|\theta)$ is always bounded below by a strictly positive function of $\boldsymbol{x}$. Hence, $\int_{-\infty}^{\infty} p(\boldsymbol{x}|\theta)\,\mathrm{d}\theta = \infty$ for all $\boldsymbol{x}$, and the "natural" objective uniform prior function $\pi(\theta) = 1$ is obviously *not* permissible, although it may be pointwise arbitrarily well approximated by a sequence of proper "flat" priors.

DEFINITION 4 (*Intrinsic association*). The *intrinsic association* α_{xy} between two random vectors $\boldsymbol{x} \in \boldsymbol{\mathcal{X}}$ and $\boldsymbol{y} \in \boldsymbol{\mathcal{Y}}$ with joint density $p(\boldsymbol{x}, \boldsymbol{y})$ and marginals $p(\boldsymbol{x})$ and $p(\boldsymbol{y})$ is the intrinsic discrepancy $\alpha_{xy} = \delta\{p_{xy}, p_x p_y\}$ between their joint density and the product of their marginals. The intrinsic coefficient of association $\rho_{xy}^2 = 1 - \exp\{-2\alpha_{xy}\}$ recales the intrinsic association to [0, 1].

The intrinsic association is a nonnegative invariant measure of association between two random vectors, which vanishes if they are independent, and tends to infinity as $\boldsymbol{y}$ and $\boldsymbol{x}$ approach a functional relationship. If their joint distribution is bivariate normal, then $\alpha_{xy} = -\frac{1}{2}\log(1 - \rho^2)$, and $\rho_{xy}^2 = \rho^2$, the square of their coefficient of correlation ρ.

The concept of intrinsic association extends that of *mutual information*; see, e.g., Cover and Thomas (1991), and references therein. Important differences arise in the context of contingency tables, where both $\boldsymbol{x}$ and $\boldsymbol{y}$ are discrete random variables which may only take a finite number of different values.

DEFINITION 5 (*Expected intrinsic information*). The *expected intrinsic information* $I\{p_\omega|\mathcal{M}\}$ from one observation of $\mathcal{M} \equiv \{p(\boldsymbol{x}|\boldsymbol{\omega}), \boldsymbol{x} \in \boldsymbol{\mathcal{X}}, \boldsymbol{\omega} \in \boldsymbol{\Omega}\}$ about the value of $\boldsymbol{\omega} \in \boldsymbol{\Omega}$ when the prior density is $p(\boldsymbol{\omega})$, is the intrinsic association $\alpha_{x\omega} = \delta\{p_{x\omega}, p_x p_\omega\}$ between $\boldsymbol{x}$ and $\boldsymbol{\omega}$, where $p(\boldsymbol{x}, \boldsymbol{\omega}) = p(\boldsymbol{x}|\boldsymbol{\omega})p(\boldsymbol{\omega})$, and $p(\boldsymbol{x}) = \int_{\boldsymbol{\Omega}} p(\boldsymbol{x}|\boldsymbol{\omega})p(\boldsymbol{\omega})\,\mathrm{d}\boldsymbol{\omega}$.

For a fixed model $\mathcal{M}$, the expected intrinsic information $I\{p_\omega|\mathcal{M}\}$ is a concave, positive functional of the prior $p(\boldsymbol{\omega})$. Under appropriate regularity conditions, in particular when data consists of a *large* random sample $\boldsymbol{x} = \{\boldsymbol{y}_1, \ldots, \boldsymbol{y}_n\}$ from some model $\{p(\boldsymbol{y}|\boldsymbol{\omega}), \boldsymbol{y} \in \boldsymbol{\mathcal{Y}}, \boldsymbol{\omega} \in \boldsymbol{\Omega}\}$, one has

$$\iint_{\boldsymbol{\mathcal{X}}\times\boldsymbol{\Omega}} \big[p(\boldsymbol{x})p(\boldsymbol{\omega}) + p(\boldsymbol{x}, \boldsymbol{\omega})\big] \log \frac{p(\boldsymbol{x})p(\boldsymbol{\omega})}{p(\boldsymbol{x}, \boldsymbol{\omega})}\,\mathrm{d}\boldsymbol{x}\,\mathrm{d}\boldsymbol{\omega} \geqslant 0 \tag{8}$$

so that $\kappa\{p_x p_\omega|p_{x\omega}\} \leqslant \kappa\{p_{x\omega}|p_x p_\omega\}$. If this is the case,

$$\begin{aligned} I\{p_\omega|\mathcal{M}\} &= \delta\{p_{x\omega}, p_x p_\omega\} = \kappa\{p_x p_\omega|p_{x\omega}\} \\ &= \iint_{\boldsymbol{\mathcal{X}}\times\boldsymbol{\Omega}} p(\boldsymbol{x}, \boldsymbol{\omega}) \log \frac{p(\boldsymbol{x}, \boldsymbol{\omega})}{p(\boldsymbol{x})p(\boldsymbol{\omega})}\,\mathrm{d}\boldsymbol{x}\,\mathrm{d}\boldsymbol{\omega} && (9) \\ &= \int_{\boldsymbol{\Omega}} p(\boldsymbol{\omega}) \int_{\boldsymbol{\mathcal{X}}} p(\boldsymbol{x}|\boldsymbol{\omega}) \log \frac{p(\boldsymbol{\omega}|\boldsymbol{x})}{p(\boldsymbol{\omega})}\,\mathrm{d}\boldsymbol{x}\,\mathrm{d}\boldsymbol{\omega} && (10) \\ &= H[p_\omega] - \int_{\boldsymbol{\mathcal{X}}} p(\boldsymbol{x})H[p_{\omega|x}]\,\mathrm{d}\boldsymbol{x}, && (11) \end{aligned}$$

where $H[p_\omega] = -\int_{\Omega} p(\boldsymbol{\omega}) \log p(\boldsymbol{\omega})\, d\boldsymbol{\omega}$ is the *entropy* of p_ω, and the expected intrinsic information reduces to the Shannon's expected information (Shannon, 1948; Lindley, 1956; Stone, 1959; de Waal and Groenewald, 1989; Clarke and Barron, 1990).

For any fixed model $\mathcal{M}$, the expected intrinsic information $I\{p_\omega|\mathcal{M}\}$ measures, as a functional of the prior p_ω, the amount of information about the value of $\boldsymbol{\omega}$ which one observation $\boldsymbol{x} \in \boldsymbol{\mathcal{X}}$ may be expected to provide. The stronger the prior knowledge described by p_ω, the smaller the information the data may be expected to provide; conversely, weak initial knowledge about $\boldsymbol{\omega}$ will correspond to large expected information from the data. This is the intuitive basis for the definition of a reference prior.

3. Reference distributions

Let $\boldsymbol{x}$ be one observation from model $\mathcal{M} \equiv \{p(\boldsymbol{x}|\boldsymbol{\omega}), \boldsymbol{x} \in \boldsymbol{\mathcal{X}}, \boldsymbol{\omega} \in \boldsymbol{\Omega}\}$, and let $\boldsymbol{\theta} = \boldsymbol{\theta}(\boldsymbol{\omega}) \in \boldsymbol{\Theta}$ be some vector of interest, whose posterior distribution is required. Notice that $\boldsymbol{x}$ represents the *complete* available data; often, but not always, this will consist of a random sample $\boldsymbol{x} = \{\boldsymbol{y}_1, \ldots, \boldsymbol{y}_n\}$ of fixed size n from some simpler model. Let $\mathcal{P}$ be the *class of candidate priors* for $\boldsymbol{\omega}$, defined as those sufficiently regular priors which are compatible with whatever agreed "objective" initial information about the value of $\boldsymbol{\omega}$ one is willing to assume. A permissible prior function $\pi^{\boldsymbol{\theta}}(\boldsymbol{\omega}|\mathcal{M}, \mathcal{P})$ is desired which may be expected to have a minimal effect (in a sense to be made precise) among all priors in $\mathcal{P}$, on the posterior inferences about $\boldsymbol{\theta} = \boldsymbol{\theta}(\boldsymbol{\omega})$ which could be derived given data generated from $\mathcal{M}$. This will be named a *reference prior function* of $\boldsymbol{\omega}$ for the quantity of interest $\boldsymbol{\theta}$, relative to model $\mathcal{M}$ and class $\mathcal{P}$ of candidate priors, and will be denoted by $\pi^{\boldsymbol{\theta}}(\boldsymbol{\omega}|\mathcal{M}, \mathcal{P})$. The reference prior function $\pi^{\boldsymbol{\theta}}(\boldsymbol{\omega}|\mathcal{M}, \mathcal{P})$ will then be used as a formal prior density to derive the required *reference posterior* distribution of the quantity of interest, $\pi(\boldsymbol{\theta}|\boldsymbol{x}, \mathcal{M}, \mathcal{P})$, via Bayes theorem and the required probability operations.

This section contains the definition and basic properties of reference distributions. The ideas are first formalized in one-parameter models, and then extended to multiparameter situations. Special attention is devoted to *restricted* reference distributions, where the class of candidate priors $\mathcal{P}$ consists of those which satisfy some set of assumed conditions. This provides a continuous collection of solutions, ranging from situations with no assumed prior information on the quantity of interest, when $\mathcal{P}$ is the class $\mathcal{P}_0$ of *all* sufficiently regular priors, to situations where accepted prior knowledge is sufficient to specify a unique prior $p_0(\boldsymbol{\omega})$, so that $\pi^{\boldsymbol{\theta}}(\boldsymbol{\omega}|\mathcal{M}, \mathcal{P}) = p_0(\boldsymbol{\theta})$, the situation commonly assumed in Bayesian subjective analysis.

3.1. One parameter models

Let $\theta \in \Theta \subset \mathbb{R}$ be a real-valued quantity of interest, and let available data $\boldsymbol{x}$ consist of one observation from model $\mathcal{M} \equiv \{p(\boldsymbol{x}|\theta), \boldsymbol{x} \in \boldsymbol{\mathcal{X}}, \theta \in \Theta\}$, so that there are no nuisance parameters. A permissible prior function $\pi(\theta) = \pi(\theta|\mathcal{M}, \mathcal{P})$ in a class $\mathcal{P}$ is desired with a minimal expected effect on the posteriors of θ which could be obtained after data $\boldsymbol{x} \in \boldsymbol{\mathcal{X}}$ generated from $\mathcal{M}$ have been observed.

Let $\boldsymbol{x}^{(k)} = \{\boldsymbol{x}_1, \ldots, \boldsymbol{x}_k\}$ consist of k conditionally independent (given θ) observations from $\mathcal{M}$, so that $\boldsymbol{x}^{(k)}$ consists of one observation from the product model

$\mathcal{M}^k = \{\prod_{j=1}^{k} p(\boldsymbol{x}_j|\theta), \boldsymbol{x}_j \in \boldsymbol{\mathcal{X}}, \theta \in \Theta\}$. Let p_θ be a prior distribution for the quantity of interest, and consider the intrinsic information about θ, $I\{p_\theta|\mathcal{M}^k\}$, which could be expected from the vector $\boldsymbol{x}^{(k)} \in \boldsymbol{\mathcal{X}}^k$. For any sufficiently regular prior p_θ, the posterior distribution of θ would concentrate on its true value as k increases and therefore, as $k \to \infty$, the true value of θ would get to be precisely known. Thus, as $k \to \infty$, the functional $I\{p_\theta|\mathcal{M}^k\}$ will approach a precise measure of the amount of *missing information* about θ which corresponds to the prior p_θ. It is natural to define the reference prior as that prior function $\pi^\theta = \pi(\theta|\mathcal{M}, \mathcal{P})$ which *maximizes the missing information* about the value of θ within the class $\mathcal{P}$ of candidate priors.

Under regularity conditions, the expected intrinsic information $I\{p_\theta|\mathcal{M}^k\}$ becomes, for large k, Shannon's expected information and hence, using (11),

$$I\{p_\theta|\mathcal{M}^k\} = H[p_\theta] - \int_{\mathcal{X}^k} p(\boldsymbol{x}^{(k)}) H[p_{\theta|\boldsymbol{x}^{(k)}}]\,\mathrm{d}\boldsymbol{x}^{(k)}, \qquad (12)$$

where $H[p_\theta] = -\int_\Theta p(\theta)\log p(\theta)\,\mathrm{d}\theta$, is the *entropy* of p_θ. It follows that, when the parameter space $\Theta = \{\theta_1, \ldots, \theta_m\}$ is finite, the missing information which corresponds to any strictly positive prior p_θ is, *for any model* $\mathcal{M}$,

$$\lim_{k\to\infty} I\{p_\theta|\mathcal{M}^k\} = H[p_\theta] = -\sum_{j=1}^{m} p(\theta_j)\log p(\theta_j), \qquad (13)$$

since, as $k \to \infty$, the discrete posterior probability function $p(\theta|\boldsymbol{x}^{(k)})$ converges to a degenerate distribution with probability one on the true value of θ and zero on all others, and thus, the posterior entropy $H[p_{\theta|\boldsymbol{x}^{(k)}}]$ converges to zero. Hence, in finite parameter spaces, the reference prior for the parameter does not depend on the precise form of the model, and it is precisely that which *maximizes the entropy* within the class $\mathcal{P}$ of candidate priors. This was the solution proposed by Jaynes (1968), and it is often used in mathematical physics. In particular, if the class of candidate priors is the class $\mathcal{P}_0$ of *all* strictly positive probability distributions, the reference prior for θ is a uniform distribution over Θ, the "noninformative" prior suggested by the old insufficient reason argument (Laplace, 1812). For further information on the concept of *maximum entropy*, see Jaynes (1968, 1982, 1985, 1989), Akaike (1977), Csiszár (1985, 1991), Clarke and Barron (1994), Grünwald and Dawid (2004), and references therein.

In the continuous case, however, $I\{p_\theta|\mathcal{M}^k\}$ typically diverges as $k \to \infty$, since an infinite amount of information is required to know exactly the value of a real number. A general definition of the reference prior (which includes the finite case as a particular case), is nevertheless possible as an appropriate limit, when $k \to \infty$, of the sequence of priors maximizing $I\{p_\theta|\mathcal{M}^k\}$ within the class $\mathcal{P}$. Notice that this limiting procedure is *not* some kind of asymptotic approximation, but an essential element of the *concept* of a reference prior. Indeed, the reference prior is defined to maximize the *missing information* about the quantity of interest which *could* be obtained by repeated sampling from $\mathcal{M}$ (not just the information expected from a finite data set), and this is precisely achieved by maximizing the expected information from the arbitrarily large data set which could be obtained by unlimited repeated sampling from the assumed model.

Since $I\{p_\theta|\mathcal{M}^k\}$ is only defined for proper priors, and $I\{p_\theta|\mathcal{M}^k\}$ is not guaranteed to attain its maximum at a proper prior, the formal definition of a reference prior is stated as a limit, as $i \to \infty$, of the sequence of solutions obtained for restrictions $\{\Theta_i\}_{i=1}^{\infty}$ of the parameter space chosen to ensure that the maximum of $I\{p_\theta|\mathcal{M}^k\}$ is actually obtained at a proper prior. The definition below (Berger et al., 2005) generalizes those in Bernardo (1979b) and Berger and Bernardo (1992c), and addresses the problems described in Berger et al. (1989).

DEFINITION 6 (*One-parameter reference priors*). Consider the one-parameter model $\mathcal{M} \equiv \{p(\boldsymbol{x}|\theta), \boldsymbol{x} \in \boldsymbol{\mathcal{X}}, \theta \in \Theta \subset \mathbb{R}\}$, and let $\mathcal{P}$ be a class of candidate priors for θ. The positive function $\pi(\theta) = \pi(\theta|\mathcal{M}, \mathcal{P})$ is a *reference prior* for model $\mathcal{M}$ given $\mathcal{P}$ if it is a permissible prior function such that, for some increasing sequence $\{\Theta_i\}_{i=1}^{\infty}$ with $\lim_{i\to\infty} \Theta_i = \Theta$ and $\int_{\Theta_i} \pi(\theta)\, \mathrm{d}\theta < \infty$,

$$\lim_{k\to\infty} \left\{ I\{\pi_i|\mathcal{M}^k\} - I\{p_i|\mathcal{M}^k\} \right\} \geqslant 0, \quad \text{for all } \Theta_i, \text{ for all } p \in \mathcal{P},$$

where $\pi_i(\theta)$ and $p_i(\theta)$ are the renormalized restrictions of $\pi(\theta)$ and $p(\theta)$ to Θ_i.

Notice that Definition 6 involves two rather different limiting processes. The limiting process of the Θ_i's towards the whole parameter space Θ is only required to guarantee the existence of the expected informations; this may often (but not always) be avoided if the parameter space is (realistically) chosen to be some finite interval $[a, b]$. On the other hand, the limiting process as $k \to \infty$ is an *essential* part of the definition. Indeed, the reference prior is *defined* as that prior function which maximizes the *missing* information, which is the expected discrepancy between prior knowledge and *perfect* knowledge; but perfect knowledge is only approached *asymptotically*, as $k \to \infty$.

Definition 6 implies that reference priors *only* depend on the *asymptotic behaviour* of the assumed model, a feature which greatly simplifies their actual derivation; to obtain a reference prior $\pi(\theta|\mathcal{M}, \mathcal{P})$ for the parameter θ of model $\mathcal{M} \equiv \{p(\boldsymbol{x}|\theta), \boldsymbol{x} \in \boldsymbol{\mathcal{X}}, \theta \in \Theta\}$, it is both necessary and sufficient to establish the asymptotic behaviour of its posterior distribution under (conceptual) repeated sampling from $\mathcal{M}$, that is the limiting form, as $k \to \infty$, of the posterior density (or probability function) $\pi(\theta|\boldsymbol{x}^{(k)}) = \pi(\theta|\boldsymbol{x}_1, \ldots, \boldsymbol{x}_k)$.

As one would hope, Definition 6 yields the maximum entropy result in the case where the parameter space is finite and the quantity of interest is the actual value of the parameter:

THEOREM 2 (Reference priors with finite parameter space). *Consider a model* $\mathcal{M} \equiv \{p(\boldsymbol{x}|\theta), \boldsymbol{x} \in \boldsymbol{\mathcal{X}}, \theta \in \Theta\}$, *with a* finite *parameter space* $\Theta = \{\theta_1, \ldots, \theta_m\}$ *and such that, for all pairs* θ_i *and* θ_j, $\delta\{p_{x|\theta_i}, p_{x|\theta_j}\} > 0$, *and let* $\mathcal{P}$ *be a class of probability distributions over* Θ. *Then the reference prior for the parameter* θ *is*

$$\pi^{\theta}(\theta|\mathcal{M}, \mathcal{P}) = \arg\max_{p_\theta \in \mathcal{P}} H\{p_\theta\},$$

where $p_\theta = \{p(\theta_1), p(\theta_2), \ldots, p(\theta_m)\}$ *and* $H\{p_\theta\} = -\sum_{j=1}^{m} p(\theta_j) \log p(\theta_j)$ *is the entropy of* p_θ. *In particular, if the class of candidate priors for* θ *is the set* $\mathcal{P}_0$ *of all* *strictly positive probability distributions over Θ, then the reference prior is the uniform distribution* $\pi^{\theta}(\theta|\mathcal{M}, \mathcal{P}_0) = \{1/m, \ldots, 1/m\}$.

Theorem 2 follows immediately from the fact that, if the intrinsic discrepancies $\delta\{p_{x|\theta_i}, p_{x|\theta_j}\}$ are all positive (and hence the m models $p(x|\theta_i)$ are all distinguishable from each other), then the posterior distribution of θ asymptotically converges to a degenerate distribution with probability one on the true value of θ (see, e.g., Bernardo and Smith, 1994, Section 5.3, and references therein). Such asymptotic posterior has zero entropy and thus, by Eq. (12), the missing information about θ when the prior is p_θ does not depend on $\mathcal{M}$, and is simply given by the prior entropy, $H\{p_\theta\}$.

Consider now a model $\mathcal{M}$ indexed by a continuous parameter $\theta \in \Theta \subset \mathbb{R}$. If the family of candidate priors consist of the class $\mathcal{P}_0$ of *all* continuous priors with support Θ, then the reference prior, $\pi(\theta|\mathcal{M}, \mathcal{P}_0)$ may be obtained as the result of an explicit limit. This provides a relatively simple procedure to obtain reference priors in models with one continuous parameter. Moreover, this analytical procedure may easily be converted into a programmable algorithm for numerical derivation of reference distributions. The results may conveniently be described in terms of any *asymptotically sufficient* statistic, i.e., a function $\boldsymbol{t}_k = \boldsymbol{t}_k(\boldsymbol{x}^{(k)})$ such that, for all θ and for all $\boldsymbol{x}^{(k)}$, $\lim_{k\to\infty}[p(\theta|\boldsymbol{x}^{(k)})/p(\theta|\boldsymbol{t}_k)] = 1$.

Obviously, the entire sample $\boldsymbol{x}^{(k)}$ is sufficient (and hence asymptotically sufficient), so there is no loss of generality in framing the results in terms of asymptotically sufficient statistics.

THEOREM 3 (Explicit form of the reference prior). *Consider the model* $\mathcal{M} \equiv \{p(\boldsymbol{x}|\theta), \boldsymbol{x} \in \mathcal{X}, \theta \in \Theta \subset \mathbb{R}\}$, *and let* $\mathcal{P}_0$ *be the class of all continuous priors with support* Θ. *Let* $\boldsymbol{x}^{(k)} = \{\boldsymbol{x}_1, \ldots, \boldsymbol{x}_k\}$ *consist of k independent observations from* $\mathcal{M}$, *so that* $p(\boldsymbol{x}^{(k)}|\theta) = \prod_{j=1}^{k} p(\boldsymbol{x}_j|\theta)$, *and let* $\boldsymbol{t}_k = \boldsymbol{t}_k(\boldsymbol{x}^{(k)}) \in \mathcal{T}$ *be any asymptotically sufficient statistic. Let* $h(\theta)$ *be a continuous strictly positive function such that, for sufficiently large k,* $\int_\Theta p(\boldsymbol{t}_k|\theta)h(\theta)\,\mathrm{d}\theta < \infty$, *and define*

$$f_k(\theta) = \exp\left\{\int_{\mathcal{T}} p(\boldsymbol{t}_k|\theta)\log\left(\frac{p(\boldsymbol{t}_k|\theta)h(\theta)}{\int_\Theta p(\boldsymbol{t}_k|\theta)h(\theta)\,\mathrm{d}\theta}\right)\mathrm{d}\boldsymbol{t}_k\right\}, \quad \textit{and} \tag{14}$$

$$f(\theta) = \lim_{k\to\infty}\frac{f_k(\theta)}{f_k(\theta_0)}, \tag{15}$$

where θ_0 *is any interior point of* Θ. *If* $f(\theta)$ *is a permissible prior function then, for any* $c > 0$, $\pi(\theta|\mathcal{M}, \mathcal{P}_0) = cf(\theta)$ *is a reference prior function.*

Intuitively, Theorem 3 states that the reference prior $\pi(\theta|\mathcal{M})$ relative to model $\mathcal{M}$ only depends on the asymptotic behaviour of the model and that, with no additional information to restrict the class of candidate priors, it has (from Eq. (14)), the form

$$\pi(\theta|\mathcal{M}, \mathcal{P}_0) \propto \exp\{E_{\boldsymbol{t}_k|\theta}[\log p(\theta|\boldsymbol{t}_k)]\}, \tag{16}$$

where $p(\theta|\boldsymbol{t}_k)$ is any asymptotic approximation to the posterior distribution of θ, and the expectation is taken with respect to the sampling distribution of the relevant asymptotically sufficient statistic $\boldsymbol{t}_k = \boldsymbol{t}_k(\boldsymbol{x}^{(k)})$. A heuristic derivation of Theorem 3 is provided below. For a precise statement of the regularity conditions and a formal proof, see Berger et al. (2005).

Under fairly general regularity conditions, the intrinsic expected information reduces to Shannon's expected information when $k \to \infty$. Thus, starting from (10), the amount of information about θ to be expected from $\mathcal{M}^k$ when the prior is $p(\theta)$ may be rewritten as $I\{p_\theta | \mathcal{M}^k\} = \int_\Theta p(\theta) \log[h_k(\theta)/p(\theta)] \, d\theta$, where $h_k(\theta) = \exp\{\int_\mathcal{T} p(\boldsymbol{t}_k|\theta) \log p(\theta|\boldsymbol{t}_k) \, d\boldsymbol{t}_k\}$. If $c_k = \int_\Theta h_k(\theta) \, d\theta < \infty$, then $h_k(\theta)$ may be renormalized to get the proper density $h_k(\theta)/c_k$, and $I\{p_\theta | \mathcal{M}^k\}$ may be rewritten as

$$I\{p_\theta | \mathcal{M}^k\} = \log c_k - \int_\Theta p(\theta) \log \frac{p(\theta)}{h_k(\theta)/c_k} \, d\theta. \tag{17}$$

But the integral in (17) is the Kullback–Leibler directed divergence of $h_k(\theta)/c_k$ from $p(\theta)$, which is nonnegative, and it is zero iff $p(\theta) = h_k(\theta)/c_k$ almost everywhere. Thus, $I\{p_\theta | \mathcal{M}^k\}$ would be maximized by a prior $\pi_k(\theta)$ which satisfies the functional equation

$$\pi_k(\theta) \propto h_k(\theta) = \exp\left\{\int_\mathcal{T} p(\boldsymbol{t}_k|\theta) \log \pi_k(\theta|\boldsymbol{t}_k) \, d\boldsymbol{t}_k\right\}, \tag{18}$$

where $\pi_k(\theta|\boldsymbol{t}_k) \propto p(\boldsymbol{t}_k|\theta)\pi_k(\theta)$ and, therefore, the reference prior should be a limiting form, as $k \to \infty$ of the sequence of proper priors given by (18). This only provides an implicit solution, since the posterior density $\pi_k(\theta|\boldsymbol{t}_k)$ in the right-hand side of (18) obviously depends on the prior $\pi_k(\theta)$; however, as $k \to \infty$, the posterior $\pi_k(\theta|\boldsymbol{t}_k)$ will approach its asymptotic form which, under the assumed conditions, is independent of the prior. Thus, the posterior density in (18) may be replaced by the posterior $\pi^0(\theta|\boldsymbol{t}_k) \propto p(\boldsymbol{t}_k|\theta)h(\theta)$ which corresponds to any fixed prior, say $\pi^0(\theta) = h(\theta)$, to obtain an explicit expression for a sequence of priors,

$$\pi_k(\theta) \propto f_k(\theta) = \exp\left\{\int_\mathcal{T} p(\boldsymbol{t}_k|\theta) \log \pi^0(\theta|\boldsymbol{t}_k) \, d\boldsymbol{t}_k\right\}, \tag{19}$$

whose limiting form will still maximize the missing information about θ. The preceding argument rests however on the assumption that (at least for sufficiently large k) the integrals in Θ of $f_k(\theta)$ are finite, but those integrals may well diverge. The problem is solved by considering an increasing sequence $\{\Theta_i\}_{i=1}^\infty$ of subsets of Θ which converges to Θ and such that, for all i and sufficiently large k, $c_{ik} = \int_{\Theta_i} f_k(\theta) \, d\theta < \infty$, so that the required integrals are finite. An appropriate limiting form of the double sequence $\pi_{ik}(\theta) = f_k(\theta)/c_{ik}$, $\theta \in \Theta_i$ will then approach the required reference prior.

Such a limiting form is easily established; indeed, let $\pi_{ik}(\theta|\boldsymbol{x})$, $\theta \in \Theta_i$, be the posterior which corresponds to $\pi_{ik}(\theta)$ and, for some interior point θ_0 of all the Θ_i's, consider the limit

$$\lim_{k\to\infty} \frac{\pi_{ik}(\theta|\boldsymbol{x})}{\pi_{ik}(\theta_0|\boldsymbol{x})} = \lim_{k\to\infty} \frac{p(\boldsymbol{x}|\theta) f_k(\theta)}{p(\boldsymbol{x}|\theta_0) f_k(\theta_0)} \propto p(\boldsymbol{x}|\theta) f(\theta), \tag{20}$$

where $f(\theta) = \lim_{k\to\infty} f_k(\theta)/f_k(\theta_0)$, which does not depend on the initial function $h(\theta)$ (and therefore $h(\theta)$ may be chosen by mathematical convenience). It follows from (20) that, for any data $\boldsymbol{x}$, the sequence of posteriors $\pi_{ik}(\theta|\boldsymbol{x})$ which maximize the missing information will approach the posterior $\pi(\theta|\boldsymbol{x}) \propto p(\boldsymbol{x}|\theta)f(\theta)$ obtained by formal use of Bayes theorem, using $f(\theta)$ as the prior. This completes the heuristic justification of Theorem 3.

3.2. *Main properties*

Reference priors enjoy many attractive properties, as stated below. For detailed proofs, see Bernardo and Smith (1994, Sections 5.4 and 5.6).

In the frequently occurring situation where the available data consist of a random sample of fixed size n from some model $\mathcal{M}$ (so that the assumed model is $\mathcal{M}^n$), the reference prior relative to $\mathcal{M}^n$ is independent of n, and may simply be obtained as the reference prior relative to $\mathcal{M}$, assuming the later exists.

THEOREM 4 (Independence of sample size). *If data $\boldsymbol{x} = \{\boldsymbol{y}_1, \ldots, \boldsymbol{y}_n\}$ consists of a random sample of size n from model $\mathcal{M} \equiv \{p(\boldsymbol{y}|\theta), \boldsymbol{y} \in \boldsymbol{\mathcal{Y}}, \theta \in \Theta\}$, with reference prior $\pi^\theta(\theta|\mathcal{M}, \mathcal{P})$ relative to the class of candidate priors $\mathcal{P}$, then, for any fixed sample size n, the reference prior for θ relative to $\mathcal{P}$ is $\pi^\theta(\theta|\mathcal{M}^n, \mathcal{P}) = \pi^\theta(\theta|\mathcal{M}, \mathcal{P})$.*

This follows from the additivity of the information measure. Indeed, for any sample size n and number of replicates k, $I\{p_\theta|\mathcal{M}^{nk}\} = nI\{p_\theta|\mathcal{M}^k\}$.

Note, however, that Theorem 4 requires $\boldsymbol{x}$ to be a random sample from the assumed model. If the model entails dependence between the observations (as in time series, or in spatial models) the reference prior may well depend on the sample size; see, for example, Berger and Yang (1994), and Berger et al. (2001).

The possible dependence of the reference prior on the sample size and, more generally, on the design of the experiment highlights the fact that a reference prior is *not* a description of (personal) prior beliefs, but a possible *consensus* prior for a particular problem of scientific inference. Indeed, genuine prior beliefs about some quantity of interest should not depend on the design of the experiment performed to learn about its value (although they will typically influence the choice of the design), but a prior function to be used as a *consensus* prior to analyse the results of an experiment may be expected to depend on its design. Reference priors, which by definition maximize the missing information which repeated observations from a *particular* experiment could possibly provide, generally depend on the design of that experiment.

As one would hope, if the assumed model $\mathcal{M}$ has a sufficient statistic $\boldsymbol{t} = \boldsymbol{t}(\boldsymbol{x})$, the reference prior relative to $\mathcal{M}$ is the same as the reference prior relative to the equivalent model derived from the sampling distribution of $\boldsymbol{t}$:

THEOREM 5 (Compatibility with sufficient statistics). *Consider a model $\mathcal{M} \equiv \{p(\boldsymbol{x}|\theta), \boldsymbol{x} \in \boldsymbol{\mathcal{X}}, \theta \in \Theta\}$ with sufficient statistic $\boldsymbol{t} = \boldsymbol{t}(\boldsymbol{x}) \in \boldsymbol{\mathcal{T}}$, and let $\mathcal{M}_t \equiv \{p(\boldsymbol{t}|\theta), \boldsymbol{t} \in \boldsymbol{\mathcal{T}}, \theta \in \Theta\}$ be the corresponding model in terms of $\boldsymbol{t}$. Then, for any class of candidate priors $\mathcal{P}$, the reference prior for θ relative to model $\mathcal{M}$ is $\pi^\theta(\theta|\mathcal{M}, \mathcal{P}) = \pi^\theta(\theta|\mathcal{M}_t, \mathcal{P})$.*

Theorem 5 follows from the fact that the expected information is invariant under such transformation, so that, for all k, $I\{p_\theta|\mathcal{M}^k\} = I\{p_\theta|\mathcal{M}_t^k\}$.

When data consist of a random sample of fixed size from some model, and there exists a sufficient statistic of fixed dimensionality, Theorems 3, 4 and 5 may be combined for an easy, direct derivation of the reference prior, as illustrated below.

EXAMPLE 5 (*Exponential model, continued*). Let $\boldsymbol{x} = \{x_1, \ldots, x_n\}$ be a random sample of size n from an exponential distribution. By Theorem 4, to obtain the corresponding reference prior it suffices to analyse the behaviour, as $k \to \infty$, of k replications of the model which corresponds to a single observation, $\mathcal{M} \equiv \{\theta\,\mathrm{e}^{-\theta y}, y > 0, \theta > 0\}$, as opposed to k replications of the actual model for data $\boldsymbol{x}$, $\mathcal{M}^n \equiv \{\prod_{j=1}^n \theta\,\mathrm{e}^{-\theta x_j}, x_j > 0, \theta > 0\}$.

Thus, consider $\boldsymbol{y}^{(k)} = \{y_1, \ldots, y_k\}$, a random sample of size k from the single observation model $\mathcal{M}$; clearly $t_k = \sum_{j=1}^k y_j$ is sufficient, and the sampling distribution of t_k has a gamma density $p(t_k|\theta) = \mathrm{Ga}(t_k|k, \theta)$. Using a constant for the arbitrary function $h(\theta)$ in Theorem 3, the corresponding posterior has a gamma density $\mathrm{Ga}(\theta|k+1, t_k)$ and, thus,

$$f_k(\theta) = \exp\left[\int_0^\infty \mathrm{Ga}(t_k|k, \theta) \log\{\mathrm{Ga}(\theta|k+1, t_k)\}\,\mathrm{d}t_k\right] = c_k \theta^{-1},$$

where c_k is a constant which does not contain θ. Therefore, using (15), $f(\theta) = \theta_0/\theta$ and, since this is a permissible prior function (see Example 3), the unrestricted reference prior (for both the single observation model $\mathcal{M}$ and the actual model $\mathcal{M}^n$) is $\pi(\theta|\mathcal{M}^n, \mathcal{P}_0) = \pi(\theta|\mathcal{M}, \mathcal{P}_0) = \theta^{-1}$.

Parametrizations are essentially arbitrary. As one would hope, reference priors are coherent under reparametrization in the sense that if $\phi = \phi(\theta)$ is a one-to-one mapping of Θ into $\Phi = \phi(\Theta)$ then, for all $\phi \in \Phi$,

(i) $\pi^\phi(\phi) = \pi^\theta\{\theta(\phi)\}$, if Θ is discrete;
(ii) $\pi^\phi(\phi) = \pi^\theta\{\theta(\phi)\}|\partial\theta(\phi)/\partial\phi|$, if Θ is continuous.

More generally, reference posteriors are coherent under piecewise invertible transformations $\phi = \phi(\theta)$ of the parameter θ in the sense that, for all $\boldsymbol{x} \in \boldsymbol{\mathcal{X}}$, the reference posterior for ϕ derived from first principles, $\pi(\phi|\boldsymbol{x})$, is precisely the same as that which could be obtained from $\pi(\theta|\boldsymbol{x})$ by standard probability calculus:

THEOREM 6 (Consistency under reparametrization). *Consider a model $\mathcal{M} \equiv \{p(\boldsymbol{x}|\theta), \boldsymbol{x} \in \boldsymbol{\mathcal{X}}, \theta \in \Theta\}$ and let $\phi(\theta)$ be a piecewise invertible transformation of θ. For any data $\boldsymbol{x} \in \boldsymbol{\mathcal{X}}$, the reference posterior density of ϕ, $\pi(\phi|\boldsymbol{x})$, is that induced by the reference posterior density of θ, $\pi(\theta|\boldsymbol{x})$.*

If $\phi(\theta)$ is one-to-one, Theorem 6 follows immediately from the fact that the expected information is also invariant under such transformation, so that, for all k, $I\{p_\theta|\mathcal{M}_\theta^k\} = I\{p_\psi|\mathcal{M}_\psi^k\}$; this may also be directly verified using Theorems 2 and 3.

Suppose now that $\phi(\theta) = \phi_j(\theta)$, $\theta \in \Theta_j$, where the Θ_j's form a partition of Θ, such that each of the $\phi_j(\theta)$'s is one-to-one in Θ_j. The reference prior for θ only depends on the asymptotic posterior of θ which, for sufficiently large samples, will concentrate on that subset Θ_j of the parameter space Θ to which the true value of θ belongs. Since $\phi(\theta)$ is one-to-one within Θ_j, and reference priors are coherent under one-to-one parametrizations, the general result follows.

An important consequence of Theorem 6 is that the reference prior of any location parameter, and the reference prior of the logarithm of any scale parameter are both uniform:

THEOREM 7 (Location models and scale models). *Consider a* location *model* $\mathcal{M}_1$, *so that for some function* f_1, $\mathcal{M}_1 \equiv \{f_1(x-\mu), x \in \mathbb{R}, \mu \in \mathbb{R}\}$, *and let* $\mathcal{P}_0$ *be the class of all continuous strictly positive priors on* $\mathbb{R}$; *then, if it exists, a reference prior for* μ *is of the form* $\pi(\mu|\mathcal{M}_1, \mathcal{P}_0) = c$. *Moreover, if* $\mathcal{M}_2$ *is a* scale *model*, $\mathcal{M}_2 \equiv \{\sigma^{-1} f_2(x/\sigma), x > 0, \sigma > 0\}$, *and* $\mathcal{P}_0$ *is the class of all continuous strictly positive priors on* $(0, \infty)$, *then a reference prior for* σ, *if it exists, is of the form* $\pi(\sigma|\mathcal{M}_2, \mathcal{P}_0) = c\sigma^{-1}$.

Let $\pi(\mu)$ be the reference prior which corresponds to model $\mathcal{M}_1$; the changes $y = x + \alpha$ and $\theta = \mu + \alpha$ produce $\{f_1(y-\theta), y \in \mathcal{Y}, \theta \in \mathbb{R}\}$, which is again model $\mathcal{M}_1$. Hence, using Theorem 6, $\pi(\mu) = \pi(\mu + \alpha)$ for all α and, therefore, $\pi(\mu)$ must be constant. Moreover, the obvious changes $y = \log x$ and $\phi = \log\sigma$ transform the scale model $\mathcal{M}_2$ into a location model; hence, $\pi(\phi) = c$ and, therefore, $\pi(\sigma) \propto \sigma^{-1}$.

EXAMPLE 6 (*Cauchy data*). Let $\boldsymbol{x} = \{x_1, \ldots, x_n\}$ be a random sample from a Cauchy distribution with unknown location μ and known scale $\sigma = 1$, so that $p(x_j|\mu) \propto [1 + (x_j - \mu)^2]^{-1}$. Since this is a location model, the reference prior is uniform and, by Bayes theorem, the corresponding reference posterior is

$$\pi(\mu|\boldsymbol{x}) \propto \prod_{j=1}^{n} [1 + (x_j - \mu)^2]^{-1}, \quad \mu \in \mathbb{R}.$$

Using the change of variable theorem, the reference posterior of (say) the one-to-one transformation $\phi = \mathrm{e}^{\mu}/(1 + \mathrm{e}^{\mu})$ mapping the original parameter space $\mathbb{R}$ into $(0, 1)$, is $\pi(\phi|\boldsymbol{x}) = \pi(\mu(\phi)|\boldsymbol{x})|\partial\mu/\partial\phi|$, $\phi \in (0, 1)$. Similarly, the reference posterior $\pi(\psi|\boldsymbol{x})$ of (say) $\psi = \mu^2$ may be derived from $\pi(\mu|\boldsymbol{x})$ using standard change of variable techniques, since $\psi = \mu^2$ is a piecewise invertible function of μ, and Theorem 6 may therefore be applied.

3.3. *Approximate location parametrization*

Another consequence of Theorem 6 is that, for any model with one continuous parameter $\theta \in \Theta$, there is a parametrization $\phi = \phi(\theta)$ (which is unique up to a largely irrelevant proportionality constant), for which the reference prior is uniform. By Theorem 6 this may be obtained from the reference prior $\pi(\theta)$ in the original parametrization as a function $\phi = \phi(\theta)$ which satisfies the differential equation $\pi(\theta)|\partial\phi(\theta)/\partial\theta|^{-1} = 1$, that is, any solution to the indefinite integral $\phi(\theta) = \int \pi(\theta)\,d\theta$. Intuitively, $\phi = \phi(\theta)$ may be expected to behave as an *approximate* location parameter; this links reference priors with the concept data translated likelihood inducing priors introduced by Box and Tiao (1973, Section 1.3). For many models, good simple approximations to the posterior distribution may be obtained in terms of this parametrization, which often yields an *exact* location model.

DEFINITION 7 (*Approximate location parametrization*). Consider the model $\mathcal{M} \equiv \{p(\boldsymbol{x}|\theta), \boldsymbol{x} \in \boldsymbol{\mathcal{X}}, \theta \in \Theta \subset \mathbb{R}\}$. An *approximate location parametrization* $\phi = \phi(\theta)$ for model $\mathcal{M}$ is one for which the reference prior is uniform. In continuous regular models, this is given by any solution to the indefinite integral $\phi(\theta) = \int \pi(\theta)\,d\theta$, where $\pi(\theta) = \pi(\theta|\mathcal{M}, \mathcal{P}_0)$ is the (unrestricted) reference prior for the original parameter.

EXAMPLE 7 (*Exponential model, continued*). Consider again the exponential model $\mathcal{M} \equiv \{\theta\,e^{-\theta x}, x > 0, \theta > 0\}$. The reference prior for θ is (see Example 5) $\pi(\theta) = \theta^{-1}$; thus an approximate location parameter is $\phi = \phi(\theta) = \int \pi(\theta)\,d\theta = \log\theta$. Using $y = -\log x$, this yields

$$\mathcal{M}_y \equiv \left\{\exp\left[-(y-\phi) + e^{-(y-\phi)}\right],\ y \in \mathbb{R}, \phi \in \mathbb{R}\right\},$$

where ϕ is an (actually exact) location parameter.

EXAMPLE 8 (*Uniform model on* $(0, \theta)$). Let $\boldsymbol{x} = \{x_1, \ldots, x_k\}$ be a random sample from the uniform model $\mathcal{M} \equiv \{p(x|\theta) = \theta^{-1}, 0 < x < \theta, \theta > 0\}$, so that $t_k = \max_{j=1}^{k} x_j$ is sufficient, and the sampling distribution of t_k is the inverted Pareto $p(t_k|\theta) = \mathrm{IPa}(t_k|k, \theta^{-1}) = k\theta^{-k}t_k^{k-1}$, if $0 < t_k < \theta$, and zero otherwise. Using a uniform prior for the arbitrary function $h(\theta)$ in Theorem 3, the corresponding posterior distribution has the Pareto density $\mathrm{Pa}(\theta|k-1, t_k) = (k-1)t_k^{k-1}\theta^{-k}$, $\theta > t_k$, and (14) becomes

$$f_k(\theta) = \exp\left[\int_0^\theta \mathrm{IPa}\left(t_k|k, \theta^{-1}\right)\log \mathrm{Pa}(\theta|k-1, t_k)\,dt_k\right] = c_k\theta^{-1},$$

where c_k is a constant which does not contain θ. Therefore, using (15), $f(\theta) = \theta_0/\theta$, $\pi(\theta|\mathcal{M}, \mathcal{P}_0) = \theta^{-1}$.

By Theorem 4, this is also the reference prior for samples of any size; hence, by Bayes theorem, the reference posterior density of θ after, say, a random sample $\boldsymbol{x} = \{x_1, \ldots, x_n\}$ of size n has been observed is

$$\pi(\theta|\boldsymbol{x}) \propto \prod_{j=1}^{n} p(x_j|\theta)\pi(\theta) = \theta^{-(n+1)}, \qquad \theta > t_n,$$

where $t_n = \max\{x_1, \ldots, x_n\}$, which is a kernel of the Pareto density $\pi(\theta|\boldsymbol{x}) = \pi(\theta|t_n) = \mathrm{Pa}(\theta|n, t_n) = n(t_n)^n\theta^{-(n+1)}$, $\theta > t_n$.

The approximate location parameter is $\phi(\theta) = \int \theta^{-1}\,d\theta = \log\theta$. The sampling distribution of the sufficient statistic $s_n = \log t_n$ in terms of the new parameter is the reversed exponential $p(s_n|n, \phi) = n\, e^{-n(\phi - s_n)}$, $s_n < \phi$, which explicitly shows ϕ as an (exact) location parameter. The reference prior of ϕ is indeed uniform, and the reference posterior after $\boldsymbol{x}$ has been observed is the shifted exponential $\pi(\phi|\boldsymbol{x}) = n\, e^{-n(\phi - s_n)}$, $\phi > s_n$, which may also be obtained by changing variables in $\pi(\theta|\boldsymbol{x})$.

3.4. Numerical reference priors

Analytical derivation of reference priors may be technically demanding in complex models. However, Theorem 3 may also be used to obtain a numerical approximation to the reference prior which corresponds to any one-parameter model $\mathcal{M} \equiv \{p(\boldsymbol{x}|\theta), \boldsymbol{x} \in \boldsymbol{\mathcal{X}}, \theta \in \Theta\}$ from which random observations may be efficiently simulated.

The proposed algorithm requires a numerical evaluation of Eq. (14). This is relatively straightforward, for simulation from the assumed model may be used to approximate by Monte Carlo the integral in (14), and the evaluation of its integrand for each simulated set of data only requires (cheap) one-dimensional numerical integration. Moderate values of k (to simulate the asymptotic posterior) are typically sufficient to obtain a good approximation to the reference prior $\pi(\theta|\mathcal{M}, \mathcal{P}_0)$ (up to an irrelevant proportionality constant). The appropriate pseudo code is quite simple:

(1) Starting values:
  Choose a moderate value for k.
  Choose an arbitrary positive function $h(\theta)$, say $h(\theta) = 1$.
  Choose the number m of samples to be simulated.
(2) For any given θ value, **repeat**, for $j = 1, \ldots, m$:
  Simulate a random sample $\{\boldsymbol{x}_{1j}, \ldots, \boldsymbol{x}_{kj}\}$ of size k from $p(\boldsymbol{x}|\theta)$.
  Compute numerically the integral $c_j = \int_\Theta \prod_{i=1}^k p(\boldsymbol{x}_{ij}|\theta)h(\theta)\, d\theta$.
  Evaluate $r_j(\theta) = \log[\prod_{i=1}^k p(\boldsymbol{x}_{ij}|\theta)h(\theta)/c_j]$.
(3) Compute $\pi(\theta) = \exp[m^{-1} \sum_{j=1}^m r_j(\theta)]$ and **store** the pair $\{\theta, \pi(\theta)\}$.
(4) **Repeat** routines (2) and (3) for all θ values for which the pair $\{\theta, \pi(\theta)\}$ is required.

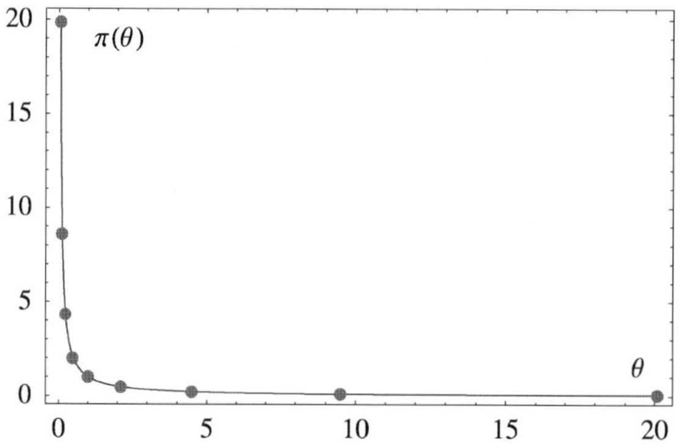

Fig. 3. Numerical reference prior for the exponential model.

EXAMPLE 9 (*Exponential data, continued*). Figure 3 represents the exact reference prior for the exponential model $\pi(\theta) = \theta^{-1}$ (continuous line) and the reference prior numerically calculated with the algorithm above for nine θ values, ranging from e^{-3} to e^3, uniformly log-spaced and rescaled to have $\pi(1) = 1$; $m = 500$ samples of $k = 25$ observations were used to compute each of the nine $\{\theta_i, \pi(\theta_i)\}$ points.

If required, a continuous approximation to $\pi(\theta)$ may easily be obtained from the computed points using standard interpolation techniques.

An educated choice of the arbitrary function $h(\theta)$ often leads to an analytical form for the required posterior, $p(\theta|\boldsymbol{x}_{1j}, \ldots, \boldsymbol{x}_{kj}) \propto \prod_{i=1}^{k} p(\boldsymbol{x}_{ij}|\theta)h(\theta)$; for instance, this is the case in Example 9 if $h(\theta)$ is chosen to be of the form $h(\theta) = \theta^a$, for some $a \geqslant -1$. If the posterior may be analytically computed, then the values of the $r_j(\theta) = \log[p(\theta|\boldsymbol{x}_{1j}, \ldots, \boldsymbol{x}_{kj})]$ are immediately obtained, and the numerical algorithm reduces to only one Monte Carlo integration for each desired pair $\{\theta_i, \pi(\theta_i)\}$.

For an alternative, MCMC based, numerical computation method of reference priors, see Lafferty and Wasserman (2001).

3.5. *Reference priors under regularity conditions*

If data consist of a random sample $\boldsymbol{x} = \{x_1, \ldots, x_n\}$ of a model with one continuous parameter θ, it is often possible to find an *asymptotically sufficient* statistic $\tilde{\theta}_n = \tilde{\theta}_n(x_1, \ldots, x_n)$ which is also a *consistent* estimator of θ; for example, under regularity conditions, the maximum likelihood estimator (mle) $\hat{\theta}_n$ is consistent and asymptotically sufficient. In that case, the reference prior may easily be obtained in terms of *either* (i) an asymptotic approximation $\pi(\theta|\tilde{\theta}_n)$ to the posterior distribution of θ, or (ii) the sampling distribution $p(\tilde{\theta}_n|\theta)$ of the asymptotically sufficient consistent estimator $\tilde{\theta}_n$.

THEOREM 8 (Reference priors under regularity conditions). *Let available data* $\boldsymbol{x} \in \mathcal{X}$ *consist of a random sample of any size from a one-parameter model* $\mathcal{M} \equiv \{p(x|\theta), x \in \mathcal{X}, \theta \in \Theta\}$. *Let* $\boldsymbol{x}^{(k)} = \{x_1, \ldots, x_k\}$ *be a random sample of size k from model* $\mathcal{M}$, *let* $\tilde{\theta}_k = \tilde{\theta}_k(\boldsymbol{x}^{(k)}) \in \Theta$ *be an asymptotically sufficient statistic which is a consistent estimator of* θ, *and let* $\mathcal{P}_0$ *be the class of all continuous priors with support* Θ. *Let* $\pi_k(\theta|\tilde{\theta}_k)$ *be any asymptotic approximation (as* $k \to \infty$*) to the posterior distribution of* θ, *let* $p(\tilde{\theta}_k|\theta)$ *be the sampling distribution of* $\tilde{\theta}_k$, *and define*

$$f_k^a(\theta) = \pi_k(\theta|\tilde{\theta}_k)\big|_{\tilde{\theta}_k=\theta}, \qquad f^a(\theta) = \lim_{k\to\infty} \frac{f_k^a(\theta)}{f_k^a(\theta_0)}, \tag{21}$$

$$f_k^b(\theta) = p(\tilde{\theta}_k|\theta)\big|_{\tilde{\theta}_k=\theta}, \qquad f^b(\theta) = \lim_{k\to\infty} \frac{f_k^b(\theta)}{f_k^b(\theta_0)}, \tag{22}$$

where θ_0 *is any interior point of* Θ. *Then, under frequently occurring additional technical conditions,* $f^a(\theta) = f^b(\theta) = f(\theta)$ *and, if* $f(\theta)$ *is a permissible prior, any function of the form* $\pi(\theta|\mathcal{M}, \mathcal{P}_0) \propto f(\theta)$ *is a reference prior for* θ.

Since $\tilde{\theta}_k$ is asymptotically sufficient, Eq. (14) in Theorem 3 becomes

$$f_k(\theta) = \exp\left\{\int_{\Theta} p(\tilde{\theta}_k \mid \theta) \log \pi_k(\theta \mid \tilde{\theta}_k)\, d\tilde{\theta}_k\right\}.$$

Moreover, since $\tilde{\theta}_k$ is consistent, the sampling distribution of $\tilde{\theta}_k$ will concentrate on θ as $k \to \infty$, $f_k(\theta)$ will converge to $f_k^a(\theta)$, and Eq. (21) will have the same limit as Eq. (15). Moreover, for any formal prior function $h(\theta)$,

$$\pi(\theta \mid \tilde{\theta}_k) = \frac{p(\tilde{\theta}_k \mid \theta) h(\theta)}{\int_{\Theta} p(\tilde{\theta}_k \mid \theta) h(\theta)\, d\theta}.$$

As $k \to \infty$, the integral in the denominator converges to $h(\tilde{\theta}_k)$ and, therefore, $f_k^a(\theta) = \pi(\theta \mid \tilde{\theta}_k)|_{\tilde{\theta}_k=\theta}$ converges to $p(\tilde{\theta}_k \mid \theta)|_{\tilde{\theta}_k=\theta} = f_k^b(\theta)$. Thus, both limits in Eqs. (21) and (22) yield the same result, and their common value provides an explicit expression for the reference prior. For details, and precise technical conditions, see Berger et al. (2005).

EXAMPLE 10 (*Exponential model, continued*). Let $\boldsymbol{x} = \{x_1, \ldots, x_n\}$ be a random sample of n exponential observations from $\mathrm{Ex}(x \mid \theta)$. The mle is $\hat{\theta}_n(\boldsymbol{x}) = 1/\bar{x}$, a sufficient, consistent estimator of θ whose sampling distribution is the inverted gamma $p(\hat{\theta}_n \mid \theta) = \mathrm{IGa}(\hat{\theta}_n \mid n\theta, n)$. Therefore, $f_n^b(\theta) = p(\hat{\theta}_n \mid \theta)|_{\hat{\theta}_n=\theta} = c_n/\theta$, where $c_n = e^{-n} n^n / \Gamma(n)$ and, using Theorem 8, the reference prior is $\pi(\theta) = \theta^{-1}$.

Alternatively, the likelihood function is $\theta^n e^{-n\theta/\hat{\theta}_n}$; hence, for any positive function $h(\theta)$, $\pi_n(\theta \mid \hat{\theta}_n) \propto \theta^n e^{-n\theta/\hat{\theta}_n} h(\theta)$ is an asymptotic approximation to the posterior distribution of θ. Taking, for instance, $h(\theta) = 1$, this yields the gamma posterior $\pi_n(\theta \mid \hat{\theta}_n) = \mathrm{Ga}(\theta \mid n+1, n/\hat{\theta}_n)$. Consequently, $f_n^a(\theta) = \pi_n(\theta \mid \hat{\theta}_n)|_{\hat{\theta}_n=\theta} = c_n/\theta$, and $\pi(\theta) = \theta^{-1}$ as before.

EXAMPLE 11 (*Uniform model, continued*). Let $\boldsymbol{x} = \{x_1, \ldots, x_n\}$ be a random sample of n uniform observations from $\mathrm{Un}(x \mid 0, \theta)$. The mle is $\hat{\theta}_n(\boldsymbol{x}) = \max\{x_1, \ldots, x_n\}$, a sufficient, consistent estimator of θ whose sampling distribution is the inverted Pareto $p(\hat{\theta}_n \mid \theta) = \mathrm{IPa}(\hat{\theta}_n \mid n, \theta^{-1})$. Therefore, $f_n^b(\theta) = p(\hat{\theta}_n \mid \theta)|_{\hat{\theta}_n=\theta} = n/\theta$ and, using Theorem 8, the reference prior is $\pi(\theta) = \theta^{-1}$.

Alternatively, the likelihood function is θ^{-n}, $\theta > \hat{\theta}_n$; hence, taking for instance a uniform prior, the Pareto $\pi^n(\theta \mid \hat{\theta}_n) = \mathrm{Pa}(\theta \mid n-1, \hat{\theta}_n)$ is found to be a particular asymptotic approximation of the posterior of θ; thus, $f_n^a(\theta) = \pi_n(\theta \mid \hat{\theta}_n)|_{\hat{\theta}_n=\theta} = (n-1)/\theta$, and $\pi(\theta) = \theta^{-1}$ as before.

The posterior distribution of the parameter is often asymptotically normal (see, e.g., Bernardo and Smith, 1994, Section 5.3, and references therein). In this case, the reference prior is easily derived. The result includes (univariate) Jeffreys (1946) and Perks (1947) rules as a particular cases:

THEOREM 9 (Reference priors under asymptotic normality). *Let data consist of a random sample from model* $\mathcal{M} \equiv \{p(\boldsymbol{y} \mid \theta), \boldsymbol{y} \in \boldsymbol{\mathcal{Y}}, \theta \in \Theta \subset \mathbb{R}\}$, *and let* $\mathcal{P}_0$ *be the class of* *all continuous priors with support Θ. If the posterior distribution of θ, $\pi(\theta|\boldsymbol{y}_1, \ldots, \boldsymbol{y}_n)$, is asymptotically normal with standard deviation $s(\tilde{\theta}_n)/\sqrt{n}$, where $\tilde{\theta}_n$ is a consistent estimator of θ, and $s(\theta)^{-1}$ is a permissible prior function, then any function of the form*

$$\pi(\theta|\mathcal{M}, \mathcal{P}_0) \propto s(\theta)^{-1} \tag{23}$$

is a reference prior. Under appropriate regularity conditions the posterior distribution of θ is asymptotically normal with variance $i(\hat{\theta}_n)^{-1}/n$, where $\hat{\theta}_n$ is the mle of θ and

$$i(\theta) = -\int_{\mathcal{Y}} p(\boldsymbol{y}|\theta) \frac{\partial^2}{\partial\theta^2} \log p(\boldsymbol{y}|\theta)\, \mathrm{d}\boldsymbol{y} \tag{24}$$

is Fisher's information function. If this is the case, and $i(\theta)^{1/2}$ is a permissible prior function, the reference prior is Jeffreys prior, $\pi(\theta|\mathcal{M}, \mathcal{P}_0) \propto i(\theta)^{1/2}$.

The result follows directly from Theorem 8 since, under the assumed conditions, $f_n^a(\theta) = \pi_n(\theta|\hat{\theta}_n)|_{\hat{\theta}_n=\theta} = c_n s(\theta)^{-1}$. Jeffreys prior is the particular case which obtains when $s(\theta) = i(\theta)^{-1/2}$.

Jeffreys (1946, 1961) prior, independently rediscovered by Perks (1947), was central in the early objective Bayesian reformulation of standard textbook problems of statistical inference (Lindley, 1965; Zellner, 1971; Press, 1972; Box and Tiao, 1973). By Theorem 9, this is also the reference prior in regular models with one continuous parameter, whose posterior distribution is asymptotically normal. By Theorem 6, reference priors are coherently transformed under one-to-one reparametrizations; hence, Theorem 9 may be typically applied with any mathematically convenient (re)parametrization. For conditions which preserve asymptotic normality under transformations see Mendoza (1994).

The posterior distribution of the exponential parameter in Example 10 is asymptotically normal; thus the corresponding reference prior may also be obtained using Theorem 9; the reference prior for the uniform parameter in Example 11 cannot be obtained however in this way, since the relevant posterior distribution is *not* asymptotically normal. Notice that, even under conditions which guarantee asymptotic normality, Jeffreys formula is not necessarily the easiest way to derive a reference prior; indeed, Theorem 8 often provides a simpler alternative.

3.6. *Reference priors and the likelihood principle*

By definition, reference priors are a function of the *entire* statistical model $\mathcal{M} \equiv \{p(\boldsymbol{x}|\theta), \boldsymbol{x} \in \boldsymbol{\mathcal{X}}, \theta \in \Theta\}$, not of the *observed* likelihood. Indeed, the reference prior $\pi(\theta|\mathcal{M})$ is a mathematical description of lack of information about θ *relative* to the information about θ which could be obtained by repeated sampling from a particular experimental design $\mathcal{M}$. If the design is changed, the reference prior may be expected to change accordingly. This is now illustrated by comparing the reference priors which correspond to direct and inverse sampling of Bernoulli observations.

EXAMPLE 12 (*Binomial and negative binomial data*). Let available data $\boldsymbol{x} = \{r, m\}$ consist of m Bernoulli trials (with m fixed in advance) which contain r successes, so that the assumed model is Binomial $\mathrm{Bi}(r|m,\theta)$:

$$\mathcal{M}_1 \equiv \left\{ p(r|m,\theta) = \binom{m}{r}\theta^r(1-\theta)^{m-r},\ r = 0, 1, \ldots, m,\ 0 < \theta < 1 \right\}.$$

Using Theorem 9, with $n = 1$, m fixed, and $\boldsymbol{y} = r$, the reference prior for θ is the (proper) prior $\pi(\theta) \propto \theta^{-1/2}(1-\theta)^{-1/2}$; Bayes theorem yields the Beta reference posterior $\pi(\theta|\boldsymbol{x}) = \mathrm{Be}(\theta|r+1/2, m-r+1/2)$. Notice that $\pi(\theta|\boldsymbol{x})$ is proper, for all values of r; in particular, if $r = 0$, the reference posterior is $\pi(\theta|\boldsymbol{x}) = \mathrm{Be}(\theta|1/2, m+1/2)$, from which sensible conclusions may be reached, even though there are no observed successes. This may be compared with the Haldane (1948) prior, also proposed by Jaynes (1968), $\pi(\theta) \propto \theta^{-1}(1-\theta)^{-1}$, which produces an improper posterior until at least one success and one failure are observed.

Consider, however, that data $\boldsymbol{x} = \{r, m\}$ consist of the sequence of Bernoulli trials observed until r successes are obtained (with $r \geqslant 1$ fixed in advance), so that the assumed model is negative binomial:

$$\mathcal{M}_2 \equiv \left\{ p(m|r,\theta) = \binom{m-1}{r-1}\theta^r(1-\theta)^{m-r}, \right.$$
$$\left. m = r, r+1, \ldots,\ 0 < \theta < 1 \right\}.$$

Using Theorem 9, with $n = 1$ and $\boldsymbol{y} = m$, the reference prior for θ is the (improper) prior $\pi(\theta) \propto \theta^{-1}(1-\theta)^{-1/2}$, and Bayes theorem yields the Beta reference posterior $\pi(\theta|\boldsymbol{x}) = \mathrm{Be}(\theta|r, m-r+1/2)$, which is proper whatever the number of observations m required to obtain r successes. Notice that $r = 0$ is *not* possible under this model: inverse binomial sampling implicitly assumes that $r \geqslant 1$ successes will occur for sure.

In reporting results, scientists are typically required to specify not only the observed data but also the conditions under which those where obtained, the *design* of the experiment, so that the data analyst has available the full specification of the model, $\mathcal{M} \equiv \{p(\boldsymbol{x}|\boldsymbol{\omega}), \boldsymbol{x} \in \boldsymbol{\mathcal{X}}, \boldsymbol{\omega} \in \boldsymbol{\Omega}\}$. To carry out a reference analysis of the data, such a full specification (that is, including the experimental design) is indeed required. The reference prior $\pi(\boldsymbol{\omega}|\mathcal{M}, \mathcal{P})$ is proposed as a *consensus* prior to analyse data *associated to a particular design* $\mathcal{M}$ (and under any agreed assumptions about the value of $\boldsymbol{\omega}$ which might be encapsulated in the choice of $\mathcal{P}$).

The *likelihood principle* (Berger and Wolpert, 1988) says that all evidence about an unknown quantity $\boldsymbol{\omega}$, which is obtained from an experiment which has produced data $\boldsymbol{x}$, is contained in the likelihood function $p(\boldsymbol{x}|\boldsymbol{\omega})$ of $\boldsymbol{\omega}$ for the *observed* data $\boldsymbol{x}$. In particular, for any *specific* prior beliefs (described by a *fixed* prior), proportional likelihoods should produce the same posterior distribution.

As Example 12 demonstrates, it may be argued that formal use of reference priors is not compatible with the likelihood principle. However, the likelihood principle applies *after* data have been observed while reference priors are derived *before* the data are observed. Reference priors are a (limiting) form of rather specific beliefs, namely those which would maximize the missing information (about the quantity or interest) *associated to a particular design*, and thus depend on the particular design considered. There is no claim that these particular beliefs describe (or even approximate) those of any particular individual; instead, they are precisely defined as possible *consensus* prior functions, presumably useful as a *reference* for scientific communication. Notice that reference prior *functions* (often improper) should *not* be interpreted as prior probability *distributions*: they are merely technical devices to facilitate the derivation of reference posteriors, and only reference posteriors support a probability interpretation.

Any statistical analysis should include an evaluation of the sensitivity of the results to accepted assumptions. In particular, any Bayesian analysis should include some discussion of the sensitivity of the results to the choice of the prior, and reference priors are better viewed as a useful tool for this important aspect of *sensitivity analysis*. The analyst is supposed to have a unique (often subjective) prior $p(\boldsymbol{\omega})$, independent of the design of the experiment, but the scientific community will presumably be interested in comparing the corresponding analyst's personal posterior with the *reference* (consensus) posterior associated to the published experimental design. To report reference posteriors (possibly for a range of alternative designs) should be seen as part of this sensitivity analysis. Indeed, reference analysis provides an answer to an important *conditional* question in scientific inference: the reference posterior encapsulates what *could* be said about the quantity of interest *if* prior information about its value were minimal *relative* to the information which repeated data from an specific experimental design $\mathcal{M}$ could possibly provide.

3.7. *Restricted reference priors*

The reference prior $\pi(\theta \mid \mathcal{M}, \mathcal{P})$ is that which maximizes the missing information about θ relative to model $\mathcal{M}$ among the priors which belong to $\mathcal{P}$, the class of all sufficiently regular priors which are compatible with available knowledge (Definition 6). By restricting the class $\mathcal{P}$ of candidate priors to those which satisfy specific restrictions (derived from assumed knowledge) one may use the reference prior algorithm as an effective tool for *prior elicitation*: the corresponding reference prior will incorporate the accepted restrictions, but no other information.

Under regularity conditions, Theorems 3, 8 and 9, make it relatively simple to obtain the unrestricted reference prior $\pi(\theta) = \pi(\theta \mid \mathcal{M}, \mathcal{P}_0)$ which corresponds to the case where the class of candidate priors is the class $\mathcal{P}_0$ of all continuous priors with support Θ. Hence, it is useful to be able to express a general reference prior $\pi(\theta \mid \mathcal{M}, \mathcal{P})$ in terms of the corresponding unrestricted reference prior $\pi(\theta \mid \mathcal{M}, \mathcal{P}_0)$, and the set of restrictions which define the class $\mathcal{P}$ of candidate priors.

If the unrestricted reference prior $\pi(\theta \mid \mathcal{M}, \mathcal{P}_0)$ is proper, then $\pi(\theta \mid \mathcal{M}, \mathcal{P})$ is the closest prior in $\mathcal{P}$ to $\pi(\theta \mid \mathcal{M}, \mathcal{P}_0)$, in the sense of minimizing the intrinsic discrepancy (see Definition 1) between them, so that

$$\pi(\theta \mid \mathcal{M}, \mathcal{P}) = \arg \inf_{p(\theta) \in \mathcal{P}} \delta\{p(\theta), \pi(\theta \mid \mathcal{M}, \mathcal{P}_0)\}.$$

If $\pi(\theta \mid \mathcal{M}, \mathcal{P}_0)$ is not proper it may be necessary to derive $\pi(\theta \mid \mathcal{M}, \mathcal{P})$ from its definition. However, in the rather large class of problems where the conditions which define $\mathcal{P}$ may all be expressed in the general form $\int_\Theta g_i(\theta)p(\theta)\,\mathrm{d}\theta = \beta_i$, for appropriately chosen functions $g_i(\theta)$ (i.e., as a collection of expected values which the prior $p(\theta)$ must satisfy), an explicit solution is available in terms of the unrestricted reference prior:

THEOREM 10 (Explicit form of restricted reference priors). *Consider a model* $\mathcal{M} \equiv \{p(\boldsymbol{x}|\theta), \boldsymbol{x} \in \boldsymbol{\mathcal{X}}, \theta \in \Theta\}$, *let* $\mathcal{P}$ *be the class of continuous proper priors with support* Θ

$$\mathcal{P} = \left\{ p_\theta;\ \int_\Theta p(\theta)\,\mathrm{d}\theta = 1,\ \int_\Theta g_i(\theta)p(\theta)\,\mathrm{d}\theta = \beta_i,\ i = 1,\ldots,m \right\}$$

which satisfies the restrictions imposed by the expected values $\mathrm{E}[g_i(\theta)] = \beta_i$, *and let* $\mathcal{P}_0$ *be the class of all continuous priors with support* Θ. *The reference prior* $\pi(\theta|\mathcal{M},\mathcal{P})$, *if it exists, is then of the form*

$$\pi(\theta|\mathcal{M},\mathcal{P}) \propto \pi(\theta|\mathcal{M},\mathcal{P}_0)\ \exp\left\{ \sum_{i=1}^m \lambda_i g_i(\theta) \right\},$$

where the λ_i*'s are constants determined by the conditions which define* $\mathcal{P}$.

Theorem 10 may be proven using a standard calculus of variations argument. If $m = 0$, so that one only has the constraint that the prior is proper, then there typically is no restricted reference prior. For details, see Bernardo and Smith (1994, p. 316).

EXAMPLE 13 (*Location models, continued*). Let $\boldsymbol{x} = \{x_1,\ldots,x_n\}$ be a random sample from a location model $\mathcal{M} \equiv \{f(x-\mu), x \in \mathcal{X}, \mu \in \mathbb{R}\}$, and suppose that the prior mean and variance of μ are restricted to be $\mathrm{E}[\mu] = \mu_0$, and $\mathrm{Var}[\mu] = \sigma_0^2$. By Theorem 7, the unrestricted reference prior $\pi(\mu|\mathcal{M},\mathcal{P}_0)$ is uniform; hence, using Theorem 10, the (restricted) reference prior must be of the form

$$\pi(\mu|\mathcal{M},\mathcal{P}) \propto \exp\{\lambda_1\mu + \lambda_2(\mu - \mu_0)^2\}$$

with $\int_{-\infty}^{\infty} \mu\pi(\mu|\mathcal{M},\mathcal{P})\,\mathrm{d}\mu = \mu_0$ and $\int_{-\infty}^{\infty} (\mu - \mu_0)^2\pi(\mu|\mathcal{M},\mathcal{P})\,\mathrm{d}\mu = \sigma_0^2$. It follows that $\lambda_1 = 0$ and $\lambda_2 = -1/(2\sigma_0^2)$ and, substituting above, the restricted reference prior is $\pi(\mu|\mathcal{M},\mathcal{P}) \propto \exp\{-(\mu - \mu_0)^2/(2\sigma_0^2)\}$, which is the *normal* distribution $\mathrm{N}(\mu|\mu_0,\sigma_0)$ with the specified mean and variance. This provides a very powerful argument for the choice of a normal density to describe prior information in location models, when prior knowledge about the location parameter is *limited* to its first two moments.

3.8. One nuisance parameter

Consider now the case where the statistical model $\mathcal{M}$ contains one nuisance parameter, so that $\mathcal{M} \equiv \{p(\boldsymbol{x}|\theta,\lambda), \boldsymbol{x} \in \boldsymbol{\mathcal{X}}, \theta \in \Theta, \lambda \in \Lambda\}$, the quantity of interest is $\theta \in \Theta \subset \mathbb{R}$, and the nuisance parameter is $\lambda \in \Lambda \subset \mathbb{R}$. To obtain the required reference posterior for θ, $\pi(\theta|\boldsymbol{x})$, an appropriate *joint* reference prior $\pi^\theta(\theta,\lambda)$ is obviously needed: by Bayes theorem, the corresponding joint posterior is $\pi^\theta(\theta,\lambda|\boldsymbol{x}) \propto p(\boldsymbol{x}|\theta,\lambda)\pi^\theta(\theta,\lambda)$

and, integrating out the nuisance parameter, the (marginal) reference posterior for the parameter of interest is

$$\pi(\theta|\boldsymbol{x}) = \int_\Lambda \pi^\theta(\theta,\lambda|\boldsymbol{x})\,\mathrm{d}\lambda \propto \int_\Lambda p(\boldsymbol{x}|\theta,\lambda)\pi^\theta(\theta,\lambda)\,\mathrm{d}\lambda.$$

The extension of the reference prior algorithm to the case of two parameters follows the usual mathematical procedure of reducing the two parameter problem to a sequential application of the established procedure for the single parameter case. Thus, the reference algorithm proceeds by combining the results obtained in two successive applications of the one-parameter solution:

(1) Conditional on θ, $p(\boldsymbol{x}|\theta,\lambda)$ only depends on the nuisance parameter λ and, hence, the one-parameter algorithm may be used to obtain the *conditional* reference prior $\pi(\lambda|\theta) = \pi(\lambda|\theta,\mathcal{M},\mathcal{P})$.

(2) If $\pi(\lambda|\theta)$ has a finite integral in Λ (so that, when normalized, yields a proper density with $\int_\Lambda \pi(\lambda|\theta)\,\mathrm{d}\lambda = 1$), the conditional reference prior $\pi(\lambda|\theta)$ may be used to integrate out the nuisance parameter and derive the one-parameter integrated model,

$$p(\boldsymbol{x}|\theta) = \int_\Lambda p(\boldsymbol{x}|\theta,\lambda)\pi(\lambda|\theta)\,\mathrm{d}\lambda, \tag{25}$$

to which the one-parameter algorithm may be applied again to obtain the *marginal* reference prior $\pi(\theta) = \pi(\theta|\mathcal{M},\mathcal{P})$.

(3) The desired θ-reference prior is then $\pi^\theta(\theta,\lambda) = \pi(\lambda|\theta)\pi(\theta)$, and the required reference posterior is

$$\pi(\theta|\boldsymbol{x}) \propto \int_\Lambda p(\boldsymbol{x}|\theta,\lambda)\pi^\theta(\theta,\lambda)\,\mathrm{d}\lambda = p(\boldsymbol{x}|\theta)\pi(\theta). \tag{26}$$

Eq. (25) suggests that conditional reference priors provides a general procedure to eliminate nuisance parameters, a major problem within the frequentist paradigm. For a review of this important topic, see Liseo (2005), in this volume.

If the conditional reference prior $\pi(\lambda|\theta)$ is *not* proper, Eq. (25) does not define a valid statistical model and, as a consequence, a more subtle approach is needed to provide a general solution; this will be described later. Nevertheless, the simple algorithm described above may be used to obtain appropriate solutions to a number of interesting problems which serve to illustrate the crucial need to identify the quantity of interest, as is the following two examples.

EXAMPLE 14 (*Induction*). Consider a finite population of (known) size N, all of whose elements may or may not have a specified property. A random sample of size n is taken without replacement, and all the elements in the sample turn out to have that property. Scientific interest often centres in the probability that all the N elements in the population have the property under consideration (natural induction). It has often been argued that for relatively large n values, this should be close to one whatever might be the population size N (typically much larger than the sample size n). Thus, if all the $n = 225$

randomly chosen turtles in an isolated volcanic island are found to show a particular difference with respect to those in the mainland, zoologists would tend to believe that all the turtles in the island share that property. Formally, if r and R respectively denote the number of elements in the sample and in the population which have the property under study, the statistical model is

$$\mathcal{M} \equiv \{p(r|n, R, N),\ r \in \{0, \dots, n\},\ R \in \{0, \dots, N\}\},$$

where R is the unknown parameter, and $p(r|n, R, N) = \binom{R}{r}\binom{N-R}{n-r}/\binom{N}{n}$ is the relevant hypergeometric distribution. The required result,

$$p(R = N|r = n, N) = \frac{p(r = n|n, R, N)p(R = N)}{\sum_{R=0}^{N} p(r = n|n, R, N)p(R)} \tag{27}$$

may immediately be obtained from Bayes theorem, once a prior $p(R)$ for the unknown number R of elements in the population which have the property has been established. If the parameter of interest were R itself, the reference prior would be uniform over its range (Theorem 2), so that $p(R) = (N+1)^{-1}$; using (27) this would lead to the posterior probability $p(R = N|r = n, N) = (n + 1)/(N + 1)$ which will be small when (as it is usually the case) the sampling fraction n/N is small. However, the quantity of interest here is *not* the value of R but whether or not $R = N$, and a reference prior is desired which maximizes the missing information about this *specific* question. Rewriting the unknown parameter as $R = (\theta, \lambda)$, where $\theta = 1$ if $R = N$ and $\theta = 0$ otherwise, and $\lambda = 1$ if $R = N$ and $\lambda = R$ otherwise (so that the quantity of interest θ is explicitly shown), and using Theorem 2 and the argument above, one gets $\pi(\lambda|\theta = 1) = 1$, $\pi(\lambda|\theta = 0) = N^{-1}$, and $\pi(\theta = 0) = \pi(\theta = 1) = 1/2$, so that the θ-reference prior is $\pi^{\theta}(R) = 1/2$ if $R = N$ and $\pi^{\theta}(R) = 1/(2N)$ if $R \neq N$. Using (27), this leads to

$$p(R = N|r = n, N) = \left[1 + \frac{1}{n+1}\left(1 - \frac{n}{N}\right)\right]^{-1} \approx \frac{n+1}{n+2} \tag{28}$$

which, as expected, clearly displays the irrelevance of the sampling fraction, and the approach to unity for large n. In the turtles example (a real question posed to the author at the Galápagos Islands in the eighties), this yields $p(R = N|r = n = 225, N) \approx 0.995$ for all large N. The *reference* result (28) does not necessarily represents any personal scientist's beliefs (although apparently it may approach actual scientist's beliefs in many situations), but the conclusions which should be reached from a situation where the missing information about the quantity of interest (whether or not $R = N$) is maximized, a situation mathematically characterized by the θ-reference prior described above. For further discussion of this problem (with important applications in philosophy of science, physical sciences and reliability), see Jeffreys (1961, pp. 128–132), Geisser (1984), Bernardo (1985b) and Singpurwalla and Wilson (2004).

EXAMPLE 15 (*Ratio of multinomial parameters*). Let data $\boldsymbol{x} = \{r_1, r_2, n\}$ consist of the result of n trinomial observations, with parameters α_1, α_2 and $\alpha_3 = 1 - \alpha_1 - \alpha_2$, so that, for $0 < \alpha_i < 1$, $\alpha_1 + \alpha_2 < 1$,

$$p(r_1, r_2|n, \alpha_1, \alpha_2) = c(r_1, r_2, n)\alpha_1^{r_1}\alpha_2^{r_2}(1 - \alpha_1 - \alpha_2)^{n-r_1-r_2},$$

where $c(r_1, r_2, n) = (n!)/(r_1!r_2!(n - r_1 - r_2)!)$, and suppose that the quantity of interest is the *ratio* $\theta = \alpha_1/\alpha_2$ of the first two original parameters. Reparametrization in terms of θ and (say) $\lambda = \alpha_2$ yields

$$p(r_1, r_2|n, \theta, \lambda) = c(r_1, r_2, n)\theta^{r_1}\lambda^{r_1+r_2}\{1 - \lambda(1+\theta)\}^{n-r_1-r_2},$$

for $\theta > 0$ and, given θ, $0 < \lambda < (1+\theta)^{-1}$. Conditional on θ, this is a model with one continuous parameter λ, and the corresponding Fisher information function is $i(\lambda|\theta) = n(1+\theta)/\{\lambda(1 - \lambda(1+\theta))\}$; using Theorem 9 the conditional reference prior of the nuisance parameter is $\pi(\lambda|\theta) \propto i(\lambda|\theta)^{1/2}$ which is the *proper* beta-like prior $\pi(\lambda|\theta) \propto \lambda^{-1/2}\{1 - \lambda(1+\theta)\}^{-1/2}$, with support on $\lambda \in [0, (1+\theta)^{-1}]$ (which depends on θ). Integration of the full model $p(r_1, r_2|n, \theta, \lambda)$ with the conditional reference prior $\pi(\lambda|\theta)$ yields $p(r_1, r_2|n, \theta) = \int_0^{(1+\theta)^{-1}} p(r_1, r_2|n, \theta, \lambda)\pi(\lambda|\theta)\, d\lambda$, the *integrated* one-parameter model

$$p(r_1, r_2|n, \theta) = \frac{\Gamma(r_1 + r_2 + \frac{1}{2})\Gamma(n - r_1 - r_2 + \frac{1}{2})}{r_1!r_2!(n - r_1 - r_2)!} \frac{\theta^{r_1}}{(1+\theta)^{r_1+r_2}}.$$

The corresponding Fisher information function is $i(\theta) = n/\{2\theta(1+\theta)^2\}$; using again Theorem 9 the reference prior of the parameter of interest is $\pi(\theta) \propto i(\theta)^{1/2}$ which is the proper prior $\pi(\theta) \propto \theta^{-1/2}(1+\theta)^{-1}, \theta > 0$. Hence, by Bayes theorem, the reference posterior of the quantity of interest is $\pi(\theta|r_1, r_2, n) \propto p(r_1, r_2|n, \theta)\pi(\theta)$; this yields

$$\pi(\theta|r_1, r_2) = \frac{\Gamma(r_1 + r_2 + 1)}{\Gamma(r_1 + \frac{1}{2})\Gamma(r_2 + \frac{1}{2})} \frac{\theta^{r_1 - 1/2}}{(1+\theta)^{r_1+r_2+1}}, \qquad \theta > 0.$$

Notice that $\pi(\theta|r_1, r_2)$ does *not* depend on n; to draw conclusions about the value of $\theta = \alpha_1/\alpha_2$ only the numbers r_1 and r_2 observed in the first two classes matter: a result $\{55, 45, 100\}$ carries precisely the same information about the *ratio* α_1/α_2 than a result $\{55, 45, 10\,000\}$. For instance, if an electoral survey of size n yields r_1 voters for party A and r_2 voters for party B, the reference posterior distribution of the *ratio* θ of the proportion of A voters to B voters in the population only depends on their respective number of voters in the sample, r_1 and r_2, whatever the size and political intentions of the other $n - r_1 - r_2$ citizens in the sample. In particular, the reference posterior probability that party A gets better results than party B is $\Pr[\theta > 1|r_1, r_2] = \int_1^\infty \pi(\theta|r_1, r_2)\, d\theta$. As one would expect, this is precisely equal to $1/2$ if, and only if, $r_1 = r_2$; one-dimensional numerical integration (or use of the incomplete beta function) is required to compute other values. For instance, whatever the total sample size n in each case, this yields $\Pr[\theta > 1|r_1 = 55, r_2 = 45] = 0.841$ (with $r_1 + r_2 = 100$) and $\Pr[\theta > 1|r_1 = 550, r_2 = 450] = 0.999$ (with the same ratio r_1/r_2, but $r_1 + r_2 = 1000$).

As illustrated by the preceding examples, in a multiparameter model, say $\mathcal{M} \equiv \{p(\boldsymbol{x}|\boldsymbol{\omega}), \boldsymbol{x} \in \boldsymbol{\mathcal{X}}, \boldsymbol{\omega} \in \boldsymbol{\Omega}\}$ the required (joint) reference prior $\pi^\theta(\boldsymbol{\omega})$ may depend on the quantity of interest, $\boldsymbol{\theta} = \boldsymbol{\theta}(\boldsymbol{\omega})$ (although, as one would certainly expect, and will later be demonstrated, this will *not* be the case if the new quantity of interest $\boldsymbol{\phi} = \boldsymbol{\phi}(\boldsymbol{\omega})$ say, is a one-to-one function of $\boldsymbol{\theta}$). Notice that this does *not* mean that the analyst's beliefs should depend on his or her interests; as stressed before, reference priors are not meant to describe the analyst's beliefs, but the mathematical formulation of a particular type of prior beliefs – those which would maximize the expected missing information about the quantity of interest – which could be adopted by consensus as a standard for scientific communication.

If the conditional reference prior $\pi(\lambda|\theta)$ is *not* proper, so that Eq. (25) does not define a valid statistical model, then integration may be performed within each of the elements of an increasing sequence $\{\Lambda_i\}_{i=1}^{\infty}$ of subsets of Λ converging to Λ over which $\pi(\lambda|\theta)$ is integrable. Thus, Eq. (25) is to be replaced by

$$p_i(\boldsymbol{x}|\theta) = \int_{\Lambda_i} p(\boldsymbol{x}|\theta,\lambda)\pi_i(\lambda|\theta)\,\mathrm{d}\lambda, \tag{29}$$

where $\pi_i(\lambda|\theta)$ is the renormalized proper restriction of $\pi(\lambda|\theta)$ to Λ_i, from which the reference posterior $\pi_i(\theta|\boldsymbol{x}) = \pi(\theta|\mathcal{M}_i,\mathcal{P})$, which corresponds to model $\mathcal{M}_i \equiv \{p(\boldsymbol{x}|\theta,\lambda), \boldsymbol{x}\in\boldsymbol{\mathcal{X}}, \theta\in\Theta, \lambda\in\Lambda_i\}$ may be derived.

The use of the sequence $\{\Lambda_i\}_{i=1}^{\infty}$ makes it possible to obtain a corresponding sequence of θ-reference posteriors $\{\pi_i(\theta|\boldsymbol{x})\}_{i=1}^{\infty}$ for the quantity of interest θ which corresponds to the sequence of integrated models (29); the required reference posterior may then be found as the corresponding intrinsic limit $\pi(\theta|\boldsymbol{x}) = \lim_{i\to\infty}\pi_i(\theta|\boldsymbol{x})$. A θ-reference prior is then defined as any positive function $\pi^{\theta}(\theta,\lambda)$ which may formally be used in Bayes' theorem to directly obtain the reference posterior, so that for all $\boldsymbol{x}\in\boldsymbol{\mathcal{X}}$, the posterior density satisfies $\pi(\theta|\boldsymbol{x}) \propto \int_{\Lambda} p(\boldsymbol{x}|\theta,\lambda)\pi^{\theta}(\theta,\lambda)\,\mathrm{d}\lambda$.

The approximating sequences should be *consistently* chosen within the same model: given a statistical model $\mathcal{M} \equiv \{p(\boldsymbol{x}|\boldsymbol{\omega}), \boldsymbol{x}\in\boldsymbol{\mathcal{X}}, \boldsymbol{\omega}\in\boldsymbol{\Omega}\}$ an appropriate approximating sequence $\{\boldsymbol{\Omega}_i\}$ should be chosen for the whole parameter space $\boldsymbol{\Omega}$. Thus, if the analysis is done in terms of $\psi = \{\psi_1,\psi_2\}\in\Psi(\boldsymbol{\Omega})$, the approximating sequence should be chosen such that $\Psi_i = \psi(\boldsymbol{\Omega}_i)$. A very natural approximating sequence in location-scale problems is $\{\mu, \log\sigma\}\in[-i,i]^2$; reparametrization to asymptotically independent parameters and approximate location reparametrizations (Definition 7) may be combined to choose appropriate approximating sequences in more complex situations. A formal definition of reference prior functions in multiparameter problems is possible along the lines of Definition 6.

As one would hope, the θ-reference prior does *not* depend on the choice of the nuisance parameter λ; thus, for any $\psi = \psi(\theta,\lambda)$ such that (θ,ψ) is a one-to-one function of (θ,λ), the θ-reference prior in terms of (θ,ψ) is simply $\pi^{\theta}(\theta,\psi) = \pi^{\theta}(\theta,\lambda)/|\partial(\theta,\psi)/\partial(\theta,\lambda)|$, the appropriate probability transformation of the θ-reference prior in terms of (θ,λ). Notice however that, as mentioned before, the reference prior *may* depend on the parameter of interest; thus, the θ-reference prior may differ from the ϕ-reference prior unless either ϕ is a one-to-one transformation of θ, or ϕ is asymptotically independent of θ. This is an expected consequence of the mathematical fact that the prior which maximizes the missing information about θ is not generally the same as the prior which maximizes the missing information about any function $\phi = \phi(\theta,\lambda)$.

The *nonexistence* of a unique "noninformative" prior for all inference problems within a given model was established by Dawid et al. (1973) when they showed that this is incompatible with *consistent marginalization*. Indeed, given a two-parameter model $\mathcal{M} \equiv \{p(\boldsymbol{x}|\theta, \lambda), \boldsymbol{x} \in \boldsymbol{\mathcal{X}}, \theta \in \Theta, \lambda \in \Lambda\}$, if the reference posterior of the quantity of interest θ, $\pi(\theta|\boldsymbol{x}) = \pi(\theta|\boldsymbol{t})$, only depends on the data through a statistic $\boldsymbol{t} = \boldsymbol{t}(\boldsymbol{x})$ whose sampling distribution, $p(\boldsymbol{t}|\theta, \lambda) = p(\boldsymbol{t}|\theta)$, only depends on θ, one would expect the reference posterior to be of the form $\pi(\theta|\boldsymbol{t}) \propto p(\boldsymbol{t}|\theta)\pi(\theta)$ for some prior $\pi(\theta)$. However, examples were found where this *cannot* be the case if a *unique* joint "noninformative" prior were to be used for all possible quantities of interest within the same statistical model $\mathcal{M}$.

By definition, a reference prior must be a *permissible* prior function. In particular (Definition 3), it must yield *proper posteriors* for all data sets large enough to identify the parameters. For instance, if data $\boldsymbol{x}$ consist of a random sample of fixed size n from a normal $\mathrm{N}(x|\mu, \sigma)$ distribution, so that, $\mathcal{M} \equiv \{\prod_{j=1}^{n} \mathrm{N}(x_j|\mu, \sigma), x_j \in \mathbb{R}, \sigma > 0\}$, the function $\pi^{\mu}(\mu, \sigma) = \sigma^{-1}$ is only a permissible (joint) prior for μ if $n \geqslant 2$ (and, without restrictions in the class $\mathcal{P}$ of candidate priors, a reference prior function *does not exist* for $n = 1$).

Under posterior asymptotic normality, reference priors are easily obtained in terms of the relevant Fisher information matrix. The following result extends Theorem 9 to models with two continuous parameters:

THEOREM 11 (Reference priors under asymptotic binormality). *Let data* $\boldsymbol{x} = \{\boldsymbol{y}_1, \ldots, \boldsymbol{y}_n\}$ *consist of* n *conditionally independent (given* θ*) observations from a model* $\mathcal{M} \equiv \{p(\boldsymbol{y}|\theta, \lambda), \boldsymbol{y} \in \boldsymbol{\mathcal{Y}}, \theta \in \Theta, \lambda \in \Lambda\}$, *and let* $\mathcal{P}_0$ *be the class of all continuous (joint) priors with support* $\Theta \times \Lambda$. *If the posterior distribution of* $\{\theta, \lambda\}$ *is asymptotically normal with dispersion matrix* $V(\hat{\theta}_n, \hat{\lambda}_n)/n$, *where* $\{\hat{\theta}_n, \hat{\lambda}_n\}$ *is a consistent estimator of* $\{\theta, \lambda\}$, *define*

$$V(\theta, \lambda) = \begin{pmatrix} v_{\theta\theta}(\theta, \lambda) & v_{\theta\lambda}(\theta, \lambda) \\ v_{\theta\lambda}(\theta, \lambda) & v_{\lambda\lambda}(\theta, \lambda) \end{pmatrix}, \qquad H(\theta, \lambda) = V^{-1}(\theta, \lambda), \quad \text{and}$$

$$\pi(\lambda|\theta) \propto h_{\lambda\lambda}^{1/2}(\theta, \lambda), \quad \lambda \in \Lambda, \tag{30}$$

and, if $\pi(\lambda|\theta)$ *is proper,*

$$\pi(\theta) \propto \exp\left\{\int_{\Lambda} \pi(\lambda|\theta) \log\left[v_{\theta\theta}^{-1/2}(\theta, \lambda)\right] \mathrm{d}\lambda\right\}, \quad \theta \in \Theta. \tag{31}$$

Then, if $\pi(\lambda|\theta)\pi(\theta)$ *is a permissible prior function, the* θ*-reference prior is*

$$\pi\left(\theta|\mathcal{M}^n, \mathcal{P}_0\right) \propto \pi(\lambda|\theta)\pi(\theta).$$

If $\pi(\lambda|\theta)$ *is not proper, integration in* (31) *is performed on elements of an increasing sequence* $\{\Lambda_i\}_{i=1}^{\infty}$ *such that* $\int_{\Lambda_i} \pi(\lambda|\theta)\,\mathrm{d}\lambda < \infty$, *to obtain the sequence* $\{\pi_i(\lambda|\theta)\pi_i(\theta)\}_{i=1}^{\infty}$, *where* $\pi_i(\lambda|\theta)$ *is the renormalization of* $\pi(\lambda|\theta)$ *to* Λ_i, *and the* θ*-reference prior* $\pi^{\theta}(\theta, \lambda)$ *is defined as its corresponding intrinsic limit.*

A heuristic justification of Theorem 11 is now provided. Under the stated conditions, given k independent observations from model $\mathcal{M}$, the *conditional* posterior distribution of λ given θ is asymptotically normal with precision $kh_{\lambda\lambda}(\theta, \hat{\lambda}_k)$, and the *marginal* posterior distribution of θ is asymptotically normal with precision $kv_{\theta\theta}^{-1}(\hat{\theta}_k, \hat{\lambda}_k)$; thus, using Theorem 9, $\pi(\lambda|\theta) \propto h_{\lambda\lambda}^{1/2}(\theta, \lambda)$, which is Eq. (30). Moreover, using Theorem 3,

$$\pi_k(\theta) \propto \exp\left\{\iint p(\hat{\theta}_k, \hat{\lambda}_k|\theta) \log\left[\mathrm{N}\{\theta|\hat{\theta}_k, k^{-1/2} v_{\theta\theta}^{1/2}(\hat{\theta}_k, \hat{\lambda}_k)\}\right] \mathrm{d}\hat{\theta}_k \, \mathrm{d}\hat{\lambda}_k\right\}, \tag{32}$$

where, if $\pi(\lambda|\theta)$ is proper, the integrated model $p(\hat{\theta}_k, \hat{\lambda}_k|\theta)$ is given by

$$p(\hat{\theta}_k, \hat{\lambda}_k|\theta) = \int_{\Lambda} p(\hat{\theta}_k, \hat{\lambda}_k|\theta, \lambda)\pi(\lambda|\theta) \, \mathrm{d}\lambda. \tag{33}$$

Introducing (33) into (32) and using the fact that $(\hat{\theta}_k, \hat{\lambda}_k)$ is a consistent estimator of (θ, λ) – so that as $k \to \infty$ integration with $p(\hat{\theta}_k, \hat{\lambda}_k|\theta, \lambda)$ reduces to substitution of $(\hat{\theta}_k, \hat{\lambda}_k)$ by (θ, λ) – directly leads to Eq. (31). If $\pi(\lambda|\theta)$ is not proper, it is necessary to integrate in an increasing sequence $\{\Lambda_i\}_{i=1}^{\infty}$ of subsets of Λ such that the restriction $\pi_i(\lambda|\theta)$ of $\pi(\lambda|\theta)$ to Λ_i is proper, obtain the sequence of reference priors which correspond to these restricted models, and then take limits to obtain the required result.

Notice that under appropriate regularity conditions (see, e.g., Bernardo and Smith, 1994, Section 5.3, and references therein) the joint posterior distribution of $\{\theta, \lambda\}$ is asymptotically normal with precision matrix $nI(\hat{\theta}_n, \hat{\lambda}_n)$, where $I(\theta)$ is Fisher information matrix; in that case, the asymptotic dispersion matrix in Theorem 11 is simply $V(\theta, \lambda) = I^{-1}(\theta, \lambda)/n$.

THEOREM 12 (Reference priors under factorization). *In the conditions of Theorem* 11, *if* (i) θ *and* λ *are variation independent – so that* Λ *does not depend on* θ *– and* (ii) *both* $h_{\lambda\lambda}(\theta, \lambda)$ *and* $v_{\theta\theta}(\theta, \lambda)$ *factorize, so that*

$$v_{\theta\theta}^{-1/2}(\theta, \lambda) \propto f_\theta(\theta) g_\theta(\lambda), \qquad h_{\lambda\lambda}^{1/2}(\theta, \lambda) \propto f_\lambda(\theta) g_\lambda(\lambda), \tag{34}$$

then the θ*-reference prior is simply* $\pi^\theta(\theta, \lambda) = f_\theta(\theta) g_\lambda(\lambda)$, *even if the conditional reference prior* $\pi(\lambda|\theta) = \pi(\lambda) \propto g_\lambda(\lambda)$ *is improper.*

If $h_{\lambda\lambda}^{1/2}(\theta, \lambda)$ factorizes as $h_{\lambda\lambda}^{1/2}(\theta, \lambda) = f_\lambda(\theta) g_\lambda(\lambda)$, then the conditional reference prior is $\pi(\lambda|\theta) \propto f_\lambda(\theta) g_\lambda(\lambda)$ and, normalizing, $\pi(\lambda|\theta) = c_1 g_\lambda(\lambda)$, which does not depend on θ. If, furthermore, $v_{\theta\theta}^{-1/2}(\theta, \lambda) = f_\theta(\theta) g_\theta(\lambda)$ and Λ does not depend on θ, Eq. (31) reduces to

$$\pi(\theta) \propto \exp\left\{\int_{\Lambda} c_1 g_\lambda(\lambda) \log\left[f_\theta(\theta) g_\theta(\lambda)\right] \mathrm{d}\lambda\right\} = c_2 f_\theta(\theta)$$

and, hence, the reference prior is $\pi^\theta(\theta, \lambda) = \pi(\lambda|\theta)\pi(\theta) = c f_\theta(\theta) g_\lambda(\lambda)$.

EXAMPLE 16 (*Inference on the univariate normal parameters*). Let data $\boldsymbol{x} = \{x_1, \ldots, x_n\}$ consist of a random sample of fixed size n from a normal distribution

$N(x|\mu,\sigma)$. The information matrix $I(\mu,\sigma)$ and its inverse matrix are respectively

$$I(\mu,\sigma) = \begin{pmatrix} \sigma^{-2} & 0 \\ 0 & 2\sigma^{-2} \end{pmatrix}, \qquad V(\mu,\sigma) = I^{-1}(\mu,\sigma) = \begin{pmatrix} \sigma^{2} & 0 \\ 0 & \frac{1}{2}\sigma^{2} \end{pmatrix}.$$

Hence, $i_{\sigma\sigma}^{1/2}(\mu,\sigma) = \sqrt{2}\sigma^{-1} = f_\sigma(\mu) g_\sigma(\sigma)$, with $g_\sigma(\sigma) = \sigma^{-1}$, so that $\pi(\sigma|\mu) = \sigma^{-1}$. Similarly, $v_{\mu\mu}^{-1/2}(\mu,\sigma) = \sigma^{-1} = f_\mu(\mu) g_\sigma(\sigma)$, with $f_\mu(\mu) = 1$, and thus $\pi(\mu) = 1$. Therefore, using Theorem 11 the μ-reference prior is $\pi^\mu(\mu,\sigma) = \pi(\sigma|\mu)\pi(\mu) = \sigma^{-1}$ for all $n \geqslant 2$. For $n = 1$ the posterior distribution is not proper, the function $h(\mu,\sigma) = \sigma^{-1}$ is *not* a permissible prior, and a reference prior does not exist. Besides, since $I(\mu,\sigma)$ is diagonal, the σ-reference prior is $\pi^\sigma(\mu,\sigma) = f_\sigma(\sigma) g_\mu(\mu) = \sigma^{-1}$, the same as $\pi^\mu(\mu,\sigma)$.

Consider now the case where the quantity of interest is *not* the mean μ or the standard deviation σ, but the *standardized* mean $\phi = \mu/\sigma$ (or, equivalently, the coefficient of variation σ/μ). Fisher's matrix in terms of the parameters ϕ and σ is $I(\phi,\sigma) = J^t I(\mu,\sigma) J$, where $J = (\partial(\mu,\sigma)/\partial(\phi,\sigma))$ is the Jacobian of the inverse transformation, and this yields

$$I(\phi,\sigma) = \begin{pmatrix} 1 & \phi\sigma^{-1} \\ \phi\sigma^{-1} & \sigma^{-2}(2+\phi^2) \end{pmatrix}, \quad V(\phi,\sigma) = \begin{pmatrix} 1+\frac{1}{2}\phi^2 & -\frac{1}{2}\phi\sigma \\ -\frac{1}{2}\phi\sigma & \frac{1}{2}\sigma^2 \end{pmatrix}.$$

Thus, $i_{\sigma\sigma}^{1/2}(\phi,\sigma) = \sigma^{-1}(2+\phi^2)^{1/2}$, and $v_{\phi\phi}^{-1/2}(\phi,\sigma) = (1+\frac{1}{2}\phi^2)^{-1/2}$. Hence, using Theorem 11, $\pi^\phi(\phi,\sigma) = (1+\frac{1}{2}\phi^2)^{-1/2}\sigma^{-1}$ ($n \geqslant 2$). In the original parametrization, this is $\pi^\phi(\mu,\sigma) = (1+\frac{1}{2}(\mu/\sigma)^2)^{-1/2}\sigma^{-2}$, which is *very* different from $\pi^\mu(\mu,\sigma) = \pi^\sigma(\mu,\sigma) = \sigma^{-1}$. The reference posterior of the quantity of interest ϕ after data $\boldsymbol{x} = \{x_1,\ldots,x_n\}$ have been observed is

$$\pi(\phi|\boldsymbol{x}) \propto \left(1+\frac{1}{2}\phi^2\right)^{-1/2} p(t|\phi), \tag{35}$$

where $t = (\sum x_j)/(\sum x_j^2)^{1/2}$, a one-dimensional statistic whose sampling distribution, $p(t|\mu,\sigma) = p(t|\phi)$, only depends on ϕ. Thus, the reference prior algorithm is seen to be consistent under marginalization.

The reference priors $\pi^\mu(\mu,\sigma) = \sigma^{-1}$ and $\pi^\sigma(\mu,\sigma) = \sigma^{-1}$ for the normal location and scale parameters obtained in the first part of Example 16 are just a particular case of a far more general result:

THEOREM 13 (Location-scale models). *If* $\mathcal{M}$ *is a* location-scale *model, so that for some function* f, $\mathcal{M} \equiv \sigma^{-1} f\{(x-\mu)/\sigma, x \in \mathcal{X}, \mu \in \mathbb{R}, \sigma > 0\}$, *and* $\mathcal{P}_0$ *is the class of all continuous, strictly positive (joint) priors for* (μ,σ), *then a reference prior for either* μ *or* σ, *if it exists, is of the form*

$$\pi^\mu(\mu,\sigma|\mathcal{M},\mathcal{P}_0) = \pi^\sigma(\mu,\sigma|\mathcal{M},\mathcal{P}_0) \propto \sigma^{-1}.$$

For a proof, which is based on the form of the relevant Fisher matrix, see Fernández and Steel (1999b).

When the quantity of interest and the nuisance parameter are *not* variation independent, derivation of the reference prior requires special care. This is illustrated in the example below:

EXAMPLE 17 (*Product of positive normal means*). Let data consist of two independent random samples $\boldsymbol{x} = \{x_1, \ldots, x_n\}$ and $\boldsymbol{y} = \{y_1, \ldots, y_m\}$ from $N(x|\alpha, 1)$ and $N(y|\beta, 1)$, $\alpha > 0$, $\beta > 0$, so that the assumed model is

$$p(\boldsymbol{x}, \boldsymbol{y}|\alpha, \beta) = \prod_{i=1}^{n} N(x_i|\alpha, 1) \prod_{j=1}^{m} N(y_j|\beta, 1), \quad \alpha > 0, \beta > 0,$$

and suppose that the quantity of interest is the product of the means, $\theta = \alpha\beta$, a frequent situation in physics and engineering. Reparametrizing in terms of the one-to-one transformation $(\theta, \lambda) = (\alpha\beta, \alpha/\beta)$, Fisher matrix $I(\theta, \lambda)$ and its inverse matrix $V(\theta, \lambda)$ are,

$$I = \begin{pmatrix} \frac{m+n\lambda^2}{4\theta\lambda} & \frac{1}{4}(n - \frac{m}{\lambda^2}) \\ \frac{1}{4}(n - \frac{m}{\lambda^2}) & \frac{\theta(m+n\lambda^2)}{4\lambda^3} \end{pmatrix}, \qquad V = \begin{pmatrix} \theta(\frac{1}{n\lambda} + \frac{\lambda}{m}) & \frac{1}{n} - \frac{\lambda^2}{m} \\ \frac{1}{n} - \frac{\lambda^2}{m} & \frac{\lambda(m+n\lambda^2)}{nm\theta} \end{pmatrix}.$$

and, therefore, using (30),

$$\pi(\lambda|\theta) \propto I_{22}(\theta, \lambda)^{1/2} \propto \theta^{1/2}\left(m + n\lambda^2\right)^{1/2}\lambda^{-3/2}. \tag{36}$$

The natural increasing sequence of subsets of the original parameter space, $\Omega_i = \{(\alpha, \beta); 0 < \alpha < i, 0 < \beta < i\}$, transforms, in the parameter space of λ, into the sequence $\Lambda_i(\theta) = \{\lambda; \theta i^{-2} < \lambda < i^2\theta^{-1}\}$. Notice that this depends on θ, so that θ and λ are *not* variation independent and, hence, Theorem 12 *cannot* be applied. Renormalizing (36) in $\Lambda_i(\theta)$ and using (31), it is found that, for large i,

$$\pi_i(\lambda|\theta) = c_i(m, n)\theta^{1/2}\left(m + n\lambda^2\right)^{1/2}\lambda^{-3/2},$$

$$\pi_i(\theta) = c_i(m, n) \int_{\Lambda_i(\theta)} \left(m + n\lambda^2\right)^{1/2}\lambda^{-3/2} \log\left(\frac{\lambda}{m} + \frac{1}{n\lambda}\right)^{-1/2} d\lambda,$$

where $c_i(m, n) = i^{-1}\sqrt{nm}/(\sqrt{m} + \sqrt{n})$, which leads to the θ-reference prior $\pi^\theta(\theta, \lambda) \propto \theta^{1/2}\lambda^{-1}(\frac{\lambda}{m} + \frac{1}{n\lambda})^{1/2}$. In the original parametrization, this corresponds to

$$\pi^\theta(\alpha, \beta) \propto \left(n\alpha^2 + m\beta^2\right)^{1/2}, \quad n \geqslant 1,\ m \geqslant 1, \tag{37}$$

which depends on the sample sizes through the ratio m/n. It has already been stressed that the reference prior depends on the experimental design. It is therefore not surprising that, if the design is unbalanced, the reference prior depends on the ratio m/n which controls the level of balance. Notice that the reference prior (37) is very different from the uniform prior $\pi^\alpha(\alpha, \beta) = \pi^\beta(\alpha, \beta) = 1$, which should be used to make reference inferences about either α or β.

It will later be demonstrated (Example 22) that the prior $\pi^\theta(\alpha, \beta)$ found above provides approximate agreement between Bayesian credible regions and frequentist confidence intervals for θ (Berger and Bernardo, 1989); indeed, this prior was originally suggested by Stein (1986) (who only considered the case $m = n$) to obtain such approximate agreement. Efron (1986) used this problem as an example in which *conventional* objective Bayesian theory encounters difficulties since, even within a fixed model $\mathcal{M} \equiv \{p(\boldsymbol{y}|\boldsymbol{\theta}), \boldsymbol{y} \in \boldsymbol{\mathcal{Y}}, \boldsymbol{\theta} \in \boldsymbol{\Theta}\}$, the "correct" objective prior depends on the particular function $\phi = \phi(\boldsymbol{\theta})$ one desires to estimate. For the reference priors associated to generalizations of the product of normal means problem, see Sun and Ye (1995, 1999).

3.9. *Many parameters*

Theorems 11 and 12 may easily be extended to any number of nuisance parameters. Indeed, let data $\boldsymbol{x} = \{\boldsymbol{y}_1, \ldots, \boldsymbol{y}_n\}$ consist of a random sample of size n from a model $\mathcal{M} \equiv \{p(\boldsymbol{y}|\boldsymbol{\omega}), \boldsymbol{y} \in \boldsymbol{\mathcal{Y}}, \boldsymbol{\omega} = \{\omega_1, \ldots, \omega_m\}, \boldsymbol{\omega} \in \boldsymbol{\Omega}\}$, let ω_1 be the quantity of interest, assume regularity conditions to guarantee that, as $n \to \infty$, the joint posterior distribution of $\boldsymbol{\omega}$ is asymptotically normal with mean $\hat{\boldsymbol{\omega}}_n$ and dispersion matrix $V(\hat{\boldsymbol{\omega}}_n)/n$, and let $H(\boldsymbol{\omega}) = V^{-1}(\boldsymbol{\omega})$. It then follows that, if $V_j(\boldsymbol{\omega})$ is the $j \times j$ upper matrix of $V(\boldsymbol{\omega})$, $j = 1, \ldots, m$, $H_j(\boldsymbol{\omega}) = V_j^{-1}(\boldsymbol{\omega})$ and $h_{jj}(\boldsymbol{\omega})$ is the lower right (j, j) element of $H_j(\boldsymbol{\omega})$, then

(1) the *conditional* posterior distribution of ω_j given $\{\omega_1, \ldots, \omega_{j-1}\}$, is asymptotically normal with precision $nh_{jj}(\hat{\boldsymbol{\omega}}_n)$ $(j = 2, \ldots, m)$ and
(2) the *marginal* posterior distribution of ω_1 is asymptotically normal with precision $nh_{11}(\hat{\boldsymbol{\omega}}_n)$.

This may be used to extend the algorithm described in Theorem 11 to sequentially derive $\pi(\omega_m|\omega_1, \ldots, \omega_{m-1})$, $\pi(\omega_{m-1}|\omega_1, \ldots, \omega_{m-2}), \ldots, \pi(\omega_2|\omega_1)$ and $\pi(\omega_1)$; their product yields the reference prior associated to the particular ordering $\{\omega_1, \omega_2, \ldots, \omega_m\}$. Intuitively, this is a mathematical description of a situation where, relative to the particular design considered $\mathcal{M}$, one maximizes the missing information about the parameter ω_1 (that of higher inferential importance), but also the missing information about ω_2 given ω_1, that of ω_3 given ω_1 and $\omega_2, \ldots$ and that of ω_m given ω_1 to ω_{m-1}. As in sequential decision theory, this must be done backwards. In particular, to maximize the missing information about ω_1, the prior which maximizes the missing information about ω_2 given ω_1 has to be derived first.

The choice of the ordered parametrization, say $\{\theta_1(\boldsymbol{\omega}), \theta_2(\boldsymbol{\omega}), \ldots, \theta_m(\boldsymbol{\omega})\}$, precisely describes the particular prior required, namely that which sequentially maximizes the missing information about the θ_j's in order of inferential interest. Indeed, "diffuse" prior knowledge about a particular sequence $\{\theta_1(\boldsymbol{\omega}), \theta_2(\boldsymbol{\omega}), \ldots, \theta_m(\boldsymbol{\omega})\}$ may be very "precise" knowledge about another sequence $\{\phi_1(\boldsymbol{\omega}), \phi_2(\boldsymbol{\omega}), \ldots, \phi_m(\boldsymbol{\omega})\}$ unless, *for all* j, $\phi_j(\boldsymbol{\omega})$ is a one-to-one function of $\theta_j(\boldsymbol{\omega})$. Failure to recognize this fact is known to produce untenable results; famous examples are the paradox of Stein (1959) (see Example 19 below) and the marginalization paradoxes (see Example 16).

THEOREM 14 (Reference priors under asymptotic normality). *Let data* $\boldsymbol{x} = \{\boldsymbol{y}_1, \ldots, \boldsymbol{y}_n\}$ *consist of a random sample of size n from a statistical model* $\mathcal{M} \equiv \{p(\boldsymbol{y}|\boldsymbol{\theta}), \boldsymbol{y} \in \boldsymbol{\mathcal{Y}}, \boldsymbol{\theta} = \{\theta_1, \ldots, \theta_m\}, \boldsymbol{\theta} \in \boldsymbol{\Theta} = \prod_{j=1}^m \Theta_j\}$, *and let* $\mathcal{P}_0$ *be the class of all continuous priors with support* $\boldsymbol{\Theta}$. *If the posterior distribution of* $\boldsymbol{\theta}$ *is asymptotically normal with dispersion matrix* $V(\hat{\boldsymbol{\theta}}_n)/n$, *where* $\hat{\boldsymbol{\theta}}_n$ *is a consistent estimator of* $\boldsymbol{\theta}$, $H(\boldsymbol{\theta}) = V^{-1}(\boldsymbol{\theta})$, V_j *is the upper* $j \times j$ *submatrix of* V, $H_j = V_j^{-1}$, *and* $h_{jj}(\boldsymbol{\theta})$ *is the lower right element of* H_j, *then the* $\boldsymbol{\theta}$*-reference prior, associated to the ordered parametrization* $\{\theta_1, \ldots, \theta_m\}$, *is*

$$\pi\big(\boldsymbol{\theta} \mid \mathcal{M}^n, \mathcal{P}_0\big) = \pi(\theta_m \mid \theta_1, \ldots, \theta_{m-1}) \times \cdots \times \pi(\theta_2 \mid \theta_1)\pi(\theta_1),$$

with $\pi(\theta_m \mid \theta_1, \ldots, \theta_{m-1}) = h_{mm}^{1/2}(\boldsymbol{\theta})$ *and, for* $i = 1, \ldots, m-1$,

$$\pi(\theta_j \mid \theta_1, \ldots, \theta_{j-1}) \propto \exp\left\{ \int_{\boldsymbol{\Theta}^{j+1}} \prod_{l=j+1}^{m} \pi(\theta_l \mid \theta_1, \ldots, \theta_{l-1}) \log\big[h_{jj}^{1/2}(\boldsymbol{\theta})\big] \, \mathrm{d}\boldsymbol{\theta}^{j+1} \right\}$$

with $\boldsymbol{\theta}^{j+1} = \{\theta_{j+1}, \ldots, \theta_m\}$, *provided* $\pi(\theta_j \mid \theta_1, \ldots, \theta_{j-1})$ *is proper for all* j.

If the conditional reference priors $\pi(\theta_j \mid \theta_1, \ldots, \theta_{j-1})$ *are not all proper, integration is performed on elements of an increasing sequence* $\{\boldsymbol{\Theta}_i\}_{i=1}^{\infty}$ *such that* $\int_{\Theta_{ij}} \pi(\theta_j \mid \theta_1, \ldots, \theta_{j-1}) \, \mathrm{d}\theta_j$ *is finite, to obtain the corresponding sequence* $\{\pi_i(\boldsymbol{\theta})\}_{i=1}^{\infty}$ *of reference priors for the restricted models. The* $\boldsymbol{\theta}$*-reference prior is then defined as their intrinsic limit.*

If, moreover, (i) Θ_j *does not depend on* $\{\theta_1, \ldots, \theta_{j-1}\}$, *and* (ii) *the functions* $h_{jj}(\theta, \lambda)$ *factorize in the form*

$$h_{jj}^{1/2}(\boldsymbol{\theta}) \propto f_j(\theta_j) g_j(\theta_1, \ldots, \theta_{j-1}, \theta_{j+1}, \ldots, \theta_m), \quad j = 1, \ldots, m,$$

then the $\boldsymbol{\theta}$*-reference prior is simply* $\pi^{\theta}(\boldsymbol{\theta}) = \prod_{j=1}^{m} f_j(\theta_j)$, *even if the conditional reference priors are improper.*

Under appropriate regularity conditions – see, e.g., Bernardo and Smith (1994, Theorem 5.14) – the posterior distribution of $\boldsymbol{\theta}$ is asymptotically normal with mean the mle $\hat{\boldsymbol{\theta}}_n$ and precision matrix $nI(\hat{\boldsymbol{\theta}}_n)$, where $I(\boldsymbol{\theta})$ is Fisher matrix,

$$i_{ij}(\boldsymbol{\theta}) = -\int_{\mathcal{Y}} p(\mathbf{y} \mid \boldsymbol{\theta}) \frac{\partial^2}{\partial\theta_i \partial\theta_j} \log\big[p(\mathbf{y} \mid \boldsymbol{\theta})\big] \, \mathrm{d}\mathbf{y};$$

in that case, $H(\boldsymbol{\theta}) = nI(\boldsymbol{\theta})$, and the reference prior may be computed from the elements of Fisher matrix $I(\boldsymbol{\theta})$. Notice, however, that in the multivariate case, the reference prior does *not* yield Jeffreys multivariate rule (Jeffreys, 1961), $\pi^J(\boldsymbol{\theta}) \propto |I(\boldsymbol{\theta})|^{1/2}$. For instance, in location-scale models, the (μ, σ)-reference prior and the (σ, μ)-reference prior are both $\pi^R(\mu, \sigma) = \sigma^{-1}$ (Theorem 13), while Jeffreys multivariate rule yields $\pi^J(\mu, \sigma) = \sigma^{-2}$. As a matter of fact, Jeffreys himself criticised his own multivariate rule. This is known, for instance, to produce both marginalization paradoxes Dawid et al. (1973), and strong inconsistencies (Eaton and Freedman, 2004). See, also, Stein (1962) and Example 23.

Theorem 14 provides a procedure to obtain the reference prior $\pi^{\theta}(\boldsymbol{\theta})$ which corresponds to any *ordered parametrization* $\boldsymbol{\theta} = \{\theta_1, \ldots, \theta_m\}$. Notice that, within any particular multiparameter model

$$\mathcal{M} \equiv \{p(\boldsymbol{x}|\boldsymbol{\theta}),\ \boldsymbol{x} \in \boldsymbol{\mathcal{X}},\ \boldsymbol{\theta} = \{\theta_1, \ldots, \theta_m\} \in \boldsymbol{\Theta} \subset \mathbb{R}^k\},$$

the reference algorithm provides a (possibly different) joint reference prior

$$\pi^{\phi}(\boldsymbol{\phi}) = \pi(\phi_m|\phi_1, \ldots, \phi_{m-1}) \times \cdots \times \pi(\phi_2|\phi_1)\pi(\phi_1),$$

for each possible ordered parametrization $\{\phi_1(\boldsymbol{\theta}), \phi_2(\boldsymbol{\theta}), \ldots, \phi_m(\boldsymbol{\theta})\}$. However, as one would hope, the results are coherent under monotone transformations of each of the $\phi_i(\boldsymbol{\theta})$'s in the sense that, in that case, $\pi^{\phi}(\boldsymbol{\phi}) = \pi^{\theta}[\boldsymbol{\theta}(\boldsymbol{\phi})]|J(\boldsymbol{\phi})|$, where $J(\boldsymbol{\phi})$ is the Jacobian of the inverse transformation $\boldsymbol{\theta} = \boldsymbol{\theta}(\boldsymbol{\phi})$, of general element $j_{ij}(\boldsymbol{\phi}) = \partial\theta_i(\boldsymbol{\phi})/\partial\phi_j$. This property of coherence under appropriate reparametrizations may be very useful in choosing a particular parametrization (for instance one with orthogonal parameters, or one in which the relevant $h_{jj}(\boldsymbol{\theta})$ functions factorize) which simplifies the implementation of the algorithm.

Starting with Jeffreys (1946) pioneering work, the analysis of the invariance properties under reparametrization of multiparameter objective priors has a very rich history. Relevant pointers include Hartigan (1964), Stone (1965, 1970), Zidek (1969), Florens (1978, 1982), Dawid (1983), Consonni and Veronese (1989), Chang and Eaves (1990), George and McCulloch (1993), Datta and J.K. Ghosh (1995b), Yang (1995), Datta and M. Ghosh (1996), Eaton and Sudderth (1999, 2002, 2004) and Severini et al. (2002). In particular, Datta and J.K. Ghosh (1995b), Yang (1995) and Datta and M. Ghosh (1996) are specifically concerned with the invariance properties of reference distributions.

EXAMPLE 18 (*Multivariate normal data*). Let data consist of a size n random sample $\boldsymbol{x} = \{\boldsymbol{y}_1, \ldots, \boldsymbol{y}_n\}$, $n \geqslant 2$, from an m-variate normal distribution with mean $\boldsymbol{\mu}$, and covariance matrix $\sigma^2\boldsymbol{I}_m$, $m \geqslant 1$, so that

$$I(\boldsymbol{\mu}, \sigma) = \begin{pmatrix} \sigma^{-2}\boldsymbol{I}_m & 0 \\ 0 & (2/m)\sigma^{-2} \end{pmatrix}.$$

It follows from Theorem 14 that the reference prior relative to the natural parametrization $\boldsymbol{\theta} = \{\mu_1, \ldots, \mu_m, \sigma\}$ is $\pi^{\theta}(\mu_1, \ldots, \mu_m, \sigma) \propto \sigma^{-1}$, and also that the result does not depend on the order in which the parametrization is taken, since their asymptotic covariances are zero. Hence, $\pi^{\theta}(\mu_1, \ldots, \mu_m, \sigma) \propto \sigma^{-1}$ is the appropriate prior function to obtain the reference posterior of any piecewise invertible function $\phi(\mu_j)$ of μ_j, and also to obtain the reference posterior of any piecewise invertible function $\phi(\sigma)$ of σ. In particular, the corresponding reference posterior for any of the μ_j's is easily shown to be the Student density

$$\pi(\mu_j|\boldsymbol{y}_1, \ldots, \boldsymbol{y}_n) = \mathrm{St}\{\mu_j|\bar{y}_j, s/\sqrt{(n-1)}, m(n-1)\}$$

with $n\bar{y}_j = \sum_{i=1}^{n} y_{ij}$, and $nms^2 = \sum_{j=1}^{m}\sum_{i=1}^{n}(y_{ij} - \bar{y}_j)^2$, which agrees with the standard argument according to which one degree of freedom should be lost by each of the unknown means. Similarly, the reference posterior of σ^2 is the inverted Gamma

$$\pi(\sigma^2|\boldsymbol{y}_1, \ldots, \boldsymbol{y}_n) = \mathrm{IGa}\{\sigma^2|n(m-1)/2, nms^2/2\}.$$

When $m = 1$, these results reduce to those obtained in Example 16.

EXAMPLE 19 (*Stein's paradox*). Let $\boldsymbol{x} \in \mathcal{X}$ be a random sample of size n from a m-variate normal $\mathrm{N}_m(\boldsymbol{x}|\boldsymbol{\mu}, I_m)$ with mean $\boldsymbol{\mu} = \{\mu_1, \ldots, \mu_m\}$ and unitary dispersion matrix. The reference prior which corresponds to any permutation of the μ_i's is uniform, and this uniform prior leads indeed to appropriate reference posterior distributions for any of the μ_j's, given by $\pi(\mu_j|\boldsymbol{x}) = \mathrm{N}(\mu_j|\bar{x}_j, 1/\sqrt{n})$. Suppose, however, that the quantity of interest is $\theta = \sum_j \mu_j^2$, the squared distance of $\boldsymbol{\mu}$ from the origin. As shown by Stein (1959), the posterior distribution of θ based on the uniform prior (or indeed any "flat" *proper* approximation) has very undesirable properties; this is due to the fact that a uniform (or nearly uniform) prior, although "noninformative" with respect to each of the individual μ_j's, is actually highly informative on the sum of their squares, introducing a severe bias towards large values of θ (Stein's paradox). However, the reference prior which corresponds to a parametrization of the form $\{\theta, \boldsymbol{\lambda}\}$ produces, for any choice of the nuisance parameter vector $\boldsymbol{\lambda} = \boldsymbol{\lambda}(\boldsymbol{\mu})$, the reference posterior for the quantity of interest $\pi(\theta|\boldsymbol{x}) = \pi(\theta|t) \propto \theta^{-1/2}\chi^2(nt|m, n\theta)$, where $t = \sum_i \bar{x}_i^2$, and this posterior is shown to have the appropriate consistency properties. For further details see Ferrándiz (1985).

If the μ_i's were known to be related, so that they could be assumed to be exchangeable, with $p(\boldsymbol{\mu}) = \prod_{i=1}^m p(\mu_i|\boldsymbol{\phi})$, for some $p(\mu|\boldsymbol{\phi})$, one would have a (very) different (hierarchical) model. Integration of the μ_i's with $p(\boldsymbol{\mu})$ would then produce a model $\mathcal{M} \equiv \{p(\boldsymbol{x}|\boldsymbol{\phi}), \boldsymbol{x} \in \mathcal{X}, \boldsymbol{\phi} \in \boldsymbol{\Phi}\}$ parametrized by $\boldsymbol{\phi}$, and only the corresponding reference prior $\pi(\boldsymbol{\phi}|\mathcal{M})$ would be required. See below (Subsection 3.12) for further discussion on reference priors in hierarchical models.

Far from being specific to Stein's example, the inappropriate behaviour in problems with many parameters of specific marginal posterior distributions derived from multivariate "flat" priors (proper or improper) is very frequent. Thus, as indicated in the introduction, uncritical use of "flat" priors (rather than the relevant reference priors), should be very strongly discouraged.

3.10. *Discrete parameters taking an infinity of values*

Due to the nonexistence of an asymptotic theory comparable to that of the continuous case, the infinite discrete case presents special problems. However, it is often possible to obtain an approximate reference posterior by embedding the discrete parameter space within a continuous one.

EXAMPLE 20 (*Discrete parameters taking an infinite of values*). In the context of capture–recapture models, it is of interest to make inferences about the population size $\theta \in \{1, 2, \ldots\}$ on the basis of data $\boldsymbol{x} = \{x_1, \ldots, x_n\}$, which are assumed to consist of a random sample from

$$p(x|\theta) = \frac{\theta(\theta+1)}{(x+\theta)^2}, \quad 0 \leqslant x \leqslant 1.$$

This arises, for instance, in software reliability, when the unknown number θ of bugs is assumed to be a continuous mixture of Poisson distributions. Goudie and Goldie (1981) concluded that, in this problem, all standard non-Bayesian methods are liable to fail;

Raftery (1988) finds that, for several plausible "diffuse looking" prior distributions for the discrete parameter θ, the corresponding posterior virtually ignores the data; technically, this is due to the fact that, for most samples, the corresponding likelihood function $p(\boldsymbol{x}|\theta)$ tends to one (rather than to zero) as $\theta \to \infty$. Embedding the discrete parameter space $\Theta = \{1, 2, \ldots\}$ into the continuous space $\Theta = (0, \infty)$ (since, for each $\theta > 0$, $p(x|\theta)$ is still a probability density for x), and using Theorem 9, the appropriate reference prior is

$$\pi(\theta) \propto i(\theta)^{1/2} \propto (\theta + 1)^{-1}\theta^{-1},$$

and it is easily verified that this prior leads to a posterior in which the data are no longer overwhelmed. If the problem requires the use of discrete θ values, the discrete approximation $\Pr(\theta = 1|\boldsymbol{x}) = \int_0^{3/2} \pi(\theta|\boldsymbol{x})\,d\theta$, and $\Pr(\theta = j|\boldsymbol{x}) = \int_{j-1/2}^{j+1/2} \pi(\theta|\boldsymbol{x})\,d\theta$, $j > 1$, may be used as an approximate discrete reference posterior, specially when interest mostly lies on large θ values, as it is typically the case.

3.11. *Behaviour under repeated sampling*

The frequentist coverage probabilities of the different types of credible intervals which may be derived from reference posterior distributions are sometimes identical, and usually very close, to their posterior probabilities; this means that, even for moderate samples, an interval with reference posterior probability q may often be interpreted as an *approximate* frequentist confidence interval with significance level $1 - q$.

EXAMPLE 21 (*Coverage in simple normal problems*). Consider again inferences about the mean μ and the variance σ^2 of a normal $\mathrm{N}(x|\mu, \sigma)$ model. Using the reference prior $\pi^{\mu}(\mu, \sigma) \propto \sigma^{1}$ derived in Example 16, the reference posterior distribution of μ after a random sample $\boldsymbol{x} = \{x_1, \ldots, x_n\}$ has been observed, $\pi(\mu|\boldsymbol{x}) \propto \int_0^{\infty} \prod_{j=1}^{n} \mathrm{N}(x_j|\mu, \sigma)\pi^{\mu}(\mu, \sigma)\,d\sigma$, is the Student density $\pi(\mu|\boldsymbol{x}) = \mathrm{St}(\mu|\bar{x}, s/\sqrt{n-1}, n-1) \propto [s^2 + (\bar{x} - \mu)^2]^{-n/2}$, where $\bar{x} = \sum_j x_j/n$, and $s^2 = \sum_j (x_j - \bar{x})^2/n$. Hence, the reference *posterior* of the standardized function of μ, $\phi(\mu) = \sqrt{n-1}(\mu - \bar{x})/s$ is standard Student with $n - 1$ degrees of freedom. But, conditional on μ, the *sampling* distribution of $t(\boldsymbol{x}) = \sqrt{n-1}(\mu - \bar{x})/s$ is *also* standard Student with $n - 1$ degrees of freedom. It follows that, for all sample sizes, posterior reference credible intervals for μ will numerically be identical to frequentist confidence intervals based on the sampling distribution of t. Similar results are obtained concerning inferences about σ: the reference posterior distribution of $\psi(\sigma) = ns^2/\sigma^2$ is a χ^2 with $n - 1$ degrees of freedom but, conditional on σ, this is also the sampling distribution of $r(\boldsymbol{x}) = ns^2/\sigma^2$.

The *exact* numerical agreement between reference posterior credible intervals and frequentist confidence intervals shown in Example 21 is however the exception, not the norm. Nevertheless, for *large* sample sizes, reference credible intervals are always *approximate* confidence intervals *approximate confidence intervals*.

More precisely, let data $\boldsymbol{x} = \{x_1, \ldots, x_n\}$ consist of n independent observations from $\mathcal{M} = \{p(x|\theta), x \in \mathcal{X}, \theta \in \Theta\}$, and let $\theta_q(\boldsymbol{x}, p_\theta)$ denote the q quantile of the posterior

$p(\theta|\boldsymbol{x}) \propto p(\boldsymbol{x}|\theta)p(\theta)$ which corresponds to the prior $p(\theta)$, so that

$$\Pr\big[\theta \leqslant \theta_q(\boldsymbol{x}, p_\theta)|\boldsymbol{x}\big] = \int_{\theta \leqslant \theta_q(\boldsymbol{x}, p_\theta)} p(\theta|\boldsymbol{x})\,\mathrm{d}\theta = q.$$

Standard asymptotic theory may be used to establish that, for any sufficiently regular pair $\{p_\theta, \mathcal{M}\}$ of prior p_θ and model $\mathcal{M}$, the *coverage* probability of the region thus defined, $R_q(\boldsymbol{x}, \theta, p_\theta) = \{\boldsymbol{x}; \theta \leqslant \theta_q(\boldsymbol{x}, p_\theta)\}$, converges to q as $n \to \infty$. Specifically, for all sufficiently regular priors,

$$\Pr\big[\theta_q(\boldsymbol{x}, p_\theta) \geqslant \theta|\theta\big] = \int_{R_q(\boldsymbol{x},\theta, p_\theta)} p(\boldsymbol{x}|\theta)\,\mathrm{d}\boldsymbol{x} = q + \mathrm{O}\big(n^{-1/2}\big).$$

It has been found however that, when there are no nuisance parameters, the reference prior π^θ typically satisfies

$$\Pr\big[\theta_q\big(\boldsymbol{x}, \pi^\theta\big) \geqslant \theta|\theta\big] = q + \mathrm{O}\big(n^{-1}\big);$$

this means that the reference prior is often a *probability matching* prior, that is, a prior for which the coverage probabilities of *one-sided* posterior credible intervals are asymptotically closer to their posterior probabilities. Hartigan (1966) showed that the coverage probabilities of *two-sided* Bayesian posterior credible intervals satisfy this type of approximation to $\mathrm{O}(n^{-1})$ for *all* sufficiently regular prior functions.

In a pioneering paper, Welch and Peers (1963) established that in the case of the one-parameter regular continuous models Jeffreys prior (which in this case, Theorem 9, is also the reference prior), is the only probability matching prior. Hartigan (1983, p. 79) showed that this result may be extended to one-parameter discrete models by using continuity corrections. Datta and Ghosh (1995a) derived a differential equation which provides a necessary and sufficient condition for a prior to be probability matching in the multiparameter continuous regular case; this has been used to verify that reference priors are typically probability matching priors.

In the nuisance parameter setting, reference priors are sometimes matching priors fro the parameter of interest, but in this general situation, matching priors may not always exist or be unique (Welch, 1965; Ghosh and Mukerjee, 1998). For a review of probability matching priors, see Datta and Sweeting (2005), in this volume.

Although the results described above only justify an *asymptotic* approximate frequentist interpretation of reference posterior probabilities, the coverage probabilities of reference posterior credible intervals derived from *relatively small samples* are also found to be typically close to their posterior probabilities. This is now illustrated within the product of positive normal means problem, already discussed in Example 17.

EXAMPLE 22 (*Product of normal means, continued*). Let available data $\boldsymbol{x} = \{x, y\}$ consist of one observation x from $\mathrm{N}(x|\alpha, 1)$, $\alpha > 0$, and another observation y from $\mathrm{N}(y|\beta, 1)$, $\beta > 0$, and suppose that the quantity of interest is the product of the means $\theta = \alpha\beta$. The behaviour under repeated sampling of the posteriors which correspond to both the conventional uniform prior $\pi^u(\alpha, \beta) = 1$, and the reference

Table 1
Coverage probabilities of 0.05-credible regions for $\theta = \alpha\beta$

$\{\alpha, \beta\}$	$\Pr[R_{0.05} \mid \theta, \pi^u]$	$\Pr[R_{0.05} \mid \theta, \pi^\theta]$
$\{1, 1\}$	0.024	0.047
$\{2, 2\}$	0.023	0.035
$\{3, 3\}$	0.028	0.037
$\{4, 4\}$	0.033	0.048
$\{5, 5\}$	0.037	0.046

prior $\pi^\theta(\alpha, \beta) = (\alpha^2 + \beta^2)^{1/2}$ (see Example 17) is analyzed by computing the coverage probabilities $\Pr[R_q \mid \theta, p_\theta] = \int_{R_q(\boldsymbol{x}, \theta, p_\theta)} p(\boldsymbol{x} \mid \theta)\, \mathrm{d}\boldsymbol{x}$ associated to the regions $R_q(\boldsymbol{x}, \theta, p_\theta) = \{\boldsymbol{x}; \theta \leqslant \theta_q(\boldsymbol{x}, p_\theta)\}$ defined by their corresponding quantiles, $\theta_q(\boldsymbol{x}, \pi^u)$ and $\theta_q(\boldsymbol{x}, \pi^\theta)$. Table 1 contains the coverage probabilities of the regions defined by the 0.05 posterior quantiles. These have been numerically computed by simulating 4000 pairs $\{x, y\}$ from $\mathrm{N}(x \mid \alpha, 1)\mathrm{N}(y \mid \beta, 1)$ for each of the $\{\alpha, \beta\}$ pairs listed in the first column of the table.

The standard error of the entries in the table is about 0.0035. It may be observed that the estimated coverages which correspond to the reference prior are appreciably closer to the nominal value 0.05 that those corresponding to the uniform prior. Notice that, although it may be shown that the reference prior *is* probability matching in the technical sense described above, the empirical results shown in the table do *not* follow from that fact, for probability matching is an *asymptotic* result, and one is dealing here with samples of size $n = 1$. For further details on this example, see Berger and Bernardo (1989).

3.12. Prediction and hierarchical models

Two classes of problems that are not specifically covered by the methods described above are hierarchical models and prediction problems. The difficulty with these problems is that the distributions of the quantities of interest must belong to specific families of distributions. For instance, if one wants to predict the value of y based on $\boldsymbol{x}$ when $(y, \boldsymbol{x})$ has density $p(y, \boldsymbol{x} \mid \boldsymbol{\theta})$, the unknown of interest is y, but its distribution is conditionally specified; thus, one needs a prior for $\boldsymbol{\theta}$, not a prior for y. Likewise, in a hierarchical model with, say, $\{\mu_1, \mu_2, \ldots, \mu_p\}$ being $\mathrm{N}(\mu_i \mid \theta, \lambda)$, the μ_i's may be the parameters of interest, but a prior is only needed for the hyperparameters θ and λ.

In hierarchical models, the parameters with conditionally known distributions may be integrated out (which leads to the so-called marginal over-dispersion models). A reference prior for the remaining parameters based on this marginal model is then required. The difficulty that arises is how to then identify parameters of interest and nuisance parameters to construct the ordering necessary for applying the reference algorithm, the real parameters of interest having been integrated out.

A possible solution to the problems described above is to define the quantity of interest to be the conditional mean of the original parameter of interest. Thus, in the prediction problem, the quantity of interest could be defined to be $\phi(\boldsymbol{\theta}) = \mathrm{E}[y \mid \boldsymbol{\theta}]$, which will be either $\boldsymbol{\theta}$ or some transformation thereof, and in the hierarchical model mentioned above the quantity of interest could be defined to be $\mathrm{E}[\mu_i \mid \theta, \lambda] = \theta$. More sophisticated choices, in terms of appropriately chosen discrepancy functions, are currently under scrutiny.

Bayesian prediction with objective priors is a very active research area. Pointers to recent suggestions include Kuboki (1998), Eaton and Sudderth (1998, 1999) and Smith (1999). Under appropriate regularity conditions, some of these proposals lead to Jeffreys multivariate prior, $\pi(\boldsymbol{\theta}) \propto |I(\boldsymbol{\theta})|^{1/2}$. However, the use of that prior may lead to rather unappealing predictive posteriors as the following example demonstrates.

EXAMPLE 23 (*Normal prediction*). Let available data consist of a random sample $\boldsymbol{x} = \{x_1, \ldots, x_n\}$ from $\mathrm{N}(x_j|\mu, \sigma)$, and suppose that one is interested in predicting a new, future observation x from $\mathrm{N}(x|\mu, \sigma)$. Using the argument described above, the quantity of interest could be defined to be $\phi(\mu, \sigma) = \mathrm{E}[x|\mu, \sigma] = \mu$ and hence (see Example 16) the appropriate reference prior would be $\pi^x(\mu, \sigma) = \sigma^{-1}$ ($n \geqslant 2$). The corresponding joint reference posterior is $\pi(\mu, \sigma|\boldsymbol{x}) \propto \prod_{j=1}^n \mathrm{N}(x_j|\mu, \sigma)\sigma^{-1}$ and the posterior predictive distribution is

$$\begin{aligned}\pi(x|\boldsymbol{x}) &= \int_0^\infty\int_{-\infty}^\infty \mathrm{N}(x|\mu, \sigma)\pi(\mu, \sigma|\boldsymbol{x})\, \mathrm{d}\mu\, \mathrm{d}\sigma \\ &\propto \left\{(n+1)s^2 + (\bar{x} - \mu)^2\right\}^{-n/2}, \\ &\propto \mathrm{St}\left(x|\bar{x}, s\left\{(n+1)/(n-1)\right\}^{1/2}, n-1\right), \quad n \geqslant 2, \end{aligned} \tag{38}$$

where, as before, $\bar{x} = n^{-1}\sum_{j=1}^n x_j$ and $s^2 = n^{-1}\sum_{j=1}^n (x_j - \bar{x})^2$. As one would expect, the reference predictive distribution (38) is proper whenever $n \geqslant 2$: in the absence of prior knowledge, $n = 2$ is the minimum sample size required to identify the two unknown parameters.

It may be verified that the predictive posterior (38) has consistent coverage properties. For instance, with $n = 2$, the reference posterior predictive probability that a third observation lies within the first two is

$$\Pr[x_{(1)} < x < x_{(2)}|x_1, x_2] = \int_{x_{(1)}}^{x_{(2)}} \pi(x|x_1, x_2)\, \mathrm{d}x = \frac{1}{3},$$

where $x_{(1)} = \min[x_1, x_2]$, and $x_{(2)} = \max[x_1, x_2]$. This is consistent with the fact that, for all μ and σ, the frequentist coverage of the corresponding region of $\mathbb{R}^3$ is precisely

$$\iiint_{\{(x_1,x_2,x_3);\, x_{(1)}<x_3<x_{(2)}\}} \prod_{i=1}^3 \mathrm{N}(x_j|\mu, \sigma)\, \mathrm{d}x_1\, \mathrm{d}x_2\, \mathrm{d}x_3 = \frac{1}{3}. \tag{39}$$

In sharp contrast, if Jeffreys multivariate rule $\pi^J(\mu, \sigma) \propto |I(\mu, \sigma)|^{1/2} = \sigma^{-2}$ were used, the posterior predictive would have been a Student t centered at $\bar{x}$, with scale $s\{(n+1)/n\}^{1/2}$, and with n degrees of freedom, which is proper whenever $n \geqslant 1$. Thus, with $\pi^J(\mu, \sigma)$ as a prior, probabilistic predictions would be possible with only *one* observation, rather unappealing when no prior knowledge is assumed. Moreover, the probability that a third observation lies within the first two which corresponds to the prior $\pi^J(\mu, \sigma)$ is $1/2$, rather than $1/3$, a less than attractive result in view of (39).

For a recent predictive probability matching approach to objective predictive posteriors, see Datta et al. (2000).

4. Reference inference summaries

From a Bayesian viewpoint, the final outcome of a problem of inference about *any* unknown quantity is nothing but the posterior distribution of that quantity. Thus, given some data $\boldsymbol{x}$ and conditions C, *all* that can be said about any function $\boldsymbol{\theta}(\boldsymbol{\omega})$ of the parameters $\boldsymbol{\omega}$ which govern the model is contained in the posterior distribution $p(\boldsymbol{\theta}\,|\,\boldsymbol{x}, C)$, and *all* that can be said about some function $\boldsymbol{y}$ of future observations from the same model is contained in its posterior predictive distribution $p(\boldsymbol{y}\,|\,\boldsymbol{x}, C)$. In fact (Bernardo, 1979a), Bayesian inference may be described as a decision problem where the space of available actions is the class of those posterior probability distributions of the quantity of interest which are compatible with accepted assumptions.

However, to make it easier for the user to assimilate the appropriate conclusions, it is often convenient to *summarize* the information contained in the posterior distribution, while retaining as much of the information as possible. This is conventionally done by (i) providing values of the quantity of interest which, in the light of the data, are likely to be "close" to its true value, and (ii) measuring the compatibility of the results with hypothetical values of the quantity of interest which might have been suggested in the context of the investigation. In this section, objective Bayesian counterparts to these traditional inference problems of *estimation* and *testing*, which are based on the joint use of intrinsic loss functions and reference analysis, are briefly considered.

4.1. Point estimation

Let $\boldsymbol{x}$ be the available data, which are assumed to consist of one observation from $\mathcal{M} \equiv \{p(\boldsymbol{x}\,|\,\boldsymbol{\omega}), \boldsymbol{x} \in \boldsymbol{\mathcal{X}}, \boldsymbol{\omega} \in \Omega\}$, and let $\boldsymbol{\theta} = \boldsymbol{\theta}(\boldsymbol{\omega}) \in \boldsymbol{\Theta}$ be the quantity of interest. Without loss of generality, the original model $\mathcal{M}$ may be written as $\mathcal{M} \equiv \{p(\boldsymbol{x}\,|\,\boldsymbol{\theta}, \boldsymbol{\lambda}), \boldsymbol{x} \in \boldsymbol{\mathcal{X}}, \boldsymbol{\theta} \in \boldsymbol{\Theta}, \boldsymbol{\lambda} \in \boldsymbol{\Lambda}\}$, in terms of the quantity of interest $\boldsymbol{\theta}$ and a vector $\boldsymbol{\lambda}$ of nuisance parameters. A *point estimate* of $\boldsymbol{\theta}$ is some value $\tilde{\boldsymbol{\theta}} \in \boldsymbol{\Theta}$ which could possibly be regarded as an appropriate proxy for the actual, unknown value of $\boldsymbol{\theta}$.

Formally, to choose a point estimate for $\boldsymbol{\theta}$ is a *decision problem*, where the action space is the class Θ of possible $\boldsymbol{\theta}$ values. From a decision-theoretic perspective, to choose a point estimate $\tilde{\boldsymbol{\theta}}$ of some quantity $\boldsymbol{\theta}$ is a *decision* to act as if $\tilde{\boldsymbol{\theta}}$ were $\boldsymbol{\theta}$, not to assert something about the value of $\boldsymbol{\theta}$ (although desire to assert something simple may well be the main reason to obtain an estimate). To solve this decision problem it is necessary to specify a *loss function* $\ell(\tilde{\boldsymbol{\theta}}, \boldsymbol{\theta})$ measuring the consequences of acting *as if* the true value of the quantity of interest were $\tilde{\boldsymbol{\theta}}$, when it is actually $\boldsymbol{\theta}$. The expected posterior loss if $\tilde{\boldsymbol{\theta}}$ were used is $l[\tilde{\boldsymbol{\theta}}\,|\,\boldsymbol{x}] = \int_{\Theta} \ell(\tilde{\boldsymbol{\theta}}, \boldsymbol{\theta})\pi(\boldsymbol{\theta}\,|\,\boldsymbol{x})\,\mathrm{d}\boldsymbol{\theta}$, and the *Bayes estimate* is that $\tilde{\boldsymbol{\theta}}$ value which minimizes $l[\tilde{\boldsymbol{\theta}}|\boldsymbol{x}]$ in $\boldsymbol{\Theta}$. The *Bayes estimator* is the function of the data $\boldsymbol{\theta}^*(\boldsymbol{x}) = \arg\min_{\tilde{\boldsymbol{\theta}}\in\boldsymbol{\Theta}} l[\tilde{\boldsymbol{\theta}}|\boldsymbol{x}]$.

For any given model and data, the Bayes estimate depends on the chosen loss function. The loss function is context specific, and should generally be chosen in terms of the anticipated uses of the estimate; however, a number of conventional loss functions have been suggested for those situations where no particular uses are envisaged. These loss functions produce estimates which may be regarded as simple descriptions of the *location* of the posterior distribution. For example, if the loss function is quadratic, so that $\ell(\tilde{\boldsymbol{\theta}}, \boldsymbol{\theta}) = (\tilde{\boldsymbol{\theta}} - \boldsymbol{\theta})^t(\tilde{\boldsymbol{\theta}} - \boldsymbol{\theta})$, then the Bayes estimate is the *posterior mean* $\boldsymbol{\theta}^* = E[\boldsymbol{\theta}|\boldsymbol{x}]$, assuming that the mean exists. Similarly, if the loss function is a zero-one function, so that $\ell(\tilde{\boldsymbol{\theta}}, \boldsymbol{\theta}) = 0$ if $\tilde{\boldsymbol{\theta}}$ belongs to a ball of radius ε centered in $\boldsymbol{\theta}$ and $\ell(\tilde{\boldsymbol{\theta}}, \boldsymbol{\theta}) = 1$ otherwise, then the Bayes estimate $\boldsymbol{\theta}^*$ tends to the *posterior mode* as the radius of the ball tends to zero, assuming that a unique mode exists.

If θ is univariate and the loss function is linear, so that $\ell(\tilde{\theta}, \theta) = c_1(\tilde{\theta} - \theta)$ if $\tilde{\theta} \geqslant \theta$, and $\ell(\tilde{\theta}, \theta) = c_2(\theta - \tilde{\theta})$ otherwise, then the Bayes estimate is the *posterior quantile* of order $c_2/(c_1 + c_2)$, so that $\Pr[\theta < \theta^*] = c_2/(c_1 + c_2)$. In particular, if $c_1 = c_2$, the Bayes estimate is the *posterior median*. The results just described for univariate linear loss functions clearly illustrate the fact that *any* possible parameter value may turn out be the Bayes estimate: it all depends on the loss function describing the consequences of the anticipated uses of the estimate.

Conventional loss functions are typically *not* invariant under reparametrization. As a consequence, the Bayes estimator ϕ^* of a one-to-one transformation $\phi = \phi(\boldsymbol{\theta})$ of the original parameter $\boldsymbol{\theta}$ is not necessarily $\phi(\boldsymbol{\theta}^*)$ (the *univariate* posterior median, which *is* coherent under reparametrization, is an interesting exception). Moreover, conventional loss functions, such as the quadratic loss, focus attention on the discrepancy between the estimate $\tilde{\boldsymbol{\theta}}$ and the true value $\boldsymbol{\theta}$, rather than on the more relevant discrepancy between the statistical *models* they label. The intrinsic discrepancy $\delta\{\mathcal{M}_{\tilde{\theta}}, p_{x|\theta,\lambda}\}$ (Definition 1) directly measures how different the probability *model* $p(\boldsymbol{x}|\boldsymbol{\theta}, \boldsymbol{\lambda})$ is from its closest approximation within the family $\mathcal{M}_{\tilde{\theta}} \equiv \{p(\boldsymbol{x}|\tilde{\boldsymbol{\theta}}, \boldsymbol{\lambda}), \boldsymbol{\lambda} \in \boldsymbol{\Lambda}\}$, and its value does not depend on the particular parametrization chosen to describe the problem.

DEFINITION 8 (*Intrinsic estimation*). Let available data $\boldsymbol{x}$ consist of one observation from $\mathcal{M} \equiv \{p(\boldsymbol{x}|\boldsymbol{\theta}, \boldsymbol{\lambda}), \boldsymbol{x} \in \boldsymbol{\mathcal{X}}, \boldsymbol{\theta} \in \boldsymbol{\Theta}, \boldsymbol{\lambda} \in \boldsymbol{\Lambda}\}$, let $\mathcal{M}_{\tilde{\theta}}$ be the restricted model $\mathcal{M}_{\tilde{\theta}} \equiv \{p(\boldsymbol{x}|\tilde{\boldsymbol{\theta}}, \boldsymbol{\lambda}), \boldsymbol{x} \in \boldsymbol{\mathcal{X}}, \boldsymbol{\lambda} \in \boldsymbol{\Lambda}\}$, and let

$$\delta\{\tilde{\boldsymbol{\theta}}, (\boldsymbol{\theta}, \boldsymbol{\lambda})\} = \delta\{\mathcal{M}_{\tilde{\theta}}, p_{x|\theta,\lambda}\} = \min_{\tilde{\boldsymbol{\lambda}}\in\boldsymbol{\Lambda}} \delta\{p(\boldsymbol{x}|\tilde{\boldsymbol{\theta}}, \tilde{\boldsymbol{\lambda}}), p(\boldsymbol{x}|\boldsymbol{\theta}, \boldsymbol{\lambda})\} \tag{40}$$

be the intrinsic discrepancy between the distribution $p(\boldsymbol{x}|\boldsymbol{\theta}, \boldsymbol{\lambda})$ and the set of distributions $\mathcal{M}_{\tilde{\theta}}$. The reference posterior expected intrinsic loss is

$$d(\tilde{\boldsymbol{\theta}}|\boldsymbol{x}) = \mathrm{E}[\delta|\boldsymbol{x}] = \int_{\boldsymbol{\Theta}}\int_{\boldsymbol{\Lambda}} \delta\{\tilde{\boldsymbol{\theta}}, (\boldsymbol{\theta}, \boldsymbol{\lambda})\}\pi^\delta(\boldsymbol{\theta}, \boldsymbol{\lambda}|\boldsymbol{x})\,\mathrm{d}\boldsymbol{\theta}\,\mathrm{d}\boldsymbol{\lambda}, \tag{41}$$

where $\pi^\delta(\boldsymbol{\theta}, \boldsymbol{\lambda}|\boldsymbol{x}) \propto p(\boldsymbol{x}|\boldsymbol{\theta}, \boldsymbol{\lambda})\pi^\delta(\boldsymbol{\theta}, \boldsymbol{\lambda})$ is the reference posterior of $(\boldsymbol{\theta}, \boldsymbol{\lambda})$ when δ is the quantity of interest. Given $\boldsymbol{x}$, the *intrinsic estimate* $\boldsymbol{\theta}^* = \boldsymbol{\theta}^*(\boldsymbol{x})$ is that value $\tilde{\boldsymbol{\theta}} \in \boldsymbol{\Theta}$

which minimizes the posterior reference expected intrinsic loss $d(\tilde{\boldsymbol{\theta}}|\boldsymbol{x})$. As a function of $\boldsymbol{x}$, $\boldsymbol{\theta}^*(\boldsymbol{x})$ is the *intrinsic estimator* of $\boldsymbol{\theta}$.

The intrinsic estimate is well defined for any dimensionality, and it is coherent under transformations, in the sense that, if $\boldsymbol{\phi}(\boldsymbol{\theta})$ is a one-to-one function of $\boldsymbol{\theta}$, then the intrinsic estimate $\boldsymbol{\phi}^*$ of $\boldsymbol{\phi}(\boldsymbol{\theta})$ is simply $\boldsymbol{\phi}(\boldsymbol{\theta}^*)$. Under broad regularity conditions (Juárez, 2004), the intrinsic estimator is admissible under the intrinsic loss. Moreover, the reference expected intrinsic loss $d(\tilde{\boldsymbol{\theta}}|\boldsymbol{x})$ is typically a convex function of $\tilde{\boldsymbol{\theta}}$ in a neighbourhood of its minimum, in which case the intrinsic estimate $\boldsymbol{\theta}^*$ is unique, and it is easily derived by either analytical or numerical methods.

EXAMPLE 24 (*Intrinsic estimation of a binomial parameter*). Consider estimation of a binomial proportion θ from r successes given n trials; the reference prior (see Example 12) is $\pi(\theta) \propto \theta^{-1/2}(1-\theta)^{-1/2}$, the corresponding reference posterior is $\pi(\theta|n,r) = \mathrm{Be}(\theta|r+\frac{1}{2}, n-r+\frac{1}{2})$, and the quadratic loss based estimator (the posterior mean) of θ is $\mathrm{E}[\theta|n,r] = (r+1/2)/(n+1)$. However, the quadratic loss based estimator of the log-odds $\phi(\theta) = \log[\theta/(1-\theta)]$, is $\mathrm{E}[\phi|n,r] = \psi(r+1/2) - \psi(n-r+1/2)$ (where $\psi(x) = d\log[\Gamma(x)]/dx$ is the *digamma* function), which is *not* equal to $\phi(\mathrm{E}[\theta|n,r])$.

On the other hand the intrinsic discrepancy between two binomial distributions with parameters θ and $\tilde{\theta}$ and the same value of n, the loss to be suffered if $\tilde{\theta}$ were used as a proxy for θ, is $\delta\{\tilde{\theta},\theta|n\} = n\delta_1\{\tilde{\theta},\theta\}$, where (see Example 1)

$$\delta_1\{\theta_i,\theta_j\} = \min\{k(\theta_i|\theta_j), k(\theta_j|\theta_i)\},$$
$$k(\theta_i|\theta_j) = \theta_j\ \log[\theta_j/\theta_i] + (1-\theta_j)\ \log[(1-\theta_j)/(1-\theta_i)].$$

The intrinsic estimator $\theta^* = \theta^*(r,n)$ is obtained by minimizing the reference expected posterior loss

$$d(\tilde{\theta}|n,r) = \int_0^1 \delta(\tilde{\theta},\theta|n)\mathrm{Be}(\theta|r+\tfrac{1}{2}, n-r+\tfrac{1}{2})\,\mathrm{d}\theta. \tag{42}$$

Since intrinsic estimation is coherent under reparametrization, the intrinsic estimator of, say, the log-odds is simply the log-odds of the intrinsic estimator of θ. The exact value of θ^* may be easily obtained by numerical methods, but a very good linear approximation, based on the reference posterior mean of the approximate location parameter (Definition 7) $\phi(\theta) = \int_0^\theta \theta^{-1/2}(1-\theta)^{-1/2}\,\mathrm{d}\theta = \frac{2}{\pi}\arcsin\sqrt{\theta}$, is

$$\theta^*(r,n) \approx \sin^2\left\{\frac{\pi}{2}E[\phi|r,n]\right\} \approx \left(r+\frac{1}{3}\right)\Big/\left(n+\frac{2}{3}\right). \tag{43}$$

The linear approximation (43) remains good even for small samples and extreme r values. For instance, the exact value of the intrinsic estimator with $r = 0$ and $n = 12$ (see Example 28 later in this section) is $\theta^* = 0.02631$, while the approximation yields 0.02632.

EXAMPLE 25 (*Intrinsic estimation of normal variance*). The intrinsic discrepancy $\delta\{p_1, p_2\}$ between two normal densities $p_1(x)$ and $p_2(x)$, with $p_i(x) = \mathrm{N}(x|\mu_i, \sigma_i)$, is $\delta\{p_1, p_2\} = \min\{k\{p_1|p_2\}, k\{p_2|p_1\}\}$, where the relevant Kullback–Leibler directed divergences are

$$\kappa\{p_i|p_j\} = \int_{\mathcal{X}} p_j(x)\log\frac{p_j(x)}{p_i(x)}\,\mathrm{d}x = \frac{1}{2}\left\{\log\frac{\sigma_i^2}{\sigma_j^2} + \frac{\sigma_j^2}{\sigma_i^2} - 1 + \frac{(\mu_i-\mu_j)^2}{\sigma_i^2}\right\}.$$

The intrinsic discrepancy between the normal $\mathrm{N}(x|\mu, \sigma)$ and the set of normals with standard deviation $\tilde{\sigma}$, $\mathcal{M}_{\tilde{\sigma}} \equiv \{\mathrm{N}(x|\tilde{\mu}, \tilde{\sigma}), \tilde{\mu} \in \mathbb{R}\}$ is achieved when $\tilde{\mu} = \mu$, and is found to be

$$\delta\{\mathcal{M}_{\tilde{\sigma}}, N(x|\mu,\sigma)\} = \delta(\theta) = \begin{cases} \frac{1}{2}[\log\theta^{-1} + \theta - 1], & \theta < 1, \\ \frac{1}{2}[\log\theta + \theta^{-1} - 1], & \theta \geqslant 1 \end{cases}$$

which only depends on the ratio $\theta = \tilde{\sigma}^2/\sigma^2$. Since, for any fixed $\tilde{\sigma}$, the intrinsic discrepancy, $\delta\{\tilde{\sigma}, (\mu, \sigma)\} = \delta(\theta)$ is a one-to-one function of σ, the reference prior when δ is the quantity of interest is $\pi^\delta(\mu, \sigma) = \sigma^{-1}$, the same as if the quantity of interest were σ (see Example 16). The corresponding posterior distribution of $\theta = \tilde{\sigma}^2/\sigma^2$, after a random sample $\boldsymbol{x} = \{x_1, \ldots, x_n\}$ of fixed size $n \geqslant 2$ has been observed, is the gamma density $\pi(\theta|\boldsymbol{x}) = \mathrm{Ga}(\theta|(n-1)/2, ns^2/\tilde{\sigma}^2)$, where $s^2 = \sum_j (x_j - \bar{x})^2/n$. The intrinsic estimate of σ is that value σ^* of $\tilde{\sigma}$ which minimizes the expected posterior loss,

$$\int_0^\infty \delta(\theta)\pi(\theta|\boldsymbol{x})\,\mathrm{d}\theta = \int_0^\infty \delta(\theta)\mathrm{Ga}\big(\theta|(n-1)/2, ns^2/\tilde{\sigma}^2\big)\,\mathrm{d}\theta.$$

The exact value of $\sigma^*(\boldsymbol{x})$ is easily obtained by one-dimensional numerical integration. However, for $n > 2$, a very good approximation is given by

$$\sigma^* = \sqrt{\frac{\sum_j (x_j - \bar{x})^2}{n-2}} \tag{44}$$

which is larger than both the mle estimate s (which divides by n the sum of squares) and the squared root of the conventional unbiased estimate of the variance (which divides by $n-1$). A good approximation for $n = 2$ is $\sigma^* = (\sqrt{5}/2)|x_1 - x_2|$. Since intrinsic estimation is coherent under one-to-one reparametrizations, the intrinsic estimator of the variance is $(\sigma^*)^2$, and the intrinsic estimator of, say, $\log\sigma$ is simply $\log\sigma^*$.

Intrinsic estimation is a very powerful, general procedure for objective, invariant point estimation. For further discussion, see Bernardo and Juárez (2003).

4.2. *Region (interval) estimation*

To describe the inferential content of the posterior distribution $\pi(\boldsymbol{\theta}|\boldsymbol{x})$ of the quantity of interest it is often convenient to quote regions $R \subset \boldsymbol{\Theta}$ of given (posterior) probability under $\pi(\boldsymbol{\theta}|\boldsymbol{x})$. Any subset of the parameter space $R_q \subset \boldsymbol{\Theta}$ such that $\int_{R_q} \pi(\boldsymbol{\theta}|\boldsymbol{x})\,\mathrm{d}\boldsymbol{\theta} = q$, $0 < q < 1$, so that, given data $\boldsymbol{x}$, the true value of $\boldsymbol{\theta}$ belongs to R_q with probability q, is said to be a (posterior) *q-credible region* of $\boldsymbol{\theta}$. Credible regions are coherent under reparametrization; thus, for any q-credible region R_q of $\boldsymbol{\theta}$ a one-to-one transformation $\boldsymbol{\phi} = \boldsymbol{\phi}(\boldsymbol{\theta})$, $\boldsymbol{\phi}(R_q)$ is a q-credible region of $\boldsymbol{\phi}$. However, for any given q there are generally infinitely many credible regions.

Sometimes, credible regions are selected to have minimum size (length, area, volume), resulting in *highest probability density* (HPD) regions, where all points in the region have larger probability density than all points outside. However, HPD regions are *not* coherent under reparametrization: the image $\boldsymbol{\phi}(R_q)$ of an HPD q-credible region R_q will be a q-credible region for $\boldsymbol{\phi}$, but will not generally be HPD; indeed, there is no compelling reason to restrict attention to HPD credible regions. In one dimension, posterior quantiles are often used to derive credible regions. Thus, if $\theta_q = \theta_q(\boldsymbol{x})$ is the $100q\%$ posterior quantile of θ, then $R_q^l = \{\theta; \theta \leqslant \theta_q\}$ is a one-sided, typically unique q-credible region, and it is coherent under reparametrization. *Probability centered q-credible regions* of the form $R_q^c = \{\theta; \theta_{(1-q)/2} \leqslant \theta \leqslant \theta_{(1+q)/2}\}$ are easier to compute, and are often quoted in preference to HPD regions. However, centered credible regions are only really appealing when the posterior density has a unique interior mode, and have a crucial limitation: they are not uniquely defined in problems with more than one dimension.

For reasonable loss functions, a typically unique credible region may be selected as a *lowest posterior loss* (LPL) region, where all points in the region have smaller (posterior) expected loss than all points outside.

DEFINITION 9 (*Intrinsic credible region*). Let available data $\boldsymbol{x}$ consist of one observation from $\mathcal{M} \equiv \{p(\boldsymbol{x}|\boldsymbol{\theta}, \boldsymbol{\lambda}), \boldsymbol{x} \in \boldsymbol{\mathcal{X}}, \boldsymbol{\theta} \in \boldsymbol{\Theta}, \boldsymbol{\lambda} \in \boldsymbol{\Lambda}\}$, let $\mathcal{M}_{\tilde{\boldsymbol{\theta}}}$ be the restricted model $\mathcal{M}_{\tilde{\boldsymbol{\theta}}} \equiv \{p(\boldsymbol{x}|\tilde{\boldsymbol{\theta}}, \boldsymbol{\lambda}), \boldsymbol{x} \in \boldsymbol{\mathcal{X}}, \boldsymbol{\lambda} \in \boldsymbol{\Lambda}\}$ and let $\delta\{\tilde{\boldsymbol{\theta}}, (\boldsymbol{\theta}, \boldsymbol{\lambda})\}$ be the intrinsic discrepancy between the distribution $p(\boldsymbol{x}|\boldsymbol{\theta}, \boldsymbol{\lambda})$ and the set $\mathcal{M}_{\tilde{\boldsymbol{\theta}}}$. An *intrinsic q-credible region* $R_q^* = R_q^*(\boldsymbol{x}) \subset \boldsymbol{\Theta}$ is a subset of the parameter space $\boldsymbol{\Theta}$ such that,

$$\int_{R_q^*(x)} \pi(\boldsymbol{\theta}|\boldsymbol{x})\,\mathrm{d}\boldsymbol{\theta} = q, \quad \forall \boldsymbol{\theta}_i \in R_q^*(\boldsymbol{x}), \forall \boldsymbol{\theta}_j \notin R_q^*(\boldsymbol{x}), d(\tilde{\boldsymbol{\theta}}_i|\boldsymbol{x}) \leqslant d(\tilde{\boldsymbol{\theta}}_j|\boldsymbol{x}),$$

where, as before, $d(\tilde{\boldsymbol{\theta}}|\boldsymbol{x}) = \mathrm{E}[\delta|\boldsymbol{x}] = \int_{\boldsymbol{\Theta}}\int_{\boldsymbol{\Lambda}} \delta\{\tilde{\boldsymbol{\theta}}, (\boldsymbol{\theta}, \boldsymbol{\lambda})\}\pi^{\delta}(\boldsymbol{\theta}, \boldsymbol{\lambda}|\boldsymbol{x})\,\mathrm{d}\boldsymbol{\theta}\,\mathrm{d}\boldsymbol{\lambda}$ is the reference posterior expected intrinsic loss.

Intrinsic credible regions are well defined for any dimensionality, and they are coherent under one-to-one transformations, in the sense that, if $\boldsymbol{\phi}\{\boldsymbol{\theta}\}$ is a one-to-one transformation of $\boldsymbol{\theta}$ and $R_q^* \subset \boldsymbol{\Theta}$ is an intrinsic q-credible region for $\boldsymbol{\theta}$, then $\boldsymbol{\phi}\{R_q^*\} \subset \boldsymbol{\Phi}$ is an intrinsic q-credible region for $\boldsymbol{\phi}$. As mentioned above, the reference expected intrinsic loss $d(\tilde{\boldsymbol{\theta}}|\boldsymbol{x})$ is often a convex function of $\tilde{\boldsymbol{\theta}}$; in that case, for each $q \in (0, 1)$ there is a unique (convex) intrinsic q-credible region.

EXAMPLE 26 (*Intrinsic binomial credible regions*). Let r be the number of successes observed in n independent Bernoulli trials with parameter θ.

As described in Example 24, the reference posterior expected intrinsic loss which corresponds to using $\tilde{\theta}$ instead of the actual (unknown) θ is the convex function

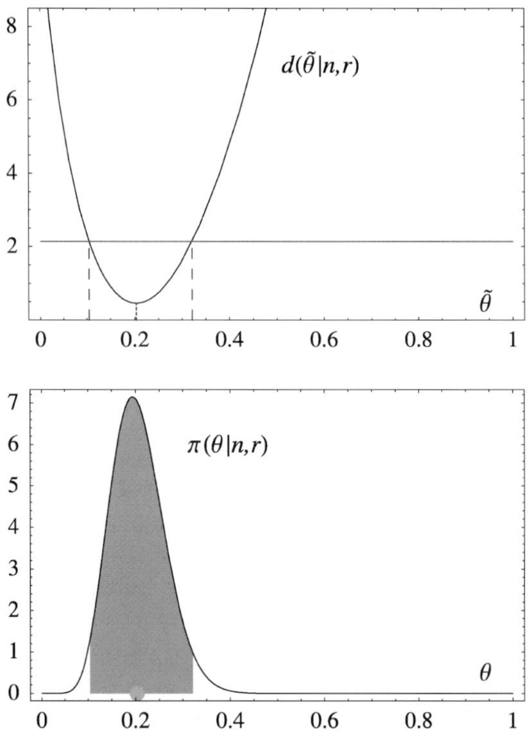

Fig. 4. Intrinsic 0.95-credible region for a binomial parameter.

$d\{\tilde{\theta}|n,r\}$ of Eq. (42), which is represented in the upper panel of Figure 4 as a function of $\tilde{\theta}$, for $r = 10$ and $n = 50$. Using the invariance of the intrinsic loss with respect to one-to-one transformations, and a normal approximation to the posterior distribution of the approximate location parameter $\phi(\theta) = 2 \arcsin \sqrt{\theta}$, it is found that

$$d\{\tilde{\theta}|n,r\} \approx \frac{1}{2} + 2n\left(\arcsin\sqrt{\tilde{\theta}} - \arcsin\sqrt{(r+\alpha_n)/(n+2\alpha_n)}\,\right)^2,$$

where $\alpha_n = (n+4)/(4n+10)$, rapidly concerging to $1/4$. A lowest posterior loss q-credible region consists of the set of $\tilde{\theta}$ points with posterior probability q and minimum expected loss. In this problem, the intrinsic q-credible region $R_q^*(r,n)$, is therefore obtained as the interval $R_q^*(r,n) = [\theta_a(r,n), \theta_b(r,n)]$ defined by the solution (θ_a, θ_b) to the system

$$\left\{ d\{\theta_a|n,r\} = d\{\theta_b|n,r\},\ \int_{\theta_a}^{\theta_b} \pi(\theta|n,r)\,\mathrm{d}\theta = q \right\}.$$

In particular, the intrinsic 0.95-credible region is the set of $\tilde{\theta}$ points with posterior expected loss smaller than 2.139 (shaded region in the lower panel of Figure 4), which is $R_{0.95}^* = \{\tilde{\theta};\, 0.105 \leqslant \tilde{\theta} \leqslant 0.321\}$. Notice that this is neither a HPD interval nor a centered interval. The point with minimum expected loss is the intrinsic estimator,

$\theta^* = 0.2034$. Since intrinsic estimation is coherent under one-to-one reparametrizations, the intrinsic estimator and the 0.95-intrinsic credible region of the log-odds, $\psi = \psi(\theta) = \log[\theta/(1-\theta)]$ are immediately derived as $\psi(\theta^*) = -1.365$ and $\psi(R^*_{0.95}) = [-2.144, -0.747]$.

It may be argued that, in practice, it is reasonable for credible regions to give privilege to the most probable values of the parameters, as HPD regions do. This is obviously incompatible with an invariance requirement, but it is interesting to notice that, in one-parameter problems, intrinsic credible regions are approximately HPD in the approximate location parametrization. Thus, in Example 26, the 0.95-credible region for the approximate location parameter, $\phi(\theta) = \frac{2}{\pi}\arcsin\sqrt{\theta}$, $\phi(R^*_{0.95}) = [0.210, 0.384]$, is nearly an HPD interval for ϕ.

4.3. Hypothesis testing

Let $\boldsymbol{x}$ be the available data, which are assumed to consist of one observation from model $\mathcal{M} \equiv \{p(\boldsymbol{x}|\boldsymbol{\theta}, \boldsymbol{\lambda}), \boldsymbol{x} \in \boldsymbol{\mathcal{X}}, \boldsymbol{\theta} \in \boldsymbol{\Theta}, \boldsymbol{\lambda} \in \boldsymbol{\Lambda}\}$, parametrized in terms of the vector of interest $\boldsymbol{\theta}$ and a vector $\boldsymbol{\lambda}$ of nuisance parameters. The posterior distribution $\pi(\boldsymbol{\theta}|\boldsymbol{x})$ of the quantity of interest $\boldsymbol{\theta}$ conveys immediate intuitive information on the values of $\boldsymbol{\theta}$ which, given $\mathcal{M}$, might be declared to be *compatible* with the observed data $\boldsymbol{x}$, namely, those with a relatively high probability density. Sometimes, a *restriction*, $\boldsymbol{\theta} \in \boldsymbol{\Theta}_0 \subset \boldsymbol{\Theta}$, of the possible values of the quantity of interest (where $\boldsymbol{\Theta}_0$ may possibly consist of a single value $\boldsymbol{\theta}_0$) is suggested in the course of the investigation as deserving special consideration, either because restricting $\boldsymbol{\theta}$ to $\boldsymbol{\Theta}_0$ would greatly simplify the model, or because there are additional, context specific arguments suggesting that $\boldsymbol{\theta} \in \boldsymbol{\Theta}_0$. Intuitively, the (null) *hypothesis* $H_0 \equiv \{\boldsymbol{\theta} \in \boldsymbol{\Theta}_0\}$ should be judged to be *compatible* with the observed data $\boldsymbol{x}$ if there are elements in $\boldsymbol{\Theta}_0$ with a relatively high posterior density. However, a more precise conclusion is typically required and this is made possible by adopting a decision-oriented approach. Formally, testing the hypothesis $H_0 \equiv \{\boldsymbol{\theta} \in \boldsymbol{\Theta}_0\}$ is a *decision problem* where the action space $\mathcal{A} = \{a_0, a_1\}$ only contains two elements: to accept (a_0) or to reject (a_1) the proposed restriction.

To solve this decision problem, it is necessary to specify an appropriate loss function, $\ell(a_i, \boldsymbol{\theta})$, measuring the consequences of accepting or rejecting H_0 as a function of the actual value $\boldsymbol{\theta}$ of the vector of interest. Notice that this requires the statement of an *alternative* a_1 to accepting H_0; this is only to be expected, for an action is taken not because it is good, but because it is better than anything else that has been imagined. Given data $\boldsymbol{x}$, the optimal action will be to reject H_0 if (and only if) the expected posterior loss of accepting the null, $\int_{\boldsymbol{\Theta}} \ell(a_0, \boldsymbol{\theta})\pi(\boldsymbol{\theta}|\boldsymbol{x})\,\mathrm{d}\boldsymbol{\theta}$, is larger than the expected posterior loss of rejecting, $\int_{\boldsymbol{\Theta}} \ell(a_1, \boldsymbol{\theta})\pi(\boldsymbol{\theta}|\boldsymbol{x})\,\mathrm{d}\boldsymbol{\theta}$, that is, if (and only if)

$$\int_{\boldsymbol{\Theta}} \big[\ell(a_0, \boldsymbol{\theta}) - \ell(a_1, \boldsymbol{\theta})\big]\pi(\boldsymbol{\theta}|\boldsymbol{x})\,\mathrm{d}\boldsymbol{\theta} = \int_{\boldsymbol{\Theta}} \Delta\ell(\boldsymbol{\theta})\pi(\boldsymbol{\theta}|\boldsymbol{x})\,\mathrm{d}\boldsymbol{\theta} > 0. \tag{45}$$

Therefore, only the loss difference $\Delta\ell(\boldsymbol{\Theta}_0, \boldsymbol{\theta}) = \ell(a_0, \boldsymbol{\theta}) - \ell(a_1, \boldsymbol{\theta})$, which measures the *advantage* of rejecting $H_0 \equiv \{\boldsymbol{\theta} \in \boldsymbol{\Theta}_0\}$ as a function of $\boldsymbol{\theta}$, has to be specified:

the hypothesis H_0 should be rejected whenever the expected advantage of rejecting is positive.

A crucial element in the specification of the loss function is a description of what is precisely meant by rejecting H_0. By assumption a_0 means to act *as if* H_0 were true, i.e., as if $\boldsymbol{\theta} \in \boldsymbol{\Theta}_0$, but there are at least two options for the alternative action a_1. This may either mean (i) the *negation* of H_0, that is to act as if $\boldsymbol{\theta} \notin \boldsymbol{\Theta}_0$ or, alternatively, it may rather mean (ii) to reject the simplification implied by H_0 and to keep the unrestricted model, $\boldsymbol{\theta} \in \boldsymbol{\Theta}$, which is true by assumption. Both options have been analyzed in the literature, although it may be argued that the problems of scientific data analysis, where hypothesis testing procedures are typically used, are better described by the second alternative. Indeed, an established model, identified by $H_0 \equiv \{\boldsymbol{\theta} \in \boldsymbol{\Theta}_0\}$, is often embedded into a more general model, $\{\boldsymbol{\theta} \in \boldsymbol{\Theta}, \boldsymbol{\Theta}_0 \subset \boldsymbol{\Theta}\}$, constructed to include promising departures from H_0, and it is then required to verify whether presently available data $\boldsymbol{x}$ are still compatible with $\boldsymbol{\theta} \in \boldsymbol{\Theta}_0$, or whether the extension to $\boldsymbol{\theta} \in \boldsymbol{\Theta}$ is really required.

The simplest loss structure has, for all values of the nuisance parameter vector $\boldsymbol{\lambda}$, a zero-one form, with $\{\ell(a_0, \boldsymbol{\theta}) = 0, \ell(a_1, \boldsymbol{\theta}) = 1\}$ if $\boldsymbol{\theta} \in \boldsymbol{\Theta}_0$, and $\{\ell(a_0, \boldsymbol{\theta}) = 1, \ell(a_1, \boldsymbol{\theta}) = 0\}$ if $\boldsymbol{\theta} \notin \boldsymbol{\Theta}_0$, so that the *advantage* $\Delta\ell\{\boldsymbol{\Theta}_0, (\boldsymbol{\theta}, \boldsymbol{\lambda})\}$ of rejecting H_0 is

$$\Delta\ell\{\boldsymbol{\Theta}_0, (\boldsymbol{\theta}, \boldsymbol{\lambda})\} = \begin{cases} 1, & \text{if } \boldsymbol{\theta} \notin \boldsymbol{\Theta}_0, \\ -1, & \text{if } \boldsymbol{\theta} \in \boldsymbol{\Theta}_0. \end{cases} \tag{46}$$

With this (rather naïve) loss function it is immediately found that the optimal action is to reject H_0 if (and only if) $\Pr(\boldsymbol{\theta} \notin \boldsymbol{\Theta}_0 | \boldsymbol{x}) > \Pr(\boldsymbol{\theta} \in \boldsymbol{\Theta}_0 | \boldsymbol{x})$. Notice that this formulation *requires* that $\Pr(\boldsymbol{\theta} \in \boldsymbol{\Theta}_0) > 0$, that is, that the (null) hypothesis H_0 has a strictly positive prior probability. If $\boldsymbol{\theta}$ is a continuous parameter and $\boldsymbol{\Theta}_0$ has zero measure (for instance, if H_0 consists of a single point $\boldsymbol{\theta}_0$), this requires the use of a nonregular "sharp" prior concentrating a positive probability mass on $\boldsymbol{\theta}_0$. With no mention to the loss structure implicit behind, this solution was early advocated by Jeffreys (1961, Chapter 5). However, this is known to lead to the difficulties associated to Lindley's paradox (Lindley, 1957; Bartlett, 1957; Bernardo, 1980; Robert, 1993; Brewer, 2002).

The intrinsic discrepancy loss may also be used to provide an attractive general alternative to Bayesian hypothesis testing, the *Bayesian reference criterion, BRC* (Bernardo, 1999a; Bernardo and Rueda, 2002). This follows from assuming that the loss structure is such that

$$\Delta\ell\{\boldsymbol{\Theta}_0, (\boldsymbol{\theta}, \boldsymbol{\lambda})\} = \delta\{\boldsymbol{\Theta}_0, (\boldsymbol{\theta}, \boldsymbol{\lambda})\} - d^*, \quad d^* > 0, \tag{47}$$

where $\delta\{\boldsymbol{\Theta}_0, (\boldsymbol{\theta}, \boldsymbol{\lambda})\}$, which describes as a function of $(\boldsymbol{\theta}, \boldsymbol{\lambda})$ the loss suffered by assuming that $\boldsymbol{\theta} \in \boldsymbol{\Theta}_0$, is the intrinsic discrepancy between the distribution $p(\boldsymbol{x} | \boldsymbol{\theta}, \boldsymbol{\lambda})$ and the set $\mathcal{M}_0 \equiv \{p(\boldsymbol{x} | \boldsymbol{\theta}_0, \boldsymbol{\lambda}), \boldsymbol{\theta}_0 \in \boldsymbol{\Theta}_0, \boldsymbol{\lambda} \in \boldsymbol{\Lambda}\}$. The function $\delta\{\boldsymbol{\Theta}_0, (\boldsymbol{\theta}, \boldsymbol{\lambda})\}$, which is invariant under one-to-one reparametrization, is nonnegative and it is zero if, and only if, $\boldsymbol{\theta} \in \boldsymbol{\Theta}_0$. The constant d^* is the (strictly positive) advantage of being able to work with the null model when it is true, measured in the same units as δ; the choice of d^*, in terms of posterior expected log-likelihood ratios, is discussed below.

DEFINITION 10 (*Intrinsic hypothesis testing: BRC*). Let available data $\boldsymbol{x}$ consist of one observation from $\mathcal{M} \equiv \{p(\boldsymbol{x} | \boldsymbol{\theta}, \boldsymbol{\lambda}), \boldsymbol{x} \in \boldsymbol{\mathcal{X}}, \boldsymbol{\theta} \in \boldsymbol{\Theta}, \boldsymbol{\lambda} \in \boldsymbol{\Lambda}\}$, let $\mathcal{M}_0$ be the restricted model $\mathcal{M}_0 \equiv \{p(\boldsymbol{x}|\boldsymbol{\theta}_0, \boldsymbol{\lambda}), \boldsymbol{\theta}_0 \in \boldsymbol{\Theta}_0, \boldsymbol{\lambda} \in \boldsymbol{\Lambda}\}$ and let $\delta\{\boldsymbol{\Theta}_0, (\boldsymbol{\theta}, \boldsymbol{\lambda})\}$ be the intrinsic discrepancy between the distribution $p(\boldsymbol{x}|\boldsymbol{\theta}, \boldsymbol{\lambda})$ and the set $\mathcal{M}_0$. The *Bayesian reference criterion* (*BRC*) rejects model $\mathcal{M}_0$ if the *intrinsic statistic* $d(\boldsymbol{\Theta}_0|\boldsymbol{x})$, defined as the reference posterior expected intrinsic loss, exceeds a critical value d^*. Formally, the null hypothesis $H_0 \equiv \{\boldsymbol{\theta} \in \boldsymbol{\Theta}_0\}$ is rejected if, and only if

$$d(\boldsymbol{\Theta}_0|\boldsymbol{x}) = \mathrm{E}[\delta|\boldsymbol{x}] = \int_{\Theta} \delta\{\boldsymbol{\Theta}_0, (\boldsymbol{\theta}, \boldsymbol{\lambda})\}\pi^{\delta}(\boldsymbol{\theta}, \boldsymbol{\lambda}|\boldsymbol{x})\, \mathrm{d}\boldsymbol{\theta}\, \mathrm{d}\boldsymbol{\lambda} > d^*,$$

where $\pi^{\delta}(\boldsymbol{\theta}, \boldsymbol{\lambda}|\boldsymbol{x}) \propto p(\boldsymbol{x}|\boldsymbol{\theta}, \boldsymbol{\lambda})\pi^{\delta}(\boldsymbol{\theta}, \boldsymbol{\lambda})$ is the reference posterior of $(\boldsymbol{\theta}, \boldsymbol{\lambda})$ when $\delta = \delta\{\boldsymbol{\Theta}_0, (\boldsymbol{\theta}, \boldsymbol{\lambda})\}$ is the quantity of interest. The conventional value $d^* = \log(100)$ may be used for scientific communication.

As the sample size increases, the expected value of $d(\boldsymbol{\Theta}_0|\boldsymbol{x})$ under sampling tends to one when H_0 is true, and tends to infinity otherwise; thus $d(\boldsymbol{\Theta}_0|\boldsymbol{x})$ may be regarded as a continuous, positive measure of the expected loss (in information units) from simplifying the model by accepting $\mathcal{M}_0$. In traditional language, $d(\boldsymbol{\Theta}_0|\boldsymbol{x})$ is a test statistic, and the BRC criterion rejects the null if this *intrinsic test statistic* $d(\boldsymbol{\Theta}_0|\boldsymbol{x})$ exceeds some *critical value* d^*. However, in sharp contrast to frequentist hypothesis testing, the critical value d^* is simply a utility constant which measures the number of *information units* which the decision maker is prepared to loose in order to be able to work with the null model H_0, *not* a function of sampling properties of the model.

The interpretation of the intrinsic discrepancy in terms of the minimum posterior expected likelihood ratio in favour of the true model (see Section 2) provides a direct *calibration* of the required critical value. Indeed, $d(\boldsymbol{\Theta}_0|\boldsymbol{x})$ is the minimum posterior expected log-likelihood ratio in favour of the unrestricted model. For instance, values around $\log[10] \approx 2.3$ should be regarded as mild evidence against H_0, while values around $\log[100] \approx 4.6$ suggest strong evidence against the null, and values larger than $\log[1000] \approx 6.9$ may be safely used to reject H_0. Notice that, in contrast to frequentist hypothesis testing, where it is hazily recommended to adjust the significance level for dimensionality and sample size, the intrinsic statistic in measured on an absolute scale which remains valid for *any* sample size and *any* dimensionality.

EXAMPLE 27 (*Testing the value of a normal mean*). Let data consist of a random sample $\boldsymbol{x} = \{x_1, \ldots, x_n\}$ from a normal $\mathrm{N}(x|\mu, \sigma)$ distribution, and consider the "canonical" problem of testing whether or not these data are compatible with some specific sharp hypothesis $H_0 \equiv \{\mu = \mu_0\}$ on the value of the mean. The intrinsic discrepancy is easily found to be

$$\delta(\mu_0, \mu|\sigma) = \frac{n}{2}\left(\frac{\mu - \mu_0}{\sigma}\right)^2, \tag{48}$$

a simple transformation of the standardized distance between μ and μ_0, which generalizes to $\delta(\boldsymbol{\mu}_0, \boldsymbol{\mu}) = (n/2)(\boldsymbol{\mu} - \boldsymbol{\mu}_0)^t\, \Sigma^{-1}(\boldsymbol{\mu} - \boldsymbol{\mu}_0)$, a linear function of the Mahalanobis distance, in the multivariate normal case.

Consider first the case where σ is assumed to be known. The reference prior for μ is then uniform; this is also the reference prior when the parameter of interest is δ, since

$\delta(\mu_0, \mu)$ is a piecewise invertible function of μ (see Theorem 6). The corresponding posterior distribution, is $\pi(\mu|\mathbf{x}) = \mathrm{N}(\mu|\bar{x}, \sigma/\sqrt{n})$ $(n \geqslant 1)$. The expected value of $\delta(\mu_0, \mu)$ with respect to this posterior yields the corresponding intrinsic statistic,

$$d(\mu_0|\mathbf{x}) = \frac{1}{2}(1 + z^2), \quad z = \frac{\bar{x} - \mu_0}{\sigma/\sqrt{n}}, \tag{49}$$

a simple function of the standardized distance between the sample mean $\bar{x}$ and μ_0. As prescribed by the general theory, the expected value of $d(\mu_0|\mathbf{x})$ under repeated sampling is one if $\mu = \mu_0$, and increases linearly with n otherwise. In this canonical example, to reject H_0 whenever $|z| > 1.96$ (the frequentist suggestion with the conventional 0.05 significance level), corresponds to rejecting H_0 whenever $d(\mu_0|\mathbf{x})$ is larger than 2.42, a rather weak evidence, since this means that the posterior expected likelihood ratio against H_0 is only about $\exp[2.42] = 11.25$. Conversely, to reject whenever posterior expected likelihood ratio against H_0 is about 100, so that $d^* = \log[100] \approx 4.6$, is to reject whenever $|z| > 2.86$, which is close to the conventional 3σ rule often used by engineers. The extreme 6σ rule, apparently popular these days, would correspond (under normality) to $d^* = 18.5 \approx \log[10^8]$.

If the scale parameter σ is also unknown, the intrinsic discrepancy is

$$\delta\{\mu_0, (\mu, \sigma)\} = \frac{n}{2} \log\left[1 + \left(\frac{\mu - \mu_0}{\sigma}\right)^2\right], \tag{50}$$

which is *not* the same as (48). The intrinsic test statistic $d(\mu_0, \mathbf{x})$ may then be found as the expected value of $\delta\{\mu_0, (\mu, \sigma)\}$ under the corresponding joint reference posterior distribution $\pi^\delta(\mu, \sigma|\mathbf{x})$ when δ is the quantity of interest. After some algebra, the exact result may be expressed in terms of hypergeometric functions (Bernardo, 1999a), but is very well approximated by the simple function

$$d(\mu_0|\mathbf{x}) \approx \frac{1}{2} + \frac{n}{2} \log\left(1 + \frac{t^2}{n}\right), \tag{51}$$

where t is the conventional statistic $t = \sqrt{n-1}(\bar{x} - \mu_0)/s$, written here in terms of the sample variance $s^2 = \sum_j (x_j - \bar{x})^2/n$. For instance, for samples sizes 5, 30 and 1000, and using the threshold $d^* = \log[100]$, the null hypothesis $H_0 \equiv \{\mu = \mu_0\}$ would be rejected whenever $|t|$ is respectively larger than 4.564, 3.073, and 2.871.

EXAMPLE 28 (*A lady tasting tea*). A lady claims that by tasting a cup of tea made with milk she can discriminate whether milk has been poured over the tea infusion or the other way round, and she is able to give the correct answer in n consecutive trials. Are these results compatible with the hypothesis that she is only guessing and has been lucky? The example, a variation suggested by Neyman (1950, Section 5.2) to a problem originally proposed by Fisher (1935, Section 2.5), has often been used to compare alternative approaches to hypothesis testing. See Lindley (1984) for a subjectivist Bayesian analysis.

The intrinsic objective Bayesian solution is immediate from the results in Examples 24 and 26. Indeed, using Definition 10, if data are assumed to consist of n Bernoulli observations and r successes have been observed, the intrinsic statistic to test the precise null $\theta = \theta_0$ is

$$d(\theta_0|r, n) = \int_0^1 \delta\{\theta_0, \theta|n\}\mathrm{Be}\left(\theta|r + \frac{1}{2}, n - r + \frac{1}{2}\right) \mathrm{d}\theta,$$

where $\delta\{\theta_0, \theta|n\}$ is given by (7). In this case, one has $r = n$ and $\theta_0 = 1/2$. For the values $n = 8$, $n = 10$ and $n = 12$ traditionally discussed, the intrinsic test statistic, $d(\theta_0|r, n)$, respectively yields the values $d(\frac{1}{2}|8, 8) \approx 4.15$, $d(\frac{1}{2}|10, 10) \approx 5.41$ and $d(\frac{1}{2}|12, 12) \approx 6.70$. Since $\log[100] \approx 4.61$, the hypothesis of pure guessing would not be rejected with $n = 8$ with the conventional threshold $d^* = \log[100]$, but would be rejected with $n = 10$ successes (and *a fortiori* with $n = 12$). Actually, the value of $d(\frac{1}{2}|8, 8)$ says that the observed data are only estimated to be about $\exp[4.15] \approx 64$ times more likely under the unrestricted model (unknown θ) that under the null model (no discrimination power, $\theta = \theta_0 = 1/2$). However, with $n = 10$ and $n = 12$ the observed data are respectively estimated to be about 224 and 811 times more likely under the unrestricted model that under the null.

The Bayesian reference criterion may also be used with nonnested problems. Thus, given two alternative models for $\boldsymbol{x} \in \mathcal{X}$, $\mathcal{M}_1 = \{p_1(\boldsymbol{x}|\boldsymbol{\theta}_1), \boldsymbol{\theta}_1 \in \boldsymbol{\Theta}_1\}$ and $\mathcal{M}_2 = \{p_2(\boldsymbol{x}|\boldsymbol{\theta}_2), \boldsymbol{\theta}_2 \in \boldsymbol{\Theta}_2\}$, one may introduce the a new parameter α to define a mixture model $p(\boldsymbol{x}|\boldsymbol{\theta}_1, \boldsymbol{\theta}_1, \alpha) = p_1(\boldsymbol{x}|\boldsymbol{\theta}_1)^{\alpha} p_2(\boldsymbol{x}|\boldsymbol{\theta}_2)^{1-\alpha}$ (with either a continuous $\alpha \in [0, 1]$ or, more simply, a discrete $\alpha \in \{0, 1\}$), and use BRC to verify whether $\mathcal{M}_1$, or $\mathcal{M}_2$, or both, are compatible with the data, assuming the mixture is. For further discussion on hypothesis testing and the development of the Bayesian reference criterion see Bernardo (1982, 1985a, 1999a), Bernardo and Bayarri (1985), Rueda (1992) and Bernardo and Rueda (2002).

5. Related work

Reference analysis already has a long history, but it still is a very active area of research. The original paper on reference analysis, (Bernardo, 1979b), is easily read and it is followed by a very lively discussion; Bernardo (1981), extends the theory to general decision problems; see also Bernardo and Smith (1994, Section 5.4.1) and Rabena (1998). Berger and Bernardo (1989, 1992c) contain crucial mathematical extensions. Bernardo (1997) is a nontechnical analysis, in a dialogue format, of the foundational issues involved, and it is followed by a discussion. A textbook level description of reference analysis is provided in Bernardo and Smith (1994, Section 5.4); Bernardo and Ramón (1998) contains a simple introduction to reference distributions. BRC, the Bayesian reference criterion for hypothesis testing, was introduced by Bernardo (1999a) and further refined in Bernardo and Rueda (2002). Intrinsic estimation was introduced in Bernardo and Juárez (2003). Berger et al. (2005) contains the last mathematical developments of reference theory at the moment of writing.

Papers which contain either specific derivations or new applications of reference analysis include, in chronological order of the first related paper by the same author(s), Bernardo (1977a, 1977b, 1978, 1980, 1982, 1985a, 1985b, 1999b), Bayarri (1981, 1985), Ferrándiz (1982, 1985), Sendra (1982), Eaves (1983a, 1983b, 1985), Armero (1985), Bernardo and Bayarri (1985), Chang and Villegas (1986), Chang and Eaves (1990), Hills (1987), Mendoza (1987, 1988, 1990), Bernardo and Girón (1988), Lindley (1988), Berger and Bernardo (1989, 1992a, 1992b, 1992c), Clarke and Barron (1990, 1994), Polson and Wasserman (1990), Phillips (1991), Severini (1991, 1993, 1995, 1999), Ye and Berger (1991), Ghosh and Mukerjee (1992), Singh and Upadhyay (1992), Stephens and Smith (1992), Berger and Sun (1993), Clarke and Wasserman (1993), Dey and Peng (1993, 1995), Kuboki (1993, 1998), Liseo (1993, 2003, 2005), Ye (1993, 1994, 1995, 1998), Berger and Yang (1994), Kubokawa and Robert (1994), Sun (1994, 1997), Sun and Berger (1994, 1998), Yang and Berger (1994, 1997), Datta and J.K. Ghosh (1995a, 1995b), Datta and M. Ghosh (1995a, 1995b), Giudici (1995), Ghosh et al. (1995), du Plessis et al. (1995), Sun and Ye (1995, 1996, 1999), de Waal et al. (1995), Yang and Chen (1995), Bernard (1996), Clarke (1996), Ghosh and Yang (1996), Armero and Bayarri (1997), Fernández et al. (1997), Garvan and Ghosh (1997, 1999), Ghosal and Samanta (1997), Ghosal (1997, 1999), Sugiura and Ishibayashi (1997), Berger et al. (1998a, 1998b, 1999, 2001) Bernardo and Ramón (1998), Chung and Dey (1998, 2002), Scholl (1998), Sun et al. (1998), Philippe and Robert (1998), Burch and Harris (1999), Brewer (1999), Scricciolo (1999), Vernotte and Zalamansky (1999), Yuan and Clarke (1999), Fernández and Steel (1998, 1999a, 1999b, 2000), Mendoza and Gutiérrez-Peña (1999), Mukerjee and Reid (1999, 2001), Aguilar and West (2000), Eno and Ye (2000, 2001), Elhor and Pensky (2000), B.H. Kim et al. (2001), D.H. Kim et al. (2000a, 2000b, 2001, 2002), van der Linde (2000), Fan (2001), Ghosh and Kim (2001), Ghosh et al. (2001), Komaki (2001, 2004), Natarajan (2001), Rosa and Gianola (2001), Aslam (2002a, 2002b), Daniels and Pourahmadi (2002), Datta et al. (2002), Millar (2002), Philippe and Rousseau (2002), Pretorius and van der Merwe (2002), Tardella (2002), Consonni and Veronese (2003), Datta and Smith (2003), Fraser et al. (2003), Ghosh and Heo (2003a, 2003b), Ghosh et al. (2003), Gutiérrez-Peña and Rueda (2003), He (2003), Leucari and Consonni (2003), Lauretto et al. (2003), Madruga et al. (2003), Ni and Sun (2003), Sareen (2003), Consonni et al. (2004), Sun and Ni (2004), Grünwald and Dawid (2004), Roverato and Consonni (2004), Stern (2004a, 2004b), van der Merwe and Chikobvu (2004) and Liseo and Loperfido (2005).

This chapter concentrates on reference analysis. It is known, however, that ostensibly different approaches to the derivation of objective priors often produce the same result, a testimony of the robustness of many solutions to the definition of what an appropriate objective prior may be in a particular problem. Many authors have proposed alternative objective priors (often comparing the resulting inferences with those obtained within the frequentist paradigm), either as general methods or as *ad hoc* solutions to specific inferential problems, and a few are openly critical with objective Bayesian methods. Many relevant papers in this very active field of Bayesian mathematical statistics are listed in the further reading section below. For reviews of many of these, see Dawid (1983), Bernardo and Smith (1994, Section 5.6.2), Kass and Wasserman (1996) and Bernardo (1997).

As suggested in Section 3.12, reference analysis in only fairly developed for relatively simple probability structures. Complicated – where most applied action is – offer challenging opportunities for advanced research in the Bayesian methodology. This chapter may be seen as an open invitation to contribute to that fascinating endeavour.

Acknowledgements

The author is indebted to many colleagues for helpful comments on an earlier draft of this paper. Special recognition is due, however, to the very detailed comments by my *maestro*, Professor Dennis Lindley.

Research supported by grant BMF 2002-2889 of the former *Ministerio de Ciencia y Tecnología*, Madrid, Spain.

References

Aguilar, O., West, M. (2000). Bayesian dynamic factor models and portfolio allocation. *J. Business Econ. Studies* **18**, 338–357.

Akaike, H. (1977). On entropy maximization principle. In: Krishnaiah, P.R. (Ed.), *Applications of Statistics*. North-Holland, Amsterdam, pp. 27–41.

Armero, C. (1985). Bayesian analysis of M/M/1/∞/fifo queues. In: Bernardo, J.M., DeGroot, M.H., Lindley, D.V., Smith, A.F.M. (Eds.), *Bayesian Statistics 2*. North-Holland, Amsterdam, pp. 613–618.

Armero, C., Bayarri, M.J. (1997). A Bayesian analysis of a queuing system with unlimited service. *J. Statist. Plann. Inference* **58**, 241–261.

Aslam, M. (2002a). Bayesian analysis for paired comparison models allowing ties and not allowing ties. *Pakistan J. Statist.* **18**, 53–69.

Aslam, M. (2002b). Reference prior for the parameters of the Rao–Kupper model. *Proc. Pakistan Acad. Sci.* **39**, 215–224.

Bartlett, M. (1957). A comment on D.V. Lindley's statistical paradox. *Biometrika* **44**, 533–534.

Bayarri, M.J. (1981). Inferencia Bayesiana sobre el coeficiente de correlación de una población normal bivariante. *Trab. Estadist.* **32**, 18–31.

Bayarri, M.J. (1985). Bayesian inference on the parameters of a Beta distribution. *Statistics and Decisions* **2**, 17–22.

Bayes, T. (1763). An essay towards solving a problem in the doctrine of chances. Published posthumously in *Philos. Trans. Roy. Soc. London* **53**, 370–418 and **54**, 296–325. Reprinted in *Biometrika* **45** (1958), 293–315, with a biographical note by G.A. Barnard.

Berger, J.O. (2000). Bayesian analysis: A look at today and thoughts of tomorrow. *J. Amer. Statist. Assoc.* **95**, 1269–1276.

Berger, J.O., Bernardo, J.M. (1989). Estimating a product of means: Bayesian analysis with reference priors. *J. Amer. Statist. Assoc.* **84**, 200–207.

Berger, J.O., Bernardo, J.M. (1992a). Ordered group reference priors with applications to a multinomial problem. *Biometrika* **79**, 25–37.

Berger, J.O., Bernardo, J.M. (1992b). Reference priors in a variance components problem. In: Goel, P.K., Iyengar, N.S. (Eds.), *Bayesian Analysis in Statistics and Econometrics*. Springer, Berlin, pp. 323–340.

Berger, J.O., Bernardo, J.M. (1992c). On the development of reference priors. In: Bernardo, J.M., Berger, J.O., Dawid, A.P., Smith, A.F.M. (Eds.), *Bayesian Statistics 4*. Oxford University Press, pp. 35–60. (With discussion.)

Bernardo, J.M., Girón, F.J. (1988). A Bayesian analysis of simple mixture problems. In: Bernardo, J.M., DeGroot, M.H., Lindley, D.V., Smith, A.F.M. (Eds.), *Bayesian Statistics 3*. Oxford University Press, pp. 67–88. (With discussion.)

Berger, J.O., Pericchi, L.R. (1996). The intrinsic Bayes factor for model selection and prediction. *J. Amer. Statist. Assoc.* **91**, 109–122.

Berger, J.O., Sun, D. (1993). Bayesian analysis for the poly-Weibull distribution. *J. Amer. Statist. Assoc.* **88**, 1412–1418.

Berger, J.O., Wolpert, R.L. (1988). *The Likelihood Principle*, second ed. IMS, Hayward, CA.

Berger, J.O., Yang, R. (1994). Noninformative priors and Bayesian testing for the AR(1) model. *Econometric Theory* **10**, 461–482.

Berger, J.O., Bernardo, J.M., Mendoza, M. (1989). On priors that maximize expected information. In: Klein, J.P., Lee, J.C. (Eds.), *Recent Developments in Statistics and their Applications*. Freedom Academy, Seoul, pp. 1–20.

Berger, J.O., Pericchi, L.R., Varshavsky, J.A. (1998a). Bayes factors and marginal distributions in invariant situations. *Sankhyā A* **60**, 307–321.

Berger, J.O., Philippe, A., Robert, C. (1998b). Estimation of quadratic functions: Noninformative priors for noncentrality parameters. *Statistica Sinica* **8**, 359–376.

Berger, J.O., Liseo, B., Wolpert, R. (1999). Integrated likelihood methods for eliminating nuisance parameters. *Statist. Sci.* **14**, 1–28.

Berger, J.O., De Oliveira, V., Sansó, B. (2001). Objective Bayesian analysis of spatially correlated data. *J. Amer. Statist. Assoc.* **96**, 1361–1374.

Berger, J.O., Bernardo, J.M., Sun, D. (2005). Reference priors from first principles: A general definition. Technical Report, SAMSI, NC, USA.

Bernard, J.-M. (1996). Bayesian interpretation of frequentist procedures for a Bernoulli process. *Amer. Statist.* **50**, 7–13.

Bernardo, J.M. (1977a). Inferences about the ratio of normal means: A Bayesian approach to the Fieller–Creasy problem. In: Barra, J.R., Brodeau, F., Romier, G., van Cutsem, B. (Eds.), *Recent Developments in Statistics*. North-Holland, Amsterdam, pp. 345–349.

Bernardo, J.M. (1977b). Inferencia Bayesiana sobre el coeficiente de variación: Una solución a la paradoja de marginalización. *Trab. Estadist.* **28**, 23–80.

Bernardo, J.M. (1978). Unacceptable implications of the left Haar measure in a standard normal theory inference problem. *Trab. Estadist.* **29**, 3–9.

Bernardo, J.M. (1979a). Expected information as expected utility. *Ann. Statist.* **7**, 686–690.

Bernardo, J.M. (1979b). Reference posterior distributions for Bayesian inference. *J. Roy. Statist. Soc., Ser. B* **41**, 113–147 (with discussion). Reprinted in Polson, N.G., Tiao, G.C. (Eds.), *Bayesian Inference*. Edward Elgar, Brookfield, VT, pp. 229–263.

Bernardo, J.M. (1980). A Bayesian analysis of classical hypothesis testing. In: Bernardo, J.M., DeGroot, M.H., Lindley, D.V., Smith, A.F.M. (Eds.), *Bayesian Statistics*. Valencia University Press, pp. 605–647. (With discussion.)

Bernardo, J.M. (1981). Reference decisions. *Symposia Mathematica* **25**, 85–94.

Bernardo, J.M. (1982). Contraste de modelos probabilísticos desde una perspectiva Bayesiana. *Trab. Estadist.* **33**, 16–30.

Bernardo, J.M. (1985a). Análisis Bayesiano de los contrastes de hipótesis paramétricos. *Trab. Estadist.* **36**, 45–54.

Bernardo, J.M. (1985b). On a famous problem of induction. *Trab. Estadist.* **36**, 24–30.

Bernardo, J.M. (1997). Non-informative priors do not exist. *J. Statist. Plann. Inference* **65**, 159–189. (With discussion.)

Bernardo, J.M. (1999a). Nested hypothesis testing: The Bayesian reference criterion. In: Bernardo, J.M., Berger, J.O., Dawid, A.P., Smith, A.F.M. (Eds.), *Bayesian Statistics 6*. Oxford University Press, pp. 101–130. (With discussion.)

Bernardo, J.M. (1999b). Model-free objective Bayesian prediction. *Rev. Acad. Ciencias de Madrid* **93**, 295–302.

Bernardo, J.M., Bayarri, M.J. (1985). Bayesian model criticism. In: Florens, J.-P., Mouchart, M., Raoult, J.-P., Simar, L. (Eds.), *Model Choice*. Pub. Fac. Univ. Saint Louis, Brussels, pp. 43–59.

Bernardo, J.M., Juárez, M. (2003). Intrinsic estimation. In: Bernardo, J.M., Bayarri, M.J., Berger, J.O., Dawid, A.P., Heckerman, D., Smith, A.F.M., West, M. (Eds.), *Bayesian Statistics 7*. Oxford University Press, pp. 465–476.

Bernardo, J.M., Ramón, J.M. (1998). An introduction to Bayesian reference analysis: Inference on the ratio of multinomial parameters. *The Statistician* **47**, 1–35.

Bernardo, J.M., Rueda, R. (2002). Bayesian hypothesis testing: A reference approach. *Internat. Statist. Rev.* **70**, 351–372.

Bernardo, J.M., Smith, A.F.M. (1994). *Bayesian Theory*. Wiley, Chichester.

Boulton, D.M., Wallace, C.S. (1970). A program for numerical classification. *The Computer J.* **13**, 63–69.

Box, G.E.P., Tiao, G.C. (1973). *Bayesian Inference in Statistical Analysis*. Addison-Wesley, Reading, MA.

Brewer, K.R.W. (1999). Testing a precise null hypothesis using reference posterior odds. *Rev. Acad. Ciencias de Madrid* **93**, 303–310.

Brewer, K.R.W. (2002). The Lindley and Bartlett paradoxes. *Pak. J. Statist.* **18**, 1–13.

Burch, B.D., Harris, I.R. (1999). Bayesian estimators of the intraclass correlation coefficient in the one-way random effects model. *Comm. Statist. Theory Methods* **28**, 1247–1272.

Casella, G. (1996). Statistical inference and Monte Carlo algorithms. *Test* **5**, 249–344. (With discussion.)

Chang, T., Eaves, D.M. (1990). Reference priors for the orbit of a group model. *Ann. Statist.* **18**, 1595–1614.

Chang, T., Villegas, C. (1986). On a theorem of Stein relating Bayesian and classical inferences in group models. *Canad. J. Statist.* **14**, 289–296.

Chung, Y., Dey, D.K. (1998). Bayesian approach to estimation in intraclass correlation using reference priors. *Comm. Statist. Theory Methods* **27**, 2241–2255.

Chung, Y., Dey, D.K. (2002). Model determination for the variance component model using reference priors. *J. Statist. Plann. Inference* **100**, 49–65.

Clarke, B. (1996). Implications of reference priors for prior information and for sample size. *J. Amer. Statist. Assoc.* **91**, 173–184.

Clarke, B., Barron, A.R. (1990). Information-theoretic asymptotics of Bayes methods. *IEEE Trans. Inform. Theory* **36**, 453–471.

Clarke, B., Barron, A.R. (1994). Jeffreys' prior is asymptotically least favourable under entropy risk. *J. Statist. Plann. Inference* **41**, 37–60.

Clarke, B., Wasserman, L. (1993). Noninformative priors and nuisance parameters. *J. Amer. Statist. Assoc.* **88**, 1427–1432.

Consonni, G., Veronese, P. (1989). A note on coherent invariant distributions as non-informative priors for exponential and location-scale families. *Comm. Statist. Theory Methods* **18**, 2883–2907.

Consonni, G., Veronese, P. (2003). Enriched conjugate and reference priors for the Wishart family on symmetric cones. *Ann. Statist.* **31**, 1491–1516.

Consonni, G., Veronese, P., Gutiérrez-Peña, E. (2004). Reference priors for exponential families with simple quadratic variance function. *J. Multivariate Anal.* **88**, 335–364.

Cover, M., Thomas, J.A. (1991). *Elements of Information Theory*. Wiley, New York.

Cox, D.R., Hinkley, D.V. (1974). *Theoretical Statistics*. Chapman and Hall, London.

Csiszár, I. (1985). An extended maximum entropy principle and a Bayesian justification. In: Bernardo, J.M., DeGroot, M.H., Lindley, D.V., Smith, A.F.M. (Eds.), *Bayesian Statistics 2*. North-Holland, Amsterdam, pp. 83–98. (With discussion.)

Csiszár, I. (1991). Why least squared and maximum entropy? An axiomatic approach to inference for linear inverse problems. *Ann. Statist.* **19**, 2032–2066.

Daniels, M.J., Pourahmadi, M. (2002). Bayesian analysis of covariance matrices and dynamic models for longitudinal data. *Biometrika* **89**, 553–566.

Datta, G.S., Ghosh, J.K. (1995a). On priors providing a frequentist validity for Bayesian inference. *Biometrika* **82**, 37–45.

Datta, G.S., Ghosh, J.K. (1995b). Noninformative priors for maximal invariant parameter in group models. *Test* **4**, 95–114.

Datta, G.S., Ghosh, M. (1995a). Some remarks on noninformative priors. *J. Amer. Statist. Assoc.* **90**, 1357–1363.

Datta, G.S., Ghosh, M. (1995b). Hierarchical Bayes estimators of the error variance in one-way ANOVA models. *J. Statist. Plann. Inference* **45**, 399–411.

Datta, G.S., Ghosh, M. (1996). On the invariance of noninformative priors. *Ann. Statist.* **24**, 141–159.

Datta, G.S., Smith, D.D. (2003). On property of posterior distributions of variance components in small area estimation. *J. Statist. Plann. Inference* **112**, 175–183.

Datta, G.S., Sweeting, T. (2005). Probability matching priors. In: Dey, D.K., Rao, C.R. (Eds.), *Bayesian Thinking: Modeling and Computation, Handbook of Statistics*, vol. 25. Elsevier, Amsterdam. This volume.

Datta, G.S., Mukerjee, R., Ghosh, M., Sweeting, T.J. (2000). Bayesian prediction with approximate frequentist validity. *Ann. Statist.* **28**, 1414–1426.

Datta, G.S., Ghosh, M., Kim, Y.-H. (2002). Probability matching priors for one-way unbalanced random effect models. *Statistics and Decisions* **20**, 29–51.

Dawid, A.P. (1980). A Bayesian look at nuisance parameters. In: Bernardo, J.M., DeGroot, M.H., Lindley, D.V., Smith, A.F.M. (Eds.), *Bayesian Statistics*. Valencia University Press, pp. 167–203. (With discussion.)

Dawid, A.P. (1983). Invariant prior distributions. In: Kotz, S., Johnson, N.L., Read, C.B. (Eds.), *Encyclopedia of Statistical Sciences, vol. 4*. Wiley, New York, pp. 228–236.

Dawid, A.P., Stone, M. (1972). Expectation consistency of inverse probability distributions. *Biometrika* **59**, 486–489.

Dawid, A.P., Stone, M. (1973). Expectation consistency and generalized Bayes inference. *Ann. Statist.* **1**, 478–485.

Dawid, A.P., Stone, M., Zidek, J.V. (1973). Marginalization paradoxes in Bayesian and structural inference. *J. Roy. Statist. Soc., Ser. B* **35**, 189–233. (With discussion.)

de Finetti, B. (1970). *Teoria delle Probabilit, vol. 1*. Einaudi, Turin. English translation as *Theory of Probability*. Wiley, Chichester, 1974.

de Waal, D.J., Groenewald, P.C.N. (1989). On measuring the amount of information from the data in a Bayesian analysis. *South African Statist. J.* **23**, 23–62. (With discussion.)

de Waal, D.J., Groenewald, P.C.N., Kemp, C.J. (1995). Perturbation of the normal model in linear regression. *South African Statist. J.* **29**, 105–130.

Dey, D.K., Peng, F. (1993). On the choice of prior for the Bayes estimation in accelerated life testing. *J. Statist. Comput. Simulation* **48**, 207–217.

Dey, D.K., Peng, F. (1995). Elimination of nuisance parameters using information measure. *Parisankhyan Samikkha* **2**, 9–29.

du Plessis, J.L., van der Merwe, A.J., Groenewald, P.C.N. (1995). Reference priors for the multivariate calibration problem. *South African Statist. J.* **29**, 155–168.

Eaton, M.L., Freedman, D.A. (2004). Dutch book against some 'objective' priors. *Bernoulli* **10**, 861–872.

Eaton, M.L., Sudderth, W.D. (1998). A new predictive distribution for normal multivariate linear models. *Sankhyā A* **60**, 363–382.

Eaton, M.L., Sudderth, W.D. (1999). Consistency and strong inconsistency of group-invariant predictive inferences. *Bernoulli* **5**, 833–854.

Eaton, M.L., Sudderth, W.D. (2002). Group invariant inference and right Haar measure. *J. Statist. Plann. Inference* **103**, 87–99.

Eaton, M.L., Sudderth, W.D. (2004). Properties of right Haar predictive inference. *Sankhyā A* **66**, 487–512.

Eaves, D.M. (1983a). On Bayesian nonlinear regression with an enzyme example. *Biometrika* **70**, 373–379.

Eaves, D.M. (1983b). Minimally informative prior analysis of a non-linear model. *The Statistician* **32**, 117.

Eaves, D.M. (1985). On maximizing the missing information about a hypothesis. *J. Roy. Statist. Soc., Ser. B* **47**, 263–266.

Efron, B. (1986). Why isn't everyone a Bayesian? *Amer. Statist.* **40**, 1–11. (With discussion.)

Elhor, A., Pensky, M. (2000). Bayesian estimation of location of lightning events. *Sankhyā* **62**, 202–206.

Eno, D.R., Ye, K. (2000). Bayesian reference prior analysis for polynomial calibration models. *Test* **9**, 191–202.

Eno, D.R., Ye, K. (2001). Probability matching priors for an extended statistical calibration problem. *Canad. J. Statist.* **29**, 19–35.

Fan, T.-H. (2001). Noninformative Bayesian estimation for the optimum in a single factor quadratic response model. *Test* **10**, 225–240.

Fernández, C., Steel, M. (1998). On Bayesian modelling of fat tails and skewness. *J. Amer. Statist. Assoc.* **93**, 359–371.

Fernández, C., Steel, M. (1999a). On the dangers of modelling through continuous distributions: A Bayesian perspective. In: Bernardo, J.M., Berger, J.O., Dawid, A.P., Smith, A.F.M. (Eds.), *Bayesian Statistics 6*. Oxford University Press, pp. 213–238.

Fernández, C., Steel, M. (1999b). Reference priors for the general location-scale model. *Statist. Probab. Lett.* **43**, 377–384.

Fernández, C., Steel, M. (2000). Bayesian regression analysis with scale mixtures of normals. *J. Econometric Theory* **16**, 80–101.

Fernández, C., Osiewalski, J., Steel, M. (1997). On the use of panel data in stochastic frontier models with improper priors. *J. Econometrics* **79**, 169–193.

Ferrándiz, J.R. (1982). Una solución Bayesiana a la paradoja de Stein. *Trab. Estadist.* **33**, 31–46.

Ferrándiz, J.R. (1985). Bayesian inference on Mahalanobis distance: An alternative approach to Bayesian model testing. In: Bernardo, J.M., DeGroot, M.H., Lindley, D.V., Smith, A.F.M. (Eds.), *Bayesian Statistics 2*. North-Holland, Amsterdam, pp. 645–654.

Fisher, R.A. (1935). *The Design of Experiments*. Oliver and Boyd, Edinburgh.

Florens, J.-P. (1978). Mesures à priori et invariance dans une expérience Bayésienne. *Pub. Inst. Statist. Univ. Paris* **23**, 29–55.

Florens, J.-P. (1982). Expériences Bayésiennes invariantes. *Ann. Inst. H. Poincaré* **18**, 309–317.

Fraser, D.A.S., Monette, G., Ng, K.W. (1985). Marginalization, likelihood and structural models. In: Krishnaiah, P.R. (Ed.), *Multivariate Analysis, vol. 6*. North-Holland, Amsterdam, pp. 209–217.

Fraser, D.A.S., Reid, N., Wong, A., Yi, G.Y. (2003). Direct Bayes for interest parameters. In: Bernardo, J.M., Bayarri, M.J., Berger, J.O., Dawid, A.P., Heckerman, D., Smith, A.F.M., West, M. (Eds.), *Bayesian Statistics 7*. Oxford University Press, pp. 529–534.

Garvan, C.W., Ghosh, M. (1997). Noninformative priors for dispersion models. *Biometrika* **84**, 976–982.

Garvan, C.W., Ghosh, M. (1999). On the property of posteriors for dispersion models. *J. Statist. Plann. Inference* **78**, 229–241.

Geisser, S. (1984). On prior distributions for binary trials. *J. Amer. Statist. Assoc.* **38**, 244–251. (With discussion.)

George, E.I., McCulloch, R. (1993). On obtaining invariant prior distributions. *J. Statist. Plann. Inference* **37**, 169–179.

Ghosal, S. (1997). Reference priors in multiparameter nonregular cases. *Test* **6**, 159–186.

Ghosal, S. (1999). Probability matching priors for non-regular cases. *Biometrika* **86**, 956–964.

Ghosal, S., Samanta, T. (1997). Expansion of Bayes risk for entropy loss and reference prior in nonregular cases. *Statistics and Decisions* **15**, 129–140.

Ghosh, J.K., Mukerjee, R. (1992). Non-informative priors. In: Bernardo, J.M., Berger, J.O., Dawid, A.P., Smith, A.F.M. (Eds.), *Bayesian Statistics 4*. Oxford University Press, pp. 195–210. (With discussion.)

Ghosh, M., Heo, J. (2003a). Default Bayesian priors for regression models with first-order autoregressive residuals. *J. Time Series Anal.* **24**, 269–282.

Ghosh, M., Heo, J. (2003b). Noninformative priors, credible sets and Bayesian hypothesis testing for the intraclass model. *J. Statist. Plann. Inference* **112**, 133–146.

Ghosh, M., Kim, Y.-H. (2001). The Behrens–Fisher problem revisited: A Bayes-frequentist synthesis. *Canad. J. Statist.* **29**, 5–17.

Ghosh, M., Mukerjee, R. (1998). Recent developments on probability matching priors. In: Ahmed, S.E., Ashanullah, M., Sinha, B.K. (Eds.), *Applied Statistical Science III*. Science Publishers, New York, pp. 227–252.

Ghosh, M., Yang, M.-Ch. (1996). Non-informative priors for the two sample normal problem. *Test* **5**, 145–157.

Ghosh, M., Carlin, B.P., Srivastava, M.S. (1995). Probability matching priors for linear calibration. *Test* **4**, 333–357.

Ghosh, M., Rousseau, J., Kim, D.H. (2001). Noninformative priors for the bivariate Fieller–Creasy problem. *Statistics and Decisions* **19**, 277–288.

Ghosh, M., Yin, M., Kim, Y.-H. (2003). Objective Bayesian inference for ratios of regression coefficients in linear models. *Statistica Sinica* **13**, 409–422.

Giudici, P. (1995). Bayes factors for zero partial covariances. *J. Statist. Plann. Inference* **46**, 161–174.

Goudie, I.B.J., Goldie, C.M. (1981). Initial size estimation for the pure death process. *Biometrika* **68**, 543–550.

Grünwald, P.D., Dawid, A.P. (2004). Game theory, maximum entropy, minimum discrepancy and robust Bayesian decision theory. *Ann. Statist.* **32**, 1367–1433.

Gutiérrez-Peña, E. (1992). Expected logarithmic divergence for exponential families. In: Bernardo, J.M., Berger, J.O., Dawid, A.P., Smith, A.F.M. (Eds.), *Bayesian Statistics 4*. Oxford University Press, pp. 669–674.

Gutiérrez-Peña, E., Rueda, R. (2003). Reference priors for exponential families. *J. Statist. Plann. Inference* **110**, 35–54.

Hadjicostas, P., Berry, S.M. (1999). Improper and proper posteriors with improper priors in a Poisson-gamma hierarchical model. *Test* **8**, 147–166.

Haldane, J.B.S. (1948). The precision of observed values of small frequencies. *Biometrika* **35**, 297–303.

Hartigan, J.A. (1964). Invariant prior distributions. *Ann. Math. Statist.* **35**, 836–845.

Hartigan, J.A. (1966). Note on the confidence prior of Welch and Peers. *J. Roy. Statist. Soc., Ser. B* **28**, 55–56.

Hartigan, J.A. (1983). *Bayes Theory*. Springer, Berlin.

He, C.Z. (2003). Bayesian modelling of age-specific survival in bird nesting studies under irregular visits. *Biometrics* **59**, 962–973.

Hills, S.E. (1987). Reference priors and identifiability problems in non-linear models. *The Statistician* **36**, 235–240.

Hobert, J.P., Casella, G. (1996a). The effect of improper priors on Gibbs sampling in hierarchical linear mixed models. *J. Amer. Statist. Assoc.* **91**, 1461–1473.

Hobert, J.P., Casella, G. (1996b). Functional compatibility, Markov chains and Gibbs sampling with improper posteriors. *J. Comput. Graph. Statist.* **7**, 42–60.

Jaynes, E.T. (1968). Prior probabilities. *IEEE Trans. Systems, Science and Cybernetics* **4**, 227–291.

Jaynes, E.T. (1982). On the rationale of maximum-entropy methods. *Proc. IEEE* **70**, 939–952.

Jaynes, E.T. (1985). Some random observations. *Synthése* **3**, 115–138.

Jaynes, E.T. (1989). *Papers on Probability, Statistics and Statistical Physics*, second ed. Kluwer Academic, Dordrecht.

Jeffreys, H. (1946). An invariant form for the prior probability in estimation problems. *Proc. Roy Soc.* **186**, 453–461.

Jeffreys, H. (1961). *Theory of Probability*, third ed. Oxford University Press.

Juárez, M. (2004). Métodos bayesianos objetivos de estimación y contraste de hipótesis. Ph.D. thesis, Universitat de València, Spain.

Ibragimov, I.A., Khasminskii, R.Z. (1973). On the information in a sample about a parameter. In: Petrov, B.N., Csaki, F. (Eds.), *Proc. 2nd Internat. Symp. Information Theory*. Akademia Kiadó, Budapest, pp. 295–309.

Kass, R.E., Wasserman, L. (1996). The selection of prior distributions by formal rules. *J. Amer. Statist. Assoc.* **91**, 1343–1370. Corr: 1998 **93**, 412.

Kim, B.H., Chang, I.H., Kang, C.K. (2001). Bayesian estimation for the reliability in Weibull stress-strength systems using noninformative priors. *Far East J. Theor. Statist.* **5**, 299–315.

Kim, D.H., Kang, S.G., Cho, J.S. (2000a). Noninformative Priors for stress–strength system in the Burr-type X model. *J. Korean Statist. Soc.* **29**, 17–28.

Kim, D.H., Lee, W.D., Kang, S.G. (2000b). Bayesian model selection for life time data under type II censoring. *Comm. Statist. Theory Methods* **29**, 2865–2878.

Kim, D.H., Kang, S.G., Lee, W.D. (2001). Noninformative priors for intraclass correlation coefficient in familial data. *Far East J. Theor. Statist.* **5**, 51–65.

Kim, D.H., Kang, S.G., Lee, W.D. (2002). Noninformative priors for the power law process. *J. Korean Statist. Soc.* **31**, 17–31.

Komaki, F. (2001). A shrinkage predictive distribution for multivariate Normal observables. *Biometrika* **88**, 859–864.

Komaki, F. (2004). Simultaneous prediction of independent Poisson observables. *Ann. Statist.* **32**, 1744–1769.

Kubokawa, T., Robert, C.P. (1994). New perspectives in linear calibration. *J. Multivariate Anal.* **51**, 178–200.

Kuboki, H. (1993). Inferential distributions for non-Bayesian predictive fit. *Ann. Inst. Statist. Math.* **45**, 567–578.

Kuboki, H. (1998). Reference priors for prediction. *J. Statist. Plann. Inference* **69**, 295–317.

Kullback, S. (1968). *Information Theory and Statistics*, second ed. Dover, New York. Reprinted in 1978, Peter Smith, Gloucester, MA.

Kullback, S. (1983). Kullback information. In: Kotz, S., Johnson, N.L., Read, C.B. (Eds.), *Encyclopedia of Statistical Sciences, vol. 4*. Wiley, New York, pp. 421–425.

Kullback, S. (1987). The Kullback–Leibler distance. *Amer. Statist.* **41**, 340–341.

Kullback, S., Leibler, R.A. (1951). On information and sufficiency. *Ann. Math. Statist.* **22**, 79–86.

Lafferty, J., Wasserman, L. (2001). Iterative Markov chain Monte Carlo computation of reference priors and minimax risk. *Uncertainty in Artificial Intelligence* (UAI, Seattle).

Lane, D., Sudderth, W.D. (1984). Coherent predictive inference. *Sankhyā A* **46**, 166–185.

Laplace, P.S. (1812). *Théorie analytique des probabilités*. Courcier, Paris. Reprinted as Oeuvres Complètes de Laplace, vol. 7, Gauthier-Villars, Paris, 1878–1912.

Laplace, P.S. (1825). *Essai philosophique sur les probabilitiés*, fifth ed. Courcier, Paris. English translation in 1952 as *Philosophical Essay on Probabilities*. Dover, New York.

Lauretto, M., Pereira, C.A.B., Stern, J.M., Zacks, S. (2003). Comparing parameters of two bivariate normal distributions using the invariant FBST, full Bayesian significance test. *Brazilian J. Probab. Statist.* **17**, 147–168.

Leucari, V., Consonni, G. (2003). Compatible priors for causal Bayesian networks. In: Bernardo, J.M., Bayarri, M.J., Berger, J.O., Dawid, A.P., Heckerman, D., Smith, A.F.M., West, M. (Eds.), *Bayesian Statistics 7*. Oxford University Press, pp. 597–606.

Lindley, D.V. (1956). On a measure of information provided by an experiment. *Ann. Math. Statist.* **27**, 986–1005.

Lindley, D.V. (1957). A statistical paradox. *Biometrika* **44**, 187–192.

Lindley, D.V. (1965). *Introduction to Probability and Statistics from a Bayesian Viewpoint*. Cambridge University Press.

Lindley, D.V. (1984). A Bayesian lady tasting tea. In: David, H.A., David, H.T. (Eds.), *Statistics: An Appraisal*. Iowa State Press, Ames, IA, pp. 455–479.

Lindley, D.V. (1988). Statistical inference concerning the Hardy–Weinberg equilibrium. In: Bernardo, J.M., DeGroot, M.H., Lindley, D.V., Smith, A.F.M. (Eds.), *Bayesian Statistics 3*. Oxford University Press, pp. 307–326. (With discussion.)

Liseo, B. (1993). Elimination of nuisance parameters with reference priors. *Biometrika* **80**, 295–304.

Liseo, B. (2003). Bayesian and conditional frequentist analyses of the Fieller's problem. A critical review. *Metron* **61**, 133–152.

Liseo, B. (2005). The problem of the elimination of nuisance parameters in a Bayesian framework. In: Dey, D.K., Rao, C.R. (Eds.), *Bayesian Thinking: Modeling and Computation, Handbook of Statistics*, vol. 25. Elsevier, Amsterdam. This volume.

Liseo, B., Loperfido, N. (2005). A note on reference priors for the scalar skew-normal distribution. *J. Statist. Plann. Inference*. In press.

Madruga, M.R., Pereira, C.A.B., Stern, J.M. (2003). Bayesian evidence test for precise hypothesis. *J. Statist. Plann. Inference* **117**, 185–198.

Mendoza, M. (1987). A Bayesian analysis of a generalized slope ratio bioassay. In: Viertl, R. (Ed.), *Probability and Bayesian Statistics*. Plenum, London, pp. 357–364.

Mendoza, M. (1988). Inferences about the ratio of linear combinations of the coefficients in a multiple regression problem. In: Bernardo, J.M., DeGroot, M.H., Lindley, D.V., Smith, A.F.M. (Eds.), *Bayesian Statistics 3*. Oxford University Press, pp. 705–711.

Mendoza, M. (1990). A Bayesian analysis of the slope ratio bioassay. *Biometrika* **46**, 1059–1069.

Mendoza, M. (1994). Asymptotic normality under transformations. A result with Bayesian applications. *Test* **3**, 173–180.

Mendoza, M., Gutiérrez-Peña, E. (1999). Bayesian inference for the ratio of the means of two normal populations with unequal variances. *Biom. J.* **41**, 133–147.

Millar, R.B. (2002). Reference priors for Bayesian fisheries models. *Canad. J. Fish. Aquat. Sci.* **59**, 1492–1502.

Mukerjee, R., Reid, N. (1999). On a property of probability matching priors: Matching the alternative coverage probabilities. *Biometrika* **86**, 333–340.

Mukerjee, R., Reid, N. (2001). Second-order probability matching priors for a parametric function with application to Bayesian tolerance limits. *Biometrika* **88**, 587–592.

Natarajan, R. (2001). On the propriety of a modified Jeffreys's prior for variance components in binary random effects models. *Statist. Probab. Lett.* **51**, 409–414.

Neyman, J. (1950). *First Course in Probability and Statistics*. Holt, New York.

Neyman, J., Scott, E.L. (1948). Consistent estimates based on partially consistent observations. *Econometrica* **16**, 1–32.
Ni, S.X., Sun, D. (2003). Noninformative priors and frequentist risks of Bayesian estimators of vector-autoregressive models. *J. Econom.* **115**, 159–197.
Palmer, J.L., Pettit, L.I. (1996). Risks of using improper priors with Gibbs sampling and autocorrelated errors. *J. Comput. Graph. Statist.* **5**, 245–249.
Pericchi, L.R. (2005). Model selection and hypothesis testing based on probabilities. In: Dey, D.K., Rao, C.R. (Eds.), *Bayesian Thinking: Modeling and Computation, Handbook of Statistics*, vol. 25. Elsevier, Amsterdam. This volume.
Perks, W. (1947). Some observations on inverse probability, including a new indifference rule. *J. Inst. Actuaries* **73**, 285–334. (With discussion.)
Philippe, A., Robert, C.P. (1998). A note on the confidence properties of reference priors for the calibration model. *Test* **7**, 147–160.
Philippe, A., Rousseau, J. (2002). Non informative priors in the case of Gaussian long-memory processes. *Bernoulli* **8**, 451–473.
Phillips, P.C.B. (1991). To criticize the critics: An objective Bayesian analysis of stochastic trends. *J. Applied Econometrics* **6**, 333–473. (With discussion.)
Polson, N.G., Wasserman, L. (1990). Prior distributions for the bivariate binomial. *Biometrika* **76**, 604–608.
Press, S.J. (1972). *Applied Multivariate Analysis: Using Bayesian and Frequentist Methods of Inference*. Krieger, Melbourne, FL. Second edition in 1982.
Pretorius, A.L., van der Merwe, A.J. (2002). Reference and probability-matching priors in Bayesian analysis of mixed linear models. *J. Anim. Breed. Genet.* **119**, 311–324.
Rabena, M.T. (1998). Deriving reference decisions. *Test* **7**, 161–177.
Raftery, A.E. (1988). Analysis of a simple debugging model. *Appl. Statist.* **37**, 12–22.
Robert, C.P. (1993). A note on Jeffreys–Lindley paradox. *Statistica Sinica* **3**, 601–608.
Robert, C.P. (1996). Intrinsic loss functions. *Theory and Decision* **40**, 192–214.
Rosa, G.J.M., Gianola, D. (2001). Inferences about the coefficient of correlation in the standard bivariate normal distribution. *South African Statist. J.* **35**, 69–93.
Roverato, A., Consonni, G. (2004). Compatible prior distributions for DAG models. *J. Roy. Statist. Soc., Ser. B* **66**, 47–61.
Rueda, R. (1992). A Bayesian alternative to parametric hypothesis testing. *Test* **1**, 61–67.
Sareen, S. (2003). Reference Bayesian inference in nonregular models. *J. Econom.* **113**, 265–288.
Savage, L.J. (1954). *The Foundations of Statistics*. Wiley, New York. Second edition in 1972, Dover, New York.
Scholl, H.R. (1998). Shannon optimal priors on independent identically distributed statistical experiments converge weakly to Jeffreys' prior. *Test* **7**, 75–94.
Scricciolo, C. (1999). Probability matching priors: A review. *J. It. Statist. Soc.* **8**, 83–100.
Sendra, M. (1982). Distribución final de referencia para el problema de Fieller–Creasy. *Trab. Estadist.* **33**, 55–72.
Severini, T.A. (1991). On the relationship between Bayesian and non-Bayesian interval estimates. *J. Roy. Statist. Soc., Ser. B* **53**, 611–618.
Severini, T.A. (1993). Bayesian interval estimates which are also confidence intervals. *J. Roy. Statist. Soc., Ser. B* **53**, 611–618.
Severini, T.A. (1995). Information and conditional inference. *J. Amer. Statist. Assoc.* **90**, 1341–1346.
Severini, T.A. (1999). On the relationship between Bayesian and non-Bayesian elimination of nuisance parameters. *Statistica Sinica* **9**, 713–724.
Severini, T.A., Mukerjee, R., Ghosh, M. (2002). On an exact probability matching property of right-invariant priors. *Biometrika* **89**, 953–957.
Shannon, C.E. (1948). A mathematical theory of communication. *Bell System Tech. J.* **27**, 379–423 and 623–656. Reprinted in Shannon, C.E., Weaver W. (Eds.), *The Mathematical Theory of Communication*. Univ. Illinois Press, Urbana, IL, 1949.
Singh, U., Upadhyay, S.K. (1992). Bayes estimators in the presence of a guess value. *Comm. Statist. Simulation Comput.* **21**, 1181–1198.

Singpurwalla, N.D., Wilson, P. (2004). When can finite testing ensure infinite trustworthiness?. *J. Iranian Statist. Soc.* **3**, 1–37. (With discussion.)

Smith, R.L. (1999). Bayesian and frequentist approaches to parametric predictive inference. In: Bernardo, J.M., Berger, J.O., Dawid, A.P., Smith, A.F.M. (Eds.), *Bayesian Statistics 6*. Oxford University Press, pp. 589–612. (With discussion).

Stein, C. (1959). An example of wide discrepancy between fiducial and confidence intervals. *Ann. Math. Statist.* **30**, 877–880.

Stein, C. (1962). Confidence sets for the mean of a multivariate normal distribution. *J. Roy. Statist. Soc., Ser. B* **24**, 265–296. (With discussion.)

Stein, C. (1986). On the coverage probability of confidence sets based on a prior distribution. In: Wetherbil, G.B., Glazebrook, K.D. (Eds.), *Sequential Methods in Statistics*. third ed. Chapman and Hall, London, pp. 485–514.

Stephens, D.A., Smith, A.F.M. (1992). Sampling–resampling techniques for the computation of posterior densities in normal means problems. *Test* **1**, 1–18.

Stern, J.M. (2004a). Paraconsistent sensitivity analysis for Bayesian significance tests. *Lecture Notes in Artificial Intelligence* **3171**, 134–143.

Stern, J.M. (2004b). Uninformative reference sensitivity in possibilistic sharp hypotheses tests. *Amer. Inst. Physics Proc.* **735**, 581–588.

Stone, M. (1959). Application of a measure of information to the design and comparison of experiments. *Ann. Math. Statist.* **30**, 55–70.

Stone, M. (1965). Right Haar measures for convergence in probability to invariant posterior distributions. *Ann. Math. Statist.* **36**, 440–453.

Stone, M. (1970). Necessary and sufficient conditions for convergence in probability to invariant posterior distributions. *Ann. Math. Statist.* **41**, 1939–1953.

Stone, M. (1976). Strong inconsistency from uniform priors. *J. Amer. Statist. Assoc.* **71**, 114–125. (With discussion.)

Stone, M., Dawid, A.P. (1972). Un-Bayesian implications of improper Bayesian inference in routine statistical problems. *Biometrika* **59**, 369–375.

Sugiura, N., Ishibayashi, H. (1997). Reference prior Bayes estimator for bivariate normal covariance matrix with risk comparison. *Comm. Statist. Theory Methods* **26**, 2203–2221.

Sun, D. (1994). Integrable expansions for posterior distributions for a two-parameter exponential family. *Ann. Statist.* **22**, 1808–1830.

Sun, D. (1997). A note on noninformative priors for Weibull distributions. *J. Statist. Plann. Inference* **61**, 319–338.

Sun, D., Berger, J.O. (1994). Bayesian sequential reliability for Weibull and related distributions. *Ann. Inst. Statist. Math.* **46**, 221–249.

Sun, D., Berger, J.O. (1998). Reference priors with partial information. *Biometrika* **85**, 55–71.

Sun, D., Ni, S. (2004). Bayesian analysis of vector-autoregressive models with noninformative priors. *J. Statist. Plann. Inference* **121**, 291–309.

Sun, D., Ye, K. (1995). Reference prior Bayesian analysis for normal mean products. *J. Amer. Statist. Assoc.* **90**, 589–597.

Sun, D., Ye, K. (1996). Frequentist validity of posterior quantiles for a two-parameter exponential family. *Biometrika* **83**, 55–65.

Sun, D., Ye, K. (1999). Reference priors for a product of normal means when variances are unknown. *Canad. J. Statist.* **27**, 97–103.

Sun, D., Ghosh, M., Basu, A.P. (1998). Bayesian analysis for a stress-strength system under noninformative priors. *Canad. J. Statist.* **26**, 323–332.

Tardella, L. (2002). A new Bayesian method for nonparametric capture-recapture models in presence of heterogeneity. *Biometrika* **89**, 807–817.

van der Linde, A. (2000). Reference priors for shrinkage and smoothing parameters. *J. Statist. Plann. Inference* **90**, 245–274.

van der Merwe, A.J., Chikobvu, D. (2004). Bayesian analysis of the process capability index C_{pk}. *South African Statist. J.* **38**, 97–117.

Vernotte, F., Zalamansky, G. (1999). A Bayesian method for oscillator stability analysis. *IEEE Trans. Ultrasonics, Ferroelec. and Freq. Controls* **46**, 1545–1550.

Villegas, C. (1977). On the representation of ignorance. *J. Amer. Statist. Assoc.* **72**, 651–654.

Villegas, C. (1990). Bayesian inference in models with Euclidean structures. *J. Amer. Statist. Assoc.* **85**, 1159–1164.

Welch, B.L. (1965). On comparisons between confidence point procedures in the case of a single parameter. *J. Roy. Statist. Soc., Ser. B* **27**, 1–8.

Welch, B.L., Peers, H.W. (1963). On formulae for confidence points based on intervals of weighted likelihoods. *J. Roy. Statist. Soc., Ser. B* **25**, 318–329.

Yang, R. (1995). Invariance of the reference prior under reparametrization. *Test* **4**, 83–94.

Yang, R., Berger, J.O. (1994). Estimation of a covariance matrix using the reference prior. *Ann. Statist.* **22**, 1195–1211.

Yang, R., Berger, J.O. (1997). A catalogue of noninformative priors. Technical Report, Duke University, ISDS 97-42.

Yang, R., Chen, M.-H. (1995). Bayesian analysis of random coefficient regression models using noninformative priors. *J. Multivariate Anal.* **55**, 283–311.

Ye, K. (1993). Reference priors when the stopping rule depends on the parameter of interest. *J. Amer. Statist. Assoc.* **88**, 360–363.

Ye, K. (1994). Bayesian reference prior analysis on the ratio of variances for the balanced one-way random effect model. *J. Statist. Plann. Inference* **41**, 267–280.

Ye, K. (1995). Selection of the reference priors for a balanced random effects model. *Test* **4**, 1–17.

Ye, K. (1998). Estimating a ratio of the variances in a balanced one-way random effects model. *Statistics and Decisions* **16**, 163–180.

Ye, K., Berger, J.O. (1991). Non-informative priors for inferences in exponential regression models. *Biometrika* **78**, 645–656.

Yuan, A., Clarke, B.S. (1999). A minimally informative likelihood for decision analysis: Illustration and robustness. *Canad. J. Statist.* **27**, 649–665.

Zellner, A. (1971). *An Introduction to Bayesian Inference in Econometrics*. Wiley, New York. Reprinted in 1987, Krieger, Melbourne, FL.

Zidek, J.V. (1969). A representation of Bayes invariant procedures in terms of Haar measure. *Ann. Inst. Statist. Math.* **21**, 291–308.

Further reading

Agliari, A., Calvi-Parisetti, C. (1988). A g-reference informative prior: A note on Zellner's g-prior. *The Statistician* **37**, 271–275.

Aitchison, J., Dunsmore, I.R. (1975). *Statistical Prediction Analysis*. Cambridge University Press.

Akaike, H. (1978). A new look at the Bayes procedure. *Biometrika* **65**, 53–59.

Akaike, H. (1980). The interpretation of improper prior distributions as limits of data dependent proper prior distributions. *J. Roy. Statist. Soc., Ser. B* **45**, 46–52.

Akaike, H. (1980). Likelihood and the Bayes procedure. In: Bernardo, J.M., DeGroot, M.H., Lindley, D.V., Smith, A.F.M. (Eds.), *Bayesian Statistics*. Valencia University Press, pp. 144–166 and 185–203 (with discussion).

Akaike, H. (1980). Ignorance prior distribution of a hyperparameter and Stein's estimator. *Ann. Inst. Statist. Math.* **32**, 171–178.

Akaike, H. (1983). On minimum information prior distributions. *Ann. Inst. Statist. Math.* **35**, 139–149.

Atwood, C.L. (1996). Constrained noninformative priors in risk assessment. *Reliability Eng. System Safety* **53**, 37–46.

Banerjee, A.K., Bhattacharyya, G.K. (1979). Bayesian results for the inverse Gaussian distribution with an application. *Technometrics* **21**, 247–251.

Barnard, G.A. (1952). The frequency justification of certain sequential tests. *Biometrika* **39**, 144–150.

Barnard, G.A. (1988). The future of statistics: Teaching and research. In: Bernardo, J.M., DeGroot, M.H., Lindley, D.V., Smith, A.F.M. (Eds.), *Bayesian Statistics 3*. Oxford University Press, pp. 17–24.

Barron, A.R. (1999). Information-theoretic characterization of Bayes performance and the choice of priors in parametric and nonparametric problems. In: Bernardo, J.M., Berger, J.O., Dawid, A.P., Smith, A.F.M. (Eds.), *Bayesian Statistics 6*. Oxford University Press, pp. 27–52. (With discussion.)

Barron, A., Rissanen, J., Yu, B. (1998). The minimum description length principle in coding and modelling. *IEEE Trans. Inform. Theory* **44**, 2743–2760.

Belzile, E., Angers, J.-F. (1995). Choice of noninformative priors for the variance components of an unbalanced one-way random model. *Comm. Statist. Theory Methods* **24**, 1325–1341.

Berger, J.O., Robert, C.P. (1990). Subjective hierarchical Bayes estimation of a multivariate mean: On the frequentist interface. *Ann. Statist.* **18**, 617–651.

Berger, J.O., Strawderman, W.E. (1996). Choice of hierarchical priors: Admissibility in estimation of normal means. *Ann. Statist.* **24**, 931–951.

Berger, R.L. (1979). Gamma minimax robustness of Bayes rules. *Comm. Statist. Theory Methods* **8**, 543–560.

Blyth, S., Smith, A.F.M. (1998). Bayesian meta-analysis of frequentist p-values. *Comm. Statist. Theory Methods* **27**, 2707–2724.

Box, G.E.P., Cox, D.R. (1964). An analysis of transformations. *J. Roy. Statist. Soc., Ser. B* **26**, 211–252. (With discussion.)

Brown, L.D., Cai, T.T., DasGupta, A. (2001). Interval estimation of a binomial proportion. *Statist. Sci.* **16**, 101–133. (With discussion.)

Brown, L.D., Cai, T.T., DasGupta, A. (2002). Confidence intervals for a binomial proportion and asymptotic expansions. *Ann. Statist.* **30**, 160–201.

Casella, G., Hwang, J.T. (1987). Employing vague prior information in the construction of confidence sets. *J. Multivariate Anal.* **21**, 79–104.

Chaloner, K. (1987). A Bayesian approach to the estimation of variance components for the unbalanced one way random model. *Technometrics* **29**, 323–337.

Chao, J.C., Phillips, P.C.B. (1998). Posterior distributions in limited information analysis of the simultaneous equations model using the Jeffreys prior. *J. Econometrics* **87**, 49–86.

Chao, J.C., Phillips, P.C.B. (2002). Jeffreys prior analysis of the simultaneous equations model in the case with $n + 1$ endogenous variables. *J. Econometrics* **111**, 251–283.

Chen, M.H., Ibrahim, J.G., Shao, Q.M. (2000). Power prior distributions for generalized linear models. *J. Statist. Plann. Inference* **84**, 121–137.

Cho, J.S., Baek, S.U. (2002). Development of matching priors for the ratio of the failure rate in the Burr model. *Far East J. Theor. Statist.* **8**, 79–87.

Cifarelli, D.M., Regazzini, E. (1987). Priors for exponential families which maximize the association between past and future observations. In: Viertl, R. (Ed.), *Probability and Bayesian Statistics*. Plenum, London, pp. 83–95.

Clarke, B., Sun, D. (1997). Reference priors under the chi-squared distance. *Sankhyā A* **59**, 215–231.

Consonni, G., Veronese, P. (1989). Some remarks on the use of improper priors for the analysis of exponential regression problems. *Biometrika* **76**, 101–106.

Consonni, G., Veronese, P. (1992). Bayes factors for linear models and improper priors. In: Bernardo, J.M., Berger, J.O., Dawid, A.P., Smith, A.F.M. (Eds.), *Bayesian Statistics 4*. Oxford University Press, pp. 587–594.

Consonni, G., Veronese, P. (1993). Unbiased Bayes estimates and improper priors. *Ann. Inst. Statist. Math.* **45**, 303–315.

Cornfield, J. (1969). The Bayesian outlook and its application. *Biometrics* **25**, 617–657. (With discussion.)

Cox, D.R., Reid, N. (1987). Parameter orthogonality and approximate conditional inference. *J. Roy. Statist. Soc., Ser. B* **49**, 1–39. (With discussion.)

Crowder, M. (1992). Bayesian priors based on a parameter transformation using the distribution function. *Ann. Inst. Statist. Math.* **44**, 405–416.

Crowder, M., Sweeting, T. (1989). Bayesian inference for a bivariate binomial distribution. *Biometrika* **76**, 599–603.

Daniels, M.J. (1999). A prior for the variance in hierarchical models. *Canad. J. Statist.* **27**, 567–578.

Daniels, M.J. (2005). A class of shrinkage priors for the dependence structure in longitudinal data. *J. Statist. Plann. Inference* **127**, 119–130.

DasGupta, A. (1985). Bayes minimax estimation in multiparameter families when the parameter space is restricted to a bounded convex set. *Sankhyā A* **47**, 326–332.

Datta, G.S. (1996). On priors providing frequentist validity for Bayesian inference of multiple parametric functions. *Biometrika* **83**, 287–298.

Datta, G.S., Mukerjee, R. (2004). *Probability Matching Priors: Higher Order Asymptotics*. Springer, Berlin.

Datta, G.S., Ghosh, M., Mukerjee, R. (2000). Some new results on probability matching priors. *Calcuta Statist. Assoc. Bull.* **50**, 179–192. Corr: **51**, 125.

Dawid, A.P. (1991). Fisherian inference in likelihood and prequential frames of reference. *J. Roy. Statist. Soc., Ser. B* **53**, 79–109. (With discussion).

de Alba, E., Mendoza, M. (1996). A discrete model for Bayesian forecasting with stable seasonal patterns. In: Hill, R. Carter (Ed.), *Advances in Econometrics II*. JAI Press, New York, pp. 267–281.

DeGroot, M.H. (1970). *Optimal Statistical Decisions*. McGraw-Hill, New York.

Delampady, M., DasGupta, A., Casella, G., Rubin, H., Strawderman, W.E. (2001). A new approach to default priors and robust Bayesian methodology. *Canad. J. Statist.* **29**, 437–450.

de Waal, D.J., Nel, D.G. (1988). A procedure to select a ML-II prior in a multivariate normal case. *Comm. Statist. Simulation Comput.* **17**, 1021–1035.

Diaconis, P., Freedman, D.A. (1998). Consistency of Bayes estimates for nonparametric regression: Normal theory. *Bernoulli* **4**, 411–444.

DiCiccio, T.J., Martin, M.A. (1991). Approximations of marginal tail probabilities for a class of smooth functions with applications to Bayesian and conditional inference. *Biometrika* **78**, 891–902.

DiCiccio, T.J., Stern, S.E. (1994). Frequentist and Bayesian Bartlett correction of test statistics based on adjusted profile likelihood. *J. Roy. Statist. Soc., Ser. B* **56**, 397–408.

Eaton, M.L. (1982). A method for evaluating improper prior distributions. In: Gupta, S.S., Berger, J.O. (Eds.), *Statistical Decision Theory and Related Topics III, vol. 1*. Academic Press, New York.

Eaton, M.L. (1992). A statistical diptych: Admissible inferences, recurrence of symmetric Markov chains. *Ann. Statist.* **20**, 1147–1179.

Eaves, D., Chang, T. (1992). Priors for ordered conditional variance and vector partial correlation. *J. Multivariate Anal.* **41**, 43–55.

Efron, B. (1993). Bayes and likelihood calculations from confidence intervals. *Biometrika* **80**, 3–26.

Erickson, T. (1989). Proper posteriors from improper priors for an unidentified error-in-variables model. *Econometrica* **57**, 1299–1316.

Evans, I.G., Nigm, A.M. (1980). Bayesian prediction for two parameter Weibull lifetime models. *Comm. Statist. Theory Methods* **9**, 649–658.

Everson, P.J., Bradlow, E.T. (2002). Bayesian inference for the Beta-binomial distribution via polynomial expansions. *J. Comput. Graph. Statist.* **11**, 202–207.

Fan, T.-H., Berger, J.O. (1992). Behaviour of the posterior distribution and inferences for a normal mean with t prior distributions. *Statistics and Decisions* **10**, 99–120.

Fatti, L.P. (1982). Predictive discrimination under the random effect model. *South African Statist. J.* **16**, 55–77.

Fernández, C., Steel, M. (1998). Reference priors for the non-normal two-sample problems. *Test* **7**, 179–205.

Fraser, D.A.S., Reid, N. (1996). Bayes posteriors for scalar interest parameters. In: Bernardo, J.M., Berger, J.O., Dawid, A.P., Smith, A.F.M. (Eds.), *Bayesian Statistics 5*. Oxford University Press, pp. 581–585.

Fraser, D.A.S., Reid, N. (2002). Strong matching of frequentist and Bayesian parametric inference. *J. Statist. Plann. Inference* **103**, 263–285.

Fraser, D.A.S., Yi, G.Y. (2002). Location reparametrization and default priors for statistical analysis. *J. Iranian Statist. Soc.* **1**, 55–78.

Fraser, D.A.S., McDunnough, P., Taback, N. (1997). Improper priors, posterior asymptotic normality, and conditional inference. In: Johnson, N.L., Balakrishnan, N. (Eds.), *Advances in the Theory and Practice of Statistics: A Volume in Honor of Samuel Kotz*. Wiley, New York, pp. 563–569.

Fraser, D.A.S., Reid, N., Wu, J. (1999). A simple general formula for tail probabilities for frequentist and Bayesian inference. *Biometrika* **86**, 249–264.

Freedman, D.A. (1966). A note on mutual singularity of priors. *Ann. Math. Statist.* **37**, 375–381.

Freedman, D.A. (1995). Some issues in the foundation of statistics. In: van Fraassen, B.C. (Ed.), *Topics in the Foundation of Statistics*. Kluwer Academic, Dordrecht, pp. 19–83. (With discussion.)
Gatsonis, C.A. (1984). Deriving posterior distributions for a location parameter: A decision theoretic approach. *Ann. Statist.* **12**, 958–970.
Geisser, S. (1965). Bayesian estimation in multivariate analysis. *Ann. Math. Statist.* **36**, 150–159.
Geisser, S. (1979). In discussion of Bernardo (1979b) *J. Roy. Statist. Soc., Ser. B* **41**, 136–137.
Geisser, S. (1980). A predictivist primer. In: Zellner, A. (Ed.), *Bayesian Analysis in Econometrics and Statistics: Essays in Honor of Harold Jeffreys*. North-Holland, Amsterdam, pp. 363–381.
Geisser, S. (1993). *Predictive Inference: An Introduction*. Chapman and Hall, London.
Geisser, S., Cornfield, J. (1963). Posterior distributions for multivariate normal parameters. *J. Roy. Statist. Soc., Ser. B* **25**, 368–376.
Ghosal, S., Ghosh, J.K., Ramamoorthi, R.V. (1997). Non-informative priors via sieves and packing numbers. In: Panchpakesan, S., Balakrisnan, N. (Eds.), *Advances in Decision Theory and Applications*. Birkhäuser, Boston, pp. 119–132.
Ghosh, J.K., Mukerjee, R. (1991). Characterization of priors under which Bayesian and frequentist Bartlett corrections are equivalent in the multivariate case. *J. Multivariate Anal.* **38**, 385–393.
Ghosh, J.K., Mukerjee, R. (1993). Frequentist validity of highest posterior density regions in the multiparameter case. *Ann. Math. Statist.* **45**, 293–302. corr: 602.
Ghosh, J.K., Mukerjee, R. (1993). On priors that match posterior and frequentist distribution functions. *Canad. J. Statist.* **21**, 89–96.
Ghosh, J.K., Mukerjee, R. (1995). Frequentist validity of highest posterior density regions in the presence of nuisance parameters. *Statistics and Decisions* **13**, 131–139.
Ghosh, J.K., Mukerjee, R. (1995). On perturbed ellipsoidal and highest posterior density regions with approximate frequentist validity. *J. Roy. Statist. Soc., Ser. B* **57**, 761–769.
Ghosh, J.K., Samanta, T. (2002). Nonsubjective Bayes testing – an overview. *J. Statist. Plann. Inference* **103**, 205–223.
Ghosh, M., Meeden, G. (1997). *Bayesian Methods for Finite Population Sampling*. Chapman and Hall, London.
Ghosh, M., Chen, M.-H., Ghosh, A., Agresti, A. (2000). Hierarchical Bayesian analysis of binary matched pairs data. *Statistica Sinica* **10**, 647–657.
Ghosh, M., Ghosh, A., Chen, M.-H., Agresti, A. (2000). Noninformative priors for one-parameter item response models. *J. Statist. Plann. Inference* **88**, 99–115.
Gokhale, D.V., Press, S.J. (1982). Assessment of a prior distribution for the correlation coefficient in a bivariate normal distribution. *J. Roy. Statist.* **145**, 237–249.
Good, I.J. (1952). Rational decisions. *J. Roy. Statist. Soc., Ser. B* **14**, 107–114.
Good, I.J. (1969). What is the use of a distribution?. In: Krishnaiah, P.R. (Ed.), *Multivariate Analysis, vol. 2*. Academic Press, New York, pp. 183–203.
Good, I.J. (1981). Flexible priors for estimating a normal distribution. *J. Statist. Comput. Simulation* **13**, 151–153.
Good, I.J. (1986). Invariant hyperpriors. *J. Statist. Comput. Simulation* **24**, 323–324.
Gutiérrez-Peña, E., Muliere, P. (2004). Conjugate priors represent strong pre-experimental assumptions. *Scand. J. Statist.* **31**, 235–246.
Hadjicostas, P. (1998). Improper and proper posteriors with improper priors in a hierarchical model with a beta-binomial likelihood. *Comm. Statist. Theory Methods* **27**, 1905–1914.
Hartigan, J.A. (1965). The asymptotically unbiased prior distribution. *Ann. Math. Statist.* **36**, 1137–1152.
Hartigan, J.A. (1971). Similarity and probability. In: Godambe, V.P., Sprott, D.A. (Eds.), *Foundations of Statistical Inference*. Holt, Rinehart and Winston, Toronto, pp. 305–313. (With discussion.)
Hartigan, J.A. (1996). Locally uniform prior distributions. *Ann. Statist.* **24**, 160–173.
Hartigan, J.A. (1998). The maximum likelihood prior. *Ann. Statist.* **26**, 2083–2103.
Hartigan, J.A. (2004). Uniform priors on convex sets improve risk. *Statist. Probab. Lett.* **67**, 285–288.
Hartigan, J.A., Murphy, T.B. (2002). Inferred probabilities. *J. Statist. Plann. Inference* **105**, 23–34.
Heath, D., Sudderth, W. (1978). On finitely additive priors, coherence, and extended admissibility. *Ann. Statist.* **6**, 333–345.

Heath, D., Sudderth, W. (1989). Coherent inference from improper priors and from finitely additive priors. *Ann. Statist.* **17**, 907–919.
Hill, B.M. (1965). Inference about variance components in the one-way model. *J. Amer. Statist. Assoc.* **60**, 806–825.
Hill, S.D., Spall, J.C. (1988). Shannon information-theoretic priors for state-space model parameters. In: Spall, J.C. (Ed.), *Bayesian Analysis of Time Series and Dynamic Models*. Marcel Dekker, New York, pp. 509–524.
Hill, S.D., Spall, J.C. (1994). Sensitivity of a Bayesian analysis to the prior distribution. *IEEE Trans. Systems, Man Cybernet.* **24**, 216–221.
Hobert, J.P., Marchev, D., Schweinsberg, J. (2004). Stability of the tail Markov chain and the evaluation of improper priors for an exponential rate parameter. *Bernoulli* **10**, 549–564.
Howlader, H.A., Weiss, G. (1988). Bayesian reliability estimation of a two parameter Cauchy distribution. *Biom. J.* **30**, 329–337.
Ibrahim, J.G. (1997). On properties of predictive priors in linear models. *Amer. Statist.* **51**, 333–337.
Ibrahim, J.G., Chen, M.-H. (1998). Prior distributions and Bayesian computation for proportional hazards models. *Sankhyā B* **60**, 48–64.
Ibrahim, J.G., Laud, P.W. (1991). On Bayesian analysis of generalized linear models using Jeffreys' prior. *J. Amer. Statist. Assoc.* **86**, 981–986.
Inaba, T. (1984). Non-informative prior distribution in a simultaneous equations model. *J. Japan Statist. Soc.* **14**, 93–103.
Jaynes, E.T. (1976). Confidence intervals vs. Bayesian intervals. In: Harper, W.L., Hooker, C.A. (Eds.), *Foundations of Probability Theory, Statistical Inference and Statistical Theories of Science, vol. 2*. Reidel, Dordrecht, pp. 175–257. (With discussion.)
Jeffreys, H. (1955). The present position in probability theory. *British J. Philos. Sci.* **5**, 275–289.
Joshi, S., Shah, M. (1991). Estimating the mean of an inverse Gaussian distribution with known coefficient of variation. *Comm. Statist. Theory Methods* **20**, 2907–2912.
Kappenman, R.F., Geisser, S., Antle, C.F. (1970). Bayesian and fiducial solutions to the Fieller Creasy problem. *Sankhyā* **25**, 330–331.
Kashyap, R.L. (1971). Prior probability and uncertainty. *IEEE Trans. Inform. Theory* **14**, 641–650.
Kass, R.E. (1989). The geometry of asymptotic inference. *Statist. Sci.* **4**, 188–235. (With discussion.)
Kass, R.E. (1990). Data-translated likelihood and Jeffreys' rule. *Biometrika* **77**, 107–114.
Keyes, T.K., Levy, M.S. (1996). Goodness of prediction fit for multivariate linear models. *J. Amer. Statist. Assoc.* **91**, 191–197.
Kim, S.W., Ibrahim, J.G. (2000). On Bayesian inference for proportional Hazards models using noninformative priors. *Lifetime Data Anal.* **6**, 331–341.
Lee, G. (1998). Development of matching priors for $P(X < Y)$ in exponential distributions. *J. Korean Statist. Soc.* **27**, 421–433.
Lee, H.K.H. (2003). A noninformative prior for neural networks. *Machine Learning* **50**, 197–212.
Lee, J.C., Chang, C.H. (2000). Bayesian analysis of a growth curve model with a general autoregressive covariance structure. *Scand. J. Statist.* **27**, 703–713.
Lee, J.C., Hwang, R.C. (2000). On estimation and prediction for temporally correlated longitudinal data. *J. Statist. Plann. Inference* **87**, 87–104.
Lee, J.J., Shin, W.S. (1990). Prior distributions using the entropy principles. *Korean J. Appl. Statist.* **3**, 91–104.
Lindley, D.V. (1958). Fiducial distribution and Bayes' theorem. *J. Roy. Statist. Soc., Ser. B* **20**, 102–107.
Lindley, D.V. (1961). The use of prior probability distributions in statistical inference and decision. In: Neyman, J., Scott, E.L. (Eds.), *Proc. Fourth Berkeley Symp. 1*. Univ. California Press, Berkeley, pp. 453–468.
Lunn, D.J., Thomas, A., Best, N., Spiegelhalter, D. (2000). WinBUGS – a Bayesian modelling framework: Concepts, structure, and extensibility. *Statist. Comput.* **10**, 325–337.
Marinucci, D., Petrella, L. (1999). A Bayesian proposal for the analysis of stationary and nonstationary AR(1) time series. In: Bernardo, J.M., Berger, J.O., Dawid, A.P., Smith, A.F.M. (Eds.), *Bayesian Statistics, vol. 6*. Oxford University Press, pp. 821–828.
Maryak, J.L., Spall, J.C. (1987). Conditions for the insensitivity of the Bayesian posterior distribution to the choice of prior distribution. *Statist. Probab. Lett.* **5**, 401–407.

McCulloch, R.E., Polson, N.G., Rossi, P.E. (2000). A Bayesian analysis of the multinomial probit model with fully identified parameters. *J. Econometrics* **99**, 173–193.

Meeden, G., Vardeman, S. (1991). A non-informative Bayesian approach to interval estimation in finite population sampling. *J. Amer. Statist. Assoc.* **86**, 972–986.

Mendoza, M., Gutiérrez-Peña, E. (2000). Bayesian conjugate analysis of the Galton–Walton process. *Test* **9**, 149–171.

Meng, X.-L., Zaslavsky, A.M. (2002). Single observation unbiased priors. *Ann. Statist.* **30**, 1345–1375.

Mengersen, K.L., Robert, C.P. (1996). Testing for mixtures: A Bayesian entropic approach. In: Bernardo, J.M., Berger, J.O., Dawid, A.P., Smith, A.F.M. (Eds.), *Bayesian Statistics 5*. Oxford University Press, pp. 255–276. (With discussion.)

Miller, R.B. (1980). Bayesian analysis of the two-parameter gamma distribution. *Technometrics* **22**, 65–69.

Molitor, J., Sun, D. (2002). Bayesian analysis under ordered functions of parameters. *Environ. Ecological Statist.* **9**, 179–193.

Moreno, E., Girón, F.J. (1997). Estimating with incomplete count data: A Bayesian approach. *J. Statist. Plann. Inference* **66**, 147–159.

Mukerjee, R., Chen, Z. (2003). On expected lengths of predictive intervals. *Scand. J. Statist.* **30**, 757–766.

Mukerjee, R., Dey, D.K. (1993). Frequentist validity of posterior quantiles in the presence of a nuisance parameter: Higher order asymptotics. *Biometrika* **80**, 499–505.

Mukerjee, R., Ghosh, M. (1997). Second-order probability matching priors. *Biometrika* **84**, 970–975.

Mukhopadhyay, S., DasGupta, A. (1997). Uniform approximation of Bayes solutions and posteriors: Frequentist valid Bayes inference. *Statistics and Decisions* **15**, 51–73.

Mukhopadhyay, S., Ghosh, M. (1995). On the uniform approximation of Laplace's prior by t-priors in location problems. *J. Multivariate Anal.* **54**, 284–294.

Natarajan, R., Kass, R.E. (2000). Reference Bayesian methods for generalized linear mixed models. *J. Amer. Statist. Assoc.* **95**, 227–237.

Natarajan, R., McCulloch, C.E. (1998). Gibbs sampling with diffuse proper priors: A valid approach to data driven inference?. *J. Comput. Graph. Statist.* **7**, 267–277.

Nicolau, A. (1993). Bayesian intervals with good frequentist behaviour in the presence of nuisance parameters. *J. Roy. Statist. Soc., Ser. B* **55**, 377–390.

Novick, M.R. (1969). Multiparameter Bayesian indifference procedures. *J. Roy. Statist. Soc., Ser. B* **31**, 29–64.

Novick, M.R., Hall, W.K. (1965). A Bayesian indifference procedure. *J. Amer. Statist. Assoc.* **60**, 1104–1117.

Ogata, Y. (1990). A Monte Carlo method for an objective Bayesian procedure. *Ann. Inst. Statist. Math.* **42**, 403–433.

Oh, M.-S., Kim, Y. (2000). Jeffreys noninformative prior in Bayesian conjoint analysis. *J. Korean Statist. Soc.* **29**, 137–153.

Paris, J.B. (1994). *The Uncertain Reasoner's Companion: A Mathematical Perspective*. Cambridge University Press.

Pauler, D.K., Wakefield, J.C., Kass, R.E. (1999). Bayes factors and approximations for variance component models. *J. Amer. Statist. Assoc.* **94**, 1242–1253.

Peers, H.W. (1965). On confidence points and Bayesian probability points in the case of several parameters. *J. Roy. Statist. Soc., Ser. B* **27**, 9–16.

Peers, H.W. (1968). Confidence properties of Bayesian interval estimates. *J. Roy. Statist. Soc., Ser. B* **30**, 535–544.

Pericchi, L.R. (1981). A Bayesian approach to transformation to normality. *Biometrika* **68**, 35–43.

Pericchi, L.R., Sansó, B. (1995). A note on bounded influence in Bayesian analysis. *Biometrika* **82**, 223–225.

Pericchi, L.R., Walley, P. (1991). Robust Bayesian credible intervals and prior ignorance. *Internat. Statist. Rev.* **59**, 1–23.

Piccinato, L. (1973). Un metodo per determinare distribuzioni iniziali relativamente non-informative. *Metron* **31**, 124–156.

Piccinato, L. (1977). Predictive distributions and non-informative priors. In: Uldrich, M. (Ed.), *Trans. 7th Prague Conf. Information Theory*. Czech. Acad. Sciences, Prague, pp. 399–407.

Poirier, D.J. (1985). Bayesian hypothesis testing in linear models with continuously induced conjugate priors across hypotheses. In: Bernardo, J.M., DeGroot, M.H., Lindley, D.V., Smith, A.F.M. (Eds.), *Bayesian Statistics 2*. North-Holland, Amsterdam, pp. 711–722.

Poirier, D.J. (1994). Jeffreys prior for logit models. *J. Econometrics* **63**, 327–339.

Pole, A., West, M. (1989). Reference analysis of the dynamic linear model. *J. Time Series Anal.* **10**, 13–147.

Polson, N.G. (1992). In discussion of Ghosh and Mukerjee (1992). In: Bernardo, J.M., Berger, J.O., Dawid, A.P., Smith, A.F.M. (Eds.), *Bayesian Statistics 4*. Oxford University Press, pp. 203–205.

Price, R.M., Bonett, D.G. (2000). Estimating the ratio of two Poisson rates. *Comput. Statist. Data Anal.* **34**, 345–356.

Raftery, A.E. (1988). Inference for the binomial *N* parameter: A hierarchical Bayes approach. *Biometrika* **75**, 223–228.

Raftery, A.E., Akman, V.E. (1986). Bayesian analysis of a Poisson process with a change-point. *Biometrika* **73**, 85–89.

Rai, G. (1976). Estimates of mean life and reliability function when failure rate is randomly distributed. *Trab. Estadist.* **27**, 247–251.

Reid, N. (1996). Likelihood and Bayesian approximation methods. In: Bernardo, J.M., Berger, J.O., Dawid, A.P., Smith, A.F.M. (Eds.), *Bayesian Statistics 5*. Oxford University Press, pp. 349–366. (With discussion.)

Rissanen, J. (1983). A universal prior for integers and estimation by minimum description length. *Ann. Statist.* **11**, 416–431.

Rissanen, J. (1986). Stochastic complexity and modelling. *Ann. Statist.* **14**, 1080–1100.

Rissanen, J. (1987). Stochastic complexity. *J. Roy. Statist. Soc., Ser. B* **49**, 223–239 and 252–265 (with discussion).

Rissanen, J. (1988). Minimum description length principle. In: Kotz, S., Johnson, N.L., Read, C.B. (Eds.), *Encyclopedia of Statistical Sciences, vol. 5*. Wiley, New York, pp. 523–527.

Rissanen, J. (1989). *Stochastic Complexity in Statistical Enquiry*. World Scientific, Singapore.

Roberts, G.O., Rosenthal, J.S. (2001). Infinite hierarchies and prior distributions. *Bernoulli* **7**, 453–471.

Rodríguez, C.C. (1991). From Euclid to entropy. In: Grandy, W.T., Schick, L.H. (Eds.), *Maximum Entropy and Bayesian Methods*. Kluwer Academic, Dordrecht, pp. 343–348.

Rousseau, J. (2000). Coverage properties of one-sided intervals in the discrete case and application to matching priors. *Ann. Inst. Statist. Math.* **52**, 28–42.

Sansó, B., Pericchi, L.R. (1992). Near ignorance classes of log-concave priors for the location model. *Test* **1**, 39–46.

Sansó, B., Pericchi, L.R. (1994). On near ignorance classes. *Rev. Brasilera Prob. Estatist.* **8**, 119–126.

Shieh, G., Lee, J.C. (2002). Bayesian prediction analysis for growth curve model using noninformative priors. *Ann. Inst. Statist. Math.* **54**, 324–337.

Singh, U., Gupta, P.K., Upadhyay, S.K. (2002). Estimation of exponentiated Weibull shape parameters under linex loss functions. *Comm. Statist. Simulation Comput.* **31**, 523–537.

Smith, R.L., Naylor, J.C. (1987). A comparison of maximum likelihood and Bayesian estimators for the three-parameter Weibull distribution. *Appl. Statist.* **36**, 358–369.

Sono, S. (1983). On a noninformative prior distribution for Bayesian inference of multinomial distribution's parameters. *Ann. Inst. Statist. Math.* **35**, 167–174.

Spall, J.C., Hill, S.D. (1990). Least informative Bayesian prior distributions for finite samples based on information theory. *IEEE Trans. Automat. Control* **35**, 580–583.

Spiegelhalter, D.J. (1985). Exact Bayesian inference on the parameters of a Cauchy distribution with vague prior information. In: Bernardo, J.M., DeGroot, M.H., Lindley, D.V., Smith, A.F.M. (Eds.), *Bayesian Statistics 2*. North-Holland, Amsterdam, pp. 743–749.

Stewart, W.E. (1987). Multiresponse parameter estimation with a new and noninformative prior. *Biometrika* **74**, 557–562.

Stone, M. (1963). The posterior *t* distribution. *Ann. Math. Statist.* **34**, 568–573.

Stone, M., Springer, B.G.F. (1965). A paradox involving quasi prior distributions. *Biometrika* **52**, 623–627.

Strachan, R.W., van Dijk, H.K. (2003). Bayesian model selection with an uninformative prior. *Oxford Bul. Econ. Statist.* **65**, 863–876.

Strawderman, W.E. (2000). Minimaxity. *J. Amer. Statist. Assoc.* **95**, 1364–1368.

Sun, D., Tsutakawa, R.K., He, Z. (2001). Propriety of posteriors with improper priors in hierarchical linear mixed models. *Statistica Sinica* **11**, 77–95.

Sweeting, T.J. (1984). On the choice of prior distribution for the Box–Cox transformed linear model. *Biometrika* **71**, 127–134.
Sweeting, T.J. (1985). Consistent prior distributions for transformed models. In: Bernardo, J.M., DeGroot, M.H., Lindley, D.V., Smith, A.F.M. (Eds.), *Bayesian Statistics 2*. North-Holland, Amsterdam, pp. 755–762.
Sweeting, T.J. (1995). A framework for Bayesian and likelihood approximations in statistics. *Biometrika* **82**, 1–24.
Sweeting, T.J. (1995). A Bayesian approach to approximate conditional inference. *Biometrika* **82**, 25–36.
Sweeting, T.J. (1996). Approximate Bayesian computation based on signed roots of log-density rations. In: Bernardo, J.M., Berger, J.O., Dawid, A.P., Smith, A.F.M. (Eds.), *Bayesian Statistics 5*. Oxford University Press, pp. 427–444. (With discussion.)
Sweeting, T.J. (2001). Coverage probability bias, objective Bayes and the likelihood principle. *Biometrika* **88**, 657–675.
Tibshirani, R. (1989). Noninformative priors for one parameter of many. *Biometrika* **76**, 604–608.
Tiwari, R.C., Chib, S., Jammalamadaka, S.R. (1989). Bayes estimation of the multiple correlation coefficient. *Comm. Statist. Theory Methods* **18**, 1401–1413.
Torgersen, E.N. (1981). Measures of information based on comparison with total information and with total ignorance. *Ann. Statist.* **9**, 638–657.
Tsutakawa, R.K. (1992). Prior distribution for item response curves. *British J. Math. Statist. Philos.* **45**, 51–74.
Upadhyay, S.K., Pandey, M. (1989). Prediction limits for an exponential distribution: A Bayes predictive distribution approach. *IEEE Trans. Reliability* **38**, 599–602.
Upadhyay, S.K., Peshwani, M. (2003). Choice between Weibull and lognormal models: A simulation-based Bayesian study. *Comm. Statist. Theory Methods* **32**, 381–405.
Upadhyay, S.K., Smith, A.F.M. (1994). Modelling complexities in reliability, and the role of simulation in Bayesian computation. *Internat. J. Continuing Eng. Education* **4**, 93–104.
Upadhyay, S.K., Agrawal, R., Smith, A.F.M. (1996). Bayesian analysis of inverse Gaussian non-linear regression by simulation. *Sankhyā* **58**, 363–378.
Upadhyay, S.K., Vasishta, N., Smith, A.F.M. (2001). Bayes inference in life testing and reliability via Markov chain Monte Carlo simulation. *Sankhyā A* **63**, 15–40.
Vaurio, J.K. (1992). Objective prior distributions and Bayesian updating. *Reliability Eng. System Safety* **35**, 55–59.
Villegas, C. (1969). On the a priori distribution of the covariance matrix. *Ann. Math. Statist.* **40**, 1098–1099.
Villegas, C. (1971). On Haar priors. In: Godambe, V.P., Sprott, D.A. (Eds.), *Foundations of Statistical Inference*. Holt, Rinehart and Winston, Toronto, pp. 409–414. (With discussion.)
Villegas, C. (1977). Inner statistical inference. *J. Amer. Statist. Assoc.* **72**, 453–458.
Villegas, C. (1981). Inner statistical inference II. *Ann. Statist.* **9**, 768–776.
Villegas, C. (1982). Maximum likelihood and least squares estimation in linear and affine functional models. *Ann. Statist.* **10**, 256–265.
Walker, S.G., Gutiérrez-Peña, E. (1999). Robustifying Bayesian procedures. In: Bernardo, J.M., Berger, J.O., Dawid, A.P., Smith, A.F.M. (Eds.), *Bayesian Statistics 6*. Oxford University Press, pp. 685–710. (With discussion.)
Walker, S., Muliere, P. (1999). A characterization of a neutral to the right prior via an extension of Johnson's sufficientness postulate. *Ann. Statist.* **27**, 589–599.
Wallace, C.S., Dowe, D.L. (1999). Minimum message length and Kolmogorov complexity. *The Computer J.* **42**, 270–283.
Wallace, C.S., Freeman, P.R. (1987). Estimation and inference by compact coding. *J. Roy. Statist. Soc., Ser. B* **49**, 240–265.
Wasserman, L. (1996). The conflict between improper priors and robustness. *J. Statist. Plann. Inference* **52**, 1–15.
Wasserman, L. (2000). Asymptotic inference for mixture models using data-dependent priors. *J. Roy. Statist. Soc., Ser. B* **62**, 159–180.
Wasserman, L., Clarke, B. (1995). Information tradeoff. *Test* **4**, 19–38.
Wolfinger, R.D., Kass, R.E. (2000). Nonconjugate Bayesian analysis of variance component models. *Biometrics* **56**, 768–774.

Zellner, A. (1977). Maximal data information prior distributions. In: Ayka, A., Brumat, C. (Eds.), *New Developments in the Applications of Bayesian Methods*. North-Holland, Amsterdam, pp. 211–232.
Zellner, A. (1983). Applications of Bayesian analysis in econometrics. *The Statistician* **32**, 23–34.
Zellner, A. (1986). On assessing prior distributions and Bayesian regression analysis with g-prior distributions. In: Goel, P.K., Zellner, A. (Eds.), *Bayesian Inference and Decision Techniques: Essays in Honor of Bruno de Finetti*. North-Holland, Amsterdam, pp. 233–243.
Zellner, A. (1986). Bayesian estimation and prediction using asymmetric loss functions. *J. Amer. Statist. Assoc.* **81**, 446–451.
Zellner, A. (1988). Optimal information processing and Bayes' theorem. *Amer. Statist.* **42**, 278–284. (With discussion.)
Zellner, A. (1991). Bayesian methods and entropy in economics and econometrics. In: Grandy, W.T., Schick, L.H. (Eds.), *Maximum Entropy and Bayesian Methods*. Kluwer Academic, Dordrecht, pp. 17–31.
Zellner, A. (1996). Models, prior information, and Bayesian analysis. *J. Econometrics* **75**, 51–68.
Zellner, A. (1997). *Bayesian Analysis in Econometrics and Statistics: The Zellner View and Papers*. Edward Elgar, Brookfield, VT.
Zellner, A., Siow, A. (1980). Posterior odds ratios for selected regression hypothesis. In: Bernardo, J.M., DeGroot, M.H., Lindkey, D.V., Smith, A.F.M. (Eds.), *Bayesian Statistics*. Valencia University Press, pp. 585–603 and 618–647 (with discussion).

Handbook of Statistics, Vol. 25
ISSN: 0169-7161
© 2005 Elsevier B.V. All rights reserved.
DOI 10.1016/S0169-7161(05)25003-4

3

Probability Matching Priors

Gauri Sankar Datta and Trevor J. Sweeting

Abstract
A probability matching prior is a prior distribution under which the posterior probabilities of certain regions coincide with their coverage probabilities, either exactly or approximately. Use of such a prior will ensure exact or approximate frequentist validity of Bayesian credible regions. Probability matching priors have been of interest for many years but there has been a resurgence of interest over the last twenty years. In this article we survey the main developments in probability matching priors, which have been derived for various types of parametric and predictive region.

1. Introduction

A *probability matching prior* (PMP) is a prior distribution under which the posterior probabilities of certain regions coincide with their coverage probabilities, either exactly or approximately. The simplest example of this phenomenon occurs when we have an observation X from the $N(\theta, 1)$ distribution, where θ is unknown. If we take an improper uniform prior π over the real line for θ then the posterior distribution of $Z = \theta - X$ is exactly the same as its sampling distribution. Therefore $\mathrm{pr}_\pi\{\theta \leqslant \theta_\alpha(X)|X\} = \mathrm{pr}_\theta\{\theta \leqslant \theta_\alpha(X)\} = \alpha$, where $\theta_\alpha(X) = X + z_\alpha$ and z_α is the α-quantile of the standard normal distribution. Thus every credible interval based on the pivot Z with posterior probability α is also a confidence interval with confidence level α. The uniform distribution is therefore a PMP. Of course, this example applies to a random sample of size n from the $N(\phi, 1)$ distribution on taking X to be the sufficient statistic $\sqrt{n}\bar{X}$ and $\theta = \sqrt{n}\phi$.

Situations in which there exist exact PMPs are very limited. Most of the literature on this topic therefore focuses on approximate PMPs, usually for large n, and usually based on the asymptotic theory of the maximum likelihood estimator. Since the posterior and sampling distributions of the pivotal quantity Z in the above example coincide, the uniform prior matches the posterior and coverage probabilities of all sets. However, one can also search for priors that approximately match the posterior probabilities of specific sets, such as likelihood regions. The answers can then be quite different to those obtained when matching for all regions. It is also possible to obtain PMPs for predictive, as opposed to parametric, regions. Other recent reviews on PMPs are Ghosh and Mukerjee (1998) and Datta and Mukerjee (2004). We also refer to Ghosh and Mukerjee (1992a), Ghosh (1994), Reid (1996), Kass and Wasserman (1996) and Sweeting (2001) for discussion on matching priors and other nonsubjective priors.

The framework in which we will be working is that of n independent and identically distributed observations from a parametric model. However, it would be possible to extend the results reviewed here to situations where the observations are not identically distributed or are dependent. Let $X^n = (X_1, \dots, X_n)$ be a vector of random variables, identically and independently distributed as the random variable X having probability density function $f(\cdot;\theta)$ with respect to some dominating σ-finite measure λ. The density $f(\cdot;\theta)$ is supposed known apart from the parameter $\theta = (\theta_1, \dots, \theta_d) \in \Omega$, an open subset of $\mathbb{R}^d$. Further suppose that a prior density $\pi(\cdot)$, which may be proper or improper, is available for θ. Then the posterior density of θ is

$$\pi\left(\theta|X^n\right) \propto \pi(\theta)L_n(\theta), \tag{1}$$

where $L_n(\theta) \propto \prod_i f(X_i;\theta)$ is the likelihood function associated with X^n.

We will assume sufficient regularity conditions on the likelihood function and prior density for the validity of the asymptotic results in this article. All statements about orders of magnitude will be in the probability sense. Unless otherwise stated we will assume the existence of a unique local maximum likelihood estimator $\hat{\theta}_n$ of θ and of Fisher's information matrix $i(\theta)$ with (i, j)th element $\kappa_{ij}(\theta) = -E_\theta\{\partial^2 \log f(X;\theta)/\partial\theta_i\partial\theta_j\}$.

The organisation of the rest of this chapter is as follows. We begin in Section 2 by giving a general discussion of the rationale behind PMPs, especially in relation to asymptotic matching. In the succeeding sections we review the major developments that have taken place over the last forty years, much of which has occurred in the last fifteen to twenty years. In Section 3 we briefly review results on exact probability matching. All the subsequent material concerns asymptotic matching. Beginning with one-parameter models in Section 4, we review the seminal paper of Welch and Peers (1963) on one-sided PMPs, various results on two-sided probability matching and probability matching in non-regular cases. We then proceed in Section 5 to the more problematic multiparameter case and consider PMPs for one or more specified interest parameters and for specified regions. The corresponding theory for matching predictive, as opposed to parametric, probability statements is reviewed in Section 6. Before finishing with some concluding remarks in Section 8, we have included in Section 7 a brief discussion on invariance of matching priors under reparameterisation.

One common feature of most of this work is the derivation of a (partial) differential equation for a PMP. In general there may be no solution, a unique solution, or many (usually an infinite number of) solutions to this equation, depending on the particular problem. Moreover, once we move away from single parameter or simple multiparameter examples, these equations cannot be solved analytically. Therefore there are important questions about implementation that need to be addressed if the theory is to be used in practice. We return to this question in Section 8.

2. Rationale

Before reviewing the main results in this area, it will be useful to consider the rationale behind PMPs. From a Bayesian point of view it can be argued that a PMP is a suitable candidate for a nonsubjective Bayesian prior, since the repeated sampling property of the associated posterior regions provides some assurance that the Bayesian results will make some inferential sense, at least on average, whatever the true value of θ. Alternatively, the matching property can be viewed as one desirable property of a proposed nonsubjective prior that has been constructed according to some other criterion. For example, the extent to which various forms of reference prior (Bernardo, 1979) are probability matching has been investigated quite extensively. For general discussions on the development of nonsubjective priors see, for example, Bernardo (1979), Berger and Bernardo (1992a), Ghosh and Mukerjee (1992a), Kass and Wasserman (1996), Bernardo and Ramón (1998), Barron (1999), Bernardo and Smith (1994, Chapter 5) and Bernardo (2005) in this volume.

Often the derivation of a frequentist property from a Bayesian property proceeds by introducing an auxiliary prior distribution that is allowed to shrink to the true parameter value, thereby producing the required frequentist probability. Although early applications of the shrinkage argument are due to Bickel and Ghosh (1990), Dawid (1991), Ghosh and Mukerjee (1991) and Ghosh (1994, Chapter 9), the argument is presented in detail in Mukerjee and Reid (2000) and also in Section 1.2 of the monograph on PMPs by Datta and Mukerjee (2004).

Suppose that the set C has posterior probability α under a given prior π and that also

$$\mathrm{pr}_\tau(\theta \in C) \equiv \int \mathrm{pr}_\theta\{\theta \in C\}\tau(\theta)\,\mathrm{d}\theta = \alpha \tag{2}$$

for every continuous prior density τ, either exactly or approximately. If the relation (2) is exact then it is equivalent to the pointwise assertion

$$\mathrm{pr}_\theta(\theta \in C) = \alpha \tag{3}$$

for every $\theta \in \Omega$. However, if (2) holds only in an asymptotic sense then further conditions are needed to deduce the 'fully frequentist' probability matching result (3). From a Bayesian standpoint it can be argued that (2) is actually sufficient for our purposes, since if there is some concern about the form of prior used, then (2) would provide a type of robustness result with respect to the prior in repeated usage, as opposed to hypothetical repeated sampling. When (2) is an asymptotic approximation we will say that the relation (3) holds *very weakly*; see Woodroofe (1986), for example. Happily this formulation also avoids a technical issue in asymptotic analysis, where (2) may hold in an asymptotic sense for every smooth prior τ but not for every point-mass prior.

From a foundational point of view, one difficulty with the notion of PMPs is that, in common with a number of other approaches to nonsubjective prior construction, probability matching involves averaging over the sample space, contravening the strong likelihood principle, and hence is a non-Bayesian property. There seems to be little one can do about this except to investigate sensitivity to the sampling rule and weigh up what is lost in this regard with what is gained through prior robustness. Some discussion of conformity of PMPs to the likelihood principle is given in Sweeting (2001).

The use of PMPs can also be justified from a frequentist point of view. A probability matching property indicates that the associated frequentist inference will possess some conditional validity, since it will be approximately the same as a direct inference based on some form of noninformative prior. Conditional properties of frequentist methods are important to ensure that the resulting inferences are relevant to the data in hand. Alternatively, PMPs may be viewed simply as a mechanism for producing approximate confidence intervals. This possibility is particularly appealing in the multiparameter case, where the construction of appropriate pivotal quantities that effectively eliminate nuisance parameters causes major difficulties in frequentist inference (cf. Stein, 1985).

3. Exact probability matching priors

In this section we review results on exact probability matching. In parametric inference, Lindley (1958) was the first to address this problem in a different set-up. He sought to provide a Bayesian interpretation of Fisher's (1956) fiducial distribution for a scalar parameter. Under the assumption of a single sufficient statistic, Lindley (1958) showed that if a suitable transformation results in a location model with a location parameter $\tau = g(\theta)$, then exact matching holds by using a uniform prior on the location parameter τ. Welch and Peers (1963) extended this work to any location family model by dispensing with the assumption of the existence of a one-dimensional sufficient statistic. Datta et al. (2000a) and Datta and Mukerjee (2004, p. 22) provided an explicit proof of this exact matching result.

Indeed the above exact matching result extends beyond location parameter and scale parameter families. Hora and Buehler (1966) studied this problem for group models. It can be shown that for credible sets satisfying a suitable invariance requirement based on a prior defined by the right Haar measure on the group models exact matching will hold; see Lemma 1 of Severini et al. (2002). In fact this exact matching holds conditionally on an ancillary statistic and hence also holds unconditionally. Welch and Peers (1963) proved exact matching of conditional coverage for a location parameter by conditioning on the successive differences of the observations. Fraser and Reid (2002) recently discussed an exact matching result for a scalar interest parameter that is a known linear combination of the location parameters in a possibly multiparameter location model.

Nearly forty years ago exact matching for a predictive distribution was considered in a binomial prediction problem by Thatcher (1964). He showed that prediction limits for the future number of successes in a binomial experiment can also be interpreted as Bayesian prediction limits. One difficulty with his solution is that one single prior does not work for both the upper and the lower prediction limits.

Datta (2000, unpublished note) obtained exact matching for prediction in a location parameter family based on a uniform prior on the location parameter. In a recent article, Severini et al. (2002) explored this problem in group models. They showed that the posterior coverage probabilities of certain invariant Bayesian predictive regions based on the right Haar measure on the group exactly match the corresponding conditional frequentist probabilities, conditioned on a certain ancillary statistic. In particular, they considered the location-scale problem, multivariate location-scale model and highest predictive density regions in elliptical contoured distributions.

4. Parametric matching priors in the one-parameter case

In the remainder of this article we will consider asymptotic probability matching. In keeping with the terminology in higher-order asymptotics, we say that an asymptotic (in n, the sample size) approximation is accurate to the kth-order, if the neglected terms in the approximation are of the order $n^{-k/2}$, $k = 1, 2, \ldots$. Thus we say a matching prior is second-order accurate if the coverage probability differs from the credible level by terms which are of order n^{-1}; it is third-order accurate if the difference is of the order $n^{-3/2}$, and so on. We note, however, that in the PMP literature priors that we call here second-order matching are often referred to as first-order matching, and priors which are defined third-order matching here are referred to as second-order matching (see, for example, Mukerjee and Ghosh, 1997; Datta and Mukerjee, 2004). We begin by investigating probability matching associated with posterior parametric statements in the case $d = 1$.

4.1. One-sided parametric intervals

Let $0 < \alpha < 1$ and suppose that π is a positive continuous prior on Ω. Let $t(\pi, \alpha)$ denote the α-quantile of the posterior distribution of θ. That is, $t(\pi, \alpha)$ satisfies

$$\mathrm{pr}_\pi\left\{\theta \leqslant t(\pi, \alpha)|X^n\right\} = \alpha. \tag{4}$$

We wish to know when is it true that, to a given asymptotic order of approximation,

$$\mathrm{pr}_\theta\left\{\theta \leqslant t(\pi, \alpha)\right\} = \alpha, \tag{5}$$

either pointwise or very weakly, as discussed in Section 2. In view of the standard first order quadratic approximation to the log-likelihood, it follows that relation (5) holds up to $\mathrm{O}(n^{-1/2})$ for every positive continuous prior π on Ω and $0 < \alpha < 1$. Thus, to the first-order of approximation *all* priors are probability matching. Welch and Peers (1963) investigated the second-order of approximation. They showed that relation (5) holds to $\mathrm{O}(n^{-1})$ pointwise for all α if and only if $\pi(\theta) \propto \{i(\theta)\}^{1/2}$. They obtained this result via an expansion of a cumulant generating function associated with the posterior density. Thus Jeffreys' invariant prior is second-order probability matching with respect to one-sided parametric regions. A conditional version of this classical result is also available. Say that a prior distribution is 'kth-order stably matching' if, conditional on any kth-order locally ancillary statistic, the very weak error in (5) is $\mathrm{O}(n^{-k/2})$. It is shown in Sweeting (1995b) that Jeffreys' prior is second-order stably matching with respect to one-sided parametric regions.

In general, it can be shown that the approximation in (5) is no better than $\mathrm{O}(n^{-1})$, unless the skewness measure $\rho_{111}(\theta) = \{i(\theta)\}^{-3/2} E_\theta\{\partial \log f(X; \theta)/\partial\theta\}^3$ is independent of θ. In that case Welch and Peers (1963) showed that the approximation in (5) is $\mathrm{O}(n^{-3/2})$ under Jeffreys' prior.

Finally we note that, although here and elsewhere in the paper a PMP will very often be improper, it is usually possible to find a suitable proper but diffuse prior that achieves matching to the given order of approximation. For example, when X is $N(\theta, 1)$, coverage matching of the one-sided credible interval will hold to $\mathrm{O}(n^{-1})$ under a $N(\mu_0, \sigma_0^2)$ prior, where $\mu_0 = \mathrm{O}(1)$, $\sigma_0^{-2} = \mathrm{O}(n^{-1/2})$. For third-order matching σ_0^{-2} needs to be $\mathrm{O}(n^{-1})$.

4.2. *Two-sided parametric intervals*

In this section we describe the construction of PMPs associated with likelihood and related regions in the case $d = 1$. For such regions matching up to $\mathrm{O}(n^{-2})$ can be achieved as a result of cancellation of directional errors. The associated family of matching priors may not contain Jeffreys' prior, however, in which case one-sided matching under a member of this family will only be to $\mathrm{O}(n^{-1/2})$.

Again let $0 < \alpha < 1$ and suppose that π is a positive continuous prior on Ω. Let $(t_1(\pi, \alpha), t_2(\pi, \alpha))$ be any interval having posterior probability α; that is,

$$\mathrm{pr}_\pi\left\{t_1(\pi, \alpha) \leqslant \theta \leqslant t_2(\pi, \alpha) | X^n\right\} = \alpha. \tag{6}$$

As before, we ask when it is also true that, to a given degree of approximation,

$$\mathrm{pr}_\theta\left\{t_1(\pi, \alpha) \leqslant \theta \leqslant t_2(\pi, \alpha)\right\} = \alpha, \tag{7}$$

either pointwise or very weakly.

Although we trivially deduce from the discussion in Section 4.1 that (7) holds to $\mathrm{O}(n^{-1/2})$ for every smooth prior π, and to $\mathrm{O}(n^{-1})$ under Jeffreys' prior, the order of approximation is usually better than this. Hartigan (1966) showed that the error in (7) is $\mathrm{O}(n^{-1})$ for every positive continuous prior π on Ω when (t_1, t_2) is an equi-tailed Bayesian region. Alternatively, if (t_1, t_2) is a likelihood region (that is, $L_n\{t_1(\pi, \alpha)\} = L_n\{t_2(\pi, \alpha)\}$) then (7) holds to $\mathrm{O}(n^{-1})$ for every positive continuous prior π. This is true since both the Bayesian and frequentist errors in approximating the likelihood ratio statistic by the chi-square distribution with 1 degree of freedom are $\mathrm{O}(n^{-1})$. Furthermore, it follows from Ghosh and Mukerjee (1991) that (7) holds to $\mathrm{O}(n^{-2})$ for all priors of the form

$$\pi(\theta) = i(\theta)^{1/2} \mathrm{e}^{-n\tau(\theta)}\left\{k_1 + k_2 \int i(\theta)^{1/2} \mathrm{e}^{n\tau(\theta)} \,\mathrm{d}\theta\right\}, \tag{8}$$

where $\tau(\theta) = \frac{1}{2}\int\{i(\theta)\}^{1/2}\rho_{111}(\theta)\,\mathrm{d}\theta$ and k_1, k_2 are arbitrary constants; see also Sweeting (1995a). That is, every prior of the form (8) is fourth-order matching with respect to likelihood regions.

Notice that the class of priors of the form (8) contains Jeffreys' prior only in the special case where the skewness $\rho_{111}(\theta)$ is independent of θ. If this is not the case then the $\mathrm{O}(n^{-2})$ two-sided matching is bought at the price of having only $\mathrm{O}(n^{-1/2})$ one-sided matching. Furthermore, it can be argued that there is nothing special about likelihood regions for the construction of objective priors. Severini (1993) showed that, by a judicious choice of interval, it is possible to have probability matching to third-order under *any* given smooth prior. In Sweeting (1999) such regions are referred to as 'Bayes-confidence regions' and they are derived using a signed-root likelihood ratio approach coupled with a Bayesian argument based on a shrinking prior, as discussed in Section 2. In the light of these results there does not seem to be a compelling case for constructing default priors via probability matching for likelihood regions, since other suitably perturbed likelihood regions could equally be used and would yield different default priors (Sweeting, 2001). One possible strategy suggested in Sweeting (2001) would be to use Jeffreys' prior to give one-sided matching to $O(n^{-1})$ and, if desired, to use perturbed likelihood regions associated with Jeffreys' prior to give two-sided matching to $O(n^{-2})$.

The special case of an exponential family model is discussed in more detail in Sweeting (2001). Here the class (8) of matching priors is shown to take a simple form. Moreover, it is shown that all members of this class are matching under a class of linear stopping rules, thereby providing some conformity with the likelihood principle. This property is generalised to perturbed likelihood regions.

4.3. Nonregular cases

The derivation of matching priors based on parametric coverage for a class of nonregular cases has been considered by Ghosal (1999). Suppose that the underlying density $f(\cdot;\theta)$ has support $S(\theta) = [a_1(\theta), a_2(\theta)]$ depending on θ with $f(\cdot;\theta)$ strictly positive on $S(\theta)$ and the family $S(\theta)$ strictly monotone in θ. Then it is shown in Ghosal (1999) that, under suitable regularity conditions, (4) and (5) hold to $O(n^{-1})$ for every positive continuous prior π on Ω and $0 < \alpha < 1$. Notice that this is an order of magnitude higher than the regular case of Section 4.1. Furthermore, it is shown that the unique PMP to $O(n^{-2})$ (very weakly) is $\pi(\theta) \propto c(\theta)$, where $c(\theta) = E_\theta\{\partial \log f(X;\theta)/\partial\theta\}$. Examples include the uniform and shifted exponential distributions. The above result is obtained via an asymptotic expansion of the posterior distribution, which has an exponential distribution as the leading term rather than a normal distribution as in regular cases, followed by a shrinking prior argument (see Section 2). Note that, like Jeffreys' prior in the regular case, $c(\theta)$ is parameterisation invariant.

As pointed out in Ghosal (1999), since highest posterior density regions tend to be one-sided intervals in these nonregular cases, it does not seem relevant to consider two-sided matching here.

5. Parametric matching priors in the multiparameter case

In this section we investigate probability matching associated with posterior parametric statements in the case $d > 1$.

As in the one-parameter case, in the multiparameter case the $O(n^{-1/2})$ equivalence of Bayesian and frequentist probability statements holds on account of the first-order equivalence of the Bayesian and frequentist normal approximations. In this section we therefore investigate higher-order probability matching in the case $d > 1$.

5.1. Matching for an interest parameter

Suppose that θ_1 is considered to be a scalar parameter of primary interest and that $(\theta_2, \dots, \theta_d)$ is regarded as a vector of nuisance parameters. Let $z_\alpha(X^n)$ be the upper α-quantile of the marginal posterior distribution of θ_1 under π; that is $z_\alpha(X^n)$ satisfies

$$\mathrm{pr}_\pi\{\theta_1 \leqslant z_\alpha(X^n) \mid X^n\} = \alpha.$$

The prior $\pi(\cdot)$ is $\mathrm{O}(n^{-1})$-probability matching with respect to θ_1 if

$$\mathrm{pr}_\theta\{\theta_1 \leqslant z_\alpha(X^n)\} = \alpha + \mathrm{O}(n^{-1}) \tag{9}$$

pointwise or very weakly for every α, $0 < \alpha < 1$.

The second-order one-parameter result of Welch and Peers (1963) was generalised to the multiparameter case by Peers (1965). Peers showed that $\pi(\theta)$ is a PMP for θ_1 if and only if it is a solution of the partial differential equation

$$D_j\{(\kappa^{11})^{-1/2}\kappa^{1j}\pi\} = 0, \tag{10}$$

where $D_j \equiv \partial/\partial\theta_j$, $\kappa^{ij}(\theta)$ is the (i, j)th element of $\{i(\theta)\}^{-1}$ and we have used the summation convention. In particular, if θ_1 and $(\theta_2, \dots, \theta_d)$ are orthogonal (cf. Cox and Reid, 1987) then $\kappa^{1j}(\theta) = 0$, $j = 2, \dots, d$ and it follows immediately from (10) that

$$\pi(\theta) \propto \kappa^{11}(\theta)^{-1/2} h(\theta_2, \dots, \theta_d) = \kappa_{11}(\theta)^{1/2} h(\theta_2, \dots, \theta_d), \tag{11}$$

where the function h is arbitrary, as indicated in Tibshirani (1989) and rigorously proved by Nicolaou (1993) and Datta and J.K. Ghosh (1995a), following earlier work by Stein (1985).

Given the arbitrary function h in (11), there is the opportunity of probability matching to an order higher than $\mathrm{O}(n^{-1})$. Mukerjee and Ghosh (1997), generalising results in Mukerjee and Dey (1993), show that it may be possible to achieve $\mathrm{o}(n^{-1})$ probability matching in (9). The priors that emerge, however, may differ depending on whether posterior quantiles or the posterior distribution function of θ_1 are considered.

Datta and J.K. Ghosh (1995a) generalised the differential equation (10) for an arbitrary parametric function. For a smooth parametric function $t(\theta)$ of interest a second-order PMP $\pi(\theta)$ satisfies the differential equation

$$D_j\{\Lambda^{-1} b^j \pi\} = 0, \tag{12}$$

where

$$b^j = \kappa^{jr} D_r t, \qquad \Lambda^2 = \kappa^{jr} D_j t D_r t. \tag{13}$$

The third-order matching results for $t(\theta)$, established by Mukerjee and Reid (2001), are more complex. One may also refer to Datta and Mukerjee (2004, Theorem 2.8.1) for this result.

EXAMPLE 5.1. Consider the location-scale model

$$f(x; \theta) = \frac{1}{\theta_2} f^*\left(\frac{x - \theta_1}{\theta_2}\right), \quad x \in \mathbb{R},$$

where $\theta_1 \in \mathbb{R}$, $\theta_2 > 0$, and $f^*(\cdot)$ is a density with support $\mathbb{R}$. It can be checked that the information matrix is $\theta_2^{-2}\Sigma$, where $\Sigma = (\sigma_{ij})$ is the covariance matrix of $U^{j-1}\mathrm{d}\log f^*(U)/\mathrm{d}U$, $j = 1, 2$, when the density of U is $f^*(u)$.

Suppose θ_1 is the interest parameter. In a class of priors of the form $g(\theta_1)h(\theta_2)$, $g(\theta_1)$ is necessarily a constant for a second-order PMP for θ_1. If an orthogonal parameterisation holds, which happens if $f^*(\cdot)$ is a symmetric density about zero, such as a standard normal or Student's t distribution, then $h(\theta_2)$ is arbitrary in this case. However, in the absence of orthogonalisation, $h(\theta_2)$ must be proportional to θ_2^{-1}. It can be checked using (2.5.15) of Datta and Mukerjee (2004) that, in the class of priors under consideration, the prior proportional to θ_2^{-1} is the unique third-order PMP with or without parametric orthogonality.

Now suppose θ_2 is the interest parameter. In the aforementioned class of priors, $g(\theta_1)/\theta_2$ is a second-order PMP. Under parametric orthogonality, again $g(\theta_1)$ is arbitrary; otherwise it is a constant. In either case, using (2.5.15) of Datta and Mukerjee (2004), it can be checked that the second-order PMP is also third-order matching for θ_2.

If PMPs are required when both θ_1 and θ_2 are interest parameters, then θ_2^{-1} is the unique prior (see Section 5.2).

Finally, consider θ_1/θ_2 as the parametric function of interest. In a class of priors of the form $\theta_2^a(\theta^{\mathrm{T}}\Sigma\theta)^b$, where a, b are constants, it can be checked from (12) that $a = -1$, b arbitrary gives the class of all second-order PMPs. In particular, $b = 0$ leads to the prior which is second-order matching simultaneously for each of θ_1, θ_2 and θ_1/θ_2. While this is an attractive choice, a negative feature of this prior is that in the special case of a normal model, it leads to a marginalisation paradox (cf. Dawid et al., 1973; Bernardo, 1979). On the other hand, for the normal model it was checked in Datta and Mukerjee (2004, p. 38) that this prior is the unique third-order PMP for θ_1/θ_2. Interestingly, $b = -1/2$ leads to an important prior, namely, the reference prior of Bernardo (1979), which avoids this paradox.

EXAMPLE 5.2. Consider the balanced one-way random effects model with t treatments and n replications for each treatment. The corresponding mixed linear model is given by

$$X_{ij} = \mu + a_i + e_{ij}, \quad 1 \leqslant i \leqslant n,\ 1 \leqslant j \leqslant t,$$

where the parameter μ represents the general mean, each random effect a_i is univariate normal with mean zero and variance λ_1, each e_{ij} is univariate normal with mean zero and variance λ_2, and the a_i's and the e_{ij}'s are all independent. Here $\mu \in \mathbb{R}$ and λ_1, λ_2 (> 0) are unknown parameters and t ($\geqslant 2$) is fixed. For $1 \leqslant i \leqslant n$, let $X_i = (X_{i1}, \ldots, X_{it})$. Then $X_1, \ldots, X_n$ are i.i.d. random variables with a multivariate normal distribution.

This example has been extensively discussed in the noninformative priors literature (cf. Box and Tiao, 1973). Berger and Bernardo (1992b) have constructed reference priors for $(\mu, \lambda_1, \lambda_2)$ under different situations when one of the three parameters is of interest and the remaining two parameters are either clustered into one group or divided into two groups according to their order of importance. Ye (1994) considered the one-to-one reparameterisation $(\mu, \lambda_2, \lambda_1/\lambda_2)$ and constructed various reference priors when

λ_1/λ_2 is the parameter of importance. Datta and M. Ghosh (1995a, 1995b) have constructed reference priors as well as matching priors for various parametric functions in this set-up.

Suppose our interest lies in the ratio λ_1/λ_2. Following Datta and Mukerjee (2004), we reparameterise as

$$\theta_1 = \lambda_1/\lambda_2, \qquad \theta_2 = \{\lambda_2^{t-1}(t\lambda_1 + \lambda_2)\}^{1/(2t)}, \qquad \theta_3 = \mu,$$

where $\theta_1, \theta_2 > 0$ and $\theta_3 \in \mathbb{R}$. It can be checked that the above is an orthogonal parameterisation with $\kappa_{11}(\theta) \propto (1+t\theta_1)^{-2}$. Hence by (11), second-order matching is achieved if and only if $\pi(\theta) = d(\theta^{(2)})/(1 + t\theta_1)$, where $\theta^{(2)} = (\theta_2, \theta_3)$ and $d(\cdot)$ is a smooth positive function. In fact, Datta and Mukerjee (2004) further showed that a subclass of this class of priors given by $\pi(\theta) = d^*(\theta_3)/\{(1 + t\theta_1)\theta_2\}$ characterises the class of all third-order matching priors, where $d^*(\theta_3)$ is a smooth positive function. In particular, taking $d^*(\theta_3)$ constant, it can be checked that the prior given by $\{(1 + t\theta_1)\theta_2\}^{-1}$ corresponds to $\{(t\lambda_1 + \lambda_2)\lambda_2\}^{-1}$ in the original parameterisation, which is one of the reference priors derived by Berger and Bernardo (1992b) and Ye (1994). This prior was also recommended by Datta and M. Ghosh (1995a) and Datta (1996) from other considerations.

An extension of this example to the unbalanced case was studied recently by Datta et al. (2002).

5.2. *Probability matching priors in group models*

We have already encountered group models in our review of exact matching in Section 3. In this section we will review various noninformative priors for a scalar interest parameter in group models. The interest parameter is maximal invariant under a suitable group of transformations G where the remaining parameters are identified with the group element g. We assume that G is a Lie group and that the parameter space Ω has a decomposition such that the space of nuisance parameters is identical with G. It is also assumed that G acts on Ω freely by left multiplication in G. Chang and Eaves (1990) derived reference priors for this model. Datta and J.K. Ghosh (1995b) used this model for a comparative study of various reference priors, including the Berger–Bernardo and Chang–Eaves reference priors. Writing the parameter vector as (ψ_1, g), Datta and J.K. Ghosh (1995b) noted that (i) κ^{11}, the first diagonal element of the inverse of the information matrix, is only a function of ψ_1, (ii) the Chang and Eaves (1990) reference prior $\pi_{CE}(\psi_1, g)$ is given by $\{\kappa^{11}(\psi_1)\}^{-1/2}h^r(g)$ and (iii) the Berger and Bernardo (1992a) reference prior for the group ordering $\{\psi_1, g\}$ is given by $\{\kappa^{11}(\psi_1)\}^{-1/2}h^l(g)$, where $h^r(g)$ and $h^l(g)$ are the right and the left invariant Haar densities on G. While the left and the right invariant Haar densities are usually different, they are identical if the group G is either commutative or compact. Typically, these reference priors are improper. It follows from Dawid et al. (1973) that $\pi_{CE}(\psi_1, g)$ will not yield any marginalisation paradox for inference for ψ_1. The same is not true for the two-group ordering Berger–Bernardo reference prior. Datta and J.K. Ghosh (1995b) illustrated this point through two examples. With regard to probability matching, this article established that while the Chang–Eaves reference prior is always second-order matching for ψ_1, this is not always the case for the other prior based on the left invariant Haar density. However these authors also noted that often the Berger–Bernardo reference prior based on a one-at-a-time parameter grouping is identical with the Chang–Eaves reference prior. We illustrate these priors through two examples given below.

EXAMPLE 5.3. Consider the location-scale family of Example 5.1. Let $\psi_1 = \theta_1/\theta_2$. Under a group of scale transformations, ψ_1 remains invariant. Since the group operation is commutative, both the left and the right invariant Haar densities are equal and given by g^{-1}. Here the group element g is identified with the nuisance parameter θ_2. It can be checked that $\kappa^{11} = (\sigma_{22} + 2\sigma_{12}\psi_1 + \sigma_{11}\psi_1^2)/|\Sigma|$, where Σ and its elements are as defined in Example 5.1. Hence the Berger–Bernardo and Chang–Eaves reference priors are given by (in the (ψ_1, g)-parameterisation) $g^{-1}(\sigma_{22} + 2\sigma_{12}\psi_1 + \sigma_{11}\psi_1^2)^{-1/2}$. It can be checked that, in the (θ_1, θ_2)-parameterisation, this prior reduces to the prior corresponding to the choices $a = -1$ and $b = -1/2$ given at the end of Example 5.1.

EXAMPLE 5.4. Consider a bivariate normal distribution with means μ_1, μ_2 and dispersion matrix $\sigma^2 I_2$. Let the parameter of interest be $\theta_1 = (\mu_1 - \mu_2)/\sigma$ and write $\mu_2 = \theta_2$ and $\sigma = \theta_3$. This problem was considered, among others, by Datta and J.K. Ghosh (1995b) and Ghosh and Yang (1996). For the group of transformations $H = \{g\colon g = (g_2, g_3), g_2 \in \mathbb{R}, g_3 > 0\}$ in the range of X_1 defined by $gX_1 = g_3X_1 + g_2 1$, the induced group of transformations on the parameter space is $G = \{g\}$ with the transformation defined by $g\theta = (\theta_1, g_3\theta_2 + g_2, g_3\theta_3)$. Here θ_1 is the maximal invariant parameter. Datta and J.K. Ghosh (1995b) obtained the Chang–Eaves reference prior and Berger–Bernardo reference prior for the group ordering $\{\theta_1, (\theta_2, \theta_3)\}$ given by

$$\pi_{CE}(\theta) \propto \left(8 + \theta_1^2\right)^{-1/2}\theta_3^{-1}, \qquad \pi_{BB}(\theta) \propto \left(8 + \theta_1^2\right)^{-1/2}\theta_3^{-2}.$$

The Chang–Eaves prior is a second-order PMP for θ_1. These two priors transform to $\sigma^{-1}\{8\sigma^2 + (\mu_1 - \mu_2)^2\}^{-1/2}$ and $\sigma^{-2}\{8\sigma^2 + (\mu_1 - \mu_2)^2\}^{-1/2}$ respectively in the original parameterisation. Datta and Mukerjee (2004) have considered priors having the structure

$$\pi^*(\mu_1, \mu_2, \sigma) = \left\{8 + \left(\frac{\mu_1 - \mu_2}{\sigma}\right)^2\right\}^{-s_1} \sigma^{-s_2},$$

where s_1 and s_2 are real numbers. They showed that such priors will be second-order matching for $(\mu_1 - \mu_2)/\sigma$ if and only if $s_2 = 2s_1 + 1$. Clearly, the Chang–Eaves prior satisfies this condition. Datta and Mukerjee (2004) have further shown that the only third-order PMP in this class is given by $s_1 = 0$ and $s_2 = 1$.

5.3. Probability matching priors and reference priors

Berger and Bernardo (1992a) have given an algorithm for reference priors. Berger (1992) has also introduced the reverse reference prior. The latter prior for the parameter grouping $\{\theta_1, \theta_2\}$, assuming $\theta = (\theta_1, \theta_2)$ for simplicity, is the prior that would result from following the reference prior algorithm for the reverse parameter grouping $\{\theta_2, \theta_1\}$. Following the algorithm with rectangular compacts, the reverse reference prior $\pi_{RR}(\theta_1, \theta_2)$ is of the form

$$\pi_{RR}(\theta_1, \theta_2) = \kappa_{11}^{1/2}(\theta) g(\theta_2),$$

where $g(\theta_2)$ is an appropriate function of θ_2. Under parameter orthogonality, the above reverse reference prior has the form of (11) and hence it is a second-order PMP for θ_1.

While the above is an interesting result for reverse reference priors, reference priors still play a dominant role in objective Bayesian inference. Datta and M. Ghosh (1995b) have provided sufficient conditions establishing the equivalence of reference and reverse reference priors. A simpler version of that result is described here. Let the Fisher information matrix be a $d \times d$ diagonal matrix with the jth diagonal element factored as $\kappa_{jj}(\theta) = h_{j1}(\theta_j) h_{j2}(\theta_{(-j)})$, where $\theta_{(-j)} = (\theta_1, \ldots, \theta_{j-1}, \theta_{j+1}, \ldots, \theta_d)$ and $h_{j1}(\cdot), h_{j2}(\cdot)$ are two positive functions. Assuming a rectangular sequence of compacts, it can be shown that for the one-at-a-time parameter grouping $\{\theta_1, \ldots, \theta_d\}$ the reference prior $\pi_R(\theta)$ and the reverse reference prior $\pi_{RR}(\theta)$ are identical and proportional to

$$\prod_{j=1}^{d} h_{j1}^{1/2}(\theta_j). \tag{14}$$

It was further shown by Datta (1996) that the above prior is second-order probability matching for each component of θ. The last result is also available in somewhat implicit form in Datta and M. Ghosh (1995b), and in a special case in Sun and Ye (1996).

EXAMPLE 5.5 (Datta and M. Ghosh 1995b). Consider the inverse Gaussian distribution with pdf

$$f(x; \mu, \sigma^2) = (2\pi\sigma^2)^{-1/2} x^{-3/2} \exp\{-(x-\mu)^2/(2\sigma^2\mu^2 x)\} I_{(0,\infty)}(x),$$

where μ (> 0) and σ^2 (> 0) are both unknown. Here $i(\mu, \sigma^2) = \mathrm{diag}(\mu^{-3}\sigma^{-2}, (2\sigma^4)^{-1})$. From the result given above it follows that $\pi_R(\mu, \sigma^2) \equiv \pi_{RR}(\mu, \sigma^2) \propto \mu^{-3/2}\sigma^{-2}$, identifying $\theta_1 = \mu$ and $\theta_2 = \sigma^2$. It further follows that it is a second-order PMP for both μ and σ^2.

5.4. *Simultaneous and joint matching priors*

Peers (1965) showed that it is not possible in general to find a single prior that is probability matching to $O(n^{-1})$ for all parameters simultaneously. Datta (1996) extended this discussion to the case of $s \leqslant d$ real parametric functions of interest, $t_1(\theta), \ldots, t_s(\theta)$. For each of these parametric functions a differential equation similar to (12) can be solved to get a second-order PMP for that parametric function. It may be possible that one or more priors exist satisfying all the s differential equations. If that happens, following Datta (1996) these priors may be referred to as *simultaneous marginal* second-order PMPs.

The above simultaneous PMPs should be contrasted with joint PMPs, which are derived via joint consideration of all these parametric functions. While simultaneous marginal PMPs are obtained by matching appropriate posterior and frequentist marginal quantiles, joint PMPs are obtained by matching appropriate posterior and frequentist joint c.d.f.'s.

A prior $\pi(\cdot)$ is said to be a joint PMP for $t_1(\theta), \ldots, t_s(\theta)$ if

$$\begin{aligned}\mathrm{pr}_\pi&\left[\frac{n^{1/2}\{t_1(\theta)-t_1(\hat\theta_n)\}}{d_1}\leqslant w_1,\ldots,\frac{n^{1/2}\{t_s(\theta)-t_s(\hat\theta_n)\}}{d_s}\leqslant w_s|X^n\right]\\ &=\mathrm{pr}_\theta\left[\frac{n^{1/2}\{t_1(\theta)-t_1(\hat\theta_n)\}}{d_1}\leqslant w_1,\ldots,\frac{n^{1/2}\{t_s(\theta)-t_s(\hat\theta_n)\}}{d_s}\leqslant w_s\right]\\ &\quad+\mathrm{o}\big(n^{-1/2}\big)\end{aligned}\tag{15}$$

for all $w_1, \ldots, w_s$ and all θ. In the above, $d_k = [\{\nabla t_k(\hat\theta_n)\}^{\mathrm{T}} C_n^{-1}\{\nabla t_k(\hat\theta_n)\}]^{1/2}$, $k = 1, \ldots, s$, and $w_1, \ldots, w_s$ are free from n, θ and X^n. Here C_n is the observed information matrix, a $d \times d$ matrix which is positive definite with pr_θ-probability $1 + \mathrm{O}(n^{-2})$. It is assumed that the $d \times s$ gradient matrix corresponding to the s parametric functions is of full column rank for all θ.

In this section we will be concerned with only second-order PMPs. Mukerjee and Ghosh (1997) showed that up to this order marginal PMPs via c.d.f. matching and quantile matching are equivalent. Thus, from the definition of joint matching it is obvious that any joint PMP for a set of parametric functions will also be a simultaneous marginal PMP for those functions.

Datta (1996) investigated joint matching by considering an indirect extension of earlier work of Ghosh and Mukerjee (1993a), in which an importance ordering as used in reference priors (cf. Berger and Bernardo, 1992a) is assumed amongst the components of θ. Ghosh and Mukerjee (1993a) considered PMPs for the entire parameter vector θ by considering a pivotal vector whose ith component can be interpreted as an approximate standardised version of the regression residual of θ_i on $\theta_1, \ldots, \theta_{i-1}$, $i = 1, \ldots, d$, in the posterior set-up. For this reason Datta (1996) referred to this approach as a regression residual matching approach. On the other hand, Datta (1996) proposed a direct extension of Datta and J.K. Ghosh (1995a) for a set of parametric functions that are of equal importance. The relationship between these two approaches has been explored in Datta (1996).

Define b_k^j and Λ_k as in (13) by replacing $t(\theta)$ by $t_k(\theta)$, $k = 1, \ldots, s$. Also define, for $k, m, u = 1, \ldots, s$,

$$\begin{aligned}\rho_{km} &= b_k^j \kappa_{jl} b_m^l/(\Lambda_k\Lambda_m) = \kappa^{jr} D_j t_k D_r t_m/(\Lambda_k\Lambda_m),\\ \zeta_{kmu} &= b_k^j D_j \rho_{mu}/\Lambda_k.\end{aligned}$$

Datta (1996) proved that a simultaneous marginal PMP for parametric functions $t_1(\theta), \ldots, t_s(\theta)$ is a joint matching prior if and only if

$$\zeta_{kmu} + \zeta_{mku} + \zeta_{ukm} = 0, \quad k, m, u = 1, \ldots, s, \tag{16}$$

hold.

Note that the conditions (16) depend only on the parametric functions and the model. Thus if these conditions fail, there would be no joint PMP even if a simultaneous marginal PMP exists. In the special case in which interest lies in the entire parameter vector, then $s = d$, $t_k(\theta) = \theta_k$. Here, if the Fisher information matrix is a $d \times d$ diagonal matrix then condition (16) holds trivially. If further the jth diagonal element factors as $\kappa_{jj}(\theta) = h_{j1}(\theta_j)h_{j2}(\theta_{(-j)})$, where $\theta_{(-j)} = (\theta_1, \ldots, \theta_{j-1}, \theta_{j+1}, \ldots, \theta_d)$ and $h_{j1}(\cdot), h_{j2}(\cdot)$ are two positive functions, then the unique second-order joint PMP $\pi_{JM}(\theta)$ is given by the prior

$$\pi_{JM}(\theta) \propto \prod_{j=1}^{d} h_{j1}^{1/2}(\theta_j), \tag{17}$$

which is the same as the reference prior given in (14). In particular, Sun and Ye's (1996) work that considered a joint PMP for the orthogonal mean and variance parameters in a two-parameter exponential family follows as a special case of the prior (17).

EXAMPLE 5.6. Datta (1996) considered the example of a p-variate normal with mean $\mu = (\mu_1, \ldots, \mu_p)$ and identity matrix as the dispersion matrix. Reparameterise as $\mu_1 = \theta_1 \cos\theta_2, \ldots, \mu_{p-1} = \theta_1 \sin\theta_2 \cdots \sin\theta_{p-1}\cos\theta_p$, $\mu_p = \theta_1 \sin\theta_2 \cdots \sin\theta_{p-1}\sin\theta_p$. Here $i(\theta) = \operatorname{diag}(1, \theta_1^2, \theta_1^2 \sin^2\theta_2, \ldots, \theta_1^2 \sin^2\theta_2 \cdots \sin^2\theta_{p-1})$ and all its diagonal elements have the desired factorisable structure. Hence by (17), $\pi(\theta) \propto 1$ is the unique joint PMP for the components of θ.

EXAMPLE 5.7. We continue Example 5.2 with a different notation. Consider the mixed linear model

$$X_{ij} = \theta_1 + a_i + e_{ij}, \quad 1 \leqslant i \leqslant n,\ 1 \leqslant j \leqslant t,$$

where the parameter θ_1 represents the general mean, each random effect a_i is univariate normal with mean zero and variance θ_2, each e_{ij} is univariate normal with mean zero and variance θ_3, and the a_i's and the e_{ij}'s are all independent. Here $\theta_1 \in \mathbb{R}$ and θ_2, θ_3 (> 0). Let $s = 3$, $t_1(\theta) = \theta_1$, $t_2(\theta) = \theta_2/\theta_3$ and $t_3(\theta) = \theta_3$. It is shown by Datta (1996) that the elements ρ_{ij} are all free from θ, and hence the condition (16) is automatically satisfied. Datta (1996) showed that $\pi(\theta) \propto \{\theta_3(\theta_3 + t\theta_2)\}^{-1}$ is the unique joint PMP for the parametric functions given above.

5.5. *Matching priors via Bartlett corrections*

Inversion of likelihood ratio acceptance regions is a standard approach for constructing reasonably optimal confidence sets. Under suitable regularity conditions, the likelihood ratio statistic is Bartlett correctable. The error incurred in approximating the distribution of a Bartlett corrected likelihood ratio statistic for the entire parameter vector by the chi-square distribution with d degrees of freedom is $O(n^{-2})$, whereas the corresponding error for the uncorrected likelihood ratio statistic is $O(n^{-1})$. Approximate confidence sets using a Bartlett corrected likelihood ratio statistic and chi-square quantiles will have coverage probability accurate to the fourth-order.

In a pioneering article Bickel and Ghosh (1990) noted that the posterior distribution of the likelihood ratio statistic is also Bartlett correctable, and the posterior distribution of the posterior Bartlett corrected likelihood ratio statistic agrees with an appropriate chi-square distribution up to $O(n^{-2})$. From this, via the shrinkage argument mentioned in Section 2, they provided a derivation of the frequentist Bartlett correction.

It follows from the above discussion that for any smooth prior one can construct approximate credible sets for θ using chi-square quantiles and posterior Bartlett corrected likelihood ratio statistics with an $O(n^{-2})$ error in approximation. Ghosh and Mukerjee (1991) utilised the existence of a posterior Bartlett correction to the likelihood ratio statistic to derive the frequentist Bartlett correction to the same and characterised priors for which these two corrections are identical up to $o(1)$. An important implication of this characterisation result is that for all such priors the resulting credible sets based on the posterior Bartlett corrected likelihood ratio statistic will also have frequentist coverage accurate to $O(n^{-2})$ and hence these priors are fourth-order PMPs.

For $1 \leqslant j, r, s \leqslant d$, let us define $V_{jr,s} = E_\theta\{D_j D_r \log f(X;\theta) D_s \log f(X;\theta)\}$ and $V_{jrs} = E_\theta\{D_j D_r D_s \log f(X;\theta)\}$. Ghosh and Mukerjee (1991) characterised the class of priors for which the posterior and the frequentist Bartlett corrections agree to $o(1)$. Any such prior $\pi(\cdot)$ is given by a solution to the differential equation

$$D_i D_j\{\pi(\theta)\kappa^{ij}\} - D_i\{\pi(\theta)\kappa^{ir}\kappa^{js}(2V_{rs,j} + V_{jrs})\} = 0. \tag{18}$$

Ghosh and Mukerjee (1992b) generalised the above result in the presence of a nuisance parameter. They considered the case of a scalar interest parameter in the presence of a scalar orthogonal nuisance parameter. DiCiccio and Stern (1993) have considered a very general adjusted likelihood for a vector interest parameter where the nuisance parameter is also vector-valued. They have shown that the posterior distribution of the resulting likelihood ratio statistic also admits posterior Bartlett correction. They subsequently utilised this fact in DiCiccio and Stern (1994), as in Ghosh and Mukerjee (1991), to characterise priors for which the posterior and the frequentist Bartlett corrections agree to $o(1)$. Such priors are obtained as solutions to a differential equation similar to the one given by (18) above. As a particular example, DiCiccio and Stern (1994) considered PMPs based on HPD regions of a vector of interest parameters in the presence of nuisance parameters (see the next subsection).

5.6. *Matching priors for highest posterior density regions*

Highest posterior density regions are very popular in Bayesian inference as they are defined for multi-dimensional interest parameters with or without a nuisance parameter, which can also be multi-dimensional. These regions have the smallest volumes for a given credible level. If such regions also have frequentist validity, they will be desirable in the frequentist set-up as well. Nonsubjective priors that lend frequentist validity to HPD regions are known in the literature as HPD matching priors. Complete characterisations of such priors are well studied in the literature. A brief account of this literature is provided below.

A prior $\pi(\cdot)$ is HPD matching for θ if and only if it satisfies the partial differential equation

$$D_u\{\pi(\theta)V_{jrs}\kappa^{jr}\kappa^{su}\} - D_j D_r\{\pi(\theta)\kappa^{jr}\} = 0. \tag{19}$$

Ghosh and Mukerjee (1993b) reported this result in a different but equivalent form. Prior to these authors Peers (1968) and Severini (1991) explored HPD matching priors for scalar θ models. Substantial simplification of Eq. (19) arises for scalar parameter models. For such models, although Jeffreys' prior is not necessarily HPD matching, it is so for location models and for scale models (cf. Datta and Mukerjee, 2004, p. 71).

For the location-scale model of Example 5.1, Ghosh and Mukerjee (1993b) obtained HPD matching priors for the parameter vector. They showed that $\pi(\theta) \propto \theta_2^{-1}$ is a solution to (19). This prior is already recommended from other considerations in Example 5.1.

For the bivariate normal model with zero means, unit variances and correlation coefficient θ, Datta and Mukerjee (2004, p. 71) checked that while Jeffreys' prior is not HPD matching, the prior $\pi(\theta) \propto (1-\theta^2)^2(1+\theta^2)^{-1}$ is HPD matching. Interestingly, this is a proper prior. See Datta and Mukerjee (2004, Section 4.3) for more examples.

Now suppose for $d \geqslant 2$ our interest lies in θ_1 while we treat the remaining parameters as nuisance parameters. Since it is possible to make the interest parameter orthogonal to the nuisance parameter vector (cf. Cox and Reid, 1987), we will present the HPD matching result under this orthogonality assumption. A prior $\pi(\cdot)$ is HPD matching for θ_1 if and only if it satisfies the partial differential equation

$$\sum_{s=2}^{d}\sum_{u=2}^{d} D_u\left\{\pi(\theta)\kappa_{11}^{-1}\kappa^{su}V_{11s}\right\} + D_1\left\{\pi(\theta)\kappa_{11}^{-2}V_{111}\right\} - D_1^2\left\{\pi(\theta)\kappa_{11}^{-1}\right\} = 0. \tag{20}$$

For a proof of this result we refer to Ghosh and Mukerjee (1995). For examples we refer to Ghosh and Mukerjee (1995) and Datta and Mukerjee (2004, Section 4.4). For the most general situation when both the interest and nuisance parameters are multidimensional we refer to DiCiccio and Stern (1994) and Ghosh and Mukerjee (1995). They derived a partial differential equation for HPD matching priors for the interest parameter vector without any assumption of orthogonality of the interest and nuisance parameters.

As a concluding remark on HPD matching priors we note that such priors are third-order matching. The resulting HPD regions based on HPD matching priors tend to have an edge over other confidence sets under the frequentist expected volume criterion (cf. Mukerjee and Reid, 1999a; Datta and DiCiccio, 2001).

5.7. Nonregular cases

Ghosal (1999) extended the single parameter case described earlier to the case $d = 2$, where the family $f(\cdot;\theta)$ is nonregular with respect to θ_1 and regular with respect to θ_2. Using an asymptotic expansion for the posterior distribution, Ghosal obtained a partial differential equation for the PMP when θ_1 is the interest parameter and the PMP when θ_2 is the interest parameter, where the latter prior has the form of Eq. (11). Two-sided matching priors for θ_2 are also of the same form as solutions to Eq. (8). Ghosal (1999) also obtained a PMP for θ_1 after integrating out θ_2 with respect to the conditional prior of θ_2 given θ_1 (cf. Condition C of Ghosh, 1994, p. 91). In this case, if $\pi(\theta) = \pi(\theta_1)\pi(\theta_2|\theta_1)$ is a PMP for θ_1, Ghosal has shown that

$$\pi(\theta_1) \propto \left\{\int \frac{\pi(\theta_2|\theta_1)}{c(\theta_1,\theta_2)}\,\mathrm{d}\theta_2\right\}^{-1},$$

where $c(\theta_1,\theta_2) = E_\theta\{D_1 \log f(X;\theta)\}$.

Define $\lambda(\theta) = \kappa_{22}^{1/2}(\theta)$, the square root of the Fisher information for θ_2. If $c(\theta)$ and $\lambda(\theta)$ each factors as $c(\theta) = c_1(\theta_1)c_2(\theta_2)$ and $\lambda(\theta) = \lambda_1(\theta_1)\lambda_2(\theta_2)$, then the prior

$$\pi(\theta) \propto c_1(\theta_1)\lambda_2(\theta_2) \tag{21}$$

is a fourth-order PMP for θ_1 and a second-order PMP for θ_2. This result holds irrespective of whether matching is done in the usual sense or in integrated sense mentioned above. There is a striking similarity of the PMP given by (21) with that given by (17) in the regular case.

EXAMPLE 5.8 (Ghosal, 1999). Let $f(x;\theta) = \theta_2^{-1} f_0((x-\theta_1)/\theta_2)$ where $f_0(\cdot)$ is a strictly positive density on $[0,\infty)$. In this case, $c(\theta) = \theta_2^{-1} f_0(0+)$, $\lambda(\theta) \propto \theta_2^{-1}$. Since the required factorisation holds, $\pi(\theta) \propto \theta_2^{-1}$ is a PMP for both θ_1 and θ_2. Note that the same prior emerges as the second-order PMP for both the location and scale parameter in the regular location-scale set-up (cf. Example 5.1).

6. Predictive matching priors

In this section we consider the construction of asymptotic PMPs for prediction. This question was discussed in Datta et al. (2000b).

6.1. One-sided predictive intervals

Let $0 < \alpha < 1$, $\pi \in \Pi_\Omega$ and let Y be a real-valued future observation from $f(\cdot;\theta)$. Let $y(\pi,\alpha)$ denote the upper α-quantile of the predictive distribution of Y, satisfying

$$\mathrm{pr}_\pi\{Y > y(\pi,\alpha) | X^n\} = \alpha. \tag{22}$$

We ask when it is also true that, to a given degree of approximation,

$$\mathrm{pr}_\theta\{Y > y(\pi,\alpha)\} = \alpha \tag{23}$$

very weakly.

It turns out that (23) holds to $O(n^{-1})$ for every positive continuous prior π on Ω and $0 < \alpha < 1$. Note that this is one order higher than the corresponding property for parametric statements. It is therefore natural to ask whether or not there exists a prior distribution for which (23) holds to a higher asymptotic order. Datta et al. (2000b) showed that

$$\mathrm{pr}_\theta\{Y > y(\pi,\alpha)\} = \alpha - \frac{1}{n\pi(\theta)} D_s\{\kappa^{st}(\theta)\mu_t(\theta,\alpha)\pi(\theta)\} + o(n^{-1}) \tag{24}$$

very weakly, where

$$\mu_t(\theta,\alpha) = \int_{q(\theta,\alpha)}^{\infty} D_t f(u;\theta)\,du$$

and $q(\theta, \alpha)$ satisfies

$$\int_{q(\theta,\alpha)}^{\infty} f(u;\theta)\,\mathrm{d}u = \alpha.$$

It follows that (23) holds to $\mathrm{o}(n^{-1})$ if and only if π satisfies the partial differential equation

$$D_s\{\kappa^{st}(\theta)\mu_t(\theta,\alpha)\pi(\theta)\} = 0. \tag{25}$$

In general, solutions to (25) depend on the level α, in which case it is not possible to obtain simultaneous probability matching for all α beyond $\mathrm{O}(n^{-1})$. On the other hand, in the case $d = 1$ it was shown in Datta et al. (2000b) that, if there does exist a prior satisfying (25) for all α, then this prior is Jeffreys' prior. Examples include all location models.

The solution to (25) in the multiparameter case was also investigated by Datta et al. (2000b). In particular, they showed that if there does exist a prior satisfying (25) that is free from α then it is not necessarily Jeffreys' prior. Consideration of particular models indicates that the prior that does emerge has other attractive properties. For example, in location-scale models the predictive approach yields the improper prior that is proportional to the inverse of the scale parameter, which is Jeffreys' right-invariant prior.

Further discussion in Datta et al. (2000b) focused on the construction of predictive regions that give rise to probability matching based on a specified prior, in the spirit of the discussion of Bayes-confidence intervals in Section 4.2. It was shown that for every positive and continuous prior on Ω it is possible to construct a predictive interval with the matching property to $\mathrm{o}(n^{-1})$.

6.2. *Highest predictive density regions*

Consider the general case where the X_i's are possibly vector-valued. The posterior quantile approach outlined above will not be applicable if the X_i's are vector-valued. No matter whether the X_i's are vector-valued or not, one may consider a highest predictive density region for predicting a future observation Y. Let $H(\pi, X^n, \alpha)$ be a highest predictive density region for Y with posterior predictive coverage probability α. Note that for each θ and $\alpha \in (0, 1)$, there exists a unique $m(\theta, \alpha)$ such that $\mathrm{pr}_\theta\{Y \in A(\theta, \alpha)\} = \alpha$ where $A(\theta, \alpha) = \{u\colon f(u;\theta) \geqslant m(\theta,\alpha)\}$. It was shown by Datta et al. (2000b) that $H(\pi, X^n, \alpha)$ is a perturbation of $A(\hat{\theta}, \alpha)$. They further showed that

$$\mathrm{pr}_\theta\{Y \in H(\pi, X^n, \alpha)\} = \alpha - \frac{1}{n\pi(\theta)} D_s\{\kappa^{st}(\theta)\xi_t(\theta,\alpha)\pi(\theta)\} + \mathrm{o}(n^{-1}), \tag{26}$$

where

$$\xi_t(\theta,\alpha) = \int_{A(\theta,\alpha)} D_t f(u;\theta)\,\mathrm{d}u.$$

It follows that $\mathrm{pr}_\theta\{Y \in H(\pi, X^n, \alpha)\} = \alpha + \mathrm{o}(n^{-1})$ if and only if π satisfies the partial differential equation

$$D_s\{\kappa^{st}(\theta)\xi_t(\theta, \alpha)\pi(\theta)\} = 0. \quad (27)$$

As before, in general solutions to (27) depend on the level α, in which case it is not possible to obtain simultaneous probability matching for all α beyond $\mathrm{O}(n^{-1})$. Various examples where such solutions exist for all α are included in Datta et al. (2000b) and Datta and Mukerjee (2004). They include bivariate normal, multivariate location, multivariate scale and multivariate location-scale models.

Before considering prediction of unobservable random effects in the next subsection, we note that following the work of Datta et al. (2000b), Datta and Mukerjee (2003) considered predictive matching priors in a regression set-up. When each observation involves a dependent variable and an independent variable, quite often one has knowledge of both the variables in the past observations and also of the independent variable in the new observation. Based on such data in regression settings, Datta and Mukerjee (2003) obtained matching priors both via the quantile approach and HPD region approach when the goal is prediction of the dependent variable in the new observation. Many examples in this case are discussed in Datta and Mukerjee (2003, 2004).

6.3. Probability matching priors for random effects

Random effects models, also known as hierarchical models, are quite common in statistics. Bayesian versions of such models are hierarchical Bayesian (HB) models. In HB analysis of these models, one often uses nonsubjective priors for the hyperparameters.

Datta et al. (2000a) considered the PMP idea to formally select suitable nonsubjective priors for hyperparameters in a simple HB model. As in Morris (1983), Datta et al. (2000a) considered a normal HB model given by: (a) conditional on $\xi_1, \ldots, \xi_n, \theta_1$ and θ_2, Y_i, $i = 1, \ldots, n$, are independent with Y_i having a normal distribution with mean ξ_i and variance σ^2; (b) conditional on θ_1 and θ_2, the ξ_i's are independent and identically distributed with mean θ_1 and variance θ_2, and (c) $\pi(\theta_1, \theta_2)$ is a suitable nonsubjective prior density on the hyperparameter $\theta = (\theta_1, \theta_2)$. Frequentist calculations for this model are based on the marginal distribution of $Y_1, \ldots, Y_n$ resulting from stages (a) and (b) of the above HB model. Here σ^2 is assumed known. Datta et al. (2000a) characterised priors $\pi(\theta_1, \theta_2)$ for which one-sided Bayesian credible intervals for ξ_i are also third-order accurate confidence intervals. In particular, they have shown that $\pi(\theta) \propto \theta_2/(\sigma^2 + \theta_2)$ is one such prior. Note that this prior is different from the standard uniform prior proposed for this problem (cf. Morris, 1983).

While Datta et al. (2000a) obtained their result from first principles, Chang et al. (2003) considered a random effects model and characterised priors that ensure approximate frequentist validity to the third-order of posterior quantiles of an unobservable random effect. Such characterisation is done, again, via a partial differential equation. For details and examples we refer to Chang et al. (2003).

7. Invariance of matching priors

It is well known that Jeffreys' prior is invariant under reparameterisation. We have mentioned in Section 4.1 that in scalar parametric models Jeffreys' prior possesses the second-order matching property for one-sided parametric intervals. It was also mentioned in Section 6.1 that in such models if a (third-order) matching prior exists for all α, then Jeffreys' prior is the unique prior possessing the matching property for one-sided predictive intervals.

Orthogonal parameterisation plays a crucial role in the study of PMPs. One could not use an orthogonal transformation without invariance of PMPs under interest parameter preserving transformations. If θ and ψ provide two alternative parameterisations of our model, and $t(\theta)$ and $u(\psi)$ are parametric functions of interest in the respective parameterisations, we say that the transformation $\theta \rightarrow \psi$ is *interest parameter preserving* if $t(\theta) = u(\psi)$. Datta and Ghosh (1996) discussed invariance of various noninformative priors, including second-order PMPs for one-sided parametric intervals, under interest parameter preserving transformations. Datta (1996) discussed invariance of joint PMPs for the multiple parametric functions reviewed in Section 5.4. While Datta and Ghosh (1996) and Datta (1996) proved such invariance results algebraically, Mukerjee and Ghosh (1997) provided an elegant argument in proving invariance of one-sided parametric intervals. Datta and Mukerjee (2004) extensively used the argument of Mukerjee and Ghosh (1997) to establish invariance of various types of PMPs (cf. Sections 2.8, 4.3 and Subsection 6.2.2 of Datta and Mukerjee, 2004).

We conclude this section with a brief remark on invariance of predictive matching priors. It should be intuitively obvious that parameterisation should play no role in prediction. Datta et al. (2000b) established invariance of predictive matching priors based on one-sided quantile matching as well as highest predictive density criteria; see also Subsection 6.2.2 of Datta and Mukerjee (2004).

8. Concluding remarks

As argued in Section 2, the probability matching property is an appealing one for a proposed nonsubjective prior, since it provides some assurance that the resulting posterior statements make some inferential sense at least from a repeated sampling point of view. However, in view of the many alternative matching criteria, and the fact that in the multiparameter case there is usually an infinite number of solutions for any one of these criteria, in the authors' view it is inappropriate to use probability matching as a general paradigm for nonsubjective Bayesian inference. Instead, probability matching should be considered as one of a number of desirable properties, such as invariance, propriety, avoidance of paradoxes, that might be investigated for a nonsubjective prior proposal. From a frequentist point of view, the fact that there are many matching solutions is not a problem in principle, although future investigation might reveal whether one of these possesses additional sampling-based optimality. Indeed the concept of alternative matching coverage probabilities due to Mukerjee and Reid (1999b) can be used to discriminate among these many matching solutions.

With the rapid advances in computational techniques for Bayesian statistics that exploit the increased computing power now available, researchers are able to adopt more realistic, and usually more complex, models. However, it is then less likely that the statistician will be able to properly elicit prior beliefs about all aspects of the model. Moreover, many parameters may not have a direct interpretation. This suggests that there is a need to develop general robust methods for prior specification that incorporate both subjective and nonsubjective components. In this case, the matching property could be recast as being the approximate equality of the posterior probability of a suitable set and the corresponding frequentist probability averaged over the parameter space with respect to any continuous prior that preserves the subjective element of the specified prior. A related idea mentioned in Sweeting (2001) would be to achieve some mixed parametric/predictive matching.

As mentioned in Section 1, important questions about implementation need to be addressed if the theory is to be used in practice. Levine and Casella (2003) present an algorithm for the implementation of matching priors for an interest parameter in the presence of a single nuisance parameter. An alternative solution is to use a suitable data-dependent prior as an approximation to a PMP. Sweeting (2005) gives a local implementation that is relatively simple to compute. There is also some prospect of computer implementation of predictive matching priors via local solutions to (25).

Acknowledgements

Datta's research was partially supported by NSF Grants DMS-0071642, SES-0241651 and NSA Grant MDA904-03-1-0016. Sweeting's research was partially supported by EPSRC Grant GR/R24210/01.

References

Barron, A.R. (1999). Information-theoretic characterization of Bayes performance and the choice of priors in parametric and nonparametric problems. In: Bernardo, J.M., Berger, J.O., Dawid, A.P., Smith, A.F.M. (Eds.), *Bayesian Statistics, vol. 6*. Oxford University Press, pp. 27–52.

Berger, J. (1992). Discussion of "Non-informative priors" by J.K. Ghosh and R. Mukerjee. In: Bernardo, J.M., Berger, J.O., Dawid, A.P., Smith, A.F.M. (Eds.), *Bayesian Statistics, vol. 4*. Oxford University Press, pp. 205–206.

Berger, J., Bernardo, J.M. (1992a). On the development of reference priors (with discussion). In: Bernardo, J.M., Berger, J.O., Dawid, A.P., Smith, A.F.M. (Eds.), *Bayesian Statistics, vol. 4*. Oxford University Press, pp. 35–60.

Berger, J., Bernardo, J.M. (1992b). Reference priors in a variance components problem. In: Goel, P.K., Iyengar, N.S. (Eds.), *Bayesian Analysis in Statistics and Econometrics*. Springer-Verlag, New York, pp. 177–194 .

Bernardo, J.M. (1979). Reference posterior distributions for Bayesian inference (with discussion). *J. Roy. Statist. Soc., Ser. B* **41**, 113–147.

Bernardo, J.M. (2005). Reference analysis. In: Dey, D.K., Rao, C.R. (Eds.), *Bayesian Thinking: Modeling and Computation, Handbook of Statistics*, vol. 25. Elsevier, Amsterdam. This volume.

Bernardo, J.M., Ramón, J.M. (1998). An introduction to Bayesian reference analysis: Inference on the ratio of multinomial parameters. *The Statistician* **47**, 101–135.

Bernardo, J.M., Smith, A.F.M. (1994). *Bayesian Theory*. Wiley, New York.

Bickel, P.J., Ghosh, J.K. (1990). A decomposition for the likelihood ratio statistic and the Bartlett correction – a Bayesian argument. *Ann. Statist.* **18**, 1070–1090.

Box, G.E.P., Tiao, G.C. (1973). *Bayesian Inference in Statistical Analysis*. Wiley, New York.

Chang, T., Eaves, D. (1990). Reference priors for the orbit in a group model. *Ann. Statist.* **18**, 1595–1614.

Chang, I.H., Kim, B.H., Mukerjee, R. (2003). Probability matching priors for predicting unobservable random effects with application to ANOVA models. *Statist. Probab. Lett.* **62**, 223–228.

Cox, D.R., Reid, N. (1987). Parameter orthogonality and approximate conditional inference (with discussion). *J. Roy. Statist. Soc., Ser. B* **53**, 79–109.

Datta, G.S. (1996). On priors providing frequentist validity of Bayesian inference for multiple parametric functions. *Biometrika* **83**, 287–298.

Datta, G.S., DiCiccio, T. (2001). On expected volumes of multidimensional confidence sets associated with the usual and adjusted likelihoods. *J. Roy. Statist. Soc., Ser. B* **63**, 691–703.

Datta, G.S., Ghosh, J.K. (1995a). On priors providing frequentist validity for Bayesian inference. *Biometrika* **82**, 37–45.

Datta, G.S., Ghosh, J.K. (1995b). Noninformative priors for maximal invariant in group models. *Test* **4**, 95–114.

Datta, G.S., Ghosh, M. (1995a). Hierarchical Bayes estimators for the error variance in one-way ANOVA models. *J. Statist. Plann. Inference* **45**, 399–411.

Datta, G.S., Ghosh, M. (1995b). Some remarks on noninformative priors. *J. Amer. Statist. Assoc.* **90**, 1357–1363.

Datta, G.S., Ghosh, M. (1996). On the invariance of noninformative priors. *Ann. Statist.* **24**, 141–159.

Datta, G.S., Mukerjee, R. (2003). Probability matching priors for predicting a dependent variable with application to regression models. *Ann. Inst. Statist. Math.* **55**, 1–6.

Datta, G.S., Mukerjee, R. (2004). *Probability Matching Priors: Higher Order Asymptotics. Lecture Notes in Statistics*. Springer, New York.

Datta, G.S., Ghosh, M., Mukerjee, R. (2000a). Some new results on probability matching priors. *Calcutta Statist. Assoc. Bull.* **50**, 179–192.

Datta, G.S., Mukerjee, R., Ghosh, M., Sweeting, T.J. (2000b). Bayesian prediction with approximate frequentist validity. *Ann. Statist.* **28**, 1414–1426.

Datta, G.S., Ghosh, M., Kim, Y.-H. (2002). Probability matching priors for one-way unbalanced random effects models. *Statist. Decis.* **20**, 29–51.

Dawid, A.P. (1991). Fisherian inference in likelihood and prequential frames of reference (with discussion). *J. Roy. Statist. Soc., Ser. B* **53**, 79–109.

Dawid, A.P., Stone, M., Zidek, J.V. (1973). Marginalization paradoxes in Bayesian and structural inference (with discussion). *J. Roy. Statist. Soc., Ser. B* **35**, 189–233.

DiCiccio, T.J., Stern, S.E. (1993). On Bartlett adjustments for approximate Bayesian inference. *Biometrika* **80**, 731–740.

DiCiccio, T.J., Stern, S.E. (1994). Frequentist and Bayesian Bartlett correction of test statistics based on adjusted profile likelihoods. *J. Roy. Statist. Soc., Ser. B* **56**, 397–408.

Fisher, R.A. (1956). *Statistical Methods and Scientific Inference*. Oliver & Boyd, Edinburgh.

Fraser, D.A.S., Reid, N. (2002). Strong matching of frequentist and Bayesian parametric inference. *J. Statist. Plann. Inference* **103**, 263–285.

Ghosal, S. (1999). Probability matching priors for non-regular cases. *Biometrika* **86**, 956–964.

Ghosh, J.K. (1994). *Higher Order Asymptotics*. Institute of Mathematical Statistics and American Statistical Association, Hayward, CA.

Ghosh, J.K., Mukerjee, R. (1991). Characterization of priors under which Bayesian and frequentist Bartlett corrections are equivalent in the multiparameter case. *J. Multivariate Anal.* **38**, 385–393.

Ghosh, J.K., Mukerjee, R. (1992a). Non-informative priors (with discussion). In: Bernardo, J.M., Berger, J.O., Dawid, A.P., Smith, A.F.M. (Eds.), *Bayesian Statistics, vol. 4*. Oxford University Press, pp. 195–210.

Ghosh, J.K., Mukerjee, R. (1992b). Bayesian and frequentist Bartlett corrections for likelihood ratio and conditional likelihood ratio tests. *J. Roy. Statist. Soc., Ser. B* **54**, 867–875.

Ghosh, J.K., Mukerjee, R. (1993a). On priors that match posterior and frequentist distribution functions. *Canad. J. Statist.* **21**, 89–96.

Ghosh, J.K., Mukerjee, R. (1993b). Frequentist validity of highest posterior density regions in multiparameter case. *Ann. Inst. Statist. Math.* **45**, 293–302.

Ghosh, J.K., Mukerjee, R. (1995). Frequentist validity of highest posterior density regions in the presence of nuisance parameters. *Statist. Decis.* **13**, 131–139.

Ghosh, M., Mukerjee, R. (1998). Recent developments on probability matching priors. In: Ahmed, S.E., Ahsanullah, M., Sinha, B.K. (Eds.), *Applied Statistical Science III*. Nova Science Publishers, New York, pp. 227–252.

Ghosh, M., Yang, M.C. (1996). Noninformative priors for the two-sample normal problem. *Test* **5**, 145–157.

Hartigan, J.A. (1966). Note on the confidence-prior of Welch and Peers. *J. Roy. Statist. Soc., Ser. B* **28**, 55–56.

Hora, R.B., Buehler, R.J. (1966). Fiducial theory and invariant estimation. *Ann. Math. Statist.* **37**, 643–656.

Kass, R.E., Wasserman, L. (1996). The selection of prior distributions by formal rules. *J. Amer. Statist. Assoc.* **91**, 1343–1370.

Levine, R.A., Casella, G. (2003). Implementing matching priors for frequentist inference. *Biometrika* **90**, 127–137.

Lindley, D.V. (1958). Fiducial distributions and Bayes' theorem. *J. Roy. Statist. Soc., Ser. B* **20**, 102–107.

Morris, D.V. (1983). Parametric empirical Bayes confidence intervals. In: Box, G.E.P., Leonard, T., Wu, C.F.J. (Eds.), *Scientific Inference, Data Analysis, and Robustness*. Academic Press, New York, pp. 25–50.

Mukerjee, R., Dey, D.K. (1993). Frequentist validity of posterior quantiles in the presence of a nuisance parameter: Higher-order asymptotics. *Biometrika* **80**, 499–505.

Mukerjee, R., Ghosh, M. (1997). Second-order probability matching priors. *Biometrika* **84**, 970–975.

Mukerjee, R., Reid, N. (1999a). On confidence intervals associated with the usual and adjusted likelihoods. *J. Roy. Statist. Soc., Ser. B* **61**, 945–954.

Mukerjee, R., Reid, N. (1999b). On a property of probability matching priors: Matching the alternative coverage probabilities. *Biometrika* **86**, 333–340.

Mukerjee, R., Reid, N. (2000). On the Bayesian approach for frequentist computations. *Brazilian J. Probab. Statist.* **14**, 159–166.

Mukerjee, R., Reid, N. (2001). Second-order probability matching priors for a parametric function with application to Bayesian tolerance limits. *Biometrika* **88**, 587–592.

Nicolaou, A. (1993). Bayesian intervals with good frequency behaviour in the presence of nuisance parameters. *J. Roy. Statist. Soc., Ser. B* **55**, 377–390.

Peers, H.W. (1965). On confidence sets and Bayesian probability points in the case of several parameters. *J. Roy. Statist. Soc., Ser. B* **27**, 9–16.

Peers, H.W. (1968). Confidence properties of Bayesian interval estimates. *J. Roy. Statist. Soc., Ser. B* **30**, 535–544.

Reid, N. (1996). Likelihood and Bayesian approximation methods (with discussion). In: Bernardo, J.M., Berger, J.O., Dawid, A.P., Smith, A.F.M. (Eds.), *Bayesian Statistics, vol. 5*. Oxford University Press, pp. 351–368.

Severini, T.A. (1991). On the relationship between Bayesian and non-Bayesian interval estimates. *J. Roy. Statist. Soc., Ser. B* **53**, 611–618.

Severini, T.A. (1993). Bayesian interval estimates which are also confidence intervals. *J. Roy. Statist. Soc., Ser. B* **55**, 533–540.

Severini, T.A., Mukerjee, R., Ghosh, M. (2002). On an exact probability matching property of right-invariant priors. *Biometrika* **89**, 952–957.

Stein, C.M. (1985). On the coverage probability of confidence sets based on a prior distribution. In: Zieliński (Ed.), *Sequential Methods in Statistics*. In: *Banach Center Publications*, vol. 16. PWN–Polish Scientific Publishers, Warsaw, pp. 485–514.

Sun, D., Ye, K. (1996). Frequentist validity of posterior quantiles for a two-parameter exponential family. *Biometrika* **83**, 55–64.

Sweeting, T.J. (1995a). A framework for Bayesian and likelihood approximations in statistics. *Biometrika* **82**, 1–23.

Sweeting, T.J. (1995b). A Bayesian approach to approximate conditional inference. *Biometrika* **82**, 25–36.

Sweeting, T.J. (1999). On the construction of Bayes-confidence regions. *J. Roy. Statist. Soc., Ser. B* **61**, 849–861.

Sweeting, T.J. (2001). Coverage probability bias, objective Bayes and the likelihood principle. *Biometrika* **88**, 657–675.

Sweeting, T.J. (2005). On the implementation of local probability matching priors for interest parameters. *Biometrika* **92**, 47–58.

Thatcher, A.R. (1964). Relationships between Bayesian and confidence limits for predictions. *J. Roy. Statist. Soc., Ser. B* **26**, 176–210.

Tibshirani, R.J. (1989). Noninformative priors for one parameter of many. *Biometrika* **76**, 604–608.

Welch, B.L., Peers, H.W. (1963). On formulae for confidence points based on integrals of weighted likelihoods. *J. Roy. Statist. Soc., Ser. B* **35**, 318–329.

Woodroofe, M. (1986). Very weak expansions for sequential confidence levels. *Ann. Statist.* **14**, 1049–1067.

Ye, K. (1994). Bayesian reference prior analysis on the ratio of variances for balanced one-way random effect model. *J. Statist. Plann. Inference* **41**, 267–280.

Handbook of Statistics, Vol. 25
ISSN: 0169-7161
© 2005 Elsevier B.V. All rights reserved.
DOI 10.1016/S0169-7161(05)25004-6

4

Model Selection and Hypothesis Testing based on Objective Probabilities and Bayes Factors

Luis Raúl Pericchi

Abstract
The basics of the Bayesian approach to model selection are first presented. Eight objective methods of developing default Bayesian approaches that have undergone considerable recent development are reviewed and analyzed in a general framework: Well Calibrated Priors (WCP), Conventional Priors (CP), Intrinsic Bayes Factor (IBF), Intrinsic Priors (IPR), Expected Posterior Priors (EP), Fractional Bayes Factor (FBF), asymptotic methods and the Bayesian Information Criterion (BIC), and Lower Bounds (LB) on Bayes Factors. These approaches will be illustrated and commented on how to use and how *not* to use them. Despite the apparent inordinate multiplicity of methods, there are important connections and similarities among different Bayesian methods. Most important, typically the results obtained by any of the methods, are closer among themselves than to results from non-Bayesian methods, and this is typically more so as the information accumulates.

Keywords: Bayes factors; Bayesian information criterion (BIC); Bayesian model selection; conventional priors; expected posterior priors; fractional Bayes factors; intrinsic Bayes factors; intrinsic priors; training samples; universal lower bound; well calibrated priors

"...the simple concept of the Bayes Factor, is basic for legal trials.
It is also basic for medical diagnosis and for the philosophy of science.
It should be taught at the pre-college level!"
I.J. Good, from the letter, "When a batterer turns murderer", *Nature* **375** (1995), 541.

1. Introduction

This article summarizes and give some guidance on the evolving subject of Objective Bayesian Bayes Factors and Posterior Model Probabilities. Here the chapter by Berger and Pericchi (2001), thereafter abbreviated as (BP01) is updated and extended, incorporating more recent developments. The reference Berger and Pericchi (1996a) will also be abbreviated as (BP96a). The reference (BP01) is recommended for an in deep study of key examples.

1.1. Basics of Bayes factors and posterior model probabilities

Suppose that we are comparing q models for the data $\boldsymbol{x}$,

$$M_i\colon\ \boldsymbol{X} \text{ has density } f_i(\boldsymbol{x}|\boldsymbol{\theta}_i), \quad i = 1, \ldots, q,$$

where the $\boldsymbol{\theta}_i$ are unknown model parameters. Suppose that we have available prior distributions, $\pi_i(\boldsymbol{\theta}_i)$, $i = 1, \ldots, q$, for the unknown parameters. Define the marginal or predictive densities of $\boldsymbol{X}$,

$$m_i(\boldsymbol{x}) = \int f_i(\boldsymbol{x}|\boldsymbol{\theta}_i)\pi_i(\boldsymbol{\theta}_i)\,\mathrm{d}\boldsymbol{\theta}_i.$$

The *Bayes factor* of M_j to M_i is given by

$$B_{ji} = \frac{m_j(\boldsymbol{x})}{m_i(\boldsymbol{x})} = \frac{\int f_j(\boldsymbol{x}|\boldsymbol{\theta}_j)\pi_j(\boldsymbol{\theta}_j)\,\mathrm{d}\boldsymbol{\theta}_j}{\int f_i(\boldsymbol{x}|\boldsymbol{\theta}_i)\pi_i(\boldsymbol{\theta}_i)\,\mathrm{d}\boldsymbol{\theta}_i}. \tag{1}$$

The Bayes factor is often interpreted as the "evidence provided by the data in favor or against model M_j vs. the alternative model M_i", but it should be remembered that Bayes Factors depend also on the priors. In fact assuming an objective Bayesian viewpoint, the priors may be thought of "weighting measures" and the Bayes Factor as the "weighted averaged likelihood ratio", which are in principle more comparable than maximized likelihood ratios, as measures of relative fit. (Notice that if M_i is nested in M_j, then logically, the maximized likelihood of M_j is a fortiori larger than the maximized likelihood of M_i.)

If prior probabilities $P(M_j)$, $j = 1, \ldots, q$, of the models are available, then one can compute the posterior probabilities of the models from the Bayes factors. Using Bayes Rule, it is easy to see that *posterior probability* of M_i, given the data $\boldsymbol{x}$, is

$$P(M_i|\boldsymbol{x}) = \frac{P(M_i)m_i(\boldsymbol{x})}{\sum_{j=1}^{q} P(M_j)m_j(\boldsymbol{x})} = \left[\sum_{j=1}^{q} \frac{P(M_j)}{P(M_i)} B_{ji}\right]^{-1}. \tag{2}$$

A particularly common choice of the prior model probabilities is $P(M_j) = 1/q$, so that each model has the same initial probability, but there are other possible choices, see, for example, (BP96a) and Section 3.

The posterior odds $(O_{ji}(\boldsymbol{x}))$ and prior odds (O_{ji}) of M_j vs. M_i, i.e. the ratio of the probabilities of M_j vs. M_i, follow the identity (directly from (2)),

$$O_{ji}(\boldsymbol{x}) = O_{ji} \times B_{ji},$$

or taking logs (and denoting by LO_{ij} the log-odds)

$$LO_{ji}(\boldsymbol{x}) = LO_{ji} + \log[B_{ji}].$$

As a direct consequence of the previous formulae, the Bayes factor is then also interpreted as the "*ratio from posterior to prior odds*" for M_j versus M_i, and the log of the Bayes Factor is defined as the "weight of the evidence" provided by the data which if added to the prior log-odds, results in the posterior log-odds (Good, 1985). Bayes Factors were introduced independently by Alan Turing, in his classified work to break the

German code during World War II, and independently by Sir Harold Jeffreys in his book *Theory of Probability*.

In principle, all inference and prediction should be based on the overall joint posterior distribution $\pi(\theta_j, M_j|\mathbf{y})$, $j = 1, \ldots, q$. Notice that the model M is explicitly recognized as a random variable. It is a distinctive advantage of the Bayesian approach that you are not forced to choose a model. For example, the prediction of a vector of future observations $\mathbf{y}_f$ is based on

$$\pi(\mathbf{y}_f|\mathbf{y}) = \sum_{j=1}^{q} \pi_j(\mathbf{y}_f|\mathbf{y}, M_j)P(M_j|\mathbf{y}). \tag{3}$$

Inferences based on (3) are referred to as Bayesian Model Averaging.

1.2. How to choose a model if you must?

However in several applications it is required to choose a model. If this is the case, a common error is to equate "*best*" model with the "*Highest Probability Model*". The highest probability model may still have very low probability! The *Median Probability Model* is introduced by Barbieri and Berger (2004) and found to be often optimal in the normal linear model. Consider models:

$$M_i\colon\ \mathbf{y} = \mathbf{X}_i\boldsymbol{\beta}_i + \boldsymbol{\varepsilon},$$

where $\boldsymbol{i} = (i_1, i_2, \ldots, i_k)$, is the model index, i_j being either 1 or 0 as covariate x_j is in or out of the model. The posterior inclusion probability is the overall posterior probability that the variable j is in the model, i.e.

$$p_j = \sum_{i:i_j=1} P(M_i|\mathbf{y}),$$

and the median probability model, if it exists, is the model formed with all variables whose posterior inclusion probability is at least $1/2$.

When there are a huge number of possible models, an intermediate compromise between Bayesian Model Averaging and choosing one model, is to do the averaging with a subset of models, neglecting the models which have probability under a small threshold. This have been called, the *Ockam's Window* (Raftery et al., 1996).

> The main *message* is that a model with low posterior probability should not be chosen as the theoretical model or for prediction, even if it is the model with highest probability.

1.3. Motivation for the Bayesian approach to model selection

Motivation 1. *Posterior probabilities are appropriate scientific responses to scientific questions*: To quote Efron (Holmes et al., 2003): "*Astronomers have stars, geologists have rocks, we have science-it's the raw material statisticians work from.*" Probability is the language of the methodology of science and the most natural basis on which to measure the evidence provided by the data. One of the main indications to recognize yourself as a Bayesian, is to find meaningful questions like: "*What is the probability that factor A influences the response B?*" Bayes Factors and posterior model probabilities answer precisely that question, as does the posterior inclusion probabilities p_j above. This is achieved however, adopting a substantial set of assumptions. Still, at least in the realm of Objective Bayesian methods, a realm that can be described as the one on which we systematically strive to avoid potentially harmful prior assumptions, even admitting that there is some influence in the prior, the influence of leaving posterior model probabilities altogether in favor of significance testing measures, is far more serious. The language of posterior probabilities is the one that addresses head on, the scientific relevant questions: the language of the science is the language of probability, and not of p-values. In a sense everything follow from this, that is why, for example, the Bayesian approach deals naturally to irksome questions for nonprobability based statistics, like *multiple-comparisons* of models or the probability of the inclusion of a variable in a model, taking into consideration *all* the interactions of that variable with other variables.

Motivation 2. Classical Significance Hypothesis Testing, whether based on fixed significance levels or p-values, needs to be replaced or at the very least complemented by alternative measures of the evidence in favor of models. See, for example, Berger and Sellke (1987), and (BP96a) for some of the reasons. The rational for searching alternatives based on Bayes Factors are not necessarily Bayesian: in fact the conditional frequentist approach may lead to the same inferences as Bayes Factors' conclusions, see Berger (2003). Other motivations are presented in (BP01) and in Berger (1999).

All this said, care has to be taken in how to formulate the choice of the conditional prior densities $\pi(\theta_j|M_j)$.

1.4. Utility functions and prediction

I.J. Good, called log-Bayes Factors *quasi-utilities*, and proposed that even though Bayes Factors were not in general necessarily obtained via an utility function, the weights of evidence (log-Bayes Factors) will always play a crucial role (Good, 1985). Time seems to have given reason to Good: other utilities might be relevant but weighs of evidence are unavoidable, since the purposes of an analysis are often multiple and variable with time, so a unique loss is not appropriate. See Key et al. (1999).

On the other hand if prediction of observations is the goal then Bayesian Model Averaging is optimal for several loss functions. If one is constrained to choose one model, then the Median Probability Model should be the choice for linear models at least. Again, for Model Averaging and for the Median Probability Model Bayes Factors and Posterior Probabilities play a crucial role.

1.5. Motivation for objective Bayesian model selection

Even though pure subjective Bayesian analysis is in principle logically invulnerable in ideal terms, it however suffers from at least two deep seated difficulties: (i) A huge number of elicitations are needed and (ii) Bayes Factors can be very nonrobust with respect to seemingly innocent priors, see Section 1.6. A different intuition is needed when more than one model is considered, than the conventional wisdom within a model.

This lack of robustness is aggravated with the requirement of comparability, that is that at least approximately two different researchers, with similar backgrounds, knowledge and data should reach similar conclusions that lack of robustness prevents.

Finally there is a profound problem with the pure subjectivist approach to Bayes Factors and Posterior Probabilities. We may call it *un-connectivity* of the conditional priors: $\pi(\theta_j)$, for models M_j: $j = 1, \ldots, q$: In the probability calculus, these priors are un-connected, they can be made to vary arbitrarily for different models. This of course does not prevent a subjective approach but suggest partially subjective approaches on which for example one prior is assessed and then it is "*projected*" towards the other models, see Section 2.5.

1.6. Difficulties in objective Bayesian model selection

The considerable scientific advantages of model comparisons based on Bayes Factors and Posterior Probabilities can not be obtained however, without considerable care about conditional priors for each model. A simple example dramatically highlights the dangers of casual model comparisons.

EXAMPLE 1 (*Exponential test*). Motivating Intrinsic and EP priors by a simple example. Assume an Exponential Likelihood $f(x|\lambda) = \lambda \exp(-\lambda \cdot x)$ and it is desired to test:

$$H_0\colon \lambda = \lambda_0 \quad \text{vs.} \quad H_1\colon \lambda \neq \lambda_0.$$

(Take $\lambda_0 = 1$ for simplicity.) The standard noninformative prior for estimating λ is the improper prior $\pi^N(\lambda) = c/\lambda$, with $c > 0$ arbitrary.

PROBLEM 1. Calculating the Bayes Factor with the π^N results in

$$B_{01}^N = \frac{(n\bar{x})^n \exp(-n\bar{x})}{c \cdot \Gamma(n)}, \tag{4}$$

and thus $B_{01} \propto 1/c$, and is thus undetermined.

PROBLEM 2. Assuming a Conjugate Prior opens several new difficulties: In the example let us assume an Exponential Prior (assuming a Gamma prior only complicates the problems exposed here) $\pi^C(\lambda) = \tau \exp(-\tau \cdot \lambda)$, leads to the Bayes Factor

$$B_{01}^C = \frac{(n\bar{x} + \tau)^{(n+1)} \exp(-n\bar{x})}{\tau \cdot \Gamma(n+1)}. \tag{5}$$

The factor $1/\tau$ dramatically highlights the extreme sensitivity (lack of robustness) of the Bayes Factor on the assessment of the hyperparameter. In particular, a very popular casual assessment is to use "vague" proper priors, priors with large variances which in this case amounts to assess $\tau \approx 0$, which will be biased (without bound!) in favor of the null producing a large B_{01} for most data. (Notice that this is a simple illustrative example. In more complex situations, the difficulties in assessment and lack of robustness are exponentiated.) In (BP01) other serious problems with conjugate priors for model comparisons (like for example with the popular "g-priors") are exposed, see Section 2.2.

The problems with undefined constants may be present also with comparisons of models of equal dimensions, an example is in Pericchi and Pérez (1994).

Recommendation. Never use arbitrary vague priors because they introduce an unbounded bias, and use of improper priors leaves the Bayes Factors determined only up to a constant. Sometimes this constant is exactly one, see "*well calibrated priors*" below. Direct assessment of the constant may give sometimes reasonable Bayes Factors, for large samples. See Spiegelhalter and Smith (1982) for specific assessment of the arbitrary constants.

Another informal model selection, very easy to implement casually with MCMC software, is to estimate the parameters with arbitrary vague priors, and if the probability intervals does not contain the null value, to reject the null hypothesis. Notice however that this procedure is not valid since the posterior of the parameters is calculated assuming the alternative hypothesis. Furthermore, this practice actually "*import*" the use of *p-values* and fixed α levels, from significance hypothesis testing, under an appearance of a Bayesian approach. The literature on the un-Bayesianity (and failure to represent weight of evidence in favor of a hypothesis) of p-values and fixed α levels is huge, see, for example, Sellke et al. (2001). However the danger of importing fixed α levels and p-values under the appearance of a Bayesian approach seems to be largely unnoticed, and we call it "*Significance Testing under the Bayesian carpet*", to paraphrase I.J Good. An example is in order.

EXAMPLE 2 (*Significance testing under the Bayesian carpet*). A basic example is a Normal random sample with mean and variance μ, σ^2 respectively (both unknown). The test is: H_0: $\mu = 0$ vs. H_1: $\mu \neq 0$. The usual confidence interval is

$$\bar{x} - t_{\alpha/2,(n-1)} \frac{S}{\sqrt{n(n-1)}} \leqslant \mu \leqslant \bar{x} + t_{\alpha/2,(n-1)} \frac{S}{\sqrt{n(n-1)}},$$

where $S^2 = \sum (y_i - \bar{y})^2$. For now, assume the usual conjugate Normal-Gamma prior, for the mean and precision $\tau = 1/\sigma^2$

$$\pi(\mu, \tau) = NG(\mu, \tau | m, r, a, b) = N\big(\mu | m, (r\tau)^{-1}\big) \cdot Ga(\tau | a, b).$$

Given a sample $\boldsymbol{y}$ of size n, the marginal density of the mean is a Student-t with $2a' = 2(\alpha + \frac{n}{2})$ degrees of freedom, and location and scale parameters given by,

$$m' = \frac{rm + n\bar{x}}{r + n}; \qquad \frac{b'}{a'r'} = \frac{b + S^2/2 + rn(\bar{x} - m)^2/(2(r+n))}{(a + n/2)(r + n)}.$$

A vague prior assignment obtained by choosing $m = 0, r = a = b = 0.01$, will give an interval virtually identical to the usual one. Taking the decision of rejecting the Null Hypothesis if and only if the Null belongs to the Highest Posterior Interval with vague priors, is then equivalent to significance testing with fix level α, even if obtained via a Bayesian route. It has been far recognized that the interval alone is not a comprehensive presentation of the evidence. It should be complemented with the posterior probability of the null, since the interval is calculated conditionally on the Alternative Hypothesis, see Barbieri and Berger (2004) and references therein.

A simple didactic illustration, still not available in close form but that requires Markov Chain Monte Carlo (MCMC) computations follows. The MCMC aspect of the example is not essential, but it is included here, since the practice of the use of vague priors and decision rule based on HPD intervals is typically performed with complicated models with no closed form analysis. Assume a Normal likelihood as before except that we now use independent priors instead of the conjugate Normal-Gamma prior, Normal for the mean and Gamma for the precision,

$$\pi(\mu, \tau) = \pi(\mu) \cdot \pi(\tau),$$

where $\pi(\mu)$ is a Normal distribution with mean 0 and precision 0.01 and $\pi(\tau)$ is a Gamma distribution with parameters $\alpha = 0.001, \beta = 0.001$, a seemingly uncontroversial vague prior for "almost" objective inference. A Normal simulated sample of $n = 25$ data points, $\bar{x} = 0.4115$ and $S^2/(n-1) = 0.8542$ yields the following 95% intervals for μ, under frequentist inference: $(0.030 < \mu < 0.793)$ and Bayesian inference: $(0.032 < \mu < 0.791)$, agreeing up to the second decimal, and both rejecting H_0: $\mu = 0$ vs. H_1: $\mu \neq 0$, for fixed level 0.05. On the other hand the Bayes Factor (under, for example, the Jeffreys' Conventional Prior, Section 2.2) can be approximated using the precise expression (21) below, obtaining $B_{01} \approx 0.701$; Odds of approximately 7 to 10, does not seem conclusive evidence against the null, and are very far from 1 to 20, as the 95% may suggest to some.

The following points summarizes the message of Examples 1 and 2:

POINT 1. The use of improper noninformative priors yields answers which depend on a arbitrary constant. The problem may be present even in comparisons of models of the same dimension.

POINT 2. The casual use of 'vague proper priors' does not solve Problem 1 and often aggravates it, giving unbounded bias in favor of one of the candidate models.

POINT 3. Probability intervals alone should not be used for rejecting null hypotheses.

It should be added that computation of Bayes Factors can be difficult, and another whole chapter may be written on computations and approximations. Recent advances on Markov Chain Monte Carlo methods, have achieved breakthroughs on Bayes Factor computations, and have made it feasible to analyze intractable or highly-dimensional problems. See, for example, Chen et al. (2000).

2. Objective Bayesian model selection methods

In this section, we introduce the eight default Bayesian approaches to model selection that will be considered in this chapter, the Well Calibrated Priors (WCP) approach, the Conventional Prior (CP) approach, the Intrinsic Bayes Factor (IBF) approach, the Intrinsic Prior (IPR) approach, the Expected Posterior Prior (EP) approach, the Fractional

Bayes Factor (FBF) approach, asymptotic methods and the Bayesian Information Criterion (BIC), and Lower Bounds on Bayes Factors and Posterior Probabilities (LB). These approaches will be illustrated by examples.

At first sight eight methods might appear as an explosion of methodology. There are however deep connections among the methods. For instance, in the exposition the concept of a minimal training sample will be a building block to several of the methods. We introduce the simplest version of the concept here.

For the q models $M_1, \dots, M_q$, suppose that (ordinary, usually improper) noninformative priors $\pi_i^N(\boldsymbol{\theta}_i)$, $i = 1, \dots, q$, are available. Define the corresponding marginal or predictive densities of $\boldsymbol{X}$,

$$m_i^N(\boldsymbol{x}) = \int f_i(\boldsymbol{x}|\boldsymbol{\theta}_i)\pi_i^N(\boldsymbol{\theta}_i)\,\mathrm{d}\boldsymbol{\theta}_i.$$

Because we will consider several training samples, we index them by l.

DEFINITION 1. A (deterministic) training sample, $\boldsymbol{x}(l)$, is a subset of the sample $\boldsymbol{x}$ which is called *proper* if $0 < m_i^N(\boldsymbol{x}(l)) < \infty$ for all M_i, and *minimal* if it is proper and no subset is proper. Minimal Training Samples will be denoted MTS.

More general versions of training samples will be introduced in the sequel. This definition is the original one in (BP96a). Real and "imaginary" training samples date back at least to Lempers (1971) and Good (1950) and references therein. In the frequentist literature, the different methods can be written as:

Likelihood Ratio × Correction Factor.

A Bayesian version for grouping the different methods is:

Un-normalized Bayes Factor × Correction Factor,

or in symbols,

$$\text{Bayes Factor}_{ij} = B_{ij} = \frac{m_i^N(\boldsymbol{x})}{m_j^N(\boldsymbol{x})} \times CF_{ji} = B_{ij}^N \times CF_{ji}.$$

Several of the methods described in this article can be written in this form.

How to judge and compare the methods? (BP96a) proposed the following (Bayesian) principle:

PRINCIPLE 1. Testing and model selection methods should correspond, in some sense, to actual Bayes factors, arising from reasonable default prior distributions.

It is natural for a Bayesian to accept that the best *discriminator between procedures* is the study of the prior distribution (if any) that give rise to the procedure or that is implied by it. Other properties like large sample size consistency, are rough as compared with the incisiveness of Principle 1. This is the case, particularly in parametric problems, where such properties follow automatically when there is correspondence of a procedure with a real Bayesian one. One of the best ways of studying any biases in a procedure is by examining the corresponding prior for biases. It is of paramount importance that we know how to interpret Bayes factors as posterior to prior odds. Indeed, other "Bayesian" measures, different from Bayes Factors and Posterior Probabilities have been put forward. But then how to interpret them? One of the main advantages of the "canonical" Bayesian approach based on Odds is its natural scientific and probabilistic interpretation. It is so natural, that far too often practitioners misinterpret p-values as posterior probabilities.

The main message is: Bayesians have to live with Bayes Factors, so we rather learn how to use this powerful measure of evidence.

2.1. *Well calibrated priors approach*

When can we compare two models using Bayes factors? The short answer is when we have (reasonable) proper priors (reasonable, excluding for example "vague" proper priors or point masses).

There is an important concept however that allows to compute exact Bayes Factors with fully improper or partially improper priors (i.e. priors which integrate infinity for some parameters but integrate finite with respect to the other parameters). The concept is "*well calibrated priors*". An important sub-class of well calibrated priors are priors which are "*predictively matched*". Other well calibrated kind of priors will be introduced in the sequel.

DEFINITION 2. For the models M_i: $f_i(\mathbf{y}|\theta_i)$ and M_j: $f_j(\mathbf{y}|\theta_j)$ the priors $\pi_i^N(\theta_i)$ and $\pi_j^N(\theta_j)$ are *predictively matched*, if for *any* minimal training sample $\mathbf{y}(l)$ holds the identity,

$$m_i^N(\mathbf{y}(l)) = \int f_i(\mathbf{y}(l)|\theta_i)\pi^N(\theta_i)\,\mathrm{d}\theta_i$$
$$= m_j^N(\mathbf{y}(l)) = \int f_j(\mathbf{y}(l)|\theta_j)\pi^N(\theta_j)\,\mathrm{d}\theta_j. \tag{6}$$

If two models are predictively matched, it is perfectly sensible to compare them with Bayes Factors that use the improper priors for which they are predictively matched, since the scaling correction given by training samples (see Eq. (14)) cancels out and the correction becomes unity. Strikingly there are large classes of well calibrated models and priors. For a general theory of predictively matched priors see Berger et al. (1998). Three substantial particular cases follow, namely location, scale and location-scale models, from the following property:

PROPERTY 1.

$$f(y_l|\mu) = f(y_l - \mu), \qquad \pi^N(\mu) = 1, \qquad \int f(y_l - \mu)\cdot 1\,\mathrm{d}\mu = 1, \tag{7}$$

$$f(y_l|\sigma) = 1/\sigma f(y_l/\sigma), \qquad \pi^N(\sigma) = 1/\sigma,$$
$$\int f(y_l/\sigma)\cdot\frac{\mathrm{d}\sigma}{\sigma^2} = c_0\cdot\frac{1}{|y_l|} \tag{8}$$

and

$$f\big((y_l-\mu)/\sigma\big),\ \pi^N(\mu,\sigma)=1/\sigma,$$

$$\int f\big((y_{l_1}-\mu)/\sigma\big)\cdot f\big((y_{l_2}-\mu)/\sigma\big)\cdot\frac{\mathrm{d}\sigma}{\sigma^3}=\frac{1}{2\cdot|y_{l_1}-y_{l_2}|},\tag{9}$$

where the minimal training samples are one data y_l in the location and scale case and two distinct observations $y_{l_1}\neq y_{l_2}$ in the location-scale case. Finally, $c_0=\int_0^\infty f(v)\,\mathrm{d}v$.

A direct application of property (9) is the "*robustification*" of the Normal distribution.

EXAMPLE 3. Let the base model M_0 be Normal, but since it is suspected the existence of a percentage of outliers, an alternative Student-t model M_ν, with ν degrees of freedom is proposed (for example $\nu=1$, i.e. the Cauchy model). Then the following Bayes Factor, of the Normal vs. Student is perfectly legal (Bayesianly) and appropriate,

$$B_{(0,\nu)}=\frac{\int f_0(\mathbf{y}|\mu,\sigma)\frac{\mathrm{d}\mu\,\mathrm{d}\sigma}{\sigma}}{\int f_\nu(\mathbf{y}|\mu,\sigma)\frac{\mathrm{d}\mu\,\mathrm{d}\sigma}{\sigma}}.\tag{10}$$

If we now predict a future observation according to the Bayesian Model averaging approach,

$$E(y_f|\mathbf{y})=E(y_f|M_0,\mathbf{y})P(M_0|\mathbf{y})+E(y_f|M_\nu,\mathbf{y})P(M_\nu|\mathbf{y}),$$

we have constructed a robustifier against outliers. See Spiegelhalter (1977) and Pericchi and Pérez (1994). It is also a powerful robustifier, switching smoothly from Normal towards Student-t inference as the presence of outliers is more pronounced. Other alternative models, e.g., asymmetrical, may be added to the location-scale set of candidate models. When a procedure is obtained in a reasonable Bayesian manner (subjective or objective) it is to be expected that the procedure is also efficient in a frequentist sense. In this case the Bayes Factor above is optimal, "*Most Powerful Invariant Test Statistics*" (Cox and Hinkley, 1974).

Do predictively matched priors obey Principle 1? To see that it does obey the Principle in a rather strong way, consider formula (14) below. Applying it to a minimal training sample (MTS), results in $B_{ij}^N(\boldsymbol{x}(l))\equiv 1$, for *any* Minimal Training Sample $\boldsymbol{x}(l)$. As a consequence, the uncorrected $B_{ij}^N(\boldsymbol{x})$ arises from *any* proper prior $\pi(\theta_k|\boldsymbol{x}(l))$, and such priors are typically sensible, since are based on the same MTS for both models. Another way to see this is that it is identical to *any* Intrinsic Bayes Factor, see Section 2.3.

The next approach, finds in the property of "well calibration" an important justification.

2.2. *Conventional prior approach*

It was Jeffreys (1961, Chapter 5) who recognized the problem of arbitrary constants arising in hypotheses testing problems, implied by the use of "Jeffreys' Rule" for choosing objective-invariant priors for estimation problems. For testing problems then a convention have to be established. His approach is based on: (i) using noninformative priors only for common parameters in the models, so that the arbitrary multiplicative constant for the priors would cancel in all Bayes factors, and (ii) using default *proper* priors for orthogonal parameters that would occur in one model but not the other. These priors are neither vague nor over-informative, but that correspond to a definite but limited amount of information. He presented arguments justifying certain default proper priors in general, but mostly on a case-by-case basis. This line of development has been successfully followed by many others (for instance, by Zellner and Siow, 1980; see (BP01), for other references). Here I revisit some examples and formally justify the use of the conventional, partially proper priors, based on the property of *well calibration* and "predictively matched" priors.

EXAMPLE 4 (*Normal mean, Jeffreys' conventional prior*). Suppose the data is $\boldsymbol{X} = (X_1, \ldots, X_n)$, where the X_i are i.i.d. $\mathcal{N}(\mu, \sigma_2^2)$ under M_2. Under M_1, the X_i are $\mathcal{N}(0, \sigma_1^2)$. Since the mean and variance are orthogonal in the sense of having diagonal expected Fisher's information matrix, Jeffreys equated $\sigma_1^2 = \sigma_2^2 = \sigma^2$. Because of this, Jeffreys suggests that the variances can be assigned the same (improper) noninformative prior $\pi^J(\sigma) = 1/\sigma$, since the indeterminate multiplicative constant for the prior would cancel in the Bayes factor. (See below for a formal justification.)

Since the unknown mean μ occurs in only M_2, it needs to be assigned a proper prior. Jeffreys, comes up with the following desiderata for such a prior that in retrospect appears as compelling: (i) it should be centered at zero (i.e. centered at the null hypothesis); (ii) have scale σ (i.e. have the information provided by one observation); (iii) be symmetric around zero, and (iv) have no moments. He then settles for the Cauchy prior Cauchy$(0, \sigma^2)$ as the simplest distribution that obeys the desiderata. In formulae, Jeffreys's conventional prior for this problem is:

$$\pi_1^J(\sigma_1) = \frac{1}{\sigma_1}, \qquad \pi_2^J(\mu, \sigma_2) = \frac{1}{\sigma_2} \cdot \frac{1}{\pi\sigma_2(1 + \mu^2/\sigma_2^2)}. \tag{11}$$

This solution is justified as a Bayesian prior, as the following property shows.

PROPERTY 2. *The priors* (11) *are predictively matched.*

The priors in (11) are improper, but one data will make them proper. Denote by y_l such a generic data point. It is clear that for M_1, $\int f((y_l - \mu)/\sigma)1/\sigma^2\, \mathrm{d}\sigma = \frac{1}{2}\frac{1}{|y_l|}$ using identity (8). For M_2, we use the fact that a Cauchy is a scale mixture of Normals, with a Gamma mixing distribution with parameters $(1/2, 2)$, $Ca(\mu|0, r^2) = \int N(\mu|0, r^2/\lambda)Ga(\lambda|1/2, 2)\, \mathrm{d}\lambda$. Thus,

$$m_2(y_l) = \int N\big(y_l|\mu, \sigma^2\big)Ca\big(\mu|0, r^2\big)\, \mathrm{d}\mu \frac{\mathrm{d}\sigma}{\sigma}$$

(expressing the Cauchy as a scale mixture and interchanging orders of integration)

$$= \int \left[\int N\big(y|0, \sigma^2(1 + 1/\lambda)\big)\frac{\mathrm{d}\sigma}{\sigma}\right] Ga(\lambda|1/2, 2)\, \mathrm{d}\lambda$$

after substitution and using identity (8)

$$\int \frac{1}{2\cdot|y_l|}\cdot Ga(\lambda|1/2,2)\,\mathrm{d}\lambda = \frac{1}{2}\cdot\frac{1}{|y_l|}.$$

In summary, had we used a training sample to "correct" the priors in (11), the correction would have exactly cancelled out, for whatever training sample $y_l \neq 0$. Moreover, choosing the scale of the prior for μ to be σ_2 (the only available nonsubjective 'scaling' in the problem) and centering it at M_1 are natural choices, and the Cauchy prior is known to be robust in various ways. It is important to recall that the property of well calibration should be checked when using a conventional, partially improper prior. It is clear that the argument given in Property 2 would work with any scale mixture of Normal priors.

EXAMPLE 5 (*Linear model, Zellner and Siow conventional priors*). It was suggested in Zellner and Siow (1980), a generalization of the above conventional Jeffreys prior for comparing two nested models within the normal linear model. Let $\boldsymbol{X} = [\mathbf{1} : \mathbf{Z}_1 : \mathbf{Z}_2]$ be the design matrix for the 'full' linear model under consideration, where $\mathbf{1}$ is the vector of 1's, and (without loss of generality) it is assumed that the regressors are measured in terms of deviations from their sample means, so that $\mathbf{1}^t\mathbf{Z}_j = 0$, $j = 1, 2$. It is also assumed that the model has been parameterized in an orthogonal fashion, so that $\mathbf{Z}_1^t\mathbf{Z}_2 = 0$. The normal linear model, M_2, for n observations $\boldsymbol{y} = (y_1, \ldots, y_n)^t$ is

$$\boldsymbol{y} = \alpha\mathbf{1} + \mathbf{Z}_1\boldsymbol{\beta}_1 + \mathbf{Z}_2\boldsymbol{\beta}_2 + \boldsymbol{\varepsilon},$$

where $\boldsymbol{\varepsilon}$ is $\mathcal{N}_n(\mathbf{0}, \sigma^2\boldsymbol{I}_n)$, the n-variate normal distribution with mean vector $\mathbf{0}$ and covariance matrix σ^2 times the identity. Here, the dimensions of $\boldsymbol{\beta}_1$ and $\boldsymbol{\beta}_2$ are $k_1 - 1$ and p, respectively.

For comparison of M_2 with the model M_1: $\boldsymbol{\beta}_2 = 0$, Zellner and Siow (1980) propose the following default conventional priors:

$$\pi_1^{ZS}(\alpha, \boldsymbol{\beta}_1, \sigma) = 1/\sigma,$$
$$\pi_2^{ZS}(\alpha, \boldsymbol{\beta}_1, \sigma, \boldsymbol{\beta}_2) = h(\boldsymbol{\beta}_2|\sigma)/\sigma,$$

where $h(\boldsymbol{\beta}_2|\sigma)$ is the $\mathrm{Cauchy}_p(\mathbf{0}, \mathbf{Z}_2^t\mathbf{Z}_2/(n\sigma^2))$ density

$$h(\boldsymbol{\beta}_2|\sigma) = c\frac{|\mathbf{Z}_2^t\mathbf{Z}_2|^{1/2}}{(n\sigma^2)^{p/2}}\left(1 + \frac{\boldsymbol{\beta}_2^t\mathbf{Z}_2^t\mathbf{Z}_2\boldsymbol{\beta}_2}{n\sigma^2}\right)^{-(p+1)/2},$$

with $c = \Gamma[(p+1)/2]/\pi^{(p+1)/2}$. Thus the improper priors of the "common" $(\alpha, \boldsymbol{\beta}_1, \sigma)$ are assumed to be the same for the two models (again justifiable by the property of being predictively matched as in Example 4), while the conditional prior of the (unique to M_2) parameter $\boldsymbol{\beta}_2$, given σ, is assumed to be the (proper) p-dimensional Cauchy distribution, with location at $\mathbf{0}$ (so that it is 'centered' at M_1) and scale matrix $\mathbf{Z}_2^t\mathbf{Z}_2/(n\sigma^2)$, "... a matrix suggested by the form of the information matrix," to quote Zellner and Siow (1980). (But see criticisms about this choice, particularly for unbalanced designs, in the sequel and in detail in Berger and Pericchi, 2004.)

Again, using the fact that a Cauchy distribution can be written as a scale mixture of normal distributions, it is possible to compute the needed marginal distributions, $m_i(\boldsymbol{y})$, with one-dimensional numerical integration, or even an involved exact calculation, is possible, as Jeffreys does. Alternatively, Laplace or another approximation can be used. In fact, perhaps the best approximation to be used in this problem is (20) below.

When there are more than two models, or the models are nonnested, there are various possible extensions of the above strategy. Zellner and Siow (1980) utilize what is often called the 'encompassing' approach (introduced by Cox, 1961), where one compares each submodel, M_i, to the *encompassing* model, M_0, that contains all possible covariates from the submodels. One then obtains, using the above priors, the pairwise Bayes factors B_{0i}, $i = 1, \ldots, q$. The Bayes factor of M_j to M_i is then *defined* to be

$$B_{ji} = B_{0i}/B_{0j}. \tag{12}$$

EXAMPLE 5 (*Continued, Linear model, conjugate g-priors*). It is perplexing (since Zellner suggestion of a g-prior was for estimation but for testing he suggested other priors), that one of the most popular choices of prior for hypothesis testing in the normal linear model is the conjugate prior, called a g-prior in Zellner (1986). For a linear model

$$M\colon\ \boldsymbol{y} = \boldsymbol{X}\boldsymbol{\beta} + \boldsymbol{\varepsilon}, \qquad \boldsymbol{\varepsilon} \sim \mathcal{N}_n\big(\boldsymbol{0}, \sigma^2 \boldsymbol{I}_n\big),$$

where σ^2 and $\boldsymbol{\beta} = (\beta_1, \ldots, \beta_k)^t$ are unknown and $\boldsymbol{X}$ is an $(n \times k)$ given design matrix of rank $k < n$, the *g-prior* density is defined by

$$\pi(\sigma) = \frac{1}{\sigma}, \qquad \pi(\boldsymbol{\beta}|\sigma) \text{ is } \mathcal{N}_k\big(\boldsymbol{0}, g\sigma^2\big(\boldsymbol{X}^t\boldsymbol{X}\big)^{-1}\big).$$

Often $g = n$ is chosen (see, also, Shively et al., 1999), while sometimes g is estimated by empirical Bayes methods (see, e.g., George and Foster, 2000; Clyde and George, 2000).

A perceived advantage of g-priors is that the marginal density, $m(\boldsymbol{y})$, is available in closed form and it is given by

$$m(\boldsymbol{y}) = \frac{\Gamma(n/2)}{2\pi^{n/2}(1+g)^{k/2}}\left(\boldsymbol{y}^t\boldsymbol{y} - \frac{g}{(1+g)}\boldsymbol{y}^t\boldsymbol{X}\big(\boldsymbol{X}^t\boldsymbol{X}\big)^{-1}\boldsymbol{X}^t\boldsymbol{y}\right)^{-n/2}.$$

Thus the Bayes factors and posterior model probabilities for comparing any two linear models are available in simple and closed form.

A major problem is that g-priors have some undesirable properties when used for model selection, as shown in (BP01): one of them may be called "*finite sample inconsistency*". Suppose one is interested in comparing the linear model above with the null model M^*: $\boldsymbol{\beta} = \boldsymbol{0}$. It can be shown that, as the least squares estimate $\hat{\boldsymbol{\beta}}$ (and the noncentrality parameter) goes to infinity, so that one becomes *certain* that M^* is wrong, the Bayes factor of M^* to M goes to the nonzero constant $(1+g)^{(k-n)/2}$. It was this defect what caused Jeffreys (1961) to reject conjugate-priors for model selection, in favor of the Cauchy priors discussed above (for which the Bayes factor will go to zero when the evidence is overwhelmingly against M^*).

The conventional prior approach is appealing in principle. However, there seems to be no general method for determining such conventional priors. Nevertheless it is a promising approach in the sense that as long as knowledge of the behavior of Bayes Factors accumulates, the acquired wisdom may be used with advantage at the moment of assuming good Conventional Priors, for substantial classes of problems. Furthermore Conventional Priors obey Principle 1, when it can be established that the priors are well calibrated, and also that are free of the "finite sample inconsistency".

The remaining methods that are discussed here, have the advantage of applying automatically to quite general situations.

2.3. *Intrinsic Bayes factor (IBF) approach*

For the q models $M_1, \ldots, M_q$, suppose that (ordinary, usually improper) noninformative priors $\pi_i^N(\boldsymbol{\theta}_i)$, $i = 1, \ldots, q$, have been chosen, preferably as 'reference priors' (see Berger and Bernardo, 1992), but other choices are possible. Define the corresponding marginal or predictive densities of $\boldsymbol{X}$,

$$m_i^N(\boldsymbol{x}) = \int f_i(\boldsymbol{x}|\boldsymbol{\theta}_i)\pi_i^N(\boldsymbol{\theta}_i)\,\mathrm{d}\boldsymbol{\theta}_i.$$

The general strategy for defining IBF's starts with the definition of a proper and minimal 'deterministic training sample,' Definition 1, above, which are simply subsets $x(l)$, $l = 1, \ldots, L$, of size m, as small as possible so that, using the training sample as a sample all the updated posteriors under each model become proper.

The advantage of employing a training sample to define Bayes factors is to use $\boldsymbol{x}(l)$ to "convert" the improper $\pi_i^N(\boldsymbol{\theta}_i)$ to proper posteriors, $\pi_i^N(\boldsymbol{\theta}_i|\boldsymbol{x}(l))$, and then use the latter to define Bayes factors for the rest of the data denoted by $\boldsymbol{x}(-l)$. The result, for comparing M_j to M_i, is the following property (BP96a).

PROPERTY 3.

$$B_{ji}(l) = \frac{\int f_j(\boldsymbol{x}(-l)|\theta_j, \boldsymbol{x}(l))\pi_j^N(\theta_j|\boldsymbol{x}(l))\,\mathrm{d}\theta_j}{\int f_i(\boldsymbol{x}(-l)|\theta_i, \boldsymbol{x}(l))\pi_j^N(\theta_i|\boldsymbol{x}(l))\,\mathrm{d}\theta_i}$$

using Bayes Rule, can be written as (assuming that all quantities involved exist)

$$B_{ji}(l) = B_{ji}^N(\boldsymbol{x}) \cdot B_{ij}^N\big(\boldsymbol{x}(l)\big),$$

where

$$B_{ji}^N = B_{ji}^N(\boldsymbol{x}) = \frac{m_j^N(\boldsymbol{x})}{m_i^N(\boldsymbol{x})} \quad \text{and} \quad B_{ij}^N(l) = B_{ij}^N\big(\boldsymbol{x}(l)\big) = \frac{m_i^N(\boldsymbol{x}(l))}{m_j^N(\boldsymbol{x}(l))} \tag{13}$$

are the Bayes factors that would be obtained for the full data $\boldsymbol{x}$ *and training sample* $\boldsymbol{x}(l)$, *respectively, if one were to use* π_i^N *and* π_j^N.

The corrected $B_{ji}(l)$ no longer depends on the scales of π_j^N and π_i^N, but it depends on the arbitrary choice of the (minimal) training sample $\boldsymbol{x}(l)$. To eliminate this dependence and to increase stability, the $B_{ji}(l)$ are averaged over all possible training samples

$\boldsymbol{x}(l)$, $l = 1, \ldots, L$. A variety of different averages are possible; here consideration is given only the *arithmetic IBF* (AIBF) and the *median IBF* (MIBF) defined, respectively, as

$$B_{ji}^{AI} = B_{ji}^{N} \cdot \frac{1}{L} \sum_{l=1}^{L} B_{ij}^{N}, \qquad B_{ji}^{MI} = B_{ji}^{N} \cdot \mathrm{Med}\big[B_{ij}^{N}(l)\big], \tag{14}$$

where "Med" denotes median. (The Geometric IBF has also received attention, Bernardo and Smith, 1994.) For the AIBF, it is typically necessary to place the more "complex" model in the numerator, i.e. to let M_j be the more complex model, and then *define* B_{ij}^{AI} by $B_{ij}^{AI} = 1/B_{ji}^{AI}$. The IBFs defined in (14) are *resampling* summaries of the evidence of the data for the comparison of models, since in the averages there is sample re-use. (Note that the Empirical EP-Prior approach, defined in Section 2.5 is also a resampling method.)

These IBFs were defined in (BP96a) along with alternate versions, such as the *encompassing* IBF and the *expected* IBF, which are recommended for certain scenarios. The MIBF is the most robust IBF (see Berger and Pericchi, 1998), although the AIBF is justified trough the existence of an implicit intrinsic prior, at least (but not only) for nested models (see next section).

EXAMPLE 1 (*Continued, Exponential model, AIBF and MIBF*). We have a minimal training sample x_l of size $m = 1$,

$$B_{01}^{N} = \frac{f(x_l|\lambda_0)}{m_1^N(x_l)} = x_l \lambda_0 \exp(-\lambda_0 x_l), \tag{15}$$

and then

$$B_{10}^{AI} = B_{10}^{N} \cdot B_{01}^{N} I(x_l) = \frac{\Gamma(n) \exp(-\lambda_0 \sum x_i)}{(\sum x_i)^n \lambda_0^n} \cdot \frac{1}{n} \sum_{l=1}^{n} x_l \lambda_0 \exp(-\lambda_0 x_l)$$

and for the Median IBF, the arithmetic average is replaced by the Median of the corrections.

EXAMPLE 4 (*Continued, Normal mean, AIBF and MIBF*). Start with the noninformative priors $\pi_1^N(\sigma_1) = 1/\sigma_1$ and $\pi_2^N(\mu, \sigma_2) = 1/\sigma_2^2$. Note that π_2^N is not the reference prior recommended; but π_2^N yields simpler expressions for illustrative purposes. It turns out that minimal training samples consist of any two distinct observations $\boldsymbol{x}(l) = (x_i, x_j)$, and integration shows that

$$m_1^N\big(\boldsymbol{x}(l)\big) = \frac{1}{2\pi(x_i^2 + x_j^2)}, \qquad m_2^N\big(\boldsymbol{x}(l)\big) = \frac{1}{\sqrt{\pi}(x_i - x_j)^2}.$$

Standard integrals yields the following (unscaled) Bayes factor for data $\boldsymbol{x}$, when using π_1^N and π_2^N directly as the priors:

$$B_{21}^{N} = \sqrt{\frac{2\pi}{n}} \cdot \left(1 + \frac{n\bar{x}^2}{s^2}\right)^{n/2}, \tag{16}$$

where $s^2 = \sum_{i=1}^{n}(x_i - \bar{x})^2$. Using (14), the AIBF is then equal to

$$B_{21}^{AI} = B_{21}^{N} \cdot \frac{1}{L}\sum_{l=1}^{L} \frac{(x_1(l) - x_2(l))^2}{2\sqrt{\pi}[x_1^2(l) + x_2^2(l)]}, \tag{17}$$

while the MIBF is given by replacing the arithmetic average by the median.

EXAMPLE 5 (*Continued, Linear models, AIBF and MIBF*). IBF's for linear and related models are developed in Berger and Pericchi (1996a, 1996b, 1997) and Rodríguez and Pericchi (2001) for Dynamic Linear Models. Suppose, for $j = 1, \ldots, q$, that model M_j for $\boldsymbol{Y}$ $(n \times 1)$ is the linear model

$$M_j\colon\ \boldsymbol{y} = \boldsymbol{X}_j \boldsymbol{\beta}_j + \boldsymbol{\varepsilon}, \quad \boldsymbol{\varepsilon}_j \sim \mathcal{N}_n\big(\boldsymbol{0}, \sigma_j^2 \boldsymbol{I}_n\big),$$

where σ_j^2 and $\boldsymbol{\beta}_j = (\beta_{j1}, \beta_{j2}, \ldots, \beta_{jk_j})^t$ are unknown, and $\boldsymbol{X}_j$ is an $(n \times k_j)$ given design matrix of rank $k_j < n$. We will consider priors of the form

$$\pi_j^N(\boldsymbol{\beta}_j, \sigma_j) = \sigma_j^{-(1+q_j)}, \quad q_j > -1.$$

Common choices of q_j are $q_j = 0$ (the reference prior), or $q_j = k_j$ (Jeffreys rule prior). When comparing model M_i nested in M_j, (BP96a) also consider a *modified Jeffreys* prior, having $q_i = 0$ and $q_j = k_j - k_i$. This is a prior between reference and Jeffreys Rule.

For these priors, a minimal training sample $\boldsymbol{y}(l)$, with corresponding design matrix $\boldsymbol{X}(l)$ (under M_j), is a sample of size $m = \max\{k_j\} + 1$ such that all $(\boldsymbol{X}_j^t \boldsymbol{X}_j)$ are nonsingular. Calculation then yields

$$B_{ji}^N = \frac{\pi^{(k_j - k_i)/2}}{2^{(q_i - q_j)/2}} \cdot \frac{\Gamma((n - k_j + q_j)/2)}{\Gamma((n - k_i + q_i)/2)} \cdot \frac{|\boldsymbol{X}_i^t \boldsymbol{X}_i|^{1/2}}{|\boldsymbol{X}_j^t \boldsymbol{X}_j|^{1/2}} \cdot \frac{R_i^{(n-k_i+q_i)/2}}{R_j^{(n-k_j+q_j)/2}}, \tag{18}$$

where R_i and R_j are the residual sums of squares under models M_i and M_j, respectively. Similarly, $B_{ij}^N(l)$ is given by the inverse of this expression with n, $\boldsymbol{X}_i$, $\boldsymbol{X}_j$, R_i and R_j replaced by m, $\boldsymbol{X}_i(l)$, $\boldsymbol{X}_j(l)$, $R_i(l)$ and $R_j(l)$, respectively; here $R_i(l)$ and $R_j(l)$ are the residual sums of squares corresponding to the training sample $\boldsymbol{y}(l)$.

Plugging these expressions in (14) results in the Arithmetic and Median IBFs for the three default priors being considered. For instance, using the modified Jeffreys prior and defining $p = k_j - k_i > 0$, the AIBF is

$$B_{ji}^{AI} = \frac{|\boldsymbol{X}_i^t \boldsymbol{X}_i|^{1/2}}{|\boldsymbol{X}_j^t \boldsymbol{X}_j|^{1/2}} \cdot \left(\frac{R_i}{R_j}\right)^{(n-k_i)/2} \cdot \frac{1}{L}\sum_{l=1}^{L} \frac{|\boldsymbol{X}_j^t(l)\boldsymbol{X}_j(l)|^{1/2}}{|\boldsymbol{X}_i^t(l)\boldsymbol{X}_i(l)|^{1/2}} \cdot \left(\frac{R_j(l)}{R_i(l)}\right)^{(p+1)/2}. \tag{19}$$

To obtain the MIBF, replace the arithmetic average by the median. Note that the MIBF does not require M_i to be nested in M_j.

When multiple linear models are being compared, IBFs can have the unappealing feature of violating the basic Bayesian coherency condition $B_{jk} = B_{ji} B_{ik}$. To avoid this (see (BP96a) for another way to solve the problem), one can utilize the *encompassing* approach, described in the paragraph preceding (12). This leads to what is called the Encompassing IBF. See Lingham and Sivaganesan (1997, 1999), Kim and Sun (2000) and Casella and Moreno (2002) for different applications.

2.4. *The intrinsic prior approach*

Byproducts of the IBF approach create related methods, like the Intrinsic Prior (this section) and the EP-Prior approach, next. As part of the general evaluation strategy of the methods, Principle 1 propose investigation of so-called *intrinsic priors* corresponding to a model selection method. Theorem 1 below shows that in nested models, intrinsic priors corresponding to the AIBF are very reasonable as *conventional priors* for model selection. Hence, the IBF approach can also be thought of as the long sought device for generation of good conventional priors for model selection in nested scenarios. The first discussion on the Intrinsic Prior Equations, was given in (BP96a).

The formal definition of an intrinsic prior, given in (BP96a), was based on an asymptotic analysis, utilizing the following approximation to a Bayes factor associated with priors π_j and π_i:

$$B_{ji} = B_{ji}^N \cdot \frac{\pi_j(\hat{\boldsymbol{\theta}}_j)\pi_i^N(\hat{\boldsymbol{\theta}}_i)}{\pi_j^N(\hat{\boldsymbol{\theta}}_j)\pi_i(\hat{\boldsymbol{\theta}}_i)}\big(1 + o(1)\big); \tag{20}$$

here $\hat{\boldsymbol{\theta}}_i$ and $\hat{\boldsymbol{\theta}}_j$ are the m.l.e.s under M_i and M_j respectively. The approximation in (20) holds in considerably greater generality than does the Schwarz approximation in (40) and it is fundamental for Bayes Factors theory and practice.

As in Section 2, several default Bayes factors can be written in the form

$$B_{ji} = B_{ji}^N \cdot CF_{ij}, \tag{21}$$

where CF_{ij} is the *correction factor*. To define intrinsic priors, equate (20) with (21), yielding,

$$\frac{\pi_j(\hat{\boldsymbol{\theta}}_j)\pi_i^N(\hat{\boldsymbol{\theta}}_i)}{\pi_j^N(\hat{\boldsymbol{\theta}}_j)\pi_i(\hat{\boldsymbol{\theta}}_i)}\big(1 + o(1)\big) = CF_{ij}. \tag{22}$$

Next we need to make some assumptions about the limiting behavior of the quantities in (22). The following are typically satisfied, and will be assumed to hold as the sample size grows to infinity:

(i) Under M_j, $\hat{\boldsymbol{\theta}}_j \to \boldsymbol{\theta}_j$, $\hat{\boldsymbol{\theta}}_i \to \psi_i(\boldsymbol{\theta}_j)$, and $CF_{ij} \to B_j^*(\boldsymbol{\theta}_j)$.

(ii) Under M_i, $\hat{\boldsymbol{\theta}}_i \to \boldsymbol{\theta}_i$, $\hat{\boldsymbol{\theta}}_j \to \psi_j(\boldsymbol{\theta}_i)$, and $CF_{ij} \to B_i^*(\boldsymbol{\theta}_i)$.

When dealing with the AIBF, it will typically be the case that, for $k = i$ or $k = j$,

$$B_k^*(\boldsymbol{\theta}_k) = \lim_{L\to\infty} E_{\boldsymbol{\theta}_k}^{M_k}\left[\frac{1}{L}\sum_{l=1}^{L} B_{ij}^N(l)\right]; \tag{23}$$

if the $\boldsymbol{X}(l)$ are exchangeable, then the limits and averages over L can be removed. For the MIBF, it will simply be the case that, for $k = i$ or $k = j$,

$$B_k^*(\boldsymbol{\theta}_k) = \lim_{L\to\infty} \text{Med}\big[B_{ij}^N(l)\big]. \tag{24}$$

Passing to the limit in (22), first under M_j and then under M_i, results in the following two equations which define the *intrinsic prior* (π_j^I, π_i^I):

$$\frac{\pi_j^I(\boldsymbol{\theta}_j)\pi_i^N(\psi_i(\boldsymbol{\theta}_j))}{\pi_j^N(\boldsymbol{\theta}_j)\pi_i^I(\psi_i(\boldsymbol{\theta}_j))} = B_j^*(\boldsymbol{\theta}_j),$$
$$\frac{\pi_j^I(\psi_j(\boldsymbol{\theta}_i))\pi_i^N(\boldsymbol{\theta}_i)}{\pi_j^N(\psi_j(\boldsymbol{\theta}_i))\pi_i^I(\boldsymbol{\theta}_i)} = B_i^*(\boldsymbol{\theta}_i). \tag{25}$$

The motivation is that priors which satisfy (25) would yield answers which are asymptotically equivalent to the use of a given default Bayes factors. We note that solutions are not necessarily unique, do not necessarily exist, and are not necessarily proper (cf. Dmochowski, 1996; Moreno et al., 1998a).

In the nested model scenario (M_i nested in M_j), and under mild assumptions, solutions to (25) are trivially given by (BP96a)

$$\pi_i^I(\boldsymbol{\theta}_i) = \pi_i^N(\boldsymbol{\theta}_i), \qquad \pi_j^I(\boldsymbol{\theta}_j) = \pi_j^N(\boldsymbol{\theta}_j)B_j^*(\boldsymbol{\theta}_j), \tag{26}$$

the last expression having a simpler expression for exchangeable observations:

$$B_j^*(\theta_j) = \int f_j\big(y(l)|\theta_j\big)\frac{m_i^N(y(l))}{m_j^N(y(l))}\,\mathrm{d}\boldsymbol{y}(l).$$

In (BP96a) an example is given for Intrinsic Priors for separate models.

EXAMPLE 1 (*Continued, Exponential test*).

$$\pi^I(\lambda) = \frac{1}{\lambda}E_\lambda^{M_1}\big[B_{01}^N(x_l)\big] = \frac{\lambda_0}{(\lambda+\lambda_0)^2}. \tag{27}$$

This prior is very sensible. It is proper with median at λ_0. It is not a particular case that this is proper. The following theorem was given in (BP96), and note that it is necessary to assume that the sampling model is absolutely continuous.

THEOREM 1. *Suppose that either the null model is simple or the prior under the null model is proper. Then the Intrinsic Prior is proper.*

If the null is composite, and its prior is not proper it was suggested in (BP96a) and implemented in Moreno et al. (1998a), to consider a sequence of compact sets in Θ_i that converges to the whole set Θ_i, and obtain the Intrinsic Prior that correspond to each set. Intrinsic Priors are seen, under conditions, to be the unique limit of a sequence of proper priors. This theorem has been generalized in Berger and Pericchi (2004) for nonabsolutely continuous likelihoods, like those under censored observations, or for discrete data.

The following is an example on which the prior under the null hypothesis is not proper.

EXAMPLE 4 (*Continued*). In (BP96a), it is shown that the intrinsic priors corresponding to the AIBF are given by the usual noninformative priors for the standard deviations ($\pi_i^I(\sigma_i) = 1/\sigma_i$) and

$$\pi_2^I(\mu|\sigma_2) = \frac{1 - \exp[-\mu^2/\sigma_2^2]}{2\sqrt{\pi}(\mu^2/\sigma_2)}.$$

This last conditional distribution is proper (integrating to one over μ) and, furthermore, is very close to the Cauchy$(0, \sigma_2)$ choice of $\pi_2(\mu|\sigma_2)$ suggested by Jeffreys (1961). Note that π_2^I also obeys all the goals of Jeffreys for choice of a good conventional prior for this situation.

That the AIBF has a conditionally proper intrinsic prior in this example of composite null hypothesis is not fortuitous, but holds for most priors $\pi^N(\cdot)$ in nested models, as discussed in (BP96a), Dmochowski (1996), and Moreno et al. (1998a). This is a rather remarkable property of AIBFs and shows that it is a powerful mechanism for generating conventional priors. It is shown in Sansó et al. (1996) that the Intrinsic Prior of Example 4 prior is well calibrated, and it is predictively matched as Jeffreys' conventional prior. But there is a more widely applicable sense on which Intrinsic Priors are well calibrated.

DEFINITION 3. Two priors are said to be *ratio normalized* if they share a common proportionality constant that cancels out in the quotient.

To be ratio-normalized is a necessary condition for priors to be reasonable but too weak a property to claim to be sufficient.

PROPERTY 4. *The Intrinsic Priors for nested models* (26) *are ratio normalized.*

PROOF. Take exchangeable observations for simplicity of notation. From (26) we may write,

$$\frac{\pi_j^I(\theta_j)}{\pi_i^N(\theta_i)} = \frac{\pi_j^N(\theta_j)}{\pi_i^N(\theta_i)} \cdot \int f_j\big(\mathbf{y}(l)|\theta_j\big) \cdot \frac{f_i(\mathbf{y}(l)|\theta_i)\pi^N(\theta_i)\,\mathrm{d}\theta_i}{\int f_j(\mathbf{y}(l)|\theta_j)\pi_j^N(\theta_j)\,\mathrm{d}\theta_j},$$

and the result follows from the fact that both, π_i^N and π_j^N appear once in the numerator and once denominator, thereby cancelling their own arbitrary constants. □

Also, an intrinsic prior can be found for the MIBF. It is very similar to π^I above, but differs by a moderate constant and hence is not conditionally proper. This constant may be interpreted as the amount of bias inherent in use of the MIBF.

EXAMPLE 5 (*Continued*). For the AIBF, pairwise intrinsic priors exist, and are discussed in Berger and Pericchi (1996b, 1997). The use of the encompassing approach when dealing with linear models,is convenient so that all models are compared with the encompassing model, M_0, and Bayes factors are computed via (12) (Pérez and Berger, 2002, discuss the alternative of comparing all models against the simplest one). The encompassing approach yields that the intrinsic prior for comparing M_1 (say) to M_0 is the usual reference prior for M_1, given by $\pi_1^I(\boldsymbol{\beta}_1, \sigma_1) = 1/\sigma_1$, while, under M_0, $\pi_0^I(\boldsymbol{\beta}_0, \boldsymbol{\beta}_1, \sigma_0) = \pi_0^I(\boldsymbol{\beta}_0|\boldsymbol{\beta}_1, \sigma_0)/\sigma_0$, where $\pi_0^I(\boldsymbol{\beta}_0|\boldsymbol{\beta}_1, \sigma_0)$ is a proper density when the AIBF is derived using reference priors (i.e. $q_1 = q_0 = 0$); is proper when the AIBF is derived using modified Jeffreys priors (i.e. $q_1 = 0$, $q_0 = k_0 - k_1 = p$); but is not proper when the AIBF is derived using the Jeffreys rule noninformative priors (i.e. $q_1 = k_1, q_0 = k_0$). Thus, the use of Jeffreys rule priors is not recommended for deriving the AIBF.

As an illustration, use of the modified Jeffreys priors, in deriving the AIBF, results in the proper conditional intrinsic prior

$$\pi_0^I(\boldsymbol{\beta}_0|\boldsymbol{\beta}_1, \sigma_0) = \frac{\sigma_0^{-p} c^*}{L} \cdot \sum_{l=1}^{L} \frac{|\boldsymbol{X}_0^t(l)\boldsymbol{X}_0(l)|^{1/2}}{|\boldsymbol{X}_1^t(l)\boldsymbol{X}_1(l)|^{1/2}} \cdot \psi\big(\lambda(l), \sigma_0\big),$$

where $c^* = (2\pi)^{-p/2}$, $\psi(\lambda(l), \sigma_0) = 2^{-p} \exp(-\lambda(l)/2) M((p+1)/2, p+1, \lambda(l)/2)$, M is the Kummer function $M(a, b, z) = \frac{\Gamma(b)}{\Gamma(a)} \cdot \sum_{j=0}^{\infty} \frac{\Gamma(a+j)}{\Gamma(b+j)} \cdot \frac{z^j}{j!}$ and $\lambda(l) = \sigma_0^{-2} \boldsymbol{\beta}_0^t \boldsymbol{X}_0^t(l)(I - \boldsymbol{X}_1(l)[\boldsymbol{X}_1^t(l)\boldsymbol{X}_1(l)]^{-1}\boldsymbol{X}_1^t(l))\boldsymbol{X}_0(l)\boldsymbol{\beta}_0$.

This (proper) conditional intrinsic prior resembles a mixture of p-variate t-densities with one degree of freedom, location $\mathbf{0}$, and scale matrices

$$\boldsymbol{\Sigma}(l) = 2\sigma_0^2\big[\boldsymbol{X}_0^t(l)\big(I - \boldsymbol{X}_1(l)\big[\boldsymbol{X}_1^t(l)\boldsymbol{X}_1(l)\big]^{-1}\boldsymbol{X}_1^t(l)\big)\boldsymbol{X}_0(l)\big]^{-1}.$$

Each individual training sample captures part of the structure of the design matrix, which thus gets reflected in the corresponding scale matrix above, but the overall mixture prior does not concentrate about the nested model to nearly as great an extent as do the earlier discussed Zellner and Siow, g-priors and we will see also Fractional Priors in Section 2.6, which effectively choose the full $\sigma_0^2(\boldsymbol{X}_0^t\boldsymbol{X}_0)^{-1}$ as the scale matrix. In this regard, the intrinsic priors seem considerably more sensible. See also Casella and Moreno (2002) for a simpler alternative, but that resembles the Intrinsic Prior, in its treatment of the covariance matrices.

The Intrinsic Prior is not analytically integrable, but use of approximation (20), gives the Expected AIBF (a very useful approximation to the AIBF),

$$B_{ji}^{EIBF} = B_{ji}^N \times \frac{1}{L} \sum_{l=1}^{L} \int f_j\big(\boldsymbol{x}(l)|\hat{\theta}_j\big) \times \frac{m_i^N(\boldsymbol{x}(l))}{m_j^N(\boldsymbol{x}(l))}\, d\boldsymbol{x}(l),$$

which involves the ratio of determinants in B_{0i}^N, which do not cancel. IBF's and EP-Priors are the only approaches studied here, on which the ratio of the determinants of the design matrices is kept in the final Bayes Factor.

2.5. *The expected posterior prior (EP) approach and the empirical EP approach*

Expected posterior prior approach. This is a promising method (related to the Intrinsic Prior Approach) developed in Pérez (1998) and Pérez and Berger (2002) (and, independently in special cases, in Schluter et al., 1999 and Neal, 2001). This method utilizes imaginary training samples to directly develop default conventional priors for use in model comparison. Letting $\boldsymbol{x}^*$ denote the imaginary training sample (usually taken to be a minimal training sample) and starting with noninformative priors $\pi_i^N(\boldsymbol{\theta}_i)$ for M_i as usual, one first defines the posterior distributions, given $\boldsymbol{x}^*$,

$$\pi_i^*(\boldsymbol{\theta}_i|\boldsymbol{x}^*) = \frac{f_i(\boldsymbol{x}^*|\boldsymbol{\theta}_i)\pi_i^N(\boldsymbol{\theta}_i)}{\int f_i(\boldsymbol{x}^*|\boldsymbol{\theta}_i)\pi_i^N(\boldsymbol{\theta}_i)\,\mathrm{d}\boldsymbol{\theta}_i}. \tag{28}$$

Since $\boldsymbol{x}^*$ is not actually observed, these posteriors are not available for use in model selection. However, we can let $m^*(\boldsymbol{x}^*)$ be a suitable *predictive measure or generator measure* for the (imaginary) data $\boldsymbol{x}^*$, and define priors for model comparison by

$$\pi_i^*(\boldsymbol{\theta}_i) = \int \pi_i^N(\boldsymbol{\theta}_i|\boldsymbol{x}^*)m^*(\boldsymbol{x}^*)\,\mathrm{d}\boldsymbol{x}^*. \tag{29}$$

These are called *expected posterior priors* (or *EP priors* for short) because this last integral can be viewed as the expectation (with respect to m^*) of the posteriors in (28). It is to be remarked that the generator measure $m^*(\boldsymbol{x}^*)$ is *common to all models*, so it does *not* change from model to model. This is the binding mechanism of the otherwise unconnected priors.

Propagation of proper prior information. It is both more feasible and appropriate to assess *one* proper prior for *one* model, typically the simplest model, and *propagate* that information to the rest of the models through Eq. (29). Since for the generator model, the prior is proper, then the predictive is so, and Eq. (29) defines proper priors for all models. This propagation of prior information constitutes a sort of mixed approach, being partly subjective (in the assessment of the prior for the generator model, but only to one model subjectivity is allowed). The rest of the models are assigned according to the rule (29). Being all proper priors, and moreover based on a common predictive data set, these priors are good candidates for model comparisons.

Improper EP-priors are ratio normalized. But what about if in Eq. (29) an improper prior $\pi^N(\cdot)$ in a fully objective approach is employed? Well at least the priors are ratio normalized since for any two models,

$$\frac{\pi_i^*(\boldsymbol{\theta}_i)}{\pi_j^*(\boldsymbol{\theta}_j)} = \frac{\int \pi_i^N(\boldsymbol{\theta}_i|\boldsymbol{x}^*)m^*(\boldsymbol{x}^*)\,\mathrm{d}\boldsymbol{x}^*}{\int \pi_j^N(\boldsymbol{\theta}_j|\boldsymbol{x}^*)m^*(\boldsymbol{x}^*)\,\mathrm{d}\boldsymbol{x}^*},$$

and the constants in $m^*(\boldsymbol{x}^*)$, cancel out. Thus we have:

PROPERTY 5. *Improper EP-priors are ratio normalized.*

EP priors can successfully deal with the difficulties discussed in Section 1.6. Of particular note is that, because they can be viewed as an expectation of training sample posteriors, it is usually possible to utilize MCMC computational schemes to compute Bayes factors. Also noteworthy is that, since the $\pi_i^*(\boldsymbol{\theta}_i)$ are actual conventional priors, they inherit the attractive properties of that approach. So in this sense, EP-priors obey Principle 1.

The key issue in the utilization of EP priors is that of appropriate choice of $m^*(\boldsymbol{x}^*)$. Two choices that are natural are (i) the empirical distribution of the actual data (so that the EP prior approach can then be viewed as a resampling approach (see below); and (ii) the (usually improper) predictive arising from an *encompassed* model (i.e. a model that is nested within all others under consideration) and under the usual noninformative prior. A key relationship of the EP priors with training samples generated from the encompassed model, is that this latter approach results in the $\pi_i^*(\boldsymbol{\theta}_i)$ being the intrinsic priors corresponding to the AIBF for nested models. Thus the EP prior approach can be viewed as a development of the IBF approach. It can also be viewed as a method of implementing the use of IBF intrinsic priors.

Relationship between arithmetic intrinsic priors and EP-priors. It turns out that the two equations in (26) can be rephrased, in the exchangeable case for simplicity of notation (using Bayes Rule) as,

$$\pi_k^I(\theta_k) = \int \pi_k\big(\theta_k|\boldsymbol{x}(l)\big)m_i^N\big(\boldsymbol{x}(l)\big)\,\mathrm{d}\boldsymbol{x}(l), \quad k = i, j. \tag{30}$$

In (30) as k equals first i and next j, we obtain the two expressions in (26). This motivated the general definition of an EP-prior as

$$\pi_k^*(\theta_k) = \int \pi_k\big(\theta_k|\boldsymbol{x}(l)\big)m_0\big(\boldsymbol{x}(l)\big)\,\mathrm{d}\boldsymbol{x}(l), \tag{31}$$

with a common "generator" of training samples, $m_0(\boldsymbol{x}(l))$ across models. This coincides with (26) when the training samples are generated by the simpler nested model. When the generation comes from the empirical distribution, that is when real data training samples are taken, we have the Empirical version, next.

The empirical EP-prior approach. This approach is not related directly with IBF's but it is the equivalent in terms of the EP-Priors to the empirical (real training samples) AIBF. Take again a (proper) training sample, but instead of averaging Bayes Factors, the idea is to (arithmetically) average posterior densities, and then use these averaged posteriors to form the Bayes Factors. Define for each model M_i the (*Empirical EP*) prior $\pi_i^{EP}(\theta_i)$ as,

$$\pi_i^{EP}(\theta_i) = \frac{1}{L}\sum_{l=1}^{L}\pi_i^N\big(\theta_i|\boldsymbol{x}(l)\big), \tag{32}$$

where

$$\pi_i^N\big(\theta_i|\boldsymbol{x}(l)\big) = \frac{f_i(x(l)|\theta_i)\cdot\pi_i^N(\theta_i)}{m_i^N(\boldsymbol{x}(l))}.$$

After the priors have been formed, then they can be used to form the Bayes Factors for any pair of models, M_i vs. M_j,

$$B_{ij}^{EP} = \frac{m_i^{EP}(\boldsymbol{x})}{m_j^{EP}(\boldsymbol{x})}, \tag{33}$$

where, for $k = i, j$,

$$m_k^{EP}(\boldsymbol{x}) = \int f_k(\boldsymbol{x}|\theta_k)\pi_k^{EP}(\theta_k)\,\mathrm{d}\theta_k.$$

This is a very general method which induce real proper priors under all models, which is a major advantage, and since the training samples are taken as minimal, the empirical priors are good candidates to obey Principle 1. Besides, there is no need to decide which is the most complex model (as in the AIBF), and can handle automatically both nested and separate model comparisons. Empirical EP-priors tend to favor more, complex models than Intrinsic Priors (Berger and Pericchi, 2004).

As a final comment, EP priors, both empirical and not, open up the possibility of employing different training sample sizes for different models which has some advantages, computational and conceptual.

2.6. *The fractional Bayes factor (FBF) approach*

The fractional Bayes factor was introduced by O'Hagan (1995) as a Bayesian alternative to the "Posterior Bayes factors" of Aitkin (1991), see O'Hagan's discussion of Aitken. Later it has been developed in series of papers by De Santis and Spezzaferri (1996, 1997, 1998a, 1998b, 1999). In short, instead of using the whole likelihood to make the improper priors proper (as Aitken's Posterior Bayes factor does), it uses a fraction of the likelihood for the same purpose. In retrospect it can also be seen as using part of the likelihood (a fraction of it), instead of using part of the sample (as a (minimal) training sample) as in the IBF. There is also a sticking connection with training samples averaging (see below).

The FBF uses a fraction, b, of each *likelihood function*, $L_i(\boldsymbol{\theta}_i) = f_i(\boldsymbol{x}|\boldsymbol{\theta}_i)$, with the remaining $1 - b$ fraction of the likelihood used for model discrimination. Using Bayes rule, it follows that the fractional Bayes factor of model M_j to model M_i is then given by

$$B_{ji}^{F(b)} = B_{ji}^N(\boldsymbol{x})\frac{\int L^b(\boldsymbol{\theta}_i)\pi_i^N(\boldsymbol{\theta}_i)\,\mathrm{d}\boldsymbol{\theta}_i}{\int L^b(\boldsymbol{\theta}_j)\pi_j^N(\boldsymbol{\theta}_j)\,\mathrm{d}\boldsymbol{\theta}_j} = B_{ji}^N(\boldsymbol{x})\frac{m_i^b(\boldsymbol{x})}{m_j^b(\boldsymbol{x})}. \tag{34}$$

The usual choice of b (as in the examples in O'Hagan, 1995, and the discussion by Berger and Mortera of O'Hagan, 1995) is $b = m/n$, where m is the "minimal training sample size," i.e. the number of observations contained in a minimal training sample. In O'Hagan (1995, 1997) are presented other possible choices.

PROPERTY 6. *From expression* (34), *since only the likelihood* (*and not the prior*) *is raised to the power b, then it follows that the FBF is* ratio normalized.

EXAMPLE 4 (*Continued, Normal mean*). Assume, as in Section 2.3, that $\pi_1^N(\sigma_1) = 1/\sigma_1$ and $\pi_2^N(\mu, \sigma_2) = 1/\sigma_2^2$. Consider $b = r/n$, where r is to be specified. Then the correction factor to B_{ji}^N in (34) can be computed to be

$$CF_{21}^{F(b)} = \frac{m_1^b(\boldsymbol{x})}{m_2^b(\boldsymbol{x})} = \left(\frac{r}{2\pi}\right)^{1/2}\left(1 + \frac{n\bar{x}^2}{s^2}\right)^{-r/2},$$

and thus

$$B_{21}^{F(b)} = B_{21}^N CF_{21}^{F(b)} = \left(\frac{r}{n}\right)^{1/2}\left(1 + \frac{n\bar{x}^2}{s^2}\right)^{(n-r)/2}, \tag{35}$$

where B_{21}^N and s^2 are as in Subsection 2.3. Since minimal training samples are of size two, the usual, and more appropriate, choice of r, as mentioned previously, is $r = m = 2$.

EXAMPLE 5 (*Continued, Linear model*). For comparison, assume that M_i is nested in M_j, use modified Jeffreys priors to derive the FBF, and assume that $b = m/n$, where $m = k_j + 1$ is the minimal training sample size.

Using the notation of Subsection 2.3, the correction factor is

$$CF_{ij}^{F(b)} = \frac{(b/2)^{(q_j - q_i)/2}}{\pi^{(k_j - k_i)/2}} \cdot \frac{\Gamma[(r - k_i + q_i)/2]}{\Gamma[(r - k_j + q_j)/2]} \cdot \frac{|\boldsymbol{X}_j^t \boldsymbol{X}_j|^{1/2}}{|\boldsymbol{X}_i^t \boldsymbol{X}_i|^{1/2}} \cdot \frac{R_j^{(r-k_j+q_j)/2}}{R_i^{(r-k_i+q_i)/2}}. \tag{36}$$

The FBF, found by multiplying B_{ji}^N by this correction factor, is thus

$$B_{ji}^{F(b)} = b^{(q_j - q_i)/2} \cdot \frac{\Gamma[(n - k_j + q_j)/2]\Gamma[(r - k_i + q_i)/2]}{\Gamma[(n - k_i + q_i)/2]\Gamma[(r - k_j + q_j)/2]} \cdot \left(\frac{R_i}{R_j}\right)^{(n-r)/2}. \tag{37}$$

Here, $r = m = \max\{k_j\} + 1$ will typically be chosen. Notice that this criterion does not depend upon the ratio of determinants of the design matrices, and hence it is assuming a prior with prior covariance matrix proportional to $Xj^t Xj$. This is an important departure from "Training sampling thinking", in the case of nonexchangeable observations.

Fractional Bayes factor as an average of training samples for exchangeable observations. It has been pointed out, by De Santis and Spezzaferri (1997) and Ghosh and Samanta (2002), that the FBF for exchangeable observations at least, can be thought of as a Bayes Factor with a correction obtained through the geometrical average of Likelihoods over all training samples, that is $m_j^b(\boldsymbol{x})$ can be obtained integrating the geometric average of the product of the likelihoods over all training samples of equal size $r = b \times n$,

$$m_j^b(\boldsymbol{x}) = \int \left[\prod f(\boldsymbol{x}(l)|\theta_j)\right]^{1/N(r)} \pi^N(\theta_j)\, \mathrm{d}\theta_j,$$

where the product is over all proper training samples of size r, and $N(r)$ is the total number of proper training samples of size $m \leqslant r < n$. This property highlights the conceptual connections between FBF's and IBF's.

Asymptotic approximation of the fractional Bayes factor. There is an interesting large sample approximation of the FBF. Performing Laplace expansions in all terms of expression (34), obtains the following remarkable simple approximation:

PROPERTY 7.

$$B_{ji}^{F(b)} \approx \frac{f_j^{(1-b)}(\mathbf{y}|\hat{\theta}_j)}{f_i^{(1-b)}(\mathbf{y}|\hat{\theta}_i)} \cdot b^{(k_j-k_i)/2}. \tag{38}$$

This approximation is potentially a substantial improvement over BIC (41) below, when $r = b \times n$ is not negligible with respect to the sample size n. The approximation (38) while very promising, is too rough for very small r. Refinements of (38) are the subject of current research (Pericchi, 2005).

Intrinsic priors of the fractional Bayes factor approach. The Intrinsic Prior equations for the FBF can be found, adapting the equations of Section 2.4, although there is no equivalent of Theorem 1 here, so there is no guarantee that the intrinsic priors integrate one. De Santis and Spezzaferri (1997) show, for the i.i.d. situation and with $b = m/n$ (where m is fixed) that, for $k = i$ or $k = j$,

$$\begin{aligned} B_k^*(\tilde{\boldsymbol{\theta}}_k) &= \lim_{n\to\infty} \frac{\int L^b(\boldsymbol{\theta}_i)\pi_i^N(\boldsymbol{\theta}_i)\,\mathrm{d}\boldsymbol{\theta}_i}{\int L^b(\boldsymbol{\theta}_j)\pi_j^N(\boldsymbol{\theta}_j)\,\mathrm{d}\boldsymbol{\theta}_j} \\ &= \frac{\int \exp\big(mE_{\tilde{\boldsymbol{\theta}}_k}^{M_k}[\log(f_i(X_l|\boldsymbol{\theta}_i))]\big)\pi_i^N(\boldsymbol{\theta}_i)\,\mathrm{d}\boldsymbol{\theta}_i}{\int \exp\big(mE_{\tilde{\boldsymbol{\theta}}_k}^{M_k}[\log(f_j(X_l|\boldsymbol{\theta}_j))]\big)\pi_j^N(\boldsymbol{\theta}_j)\,\mathrm{d}\boldsymbol{\theta}_j}, \end{aligned} \tag{39}$$

where X_l stands for a single observation and $B_k^*(\tilde{\boldsymbol{\theta}}_k)$ is as in Section 2.4.

EXAMPLE 5 (*Continued, Linear model*). Using the expressions above, for the *modified Jeffreys priors*, it turns out that the FBF has an intrinsic prior of the Zellner and Siow (1980) form, but unfortunately the conditional prior of the parameters under test does not integrate to one. The bias factor is

$$\left(\frac{m}{2}\right)^{p/2} \cdot \frac{\sqrt{\pi}}{\Gamma[(p+1)/2]},$$

which is potentially quite large. Thus, although the FBF has an intrinsic prior with a sensible centering and Cauchy form, it has a potentially large (constant) bias. There are at least two ways out of this problem. One is to re-scale the Bayes Factors multiplying by the reciprocal of the bias. A second way out is to change the initial priors, and in fact assuming the usual Jeffreys independent priors $\pi(\sigma_j) = 1/\sigma_j$, $j = 1, 2$, largely solve the bias for the Linear Model. On the other hand, the concerns regarding the scale matrix in the Zellner and Siow (1980) prior also clearly apply to the intrinsic prior for the FBF. See Berger and Pericchi (2004) for discussion and examples on this matter.

In (BP01), two examples are given on which the FBF turns out to be inconsistent. One is an example on which the support of the likelihood depends on the value of the parameters, usually called an *irregular likelihood*. In this situation no solution appears to exist. The other example is the familiar Neyman–Scott problem, on which there is an increasing multiplicity of parameters as the sample size grows; in this case the number of means grows but keeping the variance unique. On such cases, as in unbalanced linear model situations, the solution appears to be to work not with one fraction but with *multiple fractions*: the information about some parameters is fixed but the information about other (the variance) grows. So the fix is to try to write the likelihood as a product of the likelihood of the means times the likelihood of the variance an take different fractions for each, see (BP01).

On the bright side, the FBF is simpler than other methods, but not as much as BIC (next section). It is very important that there is an asymptotic expression (38), and its refinements, that can compete with advantage with BIC and has the same level of simplicity. In several regular balanced problems with fixed number of parameters, the FBF is found to be close to IBF's for comparison of even separate nonnested models (Araújo and Pereira, 2003).

In retrospect, the insight given by training samples has influenced the FBF, at least in choice of the fraction b, and in the need to employ multiple fractions for different parts of the likelihood. Also the existence of Intrinsic Priors for the FBF, possibly after re-scaling, opens up a road for a synthesis of Bayes Factors (Pericchi, 2005).

2.7. *Asymptotic methods and BIC*

Laplace's asymptotic method yields as an approximation to a Bayes factor, B_{ji}, with respect to two priors $\pi_j(\boldsymbol{\theta}_j)$ and $\pi_i(\boldsymbol{\theta}_i)$,

$$B_{ji}^L = \frac{f_j(\boldsymbol{x}|\hat{\boldsymbol{\theta}}_j)|\hat{\boldsymbol{I}}_j|^{-1/2}}{f_i(\boldsymbol{x}|\hat{\boldsymbol{\theta}}_i)|\hat{\boldsymbol{I}}_I|^{-1/2}} \cdot \frac{(2\pi)^{k_j/2}\pi_j(\hat{\boldsymbol{\theta}}_j)}{(2\pi)^{k_i/2}\pi_i(\hat{\boldsymbol{\theta}}_i)}, \tag{40}$$

where $\hat{\boldsymbol{I}}_i$ and $\hat{\boldsymbol{\theta}}_i$ are the observed information matrix and m.l.e., respectively, under model M_i, and k_i is the dimension of $\boldsymbol{\theta}_i$ (cf. Haughton, 1988; Gelfand and Dey, 1994; Kass and Raftery, 1995; Dudley and Haughton, 1997; Pauler, 1998). As the sample size goes to infinity, the first factor of B_{ji}^L typically goes to 0 or ∞, while the second factor stays bounded. The BIC criterion of Schwarz (1978) arises from choosing an appropriate constant for this second term, leading to

$$B_{ji}^S = \frac{f_j(\boldsymbol{x}|\hat{\boldsymbol{\theta}}_j)}{f_i(\boldsymbol{x}|\hat{\boldsymbol{\theta}}_i)} \cdot n^{(k_i-k_j)/2}. \tag{41}$$

In Kass and Wasserman (1995) and Pauler (1998) this choice is analyzed.

The BIC approximation has the advantages of simplicity and an (apparent) freedom from prior assumptions. However, it is valid only for very regular likelihoods. Among the problems for which it does not apply directly are situations with increasing number of parameters as sample sizes grows, models with irregular asymptotics (see BP01) and problems in which the likelihood can concentrate at the boundary of the parameter space for one of the models. Dudley and Haughton (1997) and Kass and Vaidyanathan (1992) give extensions of (40) to such situations. Recall that the approximation (20) is both more accurate and widely applicable than (40).

EXAMPLE 4 (*Continued, Normal mean*). Application of (41) yields,

$$B_{21}^S = \frac{B_{21}^N}{\sqrt{2\pi}} = \frac{1}{\sqrt{n}}\left(1 + \frac{n\bar{x}^2}{s^2}\right)^{n/2}.$$

EXAMPLE 5 (*Continued, Linear model*). The usual BIC for linear models has the simple expression

$$B_{21}^S = \left(\frac{R_1}{R_2}\right)^{n/2} \cdot n^{(k_1-k_2)/2}.$$

Considering the situation of M_i nested in M_j, the bias (with respect to the Jeffreys–Zellner and Siow approach) of the BIC Bayes factor can be computed to be

$$\frac{\sqrt{\pi}}{2^{p/2}\Gamma[(p+1)/2]} \cdot \left(\frac{R_i}{R_j}\right)^{(k_j+1)/2}.$$

This is a potentially large bias. More seriously, it is *not a constant bias*, depending strongly on the ratio of the residual sum of squares for the two models.

Priors can clearly be found for which the BIC Bayes factor equals the actual Bayes factor when the m.l.e. of the 'extra parameters' in the more complex model is at **0**. But, as this m.l.e. departs from **0**, the BIC Bayes factor will have an increasingly dramatic bias. This bias becomes even more severe as the dimensional difference between the two models increases. See Pauler (1998) for a related discussion.

It should be remembered that BIC arises as a version of the Laplace asymptotic method. When it is not valid, the general expression (41) is not valid either, and can not be justified as an approximate Bayesian method. One such situation is the example shown in Stone (1979), of a version of Neyman–Scott problem with increasing number of parameters (with the sample size) where simply the formula (41) is *not* a valid approximation of a Bayes Factor. See also Berger et al. (2003) for related discussions.

Still, BIC is the simplest of all the methods for computing Bayes Factors, and even though is the roughest approximation, is still the one that has catched, so far, more practical ground, and the one that has been more apt in hiding the prior assumptions. Formula (38) and refinements has a good chance to become a strong competitor to BIC.

2.8. *Lower bounds on Bayes factors*

In Sellke et al. (2001) (see also Bayarri and Berger, 1998), a "universal" lower bound is proposed for calibrating p-values, p_{val}. The proposal for calibrating a p-value is to compute, when $p_{\text{val}} < \mathrm{e}^{-1}$,

$$B_{01} \geqslant \underline{B}(p_{\text{val}}) = \underline{B}_{01} = -\mathrm{e} \cdot p_{\text{val}} \cdot \log_{\mathrm{e}}(p_{\text{val}}). \tag{42}$$

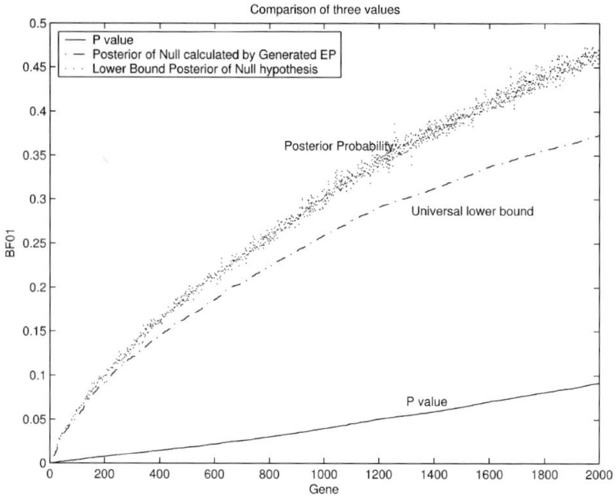

Fig. 1. Image visualising the results of a microarray experiment on mice DNA, with genes ordered according to their p-values. The figure shows the huge number of significant responses given by (the uncorrected for multiple comparisons) p-values, the improvement by the Universal Bound, and its growing difference with real posterior probabilities given by Intrinsic-EP Priors.

EXAMPLE 4 (*Microarray analyses*, Yang, 2003). The Analysis of microarray data is a good test case for the usefulness of Bayesian posterior probabilities and the inadequacy of p-values. This is not only due to the fact of the multiple comparisons involved but to the need of a re-scaling of the p-values in order to measure evidence that the individual gene has been expressed after a treatment. In our view the correct scaling are the probabilities induced by real Bayes Factors. The universal bound (42), which is valid for *any* prior for the parameters (Sellke et al., 2001) can be used as an approximate re-scaling of p-values.

The lower bound has been demonstrated for simple null hypothesis. For particular composite hypothesis, a sufficient condition in Sellke et al. (2001) has to be verified numerically.

This "Universal Lower Bound" helps calibrating the p-values, but it is typically much smaller than actual posterior probabilities. In Figure 1 an example of microarray data is shown, on which 6000 genes are evaluated pre and post treatment to find out which changed their means (we assume here a Normal likelihood). For (uncorrected) p-values, the number of significant genes is unrealistically large. The situation is improved by the Universal bound, but much more so, by the actual posterior probabilities.

The Universal Lower Bound, which is the simplest of all measures of evidence exposed here has a huge potential; First of all, for educational purposes: the bound can be introduced in elementary courses. Secondly, the bound can be implemented overnight in computer packages.

3. More general training samples

The article Berger and Pericchi (2004) is devoted to enlarge the concept of deterministic training samples in Definition 1, for more general MTS like randomized or weighted MTS's. When data vary widely in terms of their information content, a deterministic MTS (Definition 1) can be inadequate. Fortunately, there is a test for such situations to be checked, and is termed Assumption 0. Call $\mathcal{X}^I$ the population set of training samples under consideration.

ASSUMPTION 0. $P_{\boldsymbol{\theta}_i}^{M_i}(\mathcal{X}^I) = 1,\ i = 0, 1.$

This assumption can be violated for the set of minimal training samples, for instance in situations involving censoring or when inappropriate initial no informative priors are utilized.

EXAMPLE 6 (*Right censored exponential*). This is like Example 1, a Hypothesis Test involving the Exponential likelihood, but now all the observations bigger than a fixed value are censored, i.e. if $X_i > r$ then it contributes to the likelihood: $P(X_i > r|\lambda) = \exp(-r\lambda)$. If we employ the same improper prior $\pi^N(\lambda) = \lambda^{-1}$ as in Example 1, they MTS is *any* uncensored data point x_i, but it can be proved that if $x_i > r$ then the posterior is still improper. (The information in censored and uncensored observations is wildly different.) The consequence is that Assumption 0 is violated, and it turns out that the Intrinsic Prior is nonintegrable! (Berger and Pericchi, 2004). How to solve it? The set of training samples should be enlarged until the assumption is satisfied.

3.1. Randomized and weighted training samples

DEFINITION 4. A *randomized training sample* with sampling mechanism $\boldsymbol{p} = (p_1, \ldots, p_{L_P})$, where $\boldsymbol{p}$ is a probability vector, is obtained by drawing a training sample from $\mathcal{X}^P$ (the set of proper training samples) according to $\boldsymbol{p}$. Alternatively, the training samples can be considered to be *weighted training samples* with weights p_i.

Example of randomized training sample. Sequential random sampling: Very useful are sequential minimal training samples (SMTS) that are each obtained by drawing observations from the collection of data $\boldsymbol{x} = \{x_1, x_2, \ldots, x_n\}$ by simple random sampling (without replacement for a given SMTS), stopping when the subset so formed, $\boldsymbol{x}^*(l) = (x(l)_1, \ldots, x(l)_{N(l)})$, is a proper training sample. Note that $N(l)$ is itself a random variable. SMTS's are obtained by sequential random sampling, but they can also be described via Definition 4, where p_i is simply the probability of obtaining the ith SMTS via sampling without replacement from the set of observations, and all other proper training samples are assigned probability 0.

However if Assumption 0 is satisfied the Intrinsic Priors are fine. Assume for simplicity that M_0 is a simple model.

LEMMA 1 (Berger and Pericchi, 2004). *If Assumption* 0 *holds,* M_0 *is a simple model (i.e.* $\boldsymbol{\theta}_0$ *is specified), and the intrinsic prior is given by* (26), *then*

$$\int \pi_1^I(\boldsymbol{\theta}_1)\,\mathrm{d}\boldsymbol{\theta}_1 = 1.$$

EXAMPLE 6 (*Continued*). A sequential minimal training sample here is simply: "Keep sampling until you reach an uncensored observation". Using Intrinsic Prior equations (26) coupled with SMTS's yield,

$$\pi^I(\lambda) = \frac{\lambda_0}{(\lambda + \lambda_0)^2},$$

that is the same prior as the one without censoring. This is a satisfying result. The bias of the deterministic MTS for censoring is thus taken care of by random MTS.

Another group of examples that need an adaptation of training samples are those on which the observations y_i contain drastically diminishing information as i increases. Two examples follow.

EXAMPLE 7 (*Dynamic linear models and times series*). This has been analyzed in Rodríguez and Pericchi (2001). As times $t = i$ elapses, the data is drastically less informative about the parameters of the model at time $t = 0$. Thus a training sample that insists in averaging all training samples for all t, as t grows becomes increasingly improper. The solution suggested is simply stop taking training samples (real or imaginary) after a finite (and small) $t = t_0$, even for the minimum t_0 on which there is a proper training sample for all models.

The next example follow similar lines.

EXAMPLE 8 (Findley, 1991). This is a classical counterexample against BIC, due to its potential inconsistency. The model is the simple regression:

$$Y_i = d_i\theta + \varepsilon_i,$$

where the errors are a random sample from $\mathcal{N}(\varepsilon_i|0, 1)$. Findley (1991) considered the case where the regressor is $d_i = 1/\sqrt{i}$, so the regressor is vanishing with i. It is being tested H_0: $\theta = 0$ vs. H_1: $\theta \neq 0$ and the standard noninformative prior $\pi^N(\theta) = 1$ is employed. It turns out that both BIC and also more sophisticated Bayes Factors are inconsistent for $\theta^2 < 1$, see Berger and Pericchi (2004). How to solve this? If the problem is the change in information between training samples we may then weight them according to its information: the suggestion is to weigh each training sample i by $d_i^2/\sum_{j=1}^n d_j^2$. This solves the problem of inconsistency, for the Arithmetic IBF. However, still it does not pass the more stringent test of Principle 1, since the Intrinsic prior tends to an improper prior. Again, truncation on the number of training samples, not taking training samples after some $i = i_0$, solves the problem.

4. Prior probabilities

The right (and necessity) to assign prior model probabilities, in order to convert Bayes Factors in Posterior Odds, open a host of possibilities. See (BP96) rules that emphasize simpler models. The assessment of equal prior model probabilities, which is the casual one, is potentially misleading, particularly when explanatory variables are highly correlated or worst, nearly equal. In such cases, the model with the repetitive explanatory variable is "overweighed", since it is counted several times, as many as that explanatory variable appears. (See Chipman et al., 2001 for further motivation.) Working with orthogonalized variables may alleviate the problem but then the interpretation of the original variables is lost. One possible remedy (among others) has been put forward in Nadal (1999) in terms of the original variables, the so-called "Orthogonalized-Regressors Prior Model Probabilities", introduced here in the context of a Linear Model:

1. Compute for each regressor $x_i, i = 1, \ldots, k$, the correlation with the response variable $\boldsymbol{y}$. Re-order the columns of the design matrix $\boldsymbol{X}$ according to the correlations in a decreasing manner. Divide each column by its norm (so that it is independent of scale).
2. Orthogonalize the matrix $\boldsymbol{X}$, by, for example, Gram–Schmith (G–S) to obtain the matrix
   (a) $z_1 = \boldsymbol{x}_1$
   $Z = [z_1]$
   For $j = 2, \ldots, k$:
   (i) $z_j = \boldsymbol{x}_j - Z(Z^t Z)^{-1} Z^t \boldsymbol{x}_j$; (ii) $Z = [Z, z_j]$
3. Calculate the vector: $\boldsymbol{p}^0 = (p^0(1), \ldots, p^0(k)) = \frac{\sqrt{\text{diagonal}(Z^t Z)}}{\text{Trace}(\sqrt{Z^t Z})}$
4. For each model M_j of the s entertained models assign prior probability according to:
$$p_j = \prod_{i=1}^{k} p^0(i)^{\delta_{ij}} \left(1 - p^0(i)\right)^{(1-\delta_{ij})},$$
   where $\delta_{ij} = 1$ if x_i is in M_j and 0 otherwise.
5. For $j = 1, \ldots, s$
$$p(M_j) = \frac{p_j}{\sum_{r=1}^{s} p_r}.$$

This assignment deals with the problem of multicolinearity. See also (BP96a) for other choices. These and other strategies are the subject of current research.

5. Conclusions

1. Here 8 methods are analyzed for calculation or approximation of Bayes Factors. These may be taken as a proof of widespread disagreement. However, there is deep source of agreements: often these methods share the same asymptotics and give results which are close to each other, apart from deep theoretical connections among

them. On the other hand, frequentist methods, like p-values or significance testing with fixed type I errors, have different asymptotics, and thus are increasingly at odds with Bayesian methods, by an increasing amount as the data accumulates.
2. There are general concepts which offer a unnifying framework: (i) Principle 1, in Section 2. (ii) The different kinds of well calibrated priors. (iii) Assumption 0, of Section 3. (iv) The concept of Intrinsic Priors for different methods. (v) The concept of training sample, real and imaginary; deterministic and aleatory.
3. It should be remembered that a unifying view is that Bayes Factors can be seen as "Unnormalized Bayes Factors $\times$ Correction Factors". The methods discussed here are mostly about the second right-hand term which is *not* the dominant asymptotic factor, but that should be given careful consideration. Still all methods considered here have the first (or an approximation of it) factor embedded in their formulae. So the methods discussed here, evolving around the Unnormalized Bayes Factor, share a dominant (asymptotically) common term.
4. Bayes Factor is a powerful probabilistic tool, which is here to stay. We rather learn how to use it at its best. It is time to emphasize the agreements among the Bayesian approaches to form and approximate Bayes Factors. The 8 approaches visited here are most of the time able to avoid potentially harmful priors. Specially in new problems, it is advisable to compare several of them to check the reassuring agreement, a sort of robustness with respect to the method.

Acknowledgements

The research was supported by the School of Natural Sciences, UPR-RRP, and by a grant of CONICIT-USB-CESMa, Venezuela. The author held a Guggenheim Fellowship during part of his research. I take opportunity to thank: A.C. Atkinson, Sir D.R. Cox, P.J. Green, E. Moreno, N. Nadal, M.E. Pérez, I. Rodriguez-Iturbe, A. Rodriguez, B. Sansò, A.F.M. Smith and specially J.O. Berger, for discussions and joint efforts on the subject of this article.

References

Aitkin, M. (1991). Posterior Bayes factors (with discussion). *J. Roy. Statist. Soc., Ser. B* **53**, 111–142.

Araújo, M.I., Pereira, B.B. (2003). A comparison among Bayes factors for separated models: Some simulation results. Elsevier Preprint.

Barbieri, M., Berger, J. (2004). Optimal predictive variable selection. *Ann. Statist.* **32** (3). (In press).

Bayarri, M.J., Berger, J.O. (1998). Quantifying surprise in the data and model verification. In: Bernardo, J.M., et al. (Eds.), *Bayesian Statistics, vol. 6*. Oxford University Press, pp. 53–82.

Berger, J. (1999). Bayes factors. In: Kotz, S., et al. (Eds.), *Encyclopedia of Statistical Sciences,* Update Volume 3. Wiley, New York, pp. 20–29.

Berger, J. (2003). Could Fisher, Jeffreys and Neyman have agreed on testing? (with discussion). *Statist. Sci.* **18** (1), 1–32.

Berger, J., Bernardo, J.M. (1992). On the development of the reference prior method. In: Bernardo, J.M., et al. (Eds.), *Bayesian Statistics, vol. IV*. Oxford University Press, London, pp. 35–60.

Berger, J., Pericchi, L. (1996a). The intrinsic Bayes factor for model selection and prediction. *J. Amer. Statist. Assoc.* **91**, 109–122.
Berger, J., Pericchi, L. (1996b). The intrinsic Bayes factor for linear models. In: Bernardo, J.M., et al. (Eds.), *Bayesian Statistics, vol. 5*. Oxford University Press, London, pp. 23–42.
Berger, J., Pericchi, L. (1997). On the justification of default and intrinsic Bayes factors. In: Lee, J.C., et al. (Eds.), *Modeling and Prediction*. Springer-Verlag, New York, pp. 276–293.
Berger, J., Pericchi, L. (1998). Accurate and stable Bayesian model selection: The median intrinsic Bayes factor. *Sankya B* **60**, 1–18.
Berger, J., Pericchi, L. (2001). Objective Bayesian model methods for model selection: Introduction and comparison (with discussion). In: Lahiri, P. (Ed.), *IMS Lecture Notes – Monograph Series*, vol. 38, pp. 135–207.
Berger, J., Pericchi, L. (2004). Training samples in objective Bayesian model selection. *Ann. Statist.* **32** (3), 841–869.
Berger, J., Sellke, T. (1987). Testing a point null hypothesis: The irreconcilability of P-values and evidence. *J. Amer. Statist. Assoc.* **82**, 112–122.
Berger, J., Pericchi, L., Varshavsky, J. (1998). Bayes factors and marginal distributions in invariant situations. *Sankya A* **60**, 307–321.
Berger, J., Ghosh, J.K., Mukhopadhyay, N. (2003). Approximations and consistency of Bayes factors as model dimension grows. *J. Statist. Plann. Inference* **112**, 241–258.
Bernardo, J.M., Smith, A.F.M. (1994). *Bayesian Theory*. Wiley, Chichester.
Casella, G., Moreno, E. (2002). Objective Bayesian variable selection. Tech. Report, University of Florida, Department of Statistics.
Chen, M.H., Shao, Q.M., Ibrahim, J.G. (2000). *Monte Carlo Methods in Bayesian Computation. Springer Series in Statistics*.
Chipman, H., George, E.I., McCulloch, R.E. (2001). The practical implementation of Bayesian model selection (with discussion). In: Lahiri, P. (Ed.), *IMS Lecture Notes – Monograph Series, vol. 38*, pp. 65–134.
Clyde, M., George, E.I. (2000). Flexible empirical Bayes estimation for wavelets. *J. Roy. Statist. Soc., Ser. B* **62**, 681–698.
Cox, D.R. (1961). Tests of separate families of hypotheses. In: *Proceedings of the Fourth Berkeley Symposium on Mathematical Statistics and Probability, vol. 1*. University of California Press, Berkeley, pp. 105–123.
Cox, D.R., Hinkley, D.V. (1974). *Theoretical Statistics*. Chapman and Hall, London.
De Santis, F., Spezzaferri, F. (1996). Comparing hierarchical models using Bayes factor and fractional Bayes factors: A robust analysis. In: Berger, J., et al. (Eds.), *Bayesian Robustness, vol. 29*. IMS Lecture Notes, Hayward.
De Santis, F., Spezzaferri, F. (1997). Alternative Bayes factors for model selection. *Canad. J. Statist.* **25**, 503–515.
De Santis, F., Spezzaferri, F. (1998a). Consistent fractional Bayes factor for linear models. *Pubblicazioni Scientifiche del Dipartimento di Statistica, Probab. e Stat. Appl.*, Serie a, n. 19, Univ. di Roma "La Sapienza".
De Santis, F., Spezzaferri, F. (1998b). Bayes factors and hierarchical models. *J. Statist. Plann. Inference* **74**, 323–342.
De Santis, F., Spezzaferri, F. (1999). Methods for robust and default Bayesian model selection: The fractional Bayes factor approach. *Internat. Statist. Rev.* **67**, 1–20.
Dmochowski, J. (1996). Intrinsic priors via Kullback–Leibler geometry. In: Bernardo, J.M., et al. (Eds.), *Bayesian Statistics, vol. 5*. Oxford University Press, London, pp. 543–549.
Dudley, R., Haughton, D. (1997). Information criteria for multiple data sets and restricted parameters. *Statistica Sinica* **7**, 265–284.
Findley, D.F. (1991). Counterexamples to parsimony and BIC. *Ann. Inst. Statist. Math.* **43**, 505–514.
Gelfand, A.E., Dey, D.K. (1994). Bayesian model choice: Asymptotics and exact calculations. *J. Roy. Statist. Soc., Ser. B* **56**, 501–514.
George, E.I., Foster, D.P. (2000). Empirical Bayes variable selection. *Biometrika* **87**, 731–747.
Ghosh, J.K., Samanta, T. (2002). Nonsubjective Bayesian testing – an overview. *J. Statist. Plann. Inference* **103** (1), 205–223.
Good, I.J. (1950). *Probability and the Weighing of Evidence*. Haffner, New York.

Good, I.J. (1985). Weight of evidence: A brief survey. In: Bernardo, J.M., DeGroot, M., Lindley, D., Smith, A.F.M. (Eds.), *Bayesian Statistics, vol. 2*. North-Holland, Amsterdam, pp. 249–270.

Haughton, D. (1988). On the choice of a model to fit data from an exponential family. *Ann. Statist.* **16**, 342–355.

Holmes, S., Morris, C., Tibshirani, R. (2003). Bradley Efron: A conversation with good friends. *Statist. Sci.* **18** (2), 268–281.

Jeffreys, H. (1961). *Theory of Probability*. Oxford University Press, London.

Kass, R.E., Raftery, A. (1995). Bayes factors. *J. Amer. Statist. Assoc.* **90**, 773–795.

Kass, R.E., Vaidyanathan, S.K. (1992). Approximate Bayes factors and orthogonal parameters, with application to testing equality of two binomial proportions. *J. Roy. Statist. Soc.* **54**, 129–144.

Kass, R.E., Wasserman, L. (1995). A reference Bayesian test for nested hypotheses and its relationship to the Schwarz criterion. *J. Amer. Statist. Assoc.* **90**, 928–934.

Key, J.T., Pericchi, L.R., Smith, A.F.M. (1999). Bayesian model choice: What and why?. In: Bernardo, J.M., et al. (Eds.), *Bayesian Statistics, vol. 6*. Oxford University Press, pp. 343–370.

Kim, S., Sun, D. (2000). Intrinsic priors for model selection using an encompassing model. *Life Time Data Anal.* **6**, 251–269.

Lempers, F.B. (1971). *Posterior Probabilities of Alternative Linear Models*. University of Rotterdam Press, Rotterdam.

Lingham, R., Sivaganesan, S. (1997). Testing hypotheses about the power law process under failure truncation using intrinsic Bayes factors. *Ann. Inst. Statist. Math.* **49**, 693–710.

Lingham, R., Sivaganesan, S. (1999). Intrinsic Bayes factor approach to a test for the power law process. *J. Statist. Plann. Inference* **77**, 195–220.

Moreno, E., Bertolino, F., Racugno, W. (1998a). An intrinsic limiting procedure for model selection and hypothesis testing. *J. Amer. Statist. Assoc.* **93**, 1451–1460.

Nadal, N., (1999). El análisis de varianza basado en los factores de Bayes intrínsecos. Ph.D. thesis, Universidad Simón Bolívar, Venezuela.

Neal, R. (2001). Transferring prior information between models using imaginary data. Tech. Report 0108, Dept. Statistics, Univ. Toronto.

O'Hagan, A. (1995). Fractional Bayes factors for model comparisons. *J. Roy. Statist. Soc., Ser. B* **57**, 99–138.

O'Hagan, A. (1997). Properties of intrinsic and fractional Bayes factors. *Test* **6**, 101–118.

Pauler, D. (1998). The Schwarz criterion and related methods for normal linear models. *Biometrika* **85** (1), 13–27.

Pérez, J.M. (1998). Development of conventional prior distributions for model comparisons. Ph.D. thesis, Purdue University.

Pérez, J.M., Berger, J. (2002). Expected posterior prior distributions for model selection. *Biometrika* **89**, 491–512.

Pericchi, L.R. (2005). Approximations and synthesis of Bayes factors. II Congreso Bayesiano Latinoamericano, Cobal II, San Juan de los Cabos, Baja California, Feb. 6–11, 2005.

Pericchi, L.R., Pérez, M.E. (1994). Posterior robustness with more than one sampling model. *J. Statist. Plann. Inference* **40**, 279–294.

Raftery, A.E., Madigan, D., Volinsky, C.T. (1996). Accounting for model uncertainty in survival analysis improves predictive performance. In: Bernardo, J.M., et al. (Eds.), *Bayesian Statistics, vol. 5*. Oxford University Press, London, pp. 323–349.

Rodríguez, A., Pericchi, L.R. (2001). Intrinsic Bayes factors for dynamic linear models. In: George, E., Nanopoulos, P. (Eds.), *Bayesian Methods, with Applications to Science, Policy and Official Statistics*. Eurostat, Luxembourg, pp. 459–468.

Sansó, B., Pericchi, L.R., Moreno, E. (1996). On the robustness of the intrinsic Bayes factor for nested models. In: Berger, J., et al. (Eds.), *Bayesian Robustness, IMS Lecture Notes*, vol. 29. Hayward, pp. 157–176.

Schluter, P.J., Deely, J.J., Nicholson, A.J. (1999). The averaged Bayes factor: A new method for selecting between competing models. Technical Report, University of Canterbury.

Schwarz, G. (1978). Estimating the dimension of a model. *Ann. Statist.* **6**, 461–464.

Sellke, T., Bayarri, M.J., Berger, J. (2001). Calibration of P-values for testing precise null hypotheses. *Amer. Statist.* **55**, 62–71.

Shively, T.S., Kohn, R., Wood, S. (1999). Variable selection and function estimation in additive nonparametric regression using a data-based prior (with discussion). *J. Amer. Statist. Assoc.* **94**, 777–806.

Spiegelhalter, D.J. (1977). A test for normality against symmetric alternatives. *Biometrika* **68** (2), 415–418.

Spiegelhalter, D.J., Smith, A.F.M. (1982). Bayes factors for linear and log-linear models with vague prior information. *J. Roy. Statist. Soc., Ser. B* **44**, 377–387.

Stone, M. (1979). Comments on model selection criteria of Akaike and Schwarz. *J. Roy. Statist. Soc., Ser. B* **41**, 276–278.

Yang, Ch. (2003). Objective Bayesian training sample based model selection methodologies with application to microarray data analysis. M.Sc. thesis, University of Puerto Rico, Department of Mathematics.

Zellner, A. (1986). On assessing prior distributions and Bayesian regression analysis with g-prior distributions. In: Goel, P.K., Zellner, A. (Eds.), *Bayesian Inference and Decision Techniques: Essays in Honor of Bruno de Finetti*. North-Holland, Amsterdam, pp. 233–243.

Zellner, A., Siow (1980). Posterior odds for selected regression hypotheses. In: Bernardo, J.M., et al. (Eds.), *Bayesian Statistics, vol. 1*. Valencia University Press, Valencia, pp. 585–603.

Handbook of Statistics, Vol. 25
ISSN: 0169-7161
© 2005 Elsevier B.V. All rights reserved.
DOI 10.1016/S0169-7161(05)25005-8

5

Role of P-values and other Measures of Evidence in Bayesian Analysis

Jayanta Ghosh[a], *Sumitra Purkayastha and Tapas Samanta*

Abstract
We review the relation between P-values and posterior probabilities, and also the Bayesian P-values for composite null hypothesis when there is no alternative.

Keywords: Bayes factors; Bayesian P-values; Pitman alternative; posterior probability

1. Introduction

A P-value is "the smallest level at which the observations are significant in a particular direction" (Gibbons and Pratt, 1975). Equivalently, it is the probability, under a simple null or the supremum of probabilities under a composite null, of deviations of the test statistic being greater than that currently observed. As a data based measure of evidence against the null hypothesis H_0, it contains more information than just a statement of significance or lack of it at a predetermined level of significance, or the Neyman–Pearsonian probability of Type I error. Therein lies its attraction as a measure of evidence in classical statistical methodology for testing.

Gibbons and Pratt (1975) argue that in "any single experiment the P-value measures the agreement or disagreement in a specific direction" and that, therefore, it serves a useful purpose. On the other hand they point out that one cannot compare P-values across sample sizes or across experiments – a point also made very effectively in Lindley (1957). It is also clear and has been noted many times that the frequentist interpretation of a P-value under a simple null is somewhat odd in that it depends on hypothetical repetition leading to new samples which have more deviant values of a test statistic than the value actually observed. Among other things it implies an apparent use of unobserved more deviant data as evidence. A fact which is both an advantage and a disadvantage is that a P-value only depends on the distribution under the null hypothesis, it ignores the alternative. This is an advantage because the null distribution is usually

[a]Corresponding author.

easier to calculate whereas the alternative distribution may be difficult to calculate or even not well-defined. This is a disadvantage because as Sellke et al. (2001) remark, "knowing that the data are 'rare' under H_0 is of little use unless one determines whether or not they are also 'rare' under H_1".

There exist two well-known alternatives to P-values which depend only on observed data and do not ignore the alternative hypothesis. One alternative is to use the likelihood ratio test statistic

$$\lambda = \sup_{\theta \in \Theta_0} f(x|\theta) / \sup_{\theta \in \Theta} f(x|\theta), \tag{1}$$

where $f(x|\theta)$ is p.m.f. or p.d.f. under the parameter θ – evaluated at observed data x, Θ is the full parameter space, and Θ_0 is the set specified by H_0. The other alternative is the Bayes factor

$$BF_{01} = \frac{\int_{\Theta_0} f(x|\theta)\pi(\theta|H_0)\,\mathrm{d}\theta}{\int_{\Theta_1} f(x|\theta)\pi(\theta|H_1)\,\mathrm{d}\theta}, \tag{2}$$

where $\Theta_1 = \Theta_0^c$ is the set specified by the alternative H_1, and $\pi(\theta|H_i)$ is the p.d.f. of θ given H_i.

Each of these has some problems. The measure λ tends to be too small, indicating a bias against H_0, vide Section 5. The second measure is a ratio of easily interpretable probabilities (unlike P-values) but requires a careful specification of the priors $\pi(\theta|H_i)$. We consider this aspect in more detail below. Note that the posterior probability of H_0 given x is

$$P(H_0|x) = \left[1 + \frac{(1-\pi_0)}{\pi_0}\frac{1}{BF_{01}}\right]^{-1}, \tag{3}$$

where π_0 is the specified prior probability of H_0. One may choose the prior probabilities of H_0 and H_1 objectively as equal. However, as noted earlier, the choice of $\pi(\theta|H_i)$ is not easy. To reduce the dependence of BF_{01} or $P(H_0|x)$ on this choice, there has been a suggestion to report lower bounds to both BF_{01} and the posterior probabilities over various plausible classes of priors. Reporting lower bounds introduces an element of objectivity. Moreover, the lower bound over a class will be closer to the P-value than $P(H_0|x)$ for any particular prior in the class. In the case of sharp null H_0: $\theta = \theta_0$, the conditional distribution of θ given H_0 is degenerate at θ_0, while the class of symmetric unimodal p.d.f.'s is the most attractive choice for $\pi(\theta|H_1)$. The same class of symmetric unimodal priors may work well in many examples with one-sided null, e.g., in the location parameter families considered by Casella and Berger (1987). One may also use noninformative priors in problems with one-sided alternatives.

A comparison of P-values and posterior probabilities or Bayes factors is the subject of our discussion in Sections 2–6. The measure in (1) is also considered implicitly in Section 2 and explicitly in Section 5.3. In Section 2 we review the work of Edwards et al. (1963), Berger and Delampady (1987), Berger and Sellke (1987) and others comparing P-values with the Bayes factors and posterior probabilities for a sharp null hypothesis. The comparison in this section indicates that the P-values for a sharp null overstate the case against the null. A similar phenomenon has been observed also in the context of sequential testing (vide Kundu, 2001). Sellke et al. (2001) have suggested an interesting way of calibrating a P-value upwards to rectify this bias. This is presented in Section 3. The comparisons reviewed in Section 2 are of special importance because it is well-known that P-values are wrongly interpreted by users as posterior probabilities. The calibration, vide Section 3, is important for the same reason. Closely related to the conflict between P-values and Bayes factors is the so-called Jeffreys–Lindley paradox. This is presented in Section 4. In Section 5 we present some asymptotic results comparing the P-value with likelihood ratio and Bayes factor and argue that a part of the conflict between them arises from different choices of the framework for asymptotics. On the whole, in the presence of well-defined alternatives, P-values come out as inadequate measures of evidence against a sharp null hypothesis. We may add here that the asymptotic results in Section 5 appear to be new.

The aforementioned conflict does not arise if the null hypothesis is one-sided, vide Casella and Berger (1987). Their comparison between P-values and posterior probabilities for the one-sided case is presented in Section 6.

It is clear that a P-value is a natural measure of evidence in significance tests where one has a precisely stated null hypothesis but no well-defined alternatives, vide Cox and Hinkley (1974, Chapter 3). Since the usual Bayesian measures of evidence are not available in this case, there has been some recent Bayesian work on what are called Bayesian P-values. A review of these contributions appears in Section 7.

A classical P-value (p) may be interpreted as providing a ranking of values of the test statistic according to the sampling distribution of the test statistic under H_0 in the same sense as a score in a test of a candidate places it among the top $100p\%$. It ranks a given observed deviation among all possible observations with respect to deviation from H_0, but the ranking is based only on the distribution under H_0. It is not a probabilistic measure of evidence against H_0 relative to H_1 even though it is liable to be misinterpreted that way. The calibrated P-values, lower bounds to posterior probabilities, Bayesian P-values, etc. seem to provide promising alternative objective measures of evidence.

2. Conflict between P-values and lower bounds to Bayes factors and posterior probabilities: Case of a sharp null

Consider the problem of testing

$$H_0\colon \theta = \theta_0 \quad \text{versus} \quad H_1\colon \theta \neq \theta_0$$

on the basis of a random observable X having a density $f(x|\theta), \theta \in \Theta$. Let $T(X)$ be a relevant statistic, large values of which indicate deviation from H_0. The P-value for observed data x, denoted p, is given by

$$p = P_{\theta_0}\bigl(T(X) \geqslant T(x)\bigr). \tag{4}$$

In the following we denote the prior $\pi(\theta|H_1)$ by $g(\theta)$ as in Berger and Delampady (1987) and Berger and Sellke (1987). For a continuous prior density $g(\theta)$ on $\{\theta \neq \theta_0\}$, the Bayes factor of H_0 to H_1 is

$$BF_{01} = \frac{f(x|\theta_0)}{\int f(x|\theta)g(\theta)\,\mathrm{d}\theta}, \tag{5}$$

and, for a specified prior probability π_0 of H_0, the posterior probability of H_0 is given by (3) above.

By way of illustration we consider the problem of testing a normal mean. Suppose that $X = (X_1, \ldots, X_n)$ is observed, where the X_i's are i.i.d. $N(\theta, \sigma^2)$, σ^2 being known. The usual test statistic is

$$T(X) = \sqrt{n}|\overline{X} - \theta_0|/\sigma,$$

where $\overline{X}$ is the sample mean. For observed data x, with $T(x) = t$, the P-value is given by

$$p = 2\big[1 - \Phi(t)\big], \tag{6}$$

where Φ is the cdf of the standard normal distribution.

Assume now that the prior p.d.f. g for θ, conditional on H_1 is given by the $N(\mu, \tau^2)$ density. It follows that the Bayes factor is given by

$$BF_{01} = \sqrt{1 + \rho^{-2}}\exp\left(-\frac{1}{2}\left[\frac{(t - \rho\eta)^2}{(1 + \rho^2)} - \eta^2\right]\right), \tag{7}$$

where $\rho := \sigma/(\sqrt{n}\tau)$ and $\eta := (\theta_0 - \mu)/\tau$.

We consider the special case with $\mu = \theta_0$ (ensuring symmetry of g about θ_0), $\tau = \sigma$, $\pi_0 = 1/2$. The corresponding prior is close to the one recommended by Jeffreys who suggested taking a Cauchy form of the prior under H_1. Berger and Delampady (1987) and Berger and Sellke (1987) have calculated (6) and (7) and $P(H_0|x)$ given by (3) for the above choice of (μ, τ, π_0) and for several choices of the observed value t of T and sample size n. This enables one to see the conflict between P-value and the Bayesian measures of evidence without much difficulty. In particular, it is demonstrated that $P(H_0|x)$ is 5 to 50 times larger than the P-value. For example, if $n = 50$ and $t = 1.960$, then $p = 0.05$, $BF_{01} = 1.08$ and $P(H_0|x) = 0.52$. Thus, in a situation like this, a classical statistician will reject H_0 at significance level 0.05, whereas a Bayesian will conclude that H_0 has probability 0.52 of being true. In fact, following the recommendation of Jeffreys (1961, p. 432), a Bayesian will conclude that the null hypothesis is supported by the data but the evidence is not strong (see also Kass and Raftery, 1995). This simple calculation in an important problem reveals practical disagreement between classical and Bayesian statistics, with special reference to testing of point null hypothesis – a conflict pointed out independently by Lindley (1957), Jeffreys (1961), and Edwards et al. (1963). The latter authors initiated calculations of this kind, with further substantial refinements, extensions and clarifications made by Berger and Sellke (1987), Berger and Delampady (1987), and Dickey (1977). In particular, it appears that the conflict is not so much between the Bayesian and frequentist paradigms as between Bayesian measures of evidence and a popular classical measure of evidence. This point is further argued in Sections 3 and 5.

Edwards et al. (1963), Berger and Delampady (1987) and Berger and Sellke (1987) investigate to what extent the above conflict between the P-value and the Bayesian measures BF_{01} and $P(H_0|x)$ depends on the choice of the prior g. Below we present results from Berger and Sellke (1987) who calculate

$$\underline{BF}_{01}(G) \stackrel{\text{def}}{=} \inf_{g\in G} BF_{01} \tag{8}$$

and

$$\underline{P}(H_0|x, G) \stackrel{\text{def}}{=} \inf_{g\in G} P(H_0|x) = \left[1 + \frac{1-\pi_0}{\pi_0} \times \frac{1}{\underline{BF}_{01}(G)}\right]^{-1} \tag{9}$$

for suitable classes G of prior densities.

Interesting choices for G include

$$\begin{aligned} G_A &= \{\text{all densities}\},\\ G_S &= \{\text{densities symmetric about } \theta_0\},\\ G_{US} &= \{\text{unimodal, symmetric densities with mode at } \theta_0\},\\ G_{NOR} &= \{\text{normal (conjugate) densities with mean } \theta_0\}. \end{aligned} \tag{10}$$

However, G should not be too large or too small. One should allow all reasonable g in G but a too large G may lead to a "uselessly small" bound. The choice $G = G_A$ seems to be highly biased toward H_1. The class G_{US} is recommended as the most sensible choice for G. In what follows we record some of the relevant findings of Berger and Sellke (1987).

• Let $G = G_A$. In this case, $\underline{BF}_{01}(G_A)$ is equal to the likelihood ratio test statistic λ. Moreover,

$$t > 1.68,\ \pi_0 = 1/2 \Longrightarrow \underline{P}(H_0|x, G_A)/pt > \sqrt{\pi/2} \cong 1.253, \tag{11}$$

and

$$\lim_{t\to\infty} \underline{P}(H_0|x, G_A)/pt = \sqrt{\pi/2}. \tag{12}$$

Thus, for large t, $P(H_0|x)$ is at least $1.25pt$, for *any* prior. In particular, when $\pi_0 = 1/2$ and $t = 1.96$,

$$p = 0.05, \qquad \underline{P}(H_0|x, G_A) = 0.128. \tag{13}$$

• Suppose now that $G = G_S$. Then

$$t > 2.28,\ \pi_0 = 1/2 \Longrightarrow \underline{P}(H_0|x, G_S)/pt > \sqrt{2\pi} \cong 2.507, \tag{14}$$

and

$$\lim_{t\to\infty} \underline{P}(H_0|x, G_S)/pt = \sqrt{2\pi}. \tag{15}$$

In particular, when $\pi_0 = 1/2$ and $t = 1.96$,

$$p = 0.05, \qquad \underline{P}(H_0|x, G_S) = 0.227. \tag{16}$$

• Consider now the case $G = G_{US}$. Here

$$t > 0,\ \pi_0 = 1/2 \Longrightarrow \underline{P}(H_0|x, G_{US})/pt^2 > 1 \tag{17}$$

and

$$\lim_{t\to\infty} \underline{P}(H_0|x, G_{US})/pt^2 = 1. \tag{18}$$

In particular, when $\pi_0 = 1/2$ and $t = 1.96$,

$$p = 0.05, \qquad \underline{P}(H_0|x, G_{US}) = 0.290. \tag{19}$$

• Consider finally the case $G = G_{NOR}$ (Edwards et al., 1963). When $\pi_0 = 1/2$ and $t = 1.96$,

$$p = 0.05, \qquad \underline{P}(H_0|x, G_{NOR}) = 0.321. \tag{20}$$

The facts stated in (11)–(20) demonstrate convincingly the conflict between P-value and its Bayesian counterparts, in the context of testing a point null hypothesis.

Berger and Delampady (1987) consider the problem of testing a k-variate normal mean, where the covariance is a multiple of the identity matrix, and calculate, for every dimension k, the lower bounds $\underline{BF}_{01}(G)$ and $\underline{P}(H_0|x, G)$ with $G = G_{US}$. They also address similar questions for a nonsymmetric and discrete situation like binomial testing. All these investigations yield results similar to those in the univariate normal testing problem. In particular, for the multivariate normal problem, when $k = 2$ and $p = 0.05$, $\underline{P}(H_0|x, G_{US})$ is obtained as 0.2582. For a fixed p, as k increases, $\underline{P}$ decreases at a slow rate but decreases to a value much larger than p. Corresponding to $p = 0.05$, $\underline{P}$ converges to 0.2180 as $k \to \infty$.

One possible reaction to the above findings is to dismiss point null hypotheses altogether on the ground that they are unrealistic. To respond to this argument and also to answer when one can model a hypothesis testing problem as a test of a point null hypothesis, the above authors have explored when one can approximate an interval null hypothesis by a point null. More precisely, they have tried to answer when the problem of testing

$$H_0\colon |\theta - \theta_0| \leqslant \varepsilon \quad \text{against} \quad H_1\colon |\theta - \theta_0| > \varepsilon \tag{21}$$

can be approximated by that of testing

$$H_0^*\colon \theta = \theta_0 \quad \text{against} \quad H_1^*\colon \theta \neq \theta_0. \tag{22}$$

To answer this question within the framework of classical (non-Bayesian) statistics, the authors have introduced

$$\alpha_\varepsilon \stackrel{\text{def}}{=} \sup_{|\theta-\theta_0|\leqslant\varepsilon} P_\theta\big(T(X) \geqslant t\big) \tag{23}$$

and in the special case of normal testing problem have, moreover, investigated for what values of ε, α_0 is close to α_ε. They have tabulated upper bounds to $\varepsilon^* := \varepsilon\sqrt{n}/\sigma$ that achieve this. In particular, they have shown that for a moderate value of t, "as long as ε is no more than 1/4 to 1/6 of a sample standard deviation, the use of a point null will cause at most 10% error in the calculated P-value". The authors have addressed this issue also within the Bayesian framework. In this situation, one considers a prior density $\pi(\theta)$ which is continuous but sharply spiked near θ_0. Let

$$\Omega = \{\theta\colon |\theta - \theta_0| \leqslant \varepsilon\}, \qquad \overline{\Omega} = \{\theta\colon |\theta - \theta_0| > \varepsilon\},$$

and

$$\pi_0 = \int_{\Omega} \pi(\theta)\,\mathrm{d}\theta, \qquad g_0(\theta) = \frac{1}{\pi_0}\pi(\theta) I_{\Omega}(\theta),$$

$$g_1(\theta) = \frac{1}{1-\pi_0}\pi(\theta) I_{\overline{\Omega}}(\theta). \tag{24}$$

Here, π_0 is the prior probability assigned to H_0, g_0 is the conditional density of θ given that H_0 is true, and g_1 is the conditional density of θ given that H_1 is true. With all these, the authors have considered the normal testing problem. In this case, one observes $\overline{X} \sim \mathrm{N}(\theta, \sigma^2/n)$ and letting $f(\bar{x}|\theta)$ denote its density, computes the exact Bayes factor for testing H_0 against H_1 in (21) by

$$B \stackrel{\text{def}}{=} \frac{\int_{\Omega} f(\bar{x}|\theta) g_0(\theta)\,\mathrm{d}\theta}{\int_{\overline{\Omega}} f(\bar{x}|\theta) g_1(\theta)\,\mathrm{d}\theta}. \tag{25}$$

On the other hand, the Bayes factor for testing H_0^* against H_1^* is given by

$$\widehat{B} \stackrel{\text{def}}{=} \frac{f(\bar{x}|\theta_0)}{\int f(\bar{x}|\theta) g(\theta)\,\mathrm{d}\theta}, \tag{26}$$

where $g(\theta)$ is the conditional prior density of θ given that H_1^* is true and satisfies

$$\begin{aligned} &\text{(i)} \quad g(\theta) \propto g_1(\theta) \quad \text{on } \overline{\Omega}, \\ &\text{(ii)} \quad \int_{\Omega} g(\theta)\,\mathrm{d}\theta \quad \text{is suitably small.} \end{aligned} \tag{27}$$

Finally, the authors have obtained a result to answer when $\widehat{B}$ provides a good approximation to B. In particular, they have shown that when g is taken to be the p.d.f. of $\mathrm{N}(\theta_0, \sigma^2)$, and $\varepsilon = \sigma/(4\sqrt{n})$, $n \geqslant 4$, the error in approximating B by $\widehat{B}$ is no more than 10%. This result improves one due to Dickey (1976).

We conclude this section by discussing briefly another interesting measure, proposed by Efron and Gous (2001). They call it a 'frequentist Bayes factor' and define it as

$$\frac{\lambda^{-1}(x)}{\lambda^{-1}(y)},$$

where λ is the likelihood ratio test statistic as defined in (1) and y is a threshold. One would reject H_0 if this is large. Efron and Gous (2001) define what they call Fisher's scale of evidence and choose their y as the point such that $\lambda^{-1}(y)$ equals the 90th percentile of the likelihood ratio test statistic $\lambda(X)$ under H_0. They show that at least for $k = 1$, there is a prior, given by $U(-4.85, 4.85)$, such that the Bayes factor corresponding to this prior is very close to the new measure (see Figure 2 of Efron and Gous, 2001).

For higher dimensions, it appears that one would have to use percentiles of order higher than 90% (see Figure 3 of Efron and Gous, 2001). In the following α_0 refers to the area to the left of the chosen percentile. Our calculation for $k = 1$ indicates α_0 need not be fixed at 0.90 but can be varied considerably without affecting very much the agreement between the Bayesian and frequentist Bayes factors. However, the corresponding priors are quite different. For example, for $\alpha_0 = 0.99$, the prior for the matching Bayes factor is $U(-34.58, 34.58)$.

In addition to the above, there is an interesting discussion in Efron and Gous (2001) on the different scales of evidence of Fisher and Jeffreys which has some points in common with the main conclusions of Section 5 of this chapter.

3. Calibration of P-values

The discussion in Section 2 indicates that the P-value for a sharp null hypothesis overstates evidence against the null; the actual evidence against the null, as measured by the posterior probability of the null, can differ substantially from the P-value. Sellke et al. (2001) have suggested an interesting way of calibrating the P-value upwards to rectify this bias against the null. This simple calibration of the P-value p is given by

$$B(p) = -\mathrm{e}p\log(p), \tag{28}$$

when $p < 1/\mathrm{e}$, and this is interpreted as a lower bound to the Bayes factor BF_{01}. We briefly indicate how the calibration of p given in (28) is developed. We consider the case with a simple null hypothesis H_0. Under H_0, the distribution of the P-value, $p = p(X)$, is uniform over (0, 1). If the alternative H_1 is expressed in terms of the distribution of p, the problem reduces to that of testing

$$H_0\colon\ p \sim U(0, 1) \quad \text{versus} \quad H_1\colon\ p \sim f(p|\xi), \tag{29}$$

where $f(p|\xi)$ is the density of p under H_1 and ξ is an unknown parameter. Sellke et al. (2001) consider the class of Beta $(\xi, 1)$ distributions with $0 < \xi \leqslant 1$ so that $f(p|\xi) = \xi p^{\xi-1}$ and H_0 corresponds to $\xi = 1$, and obtain (28) as infimum of Bayes factors BF_{01} over all priors $\pi(\xi)$.

Sellke et al. (2001) show that (28) can be interpreted as a lower bound to the Bayes factor BF_{01} in a more general set up. They also show that in the problem of testing a normal mean there is close agreement between the proposed calibration (28) and the lower bound to BF_{01} for the class G_{US} defined in Section 2.

Another calibration of p in terms of a frequentist error probability α (in rejecting H_0) proposed by Sellke et al. (2001) is

$$\alpha(P) = \left[1 + \left(-\mathrm{e}p\log(p)\right)^{-1}\right]^{-1}. \tag{30}$$

Another way of avoiding a conflict between P-values and Bayes factors is to take a conditional frequentist point of view and use conditional frequentist measures of evidence instead of P-values, vide Berger et al. (1994) and Berger (2003).

4. Jeffreys–Lindley paradox

If instead of taking a lower bound as in the previous sections, we take a fixed prior $\pi(\theta | H_1)$ but let n go to ∞, then the conflict between P-values and posterior probabilities is further enhanced. Historically this phenomenon was noted earlier than the conflict with the lower bound, vide Jeffreys (1961) and Lindley (1957).

We consider the set-up of Section 2 and the problem of testing H_0: $\theta = \theta_0$ versus H_1: $\theta \neq \theta_0$, where θ is a normal mean. Let g be a uniform prior density over some interval $(\theta_0 - a, \theta_0 + a)$ containing θ_0. The posterior probability of H_0 given $x = (x_1, \ldots, x_n)$ is

$$P(H_0|x) = \pi_0 \exp\bigl[-n(\bar{x} - \theta_0)^2/\bigl(2\sigma^2\bigr)\bigr]/K, \tag{31}$$

where $K = \pi_0 \exp[-n(\bar{x} - \theta_0)^2/(2\sigma^2)] + \frac{1-\pi_0}{2a}\int_{\theta_0-a}^{\theta_0+a} \exp[-n(\bar{x} - \theta)^2/(2\sigma^2)]\,\mathrm{d}\theta$.

Suppose x is such that $\bar{x} = \theta_0 + z_\alpha \sigma/\sqrt{n}$ where z_α is the upper α-point of N(0, 1). Then $\bar{x}$ is significant at level α. Since $\bar{x} - \theta_0$ tends to zero as n increases, $\bar{x}$ is well within $(\theta_0 - a, \theta_0 + a)$ for sufficiently large n. We then have

$$\int_{\theta_0-a}^{\theta_0+a} \exp\bigl[-n(\bar{x} - \theta)^2/\bigl(2\sigma^2\bigr)\bigr]\,\mathrm{d}\theta \cong \sigma\sqrt{(2\pi/n)}$$

and hence

$$P(H_0|x) \cong \pi_0\,\mathrm{e}^{-\frac{1}{2}z_\alpha^2} \Big/ \left[\pi_0\,\mathrm{e}^{-\frac{1}{2}z_\alpha^2} + \frac{(1-\pi_0)}{2a}\sigma\sqrt{(2\pi/n)}\right]. \tag{32}$$

Thus $P(H_0|x) \to 1$ as $n \to \infty$ whereas the P-value is equal to α for all n.

As pointed out earlier, P-values cannot be compared across sample sizes or across experiments. Lindley (1957) remarks that the posterior probability "varies strikingly with n for fixed significance level, in an extreme case producing a result in direct conflict with the significance level, the degree of conviction is not even approximately the same in two situations with equal significance levels. 5% in today's small sample does not mean the same as 5% in tomorrow's large one". In fact even a frequentist tends to agree that the conventional value of the significance level α like $\alpha = 0.05$ or 0.01 are too large for large sample sizes. This point is further discussed in Section 5.

5. Role of the choice of an asymptotic framework

5.1. General observations

Suppose $X_1, \ldots, X_n$ are i.i.d. $N(\theta, \sigma^2)$, σ^2 known. The Jeffreys–Lindley paradox shows that for inference about θ, P-values and Bayes factors may provide contradictory evidence and hence can lead to opposite decisions. Once again, as in Section 2, the evidence against H_0 contained in P-values seems unrealistically high. We argue in this section that part of this conflict arises from the fact that different types of asymptotics are being used for the Bayes factors and the P-values. We begin with a quick review of the two relevant asymptotic frameworks in classical statistics for testing a sharp null hypothesis.

The standard asymptotics of classical statistics is based on what are called Pitman alternatives, namely, $\theta_n = \theta_0 + d/\sqrt{n}$ at $1/\sqrt{n}$-distance from the null. The Pitman alternatives are also called contiguous in the very general asymptotic theory developed by Le Cam (vide Roussas, 1972; Le Cam and Yang, 2000; Hájek and Sidák, 1967). The log-likelihood ratio of a contiguous alternative with respect to the null is stochastically bounded as $n \to \infty$. On the other hand, for a fixed alternative, the log-likelihood ratio tends to $-\infty$ (under the null) or ∞ (under the fixed alternative). If the probability of Type I error is $0 < \alpha < 1$, then the behavior of the likelihood ratio has the following implication. The probability of Type II error will converge to $0 < \beta < 1$ under a contiguous alternative θ_n and to zero if θ is a fixed alternative. This means the fixed alternatives are relatively easy to detect. So in this framework it is assumed that the alternatives of importance are the contiguous alternatives. Let us call this theory Pitman type asymptotics.

There are several other frameworks in classical statistics of which Bahadur's (Bahadur, 1971; Serfling, 1980, pp. 332–341) has been studied most. We focus on Bahadur's approach but the one due to Rubin and Sethuraman (1965), based on moderate deviations, seems also relevant. In Bahadur's theory, the alternatives of importance are fixed and do not depend on n. Given a test statistic, Bahadur evaluates its performance at a fixed alternative by the limit (in probability or a.s.) of $\frac{1}{n}(\log \text{P-value})$ when the alternative is true. Bahadur's exact slope is negative of twice this limit. An equivalent description of the limiting value is to fix the probability of Type II error at the fixed alternative and calculate the limit of $\frac{1}{n}$ log(probability of Type I error). This limit and the above limit of $\frac{1}{n}(\log \text{P-value})$ coincide. Thus the limit measures how significant is the test, on an average, under the alternative. It also measures how fast the probability of Type I error tends to zero. It may seem a little odd that the probability of Type II error is fixed rather than that of Type I error. But if the probability of Type I error is fixed and the other error at a fixed alternative is studied, one runs into a somewhat anomalous situation where the probability of Type II error gets much smaller than the probability of Type I error even though Type I error is supposed to be more important. Thus fixing α (at any conventional value) and letting $n \to \infty$ violates the usual convention of treating Type I error as the more important of the two. One may also ask what value of α should one choose. Unfortunately, Bahadur's theory does not answer this question but it does suggest α should tend to zero very fast and not be fixed at the conventional values.

Which of these two asymptotics is appropriate in a given situation should depend on how the sample size is chosen. If the sample size is chosen, as is often the case, such that the probability of Type I error is α and probability of Type II error at a given alternative θ is β, where $0 < \alpha, \beta < 1$ are preassigned, then it is easy to verify that $|\theta - \theta_0| \sim 1/\sqrt{n}$, i.e., θ is essentially a contiguous alternative. On the other hand, if there is a fixed alternative θ of importance where protection against Type II error is desired, and n is chosen to be very large, much larger than the sample size in the previous case, θ is not a contiguous alternative. In this framework, fixing α at a conventional value of 0.05 or 0.01 would clearly lead to a β smaller than α. In this kind of problem, the limiting P-value of Bahadur is a good measure of performance.

5.2. *Comparison of decisions via P-values and Bayes factors in Bahadur's asymptotics*

In this and the next subsection, essentially Bahadur's approach is followed both for P-values and Bayes factors. In Subsection 5.4, a Pitman type asymptotics is used for both. We first show that if the P-value is sufficiently small, as small as it is typically in Bahadur's theory, BF_{01} will tend to zero, calling for rejection of H_0, i.e., the evidence in the P-value points in the same direction as that in the Bayes factor or posterior probability, removing the sense of paradox in the result of Jeffreys and Lindley. One could, therefore, argue that the P-values or the significance level α assumed in Section 4 are not small enough. The asymptotic framework chosen is not appropriate when contiguous alternatives are not singled out as alternatives of importance.

We now verify the claim about limit of BF_{01}. Without loss of generality, take $\theta_0 = 0$, $\sigma^2 = 1$. First note that

$$\log BF_{01} = -\frac{n}{2}\overline{X}^2 + \frac{1}{2}\log n + R_n, \tag{33}$$

where

$$R_n = -\log \pi\big(\overline{X}|H_1\big) - \frac{1}{2}\log(2\pi) + \mathrm{o}(1)$$

provided $\pi(\theta|H_1)$ is a continuous function of θ and is positive at all θ. If we omit R_n from the right-hand side of (33), we have Schwarz's (1978) approximation to the Bayes factor via BIC.

The logarithm of P-value corresponding to observed $\overline{X}$ is

$$\log p = \log 2\big[1 - \Phi\big(\sqrt{n}|\overline{X}|\big)\big] = -\frac{n}{2}\overline{X}^2\big(1 + \mathrm{o}(1)\big) \tag{34}$$

by standard approximation to a normal tail (vide Feller, 1973, p. 175 or Bahadur, 1971, p. 1). Thus $\frac{1}{n}\log p \to -\theta^2/2$ and by (33), $\log BF_{01} \to -\infty$. This result is true as long as $|\overline{X}| > c(\log n/n)^{1/2}$, $c > \sqrt{2}$. Such deviations are called moderate deviations, vide Rubin and Sethuraman (1965). Of course, even for such P-values, $p \sim (BF_{01}/n)$ so that P-values are smaller by an order of magnitude. The conflict in measuring evidence remains but the decisions are the same.

In the next subsection we pursue the comparison of the three measures of evidence based on the likelihood ratio λ, the P-value based on the likelihood ratio test and the Bayes factor BF_{01} under general regularity conditions.

5.3. *Comparison of P-value with likelihood ratio and Bayes factor in Bahadur's asymptotics*

In this subsection, we consider i.i.d. observations with a density satisfying certain regularity conditions (vide Woodroofe, 1978).

In Bahadur's theory, the smaller the limit of the P-values (in an appropriate scale as mentioned above), the better the test statistic. Bahadur and Raghavachari (1972) showed the likelihood ratio test is optimal under regularity conditions, i.e., smallest limits of P-values are obtained when likelihood ratio test statistics are used. So we focus on such tests. We need the following large deviation result for likelihood ratio test statistic (vide

Woodroofe, 1978) which is valid under appropriate regularity conditions:

$$P_{H_0}(-\log\lambda \geqslant n\varepsilon) \sim (n\varepsilon)^{k/2-1}\,\mathrm{e}^{-n\varepsilon}/\,\Gamma(k/2) \tag{35}$$

for $\varepsilon = \varepsilon_n \to 0$ with $n\varepsilon_n \to \infty$ as $n \to \infty$ where k is the difference between numbers of parameters under H_0 and H_1. To understand such results heuristically, notice that $-2\log\lambda$ is asymptotically a χ^2 with k degrees of freedom. By a simple integration by parts (as for the normal tail),

$$P\left(\frac{1}{2}\chi_k^2 > n\varepsilon\right) \sim (n\varepsilon)^{k/2-1}\,\mathrm{e}^{-n\varepsilon}/\,\Gamma(k/2).$$

Eq. (35) can be interpreted as

$$p = \exp(\log\lambda) n^{k/2-1} \mathrm{O}(1)$$

if $(\log\lambda)/n$ is at most $\mathrm{O}(1)$. So

$$\log p = \log\lambda + \left(\frac{k}{2} - 1\right)\log n + \mathrm{O}(1).$$

Using Schwarz's approximation by BIC (Schwarz, 1978; Ghosh and Ramamoorthi, 2003, Chapter 1), the Bayes factor can be expressed as

$$\log BF_{01} = \log\lambda + \frac{k}{2}\log n + \mathrm{O}(1). \tag{36}$$

From (35) and (36), it follows that when the value of the test statistic $-2\log\lambda$ is large, of the three measures of evidence against H_0, λ is smaller (larger) than the P-value for $k > 2$ ($k < 2$) and the P-value in turn is smaller than BF_{01} since λ can be at most $\mathrm{O}(n)$ whether the null or any (fixed) alternative is true. Incidentally, while the alternative hypothesis has a bearing on both λ and BF_{01}, λ is obtained by maximizing with respect to θ whereas BF_{01} is obtained by integration.

We explore below if any simple reasonable adjustment of λ or p can reduce the conflict with BF_{01}. The mathematical similarity of these three quantities is quite striking specially because they are conceptually so different.

One reason the P-value tends to be small is that as a tail probability it depends on deviations from H_0 which are larger than those observed. It is tempting to think this can be corrected by using instead the density of what is observed. Since $-2\log\lambda$ is a χ^2 variable with k degrees of freedom (approximately – under regularity conditions, exactly – under normality assumption), the density agrees with the P-value up to $\mathrm{O}(1)$, so nothing is gained. The density may be unsatisfactory since, unlike the P-value, it is not invariant under smooth one-one transformations.

A way out is to calculate the probability that $-\log\lambda$ lies between $n\varepsilon$ and $n(\varepsilon + d\varepsilon)$. This would agree with BF_{01} up to $\mathrm{O}(1)$. A more fruitful approach is to replace $\log\lambda$ by the penalized log likelihood ratio, namely, Schwarz's BIC difference (ΔBIC), given by

$$\Delta\mathrm{BIC} = \log\lambda + \frac{k}{2}\log n.$$

As pointed out by Schwarz (1978), ΔBIC agrees with $\log BF_{01}$ up to $\mathrm{O}(1)$ but does not depend on the prior. For the testing problem considered in detail in Section 2, ΔBIC tends to be bigger than logarithm of the calibrated P-value for n not too small, say, $n > 20$. The calibrated P-value is a sort of nonparametric lower bound to a Bayes Factor, while ΔBIC may be bigger or smaller than logarithm of a Bayes Factor depending on the prior, the difference being negligible and of order o(1), if the Jeffreys prior is used as the conditional prior given the alternative hypothesis.

5.4. Pitman alternative and rescaled priors

We consider once again the problem of testing H_0: $\theta = 0$ versus H_1: $\theta \neq 0$ on the basis of a random sample from $\mathrm{N}(\theta, 1)$. Suppose that the Pitman alternatives are the most important ones and the prior $\pi(\theta|H_1)$ puts most of the mass on Pitman alternatives. One such prior is $\mathrm{N}(0, \delta/n)$. Then

$$BF_{01} = \sqrt{\delta + 1}\exp\left[-\frac{n}{2}\left(\frac{\delta}{\delta + 1}\right)\overline{X}^2\right].$$

If the P-value is close to zero, $\sqrt{n}|\overline{X}|$ is large and therefore, BF_{01} is also close to zero, i.e., for these priors there is no paradox. All three measures are of the same order but the result of Berger and Sellke (1987) for symmetric unimodal priors still implies that P-value is smaller than the Bayes factor. In particular, the comments of Sellke et al. (2001, Section 2) on the weak evidential value of $P = 0.05$ are still true.

The points made above are quite different from those in Andrews (1994). Andrews (1994) essentially shows that for priors of this kind and for fairly general parametric families satisfying usual regularity conditions, the Bayesian criterion is approximately a monotone function (with monotonicity in the right direction) of the classical test statistics. In the present case, it is clear from the above expression for BF_{01} that the monotonicity is exact. Such monotonicity does not establish that the scales of evidence corresponding to standard P-values suggested by Fisher and those suggested by Jeffreys are very similar as claimed in this subsection. It may be pointed out that exact monotonicity would hold for several standard cases.

6. One-sided null hypothesis

While the P-values and the posterior probabilities can greatly differ in testing a sharp null hypothesis H_0, in the one-sided testing problem they may be approximately and asymptotically equal. Indeed, Casella and Berger (1987) have shown that for many classes of priors the infimum of the posterior probability of H_0 is equal to the P-value and in some cases the infimum is even less than the P-value. The set up considered by Casella and Berger (1987) is the following. Suppose X having a density $f(x - \theta)$ is observed, the purpose being to test

$$H_0\colon \theta \leqslant 0 \quad \text{against} \quad H_1\colon \theta > 0.$$

It is assumed that

$$\begin{aligned} &\text{(i)} \quad f(\cdot) \text{ is symmetric around zero,} \\ &\text{(ii)} \quad f(x - \theta) \text{ has monotone likelihood ratio (MLR) in } X. \end{aligned} \tag{37}$$

It is known (vide Lehmann, 1986, p. 509) that when $f(x) > 0$ for all x, (37) is equivalent to $f(\cdot)$ being symmetric around zero, and $-\log f$ being convex. Thus, two possible choices for f are normal and logistic. Let the relevant test statistic be X. When the observed value of X is x, the P-value is given by

$$p(x) = P(X \geqslant x|\theta = 0) = \int_x^\infty f(t)\,\mathrm{d}t \tag{38}$$

and for a prior density $\pi(\theta)$ for θ, the posterior probability of H_0 is

$$P(H_0|X = x) = P(\theta \leqslant 0|X = x) = \frac{\int_{-\infty}^0 f(x-\theta)\pi(\theta)\,\mathrm{d}\theta}{\int_{-\infty}^\infty f(x-\theta)\pi(\theta)\,\mathrm{d}\theta}. \tag{39}$$

Casella and Berger (1987) have explored under what kind of circumstances (38) and (39) can be reconciled, i.e., (38) can, in some sense, be recognized as (39). The classes of prior distributions that the authors have considered are as follows:

$$\begin{aligned} &\Gamma_S = \{\pi\colon \pi \text{ is symmetric about zero}\},\\ &\Gamma_{US} = \{\pi\colon \pi \text{ is symmetric about zero and has unimodal p.d.f.}\},\\ &\Gamma_{NOR} = \{\pi\colon \pi \text{ is normal with mean zero and variance } \tau^2, \tau > 0\}. \end{aligned} \tag{40}$$

The authors have also considered the class

$$\Gamma_A = \bigl\{\pi\colon \pi\bigl((-\infty, 0]\bigr) = \pi\bigl((0,\infty)\bigr) = 1/2\bigr\} \tag{41}$$

and have addressed the same issue of reconciliation. The reason for considering these classes of priors is to ensure that the prior distribution gives equal weight to both the null and alternative hypotheses. These are called impartial prior distributions by the authors. Accordingly, the class Γ_A is the broadest possible class in this context.

For each class of distributions listed in (40) above, the authors have calculated infimum of $P(H_0|X = x)$ over the corresponding class. In particular, under the assumptions listed in (37), they have established the following:

$$\begin{aligned} &\inf_{\pi\in\Gamma_S} P(H_0|X = x) \leqslant p(x), \quad x > 0,\\ &\inf_{\pi\in\Gamma_{US}} P(H_0|X = x) = p(x), \quad x > 0,\\ &\inf_{\pi\in\Gamma_{NOR}} P(H_0|X = x) = p(x), \quad x > 0. \end{aligned} \tag{42}$$

For $x < 0$, $p(x) > 1/2$, and each of the infima above is exactly equal to $1/2$. Indeed, the information $x < 0$ will not provide any evidence against H_0 either to a Bayesian or to a frequentist. One needs also to say in this context that the last equality in (42) will hold if the class of normal distributions is replaced by any class of distributions that consists of scale transformations of a bounded, symmetric (about zero) and unimodal density. The result is as follows.

Consider the class of priors

$$\Gamma^\sigma(g) = \bigl\{\pi_\sigma\colon \pi_\sigma(\theta) = g(\theta/\sigma)/\sigma, \sigma > 0\bigr\},$$

where $g(\theta)$ is any bounded, symmetric, and unimodal prior density. For the hypothesis H_0: $\theta \leqslant 0$ against H_1: $\theta > 0$, if f is symmetric and has MLR and if $x > 0$, then

$$\inf_{\pi \in \Gamma^{\sigma}(g)} P(H_0|X = x) = p(x). \tag{43}$$

Finally, the authors have demonstrated through counter-examples that while sampling from a Cauchy location density, which is symmetric but does not have MLR, or from an exponential location density which has MLR but is asymmetric, one has the following:

$$\inf_{\pi \in \Gamma_{US}} P(H_0|X = x) = \inf_{\pi \in U_S} P(H_0|X = x) < p(x), \quad x > 0.$$

For the class in (41) above, the authors have shown that if the density f from which observations are being taken is arbitrary,

$$\inf_{\pi \in \Gamma_A} P(H_0|X = x) = 0.$$

Thus, the class Γ_A is too broad to be of any practical interest.

In the cases considered by Casella and Berger (1987), the P-value is an upper bound to the infimum of $P(H_0|X = x)$ and hence lies within or at the boundary of a set of Bayesian measures of evidence. In particular, they have shown that for the Cauchy (non-MLR) and exponential (asymmetric) sampling densities, the infimum is strictly less than $p(x)$ so that $p(x)$ is, in fact, equal to $P(H_0|X = x)$ for some prior (depending on x) in the class under consideration.

7. Bayesian P-values

Consider the problem of testing

$$H_0\colon X_1, \ldots, X_n \sim \text{i.i.d.} f(x|\theta),$$

$\theta \in \Theta$ being an unknown parameter. Let $X = (X_1, \ldots, X_n)$ and $T(X)$ be a statistic, large values of which indicate deviation from H_0. For an observed value x_{obs} of X, the classical P-value is defined as

$$p_{\text{sup}} = \sup_{\theta} P_\theta\big(T(X) \geqslant T(x_{\text{obs}})\big). \tag{44}$$

An alternative classical P-value mentioned in Bayarri and Berger (1999, 2000) is

$$p_{\text{plug}} = P_{\hat{\theta}_{\text{obs}}}\big(T(X) \geqslant T(x_{\text{obs}})\big), \tag{45}$$

where $\hat{\theta}_{\text{obs}}$ maximizes $f(x_{\text{obs}}|\theta)$. Robins et al. (2000) studied the conditional plug-in P-value p_{cplug}, which uses the maximizer $\hat{\theta}_{\text{cMLE.obs}}$ of the conditional likelihood $f(x_{\text{obs}}|T(x_{\text{obs}}); \theta)$ as a plug-in in place of $\hat{\theta}_{\text{obs}}$ in (45).

To fix ideas, we consider two simple examples: (i) $f(x|\theta) \equiv \text{N}(0, 1)$ and (ii) $f(x|\theta) = \text{N}(0, \sigma^2)$, i.e., $\theta = \sigma^2$ in the second example. In each example, we may take $T(x) = |\bar{x}|$.

Several people, notably Box (1980), Guttman (1967), Rubin (1984), Meng (1994), Gelman et al. (1996), and Bayarri and Berger (1999, 2000), have explored how a Bayesian may test for the goodness of fit to the model specified by H_0 in the absence of alternative H_1. In fact, the analysis of goodness of fit may even suggest a plausible alternative. See in this connection Bayarri and Berger (1999).

The natural analogue of the classical P-value is based on the prior predictive distribution

$$m(x) = \int \prod_{i=1}^{n} f(x_i|\theta)\pi(\theta)\,\mathrm{d}\theta, \tag{46}$$

where π is the prior density for θ under H_0. The prior predictive P-value (Box, 1980) is defined as

$$P_{\text{prior}} = P^m\big(T(X) \geqslant T(x)\big) \tag{47}$$

which gives the probability of observing data at least as deviant as that observed. If this is small, one would have doubts about goodness of fit. It is clear that $p_{\text{prior}} = p_{\text{sup}}$ if Θ is a singleton as in our example (i) above. In general $p_{\text{prior}} \leqslant p_{\text{sup}}$, i.e., in the context of testing goodness of fit, the Bayesian P-value always offers more evidence against H_0 than the classical P-value. In view of the fact, discussed in earlier sections, that P-values tend to overstate the case against H_0, it seems Bayesian P-values will do this too. Bayarri and Berger (1999) argue in favor of a cautious attitude to interpreting small Bayesian P-values automatically as implying conflict between data and null hypothesis.

The Bayesian P-value p_{prior} is usually highly influenced by the choice of the prior π. One may attempt to use noninformative priors but these priors being typically improper, yield improper m which are defined only up to an arbitrary multiplicative constant and hence cannot be used in computation of (47). This led many Bayesians, including Guttman (1967), Rubin (1984), Meng (1994) and Gelman et al. (1996) to suggest modifications by replacing π in m by the posterior $\pi(\theta|x_{\text{obs}})$:

$$m(x|x_{\text{obs}}) = \int f(x|\theta)\pi(\theta|x_{\text{obs}})\,\mathrm{d}\theta.$$

The resulting Bayesian P-value, called the posterior predictive P-value, is given by

$$p_{\text{post}} = P^{m(\cdot|x_{\text{obs}})}\big(T(X) \geqslant T(x_{\text{obs}})\big). \tag{48}$$

This rectifies, to a large extent, the difficulties mentioned above as $\pi(\theta|x_{\text{obs}})$ is typically proper even for an improper π and $m(\cdot|x_{\text{obs}})$ is influenced much more by the model than by the prior. However, this modified version of Bayesian P-value is not wholly satisfactory. There is a "double use" of data in (48) – the same data x_{obs} is used first to convert the prior $\pi(\theta)$ into the posterior $\pi(\theta|_{\text{obs}})$ and then again to compute the tail probability corresponding to $T(x_{\text{obs}})$.

A double use of data usually leads to overfitting, i.e., evidence in favor of H_0 even when a close look at the data indicates otherwise. There is an illuminating discussion of Example (ii) from this point of view in Bayarri and Berger (1999). A brief remark may be in order as to why this sort of phenomenon may be expected. In many cases,

$\hat{\theta}_{obs}$ would be close to the maximum likelihood estimate of θ based on maximizing the likelihood of $T(x_{obs})$. The latter will usually make the second classical P-value p_{plug} relatively large.

In an attempt to achieve the advantages of the prior predictive P-value and the posterior predictive P-value and eliminate the disadvantages, Bayarri and Berger (1999) introduce the conditional predictive P-value. To obtain this criterion, an appropriate statistic $U(X)$ is chosen and the conditional predictive distribution $m(t|u)$ is derived which is used to compute the P-value. The conditional predictive P-value, denoted p_{cpred}, is thus defined as

$$p_{cpred} = P^{m(\cdot|u_{obs})}\big(T(X) \geqslant T(x_{obs})\big), \tag{49}$$

where

$$\begin{aligned} m(t|u) &= \int f(t|u,\theta)\pi(\theta|u)\,\mathrm{d}\theta, \\ \pi(\theta|u) &= \frac{f(u|\theta)\pi(\theta)}{\int f(u|\theta)\pi(\theta)\,\mathrm{d}\theta}, \\ u_{obs} &= U(x_{obs}). \end{aligned}$$

Here $f(t|u,\theta)$ and $f(u|\theta)$ denote respectively the conditional and marginal densities of T and U under H_0 and $\pi(\theta|u)$ is assumed to be proper. Bayarri and Berger (1999) argue that, with an appropriate choice of U, this criterion is a satisfactory Bayesian measure of evidence against H_0; it does not involve a double use of data and also allows use of noninformative priors as long as $\pi(\theta|u)$ is proper. Bayarri and Berger (1999) illustrate this with Example (ii).

An attractive alternative way of avoiding double use of data is to use the conditional sampling distribution of X given $T(X)$ and θ, instead of $f(x|\theta)$ in calculating the posterior. This leads to the partial posterior predictive P-value

$$p_{ppost} = P^{m^*(\cdot)}\big(T(X) \geqslant T(x_{obs})\big), \tag{50}$$

where m^* and the partial posterior π^* are given by

$$\begin{aligned} m^*(t) &= \int f(t|\theta)\pi^*(\theta)\,\mathrm{d}\theta, \\ \pi^*(\theta) &\propto f\big(x_{obs}|T(x_{obs})\big)\pi(\theta) \propto \frac{f(x_{obs}|\theta)\pi(\theta)}{f_T(T(x_{obs})|\theta)}. \end{aligned}$$

EXAMPLE (ii). Assume that X_i's are i.i.d. $N(0,\sigma^2)$, σ^2 unknown, under H_0. We choose $T(X) = |\overline{X}|$, as a measure of departure from the model. Consider the usual noninformative prior for σ^2: $\pi(\sigma^2) = 1/\sigma^2$. We write $\theta = \sigma^2$. It is easy to verify that

$$\pi^*(\theta) \propto \theta^{-(n+1)/2}\exp\big(-ns^2_{obs}/2\theta\big), \quad \theta > 0, \tag{51}$$

where $ns^2 = \sum_{i=1}^n (X_i - \overline{X})^2$, and that

$$m^*(t) \propto \frac{1}{(t^2 + s^2_{obs})^{n/2}}, \quad t > 0. \tag{52}$$

It follows now from (51) and (52) that in this situation

$$p_{\text{ppost}} = 2\left[1 - F_{n-1}\left(\frac{\sqrt{n-1}t_{\text{obs}}}{s_{\text{obs}}}\right)\right],$$

where F_d is the distribution function of the t-distribution with d degrees of freedom. It should be mentioned here that this expression coincides with that of the conditional predictive P-value (vide Bayarri and Berger, 1999). On the other hand, the prior being improper, the prior predictive P-value cannot be computed for this example. However, the posterior predictive P-value equals (Bayarri and Berger, 1999)

$$p_{\text{post}} = 2\left[1 - F_n\left(\frac{\sqrt{n}t_{\text{obs}}}{s_{\text{obs}}^2 + t_{\text{obs}}^2}\right)\right].$$

Notice that when $t_{\text{obs}} \to \infty$, $p_{\text{post}} \to 2[1 - F_n(\sqrt{n})]$, a positive constant. Hence, p_{post} is unsatisfactory.

Of all the Bayesian P-values, the last two, namely, p_{cpred} and p_{ppost}, seem to be the most satisfactory. They are not sensitive to the choice of prior and avoid double use of the same data. Possibly because of this, they have the desirable property of being asymptotically uniformly distributed. None of the other P-values, including the frequentist P-values have this property, vide Robins et al. (2000). Indeed these authors observe – "in proposing p_{ppost} and p_{cpred}, both motivated through Bayesian arguments, Bayarri and Berger have solved the 'frequentist' math problem of finding a reference density $m(\cdot|x_{\text{obs}})$ such that the P-value $p = P^{m(\cdot|x_{\text{obs}})}(T(X) \geqslant T(x_{\text{obs}}))$, based on an arbitrary statistic $T(X)$ with a limiting normal distribution, is an asymptotic frequentist P-value". Further, it is to be noted that the frequentist measures p_{sup}, p_{plug} and p_{cplug} mentioned above, fail to achieve this. Presumably, because of computational difficulty the measures p_{cpred} and p_{ppost} are still not as popular as they should be. One of the computational difficulties is that the posterior π^* may not be proper. Another computational difficulty is the identification of a suitable conditioning statistic U. These points are discussed in detail in Bayarri and Berger (1999).

8. Concluding remarks

When testing a sharp null about a normal mean against usual (two-sided) alternative, P-values tend to overstate the evidence against H_0 in the sense of being smaller than a lower bound to the posterior probability for various plausible priors, like symmetric unimodal priors. A similar conflict between P-values and posterior probability (the Jeffreys–Lindley paradox) appears asymptotically under general regularity conditions. By choosing an appropriate asymptotic framework, the conflict can be reduced but not eliminated. Such conflicts do not seem to exist for one-sided alternatives.

If the null is composite and there are no alternatives, as in a problem of testing goodness of fit of a model, it is unclear how a Bayesian should measure evidence against the null. Various Bayesian P-values have been proposed of which two, namely, p_{cpred} and p_{ppost}, stand out as specially attractive. However, there remains the possibility that in some sense even Bayesian P-values may overstate the evidence against H_0. To some extent this issue is recognized and addressed well in the following remark of Bayarri and Berger (1999): "a model should really only be rejected if an alternative and *apriori believable* model can be found and shown to be superior to H_0 by a full Bayesian analysis. Note the requirement that the alternative model must be apriori believable".

References

Andrews, D. (1994). The large sample correspondence between classical hypothesis tests and Bayesian posterior odds tests. *Econometrica* **62**, 1207–1232.

Bahadur, R.R. (1971). *Some Limit Theorems in Statistics. Conference Board of the Mathematical Sciences Regional Conference Series in Applied Mathematics*, vol. 4. Society for Industrial and Applied Mathematics, Philadelphia, PA.

Bahadur, R.R., Raghavachari, M. (1972). Some asymptotic properties of likelihood ratios on general sample spaces. In: *Proceedings of the Sixth Berkeley Symposium on Mathematical Statistics and Probability (Univ. California, Berkeley, 1970/1971), vol. I. Theory of Statistics*. Univ. California Press, Berkeley, CA, pp. 129–152.

Bayarri, M.J., Berger, J.O. (1999). Quantifying surprise in the data and model verification. In: Bernardo, J.M., et al. (Eds.), *Bayesian Statistics 6*. Oxford University Press, Oxford, pp. 53–82.

Bayarri, M.J., Berger, J.O. (2000). P values in composite null models. *J. Amer. Statist. Assoc.* **95**, 1127–1142.

Berger, J.O. (2003). Could Fisher, Jeffreys, and Neyman have agreed on testing? With comments and a rejoinder by the author. *Statist. Sci.* **18**, 1–32.

Berger, J.O., Delampady, M. (1987). Testing precise hypotheses. With comments and a rejoinder by the authors. *Statist. Sci.* **2**, 317–352.

Berger, J.O., Sellke, T. (1987). Testing a point null hypothesis: The irreconcilability of *P* values and evidence. With comments and a rejoinder by the authors. *J. Amer. Statist. Assoc.* **82**, 112–122.

Berger, J.O., Brown, L.D., Wolpert, R.L. (1994). A unified conditional frequentist and Bayesian test for fixed and sequential simple hypothesis testing. *Ann. Statist.* **22**, 1787–1807.

Box, G.E.P. (1980). Sampling and Bayes inference in scientific modeling and robustness. *J. Roy. Statist. Soc., Ser. A* **143**, 383–430.

Casella, G., Berger, R.L. (1987). Reconciling Bayesian and frequentist evidence in the one-sided testing problem. With comments and a rejoinder by the authors. *J. Amer. Statist. Assoc.* **82**, 106–111.

Cox, D.R., Hinkley, D.V. (1974). *Theoretical Statistics*. Chapman and Hall, London.

Dickey, J.M. (1976). Approximate posterior distributions. *J. Amer. Statist. Assoc.* **71**, 680–689.

Dickey, J.M. (1977). Is the tail area useful as an approximate Bayes factor?. *J. Amer. Statist. Assoc.* **72**, 138–142.

Edwards, W., Lindman, H., Savage, L.J. (1963). Bayesian statistical inference for psychological research. *Psychological Review* **70**, 193–242. Reprinted in Kadane, J.B. (Ed.), *Robustness of Bayesian Analyses*. North-Holland, Amsterdam, 1984, pp. 1–62.

Efron, B., Gous, A. (2001). Scales of evidence for model selection: Fisher versus Jeffreys. In: Lahiri, P. (Ed.), *Model Selection*. In: *Lecture Notes – Monograph Series*, vol. 38. Institute of Mathematical Statistics, Beachwood, OH, pp. 208–256.

Feller, W. (1973). *Introduction to Probability Theory and Its Applications 1*, third ed. Wiley, New York.

Gelman, A., Meng, X., Stern, H. (1996). Posterior predictive assessment of model fitness via realized discrepancies. *Statistica Sinica* **6**, 733–807. (With discussion).

Ghosh, J.K., Ramamoorthi, R.V. (2003). Bayesian Nonparametrics. Springer-Verlag, New York.

Gibbons, J.D., Pratt, J.W. (1975). P-values: Interpretation and methodology. *Amer. Statist.* **29**, 20–25.

Guttman, I. (1967). The use of the concept of a future observation in goodness-of-fit problems. *J. Roy. Statist. Soc., Ser. B* **29**, 83–100.

Hajek, J., Sidak, Z.V. (1967). *Theory of Rank Tests*. Academic Press, New York.

Jeffreys, H. (1961). *Theory of Probability*, third ed. Oxford Univ. Press, London.

Kass, R.E., Raftery, A.E. (1995). Bayes factors. *J. Amer. Statist. Assoc.* **90**, 773–795.

Kundu, S. (2001). A short note on repeated significance tests and the unified Bayesian-frequentist measure. *Sequential Anal.* **20**, 77–86.

Le Cam, L., Yang, G.L. (2000). *Asymptotics in Statistics: Some Basic Concepts*, second ed. *Springer Series in Statistics*. Springer-Verlag, New York.

Lehmann, E.L. (1986). *Testing Statistical Hypotheses*, second ed. Wiley, New York.

Lindley, D.V. (1957). A statistical paradox. *Biometrika* **44**, 187–192.

Meng, X.L. (1994). Posterior predictive p-values. *Ann. Statist.* **22**, 1142–1160.

Robins, J.M., van der Vaart, A., Ventura, V. (2000). Asymptotic distribution of P values in composite null models. *J. Amer. Statist. Assoc.* **95**, 1143–1156.

Roussas, G.G. (1972). *Contiguity of Probability Measures: Some Applications in Statistics. Cambridge Tracts in Mathematics and Mathematical Physics*, vol. 63. Cambridge University Press, London.

Rubin, D.B. (1984). Bayesianly justifiable and relevant frequency calculations for the applied statistician. *Ann. Statist.* **12**, 1151–1172.

Rubin, H., Sethuraman, J. (1965). Probabilities of moderate deviations. *Sankhya, Ser. A* **27**, 325–346.

Schwarz, G. (1978). Estimating the dimension of a model. *Ann. Statist.* **6**, 461–464.

Sellke, T., Bayarri, M.J., Berger, J.O. (2001). Calibration of P-values for testing precise null hypotheses. *Amer. Statist.* **55**, 62–71.

Serfling, R.J. (1980). *Approximation Theorems of Mathematical Statistics*. Wiley, New York.

Woodroofe, M. (1978). Large deviations of likelihood ratio statistics with applications to sequential testing. *Ann. Statist.* **6**, 72–84.

Handbook of Statistics, Vol. 25
ISSN: 0169-7161
© 2005 Elsevier B.V. All rights reserved.
DOI 10.1016/S0169-7161(05)25006-X

6

Bayesian Model Checking and Model Diagnostics

Hal S. Stern and Sandip Sinharay

1. Introduction

Model checking, or assessing the fit of a model, is a crucial part of any statistical analysis. Before drawing any firm conclusions from the application of a statistical model to a data set, an investigator should assess the model's fit to make sure that the important features of the data set are adequately captured. Serious misfit (failure of the model to explain a number of aspects of the data that are of practical interest) should result in the replacement or extension of the model, if possible. Even if a model has been appointed as the final model for an application, it is important to assess its fit in order to be aware of its limitations before making any inferences. The goal is not to determine if a model is true or false, because a model is rarely perfect; rather, the interest is in determining if the limitations of the model have a noticeable effect on the substantive inferences. Statistical models are often chosen, at least in part, for convenience. Judging when assumptions of convenience can be made safely is the major goal in model checking.

In Bayesian statistics, a researcher can check the fit of the model using a variety of strategies (Gelman et al., 2003): (1) checking that the posterior inferences are reasonable, given the substantive context of the model; (2) examining the sensitivity of inferences to reasonable changes in the prior distribution and the likelihood; and (3) checking that the model can explain the data, or in other words, that the model is capable of generating data like the observed data. Each of these strategies is discussed in this chapter with the greatest emphasis placed on the last.

Model checking is related to, but still quite different from, *model selection* (which is discussed in a separate chapter of this volume). In the model selection context an investigator will typically develop a set of plausible models and then choose the best fitting model for analyzing the data. Model checking and model selection are complementary in practice. Iterative fitting and checking of models can be used to suggest a range of plausible models to serve as input to model selection. Also, once the best fitting model or combinations of models is selected, the fit of that model must still be checked. It is possible that the best fitting model does not fit the data or has a limitation sufficiently serious to require serious model changes or expansions.

2. Model checking overview

2.1. Checking that the posterior inferences are reasonable

The first and most natural form of model checking is to check that the posterior inferences are consistent with any information that was not used in the analysis. This could be data from another data source that is not being considered in the present analysis or prior information that (for one reason or another) was not incorporated in the model. There are important reasons for performing informal checks of this type. Nonsensical or paradoxical parameter values in the posterior distribution may indicate a problem in the computer program used to carry out the analysis. Implausible results may also indicate that a prior distribution chosen for mathematical or computational convenience is inappropriate. The very fact that a posterior inference seems unreasonable suggests the presence of prior information that could (and likely should) be used in the analysis (Gelman, 2004). Though this form of model checking is important, giving general advice is difficult because the types of information available depend heavily on the application context.

A recent example involved a variance components model fit to functional magnetic resonance imaging (fMRI) brain activation data. The posterior distribution was concentrated on variance component values that were larger than expected given the scale of the response variable (and larger than point estimates obtained from traditional method of moments estimates). Investigation detected an error; the prior distribution was specified as if the parameters of interest were variance parameters whereas the model parameters were actually precision parameters. Correcting the error improved things but the values were still too high which suggested that automatic application of an inverse gamma prior distribution was not appropriate for the data.

2.2. Sensitivity to choice of prior distribution and likelihood

Sensitivity analysis is important because any single model will tend to underestimate uncertainty in the inferences drawn. Other reasonable models could have fit the data equally well yet yielded different inferences. Model averaging is one approach to taking account of this information (see the chapter on model averaging in this volume). From the model checking perspective the existence of other reasonable models points to the need for sensitivity analysis. The basic technique of sensitivity analysis is to fit several probability models to the same data set, altering either the prior distribution, the likelihood, or both, and studying how the primary inferences for the problem at hand change.

Occasionally it is possible to perform a sensitivity analysis by embedding a given model in a larger family. A standard example is the use of Student's t-distribution in place of a normal distribution; inference for different values of the degrees of freedom parameter allows one to assess sensitivity to the strong normal assumption about the tail of the distribution. Even this approach is somewhat narrow as one may also wish to consider sensitivity of the results to nonsymmetric alternatives as well. Some applications of sensitivity analysis can be found in Gelman et al. (2003), Weiss (1994), and Smith et al. (1995).

There are a number of issues related to the computational efficiency with which a sensitivity analysis is carried out. It can be quite time consuming and computationally costly to obtain simulations from the posterior distribution of interest which makes the prospect of fitting a number of alternatives rather unappealing. Gelman et al. (1996a) describe approaches to assessing sensitivity that can be carried out without additional computation. The sensitivity of inferences for a particular parameter can be investigated by comparing the posterior variance of the parameter to the prior variance. If they are comparable, then the data evidently provides little information about the parameter and inferences are likely to be quite sensitive to the choice of the prior distribution. The sensitivity of inferences for a more complex quantity of inference, say a function of several parameters $Q(\boldsymbol{\omega})$, can be judged by examining scatter plots of $Q(\boldsymbol{\omega})$ against individual elements of $\boldsymbol{\omega}$. If Q is correlated with a particular parameter and posterior inference on that parameter is sensitive to the choice of prior distribution, then we expect that Q will also be sensitive to the choice of prior distribution of the relevant parameter. Another approach to sensitivity analysis that reduces the computational burden is to use importance sampling or sampling-importance-resampling (Rubin, 1987; Gelman et al., 2003) to draw approximate inferences under alternative models.

2.3. Checking that the model can explain the data adequately

It seems that a minimum requirement for a good probability model would be that it is able to explain the key features in the data set adequately, or put differently, that data generated by the model should look like the observed data. This is a self-consistency check. This intuitive idea underlies most model checking approaches including traditional residual analysis for linear models. Checks of this kind can detect patterns in the data that invalidate some aspect of the model or perhaps the entire model. Model checks can also identify individual observations that are not consistent with the model (outliers or influential points in linear models).

There are a number of approaches for carrying out a self-consistency check of the model: (1) Bayesian residual analysis; (2) cross-validatory predictive checks; (3) prior predictive checks; (4) posterior predictive checks; (5) partial posterior predictive checks; and (6) repeated data generation and analysis. Classifying model checking techniques into these categories does not in fact yield a disjoint partition as several model checking ideas can be viewed as falling into more than one of the categories. The categories are helpful however in identifying key features (e.g., use of prior information, range of diagnostics permitted) that distinguish different model checking methods. The six approaches, including relevant references, are described briefly in the next section. A thorough discussion of posterior predictive checks, the method that appears to be generating the broadest interest in scientific applications is provided in Section 4.

3. Approaches for checking if the model is consistent with the data

3.1. Bayesian residual analysis

In linear models, analysis of residuals is a popular and intuitive tool for diagnosing problems with the model fit. Plot of residuals against predicted values can identify failure of the statistical model. There is of course a natural Bayesian analogue to the traditional residual analysis (see, for example, Chaloner and Brant, 1988; Albert and Chib, 1995). Suppose y_i denotes the observation for individual i. Suppose further that $E(y_i|\boldsymbol{\omega}) = \mu_i$, where $\boldsymbol{\omega}$ denotes the vector of parameters in the model. The residual $\varepsilon_i = y_i - \mu_i$ is a function of the model parameters through μ_i. Thus there is a posterior distribution of the residuals and one can imagine a posterior distribution of residual plots or diagnostics. If a Bayesian model has been fit to the data, then y_i is considered outlying if the posterior distribution for the residual ε_i is located far from zero (Chaloner and Brant, 1988).

Examining residual plots based on posterior means of the residuals or randomly chosen posterior draws may help identify patterns that call the model assumptions into question. Residuals for discrete data can be difficult to interpret because only a few possible values of the residual are possible for a given value of μ_i. In this case it can be helpful to bin the residuals according to the value of μ_i (or other variable of interest) as in Gelman et al. (2000).

3.2. *Cross-validatory predictive checks*

Cross-validation is the technique of partitioning observed data into two parts, a training or estimation set and a test set. The model is fit to the training set and applied to the test set. This can be done with a single observation serving as the test set to detect outlying or unusual observations or groups of observations. Alternatively, for an overall measure of fit we might split the data into K subsets and then cycle through deleting one subset at a time and using the remaining $K-1$ subsets to generate a predictive distribution for the test set. The predictive distributions for the observations in the test set are evaluated relative to the observed values. This evaluation can be done by comparing the predictive mean (or median) to the observed values or by calculating the predictive density at the observed value. Poor performance in the test set may be evidence of a poorly fitting or poorly identified model. Stone (1974), Gelfand et al. (1992), and Marshall and Spiegelhalter (2003) discuss Bayesian approaches to cross-validation.

3.3. *Prior predictive checks*

Let $\boldsymbol{y}^{\text{rep}}$ denote replicate data that we might observe if the experiment that generated $\boldsymbol{y}$ is replicated and let $\boldsymbol{\omega}$ denote all the parameters of the model. Box (1980) suggested checking Bayesian models using the marginal or prior predictive distribution $p(\boldsymbol{y}^{\text{rep}}) = \int p(\boldsymbol{y}^{\text{rep}}|\boldsymbol{\omega})p(\boldsymbol{\omega})\,\mathrm{d}\boldsymbol{\omega}$ as a reference distribution for the observed data $\boldsymbol{y}$. In practice, a diagnostic measure $D(\boldsymbol{y})$ is defined and the observed value $D(\boldsymbol{y})$ compared to the reference distribution of $D(\boldsymbol{y}^{\text{rep}})$ with any significant difference between them indicating a model failure.

This method is a natural one to use for Bayesians because it compares observables to their predictive distribution under the model. A significant feature of the prior predictive approach is the important role played by the prior distribution in defining the reference distribution. Lack of fit may be due to misspecification of the prior distribution or the likelihood. The dependence on the prior distribution also means that the prior predictive distribution is undefined under improper prior distributions (and can be quite sensitive to the prior distribution if vague prior distributions are used). As the influence of Bayesian methods spreads to many application fields it appears that vague or improper priors are often used as a "noninformative" starting point in many data analyses. Thus the requirement of a proper prior distribution can be quite restrictive.

3.4. Posterior predictive checks

A slight variation on the prior predictive checks is to define $\boldsymbol{y}^{\text{rep}}$ as replicate data that we might observe if the experiment that generated $\boldsymbol{y}$ is repeated with the same underlying parameter. (For the prior predictive checks, the replicate data is based on a different parameter value drawn independently from the prior distribution.) Essentially $\boldsymbol{y}^{\text{rep}}$ is a data set that is exchangeable with the observed data if the model is correct. This leads to the use of the posterior predictive distribution $p(\boldsymbol{y}^{\text{rep}}) = \int p(\boldsymbol{y}^{\text{rep}}|\boldsymbol{\omega})p(\boldsymbol{\omega}|\boldsymbol{y})\,\mathrm{d}\boldsymbol{\omega}$ as a reference distribution for the observed data $\boldsymbol{y}$. Posterior predictive checks are concerned with the prior distribution only if the prior is sufficiently misspecified to yield a poor fit to the observed data. Also, posterior predictive checks can be applied even if the prior distribution is improper as long as the posterior distribution is a proper distribution. Posterior predictive checks are the primary focus of the remainder of the chapter. They are discussed in more detail in the next section and then illustrated in two applications.

3.5. Partial posterior predictive checks

Bayarri and Berger (2000) develop model checking strategies that can accommodate improper prior distributions but which avoid the conservatism of posterior predictive checks (this conservatism is discussed in the next section). They propose a compromise approach, known as partial posterior predictive model checks, that develops a reference distribution conditional on some of the data (but not all). Let $T(\boldsymbol{y})$ denote a test statistic with observed value t_{obs}. The partial posterior predictive checks use as a reference distribution the predictive distribution

$$p_{\text{ppost}}\left(\boldsymbol{y}^{\text{rep}}|\boldsymbol{y}\right) = \int p\left(\boldsymbol{y}^{\text{rep}}|\boldsymbol{\omega}\right)p_{\text{ppost}}(\boldsymbol{\omega}|\boldsymbol{y}, t_{\text{obs}})\,\mathrm{d}\boldsymbol{\omega},$$

where $p_{\text{ppost}}(\boldsymbol{\omega}|\boldsymbol{y}, t_{\text{obs}})$ is a partial posterior distribution conditional on the observed value of the test statistic. The partial posterior distribution is computed as $p_{\text{ppost}}(\boldsymbol{\omega}|\boldsymbol{y}, t_{\text{obs}}) \propto p(\boldsymbol{y}|t_{\text{obs}}, \boldsymbol{\omega})p(\boldsymbol{\omega})$. Model checks based on the partial posterior predictive distribution (and other alternatives described by Bayarri and Berger, 2000) have good small-sample and large-sample properties, as demonstrated by Bayarri and Berger (2000) and Robins et al. (2000). However, as Johnson (2004) comments, the p-values associated with these methods can seldom be defined and calculated in realistically complex models. Moreover the reference distribution changes for each test statistic that is conditioned on.

3.6. Repeated data generation and analysis

Dey et al. (1998) suggested a variation on the prior predictive checks that considers the entire posterior distribution as the "test statistic" (at least conceptually) rather than just a particular function. Here, as in the prior predictive method, many replicates of the data are generated using the model (i.e., using the prior distribution and the likelihood). For each replicate data set, rather than just compute a particular test statistic to compare with the observed test statistics, the posterior distribution of the parameters is determined. Each posterior distribution is known as a replicate posterior distribution. The key to model checking is then to compare the posterior distribution obtained by conditioning on the observed data with the replicate posterior distributions.

As described in Dey et al. (1998), let $y^{(r)}$ denote the rth replicate data set and say, for convenience, that $y^{(0)}$ denotes the observed data. With this approach it is possible to consider discrepancies that are functions of just the data (in which case the method is identical to the prior predictive check), functions of the parameters, or most generally functions of both, say $T(y, \boldsymbol{\omega})$. The posterior distribution $p(T|y^{(0)})$ is compared with the set $p(T|y^{(r)}), r = 1, 2, \ldots, R$. Comparing posterior distributions is a complicated problem, one possibility is to compare samples from each distribution $p(T|y^{(r)})$ with a single sample drawn from the posterior distribution based on the observed data. Dey et al. (1998) propose one approach for carrying out the comparison; they summarize each posterior sample with a vector of quantiles and then define a test statistic on this vector.

Like the prior predictive approach, this approach can not be applied when the prior is improper and can be quite sensitive to the use of vague prior distributions. The computational burden of this method is substantial; a complete Bayesian analysis is required for each replication which likely involves a complex simulation. In addition, a large amount of information needs to be stored for each replication.

4. Posterior predictive model checking techniques

Guttman (1967) first suggested the idea of the posterior predictive distribution; he used the terminology *density of a future observation* to describe the concept. Rubin applied the idea of the posterior predictive distribution to model checking and assessment (Rubin, 1981) and gave a formal Bayesian definition of the technique (Rubin, 1984). Gelman et al. (1996b) provided additional generality by broadening the class of diagnostics measures that are used to assess the discrepancy between the data and the posited model.

4.1. *Description of posterior predictive model checking*

Let $p(\mathbf{y}|\boldsymbol{\omega})$ denote the sampling or data distribution for a statistical model, where $\boldsymbol{\omega}$ denotes the parameters in the model. Let $p(\boldsymbol{\omega})$ be the prior distribution on the parameters. Then the posterior distribution of $\boldsymbol{\omega}$ is $p(\boldsymbol{\omega}|\mathbf{y}) \equiv \frac{p(\mathbf{y}|\boldsymbol{\omega})p(\boldsymbol{\omega})}{\int_{\boldsymbol{\omega}} p(\mathbf{y}|\boldsymbol{\omega})p(\boldsymbol{\omega})\,\mathrm{d}\boldsymbol{\omega}}$. Let $\mathbf{y}^{\mathrm{rep}}$ denote replicate data that one might observe if the process that generated the data $\mathbf{y}$ is replicated with the same value of $\boldsymbol{\omega}$ that generated the observed data. Then $\mathbf{y}^{\mathrm{rep}}$ is governed by the *posterior predictive distribution* (or the predictive distribution of replicated data conditional on the observed data),

$$p(\mathbf{y}^{\mathrm{rep}}|\mathbf{y}) = \int p(\mathbf{y}^{\mathrm{rep}}|\boldsymbol{\omega})\, p(\boldsymbol{\omega}|\mathbf{y})\,\mathrm{d}\boldsymbol{\omega}. \tag{1}$$

The posterior predictive approach to model checking proposes using this as a reference distribution for assessing whether various characteristics of the observed data are unusual. Gelman (2003) characterizes the inclusion of replicate data as a generalization of the model building process. Bayesian inference generalizes a probability model from the likelihood $p(\boldsymbol{y}|\boldsymbol{\omega})$ to the joint distribution $p(\boldsymbol{y}, \boldsymbol{\omega}) \propto p(\boldsymbol{y}|\boldsymbol{\omega})p(\boldsymbol{\omega})$. Model checking further generalizes to $p(\boldsymbol{y}, \boldsymbol{y}^{\text{rep}}, \boldsymbol{\omega})$ with the posterior predictive approach using the factorization $p(\boldsymbol{\omega})p(\boldsymbol{y}|\boldsymbol{\omega})p(\boldsymbol{y}^{\text{rep}}|\boldsymbol{\omega})$ (in which $\boldsymbol{y}$, $\boldsymbol{y}^{\text{rep}}$ are two exchangeable draws from the sampling distribution).

To carry out model checks *test quantities* or *discrepancy measures* $D(\boldsymbol{y}, \boldsymbol{\omega})$ are defined (Gelman et al., 1996b), and the posterior distribution of $D(\boldsymbol{y}, \boldsymbol{\omega})$ compared to the posterior predictive distribution of $D(\boldsymbol{y}^{\text{rep}}, \boldsymbol{\omega})$, with any significant difference between them indicating a model failure. If $D(\boldsymbol{y}, \boldsymbol{\omega}) = D(\boldsymbol{y})$, then the discrepancy measure is a test statistic in the usual sense.

Model checking can be carried out by graphically examining the replicate data and the observed data, by graphically examining the joint distribution of $D(\boldsymbol{y}, \boldsymbol{\omega})$ and $D(\boldsymbol{y}^{\text{rep}}, \boldsymbol{\omega})$ (possibly for several different discrepancy measures), or by calculating a numerical summary of such distributions. Gelman et al. (2003) provide a detailed discussion of graphical posterior predictive checks. One numerical summary of the model diagnostic's posterior distribution is the tail-area probability or as it is sometimes known, the posterior predictive p-value:

$$\begin{aligned} p_b &= P\big(D\big(\boldsymbol{y}^{\text{rep}}, \boldsymbol{\omega}\big) \geqslant D(\boldsymbol{y}, \boldsymbol{\omega})|\boldsymbol{y}\big) \\ &= \iint I_{[D(\boldsymbol{y}^{\text{rep}},\boldsymbol{\omega}) \geqslant D(\boldsymbol{y},\boldsymbol{\omega})]}\, p\big(\boldsymbol{y}^{\text{rep}}|\boldsymbol{\omega}\big)\, p(\boldsymbol{\omega}|\boldsymbol{y})\, \mathrm{d}\boldsymbol{y}^{\text{rep}}\, \mathrm{d}\boldsymbol{\omega}, \end{aligned} \tag{2}$$

where $I_{[A]}$ denotes the indicator function for the event A. Small or large tail area probabilities indicate that D identifies an aspect of the observed data for which replicate data generated under the model does not match the observed data.

Because of the difficulty in dealing with (1) or (2) analytically for all but the most simple problems, Rubin (1984) suggests simulating replicate data sets from the posterior predictive distribution. One draws L simulations $\boldsymbol{\omega}^1, \boldsymbol{\omega}^2, \ldots, \boldsymbol{\omega}^L$ from the posterior distribution $p(\boldsymbol{\omega}|\boldsymbol{y})$ of $\boldsymbol{\omega}$, and then draws $\boldsymbol{y}^{\text{rep},l}$ from the sampling distribution $p(\boldsymbol{y}|\boldsymbol{\omega}^l)$, $l = 1, 2, \ldots, L$. The process results in L draws from the joint posterior distribution $p(\boldsymbol{y}^{\text{rep}}, \boldsymbol{\omega}|\boldsymbol{y})$. Then graphical or numerical model checks are carried out using these sampled values.

4.2. *Properties of posterior predictive p-values*

Though the tail area probability is only one possible summary of the model check it has received a great deal of attention. The inner integral in (2) can be interpreted as a traditional p-value for assessing a hypothesis about a fixed value of $\boldsymbol{\omega}$ given the test measure D. If viewed in this way, the various model checking approaches represent different ways of handling the parameter $\boldsymbol{\omega}$. The posterior predictive p-value is an average of the classical p-value over the posterior uncertainty about the true $\boldsymbol{\omega}$. The Box prior predictive approach averages over the prior distribution for the parameter.

Meng (1994) provides a theoretical comparison of classical and Bayesian p-values. One unfortunate result is that posterior predictive p-values do not share some of the features of the classical p-values that dominate traditional significance testing. In particular, they do not have a uniform distribution when the assumed model is true, instead they are more concentrated around 0.5 than a uniform distribution. This is a result of using the same data to define the reference distribution and the tail event measured by the p-value. For some (see, e.g., Bayarri and Berger, 2000; Robins et al., 2000) the resulting conservatism of the p-values is a major disadvantage; this has motivated some of the alternative model checking strategies described earlier. Posterior predictive checks remain popular because the posterior predictive distributions of suitable discrepancy measures (not just the p-values) are easy to calculate and relevant to assessing model fitness.

4.3. *Effect of prior distributions*

Because posterior predictive checks are Bayesian by nature, a question arises about the sensitivity of the results obtained to the prior distribution on the model parameters. Because posterior predictive checks are based on the posterior distribution they are generally less sensitive to the choice of prior distribution than are prior predictive checks. Model failures are detected only if the posterior inferences under the model seem flawed. Unsuitable prior distributions may still be judged acceptable if the posterior inferences are reasonable. For example, with large sample sizes the prior distribution has little effect on the posterior distribution, and hence on posterior predictive checks. Gelman et al. (1996a) comment that if the parameters are well-estimated from the data, posterior predictive checks give results similar to classical model checking procedures for reasonable prior distributions. In such situations the focus is on assessing the fit of the likelihood part of the model.

Strongly informative prior distributions may of course have a large impact on the results of posterior predictive model checks. The replicated data sets obtained under strong incorrect prior specifications may be quite far from the observed data. In this way posterior predictive checks maintain the capability of rejecting a probability model if the prior distribution is sufficiently poorly chosen to negatively impact the fit of the model to the data. Gelman et al. (1996a, p. 757) provide one such example. On the contrary, a strong prior distribution, if trustworthy, can help a researcher assess the fit of the likelihood part of the data more effectively.

4.4. *Definition of replications*

In the description thus far the replications have been defined as data sets that are exchangeable with the original data under the model. This is implemented as an independent draw from the sampling distribution conditional on the model parameters. This is a natural definition for models where the data $\boldsymbol{y}$ depend on a parameter vector $\boldsymbol{\omega}$ which is given a (possibly noninformative) prior distribution. In hierarchical models, where the distribution of the data $\boldsymbol{y}$ depends on parameters $\boldsymbol{\omega}$ and the parameters $\boldsymbol{\omega}$ are given a prior or population distribution that depends on parameters $\boldsymbol{\alpha}$, there can be more than one possible definition of replications. To illustrate we consider a commonly occurring situation involving a hierarchical model.

Suppose that $\mathbf{y}$ are measurements (perhaps weights) with the measurement y_i of object i having a Gaussian distribution with mean equal to the true measurement ω_i and known variance. If a number of related objects are measured, then it is natural to model elements of $\boldsymbol{\omega}$ as independent Gaussian random variables conditional on parameters $\boldsymbol{\alpha}$ (the population mean and variance). Then one possible definition of replicate data sets corresponds to taking new measurements of the same objects. This corresponds to the joint distribution $p(\boldsymbol{\alpha}, \boldsymbol{\omega}, \mathbf{y}, \mathbf{y}^{\text{rep}})$ having factorization $p(\boldsymbol{\alpha})p(\boldsymbol{\omega}|\boldsymbol{\alpha})p(\mathbf{y}|\boldsymbol{\omega})p(\mathbf{y}^{\text{rep}}|\boldsymbol{\omega})$. The final term in the factorization reflects the assumption that, conditional on $\mathbf{y}$ and the parameters, $\mathbf{y}^{\text{rep}}$ depends only on the true measurements of the objects ($\boldsymbol{\omega}$). In practice the replicate data are obtained by simulating from the posterior distribution $p(\boldsymbol{\alpha}, \boldsymbol{\omega}|y)$ and then $p(\mathbf{y}^{\text{rep}}|\boldsymbol{\omega})$.

An alternative definition in this case would take the replications as corresponding to measurements of new objects from the same population. Thiscorresponds to the joint distribution $p(\boldsymbol{\alpha}, \boldsymbol{\omega}, \mathbf{y}, \mathbf{y}^{\text{rep}})$ having factorization $p(\boldsymbol{\alpha})p(\boldsymbol{\omega}|\boldsymbol{\alpha})p(\mathbf{y}|\boldsymbol{\omega})p(\mathbf{y}^{\text{rep}}|\boldsymbol{\alpha})$. The final term in the factorization now reflects the assumption that, conditional on $\mathbf{y}$ and the parameters, $\mathbf{y}^{\text{rep}}$ depends only on the population parameters ($\boldsymbol{\alpha}$) (because we have new objects from that population). In practice the replicate data for this case are obtained by simulating from the posterior distribution $p(\boldsymbol{\alpha}|y)$, then simulating new "true" measurements from $p(\boldsymbol{\omega}|\boldsymbol{\alpha})$, and finally simulating replicated data sets based on these new "true" measurements $p(\mathbf{y}^{\text{rep}}|\boldsymbol{\omega})$.

Diagnostics geared at assessing the likelihood, e.g., assessing whether symmetry is a reasonable model, would likely use the first definition corresponding to repeated measurements on the same objects. Diagnostics geared at assessing the appropriateness of the population distribution would likely use the alternative definition to ask whether measurements of objects from the assumed population would look like our sample. This one example is intended to demonstrate that, for sophisticated modeling efforts, it is important to carefully define the replicate data that serve as a reference distribution for the observed data.

4.5. Discrepancy measures

Technically, any function of the data and the parameters can play the role of a discrepancy measure in posterior predictive checks. The choice of discrepancy measures is very important. Virtually all models are wrong, and a statistical model applied to a data set usually explains certain aspects of the data adequately and some others inadequately. The challenge to the researcher in model checking is to develop discrepancy measures that have the power to detect the aspects of the data that the model cannot explain satisfactorily. A key point is that discrepancy measures corresponding to features of the data that are directly addressed by model parameters will never detect a lack of fit. For example, asking whether a normal model produces data with the same mean as the observed data, that is choosing $T(\mathbf{y}; \boldsymbol{\omega}) = \text{mean}\{y_i\}$, is sure to conclude that the model is adequate because the location parameter in the normal model will be fit to the observed data and the model will then generate replicate data centered in the same place. The particular model at hand may in fact fit quite poorly in the tails of the distribution but that will not be uncovered by a poor choice of discrepancy. Discrepancy measures that relate to features of the data not directly addressed by the probability model are better able to detect model failures (Gelman et al., 2003). Thus a measure of tail size (perhaps based on quantiles of the observed and replicate data) is more likely to detect the lack of fit in the situation described above. Failure to develop suitable discrepancy measures may lead to the incorrect conclusion that the model fits the data adequately. For a practical problem, a good strategy is to examine a number of discrepancies corresponding to aspects of practical interest as well as some standard checks on overall fitness (Gelman et al., 1996a).

4.6. Discussion

This section has outlined the theory behind posterior predictive model checks and some of the issues to be addressed before implementation. The fact that posterior predictive checks tend to be somewhat conservative, and the related fact that posterior predictive p-values are not uniformly distributed under the true model, have been cited as arguments against the use of posterior predictive checks. Despite this a number of advantages have made them one of the more practical tools available for practicing Bayesian statisticians. Posterior predictive model checks are straightforward to carry out once the difficult task of generating simulations from the posterior distribution of the model parameters is done. One merely has to take the simulated parameter values and then simulate data according to the model's sampling distribution (often a common probability distribution) to obtain replicate data sets. The important conceptual tasks of defining discrepancy measures and replications will typically require interaction with subject matter experts. This too can be thought of as an advantage in that the probing of the model required to define suitable discrepancies is a useful exercise. Finally posterior predictive checks lead to intuitive graphical and probabilistic summaries of the quality of the model fit.

The succeeding sections discuss two applications of posterior predictive model checks. Successful applications in the literature include Belin and Rubin (1995), Rubin and Wu (1997), Glickman and Stern (1998), Gelman et al. (2000, 2003), Fox and Glas (2003), Sorensen and Waagepetersen (2003), Gelman (2004), and Sinharay (in press, b).

5. Application 1

The preceding sections provide the theoretical overview of diagnostic methods. The chapter concludes with two applications. The first, described in the remainder of this section, is a modest application using a small data set and a simple Gaussian model. Though not particularly impressive the application serves to illustrate the posterior predictive approach without requiring a major investment in a particular application area. The second application, in the following section, uses a more complex probability model and a larger data set.

The small data set taken from Jeffreys (1961) consists of 20 observations of the mass of nitrogen (in grams). These data, collected by Rayleigh, include 12 observations in which measurements are based on atmospheric nitrogen and 8 observations in which measurements are derived using chemical reactions involving compounds like nitrous oxide that contain nitrogen. Because the observations all attempt to measure the same quantity and the measurement error is the only possible source of uncertainty, one might expect a simple model like the $\mathcal{N}(\mu, \sigma^2)$-model to be reasonable for the data set. A good statistical consultant would of course counsel that measurements from different paradigms not be pooled without first assessing their comparability. Assessing the fit of a pooled analysis under a Gaussian model is not too unrealistic though because it there are often several dimensions on which data samples are not exchangeable and all of these are not always incorporated in our model.

Suppose that the data are analyzed as a sample of 20 observations conditionally independent (given the mean and variance) under the $\mathcal{N}(\mu, \sigma^2)$ model and that the standard noninformative prior $p(\mu, \sigma^2) \propto 1/\sigma^2$ is used. As part of model checking it is natural to ask whether a random selection of 12 observations from a posterior predictive replicate data set could by chance differ from the remaining 8 observations by as much as we have observed in the nitrogen data. Figure 1 plots the observed data (at the top left) and eight replicated data sets under the model and the standard prior. The replications

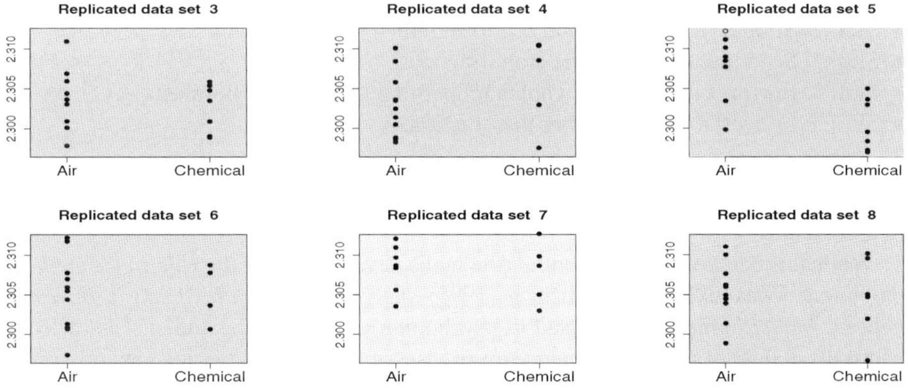

Fig. 1. The observed data and eight replicated data sets for the nitrogen example. The difference between one group of 12 measurements (labelled "air") and a second group of 8 measurements (labelled "chemical") under the model is unlikely to be as large as the difference in the observed data.

are obtained from the posterior predictive distribution of the data, which is given by

$$p(y^{\text{rep}}|\mathbf{y}) = t_{n-1}\left(\bar{y}, s\left(1 + \frac{1}{n}\right)^{1/2}\right),$$

where $\bar{y}$ is the observed data sample mean, s is the observed data sample standard deviation, $n = 20$ is the sample size, and t_{n-1} refers to the Student's t-distribution probability density function with $n - 1$ degrees of freedom.

The figure provides a clear indication that the model is not consistent with the data. In the observed data, the average for "air" is larger than that for "chemical" by a much greater amount than in any of the eight replicate data sets. One does not have to explore any further to conclude a clear misfit of the model. What next? In practice one would likely report inferences separately for the two data collection methods and then set out on the difficult task of determining why there is a difference when none was expected. An interesting historical note is that the discrepancy between the two types of samples led to the discovery of argon, an element heavier than nitrogen that contaminated the observations from air.

6. Application 2

In reality the failure of the previous model would be easily detected in any number of ways, including a simple t-test to compare the observations from the two measurement protocols. The greater question is how to carry out model checking in the context of the more complex models that are increasingly common in many fields. As an example, consider part of a data set from the field of educational testing (Tatsuoka, 1984, 1990; Tatsuoka et al., 1988). The data are responses of 325 middle-school students to 15 mixed number subtraction problems for which the fractions have a common denominator. The item writers designed these items to diagnose student mastery on five skills (e.g., Tatsuoka et al., 1988)

- Skill 1: Basic fraction subtraction.
- Skill 2: Simplify/reduce fraction or mixed number.
- Skill 3: Separate whole number from fraction.
- Skill 4: Borrow one from the whole number in a given mixed number.
- Skill 5: Convert a whole number to a fraction.

Table 1 lists the 15 items considered, characterized by the skills they require. Horizontal lines in the table separate the items into groups of items with common skill requirements.

While this data set played a central role in the development of the *rule space methodology* (e.g., Tatsuoka, 1990), Mislevy (1995) analyzed the data set using a *Bayesian network model*, a model often used in psychometrics to measure students' knowledge and skills in light of their performance on an assessment. In the Bayesian network model $\boldsymbol{\theta}_i = \{\theta_{i1}, \ldots, \theta_{i5}\}'$ denotes a vector of latent variables describing the attribute/skill vector for examinee i with

$$\theta_{ik} = \begin{cases} 1 & \text{if examinee } i \text{ has skill } k, \\ 0 & \text{otherwise.} \end{cases} \tag{3}$$

Table 1
Skill requirements for the mixed number subtraction problems

Item no.	Text of the item	Skills required 1	2	3	4	5	Proportion correct
2	$\frac{6}{7}-\frac{4}{7}$	x					0.79
4	$\frac{3}{4}-\frac{3}{4}$	x					0.70
8	$\frac{11}{8}-\frac{1}{8}$	x	x				0.71
9	$3\frac{4}{5}-3\frac{2}{5}$	x		x			0.75
11	$4\frac{5}{7}-1\frac{4}{7}$	x		x			0.74
5	$3\frac{7}{8}-2$	x		x			0.69
1	$3\frac{1}{2}-2\frac{3}{2}$	x		x	x		0.37
7	$4\frac{1}{3}-2\frac{4}{3}$	x		x	x		0.37
12	$7\frac{3}{5}-\frac{4}{5}$	x		x	x		0.34
15	$4\frac{1}{3}-1\frac{5}{3}$	x		x	x		0.31
13	$4\frac{1}{10}-2\frac{8}{10}$	x		x	x		0.41
10	$2-\frac{1}{3}$	x		x	x	x	0.38
3	$3-2\frac{1}{5}$	x		x	x	x	0.33
14	$7-1\frac{4}{3}$	x		x	x	x	0.26
6	$4\frac{4}{12}-2\frac{7}{12}$	x	x	x	x		0.31

The θ_{ik}'s are called *proficiency variables* and a Bayesian network model attempts to draw inferences about these variables based on the individual's responses. Prior analyses (see, e.g., Mislevy et al., 2001) revealed that skill 3 is a prerequisite to skill 4. A three-level auxiliary variable θ_{WN} (WN for whole number) incorporates this constraint. Level 0 of θ_{WN} corresponds to the participants who have mastered neither skill (in other words, $\theta_{i,WN} = 0$ indicates $\theta_{i3} = \theta_{i4} = 0$); Level 1 represents participants who have mastered skill 3 but not skill 4 ($\theta_{i,WN} = 1$ implies $\theta_{i3} = 1, \theta_{i4} = 0$); Level 2 represents participants who mastered both skills ($\theta_{i,WN} = 2$ implies $\theta_{i3} = \theta_{i4} = 1$).

Let X_{ij} denote the score (1 for correct answer, 0 otherwise) of the ith examinee for the jth item, $i = 1, 2, \ldots, I \equiv 325$, $j = 1, 2, \ldots, J \equiv 15$. In the Bayesian network model, the responses are modeled as Bernoulli trials with item-specific success probabilities depending on the skill set of the individual,

$$P(X_{ij} = 1|\boldsymbol{\theta}_i, \boldsymbol{\pi}_j) = \begin{cases} \pi_{j1} & \text{if examinee } i \text{ mastered all the skills needed to solve item } j, \\ \pi_{j0} & \text{otherwise.} \end{cases} \tag{4}$$

The conditioning on $\boldsymbol{\theta}_i$ above is somewhat subtle; the success probability takes one (higher) level if all of the skills required (see Table 1) are present and a second (lower) level if one or more of the required skills is missing. It is common to assume local independence, i.e., given the proficiency vector $\boldsymbol{\theta}_i$, the responses of an examinee to the different items are assumed independent. The probability π_{j1} represents a "true-positive" probability for the item, i.e., it is the probability of getting the item right for students who have mastered all of the required skills. The probability π_{j0} represents a "false-positive" probability; it is the probability of getting the item right for students who have yet to master at least one of the required skills. Mislevy et al. (2001) assume the same prior distributions for the two success probabilities for all items:

$$\pi_{j0} \sim \text{Beta}(6, 21), \qquad \pi_{j1} \sim \text{Beta}(21, 6), \quad \text{for } j = 1, \dots, J.$$

The distribution of proficiency vectors over the population of examinees is described in terms of a series of lower-dimensional distributions. The marginal probability of an examinee possessing the first, most basic, skill is taken as λ_1 and then the probability of being proficient on more sophisticated skills is given conditional on lower order skills.

$$\begin{aligned}
\lambda_1 &= P(\theta_1 = 1),\\
\lambda_{2,m} &= P(\theta_2 = 1|\theta_1 = m) \quad \text{for } m = 0, 1,\\
\lambda_{5,m} &= P(\theta_5 = 1|\theta_1 + \theta_2 = m) \quad \text{for } m = 0, 1, 2,\\
\lambda_{WN,m,n} &= P(\theta_{WN} = n|\theta_1 + \theta_2 + \theta_5 = m)\\
&\quad \text{for } m = 0, 1, 2, 3 \text{ and } n = 0, 1, 2.
\end{aligned}$$

The proficiency probabilities λ_1, $\boldsymbol{\lambda}_2$, $\boldsymbol{\lambda}_5$, and $\boldsymbol{\lambda}_{WN}$ are assumed to be *a priori* independent. The natural conjugate priors for the components of $\boldsymbol{\lambda}$ are Beta or Dirichlet distributions. The hyper-parameters of the Beta and Dirichlet distributions are chosen so that they sum to 27; this gives the prior information for these quantities weight equivalent to data from 27 examinees (relatively strong prior information given the sample size of 325). The specific values chosen are:

$$\begin{aligned}
&\lambda_1 \sim \text{Beta}(23.5, 3.5),\\
&\lambda_{2,0} \sim \text{Beta}(3.5, 23.5); \qquad \lambda_{2,1} \sim \text{Beta}(23.5, 3.5),\\
&\lambda_{5,0} \sim \text{Beta}(3.5, 23.5); \qquad \lambda_{5,1} \sim \text{Beta}(13.5, 13.5);\\
&\lambda_{5,2} \sim \text{Beta}(23.5, 3.5),\\
&\boldsymbol{\lambda}_{WN,0,\cdot} = (\boldsymbol{\lambda}_{WN,0,0}, \boldsymbol{\lambda}_{WN,0,1}, \boldsymbol{\lambda}_{WN,0,2}) \sim \text{Dirichlet}(15, 7, 5),\\
&\boldsymbol{\lambda}_{WN,1,\cdot} \sim \text{Dirichlet}(11, 9, 7); \qquad \boldsymbol{\lambda}_{WN,2,\cdot} \sim \text{Dirichlet}(7, 9, 11);\\
&\boldsymbol{\lambda}_{WN,3,\cdot} \sim \text{Dirichlet}(5, 7, 15).
\end{aligned}$$

The prior probabilities for λ_1, λ_2, and λ_5 give a chance of 87% (23.5/27) of acquiring a skill when the previous skills are mastered and a 13% chance of acquiring the skill when the previous skills are not mastered. The structure for λ_{WN} is more complex. Mislevy et al. (2001) argue that with the complex latent structure in the model, strong priors such as the ones used here are required to prevent problems with identifiability. Specifically since we don't observe the $\boldsymbol{\theta}_i$'s directly it can be hard to learn about the $\boldsymbol{\lambda}$ parameters.

As in Mislevy et al. (2001), the BUGS software (Spiegelhalter et al., 1995) is used to obtain samples from the posterior distribution of the $\boldsymbol{\theta}_i$'s and the components of $\boldsymbol{\lambda}$, and subsequently to generate posterior predictive data sets. The posterior predictive data $\boldsymbol{y}^{\text{rep}}$ is generated as a replicate for the same examinees, i.e., we simulate $\boldsymbol{y}^{\text{rep}}$ conditional on the proficiencies $\boldsymbol{\theta}_i$, $i = 1, \dots, I$, for these 325 examinees.

6.1. Direct data display

An initial, rather unfocused model check, is to examine the observed data matrix and the replicate data graphically. Consider Figure 2, which shows the observed data and 10 replicated data sets. In each display the examinees are shown along the vertical axis with the lowest ability examinees on the bottom and the items are shown along the horizontal axis with the easiest items on the left. A dark square indicates a correct response. The plot is somewhat similar to Figures 6.6 and 6.7 of Gelman et al. (2003) that are used to assess the fit of a logistic regression model. Sinharay (in press, a) has a similar plot as well.

There are certain patterns in the observed data that are not present in the replicated data sets. For example, the observed examinees with the lowest scores (at the bottom) could answer only item 4 or item 5 correctly (interestingly, these are items which can be solved without any knowledge of mixed number subtraction; e.g., item 4 can be solved by arguing that any quantity minus itself is zero). In the replicated data sets, however, the weaker examinees sometimes get other items correct as well. The more successful examinees get all items correct in the observed data, which is not true for the replicated data sets. Further, the weakest half of the examinees rarely get any hard items correct in the observed data, which is not true in the replicated data sets. These discrepancies suggest that the fitted model is not entirely satisfactory for the data set. Having reached

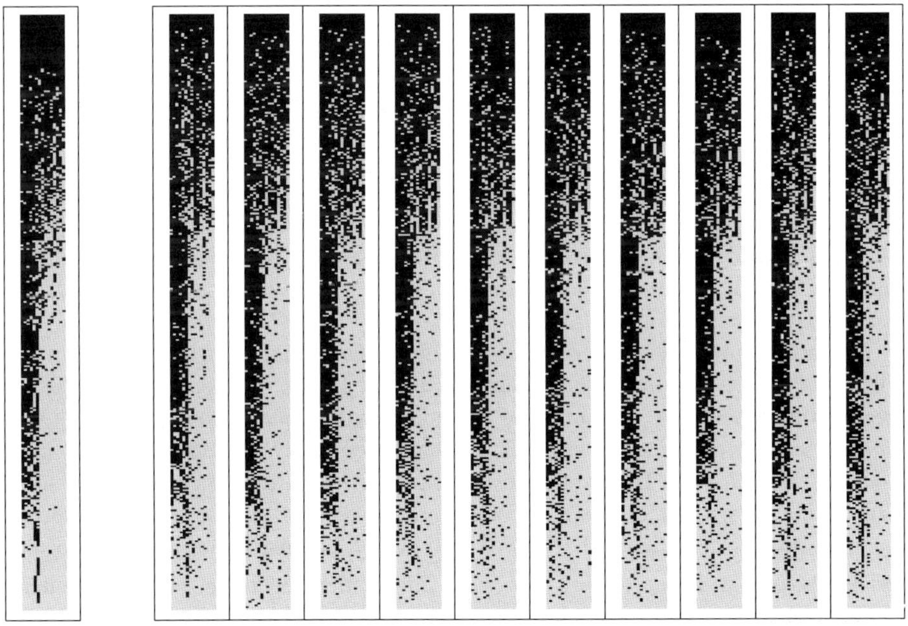

Fig. 2. Left column shows all observed responses, with a little black box indicating a correct response. The items, sorted according to increasing difficulty, are shown along the x-axis; the examinees, sorted according to increasing raw scores, are shown along the y-axis. Right columns show 10 replicated data sets from the model. There are some clear patterns in the observed data set that the model fails to predict.

this conclusion it is natural to probe further to identify specific aspects of the model that might be improved.

6.2. *Item fit*

In educational test analyses like this, researchers are often interested in examining if the model explains the responses for individual items adequately. Misfitting items may be discarded from the item pool to ensure that the inferences on the examinees (or at least future examinees) are not biased by the inferior items. The goal here then is to find a discrepancy measure that will be effective to detect lack of fit of the model for an individual item.

The proportion correct for an item would be an obvious, but unhelpful, suggestion for a possible discrepancy measure. The fitted model explains the proportion correct for each item adequately with posterior predictive p-values ranging from 0.23 to 0.73. This result is not unexpected (and the choice of diagnostic clearly flawed) because the model has parameters π_{j0} and π_{j1} for each item that directly address the proportion of correct responses for that item. That is to say, these parameters are estimated from the data and consequently using the estimated parameter values to generate replicate data will tend to produce overall performance on the items that is quite similar to that found in the original data. This reinforces the notion that design of appropriate discrepancy measures is crucial in implementing posterior predictive model checks. The failure to find any model failures may just be a reflection of an inferior choice of discrepancy measure.

Yan et al. (2003) try to avoid this problem by defining a standardized residual for each examinee-item combination as $\frac{X_{ij}-E(X_{ij}|\boldsymbol{\pi}_j,\boldsymbol{\theta}_i)}{V(X_{ij}|\boldsymbol{\pi}_j,\boldsymbol{\theta}_i)^{1/2}}$ and then add the squares of these residuals up over all individuals (i.e., over all i) to obtain a sum-of-squared-error discrepancy for item j. Unfortunately this measure has little power to detect item fit as well (posterior predictive p-values range from .22 to .74). In this case the problem is the latent variable parameter vector $\boldsymbol{\theta}_i$ which ensures that residuals for an individual cannot be too unusual; the same phenomenon is observed by Gelman et al. (2000) in a discrete data regression model.

The key is to define a discrepancy that identifies a feature of the model not explicitly addressed by any of the parameters in the model. We apply as discrepancy measures for item j the proportion of examinees with *number-correct score* (or *raw score*) k on the other items who answer item j correctly, for each possible value $k = 1, 2, \ldots, (J-1)$. Denote the observed proportion of examinees with raw score k answering item j correctly as p_{jk}. For each replicated data set, there is a corresponding proportion correct, denoted p_{jk}^{rep}. For each item j, comparison of the values of p_{jk} and p_{jk}^{rep}'s, $k = 1, 2, \ldots, (J-1)$, provides information regarding the fit of the item. One way to make the comparison is to compute the posterior predictive p-values for each item j and raw score k. We use a graphical approach to make the comparison.

Figure 3, like a plot in Sinharay (in press, a) provides item fit plots for all items for this example. The horizontal axis of each plot denotes the raw scores of the examinees. The vertical axis represents the proportion correct on the item being assessed for the

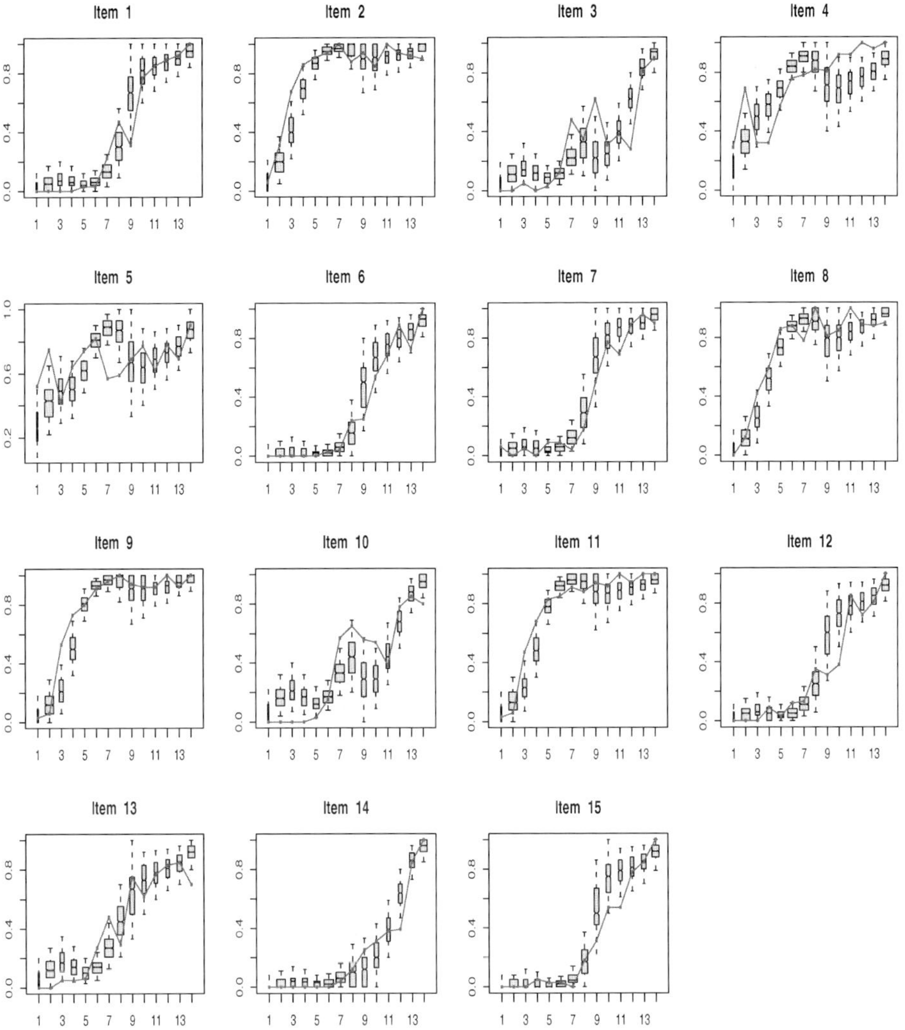

Fig. 3. Item fit plots for the mixed number subtraction data. The horizontal axis in each plot is the number of correct responses. The vertical axis is the proportion of correct responses on the item being considered. The side-by-side box plots in each plot given the posterior predictive distributions for the proportion of correct responses for each raw score subpopulation. The observed proportion correct seems to lie outside the posterior predictive distribution often, especially for items 5 and 10.

subgroup with each given raw score. For any raw score, a point on the solid line denotes the observed proportion correct and a box plot represents the posterior predictive distribution of the replicated proportions correct for that score. The observed proportions correct are connected by a line for ease of viewing only. The whiskers of the box stretch from the 5th to the 95th percentile of the empirical posterior predictive distribution and a notch near the middle of the box denotes the median. The width of a box is proportional to the observed number of examinees with the corresponding raw score.

(The observed number with a particular raw score provides some idea about the importance of seeing a difference between the observed and replicated proportions correct; a substantial difference for a large subgroup would be considered more important than a similar difference for a small subgroup.) The replicate data produces a natural monotone pattern; as we move from left to right the examinees are performing better overall and we expect them to be more likely to get the item correct. For any item whose fit in the model is being examined, too many observed proportions that are discrepant relative to their posterior predictive distributions indicate a failure of the model to explain the response pattern for the item. Many of the items provide some evidence for lack of fit. There are several items for which the intuitive monotone pattern in the replicate data looks very different than the pattern in the observed data (see, for example, items 5 and 10). More generally, the plots for most of the items (other than 2, 6, 7, 14, and 15) have a number of observed proportions lying outside the corresponding 90% posterior predictive intervals. In addition to providing an overall idea about the fit of the model for an individual item, the suggested item fit plot also provides some idea about the region where the misfit occurs (i.e., for low or high-scoring individuals) and the direction of the misfit (i.e., whether the model overestimates/underestimates performance in the discrepant regions). Sinharay (in press, a) develops a single χ^2-type summary measure based on each item's plot to use as a discrepancy measure.

6.3. Studying the association among the items

A number of researchers (e.g., van den Wollenberg, 1982; Reiser, 1996) show that the second-order marginals, which is the proportion of individuals answering a pair of items correctly, are useful quantities to examine in the context of models for item response data. Such quantities allow one to assess whether the model can adequately explain the observed associations among items.

Consider the item pair consisting of items i and j on a test. Let $n_{kk'}$ denote the number of individuals obtaining a score k on item i and k' on item j with $k, k' = 0, 1$. We examine the empirical odds ratio

$$OR_{ij} = \frac{n_{11}n_{00}}{n_{10}n_{01}}$$

which estimates the population odds ratio:

$$\frac{P(\text{item } i \text{ correct}|\text{item } j \text{ correct})/P(\text{item } i \text{ wrong}|\text{item } j \text{ correct})}{P(\text{item } i \text{ correct}|\text{item } j \text{ wrong})/P(\text{item } i \text{ wrong}|\text{item } j \text{ wrong})}.$$

The odds ratio is a measure of association and therefore can be used as a discrepancy measure (actually a test statistic) to detect if the fitted model can adequately explain the association among the test items.

For each pair of items the posterior predictive distribution of the odds ratio provides a reference distribution for the observed odds ratio. For example, Figure 4 shows the observed odds ratio and the corresponding posterior predictive distribution (using a dot and a box respectively) for all item pairs involving item 4. It is evident that the observed odds ratio for most of these pairs is much greater than would be expected under the posterior predictive distribution using the Bayesian network model. With 15 items there

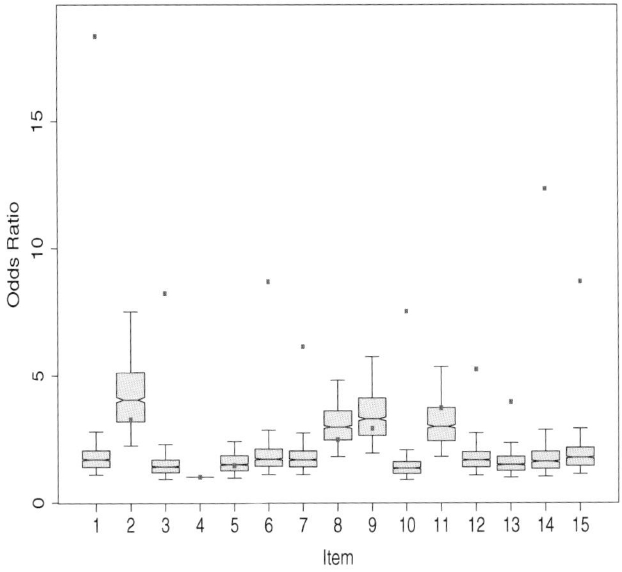

Fig. 4. Observed and predicted odds ratios for all item pairs involving item 4.

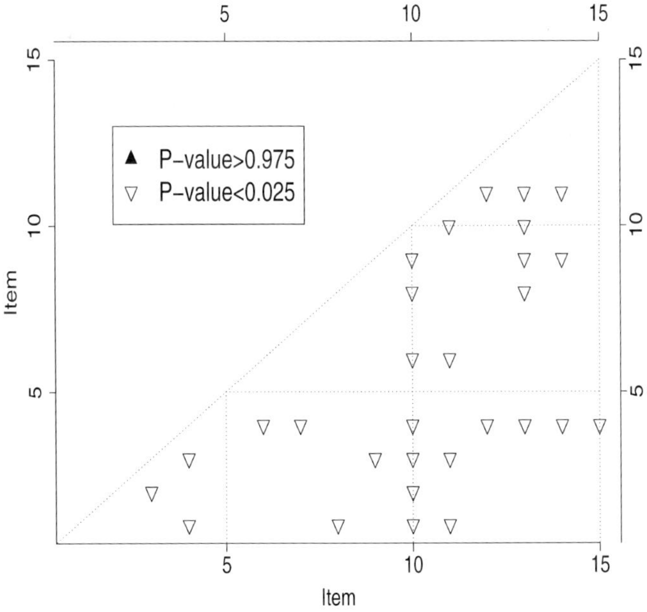

Fig. 5. Summary of posterior predictive p-values for odds ratios for the mixed number subtraction example. The dashed lines are to assist in identifying locations. The are many pairs of items with observed associations larger than expected under the model (p-values less than .025).

are 15 such plots. As a summary Figure 5 identifies pairs of items with unusually high or low odds ratios compared to the posterior predictive distribution. A triangle indicates a significant over-prediction by the model indicating that the model predicts greater association than is actually observed; there are no such results. An inverted triangle indicates a significant under-prediction wherein the replicate data demonstrates consistently smaller degrees of association than we observe. Horizontal and vertical grid-lines at multiples of 5 in the figure, as well as vertical axis-labels on the right side of the figure and a line along the main diagonal are there for convenience in reading the plot. The plot had many inverted triangles, indicating that the observed associations among items are much higher than expected under the model. The model especially performs poorly for item pairs involving items 3, 4, 10, 11, 13. This information calls into question the local independence assumption. It appears that even after accounting for the skills required, some items are more alike than others.

6.4. Discussion

The Bayes network model is found to perform rather poorly for this data set. A natural next step is to ask how the model's deficits could be addressed in a new or expanded model. For example, Figure 2 suggests that the model fails to fit the weakest and strongest examinees; one reason could be the use of a dichotomous distribution for the probability of a correct response to an item; i.e., the model divides the examinees into two groups based on whether or not they have mastered all the necessary skills for solving an item, and assigns the same success probabilities to all examinees within a group. It may be that it is easier to compensate for the lack of one skill than for the lack of many skills. This would suggest more than two different probabilities of success are needed. The model also does not incorporate any special indication of overall mathematical proficiency. It is possible that such proficiency would be related to both how quickly a student could master the skills and how readily the participant could apply them to a given problem. Sinharay (in press, a) expands the Bayes network model by adding factors to account for overall proficiency and to allow for the possibility of one or more key skills. The expanded model appears to fit the data more satisfactorily according to the model diagnostics described here. Another completely different model, the three-parameter logistic model (3PL; Lord, 1980) which allows the proficiency of an individual to be a continuous variable, also performs better than the Bayes network model based on the diagnostics considered here.

7. Conclusions

It is self-evident that model checking is an essential part of the modeling process. Despite this one still sees data analyses involving sophisticated probability models that stop with the posterior distribution and do not include any model checking. Model checking offers several important benefits. It can play an important role in model development, when failures of initial models suggest the nature of refinements that are needed to improve the model. Model checking can also help identify particular observations or data features that are not consistent with the model even when the overall fit is adequate.

There are a large number of approaches to model checking and investigators should call on whatever is helpful in a particular problem. The two examples here demonstrate how posterior predictive model checking methods can be useful in detecting model misfit. This includes gross model inadequacy as in the case of Rayleigh's nitrogen data and more subtle failures of particular assumptions in the more sophisticated Bayesian network model for educational testing data. Posterior predictive checks can usually be added to an existing data analysis with relatively little additional computational work. Careful thought about the underlying application area is required to define informative discrepancy measures – the interactions between statisticians and their collaborators required to construct such discrepancy measures is yet another positive aspect of model checking.

References

Albert, J., Chib, S. (1995). Bayesian residual analysis for binary response regression models. *Biometrika* **82**, 747–759.

Bayarri, S., Berger, J. (2000). P-values for composite null models. *J. Amer. Statist. Assoc.* **95**, 1127–1142.

Belin, T.R., Rubin, D.B. (1995). The analysis of repeated-measures data on schizophrenic reaction times using mixture models. *Statistics in Medicine* **14**, 747–768.

Box, G.E.P. (1980). Sampling and Bayes inference in scientific modelling and robustness. *J. Roy. Statist. Soc., Ser. A* **143**, 383–430.

Chaloner, K., Brant, R. (1988). A Bayesian approach to outlier detection and residual analysis. *Biometrika* **75**, 651–659.

Dey, D.K., Gelfand, A.E., Swartz, T.B., Vlachos, P.K. (1998). A simulation-intensive approach for checking hierarchical models. *Test* **7**, 325–346.

Fox, J.P., Glas, C.A.W. (2003). Bayesian modeling of measurement error in predictor variables using item response theory. *Psychometrika* **68**, 169–191.

Gelfand, A.E., Dey, D.K., Chang, H. (1992). Model determination using predictive distributions, with implementation via sampling-based methods. In: *Bayesian Statistics, vol. 4. Proceedings of the Fourth Valencia International Meeting*, pp. 147–159.

Gelman, A. (2003). A Bayesian formulation of exploratory data analysis and goodness-of-fit testing. *Internat. Statist. Rev.* **71**, 369–382.

Gelman, A. (2004). Exploratory data analysis for complex problems. *J. Comput. Graph. Statist.* **13**, 755–779.

Gelman, A., Bois, F.Y., Jiang, J. (1996a). Physiological pharmacokinetic analysis using population modeling and informative prior distributions. *J. Amer. Statist. Assoc.* **91**, 1400–1412.

Gelman, A., Meng, X.-L., Stern, H. (1996b). Posterior predictive assessment of model fitness via realized discrepancies. *Statistica Sinica* **6**, 733–807.

Gelman, A., Goegebeur, Y., Tuerlinckx, F., van Mechelen, I. (2000). Diagnostic checks for discrete-data regression models using posterior predictive simulations. *Appl. Statist.* **49**, 247–268.

Gelman, A., Carlin, J.B., Stern, H.S., Rubin, D.B. (2003). *Bayesian Data Analysis*, second ed. Chapman & Hall/CRC, Boca Raton.

Glickman, M.E., Stern, H.S. (1998). A state-space model for National Football League scores. *J. Amer. Statist. Assoc.* **93**, 25–35.

Guttman, I. (1967). The use of the concept of a future observation in goodness-of-fit problems. *J. Roy. Statist. Soc., Ser. B* **29**, 83–100.

Jeffreys, H. (1961). *Theory of Probability*. Oxford University Press.

Johnson, V. (2004). A Bayesian χ^2 test for goodness-of-fit. *Ann. Statist.* **32**, 2361–2384.

Lord, F.M. (1980). *Applications of Item Response Theory to Practical Testing Problems*. Lawrence Erlbaum Associates, Hillsdale, NJ.

Marshall, E.C., Spiegelhalter, D.J. (2003). Approximate cross-validatory predictive checks in disease-mapping model. *Statistics in Medicine* **22**, 1649–1660.
Meng, X.-L. (1994). Posterior predictive p-values. *Ann. Statist.* **22**, 1142–1160.
Mislevy, R.J. (1995). Probability-based inference in cognitive diagnosis. In: Nichols, P., Chipman, S., Brennan, R. (Eds.), *Cognitively Diagnostic Assessment*. Erlbaum, Hillsdale, NJ, pp. 43–71.
Mislevy, R.J., Almond, R.G., Yan, D., Steinberg, L.S. (2001). Bayes nets in educational assessment: Where the numbers come from. In: *Proceedings of the Fifteenth Conference on Uncertainty in Artificial Intelligence*, pp. 437–446.
Reiser, M. (1996). Analysis of residuals for the multinomial item response models. *Psychometrika* **61**, 509–528.
Robins, J.M., van der Vaart, A., Ventura, V. (2000). The asymptotic distribution of p-values in composite null models. *J. Amer. Statist. Assoc.* **95**, 1143–1172.
Rubin, D.B. (1981). Estimation in parallel randomized experiments. *J. Educational Statist.* **6**, 377–401.
Rubin, D.B. (1984). Bayesianly justifiable and relevant frequency calculations for the applied statistician. *Ann. Statist.* **12**, 1151–1172.
Rubin, D.B. (1987). A noniterative sampling/importance resampling alternative to the data augmentation algorithm for creating a few imputations when fractions of missing information are modest: The SIR algorithm. Discussion of Tanner and Wong (1987). *J. Amer. Statist. Assoc.* **82**, 543–546.
Rubin, D.B., Wu, Y.N. (1997). Modelling schizophrenic behavior using general mixture components. *Biometrics* **53**, 243–261.
Sinharay, S. (in press, a). Assessing fit of Bayesian networks using the posterior predictive model checking method. *J. Educational and Behavioral Statist.* Submitted for publication.
Sinharay, S. (in press, b). Practical applications of posterior predictive model checking for assessing fit of common item response theory models. *J. Educational Measurement.* Submitted for publication.
Smith, A.F.M., Spiegelhalter, D.J., Thomas, A. (1995). Bayesian approaches to random-effects meta-analysis: A comparative study. *Statistics in Medicine* **14**, 2685–2699.
Sorensen, D., Waagepetersen, R. (2003). Normal linear models with genetically structured residual variance heterogeneity: A case study. *Genetical Research Cambridge* **82**, 207–222.
Spiegelhalter, D.J., Thomas, A., Best, N.G., Gilks, W.R. (1995). BUGS: Bayesian inference using Gibbs sampling, Version 0.50. MRC Biostatistics Unit, Cambridge.
Stone, M. (1974). Cross-validatory choice and assessment of statistical predictions. *J. Roy. Statist. Soc., Ser. B, Methodological* **36**, 111–147.
Tatsuoka, K.K. (1984). Analysis of errors in fraction addition and subtraction problems (NIE Final Rep. for Grant No. NIE-G-81-002). Computer-Based Education Research, University of Illinois, Urbana, IL.
Tatsuoka, K.K. (1990). Toward an integration of item response theory and cognitive error diagnosis. In: Frederiksen, N., Glaser, R., Lesgold, A., Shafto, M.G. (Eds.), *Diagnostic Monitoring of Skill and Knowledge Acquisition*. Erlbaum, Hillsdale, NJ, pp. 453–488.
Tatsuoka, K.K., Linn, R.L., Tatsuoka, M.M., Yamamoto, K. (1988). Differential item functioning resulting from the use of different solution strategies. *J. Educational Measurement* **25** (4), 301–319.
van den Wollenberg, A.L. (1982). Two new test statistics for the Rasch model. *Psychometrika* **47**, 123–140.
Weiss, R.E. (1994). Pediatric pain, predictive inference, and sensitivity analysis. *Evaluation Review* **18**, 651–678.
Yan, D., Mislevy, R.J., Almond, R.G. (2003). Design and analysis in a cognitive assessment. ETS Research Report 03-32, Educational Testing Service, Princeton, NJ.

Handbook of Statistics, Vol. 25
ISSN: 0169-7161
© 2005 Elsevier B.V. All rights reserved.
DOI 10.1016/S0169-7161(05)25007-1

7

The Elimination of Nuisance Parameters[1]

Brunero Liseo

Abstract

We review the Bayesian approach to the problem of the elimination of nuisance parameters from a statistical model. Many Bayesian statisticians feel that the framework of Bayesian statistics is so clear and simple that the elimination of nuisance parameters should not be considered a problem: one has simply to compute the marginal posterior distribution of the parameter of interest. However we will argue that this exercise need not be so simple from a practical perspective.

The paper is divided in two main parts: the first deals with regular parametric models whereas the second will focus on non regular problem, including the so-called Neyman and Scott's class of models and semiparametric models where the nuisance parameter lies in an infinite dimensional space. Finally we relate the issues of the elimination of nuisance parameters to other, apparently different, problems. Occasionally, we will mention non-Bayesian treatment of nuisance parameters, mainly for comparative analyses.

Keywords: binary matched pairs; conditional likelihood; Kullback–Leibler divergence; integrated likelihood; marginal likelihood; modified profile likelihood; profile likelihood; Jeffreys' prior; Neyman–Scott problem; nonparametric likelihood; reference priors

1. Introduction

1.1. Synopsis

We review the fundamental issue of the elimination of nuisance parameters from a statistical model. This is one of the statistical issues that epitomize the differences among inferential paradigms. We will illustrate the Bayesian approach, and we will only sporadically compare it with the methods proposed by other schools: these alternative methods will be briefly summarized in Section 1.2.

We will firstly consider the regular parametric framework where one observes a vector $\boldsymbol{Y} = (Y_1, \ldots, Y_n)$, which, conditionally on some vector-valued parameter $\boldsymbol{\omega} \in \Omega$,

[1] This research has been partially supported by MIUR-Italy, grant no. 2002134833-001.

consists of independent and identically distributed random variables.[2] With an abuse of notation, for $i = 1, \ldots, n$,

$$Y_i \sim f(y; \boldsymbol{\omega}).$$

Rarely the entire parameter vector $\boldsymbol{\omega}$ is of interest and one often wants to make inference on some lower-dimensional, possibly real, function $\theta = t_1(\omega)$. We will denote by $\lambda = t_2(\omega)$ the complementary transformation in order to get $\boldsymbol{\omega} = (\theta, \lambda)$. In this context, the problem of eliminating the nuisance parameter λ corresponds to provide inference on θ which does not depend on λ. One may notice that, in general, there is no a unique way to define the nuisance parameter λ and rarely the choice can be based on the ground of a physical interpretation of the quantities in the model.

We will then discuss a special class of problems where the number of parameters increases with the sample size, which are known as the Neyman and Scott's problem and which still poses crucial questions both from a theoretical perspective and from an applied one: see, for example, the recent review on the role of Neyman and Scott problem in Econometrics by Lancaster (2000).

1.2. Preliminaries

In the frequentist approach the problem of eliminating the nuisance parameters, although present, is not clearly formalized, and, sometimes, not even mentioned, at least in traditional textbooks (Mood et al., 1974). The goal is to produce statistical inferences whose long run properties depend only on θ. Consider the estimation of a normal mean – from an i.i.d. sample – when the standard deviation σ is unknown: the procedure based on the Student's t, provides an example of a technique specifically set up to eliminating the influence of σ upon the conclusions. In general, this problem might admit no solution, as illustrated by the following well known example.

EXAMPLE 1 (*The Behrens–Fisher problem*). We observe $n_1 + n_2$ independent Normal random variables, the first n_1 drawn from a $N(\mu_1, \sigma_1^2)$ distribution, and the other n_2 from a $N(\mu_2, \sigma_2^2)$ distribution. Here the aim is at constructing a confidence interval for the parameter of interest $\theta = \mu_1 - \mu_2$. Let $\overline{X}_1$, $\overline{X}_2$, S_1 and S_2, with obvious meaning of the symbols, be the jointly sufficient statistics. To achieve the goal one needs to find a function of the sample standard deviations, say $h(s_1, s_2)$, not depending on the unknown parameters, such that

$$1 - \alpha = \mathrm{P}\big(\big|\overline{X}_1 - \overline{X}_2 - \theta\big| \leqslant h(S_1, S_2); \delta, \sigma_1, \sigma_2\big).$$

Linnik (1968) has proved that such a function does not exist, and this is a clear illustration of the potential difficulties in eliminating nuisance parameters in exact frequentist inference (Welsh, 1996).

Another aspect to remark is that, in the Neyman–Pearson framework, the nuisance parameter λ need not be explicitly defined, although the strength of the conclusions

[2] The last two assumptions are introduced with the aim to ease the illustration of the theory, although in a Bayesian framework, they do not play a crucial role, and could be removed.

may heavily depend on the values of some component of λ: here is an example of this phenomenon.

EXAMPLE 2 (*Fieller's problem*). In its simplest form, the problem can be stated as follows: let x_1 and x_2 be two sample means obtained from n replications of a bivariate normal random variable with mean $\omega = (\omega_1, \omega_2)$ and covariance matrix $\sigma^2 I_2$, where I_2 denotes the two-dimensional identity matrix. The Fieller's problem amounts to construct a confidence region for the mean ratio $\theta = \omega_1/\omega_2$. For mathematical convenience we may adopt, as the complementary nuisance parameter, either $\lambda = \omega_2$, the mean of X_2, or $\xi = \omega_1^2 + \omega_2^2$. The latter choice has the attractive feature of being orthogonal to θ in terms of the expected Fisher information matrix. Fieller (1954) proposed to construct a confidence set of level $\gamma = 1 - \alpha$, $\alpha > 0$, based on the pivotal quantity

$$\frac{\sqrt{n}(X_1 - \theta X_2)}{\sigma\sqrt{1+\theta^2}} \sim N(0, 1). \tag{1}$$

It is well known that this procedure provides regions of infinite size with positive sampling probability, for each value of θ and for each positive confidence level α. Interestingly, the frequentist probability of obtaining a (useless) confidence set which coincides with the real line can be shown to be

$$P_\omega\left(X_1^2 + X_2^2 \leqslant \frac{\sigma^2 z_{\alpha/2}^2}{n}\right) = \chi_2^2\left(\frac{n}{\sigma^2}\xi, z_{\alpha/2}^2\right), \tag{2}$$

where $\chi_\nu^2(\kappa, t)$ is the cdf of a χ^2 random variable with ν degrees of freedom and non-centrality parameter κ, evaluated at t. This means that Fieller's solution only apparently eliminates the dependence on the nuisance parameter $\xi = \omega_1^2 + \omega_2^2$. Its value actually determines the frequentist probability that $C_F(x)$ will be completely uninformative.

In the likelihood approach, the presence of nuisance parameters is made explicit. In order to make inference on the parameter of interest, one must treat them, in some way. The nuisance parameters are often introduced with the aim to creating more flexible models that might better interpret data at hand. Even if some of the additional parameters are not particularly important in terms of estimation – they often lack a precise physical meaning – their estimates can affect enormously the inference on the parameter of interest. Ideally, the ultimate goal would be to construct a likelihood function for the parameter of interest only. In some special circumstances, it may happen that one or more components of the sufficient statistics have marginal or conditional distributions which depend on θ, but not on λ. Such distributions are then used as pseudo-likelihood for θ: they are called "marginal likelihood" and "conditional likelihood" respectively. Basu (1977) gives an example where it is possible to determine a marginal likelihood and a conditional likelihood for the same parameter, which provide conflicting evidence; this counterexample indicates that use of such techniques is likely to remain somewhat arbitrary. Chapter 8 of Severini (2000) provides a comprehensive discussion of these approaches, along with a series of examples about the construction and the use of these pseudo-likelihoods. Marginal and conditional likelihoods are special cases of the more general *partial likelihood* (Cox, 1975). If there exists a partition (y, z) of the data x, such that

$$f(x|\theta, \lambda) = h(x) f_1(y|\theta, \lambda) f_2(z|y, \theta) \tag{3}$$

or

$$f(x|\theta, \lambda) = h(x) f_1(y|\theta) f_2(z|y, \theta, \lambda), \tag{4}$$

then the terms hf_1 in (3) or hf_2 in (4) are ignored, and the remaining factor is taken as the partial likelihood for θ. Since the ignored term depends on θ, there is always a loss of information. Supporters of the approach suggest, however, that one loses only the information about θ which is inextricably tied with the unknown parameter λ; it does not seem possible, however, to *quantify* the loss. Moreover, many authors have criticized the idea of partial likelihood on the ground that it usually cannot be interpreted in terms of sampling distribution of some suitable observable statistics; in other words the partial likelihood is not a *genuine* likelihood, and one is left only with the possibility of exploring the shape of the particular observed partial likelihood.

A more general and common method is to eliminate λ via conditional maximization. Let $L(\theta, \lambda) = f(\boldsymbol{y}; \theta, \lambda)$ be the likelihood function; for fixed θ, the nuisance parameter λ is replaced by its conditional maximum likelihood estimate $\hat{\lambda}_\theta$, and the analysis is based on the so-called profile likelihood

$$\hat{L}(\theta) = L(\theta, \hat{\lambda}_\theta). \tag{5}$$

As illustrated in Pawitan (2001), the use of profile likelihood induce a sort of bias which can be appreciated by noticing that, for example, the mean value of the score function based on profile likelihood need not be zero, as with regular likelihoods. In standard problems, however, this bias is generally small relative to the standard error of the estimates; this holds, for example, when the number of nuisance parameters is small compared to the sample size. The situation is likely to be more complicated when the number of nuisance parameters is of the same order of magnitude as the sample size. In these situations the use of the profile likelihood may be highly misleading and several corrections have been proposed in literature. We will consider a specific example of this sort in Section 5. Finally, note that, in the likelihood approaches – with the only exception of profile likelihood – the choice of the parameterization of the nuisance parameters may be very important. The interested reader may refer to Severini (2000) or Pace and Salvan (1997) for a comprehensive discussion on the subject.

2. Bayesian elimination of nuisance parameters

Bayesian techniques for eliminating λ seem to be simple and intuitive. Let $\pi(\theta, \lambda) = \pi(\theta)\pi(\lambda|\theta)$ be a prior distribution; under general regularity conditions, inference will be based on the marginal posterior distribution of θ given the observations $\boldsymbol{Y} = \boldsymbol{y}$

$$\pi(\theta|\boldsymbol{y}) = \frac{\pi(\theta) \int_\Lambda L(\theta, \lambda)\pi(\lambda|\theta)\,\mathrm{d}\lambda}{\int_\Omega L(\theta, \lambda)\pi(\theta)\pi(\lambda|\theta)\,\mathrm{d}\lambda\,\mathrm{d}\theta}, \tag{6}$$

where $L(\theta, \lambda) = f(\mathbf{y}|\theta, \lambda)$ is the likelihood function associated with the experiment. Note also that there are no compelling reasons, other than computational ones, of choosing a particular parametrization of λ. Any choice of $\eta = g(\lambda, \theta)$ will be equivalent, provided that the Jacobian $|\partial\lambda/\partial\eta|$ is included in the new posterior. This point is mathematically trivial but crucial because, out of the Bayesian approach, it is difficult and, in general, not possible, to account for the Jacobian if one switches from λ to η. In other words, what we are really using is a *marginal* model $f(\mathbf{y}|\theta)$ together with the marginal prior for θ. As in the frequentist case, the way in which the nuisance parameters are expressed is not important; but now their effect is marginalized out and their value does not affect $\pi(\theta|\mathbf{y})$.

EXAMPLE 1 (*Continued*). Suppose for the sake of simplicity, that we elicit conjugate Normal priors for the parameters of the model, so that

$$\mu_1|\sigma_1 \sim N\left(\alpha_1, \frac{\tau_1^2}{\sigma_1^2}\right), \qquad \frac{1}{\sigma_1^2} \sim \text{Gamma}(r_1, m_1)$$

and

$$\mu_2|\sigma_2 \sim N\left(\alpha_2, \frac{\tau_2^2}{\sigma_2^2}\right), \qquad \frac{1}{\sigma_2^2} \sim \text{Gamma}(r_2, m_2);$$

then it is easy to see that the conditional distribution of each single parameter given the others and the observed data belong to a known family. In fact, for $i = 1, 2$, and denoting with $[\psi]^c$ the parameter vector ω minus the ψ component,

$$\mu_i|\mathbf{y}, [\mu_i]^c \sim N\left(\frac{n_i\bar{x}_i + \tau_i\alpha_i}{n + \tau_i}, \frac{2\sigma_i^2}{n + \tau_i}\right)$$

and

$$\frac{1}{\sigma_i^2}|\mathbf{y}, [\sigma_i^2]^c$$
$$\sim \text{Gamma}\left(r_1 + \frac{1}{2}\left(ns_i^2 + \tau_i(\mu_i - \alpha_i)^2 + n(\mu_i - \bar{x}_i)^2\right), m_1 + \frac{n_i + 1}{2}\right).$$

A simple Gibbs Sampling can then be performed to obtain a sample from the posterior distribution of (μ_1, μ_2), and consequently, from the posterior of θ.

EXAMPLE 3 (*Poisson regression*, Severini, 2000). Let $Y_1, Y_2, \ldots, Y_n$ denote independent random variables such that Y_j has a Poisson distribution with mean $\lambda + \theta x_j$, where $x_1, \ldots, x_n$ are known constants. The likelihood function is

$$L(\theta, \lambda) \propto \exp\left\{-\sum_j e^{\lambda+\theta x_j}\right\} \exp\left\{\lambda \sum_j y_j + \theta \sum_j x_j y_j\right\}.$$

An initial and simple Bayesian approach to the problem is the following: let η be equal to $\exp\{\lambda\}$ and consider a Gamma prior on η, independently of θ, namely

$$\pi(\eta) \propto \eta^{c-1} \exp\{-d\eta\}; \tag{7}$$

then the integrated likelihood for θ is

$$\tilde{L}(\theta) \propto \exp\left\{\theta \sum_j x_j y_j\right\} \int_0^\infty \eta^{\sum_j y_j + c - 1} \exp\left\{-\eta\left(d + \sum_j e^{\theta x_j}\right)\right\}$$

$$\propto \frac{\exp\{\theta \sum_j x_j y_j\}}{(\sum_j e^{\theta x_j} + d)^{\sum_j y_j + c}}. \tag{8}$$

This is in no way the suggested default prior for the nuisance parameter: we will return to this example in Section 3.1.

From a purely theoretical perspective there seems to be no problem in this program. One can build a very sophisticated and flexible model and then eliminate the influence of the nuisance parameters simply by marginalizing out them as in (6). However, there are no free lunches and the use of models with a large-dimensional nuisance parameter λ implies the elicitation of a multivariate conditional prior $\pi(\lambda|\theta)$ which is almost never easy to assess. For these reasons, even a *pucca* – fully baked, in the terminology of Basu (1977) – Bayesian would agree to consider a half-baked Bayesian strategy which can be either

- use a default prior for (θ, λ), say $\pi^N(\theta, \lambda)$ and, following the Bayesian paradigm, base the conclusions upon (6)

or, even less-baked,

- use only the conditional default prior of $\lambda|\theta$ and produce the so-called integrated likelihood function for θ, namely

$$\tilde{L}(\theta) = \int_\Lambda L(\theta, \lambda)\pi^N(\lambda|\theta)\, d\lambda. \tag{9}$$

Of course an integrated likelihood could be also calculated using a subjective conditional prior but this is less common in practice: a subjective Bayesian would go one step further and he would use his subjective $\pi(\theta)$ too! We will return on the legitimate use of an integrated likelihood. These two approaches will be discussed in some details in Sections 3.1 and 3.2.

The use of default improper priors, or of integrated likelihoods, however, poses some problems. It is well recognized – see, for example, Berger and Bernardo (1992a) and Berger et al. (1999), Example 5 – that the correct derivation of reference priors depends on what the parameter of interest is and the use of Jeffreys' invariant priors can be misleading in high-dimensional problems. Also, default priors are almost always improper. This implies two important remarks, among others:

- the simulation approach followed in Example 1 cannot be directly used in the presence of improper priors; there is simply no guarantee that the resulting posterior is proper and this must be checked, particularly in the presence of nonstandard models.

- the use of improper priors induce improper marginal models, that is, $f(\boldsymbol{y}|\theta)$ is not a probability distribution: consequently, the resulting integrated likelihood is not a genuine likelihood and care must be taken in dealing with it.

However, it is a matter of fact that, when the dimension of the parameter space is large, a fully subjective Bayesian analysis of the problem is hard to perform; in these cases the choice a good *default* prior at least for the nuisance parameters, can be a valid alternative. Among the many reasons why even a Bayesian would consider to base his inference upon the integrated likelihood, Berger et al. (1999) mention

(i) *Scientific reporting*: It is usually considered good form to report separately $L(\theta)$ and $\pi(\theta|x)$ in order to make explicit the effect of the prior distribution. This also allows others to utilize their own prior distributions for θ.
(ii) *Sensitivity analysis*: It is important to consider sensitivity with respect to $\pi(\theta)$, and having $L(\theta)$ available for this purpose is valuable. (Of course, sensitivity to $\pi(\lambda|\theta)$ is also a potential concern, but frequently this is of less importance.)
(iii) *Elicitation cost*: It is typically very expensive (in terms of time and effort) to obtain subjective prior distributions, especially when the nuisance parameter are many and they cannot be safely considered independent. Under the frequent cost limitations with which we operate, elicitation efforts may have to be limited. It is often cost effective to eliminate nuisance parameters in a default fashion, resulting in $L(\theta)$, and concentrate subjective elicitation efforts on $\pi(\theta)$.
(iv) *Objectivity*: Although most statisticians (including myself) are justifiably skeptical of the possibility of truly "objective" analysis, there is an undeniable need, in some applications, for an analysis that appears, as much as possible, objective. Using $L(\theta)$, with default $\pi(\lambda|\theta)$, can satisfy this need.
(v) *Combining likelihoods*: If one obtains information about θ from different independent sources, and the information arrives in the form of several likelihood functions, say $L_i(\theta)$, then one can summarize the information by $\prod_i L_i(\theta)$ and proceed, coherently with Bayes' theorem, obtaining his own personal posterior from his prior distribution. This is the basis of many important meta-analysis techniques. One could not, of course, simply multiply posteriors in this way.
(vi) *Improper priors*: Focusing on integrated likelihood seems to reduce some of the dangers of using improper priors.

3. Objective Bayes analysis

In this section we describe how to deal with the nuisance parameters in absence of genuine prior information on them. In Section 3.1 we illustrate the use of integrated likelihood to produce, in a default way, a likelihood function for the parameter of interest; that section will be mainly based on Berger et al. (1999). The obvious objective Bayesian way to deal with nuisance parameters is, however, to base conclusions on the marginal posterior distribution of the parameter of interest based on some default prior. In Sections 3.2 and 5 we will discuss the use of reference priors and we will try to show how this approach is superior to the use of the Jeffreys's invariant prior.

3.1. Integrated likelihood

Let $q(\lambda)$ be a nonnegative weight function defined on Λ and such that $\int q(\lambda)\,\mathrm{d}\lambda < \infty$. An integrated likelihood function for the parameter of interest θ is then

$$\tilde{L}_q(\theta) \propto \int_{\Lambda} f(\mathbf{y}, \theta, \lambda) q(\lambda)\,\mathrm{d}(\lambda) = f_q(\mathbf{y}; \theta). \tag{10}$$

It is trivial to see (and clearly intuitive) that a Bayesian would select $q(\lambda)$ as his/her conditional prior distribution for $\lambda|\theta$, say $\pi(\lambda|\theta)$. As long as π is a proper distribution this correspond to use the marginal model $f_q(\mathbf{y}; \theta)$: in this sense $\tilde{L}_q(\theta)$ is a genuine likelihood function. If π is improper, as it is often the case in the default approach, this is not guaranteed and care must be taken in the choice of π itself.

In the effort of being noninformative, a first natural choice here seems the use of a uniform prior over the nuisance parameter λ. Although this choice can be reasonable from several perspectives, it cannot be considered a conclusive solution. In fact, uniformity is not compatible with invariance and a uniform prior for λ will not be uniform over a new parameter $\eta = g(\lambda, \theta)$.

EXAMPLE 3 (*Continued*). The uniform prior on $\eta = \exp\{\lambda\}$ will give here

$$\tilde{L}_1(\theta) = \frac{\exp\{\theta \sum_j x_j y_j\}}{(\sum_j \mathrm{e}^{\theta x_j})^{\sum_j y_j + 1}},$$

which correspond to the use of (7) with $c = 1$ and $d = 0$. However, a uniform prior on λ would correspond to use a prior for η which is proportional to η^{-1}: this would give an integrated likelihood $\tilde{L}_2(\theta)$ again as in (7), with $c = d = 0$.

These kinds of incoherency will have minor consequences in regular problems but they can lead to strongly misleading results when the number of nuisance parameters is large: see Section 5.

Let $I(\theta, \lambda)$ represent the expected Fisher information. The Jeffreys's invariant method for deriving default priors is based on the use of

$$\pi^J(\theta, \lambda) = |\det I(\theta, \lambda)|^{1/2}.$$

This approach is explicitly tailored to obtain a default prior when the whole parameter vector is of interest. Within this approach one cannot distinguish the roles of θ and λ. There are many cases where, being noninformative on both θ and λ is equivalent to introduce, unwittingly, a quite informative prior over some specific sub-parameter $\tau = h(\theta, \lambda)$: see Example 7 below. Another aspect of the same problem is that, whatever definition of *noninformativeness* is chosen, one cannot be, at the same time, noninformative on the parameter of interest θ and on the whole vector ω. The following example is adapted from Berger and Bernardo (1992b).

EXAMPLE 4 (*Binomial–Trinomial model*). A medical doctor observes n patients which can answer positively or not to a specific test; he models the problem as n Binomial trials, $X_1, X_2, \ldots, X_n$ i.i.d. $\mathrm{Be}(\theta)$, where θ is the probability of a negative answer: in this case the obvious Jeffreys' prior is the Beta($\frac{1}{2}, \frac{1}{2}$) distribution with mean 0.5. Now suppose the doctor feels that the dichotomous answer is too coarse and he decides to classify the patients into three groups, say, for example, *strongly positive*, *moderate positively*, and *negative*. The appropriate model is now the Trinomial(θ, λ), where θ is still the probability of a negative answer and λ is the probability of a moderately positive answer. The Jeffreys' prior for the trinomial model is given by

$$\pi^J(\theta, \lambda) = \frac{1}{2\pi}(\theta\lambda)^{-1/2}(1-\theta-\lambda)^{-1/2}. \tag{11}$$

The marginal prior for θ is $\pi_1^J(\theta) = (2\sqrt{\theta})^{-1}$ which is quite different from the Jeffreys' prior in the Binomial model: for example the resulting prior mean is now $1/3$. The gist of this example is to clarify that, in a default Bayes analysis, one should *declare* what is the main goal of the inference. The reference prior method, which we illustrate below, tries to provide an answer to these problems.

A final difficulty for a practical use of the Jeffreys' prior is that there is no guidance on how to factorize $\pi^J(\theta, \lambda)$ into $\pi^J(\theta)\pi^J(\lambda|\theta)$ in order to use the last factor to produce an integrated likelihood.

The reference prior approach, proposed by Berger and Bernardo in a series of papers (1989, 1992a, 1992b, 1992c), typically provides the same priors as the Jeffreys approach in the one-dimensional case; when the parameter space is multidimensional, the reference algorithm takes into account the order of inferential importance of the parameters, by partitioning the parameter vector $\omega = (\omega_1, \omega_2, \ldots, \omega_p)$ into several blocks of decreasing interest to the investigator. Berger and Bernardo suggest using the one-at-time reference prior which corresponds to partitioning ω into p one-dimensional blocks.

In the standard Bayesian approach, the reference prior is used to produce the joint posterior distribution for (θ, λ); then λ is integrated out to obtain the marginal posterior for θ. In this approach, there is no need to develop the notion of a likelihood for θ. In order to produce a reference integrated likelihood, we need to slightly modify the reference prior algorithm. For the sake of simplicity, we will simply partition ω in two blocks, where the parameter vector is split into the parameters of interest, θ, and the nuisance parameters, λ; extensions to several groups is immediate. For simplicity of notation, we will take θ and λ to be scalars.

Under certain regularity conditions, basically the existence of a consistent and asymptotically normal estimator of the parameters, the reference prior for λ, when θ is known, is defined to be

$$\pi^*(\lambda|\theta) \propto \sqrt{I_{22}(\theta, \lambda)}, \tag{12}$$

where $I_{22}(\theta, \lambda)$ is the lower right corner of the expected Fisher information matrix. The direct use of (12) to obtain an integrated likelihood is problematical due to its typical impropriety; a possible correction (coherent with the spirit of the reference prior algorithm) is then to consider a sequence of nested subsets of $\Omega = \Theta \times \Lambda$, $\Omega_1, \Omega_2, \ldots,$ increasing to Ω and over which (12) can be normalized; then the conditional reference prior, and associated integrated likelihood, will be defined as the appropriate limit. Defining $\Lambda_m(\theta) = \{\lambda\colon (\theta, \lambda) \in \Omega_m\}$ and the normalizing constant over this compact set as $K_m(\theta)^{-1} = \int_{\Lambda_m(\theta)} \pi^*(\lambda|\theta)\,\mathrm{d}\lambda$, the *conditional reference prior* is defined as

$$\pi^R(\lambda|\theta) = h(\theta)\pi^*(\lambda|\theta),$$

where

$$h(\theta) = \lim_{m\to\infty} \frac{K_m(\theta)}{K_m(\theta_0)}, \tag{13}$$

assuming the limit is unique up to a proportionality constant for any θ_0 in the interior of Θ. The corresponding integrated likelihood is then given by

$$L^R(\theta) = \int_\Lambda f(x|\theta,\lambda)\pi^R(\lambda|\theta)\,\mathrm{d}\lambda,$$

which is in the "standard" form of an integrated likelihood.

This definition of conditional reference prior can be shown to be consistent with the definition of the joint reference prior $\pi^R(\theta,\lambda)$ in Berger and Bernardo (1992a), providing the joint reference prior exists and (13) holds, in the sense that

$$\pi^R(\theta,\lambda) = \pi^R(\lambda|\theta)\pi^R(\theta)$$

for some function $\pi^R(\theta)$. We will then define $\pi^R(\theta)$ to be the *marginal reference prior* for θ. The conditional reference prior can be shown to share many of the desirable properties of the joint reference prior, such as invariance to reparameterization of the nuisance parameters.

EXAMPLE 3 (*Continued*). Here the Fisher matrix, in the (θ,η) parameterization, is given by

$$I(\theta,\eta) = \begin{pmatrix} \sum_j x_j^2\,\mathrm{e}^{\theta x_j} & \sum_j x_j\,\mathrm{e}^{\theta x_j} \\ \sum_j x_j\,\mathrm{e}^{\theta x_j} & \frac{1}{\eta}\sum_j \mathrm{e}^{\theta x_j} \end{pmatrix}.$$

From a corollary of Proposition 5.30 in Bernardo and Smith (1994), it is easy to see that the conditional prior for $\eta|\theta$ is given by $\pi^R(\eta|\theta) \propto \eta^{-1}$ and the corresponding integrated likelihood corresponds to the already obtained $\tilde{L}_2(\theta)$.

3.2. *Reference prior approach*

The most obvious way to make inference, in an *objective* Bayesian way, on a given parameter of interest θ is of course to produce an objective marginal posterior distribution for θ. The most popular and general approach to this problem are the Jeffreys' invariant approach (Kass and Wasserman, 1996) and the reference prior approach (Berger and Bernardo, 1992a).

EXAMPLE 1 (*Continued*). Using the parameterization $\theta = \mu_1 - \mu_2$ and $\lambda = \mu_1 + \mu_2$, the expected Fisher matrix for this problem is given by

$$H(\theta, \lambda, \sigma_1, \sigma_2) = \begin{pmatrix} \frac{1}{4}\left(\frac{1}{\sigma_1^2} + \frac{1}{\sigma_2^2}\right) & \frac{1}{4}\left(\frac{1}{\sigma_1^2} - \frac{1}{\sigma_2^2}\right) & 0 & 0 \\ \frac{1}{4}\left(\frac{1}{\sigma_1^2} - \frac{1}{\sigma_2^2}\right) & \frac{1}{4}\left(\frac{1}{\sigma_1^2} + \frac{1}{\sigma_2^2}\right) & 0 & 0 \\ 0 & 0 & \frac{2}{\sigma_1^2} & 0 \\ 0 & 0 & 0 & \frac{2}{\sigma_2^2} \end{pmatrix}.$$

It is simple to see that the Jeffreys' prior for the Behrens–Fisher problem is

$$\pi^J(\theta, \lambda, \sigma_1, \sigma_2) \propto \frac{1}{\sigma_1^2 \sigma_2^2}$$

and the reference prior, when the parameter of interest is θ and for every possible ordering of the nuisance parameters, is

$$\pi^R(\theta, \lambda, \sigma_1, \sigma_2) \propto \frac{1}{\sigma_1 \sigma_2}.$$

The difference among the two prior here is modest, and for moderate to large sample sizes, the resulting inferences are practically identical; however the discrepancy among the two priors is typical of the fact that the reference algorithm explicitly recognizes the ordering of interest among parameters, and π^R is actually the prior which asymptotically maximizes a particular version of the Kullback–Leibler distance among the marginal prior and the marginal posterior from the parameter of interest θ; see Ghosh and Mukerjee (1992) and the paper by Bernardo in this volume for more details on this issue.

One should also notice that, in this example, Jeffreys itself would have used the reference prior π^R instead of π^J: as illustrated in Kass and Wasserman (1996), in fact, the Jeffreys' rule should be modified in the presence of location and scale parameters.

EXAMPLE 2 (*Continued*). The complete reference prior approach here will provide a marginal prior for θ given by

$$\pi^R(\theta) \propto \frac{1}{\sqrt{1+\theta^2}}$$

and it is possible to show that the resulting marginal posterior for θ is proper and it can be written in a closed form as

$$\pi^R(\theta|x_1, x_2) = \frac{1}{\pi(1+\theta^2)} \exp\left\{-\frac{n(x_1 - \theta x_2)^2}{2\sigma^2(1+\theta^2)}\right\} \frac{\exp\left\{\frac{n(x_1^2+x_2^2)}{4\sigma^2}\right\}}{I_0\left(\frac{n(x_1^2+x_2^2)}{4\sigma^2}\right)}, \tag{14}$$

where I_0 denotes the first-type Bessel function of zero order. This posterior distribution is a legitimate Bayesian summary for inference on θ; however, unlike more regular problems, it is not well calibrated from the frequentist coverage point of view. Because of the propriety of the posteriors, the induced high posterior densities credible regions do turn out to be finite, although sometimes they are disconnected; then the practical results from a default Bayesian analysis are quite different from those obtained in a frequentist way. The Fieller' problem is an example which shows that frequentist and conditional approaches are not always compatible. Whatever the nuisance parameters are eliminated, it is possible to observe samples which produce a weakly informative likelihood for θ: consequently, a confidence interval may be infinite. On the other hand a Bayesian (even noninformative) analysis introduces some prior information which makes the posterior proper.

4. Comparison with other approaches

The large amount of literature on the problem of eliminating nuisance parameters is in sharp contrast with the small number of papers which try to make comparisons among different approaches. One reason to that could be the fact that each method, either based on maximization, integration, or marginalization is well suited within a theory and it is not considered within another. So, in the last years, we have seen dramatic improvements both within the likelihood and the Bayesian schools, in terms of accuracy of the approximations for the modified profile likelihoods and in terms of the choice of the prior for the integrated likelihood. On the other hand, this philosophical separation has produced no systematic comparisons among different schools and there exist only a few attempts to make some partial comparison. Liseo (1993) consider a series of examples where the marginal posterior distribution for the parameter of interest is compared in a frequentist sense, with the available likelihoods for that problem, pretending that likelihoods could be treated as probability distributions: this approach may be biased in favor of the Bayesian approach. See also Reid (1996), where comparisons are based in terms of p-values, introducing then a bias in favor of the likelihood approaches. Another possibility is depicted in an unpublished technical report (Liseo and Sun, 1999), where a simple and general method of comparing two different pseudo-likelihoods for the same model is presented; comparisons are done in a robust Bayesian perspective, by computing, for each prior distribution in a given class Γ, the Kullback–Leibler divergence between each single pseudo-marginal posterior (induced by the pseudo-likelihood) and the true marginal posterior distribution of the parameter of interest. A pseudo-likelihood will be superior to another in terms of Kullback–Leibler divergence, if the distance which induces from the true marginal posterior is smaller compared to the other.

In general, the different rationale behind the different approaches makes the comparison very difficult and one can learn more from specific examples rather than attempting a systematic comparison.

EXAMPLE 2 (*Continued*). The likelihood function for this problem is

$$L(\theta, \lambda) \propto \exp\left\{-\frac{n}{2\sigma^2}\left[(x_1 - \lambda\theta)^2 + (x_2 - \lambda)^2\right]\right\}$$

$$= \exp\left\{-\frac{n}{2\sigma^2}\left[\left(1+\theta^2\right)\left(\lambda - \frac{x_1\theta + x_2}{1+\theta^2}\right)^2 + \frac{(x_1 - \theta x_2)^2}{1+\theta^2}\right]\right\}; \quad (15)$$

from (15), it is immediate to see that $\hat{\lambda}_\theta = (x_1\theta + x_2)/(1+\theta)^2$ and the profile likelihood is

$$\hat{L}(\theta) \propto \exp\left\{-\frac{n}{2\sigma^2}\frac{(x_1 - \theta x_2)^2}{1+\theta^2}\right\}, \tag{16}$$

which is bounded away from zero, since, as $|\theta| \to \infty$, $\hat{L}$ converges to $\exp\{-nx_2^2/2\}$. Although this can be reasonable, it is hard to use this likelihood, especially when the observed value of $|x_1/x_2|$ is large: in fact, in these cases, any reasonable likelihood set, that is $\{\theta\colon \hat{L}(\theta) > k\}$, for some $k > 0$, would have infinite size. This behavior is typical of a larger family of statistical problems, known as the Gleser and Hwang's class (Gleser and Hwang, 1987; see also Berger et al., 1999).

This problem can be also handled in a pure likelihood set up, using the so-called modified profile likelihood approach (Barndorff-Nielsen, 1983), which can be seen as an approximated marginal likelihood based on the sampling distribution of the maximum likelihood estimator of the parameter of interest, conditionally on some ancillary statistics. It can be shown that in this problem the correction induced by the modified profile likelihood approach is given by the multiplicative factor

$$\left|j_\lambda(\theta, \hat{\lambda}_\theta)\right|^{1/2}\left|\frac{\lambda}{\hat{\lambda}_\theta}\right| = \sqrt{1+\theta^2},$$

where $|j_\lambda(\theta, \lambda)|$ is the observed Fisher information for λ. The tail behavior is then even worse than (16). A direct use of the conditional reference prior algorithm here gives that

$$\pi(\lambda|\theta) \propto 1$$

and, consequently,

$$\tilde{L}(\theta) \propto \hat{L}(\theta)\frac{1}{\sqrt{1+\theta^2}}.$$

This likelihood eliminates the problem on the tails and allows to produce finite likelihood sets, which, because of the Gleser and Hwang's phenomenon, cannot have good frequentist properties for finite sample sizes.

The following example is relatively less known but it shows clearly the inherent consequences of the various elimination philosophies.

EXAMPLE 5 (*The broken sample*, DeGroot and Goel, 1980; Goel and Ramalingam, 1989). Suppose that a random sample of size n say $(X_i, Z_i), i = 1, \ldots, n$, is drawn from a bivariate normal distribution. For the sake of simplicity we assume that all the parameters but the correlation coefficient ρ are known: in particular we assume that $\mu_x = \mu_z = 0$ and $\sigma_x = \sigma_z = 1$. Suppose also, however, that before the sample values are recorded, each observation (x_i, z_i) gets broken into two separate components x_i and z_i. As a consequence, the available information comes into the form of $\boldsymbol{x} = (x_1, x_2, \ldots, x_n)$ and $\boldsymbol{y} = (y_1, y_2, \ldots, y_n)$ where $\boldsymbol{y}$ is some permutation of the original values $z_1, z_2, \ldots, z_n$.

In this context ρ is our parameter of interest while the permutation of the z_i's can be considered a nuisance parameter; this is actually a typical example where different inferential schools give different names to the same mathematical object: in a classical or likelihood framework the permutation would be called "latent structure" or, maybe, a vector of missing items.

Let ϕ denote the generic permutation of the z_i's; it is easily seen that the log-likelihood function for this problem is given by

$$\ell(\rho, \phi) = -\frac{n}{2}\log(1-\rho^2) + \frac{\rho}{1-\rho^2}\sum_{i=1}^{n} x_i y_{\phi(i)} - \frac{1}{2(1-\rho^2)}\left(\sum_{i=1}^{n}(x_i^2 + y_i^2)\right); \tag{17}$$

the profile likelihood is then obtained by choosing the permutation $\hat{\phi}_\rho$ which maximizes the second term of (17). Let

$$r(\phi) = \frac{1}{n}\sum_{i=1}^{n} x_i y_{\phi(i)}$$

be the sample correlation coefficient for a given permutation; also let $x_{(1)} \leqslant x_{(2)} \leqslant \cdots \leqslant x_{(n)}$ and $y_{(1)} \leqslant y_{(2)} \leqslant \cdots \leqslant y_{(n)}$ denote the order statistics of the sample values. Finally let us define the quantities

$$\phi^{(2)} = \arg\max_{\phi} r(\phi); \qquad \phi^{(1)} = \arg\min_{\phi} r(\phi).$$

DeGroot and Goel (1980) show that $\phi^{(2)}$ is given by the maximal concordant permutation, that is the one which matches $x_{(i)}$ to $y_{(i)}$, for $i = 1, \dots, n$. Analogously, one can see that $\phi^{(1)}$ is equal to the maximal discordant permutation, that is when $x_{(i)}$ is paired to $y_{(n-i+1)}$, for $i = 1, \dots, n$. They also show that $\phi^{(1)} < 0 < \phi^{(2)}$. All together, it is easy to see that the conditional maximum likelihood estimate of ϕ is given by

$$\hat{\phi}_\rho = \begin{cases} \text{maximal concordant permutation}, & \rho > 0, \\ \text{maximal discordant permutation}, & \rho \leqslant 0. \end{cases}$$

The resulting profile log-likelihood is then

$$\hat{L}(\rho) = \begin{cases} -\frac{n}{2}\log(1-\rho^2) + \frac{\rho}{1-\rho^2} n r(\phi^{(2)}) \\ \quad - \frac{1}{2(1-\rho^2)}(\sum_{i=1}^{n}(x_i^2 + y_i^2)), & \rho > 0, \\ -\frac{n}{2}\log(1-\rho^2) + \frac{\rho}{1-\rho^2} n r(\phi^{(1)}) \\ \quad - \frac{1}{2(1-\rho^2)}(\sum_{i=1}^{n}(x_i^2 + y_i^2)), & \rho \leqslant 0. \end{cases} \tag{18}$$

The profile likelihood is not a reasonable likelihood for ρ. For example, the resulting maximum likelihood estimator for ρ is always given by either the maximum or the minimum sample correlation coefficient. There is no point in replacing the uncertainty about the "true" permutation of the z_i's with a single one, which maximizes (or minimizes) the correlation among $\boldsymbol{x}$ and $\boldsymbol{y}_{\phi}$. Note also that the discreteness of the nuisance parameter makes impractical the adoption of the usual corrections to the profile likelihood; no solutions based on conditional or marginal experiments are known.

The integrated likelihood for this problem is obtained by averaging with respect the uniform distribution over the set Φ of all the permutations,

$$\tilde{L}(\rho) = \frac{1}{n!}\left(1-\rho^2\right)^{-n/2} \exp\left\{-\frac{1}{2(1-\rho^2)} \sum_i \left(x_i^2 + y_i^2\right)\right\} \times \sum_{\phi\in\Phi}\left[\exp\left\{\frac{n\rho}{1-\rho^2} r(\phi)\right\}\right]. \tag{19}$$

The value of ρ which maximizes (19) must be found numerically; it can be proved that there are at least two local maxima and they have always different signs.

This particular problem clarifies the different ways in which a Bayesian analysis can be useful in eliminating nuisance parameters. In the above set-up, it is difficult to disagree that the conditional prior for ϕ should be uniform over the set of all the $n!$ possible permutations of the z_i's: this is an example of "objective" elimination of the nuisance parameter. Similar problems emerge in a special class of problems, which are known as record linkage; see, for example, Fellegi and Sunter (1969) and Belin and Rubin (1995). In record linkage problems the matching step is based on some additional variables which are observed twice (once together with $\boldsymbol{x}$ and once together with $\boldsymbol{y}$); this extra information modifies the likelihood of each single pair of being a true match or not: in cases like that it is natural to compute (19) using the posterior (given the additional observed variables) distribution of ϕ. See Lahiri and Larsen (2005) for a classical treatment of this problem and Liseo and Tancredi (2004) for a Bayesian analysis.

5. The Neyman and Scott class of problems

In this section we consider the case where the number of nuisance parameters increase at the same rate as the sample size. These problems are well known in statistical, econometric and bio-statistical literatures as the Neyman and Scott's type of problems (Neyman and Scott, 1948). Here we address the following general set-up: consider independent stratified observations in the form

$$Y_{ij} \sim p_{ij}(y_{ij};\theta,\lambda_i),$$

where $i = 1,\ldots,k$ and $j = 1,\ldots,m$, and the sample size is $n = km$. The parameter of interest is θ and the nuisance parameters, often called *incidental*, are given by the strata parameters $\lambda_1,\ldots,\lambda_k$. If we take m fixed and let k increase, then the sample size and the number of nuisance parameters increase at the same rate; in cases like this the profile likelihood does not provide good results, the maximum profile likelihood estimates being often inconsistent, and modifications of the likelihood theory are necessary: see, for example, Pace and Salvan (1997), and Sartori (2003). For the sake of simplicity and in order to make practical comparisons, we restrict our analysis to the following restricted scenario.

EXAMPLE 6 (*Common odds-ratio*). Suppose Y_{ij} are binomial random variables with parameters n_{ij} and p_{ij} ($j = 0, 1$) which follow a logistic model

$$\operatorname{logit} p_{ij} = \lambda_i + \theta_j,$$

with $i = 1, \ldots, k$; for identifiability reasons θ_0 is set equal to zero and $\theta_1 = \theta$, which is considered constant across the k strata, is the parameter of interest. It can happen that the n_{ij}'s are small compared to k; in these cases the maximum likelihood estimate $\hat{\theta}$ of θ is seriously biased and, in the extreme case of matched pairs of Bernoulli observations – e.g., $n_{ij} = 1$ – one can see that $\hat{\theta} \to 2\theta$, as $k \to \infty$.

We now consider all the existing likelihood and Bayesian solutions for this problem, and we argue that the difference among the consequences of integrating out or profiling the nuisance parameters gets sharper as the dimension of the nuisance parameter increases.

Following Davison (2003), we formalize the problem in the following way: let (R_{i0}, R_{i1}) ($i = 1, \ldots, k$) be k independent matched pairs such that, for all i,

$$R_{i0} \sim \mathrm{Be}\left(\frac{e^{\lambda_i}}{1 + e^{\lambda_i}}\right), \qquad R_{i1} \sim \mathrm{Be}\left(\frac{e^{\lambda_i + \theta}}{1 + e^{\lambda_i + \theta}}\right),$$

where the λ_i's are the (individual specific) nuisance parameters. The complete likelihood function for $\boldsymbol{\lambda} = (\lambda_1, \ldots, \lambda_k)$ and θ is

$$\begin{aligned} L(\theta, \boldsymbol{\lambda}) &= \prod_{i=1}^{k} \left[\frac{e^{\lambda_i R_{i0}}}{1 + e^{\lambda_i}} \frac{e^{(\lambda_i + \theta) R_{i1}}}{1 + e^{\lambda_i + \theta}} \right] \\ &= \frac{\exp\{\sum_{i=1}^{k} \lambda_i S_i + \theta T\}}{\prod_{i=1}^{k} (1 + e^{\lambda_i})(1 + e^{\lambda_i + \theta})}, \end{aligned}$$

where $S_i = R_{i0} + R_{i1}$, $i = 1, \ldots, k$, and $T = \sum_{i=1}^{k} R_{i1}$ is the number of successes among the *cases*. S_i can be either 0, 1 or 2. It is readily seen that, when $S_i = 0$ or 2, the conditional maximum likelihood estimate of the corresponding λ_i is infinite. Because of this result, classical and likelihood approaches to this problem eliminate the matched pairs for which $S_i = 0$ or 2 from the analysis, since *they are noninformative on* θ. A Bayesian would not agree on this point: if a matched pair gives the same answer to both *treatment* and *control*, this may be an indication that the difference among *treatment* and *control* is not dramatic, at least on that specific pair! This point is deeply discussed in Altham (1971): she performed a fully Bayesian conjugate analysis of the problem and showed that, while the posterior probability of the event $\theta > 0$ is not influenced by the pairs with $S_i = 0$ or 2, the complete distribution of θ does!

The likelihood analysis goes on, restricted to the h units with $S_i = 1$, with the computation of the profile likelihood. Since $\hat{\lambda}_{i,\theta} | (S_i = 1) = -\frac{1}{2}\theta$, one has

$$\hat{L}(\theta) \propto \frac{e^{(T - h/2)\theta}}{(1 + e^{-\theta/2})^h (1 + e^{\theta/2})^h} = \frac{e^{\theta T}}{(1 + e^{\theta/2})^{2h}}.$$

This likelihood is useless: the maximum profile likelihood is $\hat{\theta} = 2\log\frac{T}{h-T}$. Since, with respect to units with $S_i = 1$,

$$T|\theta \sim \text{Binomial}\left(h, \frac{e^{\theta}}{1+e^{\theta}}\right), \tag{20}$$

it is easy to see that $\hat{\theta}$ is inconsistent, because, as $h \to \infty$, $\hat{\theta} \to 2\theta$.

An alternative likelihood approach is based on a conditional argument. Because of (20), one can use the conditional distribution of T given $S_1 = S_2 = \cdots = S_h = 1$ to get a conditional likelihood

$$L_C(\theta) \propto \binom{h}{T}\frac{e^{\theta T}}{(1+e^{\theta})^h}.$$

This likelihood solves the inconsistency problem, since the corresponding estimate for θ is

$$\hat{\theta}_C = \log\frac{T}{h-T} \to \theta.$$

One can also try to *correct* the profile likelihood by using the modified profile likelihood (Barndorff-Nielsen, 1983), as illustrated in Davison (2003). In this case the correction amounts to multiply the profile likelihood by a term

$$M(\theta) = \left|J_{\lambda\lambda}(\theta, \hat{\lambda}_\theta)\right|^{-1/2}\left|\frac{\partial\hat{\lambda}}{\partial\hat{\lambda}_\theta^T}\right| = \frac{e^{h\theta/4}}{(1+e^{\theta/2})^h},$$

where $\hat{\lambda}_\theta$ is the maximum likelihood estimate of λ for fixed θ and $J_{\lambda\lambda}(\theta, \lambda)$ represents the lower right corner of the observed Fisher information matrix. Intuitively, the first factor of the correction penalizes the profile likelihood by an amount that depends on the available information on λ; the second term is a Jacobian which renders the modified profile likelihood invariant to interest-preserving reparameterization of the nuisance parameters: see Pace and Salvan (1997) and Davison (2003) for details. Notice that the correction $M(\theta)$, in this example, does not depend on the observed sample but only on the sample size. Then the resulting maximum modified profile likelihood estimate of θ turns out to be

$$\hat{\theta}_{mp} = 2\log\left(\frac{h+4T}{5h-4T}\right),$$

which is still inconsistent although the problem is not so serious compared with the profile likelihood. Puzzling facts occur for small sample sizes: in fact, in these cases, h can be very small and all the units included in the analysis have the same pattern (e.g., $R_{i0} = 0$ and $R_{i1} = 1$ or vice versa). In these cases $T = 0$ or $T = h$ and, in the last case, for example,

- the conditional likelihood and the profile likelihood are increasing functions of θ,
- the modified profile likelihood attains its maximum *always* at $\theta_0 = 2\log 5 = 3.219$.

In these extreme cases the modified profile likelihood provides reasonable results as long as the true value of $|\theta|$ is less than 3 (Davison, 2003).

A Bayesian analysis based on default priors can be easily performed here. Firstly, one needs not to discard the pairs with $S_i = 0$ or 2; the likelihood contribution of the ith pair is

$$L_i(\lambda_i, \theta) = \frac{\mathrm{e}^{\lambda_i S_i + \theta R_{i1}}}{(1+\mathrm{e}^{\lambda_i})(1+\mathrm{e}^{\lambda_i+\theta})}.$$

Setting $\omega_i = \mathrm{e}^{\lambda_i}/(1+\mathrm{e}^{\lambda_i})$, one can easily see that the conditional reference prior for ω_i is the usual (proper) Jeffreys' prior

$$\pi(\omega_i|\theta) \propto \omega_i^{-1/2}(1-\omega_i)^{-1/2}.$$

After some algebra, the resulting integrated likelihood contribution of the ith pair is

$$\tilde{L}_i(\theta) = \mathrm{e}^{\theta R_{i1}} \int_0^1 \frac{\omega_i^{S_i - 1/2}(1-\omega_i)^{3/2 - S_i}}{(1-\omega_i(1-\mathrm{e}^{\theta}))}\,\mathrm{d}\omega_i$$

which can be shown (Abramowitz and Stegun, 1964, formula 15.3.1, p. 558) to be proportional to

$$\tilde{L}_i(\theta) \propto \mathrm{e}^{\theta R_{i1}}\, {}_2F_1\left(1, S_i + \frac{1}{2}, 3, 1-\mathrm{e}^{\theta}\right), \tag{21}$$

where ${}_2F_1$ is the Hypergeometric or Gauss Geometric series defined as

$${}_2F_1(a,b,c,z) = \frac{\Gamma(c)}{\Gamma(a)\Gamma(b)} \sum_{j=0}^{\infty} \frac{\Gamma(a+j)\Gamma(b+j)}{\Gamma(c+j)} \frac{z^j}{j!}.$$

Denoting with $\tilde{L}_{jk}(\theta)$ the integrated likelihood contribution for a pair such that $(R_{i0}, R_{i1}) = (j,k)$, and with $n_{j,k}$ the number of pairs with $(R_{i0}, R_{i1}) = (j,k)$, the resulting likelihood is then

$$\tilde{L}(\theta) \propto \prod_{j,k=0,1} \tilde{L}_{jk}(\theta)^{n_{jk}}.$$

This likelihood always provides a reasonable behavior; it is not directly comparable with conditional and profile likelihoods since it uses the units with $j = k$, also, and situations with $T = m$ are, in a frequentist sense, less common.

Figures 1 and 2 show two examples with simulated data. Let $F = (F_0, F_1, F_2)$ the vector of absolute frequencies of pairs with the value of S_i equal to 0, 1, 2. In the former example, for a fixed value of $\theta = 1$ (quite common in real applications) and stratum parameters $\lambda_i = \log(\xi_i/(1-\xi_i))$ generated from a $U(0,1)$ distribution for the ξ_i, we have generated $m = 30$ matched pairs, with $F = (10, 13, 7)$; also $T = 10$ in the reduced sample and $T = 17$ in the computation of $\tilde{L}$. All the competing likelihoods have here a reasonable behavior although the profile likelihood and the modified profile likelihood seem to be biased toward larger values of θ. In the latter example, the true θ

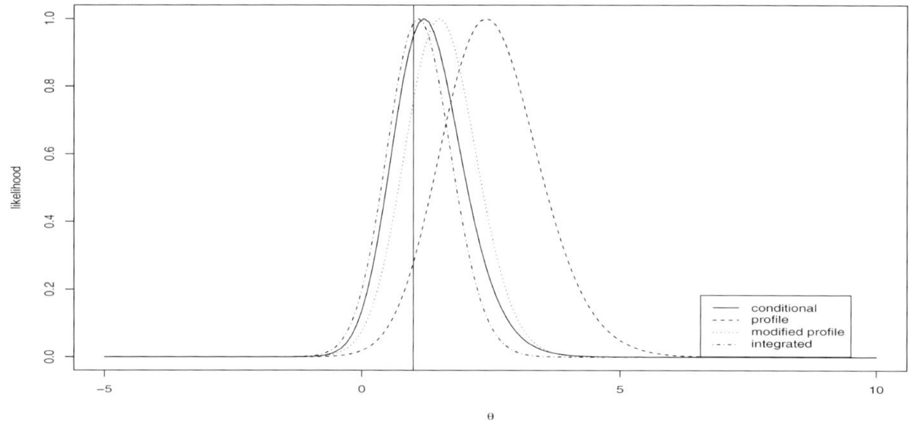

Fig. 1. Various likelihood functions for the binary matched problem: $[m = 30, F = (10, 13, 7)]$.

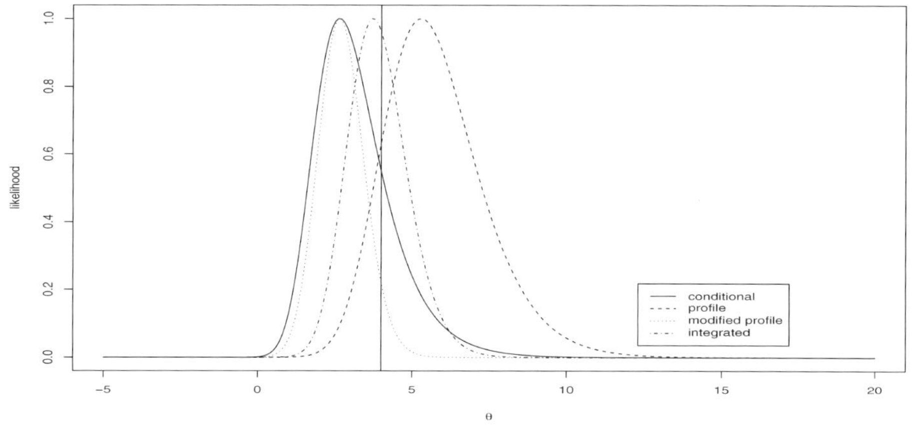

Fig. 2. Various likelihood functions for the binary matched problem: $[m = 30, F = (3, 15, 12)]$.

is equal to 4, less frequent in applications; also, $m = 30$, $F = (3, 15, 12)$ and $T = 14$ in the reduced sample and $T = 26$ in the computation of $\tilde{L}$. In this case, the profile still overestimates θ, while $L_C(\theta)$ and the modified profile likelihood are biased toward smaller values of θ.

Another example belonging to this category, which also sheds light on the importance of a correct default analysis is the following.

EXAMPLE 7 (*The norm of a normal mean vector*, Berger et al., 1999). Suppose that, independently for $i = 1, \ldots, p$, $X_i \sim N(\mu_i, 1)$. The parameter of interest is

$$\theta = \frac{1}{p}\|\mu\|^2 = \frac{1}{p}\sum_{i=1}^{p} \mu_i^2.$$

The usual choice of nuisance parameter here is $\lambda = \boldsymbol{\mu}/\|\boldsymbol{\mu}\|$, i.e., the "direction" of $\boldsymbol{\mu}$ in R^p. Note that λ can also be viewed as a point on the surface of the unit ball in R^p; hence it is natural to assign λ the uniform prior on this surface. The resulting uniform-integrated likelihood (see Chang and Eaves, 1990 or Berger et al., 1990) is

$$\tilde{L}^U(\theta) \propto \theta^{-(p-2)/4}\, e^{-p\theta/2}\, I_{(p-2)/2}\big(\sqrt{p\theta}\,\|x\|\big), \tag{22}$$

where I_ν is the modified Bessel function of the first type of order ν.

For large p and assuming that θ stays bounded as $p \to \infty$, it can be shown that the mode of (22) is approximately

$$\hat{\theta} \approx \frac{1}{p}\|x\|^2 - \frac{p-1}{p} - \frac{1}{p}\left(\frac{2\|x\|^2}{p-2} - 1\right)^{-1}.$$

This is a sensible estimate. For instance, since $|x|^2/p - E|X|^2/p \to 0$ as $p \to \infty$ by the law of large numbers, it is immediate that $\hat{\theta} - \theta \overset{p\to\infty}{\longrightarrow} 0$. Thus $\hat{\theta}$ is consistent for θ.

In contrast, the profile likelihood does not yield a consistent estimator. Indeed, the profile likelihood for this problem can easily be seen to be

$$\hat{L}(\theta) \propto e^{-(\|x\| - \sqrt{p\theta})^2},$$

which has mode $\hat{\theta} = \|x\|^2/p$. Clearly $\hat{\theta} - \theta \to 1$, establishing the inconsistency. There exist "classical" methods which would yield reasonable answers for this problem. One such is to look at the marginal distribution of $\sum_{i=1}^p X_i^2$, which is a noncentral chi-square distribution with parameter $p\theta$; the resulting likelihood, though complicated, will behave reasonably.

The common "naive" default Bayesian approach to this problem would be to base inference on the noninformative prior $\pi(\boldsymbol{\mu}) = 1$ for $\boldsymbol{\mu} = (\mu_1, \ldots, \mu_p)$; the resulting posterior distribution, given $\boldsymbol{x} = (x_1, \ldots, x_p)$, is

$$\pi(\boldsymbol{\mu}|\boldsymbol{x}) = (2\pi)^{-p/2} \exp\left\{-\sum_{i=1}^{p} (\mu_i - x_i)^2/2\right\}.$$

Since the parameter of interest is $\theta = (\sum_{i=1}^p \mu_i^2)/p$, it is easy to see that $p\theta$ has a posterior marginal noncentral chi-square distribution with p degrees of freedom and noncentrality parameter $\|\boldsymbol{x}\|^2$.

This is not a "good posterior" for θ. For instance, the posterior mean here would be $\hat{\theta} = \|\boldsymbol{x}\|^2/p + 1$. This is inconsistent as $p \to \infty$, in the sense that $\hat{\theta} - \theta \to 2$. The posterior median and posterior mode also exhibit this behavior.

Thinking in terms of θ and the nuisance parameters λ in this example allows to avoid this problem. It was "obvious" to use the uniform distribution on λ to integrate out the nuisance parameters, and the ensuing integrated likelihood will work well with any reasonable $\pi(\theta)$. The "error" in use of $\pi(\boldsymbol{\mu}) = 1$ above can be seen by changing variables to (θ, λ). Then $\pi(\boldsymbol{\mu}) = 1$ is transformed into

$$\pi(\theta, \lambda) = \theta^{-(p-2)/2}\, \pi(\lambda|\theta),$$

where $\pi(\lambda|\theta)$ is the uniform distribution on the surface of the unit ball, as before. Thus, $\pi(\boldsymbol{\mu}) = 1$ has unwittingly introduced a drastic and unreasonable prior distribution on θ.

Separately considering λ and θ, through integrated likelihood, can help to avoid this type of error. More advanced default Bayesian approaches, such as the reference prior approach, automatically avoid this type of error.

6. Semiparametric problems

There is no clearcut definition of what a semiparametric model is (Van der Vaart, 2000). One can say that a "semiparametric model" is one that is in-between the two extremes of the classical parametric model we have considered so far and a nonparametric model, which consists of all the probability distributions on the sample space or, at least, all the distributions satisfying some conditions (smooth densities, existence of the second moment, etc.).

To fix the ideas we can think of a semiparametric model as one in which the parameter space Ω can be written as $\Theta \times \Lambda$ with $\Theta \subset \mathrm{R}_d$ for some d while Λ is infinite-dimensional. If θ is our parameter of interest then the analysis of such a model can be considered a problem of elimination of the infinite-dimensional nuisance parameters λ.

Murphy and Van der Vaart (2000) establish a first-order asymptotic likelihood theory for general classes of semiparametric models. Their approach are based on the construction of a least-favorable parametric family, with respect to which the usual asymptotic results are considered. In their framework the profile likelihood inherits many of its asymptotic properties for the parametric case. A full description of these issues is not the central issues of this chapter; we will limit ourselves to the discussion of the relatively simple, although meaningful, example of the construction of a semiparametric likelihood for the median of a distribution.

EXAMPLE 8 (*Semiparametric estimation of the median*). Conditionally on $\theta \in \mathrm{R}$ and for a given continuous probability density λ, with median equal to 0 defined on R, suppose we observe n i.i.d. random variables $X_i = \theta + \varepsilon_i$, $i = 1, \ldots, n$, with $\varepsilon_i \sim \lambda(\cdot)$. The parameter of interest is θ, the median of the X_i's, while λ is the infinite-dimensional nuisance parameter. Let $\mathcal{F}$ be the set of all continuous, 0-median, probability density functions. This problem is a truly basic one, and it has been considered in several papers, some of which are referenced in the sequel.

The nonparametric nature of the model makes it difficult to work with the "likelihood"[3]

$$L(\theta, \lambda) \propto \prod_{i=1}^{n} \lambda(x_i - \theta),$$

[3] This can be considered a likelihood only under the assumption that each $\lambda \in \mathcal{F}$ has a density w.r.t. a common σ-finite measure; without this condition we would not have a dominated model and we could not even define a likelihood function.

and non-Bayesian inference is usually performed via the construction of some pseudo-likelihood for θ. The original idea dates back to Jeffreys (1961, p. 211); let

$$K(\theta) = \sum_{i=1}^{n} I_{(-\infty,\theta]}(X_i)$$

be the number of observations in the sample which are not greater than θ. Then, for fixed θ, $K(\theta) \sim \text{Bin}(n, 0.5)$ and a possible likelihood for θ is

$$L_J(\theta) \propto \binom{n}{k(\theta)}. \tag{23}$$

This is not a true likelihood, since $k(\theta)$ is not observable; also it is easy to see that, for large values of $|\theta|$, L_J converges to a constant larger than 0: for this reason, it is difficult to use $L_J(\theta)$ in default Bayes analysis, since the default prior for the location parameter θ will be uniform, that is improper. On the other hand, L_J can be used as a proper likelihood for testing hypotheses such as, for example,

$$H_0\colon \theta \leqslant \theta_0, \quad \text{vs.} \quad H_1\colon \theta > \theta_0.$$

In this case, the likelihood ratio test procedure, based on $L_J(\theta_0)/L_J(\hat{\theta})$ gives results equivalent to those obtained with the sign test.

Alternatively one can maximize $L(\theta, \lambda)$ over the set $\mathcal{F}$, for fixed θ, thus obtaining the profile likelihood for θ (Wasserman, 1988)

$$\hat{L}(\theta) \propto k(\theta)^{-k(\theta)}\big[n - k(\theta)\big]^{-[n-k(\theta)]}. \tag{24}$$

This profile likelihood can be shown to be equal to the empirical likelihood derived by Owen (1988): see also Chen and Hall (1993) and Adimari (1998). Criticism expressed apropos of $L_J(\theta)$ applies for $\hat{L}(\theta)$ also. In spite of that, the methods which lead to (23) and (24) are both very simple and, still, nonparametric. The main drawbacks of such analyses is that they can make no use at all of any further information that could be available concerning the shape of $\lambda(\cdot)$; this last remark also explains why this is an example where one would like to be able (at least potentially) to introduce an informative prior on the nuisance parameter.

The Bayesian analysis of this problem is not straightforward and requires some care; one takes the parameter space to be $\Pi = \mathcal{F} \times \mathrm{R}$ and puts a prior distribution over Π; then the inference on θ is based on its marginal posterior distribution given $X_1, \ldots, X_n$. Doss (1985a, 1985b) considered θ and λ independent; he assumed a symmetrized Dirichlet process prior on λ with mean $\alpha(\cdot)$, and a prior $\nu(\cdot)$, for θ. He shows that his approach does not guarantee the consistency of the estimates. In particular, Doss (1985b) shows that, when the ε_i are i.i.d. from a discrete distribution then the posterior Bayes estimate of θ – under squared error loss –, say $\hat{\theta}_B$, is consistent, but when the ε_i are i.i.d. from a continuous distribution, then inconsistency may occur: for example $\hat{\theta}_B$ may converge to a wrong value, or it may oscillate among two values a and b such that the true value θ_0 is in (a, b). Diaconis and Friedamn (1986) and Ghosal et al. (1999) consider a similar problem, where the support of λ is restricted to the symmetric densities on the real line. Ghosal et al. (1999) adopt a Polya tree prior for λ and show that, under some constraints on the choice of the parameters of the Polya tree, the posterior of θ will be consistent for a large class of true symmetric densities.

The results of Doss (1985a) can also be used to compute an integrated likelihood for θ. His Theorem 1, p. 1436, contains the expression of the marginal posterior for θ (formulae 3.2). The factor $M(X,\theta)$, namely

$$M(X,\theta) = \frac{1}{\Gamma\left(\frac{\alpha(\mathrm{R})}{2} + nF_n(\theta)\right)\Gamma\left(\frac{\alpha(\mathrm{R})}{2} + n(1 - F_n(\theta))\right)}$$

which appears in the posterior is exactly the integrated likelihood when the nuisance parameter λ is given a Dirichlet process prior centered at $\alpha(\cdot)$. In the above formula, $\alpha(\mathrm{R})$ represents the dispersion parameter for the Dirichlet process prior and F_n is the empirical c.d.f. For a noninformative choice of α (small values of $\alpha(\mathrm{R})$), $M(X,\theta)$ is practically indistinguishable from (23) and (24). This can be seen as an indirect validation of the Jeffreys' approach. See also Lavine (1995) for other properties of (23).

7. Related issues

In the last section of the chapter, it is time to tentatively give an at least vague definition of what the nuisance parameters are in a statistical analysis. In my opinion, we can call nuisance parameter any aspect of the statistical model which is, at the same time,

- not completely known;
- such that, its partial or complete knowledge might not be important *per sé*, rather for improving the accuracy of our inferential procedures.

This definition broadens the class of statistical problems in which we have practically to deal with the elimination of nuisance parameters. Here we give a brief and by no means exhaustive list.

7.1. Prediction and model averaging

In a prototypical prediction problem, one observes a vector $\boldsymbol{y}$ of observations from the model

$$Y_i \sim f(y; \boldsymbol{\omega})$$

and try to predict the value of a future observation y^*. The obvious Bayesian way to face this problem is to elicit a prior distribution $\pi(\boldsymbol{\omega})$ and compute the posterior predictive distribution of y^*:

$$f(y^*|\boldsymbol{y}) = \int_{\boldsymbol{\omega}} f(y^*|\boldsymbol{\omega}, \boldsymbol{y})\pi(\boldsymbol{\omega}|\boldsymbol{y})\,\mathrm{d}\boldsymbol{\omega}; \tag{25}$$

it is obvious that, in these cases, the whole parameter vector ω is a nuisance parameter, and concepts like the order of interest among the coordinates of $\boldsymbol{\omega}$ make no sense. Pursuing this issue further, it can happen that the entire model is a nuisance parameter, when the ultimate goal of the analysis is the prediction of a future value. Consider the following example.

EXAMPLE 9 (*Linear models for prediction*, Barbieri and Berger, 2004). Suppose we have the usual normal linear model:

$$y = \boldsymbol{X}\beta + \varepsilon, \tag{26}$$

where y is the n-vector of observed values of the response variable, $\boldsymbol{X}$ is the $n \times k$ full rank design matrix of covariates, and β is a k-vector of unknown coefficients. For simplicity assume that the coordinates of the random error vector ε are independent, each with a $N(0, \sigma^2)$ distribution, with σ^2 known. Eq. (26) is the *full* model, and possible alternative and less expensive models take the form

$$M_\ell\colon\ y_\ell = \boldsymbol{X}_\ell \beta_\ell + \varepsilon, \tag{27}$$

where $\ell = (l_1, l_2, \ldots, l_k)$ is the model index, l_i being either 1 or 0 as covariate x_i is in or out of the model; $\boldsymbol{X}_\ell$ contains the columns of $\boldsymbol{X}$ corresponding to the nonzero coordinates of ℓ, and β_ℓ is the corresponding vector of regression coefficients. Suppose that the ultimate goal of the analysis is to predict a future observation

$$y^* = x^*\beta + \varepsilon,$$

where $x^* = (x_1^*, \ldots, x_k^*)$ is the vector of covariates at which the prediction is to be performed. In this set-up the choice of the particular subset ℓ of covariates is a nuisance parameter that we need not to estimate: in fact, choosing a specific model makes little sense and one should, rather, base the prediction on Bayesian model averaging: see, for example, Clyde (1999) and Hoeting et al. (1999).

7.2. *Significance tests*

Sometimes it is necessary to check the compatibility of a given working model, say M_0: $X \sim f(x; \omega)$, with the observed data, and there are cases where there is no explicit alternative to M_0 (Bayarri and Berger, 2000). This operation is usually done, in a non-Bayesian way, via the computation of a p-value based on a test statistic $T(X)$, namely

$$p = \mathrm{P}\big(T(X) \geqslant T(x_{\text{obs}}|\omega)\big).$$

Since the distribution of T usually depends on ω, the actual calculation of p is nothing but another statistical issue which can be rephrased in terms of elimination of nuisance parameters. Bayarri and Berger (2000) review several different ways of eliminating ω and discuss some new proposals. All the existing methods can be roughly classified into three categories:

(i) *plug-in* p-values, where ω is replaced by some point estimate $\hat{\omega}$, usually the m.l.e.,
(ii) p-values based on the conditional distribution of T given a sufficient statistics U,
(iii) *Bayesian* p-values, based on the marginal – with respect to ω – distribution of X, namely

$$m(x) = \int_\Omega f(x|\omega)\pi(\omega)\,\mathrm{d}\omega,$$

where $\pi(\omega)$ is a suitable distribution on the parameter space.

It is quite interesting to note that the three above classes correspond to the already discussed methods of eliminating the nuisance parameters, namely, (i) profile likelihood, (ii) conditional likelihood and (iii) integrated likelihood.

However, here the problem is more complicated, since $\pi(\omega)$ *must* be a proper distribution, in order to obtain a proper and usable $m(x)$. Although desirable, this requirement is not strictly necessary for producing an integrated likelihood function. Monahan and Boos (1992) and Lavine (1995) (see also Severini, 2000) discuss the role and the properties of a *proper* likelihood function, that is one which can be interpreted as the sampling distribution of some observable quantity. In this respect, the proposal in Bayarri and Berger (2000), based on the U-conditional predictive p-value, may play a role also in the problem of producing a *proper* integrated likelihood. This connection will be explored elsewhere.

Acknowledgements

The author kindly thanks Silvia Polettini for patiently reading the paper and for solving several Lateχ and R problems.

References

Abramowitz, M., Stegun, I.A. (1964). *Handbook of Mathematical Functions with Formulas, Graphs, and Mathematical Tables. National Bureau of Standards Applied Mathematics Series*, vol. 55. U.S. Government Printing Office, Washington, DC.

Adimari, G. (1998). An empirical likelihood statistic for quantiles. *J. Statist. Comput. Simulation* **60**, 85–95.

Altham, P. (1971). The analysis of matched proportions. *Biometrika* **58** (3), 561–576.

Barbieri, M., Berger, J. (2004). Optimal predictive model selection. *Ann. Statist.* **32** (3), 870–897.

Barndorff-Nielsen, O. (1983). On a formula for the distribution of the maximum likelihood estimator. *Biometrika* **70** (2), 343–365.

Basu, D. (1977). On the elimination of nuisance parameters. *J. Amer. Statist. Assoc.* **72** (358), 355–366.

Bayarri, M.J., Berger, J.O. (2000). p values for composite null models (with discussion). *J. Amer. Statist. Assoc.* **95** (452), 1127–1142.

Belin, T., Rubin, D. (1995). A method for calibrating false – match rates in record linkage. *J. Amer. Statist. Assoc.* **90**, 694–707.

Berger, J.O., Bernardo, J.-M. (1989). Estimating a product of means: Bayesian analysis with reference priors. *J. Amer. Statist. Assoc.* **84** (405), 200–207.

Berger, J., Bernardo, J. (1992a). On the development of reference priors. In: Berger, J.O., Bernardo, J.M., Dawid, A.P., Smith, A.F.M. (Eds.), *Bayesian Statistics, vol. 4*. Oxford University Press, London, pp. 35–60.

Berger, J.O., Bernardo, J.M. (1992b). Ordered group reference priors with application to the multinomial problem. *Biometrika* **79** (1), 25–37.

Berger, J.O., Bernardo, J.M. (1992c). Reference priors in a variance components problem. In: *Bayesian Analysis in Statistics and Econometrics*. Bangalore, 1988, In: *Lecture Notes in Statist.*, vol. 75. Springer, New York, pp. 177–194.

Berger, J.O., Philippe, A., Robert, C. (1990). Estimation of quadratic functions: Noninformative priors for non-centrality parameters. *Statistica Sinica* **18**, 1595–1614.

Berger, J.O., Liseo, B., Wolpert, R.L. (1999). Integrating likelihood methods for eliminating nuisance parameters. *Statist. Sci.* **14**, 1–28.

Bernardo, J., Smith, A. (1994). *Bayesian Theory*. Wiley, New York.
Chang, T., Eaves, D. (1990). Reference priors for the orbit in a group model. *Ann. Statist.* **18**, 1595–1614.
Chen, S., Hall, P. (1993). Smoothed empirical likelihood confidence intervals for quantiles. *Ann. Statist.* **21**, 1166–1181.
Clyde, M. (1999). Bayesian model averaging and model search strategies. In: Berger, J.O., Bernardo, J.M., Dawid, A.P., Smith, A.F.M. (Eds.), *Bayesian Statistics, vol. 6*. Oxford University Press, London, pp. 157–185.
Cox, D.R. (1975). Partial likelihood. *Biometrika* **62** (2), 269–276.
Davison, A.C. (2003). *Statistical Models. Cambridge Series in Statistical and Probabilistic Mathematics*. Cambridge University Press, Cambridge.
DeGroot, M.H., Goel, P.K. (1980). Estimation of the correlation coefficient from a broken random sample. *Ann. Statist.* **8** (2), 264–278.
Diaconis, P., Friedamn, D. (1986). On inconsistent Bayes estimates of location. *Ann. Statist.* **14**, 68–87.
Doss, H. (1985a). Bayesian nonparametric estimation of the median. Part i: Computation of the estimates. *Ann. Statist.* **13**, 1432–1444.
Doss, H. (1985b). Bayesian nonparametric estimation of the median. Part ii: Asymptotic properties of the estimates. *Ann. Statist.* **13**, 1445–1464.
Fellegi, I., Sunter, A. (1969). A theory of record linkage. *J. Amer. Statist. Assoc.* **64**, 1183–1210.
Fieller, E.C. (1954). Symposium on interval estimation: Some problems in interval estimation. *J. Roy. Statist. Soc., Ser. B* **16**, 175–185.
Ghosal, S., Ghosh, J., Ramamoorthi, R. (1999). Consistent semiparametric Bayesian inference about a location parameter. *J. Statist. Plann. Inference* **77**, 181–193.
Ghosh, J., Mukerjee, R. (1992). Non-informative priors. In: Berger, J.O., Bernardo, J.M., Dawid, A.P., Smith, A.F.M. (Eds.), *Bayesian Statistics, vol. 4*. Oxford University Press, New York, pp. 195–210.
Gleser, L., Hwang, J. (1987). The nonexistence of $100(1-\alpha)$ confidence sets of finite expected diameter in errors-in-variable and related models. *Ann. Statist.* **15**, 1351–1362.
Goel, P., Ramalingam, T. (1989). *The Matching Methodology: Some Statistical Properties. Lecture Notes in Statistics*, vol. 52. Springer-Verlag, Berlin.
Hoeting, J., Madigan, D., Raftery, A., Volinsky, C. (1999). Bayesian model averaging: A tutorial. *Statist. Sci.* **14** (1), 382–417.
Jeffreys, H. (1961). *Theory of Probability*, third ed. Oxford University Press, Oxford.
Kass, R., Wasserman, L. (1996). Formal rules for selecting prior distributions: A review and annotated bibliography. *J. Amer. Statist. Assoc.* **91**, 1343–1370.
Lahiri, P., Larsen, M. (2005). Regression analysis with linked data. *J. Amer. Statist. Assoc.* **100** (469), 222–230.
Lancaster, T. (2000). The incidental parameter problem since 1948. *J. Econometrics* **95** (2), 391–413.
Lavine, M. (1995). On an approximate likelihood for quantiles. *Biometrika* **82** (1), 220–222.
Linnik, J.V. (1968). *Statistical Problems with Nuisance Parameters. Translations of Mathematical Monographs*, vol. 20. Amer. Math. Soc., Providence, RI. Translated from the Russian by Scripta Technica.
Liseo, B. (1993). Elimination of nuisance parameters with reference priors. *Biometrika* **80** (2), 295–304.
Liseo, B., Sun, D. (1999). A general method of comparison for likelihood. Technical Report, Università di Roma La Sapienza.
Liseo, B., Tancredi, A. (2004). Statistical inference with linked data. In: *Proceedings of International Workshop on Statistical Modelling*. Firenze Univ. Press, pp. 224–228.
Monahan, J., Boos, D. (1992). Proper likelihoods for Bayesian analysis. *Biometrika* **79** (3), 271–278.
Mood, A., Graybill, F., Boes, D. (1974). *Introduction to the Theory of Statistics*, third ed. McGraw-Hill, Singapore. A Wiley-Interscience publication.
Murphy, S.A., Van der Vaart, A.W. (2000). On profile likelihood. *J. Amer. Statist. Assoc.* **95** (450), 449–485. (With discussion.)
Neyman, J., Scott, E. (1948). Consistent estimates based on partially consistent observations. *Econometrica* **16**, 1–32.
Owen, A. (1988). Empirical likelihood ratio confidence intervals for a single functional. *Biometrika* **75**, 237–249.

Pace, L., Salvan, A. (1997). *Principles of Statistical Inference. From a Neo-Fisherian Perspective*. World Scientific Publishing Co. Inc., River Edge, NJ.

Pawitan, Y. (2001). *In All Likelihood*. Clarendon Press, Oxford.

Reid, N. (1996). Likelihood and Bayesian approximation methods. In: Berger, J.O., Bernardo, J.M., Dawid, A.P., Smith, A.F.M. (Eds.), *Bayesian Statistics, vol. 5*. Oxford University Press, New York, pp. 351–369.

Sartori, N. (2003). Modified profile likelihoods in models with stratum nuisance parameters. *Biometrika* **90** (3), 533–549.

Severini, T.A. (2000). *Likelihood Methods in Statistics*. Oxford University Press, Oxford.

Van der Vaart, A. (2000). *Semiparametric Statistics*. School of St. Flour, Lecture Notes.

Wasserman, L. (1988). Some applications of belief functions to statistical inference. Doctoral dissertation, Univ. of Toronto.

Welsh, A.H. (1996). *Aspects of Statistical Inference*. Wiley, New York.

Handbook of Statistics, Vol. 25
ISSN: 0169-7161
© 2005 Elsevier B.V. All rights reserved.
DOI 10.1016/S0169-7161(05)25008-3

8

Bayesian Estimation of Multivariate Location Parameters

Ann Cohen Brandwein and William E. Strawderman

Abstract

This paper presents an expository development of Bayesian estimation with substantial emphasis on exact results for the multivariate normal location models with respect to squared error loss. From the time Stein, in 1956, showed the inadmissibility of the best invariant estimator when sampling from a multivariate normal distribution in 3 or more dimensions, there has been an outpouring of improved estimators with a Bayesian flavor, encouraged largely by the connections between Bayes estimation, admissibility and minimaxity. In this chapter, we attempt to give a coherent presentation of numerous Bayesian results (proper, generalized, empirical) for this case. Generalizations for the location parameter of multivariate normal distributions with unknown covariance matrices and general quadratic loss are also presented.

1. Introduction

This paper presents an expository development of Bayesian estimation with substantial emphasis on exact results for the multivariate normal location models in 3 or more dimensions. Specifically, consider the problem of estimating the mean, θ of a p-dimensional multivariate normal distribution with covariance matrix equal to the identity ($X \sim \mathrm{MVN}_p(\theta, I)$) and loss equal to the sum of squared errors loss. From the time Stein (1956) showed the inadmissibility of the best invariant estimator, $\overline{X}$, when sampling from a $\mathrm{MVN}_p(\theta, I)$ in 3 or more dimensions, the floodgates opened and there was an outpouring of improved estimators with a Bayesian flavor, encouraged largely by the connections between Bayes estimation, admissibility and minimaxity. Throughout this chapter, we will attempt to give a coherent presentation of numerous Bayesian results (proper, generalized, empirical) for this case. Generalizations for the location parameter of multivariate normal distributions with unknown covariance matrices and general quadratic loss will also be presented.

Section 2 gives an outline of some of the basic results of Decision Theory such as admissibility, minimaxity, and Bayes estimation. Section 3 gives an intuitive introduction to Stein estimation and the basic theoretical results concerning the James–Stein estimator and Baranchik-type estimators for the case where the covariance matrix is the equal to identity. Section 4 discusses the connection between Bayes estimation and Stein estimation when the covariance is the identity. Section 5 discusses extensions of the results in Sections 3 and 4 to the case of a general known covariance matrix and general quadratic loss. Section 6 contains a brief discussion of some developments not covered in this paper together with some references to this material.

2. Bayes, admissible and minimax estimation

This section will present the basic ideas and notation for Bayes, admissible and minimax estimation. For the most part, we will incorporate the notation found in Lehmann and Casella (1998) and borrow some of their definitions and theorems where appropriate.

Given an observation, X, on a p-dimensional multivariate normal distribution with unknown mean, θ, and covariance matrix known to be the identity, I ($X \sim \mathrm{MVN}_p(\theta, I)$) then the density of X is given (for $x \in \mathrm{R}^p$) by

$$f(x;\theta) = (2\pi)^{-p/2} \exp\bigl(-\|x-\theta\|^2/2\bigr). \tag{2.1}$$

We consider the problem of estimating θ subject to the squared error loss function, i.e. $\delta(X)$ is a measurable (vector-valued) function, and the loss is given by

$$L(\delta,\theta) = \|\delta-\theta\|^2 = \sum_{i=1}^{p} (\delta_i - \theta)^2, \tag{2.2}$$

where $\delta = [\delta_1, \delta_2, \ldots, \delta_p]_-$ and $\theta = [\theta_1, \theta_2, \ldots, \theta_p]_-$.

The more general case of a sample of size n from a multivariate normal population with unknown mean θ, and known covariance matrix $\sigma^2 I$ can be reduced to the foregoing by sufficiency (and scale transformation).

We will begin by first defining admissibility. If $\delta(X)$ is a decision procedure, for any loss function $L(\delta,\theta)$ *the risk function of* δ is defined as

$$R(\delta,\theta) = E_\theta L\bigl(\delta(X),\theta\bigr). \tag{2.3}$$

An estimator δ_{a} is an *admissible estimator* if there does not exist another statistical decision procedure δ such that $R(\delta,\theta) \leqslant R(\delta_{\mathrm{a}},\theta)$ for all $\theta \in \Theta$ with strict inequality for some θ.

A minimax statistical decision procedure, is by definition, one that minimizes (over the class of procedures) the maximum (over the parameter space Θ) of the risk function (2.3). Thus, δ^{M} is a *minimax estimator* of θ if $\sup_\theta R(\delta^{\mathrm{M}},\theta) = \inf_\delta \sup_\theta R(\delta,\theta)$ for all possible decision rules δ.

As is pointed out by Strawderman (2000), minimaxity is considered by some to be a principle that should always be used in a statistical problem. Though more typically it is put forward mainly for the purpose of refuting it. However, the study of minimaxity is an important part of theoretical statistics which has lead to a deeper understanding of various areas of statistics, and has provided useful new procedures.

There is no universally successful method for finding a minimax estimator, however, understanding Bayesian estimation and the relationship between admissibility and a Bayes estimator and a minimax estimator can lead to such optimal procedures.

Suppose we view our parameter θ as a random variable with a probability distribution Π ($\Theta \sim \Pi$) and consider the weighted risk $r(\delta, \Pi)$ defined by,

$$r(\delta, \Pi) = \int R(\delta, \theta)\, \mathrm{d}\Pi(\theta). \tag{2.4}$$

Then the estimator δ_Π that minimizes the risk (2.4) is called the *Bayes estimator* (*proper Bayes estimator*) with respect to Π.

In lieu of this new treatment of our unknown parameter θ, Bayesian decision theorists now view the density of X, defined in (2.1) as $f(x; \theta)$ to be a conditional density, $f(x|\theta)$. The distribution Π is called the *prior distribution* of Θ and the density of the prior distribution is typically denoted by $\pi(\theta)$. Once data is collected, the prior distribution is revised (updated) and replaced by the posterior distribution, $\Pi(\theta|x)$. The Bayes estimator now becomes the estimator that minimizes the expected value of the posterior loss. That is, the Bayes estimator is that $\delta_\Pi(X)$ which minimizes $E_\theta(L(\delta(X), \Theta|X = x))$. It is worthwhile to point out, that for estimating the mean of a $\mathrm{MVN}_p(\theta, I)$ distribution, with respect to quadratic loss defined in (2.2), the Bayes estimator simply reduces to the mean of the posterior distribution, i.e.

$$\delta_\Pi(X) = E(\Theta|X). \tag{2.5}$$

Thus, with respect to quadratic loss, if $X \sim \mathrm{MVN}_p(\theta, I)$, we can express the ith coordinator of the Bayes estimator $\delta_\Pi(X)$ uniquely as

$$\delta_{\Pi_i}(X) = E(\theta_i|X) = \frac{\int \theta_i \exp(-(1/2)\|X - \theta\|^2)\, \mathrm{d}\Pi(\theta)}{\int \exp(-(1/2)\|X - \theta\|^2)\, \mathrm{d}\Pi(\theta)}. \tag{2.6}$$

The following important relationships between admissibility, minimaxity and Bayes estimation can all be found (along with their proofs) as either theorems, corollaries and lemmas in Lehmann and Casella (1998). We gather them together in the following Proposition 2.1.

PROPOSITION 2.1. (a) *Any unique Bayes estimator is admissible.*

(b) *If the risk of a Bayes estimator $\delta_\Pi(X)$, $r(\delta, \Pi) = \sup_\theta R(\delta_\Pi, \theta)$ then*:

(i) *$\delta_\Pi(X)$ is minimax.*
(ii) *If $\delta_\Pi(X)$ is the unique Bayes solution with respect to Π, it is also the unique minimax procedure.*

(c) *If a Bayes solution $\delta_\Pi(X)$ has a constant risk, then it is minimax.*
(d) *If an estimator has constant risk and is admissible, it is minimax.*
(e) *If an estimator is unique minimax, it is admissible.*

By Proposition 2.1(a), since any unique Bayes estimator is admissible, the unique Bayes estimator defined by (2.6) will be admissible. Moreover, we begin to see how Bayes estimation can lead to admissible estimators. Additionally, linking these ideas to minimaxity as in Proposition 2.1, we see the possibility of finding procedures which are admissible proper Bayes minimax procedures.

To show that $\overline{X}$ is a minimax procedure in the normal case for any p requires the following extension of Proposition 2.1 (see Lehmann and Casella, 1998, for a proof).

PROPOSITION 2.2. *Let $\delta(X)$ be an estimator. If Π_n is a sequence of prior distributions with corresponding Bayes risks r_n such that* $\lim_{n\to\infty} r_n = \sup_\theta R(\delta, \theta)$, *then $\delta(X)$ is minimax.*

If $X \sim \mathrm{MVN}_p(\theta, I)$ and the loss is $\|\delta - \theta\|^2$, then it follows that $[\tau^2/(1+\tau^2)]X$ is the Bayes estimator of θ for the prior distribution $\Theta \sim \mathrm{MVN}_p(0, \tau^2 I)$. The Bayes risk is easily seen to be $r_\tau = p\tau^2/(1+\tau^2)$ (see also Section 4). Hence

$$\lim_{\tau\to\infty} r_\tau = \lim_{\tau\to\infty}\left[p\tau^2/\left(1+\tau^2\right)\right] = p \equiv R(X,\theta) = \sup_\theta R(X,\theta)$$

and we have the following result.

PROPOSITION 2.3. *If $X \sim \mathrm{MVN}_p(\theta, I)$ and the loss is $\|\delta - \theta\|^2$, then X is a minimax estimator for any p. The minimax risk is equal to p.*

The concept of a generalized Bayes estimator is similar to that of a proper Bayes estimator as defined above, but allows the prior $\Pi(\theta)$ (and thus the posterior distribution $\Pi(\theta|X)$) to be a measure that is not necessarily a proper probability distribution. So given a (not necessarily finite) measure $\Pi(\theta)$ an estimator $\delta^\Pi(X)$ is said to be a *generalized Bayes estimator* if the expected posterior loss, $E_\Theta(L(\delta(X), \Theta|X))$ is minimized when $\delta = \delta^\Pi$. As discussed in Strawderman and Cohen (1971), it is easy to see that when $\Pi(\theta)$ is a proper posterior distribution then the generalized Bayes estimator just reduces to the proper Bayes estimator and for $X \sim \mathrm{MVN}_p(\theta, I)$, with respect to quadratic loss, it is computed using expression (2.6).

Interestingly, such a unique generalized Bayes estimator will not necessarily be admissible. However, in this multivariate normal case, as shown by Sacks (2003) and then later by others, any admissible estimator must be generalized Bayes.

Finally, an *empirical Bayes estimator* is conceptually similar to that of a proper Bayes estimator as defined above, but the parameters of the prior distribution are themselves estimated using the data. Often the parameters of the prior distribution (these prior distribution parameters are typically referred to as the *hyperparameters*) will be estimated using the maximum likelihood estimate. Once the prior parameters have been estimated, the procedure is the same as the proper Bayes procedure and the resulting estimator is the empirical Bayes estimator. As pointed out in Lehmann and Casella (1998), empirical Bayes estimation is an effective technique for constructing estimators that perform well under both Bayesian and frequentist criteria. When the prior distribution is not necessarily specified correctly, empirical Bayes estimators tend to be more robust than these other estimators.

3. Stein estimation and the James–Stein estimator

In this section we look more closely at the problem of estimating the multivariate normal mean and start to pursue the connections between Bayes, minimaxity and admissibility introduced in Section 2. This section will concentrate mostly on minimaxity and domination results, while in Section 4 we will continue with the Bayesian connections. Charles Stein investigated the question of the admissibility of the usual procedure, the sample mean, when sampling from a $\text{MVN}_p(\theta, I)$ distribution with respect to quadratic loss (2.2). The admissibility of the best equivariant procedure, the sample mean, in the univariate case was well-known and though Stein's objective was to prove its admissibility in 2 or more dimension, he could in fact only prove admissibility when $p = 2$. He actually proved the inadmissibility of the sample mean for $p \geqslant 3$. This result was quite astonishing at the time and has lead to an enormous and rich literature of substantial importance in statistical theory and practice. It can indeed be argued that much of the impetus behind the recent explosion of interest in Bayesian methods was due to this discovery and the related results which followed.

Specifically, for $X \sim \text{MVN}_p(\theta, I)$ ($p \geqslant 3$), Stein (1956) showed that estimators of the form

$$\delta_{a,b}(X) = \left(1 - \frac{a}{b + \|X\|^2}\right)X \tag{3.1}$$

dominate X for sufficiently small a and sufficiently large b. Stein (1956) and Brown (1966) have shown that the inadmissibility of the best equivariant estimators of a location parameter in 3 or more dimensions is a general phenomenon and not just restricted to the multivariate normal distribution. Note that any procedure that dominates X will be minimax by Proposition 2.3.

James and Stein (1961) sharpened Stein's result and gave an explicit class of dominating estimators,

$$\delta_a(X) = \left(1 - \frac{a}{\|X\|^2}\right)X \quad \text{for } 0 < a < 2(p-2). \tag{3.2}$$

A rough geometric argument that is basically due to Stein (1962) (see also Brandwein and Strawderman, 1990), will help to provide some intuition for the James–Stein estimators. Consider an observation vector X in p dimensions with mean vector θ and independent (uncorrelated) components. Assume that the components have variances all equal to 1. The situation is depicted in Figure 1.

Since $E(X - \theta)'\theta = 0$, we expect $X - \theta$ and θ to be nearly orthogonal, especially for large $\|\theta\|$. Since $E\|X\|^2 = p + \|\theta\|^2$, it appears that X as an estimator of θ might be too long and that the projection of θ on X or something close to it might be a better estimator. This projection, of course, depends on θ and therefore is not a valid estimator, but perhaps we can approximate it. If we denote this projection by $(1-a)X$, the problem is to approximate a. That is, if we believe that we can improve on X by

$$\delta_a(X) = (1-a)X \tag{3.3}$$

then what form must a take?

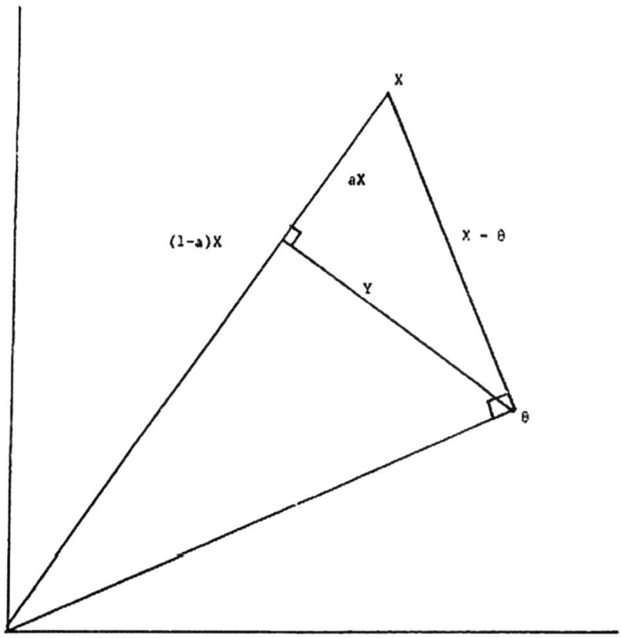

Fig. 1. Observation vector X in p dimensions with mean vector θ and independent (or uncorrelated) components.

A geometric approach to finding the form of a is to assume that the angle between θ and $X - \theta$ is exactly a right angle and to assume that $\|X\|^2$ is exactly equal to its expected value $\theta'\theta + p$. In this case, we have (using the two small triangles in Figure 1)

$$\|Y\|^2 = \|X - \theta\|^2 - a^2\|X\|^2 \approx p - a^2\|X\|^2$$

and

$$\begin{aligned}\|Y\|^2 &= \|\theta\|^2 - (1-a)^2\|X\|^2 \\ &\approx \|X\|^2 - p - (1-a)^2\|X\|^2.\end{aligned}$$

Equating the two approximations for $\|Y\|^2$ we have

$$p - a^2\|X\|^2 = \|X\|^2 - p - (1-a)^2\|X\|^2$$

or equivalently

$$(1 - 2a)\|X\|^2 = \|X\|^2 - 2p.$$

Solving for a gives $a = p/\|X\|^2$ and so the suggested estimator is

$$\delta_a(X) = (1-a)X = \left(1 - \frac{p}{\|X\|^2}\right)X. \tag{3.4}$$

This estimator is just the estimator given by (3.3) when $a = p$.

Note that the above development does not particularly depend on X being normally distributed or even that θ is a location vector. It also fails to be a proof of the inadmissibility of X and does not distinguish for which values of p, X is inadmissible. However, it is suggestive that the possibility of improving on the unbiased vector X by shrinkage toward the origin may be quite general.

Another intuitive approach to finding the form of a in $\delta_a(X) = (1-a)X$ is the decision theoretic approach. With respect to the loss $\|\delta - \theta\|^2$ the risk of δ_a given by (3.3) can be written as

$$\begin{aligned} R(\delta_a, \theta) &= E_\theta L\big(\delta(X), \theta\big) = E_\theta \big\|(1-a)X - \theta\big\|^2 \\ &= \sum_{i=1}^{p} \mathrm{Var}(\delta_i) + \sum_{i=1}^{p} \big(\mathrm{Bias}(\delta_i)\big)^2 \\ &= p(1-a)^2 + \sum_{i=1}^{p} a^2\theta_i^2 = p(1-a)^2 + a^2 p\|\theta\|^2. \end{aligned} \tag{3.5}$$

To find the a that minimizes this risk, take the derivative with respect to a, of $R(\delta_a, \theta)$ given by (3.5) and set it equal to 0. Using the expression in the last line of (3.5), we have

$$\frac{\mathrm{d}}{\mathrm{d}a} R(\delta_a, \theta) = -2p(1-a) + 2ap\|\theta\|^2 = 0. \tag{3.6}$$

Solving Eq. (3.6) for a we find again

$$a = \frac{p}{p + \|\theta\|^2}. \tag{3.7}$$

Of course an estimator cannot depend on the unknown parameter θ. But since, $E\|X\|^2 = p + \|\theta\|^2$, substituting $\|X\|^2$ for $p + \|\theta\|^2$ in expression (3.7) we have

$$a = \frac{p}{\|X\|^2} \quad \text{and thus} \quad \delta_a(X) = (1-a)X = \left(1 - \frac{p}{\|X\|^2}\right)X.$$

This is, of course the same estimator arrived at via the geometric approach in expression (3.4).

Baranchik (1964, 1970) exhibited classes of minimax estimator for the mean θ of a $\mathrm{MVN}_p(\theta, I)$ ($p \geqslant 3$) with respect to quadratic loss (2.2) which includes the James–Stein class. These estimators are of the form

$$\delta_r(X) = \left(1 - \frac{(p-2)r(\|X\|^2)}{\|X\|^2}\right)X, \tag{3.8}$$

where $r(\|X\|^2)$ is a nondecreasing function of $\|X\|$ and $0 \leqslant r(\cdot) \leqslant 2$.

More modern developments of the theory of Stein estimation depend on the following basic result due to Stein (1973, 1981) which leads to an unbiased estimate of the risk for a general estimator of the form

$$\delta_g(X) = X + g(X). \tag{3.9}$$

We use the following notation,

$$\nabla g = \sum \nabla_i g_i, \quad \text{where } \nabla_i = \frac{\partial}{\partial X_i}. \tag{3.10}$$

The basic idea is quite easy and is presented in the following result for the one-dimensional case.

LEMMA 3.1. *Let $X \sim N(\theta, 1)$ and let $h(X)$ be a continuously differentiable such that $E[h(X)]$ is finite. Then $E[(X-\theta)h(X)] = E[h'(X)]$.*

PROOF. Using integration by parts, it is straightforward to obtain

$$E\big[(X-\theta)'h(X)\big] = E\big[h'(X)\big].$$

□

The multivariate extension given next follows from Lemma 3.1 if stringent conditions are placed on $h(\cdot)$, but it follows under the much weaker stated conditions using Stoke's theorem. Stein's development is slightly different but essentially equivalent.

LEMMA 3.2. *If $h : \mathrm{R}^p \to \mathrm{R}^p$ is an almost differentiable function with $E_\theta \|h(X)\| < \infty$, then*

$$E_\theta\big(\nabla h(X)\big) = E_\theta\big\{(X-\theta)'h(X)\big\}. \tag{3.11}$$

The following Theorem 3.1 is Theorem 1 from Section 2 of Stein's (1981) paper.

THEOREM 3.1. *Suppose $X \sim \mathrm{MVN}_p(\theta, I)$, and consider the estimator $\delta(X) = X + g(X)$ for θ. Assume $g : \mathrm{R}^p \to \mathrm{R}^p$ is an almost differentiable function for which*

$$E_\theta \sum \big|\nabla_i g_i(X)\big| < \infty.$$

Then, for each $i \in (1, \ldots, p)$,

$$E_\theta\big(X_i + g_i(X) - \theta_i\big)^2 = 1 + E_\theta\big\{g_i^2(X) + 2\nabla_i g_i(X)\big\}, \tag{3.12}$$

and consequently

$$E_\theta\big\|X + g(X) - \theta\big\|^2 = p + E_\theta\big\{\big\|g(X)\big\|^2 + 2\nabla g(X)\big\}. \tag{3.13}$$

PROOF. From formula (3.11) with $h = g_i$, it follows that

$$\begin{aligned} E_\theta\big(X_i + g_i(X) - \theta_i\big)^2 &= E_\theta\big\{(X_i-\theta_i)^2 + 2(X_i-\theta_i)g_i(X) + g_i^2(X)\big\} \\ &= 1 + 2E_\theta \nabla_i g_i(X) + E_\theta g_i^2(X) \end{aligned}$$

and this last expression is just (3.12). Summing over i, we immediately obtain (3.13). Notice, that (3.13) asserts that $p + E_\theta\{\|g(X)\|^2 + 2\nabla g(X)\}$ is an unbiased estimator of the risk of the general estimator $\delta_g(X) = X + g(X)$ for location parameter θ. □

As Stein (1981) points out here, when the dimension is 3 or more, by specializing this theorem to a class of estimators which contains all formal Bayes estimators, we will be able to find a large class of dominating minimax estimators for θ. We develop these results in Section 4.

To see the utility of this result now, note the following corollary which immediately follows from Theorem 3.1.

COROLLARY 3.1. *If* $X \sim \mathrm{MVN}_p(\theta, I)$ *and* $\delta_g(X) = X + g(X)$ *and* g *is such that* $E(\|g(X)\|^2) < \infty$ *and* $E_\theta\{\|g(X)\|^2 + 2\nabla g(X)\} \leqslant 0$ *then with respect to quadratic loss* (2.2), $R(\delta_g, \theta) \leqslant R(X, \theta) = p$. *In particular this will hold if* $\|g(X)\|^2 + 2\nabla g(X) \leqslant 0$ *for all* X *with strict inequality on a set of positive measure. Hence if any such estimator exits, then* X *is inadmissible and* $\delta_g(X) = X + g(X)$ *is minimax.*

Moreover, by the proper choice of g (namely, $g(X) = -(\frac{r(\|X\|^2)}{\|X\|^2})X$ provided $r(\cdot)$ is nondecreasing and $0 \leqslant r(\cdot) \leqslant 2(p-2)$) we obtain Baranchik's result (and thus the James–Stein result) given by (3.8). We summarize this last statement in Corollary 3.2 and provide a proof.

COROLLARY 3.2. *Let* $X \sim \mathrm{MVN}_p(\theta, I)$ $(p \geqslant 3)$, *and* $\delta_g(X) = X + g(X)$ *where* $g(X) = -(\frac{r(\|X\|^2)}{\|X\|^2})X$ *for* $r(\cdot)$ *a nondecreasing function such that* $0 < r(\cdot) \leqslant 2(p-2)$ *with strict inequality on a set of positive measure. Then, for quadratic loss* (2.2), $\delta_g(X)$ *has smaller risk than* X *and is thus minimax.*

PROOF. To prove this result, from Corollary 3.1, we need only show that $E_\theta\{\|g(X)\|^2 + 2\nabla g(X)\} \leqslant 0$. By Lemma 3.2,

$$
\begin{aligned}
&E_\theta\{\|g(X)\|^2 + 2\nabla g(X)\} \\
&\quad = E_\theta\left\{\left\|-\left(\frac{r(\|X\|^2)}{\|X\|^2}\right)X\right\|^2 + 2\nabla\left(\frac{-r(\|X\|^2)}{\|X\|^2}\right)X\right\}.
\end{aligned}
$$

Now

$$
\begin{aligned}
E_\theta\left\{2\nabla\left(\frac{r(\|X\|^2)}{\|X\|^2}\right)X\right\} &= (p-2)E_\theta\left(\frac{r(\|X\|^2)}{\|X\|^2}\right) + 2E_\theta\left(\frac{r'(\|X\|^2)}{\|X\|^2}\right) \\
&\geqslant (p-2)E_\theta\left(\frac{r(\|X\|^2)}{\|X\|^2}\right)
\end{aligned}
$$

since r is a nondecreasing function. Further since $0 < r(\cdot) \leqslant 2(p-2)$ with strict inequality on a set of positive measure,

$$
E_\theta\left\{\left\|-\left(\frac{r(\|X\|^2)}{\|X\|^2}\right)X\right\|^2\right\} = E_\theta\left(\frac{r^2(\|X\|^2)}{\|X\|^2}\right) < 2(p-2)E_\theta\left(\frac{r(\|X\|^2)}{\|X\|^2}\right).
$$

Thus

$$
\begin{aligned}
&E_\theta\{\|g(X)\|^2 + 2\nabla g(X)\} \\
&\quad = E_\theta\left\{\left\|-\left(\frac{r(\|X\|^2)}{\|X\|^2}\right)X\right\|^2 + 2\nabla\left(\frac{-r(\|X\|^2)}{\|X\|^2}\right)X\right\}
\end{aligned}
$$

$$< 2(p-2)E_\theta\left(\frac{r(\|X\|^2)}{\|X\|^2}\right) - 2(p-2)E_\theta\left(\frac{r(\|X\|^2)}{\|X\|^2}\right) = 0. \qquad \square$$

We will see in Section 5, that many of the results we have presented here, can be modified for the $\mathrm{MVN}_p(\theta, \sigma^2 I)$.

4. Bayes estimation and the James–Stein estimator for the mean of the multivariate normal distribution with identity covariance matrix

It is clear from the discussions in the previous sections, that there is a strong connection between minimaxity, Bayesian estimation and the James–Stein estimator. Efron and Morris (1973) brought these ideas together, and actually came up with an estimator which is (empirical) Bayes, minimax and improves on the James–Stein estimator.

To start, we begin with the empirical Bayesian derivation of the James–Stein estimator which can be found in numerous sources including Efron and Morris (1972, 1973), Brandwein and Strawderman (1990) and Lehmann and Casella (1998). Our development will be similar to that in Lehmann and Casella (1998). Let $X \sim \mathrm{MVN}_p(\theta, I)$ and let the loss be quadratic loss $L(\delta, \theta) = \|\delta - \theta\|^2 = \sum_{i=1}^p (\delta_i - \theta)^2$, as in (2.2), where $\delta = [\delta_1, \delta_2, \ldots, \delta_p]'$ and $\theta = [\theta_1, \theta_2, \ldots, \theta_p]'$. Consider the prior distribution $\Theta \sim N(0, \tau^2 I)$, where τ^2 is assumed to be known. It is straight forward to show that the Bayes estimator, will simply be the mean of the posterior distribution, $E(\Theta|X)$, which, with respect to this normal prior, is

$$\delta^\tau(X) = \left(\frac{\tau^2}{1+\tau^2}\right)X = \left(1 - \frac{1}{1+\tau^2}\right)X \tag{4.1}$$

and the Bayes risk for $\delta^\tau(X)$ is

$$r(\delta_{\mathrm{B}}, \tau) = \left(\frac{p\tau^2}{1+\tau^2}\right). \tag{4.2}$$

If the parameter τ (referred to as a hyperparameter as mentioned in Section 2) is considered to be unknown and is estimated by the data, we obtain an empirical Bayes estimator for θ. One way to proceed is by finding an unbiased estimator of $1/(1+\tau^2)$ based on the marginal distribution of X (though, perhaps more typically, the hyperparameter will be estimated using the method of maximum likelihood). Since $X - \Theta$, conditional on Θ has a normal distribution with mean vector 0 and covariance I, $X - \Theta$ is independent of Θ. Hence $X = (X - \Theta) + \Theta$ is marginally distributed as a p-variate normal distribution with mean vector 0 and covariance matrix $(1+\tau^2)I$. Therefore $\|X\|^2/(1+\tau^2)$ has a central chi-square distribution with p degrees of freedom. It follows that marginally,

$$E\left(\frac{p-2}{\|X\|^2}\right) = \frac{1}{1+\tau^2}. \tag{4.3}$$

Thus $(\frac{p-2}{\|X\|^2})$ is an unbiased estimator of $1/(1+\tau^2)$. Substituting this estimator into the second expression of (4.1), the empirical Bayes estimator is

$$\delta_{\mathrm{EB}} = \left(1 - \frac{p-2}{\|X\|^2}\right)X \tag{4.4}$$

which is just the James–Stein estimator defined by (3.2) when $a = p - 2$.

To see how well the James–Stein (EB) estimator given by (4.4) does in comparison with the Bayes estimator defined by (4.1) it is interesting to compare their average (Bayes) risks with respect to the assumed normal prior. For δ_{EB}, $g(X) = (-\frac{p-2}{\|X\|^2})X$. Hence, by Theorem 3.1, the risk of $\delta_{\text{EB}} = (1 - \frac{p-2}{\|X\|^2})X$ is

$$\begin{aligned} R(\delta_{\text{EB}}, \theta) &= E_\theta \big\|\delta_{\text{EB}}(X - \theta)\big\|^2 = p + E_\theta\big\{\big\|g(X)\big\|^2 + 2\nabla g(X)\big\} \\ &= p + E_\theta\left\{\left\|\left(-\frac{p-2}{\|X\|^2}\right)X\right\|^2\right\} - 2E_\theta\nabla\left(\frac{p-2}{\|X\|^2}\right)X \\ &= p + (p-2)^2 E_\theta\left(\frac{1}{\|X\|^2}\right) - 2(p-2)^2 E_\theta\left(\frac{1}{\|X\|^2}\right) \\ &= p - (p-2)^2 E_\theta\left(\frac{1}{\|X\|^2}\right). \end{aligned} \tag{4.5}$$

Combining (4.3) and (4.5) and by (4.2), the Bayes risk for δ_{EB} will be

$$\begin{aligned} r(\delta_{\text{EB}}, \tau) &= p - (p-2)\left(\frac{1}{1+\tau^2}\right) = \left(\frac{p\tau^2}{1+\tau^2}\right) + \frac{2}{1+\tau^2} \\ &= r(\delta_{\text{B}}, \tau) + \frac{2}{1+\tau^2}. \end{aligned}$$

Hence, by estimating the hyperparameter, the empirical Bayes estimator has an increased Bayes risk of $2/(1+\tau^2)$.

If, instead of using the unbiased estimator for the hyperparameter, we chose to use the maximum likelihood estimator of $1 + \tau^2$ (which is $\max(1, \|X\|^2/p)$, the resulting empirical Bayes estimator would be

$$\delta_{\text{EB}}^+ = \left(1 - \frac{p}{\|X\|^2}\right)^+ X. \tag{4.6}$$

Generally,

$$\delta_a^+ = \left(1 - \frac{a}{\|X\|^2}\right)^+ X, \tag{4.7}$$

where $(a)^+ = \max(0, a)$ is known as a *positive part James–Stein estimator*. This estimator appeared and was developed in Baranchik (1964, 1970). It improves on the usual James–Stein estimator because it does not cause the estimator to change sign when $\|X\|^2$ becomes too small (i.e. $< a$). In this case, the positive part estimator shrinks toward the origin, and for this reason these estimators are often referred to as *shrinkage estimators*.

The following theorem, due to Efron and Morris (1973), proves a more general result, which includes the James–Stein and positive part James–Stein estimators. We basically follow Lehmann and Casella (1998).

THEOREM 4.1. *Let* $X \sim \mathrm{MVN}_p(\theta, I)$ *and* $\Theta \sim \mathrm{MVN}_p(0, \tau^2 I)$, *with loss quadratic loss* (2.2). *If* $\delta_{\mathrm{B}}(X)$ *is an estimator of the form*

$$\delta_{\mathrm{B}}(X) = \big(1 - B(\|X\|^2)\big)X \tag{4.8}$$

and if

$$\delta_{\mathrm{B}}^{+}(X) = \big(1 - B(\|X\|^2)\big)^{+}X, \tag{4.9}$$

then $R(\delta_{\mathrm{B}}, \theta) \geqslant R(\delta_{\mathrm{B}}^{+}, \theta)$ *(i.e. the risk of* δ_{B} *is at least as great as the risk of* δ_{B}^{+}*), with strict inequality if* $P_\theta(\delta_{\mathrm{B}}(X)\delta_{\mathrm{B}}^{+}(X)) > 0$. *It is apparent that the Bayes risks will also satisfy an inequality in the same direction.*

Clearly, when $B(X) = (\frac{a}{\|X\|^2})$ for $0 < a < 2(p-2)$ (4.8) and (4.9) reduce to the James–Stein and positive part James–Stein estimators. Moreover, when $a = p - 2$, that estimator has smaller risk than all the other James–Stein estimators, $\delta_a = (1 - \frac{a}{\|X\|^2})X$, $0 < a < 2(p-2)$. However, $\delta_{a=p-2}$ is inadmissible since, by Theorem 4.1 its positive part version dominates it. Though it is difficult to find an explicit estimator dominating the positive part estimator, it too is inadmissible since it can be shown that it is not generalized Bayes (Brown, 1971). Shao and Strawderman (1994) found explicit estimators which dominate the positive part James–Stein estimator which are partly based on δ_a^{+} given by (4.7) when $a = p - 2$.

We know from Proposition 2.1 that proper Bayes estimators, under certain conditions, can lead to admissible minimax estimators. Strawderman (1971) considered Baranchik's class of estimators given by

$$\delta_r(X) = \left(1 - \frac{r(\|X\|^2)}{\|X\|^2}\right)X \tag{4.10}$$

for $X \sim \mathrm{MVN}_p(\theta, I)$ and found, by the appropriate choice of $r(\|X\|^2)$, admissible proper Bayes minimax estimators for dimensions $p \geqslant 5$. In particular he considered the following hierarchical prior distribution. Consider a two-stage prior for θ where at the first stage $\Theta|\lambda \sim \mathrm{MVN}_p(0, \lambda^{-1}(1-\lambda)I)$ and at the second stage $\Lambda \sim (1-b)\lambda^{-b}$ $(b < 1)$ for $0 < \lambda < 1$. The proper Bayes estimator is given by

$$\begin{aligned} E(\Theta|X) &= E\big(E(\Theta|X,\lambda)|X\big) \\ &= E\left[\left(1 - \frac{1}{1 + [(1-\lambda)/\lambda]}\right)X \middle| X\right] = \big[1 - E[\Lambda|X]\big]X. \end{aligned} \tag{4.11}$$

In the following theorem, we state the conditions under which the Bayes estimator (4.11) is minimax.

THEOREM 4.2. (a) *For* $p \geqslant 5$ *the proper Bayes estimator* (4.11) *is minimax provided* $b \geqslant \frac{6-p}{2}$.

(b) *For* $p \geqslant 3$ *the estimator* (4.11) *is generalized Bayes and minimax provided*

$$\frac{6-p}{2} \leqslant b < \frac{p+2}{2}.$$

PROOF. (a) Since the distribution of $\Lambda|X$ depends only on $\|X\|^2$, the estimator (4.11) can be expressed in the form (4.10) by setting $E[\Lambda|X] = r(\|X\|^2)/\|X\|^2$. A straightforward calculation gives

$$E[\Lambda|X] = \frac{1}{\|X\|^2}\left[p + 2 - 2b - \frac{2\,\mathrm{e}^{-\|X\|^2/2}}{\int_0^1 \lambda^{p/2-b}\,\mathrm{e}^{-\lambda\|X\|^2/2}\,\mathrm{d}\lambda}\right], \tag{4.12}$$

where

$$\begin{aligned} r(\|X\|^2) &= \left[p + 2 - 2b - \frac{2\,\mathrm{e}^{-\|X\|^2/2}}{\int_0^1 \lambda^{p/2-b}\,\mathrm{e}^{-\lambda\|X\|^2/2}\,\mathrm{d}\lambda}\right] \\ &= \left[p + 2 - 2b - \frac{2}{\int_0^1 \lambda^{p/2-b}\,\mathrm{e}^{(1-\lambda)\|X\|^2/2}\,\mathrm{d}\lambda}\right]. \end{aligned} \tag{4.13}$$

Hence, we see that $r(\|X\|^2) < p+2-2b$ and is clearly an increasing function of $\|X\|^2$. Thus the conditions of Corollary 3.2 are met provided $p + 2 - 2b \leqslant 2(p-2)$ or equivalently, $b \geqslant (6-p)/2$. Since in order for the second stage prior to be integrable we must have $b < 1$, the estimator (4.11) is a proper Bayes (and therefore admissible) minimax estimators provided $p \geqslant 5$ and $1 > b \geqslant (6-p)/2$. This proves part (a).

(b) Noting that (4.12) makes sense (i.e. the generalized Bayes estimator exits) provided $p+2-2b > 1$, which is equivalent to $b < (p+2)/2$, we see that the generalized Bayes estimator (4.11) is minimax provided $(6-p)/2 \leqslant b < (p+2)/2$ (note this inequality can only hold when $p > 2$). Thus part (b) follows. □

Strawderman (1973) showed that no proper Bayes minimax estimators exist for $p < 5$.

Stein (1981) gives a very general approach to the problem of finding Bayes minimax estimators. Recall from Section 2, that if $\pi(\theta)$ is the (generalized) prior density, then with respect to quadratic loss (2.2) the (generalized) Bayes estimator is given by (2.6).

However,

$$\begin{aligned} \delta_\Pi(X) = E(\theta|X) &= \frac{\int \theta\,\mathrm{e}^{(-(1/2)\|X-\theta\|^2)}\,\mathrm{d}\Pi(\theta)}{\int \mathrm{e}^{(-(1/2)\|X-\theta\|^2)}\,\mathrm{d}\Pi(\theta)} \\ &= X + \frac{\int (\theta - X)\,\mathrm{e}^{(-(1/2)\|X-\theta\|^2)}\,\mathrm{d}\Pi(\theta)}{\int \mathrm{e}^{(-(1/2)\|X-\theta\|^2)}\,\mathrm{d}\Pi(\theta)} = X + \frac{\nabla m(X)}{m(X)}, \end{aligned} \tag{4.14}$$

where $m(X)$ is the marginal density of X.

Returning to the two-stage prior for θ where at the first stage $\Theta|\lambda \sim \mathrm{MVN}_p(0, \lambda^{-1}(1-\lambda)I)$ and at the second stage $\Lambda \sim (1-b)\lambda^{-b}$ $(b < 1)$ for $0 < \lambda < 1$.

$$m(x|\lambda) \propto \lambda^{p/2}\exp\left(-\frac{\lambda\|x\|^2}{2}\right) \quad \text{and} \quad m(x) \propto \int \lambda^{p/2-b}\exp\left(-\frac{\lambda\|x\|^2}{2}\right)\mathrm{d}\lambda.$$

$$\frac{\nabla m(X)}{m(X)} = -X\left(\frac{\int \lambda^{p/2-b+1}\exp(-\frac{\lambda\|x\|^2}{2})\,\mathrm{d}\lambda}{\int \lambda^{p/2-b}\exp(-\frac{\lambda\|x\|^2}{2})\,\mathrm{d}\lambda}\right) = -\frac{r(\|X\|^2)}{\|X\|}X.$$

So, we see that the Baranchik estimator given by (4.10) can take the form of (4.14).

Recall that a function $f(X)$ is superharmonic if $\nabla^2 f(X) \leqslant 0\ \forall X$. Combining expression (4.14) with Stein's theorem (our Theorem 3.1), we obtain the following neat result which leads to a broad class of minimax (generalized) Bayes estimators.

THEOREM 4.3. *If* $X \sim \mathrm{MVN}_p(\theta, I)$ *and* $\pi(\theta)$ *is superharmonic, then the estimator* (4.14) *is minimax with respect to quadratic loss* (2.2).

PROOF. If we write the estimator (4.14) as $\delta_\Pi(X) = X + g(X)$ then clearly $g(X) = \frac{\nabla m(X)}{m(X)}$. First note that convex combinations of superharmonic functions are superharmonic. So, since $\pi(\theta)$ is a superharmonic prior and

$$m(X) = \frac{\int \mathrm{e}^{(-(1/2)\|X-\theta\|^2)}\,\mathrm{d}\Pi(\theta)}{(2\pi)^p}$$

then $m(X)$ is also superharmonic, and hence $\nabla^2 m(X) \leqslant 0$. According to Theorem 3.1, the risk of $\delta_\Pi(X)$ with respect to quadratic loss (2.2) is

$$\begin{aligned} R(\delta_\Pi, \theta) &= p + E_\theta\left\{\left\|g(X)\right\|^2 + 2\nabla g(X)\right\} \\ &= p + E_\theta\left\{\left\|\frac{\nabla m(X)}{m(X)}\right\|^2 + 2\frac{m(X)\nabla^2 m(X) - \|\nabla m(X)\|^2}{m^2(X)}\right\} \\ &\leqslant p + 2E_\theta\left\{\frac{\nabla^2 m(X)}{m(X)}\right\} \leqslant p = R(X, \theta). \end{aligned} \tag{4.15}$$

Thus the desired minimax result for the Bayes estimator $\delta_\Pi(X)$ follows.

Notice that we just needed the superharmonicity of $m(X)$ to get the desired result. Moreover, note using (4.15) we can re-express the risk of the Bayes estimator, $\delta_\Pi(X)$ of (4.14), in another useful form. Specifically,

$$\begin{aligned} R(\delta_\Pi, \theta) &= p + E_\theta\left\{\left\|\frac{\nabla m(X)}{m(X)}\right\|^2 + 2\frac{m(X)\nabla^2 m(X) - \|\nabla m(X)\|^2}{m^2(X)}\right\} \\ &= p + 4E_\theta\left[\frac{\nabla^2\sqrt{m(X)}}{\sqrt{m(X)}}\right]. \end{aligned} \tag{4.16}$$

This implies that if $\sqrt{m(X)}$ is superharmonic, then the Bayes estimator is minimax. □

We can summarize the observations about the superharmonicity of $m(X)$ and $\sqrt{m(X)}$ in the following corollary:

COROLLARY 4.1. *If* $X \sim \mathrm{MVN}_p(\theta, I)$ *and the loss is quadratic loss* (2.2) *then the estimator* (4.14) *is minimax provided* $E\|\frac{\nabla m(X)}{m(X)}\|^2 < \infty$, *and any of the following hold*:

(a) $\pi(\theta)$ *is superharmonic,*
(b) $m(X)$ *is superharmonic,*
(c) $\sqrt{m(X)}$ *is a superharmonic.*

Further (a) *implies* (b) *implies* (c).

5. Generalizations for Bayes and the James–Stein estimation or the mean for the multivariate normal distribution with known covariance matrix Σ

In Sections 1 through 4 we have focused our attention on the case where $X \sim \mathrm{MVN}_p(\theta, I)$. In this section, we will generalize many of these results to the case $X \sim \mathrm{MVN}_p(\theta, \Sigma)$ where Σ is known and the loss is now general quadratic loss defined by

$$L(\delta, \theta) = (\delta - \theta)' Q(\delta - \theta), \tag{5.1}$$

and where Q is a positive definite matrix.

Consider first the case when $\Sigma = \sigma^2 I$ and the loss is still equal to squared error loss $\|\delta - \theta\|^2$ as given by (2.2). It is fairly straightforward to generalize many of the results that were presented in Sections 3 and 4 for the problem of estimating the mean vector of a multivariate normal distribution.

First, we take as the generalization of the James–Stein estimator given by (3.2) and generalize it to the estimator

$$\delta_a(X) = \left(1 - \frac{a\sigma^2}{\|X\|^2}\right)X. \tag{5.2}$$

When $p \geqslant 3$ this estimator will have smaller risk than the usual estimator X with respect to the general quadratic loss (5.1) and hence will be minimax for $0 < a < 2(p-2)$ (this is proven later on in this section). Additionally Baranchik's estimator given by (3.8) takes on the slightly amended form

$$\delta_r(X) = \left(1 - \frac{(p-2)\sigma^2 r(\|X\|^2)}{\|X\|^2}\right)X, \tag{5.3}$$

where $r(\|X\|^2)$ is a nondecreasing function of $\|X\|$ and $0 < r(\cdot) < 2$.

Strawderman (2003) summarizes the results for this case for an estimator of the form

$$\delta(X) = X + \sigma^2 g(X) \tag{5.4}$$

assuming $g(X)$ is weakly differentiable and $E(\|g(X)\|^2) < \infty$. The results we present are fairly straightforward based on those in Sections 3 and 4. We gather them, as Strawderman (2003) did in the following theorem.

THEOREM 5.1. *Let $X \sim \mathrm{MVN}_p(\theta, \sigma^2 I)$ and suppose that the loss is $\|\delta - \theta\|^2$.*

(a) $E_\theta\big[(X-\theta)' g(X)\big] = \sigma^2 E_\theta \nabla \cdot g(X)$ *(generalization of Lemma* 3.2).

(b) *For $\delta(X) = X + \sigma^2 g(X)$, the risk of δ,*

$$\begin{aligned} R(\delta, \theta) &= E_\theta\big(\|\delta - \theta\|^2\big) = E_\theta \big\|X + \sigma^2 g(X) - \theta\big\|^2 \\ &= p\sigma^2 + \sigma^4 E_\theta\big\{\big\|g(X)\big\|^2 + 2\nabla g(X)\big\} \end{aligned}$$

(generalization of (3.13) *in Theorem* 3.1).

(c) *If $\|g(X)\|^2 + 2\nabla g(X) \leqslant 0\ \forall X$, then $\delta(X)$ defined by* (5.4) *is minimax.* *(Generalization of Corollary* 3.1.)

(d) *If* $\theta \sim \Pi(\theta)$, *then the Bayes estimator is given by*

$$\delta_\Pi(X) = X + \sigma^2 \frac{\nabla m(X)}{m(X)}, \quad \textit{where} \tag{5.5}$$

$$m(X) = \frac{\int e^{(-(1/2\sigma^2)\|X-\theta\|^2)}\, d\Pi(\theta)}{(2\pi\sigma)^p} \quad \textit{is the marginal distribution of } X.$$

(*Generalization of* (4.14).)

(e) *If* $\theta \sim \Pi(\theta)$, *and* $\delta_\Pi(X)$ *is the Bayes estimator* (5.5) *then the risk of* δ_Π *is*

$$\begin{aligned} R(\delta_\Pi, \theta) &= p\sigma^2 + \sigma^4 E_\theta\{\|g(X)\|^2 + 2\nabla g(X)\} \\ &= p\sigma^2 + \sigma^4 E_\theta \left\{ \left[\left\| \frac{\nabla m(X)}{m(X)} \right\|^2 \right.\right. \\ &\quad \left.\left. + 2 \frac{m(X)\nabla^2 m(X) - \|\nabla m(X)\|^2}{m^2(X)} \right] \right\} \\ &= p\sigma^2 + 4\sigma^4 E_\theta \left[\frac{\nabla^2 \sqrt{m(X)}}{\sqrt{m(X)}} \right]. \end{aligned}$$

Hence δ_Π *is minimax provided* $\sqrt{m(X)}$ *is superharmonic.*
(*Generalization of* (4.16) *and Corollary* 4.1.)

Since superharmonicity of $\pi(\theta)$ *implies superharmonicity of* $m(X)$ *which in turn implies superharmonicity of* $\sqrt{m(X)}$, *then if either* $\pi(\theta)$, $m(X)$, *or* $\sqrt{m(X)}$ *is superharmonic* $\delta_\Pi(X) = X + \sigma^2 \frac{\nabla m(X)}{m(X)}$ *is minimax.* (*Generalization of Theorem* 4.3 *and Corollary* 4.1.)

As Strawderman (2003) discusses, following Bock (1988) if an estimator takes the form of the Bayes estimator (5.5) but $m(X)$ may not be a true marginal distribution resulting from a (generalized) prior $\pi(\theta)$, it is called a *pseudo-Bayes* estimator (and will denote it by $\delta_m(X)$, to distinguish it from the proper Bayes estimator in (5.5)). Moreover, provided $\sqrt{m(X)}$ is superharmonic, this pseudo-Bayes estimator will be minimax.

Lastly, by the right choice of $m(X)$, the resulting pseudo-Bayes estimator is a James–Stein type estimator. We demonstrate this in the following example.

EXAMPLE 5.1. If $m(X) = 1/\|X\|^{2b}$, then the pseudo-Bayes estimator

$$\delta_m(X) = \left(1 - \frac{2b\sigma^2}{\|X\|^2}\right) X. \tag{5.6}$$

Moreover, if $m(X) = 1/\|X\|^{2b}$ then $\sqrt{m(X)} = \|X\|^{-b}$ and clearly, $\nabla\sqrt{m(X)} = -bX/\|X\|^{b+2}$ and

$$\begin{aligned} \nabla^2 \sqrt{m(X)} &= -b\left(\frac{p\|X\|^{b+2} - (b+2)\|X\|^{b+2}}{\|X\|^{2b+4}}\right) \\ &= -\left(\frac{b((p-2)-b)}{\|X\|^{b+2}}\right) \leqslant 0, \end{aligned}$$

i.e. $\sqrt{m(X)}$ is superharmonic for $0 \leqslant b \leqslant (p-2)$. So, the James–Stein estimator given by (5.2) is minimax for $0 \leqslant a \leqslant 2(p-2)$. It is a simple matter to extend the minimaxity result to the Baranchik estimator (5.3), as we did in Section 3.

By Theorem 5.1 part (e)

$$R(\delta_m, \theta) = p\sigma^2 + 4\sigma^4 E_\theta\left[\frac{\nabla^2\sqrt{m(X)}}{\sqrt{m(X)}}\right]$$
$$= p\sigma^2 - 4\sigma^4 E_\theta\left(\frac{b((p-2)-b)}{\|X\|^2}\right). \tag{5.7}$$

Fourdrinier, Strawderman and Wells (FSW) (1998) point out the importance in the distinction between the superharmonicity of $m(X)$ and $\sqrt{m(X)}$. They note that it is impossible for $m(X)$ to be both superharmonic and lead to a proper Bayes estimator, however, this is not the case when $\sqrt{m(X)}$ is superharmonic. In fact, they show that when $p \geqslant 5$ that propriety of $m(X)$ and superharmonicity of $\sqrt{m(X)}$ are possible. A main result from the FSW paper includes a generalization of Theorem 4.2.

THEOREM 5.2. *Suppose Θ has a prior distribution with the following hierarchical structure*

$$\Theta|\lambda \sim \mathrm{MVN}_p\big(0, \sigma^2(1-\lambda)\lambda^{-1}I\big) \quad \text{and} \quad \Lambda \sim h(\lambda)\ (b < 1),\ 0 < \lambda < 1.$$

(a) *The marginal distribution of X conditional on λ is*

$$X|\lambda \sim \mathrm{MVN}_p\big(0, \sigma^2\lambda^{-1}I\big).$$

(b) $m(x) \propto \displaystyle\int_0^1 \lambda^{p/2}\exp\left[-\frac{\lambda\|x\|^2}{2\sigma^2}\right]h(\lambda)\,d\lambda.$

(c) *$\sqrt{m(X)}$ is superharmonic (and hence $\delta_\Pi(X)$ is minimax) provided $h(\lambda)$ satisfies $\frac{\lambda h'(\lambda)}{h(\lambda)} = l_1(\lambda) + l_2(\lambda)$, where $l_1(\lambda) \leqslant A$, is nonincreasing in λ, and $0 \leqslant l_2(\lambda) \leqslant B$ where $\frac{1}{2}A + B \leqslant (p-6)/4$ and $\lim_{\lambda\to 0}\lambda^{p/2}h(\lambda) = 0$ and $\lim_{\lambda\to 1}h(\lambda) < \infty$.*

EXAMPLE 5.2. For the obvious generalization of the 2-stage prior in Section 4, the conditions of Theorem 5.2 are satisfied. That is, consider a two-stage prior for θ where at the first stage $\Theta|\lambda \sim \mathrm{MVN}_p(0, \sigma^2(1-\lambda)\lambda^{-1}I)$ for $0 < \lambda < 1$ and at the second stage $\Lambda \sim (1-b)\lambda^{-b}$ for $0 < \lambda < 1$, then the conditions of Theorem 5.2 are satisfied for $(6-p)/2 \leqslant b < 1$ and this property gives proper Bayes minimax estimators for $p \geqslant 5$. For $p \geqslant 3$ the above prior gives admissible minimax estimators provided $(6-p)/2 \leqslant b < 2$.

George (1986) studied *multiple shrinkage estimators*, that is, estimators for which there are several possible points ν_i, $i = 1, 2, \ldots, k$, towards which to shrink. The desired result is to shrink "adaptively" toward one of the points so that the resulting procedure will be minimax. The following theorem is a version of George's result and is presented without proof.

THEOREM 5.3. *Suppose $\pi_\alpha(\theta), \alpha \in A$, are a collection of superharmonic (generalized) priors and $m_\alpha(X)$ is the corresponding collection of marginal distributions. Let $\lambda(\alpha)$ be any finite mixing distribution on $\alpha \in A$. Then*

(a) $\int \pi_\alpha(\theta)h(\alpha)\,\mathrm{d}\alpha$ *is superharmonic with corresponding superharmonic marginal*

$$\int m_\alpha(X)h(\alpha)\,\mathrm{d}\alpha.$$

(b) *The resulting generalized (or pseudo, if $m_\alpha(X)$ are given) Bayes estimator is minimax.*

To see how this is applied in Example 5.3, first note that if each π_i is superharmonic, then so is $\Sigma\alpha_i\pi_i(\theta)$ (and more generally for an arbitrary mixture of superharmonic functions).

EXAMPLE 5.3. Suppose $\nu_i, i = 1, 2, \ldots, k$, are vectors in R^p, and $m_i(X) = 1/\|X - \nu_i\|^{2b}$ for $0 < b < (p-2)/2$. Let $m(X) = (1/k)(\sum_{i=1}^k m_i(X))$. Then $m_i(X)$ and $m(X)$ are superharmonic and the resulting pseudo-Bayes estimator given by

$$\delta_m(X) = X - \frac{2b\sigma^2 \sum_{i=1}^k (X - \nu_i)/(\|X - \nu_i\|^{2b+2})}{\sum_{i=1}^k (1/(\|X - \nu_i\|^{2b}))} \tag{5.8}$$

is minimax. This estimator given by (5.8) "adaptively" shrinks X toward the "closest" ν_i or alternatively, $\delta_m(X)$ given by (5.8) is a weighted combination of James–Stein like estimators shrinking toward the ν_i with greater weights ($\|X - \nu_i\|^{2b}$) put on the ν_i closest to X.

To further extend the results in this section as well as Sections 3 and 4, consider now $X \sim \mathrm{MVN}_p(\theta, \Sigma)$ where $\Sigma = \mathrm{diag}(\sigma_1^2, \sigma_2^2, \ldots, \sigma_p^2)$ and in the general quadratic loss defined by (5.2) let $Q = \mathrm{diag}(d_1, d_2, \ldots, d_p) = D$. The extension of Theorem 5.1 (whose proof is similar to Theorem 5.1 and can be found in Stein, 1981) is the following theorem.

THEOREM 5.4. *Let $X \sim \mathrm{MVN}_p(\theta, \Sigma)$ and suppose that the loss is $L(\delta, \theta) = (\delta - \theta)'Q(\delta - \theta)$, where Σ and D are defined as above.*

(a) *If $\delta(X) = X + \Sigma g(X)$, where $g(X)$ is weakly differentiable and $E\|g\|^2 < \infty$, then the risk of δ,*

$$\begin{aligned} R(\delta, \theta) &= E_\theta\big((\delta - \theta)'Q(\delta - \theta)\big) \\ &= \mathrm{tr}\,\Sigma D + E_\theta\left\{\left[\sum_{i=1}^p \sigma_i^4 d_i\left(g_i^2(X) + 2\frac{\partial g_i(X)}{\partial X_i}\right)\right]\right\}. \end{aligned}$$

(b) *If $\theta \sim \Pi(\theta)$, then the Bayes estimator of θ is*

$$\delta_\Pi(X) = X + \Sigma\frac{\nabla m(X)}{m(X)},$$

where $m(X)$ is the marginal distribution of X.

(c) *If* $\theta \sim \Pi(\theta)$ *then the risk of a proper (generalized, pseudo-) Bayes estimator of the form* $\delta_m(X) = X + \Sigma \frac{\nabla m(X)}{m(X)}$ *is given by*

$$R(\delta_m, \theta) = \operatorname{tr} \Sigma D + E_\theta\left[\frac{2m(X)\left(\sum_{i=1}^{p} \sigma_i^4 d_i (\partial m^2(X)/\partial^2 X_i)\right)}{m^2(X)} - \frac{\left(\sum_{i=1}^{p} \sigma_i^4 d_i (\partial m(X)/\partial X_i)^2\right)}{m^2(X)}\right]$$

$$= \operatorname{tr} \Sigma D + 4E_\theta\left[\frac{\left(\sum_{i=1}^{p} \sigma_i^4 d_i (\partial^2 \sqrt{m(X)}/\partial^2 X_i)\right)}{\sqrt{m(X)}}\right].$$

(d) *If* $[(\sum_{i=1}^{p} \sigma_i^4 d_i (\partial^2 \sqrt{m(X)}/\partial^2 X_i))/\sqrt{m(X)}]$ *is nonpositive, the proper (generalized, pseudo-) Bayes* $\delta_m(X)$ *is minimax.*

What now follows is the key observation that allows us to construct Bayes minimax procedures for the generalized situation, based on the procedures when $\Sigma = D = I$.

LEMMA 5.1. *Suppose* $\eta(X)$ *is such that* $\nabla^2 \eta(X) = \sum_{i=1}^{p} (\partial^2 \eta(X)/\partial^2 X_i^2) \leqslant 0$ *(i.e.* $\eta(X)$ *is superharmonic) then* $\eta^*(X) = \eta(\Sigma^{-1} D^{-1/2} X)$ *is such that* $\sum_{i=1}^{p} \sigma_i^4 d_i (\partial^2 \eta^*(X)/\partial^2 X_i) \leqslant 0$.

The proof is a straightforward calculation and can be found in Strawderman (2003). Note that for any scalar, a, if $\eta(X)$ is superharmonic, then so is $\eta(aX)$. This all leads to the following main result.

THEOREM 5.5. *Let* $X \sim \mathrm{MVN}_p(\theta, \Sigma)$ *and suppose that the loss is* $L(\delta, \theta) = (\delta - \theta)' Q(\delta - \theta)$, *where* Σ *and* D *are defined diagonal matrices.*

(a) *Suppose* $\sqrt{m(X)}$ *is superharmonic (*$m(X)$ *is a proper, generalized, or pseudo-marginal for the case* $\Sigma = D = I$*). Then*

$$\delta_m(X) = X + \Sigma\left(\frac{\nabla m(\Sigma^{-1} D^{-1/2} X)}{m(\Sigma^{-1} D^{-1/2} X)}\right)$$

is a proper, generalized, or pseudo-Bayes minimax estimator.

(b) *If* $\sqrt{m(\|X\|^2)}$ *is spherically symmetric and superharmonic, then*

$$\delta_m(X) = X + \Sigma \frac{2m'(X'\Sigma^{-1} D^{-1} \Sigma^{-1} X) D^{-1} \Sigma^{-1} X}{m(X' \Sigma^{-1} D^{-1} \Sigma^{-1} X)}$$

is minimax.

(c) *Suppose the prior distribution* $\Pi(\theta)$ *has the hierarchical structure* $\Theta | \lambda \sim \mathrm{MVN}_p(0, A_\lambda)$ *for* $\lambda \sim h(\lambda)$, $0 < \lambda < 1$, *where* $A_\lambda = (c/\lambda) \Sigma D \Sigma - \Sigma$ *where* c *is such that* A_1 *is positive definite and* $h(\lambda)$ *satisfies the condition in part (c) of Theorem 5.2. Then* $\delta_\Pi(X) = X + \Sigma \frac{\nabla m(X)}{m(X)}$ *is minimax.*

(d) *Suppose $m_i(X)$, $i = 1, 2, \dots, k$, are superharmonic, then the multiple shrinkage estimator*

$$\delta_m(X) = X + \Sigma\left[\frac{\sum_{i=1}^k \nabla m_i(\Sigma^{-1}D^{-1/2}X)}{\sum_{i=1}^k m_i(\Sigma^{-1}D^{-1/2}X)}\right] \tag{5.9}$$

is a minimax multiple shrinkage estimator.

PROOF. (a) This follows directly from Theorem 5.4 parts (c) and (d) and Lemma 5.1.

(b) This follows from part (a) and Theorem 5.4 part (b) with a straightforward calculation.

(c) First note that $\Theta|\lambda \sim \mathrm{MVN}_p(0, A_\lambda)$ and $X - \Theta|\lambda \sim \mathrm{MVN}_p(0, \Sigma)$. So, $X - \Theta$ and Θ are therefore conditionally independent given λ. Hence $X|\lambda \sim \mathrm{MVN}_p(0, A_\lambda + \Sigma)$. It follows that

$$m(X) \propto \int_0^1 \lambda^{p/2} \exp\left[-\frac{\lambda}{c}\left(X'\Sigma^{-1}D^{-1}\Sigma^{-1}X\right)\right] h(\lambda)\, \mathrm{d}\lambda$$

but $m(X) = \eta(X'\Sigma^{-1}D^{-1}\Sigma^{-1}X/c)$, where $\sqrt{\eta(X'X)}$ is superharmonic by Theorem 5.2 part (c). Hence by part (b) $\delta_\Pi(X)$ is minimax (and proper or generalized Bayes depending on whether $h(\lambda)$ is integrable or not).

(d) Since $\eta(X)$ superharmonic implies that $\sqrt{\eta(X)}$ is superharmonic, this part follows from part (a). □

EXAMPLE 5.4 (*Pseudo-Bayes minimax estimators*). When $\Sigma = D = I$ we saw in Example 5.1, that by choosing $m(X) = 1/\|X\|^{2b}$, the pseudo-Bayes estimator was the James–Stein estimator $\delta_m(X) = (1 - 2b\sigma^2/\|X\|^2)X$ given by (5.6). It now follows from this and part (b) of Theorem 5.5 that $m(X'\Sigma^{-1}D^{-1}\Sigma^{-1}X) = (1/X'\Sigma^{-1}D^{-1}\Sigma^{-1}X)^b$ has associated with it the pseudo-Bayes estimator $\delta_m(X) = (1 - 2bD^{-1}\Sigma^{-1}/(X'\Sigma^{-1}D^{-1}\Sigma^{-1}X))X$ and that estimator is minimax for $0 < b \leqslant 2(p-2)$.

EXAMPLE 5.5 (*Hierarchical (proper) Bayes minimax estimator*). As suggested by Berger (1976) suppose the prior distribution $\Pi(\theta)$ has the hierarchical structure $\Theta|\lambda \sim \mathrm{MVN}_p(0, A_\lambda)$ where $A_\lambda = c\Sigma D\Sigma - \Sigma$ and $c > (1/\min(\sigma_i^2 d_i)$ and $h(\lambda) \propto {\sim}\lambda^{-b}$, $0 < \lambda < 1$, for $(6-p)/2 \leqslant b < 1$. The resulting proper Bayes estimator will be minimax for $p \geqslant 5$ by Theorem 5.5 part (c) and Example 5.2. For $p \geqslant 3$ the estimator $\delta_\Pi(X)$ given in part (c) of Theorem 5.5 is a generalized Bayes admissible minimax provided $(6-p)/2 \leqslant b < 2$.

EXAMPLE 5.6 (*Multiple shrinkage minimax estimator*). It follows from Example 5.3 and Theorem 5.5 that

$$m(X) = \sum_{i=1}^k \left[\frac{1}{(X - \nu_i)'\Sigma^{-1}D^{-1}\Sigma^{-1}(X - \nu_i)}\right]^b$$

satisfies the conditions of Theorem 5.5(d) for $0 < b \leqslant (p-2)/2$ and hence that

$$\delta_m(X) = X - \frac{2b\sum_{i=1}^{k}[D^{-1}\Sigma^{-1}(X-\nu_i)]/[(X-\nu_i)'\Sigma^{-1}D^{-1}\Sigma^{-1}(X-\nu_i)]^{b+1}}{\sum_{i=1}^{k} 1/[(X-\nu_i)'\Sigma^{-1}D^{-1}\Sigma^{-1}(X-\nu_i)]^{b}} \tag{5.10}$$

is a minimax multiple shrinkage (pseudo-Bayes) estimator.

If in Example 5.3 and this example, we used the following generalized prior

$$\pi(\theta) = \sum_{i=1}^{k}\left[\frac{1}{(\theta-\nu_i)'\Sigma^{-1}D^{-1}\Sigma^{-1}(\theta-\nu_i)}\right]^{b}$$

the resulting estimators would be generalized Bayes minimax estimators (as opposed to pseudo-Bayes) for $0 < b \leqslant (p-2)/2$.

We are now ready to consider the general situation stated at the beginning of this section, $X \sim \mathrm{MVN}_p(\theta, \Sigma)$ where Σ is known and the loss is now general quadratic loss defined by $L(\delta, \theta) = (\delta - \theta)'Q(\delta - \theta)$, where Q is a positive definite matrix. We will see that this case can be reduced to the canonical form $\Sigma = I$ and $Q = \mathrm{diag}(d_1, d_2, \ldots, d_p) = D$. We continue to follow the development in Strawderman (2003).

The following well known fact will be used repeatedly to generalize our results.

LEMMA 5.2. *For any pair of positive definite matrices, Σ and Q, there exits a nonsingular matrix A such that $A\Sigma A' = I$ and $(A')^{-1}QA^{-1} = D$ where D is diagonal.*

Using this fact we can now present the canonical form of the estimation problem.

THEOREM 5.6. *Let $X \sim \mathrm{MVN}_p(\theta, \Sigma)$ and suppose that the loss is $L_1(\delta, \theta) = (\delta - \theta)'Q(\delta - \theta)$. Let A and D be as in Lemma 5.2, $Y = AX \sim \mathrm{MVN}_p(\nu, I)$, where $\nu = A\theta$, and $L_2(\delta, \nu) = (\delta - \nu)'D(\delta - \nu)$.*

(a) *If $\delta_1(X)$ is an estimator with risk function $R_1(\delta_1, \theta) = E_\theta L_1(\delta_1(X), \theta)$, then the estimator $\delta_2(Y) = A\delta_1(A^{-1}Y)$ has risk function $R_2(\delta_2, \nu) = R_1(\delta_1, \theta) = E_\theta L_2(\delta_2(Y), \nu)$.*

(b) *$\delta_1(X)$ is proper or generalized Bayes with respect to the proper prior distribution $\Pi_1(\theta)$ (or pseudo-Bayes with respect to the pseudo-marginal $m_1(X)$) under loss L_1 if and only if $\delta_2(Y) = A\delta_1(A^{-1}Y)$ is proper or generalized Bayes with respect to $\Pi_2(\nu) = \pi_1(A^{-1}\nu)$ (or pseudo-Bayes with respect to the pseudo-marginal $m_2(Y) = m_1(A^{-1}Y)$).*

(c) *$\delta_1(X)$ is admissible (or minimax or dominates $\delta_1^*(X) = \delta(\Sigma^{-1}D^{-1/2}X)$) under L_1 if and only if $\delta_2(Y) = A\delta_1(A^{-1}Y)$ is admissible (or minimax or dominates $\delta_2^*(Y) = A\delta_1^*(A^{-1}Y)$ under L_2).*

PROOF.

(a) $$R_2(\delta_2, \nu) = E_\theta L_2[\delta_2(Y), \nu] = E_\theta L_2\big[\big(\delta_2(Y) - \nu\big)'D\big(\delta_2(Y) - \nu\big)\big]$$

$$= E_\theta L_2\big[\big(A\delta_1\big(A^{-1}(AX)\big) - A\theta\big)' D\big(A\delta_1\big(A^{-1}(AX)\big) - A\theta\big)\big]$$
$$= E_\theta L_2\big[\big(\delta_1\big((X) - \theta\big)' A' D A\big(\delta_1(X) - \theta\big)\big]$$
$$= E_\theta L_2\big[\big(\delta_1\big((X) - \theta\big)' Q\big(\delta_1(X) - \theta\big)\big] = R_1(\delta_1, \theta).$$

(b) The Bayes estimator for any quadratic loss is the posterior mean. Hence, since $\Theta \sim \Pi_1(\theta)$ and $N = A\Theta \sim \Pi_2(\nu) = \Pi_1(A^{-1}\nu)$ (ignoring constants) then

$$\delta_2(Y) = E[N|Y] = E[A\Theta|Y] = E\big[A(\Theta|AX)\big]$$
$$= AE[\Theta|X] = A\delta_1(X) = A\delta_1\big(A^{-1}Y\big).$$

(c) This follows directly from part (a). □

If $\Sigma^{1/2}$ is the positive definite square root of Σ and $A = P\Sigma^{-1}$ where P is orthogonal and such that it diagonalizes $\Sigma^{1/2}Q\Sigma^{1/2}$ then Theorem 5.6 works for this case.

EXAMPLE 5.7. Proceeding as we did in Example 5.4 and applying Theorem 5.6,

$$m\big(X'\Sigma^{-1}Q^{-1}\Sigma^{-1}X\big) = \big(X'\Sigma^{-1}Q^{-1}\Sigma^{-1}X\big)^{-b}$$

has associated with it, for $0 < b \leqslant 2(p-2)$, the pseudo-Bayes minimax James–Stein estimators

$$\delta_m(X) = \left(1 - \frac{2bQ^{-1}\Sigma^{-1}}{(X'\Sigma^{-1}Q^{-1}\Sigma^{-1}X)}\right)X.$$

Note that generalizations of Examples 5.5 to hierarchical Bayes minimax estimators and 5.6 to Multiple shrinkage estimators are straightforward. We omit the details.

6. Conclusion and extensions

We have attempted to give a coherent review of much of the development of Bayes Minimax estimation of the mean vector of a multivariate normal distribution under quadratic loss. We have restricted the discussion to the case where the covariance is known. Our approach has been more decision theoretic than purely Bayesian in orientation and has followed the development primarily from the perspective of its connections with Stein-type or shrinkage estimation.

There is a vast literature on the general area of shrinkage estimation that we have not covered at all. We mention below some of the areas we have not covered together with a few references so that the interested reader may learn more.

The unknown covariance case. Strawderman (1973) found proper Bayes minimax estimators for the multivariate normal case when $\Sigma = \sigma^2 I$ and the loss is normed squared error loss. Berger et al. (1977) (see also Gleser, 1979 and Lin and Tsai, 1973) found minimax Stein-like estimators for the case of completely unknown Σ and general quadratic loss.

The nonnormal case. Strawderman (1974) developed Baranchik-type and also generalized Bayes minimax estimators of the mean vector of a scale mixture of multivariate normal distributions under squared error loss. Brandwein and Strawderman (1978) and Brandwein (1979) gave Baranchik-type minimax estimators for spherically symmetric unimodal and spherically symmetric distributions respectively under squared error loss. Berger (1975) also gave similar results but with different methods of proof in the spherically symmetric case. See also Shinozaki (1994), Brandwein and Strawderman (1991), Cellier and Fourdrinier (1995), Fourdrinier and Strawderman (1996), and Maruyama (2003a, 2003b).

Nonquadratic loss. Brandwein and Strawderman (1978) gave minimax Baranchik-type estimators for spherically symmetric distributions when the loss function is a convex function of quadratic loss.

Admissibility. Brown (1966) establishes admissibility (for $p < 3$) and inadmissibility conditions (for $p \geqslant 3$), for the best equivariant estimator of a location parameter under a very general loss function. Brown (1971) gave conditions on the prior distribution under which a Bayes estimator is admissible or inadmissible for the known covariance, quadratic loss case. See also Berger and Strawderman (1996).

References

Baranchik, A.J. (1964). Multiple regression and estimation of the mean of a multivariate normal distribution. Stanford University Technical Report No. 51.

Baranchik, A.J. (1970). A family of minimax estimators of the mean of a multivariate normal distribution. *Ann. Math. Statist.* **41**, 642–645.

Berger, J. (1975). Minimax estimation of location vectors for a wide class of densities. *Ann. Statist.* **3**, 1318–1328.

Berger, J. (1976). Admissible minimax estimation of multivariate normal mean with arbitrary quadratic loss. *Ann. Statist.* **4**, 223–226.

Berger, J., Strawderman, W.E. (1996). Choice of hierarchical priors: Admissibility in estimation of normal means. *Ann. Statist.* **24**, 931–951.

Berger, J., Bock, M.E., Brown, L., Casella, G., Gleser, L. (1977). Minimax estimation of a normal mean vector for arbitrary quadratic loss and unknown covariance matrix. *Ann. Statist.* **5**, 763–771.

Bock, M.E. (1988). Shrinkage estimators: Pseudo Bayes estimators for normal mean vectors. In: Gupta, S.S., Berger, J.O. (Eds.), *Statistical Decision Theory IV*. Springer-Verlag, New York, pp. 281–298.

Brandwein, A.C. (1979). Minimax estimation of the mean of spherically symmetric distributions under quadratic loss. *J. Multivariate Anal.* **9**, 579–588.

Brandwein, A.C., Strawderman, W.E. (1990). Stein estimation: The spherically symmetric case. *Statist. Sci.* **5**, 356–369.

Brandwein, A.C., Strawderman, W.E. (1991). Generalizations of James–Stein estimators under spherical symmetry. *Ann. Statist.* **19**, 1639–1650.

Brown, L.D. (1966). On the admissibility of invariant estimators of one or more location parameters. *Ann. Math. Statist.* **37**, 1087–1136.

Brown, L.D. (1971). Admissible estimators, recurrent diffusions, and insoluble boundary value problems. *Ann. Math. Statist.* **42**, 855–903.

Cellier, D., Fourdrinier, D. (1995). Shrinkage estimators under spherical symmetry for the general linear model. *J. Multivariate Anal.* **52**, 338–351.

Efron, B., Morris, C. (1972). Empirical Bayes on vector observations: An extension of Stein's method. *Biometrika* **59**, 335–347.

Efron, B., Morris, C. (1973). Stein's estimation rule and its competitors – an empirical Bayes approach. *J. Amer. Statist. Assoc.* **68**, 117–130.
Fourdrinier, D., Strawderman, W.E. (1996). A paradox concerning shrinkage estimators: Should a known scale parameter be replaced by an estimated value in the shrinkage factor. *J. Multivariate Anal.* **59**, 109–140.
Fourdrinier, D., Strawderman, W.E., Wells, M. (1998). On the construction of Bayes minimax estimators. *Ann. Statist.* **26**, 660–671.
George, E.I. (1986). Minimax multiple shrinkage estimators. *Ann. Statist.* **14**, 188–205.
Gleser, L. (1979). Minimax estimation of a normal mean vector when the covariance matrix is unknown. *Ann. Statist.* **7**, 838–846.
James, W., Stein, C. (1961). Estimation of quadratic loss. In: *Proc. 4th Berkeley Symposium on Mathematical Statistics and Probability*. Univ. of California Press, pp. 311–319.
Lehmann, E.L., Casella, G. (1998). *The Theory of Point Estimation*, second ed. Springer-Verlag, New York.
Lin, P., Tsai, H. (1973). Generalize Bayes minimax estimators of the multivariate normal mean with unknown covariance matrix. *Ann. Statist.* **1**, 142–145.
Maruyama, Y. (2003a). Admissible minimax estimators of a mean vector of scale mixtures of multivariate normal distributions. *J. Multivariate Anal.* **84**, 274–283.
Maruyama, Y. (2003b). A robust generalized Bayes estimator improving on the James–Stein estimator for spherically symmetric distributions. *J. Multivariate Anal.* **84**, 274–283.
Sacks, J. (2003). Generalized Bayes solutions in estimation problems. *Statistics and Decisions* **21**, 69–78.
Shao, P., Strawderman, W.E. (1994). Improving on the James–Stein positive part estimator. *Ann. Statist.* **22**, 1517–1539.
Shinozaki, N. (1994). Simultaneous estimation of location parameters under quadratic loss. *Ann. Statist.* **22**, 233–235.
Stein, C. (1956). Inadmissibility of the usual estimator for the mean of a multivariate normal distribution. In: *Proc. Third Berkeley Symposium on Mathematical Statistics and Probability*. Univ. of California Press, pp. 197–206.
Stein, C. (1962). Confidence sets for the mean of a multivariate normal distribution. *J. Roy. Statist. Soc., Ser. B* **24**, 265–296.
Stein, C. (1973). Estimation of the mean of a multivariate normal distribution. In: *Proc. Prague Symposium Asymptotic Statistics*, pp. 345–381.
Stein, C. (1981). Estimation of the mean of a multivariate normal distribution. *Ann. Statist.* **9**, 1131–1151.
Strawderman, W.E. (1971). Proper Bayes minimax estimators of multivariate normal mean. *Ann. Math. Statist.* **42**, 385–388.
Strawderman, W.E. (1973). Proper Bayes minimax estimators of multivariate normal mean for the case of common unknown variances. *Ann. Statist.* **1**, 1189–1194.
Strawderman, W.E. (1974). Minimax estimation of location parameters for certain spherically symmetric distributions. *J. Multivariate Anal.* **4**, 255–264.
Strawderman, W.E. (2000). Minimaxity. *J. Amer. Statist. Assoc.* **95**, 1364–1368.
Strawderman, W.E. (2003). On minimax estimation of a normal mean vector for general quadratic loss. In: Moore, M., Froda, S., Leger, C. (Eds.), *Mathematical Statistics & Applications: A Festschrift for Constance van Eeden*. In: *IMS Monograph Series*, vol. 42, pp. 3–14.
Strawderman, W.E., Cohen, A. (1971). Admissibility of estimators of the mean vector of a multivariate normal distribution with quadratic loss. *Ann. Math. Statist.* **42**, 270–296.

Handbook of Statistics, Vol. 25
ISSN: 0169-7161
© 2005 Elsevier B.V. All rights reserved.
DOI 10.1016/S0169-7161(05)25009-5

9

Bayesian Nonparametric Modeling and Data Analysis: An Introduction

Timothy E. Hanson, Adam J. Branscum and Wesley O. Johnson

Abstract
Statistical models are developed for the purpose of addressing scientific questions. For each scientific question for which data are collected, the truth is sought by developing statistical models that are useful in this regard. Despite the fact that restrictive parametric models have been shown to be extraordinarily effective in many instances, there is and has been much scope for developing statistical inferences for models that allow for greater flexibility. It would seem that just about any statistical modeling endeavor can be expanded and approached, at least conceptually, as a nonparametric problem. The purpose of this chapter is to give a brief discussion of, and introduction to, one of the two major approaches to the whole of statistics as it were, Bayesian nonparametrics.

1. Introduction to Bayesian nonparametrics

The term 'nonparametric' is somewhat of a misnomer. It literally connotes the absence of parameters. But it is usually the case that the goals of a data analysis include making inferences about functionals of an unknown probability measure, F, which are themselves parameters, regardless of whether the class of probability measures under consideration is quite broad (e.g., not indexed by parameters). Nonetheless, the spirit of the term 'nonparametric' is to be free of restrictive, inappropriate, or unrealistic constraints that are implied by particular parametric models. For example, it is often necessary to consider models that allow for unspecified multimodality, asymmetry and nonlinearity. This can be accomplished by considering a broad class of distributions and by making statistical inferences within that context. Semiparametric modeling involves incorporating parametric and nonparametric components into a single model, an example being a linear regression where the error distribution is allowed to be arbitrary subject to having median zero. Hundreds of frequentist nonparametric and semiparametric papers have been published. Classic methods were typically based on permutations and ranking, while with increases in computational capabilities, jackknifing and resampling methods have more recently played a major role. Bayesian and frequentist nonparametric regression modeling, density estimation, and smoothing remains an active area of research.

Parametric modeling has dominated the Bayesian landscape for many years. In the parametric setting, data are modeled according to a family of probability measures $\{F_\theta: \theta \in \Theta\}$ with corresponding probability density functions (pdf) $\{p(\cdot|\theta): \theta \in \Theta\}$. Scientific evidence for θ, which is obtained independently of the current data, is used to construct a parametric "prior" pdf, $p(\mathrm{d}\theta)$. As a first step, the posterior pdf of θ, $p(\mathrm{d}\theta|\text{data})$, is obtained. The next steps usually involve finding various posterior characteristics such as medians or means, standard deviations, and probability intervals. Prediction is accomplished by integrating the sampling pdf for the future observation given the data against the posterior.

Nonparametric modeling begins with the specification of a broad class of models for the data at hand. For example, consider a single sample of data from an unknown distribution F. The goal is to make inferences about functionals of F, or possibly the pdf corresponding to F. We could simply assert that F belongs to $\mathcal{F}$, the class of all continuous distributions on the real line. Alternatively, standard regression data, $\{(y_i, x_i): i = 1, \ldots, n\}$, can be modeled as $y_i|(x_i, f, \theta) \overset{\perp}{\sim} N(f(x_i), \theta)$, where $f \in \mathcal{F}^*$, a broad class of possible regression functions, and where $\theta \in (0, \infty)$. Bayesian approaches to these problems require specifying probability measures, $\mathcal{P}(\mathrm{d}F)$ and $\mathcal{P}^*(\mathrm{d}f)$ on $\mathcal{F}$ and $\mathcal{F}^*$, respectively, as well as a suitable parametric probability measure for θ. In general, constructing suitable $\mathcal{P}$'s on function spaces has been accomplished by a number of authors. Data analysis and applications involving these models were limited at first due to analytical intractability. However, the last fifteen years has seen a dramatic increase in nonparametric and semiparametric Bayesian modeling due to remarkable improvements in computational techniques and capabilities.

Müller and Quintana (2004) noted that Bayesian nonparametric models are also used to "robustify" parametric models and to perform sensitivity analyses. For example, the above regression problem includes standard parametric linear regression as a special case. Bayesian modeling can take specific account of this by constructing a prior $\mathcal{P}^*(\mathrm{d}f)$ that is centered on the parametric regression function. Along these lines, Ibrahim and Kleinman (1998) embedded the family of zero-mean normal models in a broader class of models for random effects in a generalized linear mixed model framework, and Berger and Gugliemi (2001) developed general Bayesian nonparametric (BNP) methodology for embedding a family of parametric models in a broader class for the purpose of determining the adequacy of parametric models.

In this chapter, we first discuss the basics of BNP modeling, e.g., the determination of suitable $\mathcal{P}$ to be defined on $\mathcal{F}$. This development begins with the Dirichlet process (DP) (Ferguson, 1973), the mixture of DP's (MDP) (Antoniak, 1974), and Dirichlet process mixture (DPM) models (Antoniak, 1974; Escobar, 1994), and then presents Polya tree (PT) (Lavine, 1992, 1994), mixtures of PT's (MPT) (Lavine, 1992) and the gamma process (GP) (Kalbfleisch, 1978). Special emphasis is given to the DPM, MPT and GP models so more details and/or illustrations are given for them. There are many other choices for $\mathcal{P}$, but we mainly focus on these. This material is like root stock, from which it is possible to grow more complex models and methods.

After this development, we present a variety of illustrations starting with an application to the independent two sample problem, and moving on to a variety of regression problems. The regression scenarios considered include (i) approaches to linear regression modeling with an unknown error distribution, which are illustrated in a survival analysis setting, (ii) nonlinear regression modeling with a parametric error distribution, which is illustrated on highly nonlinear data, and (iii) a fully nonparametric model where the regression function and the error distribution are modeled nonparametrically. Our presentation of nonparametric regression modeling of a mean function involves the representation of the mean function as an infinite linear combination of known basis functions (the coefficients are unknown). Bayesian modeling in this setting involves the truncation of the infinite series, resulting in a regression function specified as a finite linear combination. This can lead to a dimension varying linear model and requires specifying a joint prior probability distribution on the corresponding basis coefficients and (possibly) the number of basis functions to be included in the model. The resulting linear model is essentially a highly flexible parametric model so that standard parametric methods are applicable in fitting the semiparametric model. For this particular application, the fundamental background material is not needed.

We also discuss a variety of other modeling situations, but in less detail. We make no attempt to present an exhaustive discussion of Bayesian nonparametrics since it is possible to discuss *all* of inferential statistics from a BNP perspective, and this would be beyond the scope of any single article. We shall instead discuss basic ideas, provide some simple illustrations, and give the reader a taste of recent progress in a few important subfields.

The computing environment WinBUGS (Spiegelhalter et al., 2003) has made Bayesian modeling available to the masses. In our discussion, we indicate how this user-friendly software can be used to fit data to a number of non/semiparametric models. Congdon (2001, Section 6.7) presents examples of DPM and PT models fit in WinBUGS.

From here on we use the notation F to mean both a probability measure and its corresponding cumulative distribution function (CDF) where we trust that the context will make clear the distinction.

2. Probability measures on spaces of probability measures

In modeling a probability measure F as $F \sim \mathcal{P}(\mathrm{d}F)$, common choices of $\mathcal{P}$ are the DP, MDP, DPM, PT, MPT and GP (a primary application of the GP is in the area of survival analysis where the GP can be used to model the cumulative hazard function and thus induces a distribution on F). For many years, emphasis was placed on the DP due to its mathematical tractability in simple situations, however the DP prior was criticized because it places prior probability one on the class of discrete distributions. The use of MDP's has been limited due to the complexity resulting from a computational explosion associated with possibilities for ties (Antoniak, 1974; Berry and Christensen, 1979; Johnson and Christensen, 1989).

The advent of modern BNP data analysis stems first from the development of Markov chain Monte Carlo (MCMC) technology starting with Gelfand and Smith (1990) and then from the observation by Escobar (1994) that these methods (in particular, Gibbs sampling) could be applied to DPM's after marginalization over the process F. There have been many papers that used DPM's for modeling and analyzing data since 1994 (for a small sampling see Dey et al., 1998). While PT priors have been discussed as early as Freedman (1963), Fabius (1964) and Ferguson (1974), the natural starting point for understanding their potential use in modeling data is Lavine (1992, 1994). The utility of using MPT's to generalize existing parametric families was illustrated by Berger and Gugliemi (2001) and Hanson and Johnson (2002). The GP model was used to model survival data in the context of the proportional hazards model by Kalbfleisch (1978) and we present a particular implementation of this model here.

Other general probability models for $\mathcal{P}$ have been developed by Freedman (1963) and Doksum (1974). In the particular area of survival analysis, there are a number of nonparametric and semiparametric models beyond the GP that have been developed that are based on modeling the hazard function and the cumulative hazard function (see, e.g., Dykstra and Laud, 1981; Ibrahim et al., 2001; Nieto-Barajas and Walker, 2002, 2004) but we do not discuss these here. Review articles by Müller and Quintana (2004), Walker et al. (1999), Gelfand (1999), Sinha and Dey (1997), the volume by Dey et al. (1998), the monograph by Ghosh and Ramamoorthi (2003), and the article in this volume by Choudhuri et al. (2005), all provide additional background and breadth beyond what we present here.

2.1. *The Dirichlet process*

Ferguson (1973) introduced the DP as a means to specify a (prior) probability measure $\mathcal{P}(\mathrm{d}F)$ on a probability measure F taking values in the space of all probability measures, $\mathcal{F}$, in the context of modeling statistical data. A random probability measure F is said to be a DP with parameter αF_0 if for all finite measurable partitions $\{A_j\}_{j=1}^J$ of the sample space, the vector $(F(A_1), F(A_2), \ldots, F(A_J))$ has a Dirichlet distribution with parameter $(\alpha F_0(A_1), \alpha F_0(A_2), \ldots, \alpha F_0(A_J))$. The parameter (αF_0) of a DP consists of a scalar precision parameter $\alpha > 0$ and a completely known base probability measure F_0. The DP is centered at F_0 in the sense that for any measurable set B, $E[F(B)] = F_0(B)$. The parameter α is referred to as a precision parameter because the prior variance for the probability of any measurable set, $\mathrm{Var}[F(B)] = \frac{F_0(B)[1-F_0(B)]}{\alpha+1}$, is small for large α. These results follow from the fact that $F(B) \sim \mathrm{Beta}(\alpha F_0(B), \alpha F_0(B^c))$. We write $F|\alpha, F_0 \sim \mathrm{DP}(\alpha F_0)$.

A key conjugacy result holds for the DP. Consider the model

$$\begin{aligned} y_1, y_2, \ldots, y_n | F &\overset{\text{i.i.d.}}{\sim} F \\ F|\alpha, F_0 &\sim \mathrm{DP}(\alpha F_0) \end{aligned}$$

and define $Y = (y_1, y_2, \ldots, y_n)$. Then the posterior distribution of F is $F|Y \sim \mathrm{DP}(\alpha^* F_0^*)$ where $\alpha^* = \alpha + n$ and $F_0^* = \frac{\alpha}{\alpha+n} F_0 + \frac{1}{\alpha+n} \sum_{i=1}^n \delta_{y_i}$; $\delta_y(\cdot)$ denotes point mass at y, e.g., $\delta_y(B) = 1$, if $y \in B$, and zero otherwise. Hence the posterior mean of the CDF $F(t)$ is given by

$$\widehat{F}(t) = E\big[F(t)|Y\big] = \frac{\alpha}{\alpha+n}F_0(t) + \frac{n}{\alpha+n}\widehat{F}_n(t),$$

where $\widehat{F}_n(t)$ is the empirical distribution function based on $(y_1, \ldots, y_n)$. This is a common occurrence in Bayesian statistics that the estimate is a weighted average of the prior mean of F and an empirical estimate, in this instance the nonparametric maximum likelihood estimate.

In addition to estimating F, inferences for functionals, $T(F)$, are of interest. For instance, the mean functional is given by $E(y|F) = \int y\,\mathrm{d}F(y)$. Inferences for $T(F)$ can be obtained using the approach of Gelfand and Kottas (2002) where $\{F^j: j = 1, \ldots, MC\}$ are simulated from the posterior distribution $F|Y$ and used to obtain the corresponding Monte Carlo sample of $T(F^j)$'s. We shall discuss this approach in detail for the DPM model.

Predictive inference for a future observation is also straightforward. The predictive distribution of a future observation y_f where $y_f|Y, F \sim F$, is F_0^*. This follows from the generalized Polya urn representation for the marginal distribution of Y (Blackwell and MacQueen, 1973).

There are two features of the DP that typically are viewed as its primary limitations. As previously indicated, the support of the DP distribution is the set of all discrete distributions (Ferguson, 1973; Blackwell, 1973). This can be visualized from the constructive definition of F (Sethuraman, 1994):

$$F = \sum_{j=1}^{\infty} V_j \delta_{\theta_j},$$

where with $W_i \overset{\text{i.i.d.}}{\sim} \text{Beta}(1, \alpha)$, the V_j's are defined as $V_1 = W_1, \ldots, V_j = W_j \prod_{r=1}^{j-1}(1 - W_r), \ldots$, and $\theta_j \overset{\text{i.i.d.}}{\sim} F_0$. This is often referred to as the "stick-breaking" representation as the weights are defined in a way that the interval [0, 1] (the stick) is successively broken up or partitioned into pieces starting with the interval $[0, w_1]$, and then adding $[w_1, w_1 + (1 - w_1)w_2]$ etc. The lengths of each of the corresponding subintervals are the weights in the Sethuraman representation of F. The second drawback of the DP is that for any disjoint measurable sets B_1 and B_2, the correlation between $F(B_1)$ and $F(B_2)$ is negative, which for ("small") adjacent sets violates a belief that these two probabilities should be positively correlated.

2.2. *Mixtures of Dirichlet processes*

Centering the DP on a fixed F_0 may be appropriate for some applications but for the majority of applied problems centering the DP on a *family* of parametric distributions is preferable. The goal then is to embed a parametric family in the broad class of models $\mathcal{F}$.

The MDP model is specified as:

$$\begin{aligned} y_1, y_2, \ldots, y_n|F &\overset{\text{i.i.d.}}{\sim} F \\ F|\alpha, F_\theta &\sim \text{DP}(\alpha F_\theta) \\ \theta &\sim p(\mathrm{d}\theta), \end{aligned}$$

where $\{F_\theta\colon \theta \in \Theta\}$ is a parametric family of probability models. The standard representation for the MDP is $F \sim \int \mathrm{DP}(\alpha F_\theta)p(\mathrm{d}\theta)$. This representation makes it clear that F is distributed as a literal mixture of DP's. Antoniak (1974) presented theoretical results for the MDP model and also gave a number of applications. In particular, Antoniak (1974) obtained the posterior pdf for θ, assuming absolutely continuous F_θ with pdf $p(\cdot|\theta)$, as:

$$p(\mathrm{d}\theta|Y) \propto p(\mathrm{d}\theta)\prod_{i=1}^{k} p(y_i^*|\theta),$$

where $\{y_i^*, i = 1,\dots,k \leqslant n\}$ are the distinct y_j's. Also, for given θ, $F|Y,\theta \sim \mathrm{DP}(\alpha^* F_\theta^*)$ where $F_\theta^* = \frac{\alpha}{\alpha+n}F_\theta + \frac{1}{\alpha+n}\sum_{i=1}^n \delta_{y_i}$. Hence, inferences for functionals $T(F)$ can be obtained by first sampling $\theta^j \overset{\text{i.i.d.}}{\sim} p(\mathrm{d}\theta|Y)$, $j = 1, 2, \dots, MC$, then (partially) sampling $F^j|Y, \theta^j$ from $\mathrm{DP}(\alpha^* F_{\theta j}^*)$, and finally computing $T(F^j)$.

The posterior mean $E[F|Y] = \int F_\theta^* p(\mathrm{d}\theta|Y)$ provides an estimate of F and can be approximated by Monte Carlo integration, e.g.,

$$E[F|Y] \doteq \frac{1}{MC}\sum_{j=1}^{MC} F_{\theta j}^*.$$

If $p(\mathrm{d}\theta)$ is conjugate to $p(y|\theta)$, $p(\mathrm{d}\theta|Y)$ is easily sampled. Otherwise, sampling from $p(\mathrm{d}\theta|Y)$ can be accomplished, for instance, using a Metropolis sampler (Tierney, 1994).

Briefly consider a BNP version of the classic empirical Bayes problem. Let $y_i|\theta_i \overset{\text{ind}}{\sim} F_{\theta_i}, \theta_i|G \overset{\text{i.i.d.}}{\sim} G$, $G \sim \mathrm{DP}(\alpha G_0)$, $i = 1,\dots,n$. This model can be represented as $y_i|F \overset{\text{i.i.d.}}{\sim} F \equiv \int F_\theta G(\mathrm{d}\theta)$, $G \sim \mathrm{DP}(\alpha G_0)$. The definition of F here corresponds to the definition of a DPM in the next subsection. Antoniak (1974) established in his Corollary 3.1 that the posterior distribution of $F|y$ (for a single y) can be represented as an MDP, namely $F|y \sim \int \mathrm{DP}((\alpha+1)(G_0+\delta_\theta))p(\mathrm{d}\theta|y)$. Thus there is a connection between the MDP and the DPM models. But aside from that, computational complexities arise using this model for the empirical Bayes problem as soon as one attempts to characterize the full posterior distribution. From Corollary 3.2 of Antoniak (1974), and with $\theta = (\theta_1,\dots,\theta_n)$, we have $F|Y \sim \int \mathrm{DP}((\alpha+n)(wF_0 + (1-w)\sum_{i=1}^n \delta_{\theta_i}/n))p(\mathrm{d}\theta|Y)$, $w = \alpha/(\alpha+n)$. It is here where Berry and Christensen (1979) and Lo (1984) realized how complicated the problem is due to the discreteness of the distribution of $\theta|Y$. A brute force approach to the problem must consider all possible combinations of ties among the θ_i's. The Monte Carlo approach of Escobar (1994) made it possible to actually analyze data modeled as a DPM.

2.3. *Dirichlet process mixture models*

The DPM model has been very popular for use in BNP inference. A standard parametric model that strives to achieve flexibility is the finite mixture model

$$y_i \overset{\text{i.i.d.}}{\sim} \sum_{j=1}^{K} p_j F_{\theta_j},$$

where $\{F_\theta: \theta \in \Theta\}$ represents a standard parametric family, and $\theta_j \in \Theta$ for $j = 1, \ldots, K$ are assumed to be distinct so the mixture is comprised of K distinct members of this family. The fixed unknown mixing probabilities $\{p_j, j = 1, \ldots, K\}$ add to one and there are additional constraints that insure identifiability (Titterington et al., 1985). Bayesian inference for this model is achieved by placing a prior distribution on K, $\{p_j, j = 1, \ldots, K\}$, and $\{\theta_j, j = 1, \ldots, K\}$. Such a model results in a varying dimensional parameter space and consequently specialized computational techniques, such as reversible jump MCMC (Green, 1995), are required.

The DPM model avoids such concerns as the data are modeled according to an infinite mixture model which, using the Sethuraman (1994) representation, is given by

$$y_i \overset{\text{i.i.d.}}{\sim} \sum_{j=1}^{\infty} V_j F_{\theta_j},$$

where the F_{θ_j}'s are parametric CDFs (the CDFs that would be used in a finite mixture model) with V_j and θ_j defined as in the DP. Here the (implied) induced prior on the θ_j's is that they are i.i.d. from the base measure (F_0) of the DP. This representation of the model makes clear that the DPM model is equivalent to selecting an infinite mixture and where the DP prior induces the specified distribution on the weights and the θ's. So while the DPM generalizes the Bayesian version of the finite mixture model above by allowing for an infinite mixture, it does so at the expense of having a particular prior for these inputs. With a small weight α selected for the DP, the DP places high probability on a few nonnegligible components. In this instance, the DPM model effectively results in a finite mixture model but where it is not necessary to specify the number of components of the mixture in advance. The data are allowed to determine the likely number of mixture components.

Alternatively, the DPM model is specified as

$$y_1, y_2, \ldots, y_n | F \overset{\text{i.i.d.}}{\sim} F(\cdot|G) = \int F_\theta(\cdot) G(\mathrm{d}\theta), \qquad G|\alpha, G_0 \sim \mathrm{DP}(\alpha G_0).$$

Because G is a random probability measure, F is a random probability measure. Note that if F_θ is continuous, then $F(\cdot|G)$ is also continuous with probability one. Thus the DPM model does not suffer the same fate as the DP in this regard.

An equivalent (and more commonly used) DPM model specification introduces latent variables as discussed at the end of the previous section:

$$\begin{aligned}
&y_i|\theta_i \overset{\perp}{\sim} F_{\theta_i}\\
&\theta_i|G \overset{\text{i.i.d.}}{\sim} G\\
&G|\alpha, G_0 \sim \mathrm{DP}(\alpha G_0).
\end{aligned}$$

Contributions related to fitting DPM models include the work of Escobar (1994), MacEachern (1994), Escobar and West (1995), Bush and MacEachern (1996), MacEachern and Müller (1998), Walker and Damien (1998), MacEachern et al. (1999), and Neal (2000). Contributions related to obtaining inferences for F and functionals $T(F)$ for DPM models have been provided by Gelfand and Mukhopadhyay (1995),

Mukhopadhyay and Gelfand (1997), Kleinman and Ibrahim (1998), Gelfand and Kottas (2002), and Regazzini et al. (2002) among many others.

We now proceed to discuss details of fitting the basic DPM model and some of its extensions since it is perhaps the single most important BNP model to date.

2.3.1. Fitting DPM models

A Monte Carlo approach to approximating the posterior distribution of $T(F)$ would involve sampling the infinite-dimensional parameter G. Such an approach cannot be implemented without introducing finite approximations. Escobar (1994) considered the DPM model obtained after marginalizing the DP. This reduces the problem to sampling only the finite-dimensional variables $(\theta_1, \ldots, \theta_n)$ as will be seen below. Using the third characterization of the DPM, Escobar obtained a numerical approximation to the posterior of the vector $\theta = (\theta_1, \ldots, \theta_n)$ using Gibbs sampling, e.g., by iteratively sampling $\theta_i|\theta_{-i}, y_i$ where θ_{-i} denotes the vector of all θ_j's excluding θ_i.

The marginalized DPM model is given by

$$y_i|\theta_i \overset{\perp}{\sim} F_{\theta_i}$$
$$p(\theta_1, \theta_2, \ldots, \theta_n) = p(\theta_1)p(\theta_2|\theta_1)p(\theta_3|\theta_1, \theta_2)\cdots p(\theta_n|\theta_{1:n-1}),$$

where $\theta_{1:i-1} = (\theta_1, \theta_2, \ldots, \theta_{i-1})$, $i = 2, \ldots, n$, and dependence of the distribution for θ on (α, G_0) has been suppressed. The generalized Polya urn scheme (Blackwell and MacQueen, 1973) is used to specify $p(\theta_1, \theta_2, \ldots, \theta_n)$ as

$$\theta_1 \sim G_0$$

$$\theta_i|\theta_{1:i-1} \begin{cases} \sim G_0 & \text{with probability } \frac{\alpha}{\alpha+i-1}, \\ = \theta_j & \text{with probability } \frac{1}{\alpha+i-1},\ j = 1, 2, \ldots, i-1. \end{cases}$$

This follows from the fact that, for an appropriate measurable set A, $\Pr(\theta_i \in A|\theta_{1:i-1}) = E[G(A)|\theta_{1:i-1}]$. For $i = 1$, we have $E[G(A)] = G_0(A)$. For $i > 1$, since $G|\theta_{1:i-1}$ is an updated DP, we have

$$G(A)|\theta_{1:i-1} \sim \text{Beta}\left(\alpha G_0(A) + \sum_{j=1}^{i-1} \delta_{\theta_j}(A), \alpha G_0(A^c) + \sum_{j=1}^{i-1} \delta_{\theta_j}(A^c)\right).$$

Hence $E[G(A)|\theta_{1:i-1}] = \frac{\alpha}{\alpha+i-1} G_0(A) + \frac{1}{\alpha+i-1} \sum_{j=1}^{i-1} \delta_{\theta_j}(A)$, which yields the above result.

Combining the distribution for $\theta_i|\theta_{1:i-1}$ with the contribution $p(y_i|\theta_i)$ and because the latent θ_j's are exchangeable, the full conditional for θ_i is:

$$\theta_i|\theta_{-i}, y_i \begin{cases} = \theta_j & \text{with probability } \frac{p(y_i|\theta_j)}{A(y_i) + \sum_{j\neq i} p(y_i|\theta_j)},\ j \neq i, \\ \sim p(\mathrm{d}\theta_i|y_i) & \text{with probability } \frac{A(y_i)}{A(y_i)+\sum_{j\neq i} p(y_i|\theta_j)}, \end{cases} \tag{1}$$

where $A(y_i) = \alpha \int p(y_i|\theta)G_0(\mathrm{d}\theta)$ and $p(\mathrm{d}\theta_i|y_i)$ is the conditional distribution for θ_i given the single observation y_i based on a parametric model with likelihood contribution $p(y_i|\theta_i)$ (the pdf corresponding to F_{θ_i}) and prior distribution G_0 on θ_i. Sampling these full conditional distributions will be straightforward if $p(y_i|\theta_i)$ and G_0 are a conjugate pair so that computing $A(y_i)$ and sampling $p(\mathrm{d}\theta_i|y_i)$ will be routine. Such models are referred to as conjugate DPM models. Escobar and West (1995) considered a generalization of this model with $y_i|\mu_i, \sigma_i^2 \sim N(\mu_i, \sigma_i^2)$ and $G_0(\mathrm{d}\mu, \mathrm{d}\sigma^2) = N(\mathrm{d}\mu|m, \tau\sigma^2)IG(\mathrm{d}\sigma^2|a, b)$, a normal/inverse gamma base measure.

Although fitting a conjugate DPM model using the Gibbs sampler above is straightforward, the Gibbs sampler will often exhibit slow convergence to the joint marginal posterior, and once convergence is achieved, subsequent sampling of the θ_i's may be very inefficient, as discussed by Neal (2000). This is due to the discreteness of the DP. The θ's will cluster at each iteration of the Gibbs sampler, namely there will be a vector of distinct values of $(\theta_1, \ldots, \theta_n)$, say $\phi = (\phi_1, \ldots, \phi_k)$ for $k \leqslant n$. The inefficiency results from ignoring this fact in the Gibbs sampler described above.

MacEachern and Müller (1998) overcome this problem by using the following sampling approach for conjugate DPM models. At a given iteration of the Gibbs sampler, let the vector $c = (c_1, c_2, \ldots, c_n)$ denote the cluster membership of y_i so that $c_i = j$ if $\theta_i = \phi_j$ for $i = 1, 2, \ldots, n$, and $j = 1, \ldots, k$. The current state of the Markov chain is (c, ϕ). The actual sampling is accomplished in two steps: (i) Sample θ_i as previously described but only for the purpose of determining the cluster membership c_i of each y_i. This involves the possibility of adding a new value of θ or sampling one of the current values in the vector ϕ. If a new value is added, the vector ϕ is augmented to include the new value and $k \to k + 1$. It is also possible that in sampling a θ_i when the current value of θ_i has only multiplicity one (e.g., $c_i = j, \sum_l \delta_j(c_l) = 1$), the new value will be one of the θ_{-i} values so that the vector ϕ must be redefined to accommodate its removal from the collection and hence $k \to k - 1$ in this instance. (ii) Then generate ϕ_j by sampling from the posterior distribution of ϕ_j based on the parametric model with likelihood $p(\cdot|\phi_j)$ and prior G_0 on ϕ_j where the posterior distribution is computed using only the y_i's that belong to cluster j. With this approach, all the θ_i's associated with a given cluster will be updated to a new value simultaneously.

MacEachern and Müller (1998) and Neal (2000) developed and discussed methods for sampling nonconjugate DPM models. Such methods are necessary, for example, if the data are assumed to be normally distributed conditional on $\theta = (\mu, \sigma^2)$ but where the DP $G(\mathrm{d}\mu, \mathrm{d}\sigma^2)$ is centered on $G_0(\mathrm{d}\mu, \mathrm{d}\sigma^2) = N(\mathrm{d}\mu|a, b)\Gamma(\mathrm{d}\sigma^2|c, \mathrm{d})$ instead of the usual conjugate normal–gamma distribution. Alternatively, let F_θ denote a Poisson(θ) distribution and assume $G(\mathrm{d}\theta)$ is centered a log-normal distribution.

The issue that remains is how to use the MC samples from the marginal posterior of θ in order to make inferences. There are some inferences that can be made and some that cannot. For example, it is not possible to obtain interval inferences for the unknown CDF $F(\cdot|G)$, or the population mean $\int yF(\mathrm{d}y|G)$ based solely on an MC sample from $p(\mathrm{d}\theta|Y)$. In general, for the marginalized DPM model, full inferences are not available for arbitrary functionals of $F(\cdot|G)$ because G is not sampled. Subsection 2.3.3 addresses these issues. However, as pointed out by Gelfand and Mukhopadhyay (1995), it is possible to obtain posterior expectations of linear functionals. For example, let $p(\cdot|\theta^*)$ denote the pdf for a sampled observation were the value of θ^* to be known. Then the modeled sampling density is $p(\cdot|G) = \int p(\cdot|\theta^*)G(\mathrm{d}\theta^*)$. Let $T(p(\cdot))$ be a linear functional of an arbitrary pdf $p(\cdot)$. Then it is not difficult to show that (see Gelfand

and Mukhopadhyay, 1995)

$$\int T\big(p(\cdot|G)\big)p(\mathrm{d}G|Y) = \int T\big(p(\cdot|\theta^*)\big)p\big(\mathrm{d}\theta^*|\theta\big)p(\mathrm{d}\theta|Y).$$

Having obtained a sample from the marginal posterior for (θ^*, θ) (using (1) to obtain the full conditional for θ^*), the above integral is easily approximated. So clearly it is possible to obtain MCMC approximations to the posterior mean of the conditional mean $E(y|\theta^*)$, and also the pdf $p(y|\theta^*)$, and corresponding CDF $F_{\theta^*}(y)$, for all y. West et al. (1994) catalogue very interesting applications of DPM's to multivariate multimodal density estimation and random coefficient growth curves. Kottas and Gelfand (2001a) modeled semiparametric survival data with DPM's and showed how to make inferences for the median time to survival functional.

2.3.2. Extensions

Three extensions of the basic DPM model include the incorporation of covariates for semiparametric regression, a prior distribution for α, and centering the DP on a family of parametric distributions G_η with a prior distribution specified for η.

Perhaps the most important extension involves the incorporation of covariates into the model. Gelfand (1999), Kleinman and Ibrahim (1998), Mukhopadhyay and Gelfand (1997), and Bush and MacEachern (1996) discussed semiparametric regression for the DPM model. The basic model is given by:

$$\begin{aligned} y_i|\theta_i, x_i, \beta &\overset{\perp}{\sim} p(y_i|\theta_i, x_i, \beta) \\ \theta_i|G &\overset{\text{i.i.d.}}{\sim} G \\ G|\alpha, G_0 &\sim \mathrm{DP}(\alpha G_0) \\ \beta &\sim p(\mathrm{d}\beta), \end{aligned}$$

where y_i denotes the response for subject i with covariate vector x_i, and the θ_i's are random effects. The model is fitted using Gibbs sampling where the θ_i's are sampled from the full (marginal) conditional distribution corresponding to $p(\mathrm{d}\theta|\beta, Y)$, which is obtained with only slight notational changes from what was previously described, and where β is sampled from the full (marginal) conditional distribution $p(\mathrm{d}\beta|\theta, Y) \propto p(\mathrm{d}\beta)\prod_{i=1}^n p(y_i|\theta_i, x_i, \beta)$. Sampling the full conditional distribution for θ will often require nonconjugate methods.

The precision parameter α can also be modeled thereby inducing a prior distribution on the number of distinct clusters. Escobar and West (1995) used a data augmentation approach to model α using a gamma prior, $\alpha|a, b \sim \Gamma(a, b)$, and introducing a clever latent variable that makes the Gibbs sampling easy. This same approach can be used in DP and MDP models.

Centering the DP on a parametric family $\{G_\eta\colon \eta \in \Omega\}$ with prior $p(\mathrm{d}\eta)$ is also possible, e.g., $G \sim \int \mathrm{DP}(\alpha G_\eta)p(\mathrm{d}\eta)$. The full conditional distribution for η is obtained in the same way the marginal conditional was obtained in the MDP model. For the normal linear mixed model with simple random effects, centering the random effects distribution on the $N(0, \sigma^2)$ family with an inverse gamma prior on σ^2 results in an inverse gamma distribution for the full conditional for σ^2 (Bush and MacEachern, 1996).

Modeling a random effects distribution with a DP prior centered on the zero-mean multivariate normal distribution with covariance matrix D, where D^{-1} is distributed Wishart, the full conditional of D^{-1} is distributed Wishart (Kleinman and Ibrahim, 1998).

2.3.3. *General inferences*

Inferences for the marginalized DPM model were discussed at the end of Section 2.3.1. The full DPM model is in the form $y_1, y_2, \ldots, y_n | F \overset{\text{i.i.d.}}{\sim} F(\cdot|G) = \int F_\theta(\cdot) G(d\theta)$, where F_θ has corresponding pdf $p(\cdot|\theta)$. We first indicate how to obtain full inferences for linear functionals and then for arbitrary functionals of F.

In the first instance, run the Gibbs sampler for the marginalized DPM. Once convergence is achieved and the "burn-in" discarded, the Gibbs sampler yields the output $\{\theta^j = (\theta_1^j, \ldots, \theta_n^j)\colon j = 1, \ldots, MC\}$. Linear functionals of $F \equiv F(\cdot|G)$ are again given by $T \equiv T(F) = \int T[p(\cdot|\theta_*)]G(d\theta_*)$. Then for each θ^j, we obtain T^j by first sampling from the updated DP for G, namely sample $G^j \sim G|\theta^j$ using the Sethuraman (1994) construct. Then for each j obtain a sample of B i.i.d. values from G^j, e.g., sample $\theta_*^i \overset{\text{i.i.d.}}{\sim} G^j$. Finally obtain,

$$T^j = \frac{1}{B} \sum_{i=1}^{B} T\big[p(\cdot|\theta_*^i)\big], \quad j = 1, \ldots, MC,$$

which yield (approximate) realizations from the posterior distribution of $T(F)|Y$. The sample $\{T^j\}_{j=1}^{MC}$ is used to obtain point and interval estimates of $T(F)$, as well as its posterior pdf.

Posterior inferences for nonlinear $T(F)$ are obtained as above by simply obtaining $F^j = \int p(\cdot|\theta_*)G^j(d\theta_*)$, and the corresponding $T(F^j)$, $j = 1, \ldots, MC$. In each instance, G^j is obtained by sampling a truncated version of the Sethuraman (1994) representation for G. Gelfand and Kottas (2002) give details.

2.4. *Polya tree and mixtures of Polya tree models*

Polya tree models were first discussed by Freedman (1963), Fabius (1964) and Ferguson (1974). Use of PT models for complicated data was historically difficult due to mathematical intractability. However, as with DPM models, modern MCMC methods have allowed data analysts to once again consider PT's for modeling data nonparametrically. Lavine (1992, 1994) and Mauldin et al. (1992) have carefully developed and catalogued much of the current theory governing PT's.

The PT is a generalization of the DP. A particular general specification of the PT places probability one on absolutely continuous F's, thus avoiding the discreteness issues associated with the DP. Here, the sample space, Ω, is successively partitioned into finer-and-finer disjoint sets using binary partitioning. At the first level of the tree, a two set partition is constructed with a single pair of corresponding branch probabilities defining the marginal probabilities of these sets. The mth level partition has 2^m sets and corresponding conditional branch probabilities (probability of being in a set in this partition, given that it is contained in the corresponding parent set in the

($m-1$)st level). Starting from the first level (i.e. the top of the tree), there is a unique path down the branches of the tree to each set at level m, and consequently to any real number in Ω if one continues as $m \to \infty$. The marginal probability of any level m set is simply the product of the corresponding conditional branch probabilities that lead to that set. Randomness is incorporated by specifying independent Dirichlet distributions on each of the pairs of conditional branch probabilities at each level of the tree.

To make this more precise, the first partition of Ω is $\{B_0, B_1\}$. Then further split B_0 into $\{B_{00}, B_{01}\}$, and split B_1 into $\{B_{10}, B_{11}\}$ yielding the 4 disjoint sets at level 2 of the tree. Continue by letting $\varepsilon = \varepsilon_1 \cdots \varepsilon_m$ be an arbitrary binary number, and split B_ε into $\{B_{\varepsilon 0}, B_{\varepsilon 1}\}$ for all ε, and continue *ad infinitum*. The schematic below conveys the splitting for $m = 2$.

B_0		B_1	
B_{00}	B_{01}	B_{10}	B_{11}

Then define the random marginal probabilities $Y_0 = F(B_0)$, $Y_1 = 1 - Y_0 = F(B_1)$, and the successive conditional probabilities $Y_{00} = F(B_{00}|B_0)$, $Y_{01} = 1 - Y_{00} = F(B_{01}|B_0)$, $Y_{10} = F(B_{10}|B_1)$, $Y_{11} = 1 - Y_{10} = F(B_{11}|B_1), \ldots, Y_{\varepsilon 0} = F(B_{\varepsilon 0}|B_\varepsilon)$, $Y_{\varepsilon 1} = 1 - Y_{\varepsilon 0} = F(B_{\varepsilon 1}|B_\varepsilon)$, etc. The marginal probability of a set in the mth partition is calculated as $F(B_{\varepsilon_1 \cdots \varepsilon_m}) = \prod_{j=1}^m Y_{\varepsilon_1 \cdots \varepsilon_j}$. The PT specification is completed by specifying $Y_{\varepsilon_1 \cdots \varepsilon_m 0} \overset{\text{ind}}{\sim} \text{Beta}(\alpha_{\varepsilon_1 \cdots \varepsilon_m 0}, \alpha_{\varepsilon_1 \cdots \varepsilon_m 1})$ (i.e. $(Y_{\varepsilon_1 \cdots \varepsilon_m 0}, Y_{\varepsilon_1 \cdots \varepsilon_m 1}) \sim \text{Dirichlet}(\alpha_{\varepsilon_1 \cdots \varepsilon_m 0}, \alpha_{\varepsilon_1 \cdots \varepsilon_m 1})$), for all sets in all partitions. The collection of partitions is denoted as Π and the collection of parameters of all the beta distributions is denoted $\mathcal{A}$. We write $F|\Pi, \mathcal{A} \sim \text{PT}(\Pi, \mathcal{A})$.

It is straightforward to establish conjugacy of the PT model, namely if $y|F \sim F$, $F \sim \text{PT}(\Pi, \mathcal{A})$, then $F|y, \Pi, \mathcal{A} \sim \text{PT}(\Pi, \mathcal{A}^*)$, $\mathcal{A}^* = \{\alpha_\varepsilon + I(y \in B_\varepsilon), \forall \varepsilon\}$.

The PT process can be centered on a particular F_0 by selecting $\Pi = \{F_0^{-1}((i-1)/2^m), F_0^{-1}(i/2^m))\colon i = 1, \ldots, 2^m, m = 1, 2, \ldots\}$. Then setting $\alpha_{\varepsilon 0} = \alpha_{\varepsilon 1}$ for all ε, we obtain $E\{F(B_{\varepsilon_1 \cdots \varepsilon_m})\} = 2^{-m} = F_0(B_{\varepsilon_1 \cdots \varepsilon_m})$. Ferguson (1974) noted that for $\gamma > 0$ and $\alpha_{\varepsilon_1 \cdots \varepsilon_{m-1} 0} = \alpha_{\varepsilon_1 \cdots \varepsilon_{m-1} 1} = \gamma m^2$, F is absolutely continuous with probability one. This has become the "standard" parameterization for α_ε. The parameter γ determines how concentrated the prior specification is about the prior guess, F_0. Large γ results in the prior being more concentrated on F_0, e.g., random F's sampled from the PT will concentrate both in terms of similarity in shape and distance from the fixed F_0, while with γ near zero, simulated CDF's often will be considerably dispersed in terms of shape and distance from the fixed F_0. From here on, we choose this standard parametrization and denote the PT distribution as $\text{PT}(\Pi, \gamma)$.

A major criticism of the PT is that, unlike the DP, inferences are somewhat sensitive to the choice of a fixed partition Π. This led Paddock et al. (2003) to consider "jittered" partitions. Hanson and Johnson (2002) instead considered MPT's, wherein inferences are obtained having mixed over a random partition Π_θ thereby alleviating the influence of a fixed partition on inferences.

The MPT is simply defined by allowing the base probability measure to depend on an unknown $\theta \in \Theta$. Thus the base measure becomes a family of probability measures,

$\{F_\theta\colon \theta \in \Theta\}$. This leads to a family of partition families $\{\Pi_\theta\colon \theta \in \Theta\}$. A prior is placed on θ, $p(\mathrm{d}\theta)$. The basic MPT model is represented as

$$y_1, \ldots, y_n | F_\theta \overset{\text{i.i.d.}}{\sim} F_\theta, \qquad F_\theta \sim \mathrm{PT}(\Pi_\theta, \gamma), \quad \theta \sim p(\mathrm{d}\theta)$$

or equivalently $y_i | F \overset{\text{i.i.d.}}{\sim} F$ where $F \sim \int \mathrm{PT}(\Pi_\theta, \gamma) p(\mathrm{d}\theta)$.

If we only specify Π or Π_θ to a finite level M, then we have defined a partially specified (or finite) PT or MPT. For a finite PT we write $F | \Pi_M, \gamma \sim \mathrm{PT}(\Pi_M, \gamma)$. Lavine (1994) detailed how such a level M can be chosen by placing bounds on the posterior predictive density at a point. Hanson and Johnson (2002) have recommended the rule of thumb $M \doteq \log_2 n$. On the sets that comprise level M of the tree, one may consider F to follow F_0 (or F_θ) restricted to this set.

Barron et al. (1999) note that the posterior predictive densities of future observations computed from Polya tree priors have noticeable jumps at the boundaries of partition sets and that a choice of centering distribution F_0 "that is particularly unlike the sample distribution of the data will make convergence of the posterior very slow." The MPT appears to mitigate some of these problems (Hanson and Johnson, 2002). In particular, with a MPT, the predictive density in a regression problem was shown to be differentiable by Hanson and Johnson (2002).

Methods of fitting Polya trees to real data are discussed by Walker and Mallick (1997, 1999), and methods for MPT's are discussed by Hanson and Johnson (2002). Berger and Gugliemi (2001) considered the problem of model fit by embedding a parametric family in a larger MPT family.

2.5. The gamma process model

The survival function for nonnegative data is defined as $S(t) = 1 - F(t), t > 0$. For continuous data, the corresponding hazard function is defined to be $\lambda(t) = -\frac{\mathrm{d}}{\mathrm{d}t}\ell n(S(t))$, and the cumulative hazard is defined to be $\Lambda(t) = \int_0^t \lambda(s)\,\mathrm{d}s$. It follows that $S(t) = \exp(-\Lambda(t))$. Thus in survival modeling for a continuous response, it is possible to place a probability distribution on the space of all probability models for nonnegative continuous data by placing a probability distribution on the family of all possible cumulative hazard functions. Kalbfleisch (1978) proposed using the gamma process (GP) to model the cumulative hazard function $\Lambda(\cdot)$ in the context of the proportional hazards model (Cox, 1972). We follow Ibrahim et al. (2001) and define the GP as follows.

On $[0, \infty)$ let $\Lambda_0(t)$ be an increasing, left-continuous function such that $\Lambda_0(0) = 0$. Let $\Lambda(\cdot)$ be a stochastic process such that (i) $\Lambda(0) = 0$, (ii) $\Lambda(\cdot)$ has independent increments in disjoint intervals, and (iii) $\Lambda(t_2) - \Lambda(t_1) \sim \Gamma(\alpha(\Lambda_0(t_2) - \Lambda_0(t_1)), \alpha)$ for $t_2 > t_1$. Then $\{\Lambda(t)\colon t \geqslant 0\}$ is said to be a GP with parameter (α, Λ_0) and denoted $\Lambda \sim \mathrm{GP}(\alpha, \Lambda_0)$.

Note that $\mathrm{E}(\Lambda(t)) = \Lambda_0(t)$ so that Λ is centered at Λ_0. Also, $\mathrm{Var}(\Lambda(t)) = \Lambda_0(t)/\alpha$ so that, similar to the DP and PT, α controls how "close" Λ is to Λ_0 and provides a measure of how certain we are that Λ is near Λ_0. It is interesting to note that Ferguson (1973, Section 4) recasts the DP as a scaled GP.

The posterior of the GP is characterized by Kalbfleisch (1978); his results for the proportional hazards model simplify when no covariates are specified. With probability one, the GP is a monotone nondecreasing step function, implying that the corresponding survivor function is a nonincreasing step function. Similar to the DP, matters are complicated by the presence of ties in the data with positive probability. When present in the observed data, such ties make the resulting computations awkward. Clayton (1991) described a Gibbs sampler for obtaining inferences in the proportional hazards model with a GP baseline.

3. Illustrations

In this section we discuss particular modeling applications and we analyze three data sets using a variety of BNP techniques. We first consider a two sample problem and apply BNP models to analyze these data. Next we discuss the rather large area of semi-parametric regression modeling and illustrate with a number of fundamental survival analysis models for data. We analyze a classic data set on time to death from diagnosis with leukemia. We then discuss nonparametric regression function estimation using a variety of basis models for representing the regression function. These methods are illustrated on a data set involving the estimation of mean response of nitric oxide and nitric dioxide in engine exhaust (using ethanol as fuel) as a function of the air to fuel ratio. Methods for the two sample problem were implemented in S-Plus while the survival analysis and the function estimation analyses were done in WinBUGS and Mathematica.

3.1. Two sample problem

A randomized comparative study was conducted to assess the association between amount of calcium intake and reduction of systolic blood pressure (SBP) in black males. Of 21 healthy black men, 10 were randomly assigned to receive a calcium supplement (group 1) over a 12 week period. The other men received a placebo during the 12 week period (group 2). The response variable was amount of decrease in systolic blood pressure. Negative responses correspond to increases in SBP. The data appear in Moore (1995, p. 439). Summary statistics for both groups are given in Table 1.

Let F_1 and F_2 denote the population distributions for decrease in SBP for groups 1 and 2, respectively. The data were fitted to the DP, MDP, DPM, PT, and MPT models. Prior distributions were constructed assuming the range of decrease of SBP for the calcium group is between -20 and 30 and that the data for the placebo group would range between -20 and 20. The midpoints were used for prior estimates of the mean change in SBP, namely 5 and 0 for groups 1 and 2. Prior estimates for the standard deviation were computed as the range/6. Hence, for the calcium group we centered the DP and PT distributions on an $F_{10} = N(5, 70)$, and we centered the placebo group on an $F_{20} = N(0, 44)$.

For the MDP model, we assume $F_1|(\mu_1, \sigma_1^2) \sim \text{DP}(\alpha N(\mu_1|\ 5, \sigma_1^2) IG(\sigma_1^2|\ 2, 70))$ and $F_2|(\mu_2, \sigma_2^2) \sim \text{DP}(\alpha N(\mu_2|\ 0, \sigma_2^2) IG(\sigma_2^2|\ 2, 44))$. Therefore $E(\sigma_1^2) = 70$ and $E(\sigma_2^2) = 44$, and both prior variances are infinite.

Table 1
Blood pressure data: summary statistics for the decrease in systolic blood pressure data for the calcium and placebo groups

	n	Mean	Median	Std. Dev.	Min	Max
Calcium	10	5.0	4	8.7	−5	18
Placebo	11	−0.27	−1	5.9	−11	12

Table 2
Blood pressure data: prior and posterior medians and 95% probability intervals for functionals $T(F)$ for the two-sample problem. The mean and median functionals are denoted by $\mu(\cdot)$ and $\eta(\cdot)$, respectively

	DP		MDP		DPM	
$T(F)$	Prior	Posterior	Prior	Posterior	Prior	Posterior
$\mu(F_1)$	5.08 (−10.4, 20.3)	4.96 (0.5, 9.9)	4.90 (−14.7, 25.9)	4.97 (0.6, 10.0)	5.05 (−5.2, 16.5)	5.08 (0.3, 9.9)
$\mu(F_2)$	−0.08 (−9.5, 9.3)	−0.31 (−3.3, 3.0)	0.02 (−16.2, 15.3)	−0.25 (−3.2, 3.1)	0.13 (−8.8, 9.6)	−0.30 (−3.3, 2.8)
$\eta(F_1)$	5.01 (−10.3, 20.3)	5.17 (−3.0, 11.0)	4.93 (−16.4, 27.6)	5.27 (−3.0, 11.0)	5.14 (−4.6, 15.9)	4.89 (0.2, 9.9)
$\eta(F_2)$	−0.10 (−12.4, 11.9)	−1.1 (−3.1, 2.9)	−0.10 (−17.8, 17.1)	−1.1 (−3.1, 2.9)	0.25 (−8.1, 8.7)	−0.35 (−3.3, 2.6)
$\mu(F_1)-\mu(F_2)$	5.12 (−9.8, 20.5)	5.23 (−0.3, 11.1)	4.86 (−19.4, 31.5)	5.23 (−0.3, 10.8)	5.08 (−8.94, 20.4)	5.24 (0.0, 10.6)
$\eta(F_1)-\eta(F_2)$	5.19 (−14.0, 24.7)	4.91 (−3.9, 14.1)	4.86 (−22.3, 34.2)	5.01 (−3.9, 14.1)	5.22 (−8.4, 18.9)	4.99 (−0.3, 10.8)

The DPM model used was, for $k = 1, 2$, and $i = 1, \ldots, n_k$ with $n_1 = 10$, $n_2 = 11$,

$$x_{ki} \mid (\mu_{ki}, \sigma_{ki}^2) \overset{\text{ind}}{\sim} N(\mu_{ki}, \tau\sigma_{ki}^2)$$
$$(\mu_{ki}, \sigma_{ki}^2) \mid G_k \overset{\text{ind}}{\sim} G_k$$
$$G_k \mid \alpha, G_{k0} \overset{\text{ind}}{\sim} \mathrm{DP}(\alpha G_{k0}).$$

Escobar and West (1995) discuss the parameter τ, which for density estimation can be interpreted as a smoothing parameter. For the current problem, we selected $G_{10}(\mathrm{d}\mu_1, \mathrm{d}\sigma_1^2) = N(\mathrm{d}\mu_1|5, \tau\,\mathrm{d}\sigma_1^2)IG(\mathrm{d}\sigma_1^2|2, 70)$, and $G_{20}(\mathrm{d}\mu_2, \mathrm{d}\sigma_2^2) = N(\mathrm{d}\mu_2|5, \tau\,\mathrm{d}\sigma_2^2)IG(\mathrm{d}\sigma_2^2|2, 70)$, where τ was selected to be either 1 or 10 in the current analysis.

For the MPT model, we centered on the family $F_{10}(\mathrm{d}\mu_1, \mathrm{d}\sigma_1) = N(\mathrm{d}\mu_1|5, 5)\,\Gamma(\mathrm{d}\sigma_1|0.64, 0.08)$ for the calcium group and $F_{20}(\mathrm{d}\mu_2, \mathrm{d}\sigma_2) = N(\mathrm{d}\mu_2|0, 5)\Gamma(\mathrm{d}\sigma_2|0.45, 0.067)$ for the placebo group. In all models with DP components, we set $\alpha = 1$ and we set $\gamma = 0.1$ for models involving Polya trees.

Table 2 contains prior and posterior medians and 95% probability intervals for functionals of F_1 and F_2 using the DP, MDP, and DPM models. The posterior estimates are similar for all 3 models, especially for the DP and MDP models. Estimates of these functionals using PT and MPT models are also readily available. For example, based on the MPT models, the population median change in SBP for the calcium group,

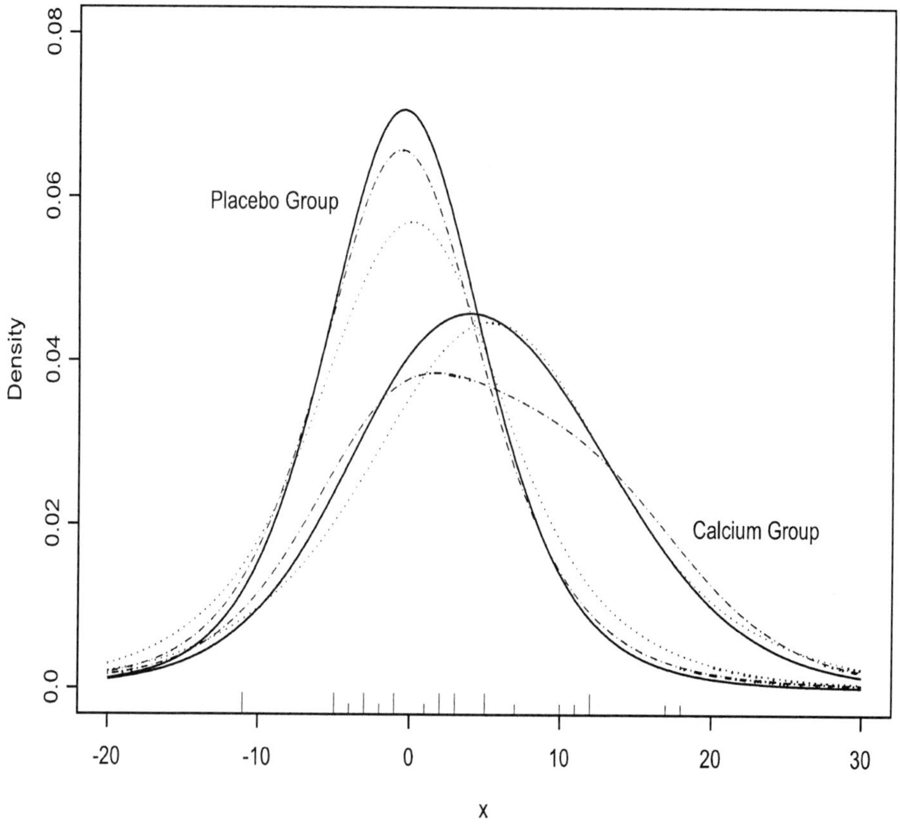

Fig. 1. Blood pressure data: prior (dotted) and posterior density estimates for both groups using the DPM model with $\tau = 1$ (solid) and $\tau = 10$ (dashed). The longer tick marks along the x-axis correspond to the observed data for the placebo group and the shorter tick marks to the observed data for the calcium group.

median(F_1), is estimated to be 3.86 (−2.45, 11.53) and for the placebo group, an estimate of *median*(F_2) is −1.0 (−3.27, 2.83). Inferences for the differences in means and medians are also given in Table 2. It appears that there would be a significant difference if 90% intervals had been considered. Observe that no attempt was made to guarantee that the priors were consistent across models and that this is clearly reflected in the induced priors for the functionals considered in Table 2.

Density estimates from DPM models with $\tau = 1$ and $\tau = 10$ are given in Figure 1. Also, the estimated CDF's for both groups using MDP, DPM, and MPT models are in Figure 2. The estimated CDF's using DP models (not shown) are essentially identical (for these data and for the given choices of α and γ) to those from the MDP models and the estimated CDF's from the PT models (not shown) were similar to those from the MPT models, but differ in that they were not as smooth due to partition effects. Finally note that the prior and posterior density estimates are quite similar. Since our prior was obtained independently of the data (from the second author), this is an indication that

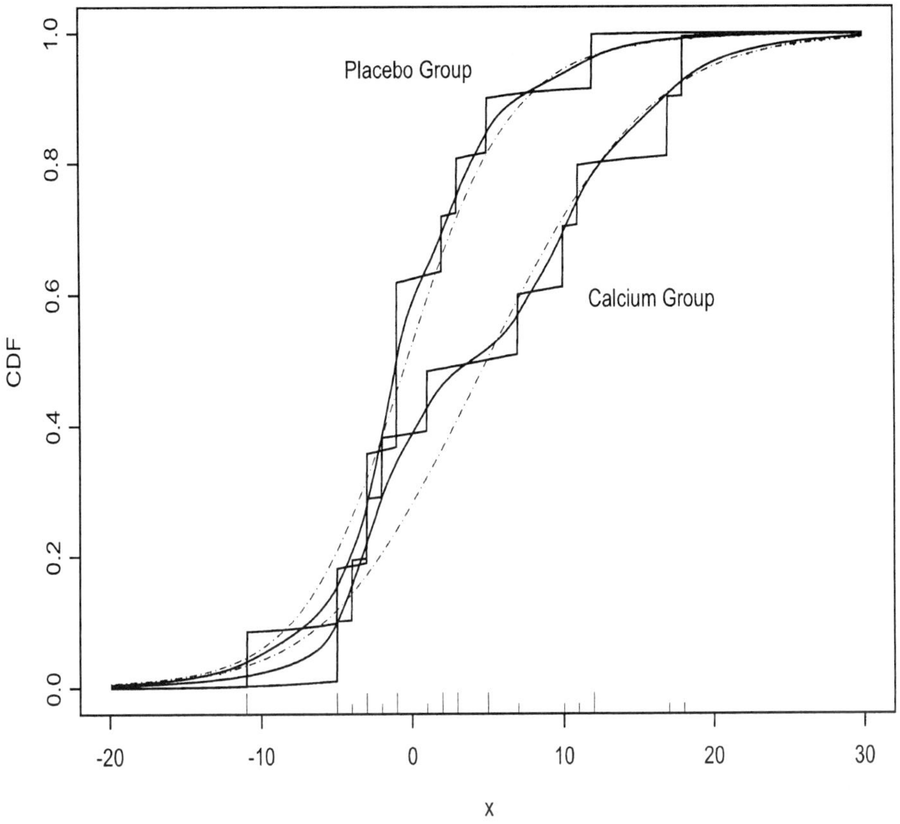

Fig. 2. Blood pressure data: posterior CDF estimates for both groups using the MDP (jagged), DPM (dashed), and MPT (solid) models. The longer tick marks along the x-axis correspond to the observed data for the placebo group and the shorter tick marks to the observed data for the calcium group.

the prior information was quite accurate. One final note is that the sample sizes are so small for this problem that the DPM model density estimates look parametric. If there were bumps in the true densities and with larger sample sizes, the DPM model would reflect this fact. Since the truth is unknown here, we are not in a position to say that any of the models are preferable.

3.2. Regression examples

Here, we mainly discuss two types of regression models. Both types can be expressed in the usual form $y = f(x) + \varepsilon$. In one instance, we consider $f(x) = x\beta$, and with $\varepsilon \sim F$, $F \in \mathcal{F}$ where $\mathcal{F}$ consists of continuous distributions with median zero, which results in $x\beta$ as the median of $y|x, \beta$ or what has been called median regression. In a second instance, we consider $f \in \mathcal{F}^*$ and where the distribution of the error is assumed to have been generated according to a parametric family. When the primary goal is estimation of the regression function, f, parametric error models may suffice, but when considering predictive inference or in estimation of certain survival models, it is desirable to estimate F nonparametrically. We discuss these two types of models in some detail and give illustrations. We also discuss the situation where both f and F are allowed to be flexible.

We want to emphasize at the outset that our purpose here is mainly to illustrate the fundamental ideas and methods. The published literature clearly goes well beyond what we present here and we make no claims to having used the best or most sophisticated method for any given problem. We emphasize the simplicity of the methods that are presented here as many of them are accomplished in WinBUGS.

Our main illustration of semiparametric regression with unknown error distribution is in the area of survival analysis, which is discussed next.

3.2.1. *Regression for survival data*

In this subsection, we first briefly discuss univariate survival data with censoring. We proceed to discuss semiparametric accelerated failure time (AFT) and proportional hazards (PH) models for censored survival data with covariates. We ultimately analyze a classic data set on leukemia remission times using BNP methodology applied to AFT and PH models. See Ibrahim et al. (2001) for descriptions of these models and for other analyses of these data. All of the modeling done here is also applicable to uncensored data and thus to standard linear regression.

Denote survival times for n independently sampled individuals as $T_1, \ldots, T_n$. Right censored data are denoted $\{(t_i, \delta_i): i = 1, \ldots, n\}$ where $\delta_i = 0$ implies that $T_i > t_i$, which corresponds to t_i being an observed censoring time, and $\delta_i = 1$ implies $T_i = t_i$. Censoring times are assumed independent of event times. With covariate information, we have data $\{(t_i, \delta_i, x_i): i = 1, \ldots, n\}$.

Let T_0 be a random survival time from a baseline distribution. The AFT model specifies that an individual with covariate vector x has the survival time $T_x = g(x'\beta)T_0$, for regression coefficients β and a monotone function g. This is equivalent to $S(t|x) = S(t/g(x'\beta))$ where $S(t) = P(T_0 > t)$ is the baseline survival function and $S(t|x) = P(T_x > t)$.

Usually, g is taken to be the exponential function and the model is then equivalent to $\log(T_x) = x'\beta + \log(T_0)$, i.e. a standard linear regression model. Standard parametric analyses further assume that $\log(T_0) = \sigma\varepsilon$ where ε is standard normal, extreme value, or logistic. If ε has median zero, a median-zero regression model is obtained.

Christensen and Johnson (1988) obtain approximate, marginal inference in the AFT model with a DP baseline S while Johnson and Christensen (1989) show that obtaining full posterior inference from an AFT model with a DP baseline is infeasible. Kuo and Mallick (1997) circumvent this difficulty by considering a DPM for S. They interpret the baseline model as a "smoothed" DP. Walker and Mallick (1999), and Hanson and Johnson (2002) considered, respectively, PT and MPT baselines in the AFT model, whereas Kottas and Gelfand (2001a) described a DPM model for the baseline in the AFT model; these models are all median regression models. Hanson and Johnson (2004) extended the MDP model of Doss (1994) to an AFT model with a MDP baseline for interval censored data.

On the other hand, the PH model has by far enjoyed the greatest success of any other statistical model for survival data with covariates. Frequentist and Bayesian statistical literature on the topic far exceed that for any other survival model. The PH model is specified using the baseline hazard function $\lambda(t)$. For baseline survival T_0, the hazard is defined as

$$\lambda(t) = \lim_{dt \to 0^+} \frac{P(t \leqslant T_0 < t + dt)}{dt},$$

or $\lambda(t)\,dt \approx P(t \leqslant T_0 < t + dt)$ for small dt. If T_0 is absolutely continuous then $\lambda(t) = f(t)/S(t)$ where f and S are the pdf and survivor function for T_0, respectively. Cox's PH model (Cox, 1972) assumes that for an individual with covariate x, $\lambda(t|x) = g(x'\beta)\lambda(t)$, where g and β are as before (except there is no intercept term here). Typically g is taken to be the exponential function yielding the interpretation of $\exp(x\beta)$ as a relative risk of "instantaneous failure" comparing an individual with covariates x to a baseline individual. Under the PH model $S(t|x) = \exp(-e^{x\beta}\Lambda(t))$. The latter expression can be used to define the PH model when Λ has jump discontinuities, e.g., when T_0 is a mixture of continuous and discrete distributions.

The success of the PH model across a wide spectrum of disciplines is in part due to the interpretability of the regression parameters and in part due to the availability of easy to use software to fit the frequentist version of the model. In statistical packages the model is fit via *partial likelihood*, involving only β, which is not a proper likelihood but which does yield estimators with desirable properties such as asymptotic normality. The infinite-dimensional parameter Λ is treated as a nuisance parameter and, if needed, is estimated following the estimation of β. Bayesian approaches to the Cox model have considered both the use of the partial likelihood in inference and the consideration of a full probability model for (β, λ). We discuss only the latter and view the full, joint modeling of parameters, as well as nonasymptotic inference as a particular benefit of the Bayesian approach. Other BNP approaches have been discussed by Sinha and Dey (1997), Laud et al. (1998) and Ibrahim et al. (2001). It should be pointed out that, despite the flexibility of the PH model due to the baseline hazard being unspecified, the PH assumption is still quite restrictive and easily fails for many data sets. The semiparametric AFT model serves as a potential alternative when this is the case.

Given all of this background, we consider here a simple application of BNP methodology to a two sample survival analysis problem. Clearly there are many possible approaches but we only consider two here for the purpose of illustration.

Data on the remission times from two groups of leukemia patients are considered by Gehan (1965) and Kalbfleisch (1978) and are reproduced in Table 3. A PT AFT model was fitted to these data with a Weibull(1.47, 19.61) base measure, estimated from a parametric fit. We set $\gamma = 0.1$. The posterior median and equal-tailed 95% PI for β is 1.62 (0.70, 1.97). The Group 2 population has a median survival time estimated to be about $e^{1.62} \approx 5$ times that of Group 1. In Figure 3, estimated survival curves are plotted for the two groups.

We now turn to the PH model. Although many stochastic processes have been used as priors for Λ in the Cox model, we focus attention on the first to be used in this context, the independent increments GP, which was discussed in Section 2.5. We now give a detailed discussion of the implementation of this model for use in WinBUGS, before discussing the BNP PH analysis of the leukemia data.

Table 3
Leukemia data: weeks of remission for leukemia patients

Group 1:	1, 1, 2, 2, 3, 4, 4, 5, 5, 8, 8, 8, 8, 11, 11, 12, 12, 15, 17, 22, 23
Group 2:	6, 6, 6, 6*, 7, 9*, 10, 10*, 11*, 13, 16, 17*, 19*, 20*, 22, 23, 25*, 32*, 32*, 34*, 35*

*Right censored observation.

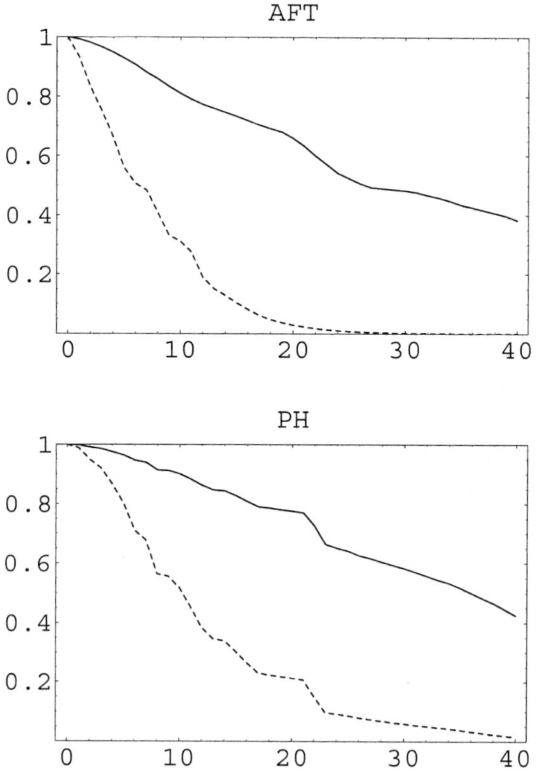

Fig. 3. Leukemia data: estimated survival curves for AFT and PH models; Group 2 (solid) and Group 1 (dashed).

Burridge (1981) and Ibrahim et al. (2001) suggest that the model as proposed by Kalbfleisch (1978) and extended by Clayton (1991) is best suited to grouped survival data. Walker and Mallick (1997) considered an approximation to the GP for continuous data that we describe here, in part because it is readily implemented in WinBUGS. Define a partition of $(0, \infty)$ by $\{(a_{j-1}, a_j]\}_{j=1}^{J} \cup (a_J, \infty)$ where $0 = a_0 < a_1 < a_2 < \cdots < a_{J+1} = \infty$. Here, a_J is taken to be equal to $\max(\{t_i\}_{i=1}^{n})$. If $\Lambda \sim \text{GP}(\alpha, \Lambda_0)$ then by definition $\Lambda(a_j) - \Lambda(a_{j-1}) \overset{\text{ind}}{\sim} \Gamma(\alpha(\Lambda_0(a_j) - \Lambda_0(a_{j-1})), \alpha)$. Walker and Mallick (1997) make this assumption *for the given partition* and further assume that $\lambda(t)$ is constant and equal to λ_j on each $(a_{j-1}, a_j]$ for $j = 1, \ldots, J$. This implies $\lambda_j \sim$

$\Gamma(\alpha\lambda_{0j}, \alpha)$ where $\lambda_{0j} = (\Lambda_0(a_j) - \Lambda_0(a_{j-1}))/(a_j - a_{j-1})$, and yields a particular piecewise exponential model.

Now given $\Lambda(\cdot)$, or equivalently $\{\lambda_j\}_{j=1}^J$, $S(t) = \exp(-\Lambda(t))$, $S(t|x) = \exp(-e^{x'\beta}\Lambda(t))$ and $f(t|x) = e^{x'\beta}\lambda(t)\exp(-e^{x'\beta}\Lambda(t))$. Assume that the event times $\{t_i\}_{i=1}^n$ are included as some of the partition points $\{a_j\}_{j=1}^J$. Let $j(i)$ be such that $t_i = a_{j(i)}$. Then $\Lambda(t_i) = \sum_{j=1}^{j(i)} \lambda_j \Delta_j$ where $\Delta_j = a_j - a_{j-1}$ and $\lambda(t_i) = \lambda_{j(i)}$. The likelihood is given by

$$\mathcal{L}(\lambda, \beta) = \prod_{i=1}^{n} \exp\bigl(-e^{x_i\beta}\Lambda(t_i)\bigr)\bigl[e^{x_i\beta}\lambda(t_i)\bigr]^{\delta_i}$$

$$= \prod_{i}\prod_{j=1}^{j(i)} \exp\bigl(-e^{x_i\beta}\lambda_j\Delta_j\bigr) \prod_{\{i:\delta_i=1\}} e^{x_i\beta}\lambda_{j(i)},$$

which is proportional to a product of Poisson kernels. Therefore, with independent gamma priors on $\{\lambda_j\}$, this model is readily fitted in WinBUGS. This likelihood is similar to that obtained by Clayton (1991) using a counting process argument (for example, see "Leuk: survival analysis using Cox regression" in Examples Volume I, WinBUGS 1.4). Clayton's approach requires sampling $\Lambda(\cdot)$ only at the $\{t_i\}$ to obtain full inference for β. The piecewise exponential model has been used to accommodate approximations to a correlated prior process (Ibrahim et al., 2001, Section 3.6) and also used in joint models accommodating a latent longitudinal marker that affects survival (Wang and Taylor, 2001; Brown and Ibrahim, 2003) due to the simple structure of the model.

To get more of the flavor of the GP from this approximation, one might take the partition to be a fine mesh. Furthermore, a mixture of gamma processes can be induced by assuming $\Lambda \sim \mathrm{GP}(\alpha, \Lambda_\theta)$, $\theta \sim f(\theta)$. For example, one might center $\Lambda(\cdot)$ at $\Lambda_\theta = \theta t$, the exponential family, and place a hyperprior on θ. This results in a mixture of GP's (MGP).

We adapted this approach and fit the MGP PH model to the leukemia data using vague hyperpriors in WinBUGS. The posterior median and 95% PI for β is 1.56 (0.84, 2.36). The hazard of expiring in Group 1 is about $e^{1.56} \approx 4.8$ times as likely as Group 2 at any time t. Estimated survival curves are given in Figure 3.

Other prior processes used in PH survival models include the beta process (Hjort, 1990), and the extended gamma process (Dykstra and Laud, 1981), which smooths the GP with a known kernel. Ishwaran and James (2004) extend this work and the work of others (notably Lo and Weng, 1989, and Ibrahim et al., 1999) to a very general setting by capitalizing on a connection between the GP and the DP. Often Bayesian semiparametric survival models are fit by partitioning $[0, \infty)$ into a fine mesh and computing grouped data likelihoods; the approach of Ishwaran and James (2004) avoids this computationally intensive approach. Kim and Lee (2003) consider the PH model with left truncated and right censored data for very general neutral to the right priors.

Ibrahim et al. (2001) also discuss the implementation of frailty, cure rate, and joint survival and longitudinal marker models. Mallick and Walker (2003) develop a frailty model that uses PTs and includes proportional odds, AFT, and PH all as special cases. The model utilizes a PT error term and a monotone transformation function modeled with a mixture of incomplete beta functions. Prior elicitation for survival models are discussed by Ibrahim et al. (2001). Methods on prior elicitation for regression coefficients in parametric survival models developed by Bedrick et al. (2000) apply to Bayesian semiparametric AFT modeling. Ishwaran and James (2004) develop weighted GP's in the multiplicative intensity model. Huzurbazar (2004) provides an extensive introduction to the use of Bayesian flowgraph models for the modeling of survival data. Mallick et al. (1999) use multivariate adaptive regression splines in a highly flexible model allowing for time-dependent covariates. Space does not permit us to discuss the extensive literature on semiparametric cure rate models, competing risks models, multivariate models, and other important areas.

In the absence of covariates, Susarla and van Ryzin (1976) assumed a DP prior for F for right-censored data and established that the Kaplan and Meier (1958) estimator is obtained as $\alpha \to 0^+$. Johnson and Christensen (1986) extended the model to grouped survival data and similarly showed that Turnbull's (1974) estimator is the corresponding limiting form. Doss (1994) and Doss and Huffer (2004) discussed fitting the MDP model to censored data and compared various algorithms based on importance sampling and MCMC to obtain inferences. They also provided user-friendly software for the statistical packages R and S-Plus to fit these models. Other related approaches include Lavine (1992), who gave an example of density estimation for survival data via PT's. Wiper et al. (2001) used a mixture of Gamma densities in the spirit of Richardson and Green (1997) to model data with support on $[0, \infty)$. The DPM model of Escobar and West (1995) can also be used for survival data or log survival data.

3.2.2. Nonparametric regression with known error distribution

Estimation of an unknown regression function is a common and extensively researched area across many disciplines. The problem is typically to estimate the mean function f from data $\{(x_i, y_i)\}_{i=1}^n$ in the model

$$y_i = f(x_i) + \varepsilon_i, \quad \varepsilon_i \text{ i.i.d.}, \ E(\varepsilon_i) = 0,$$

but in some applications the shape of the error distribution ε_i is of interest as well. We initially assume x_i is univariate but later discuss the case when x_i is a vector of predictors.

Denison et al. (2002) provide an introduction to Bayesian semiparametric regression methods focusing primarily on splines. Müller and Quintana (2004) review advances in Bayesian regression and additionally discuss neural networks. Müller and Vidakovic (1999) discuss Bayesian models incorporating wavelets.

One successful approach borrows from the field of *harmonic analysis* and assumes f can be represented as a weighted sum of basis functions. For f sufficiently smooth, and given an orthonormal basis $\{\phi_j\}_{j=1}^\infty$ of the function space of square-integrable functions on some region R, $\mathcal{L}^2(R)$, one can write the Fourier representation of f as $f(x) = \sum_{j=0}^\infty \beta_j\phi_j(x)$ where $\beta_j = \int_R f(x)\phi_j(x)\,\mathrm{d}x$. The basis is said to be orthonormal if $\int_R \phi_i(x)\phi_j(x)\,\mathrm{d}x = \delta_{ij}$ where $\delta_{ij} = 1$ if $i = j$ and zero otherwise. Orthonormal bases make certain common calculations trivial in some problems, but are not required of this approach. Popular choices for $\{\phi_j\}$ are the Fourier series (sines and cosines), spline bases, polynomials, and wavelet bases.

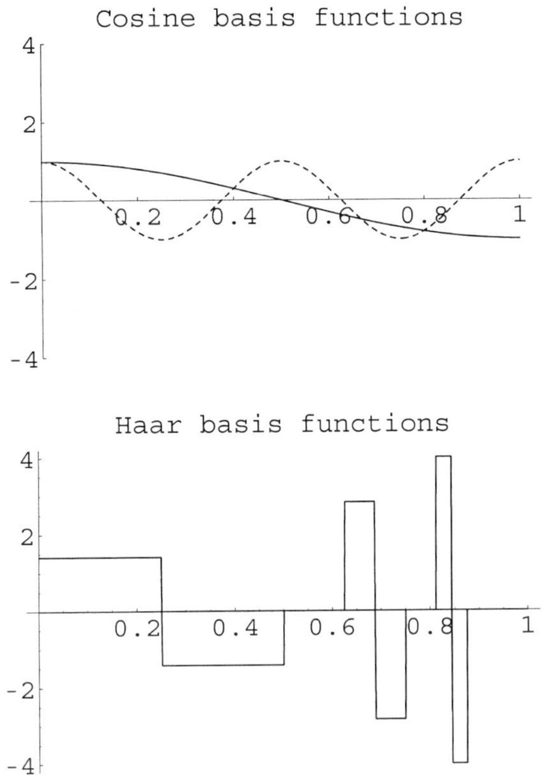

Fig. 4. Cosine basis functions $\cos(x\pi)$ (solid) and $\cos(x4\pi)$ (dashed); Haar basis functions $\phi_{2,1}$, $\phi_{4,6}$ and $\phi_{5,14}$ from left to right.

It is impossible to estimate $\{\beta_j\}_{j=0}^{\infty}$ with finite data. All but a finite number of these coefficients must be set to zero for estimation to proceed and therefore f is approximated by a finite number of basis functions $f(x) = \sum_{j=0}^{J} \beta_j \phi_j(x)$ in practice. The basis functions are often ordered in some fashion from broad functions that indicate a rough trend to functions that are highly oscillatory over R. A statistical problem is to determine at which point noise is essentially being modeled by the more oscillatory functions, or equivalently at which point J to "cutoff" the basis functions. In Figure 4 we see two of the cosine basis functions $\{\cos(xj\pi)\}_{j=0}^{\infty}$ and three Haar basis functions (described later) on $R = [0, 1]$.

Traditionally, the choice of J is an interesting problem with many reasonable, typically *ad hoc*, solutions. This choice deals intimately with the issue of separating signal from the noise. It is well-known that an $(n-1)$-degree polynomial fits data $\{(x_i, y_i)\}_{i=1}^{n}$ perfectly, an example of overfitting, or the inclusion of too many basis functions. Efromovich (1999) overviews common bases used in regression function estimation and addresses choosing J in small and large samples.

If one fixes J and assumes i.i.d. Gaussian errors then the standard linear model is obtained:

$$y_i = \beta_0 + \beta_1\phi_1(x_i) + \cdots + \beta_J\phi_J(x_i) + \varepsilon_i, \quad \varepsilon_i \overset{\text{i.i.d.}}{\sim} N(0, \sigma^2).$$

Placing a prior on β and σ^{-2} yields the Bayesian linear model (Lindley and Smith, 1972), which is easily implemented in WinBUGS. In Figure 5 we examine orthonormal series fits to data on the amount of nitric oxide and nitric dioxide in the exhaust of a single-cylinder test engine using ethanol as fuel (Brinkman, 1981). The response is in μg per joules and the predictor is a measure of the air to fuel ratio. These data are part of a larger data set used throughout the S-Plus Guide to Statistics (MathSoft, 1999) to illustrate various smoothing techniques, including locally weighted regression smoothing, kernel smoothers, and smoothing splines. The cosine, $\phi_i(x) = \cos(i\pi(x - 0.5)/0.8)$, and Legendre polynomial bases were used for illustration with $R = [0.5, 1.3]$ and fixed $J = 5$. Independent $N(0, 1000)$ priors were placed on the regression coefficients and the precision σ^{-2} was assumed to be distributed $\Gamma(0.001, 0.001)$ as an approximation to Jeffreys' prior. The prior of Bedrick et al. (1996) can be used to develop an informative prior on β. The choice of basis functions, cutoff J, and region R will all affect posterior inference.

Multivariate predictors $x_i = (x_{i1}, \ldots, x_{ip})$ can be accommodated via series expansions by considering products of univariate basis functions. For example, in the plane, simple products are formed as $\phi_{jk}(x_1, x_2) = \phi_j(x_1)\phi_k(x_2)$. The regression model is then $y_i = \sum_{j=1}^{J}\sum_{k=1}^{J}\beta_{jk}\phi_{jk}(x_{i1}, x_{i2}) + \varepsilon_i$. Additive models are an alternative where the mean response is the sum of curves in each predictor, e.g., $E(y_i) = \sum_{j=1}^{J_1}\beta_{j1}\phi_j(x_{i1}) + \sum_{j=1}^{J_2}\beta_{j2}\phi_j(x_{i2})$.

A popular Bayesian alternative to fixing the number of components is to place a prior on J and implement the reversible jump algorithm of Green (1995). Reversible jump MCMC approximates posterior inference over a model space where each model has a parameter vector of possibly different dimension. A prior probability is placed on each of $J = 1, 2, \ldots, J_0$, where J_0 is some natural upper bound chosen such that consideration of $J > J_0$ would be superfluous. Reversible jump for the regression problem in the context of a spline basis is discussed in Denison et al. (2002) and used, for example, by Mallick et al. (1999) and Holmes and Mallick (2001). Many spline bases are built from truncated polynomials. For example $\{(x - a_j)_+^3\}_{j=1}^J$ is a subset of a cubic spline basis, where $\{a_j\}_{j=1}^J$ are termed *knots* and $(x)_+$ is equal to x when $x > 0$ and equal to zero otherwise.

Another approach is to fix J quite large and allow some of the $\{\beta_j\}_{j=1}^J$ to be zero with positive probability. This approach, advocated by Smith and Kohn (1996), can be formulated as $\beta_j \sim \gamma_j\beta_j^*$ where $\gamma_j \sim \text{Bernoulli}(\theta_j)$ independent of $\beta_j^* \sim N(b_j, \eta_j^2)$, and for moderate J and independent β_j priors can be programmed in WinBUGS. For the ethanol data using the cosine basis, we consider the rather naive, data-driven prior $\gamma_j \overset{\text{i.i.d.}}{\sim} \text{Bernoulli}(0.5)$, $\beta_j^*|\sigma^2 \sim N(b_j, 10\sigma^2 v_j)$. Where X is the design matrix from the model with all basis functions up to J, i.e., $\gamma_1 = \cdots = \gamma_J = 1$, v_j are the diagonal elements of $(X'X)^{-1}$ and $(b_1, \ldots, b_J)$ are the least squares estimates taken from $(X'X)^{-1}X'y$. Figure 6 shows the resulting estimate of the regression function. Five of

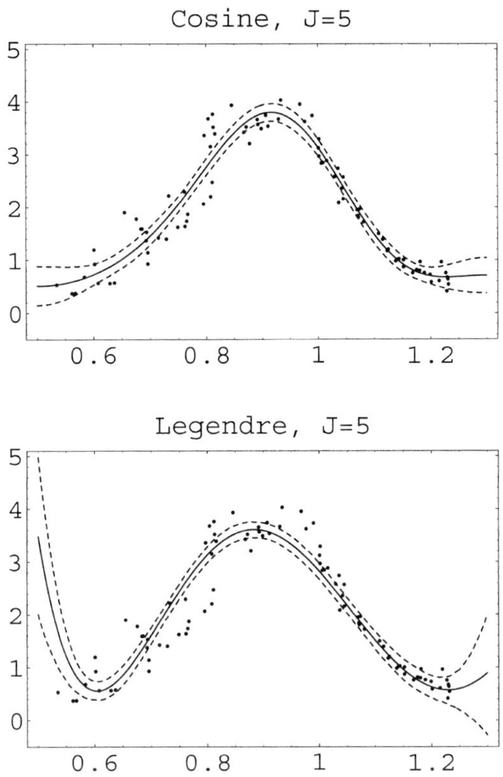

Fig. 5. Ethanol data: mean function estimate using cosine and Legendre bases, $J = 5$.

the ten basis functions have posterior probability $P(\gamma_j = 0|y)$ less than the prior value of 0.5. Clyde and George (2004) discuss priors of this type, specifically the g-prior, in more detail.

Crainiceanu et al. (2004) outline a strategy for fitting penalized spline models in WinBUGS. They capitalize on an equivalence between fitting penalized spline and mixed effect models and the resulting WinBUGS implementation is straightforward. They illustrate the possibilities by fitting nonparametric regression, binomial regression, and nonparametric longitudinal ANOVA models, all in WinBUGS. An advantage of the Bayesian approach over the frequentist approach is that it obviates the use of "plug-in" estimates when computing interval estimates. We apply the approach of Crainiceanu et al. (2004) by fitting a penalized quadratic spline model to the ethanol data. Specifically, the model is

$$y_i = \beta_0 + \beta_1 x_i + \beta_2 x_i^2 + \sum_{k=1}^{10} b_k (x_i - \kappa_k)_+^2 + \varepsilon_i,$$

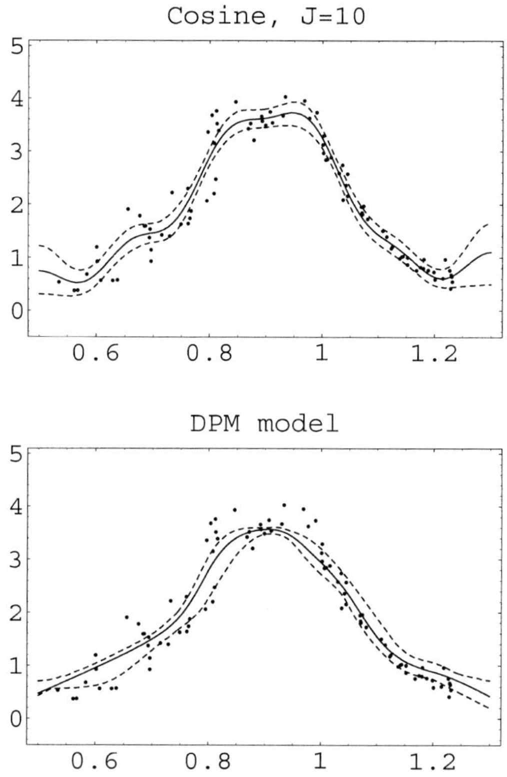

Fig. 6. Ethanol data: estimates of regression mean functions using a cosine basis, and a DPM model (see Section 3.2.3).

where $b_k|\sigma_b \overset{\text{i.i.d.}}{\sim} N(0, \sigma_b^2)$ independent of $\varepsilon_i \overset{\text{i.i.d.}}{\sim} N(0, \sigma_\varepsilon^2)$. Here, the knots $\{\kappa_k\}_{k=1}^{10}$ are defined as $\kappa_i = 0.4 + 0.1i$, evenly spaced over the range of the predictor. In Figure 7 we see the penalized spline estimate along with 95% pointwise probability intervals.

A unique class of orthonormal bases are wavelet bases. Wavelets are useful for modeling functions whose behavior changes dramatically at different locations and scales, often termed "spatially inhomogeneous." Think of a grayscale photograph of the Rocky mountains. Much of the photograph will be flat, rocky homogeneous areas where the grayscale changes little. At the edges of a mountain leading to sky, however, the scale changes abruptly. Also, foliage around the base of the mountain will have highly varying grayscale in a small area relative to the mountainous part. Wavelets can capture these sorts of phenomena and for this reason are extensively used in image processing.

The simplest wavelet basis is the Haar basis (Haar, 1910). The Haar basis is also the only wavelet basis with basis functions that have a closed form. On the interval $R = [0, 1]$ the Haar basis (as well as other wavelet bases) is managed conveniently by

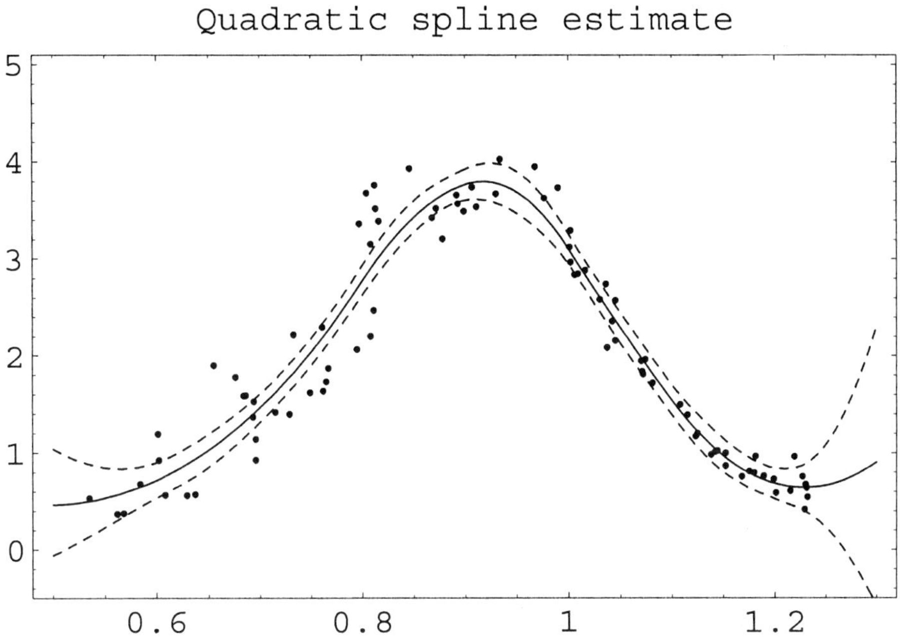

Fig. 7. Ethanol data: estimate of regression mean function using a penalized spline.

a double index and is derived from the Haar mother wavelet

$$\phi(x) = \left\{ \begin{array}{ll} 1, & 0 \leqslant x < 0.5, \\ -1, & 0.5 \leqslant x \leqslant 1, \\ 0, & \text{otherwise} \end{array} \right\}$$

through the relation $\phi_{ij}(x) = \phi(2^{(i-1)}x - j + 1)2^{(i-1)/2}$ for $i = 1, \ldots, \infty$, and $j = j(i) = 1, \ldots, 2^{(i-1)}$. The set $\{I_{[0,1]}(x)\} \cup \{\phi_{ij}\}$ forms an orthonormal basis of $[0, 1]$. Figure 4 shows three of the Haar basis functions; the i indexes the scale of the basis function whereas the j indexes location. For large i, wavelet basis functions can model very localized behavior. Contrast the Haar basis to the cosine basis where basis functions oscillate over the entire region R. For this reason wavelets can model highly inhomogeneous functions but also require special tools to ensure that mean estimates do not follow the data too closely. These tools, broadly termed "thresholding," require that there is substantial data-driven evidence that a wavelet basis function belongs in the model, and more evidence is required for larger i. Bayesian thresholding typically places mixture priors on basis coefficients in the wavelet domain after transforming data using the discrete wavelet transform. These priors place positive probability that some coefficients are very small (or zero). Müller and Vidakovic (1999) discuss Bayesian wavelet modeling in detail. A nice, short introduction to Bayesian wavelets and thresholding is provided by Vidakovic (1998).

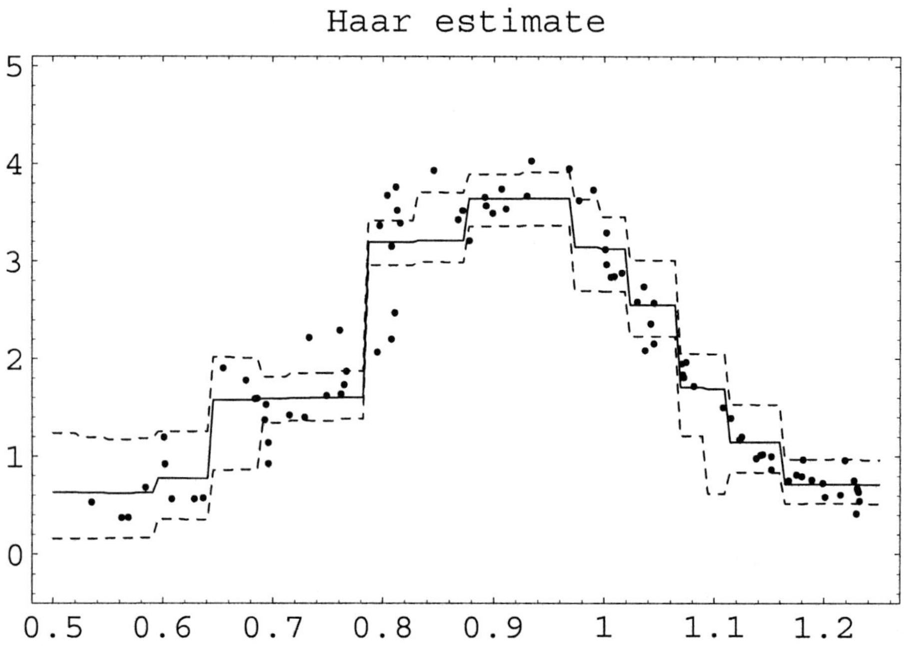

Fig. 8. Ethanol data: estimates of regression mean functions using Haar wavelets.

For illustrative purposes, we fit the following Haar wavelet model to the ethanol data in WinBUGS:

$$y_k = \beta_0 + \sum_{i=1}^{4} \sum_{j=1}^{2^{i-1}} \gamma_{ij} \beta_{ij}^* \phi_{ij}(x_k) + \varepsilon_k, \quad \varepsilon_k \overset{\text{i.i.d.}}{\sim} N\bigl(0, \sigma^2\bigr).$$

A simple data-driven prior was constructed in the same manner as for the cosine basis except $\gamma_{ij} \sim \text{Bernoulli}(2^{-i})$, ensuring that the prior probability of including a basis function decreases with how "localized" the function is. Figure 8 shows the resultant mean function estimate. Four of the 15 basis functions considered had posterior probabilities of being included in the model less than 0.1.

3.2.3. Nonparametric regression with unknown error distribution

Combinations of the approaches discussed so far yield very rich, highly flexible models for both the regression function f and the error ε. Alternatively, a highly flexible model that has ties to kernel regression and local linear regression is the use of DPM's for multivariate data.

To obtain inference in the general model $y(x) = f(x) + e(x)$, Müller et al. (1996) suggest modeling data $\{z_i = (x_i, y_i)\}_{i=1}^n$ as arising from a DPM of multivariate Gaussian densities. As is typical, inference is obtained with the DP integrated out and the model reduces to a particular finite mixture model. The model is given hierarchically by

$$z_i|\mu_i, \Sigma_i \overset{\text{ind}}{\sim} N(\mu_i, \Sigma_i), \quad (\mu_i, \Sigma_i)|G \overset{\text{i.i.d.}}{\sim} G, \quad G \sim \text{DP}(\alpha G_0).$$

The authors consider the prior $g_0(\mu, \Sigma^{-1}) = N_p(\mu|m, B)W_p(\Sigma^{-1}; \nu, (S\nu)^{-1})$ where p is the dimension of $z_i = (x_i, y_i)$, g_0 is the density of G_0, $N_p(x|\mu, \Sigma)$ is the pdf of a multivariate normal variate with mean μ and covariance Σ, and $W_q(\nu, \Sigma)$ is the pdf of a Wishart variate with degrees of freedom ν and mean $\nu\Sigma$. Hyperpriors can be further placed on m, B, S, and α.

An estimate of $f(x_0)$ is provided by $E(y_{n+1}|x_{n+1} = x_0, x, y)$ and is obtained using conditioning arguments. This estimate is essentially a locally-weighted piecewise linear estimate averaged over the MCMC iterates. We consider a simple version of this model for the ethanol data by taking m to be the sample mean $\bar{z}$, B as 10 times the sample covariance of $\{z_i\}_{i=1}^n$, $\alpha = 2$, $S = \text{diag}(0.05^2, 0.25^2)$, and $\nu = 2$.

The prior expected number of components is about 8. Let k denote the number of distinct components in the model. A posteriori, we find $P(k \leqslant 3|z) \approx 0$, $P(4 \leqslant k \leqslant 6|z) \approx 0.94$, and $P(k \geqslant 7|z) \approx 0.04$. The estimated regression function and pointwise 95% probability intervals are in Figure 6, assuming that the marginal finite mixture model (induced by the DPM) is the full probability model. Although the example illustrates regression with one predictor, an attractive feature of the DPM model is that it is readily extended to many predictors, as long as modeling assumptions are reasonable.

4. Concluding remarks

The field of Bayesian nonparametrics relies on an interesting combination of the (sometimes abstract) development of probability models on large spaces and modern Markov chain Monte Carlo technology. The former is necessary for the application of Bayes theorem and the latter for its implementation. Analysis of complex and interesting data using BNP methodology was made to wait for the recent development of MCMC methods. This chapter has attempted to give a flavor of what is now possible due to the merger of these areas. We remind the reader that our goal was to present fundamental ideas and to illustrate them with relatively simple methods, rather than the most sophisticated ones.

There is a long list of methods and models that have been left out, too long to mention all. We simply mention a few. First, we have not discussed nonparametric dependent data modeling. MacEachern (2000) invented the dependent Dirichlet Process (DDP), which builds in dependence among a collection of random probability measures. The DDP has recently been used by De Iorio et al. (2004), who used ANOVA structure in modeling dependence, and by Gelfand et al. (2004) for modeling spatial data. Longitudinal modeling using the DDP should be straightforward given their development for spatial data. Dependent nonparametric processes were also considered by Gelfand and Kottas (2001) and Kottas and Gelfand (2001b), and Hoff (2003) in the context of modeling stochastic order. Another area that is ripe for development is the application of BNP methods to bioinformatics and proteomics, see for example Do et al. (2004). Areas that, to our knowledge, still require attention are (i) the development of mixtures of Polya tree priors for multivariate data and (ii) methods for model selection and model fit, for example how can one formally choose between semiparametric PH and AFT models and also assess their goodness of fit.

Throughout this chapter, very little has been said about theory since our goal was to present basic modeling techniques and to give a flavor for their application to data. There are of course many articles that develop theoretical aspects of BNP models. See for example Diaconis and Freedman (1986) for a BNP model and method based on DP's that fails. However, there is much theoretical work that establishes that BNP methods are valid. For example, Ghosal et al. (1999) established consistency of density estimates based on an DPM. Regazzini et al. (2002) recently presented results for exact distributions of functionals of a DP. Choudhuri et al. (2005) discuss asymptotic properties of BNP function estimates and give many references. Also see the monograph of Ghosh and Ramamoorthi (2003) for additional theoretical background material and references.

References

Antoniak, C.E. (1974). Mixtures of Dirichlet processes with applications to Bayesian nonparametric problems. *Ann. Statist.* **2**, 1152–1174.

Barron, A., Schervish, M., Wasserman, L. (1999). The consistency of posterior distributions in nonparametric problems. *Ann. Statist.* **27**, 536–561.

Bedrick, E.J., Christensen, R., Johnson, W.O. (1996). A new perspective on priors for generalized linear models. *J. Amer. Statist. Assoc.* **91**, 1450–1460.

Bedrick, E.J., Christensen, R., Johnson, W.O. (2000). Bayesian accelerated failure time analysis with application to veterinary epidemiology. *Statist. in Med.* **19**, 221–237.

Berger, J.O., Gugliemi, A. (2001). Bayesian testing of a parametric model versus nonparametric alternatives. *J. Amer. Statist. Assoc.* **96**, 174–184.

Berry, D., Christensen, R. (1979). Empirical Bayes estimation of a binomial parameter via mixture of Dirichlet processes. *Ann. Statist.* **7**, 558–568.

Blackwell, D. (1973). Discreteness of Ferguson selections. *Ann. Statist.* **1**, 356–358.

Blackwell, D., MacQueen, J.B. (1973). Ferguson distributions via Polya urn schemes. *Ann. Statist.* **2**, 353–355.

Brinkman, N.D. (1981). Ethanol fuel – a single-cylinder engine study of efficiency and exhaust emissions. *SAE Transactions* **90**, 1410–1424.

Brown, E.R., Ibrahim, J.G. (2003). A Bayesian semiparametric joint hierarchical model for longitudinal and survival data. *Biometrics* **59**, 221–228.

Burridge, J. (1981). Empirical Bayes analysis of survival time data. *J. Roy. Statist. Soc., Ser. B* **43**, 65–75.

Bush, C.A., MacEachern, S.N. (1996). A semi-parametric Bayesian model for randomized block designs. *Biometrika* **83**, 275–286.

Choudhuri, N., Ghosal, S., Roy, A. (2005). Bayesian methods for function estimation. In: Dey, D.K., Rao, C.R. (Eds.), *Bayesian Thinking: Modeling and Computation, Handbook of Statistics*, vol. 25. Elsevier, Amsterdam, pp. 373–414. This volume.

Christensen, R., Johnson, W.O. (1988). Modeling accelerated failure time with a Dirichlet process. *Biometrika* **75**, 693–704.

Clayton, D.G. (1991). A Monte Carlo method for Bayesian inference in frailty models. *Biometrics* **47**, 467–485.

Clyde, M., George, E.I. (2004). Model uncertainty. *Statist. Sci.* **19**, 81–94.

Congdon, P. (2001). *Bayesian Statistical Modeling*. Wiley, Chichester.

Cox, D.R. (1972). Regression models and life-tables (with discussion). *J. Roy. Statist. Soc., Ser. B Methodological* **34**, 187–220.

Crainiceanu, C.M., Ruppert, D., Wand, M.P. (2004). Bayesian analysis for penalized spline regression using WinBUGS. Johns Hopkins University, Dept. of Biostatistics Working Papers, Working Paper 40. http://www.bepress.com/jhubiostat/paper40.

Diaconis, P., Freedman, D. (1986). On inconsistent Bayes estimates of location. *Ann. Statist.* **14**, 68–87.

De Iorio, M., Müller, P., Rosner, G.L., MacEachern, S.N. (2004). An ANOVA model for dependent random measures. *J. Amer. Statist. Assoc.* **99**, 205–215.

Denison, D.G.T., Holmes, C.C., Mallick, B.K., Smith, A.F.M. (2002). *Bayesian Methods for Nonlinear Classification and Regression*. Wiley, Chichester.

Dey, D., Müller, P., Sinha, D. (1998). *Practical Nonparametric and Semiparametric Bayesian Statistics*. Springer Lecture Notes, New York.

Do, K.-A., Müller, P., Tang, F. (2004). A Bayesian mixture model for differential gene expression. Preprint.

Doksum, K.A. (1974). Tailfree and neutral random probabilities and their posterior distributions. *Ann. Probab.* **2**, 183–201.

Doss, H. (1994). Bayesian nonparametric estimation for incomplete data via successive substitution sampling. *Ann. Statist.* **22**, 1763–1786.

Doss, H., Huffer, F.W. (2004). Monte Carlo methods for Bayesian analysis of survival data using mixtures of Dirichlet process priors. *J. Comput. Graph. Statist.* **12**, 282–307.

Dykstra, R.L., Laud, P.W. (1981). A Bayesian nonparametric approach to reliability. *Ann. Statist.* **9**, 356–367.

Efromovich, S. (1999). *Nonparametric Curve Estimation*. Springer-Verlag, New York.

Escobar, M.D. (1994). Estimating normal means with a Dirichlet process prior. *J. Amer. Statist. Assoc.* **89**, 268–277.

Escobar, M.D., West, M. (1995). Bayesian density estimation and inference using mixtures. *J. Amer. Statist. Assoc.* **90**, 577–588.

Fabius, J. (1964). Asymptotic behavior of Bayes' estimates. *Ann. Math. Statist.* **35**, 846–856.

Ferguson, T.S. (1973). A Bayesian analysis of some nonparametric problems. *Ann. Statist.* **1**, 209–230.

Ferguson, T.S. (1974). Prior distributions on spaces of probability measures. *Ann. Statist.* **2**, 615–629.

Freedman, D.A. (1963). On the asymptotic behavior of Bayes' estimates in the discrete case. *Ann. Math. Statist.* **34**, 1194–1216.

Gehan, E.A. (1965). A generalized Wilcoxin test for comparing arbitrarily single-censored samples. *Biometrika* **52**, 203–224.

Gelfand, A.E. (1999). Approaches for semiparametric Bayesian regression. In: Ghosh, S. (Ed.), *Asymptotics, Nonparametrics and Time Series*. Marcel Dekker, New York, pp. 615–638.

Gelfand, A.E., Kottas, A.A. (2001). Nonparametric Bayesian modeling for stochastic order. *Ann. Inst. Statist. Math.* **53**, 865–876.

Gelfand, A.E., Kottas, A.A. (2002). Computational approach for full nonparametric Bayesian inference under Dirichlet process mixture models. *J. Comput. Graph. Statist.* **11**, 289–305.

Gelfand, A.E., Mukhopadhyay, S. (1995). On nonparametric Bayesian inference for the distribution of a random sample. *Canad. J. Statist.* **23**, 411–420.

Gelfand, A.E., Smith, A.F.M. (1990). Sampling based approaches to calculating marginal densities. *J. Amer. Statist. Assoc.* **85**, 398–409.

Gelfand, A.E., Kottas, A.A., MacEachern, S.N. (2004). Bayesian nonparametric spatial modeling with Dirichlet process mixing. *J. Amer. Statist. Assoc.*, submitted for publication.

Ghosh, J.K., Ramamoorthi, R. (2003). *Bayesian Nonparametrics*. Springer-Verlag, New York.

Ghosal, S., Ghosh, J.K., Ramamoorthi, R. (1999). Posterior consistency of Dirichlet mixtures in density estimation. *Ann. Statist.* **17**, 143–158.

Green, P.J. (1995). Reversible jump Markov chain Monte Carlo computation and Bayesian model determination. *Biometrika* **82**, 711–732.

Haar, A. (1910). Zur theorie der orthogonalen funktionensysteme. *Math. Ann.* **69**, 331–371.

Hanson, T.E., Johnson, W.O. (2002). Modeling regression error with a mixture of Polya trees. *J. Amer. Statist. Assoc.* **97**, 1020–1033.

Hanson, T.E., Johnson, W.O. (2004). A Bayesian semiparametric AFT model for interval censored data. *J. Comput. Graph. Statist.* **13**, 341–361.

Hjort, N.L. (1990). Nonparametric Bayes estimators based on beta processes in models of life history data. *Ann. Statist.* **18**, 1259–1294.

Hoff, P.D. (2003). Bayesian methods for partial stochastic orderings. *Biometrika* **90**, 303–317.

Holmes, C.C., Mallick, B.K. (2001). Bayesian regression with multivariate linear splines. *J. Roy. Statist. Soc., Ser. B* **63**, 3–17.

Huzurbazar, A.V. (2004). *Flowgraph Models for Multistate Time-to-Event Data*. Wiley, New York.

Ibrahim, J.G., Kleinman, K.P. (1998). Semiparametric Bayesian methods for random effects models. In: *Practical Nonparametric and Semiparametric Bayesian Statistics*. In: *Lecture Notes in Statistics*, vol. 133. Springer-Verlag, New York.

Ibrahim, J.G., Chen, M.-H., MacEachern, S.N. (1999). Bayesian variable selection for proportional hazards models. *Canad. J. Statist.* **37**, 701–717.

Ibrahim, J.G., Chen, M.-H., Sinha, D. (2001). *Bayesian Survival Analysis*. Springer-Verlag, New York.

Ishwaran, H., James, L.F. (2004). Computational methods for multiplicative intensity models using weighted gamma processes: Proportional hazards, marked point processes and panel count data. *J. Amer. Statist. Assoc.* **99**, 175–190.

Johnson, W.O., Christensen, R. (1986). Bayesian nonparametric survival analysis for grouped data. *Canad. J. Statist.* **14**, 307–314.

Johnson, W.O., Christensen, R. (1989). Nonparametric Bayesian analysis of the accelerated failure time model. *Statist. Probab. Lett.* **8**, 179–184.

Kalbfleisch, J.D. (1978). Non-parametric Bayesian analysis of survival time data. *J. Roy. Statist. Soc., Ser. B* **40**, 214–221.

Kaplan, E.L., Meier, P. (1958). Nonparametric estimation from incomplete observations. *J. Amer. Statist. Assoc.* **53**, 457–481.

Kim, Y., Lee, J. (2003). Bayesian analysis of proportional hazard models. *Ann. Statist.* **31**, 493–511.

Kleinman, K., Ibrahim, J. (1998). A semi-parametric Bayesian approach to generalized linear mixed models. *Statist. in Med.* **17**, 2579–2596.

Kottas, A.A., Gelfand, A.E. (2001a). Bayesian semiparametric median regression modeling. *J. Amer. Statist. Assoc.* **95**, 1458–1468.

Kottas, A.A., Gelfand, A.E. (2001b). Modeling variability order: A semiparametric Bayesian approach. *Methodology and Computing in Applied Probability* **3**, 427–442.

Kuo, L., Mallick, B. (1997). Bayesian semiparametric inference for the accelerated failure-time model. *Canad. J. Statist.* **25**, 457–472.

Laud, P., Damien, P., Smith, A.F.M. (1998). Bayesian nonparametric and covariate analysis of failure time data. In: Dey, D., Müller, P., Sinha, D. (Eds.), *Practical Nonparametric and Semiparametric Bayesian Statistics*. Springer, New York, pp. 213–226.

Lavine, M. (1992). Some aspects of Polya tree distributions for statistical modeling. *Ann. Statist.* **20**, 1222–1235.

Lavine, M. (1994). More aspects of Polya tree distributions for statistical modeling. *Ann. Statist.* **22**, 1161–1176.

Lindley, D.V., Smith, A.F.M. (1972). Bayes estimates for the linear model (with discussion). *J. Roy. Statist. Soc., Ser. B* **34**, 1–41.

Lo, A.Y. (1984). On a class of Bayesian nonparametric estimates: I. Density estimates. *Ann. Statist.* **12**, 351–357.

Lo, A.Y., Weng, C.S. (1989). On a class of Bayesian nonparametric estimates: II. Hazard rate estimates. *Ann. Inst. Statist. Math.* **41**, 227–245.

MacEachern, S.N. (1994). Estimating normal means with a conjugate style Dirichlet process prior. *Comm. Statist. Simulation Comput.* **23**, 727–741.

MacEachern, S.N. (2000). Dependent Dirichlet processes. Technical Report, Dept. of Statistics, The Ohio State University.

MacEachern, S.N., Clyde, M., Liu, J. (1999). Sequential importance sampling for nonparametric Bayes models: The next generation. *Canad. J. Statist.* **27**, 251–267.

MacEachern, S.N., Müller, P. (1998). Estimating mixture of Dirichlet process models. *J. Comput. Graph. Statist.* **7**, 223–238.

Mallick, B.K., Denison, D.G.T., Smith, A.F.M. (1999). Bayesian survival analysis using a MARS model. *Biometrics* **55**, 1071–1077.

Mallick, B.K., Walker, S.G. (2003). A Bayesian semiparametric transformation model incorporating frailties. *J. Statist. Plann. Inference* **112**, 159–174.

MathSoft, Inc. (1999). *S-Plus 5 for UNIX Guide to Statistics*. Data Analysis Products Division, MathSoft, Seattle.

Mauldin, R.D., Sudderth, W.D., Williams, S.C. (1992). Polya trees and random distributions. *Ann. Statist.* **20**, 1203–1221.

Moore, D.S. (1995). *The Basic Practice of Statistics*, first ed. W.H. Freeman and Company.

Mukhopadhyay, S., Gelfand, A.E. (1997). Dirichlet process mixed generalized linear models. *J. Amer. Statist. Assoc.* **92**, 633–639.

Müller, P., Erkanli, A., West, M. (1996). Bayesian curve fitting using multivariate normal mixtures. *Biometrika* **83**, 67–79.

Müller, P., Quintana, F.A. (2004). Nonparametric Bayesian data analysis. *Statist. Sci.* **19**, 95–110.

Müller, P., Vidakovic, B. (1999). *Bayesian Inference in Wavelet-Based Models*. Springer-Verlag, New York.

Neal, R.M. (2000). Markov chain sampling methods for Dirichlet process mixture models. *J. Comput. Graph. Statist.* **9**, 249–265.

Nieto-Barajas, L., Walker, S. (2002). Markov beta and gamma processes for modeling hazard rates. *Scand. J. Statist.* **29**, 413–424.

Nieto-Barajas, L., Walker, S. (2004). Bayesian nonparametric survival analysis via Lévy driven Markov processes. *Statistica Sinica* **14**, 1127–1146.

Paddock, S.M., Ruggeri, F., Lavine, M., West, M. (2003). Randomized Polya tree models for nonparametric Bayesian inference. *Statistica Sinica* **13** (2), 443–460.

Regazzini, E., Guglielmi, A., Di Nunno, G. (2002). Theory and numerical analysis for exact distributions of functionals of a Dirichlet process. *Ann. Statist.* **30**, 1376–1411.

Richardson, S., Green, P. (1997). On Bayesian analysis of mixtures with an unknown number of components. *J. Roy. Statist. Soc., Ser. B* **59**, 731–792.

Sethuraman, J. (1994). A constructive definition of the Dirichlet prior. *Statistica Sinica* **4**, 639–650.

Sinha, D., Dey, D.K. (1997). Semiparametric Bayesian analysis of survival data. *J. Amer. Statist. Assoc.* **92**, 1195–1212.

Smith, M., Kohn, R. (1996). Nonparametric regression using Bayesian variable selection. *J. Econometrics* **75**, 317–343.

Spiegelhalter, D., Thomas, A., Best, N., Lunn, D. (2003). WinBUGS 1.4 User Manual. MRC Biostatistics Unit, Cambridge. www.mrc-bsu.cam.ac.uk/bugs/winbugs/contents.shtml.

Susarla, J., van Ryzin, J. (1976). Nonparametric Bayesian estimation of survival curves from incomplete observations. *J. Amer. Statist. Assoc.* **71**, 897–902.

Tierney, L. (1994). Markov chains for exploring posterior distributions. *Ann. Statist.* **22**, 1701–1762.

Titterington, D.M., Smith, A.F.M., Makov, U.E. (1985). *Statistical Analysis of Finite Mixture Distributions*. Wiley, New York.

Turnbull, B.W. (1974). Nonparametric estimation of a survivorship function with doubly censored data. *J. Amer. Statist. Soc.* **69**, 169–173.

Vidakovic, B. (1998). Wavelet-based nonparametric Bayes methods. In: Dey, D., Müller, P., Sinha, D. (Eds.), *Practical Nonparametric and Semiparametric Bayesian Statistics*. In: *Lecture Notes in Statistics*, vol. 133. Springer-Verlag, New York, pp. 133–155.

Walker, S., Damien, P. (1998). Sampling methods for Bayesian nonparametric inference involving stochastic processes. In: D. Dey, P. Mueller, D. Sinha, (Eds.), *Practical Nonparametric and Semiparametric Bayesian Statistics*, Springer Lecture Notes, pp. 243–254.

Walker, S.G., Mallick, B.K. (1997). Hierarchical generalized linear models and frailty models with Bayesian nonparametric mixing. *J. Roy. Statist. Soc., Ser. B* **59**, 845–860.

Walker, S.G., Mallick, B.K. (1999). Semiparametric accelerated life time model. *Biometrics* **55**, 477–483.

Walker, S.G., Damien, P., Laud, P., Smith, A.F.M. (1999). Bayesian nonparametric inference for random distributions and related functions (with discussion). *J. Roy. Statist. Soc., Ser. B* **61**, 485–527.

Wang, Y., Taylor, J.M.G. (2001). Jointly modeling longitudinal and event time data with application to acquired immunodeficiency syndrome. *J. Amer. Statist. Assoc.* **96**, 895–905.

West, M., Müller, P., Escobar, M.D. (1994). Hierarchical priors and mixture models with applications in regression and density estimation. In: Smith, A.F.M., Freeman, P.R. (Eds.), *Aspects of Uncertainty: A Tribute D. Lindley*. Wiley, London, pp. 363–386.

Wiper, M.P., Ríos Insua, D., Ruggeri, F. (2001). Mixtures of gamma distributions with applications. *J. Comput. Graph. Statist.* **10**, 440–454.

Handbook of Statistics, Vol. 25
ISSN: 0169-7161
© 2005 Elsevier B.V. All rights reserved.
DOI 10.1016/S0169-7161(05)25010-1

10

Some Bayesian Nonparametric Models

Paul Damien

1. Introduction

The aim of this article is to describe some popular Bayesian nonparametric models that have been studied extensively in the literature, and which have found some use in practice as well. The style of the paper is to highlight the salient features of such models with a view to implementing them.

At the outset it is noted that emphasis will be placed on the stochastic process approach to Bayesian nonparametrics, in particular the Dirichlet process. Tens of papers have been published using stochastic process formulations. This in no way implies other approaches are less important; copious references to alternative approaches will be provided in the last section. It is safe to say that, at the present time, the Dirichlet process is to Bayesian nonparametrics what the proportional hazards models is to biostatistics, or, more generally, what the normal distribution is to statistics. This is because, like the proportional hazards model and the normal distribution, the Dirichlet process has appealing theoretical properties and is remarkably easy to implement in a variety of contexts: in short, it works. Therefore, unless otherwise specified, henceforward the phrase "Bayesian nonparametrics" should be interpreted as "the stochastic process approach to Bayesian nonparametrics".

Also at the outset it should be said that nonparametric methods might not be viable in certain applications due to paucity of data; this is particularly true in some stylized biomedical contexts. However, in most contexts voluminous data are readily available in the Internet era, and hence nonparametric methods could very well be attempted, but keeping in mind that a parametric model might be adequate to resolve the problem under consideration.

Lastly at the outset, there are five papers fundamental to understanding and implementing Bayesian nonparametrics. Chronologically these are Lindley and Smith (1972), Ferguson (1973), Antoniak (1974), Gelfand and Smith (1990), and Escobar (1994). Ferguson developed the Dirichlet process; Antoniak developed mixtures of Dirichlet processes; Lindley and Smith developed hierarchical models, which, in conjunction with Antoniak's work, leads to Bayesian nonparametric hierarchical mixture models; Gelfand and Smith introduced the Gibbs sampler (due to Hastings, 1970 and Geman and

Geman, 1984) in Bayesian applications, and which is now the standard tool to implement Bayesian nonparametrics in virtually every context. Finally, Escobar was the first to implement the Dirichlet process using Markov chain Monte Carlo (MCMC) methods, setting the stage for all future applied work in this context. In fairness, it should be mentioned that Kuo (1986) developed an importance sampler for Dirichlet process models; however, this approach is not as efficient and general as the ideas described in Escobar.

1.1. Parametric Bayes

In parametric Bayesian inference, a class of distribution functions, $F_\theta(x)$ say, with density, $f_\theta(x)$ is indexed by a parameter $\theta \in \Theta$ where $\Theta \subseteq \Re^n$, for some finite $n \geqslant 1$. You then choose a prior distribution for the indexing parameter θ from a suitable class of prior distributions. Inferences about the random quantity θ follows from the posterior distribution of θ given the observed data. Formally, given an observation vector, x from $f_\theta(x)$, and a prior distribution $p(\theta)$, the posterior distribution of θ given x, by Bayes Theorem, is

$$p(\theta|x) = \frac{p(\theta) f_\theta(x)}{\int_{\theta \in \Theta} p(\theta) f_\theta(x)\, d\theta}. \tag{1}$$

1.2. Nonparametric Bayes

If the class of distribution functions cannot be indexed by a finite-dimensional parameter θ, parametric inference becomes nonparametric inference. While a distribution function on an n-dimensional space is used to define the prior in parametric inference, the idea now is to define a prior probability measure on the indexing set with an appropriate σ-algebra. The n-dimensional distribution function achieves this when the indexing parameter belongs to $\Re^n$.

Typically, although not exclusively, in the nonparametric case, the distribution of a stochastic process is specified which then indexes the family of distribution functions. Thus, the distribution of the process serves as a prior distribution over the indexed family. Given a sample path of a stochastic process, the observation is assumed to have the distribution function in the family indexed by that sample path. The posterior distribution of the stochastic process is the conditional distribution of the process given the observed data. Conceptually, this parallels, exactly, the parametric set-up. However, unlike the parametric framework, in this case, the simple density form in (1) is typically not available. One reason for this is that the posterior process is usually a complex mixture of processes in a high-dimensional space, thus rendering it quite difficult to obtain an analytical form for the distribution of the process.

Regardless of the mathematical sophistication involved, note that the word "nonparametrics" is somewhat of a misnomer. Even in the rigorous stochastic process formulation of the problem, it is clear that the parameter space is now a function space, where each element is an infinite-dimensional *parameter*. Of theoretical interest here is the general form of de Finetti's representation theorem that does not distinguish between parametric and nonparametric priors/models; see, for example, Bernardo and Smith (1994) who discuss this in some detail.

There are many approaches to nonparametric Bayes; see, for example, the collected papers in the volume edited by Dey et al. (1998). However, historically and currently the most popular approach to Bayesian nonparametric inference is to use the sample paths of a stochastic process to index families or subclasses of distribution functions. Then, given observations from some member in the indexed family, a posterior process is calculated. There are two main reasons for the popularity of the stochastic process approach: first, theoretical properties underlying stochastic process formulations are appealing and have been studied extensively; indeed the seminal paper by Ferguson (1973) on Dirichlet processes did set the research agenda for the last three decades. Second, many researchers have tackled the implementation aspects of stochastic process models extensively, leading to their widespread acceptance.

In recent years, Bayesian models are being widely used because MCMC methods have obviated the need for closed-form posterior distributions; you merely sample random variates from the target posterior distribution of interest, which could then be used to reconstruct useful features of the distribution, enabling inference; see Gelfand and Smith (1990). From an MCMC perspective, nonparametric procedures can be roughly described as following. You construct a (perhaps high-dimensional) grid on which the posterior process lives, and then simulate the process based on the properties of the particular process. The griding is intended purely to reconstruct the posterior process using MCMC. This in no way violates any aspect of the underlying process theory. Stated differently, formal stochastic process theory leads to identifying the posterior process. A finite-dimensional approximation to such a posterior process is required if one wishes to employ MCMC to generate random variates from the posterior. The notion of a grid-based approach to estimating the underlying distribution function implies that the "smoothness" of the estimates can be improved by choosing a highly refined grid; i.e., a very fine partition of the domain of the function. The trade-off is the capability of computing resources available to the practitioner, as well as the amount of data used in the estimation.

The discussion on Bayesian nonparametrics will be divided into three parts. First, models for random distribution functions; second, models for context-specific random functions, such as the hazard rate or the cumulative hazard rate; and third, another popular approach to Bayesian nonparametrics, namely, the Polya tree, will be described and exemplified.

2. Random distribution functions

Before proceeding with the mathematics, a brief overview of some relevant theoretical literature is provided.

Ferguson (1973) defines the Dirichlet process such that the support of the prior is the class of all distribution functions. In addition to having a large support, the Dirichlet process is also a conjugate prior; i.e., the posterior distribution of the process given the data is again a Dirichlet process. The posterior distribution of the process is then used by Ferguson to estimate the random distribution function and its moments under squared error loss.

Antoniak (1974) considers mixtures of Dirichlet processes within the context of bioassay and regression analysis and demonstrates that the posterior distribution of these mixtures is also a mixture of Dirichlet processes.

Doksum (1974) defines tailfree processes and processes neutral to the right. He succeeds in obtaining the Dirichlet process as a special case. Doksum also proves that if the prior is neutral to the right, then so is the posterior distribution. Susarla and Van Ryzin (1976), using Ferguson's Dirichlet process, model censored observations in survival analysis. Ferguson and Phadia (1979) combine the priors that are neutral to the right along with right censored data. They also show the connection between a survival and bioassay likelihood; thus, their results also apply to problems that arise in bioassay.

2.1. *The Dirichlet process*

Practitioners sometimes have a hard time understanding the Dirichlet process. And so, before delving into the Dirichlet Process, consider a simple example to motivate the general idea underlying this process. Also, the example naturally transits from parametrics to nonparametrics; see Wakefield and Walker (1997).

Suppose X is a random variable which takes the value 1 with probability p and the value 2 with probability $1 - p$. Uncertainty about the unknown distribution function F is equivalent to uncertainty about (p_1, p_2), where $p_1 = p$ and $p_2 = 1 - p$. A Bayesian would put a prior distribution over the two unknown probabilities p_1 and p_2. Of course here, essentially, there is only one unknown probability since p_1 and p_2 must sum to one. A convenient prior distribution is the Beta distribution given, up to proportionality, by:

$$f(p_1, p_2) \propto p_1^{\alpha_1 - 1} p_2^{\alpha_2 - 1},$$

where $p_1, p_2, \alpha_1, \alpha_2 \geqslant 0$, and $p_1 + p_2 = 1$. It is denoted $\text{Beta}(\alpha_1, \alpha_2)$. Different prior opinions can be expressed by different choices of α_1 and α_2. Set $\alpha_i = cq_i$ with $q_i \geqslant 0$ and $q_1 + q_2 = 1$. Then,

$$E[p_i] = q_i \quad \text{and} \quad \text{var}(p_i) = \frac{q_i(1 - q_i)}{c + 1}. \tag{2}$$

If $q_i = 0.5$ and $c = 2$, a noninformative prior results. Denote the prior guess (q_1, q_2) by F_0. The interpretation is that the q_i center the prior and c reflects your degree of belief in the prior: a large value of c implies a small variance, and hence strong prior beliefs.

The Beta prior is convenient in the example above. Why? Suppose you obtain a random sample of size n from the distribution F. This is a binomial experiment with the value $X = 1$ occurring n_1 times (say) and the value $X = 2$ occurring n_2 times, where $n_2 = n - n_1$. The posterior distribution of (p_1, p_2) is once again a Beta distribution with parameters updated to $\text{Beta}(\alpha_1 + n_1, \alpha_2 + n_2)$. Since the posterior distribution belongs to the same family as the prior distribution, namely, the Beta distribution, such a prior to posterior analysis is called a *conjugate update*, with the prior being referred to as a *conjugate prior*. The above example is called the Beta–Binomial model for p.

Next, generalize the conjugate Beta–Binomial model to the conjugate Multinomial–Dirichlet model. Now the random variable X can take the value X_i with probability p_i,

$i = 1, \ldots, K$, with $p_i \geqslant 0$, and $\sum_{i=1}^{K} p_i = 1$. So now uncertainty about the unknown distribution function F is equivalent to uncertainty about $p = (p_1, \ldots, p_K)$. The conjugate prior distribution in this case is the Dirichlet *distribution* (not to be confused with the Dirichlet *Process*), given, up to proportionality, by:

$$f(p_1, \ldots, p_K) \propto p_1^{\alpha_1 - 1} \times \cdots \times p_K^{\alpha_K - 1},$$

where $\alpha_i, p_i \geqslant 0$ and $\sum_{i=1}^{K} p_i = 1$. Setting $\alpha_i = cq_i$ results in the same interpretation of the prior, and, in particular, the mean and variance are again given by Eq. (2). As before, $(q_1, \ldots, q_K)$ represents your prior guess (F_0) and c the certainty in this guess. A random sample from F now constitutes a Multinomial experiment and when this likelihood is combined with the Dirichlet prior, the posterior distribution for p is once again a Dirichlet distribution with parameters $\alpha_i + n_i$, where n_i is the number of observations in the ith category.

Now transit from a discrete X to a continuous X by letting $K \to \infty$ in the above discussion. In traditional parametric Bayesian analysis, the distribution of $X(F)$ would be assumed to belong to a particular family of probability density functions. For example, if X can take on any real value the family of distributions is often assumed to be the Normal distribution, denoted $N(\mu, \sigma^2)$. A Bayesian analysis would then proceed by first placing a prior distribution on μ and σ^2, and then obtaining the resultant posterior distributions of these two *finite-dimensional* parameters.

The realm of Bayesian nonparametrics is entered when F (in the last paragraph) itself is treated as a random quantity; that is, you must now assign a prior distribution to F. Since F is infinite-dimensional, a stochastic process whose sample paths index the entire space of distribution functions is required.

Now for the Dirichlet Process, but first some notation. A *partition* $B_1, \ldots, B_k$ of the sample space Ω is such that $\bigcup_{i=1}^{K} B_i = \Omega$ and $B_i \cap B_j = \emptyset$ for all $i \neq j$; that is, a group of sets that are disjoint and exhaustive, covering the whole sample space.

DEFINITION 1. A Dirichlet Process prior with parameter α generates random probability distributions F such that for every $k = 1, 2, 3, \ldots$ and partition $B_1, \ldots, B_k$ of Ω, the distribution of $(F(B_1), \ldots, F(B_k))$ is the Dirichlet distribution with parameter $(\alpha(B_1), \ldots, \alpha(B_k))$. Here α is a finite measure on Ω and so we can put $\alpha(\cdot) = cF_0(\cdot)$, where $c > 0$ and F_0 is a probability distribution on Ω.

EXAMPLE 1. Consider a random variable X with distribution function F, defined on the real line. Now consider the probability $p = \mathrm{pr}(X < a)$ and specify a Dirichlet process prior with parameter α for F. Put $B_1 = (-\infty, a]$ and $B_2 = (a, \infty)$. Then from the above definition, *a priori*, p has a Beta distribution with parameters $\alpha_1 = \alpha(B_1) = cF_0(a)$ and $\alpha_2 = \alpha(B_2) = c(1 - F_0(a))$, where $c = \alpha(-\infty, \infty)$. This prior is such that $E(p) = F_0(a)$ and $\mathrm{var}(p) = F_0(a)(1 - F_0(a))/(c + 1)$, thus showing the link to Eq. (2). The variance of p and hence the level of fluctuation of p about $F_0(a)$ depends on c. In particular, large values of c implies strong belief in $F_0(a)$ and $\mathrm{var}(p)$ is small. Note that this is true for *all* partitions B_1, B_2 of the real line, or, equivalently, all values of a.

Next consider observations $X_1, \ldots, X_n$ from F. Let F be assigned a Dirichlet process prior, denoted Dir(c, F_0). Then Ferguson showed that the posterior process F has parameters given by

$$c + n \quad \text{and} \quad \frac{cF_0 + nF_n}{c + n}.$$

F_n is the empirical distribution function for the data; i.e., the step function with jumps of $1/n$ at each X_i. The classical maximum likelihood estimator is given by F_n. The posterior mean of $F(A)$ for any set $A \in (-\infty, \infty)$ is given by

$$E\big[F(A)|\text{data}\big] = p_n F_0(A) + (1 - p_n) F_n(A),$$

where $p_n = c/(c + n)$. Thus $c = 1$ suggests a weight of $1/(n + 1)$ on the prior F_0 and $n/(n + 1)$ on the "data" F_n. As c increases, the posterior mean is influenced more by the prior, that is, you have more faith in your prior choice, F_0.

3. Mixtures of Dirichlet processes

One limitation of the Dirichlet process is that it indexes the class of discrete (random) distributions with probability one. While this is not such a glaring difficulty, it would be fruitful to extend the indexing to include continuous distributions. A mixture of Dirichlet processes (MDPs) achieves this. In addition, MDPs are also valuable from a modeling perspective.

Mixture models are essentially hierarchical models, and dates back to Lindley and Smith (1972) who consider parametric mixtures. As an example, consider the following parametric set-up, where N and IG denote the Normal and Inverse-Gamma distributions, respectively, and Y_i are the observed data.

$$\begin{aligned} &Y_i|\theta_i \sim F(\cdot), \\ &\theta_i|\lambda \sim G(\cdot), \\ &\lambda \sim H_\lambda(\cdot). \end{aligned} \tag{3}$$

Set $F = N(\theta_i, \sigma^2)$, and define $\lambda = (\mu, \tau^2)$, thus $G = N(\mu, \sigma^2)$. For a conjugate hierarchical model, you would take $H = N(\mu_0, \sigma_0^2) \times IG(\alpha_0, \beta_0)$. The fixed hyperparameters, depicted with the subscript 0, can be chosen to reflect less intrusive prior beliefs, a so-called noninformative choice.

Extending (3) to include a nonparametric component using the Dirichlet process is easy. A hierarchical mixture of Dirichlet processes (Antoniak, 1974; Escobar, 1995) is given by:

$$\begin{aligned} &Y_i|\theta_i \sim F(\cdot), \\ &\theta_i|G \sim G, \\ &G|c, \lambda \sim \mathit{Dirichlet}\big(c, G_0(\lambda)\big), \\ &\lambda \sim H_\lambda(\cdot), \\ &c \sim H_c(\cdot). \end{aligned} \tag{4}$$

The set of equations labeled (4) is also called a "mixture of Dirichlet processes (MDPs)" because G is a discrete random distribution (taken from the Dirichlet process) conditional on which the mixing parameters, $\{\theta_i\}$, are independent and identically distributed (i.i.d.).

The parameters of the Dirichlet process are the distribution G_0 (also referred to as the base measure) that centers the process, in that $E(G) = G_0$, and $c > 0$, which loosely represents your strength of belief in G_0 via a prior sample size; note also that the marginal prior distribution of θ_i is G_0.

If the sampling distribution of Y_i forms a conjugate pair with the base measure G_0 you obtain what is called a *conjugate MDP*, else you obtain a *nonconjugate MDP*. The distinction is critical from a computational perspective with the obvious deduction that nonconjugate MDPs are generally more difficult to implement.

It is immediately clear from the MDP model that the nonparametric component, G, provides additional flexibility in the modeling process since it is random. Whereas in the hierarchical parametric mixture model the θ_i are drawn from a fixed class of (normal) distributions indexed by λ, under the Dirichlet process mixture model the θ_is could be drawn from distributions that exhibit skewness, higher degrees of kurtosis than the normal, multimodality, etc. The user only needs to specify a distribution for the prior shape of G, G_0. This, of course, could be taken to be any well-known distribution such as the normal, beta, and so on, keeping in mind conjugate versus nonconjugate choices. Likewise, uncertainty regarding the scale parameter, c, of the Dirichlet process could also be modeled by assigning it a prior distribution; thus, for example, in (4), you could set $H_c(\cdot)$ to be a gamma distribution. Once again, as in the parametric model, less intrusive prior beliefs can be entertained by an appropriate choice of the hyperparameters which define $H_\lambda(\cdot)$ and $H_c(\cdot)$: of course, the least intrusive prior is the one based solely on your subjective judgments about the observed data, resulting in your choice of an appropriate likelihood function, leading to the limiting special case of likelihood-based inference.

A key point to note in the MDP model is the use of the Dirichlet process in the hierarchy. Earlier an analogy between the Dirichlet process and the proportional hazards model and the normal distribution was made. It is easy to see why. If instead of the Dirichlet process some other stochastic process prior is imposed in the hierarchy, implementation tends to get nasty very quickly. Also, note that the Dirichlet process mixture-modeling framework provides an incredibly large class of models, which should prove quite adequate in a vast range of applied problems. It is for these reasons that statistical and interdisciplinary literature consist of a plethora of papers which employ some variant of the MDP model.

EXAMPLE 2 (*Poisson–Gamma MDP*). Let Y_i be distributed *Poisson*(θ_i). Set G_0 to be *gamma*(a, b). A parametric conjugate Bayesian analysis results in the posterior distribution for $\theta_i | Y_i, a, b$ equaling *gamma*$(a + Y_i, b + 1)$. A conjugate MDP model is given by:

$$Y_i|\theta_i \sim Poisson(Y_i|\theta_i),$$
$$\theta_i|G \sim G(\theta_i),$$

$$
\begin{aligned}
G|c,a,b &\sim \mathit{Dirichlet}\big(c, G_0(a,b)\big),\\
G_0 &\sim \mathit{gamma}(a,b),\\
c &\sim \mathit{gamma}(e,f).
\end{aligned}
\tag{5}
$$

See Escobar and West (1998) for the Gibbs sampling details pertaining to the MDP model.

EXAMPLE 3 (*Beta–Binomial MDP*). Let Y_i be distributed $\mathit{Binomial}(x_i, \theta_i)$. Set G_0 to be $\mathit{beta}(a,b)$. A conjugate MDP model is given by:

$$
\begin{aligned}
Y_i|\theta_i &\sim \mathit{Binomial}(x_i,\theta_i),\\
\theta_i|G &\sim G(\theta_i),\\
G|c,a,b &\sim \mathit{Dirichlet}\big(c, G_0(a,b)\big),\\
G_0 &\sim \mathit{beta}(a,b),\\
\frac{a}{a+b} &\sim \mathit{beta}(e_0, f_0),\\
c &\sim \mathit{lognormal}\big(\mu_0, \sigma_0^2\big).
\end{aligned}
\tag{6}
$$

MacEachern (1998) provides Gibbs sampling details pertaining to this MDP model.

EXAMPLE 4 (*Normal linear random effects model*). For subject i with n_i repeated measurements, the Normal linear random effects model for observed data y_i is given by:

$$
y_i = X_i\beta + Z_i b_i + e_i, \quad i = 1,\ldots,N, \tag{7}
$$

where X_i (β_i) and Z_i (b_i) correspond to fixed effects covariates (regression coefficients) and random effects covariates (regression coefficients), respectively. Assume e_i and b_i are independent and normally distributed with $e_i \sim N(0, \sigma^2 I)$ and $b_i \sim N(0, D)$, where I is the identity matrix of appropriate dimension. Then a parametric normal linear random effects model is given by:

$$
y_i|\beta, b_i \sim N\big(X_i\beta + Z_i b_i, \sigma^2 I\big). \tag{8}
$$

Relaxing the normality assumption on the b_i by assigning it a random distribution with a Normal base measure results in a conjugate MDP model given as following, with $\tau = \sigma^{-2}$.

$$
\begin{aligned}
y_i|\beta, b_i &\sim N\big(X_i\beta + Z_i b_i, \sigma^2 I\big),\\
\tau &\sim \mathit{Gamma}(a/2, b/2),\\
\beta &\sim N(\mu_0, \Sigma_0),\\
b_i &\sim G,\\
G &\sim \mathit{Dirichlet}\big(c, N(0,D)\big).
\end{aligned}
\tag{9}
$$

As in the Poisson–Gamma example, you could specify a prior for c and D (the variance–covariance matrix). Typically, D is assigned a Wishart prior.

If the distribution of the y_is is taken to be a member of the exponential family other than the Normal, you obtain the family of parametric generalized linear mixed models. In this instance, a nonconjugate MDP is obtained by leaving the baseline measure to be Normal. Kleinman and Ibrahim (1998a, 1998b) provide details on the implementation aspects of these models using the Gibbs sampler.

4. Random variate generation for NTR processes

In Bayesian nonparametric inference problems, density representations are typically not available and the potential for implementation by simulation methods is not so obvious; see, for example, Bondesson (1982), Damien et al. (1995), and Walker and Damien (1998a, 1998b). Focus in this paper, thus far, has been on the Dirichlet process, which is also more generally referred to as a *neutral to the right process* (NTR). (The abbreviation NTR is somewhat odd in that NTTR seems more appropriate. But the former convention has stuck and is continued here.) The aim of this section is to provide a canonical scheme to sample any NTR process. It should be noted that no claim is made here about the efficiency of this scheme compared to other sampling approaches to the problem. The key point is the sampling scheme works quite nicely in practice.

4.1. Neutral to the right processes

The goal of this section is to conclude with a theorem, which, from the perspective of random variate generation, is simply this: any NTR process can be broken up into two components: a *jump component* that is straightforward to sample since its distribution is usually amenable to standard random variate generation; and a *continuous component* that is difficult to sample since its distribution is an infinitely divisible law with unknown density representation. Once this general characterization of a posterior NTR process is achieved, a procedure for sampling the posterior process is developed in the next sub-subsection.

DEFINITION 2 (Ferguson, 1974). A random distribution function $F(t)$ on $(0, \infty)$ is neutral to the right (NTR) if for every m and $0 < t_1 < \cdots < t_m$ there exist independent random variables $V_1, \ldots, V_m$ such that $\mathcal{L}(1 - F(t_1), \ldots, 1 - F(t_m)) = \mathcal{L}(V_1, V_1V_2, \ldots, \prod_{j=1}^{m} V_j)$.

If F is NTR then $Z(t) = -\log(1 - F(t))$ has independent increments. The converse is also true: if $Z(t)$ is an independent increments process, nondecreasing almost surely (a.s.), right continuous a.s., $Z(0) = 0$ a.s. and $\lim_{t\to\infty} Z(t) = +\infty$ a.s. then $F(t) = 1 - \exp(-Z(t))$ is NTR.

For the process $Z(\cdot)$ there exist at most countably many fixed points of discontinuity at $t_1, t_2, \ldots$ with jumps $J_1, J_2, \ldots$ (independent) having densities $f_{t_1}, f_{t_2}, \ldots$.

Then $Z_c(t) = Z(t) - \sum_{t_j \leqslant t} J_j$ has no fixed points of discontinuity and therefore has a Lévy representation with log-Laplace transform given by

$$-\log E\exp\bigl(-\phi Z_c(t)\bigr) = \int_0^\infty \bigl(1-\exp(-\phi z)\bigr)\,\mathrm{d}N_t(z),$$

where $N_t(\cdot)$, a Lévy measure, satisfies: $N_t(B)$ is nondecreasing and continuous for every Borel set B; $N_t(\cdot)$ is a measure on $(0,\infty)$ for each t and $\int_{(0,\infty)} z(1+z)^{-1}\,\mathrm{d}N_t(z) < \infty$. In short, $F(t) = 1 - \exp[-Z(t)]$, with Z an independent increments (Lévy) process.

THEOREM 1 (Ferguson, 1974). *If F is NTR and $X_1, \ldots, X_n$ is a sample from F, including the possibility of right censored samples (where X_i represents the censoring time if applicable), then the posterior distribution of F given $X_1, \ldots, X_n$ is NTR.*

4.2. Specifying prior distributions

The prior distribution for $Z(\cdot)$ is characterized by $[M = \{t_1, t_2, \ldots\}, \{f_{t_1}, f_{t_2}, \ldots\}]$, the set of fixed points of discontinuity with corresponding densities for the jumps, and $N_t(\cdot)$, the Lévy measure for the part of the process without fixed points of discontinuity.

In the following, assume the Lévy measure to be of the type $\mathrm{d}N_t(z) = \mathrm{d}z \int_{(0,t]} K(z,s)\,\mathrm{d}s$; i.e., a beta-Stacy process (a generalization of the Dirichlet process) with parameters $\alpha(\cdot)$ and $\beta(\cdot)$ when

$$K(z,s)\,\mathrm{d}s = \bigl(1-\mathrm{e}^{-z}\bigr)^{-1}\exp\bigl[-z\beta(s)\bigr]\,\mathrm{d}\alpha(s);$$

see Walker and Muliere (1997a). The Dirichlet process is obtained when α is a finite measure and $\beta(s) = \alpha(s,\infty)$. Also, the simple homogeneous process of Ferguson and Phadia (1979) results when β is constant.

The beta-Stacy process is used as the prior process due to the following facts.

Fact 1. The Dirichlet process is not conjugate with respect to right censored data. The beta-Stacy process is conjugate with respect to right censored data. If the prior is a Dirichlet process, then the posterior, given censored data, is a beta-Stacy process.

Fact 2. After a suitable transformation and selection of K, other processes such as the extended gamma (Dykstra and Laud, 1981) and the beta process (Hjort, 1990) can be obtained; these processes will be discussed later.

Walker and Damien (1998b) provide a way of specifying the mean and variance of the distribution function based on the beta-Stacy process. This method has the merit that the practitioner can model the prior mean and variance via a Bayesian parametric model. Let

$$\mu(t) = -\log\bigl\{ES(t)\bigr\} = \int_0^\infty \bigl(1-\mathrm{e}^{-v}\bigr)\,\mathrm{d}N_t(v)$$

and

$$\lambda(t) = -\log\{E[S^2(t)]\} = \int_0^\infty (1 - e^{-2v})\, dN_t(v),$$

where $S(t) = 1 - F(t)$. Recalling the Lévy measure for the beta-Stacy process, and assuming for simplicity that there are no fixed points of discontinuity in the prior process, Walker and Damien (1998b) show that there exist $\alpha(\cdot)$ and $\beta(\cdot)$ that provide an explicit solution satisfying the above two conditions for arbitrary μ and λ, satisfying $\mu < \lambda < 2\mu$, which corresponds to $[ES]^2 < E[S^2] < ES$. These are given by $d\alpha(t) = d\mu(t)\beta(t)$ and $d\lambda(t)/d\mu(t) = 2 - (1 + \beta(t))^{-1}$.

Suppose, for example, you wish to center, up to and including second moments, the nonparametric model on the parametric Bayesian model given by $S(t) = \exp(-at)$ with $a \sim gamma(p, q)$. Then $\mu(t) = p\log(1+t/q)$ and $\lambda(t) = p\log(1+2t/q)$ giving

$$\beta(t) = q/(2t) \quad \text{and} \quad d\alpha(t) = pq\, dt/[2t(q+t)].$$

In the absence of strong prior information, this provides a flexible form of prior specification. You can specify a p and q to reflect beliefs concerning the "likely" position of S; that is, a region of high probability in which S is thought most likely to be. The unrestricted nature of the prior will then allow S to "find" its correct shape within this specified region, given sufficient data.

4.3. The posterior distributions

The posterior distribution is now given for a single observation X. The case for n observations can then be obtained by repeated application.

THEOREM 2 (Ferguson, 1974; Ferguson and Phadia, 1979). *Let F be NTR and let X be a random sample from F.*

(i) *Given $X > x$ the posterior parameters (denoted by a^*) are*

$$M^* = M,$$
$$f^*_{t_j}(z) = \begin{cases} c \cdot e^{-z} f_{t_j}(z) & \text{if } t_j \leqslant x, \\ f_{t_j}(z) & \text{if } t_j > x \end{cases}$$

and $K^*(z, s) = \exp[-zI(x \geqslant s)]K(z, s)$ (here c is the normalizing constant).

(ii) *Given $X = x \in M$ the posterior parameters are*

$$M^* = M,$$
$$f^*_{t_j}(z) = \begin{cases} c \cdot e^{-z} f_{t_j}(z) & \text{if } t_j < x, \\ c \cdot (1 - e^{-z}) f_{t_j}(z) & \text{if } t_j = x, \\ f_{t_j}(z) & \text{if } t_j > x \end{cases}$$

and, again, $K^*(z, s) = \exp[-zI(x \geqslant s)]K(z, s)$.

(iii) *Given* $X = x \notin M$ *the posterior parameters are* $M^* = M \cup \{x\}$, *with* $f_x(z) = c \cdot (1 - \mathrm{e}^{-z})K(z, x)$,

$$f^*_{t_j}(z) = \begin{cases} c \cdot \mathrm{e}^{-z} f_{t_j}(z) & \text{if } t_j < x, \\ f_{t_j}(z) & \text{if } t_j > x \end{cases}$$

and, again, $K^*(z, s) = \exp[-zI(x \geqslant s)]K(z, s)$.

4.4. Simulating the posterior process

Focus is now on simulating from the posterior distribution of $F[a, b)$. Since $F(t) = 1 - \exp[-Z(t)]$, you have

$$F[a, b) = \exp\bigl[-Z(0, a)\bigr]\bigl[1 - \exp\bigl(-Z[a, b)\bigr)\bigr],$$

and note that $Z(0, a)$ and $Z[a, b)$ are independent. Therefore you only need to consider sampling random variables of the type $Z[a, b)$. This will involve sampling from densities $\{f^*_{t_j}\colon t_j \in [a, b)\}$, corresponding to the jumps in the process, and secondly, the continuous component, the random variable $Z^*_c[a, b)$. The type of densities corresponding to points in M^*, the jumps, are defined, up to a constant of proportionality, by

$$f^*(z) \propto \bigl(1 - \exp(-z)\bigr)^{\lambda} \exp(-\mu z) f(z), \quad \lambda, \mu \geqslant 0 \text{ (integers)}, \tag{10}$$

and

$$f^*(z) \propto \bigl(1 - \exp(-z)\bigr)^{\lambda} K(z, x), \quad \lambda > 0 \text{ (integer)}, \tag{11}$$

where the subscripts t_j have been omitted.

4.4.1. Simulating the jump component

The types of densities for the random variable corresponding to the jump in the process given in Eq. (11) will most likely be a member of the Smith–Dykstra (SD) family of densities (Damien et al., 1995) and algorithms for sampling from such densities are provided in that paper. For the types of densities appearing in Eq. (10), you can proceed via the introduction of latent variables. Define the joint density of z, $u = (u_1, \ldots, u_\lambda)$, and v by

$$f(z, u, v) \propto \mathrm{e}^{-\mu v} I(v > z) \left\{ \prod_{l=1}^{\lambda} \mathrm{e}^{-u_l} I(u_l < z) \right\} f(z).$$

Clearly the marginal density of z is as required. The full conditional densities for implementing Gibbs sampling (Gelfand and Smith, 1990) are given by

$$f(u_l | u_{-l}, v, z) \propto \mathrm{e}^{-u_l} I(u_l < z), \quad l = 1, \ldots, \lambda,$$
$$f(v | u, z) \propto \mathrm{e}^{-\mu v} I(v > z)$$

and

$$f(z | u, v) \propto f(z) I\Bigl(z \in \bigl(\max_l \{u_l\}, v\bigr)\Bigr).$$

Provided it is possible to sample from $f(z)$, this algorithm is easy to implement; see, for example, Damien et al. (1999).

4.4.2. Simulating the continuous component

The random variable $Z = Z_c^*[a, b)$, corresponding to the continuous component, is *infinitely divisible* (id) with log-Laplace transform given by

$$-\log E\exp(-\phi Z) = \int_0^\infty \bigl(1-\exp(-\phi z)\bigr)\int_a^b \exp\Bigl(-z\sum_{i=1}^n I(X_i \geqslant s)\Bigr) K(z,s)\,\mathrm{d}s\,\mathrm{d}z.$$

Damien et al. (1995) introduced an algorithm for sampling from id laws, which relies on the characterization of an id random variable as the limit of sequences of sums of Poisson types. Here, based on the fact that $\mathcal{L}Z = \mathcal{L}\int_0^\infty z\,\mathrm{d}P(z)$, where $P(\cdot)$ is a Poisson process with intensity measure

$$\mathrm{d}z \int_{[a,b)} K^*(z,s)\,\mathrm{d}s,$$

an alternative approach to simulating an id distribution is given. This method is an improvement to the Damien et al. method.

Simulating an id distribution. Consider sampling an infinitely divisible variable Z with log-Laplace transform given by

$$\log E\exp(-\phi Z) = \int_0^\infty \bigl(\mathrm{e}^{-z\phi}-1\bigr)\,\mathrm{d}N_\Delta(z),$$

where

$$\mathrm{d}N_\Delta(z) = \frac{\mathrm{d}z}{1-\mathrm{e}^{-z}}\int_\Delta \exp\bigl(-z\beta(s)\bigr)\,\mathrm{d}\alpha(s).$$

Recall $\mathcal{L}Z = \mathcal{L}(\int_{(0,\infty)} z\,\mathrm{d}P(z))$ where $P(\cdot)$ is a Poisson process with intensity measure $\mathrm{d}N_\Delta(\cdot)$. For a fixed $\varepsilon > 0$, let $Z = X_\varepsilon + Y_\varepsilon$ (X_ε, Y_ε independent) where X_ε is the sum of the jumps of $P(\cdot)$ in (ε,∞) and Y_ε is the sum of the jumps in $(0,\varepsilon]$; note that these are the sizes of the jumps in the Lévy process bigger and smaller than ε, respectively, with log-Laplace transform given by $\int_{(0,\varepsilon]}[\exp(-\phi z)-1]\,\mathrm{d}N_\Delta(z)$.

The aim is to approximate a random draw Z by sampling an X_ε. The error variable Y_ε has mean $EY_\varepsilon = \int_{(0,\varepsilon]} z\,\mathrm{d}N_\Delta(z)$ and variance $\operatorname{var} Y_\varepsilon = \int_{(0,\varepsilon]} z^2\,\mathrm{d}N_\Delta(z)$. ε can be taken as small as one wishes and it is not difficult to show that $EY_\varepsilon < \varepsilon$ and $\operatorname{var} Y_\varepsilon < \varepsilon^2/2$ for very small ε. Let $\lambda_\varepsilon = N_\Delta(\varepsilon,\infty) < \infty$ and take $\nu \sim \mathit{Poisson}(\lambda_\varepsilon)$. Here ν is the (random) number of jumps in (ε,∞) and therefore the jumps $\tau_1,\dots,\tau_\nu$ are taken i.i.d. from $G_\varepsilon(s) = \lambda_\varepsilon^{-1}N_\Delta(\varepsilon,s]$, for $s\in(\varepsilon,\infty)$. Then X_ε is given by $X_\varepsilon = \sum_{k=1}^\nu \tau_k$.

To implement the algorithm it is required to:

(A): simulate from $G_\varepsilon(\cdot)$;
(B): calculate λ_ε.

A: simulating from $G_\varepsilon(\cdot)$. For the beta-Stacy process the density corresponding to G_ε is given, up to a constant of proportionality, by

$$g_\varepsilon(z) \propto \left[1 - \exp(-z)\right]^{-1} I(z > \varepsilon) \int_\Delta \exp\bigl(-z\beta(s)\bigr)\, \mathrm{d}\alpha(s).$$

To sample from g_ε define the joint density function of z, s and u by

$$f(z, s, u) \propto I\bigl(u < \left[1 - \exp(-z)\right]^{-1}\bigr) \exp\bigl(-z\beta(s)\bigr)\alpha(s) I(z > \varepsilon, s \in \Delta),$$

where u is a random variable defined on the interval $(0, [1 - \exp(-\varepsilon)]^{-1})$, s is a random variable defined on $(0, \infty)$ and $\alpha(s)\,\mathrm{d}s = \mathrm{d}\alpha(s)$. Clearly the marginal distribution for z is given by g_ε. Consider now a hybrid Gibbs/Metropolis sampling algorithm for obtaining random variates $\{z^{(l)}\}$ from g_ε. The algorithm involves simulating a Markov chain $\{z^{(l)}, s^{(l)}, u^{(l)}\}$ such that as $l \to \infty$ then $z^{(l)}$ can be taken as a random draw from $g_\varepsilon(z)$. The algorithm is given, with starting values $\{z^{(1)}, s^{(1)}, u^{(1)}\}$, by the following.

(i) Take $z^{(l+1)}$ from the exponential distribution with mean value $1/\beta(s^{(l)})$ restricted to the interval $(\varepsilon, -\log[1 - 1/u^{(l)}])$ if $u^{(l)} > 1$ and the interval (ε, ∞) if $u^{(l)} \leqslant 1$.
(ii) Take $u^{(l+1)}$ from the uniform distribution on the interval

$$\bigl(0, \left[1 - \exp\bigl(-z^{(l+1)}\bigr)\right]^{-1}\bigr)$$

and
(iii) for example, take s^* from $\alpha(\cdot)/\alpha(\Delta)$ on Δ and ξ from the uniform distribution on the interval $(0, 1)$. If

$$\xi < \{\min\}\bigl\{1, \exp\bigl(-z^{(l+1)}\bigl[\beta(s^*) - \beta\bigl(s^{(l)}\bigr)\bigr]\bigr)\bigr\}$$

then $s^{(l+1)} = s^*$ else $s^{(l+1)} = s^{(l)}$.

B: Calculating λ_ε Define the joint density function of z and s by

$$f(z, s) \propto \exp\bigl(-z\beta(s)\bigr)\alpha(s) I(z > \varepsilon, s \in \Delta)$$

and let the right-hand side of this expression be $h(z, s)$. Then

$$\lambda_\varepsilon = \int_0^\infty \int_0^\infty \left[1 - \exp(-z)\right]^{-1} h(z, s)\, \mathrm{d}z\, \mathrm{d}s,$$

so

$$\lambda_\varepsilon = E_f\bigl\{\left[1 - \exp(-z)\right]^{-1}\bigr\} \times \int_0^\infty \int_0^\infty h(z, s)\, \mathrm{d}z\, \mathrm{d}s.$$

Now $E_f\{[1-\exp(-z)]^{-1}\}$ can be obtained from $L^{-1}\sum_{l=1}^{L}[1-\exp(-z^{(l)})]^{-1}$ where $\{z^{(l)}: l=1,2,\ldots\}$ are obtained via a Gibbs sampling algorithm which involves sampling from $f(z|s)$, the exponential distribution with mean value $1/\beta(s)$ restricted to the interval (ε,∞), and sampling from $f(s|z)$, done as in (iii) under (A).

Finally,

$$\int_0^\infty\int_0^\infty h(z,s)\,\mathrm{d}z\,\mathrm{d}s = \int_\Delta \exp\bigl(-\varepsilon\beta(s)\bigr)\,\mathrm{d}\alpha(s)/\beta(s),$$

and it should be adequate to take this to be $\int_\Delta \mathrm{d}\alpha(s)/\beta(s) - \varepsilon\alpha(\Delta)$. The error involved in the algorithm is in the taking of X_ε for X and is contained in the variable Y_ε. Now

$$EY_\varepsilon = \int_0^\varepsilon\int_\Delta z\bigl[1-\exp(-z)\bigr]^{-1}\exp\bigl(-z\beta(s)\bigr)\,\mathrm{d}\alpha(s)\,\mathrm{d}z,$$

which can easily be shown to be less than ε. The variance of Y_ε can also be shown to be less than $\varepsilon^2/2$.

Sampling MDP models has been discussed extensively in the literature, and is omitted here; see, as examples, Escobar (1994), MacEachern (1998), MacEachern and Mueller (1998), Walker and Damien (1998a, 1998b).

5. Sub-classes of random distribution functions

The Dirichlet process is a prior on the space of distribution functions. However, in many applications such as biomedicine, context-specific knowledge might be available about some other aspect of the data generating process, such as, for instance, the hazard rate or cumulative hazard rate. In this regard, Kalbfleisch (1978) and Clayton (1991) provide examples of the use of gamma process priors in the context of Cox regressions. Hjort (1990) developed the beta process to model a cumulative hazard function. Simulation algorithms for carrying out prior to posterior analysis for the beta process appear in Damien et al. (1996). Dykstra and Laud (1981) consider modeling monotone hazard rates nonparametrically by developing a class of processes called the extended-gamma process. Laud et al. (1996) developed simulation methods for the extended-gamma process. Ammann (1984) enhanced the extended-gamma process to model bathtub hazard rates. Arjas and Gasbarra (1994) developed stylized processes to model the hazard rate piecewise.

Since biostatistical research is popular these days, it would be timely to revisit the well-known Cox regression model using nonparametric priors on the cumulative hazard and hazard rate functions. See Sinha and Dey (1997), and the many references therein, for a comprehensive review of Bayesian semiparametric survival analysis.

5.1. The Beta process

The NTR processes considered earlier sought to place a prior on the space of CDFs. Hjort (1990) develops Bayesian theory by placing a prior on the space of cumulative hazard functions (CHFs). He constructs prior distributions that correspond to cumulative hazard rate processes with nonnegative independent increments. A particular class of processes, termed Beta processes, is introduced and is shown to constitute a conjugate class. Here only the time-continuous version of the Beta process is considered; see Sinha (1997) for a discussion of the discrete version of the Beta process in survival analysis under interval censoring.

Let T be a random variable with CDF $F(t) = \Pr(T \leqslant t)$ on $[0, \infty)$ and $F(0) = 0$. The cumulative hazard function (CHF) for F or T is a nonnegative, nondecreasing, right continuous function A on $[0, \infty)$ satisfying

$$\mathrm{d}A(s) = A[s, s + \mathrm{d}s) = \Pr\{T \in [s, s + \mathrm{d}s) | T \geqslant s\} = \mathrm{d}F(s)/F[s, \infty).$$

Define

$$A[a, b) = \int_{[a,b)} \frac{\mathrm{d}F(s)}{F[s, \infty)},$$

and require

$$F[a, b) = \int_{[a,b)} F[s, \infty)\,\mathrm{d}A(s), \quad 0 \leqslant a \leqslant b < \infty.$$

Using product integral techniques (Gill and Johansen, **?**), Hjort shows that F is restorable from A as

$$F(t) = 1 - \prod_{[0,t]} \{1 - \mathrm{d}A(s)\}, \quad t \geqslant 0.$$

Hjort constructs prior distributions for A as follows. Let $\boldsymbol{F}$ be the set of all CDFs F on $[0, \infty)$ having $F(0) = 0$ and let $\boldsymbol{B}$ be the set of all nondecreasing, right continuous functions B on $[0, \infty)$ having $B(0) = 0$. Define the space of cumulative hazard rates as

$$\boldsymbol{A} = \{A \in \boldsymbol{B}\ F \in \boldsymbol{F}\}.$$

The idea is to place a probability distribution on $(\boldsymbol{A}, \Sigma_{\boldsymbol{A}})$, where $\Sigma_{\boldsymbol{A}}$ is the σ-algebra generated by the Borel cylinder sets. Such a probability distribution, say $\boldsymbol{P}$, is determined if the distribution of every finite set of increments $A[a_{j-1}, a_j)$ is specified, provided the Kolmogorov consistency condition is satisfied.

As in the construction of priors on spaces of CDFs, it is natural to consider nonnegative, nondecreasing processes on $[0, \infty)$ that start at zero and have independent increments. These processes are the Lévy processes in this context. Note that if F is Dirichlet, then $-\log(1 - F)$ is Lévy. Moreover, if B is a Lévy process, then given possibly censored data, the posterior B process is still Lévy. Hjort considers $A = -\log(1-F)$ rather than B because

- A is easier to generalize to different applications;
- it has a natural interpretation in terms of hazard functions; and
- neutral to the right processes such as the Dirichlet can be generalized easily.

The difficulty in starting with CHFs is that not every Lévy process can be used as a prior for A. The gamma process, for example, which has independent increments of the form $G[s, s+\varepsilon) \sim Gamma\{cG_0[s, s+\varepsilon), c\}$, does *not* have paths that a.s. produce proper CDFs $F = 1 - \prod_{[0,.]}\{1 - \mathrm{d}G(s)\}$, i.e., the subset $\boldsymbol{A}$ of $\boldsymbol{B}$ does not have outer measure 1.

In order to address this limitation, Hjort constructs a rich class of priors, the Beta process, whose sample paths lie in $\boldsymbol{A}$. He constructs a Beta process on $[0, \infty)$ with paths in $\boldsymbol{A}$ so that it has independent increments of the type

$$\mathrm{d}A(s) \sim Beta\{c(s)\,\mathrm{d}A_0(s), c(s)[1 - \mathrm{d}A_0(s)]\}. \tag{12}$$

A_0 is the prior guess at the cumulative hazard, and as will be shown later, can be used to model families such as the Weibull quite easily. $c(\cdot)$ can be interpreted as the number at risk at s in an imagined prior sample with intensity or hazard rate corresponding to A_0.

Let A be any Lévy process with jumps $S_j = A\{t_j\} = A(t_j) - A(t_j^-)$. Then A has Lévy representation given as

$$E\left(\mathrm{e}^{-\theta A(t)}\right) = \left[\prod_{j:\, t_j \leqslant t} E\left(\mathrm{e}^{-\theta S_j}\right)\right] \exp\left[\int_0^\infty \left(1 - \mathrm{e}^{-\theta s}\right) \mathrm{d}L_t(s)\right],$$

where $\{L_t\ t \geqslant 0\}$ is a continuous Lévy measure. This means that L_t for each t is a measure on $(0, \infty)$, $L_t(D)$ is nondecreasing and continuous in t for each Borel set D in $(0, \infty)$ and $L_0(D) = 0$. From this it follows that $A(t)$ is finite a.s. whenever

$$\int_0^\infty \frac{s}{1+s}\,\mathrm{d}L_t(s) < \infty.$$

Assume that A contains no nonrandom part. Thus the distribution of a prior P is specified by $\{t_1, t_2, \ldots\}$, the distributions of $S_1, S_2, \ldots$ and $\{L_t;\ t \geqslant 0\}$. Hjort's Theorem 3.1 yields a useful result.

Fact 3. For a time continuous Beta process, the accompanying Lévy measures are concentrated on [0, 1].

Thus, Hjort's Lévy process A is a Beta process with parameters $c(\cdot)$ and $A_0(\cdot)$. Note the similarity between this and the Dirichlet process, which too has a scale parameter and a centering parameter.

In the description so far, there is an implicit assumption that the prior guess A_0 is continuous. If $A \sim Beta\{c, A_0\}$ and a sample is drawn from $F = F_A$, then the posterior distribution of A is such that the jump $A\{x\}$ is positive, whenever x is a point at which an observation occurs. Thus the posterior process has fixed points of discontinuity. To account for this, the above description of A must be extended.

The motivation for the following is to mimic exactly the NTR process development, in that the Beta process should be split up into a *continuous* and a *discrete* component, with the aim of being able to sample the posterior process easily.

DEFINITION 3 (Hjort, 1990). Let A_0 be a CHF with a finite number of jumps taking place at $t_1, t_2, \ldots,$ and let $c(\cdot)$ be a piecewise continuous, nonnegative function on

$[0, \infty)$. Suppose A is a Lévy process that is a Beta process with parameters $c(\cdot)$, $A_0(\cdot)$, and write

$$A \sim Beta\{c(\cdot), A_0(\cdot)\}$$

to denote this, if the following holds: A has Lévy representation with

$$S_j = A\{t_j\} \sim Beta\{c(t_j)A_0\{t_j\}, c(t_j)(1 - A_0\{t_j\})\},$$

and

$$\mathrm{d}L_t(s) = \int_0^t c(z)s^{-1}(1-s)^{c(z)-1}\,\mathrm{d}A_{0,c}(z)\,\mathrm{d}s,$$

for $t \geqslant 0$ and $0 < s < 1$, in which $A_{0,c}(t) = A_0(t) - \sum_{t_j \leqslant t} A_0\{t_j\}$ is A_0 with its jumps removed.

The last part of the definition may be restated as

$$A(t) = \sum_{t_j \leqslant t} S_j + A_c(t),$$

where the jumps, S_j, are independent and distributed as above, and A_c is distributed $Beta\{c(\cdot), A_{0,c}(\cdot)\}$ as defined earlier. The general class of possible prior distributions for A can now be stated.

Let A have fixed points of discontinuity $M = \{t_1, \ldots, t_m\}$, with jumps $S_j = A\{t_j\}$ that have densities $f_j(s)\,\mathrm{d}s$ on $[0, 1]$. The process

$$A_c(t) = A(t) - \sum_{t_j \leqslant t} S_j$$

is free of fixed points of discontinuity and has Lévy formula

$$E\left[\mathrm{e}^{-\theta A_c(t)}\right] = \exp\left[-\int_0^1 \left(1 - \mathrm{e}^{-\theta s}\right)\mathrm{d}L_t(s)\right],$$

where the continuous family of Lévy measures $\{L_t;\ t \geqslant 0\}$ is assumed to be of the form

$$\mathrm{d}L_t(s) = \begin{cases} \int_0^t a(s,z)\,\mathrm{d}H(z)\,\mathrm{d}s & \text{if } t \geqslant 0,\ s \in (0,1), \\ 0 & \text{if } s \geqslant 1. \end{cases}$$

H is a continuous, nondecreasing function having $H(0) = 0$ and $a(s, z)$ is some nonnegative function, assumed to be continuous in (s, z) except possibly on the line segments where $z \in M$ and chosen such that $\int_0^1 s\,\mathrm{d}L_t(s)$ is finite. The case $A \sim Beta(c, A_0)$ corresponds to the case where the $f_j(s)$'s are Beta densities and $a(s,z) = c(z)s^{-1}(1-s)^{c(z)-1}$, $H = A_0$.

Hjort proves if A is Lévy, then given the data A is still Lévy. Thus the property of being Lévy (i.e., having independent, nonnegative increments) is preserved. However, given the different form of the Lévy measure for A, the description of the posterior process is quite different from the NTR processes considered earlier; recall Theorem 2 in Section 4.3.

Let $X_1, \dots, X_n$ be i.i.d. given A, and assume that $(T_1, \delta_1), \dots, (T_n, \delta_n)$ is observed, where $T_i = \min(X_i, c_i)$, $\delta_i = I\ \{X_i \leqslant c_i\}$, and $c_1, \dots, c_n$ are censoring times. Define the counting process N and the left-continuous at-risk process Y as

$$N(t) = \sum_{i=1}^{n} I\{T_i \leqslant t, \delta_i = 1\} \quad \text{and} \quad Y(t) = \sum_{i=1}^{n} I\{T_i \geqslant t\}.$$

In particular, note that $\mathrm{d}N(t) = N(t)$ is the number of observed X_i's at the spot t.

THEOREM 3 (Hjort, 1990). *Let the Lévy process A have prior distribution as in Definition 3, but with no fixed points of discontinuity, i.e., M is empty. Let $\{u_1, \dots, u_p\}$ be the distinct points at which noncensored observations occur. Then the posterior distribution of A is a Lévy process with parameters $M^* = \{u_1, \dots, u_p\}$, $H^* = H$, $a^*(s, z) = (1-s)^{Y(z)} a(s, z)$ and $A\{u_j\}$ has density*

$$f_j^*(s) \propto s^{\mathrm{d}N(u_j)} (1-s)^{Y(u_j) - \mathrm{d}N(u_j)} a(s, u_j).$$

In a typical application, one begins with a continuous prior guess for A_0. In the above, suppose M is not empty. Let $t \in M$ be a point with prior density $f_t(s)$ for the jump $S = A\{t\}$. The posterior density of S is given by

$$f_j^*(s) \propto s^{\mathrm{d}N(t)} (1-s)^{Y(t) - \mathrm{d}N(t)} f_t(s).$$

Fact 4. Let $A \sim \mathit{Beta}\{c(\cdot), A_0(\cdot)\}$ be as defined earlier. Then, given the data $(T_1, \delta_1), \dots, (T_n, \delta_n)$,

$$A \sim \mathit{Beta}\left\{ c(\cdot) + Y(\cdot), \int_0^{(\cdot)} \frac{c\,\mathrm{d}A_0 + \mathrm{d}N}{c + Y} \right\}.$$

Fact 5. The distribution of the *jumps* in the posterior is given by

$$A\{t\}|\text{data} \sim \mathit{Beta}\big(c(t) A_0\{t\} + \mathrm{d}N(t),\ c(t)\big(1 - A_0\{t\}\big) + Y(t) - \mathrm{d}N(t)\big). \tag{13}$$

Fact 6. The distribution of the *continuous* component, namely, the intervals between the jump sites, have Lévy formula

$$\mathrm{d}L_t^*(s) = \int_0^t \big[c(z) + Y(z)\big] s^{-1} \big[(1-s)^{c(z)+Y(z)-1}\big] \frac{c(z)}{c(z) + Y(z)}\, \mathrm{d}A_{0,c}(z). \tag{14}$$

Thus, the development so far, in a sense, mimics the discussion in Section 4.3, with the focus being on splitting the posterior Beta process into a *jump* and a *continuous* component. Sampling these separately and putting them together results in a sample from the posterior Beta process.

5.1.1. Some insights into the Beta process

Simulating the Beta process is actually quite trivial. Details can be found in Laud et al. (1996). Here it is simply noted that the continuous component, Eq. (14), can be well approximated by standard beta distributions; see Hjort (1990) for further details. The jump component, Eq. (13), is a standard beta distribution. Thus, unlike the Dirichlet process or other NTR processes, the Beta process simulation is straightforward.

Among its many attractive properties, the Beta process generalizes Ferguson's Dirichlet processes. To see this, let A be a Beta process with parameters $c(\cdot)$ and $A_0(\cdot)$ and consider the random CDF $F(z) = 1 - \prod_{[0,z]}\{1 - \mathrm{d}A_0(s)\}$. Choosing $c(z) = rF_0[z, \infty)$, $r > 0$, Hjort shows that F is a Dirichlet process with parameter $rF_0(\cdot)$. Hjort refers to these as Generalized Dirichlet processes with parameters $c(\cdot)$ and $F_0(\cdot)$.

EXAMPLE 5. Following Ferguson and Phadia (1979), set

$$F_0[z, \infty) = \mathrm{e}^{-az}.$$

Then to express this prior choice in terms of the Beta process, set $c(z) = rF_0[z, \infty)$, $r = 1$. Letting $Y(z) = \sum_{i=1}^{n} I(T_i \geqslant z)$ denote the number at risk at time z provides an illustration that uses the standard Dirichlet process formulation with a constant failure rate.

Another popular model in biomedical contexts is the following. Specify the underlying prior mean cumulative hazard to have a Weibull distribution; i.e., $\mathrm{d}A_0(t) = at^{\alpha-1}\,\mathrm{d}t$. This translates to $c(z) = \mathrm{e}^{-az^{\alpha}/\alpha}$. For illustration, set the mean and variance of the hazard function to equal 10 and 100, respectively. This results in $\alpha = 2$ and $a = 0.025$. This illustrates the use of a generalized Dirichlet process in addition to a specific parametric choice for the prior CHF.

EXAMPLE 6 (*Time-Inhomogeneous Markov Chains*). Suppose that $X_a = \{X_{a(t)};\ t \geqslant 0\}$ are Markov chains for $a = 1, \ldots, n$, moving around in the state space $\{1, \ldots, k\}$ independently of each other and each with the same set of CHFs A_{ij}. Assume that X_a is followed over the interval $[0, w_a]$ and introduce

$$N_{ij}(t) = \text{number of observed transitions } i \to j \text{ during } [0, t],$$
$$Y_i(t) = \text{number at risk in state } i \text{ just before time } t.$$

THEOREM 4 (Hjort, 1990). *Let A_{ij}, $i \neq j$, be independent Beta$(c_{ij}, A_{0,ij})$ processes and assume that the prior guesses $A_{0,ij}$ and $A_{0,il}$ have disjoint sets of discontinuity points when $j \neq l$ (they will often be chosen continuous). Then, given data collected from the n individual Markov chains, the A_{ij}'s are still independent and*

$$A_{ij}|\text{data} \sim \mathit{Beta}\left(c_{ij} + Y_i,\ \frac{\int_0^{\cdot} c_{ij}\,\mathrm{d}A_{0,ij} + \mathrm{d}N_{ij}}{\{c_{ij} + Y_i\}}\right).$$

Note that the posterior distribution of the CHFs for each of the A_{ij} processes will have Lévy form as in Eq. (14), and the distribution of the jumps as in Eq. (13). Thus, given the data, one simply processes, in parallel, the computations of a single Beta process.

6. Hazard rate processes

Kalbfleisch (1978) considered the following problem. Let $T > 0$ represent the failure time of an individual for whom $z = (z_1, \ldots, z_q)$ are q measured covariables. Following Cox (1972), you can model the distribution of T by

$$P(T \geqslant s|z) = \exp\bigl(-H(s)\exp\bigl(z'\boldsymbol{\beta}\bigr)\bigr), \tag{15}$$

where $H(s) = \int_0^s h(t)\,dt$. $h(t)$ is the hazard rate, and correspondingly, $H(t)$ is the cumulative hazard function. The Beta process places a nonparametric prior distribution on $H(\cdot)$. In this section, nonparametric priors are imposed on $h(\cdot)$.

Kalbfleisch (1978) and Clayton (1991) offer an excellent description of the problem from a Bayesian perspective. Both authors model the $H(\cdot)$ process as a gamma process. However, as Kalbfleisch notes, "One deficiency of the gamma hazard process or the Dirichlet process is that with probability 1 the resulting cumulative hazard function $H(t)$ is discrete...". The implications of this are described in detail by Kalbfleisch. The Extended-gamma process of Dykstra and Laud (1981) used to model the hazard rate overcomes this deficiency. An alternative approach to model the hazard rate has been proposed by Arjas and Gasbarra (1994), and they use the Gibbs Sampler to carry out the posterior calculations. Ammann (1984) models bathtub hazard rates, extending the work of Dykstra and Laud (1981). Sinha and Dey (1997), and Dey et al. (1998) provide many other references.

6.1. *Extended-gamma process*

Dykstra and Laud (1981) consider a subclass of the space of distribution functions within the context of reliability. They consider the class of distribution functions with nondecreasing/nonincreasing hazard rates. They then proceed to define a stochastic process called the Extended-gamma (EG) process whose sample paths index the class of nondecreasing/nonincreasing hazard rates with probability one. Observations are then assumed to arise from the distribution function associated with the random hazard rate, and posterior distributions for censored and exact observations are calculated.

Among the key features of the EG process are that it forms a conjugate prior when one has right censored data; it indexes families of absolutely continuous distribution functions with probability one; and, in the case of exact observations, the posterior distribution of the process involves a mixture of EG processes. Also, in the latter instance, Dykstra and Laud define a class of probability distributions called D-distributions which contains the gamma distribution as a special case. Dykstra and Laud (1981) point out that while the EG process has nice properties, it is computationally quite off-putting to calculate the posterior distribution in the presence of left-censored data and/or a large number of exact data. Ammann (1984) combines two EG processes thus allowing for models where both the class of decreasing and increasing hazard rates can be indexed by the sample paths of the combined process. Computations are limited to calculating posterior expectations under squared error loss. Lévy forms for the moment generating function of the posterior distribution of the process are derived.

Given a vector of covariates z, the corresponding vector of parameters $\boldsymbol{\beta}$, and the hazard rate $h(\cdot)$, rewrite Eq. (15) as

$$F(s) = P(T \leqslant s) = 1 - \exp\left(- \exp(z'\boldsymbol{\beta}) \int_0^s h(t)\,\mathrm{d}t \right). \tag{16}$$

Following Dykstra and Laud (1981) and Laud et al. (1996), a prior distribution for *increasing* hazard rates $h(t)$ is constructed.

First, note that, given the data, the probability statement in (16) is conditional on $h(\cdot)$ and $\boldsymbol{\beta}$, the parameters in the model. $\boldsymbol{\beta}$ is finite-dimensional, and hence specifying a prior for $\boldsymbol{\beta}$ is straightforward; you could take it be normally distributed, say. To motivate a prior distribution for $h(\cdot)$, start by treating $h(\cdot)$ as the realization of a stochastic process. To this end, consider a grid of m points $0 = S_0 < S_1 < S_2 < \cdots < S_m < S_{m+1} = \infty$. Let $\delta_j = h(S_j) - h(S_{j-1})$, where δ_j is the increment in the hazard rate in the jth interval. The model in (16) considered on this grid can thus be written as

$$\begin{aligned} p_j &= F(S_j) - F(S_{j-1}) \\ &= \exp\left(- \exp(z'\beta) \sum_{i=1}^{j-1} \delta_i (S_{j-1} - S_{i-1}) \right) \\ &\quad \times \left(1 - \exp\left(- \exp(z'\beta)(S_j - S_{j-1}) \sum_{i=1}^{j} \delta_i \right) \right), \end{aligned}$$

for $j = 1, \ldots, m+1$, and where $\delta_{m+1} = \infty$ and $\sum_{i=1}^{0} = 0$ by definition. The transition from the continuous to the discrete follows by noting that $\int_0^s h(t)\,\mathrm{d}t$ can be approximated by $\sum_{i=1}^{k} (S_k - S_{i-1})\delta_i$.

To define a prior for $h(\cdot)$, let $G(\alpha, \lambda)$ denote the gamma distribution with shape parameter $\alpha \geqslant 0$ and scale parameter $1/\lambda > 0$. For $\alpha = 0$, define this distribution to be degenerate at zero. For $\alpha > 0$, its density is

$$g(x|\alpha, \lambda) = \begin{cases} \frac{\lambda^{\alpha} x^{\alpha-1} \exp(-\lambda x)}{\Gamma(\alpha)} & \text{if } x > 0; \\ 0 & \text{if } x \leqslant 0. \end{cases}$$

Let $\alpha(t)$, $t \geqslant 0$, be a nondecreasing left continuous real valued function such that $\alpha(0) = 0$. Let $\lambda(t), t \geqslant 0$, be a positive right continuous real valued function with left-hand limits existing and bounded away from 0 and ∞. Let $Z(t), t \geqslant 0$, be a gamma process with the parameter function $\alpha(\cdot)$. That is, $Z(0) = 0$, $Z(t)$ has independent increments and for $t > s$, $Z(t) - Z(s)$ is $G(\alpha(t) - \alpha(s), 1)$. Consider a version of this process such that all its sample paths are nondecreasing and left continuous and let

$$h(t) = \int_{[0,t)} \left[\lambda(s)\right]^{-1} \mathrm{d}Z(s). \tag{17}$$

This process $\{h(t), t \geqslant 0\}$ is called the EG process and is denoted

$$h(t) \sim \Gamma\big(\alpha(\cdot), \lambda(\cdot)\big).$$

Some practical implications and useful features of the EG process within the context of the discrete model in (16) are considered.

(i) A realization of the stochastic process in (17) corresponds to an underlying random hazard rate for the model in (16).

(ii) Following Dykstra and Laud (1981), for a fixed t, the distribution of $h(t)$ does not have a closed form; however, its Laplace transform is given by

$$L_{h_t}(u) = E\left[\mathrm{e}^{-uh(t)}\right] = \exp\left\{-\int_0^t \log\left(\frac{1+u}{\{\lambda(s)\}}\right) \mathrm{d}\alpha(s)\right\}. \tag{18}$$

In particular, replace the limits of integration in (16) by S_{j-1} and S_j, and denote the corresponding density by f_{δ_j}.

(iii) Each increment δ_j has prior distribution determined by (17).

(iv) A prior distribution on the δ_js will imply a prior distribution on the p_js.

(v) In assigning a prior probability by this method, you need to input the functions $\alpha(\cdot)$ and $\lambda(\cdot)$. Dykstra and Laud provide details for choosing these functions. Here simply note that you can "control" the variance in the process by choosing $\lambda_i(\cdot)$ large (small) to reflect sharp (vague) beliefs at various time intervals.

(vi) Using the EG prior for $h(\cdot)$ guarantees that the distribution function in (16) is absolutely continuous. This bypasses the difficulties encountered when one models the hazard function process as gamma processes; see, also, Kalbfleisch (1978) and Doksum (1974).

(vii) Unlike the gamma process whose scale parameter is fixed, the scale parameter in the EG process is allowed to vary with time. From a modeling perspective, this offers the practitioner more freedom in accounting for uncertainty in the behavior of the hazard rate in each interval of the time axis.

(viii) The EG process implies that a model based on the hazard rate does *not* lead to the (unrealistic) independent increments property in the cumulative hazard. Stated another way, methods using priors on hazard functions generally imply the undesirable assumption of independent increments in the distribution function. Although the EG model assumes independent increments in the hazard rate, this does not carry over to the corresponding distribution function (Dykstra and Laud, 1981).

The likelihood function

Consider for the moment only exact observations $x_1, x_2, \ldots, x_n$ from the model implied by the development so far, with covariate vectors $z_1, \ldots, z_n$. The likelihood on the grid $0 = S_0 < S_1 < \cdots < S_m < S_{m+1} = \infty$ can be written as

$$L(\Delta; \boldsymbol{x}, \boldsymbol{Z}) \propto \prod_{k=1}^{n} \prod_{j=1}^{m} \exp\left(- \exp\left(z_k' \beta\right) I_{jk} \sum_{i=1}^{j-1} \delta_i (S_{j-1} - S_{i-1}) \right)$$
$$\times \left(1 - \exp\left(- \exp\left(z_k' \beta\right) (S_j - S_{j-1}) \sum_{i=1}^{j} \delta_i \right) \right)^{I_{jk}}, \qquad (19)$$

where I_{jk} is an indicator equaling 1 if x_k is in $(S_{j-1}, S_j]$ and 0 otherwise. Also, Δ, $\boldsymbol{x}$ and $\boldsymbol{Z}$ denote, respectively, the collections of δ_js, $\boldsymbol{x}_k$'s and z_k's.

The posterior distribution

Based on the development so far, the posterior joint distribution of the (discretized) random hazard rate, Δ, and the regression parameter vector, $\boldsymbol{\beta}$, is given by

$$\Delta, \boldsymbol{\beta} | \boldsymbol{x}, \boldsymbol{Z}, \lambda \propto L(\Delta, \beta; \boldsymbol{x}, \boldsymbol{Z})[\boldsymbol{\beta}] \prod_{j=1}^{m} f_{\delta_j}, \qquad (20)$$

where $[\boldsymbol{\beta}]$ is the prior distribution for $\boldsymbol{\beta}$, and f_{δ_j} is defined under (iii) in Eq. (18). The motivation for conditioning the posterior distribution on λ, the scale parameter in the EG process, is now explained.

Right censored data

Theorem 3.2 of Dykstra and Laud (1981) states that given right censored data, the posterior process is once again an EG process; i.e., the EG process serves as a conjugate prior for such data. In particular, they provide a very simple way of updating the parameters of the process given right censored data. While the $\alpha(\cdot)$ parameter remains unchanged, $\lambda(\cdot)$ is updated by

$$\lambda(s) + \sum_{j=1}^{n_r} \exp\left(z_j' \beta\right) \left(x_j^{(r)} - s\right)^+, \qquad (21)$$

where $x_1^{(r)}, \ldots, x_{n_r}^{(r)}$ are right censored observations, and t^+ equals t if $t > 0$ and 0 otherwise. Thus, the $\lambda(\cdot)$ parameter is updated as in (21). Alas, however, note that (21) depends on the covariate vector $\boldsymbol{\beta}$. In the absence of covariates, $\lambda(\cdot)$ is updated just *once* using the right censored data. In the presence of covariates, while using the Gibbs Sampler, (21) gets updated through *each* pass of the Gibbs Sampler via $\boldsymbol{\beta}$. In other words, $\lambda(\cdot)$ itself is *not* a random quantity from a simulation perspective.

Left censored data

The contribution to the likelihood function in the jth interval from left censored data is handled using a simple data augmentation scheme described in Kuo and Smith (1992).

The computational model

Based on the model developed above, Bayesian inference using the EG process as a prior on hazard rates would require the calculation of the posterior distribution of $\boldsymbol{p} = (p_1, \ldots, p_m)$, or any function of $\boldsymbol{p}$ such as the survival probability at some fixed time. (Recall from (16), the one-to-one relationship between $\Delta = (\delta_1, \ldots, \delta_m)$ and $\boldsymbol{p}$.) A Gibbs Sampler will thus have to iterate between the two sets of conditional distributions,

$$\delta_j | \delta_i, \beta, x, Z, \quad j \neq i, j = 1, \ldots, m,$$
$$\beta_u | \beta_v, \Delta, x, Z, \quad u \neq v, u = 1, \ldots, q.$$

Abbreviate the above as $\Delta|\cdot$ and $\boldsymbol{\beta}|\cdot$, or, $\delta_j|\cdot$ and $\beta_u|\cdot$, and use either forms interchangeably where the meaning is clear within the context.

Following some algebra, one obtains,

$$\Delta|\cdot \propto \prod_{j=1}^{m} f_{\delta_j} \exp\left(-\left\{\sum_{k=1}^{n} e^{z'_k \beta} I_{jk} \sum_{i=1}^{j-1} \delta_i (S_{j-1} - S_{i-1})\right\}\right)$$
$$\times \prod_{k=1}^{n} \prod_{j=1}^{m} \left(1 - \exp\left(-(S_j - S_{j-1}) e^{z'_k \beta} \sum_{i=1}^{j} \delta_i\right)\right)^{I_{jk}},$$

and

$$\beta|\cdot \propto [\beta] \prod_{k=1}^{n} \exp\left(-c_k e^{z'_k \beta}\right)\left(1 - \exp\left(-b_k e^{z'_k \beta}\right)\right), \tag{22}$$

where $c_k = \sum_{j=1}^{m} I_{jk} \sum_{i=1}^{j-1} \delta_i (S_{j-1} - S_{i-1})$, $b_k = \sum_{j=1}^{m} (S_j - S_{j-1}) (\sum_{i=1}^{j} \delta_i) I_{jk}$, and $[\beta]$ is the prior distribution for β.

The above description completes the modeling and implementation aspect of Cox regressions via a random hazard rate model based on EG processes. Unlike the NTR and Beta processes, the EG process development doesn't need to be split up into discrete and continuous components; the simulation proceeds in one block as described above. Data analyses using the above model is provided in Laud et al. (1996).

7. Polya trees

The phrase "Bayesian nonparametrics" which until now meant "Bayesian nonparametrics using stochastic processes" is extended to include a class of nonparametric priors called Polya Trees. Detailed background to the material in this section can be found in Ferguson (1974), Lavine (1992, 1994), Mauldin et al. (1992), Muliere and Walker (1998), Walker et al. (1999), Hanson and Johnson (2002).

The Polya tree prior relies on a binary tree partitioning of the sample space Ω. There are two aspects to a Polya tree: a binary tree partition of Ω and a nonnegative parameter associated with each set in the binary partition. The binary tree partition is given by $\Pi = \{B_\varepsilon\}$ where ε is a binary sequence which "places" the set B_ε in the tree. Denote the sets at level 1 by (B_0, B_1), a measurable partition of Ω; denote by (B_{00}, B_{01}) the "offspring" of B_0, so that $B_{00}, B_{01}, B_{10}, B_{11}$ denote the sets at level 2, and so on. The number of partitions at the mth level is 2^m. In general, B_ε splits into $B_{\varepsilon 0}$ and $B_{\varepsilon 1}$ where $B_{\varepsilon 0} \cap B_{\varepsilon 1} = \emptyset$ and $B_{\varepsilon 0} \cup B_{\varepsilon 1} = B_\varepsilon$.

A helpful image is that of a particle cascading through these partitions. It starts in Ω and moves into B_0 with probability C_0, or into B_1 with probability $1 - C_0$. In general, on entering B_ε the particle could either move into $B_{\varepsilon 0}$ or into $B_{\varepsilon 1}$. Let it move into the former with probability $C_{\varepsilon 0}$ and into the latter with probability $C_{\varepsilon 1} = 1 - C_{\varepsilon 0}$. For Polya trees, these probabilities are random, beta variables, $(C_{\varepsilon 0}, C_{\varepsilon 1}) \sim \text{beta}(\alpha_{\varepsilon 0}, \alpha_{\varepsilon 1})$ with nonnegative $\alpha_{\varepsilon 0}$ and $\alpha_{\varepsilon 1}$. If you denote the collection of αs by $\mathcal{A} = \{\alpha_\varepsilon\}$, a particular Polya tree distribution is completely defined by Π and $\mathcal{A}$.

DEFINITION 4 (Lavine, 1992). A random probability measure F on Ω is said to have a Polya tree distribution, or a Polya tree prior, with parameters $(\Pi, \mathcal{A})$, written $F \sim PT(\Pi, \mathcal{A})$, if there exist nonnegative numbers $\mathcal{A} = (\alpha_0, \alpha_1, \alpha_{00}, \ldots)$ and random variables $\mathcal{C} = (C_0, C_{00}, C_{10}, \ldots)$ such that the following hold:

(i) all the random variables in $\mathcal{C}$ are independent;
(ii) for every ε, $C_{\varepsilon 0} \sim \beta(\alpha_{\varepsilon 0}, \alpha_{\varepsilon 0})$; and
(iii) for every $m = 1, 2, \ldots$ and every $\varepsilon = \varepsilon_1 \cdots \varepsilon_m$ define

$$F(B_{\varepsilon_1 \cdots \varepsilon_m}) = \left(\prod_{j=1}^{m} {}_{\varepsilon_j = 0} C_{\varepsilon_1 \cdots \varepsilon_{j-1} 0}\right)\left(\prod_{j=1}^{m} {}_{\varepsilon_j = 1}(1 - C_{\varepsilon_1 \cdots \varepsilon_{j-1} 0})\right),$$

where the first terms (i.e., for $j = 1$) are interpreted as C_0 and $1 - C_0$.

Fact 7. The Polya tree class indexes priors which assign probability 1 to the set of continuous distributions, unlike the Dirichlet process which has sample distribution functions that are discrete with probability 1. For example, the choice $\alpha_\varepsilon = m^2$, whenever ε defines a set at level m, leads to an F which is absolutely continuous (Ferguson, 1974).

Fact 8. It is easy to show that the discrete versions of the beta process (Hjort, 1990), the beta-Stacy process, and hence the Dirichlet process can all be characterized as Polya trees; see, for example, Muliere and Walker (1997).

7.1. *Prior specifications and computational issues*

Probably the main difficulty in implementing the Polya tree is to limit the infinite collection $\mathcal{C}$ in Definition 4 to some manageable (finite) number of random variables that is practically sensible. Once this is accomplished, the rest of the process is utterly simple.

7.2. *Specifying the Polya tree*

Polya trees require simulating a random probability measure $F \sim PT(\Pi, \mathcal{A})$. This is done by sampling $\mathcal{C}$ using the constructive form given in Definition 4. Since $\mathcal{C}$ is an infinite set an approximate probability measure from $PT(\Pi, \mathcal{A})$ is sampled by terminating the process at a finite level M. Let this finite set be denoted by $\mathcal{C}_M$ and denote by F_M the resulting random measure constructed to level M (which Lavine, 1992, refers to as a "partially specified Polya tree"). From the sampled variates of $\mathcal{C}_M$ define F_M by

$F(B_{\varepsilon_1\cdots\varepsilon_M})$ for each $\varepsilon = \varepsilon_1 \cdots \varepsilon_M$ according to (iii) under Definition 4. So, for example, if $M = 8$, you obtain a random distribution which assigns random mass to $r = 2^8$ sets.

It is possible to center the Polya tree prior on a particular probability measure F_0 on Ω by taking the partitions to coincide with percentiles of F_0 and then to take $\alpha_{\varepsilon 0} = \alpha_{\varepsilon 1}$ for each ε. This involves setting $B_0 = (-\infty, F_0^{-1}(1/2))$, $B_1 = [F_0^{-1}(1/2), \infty)$ and, at level m, setting, for $j = 1, \ldots, 2^m$, $B_j = [F_0^{-1}((j-1)/2^m), F_0^{-1}(j/2^m))$, with $F_0^{-1}(0) = -\infty$ and $F_0^{-1}(1) = +\infty$, where $(B_j\colon j = 1, \ldots, 2^m)$ correspond, in order, to the 2^m partitions of level m. It is then straightforward to show that $EF(B_\varepsilon) = F_0(B_\varepsilon)$ for all ε.

In practice, you may not wish to assign a separate α_ε for each ε. It may be convenient, therefore, to take $\alpha_\varepsilon = c_m$ whenever ε defines a set at level m. For the top levels (m small) it is not necessary for $F(B_{\varepsilon 0})$ and $F(B_{\varepsilon 1})$ to be “close”; on the contrary, a large amount of variability is desirable. However, as you move down the levels (m large) you will increasingly want $F(B_{\varepsilon 0})$ and $F(B_{\varepsilon 1})$ to be close, if you believe in the underlying continuity of F. This can be achieved by allowing c_m to be small for small m and allowing c_m to increase as m increases; choosing, for example, $c_m = cm^2$ for some $c > 0$. According to Ferguson (1974), $c_m = m^2$ implies that F is absolutely continuous with probability 1 and therefore according to Lavine (1992) this “would often be a sensible canonical choice”. In what follows, set cm^2 for the αs. Note that the Dirichlet process arises when $c_m = c/2^m$, which means that $c_m \to 0$ as $m \to \infty$ (the wrong direction as far as the continuity of F is concerned) and F is discrete with probability 1 (Blackwell and MacQueen, 1973).

The model can be extended by assigning a prior to c, but in most applications it might suffice to report the analyses based on a range of judiciously specified choices of c. An alternative idea is to understand and assign the c_ms in terms of the variance of the probabilities associated with the sets on level m. These variances are all equal to

$$\frac{1}{2^m}\left\{\prod_{k=1}^{m}\frac{c_k+1}{2c_k+1}-\frac{1}{2^m}\right\}$$

which gives a procedure for assigning the c_ms based on uncertainty in the centering of $F(B_\varepsilon)$. For example, if you want $\operatorname{var} F(B_\varepsilon) = v_m$ whenever ε defines a set at level m, then you would need

$$\frac{c_m+1}{2c_m+1}=\frac{4^m v_m+1}{2(4^{m-1}v_{m-1}+1)}.$$

Note that this imposes a constraint on the v_ms given by $v_{m-1}/4 < v_m < v_{m-1}/2 + 1/4^m$. More generally, you could define the α_ε to match $E_{PT}F(B_\varepsilon)$ and $E_{PT}[F^2(B_\varepsilon)]$ with those obtained from a parametric model; Walker and Damien (1998a, 1998b) provide examples of these.

7.3. Posterior distributions

Consider a Polya tree prior $PT(\Pi, \mathcal{A})$. Following Lavine (1992), given an observation Y_1, the posterior Polya tree distribution is easily obtained. Write $(F|Y_1) \sim$

$PT(\Pi, \mathcal{A}|Y_1)$ with $(\mathcal{A}|Y_1)$ given by

$$\alpha_\varepsilon|Y_1 = \begin{cases} \alpha_\varepsilon + 1 & \text{if } Y_1 \in B_\varepsilon, \\ \alpha_\varepsilon & \text{otherwise.} \end{cases}$$

If Y_1 is observed exactly, then an α needs to be updated at each level, whereas in the case of censored data (in one of the sets B_ε), only a finite number require to be updated. For n observations, let $\mathcal{Y} = (Y_1, \ldots, Y_n)$, with $(\mathcal{A}|\mathcal{Y})$ given by $(\alpha_\varepsilon|\mathcal{Y}) = \alpha_\varepsilon + n_\varepsilon$, where n_ε is the number of observations in B_ε. Let $q_\varepsilon = P(Y_{n+1} \in B_\varepsilon|\mathcal{Y})$, for some ε, denote the posterior predictive distribution, and let $\varepsilon = \varepsilon_1 \cdots \varepsilon_m$; then, in the absence of censoring,

$$q_\varepsilon = \frac{\alpha_{\varepsilon_1} + n_{\varepsilon_1}}{\alpha_0 + \alpha_1 + n} \frac{\alpha_{\varepsilon_1\varepsilon_2} + n_{\varepsilon_1\varepsilon_2}}{\alpha_{\varepsilon_1 0} + \alpha_{\varepsilon_1 1} + n_{\varepsilon_1}} \cdots \frac{\alpha_{\varepsilon_1\cdots\varepsilon_m} + n_{\varepsilon_1\cdots\varepsilon_m}}{\alpha_{\varepsilon_1\cdots\varepsilon_{m-1}0} + \alpha_{\varepsilon_1\cdots\varepsilon_{m-1}1} + n_{\varepsilon_1\cdots\varepsilon_{m-1}}}.$$

For censored data,

$$q_\varepsilon = \frac{\alpha_{\varepsilon_1} + n_{\varepsilon_1}}{\alpha_0 + \alpha_1 + n} \cdots \frac{\alpha_{\varepsilon_1\cdots\varepsilon_m} + n_{\varepsilon_1\cdots\varepsilon_m}}{\alpha_{\varepsilon_1\cdots\varepsilon_{m-1}0} + \alpha_{\varepsilon_1\cdots\varepsilon_{m-1}1} + n_{\varepsilon_1\cdots\varepsilon_{m-1}} - s_{\varepsilon_1\cdots\varepsilon_{m-1}}},$$

where s_ε is the number of observations censored in B_ε.

EXAMPLE 7 (*The Kaplan–Meier data*). The NTR process approach to the analysis of survival data is mathematically involved. Polya trees can simplify this complexity a great deal. However, what should the partitions be? Recall that with the NTR analysis the observations effectively partitioned the time axis with each partition having to be treated separately (and independently), all the different parts subsequently being "put together": that is, the jump and continuous components were considered separately and then put together. It seems obvious that the partitions of the Polya tree should coincide with at least some of the observations. In fact the only way to make progress is to have the censoring sets, e.g., $[1.0, \infty)$, coinciding with partition sets; see below. Consider the Kaplan–Meier data, with exact observations at 0.8, 3.1, 5.4 and 9.2 months, and right-censored data at 1.0, 2.7, 7.0 and 12.1 months. Next, consider the following partitions of these data.

$$B_0 = [0, 1.0), \qquad B_1 = [1.0, \infty),$$
$$B_{10} = [1.0, 2.7), \qquad B_{11} = [2.7, \infty),$$
$$B_{110} = [2.7, 7.0), \qquad B_{111} = [7.0, \infty),$$

and

$$B_{1110} = [7.0, 12.1), \qquad B_{1111} = [12.1, \infty).$$

The inclusion of these partitions is compulsory; the choice of the remainder is somewhat arbitrary.

In the above, a particular partition could be selected in the light of what specific questions need to be answered; for example, for a future observation, Y_{n+1}, what is $P(Y_{n+1} > 6.0|\text{data})$? This can be answered by partitioning B_{110} appropriately. Thus an estimate of the survival curve $S(t) = 1 - F(t)$ can be found for any $t \in (0, \infty)$ and its computation for a range of ts will be sufficient to provide a clear picture of S.

In summary, using Polya trees for analyzing survival data is more straightforward than using NTR processes. Posterior distributions are more tractable and there is little need for computer intensive simulations. An additional advantage is that you can select Polya trees which are continuous, whereas NTR processes are discrete, unless you model the hazard rate process using an extended-gamma process. Several papers have appeared in the literature which use Polya trees. A comprehensive lists of such papers, along with illustrations, are provided in Walker et al. (1999).

8. Beyond NTR processes and Polya trees

The NTR and Polya tree priors are just two of many approaches to Bayesian nonparametrics. The reason these approaches have found broad acceptance is two-fold: they are easy to implement; and the theoretical properties of these priors are very appealing. But in a world of fast computers, the first merit is not that critical.

Broadly speaking, other approaches to Bayesian nonparametrics such as density estimation, wavelets, splines, bootstrapping, and nonparametric empirical Bayes estimation methods (Deely and Kruse, 1968) may be classified as "curve-fitting". Of course, this is not entirely a fair characterization since the methods described in this paper are also types of smoothing and curve-fitting techniques, and indeed theoretical intersections among all these approaches can be found.

As an illustration, if F is a beta-Stacy process with parameters α and β then, given an iid sample from F, with possible right censoring, the Bayes estimate of $F(t)$, under a quadratic loss function, is given by

$$\widehat{F}(t) = 1 - \prod_{[0,t]}\left\{1 - \frac{\mathrm{d}\alpha(s) + \mathrm{d}N(s)}{\beta(s) + M(s)}\right\},$$

where $N(t) = \sum_i I(Y_i \leqslant t)$, $M(t) = \sum_i I(Y_i \geqslant t)$ and $\prod_{[0,t]}$ represents a product integral. The Kaplan–Meier estimate is obtained as $\alpha, \beta \to 0$, which is also the basis for both the censored data Bayesian bootstrap (Lo, 1984) and the finite population censored data Bayesian bootstrap (Muliere and Walker, 1997). Also, the underpinning for Rubin's (1981) bootstrap technique is the Dirichlet process.

Regardless of the approach to Bayesian nonparametrics, several researchers have also studied some general nonparametric theoretical issues, the central ones being consistency or robustness properties of nonparametric priors. The motivation for this is to find answers to the following types of questions. As more and more data accumulate, are Bayesian nonparametric procedures consistent? What *is* consistency in Bayesian nonparametrics? Are there mathematical tools that are unique to studying Bayesian nonparametric procedures from a large sample perspective? This research is not without controversy because not all Bayesians are agreed on the need for examining asymptotic properties of estimators, an activity that is popular among non-Bayesians. Indeed, some Bayesians would dismiss outright such an activity, claiming it goes against Bayesian thinking altogether. Be that as it may, considerable progress has been made on this front.

In addition to the references provided throughout this paper, this section also lists research done by several statisticians who have contributed to other approaches to

Bayesian nonparametrics, as well as to theoretical issues. Furthermore, additional references on the topics covered in this review are provided. The list is by no means exhaustive, and the reader is invited to seek out references contained in the papers and the books cited here. Any omission is completely unintentional since the field is growing at such a rapid pace as to render it impossible to closely monitor all the advances. It is hoped that the list, arranged alphabetically, will prove a useful starting point for those who wish to understand Bayesian nonparametrics as a whole, not just the subset that was described in some detail in this review.

Consistency issues. Barron et al. (1998), Diaconis and Freedman (1986), Ghosal et al. (1999), Ghosh and Ramamoorthi (2002), Petrone and Wassserman (2002), Salinetti (2003), Walker and Hjort (2001) and Walker (2003).

Wavelets, splines, density estimation, etc. Berger and Guglielmi (2001), Berry and Christensen (1979), Bhattacharya (1981), Binder (1982), Blackwell (1973), Breiman et al. (1984), Breth (1979), Brunk (1978), Brunner (1992), Brunner and Lo (1989, 1994), Bush and MacEachern (1996), Campbell and Hollander (1978), Chipman et al. (1997, 1998), Christensen and Johnson (1988), Cifarelli and Regazzini (1990), Clyde et al. (1996), Dalal (1979), Dalal and Hall (1980), Denison et al. (1998), Diaconis and Freedman (1993), Diaconis and Kemperman (1996), Doksum (1974), Donoho and Johnstone (1994, 1995), Doss (1985a, 1985b, 1994), Doss and Sellke (1982), Escobar and West (1995, 1998), Ferguson et al. (1992), Florens et al. (1992), Friedman (1991), Gelfand (1999), Gelfand and Kuo (1991), Gopalan and Berry (1998), Hall and Patil (1995, 1996), Hannum et al. (1981), Hernandez and Weiss (1996), Hettmansperger and McKean (1998), Hjort and Walker (2001), Ishwaran and Zarepour (2000), Kaplan and Meier (1958), Kottas and Gelfand (2001), Kuo and Smith (1992), Lenk (1988, 1991, 1999), Laud et al. (1998), Leonard (1978), Lo (1984, 1986, 1988, 1991), MacEachern (1998), MacEachern et al. (1999), MacEachern and Mueller (1998), Mallick et al. (2000), Mallick and Gelfand (1996), Mallick and Walker (1997), Mauldin et al. (1992), Mukhopadhyay (2000), Mukhopadhyay and Gelfand (1997), Muliere and Tardella (1998), Muliere and Walker (1998), Mueller et al. (1996, 1997), Mueller and Roeder (1997), Mueller and Rosner (1997), Neal (2000), Newton et al. (1996), Newton and Zhang (1999), Ogden (1997), Petrone (1999a, 1999b), Quintana (1998), Richardson and Green (1997), Robbins (1964), Schwarz (1978), Sethuraman (1994), Sethuraman and Tiwari (1982), Shaked (1980), Shively et al. (1999), Silverman (1986), Susarla and Van Ryzin (1978), Sinha (1993), Smith and Kohn (1996, 1997, 1998), Stone (1994), Thorburn (1986), Vidakovic (1998), Wahba (1990, 1994), Walker and Mallick (1996), Walker and Muliere (1997a, 1997b), Walter (1994), Wolpert and Ickstadt (1998a, 1998b), West et al. (1994) and Yamato (1984).

References

Ammann, L. (1984). Bayesian nonparametric inference for quantal response data. *Ann. Statist.* **2**, 636–645.

Antoniak, C.E. (1974). Mixtures of Dirichlet processes with applications to Bayesian nonparametric problems. *Ann. Statist.* **2**, 1152–1174.

Arjas, E., Gasbarra, D. (1994). Nonparametric Bayesian inference from right censored survival data, using the Gibbs sampler. *Statistica Sinica* **4**, 505–524.

Barron, A., Schervish, M., Wasserman, L. (1998). The consistency of posterior distributions in nonparametric problems. *Ann. Statist.* **27**, 536–561.

Berger, J.O., Guglielmi, A. (2001). Bayesian testing of a parametric model versus nonparametric alternatives. *J. Amer. Statist. Assoc.* **96**, 174–184.

Bernardo, J.M., Smith, A.F.M. (1994). *Bayesian Theory*. Wiley, Chichester.

Berry, D.A., Christensen, R. (1979). Empirical Bayes estimation of a binomial parameter via mixtures of Dirichlet processes. *Ann. Statist.* **7**, 558–568.

Bhattacharya, P.K. (1981). Posterior distribution of a Dirichlet process from quantal response data. *Ann. Statist.* **9**, 803–811.

Binder, D.A. (1982). Nonparametric Bayesian models for samples from finite populations. *J. Roy. Statist. Soc., Ser. B* **44**, 388–393.

Blackwell, D. (1973). Discreteness of Ferguson selections. *Ann. Statist.* **1**, 356–358.

Blackwell, D., MacQueen, J.B. (1973). Ferguson distributions via Polya urn schemes. *Ann. Statist.* **1**, 353–355.

Bondesson, L. (1982). On simulation from infinitely divisible distributions. *Adv. Appl. Probab.* **14**, 855–869.

Breiman, L., Friedman, J.H., Olshen, R.A., Stone, C.J. (1984). *Classification and Regression Trees*. Wadsworth, Belmont, CA.

Breth, M. (1979). Nonparametric Bayesian interval estimation. *Biometrika* **66**, 641–644.

Brunk, H. (1978). Univariate density estimation by orthogonal series. *Biometrika* **65**, 521–528.

Brunner, L.J. (1992). Bayesian nonparametric methods for data from a unimodal density. *Statist. Probab. Lett.* **14**, 195–199.

Brunner, L.J., Lo, A.Y. (1989). Bayes methods for a symmetric unimodal density and its mode. *Ann. Statist.* **17**, 1550–1566.

Brunner, L.J., Lo, A.Y. (1994). Nonparametric Bayes methods for directional data. *Canad. J. Statist.* **22**, 401–412.

Bush, C.A., MacEachern, S.N. (1996). A semiparametric Bayesian model for randomized block designs. *Biometrika* **83**, 275–285.

Campbell, G., Hollander, M. (1978). Rank order estimation with the Dirichlet prior. *Ann. Statist.* **6**, 142–153.

Chipman, H., McCulloch, R., Kolaczyk, E. (1997). Adaptive Bayesian wavelet shrinkage. *J. Amer. Statist. Assoc.* **92**, 1413–1421.

Chipman, H., George, E.I., McCulloch, R. (1998). Bayesian CART model search (with discussion). *J. Amer. Statist. Assoc.* **93**, 935–960.

Christensen, R., Johnson, W. (1988). Modeling accelerated failure time with a Dirichlet process. *Biometrika* **75**, 693–704.

Cifarelli, D.M., Regazzini, E. (1990). Distribution functions of means of a Dirichlet process. *Ann. Statist.* **18**, 429–442.

Clayton, D.G. (1991). A Monte Carlo method for Bayesian inference in frailty models. *Biometrics* **47**, 467–485.

Clyde, M., Desimone, H., Parmigian, G. (1996). Prediction via orthogonalized model mixing. *J. Amer. Statist. Assoc.* **91**, 1197–1208.

Cox, D.R. (1972). Regression models and life tables (with discussion). *J. Roy. Statist. Soc., Ser. B* **34**, 187–220.

Dalal, S.R. (1979). Dirichlet invariant processes and applications to nonparametric estimation of symmetric distribution functions. *Stochastic Process. Appl.* **9**, 99–107.

Dalal, S.R., Hall, G.J. Jr. (1980). On approximating parametric Bayes models by nonparametric Bayes models. *Ann. Statist.* **8**, 664–672.

Damien, P., Laud, P.W., Smith, A.F.M. (1995). Random variate generation approximating infinitely divisible distributions with application to Bayesian inference. *J. Roy. Statist. Soc., Ser. B* **57**, 547–564.

Damien, P., Laud, P.W., Smith, A.F.M. (1996). Implementation of Bayesian nonparametric inference based on beta processes. *Scand. J. Statist.* **23**, 27–36.

Damien, P., Wakefield, J.C., Walker, S.G. (1999). Gibbs sampling for Bayesian nonconjugate and hierarchical models using auxiliary variables. *J. Roy. Statist. Soc., Ser. B* **61**, 331–344.

Deely, J.J., Kruse, R.L. (1968). Construction of sequences estimating the mixing distribution. *Ann. Statist.* **39**, 286–288.

Denison, D.G.T., Mallick, B.K., Smith, A.F.M. (1998). Automatic Bayesian curve fitting. *J. Roy. Statist. Soc., Ser. B* **60**, 333–350.

Dey, D.K., Mueller, P.M., Sinha, D. (1998). *Practical Nonparametric and Semiparametric Bayesian Statistics*. Springer-Verlag, New York.

Diaconis, P., Freedman, D. (1986). On the consistency of Bayes estimates (with discussion). *Ann. Statist.* **14**, 1–67.

Diaconis, P., Freedman, D. (1993). Nonparametric binary regression: A Bayesian approach. *Ann. Statist.* **21**, 2108–2137.

Diaconis, P., Kemperman, J. (1996). Some new tools for Dirichlet priors. In: Bernardo, J.M., Berger, J.O., Dawid, A.P., Smith, A.F.M. (Eds.), *Bayesian Statistics, vol. 5, Proceedings of the Fifth Valencia International Meeting*. Clarendon Press, Oxford, pp. 97–106.

Doksum, K. (1974). Tailfree and neutral random probabilities and their posterior distributions. *Ann. Probab.* **2**, 183–201.

Donoho, D., Johnstone, I. (1994). Ideal spatial adaptation by wavelet shrinkage. *Biometrika* **81**, 425–455.

Donoho, D., Johnstone, I. (1995). Adapting to unknown smoothness via wavelet shrinkage. *J. Roy. Statist. Soc., Ser. B* **57**, 301–370.

Doss, H. (1985a). Bayesian nonparametric estimation of the median; Part I: Computation of the estimates. *Ann. Statist.* **13**, 1432–1444.

Doss, H. (1985b). Bayesian nonparametric estimation of the median; Part II: Asymptotic properties of the estimates. *Ann. Statist.* **13**, 1445–1464.

Doss, H. (1994). Bayesian nonparametric estimation for incomplete data via successive substitution sampling. *Ann. Statist.* **22**, 1763–1786.

Doss, H., Sellke, T. (1982). The tails of probabilities chosen from a Dirichlet prior. *Ann. Statist.* **10**, 1302–1305.

Dykstra, R.L., Laud, P.W. (1981). A Bayesian nonparametric approach to reliability. *Ann. Statist.* **9**, 356–367.

Escobar, M.D. (1994). Estimating normal means with a Dirichlet process prior. *J. Amer. Statist. Assoc.* **89**, 268–277.

Escobar, M.D. (1995). Nonparametric Bayesian methods in hierarchical models. *J. Statist. Plann. Inference* **43**, 97–106.

Escobar, M.D., West, M. (1995). Bayesian density estimation and inference using mixtures. *J. Amer. Statist. Assoc.* **90**, 577–588.

Escobar, M.D., West, M. (1998). Computing nonparametric hierarchical models. In: Dey, D., Mueller, P., Sinha, D. (Eds.), *Practical Nonparametric and Semiparametric Bayesian Statistics*. Springer, New York, pp. 1–22.

Ferguson, T.S. (1973). A Bayesian analysis of some nonparametric problems. *Ann. Statist.* **1**, 209–230.

Ferguson, T.S. (1974). Prior distributions on spaces of probability measures. *Ann. Statist.* **2**, 615–629.

Ferguson, T.S., Phadia, E.G. (1979). Bayesian nonparametric estimation based on censored data. *Ann. Statist.* **7**, 163–186.

Ferguson, T.S., Phadia, E.G., Tiwari, R.C. (1992). Bayesian nonparametric inference. In: *Current Issues in Statistical Inference: Essays in Honor of D. Basu*, pp. 127–150.

Florens, J.P., Mouchart, M., Rolin, J.M. (1992). Bayesian analysis of mixtures: Some results on exact estimability and identification (with discussion). In: Bernardo, J.M., Berger, J.O., Dawid, A.P., Smith, A.F.M. (Eds.), *Bayesian Statistics, vol. 4*. Oxford University Press, Oxford, pp. 127–145.

Friedman, J. (1991). Multivariate adaptive regression splines (with discussion). *Ann. Statist.* **19**, 1–141.

Gelfand, A.E. (1999). Approaches for semiparametric Bayesian regression. In: Ghosh, S. (Ed.), *Asymptotics, Nonparametrics and Time Series*. Marcel Dekker, Inc., New York, pp. 615–638.

Gelfand, A.E., Kuo, L. (1991). Nonparametric Bayesian bioassay including ordered polytomous response. *Biometrika* **78**, 657–666.

Gelfand, A.E., Smith, A.F.M. (1990). Sampling-based approaches to calculating marginal densities. *J. Amer. Statist. Assoc.* **85**, 398–409.

Geman, S., Geman, D. (1984). Stochastic relaxation, Gibbs distributions and the Bayesian restoration of images. *IEEE Trans. Pattern Analysis and Machine Intelligence* **6**, 721–741.

Ghosal, S., Ghosh, J.K., Ramamoorthi, R.V. (1999). Posterior consistency of Dirichlet mixtures in density estimation. *Ann. Statist.* **27**, 143–158.

Ghosh, J.K., Ramamoorthi, R.V. (2002). *Bayesian Nonparametrics*. Springer-Verlag, New York.

Gopalan, R., Berry, D.A. (1998). Bayesian multiple comparisons using Dirichlet process priors. *J. Amer. Statist. Assoc.* **93**, 1130–1139.

Hall, P., Patil, P. (1995). Formulae for the mean integrated square error of non-linear wavelet based density estimators. *Ann. Statist.* **23**, 905–928.

Hall, P., Patil, P. (1996). On the choice of the smoothing parameter, threshold and truncation in nonparametric regression by nonlinear wavelet methods. *J. Roy. Statist. Soc., Ser. B* **58**, 361–377.

Hanson, T., Johnson, W. (2002). Modeling regression errors with a mixture of Polya trees. *J. Amer. Statist. Assoc.* **97**, 1020–1033.

Hannum, R.C., Hollander, M., Langberg, N.A. (1981). Distributional results for random functionals of a Dirichlet process. *Ann. Probab.* **9**, 665–670.

Hastings, W.K. (1970). Monte Carlo sampling methods using Markov chains and their applications. *Biometrika* **57**, 97–109.

Hernandez, E., Weiss, G. (1996). *A First Course on Wavelets*. CRC Press, Boca Raton, FL.

Hettmansperger, T.P., McKean, J.W. (1998). *Robust Nonparametric Statistical Methods*. Arnold, London.

Hjort, N.L. (1990). Nonparametric Bayes estimators based on beta processes in models for life history data. *Ann. Statist.* **18**, 1259–1294.

Hjort, N.L., Walker, S.G. (2001). A note on kernel density estimators with optimal bandwidths. *Statist. Probab. Lett.* **54**, 153–159.

Ishwaran, H., Zarepour, M. (2000). Markov chain Monte Carlo in approximate Dirichlet and beta two-parameter process hierarchical models. *Biometrika* **87**, 371–390.

Kalbfleisch, J.D. (1978). Nonparametric Bayesian analysis of survival time data. *J. Roy. Statist. Soc., Ser. B* **40**, 214–221.

Kaplan, E.L., Meier, P. (1958). Nonparametric estimation from incomplete observations. *J. Amer. Statist. Assoc.* **53**, 457–481.

Kleinman, K.P., Ibrahim, J.G. (1998a). A semiparametric Bayesian approach to the random effects model. *Biometrics* **54**, 921–938.

Kleinman, K.P., Ibrahim, J.G. (1998b). A semiparametric Bayesian approach to generalized linear mixed models. *Statistics in Medicine* **17**, 2579–2976.

Kottas, A., Gelfand, A.E. (2001). Bayesian semiparametric median regression modeling. *J. Amer. Statist. Assoc.* **96**.

Kuo, L. (1986). Computations of mixtures of Dirichlet processes. *SIAM J. Sci. Statist. Comput.* **7**, 60–71.

Kuo , L., Smith, A.F.M. (1992). Bayesian computations in survival models via the Gibbs sampler. In: Klein, J.P., Goel, P.K. (Eds.), *Survival Analysis: State of the Art*. Kluwer Academic, Dordrecht, pp. 11–24.

Laud, P.W., Smith, A.F.M., Damien, P. (1996). Monte Carlo methods approximating a posterior hazard rate process. *Statist. Comput.* **6**, 77–84.

Laud, P.W., Damien, P., Smith, A.F.M. (1998). Bayesian nonparametric and covariate analysis of failure time data. In: Dey, D., Mueller, P., Sinha, D. (Eds.), *Practical Nonparametric and Semiparametric Bayesian Statistics*. Springer-Verlag, New York, pp. 213–225.

Lavine, M. (1992). Some aspects of Polya tree distributions for statistical modeling. *Ann. Statist.* **20**, 1222–1235.

Lavine, M. (1994). More aspects of Polya trees for statistical modeling. *Ann. Statist.* **22**, 1161–1176.

Lenk, P. (1988). The logistic normal distribution for Bayesian, nonparametric predictive densities. *J. Amer. Statist. Assoc.* **83**, 509–516.

Lenk, P.J. (1991). Towards a practicable Bayesian nonparametric density estimator. *Biometrika* **78**, 531–543.

Lenk, P. (1999). Bayesian inference for semiparametric regression using a Fourier representation. *J. Roy. Statist. Soc., Ser. B* **61**, 863–879.

Lindley, D.V., Smith, A.F.M. (1972). Bayes estimates for the linear model (with discussion). *J. Roy. Statist. Soc., Ser. B* **34**, 1–42.

Leonard, T. (1978). Density estimation, stochastic processes and prior information. *J. Roy. Statist. Soc., Ser. B* **40**, 113–146.

Lo, A.Y. (1984). On a class of Bayesian nonparametric estimates: I Density estimates. *Ann. Statist.* **12**, 351–357.

Lo, A.Y. (1986). Bayesian statistical inference for sampling a finite population. *Ann. Statist.* **14**, 1226–1233.

Lo, A.Y. (1988). A Bayesian bootstrap for a finite population. *Ann. Statist.* **16**, 1684–1695.

Lo, A.Y. (1991). A characterization of the Dirichlet process. *Statist. Probab. Lett.* **12**, 185–187.

MacEachern, S.N. (1998). Computational methods for mixture of Dirichlet process models. In: Dey, D., Mueller, P., Sinha, D. (Eds.), *Practical Nonparametric and Semiparametric Bayesian Statistics*. Springer, New York, pp. 23–43.

MacEachern, S.N., Clyde, M., Liu, J.S. (1999). Sequential importance sampling for nonparametric Bayes models: The next generation. *Canad. J. Statist.* **27**, 251–267.

MacEachern, S.N., Mueller, P. (1998). Estimating mixture of Dirichlet process models. *J. Comput. Graph. Statist.* **7**, 223–238.

Mallick, B.K., Gelfand, A.E. (1996). Semiparametric errors-in-variables models: A Bayesian approach. *J. Statist. Plann. Inference* **52**, 307–321.

Mallick, B.K., Walker, S.G. (1997). Combining information from several experiments with nonparametric priors. *Biometrika* **84**, 697–706.

Mallick, B., Denison, T., Smith, A.F.M. (2000). *Bayesian Functional Estimation*. Wiley, Chichester.

Mauldin, R.D., Sudderth, W.D., Williams, S.C. (1992). Polya trees and random distributions. *Ann. Statist.* **20**, 1203–1221.

Mueller, P., Roeder, K. (1997). A Bayesian semiparametric model for case-control studies with errors in variables. *Biometrika* **84**, 523–537.

Mueller, P., Rosner, G.L. (1997). A Bayesian population model with hierarchical mixture priors applied to blood count data. *J. Amer. Statist. Assoc.* **92**, 1279–1292.

Mueller, P., Erkanli, A., West, M. (1996). Bayesian curve fitting using multivariate normal mixtures. *Biometrika* **83**, 67–79.

Mueller, P., West, M., MacEachern, S. (1997). Bayesian models for non-linear autoregressions. *J. Time Series Anal.* **18**, 593–614.

Mukhopadhyay, S. (2000). Bayesian nonparametric inference on the dose level with specified response rate. *Biometrics* **56**, 220–226.

Mukhopadhyay, S., Gelfand, A.E. (1997). Dirichlet process mixed generalized linear models.

Muliere, P., Tardella, L. (1998). Approximating distributions of random functionals of Ferguson Dirichlet priors. *Canad. J. Statist.* **26**, 283–297.

Muliere, P., Walker, S.G. (1997). A Bayesian nonparametric approach to survival analysis using Polya trees. *Scand. J. Statist.* **24**, 331–340.

Muliere, P., Walker, S.G. (1998). Extending the family of Bayesian bootstraps and exchangeable urn schemes. *J. Roy. Statist. Soc., Ser. B* **60**, 175–182.

Neal, R.M. (2000). Markov chain sampling methods for Dirichlet process mixture models. *J. Comput. Graph. Statist.* **9**, 249–265.

Newton, M.A., Zhang, Y. (1999). A recursive algorithm for nonparametric analysis with missing data. *Biometrika* **86**, 15–26.

Newton, M.A., Czado, C., Chappell, R. (1996). Bayesian inference for semiparametric binary regression. *J. Amer. Statist. Assoc.* **91**, 142–153.

Ogden, T. (1997). *Essential Wavelets for Statistical Applications and Data Analysis*. Birkhäuser, Boston.

Petrone, S. (1999a). Bayesian density estimation using Bernstein polynomials. *Canad. J. Statist.* **27**, 105–126.

Petrone, S. (1999b). Random Bernstein polynomials. *Scand. J. Statist.* **26**, 373–393.

Petrone, S., Wassserman, L. (2002). Consistency of Bernstein polynomial posteriors. *J. Roy. Statist. Soc., Ser. B* **64**, 79–100.

Quintana, F.A. (1998). Nonparametric Bayesian analysis for assessing homogeneity in $k\Theta$ contingency tables with fixed right margin totals. *J. Amer. Statist. Assoc.* **93**, 1140–1149.

Richardson, S., Green, P.J. (1997). On Bayesian analysis of mixtures with an unknown number of components (with discussion). *J. Roy. Statist. Soc., Ser. B* **59**, 731–792.

Robbins, H. (1964). The empirical Bayes approach to statistical decision problems. *Ann. Statist.* **35**, 1–20.

Rubin, D.B. (1981). The Bayesian bootstrap. *Ann. Statist.* **9**, 130–134.

Salinetti, G. (2003). New tools for consistency (with discussion). In: Bernardo, J.M., Bayarri, M.J., Berger, J.O., Dawid, A.P., Heckerman, D., Smith, A.F.M., West, M. (Eds.), *Bayesian Statistics, vol. 7*. Oxford University Press.

Schwarz, G. (1978). Estimating the dimension of a model. *Ann. Statist.* **6**, 461–464.

Sethuraman, J. (1994). Constructive definition of Dirichlet priors. *Statistica Sinica* **4**, 639–650.

Sethuraman, J., Tiwari, R.C. (1982). Convergence of Dirichlet measures and the interpretation of their parameter. In: Gupta, S., Berger, J.O. (Eds.), *Statistical Decision Theory and Related Topics III, vol. 2*. Springer-Verlag, New York, pp. 305–315.

Shaked, M. (1980). On mixtures from exponential families. *J. Roy. Statist. Soc., Ser. B* **42**, 192–198.

Shively, T.S., Kohn, R., Wood, S. (1999). Variable selection and function estimation in additive nonparametric regression using a data-based prior (with discussion). *J. Amer. Statist. Assoc.* **94**, 777–806.

Silverman, B.W. (1986). *Density Estimation for Statistics and Data Analysis*. Chapman and Hall, New York.

Sinha, D. (1993). Semiparametric Bayesian analysis of multiple time data. *J. Amer. Statist. Assoc.* **88**, 979–983.

Sinha, D. (1997). Time-discrete beta process model for interval censored survival data. *Canad. J. Statist.* **25**, 445–456.

Sinha, D., Dey, D.K. (1997). Semiparametric Bayesian analysis of survival data. *J. Amer. Statist. Assoc.* **92**, 1195–1212.

Smith, M., Kohn, R. (1996). Nonparametric regression using Bayesian variable selection. *J. Econometrics* **75**, 317–344.

Smith, M., Kohn, R. (1997). A Bayesian approach to nonparametric bivariate regression. *J. Amer. Statist. Assoc.* **92**, 1522–1535.

Smith, M., Kohn, R. (1998). Additive nonparametric regression with autocorrelated errors. *J. Roy. Statist. Soc., Ser. B*.

Stone, C. (1994). The use of polynomial splines and their tensor products in multivariate function estimation (with discussion). *Ann. Statist.* **22**, 118–184.

Susarla, V., Van Ryzin, J. (1976). Nonparametric Bayesian estimation of survival curves from incomplete observations. *J. Amer. Statist. Assoc.* **71**, 897–902.

Susarla, V., Van Ryzin, J. (1978). Large sample theory for a Bayesian nonparametric survival curve estimator based on censored samples. *Ann. Statist.* **6**, 755–768. Addendum (1980), *Ann. Statist.* **8**, 693.

Thorburn, D. (1986). A Bayesian approach to density estimation. *Biometrika* **73**, 65–75.

Vidakovic, B. (1998). Nonlinear wavelet shrinkage Bayes rules and Bayes factors. *J. Amer. Statist. Assoc.* **93**, 173–179.

Wahba, G. (1990). *Spline Models for Observational Data*. SIAM, Philadelphia.

Wahba, G. (1994). Data-based optimal smoothing of orthogonal series density estimates. *Ann. Statist.* **9**, 146–156.

Wakefield, J.C., Walker, S.G. (1997). Bayesian nonparametric population models: Formulation and comparison with likelihood approaches. *J. Pharmacokinetics & Biopharmaceutics* **25**, 235–253.

Walker, S.G. (2003). On sufficient conditions for Bayesian consistency. *Biometrika* **90**, 482–488.

Walker, S.G., Damien, P. (1998a). Sampling methods for Bayesian nonparametric inference involving stochastic processes. In: Dey, D., Mueller, P., Sinha, D. (Eds.), *Practical Nonparametric and Semiparametric Bayesian Statistics*. Springer-Verlag, New York.

Walker, S.G., Damien, P. (1998b). A full Bayesian nonparametric analysis involving a neutral to the right process. *Scand. J. Statist.* **25**, 669–680.

Walker, S.G., Hjort, N.L. (2001). On Bayesian consistency. *J. Roy. Statist. Soc., Ser. B* **63**, 811–821.

Walker, S.G., Mallick, B.K. (1996). Hierarchical generalized linear models and frailty models with Bayesian nonparametric mixing. *J. Roy. Statist. Soc., Ser. B* **59**, 845–860.

Walker, S.G., Muliere, P. (1997a). Beta-Stacy processes and a generalization of the Polya-urn scheme. *Ann. Statist.* **25**, 1762–1780.

Walker, S.G., Muliere, P. (1997b). A characterization of Polya tree distributions. *Statist. Probab. Lett.* **31**, 163–168.

Walker, S.G., Damien, P., Laud, P.W., Smith, A.F.M. (1999). Bayesian inference for random distributions and related functions (with discussion). *J. Roy. Statist. Soc., Ser. B* **34**, 1–42.

Walter, G.G. (1994). *Wavelets and Other Orthogonal Systems with Applications*. CRC Press, Boca Raton, FL.

West, M., Mueller, P., Escobar, M.D. (1994). Hierarchical priors and mixture models with application in regression and density estimation. In: Smith, A.F.M., Freeman, P. (Eds.), *Aspects of Uncertainty: A tribute to D.V. Lindley*. Wiley, New York.

Wolpert, R.L., Ickstadt, K. (1998a). Poisson/Gamma random field models for spatial statistics. *Biometrika* **85**, 251–267.

Wolpert, R.L., Ickstadt, K. (1998b). Simulation of Lévy random fields. In: Dey, D., Mueller, P., Sinha, D. (Eds.), *Practical Nonparametric and Semiparametric Bayesian Statistics*. Springer, New York, pp. 227–242.

Yamato, H. (1984). Characteristic functions of means of distributions chosen from a Dirichlet process. *Ann. Probab.* **12**, 262–267.

Handbook of Statistics, Vol. 25
ISSN: 0169-7161
© 2005 Elsevier B.V. All rights reserved.
DOI 10.1016/S0169-7161(05)25011-3

11

Bayesian Modeling in the Wavelet Domain

Fabrizio Ruggeri and Brani Vidakovic

Abstract
Wavelets are the building blocks of wavelet transforms the same way that the functions e^{inx} are the building blocks of the ordinary Fourier transform. But in contrast to sines and cosines, wavelets can be (or almost can be) supported on an arbitrarily small closed interval. This feature makes wavelets a very powerful tool in dealing with phenomena that change rapidly in time. In many statistical applications, there is a need for procedures to (i) adapt to data and (ii) use prior information. The interface of wavelets and the Bayesian paradigm provides a natural terrain for both of these goals. In this chapter, the authors provide an overview of the current status of research involving Bayesian inference in wavelet nonparametric problems. Two applications, one in functional data analysis (FDA) and the second in geoscience are discussed in more detail.

1. Introduction

Wavelet-based tools are now indispensable in many areas of modern statistics, for example in regression, density and function estimation, factor analysis, modeling and forecasting of time series, functional data analysis, data mining and classification, with ranges of application areas in science and engineering. Wavelets owe their initial popularity in statistics to shrinkage, a simple and yet powerful procedure in nonparametric statistical modeling. It can be described by the following three steps: (i) observations are transformed into a set of wavelet coefficients; (ii) a shrinkage of the coefficients is performed; and (iii) the processed wavelet coefficients are back transformed to the domain of the original data.

Wavelet domains form desirable modeling environments; several supporting arguments are listed below.

Discrete wavelet transforms tend to "disbalance" the data. Even though the orthogonal transforms preserve the ℓ_2 norm of the data (the square root of sum of squares of observations, or the "energy" in engineering terms), most of the ℓ_2 norm in the transformed data is concentrated in only a few wavelet coefficients. This concentration narrows the class of plausible models and facilitates the thresholding. The disbalancing property also yields a variety of criteria for the selection of best basis.

Wavelets, as building blocks in modeling, are localized well in both time and scale (frequency). Signals with rapid local changes (signals with discontinuities, cusps, sharp spikes, etc.) can be well represented with only a few wavelet coefficients. This parsimony does not, in general, hold for other standard orthonormal bases which may require many "compensating" coefficients to describe discontinuity artifacts or local bursts.

Heisenberg's principle states that time-frequency models not be precise in the time and frequency domains simultaneously. Wavelets adaptively distribute the time-frequency precision by their innate nature. The economy of wavelet transforms can be attributed to their ability to confront the limitations of Heisenberg's principle in a data-dependent manner.

An important feature of wavelet transforms is their whitening property. There is ample theoretical and empirical evidence that wavelet transforms simplify the dependence structure in the original signal. For example, it is possible, for any given stationary dependence in the input signal, to construct a biorthogonal wavelet basis such that the corresponding in the transform are uncorrelated (a wavelet counterpart of Karhunen–Loève transform). For a discussion and examples see Walter and Shen (2001).

We conclude this incomplete list of wavelet transform features by pointing out their sensitivity to self-similar data. The scaling laws are distinctive features of self-similar data. Such laws are clearly visible in the wavelet domain in the so called wavelet spectra, wavelet counterparts of the Fourier spectra.

More arguments can be provided: computational speed of the wavelet transform, easy incorporation of prior information about some features of the signal (smoothness, distribution of energy across scales), etc.

Basics on wavelets can be found in many texts, monographs, and papers at many different levels of exposition. The interested reader should consult monographs by Daubechies (1992), Ogden (1997), Vidakovic (1999), and Walter and Shen (2001), among others. An introductory article is Vidakovic and Müller (1999).

With self-containedness of this chapter in mind, we provide a brief overview of the discrete wavelet transforms (DWT).

1.1. Discrete wavelet transforms and wavelet shrinkage

Let $\boldsymbol{y}$ be a data-vector of dimension (size) n, where n is a power of 2, say 2^J. We assume that measurements $\boldsymbol{y}$ belong to an interval and consider periodized wavelet bases. Generalizations to different sample sizes and general wavelet and wavelet-like transforms are straightforward.

Suppose that the vector $\boldsymbol{y}$ is wavelet-transformed to a vector $\boldsymbol{d}$. This linear and orthogonal transform can be fully described by an $n \times n$ orthogonal matrix $\boldsymbol{W}$. In practice, one performs the DWT without exhibiting the matrix $\boldsymbol{W}$ explicitly, but by using fast filtering algorithms. The filtering procedures are based on so-called quadrature mirror filters which are uniquely determined by the wavelet of choice and fast Mallat's algorithm (Mallat, 1989). The wavelet decomposition of the vector $\boldsymbol{y}$ can be written as

$$\boldsymbol{d} = \left(H^\ell \boldsymbol{y}, GH^{\ell-1}\boldsymbol{y}, \ldots, GH^2\boldsymbol{y}, GH\boldsymbol{y}, G\boldsymbol{y}\right). \tag{1.1}$$

Note that in (1.1), $\boldsymbol{d}$ has the same length as $\boldsymbol{y}$ and ℓ is any fixed number between 1 and $J = \log_2 n$. The operators G and H are defined coordinate-wise via

$$(Ha)_k = \sum_{m \in \mathbf{Z}} h_{m-2k} a_m, \quad \text{and} \quad (Ga)_k = \sum_{m \in \mathbf{Z}} g_{m-2k} a_m, \quad k \in \mathbf{Z},$$

where g and h are high- and low-pass filters corresponding to the wavelet of choice. Components of g and h are connected via the *quadrature mirror* relationship $g_n = (-1)^n h_{1-n}$. For all commonly used wavelet bases, the taps of filters g and h are readily available in the literature or in standard wavelet software packages.

The elements of $\boldsymbol{d}$ are called "wavelet coefficients." The sub-vectors described in (1.1) correspond to detail levels in a levelwise organized decomposition. For instance, the vector $G\boldsymbol{y}$ contains $n/2$ coefficients representing the level of the finest detail. When $\ell = J$, the vectors $GH^{J-1}\boldsymbol{y} = \{d_{00}\}$ and $H^J\boldsymbol{y} = \{c_{00}\}$ contain a single coefficient each and represent the coarsest possible level of detail and the smooth part in wavelet decomposition, respectively.

In general, jth detail level in the wavelet decomposition of $\boldsymbol{y}$ contains 2^j elements, and is given as

$$GH^{J-j-1}\boldsymbol{y} = (d_{j,0}, d_{j,1}, \ldots, d_{j,2^j-1}). \tag{1.2}$$

Wavelet shrinkage methodology consists of shrinking wavelet coefficients. The simplest wavelet shrinkage technique is thresholding. The components of $\boldsymbol{d}$ are replaced by 0 if their absolute value is smaller than a fixed threshold λ.

The two most common thresholding policies are *hard* and *soft* thresholding with corresponding rules given by:

$$\theta^h(d, \lambda) = d\mathbf{1}(|d| > \lambda),$$
$$\theta^s(d, \lambda) = (d - \operatorname{sign}(d)\lambda)\mathbf{1}(|d| > \lambda),$$

where $\mathbf{1}(A)$ is the indicator of relation A, i.e., $\mathbf{1}(A) = 1$ if A is true and $\mathbf{1}(A) = 0$ if A is false.

In the next section we describe how the Bayes rules, resulting from the models on wavelet coefficients, can act as shrinkage/thresholding rules.

2. Bayes and wavelets

Bayesian paradigm has become very popular in wavelet data processing since Bayes rules are shrinkers. This is true in general, although examples of Bayes rules that expand can be constructed, see Vidakovic and Ruggeri (1999). The Bayes rules can be constructed to mimic the thresholding rules: to slightly shrink the large coefficients and heavily shrink the small coefficients. Bayes shrinkage rules result from realistic statistical models on wavelet coefficients and such models allow for incorporation of prior information about the *true* signal. Furthermore, most Practicable Bayes rules should be easily computed by simulation or expressed in a closed form. Reviews on early Bayesian approaches can be found in Abramovich et al. (2000) and Vidakovic (1998b, 1999). An edited volume on Bayesian modeling in the wavelet domain was edited by Muller and Vidakovic and appeared more than 5 years ago (Müller and Vidakovic, 1999c).

One of the tasks in which the wavelets are typically applied is recovery of an unknown signal $\boldsymbol{f}$ affected by noise $\boldsymbol{\varepsilon}$. Wavelet transforms $\boldsymbol{W}$ are applied to noisy measurements $y_i = f_i + \varepsilon_i$, $i = 1, \ldots, n$, or, in vector notation, $\boldsymbol{y} = \boldsymbol{f} + \boldsymbol{\varepsilon}$. The linearity of $\boldsymbol{W}$ implies that the transformed vector $\boldsymbol{d} = \boldsymbol{W}(\boldsymbol{y})$ is the sum of the transformed signal $\boldsymbol{\theta} = \boldsymbol{W}(\boldsymbol{f})$ and the transformed noise $\boldsymbol{\eta} = \boldsymbol{W}(\boldsymbol{\varepsilon})$. Furthermore, the orthogonality of $\boldsymbol{W}$ and normality of the noise vector $\boldsymbol{\varepsilon}$ implies the noise vector $\boldsymbol{\eta}$ is also normal.

Bayesian methods are applied in the wavelet domain, i.e., after the data have been transformed. The wavelet coefficients can be modeled in totality, as a single vector, or one by one, due to decorrelating property of wavelet transforms. Several authors consider modeling blocks of wavelet coefficients, as an intermediate solution (e.g., Abramovich et al., 2002; De Canditiis and Vidakovic, 2004).

We concentrate on typical wavelet coefficient and model: $d = \theta + \varepsilon$. Bayesian methods are applied to estimate the location parameter θ, which will be, in the sequel, retained as the shrunk wavelet coefficient and back transformed to the data domain. Various Bayesian models have been proposed. Some models have been driven by empirical justifications, others by pure mathematical considerations; some models lead to simple, close-form rules the other require extensive Markov Chain Monte Carlo simulations to produce the estimate.

2.1. *An illustrative example*

As an illustration of the Bayesian approach we present BAMS (Bayesian Adaptive Multiresolution Shrinkage). The method, due to Vidakovic and Ruggeri (2001), is motivated by empirical considerations on the coefficients and leads to easily implementable Bayes estimates, available in closed form.

The BAMS originates from the observation that a realistic Bayes model should produce prior predictive distributions of the observations which "agree" with the observations. Other authors were previously interested in the empirical distribution of the wavelet coefficients; see, for example, Leporini and Pesquet (1998, 1999), Mallat (1989), Ruggeri (1999), Simoncelli (1999), and Vidakovic (1998b). Quoting Vidakovic and Ruggeri (2001), their common argument can be summarized by the following statement:

> For most of the signals and images encountered in practice, the empirical distribution of a typical detail wavelet coefficient is notably centered about zero and peaked at it.

In the spirit of this statement, Mallat (1989) suggested an empirically justified model for typical wavelet coefficients, the exponential power distribution

$$f(d) = C \cdot \mathrm{e}^{-(|d|/\alpha)^{\beta}}, \qquad \alpha, \beta > 0,$$

with $C = \frac{\beta}{2\alpha\Gamma(1/\beta)}$.

Following the Bayesian paradigm, prior distributions should be elicited on the parameters of the model $d|\theta, \sigma^2 \sim \mathcal{N}(\theta, \sigma^2)$ and Bayesian estimators (namely, posterior means under squared loss functions) computed. In BAMS, prior on σ^2 is sought so that the marginal likelihood of the wavelet coefficient is a double exponential distribution $\mathcal{DE}$. The double exponential distribution can be obtained as scale mixture of normals with exponential as the mixing distribution. The choice of the exponential prior is can be additionally justified by its *maxent* property, i.e. the exponential distribution is the entropy maximizer in the class of all distributions supported on $(0, \infty)$ with a fixed first moment.

Thus, BAMS uses the exponential prior $\sigma^2 \sim \mathcal{E}(\mu),\ \mu > 0$, that leads to the marginal likelihood

$$d|\theta \sim \mathcal{DE}\left(\theta, \frac{1}{\sqrt{2\mu}}\right), \qquad \text{with density } f(d|\theta) = \frac{1}{2}\sqrt{2\mu}\,\mathrm{e}^{-\sqrt{2\mu}|d-\theta|}.$$

The next step is elicitation of a prior on θ. Vidakovic (1998a) considered the previous model and proposed t distribution as a prior on θ. He found the Bayes rules with respect to the squared error loss exhibit desirable shape. In personal communication with the second author, Jim Berger and Peter Müller suggested in 1993 the use of ε-contamination priors in the wavelet context pointing out that such priors would lead to rules which are smooth approximations to a thresholding.

The choice

$$\pi(\theta) = \varepsilon\delta(0) + (1-\varepsilon)\xi(\theta) \tag{2.1}$$

is now standard, and reflects prior belief that some locations (corresponding to the signal or function to be estimated) are 0 and that there is nonzero spread component ξ describing "large" locations. In addition to this prior sparsity of the signal part, this prior leads to desirable shapes of their resulting Bayes rules.

In BAMS, the spread part ξ is chosen as

$$\theta \sim \mathcal{DE}(0, \tau),$$

for which the predictive distribution for d is

$$m_\xi(d) = \frac{\tau\,\mathrm{e}^{-|d|/\tau} - \frac{1}{\sqrt{2\mu}}\,\mathrm{e}^{-\sqrt{2\mu}|d|}}{2\tau^2 - 1/\mu}.$$

The Bayes rule with respect to the prior ξ, under the squared error loss, is

$$\delta_\xi(d) = \frac{\tau(\tau^2 - 1/(2\mu))d\,\mathrm{e}^{-|d|/\tau} + \tau^2(\mathrm{e}^{-|d|\sqrt{2\mu}} - \mathrm{e}^{-|d|/\tau})/\mu}{(\tau^2 - 1/(2\mu))(\tau\,\mathrm{e}^{-|d|/\tau} - (1/\sqrt{2\mu})\,\mathrm{e}^{-|d|\sqrt{2\mu}})}. \tag{2.2}$$

When

$$\pi(\theta) = \varepsilon\delta_0 + (1-\varepsilon)\mathcal{DE}(0,\tau), \tag{2.3}$$

we use the marginal and rule under ξ to express the marginal and rule under the prior π. The marginal under π is

$$m_\pi(d) = \varepsilon\mathcal{DE}\left(0, \frac{1}{\sqrt{2\mu}}\right) + (1-\varepsilon)m_\xi(d)$$

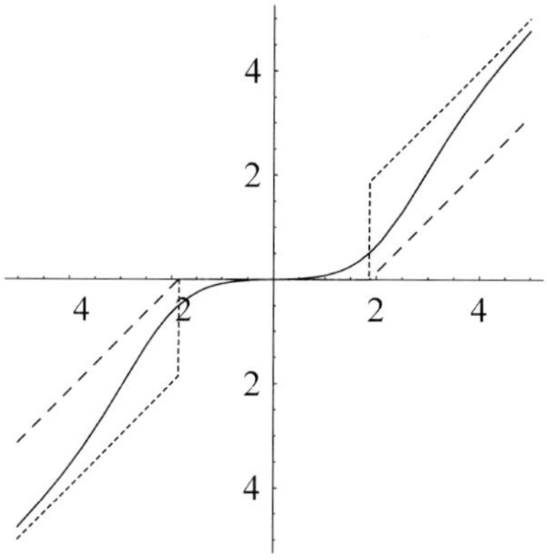

Fig. 1. BAMS rule (2.4) and comparable hard and soft thresholding rules.

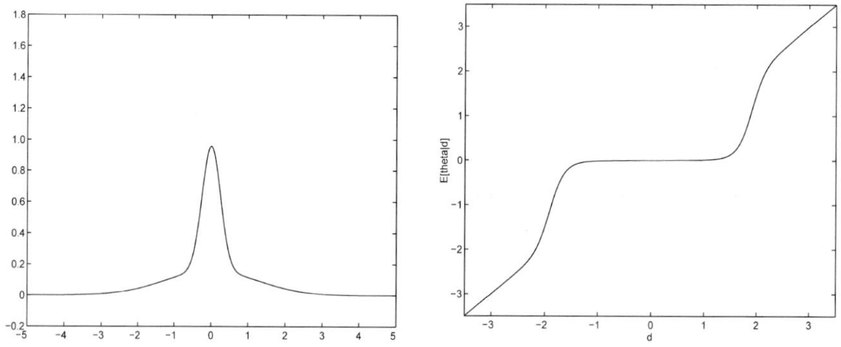

(a) (b)

Fig. 2. (a) Prior on θ as a mixture of two normal distributions with different variances. (b) Bayes rule (2.5) in Chipman et al. (1997).

while the Bayes rule is

$$\delta_\pi(d) = \frac{(1-\varepsilon)m_\xi(d)\delta_\xi(d)}{(1-\varepsilon)m_\xi(d) + \varepsilon\mathcal{DE}\big(0, \frac{1}{\sqrt{2\mu}}\big)}. \tag{2.4}$$

The rule (2.4) is the BAMS rule. As shown in Figure 1, the BAMS rule falls between comparable hard and soft thresholding rules.

Tuning of hyperparameters is an important implementational issue and it is thoroughly discussed in Vidakovic and Ruggeri (2001), who suggest an automatic selection based on the nature of the data.

2.2. *Regression problems*

In the context of wavelet regression, we will discuss two early approaches in more detail. The first one is Adaptive Bayesian Wavelet Shrinkage (ABWS) proposed by Chipman et al. (1997). Their approach is based on the stochastic search variable selection (SSVS) model proposed by George and McCulloch (1997), with the assumption that σ is known.

The likelihood in ABWS is

$$[d|\theta] \sim \mathcal{N}(\theta, \sigma^2).$$

The prior on θ is defined as a mixture of two normals (Figure 2(a))

$$[\theta|\gamma_j] \sim \gamma_j \mathcal{N}(0, (c_j\tau_j)^2) + (1-\gamma_j)\mathcal{N}(0, \tau_j^2),$$

where

$$[\gamma_j] \sim Ber(p_j).$$

Because the hyperparameters p_j, c_j, and τ_j depend on the level j to which the corresponding θ (or d) belongs, and can be level-wise different, the method is date-adaptive.

The Bayes rule under squared error loss for θ (from the level j) has an explicit form,

$$\delta(d) = \left[P(\gamma_j = 1|d)\frac{(c_j\tau_j)^2}{\sigma^2 + (c_j\tau_j)^2} + P(\gamma_j = 0|d)\frac{\tau_j^2}{\sigma^2 + \tau_j^2}\right]d, \tag{2.5}$$

where

$$P(\gamma_j = 1|d) = \frac{p_j\pi(d|\gamma_j = 1)}{(1-p_j)\pi(d|\gamma_j = 0)}$$

and

$$\pi(d|\gamma_j = 1) \sim \mathcal{N}(0, \sigma^2 + (c_j\tau_j)^2) \quad \text{and} \quad \pi(d|\gamma_j = 0) \sim \mathcal{N}(0, \sigma^2 + \tau_j^2).$$

The shrinkage rule ((2.5), Figure 2(b)) can be interpreted as a smooth interpolation between two linear shrinkages with slopes $\tau_j^2/(\sigma^2 + \tau_j^2)$ and $(c_j\tau_j)^2/(\sigma^2 + (c_j\tau_j)^2)$. The authors utilize empirical Bayes argument for tuning the hyperparameters level-wise. We note that most popular way to specify hyperparameters in Bayesian shrinkage is via empirical Bayes, see, for example, Abramovich et al. (2002), Angelini and Sapatinas (2004), Clyde and George (1999, 2000), and Huang and Cressie (1999). De Canditiis and Vidakovic (2004) extend the ABWS method to multivariate case and unknown σ^2 using a mixture of normal–inverse Gamma priors.

The approach used by Clyde et al. (1998) is based on a limiting form of the conjugate SSVS prior in George and McCulloch (1997). A similar model was used before in Müller and Vidakovic (1999a) but in the context of density estimation.

Clyde et al. (1996) consider a prior for $\boldsymbol{\theta}$ which is a mixture of a point mass at 0 if the variable is excluded from the wavelet regression and a normal distribution if it is included,

$$[\theta|\gamma_j, \sigma^2] \sim \mathcal{N}(0, (1-\gamma_j) + \gamma_j c_j \sigma^2).$$

The γ_j are indicator variables that specify which basis element or column of W should be selected. As before, the subscript j points to the level to which θ belongs. The set of all possible vectors $\boldsymbol{\gamma}$'s will be referred to as the subset space. The prior distribution for

σ^2 is inverse χ^2, i.e.

$$[\lambda\nu/\sigma^2] \sim \chi^2_\nu,$$

where λ and ν are fixed hyperparameters and the γ_j's are independently distributed as Bernoulli $Ber(p_j)$ random variables.

The posterior mean of $\boldsymbol{\theta}|\boldsymbol{\gamma}$ is

$$E(\boldsymbol{\theta}|\boldsymbol{d},\boldsymbol{\gamma}) = \Gamma\left(I_n + C^{-1}\right)^{-1}\boldsymbol{d}, \tag{2.6}$$

where Γ and C are diagonal matrices with γ_{jk} and c_{jk}, respectively, on the diagonal and 0 elsewhere. For a particular subset determined by the ones in $\boldsymbol{\gamma}$, (2.6) corresponds to thresholding with linear shrinkage.

The posterior mean is obtained by averaging over all models. Model averaging leads to a multiple shrinkage estimator of θ:

$$E(\boldsymbol{\theta}|\mathbf{d}) = \sum_{\gamma} \pi(\boldsymbol{\gamma}|\boldsymbol{d})\Gamma\left(I_n + C^{-1}\right)^{-1}\boldsymbol{d},$$

where $\pi(\boldsymbol{\gamma}|\boldsymbol{d})$ is the posterior probability of a particular subset $\boldsymbol{\gamma}$.

An additional nonlinear shrinkage of the coefficients to 0 results from the uncertainty in which subsets should be selected.

Calculating the posterior probabilities of $\boldsymbol{\gamma}$ and the mixture estimates for the posterior mean of $\boldsymbol{\theta}$ above involve sums over all 2^n values of $\boldsymbol{\gamma}$. The calculational complexity of the mixing is prohibitive even for problems of moderate size, and either approximations or stochastic methods for selecting subsets $\boldsymbol{\gamma}$ possessing high posterior probability must be used.

In the orthogonal case, Clyde et al. (1996) obtain an approximation to the posterior probability of $\boldsymbol{\gamma}$ which is adapted to the wavelet setting in Clyde et al. (1998). The approximation can be achieved by either conditioning on σ (plug-in approach) or by assuming independence of the elements in $\boldsymbol{\gamma}$.

The approximate model probabilities, for the conditional case, are functions of the data through the regression sum of squares and are given by:

$$\pi(\boldsymbol{\gamma}|\boldsymbol{d}) \approx \prod_{j,k} \rho_{jk}^{\gamma_{jk}}(1-\rho_{jk})^{1-\gamma_{jk}},$$

$$\rho_{jk}(\boldsymbol{d},\sigma) = \frac{a_{jk}(\boldsymbol{d},\sigma)}{1+a_{jk}(\boldsymbol{d},\sigma)},$$

where

$$a_{jk}(\boldsymbol{d},\sigma) = \frac{p_{jk}}{1-p_{jk}}(1+c_{jk})^{-1/2}\cdot\exp\left\{\frac{1}{2}\frac{S_{jk}^2}{\sigma^2}\right\},$$

$$S_{jk}^2 = d_{jk}^2/\left(1+c_{jk}^{-1}\right).$$

The p_{jk} can be used to obtain a direct approximation to the multiple shrinkage Bayes rule. The independence assumption leads to more involved formulas. Thus, the posterior

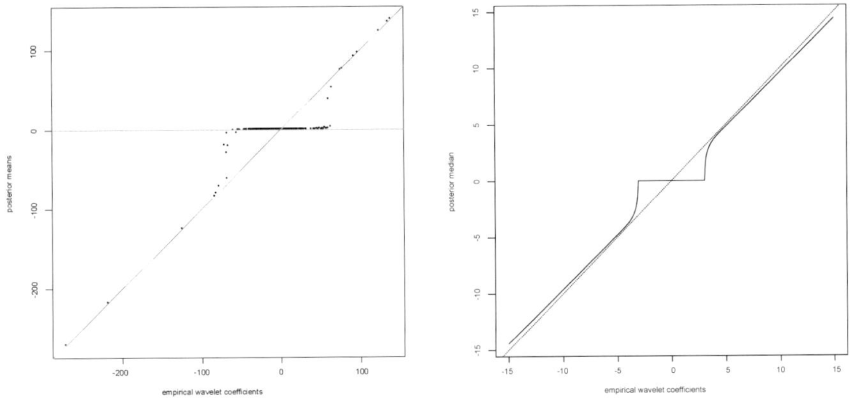

(a) (b)

Fig. 3. (a) Shrinkage rule from Clyde et al. (1998) based on independence approximation (2.7). (b) Posterior median thresholding rule (2.9) from Abramovich et al. (1998).

mean for θ_{jk} is approximately

$$\rho_{jk}\left(1 + c_{jk}^{-1}\right)^{-1} d_{jk}. \tag{2.7}$$

Eq. (2.7) can be viewed as a level dependent wavelet shrinkage rule, generating a variety of nonlinear rules. Depending on the choice of prior hyperparameters, shrinkage may be monotonic, if there are no level dependent hyperparameters, or nonmonotonic; see Figure 3(a).

Clyde and George (1999, 2000) propose a model in which the distributions for the error ε and θ are scale mixtures of normals, thus justifying Mallat's paradigm (Mallat, 1989). They use an empirical Bayes approach to estimate the prior hyperparamters, and provide analytic expressions for the shrinkage estimator based on Bayesian model averaging. They report an excellent denoising performance of their shrinkage method for a range of noise distributions.

2.3. Bayesian thresholding rules

Bayes rules under the squared error loss and regular models are never thresholding rules. We discuss two possible approaches for obtaining *bona fide* thresholding rules in a Bayesian manner. The first one is via hypothesis testing, while the second one uses weighted absolute error loss.

Donoho and Johnstone (1994, 1995) gave a heuristic for the selection of the universal threshold via rejection regions of suitable hypotheses tests. Testing a precise hypothesis in Bayesian fashion requires a prior which has a point mass component. A method based on Bayes factors is discussed first. For details, see Vidakovic (1998a).

Let

$$[d|\theta] \sim f(d|\theta).$$

After observing the coefficient d, the hypothesis H_0: $\theta = 0$, versus H_1: $\theta \neq 0$ is tested. If the hypothesis H_0 is rejected, θ is estimated by d. Let

$$[\theta] \sim \pi(\theta) = \pi_0\delta_0 + \pi_1\xi(\theta), \tag{2.8}$$

where $\pi_0 + \pi_1 = 1$, δ_0 is a point mass at 0, and $\xi(\theta)$ is a prior that describes distribution of θ when H_0 is false.

The resulting Bayesian procedure is:

$$\hat{\theta} = d\mathbf{1}\left(P(H_0|d) < \frac{1}{2}\right),$$

where

$$P(H_0|d) = \left(1 + \frac{\pi_1}{\pi_0}\frac{1}{B}\right)^{-1},$$

is the posterior probability of the H_0 hypothesis, and $B = \frac{f(d|0)}{\int_{\theta\neq 0} f(d|\theta)\xi(\theta)\,\mathrm{d}\theta}$ is the Bayes factor in favor of H_0. The optimality of Bayes Factor shrinkage was recently explored by Pensky and Sapatinas (2004) and Abramovich et al. (2004). They show that Bayes Factor shrinkage rule is optimal for wide range smoothness spaces and can outperform the posterior mean and the posterior median.

Abramovich et al. (1998) use weighted absolute error loss and show that for a prior on θ

$$[\theta] \sim \pi_j\mathcal{N}\left(0, \tau_j^2\right) + (1 - \pi_j)\delta(0)$$

and normal $\mathcal{N}(\theta, \sigma^2)$ likelihood, the posterior median is

$$Med(\theta|d) = \mathrm{sign}(d)\max(0, \zeta). \tag{2.9}$$

Here

$$\zeta = \frac{\tau_j^2}{\sigma^2 + \tau_j^2}|d| - \frac{\tau_j\sigma}{\sqrt{\sigma^2 + \tau_j^2}}\Phi^{-1}\left(\frac{1 + \min(\omega, 1)}{2}\right), \quad \text{and}$$

$$\omega = \frac{1 - \pi_j}{\pi_j}\frac{\sqrt{\tau_j^2 + \sigma^2}}{\sigma}\exp\left\{-\frac{\tau_j^2 d^2}{2\sigma^2(\tau_j^2 + \sigma^2)}\right\}.$$

The index j, as before, points to the level containing θ (or d). The plot of the thresholding function (2.9) is given in Figure 3(b).

The authors compare the rule (2.9), they call BayesThresh, with several methods (Cross-Validation, False Discovery Rate, VisuShrink and GlobalSure) and report very good MSE performance.

2.4. *Bayesian wavelet methods in functional data analysis*

Recently wavelets have been used in functional data analysis as a useful tool for dimension reduction in the modeling of multiple curves. This has also led to important contributions in interdisciplinary fields, such as chemometrics, biology and nutrition.

Brown et al. (2001) and Vannucci et al. (2001) considered regression models that relate a multivariate response to functional predictors, applied wavelet transforms to the curves, and used Bayesian selection methods to identify features that best predict the responses. Their model in the data domain is

$$\boldsymbol{Y} = \mathbf{1}_n \boldsymbol{\alpha}' + \boldsymbol{X}\boldsymbol{B} + \boldsymbol{E}, \tag{2.10}$$

where $\boldsymbol{Y}(n \times q)$ are q-variate responses and $\boldsymbol{X}(n \times p)$ the functional predictors data, each row of $\boldsymbol{X}$ being a vector of observations of a curve at p equally spaced points. In the practical context considered by the authors, the responses are given by the composition (by weight) of the $q = 4$ constituents of 40 biscuit doughs made with variations in quantities of fat, flour, sugar and water in a recipe. The functional predictors are near infrared spectral data measured at $p = 700$ wavelengths (from 1100 nm to 2498 nm in steps of 2 nm) for each dough piece. The goal is to use the spectral data to predict the composition. With $n \ll p$ and p large, wavelet transforms are employed as an effective tool for dimension reduction that well preserves local features.

When a wavelet transform is applied to each row of $\boldsymbol{X}$, the model (2.10) becomes

$$\boldsymbol{Y} = \mathbf{1}_n \boldsymbol{\alpha}' - \boldsymbol{Z}\widetilde{\boldsymbol{B}} + \boldsymbol{E}$$

with $\boldsymbol{Z} = \boldsymbol{X}\boldsymbol{W}'$ a matrix of wavelet coefficients and $\widetilde{\boldsymbol{B}} = \boldsymbol{W}\boldsymbol{B}$ the transformed regression coefficients. Shrinkage mixture priors are imposed on the regression coefficients. A latent vector with p binary entries serves to identify one of two types of regression coefficients, those close to zero and those not. The authors use results from Vannucci and Corradi (1999) to specify suitable prior covariance structures in the domain of the data that nicely transform to modified priors on the wavelet coefficients domain. Using a natural conjugate Gaussian framework, the marginal posterior distribution of the binary latent vector is derived. Fast algorithms aid its direct computation, and in high dimensions these are supplemented by a Markov Chain Monte Carlo approach to sample from the known posterior distribution. Predictions are then based on the selected coefficients. The authors investigate both model averaging strategies and single models predictions.

In Vannucci et al. (2003) an alternative decision theoretic approach is investigated where variables have genuine costs and a single subset is sought. The formulation they adopt assumes a joint normal distribution of the q-variate response and the full set of p regressors. Prediction is done assuming quadratic losses with an additive cost penalty nondecreasing in the number of variables. Simulated annealing and genetic algorithms are used to maximize the expected utility.

Morris et al. (2003) extended wavelet regression to the nested functional framework. Their work was motivated by a case study investigating the effect of diet on *O^6-methylguanine-DNA-methyltransferase* (MGMT), an important biomarker in early colon carcinogenesis. Specifically, two types of dietary fat (fish oil or corn oil) were investigated as potentially important factors that affect the initiation stage of carcinogenesis, i.e., the first few hours after the carcinogen exposure. In the experiment 30 rats were fed one of the 2 diets for 14 days, exposed to a carcinogen, then sacrificed at one of 5 times after exposure (0, 3, 6, 9, or 12 hours). Rat's colons were removed and dissected, and measurements of various biomarkers, including MGMT, were obtained. Each biomarker was measured on a set of 25 crypts in the distal and proximal

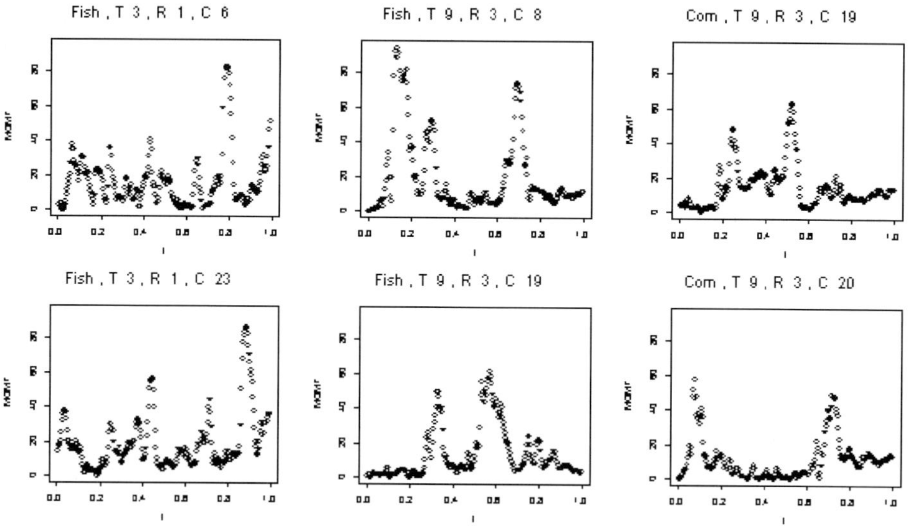

Fig. 4. DNA repair enzyme for selected crypts.

regions of each rat's colon. Crypts are fingerlike structures that extend into the colon wall. The procedure yielded observed curves for each crypt consisting of the biomarker quantification as a function of relative cell position within the crypt, the position being related to cell age and stage in the cell cycle. Due to the image processing used to quantify the measurements, these functions may be very irregular, with spikes presumably corresponding to regions of the crypt with high biomarker levels, see Figure 4.

The primary goal of the study was to determine whether diet has an effect on MGMT levels, and whether this effect depends on time and/or relative depth within the crypt. Another goal was to assess the relative variability between crypts and between rats. The authors model the curves in the data domain using a nonparametric hierarchical model of the type

$$
\begin{aligned}
Y_{abc} &= g_{abc}(t) + \varepsilon_{abc}, \\
g_{abc}(t) &= g_{ab}(t) + \eta_{abc}(t), \\
g_{ab}(t) &= g_{a}(t) + \zeta_{ab}(t),
\end{aligned}
$$

with Y_{abc} the response vector for crypt c, rat b and treatment a, $g_{\cdot}(t)$ the true crypt, rat or treatment profile, ε_{abc} the measurement error and $\eta_{abc}(t)$, $\zeta_{ab}(t)$ the crypt/rat level error.

The wavelet-based Bayesian method suggested by the authors leads to adaptively regularized estimates and posterior credible intervals for the mean function and random effects functions, as well as the variance components of the model. The approach first applies DWT to each observed curve to obtain the corresponding wavelet coefficients. This step results in the projection of the original curves into a transformed domain, where modeling can be done in a more parsimonious way. A Bayesian model is then fit

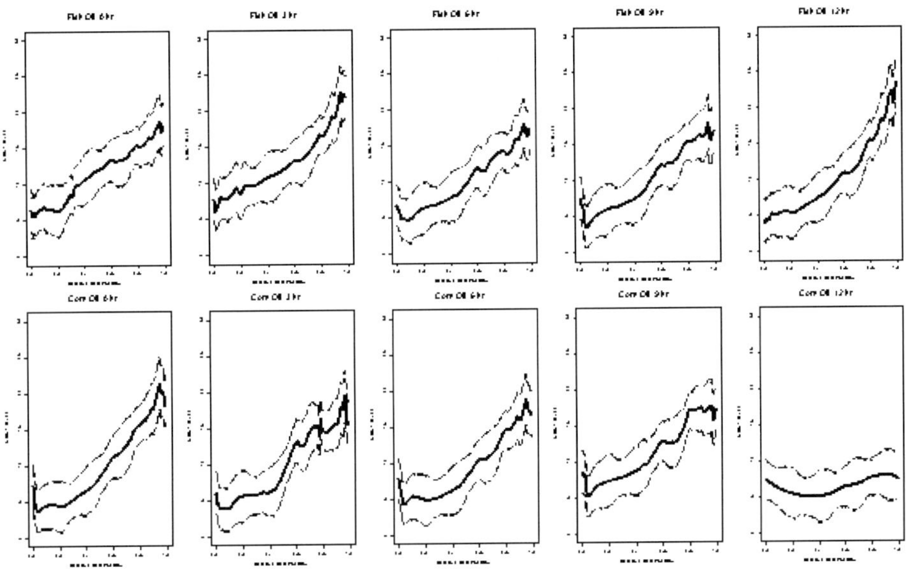

Fig. 5. Estimated mean profiles by diet/time with 90% posterior bounds.

to each wavelet coefficient across curves using an MCMC procedure to obtain posterior samples of the wavelet coefficients corresponding to the functions at each hierarchical level and the variance components. The inverse DWT is applied to transform the obtained estimates back to the data domain.

The Bayesian modeling adopted is such that the function estimates at all levels are adaptively regularized using a multiple shrinkage prior imposed at the top level of the hierarchy. The treatment level functions are directly regularized by this shrinkage, while the functions at the lower levels of the hierarchy are subject to some regularization induced by the higher levels, as modulated by the variance components. The authors provide guidelines for selecting these regularization parameters, together with empirical Bayes estimates, introduced during the rejoinder to the discussion.

Results from the analysis of the case study reveal that there is more MGMT expressed at the lumenal surface of the crypt, and suggest a diet difference in the MGMT expression at this location 12 hours after exposure to the carcinogen, see Figure 5. Also, the multiresolution wavelet analysis highlights features present in the crypt-level profiles that may correspond to individual cells, suggesting the hypothesis that MGMT operates on a largely cell-by-cell basis.

2.5. *The density estimation problem*

Donoho et al. (1996), Hall and Patil (1995), and Walter and Shen (2001), among others, applied wavelets in density estimation from a classical and data analytic perspective.

Chencov (1962) proposed projection type density estimators in terms of an arbitrary orthogonal basis. In the case of a wavelet basis, Chencov's estimator has the form

$$\hat{f}(x) = \sum_k c_{j_0k}\phi_{j_0k}(x) + \sum_{j_0\leqslant j\leqslant j_1}\sum_k d_{jk}\psi_{jk}(x), \tag{2.11}$$

where the coefficients c_{jk} and d_{jk} constituting the vector d are defined via the standard empirical counterparts of $\langle f, \phi_{jk}\rangle$ and $\langle f, \psi_{jk}\rangle$. Let $X_1, \ldots, X_n$ be a random sample from f. Then

$$c_{jk} = \frac{1}{n}\sum_{i=1}^{n}\phi_{jk}(X_i), \quad \text{and} \quad d_{jk} = \frac{1}{n}\sum_{i=1}^{n}\psi_{jk}(X_i). \tag{2.12}$$

Müller and Vidakovic (1999a, 1999b) parameterize an unknown density $f(x)$ by a wavelet series on its logarithm, and propose a prior model which explicitly defines geometrically decreasing prior probabilities for nonzero wavelet coefficients at higher levels of detail.

The unknown probability density function $f(\cdot)$ is modeled by:

$$\log f(x) = \sum_{k\in Z}\xi_{j_0k}\phi_{j_0,k}(x) + \sum_{j\geqslant j_0,k\in Z} s_{jk}\theta_{jk}\psi_{jk}(x) - \log K, \tag{2.13}$$

where $K = \int f(x)\,\mathrm{d}x$ is the normalization constant and $s_{jk} \in \{0, 1\}$ is an indicator variable that performs model induced thresholding.

The dependence of $f(x)$ on the vector $\theta = (\xi_{j_0,k}, s_{jk}, \theta_{jk},\ j = j_0, \ldots, j_1,\ k \in Z)$ of wavelet coefficients and indicators is expressed by $f(x) = p(x|\theta)$. The sample $X = \{X_1, \ldots, X_n\}$ defines a likelihood function $p(X|\theta) = \prod_{i=1}^{n} p(X_i|\theta)$.

The model is completed by a prior probability distribution for θ. Without loss of generality, $j_0 = 0$ can be assumed. Also, any particular application will determine the finest level of detail j_1.

$$\begin{aligned}
[\xi_{0k}] &\sim \mathcal{N}(0, \tau r_0),\\
[\theta_{jk}|s_{jk} = 1] &\sim \mathcal{N}(0, \tau r_j), \quad r_j = 2^{-j},\\
[s_{jk}] &\sim \mathit{Bernoulli}\big(\alpha^j\big),\\
[\alpha] &\sim \mathit{Beta}(a, b),\\
[1/\tau] &\sim \mathit{Gamma}(a_\tau, b_\tau).
\end{aligned} \tag{2.14}$$

The wavelet coefficients θ_{jk} are nonzero with geometrically decreasing probabilities. Given that a coefficient is nonzero, it is generated from a normal distribution. The parameter vector θ is augmented in order to include all model parameters, i.e., $\theta = (\theta_{jk}, \xi_{jk}, s_{jk}, \alpha, \tau)$.

The scale factor r_j contributes to the adaptivity of the method. Wavelet shrinkage is controlled by both: the factor r_j and geometrically decreasing prior probabilities for nonzero coefficient, α^j.

The conditional prior $p(\theta_{jk}|s_{jk} = 0) = h(\theta_{jk})$ is a pseudo-prior as discussed in Carlin and Chib (1995). The choice of $h(\cdot)$ has no bearing on the inference about $f(x)$.

In fact, the model could be alternatively formulated by dropping θ_{jk} under $s_{jk} = 0$. However, this model would lead to a parameter space of varying dimension. Carlin and Chib (1995) argue that the pseudo-prior $h(\theta_{jk})$ should be chosen to produce values for θ_{jk} which are consistent with the data.

The particular MCMC simulation scheme used to estimate the model (2.13), (2.14) is described. Starting with some initial values for θ_{jk}, $j = 0, \ldots, j_1$, ξ_{00}, α, and τ, the following Markov chain was implemented.

1. For each $j = 0, \ldots, j_1$ and $k = 1, \ldots, 2^j - 1$ go over the steps 2 and 3.
2. Update s_{jk}. Let θ_0 and θ_1 indicate the current parameter vector θ with s_{jk} replaced by 0 and 1, respectively. Compute $p_0 = p(y|\theta_0) \cdot (1-\alpha^j)h(\theta_{jk})$ and $p_1 = p(y|\theta_0) \cdot \alpha^j p(\theta_{jk}|s_{jk} = 1)$. With probability $p_1/(p_0 + p_1)$ set $s_{jk} = 1$, else $s_{jk} = 0$.

3a. Update θ_{jk}. If $s_{jk} = 1$, generate $\tilde{\theta}_{jk} \sim g(\tilde{\theta}_{jk}|\theta_{jk})$. Use, for example, $g(\tilde{\theta}_{jk}|\theta_{jk}) = \mathcal{N}(\theta_{jk}, 0.25\sigma_{jk})$, where σ_{jk} is some rough estimate of the posterior standard deviation of θ_{jk}. We will discuss alternative choices for the probing distribution $g(\cdot)$ below.

Compute

$$a(\theta_{jk}, \tilde{\theta}_{jk}) = \min\left[1, \frac{p(y|\tilde{\theta})p(\tilde{\theta}_{jk})}{p(y|\theta)p(\theta_{jk})}\right],$$

where $\tilde{\theta}$ is the parameter vector θ with θ_{jk} replaced by $\tilde{\theta}_{jk}$, and $p(\theta_{jk})$ is the p.d.f. of the normal prior distribution given in (2.14).

With probability $a(\theta_{jk}, \tilde{\theta}_{jk})$ replace θ_{jk} by $\tilde{\theta}_{jk}$; else keep θ_{jk} unchanged.

3b. If $s_{jk} = 0$, generate θ_{jk} from the full conditional posterior $p(\theta_{jk}|\ldots, X) = p(\theta_{jk}|s_{jk} = 0) = h(\theta_{jk})$.

4. Update ξ_{00}. Generate $\tilde{\xi}_{00} \sim g(\tilde{\xi}_{00}|\xi_{00})$. Use, for example, $g(\tilde{\xi}_{00}|\xi_{00}) = \mathcal{N}(\xi_{00}, 0.25\rho_{00})$, where ρ_{00} is some rough estimate of the posterior standard deviation of ξ_{00}. Analogously to step 3a, compute an acceptance probability a and replace ξ_{00} with probability a.
5. Update α. Generate $\tilde{\alpha} \sim g_\alpha(\tilde{\alpha}|\alpha)$ and compute

$$a(\alpha, \tilde{\alpha}) = \min\left[1, \frac{\prod_{jk} \tilde{\alpha}^{js_{jk}}(1 - \tilde{\alpha}^j)^{s_{jk}}}{\prod_{jk} \alpha^{js_{jk}}(1 - \alpha^j)^{s_{jk}}}\right].$$

   With probability $a(\alpha, \tilde{\alpha})$ replace α by $\tilde{\alpha}$, else keep α unchanged.
6. Update τ. Resample τ from the complete inverse Gamma conditional posterior.
7. Iterate over steps 1 through 6 until the chain is judged to have practically converged.

The algorithm implements a Metropolis chain changing one parameter at a time in the parameter vector. See, for example, Tierney (1994) for a description and discussion of Metropolis chains for posterior exploration. For a practical implementation, g should be chosen such that the acceptance probabilities a are neither close to zero, nor close to one. In the implementations, $g(\tilde{\theta}_{jk}|\theta_{jk}) = \mathcal{N}(\theta_{jk}, 0.25\sigma_{jk})$ with $\sigma_{jk} = 2^{-j}$, was used.

Described wavelet based density estimation model is illustrated on the galaxy data set (Roeder, 1990). The data is rescaled to the interval [0, 1]. The hyperparameters were

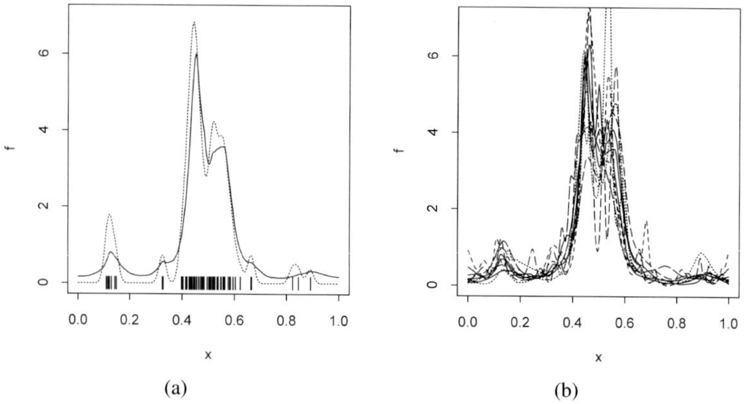

Fig. 6. (a) The estimated p.d.f. $\hat{f}(x) = \int p(x|\theta)\,\mathrm{d}p(\theta|X)$. The dotted line plots a conventional kernel density estimate for the same data. (b) The posterior distribution of the unknown density $f(x) = p(x|\theta)$ induced by the posterior distribution $p(\theta|X)$. The lines plot $p(x|\theta_i)$ for ten simulated draws posterior $\theta_i \sim p(\theta|X), i = 1, \ldots, 10$.

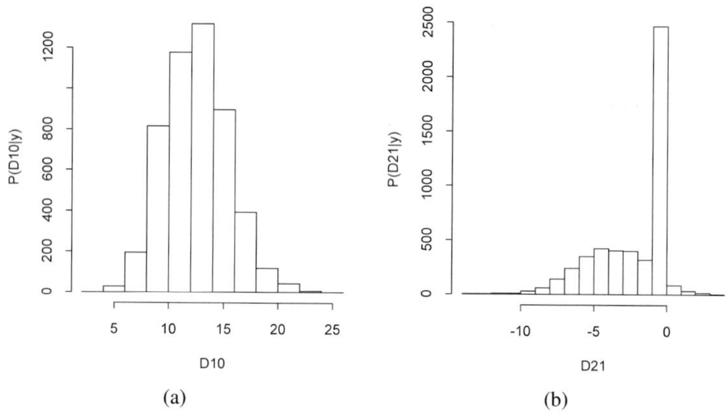

Fig. 7. Posterior distributions $p((s_{10}\theta_{10})|X)$ and $p((s_{21}\theta_{21})|X)$. While $s_{10}\theta_{10}$ is nonzero with posterior probability close to one, the posterior distribution $p((s_{21}\theta_{21})|X)$ is a mixture of a point mass at zero and a continuous part.

fixed as $a = 10$, $b = 10$, and $a_\tau = b_\tau = 1$. The $Beta(10, 10)$ prior distribution on α is reasonably noninformative compared to the likelihood based on $n = 82$ observations.

Initially, all s_{jk} are set to one, and α to its prior mean $\alpha = 0.5$. The first 10 iterations as burn-in period were discarded, then 1000 iterations of steps 1 through 6 were simulated. For each j, k, step 3 was repeated three times. The maximum level of detail selected was $j_1 = 5$.

Figures 6 and 7 describe some aspects of the analysis.

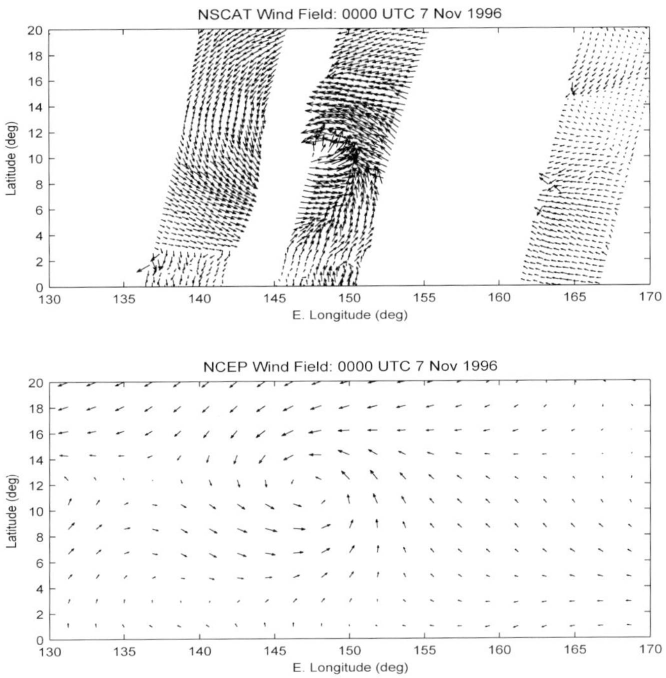

Fig. 8. (Top) Incomplete satellite-derived estimates of near-surface winds over the ocean. (Bottom) The "analysis" winds from the National Centers for Environmental Prediction (NCEP).

2.6. An application in geoscience

Wikle et al. (2001) considered a problem common in the atmospheric and ocean sciences, in which there are two measurement systems for a given process, each of which is imperfect. Satellite-derived estimates of near-surface winds over the ocean are available, and although they have very high spatial resolution, their coverage is incomplete (e.g., top panel of Figure 8; the direction of the wind is represented by the direction of the arrow and the wind speed is proportional to the length of the arrow). In addition, wind fields from the major weather centers are produced through combinations of observations and deterministic weather models (so-called "analysis" winds). These winds provide complete coverage, but are have relatively low spatial resolution (e.g., bottom panel of Figure 8 shows such winds from the National Centers for Environmental Prediction (NCEP)). Wikle et al. (2001) were interested in predicting spatially distributed wind fields at intermediate spatial resolutions and regular time intervals (every six hours) over the tropical oceans. Thus, they sought to combine these data sets (over multiple time periods) in such a way as to incorporate the space-time dynamics inherent in the surface wind field. They utilized the fact that to a first approximation, tropical winds can be described by so-called "linear shallow-water" dynamics. In addition, previous studies (e.g., Wikle et al., 1999) showed that tropical near surface winds exhibit turbulent scaling behavior in space. That is, the system can be modeled as a fractal

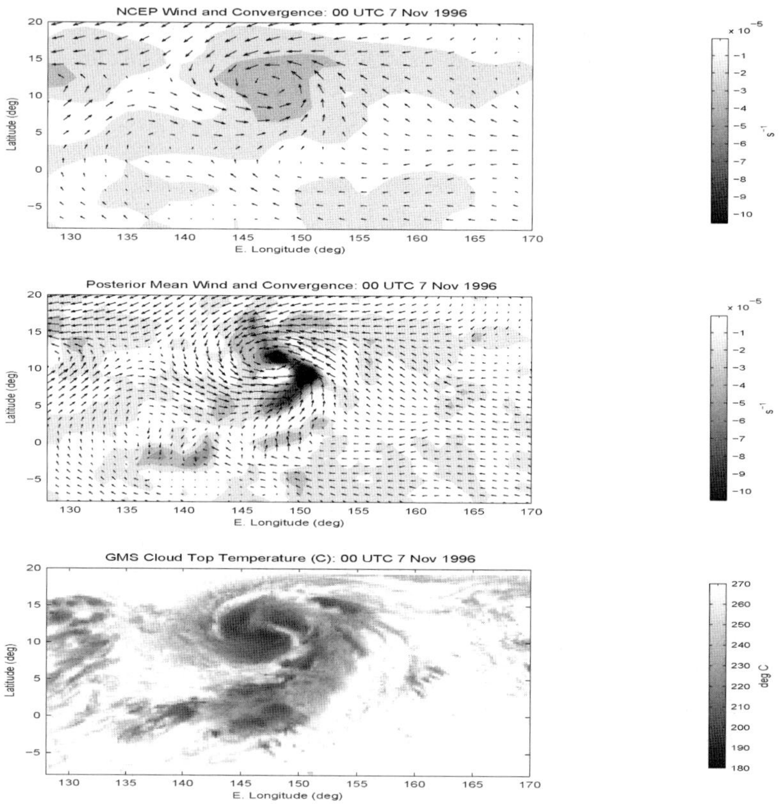

Fig. 9. (Top) The wind field from the NCEP weather center analysis winds (arrows) along with the implied convergence/divergence field (shaded area). More intense convergence (darker shading) often implies strong upward vertical motion and thus is suggestive of higher clouds and stronger precipitation development. (Middle) The posterior mean wind and convergence fields from the model of Wikle et al. (2001). (Bottom) Satellite cloud image (cloud top temperature) for the same period (darker colors imply higher clouds and thus more intense precipitation).

self-similar process that is nonstationary in space, but when considered through certain spatially-invariant filters, appears stationary (i.e., a $1/f$ process). In the case of tropical near-surface winds, the energy spectrum is proportional to the inverse of the spatial frequency taken to the 5/3 power (Wikle et al., 1999). Thus, the model for the underlying wind process must consider the shallow-water dynamics and the spectral scaling relationship.

Wikle et al. (2001) considered a Bayesian hierarchical approach that considered models for the data conditioned on the desired wind spatio-temporal wind process and parameters, models for the process given parameters and finally models for the parameters. The data models considered the change of support issues associated with the two disparate data sets. More critically, the spatio-temporal wind process was decomposed into three components, a spatial mean component representative of climatological winds over the area, a large-scale dynamical component representative of linear shallow-water dynamics, and a multiresolution (wavelet) component representative of medium to fine-scale processes in the atmosphere. The dynamical evolution models for these latter two components then utilized quite informative priors which made use of the aforementioned prior theoretical and empirical knowledge about the shallow-water and multiresolution processes. In this way, information from the two data sources, regardless of level of resolution, could impact future prediction times through the evolution equations. An example of the output from this model is shown in Figure 9.

The top panel shows the wind field from the NCEP weather center analysis winds (arrows) along with the implied convergence/divergence field (shaded area). More intense convergence (darker shading) often implies strong upward vertical motion and thus is suggestive of higher clouds and stronger precipitation development. This location and time corresponds to be a period when tropical cyclone "Dale" (centered near 146 E, 12 N) was the dominant weather feature. Note that the this panel does not show much definition to the storm, relative to panel 3 which shows a satellite cloud image (cloud top temperature) for the same period (darker colors imply higher clouds and thus more intense precipitation). The posterior mean wind and convergence fields from the model of Wikle et al. (2001) are shown in the middle panel of Figure 9. Note in this case that the convergence fields match up more closely with features in the satellite cloud image. In addition, the multiresolution (turbulent) nature of the cloud image is better represented in the posterior mean fields as well. Although not shown here, another advantage of this approach is that one can quantify the uncertainty in the predictions of the winds and derivative fields (see Wikle et al., 2001).

3. Other problems

The field of Bayesian wavelet modeling deserves an extensive monograph and this chapter is only highlighting the field. In this section we add some more references on various Bayesian approaches in wavelet data processing.

Lina and MacGibbon (1997) apply a Bayesian approach to wavelet regression with complex valued Daubechies wavelets. To some extent, they exploit redundancy in the representation of real signals by the complex wavelet coefficients. Their shrinkage technique is based on the observation that the modulus and the phase of wavelet coefficients encompass very different information about the signal. A Bayesian shrinkage model is constructed for the modulus, taking into account the corresponding phase.

Simoncelli and Adelson (1996) discuss Bayes "coring" procedure in the context of image processing. The prior on the signal is Mallat's model, see Mallat (1989), while the noise is assumed normal. They implement their noise reduction scheme on an oriented multiresolution representation – known as the *steerable pyramid*. They report that Bayesian coring outperforms classical Wiener filtering. See also Simoncelli (1999) and Portilla et al. (2003) for related research. A comprehensive comparison of Bayesian and non-Bayesian wavelet models applied to neuroimaging can be found in Fadili and Bullmore (2004).

Crouse et al. (1998) consider hidden Markov fields in a problems of image denoising. They develop the Efficient Expectation Maximization (EEM) algorithm to fit their model. See also Figueiredo and Nowak (2001). Shrinkage induced by Bayesian models in which the hyperparameters of the prior are made time dependent in an empirical Bayes fashion is considered in Vidakovic and Bielza Lozoya (1998). Kolaczyk (1999) and Nowak and Kolaczyk (2000) apply Bayesian modeling in the 2-D wavelet domains where Poisson counts are of interest.

Leporini and Pesquet (1998) explore cases for which the prior is an exponential power distribution [$\mathcal{EPD}(\alpha, \beta)$]. If the noise also has an $\mathcal{EPD}(a, b)$ distribution with $0 < \beta < b \leqslant 1$, the maximum aposteriori (MAP) solution is a hard-thresholding rule. If $0 < \beta \leqslant 1 < b$ then the resulting MAP rule is

$$\delta(d) = d - \left(\frac{\beta a^b}{b\alpha^\beta}\right)^{1/(b-1)} |d|^{(\beta-1)/(b-1)} + \mathrm{o}\bigl(|d|^{(\beta-1)/(b-1)}\bigr).$$

The same authors consider the Cauchy noise as well and explore properties of the resulting MAP rules. When the priors are hierarchical (mixtures) Leporini et al. (1999) demonstrated that the MAP solution can be degenerated and suggested Maximum Generalized Marginal Likelihood method. Some related derivations can be found in Chambolle et al. (1998) and Leporini and Pesquet (1999).

Pesquet et al. (1996) develop a Bayesian-based approach to the best basis problem, while preserving the classical tree search efficiency in wavelet packets and local trigonometric bases. Kohn et al. (2000) use a model similar to one in Chipman et al. (1997) but in the context of the best basis selection.

Ambler and Silverman (2004a, 2004b) allow for the possibility that the wavelet coefficients are locally correlated in both location (time) and scale (frequency). This leads to an analytically intractable prior structure. However, they show that it is possible to draw independent samples from a close approximation to the posterior distribution by an approach based on *Coupling From The Past*, making it possible to take a simulation-based approach to wavelet shrinkage.

Angelini and Vidakovic (2004) show that Γ-minimax shrinkage rules are Bayes with respect to a least favorable contamination prior with a uniform spread distribution ξ. Their method allows for incorporation of information about the energy in the signal of interest.

Ogden and Lynch (1999) describe a Bayesian wavelet method for estimating the location of a change-point. Initial treatment is for the standard change-point model (that is, constant mean before the change and constant mean after it) but extends to the case of detecting discontinuity points in an otherwise smooth curve. The conjugate prior distribution on the change point τ is given in the wavelet domain, and it is updated by observed empirical wavelet coefficients.

Ruggeri and Vidakovic (1999) discuss Bayesian decision theoretic thresholding. In the set of all hard thresholding rules, they find restricted Bayes rules under a variety of models, priors, and loss functions. When the data are multivariate, Chang and Vidakovic (2002) propose a wavelet-based shrinkage estimation of a single data component of interest using information from the rest of multivariate components. This incorporation of information is done via Stein-type shrinkage rule resulting from an empirical Bayes standpoint. The proposed shrinkage estimators maximize the predictive density under appropriate model assumptions on the wavelet coefficients.

Huang and Lu (2000) and Lu et al. (2003) suggested linear Bayesian wavelet shrinkage in a nonparametric mixed-effect model. Their formulation is conceptually inspired by the duality between reproducing kernel Hilbert spaces and random processes as well as by connections between smoothing splines and Bayesian regressions. The unknown function f in the standard nonparametric regression formulation ($y = f(x_i)+\sigma\varepsilon_i$, $i = 1,\dots,n$; $0 \leqslant x \leqslant 1$; $\sigma > 0$; $\mathrm{Cov}(\varepsilon_1,\dots,\varepsilon_n) = R$) is given a prior of the form $f(x) = \Sigma_k \alpha_{Jk}\phi_{Jk}(x) + \delta Z(x)$; $Z(x) \sim \Sigma_{j\geqslant J}\Sigma_k \theta_{jk}\psi_{jk}(x)$, where θ_{jk} are uncorrelated random variables such that $E\theta_{jk} = 0$ and $E\theta_{jk}^2 = \lambda_j$. The authors propose a linear, empirical Bayes estimator $\hat{f}$ of f that enjoys Gauss–Markov type of optimality. Several nonlinear versions of the estimator are proposed, as well. Independently, and by using different techniques, Huang and Cressie (1999, 2000) consider the same problem and derive a Bayesian estimate.

Acknowledgements

We would like to thank Dipak Dey for the kind invitation to compile this chapter. This work was supported in part by NSA Grant E-24-60R at the Georgia Institute of Technology.

References

Abramovich, F., Sapatinas, T., Silverman, B.W. (1998). Wavelet thresholding via Bayesian approach. *J. Roy. Statist. Soc., Ser. B* **60**, 725–749.

Abramovich, F., Bailey, T.C., Sapatinas, T. (2000). Wavelet analysis and its statistical applications. *The Statistician* **49**, 1–29.

Abramovich, F., Besbeas, P., Sapatinas, T. (2002). Empirical Bayes approach to block wavelet function estimation. *Comput. Statist. Data Anal.* **39**, 435–451.

Abramovich, F., Amato, U., Angelini, C. (2004). On optimality of Bayesian wavelet estimators. *Scand. J. Statist.* **31**, 217–234.

Ambler, G.K., Silverman, B.W. (2004a). Perfect simulation of spatial point processes using dominated coupling from the past with application to a multiscale area-interaction point process. Manuscript, Department of Mathematics, University of Bristol.

Ambler, G.K., Silverman, B.W. (2004b). Perfect simulation for wavelet thresholding with correlated coefficients. Technical Report 04:01, Department of Mathematics, University of Bristol.

Angelini, C., Sapatinas, T. (2004). Empirical Bayes approach to wavelet regression using ε-contaminated priors. *J. Statist. Comput. Simulation* **74**, 741–764.

Angelini, C., Vidakovic, B. (2004). Γ-Minimax wavelet shrinkage: A robust incorporation of information about energy of a signal in denoising applications. *Statistica Sinica* **14**, 103–125.

Brown, P.J., Fearn, T., Vannucci, M. (2001). Bayesian wavelet regression on curves with application to a spectroscopic calibration problem. *J. Amer. Statist. Assoc.* **96**, 398–408.

Carlin, B., Chib, S. (1995). Bayesian model choice via Markov chain Monte Carlo. *J. Roy. Statist. Soc., Ser. B* **57**, 473–484.

Chambolle, A., DeVore, R.A., Lee, N.-Y., Lucier, B.J. (1998). Nonlinear wavelet image processing: Variational problems, compression and noise removal through wavelet shrinkage. *IEEE Trans. Image Processing* **7**, 319–335.

Chang, W., Vidakovic, B. (2002). Wavelet estimation of a baseline signal from repeated noisy measurements by vertical block shrinkage. *Comput. Statist. Data Anal.* **40**, 317–328.

Chencov, N.N. (1962). Evaluation of an unknown distribution density from observations. *Doklady* **3**, 1559–1562.

Chipman, H., McCulloch, R., Kolaczyk, E. (1997). Adaptive Bayesian wavelet shrinkage. *J. Amer. Statist. Assoc.* **92**, 1413–1421.

Clyde, M., George, E. (1999). Empirical Bayes estimation in wavelet nonparametric regression. In: Müller, P., Vidakovic, B. (Eds.), *Bayesian Inference in Wavelet Based Models*. In: *Lecture Notes in Statistics*, vol. 141. Springer-Verlag, New York, pp. 309–322.

Clyde, M., George, E. (2000). Flexible empirical Bayes estimation for wavelets. *J. Roy. Statist. Soc., Ser. B* **62**, 681–698.

Clyde, M., DeSimone, H., Parmigiani, G. (1996). Prediction via orthogonalized model mixing. *J. Amer. Statist. Assoc.* **91**, 1197–1208.

Clyde, M., Parmigiani, G., Vidakovic, B. (1998). Multiple shrinkage and subset selection in wavelets. *Biometrika* **85**, 391–402.

Crouse, M., Nowak, R., Baraniuk, R. (1998). Wavelet-based statistical signal processing using hidden Markov models. *IEEE Trans. Signal Processing* **46**, 886–902.

Daubechies, I. (1992). *Ten Lectures on Wavelets*. SIAM, Philadelphia.

De Canditiis, D., Vidakovic, B. (2004). Wavelet Bayesian block shrinkage via mixtures of normal-inverse gamma priors. *J. Comput. Graph. Statist.* **13**, 383–398.

Donoho, D., Johnstone, I. (1994). Ideal spatial adaptation by wavelet shrinkage. *Biometrika* **81**, 425–455.

Donoho, D., Johnstone, I. (1995). Adapting to unknown smoothness via wavelet shrinkage. *J. Amer. Statist. Assoc.* **90**, 1200–1224.

Donoho, D., Johnstone, I., Kerkyacharian, G., Pickard, D. (1996). Density estimation by wavelet thresholding. *Ann. Statist.* **24**, 508–539.

Fadili, M.J., Bullmore, E.T. (2004). A comparative evaluation of wavelet-based methods for hypothesis testing of brain activation maps. *NeuroImage* **23**, 1112–1128.

Figueiredo, M., Nowak, R. (2001). Wavelet-based image estimation: An empirical Bayes approach using Jeffreys' noninformative prior. *IEEE Trans. Image Processing* **10**, 1322–1331.

George, E.I., McCulloch, R. (1997). Approaches to Bayesian variable selection. *Statistica Sinica* **7**, 339–373.

Hall, P., Patil, P. (1995). Formulae for the mean integrated square error of non-linear wavelet based density estimators. *Ann. Statist.* **23**, 905–928.

Huang, H.-C., Cressie, N. (1999). Empirical Bayesian spatial prediction using wavelets. In: Müller, P., Vidakovic, B. (Eds.), *Bayesian Inference in Wavelet Based Models*. In: *Lecture Notes in Statistics*, vol. 141. Springer-Verlag, New York, pp. 203–222.

Huang, H.-C., Cressie, N. (2000). Deterministic/stochastic wavelet decomposition for recovery of signal from noisy data. *Technometrics* **42**, 262–276. (Matlab code is available from the Wavelet Denoising software written by Antoniadis, Bigot, and Sapatinas, 2001).

Huang, S.Y., Lu, H.S. (2000). Bayesian wavelet shrinkage for nonparametric mixed-effects models. *Statistica Sinica* **10**, 1021–1040.

Kohn, R., Marron, J.S., Yau, P. (2000). Wavelet estimation using Bayesian basis selection and basis averaging. *Statistica Sinica* **10**, 109–128.

Kolaczyk, E.D. (1999). Bayesian multi-scale models for Poisson processes. *J. Amer. Statist. Assoc.* **94**, 920–933.

Leporini, D., Pesquet, J.-C. (1998). Wavelet thresholding for a wide class of noise distributions. In: *EUSIPCO'98*, Rhodes, Greece, pp. 993–996.

Leporini, D., Pesquet, J.-C. (1999). Bayesian wavelet denoising: Besov priors and non-Gaussian noises. *Signal Process.* **81**, 55–67.

Leporini, D., Pesquet, J.-C., Krim, H. (1999). Best basis representations with prior statistical models. In: Müller, P., Vidakovic, B. (Eds.), *Bayesian Inference in Wavelet Based Models*. In: *Lecture Notes in Statistics*, vol. 141. Springer-Verlag, New York, pp. 109–113.

Lina, J-M., MacGibbon, B., (1997). Non-linear shrinkage estimation with complex Daubechies wavelets. In: *Proceedings of SPIE, vol. 3169*. Wavelet Applications in Signal and Image Processing V, pp. 67–79.

Lu, H.S., Huang, S.Y., Lin, F.J. (2003). Generalized cross-validation for wavelet shrinkage in nonparametric mixed-effects models. *J. Comput. Graph. Statist.* **12**, 714–730.

Mallat, S. (1989). A theory for multiresolution signal decomposition: The wavelet representation. *IEEE Trans. Pattern Anal. Machine Intell.* **11**, 674–693.

Morris, J.S., Vannucci, M., Brown, P.J., Carroll, R.J. (2003). Wavelet-based nonparametric modeling of hierarchical functions in colon carcinogenesis (with discussion). *J. Amer. Statist. Assoc.* **98**, 573–597.

Müller, P., Vidakovic, B. (1999a). Bayesian inference with wavelets: Density estimation. *J. Comput. Graph. Statist.* **7**, 456–468.

Müller, P., Vidakovic, B. (1999b). MCMC methods in wavelet shrinkage: Non-equally spaced regression, density and spectral density estimation. In: Müller, P., Vidakovic, B. (Eds.), *Bayesian Inference in Wavelet Based Models*. In: *Lecture Notes in Statistics*, vol. 141. Springer-Verlag, New York, pp. 187–202.

Müller, P., Vidakovic, B. (Eds.) (1999c). *Bayesian Inference in Wavelet Based Models. Lecture Notes in Statistics*, vol. 141. Springer-Verlag, New York.

Nowak, R., Kolaczyk, E. (2000). A Bayesian multiscale framework for Poisson inverse problems. *IEEE Trans. Inform. Theory, Special Issue on Information-Theoretic Imaging* **46**, 1811–1825.

Ogden, T. (1997). *Essential Wavelets for Statistical Applications and Data Analysis*. Birkhäuser, Boston.

Ogden, R.T., Lynch, J.D. (1999). Bayesian analysis of change-point models. In: Müller, P., Vidakovic, B. (Eds.), *Bayesian Inference in Wavelet Based Models*. In: *Lecture Notes in Statistics*, vol. 141. Springer-Verlag, New York, pp. 67–82.

Pensky, M., Sapatinas, T. (2004). Frequentist optimality of Bayes factor thresholding estimators in wavelet regression models. Technical Report at University of Cyprus.

Pesquet, J., Krim, H., Leporini, D., Hamman, E. (1996). Bayesian approach to best basis selection. In: *IEEE International Conference on Acoustics, Speech, and Signal Processing, vol. 5*, pp. 2634–2637.

Portilla, J., Strela, V., Wainwright, M., Simoncelli, E. (2003). Image denoising using scale mixtures of Gaussians in the wavelet domain. *IEEE Trans. Image Process.* **12**, 1338–1351.

Roeder, K. (1990). Density estimation with confidence sets exemplified by superclusters and voids in the galaxies. *J. Amer. Statist. Assoc.* **85**, 617–624.

Ruggeri, F. (1999). Robust Bayesian and Bayesian decision theoretic wavelet shrinkage. In: Müller, P., Vidakovic, B. (Eds.), *Bayesian Inference in Wavelet Based Models*. In: *Lecture Notes in Statistics*, vol. 141. Springer-Verlag, New York, pp. 139–154.

Ruggeri, F., Vidakovic, B. (1999). A Bayesian decision theoretic approach to the choice of thresholding parameter. *Statistica Sinica* **9**, 183–197.

Simoncelli, E. (1999). Bayesian denoising of visual images in the wavelet domain. In: Müller, P., Vidakovic, B. (Eds.), *Bayesian Inference in Wavelet Based Models*. In: *Lecture Notes in Statistics*, vol. 141. Springer-Verlag, New York, pp. 291–308.

Simoncelli, E., Adelson, E. (1996). Noise removal via Bayesian wavelet coring. Presented at: 3rd IEEE International Conference on Image Processing, Lausanne, Switzerland.

Tierney, L. (1994). Markov chains for exploring posterior distributions. *Ann. Statist.* **22**, 1701–1728.

Vannucci, M., Corradi, F. (1999). Covariance structure of wavelet coefficients: Theory and models in a Bayesian perspective. *J. Roy. Statist. Soc., Ser. B* **61**, 971–986.

Vannucci, M., Brown, P.J., Fearn, T. (2001). Predictor selection for model averaging. In: George, E.I., Nanopoulos, P. (Eds.), *Bayesian Methods with Applications to Science, Policy and Official Statistics*. Eurostat, Luxemburg, pp. 553–562.

Vannucci, M., Brown, P.J., Fearn, T. (2003). A decision theoretical approach to wavelet regression on curves with a high number of regressors. *J. Statist. Plann. Inference* **112**, 195–212.

Vidakovic, B. (1998a). Nonlinear wavelet shrinkage with Bayes rules and Bayes factors. *J. Amer. Statist. Assoc.* **93**, 173–179.

Vidakovic, B. (1998b). Wavelet-based nonparametric Bayes methods. In: Dey, D., Müller, P., Sinha, D. (Eds.), *Practical Nonparametric and Semiparametric Bayesian Statistics*. In: *Lecture Notes in Statistics*, vol. 133. Springer-Verlag, New York, pp. 133–155.

Vidakovic, B. (1999). *Statistical Modeling by Wavelets*. Wiley, New York. 384 pp.

Vidakovic, B., Bielza Lozoya, C. (1998). Time-adaptive wavelet denoising. *IEEE Trans. Signal Process.* **46**, 2549–2554.

Vidakovic, B., Müller, P. (1999). An introduction to wavelets. In: Müller, P., Vidakovic, B. (Eds.), *Bayesian Inference in Wavelet Based Models*. In: *Lecture Notes in Statistics*, vol. 141. Springer-Verlag, New York, pp. 1–18.

Vidakovic, B., Ruggeri, F. (1999). Expansion estimation by Bayes rules. *J. Statist. Plann. Inference* **79**, 223–235.

Vidakovic, B., Ruggeri, F. (2001). BAMS method: Theory and simulations. *Sankhyā, Ser. B* **63**, 234–249. (Matlab code is available from the Wavelet Denoising software written by Antoniadis, Bigot, and Sapatinas, 2001).

Walter, G.G., Shen, X. (2001). *Wavelets and Others Orthogonal Systems*, second ed. Chapman & Hall/CRC, Boca Raton.

Wikle, C.K., Milliff, R.F., Large, W.G. (1999). Observed wavenumber spectra from 1 to 10,000 km of near-surface winds during the TOGA COARE IOP. *J. Atmospheric Sci.* **56**, 2222–2231.

Wikle, C.K., Milliff, R.F., Nychka, D., Berliner, L.M. (2001). Spatiotemporal hierarchical Bayesian modeling: Tropical ocean surface winds. *J. Amer. Statist. Assoc.* **96**, 382–397.

Handbook of Statistics, Vol. 25
ISSN: 0169-7161
© 2005 Elsevier B.V. All rights reserved.
DOI 10.1016/S0169-7161(05)25012-5

12

Bayesian Nonparametric Inference

Stephen Walker

1. Introduction

The correct interpretation of the word "nonparametric" in the title Bayesian Nonparametrics, is "infinite-dimensional". The traditional view of Bayesian statistics is of the parametric family of densities, say $f(y;\theta)$, with a prior distribution, say $\pi(\theta)$, assigned to the parameter space Θ. And Θ is finite-dimensional, that is, a subset of $\mathbb{R}^m$ with $m < \infty$. I have often heard mentioned from eminent statisticians that the problem for the Bayesian is two-fold; find both likelihood and prior. However, in Bayesian nonparametric problems there is one task – find a probability measure on a space of density functions (or related functions such as distribution functions or hazard functions). I actually believe this is the objective for all types of Bayesians and the parametric Bayesian is using both $f(\cdot;\theta)$ and $\pi(\theta)$ to construct a probability measure on densities.

If a statistician is willing to define relevant unknowns as parameters of a parametric family of densities then the underlying assumption is that the chosen family possesses some unique relationship with the data, $Y_1, Y_2, \ldots$. Presumably that there exists a $\theta_0 \in \Theta$ such that, for example, $Y_1, Y_2, \ldots$ are independent and identically distributed from $f(y;\theta_0)$. The prior then formulates what is believed about θ_0 prior to observing the data. There are obviously deeper insights than this and even an axiomatic development of Bayesian procedures (see, for example, Bernardo and Smith, 1994).

The nonparametric point of view is that the real unknown is the density (or distribution) function. In this case the emphasis shifts from priors on parameter spaces to priors on spaces of density functions. Now it has never been thought that constructing probability measures on spaces of density functions, let alone deriving posteriors, is easy. But one can always exploit the parametric model. If A is a set of densities, one can achieve

$$\mathrm{pr}\{f \in A\} \equiv \Pi(A) = \int_{\{\theta: f(\cdot;\theta)\in A\}} \pi(\theta)\,\mathrm{d}\theta.$$

This does pose the interesting question as to what exactly is the prior. Is it π or Π?

For a parametric model let Ω be the smallest set of densities which are allocated probability one. There are technical details about sets of densities which are being ignored here and we will do this for the moment. One is then declaring that

$\mathrm{pr}\{f \in \Omega\} = 1$ and this is clear as clear can be; probability one. However, more often than not, a parametric family is checked off with the data once it has been observed to see if the two are compatible. This is a blatant violation of the declaration of probability one and all that follows, including the check, are based on this declaration. See Lindsey (1999) for more words on this aspect of Bayesian inference. To cut a long story short, it is the responsibility of the Bayesian to select Ω large enough to make any such checks redundant. If this means having Ω to be the set of all densities, so be it.

If Ω can be worked through from a parametric family then we can consider the prior to be π. If we can not do this, and then necessarily the model is nonparametric, the prior can be considered as Π and we are required to construct the probability measure directly onto the space of all density functions without help from a parametric family. It is the latter scenario that is the subject of this chapter. Hence, we are concerned with the construction of probability measures on spaces of density (or related) functions.

There is now a substantial amount of literature on Bayesian nonparametric methods. Recent reviews are provided by Walker et al. (1999) and Dey et al. (1998). For Bayesian nonparametric survival analysis there is the book by Ibrahim et al. (2001). More recent reviews are given by Hjort (2003), a chapter in the HSSS book, and a forthcoming review by Müller and Quintana (2004) for a special edition of Statistical Science. The book by Ghosh and Ramamoorthi (2003) also provides a (technical) review which concentrates on consistency issues. On the other hand, the book by Denison et al. (2002) provides a review of the authors' work on Bayesian nonparametric regression and curve fitting.

1.1. Notation

Unfortunately, when dealing with probability measures and Bayes theorem on spaces of density functions, which are infinite-dimensional, the mathematics becomes necessarily more difficult. The first thing is that we need to work in a topology of sets of densities. This is most conveniently done via neighbourhoods of densities using a distance, or metric, between density functions. Note that we could as equally well do with a distance measure between distribution functions but for now we will stick with density functions. A neighbourhood of the density g with respect to the distance $d(f, g)$ and of size $\varepsilon > 0$ is defined to be the set

$$N_\varepsilon(g) = \{f\colon d(f, g) < \varepsilon\}.$$

The support of the probability measure Π is then defined to include all those densities g for which

$$\Pi(N_\varepsilon(g)) > 0$$

for all $\varepsilon > 0$. Distances between densities, which will be appearing later on, include the L_1 distance

$$d_1(f, g) = \int |f - g|,$$

the Hellinger distance

$$d_H(f, g) = \left\{\int \left(\sqrt{f} - \sqrt{g}\right)^2\right\}^{1/2},$$

and we will be making use of $h(f, g) \equiv \frac{1}{2} d_H^2(f, g) = 1 - \int \sqrt{fg}$, and the Kullback–Leibler divergence

$$d_K(f, g) = \int f \log(f/g).$$

The likelihood function for a sequence of independent and identically distributed samples $Y_1, \ldots, Y_n$ is given by

$$\mathscr{L}_n(f) = \prod_{i=1}^{n} f(Y_i)$$

and so the posterior mass assigned to a set A is given by

$$\Pi_n(A) = \Pi(A|Y_1, \ldots, Y_n) = \frac{\int_A \mathscr{L}_n(f)\Pi(\mathrm{d}f)}{\int_\Omega \mathscr{L}_n(f)\Pi(\mathrm{d}f)}.$$

We will use f and F to denote density and distribution functions, respectively. While constructing probability measures on spaces of density functions is now relatively straightforward, problems with inference remain. In particular, the expectations appearing in the numerator and denominator of Π_n are difficult to compute analytically and usually simulation is the preferred choice.

1.2. History

While there are papers on Bayesian nonparametrics prior to the seminal paper by Ferguson (1973), including descriptions of tail-free priors and consistency (see, for example, Freedman, 1963, 1965), for the purposes of getting to grips with the current day literature and developments we only need to start with Ferguson. Almost twenty years then elapsed before sampling algorithms became routine and this could not have come quickly enough. Bayesian nonparametrics was unable to demonstrate it was a useful area of statistics with only a handful of problems being tackled and with only one or two type of prior being analytically tractable. Nowadays, many types of prior can be employed. Sampling has given Bayesian nonparametrics a new lease of life. And with this has come many new theoretical contributions.

1.3. Outline

We will obviously start by looking at some examples of probability measures on spaces of density functions, and related functions such as distribution and hazard functions. This will also include a look at properties and characterizations of such priors. Priors include the Dirichlet process, neutral to the right processes, the mixture of Dirichlet process model, Gaussian process priors, Pólya trees among others.

Of concern is the derivation of posterior distributions. This is typically a nontrivial exercise. Even once this has been done, obtaining posterior summaries is also often a difficult proposition. Fortunately, with the advent of Markov chain Monte Carlo methods (Escobar, 1988, Tierney, 1994), sampling from posterior probability measures has become quite standard and highly feasible. In fact, apparently the first such algorithm appeared in the Ph.D. thesis of Escobar (1988) who provided a Gibbs sampler (Smith and Roberts, 1993) for sampling a posterior distribution associated with the well known Dirichlet process prior (Ferguson, 1973).

Moving on, large sample properties of posterior distributions are also of interest and so we will take a look at Bayesian consistency. We will also take a look at Bayesian nonparametric regression models and, finally, the idea of reinforcement and its connections with exchangeable sequences.

2. The Dirichlet process

A lot has now been written about the Dirichlet process and so an exhaustive account of this particular prior is completely unnecessary. However, the best reference and source of information remains Ferguson (1973). It is first to be noted that the Dirichlet process prior is a probability measure on spaces of distribution functions. Furthermore, the prior is supported by discrete distribution functions (see Blackwell, 1973). Whereas each distribution generated from the prior will have a probability mass function, there is no common dominating measure for all such mass functions. This has implications when it comes to finding the posterior of a Dirichlet process prior.

Without doubt the best way to think of the Dirichlet process prior is as a probability measure governing a stochastic process. If we are interested in the measurable space $(\mathscr{Y}, \mathscr{B})$, with observations $Y \in \mathscr{Y}$, then the stochastic process provides a value for the random mass $F(B)$ allocated to each set $B \in \mathscr{B}$. That is, we have a stochastic process indexed by the measurable sets; $\{B\colon\ B \in \mathscr{B}\}$. Of course, certain conditions need to be met for F to be a random probability distribution, such as $F(\mathscr{Y}) = 1$ almost surely, $0 \leqslant F(B) \leqslant 1$ almost surely and, for example, when $B_1 \cap B_2 = \emptyset$,

$$F(B_1 \cup B_2) = F(B_1) + F(B_2) \quad \text{almost surely.}$$

For stochastic processes it is the finite-dimensional distributions which essentially characterize the process, and for the Dirichlet process these turn out to be Dirichlet distributions. That is, for any measurable partition, $(B_1, \ldots, B_M)$, $M = 2, 3, \ldots$, it is that

$$\mathrm{Law}\big(F(B_1), \ldots, F(B_M)\big) = \mathrm{Dir}\big(\alpha(B_1), \ldots, \alpha(B_M)\big),$$

where Dir denotes the Dirichlet distribution and α is a finite, nonnull measure. The density function for the Dirichlet distribution is given by

$$\begin{aligned} &p(x_1, \ldots, x_M; \alpha_1, \ldots, \alpha_M) \\ &\quad = \prod_{j=1}^{M} \Gamma(\alpha_j) x^{\alpha_j - 1} \mathbf{1}\left(\sum_{j=1}^{M} x_j = 1\right) \Big/ \Gamma\left(\sum_{j=1}^{M} \alpha_j\right). \end{aligned}$$

We will write $F \sim \mathscr{D}(\alpha)$ to indicate that F is a Dirichlet process with parameter α. Properties of the Dirichlet process are still being discovered; the important ones are that

$$\mathrm{E}\big\{F(B)\big\} = \alpha(B)/\alpha(\mathscr{Y}) \quad \text{and} \quad \mathrm{Var}\big\{F(B)\big\} = \frac{\alpha(B)\{\alpha(\mathscr{Y}) - \alpha(B)\}}{\alpha(\mathscr{Y})^2\{1 + \alpha(\mathscr{Y})\}}.$$

With these one can attempt to understand the parameter α. Since α is a finite measure one can write $\alpha(\cdot) = c\,G(\cdot)$ where G is a probability measure and $c = \alpha(\mathscr{Y})$ is a positive real. Then

$$\mathrm{E}\{F(B)\} = G(B),$$

giving obvious interpretation for G, and

$$\mathrm{Var}\{F(B)\} = G(B)\{1 - G(B)\}/(1 + c).$$

This expression should provide interpretation for c. That is, for large variance one would take c close to zero. But this is not clear cut; a c close to zero puts a lot of mass on distributions having a large jump – highly discrete distributions. See Sethuraman and Tiwari (1982) for more on this. Walker and Mallick (1997) discuss the interpretation of c if the aim is to estimate a functional of F; such as

$$\phi(F) = \int k(y)\,\mathrm{d}F(y)$$

for some suitable function k. The corresponding mean and variance terms are given by

$$\mathrm{E}(\phi) = \int k(y)\,\mathrm{d}G(y)$$

and

$$\mathrm{Var}(\phi) = \frac{\int k^2(y)\,\mathrm{d}G(y) - \{\int k(y)\,\mathrm{d}G(y)\}^2}{1 + c}.$$

Hence, one can assign a value to c based on assigning a value to the prior variance for ϕ.

A lot of recent work on the Dirichlet process has focused on the distribution of means of Dirichlet processes. This work was pioneered by Cifarelli and Regazzini (1990). More recently by Regazzini et al. (2002) and Lijoi and Regazzini (2004).

For simulating from functionals of Dirichlet processes see Muliere and Tardella (1998), Guglielmi et al. (2002) and Guglielmi and Tweedie (2001).

A useful representation in distribution of a Dirichlet process is given by

$$F(\cdot) = \sum_{k=1}^{\infty} w_k \mathbf{1}(\theta_k \in \cdot),$$

where

$$w_1 = v_1, \qquad w_k = v_k \prod_{j<k}(1 - v_j), \quad k = 2, 3, \ldots,$$

the v_k are independent be$(1, c)$ variables and the θ_k are independent variables with distribution function G. See Sethuraman (1994). It is this representation which proved useful in the work of Muliere and Tardella (1998).

The Dirichlet process is known as a nonparametric prior. It is clearly an infinite-dimensional prior as a random distribution can not be pinned down with the knowledge of a finite set of parameters. But the infinite dimensionality should be providing large support with respect to some suitable topology. This is the case. For any positive integer

M and sets $(B_1, \ldots, B_M)$ and any $\varepsilon > 0$ it is that

$$\mathscr{D}\{|F(B_j) - Q(B_j)| < \varepsilon \text{ for } j = 1, \ldots, M\} > 0$$

for any distribution Q absolutely continuous with respect to G.

2.1. Posterior distribution

Obtaining the posterior Dirichlet process given observations $Y^n = (Y_1, \ldots, Y_n)$ from F is not a straightforward proposition. Without going into irrelevant technical details, the basic idea is to "find" the probability law $\mathscr{D}(\mathrm{d}F|Y^n)$ such that

$$\int_{F\in A, Y^n \in B^n} \mathscr{D}(\mathrm{d}F|Y^n)\, p(Y^n)\, \mathrm{d}Y^n \equiv \int_{F\in A} \left\{\prod_{i=1}^n F(B_i)\right\} \mathscr{D}(\mathrm{d}F),$$

for all $B^n = (B_1, \ldots, B_n)$ and A. This is two ways of writing the joint probability

$$\mathrm{pr}(Y_i \in B_i,\ i = 1, \ldots, n,\ F \in A).$$

Ferguson (1973) showed that this is satisfied when

$$\mathscr{D}(F|Y^n) \equiv \mathscr{D}(\alpha + nF_n),$$

where

$$F_n = \frac{1}{n} \sum_{i=1}^n \delta_{Y_i}$$

is the empirical distribution function. Here, δ_Y is the probability measure putting mass 1 on the point Y. Hence,

$$\mathrm{E}(F|Y^n) = \frac{cG + nF_n}{c+n},$$

which is an appealing mixture of the empirical distribution and the prior expectation. From this perspective, the original and popular interpretation of c is as a prior sample size. The Bayesian bootstrap arises by taking the "posterior" $\mathscr{D}(nF_n)$. See Rubin (1981).

A special mention must be made for $p(Y^n)$ which appears in the identity involving $\mathscr{D}(F|Y^n)$. It is the marginal distribution of Y^n, having integrated out the random distribution function F;

$$p(Y^n) = \int p(Y^n|F)\mathscr{D}(\mathrm{d}F)$$

and for $p(Y^n|F)$ we have that the Y_i are independent and identically distributed with distribution function F. It is clear that for each n the $(Y_1, \ldots, Y_n)$ are *exchangeable*. The exchangeable sequence when F is a Dirichlet process is a general Pólya-urn scheme introduced by Blackwell and MacQueen (1973). The sequence can be generated as follows. Take Y_1 from the distribution G and, in general, take Y_n to be distributed from G with probability $c/(c+n-1)$ or take Y_n to be Y_j, $j \in \{1, \ldots, n-1\}$, with probability $1/(c+n-1)$. This scheme has been studied extensively. The joint density $p(Y^n)$ can be constructed from the predictive densities

$$\mathrm{pr}(Y_n \in B|Y_1, \ldots, Y_{n-1}) = c_n G(B) + (1 - c_n) \sum_{i=1}^{n-1} \mathbf{1}(Y_i \in B)/(n-1),$$

where $c_n = c/(c+n-1)$. A result of Antoniak (1974) is that the number of samples taken from G climbs to infinity.

The general Pólya-urn scheme is extremely important and useful for studying the mixture of Dirichlet process (MDP) model introduced by Lo (1984). See Section 2.2.

Also, Blackwell and MacQueen (1973) show that if $Y_1, Y_2, \ldots$ follow a Pólya urn scheme then

$$\frac{1}{n} \sum_{i=1}^{n} \delta_{Y_i}$$

converges with probability one to a Dirichlet process random distribution function.

Finally in this section, we mention Antoniak (1974) who generalized the Dirichlet process via a mixture. The prior is written as

$$F \sim \int \mathscr{D}(\alpha_u)\, \mathrm{d}H(u)$$

which means that

$$F|u \sim \mathscr{D}(\alpha_u)$$

with

$$u \sim H.$$

The main result of Antoniak (1974) is that if $Y^n = (Y_1, \ldots, Y_n)$ is an independent and identically distributed sample from F, then

$$F|Y^n \sim \int \mathscr{D}\left(\alpha_u + \sum_{i=1}^{n} \delta_{Y_i}\right) \mathrm{d}H^*(u),$$

where

$$\mathrm{d}H^*(u) = \frac{\frac{c}{c^{(n)}} \prod_{i=1}^{k} g(Y_i^*; \phi)\, \mathrm{d}H(u)}{\int \frac{c}{c^{(n)}} \prod_{i=1}^{k} g(Y_i^*; \phi)\, \mathrm{d}H(u)}.$$

Here we have assumed that $u = (c, \phi)$, $\alpha_u(\cdot) = c\, G(\cdot; \phi)$ snd g is the density function corresponding to G. Also $c^{(n)} = c(c+1) \cdots (c+n-1)$ and k is the number of distinct $\{Y_i\}_{i=1}^n$ with $\{Y_i^*\}_{i=1}^k$ being the distinct set.

This model has been used by Doss (1995) and Doss and Huffer (2003) to model independent and identically distributed data subject to random censoring; so that the observed data consists of $\{Y_i \in B_i\colon i = 1, \ldots, n\}$. Due to the censoring, there is no analytic update or posterior summary of any use and hence sampling algorithms are required.

2.2. *The MDP model*

The purpose of the MDP model is to achieve a probability measure supported by densities. A random density function is chosen as

$$f_P(y) = \int k(y;\theta)\,\mathrm{d}P(\theta),$$

where P is chosen as a Dirichlet process. Provided $k(y;\theta)$ is a density for each θ then f_P will be a density. This model was first considered by Lo (1984). The model was left on the shelf for a considerable period before it was understood how to use it. The problem is obtaining posterior summaries. There is actually only one way to do this and that is via computer simulation from the posterior distribution. The best simulation methods are those based on Markov chain Monte Carlo (MCMC) algorithms which first started to appear in the statistics literature in the late eighties; see Escobar (1988).

The most common form of the mixture of Dirichlet process model is as a hierarchical model, and is given here;

$$\begin{aligned}
&Y_i|\theta_i \sim k(\cdot;\theta_i), \quad i = 1,\ldots,n,\\
&\theta_1,\ldots,\theta_n|P \sim_{\text{i.i.d.}} P,\\
&P \sim \mathscr{D}\{cG(\cdot;\phi)\},\\
&c \sim \pi_c(c) \quad \text{and} \quad \phi \sim \pi_\phi(\phi),
\end{aligned}$$

where ϕ are the parameters of G. Hence the prior for P is as described in Antoniak (1974). This can be viewed as a more flexible version of the parametric hierarchical model, introduced by Lindley and Smith (1972);

$$\begin{aligned}
&Y_i|\theta_i \sim k(\cdot;\theta_i), \quad i = 1,\ldots,n,\\
&\theta_1,\ldots,\theta_n|\phi \sim_{\text{i.i.d.}} G(\cdot;\phi),\\
&\phi \sim \pi_\phi(\phi).
\end{aligned}$$

It is well known that P can be integrated out of the nonparametric model, see, for example, Blackwell and MacQueen (1973), leading to the revised, but equivalent, hierarchical model;

$$\begin{aligned}
&Y_i|\theta_i \sim k(\cdot;\theta_i), \quad i = 1,\ldots,n,\\
&p(\theta_1,\ldots,\theta_n) = g(\theta_1;\phi)\prod_{i=2}^{n}\frac{cg(\theta_i;\phi) + \sum_{j=1}^{i-1}\delta_{\theta_j}(\theta_i)}{c+i-1},\\
&c \sim \pi_c(c) \quad \text{and} \quad \phi \sim \pi_\phi(\phi).
\end{aligned}$$

Here g is the density function corresponding to G.

The posterior distribution is mathematically intractable and only sampling strategies are going to be able to obtain posterior summaries of interest. Kuo (1986) tried out an importance sampling scheme but the efficient algorithms that are currently in use are all based on MCMC methods.

The idea of Escobar (1988) was to sample the posterior distribution of the parameters $(\theta_1,\ldots,\theta_n)$ using a Markov chain, which he constructed by successively sampling from the full conditional densities,

$$p(\theta_i|\theta_{-i}, y_1, \dots, y_n), \quad i = 1, \dots, n,$$

where $\theta_{-i} = (\theta_1, \dots, \theta_{i-1}, \theta_{i+1}, \dots, \theta_n)$ and, for example,

$$p(\theta_1|\theta_{-1}, y_1, \dots, y_n) \propto k(y_1; \theta_1)\left\{cg(\theta_1; \phi) + \sum_{i=2}^{n} \delta_{\theta_i}(\theta_1)\right\}.$$

This algorithm is what we know now to be a Gibbs sampler. Escobar (1988) used this idea to estimate the normal means. MacEachern (1994) developed an alternative sampling strategy which was still based on the Gibbs sampler. Current mixture of Dirichlet process sampling strategies are based on the idea of MacEachern (1994). A sample from $p(\theta_1, \dots, \theta_n)$ defines a random variable S_n, defined on $\mathcal{S}_n = \{1, \dots, n\}$; S_n is the number of clusters or the number of distinct elements in $\mathcal{S}_n$. The distinct θ_i are independent and identically distributed from g and denote these distinct elements by θ_j^*, $j = 1, \dots, S_n$. The scheme also defines a random variable $s = (s_1, \dots, s_n)$ which indicates the cluster in which θ_i lies. That is, $s_i = j$ means that θ_i is in the jth cluster and $\theta_i = \theta_j^*$. Let k denote the number of clusters, and n_j, for $j = 1, \dots, k$, the number of cases in each cluster. Then let n_j^- and k^- denote the same things with case i removed. The basic algorithm is as follows:

(1) Generate $(s_i, \theta_i)|s_{-i}, \theta_{-i}, y_i, c, \phi$ by taking $s_i = j$ with probability proportional to

$$q_j = n_j^- k\big(y_i, \theta_j^*\big)$$

or taking $s_i = k^- + 1$ with probability proportional to

$$q_0 = c \int k(y_i, \theta) g(\theta; \phi)\, \mathrm{d}\theta.$$

If $s_i = k^- + 1$ then a new location $\theta_i = \theta_k^*$ for the new cluster, with 1 element in it, is sampled from the density proportional to

$$k(y_i; \theta) g(\theta; \phi).$$

(2) Generate $\theta_j^*|y_1, \dots, y_n, k$ by taking θ_j^* from the density proportional to

$$\left\{\prod_{s_i=j} k(y_i; \theta)\right\} g(\theta; \phi).$$

Step (1) is repeated for $i = 1, \dots, n$ and step (2) for $j = 1, \dots, k$.

(3) Generate $\phi|\theta_1^*, \dots, \theta_k^*, k$ from the density proportional to

$$\left\{\prod_{j=1}^{k} g\big(\theta_j^*; \phi\big)\right\} \pi_\phi(\phi).$$

(4) Generate $c|k$ from the density proportional to

$$c^k \frac{\Gamma(c)}{\Gamma(c+k)} \pi_c(c).$$

See also Escobar (1994) and Escobar and West (1995) for sampling the MDP model. A recent review is given in MacEachern (1998). Difficulties arise with this Gibbs sampler if $k(y;\theta)$ and $g(\theta;\phi)$ do not form a conjugate pair. The integration needed to find q_0 will not be possible. Modifications to the algorithm required to cope with this problem are given in MacEachern and Müller (1998). See also Müller et al. (1996) who use the MDP model for curve fitting.

More recent work on simulation of the MDP model has been done by MacEachern et al. (1999), Neal (2000), Kottas and Gelfand (2002) and approximate simulation by Ishwaran and Zarepour (2000) and Ishwaran and James (2002).

3. Neutral to the right processes

A way of representing a Dirichlet process on the positive real line is as an independent increment gamma process. Such a process, which we will refer to as $Z(t)$, for $t>0$, starts at the origin, that is $Z(0)=0$, and jumps at random locations, though a countable number of jumps can have fixed locations, with random height. The height of the jumps are independent gamma random variables. Formally, for some finite measure α, and for any partition $0\equiv t_0<t_1<\cdots<t_M<t_{M+1}\equiv\infty$, with a slight abuse of notation,

$$\mathrm{Law}\{Z(t_1), Z(t_2)-Z(t_1),\ldots,Z(\infty)-Z(t_M)\}=\prod_{j=1}^{M+1}\mathrm{ga}\big(\alpha(t_{j-1},t_j),1\big).$$

In particular, from a marginal point of view, $Z(\infty)\sim \mathrm{ga}(\alpha(0,\infty),1)$ and so $Z(\infty)<\infty$ with probability one. Therefore,

$$F(t)=Z(t)/Z(\infty)$$

behaves almost surely as a probability distribution function and is actually a Dirichlet process. This is quite easy to understand given the connection between ratios of independent gamma variables and the Dirichlet distribution. Regazzini et al. (2003) investigate random distribution functions generated with alternative independent increment processes and in particular consider the distributions of functionals.

Taking Z now to be a general independent increment process, if instead of forcing $Z(\infty)$ to be finite almost surely we force $Z(t)\to\infty$ as $t\to\infty$ then another transform

$$F(t)=1-\exp\{-Z(t)\}$$

also behaves as a distribution function. Preferably, we can write

$$S(t)=\exp\{-Z(t)\}$$

as the corresponding survival function. In order to derive posterior distributions, including not just exact observations but also random right censored observations, it is necessary to consider the characterization of an independent increment process. Such a process can be characterized as having independent increments and the form of the log-Laplace transform is given by

$$-\log \mathrm{E}\big[\exp\{-\phi Z(t)\}\big] = \int_0^\infty \{1 - \exp(-\phi u)\}\, \mathrm{d}N_t(u),$$

where $N_t(\cdot)$ is a measure on the Borel sets of $(0, \infty)$ for each t, and is sometimes referred to as a Lévy measure, satisfying

$$\int_0^\infty \min\{1, u\}\, \mathrm{d}N_t(u) < \infty \quad \text{for all } t$$

and

$$N_t(B) \text{ is nondecreasing and continuous in } t \text{ for every set } B.$$

The term neutral to the right follows from the fact that for any $t_1 < t_2 < \cdots < t_m$, we have

$$S(t_1),\ S(t_2)/S(t_1), \ldots, S(t_m)/S(t_{m-1})$$

are mutually independent random variables. See Doksum (1974) and Ferguson (1974).

Expressions for the prior mean and variance of $S(t)$ are easy to obtain. We have

$$\mathrm{E}\{S(t)\} = \int_0^\infty u\, \mathrm{d}N_t(u), \qquad \mathrm{Var}\{S(t)\} = \int_0^\infty u^2\, \mathrm{d}N_t(u).$$

For the Dirichlet process the Lévy measure is given by

$$\mathrm{d}N_t(u) = \frac{\mathrm{d}u}{1 - \mathrm{e}^{-u}} \int_0^t \mathrm{e}^{-u\alpha(s,\infty)}\, \mathrm{d}\alpha(s).$$

This can be made more general by allowing the $\alpha(s, \infty)$ to be replaced by a general positive function $\beta(s)$. See, for example, Walker and Muliere (1997) who call it a beta-Stacy process. The simple homogeneous process of Ferguson and Phadia (1979) then arises by taking β to be a constant. Dey et al. (2003) provide a comprehensive review of neutral to the right processes.

A distinct advantage in allowing both α and β to be arbitrarily is that both $\mathrm{E}\{S(t)\}$ and $\mathrm{Var}\{S(t)\}$ can be specified to match with prior knowledge. Hence, for example, we have

$$\mathrm{E}\{S(t)\} = \exp\left\{-\int_0^t \mathrm{d}\alpha(s)/\beta(s)\right\}.$$

Finally, in this part we mention the work of Epifani et al. (2003) who consider existence of and the law of means of neutral to the right processes. The mean of a neutral to the right process is given by

$$\int_0^\infty \exp\{-Z(t)\}\,\mathrm{d}t,$$

where Z is the Lévy process.

3.1. Posterior distribution

The update from prior to posterior for neutral to the right processes is quite tricky. However, it is well known that the posterior is also a neutral to the right process (Doksum, 1974). To build up the story we need to consider more general Lévy processes which allow jumps of random height to occur at a possible countable number of fixed locations. For simplicity we will assume there are no such jumps in the prior process. However, the posterior process will have such jumps.

Given an observation T we can characterize the posterior via the updated Lévy measure and the height of the jump at the fixed location. The new Lévy measure for the beta-Stacy process is the same as for the prior except the β has changed to

$$\beta(t) \to \beta(t) + \mathbf{1}(T \geqslant t).$$

There will also be a jump in the process at the location T and the density function for the height of the jump is given by

$$p(v) \propto \mathrm{e}^{-v\beta(T)}.$$

For a set of data $T_1, \dots, T_n$, which may include some random right censored observations, it remains

$$\beta(t) \to \beta(t) + \sum_{i=1}^{n} \mathbf{1}(T_i \geqslant t).$$

However, only the uncensored observations contribute to the jumps at fixed locations. The density function for the height at T is given by

$$p(v) \propto \left(1 - \mathrm{e}^{-v}\right)^{N\{T\}-1} \mathrm{e}^{-v(\beta(T)+M(T)-N\{T\})},$$

where $N\{t\} = N(t) - N(t-)$ and

$$N(t) = \sum_{i=1}^{n} \mathbf{1}(T_i \leqslant t)$$

and $M(t) = \sum_{i=1}^{n} \mathbf{1}(T_i \geqslant t)$. The posterior Bayes estimate is given by

$$\mathrm{E}\{S(t)|T_1, \dots, T_n\} = \mathrm{PI}_{[0,t]}\left\{1 - \frac{\mathrm{d}\alpha(s) + \mathrm{d}N(s)}{\beta(s) + M(s)}\right\},$$

where PI denotes a product integral (Gill and Johansen, 1990). This estimator, which generalises the Kaplan–Meier estimator, was first found by Hjort (1990). Note that the

Kaplan–Meier estimator is given by

$$\widehat{S}(t) = \mathrm{PI}_{[0,t]}\left\{1 - \frac{\mathrm{d}N(s)}{M(s)}\right\}.$$

3.2. Alternative representation

The cumulative hazard function corresponding to S can be written as

$$A(t) = \int_0^t \frac{\mathrm{d}F(s)}{S(s-)}.$$

From this we can re-arrange to get

$$S(t) = \mathrm{PI}_{[0,t]}\{1 - \mathrm{d}A(s)\}.$$

Hjort (1990) chose to model the cumulative hazard A using a Lévy process with jumps in $(0, 1)$. If A is a Lévy process then $\mathrm{d}Z = -\log(1 - \mathrm{d}A)$ is also a Lévy process and hence the corresponding S will be a neutral to the right process.

Hjort (1990) specifically used a beta process which has Lévy measure given by

$$\mathrm{d}N_t(u) = \frac{\mathrm{d}u}{u} \int_0^t \beta(s)(1-u)^{\beta(s)-1}\,\mathrm{d}H(s)$$

with $u \in (0, 1)$. Here H is a continuous nondecreasing function with $H(0) = 0$ and β is a nonnegative piecewise continuous function. The posterior distribution can also be characterized via the Lévy measure and jumps at fixed locations. The posterior Lévy measure is given by

$$\mathrm{d}N_t(u) = \frac{\mathrm{d}u}{u} \int_0^t \beta(s)(1-u)^{\beta(s)+M(s)-1}\,\mathrm{d}H(s)$$

and at each of the exact observations T, there is a jump with density function for the height of the jump being

$$f(v) \propto v^{N\{T\}-1}(1-v)^{M(T)-N\{T\}+\beta(T)-1}.$$

Consequently, the Bayes estimate for the cumulative hazard function is

$$\mathrm{E}\{A(t) | T_1, \ldots, T_n\} = \int_0^t \frac{\beta(s)\,\mathrm{d}H(s) + \mathrm{d}N(s)}{\beta(s) + M(s)}.$$

Note that the Nelson–Aalen estimator is given by

$$\widehat{A}(t) = \int_0^t \frac{\mathrm{d}N(s)}{M(s)}.$$

The modelling of the cumulative hazard function is important in survival analysis and the beta process approach allows advances to more complex models such as the proportional hazards model. See, for example, Kim and Lee (2003).

3.3. Simulation

Sampling from the posterior distribution is a useful way to do inference. In the case of a neutral to the right process this will involve the simulation of a Lévy process. For a pure jump Lévy process it is possible to arrange the countably many jumps in decreasing order of magnitude. From a simulation perspective this is quite attractive since a truncation of the infinite number of jumps must be exercised at some point. It is therefore convenient to make sure the largest jumps are being taken. Hence, the representation of such a process given in Ferguson and Klass (1972) which does indeed develop the process from the largest to the smallest jumps should be a good way of simulating the process.

Suppose the Lévy measure is given by N_t and the limiting measure is N_∞. Let $M(x) = N_\infty[x, \infty)$, we assume that $M(0) = \infty$, and define the decreasing sequence of random variables $J_1, J_2, \ldots$ via

$$\mathrm{pr}(J_1 \leqslant x_1) = \exp\{-M(x_1)\}$$

and

$$\mathrm{pr}(J_i \leqslant x_i | J_{i-1} = x_{i-1}) = \exp\{-M(x_1) + M(x_{i-1})\}, \quad 0 < x_i < x_{i-1}.$$

It is possible to obtain the J_i directly via $\tau_i = M(J_i)$ and the τ_i are the event times of a standard Poisson process. The Ferguson and Klass (1972) representation of a Lévy process with Lévy measure N_t is given by

$$Z(t) = \sum_i J_i \mathbf{1}\{u_i \leqslant n_t(J_i)\},$$

where the u_i are independent uniform random variables from [0, 1] and

$$n_t(v) = \mathrm{d}N_t(v)/\mathrm{d}N_\infty(v).$$

For example, if $\mathrm{d}N_t(u) = \mathrm{d}u \int_0^t K(u, s)\,\mathrm{d}s$, for some suitable function K, then

$$n_t(v) = \int_0^t K(v, s)\,\mathrm{d}s \bigg/ \int_0^\infty K(v, s)\,\mathrm{d}s.$$

A re-arrangement of the ordering of the J_i's will result in the alternative representation, and hence simulation strategy, provided by Wolpert and Ickstadt (1998). See Walker and Damien (2000) for further details.

A homogeneous version of the process arises when $K(u, s) = K(u)\eta(s)$ and simulation of such processes using the Ferguson and Klass (1972) approach was undertaken in Barndorff-Nielsen and Shephard (2001).

An alternative simulation strategy was developed by Damien et al. (1995) and was used in an analysis of survival data based on the beta process in Damien et al. (1996).

4. Other priors

The nonparametric priors described in Sections 2 and 3, except for the MDP model, are based on probability measures on spaces of distribution functions. No corresponding probability measure exists on any space of density functions. In this section we will briefly look at priors which are based on probability measures on spaces of density functions.

4.1. Log-Gaussian prior

The idea here is the normalization of a nonnegative valued stochastic process, say $Z(y)$, for $y \in I$, where I is a closed interval, and to define a random density function accordingly,

$$f(y) = \frac{Z(y)}{\int_I Z(y)\,\mathrm{d}y}.$$

The most popular choice here is to take $Z = \exp W$ with W a Gaussian process.

The process W is said to be Gaussian with mean μ and covariance function σ, a positive definite function on $\mathbb{R} \times \mathbb{R}$, if for any $x_1, \ldots, x_m$, $W(x_1), \ldots, W(x_m)$ has a multivariate normal distribution with mean $\mu(x_1), \ldots, \mu(x_m)$ and covariance matrix with (i, j)th element given by $\sigma(x_i, x_j)$. According to Lenk (1988), if there exist positive constants C, β and ε and a nonnegative integer r so that

$$\mathrm{E}\big|W(x) - W(y)\big| \leqslant C|x - y|^{1+r+\varepsilon}$$

then the sample paths of W are r times continuously differentiable with probability one. The log-normal process is defined to be $Z = \exp W$ and it was this model that was used by Leonard (1978).

Lenk (1988) generalized this process. Let Λ denote the law of the log-normal process. The law of the general log-Gaussian process is given by

$$\Lambda_\xi(A) = \int_A \left(\int_I W(x)\,\mathrm{d}x\right)^\xi \mathrm{d}\Lambda(W)/C(\xi, \mu),$$

where

$$C(\xi, \mu) = \mathrm{E}\left[\left(\int_I W(x)\,\mathrm{d}x\right)^\xi\right].$$

Clearly, if $\xi = 1$ then we recover the log-normal process prior. Denote the general log-Gaussian by $\mathrm{LN}(\mu, \sigma, \xi)$. The posterior, given independent and identically distributed samples $y_1, \ldots, y_n$, is also a general log-Gaussian process with updates given by

$$\mu(x) \to \mu(x) + \sum_{i=1}^n \sigma(x, y_i),$$

$$\sigma \to \sigma,$$

$$\xi \to \xi - n.$$

Lenk (1988) goes on to provide expressions for the predictive density. See also related work in Lenk (1991).

4.2. *Infinite-dimensional exponential family*

Let us consider such a class of random density functions defined on [0, 1]. Let β_j be a set of independent random variables, typically $\beta_j \sim \mathrm{N}(0, \sigma_j^2)$. Also, let $\{\phi_j(y)\}$ be a sequence of orthogonal polynomials on [0, 1] and define the random density

$$f(y) = \frac{\exp\{\sum_{j=1}^{\infty} \beta_j\, \phi_j(y)\}}{\int_0^1 \exp\{\sum_{j=1}^{\infty} \beta_j\, \phi_j(y)\}\, \mathrm{d}y}.$$

To ensure that f is a density with probability one it is sufficient that

$$\sum_{j=1}^{\infty} \sigma_j \sup_{0 \leqslant y \leqslant 1} \phi_j(y) < \infty.$$

The posterior is quite difficult to manage and sampling method are used.

For such models and similar models involving sums of infinite terms, it is quite usual to truncate the sum at some random integer, say K, and to model K with a prior distribution. To jump between dimensions it is possible to use a reversible jump MCMC algorithm (Green, 1995). See Richardson and Green (1997) for an illustration of such a model (mixtures of normals) and algorithm, and Petrone (1999) for a mixture of beta distributions.

4.3. *Pólya trees*

The original partition models were the tailfree process, studied in the early sixties, mainly by Freedman (1963). Ferguson (1974) reviewed them briefly but they were largely forgotten until the early nineties when they made a return under a different name – Pólya trees. Pólya trees are a special kind of tailfree prior, and their importance is based on a desirable conjugacy property.

Here we describe how to construct a Pólya tree; the main ingredient of which is a partition of the relevant sample space. So, let Ω be this space and let (B_0, B_1) be a partition of Ω. Then let (B_{00}, B_{01}) be a partition of B_0 and (B_{10}, B_{11}) be a partition of B_1. Continue in this fashion ad infinitum and call the set of partitions a binary tree partition of Ω. Let, for some m, $\epsilon = \epsilon_1 \ldots \epsilon_m$ with each $\epsilon_k \in \{0, 1\}$, $k = 1, \ldots, m$, so that each ϵ defines a unique set B_ϵ. The number of partitions at the mth level is 2^m. Thus, in general B_ϵ splits into $B_{\epsilon 0}$ and $B_{\epsilon 1}$.

To obtain the randomness of the mass assigned to the partitions, it is helpful to think of a particle cascading through these partitions. It starts in Ω and moves into B_0 with probability $C_\emptyset$, or B_1 with probability $1 - C_\emptyset$. In general, on entering B_ϵ, the particle could either move into $B_{\epsilon 0}$ or $B_{\epsilon 1}$. Let it move into the former with probability C_ϵ and into the latter with probability $1 - C_\epsilon$. For Pólya trees these probabilities are random and are beta variables; that is, $C_\epsilon \sim \mathrm{beta}(\alpha_{\epsilon 0}, \alpha_{\epsilon 1})$ with nonnegative $\alpha_{\epsilon 0}$ and $\alpha_{\epsilon 1}$. Let

$\mathscr{B} = (B_0, B_1, B_{00}, \ldots)$ and $\mathscr{A} = (\alpha_0, \alpha_1, \alpha_{00}, \ldots)$. In order to define a particular Pólya tree distribution both $\mathscr{B}$ and $\mathscr{A}$ need to be specified.

Random mass is then allocated to the sets in $\mathscr{B}$, assumed to generate the Borel sets, such that all finite-dimensional distributions are defined as follows;

$$F(B_0) = C_\emptyset, \qquad F(B_1) = 1 - C_\emptyset,$$

$$F(B_{00}) = C_0 C_\emptyset, \qquad F(B_{01}) = (1 - C_0)C_\emptyset,$$

$$F(B_{10}) = C_1(1 - C_\emptyset), \qquad F(B_{11}) = (1 - C_1)(1 - C_\emptyset)$$

and so on down the partitions. In general, if $F(B_\epsilon)$ has been assigned then

$$F(B_{\epsilon 0}) = C_\epsilon F(B_\epsilon), \qquad F(B_{\epsilon 1}) = (1 - C_\epsilon)F(B_\epsilon).$$

It is not difficult to verify that the conditions for the existence of a probability measure on the set of distributions on Ω are met. We write $F \sim \mathrm{PT}(\mathscr{B}, \mathscr{A})$.

We can obtain a Dirichlet process prior by taking $\alpha_\epsilon = \alpha(B_\epsilon)$, for some finite measure $\alpha(\cdot)$. We obviously do not need to do this and can look for alternative prior specifications. The problem in doing so is that necessarily the prior then depends on the partition $\mathscr{B}$.

A popular idea for specifying $\mathscr{A}$ is to take $\alpha_\epsilon = m^2$ whenever ϵ represents a partition at level m. The reason for this choice is that F is then absolutely continuous with respect to the Lebesgue measure with probability one and so there exists a density.

The prior predictive for a first observation Y_1, that is, $\mathrm{pr}(Y_1 \in B)$, for some $B \in \mathscr{B}$, is given by

$$\mathrm{pr}(Y_1 \in B) = \int F(B)\, \mathrm{dPT}(F).$$

Given an observation Y_1 the posterior is also a Pólya tree distribution and the updated parameter $\mathscr{A}$ easily obtained. Note that the partition $\mathscr{B}$ remains fixed. We write $F|Y_1 \sim \mathrm{PT}(\mathscr{B}, \mathscr{A}|Y_1)$ with $[\mathscr{A}|Y_1]$ given by

$$[\alpha_\epsilon|Y_1] = \begin{cases} \alpha_\epsilon + 1 & \text{if } Y_1 \in B_\epsilon, \\ \alpha_\epsilon & \text{otherwise.} \end{cases}$$

For Y_1, an α from each level needs to be updated. For n observations, $Y^n = (Y_1, \ldots, Y_n)$, then $[\mathscr{A}|Y^n]$ is given by

$$\left[\alpha_\epsilon|Y^n\right] = \alpha_\epsilon + n_\epsilon,$$

where n_ϵ is the number of observations which lie in B_ϵ.

While Pólya trees are quite easy to work with, it is the partition which is not popular and hence not many researchers have championed its cause. For applications of Pólya trees and other aspects see Lavine (1992, 1994). See also Mauldin et al. (1992) for a recent survey of Pólya trees.

4.4. Lévy driven processes

The main drawback of survival models based on independent increment processes is that with probability one the survival function is discrete. In order to overcome this problem, Dystra and Laud (1981) considered an independent increment process to model the hazard rate function. The corresponding cumulative hazard function, and hence survival function, are continuous. A disadvantage is that the model only allows monotone hazard rate functions. Here we describe a piecewise continuous Markov process to model the hazard rate function and by integration obtain a continuous process to model cumulative hazard and survival functions. The process allows us to model nonmonotone hazard rate functions and in a special limiting case we do arrive at the model of Dystra and Laud (1981). Hence, we retain the desirable continuous models without the limitations of a monotone hazard rate function.

Let T be a continuous random variable defined on $(0, \infty)$ with (conditional) cumulative distribution function given by

$$F(t|Z) = 1 - \exp\{-Z(t)\},$$

and density function

$$f(t|Z) = x(t)\exp\{-Z(t)\},$$

where $Z(\cdot)$ is the cumulative hazard function given by

$$Z(t) = \int_0^t x(s)\,\mathrm{d}s,$$

with $x(t)$ the hazard rate function.

Let $L(t)$ be an independent increments (Lévy) process defined on $[0, \infty)$ without Gaussian components, so $L(t)$ is a pure jump process (Ferguson and Klass, 1972). For the neutral to the right process,

$$Z_{NTR}(t) = \int_0^t \mathrm{d}L(v) = L(t).$$

Dystra and Laud (1981) considered

$$Z_{DL}(t) = \int_0^t (t - v)\,\mathrm{d}L(v),$$

and in particular they took $L(\cdot)$ to be an extended gamma process. Here $Z_{DL}(\cdot)$ has continuous sample paths but it is easy to see that in this case the hazard rate function is given by $x_{DL}(t) = \int_0^t \mathrm{d}L(v)$, and so is monotone.

Nieto-Barajas and Walker (2004) consider

$$Z(t) = \int_0^t \frac{1}{a}\{1 - \mathrm{e}^{-a(t-v)}\}\,\mathrm{d}L(v)$$

for some $a > 0$. In this case, the hazard rate function is given by

$$x(t) = \int_0^t \exp\{-a(t-v)\}\, \mathrm{d}L(v).$$

Due to the fact that for $a > 0$ and $t > v$, the factors $\{\exp^{-a(t-v)}\}$ are decreasing functions of t for each v, the sample paths of are nonmonotone functions. Moreover, the sample paths of the hazard rate process $x(t)$ are piece-wise continuous functions. Therefore, the cumulative hazard process $Z(t)$ is continuous with probability one.

Other authors have considered kernel mixture models of the type

$$x(t) = \int k(t,u)\, \mathrm{d}L(u).$$

Lo and Weng (1989) take $L(u)$ to be a weighted gamma process and k any suitable kernel function. Brix (1999) takes k to be an arbitrary kernel and $L(u)$ to be a generalized gamma process. James (2002, Section 4) also considers models of this type taking $L(u)$ to be a general random measure. This general measure contains the generalized gamma processes and the weighted gamma processes as particular cases.

To consider posterior distributions, we will write the process $Z(\cdot)$ as

$$Z(t) = \int_0^\infty k(t,v)\, \mathrm{d}L(v),$$

where

$$k(t,v) = \frac{1}{a}\left[1 - \exp\{-a(t-v)^+\}\right].$$

We will take the Lévy measure to be

$$\mathrm{d}N_t(u) = \mathrm{d}u \int_0^t \exp\{-u\beta(s)\}\, \mathrm{d}\alpha(s),$$

where $\alpha(\cdot)$ is a continuous measure and $\beta(\cdot)$ is a nonnegative piece-wise continuous function. The following results are taken from Nieto-Barajas and Walker (2004).

THEOREM 1. *The posterior Lévy measure given an observation $T > t$ is given by*

$$\mathrm{d}N_\tau(v) = \mathrm{d}v \int_0^\tau \exp\{-v\beta'(u)\}\, \mathrm{d}\alpha(u),$$

where $\beta'(u) = \beta(u) + k(t,u)$.

Consequently, given n observations $T_1 > t_1, \ldots, T_n > t_n$, the posterior Lévy measure becomes

$$\mathrm{d}N'_\tau(v) = \mathrm{d}v \int_0^\tau \exp\{-v\beta'(u)\}\, \mathrm{d}\alpha(u),$$

where now $\beta'(u) = \beta(u) + \sum_{1 \leqslant i \leqslant n} k(t_i, u)$.

It is more difficult to find the posterior distribution given an observation $T = t$. In order to do this we need to introduce a latent parameter or observation. We assume we have witnessed, along with $T = t$, the observation $S = s$, where the random mass function for S is given by

$$f(s|t, L) = \mathrm{pr}(S = s|t, L) \propto \mathrm{e}^{-a(t-s)} L\{s\} I(0 \leqslant s \leqslant t),$$

coming from the joint density

$$f(t, s|L) = \mathrm{e}^{-a(t-s)} L\{s\} \exp\left\{-\int k(t, v)\, \mathrm{d}L(v)\right\} \mathbf{1}(0 \leqslant s \leqslant t),$$

where $L\{s\} = L(s) - L(s-)$. Of course, this ensures that

$$f(t|L) = x(t) \exp\left\{-\int_0^t x(s)\, \mathrm{d}s\right\}.$$

THEOREM 2. *The posterior Lévy measure given an observation $T = t$ and $S = s$ is given by*

$$\mathrm{d}N_t(u) = \mathrm{d}u \int_0^t \exp\{-u\beta'(s)\}\, \mathrm{d}\alpha(s),$$

where $\beta'(s) = \beta(s) + k(t, s)$ and the updated Lévy process has a fixed point of discontinuity at s with the distribution of $L'\{s\}$ being $\mathrm{ga}(2, \beta(s) + k(t, s))$.

Therefore, we are now in a position to write down the full posterior distributions, allowing the process to have prior fixed jumps.

THEOREM 3. *Let $L(\cdot)$ be a Lévy process such that*

$$L(t) = L_c(t) + \sum_j L\{\tau_j\} \mathbf{1}(\tau_j \leqslant t)$$

with Lévy measure for L_c given by

$$\mathrm{d}N_t(u) = \mathrm{d}u \int_0^t \exp\{-u\beta(s)\}\, \mathrm{d}\alpha(s),$$

and let $M = \{\tau_1, \tau_2, \ldots\}$ be the set of prior fixed points of discontinuity of $L(\cdot)$ and the density function of $L\{\tau_j\}$ is f_j. We will deal with the updates for a single observation of both types and we will use a $'$ (prime) to denote an updated parameter/function.

(i) *Given an observation $T > t$, the posterior Lévy measure is given by*

$$\mathrm{d}N_t'(u) = \mathrm{d}u \int_0^t \exp\{-u\beta'(s)\}\, \mathrm{d}\alpha(s),$$

where $\beta'(s) = \beta(s) + k(t, s)$*. Furthermore,* $M' = M$ *and*

$$f_j'(x) \propto \begin{cases} e^{-k(t,\tau_j)x} f_j(x) & \text{if } t > \tau_j, \\ f_j(x) & \text{otherwise.} \end{cases}$$

(ii) *Given an observation* $T = t$ *and* $S = s$*, the posterior Lévy measure is given by*

$$dN_t'(u) = du \int_0^t \exp\{-u\beta'(s)\}\, d\alpha(s),$$

where $\beta'(s) = \beta(s) + k(t, s)$.

(a) *If* $s \notin M$ *then* $M' = M \cup \{s\}$ *and* f_s' *is* ga$(2, \beta(s) + k(t, s))$*. Furthermore,*

$$f_j'(x) \propto \begin{cases} e^{-k(t,\tau_j)x} f_j(x) & \text{if } t > \tau_j, \\ f_j(x) & \text{otherwise.} \end{cases}$$

(b) *If* $s \in M$ *then* $M' = M$ *and*

$$f_s'(x) \propto x\, e^{-k(t,s)x} f_s(x).$$

Furthermore, for $\tau_j \neq s$*,*

$$f_j'(x) \propto \begin{cases} e^{-k(t,\tau_j)x} f_j(x) & \text{if } t > \tau_j, \\ f_j(x) & \text{otherwise.} \end{cases}$$

With this it is possible to build the full posterior based on a sample of size n. Recall the s are latent and are required whenever an uncensored observation is witnessed. In practice, a Gibbs sampler (Smith and Roberts, 1993) would be needed to sample from the joint posterior distribution of $L(\cdot)$ and s.

A similar idea as the above was used by Nieto-Barajas et al. (2004) to provide a class of random density functions driven by a Lévy process. A random density is generated via

$$f(y) = \frac{\int k(y; x)\, dL(x)}{Z},$$

where L is the pure jump Lévy process, k a suitable kernel function and

$$Z = \iint k(y; x)\, dL(x)\, dy.$$

Conditions under which $Z < \infty$ almost surely are provided by Nieto-Barajas et al. (2004). These authors also provide distributions of means of both prior and posterior and also provide a simulation algorithm for posterior inference.

5. Consistency

Let $Y_1, Y_2, \ldots,$ taking values in $(\mathcal{Y}, \mathcal{B})$, be independent and identically distributed random variables from some fixed but unknown (the true) density function f_0, with corresponding distribution function F_0.

Bayesian asymptotics is concerned with the sequence of posterior distributions and what happens to them as the sample size grows to infinity. For consistency the requirement is that the posteriors accumulate in suitable neighbourhoods of f_0. Here we will consider neighbourhoods generated by the Hellinger distance between density functions; $h(f, g)$, see Section 1.1. The posterior distribution Π_n assigns mass

$$\Pi_n(A) = \frac{\int_A R_n(f)\Pi(\mathrm{d}f)}{\int R_n(f)\Pi(\mathrm{d}f)}$$

to the set of densities A, where

$$R_n(f) = \prod_{i=1}^{n} f(Y_i)/f_0(Y_i).$$

Consistency is characterized as

$$\Pi_n\big(\{f\colon h(f, f_0) > \varepsilon\}\big) \to 0 \quad \text{almost surely } \big[F_0^\infty\big]$$

for all $\varepsilon > 0$.

Previous studies of Hellinger consistency, see, for example, Barron (1988), Barron et al. (1999) and Ghosal et al. (1999), deal with the numerator and denominator in the expression for $\Pi_n(A)$ separately. Briefly, if Π puts positive mass on all Kullback–Leibler neighbourhoods of f_0 (which will be referred to as the Kullback–Leibler property for Π) then the denominator is eventually bounded below by $\exp(-nc)$ for all $c > 0$. With constraints on the prior, ensuring sufficiently low mass on densities which track the data too closely, the numerator can be eventually bounded above by $\exp(-n\delta)$, for some $\delta > 0$, when A is the set $A = \{f\colon h(f, f_0) > \varepsilon\}$, for any $\varepsilon > 0$. Consequently, with the appropriate conditions in place, $\Pi_n(A) \to 0$ almost surely $[F_0^\infty]$, with exponential rate, for all $\varepsilon > 0$.

To be more explicit, the basic ideas of current approaches are based on the introduction of an increasing sequence of sets $\mathcal{G}_n$, a sieve, and to consider

$$\Pi_n(A) = \Pi_n(\mathcal{G}_n \cap A) + \Pi_n\big(\mathcal{G}_n^c \cap A\big).$$

Putting sufficiently low mass on densities which track the data too closely, that is the densities in $\mathcal{G}_n^c$, involves ensuring that $\Pi(\mathcal{G}_n^c) < \exp(-n\xi)$ for all large n and for some $\xi > 0$. This results in $\Pi_n(\mathcal{G}_n^c) < \exp(-n\xi^*)$ almost surely for all large n for some $\xi^* > 0$. The aim then is to find $\mathcal{G}_n$ such that

$$\int_{A\cap\mathcal{G}_n} R_n(f)\Pi(\mathrm{d}f) < \exp(-n\delta) \quad \text{almost surely}$$

for all large n for some $\delta > 0$. Approaches differ in the precise form of $\mathcal{G}_n$ which guarantees the above. For example, Ghosal et al. (1999) have $J(\eta, \mathcal{G}_n) < n\beta_\eta$ for all large n, for some $\beta_\eta > 0$ for all $\eta > 0$, where J is the L_1 metric entropy.

To fix the notation, define the predictive density as

$$f_n(x) = \int f(x)\Pi_n(\mathrm{d}f)$$

and

$$f_{nA}(x) = \int f(x)\Pi_{nA}(\mathrm{d}f)$$

to be the predictive density with posterior distribution restricted to the set A. Also define $I_n = \int R_n(f)\Pi(\mathrm{d}f)$. The Kullback–Leibler property is given by

$$\Pi\big(\{f\colon d_K(f, f_0) < \varepsilon\}\big) > 0$$

for all $\varepsilon > 0$. Since f_0 is unknown, the condition is

$$\Pi\big(\{f\colon d_K(f, g) < \varepsilon\}\big) > 0$$

for all densities g and all $\varepsilon > 0$. This is possible to achieve using nonparametric priors. See Barron et al. (1999) and Ghosal et al. (1999). The following results are taken from Walker (2004).

THEOREM 4. *If Π has the Kullback–Leibler property then*

$$\Pi_n(A) \to 0 \quad \textit{almost surely}$$

for all sets A for which $\liminf_n h(f_{nA}, f_0) > 0$ *almost surely.*

PROOF. Assume that $\Pi(A) > 0$ and define $L_n = I_n\Pi_n(A) = \int_A R_n(f)\Pi(\mathrm{d}f)$ with $L_0 = \Pi(A)$. The following identity, which can be easily checked, holds;

$$L_{n+1}/L_n = f_{nA}(X_{n+1})/f_0(X_{n+1}).$$

Hence,

$$\mathrm{E}\big(\sqrt{L_{n+1}/L_n}\,\big|\mathscr{F}_n\big) = 1 - h(f_{n\,A}, f_0),$$

where $\mathscr{F}_n = \sigma(Y_1, \ldots, Y_n)$. Now consider the Martingale sequence $\{S_N\}$ given by

$$S_N = \sum_{n=1}^{N}\big\{\sqrt{L_n/L_{n-1}} - \mathrm{E}\big(\sqrt{L_n/L_{n-1}}\,\big|\mathscr{F}_{n-1}\big)\big\}.$$

The stability theorem (see, for example, Loève, 1963) yields $S_N/N \to 0$ almost surely if

$$\sum_n n^{-2}\,\mathrm{Var}\sqrt{L_n/L_{n-1}} < \infty.$$

Now $\mathrm{Var}(\sqrt{L_{n+1}/L_n}) \leqslant \mathrm{E}(L_{n+1}/L_n) = 1$ and so $S_N/N \to 0$ almost surely. Also,

$$\frac{1}{N}\sum_{n=1}^{N}\sqrt{L_n/L_{n-1}} \geqslant (L_N/L_0)^{1/(2N)}$$

and therefore

$$(L_N/L_0)^{1/(2N)} - 1 + \frac{1}{N}\sum_{n=1}^{N} h(f_{n-1\,A}, f_0) \leqslant S_N/N.$$

If $\liminf_n h(f_{nA}, f_0) > 0$ then

$$\limsup_N (L_N/L_0)^{1/(2N)} < 1.$$

That is, there exists a $\delta > 0$ such that $L_N < L_0 \exp(-N\delta)$ almost surely for all large N. If Π has the Kullback–Leibler property then $I_n > \exp(-nc)$ almost surely for all large n, for any $c > 0$; see Barron et al. (1999). Thus,

$$\Pi_n(A) < \Pi(A)\exp\{-n(\delta - c)\} \quad \text{almost surely}$$

for all large n and for any $c > 0$. Choosing $c < \delta$ implies that $\Pi_n(A) \to 0$ almost surely. This completes the proof. □

If A^c is a weak neighbourhood of f_0 then $\liminf_n h(f_{nA}, f_0) > 0$ automatically so Theorem 4 provides an alternative to the Schwartz (1965) proof for weak consistency. Theorem 4 can also be used to derive the Hellinger consistency result of Ghosal et al. (1999). See Walker (2003) for further details.

Now Ω is separable; that is we can cover Ω with a countable set of Hellinger balls of radius δ for any $\delta > 0$. Therefore,

$$A = \{f\colon h(f, f_0) > \varepsilon\}$$

can be covered by the countable union of disjoint sets A_j, where $A_j \subseteq A_j^* = \{f\colon h(f, f_j) < \delta\}$, and f_j are a set of densities such that $h(f_j, f_0) > \varepsilon$. We can take $\delta < \varepsilon$ so that $h(f_{nA_j}, f_0) > \gamma > 0$, where $\gamma = \varepsilon - \delta$. This follows since

$$h(f_{nA_j}, f_0) \geqslant h(f_j, f_0) - h(f_{nA_j}, f_j)$$

and $h(f_{nA_j}, f_j) \leqslant \delta$.

THEOREM 5. *If Π has the Kullback–Leibler property and*

$$\sum_j \sqrt{\Pi(A_j)} < \infty,$$

then

$$\Pi_n(A) \to 0 \quad \textit{almost surely}.$$

PROOF. Now

$$\Pi_n(A) = \sum_j \Pi_n(A_j) \leqslant \sum_j \sqrt{\Pi_n(A_j)}$$
$$= \sum_j \sqrt{\int_{A_j} R_n(f)\Pi(\mathrm{d}f)/I_n}.$$

If

$$L_{nj} = \sqrt{\int_{A_j} R_n(f)\Pi(\mathrm{d}f)}$$

then

$$L_{n+1\,j} = L_{nj}\sqrt{f_{nA_j}(X_{n+1})/f_0(X_{n+1})}$$

which includes the case when $n = 0$ and $L_{0j} = \sqrt{\Pi(A_j)}$. So,

$$\mathrm{E}(L_{n+1\,j}|\mathscr{F}_n) = L_{nj}\big\{1 - h(f_{nA_j}, f_0)\big\} < (1-\gamma)L_{nj}$$

and hence $\mathrm{E}(L_{nj}) < (1-\gamma)^n\sqrt{\Pi(A_j)}$. Therefore,

$$\mathrm{pr}\left\{\sum_j L_{nj} > \exp(-nd)\right\} < \exp(nd)(1-\gamma)^n\sum_j\sqrt{\Pi(A_j)}$$

and so if

$$\sum_j\sqrt{\Pi(A_j)} < \infty$$

then

$$\sum_j L_{nj} < \exp(-nd) \quad \text{almost surely}$$

for all large n, for any $d < -\log(1-\gamma)$. The Kullback–Leibler property for Π ensures that $I_n > \exp(-nc)$ almost surely for all large n, for any $c > 0$. This completes the proof. □

In fact, it is possible to extend this theorem quite easily. The second condition could as well be

$$\sum_k \Pi^{1-\delta}(A_k) < \infty$$

for any $0 < \delta < 1$.

5.1. *Illustration*

Here we consider an example. We have Ω being covered by $\{A_1, A_2, \ldots\}$ which are mutually disjoint Hellinger balls of radius δ. The aim then is to show that

$$\sum_k\sqrt{\Pi(A_k)} < \infty$$

and that this holds for all $\delta > 0$.

A popular type of nonparametric prior consists of a mixture of priors;

$$\Pi = \sum_N p_N\,\Pi_N,$$

where $\sum_N p_N = 1$ and the $\{p_N\}$ are fixed. Here Π_N is supported by densities in $C_N \subseteq \Omega$, so that $\Pi_N(C_N) = 1$. Typically, C_N is totally bounded, that is C_N can be covered by a finite number of Hellinger balls of size δ, for any $\delta > 0$. The number of such balls will be denoted by $I_N(\delta)$. More often than not, $C_N \subseteq C_{N+1}$ and C_N converges to Ω. See Petrone and Wasserman (2002), for example, who consider random densities generated via Bernstein polynomials.

Following the above specifications, if $\bigcup_k A_k$ covers Ω then we assume that C_N is covered by $\{A_1, \ldots, A_{I_N}\}$ and therefore $\Pi_N(A_k) = 0$ for $k > I_N$. Hence,

$$\sum_k \sqrt{\Pi(A_k)} \leqslant \sum_k \sqrt{\sum_{I_N \geqslant k} p_N}.$$

Consequently, if

$$\sum_k \sqrt{\overline{P}(M_k)} < \infty,$$

where $\overline{P}(M_k) = \sum_{N \geqslant M_k} p_N$ and $M_k = \min\{N\colon\ I_N \geqslant k\}$, then Hellinger consistency holds.

For example, if $I_N(\delta) = (c/\delta)^N$, for some $c > 0$ not depending on δ, as is the case with Bernstein polynomials, then

$$M_k(\delta) = \lfloor \log k / \log(c/\delta) \rfloor$$

and hence we would wish that

$$\overline{P}\bigl(M_k(\delta)\bigr) < ak^{-2-r},$$

for some $r > 0$ and $a > 0$, for all large k and for all $\delta > 0$. This holds if $\overline{P}(N) < a\exp(-N\psi)$ for all $\psi > 0$ for all large N, which holds if $N^{-1} \log \overline{P}(N) \to -\infty$ as $N \to \infty$.

Recent work on consistency has involved establishing rates of convergence of posterior distributions. See Shen and Wasserman (2001) and Ghosal et al. (2000).

6. Nonparametric regression

There are essentially two types of regression problem that are studied; both are semi-parametric models rather than being fully nonparametric.

1. The linear regression model $Y_i = X_i\beta + \varepsilon_i$ where the ε_i have an unknown distribution, centered at zero, which is modelled nonparametrically.
2. The regression model is given by $Y_i = \psi(X_i) + \varepsilon_i$, $X \in \mathbb{R}^d$, for some finite integer d, where the regression function ψ is modelled nonparametrically and the ε_i are typically assumed to be normal distributed with zero mean and variance σ^2.

6.1. Case 1

In most situations here, the distribution of the errors ε_i is modelled symmetrically about zero so that $\mathrm{E}(Y_i) = X_i\beta$. The most popular models are based on symmetrized Dirichlet processes. The idea from Brunner and Lo (1989) was to characterize unimodal and symmetric densities via nonincreasing densities on $\mathbb{R}_+$ which themselves can be characterized via scale mixtures of uniform distributions;

$$f_P(y) = \int u^{-1}\mathbf{1}(y < u)\,\mathrm{d}P(u).$$

In particular they took P to be a Dirichlet process. This idea was implemented via MCMC methods in Brunner (1995). More recently Kottas and Gelfand (2001) used the Dirichlet process mixture model for P and employed a different P for $\mathbb{R}_-$ and $\mathbb{R}_+$. Consequently, the distribution for the errors is not symmetric in this case but the unimodality and median at zero were maintained.

On the other hand, Mallick and Walker (2003) use a Pólya tree prior for P. In order to obtain a median at zero for the random distribution functions it is possible, without encountering any technical difficulties, to fix $F(-\infty, 0] = F(0, \infty) = 1/2$. Essentially, this would involve taking the partition at the top level to be $B_0 = (-\infty, 0]$ and $B_1 = (0, \infty)$ and taking the parameters of the top beta distribution, that is α_0 and α_1, to be infinite.

It is perhaps in order to make special mention of the binary regression problem for which a Bernoulli observations Y has probability of success p_i, which are modelled via

$$p_i = F(a + bX_i),$$

where F, a distribution function, is modelled nonparametrically. Here the X_i could represent dose levels of a drug. Constraints on F are necessary in order to ensure the identifiability of the model. See, for example, Newton et al. (1996).

6.2. *Case 2*

The basic idea here is to use a set of basis functions, say $\phi_j(x)$, and to assign

$$\psi(x) = \sum_{j=1}^{M} \beta_j \phi_j(x).$$

Here the unknowns are the β_j and possibly M, and to be estimated from the data. There are a number of choices for the ϕ_j, including polynomials, splines and wavelets. There is a vast amount of literature on this area of research and the reader is referred to Smith and Kohn (1998), Denison et al. (2002) and Yau et al. (2003) for further information and references.

7. Reinforcement and exchangeability

If $Y_1, Y_2, \ldots$ are assumed to be exchangeable on some separable metric space then de Finetti's theorem (de Finetti, 1937; Hewitt and Savage, 1955) guarantees the existence of a probability measure on the space of relevant distribution functions. This probability measure is the prior in Bayesian problems. For example, if the sequence is from a Pólya urn scheme then the de Finetti measure or prior is the Dirichlet process.

To see how reinforcement plays a key role in Bayesian nonparametric problems, let us consider a discrete time beta process on $\{1, 2, \dots\}$. We use positive (α_t, β_t) as parameters and a walk along the integers involves sampling $Z_1, Z_2, \dots$ until $Z_t = 1$, where the Z_t are independent and

$$\mathrm{pr}(Z_t = 1) = \frac{\alpha_t}{\alpha_t + \beta_t}.$$

We say the walk ends at $Y = t$ when the event $\{Z_1 = 0, \dots, Z_{t-1} = 0, Z_t = 1\}$ occurs. Along the walk, whenever $Z_t = 0$ we update $\beta_t \to \beta_t + 1$ and whenever $Z_t = 1$ we update $\alpha_t \to \alpha_t + 1$. When the first walk is completed, a second one can begin, which follows the same mechanism and use the updated (α_t, β_t) from the first walk. Continue in this fashion, updating the parameters again and continuing with the walks. The reinforcement is obvious to see as we add 1's to the β's and α's. This increases the probability of the second walk replicating the first walk. The idea of reinforcing walks in this way is due to Coppersmith and Diaconis (1987) though their ideas were not specifically targeted at exchangeable sequences.

After n walks it is clear that $\alpha_t \to \alpha_t + n_t$, where n_t is the number of walks that are completed at time t, and $\beta_t \to \beta_t + m_t$, where m_t is the number of walks which are completed at a time after time t. It is shown in Walker and Muliere (1997) that the completion times $Y_1, Y_2, \dots$ are exchangeable and the de Finetti measure is a discrete time beta-Stacy process.

It is clear that this idea of generating exchangeable observations via reinforcement is generalizable. This was done in Muliere et al. (2000). The extension to continuous time models was undertaken in Muliere et al. (2003a) where, in particular, an exchangeable sequence of semi-Markov processes was characterized. In many of the cases introduced in this paper, the de Finetti measure remains unidentified; rather it is the ability to simulate from the posterior and obtain predictive distributions which provide the key to inference.

Developing the idea of reinforcement to construct partially exchangeable sequences is investigated in Muliere et al. (2003b). The idea here is to construct branching processes of the type considered in Muliere et al. (2000) and Muliere et al. (2003a). To provide a brief explanation, we will consider a bivariate, single branching, example described in Walker and Muliere (2003). Consider a Pólya urn sequence, say $Y_1, Y_2, \dots,$ the probability distributions for which were described in Section 2.1. Suppose at r, the sequence branches into two conditionally independent sequences. So, we have

$$Y_1, Y_2, \dots, Y_r \begin{cases} Y_{1r+1}, Y_{1,r+2}, \dots \\ Y_{2r+1}, Y_{2,r+2}, \dots \end{cases}$$

Marginaliy, both sequences are Pólya urn sequences with the same marginal distributions. The sequences are also dependent on each other with the degree of dependence determined by the size of r. As such, a partially exchangeable sequence has been constructed which actually in this case determines a bivariate Dirichlet process. If (F_1, F_2) represent a random selection then the key to the dependence is

$$\mathrm{corr}\big(F_1(B), F_2(B)\big) = \frac{r}{c + r}$$

for all sets B. Here $c = \alpha(\mathscr{Y})$ is the scale parameter of the marginal Dirichlet process; that is $F_j \sim \mathscr{D}(\alpha)$ for both $j = 1, 2$.

The generalization to more complex partially exchangeable sequences is obtained by then branching the sequences Y_{1t} and Y_{2t} and so on.

8. Discussion

Bayesian nonparametric methods have a lot to offer Bayesians. Finite-dimensional parametric models have their place but their limitations are becoming increasingly exposed as nonparametric models become better understood and useable. This is from a practical perspective. From a theoretical perspective, there is something unsatisfactory about putting probability one on a restricted set of densities, as is done in the case of a parametric model. In such cases many densities are excluded, that is have probability zero, and while in practice they can be returned once the data has been viewed, this does not lead to good statistics.

One simple idea of working with parametric models is to be a Bayesian in the nonparametric framework, that is to construct nonparametric priors and posteriors and to make any finite-dimensional summaries the solutions to well defined decision problems set up in the nonparametric framework. For example, consider the problem of constructing a parametric predictive density, $p(y;\theta)$. The use of the logarithmic score utility function combined with the nonparametric posterior expectation leads to the best θ as that one maximizing

$$U(\theta; n) = \int \log p(y;\theta)\,\mathrm{d}F_n(y),$$

where F_n is the predictive distribution function from the nonparametric model. An interesting case arises when the nonparametric posterior is taken to be the Bayesian bootstrap since then F_n is the empirical distribution function and so the optimal θ is precisely the maximum likelihood estimator.

Remaining problems in Bayesian nonparametrics include developing a fully nonparametric regression model. This is difficult. For any covariate x one requires a random density function model $f(\cdot; x)$. These models need to be suitably related with each other for the overall model to make sense. If the x belong to a finite or countable set then this may be achievable. On the other hand, if x belongs to the real line or even higher dimensions then things are not so easy.

Consistency is also a fast moving area of current research. Theorems to date are restricted to a large extent to independent data. Rates of convergence of posterior distributions are also being developed.

Acknowledgement

The author's work is supported by an EPSRC Advanced Research Fellowship.

References

Antoniak, C.E. (1974). Mixtures of Dirichlet processes with applications to Bayesian nonparametric problems. *Ann. Statist.* **2**, 1152–1174.

Barndorff-Nielsen, O.E., Shephard, N. (2001). Non-Gaussian Orstein–Uhlenbeck based models and some of their uses in financial economics. *J. Roy. Statist. Soc., Ser. B* **63**, 167–241.

Barron, A. (1988). The exponential convergence of posterior probabilities with implications for Bayes estimators of density functions. Unpublished manuscript.

Barron, A., Schervish, M.J., Wasserman, L. (1999). The consistency of distributions in nonparametric problems. *Ann. Statist.* **27**, 536–561.

Bernardo, J.M., Smith, A.F.M. (1994). *Bayesian Theory*. Wiley, New York.

Blackwell, D. (1973). The discreteness of Ferguson selections. *Ann. Statist.* **1**, 356–358.

Blackwell, D., MacQueen, J.B. (1973). Ferguson distributions via Pólya-urn schemes. *Ann. Statist.* **1**, 353–355.

Brix, A. (1999). Generalized gamma measures and shot-noise Cox processes. *Adv. Appl. Probab.* **31**, 929–953.

Brunner, L.J. (1995). Bayesian linear regression with error terms that have symmetric unimodal densities. *J. Nonparametr. Statist.* **4**, 335–348.

Brunner, L.J., Lo, A.Y. (1989). Bayes methods for a symmetric unimodal density and its mode. *Ann. Statist.* **17**, 1550–1566.

Cifarelli, D.M., Regazzini, E. (1990). Distribution functions of means of a Dirichlet process. *Ann. Statist.* **18**, 429–442. (Correction in *Ann. Statist.* **22** (1994), 1633–1634).

Coppersmith, D., Diaconis, P. (1987). Random walk with reinforcement. Unpublished manuscript.

Damien, P., Laud, P.W., Smith, A.F.M. (1995). Approximate random variate generation from infinitely divisible distributions with applications to Bayesian inference. *J. Roy. Statist. Soc., Ser. B* **57**, 547–563.

Damien, P., Laud, P.W., Smith, A.F.M. (1996). Implementation of Bayesian nonparametric inference based on beta processes. *Scand. J. Statist.* **23**, 27–36.

de Finetti, B. (1937). La prevision: Ses lois logiques, ses sources subjectives. *Ann. Inst. H. Poincaré* **7**, 1–68.

Denison, D.G.T., Holmes, C.C., Mallick, B.K., Smith, A.F.M. (2002). *Bayesian Methods for Nonlinear Classification and Regression*. Wiley, New York.

Dey, D., Sinha, D., Müller, P. (1998). *Practical Nonparametric and Semiparametric Bayesian Statistics. Lecture Notes in Statistics*. Springer, New York.

Dey, J., Erickson, R.V., Ramamoorthi, R.V. (2003). Some aspects of neutral to right priors. *Internat. Statist. Rev.* **71**, 383–401.

Doksum, K.A. (1974). Tailfree and neutral random probabilities and their posterior distributions. *Ann. Probab.* **2**, 183–201.

Doss, H. (1995). Bayesian nonparametric estimation for incomplete data via substitution sampling. *Ann. Statist.* **22**, 1763–1786.

Doss, H., Huffer, F.W. (2003). Monte Carlo methods for Bayesian analysis of survival data using mixtures of Dirichlet process priors. *J. Comput. Graph. Statist.* **12**, 282–307.

Dystra, R.L., Laud, P.W. (1981). A Bayesian nonparametric approach to reliability. *Ann. Statist.* **9**, 356–367.

Epifani, I., Lijoi, A., Prünster, I. (2003). Exponential functionals and means of neutral to the right priors. *Biometrika* **90**, 791–808.

Escobar, M.D. (1988). Estimating the means of several normal populations by nonparametric estimation of the distribution of the means. Unpublished Ph.D. dissertation, Department of Statistics, Yale University.

Escobar, M.D. (1994). Estimating normal means with a Dirichlet process prior. *J. Amer. Statist. Assoc.* **89**, 268–277.

Escobar, M.D., West, M. (1995). Bayesian density estimation and inference using mixtures. *J. Amer. Statist. Assoc.* **90**, 577–588.

Ferguson, T.S. (1973). A Bayesian analysis of some nonparametric problems. *Ann. Statist.* **1**, 209–230.

Ferguson, T.S. (1974). Prior distributions on spaces of probability measures. *Ann. Statist.* **2**, 615–629.

Ferguson, T.S., Klass, M.J. (1972). A representation of independent increment processes without Gaussian components. *Ann. Math. Statist.* **43**, 1634–1643.

Ferguson, T.S., Phadia, E.G. (1979). Bayesian nonparametric estimation based on censored data. *Ann. Statist.* **7**, 163–186.

Freedman, D.A. (1963). On the asymptotic behaviour of Bayes estimates in the discrete case I. *Ann. Math. Statist.* **34**, 1386–1403.

Freedman, D.A. (1965). On the asymptotic behaviour of Bayes estimates in the discrete case II. *Ann. Math. Statist.* **36**, 454–456.

Ghosal, S., Ghosh, J.K., Ramamoorthi, R.V. (1999). Posterior consistency of Dirichlet mixtures in density estimation. *Ann. Statist.* **27**, 143–158.

Ghosal, S., Ghosh, J.K., van der Vaart, A. (2000). Convergence rates of posterior distributions. *Ann. Statist.* **28**, 500–531.

Ghosh, J.K., Ramamoorthi, R.V. (2003). *Bayesian Nonparametrics*. Springer-Verlag, New York.

Gill, R.D., Johansen, S. (1990). A survey of product integration with a view toward application in survival analysis. *Ann. Statist.* **18**, 1501–1555.

Green, P.J. (1995). Reversible jump Markov chain Monte Carlo computation and Bayesian model determination. *Biometrika* **82**, 711–732.

Guglielmi, A., Tweedie, R. (2001). MCMC estimation of the law of the mean of a Dirichlet process. *Bernoulli* **7**, 573–592.

Guglielmi, A., Holmes, C.C., Walker, S.G. (2002). Perfect simulation involving functionals of a Dirichlet process. *J. Comput. Graph. Statist.* **11**, 306–310.

Hewitt, E., Savage, L.J. (1955). Symmetric measures on Cartesian products. *Trans. Amer. Math. Soc.* **80**, 470–501.

Hjort, N.L. (1990). Nonparametric Bayes estimators based on beta processes in models for life history data. *Ann. Statist.* **18**, 1259–1294.

Hjort, N.L (2003). Topics in nonparametric Bayesian statistics. In: Green, P.J., Hjort, N.L., Richardson, S. (Eds.), *Highly Structured Stochastic Systems*. Oxford University Press, pp. 455–478.

Ibrahim, J.G., Chen, M.H., Sinha, D. (2001). *Bayesian Survival Analysis*. Springer-Verlag, New York.

Ishwaran, H., James, L.F. (2002). Approximate Dirichlet process computing in finite normal mixtures: Smoothing and prior information. *J. Comput. Graph. Statist.* **11**, 508–533.

Ishwaran, H., Zarepour, M. (2000). Markov chain Monte Carlo in approximate Dirichlet and beta two parameter process hierarchical models. *Biometrika* **87**, 371–390.

James, L.F. (2002). Poisson process partition calculus with applications to exchangeable models and Bayesian nonparametrics. Available at http://arXiv-.org/abs/math/0205093.

Kim, Y., Lee, J. (2003). Bayesian analysis of proportional hazards model. *Ann. Statist.* **31**, 493–511.

Kottas, A., Gelfand, A.E. (2001). Bayesian semiparametric median regression modelling. *J. Amer. Statist. Assoc.* **96**, 1458–1468.

Kottas, A., Gelfand, A.E. (2002). A computational approach for full nonparametric Bayesian inference under Dirichlet process mixture models. *J. Comput. Graph. Statist.* **11**, 289–305.

Kuo, L. (1986). Computations of mixtures of Dirichlet processes. *SIAM J. Sci. Statist. Comput.* **7**, 60–71.

Lavine, M. (1992). Some aspects of Pólya tree distributions for statistical modelling. *Ann. Statist.* **20**, 1203–1221.

Lavine, M. (1994). More aspects of Pólya trees for statistical modelling. *Ann. Statist.* **22**, 1161–1176.

Lenk, P.J. (1988). The logistic normal distribution for Bayesian, nonparametric, predictive densities. *J. Amer. Statist. Assoc.* **83**, 509–516.

Lenk, P.J. (1991). Towards a practicable Bayesian nonparametric density estimator. *Biometrika* **78**, 531–543.

Leonard, T. (1978). Density estimation, stochastic processes and prior information. *J. Roy. Statist. Soc., Ser. B* **40**, 113–146.

Lijoi, A., Regazzini, E. (2004). Means of a Dirichlet process and multiple hypergeometric functions. *Ann. Probab.* **32**, 1469–1495.

Lindley, D.V., Smith, A.F.M. (1972). Bayes estimates for the linear model. *J. Roy. Statist. Soc., Ser. B* **34**, 1–41.

Lindsey, J.K. (1999). Some statistical heresies. *The Statistician* **48**, 1–40.

Lo, A.Y. (1984). On a class of Bayesian nonparametric estimates I. Density estimates. *Ann. Statist.* **12**, 351–357.

Lo, A.Y., Weng, C.S. (1989). On a class of Bayesian nonparametric estimates: II. Hazard rate estimates. *Ann. Inst. Statist. Math.* **41**, 227–245.

Loève, M. (1963). *Probability Theory*, third ed. D. Van Nostrand Company Canada Ltd.

MacEachern, S.N. (1994). Estimating normal means with a conjugate style Dirichlet process prior. *Comm. Statist. Simulation Comput.* **23**, 727–741.
MacEachern, S.N. (1998). Computational methods for mixture of Dirichlet process models. In: Dey, D., Müller, P., Sinha, D. (Eds.), *Practical Nonparametric and Semiparametric Bayesian Statistics*. Springer, New York, pp. 23–43.
MacEachern, S.N., Müller, P. (1998). Estimating mixtures of Dirichlet process models. *J. Comput. Graph. Statist.* **7**, 223–238.
MacEachern, S.N., Clyde, M., Liu, J.S. (1999). Sequential importance sampling for nonparametric Bayes models: The next generation. *Canad. J. Statist.* **27**, 251–267.
Mallick, B.K., Walker, S.G. (2003). A Bayesian semiparametric transformation model incorporating frailties. *J. Statist. Plann. Inference* **112**, 159–174.
Mauldin, R.D., Sudderth, W.D., Williams, S.C. (1992). Pólya trees and random distributions. *Ann. Statist.* **20**, 1203–1221.
Muliere, P., Tardella, L. (1998). Approximating distributions of random functionals of Ferguson–Dirichlet priors. *Canad. J. Statist.* **26**, 283–297.
Muliere, P., Secchi, P., Walker, S.G. (2000). Urn schemes and reinforced random walks. *Stochastic Process. Appl.* **88**, 59–78.
Muliere, P., Secchi, P., Walker, S.G. (2003a). Reinforced random processes in continuous time. *Stochastic Process. Appl.* **104**, 117–130.
Muliere, P., Secchi, P., Walker, S.G. (2003b). Partially exchangeable processes indexed by the vertices of a k-tree constructed via reinforcement. Submitted for publication.
Müller, P., Quintana, F.A. (2004). Nonparametric Bayesian data analysis. *Statist. Sci.* **19**, 95–110.
Müller, P., Erkanli, A., West, M. (1996). Bayesian curve fitting using multivariate normal mixtures. *Biometrika* **83**, 67–79.
Neal, R.M. (2000). Markov chain sampling methods for Dirichlet process mixture models. *J. Comput. Graph. Statist.* **9**, 249–265.
Newton, M.A., Czado, C., Chappell, R. (1996). Bayesian inference for semiparametric binary regression. *J. Amer. Statist. Assoc.* **91**, 142–153.
Nieto-Barajas, L.E., Walker, S.G. (2004). Bayesian nonparametric survival analysis via Lévy driven Markov processes. *Statistica Sinica* **14**, 1127–1146.
Nieto-Barajas, L.E., Prünster, I., Walker, S.G. (2004). Normalized random measures driven by increasing additive processes. *Ann. Statist.* **32**, 2343–2360.
Petrone, S. (1999). Random Bernstein polynomials. *Scand. J. Statist.* **26**, 273–393.
Petrone, S., Wasserman, L. (2002). Consistency of Bernstein polynomial posteriors. *J. Roy. Statist. Soc., Ser. B* **64**, 79–100.
Regazzini, E., Guglielmi, A., Di Nunno, G. (2002). Theory and numerical analysis for exact distribution of functionals of a Dirichlet process. *Ann. Statist.* **30**, 1376–1411.
Regazzini, E., Lijoi, A., Prünster, I. (2003). Distributional results for means of normalized random measures with independent increments. *Ann. Statist.* **31**, 560–585.
Richardson, S., Green, P.J. (1997). On Bayesian analysis of mixtures with an unknown number of components. *J. Roy. Statist. Soc., Ser. B* **59**, 731–792.
Rubin, D.B. (1981). The Bayesian bootstrap. *Ann. Statist.* **9**, 130–134.
Schwartz, L. (1965). On Bayes procedures. *Z. Wahrsch. Verw. Gebiete* **4**, 10–26.
Sethuraman, J. (1994). A constructive definition of Dirichlet priors. *Statistica Sinica* **4**, 639–650.
Sethuraman, J., Tiwari, R. (1982). Convergence of Dirichlet measures and the interpretation of their parameter. In: Gupta, S.S., Berger, J.O. (Eds.), *Proceedings of the Third Purdue Symposium on Statistical Decision Theory and Related Topics*. Academic Press, New York.
Shen, X., Wasserman, L. (2001). Rates of convergence of posterior distributions. *Ann. Statist.* **29**, 687–714.
Smith, A.F.M., Roberts, G.O. (1993). Bayesian computations via the Gibbs sampler and related Markov chain Monte Carlo methods. *J. Roy. Statist. Soc., Ser. B* **55**, 3–23.
Smith, M., Kohn, R. (1998). Nonparametric estimation of irregular functions with independent or autocorrelated errors. In: Dey, D., Müller, P., Sinha, D. (Eds.), *Practical Nonparametric and Semiparametric Bayesian Statistics*. Springer, New York, pp. 157–179.
Tierney, L. (1994). Markov chains for exploring posterior distributions. *Ann. Statist.* **22**, 1701–1722.

Walker, S.G. (2003). On sufficient conditions for Bayesian consistency. *Biometrika* **90**, 482–488.

Walker, S.G. (2004). New approaches to Bayesian consistency. *Ann. Statist.* **32**, 2028–2043.

Walker, S.G., Damien, P. (2000). Representations of Lévy processes with Gaussian components. *Biometrika* **87**, 477–483.

Walker, S.G., Mallick, B.K. (1997). A note on the scale parameter of the Dirichlet process. *Canad. J. Statist.* **25**, 473–479.

Walker, S.G., Muliere, P. (1997). Beta-Stacy processes and a generalization of the Pólya-urn scheme. *Ann. Statist.* **25**, 1762–1780.

Walker, S.G., Muliere, P. (2003). A bivariate Dirichlet process. *Statist. Probab. Lett.* **64**, 1–7.

Walker, S.G., Damien, P., Laud, P.W., Smith, A.F.M. (1999). Bayesian nonparametric inference for random distributions and related functions (with discussion). *J. Roy. Statist. Soc., Ser. B* **61**, 485–527.

Wolpert, R.L., Ickstadt, K. (1998). Poisson/gamma random field models for spatial statistics. *Biometrika* **85**, 251–267.

Yau, P., Kohn, R., Wood, S. (2003). Bayesian variable selection and model averaging in high-dimensional multinomial nonparametric regression. *J. Comput. Graph. Statist.* **12**, 23–54.

Handbook of Statistics, Vol. 25
ISSN: 0169-7161
© 2005 Elsevier B.V. All rights reserved.
DOI 10.1016/S0169-7161(05)25013-7

13

Bayesian Methods for Function Estimation

Nidhan Choudhuri, Subhashis Ghosal and Anindya Roy

Keywords: consistency; convergence rate; Dirichlet process; density estimation; Markov chain Monte Carlo; posterior distribution; regression function; spectral density; transition density

1. Introduction

Nonparametric and semiparametric statistical models are increasingly replacing parametric models, for the latter's lack of sufficient flexibility to address a wide variety of data. A nonparametric or semiparametric model involves at least one infinite-dimensional parameter, usually a function, and hence may also be referred to as an infinite-dimensional model. Functions of common interest, among many others, include the cumulative distribution function, density function, regression function, hazard rate, transition density of a Markov process, and spectral density of a time series. While frequentist methods for nonparametric estimation have been flourishing for many of these problems, nonparametric Bayesian estimation methods had been relatively less developed.

Besides philosophical reasons, there are some practical advantages of the Bayesian approach. On the one hand, the Bayesian approach allows one to reflect ones prior beliefs into the analysis. On the other hand, the Bayesian approach is straightforward in principle where inference is based on the posterior distribution only. Subjective elicitation of priors is relatively simple in a parametric framework, and in the absence of any concrete knowledge, there are many default mechanisms for prior specification. However, the recent popularity of Bayesian analysis comes from the availability of various Markov chain Monte Carlo (MCMC) algorithms that make the computation feasible with today's computers in almost every parametric problem. Prediction, which is sometimes the primary objective of a statistical analysis, is solved most naturally if one follows the Bayesian approach. Many non-Bayesian methods, including the maximum likelihood estimator (MLE), can have very unnatural behavior (such as staying on the boundary with high probability) when the parameter space is restricted, while a Bayesian estimator does not suffer from this drawback. Besides, the optimality of a parametric Bayesian procedure is often justified through large sample as well as finite sample admissibility properties.

The difficulties for a Bayesian analysis in a nonparametric framework is threefold. First, a subjective elicitation of a prior is not possible due to the vastness of the parameter space and the construction of a default prior becomes difficult mainly due to the absence of the Lebesgue measure. Secondly, common MCMC techniques do not directly apply as the parameter space is infinite-dimensional. Sampling from the posterior distribution often requires innovative MCMC algorithms that depend on the problem at hand as well as the prior given on the functional parameter. Some of these techniques include the introduction of latent variables, data augmentation and reparametrization of the parameter space. Thus, the problem of prior elicitation cannot be separated from the computational issues.

When a statistical method is developed, particular attention should be given to the quality of the corresponding solution. Of the many different criteria, asymptotic consistency and rate of convergence are perhaps among the least disputed. Consistency may be thought of as a validation of the method used by the Bayesian. Consider an imaginary experiment where an experimenter generates observations from a given stochastic model with some value of the parameter and presents the data to a Bayesian without revealing the true value of the parameter. If enough information is provided in the form of a large number of observations, the Bayesian's assessment of the unknown parameter should be close to the true value of it. Another reason to study consistency is its relationship with robustness with respect to the choice of the prior. Due to the lack of complete faith in the prior, we should require that at least eventually, the data overrides the prior opinion. Alternatively two Bayesians, with two different priors, presented with the same data eventually must agree. This large sample "merging of opinions" is equivalent to consistency (Blackwell and Dubins, 1962; Diaconis and Freedman, 1986a, 1986b; Ghosh et al., 1994). For virtually all finite-dimensional problems, the posterior distribution is consistent (Ibragimov and Has'minskii, 1981; Le Cam, 1986; Ghosal et al., 1995) if the prior does not rule out the true value. This is roughly a consequence of the fact that the likelihood is highly peaked near the true value of the parameter if the sample size is large. However, for infinite-dimensional problems, such a conclusion is false (Freedman, 1963; Diaconis and Freedman, 1986a, 1986b; Doss, 1985a, 1985b; Kim and Lee, 2001). Thus posterior consistency must be verified before using a prior.

In this chapter, we review Bayesian methods for some important curve estimation problems. There are several good reviews available in the literature such as Hjort (1996, 2003), Wasserman (1998), Ghosal et al. (1999a), the monograph of Ghosh and Ramamoorthi (2003) and several chapters in this volume. We omit many details which may be found from these sources. We focus on three different aspects of the problem: prior specification, computation and asymptotic properties of the posterior distribution. In Section 2, we describe various priors on infinite-dimensional spaces. General results on posterior consistency and rate of convergence are reviewed in Section 3. Specific curve estimation problems are addressed in the subsequent sections.

2. Priors on infinite-dimensional spaces

A well accepted criterion for the choice of a nonparametric prior is that the prior has a large or full topological support. Intuitively, such a prior can reach every corner of the parameter space and thus can be expected to have a consistent posterior. More flexible models have higher complexity and hence the process of prior elicitation becomes more complex. Priors are usually constructed from the consideration of mathematical tractability, feasibility of computation, and good large sample behavior. The form of the prior is chosen according to some default mechanism while the key hyper-parameters are chosen to reflect any prior beliefs. A prior on a function space may be thought of as a stochastic process taking values in the given function space. Thus, a prior may be specified by describing a sampling scheme that generate random function with desired properties or they can be specified by describing the finite-dimensional laws. An advantage of the first approach is that the existence of the prior measure is automatic, while for the latter, the nontrivial proposition of existence needs to be established. Often the function space is approximated by a sequence of sieves in such a way that it is easier to put a prior on these sieves. A prior on the entire space is then described by letting the index of the sieve vary with the sample size, or by putting a further prior on the index thus leading to a hierarchical mixture prior. Here we describe some general methods of prior construction on function spaces.

2.1. Dirichlet process

Dirichlet processes were introduced by Ferguson (1973) as prior distributions on the space of probability measures on a given measurable space $(\mathfrak{X}, \mathcal{B})$. Let $M > 0$ and G be a probability measure on $(\mathfrak{X}, \mathcal{B})$. A Dirichlet process on $(\mathfrak{X}, \mathcal{B})$ with parameters (M, G) is a random probability measure P which assigns a number $P(B)$ to every $B \in \mathcal{B}$ such that

(i) $P(B)$ is a measurable $[0, 1]$-valued random variable;
(ii) each realization of P is a probability measure on $(\mathfrak{X}, \mathcal{B})$;
(iii) for each measurable finite partition $\{B_1, \ldots, B_k\}$ of $\mathfrak{X}$, the joint distribution of the vector $(P(B_1), \ldots, P(B_k))$ on the k-dimensional unit simplex has Dirichlet distribution with parameters $(k; MG(B_1), \ldots, MG(B_k))$.

(We follow the usual convention for the Dirichlet distribution that a component is a.s. 0 if the corresponding parameter is 0.) Using Kolmogorov's consistency theorem, Ferguson (1973) showed that a process with the stated properties exists. The argument could be made more elegant and transparent by using a countable generator of $\mathcal{B}$ as in Blackwell (1973). The distribution of P is also uniquely defined by its specified finite-dimensional distributions in (iii) above. We shall denote the process by $\mathrm{Dir}(M, G)$. If $(M_1, G_1) \neq (M_2, G_2)$ then the corresponding Dirichlet processes $\mathrm{Dir}(M_1, G_1)$ and $\mathrm{Dir}(M_2, G_2)$ are different, unless both G_1 and G_2 are degenerate at the same point. The parameter M is called the precision, G is called the center measure, and the product MG is called the base measure of the Dirichlet process. Note that

$$\mathrm{E}\big(P(B)\big) = G(B), \qquad \mathrm{var}\big(P(B)\big) = \frac{G(B)(1 - G(B))}{1 + M}. \tag{2.1}$$

Therefore, if M is large, P is tightly concentrated about G justifying the terminology. The relation (2.1) easily follows by the observation that each $P(B)$ is distributed as beta with parameters $MG(B)$ and $M(1-G(B))$. By considering finite linear combinations of indicator of sets and passing to the limit, it readily follows that (2.1) could be extended to functions, that is, $\mathrm{E}(\int \psi \, \mathrm{d}P) = \int \psi \, \mathrm{d}G$, and $\mathrm{var}(\int \psi \, \mathrm{d}P) = \mathrm{var}_G(\psi)/(1+M)$.

As $P(A)$ is distributed as beta $(MG(A), MG(A^c))$, it follows that $P(A) > 0$ a.s. if and only if $G(A) > 0$. However, this does not imply that P is a.s. mutually absolutely continuous with G, as the null set could depend on A. As a matter of fact, the two measures are often a.s. mutually singular.

If $\mathfrak{X}$ is a separable metric space, the topological support of a measure on $\mathfrak{X}$ and the weak[1] topology on the space $\mathfrak{M}(\mathfrak{X})$ of all probability measures on $\mathfrak{X}$ may be defined. The support of $\mathrm{Dir}(M, G)$ with respect to the weak topology is given by $\{P \in \mathfrak{M}(\mathfrak{X})\colon \mathrm{supp}(P) \subset \mathrm{supp}(G)\}$. In particular, if the support of G is $\mathfrak{X}$, then the support of $\mathrm{Dir}(M, G)$ is the whole of $\mathfrak{M}(\mathfrak{X})$. Thus the Dirichlet process can be easily chosen to be well spread over the space of probability measures. This may however look apparently contradictory to the fact that a random P following $\mathrm{Dir}(M, G)$ is a.s. discrete. This important (but perhaps somewhat disappointing) property was observed in Ferguson (1973) by using a gamma process representation of the Dirichlet process and in Blackwell (1973) by using a Polya urn scheme representation. In the latter case, the Dirichlet process arises as the mixing measure in de Finetti's representation in the following continuous analogue of the Polya urn scheme: $X_1 \sim G$; for $i = 1, 2, \ldots$, $X_i = X_j$ with probability $1/(M+i-1)$ for $j = 1, \ldots, i-1$ and $X_i \sim G$ with probability $M/(M+i-1)$ independently of the other variables. This representation is extremely crucial for MCMC sampling from a Dirichlet process. The representation also shows that ties are expected among $X_1, \ldots, X_n$. The expected number of distinct X's, as $n \to \infty$, is $M \log \frac{n}{M}$, which asymptotically much smaller than n. A simple proof of a.s. discreteness of Dirichlet random measure, due to Savage, is given in Theorem 3.2.3 of Ghosh and Ramamoorthi (2003).

Sethuraman (1994) gave a constructive representation of the Dirichlet process. If $\theta_1, \theta_2, \ldots$ are i.i.d. G_0, $Y_1, Y_2, \ldots$ are i.i.d. beta$(1, M)$, $V_i = Y_i \prod_{j=1}^{i-1}(1-Y_j)$ and

$$P = \sum_{i=1}^{\infty} V_i \delta_{\theta_i}, \tag{2.2}$$

then the above infinite series converges a.s. to a random probability measure that is distributed as $\mathrm{Dir}(M, G)$. It may be noted that the masses V_i's are obtained by successive "stick-breaking" with $Y_1, Y_2, \ldots$ as the corresponding stick-breaking proportions, and allotted to randomly chosen points $\theta_1, \theta_2, \ldots$ generated from G. Sethuraman's representation has made it possible to use the Dirichlet process in many complex problem using some truncation and Monte Carlo algorithms. Approximations of this type are discussed by Muliere and Tardella (1998) and Iswaran and Zarepour (2002a, 2002b). Another consequence of the Sethuraman representation is that if $P \sim \mathrm{Dir}(M, G)$, $\theta \sim G$ and $Y \sim \mathrm{beta}(1, M)$, and all of them are independent, then $Y\delta_\theta + (1-Y)P$ also has $\mathrm{Dir}(M, G)$ distribution. This property leads to important distributional equations for

[1] What we call weak is termed as weak star in functional analysis.

functionals of the Dirichlet process, and could also be used to simulate a Markov chain on $\mathfrak{M}(\mathfrak{X})$ with $\mathrm{Dir}(M, G)$ as its stationary distribution.

The Dirichlet process has a very important conditioning property. If A is set with $G(A) > 0$ (which implies that $P(A) > 0$ a.s.), then the random measure $P|_A$, the restriction of P to A defined by $P|_A(B) = P(B|A) = P(B \cap A)/P(A)$, is distributed as Dirichlet process with parameters $MG(A)$ and $G|_A$ and is independent of $P(A)$. The argument can be extended to more than one set. Thus the Dirichlet process locally splits into numerous independent Dirichlet processes.

A peculiar property of the Dirichlet process is that any two Dirichlet processes $\mathrm{Dir}(M_1, G_1)$ and $\mathrm{Dir}(M_2, G_2)$ are mutually singular if G_1, G_2 are nonatomic and $(M_1, G_1) \neq (M_2, G_2)$.

The distribution of a random mean functional $\int \psi \, \mathrm{d}P$, where ψ is a measurable function, is of some interest. Although, $\int \psi \, \mathrm{d}P$ has finite mean if and only if $\int |\psi| \, \mathrm{d}G < \infty$, P has a significantly shorter tail than that of G. For instance, the random P generated by a Dirichlet process with Cauchy base measure has all moments. Distributions of the random mean functional has been studied in many articles including Cifarelli and Regazzini (1990) and Regazzini et al. (2002). Interestingly the distribution of $\int x \, \mathrm{d}P(x)$ is G if and only if G is Cauchy.

The behavior of the tail probabilities of a random P obtained from a Dirichlet process is important for various purposes. Fristedt (1967) and Fristedt and Pruitt (1971) characterized the growth rate of a gamma process and using their result, Doss and Sellke (1982) obtained analogous results for the tail probabilities of P.

Weak convergence properties of the Dirichlet process are controlled by the convergence of its parameters. Let G_n weakly converge to G. Then

(i) if $M_n \to M > 0$, then $\mathrm{Dir}(M_n, G_n)$ converges weakly to $\mathrm{Dir}(M, G)$;
(ii) if $M_n \to 0$, then $\mathrm{Dir}(M_n, G_n)$ converges weakly to a measure degenerated at a random $\theta \sim G$;
(iii) if $M_n \to \infty$, then $\mathrm{Dir}(M_n, G_n)$ converges weakly to random measure degenerate at G.

2.2. *Processes derived from the Dirichlet process*

2.2.1. *Mixtures of Dirichlet processes*

The mixture of Dirichlet processes was introduced by Antoniak (1974). While eliciting the base measure using (2.1), it may be reasonable to guess that the prior mean measure is normal, but it may be difficult to specify the values of the mean and the variance of this normal distribution. It therefore makes sense to put a prior on the mean and the variance. More generally, one may propose a parametric family as the base measure and put hyper-priors on the parameters of that family. The resulting procedure has an intuitive appeal in that if one is a weak believer in a parametric family, then instead of using a parametric analysis, one may use the corresponding mixture of Dirichlet to robustify the parametric procedure. More formally, we may write the hierarchical Bayesian model $P \sim \mathrm{Dir}(M_\theta, G_\theta)$, where the indexing parameter $\theta \sim \pi$.

In semiparametric problems, mixtures of Dirichlet priors appear if the nonparametric part is given a Dirichlet process. In this case, the interest is usually in the posterior distribution of the parametric part, which has a role much bigger than that of an indexing parameter.

2.2.2. *Dirichlet mixtures*

Although the Dirichlet process cannot be used as a prior for estimating a density, convoluting it with a kernel will produce smooth densities. Such an approach was pioneered by Ferguson (1983) and Lo (1984). Let Θ be a parameter set, typically a Euclidean space. For each θ, let $\psi(x,\theta)$ be a probability density function. A nonparametric mixture of $\psi(x,\theta)$ is obtained by considering $p_F(x) = \int \psi(x,\theta)\,\mathrm{d}F(\theta)$. These mixtures can form a very rich family. For instance, the location and scale mixture of the form $\sigma^{-1}k((x-\mu)/\sigma)$, for some fixed density k, may approximate any density in the L_1-sense if σ is allowed to approach to 0. Thus, a prior on densities may be induced by putting a Dirichlet process prior on the mixing distribution F and a prior on σ.

The choice of an appropriate kernel depends on the underlying sample space. If the underlying density function is defined on the entire real line, a location-scale kernel is appropriate. If on the unit interval, beta distributions form a flexible two parameter family. If on the positive half line, mixtures of gamma, Weibull or lognormal may be used. The use of a uniform kernel leads to random histograms. Petrone and Veronese (2002) motivated a canonical way of viewing the choice of a kernel through the notion of the Feller sampling scheme, and call the resulting prior a Feller prior.

2.2.3. *Invariant Dirichlet process*

The invariant Dirichlet process was considered by Dalal (1979). Suppose that we want to put a prior on the space of all probability measures symmetric about zero. One may let P follow Dir(M,G) and put $\overline{P}(A) = (P(A) + P(-A))/2$, where $-A = \{x\colon -x \in A\}$.[2] More generally, one can consider a compact group $\mathfrak{G}$ acting on the sample space $\mathfrak{X}$ and consider the distribution of $\overline{P}$ as the invariant Dirichlet process where $\overline{P}(A) = \int P(gA)\,\mathrm{d}\mu(g)$, μ stands for the Haar probability measure on $\mathfrak{G}$ and P follows the Dirichlet process.

The technique is particularly helpful for constructing priors on the error distribution F for the location problem $X = \theta + \epsilon$. The problem is not identifiable without some restriction on F, and symmetry about zero is a reasonable condition on F ensuring identifiability. The symmetrized Dirichlet process prior was used by Diaconis and Freedman (1986a, 1986b) to present a striking example of inconsistency of the posterior distribution.

2.2.4. *Pinned-down Dirichlet*

If $\{B_1,\dots,B_k\}$ is a finite partition, called control sets, then the conditional distribution of P given $\{P(B_j) = w_j,\ j = 1,\dots,k\}$, where P follows Dir(M,G) and $w_j \geqslant 0$, $\sum_{j=1}^k w_j = 1$, is called a pinned-down Dirichlet process. By the conditioning property of the Dirichlet process mentioned in the last subsection, it follows that the above

[2] Another way of randomly generating symmetric probabilities is to consider a Dirichlet process P on $[0,\infty)$ and unfold it to $\widetilde{P}$ on $\mathbb{R}$ by $\widetilde{P}(-A) = \widetilde{P}(A) = \frac{1}{2}P(A)$.

process may be written as $P = \sum_{j=1}^{k} w_j P_j$, where each P_j is a Dirichlet process on B_j. Consequently P is a countable mixture of Dirichlet (with orthogonal supports).

A particular case of pinned-down Dirichlet is obtained when one puts the restriction that P has median 0. Doss (1985a, 1985b) used this idea to put a prior for the semiparametric location problem and showed an inconsistency result similar to Diaconis and Freedman (1986a, 1986b) mentioned above.

2.3. *Generalizations of the Dirichlet process*

While the Dirichlet process is a prior with many fascinating properties, its reliance on only two parameters may sometimes be restrictive. One drawback of Dirichlet process is that it always produces discrete random probability measures. Another property of Dirichlet which is sometimes embarrassing is that the correlation between the random probabilities of two sets is always negative. Often, random probabilities of sets that are close enough are expected to be positively related if some smoothness is present. More flexible priors may be constructed by generalizing the way the prior probabilities are assigned. Below we discuss some of the important generalizations of a Dirichlet process.

2.3.1. *Tail-free and neutral to the right process*

The concept of a tail-free process was introduced by Freedman (1963) and chronologically precedes that of the Dirichlet process. A tail-free process is defined by random allocations of probabilities to sets in a nested sequence of partitions. Let $E = \{0, 1\}$ and E^m be the m-fold Cartesian product $E \times \cdots \times E$ where $E^0 = \emptyset$. Further, set $E^* = \bigcup_{m=0}^{\infty} E^m$. Let $\pi_0 = \{\mathfrak{X}\}$ and for each $m = 1, 2, \ldots$, let $\pi_m = \{B_\varepsilon\colon \varepsilon \in E^m\}$ be a partition of $\mathfrak{X}$ so that sets of π_{m+1} are obtained from a binary split of the sets of π_m and $\bigcup_{m=0}^{\infty} \pi_m$ be a generator for the Borel sigma-field on $\mathbb{R}$. A probability P may then be described by specifying all the conditional probabilities $\{V_\varepsilon = P(B_{\varepsilon 0} | B_\varepsilon)\colon \varepsilon \in E^*\}$. A prior for P may thus be defined by specifying the joint distribution of all V_ε's. The specification may be written in a tree form. The different hierarchy in the tree signifies prior specification of different levels. A prior for P is said to be tail-free with respect to the sequence of partitions $\{\pi_m\}$ if the collections $\{V_\emptyset\}, \{V_0, V_1\}, \{V_{00}, V_{01}, V_{10}, V_{11}\}, \ldots$, are mutually independent. Note that, variables within the same hierarchy need not be independent; only the variables at different levels are required to be so. Partitions more general than binary partitions could be used, although that will not lead to more general priors.

A Dirichlet process is tail-free with respect to any sequence of partitions. Indeed, the Dirichlet process is the only prior that has this distinguished property; see Ferguson (1974) and the references therein. Tail-free priors satisfy some interesting zero-one laws, namely, the random measure generated by a tail-free process is absolutely continuous with respect to a given finite measure with probability zero or one. This follows from the fact that the criterion of absolute continuity may be expressed as tail event with respect to a collection of independent random variables and Kolmogorov's zero-one law may be applied; see Ghosh and Ramamoorthi (2003) for details. Kraft (1964) gave a very useful sufficient condition for the almost sure absolute continuity of a tail-free process.

Neutral to the right processes, introduced by Doksum (1974), are also tail-free processes, but the concept is applicable only to survival distribution functions. If F is a random distribution function on the positive half line, then F is said to follow a neutral to the right process if for every k and $0 < t_1 < \cdots < t_k$, there exists independent random variables $V_1, \dots, V_k$ such that the joint distribution of $(1 - F(t_1), 1 - F(t_2), \dots, 1 - F(t_k))$ is same as that of the successive products $(V_1, V_1 V_2, \dots, \prod_{j=1}^{k} V_j)$. Thus a neutral to the right prior is obtained by stick breaking. Clearly the process is tail-free with respect to the nested sequence $\{[0, t_1], (t_1, \infty)\}, \{[0, t_1], (t_1, t_2], (t_2, \infty)\}, \dots$ of partitions. Note that $F(x)$ may be written as $\mathrm{e}^{-H(x)}$, where $H(\cdot)$ is a process of independent increments.

2.3.2. *Polya tree process*

A Polya tree process is a special case of a tail-free process, where besides across row independence, the random conditional probabilities are also independent within row and have beta distributions. To elaborate, let $\{\pi_m\}$ be a sequence of binary partition as before and $\{\alpha_\varepsilon \colon \varepsilon \in E^*\}$ be a collection of nonnegative numbers. A random probability measure P on $\mathbb{R}$ is said to possess a Polya tree distribution with parameters $(\{\pi_m\}, \{\alpha_\varepsilon \colon \varepsilon \in E^*\})$, if there exist a collection $\mathcal{Y} = \{Y_\varepsilon \colon \varepsilon \in E^*\}$ of random variables such that the following hold:

(i) The collection $\mathcal{Y}$ consists of mutually independent random variables;
(ii) For each $\varepsilon \in E^*$, Y_ε has a beta distribution with parameters $\alpha_{\varepsilon 0}$ and $\alpha_{\varepsilon 1}$;
(iii) The random probability measure P is related to $\mathcal{Y}$ through the relations

$$P(B_{\varepsilon_1 \cdots \varepsilon_m}) = \left(\prod_{j=1; \varepsilon_j = 0}^{m} Y_{\varepsilon_1 \cdots \varepsilon_{j-1}} \right) \left(\prod_{j=1; \varepsilon_j = 1}^{m} (1 - Y_{\varepsilon_1 \cdots \varepsilon_{j-1}}) \right),$$

$m = 1, 2, \dots,$

where the factors are $Y_\emptyset$ or $1 - Y_\emptyset$ if $j = 1$.

The concept of a Polya tree was originally considered by Ferguson (1974) and Blackwell and MacQueen (1973), and later studied thoroughly by Mauldin et al. (1992) and Lavine (1992, 1994). The prior can be seen as arising as the de Finetti measure in a generalized Polya urn scheme; see Mauldin et al. (1992) for details.

The class of Polya trees contain all Dirichlet processes, characterized by the relation that $\alpha_{\varepsilon 0} + \alpha_{\varepsilon 1} = \alpha_\varepsilon$ for all ε. A Polya tree can be chosen to generate only absolutely continuous distributions. The prior expectation of the process could be easily written down; see Lavine (1992) for details. Below we consider an important special case for discussion, which is most relevant for statistical use. Consider $\mathfrak{X}$ to be a subset of the real line and let G be a probability measure. Let the partitions be obtained successively by splitting the line at the median, the quartiles, the octiles, and in general, binary quantiles of G. If $\alpha_{\varepsilon 0} = \alpha_{\varepsilon 1}$ for all $\varepsilon \in E^*$, then it follows that $\mathrm{E}(P) = G$. Thus G will have the role similar to that of the center measure of a Dirichlet process, and hence will be relatively easy to elicit. Besides, the Polya tree will have infinitely many more parameters which may be used to describe one's prior belief. Often, to avoid specifying too many parameters, a default method is adopted, where one chooses α_ε depending only on the length of the finite string ε. Let a_m stand for the value of α_ε when ε has length m. The growth rate of a_m controls the smoothness of the Polya tree process. For instance, if $a_m = c2^{-m}$, we obtain the Dirichlet process, which generate discrete probabilities. If $\sum_{m=1}^{\infty} a_m^{-1} < \infty$ (for instance, if $a_m = cm^2$), then it follows from Kraft's (1964) result that the random P is absolutely continuous with respect to G. The choice $a_m = c$ leads to singular continuous distributions almost surely; see Ferguson (1974). This could guide one to choose the sequence a_m. For smoothness, one should choose rapidly growing a_m. One may actually like to choose according to one's prior belief in the beginning of the tree deviating from the above default choice, and let a default method choose the parameters at the later stages where practically no prior information is available. An extreme form of this will lead to partially specified Polya trees, where one chooses a_m to be infinity after a certain stage (which is equivalent to uniformly spreading the mass inside a given interval).

Although the prior mean distribution function may have a smooth Lebesgue density, the randomly sampled densities from a Polya tree are very rough, being nowhere differentiable. To overcome this difficulty, mixtures of a Polya tree, where the partitioning measure G involves some additional parameter θ with some prior, may be considered. The additional parameter will average out jumps to yield smooth densities; see Hanson and Johnson (2002). However, then the tail-freeness is lost and the resulting posterior distribution could be inconsistent. Berger and Guglielmi (2001) considered a mixture where the partition remains fixed and the α-parameters depend on θ, and applied the resulting prior to a model selection problem.

2.3.3. Generalized Dirichlet process

The k-dimensional Dirichlet distribution may be viewed as the conditional distribution of $(p_1, \dots, p_k)$ given that $\sum_{j=1}^{k} p_j = 1$, where $p_j = \mathrm{e}^{-Y_j}$ and Y_j's are independent exponential variables. In general, if Y_j's have a joint density $h(y_1, \dots, y_k)$, the conditional joint density of $(p_1, \dots, p_{k-1})$ is proportional to $h(-\log p_1, \dots, -\log p_k) p_k^{-1} \cdots p_k^{-1}$, where $p_k = 1 - \sum_{j=1}^{k-1} p_j$. Hjort (1996) considered the joint density of Y_j's to be proportional to $\prod_{j=1}^{k} \mathrm{e}^{-\alpha_j y_j} g_0(y_1, \dots, y_k)$, and hence the resulting (conditional) density of $p_1, \dots, p_{k-1}$ is proportional to $p_1^{\alpha_1 - 1} \cdots p_k^{\alpha_k - 1} g(p_1, \dots, p_k)$, where $g(p_1, \dots, p_k) = g_0(-\log p_1, \dots, -\log p_k)$. We may put $g(p) = \mathrm{e}^{-\lambda \Delta(p)}$, where $\Delta(p)$ is a penalty term for roughness such as $\sum_{j=1}^{k-1}(p_{j+1} - p_j)^2$, $\sum_{j=2}^{k-1}(p_{j+1} - 2p_j + p_{j-1})^2$ or $\sum_{j=1}^{k-1}(\log p_{j+1} - \log p_j)^2$. The penalty term helps maintain positive correlation and hence "smoothness". The tuning parameter λ controls the extent to which penalty is imposed for roughness. The resulting posterior distribution is conjugate with mode equivalent to a penalized MLE. Combined with random histogram or passing through the limit as the bin width goes to 0, the technique could also be applied to continuous data.

2.3.4. Priors obtained from random series representation

Sethuraman's (1994) infinite series representation creates a lot of possibilities of generalizing the Dirichlet process by changing the distribution of the weights, the support points, or even the number of terms. Consider a random probability measure given by

$P = \sum_{i=1}^{N} V_i \delta_{\theta_i}$, where $1 \leqslant N \leqslant \infty$, $\sum_{i=1}^{N} V_i = 1$ and N may be given a further prior distribution. Note that the resulting random probability measure is almost surely discrete. Choosing $N = \infty$, θ_i's as i.i.d. G as in the Sethuraman representation, $V_i = Y_i \prod_{j=1}^{i-1}(1 - Y_j)$, where $Y_1, Y_2, \ldots$ are i.i.d. beta(a, b), Hjort (2000) obtained an interesting generalization of the Dirichlet process. The resulting process admits, as in the case of a Dirichlet process, explicit formulae for the posterior mean and variance of a mean functional.

From computational point of view, a prior is more tractable if N is chosen to be finite. To be able to achieve reasonable large sample properties, either N has to depend on the sample size n, or N must be given a prior which is infinitely supported. Given $N = k$, the prior on $(V_1, \ldots, V_k)$ is taken to be k-dimensional Dirichlet distribution with parameters $(\alpha_{1,n}, \ldots, \alpha_{k,n})$. The parameters θ_i's are usually chosen as in the Sethuraman's representation, that is i.i.d. G. Iswaran and Zarepour (2002a) studied convergence properties of these random measures. For the choice $\alpha_{j,k} = M/k$, the limiting measure is $\mathrm{Dir}(M, G)$. However, the commonly advocated choice $\alpha_{j,k} = M$ leads essentially to a parametric prior, and hence to an inconsistent posterior.

2.4. *Gaussian process*

Considered first by Leonard (1978), and then by Lenk (1988, 1991) in the context of density estimation, a Gaussian process may be used in a wider generality because of its ability to produce arbitrary shapes. The method may be applied to nonparametric regression where only smoothness is assumed for the regression function. The mean function reflects any prior belief while the covariance kernel may be tuned to control the smoothness of the sample paths as well as to reflect the confidence in the prior guess. In a generalized regression, where the function of interest has restricted range, a link function is used to map the unrestricted range of the Gaussian process to the desired one. A commonly used Gaussian process in the regression context is the integrated Wiener process with some random intercept term as in Wahba (1978). Choudhuri et al. (2004b) used a general Gaussian process prior for binary regression.

2.5. *Independent increment process*

Suppose that we want to put a prior on survival distribution functions, that is, distribution functions on the positive half line. Let $Z(t)$ be a process with independent nonnegative increments such that $Z(\infty)$, the total mass of Z, is a.s. finite. Then a prior on F may be constructed by the relation $F(t) = Z(t)/Z(\infty)$. Such a prior is necessarily neutral to the right. When $Z(t)$ is the gamma process, that is an independent increment process with $Z(t) \sim \mathrm{gamma}(MG(t), 1)$, then the resulting distribution of P is Dirichlet process $\mathrm{Dir}(M, G)$.

For estimating a survival function, it is often easier to work with the cumulative hazard function. If $Z(t)$ is a process such that $Z(\infty) = \infty$ a.s., then $F(t) = 1 - \mathrm{e}^{-Z(t)}$ is a distribution function. The process $Z(t)$ may be characterized in terms of its Lévy measure $N_t(\cdot)$, and is called a Lévy process. Unfortunately, as $Z(t)$ necessarily increases by jumps only, $Z(t)$ is not the cumulative hazard function corresponding to $F(t)$. Instead, one may define $F(t)$ by the relation $Z(t) = \int_0^t \mathrm{d}F(s)/(1 - F(s-))$. The expressions of prior mean and variance, and posterior updating are relatively straightforward in terms of the Lévy measure; see Hjort (1990) and Kim (1999). Particular choices of the Lévy measure lead to special priors such as the Dirichlet process, completely homogeneous process (Ferguson and Phadia, 1979), gamma process (Lo, 1982), beta process (Hjort, 1990), beta-Stacy process (Walker and Muliere, 1997) and extended beta process (Kim and Lee, 2001). Kim and Lee (2001) settled the issue of consistency, and provided an interesting example of inconsistency.

A disadvantage of modeling the process $Z(t)$ is that the resulting F is discrete. Dykstra and Laud (1981) considered a Lévy process to model the hazard rate. However, this approach leads only to monotone hazard functions. Nieto-Barajas and Walker (2004) replaced the independent increments process by a Markov process and obtained continuous sample paths.

2.6. *Some other processes*

One approach to putting a prior on a function space is to decompose a function into a basis expansion of the form $\sum_{j=1}^{\infty} b_j \psi_j(\cdot)$ for some fixed basis functions and then putting priors on b_j's. An orthogonal basis is very useful if the function space of interest is a Hilbert space. Various popular choices of such basis include polynomials, trigonometric functions, splines and wavelets among many others. If the coefficients are unrestricted, independent normal priors may used. Interestingly, when the coefficients are normally distributed, the prior on the random function is a Gaussian process. Conversely, a Gaussian process may be represented in this way by virtue of the Karhunen–Loévé expansion. When the function values are restricted, transformations should be used prior to a basis expansion. For instance, for a density function, an expansion should be raised to the exponential and then normalized. Barron et al. (1999) used polynomials to construct an infinite-dimensional exponential family. Hjort (1996) discussed a prior on a density induced by the Hermite polynomial expansion and a prior on the sequence of cumulants.

Instead of considering an infinite series representation, one may consider a series based on the first k terms, where k is deterministically increased to infinity with the sample size, or is itself given a prior that has infinite support. The span of the first k functions, as k tends to infinity, form approximating sieves in the sense of Grenander (1981). The resulting priors are recommended as default priors in infinite-dimensional spaces by Ghosal et al. (1997). In Ghosal et al. (2000), this idea was used with a spline basis for density estimation. They showed that with a suitable choice of k, depending on the sample size and the smoothness level of the target function, optimal convergence rates could be obtained.

If the domain is a bounded interval then the sequence of moments uniquely determines the probability measure. Hence a prior on the space of probability measures could be induced from that on the sequence of moments. One may control the location, scale, skewness and kurtosis of the random probability by using subjective priors on the first four moments. Priors for the higher-order moments are difficult to elicit, and some default method should be used.

Priors for quantiles are much easier to elicit than that for moments. One may put priors on all dyadic quantiles honoring the order restrictions. Conceptually, this operation is opposite to that of specifying a tree based prior such as the Polya tree or a tail-free process. Here masses are predetermined and the partitions are chosen randomly. In practice, one may put priors only for a finite number of quantiles, and then distribute the remaining masses uniformly over the corresponding interval. Interestingly, if the prior on the quantile process is induced from a Dirichlet process on the random probability, then the posterior expectation of a quantile (in the noninformative limit $M \to 0$) is seen to be a Bernstein polynomial smoother of the empirical quantile process. This leads to a quantile density estimator, which, upon inversion, leads to an automatically smoothed empirical density estimator; see Hjort (1996) for more details.

3. Consistency and rates of convergence

Let $\{(\mathfrak{X}^{(n)}, \mathcal{A}^{(n)}, P_\theta^{(n)})\colon \theta \in \Theta\}$ be a sequence of statistical experiments with observations $X^{(n)}$, where the parameter set Θ is an arbitrary topological space and n is an indexing parameter, usually the sample size. Let $\mathcal{B}$ be the Borel sigma-field on Θ and Π_n be a probability measure on $(\Theta, \mathcal{B})$, which, in general, may depend on n. The posterior distribution is defined to be a version of the regular conditional probability of θ given $X^{(n)}$, and is denoted by $\Pi_n(\cdot|X^{(n)})$.

Let $\theta_0 \in \Theta$. We say that the posterior distribution is consistent at θ_0 (with respect to the given topology on Θ) if $\Pi_n(\cdot|X^{(n)})$ converges weakly to δ_{θ_0} as $n \to \infty$ under $P_{\theta_0}^{(n)}$-probability, or almost surely under the distribution induced by the parameter value θ_0. If the latter makes sense, it is a more appealing concept.

The above condition (in the almost sure sense) is equivalent to checking that except on a θ_0-induced null set of sample sequences, for any neighborhood U of θ_0, $\Pi_n(U^c|X^{(n)}) \to 0$. If the topology on Θ is countably generated (as in the case of a separable metric space), this reduces to $\Pi_n(U^c|X^{(n)}) \to 0$ a.s. under the distribution induced by θ_0 for every neighborhood U. An analogous conclusion holds for consistency in probability. Henceforth we work with the second formulation.

Consistency may be motivated as follows. A (prior or posterior) distribution stands for one's knowledge about the parameter. Perfect knowledge implies a degenerate prior. Thus consistency means weak convergence of knowledge towards the perfect knowledge as the amount of data increases.

Doob (1948) obtained a very general result on posterior consistency. Let the prior Π be fixed and the observations be i.i.d. Under some mild measurability conditions on the sample space (a standard Borel space will suffice) and model identifiability, Doob (1948) showed that the set of all $\theta \in \Theta$ where consistency does not hold is Π-null. This follows by the convergence of the martingale $\mathrm{E}I(\theta \in B|X_1, \ldots, X_n)$ to $\mathrm{E}I(\theta \in B|X_1, X_2, \ldots) = I(\theta \in B)$. The condition of i.i.d. observations could be replaced by the assumption that in the product space $\Theta \times \mathfrak{X}^\infty$, the parameter θ is $\mathcal{A}^\infty$-measurable. Statistically speaking, the condition holds if there is a consistent estimate of some bimeasurable function of θ.

The above result should not however create a false sense of satisfaction as the Π-null set could be very large. It is important to know at which parameter values consistency holds. Indeed, barring a countable parameter space, Doob's (1948) is of little help. On the other hand, Doob's (1948) theorem implies that consistency holds at a parameter point whenever there is a prior point mass there.

Freedman (1963) showed that merely having positive Π-probability in a neighborhood of θ_0 does not imply consistency at that point.

EXAMPLE 1. Let $\Theta = \mathfrak{M}(\mathbb{Z}_+)$, the space of all discrete distribution on positive integers with the total variation distance on Θ. Let θ_0 be the geometric distribution with parameter 1/4. There exists a prior Π such that every neighborhood of θ_0 has positive probability under Π, yet

$$\Pi(\theta \in U | X_1, \ldots, X_n) \to 1 \quad \text{a.s. } [\theta_0^\infty], \tag{3.1}$$

where U is any neighborhood of θ_1, the geometric distribution with parameter 3/4.

Indeed, the following result of Freedman (1963) shows that the above example of inconsistency is somewhat generic in a topological sense.

THEOREM 1. *Let $\Theta = \mathfrak{M}(\mathbb{Z}_+)$ with the total variation distance on it, and let $\mathfrak{M}(\Theta)$ be the space of all priors on Θ with the weak topology. Put the product topology on $\Theta \times \mathfrak{M}(\Theta)$. Then*

$$\Big\{(\theta, \Pi) \in \Theta \times \mathfrak{M}(\Theta) \colon \limsup_{n\to\infty} \Pi(\theta \in U | X_1, \ldots, X_n) = 1 \quad \forall U \text{ open, } U \neq \emptyset\Big\} \tag{3.2}$$

is the complement of a meager set.[3]

Thus, Freedman's (1963) result tells us that except for a relatively small collection of pairs of (θ, Π), the posterior distribution wanders aimlessly around the parameter space. In particular, consistency will not hold at any given θ. While this result cautions us about naive uses of Bayesian methods, it does not mean that Bayesian methods are useless. Indeed, a pragmatic Bayesian's only aim might be to just be able to find a reasonable prior complying with one's subjective belief (if available) and obtaining consistency at various parameter values. There could be plenty of such priors available even though there will be many more that are not appropriate. The situation may be compared with the role of differentiable functions among the class of all continuous functions. Functions that are differentiable at some point form a small set in the same sense while nowhere differentiable functions are much more abundant.

From a pragmatic point of view, useful sufficient conditions ensuring consistency at a given point is the most important proposition. Freedman (1963, 1965) showed that for estimation of a probability measure, if the prior distribution is tail-free, then (a suitable version of) the posterior distribution is consistent at any point with respect to the weak topology. The idea behind this result is reducing every weak neighborhood to a Euclidean neighborhood in some finite-dimensional projection using the tail-free property.

[3] A meager set is one which can be written as a countable union of closed sets without any interior points, and is considered to be topologically small.

Schwartz (1965), in a celebrated paper, obtained a general result on consistency. Schwartz's (1965) theorem requires a testing condition and a condition on the support of the prior.

Consider i.i.d. observations generated by a statistical model indexed by an abstract parameter space Θ admitting a density $p(x, \theta)$ with respect to some sigma-finite measure μ. Let $K(\theta_1, \theta_2)$ denote the Kullback–Leibler divergence $\int p(x, \theta_1) \log(p(x, \theta_1)/p(x, \theta_2))\, \mathrm{d}\mu(x)$. We say that $\theta_0 \in \Theta$ is in the Kullback–Leibler support of Π, we write $\theta_0 \in \mathrm{KL}(\Pi)$, if for every $\varepsilon > 0$, $\Pi\{\theta\colon K(\theta_0, \theta) < \varepsilon\}$. As the Kullback–Leibler divergence is asymmetric and not a metric, the support may not be interpreted in a topological sense. Indeed, a prior may have empty Kullback–Leibler support even on a separable metric space.

THEOREM 2. *Let $\theta_0 \in U \subset \Theta$. If there exists $m \geqslant 1$, a test function $\phi(X_1, \ldots, X_m)$ for testing $H_0\colon \theta = \theta_0$ against $H\colon \theta \in U^c$ with the property that* $\inf\{\mathrm{E}_\theta \phi(X_1, \ldots, X_m)\colon \theta \in U^c\} > \mathrm{E}_{\theta_0} \phi(X_1, \ldots, X_m)$ *and $\theta_0 \in \mathrm{KL}(\Pi)$, then $\Pi\{\theta \in U^c | X_1, \ldots, X_n\} \to 0$ a.s. $[P_{\theta_0}^\infty]$.*

The importance of Schwartz's theorem cannot be overemphasized. It forms the basic foundation of Bayesian asymptotic theory for general parameter spaces. The first condition requires existence of a strictly unbiased test for testing the hypothesis $H_0\colon \theta = \theta_0$ against the complement of a neighborhood U. The condition implies the existence of a sequence of tests $\Phi_n(X_1, \ldots, X_n)$ such that probabilities of both the type I error $\mathrm{E}_{\theta_0} \Phi_n(X_1, \ldots, X_n)$ and the (maximum) type II error $\sup_{\theta \in U^c} \mathrm{E}_\theta(1 - \Phi_n(X_1, \ldots, X_n))$ converges to zero exponentially fast. This existence of test is thus only a size restriction on the model and not a condition on the prior. Writing

$$\Pi\big(\theta \in U^c | X_1, \ldots, X_n\big) = \frac{\int_{U^c} \prod_{i=1}^n \frac{p(X_i, \theta)}{p(X_i, \theta_0)}\, \mathrm{d}\Pi(\theta)}{\int_\Theta \prod_{i=1}^n \frac{p(X_i, \theta)}{p(X_i, \theta_0)}\, \mathrm{d}\Pi(\theta)}, \tag{3.3}$$

this condition is used to show that for some $c > 0$, the numerator in (3.3) is smaller than e^{-nc} for all sufficiently large n a.s. $[P_{\theta_0}^\infty]$. The condition on Kullback–Leibler support is a condition on the prior as well as the model. The condition implies that for all $c > 0$, $\mathrm{e}^{nc} \int_\Theta \prod_{i=1}^n \frac{p(X_i, \theta)}{p(X_i, \theta_0)}\, \mathrm{d}\Pi(\theta) \to \infty$ a.s. $[P_{\theta_0}^\infty]$. Combining these two assertions, we obtain the result of the theorem. The latter assertion follows by first replacing Θ by the subset $\{\theta\colon K(\theta_0, \theta) < \varepsilon\}$, applying the strong law of large numbers to the integrand and invoking Fatou's lemma. It may be noted that θ_0 needs to be in the Kullback–Leibler support, not merely in the topological support of the prior for this argument to go through. In practice, the condition is derived from the condition that θ_0 is in the topological support of the prior along with some conditions on "nicety" of $p(x, \theta_0)$.

The testing condition is usually more difficult to satisfy. In finite dimension, the condition usually holds. On the space of probability measures with the weak topology on it, it is also not difficult to show that the required test exists; see Theorem 4.4.2 of Ghosh and Ramamoorthi (2003). However, in more complicated problems or for stronger topologies on densities (such as the variation or the Hellinger distance), the required tests do not exist without an additional compactness condition. Le Cam (1986) and Birgé (1983) developed an elegant theory of existence of uniformly exponentially powerful tests. However, the theory applies provided that the two hypotheses are convex. It is therefore helpful to split U^c into small balls for which required tests exist. If Θ is compact, the number of balls needed to cover U^c will be finite, and hence by taking the maximum of the resulting tests, the required test for testing $\theta = \theta_0$ against $\theta \in U^c$ may be obtained. However, the compactness condition imposes a severe restriction.

By a simple yet very useful observation, Barron (1988) concluded that it suffices that Φ_n satisfy

$$\sup_{\theta \in U^c \cap \Theta_n} \mathrm{E}_\theta\big(1 - \Phi_n(X_1, \ldots, X_n)\big) < a\,\mathrm{e}^{-bn} \tag{3.4}$$

for some constants $a, b > 0$ and some "sieve" $\Theta_n \subset \Theta$, provided that it can be shown separately that

$$\Pi\big(\theta \in \Theta_n^c | X_1, \ldots, X_n\big) \to 0 \quad \text{a.s.}\ \big[P_{\theta_0}^\infty\big]. \tag{3.5}$$

By a simple application of Fubini's theorem, Barron (1988) concluded that (3.5) is implied by a condition only on the prior probability, namely, for some $c, d > 0$, $\Pi(\theta \in \Theta_n^c) \leqslant c\,\mathrm{e}^{-nd}$. Now one may choose each Θ_n to be compact. However, because of dependence on n, one needs to estimate the number of balls required to cover Θ_n. From the same arguments, it follows that one needs to cover the sieve Θ_n with a maximum of e^{nc} balls, which is essentially a restriction on the covering number of the sieve Θ_n. The remaining part Θ_n^c, which may be topologically much bigger receives only a negligible prior probability by the given condition. It is interesting to note that unlike in sieve methods in non-Bayesian contexts, the sieve is merely a technical device for establishing consistency; the prior and the resulting Bayes procedure is not influenced by the choice of the sieve. Moreover, the sieve can be chosen depending on the accuracy level defined by the neighborhood U.

Barron's (1988) useful observation made it possible to apply Schwartz's ideas to prove posterior consistency in noncompact spaces as well. When the observations are i.i.d., one may take the parameter θ to be the density p itself. Let p_0 stand for the true density of each observation. Exploiting this idea, for a space $\mathcal{P}$ of densities, Barron et al. (1999) gave a sufficient condition for posterior consistency in Hellinger distance $d_H(p_1, p_2) = (\int (p_1^{1/2} - p_2^{1/2})^2)^{1/2}$ in terms of a condition on bracketing Hellinger entropy[4] a sieve $\mathcal{P}_n \subset \mathcal{P}$. Barron et al. (1999) used brackets to directly bound the likelihood ratios uniformly in the numerator of (3.4). The condition turns out to be considerably stronger than necessary in that we need to bound only an average likelihood ratio. Following Schwartz's (1965) original approach involving test functions, Ghosal et al. (1999b) constructed the required tests using a much weaker condition on metric entropies. These authors considered the total variation distance $d_V(p_1, p_2) = \int |p_1 - p_2|$ (which is equivalent to d_H), constructed a test directly for a point null against a small variation ball using Hoeffding's inequality, and combined the resulting tests using the condition on the metric entropy.

[4] The ε-bracketing Hellinger entropy of a set is the logarithm of the number ε-brackets with respect to the Hellinger distance needed to cover the set; see van der Vaart and Wellner (1996) for details on this and the related concepts.

For a subset S of a metric space with a metric d on it, let $N(\varepsilon, S, d)$, called the ε-covering number of S with respect to the metric d, stand for the minimum number of ε-balls needed to cover S. The logarithm of $N(\varepsilon, S, d)$ is often called the ε-entropy.

Assume that we have i.i.d. observations from a density $p \in \mathcal{P}$, a space of densities. Let p_0 stand for the true density and consider the variation distance d_V on $\mathcal{P}$. Let Π be a prior on $\mathcal{P}$.

THEOREM 3. *Suppose that* $p_0 \in \mathrm{KL}(\Pi)$. *If given any* $\varepsilon > 0$, *there exist* $\delta < \varepsilon/4$, $c_1, c_2 > 0$, $\beta < \varepsilon^2/8$ *and* $\mathcal{P}_n \subset \mathcal{P}$ *such that* $\Pi(\mathcal{P}_n^c) \leqslant c_1 \mathrm{e}^{-nc_2}$ *and* $\log N(\delta, \mathcal{P}_n, d_V) \leqslant n\beta$, *then* $\Pi(P\colon d_V(P, P_0) > \varepsilon | X_1, \ldots, X_n) \to 0$ *a.s.* $[P_0^\infty]$.

Barron (1999) also noted that the testing condition in Schwartz's theorem is, in a sense, also necessary for posterior consistency to hold under Schwartz's condition on Kullback–Leibler support.

THEOREM 4. *Let* $\mathcal{P}$ *be a space of densities,* $p_0 \in \mathcal{P}$ *be the true density and* P_0 *be the probability measure corresponding to* p_0. *Let* $p_0 \in \mathrm{KL}(\Pi)$. *Then the following conditions are equivalent*:

(1) *There exists a* β_0 *such that* $P_0\{\Pi(U^c | X_1, \ldots, X_n) > \mathrm{e}^{-n\beta_0}$ *infinitely often*$\} = 0$.
(2) *There exist subsets* $V_n, W_n \subset \mathcal{P}$, $c_1, c_2, \beta_1, \beta_2 > 0$ *and a sequence of test functions* $\Phi_n(X_1, \ldots, X_n)$ *such that*
  (a) $U^c \subset V_n \cup W_n$,
  (b) $\Pi(W_n) \leqslant c_1 \mathrm{e}^{-nc_2}$,
  (c) $P_0\{\Phi_n > 0$ *infinitely often*$\} = 0$ *and* $\sup\{\mathrm{E}_p(1 - \Phi_n)\colon p \in V_n\} \leqslant c_2 \mathrm{e}^{-n\beta_2}$.

In a semiparametric problem, an additional Euclidean parameter is present apart from an infinite-dimensional parameter, and the Euclidean parameter is usually of interest. Diaconis and Freedman (1986a, 1986b) demonstrated that putting a prior that gives consistent posterior separately for the nonparametric part may not lead to a consistent posterior when the Euclidean parameter is incorporated in the model. The example described below appeared to be counter-intuitive when it first appeared.

EXAMPLE 2. Consider i.i.d. observations from the location model $X = \theta + \epsilon$, where $\theta \in \mathbb{R}$, $\epsilon \sim F$ which is symmetric. Put any nonsingular prior density on θ and the symmetrized Dirichlet process prior on F with a Cauchy center measure. Then there exists a symmetric distribution F_0 such that if the X observations come from F_0, then the posterior concentrates around two wrong values $\pm\gamma$ instead of the true value $\theta = 0$.

A similar phenomenon was observed by Doss (1985a, 1985b). The main problem in the above is that the posterior distribution for θ is close to the parametric posterior with a Cauchy density, and hence the posterior mode behaves like the M-estimator based on the criterion function $m(x, \theta) = \log(1 + (x - \theta)^2)$. The lack of concavity of m leads to undesirable solutions for some peculiar data generating distribution like F_0. Consistency however does obtain for the normal base measure since $m(x, \theta) = (x - \theta)^2$ is convex, or even for the Cauchy base measure if F_0 has a strongly unimodal density. Here, addition of the location parameter θ to the model destroys the delicate tail-free structure, and hence Freedman's (1963, 1965) consistency result for tail-free processes cannot be applied. Because the Dirichlet process selects only discrete distribution, it is also clear that Schwartz's (1965) condition on Kullback–Leibler support does not hold. However, as shown by Ghosal et al. (1999c), if we start with a prior on F that satisfies Schwartz's (1965) condition in the nonparametric model (that is, the case of known $\theta = 0$), then the same condition holds in the semiparametric model as well. This leads to weak consistency in the semiparametric model (without any additional testing condition) and hence consistency holds for the location parameter θ. The result extends to more general semiparametric problems. Therefore, unlike the tail-free property, Schwartz's condition on Kullback–Leibler support is very robust which is not altered by symmetrization, addition of a location parameter or formation of mixtures. Thus Schwartz's theorem is the right tool for studying consistency in semiparametric models.

Extensions of Schwartz's consistency theorem to independent, nonidentically distributed observations have been obtained by Amewou-Atisso et al. (2003) and Choudhuri et al. (2004a). The former does not use sieves and hence is useful only when weak topology is put on the infinite-dimensional part of the parameter. In semiparametric problems, this topology is usually sufficient to derive posterior consistency for the Euclidean part. However, for curve estimation problems, stronger topologies need to be considered and sieves are essential. Consistency in probability instead of that in the almost sure sense allows certain relaxations in the condition to be verified. Choudhuri et al. (2004a) considered such a formulation which is described below.

THEOREM 5. *Let $Z_{i,n}$ be independently distributed with density $p_{i,n}(\cdot;\theta)$, $i = 1,\dots,r_n$, with respect to a common σ-finite measure, where the parameter θ belongs to an abstract measurable space Θ. The densities $p_{i,n}(\cdot,\theta)$ are assumed to be jointly measurable. Let $\theta_0 \in \Theta$ and let $\overline{\Theta}_n$ and $\mathcal{U}_n$ be two subsets of Θ. Let θ have prior Π on Θ. Put $K_{i,n}(\theta_0,\theta) = \mathrm{E}_{\theta_0}(\Lambda_i(\theta_0,\theta))$ and $V_{i,n}(\theta_0,\theta) = \mathrm{var}_{\theta_0}(\Lambda_i(\theta_0,\theta))$, where $\Lambda_i(\theta_0,\theta) = \log\frac{p_{i,n}(Z_{i,n};\theta_0)}{p_{i,n}(Z_{i,n};\theta)}$.*

(A1) *Prior positivity of neighborhoods.*
Suppose that there exists a set B with $\Pi(B) > 0$ such that

(i) $\displaystyle \frac{1}{r_n^2}\sum_{i=1}^{r_n} V_{i,n}(\theta_0,\theta) \to 0$ *for all* $\theta \in B$,

(ii) $\displaystyle \liminf_{n\to\infty} \Pi\left(\left\{\theta \in B\colon \frac{1}{r_n}\sum_{i=1}^{r_n} K_{i,n}(\theta_0,\theta) < \varepsilon\right\}\right) > 0$ *for all* $\varepsilon > 0$.

(A2) *Existence of tests.*
Suppose that there exists test functions $\{\Phi_n\}$, $\Theta_n \subset \overline{\Theta}_n$ and constants $C_1, C_2, c_1, c_2 > 0$ such that

(i) $\mathrm{E}_{\theta_0}\Phi_n \to 0$,

(ii) $\sup_{\theta\in\mathcal{U}_n^c\cap\Theta_n} \mathrm{E}_\theta(1-\Phi_n) \leqslant C_1\,\mathrm{e}^{-c_1 r_n}$,

(iii) $\Pi(\overline{\Theta}_n \cap \Theta_n^c) \leqslant C_2\,\mathrm{e}^{-c_2 r_n}$.

Then $\Pi(\theta \in \mathcal{U}_n^c \cap \overline{\Theta}_n | Z_{1,n}, \ldots, Z_{r_n,n}) \to 0$ *in* $P_{\theta_0}^n$*-probability.*

Usually, the theorem will be applied to $\overline{\Theta}_n = \Theta$ for all n. If, however, condition (A2) could be verified only on a part of Θ which may possibly depend on n, the above formulation could be useful. However, the final conclusion should then be complemented by showing that $\Pi(\overline{\Theta}_n^c | Z_1, \ldots, Z_{r_n}) \to 0$ in $P_{\theta_0}^n$-probability by some alternative method.

The first condition (A1) asserts that certain sets, which could be thought of as neighborhoods of the true parameter θ_0, have positive prior probabilities. This condition ensures that the true value of the parameter is not excluded from the support of the prior. The second condition (A2) asserts that the hypothesis $\theta = \theta_0$ can be tested against the complement of a neighborhood for a topology of interest with a small probability of type I error and a uniformly exponentially small probability of type II error on most part of the parameter space in the sense that the prior probability of the remaining part is exponentially small.

The above theorem is also valid for a sequence of priors Π_n provided that (A1)(i) is strengthened to uniform convergence.

It should be remarked that Schwartz's condition on the Kullback–Leibler support is not necessary for posterior consistency to hold. This is clearly evident in parametric nonregular cases, where Kullback–Leibler divergence to some direction could be infinity. For instance, as in Ghosal et al. (1999a), for the model $p_\theta = \text{Uniform}(0, \theta)$ density, $0 < \theta \leqslant 1$, the Kullback–Leibler numbers $\int p_1 \log(p_1/p_\theta) = \infty$. However, the posterior is consistent at $\theta = 1$ if the prior Π has 1 in its support. Modifying the model to uniform$(\theta - 1, \theta + 1)$, we see that the Kullback–Leibler numbers are infinite for every pair. Nevertheless, consistency for a general parametric family including such nonregular cases holds under continuity and positivity of the prior density at θ_0 provided that the general conditions of Ibragimov and Has'minskii (1981) can be verified; see Ghosal et al. (1995) for details. For infinite-dimensional models, consistency may hold without Schwartz's condition on Kullback–Leibler support by exploiting special structure of the posterior distribution as in the case of the Dirichlet or a tail-free process. For estimation of a survival distribution using a Lévy process prior, Kim and Lee (2001) concluded consistency from the explicit expressions for pointwise mean and variance and monotonicity. For densities, consistency may also be shown by using some alternative conditions. One approach is by using the so-called Le Cam's inequality: For any two disjoint subsets $U, V \subset \mathfrak{M}(\mathfrak{X})$, test function Φ, prior Π on $\mathfrak{M}(\mathfrak{X})$ and probability measure P_0 on $\mathfrak{X}$,

$$\int \Pi(V|x)\,\mathrm{d}P_0(x) \leqslant d_V(P_0, \lambda_U) + \int \Phi\,\mathrm{d}P_0 + \frac{\Pi(V)}{\Pi(U)} \int (1 - \Phi)\,\mathrm{d}\lambda_V, \tag{3.6}$$

where $\lambda_U(B) = \int_U P(B)\,\mathrm{d}\Pi(P)/\Pi(U)$, the conditional expectation of $P(B)$ with respect to the prior Π restricted to the set U. Applying this inequality to V the complement of a neighborhood of P_0 and n i.i.d. observations, it may be shown that posterior consistency in the weak sense holds provided that for any $\beta, \delta > 0$,

$$\mathrm{e}^{n\beta}\,\Pi\big(P\colon d_V(P, P_0) < \delta/n\big) \to \infty. \tag{3.7}$$

Combining with appropriate testing conditions, stronger notions of consistency could be derived. The advantage of using this approach is that one need not control likelihood ratios now, and hence the result could be potentially used for undominated families as well, or at least can help reduce some positivity condition on the true density p_0. On the other hand, (3.7) is a quantitative condition on the prior unlike Schwartz's, and hence is more difficult to verify in many examples.

Because the testing condition is a condition only on a model and is more difficult to verify, there have been attempts to prove some assertion on posterior convergence using Schwartz's condition on Kullback–Leibler support only. While Theorem 4 shows that the testing condition is needed, it may be still possible to show some useful results by either weakening the concept of convergence, or even by changing the definition of the posterior distribution! Barron (1999) showed that if $p_0 \in \mathrm{KL}(\Pi)$, then

$$n^{-1}\sum_{i=1}^{n} \mathrm{E}_{p_0}\left(\log \frac{p_0(X_i)}{p(X_i|X_1,\ldots,X_{i-1})}\right) \to 0, \tag{3.8}$$

where $p(X_i|X_1,\ldots,X_{i-1})$ is the predictive density of X_i given $X_1,\ldots,X_{i-1}$. It may be noted that the predictive distribution is equal to the posterior mean of the density function. Hence in the Cesàro sense, the posterior mean density converges to the true density with respect to Kullback–Leibler neighborhoods, provided that the prior puts positive probabilities on Kullback–Leibler neighborhoods of p_0. Walker (2003), using a martingale representation of the predictive density, showed that the average predictive density converges to the true density almost surely under d_H. Walker and Hjort (2001) showed that the following pseudo-posterior distribution, defined by

$$\Pi_\alpha(p \in B|X_1,\ldots,X_n) = \frac{\int_B \prod_{i=1}^{n} p^\alpha(X_i)\,\mathrm{d}\Pi(p)}{\int_B \prod_{i=1}^{n} p^\alpha(X_i)\,\mathrm{d}\Pi(p)} \tag{3.9}$$

is consistent at any $p_0 \in \mathrm{KL}(\Pi)$, provided that $0 < \alpha < 1$.

Walker (2004) obtained another interesting result using an idea of restricting to a subset and looking at the predictive distribution (in this case, in the posterior) somewhat similar to that in Le Cam's inequality. If V is a set such that $\liminf_{n\to\infty} d_H(\lambda_{n,V}, p_0) > 0$, where $\lambda_{n,V}(B) = (\Pi(V|X_1,\ldots,X_n))^{-1}\int_V p(B)\,\mathrm{d}\Pi(p|X_1,\ldots,X_n)$, then $\Pi(V|X_1,\ldots,X_n) \to 0$ a.s. under P_0. A martingale property of the predictive distribution is utilized to prove the result. If V is the complement of a suitable weak neighborhood of p_0, then $\liminf_{n\to\infty} d_H(\lambda_{n,V}, p_0) > 0$, and hence the result provides an alternative way of proving the weak consistency result without appealing to Schwartz's theorem. Walker (2004) considered other topologies also.

The following is another result of Walker (2004) proving sufficient conditions for posterior consistency in terms of a suitable countable covering.

THEOREM 6. *Let* $p_0 \in \mathrm{KL}(\Pi)$ *and* $V = \{p\colon d_H(p,p_0) > \varepsilon\}$. *Let there exists* $0 < \delta < \varepsilon$ *and* $V_1, V_2, \ldots$ *a countable disjoint cover of* V *such that* $d_H(p_1,p_2) < 2\delta$ *for all* $p_1, p_2 \in V_j$ *and for all* $j = 1, 2, \ldots$, *and* $\sum_{j=1}^{\infty}\sqrt{\Pi(V_j)} < \infty$. *Then* $\Pi(V|X_1,\ldots,X_n) \to 0$ *a.s.* $[p_0^\infty]$.

While the lack of consistency is clearly undesirable, consistency itself is a very weak requirement. Given a consistency result, one would like to obtain information on the rates of convergence of the posterior distribution and see whether the obtained rate matches with the known optimal rate for point estimators. In finite-dimensional problems, it is well known that the posterior converges at a rate of $n^{-1/2}$ in the Hellinger distance; see Ibragimov and Has'minskii (1981) and Le Cam (1986).

Conditions for the rate of convergence given by Ghosal et al. (2000) and described below are quantitative refinement of conditions for consistency. A similar result, but under a much stronger condition on bracketing entropy numbers, was given by Shen and Wasserman (2001).

THEOREM 7. *Let $\varepsilon_n \to 0$, $n\varepsilon_n^2 \to \infty$ and suppose that there exist $\mathcal{P}_n \subset \mathcal{P}$, constants $c_1, c_2, c_3, c_4 > 0$ such that*

(i) $\log D(\varepsilon_n, \mathcal{P}_n, d) \leqslant c_1 n\varepsilon_n^2$, *where D stands for the packing number*;
(ii) $\Pi(\mathcal{P} \setminus \mathcal{P}_n) \leqslant c_2\,\mathrm{e}^{-(c_3+4)n\varepsilon_n^2}$;
(iii) $\Pi(p\colon \int p_0 \log \frac{p_0}{p} < \varepsilon_n^2, \int p_0 \log^2 \frac{p_0}{p} < \varepsilon_n^2) \geqslant c_4\,\mathrm{e}^{-c_3 n\varepsilon_n^2}$.

Then for some M, $\Pi(d(p, p_0) > M\varepsilon_n | X_1, X_2, \dots, X_n) \to 0$.

More generally, the entropy condition can be replaced by a testing condition, though, in most applications, a test is constructed from entropy bounds. Some variations of the theorem are given by Ghosal et al. (2000), Ghosal and van der Vaart (2001) and Belitser and Ghosal (2003).

While the theorems of Ghosal et al. (2000) satisfactorily cover i.i.d. data, major extensions are needed to cover some familiar situations such as regression with a fixed design, dose response study, generalized linear models with an unknown link, Whittle estimation of a spectral density and so on. Ghosal and van der Vaart (2003a) considered the issue and showed that the basic ideas of the i.i.d. case work with suitable modifications. Let d_n^2 be the average squared Hellinger distance defined by $d_n^2(\theta_1, \theta_2) = n^{-1}\sum_{i=1}^n d_H^2(p_{i,\theta_1}, p_{i,\theta_2})$. Birgé (1983) showed that a test for θ_0 against $\{\theta\colon d_n(\theta, \theta_1) < d_n(\theta_0, \theta_1)/18\}$ with error probabilities at most $\exp(-n d_n^2(\theta_0, \theta_1)/2)$ may be constructed. To find the intended test for θ_0 against $\{\theta\colon d_n(\theta, \theta_0) > \varepsilon\}$, one therefore needs to cover the alternative by d_n balls of radius $\varepsilon/18$. The number of such balls is controlled by the d_n-entropy numbers. Prior concentration near θ_0 controls the denominator as in the case of i.i.d. observations. Using these ideas, Ghosal and van der Vaart (2003a) obtained the following theorem on convergence rates that is applicable to independent, nonidentically distributed observations, and applied the result to various non-i.i.d. models.

THEOREM 8. *Suppose that for a sequence $\varepsilon_n \to 0$ such that $n\varepsilon_n^2$ is bounded away from zero, some $k > 1$, every sufficiently large j and sets $\Theta_n \subset \Theta$, the following conditions are satisfied*:

$$\sup_{\varepsilon > \varepsilon_n} \log N\big(\varepsilon/36, \{\theta \in \Theta_n\colon d_n(\theta, \theta_0) < \varepsilon\}, d_n\big) \leqslant n\varepsilon_n^2, \tag{3.10}$$

$$\Pi_n(\Theta \setminus \Theta_n)/\Pi_n\big(B_n^*(\theta_0, \varepsilon_n; k)\big) = \mathrm{o}\big(\mathrm{e}^{-2n\varepsilon_n^2}\big), \tag{3.11}$$

$$\frac{\Pi_n(\theta \in \Theta_n\colon\ j\varepsilon_n < d_n(\theta, \theta_0) \leqslant 2j\varepsilon_n)}{\Pi_n(B_n^*(\theta_0, \varepsilon_n; k))} \leqslant \mathrm{e}^{n\varepsilon_n^2 j^2/4}. \tag{3.12}$$

Then $P_{\theta_0}^{(n)} \Pi_n(\theta\colon\ d_n(\theta, \theta_0) \geqslant M_n\varepsilon_n | X^{(n)}) \to 0$ *for every* $M_n \to \infty$.

Ghosal and van der Vaart (2003a) also considered some dependent cases such as Markov chains, autoregressive model and signal estimation in presence of Gaussian white noise.

When one addresses the issue of optimal rate of convergence, one considers a smoothness class of the involved functions. The method of construction of the optimal prior with the help of bracketing or spline functions, as in Ghosal et al. (2000) requires the knowledge of the smoothness index. In practice, such information is not available and it is desirable to construct a prior that is adaptive. In other words, we wish to construct a prior that simultaneously achieves the optimal rate for every possible smoothness class under consideration. If only countably many models are involved, a natural and elegant method would be to consider a prior that is a mixture of the optimal priors for different smoothness classes. Belitser and Ghosal (2003) showed that the strategy works for an infinite-dimensional normal. Ghosal et al. (2003) and Huang (2004) obtained similar results for the density estimation problem.

Kleijn and van der Vaart (2002) considered the issue of misspecification, where p_0 may not lie in the support of the prior. In such a case, consistency at p_0 cannot hold, but it is widely believed that the posterior concentrates around the Kullback–Leibler projection p^* of p_0 to the model; see Berk (1966) for some results for parametric exponential families. Under suitable conditions which could be regarded as generalizations of the conditions of Theorem 7, Kleijn and van der Vaart (2002) showed that the posterior concentrates around p^* at a rate described by a certain entropy condition and concentration rate of the prior around p^*. Kleijn and van der Vaart (2002) also defined a notion of covering number for testing under misspecification that turns out to be the appropriate way of measuring the size of the model in the misspecified case. A weighted version of the Hellinger distance happens to be the proper way of measuring distance between densities that leads to a fruitful theorem on rates in the misspecified case. A useful theorem on consistency (in the sense that the posterior distribution concentrates around p^*) follows as a corollary.

When the posterior distribution converges at a certain rate, it is also important to know whether the posterior measure, after possibly a random centering and scaling, converges to a nondegenerate measure. For smooth parametric families, convergence to a normal distribution holds and is popularly known as the Bernstein–von Mises theorem; see Le Cam and Yang (2000) and van der Vaart (1998) for details. For a general parametric family which need not be smooth, a necessary and sufficient condition in terms of the limiting likelihood ratio process for convergence of the posterior (to some nondegenerate distribution using some random centering) is given by Ghosh et al. (1994) and Ghosal et al. (1995). For infinite-dimensional cases, results are relatively rare. Some partial results were obtained by Lo (1983, 1986) for Dirichlet process, Shen (2002) for certain semiparametric models, Susarla and Van Ryzin (1978) and Kim and Lee (2004) for certain survival models respectively with the Dirichlet process and Lévy process priors. However, it appears from the work of Cox (1993) and Freedman (1999) that Bernstein–von Mises theorem does not hold in most cases when the convergence rate is slower than $n^{-1/2}$. Freedman (1999) indeed showed that for the relatively simple problem of the estimation of the mean of an infinite-dimensional normal distribution with independent normal priors, the frequentist and the Bayesian distribution of L_2-norm of the difference of the Bayes estimate and the parameter differ by an amount equal to the scale of interest, and the frequentist coverage probability of a Bayesian credible set for the parameter is asymptotically zero. However, see Ghosal (2000) for a partially positive result.

4. Estimation of cumulative probability distribution

4.1. *Dirichlet process prior*

One of the nicest properties of the Dirichlet distribution, making it hugely popular, is its conjugacy for estimating a distribution function (equivalently, the probability law) with i.i.d. observations. Consider $X_1, \ldots, X_n$ are i.i.d. samples from an unknown cumulative distribution function (cdf) F on $\mathbb{R}^d$. Suppose F is given a Dirichlet process prior with parameters (M, G). Then the posterior distribution is again a Dirichlet process with the two parameters updated as

$$M \mapsto M + n \quad \text{and} \quad G \mapsto (MG + n\mathbb{F}_n)/(M + n), \tag{4.1}$$

where $\mathbb{F}_n$ is the empirical cdf. This may be easily shown by reducing the data to counts of sets from a partition, using the conjugacy of the finite-dimensional Dirichlet distribution for the multinomial distribution and passing to the limit with the aid of the martingale convergence theorem. Combining with (2.1), this implies that the posterior expectation and variance of $F(x)$ are given by

$$\begin{aligned} \widetilde{\mathbb{F}}_n(x) = \mathrm{E}\big(F(x)|X_1, \ldots, X_n\big) &= \frac{M}{M+n}G(x) + \frac{n}{M+n}\mathbb{F}_n(x), \\ \mathrm{var}\big(F(x)|X_1, \ldots, X_n\big) &= \frac{\widetilde{\mathbb{F}}_n(x)(1 - \widetilde{\mathbb{F}}_n(x))}{1 + M + n}. \end{aligned} \tag{4.2}$$

Therefore the posterior mean is a convex combination of the prior mean and the empirical cdf. As the sample size increases, the behavior of the posterior mean is inherited from that of the empirical probability measure. Also M could be interpreted as the strength in the prior or the "prior sample size".

The above discussion may lull us to interpret the limiting case $M \to 0$ as noninformative. Indeed, Rubin (1981) proposed $\mathrm{Dir}(n, \mathbb{F}_n)$ as the Bayesian bootstrap, which corresponds to the posterior obtained from the Dirichlet process by letting $M \to 0$. However, some caution is needed while interpreting the case $M \to 0$ as noninformative because of the role of M in also controlling the number of ties among samples drawn from P, where P itself is drawn from the Dirichlet process. Sethuraman and Tiwari

(1982) pointed out that as $M \to 0$, the Dirichlet process converges weakly to the random measure which is degenerate at some point θ distributed as G by property (ii) of convergence of Dirichlet measures mentioned in Section 2.1. Such a prior is clearly "very informative", and hence is unsuitable as a noninformative prior.

To obtain posterior consistency, note that (4.1) converges a.s. to the true cdf generating data. An important consequence of the above assertions is that the posterior distribution based on the Dirichlet process, not just the posterior mean, is consistent for the weak topology. Thus, by the weak convergence property of Dirichlet process, the posterior is consistent with respect to the weak topology. It can also be shown that, the posterior is consistent in the Kolmogorov–Smirnov distance defined as $d_{KS}(F_1, F_2) = \sup_x |F_1(x) - F_2(x)|$. The space of cdf's under d_{KS} is however not separable.

If the posterior distribution of F is given a prior that is a mixture of Dirichlet process, the posterior distribution is still a mixture of Dirichlet processes; see Theorem 3 of Antoniak (1974). However, mixtures may lead to inconsistent posterior distribution, unlike a single Dirichlet process. Nevertheless, if M_θ is bounded in θ, then posterior consistency holds.

4.2. Tail-free and Polya tree priors

Tail-free priors are extremely flexible, yet have some interesting properties. If the distribution function generating the i.i.d. data is given a tail-free prior, the posterior distribution is also tail-free. Further, as mentioned in Section 3, Freedman (1963, 1965) showed that the posterior obtained from a tail-free process prior is weakly consistent. The tail-free property helps reduce a weak neighborhood to a neighborhood involving only finitely many variables in the hierarchical representation, and hence the problem reduces to a finite-dimensional multinomial distribution, where consistency holds. Indeed Freedman's original motivation was to avoid pitfall as in Example 1.

A Polya tree prior may be used if one desires some smoothness of the random cdf. The most interesting property of a Polya tree process is its conjugacy. Conditional on the data $X_1, \dots, X_n$, the posterior distribution is again a Polya tree with respect to the same partition and α_ε updated to $\alpha^*_\varepsilon = \alpha_\varepsilon + \sum_{i=1}^n I\{X_i \in B_\varepsilon\}$. Besides, they lead to a consistent posterior in the weak topology as Polya trees are also tail-free processes.

4.3. Right censored data

Let X be a random variable of interest that is right censored by another random variable Y. The observation is (Z, Δ), where $Z = \min(X, Y)$ and $\Delta = I(X > Y)$. Assume that X and Y are independent with corresponding cdf F and H, where both F and H are unknown. The problem is to estimate F. Susarla and Van Ryzin (1976) put a Dirichlet process prior on F. Blum and Susarla (1977) found that the posterior distribution for i.i.d. data can be written as a mixture of Dirichlet processes. Using this idea, Susarla and Van Ryzin (1978) obtained that the posterior is mean square consistent with rate $O(n^{-1})$, almost surely consistent with rate $O(\log n/n^{1/2})$, and that the posterior distribution of $\{F(u)\colon 0 < u < T\}$, $T < \infty$, converges weakly to a Gaussian process whenever F and H are continuous and that $P(X > u)P(Y > u) > 0$. The mixture representation is however cumbersome. Ghosh and Ramamoorthi (1995) showed that the posterior distribution can also be written as a Polya tree process (with partitions dependent on the uncensored samples). They proved consistency by an elegant argument.

Doksum (1974) found that the neutral to right process for F form a conjugate family for the right censored data. Viewed as a prior on the cumulative hazard process, the prior can be identified with an independent increment process. An updating mechanism is described by Kim (1999) using a counting process approach. Beta processes, introduced by Hjort (1990), also form a conjugate family. Kim and Lee (2001) obtained sufficient conditions for posterior consistency for a Lévy process prior, which includes Dirichlet processes and beta processes. Under certain conditions, the posterior also converges at the usual $n^{-1/2}$ rate and admits a Bernstein–von Mises theorem; see Kim and Lee (2004).

5. Density estimation

Density estimation is one of the fundamental problems of nonparametric inference because of its applicability to various problems including cluster analysis and robust estimation. A common approach to constructing priors on the space of probability densities is to use Dirichlet mixtures where the kernels are chosen depending on the sample space. The posterior distributions are analytically intractable and the MCMC techniques are different for different kernels. Other priors useful for this problem are Polya tree processes and Gaussian processes. In this section, we discuss some of the computational issues and conditions for consistency and convergence rates of the posterior distribution.

5.1. Dirichlet mixture

Consider that the density generating the data is a mixture of densities belonging to some parametric family, that is, $p_F(x) = \int \psi(x, \theta)\, \mathrm{d}F(\theta)$. Let the mixing distribution F be given a $\mathrm{Dir}(M, G)$ prior. Viewing $p_F(x)$ as a linear functional of F, the prior expectation of $p_F(x)$ is easily found to be $\int \psi(x, \theta)\, \mathrm{d}G(\theta)$. To compute the posterior expectation, the following hierarchical representation of the above prior is often convenient:

$$X_i \overset{\text{ind}}{\sim} \psi(\cdot, \theta_i), \qquad \theta_i \overset{\text{i.i.d.}}{\sim} F, \qquad F \sim \mathrm{Dir}(M, G). \tag{5.1}$$

Let $\Pi(\boldsymbol{\theta}|X_1, \dots, X_n)$ stand for the distribution of $(\theta_1, \dots, \theta_n)$ given $(X_1, \dots, X_n)$. Observe that given $\boldsymbol{\theta} = (\theta_1, \dots, \theta_n)$, the posterior distribution of F is Dirichlet with base measure $MG + n\mathbb{G}_n$, where $\mathbb{G}_n(\cdot, \boldsymbol{\theta}) = n^{-1} \sum_{i=1}^n \delta_{\theta_i}$, the empirical distribution of $(\theta_1, \dots, \theta_n)$. Hence the posterior distribution of F may be written as a mixture of Dirichlet processes. The posterior mean of $F(\cdot)$ may be written as

$$\frac{M}{M+n} G(\cdot) + \frac{n}{M+n} \int \mathbb{G}_n(\cdot, \boldsymbol{\theta}) \Pi(\mathrm{d}\boldsymbol{\theta}|X_1, \dots, X_n) \tag{5.2}$$

and the posterior mean of the density at x becomes

$$\frac{M}{M+n} \int \psi(x, \theta)\, \mathrm{d}G(\theta) + \frac{n}{M+n} \frac{1}{n} \sum_{i=1}^n \int \psi(x, \theta_i) \Pi(\mathrm{d}\boldsymbol{\theta}|X_1, \dots, X_n).$$

(5.3)

The Bayes estimate is thus composed of a part attributable to the prior and a part due to observations. Ferguson (1983) remarks that the factor $n^{-1}\sum_{i=1}^{n}\int\psi(x,\theta_i)\Pi(\mathrm{d}\boldsymbol{\theta}|X_1,\ldots,X_n)$ in the second term of (5.3) can be viewed as a partially Bayesian estimate with the influence of the prior guess reduced. The evaluation of the above quantities depend on $\Pi(\mathrm{d}\boldsymbol{\theta}|X_1,\ldots,X_n)$. The joint prior for $(\theta_1,\theta_2,\ldots,\theta_n)$ is given by the generalized Polya urn scheme

$$G(\mathrm{d}\theta_1)\times\frac{(MG(\mathrm{d}\theta_2)+\delta_{\theta_1})}{M+1}\times\cdots\times\frac{\left(MG(\mathrm{d}\theta_n)+\sum_{i=1}^{n-1}\delta_{\theta_i}\right)}{M+n}. \tag{5.4}$$

Further, the likelihood given $(\theta_1,\theta_2,\ldots,\theta_n)$ is $\prod_{i=1}^{n}\psi(X_i,\theta_i)$. Hence H can be written down using the Bayes formula. Using the above equations and some algebra, Lo (1984) obtained analytical expressions of the posterior expectation of $f(x)$. However, the formula is of marginal use because the number of terms grows very fast with the sample size. Computations are thus done via MCMC techniques as in the special case of normal mixtures described in the next subsection; see the review article Escobar and West (1998) for details.

5.1.1. *Mixture of normal kernels*

Suppose that the unknown density of interest is supported on the entire real line. Then a natural choice of the kernel is $\phi_\sigma(x-\mu)$, the normal density with mean μ and variance σ^2. The mixture distribution F is given Dirichlet process prior with some base measure MG, while G is often given a normal/inverse-gamma distribution to achieve conjugacy. Thus, under G, $\sigma^{-2}\sim\mathrm{Gamma}(s,\beta)$, a gamma distribution with shape parameter s and scale parameter β, and $(\mu|\sigma)\sim N(m,\sigma^2)$. Let $\theta=(\mu,\sigma)$. Then the hierarchical model is

$$X_i|\theta_i\overset{\mathrm{ind}}{\sim}N\left(\mu_i,\sigma_i^2\right),\qquad \theta_i\overset{\mathrm{i.i.d.}}{\sim}F,\qquad F\sim\mathrm{Dir}(M,G). \tag{5.5}$$

Given $\boldsymbol{\theta}=(\theta_1,\ldots,\theta_n)$, the distribution of F may be updated analytically. Thus, if one can sample from the posterior distribution of $\boldsymbol{\theta}$, Monte Carlo averages may be used to find the posterior expectation of F and thus the posterior expectation of $p(x)=\int\phi_\sigma(x-\mu)\,\mathrm{d}F(x)$. Escobar (1994) and Escobar and West (1995) provided an algorithm for sampling from the posterior distribution of $\boldsymbol{\theta}$. Let $\theta_{-i}=\{\theta_1,\ldots,\theta_{i-1},\theta_{i+1},\ldots,\theta_n\}$. Then

$$(\theta_i|\theta_{-i},x_1,\ldots,x_n)\sim q_{i0}G_i(\theta_i)+\sum_{j=1,j\neq i}^{n}q_{ij}\delta_{\theta_j}(\theta_i), \tag{5.6}$$

where $G_i(\theta_i)$ is the bivariate normal/inverse-gamma distribution under which

$$\begin{aligned}&\sigma_i^{-2}\sim\mathrm{Gamma}\left(s+1/2,\beta+(x_i-m)^2/2\right),\\&(\mu_i|\sigma_i)\sim N\left(m+x_i,\sigma_i^2\right)\end{aligned} \tag{5.7}$$

and the weights q_{ij}'s are defined by $q_{i0}\propto M\Gamma(s+1/2)(2\beta)^s\Gamma(s)^{-1}\{2\beta+(x_i-m)^2\}^{-(s+1/2)}$ and $q_{ij}\propto\sqrt{\pi}\phi_{\sigma_i}(x_i-\mu_i)$ for $j\neq i$. Thus a Gibbs sampler algorithm is described by updating $\boldsymbol{\theta}$ componentwise through the conditional distribution in (5.6). The initial values of θ_i could be a sample from G_i.

The bandwidth parameter σ is often kept constant depending on the sample size, say σ_n. This leads to only the location mixture. In that case a Gibbs sampler algorithm is obtained by keeping σ_i fixed at σ_n in the earlier algorithm and updating only the location components μ_i.

Consistency of the posterior distribution for Dirichlet mixture of normals was studied by Ghosal et al. (1999b). Let p_0 stand for the true density.

THEOREM 9. *If $p_0 = \int \phi_\sigma(x - \mu)\,\mathrm{d}F_0(\mu, \sigma)$, where F_0 is compactly supported and in the weak support of Π, then $p_0 \in \mathrm{KL}(\Pi)$.*

If p_0 is not a mixture of normals but is compactly supported, 0 is in the support of the prior for σ, and $\lim_{\sigma\to 0} \int p_0 \log(p_0/p_0 * \phi_\sigma) = 0$, then $p_0 \in \mathrm{KL}(\Pi)$.

If $p_0 \in \mathrm{KL}(\Pi)$, the base measure G of the underlying Dirichlet process is compactly supported and $\Pi(\sigma < t) \leqslant c_1 \mathrm{e}^{-c_2/t}$, then the posterior is consistent at p_0 for the total variation distance d_V. If the condition of compactly support G is replaced by the condition that for every $\varepsilon > 0$, there exist a_n, σ_n with $a_n/\sigma_n < \varepsilon n$ satisfying $G[-a_n, a_n] < \mathrm{e}^{-n\beta_1}$ and $\Pi(\sigma < \sigma_n) \leqslant \mathrm{e}^{-n\beta_2}$ for $\beta_1, \beta_2 > 0$, then also consistency for d_V holds at any $p_0 \in \mathrm{KL}(\Pi)$.

The condition $p_0 \in \mathrm{KL}(\Pi)$ implies weak consistency by Schwartz's theorem. The condition for $p_0 \in \mathrm{KL}(\Pi)$ when p_0 is neither a normal mixture nor compactly supported, as given by Theorem 5 of Ghosal et al. (1999b) using estimates of Dirichlet tails, is complicated. However, the conditions holds under strong integrability conditions on p_0. The base measure for the Dirichlet could be normal and the prior on σ could be a truncated inverse gamma possibly involving additional parameters. Better sufficient condition for $p_0 \in \mathrm{KL}(\Pi)$ is given by Tokdar (2003). Consider a location-scale mixture of normal with a prior Π on the mixing measure. If p_0 is bounded, nowhere zero, $\int p_0|\log p_0| < \infty$, $\int p_0 \log(p_0/\psi) < \infty$ where $\psi(x) = \inf\{p_0(t)\colon x-1 \leqslant t \leqslant x+1\}$, $\int |x|^{2+\delta} p_0(x)\,\mathrm{d}x < \infty$, and every compactly supported probability lies in $\mathrm{supp}(\Pi)$, then $p_0 \in \mathrm{KL}(\Pi)$. The moment condition can be weakened to only δ-moment if Π is Dirichlet. In particular, the case that p_0 is Cauchy could be covered.

Convergence rates of the posterior distribution were obtained by Ghosal and van der Vaart (2001, 2003b) respectively the "super smooth" and the "smooth" cases. We discuss below the case of location mixtures only, where the scale gets a separate independent prior.

THEOREM 10. *Assume that $p_0 = \phi_{\sigma_0} * F_0$, and the prior on σ has a density that is compactly supported in $(0, \infty)$ but is positive and continuous at σ_0. Suppose that F_0 has compact support and the base measure G has a continuous and positive density on an interval containing the support of F_0 and has tails $G(|z| > t) \lesssim \mathrm{e}^{-b|t|^\delta}$. Then the posterior converges at a rate $n^{-1/2}(\log n)^{\max(\frac{2}{\delta},\frac{1}{2})+\frac{1}{2}}$ with respect to d_H. The condition of compact support of F_0 could be replaced by that of sub-Gaussian tails if G is normal, in which case the rate is $n^{-1/2}(\log n)^{3/2}$.*

If instead p_0 *is compactly supported, twice continuously differentiable and* $\int (p_0''/p_0)^2 p_0 < \infty$ *and* $\int (p_0'/p_0)^4 p_0 < \infty$, *and the prior on* (σ/σ_n) *has a density that is compactly supported in* $(0, \infty)$, *where* $\sigma_n \to 0$, *then the posterior converges at a rate* $\max((n\sigma_n)^{-1/2}(\log n), \sigma_n^2 \log n)$. *In particular, the best rate* $\varepsilon_n \sim n^{-2/5}(\log n)^{-4/5}$ *is obtained by choosing* $\sigma_n \sim n^{-1/5}(\log n)^{-2/5}$.

The proofs are the result of some delicate estimates of the number of components a discrete mixing distribution must have to approximate a general normal mixture. Some further results are given by Ghosal and van der Vaart (2003b) when p_0 does not have compact support.

5.1.2. Uniform scale mixtures

A nonincreasing density on $[0, \infty)$ may be written as a mixture of the form $\int \theta^{-1} I\{0 \leqslant x \leqslant \theta\} F(\mathrm{d}\theta)$ by a well known representation theorem of Khinchine and Shepp. This lets us put a prior on this class from that on F. Brunner and Lo (1989) considered this idea and put a Dirichlet prior for F. Coupled with a symmetrization technique as in Section 2.2.3, this leads to a reasonable prior for the error distribution. Brunner and Lo (1989) used this approach for the semiparametric location problem. The case of asymmetric error was treated by Brunner (1992) and that of semiparametric linear regression by Brunner (1995).

5.1.3. Mixtures on the half line

Dirichlet mixtures of exponential distributions may be considered as a reasonable model for a decreasing, convex density on the positive half line. More generally, mixtures of gamma densities, which may be motivated by Feller approximation procedure using a Poisson sampling scheme in the sense of Petrone and Veronese (2002), may be considered to pick up arbitrary shapes. Such a prior may be chosen to have a large weak support. Mixtures of inverse gammas may be motivated similarly by Feller approximation using a gamma sampling scheme. In general, a canonical choice of a kernel function could be made once a Feller sampling scheme appropriate for the domain could be specified. For a general kernel, weak consistency may be shown exploiting Feller approximation property as in Petrone and Veronese (2002).

Mixtures of Weibulls or lognormals are dense in the stronger sense of total variation distance provided that we let the shape parameter of the Weibull to approach infinity or that of the lognormal to approach zero. To see this, observe that these two kernels form location-scale families in the log-scale, and hence are approximate identities. Kottas and Gelfand (2001) used these mixtures for median regression, where asymmetry is an important aspect. The mixture of Weibull is very useful to model observations of censored data because its survival function has a simpler expression compared to that for the mixtures of gamma or lognormal. Ghosh and Ghosal (2003) used these mixtures to model a proportional mean structure censored data. The posterior distribution was computed using an MCMC algorithm for Dirichlet mixtures coupled with imputation of censored data. Posterior consistency can be established by reducing the original model to a standard regression model with unknown error for which the results of Amewou-Atisso et al. (2003) apply. More specifically, consistency holds if the true baseline density is in the Kullback–Leibler support of the Dirichlet mixture prior. The last condition can be established under reasonable conditions using the ideas of Theorem 9 and its extension by Tokdar (2003).

5.1.4. Bernstein polynomials

On the unit interval, the family of beta distributions form a flexible two-parameter family of densities and their mixtures form a very rich class. Indeed, mixtures of beta densities with integer parameters are sufficient to approximate any distribution. For a continuous probability distribution function F on $(0, 1]$, the associated Bernstein polynomial $B(x; k, F) = \sum_{j=0}^{k} F(j/k)\binom{k}{j}x^j(1-x)^{k-j}$, which is a mixture of beta distributions, converges uniformly to F as $k \to \infty$. Using an idea of Diaconis that this approximation property may be exploited to construct priors with full topological support, Petrone (1999a, 1999b) proposed the following hierarchical prior called the Bernstein polynomial prior:

- $f(x) = \sum_{j=1}^{k} w_{j,k}\beta(x; j, k-j+1)$,
- $k \sim \rho(\cdot)$,
- $(\boldsymbol{w}_k = (w_{1,k}, \ldots, w_{k,k})|k) \sim H_k(\cdot)$, a distribution on the k-dimensional simplex.

Petrone (1999a) showed that if for all k, $\rho(k) > 0$ and $\boldsymbol{w}_k$ has full support on Δ_k, then every distribution on $(0, 1]$ is in the weak support of the Bernstein polynomial prior, and every continuous distribution is in the topological support of the prior defined by the Kolmogorov–Smirnov distance.

The posterior mean, given k, is

$$E\big(f(x)|k, x_1, \ldots, x_n\big) = \sum_{j=1}^{k} E(w_{j,k}|x_1, \ldots, x_n)\beta(x; j, k-j+1), \tag{5.8}$$

and the distribution of k is updated to $\rho(k|x_1, \ldots, x_n)$. Petrone (1999a, 1999b) discussed MCMC algorithms to compute the posterior expectations and carried out extensive simulations to show that the resulting density estimates work well.

Consistency is given by Petrone and Wasserman (2002). The corresponding results on convergence rates are obtained by Ghosal (2001).

THEOREM 11. *If p_0 is continuous density on* $[0, 1]$, *the base measure G has support all of* $[0, 1]$ *and the prior probability mass function $\rho(k)$ for k has infinite support, then* $p_0 \in \mathrm{KL}(\Pi)$. *If further $\rho(k) \lesssim \mathrm{e}^{-\beta k}$, then the posterior is consistent for d_H.*

If p_0 is itself a Bernstein polynomial, then the posterior converges at the rate $n^{-1/2}\log n$ *with respect to d_H.*

If p_0 is twice continuously differentiable on $[0, 1]$ *and bounded away from zero, then the posterior converges at the rate* $n^{-1/3}(\log n)^{5/6}$ *with respect to d_H.*

5.1.5. Random histograms

Gasparini (1996) used the Dirichlet process to put a prior on histograms of different bin width. The sample space is first partitioned into (possibly an infinite number of) intervals of length h, where h is chosen from a prior. Mass is distributed to the intervals according to a Dirichlet process, whose parameters $M = M_h$ and $G = G_h$ may depend on h. Mass assigned to any interval is equally distributed over that interval. The method corresponds to Dirichlet mixtures with a uniform kernel $\psi(x, \theta, h) = h^{-1}$, $x, \theta \in (jh, (j+1)h)$ for some j.

If $n_j(h)$ is the number of X_i's in the bin $[jh, (j+1)h)$, it is not hard to see that the posterior is of the same form as the prior with $M_h G_h$ updated to $M_h G_h + \sum_j n_j(h) I[jh, (j+1)h)$ and the prior density $\pi(h)$ of h changed to

$$\pi^*(h) = \frac{\pi(h) \prod_{j=1}^{\infty} (M_h G_h([jh, (j+1)h)))^{(n_j(h)-1)}}{M_h + n}. \tag{5.9}$$

The predictive density with no observations is given by $\int f_h(x)\pi(h)\,\mathrm{d}h$, where $f_h(x) = h^{-1} \sum_{j=-\infty}^{\infty} G_h([jh, (j+1)h)) I_{[jh,(j+1)h]}(x)$. In view of the conjugacy property, the predictive density given n observations can be easily written down. Let P_h stand for the histogram of bin-width h obtained from the probability measure P. Assume that $G_h(j)/G_h(j-1) \leqslant K_h$. If $\int x^2 p_0(x)\,\mathrm{d}x < \infty$ and $\lim_{h\to 0} \int p_0(x) \log \frac{p_{0,h}}{p_0} = 0$, then the posterior is weakly consistent at p_0. Gasparini (1996) also gave additional conditions to ensure consistency of the posterior mean of p under d_H.

5.2. *Gaussian process prior*

For density estimation on a bounded interval I, Leonard (1978) defined a random density on I through $f(x) = \frac{\mathrm{e}^{Z(x)}}{\int_I \mathrm{e}^{Z(t)}\,\mathrm{d}t}$, where $Z(x)$ is a Gaussian process with mean function $\mu(x)$ and covariance kernel $\sigma(x, x')$. Lenk (1988) introduces an additional parameter ξ to obtain a conjugate family. It is convenient to introduce the intermediate lognormal process $W(x) = \mathrm{e}^{Z(x)}$. Denote the distribution of W by $LN(\mu, \sigma, 0)$. For each ξ define a positive valued random process $LN(\mu, \sigma, \xi)$ on I whose Radon–Nikodym derivative with respect to $LN(\mu, \sigma, 0)$ is $(\int_I W(x, \omega)\,\mathrm{d}x)^{\xi}$. The normalization $f(x, \omega) = \frac{W(x)}{\int W(t)\,\mathrm{d}t}$ gives a random density and the distribution of this density under $LN(\mu, \sigma, \xi)$ is denoted by $LNS(\mu, \sigma, \xi)$. If $X_1, \ldots, X_n$ are i.i.d. f and $f \sim LNS(\mu, \sigma, \xi)$, then the posterior is $LNS(\mu^*, \sigma, \xi^*)$, where $\mu^*(x) = \mu(x) + \sum_{i=1}^{n} \sigma(x_i, x)$ and $\xi^* = \xi - n$.

The interpretation of the parameters are somewhat unclear. Intuitively, for a stationary covariance kernel, a higher value of $\sigma(0)$ leads to more fluctuations in $Z(x)$ and hence more noninformative. Local smoothness is controlled by $-\sigma''(0)$ – smaller value implying a smoother curve. The parameter ξ, introduced somewhat unnaturally, is the least understood. Apparently, the expression for the posterior suggests that $-\xi$ may be thought of as the "prior sample size".

5.3. *Polya tree prior*

A Polya tree prior satisfying $\sum_{m=1}^{\infty} a_m^{-1} < \infty$ admits densities a.s. by Kraft (1964) and hence may be considered for density estimation. The posterior expected density is given by

$$\mathrm{E}\big(f(x)|X_1, \ldots, X_n\big) = \alpha(x) \prod_{m=1}^{\infty} \frac{2a_m + 2N(B_{\epsilon(m)})}{2a_m + N(B_{\epsilon(m)}) + N(B'_{\epsilon(m)})}, \tag{5.10}$$

where $N(B_{\epsilon(m)})$ stand for the number of observations falling in $B_{\epsilon(m)}$, the set in the m-level partition which contains x and $N(B'_{\epsilon(m)})$ is the number of observations falling in its sibling $B'_{\epsilon(m)}$. From Theorem 3.1 of Ghosal et al. (1999c), it follows that under the condition $\sum_{m=1}^{\infty} a_m^{-1/2} < \infty$, any p_0 with $\int p_0 \log(p_0/\alpha) < \infty$ satisfies $p_0 \in \mathrm{KL}(\Pi)$ and hence the weak consistency holds. Consistency under d_H has been obtained by Barron et al. (1999) under the rather strong condition that $a_m = 8^m$. This high value of 8^m appears to be needed to control the roughness of the Polya trees. Using the pseudo-posterior distribution as described in Section 3, Walker and Hjort (2001) showed that the posterior mean converges in d_H solely under the condition $\sum_{m=1}^{\infty} a_m^{-1/2} < \infty$. Interestingly, they identify the posterior mean with the mean of a pseudo-posterior distribution that also comes from a Polya tree prior with a different set of parameters.

6. Regression function estimation

Regression is one of the most important and widely used tool in statistical analysis. Consider a response variable Y measured with some covariate X that may possibly be multivariate. The regression function $f(x) = \mathrm{E}(Y|X = x)$ describes the overall functional dependence of Y on X and thus becomes very useful in prediction. Spatial and geostatistical problems can also be formulated as regression problems. Classical parametric models such as linear, polynomial and exponential regression models are increasingly giving way to nonparametric regression model. Frequentist estimates of the regression functions such as the kernel estimate, spline or orthogonal series estimators have been in use for a long time and their properties have been well studied. Some nonparametric Bayesian methods have also been developed recently. The Bayesian analysis depends on the dependence structure of Y on X and are handled differently for different regression models.

6.1. Normal regression

For continuous response, a commonly used regression model is $Y_i = f(X_i) + \epsilon_i$, where ϵ_i are assumed to be i.i.d. mean zero Gaussian errors with unknown variance and be independent of X_i's. Leading nonparametric Bayesian techniques, among some others, include those based on (i) Gaussian process prior, (ii) orthogonal basis expansion, and (iii) free-knot splines.

Wahba (1978) considered a Gaussian process prior for f. The resulting Bayes estimator is found to be a smoothing spline with the appropriate choice of the covariance kernel of the Gaussian process. A commonly used prior for f is defined through the stochastic differential equation $\frac{\mathrm{d}^2 f(x)}{\mathrm{d}x^2} = \tau \frac{\mathrm{d}W(x)}{\mathrm{d}x}$, where $W(x)$ is a Wiener process. The scale parameter τ is given an inverse gamma prior while the intercept term $f(0)$ is given an independent Gaussian prior. Ansley et al. (1993) described an extended state-space representation for computing the Bayes estimate. Barry (1986) used a similar prior for multiple covariates and provided asymptotic result for the Bayes estimator.

Another approach to putting a nonparametric prior on f is through an orthogonal basis expansion of the form $f(x) = \sum_{j=1}^{\infty} b_j \psi_j(x)$ and then putting a prior on the coefficients b_j's. Smith and Kohn (1997) considered such an approach when the infinite series is truncated at some predetermined finite stage k. Zhao (2000) considered a sieve prior putting an infinitely supported prior on k. Shen and Wasserman (2001) investigated the asymptotic properties for this sieve prior and obtained a convergence rate $n^{-q/(2q+1)}$ under some restriction on the basis function and for a Gaussian prior on the b_j's. Variable selection problem is considered in Shively et al. (1999) and Wood et al. (2002a). Wood et al. (2002b) extended this approach to spatially adaptive regression, while Smith et al. (1998) extended the idea to autocorrelated errors.

A free-knot spline approach is considered by Denison et al. (1998) and DiMatteo et al. (2001). They modeled f as a polynomial spline of fixed order (usually cubic), while putting a prior on the number of the knots, the location of the knots and the coefficients of the polynomials. Since the parameter space is canonical, computations are done through Monte Carlo averages while samples from the posterior distribution is obtained by reversible jump MCMC algorithm of Green (1995).

6.2. *Binary regression*

In this case, $Y|X = x \sim \text{binom}(1, f(x))$ so that $f(x) = \mathrm{P}(Y = 1|X = x) = \mathrm{E}(Y|X = x)$. Choudhuri et al. (2004b) induced a prior on $f(x)$ by using a Gaussian process $\eta(x)$ and mapping $\eta(x)$ into the unit interval as $f(x) = H(\eta(x))$ for some strictly increasing continuous chosen "link function" H. The posterior distribution of $f(x)$ is analytically intractable and the MCMC procedure depends on the choice of link function. The most commonly used link function is the probit link in which H is the standard normal cdf. In this case, an elegant Gibbs sampler algorithm is obtained by introducing some latent variables following an idea of Albert and Chib (1993).

Let $\boldsymbol{Y} = (Y_1, \dots, Y_n)^{\mathrm{T}}$ be the random binary observations measured along with the corresponding covariate values $\boldsymbol{X} = (X_1, \dots, X_n)^{\mathrm{T}}$. Let $\boldsymbol{Z} = (Z_1, \dots, Z_n)^{\mathrm{T}}$ be some unobservable latent variables such that conditional on the covariate values $\boldsymbol{X}$ and the functional parameter η, Z_i's are independent normal random variables with mean $\eta(X_i)$ and variance 1. Assume that the observations Y_i's are functions of these latent variables defined as $Y_i = I(Z_i > 0)$. Then, conditional on $(\eta, \boldsymbol{X})$, Y_i's are independent Bernoulli random variables with success probability $\Phi(\eta(X_i))$ and thus leads to the probit link model. Had we observed Z_i's, the posterior distribution of η could have been obtained analytically, which is also a Gaussian process by virtue of the conjugacy of the Gaussian observation with a Gaussian prior for the mean. However, $\boldsymbol{Z}$ is unobservable. Given the data $(\boldsymbol{Y}, \boldsymbol{X})$ and the functional parameter η, Z_i's are conditionally independent and their distributions are truncated normal with mean $\eta(X_i)$ and variance 1, where Z_i is right truncated at 0 if $Y_i = 0$, while Z_i is right truncated at 0 if $Y_i = 1$, then Z_i is taken to be positive. Now, using the conditional distributions of $(\boldsymbol{Z}|\eta, \boldsymbol{Y}, \boldsymbol{X})$ and $(\eta|\boldsymbol{Z}, \boldsymbol{Y}, \boldsymbol{X})$, a Gibbs sampler algorithm is formulated for sampling from the distribution of $(\boldsymbol{Z}, \eta|\boldsymbol{Y}, \boldsymbol{X})$. Choudhuri et al. (2004b) also extended this Gibbs sampler algorithm to the link function that is a mixture of normal cdfs. These authors also showed that the posterior distribution is consistent under mild conditions, as stated below.

THEOREM 12. *Let the true response probability function $f_0(x)$ be continuous, $(d+1)$-times differentiable and bounded away form* 0 *and* 1, *and that the underlying Gaussian* *process has mean function and covariance kernel $(d+1)$-times differentiable, where d is the dimension of the covariate X. Assume that the range of X is bounded.*

If the covariate is random having a nonsingular density $q(x)$, then for any $\varepsilon > 0$, $\Pi(f\colon \int |f(x) - f_0(x)| q(x)\,\mathrm{d}x > \varepsilon | X_1, Y_1, \ldots, X_n, Y_n) \to 0$ in P_{f_0}-probability.

If the covariates are nonrandom, then for any $\varepsilon > 0$, $\Pi(f\colon n^{-1}\sum_{i=1}^n |f(X_i) - f_0(X_i)| > \varepsilon | Y_1, \ldots, Y_n) \to 0$ in P_{f_0}-probability.

To prove the result, conditions of Theorem 3 and Theorem 5 respectively for random and nonrandom covariates, are verified. The condition on the Kullback–Leibler support is verified by approximating the function by a finite Karhunene–Loèvè expansion and by the nonsingularity of the multivariate normal distributions. The testing condition is verified on a sieve that is given by the maximum of f and its $(d+1)$ derivatives bounded by some $M_n = \mathrm{o}(n)$. The complement of the sieve has exponentially small prior probability if M_n is not of smaller order than $n^{1/2}$.

Wood and Kohn (1998) considered the integrated Wiener process prior for the probit transformation of f. The posterior is computed via Monte Carlo averages using a data augmentation technique as above. Yau et al. (2003) extended the idea to multinomial problems. Holmes and Mallick (2003) extended the free-knot spline approach to generalized multiple regression treating binary regression as a particular case.

A completely different approach to semiparametric estimation of f is to nonparametrically estimate the link function H while using a parametric form, usually linear, for $\eta(x)$. Observe that H is a nondecreasing function with range $[0, 1]$ and this is an univariate distribution function. Gelfand and Kuo (1991), and Newton et al. (1996) used a Dirichlet process prior for H. Mallick and Gelfand (1994) modeled H as a mixture of beta cdf's with a prior probability on the mixture weights, which resulted in smoother estimates. Basu and Mukhopadhyay (2000) modeled the link function as Dirichlet scale mixture of truncated normal cdf's. Posterior consistency results for these procedures were obtained by Amewou-Atisso et al. (2003).

7. Spectral density estimation

Let $\{X_t\colon t = 1, 2, \ldots\}$ be a stationary time series with autocovariance function $\gamma(\cdot)$ and spectral density $f^*(\omega^*) = (2\pi)^{-1}\sum_{r=-\infty}^{\infty} \gamma(r)\,\mathrm{e}^{-ir\omega^*}$, $-\pi < \omega^* \leqslant \pi$. To estimate f^*, it suffices to consider the function $f(\omega) = f^*(\pi\omega)$, $0 \leqslant \omega \leqslant 1$, by the symmetry of f^*. Because the actual likelihood of f is difficult to handle, Whittle (1957, 1962) proposed a "quasi-likelihood"

$$L_n(f|X_1, \ldots, X_n) = \prod_{l=1}^{\nu} \frac{1}{f(\omega_l)}\,\mathrm{e}^{-I_n(\omega_l)/f(\omega_l)}, \tag{7.1}$$

where $\omega_l = 2l/n$, ν is the greatest integer less than or equal to $(n-1)/2$, and $I_n(\omega) = |\sum_{t=1}^n X_t\,\mathrm{e}^{-it\pi\omega}|^2/(2\pi n)$ is the periodogram. A pseudo-posterior distribution may be obtained by updating the prior using this likelihood.

7.1. Bernstein polynomial prior

Normalizing f to $q = f/\tau$ with the normalizing constant $\tau = \int f$, Choudhuri et al. (2004a) induced a prior on f by first putting a Bernstein polynomial prior on q and then putting an independent prior on τ. Thus, the prior on f is described by the following hierarchical scheme:

- $f(\omega) = \tau \sum_{j=1}^{k} F((j-1)/k, j/k]\beta(\omega; j, k-j+1)$;
- $F \sim \mathrm{Dir}(M, G)$, where G has a Lebesgue density g;
- k has probability mass function $\rho(k) > 0$ for $k = 1, 2, \ldots$;
- The distribution of τ has Lebesgue density π on $(0, \infty)$;
- F, k, and τ are a priori independent.

The pseudo-posterior distribution is analytically intractable and hence is computed by an MCMC method. Using the Sethuraman representation for F as in (2.2), (f, k, τ) may be reparameterized as $(\theta_1, \theta_2, \ldots, Y_1, Y_2, \ldots, k, \tau)$. Because the infinite series in (2.2) is almost surely convergent, it may be truncated at some large L. Then one may represent F as $F = \sum_{l=1}^{L} V_l \delta_{\theta_l} + (1 - V_1 - \cdots - V_L)\delta_{\theta_0}$, where $\theta_0 \sim G$ and is independent of the other parameters. The last term is added to make F a distribution function even after the truncation. Now the problem reduces to a parametric one with finitely many parameters $(\theta_0, \theta_1, \ldots, \theta_L, Y_1, \ldots, Y_L, k, \tau)$. The functional parameter f may be written as a function of these univariate parameters as

$$f(\omega) = \tau \sum_{j=1}^{k} w_{j,k} \beta(\omega; j, k-j+1), \tag{7.2}$$

where $w_{j,k} = \sum_{l=0}^{L} V_l I\{\frac{j-1}{k} < \theta_l \leqslant \frac{j}{k}\}$ and $V_0 = 1 - V_1 - \cdots - V_L$. The posterior distribution of $(\theta_0, \theta_1, \ldots, \theta_L, Y_1, \ldots, Y_L, k, \tau)$ is proportional to

$$\left[\prod_{m=1}^{\nu} \frac{1}{f(2m/n)} \mathrm{e}^{-U_m/f(2m/n)}\right]\left[\prod_{l=1}^{L} M(1-y_l)^{M-1}\right]\left[\prod_{l=0}^{L} g(\theta_l)\right]\rho(k)\pi(\tau). \tag{7.3}$$

The discrete parameter k may be easily simulated from its posterior distribution given the other parameters. If the prior on τ is an inverse gamma distribution, then the posterior distribution of τ conditional on the other parameters is also inverse gamma. To sample from the posterior density of θ_i's or Y_i's conditional on the other parameters, Metropolis algorithm is within the Gibbs sampling step is used. The starting values of τ may be set to the sample variance divided by 2π, while the starting value of k may be set to some large integer K_0. The approximate posterior mode of θ_i's and Y_i's given the starting values of τ and k may be considered as the starting values for the respective variables.

Let f_0^* be the true spectral density. Assume that the time series satisfies the conditions

(M1) the time series is Gaussian with $\sum_{r=0}^{\infty} r^{\alpha}\gamma(r); < \infty$ for some $\alpha > 0$;
(M2) for all ω^*, $f_0^*(\omega^*) > 0$;

and the prior satisfies

(P1) for all k, $0 < \rho(k) \leqslant C\,\mathrm{e}^{-ck(\log k)^{1+\alpha'}}$ for some constants $C, c, \alpha' > 0$;
(P2) g is bounded, continuous, and bounded away from zero;
(P3) the prior on τ is degenerate at the true value $\tau_0 = \int f_0$.

Using the contiguity result of Choudhuri et al. (2004c), the following result was shown by Choudhuri et al. (2004a) under the above assumptions.

THEOREM 13. *For any* $\varepsilon > 0$, $\Pi_n\{f^*: \|f^* - f_0^*\|_1 > \varepsilon\} \to 0$ *in* $P^n_{f_0^*}$*-probability, where* Π_n *is the pseudo-posterior distribution computed using the Whittle likelihood of and* $P^n_{f_0^*}$ *is the actual distribution of the data* $(X_1, \dots, X_n)$.

REMARK 1. The conclusion of Theorem 13 still holds if the degenerated prior on τ is replaced by a sequence of priors distribution that asymptotically bracket the true value, that is, the prior support of τ is in $[\tau_0 - \delta_n, \tau_0 + \delta_n]$ for some $\delta_n \to 0$. A two-stage empirical Bayes method, by using one part of the sample to consistently estimate τ and the other part to estimate q, may be considered to construct the above asymptotically bracketing prior.

7.2. *Gaussian process prior*

Since the spectral density is nonnegative valued function, a Gaussian process prior may be assigned to $g(\omega) = \log(f(\omega))$. Because the Whittle likelihood in (7.1) arises by assuming that $I_n(\omega_l)$'s are approximately independent exponential random variables with mean $f(\omega_l)$, one may obtain a regression model of the form $\log(I_n(\omega_l)) = g(\omega_l) + \epsilon_l$, where the additive errors ϵ_l's are approximately i.i.d. with the Gumbel distribution.

Carter and Kohn (1997) considered an integrated Wiener process prior for g. They described an elegant Gibbs sampler algorithm for sampling from the posterior distribution. Approximating the distribution of ϵ_l's as a mixture of five known normal distribution, they introduced latent variables indicating the mixture components for the corresponding errors. Given the latent variables, conditional posterior distribution of g is obtained by a data augmentation technique. Given g, the conditional posterior distribution of the latent variables are independent and samples are easily drawn from their finite support.

Gangopadhyay et al. (1998) considered the free-knot spline approach to modeling g. In this case, the posterior is computed by the reversible jump algorithm of Green (1995). Liseo et al. (2001) considered a Brownian motion process as prior on g. For sampling from the posterior distribution, they considered the Karhunen–Loévé series expansion for the Brownian motion and then truncated the infinite series to a finite sum.

8. Estimation of transition density

Estimation of the transition density of a discrete-time Markov process is an important problem. Let Π be a prior on the transition densities $p(y|x)$. Then the predictive density of a future observation X_{n+1} given the data $X_1, \dots, X_n$ equals to

$E(p(\cdot|X_n)|X_1, \ldots, X_n)$, which is the Bayes estimate of the transition density p at X_n. The prediction problem thus directly relates to the estimation of the transition density.

Tang and Ghosal (2003) considered a mixture of normal model

$$p(y|x) = \int \phi_\sigma\big(y - H(x;\theta)\big)\,dF(\theta,\sigma), \tag{8.1}$$

where θ is possibly vector valued and $H(x;\theta)$ is a known function. Such models are analogous to the normal mixture models in the density estimation where the unknown probability density is modeled as $p(y) = \int \phi_\sigma(y-\mu)\,dF(\mu,\sigma)$. A reasonable choice for the link function H in (8.1) could be of the form $\tau + \gamma\psi(\delta + \beta x)$ for some known function ψ.

As in density estimation, this mixture model may be represented as

$$X_i \sim N\big(H(X_{i-1};\theta_i), \sigma_i^2\big), \qquad (\theta_i, \sigma_i) \overset{\text{i.i.d.}}{\sim} F. \tag{8.2}$$

Here, unlike a parametric model, the unknown parameters are varying along with the index of the observation, and are actually drawn as i.i.d. samples from an unknown distribution. Hence the model is "dynamic" as opposed to a "static" parametric mixture model.

Tang and Ghosal (2003) let the mixing distribution F have a Dirichlet process prior $\mathrm{Dir}(M, G)$. As in density estimation, the hierarchical representation (8.2) helps develop Gibbs sampler algorithms for sampling from the posterior distribution. However, because of the nonstandard forms of the conditionals, special techniques, such as the "no gaps" algorithm of MacEachern and Muller (1998) need to be implemented.

To study the large sample properties of the posterior distribution, Tang and Ghosal (2003) extended Schwartz's (1965) theorem to the context of an ergodic Markov processes. For simplicity, X_0 is assumed to be fixed below, although the conclusion extends to random X_0 also.

THEOREM 14. *Let $\{X_n, n \geqslant 0\}$ be an ergodic Markov process with transition density $p \in \mathcal{P}$ and stationary distribution π. Let Π be a prior on $\mathcal{P}$. Let $p_0 \in \mathcal{P}$ and π_0 be respectively the true values of p and π. Let U_n be a sequence of subsets of $\mathcal{P}$ containing p_0.*

Suppose that there exist a sequence of tests Φ_n, based on $X_0, X_1, \ldots, X_n$ for testing the pair of hypotheses H_0: $p = p_0$ against H: $p \in U_n^c$, and subsets $V_n \subset \mathcal{P}$ such that

(i) *p_0 is in the Kullback–Leibler support of Π, that is $\Pi\{p\colon K(p_0, p) < \varepsilon\} > 0$, where*

$$K(p_0, p) = \iint \pi_0(x) p_0(y|x) \log \frac{p_0(y|x)}{p(y|x)}\,dy\,dx,$$

(ii) *$\Phi_n \to 0$ a.s. $[P_{f_0}^\infty]$,*

(iii) *$\sup_{p\in U_n^c \cap V_n} E_p(1 - \Phi_n) \leqslant C_1 e^{-n\beta_1}$ for some constants C_1 and β_1,*

(iv) *$\Pi(p \in V_n^c) \leqslant C_2 e^{-n\beta_2}$ for some constants C_2 and β_2.*

Then $\Pi(p \in U_n | X_0, X_1, \ldots, X_n) \to 1$ a.s. $[P_0^\infty]$, where $[P_0^\infty]$ denote the distribution of the infinite sequence $(X_0, X_1, \ldots)$.

Assume that $p_0(y|x)$ is of the form (8.1). Let F_0 denote the true mixing distribution, and π_0 denote the corresponding invariant distribution. Let the sup-L_1 distance on the space of transition probabilities be given by $d(p_1, p_2) = \sup_x \int |p_1(y|x) - p_2(y|x)|\,dy$. Let H be uniformly equicontinuous in x and the support of G be compact containing the support of F_0. Tang and Ghosal (2003) showed that (i) the test $I\{\sum_{i=1}^k \log \frac{p_1(X_{2i}|X_{2i-1})}{p_0(X_{2i}|X_{2i-1})} > 0\}$, where $n = 2k$ or $2k+1$, for testing p_0 against a small ball around p_1, has exponentially small error probabilities, (ii) the space of transition probabilities supported by the prior is compact under the sup-L_1 distance, and (iii) the Kullback–Leibler property holds at p_0. By the compactness property, a single test can be constructed for the entire alternative having exponentially small error probabilities. It may be noted that because of the compactness of $\mathcal{P}$, it is not necessary to consider sieves. Thus by Theorem 14, the posterior distribution is consistent at p_0 with respect to the sup-L_1 distance.

The conditions assumed in the above result are somewhat stringent. For instance $H(x, \beta, \delta, \tau) = \tau + \gamma\psi(\delta + \beta x)$, then ψ is necessarily bounded, ruling out the linear link. If a suitable weaker topology is employed, Tang and Ghosal (2003) showed that consistency can be obtained under weaker conditions by extending Walker's (2004) approach to Markov processes. More specifically, the Kullback–Leibler property holds if H satisfies uniform equicontinuity on compact sets only. If now a topology is defined by the neighborhood base $\{f\colon \int |\int g_i(y) f(y|x)\,dy - \int g_i(y) f_0(y|x)\,dy| \nu(x)\,dx < \varepsilon, i = 1, \dots, k\}$, where ν is a probability density, then consistency holds if σ is bounded below and contains σ_0 in its support. If further σ is also bounded above and the θ is supported on a compact set, then consistency also holds in the integrated-L_1 distance integrated with respect to ν. For a linear link function $H(x, \rho, b) = \rho x + b$, $|\rho| < 1$, the compactness condition can be dropped, for instance, if the distribution of b under G is normal.

9. Concluding remarks

In this article, we have reviewed Bayesian methods for the estimation of functions of statistical interest such as the cumulative distribution function, density function, regression function, spectral density of a time series and the transition density function of a Markov process. Function estimation can be viewed as a problem of the estimation of one or more infinite-dimensional parameter arising in a statistical model. It has been argued that the Bayesian approach to function estimation, commonly known as Bayesian nonparametric estimation, can provide an important, coherent alternative to more familiar classical approaches to function estimation. We have considered the problems of construction of appropriate prior distributions on infinite-dimensional spaces. It has been argued that, because of the lack of subjective knowledge about every detail of a distribution in an infinite-dimensional space, some default mechanism of prior specification needs to be followed. We have discussed various important priors on infinite-dimensional spaces, and their merits and demerits. While certainly not exhaustive, these priors and their various combinations provide a large catalog of priors in a statistician's toolbox, which may be tried and tested for various curve estimation problems including, but not restricted to, the problems we discussed. Due to the vastness of the relevant literature and the rapid growth of the subject, it is impossible to even attempt to mention all the problems of Bayesian curve estimation. The material presented here is mostly a reflection of the authors' interest and familiarity. Computation of the posterior distribution is an important issue. Due to the lack of useful analytical expressions for the posterior distribution in most curve estimation problems, computation has to be done by some numerical technique, usually by the help of Markov chain Monte Carlo methods. We described computing techniques for the curve estimation problems considered in this chapter. The simultaneous development of innovative sampling techniques and computing devices has brought tremendous computing power to nonparametric Bayesians. Indeed, for many statistical problems, the computing power of a Bayesian now exceeds that of a non-Bayesian. While these positive developments are extremely encouraging, one should however be extremely cautious about naive uses of Bayesian methods for nonparametric problems to avoid pitfalls. We argued that it is important to validate the use of a particular prior by using some benchmark criterion such as posterior consistency. We discussed several techniques of proving posterior consistency and mentioned some examples of inconsistency. Sufficient conditions for posterior consistency are discussed in the problems we considered. Convergence rates of posterior distributions have also been discussed, together with the related concepts of optimality, adaptation, misspecification and Berntsein–von Mises theorem.

The popularity of Bayesian nonparametric methods is rapidly growing among practitioners as theoretical properties are increasingly better understood and the computational hurdles are being removed. Innovative Bayesian nonparametric methods for complex models arising in biomedical, geostatistical, environmental, econometric and many other applications are being proposed. Study of theoretical properties of nonparametric Bayesian beyond the traditional i.i.d. set-up has started to receive attention recently. Much more work will be needed to bridge the gap. Developing techniques of model selection, the Bayesian equivalent of hypothesis testing, as well as the study of their theoretical properties will be highly desirable.

References

Albert, J., Chib, S. (1993). Bayesian analysis of binary and polychotomous response data. *J. Amer. Statist. Assoc.* **88**, 669–679.

Amewou-Atisso, M., Ghosal, S., Ghosh, J.K., Ramamoorthi, R.V. (2003). Posterior consistency for semiparametric regression problems. *Bernoulli* **9**, 291–312.

Ansley, C.F., Kohn, R., Wong, C. (1993). Nonparametric spline regression with prior information. *Biometrika* **80**, 75–88.

Antoniak, C. (1974). Mixtures of Dirichlet processes with application to Bayesian non-parametric problems. *Ann. Statist.* **2**, 1152–1174.

Barron, A.R. (1988). The exponential convergence of posterior probabilities with implications for Bayes estimators of density functions. Unpublished manuscript.

Barron, A.R. (1999). Information-theoretic characterization of Bayes performance and the choice of priors in parametric and nonparametric problems. In: Bernardo, J.M., et al. (Eds.), *Bayesian Statistics, vol. 6*. Oxford University Press, New York, pp. 27–52.

Barron, A., Schervish, M., Wasserman, L. (1999). The consistency of posterior distributions in nonparametric problems. *Ann. Statist.* **27**, 536–561.

Barry, D. (1986). Nonparametric Bayesian regression. *Ann. Statist.* **14**, 934–953.

Basu, S., Mukhopadhyay, S. (2000). Bayesian analysis of binary regression using symmetric and asymmetric links. *Sankhyā, Ser. B* **62**, 372–387.

Belitser, E.N., Ghosal, S. (2003). Adaptive Bayesian inference on the mean of an infinite-dimensional normal distribution. *Ann. Statist.* **31**, 536–559.

Berger, J.O., Guglielmi, A. (2001). Bayesian and conditional frequentist testing of a parametric model versus nonparametric alternatives. *J. Amer. Statist. Assoc.* **96** (453), 174–184.

Berk, R. (1966). Limiting behavior of the posterior distribution when the model is incorrect. *Ann. Math. Statist.* **37**, 51–58.

Birgé, L. (1983). Robust testing for independent non-identically distributed variables and Markov chains. In: Florens, J.P., et al. (Eds.), *Specifying Statistical Models. From Parametric to Non-Parametric. Using Bayesian or Non-Bayesian Approaches*. In: *Lecture Notes in Statistics*, vol. 16. Springer-Verlag, New York, pp. 134–162.

Blackwell, D. (1973). Discreteness of Ferguson selection. *Ann. Statist.* **1**, 356–358.

Blackwell, D., Dubins, L.E. (1962). Merging of opinions with increasing information. *Ann. Math. Statist.* **33**, 882–886.

Blackwell, D., MacQueen, J.B. (1973). Ferguson distributions via Polya urn schemes. *Ann. Statist.* **1**, 353–355.

Blum, J., Susarla, V. (1977). On the posterior distribution of a Dirichlet process given randomly right censored observations. *Stochastic Process. Appl.* **5**, 207–211.

Brunner, L.J. (1992). Bayesian nonparametric methods for data from a unimodal density. *Statist. Probab. Lett.* **14**, 195–199.

Brunner, L.J. (1995). Bayesian linear regression with error terms that have symmetric unimodal densities. *J. Nonparametr. Statist.* **4**, 335–348.

Brunner, L.J., Lo, A.Y. (1989). Bayes methods for a symmetric unimodal density and its mode. *Ann. Statist.* **17**, 1550–1566.

Carter, C.K., Kohn, R. (1997). Semiparametric Bayesian inference for time series with mixed spectra. *J. Roy. Statist. Soc., Ser. B* **59**, 255–268.

Choudhuri, N., Ghosal, S., Roy, A. (2004a). Bayesian estimation of the spectral density of a time series. *J. Amer. Statist. Assoc.* **99**, 1050–1059.

Choudhuri, N., Ghosal, S., Roy, A. (2004b). Bayesian nonparametric binary regression with a Gaussian process prior. Preprint.

Choudhuri, N., Ghosal, S., Roy, A. (2004c). Contiguity of the Whittle measure in a Gaussian time series. *Biometrika* **91**, 211–218.

Cifarelli, D.M., Regazzini, E. (1990). Distribution functions of means of a Dirichlet process. *Ann. Statist.* **18**, 429–442.

Cox, D.D. (1993). An analysis of Bayesian inference for nonparametric regression. *Ann. Statist.* **21**, 903–923.

Dalal, S.R. (1979). Dirichlet invariant processes and applications to nonparametric estimation of symmetric distribution functions. *Stochastic Process. Appl.* **9**, 99–107.

Denison, D.G.T., Mallick, B.K., Smith, A.F.M. (1998). Automatic Bayesian curve fitting. *J. Roy. Statist. Soc., Ser. B Stat. Methodol.* **60**, 333–350.

Diaconis, P., Freedman, D. (1986a). On the consistency of Bayes estimates (with discussion). *Ann. Statist.* **14**, 1–67.

Diaconis, P., Freedman, D. (1986b). On inconsistent Bayes estimates. *Ann. Statist.* **14**, 68–87.

DiMatteo, I., Genovese, C.R., Kass, R.E. (2001). Bayesian curve-fitting with free-knot splines. *Biometrika* **88**, 1055–1071.

Doksum, K.A. (1974). Tail free and neutral random probabilities and their posterior distributions. *Ann. Probab.* **2**, 183–201.

Doob, J.L. (1948). Application of the theory of martingales. Coll. Int. du CNRS, Paris, pp. 22–28.

Doss, H. (1985a). Bayesian nonparametric estimation of the median. I. Computation of the estimates. *Ann. Statist.* **13**, 1432–1444.

Doss, H. (1985b). Bayesian nonparametric estimation of the median. II. Asymptotic properties of the estimates. *Ann. Statist.* **13**, 1445–1464.

Doss, H., Sellke, T. (1982). The tails of probabilities chosen from a Dirichlet prior. *Ann. Statist.* **10**, 1302–1305.

Dykstra, R.L., Laud, P.W. (1981). A Bayesian nonparametric approach to reliability. *Ann. Statist.* **9**, 356–367.

Escobar, M. (1994). Estimating normal means with a Dirichlet process prior. *J. Amer. Statist. Assoc.* **89**, 268–277.

Escobar, M., West, M. (1995). Bayesian density estimation and inference using mixtures. *J. Amer. Statist. Assoc.* **90**, 577–588.

Escobar, M., West, M. (1998). Computing nonparametric hierarchical models. In: *Practical Nonparametric and Semiparametric Bayesian Statistics*. In: *Lecture Notes in Statistics*, vol. 133. Springer, New York, pp. 1–22.

Ferguson, T.S. (1973). A Bayesian analysis of some nonparametric problems. *Ann. Statist.* **1**, 209–230.

Ferguson, T.S. (1974). Prior distribution on the spaces of probability measures. *Ann. Statist.* **2**, 615–629.

Ferguson, T.S. (1983). Bayesian density estimation by mixtures of Normal distributions. In: Rizvi, M., Rustagi, J., Siegmund, D. (Eds.), *Recent Advances in Statistics*, pp. 287–302.

Ferguson, T.S., Phadia, E.G. (1979). Bayesian nonparametric estimation based on censored data. *Ann. Statist.* **7**, 163–186.

Freedman, D. (1963). On the asymptotic distribution of Bayes estimates in the discrete case I. *Ann. Math. Statist.* **34**, 1386–1403.

Freedman, D. (1965). On the asymptotic distribution of Bayes estimates in the discrete case II. *Ann. Math. Statist.* **36**, 454–456.

Freedman, D. (1999). On the Bernstein–von Mises theorem with infinite-dimensional parameters. *Ann. Statist.* **27**, 1119–1140.

Fristedt, B. (1967). Sample function behavior of increasing processes with stationary independent increments. *Pacific J. Math.* **21**, 21–33.

Fristedt, B., Pruitt, W.E. (1971). Lower functions for increasing random walks and subordinators. *Z. Wahsch. Verw. Gebiete* **18**, 167–182.

Gangopadhyay, A.K., Mallick, B.K., Denison, D.G.T. (1998). Estimation of spectral density of a stationary time series via an asymptotic representation of the periodogram. *J. Statist. Plann. Inference* **75**, 281–290.

Gasparini, M. (1996). Bayesian density estimation via Dirichlet density process. *J. Nonparametr. Statist.* **6**, 355–366.

Gelfand, A.E., Kuo, L. (1991). Nonparametric Bayesian bioassay including ordered polytomous response. *Biometrika* **78**, 657–666.

Ghosal, S. (2000). Asymptotic normality of posterior distributions for exponential families with many parameters. *J. Multivariate Anal.* **74**, 49–69.

Ghosal, S. (2001). Convergence rates for density estimation with Bernstein polynomials. *Ann. Statist.* **29**, 1264–1280.

Ghosal, S., van der Vaart, A.W. (2001). Entropies and rates of convergence for maximum likelihood and Bayes estimation for mixtures of normal densities. *Ann. Statist.* **29**, 1233–1263.

Ghosal, S., van der Vaart, A.W. (2003a). Convergence rates for non-i.i.d. observations. Preprint.

Ghosal, S., van der Vaart, A.W. (2003b). Posterior convergence rates of Dirichlet mixtures at smooth densities. Preprint.

Ghosal, S., Ghosh, J.K., Samanta, T. (1995). On convergence of posterior distributions. *Ann. Statist.* **23**, 2145–2152.

Ghosal, S., Ghosh, J.K., Ramamoorthi, R.V. (1997). Noniformative priors via sieves and consistency. In: Panchapakesan, S., Balakrishnan, N. (Eds.), *Advances in Statistical Decision Theory and Applications*. Birkhäuser, Boston, pp. 119–132.

Ghosal, S., Ghosh, J.K., Ramamoorthi, R.V. (1999a). Consistency issues in Bayesian nonparametrics. In: Ghosh, S. (Ed.), *Asymptotics, Nonparametrics and Time Series: A Tribute to Madan Lal Puri*. Marcel Dekker, New York, pp. 639–668.

Ghosal, S., Ghosh, J.K., Ramamoorthi, R.V. (1999b). Posterior consistency of Dirichlet mixtures in density estimation. *Ann. Statist.* **27**, 143–158.

Ghosal, S., Ghosh, J.K., Ramamoorthi, R.V. (1999c). Consistent semiparametric Bayesian inference about a location parameter. *J. Statist. Plann. Inference* **77**, 181–193.

Ghosal, S., Ghosh, J.K., van der Vaart, A.W. (2000). Convergence rates of posterior distributions. *Ann. Statist.* **28**, 500–531.

Ghosal, S., Lember, Y., van der Vaart, A.W. (2003). On Bayesian adaptation. *Acta Appl. Math.* **79**, 165–175.

Ghosh, J.K., Ramamoorthi, R.V. (1995). Consistency of Bayesian inference for survival analysis with or without censoring. In: Koul, H. (Ed.), *Analysis of Censored Data*. In: *IMS Lecture Notes Monograph Series*, vol. 27. Inst. Math. Statist., Hayward, CA, pp. 95–103.

Ghosh, J.K., Ramamoorthi, R.V. (2003). *Bayesian Nonparametrics*. Springer-Verlag, New York.

Ghosh, S.K., Ghosal, S. (2003). Proportional mean regression models for censored data. Preprint.

Ghosh, J.K., Ghosal, S., Samanta, T. (1994). Stability and convergence of posterior in non-regular problems. In: Gupta, S.S., Berger, J.O. (Eds.), *Statistical Decision Theory and Related Topics V*. Springer-Verlag, New York, pp. 183–199.

Green, P. (1995). Reversible jump Markov chain Monte Carlo computation and Bayesian model determination. *Biometrika* **82**, 711–732.

Grenander, U. (1981). *Abstract Inference*. Wiley, New York.

Hanson, T., Johnson, W.O. (2002). Modeling regression error with a mixture of Polya trees. *J. Amer. Statist. Assoc.* **97**, 1020–1033.

Hjort, N.L. (1990). Nonparametric Bayes estimators based on beta processes in models for life history data. *Ann. Statist.* **18**, 1259–1294.

Hjort, N.L. (1996). Bayesian approaches to non- and semiparametric density estimation. In: Bernardo, J., et al. (Eds.), *Bayesian Statistics, vol. 5*, pp. 223–253.

Hjort, N.L. (2000). Bayesian analysis for a generalized Dirichlet process prior. Preprint.

Hjort, N.L. (2003). Topics in nonparametric Bayesian statistics (with discussion). In: Green, P.J., Hjort, N., Richardson, S. (Eds.), *Highly Structured Stochastic Systems*. Oxford University Press, pp. 455–487.

Holmes, C.C., Mallick, B.K. (2003). Generalized nonlinear modeling with multivariate free-knot regression splines. *J. Amer. Statist. Assoc.* **98**, 352–368.

Huang, T.Z. (2004). Convergence rates for posterior distributions and adaptive estimation. *Ann. Statist.* **32**, 1556–1593.

Ibragimov, I.A., Has'minskii, R.Z. (1981). *Statistical Estimation: Asymptotic Theory*. Springer-Verlag, New York.

Iswaran, H., Zarepour, M. (2002a). Exact and approximate sum representation for the Dirichlet process. *Canad. J. Statist.* **26**, 269–283.

Iswaran, H., Zarepour, M. (2002b). Dirichlet prior sieves in finite normal mixture models. *Statistica Sinica*, 269–283.

Kim, Y. (1999). Nonparametric Bayesian estimators for counting processes. *Ann. Statist.* **27**, 562–588.

Kim, Y., Lee, J. (2001). On posterior consistency of survival models. *Ann. Statist.* **29**, 666–686.

Kim, Y., Lee, J. (2004). A Bernstein–von Mises theorem in the nonparametric right-censoring model. *Ann. Statist.* **32**, 1492–1512.

Kleijn, B., van der Vaart, A.W. (2002). Misspecification in infinite-dimensional Bayesian statistics. Preprint.

Kottas, A., Gelfand, A.E. (2001). Bayesian semiparametric median regression modeling. *J. Amer. Statist. Assoc.* **96**, 1458–1468.

Kraft, C.H. (1964). A class of distribution function processes which have derivatives. *J. Appl. Probab.* **1**, 385–388.

Lavine, M. (1992). Some aspects of Polya tree distributions for statistical modeling. *Ann. Statist.* **20**, 1222–1235.

Lavine, M. (1994). More aspects of Polya tree distributions for statistical modeling. *Ann. Statist.* **22**, 1161–1176.

Le Cam, L. (1986). *Asymptotic Methods in Statistical Decision Theory*. Springer-Verlag, New York.

Le Cam, L., Yang, G.L. (2000). *Asymptotics in Statistics*, second ed. Springer-Verlag.

Lenk, P.J. (1988). The logistic normal distribution for Bayesian, nonparametric, predictive densities. *J. Amer. Statist. Assoc.* **83**, 509–516.

Lenk, P.J. (1991). Towards a practicable Bayesian nonparametric density estimator. *Biometrika* **78**, 531–543.

Leonard, T. (1978). Density estimation, stochastic processes, and prior information. *J. Roy. Statist. Soc., Ser. B* **40**, 113–146.

Liseo, B., Marinucci, D., Petrella, L. (2001). Bayesian semiparametric inference on long-range dependence. *Biometrika* **88**, 1089–1104.
Lo, A.Y. (1982). Bayesian nonparametric statistical inference for Poisson point process. *Z. Wahsch. Verw. Gebiete* **59**, 55–66.
Lo, A.Y. (1983). Weak convergence for Dirichlet processes. *Sankhyā, Ser. A* **45**, 105–111.
Lo, A.Y. (1984). On a class of Bayesian nonparametric estimates I: Density estimates. *Ann. Statist.* **12**, 351–357.
Lo, A.Y. (1986). A remark on the limiting posterior distribution of the multiparameter Dirichlet process. *Sankhyā, Ser. A* **48**, 247–249.
MacEachern, S.N., Muller, P. (1998). Estimating mixture of Dirichlet process models. *J. Comput. Graph. Statist.* **7**, 223–228.
Mallick, B.K., Gelfand, A.E. (1994). Generalized linear models with unknown link functions. *Biometrika* **81**, 237–245.
Mauldin, R.D., Sudderth, W.D., Williams, S.C. (1992). Polya trees and random distributions. *Ann. Statist.* **20**, 1203–1221.
Muliere, P., Tardella, L. (1998). Approximating distributions of functionals of Ferguson–Dirichlet priors. *Canad. J. Statist.* **30**, 269–283.
Newton, M.A., Czado, C., Chappell, R. (1996). Bayesian inference for semiparametric binary regression. *J. Amer. Statist. Assoc.* **91**, 142–153.
Nieto-Barajas, L.E., Walker, S.G. (2004). Bayesian nonparametric survival analysis via Lévy driven Markov process. *Statistica Sinica* **14**, 1127–1146.
Petrone, S. (1999a). Random Bernstein polynomials. *Scand. J. Statist.* **26**, 373–393.
Petrone, S. (1999b). Bayesian density estimation using Bernstein polynomials. *Canad. J. Statist.* **26**, 373–393.
Petrone, S., Veronese, P. (2002). Nonparametric mixture priors based on an exponential random scheme. *Statist. Methods Appl.* **11**, 1–20.
Petrone, S., Wasserman, L. (2002). Consistency of Bernstein polynomial posteriors. *J. Roy. Statist. Soc., Ser. B* **64**, 79–100.
Regazzini, E., Guglielmi, A., Di Nunno, G. (2002). Theory and numerical analysis for exact distributions of functionals of a Dirichlet process. *Ann. Statist.* **30**, 1376–1411.
Rubin, D. (1981). The Bayesian bootstrap. *Ann. Statist.* **9**, 130–134.
Schwartz, L. (1965). On Bayes procedures. *Z. Wahsch. Verw. Gebiete* **4**, 10–26.
Sethuraman, J. (1994). A constructive definition of Dirichlet priors. *Statistica Sinica* **4**, 639–650.
Sethuraman, J., Tiwari, R. (1982). Convergence of Dirichlet measures and interpretation of their parameters. In: Gupta, S.S., Berger, J.O. (Eds.), *Statistical Decision Theory and Related Topics. III, vol. 2*. Academic Press, New York, pp. 305–315.
Shen, X. (2002). Asymptotic normality of semiparametric and nonparametric posterior distributions. *J. Amer. Statist. Assoc.* **97**, 222–235.
Shen, X., Wasserman, L. (2001). Rates of convergence of posterior distributions. *Ann. Statist.* **29**, 687–714.
Shively, T.S., Kohn, R., Wood, S. (1999). Variable selection and function estimation in additive nonparametric regression using a data-based prior (with discussions). *J. Amer. Statist. Assoc.* **94**, 777–806.
Smith, M., Kohn, R. (1997). A Bayesian approach to nonparametric bivariate regression. *J. Amer. Statist. Assoc.* **92**, 1522–1535.
Smith, M., Wong, C., Kohn, R. (1998). Additive nonparametric regression with autocorrelated errors. *J. Roy. Statist. Soc., Ser. B* **60**, 311–331.
Susarla, V., Van Ryzin, J. (1976). Nonparametric Bayesian estimation of survival curves from incomplete observations. *J. Amer. Statist. Assoc.* **71**, 897–902.
Susarla, V., Van Ryzin, J. (1978). Large sample theory for a Bayesian nonparametric survival curve estimator based on censored samples. *Ann. Statist.* **6**, 755–768.
Tang, Y., Ghosal, S. (2003). Posterior consistency of Dirichlet mixtures for estimating a transition density. Preprint.
Tokdar, S.T. (2003). Posterior consistency of Dirichlet location-scale mixtures of normals in density estimation and regression. Preprint.
van der Vaart, A.W. (1998). *Asymptotic Statistics*. Cambridge University Press.

van der Vaart, A.W., Wellner, J.A. (1996). *Weak Convergence and Empirical Processes*. Springer-Verlag, New York.

Wahba, G. (1978). Improper priors, spline smoothing and the problem of guarding against model errors in regression. *J. Roy. Statist. Soc., Ser. B* **40**, 364–372.

Walker, S.G. (2003). On sufficient conditions for Bayesian consistency. *Biometrika* **90**, 482–490.

Walker, S.G. (2004). New approaches to Bayesian consistency. *Ann. Statist.* **32**, 2028–2043.

Walker, S.G., Hjort, N.L. (2001). On Bayesian consistency. *J. Roy. Statist. Soc., Ser. B* **63**, 811–821.

Walker, S.G., Muliere, P. (1997). Beta-Stacy processes and a generalization of the Polya-urn scheme. *Ann. Statist.* **25**, 1762–1780.

Wasserman, L. (1998). Asymptotic properties of nonparametric Bayesian procedures. In: Dey, D., et al. (Eds.), *Practical Nonparametric and Semiparametric Bayesian Statistics*. In: *Lecture Notes in Statistics*, vol. 133. Springer-Verlag, New York, pp. 293–304.

Whittle, P. (1957). Curve and periodogram smoothing. *J. Roy. Statist. Soc., Ser. B* **19**, 38–63.

Whittle, P. (1962). Gaussian estimation in stationary time series. *Bull. Int. Statist. Inst.* **39**, 105–129.

Wood, S., Kohn, R. (1998). A Bayesian approach to robust binary nonparametric regression. *J. Roy. Statist. Soc., Ser. B* **93**, 203–213.

Wood, S., Kohn, R., Shively, T., Jiang, W. (2002a). Model selection in spline nonparametric regression. *J. Roy. Statist. Soc., Ser. B* **64**, 119–139.

Wood, S., Jiang, W., Tanner, M. (2002b). Bayesian mixture of splines for spatially adaptive nonparametric regression. *Biometrika* **89**, 513–528.

Yau, P., Kohn, R., Wood, S. (2003). Bayesian variable selection and model averaging in high-dimensional multinomial nonparametric regression. *J. Comput. Graph. Statist.* **12**, 23–54.

Zhao, L.H. (2000). Bayesian aspects of some nonparametric problems. *Ann. Statist.* **28**, 532–552.

Handbook of Statistics, Vol. 25
ISSN: 0169-7161
© 2005 Elsevier B.V. All rights reserved.
DOI 10.1016/S0169-7161(05)25014-9

14

MCMC Methods to Estimate Bayesian Parametric Models

Antonietta Mira

1. Motivation

Integration plays a fundamental role in Bayesian statistics (see Section 3). The integrals needed are, in most real applications, complicated and cannot be solved analytically (see Section 4). This is the reason why Bayesian inference remained an elegant theoretical construction with little practical application until the introduction of efficient approximations techniques and in particular of Markov chain Monte Carlo (MCMC) simulation, a way to approximate complex integrals that originated in the physics literature (Metropolis et al., 1953), and was brought into the mainstream statistical literature by Besag (1974), Geman and Geman (1984), and Gelfand and Smith (1990).

Before MCMC become routine and related approximation or numerical methods where developed, A.P. Dempster (1980, p. 273) writes:

> The application of inference techniques is held back by conceptual factors and computational factors. I believe that Bayesian inference is conceptually much more straightforward than non-Bayesian inference, one reason being that Bayesian inference has a unified methodology for coping with nuisance parameters, whereas non-Bayesian inference has only a multiplicity of ad hoc rules. Hence, I believe that the major barrier to much more widespread application of Bayesian methods is computational.... The development of the field depends heavily on the preparation of effective computer programs.

Approximately 10 years later, S.I. Press (1989, p. 69), quoting Dempster, writes:

> Bayesian paradigm is conceptually simple, intuitively plausible, and probabilistically elegant. Its numerical implementation is not always easy and straightforward, however. Posterior distributions are often expressible only in terms of complicated analytical functions; we often know only the kernel of the posterior density (and not the normalizing constant); we often can't readily compute the marginal densities and moments of the posterior distribution in terms of exact closed form explicit expressions; and it is sometimes difficult to find numerical fractiles of the posterior cdf.

Approximately 10 years since then, we can say that, today, Bayesian statistics has become widely applicable thanks to effective computational factors (reliable and fairly flexible freeware software and cheap and massive computational power), as Dempster envisioned 20 years ago. Thanks to approximation techniques the implementation of the Bayesian paradigm is nowadays much more easy and straightforward than at Press time.

This chapter is about MCMC methods and their use within the Bayesian framework. We point out that MCMC can be quite helpful also within the frequentist approach. We also stress that stochastic simulation (as Monte Carlo and MCMC) is not the only way to approximate integrals, we could also resort to deterministic simulation such as numerical integration (that scale poorly as the number of parameters increases) or analytic approximation techniques such as Laplace approximations. Still, MCMC has become the approximation method of choice for a wide range of Bayesian problems thanks to its flexibility also for high-dimensional posteriors and we agree with Roberts (2004) when he writes that "*its success has even inspired new and developing areas of statistical modeling, due to the existence of algorithms to perform inference on these new models*".

2. Bayesian ingredients

The basic ingredients in the Bayesian parametric approach are:

- a parameter, θ, that takes value in a parameter space $\Theta \subset R^d$;
- a prior distribution on the parameter: $\pi(\theta)$;
- some data, X, that take value in the sample space, $\mathcal{X} \subset R^N$;
- a model for the data, given the parameter: $l(X|\theta)$;
- a loss function $L(\theta, d)$ that compares the decision d with the true value of the parameter θ.

In the Bayesian parametric approach, the parameter is treated as the unknown and a distribution is elicited on its plausible values, based on subjective knowledge, prior of observing the data. Unlike in the frequentist approach, data are treated as fixed, once observed, and inference does not take into account possible experimental outcomes that could have occurred. The result is a unified and coherent paradigm that avoids some classical paradoxes.

3. Bayesian recipe

The Bayes theorem yields the posterior distribution of the parameter given the observed data, $X = x$:

$$\pi(\theta|x) \propto l(x|\theta)\pi(\theta).$$

Inference is based on such distribution and, in particular, the solution of the following minimization problem is sought:

$$\min_d \int_\Theta \pi(\theta|x)L(\theta, d)\,\mathrm{d}\theta. \tag{1}$$

If, for example, a squared error loss function is used and we are interested in point estimation of θ, the minimum of (1) is achieved when d is the mean of the posterior distribution:

$$\int_{\Theta} \theta \pi(\theta|x)\,d\theta = \frac{\int_{\Theta} \theta l(x|\theta)\pi(\theta)\,d\theta}{\int_{\Theta} l(x|\theta)\pi(\theta)\,d\theta}.$$

The normalizing constant of the posterior distribution is given by the following integral over the parameter space:

$$\int_{\Theta} l(x|\theta)\pi(\theta)\,d\theta.$$

Furthermore, suppose the parameter is multivariate, $\theta = (\theta^{(1)}, \theta^{(2)})$ and that, for example, $\theta^{(2)}$ is a nuisance parameter and we are interested only in $\theta^{(1)}$. We would then base our inference on the following marginal posterior distribution:

$$\pi\left(\theta^{(1)}|x\right) = \int \pi\left(\theta^{(1)}\theta^{(2)}|x\right) d\theta^{(2)}.$$

In the Bayesian context, the most widespread model choice criterion is the Bayes factor (Kass and Raftery, 1995) defined as the ratio of the marginal likelihoods for a pair of models, which represents the evidence provided by the data in favor of a certain model. Exact evaluation of the Bayes factor is possible only in certain restricted examples, since it requires the analytical tractability of generally complicated integrals. Therefore, its estimation has attracted considerable interest in the recent MCMC literature. The Bayes factor between two models, say $\mathcal{M}_l$ and $\mathcal{M}_k$, is

$$B_{lk} = \frac{p(y|l)}{p(y|k)},$$

where

$$p(y|k) = \int_{\Theta_k} p(y|\theta_k, k)p(\theta_k|k)\,d\theta_k$$

is the marginal likelihood for model $\mathcal{M}_k$, characterized by the parameter vector $\theta_k \in \Theta_k$ and likelihood $p(y|\theta_k, k)$. The larger is B_{lk}, the greater is the evidence provided by the data in favor of $\mathcal{M}_l$ with respect to $\mathcal{M}_k$ (see Kass and Raftery, 1995). Estimation of Bayes factors is treated in Chapter 4 in this volume.

4. How can the Bayesian pie burn

As we have pointed out in the previous section, the Bayesian approach typically requires the evaluation of complicated integrals or sums over state spaces too big to be able to deal with them in a reasonable amount of time, as in image analysis applications for example, where the space of all possible black and white images on a 256×256 pixel grid is $2^{256\times256}$. In particular each one of the ingredients needed to cook the Bayesian pie can cause problems at this stage:

- the parameter space Θ can be so complicated (constrained for example) to make the integration required to solve (1) difficult;
- the prior distribution can be complex due to elicitation procedures or sequential schemes (where the posterior of one stage becomes the prior for the next stage);
- some data may be missing or unobservable or the dataset can be huge, as in data mining;
- the model can be complicated requiring a hierarchical structure with interactions among the variables or hyper prior distributions.

5. MCMC methods

A powerful tool to address the computational challenges posed by the Bayesian paradigm is MCMC simulation. The two building blocks of MCMC are, as the name itself suggests, Monte Carlo simulation and Markov chains. Before introducing MCMC algorithms let us briefly review Monte Carlo integration and general theory on Markov chains. For a more detailed introduction to Markov chains (MC) we refer the reader to Meyn and Tweedie (1993).

5.1. Monte Carlo integration and Markov chains

Suppose, in general, we need to evaluate the integral of a function $f(\theta)$ with respect to a distribution, $\pi(\theta)$:

$$\mu_\pi(f) = \int_\Theta f(\theta)\pi(\theta)\,\mathrm{d}\theta. \tag{2}$$

As emerged from the previous sections the distribution of interest is typically the posterior of the parameter while the function can be the identity (we thus recover the posterior mean) or the indicator function of some set (for confidence regions) or any other (integrable) function of the parameter of interest.

If a random sample of i.i.d. observations, $(\theta_1, \dots, \theta_n)$, from π is obtainable, either by direct simulation, rejection sampling (see Ripley, 1987), or perfect simulation (see Section 6), we can resort to *Monte Carlo simulation* and estimate $\mu_\pi(f)$ by

$$\hat{\mu}_n(f) = \frac{1}{n}\sum_{i=1}^{n} f(\theta_i).$$

Assuming that f has finite variance under π, the law of large numbers guaranties that $\hat{\mu}_n(f)$ is an asymptotically unbiased estimator of $\mu_\pi(f)$ and the central limit theorem guaranties that the limiting distribution of the Monte Carlo estimator, properly normalized, $\sqrt{n}(\hat{\mu}_n(f) - \mu_\pi(f))$, is Normal with variance given by $\sigma_\pi^2(f)$, the variance of f with respect to π. Thus the error term is of order $\sqrt{n}$ regardless of the dimensionality of θ.

The second building block of MCMC are Markov chains. A MC is a stochastic process $\{X_0, X_1, \dots\}$ that evolves in time (typically MC used for MCMC purposes are discrete time MC) with the property that the future is independent from the past given the present:

$$P\{X_t \in A|X_0, X_1, \ldots, X_{t-1}\} = P\{X_t \in A|X_{t-1}\}$$

for any set A in the state space E. Typically, MC used for MCMC purposes, have transition probabilities that are invariant over time (time-homogeneous) or are constructed combining time-homogeneous transition probabilities in a, possibly not time-homogeneous, way (as in the random scan Gibbs sampler, see Section 5.3).

We identify a MC with the corresponding transition kernel, P defined as:

$$P(x, A) = P\{X_1 \in A|X_0 = x\}.$$

Different MCMC strategies (such as the Gibbs sampler or the Metropolis–Hastings algorithm) give rise to different transition kernels. The n-step transition kernel is given by

$$P^n(x, A) = P\{X_n \in A|X_0 = x\}.$$

We denote by P_x and by P_π probabilities for a MC started with $X_0 = x$ or with initial distribution, $X_0 \sim \pi$, respectively. A MC has invariant (or stationary) distribution π if

$$\pi P(A) = \int P(x, A)\pi(x)\,\mathrm{d}x = \pi(A), \quad \forall A \subset E.$$

Not all MC have an invariant distribution and even when an invariant distribution exists it may not be unique. The basic principle behind MCMC is that certain Markov chains converge to a unique invariant distribution and can thus be used to estimate expectations with respect to this distribution. In the sequel some properties are given that help identifying Markov chains that have a unique invariant distribution that is also the limiting distribution of the process.

A MC is said to be *ϕ-irreducible* for a probability distribution ϕ on E, if $\phi(A) > 0$ implies that

$$P_x\{\text{first return time to } A < \infty\} > 0.$$

A chain is *irreducible* if it is ϕ-irreducible for some ϕ. In other words an irreducible chain has nonzero probability to move from any one position in the state space to any other position, in a finite number of steps. This guaranties that all interesting portions of the state space will be visited. *Recurrence*, on the other hand, guaranties that all interesting subsets of the state space will be visited infinitely often, at least from almost all starting points. If a MC is irreducible and has a proper invariant distribution, π, then it must be positive recurrent and π is also the unique invariant distribution. Since all MC used for MCMC purposes are constructed to have an invariant distribution, π, we only need to check ϕ-irreducibility (or π-irreducibility) which is often straightforward. A sufficient condition for a MC to be irreducible with respect to a distribution ϕ is that the n-step transition kernel has a positive density with respect to ϕ for some $n \geqslant 1$. This is often the case for the Gibbs sampler (see Section 5.3). For a MC that has both continuous and discrete components, as often happens for a Metropolis–Hastings sampler, then it is sufficient for the continuous component of the transition kernel to have a positive density with respect to ϕ. If a MC is irreducible then it has many irreducibility distributions. However there is a *maximal irreducibility distribution* in the sense that all other irreducibility distributions are absolutely continuous with respect to it.

Given an irreducible MC with invariant distribution π and a real-valued function such that $\int |f(x)|\pi(x)\,\mathrm{d}x < \infty$ we have a *strong law of large numbers*, that is:

$$P_x\{\hat{\mu}_n(f) \to \mu_\pi(f)\}$$

for π-almost all starting values x. In other words, the observed and expected time spent in a set A converges to $\pi(A)$.

To get stronger results, such as convergence is total variation distance of the n-step transition kernel to the stationary distribution, we need to rule out periodic or cyclic behavior. A Markov chain is said *aperiodic* if the maximum common divider of the number of steps it takes for the chain to come back to the starting point (no matter what this is) is one.

To rule out null sets of initial starting points where the law of large numbers may fail, *Harris recurrent* MC are considered, that is, ϕ-irreducible chains, where ϕ is the maximal irreducibility distribution, and such that, for any $A \subset E$, with $\phi(A) > 0$, we have:

$$P_x\{X_n \in A \text{ infinitely often}\} = 1$$

for all $x \in E$. A MC is *ergodic* if it is irreducible, aperiodic and positive Harris-recurrent. In most MCMC applications ergodicity is of little importance since we are typically interested in results concerning sample path averages.

Uniform and geometric ergodicity conditions are related to the rate at which a MC converges to stationarity. In particular a MC with invariant distribution π, is geometrically ergodic if there exist a nonnegative extended real valued function, M, such that $\int M(x)\pi(x)\,\mathrm{d}x < \infty$, and a positive constant, $\rho < 1$, such that

$$\left\| P^n(x,\cdot) - \pi(\cdot) \right\| \leqslant M(x)\rho^n, \quad \forall x, n;$$

if, furthermore, M is constant (finite and positive) then the chain is uniformly ergodic. A geometrically ergodic MC started in stationarity is α-mixing at a geometric rate:

$$\alpha(n) = \sup_{A,B\subset E} \left| P_\pi\{X_0 \in A, X_n \in B\} - \pi(A)\pi(B) \right| = \mathrm{O}(r^n).$$

Uniform ergodicity is equivalent to Doeblin's condition and to exponential ϕ-mixing:

$$\phi(n) = \sup_{A,B\subset E} \left| P_\pi\{X_n \in B | X_0 \in A\} - \pi(B) \right| = \mathrm{O}(\rho^n).$$

There are only a few results on general geometric or uniform convergence of MCMC samplers. We recall here that, in Liu et al. (1994), the Gibbs sampler is proved to converge geometrically under some regularity conditions and the rate of convergence is related to how the variables are correlated with each other, thus the suggestion of updating together (blocking) correlated components (see Section 5.3).

If an ergodic MC, P, with invariant distribution π and a real valued function f are such that one of the following conditions holds:

- the MC is geometrically ergodic and $\int |f|^{2+\varepsilon}\pi(x)\,\mathrm{d}x < \infty$, for some $\varepsilon > 0$;
- the MC is uniformly ergodic and $\int f^2\pi(x)\,\mathrm{d}x < \infty$;

then

$$0 < V(f, P) = \sigma_\pi^2(f) + 2\sum_{k=1}^{\infty} \mathrm{Cov}_\pi\big[f(X_0), f(X_k)\big] < \infty$$

and $\sqrt{n}(\hat{\mu}_n(f) - \mu_\pi(f))$ converges in distribution to a $N(0, V(f, P))$ random variable. The above theorem is stated and commented in Tierney (1996).

In the article "MC's for MCMC'ists", Nummelin (2002) gives the minimum amount of theory of Markov chains to prove the law of large numbers and the central limit theorem for standard MCMC algorithms.

In Roberts (2003) different statements of the central limit theorem for Markov chains are given in terms of drift, minorization conditions and small sets.

5.2. *The Metropolis–Hastings algorithm*

A very general recipe to construct a Markov chain stationary with respect to a specified distribution, π, is the Metropolis–Hastings algorithm (Metropolis et al., 1953; Hastings, 1970): given the current position of the MC, $\theta_t = \theta$, a move to θ' is proposed using the distribution $q(\theta, \theta')$, that may depend on the current position. Such move is accepted with probability:

$$\alpha(\theta, \theta') = \min\left\{1, \frac{\pi(\theta')q(\theta', \theta)}{\pi(\theta)q(\theta, \theta')}\right\}, \tag{3}$$

where we set $\alpha(\theta, \theta') = 1$ if $\pi(\theta) = 0$. If the move is rejected the current position is retained. Notice that, in the acceptance probability, the target distribution enters only as a ratio: $\pi(\theta')/\pi(\theta)$. This means that, the possibly unknown normalizing constant of the posterior, cancels and we can thus easily implement MCMC in a Bayesian setting as long as we can evaluate the product $l(x|\theta)\pi(\theta)$ for any given value of θ up to a constant of proportionality. It is important to stress that, if improper priors are used, the algorithm, as described above, is still well defined even if the posterior might be improper and we might not realize this by running our simulation. A good practice is thus to check that the posterior distribution is proper when improper priors are used. In Gelfand and Sahu (1999) rather general results are given for checking that the resulting posterior is proper for generalized linear models with improper priors. However, the same Authors also show that, if an MCMC sampler is run with an improper posterior, it may still be possible to use the output to obtain meaningful inference for certain model unknowns.

The MH algorithm ensures, not only that the resulting MC is stationary, but also *reversible* with respect to π that is:

$$\int_{(\theta,\theta')\in A\times B} \pi(\theta)q\big(\theta,\theta'\big)\alpha\big(\theta,\theta'\big)\,\mathrm{d}\theta\,\mathrm{d}\theta'$$
$$= \int_{(\theta,\theta')\in A\times B} \pi\big(\theta'\big)q\big(\theta',\theta\big)\alpha\big(\theta',\theta\big)\,\mathrm{d}\theta'\,\mathrm{d}\theta,$$

for all $A, B \subset E$. It is easy to verify that reversibility implies stationarity. Conditions that ensure irreducibility and aperiodicity (and hence ergodicity) of the Metropolis–Hastings sampler are given, for example, in Roberts and Smith (1994).

There are some special cases of the Metropolis–Hastings algorithm such as the Gibbs sampler (see Section 5.3); the *independence Metropolis–Hastings* where the proposal distribution does not depend on the current position of the Markov chain; the *random walk Metropolis–Hastings* where the prosed move is obtained by randomly perturbating the current position using a spherically symmetric distribution (Gaussian or uniform, for example): $\theta' = \theta + \varepsilon$ where $\varepsilon \sim q_\sigma$ and σ represents the "range" of the proposal exploration and is controlled by the user.

Geometric convergence and central limit theorems for general multidimensional Metropolis–Hastings algorithms are given in Roberts and Tweedie (1996b). General conditions for geometric ergodicity of the random walk sampler can be found in Mengersen and Tweedie (1996).

A suitable *initial distribution* is chosen to start the MCMC process and simulation proceeds until the stationary regime is reached. In a Bayesian setting the initial distribution can often coincide with the prior. This is a good choice especially if the number of observations do not make the likelihood overcome the prior influence in the posterior. In any case, if the chain is irreducible the choice of the starting values will not affect the stationary distribution, still a good starting point can help avoiding long burn-in phases, that is, time before the stationary regime is reached. On the other hand beginning your simulation from "difficult" starting point can help highlighting possible problems with your sampler. Thus, from a practical point of view, I believe one should not spent too much time looking for "good" starting points because, if the sampler is well designed, it will converge no matter what (you might have to wait a little longer to reach the stationary regime), on the other hand, if the sampler is not a good one, then you want to know about it as soon as possible and initializing it poorly can help detect failures.

Calibrating the *proposal distribution* is one of the main issues of MCMC simulation (the other one is detecting stationarity). In theory any proposal will work as long as it guaranties irreducibility that is, as long as, no matter where we are, we can reach any part of the state space if we wait long enough. This may sound like magic but the accept–reject mechanism is there to correct for bad proposal. To get a feeling of this, consider the simple case of a proposal distributions that is symmetric in the sense that $q(x, y) = q(y, x)$. In this setting the acceptance probability reduces to the minimum between one and the ratio of the target at the current position ($\pi(\theta)$, in the denominator) and at the proposed position ($\pi(\theta')$, in the numerator). Thus a proposed move with π value greater than the current one, will always be accepted. On the other hand, if $\pi(\theta')$ is small (bad proposal), the move is likely rejected. If the target is reasonably smooth, a proposal distribution with small range of exploration will likely suggest moves with π

value close to the current one and thus with high acceptance probability. But if the range of exploration is too small compared to the support of the target distribution and its variability, the sampler will be highly inefficient because it will take too long to explore the relevant state space (poor mixing). Furthermore the autocorrelation of the sampled path of the chain will likely increase thus increasing the variance of the resulting estimators (compare with formula (5) of the asymptotic variance of MCMC estimators). Large range of exploration is typically obtained by proposals with big variances that will allow to perform big jumps if the resulting move is accepted but have the opposite risk of proposing moves that are in the tail of the target distribution and thus have low chance of being accepted. Thus proper calibration of the proposal is crucial for the success of the resulting sampler. Off line tuning of the proposal via pilot runs is what is typically done. Still, this might not solve all problems since there might be portions of the state space where proposals with big variances are recommended while small variances are better elsewhere. Thus a single proposal might not be optimal for the whole state space and local calibration would be preferred. In these situation one could combine different proposals either by mixture, cycles (Tierney, 1994), or delayed rejection (Tierney and Mira, 1998).

As a general advice, search for proposals that, despite making big jumps, change the likelihood or the prior (or the posterior) by little so that the acceptance probability is likely to be high. This amounts to finding transformations that possibly leave the posterior invariant. These type of moves alone might not produce an irreducible MC and need to be combined with, say, random walk type moves. A general discussion of proposals that leave the target distribution invariant, as in the Gibbs sampler, is given by Liu and Sabbati (2000) where a generalized version of the Gibbs sampler is given.

5.3. The Gibbs sampler

The Gibbs sampler is a special case of the MH algorithm. Suppose θ is high-dimensional and we can partition its coordinates into subsets for which the full conditional distributions are available, that is $\theta = (\theta^1, \theta^2, \ldots, \theta^k)$ and the distribution of $\theta^i|\theta^{-i}$ (we indicate with $\theta^{-i} = \{\theta^j, j \neq i\}$) is known and we can simulate from it. Typically in the partition, single coordinates are considered separately and can be updated in a random or fixed scan. When a fixed scan is used the coordinates are updated sequentially in a predefined order and resulting sampler is not reversible. To retain reversibility a random scan can be adopted, that is, to decide which coordinate to update, a random number is generated between 1 and k and the corresponding coordinate is moved. There is no need for the random number to be uniformly distributed on the interval: coordinates that present higher variability can be updated more often that others as long as, eventually, all coordinates are moved. If we were to use such full conditional distributions as proposals (with, therefore, no need of tuning them) in a Metropolis–Hastings algorithm, then the acceptance probability would always be equal to one. This is an advantage, from a computational point of view, that is compensated by the fact the full conditionals are often not easy to obtain and sample in an i.i.d. fashion. In this case we could resort to a *Metropolis within Gibbs* algorithm in which the easy to sample full conditionals are used as proposals while the other ones are substituted with different proposals and the corresponding acceptance probability is computed. Even if all full conditionals are available, the Gibbs sampler may not be the most efficient MCMC algorithm especially if the coordinates are highly correlated in the posterior distribution since this causes very slow mixing of the sampler: the full conditionals force the sampler to make moves parallel to the main axis of the space and the legal moves might have very limited range due to high correlation. This possible problem can be identified by plotting the autocorrelation functions of, say $f(\theta^i) = \theta^i, \forall i$. If the autocorrelations remain significantly different from zero at high lags (compared to the length of the simulation) a good suggestion is to group, in the partition, coordinates that are highly correlated in the posterior and then update them together (Liu et al., 1994). Groups of correlated coordinates can be identified via pilot runs.

Conditions that ensure irreducibility and aperiodicity (and hence ergodicity) of the Gibbs sampler are given, for example, in Roberts and Smith (1994), while Roberts and Polson (1994) gives results on geometric convergence.

How powerful Gibbs sampling can be, when applied to univariate full conditional distributions, is demonstrated by the freely available software, WinBUGS, Bayesian Inference Using Gibbs Sampling (Spiegelhalter et al., 1995). Within this software the model is specified by mean of a directed acyclic graph and its conditional independence structure is exploited to determine the full conditional distributions that are then sampled by adaptive rejection Metropolis sampling, ARMS (Gilks et al., 1995) or by slice sampling (see Section 5.4).

A variety of Gibbs samplers for mixture models are illustrated in details in Chapter 16, this volume.

5.4. *Auxiliary variables in MCMC*

Auxiliary variables can be introduced in an MCMC simulation to obtain better mixing or faster convergence to stationarity of the process, or to make the implemention of the sampler easier or faster in terms of CPU time.

The idea of introducing auxiliary variables in MCMC sampling arose in statistical physics (Swendson and Wang, 1987; Edwards and Sokal, 1988) and was brought into the mainstream statistical literature by Besag and Green (1993).

Auxiliary variable techniques exploit the general principle that a complicated posterior distribution may become more tractable if it is embedded in a higher-dimensional state space. Once the higher-dimensional MCMC sample is obtained, it is projected back on the original state space by discarding the auxiliary variables.

The simple *slice sampler* is probably the most popular auxiliary variable MCMC method, where a single auxiliary variable is introduced, γ, and the joint distribution of θ and γ is specified by taking the marginal distribution of θ to be the target of interest and specifying the distribution of $\gamma|\theta$ to be uniform on the interval $[0, \pi(\theta)]$:

$$\pi(\gamma, \theta) = \pi(\theta)\pi(\gamma|\theta) = \pi(\theta)\frac{1}{\pi(\theta)}\mathbf{1}_{\{0 \leqslant \gamma \leqslant \pi(\theta)\}}(\gamma), \tag{4}$$

where $\mathbf{1}_A(\gamma)$ is the indicator function of the set A. The joint distribution over the enlarged state space is then explored via Gibbs Sampler. This amounts to update $\gamma|\theta$ by generating a uniform random variable over the interval specified above and then sample

$\theta|\gamma$ from a uniform distribution over the set $\{\theta\colon \pi(\theta) \geqslant \gamma\}$. While the first updating step is straightforward, the second one might be hard since it requires the identification of the level sets of the target distribution. For a clever way to perform this second update see Neal (2003) where an adaptive rejection mechanism is proposed to this aim.

Easy to check conditions for uniform and geometric ergodicity of the slice sampler are given in Mira and Tierney (2002) and Roberts and Rosenthal (1999) respectively. In Roberts and Rosenthal (2002) an interesting variation of the slice sampler, the polar slice sampler, is suggested and, under some circumstances, the convergence properties of the algorithm are proved to be essentially independent of the dimension of the problem. In the same paper, computable bounds to the total variation distance to stationarity for log-concave densities are given.

The simple slice sampler can be extended by introducing more than one auxiliary variable and this is of particular interest in a Bayesian context where the posterior distribution is proportional to the product of the prior and the likelihood. The latter is, in most applications, obtained as a product, $\prod_{j=1}^{N} l(X_j|\theta)$. In this setting N independent auxiliary variables can be introduced: $U_j \sim \text{Uniform}[0, l(X_j|\theta)]$ thus obtaining a joint target distribution which is again uniform on the region under the plot of the posterior density function. The clever introduction of multiple auxiliary variables in a variety of Bayesian nonconjugate and hierarchical models is illustrated in Damien et al. (1999).

In Mira and Tierney (2002) it is proved that an independence Metropolis–Hastings can be turned into a corresponding slice sampler which produces MCMC estimators with a smaller asymptotic variance. In other words the independence Metropolis–Hastings is dominated in the Peskun ordering (Peskun, 1973), by the corresponding slice sampler. One can also easily devise overrelaxed versions of slice sampling (Neal, 2003), which sometimes greatly improve efficiency by suppressing random walk behavior or by using 'reflective' methods, which bounce off the edges of the slice.

Furthermore, in Mira et al. (2001), the slice sampler is proved to be stochastically monotone with respect to the ordering given on the state space by the target: $X \prec Y$ if $\pi(X) \leqslant \pi(Y)$. This means that, if $X_t \prec Y_t$ then $X_{t+1} \prec Y_{t+1}$, when the update from time t to time $t+1$ is done using the same random bits for both chains. This is the condition that is needed to turn the slice sampler into a perfect simulation algorithm (see Section 6). Another version of perfect slice sampler for mixture models is given in Casella et al. (2002).

A different interesting way of enlarging the state space by introducing auxiliary variables is given by *hybrid Monte Carlo* algorithms (Duane et al., 1987; Neal, 1994), where a momentum vector (with the same dimension as the original variable of interest, θ) is sampled from a Gaussian distribution and then a deterministic time-reversible and volume-preserving algorithm is run on the enlarged state space for some steps to obtain the proposed move which is then either retained or rejected, based on some acceptance probability that preserves reversibility over the enlarged state space.

Finally, auxiliary variables are introduced in data augmentation and EM algorithms (Tanner and Wong, 1987; Wei and Tanner, 1990; Meng and van Dyk, 1997). Use of these algorithms for mixture models is illustrated in Chapter 16, this volume.

5.5. *Convergence diagnostics*

The issue of checking if the MC has reached its stationary regime from the path of the MC itself is an impossible task. All convergence diagnostics based on statistics evaluated on the output of the simulation can, at most, give warning signals and motivate longer runs but will never be able to guarantee that stationarity has been reached. Still, it is a good practice to perform some convergence diagnostics. Of course, since different diagnostics aim at detecting failure to convergence in different ways, the more you perform the safer you are (but recall, you will never be sure!). Software like CODA (Best et al., 1995), and BOA (http://www.public-health.uiowa.edu/boa) are available at this end. Good reviews of convergence diagnostic tools such as (Cowles and Carlin, 1996; Mengersen et al., 1999) are also recommended.

A first class of convergence diagnostic is based on the output of a single chain that is typically partitioned into smaller portions and statistics on these portions are computed and compared. A second class is based on statistics computed on more chains initialized at different starting points. The underlying idea is that the stationary regime is reached once the starting point is forgotten and since all chains are designed to converge to the same distribution, when they "settle down" to a common path it is likely that stationarity is achieved. The average of with-in chain variances and the variance between the means of some statistics are compared. This is the rationale behind Gelman and Rubin (1992), convergence diagnostic.

We concur with Cowles and Carlin (1996) and suggest the use of a combination of Gelman and Rubin (1992) and Geyer (1992) convergence diagnostics that usually provide an effective, yet simple, method for monitoring convergence.

5.6. *Estimating the variance of MCMC estimators*

Once the MCMC estimator is obtained it is a good practice to also provide a measure of its efficiency such as its asymptotic mean squared error. The latter coincides with the asymptotic variance thanks to the law of large numbers that guaranties asymptotic unbiasedness.

The limit, as n tends to infinity, of n times the variance of the estimator, $\hat{\mu}_n(f)$, obtained by running a stationary MC having transition kernel P will be indicated by $V(f, P)$. Assuming f is square integrable with respect to π, $f \in L^2(\pi)$, we have that

$$V(f, P) = \lim_{n\to\infty} n \operatorname{Var}_\pi\left[\hat{\mu}_n\right] = \sigma_\pi^2(f)\left\{1 + 2\sum_{k=1}^{\infty} \rho_k\right\}, \tag{5}$$

where $\rho_k = \operatorname{Cov}_\pi[f(X_0), f(X_k)]/\sigma_\pi^2(f)$.

The right-hand side of the last expression in (5) stresses that the asymptotic variance is made of two components, the first one is equal to the variance we would have if we were doing Monte Carlo simulation. The second part is the "price" we pay for not being able to get i.i.d. samples from the posterior. Although the second component is typically positive, we could in theory be very clever in constructing our MC and induce negative correlation along the sampled path of the chain.

The quantity:

$$\tau = 1 + 2\sum_{k=1}^{\infty} \rho_k \tag{6}$$

is known as the *integrated autocorrelation time* and can be interpreted as the number of correlated samples with the same variance-reducing power as one independent sample. The integrated autocorrelation time can be estimated using Sokal's (1998) adaptive truncated periodogram estimator $\hat{\tau} = \sum_{|k| \leqslant M} \hat{\rho}_k$, with the window width, M, chosen adaptively as the minimum integer with $M \geqslant 3\hat{\tau}$. A fair comparison among competing MCMC algorithms can be based on $\hat{\tau} \times T$ where T is the time needed to run the MC for a fixed number of sweeps.

There exist other ways of estimating $V(f, P)$ using batch means (Roberts, 1996) or regeneration times (Mykland et al., 1995).

The asymptotic variance of MCMC estimators can be reduced either post processing the chain, by subsampling or Rao–Blackwellization, or by carefully designing the sampler adopting, for example, the delaying rejection strategy (see Section 5.9).

When subsampling, typically one every $\hat{\tau}$ observations is retained to build our MCMC estimators on almost independent samples, thus reducing the covariance term in (5).

Rao–Blackwellization relies on the well known principle that conditioning reduces the variance. The idea is to replace $f(x_i)$ in $\hat{\mu}_n(f)$ by a conditional expectation, $E_\pi[f(x_i)|h(x_i)]$, for some function h or to condition on the previous value of the chain thus using $E[f(x_i)|x_{i-1} = x]$ instead (Gelfand and Smith, 1990; Liu et al., 1994; Casella and Robert, 1996; McKeague and Wefelmeyer, 2000).

5.7. Reversible jump MCMC

In many settings the dimension of the state space where the posterior distribution is defined is not a-priori known. This happens, for example, for mixture models with an unknown number of components; change point problems with an unknown number of change points; regression models with an unknown number of regressors and graphical models with an unknown number of nodes. In all these settings the number of parameters of interest is itself object of our inference. The Metropolis–Hastings algorithm can be reformulated to allow reversible jumps among state spaces of different dimensions so that the MC can freely move, as suggested by Green (1995). The acceptance probability given in (3) is then adjusted by means of the Jacobian of the transformation that takes as inputs the current position of the MC, say $\theta \in R^d$, together with some auxiliary random numbers and gives as output the proposed move, say $\theta' \in R^{d'}$. If, for example, $d' > d$, we need at least $d' - d$ auxiliary random variables, U, to make the transformation $\theta' = h(\theta, U)$ a diffeomorphism (the transformation and its inverse need to be differentiable). If more random bits are used to obtain the proposed move, say m, we need to balance the dimensions and construct the reverse path from θ' back to θ using m' random bits and take m' so that $d + m = d' + m'$. This simple strategy is called "matching dimensions" and guaranties that the Jacobian of the transformation involved is not zero and thus moves among different state spaces have a positive chance of being accepted. Still the acceptance probability can be quite small if the proposal distribution is not well calibrated. The main implementational issue in reversible jump MCMC is that there is commonly no natural way to calibrate the proposal since "there is no Euclidean structure in the parameter space to guide the choice" as pointed out by Brooks et al. (2003) where various strategies are suggested to circumvent this issue.

As a general suggestion, sometimes, even in a fixed dimensional setting, it is useful to specify the proposed candidate move indirectly as a deterministic function of some underlying random bits and the current position of the MC and compute the acceptance probability using the Jacobian of the resulting transformation. My feeling is that this formalism allows you to use your intuition more freely and to take advantage of your understanding of the problem structure. Of course there are various levels where you can take this reasoning. If, for example, you pick your proposal to be Gaussian centered at the current position you could express this, indirectly, in terms of standard Gaussian random bits or go further down, to a lower level, and use standard uniform random bits.

5.8. *Langevin algorithms*

If the gradient of the posterior distribution is available it can be used to set up a Langevin algorithm. The simplest form of this type of algorithms uses as proposal density:

$$q(\theta, \theta') = \frac{1}{(2\pi\sigma)^{d/2}} \exp\left\{\frac{-\|\theta' - \theta - \sigma^2 \nabla \log \pi(x)/2\|^2}{2\sigma^2}\right\} \tag{7}$$

for some suitable scaling parameter σ. The only difference with respect to a regular MH algorithm is that the additional information contained in the gradient is used to push the mean of the proposal towards high posterior regions which, intuitively, is a sensible thing to do. The only caveat is that care should be taken in avoiding overshooting that causes the MC to jump from one tail-region of the posterior to another. Theoretical study of convergence properties of Langevin algorithm is given in Roberts and Tweedie (1996a), Stramer and Tweedie (1999a, 1999b). Unlike in the physics literature where the original idea comes from, Langevin algorithms have had little success among statisticians. The reason, in my opinion, is related to the fact that the gradient of the target distribution is typically not available in closed form. Still, within an MCMC simulation, the gradient at any point can be successfully numerically evaluated and used to construct a better proposal distribution as suggested, for example, in (7).

5.9. *Adaptive MCMC and particle filters*

Recent research lines in MCMC aim at finding almost automatic algorithms that need little off-line pilot tuning of the proposal. To this scope a few adaptive MCMC strategies have been recently proposed in the literature. There are mainly two ways of learning from the past history of the simulation to better design and calibrate the proposal. One can either preserve the Markovian structure of the sampler and this is obtained by performing only partial adaption as in the delayed rejection strategy (Tierney and Mira, 1998; Green and Mira, 2001) or by adapting at regeneration times as in Gilks et al. (1998). Or one can adapt to the whole history thus obtaining a non-Markovian sampler but then has to prove ergodicity of the resulting algorithm from first principles as in Gilks et al. (1998), Haario et al. (1999, 2001), Holden (1998), Warnes (2001).

In the sequel we will focus on two adaptive strategies, a local and a global one. We will then indicate ways of combining them. The intuition behind the *delayed rejection strategy* is that, upon rejection of a candidate move, instead of advancing simulation time and retaining the same position, a second stage proposal can be used to generate a new candidate. The acceptance probability of this second stage move is computed to preserve reversibility with respect to the posterior (Mira, 2002). The rationale of this second stage proposal is twofold: by trying harder to move away from the current position we aid better exploration of the state space. This improves the mixing properties of the sampler as can be formally proved by showing that the resulting algorithm dominates in terms of asymptotic efficiency (that is relative to the Peskun, 1973, ordering) the corresponding estimators obtained from the basic Metropolis–Hastings sampler. The second reason that makes the DR strategy interesting is that the second stage proposal can be different from the first stage one and can depend, not only on the current position of the MC, as in a regular Metropolis–Hastings sampler, but also on the first stage proposal. This allows partial adaptation and local calibration of the proposal to the posterior.

In the *adaptive Metropolis* algorithm a Gaussian proposal distribution centered at the current position and with a covariance matrix calibrated using the whole sample path of the Markov chain is constructed and adapted as the simulation proceeds. After an initial nonadaptation period of length t_0, the covariance matrix is obtained as

$$C_t = s_d \operatorname{Cov}(X_0, \dots, X_{t-1}) + s_d \varepsilon I_d, \quad t > t_0, \tag{8}$$

where s_d is a parameter that depends only on the dimension d of the state space where π is defined, $\varepsilon > 0$ is a constant that we may choose very small and I_d denotes the d-dimensional identity matrix. In Haario et al. (2001) ergodicity of the resulting algorithm is proved if the target is bounded and supported on a bounded set.

Many important applications involve real-time data that are collected in a sequential manner. Examples include signal processing problems, target tracking, on-line medical monitoring, computer vision, financial applications, speech recognition and communications, among others. When new data becomes available the posterior distribution is updated and thus the target distribution of the MCMC simulation evolves as we analyze it. In this setting there is little time to run, until convergence, a new simulation between two successive data inputs. Still we can rely on the fact that the posterior distribution should not change much between two successive time intervals and thus, instead of running a totally new MCMC sampler, one could re-weight the current realized Markov chain path. Furthermore, instead of running a single chain we can work we a set of "particles" that approximate the target at any given time point.

More formally, suppose $\pi_t(\theta_t)$ is the posterior distribution at time t. *Particle filters* rely upon the idea that the target can be approximated by a set of N points $(\theta_{t,1}, \dots, \theta_{t,N})$ and associated weights $(w_{t,1}, \dots, w_{t,N})$. The weighted points are called particles, thus the name. This set of particles properly approximates π_t if, for any bounded function of the state space, f_t we have:

$$\lim_{N\to\infty} \frac{\sum_{i=1}^N w_{t,i} f_t(\theta_{t,i})}{\sum_{i=1}^N w_{t,i}} = E_{\pi_t} f_t(\theta_t).$$

A particle filter is any algorithm that takes the current set of particles

$$S_t = (\theta_{t,i}, w_{t,i}), \quad i = 1, \dots, N;$$

and updates it into a new set, S_{t+1}, that properly approximates π_{t+1}. Particle filters algorithms are constructed by combining steps that involve updating the weights, resampling with probabilities proportional to the weights to eliminate unrepresentative particles and changing the particle position by MCMC moves, among others. Typically a fixed number of particles, N, is maintained across the simulation but algorithms with a random number of particles have also been suggested (with the problem of preventing the number from either increasing or decreasing too much).

We refer the interested researcher to the web site of sequential Monte Carlo methods and particle filtering: http://awl-11.eng.cam.ac.uk/oldsite/smc/ and to the book by Doucet et al. (2001), for further reading on the topic.

To conclude we stress that particle filters can be helpful also in static contexts where standard MCMC algorithms show poor mixing. Furthermore, a given target distribution can artificially be turned into a dynamic one by splitting the available data into batches and artificially constructing a posterior distribution that evolves in time as suggested in Chopin (2002).

5.10. Importance sampling and population Monte Carlo

A direct competitor of MCMC is importance sampling. An importance distribution is selected, say g, and samples are generate from it. Since the quantity $\mu_f(\pi)$ can be rewritten as

$$\mu_f(\pi) = \int f(x)\frac{\pi(x)}{g(x)} g(x)\,\mathrm{d}x,$$

an asymptotically unbiased estimator of μ is

$$\hat{\mu}^2 = \frac{1}{n}\sum_{i=1}^n f(X_i)\frac{\pi(X_i)}{g(X_i)},$$

where X_i are i.i.d. from g. The quantities $w_i = \frac{\pi(X_i)}{g(X_i)}$ are the importance weights. The importance distribution, g, plays the same role as the proposal distribution, q, in the Metropolis–Hastings sampler and the weights, w, play a role similar to the one of the acceptance probabilities, α. If the ratio π/g is only known up to a constant of proportionality the estimator:

$$\hat{\mu}^3 = \frac{\sum_{i=1}^n f(X_i) w_i}{\sum_{i=1}^n w_i}$$

can be used instead. Notice that the same samples and, likewise, the same Markov chain path, can be re-used to estimate different expectations. The importance sampling estimator has finite variance only if

$$E_\pi\left[f^2(x)\frac{\pi(x)}{g(x)}\right] < \infty.$$

Thus, if g has tails lighter than π, i.e. if $\sup \pi/g < \infty$, the weights vary widely, giving too much importance to few X_i thus increasing the variance of the resulting estimator.

We point out that if a distribution is available such that

$$\sup \pi/g < \infty, \tag{9}$$

then, that distribution, could be used to set up a uniformly ergodic slice sampler. Alternatively, g can be an importance sampling distribution, a proposal in an independence Metropolis–Hastings sampler or, finally, can be adopted to perform rejection sampling. In Mengersen and Tweedie (1996), the condition given in (9) is showed to be a necessary and sufficient for uniform ergodicity of the independence Metropolis–Hastings sampler. But if (9) does not hold the resulting algorithm is not even geometrically ergodic. On the other hand, in Mira and Tierney (2002) it is shown that the slice sampler can still be geometrically ergodic in this setting. Liu (1996) compares the performance of the independence Metropolis–Hastings, rejection and importance sampling in terms of variance of the resulting estimators.

An interesting combination of importance sampling, adaptive Metropolis–Hastings algorithms and particle filters, gives rise to Population Monte Carlo. For a detailed description of Population Monte Carlo see Chapter 16, this volume.

6. The perfect Bayesian pie: How to avoid "burn-in" issues

Perfect simulation is a way of taking a Markov chain stationary with respect to the posterior and running it in an unconventional way in order to avoid the difficult task of assessing its convergence to the target. The "burn-in" issue is thus circumvented by turning a problem that is "infinite" on the simulation time axis into a problem that is "infinite" on the state space dimension. In other words, instead of ideally running a single MC for a theoretically infinite amount of time (until convergence), we ideally run an infinite number of MC (each one initialized in any of the possible starting point) until they coalesce. Unfortunately the coalescence time is not a stopping time, we thus need to run the MC from, say, time $-T$ until time zero. Furthermore we do this by using the same random numbers to update each one of the chains so that, when two paths meet, they proceed together. Since all Markov chains have the same stationary distribution it is possible to show that (Propp and Wilson, 1996) if they all couple, the common value they have at time zero, say X_0, is in stationarity. The intuition behind the proof of this result is the following: imagine a MC ideally started at time $-\infty$, updated using the transition kernel we have chosen and thus having the posterior of interest as its unique limiting distribution. By time $-T$ this MC will necessarily go trough one of the possible states. From time $-T$ on we could, in theory, update such MC using the random numbers that we have already used to update our processes and thus, the ideal MC, at time zero, will be in state X_0 and, since this MC has ideally run for an infinite amount of time, X_0 is a sample exactly from the posterior of interest. If the chains do not all couple in T steps we give them some more time by restarting our simulation say from time $-2T$, and so on, until complete coalescence, having the caution of reusing the same random numbers that we have already generated all the times we find ourselves having to simulate in a time interval that we have already visited.

If the structure of the model and thus of the posterior, allows to identify orderings on the state space that are preserved by the updating mechanism of the MC, the impossible task of running infinite MC is turned into the issue of identifying a minimal and a maximal state (or process, see Kendall and Møller, 2000) and suggesting ways of simulating the MC so that coupling of two or more of them is possible also in a continuous state space (see Murdoch and Green, 1998; Green and Murdoch, 1999; Mira et al., 2001). Once the minimal and the maximal chain have coupled we are guarantied, by a sandwiching mechanism, that all possible Markov chains started in any other point of the state space will be "trapped" between the two and thus will also have coalesced.

The way we presented perfect simulation makes it quite intuitive and corresponds to the original formulation given in the seminal paper by Propp and Wilson (1996). Since then many variations have been proposed that, for example, do not require the book-keeping of the random numbers nor the inversion of simulation time, see, for example, Wilson (2000). Also, interruptible perfect samplers have been proposed following the seminal paper by Fill (1998). These algorithms do not suffer of the bias induced by the impatient user who is tempted to interrupt the simulation if coalescence has not happened, say, by coffee break time (by that time we would like our Bayesian pie to be well cooked and not burned!).

Both the independence sampler (Corcoran and Tweedie, 2002) and the slice sampler (Mira et al., 2001) can be turned into perfect samplers under some regularity conditions that are the same conditions that guarantee uniform ergodicity. We do not go further into the details of perfect simulation because we believe that, although it has been a quite successful area of research and interesting perfect algorithms have emerged for certain classes of models, its general implementation is still difficult in problems that do not present any special structure that can be exploited to identify the ordering, the maximal and minimal state and the coupling mechanism. We refer the interested reader to the following web site: http://dimacs.rutgers.edu/~dbwilson/exact.html that contains numerous interesting examples, running Java-applets and an annotated bibliography.

7. Conclusions

To summarize the important aspects of MCMC that have emerged, let us try to identify the relative pros and contras. The contras mostly coincide with open questions, some of which arise every time we practically set up an MCMC algorithm. The pros are partial answers to the above mentioned open questions.

CONTRAS (open questions):

- How do we scale (set the range of exploration of) the proposal distribution?
- How long should the burn-in be?
- Writing and debugging the code can be very time consuming;
- Even apparently very simple algorithms posses quite subtle pitfalls that are difficult to detect empirically such as lack of geometric ergodicity (see Roberts, 2004).

PROS:

- Adaptive MCMC can help in scaling the proposal distribution;
- Free software for convergence diagnostics can help assessing the length of the burn-in (no certainty);
- Perfect simulation avoids the burn-in issues (not always available);
- MCMC does not suffer of the course of dimensionality like many other deterministic approximation methods.

From a research point of view we refer the reader to the following web site: http://www.statslab.cam.ac.uk/MCMC where preprint of many papers related to MCMC referring to both theoretical and applied aspects, are collected.

To conclude, we point out that there remains much to be done towards automation of MCMC simulations. Still a caveat is to blind use of MCMC without first trying to analytically integrate out some of the parameters, whenever possible, or checking that the posterior of interest is proper. Another area that would require more research is error assessment of MCMC estimators.

Finally, as mentioned in the introduction, MCMC is the method of choice to estimate many Bayesian models (both parametric and nonparametric) and often it is the only viable alternative. For a more detailed discussion of all the algorithms presented here and their connections with Bayesian statistics, we refer the reader to the book "Monte Carlo Methods in Bayesian Computation" (Chen et al., 2000).

References

Besag, J. (1974). Spatial interaction and the statistical analysis of lattice systems (with discussion). *J. Roy. Statist. Soc., Ser. B* **36**, 192–326.

Besag, J., Green, P.J. (1993). Spatial statistics and Bayesian computation (with discussion). *J. Roy. Statist. Soc., Ser. B* **55**, 25–38.

Best, N.G., Cowles, M.K., Vines, K. (1995). CODA: Convergence diagnosis and output analysis software for Gibbs sampling output, version 0.30. Technical Report, Univ. of Cambridge.

Brooks, S.P., Giudici, P., Roberts, G.O. (2003). Efficient construction of reversible jump Markov chain Monte Carlo proposal distributions (with discussion). *J. Roy. Statist. Soc., Ser. B* **65** (1), 3–55.

Casella, G., Robert, C.P. (1996). Rao–Blackwellisation of sampling schemes. *Biometrika* **83** (1), 81–94.

Casella, G., Mengersen, K.L., Robert, C.P., Titterington, D.M. (2002). Perfect slice samplers for mixtures of distributions. *J. Roy. Statist. Soc., Ser. B* **64** (4), 777–790.

Chen, M.M., Shao, Q., Ibrahim, J.G. (2000). *Monte Carlo Methods in Bayesian Computation*. Springer-Verlag, New York.

Chopin, N. (2002). A sequential particle filter method for static models. *Biometrika* **89**, 539–552.

Corcoran, J., Tweedie, R.T. (2002). Perfect sampling from independent Metropolis–Hastings chains. *J. Statist. Plann. Inference* **104** (2), 297–314.

Cowles, M.K., Carlin, B.P. (1996). Markov chain Monte Carlo convergence diagnostics: A comparative study. *J. Amer. Statist. Assoc.* **91**, 883–904.
Damien, P., Wakefield, J., Walker, S. (1999). Gibbs sampling for Bayesian non-conjugate and hierarchical models by using auxiliary variables. *J. Roy. Statist. Soc., Ser. B* **61** (2), 331–344.
Dempster, A.P. (1980). Bayesian inference in applied statistics. In: Bernardo, J.M., De Groot, M.H., Lindley, D.V., Smith, A.F.M. (Eds.), *Bayesian Statistics*. Valencia University Press, Valencia, Spain, pp. 266–291.
Doucet, A., de Freitas, N., Gordon, N. (2001). *Sequential Monte Carlo Methods in Practice*. Springer-Verlag.
Duane, S., Kennedy, A.D., Pendleton, B.J., Roweth, D. (1987). Hybrid Monte Carlo. *Phys. Lett. B* **195**, 216–222.
Edwards, R.G., Sokal, A.D. (1988). Generalization of the Fortium–Kasteleyn–Swendsen–Wang representation and Monte Carlo algorithm. *Phys. Rev. D* **38**, 2009–2012.
Fill, J.A. (1998). An interruptible algorithm for exact sampling via Markov chains. *Ann. Appl. Probab.* **8**, 131–162.
Gelfand, A.E., Sahu, S.K. (1999). Identifiability, improper priors, and Gibbs sampling for generalized linear models. *J. Amer. Statist. Assoc.* **94**, 247–253.
Gelfand, A.E., Smith, A.F.M. (1990). Sampling based approaches to calculating marginal densities. *J. Amer. Statist. Assoc.* **85**, 398–409.
Gelman, A., Rubin, D.B. (1992). Inference from iterative simulation using multiple sequences (with discussion). *Statist. Sci.* **7**, 457–511.
Geman, S., Geman, D. (1984). Stochastic relaxation, Gibbs distributions and the Bayesian restoration of images. *IEEE Trans. Pattern Anal. Mach. Intell.* **6**, 721–741.
Geyer, C.J. (1992). Practical Monte Carlo Markov chain (with discussion). *Statist. Sci.* **7**, 473–511.
Gilks, W.R., Best, N.G., Tan, K.K.C. (1995). Adaptive rejection Metropolis sampling within Gibbs sampling. *Appl. Statist. (Ser. C)* **44**, 455–472.
Gilks, W.R., Roberts, G.O., Sahu, S.K. (1998). Adaptive Markov chain Monte Carlo. *J. Amer. Statist. Assoc.* **93**, 1045–1054.
Green, P.J. (1995). Reversible jump MCMC computation and Bayesian model determination. *Biometrika* **82** (4), 711–732.
Green, P.J., Mira, A. (2001). Delayed rejection in reversible jump Metropolis–Hastings. *Biometrika* **88**, 1035–1053.
Green, P.J., Murdoch, D. (1999). Exact sampling for Bayesian inference: Towards general purpose algorithms. In: Berger, J.O., Bernardo, J.M., Dawid, A.P., Lindley, V.F., Smith, A.F.M. (Eds.), *Bayesian Statistics, vol. 6*. Oxford University Press, Oxford, pp. 302–321.
Haario, H., Saksman, E., Tamminen, J. (1999). Adaptive proposal distribution for random walk Metropolis algorithm. *Comput. Statist.* **14** (3), 375–395.
Haario, H., Saksman, E., Tamminen, J. (2001). An adaptive Metropolis algorithm. *Bernoulli* **7** (2), 223–242.
Hastings, W.K. (1970). Monte Carlo sampling methods using Markov chains and their application. *Biometrika* **57**, 97–109.
Holden, L. (1998). Adaptive chains. Technical Report, Norwegian Computing Center.
Kass, R.E., Raftery, A.E. (1995). Bayes factors. *J. Amer. Statist. Assoc.* **90**, 773–795.
Kendall, W.S., Møller, J. (2000). Perfect simulation using dominating processes on ordered spaces, with application to locally stable point processes. *Adv. Appl. Probab.* **32**, 844–865.
Liu, J.S. (1996). Metropolized independent sampling with comparisons to rejection sampling and importance sampling. *Statist. Comput.* **6**, 113–119.
Liu, J.S., Sabatti, C. (2000). Generalized Gibbs sample and multigrid Monte Carlo for Bayesian computation. *Biometrika* **87**, 353–369.
Liu, J.S., Wong, W.H., Kong, A. (1994). Covariance structure of the Gibbs sampler with applications to the comparisons of estimators and sampling schemes. *Biometrika* **81**, 27–40.
McKeague, I.W., Wefelmeyer, W. (2000). Markov chain Monte Carlo and Rao–Blackwellisation. *J. Statist. Plann. Inference* **85**, 171–182.
Meng, X.L., van Dyk, D. (1997). The EM algorithm – an old folk-song sung to a new tune (with discussion). *J. Roy. Statist. Soc., Ser. B* **59**, 511–568.
Mengersen, K.L., Robert, C.P., Guihenneuc-Jouyaux, C. (1999). MCMC convergence diagnostics: A "reviewww". In: Berger, J.O., Bernardo, J.M., Dawid, A.P., Lindley, V.F., Smith, A.F.M. (Eds.), *Bayesian Statistics, vol. 6*. Oxford University Press, Oxford, pp. 415–440.

Mengersen, K.L., Tweedie, R.L. (1996). Rates of convergence of the Hastings and Metropolis algorithms. *Ann. Statist.* **24**, 101–121.

Metropolis, N., Rosenbluth, A.W., Rosenbluth, M.N., Teller, A.H., Teller, E. (1953). Equations of state calculations by fast computing machines. *J. Chem. Phys.* **21**, 1087–1092.

Meyn, S.P., Tweedie, R.L. (1993). *Markov Chains and Stochastic Stability*. Springer-Verlag.

Mira, A. (2002). On Metropolis–Hastings algorithms with delayed rejection. *Metron* **59**, 231–241.

Mira, A., Tierney, L. (2002). Efficiency and convergence properties of slice samplers. *Scand. J. Statist.* **29** (1), 1–12.

Mira, A., Møller, J., Roberts, G.O. (2001). Perfect slice samplers. *J. Roy. Statist. Soc., Ser. B* **63**, 583–606.

Murdoch, D.J., Green, P.J. (1998). Exact sampling for a continuous state. *Scand. J. Statist.* **25** (3), 483–502.

Mykland, P., Tierney, L., Yu, B. (1995). Regeneration in Markov chain samplers. *J. Amer. Statist. Assoc.* **90**, 233–241.

Neal, R.M. (1994). An improved acceptance procedure for the hybrid Monte Carlo algorithm. *J. Comput. Phys.* **111**, 194–203.

Neal, R.M. (2003). Slice sampling (with discussion). *Ann. Statist.* **31**, 705–767.

Nummelin, E. (2002). MC's for MCMC'ist. *Internat. Statist. Rev.* **70**, 215–240.

Peskun, P.H. (1973). Optimum Monte Carlo sampling using Markov chains. *Biometrika* **60**, 607–612.

Press, S.I. (1989). *Bayesian Statistics: Principles, Models and Applications*. Wiley, New York.

Propp, J.G., Wilson, D.B. (1996). Exact sampling with coupled Markov chains and applications to statistical mechanics. *Random Structures Algorithms* **9**, 223–252.

Ripley, B.D. (1987). *Stochastic Simulation*. Wiley, New York.

Roberts, G.O. (1996). Markov chain concepts related to sampling algorithms. In: Gilks, W.R., Richardson, S.T., Spiegelhalter, D.J. (Eds.), *Markov Chain Monte Carlo in Practice*. Chapman and Hall, pp. 45–57.

Roberts, G.O. (2003). Linking theory and practice of MCMC. In: Green, P.J., Hjort, N.L., Richardson, S.T. (Eds.), *Highly Structured Stochastic Systems*. Oxford University Press, pp. 145–166.

Roberts, G.O. (2004). Linking theory and practice of MCMC. In: Green, P.J., Hjort, N.L., Richardson, S.T. (Eds.), *Highly Structured Stochastic Systems*. Oxford University Press, pp. 145–166.

Roberts, G.O., Polson, N. (1994). A note on the geometric convergence of the Gibbs sampler. *J. Roy. Statist. Soc., Ser. B* **56**, 377–384.

Roberts, G.O., Rosenthal, J.S. (1999). Convergence of slice sampler Markov chains. *J. Roy. Statist. Soc., Ser. B* **61**, 643–660.

Roberts, G.O., Rosenthal, J.S. (2002). The polar slice sampler. *Stochastic Models* **18**, 257–280.

Roberts, G.O., Smith, A.F.M. (1994). Simple conditions for convergence of the Gibbs sampler and Hastings–Metropolis algorithms. *Stochastic Process. Appl.* **49**, 207–216.

Roberts, G.O., Tweedie, R.L. (1996a). Exponential convergence of Langevin diffusions and their discrete approximations. *Bernoulli* **2**, 341–364.

Roberts, G.O., Tweedie, R.L. (1996b). Geometric convergence and central limit theorems for multidimensional Hastings and Metropolis algorithms. *Biometrika* **83**, 95–110.

Sokal, A.D. (1998). Monte Carlo methods in statistical mechanics: Foundations and new algorithms. *Cours de Troisième Cycle de la Physique en Suisse Romande*, Lausanne.

Spiegelhalter, D.J., Thomas, A., Best, N., Gilks, W.R. (1995). BUGS: Bayesian inference using Gibbs sampling. Technical Report, Medical Research Council Biostatistics Unit, Institute of Public Health, Cambridge University.

Stramer, O., Tweedie, R.L. (1999a). Langevin-type models i: Diffusions with given stationary distributions, and their discretizations. *Methodology and Computing in Applied Probability* **1**, 283–306.

Stramer, O., Tweedie, R.L. (1999b). Langevin-type models ii: Self-targeting candidates for Hastings–Metropolis algorithms. *Methodology and Computing in Applied Probability* **1**, 307–328.

Swendson, R.H., Wang, J.S. (1987). Nonuniversal critical dynamics in Monte Carlo simulations. *Phys. Rev. Lett.* **58**, 86–88.

Tanner, M., Wong, W. (1987). The calculation of posterior distributions by data augmentation. *J. Amer. Statist. Assoc.* **82**, 528–550.

Tierney, L. (1994). Markov chains for exploring posterior distributions (with discussion). *Ann. Statist.* **22**, 1701–1786.

Tierney, L. (1996). Introduction to general state-space Markov chain theory. In: Gilks, W.R., Richardson, S.T., Spiegelhalter, D.J. (Eds.), *Markov Chain Monte Carlo in Practice*. Chapman and Hall, pp. 59–74.

Tierney, L., Mira, A. (1998). Some adaptive Monte Carlo methods for Bayesian inference. *Statistics in Medicine* **18**, 2507–2515.

Warnes, G.R. (2001). The Normal kernel coupler: An adaptive Markov Chain Monte Carlo method for efficiently sampling from multi-modal distributions. Technical Report 395, University of Washington.

Wei, G.C.G., Tanner, M.A. (1990). A Monte Carlo implementation of the EM algorithm and the poor man's data augmentation algorithm. *J. Amer. Statist. Assoc.* **85**, 699–704.

Wilson, D. (2000). How to couple from the past using a read-once source of randomness. *Random Structures Algorithms* **16** (1), 85–113.

Handbook of Statistics, Vol. 25
ISSN: 0169-7161
© 2005 Elsevier B.V. All rights reserved.
DOI 10.1016/S0169-7161(05)25015-0

15

Bayesian Computation: From Posterior Densities to Bayes Factors, Marginal Likelihoods, and Posterior Model Probabilities

Ming-Hui Chen

1. Introduction

In Bayesian inference, a joint posterior distribution is available through the likelihood function and a prior distribution. One way to summarize a posterior distribution is to calculate and display marginal posterior densities because the marginal posterior densities provide complete information about parameters of interest. Posterior density estimation is one of the most important topics in Bayesian computation, and its usefulness and applications have not been fully explored in most of the Bayesian computational literature. In this chapter, we will summarize the current state of the art in the area of estimating marginal and/or full posterior densities and various applications of the posterior density estimation in computing Bayes factors, marginal likelihoods, and posterior model probabilities.

Section 2 provides a most updated overview on various Monte Carlo methods for computing marginal or full posterior densities, including the kernel density estimation, the conditional marginal density estimator (CMDE) of Gelfand et al. (1992), the importance weighted marginal density estimation (IWMDE) of Chen (1994), the Gibbs stopper approach (Yu and Tanner, 1999; Oh, 1999), and an approach based on the Metropolis–Hastings output (Chib and Jeliazkov, 2001; Mira and Nicholls, 2004). The properties of each method and potential connections among those methods are discussed in details. The generalized linear model (GLM) is introduced and implementational issues for computing marginal posterior densities for the GLM are addressed in Section 3. In the later section, posterior model probabilities via informative priors will be constructed for the GLM in the context of Bayesian variable selection.

Sections 4 to 6 are devoted to various applications of the marginal posterior density estimation in several well-known or recently developed Monte Carlo methods for computing Bayes factors, marginal likelihoods, and posterior model probabilities. The methods being discussed include the Savage–Dickey density ratio (Dickey, 1971) and its generalization (Verdinelli and Wasserman, 1995), the methods for computing marginal likelihoods (Chib, 1995; Chib and Jeliazkov, 2001; Yu and Tanner, 1999), the methods proposed by Chen et al. (1999), Ibrahim et al. (2000), and Chen et al. (2000a) for computing the analytically intractable prior and/or posterior model probabilities in the context of Bayesian variable selection. Finally, brief concluding remarks are provided in Section 7.

2. Posterior density estimation

2.1. Marginal posterior densities

Let $\boldsymbol{\theta}$ be a p-dimensional column vector of parameters. Assume that the joint posterior density, $\pi(\boldsymbol{\theta}|D)$, is of the form

$$\pi(\boldsymbol{\theta}|D) = \frac{L(\boldsymbol{\theta}|D)\pi(\boldsymbol{\theta})}{m(D)}, \tag{2.1}$$

where D denotes *data*, $L(\boldsymbol{\theta}|D)$ is the likelihood function given data D, $\pi(\boldsymbol{\theta})$ is the prior, and $m(D)$ is the unknown normalizing constant.

Let Ω denote the support of the joint posterior density $\pi(\boldsymbol{\theta}|D)$. Also let $\boldsymbol{\theta}^{(j)}$ denote a $p_j \times 1$ subvector of $\boldsymbol{\theta}$ and let $\boldsymbol{\theta}^{(-j)}$ be $\boldsymbol{\theta}$ with $\boldsymbol{\theta}^{(j)}$ deleted. Thus, $\boldsymbol{\theta}^{(-j)}$ is a $(p - p_j) \times 1$ vector. Write $\boldsymbol{\theta} = ((\boldsymbol{\theta}^{(j)})', (\boldsymbol{\theta}^{(-j)})')'$. Then the marginal posterior density of $\boldsymbol{\theta}^{(j)}$ evaluated at $\boldsymbol{\theta}^{*(j)}$ has the form

$$\pi\left(\boldsymbol{\theta}^{*(j)}|D\right) = \int_{\Omega_{-j}(\boldsymbol{\theta}^{*(j)})} \pi\left(\boldsymbol{\theta}^{*(j)}, \boldsymbol{\theta}^{(-j)}|D\right) \mathrm{d}\boldsymbol{\theta}^{(-j)}, \tag{2.2}$$

where

$$\Omega_{-j}\left(\boldsymbol{\theta}^{*(j)}\right) = \left\{\boldsymbol{\theta}^{(-j)}\colon \left(\left(\boldsymbol{\theta}^{*(j)}\right)', \left(\boldsymbol{\theta}^{(-j)}\right)'\right)' \in \Omega\right\}.$$

There are several special features in Bayesian marginal density computation. First, the joint posterior density is known up to an unknown normalizing constant, namely, $m(D)$. Second, the integral given by (2.2) is analytically intractable for most Bayesian models. Third, we can generate a Markov chain Monte Carlo (MCMC) sample from $\pi(\boldsymbol{\theta}|D)$, which does not require knowing the unknown normalizing constant $m(D)$. Fourth, it is relatively inexpensive to obtain an MCMC sample with a large size, since the MCMC sample is drawn on a computer and not obtained from a real costly probability experiment. Due to these unique features, computing Bayesian marginal densities is very different than the traditional density estimation, in which a data set from some unknown population has already been obtained and the sample size is usually small.

Now, suppose $\{\boldsymbol{\theta}_i, i = 1, 2, \ldots, n\}$ is an MCMC sample from $\pi(\boldsymbol{\theta}|D)$, which may be obtained via a very popular Bayesian software `WinBUGS`. Then, a traditional nonparametric method, such as the kernel density estimation, can be directly used for computing the marginal density $\pi(\boldsymbol{\theta}^{*(j)}|D)$. Although the kernel estimation is very easy to implement, it may not be efficient, since it does not use the known structure of a posterior density. Alternatively, a number of authors (e.g., Gelfand et al., 1992; Johnson, 1992; Chen, 1994; Chen and Shao, 1997a; Chib, 1995; Verdinelli and Wasserman, 1995; Yu and Tanner, 1999; Chib and Jeliazkov, 2001; Mira and Nicholls, 2004) have proposed several efficient parametric full or marginal posterior density estimators based on the MCMC sample. We will briefly discuss some of those methods in the subsequent subsections.

2.2. *Kernel methods*

A widely used nonparametric density estimator is the kernel density estimator, which has the form

$$\hat{\pi}_{\text{kernel}}\big(\boldsymbol{\theta}^{*(j)}|D\big) = \frac{1}{nh_n^j}\sum_{i=1}^{n}\mathcal{K}\bigg(\frac{\boldsymbol{\theta}^{*(j)} - \boldsymbol{\theta}_i^{(j)}}{h_n}\bigg), \tag{2.3}$$

where $\boldsymbol{\theta}_i = (\boldsymbol{\theta}_i^{(j)}, \boldsymbol{\theta}_i^{(-j)})$, the kernel $\mathcal{K}$ is a bounded density on a j-dimensional Euclidean space R^j, and h_n is the bandwidth.

Assume that $j = 1$ and $\{\boldsymbol{\theta}_i, i = 1, 2, \dots, n\}$ is a random sample. Then, if $\pi(\boldsymbol{\theta}^{(j)}|D)$ is uniformly continuous on R and as $n \to \infty$, $h_n \to 0$, and $nh_n(\log n)^{-1} \to \infty$, we have $\lim_{n\to\infty}\hat{\pi}_{\text{kernel}}(\boldsymbol{\theta}^{*(j)}|D) \overset{\text{a.s.}}{=} \pi(\boldsymbol{\theta}^{*(j)}|D)$ (see Silverman, 1986, p. 72). Under slightly stronger regularity conditions, similar consistent results can be obtained for $j > 1$ (see Devroye and Wagner, 1980). However, the consistency of the kernel estimator is not so clear when $\{\boldsymbol{\theta}_i, i = 1, 2, \dots, n\}$ is an MCMC sample.

Silverman (1986) provides a detailed discussion on choosing the smoothing parameter. For example, when $j = 1$, if a Gaussian kernel, i.e., $\mathcal{K}(t) = (1/\sqrt{2\pi})\,\mathrm{e}^{-t^2/2}$, is used, the optimal choice for h_n is $1.06\sigma^* n^{-1/5}$, where σ^* is the sample standard deviation of the $\boldsymbol{\theta}_i^{(j)}$'s, provided that the $\boldsymbol{\theta}_i^{(j)}$'s are independent.

Although there is a consistency issue with the kernel estimator when a dependent MCMC sample is used, the kernel method does not require knowing the specific form of the distribution from which the sample is generated. Thus, it can be served as a "black-box" density estimator. It may be for this reason that the popular Bayesian software, `WinBUGS`, uses the kernel method for computing and then displaying univariate marginal posterior densities. However, caution must be taken in using the kernel method if the density value at a given point $\boldsymbol{\theta}^{*(j)}$ is of primary interest, since the kernel estimator may not be efficient. We elaborate this point more in the subsequent subsections.

2.3. *Conditional marginal density estimation*

Assume that the analytical evaluation of the conditional posterior density, $\pi(\boldsymbol{\theta}^{(j)}|\boldsymbol{\theta}^{(-j)}, D)$, is available in closed form. It is easy to show that (2.2) can be rewritten as

$$\pi\big(\boldsymbol{\theta}^{*(j)}|D\big) = \int_{\Omega} \pi\big(\boldsymbol{\theta}^{*(j)}|\boldsymbol{\theta}^{(-j)}, D\big)\pi(\boldsymbol{\theta}|D)\,\mathrm{d}\boldsymbol{\theta}. \tag{2.4}$$

Using the above identity, Gelfand et al. (1992) propose the conditional marginal density estimator (CMDE) of $\pi(\boldsymbol{\theta}^{*(j)}|D)$, which has the form

$$\hat{\pi}_{\text{CMDE}}\big(\boldsymbol{\theta}^{*(j)}|D\big) = \frac{1}{n}\sum_{i=1}^{n}\pi\big(\boldsymbol{\theta}^{*(j)}|\boldsymbol{\theta}_i^{(-j)}, D\big), \tag{2.5}$$

where $\{\boldsymbol{\theta}_i = (\boldsymbol{\theta}_i^{(j)}, \boldsymbol{\theta}_i^{(-j)}),\ i = 1, 2, \ldots, n\}$ is an MCMC sample from the joint posterior distribution $\pi(\boldsymbol{\theta}|D)$. It can be shown that under some minor regularity conditions, $\hat{\pi}_{\mathrm{CMDE}}(\boldsymbol{\theta}^{*(j)}|D)$ is an unbiased and consistent estimator of $\pi(\boldsymbol{\theta}^{*(j)}|D)$, i.e.,

$$E\big(\hat{\pi}_{\mathrm{CMDE}}\big(\boldsymbol{\theta}^{*(j)}|D\big)\big) = \pi\big(\boldsymbol{\theta}^{*(j)}|D\big),$$

where the expectation is taken with respect to the joint posterior distribution $\pi(\boldsymbol{\theta}|D)$, and

$$\lim_{n\to\infty} \hat{\pi}_{\mathrm{CMDE}}\big(\boldsymbol{\theta}^{*(j)}|D\big) \overset{\text{a.s.}}{=} \pi\big(\boldsymbol{\theta}^{*(j)}|D\big).$$

When a closed-form expression of $\pi(\boldsymbol{\theta}^{*(j)}|\boldsymbol{\theta}_i^{(-j)}, D)$ is available, the CMDE method is most efficient since it is a Rao–Blackwellized estimator. As mentioned in Gelfand et al. (1992), the CMDE is better than the kernel density estimator under a wide range of loss functions.

2.4. *Importance weighted marginal density estimation*

Although the CMDE method is simple and efficient, it requires knowing the closed form of the conditional posterior density. Unfortunately, for many Bayesian models, especially when the parameter space is constrained, the conditional posterior densities are known up only to a normalizing constant. To overcome this difficulty, Chen (1994) proposes an importance weighted marginal density estimation (IWMDE) method, which is a natural extension of the CMDE.

Consider the following identity:

$$\pi\big(\boldsymbol{\theta}^{*(j)}|D\big) = \int_{\Omega} \frac{w(\boldsymbol{\theta}^{(j)}|\boldsymbol{\theta}^{(-j)})L(\boldsymbol{\theta}^{*(j)}, \boldsymbol{\theta}^{(-j)}|D)\pi(\boldsymbol{\theta}^{*(j)}, \boldsymbol{\theta}^{(-j)})}{L(\boldsymbol{\theta}^{(j)}, \boldsymbol{\theta}^{(-j)}|D)\pi(\boldsymbol{\theta}^{(j)}, \boldsymbol{\theta}^{(-j)})} \pi(\boldsymbol{\theta}|D)\, \mathrm{d}\boldsymbol{\theta}, \tag{2.6}$$

where $w(\boldsymbol{\theta}^{(j)}|\boldsymbol{\theta}^{(-j)})$ is a completely known conditional density whose support is contained in, or equal to, the support, $\Omega_j(\boldsymbol{\theta}^{(-j)})$, of the conditional density $\pi(\boldsymbol{\theta}^{(j)}|\boldsymbol{\theta}^{(-j)}, D)$. Here, "*completely* known" means that $w(\boldsymbol{\theta}^{(j)}|\boldsymbol{\theta}^{(-j)})$ can be evaluated at any point of $(\boldsymbol{\theta}^{(j)}, \boldsymbol{\theta}^{(-j)})$. In other words, the kernel *and* the normalizing constant of this conditional density are available in closed form. The identity (2.6) can be derived as follows:

$$\begin{aligned}
&\pi\big(\boldsymbol{\theta}^{*(j)}|D\big)\\
&= \int \pi\big(\boldsymbol{\theta}^{*(j)}, \boldsymbol{\theta}^{(-j)}|D\big)\, \mathrm{d}\boldsymbol{\theta}^{(-j)}\\
&= \int \frac{L(\boldsymbol{\theta}^{*(j)}, \boldsymbol{\theta}^{(-j)}|D)\pi(\boldsymbol{\theta}^{*(j)}, \boldsymbol{\theta}^{(-j)})}{m(D)}\, \mathrm{d}\boldsymbol{\theta}^{(-j)}\\
&= \iint w\big(\boldsymbol{\theta}^{(j)}|\boldsymbol{\theta}^{(-j)}\big) \frac{L(\boldsymbol{\theta}^{*(j)}, \boldsymbol{\theta}^{(-j)}|D)\pi(\boldsymbol{\theta}^{*(j)}, \boldsymbol{\theta}^{(-j)})}{m(D)}\, \mathrm{d}\boldsymbol{\theta}^{(j)}\, \mathrm{d}\boldsymbol{\theta}^{(-j)}\\
&= \int w\big(\boldsymbol{\theta}^{(j)}|\boldsymbol{\theta}^{(-j)}\big) \frac{L(\boldsymbol{\theta}^{*(j)}, \boldsymbol{\theta}^{(-j)}|D)\pi(\boldsymbol{\theta}^{*(j)}, \boldsymbol{\theta}^{(-j)})}{L(\boldsymbol{\theta}^{(j)}, \boldsymbol{\theta}^{(-j)}|D)\pi(\boldsymbol{\theta}^{(j)}, \boldsymbol{\theta}^{(-j)})} \pi(\boldsymbol{\theta}|D)\, \mathrm{d}\boldsymbol{\theta}.
\end{aligned}$$

Using the identity (2.6), the IWMDE of $\pi(\boldsymbol{\theta}^{*(j)}|D)$ is defined as

$$\hat{\pi}_{\mathrm{IWMDE}}\big(\boldsymbol{\theta}^{*(j)}|D\big) = \frac{1}{n}\sum_{i=1}^{n} w\big(\boldsymbol{\theta}_i^{(j)}|\boldsymbol{\theta}_i^{(-j)}\big)\frac{L(\boldsymbol{\theta}^{*(j)},\boldsymbol{\theta}_i^{(-j)}|D)\pi(\boldsymbol{\theta}^{*(j)},\boldsymbol{\theta}_i^{(-j)})}{L(\boldsymbol{\theta}_i|D)\pi(\boldsymbol{\theta}_i)}. \tag{2.7}$$

Here, w plays the role of a weight function.

The IWMDE has several nice properties:

(i) $\hat{\pi}_{\mathrm{IWMDE}}(\boldsymbol{\theta}^{*(j)}|D)$ is always unbiased.

(ii) Under the ergodicity condition, $\hat{\pi}_{\mathrm{IWMDE}}(\boldsymbol{\theta}^{*(j)}|D)$ is consistent.

(iii) Under the uniform ergodicity and the finite second posterior moment of

$$w\big(\boldsymbol{\theta}^{*(j)}|\boldsymbol{\theta}^{(-j)}\big)\frac{L(\boldsymbol{\theta}^{*(j)},\boldsymbol{\theta}^{(-j)}|D)\pi(\boldsymbol{\theta}^{*(j)},\boldsymbol{\theta}^{(-j)})}{L(\boldsymbol{\theta}|D)\pi(\boldsymbol{\theta})},$$

the central limit theorem holds (see, for example, Verdinelli and Wasserman, 1995).

(iv) The IWMDE is a generalization of the CMDE.

If we choose

$$w = w\big(\boldsymbol{\theta}^{(j)}|\boldsymbol{\theta}^{(-j)}\big) = \pi\big(\boldsymbol{\theta}^{(j)}|\boldsymbol{\theta}^{(-j)}, D\big),$$

then the IWMDE reduces to

$$\begin{aligned}
&\hat{\pi}_{\mathrm{IWMDE}}\big(\boldsymbol{\theta}^{*(j)}|D\big)\\
&\quad= \frac{1}{n}\sum_{i=1}^{n} w\big(\boldsymbol{\theta}_i^{(j)}|\boldsymbol{\theta}_i^{(-j)}\big)\frac{L(\boldsymbol{\theta}^{*(j)},\boldsymbol{\theta}_i^{(-j)}|D)\pi(\boldsymbol{\theta}^{*(j)},\boldsymbol{\theta}_i^{(-j)})}{L(\boldsymbol{\theta}_i|D)\pi(\boldsymbol{\theta}_i)}\\
&\quad= \frac{1}{n}\sum_{i=1}^{n} \frac{\pi(\boldsymbol{\theta}^{(j)},\boldsymbol{\theta}^{(-j)}|D)}{\pi(\boldsymbol{\theta}^{(-j)}|D)}\frac{L(\boldsymbol{\theta}^{*(j)},\boldsymbol{\theta}_i^{(-j)}|D)\pi(\boldsymbol{\theta}^{*(j)},\boldsymbol{\theta}_i^{(-j)})}{L(\boldsymbol{\theta}_i|D)\pi(\boldsymbol{\theta}_i)}\\
&\quad= \frac{1}{n}\sum_{i=1}^{n} \frac{1}{\pi(\boldsymbol{\theta}^{(-j)}|D)}\frac{L(\boldsymbol{\theta}^{*(j)},\boldsymbol{\theta}_i^{(-j)}|D)\pi(\boldsymbol{\theta}^{*(j)},\boldsymbol{\theta}_i^{(-j)})}{m(D)}\\
&\quad= \frac{1}{n}\sum_{i=1}^{n} \frac{\pi(\boldsymbol{\theta}^{*(j)},\boldsymbol{\theta}_i^{(-j)}|D)}{\pi(\boldsymbol{\theta}^{(-j)}|D)} = \frac{1}{n}\sum_{i=1}^{n}\pi\big(\boldsymbol{\theta}^{*(j)}|\boldsymbol{\theta}_i^{(-j)}|D\big),
\end{aligned}$$

which is $\hat{\pi}_{\mathrm{CMDE}}(\boldsymbol{\theta}^{*(j)}|D)$.

(v) The IWMDE is better than the kernel method in terms of the Kullback–Leibler divergence (Chen and Shao, 1997a).

(vi) The CMDE is the best among all IWMDEs.

Let

$$\Pi_{i,j}\big(\boldsymbol{\theta}^{*(j)}|D\big) = w\big(\boldsymbol{\theta}_i^{(j)}|\boldsymbol{\theta}_i^{(-j)}\big)\frac{L(\boldsymbol{\theta}^{*(j)},\boldsymbol{\theta}_i^{(-j)}|D)\pi(\boldsymbol{\theta}^{*(j)},\boldsymbol{\theta}_i^{(-j)})}{L(\boldsymbol{\theta}_i|D)\pi(\boldsymbol{\theta}_i)}$$

and

$$V_w = \mathrm{Var}\big(\Pi_{i,j}\big(\boldsymbol{\theta}^{*(j)}|D\big)\big).$$

Then, it can be shown that

$$V_w \leqslant V_{w_{\text{opt}}},$$

where

$$w_{\text{opt}} = w\big(\boldsymbol{\theta}^{(j)}|\boldsymbol{\theta}^{(-j)}\big). \tag{2.8}$$

Note that unlike the IWMDE $\hat{\pi}_{\text{IWMDE}}(\boldsymbol{\theta}^{*(j)}|D)$, the kernel estimator $\hat{\pi}_{\text{kernel}}(\boldsymbol{\theta}^{*(j)}|D)$ is not unbiased.

2.5. *The Gibbs stopper approach*

The Gibbs stopper is originally proposed to assess the distributional convergence of draws from the Gibbs sampler by Ritter and Tanner (1992). It can also be applied to calculation of the posterior density.

Let $\{\boldsymbol{\theta}_i = (\theta_{1i}, \theta_{2i}, \ldots, \theta_{pi})', i = 1, 2, \ldots, n\}$ denote a realization of a Markov chain with transition probability from $\boldsymbol{\theta}$ to $\boldsymbol{\theta}^*$ as

$$\begin{aligned} K\big(\boldsymbol{\theta}, \boldsymbol{\theta}^*\big) = {} & \pi\big(\theta_1^*|\theta_2, \ldots, \theta_p, D\big)\pi\big(\theta_2^*|\theta_1^*, \theta_3, \ldots, \theta_p, D\big) \\ & \cdots \pi\big(\theta_p^*|\theta_1^*, \ldots, \theta_{p-1}^*, D\big), \end{aligned} \tag{2.9}$$

where $\pi(\theta_j|\theta_1, \ldots, \theta_{j-1}, \theta_{j+1}, \ldots, \theta_p, D)$ is the conditional posterior density of θ_j given $\theta_1, \ldots, \theta_{j-1}, \theta_{j+1}, \ldots, \theta_p$ and D. Note that $K(\boldsymbol{\theta}, \boldsymbol{\theta}^*)$ given in (2.9) is a typical transition kernel resulting from the Gibbs sampling algorithm. After the chain has converged, we have

$$\pi\big(\boldsymbol{\theta}^*|D\big) = \int K\big(\boldsymbol{\theta}, \boldsymbol{\theta}^*\big)\pi(\boldsymbol{\theta}|D)\, \mathrm{d}\boldsymbol{\theta}.$$

This identity directly leads to the Gibbs stopper estimator (Yu and Tanner, 1999) of $\pi(\boldsymbol{\theta}^*|D)$ given by

$$\hat{\pi}_{\text{GS}}\big(\boldsymbol{\theta}^*|D\big) = \frac{1}{n}\sum_{i=1}^{n} K\big(\boldsymbol{\theta}_i, \boldsymbol{\theta}^*\big). \tag{2.10}$$

The Gibbs stopper estimator has several attractive features:

(i) After the chain has converged, $\hat{\pi}_{\text{GS}}(\boldsymbol{\theta}^*|D)$ is unbiased and consistent.
(ii) If the one-dimensional conditional posterior distribution $\pi(\theta_j|\theta_1, \ldots, \theta_{j-1}, \theta_{j+1}, \ldots, \theta_p, D)$ can be either analytically available or easily computed numerically, the computation of $\hat{\pi}_{\text{GS}}(\boldsymbol{\theta}^*|D)$ is straightforward.
(iii) $\hat{\pi}_{\text{GS}}(\boldsymbol{\theta}^*|D)$ estimates the joint posterior $\pi(\boldsymbol{\theta}|D)$ at a fixed point $\boldsymbol{\theta}^*$ and it does not require knowing the normalizing constant $m(D)$.

Oh (1999) proposes the following approach to estimate the marginal posterior density $\pi(\boldsymbol{\theta}^{*(j)}|D)$. Write $\boldsymbol{\theta}^{*(j)} = ((\boldsymbol{\theta}_1^*)', (\boldsymbol{\theta}_2^*)')'$. Then, an estimator of $\pi(\boldsymbol{\theta}^{*(j)}|D)$ proposed by Oh (1999) is given by

$$\hat{\pi}_{\text{Oh}}\big(\boldsymbol{\theta}^{*(j)}|D\big) = \frac{1}{n}\sum_{i=1}^{n} \pi\big(\boldsymbol{\theta}_1^*|\boldsymbol{\theta}_2^*, \boldsymbol{\theta}_i^{(-j)}, D\big)\pi\big(\boldsymbol{\theta}_2^*|\boldsymbol{\theta}_{1,i}, \boldsymbol{\theta}_i^{(-j)}, D\big), \tag{2.11}$$

where $\{\boldsymbol{\theta}_i,\ i = 1, 2, \ldots, n\}$ is an MCMC sample from $\pi(\boldsymbol{\theta}|D)$. The motivation of the Oh's approach is clear, since the dimensions of $\boldsymbol{\theta}_1^*$ and $\boldsymbol{\theta}_2^*$ are smaller than $\boldsymbol{\theta}^{*(j)}$ and every term in (2.11) is written in terms of the complete conditional posterior densities. It is easy to show that $\hat{\pi}_{\mathrm{Oh}}(\boldsymbol{\theta}^{*(j)}|D)$ is unbiased and consistent if $\pi(\boldsymbol{\theta}_1^*|\boldsymbol{\theta}_2^*, \boldsymbol{\theta}_i^{(-j)}, D)$ and $\pi(\boldsymbol{\theta}_2^*|\boldsymbol{\theta}_{1,i}, \boldsymbol{\theta}_i^{(-j)}, D)$ are analytically available or they can be computed numerically. Clearly, the construction of $\hat{\pi}_{\mathrm{Oh}}(\boldsymbol{\theta}^{*(j)}|D)$ is closely related to the Gibbs stopper. However, the sample $\{\boldsymbol{\theta}_i,\ i = 1, 2, \ldots, n\}$ used in $\hat{\pi}_{\mathrm{Oh}}(\boldsymbol{\theta}^{*(j)}|D)$ can be obtained by any MCMC sampling algorithm, which may not necessarily be the Gibbs sampling algorithm.

We note that $\hat{\pi}_{\mathrm{Oh}}(\boldsymbol{\theta}^{*(j)}|D)$ is a special case of IWMDE. To see this, we take

$$w_{\mathrm{Oh}} = w\big(\boldsymbol{\theta}_1, \boldsymbol{\theta}_2|\boldsymbol{\theta}^{(-j)}\big) = \pi\big(\boldsymbol{\theta}_2|\boldsymbol{\theta}_1, \boldsymbol{\theta}^{(-j)}, D\big)\pi\big(\boldsymbol{\theta}_1|\boldsymbol{\theta}_2^*, \boldsymbol{\theta}^{(-j)}, D\big).$$

Then, from IWMDE, we have

$$\begin{aligned}
w\big(\boldsymbol{\theta}^{(j)}|\boldsymbol{\theta}^{(-j)}\big)&\frac{L(\boldsymbol{\theta}_1^*, \boldsymbol{\theta}_2^*, \boldsymbol{\theta}^{(-j)}|D)\pi(\boldsymbol{\theta}_1^*, \boldsymbol{\theta}_2^*, \boldsymbol{\theta}^{(-j)})}{L(\boldsymbol{\theta}_1, \boldsymbol{\theta}_2, \boldsymbol{\theta}^{(-j)}|D)\pi(\boldsymbol{\theta}_1, \boldsymbol{\theta}_2, \boldsymbol{\theta}^{(-j)})}\\
&= \frac{\pi(\boldsymbol{\theta}_1, \boldsymbol{\theta}_2, \boldsymbol{\theta}^{(-j)}|D)}{\pi(\boldsymbol{\theta}_1, \boldsymbol{\theta}^{(-j)}|D)} \times \frac{\pi(\boldsymbol{\theta}_1, \boldsymbol{\theta}_2^*, \boldsymbol{\theta}^{(-j)}|D)}{\pi(\boldsymbol{\theta}_2^*, \boldsymbol{\theta}^{(-j)}|D)}\\
&\quad \times \frac{L(\boldsymbol{\theta}_1^*, \boldsymbol{\theta}_2^*, \boldsymbol{\theta}^{(-j)}|D)\pi(\boldsymbol{\theta}_1^*, \boldsymbol{\theta}_2^*, \boldsymbol{\theta}^{(-j)})}{L(\boldsymbol{\theta}_1, \boldsymbol{\theta}_2, \boldsymbol{\theta}^{(-j)}|D)\pi(\boldsymbol{\theta}_1, \boldsymbol{\theta}_2, \boldsymbol{\theta}^{(-j)})}\\
&= \frac{\pi(\boldsymbol{\theta}_1, \boldsymbol{\theta}_2^*, \boldsymbol{\theta}^{(-j)}|D)\pi(\boldsymbol{\theta}_1^*, \boldsymbol{\theta}_2^*, \boldsymbol{\theta}^{(-j)}|D)}{\pi(\boldsymbol{\theta}_1, \boldsymbol{\theta}^{(-j)}|D)\pi(\boldsymbol{\theta}_2^*, \boldsymbol{\theta}^{(-j)}|D)}\\
&= \pi\big(\boldsymbol{\theta}_1^*|\boldsymbol{\theta}_2^*, \boldsymbol{\theta}^{(-j)}|D\big)\pi\big(\boldsymbol{\theta}_2^*|\boldsymbol{\theta}_1, \boldsymbol{\theta}^{(-j)}, D\big).
\end{aligned}$$

Therefore, with $w = w_{\mathrm{Oh}}$, the IWMDE automatically reduces to $\hat{\pi}_{\mathrm{Oh}}(\boldsymbol{\theta}^{*(j)}|D)$. Since w_{Oh} is not w_{opt} given by (2.8), $\hat{\pi}_{\mathrm{Oh}}(\boldsymbol{\theta}^{*(j)}|D)$ may not be most efficient. Although $\hat{\pi}_{\mathrm{Oh}}(\boldsymbol{\theta}^{*(j)}|D)$ is not optimal, w_{Oh} in $\hat{\pi}_{\mathrm{Oh}}(\boldsymbol{\theta}^{*(j)}|D)$ does provide an interesting alternative in constructing w in IWMDE.

2.6. Estimating posterior densities from the Metropolis–Hastings output

Suppose that $\{\boldsymbol{\theta}_i,\ i = 1, 2, \ldots, n\}$ is a posterior sample from $\pi(\boldsymbol{\theta}|D)$ given by (2.1) by the Metropolis–Hastings (MH) algorithm. Let $q(\boldsymbol{\theta}, \boldsymbol{\theta}^*|D)$ denote the proposal density for the transition from $\boldsymbol{\theta}$ to $\boldsymbol{\theta}^*$. Let

$$\alpha\big(\boldsymbol{\theta}, \boldsymbol{\theta}^*|D\big) = \min\left\{1, \frac{L(\boldsymbol{\theta}^*|D)\pi(\boldsymbol{\theta}^*)}{L(\boldsymbol{\theta}|D)\pi(\boldsymbol{\theta})}\frac{q(\boldsymbol{\theta}^*, \boldsymbol{\theta}|D)}{q(\boldsymbol{\theta}, \boldsymbol{\theta}^*|D)}\right\}$$

denote the probability of move. Chib and Jeliazkov (2001) show that the posterior density at a fixed point $\boldsymbol{\theta}^*$ is given by

$$\begin{aligned}
\pi\big(\boldsymbol{\theta}^*|D\big) &= \frac{\int \alpha(\boldsymbol{\theta}, \boldsymbol{\theta}^*|D)q(\boldsymbol{\theta}, \boldsymbol{\theta}^*|D)\pi(\boldsymbol{\theta}|D)\,\mathrm{d}\boldsymbol{\theta}}{\int \alpha(\boldsymbol{\theta}^*, \boldsymbol{\theta}|D)q(\boldsymbol{\theta}^*, \boldsymbol{\theta}|D)\,\mathrm{d}\boldsymbol{\theta}}\\
&= \frac{E_1[\alpha(\boldsymbol{\theta}, \boldsymbol{\theta}^*|D)q(\boldsymbol{\theta}, \boldsymbol{\theta}^*|D)]}{E_2[\alpha(\boldsymbol{\theta}^*, \boldsymbol{\theta}|D)]},
\end{aligned} \tag{2.12}$$

where the expectation E_1 is with respect to the joint posterior distribution $\pi(\boldsymbol{\theta}|D)$ and the expectation E_2 is with respect to $q(\boldsymbol{\theta}^*, \boldsymbol{\theta}|D)$. Using (2.12), the estimator of the full posterior density at a fixed point $\boldsymbol{\theta}^*$ proposed by Chib and Jeliazkov (2001) is given by

$$\hat{\pi}_{\mathrm{CJ}}\big(\boldsymbol{\theta}^*|D\big) = \frac{\frac{1}{n}\sum_{i=1}^{n} \alpha(\boldsymbol{\theta}_i, \boldsymbol{\theta}^*|D) q(\boldsymbol{\theta}_i, \boldsymbol{\theta}^*|D)}{\frac{1}{m}\sum_{j=1}^{m} \alpha(\boldsymbol{\theta}^*, \tilde{\boldsymbol{\theta}}_j|D)}, \tag{2.13}$$

where $\{\tilde{\boldsymbol{\theta}}_j,\ j = 1, 2, \ldots, m\}$ is a random (possibly MCMC) sample from $q(\boldsymbol{\theta}^*, \boldsymbol{\theta}|D)$. Note that (a) $\hat{\pi}_{\mathrm{CJ}}(\boldsymbol{\theta}^*|D)$ is consistent but not unbiased and (b) it requires two samples, one from $\pi(\boldsymbol{\theta}|D)$ and another from $q(\boldsymbol{\theta}^*, \boldsymbol{\theta}|D)$, while the second sample depends on the choice of $\boldsymbol{\theta}^*$.

Mira and Nicholls (2004) observe a connection between (2.12) and the ratio of two normalizing constants. Specifically, let

$$\pi_1(\boldsymbol{\theta}|D) = c_1 \pi_1^*(\boldsymbol{\theta}|D),$$

where $c_1 = 1/m(D)$ and $\pi_1^*(\boldsymbol{\theta}|D) = L(\boldsymbol{\theta}|D)\pi(\boldsymbol{\theta})$, which is the unnormalized full posterior density, and

$$\pi_2(\boldsymbol{\theta}|D) = c_2 \pi_2^*(\boldsymbol{\theta}|D),$$

where $c_2 = 1/\pi_1^*(\boldsymbol{\theta}^*|D)$ and $\pi_2^*(\boldsymbol{\theta}|D) = \pi_1^*(\boldsymbol{\theta}^*|D) q(\boldsymbol{\theta}^*, \boldsymbol{\theta}|D)$. Then,

$$r = \frac{c_1}{c_2} = \frac{\pi_1^*(\boldsymbol{\theta}^*|D)}{m(D)} = \pi\big(\boldsymbol{\theta}^*|D\big). \tag{2.14}$$

Following Meng and Wong (1996), for any $h(\boldsymbol{\theta})$ such that

$$0 < \left| \int h(\boldsymbol{\theta}) \pi_1(\boldsymbol{\theta}|D) \pi_2(\boldsymbol{\theta}|D)\, \mathrm{d}\boldsymbol{\theta} \right| < \infty, \tag{2.15}$$

we have

$$r = \frac{E_1[\pi_2^*(\boldsymbol{\theta}|D) h(\boldsymbol{\theta})]}{E_2[\pi_1^*(\boldsymbol{\theta}|D) h(\boldsymbol{\theta})]}, \tag{2.16}$$

where the expectation E_k is taken with respect to $\pi_k(\boldsymbol{\theta}|D)$ for $k = 1, 2$. In particular, when

$$h(\boldsymbol{\theta}) = \frac{\alpha(\boldsymbol{\theta}^*, \boldsymbol{\theta}|D)}{\pi_1^*(\boldsymbol{\theta}|D)}, \tag{2.17}$$

we have

$$\begin{aligned}
&E_1\left[\pi_2^*(\boldsymbol{\theta}|D) \frac{\alpha(\boldsymbol{\theta}^*, \boldsymbol{\theta}|D)}{\pi_1^*(\boldsymbol{\theta}|D)}\right] \\
&\quad = \int L\big(\boldsymbol{\theta}^*|D\big)\pi\big(\boldsymbol{\theta}^*\big) q\big(\boldsymbol{\theta}^*, \boldsymbol{\theta}|D\big) \frac{\alpha(\boldsymbol{\theta}^*, \boldsymbol{\theta}|D)}{L(\boldsymbol{\theta}|D)\pi(\boldsymbol{\theta})} \pi(\boldsymbol{\theta}|D)\, \mathrm{d}\boldsymbol{\theta} \\
&\quad = \int \alpha\big(\boldsymbol{\theta}, \boldsymbol{\theta}^*|D\big) q\big(\boldsymbol{\theta}, \boldsymbol{\theta}^*|D\big) \pi(\boldsymbol{\theta}|D)\, \mathrm{d}\boldsymbol{\theta}
\end{aligned}$$

and

$$E_2\big[\pi_1^*(\boldsymbol{\theta}|D)h(\boldsymbol{\theta})\big] = \int \pi_1^*(\boldsymbol{\theta}|D)\frac{\alpha(\boldsymbol{\theta}^*, \boldsymbol{\theta}|D)}{\pi_1^*(\boldsymbol{\theta}|D)} q\big(\boldsymbol{\theta}^*, \boldsymbol{\theta}|D\big)\,\mathrm{d}\boldsymbol{\theta}$$
$$= \int \alpha\big(\boldsymbol{\theta}^*, \boldsymbol{\theta}|D\big) q\big(\boldsymbol{\theta}^*, \boldsymbol{\theta}|D\big)\,\mathrm{d}\boldsymbol{\theta}.$$

Thus, (2.16) with an h given by (2.17) reduces to (2.12). Also note that for h given by (2.17),

$$\int h(\boldsymbol{\theta})\pi_1(\boldsymbol{\theta}|D)\pi_2(\boldsymbol{\theta}|D)\,\mathrm{d}\boldsymbol{\theta} = \int \frac{\alpha(\boldsymbol{\theta}^*, \boldsymbol{\theta}|D)}{\pi_1^*(\boldsymbol{\theta}|D)} c_1\pi_1^*(\boldsymbol{\theta}|D)\pi_2(\boldsymbol{\theta}|D)\,\mathrm{d}\boldsymbol{\theta}$$
$$= c_1 \int \alpha\big(\boldsymbol{\theta}^*, \boldsymbol{\theta}|D\big)\pi_2(\boldsymbol{\theta}|D)\,\mathrm{d}\boldsymbol{\theta},$$

and, hence, (2.15) holds since $\alpha(\boldsymbol{\theta}^*, \boldsymbol{\theta}|D) \leqslant 1$ and $\alpha(\boldsymbol{\theta}^*, \boldsymbol{\theta}|D) > 0$ for $\boldsymbol{\theta}$ in the support of $\pi_2(\boldsymbol{\theta}|D)$ as long as $q(\boldsymbol{\theta}^*, \boldsymbol{\theta})$ is appropriately constructed.

As discussed in Mira and Nicholls (2004), $\hat{\pi}_{\mathrm{CJ}}(\boldsymbol{\theta}^*|D)$ is in the class considered by Meng and Wong (1996). However, $\hat{\pi}_{\mathrm{CJ}}(\boldsymbol{\theta}^*|D)$ is not optimal in the sense of minimizing the relative mean square error. The "optimal" estimator of r and, hence, $\pi(\boldsymbol{\theta}^*|D)$ is the optimal bridge sampling estimator of Meng and Wong (1996), which is the unique solution to the following "score" equation:

$$S(r) = \sum_{i=1}^{n} \frac{s_2 r\pi_2^*(\boldsymbol{\theta}_i|D)}{s_1\pi_1^*(\boldsymbol{\theta}_i|D) + s_2 r\pi_2^*(\boldsymbol{\theta}_i|D)} - \sum_{j=1}^{m} \frac{s_1\pi_1^*(\tilde{\boldsymbol{\theta}}_j|D)}{s_1\pi_1^*(\tilde{\boldsymbol{\theta}}_j|D) + s_2 r\pi_2^*(\tilde{\boldsymbol{\theta}}_j)|D}$$
$$= 0, \tag{2.18}$$

where $s_1 = n/(n+m)$ and $s_2 = m/(n+m)$. This resulting estimator of $\pi(\boldsymbol{\theta}^*|D)$ is denoted by $\hat{\pi}_{\mathrm{MN}}(\boldsymbol{\theta}^*|D)$ due to Mira and Nicholls (2004). We note that Meng and Schilling (2002) also discuss the connection between (2.12) and the ratio of two normalizing constants given by (2.14). The early version of Mira and Nicholls (2004) was cited in Meng and Schilling (2002). Furthermore, Meng and Schilling (2002) provide empirical comparisons of several warp bridge sampling estimators of r under the nonlinear mixed-effect model using the data from the Fort Bragg evaluation project (Bickman et al., 1995).

Chib and Jeliazkov (2001) also propose a Monte Carlo method for estimating the marginal posterior density $\pi(\boldsymbol{\theta}^{*(j)}|D)$ from the Metropolis–Hastings output. The main goal in Chib and Jeliazkov (2001) is to compute the marginal likelihood. However, here our objective is to compute the marginal posterior density. The two parameter blocks of Chib and Jeliazkov (2001) serve very well for our purposes.

Following the notation given in Chib and Jeliazkov (2001), suppose there is a vector z of latent variables so that

$$\pi(\boldsymbol{\theta}|D) = \int \pi(\boldsymbol{\theta}, z|D)\,\mathrm{d}z,$$

where

$$\pi(\boldsymbol{\theta}, z|D) = \frac{L(\boldsymbol{\theta}, z|D)\pi(\boldsymbol{\theta})}{m(D)}$$

and $L(\boldsymbol{\theta}, z|D)$ is the joint likelihood function so that $\int L(\boldsymbol{\theta}, z|D)\,\mathrm{d}z = L(\boldsymbol{\theta}|D)$.

Let $q(\boldsymbol{\theta}^{(j)}, \boldsymbol{\theta}^{*(j)}|\boldsymbol{\theta}^{(-j)}, z, D)$ denote the proposal density for transition from $\boldsymbol{\theta}^{(j)}$ to $\boldsymbol{\theta}^{*(j)}$. Also let

$$\alpha\bigl(\boldsymbol{\theta}^{(j)}, \boldsymbol{\theta}^{*(j)}|\boldsymbol{\theta}^{(-j)}, z, D\bigr) = \min\left\{1, \frac{L(\boldsymbol{\theta}^{*(j)}, \boldsymbol{\theta}^{(-j)}, z|D)\pi(\boldsymbol{\theta}^{*(j)}, \boldsymbol{\theta}^{(-j)})}{L(\boldsymbol{\theta}^{(j)}, \boldsymbol{\theta}^{(-j)}, z|D)\pi(\boldsymbol{\theta}^{(j)}, \boldsymbol{\theta}^{(-j)})} \frac{q(\boldsymbol{\theta}^{*(j)}, \boldsymbol{\theta}^{(j)}|\boldsymbol{\theta}^{(-j)}, z, D)}{q(\boldsymbol{\theta}^{(j)}, \boldsymbol{\theta}^{*(j)}|\boldsymbol{\theta}^{(-j)}, z, D)}\right\}.$$

Chib and Jeliazkov (2001) establish the following identity:

$$\pi\bigl(\boldsymbol{\theta}^{*(j)}|D\bigr) = \frac{E_1[\alpha(\boldsymbol{\theta}^{(j)}, \boldsymbol{\theta}^{*(j)}|\boldsymbol{\theta}^{(-j)}, z, D)q(\boldsymbol{\theta}^{(j)}, \boldsymbol{\theta}^{*(j)}|\boldsymbol{\theta}^{(-j)}, z, D)]}{E_2[\alpha(\boldsymbol{\theta}^{*(j)}, \boldsymbol{\theta}^{(j)}|\boldsymbol{\theta}^{(-j)}, z, D)]}, \tag{2.19}$$

where the expectation E_1 is with respect to the full posterior distribution $\pi(\boldsymbol{\theta}, z|D)$ and the expectation E_2 is with respect to the distribution $\pi(\boldsymbol{\theta}^{(-j)}, z|\boldsymbol{\theta}^{*(j)}, D)q(\boldsymbol{\theta}^{*(j)}, \boldsymbol{\theta}^{(j)}|\boldsymbol{\theta}^{(-j)}, z, D)$.

Suppose that $\{(\boldsymbol{\theta}_i^{(j)}, \boldsymbol{\theta}_i^{(-j)}, z_i),\ i = 1, 2, \ldots, n\}$ is an MH sample from $\pi(\boldsymbol{\theta}, z|D)$ and

$$\bigl\{\bigl(\tilde{\boldsymbol{\theta}}_k^{(j)}, \tilde{\boldsymbol{\theta}}_k^{(-j)}, \tilde{z}_k\bigr),\ k = 1, 2, \ldots, m\bigr\}$$

is an MH sample from the distribution $\pi(\boldsymbol{\theta}^{(-j)}, z|\boldsymbol{\theta}^{*(j)}, D)q(\boldsymbol{\theta}^{*(j)}, \boldsymbol{\theta}^{(j)}|\boldsymbol{\theta}^{(-j)}, z, D)$. As discussed in Chib and Jeliazkov (2001), the second MH sample can be obtained by the "reduced" MH runs, namely, we generate $(\tilde{\boldsymbol{\theta}}_k^{(j)}, \tilde{\boldsymbol{\theta}}_k^{(-j)}, \tilde{z}_k)$ from the following conditional distributions in turn: (a) $\pi(\boldsymbol{\theta}^{(-j)}|\boldsymbol{\theta}^{*(j)}, z, D)$, (b) $\pi(z, |\boldsymbol{\theta}^{*(j)}, \boldsymbol{\theta}^{(-j)}, D)$, and (c) $q(\boldsymbol{\theta}^{*(j)}, \boldsymbol{\theta}^{(j)}|\boldsymbol{\theta}^{(-j)}, z, D)$. Using (2.19), a simulation-consistent estimator of $\pi(\boldsymbol{\theta}^{*(j)}|D)$ is given by

$$\hat{\pi}_{\mathrm{CJ}}\bigl(\boldsymbol{\theta}^{*(j)}|D\bigr) = \frac{\frac{1}{n}\sum_{i=1}^{n} \alpha(\boldsymbol{\theta}_i^{(j)}, \boldsymbol{\theta}^{*(j)}|\boldsymbol{\theta}_i^{(-j)}, z_i, D)q(\boldsymbol{\theta}_i^{(j)}, \boldsymbol{\theta}^{*(j)}|\boldsymbol{\theta}_i^{(-j)}, z_i, D)}{\frac{1}{m}\sum_{k=1}^{m} \alpha(\boldsymbol{\theta}^{*(j)}, \tilde{\boldsymbol{\theta}}_k^{(j)}|\tilde{\boldsymbol{\theta}}_k^{(-j)}, \tilde{z}_k, D)}. \tag{2.20}$$

Mira and Nicholls (2004) mainly focus on how to compute the joint posterior density at a fixed point. However, their idea can still be applied to the marginal posterior density estimation. Let

$$\pi_1(\boldsymbol{\theta}, z|D) = c_1\pi_1^*(\boldsymbol{\theta}, z|D),$$

where $c_1 = 1/m(D)$ and $\pi_1^*(\boldsymbol{\theta}, z|D) = L(\boldsymbol{\theta}, z|D)\pi(\boldsymbol{\theta})$, and

$$\pi_2(\boldsymbol{\theta}, z|D) = c_2\pi_2^*(\boldsymbol{\theta}, z|D),$$

where $c_2 = /[m(D)\pi(\boldsymbol{\theta}^{*(j)}|D)]$ and

$$\pi_2^*(\boldsymbol{\theta}, z|D) = \pi_1^*\bigl(\boldsymbol{\theta}^{*(j)}, \boldsymbol{\theta}^{(-j)}, z|D\bigr)q\bigl(\boldsymbol{\theta}^{*(j)}, \boldsymbol{\theta}^{(j)}|\boldsymbol{\theta}^{(-j)}, z, D\bigr).$$

Then, we have

$$r = \frac{c_1}{c_2} = \pi\bigl(\boldsymbol{\theta}^{*(j)}|D\bigr).$$

Following Meng and Wong (1996), for any $h(\boldsymbol{\theta})$ such that $0 < |\int h(\boldsymbol{\theta}, z)\pi_1(\boldsymbol{\theta}, z|D)$ $\pi_2(\boldsymbol{\theta}, z|D)\, d\boldsymbol{\theta}\, dz| < \infty$, we have

$$r = \frac{E_1[\pi_2^*(\boldsymbol{\theta}, z|D)h(\boldsymbol{\theta}, z)]}{E_2[\pi_1^*(\boldsymbol{\theta}, z|D)h(\boldsymbol{\theta}, z)]}, \tag{2.21}$$

where the expectation E_k is taken with respect to the distribution $\pi_k(\boldsymbol{\theta}, z|D)$ for $k = 1, 2$. By taking

$$h(\boldsymbol{\theta}, z) = \frac{\alpha(\boldsymbol{\theta}^{*(j)}, \boldsymbol{\theta}^{(j)}|\boldsymbol{\theta}^{(-j)}, z, D)}{\pi_1^*(\boldsymbol{\theta}, z|D)},$$

we can show that (2.21) reduces to (2.19) and again $h(\boldsymbol{\theta}, z)$ satisfies the condition for (2.21). Thus, the optimal bridge sampling estimator, denoted by $\hat{\pi}_{\mathrm{BS}}(\boldsymbol{\theta}^{*(j)}|D)$, of Meng and Wong (1996) can be better than $\hat{\pi}_{\mathrm{CJ}}(\boldsymbol{\theta}^{*(j)}|D)$ given by (2.20). The estimator, $\hat{\pi}_{\mathrm{BS}}(\boldsymbol{\theta}^{*(j)}|D)$, is the unique solution to the "score" equation similar to (2.18).

Unlike $\hat{\pi}_{\mathrm{Oh}}(\boldsymbol{\theta}^{*(j)}|D)$ given by (2.11), $\hat{\pi}_{\mathrm{CJ}}(\boldsymbol{\theta}^{*(j)}|D)$ is not a special case of IWMDE. Thus, $\hat{\pi}_{\mathrm{CJ}}(\boldsymbol{\theta}^{*(j)}|D)$ or $\hat{\pi}_{\mathrm{BS}}(\boldsymbol{\theta}^{*(j)}|D)$ is a nice alternative to IWMDE when the MH algorithm is used to sample from the joint posterior distribution. In IWMDE, we need to construct the weight function $w(\boldsymbol{\beta}^{(j)}|\boldsymbol{\beta}^{(-j)})$, while $\hat{\pi}_{\mathrm{CJ}}(\boldsymbol{\theta}^{*(j)}|D)$ requires knowing the closed-form expression of the proposal density $q(\boldsymbol{\theta}^{(j)}, \boldsymbol{\theta}^{*(j)}|\boldsymbol{\theta}^{(-j)}, z, D)$. However, if one is interested in estimating the whole marginal density curve, both $\hat{\pi}_{\mathrm{CJ}}(\boldsymbol{\theta}^{*(j)}|D)$ and $\hat{\pi}_{\mathrm{BS}}(\boldsymbol{\theta}^{*(j)}|D)$ are quite computationally expensive, since an MH sample is required from the distribution $\pi(\boldsymbol{\theta}^{(-j)}, z|\boldsymbol{\theta}^{*(j)}, D)$ for every point $\boldsymbol{\theta}^{*(j)}$ on the curve. In this regard, the IWMDE may be more preferable, since it requires only one MCMC sample from the full posterior density $\pi(\boldsymbol{\theta}|D)$.

3. Marginal posterior densities for generalized linear models

Consider the generalized linear model (see, for example, McCullagh and Nelder, 1989) given by

$$f(y_i|g_i, \tau) = \exp\{a_i^{-1}(\tau)(y_i g_i - b(g_i)) + c(y_i, \tau)\}, \quad i = 1, \ldots, n, \tag{3.1}$$

indexed by the canonical parameter g_i and the scale parameter τ. The functions b and c determine a particular family in the class, such as the binomial, normal, Poisson, etc. The functions $a_i(\tau)$ are commonly of the form $a_i(\tau) = \tau^{-1}w_i^{-1}$, where the w_i's are known weights. For ease of exposition, we assume that $w_i = 1$ and τ is known throughout.

We further assume the g_i's satisfy the equations

$$g_i = g(\eta_i), \quad i = 1, 2, \ldots, n,$$

and

$$\boldsymbol{\eta} = X\boldsymbol{\beta},$$

where $\boldsymbol{\eta} = (\eta_1, \ldots, \eta_n)$, $X\boldsymbol{\beta} = (\boldsymbol{x}_1'\boldsymbol{\beta}, \boldsymbol{x}_2'\boldsymbol{\beta}, \ldots, \boldsymbol{x}_n'\boldsymbol{\beta})'$, $\boldsymbol{x}_i$ is a $p \times 1$ vector of covariates, and $\boldsymbol{\beta}$ is a $p \times 1$ vector of regression coefficients. The function $g(\eta_i)$ is called the g-link, and for a canonical link, $g_i = \eta_i$.

Let $D = (\boldsymbol{y}, X, n)$ denote the observed data. In vector notation, the likelihood function for the generalized linear model (GLM) can be written as

$$L(\boldsymbol{\beta}|D) = \exp\{\tau(\boldsymbol{y}'\boldsymbol{g}(X\boldsymbol{\beta}) - b[\boldsymbol{g}(X\boldsymbol{\beta})]) + c(\boldsymbol{y}, \tau)\}, \tag{3.2}$$

where $\boldsymbol{g}(X\boldsymbol{\beta})$ is a function of $X\boldsymbol{\beta}$ that depends on the link. When a canonical link is used, $\boldsymbol{g}(X\boldsymbol{\beta}) = X\boldsymbol{\beta}$. Suppose that we take a normal prior for $\boldsymbol{\beta}$, i.e.,

$$\boldsymbol{\beta} \sim N_p(\boldsymbol{\beta}_0, \Sigma_0).$$

Then, the posterior is of the form

$$\pi(\boldsymbol{\beta}|D) \propto L(\boldsymbol{\beta}|D)\pi(\boldsymbol{\beta}).$$

Let $\boldsymbol{\beta} = (\boldsymbol{\beta}^{(j)}, \boldsymbol{\beta}^{(-j)})$. We are interested in computing the marginal posterior density of $\boldsymbol{\beta}^{(j)}$.

When the Gibbs sampling algorithm is used to sample from the posterior distribution $\pi(\boldsymbol{\beta}|D)$, the Gibbs stopper estimators, such as $\hat{\pi}_{\mathrm{GS}}(\boldsymbol{\beta}^*|D)$ given by (2.10) and $\hat{\pi}_{\mathrm{Oh}}(\boldsymbol{\beta}^{*(j)}|D)$ given by (2.11), may not be applicable or computationally expensive, because for the GLM, the closed-form expression of the one-dimensional conditional posterior distribution $\pi(\beta_k|\beta_1, \ldots, \beta_{k-1}, \beta_k, \ldots, \beta_p, D)$ is not available. When the MH algorithm is used, $\hat{\pi}_{\mathrm{CJ}}(\boldsymbol{\beta}^{*(j)}|D)$ given by (2.20) or $\hat{\pi}_{\mathrm{BS}}(\boldsymbol{\beta}^{*(j)}|D)$ are ready to be used. However, when the MCMC draws are obtained by using some available free or commercial softwares such as BUGS or SBayes in Splus from the posterior distribution, the proposal density (q) and the probability of move (α) may not be known. Hence, the posterior density estimators based on the MH output may be difficult to apply. In this case, the kernel density estimator and the IWMDE may be more desirable.

As we discussed earlier, the performance of an IWMDE depends on the choice of the weight function $w(\boldsymbol{\beta}^{(j)}|\boldsymbol{\beta}^{(-j)})$. Fortunately, for the GLM, a good $w(\boldsymbol{\beta}^{(j)}|\boldsymbol{\beta}^{(-j)})$, which is close to the optimal choice, can be constructed based on the asymptotic approximation to the joint posterior proposed by Chen (1985).

Following Chen (1985) and ignoring constants that are free of the parameters, we have

$$L(\boldsymbol{\beta}|D) \approx \exp\left\{-\frac{1}{2}(\boldsymbol{\beta} - \hat{\boldsymbol{\beta}})'\widehat{\Sigma}^{-1}(\boldsymbol{\beta} - \hat{\boldsymbol{\beta}})\right\}, \tag{3.3}$$

where $\hat{\boldsymbol{\beta}}$ is the maximum likelihood estimator (MLE) of $\boldsymbol{\beta}$ based on the data D, and

$$\widehat{\Sigma} = \left\{-\frac{\partial^2 \ln L(\boldsymbol{\beta}|D)}{\partial\boldsymbol{\beta}\partial\boldsymbol{\beta}'}\bigg|_{\boldsymbol{\beta}=\hat{\boldsymbol{\beta}}}\right\}^{-1}.$$

Note that $\hat{\boldsymbol{\beta}}$ and $\widehat{\Sigma}$ can be obtained using standard statistical software packages, such as SAS and S-Plus for the GLM. Thus, these are almost the "computationally free" quantities. Using (3.3), the joint posterior distribution can be approximated by

$$\boldsymbol{\beta} \sim N_p\left([\widehat{\Sigma}^{-1} + \Sigma_0^{-1}]^{-1}(\widehat{\Sigma}^{-1}\hat{\boldsymbol{\beta}} + \Sigma_0^{-1}\boldsymbol{\beta}_0), [\widehat{\Sigma}^{-1} + \Sigma_0^{-1}]^{-1}\right). \tag{3.4}$$

Then, we can simply take $w(\boldsymbol{\beta}^{(j)}|\boldsymbol{\beta}^{(-j)})$ to be the conditional density of $\boldsymbol{\beta}^{(j)}$ given $\boldsymbol{\beta}^{(-j)}$ with respect to the approximate posterior distribution given by (3.4). It is a well-known fact that the conditional distribution of a multivariate normal distribution is still normal and thus the closed form is available.

One of the nice features of the IWMDE is that we are able to estimate all possible marginal posterior densities of $\boldsymbol{\beta}^{(-j)}$ using *only* a single MCMC output from the GLM. It is expected that when n is reasonably large, the IWMDE will perform well, since in this case, the approximation given by (3.4) is close to the joint posterior distribution for $\boldsymbol{\beta}$. Also note that the CMDE may be difficult to be implemented for the GLM, since the closed-form expression of the conditional density $\pi(\boldsymbol{\beta}^{(j)}|\boldsymbol{\beta}^{(j)}, D)$ is not available. A similar procedure for choosing w can be developed even for the unknown τ case if our primary interest is in computing marginal posterior densities for $\boldsymbol{\beta}$.

4. Savage–Dickey density ratio

Suppose that the posterior $\pi(\boldsymbol{\theta}, \boldsymbol{\psi}|D)$ is proportional to $L(\boldsymbol{\theta}, \boldsymbol{\psi}|D) \times \pi(\boldsymbol{\theta}, \boldsymbol{\psi})$, where $(\boldsymbol{\theta}, \boldsymbol{\psi}) \in \Omega \times \boldsymbol{\psi}$, $L(\boldsymbol{\theta}, \boldsymbol{\psi}|D)$ is the likelihood function given data D, and $\pi(\boldsymbol{\theta}, \boldsymbol{\psi})$ is the prior. We wish to test H_0: $\boldsymbol{\theta} = \boldsymbol{\theta}_0$ versus H_1: $\boldsymbol{\theta} \neq \boldsymbol{\theta}_0$. The Bayes factor is

$$B = m_0/m,$$

where $m_0 = \int_\Psi L(\boldsymbol{\theta}_0, \boldsymbol{\psi}|D)\pi_0(\boldsymbol{\psi})\,\mathrm{d}\boldsymbol{\psi}$, $m = \int_{\Omega\times\Psi} L(\boldsymbol{\theta}, \boldsymbol{\psi}|D)\pi(\boldsymbol{\theta}, \boldsymbol{\psi})\,\mathrm{d}\boldsymbol{\theta}\,\mathrm{d}\boldsymbol{\psi}$, and $\pi_0(\boldsymbol{\psi})$ is the prior under H_0.

Dickey (1971) shows that if

$$\pi(\boldsymbol{\psi}|\boldsymbol{\theta}_0) = \pi_0(\boldsymbol{\psi}),$$

then

$$B = \frac{\pi(\boldsymbol{\theta}_0|D)}{\pi(\boldsymbol{\theta}_0)}, \tag{4.1}$$

where $\pi(\boldsymbol{\theta}_0|D) = \int_\Psi \pi(\boldsymbol{\theta}, \boldsymbol{\psi}|D)\,\mathrm{d}\boldsymbol{\psi}$ and $\pi(\boldsymbol{\theta}) = \int_\Psi \pi(\boldsymbol{\theta}, \boldsymbol{\psi})\,\mathrm{d}\boldsymbol{\psi}$. The reduced form of the Bayes factor B is called the "Savage–Dickey density ratio." Observe that

$$\begin{aligned} m &= \frac{L(\boldsymbol{\theta}_0, \boldsymbol{\psi}|D)\pi(\boldsymbol{\theta}_0, \boldsymbol{\psi})}{\pi(\boldsymbol{\theta}_0, \boldsymbol{\psi}|D)} = \frac{L(\boldsymbol{\theta}_0, \boldsymbol{\psi}|D)\pi(\boldsymbol{\psi}|\boldsymbol{\theta}_0)\pi(\boldsymbol{\theta}_0)}{\pi(\boldsymbol{\psi}|\boldsymbol{\theta}_0, D)\pi(\boldsymbol{\theta}_0|D)} \\ &= \frac{L(\boldsymbol{\theta}_0, \boldsymbol{\psi}|D)\pi_0(\boldsymbol{\psi})}{\pi(\boldsymbol{\psi}|\boldsymbol{\theta}_0, D)} \times \frac{\pi(\boldsymbol{\theta}_0)}{\pi(\boldsymbol{\theta}_0|D)} = m_0 \times \frac{\pi(\boldsymbol{\theta}_0)}{\pi(\boldsymbol{\theta}_0|D)}. \end{aligned}$$

Thus, the derivation of the Savage–Dickey density ratio is straightforward.

The Savage–Dickey density ratio indicates that the Bayes factor can be reduced to the ratio of the marginal posterior density and the prior density evaluated at $\boldsymbol{\theta}_0$. Therefore, the marginal density estimation can be used for computing Bayes factors.

In the cases where $\pi(\boldsymbol{\psi}|\boldsymbol{\theta}_0)$ depends on $\boldsymbol{\theta}_0$, Verdinelli and Wasserman (1995) obtain a generalized version of the Savage–Dickey density ratio. Assume that $0 < \pi(\boldsymbol{\theta}_0|D) < \infty$ and $0 < \pi(\boldsymbol{\theta}_0, \boldsymbol{\psi}) < \infty$ for almost all $\boldsymbol{\psi}$. Then the generalized Savage–Dickey density ratio is given by

$$B = \pi(\boldsymbol{\theta}_0|D)E\left[\frac{\pi_0(\boldsymbol{\psi})}{\pi(\boldsymbol{\theta}_0, \boldsymbol{\psi})}\right] = \frac{\pi(\boldsymbol{\theta}_0|D)}{\pi(\boldsymbol{\theta}_0)}E\left[\frac{\pi_0(\boldsymbol{\psi})}{\pi(\boldsymbol{\psi}|\boldsymbol{\theta}_0)}\right],$$

where the expectation is taken with respect to $\pi(\boldsymbol{\psi}|\boldsymbol{\theta}_0, D)$ (the conditional posterior density of $\boldsymbol{\psi}$ given $\boldsymbol{\theta} = \boldsymbol{\theta}_0$).

To evaluate the generalized density ratio, we must compute $\pi(\boldsymbol{\theta}_0|D)$ and $E[\pi_0(\boldsymbol{\psi})/\pi(\boldsymbol{\theta}_0, \boldsymbol{\psi})]$. To do this, we need two MCMC samples, one from $\pi(\boldsymbol{\theta}, \psi|D)$ and another from $\pi(\boldsymbol{\psi}|\boldsymbol{\theta}_0, D)$. Then an estimate of $\pi(\boldsymbol{\theta}_0|D)$ can be obtained via the posterior density estimation methods discussed in Section 2, while computing $E[\pi_0(\boldsymbol{\psi})/\pi(\boldsymbol{\theta}_0, \boldsymbol{\psi})]$ is straightforward when an MCMC draw is available from $\pi(\boldsymbol{\psi}|\boldsymbol{\theta}_0, D)$.

5. Computing marginal likelihoods

To compute the marginal likelihood $m(D)$ given in (2.1), Chib (1995) observes the following novel identity:

$$m(D) = \frac{L(\boldsymbol{\theta}^*|D)\pi(\boldsymbol{\theta}^*)}{\pi(\boldsymbol{\theta}^*|D)} \quad \text{or}$$

$$\log\bigl[m(D)\bigr] = \log L\bigl(\boldsymbol{\theta}^*|D\bigr) + \log\pi\bigl(\boldsymbol{\theta}^*\bigr) - \log\pi(\boldsymbol{\theta}^*|D), \tag{5.1}$$

where $\boldsymbol{\theta}^*$ be a fixed value (normally, the posterior mean or the posterior mode). Assume the values of $L(\boldsymbol{\theta}^*|D)$ and $\pi(\boldsymbol{\theta}^*)$ are available directly. From (5.1), it is clear that the problem for computing the marginal likelihood $m(D)$ reduces to the problem of estimating posterior density of $\boldsymbol{\theta}$ evaluated at $\boldsymbol{\theta}^*$.

Suppose we can write

$$\pi(\boldsymbol{\theta}|D) = \int \pi(\boldsymbol{\theta}, z|D)\,\mathrm{d}z = \int \pi(\boldsymbol{\theta}|z, D)\pi(z|D),$$

where z is the vector of latent variables, $\pi(\boldsymbol{\theta}|z, D)$ is the conditional posterior density of $\boldsymbol{\theta}$ given z and D, and $\pi(z|D)$ is the marginal posterior density. If the closed-form for $\pi(\boldsymbol{\theta}|z, D)$ is available, CMDE is ready to be used for computing $\pi(\boldsymbol{\theta}^*|D)$ and if the conditional posterior density is not available in closed-form, the other methods discussed in Section 2 can be considered.

Suppose that the parameter vector is split into two blocks so that $\boldsymbol{\theta} = (\boldsymbol{\theta}^{(j)}, \boldsymbol{\theta}^{(-j)})$, and the Gibbs sampler is applied to the complete conditional densities

$$\pi\bigl(\boldsymbol{\theta}^{(j)}|\boldsymbol{\theta}^{(j)}, z, D\bigr),\ \pi\bigl(\boldsymbol{\theta}^{(j)}|\boldsymbol{\theta}^{(j)}, z, D\bigr),\ \pi(z|\boldsymbol{\theta}, D),$$

where z denotes the vector of latent variables. By the law of total probability, we have

$$\pi\bigl(\boldsymbol{\theta}^*|D\bigr) = \pi\bigl(\boldsymbol{\theta}^{*(j)}|D\bigr)\pi\bigl(\boldsymbol{\theta}^{*(-j)}|\boldsymbol{\theta}^{*(j)}, D\bigr). \tag{5.2}$$

Chib (1995) proposes to obtain a Gibbs sample from the joint posterior $\pi(\boldsymbol{\theta}, z|D)$ for computing $\pi(\boldsymbol{\theta}^{*(j)}|D)$ and a reduced Gibbs run from $\pi(\boldsymbol{\theta}^{(-j)}, z|\boldsymbol{\theta}^{*(j)}, D)$ for computing $\pi(\boldsymbol{\theta}^{*(-j)}|\boldsymbol{\theta}^{*(j)}, D)$. The extension can be made for more than two blocks, however, more reduced Gibbs runs are also required. This method works very well for many practical problems.

Chib and Jeliazkov (2001) extend the approach of Chib (1995) when $\boldsymbol{\theta}$ is sample from the posterior $\pi(\boldsymbol{\theta}|D)$ using the Metropolis–Hastings algorithm with a proposal density, $q(\boldsymbol{\theta}, \boldsymbol{\theta}^*|D)$, and the probability of move, $\alpha(\boldsymbol{\theta}, \boldsymbol{\theta}^*|D)$, given in Section 2.6. Using the balanced equation or the reversibility of the Metropolis–Hastings chain,

$$q(\boldsymbol{\theta}, \boldsymbol{\theta}^*|D)\alpha(\boldsymbol{\theta}, \boldsymbol{\theta}^*)\pi(\boldsymbol{\theta}|D) = q(\boldsymbol{\theta}^*, \boldsymbol{\theta}|D)\alpha(\boldsymbol{\theta}^*, \boldsymbol{\theta})\pi(\boldsymbol{\theta}^*|D),$$

and integrating out $\boldsymbol{\theta}$ from both sides, Chib and Jeliazkov (2001) obtain the identity (2.12). Using (5.1), $\hat{\pi}_{\mathrm{CJ}}(\boldsymbol{\theta}^*|D)$, or $\hat{\pi}_{\mathrm{MN}}(\boldsymbol{\theta}^*|D)$, $m(D)$ is ready to be computed. This method is very attractive when an MH algorithm is used to sample from the joint posterior distribution and the proposal density $q(\boldsymbol{\theta}, \boldsymbol{\theta}^*|D)$ is known in a closed-form.

Yu and Tanner (1999) compare the Gibbs stopper estimator $\hat{\pi}_{\mathrm{GS}}(\boldsymbol{\theta}^*|D)$ and the Chib's estimator via (5.2) for the two-blocks case without involving any latent variables in estimating the marginal likelihood $m(D)$ when the Gibbs sampling algorithm is used to sample from the posterior distribution. They empirically show that Chib's estimator has a smaller variance than the Gibbs stopper estimator. But, they further note that the Chib's estimator requires more computing time, and, for the same amount of work, the estimator from Gibbs stopper estimator actually has a smaller variance for the cases they have been investigating.

6. Computing posterior model probabilities via informative priors

Suppose that we consider Bayesian variable selection for the GLM given by (3.2) with a known τ. Let $\mathcal{M}$ denote the model space. We enumerate the models in $\mathcal{M}$ by $k = 1, 2, \ldots, \mathcal{K}$, where $\mathcal{K}$ is the dimension of $\mathcal{M}$ and model $\mathcal{K}$ denotes the full model. Also, let $\boldsymbol{\beta}^{(\mathcal{K})} = (\beta_1, \ldots, \beta_p)'$ denote the regression coefficients for the full model and let $\boldsymbol{\beta}^{(k)}$ denote a $p_k \times 1$ vector of regression coefficients for model k. We write $\boldsymbol{\beta}^{(\mathcal{K})} = (\boldsymbol{\beta}^{(k)\prime}, \boldsymbol{\beta}^{(-k)\prime})'$, where $\boldsymbol{\beta}^{(-k)}$ is $\boldsymbol{\beta}^{(\mathcal{K})}$ with $\boldsymbol{\beta}^{(k)}$ deleted. The likelihood function under model k for (3.2) based on the data D (called the current data) can be written as

$$L(\boldsymbol{\beta}^{(k)}|k, D) = \exp\{\tau(\boldsymbol{y}'\boldsymbol{g}(X^{(k)}\boldsymbol{\beta}^{(k)}) - b[\boldsymbol{g}(X^{(k)}\boldsymbol{\beta}^{(k)})]) + c(\boldsymbol{y}, \tau)\}, \quad (6.1)$$

where $X^{(k)}$ is an $n \times p_k$ matrix of fixed covariates of rank p_k.

Suppose that there exists a previous study that measures the same response variable and covariates as the current study. Let n_0 denote sample size for the previous study, let $\boldsymbol{y}_0$ be an $n_0 \times 1$ response vector for the previous study, let $X_0^{(k)}$ be an $n_0 \times p_k$ matrix of covariates corresponding to $\boldsymbol{y}_0$. Let D_0 denote the (historical) data from the previous study. The power prior for $\boldsymbol{\beta}^{(k)}$ (Ibrahim and Chen, 2000; Chen et al., 2000b) takes the form

$$\begin{aligned}\pi(\boldsymbol{\beta}^{(k)}, a_0|k, D_0) &\propto \pi^*(\boldsymbol{\beta}^{(k)}, a_0|k, D_0)\\ &\equiv \{L(\boldsymbol{\beta}^{(k)}|k, D_0)\}^{a_0}\pi_0(\boldsymbol{\beta}^{(k)}|c_0)a_0^{\delta_0-1}(1-a_0)^{\lambda_0-1},\end{aligned} \quad (6.2)$$

where $L(\boldsymbol{\beta}^{(k)}|k, D_0)$ is similar to $L(\boldsymbol{\beta}^{(k)}|k, D)$ in (6.1) with the replacement of the current data D by the historical data D_0, (δ_0, λ_0) and c_0 are specified hyperparameters, while c_0 controls the impact of initial prior $\pi_0(\boldsymbol{\beta}^{(k)}|c_0)$ on the entire prior. Here, a_0 is a scalar prior parameter that controls the influence of historical data on the current data. We further take $\pi_0(\boldsymbol{\beta}^{(k)} \mid c_0)$ to be a normal density with mean 0 and covariance matrix $c_0 W_0^{(k)}$, i.e.,

$$N_{p_k}\big(0, c_0 W_0^{(k)}\big),$$

where $W_0^{(k)}$ to be the submatrix of $W_0^{(\mathcal{K})}$ corresponding to model k, where $W_0^{(\mathcal{K})}$ is a $p \times p$ matrix for $\boldsymbol{\beta}^{(\mathcal{K})}$ under the full model.

Let

$$p_0^*\big(\boldsymbol{\beta}^{(k)}|k, D_0\big) = L\big(\boldsymbol{\beta}^{(k)}|k, D_0\big)\pi_0\big(\boldsymbol{\beta}^{(k)}|d_0\big),$$

where $\pi_0(\boldsymbol{\beta}^{(k)}|d_0)$ is the same density as $\pi_0(\boldsymbol{\beta}^{(k)}|c_0)$ with c_0 replaced by d_0. Chen et al. (1999) propose to take the prior probability of model k as

$$p(k) \equiv p(k|D_0) = \frac{\int p_0^*(\boldsymbol{\beta}^{(k)}|k, D_0)\, \mathrm{d}\boldsymbol{\beta}^{(k)}}{\sum_{j=1}^{\mathcal{K}} \int p_0^*(\boldsymbol{\beta}^{(j)}|j, D_0)\, \mathrm{d}\boldsymbol{\beta}^{(j)}}. \tag{6.3}$$

The parameter d_0 is a scalar prior parameter that controls the impact of $\pi_0(\boldsymbol{\beta}^{(k)}|d_0)$ on the prior model probability $p(k)$.

For the GLM, the posterior distribution with the power prior (6.2) for $\boldsymbol{\beta}^{(k)}$ under model k is given by

$$\pi\big(\boldsymbol{\beta}^{(k)}, a_0|k, D, D_0\big) \propto L\big(\boldsymbol{\beta}^{(k)}|k, D\big)\pi\big(\boldsymbol{\beta}^{(k)}, a_0|k, D_0\big),$$

where

$$\pi\big(\boldsymbol{\beta}^{(k)}, a_0|k, D_0\big) = \frac{\pi^*(\boldsymbol{\beta}^{(k)}, a_0|k, D_0)}{m(D_0|k)},$$

$\pi^*(\boldsymbol{\beta}^{(k)}, a_0|k, D_0)$ is an unnormalized prior distribution defined in (6.2), and

$$m(D_0|k) = \int \pi^*\big(\boldsymbol{\beta}^{(k)}, a_0|k, D_0\big)\, \mathrm{d}\boldsymbol{\beta}^{(k)}\, \mathrm{d}a_0.$$

The marginal likelihood of the data D under model k is, thus, given by

$$m(D|k) = \iint L\big(\boldsymbol{\beta}^{(k)}|k, D\big)\pi\big(\boldsymbol{\beta}^{(k)}, a_0|k, D_0\big)\, \mathrm{d}\boldsymbol{\beta}^{(k)}\, \mathrm{d}a_0. \tag{6.4}$$

Since the normalizing constant of the posterior distribution is given by

$$m^*(D|k) = \int L\big(\boldsymbol{\beta}^{(k)}|k, D\big)\pi^*\big(\boldsymbol{\beta}^{(k)}, a_0|k, D_0\big)\, \mathrm{d}\boldsymbol{\beta}^{(k)}\, \mathrm{d}a_0,$$

we have

$$m(D|k) = \frac{m^*(D|k)}{m(D_0|k)},$$

which is a ratio of two normalizing constants. For each $m(D|k)$, we need to compute $m^*(D|k)$ and $m(D_0|k)$. Thus, computing $m(D|k)$ with the power prior is much more challenging than the one discussed in Section 5.

Using the Bayes theorem, the posterior probability of model k is given by

$$p(k|D) = \frac{m(D|k)p(k)}{\sum_{j=1}^{\mathcal{K}} m(D|j)p(j)}, \tag{6.5}$$

where $m(D|k)$ is defined by (6.4), which is the marginal distribution of the data D for the current study under model k, and $p(k)$ is the prior probability of model k given by (6.3). Chen et al. (1999) obtain a key theoretical result, which essentially makes the computation of $p(k|D)$ feasible. We summarize this result in the following theorem and the proof of the theorem can be found in Chen et al. (1999).

THEOREM 6.1. *Let* $\pi(\boldsymbol{\beta}^{(-k)}|\mathcal{K}, D_0)$ *and* $\pi(\boldsymbol{\beta}^{(-k)}|\mathcal{K}, D)$ *denote the respective marginal prior and posterior densities of* $\boldsymbol{\beta}^{(-k)}$ *under the full model* $\mathcal{K}$. *Then the posterior probability of model k given by* (6.5) *reduces to*

$$p(k|D) = \frac{(\pi(\boldsymbol{\beta}^{(-k)} = 0|\mathcal{K}, D)/\pi(\boldsymbol{\beta}^{(-k)} = 0|\mathcal{K}, D_0))p(m)}{\sum_{j=1}^{\mathcal{K}}(\pi(\boldsymbol{\beta}^{(-j)} = 0|\mathcal{K}, D)/\pi(\boldsymbol{\beta}^{(-j)} = 0|\mathcal{K}, D_0))p(j)}, \tag{6.6}$$

for $k = 1, 2, \ldots, \mathcal{K}$, *where* $\pi(\boldsymbol{\beta}^{(-k)} = 0|\mathcal{K}, D_0)$ *and* $\pi(\boldsymbol{\beta}^{(-k)} = 0|\mathcal{K}, D)$ *are the marginal prior and posterior densities of* $\boldsymbol{\beta}^{(-k)}$ *evaluated at* $\boldsymbol{\beta}^{(-k)} = 0$ *under the full model.*

The result given in Theorem 6.1 is very attractive since it shows that the posterior probability $p(k|D)$ is simply a function of the prior model probabilities $p(k)$ and the marginal prior and posterior density functions of $\boldsymbol{\beta}^{(-k)}$ under the full model evaluated at $\boldsymbol{\beta}^{(-k)} = 0$. Therefore, the computational problem for the posterior model probabilities for all $k \in \mathcal{M}$ reduces to the one for computing the marginal prior and posterior density functions under the full model. To compute $p(k)$ in (6.3), the Monte Carlo method of Chen and Shao (1997b) can be directly applied. To compute the posterior model probability $p(k|D)$, each of the density estimation methods given in Section 2 can be used in principle. However, the IWMDE may be most applicable. With the IWMDE, we need only one MCMC sample from each of prior and posterior distributions $\pi(\boldsymbol{\beta}^{(\mathcal{K})}, a_0|\mathcal{K}, D_0)$ and $\pi(\boldsymbol{\beta}^{(\mathcal{K})}, a_0|\mathcal{K}, D)$ under the full model to compute $\pi(\boldsymbol{\beta}^{(-k)} = 0|\mathcal{K}, D_0)$ and $\pi(\boldsymbol{\beta}^{(-k)} = 0|\mathcal{K}, D)$ for all k. In addition, the IWMDE does not require knowing the specific structure or the form of the MCMC sampling algorithm which is used to generate the MCMC sample.

To examine the performance of the above method, we conduct a comprehensive simulation study.

Simulation study. We simulate two data sets. The first data set (D) represents the current study and the second dataset (D_0) represents the previous study.

For the current study, $n = 200$ independent Bernoulli observations are simulated with success probability

$$p_i = \frac{\exp\{-1.0 - 0.5x_{i2} - 2.0x_{i4}\}}{1 + \exp\{-1.0 - 0.5x_{i2} - 2.0x_{i4}\}}, \quad i = 1, \dots, n,$$

where x_{i2} and x_{i4} are i.i.d. $N_2((1.0, 0.8)'\text{Diag}(1, 0.8))$. Thus, the true model contains the covariates (x_2, x_4). Two additional covariates (x_{i3}, x_{i5}) are randomly generated, such that the joint distribution of $(x_{i2}, x_{i3}, x_{i4}, x_{i5})'$ is $N_4(\boldsymbol{\mu}, \Sigma)$, where $\boldsymbol{\mu} = (1.0, 0.5, 0.8, 1.4)$ and

$$\Sigma = \begin{pmatrix} 1.0 & 0.353 & 0 & 0 \\ 0.353 & 0.5 & 0 & 0 \\ 0 & 0 & 0.8 & 0.588 \\ 0 & 0 & 0.588 & 1.2 \end{pmatrix}.$$

Thus the full model consists of the five covariates $(1, x_2, \dots, x_5)$ including an intercept.

For the previous study, $n_0 = 400$ Bernoulli observations are generated with success probability

$$p_{0i} = \frac{\exp\{-1.0 - 1.5x_{0i2} - 0.8x_{0i4}\}}{1 + \exp\{-1.0 - 1.5x_{0i2} - 0.8x_{0i5}\}}, \quad i = 1, \dots, n_0,$$

where (x_{0i2}, x_{0i4}) have the same distribution as (x_{i2}, x_{i4}). In addition, two additional covariates, (x_{0i3}, x_{0i5}), are generated such that $(x_{0i2}, x_{0i3}, x_{0i4}, x_{0i5})'$ has the same distribution as $(x_{i2}, x_{i3}, x_{i4}, x_{i5})'$.

Suppose that we consider all possible subset variable models. An intercept is included in all models, and thus the model space $\mathcal{M}$ consists of 16 models. Using the current data along, the top two models based on the Akaike Information Criterion (AIC) (Akaike, 1973) and Bayesian Information Criterion (BIC) (Schwarz, 1978) are (x_4) with AIC $= 99.2$ and BIC $= 105.8$ and (x_2, x_4) with AIC $= 99.6$ and BIC $= 109.5$. Note that for model k, AIC and BIC are given by

$$\text{AIC}_k = -2 \log L(\hat{\boldsymbol{\beta}}^{(k)} | k, D) + 2p_k \quad \text{and}$$
$$\text{BIC}_k = -2 \log L(\hat{\boldsymbol{\beta}}^{(k)} | k, D) + \log(n) p_k,$$

where $\hat{\boldsymbol{\beta}}^{(k)}$ is the maximum likelihood estimate of $\boldsymbol{\beta}^{(k)}$ under model k.

Now we consider the fully Bayesian approach using the prior (6.2). To compute the prior and posterior model probabilities, 50,000 Gibbs iterations were used in all the following computations. Also, the IWMDE was used in estimating marginal prior and posterior densities.

To obtain a result similar to AIC and BIC, we let $d_0 \to 0$. With $d_0 = .001$, and $c_0 = 50$, the top model is x_4 with posterior model probability of 0.22. Table 1 gives the model with the largest posterior probability using $\delta_0 = \lambda_0 = 15$ for several values of c_0. For each value of c_0 in Table 1, the true model, (x_2, x_4), obtains the largest posterior probability, and thus model choice is not sensitive to these values.

From Table 2, we see how the prior model probability is affected as d_0 is changed. In each case, the true model obtains the largest posterior probability. Under the settings

Table 1
Posterior model probabilities for $(\delta_0, \lambda_0) = (15, 15)$, $d_0 = 5$ and various choices of c_0

c_0	k	$p(k)$	$m(D\|k)$	$p(k\|D)$
5	(x_2, x_4)	0.17	0.22	0.31
10	(x_2, x_4)	0.17	0.19	0.35
50	(x_2, x_4)	0.17	0.15	0.38
100	(x_2, x_4)	0.17	0.14	0.39

Table 2
The posterior model probabilities for $(\delta_0, \lambda_0) = (15, 15)$, $c_0 = 5$ and various choices of d_0

d_0	k	$p(k)$	$m(D\|k)$	$p(k\|D)$
10	(x_2, x_4)	0.29	0.22	0.45
50	(x_2, x_4)	0.54	0.22	0.69
100	(x_2, x_4)	0.59	0.22	0.76

Table 3
The posterior model probabilities for $c_0 = 10$, $d_0 = 5$ and various choices of (δ_0, λ_0)

(δ_0, λ_0)	k	$p(k)$	$m(D\|k)$	$p(k\|D)$
(1, 1)	(x_2, x_4)	0.17	0.23	0.31
(5, 5)	(x_2, x_4)	0.17	0.24	0.34
(15, 15)	(x_2, x_4)	0.17	0.19	0.35
(50, 1)	(x_2, x_4)	0.17	0.08	0.36

of Table 2, the true model also obtains the largest prior probability when $d_0 \geqslant 50$. Table 2 indicates a monotonic increase in the prior (and posterior) model probability as d_0 increases.

Table 3 shows a sensitivity analysis with respect to (δ_0, λ_0). The true model obtains the largest posterior probability in each case. Under these settings, model choice is not sensitive to the choice of (δ_0, λ_0). There is a monotonic increase in the posterior model probability as more weight is given to the historical data.

In summary, Tables 1, 2, and 3 show that if c_0 and d_0 are moderately large, then model choice is not sensitive to their choices. However, model choice can be sensitive if c_0 and/or d_0 are taken to be small. Moreover, we see that monotonic increases in the posterior model probability as c_0 or d_0 are increased.

Finally, we note that Ibrahim et al. (2000) and Chen et al. (2000a) have also established a nice connection between the posterior model probabilities and the marginal posterior densities in the context of Bayesian variable selection for time series count data and multivariate mortality data, respectively.

7. Concluding remarks

In this chapter, we provide an overview of various Monte Carlo methods for computing full or marginal posterior densities. The density estimation methods discussed in Sections 2.3 to 2.6 are generally more efficient than the kernel method, since the known structure or the information of the joint posterior density or the MCMC sampling algorithms is completely ignored in the kernel estimator. In addition, we demonstrate that Bayesian computational problems such as computing marginal likelihoods, Bayes factors, and posterior model probabilities essentially reduce to the problem of full or marginal posterior density estimation. In the Bayesian computational literature, there are also several other popular Monte Carlo methods, such as importance sampling and the methods for computing a single normalizing constant or the ratios of two normalizing constants. In fact, if we rewrite (2.6) as

$$\pi\left(\boldsymbol{\theta}^{*(j)}|D\right) = \int \frac{w(\boldsymbol{\theta}^{(j)}|\boldsymbol{\theta}^{(-j)})}{\pi(\boldsymbol{\theta}|D)}\pi\left(\boldsymbol{\theta}^{*(j)}, \boldsymbol{\theta}^{(-j)}|D\right)\pi(\boldsymbol{\theta}|D)\,\mathrm{d}\boldsymbol{\theta}.$$

Then, the IWMDE is an importance sampling estimator with the importance sampling weight as $w(\boldsymbol{\theta}^{(j)}|\boldsymbol{\theta}^{(-j)})/\pi(\boldsymbol{\theta}|D)$.

Due to the Chib's identity (5.1), computing a single normalizing constant can be converted to computing the full posterior density at a fixed point. On the other hand, the Monte Carlo method for computing ratios of normalizing constants can be used to derive an improved the density estimator for computing the full or marginal posterior densities. The construction of $\hat{\pi}_{\mathrm{CJ}}(\boldsymbol{\theta}^{*(j)}|D)$ or $\hat{\pi}_{\mathrm{BS}}(\boldsymbol{\theta}^{*(j)}|D)$ via the bridge sampling approach of Meng and Wong (1996) is one of these examples. The Savage–Dickey density ratio and the Monte Carlo method developed by Chen et al. (1999) illustrate that the posterior density estimation can be very useful in computing the Bayes factor and the posterior model probability.

Perhaps one of the most challenging Bayesian computational problems is to carry out Bayesian variable selection when the number of covariates is large. The main reason is that it is impossible to enumerate all possible models in this case. One possible solution to this computational problem is to use the reversible jump MCMC algorithm (Green, 1995). However, for efficiently estimating posterior model probabilities, the reversible jump MCMC algorithm may not be most desirable as it requires a very large MCMC sample size in order to visit all probable models. The development of an efficient and practically useful Monte Carlo method for this problem is a very challenging and important future project.

References

Akaike, H. (1973). Information theory and an extension of the maximum likelihood principle. In: Petrov, B.N., Csaki, F. (Eds.), *International Symposium on Information Theory*. Akademia Kiado, Budapest, pp. 267–281.

Bickman, L., Guthrie, P.R., Foster, E.M., Lambert, E.W., Summerfelt, W.T., Breda, C.S., Helfinger, C.A. (1995). *Evaluating Managed Mental Health Services: Fort Bragg Experiment*. Plenum, New York, pp. 245–268.

Chen, C.F. (1985). On asymptotic normality of limiting density functions with Bayesian implications. *J. Roy. Statist. Soc., Ser. B* **97**, 540–546.
Chen, M.-H. (1994). Importance-weighted marginal Bayesian posterior density estimation. *J. Amer. Statist. Assoc.* **89**, 818–824.
Chen, M.-H., Shao, Q.-M. (1997a). Performance study of marginal posterior density estimation via Kullback–Leibler divergence. *Test* **6**, 321–350.
Chen, M.-H., Shao, Q.-M. (1997b). Estimating ratios of normalizing constants for densities with different dimensions. *Statistica Sinica* **7**, 607–630.
Chen, M.-H., Ibrahim, J.G., Yiannoutsos, C. (1999). Prior elicitation and Bayesian computation for logistic regression models with applications to variable selection. *J. Roy. Statist. Soc., Ser. B* **61**, 223–242.
Chen, M.-H., Dey, D.K., Sinha, D. (2000a). Bayesian analysis of multivariate mortality data with large families. *Appl. Statist.* **49**, 129–144.
Chen, M.-H., Ibrahim, J.G., Shao, Q.-M. (2000b). Power prior distributions for generalized linear models. *J. Statist. Plann. Inference* **84**, 121–137.
Chib, S. (1995). Marginal likelihood from the Gibbs output. *J. Amer. Statist. Assoc.* **90**, 1313–1321.
Chib, S., Jeliazkov, I. (2001). Marginal likelihood from the Metropolis–Hastings output. *J. Amer. Statist. Assoc.* **96**, 270–281.
Devroye, L., Wagner, T.J. (1980). The strong uniform consistency of kernel density estimates. In: Krishnaiah, P.K. (Ed.), *Multivariate Analysis V*. North-Holland, Amsterdam, pp. 59–77.
Dickey, J. (1971). The weighted likelihood ratio, linear hypotheses on normal location parameters. *Ann. Statist.* **42**, 204–223.
Gelfand, A.E., Smith, A.F.M., Lee, T.M. (1992). Bayesian analysis of constrained parameter and truncated data problems using Gibbs sampling. *J. Amer. Statist. Assoc.* **87**, 523–532.
Green, P.J. (1995). Reversible jump Markov chain Monte Carlo computation and Bayesian model determination. *Biometrika* **82**, 711–732.
Ibrahim, J.G., Chen, M.-H. (2000). Power prior distributions for regression models. *Statist. Sci.* **15**, 46–60.
Ibrahim, J.G., Chen, M.-H., Ryan, L.-M. (2000). Bayesian variable selection for time series count data. *Statistica Sinica* **10**, 971–987.
Johnson, V.E. (1992). A technique for estimating marginal posterior densities in hierarchical models using mixtures of conditional densities. *J. Amer. Statist. Assoc.* **87**, 852–860.
McCullagh, P., Nelder, J.A. (1989). *Generalized Linear Models*, second ed. Chapman and Hall, London.
Meng, X.-L., Schilling, S. (2002). Warp bridge sampling. *J. Comput. Graph. Statist.* **11**, 552–586.
Meng, X.-L., Wong, W.H. (1996). Simulating ratios of normalizing constants via a simple identity: A theoretical exploration. *Statistica Sinica* **6**, 831–860.
Mira, A., Nicholls, G. (2004). Bridge estimation of the probability density at a point. *Statistica Sinica* **14**, 603–612.
Oh, M.-S. (1999). Estimation of posterior density functions from a posterior sample. *Comput. Statist. Data Anal.* **29**, 411–427.
Ritter, C., Tanner, M.A. (1992). The Gibbs stopper and the griddy-Gibbs sampler. *J. Amer. Statist. Assoc.* **87**, 861–868.
Schwarz, G. (1978). Estimating the dimension of a model. *Ann. Statist.* **6**, 461–464.
Silverman, B.W. (1986). *Density Estimation for Statistics and Data Analysis*. Chapman & Hall, London.
Verdinelli, I., Wasserman, L. (1995). Computing Bayes factors using a generalization of the Savage–Dickey density ratio. *J. Amer. Statist. Assoc.* **90**, 614–618.
Yu, J.Z., Tanner, M.A. (1999). An analytical study of several Markov chain Monte Carlo estimators of the marginal likelihood. *J. Comput. Graph. Statist.* **8**, 839–853.

Handbook of Statistics, Vol. 25
ISSN: 0169-7161
© 2005 Elsevier B.V. All rights reserved.
DOI 10.1016/S0169-7161(05)25016-2

16

Bayesian Modelling and Inference on Mixtures of Distributions

Jean-Michel Marin, Kerrie Mengersen and Christian P. Robert[1]

'But, as you have already pointed out, we do not need any more disjointed clues,' said Bartholomew. 'That has been our problem all along: we have a mass of small facts and small scraps of information, but we are unable to make any sense out of them. The last thing we need is more.'
Susanna Gregory, *A Summer of Discontent*

1. Introduction

Today's data analysts and modellers are in the luxurious position of being able to more closely describe, estimate, predict and infer about complex systems of interest, thanks to ever more powerful computational methods but also wider ranges of modelling distributions. Mixture models constitute a fascinating illustration of these aspects: while within a parametric family, they offer malleable approximations in nonparametric settings; although based on standard distributions, they pose highly complex computational challenges; and they are both easy to constrain to meet identifiability requirements and fall within the class of ill-posed problems. They also provide an endless benchmark for assessing new techniques, from the EM algorithm to reversible jump methodology. In particular, they exemplify the formidable opportunity provided by new computational technologies like Markov chain Monte Carlo (MCMC) algorithms. It is no coincidence that the Gibbs sampling algorithm for the estimation of mixtures was proposed *before* (Tanner and Wong, 1987) and *immediately after* (Diebolt and Robert, 1990c) the seminal paper of Gelfand and Smith (1990): before MCMC was popularised, there simply was no satisfactory approach to the computation of Bayes estimators for mixtures of distributions, even though older importance sampling algorithms were later discovered to apply to the simulation of posterior distributions of mixture parameters (Casella et al., 2002).

[1]Jean-Michel Marin is lecturer in Université Paris Dauphine, Kerrie Mengersen is professor in the University of Newcastle, and Christian P. Robert is professor in Université Paris Dauphine and head of the Statistics Laboratory of CREST.

Mixture distributions comprise a finite or infinite number of components, possibly of different distributional types, that can describe different features of data. They thus facilitate much more careful description of complex systems, as evidenced by the enthusiasm with which they have been adopted in such diverse areas as astronomy, ecology, bioinformatics, computer science, economics, engineering, robotics and biostatistics. For instance, in genetics, location of quantitative traits on a chromosome and interpretation of microarrays both relate to mixtures, while, in computer science, spam filters and web context analysis (Jordan, 2004) start from a mixture assumption to distinguish spams from regular emails and group pages by topic, respectively.

Bayesian approaches to mixture modelling have attracted great interest among researchers and practitioners alike. The Bayesian paradigm (Berger, 1985; Besag et al., 1995; Robert, 2001, see, e.g.) allows for probability statements to be made directly about the unknown parameters, prior or expert opinion to be included in the analysis, and hierarchical descriptions of both local-scale and global features of the model. This framework also allows the complicated structure of a mixture model to be decomposed into a set of simpler structures through the use of hidden or latent variables. When the number of components is unknown, it can well be argued that the Bayesian paradigm is the only sensible approach to its estimation (Richardson and Green, 1997).

This chapter aims to introduce the reader to the construction, prior modelling, estimation and evaluation of mixture distributions in a Bayesian paradigm. We will show that mixture distributions provide a flexible, parametric framework for statistical modelling and analysis. Focus is on methods rather than advanced examples, in the hope that an understanding of the practical aspects of such modelling can be carried into many disciplines. It also stresses implementation via specific MCMC algorithms that can be easily reproduced by the reader. In Section 2, we detail some basic properties of mixtures, along with two different motivations. Section 3 points out the fundamental difficulty in doing inference with such objects, along with a discussion about prior modelling, which is more restrictive than usual, and the constructions of estimators, which also is more involved than the standard posterior mean solution. Section 4 describes the completion and noncompletion MCMC algorithms that can be used for the approximation to the posterior distribution on mixture parameters, followed by an extension of this analysis in Section 5 to the case in which the number of components is unknown and may be estimated by Green's (1995) reversible jump algorithm and Stephens' 2000 birth-and-death procedure. Section 6 gives some pointers to related models and problems like mixtures of regressions (or conditional mixtures) and hidden Markov models (or dependent mixtures), as well as Dirichlet priors.

2. The finite mixture framework

2.1. *Definition*

The description of a mixture of distributions is straightforward: any convex combination

$$\sum_{i=1}^{k} p_i f_i(x), \qquad \sum_{i=1}^{k} p_i = 1, \quad k > 1, \tag{1}$$

of other distributions f_i is a *mixture*. While continuous mixtures

$$g(x) = \int_{\Theta} f(x|\theta)h(\theta)\,\mathrm{d}\theta$$

are also considered in the literature, we will not treat them here. In most cases, the f_i's are from a parametric family, with unknown parameter θ_i, leading to the parametric mixture model

$$\sum_{i=1}^{k} p_i f(x|\theta_i). \tag{2}$$

In the particular case in which the $f(x|\theta)$'s are all normal distributions, with θ representing the unknown mean and variance, the range of shapes and features of the mixture (2) can widely vary, as shown[2] by Figure 1.

Since we will motivate mixtures as approximations to unknown distributions (Section 2.3), note at this stage that the tail behaviour of a mixture is always described by one or two of its components and that it therefore reflects the choice of the parametric

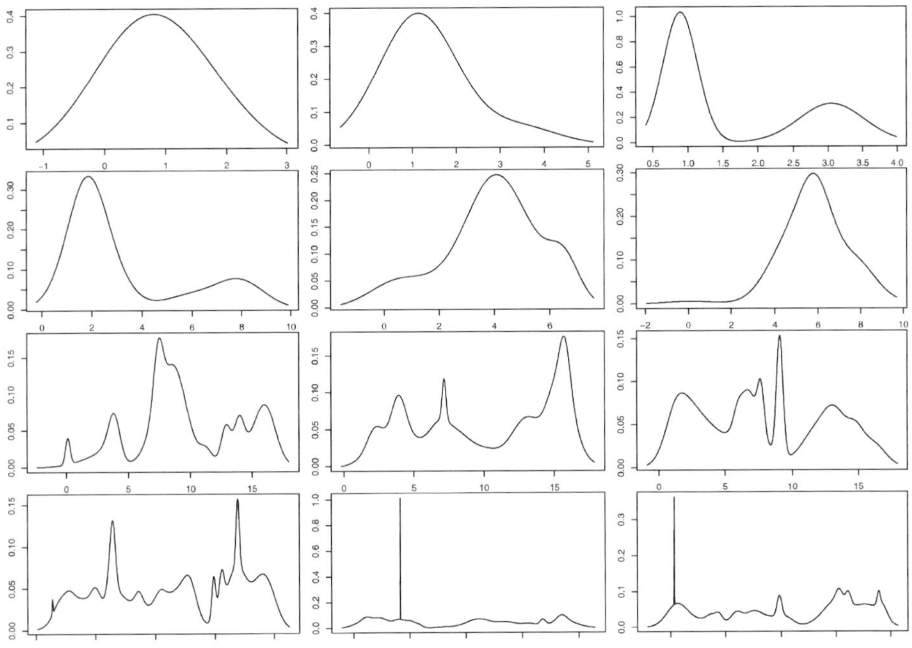

Fig. 1. Some normal mixture densities for $K = 2$ (*first row*), $K = 5$ (*second row*), $K = 25$ (*third row*) and $K = 50$ (*last row*).

[2] To draw this set of densities, we generated the weights from a Dirichlet $\mathscr{D}(1, \ldots, 1)$ distribution, the means from a uniform $\mathscr{U}[0, 5\log(k)]$ distribution, and the variances from a Beta $\mathscr{B}e(1/(0.5+0.1\log(k)), 1)$, which means in particular that the variances are all less than 1. The resulting shapes reflect this choice, as the reader can easily check by running her or his own simulation experiment.

family $f(\cdot|\theta)$. Note also that the representation of mixtures as convex combinations of distributions implies that the moments of (1) are convex combinations of the moments of the f_j's:

$$\mathbb{E}[X^m] = \sum_{i=1}^{k} p_i \mathbb{E}^{f_i}[X^m].$$

This fact was exploited as early as 1894 by Karl Pearson to derive a moment estimator of the parameters of a normal mixture with two components,

$$p\varphi(x; \mu_1, \sigma_1) + (1 - p)\varphi(x; \mu_2, \sigma_2), \tag{3}$$

where $\varphi(\cdot; \mu, \sigma)$ denotes the density of the $\mathscr{N}(\mu, \sigma^2)$ distribution.

Unfortunately, the representation of the mixture model given by (2) is detrimental to the derivation of the maximum likelihood estimator (when it exists) and of Bayes estimators. To see this, consider the case of n i.i.d. observations $\underline{x} = (x_1, \ldots, x_n)$ from this model. Defining $\underline{p} = (p_1, \ldots, p_k)$ and $\underline{\theta} = (\theta_1, \ldots, \theta_k)$, we see that even though conjugate priors may be used for each component parameter (p_i, θ_i), the explicit representation of the corresponding posterior expectation involves the expansion of the likelihood

$$\mathbb{L}(\underline{\theta}, \underline{p}|\underline{x}) = \prod_{i=1}^{n} \sum_{j=1}^{k} p_j f(x_i|\theta_j) \tag{4}$$

into k^n terms, which is computationally too expensive to be used for more than a few observations (see Diebolt and Robert, 1990a, 1990b, and Section 3.1). Unsurprisingly, one of the first occurrences of the Expectation–Maximization (EM) algorithm of Dempster et al. (1977) addresses the problem of solving the likelihood equations for mixtures of distributions, as detailed in Section 3.2. Other approaches to overcoming this computational hurdle are described in the following sections.

2.2. *Missing data approach*

There are several motivations for considering mixtures of distributions as a useful extension to "standard" distributions. The most natural approach is to envisage a dataset as constituted of several strata or subpopulations. One of the early occurrences of mixture modelling can be found in Bertillon (1895) where the bimodal structure on the height of (military) conscripts in central France can be explained by the mixing of two populations of young men, one from the plains and one from the mountains (or hills). The mixture structure appears because the origin of each observation, that is, the allocation to a specific subpopulation or stratum, is lost. Each of the X_i's is thus *a priori* distributed from either of the f_j's with probability p_j. Depending on the setting, the inferential goal may be either to reconstitute the groups, usually called *clustering*, to provide estimators for the parameters of the different groups or even to estimate the number of groups.

While, as seen below, this is not always the reason for modelling by mixtures, the missing structure inherent to this distribution can be exploited as a technical device to facilitate estimation. By a demarginalisation argument, it is always possible to associate to a random variable X_i from a mixture of k distributions (2) another random variable Z_i such that

$$X_i|Z_i = z \sim f(x|\theta_z), \qquad Z_i \sim \mathscr{M}_k(1; p_1, \ldots, p_k), \tag{5}$$

where $\mathscr{M}_k(1; p_1, \ldots, p_k)$ denotes the multinomial distribution with k modalities and a single observation. This auxiliary variable identifies to which component the observation x_i belongs. Depending on the focus of inference, the Z_i's will or will not be part of the quantities to be estimated.[3]

2.3. *Nonparametric approach*

A different approach to the interpretation and estimation mixtures is semi-parametric. Noticing that very few phenomena obey the most standard distributions, it is a trade-off between fair representation of the phenomenon and efficient estimation of the underlying distribution to choose the representation (2) for an unknown distribution. If k is large enough, there is support for the argument that (2) provides a good approximation to most distributions. Hence a mixture distribution can be approached as a type of basis approximation of unknown distributions, in a spirit similar to wavelets and such, but with a more intuitive flavour. This argument will be pursued in Section 3.5 with the construction of a new parameterisation of the normal mixture model through its representation as a sequence of perturbations of the original normal model.

Note first that the most standard nonparametric density estimator, namely the *Nadaraya–Watson* kernel (Hastie et al., 2001) estimator, is based on a (usually Gaussian) mixture representation of the density,

$$\hat{k}_n(x|\underline{x}) = \frac{1}{nh_n}\sum_{i=1}^{n} \varphi(x; x_i, h_n),$$

where $\underline{x} = (x_1, \ldots, x_n)$ is the sample of i.i.d. observations. Under weak conditions on the so-called *bandwidth* h_n, $\hat{k}_n(x)$ does converge (in L_2 norm and pointwise) to the true density $f(x)$ (Silverman, 1986).[4]

The most common approach in Bayesian nonparametric Statistics is to use the so-called *Dirichlet process distribution*, $\mathscr{D}(F_0, \alpha)$, where F_0 is a cdf and α is a precision parameter (Ferguson, 1974). This prior distribution on the space of distributions enjoys the coherency property that, if $F \sim \mathscr{D}(F_0, \alpha)$, the vector $(F(A_1), \ldots, F(A_p))$ is distributed as a Dirichlet variable in the usual sense

$$\mathscr{D}_p\big(\alpha F_0(A_1), \ldots, \alpha F_0(A_p)\big)$$

---

[3] It is always awkward to talk of the Z_i's as parameters because, on the one hand, they may be purely artificial, and thus not pertain to the distribution of the observables, and, on the other hand, the fact that they increase in dimension at the same speed as the observables creates a difficulty in terms of asymptotic validation of inferential procedures (Diaconis and Freedman, 1986). We thus prefer to call them *auxiliary variables* as in other simulation setups.

[4] A remark peripheral to this chapter but related to footnote 3 is that the Bayesian *estimation* of h_n does not produce a consistent estimator of the density.

for every partition $(A_1, \ldots, A_p)$. But, more importantly, it leads to a mixture representation of the posterior distribution on the unknown distribution: if $x_1, \ldots, x_n$ are distributed from F and $F \sim \mathscr{D}(F_0, \alpha)$, the marginal conditional cdf of x_1 given $(x_2, \ldots, x_n)$ is

$$\left(\frac{\alpha}{\alpha+n-1}\right) F_0(x_1) + \left(\frac{1}{\alpha+n-1}\right) \sum_{i=2}^{n} \mathbb{I}_{x_i \leqslant x_1}.$$

Another approach is to be found in the Bayesian nonparametric papers of Verdinelli and Wasserman (1998), Barron et al. (1999) and Petrone and Wasserman (2002), under the name of *Bernstein polynomials*, where bounded continuous densities with supports on [0, 1] are approximated by (infinite) Beta mixtures

$$\sum_{(\alpha_k, \beta_k) \in \mathbb{N}_+^2} p_k \mathscr{B}e(\alpha_k, \beta_k),$$

with integer parameters (in the sense that the posterior and the predictive distributions are consistent under mild conditions). More specifically, the prior distribution on the distribution is that it is a Beta mixture

$$\sum_{j=1}^{k} \omega_{kj}\, \mathscr{B}e(j, k+1-j)$$

with probability $p_k = \mathbb{P}(K = k)$ $(k = 1, \ldots)$ and $\omega_{kj} = F(j/k) - F((j-1)/k)$ for a certain cdf F. Given a sample $\underline{x} = (x_1, \ldots, x_n)$, the associated predictive is then

$$\hat{f}_n(x|\underline{x}) = \sum_{k=1}^{\infty} \sum_{j=1}^{k} \mathbb{E}^{\pi}[\omega_{kj}|\underline{x}]\mathscr{B}e(j, k+1-j)\, \mathbb{P}(K = k|\underline{x}).$$

The sum is formally infinite but for obvious practical reasons it needs to be truncated to $k \leqslant k_n$, with $k_n \propto n^{\alpha}$, $\alpha < 1$ (Petrone and Wasserman, 2002). Figure 2 represents a few simulations from the Bernstein prior when K is distributed from a Poisson $\mathscr{P}(\lambda)$ distribution and F is the $\mathscr{B}e(\alpha, \beta)$ cdf.

As a final illustration, consider the goodness of fit approach proposed by Robert and Rousseau (2002). The central problem is to test whether or not a given parametric model is compatible with the data at hand. If the null hypothesis holds, the cdf distribution of the sample is $\mathscr{U}(0, 1)$. When it does not hold, the cdf can be any cdf on [0, 1]. The choice made in Robert and Rousseau (2002) is to use a general mixture of Beta distributions,

$$p_0 \mathscr{U}(0, 1) + (1 - p_0) \sum_{k=1}^{K} p_k \mathscr{B}e(\alpha_k, \beta_k), \tag{6}$$

to represent the alternative by singling out the $\mathscr{U}(0, 1)$ component, which also is a $\mathscr{B}e(1, 1)$ density. Robert and Rousseau (2002) prove the consistency of this approximation for a large class of densities on [0, 1], a class that obviously contains the continuous bounded densities already well-approximated by Bernstein polynomials. Given that this

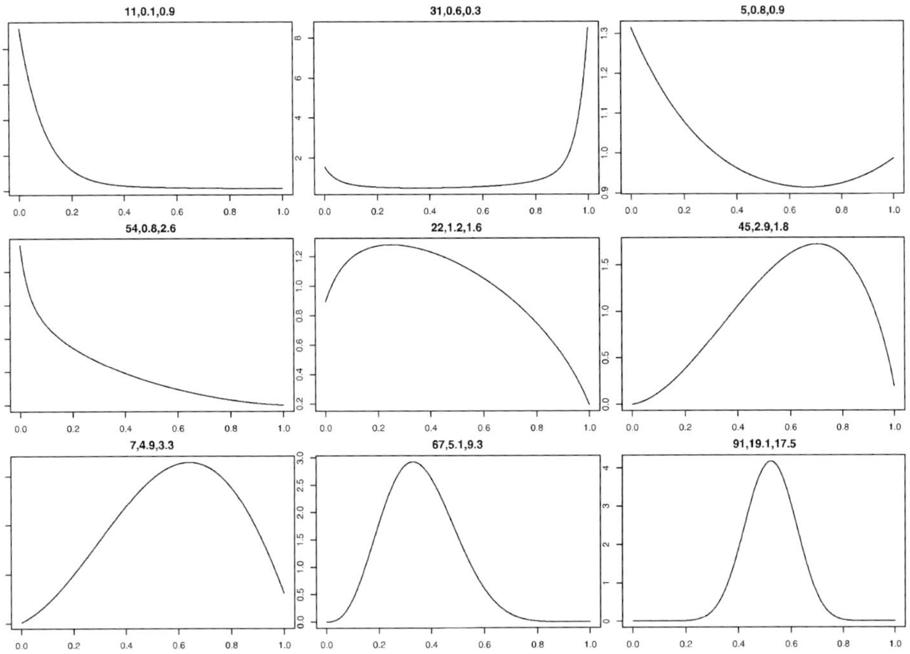

Fig. 2. Realisations from the Bernstein prior when $K \sim \mathscr{P}(\lambda)$ and F is the $\mathscr{B}e(\alpha, \beta)$ cdf for various values of (λ, α, β).

is an approximation of the true distribution, the number of components in the mixture is unknown and needs to be estimated. Figure 3 shows a few densities corresponding to various choices of K and p_k, α_k, β_k. Depending on the range of the (α_k, β_k)'s, different behaviours can be observed in the vicinities of 0 and 1, with much more variability than with the Bernstein prior which restricts the (α_k, β_k)'s to be integers.

An alternative to mixtures of Beta distributions for modelling unknown distributions is considered in Perron and Mengersen (2001) in the context of nonparametric regression. Here, mixtures of triangular distributions are used instead and compare favourably with Beta equivalents for certain types of regression, particularly those with sizeable jumps or changepoints.

2.4. *Reading*

Very early references to mixture modelling start with Pearson (1894), even though earlier writings by Quetelet and other 19th century statisticians mention these objects and sometimes try to recover the components. Early (modern) references to mixture modelling include Dempster et al. (1977), who considered maximum likelihood for incomplete data via the EM algorithm. In the 1980's, increasing interest in mixtures included Bayesian analysis of simple mixture models (Bernardo and Giron, 1988), stochastic EM derived for the mixture problem (Celeux and Diebolt, 1985), and approximation of priors by mixtures of natural conjugate priors (Redner and Walker, 1984). The 1990's

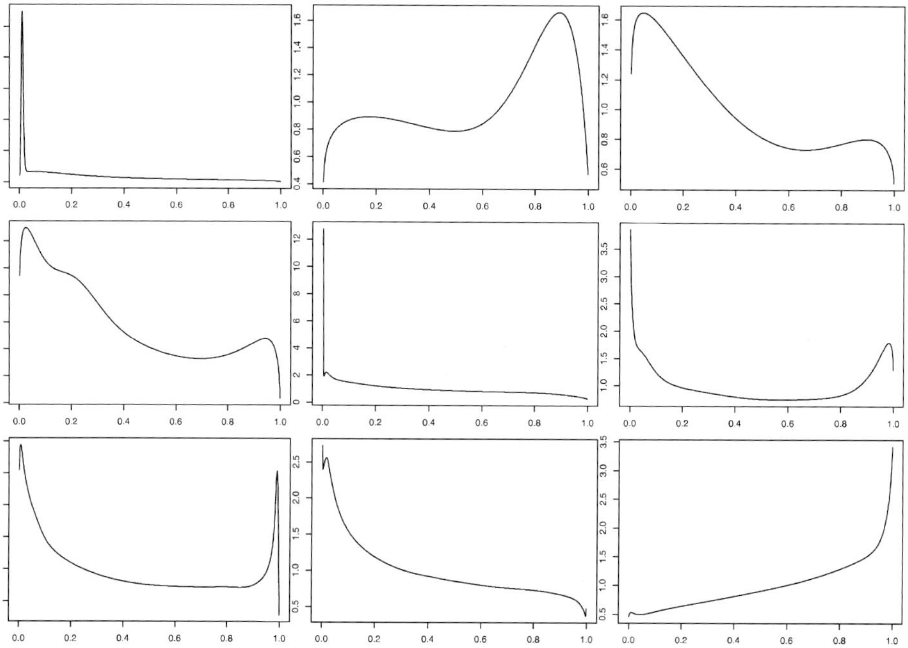

Fig. 3. Some beta mixture densities for $K = 10$ (*upper row*), $K = 100$ (*central row*) and $K = 500$ (*lower row*).

saw an explosion of publications on the topic, with many papers directly addressing mixture estimation and many more using mixtures of distributions as in, e.g., Kim et al. (1998). Seminal texts for finite mixture distributions include Titterington et al. (1985), MacLachlan and Basford (1988), and MacLachlan and Peel (2000).

3. The mixture conundrum

If these finite mixture models are so easy to construct and have such widely recognised potential, then why are they not universally adopted? One major obstacle is the difficulty of estimation, which occurs at various levels: the model itself, the prior distribution and the resulting inference.

EXAMPLE 1. To get a first impression of the complexity of estimating mixture distributions, consider the simple case of a two component normal mixture

$$p\mathcal{N}(\mu_1, 1) + (1 - p)\mathcal{N}(\mu_2, 1), \tag{7}$$

where the weight $p \neq 0.5$ is known. The parameter space is then $\mathbb{R}^2$ and the parameters are identifiable: the switching phenomenon presented in Section 3.4 does not occur because μ_1 cannot be confused with μ_2 when p is known and different from 0.5. Nonetheless, the log-likelihood surface represented in Figure 4 exhibits two modes: one close to the true value of the parameters used to simulate the corresponding dataset and one being a "spurious" mode that does not mean much in terms of the true values of the parameters, but is always present. Obviously, if we plot the likelihood, only one mode is visible because of the difference in the magnitudes.

3.1. Combinatorics

As noted earlier, the likelihood function (4) leads to k^n terms when the inner sums are expanded. While this expansion is not necessary to compute the likelihood at a given value $(\underline{\theta}, \underline{p})$, which is feasible in $O(nk)$ operations as demonstrated by the representation in Figure 4, the computational difficulty in using the expanded version of (4) precludes analytic solutions via maximum likelihood or Bayes estimators (Diebolt and Robert, 1990b). Indeed, let us consider the case of n i.i.d. observations from model (2) and let us denote by $\pi(\underline{\theta}, \underline{p})$ the prior distribution on $(\underline{\theta}, \underline{p})$. The posterior distribution is then

$$\pi(\underline{\theta}, \underline{p}|\underline{x}) \propto \left(\prod_{i=1}^{n}\sum_{j=1}^{k} p_j f(x_i|\theta_j)\right)\pi(\underline{\theta}, \underline{p}). \tag{8}$$

EXAMPLE 2. As an illustration of this frustrating combinatoric explosion, consider the case of n observations $\underline{x} = (x_1, \ldots, x_n)$ from a normal mixture

$$p\varphi(x; \mu_1, \sigma_1) + (1-p)\varphi(x; \mu_2, \sigma_2) \tag{9}$$

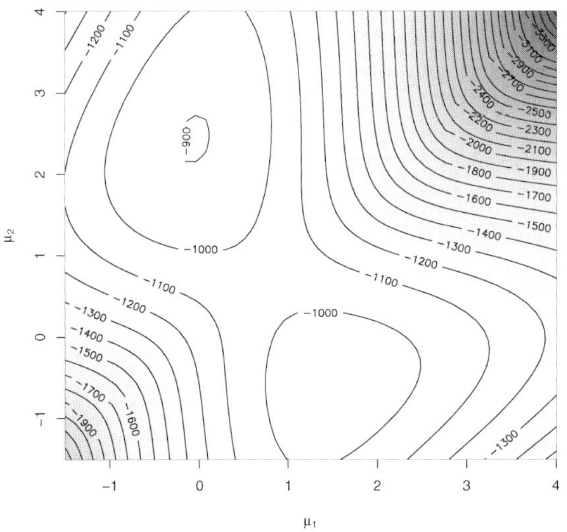

Fig. 4. R image representation of the log-likelihood of the mixture (7) for a simulated dataset of 500 observations and true value $(\mu_1, \mu_2, p) = (0, 2.5, 0.7)$. For a color reproduction of this figure see the color figures section, page 993.

under the pseudo-conjugate priors ($i = 1, 2$)

$$\mu_i|\sigma_i \sim \mathscr{N}(\zeta_i, \sigma_i^2/\lambda_i), \qquad \sigma_i^{-2} \sim \mathscr{G}a(\nu_i/2, s_i^2/2), \qquad p \sim \mathscr{B}e(\alpha, \beta),$$

where $\mathscr{G}a(\nu, s)$ denotes the Gamma distribution. Note that the hyperparameters $\zeta_i, \sigma_i, \nu_i, s_i, \alpha$ and β need to be specified or endowed with an hyperprior when they cannot be specified. In this case $\underline{\theta} = (\mu_1, \mu_2, \sigma_1^2, \sigma_2^2)$, $\underline{p} = p$ and the posterior is

$$\pi(\underline{\theta}, p|\underline{x}) \propto \prod_{j=1}^{n} \{p\varphi(x_j; \mu_1, \sigma_1) + (1-p)\varphi(x_j; \mu_2, \sigma_2)\}\pi(\underline{\theta}, p).$$

This posterior could be computed at a given value $(\underline{\theta}, p)$ in $\mathrm{O}(2n)$ operations. Unfortunately, the computational burden is that there are 2^n terms in this sum and it is impossible to give analytical derivations of maximum likelihood and Bayes estimators.

We will now present another decomposition of expression (8) which shows that only very few values of the k^n terms have a nonnegligible influence. Let us consider the auxiliary variables $\underline{z} = (z_1, \ldots, z_n)$ which identify to which component the observations $\underline{x} = (x_1, \ldots, x_n)$ belong. Moreover, let us denote by $\mathcal{Z}$ the set of all k^n allocation vectors $\underline{z}$. The set $\mathcal{Z}$ has a rich and interesting structure. In particular, for k labelled components, we can decompose $\mathcal{Z}$ into a partition of sets as follows. For a given allocation size $(n_1, \ldots, n_k)$, where $n_1 + \cdots + n_k = n$, let us define the set

$$\mathcal{Z}_j = \left\{\underline{z}\colon \sum_{i=1}^{n} \mathbb{I}_{z_i=1} = n_1, \ldots, \sum_{i=1}^{n} \mathbb{I}_{z_i=k} = n_k\right\}$$

which consists of all allocations with the given allocation size $(n_1, \ldots, n_k)$, relabelled by $j \in \mathbb{N}$, using for instance the lexicographical ordering on $(n_1, \ldots, n_k)$. The number of nonnegative integer solutions of the decomposition of n into k parts such that $n_1 + \cdots + n_k = n$ is equal to

$$r = \binom{n+k-1}{n}.$$

Thus, we have the partition $\mathcal{Z} = \bigcup_{i=1}^{r} \mathcal{Z}_i$. Although the total number of elements of $\mathcal{Z}$ is the typically unmanageable k^n, the number of partition sets is much more manageable since it is of order $n^{k-1}/(k-1)!$. The posterior distribution can be written as

$$\pi(\underline{\theta}, \underline{p}|\underline{x}) = \sum_{i=1}^{r} \sum_{\underline{z}\in\mathcal{Z}_i} \omega(\underline{z})\pi(\underline{\theta}, \underline{p}|\underline{x}, \underline{z}), \tag{10}$$

where $\omega(\underline{z})$ represents the posterior probability of the given allocation $\underline{z}$. Note that with this representation, a Bayes estimator of $(\underline{\theta}, \underline{p})$ could be written as

$$\sum_{i=1}^{r} \sum_{\underline{z}\in\mathcal{Z}_i} \omega(\underline{z})\mathbb{E}^{\pi}[\underline{\theta}, \underline{p}|\underline{x}, \underline{z}]. \tag{11}$$

This decomposition makes a lot of sense from an inferential point of view: the Bayes posterior distribution simply considers each possible allocation $\underline{z}$ of the dataset, allocates a posterior probability $\omega(\underline{z})$ to this allocation, and then constructs a posterior distribution for the parameters conditional on this allocation. Unfortunately, as for the likelihood, the computational burden is that there are k^n terms in this sum. This is even more frustrating given that the overwhelming majority of the posterior probabilities $\omega(\underline{z})$ will be close to zero. In a Monte Carlo study, Casella et al. (2000) have showed that the nonnegligible weights correspond to very few values of the partition sizes. For instance, the analysis of a dataset with $k = 4$ components, presented in Example 4 below, leads to the set of allocations with the partition sizes $(n_1, n_2, n_3, n_4) = (7, 34, 38, 3)$ with probability 0.59 and $(n_1, n_2, n_3, n_4) = (7, 30, 27, 18)$ with probability 0.32, with no other size group getting a probability above 0.01.

EXAMPLE 1 (*Continued*). In the special case of model (7), if we take *the same* normal prior on both μ_1 and μ_2, $\mu_1, \mu_2 \sim \mathcal{N}(0, 10)$, the posterior weight associated with an allocation $\underline{z}$ for which l values are attached to the first component, i.e. such that $\sum_{i=1}^{n} \mathbb{I}_{z_i=1} = l$, will simply be

$$\omega(\underline{z}) \propto \sqrt{(l + 1/4)(n - l + 1/4)}\, p^l (1 - p)^{n-l},$$

because the marginal distribution of $\underline{x}$ is then independent of $\underline{z}$. Thus, when the prior does *not* discriminate between the two means, the posterior distribution of the allocation $\underline{z}$ only depends on l and the repartition of the partition size l simply follows a distribution close to a binomial $\mathscr{B}(n, p)$ distribution. If, instead, we take two different normal priors on the means,

$$\mu_1 \sim \mathcal{N}(0, 4), \qquad \mu_2 \sim \mathcal{N}(2, 4),$$

the posterior weight of a given allocation $\underline{z}$ is now

$$\begin{aligned}\omega(\underline{z}) \propto{}& \sqrt{(l + 1/4)(n - l + 1/4)}\, p^l (1 - p)^{n-l} \\ &\times \exp\left\{-\left[(l + 1/4)\hat{s}_1(\underline{z}) + l\left\{\bar{x}_1(\underline{z})\right\}^2/4\right]/2\right\} \\ &\times \exp\left\{-\left[(n - l + 1/4)\hat{s}_2(\underline{z}) + (n - l)\left\{\bar{x}_2(\underline{z}) - 2\right\}^2/4\right]/2\right\},\end{aligned}$$

where

$$\bar{x}_1(\underline{z}) = \frac{1}{l}\sum_{i=1}^{n} \mathbb{I}_{z_i=1} x_i, \qquad \bar{x}_2(\underline{z}) = \frac{1}{n-l}\sum_{i=1}^{n} \mathbb{I}_{z_i=2} x_i,$$

$$\hat{s}_1(\underline{z}) = \sum_{i=1}^{n} \mathbb{I}_{z_i=1}\left(x_i - \bar{x}_1(\underline{z})\right)^2, \qquad \hat{s}_2(\underline{z}) = \sum_{i=1}^{n} \mathbb{I}_{z_i=2}\left(x_i - \bar{x}_2(\underline{z})\right)^2.$$

This distribution obviously depends on both $\underline{z}$ and the dataset. While the computation of the weight of all partitions of size l by a complete listing of the corresponding $\underline{z}$'s is impossible when n is large, this weight can be approximated by a Monte Carlo experiment, when drawing the $\underline{z}$'s at random. For instance, a sample of 45 points simulated from (7) when $p = 0.7$, $\mu_1 = 0$ and $\mu_2 = 2.5$ leads to $l = 23$ as the most

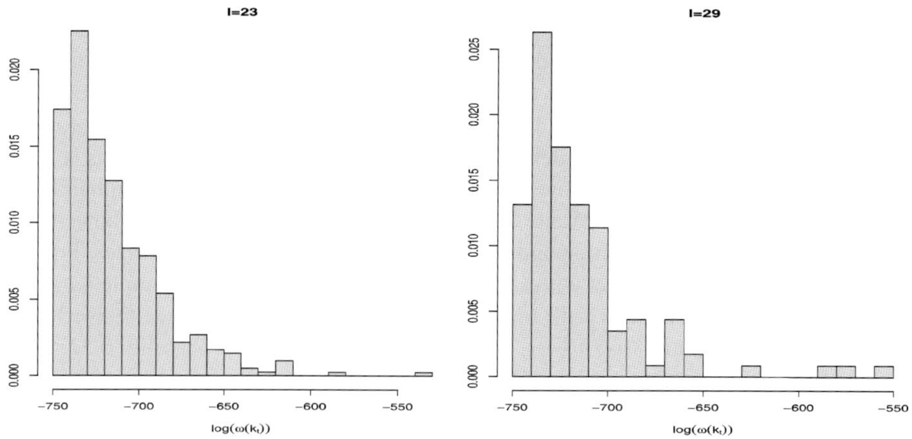

Fig. 5. Comparison of the distribution of the $\omega(\underset{\sim}{z})$'s (up to an additive constant) when $l = 23$ and when $l = 29$ for a simulated dataset of 45 observations and true values $(\mu_1, \mu_2, p) = (0, 2.5, 0.7)$.

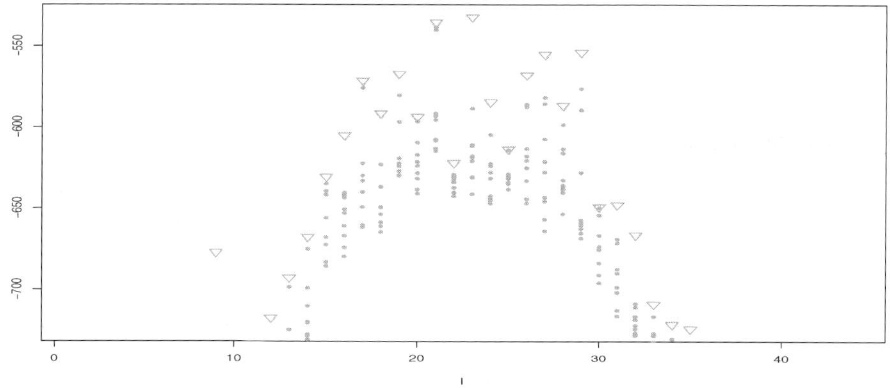

Fig. 6. Ten highest log-weights $\omega(\underset{\sim}{z})$ (up to an additive constant) found in the simulation of random allocations for each partition size l for the same simulated dataset as in Figure 5. (Triangles represent the highest weights.) For a color reproduction of this figure see the color figures section, page 993.

likely partition, with a weight approximated by 0.962. Figure 5 gives the repartition of the $\log \omega(\underset{\sim}{z})$'s in the cases $l = 23$ and $l = 27$. In the latter case, the weight is approximated by $4.56\,10^{-11}$. (The binomial factor $\binom{n}{l}$ that corresponds to the actual number of different partitions with l allocations to the first component was taken into account for the approximation of the posterior probability of the partition size.) Note that both distributions of weights are quite concentrated, with only a few weights contributing to the posterior probability of the partition. Figure 6 represents the 10 highest weights associated with each partition size ℓ and confirms the observation by Casella et al. (2000) that the number of likely partitions is quite limited. Figure 7 shows how observations

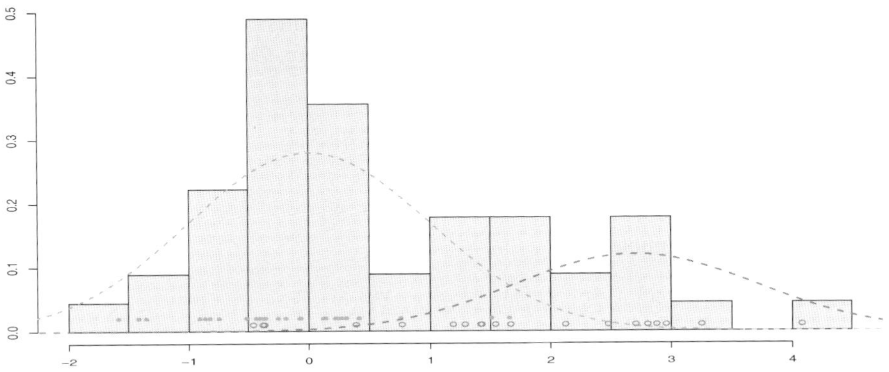

Fig. 7. Histogram, true components, and most likely allocation found over 440,000 simulations of $\underline{z}$'s for a simulated dataset of 45 observations and true values as in Figure 5. Full dots are associated with observations allocated to the first component and empty dots with observations allocated to the second component. For a color reproduction of this figure see the color figures section, page 994.

are allocated to each component in an occurrence where a single[5] allocation $\underline{z}$ took all the weight in the simulation and resulted in a posterior probability of 1.

3.2. The EM algorithm

For maximum likelihood computations, it is possible to use numerical optimisation procedures like the *EM algorithm* (Dempster et al., 1977), but these may fail to converge to the major mode of the likelihood, as illustrated below. Note also that, for location-scale problems, it is most often the case that the likelihood is unbounded and therefore the resultant likelihood estimator is only a local maximum For example, in (3), the limit of the likelihood (4) is infinite if σ_1 goes to 0.

Let us recall here the form of the EM algorithm, for later connections with the Gibbs sampler and other MCMC algorithms. This algorithm is based on the missing data representation introduced in Section 2.2, namely that the distribution of the sample $\underline{x}$ can be written as

$$f(\underline{x}|\theta) = \int g(\underline{x}, \underline{z}|\theta)\,\mathrm{d}\underline{z} = \int f(\underline{x}|\theta)k(\underline{z}|\underline{x}, \theta)\,\mathrm{d}\underline{z} \tag{12}$$

leading to a *complete* (unobserved) log-likelihood

$$\mathbb{L}^c(\theta|\underline{x}, \underline{z}) = \mathbb{L}(\theta|\underline{x}) + \log k(\underline{z}|\underline{x}, \theta),$$

[5] Note however that, given this extreme situation, the output of the simulation experiment must be taken with a pinch of salt: while we simulated a total of about 450,000 permutations, this is to be compared with a total of 2^{45} permutations many of which could have a posterior probability at least as large as those found by the simulations.

where $\mathbb{L}$ is the observed log-likelihood. The EM algorithm is then based on a sequence of completions of the missing variables $\underline{z}$ based on $k(\underline{z}|\underline{x}, \theta)$ and of maximisations of the expected complete log-likelihood (in θ):

GENERAL EM ALGORITHM.

0. Initialization: choose $\theta^{(0)}$,
1. Step t. For $t = 1, \ldots$

1.1 The E-step, compute

$$Q\left(\theta|\theta^{(t-1)}, \underline{x}\right) = \mathbb{E}_{\theta^{(t-1)}}\left[\log \mathbb{L}^c(\theta|\underline{x}, \underline{Z})\right],$$

where $\underline{Z} \sim k(\underline{z}|\theta^{(t-1)}, \underline{x})$.

1.2 The M-step, maximize $Q(\theta|\theta^{(t-1)}, \underline{x})$ in θ and take

$$\theta^{(t)} = \arg\max_{\theta} Q\left(\theta|\theta^{(t-1)}, \underline{x}\right).$$

The result validating the algorithm is that, at each step, the *observed* $\mathbb{L}(\theta|\underline{x})$ increases.

EXAMPLE 1 (*Continued*). For an illustration in our setup, consider again the special mixture of normal distributions (7) where all parameters but $\underline{\theta} = (\mu_1, \mu_2)$ are known. For a simulated dataset of 500 observations and true values $p = 0.7$ and $(\mu_1, \mu_2) = (0, 2.5)$, the log-likelihood is still bimodal and running the EM algorithm on this model means, at iteration t, computing the expected allocations

$$z_i^{(t-1)} = \mathbb{P}\left(Z_i = 1|\underline{x}, \underline{\theta}^{(t-1)}\right)$$

in the E-step and the corresponding posterior means

$$\mu_1^{(t)} = \sum_{i=1}^n \left(1 - z_i^{(t-1)}\right)x_i \Big/ \sum_{i=1}^n \left(1 - z_i^{(t-1)}\right),$$

$$\mu_2^{(t)} = \sum_{i=1}^n z_i^{(t-1)} x_i \Big/ \sum_{i=1}^n z_i^{(t-1)}$$

in the M-step. As shown on Figure 8 for five runs of EM with starting points chosen at random, the algorithm always converges to a mode of the likelihood but only two out of five sequences are attracted by the higher and more significant mode, while the other three go to the lower spurious mode (even though the likelihood is considerably smaller). This is because the starting points happened to be in the domain of attraction of the lower mode.

3.3. *An inverse ill-posed problem*

Algorithmically speaking, mixture models belong to the group of *inverse problems*, where data provide information on the parameters only indirectly, and, to some extent,

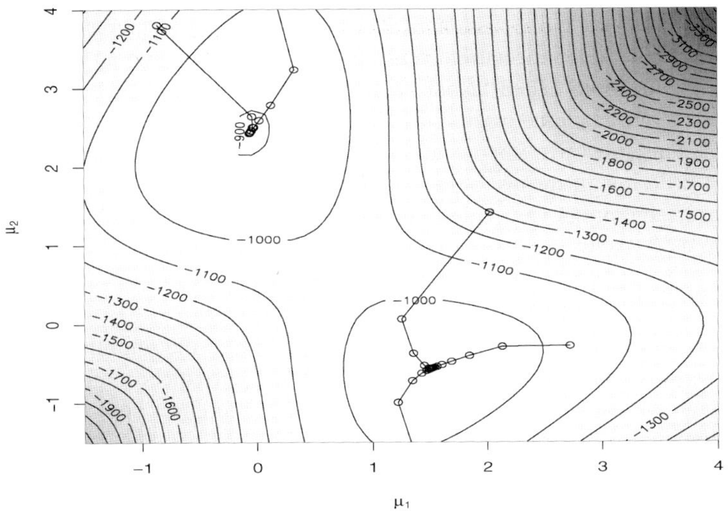

Fig. 8. Trajectories of five runs of the EM algorithm on the log-likelihood surface, along with R contour representation. For a color reproduction of this figure see the color figures section, page 994.

to the class of *ill-posed problems*, where small changes in the data may induce large changes in the results. In fact, when considering a sample of size n from a mixture distribution, there is a nonzero probability $(1 - p_i)^n$ that the ith component is empty, holding none of the random variables. In other words, there always is a nonzero probability that the sample brings no information[6] about the parameters of one or more components! This explains why the likelihood function may become unbounded and also why improper priors are delicate to use in such settings (see below).

3.4. *Identifiability*

A basic feature of a mixture model is that it is invariant under permutation of the indices of the components. This implies that the component parameters θ_i are not identifiable *marginally*: we cannot distinguish component 1 (or θ_1) from component 2 (or θ_2) from the likelihood, because they are exchangeable. While identifiability is not a strong issue in Bayesian statistics,[7] this particular identifiability feature is crucial for both Bayesian inference and computational issues. First, in a k component mixture, the number of modes is of order $O(k!)$ since, if $(\theta_1, \ldots, \theta_k)$ is a local maximum, so is $(\theta_{\sigma(1)}, \ldots, \theta_{\sigma(k)})$ for every permutation $\sigma \in \mathfrak{S}_n$. This makes maximisation and even exploration of the posterior surface obviously harder. Moreover, if an exchangeable prior is used on $\underline{\theta} = (\theta_1, \ldots, \theta_k)$, all the marginals on the θ_i's are identical, which means for instance that

[6] This is not contradictory with the fact that the Fisher information of a mixture model is well defined (Titterington et al., 1985).

[7] This is because it can be either imposed at the level of the prior distribution or bypassed for prediction purposes.

the posterior expectation of θ_1 is identical to the posterior expectation of θ_2. Therefore, alternatives to posterior expectations must be constructed as pertinent estimators.

This problem, often called "label switching", thus requires either a specific prior modelling or a more tailored inferential approach. A naïve answer to the problem found in the early literature is to impose an *identifiability constraint* on the parameters, for instance by ordering the means (or the variances or the weights) in a normal mixture (3). From a Bayesian point of view, this amounts to truncating the original prior distribution, going from $\pi(\underline{\theta}, \underline{p})$ to

$$\pi(\underline{\theta}, \underline{p})\,\mathbb{I}_{\mu_1 \leqslant \cdots \leqslant \mu_k}$$

for instance. While this seems innocuous (because indeed the sampling distribution is the same with or without this indicator function), the introduction of an identifiability constraint has severe consequences on the resulting inference, both from a prior and from a computational point of view. When reducing the parameter space to its constrained part, the imposed truncation has no reason to respect the topology of either the prior or of the likelihood. Instead of singling out one mode of the posterior, the constrained parameter space may then well include parts of several modes and the resulting posterior mean may for instance lay in a very low probability region, while the high posterior probability zones are located at the boundaries of this space. In addition, the constraint may radically modify the prior modelling and come close to contradicting the prior information. For instance, Figure 9 gives the marginal distributions of the ordered random variables $\theta_{(1)}$, $\theta_{(10)}$, and $\theta_{(19)}$, for a $\mathcal{N}(0, 1)$ prior on $\theta_1, \ldots, \theta_{19}$. The comparison of the observed distribution with the original prior $\mathcal{N}(0, 1)$ clearly shows the impact of the ordering. For large values of k, the introduction of a constraint also has a consequence on posterior inference: with many components, the ordering of components in terms of one of its parameters is unrealistic. Some components will be close in mean while others will be close in variance or in weight. As demonstrated in Celeux et al. (2000), this may lead to very poor estimates of the distribution in the end. One

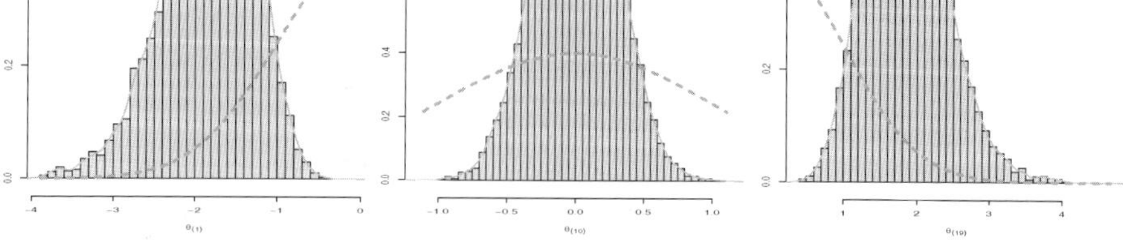

Fig. 9. Distributions of $\theta_{(1)}$, $\theta_{(10)}$, and $\theta_{(19)}$, compared with the $\mathcal{N}(0, 1)$ prior. For a color reproduction of this figure see the color figures section, page 995.

alternative approach to this problem includes reparametrisation, as discussed below in Section 3.5. Another one is to select one of the $k!$ modal regions of the posterior distribution and do the relabelling in terms of proximity to this region, as in Section 4.1.

If the index identifiability problem is solved by imposing an identifiability constraint on the components, most mixture models are identifiable, as described in detail in both Titterington et al. (1985) and MacLachlan and Peel (2000).

3.5. *Choice of priors*

The representation of a mixture model as in (2) precludes the use of independent improper priors,

$$\pi(\underline{\theta}) = \prod_{i=1}^{k} \pi_i(\theta_i),$$

since, if

$$\int \pi_i(\theta_i)\, \mathrm{d}\theta_i = \infty$$

then for every n,

$$\int \pi(\underline{\theta}, \underline{p}|\underline{x})\, \mathrm{d}\underline{\theta}\, \mathrm{d}\underline{p} = \infty$$

because, among the k^n terms in the expansion of $\pi(\underline{\theta}, \underline{p}|\underline{x})$, there are $(k-1)^n$ with *no* observation allocated to the ith component and thus a conditional posterior $\pi(\theta_i|\underline{x}, \underline{z})$ equal to the prior $\pi_i(\theta_i)$.

The inability to use improper priors can be seen by some as a *marginalia*, that is, a fact of little importance, since proper priors with large variances can be used instead.[8] However, since mixtures are ill-posed problems, this difficulty with improper priors is more of an issue, given that the influence of a particular proper prior, no matter how large its variance, cannot be truly assessed.

There is still a possibility of using improper priors in mixture models, as demonstrated by Mengersen and Robert (1996), simply by adding some degree of dependence between the components. In fact, it is quite easy to argue *against* independence in mixture models, because the components are only defined in relation with one another. For the very reason that exchangeable priors lead to identical marginal posteriors on all components, the relevant priors must contain the information that components are *different* to some extent and that a mixture modelling *is* necessary.

The proposal of Mengersen and Robert (1996) is to introduce first a common reference, namely a scale, location, or location–scale parameter. This reference parameter θ_0 is related to the global size of the problem and thus can be endowed with a improper prior: informally, this amounts to first standardising the data before estimating the component parameters. These parameters θ_i can then be defined in terms of *departure* from θ_0, as for instance in $\theta_i = \theta_0 + \vartheta_i$. In Mengersen and Robert (1996), the θ_i's are more

[8] This is the stance taken in the Bayesian software winBUGS where improper priors cannot be used.

strongly tied together by the representation of each θ_i as a perturbation of θ_{i-1}, with the motivation that, if a k component mixture model is used, it is because a $(k-1)$ component model would not fit, and thus the $(k-1)$th component is not sufficient to absorb the remaining variability of the data but must be split into two parts (at least). For instance, in the normal mixture case (3), we can consider starting from the $\mathscr{N}(\mu, \tau^2)$ distribution, and creating the two component mixture

$$p\mathscr{N}(\mu, \tau^2) + (1-p)\mathscr{N}(\mu + \tau\theta, \tau^2\varpi^2).$$

If we need a three component mixture, the above is modified into

$$\begin{aligned} &p\mathscr{N}(\mu, \tau^2) + (1-p)q\mathscr{N}(\mu + \tau\vartheta, \tau^2\varpi_1^2) \\ &\quad + (1-p)(1-q)\mathscr{N}(\mu + \tau\vartheta + \tau\sigma\varepsilon, \tau^2\varpi_1^2\varpi_2^2). \end{aligned}$$

For a k component mixture, the ith component parameter will thus be written as

$$\begin{aligned} \mu_i &= \mu_{i-1} + \tau_{i-1}\vartheta_i = \mu + \cdots + \sigma_{i-1}\vartheta_i, \\ \sigma_i &= \sigma_{i-1}\varpi_i = \tau \cdots \varpi_i. \end{aligned}$$

If, notwithstanding the warnings in Section 3.4, we choose to impose identifiability constraints on the model, a natural version is to take

$$1 \geqslant \varpi_1 \geqslant \cdots \geqslant \varpi_{k-1}.$$

A possible prior distribution is then

$$\begin{aligned} &\pi(\mu, \tau) = \tau^{-1}, \qquad p, q_j \sim \mathscr{U}_{[0,1]}, \qquad \varpi_j \sim \mathscr{U}_{[0,1]}, \\ &\vartheta_j \sim \mathscr{N}(0, \zeta^2), \end{aligned} \tag{13}$$

where ζ is the only hyperparameter of the model and represents the amount of variation allowed between two components. Obviously, other choices are possible and, in particular, a nonzero mean could be chosen for the prior on the ϑ_j's. Figure 10 represents a few mixtures of distributions simulated using this prior with $\zeta = 10$: as k increases, higher-order components are more and more concentrated, resulting in the spikes seen in the last rows. The most important point, however, is that, with this representation, we can use an improper prior on (μ, τ), as proved in Robert and Titterington (1998). See Wasserman (2000), Pérez and Berger (2002), Moreno and Liseo (2003) for different approaches to the use of default or noninformative priors in the setting of mixtures.

These reparametrisations have been developed for Gaussian mixtures (Roeder and Wasserman, 1997), but also for exponential (Gruet et al., 1999) and Poisson mixtures (Robert and Titterington, 1998). However, these alternative representations do require the artificial identifiability restrictions criticised above, and can be unwieldy and less directly interpretable.[9]

In the case of mixtures of Beta distributions used for goodness of fit testing mentioned at the end of Section 2.3, a specific prior distribution is used by Robert and Rousseau (2002) in order to oppose the uniform component of the mixture (6) with the

[9] It is actually possible to generalise the $\mathscr{U}_{[0,1]}$ prior on ϖ_j by assuming that either ϖ_j or $1/\varpi_j$ are uniform $\mathscr{U}_{[0,1]}$, with equal probability. This was tested in Robert and Mengersen (1999).

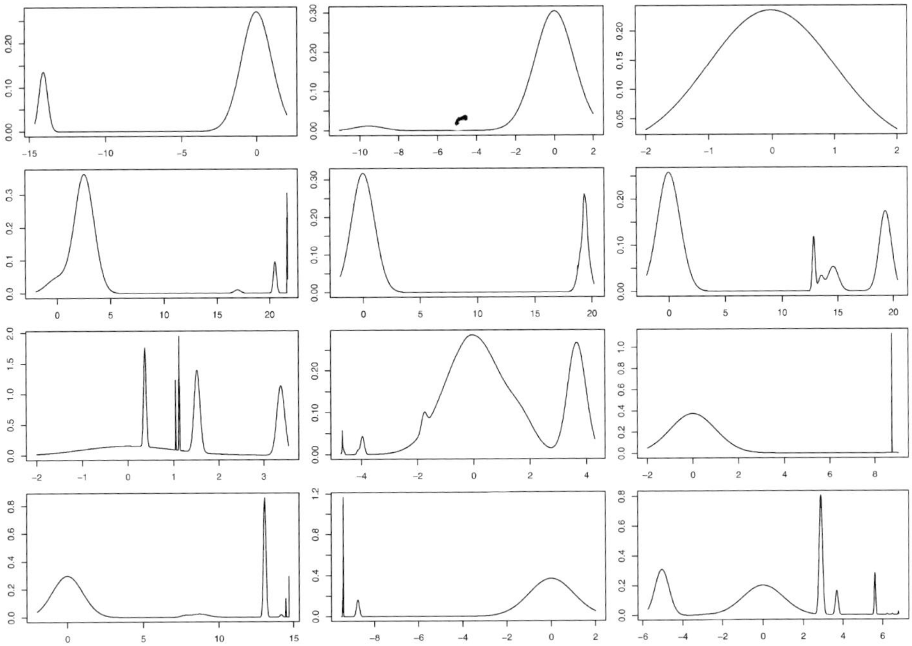

Fig. 10. Normal mixtures simulated using the Mengersen and Robert (1996) prior for $\zeta = 10$, $\mu = 0$, $\tau = 1$ and $k = 2$ (*first row*), $k = 5$ (*second row*), $k = 15$ (*third row*) and $k = 50$ (*last row*).

other components. For the uniform weight,

$$p_0 \sim \mathscr{B}e(0.8, 1.2),$$

favours small values of p_0, since the distribution $\mathscr{B}e(0.8, 1.2)$ has an infinite mode at 0, while p_k is represented as $(k = 1, \ldots, K)$

$$p_k = \frac{\omega_k}{\sum_{i=1}^{K} \omega_i}, \qquad \omega_k \sim Be(1, k),$$

for parsimony reasons (so that higher-order components are less likely) and the prior

$$(\alpha_k, \varepsilon_k) \sim \left\{1 - \exp\left[-\theta\left\{(\alpha_k - 2)^2 + (\varepsilon_k - .5)^2\right\}\right]\right\} \\ \times \exp\left[-\zeta/\left\{\alpha_k^2 \varepsilon_k (1 - \varepsilon_k)\right\} - \kappa\alpha_k^2/2\right] \tag{14}$$

is chosen for the $(\alpha_k, \varepsilon_k)$'s, where (θ, ζ, κ) are hyperparameters. This form[10] is designed to avoid the $(\alpha, \varepsilon) = (2, 1/2)$ region for the parameters of the other components.

3.6. *Loss functions*

As noted above, if no identifying constraint is imposed in the prior or on the parameter space, it is impossible to use the standard Bayes estimators on the parameters, since they

[10] The reader must realise that there is a lot of arbitrariness involved in this particular choice, which simply reflects the limited amount of prior information available for this problem.

Table 1

Estimates of the parameters of a three component normal mixture, obtained for a simulated sample of 500 points by re-ordering according to one of three constraints, p: $p_1 < p_2 < p_3$, μ: $\mu_1 < \mu_2 < \mu_3$, or σ: $\sigma_1 < \sigma_2 < \sigma_3$. (*Source:* Celeux et al., 2000)

Order	p_1	p_2	p_3	θ_1	θ_2	θ_3	σ_1	σ_2	σ_3
p	0.231	0.311	0.458	0.321	−0.55	2.28	0.41	0.471	0.303
θ	0.297	0.246	0.457	−1.1	0.83	2.33	0.357	0.543	0.284
σ	0.375	0.331	0.294	1.59	0.083	0.379	0.266	0.34	0.579
true	0.22	0.43	0.35	1.1	2.4	−0.95	0.3	0.2	0.5

are identical for all components. As also pointed out, using an identifying constraint has some drawbacks for exploring the parameter space and the posterior distribution, as the constraint may well be at odds with the topology of this distribution. In particular, stochastic exploration algorithms may well be hampered by such constraints if the region of interest happens to be concentrated on boundaries of the constrained region.

Obviously, once a sample has been produced from the unconstrained posterior distribution, for instance by an MCMC sampler (Section 4), the ordering constraint can be imposed *ex post*, that is, after the simulations have been completed, for estimation purposes (Stephens, 1997). Therefore, the simulation hindrance created by the constraint can be completely bypassed. However, the effects of different possible ordering constraints on the *same* sample are not innocuous, since they lead to very different estimations. This is not absolutely surprising given the preceding remark on the potential clash between the topology of the posterior surface and the shape of the ordering constraints: computing an average under the constraint may thus produce a value that is unrelated to the modes of the posterior. In addition, imposing a constraint on *one* and only one of the different types of parameters (weights, locations, scales) may fail to discriminate between *some* components of the mixture.

This problem is well-illustrated by Table 1 of Celeux et al. (2000). Depending on which order is chosen, the estimators vary widely and, more importantly, so do the corresponding plug-in densities, that is, the densities in which the parameters have been replaced by the estimate of Table 1, as shown by Figure 11. While *one* of the estimations is close to the true density (because it happens to differ widely enough in the means), the two others are missing one of the three modes altogether!

Empirical approaches based on clustering algorithms for the parameter sample are proposed in Stephens (1997) and Celeux et al. (2000), and they achieve some measure of success on the examples for which they have been tested. We rather focus on another approach, also developed in Celeux et al. (2000), which is to call for new Bayes estimators, based on appropriate loss functions.

Indeed, if $L((\underline{\theta}, \underline{p}), (\hat{\underline{\theta}}, \hat{\underline{p}}))$ is a loss function for which the labelling is immaterial, the corresponding Bayes estimator $(\hat{\underline{\theta}}, \hat{\underline{p}})^*$

$$(\hat{\underline{\theta}}, \hat{\underline{p}})^* = \arg\min_{(\hat{\underline{\theta}}, \hat{\underline{p}})} \mathbb{E}_{(\underline{\theta}, \underline{p})|\underline{x}}\left[L\big((\underline{\theta}, \underline{p}), (\hat{\underline{\theta}}, \hat{\underline{p}})\big)\right] \tag{15}$$

will not face the same difficulties as the posterior average.

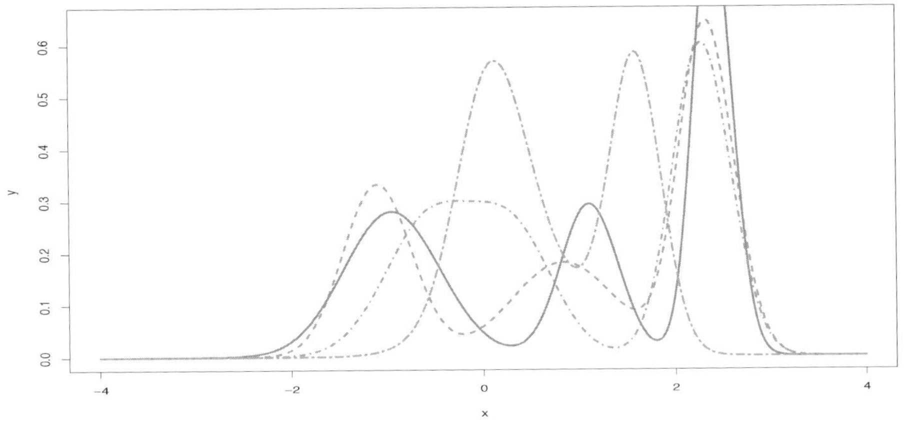

Fig. 11. Comparison of the plug-in densities for the estimations of Table 1 and of the true density (full line). For a color reproduction of this figure see the color figures section, page 995.

A first loss function for the estimation of the parameters is based on an image representation of the parameter space for one component, like the (p, μ, σ) space for normal mixtures. It is loosely based on the Baddeley Δ metric (Baddeley, 1992). The idea is to have a collection of reference points in the parameter space, and, for each of these to calculate the distance to the closest parameter point for both sets of parameters. If $t_1, \ldots, t_n$ denote the collection of reference points, which lie in the same space as the θ_i's, and if $d(t_i, \theta)$ is the distance between t_i and the closest of the θ_i's, the (L_2) loss function reads as follows:

$$L\big((\underline{\theta}, \underline{p}), (\hat{\underline{\theta}}, \hat{\underline{p}})\big) = \sum_{i=1}^{n} \big(d\big(t_i, (\underline{\theta}, \underline{p})\big) - d\big(t_i, (\hat{\underline{\theta}}, \hat{\underline{p}})\big)\big)^2. \tag{16}$$

That is, for each of the fixed points t_i, there is a contribution to the loss if the distance from t_i to the nearest θ_j is not the same as the distance from t_i to the nearest $\hat{\theta}_j$.

Clearly the choice of the t_i's plays an important role since we want $L((\underline{\theta}, \underline{p}), (\hat{\underline{\theta}}, \hat{\underline{p}})) = 0$ only if $(\underline{\theta}, \underline{p}) = (\hat{\underline{\theta}}, \hat{\underline{p}})$, and for the loss function to respond appropriately to changes in the two point configurations. In order to avoid the possibility of zero loss between two configurations which actually differ, it must be possible to determine $(\underline{\theta}, \underline{p})$ from the $\{t_i\}$ and the corresponding $\{d(t_i, (\underline{\theta}, \underline{p}))\}$. For the second desired property, the t_i's are best positioned in high posterior density regions of the (θ_j, p_j)'s space. Given the complexity of the loss function, numerical maximisation techniques like simulated annealing must be used (see Celeux et al., 2000).

When the object of inference is the predictive distribution, more global loss functions can be devised to measure distributional discrepancies. One such possibility is the integrated squared difference

$$L\big((\underline{\theta}, \underline{p}), (\hat{\underline{\theta}}, \hat{\underline{p}})\big) = \int_{\mathcal{R}} \big(f_{(\underline{\theta}, \underline{p})}(y) - f_{(\hat{\underline{\theta}}, \hat{\underline{p}})}(y)\big)^2 \,\mathrm{d}y, \tag{17}$$

where $f_{(\underline{\theta},\underline{p})}$ denotes the density of the mixture (2). Another possibility is a symmetrised Kullback–Leibler distance

$$L\big((\underline{\theta},\underline{p}),(\underline{\hat{\theta}},\underline{\hat{p}})\big)$$
$$= \int_{\mathcal{R}} \left\{ f_{(\underline{\theta},\underline{p})}(y) \log \frac{f_{(\underline{\theta},\underline{p})}(y)}{f_{(\underline{\hat{\theta}},\underline{\hat{p}})}(y)} + f_{(\underline{\hat{\theta}},\underline{\hat{p}})}(y) \log \frac{f_{(\underline{\hat{\theta}},\underline{\hat{p}})}(y)}{f_{(\underline{\theta},\underline{p})}(y)} \right\} \mathrm{d}y, \tag{18}$$

as in Mengersen and Robert (1996). We refer again to Celeux et al. (2000) for details on the resolution of the minimisation problem and on the performance of both approaches.

4. Inference for mixtures models with known number of components

Mixture models have been at the source of many methodological developments in computational Statistics. Besides the seminal work of Dempster et al. (1977), see Section 3.2, we can point out the Data Augmentation method proposed by Tanner and Wong (1987) which appears as a forerunner of the Gibbs sampler of Gelfand and Smith (1990). This section covers three Monte Carlo or MCMC (Markov chain Monte Carlo) algorithms that are customarily used for the approximation of posterior distributions in mixture settings, but it first discusses in Section 4.1 the solution chosen to overcome the label-switching problem.

4.1. Reordering

For the k-component mixture (2), with n i.i.d. observations $\underline{x} = (x_1, \ldots, x_n)$, we assume that the densities $f(\cdot|\theta_i)$ are known up to a parameter θ_i. In this section, the number of components k is known. (The alternative situation in which k is unknown will be addressed in the next section.)

As detailed in Section 3.1, the fact that the expansion of the likelihood (2) is of complexity $\mathrm{O}(k^n)$ prevents an analytical derivation of Bayes estimators: Eq. (11) shows that a posterior expectation is a sum of k^n terms which correspond to the different allocations of the observations x_i and, therefore, is never available in closed form.

Section 3.4 discussed the drawbacks of imposing identifiability ordering constraints on the parameter space. We thus consider an unconstrained parameter space, which implies that the posterior distribution has a multiple of $k!$ different modes. To derive proper estimates of the parameters of (2), we can thus opt for one of two strategies: either use a loss function as in Section 3.6, for which the labelling is immaterial or impose a reordering constraint *ex-post*, that is, after the simulations have been completed, and then use a loss function depending on the labelling.

While the first solution is studied in Celeux et al. (2000), we present the alternative here, mostly because the implementation is more straightforward: once the simulation output has been reordered, the posterior mean is approximated by the empirical average. Reordering schemes that do not face the difficulties linked to a forced ordering of the means (or other quantities) can be found in Stephens (1997) and Celeux et al. (2000), but we use here a new proposal that is both straightforward and very efficient.

For a permutation $\tau \in \mathfrak{S}_k$, set of all permutations of $\{1, \ldots, k\}$, we denote by

$$\tau(\underline{\theta}, \underline{p}) = \{(\theta_{\tau(1)}, \ldots, \theta_{\tau(k)}), (p_{\tau(1)}, \ldots, p_{\tau(k)})\}$$

the corresponding permutation of the parameter $(\underline{\theta}, \underline{p})$ and we implement the following reordering scheme, based on a simulated sample of size M,

(i) compute the pivot $(\underline{\theta}, \underline{p})^{(i^*)}$ such that

$$i^* = \arg\max_{i=1,\ldots,M} \pi\big((\underline{\theta}, \underline{p})^{(i)} | \underline{x}\big),$$

that is, a Monte Carlo approximation of the Maximum a Posteriori (MAP) estimator[11] of $(\underline{\theta}, \underline{p})$.

(ii) For $i \in \{1, \ldots, M\}$:

1. Compute

$$\tau_i = \arg\min_{\tau \in \mathfrak{S}_k} \big\langle \tau\big((\underline{\theta}, \underline{p})^{(i)}\big), (\underline{\theta}, \underline{p})^{(i^*)} \big\rangle_{(\kappa+1)k},$$

where κ is the dimension of the θ_i's and $\langle \cdot, \cdot \rangle_l$ denotes the canonical scalar product of $\mathbb{R}^l$.

2. Set $(\underline{\theta}, \underline{p})^{(i)} = \tau_i((\underline{\theta}, \underline{p})^{(i)})$.

The step (ii) chooses the reordering that is the closest to the approximate MAP estimator and thus solves the identifiability problem without requiring a preliminary and most likely unnatural ordering on one of the parameters of the model. Then, after the reordering step, the Monte Carlo estimation of the posterior expectation of θ_i, $\mathbb{E}^{\pi}_{\underline{x}}(\theta_i)$, is given by $\sum_{j=1}^{M} (\theta_i)^{(j)}/M$.

4.2. *Data augmentation and Gibbs sampling approximations*

The Gibbs sampler is the most commonly used approach in Bayesian mixture estimation (Diebolt and Robert, 1990a, 1994; Lavine and West, 1992; Verdinelli and Wasserman, 1992; Chib, 1995; Escobar and West, 1995). In fact, a solution to the computational problem is to take advantage of the missing data introduced in Section 2.2, that is, to associate with each observation x_j a missing multinomial variable $z_j \sim \mathscr{M}_k(1; p_1, \ldots, p_k)$ such that $x_j | z_j = i \sim f(x|\theta_i)$. Note that in heterogeneous populations made of several homogeneous subgroups, it makes sense to interpret z_j as the index of the population of origin of x_j, which has been lost in the observational process. In the alternative nonparametric perspective, the components of the mixture and even the number k of components in the mixture are often meaningless for the problem to be analysed. However, this distinction between natural and artificial completion is lost to the MCMC sampler, whose goal is simply to provide a Markov chain that converges to the posterior distribution. Completion is thus, from a simulation point of view, a means to generate such a chain.

Recall that $\underline{z} = (z_1, \ldots, z_n)$ and denote by $\pi(\underline{p}|\underline{z}, \underline{x})$ the density of the distribution of $\underline{p}$ given $\underline{z}$ and $\underline{x}$. This distribution is in fact independent of $\underline{x}$, $\pi(\underline{p}|\underline{z}, \underline{x}) = \pi(\underline{p}|\underline{z})$.

[11] Note that the pivot is itself a good estimator.

In addition, denote $\pi(\underline{\theta}|\underline{z}, \underline{x})$ the density of the distribution of $\underline{\theta}$ given $(\underline{z}, \underline{x})$. The most standard Gibbs sampler for mixture models (2) (Diebolt and Robert, 1994) is based on the successive simulation of $\underline{z}$, $\underline{p}$ and $\underline{\theta}$ conditional on one another and on the data:

GENERAL GIBBS SAMPLING FOR MIXTURE MODELS.

0. **Initialization:** choose $\underline{p}^{(0)}$ and $\underline{\theta}^{(0)}$ arbitrarily
1. **Step t.** For $t = 1, \ldots$

1.1 Generate $z_i^{(t)}$ $(i = 1, \ldots, n)$ from $(j = 1, \ldots, k)$

$$\mathbb{P}\big(z_i^{(t)} = j|p_j^{(t-1)}, \theta_j^{(t-1)}, x_i\big) \propto p_j^{(t-1)} f\big(x_i|\theta_j^{(t-1)}\big)$$

1.2 Generate $\underline{p}^{(t)}$ from $\pi(\underline{p}|\underline{z}^{(t)})$

1.3 Generate $\underline{\theta}^{(t)}$ from $\pi(\underline{\theta}|\underline{z}^{(t)}, \underline{x})$.

Given that the density f most often belongs to an exponential family,

$$f(x|\theta) = h(x)\exp\big(\langle r(\theta), t(x)\rangle_k - \phi(\theta)\big), \tag{19}$$

where h is a function from $\mathbb{R}$ to $\mathbb{R}_+$, r and t are functions from Θ and $\mathbb{R}$ to $\mathbb{R}^k$, the simulation of both $\underline{p}$ and $\underline{\theta}$ is usually straightforward. In this case, a conjugate prior on θ (Robert, 2001) is given by

$$\pi(\theta) \propto \exp\big(\langle r(\theta), \alpha\rangle_k - \beta\phi(\theta)\big), \tag{20}$$

where $\alpha \in \mathbb{R}^k$ and $\beta > 0$ are given hyperparameters. For a mixture of distributions (19), it is therefore possible to associate with each θ_j a conjugate prior $\pi_j(\theta_j)$ with hyperparameters α_j, β_j. We also select for $\underline{p}$ the standard Dirichlet conjugate prior, $\underline{p} \sim \mathscr{D}(\gamma_1, \ldots, \gamma_k)$. In this case, $\underline{p}|\underline{z} \sim \mathscr{D}(n_1 + \gamma_1, \ldots, n_k + \gamma_k)$ and

$$\pi(\underline{\theta}|\underline{z}, \underline{x}) \propto \prod_{j=1}^{k} \exp\left(\left\langle r(\theta_j), \alpha + \sum_{i=1}^{n} \mathbb{I}_{z_i=j} t(x_i)\right\rangle_k - \phi(\theta_j)(n_j + \beta)\right),$$

where $n_j = \sum_{l=1}^{n} \mathbb{I}_{z_l=j}$. The two steps of the Gibbs sampler are then:

GIBBS SAMPLING FOR EXPONENTIAL FAMILY MIXTURES.

0. **Initialization.** Choose $\underline{p}^{(0)}$ and $\underline{\theta}^{(0)}$
1. **Step t.** For $t = 1, \ldots$

1.1 Generate $z_i^{(t)}$ $(i = 1, \ldots, n,\ j = 1, \ldots, k)$ from

$$\mathbb{P}\big(z_i^{(t)} = j|p_j^{(t-1)}, \theta_j^{(t-1)}, x_i\big) \propto p_j^{(t-1)} f\big(x_i|\theta_j^{(t-1)}\big)$$

1.2 Compute $n_j^{(t)} = \sum_{i=1}^{n} \mathbb{I}_{z_i^{(t)}=j}$, $s_j^{(t)} = \sum_{i=1}^{n} \mathbb{I}_{z_i^{(t)}=j} t(x_i)$

1.3 Generate $\underline{p}^{(t)}$ from $\mathscr{D}(\gamma_1 + n_1, \ldots, \gamma_k + n_k)$

1.4 Generate $\theta_j^{(t)}$ $(j = 1, \ldots, k)$ from

$$\pi\big(\theta_j|\underline{z}^{(t)}, \underline{x}\big) \propto \exp\big(\langle r(\theta_j), \alpha + s_j^{(t)}\rangle_k - \phi(\theta_j)(n_j + \beta)\big).$$

As with all Monte Carlo methods, the performance of the above MCMC algorithms must be evaluated. Here, performance comprises a number of aspects, including the autocorrelation of the simulated chains (since high positive autocorrelation would require longer simulation in order to obtain an equivalent number of independent samples and 'sticky' chains will take much longer to explore the target space) and Monte Carlo variance (since high variance reduces the precision of estimates). The integrated autocorrelation time provides a measure of these aspects. Obviously, the convergence properties of the MCMC algorithm will depend on the choice of distributions, priors and on the quantities of interest. We refer to Mengersen et al. (1999) and Robert and Casella (2004, Chapter 12), for a description of the various convergence diagnostics that can be used in practise.

It is also possible to exploit the latent variable representation (5) when evaluating convergence and performance of the MCMC chains for mixtures. As detailed by Robert (1998a), the 'duality' of the two chains $(\underline{z}^{(t)})$ and $(\underline{\theta}^{(t)})$ can be considered in the strong sense of data augmentation (Tanner and Wong, 1987; Liu et al., 1994) or in the weaker sense that $\underline{\theta}^{(t)}$ can be derived from $z^{(t)}$. Thus probabilistic properties of $(\underline{z}^{(t)})$ transfer to $\underline{\theta}^{(t)}$. For instance, since $\underline{z}^{(t)}$ is a finite state space Markov chain, it is uniformly geometrically ergodic and the Central Limit Theorem also applies for the chain $\underline{\theta}^{(t)}$. Diebolt and Robert (1993, 1994) termed this the 'Duality Principle'.

In this respect, Diebolt and Robert (1990b) have shown that the naïve MCMC algorithm that employs Gibbs sampling through completion, while appealingly straightforward, does not necessarily enjoy good convergence properties. In fact, the very nature of Gibbs sampling may lead to "trapping states", that is, concentrated local modes that require an enormous number of iterations to escape from. For example, components with a small number of allocated observations and very small variance become so tightly concentrated that there is very little probability of moving observations in or out of them. So, even though the Gibbs chain $(\underline{z}^{(t)}, \underline{\theta}^{(t)})$ is formally irreducible and uniformly geometric, as shown by the above duality principle, there may be no escape from this configuration. At another level, as discussed in Section 3.1, Celeux et al. (2000) show that most MCMC samplers, including Gibbs, fail to reproduce the permutation invariance of the posterior distribution, that is, do not visit the $k!$ replications of a given mode.

EXAMPLE 1 (*Continued*). For the mixture (7), the parameter space is two-dimensional, which means that the posterior surface can be easily plotted. Under a normal prior $\mathscr{N}(\delta, 1/\lambda)$ ($\delta \in \mathbb{R}$ and $\lambda > 0$ are known hyper-parameters) on both μ_1 and μ_2, with $s_j^x = \sum_{i=1}^n \mathbb{I}_{z_i=j} x_i$, it is easy to see that μ_1 and μ_2 are independent, given $(\underline{z}, \underline{x})$, with conditional distributions

$$\mathscr{N}\left(\frac{\lambda\delta + s_1^x}{\lambda + n_1}, \frac{1}{\lambda + n_1}\right) \quad \text{and} \quad \mathscr{N}\left(\frac{\lambda\delta + s_2^x}{\lambda + n_2}, \frac{1}{\lambda + n_2}\right)$$

respectively. Similarly, the conditional posterior distribution of $\underline{z}$ given (μ_1, μ_2) is easily seen to be a product of Bernoulli rv's on $\{1, 2\}$, with $(i = 1, \ldots, n)$

$$\mathbb{P}(z_i = 1|\mu_1, x_i) \propto p\exp\bigl(-0.5(x_i - \mu_1)^2\bigr).$$

GIBBS SAMPLING FOR THE MIXTURE (7).

0. **Initialization.** Choose $\mu_1^{(0)}$ and $\mu_2^{(0)}$
1. **Step t.** For $t = 1, \ldots$

1.1 Generate $z_i^{(t)}$ $(i = 1, \ldots, n)$ from

$$\mathbb{P}\big(z_i^{(t)} = 1\big) = 1 - \mathbb{P}\big(z_i^{(t)} = 2\big) \propto p \exp\left(-\frac{1}{2}\big(x_i - \mu_1^{(t-1)}\big)^2\right)$$

1.2 Compute

$$n_j^{(t)} = \sum_{i=1}^{n} \mathbb{I}_{z_i^{(t)}=j} \quad \text{and} \quad \big(s_j^x\big)^{(t)} = \sum_{i=1}^{n} \mathbb{I}_{z_i^{(t)}=j} x_i$$

1.3 Generate

$$\mu_j^{(t)} \ (j = 1, 2) \quad \text{from} \quad \mathscr{N}\left(\frac{\lambda\delta + (s_j^x)^{(t)}}{\lambda + n_j^{(t)}}, \frac{1}{\lambda + n_j^{(t)}}\right).$$

Figure 12 illustrates the behaviour of this algorithm for a simulated dataset of 500 points from $.7\mathscr{N}(0, 1) + .3\mathscr{N}(2.5, 1)$. The representation of the Gibbs sample over 10,000 iterations is quite in agreement with the posterior surface, represented here by grey levels and contours.

This experiment gives a false sense of security about the performances of the Gibbs sampler, however, because it does not indicate the structural dependence of the sampler on the initial conditions. Because it uses conditional distributions, Gibbs sampling is often restricted in the width of its moves. Here, conditioning on $\underline{z}$ implies that the

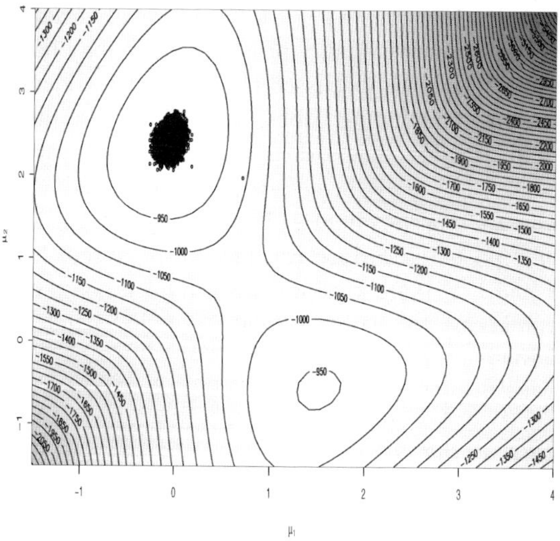

Fig. 12. Log-posterior surface and the corresponding Gibbs sample for the model (7), based on 10,000 iterations. For a color reproduction of this figure see the color figures section, page 996.

proposals for (μ_1, μ_2) are quite concentrated and do not allow for drastic changes in the allocations at the next step. To obtain a significant modification of $\underline{z}$ does require a considerable number of iterations once a stable position has been reached. Figure 13 illustrates this phenomenon for the same sample as in Figure 12: a Gibbs sampler initialised close to the spurious second mode (described in Figure 4) is unable to leave it, even after a large number of iterations, for the reason given above. It is quite interesting to see that this Gibbs sampler suffers from the same pathology as the EM algorithm, although this is not surprising given that it is based on the same completion.

This example illustrates quite convincingly that, while the completion is natural from a model point of view (since it is somehow a part of the definition of the model), the utility does not necessarily transfer to the simulation algorithm.

EXAMPLE 3. Consider a mixture of 3 univariate Poisson distributions, with an i.i.d. sample $\underline{x}$ from $\sum_{j=1}^{3} p_j \mathscr{P}(\lambda_j)$, where, thus, $\underline{\theta} = (\lambda_1, \lambda_2, \lambda_3)$ and $\underline{p} = (p_1, p_2, p_3)$. Under the prior distribution $\lambda_j \sim \mathscr{G}a(\alpha_j, \beta_j)$ and $p \sim \mathscr{D}(\gamma_1, \gamma_2, \gamma_3)$, where $(\alpha_j, \beta_j, \gamma_j)$ are known hyperparameters, $\lambda_j | \underline{x}, \underline{z} \sim \mathscr{G}a(\alpha_j + s_j^x, \beta_j + n_j)$ and we derive the corresponding Gibbs sampler as follows:

GIBBS SAMPLING FOR A POISSON MIXTURE.

0. **Initialization.** Choose $\underline{p}^{(0)}$ and $\underline{\theta}^{(0)}$
1. **Step t.** For $t = 1, \ldots$
   1.1 Generate $z_i^{(t)}$ $(i = 1, \ldots, n)$ from $(j = 1, 2, 3)$

$$\mathbb{P}(z_i^{(t)} = j) \propto p_j^{(t-1)} \big(\lambda_j^{(t-1)}\big)^{x_i} \exp\big(-\lambda_j^{(t-1)}\big)$$

   Compute

$$n_j^{(t)} = \sum_{i=1}^{n} \mathbb{I}_{z_i^{(t)}=j} \quad \text{and} \quad (s_j^x)^{(t)} = \sum_{i=1}^{n} \mathbb{I}_{z_i^{(t)}=j} x_i$$

   1.2 Generate $\underline{p}^{(t)}$ from $\mathscr{D}(\gamma_1 + n_1^{(t)}, \gamma_2 + n_2^{(t)}, \gamma_3 + n_3^{(t)})$
   1.3 Generate $\lambda_j^{(t)}$ from $\mathscr{G}a(\alpha_j + (s_j^x)^{(t)}, \beta_j + n_j^{(t)})$.

The previous sample scheme has been tested on a simulated dataset with $n = 1000$, $\underline{\lambda} = (2, 6, 10)$, $p_1 = 0.25$ and $p_2 = 0.25$. Figure 14 presents the results. We observe that the algorithm reaches very quickly one mode of the posterior distribution but then remains in its vicinity.

EXAMPLE 4. This example deals with a benchmark of mixture estimation, the galaxy dataset of Roeder (1992), also analysed in Richardson and Green (1997) and Roeder and Wasserman (1997), among others. It consists of 82 observations of galaxy velocities. All authors consider that the galaxies velocities are realisations of i.i.d. random variables distributed according to a mixture of k normal distributions. The evaluation of the number k of components for this dataset is quite delicate,[12] since the estimates

[12] In a talk at the 2001 ICMS Workshop on mixtures, Edinburgh, Radford Neal presented convincing evidence that, from a purely astrophysical point of view, the number of components was at least 7. He also argued against the use of a mixture representation for this dataset!

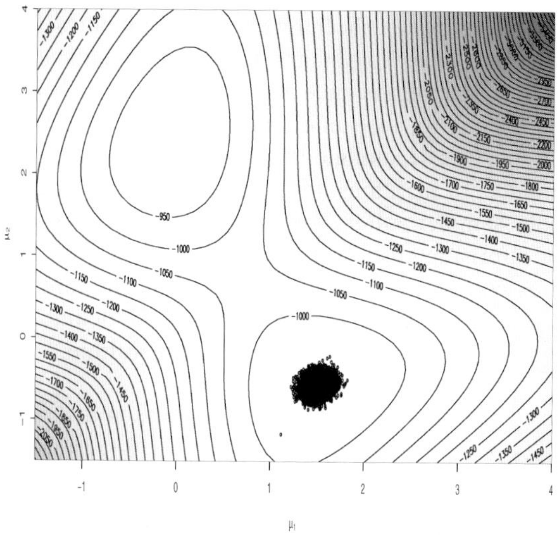

Fig. 13. Same graph, when initialised close to the second and lower mode, based on 10,000 iterations. For a color reproduction of this figure see the color figures section, page 996.

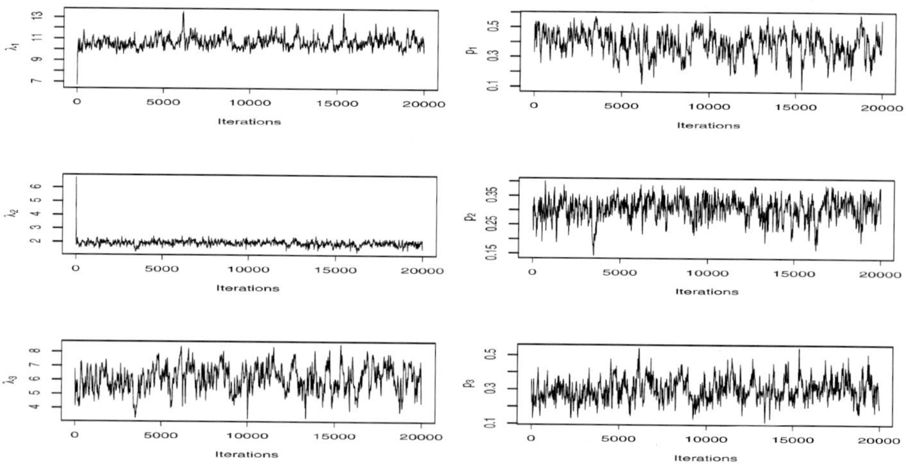

Fig. 14. Evolution of the Gibbs chains over 20,000 iterations for the Poisson mixture model.

range from 3 for Roeder and Wasserman (1997) to 5 or 6 for Richardson and Green (1997) and to 7 for Escobar and West (1995), Phillips and Smith (1996). For illustration purposes, we follow Roeder and Wasserman (1997) and consider 3 components, thus modelling the data by

$$\sum_{j=1}^{3} p_j \mathcal{N}\left(\mu_j, \sigma_j^2\right).$$

In this case, $\underline{\theta} = (\mu_1, \mu_2, \mu_3, \sigma_1^2, \sigma_2^3, \sigma_3^2)$. As in Casella et al. (2000), we use conjugate priors

$$\sigma_j^2 \sim \mathscr{IG}(\alpha_j, \beta_i),\ \mu_j|\sigma_j^2 \sim \mathscr{N}(\lambda_j, \sigma_j^2/\tau_j),$$
$$(p_1, p_2, p_3) \sim \mathscr{D}(\gamma_1, \gamma_2, \gamma_3),$$

where $\mathscr{IG}$ denotes the inverse gamma distribution and $\eta_j, \tau_j, \alpha_j, \beta_j, \gamma_j$ are known hyperparameters. If we denote

$$s_j^v = \sum_{i=1}^n \mathbb{I}_{z_i=j}(x_i - \mu_j)^2,$$

then

$$\mu_j|\sigma_j^2, \underline{x}, \underline{z} \sim \mathscr{N}\left(\frac{\lambda_j\tau_j + s_j^x}{\tau_j + n_j}, \frac{\sigma_j^2}{\tau_j + n_j}\right),$$
$$\sigma_j^2|\mu_j, \underline{x}, \underline{z} \sim \mathscr{IG}\big(\alpha_j + 0.5(n_j + 1), \beta_j + 0.5\tau_j(\mu_j - \lambda_j)^2 + 0.5s_j^v\big).$$

GIBBS SAMPLING FOR A GAUSSIAN MIXTURE.

0. **Initialization.** Choose $\underline{p}^{(0)}, \underline{\theta}^{(0)}$
1. **Step t.** For $t = 1, \ldots$

   1.1 Generate $z_i^{(t)}$ $(i = 1, \ldots, n)$ from $(j = 1, 2, 3)$

$$\mathbb{P}\big(z_i^{(t)} = j\big) \propto \frac{p_j^{(t-1)}}{\sigma_j^{(t-1)}} \exp\big(-\big(x_i - \mu_j^{(t-1)}\big)^2/2\big(\sigma_j^2\big)^{(t-1)}\big)$$

   Compute

$$n_j^{(t)} = \sum_{l=1}^n \mathbb{I}_{z_l^{(t)}=j}, \qquad (s_j^x)^{(t)} = \sum_{l=1}^n \mathbb{I}_{z_l^{(t)}=j} x_l$$

   1.2 Generate $\underline{p}^{(t)}$ from $\mathscr{D}(\gamma_1 + n_1, \gamma_2 + n_2, \gamma_3 + n_3)$

   1.3 Generate $\mu_j^{(t)}$ from

$$\mathscr{N}\left(\frac{\lambda_j\tau_j + (s_j^x)^{(t)}}{\tau_j + n_j^{(t)}}, \frac{(\sigma_j^2)^{(t-1)}}{\tau_j + n_j^{(t)}}\right)$$

   Compute

$$(s_j^v)^{(t)} = \sum_{l=1}^n \mathbb{I}_{z_l^{(t)}=j}(x_l - \mu_j^{(t)})^2$$

   1.4 Generate $(\sigma_j^2)^{(t)}$ $(j = 1, 2, 3)$ from

$$\mathscr{IG}\left(\alpha_j + \frac{n_j + 1}{2}, \beta_j + 0.5\tau_j\big(\mu_j^{(t)} - \lambda_j\big)^2 + 0.5\big(s_j^v\big)^{(t)}\right).$$

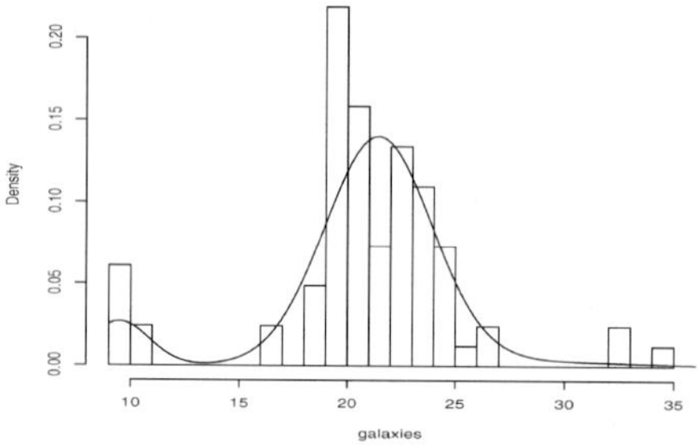

Fig. 15. Histogram of the velocity of 82 galaxies against the plug-in estimated 3 component mixture, using a Gibbs sampler.

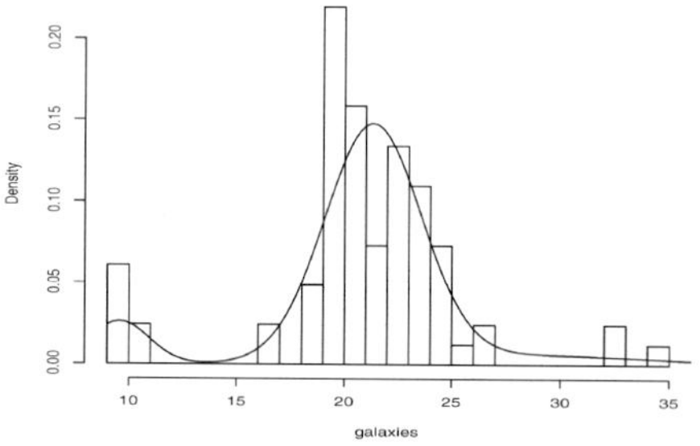

Fig. 16. Same graph, when using a Metropolis–Hastings algorithm with $\zeta^2 = .01$.

After 20,000 iterations, the Gibbs sample is quite stable (although more detailed convergence assessment is necessary and the algorithm fails to visit the permutation modes) and, using the 5,000 last reordered iterations, we find that the posterior mean estimations of μ_1, μ_2, μ_3 are equal to 9.5, 21.4, 26.8, those of $\sigma_1^2, \sigma_2^2, \sigma_3^2$ are equal to 1.9, 6.1, 34.1 and those of p_1, p_2, p_3 are equal to 0.09, 0.85, 0.06. Figure 15 shows the histogram of the data along with the estimated (plug-in) density.

4.3. Metropolis–Hastings approximations

As shown by Figure 13, the Gibbs sampler may fail to escape the attraction of the local mode, even in a well-behaved case as Example 1 where the likelihood and the posterior distributions are bounded and where the parameters are identifiable. Part of the difficulty is due to the completion scheme that increases the dimension of the simulation space and reduces considerably the mobility of the parameter chain. A standard alternative that does not require completion and an increase in the dimension is the Metropolis–Hastings algorithm. In fact, the likelihood of mixture models is available in closed form, being computable in $O(kn)$ time, and the posterior distribution is thus available up to a multiplicative constant.

GENERAL METROPOLIS–HASTINGS ALGORITHM FOR MIXTURE MODELS.

0. **Initialization.** Choose $\underline{p}^{(0)}$ and $\underline{\theta}^{(0)}$
1. **Step t.** For $t = 1, \ldots$

   1.1 Generate $(\underline{\tilde{\theta}}, \underline{\tilde{p}})$ from $q(\underline{\theta}, \underline{p}|\underline{\theta}^{(t-1)}, \underline{p}^{(t-1)})$

   1.2 Compute

$$r = \frac{f(\underline{x}|\underline{\tilde{\theta}}, \underline{\tilde{p}})\pi(\underline{\tilde{\theta}}, \underline{\tilde{p}})q(\underline{\theta}^{(t-1)}, \underline{p}^{(t-1)}|\underline{\tilde{\theta}}, \underline{\tilde{p}})}{f(\underline{x}|\underline{\theta}^{(t-1)}, \underline{p}^{(t-1)})\pi(\underline{\theta}^{(t-1)}, \underline{p}^{(t-1)})q(\underline{\tilde{\theta}}, \underline{\tilde{p}}|\underline{\theta}^{(t-1)}, \underline{p}^{(t-1)})}$$

   1.3 Generate $u \sim \mathscr{U}_{[0,1]}$

   If $r < u$ then $(\underline{\theta}^{(t)}, \underline{p}^{(t)}) = (\underline{\tilde{\theta}}, \underline{\tilde{p}})$

   else $(\underline{\theta}^{(t)}, \underline{p}^{(t)}) = (\underline{\theta}^{(t-1)}, \underline{p}^{(t-1)})$.

The major difference with the Gibbs sampler is that we need to choose the proposal distribution q, which can be *a priori* anything, and this is a mixed blessing! The most generic proposal is the random walk Metropolis–Hastings algorithm where each unconstrained parameter is the mean of the proposal distribution for the new value, that is,

$$\tilde{\theta}_j = \theta_j^{(t-1)} + u_j,$$

where $u_j \sim \mathscr{N}(0, \zeta^2)$. However, for constrained parameters like the weights and the variances in a normal mixture model, this proposal is not efficient.

This is the case for the parameter $\underline{p}$, due to the constraint that $\sum_{i=1}^{k} p_k = 1$. To solve the difficulty with the weights (since $\underline{p}$ belongs to the simplex of $\mathbb{R}^k$), Cappé et al. (2002) propose to overparameterise the model (2) as

$$p_j = w_j \bigg/ \sum_{l=1}^{k} w_l, \quad w_j > 0,$$

thus removing the simulation constraint on the p_j's. Obviously, the w_j's are not identifiable, but this is not a difficulty from a simulation point of view and the p_j's remain identifiable (up to a permutation of indices). Perhaps paradoxically, using overparameterised representations often helps with the mixing of the corresponding MCMC algorithms since they are less constrained by the dataset or the likelihood. The proposed move on the w_j's is $\log(\tilde{w}_j) = \log(w_j^{(t-1)}) + u_j$ where $u_j \sim \mathscr{N}(0, \zeta^2)$.

EXAMPLE 1 (*Continued*). For the posterior associated with (7), the Gaussian random walk proposal is

$$\tilde{\mu}_1 \sim \mathscr{N}\big(\mu_1^{(t-1)}, \zeta^2\big), \quad \text{and} \quad \tilde{\mu}_2 \sim \mathscr{N}\big(\mu_2^{(t-1)}, \zeta^2\big)$$

associated with the following algorithm:

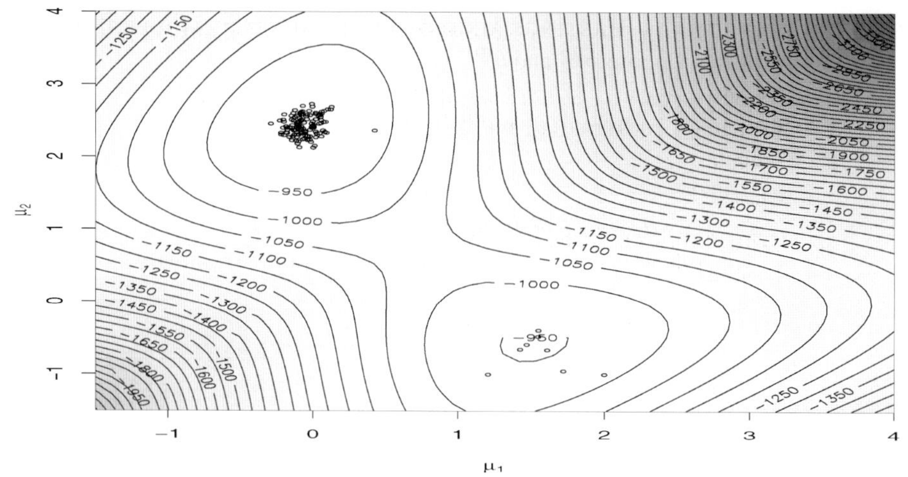

Fig. 17. Track of a 10,000 iterations random walk Metropolis–Hastings sample on the posterior surface, the starting point is equal to $(2, -1)$. The scale of the random walk ζ^2 is equal to 1. For a color reproduction of this figure see the color figures section, page 997.

METROPOLIS–HASTINGS ALGORITHM FOR MODEL (7).

0. **Initialization.** Choose $\mu_1^{(0)}$ and $\mu_2^{(0)}$
1. **Step t.** For $t = 1, \ldots$
   1.1 Generate $\tilde{\mu}_j$ $(j = 1, 2)$ from $\mathscr{N}(\mu_j^{(t-1)}, \zeta^2)$
   1.2 Compute
$$r = \frac{f(x|\tilde{\mu}_1, \tilde{\mu}_2,)\pi(\tilde{\mu}_1, \tilde{\mu}_2)}{f(x|\mu_1^{(t-1)}, \mu_2^{(t-1)})\pi(\mu_1^{(t-1)}, \mu_2^{(t-1)})}$$
   1.3 Generate $u \sim \mathscr{U}_{[0,1]}$
   If $r < u$ then $(\mu_1^{(t)}, \mu_2^{(t)}) = (\tilde{\mu}_1, \tilde{\mu}_2)$
   else $(\mu_1^{(t)}, \mu_2^{(t)}) = (\mu_1^{(t-1)}, \mu_2^{(t-1)})$.

On the same simulated dataset as in Figure 12, Figure 17 shows how quickly this algorithm escapes the attraction of the spurious mode: after a few iterations of the algorithm, the chain drifts over the poor mode and converges almost deterministically to the proper region of the posterior surface. The Gaussian random walk is scaled as $\tau^2 = 1$, although other scales would work as well but would require more iterations to reach the proper model regions. For instance, a scale of 0.01 needs close to 5,000 iterations to attain the main mode. In this special case, the Metropolis–Hastings algorithm seems to overcome the drawbacks of the Gibbs sampler.

EXAMPLE 3 (*Continued*). We have tested the behaviour of the Metropolis–Hastings algorithm (same dataset as the Gibbs), with the following proposals:

$$\tilde{\lambda}_j \sim \mathscr{LN}\big(\log(\lambda_j^{(t-1)}), \zeta^2\big), \qquad \tilde{w}_j \sim \mathscr{LN}\big(\log(w_j^{(t-1)}), \zeta^2\big),$$

where $\mathscr{LN}(\mu, \sigma^2)$ refers to the log-normal distribution with parameters μ and σ^2.

METROPOLIS–HASTINGS ALGORITHM FOR A POISSON MIXTURE.

0. **Initialization.** Choose $\underline{w}^{(0)}$ and $\underline{\theta}^{(0)}$
1. **Step t.** For $t = 1, \ldots$
   1.1 Generate $\tilde{\lambda}_j$ from $\mathscr{LN}(\log(\lambda_j^{(t-1)}), \zeta^2)$
   1.2 Generate $\tilde{w}_j$ from $\mathscr{LN}(\log(w_j^{(t-1)}), \zeta^2)$
   1.3 Compute

$$r = \frac{f(x|\tilde{\underline{\theta}}, \tilde{\underline{w}})\pi(\tilde{\underline{\theta}}, \tilde{\underline{w}}) \prod_{j=1}^{3} \tilde{\lambda}_j \tilde{w}_j}{f(x|\underline{\theta}^{(t-1)}, \underline{w}^{(t-1)})\pi(\underline{\theta}^{(t-1)}, \underline{w}^{(t-1)}) \prod_{j=1}^{3} \lambda_j^{(t-1)} w_j^{(t-1)}}$$

   1.4 Generate $u \sim \mathscr{U}_{[0,1]}$
   If $u \leqslant r$ then $(\underline{\theta}^{(t)}, \underline{w}^{(t)}) = (\tilde{\underline{\theta}}, \tilde{\underline{w}})$
   else $(\underline{\theta}^{(t)}, \underline{w}^{(t)}) = (\underline{\theta}^{(t-1)}, \underline{w}^{(t-1)})$.

Figure 18 shows the evolution of the Metropolis–Hastings sample for $\zeta^2 = 0.1$. Contrary to the Gibbs sampler, the Metropolis–Hastings samples visit more than one mode of the posterior distribution. There are three moves between two labels in 20,000 iterations, but the bad side of this mobility is the lack of local exploration of the sampler: as a result, the average acceptance probability is very small and the proportions p_j are very badly estimated. Figure 18 shows the effect of a smaller scale, $\zeta^2 = 0.05$, over the evolution of the Metropolis–Hastings sample. There is no move between the different modes but all the parameters are well estimated. If $\zeta^2 = 0.05$, the algorithm has the same behaviour as for $\zeta^2 = 0.01$. This example illustrates both the sensitivity of the random walk sampler to the choice of the scale parameter and the relevance of using several scales to allow both for local and global explorations,[13] a fact exploited in the alternative developed in Section 4.4.

EXAMPLE 4 (*Continued*). We have tested the behaviour of the Metropolis–Hastings algorithm with the following proposals:

$$\begin{aligned}
\tilde{\mu}_j &\sim \mathscr{N}\big(\mu_j^{(t-1)}, \zeta^2\big),\\
\tilde{\sigma}_j^2 &\sim \mathscr{LN}\big(\log\big((\sigma_j^2)^{(t-1)}\big), \zeta^2\big),\\
\tilde{w}_j &\sim \mathscr{LN}\big(\log\big(w_j^{(t-1)}\big), \zeta^2\big).
\end{aligned}$$

After 20,000 iterations, the Metropolis–Hastings algorithm seems to converge (in that the path is stable) and, by using the 5,000 last reordered iterations, we find that the posterior means of μ_1, μ_2, μ_3 are equal to 9.6, 21.3, 28.1, those of $\sigma_1^2, \sigma_2^2, \sigma_3^2$ are equal to 1.9, 5.0, 38.6 and those of p_1, p_2, p_3 are equal to 0.09, 0.81, 0.1. Figure 16 shows the histogram along with the estimated (plug-in) density.

[13] It also highlights the paradox of label-switching: when it occurs, inference gets much more difficult, while, if it does not occur, estimation is easier but based on a sampler that has not converged!

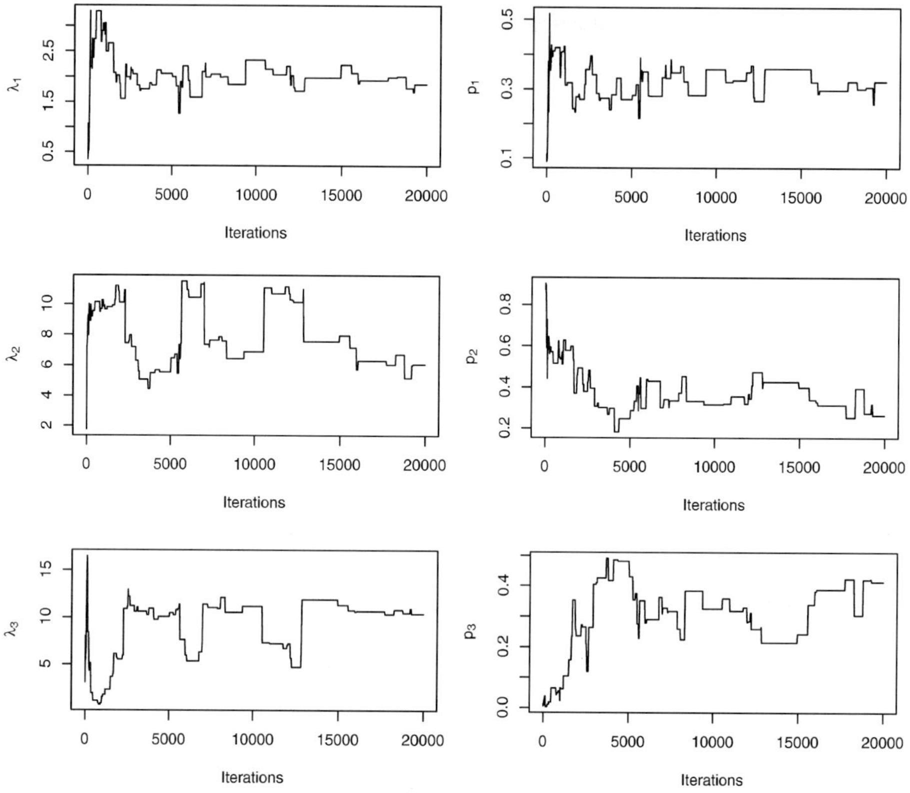

Fig. 18. Evolution of the Metropolis–Hastings sample over 20,000 iterations. (*The scale* ζ^2 *of the random walk is equal to* 0.1.)

4.4. Population Monte Carlo approximations

As an alternative to MCMC, Cappé et al. (2004a) have shown that the importance sampling technique (Robert and Casella, 2004, Chapter 3) can be generalised to encompass much more adaptive and local schemes than thought previously, without relaxing its essential justification of providing a correct discrete approximation to the distribution of interest. This leads to the Population Monte Carlo (PMC) algorithm, following Iba's (2000) denomination. The essence of the PMC scheme is to learn from experience, that is, to build an importance sampling function based on the performances of earlier importance sampling proposals. By introducing a temporal dimension to the selection of the importance function, an adaptive perspective can be achieved at little cost, for a potentially large gain in efficiency. Celeux et al. (2003) have shown that the PMC scheme is a viable alternative to MCMC schemes in missing data settings, among others for the stochastic volatility model (Shephard, 1996). Even with the standard choice of the full conditional distributions, this method provides an accurate representation of the distribution of interest in a few iterations. In the same way, Guillin et al. (2003) have illustrated

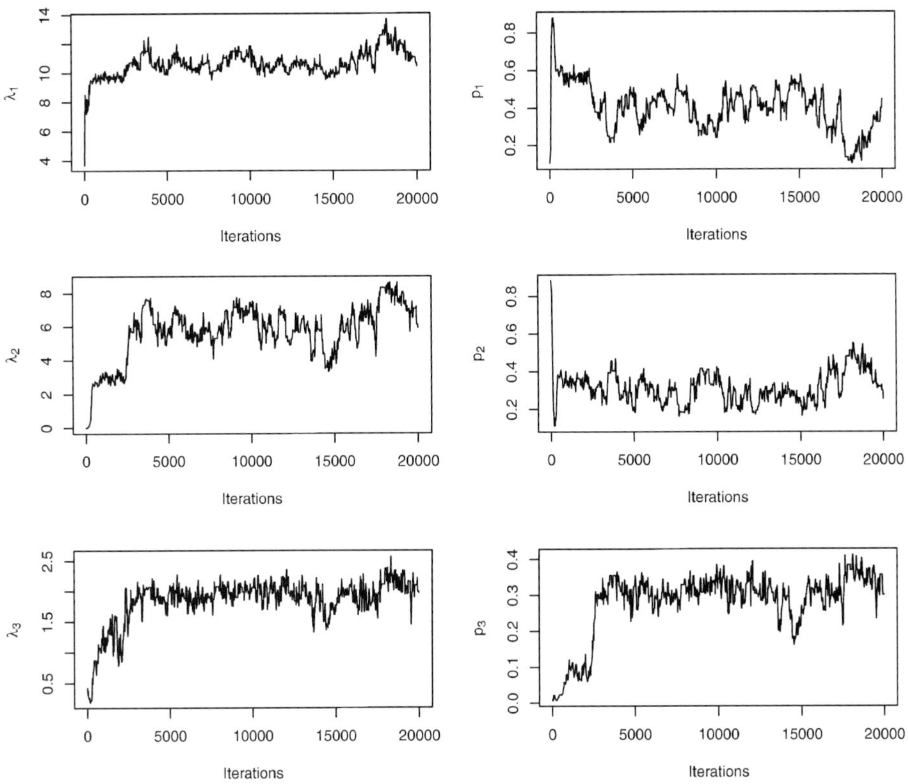

Fig. 19. Same graph with a scale ζ^2 equal to 0.01.

the good properties of this scheme on a switching ARMA model (Hamilton, 1988) for which the MCMC approximations are less satisfactory.

To construct acceptable adaptive algorithms, while avoiding an extended study of their theoretical properties, a better alternative is to leave the setting of Markov chain algorithms and to consider *sequential* or *population* Monte Carlo methods that have much more in common with importance sampling than with MCMC. They are inspired from *particle systems* that were introduced to handle rapidly changing target distributions like those found in signal processing and imaging (Gordon et al., 1993; Shephard and Pitt, 1997; Doucet et al., 2001) but they primarily handle fixed but complex target distributions by building a sequence of increasingly better proposal distributions. Each iteration of the population Monte Carlo (PMC) algorithm thus produces a sample approximately simulated from the target distribution but the iterative structure allows for adaptivity toward the target distribution. Since the validation is based on importance sampling principles, dependence on the past samples can be arbitrary *and* the approximation to the target is valid (unbiased) at *each iteration* and does not require convergence times or stopping rules.

If t indexes the iteration and i the sample point, consider proposal distributions q_{it} that simulate the $\underline{\theta}_{(t)}^{(i)}, \underline{p}_{(t)}^{(i)}$'s and associate to each an importance weight

$$\rho_{(t)}^{(i)} = \frac{f(x|\underline{\theta}_{(t)}^{(i)}, \underline{p}_{(t)}^{(i)})\pi(\underline{\theta}_{(t)}^{(i)}, \underline{p}_{(t)}^{(i)})}{q_{it}(\underline{\theta}_{(t)}^{(i)}, \underline{p}_{(t)}^{(i)}|\boldsymbol{\theta}^{(t-1)}, \mathbf{P}^{(t-1)})}, \qquad i = 1, \dots, M,$$

where

$$\boldsymbol{\theta}^{(t)} = (\underline{\theta}_{(0)}^{(1)}, \dots, \underline{\theta}_{(0)}^{(M)}, \dots, \underline{\theta}_{(t)}^{(M)}), \qquad \mathbf{P}^{(t)} = (\underline{p}_{(0)}^{(1)}, \dots, \underline{p}_{(0)}^{(M)}, \dots, \underline{p}_{(t)}^{(M)}).$$

Approximations of the form

$$\frac{1}{M}\sum_{i=1}^{M} \rho_{(t)}^{(i)} h(\underline{\theta}_{(t)}^{(i)}, \underline{p}_{(t)}^{(i)}) \Big/ \sum_{l=1}^{M} \rho_{(t)}^{(l)}$$

are then approximate unbiased estimators of $\mathbb{E}_x^{\pi}[h(\underline{\theta}, \underline{p})]$, even when the importance distribution q_{it} depends on the entire past of the experiment. Since the above establishes that a simulation scheme based on sample dependent proposals is fundamentally a specific kind of importance sampling, the following algorithm is validated by the same principles as regular importance sampling:

General Population Monte Carlo scheme.

0. **Initialization.** Choose $\underline{\theta}_{(0)}^{(1)}, \dots, \underline{\theta}_{(0)}^{(M)}$ and $\underline{p}_{(0)}^{(1)}, \dots, \underline{p}_{(0)}^{(M)}$
1. **Step t.** For $t = 1, \dots, T$
   1.1 For $i = 1, \dots, M$
      1.1.1 Generate $(\underline{\theta}_{(t)}^{(i)}, \underline{p}_{(t)}^{(i)})$ from $q_{it}(\theta, p)$
      1.1.2 Compute

$$\rho^{(i)} = \frac{f(x|\underline{\theta}_{(t)}^{(i)}, \underline{p}_{(t)}^{(i)})\pi(\underline{\theta}_{(t)}^{(i)}, \underline{p}_{(t)}^{(i)})}{q_{it}(\underline{\theta}_{(t)}^{(i)}, \underline{p}_{(t)}^{(i)}|\boldsymbol{\theta}^{(t-1)}, \mathbf{P}^{(t-1)})}$$

   1.2 Compute

$$\omega^{(i)} = \rho^{(i)} \Big/ \sum_{l=1}^{M} \rho^{(l)}$$

   1.3 Resample M values with replacement from the $(\underline{\theta}_{(t)}^{(i)}, \underline{p}_{(t)}^{(i)})$'s using the weights $\omega^{(i)}$.

Adaptivity can be extended to the individual level and the q_{it}'s can be chosen based on the performances of the previous $q_{i(t-1)}$'s or even on all the previously simulated samples, if storage allows. For instance, the q_{it}'s can include large tail proposals as in the *defensive sampling* strategy of Hesterberg (1998), to ensure finite variance. Similarly, Warnes' (2001) nonparametric Gaussian kernel approximation can be used as a proposal.

The generality in the choice of the proposal distributions q_{it} is obviously due to the abandonment of the MCMC framework. This is not solely a theoretical advantage:

proposals based on the whole past of the chain do not often work. Even algorithms validated by MCMC steps may have difficulties: in one example of Cappé et al. (2004a), a Metropolis–Hastings scheme fails to converge, while a PMC algorithm based on the same proposal produces correct answers.

EXAMPLE 1 (*continued*). In the case of the normal mixture (7), a PMC sampler can be efficiently implemented *without* the (Gibbs) augmentation step, using normal random walk proposals based on the previous sample of (μ_1, μ_2)'s. Moreover, the difficulty inherent to random walks, namely the selection of a "proper" scale, can be bypassed by the adaptivity of the PMC algorithm. Indeed, several proposals can be associated with a range of variances v_k, $k = 1, \ldots, K$. At each step of the algorithm, new variances can be selected proportionally to the performances of the scales v_k on the previous iterations. For instance, a scale can be chosen proportionally to its *nondegeneracy rate* in the previous iteration, that is, the percentage of points generated with the scale v_k that survived after resampling. When the survival rate is null, in order to avoid the complete removal of a given scale v_k, the corresponding number r_k of proposals with that scale is set to a positive value, like 1% of the sample size.

Compared with MCMC algorithms, this algorithm can thus deal with multiscale proposals in an unsupervised manner. We use four different scales, $v_1 = 1$, $v_2 = 0.5$, $v_3 = 0.1$ and $v_4 = 0.01$. We have iterated the PMC scheme 10 times with $M = 1000$ and, after 3 iterations, the two largest variances v_1 and v_2 most often have a zero survival rate, with, later, episodic bursts of survival (due to the generation of values near a posterior mode and corresponding large weights).

POPULATION MONTE CARLO FOR A GAUSSIAN MIXTURE.

0. **Initialization.** Choose $(\mu_1)^{(1)}_{(0)}, \ldots, (\mu_1)^{(M)}_{(0)}$ and $(\mu_2)^{(1)}_{(0)}, \ldots, (\mu_2)^{(M)}_{(0)}$
1. **Step t.** For $t = 1, \ldots, T$
   1.1 For $i = 1, \ldots, M$
      1.1.1 Generate k from $\mathscr{M}(1; r_1, \ldots, r_K)$
      1.1.2 Generate $(\mu_j)^{(i)}_{(t)}$ $(j = 1, 2)$ from $\mathscr{N}((\mu_j)^{(i)}_{(t-1)}, v_k)$
      1.1.4 Compute

$$\rho^{(i)} = \frac{f(x|(\mu_1)^{(i)}_{(t)}, (\mu_2)^{(i)}_{(t)})\pi((\mu_1)^{(i)}_{(t)}, (\mu_2)^{(i)}_{(t)})}{\sum_{l=1}^{K}\prod_{j=1}^{2}\varphi((\mu_j)^{(i)}_{(t)}; (\mu_1)^{(i)}_{(t-1)}, v_l)}$$

   1.2 Compute

$$\omega^{(i)} = \rho^{(i)} \Bigg/ \sum_{l=1}^{M} \rho^{(l)}$$

   1.3 Resample the $(\mu_1)^{(i)}_{(t)}, (\mu_2)^{(i)}_{(t)}$'s using the weights $\omega^{(i)}$

   1.4 Update the r_l's: r_l is proportional to the number of $(\mu_1)^{(i)}_{(t)}, (\mu_2)^{(i)}_{(t)}$'s with variance v_l resampled.

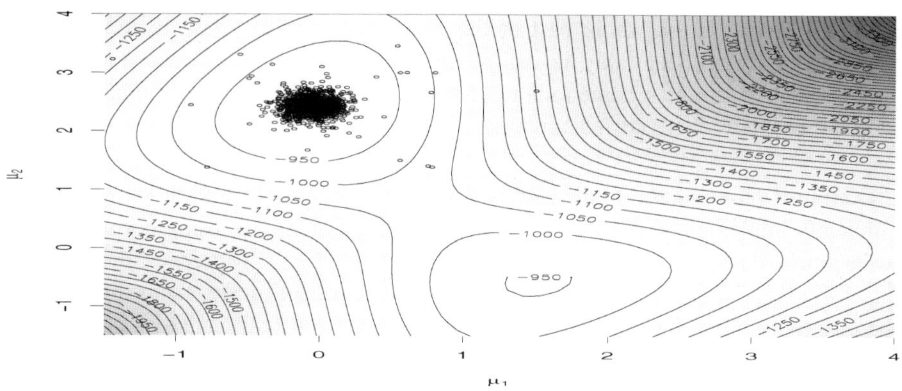

Fig. 20. Representation of the log-posterior distribution with the PMC weighted sample after 10 iterations (the weights are proportional to the circles at each point). For a color reproduction of this figure see the color figures section, page 997.

Figure 20 shows that the sample produced by the PMC algorithm is quite in agreement with the (significant) modal zone of the posterior distribution, while the spurious mode is not preserved after the first iteration.

EXAMPLE 3 (*Continued*). Based on the same multi-scale approach as earlier, the PMC scheme for the reparameterised mixture of 3 Poisson distributions can use the same proposals as in the Metropolis–Hastings setup. With $K = 3$, $\upsilon_1 = 1$, $\upsilon_2 = 0.5$, $\upsilon_3 = 0.1$, $T = 10$ and $M = 2000$, we obtain excellent results.

4.5. *Perfect sampling*

Perfect sampling (see, e.g., Robert and Casella, 2004, Chapter 13) removes the requirement for a burn-in, since samples are guaranteed to come exactly from the target distribution. Perfect sampling for mixtures of distributions has been considered by Hobert et al. (1999), who show that perfect sampling in the mixture context is 'delicate'. However, Casella et al. (2002) achieve a modicum of success by focusing on exponential families and conjugate priors, and using a perfect slice sampler in the spirit of Mira et al. (2001). See Robert and Casella (2004, Chapter 13) for a detailed coverage of perfect sampling simulation for mixtures of distributions.

5. Inference for mixture models with unknown number of components

Estimation of k, the number of components in (2), is a special kind of model choice problem, for which there is a number of possible solutions:

(i) Bayes factors (Kass and Raftery, 1995; Richardson and Green, 1997);
(ii) entropy distance or K–L divergence (Mengersen and Robert, 1996; Sahu and Cheng, 2003);

(iii) reversible jump MCMC (Richardson and Green, 1997; Gruet et al., 1999);
(iv) birth-and-death processes (Stephens, 2000a; Cappé et al., 2002)

depending on whether the perspective is on testing or estimation. We will focus on the latter, because it exemplifies more naturally the Bayesian paradigm and offers a much wider scope for inference, including model averaging in the nonparametric approach to mixture estimation.[14]

The two first solutions above pertain more strongly to the testing perspective, the entropy distance approach being based on the Kullback–Leibler divergence between a k component mixture and its projection on the set of $k-1$ mixtures, in the same spirit as Dupuis and Robert (2003).

5.1. Reversible jump algorithms

When the number of components k is unknown, we have to simultaneously consider several models $\mathfrak{M}_k$, with corresponding parameter sets Θ_k. We thus face a collection of models with a possibly infinite parameter space (and a corresponding prior distribution on this space), for which the computational challenge is higher than in the previous section.

The MCMC solution proposed by Green (1995) is called *reversible jump MCMC* (RJMCMC), because it is based on a *reversibility* constraint on the dimension-changing moves that bridge the sets Θ_k. In fact, the only real difficulty compared with previous developments is to validate moves (or *jumps*) between the Θ_k's, since proposals restricted to a given Θ_k follow from the usual (fixed-dimensional) theory. Furthermore, *reversibility* can be processed at a local level: since the model indicator μ is a integer-valued random variable, we can impose reversibility for each pair (k_1, k_2) of possible values of μ. The idea at the core of reversible jump MCMC is then to supplement each of the spaces Θ_{k_1} and Θ_{k_2} with adequate artificial spaces in order to create a *bijection* between them, most often by augmenting the space of the smaller model. For instance, if $\dim(\Theta_{k_1}) > \dim(\Theta_{k_2})$ and if the move from Θ_{k_1} to Θ_{k_2} is chosen to be a *deterministic* transformation of $\theta^{(k_1)}$

$$\theta^{(k_2)} = T_{k_1 \to k_2}\big(\theta^{(k_1)}\big),$$

Green (1995) imposes a reversibility condition which is that the opposite move from Θ_{k_2} to Θ_{k_1} is concentrated on the curve

$$\big\{\theta^{(k_1)}\colon\ \theta^{(k_2)} = T_{k_1 \to k_2}\big(\theta^{(k_1)}\big)\big\}.$$

In the general case, if $\theta^{(k_1)}$ is completed by a simulation $u_1 \sim g_1(u_1)$ into $(\theta^{(k_1)}, u_1)$ and $\theta^{(k_2)}$ by $u_2 \sim g_2(u_2)$ into $(\theta^{(k_2)}, u_2)$ so that the mapping between $(\theta^{(k_1)}, u_1)$ and $(\theta^{(k_2)}, u_2)$ is a bijection,

$$\big(\theta^{(k_2)}, u_2\big) = T_{k_1 \to k_2}\big(\theta^{(k_1)}, u_1\big), \tag{21}$$

[14] In addition, the unusual topology of the parameter space invalidates standard asymptotic approximations of testing procedures (Lindsay, 1995).

the probability of acceptance for the move from model $\mathfrak{M}_{k_1}$ to model $\mathfrak{M}_{k_2}$ is then

$$\min\left(\frac{\pi(k_2, \theta^{(k_2)})}{\pi(k_1, \theta^{(k_1)})}\frac{\pi_{21}}{\pi_{12}}\frac{g_2(u_2)}{g_1(u_1)}\left|\frac{\partial T_{k_1\to k_2}(\theta^{(k_1)}, u_1)}{\partial(\theta^{(k_1)}, u_1)}\right|, 1\right),$$

involving the Jacobian of the transform $T_{k_1\to k_2}$, the probability π_{ij} of choosing a jump to $\mathcal{M}_{k_j}$ while in $\mathcal{M}_{k_i}$, and g_i, the density of u_i. The acceptance probability for the reverse move is based on the inverse ratio if the move from $\mathfrak{M}_{k_2}$ to $\mathfrak{M}_{k_1}$ also satisfies (21) with $u_2 \sim g_2(u_2)$. The pseudo-code representation of Green's algorithm is thus as follows:

GREEN REVERSIBLE JUMP ALGORITHM.

0. At iteration t, if $x^{(t)} = (m, \theta^{(m)})$.
1. Select model $\mathfrak{M}_n$ with probability π_{mn}.
2. Generate $u_{mn} \sim \varphi_{mn}(u)$.
3. Set $(\theta^{(n)}, v_{nm}) = T_{m\to n}(\theta^{(m)}, u_{mn})$.
4. Take $x^{(t+1)} = (n, \theta^{(n)})$ with probability

$$\min\left(\frac{\pi(n, \theta^{(n)})}{\pi(m, \theta^{(m)})}\frac{\pi_{nm}\varphi_{nm}(v_{nm})}{\pi_{mn}\varphi_{mn}(u_{mn})}\left|\frac{\partial T_{m\to n}(\theta^{(m)}, u_{mn})}{\partial(\theta^{(m)}, u_{mn})}\right|, 1\right),$$

and take $x^{(t+1)} = x^{(t)}$ otherwise.

EXAMPLE 4 (*Continuation*). If model $\mathfrak{M}_k$ is the k component normal mixture distribution,

$$\sum_{j=1}^{k} p_{jk}\mathcal{N}(\mu_{jk}, \sigma_{jk}^2),$$

as in Richardson and Green (1997), we can restrict the moves from $\mathfrak{M}_k$ to only neighbouring models $\mathfrak{M}_{k+1}$ and $\mathfrak{M}_{k-1}$. The simplest solution is to use birth and death moves: The *birth step* consists in adding a new normal component in the mixture generated from the prior and the *death step* is the opposite, namely removing one of the k components at random. In this case, the birth acceptance probability is

$$\min\left(\frac{\pi_{(k+1)k}}{\pi_{k(k+1)}}\frac{(k+1)!}{k!}\frac{\pi_{k+1}(\theta_{k+1})}{\pi_k(\theta_k)\,(k+1)\varphi_{k(k+1)}(u_{k(k+1)})}, 1\right)$$

$$= \min\left(\frac{\pi_{(k+1)k}}{\pi_{k(k+1)}}\frac{\varrho(k+1)}{\varrho(k)}\frac{\ell_{k+1}(\theta_{k+1})(1-p_{k+1})^{k-1}}{\ell_k(\theta_k)}, 1\right),$$

where ℓ_k denotes the likelihood of the k component mixture model $\mathfrak{M}_k$, p_{k+1} the weight of the new component and $\varrho(k)$ is the prior probability of model $\mathfrak{M}_k$. (And the death acceptance probability simply is the opposite.)

While this proposal can work well in some settings, as in Richardson and Green (1997) when the prior is calibrated against the data, it can also be inefficient, that is, leading to a high rejection rate, if the prior is vague, since the birth proposals are not tuned properly. A second proposal, central to the solution of Richardson and Green (1997), is to devise more local jumps between models, called *split* and *combine* moves,

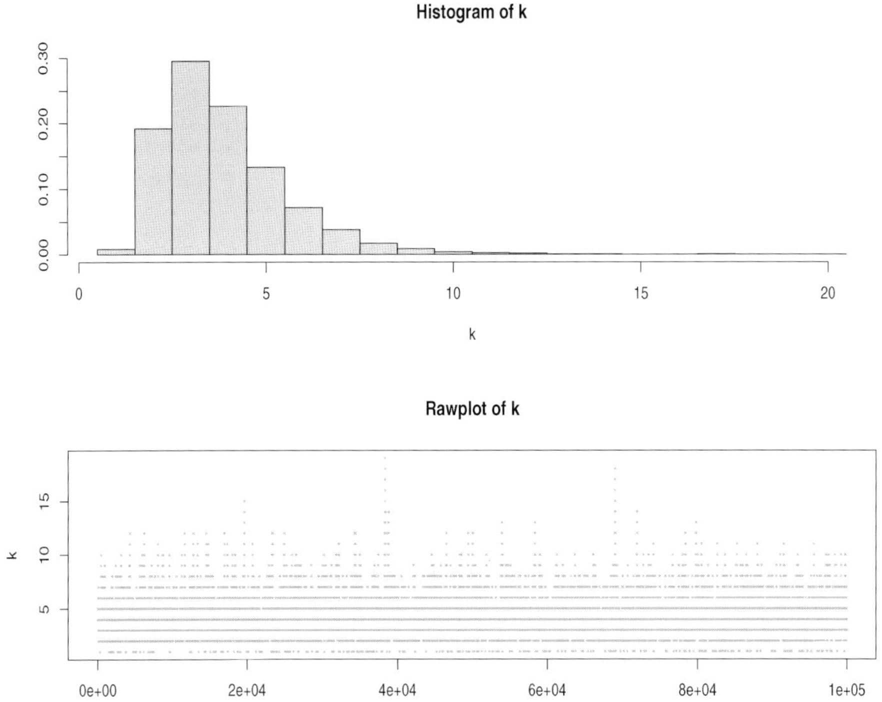

Fig. 21. Histogram and raw plot of 100,000 k's produced by a reversible jump MCMC algorithm for the Galaxy dataset. (*Source:* Robert and Casella, 2004) For a color reproduction of this figure see the color figures section, page 998.

since a new component is created by splitting an existing component into two, under some moment preservation conditions, and the reverse move consists in combining two existing components into one, with symmetric constraints that ensure reversibility.

Figures 21–23 illustrate the implementation of this algorithm for the Galaxy dataset. On Figure 21, the MCMC output on the number of components k is represented as a histogram on k, and the corresponding sequence of k's. The prior used on k is a uniform distribution on $\{1, \ldots, 20\}$: as shown by the lower plot, most values of k are explored by the reversible jump algorithm, but the upper bound does not appear to be restrictive since the $k^{(t)}$'s hardly ever reach this upper limit. Figure 22 illustrates the fact that conditioning the output on the most likely value of k (3 here) is possible. The nine graphs in this figure show the joint variation of the parameters of the mixture, as well as the stability of the Markov chain over the 1,000,000 iterations: the cumulated averages are quite stable, almost from the start. The density plotted on top of the histogram in Figure 23 is another good illustration of the inferential possibilities offered by reversible jump algorithms in that it provides an average of *all* the mixture densities corresponding to the iterations of this MCMC sampler, with higher efficiency properties than a plug-in estimator that would necessitate to condition on k.

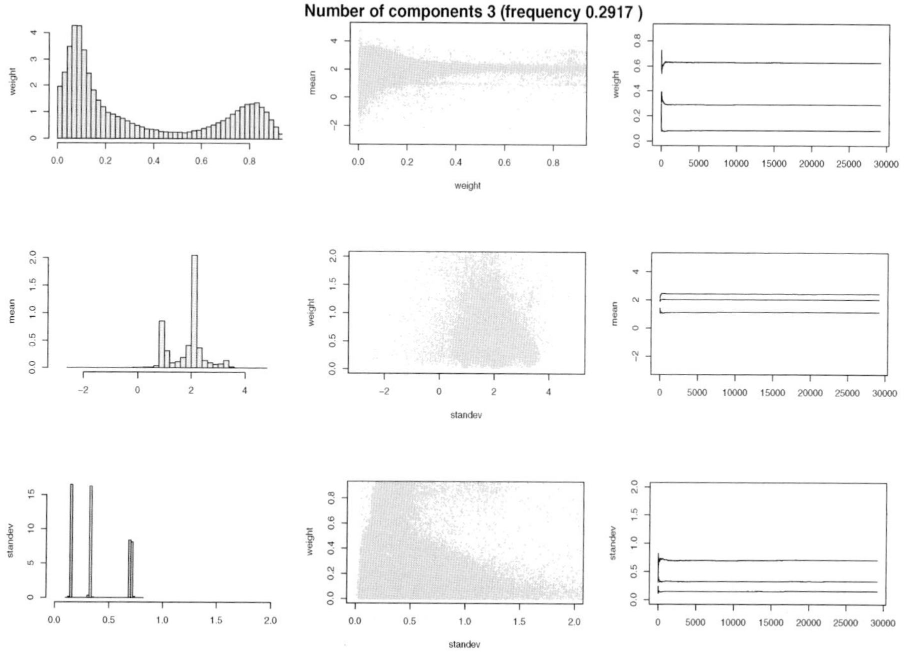

Fig. 22. Reversible jump MCMC output on the parameters of the model $\mathcal{M}_3$ for the Galaxy dataset, obtained by conditioning on $k = 3$. *The left column gives the histogram of the weights, means, and variances; the middle column the scatterplot of the pairs weights–means, means–variances, and variances–weights; the right column plots the cumulated averages (over iterations) for the weights, means, and variances.* (*Source:* Robert and Casella, 2004) For a color reproduction of this figure see the color figures section, page 999.

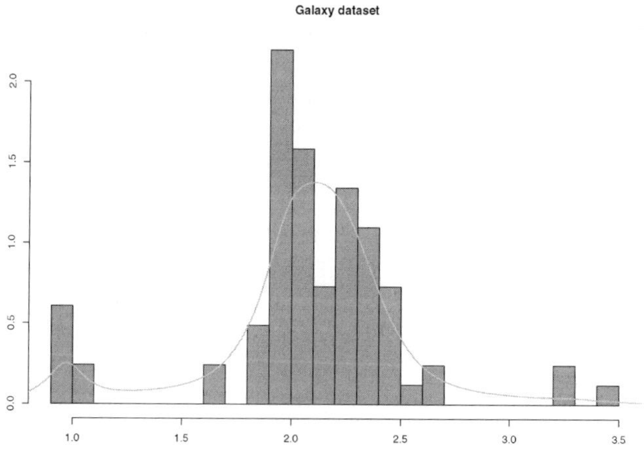

Fig. 23. Fit of the dataset by the averaged density. (*Source:* Robert and Casella, 2004) For a color reproduction of this figure see the color figures section, page 1000.

5.2. *Birth-and-death processes*

The Birth-and-death MCMC (BDMCMC) approach of Stephens (2000a) is already found in Ripley (1987), Grenander and Miller (1994), Phillips and Smith (1996). The algorithm can be described as follows: new components are created (born) in continuous time at a rate $\beta(\alpha)$, where α refers to the current state of the sampler. Whenever a new component is born, its weight ω and parameters θ are drawn from a joint density $f(\alpha; \omega, \theta)$, and old component weights are scaled down proportionally to make all of the weights, including the new one, sum to unity. The new component pair (ω, θ) is then added to α. Components die at a death rate such that the stationary distribution of the process is the posterior distribution as in Section 5.1. The continuous time jump process is thus associated with the birth and death rates: whenever a jump occurs, the corresponding move is always accepted. The acceptance probability of usual MCMC methods is replaced by differential holding times. In particular, implausible configurations die quickly. An extension by Cappé et al. (2002) to BDMCMC introduces split and combine moves, replacing the marked point process framework used by Stephens with a Markov jump process framework.

Cappé et al. (2002) compare the RJMCMC and BDMCMC algorithms and their properties. The authors notice that the reversible jump algorithm, when restricted to birth and death moves with birth proposals based on the prior distribution, enjoys similar properties to BDMCMC. They also show that for any BDMCMC process satisfying some weak regularity conditions, there exists a sequence of RJMCMC processes that converges to the BDMCMC process. More pragmatic comparisons are also to be found in Cappé et al. (2002) for the Gaussian mixture model: The numerical comparison of RJMCMC and BDMCMC revealed that when only birth and death moves are used in addition to moves that do not modify the number of components, there is no significant difference between the samplers. However, when split and combine moves are included there is a small advantage for the BDMCMC sampler. Ranking all techniques on computation time, Cappé et al. (2002) report that "the optimal choice was the RJMCMC with birth and death only, very closely followed by the equivalent BDMCMC sampler, then at some distance, RJMCMC with both types of dimension changing moves enabled and finally BDMCMC in the same conditions".

6. Extensions to the mixture framework

The mixture model (2) can be readily extended to accommodate various complexities. For example, Robert (1998b) gives an example of a hidden Markov model (HMM) that assumes a Markov dependence between the latent variables of (5). For instance, in the Gaussian case,

$$P(z_t = u|z_j, j < t) = p_{z_{t-1}u}; \qquad x_t|z, x_j, j \neq t \sim \mathcal{N}\left(\mu_{z_t}, \sigma^2_{z_t}\right).$$

Such HMMs are commonly used in signal processing and econometrics; see, for example, Hamilton (1989) and Archer and Titterington (1995). Robert and Titterington (1998) also showed that reparameterisation and noninformative prior distributions are also valid in this setting. Convergence for MCMC on HMMs, including nonparametric tests, are described by Robert et al. (1999). See Cappé et al. (2004b) for a complete entry to HMMs. This additional dependency in the observed variables either as another Markov chain or through the model structure is further described by Robert and Casella (2004) through the study of a switching ARMA model. Celeux et al. (2003) also exemplifies the extension of both MCMC and population Monte Carlo techniques to more complex (continuous) latent variables in the study of stochastic volatility.

Extensions to mixtures of regression are examined by Hurn et al. (2003). Here, the switching regression, which is well known in econometrics and chemometrics, may be written as $y = x'\beta_i + \sigma_i\varepsilon$, $\varepsilon \sim g(\varepsilon)$, where the (β_i, σ_i)'s $(i = 1, \ldots, k)$ vary among a set of k possible values with probabilities $p_1, \ldots, p_k$. Thus, if the ε are Gaussian, $y|x \sim p_1\mathscr{N}(x'\beta_1, \sigma_1^2) + \cdots + p_1\mathscr{N}(x'\beta_1, \sigma_1^2)$. Extensions cover both modifications of the model to accommodate the time-dependency encountered in HMMs and nonlinear switching regressions, and modification of the MCMC algorithm to obtain Monte Carlo confidence bands. Mixtures of logistic regressions and of Poisson regressions, and corresponding Gibbs and Metropolis–Hastings algorithms are also detailed. The authors point out that despite the identifiability problem, the MCMC output contains sufficient information on the regression lines and the parameters of the model to enable inference. They formalise this by considering loss-based inference, in which loss functions are specified for the various inferential questions.

Some authors (see, for example, Fernandez and Green, 2002) describe the analysis of spatially correlated Poisson data by a Poisson mixture model in which the weights of the mixture model vary across locations and the number of components is unknown. A missing data structure is detailed for the more complex models of qualitative regression and censored or grouped data.

Further interest in mixture models, their methodology, the associated computational tools, and their application in diverse fields, is evidenced by the wealth of references to Bayesian mixtures in the *Current Index to Statistics*. Since 2000 alone, they have been adopted in mixture hazard models (Louzada-Neto et al., 2002), spatio-temporal models (Stroud et al., 2001), structural equation models (Zhu and Lee, 2001), disease mapping (Green and Richardson, 2002), analysis of proportions (Brooks, 2001), correlated data and clustered models (Chib and Hamilton, 2000; Dunson, 2000; Chen and Dey, 2000), classification and discrimination (Wruck et al., 2001), experimental design and analysis (Nobile and Green, 2000; Sebastiani and Wynn, 2000), random effects generalised linear models (Lenk and DeSarbo, 2000) and binary data (Basu and Mukhopadhyay, 2000). Mixtures of Weibulls (Tsionas, 2002) and Gammas (Wiper et al., 2001) have been considered, along with computational issues associated with MCMC methods (Liang and Wong, 2001), issues of convergence (Liang and Wong, 2001), the display of output (Fan and Berger, 2000), model selection (Ishwaran et al., 2001; Stephens, 2000a) and inference (Lauritzen and Jensen, 2001; Gerlach et al., 2000; Aitkin, 2001; Humphreys and Titterington, 2000; Stephens, 2000b). Nonparametric approaches were popular, exhibited through Dirichlet process mixtures (Gelfand and Kottas, 2002; Green and Richardson, 2001), multiple comparisons (Cho et al., 2001), density estimation (Ghosal, 2001) and regression (Perron and Mengersen, 2001). Mixtures in regression were also identified in changepoint analysis (Skates et al., 2001; Pievatolo and Rotondi, 2000), switching models (Frühwirth-Schnatter, 2001; Hurn et al., 2003) and wavelets (Brown et al., 2001). Bayesian mixtures were also applied to a rich diversity of problems, including earthquake analysis (Walshaw, 2000), biostatistics (Dunson and Weinberg, 2000; Dunson and Dinse, 2000; Qu and Qu, 2000; Dunson and Zhou, 2000), finance (Watanabe, 2000), ecology (Leite et al., 2000) and industrial quality control (Kvam and Miller, 2002; Nair et al., 2001).

This literature, along with the challenges and solutions described in the earlier sections of this chapter, demonstrate the exciting potential for Bayesian mixture modelling in the 21st century.

Acknowledgements

K. Mengersen acknowledges support from an Australian Research Council Discovery Project. Part of this chapter was written while C. Robert was visiting the Australian Mathematical Science Institute, Melbourne, for the Australian Research Council Center of Excellence for Mathematics and Statistics of Complex Systems workshop on Monte Carlo, whose support he most gratefully acknowledges. The authors are also grateful to Agnostino Nobile for his comments on this chapter.

References

Aitkin, M. (2001). Likelihood and Bayesian analysis of mixtures. *Statistical Modelling* **1** (4), 287–304.

Archer, G., Titterington, D. (1995). Parameter estimation for hidden Markov chains. *J. Statist. Plann. Inference*.

Baddeley, A. (1992). Errors in binary images and a l^p version of the Hausdorff metric. *Nieuw Archief voor Wiskunde* **10**, 157–183.

Barron, A., Schervish, M., Wasserman, L. (1999). The consistency of posterior distributions in nonparametric problems. *Ann. Statist.* **27**, 536–561.

Basu, S., Mukhopadhyay, S. (2000). Bayesian analysis of binary regression using symmetric and asymmetric links. *Sankhya, Ser. B* **62** (2), 372–387.

Berger, J. (1985). *Statistical Decision Theory and Bayesian Analysis*, second ed. Springer-Verlag.

Bernardon, J.M., Girón, F.J. (1988). A Bayesian analysis of simple mixture problems. In: Bernardo, J.M., DeGrooh, M.H., Lindley, D.V. (Eds.), *Bayesian Statistics, vol. 3*. Oxford University Press.

Bertillon, J. (1895). *Cours élémentaire de statistique administrative*. SES, Paris.

Besag, J., Green, E., Higdon, D., Mengersen, K. (1995). Bayesian computation and stochastic systems (with discussion). *Statist. Sci.* **10**, 3–66.

Brooks, S. (2001). On Bayesian analyses and finite mixtures for proportions. *Statist. Comput.* **11** (2), 179–190.

Brown, P., Fearn, T., Vannucci, M. (2001). Bayesian wavelet regression on curves with application to a spectroscopic calibration problem. *J. Amer. Statist. Assoc.* **96** (454), 398–408.

Cappé, O., Robert, C., Rydén, T. (2002). Reversible jump MCMC converging to birth-and-death MCMC and more general continuous time samplers. *J. Roy. Statist. Soc., Ser. B* **65** (3), 679–700.

Cappé, O., Guillin, A., Marin, J., Robert, C. (2004a). Population Monte Carlo. *J. Comput. Graph. Statist.* **13**, 907–929.

Cappé, O., Moulines, E., Rydén, T. (2004b). *Hidden Markov Models*. Springer-Verlag.

Casella, G., Mengersen, K., Robert, C., Titterington, D. (2002). Perfect slice samplers for mixtures of distributions. *J. Roy. Statist. Soc., Ser. B* **64** (4), 777–790.

Casella, G., Robert, C., Wells, M. (2000). Mixture models, latent variables and partitioned importance sampling. Technical Report 2000-03, CREST, INSEE, Paris.

Celeux, G., Diebolt, J. (1985). The SEM algorithm: A probabilistic teacher algorithm derived from the EM algorithm for the mixture problem. *Comput. Statist. Quaterly* **2**, 73–82.

Celeux, G., Hurn, M., Robert, C. (2000). Computational and inferential difficulties with mixtures posterior distribution. *J. Amer. Statist. Assoc.* **95** (3), 957–979.

Celeux, G., Marin, J.M., Robert, C.P. (2003). Iterated importance sampling in missing data problems. Technical Report, Université Paris Dauphine.

Chen, M.-H., Dey, D. (2000). A unified Bayesian approach for analyzing correlated ordinal response data. *Rev. Brasil. Probab. Estatist.* **14** (1), 87–111.

Chib, S. (1995). Marginal likelihood from the Gibbs output. *J. Amer. Statist. Assoc.* **90**, 1313–1321.

Chib, S., Hamilton, B. (2000). Bayesian analysis of cross-section and clustered data treatment models. *J. Econometrics* **97** (1), 25–50.

Cho, J., Kim, D., Kang, S. (2001). Nonparametric Bayesian multiple comparisons for the exponential populations. *Far East J. Theoret. Statist.* **5** (2), 327–336.

Dempster, A., Laird, N., Rubin, D. (1977). Maximum likelihood from incomplete data via the EM algorithm (with discussion). *J. Roy. Statist. Soc., Ser. B* **39**, 1–38.

Diaconis, P., Freedman, D. (1986). On the consistency of Bayes estimates. *Ann. Statist.* **14**, 1–26.

Diebolt, J., Robert, C. (1990a). Bayesian estimation of finite mixture distributions, Part i: Theoretical aspects. Technical Report 110, LSTA, Université Paris VI, Paris.

Diebolt, J., Robert, C. (1990b). Bayesian estimation of finite mixture distributions, Part ii: Sampling implementation. Technical Report 111, LSTA, Université Paris VI, Paris.

Diebolt, J., Robert, C. (1990c). Estimation des paramètres d'un mélange par échantillonnage bayésien. *Notes aux Comptes–Rendus de l'Académie des Sciences I* **311**, 653–658.

Diebolt, J., Robert, C. (1993). Discussion of Bayesian computations via the Gibbs sampler by A.F.M. Smith and G. Roberts. *J. Roy. Statist. Soc., Ser. B* **55**, 71–72.

Diebolt, J., Robert, C. (1994). Estimation of finite mixture distributions by Bayesian sampling. *J. Roy. Statist. Soc., Ser. B* **56**, 363–375.

Doucet, A., de Freitas, N., Gordon, N. (2001). *Sequential Monte Carlo Methods in Practice*. Springer-Verlag.

Dunson, D. (2000). Bayesian latent variable models for clustered mixed outcomes. *J. Roy. Statist. Soc., Ser. B* **62** (2), 355–366.

Dunson, D., Dinse, G. (2000). Distinguishing effects on tumor multiplicity and growth rate in chemoprevention experiments. *Biometrics* **56** (4), 1068–1075.

Dunson, D., Weinberg, C. (2000). Modeling human fertility in the presence of measurement error. *Biometrics* **56** (1), 288–292.

Dunson, D., Zhou, H. (2000). A Bayesian model for fecundability and sterility. *J. Amer. Statist. Assoc.* **95** (452), 1054–1062.

Dupuis, J., Robert, C. (2003). Model choice in qualitative regression models. *J. Statist. Plann. Inference* **111**, 77–94.

Escobar, M., West, M. (1995). Bayesian prediction and density estimation. *J. Amer. Statist. Assoc.* **90**, 577–588.

Fan, T.-H., Berger, J. (2000). Robust Bayesian displays for standard inferences concerning a normal mean. *Comput. Statist. Data Anal.* **33** (1), 381–399.

Ferguson, T. (1974). Prior distributions in spaces of probability measures. *Ann. Statist.* **2**, 615–629.

Fernandez, C., Green, P. (2002). Modelling spatially correlated data via mixtures: A Bayesian approach. *J. Roy. Statist. Soc., Ser. B* **64**, 805–826.

Frühwirth-Schnatter, S. (2001). Markov chain Monte Carlo estimation of classical and dynamic switching and mixture models. *J. Amer. Statist. Assoc.* **96** (453), 194–209.

Gelfand, A., Kottas, A. (2002). A computational approach for full nonparametric Bayesian inference under Dirichlet process mixture models. *J. Comput. Graph. Statist.* **11** (2), 289–305.

Gelfand, A., Smith, A. (1990). Sampling based approaches to calculating marginal densities. *J. Amer. Statist. Assoc.* **85**, 398–409.

Gerlach, R., Carter, C., Kohn, R. (2000). Efficient Bayesian inference for dynamic mixture models. *J. Amer. Statist. Assoc.* **95** (451), 819–828.

Ghosal, S. (2001). Convergence rates for density estimation with Bernstein polynomials. *Ann. Statist.* **29** (5), 1264–1280.

Gordon, N., Salmon, J., Smith, A. (1993). A novel approach to non-linear/non-Gaussian Bayesian state estimation. *IEEE Proc. Radar and Signal Process.* **140**, 107–113.

Green, P. (1995). Reversible jump MCMC computation and Bayesian model determination. *Biometrika* **82** (4), 711–732.

Green, P., Richardson, S. (2001). Modelling heterogeneity with and without the Dirichlet process. *Scand. J. Statist.* **28** (2), 355–375.

Green, P., Richardson, S. (2002). Hidden Markov models and disease mapping. *J. Amer. Statist. Assoc.* **97** (460), 1055–1070.

Grenander, U., Miller, M. (1994). Representations of knowledge in complex systems (with discussion). *J. Roy. Statist. Soc., Ser. B* **56**, 549–603.

Gruet, M., Philippe, A., Robert, C. (1999). MCMC control spreadsheets for exponential mixture estimation. *J. Comput. Graph. Statist.* **8**, 298–317.

Guillin, A., Marin, J., Robert, C. (2003). Estimation bayésienne approximative par échantillonnage préférentiel. Technical Report 0335, Cahiers du Ceremade, Université Paris Dauphine.

Hamilton, J.D. (1988). Rational-expectations econometric analysis of changes in regime: An investigation of the term structure of interest rates. *J. Economic Dynamics and Control* **12**, 385–423.

Hamilton, J.D. (1989). A new approach to the economic analysis of nonstationary time series and the business cycles. *Econometrica* **57**, 357–384.

Hastie, T., Tibshirani, R., Friedman, J.H. (2001). *The Elements of Statistical Learning: Data Mining, Inference, and Prediction*. Springer-Verlag, New York.

Hesterberg, T. (1998). Weighted average importance sampling and defensive mixture distributions. *Technometrics* **37**, 185–194.

Hobert, J., Robert, C., Titterington, D. (1999). On perfect simulation for some mixtures of distributions. *Statist. Comput.* **9** (4), 287–298.

Humphreys, K., Titterington, D. (2000). Approximate Bayesian inference for simple mixtures. In: *COMPSTAT – Proceedings in Computational Statistics*, pp. 331–336.

Hurn, M., Justel, A., Robert, C. (2003). Estimating mixtures of regressions. *J. Comput. Graph. Statist.* **12**, 1–25.

Iba, Y. (2000). Population-based Monte Carlo algorithms. *Trans. Japan. Soc. Artificial Intelligence* **16** (2), 279–286.

Ishwaran, H., James, L., Sun, J. (2001). Bayesian model selection in finite mixtures by marginal density decompositions. *J. Amer. Statist. Assoc.* **96** (456), 1316–1332.

Jordan, M. (2004). Graphical models. *Statist. Sci.* Submitted for publication.

Kass, R., Raftery, A. (1995). Bayes factors. *J. Amer. Statist. Assoc.* **90**, 773–795.

Kim, S., Shephard, N., Chib, S. (1998). Stochastic volatility: Likelihood inference and comparison with ARCH models. *Rev. Econom. Stud.* **65**, 361–393.

Kvam, P., Miller, J. (2002). Discrete predictive analysis in probabilistic safety assessment. *J. Quality Technology* **34** (1), 106–117.

Lauritzen, S., Jensen, F. (2001). Stable local computation with conditional Gaussian distributions. *Statist. Comput.* **11** (2), 191–203.

Lavine, M., West, M. (1992). A Bayesian method for classification and discrimination. *Canad. J. Statist.* **20**, 451–461.

Leite, J., Rodrigues, J., Milan, L. (2000). A Bayesian analysis for estimating the number of species in a population using nonhomogeneous Poisson process. *Statist. Probab. Lett.* **48** (2), 153–161.

Lenk, P., DeSarbo, W. (2000). Bayesian inference for finite mixtures of generalized linear models with random effects. *Psychometrika* **65** (1), 93–119.

Liang, F., Wong, W. (2001). Real-parameter evolutionary Monte Carlo with applications to Bayesian mixture models. *J. Amer. Statist. Assoc.* **96** (454), 653–666.

Lindsay, B. (1995). *Mixture Models: Theory, Geometry and Applications*. IMS Monographs, Hayward, CA.

Liu, J., Wong, W., Kong, A. (1994). Covariance structure of the Gibbs sampler with applications to the comparisons of estimators and sampling schemes. *Biometrika* **81**, 27–40.

Louzada-Neto, F., Mazucheli, J., Achcar, J. (2002). Mixture hazard models for lifetime data. *Biometrical J.* **44** (1), 3–14.

MacLachlan, G., Basford, K. (1988). *Mixture Models: Inference and Applications to Clustering*. Marcel Dekker, New York.

MacLachlan, G., Peel, D. (2000). *Finite Mixture Models*. Wiley, New York.

Mengersen, K., Robert, C. (1996). Testing for mixtures: A Bayesian entropic approach (with discussion). In: Berger, J., Bernardo, J., Dawid, A., Lindley, D., Smith, A. (Eds.), *Bayesian Statistics, vol. 5*. Oxford University Press, Oxford, pp. 255–276.

Mengersen, K., Robert, C., Guihenneuc-Jouyaux, C. (1999). MCMC convergence diagnostics: a "reviewww". In: Berger, J., Bernardo, J., Dawid, A., Lindley, D., Smith, A. (Eds.), *Bayesian Statistics, vol. 6*. Oxford University Press, Oxford, pp. 415–440.

Mira, A., Møller, J., Roberts, G. (2001). Perfect slice samplers. *J. Roy. Statist. Soc., Ser. B* **63**, 583–606.

Moreno, E., Liseo, B. (2003). A default Bayesian test for the number of components in a mixture. *J. Statist. Plann. Inference* **111** (1–2), 129–142.

Nair, V., Tang, B., Xu, L.-A. (2001). Bayesian inference for some mixture problems in quality and reliability. *J. Quality Technology* **33**, 16–28.

Nobile, A., Green, P. (2000). Bayesian analysis of factorial experiments by mixture modelling. *Biometrika* **87** (1), 15–35.

Pearson, K. (1894). Contribution to the mathematical theory of evolution. *Proc. Trans. Roy. Soc. A* **185**, 71–110.

Pérez, J., Berger, J. (2002). Expected-posterior prior distributions for model selection. *Biometrika* **89** (3), 491–512.

Perron, F., Mengersen, K. (2001). Bayesian nonparametric modelling using mixtures of triangular distributions. *Biometrics* **57**, 518–528.

Petrone, S., Wasserman, L. (2002). Consistency of Bernstein polynomial posteriors. *J. Roy. Statist. Soc., Ser. B* **64**, 79–100.

Phillips, D., Smith, A. (1996). Bayesian model comparison via jump diffusions. In: Gilks, W., Richardson, S., Spiegelhalter, D. (Eds.), *Markov Chain Monte Carlo in Practice*. Chapman and Hall, pp. 215–240.

Pievatolo, A., Rotondi, R. (2000). Analysing the interevent time distribution to identify seismicity phases: A Bayesian nonparametric approach to the multiple-changepoint problem. *Appl. Statist.* **49** (4), 543–562.

Qu, P., Qu, Y. (2000). A Bayesian approach to finite mixture models in bioassay via data augmentation and Gibbs sampling and its application to insecticide resistance. *Biometrics* **56** (4), 1249–1255.

Redner, R.A., Walker, H.F. (1984). Mixture densities, maximum likelihood and the EM algorithm. *SIAM Review* **26** (2), 195–239.

Richardson, S., Green, P. (1997). On Bayesian analysis of mixtures with an unknown number of components (with discussion). *J. Roy. Statist. Soc., Ser. B* **59**, 731–792.

Ripley, B. (1987). *Stochastic Simulation*. Wiley, New York.

Robert, C. (1998a). *Discretization and MCMC Convergence Assessment. Lecture Notes in Statistics*, vol. 135. Springer-Verlag.

Robert, C. (1998b). MCMC specifics for latent variable models. In: Payne, R., Green, P. (Eds.), COMPSTAT 1998, Physica-Verlag, Heidelberg, pp. 101–112.

Robert, C. (2001). *The Bayesian Choice*, second ed. Springer-Verlag.

Robert, C., Casella, G. (2004). *Monte Carlo Statistical Methods*, second ed. Springer-Verlag, New York.

Robert, C., Mengersen, K. (1999). Reparametrization issues in mixture estimation and their bearings on the Gibbs sampler. *Comput. Statist. Data Anal.* **29**, 325–343.

Robert C., Rousseau, J. (2002). A mixture approach to Bayesian goodness of fit. Technical Report, Cahiers du CEREMADE, Université Paris Dauphine.

Robert, C., Titterington, M. (1998). Reparameterisation strategies for hidden Markov models and Bayesian approaches to maximum likelihood estimation. *Statist. Comput.* **8** (2), 145–158.

Robert, C., Rydén, T., Titterington, D. (1999). Convergence controls for MCMC algorithms, with applications to hidden Markov chains. *J. Statist. Comput. Simulation* **64**, 327–355.

Roeder, K. (1992). Density estimation with confidence sets exemplified by superclusters and voids in galaxies. *J. Amer. Statist. Assoc.* **85**, 617–624.

Roeder, K., Wasserman, L. (1997). Practical Bayesian density estimation using mixtures of normals. *J. Amer. Statist. Assoc.* **92**, 894–902.

Sahu, S., Cheng, R. (2003). A fast distance based approach for determining the number of components in mixtures. *Canad. J. Statist.* **31**, 3–22.

Sebastiani, P., Wynn, H. (2000). Maximum entropy sampling and optimal Bayesian experimental design. *J. Roy. Statist. Soc., Ser. B* **62** (1), 145–157.

Shephard, N. (1996). Statistical aspects of ARCH and stochastic volatility. In: Cox, D.R., Barndorff-Nielsen, O.E., Hinkley, D.V. (Eds.), *Time Series Models in Econometrics, Finance and Other Fieds*. Chapman and Hall.

Shephard, N., Pitt, M. (1997). Likelihood analysis of non-Gaussian measurement time series. *Biometrika* **84**, 653–668.

Silverman, B. (1986). *Density Estimation for Statistics and Data Analysis*. Chapman and Hall.

Skates, S., Pauler, D., Jacobs, I. (2001). Screening based on the risk of cancer calculation from Bayesian hierarchical changepoint and mixture models of longitudinal markers. *J. Amer. Statist. Assoc.* **96** (454), 429–439.

Stephens, M. (1997). Bayesian methods for mixtures of normal distributions. Ph.D. thesis, University of Oxford.

Stephens, M. (2000a). Bayesian analysis of mixture models with an unknown number of components – an alternative to reversible jump methods. *Ann. Statist.* **28**, 40–74.

Stephens, M. (2000b). Dealing with label switching in mixture models. *J. Roy. Statist. Soc., Ser. B* **62** (4), 795–809.

Stroud, J., Müller, P., Sansó, B. (2001). Dynamic models for spatiotemporal data. *J. Roy. Statist. Soc., Ser. B* **63** (4), 673–689.

Tanner, M., Wong, W. (1987). The calculation of posterior distributions by data augmentation. *J. Amer. Statist. Assoc.* **82**, 528–550.

Titterington, D., Smith, A., Makov, U. (1985). *Statistical Analysis of Finite Mixture Distributions*. Wiley, New York.

Tsionas, E. (2002). Bayesian analysis of finite mixtures of Weibull distributions. *Comm. Statist. Part A – Theory and Methods* **31** (1), 37–48.

Verdinelli, I., Wasserman, L. (1992). Bayesian analysis of outliers problems using the Gibbs sampler. *Statist. Comput.* **1**, 105–117.

Verdinelli, I., Wasserman, L. (1998). Bayesian goodness-of-fit testing using infinite-dimensional exponential families. *aos* **26**, 1215–1241.

Walshaw, D. (2000). Modelling extreme wind speeds in regions prone to hurricanes. *Appl. Statist.* **49** (1), 51–62.

Warnes, G. (2001). The Normal kernel coupler: An adaptive Markov Chain Monte Carlo method for efficiently sampling from multi-modal distributions. Technical Report 395, University of Washington.

Wasserman, L. (2000). Asymptotic inference for mixture models using data dependent priors. *J. Roy. Statist. Soc., Ser. B* **62**, 159–180.

Watanabe, T. (2000). A Bayesian analysis of dynamic bivariate mixture models: Can they explain the behavior of returns and trading volume?. *J. Business and Economic Statistics* **18** (2), 199–210.

Wiper, M., Insua, D., Ruggeri, F. (2001). Mixtures of Gamma distributions with applications. *J. Comput. Graph. Statist.* **10** (3), 440–454.

Wruck, E., Achcar, J., Mazucheli, J. (2001). Classification and discrimination for populations with mixture of multivariate normal distributions. *Rev. Mat. Estatist.* **19**, 383–396.

Zhu, H.-T., Lee, S.-Y. (2001). A Bayesian analysis of finite mixtures in the LISREL model. *Psychometrika* **66** (1), 133–152.

Handbook of Statistics, Vol. 25
ISSN: 0169-7161
© 2005 Elsevier B.V. All rights reserved.
DOI 10.1016/S0169-7161(05)25017-4

17

Simulation Based Optimal Design

Peter Müller

Abstract
We review simulation based approaches to optimal design, with an emphasis on problems that cast optimal design as formal decision problems. Under this perspective we approach optimal design problems as expected utility maximization, characterized by an action space, a probability model and a utility function. We first discuss the use of independent Monte Carlo simulation to evaluate expected utility. A variation of such approaches replaces the maximization in the design problem by a simulation problem. This is achieved by augmenting the original probability model on data and parameters to a probability model on data, parameters and actions in such a way that the marginal distribution on the actions is proportional to expected utility. Using power transformations lets us replace the typically flat expected utility function by a more peaked target function, without changing the solution of the optimization. This leads to simulated annealing type algorithms.

A interesting class of decision problems are sequential design problems. We discuss simulation based approaches to solve these typically computation intensive decision problems. Another interesting set of decision problems are found in massive multiple comparisons. We discuss the use of posterior simulation to solve such problems. Finally, we show some examples of important alternative, nondecision theoretic approaches.

Keywords: decision problem; expected utility maximization; Markov chain Monte Carlo; optimal design

1. Introduction

We discuss simulation based approaches to expected utility optimization. Consider a decision problem described by a utility function $u(d, \theta, y)$ and a probability model $p_d(\theta, y)$ on all unknowns. Here d is the decision, θ are unknown parameters and y are the observable data. The utility function models preferences over consequences and $p_d(\theta, y)$ models beliefs, possibly influenced by actions. Decision theoretic approaches assume that the investigator is a rational decision maker choosing an action that maximizes the utility of the possible consequences, averaging with respect to all relevant unknowns (Raiffa and Schlaifer, 1961; DeGroot, 1970). The average is defined as expectation with respect to all unknowns at the time of decision making, including future data that will only be observed after making the decision, as well as unknown parameters that will never be observed. In contrast, the place of data that is known at the time of decision making is on the right-hand side of the conditioning bar. The expected utility average is conditional on all relevant information available at the time of decision making.

Many decision problems consider utility functions that involve inference loss. See, for example, Chaloner and Verdinelli (1995) for a review of Bayesian approaches to such decision problems traditionally known as optimal design problems. Here we focus instead on problems with more general utility functions. We propose an approach that is suitable when the expected utility is an analytically intractable integral, possibly due to a complex probability model, the nature of the utility function or the nature of the action space. For example, the decision about shrinking a network of monitoring stations requires maximization over all possible patterns of included and excluded candidate sites, typically allowing neither analytical maximization nor full enumeration. Such problems are discussed, for example, in Clayton et al. (1999), Nychka and Saltzman (1998) and Zidek et al. (2000). Decision problems that require simulation based solutions also frequently arise in biomedical research problems. Berry (1993), Spiegelhalter et al. (1994) and Berry and Stangl (1996) review related problems. See Parmigiani (2002) and Spiegelhalter (2004) for an extensive discussion of medical decision problems.

Simulation based approaches to decision problems have been successfully used in many contexts. Carlin et al. (1998), Brockwell and Kadane (2003) and Berry et al. (2001) propose simulation based solutions to sequential decision problems. In these papers simulation is used to evaluate expected utility integrals. Müller and Parmigiani (1996) propose a similar strategy for general nonsequential problems, exploiting the continuity of $U(d)$ to reduce computational effort. Extensive literature in operations research deals with simulation based approaches to solve decision problems represented in so-called influence diagrams. See, for example, Bielza et al. (1999) for a discussion and references.

Expected utility integration is only one of the computational challenges in optimal design. The other computational challenge is the maximization over the design space. One approach to implement simulation based maximization are simulated annealing methods. See van Laarhoven and Aarts (1987) for a review of simulated annealing. Geman and Geman (1984) use simulated annealing for maximum *a posteriori* estimation. Fei and Berliner (1991) and Laud et al. (1992) discuss convergence issues and alternative probing distributions. The paradigm of simulated annealing is nonhomogeneous Markov chain simulation with a sequence of stationary distributions which increasingly concentrate around the point of maximum (or minimum). The use of such algorithms for expected utility maximization is explored in Müller et al. (2004a) and in Amzal et al. (2003).

2. Monte Carlo evaluation of expected utility

Consider the generic decision problem defined by a utility function $u(d, \theta, y)$ and a probability model $p_d(\theta, y)$. Typically the probability model factors as $p_d(\theta, y|d) = p(\theta)\, p_d(y|\theta)$ into a prior $p(\theta)$ which is invariant under the decision d and a sampling distribution $p_d(y|\theta)$. The optimal decision problem is formally described as

$$d^* = \arg\max_{d \in \mathcal{D}} U(d) \quad \text{with } U(d) = \int u(d, \theta, y)\, \mathrm{d}p_d(y|\theta)\, \mathrm{d}p(\theta). \tag{1}$$

The assumptions about the probability model are not critical. Little changes in the following discussion if $p_d(\theta, y)$ does not factor as assumed in (1), or if the prior probability model $p_d(\theta)$ depends on d. When the target function $U(d)$ is not available in closed form standard optimization approaches are difficult to implement. However, $U(d)$ can easily be approximated by appropriate Monte Carlo integration. This is generally possible by independent Monte Carlo simulation since both, prior $p(\theta)$ and sampling model $p_d(y|\theta)$, are typically available for efficient computer simulation.

We can generate a Monte Carlo sample $(\theta_j, y_j) \sim p(\theta_j) p_d(y_j|\theta_j)$, $j = 1, \ldots, M$, to obtain an approximation

$$\widehat{U}(d) = 1/M \sum u(d, \theta_j, y_j). \tag{2}$$

The use of such approximations is a common technique in many simulation based optimal design approaches, including, for example, Sun et al. (1996), Carlin et al. (1998), or Müller and Parmigiani (1996).

Figure 1 shows a typical example. The figure includes a simple, but very powerful extension of (2). Exploiting a notion of continuity of $U(d)$ we replace the sample average $\widehat{U}(d)$ by a smooth fit of a scatterplot of pairs (d, u), where $u = u(d, \theta_j, y_j)$ for simulated experiments (θ_j, y_j) under design d, with d varying over a grid of possible design choices, in the example shown in Figure 1 the grid is from $d = 2$ through 120, with $M = 50$ simulated experiments (θ_j, y_j) for each design d. The fitted curve formalizes the idea of borrowing strength across different design choices d.

3. Augmented probability simulation

Evaluation and maximization of (2) is practicable only for low-dimensional decision spaces, say for up to four-dimensional d. Also, simulation across the entire grid on d potentially wastes major simulation effort for uninteresting designs. For example, in Figure 1 simulations for $J = 2$ and $J = 3$ are not needed for the final maximization. The problem is exacerbated in higher dimensions. Finally, maximization of $\widehat{U}(d)$ is only reliable for very peaked expected utility functions. For relatively flat $U(d)$ the numerical errors inherent in the evaluation of $\widehat{U}$ might overwhelm a relatively small difference in $U(d)$ between the optimal and worst designs. For these reasons, alternative methods are required for the evaluation and maximization of $U(d)$.

An interesting approach is to recast the stochastic optimization problem (1) as a simulation problem from a sequence of augmented probability models. The central idea

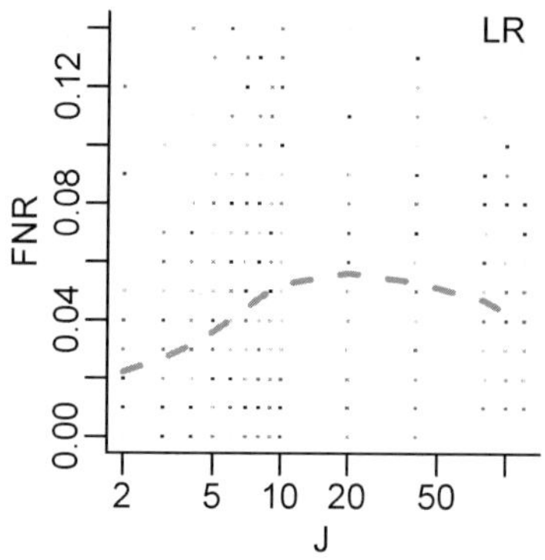

Fig. 1. A typical expected utility function $U(d)$. The figure shows estimated expected utility $\widehat{U}(d)$ (labeled FNR) against decisions d (labeled J). The dots show realized utilities $u(d, \theta_j, y_j)$. The dashed line shows a fitted smooth curve $\widehat{U}(d)$. The example is taken from Müller et al. (2004a). The decision is the number of arrays in a microarray experiment. The utility function is a combination of false discovery and false negative rate. For a color reproduction of this figure see the color figures section, page 1000.

is to define an auxiliary distribution $h_J(d, \cdot)$ such that the marginal distribution in d is proportional to $U(d)^J$. Assume that $u(d, \theta, y)$ is nonnegative and bounded. Consider an artificial distribution h_J as

$$h_J(d, y_1, \theta_1, \ldots, y_J, \theta_J) \propto \prod_{j=1}^{J} u(d, y_j) p_d(y_j, \theta_j). \tag{3}$$

Marginalizing over $(y_1, \ldots, \theta_J)$ we find that

$$h_J(d) \propto \left[\int u \, \mathrm{d}p_d(y, \theta) \right]^J = U(d)^J,$$

as desired. In (3) we have augmented the original probability model on parameters and data to include the design parameter d as a random variable.

Fixing $J = 1$ the described procedure defines a probability model that randomly generates designs d proportional to expected utility. This is exploited in Bielza et al. (1999) by defining a Markov chain Monte Carlo (MCMC) simulation to generate from $h(d, \theta, y) \equiv h_1(d, \theta, y)$. This can be implemented using any of the MCMC schemes as reviewed, for example, in Tierney (1994).

The fact that the target distribution is $h(\theta, y, \theta)$, instead of a posterior distribution $p(\theta|y)$ as in traditional posterior MCMC simulation, does not hinder application of the same simulation strategies. Appropriate summaries of the simulation output in d allow inference about the optimal design d^*. For example, for univariate d one can find the optimal design as the mode of the histogram of simulated d values. However, such strategies are limited to low-dimensional design vectors d, say up to $\dim(d) = 4$. For higher-dimensional design spaces it becomes impracticable to estimate the mode of $h(d)$ from the simulation output. An exception is the special case when $U(d)$ is very peaked and gives only negligible probability to designs d significantly away from d^*. The optimum can then be inferred as the sample average of simulated d values.

This special case motivates an algorithm proposed in Müller et al. (2004a) and Amzal et al. (2003). We start by simulating from h. In subsequent iterations of the MCMC algorithm, $n = 1, 2, \ldots$, we change the target distribution to h_J, incrementing $J = J_n$ as we proceed. For sufficiently large J the marginal distribution $h_J(d)$ is tightly concentrated on the set of optimal designs d^*.

The outlined algorithm is related to simulated annealing for a target function $f(d) = \log U(d)$. Central to simulated annealing is a sequence of equilibrium distributions $\pi_n(d) \propto \exp\{f(d)/T_n\}$. Defining $T_n = 1/J_n$ we find $\pi_n(d) = h_{J_n}(d)$, highlighting the similarity to simulated annealing. However, standard simulated annealing requires that the target function $\log U(d)$ be available for direct evaluation. But it is exactly this evaluation of the expected utility integral which complicates the optimal design problem (1) in the first place.

In Amzal et al. (2003) the same idea is approached from a different perspective. The Monte Carlo simulation over increasing J is considered as an instance of a particle filter. Particle filters are a popular method for Monte Carlo simulation in state space time series models. Let y_t and x_t denote the observed data at time t and a latent state parameter at time t. Particle filter methods are Monte Carlo simulation methods that provide Monte Carlo posterior samples from a sequence of posterior distributions $p_t = p(x_t | y_1, \ldots, y_t)$. Amzal et al. (2003) recognize the formal similarity of the particle filtering problem and simulation from the sequence of distributions h_J.

4. Sequential design

In many applications, decisions are made sequentially. Let $d = (d_1, \ldots, d_T)$, and $y = (y_1, \ldots, y_T)$. Assume that y_t depends on $d_1, \ldots, d_t$ only and that decision d_t is made after decision d_{t-1} and after observing y_{t-1}. Then $d_t = D_t(d_1, \ldots, d_{t-1}, y_1, \ldots, y_{t-1})$ is a policy depending on the observed outcomes from the first $t-1$ decision periods. Such sequential decision problems where later decisions depend on earlier outcomes are notoriously difficult to optimize. A complete solution requires, in general, backward induction involving an exponentially increasing number of possible scenarios. See, for example, (Berger, 1985, Chapter 7). The paradigm of backward induction is also known as Bellman's equation or as dynamic programming (Bellman, 1957).

In some problems the optimal policy $D = (D_1, \ldots, D_T)$ is characterized by a low-dimensional summary vector of control parameters. For example, Carlin et al. (1998) address stopping a clinical trial at each of K interim analyses. In their setup they show that the optimal Bayes sequential procedure can be characterized by a set of $2K$ critical cutoffs for the posterior mean of a parameter that quantifies the advantage of a treatment over placebo. The problem is reduced to choosing these cutoffs. Similar simplifications occur whenever the loss function and the posterior distribution depend on the current history only indirectly through a low dimensional statistic. Typically this summary statistic is a sufficient statistic for the posterior distribution, i.e., it summarizes the current state of information about the unknown parameters. For example, (Lewis and Berry, 1994) discuss optimal sequential design under binomial sampling and a 0–1 type inference loss related to a hypothesis test about the unknown success probabilities. In their setup, the state of information at any given time is given by the number of patients treated and the parameters of the beta posterior distribution corresponding to the binomial success probabilities. Christen and Nakamura (2001) discuss sequential stopping rules for species accumulation. They show that the optimal rule depends on the current history only indirectly through a bivariate summary, namely, the number of species accumulated and the area under the accumulation curve. Hsiao and Clayton (1999) show that truncated repeated significance boundaries (Lan and DeMets, 1983) can be optimal from a Bayesian perspective. They assume a decision problem related to testing a hypothesis about the drift parameter in a Brownian motion with drift. Hsiao and Clayton (1999) demonstrate how repeated significance test boundaries can be optimal from a Bayesian perspective under an appropriate choice of loss function and sampling cost.

These special cases motivate an approximate simulation-based solution for the general sequential decision problem. Consider a sufficient statistic S_t for the posterior distribution conditional on data at time t and record expected utilities and optimal decisions for a grid of values for S_t. For low-dimensional S_t, this strategy avoids the need to record solutions for an exponentially increasing number of histories. If necessary, Monte Carlo simulation can be used to approximate expected utility integrals for each possible value of S_t (on a finite grid). This strategy is used in Brockwell and Kadane (2003). The same strategy is developed in Berry et al. (2001) and Müller et al. (1999), with the additional twist of replacing the sufficient statistic by a low-dimensional summary statistic S_t that captures most of the relevant information, but is not necessarily a sufficient statistic. This allows to keep the computational effort manageable even in the absence of low-dimensional sufficient summaries. Forward simulation is used to evaluate expected utilities on the grid over S_t.

Another traditional approximate solution to sequential decision problems are m-step lookahead solutions. The idea in a one-step lookahead approach is to proceed as if the next period were the final horizon, freeing the implementation of the need to record solutions for all possible histories to a possibly distant horizon. For example, in a clinical trial we would act as if the next patient were the last patient in the trial. The general m-step lookahead approach proceeds as if the final horizon were m steps away. The resulting decision is sub-optimal. See, for example, (Berger, 1985, Chapter 7) for a description of m-step lookahead procedures and variations. Christen et al. (2003) describe a simulation-based m-step lookahead designs. The decision problem is adaptive allocation and stopping in a k-arm clinical trial. The utility function is general and not constrained to analytic evaluation of expected utility integrals. Instead simulation is used to approximate expected utilities.

5. Multiple comparisons

A challenging class of decision problems arises in massive multiple comparison problems. Typical examples are experiments with high throughput gene expression data, requiring decisions about differential expression for many, often thousands, of genes simultaneously. From a frequentist perspective, ignoring the multiplicity problem would inflate type I error rates. Using traditional adjustments would lead to hopelessly conservative rules. Similar problems arise in a formal decision-theoretic solution. A commonly used approach to address the massive multiple comparison problem is the control of false discovery rates (FDR). This approach is developed in Benjamini and Hochberg (2002), and further developed, among many others, in Genovese and Wasserman (2002a, 2002b), Efron and Tibshirani (2002) or Storey (2003).

For a formal definition of FDR consider a sequence of comparisons $\delta_i \in \{0, 1\}$, $i = 1, \ldots, n$. For example, δ_i might be an indicator for gene i being differentially expressed under two biologic conditions of interest. In other words, δ_i is the unknown truth of the ith comparison. In the context of the terminology introduced earlier δ_i is an unknown parameter. Let $d_i \in \{0, 1\}$ denote the decision for the ith comparison. The desired decision $d = (d_1, \ldots, d_n)$ is an n-dimensional binary vector. We refer to $d_i = 1$ as a rejection or discovery (because we have potentially "discovered" an important gene). FDR is defined as the number of false rejections relative to the number of rejections. Let $D = \sum_{i=1}^n d_i$ denote the total number of rejections.

$$\mathrm{FDR} = \frac{\sum_{i=1}^n d_i(1 - \delta_i)}{D}, \tag{4}$$

$\mathrm{FDR}(d, \delta)$ is a function of the decision d and the unknown parameters δ. The analogous notion of false negative rate (FNR) defines

$$\mathrm{FNR} = \frac{\sum_{i=1}^n (1 - d_i)\delta_i}{n - D}. \tag{5}$$

Let Y denote the data and let $v_i = E(\delta_i \mid Y)$ denote the marginal posterior probability of the ith comparison. It can be shown (Müller et al., 2004b) that for a variety of reasonable loss functions combining FDR and FNR the optimal decision is of the type $d_i = I(v_i > t)$. The value of the threshold t depends on the specific loss function. Usual posterior Markov chain Monte Carlo simulation is used to evaluate v_i, thus making the use of FDR based adjustment for multiplicities natural for simulation based Bayesian optimal design.

While FDR-based approaches are traditionally presented in the context of high throughput gene expression data analysis, the problem is more general and also arises in many other application areas. A good example is the record linkage problem (Fortini et al., 2001).

6. Calibrating decision rules by frequentist operating characteristics

So far we have restricted the discussion to strictly decision theoretic approaches, based on casting the design problem as expected utility maximization with a given probability model and utility function. While this can be argued to be a coherent approach for a rational decision maker, there are many impediments for the practical implementation of such approaches. Often elicitation of a utility function requires to trade off multiple competing goals, related to, for example, sampling cost, inference on treatment effects, prediction for individual patient counseling, optimal treatment of patients enrolled in the current study and optimal treatment for future patients. The specific functional form of combining these competing goals into one encompassing utility function is often an arbitrary decision. But the solution of the decision problem critically hinges upon the statement of this utility function.

This leads many authors to using pragmatic alternatives to a formal decision theoretic method. A commonly used approach is to consider a set of intuitively reasonable decision rules. For example, one could consider sequential stopping rules for a clinical trial by proceeding as follows. After each cohort of patients compute the posterior probability of clinically relevant events. These events might be, for example, the event that the experimental treatment is more effective than the current standard of care (or placebo), and the event that the experimental treatment has toxicity beyond some preset acceptable bound. Denote the two events by E and T, respectively. Let $p(E \mid y_n)$ and $p(T \mid y_n)$ denote the posterior probabilities after the nth cohort. In this context we could then consider a decision rule that stops when $p(E \mid y_n)$ or $p(T \mid y_n)$ are beyond a certain pre-set boundaries p_E and p_T, respectively. If we stop for large $p(E \mid y_n)$ we recommend the treatment for further development. If we stop for large $p(T \mid y_n)$ we recommend against the treatment. For an assumed true scenario of efficacy and toxicity we can then evaluate frequentist type-I error probability, power and average sample size. By appropriately adjusting p_E and p_T one can achieve desired frequentist operating characteristics.

We refer to such approaches as stylized Bayes. Inference on the events of interest is implemented as proper posterior inference in a Bayes model. But the approach avoids the complications arising from specifying a utility function. Control of frequentist operating characteristics is an attractive feature. In many applications, even if one proceeds with an entirely decision theoretic approach, computation of frequentist operating characteristics are required for purposes of communication and validation. This is in particular true for clinical trial designs.

In the context of clinical trials this approach is developed, among others, in Thall et al. (1995). See Estey and Thall (2003) for a recent review of related approaches.

An alternative resolution of the problem posed by the need to accommodate many possibly competing goals is the use of suitable heuristic criteria. For example, Zidek et al. (2000) argues for the use of entropy based designs for the network design problem, arguing that the entropy-based approach captures the spirit of all the objectives without emphasizing any one of them. A similar consideration leads Nychka and Saltzman (1998) to propose space-filling designs. Both approaches are not necessarily simulation-based methods.

7. Discussion

We have described the use of simulation based methods to solve expected utility maximization problems. The main features of the reviewed approaches are generality and flexibility. Most simulation based approaches only require that under a given decision and assuming hypothesized future data y and parameter values θ the decision maker be able to evaluate utilities $u(d, \theta, y)$. Similarly the probability model can be quite general. As in posterior MCMC simulation, essentially any probability model that allows pointwise evaluation of posterior distributions for given parameter values can be used.

As in any complex MCMC simulation the main limitation of the proposed method is the computation intensive implementation. The approach strictly assumes a decision theoretic setup, and thus inherits any limitations related to this paradigm. In particular, we assume that there is one single utility function, with known tradeoffs between possibly competing goals. Also, the need for pointwise evaluation of utilities limits applicability of the method for problems with traditional inference loss involving often analytically intractable posterior variances.

In summary, the reviewed algorithms are likely to be useful for problems with complicated design spaces, like in the network design example, and problems with utility functions that involve complicated functions of future data and parameters, like life histories of future patients undergoing screening for chronic disease.

References

Amzal, B., Bois, F.Y., Parent, E., Robert, C.P. (2003). Bayesian optimal design via interacting MCMC. Technical Report 03/0348, CEREMADE, Université de Paris IX, Paris.

Bellman, R. (1957). *Dynamic Programming*. Princeton University Press.

Benjamini, Y., Hochberg, Y. (2002). Controlling the false discovery rate: A practical and powerful approach to multiple testing. *J. Roy. Statist. Soc., Ser. B (Methodological)* **57**, 289–300.

Berger, J.O. (1985). *Statistical Decision Theory and Bayesian Analysis*, second ed. Springer-Verlag, New York.

Berry, D. (1993). A case for Bayesianism in clinical trials (with discussion). *Statistics in Medicine* **12**, 1377–1404.

Berry, D.A., Stangl, D.K. (Eds.) (1996). *Bayesian Biostatistics. Statistics: Textbooks and Monographs*, vol. 151. Marcel Dekker, New York.

Berry, D.A., Mueller, P., Grieve, A.P., Smith, M., Parke, T., Blazek, R., Mitchard, M., Krams, M. (2001). Adaptive Bayesian designs for dose-ranging drug trials. In: Gatsonis, C., Kass, R.E., Carlin, B., Carriquiry, A., Gelman, A., Verdinelli, I., West, M. (Eds.), *Case Studies in Bayesian Statistics, vol. V*. In: *Lecture Notes in Statistics*. Springer-Verlag, New York.

Bielza, C., Müller, P., Insua, D.R. (1999). Monte Carlo methods for decision analysis with applications to influence diagrams. *Management Sci.* **45**, 995–1007.

Brockwell, A.E., Kadane, J.B. (2003). Sequential analysis by griding sufficient statistics. *J. Comput. Graph. Statist.* **12**, 566–584.

Carlin, B., Kadane, J., Gelfand, A. (1998). Approaches for optimal sequential decision analysis in clinical trials. *Biometrics* **54**, 964–975.

Chaloner, K., Verdinelli, I. (1995). Bayesian experimental design: A review. *Statist. Sci.* **10**, 273–304.

Christen, J.A., Nakamura, M. (2001). Sequential stopping rules for species accumulation. Technical Report, CIMAT, Guanajuato, Mexico.

Christen, J., Müller, P., Wathen, K. (2003). Robust Bayesian randomized clinical trials. Technical Report, U.T. M.D. Anderson Cancer Center, Houston.

Clayton, M.K., Shepherd, J., Hsia, C.-F., Li, S.-H. (1999). Bayesian sequential design of a network of sensors. Technical Report, University of Wisconsin.

DeGroot, M. (1970). *Optimal Statistical Decisions*. McGraw-Hill.

Efron, B., Tibshirani, R. (2002). Empirical Bayes methods and false discovery rates for microarrays. *Genetic Epidemiology* **23**, 70–86.

Estey, E.H., Thall, P.F. (2003). New designs for phase 2 clinical trials. *Blood* **102**, 442–448.

Fei, L., Berliner, L.M. (1991). Asymptotic properties of stochastic probing for global optimization: The finite case. Technical Report, Ohio State University.

Fortini, M., Liseo, B., Nuccitelli, A., Scanu, M. (2001). On Bayesian record linkage. *Research in Official Statistics* **4**, 185–198.

Geman, S., Geman, A. (1984). Stochastic relaxation, Gibbs distributions and the Bayesian restoration of images. *IEEE Trans. Patt. Anal. Mac. Int.* **6**, 721–741.

Genovese, C., Wasserman, L. (2002a). Bayesian and frequentist multiple testing. In: Bernardo, A.P.D.J.M., Berger, J.O., Smith, A.F.M. (Eds.), *Bayesian Statistics, vol. 7*. Oxford University Press, Oxford.

Genovese, C., Wasserman, L. (2002b). Operating characteristics and extensions of the false discovery rate procedure. *J. Roy. Statist. Soc., Ser. B (Methodological)* **64**, 499–518.

Hsiao, C.-F., Clayton, M.K. (1999). Bayes discrete sequential boundaries for clinical trials. Technical Report, University of Wisconsin.

Lan, K.K.G., DeMets, D.L. (1983). Bayesian approaches to randomized trials. *Biometrika* **157**, 357–416.

Laud, P.W., Berliner, L.M., Goel, P.K. (1992). A stochastic probing algorithm for global optimization. *J. Global Optim.* **2**, 209–224.

Lewis, R., Berry, D. (1994). Group sequential clinical trials: A classical evaluation of Bayesian decision-theoretic designs. *J. Amer. Statist. Assoc.* **89**, 1528–1534.

Müller, P., Parmigiani, G. (1996). Optimal design via curve fitting of Monte Carlo experiments. *J. Amer. Statist. Assoc.* **90**, 1322–1330.

Müller, P., Berry, D.A., Grieve, A., Smith, M., Krams, M. (1999). Simulation-based sequential Bayesian design. Technical Report, Duke University ISDS, USA.

Müller, P., Sansó, B., de Iorio, M. (2004a). Optimal Bayesian design by inhomogeneous Markov chain simulation. *J. Amer. Statist. Assoc.* (Submitted for publication).

Müller, P., Parmigiani, G., Robert, C., Rousseau, J. (2004b). Optimal sample size for multiple testing: The case of gene expression microarrays. *J. Amer. Statist. Assoc.* (Submitted for publication).

Nychka, D., Saltzman, N. (1998). Design of air-quality monitoring networks. In: Nychka, D., Piegorsch, W., Cox, L. (Eds.), *Case Studies in Environmental Statistics*. New York.

Parmigiani, G. (2002). *Modeling in Medical Decision Making: A Bayesian Approach*. Wiley, New York.

Raiffa, H., Schlaifer, R. (1961). *Applied Statistical Decision Theory*. Harvard University Press, Boston.

Spiegelhalter, D. (2004). *Bayesian Approaches to Clinical Trials and Health-Care Evaluation*. Wiley, New York.

Spiegelhalter, D.J., Freedman, L.S., Parmar, M.K.B. (1994). Bayesian approaches to randomized trials. *J. Roy. Statist. Soc., Ser. A, General* **157**, 357–387.

Storey, J. (2003). The positive false discovery rate: A Bayesian interpretation and the q-value. *Ann. Statist.* **31**, 2013–2035.

Sun, D., Tsutakawa, R.K., Lu, W.-S. (1996). Bayesian design of experiment for quantal responses: What is promised versus what is delivered. *J. Statist. Plann. Inference* **52**, 289–306.

Thall, P.F., Simon, R.M., Estey, E.H. (1995). Bayesian sequential monitoring designs for single-arm clinical trials with multiple outcomes. *Statistics in Medicine* **14**, 357–379.

Tierney, L. (1994). Markov chains for exploring posterior distributions. *Ann. Statist.* **22**, 1701–1728.

van Laarhoven, P., Aarts, E. (1987). *Simulated Annealing: Theory and Applications*. D. Reidel.

Zidek, J.V., Sun, W., Le, N.D. (2000). Designing and integrating composite networks for monitoring multivariate Gaussian pollution fields. *Appl. Statist.* **49**, 63–79.

Handbook of Statistics, Vol. 25
ISSN: 0169-7161
© 2005 Elsevier B.V. All rights reserved.
DOI 10.1016/S0169-7161(05)25018-6

18

Variable Selection and Covariance Selection in Multivariate Regression Models

Edward Cripps, Chris Carter and Robert Kohn

Abstract

This article provides a general framework for Bayesian variable selection and covariance selection in a multivariate regression model with Gaussian errors. By variable selection we mean allowing certain regression coefficients to be zero. By covariance selection we mean allowing certain elements of the inverse covariance matrix to be zero. We estimate all the model parameters by model averaging using a Markov chain Monte Carlo simulation method. The methodology is illustrated by applying it to four real data sets. The effectiveness of variable selection and covariance selection in estimating the multivariate regression model is assessed by using four loss functions and four simulated data sets. Each of the simulated data sets is based on parameter estimates obtained from a corresponding real data set.

Keywords: cross-sectional regression; longitudinal data; model averaging; Markov chain Monte Carlo

1. Introduction

This article provides a general framework for estimating a multivariate regression model. The methodology is Bayesian and allows for variable selection and covariance selection, as well as allowing some of the dependent variables to be missing. By variable selection we mean that the regression model allows for some of the regression coefficients to be identically zero. By covariance selection we mean that the model allows the off-diagonal elements of the inverse of the covariance matrix to be identically zero. We estimate all functionals of the parameters by model averaging, i.e., by taking a weighted average of the values of the functional, where the average is over the allowable configurations of the regression coefficients and the covariance matrix and the weights are the posterior probabilities of the configurations. The computation is carried out using a Markov chain Monte Carlo simulation method.

There is an extensive literature on Bayesian variable selection. In the univariate case, see Mitchell and Beauchamp (1988), George and McCulloch (1993, 1997), Raftery et al. (1997), Smith and Kohn (1996), Kohn et al. (2001) and Hoeting et al. (1999) for further discussions and citations. Raftery et al. (1997) argue that if prediction is the goal of the analysis, then it may be better to use model averaging rather than trying to find the "optimal" subset of variables by variable selection. Further support for model averaging is given by Breiman (1996) who argues that subset selection is unstable in the univariate linear regression case. In a series of papers, Brown et al. (1998, 1999, 2002) consider variable selection and model averaging for multivariate regression models. These papers assume that a variable is either in all equations or in none of them. This is a special case of the variable selection approach taken in our article, which is based on Smith and Kohn (1996) and Kohn et al. (2001).

Efficiently estimating a covariance matrix is a difficult statistical problem, especially when the dimension of the covariance matrix is large relative to the sample size (e.g., see Stein, 1956; Dempster, 1969) because the number of unknown parameters in the covariance matrix increases quadratically with dimension and because it is necessary to keep the estimate of the covariance matrix positive definite. Early work on the efficient estimation of covariance matrices is by Stein (see Stein, 1956, and other unpublished papers by Stein that are cited by Yang and Berger, 1994) and Efron and Morris (1976). For more recent work see Leonard and Hsu (1992) and Chiu et al. (1996) who modeled the matrix logarithm of the covariance matrix. Yang and Berger (1994) used a Bayesian approach based on a spectral decomposition of the covariance matrix. Pourahmadi (1999, 2000) estimated the covariance matrix by parameterizing the Cholesky decomposition of its inverse. Smith and Kohn (2002) used a prior that allows for zero elements in the strict lower triangle of the Cholesky decomposition of the inverse of the covariance matrix to obtain a parsimonious representation of the covariance matrix. Although the Cholesky decomposition applies to a general covariance matrix, it is most useful and interpretable for longitudinal data. Barnard et al. (2000) modeled the covariance matrix in terms of standard deviations and correlations and proposed several shrinkage estimators. Further results and simulation comparisons are given by Daniels and Kass (1999).

Dempster (1972) proposed estimating the covariance matrix parsimoniously by identifying zero elements in its inverse. He called models for the covariance matrix obtained in this way covariance selection models. His idea was that in many statistical problems the inverse of the covariance matrix has a large number of zeros in its off-diagonal elements and these should be exploited in estimating the covariance matrix. There is a natural interpretation of such zeros: the i, jth element of the inverse is zero if and only if the partial correlation between the ith and j variables is 0 (e.g., Whittaker, 1990). This means that a covariance selection model can be interpreted as a Gaussian graphical model (e.g., Lauritzen, 1996). Giudici and Green (1999) gave a Bayesian approach for estimating the structure of a decomposable graphical model. Their approach can be used to efficiently estimate a covariance matrix with a decomposable graphical structure, and possibly more general covariance matrices. Wong et al. (2003) give a Bayesian approach for estimating a general covariance selection model and it is their approach that we use in the chapter.

Our chapter makes the following contributions to the literature. First, it combines model averaging over the regression coefficients with model averaging over the inverse covariance matrix in the multivariate normal linear regression model. Second, it presents a more general approach to variable selection in the mean of the regression model than that given by Brown et al. (1998, 1999, 2002). Third, it illustrates the methodology using four real examples. Fourth, it studies whether model averaging based on variable selection or covariance selection or both improves the estimation of the multivariate regression model. The assessment is based on a study of performance of four simulated examples using four loss functions. Each of the simulated examples is based on the estimates of one of the four real examples. The four loss functions consider separately the estimates of the predictive distribution, the estimates of the covariance matrix only, the estimates of the regression coefficients and the estimates of the fitted values.

The article is organized as follows. Section 2 describes the multivariate model and the priors for variable and covariance selection. Section 3 discusses the sampling scheme and computational issues. Section 4 describes the real data sets and reports on the analysis of the real data. Section 5 describes the simulated data based on the real examples and presents the results of the simulation. Section 6 summarizes the chapter.

2. Model description

2.1. Introduction

For $t = 1, \ldots, n$, let Y_t be a $p \times 1$ vector of responses, X_t a $p \times q$ matrix of covariates and β the $q \times 1$ vector of regression coefficients. We assume the model

$$Y_t = X_t\beta + e_t, \quad e_t \sim N(0, \Sigma). \tag{2.1}$$

Let $\gamma = (\gamma_1, \ldots, \gamma_q)$ be a vector of binary variables such that the ith column of X_t is included in the regression if $\gamma_i = 1$ and is excluded if $\gamma_i = 0$. We write $X_{t,\gamma}$ for the matrix that contains all columns of X_t for which $\gamma_i = 1$ and β_γ for the corresponding subvector of regression coefficients. Therefore, the vector γ indexes all the mean functions for the regression model (2.1). Conditional on γ, (2.1) becomes

$$Y_t = X_{t,\gamma}\beta_\gamma + e_t, \quad e_t \sim N(0, \Sigma). \tag{2.2}$$

Model (2.1) contains as a special case the multivariate model

$$Y_t = Bx_t + e_t, \tag{2.3}$$

where B is a matrix of regression coefficients and x_t is a vector of covariates. It is clear that model (2.3) is a special case of model (2.1) by taking $X_t = x_t' \otimes I_p$ and $\beta = \text{vec}(B)$. We note that $\otimes$ means Kronecker product and $\text{vec}(B)$ is the vector obtained by stacking the columns of B beneath each other. The model (2.3) is used extensively in multivariate regression analysis, e.g., Mardia et al. (1979, p. 157) and in particular Brown et al. (1998, 1999, 2002). We note that Brown et al. (1998, 1999, 2002) do variable selection on x_t which means that when they drop a covariate they drop a whole column of the matrix B. We show in Section 2.8 how this can be done in general for the model (2.1).

We follow Wong et al. (2003) and parameterise $\Sigma^{-1} = \Omega$ as

$$\Omega = TCT, \tag{2.4}$$

where T is a diagonal matrix with $T_i^2 = \Omega_{ii}$ which is the inverse of the partial variance of $Y_{i,t}$ and C is a correlation matrix. The partial correlation coefficients ρ^{ij} of Σ are given by

$$\rho^{ij} = \frac{-\Omega_{ij}}{(\Omega_{ii}\Omega_{jj})^{1/2}} = -C_{ij}, \tag{2.5}$$

so that the off-diagonal elements of C are the negative of the partial correlation coefficients.

Let $Y = (Y_1', \ldots, Y_n')'$. From (2.2) the likelihood is

$$\begin{aligned} p(Y|\beta, \gamma, \Sigma) &= \left|2\pi\Omega^{-1}\right|^{-n/2} \exp\left\{-\frac{1}{2}\sum_{t=1}^{n}(Y_t - X_{t,\gamma}\beta_\gamma)'\Omega(Y_t - X_{t,\gamma}\beta_\gamma)\right\} \\ &\propto |\Omega|^{n/2} \exp\left\{-\frac{1}{2}\text{trace}\big(\Omega(\text{SRR}_\gamma)\big)\right\} \\ &\propto |T|^n |C|^{n/2} \exp\left\{-\frac{1}{2}\text{trace}\Big(TCT(\text{SRR}_\gamma)\Big)\right\}, \end{aligned} \tag{2.6}$$

where $(\text{SRR}_\gamma) = \sum_{t=1}^{n}(Y_t - X_{t,\gamma}\beta_\gamma)(Y_t - X_{t,\gamma}\beta_\gamma)'$.

2.2. *Prior for the regression coefficients*

Similarly to Smith and Kohn (1996), we take the prior for the regression coefficients as noninformative with respect to the likelihood and having a mode at zero. To motivate the prior, it is useful to rewrite the likelihood as follows.

$$\begin{aligned} &p(Y|\beta, \gamma, \Omega) \\ &= |2\pi\Omega|^{n/2} \exp\left\{-\frac{1}{2}\sum_{t=1}^{n}(Y_t - X_{t,\gamma}\beta_\gamma)'\Omega(Y_t - X_{t,\gamma}\beta_\gamma)\right\} \\ &= |2\pi\Omega|^{n/2} \exp\left\{-\frac{1}{2}\sum_{t=1}^{n} Y_t'\Omega Y_t - 2\beta_\gamma'\sum_{t=1}^{n} X_{t,\gamma}'\Omega Y_t \right. \\ &\quad \left. + \beta_\gamma'\sum_{t=1}^{n} X_{t,\gamma}'\Omega X_{t,\gamma}\right\} \\ &= |2\pi\Omega|^{n/2} \exp\left\{-\frac{1}{2}(\text{SYY}) - 2\beta_\gamma'(\text{SXY}_\gamma) + \beta_\gamma'(\text{SXX}_\gamma)\beta_\gamma\right\}, \end{aligned} \tag{2.7}$$

where

$$\text{SYY} = \sum_{t=1}^{n} Y_t'\Omega Y_t, \qquad \text{SXY}_\gamma = \sum_{t=1}^{n} X_{t,\gamma}'\Omega Y_t, \quad \text{and}$$

$$\text{SXX}_\gamma = \sum_{t=1}^{n} X_{t,\gamma}'\Omega X_{t,\gamma}.$$

As a function of β_γ, the likelihood is Gaussian with a mean of $(\text{SXX}_\gamma)^{-1}(\text{SXY}_\gamma)$ and covariance matrix $(\text{SXX}_\gamma)^{-1}$.

Conditional on the binary indicator vector and the covariance matrix we take the prior for β_γ as

$$\beta_\gamma|\Sigma, \gamma \sim N\big(0, c(\text{SXX}_\gamma)^{-1}\big) \tag{2.8}$$

and set $c = n$ such that the prior variance of β_γ stays approximately the same as n increases.

From (2.7) and (2.8) we can write the density of β_γ conditional on Y, Σ and γ as

$$\beta_\gamma|Y, \gamma, \Sigma \sim N\left\{\frac{c}{1+c}(\text{SXX}_\gamma)^{-1}(\text{SXY}_\gamma), \frac{c}{1+c}(\text{SXX}_\gamma)^{-1}\right\}. \tag{2.9}$$

2.3. *Prior for the vector of binary indicator variables*

We first define

$$q_\gamma = \sum_{i=1}^{q} \gamma_i,$$

which is the number of columns contained in X_t specified by $\gamma_i = 1$.

We assume that γ and Σ are independent apriori and as in Kohn et al. (2001) we specify the prior for γ as

$$p(\gamma|\pi) = \pi^{q_\gamma}(1-\pi)^{q-q_\gamma}, \quad \text{with } 0 \leqslant \pi \leqslant 1. \tag{2.10}$$

We set the prior for π as uniform, i.e., $p(\pi) = 1$ for $0 \leqslant \pi \leqslant 1$, so that

$$\begin{aligned} p(\gamma) &= \int p(\gamma|\pi)p(\pi)\,\mathrm{d}\pi \\ &= \int \pi^{q_\gamma}(1-\pi)^{q-q_\gamma}\,\mathrm{d}\pi \\ &= \mathrm{B}(q_\gamma + 1, q - q_\gamma + 1), \end{aligned}$$

where B is the beta function defined by

$$\mathrm{B}(\alpha, \beta) = \frac{\Gamma(\alpha)\Gamma(\beta)}{\Gamma(\alpha+\beta)}.$$

The likelihood for γ and Σ, with β_γ integrated out is

$$\begin{aligned} p(Y|\gamma, \Sigma) &= \int p(Y|\beta, \Sigma, \gamma)p(\beta_\gamma|\gamma, \Sigma)\,\mathrm{d}\beta_\gamma \\ &\propto (1+c)^{-q_\gamma/2} \\ &\quad \times \exp\left\{-\frac{1}{2}\left((\text{SYY}) - \frac{c}{1+c}(\text{SXY}_\gamma)'(\text{SXX}_\gamma)^{-1}(\text{SXY}_\gamma)\right)\right\}. \end{aligned} \tag{2.11}$$

We can write the density of γ conditional on Y and Σ as

$$p(\gamma|Y, \Sigma) \propto p(Y|\Sigma, \gamma)p(\gamma),$$

and we use this density to update γ in the Markov chain Monte Carlo simulation.

2.4. Prior for Ω_{ii}

Following Wong et al. (2003), we take the prior for Ω_{ii} as a gamma distribution such that

$$\Omega_{ii} \propto \Omega_{ii}^{\tau-1} \exp\{-\nu\Omega_{ii}\},$$

which means that the prior for T_i is

$$p(T_i) \propto p(\Omega_{ii})\frac{\mathrm{d}\Omega_{ii}}{\mathrm{d}T_i}$$
$$\propto T_i^{2\tau-1} \exp\{-\nu T_i^2\}. \tag{2.12}$$

To ensure the prior is noninformative we follow Wong et al. (2003) and set $\tau = 10^{-10}$ and $\nu = 10^{-10}$ in the rest of the chapter.

From (2.6) and (2.7) we have

$$p(T_i|Y, T_{\{-i\}}, C, \beta, \gamma)$$
$$\propto P(Y|T, C, \beta, \gamma)p(T_i)$$
$$\propto T_i^n \exp\left[-\frac{1}{2}\left\{T_i^2(\mathrm{SRR}_\gamma)_{ii} + 2T_i\sum_{j\neq i}^{p}(\mathrm{SRR}_\gamma)_{ij}C_{ij}T_j\right\}\right]$$
$$\propto T^{n_\tau} \exp\{-aT_i^2 - 2bT_i\}, \tag{2.13}$$

where $n_\tau = n + 2\tau - 1$, $a = (\mathrm{SRR}_\gamma)_{ii}/2 + \nu$ and $b = (1/2)\sum_{j\neq i}^{p}(\mathrm{SRR}_\gamma)_{ij}C_{ij}T_j$. This is the conditional density we use to generate T_i. Wong et al. (2003) show that this conditional density of T_i tends to normality as $n \to \infty$.

2.5. Prior for the partial correlation matrix C

We use the covariance selection prior for C in Wong et al. (2003), which allows the off-diagonal elements of C to be identically zero. This prior is similar in intention to the variable selection prior used in Section 2.3, except that it is now necessary to keep the matrix C positive definite. For $j = 1, \ldots, p$ and $i < j$, we define the binary variable $J_{ij} = 0$ if C_{ij} is identically zero and $J_{ij} = 1$ otherwise. Let $J = \{J_{ij}, i < j, j = 1, \ldots, p\}$. These binary variables are analogous to the γ_i binary variables that we use for variable selection. Let

$$S(J) = \sum_{ij} J_{ij}, \quad i < j,$$

and let $r = p(p-1)/2$ making $0 \leqslant S(J) \leqslant r$. Let C_p be the set of $p \times p$ positive definite correlation matrices. Let

$$V(J^*) = \int_{C\in C_p:J(C)=J^*} \left(\prod_{i\leqslant j, J_{ij}=1} \mathrm{d}C_{ij}\right)$$

be the volume of the positive definite region for C, given the constraints imposed by J^*, and let

$$\overline{V}(l) = \binom{r}{l}^{-1} \Big(\sum_{J:S(J)=l} V(J) \Big)$$

be the average volume for regions with size l.

The hierarchical prior for C is given by

$$p(\mathrm{d}C|J) = V(J)^{-1}\,\mathrm{d}C_{J=1} I(C \in C_p), \tag{2.14}$$

$$p(J|S(J) = l) = \binom{r}{l}^{-1} \frac{V(J)}{\overline{V}(l)}, \tag{2.15}$$

$$p(S(J) = l|\psi) = \binom{r}{l} \psi^l (1-\psi)^{r-l}, \tag{2.16}$$

where $0 \leqslant \psi \leqslant 1$, $I(C \in C_p) = 1$ if C is a correlation matrix and 0 otherwise and $C_{J=1} = \{C_{ij}\colon C_{ij} \neq 0\}$. The parameter ψ is the probability that $J_{ij} = 1$. For the remainder of this chapter we take $p(\psi) = 1$. For a more extensive discussion of this prior see Wong et al. (2003).

2.6. *Missing values*

The Bayesian methodology coupled with the Markov chain Monte Carlo simulation method make it straightforward to handle missing values in the dependent variable as part of the estimation problem, (e.g., Gelman et al., 2000, pp. 443–447). We assume that the observations are missing at random. Suppose that Y_t^m is the subvector of Y_t that is missing and Y_t^o is the subvector that is observed. Then, $p(Y_t^m|Y_t^o, \beta, \Sigma)$ is Gaussian and it is straightforward to obtain $E(Y_t^m|Y_t^o, \beta, \Sigma)$ and $\mathrm{var}(Y_t^m|Y_t^o, \beta, \Sigma)$, and hence to generate Y_t^m.

2.7. *Permanently selected variables*

We frequently wish to permanently retain some variables in the regression. For example, we may wish to retain all the intercept terms in the regression. We do so by setting the indicators γ for these variables to be identically one and setting q_γ in Section 2.3 to be the sum of the γ_i, excluding those γ_i that are identically 1.

If we wish to estimate the model with no variable selection, then we would set the whole γ vector to be identically 1.

2.8. *Selecting variables in groups*

In some problems it is useful to add or delete a group of variables rather than a single variable. For example, suppose that in the model (2.3) we wish to add or drop elements of the vector x_t, let's say the second element of x_t. This is equivalent to dropping columns $p+1, \ldots, 2p$ of the matrix $X_t = x_t' \otimes I$ in (2.1) or, equivalently, setting the second column of the coefficient matrix B to zero in (2.3).

The binary indicator vector γ will now refer to groups of columns of X_t rather than individual columns. Similarly to Section 2.7, we can choose to retain some groups permanently, e.g., we may wish to retain all the intercepts in the model (2.3).

2.9. *Noninformative prior on Σ*

In Sections 4 and 5 we compare the effect of covariance selection with a prior for Σ that does not allow for covariance selection. One way to do so is to use the prior for Σ given in Sections 2.4 and 2.5, but with the C_{ij} always generated, i.e., no covariance selection but same shrinkage prior. The effect of the prior on the estimation of Σ was reported in Wong et al. (2003).

In our chapter we use the following prior for Σ when we do not wish to perform covariance selection,

$$p(\Sigma) \propto \det(\Sigma)^{-(p+1)/2}, \tag{2.17}$$

which implies that the prior for Ω is also of this form. The prior (2.17) is improper and uninformative.

We now show that the posterior distribution of Ω is proper. For conciseness, we do so when all regressors are in the model, but the extension to the case when there is variable selection is straightforward. It is not difficult to show that

$$\begin{aligned} p(\Omega|Y) &\propto p(Y|\Omega)p(\Omega) \\ &\propto \int p(Y|\Omega,\beta)p(\beta|\Omega)p(\Omega)\,\mathrm{d}\beta \\ &\propto \det(\Omega)^{(n-p-1)/2} \\ &\quad\times \exp\left[-\frac{1}{2}\left\{(\mathrm{SYY}) - \frac{c}{1+c}(\mathrm{SXY}_\gamma)'(\mathrm{SXX}_\gamma)^{-1}(\mathrm{SXY}_\gamma)\right\}\right]. \end{aligned} \tag{2.18}$$

With a little algebra we can show that

$$(\mathrm{SYY}) - \frac{c}{1+c}(\mathrm{SXY}_\gamma)'(\mathrm{SXX}_\gamma)^{-1}(\mathrm{SXY}_\gamma) \geqslant \frac{1}{c+1}(\mathrm{SYY})$$

which implies that

$$p(\Omega|Y) \leqslant \det(\Omega)^{(n-p-1)/2}\exp\left\{-\frac{1}{2(c+1)}(\mathrm{SYY})\right\},$$

with the right side of the expression above being a proper Wishart distribution in Ω for $n \geqslant p$. It follows that the posterior distribution of Ω, and hence Σ, is proper.

3. Sampling scheme

Let Y_{miss} be the vector of missing values of Y, Y_{obs} the vector of observed values of Y. We generate $\beta, \gamma_i, i = 1, \ldots, q, T_i, i = 1, \ldots, p, C_{ij}, i < j$, and Y_{miss} using the following Markov chain Monte Carlo scheme.

1. $\gamma_i | Y_{\text{obs}}, T, C, Y_{\text{miss}}, \gamma_{\{-i\}}$ for $i = 1, \ldots, q$;
2. $\beta_\gamma | Y_{\text{obs}}, T, C, \gamma, Y_{\text{miss}}$;
3. $C_{ij} | Y_{\text{obs}}, T, \gamma, \beta_\gamma, Y_{\text{miss}}, C_{\{-ij\}}$ for $i = 1, p-1, j < i$;
4. $T_i | Y_{\text{obs}}, C, \gamma, \beta_\gamma, Y_{\text{miss}}, T_{\{-i\}}$ for $i = 1, \ldots, p$;
5. $Y_{\text{miss}} | Y_{\text{obs}}, \gamma, T, C, \beta_\gamma$.

We generate the elements of γ one at a time by calculating

$$\begin{aligned} p(\gamma_i &= 1 | Y, T, C, \gamma_{\{-i\}}) \\ &= \frac{p(Y|T, C, \gamma_i = 1, \gamma_{\{-i\}}) p(\gamma_i = 1 | \gamma_{\{-i\}})}{\sum_{k=0}^{1} p(Y|T, C, \gamma_i = k, \gamma_{\{-k\}}) p(\gamma_i = k | \gamma_{\{-i\}})}; \end{aligned} \tag{3.1}$$

see Kohn et al. (2001) for details. β_γ and Y_{miss} are generated from their conditionals as described above. The C_{ij} and T_i are generated one element at a time using a Metropolis–Hastings step. Details are given by Wong et al. (2003).

4. Real data

This section studies both variable selection and covariance selection on real data by using two models to analyze four data sets. The first model carries out variable selection but not covariance selection and uses the prior (2.17) for Σ. We call this the *NCSVS* model. The second model does both variable selection and covariance selection and we call it the *CSVS* model. We report the posterior means and standard errors of the regression coefficients, the posterior probabilities of including a predictor variable in the regression and the image plots of the estimated partial correlation matrix. For the *CSVS* model we also include the image plots for the posterior probabilities that the elements of the partial correlation matrix are nonzero. The image plots are lighter where the matrix is sparser.

4.1. Cow milk protein data

This is a longitudinal data set described in Diggle et al. (2002, p. 5) who analyzed it to determine the effect of diet on the protein content in cow's milk.[1] The data was collected weekly for 79 cows. Each cow was assigned to one of three diets: barley (25 cows), mixture of barley and lupins (27 cows), or lupins (27 cows). Time was measured in weeks since calving and the experiment was terminated 19 weeks after the earliest calving, resulting in 38 observations with some incomplete measurements. Diggle et al. (2002, p. 6) note that calving may be associated with the milk protein content and as such the incomplete data should not be ignored. There are 11 other missing data values. We treat all the missing values as described in Section 2.6.

Exploratory analysis by Diggle et al. (2002, pp. 5–9) suggests that the barley diet yields the highest mean protein content and the mixture diet yields the second highest protein content. Diggle et al. (2002, p. 99) also point out that the mean response shows an initial drop in the protein content in cow's milk followed by a constant mean

[1] The data can be obtained from http://www.maths.lancs.ac.uk/~diggle/lda/Datasets/.

Table 1
Posterior means, standard errors and probabilities of being nonzero for the regression coefficients using model *NCSVS* for the cow milk protein data. NA means not available as the coefficient is always included

	$\beta_{1,0}$	$\beta_{1,1}$	$\beta_{2,0}$	$\beta_{2,1}$	$\beta_{3,0}$	$\beta_{3,1}$
Post. mean	3.2468	0.0022	3.1972	−0.0007	3.1073	−0.0042
Post. std. error	0.0663	0.0043	0.0621	0.0031	0.0711	0.0056
Post. prob.	NA	0.33450	NA	0.2240	NA	0.4725

response over the majority of the experiment with a slight rise towards the end of the experiment. Diggle et al. (2002, pp. 99–103) use a model to determine whether diet affects the protein content in cow's milk and conclude that diet does affect the mean response but that there is no significant rise in the mean response towards the end of the experiment. We model the mean as a linear function of time, allowing for different regression coefficients for different diets and do not explore the presence of a nonlinear trend. For each cow t denote the vector of the milk protein content Y_t such that

$$Y_t = Z\beta_d + e_t, \quad e_t \sim N(0, \Sigma), \tag{4.1}$$

for $d = 1, 2, 3$, corresponding to the diets of barley, the mixture of barley and lupin, and lupin respectively. In (4.1) the predictor matrix Z and the vector of regression coefficients are

$$Z = \begin{bmatrix} 1 & 1 \\ 1 & 2 \\ \vdots & \vdots \\ 1 & 19 \end{bmatrix} \quad \text{and} \quad \beta_d = \begin{bmatrix} \beta_{d,0} \\ \beta_{d,1} \end{bmatrix},$$

where the second column in Z indexes the time in weeks since calving. Writing (4.1) in the more general form of (2.1) we have

$$Y_t = X_t\beta + e_t, \quad e_t \sim N(0, \Sigma), \tag{4.2}$$

where

$$\beta = (\beta_{1,0}, \beta_{1,1}, \beta_{2,0}, \beta_{2,1}, \beta_{3,0}, \beta_{3,1})',$$

and

$$X_t = \begin{cases} [Z\ 0_{19\times 2}\ 0_{19\times 2}] & \text{if cow } t \text{ receives diet 1,} \\ [0_{19\times 2}\ Z\ 0_{19\times 2}] & \text{if cow } t \text{ receives diet 2,} \\ [0_{19\times 2}\ 0_{19\times 2}\ Z] & \text{if cow } t \text{ receives diet 3,} \end{cases}$$

where $0_{19\times 2}$ denotes a matrix of zeros of size 19×2.

The estimated posterior means and standard errors for the regression coefficients and the posterior probabilities that the regression coefficients are nonzero are recorded in Table 1 for model *NCSVS* and in Table 2 for model *CSVS*. As in Diggle et al. (2002), the magnitude of the intercept terms for both models for each of the diet groups agrees with the initial exploratory analysis. That is, the intercept for the barley diet is greater than the mixture diet which in turn is greater than the lupin diet. Tables 1 and 2 suggest

Table 2
Posterior means, standard errors and probabilities of being nonzero for the regression coefficients using model *CSVS* for the cow milk protein data. NA means not available as the coefficient is always included

	$\beta_{1,0}$	$\beta_{1,1}$	$\beta_{2,0}$	$\beta_{2,1}$	$\beta_{3,0}$	$\beta_{3,1}$
Post. mean	3.3093	0.0005	3.2200	−0.0003	3.1389	−0.0032
Post. std. error	0.0496	0.0023	0.0475	0.0023	0.0597	0.0049
Post. prob.	NA	0.1775	NA	0.1485	NA	0.3720

Table 3
Posterior means and standard errors of the difference in the intercepts of the three diets using model *NCSVS* for the cow milk protein data

	$\beta_{1,0} - \beta_{2,0}$	$\beta_{1,0} - \beta_{3,0}$	$\beta_{2,0} - \beta_{3,0}$
Post. mean	0.0495	0.1394	0.0899
Post. std. error	0.0639	0.0763	0.0683

Table 4
Posterior means and standard errors of the difference in the intercepts of the three diets using model *CSVS* for the cow milk protein data

	$\beta_{1,0} - \beta_{2,0}$	$\beta_{1,0} - \beta_{3,0}$	$\beta_{2,0} - \beta_{3,0}$
Post. mean	0.0893	0.1704	0.0811
Post. std. error	0.0535	0.0649	0.0618

the coefficients of the time trend are not significant, statistically or practically for any of the three diets.

The posterior means and standard errors of the difference between the intercepts for each diet are contained in Tables 3 and 4. The tables suggest that the difference for model *CSVS* between the barley and lupin diet are statistically significant, but insignificant for the remaining differences. The results for model *NCSVS* suggest that none of the intercepts are statistically different from each other.

Figure 1 shows the image plots of the posterior means of the partial correlations and posterior probabilities that the elements in the partial correlation matrix are nonzero. Model *CSVS* estimates a sparser partial correlation matrix than model *NCSVS* and the image plots of model *CSVS* suggest that the matrix of partial correlations is banded.

4.2. *Hip replacement data*

This data set contains observations on 30 patients who had hip replacements and is in Crowder and Hand (1990, p. 79). Each patient had their haematocrit levels measured four times, once before the operation and three times afterward. The main goals of this analysis are to examine differences in haematocrit levels between men and women who experienced hip replacements and to investigate whether an age effect is present. In addition to each patient's haematocrit levels their gender and age were recorded. We denote the age of individual t as a_t and take $s_t = 1$ if individual t is male and $s_t = -1$ if individual t is female.

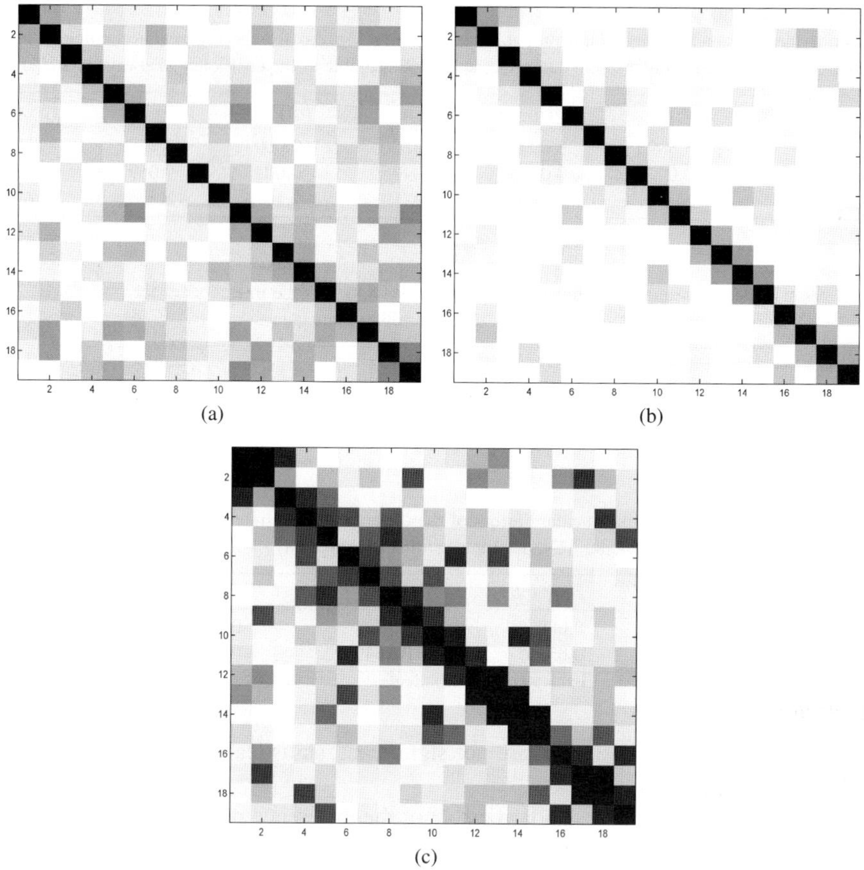

Fig. 1. Image plots for the cow milk protein data. Panel (a) is the image plot of the partial correlation matrix estimated by model *NCSVS*. Panel (b) is the image plot of the partial correlation matrix estimated by model *CSVS*. Panel (c) is the image plot of the probabilities of the elements of the correlation matrix being nonzero as estimated by model *CSVS*.

The third time measurement had 19 out of 30 values missing which appeared to distort our results. Following Crowder and Hand (1990) we omit all the measurements at the third time point. There are two other missing values. One is patient 8 at time 4 and the other is patient 15 at time 1. Section 2.6 describes how we deal with such missing values.

We assume the same mean structure as Crowder and Hand (1990) and allow for a different intercept for each time point and also include covariates for gender and age. Denote the heamatocrit levels as Y_{ti} for individual t at time i. We write the regression for observation Y_{ti} as

$$Y_{ti} = \alpha_i + \lambda_1 s_t + \lambda_2 a_t + e_{ti}, \quad \text{for } i = 1, 2, 3. \tag{4.3}$$

Table 5
Posterior means, standard errors and probabilities of being nonzero for the regression coefficients using model *NCSVS* for the hip replacement data. NA means not available as the coefficient is always included

	β_1	β_2	β_3	β_4	β_5
Post. mean	38.8217	29.4162	31.6827	0.1674	0.0032
Post. std. error	1.9205	1.8555	1.9003	0.4208	0.0257
Post. prob.	NA	NA	NA	0.2165	0.1425

Table 6
Posterior means, standard errors and probabilities of being nonzero for the regression coefficients using model *CSVS* for the hip replacement data. NA means not available as the coefficient is always included

	β_1	β_2	β_3	β_4	β_5
Post. mean	38.6631	29.2353	31.5143	0.3098	0.0060
Post. std. error	1.9948	1.9397	1.9946	0.5097	0.0269
Post. prob.	NA	NA	NA	0.3545	0.1770

Writing (4.3) in the notation of (2.1) and denoting Y_t as the 3×1 vector of responses across time for individual t we have

$$Y_t = X_t\beta + e_t, \quad e_t \sim N(0, \Sigma), \tag{4.4}$$

where the predictor matrix for each individual X_t and the vector of regression coefficients are

$$X_t = \begin{bmatrix} 1 & 0 & 0 & s_t & a_t \\ 0 & 1 & 0 & s_t & a_t \\ 0 & 0 & 1 & s_t & a_t \end{bmatrix} \quad \text{and} \quad \beta = \begin{bmatrix} \alpha_1 \\ \alpha_2 \\ \alpha_3 \\ \lambda_1 \\ \lambda_2 \end{bmatrix}.$$

The estimated posterior means and standard errors for the regression coefficients and the posterior probabilities that the regression coefficients are nonzero are recorded in Tables 5 and 6. The estimated posterior probabilities are similar for model *NCSVS* and model *CSVS* when estimating (4.4). The estimated regression coefficients for the first three predictors for both models *NCSVS* and *CSVS* are comparable to those in Crowder and Hand (1990). The last two regression coefficients were estimated by Crowder and Hand (1990, p. 81) as 0.807 and 0.031 respectively, but do not greatly exceed their standard errors. Models *NCSVS* and *CSVS* also suggest these coefficients are not significantly different from zero and are therefore comparable with Crowder and Hand (1990, p. 81). Tables 7 and 8 show the posterior means for the differences in the intercepts for haemocratic levels across each time period and show that the haemocratic levels are more significantly different when comparing the first time period with the second and third time periods than when comparing the difference between the second time period and the third time period.

Figure 2 shows the image plots for the posterior means for the partial correlation matrices for both models and the posterior probabilities that the elements in the partial correlation matrix are nonzero for model *CSVS*. The patterns in the first two image plots

Table 7
Posterior means and standard errors of the difference in the intercepts of the three time points using model *NCSVS* for the hip replacement data

	$\beta_1 - \beta_2$	$\beta_1 - \beta_3$	$\beta_2 - \beta_3$
Post. mean	9.4056	7.1390	−2.2665
Post. std. error	1.0226	0.9727	1.0788

Table 8
Posterior means and standard errors of the difference in the intercepts of the three time points using Model *CSVS* for the hip replacement data

	$\beta_1 - \beta_2$	$\beta_1 - \beta_3$	$\beta_2 - \beta_3$
Post. mean	9.4278	7.1487	−2.2791
Post. std. error	1.0861	1.1244	1.1102

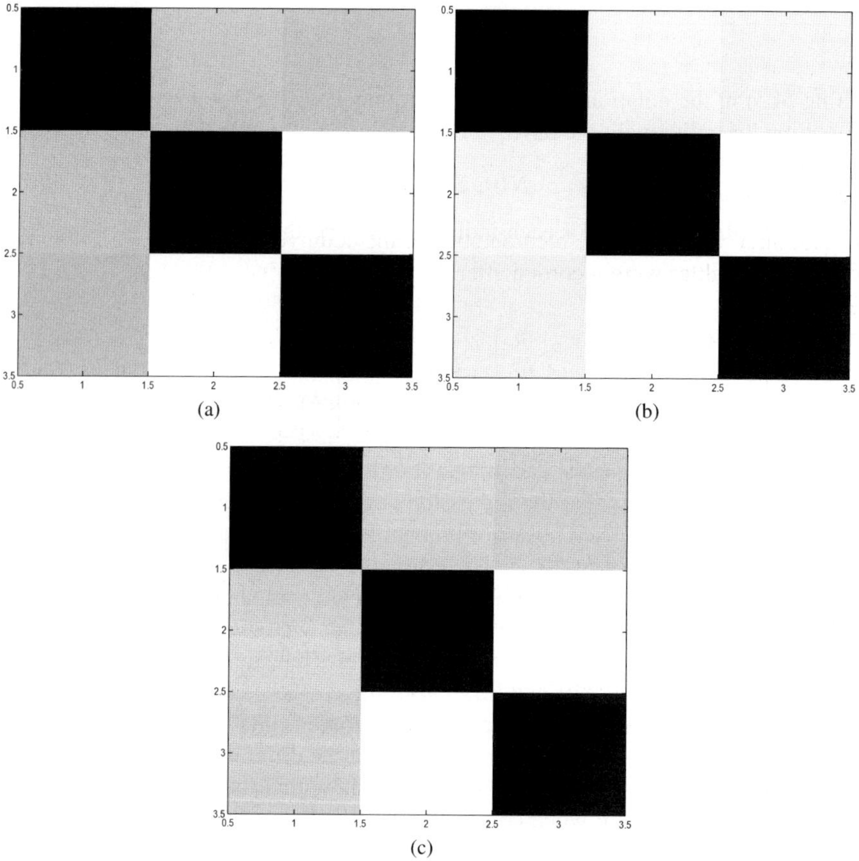

Fig. 2. Image plots for the hip replacement data. Panel (a) is the image plot of the partial correlation matrix estimated by model *NCSVS*. Panel (b) is the image plot of the partial correlation matrix estimated by model *CSVS*. Panel (c) is the image plot of the probabilities of the elements of the correlation matrix being nonzero as estimated by model *CSVS*.

are similar but, as expected, the estimated partial correlation matrix for model *CSVS* is sparser than the estimated partial correlation matrix for model *NCSVS*.

4.3. Cow diet data

This data set consists of observations on 50 cows that are subjected to a diet additive. The data is cross-sectional and is described in Gelman et al. (2000, pp. 213–215).[2] The diet additive is methionine hydroxy analog and each cow is assigned to one of four different levels: 0% for the first 12 cows, 0.1% cows 13–25, 0.2% for cows 26–38 and 0.3% for the remaining 12 cows. The following variables were also recorded for each cow:

1. Lactation
2. Age (mos)
3. Initial weight (lb)
4. Mean daily dry matter consumed (kg)
5. Mean daily milk product (lb)
6. Milk fat (%)
7. Milk solids nonfat (%)
8. Final weight (lb)
9. Milk protein (%)

The first three variables were recorded before the additive was included in the diet and the last six variables were recorded after the additive was included in the diet. We treat the six post-diet additive variables as the multivariate response and the diet additive and the three pre-diet additive variables as the predictors. We model the data as in (2.3) which allows the same covariates to have different regression coefficients for each element in the vector of the responses. An interesting feature of this data is the high correlation amongst some of the predictor variables. In particular, the correlation between lactation and age is 0.9624, the correlation between lactation and initial weight is 0.7504 and the correlation between age and initial weight is 0.7808.

The response vector for the tth cow is $Y_t = (Y_{t1}, Y_{t2}, \ldots, Y_{t6})'$ where Y_t contains, in the following order, mean daily dry matter consumed, mean daily milk product, milk fat, milk solids nonfat, final weight and milk protein respectively. The predictor vector for the ith cow is $x_t = (x_{t0}, x_{t1}, x_{t2}, x_{t3}, x_{t4})'$, where x_t contains, in the following order, an intercept, diet additive, lactation, age and initial weight. The matrix of regression coefficients in (2.3) for this example is,

$$B = \begin{bmatrix} \beta_{1,0} & \beta_{1,1} & \beta_{1,2} & \beta_{1,3} & \beta_{1,4} \\ \beta_{2,0} & \beta_{2,1} & \beta_{2,2} & \beta_{2,3} & \beta_{2,4} \\ \vdots & \vdots & \vdots & \vdots & \vdots \\ \beta_{6,0} & \beta_{6,1} & \beta_{6,2} & \beta_{6,3} & \beta_{6,4} \end{bmatrix}.$$

We specify our model such that if a variable is dropped from x_t, then it is dropped from all equations. To allow for this structure in we write the regression equation in terms of

---

[2] The data is available from http://www.stat.columbia.edu/~gelman/book/data/.

(2.1) as

$$Y_t = X_t\beta + e_t, \quad e_t \sim N(0, \Sigma), \tag{4.5}$$

where

$$X_t = x_t' \otimes I_6 \quad \text{and} \quad \beta = vec(B),$$

and I_6 is a 6×6 identity matrix.

To enable variables to be dropped from all regression models we group the variables into 5 groups as outlined in Section 2.8. The first group contains the intercepts for each equation and is always included.

Tables 9 and 10 contain the posterior means, standard errors and probabilities of being nonzero of the regression coefficients for models *NCSVS* and *CSVS* for the grouped variables case. The predictor variables diet additive and initial weight have the highest posterior probabilities of inclusion. Model *NCSVS* estimates the probabilities of inclusion for diet additive and initial weight as 0.4250 and 1.0000. Model *CSVS* estimates the probabilities of inclusion for diet additive and initial weight as 0.4390 and 1.000. The predictor variables lactation and age both have estimated posterior probabilities close to zero for models *NCSVS* and *CSVS*.

Figure 3 shows the image plots for the estimated partial correlations matrix for models *NCSVS* and *CSVS* and the image plot of the estimated posterior probabilities that the element of the partial correlation matrix is nonzero for model *CSVS* in the grouped variable case. The plots indicate the difference in the sparsity of the estimated partial correlation matrix between models *NCSVS* and *CSVS* is negligible.

We re-estimate (4.5) but now without grouping the predictor variables. Each predictor variable now has different posterior probabilities of inclusion for different equations in the response. Tables 11 and 12 are similar to Tables 9 and 10 but for the nongrouped variables case. Recall the posterior probabilities of including initial weight is 1.0000 for models *NCSVS* and *CSVS*. Tables 11 and 12 show that when initial weight is allowed to be in or out of individual equations, the probabilities of inclusion can be as low as 0.1046 (model *CSVS*, equation 4 of the response) and as high as 1.0000 (models *NCSVS* and *CSVS*, equation 5 of the response). When grouping the predictor variable diet additive the posterior probability of inclusion for model *NCSVS* is 0.4250 and 0.4390 for model *CSVS*. Relaxing the grouping assumption results in posterior probabilities as low as 0.0663 (model *CSVS*, equation 4 of the response) and as high as 0.9833 (model *CSVS*, equation 2 of the response). The posterior probabilities for the predictors lactation and age increase across every equation for the nongrouped variables case compared to the grouped variables.

Figure 4 is similar to Figure 3, but for the nongrouped variables. The estimates of the partial correlation coefficients are similar for models *NCSVS* and *CSVS*. The posterior probabilities that the partial correlations are nonzero are also similar for model *CSVS* for the grouped and ungrouped cases.

4.4. *Pig bodyweight data*

This longitudinal data set contains observations on 48 pigs measured over 9 successive weeks. It is described in Diggle et al. (2002, pp. 34–35) who analyzed it to examine

Table 9
Posterior means, standard errors and probabilities of being nonzero for the regression coefficients using model *NCSVS* for the grouped cow diet data. NA means not available as the coefficient is always included

	$\beta_{1,0}$	$\beta_{1,1}$	$\beta_{1,2}$	$\beta_{1,3}$	$\beta_{1,4}$	$\beta_{2,0}$	$\beta_{2,1}$	$\beta_{2,2}$	$\beta_{2,3}$	$\beta_{2,4}$
Post. mean	8.1754	0.4743	0.0000	0.0000	0.0062	24.2911	−0.1390	0.0001	0.0000	0.0271
Post. std. error	2.3647	2.0404	0.0021	0.0002	0.0018	8.7645	7.2526	0.0084	0.0004	0.0069
Post. prob.	NA	0.4250	0.0001	0.0001	1.0000	NA	0.4250	0.0001	0.0001	1.0000
	$\beta_{3,0}$	$\beta_{3,1}$	$\beta_{3,2}$	$\beta_{3,3}$	$\beta_{3,4}$	$\beta_{4,0}$	$\beta_{4,1}$	$\beta_{4,2}$	$\beta_{4,3}$	$\beta_{4,4}$
Post. mean	2.6561	0.8378	0.0000	−0.0000	0.0006	8.5527	−0.0891	−0.0000	−0.0000	−0.0000
Post. std. error	0.5166	1.0481	0.0012	0.0001	0.0004	0.3677	0.3236	0.0013	0.0001	0.0003
Post. prob.	NA	0.4250	0.0001	0.0001	1.0000	NA	0.4250	0.0001	0.0001	1.0000
	$\beta_{5,0}$	$\beta_{5,1}$	$\beta_{5,2}$	$\beta_{5,2}$	$\beta_{5,4}$	$\beta_{6,0}$	$\beta_{6,1}$	$\beta_{6,2}$	$\beta_{6,3}$	$\beta_{6,4}$
Post. mean	218.3280	−95.6234	0.0004	0.0001	0.8076	3.3429	−0.0735	0.0000	−0.0000	−0.0001
Post. std. error	87.1405	130.8774	0.2096	0.0102	0.0668	0.2607	0.2326	0.0002	0.0000	0.0002
Post. prob.	NA	0.4250	0.0001	0.0001	1.0000	NA	0.4250	0.0001	0.0001	1.0000

Table 10
Posterior means, standard errors and probabilities of being nonzero for the regression coefficients using model *CSVS* for the grouped cow diet data. NA means not available as the coefficient is always included

	$\beta_{1,0}$	$\beta_{1,1}$	$\beta_{1,2}$	$\beta_{1,3}$	$\beta_{1,4}$	$\beta_{2,0}$	$\beta_{2,1}$	$\beta_{2,2}$	$\beta_{2,3}$	$\beta_{2,4}$
Post. mean	8.1690	0.4862	0.0000	−0.0000	0.0063	24.3082	−0.1750	0.0001	−0.0000	0.0271
Post. std. error	2.2344	1.9875	0.0015	0.0005	0.0017	8.3744	7.1448	0.0104	0.0023	0.0065
Post. prob.	NA	0.4390	0.0001	0.0003	1.0000	NA	0.4390	0.0001	0.0003	1.0000
	$\beta_{3,0}$	$\beta_{3,1}$	$\beta_{3,2}$	$\beta_{3,3}$	$\beta_{3,4}$	$\beta_{4,0}$	$\beta_{4,1}$	$\beta_{4,2}$	$\beta_{4,3}$	$\beta_{4,4}$
Post. mean	2.6559	0.8612	0.0000	0.0000	0.0006	8.5518	−0.0904	−0.0000	−0.0000	−0.0000
Post. std. error	0.4951	1.0446	0.0009	0.0001	0.0004	0.3454	0.3102	0.0010	0.0001	0.0003
Post. prob.	NA	0.4390	0.0001	0.0003	1.0000	NA	0.4390	0.0001	0.0003	1.0000
	$\beta_{5,0}$	$\beta_{5,1}$	$\beta_{5,2}$	$\beta_{5,2}$	$\beta_{5,4}$	$\beta_{6,0}$	$\beta_{6,1}$	$\beta_{6,2}$	$\beta_{6,3}$	$\beta_{6,4}$
Post. mean	218.6719	−98.9527	−0.0008	0.0001	0.0636	0.2448	0.2242	0.0002	0.0001	0.0002
Post. std. error	83.0269	130.3103	0.0893	0.0087	0.0636	0.2448	0.2242	0.0002	0.0001	0.0002
Post. prob.	NA	0.4390	0.0001	0.0003	1.0000	NA	0.4390	0.0001	0.0003	1.0000

Table 11
Posterior, means, standard errors and probabilities of being nonzero for the regression coefficients using model *NCSVS* for the cow diet data. NA means not available as the coefficient is always included

	$\beta_{1,0}$	$\beta_{1,1}$	$\beta_{1,2}$	$\beta_{1,3}$	$\beta_{1,4}$	$\beta_{2,0}$	$\beta_{2,1}$	$\beta_{2,2}$	$\beta_{2,3}$	$\beta_{2,4}$
Post. mean	10.2406	0.6445	0.0606	0.0018	0.0044	34.2230	−0.6413	0.3325	0.0094	0.0183
Post. std. error	3.6447	1.7309	0.2107	0.0118	0.0031	14.6015	3.6768	1.0009	0.0534	0.0127
Post. prob	NA	0.1815	0.1456	0.1155	0.7386	NA	0.0886	0.1679	0.1182	0.7436
	$\beta_{3,0}$	$\beta_{3,1}$	$\beta_{3,2}$	$\beta_{3,3}$	$\beta_{3,4}$	$\beta_{4,0}$	$\beta_{4,1}$	$\beta_{4,2}$	$\beta_{4,3}$	$\beta_{4,4}$
Post. mean	2.9680	2.0439	0.0302	−0.0000	0.0001	8.5329	−0.0052	−0.0144	−0.0005	0.0000
Post. std. error	0.3811	0.5958	0.0699	0.0039	0.0003	0.1885	0.1057	0.0351	0.0021	0.0002
Post. prob	NA	0.9780	0.2732	0.1499	0.2199	NA	0.0691	0.2240	0.1574	0.1163
	$\beta_{5,0}$	$\beta_{5,1}$	$\beta_{5,2}$	$\beta_{5,2}$	$\beta_{5,4}$	$\beta_{6,0}$	$\beta_{6,1}$	$\beta_{6,2}$	$\beta_{6,3}$	$\beta_{6,4}$
Post. mean	238.8083	−201.2394	0.4431	−0.0305	0.8041	3.3002	−0.0134	0.0011	0.0001	−0.0000
Post. std. error	87.1000	117.7308	4.7685	0.3408	0.0704	0.1420	0.0877	0.0110	0.0009	0.0001
Post. prob	NA	0.8181	0.0754	0.0759	1.0000	NA	0.0809	0.0867	0.0916	0.1403

Table 12
Posterior, means, standard errors and probabilities of being nonzero for the regression coefficients using model *CSVS* for the cow diet data. NA means not available as the coefficient is always included

	$\beta_{1,0}$	$\beta_{1,1}$	$\beta_{1,2}$	$\beta_{1,3}$	$\beta_{1,4}$	$\beta_{2,0}$	$\beta_{2,1}$	$\beta_{2,2}$	$\beta_{2,3}$	$\beta_{2,4}$
Post. mean	9.5310	0.5236	0.0623	0.0010	0.0050	30.7891	−0.4925	0.2332	0.0084	0.0213
Post. std.	3.3086	1.5375	0.2179	0.0119	0.0029	13.0697	3.1737	0.8675	0.0527	0.0116
Post. prob	NA	0.1598	0.1453	0.1027	0.8208	NA	0.0816	0.1340	0.1116	0.8339
	$\beta_{3,0}$	$\beta_{3,1}$	$\beta_{3,2}$	$\beta_{3,3}$	$\beta_{3,4}$	$\beta_{4,0}$	$\beta_{4,1}$	$\beta_{4,2}$	$\beta_{4,3}$	$\beta_{4,4}$
Post. mean	2.9592	2.0622	0.0316	0.0001	0.0001	8.5256	−0.0074	−0.0109	−0.0004	0.0000
Post. std.	0.3669	0.5725	0.0688	0.0040	0.0003	0.1730	0.1014	0.0310	0.0019	0.0002
Post. prob	NA	0.9833	0.2918	0.1614	0.2203	NA	0.0663	0.1848	0.1436	0.1046
	$\beta_{5,0}$	$\beta_{5,1}$	$\beta_{5,2}$	$\beta_{5,2}$	$\beta_{5,4}$	$\beta_{6,0}$	$\beta_{6,1}$	$\beta_{6,2}$	$\beta_{6,3}$	$\beta_{6,4}$
Post. mean	232.9853	−191.9362	0.3351	−0.0346	0.8080	3.2934	−0.0119	0.0009	0.0001	−0.0000
Post. std.	82.8676	120.2146	4.5379	0.3559	0.0671	0.1259	0.0848	0.0103	0.0008	0.0001
Post. prob	NA	0.7952	0.0717	0.0746	1.0000	NA	0.0780	0.0809	0.0819	0.1212

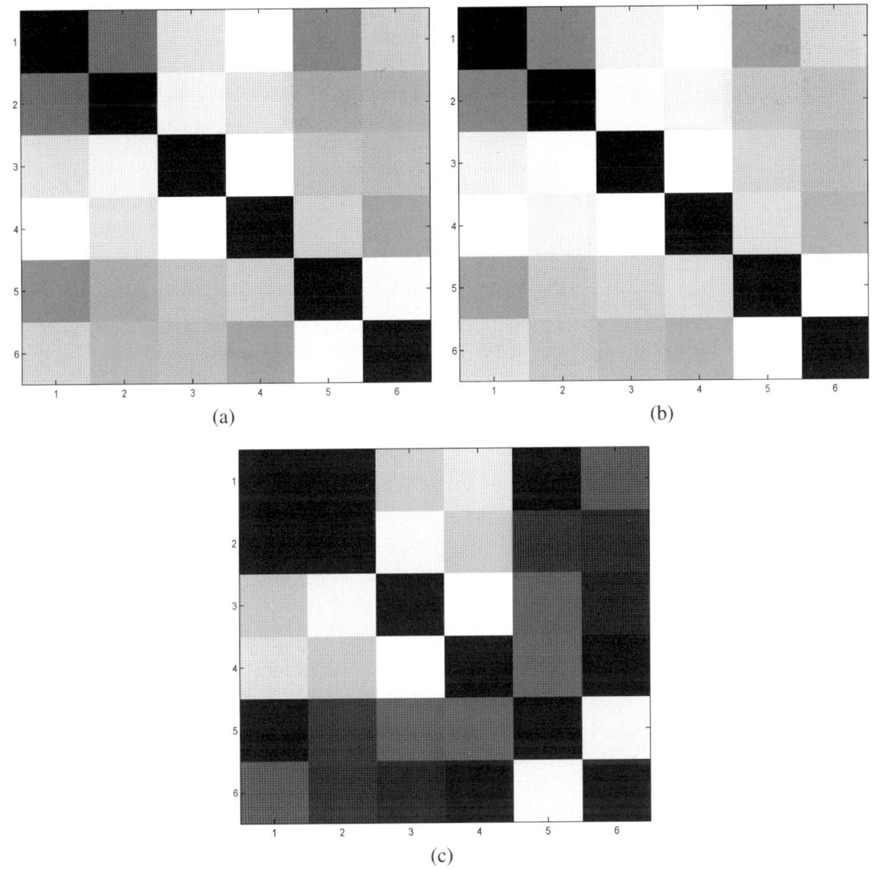

Fig. 3. Image plots for the grouped cow milk protein data. Panel (a) is the image plot of the partial correlation matrix estimated by model *NCSVS*. Panel (b) is the image plot of the partial correlation matrix estimated by model *CSVS*. Panel (c) is the image plot of the probabilities of the elements of the correlation matrix being nonzero as estimated by model *CSVS*.

the growth rates of pigs.[3] Diggle et al. (2002, p. 34) note that the trend in pig growth rates is approximately linear but each individual pig varies in both initial weight and in its growth rate. As a result Diggle et al. (2002, pp. 76–77) structure the pig growth rate data as a random effects model. In this article we model the mean function of the pigs growth rate as a piecewise linear trend such that at each time point we allow the slope to change. Diggle et al. (2002, p. 35) contains a plot of the pig bodyweights across time. The plot reveals that while time periods prior to the fourth week appear have a constant slope, after the fourth period the individual trajectories exhibit more variation and hence perhaps have different slopes for different time periods.

[3] The data can be obtained from http://www.maths.lancs.ac.uk/~diggle/lda/Datasets/.

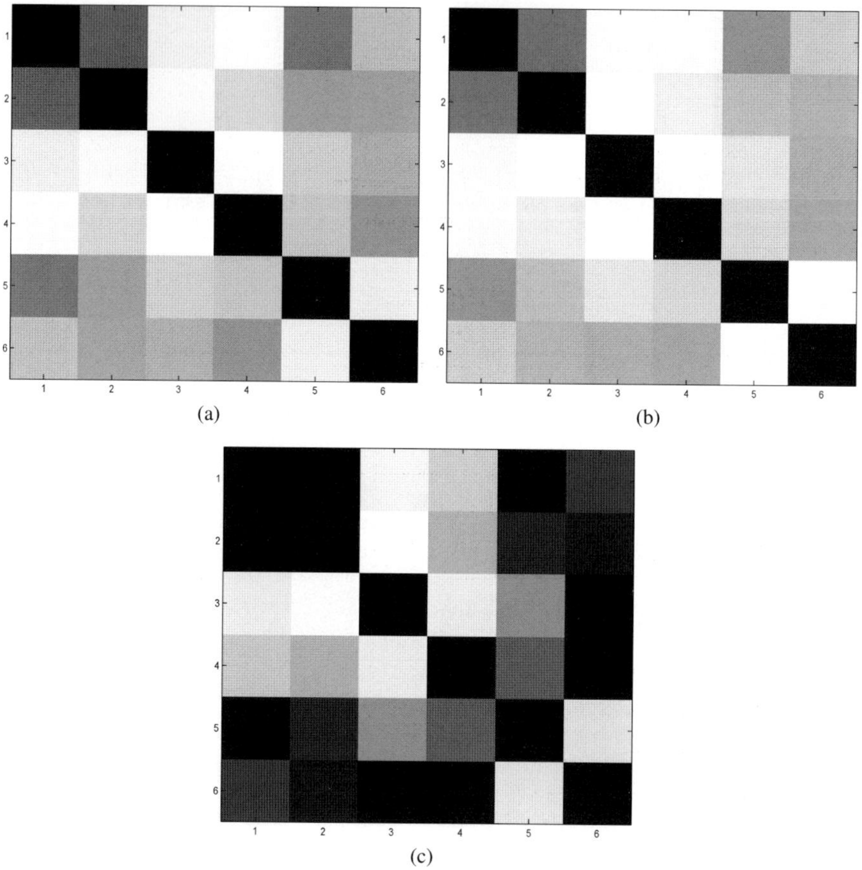

Fig. 4. Image plots for the cow milk protein data. Panel (a) is the image plot of the partial correlation matrix estimated by model *NCSVS*. Panel (b) is the image plot of the partial correlation matrix estimated by model *CSVS*. Panel (c) is the image plot of the probabilities of the elements of the correlation matrix being nonzero as estimated by model *CSVS*.

Table 13
Posterior means, standard errors and probabilities of being nonzero for the regression coefficients using model *NCSVS* for the pig growth rate data. NA means not available as the coefficient is always included

	β_0	β_1	β_2	β_3	β_4	β_5	β_6	β_7	β_8
Post. mean	17.7743	6.7040	0.0643	−1.2002	0.1451	0.2608	−0.0396	0.4606	−0.5418
Post. std. error	0.4323	0.1807	0.1902	0.3035	0.2357	0.3617	0.1983	0.4190	0.4328
Post. prob.	NA	1.0000	0.2950	0.9980	0.4455	0.5065	0.3040	0.6720	0.7295

Let Y_{ti} be the response for pig t at time i and write the piecewise linear time trend as

$$\beta_0 + \beta_1 i + \beta_2(i-2)_+ + \beta_3(i-3)_+ + \cdots + \beta_8(i-8)_+,$$
$$\text{for} \quad i = 1, \dots, 9, \tag{4.6}$$

Table 14
Posterior means, standard errors and probabilities of being nonzero for the regression coefficients using model *CSVS* the pig growth rate data. NA means not available as the coefficient is always included

	β_1	β_2	β_3	β_4	β_5	β_6	β_7	β_8	β_9
Post. mean	17.8010	6.6829	0.1251	−1.2903	0.1648	0.3164	−0.0604	0.6258	−0.7225
Post. std	0.3903	0.1630	0.2238	0.2920	0.2357	0.3759	0.2233	0.4031	0.4614
Post. prob.	1.0000	1.0000	0.3950	1.0000	0.5185	0.5905	0.3465	0.8180	0.8145

where

$$x_+ = \begin{cases} x & \text{if } x \geqslant 0, \\ 0 & \text{if } x < 0. \end{cases}$$

Writing (4.6) in the notation of (2.1) the matrix X_t of covariates is

$$X_t = \begin{bmatrix} 1 & 1 & 0 & 0 & 0 & 0 & 0 & 0 & 0 \\ 1 & 2 & 0 & 0 & 0 & 0 & 0 & 0 & 0 \\ 1 & 3 & 1 & 0 & 0 & 0 & 0 & 0 & 0 \\ 1 & 4 & 2 & 1 & 0 & 0 & 0 & 0 & 0 \\ 1 & 5 & 3 & 2 & 1 & 0 & 0 & 0 & 0 \\ 1 & 6 & 4 & 3 & 2 & 1 & 0 & 0 & 0 \\ 1 & 7 & 5 & 4 & 3 & 2 & 1 & 0 & 0 \\ 1 & 8 & 6 & 5 & 4 & 3 & 2 & 1 & 0 \\ 1 & 9 & 7 & 6 & 5 & 4 & 3 & 2 & 1 \end{bmatrix}$$

and β is the corresponding 9×1 vector of regression coefficients.

The estimated posterior means and standard errors for the regression coefficients and the posterior probabilities that the regression coefficients are nonzero are recorded in Table 13 for model *NCSVS* and in Table 14 for model *CSVS*. For both models the coefficient β_2 is significant but the remaining regression coefficients contained in Table 13 suggest that the change in slope is only significant at the third time point. Figure 5 shows the estimated posterior mean of the pig growth rate data with 95% credible regions. The time trend is approximately linear but changes slope slightly after the third time point.

Figure 6 shows the image plots of the posterior means of the partial correlations and the posterior probabilities that the elements in the partial correlation matrix are nonzero. The plots suggest that the partial correlations have an autoregressive type structure.

5. Simulation study

This section uses four different loss functions to study the performance of variable selection and covariance selection. The Kullback–Liebler loss function looks at the effect on the whole predictive density, the L_1 loss function looks at the effect on the covariance matrix, the beta loss function looks at the effect on the regression coefficients, and the fit loss function looks at the effect on the fitted values.

Our study uses four simulated data sets. Each data set is based on the corresponding model that was estimated for one of the four real data sets considered in Section 4.

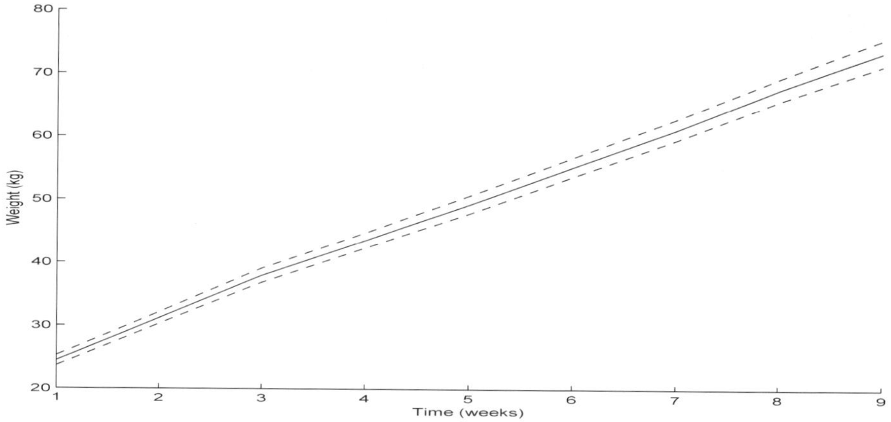

Fig. 5. Posterior mean (solid line) with a 95 percent credible region (dotted lines) of the time trend for pig growth rate data estimated by Model *NCSVS*.

Unless stated otherwise, each of the real data sets is first estimated using the *NCSVS* model and the parameter estimates are treated as the 'true model' parameters for the simulated data. Fifty replicates of data from this 'true model' are then constructed using the same values of the covariates as in the real data.

We now describe the four loss functions in detail.

1. *The Kullback–Liebler loss function.* This is our empirical version of the Kullback–Liebler distance between the true predictive density and the estimated predictive density. We use this loss function to assess the effect of variable and covariance selection on the estimation of the whole of the predictive distribution.

   Let $p(Y|Y_{\text{data}}, X)$ be the estimated predictive density of Y given X and the observed data Y_{data} and let $p_T(Y|X)$ be the true density of Y given X. For a given value of X, the Kullback–Liebler divergence between $p(Y|Y_{\text{data}}, X)$ and $p_T(Y|X)$ is (Gelman et al., 2000, p. 485)

$$\int p_T(Y|X) \log\left(\frac{p(Y|Y_{\text{data}}, X)}{p_T(Y|X)}\right) \mathrm{d}Y \tag{5.1}$$

   and it can be shown that the integral in (5.1) is less than or equal to 0, with strict inequality unless $p(Y|Y_{\text{data}}, X) = p_T(Y|X)$ for all Y.

   We cannot compute the integral in (5.1) analytically because Y is multivariate and because $p(Y|Y_{\text{data}}, X)$ is estimated by simulation. To approximate (5.1) for a given X, we generate K values of Y from $p_T(Y|X)$, which we call $Y_{k,X}, k = 1, \ldots, K$, and define the empirical Kullback–Liebler divergence at X as

$$\frac{1}{K}\sum_{k=1}^{K} \log\left(\frac{p(Y_{k,X}|Y_{\text{data}}, X)}{A_X p_T(Y_{k,X}|X)}\right), \tag{5.2}$$

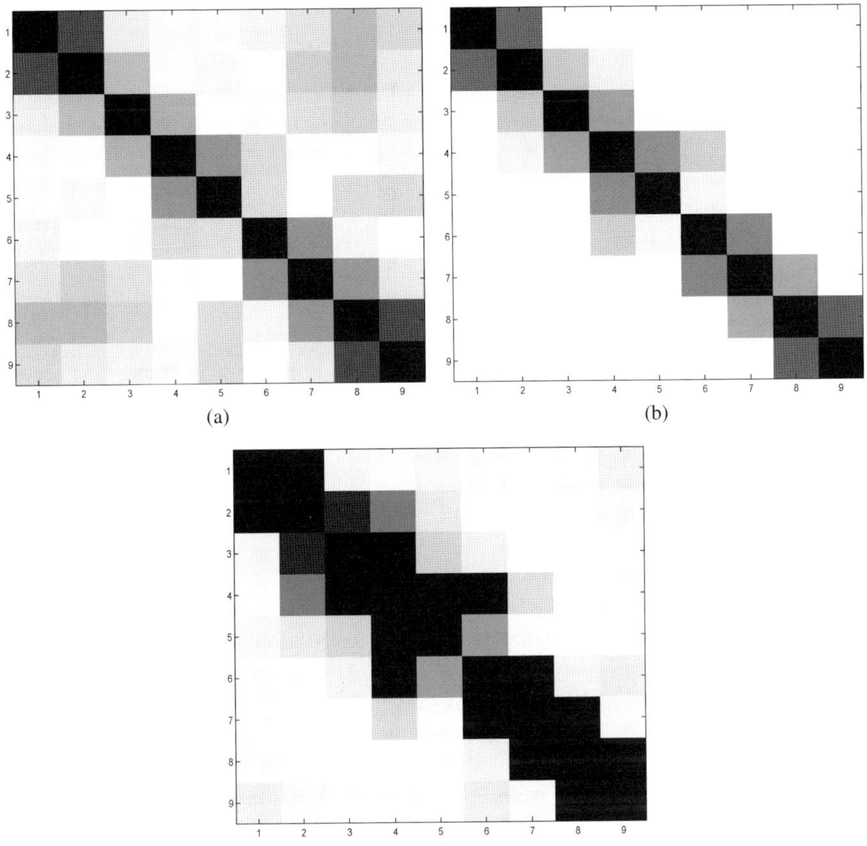

Fig. 6. Image plots for pig growth rate data. Panel (a) is the image plot of the partial correlation matrix estimated by model *NCSVS*. Panel (b) is the image plot of the partial correlation matrix estimated by Model *CSVS*. Panel (c) is the image plot of the probabilities of the elements of the correlation matrix being nonzero as estimated by model *CSVS*.

where

$$A_X = \frac{1}{K}\sum_{k=1}^{K} \frac{p(Y_{k,X}|Y_{\text{data}}, X)}{p_T(Y_{k,X}|X)}.$$

It is straightforward to show using Jensen's inequality (Gradshteyn and Ryzhik, 2000, p. 1101) that the sum in (5.2) is always less than or equal to 0 and it is strictly less than 0 unless $p(Y_{k,X}|Y_{\text{data}}, X) = p_T(Y_{k,X}|X)$ for all $k = 1, \ldots, K$. To get a representative set of values of X, we choose L values $X_l, l = 1, \ldots, L$, of X at random from the observed covariate matrices and define the empirical Kullback–Liebler

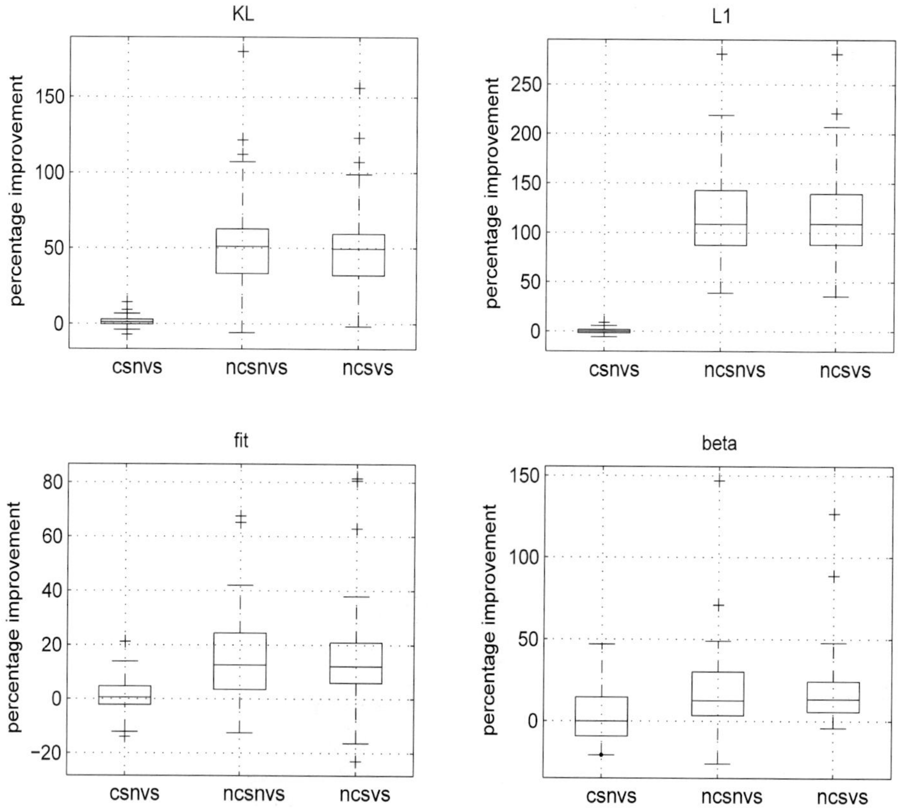

Fig. 7. Longitudinal cow data. The boxplots represent the percentage change in going from model *CSVS* to the models *CSNVS*, *NCSNVS* and *NCSVS*. From top left and reading clockwise the boxplots show the Kullback–Liebler, L_1, fit, and beta loss functions.

distance as

$$KL\{p_{\text{pred}}, p_T\} = \frac{1}{KL}\sum_{l=1}^{L}\sum_{k=1}^{K}\log\left(\frac{\hat{p}(Y_{k,l}|Y_{\text{data}}, X_l)}{A_l\, p_T(Y_{k,l}|X_l)}\right), \tag{5.3}$$

where $Y_{k,l}$ and A_l are $Y_{k,X}$ and A_X evaluated at $X = X_l$.

The predictive density $p(Y|X, Y_{\text{data}})$ is estimated using the output of the Markov chain Monte Carlo simulation. Let $\beta^{[j]}, \Sigma^{[j]}, j = 1, \ldots, J$, be iterates of β and Σ generated from the posterior distribution. We take

$$\hat{p}(Y|Y_{\text{data}}, X) = \frac{1}{J}\sum_{j=1}^{J} p\big(Y|X, \beta^{[j]}, \Sigma^{[j]}\big),$$

using the output of the Markov chain Monte Carlo simulation scheme.

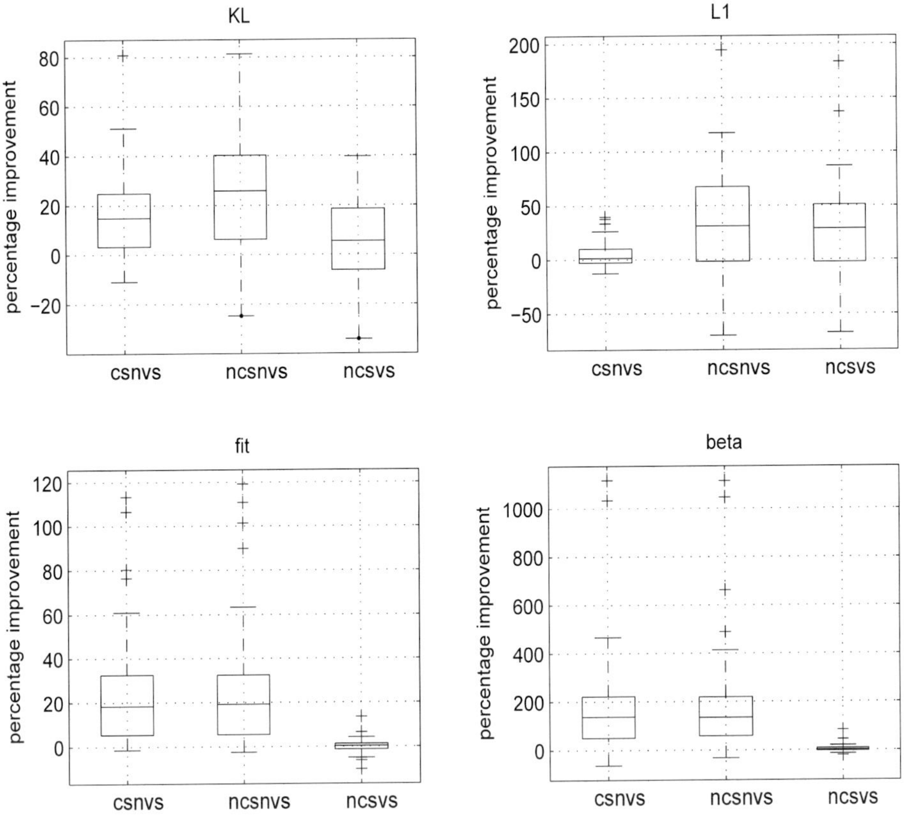

Fig. 8. Longitudinal hip replacement data. The boxplots represent the percentage change in going from the *CSVS* model to the *CSNVS*, *NCSNVS* and *NCSVS* models. From top left and reading clockwise, the boxplots show the Kullback–Liebler, L_1, fit, and beta loss functions.

In the simulations we use $K = 400$ and $L = 10$, with values of K greater than 400 giving the same numerical results (to 3 decimal places) of the Kullback–Liebler loss.

2. *The L_1 loss function.* Let $\widehat{\Sigma}$ be an estimate of the true covariance matrix Σ. The L_1 loss function for the covariance matrix is given by Yang and Berger (1994) as

$$L_1(\widehat{\Sigma}, \Sigma) = \operatorname{tr}(\widehat{\Sigma}\Sigma^{-1}) - \log|\widehat{\Sigma}\Sigma^{-1}| - p.$$

This loss function assesses the effect of variable and covariance selection on the estimation of Σ.

3. *The beta loss function.* We write the beta loss function as

$$\operatorname{beta}(\hat{\beta}, \beta) = \left\{\sum_{i=1}^{q}(\hat{\beta}_i - \beta_i)^2/q\right\}^{1/2},$$

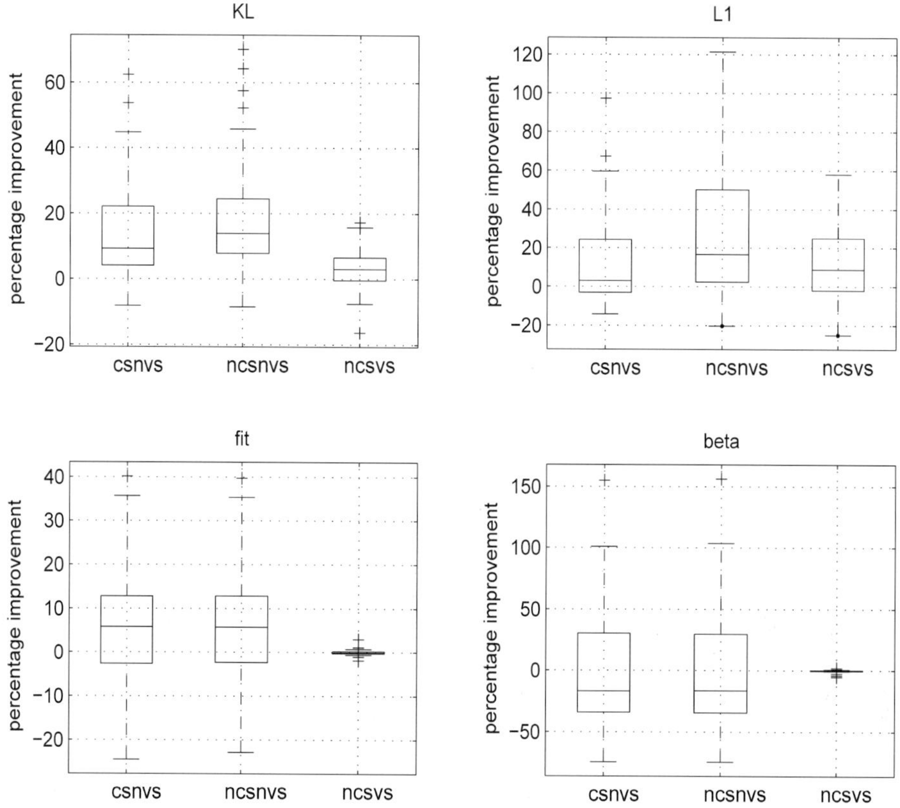

Fig. 9. Cow diet data generated from estimates using grouped variables. The boxplots represent the percentage change in going from *CSVS* model to the *CSNVS*, *NCSNVS*, and *NCSVS* models. From top left and reading clockwise, the boxplots show the Kullback–Liebler, L_1, fit, and beta loss functions.

where q is the length of β. This loss function assesses the effect of variable and covariance selection on the estimation of the regression coefficients.

4. *The fit loss function.* Let $f_l = X_l\beta$ be the actual fit and $\hat{f}_l = X_l\beta$ be the fitted vector at the covariate matrices X_l, $l = 1, \ldots, L$. These are the same covariate matrices that are used for the Kullback–Liebler loss function. We write the fit loss function as

$$\mathrm{fit}(f, \hat{f}) = \left\{ \sum_{l=1}^{L} (f_l - \hat{f}_l)'(f_l - \hat{f}_l)/(L \times p) \right\}^{1/2}.$$

This loss function measures of the effect of variable selection and covariance selection on the fitted values.

We estimate four different models for each data set. The first two models are *NCSVS* and *CSVS* and are introduced in Section 4. The third model carries out covariance selection but no variable selection and we call this model *CSNVS*. The fourth model does not carry out either variable selection nor covariance selection and we call this model

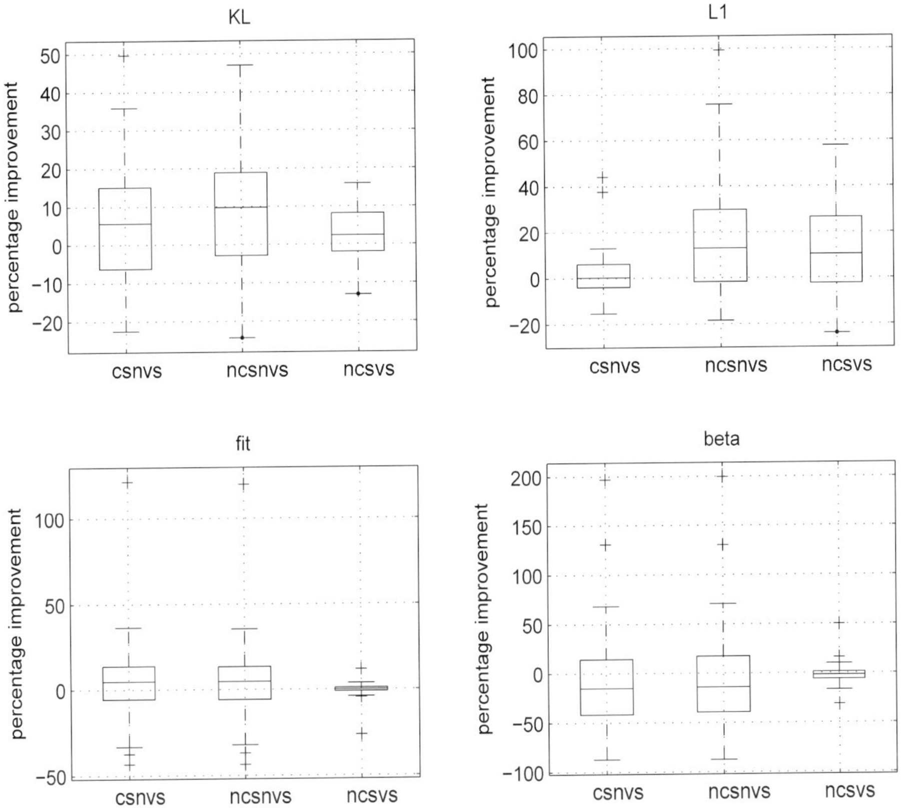

Fig. 10. Cow diet data generated from estimates using nongrouped variables. The boxplots represent the percentage change in going from the *CSVS* model to the *CSNVS*, *NCSNVS*, and *NCSVS* models. From top left and reading clockwise, the boxplots show the Kullback–Liebler, L_1, fit, and beta loss functions.

NCSNVS. We compare model *CSNVS* to model *CSVS* for a given loss function LOSS by the percentage increase (or decrease) in LOSS in going from model *CSVS* to say model *CSNVS*, i.e.,

$$D(CSNVS, CSVS) = \frac{\text{LOSS}(CSNVS) - \text{LOSS}(CSVS)}{\text{LOSS}(CSVS)} \times 100. \tag{5.4}$$

If model *CSVS* outperforms model *CSNVS* then $D(CSNVS, CSVS) > 0$, and conversely if model *CSNVS* outperforms *CSVS* then $D(CSNVS, CSVS) < 0$. We carried out similar comparisons for models *NCSNVS* and *NCSVS* with respect to model *CSVS*.

5.1. Cow milk protein data

Figure 7 reports the percentage change in going from model *CSVS* to the other three models for the four loss functions. The figure shows that model *CSVS* outperforms model *NCSNVS* and model *NCSVS* under all four loss functions. The improvement is

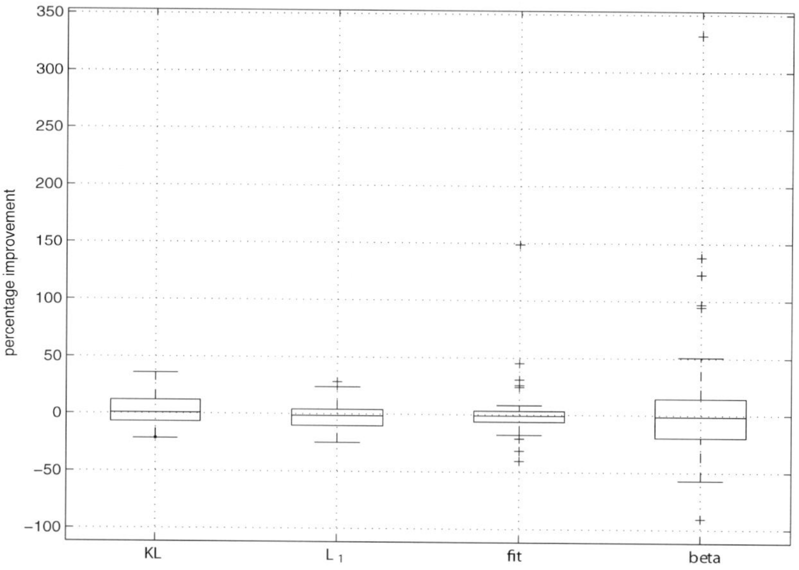

Fig. 11. Boxplots of the percentage change in going from nongrouped predictors to grouped predictors for the cross-sectional cow data using the *CSVS* model. The boxplots represent from left to right the Kullback–Liebler, L_1, fit, and beta loss functions.

particularly pronounced for the L_1 loss function, probably due to the sparsity in Ω. There is no improvement of model *CSVS* over model *CSNVS* across all loss functions. The results imply that for this example covariance selection is useful but variable selection is not.

5.2. *Hip replacement data*

The simulation results are summarised in Figure 8, which has a similar interpretation to Figure 7. The figure shows that covariance selection improves performance on the Kullback–Liebler and L_1 loss functions and that variable selection improves performance for the Kullback–Liebler, beta and fitted loss functions. The results suggest that both variable selection and covariance selection improve performance in this example.

5.3. *Cow diet data*

Figure 9 has a similar interpretation to Figures 7 and 8 and comes from data simulated from the estimated parameters in Section 4 when the predictor variables are grouped. The figure shows that covariance selection improves performance for the Kullback–Liebler and L_1 loss functions, while variable selection does not seem to improve performance for any of the loss functions.

Figure 10 shows the same results for the data generated when the predictor variables are not grouped. The results are similar to the case where the variables are grouped.

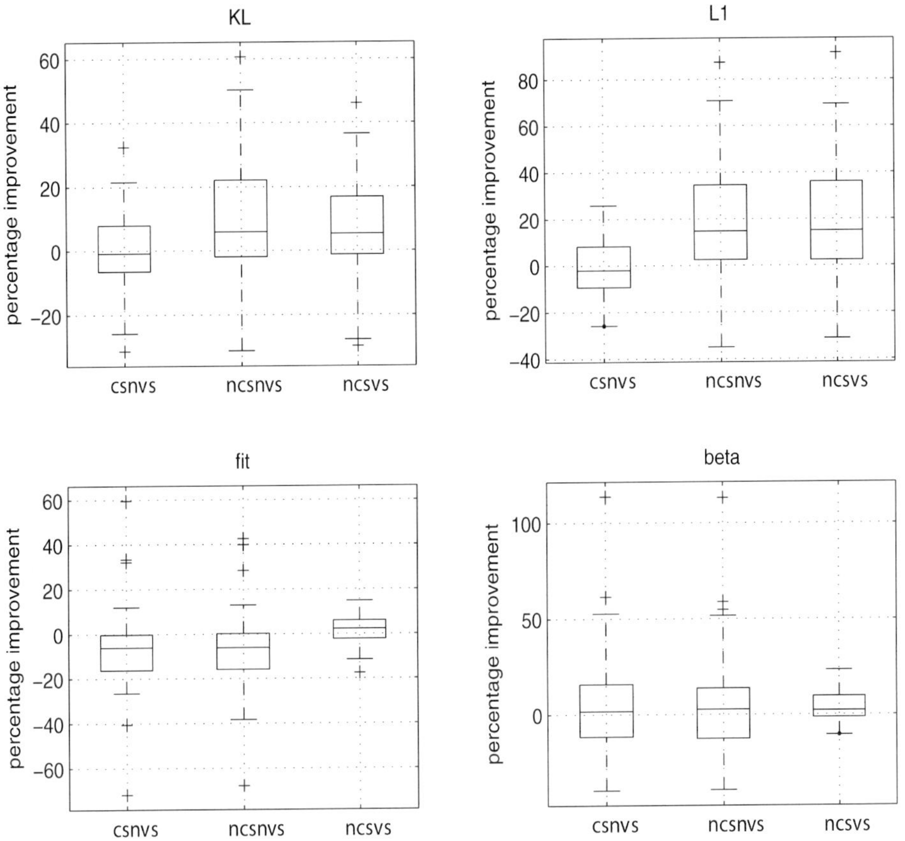

Fig. 12. Longitudinal pig data with a full inverse covariance matrix. The boxplots represent the percentage change in going from the *CSVS* model to the *CSNVS*, *NCSNVS*, and *NCSVS* models. From top left and reading clockwise the boxplots show the Kullback–Liebler, L_1, fit, and beta loss functions.

Figure 11 shows the results of going from model *CSVS* not using grouped predictor variables to model *CSVS* using grouped predictor variables. The data is simulated using parameters estimated from Section 4 using grouped predictor variables and model *CSVS*. For each loss function there seems to be no benefit in using grouped variables.

5.4. *Pig bodyweight data*

For the pig bodyweight data we generated 50 replications using the original predictor matrix and the estimated parameters from the *NCSVS* model reported in Section 4. The results are presented in Figure 12. We then generated another 50 replications using the parameters estimated by model *CSVS*. The results are reported in Figure 13.

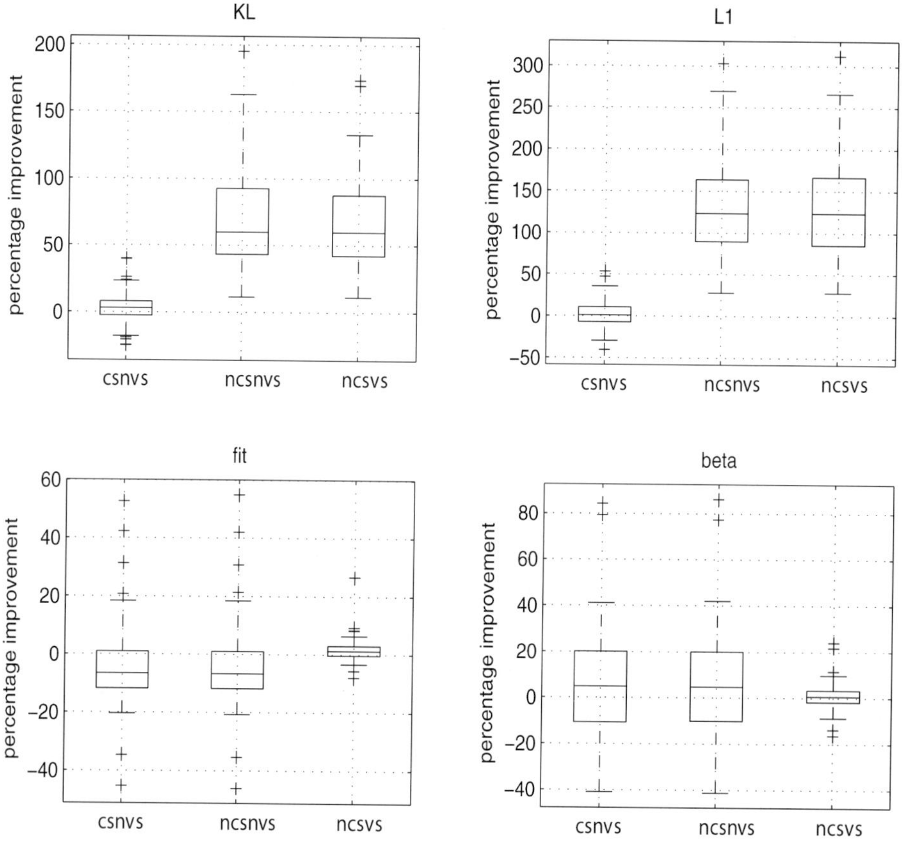

Fig. 13. Longitudinal pig data with sparse inverse covariance matrix. The boxplots represent the percentage change in going from the *CSVS* model to the *CSNVS*, *NCSNVS*, and *NCSVS* models. From top left and reading clockwise the boxplots show the Kullback–Liebler, L_1, fit, and beta loss functions.

The results in Figure 12 suggest that neither covariance selection nor variable selection affect performance, except for the L_1 loss function where covariance selection slightly improves performance.

Figure 13 shows similar results to those in Figure 12 for the beta and fit loss functions. However, covariance selection improves performance appreciably for the Kullback–Liebler and L_1 loss functions. The reason for the improved performance is that the estimated covariance matrix has a very sparse inverse when it was estimated using covariance selection.

6. Summary

The chapter presents a unified Bayesian methodology for variable and covariance selection in multivariate regression models. The methodology can be applied to both longitudinal and cross-sectional data. The simulation results suggest that when the inverse of the covariance matrix is sparse, implementing covariance selection leads to more efficient estimates of the covariance matrix as well as the predictive distribution. Similarly, when there are redundant variables in the model, variable selection will give more efficient estimators of the regression parameters and the predictive distribution.

References

Barnard, J., McCulloch, R., Meng, X. (2000). Modeling covariance matrices in terms of standard deviations and correlations, with application to shrinkage. *Statistica Sinica* **10**, 1281–1311.

Breiman, L. (1996). Heuristics of instability and stabilisation in model selection. *Ann. Statist.* **24**, 2350–2383.

Brown, P., Vannucci, M., Fearn, T. (1998). Multivariate Bayesian variable selection and prediction. *J. Roy. Statist. Soc., Ser. B* **60**, 627–641.

Brown, P., Fearn, T., Vannucci, M. (1999). The choice of variables in multivariate regression: A non-conjugate Bayesian decision theory approach. *Biometrika* **86**, 635–648.

Brown, P., Vanucci, M., Fearn, T. (2002). Bayes model averaging with selection of regressors. *J. Roy. Statist. Soc., Ser. B* **64**, 519–536.

Chiu, T., Leonard, T., Tsui, K. (1996). The matrix-logarithm covariance model. *J. Amer. Statist. Assoc.* **81**, 310–320.

Crowder, M., Hand, D. (1990). *Analysis of Repeated Measures*. Chapman and Hall, London.

Daniels, M., Kass, R. (1999). Nonconjugate Bayesian estimation of covariance matrices. *J. Amer. Statist. Assoc.* **94**, 1254–1263.

Dempster, A. (1972). Covariance selection. *Biometrics* **28**, 157–175.

Dempster, A.P. (1969). *Elements of Continuous Multivariate Analysis*. Addison-Wesley, Reading, MA.

Diggle, P., Heagerty, P., Liang, K., Zeger, S. (2002). *Analysis of Longitudinal Data*. Oxford University Press.

Efron, B., Morris, C. (1976). Multivariate empirical Bayes and estimation of covariance matrices. *Ann. Statist.* **4**, 22–32.

Gelman, A., Carlin, J., Stern, H., Rubin, D. (2000). *Bayesian Data Analysis*. Chapman and Hall/CRC.

George, E., McCulloch, R. (1993). Variable selection via Gibbs sampling. *JASA* **88**, 881–889.

George, E., McCulloch, R. (1997). Approaches for Bayesian variable selection. *Statistica Sinica* **7**, 339–373.

Giudici, P., Green, P. (1999). Decomposable graphical Gaussian model determination. *Biometrika* **86**, 785–801.

Gradshteyn, I., Ryzhik, I. (2000). *Tables of Intergals, Series, Products*, sixth ed. Academic Press, San Diego, CA.

Hoeting, J., Madigan, D., Raftery, A., Volinsky, C. (1999) Bayesian model averaging: A tutorial (with discussion). *Statist. Sci.* **14**, 382–417. Corrected version at http://www.stat.washington.edu/www/research/online/hoeting1999.pdf.

Kohn, R., Smith, M., Chan, D. (2001). Nonparametric regression using linear combinations of basis functions. *Statist. Comput.* **11**, 313–322.

Lauritzen, S. (1996). *Graphical Models*. Oxford University Press, Oxford.

Leonard, T., Hsu, J.S.J. (1992). Bayesian inference for a covariance matrix. *Ann. Statist.* **20**, 1669–1696.

Mardia, K.V., Kent, J.T., Bibby, J.M. (1979). *Multivariate Analysis*. Academic Press, London.

Mitchell, T., Beauchamp, J. (1988). Bayesian variable selection in linear regression. *J. Amer. Statist. Assoc.* **83**, 1023–1036.

Pourahmadi, M. (1999). Joint mean-covariance models with application to longitudinal data: Unconstrained parameterisation. *Biometrika* **86**, 677–690.

Pourahmadi, M. (2000). Maximum likelihood estimation of generalised linear models for multivariate normal covariance matrix. *Biometrika* **87**, 425–435.

Raftery, A., Madigan, D., Hoeting, J. (1997). Bayesian model averaging for linear regression models. *J. Amer. Statist. Assoc.* **94**, 179–191.

Smith, M., Kohn, R. (1996). Nonparametric regression using Bayesian variable selection. *J. Econometrics* **75**, 317–342.

Smith, M., Kohn, R. (2002). Bayesian parsimonious covariance matrix estimation for longitudinal data. *J. Amer. Statist. Assoc.* **87**, 1141–1153.

Stein, C. (1956). Some problems in multivariate analysis. Part I. Technical Report 6, Dept. Statistics, Stanford University.

Whittaker, J. (1990). *Graphical Models in Applied Mathematical Analysis*. Wiley, New York.

Wong, F., Carter, C., Kohn, R. (2003). Efficient estimation of covariance selection models. *Biometrika* **90**, 809–830.

Yang, R., Berger, J. (1994). Estimation of a covariance matrix using the reference prior. *Ann. Statist.* **22**, 1195–1211.

Handbook of Statistics, Vol. 25
ISSN: 0169-7161
© 2005 Elsevier B.V. All rights reserved.
DOI 10.1016/S0169-7161(05)25019-8

Dynamic Models

Helio S. Migon, Dani Gamerman, Hedibert F. Lopes and Marco A.R. Ferreira

1. Model structure, inference and practical aspects

Dynamic Bayesian modelling and forecasting of time series is one of the most important areas emerged in Statistics at the end of the last century. Taking as starting point the regression model, an extension is provided by the introduction of an equation governing the regression coefficients evolution through time. Many of the most important problems in Statistics will be encapsulated in this structure. Some special characteristics of Bayesian Forecasting include: (i) all relevant information sources are used, including history, factual or subjective experiences, and knowledge of forthcoming events. (ii) routine forecasting is produced by a statistical model and exceptions can be considered as an anticipation or in a retrospective base, (iii) prospective (what happened) and retrospective (what if) analysis are easily accommodate, (iv) model decomposition: a full Bayesian forecasting model may be decomposed into independent dynamic linear models (DLM), each one describing particular features of the process under analysis. We present in this chapter an overview of dynamic Bayesian models. A more comprehensive treatment of the subject can be found in books such as West and Harrison (1997), Pole et al. (1994), Durbin and Koopman (2002) and Harvey (1989).

In Section 1.1 we describe the class of DLM both to set the notation and to introduce important arguments of Bayesian dynamic models such as model superposition. Inference and practical aspects of Bayesian forecasting are discussed in Sections 1.2 and 1.3, respectively. The nonlinear and nonnormal models are discussed in Section 1.4, giving particular attention to the class of dynamic generalized linear models, an extension of dynamic normal linear models to the exponential family. Section 1.5 briefly summarizes the class of dynamic hierarchical models.

1.1. Dynamic linear models: General notation

Dynamic linear models are a broad class of models with time varying parameters, useful to modeling time series data and regression. It was introduced by Harrison and Stevens (1976) and is very well documented in the book by West and Harrison (1997). In this section some fundamental aspects of dynamic models will be introduced and some examples in time series as well as in regression will be addressed.

Dynamic linear models are parametric models where the parameter variation and the available data information are described probabilistically. They are characterized by a pair of equations, named observational equation and parameter evolution or system equation. The DLM can be seen as a generalization of the regression models allowing changes in parameters values throughout time. The observational and system equations are respectively given by

$$y_t = F_t'\theta_t + \varepsilon_t, \quad \varepsilon_t \sim N(0, V_t), \tag{1.1}$$

$$\theta_t = G_t\theta_{t-1} + \omega_t, \quad \omega_t \sim N(0, W_t), \tag{1.2}$$

where y_t is a time sequence of scalar observations, conditionally independent given the sequence of parameters θ_t, F_t is a $p \times 1$ vector of explanatory variables, θ_t is a $p \times 1$ vector of parameters, G_t is a $p \times p$ matrix describing the parameter evolution and, finally, V_t and W_t are the variances of the errors associated with the unidimensional observation and with the p-dimensional vector of parameters, respectively. This class includes many of the models found in the statistical literature. For example, if $G = I_p$, the identity matrix of order p and $\omega_t = 0$, $\forall t$, all linear regression models can be represented. On the other hand if F_t, V_t and W_t are constant $\forall t$, then the model covers the linear time series models such as ARIMA processes of Box and Jenkins (1976).

Summarizing, a dynamic linear model is completely specified by the quadruple $\{F_t, G_t, V_t, W_t\}$. Two special cases are, respectively, time series models characterized by $F_t = F$ and $G_t = G$, $\forall t$, and dynamic regression models, described by $G_t = I_p$.

EXAMPLE 1.1 (*1st-order polynomial model*). The simplest model in time series is the 1st-order polynomial model, which corresponds to a 1st-order Taylor series approximation of a smooth time function, named the time series trend. This model is completely defined by the quadruple $\{1, 1, V_t, W_t\}$. The above equations specialize to $y_t = \theta_t + \varepsilon_t$, $\varepsilon_t \sim N(0, V_t)$, $\theta_t = \theta_{t-1} + \omega_t$, $\omega_t \sim N(0, W_t)$, where θ_t is unidimensional and describes the underlying trend of the process. Although this model is very simple, it can be applied in many short-term forecasting systems involving a large number of time series such as in stock control or production planning. The observational and parameters evolution variance can also evolve in time, offering a broad scope for modeling.

A slightly more elaborated model, named linear growth model (LGM, in short), is derived after including an extra parameter $\theta_{2,t}$ to describe the underlying growth of the process. Then, after some minor modifications in state space equations, it follows that $\theta_{1,t} = \theta_{1,t-1} + \theta_{2,t-1} + \omega_{1,t}$, $\theta_{2,t} = \theta_{2,t-1} + \omega_{2,t}$ $\quad \omega_t \sim N(0, W_t)$. The parameter $\theta_{1,t}$ is interpreted as the current level of the process and it is easy to verify that $F_t = (1, 0)'$ and $G_t = \left(\begin{smallmatrix} 1 & 1 \\ 0 & 1 \end{smallmatrix}\right)$, $\forall t$, characterizing a time series model.

EXAMPLE 1.2 (*Simple dynamic linear regression*). Suppose, in this example, that pairs of values (x_t, y_t) are observed through time and that it is wished to model the existing relationship between x_t and y_t. Assuming that the linear model is a good approximation for the relationship between these values, a simple linear regression model can be set. Since the linear relationship is only a local approximation for the true functional dependence involving x and y, a model with varying parameters is appropriate. For example, the omission of some variables, the nonlinearity of the functional relationship connecting x and y or some structural changes occurring in the process under investigation, can be responsible for the parameter instability. These situations can be modeled as $y_t = F_t'\theta_t + \varepsilon_t$, $\theta_t = \theta_{t-1} + \omega_t$ where $F_t = (1, x_t)'$ and $\omega_t \sim N(0, W_t)$. Note that, in this case, $G_t = I_2$.

As we can observe, the choice of F_t and G_t depends on the model and the nature of the data that is being analyzed. To complete the model specification the variances V_t and W_t must be set. The latter describes the speed of the parameters evolution. In applications V_t is often larger than the elements of W_t. To make the parameter estimation method easier for the conjugate analysis, W_t is scaled by V_t and the conditional variances of the ω_t becomes $V_t W_t$. Therefore, the matrix W_t can be interpreted as a matrix of relative weights with respect to the observational variance. The parameter evolution variance matrix must be assessed subjectively by the user of the method and, in order to do that, the notion of discount factor will be useful (see Ameen and Harisson, 1985). Alternatively, it can be estimated by one of the methods described in the next sections. Therefore the equations presented before can be rewritten as

$$y_t|\theta_t \sim N(F_t'\theta_t, V_t), \tag{1.3}$$

$$\theta_t|\theta_{t-1} \sim N(G_t\theta_{t-1}, V_t W_t). \tag{1.4}$$

EXAMPLE 1.3 (*Component models*). It is a good practice to build a model step by step. A component model is structured as a composition of say r DLM's $\{F_i, G_i, 0, W_i\}_t$, $i = 1, \dots, r$, and a noise model $\{0, 0, V, 0\}_t$. Then the model elements are expressed as $\theta' = (\theta_1', \dots, \theta_r')$, $F_t' = (F_1', \dots, F_r')_t$, $G_t = \mathrm{diag}(G_1, \dots, G_r)_t$ and $W_t = \mathrm{diag}(W_1, \dots, W_r)_t$. An example, assuming that the data is observed quarterly, with $r = 3$ components includes the *linear trend* and the *regression component* (the 2 examples above) and the quarterly *seasonal component* $\{(1, 0, 0, 0), G_{3,t}, 0, W_{t,3}\}$, where $G_{3,t} = \left(\begin{smallmatrix} 0 & I_3 \\ 1 & 0 \end{smallmatrix}\right)$, named also form free seasonal component model in opposition to the Fourier form representation of seasonality (West and Harrison, 1997).

It is worth noting that it is assumed that for any time t, the current observation y_t is independent of the past observations given the knowledge of θ_t. This means that the temporal dynamics are summarized in the state parameters evolution. This linear structure for modeling data observed through time combines very well with the principles of Bayesian inference because it is possible to describe subjectively the involved probabilities and because of its sequential nature. Therefore, subjective information is coherently combined with past information to produce convenient inference.

1.2. *Inference in DLM*

The inference in DLM follows the usual steps in Bayesian inference. It explores the sequential aspects of Bayesian inference combining two main operations: *evolution* to build up the prior and *updating* to incorporate the new observation arrived at time t. Let $D_t = D_{t-1} \cap \{y_t\}$ denote the information until time t, including the values of x_t and G_t, $\forall t$, which are supposed to be known, with D_0 representing the prior information. Then for each time t the prior, predictive and posterior distribution are respectively given by

$$p(\theta_t|D_{t-1}) = \int p(\theta_t|\theta_{t-1})p(\theta_{t-1}|D_{t-1})\,d\theta_{t-1}, \tag{1.5}$$

$$p(y_t|D_{t-1}) = \int p(y_t|\theta_t)p(\theta_t|D_{t-1})\,d\theta_t, \tag{1.6}$$

$$p(\theta_t|D_t) \propto p(\theta_t|D_{t-1})p(y_t|D_{t-1}), \tag{1.7}$$

where the last one is obtained via Bayes theorem. The constant of integration in the above specification is sometimes easily obtained. This is just the case when $(F, G, V, W)_t$ are all known and normality is assumed. The resulting algorithm in this very narrow case is known as Kalman filter (Anderson and Moore, 1979). Usually the above matrices depend on some unknown parameters denoted generically by ψ.

The forecasting function and DLM design

In time series model it is very useful to obtain the mean response forecasting function $E\{E[y_{t+h}|\theta_{t+h}]|D_t\} = F'E[\theta_{t+h}|D_t] = F'G^h m_t$ where h is the forecasting horizon, $m_t = E[\theta_t|D_t]$. The structure of this function depends mainly on the eigenvalues of the design matrix G. The interested reader must consult West and Harrison (1997) to learn how to apply these theoretical results in model design.

Evolution and updating equations

The equations described before enable a joint description of (y_t, θ_t) given the past observed data D_{t-1} via $p(y_t, \theta_t|D_{t-1}) = p(y_t|\theta_t)p(\theta_t|D_{t-1})$. This leads to the predictive distribution after integrating out θ_t.

One of the main characteristics of the dynamic linear model is that, at each instant of time, all the information available is used to describe the posterior distribution of the state vector. The theorem that follows shows how to evolve from the posterior distribution at time $t-1$ to the posterior at t.

THEOREM 1.1. *Consider a normal dynamic linear model with $V_t = V, \forall t$. Denote the posterior distribution at $t-1$ by $(\theta_{t-1}|D_{t-1}, V) \sim N(m_{t-1}, VC_{t-1})$ and the marginal posterior distribution of $\phi = V^{-1}$ as $\phi|D_{t-1} \sim G(n_{t-1}/2, n_{t-1}s_{t-1}/2)$. Then,*

(1) *Conditionally on V, it follows*
   (a) *Evolution – the prior distribution at t will be $\theta_t|V, D_{t-1} \sim N(a_t, VR_t)$, with $a_t = G_t m_{t-1}$ and $R_t = G_t C_{t-1} G_t' + W_t$.*
   (b) *the one step ahead predictive distribution will be $y_t|V, D_{t-1} \sim N(f_t, VQ_t)$, with $f_t = \mathrm{F}_t' a_t$ and $Q_t = \mathrm{F}_t' R_t \mathrm{F}_t + 1$.*
   (c) *Updating – the posterior distribution at t will be $\theta_t|V, D_t \sim N(m_t, VC_t)$, with $m_t = a_t + A_t e_t$ and $C_t = R_t - A_t A_t' Q_t$, where $A_t = R_t' \mathrm{F}_t / Q_t$ and $e_t = y_t - f_t$.*

(2) *The precision ϕ is updated by the relation $\phi|D_t \sim G(n_t/2, n_t s_t/2)$, with $n_t = n_{t-1} + 1$ and $n_t s_t = n_{t-1}s_{t-1} + e_t^2/Q_t$.*

(3) *Unconditionally on V, we will have:* (a) $\theta_t|D_{t-1} \sim t_{n_{t-1}}(a_t, s_{t-1}R_t)$, (b) $y_t|D_{t-1} \sim t_{n_{t-1}}(f_t, Q_t^*)$, *with $Q_t^* = s_{t-1}Q_t$ and* (c) $\theta_t|D_t \sim t_{n_{t-1}}(m_t, s_t C_t)$.

PROOF. Item (1)(a) follows immediately using the parameter evolution equation and standard facts from the normal theory. With respect to (1)(b), using the prior distribution in (a), it follows that $f_t = E[E(y_t|\theta_t)|V, D_{t-1}] = \mathrm{F}_t' a_t$, $Q_t = \mathrm{VAR}[E(y_t|\theta_t)|V, D_{t-1}] + E[\mathrm{VAR}(y_t|\theta_t)|V, D_{t-1}] = V(\mathrm{F}_t' R_t \mathrm{F}_t + 1)$ and the normality is a consequence of the fact that all the distributions involved are normal. To prove part (1)(c), suppose that the posterior distribution at $t-1$ is as given in the theorem. We wish to show that (c) follows from the application of Bayes theorem, that is, $p(\theta_t|V, D_t) \propto p(\theta_t|V, D_{t-1})p(y_t|\theta_t, V)$. To show that it is sufficient to use Theorem 2.1 in Migon and Gamerman (1999) and the identity $C_t^{-1} = R_t^{-1} + \phi \mathrm{F}_t \mathrm{F}_t'$.

If V is unknown, it will follow

- By hypothesis, $\phi|D_{t-1} \sim G(n_{t-1}/2, n_{t-1}s_{t-1}/2)$, and $y_t|\phi, D_{t-1} \sim N(f_t, Q_t/\phi)$. Then, by Bayes theorem, $p(\phi|D_t) \propto \phi^{(n_{t-1}+1)/2-1} \exp\{-\frac{\phi}{2}(n_{t-1}s_{t-1} + \frac{e_t^2}{Q_t})\}$ and therefore, $\phi|D_t \sim G(n_t/2, n_t s_t/2)$.
- Finally, for part (3) of the theorem, the proofs of items (a)–(c) follow from the results about conjugacy of the normal–gamma to the normal model and from the marginal distributions obtained after integrating out V.

□

1.3. Practical aspects of Bayesian forecasting

In this subsection, some special aspects involved in dynamic Bayesian modeling will be briefly discussed, including variance law, discount factor, smoothing, intervention and monitoring.

Variance law

The possibility of modeling the observational variance deserves special attention among the special aspects involved in the dynamic Bayesian modeling. For example, the observational variance can be modeled as a power law, $V_t = Vv(\mu_t)$, where $v(\mu_t)$ is predictable, as for example: $v(\mu_t) = \mu_t^{b_1}(1-\mu_t)^{b_2}$ where $\mu_t = \mathrm{F}_t'\theta_t$ is the process mean level. With $b_1 = b_2 = 1$, this mimics the binomial law and fixing $b_1 = 1, b_2 = 0$ it follows a Poisson law. An explanation of this law in a commercial environment is that the number of orders is Poisson ($b_1 = 1, b_2 = 0$) but the amounts per order vary so that demand follows a compound Poisson distribution. For economic indices, the log transformation is often applied, which is equivalent to using $b_1 = 2, b_2 = 0$. The constant b_1 can be chosen in parallel to the well-known Box–Cox family of transformation. The scale factor V can be sequentially estimated as stated in the theorem or more generally assuming a dynamic evolution governed by the transformation $\phi_t = \gamma_t \phi_{t-1}$, where $\gamma_{t-1} \sim Ga(n_{t-1}/2, n_{t-1}s_{t-1}/2)$, $\gamma_t \sim Be(\delta_v n_{t-1}/2, (1-\delta_v)n_{t-1}/2)$, $\delta_v \in (0, 1)$ is a (variance) discount factor and s_t is an approximate point estimate of the variance. The main advantage in this form of modeling is to avoid the transformation of the original data, keeping in this way, the interpretation of the parameters, which is very useful, for example, when one wishes to perform some subjective intervention.

Discount factor
The use of discount factor is recommended to avoid the difficult task of directly setting the state parameters evolution matrix. These are fixed numbers between zero and one describing subjectively the loss of information through time. Remember that the prior variance of the state vector is obtained as $R_t = P_t + W_t$ where $P_t = G_t C_{t-1} G'_t$. Denoting the discount factor by δ, we can rewrite $R_t = P_t/\delta$, showing clearly that there is a relationship between W_t and δ. This is given by $W_t = (\delta^{-1} - 1)P_{t-1}$, showing that the loss of information is proportional to the posterior variance of the state parameters. For example, if $\delta = 0.9$, only about 90% of the information passes through time. The case of multiple discount factors is easily incorporated in the DLM and is very useful in practice since DLM components typically lose information at different rates.

Missing observation
Other relevant aspects of dynamic linear models are to easily take care of missing observations and to automatically implement subjective interventions. In the first case, it suffices not to use the updating equations at the time the observations are missing. In this way, the uncertainties increase with the evaluation of the new prior distribution and the recurrence equation continues to be valid without any additional problem. From the intervention point of view, the simplest proposal is to use a small discount factor, close to zero, at the time of announced structural changes in the data generation process. In this way the more recent observations will be strongly considered in the updating of the prior distribution and the system can be more adaptive to possible changes. An application of this idea in financial data analysis can be found in Reyna et al. (1999).

Retrospective analysis
It is worth mentioning that parameters distribution at any time t can be revised with the arrival of new observations. We can generically obtain the parameter distributions $p(\theta_t|D_{t+k})$, $\forall k$ integer. If $k > 0$, this is named the smoothed distribution, if $k = 0$, it is just the posterior and if $k < 0$ it is the prior distribution. In dynamic modeling it is common to use the distributions $p(\theta_t|D_T)$, $\forall t = 1, \ldots, T$, where T is the size of the series, to retrospectively analyze the parameter behavior. For example, one may want to quantify the change in behavior induced by some measure of economic policy. The future data would inform about the change occurred in any particular parameter of the model describing the behavior of the involved economic agents.

Monitoring and interventions
A continual assessment of the performance of any forecast system is vital for its effective use. Model monitoring is concerned with detecting inadequacy in the current model due to any major unanticipated event. One of the most important features of the Bayesian forecasting methodology is the sequential monitoring (West and Harrison, 1986) and intervention process described by West and Harrison (1997), which is based on the analysis of the standardized one step ahead forecasting error, $e_t = (y_t - f_t)/Q_t^{1/2}$, and allows them to identify the existence of level and variance changes. Suppose that at time t a suitable alternative model exists and is denoted by A. Based on the cumulative Bayes factor a sequential monitoring algorithm, like a Cusum test, follows as: (i) the

model A is accepted if $H_t(1) \geqslant \tau^{-1}$, the current model is accepted if $H_t(1) \leqslant 1$ and otherwise go on cumulating some more evidence, where $H_t(r) = \frac{p(e_t,\ldots,e_{t-r+1}|D_{t-r})}{p_A(e_t,\ldots,e_{t-r+1}|D_{t-r})}$ and τ and r are constants chosen in advance.

An implicit decision problem is associated with the above algorithm, where the state of the nature is $M = \{A_0, A\}$, with A_0 representing the current model and A an alternative one. The decision space is given by $D = \{d_0, d_1, d_2\}$, with d_0 indicating that the standard model is accepted, d_1, the alternative model is accepted and d_2 standing for the fact that the available information is not sufficient to decide between A_0 and A. Finally, assume that the loss function is proportional to the following payoffs: $l(d_0, A_0) = l(d_1, A) = 1 - l(d_0, A) = 0$, and $l(d_1, A_0) = l_1$ and $l(d_2, A_0) = l(d_2, A) = l_2$, with $l_2 < 1 < l_1$. Letting π_0 be the prior probability of the current model, it follows that the West and Harrison sequential monitoring algorithm corresponds to the solution of this decision problem with $l_1 = \pi_1(1 + \pi_1/(\tau\pi_0))$, $l_2 = \pi_1$ and $\pi_1 > \tau^{1/2}/(1 + \tau^{1/2})$, where $\pi_1 = 1 - \pi_0$.

In a recent paper, Harrison and Lai (1999) discuss a sequential Bayesian decision monitor for application in quality control, allowing to monitor the natural parameters of many exponential family models with conjugate prior. This proposal is able to discern whether the deterioration is due to change in level, slope, regression coefficient or seasonal pattern, individually. An extended version of the West and Harrison monitoring algorithm, allowing for, jointly, monitoring of several types of shocks, is proposed in Gargallo and Salvador (2002), where a frequentist evaluation of the proposed procedure is investigated and some guidelines as to how to choose its parameters are given.

Multiprocess models

The multi-process was introduced by Harrison and Stevens (1976) and involves a set of sub-models which together provide individual component models for outlying observations and changes in any or all of the components of the parameter vector. This is an alternative approach for model monitoring and intervention, where the alternative sub-models are fully stated. The multi-process approach is quite demanding computationally but proves extremely effective in identifying and adapting to changes with minimum loss of predictive performance. The model structure involves k states, $S_t(i)$, $i = 1, \ldots, k$. The first state $S_t(1)$ refers to the no change, nonoutlier model, the other one $S_t(2)$ allows for an outlying observation and the remaining ones, $S_t(i)$, $i = 3, \ldots, k-2$, to changes in particular components of the parameter vector, θ_t. At each time the probability of any particular state is assumed to be independent of what has occurred up to time $t - 1$ and are denoted by $\pi_i = P[S_t(i)]$. A very useful example, in short term forecasting, includes a four states multi-process model. The first state is a linear growth model, the others represent the outlier model and changes in level and growth respectively.

The posterior distribution of the parameters at time $t - 1$ is defined conditionally on the states at that time, that is: $p(\theta_{t-1}|S_{t-1}(i), D_{t-1})$. The unconditional posterior of the parameters at time $t - 1$ can be easily obtained as a mixture since the posterior probability over the states is known.

The prior distribution for the parameters is evaluated conditional on the states $S_t(j)$ and $S_{t-1}(i)$, that is: $p(\theta_t|S_{t-1}(i), S_t(j), D_{t-1})$. As soon as a new observation is obtained, k^2 posterior distributions are obtained. The full conditional posterior distribution of θ_t is given by the k^2 component mixture. The transitions probabilities involved in the mixture are obtained from Bayes theorem as: $p_t(i, j) \propto \pi_j p_{t-1}(i) p(y_t | S_t(j), S_{t-1}(i), D_{t-1})$, where $p_t(i, j) = P[S_t(j), S_{t-1}(i) | D_t]$. From this, $p_t(j) = \sum_{i=1}^k p_t(i, j)$ and $p_{t-1}(i) = \sum_{j=1}^k p_t(i, j)$. These probabilities prove useful in diagnosing the particular type of change or outlier after the event. Then, to avoid getting involved with an explosive number of alternative distributions, a collapsing procedure is in order.

The specification of the alternative models and many practical aspects involved in the multi-process implementation can be found in West and Harrison (1997), Chapter 12. The original multi-process model of Harrison and Stevens was developed for normal linear processes. A version based on the ideas of discounting was successfully implemented in the eighties. The extension to nonnormal and nonlinear models was developed in West (1986). An alternative method, computationally more efficient, is to use a simple monitoring scheme coupled with a multi-process approach, that is applied only when the monitor detects model breakdown (Ameen and Harrison, 1983).

1.4. Dynamic nonlinear/nonnormal models

There are many practical situations where nonlinearity is implicit in the modeling process. An initial very simple example in time series is the seasonal multiplicative model. The mean effect is defined as $\mu_t = \alpha_t(1 + \phi_t)$ where α_t is the trend component and ϕ_t the multiplicative seasonal effect. As a second example let us consider the first-order transfer response. This is useful to describe the dynamic relationship involved in a pulse response random variable x_t. This response function is given by $E_t = \phi E_{t-1} + \gamma x_t$, where $\gamma > 0$ represents the instantaneous "gain" and $|\phi| < 1$ models the decay rate. The mean response can be phrased as $\mu_t = \alpha_t + E_t$. Extensive applications of nonlinear and nonnormal models is reported on Migon and Harrison (1985), where the impact of consumer awareness of the advertising of various products is assessed.

The case of nonnormal models is often met in the practice of dynamic modeling. Some examples include count data and nonnegative observations and also the desire to use heavy tails density to protect oneself against spurious observations.

There are many different ways to approach both the above questions. The simplest one is to use some sort of transformation, with the undesirable cost of losing interpretability. Moreover the nonlinearity can be handled through linearization, which consists in the use of a low-order Taylor approximation.

Let the model be $y_t = F_t(\theta_t) + \varepsilon_t$, with $\theta_t = g(\theta_{t-1}) + \omega_t$. Assuming that the terms other than the linear one are negligible, the evolution becomes $g_t(\theta_{t-1}) \simeq g_t(m_{t-1}) + G_t(\theta_{t-1} - m_{t-1}) + \omega_t = h_t + G_t\theta_{t-1} + \omega_t$, with $G_t = \frac{\partial}{\partial \theta_{t-1}} g_t(\theta_{t-1})|_{m_{t-1}}$, where $m_{t-1} = E[\theta_{t-1} | D_{t-1}]$ and $h_t = g_t(m_{t-1}) - G_t(m_{t-1})$. The same sort of first-order approximation can also be applied to $F(\theta_t) \simeq F_t(a_t) + F_t'(\theta_t - a_t)$, where $F_t' = \frac{\partial}{\partial \theta_t} F_t(\theta_t)|_{a_t}$. Then the observation equation simplifies to $y_t = (F_t(a_t) - F_t' a_t) + F_t'\theta_t + \varepsilon_t$, where $a_t = E[\theta_t | D_{t-1}]$. The inference follows using the theorem stated for the linear case.

The interested reader is referred to West and Harrison (1997), Chapter 13, and Durbin and Koopman (2001), Chapter 10.

EXAMPLE 1.4. A broad class of nonlinear growth curve models, including the modified exponential ($\lambda = 1$), the logistic ($\lambda = -1$) and the Gompertz ($\lambda \to 0$), was introduced in Migon and Gamerman (1993). The global mean function of the growth process is defined by $\mu_t = [a + b\exp(\gamma t)]^{1/\lambda}$, with parameterization (a, b, γ, λ). The major advantage of this approach is to keep the measurements in the original scale, making the interpretation easier. A similar local model is easily obtained and can be represented as a dynamic nonlinear model with $F_t(\theta_t) = \theta_{1,t}$ and $G(\theta_t) = (\theta_1 + \theta_2, \theta_2\theta_3, \theta_3)'$, the link function is $g(\mu_t) = F_t(\theta_t)$, where g is an invertible function and a variance law $v(\mu_t)$.

The non-Gaussian state space model is characterized by the same structure as in the linear Gaussian case, with $p(y_t|\mu_t)$, $\mu_t = F_t'\theta_t$, and the states evolving as $\theta_t = G_t\theta_t + \omega_t$, $\omega_t \sim p(\omega_t)$, where $\omega_t, t = 1, \ldots, n$, are serially independent and either $p(y_t|\mu_t)$ or $p(\omega_t)$ or both can be non-Gaussian. Two special cases are worth considering: (a) observation from the exponential family, and (b) observation generated by the relation $y_t = \mu + \varepsilon_t$, with $\varepsilon_t \sim p(\varepsilon_t)$ non-Gaussian. The first case will be considered in the next subsection and follows West et al. (1985), while the case (b) will be considered in the next sections of this chapter. The case where only $p(\varepsilon_t)$ is nonnormal will be ilustrated in the following example.

EXAMPLE 1.5 (*A non-Gaussian model*). In finance V_t may depend upon the time interval between trading days, the time to a bond's maturity, trading volume and so forth. A broad class of models was presented in the literature of the end of the last century. An excellent example is the stochastic volatility model. Let us denote the first differences of a particular series of asset log prices by y_t. A basic stochastic volatility model is given by: $y_t = V_t^{1/2}\varepsilon_t$, where $\varepsilon_t \sim N[0, 1]$. Taking the square and the logarithm we obtain: $\log(y_t^2) = h_t + \log(\varepsilon_t^2)$, where $h_t = \log(V_t)$. Assuming that $h_t = \alpha + \beta h_{t-1} + \omega_t$ we obtain a non-Gaussian first-order dynamic model named, in the recent literature of finance, as the log-stochastic volatility model. The estimation can be done using Bayes' linear method, to be described in the next subsection, since $\log(\varepsilon_t^2)$ is $\log\text{-}\chi^2$ distribution. This approach is similar to the quasi-likelihood method proposed in Harvey et al. (1994). Various extensions of the SV models can be considered. For example the normality assumption can be replaced by some heavy tail distribution such as the t-distribution, as supported by many empirical studies.

Triantafyllopoulos and Harrison (2002) discuss stochastic volatility forecasting in the context of dynamic Bayesian models. They criticize the GARCH models showing some of their limitations and presenting dynamic models for forecasting the unknown stochastic volatility. The dynamic variance law produces a volatility estimate in the form of a discount weight moving average. A zero drift model ($F_t = 1, G_t = 0$ and $W_t = 0$) produces the volatility estimate $n_t\widehat{V}_t = \beta n_{t-1}\widehat{V}_{t-1} + y_t^2$, which looks like a GARCH recursion. Previous application of Bayesian forecasting in financial time series is presented in Migon and Mazuchelli (1999). A dynamic Bayesian GARCH model was introduced and the volatility of four different kinds of assets were predicted.

Dynamic generalized linear models
The extension of DLMs to allow observations in the exponential family was introduced by West et al. (1985) based on the generalized linear models of Nelder and Wedderburn (1972). The observation Eq. (1.1) is replaced by

$$p(y_t|\eta_t) \propto \exp\big[\big(y_t\eta_t - b(\eta_t)\big)/\phi_t\big] \tag{1.8}$$

and, in addition, a suitable link function is introduced, relating the mean $\mu_t = E[y_t|\eta_t] = b'(\eta_t)$ to the regressors F_t through $g(\mu_t) = \lambda_t = F_t'\theta_t$. A conjugate prior for the (1.8) is given as $p(\eta_t|D_{t-1}) \propto \exp[(r_t\eta_t - b(\eta_t))/s_t]$. The integral in Eqs. (1.5)–(1.7) cannot be obtained in closed form, and so the inference must be done in an approximate way. A procedure allowing the sequential analysis of DGLM was implement in West et al. (1985) using linear Bayes estimation.

The evolution equation (1.2) is only partially specified. This means that the distributions of $\theta_{t-1}|D_{t-1}$ and ω_t are only specified by the first- and second-order moments, that is: $\theta_{t-1}|D_{t-1} \sim [m_{t-1}, C_{t-1}]$ and $\omega_t \sim [0, W_t]$. Then the prior distribution for the state parameters is also partially specified as $\theta_t|D_t \sim [a_t, R_t]$ as given before. Then the prior distribution for $\lambda_t = g(\mu_t)$ is $\lambda_t|D_{t-1} \sim [f_t, q_t]$ where $f_t = F_t'a_t$ and $q_t = F_t'R_tF_t$. The parameters (r_t, s_t) in the prior distribution for η_t must be related to f_t and q_t through the $E[g(b'(\eta_t))|D_{t-1}] = f_t$ and $\mathrm{var}[g(b'(\eta_t))|D_{t-1}] = q_t$. Then the posterior for η_t is in the same form of its (conjugate) prior distribution with parameters $(r_t/s_t + y_t/\phi_t, 1/s_t + 1/\phi_t)$. The posterior distribution of the linear predictor is $\lambda_t|D_t \sim [f_t^*, q_t^*]$, where, again, $f_t^* = E[g(b'(\eta_t))|D_t]$ and $q_t^* = \mathrm{var}[g(b'(\eta_t))|D_t]$. Moreover, to complete the analysis, the posterior distribution of the state parameters must be obtained. The linear Bayes estimation is used to approximate the first- and second-order moments of this distribution, leading to: $\widehat{E}[\theta_t|\eta_t, D_{t-1}] = a_t + R_tF_t[\eta_t - f_t]/q_t$ and $\widehat{\mathrm{var}}[\theta_t|\eta_t, D_{t-1}] = R_t - R_tF_tF_t'R_t/q_t$. The moments of $\theta_t|D_t$ are calculated using the iterated expectation law given $\theta_t|D_t \sim [m_t, C_t]$, where $m_t = a_t + R_tF_t[f_t^* - f_t]/q_t$ and $C_t = R_t - R_tF_tF_t'R_t(1 - q_t^*/q_t)/q_t$.

A practical example
This example describes a model to predict the value of the Brazilian industrialized exports (y_t); for more details the interested reader should see Migon (2000). The model is built up step by step starting with a supply and demand equilibrium model. Dynamics are introduced via the general adaptive expectations hypothesis. After some simplifications we end up with a dynamic nonlinear model: $\log y_t = \eta_t + \varepsilon_t$, where, $\varepsilon_t \sim N[0, \sigma_t^2]$ and $\eta_t = \mu_t + E_t$. The parameters evolution is given by: $\theta_t = \mathrm{diag}(G_1, G_2, G_3)\theta_{t-1} + \omega_t$, where $\theta_t = (\mu, \beta, E, \gamma, \lambda)_t$, with $G_1 = \left(\begin{smallmatrix}1 & 1\\ 0 & 1\end{smallmatrix}\right)$, $G_2 = \left(\begin{smallmatrix}\lambda_{t-1} & \log r_t\\ 0 & 1\end{smallmatrix}\right)$ and $G_3 = 1$. The scale factor $\phi_t = \sigma_t^{-2}$ evolves through time keeping its mean and increasing the uncertainty via a discount factor. Next we will present a summary description of the data used and the main findings. The period of the data used in this application, Jun./79 up to Dec./84, is characterized by a deep recession in the world trade and many devaluations of the Brazilian exchange rate (r_t), which make it attractive to illustrate the capabilities of the models developed. In Figure 1(a), the one step ahead point forecasting, posterior mode, obtained using a simple linear growth model *with and without* subjective

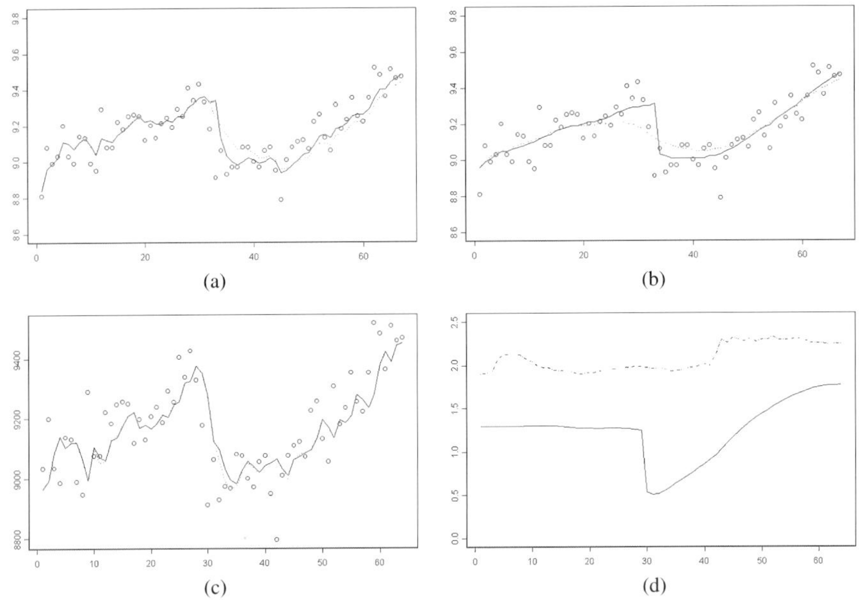

Fig. 1. (a) On-line fitting of exports linear growth model without and with intervention; (b) smoothed fitting of exports linear growth model without and with intervention; (c) on line fitting of exports transfer function model without and with intervention; (d) exchange rate effect.

intervention can be observed and also the smoothed mean of the level is plotted. The interventions improve considerably the predictive performance. The mean absolute error (MAD) decreases from .12 to .09 and the logarithm of the predictive likelihood (LPL) raises from 35 to 47, supporting the need of intervention. It is worth pointing out that the interventions were necessary to cope with change in the growth at Feb./82 and Jun./82 as is evident from Figure 1(b), where the on line fitting ($E[\mu_t|D_t]$) is shown. Examining the residuals of those fittings two alternatives can take place: to include a seasonal component or some of the omitted variables. In this study the exchange rate r_t is the candidate suggested by the theoretical model developed before. In fact seasonality is not supported by the data (MAD $=$ .12 and LPL $=$ 37). On the other hand, inclusion of the control variable, exchange rate, improves considerably the results. The best prediction results were obtained with the first-order transfer response model, which is simple and sophisticated. In the graph below, Figure 1(c), one can appreciate the predictive capability of the model and in Figure 1(d), the smoothed effect of the exchange rate is plotted.

An application of transfer response model in hydrology can be found in Monteiro and Migon (1997) where the dynamic relationship between rainfall and runoff is modeled.

1.5. Dynamic hierarchical models

This is a very important class of models useful for modeling several concomitant time series of observations. The methodology was developed in Gamerman and Migon

(1993), combining the dynamic models previously presented and the hierarchical models of Lindley and Smith (1972). These models appear in the Statistical literature under the name of panel data or longitudinal studies (Jones, 1993). The dynamic evolution provides a general framework for analysis of multivariate time series. The model specification includes a set of structural equations besides the usual observation and evolution equations. The structural equations progressively reduce the dimension of the parameter space as the level becomes higher.

The dynamic hierarchical model, with three levels, can be written as: $y_t = F_{1,t}\theta_{1,t} + \nu_{1,t}$ where $\nu_{i,t} \sim N[0, V_{1,t}]$, $\theta_{i,t} = F_{i+1,t}\theta_{i,t-1} + \nu_{i,t}$, $i = 1, 2$, and, finally, $\theta_{3,t} = G_t\theta_{3,t-1} + \omega_t$ where $\nu_{i,t} \sim N[0, V_{1,t}]$, $\omega_t \sim N[0, W_t]$, all the disturbance terms are independent with known variance and $F_{i,t}$, $i = 1, 2, 3$, known full rank matrices. It is mandatory that the evolution equation is applied to the higher level of the hierarchy.

The example includes cross-section of random samples with linear growing exchangeable means and of regression models with steady parameters, where all the observations at time t are explained by the same regressors with known values for each observation i. Inference for dynamic hierarchical models can be found in Gamerman and Migon (1993). Extension to the case of multivariate time series in presented in Landim and Gamerman (2000). A very interesting application for modeling animal growth curves is presented in Barbosa and Migon (1996).

2. Markov Chain Monte Carlo

Dynamic models introduced in the previous section allow for full inference only when the F_t's, G_t's and W_t's are entirely known and a conjugate form is imposed on the $V = V_t$, $\forall t$. In general, these quantities or other quantities used in their definition are unknown and inference about them must be based on their posterior distribution. This distribution is generally not analytically tractable.

In this section, the problem of making full (about all model parameters) inference about DLM's is considered. We start by considering normal DLM's and then generalize ideas to nonnormal DLM's. Ideally analytically unsolved problems in Bayesian inference can be approximately solved by sampling from the relevant posterior distribution. The distributions involved here are too complicated for directly drawing samples from. The tool presented in this section to solve this problem is MCMC, a powerful collection of sampling techniques that has revolutionized Bayesian inference for complex models in the last decades. Space constraints allow only a few descriptive lines here about MCMC. Full consideration of MCMC at an expository level can be seen in Gamerman (1997).

MCMC stands for Markov chain Monte Carlo and deals with sampling from a complicated distribution (in our case, the joint posterior of all model parameters) when direct sampling is not available. It is a two step procedure. In the first step, a Markov chain with a transition kernel such that the limiting distribution is given by the joint posterior distribution is constructed. In the second step, a trajectory is sampled from this chain. For a suitably large iteration, values are virtually sampled from the posterior distribution.

The most used transition kernels are derived from the Gibbs sampling (Geman and Geman, 1984) and componentwise Metropolis–Hastings steps (Metropolis et al., 1953; Hastings, 1970). In the first group, the kernel is composed of product of the posterior full conditionals of parameters divided in blocks. In the second group, the kernel is still composed of products of proposal full conditionals, not necessarily derived from the posterior. A correction term must then be imposed to blocks not drawn from the posterior to ensure convergence of the chain to the posterior. This scheme can also be applied if all parameters are gathered into a single block. In this case, the proposal replaces the joint posterior. In theory, any partition of the parameters into blocks ensures convergence. In practice, however, this partition choice plays an important role in providing a computationally feasible solution.

2.1. *Normal DLM*

This subsection considers the solution to full inference in normal DLM by sampling. Alternative sampling schemes are presented and compared in theoretical and empirical terms. The basic difference between them lies in the specification of blocks of parameters. Schemes are presented in the order of increasing dimension of the blocks, or in other words, decreasing number of blocks.

The normal dynamic models considered here are given by (1.1)–(1.2) with known F_t and G_t and constant variances of the observations and system disturbances, i.e., $V_t = V$ and $W_t = W$, for all t. This restriction is aimed mainly at presentation clarity. It also provides for a more parsimonious model. The extension to the general case of unequal variances is not difficult but can hinder meaningful inference unless substantial prior information is available for the different variance parameters. Extensions to cases where unknown quantities are present in F_t and G_t are not difficult to implement with the methodology described below.

The previous section showed how to perform exact inference in normal DLM when the evolution variance matrix W was known. It also showed a few simple alternatives to inference with respect to an unknown W, namely comparison of a few values for W and its specification through the use of discount factors, based on scaling on V.

Componentwise sampling schemes

Now, a sample from the posterior of $(\theta_1, \ldots, \theta_T, V, W)$ after observing the series up to time n is drawn. The joint posterior is

$$\pi(\theta, V, W|D_T) \propto \prod_{i=1}^{T} p(y_t|\theta_t, V) \prod_{i=2}^{T} p(\theta_t|\theta_{t-1}, W) p(\theta_1) p(V, W),$$

where $\theta = (\theta_1, \ldots, \theta_T)$. The specification of the model can be taken as a basis for blocking parameters. So, an initial blocking choice is $\theta_1, \ldots, \theta_T$, V and W. The full conditional of the θ_t's, denoted by $\pi(\theta_t|\theta_{-t}, V, W, D_T)$, are normal distributions that are easily obtained. If the joint prior for (V, W) is a product of an inverse Gamma for V and inverse Wishart for W, then their full conditional will also be so. So, sampling from their full conditionals is easily available.

These full conditional distributions complete a cycle of the Gibbs sampler. They are all easy to sample from and a MCMC-based Bayesian inference can be performed. This approach was introduced by Carlin et al. (1992) in the context on nonlinear and scale mixture of normal models and is summarized in Algorithm 1.

The prior for θ is highly dependent on the value of W. For small values of W, large correlations are induced and preserved through to the posterior. Unfortunately, this is usually the case observed in practice with system parameters typically experiencing small disturbances through time. This means that sampling the θ components separately may imply a kernel that will be very slow and may only achieve equilibrium after an unnecessarily large number of iterations was drawn. Figure 2(b) illustrates the convergence of state parameters with this sampling scheme.

ALGORITHM 1 (*Initial MCMC for DLM*).

a. Initialization: *set initial values* $(\theta^{(0)}, V^{(0)}, W^{(0)})$ *and iteration counter* $j = 1$;
b. Sampling θ**:** $\theta^{(j)}$ *is sampled componentwise as follows*

1. *set* $t = 1$;
2. *sample* $\theta_t^{(j)}$ *from* $\pi(\theta_t|\theta_1^{(j)}, \ldots, \theta_{t-1}^{(j)}, \theta_{t+1}^{(j-1)}, \ldots, \theta_T^{(j-1)}, V^{(j-1)}, W^{(j-1)}, D_T)$;
3. *set* $t \to t+1$ *and return to* **b2**, *if* $t < T$;

c. Sampling V **and** W**:** $V^{(j)}$ *and* $W^{(j)}$ *are sampled successively from the respective full conditionals* $\pi(V|\theta^{(j)}, W^{(j-1)}, D_T)$ *and* $\pi(W|\theta^{(j)}, V^{(j)}, D_T)$;
d. Updating: *set* $j \to j+1$ *and return to* **b** *until convergence.*

Block sampling schemes

One alternative to avoid this computational problem is to sample θ is a single block. There are a number of ways to obtain a sample for the posterior full conditional of $(\theta|V, W, D_T)$. A statistical way explores the properties of the model by noting that

$$\pi(\theta|V, W, D_T) = p(\theta_T|V, W, D_T)\prod_{t=1}^{T} p(\theta_t|\theta_{t+1}, V, W, D_t).$$

The distributions of $(\theta_t|\theta_{t+1}, V, W, D_t)$ are obtained from $p(\theta_{t+1}|\theta_t, V, W, D_t)$ and $p(\theta_t|V, W, D_t)$ and are given by

$$\begin{aligned}(\theta_t|\theta_{t+1}, V, W, D_t) \sim N\big[&(G_t'W^{-1}G_t + C_t^{-1})^{-1}(G_t'W^{-1}\theta_{t+1} + C_t^{-1}m_t),\\ &(G_t'W^{-1}G_t + C_t^{-1})^{-1}\big]\end{aligned} \tag{2.1}$$

for $t = 1, \ldots, n-1$. So, a scheme for sampling from the full conditional of θ is given by incorporation of this backward (in time) sampling step in the forward (in time) Kalman filter presented in the previous section. The above sampling scheme was independently proposed by Carter and Kohn (1994) and Frühwirth-Schnatter (1994), that appropriately named it forward filtering backward smoothing (FFBS). The sampling algorithm is summarized as:

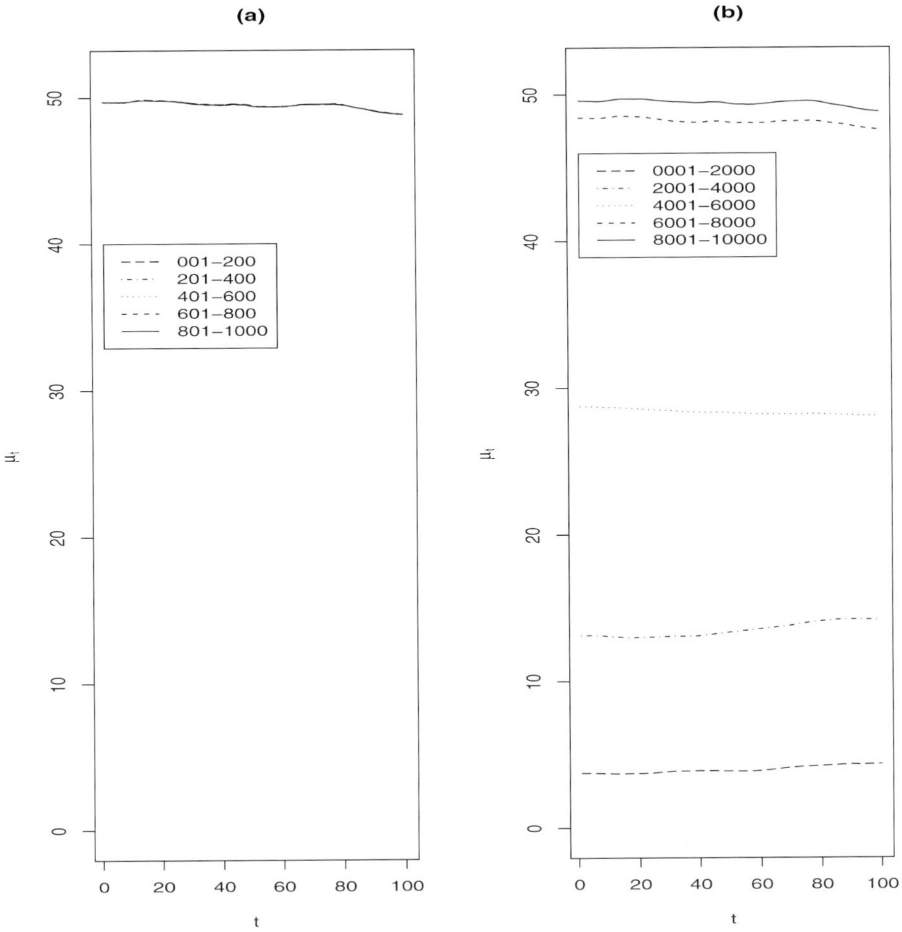

Fig. 2. Average trajectory of μ_t, $t = 1, \ldots, 100$, over batches of M successive iterations under different sampling schemes: (a) – sampling $(\mu_1, \ldots, \mu_{100})$ in a block ($M = 200$); (b) – sampling each μ_t separately ($M = 2000$). Data was generated from a first-order model with $V = 1$ and $W = 0.01$. Chains were started from 0.

ALGORITHM 1* (*FFBS*).

Exactly as Algorithm 1 *but for replacement of* θ *sampling* (*step* **b**) *by:*

b1*. *sample* θ_T *from its updated distribution (given in previous section) and set* $t = T - 1$.

b2*. *Sample* θ_t *from the distribution* (2.1).

b3*. *Decrease* t *to* $t - 1$ *and return to step* **b2*** *until* $t = 1$.

Step **b1*** is obtained by running the Kalman filter from $t = 1$ to $t = T$ with given values of V and W. When running the filter, the updated means m_t and variances C_t, $t = 1, \ldots, T$, are stored for use in step **b2***.

A more numerical way to obtain a sample from θ is given by direct evaluation of the prior full conditional of $(\theta|V, W) \sim N(A, P^{-1})$ with

$$A = \begin{pmatrix} I \\ G_2 \\ G_3G_2 \\ \vdots \\ \prod_{t=2}^{T} G_t \end{pmatrix} a \quad \text{and}$$

$$P = \begin{pmatrix} P_{11} & P_{12} & 0 & \cdots & \cdots & \cdots & 0 \\ P_{21} & P_{22} & P_{23} & \ddots & & & \vdots \\ 0 & \ddots & \ddots & \ddots & \ddots & & \vdots \\ \vdots & \ddots & \ddots & \ddots & \ddots & \ddots & \vdots \\ \vdots & & \ddots & \ddots & \ddots & \ddots & 0 \\ \vdots & & & \ddots & \ddots & \ddots & P_{T-1,T} \\ 0 & \cdots & \cdots & \cdots & 0 & P_{T,T-1} & P_{TT} \end{pmatrix}.$$

The precision matrix P is a symmetric, block tridiagonal matrix with main diagonal elements $P_{11} = R^{-1} + G_2' W^{-1} G_2$, $P_{TT} = W^{-1}$ and $P_{tt} = W^{-1} + G_{t+1}' W^{-1} G_{t+1}$, for $t = 2, \ldots, n-1$ and secondary diagonal elements given by $P_{t,t+1}' = P_{t+1,t} = W^{-1} G_t$, for $t = 1, \ldots, T-1$.

The likelihood is now given in matrix form by $y|\theta, V \sim N(F\theta, VI_T)$, where $y = (y_1, \ldots, y_T)$ and $F = \text{diag}(F_1', \ldots, F_T')$. Combining prior with likelihood leads to the posterior $\theta|V, W, D_T \sim N(M, Q^{-1})$ where $M = Q^{-1}(V^{-1}Fy + PA)$ and $Q = P + V^{-1}F'F$. Note that, like P, Q is also block tridiagonal since $F'F$ is block diagonal.

Great computational advantages can be obtained from the sparseness of Q. In particular, fast inversion algorithms can be used ensuring that samples from θ are quickly drawn. Thus, an alternative algorithm is given by

ALGORITHM $1^\dagger$ (*Block sampling*).
Exactly as Algorithm 1^* *but for replacement of* θ *sampling* (*step* $\mathbf{b}^*$) *by:*

$\mathbf{b}^\dagger$. *sample* $\theta^{(j)}$ *from its joint full conditional* $N(M, Q^{-1})$, *using a fast inversion algorithm.*

Similar comments carry over to models with dependence of k lags in the system equation (1.2). Instead of bandwidth of 3 $(= 1 + 2.1)$ blocks in Q, one would have bandwidth of $(1 + 2.k)$ blocks but algorithms are still fast, provided k is orders of magnitude smaller than T. Results in this direction are more common in the context of Markov random fields (MRF) used in spatial models (Gamerman et al., 2003) but can be equally applied in dynamic modeling settings. Connections between MRF and DLM are described in Section 4.3.

A comparison between the sampling schemes is shown in Figure 2. The computational advantages are clearly seen from this figure. Blocking θ turns the convergence issue into a much faster task. The computing time required for each iteration is higher but it is compensated by the smaller number of iterations required.

So far, we concentrated on θ sampling, with the hyperparameters V and W still sampled separately (from their full conditionals). It is possible however to sample all parameters in a single block. This task is achieved by use of

$$\pi(\theta, V, W|D_T) = \pi(\theta|V, W, D_T)\pi(V, W|D_T).$$

The joint density of all model parameters is known up to a proportionality constant and the distribution of $(\theta|V, W, D_T)$ was derived above. Therefore, the marginal density $\pi(V, W|D_T)$ is also known up to a proportionality constant. It does not have a known form and direct sampling becomes difficult. MCMC sampling can be used instead with the Metropolis–Hastings algorithm. Subsequent sampling of θ is done through its full conditional. This sampling strategy was discussed in Gamerman and Moreira (2002) and is summarized in Algorithm 2.

Algorithm 2 avoids MCMC convergence problems associated with the posterior correlation between θ and (V, W). Note that the algorithm uses $\pi(V, W|D_T)$ only in ratio form. Therefore, the unknown proportionality constant of this density is not required. The algorithm requires suitable choices of the proposal density q in **c1** to yield large enough values of α, thus ensuring computational efficiency. This may become a potentially important drawback, specially if the dimensionality of W becomes large. Usual choices of proposals are provided by a product of log random walk forms where $q(V, W|\theta^{(j-1)}, V^{(j-1)}, W^{(j-1)}) = q_1(V|\theta^{(j-1)}, V^{(j-1)})q_2(W|\theta^{(j-1)}, W^{(j-1)})$, with q_1 given by an inverse Gamma density centered around $V^{(j-1)}$ and q_2 given by an inverse Wishart density centered around $W^{(j-1)}$.

ALGORITHM 2 (*Joint sampling of all parameters*).
Exactly as Algorithm $1^\dagger$ *but for replacement of* (V, W) *sampling* (*step* **c**) *by:*

c1. *Sample* (V^*, W^*) *from* $q(V, W|\theta^{(j-1)}, V^{(j-1)}, W^{(j-1)})$;
c2. *set* $(V^{(j)}, W^{(j)}) = (V^*, W^*)$, *with probability* α *and* $(V^{(j-1)}, W^{(j-1)})$ *with probability* $1 - \alpha$, *where*

$$\alpha = \min\left\{1, \frac{\pi(V^*, W^*|D_T)}{\pi(V^{(j-1)}, W^{(j-1)}|D_T)} \frac{q(V^{(j-1)}, W^{(j-1)}|\theta^{(j-1)}, V^*, W^*)}{q(V^*, W^*|\theta^{(j-1)}, V^{(j-1)}, W^{(j-1)})}\right\}.$$

The case where the evolution disturbance matrix W is scaled by V can be easily accommodated within the framework here. Algorithms 1^*, $1^\dagger$ and 2 would sample (θ, V) jointly as now their full conditional is known to be in Normal-inverse Gamma form. Algorithm 2 would still consider sampling W via a proposal while Algorithms 1^* and $1^\dagger$ would sample W from its inverse Wishart full conditional.

The case where hyperparameters include also unknowns in the expression of F_t and G_t (examples of such models are given in Section 2) would require an additional step in the algorithms to handle sampling from these unknowns. Their full conditional will typically not be tractable for direct sampling and Metropolis proposals should be used. Unfortunately, the diversity of options available make it very hard to design multipurpose proposals that would fit any model. Again, usual choices for proposals are random walk forms centered around previous chain values on some suitable transformation leading to real-valued parameters and rescaled priors. Either way, some tuning of the precision of this proposal is needed to allow appropriate chain movement. This in turn, will hopefully lead to fast convergence.

2.2. Nonnormal models

This subsection considers the solution to full inference in nonnormal DLM by sampling. Presentation moves progressively away from normality, starting with (scale) mixture of normals, then moving to close approximation to scale mixture of normals and finally to exponential-family models.

Mixture of normals

The first case of nonnormality of the time series is associated with deviations from Gaussianity due to mixtures in the scale. In these cases, auxiliary variables ψ_t can be introduced such that $y_t|\psi_t, \theta, V \sim N(\mu_t, \psi_t V)$. The observation model is completed with specification of the mixing distribution for the scales ψ_t. These classes of models were studied in the context of DLM by Carlin et al. (1992) and include the important family of t-Student distribution obtained when ψ_t has inverse Gamma prior. This family is possibly the most common choice made by practitioners wanting to guard against marked observational departures from the standard or outliers.

An even simpler form of mixture of normals is provided by discrete mixture of normals where the observational density is given by $\sum_{i=1}^{k} p_i f_i(y|\mu_i, V_i)$ with the nonnegative weights p_i summing to 1 and $f_i(\cdot|\mu, V)$ denoting the $N(\mu, V)$ density. Defining mixture indicators z_t having multinomial distributions leads to normal models for $y_t|z_t$. An application of this setting can be presented in the context of Example 6. Kim et al. (1998) suggested that the $\log -\chi^2$ for the squared residual observation can be well approximated by a finite mixture of normals.

Another source of data nonnormality is provided by time series that can become normal after suitable transformations are performed to observations. Among normalizing transformations, the most common ones are provided by the Box–Cox family $g_\lambda(y) = (y^\lambda - 1)/\lambda$, for $\lambda \neq 0$ and $g_\lambda(y) = \log y$, for $\lambda = 0$. The model is completed with specification of a prior distribution for the λ_t's.

These three classes are qualitatively similar since they are examples of distributions that are conditionally normalized. Conditional on the additional parameters Ψ respectively given above by the ψ_t's, z_t's or λ_t's, the sampling schemes for all the other model parameters proceed as in the previous subsection. Additional sampling steps are required for the extra parameters.

Exponential-family models

The extensions above imply a MCMC solution that is based on extra steps in the sampling algorithms for normal DLMs. The observation models considered here in more detail are provided by the class of DGLM and do not fall into the categories above. The main problem of the previous section was the computational task of sampling θ efficiently but its full conditional was always directly available.

In the case of DGLM, the full conditional of θ does not fall into any recognizable form and can not be directly sampled from or integrated out. Not even the full conditional of θ_t is recognizable. Therefore, sampling from them is ruled out and proposals must be employed. Suitable proposals for θ with efficient correction terms are extremely hard to find due to the dimension of θ. Suitable proposals for θ_t are easier to find and can be constructed. However, they lead to slow convergence of the chain for the same reasons presented for normal models.

Shephard and Pitt (1997) and Knorr-Held (1999) suggested the use of random blocks containing some of the components of θ. Their proposals were based on normal approximations to the likelihood and random walk moves, respectively. Gamerman (1998) suggested the use of time-specific blocks but used a reparametrization in terms of the disturbances w_t's to avoid the strong correlation between θ_t's. The former can be rewritten as $w_t = \theta_t - G_t\theta_{t-1}$, $t = 2, \ldots, T$, in terms of the system parameters θ_t, with $w_1 = \theta_1$. The system parameters θ_t are easily recovered by the inverse relation

$$\theta_t = \sum_{l=1}^{t}\left(\prod_{k=1}^{t-l} G_{t-k+1}\right) w_l$$

for $t = 2, \ldots, n$ and $\theta_1 = w_1$. Note that $\pi(w|W, D_T) = \pi(\theta(w)|W, D_T)$ for $w = (w_1, \ldots, w_T)$, since the Jacobian of the transformation is 1. The sampling algorithm is given by

ALGORITHM 3 (*MCMC for DGLM based on state reparametrization*).

a. Initialization: *set initial values* $(\theta^{(0)}, W^{(0)})$ *and iteration counter* $j = 1$;
b. Sampling θ**:** $\theta^{(j)}$ *is sampled componentwise as follows*

1. *set* $t = 1$;
2. *Sample* w_t^* *from* $q_t(w_t|w_t^{(j-1)}, w_{-t}^{(j-1)}, W^{(j-1)}, D_T)$, *with* $w_{-t}^{(j-1)} = (w_1^{(j)}, \ldots, w_{t-1}^{(j)}, w_{t+1}^{(j-1)}, \ldots, w_T^{(j-1)})$;
3. *set* $w_t^{(j)} = w_t^*$ *with probability* α_t *and* $w_t^{(j)} = w_t^{(j-1)}$ *with probability* $1 - \alpha_t$, *where*

$$\alpha_t = \min\left\{1, \frac{\pi(w_t^*|w_{-t}^{(j)}, W^{(j-1)}, D_T)}{\pi(w_t^{(j-1)}|w_{-t}^{(j)}, W^{(j-1)}, D_T)} \times \frac{q_t(w_t^{(j-1)}|w_t^*, w_{-t}^{(j-1)}, W^{(j-1)}, D_T)}{q_t(w_t^*|w_t^{(j-1)}, w_{-t}^{(j-1)}, W^{(j-1)}, D_T)}\right\};$$

4. *reconstruct* $\theta_t^{(j)} = G_t\theta_{t-1}^{(j)} + w_t^{(j)}$;
5. *set* $t \to t + 1$ *and return to* ***b2***, *if* $t < T$;

c. Sampling W**:** $W^{(j)}$ *is sampled from its full conditional* $\pi(W|\theta^{(j)}, D_T)$;
d. Updating: *set* $j \to j + 1$ *and return to* ***b*** *until convergence.*

The full conditional densities of w_t required in the expression of α_t are simply obtained from the joint density $\pi(w|W, D_T)$ after eliminating terms not involving w_t and

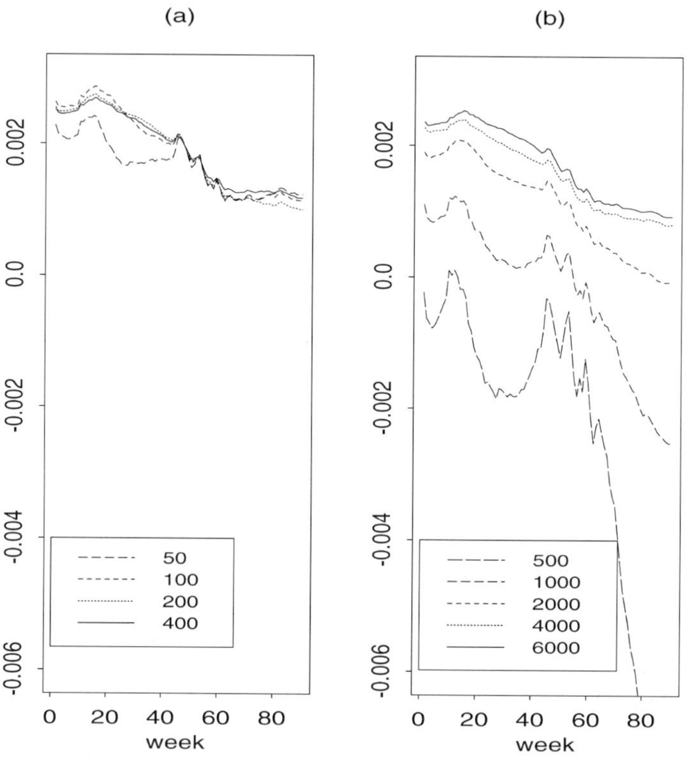

Fig. 3. Gamerman (1998) Average trajectory of a regression coefficient in 500 parallel chains with number of iterations for sampling from: (a) – w_t's; (b) – θ_t's. The data application is a dynamic logistic regression of expenditure on counts of advertising awareness.

are known up to a proportionality constant. As it appears in ratio form, knowledge of the constant is unnecessary. The proposal densities q_t used by Gamerman (1998) are based on the working variables used by Singh and Roberts (1992) for mode evaluation is DGLM. They are given by the (normal) full conditional of w_t in a DLM with same system equation and modified observation equation $\tilde{y}_t \sim N(F_t'\theta_t, \widetilde{V}_t)$, with $\tilde{y}_t = g(\mu_t) + g'(\mu_t)(y_t - \mu_t)$ and $\widetilde{V}_t = [g'(\mu_t)]^2 \operatorname{Var}(y_t|\theta_t)$. These variables are an extension of the working variables used in maximum likelihood evaluation of generalized linear models. Ferreira and Gamerman (2000) detail and illustrate the use of this sampling scheme.

The iterations are more costly here because of the reconstruction equations above but provide savings in computing time. Figure 3 shows a comparison between convergence with and without this reparametrization. It shows a significant improvement of the reparametrization in terms of number of iterations to convergence. The savings observed are comparable to the ones observed in Section 2.1 in the comparison between sampling θ_t's separately or in a single block.

3. Sequential Monte Carlo

Let us start this section with a fairly general dynamic model:

$$\text{Observational Equation: } p(y_t|\theta_t, \psi), \tag{3.1}$$

$$\text{System Equation: } p(\theta_t|\theta_{t-1}, \psi), \tag{3.2}$$

where y_t is the vector of observable time series, θ_t is the vector of unobservable state parameters and ψ is the vector of static parameters, sometimes referred to as the *hyper-parameters*. For the moment suppose that ψ is known and omitted from the notation. Later on we will show how to include ψ in the analysis. Therefore, the *evolution equation* at each time t is given by (1.5) while the *updating equation* at each time t is given by (1.7). Apart from the general normal DLM case (see Section 1), the filtering densities cannot be obtained analytically and numerical methods must be used.

The advent of MCMC methods generated a number of algorithms to sample from $p(\theta_1, \ldots, \theta_T|D_T)$, described in the previous section. Therefore, in principle, one could simply rerun the MCMC algorithm as data become available. However, in many practical situations, the cost of rerunning MCMC routines are prohibitive and real-time sequential algorithms play a crucial role in assessing $p(\theta_t|D_t)$. Doucet (2003) illustrates this issue pointing to a situation in wireless communication where you do not want to wait for a MCMC chain to converge in order to listen to your correspondent's next bit of conversation. Another common situation is when a financial analyst needs to forecast a few steps ahead, say for the next five hours, the volatilities for several assets that compound her portfolio.

Analytical approximations have been proposed in the literature. Kitagawa (1987), for instance, has introduced an algorithm where piecewise linear functions approximates the densities in both Eqs. (1.5) and (1.7). Similarly, Pole and West (1990) apply quadrature techniques to the class of conditionally conjugate models where one, or more, of the four elements that defines the general normal linear dynamic model is function of unknowns, say ψ. Quadrature techniques has recently reappeared in Bolviken and Storvik (2001). Both algorithms, based on linear splines and Gaussian quadratures, become computationally prohibitive as either the sample size or the dimension of the state vector increases.

The remainder of this section is dedicated to introducing several *particle filters*, as most sequential Monte Carlo algorithms are commonly referred to (Kitagawa, 1996). We start with the *bootstrap filter* or simply the *SIR filter*, in Section 3.1. Sequential importance sampling is also introduced, but immediately discarded because of an inherent degeneracy problem as t increases. In Section 3.2, Pitt and Shephard's (1999) extensively used *auxiliary particle filter* is introduced. Section 3.3 presents an extension of Pitt and Shephard's filter, proposed by Liu and West (2001), to deal with the sequential estimation of fixed parameters, ψ in Eqs. (3.1) and (3.2). Alternative schemes for parameter estimation are briefly discussed.

Before we proceed, let us introduce some notation that will facilitate algorithmic expositions. $\{(\theta_t^{(1)}, w_t^{(1)}), \ldots, (\theta_t^{(N)}, w_t^{(N)})\} \overset{a}{\sim} p(\theta_t|D_t)$ is used to denote that the probability density function, $p(\theta_t|D_t)$, of the continuous random variable, θ_t, is approximated by a discrete variable with random support. Therefore, if one is interested, for instance, in computing $E(g(\theta_t)|D_t)$, an approximation based on the set of points $\theta_t^{(1)}, \ldots, \theta_t^{(N)}$ is $\sum_{i=1}^N w_t^{(i)} g(\theta_t^{(i)})$.

3.1. SIR- and SIS-based filters

Simultaneous rediscovery of related approaches can be tracked back to Gordon et al. (1993), West (1993a, 1993b), and Kitagawa (1993, 1996) with the first becoming a seminal paper in the field of sequential Monte Carlo methods, at least for (Bayesian) statisticians. Doucet (2003) argues that Hetherington (1984) was the first to introduce, in physics, a multinomial resampling step. Gordon et al. (1993) use iterative sampling importance resampling (SIR, in short) (Rubin, 1988; Smith and Gelfand, 1993) in what turned out to be worldly known as the *bootstrap filter*. For consistency with the other filters, we will simply call it the *SIR filter*.

More specifically, if $\theta_{t-1}^{(1)}, \ldots, \theta_{t-1}^{(N)}$ represents a sample, commonly called *particles*, from $p(\theta_{t-1}|D_{t-1})$ a natural step is to use the evolution equation (3.2) to sample a new set of particles, $\tilde{\theta}_t^{(1)}, \ldots, \tilde{\theta}_t^{(N)}$. The new set of particles represent a sample from $p(\theta_t|D_{t-1})$, which is the state prior distribution at time t. Finally, the particles $\tilde{\theta}_t^{(i)}$ are resampled, for simplicity N times, with weights proportional to $p(y_t|\tilde{\theta}_t^{(i)})$ to produce a new set of particles $\theta_t^{(1)}, \ldots, \theta_t^{(M)}$ that are approximately distributed as $p(\theta_t|D_t)$. This is schematically represented in Algorithm 4.

The evolution and updating steps are also known as importance sampling and selection steps, respectively. It can be easily seen that $I_1 = \sum_{i=1}^N \omega_t^{(i)} g(\tilde{\theta}_t^{(i)})$ converges to $I = \int g(\theta_t)p(\theta_t|D_t)\, \mathrm{d}\theta_t$ as $N \to \infty$. Theoretically, the updating step in the previous algorithm is unnecessary and, in fact, $\mathrm{Var}(I_1) \leqslant \mathrm{Var}(I_2)$, where $I_2 = \sum_{i=1}^N g(\theta_t^{(i)})/N$ is also a consistent estimator of I (Geweke, 1989). Lopes et al. (1999) and Schmidt et al. (1999) investigate empirically the performance of both procedures in estimating fixed parameters in a class of nonlinear/nonnormal dynamic model.

ALGORITHM 4 (*SIR filter*).

Posterior at $t-1$**:** $\{(\theta_{t-1}^{(1)}, \frac{1}{N}), \ldots, (\theta_{t-1}^{(N)}, \frac{1}{N})\} \overset{a}{\sim} p(\theta_{t-1}|D_{t-1})$
Evolution: *For* $i = 1, \ldots, n$, *sample* $\tilde{\theta}_t^{(i)}$ *from* $p(\theta_t|\theta_{t-1}^{(i)})$
Weights: *For* $i = 1, \ldots, n$, *compute* $\omega_t^{(i)} \propto p(y_t|\tilde{\theta}_t^{(i)})$
Updating: *For* $i = 1, \ldots, n$, *sample* $\theta_t^{(i)}$ *from* $\{(\tilde{\theta}_t^{(1)}, \omega_t^{(i)}), \ldots, (\theta_t^{(N)}, \omega_t^{(i)})\}$
Posterior at t**:** $\{(\theta_t^{(1)}, \frac{1}{N}), \ldots, (\theta_t^{(N)}, \frac{1}{N})\} \overset{a}{\sim} p(\theta_t|D_t)$

If the updating step is deleted from the SIR filter, then sample $\theta_{t-1}^{(1)}, \ldots, \theta_{t-1}^{(N)}$ must be accompanied by the weights $\omega_{t-1}^{(1)}, \ldots, \omega_{t-1}^{(N)}$, and the previous algorithm becomes Algorithm 5.

ALGORITHM 5 (*SIS filter*).

Posterior at $t-1$**:** $\{(\theta_{t-1}^{(1)}, \omega_{t-1}^{(1)}), \ldots, (\theta_{t-1}^{(N)}, \omega_{t-1}^{(N)})\} \overset{a}{\sim} p(\theta_{t-1}|D_{t-1})$

Evolution: *For* $i = 1, \ldots, n$, *sample* $\theta_t^{(i)}$ *from* $p(\theta_t|\theta_{t-1}^{(i)})$
Weights: *For* $i = 1, \ldots, n$, *compute* $\omega_t^{(i)} \propto \omega_{t-1}^{(i)} p(y_t|\theta_t^{(i)})$
Posterior at t: $\{(\theta_t^{(1)}, \omega_t^{(1)}), \ldots, (\theta_t^{(N)}, \omega_t^{(N)})\} \overset{a}{\sim} p(\theta_t|D_t)$

At any given time t, the importance function in the SIS filter is the prior distribution for the whole state vector up to time t, i.e., $p(\theta_0)\prod_{j=1}^{t} p(\theta_j|\theta_{j-1})$. The dimension of $(\theta_1, \ldots, \theta_t)$ increases with t and, as it is well known, importance sampling techniques become practically infeasible even for moderately large t (Geweke, 1989). It is not unusual, after a few step in the SIS algorithm, to wind up with only a handful of particles with nonnegligible weights. One of the neatest aspects of the SIR filter is that at each time point only particles with high weights are kept, or using *genetic algorithm* terminology, at each generation only strong strings are selected. See Higuchi (1997, 2001) for further comparisons between *Monte Carlo filters*, as he calls sequential Monte Carlo methods, and genetic algorithms.

Finally, since at any given time t what we are effectively trying to do is to sample from the posterior distribution, $p(\theta_t|D_t)$, rejection sampling and MCMC methods could be alternatively used. Their main drawback is the mostly invariable need of evaluation the prior distribution, $p(\theta_t|D_{t-1})$, which can be troublesome. Müller (1991, 1992) proposes a Metropolis algorithm where $p(\theta_t|D_{t-1})$ is reconstructed by a mixture of Dirichlet process model based on the prior sample, $(\tilde{\theta}_t^{(1)}, \ldots, \tilde{\theta}_t^{(N)})$. West (1993a, 1993b) uses mixture of multivariate normals to reconstructed the prior and posterior distributions of the state vector at each time.

Below we introduce yet another filter, the *auxiliary particle filter*, introduced by Pitt and Shephard (1999) to tackle the main weakness of the SIR filter, which is sampling from the prior, $p(\theta_t|D_t)$ and perform poorly when the next observation, y_t, is on the tail of $p(y_t|\theta_t)$.

3.2. Auxiliary particle filter

Assuming that $\{(\theta_{t-1}^{(1)}, \omega_{t-1}^{(1)}), \ldots, (\theta_{t-1}^{(N)}, \omega_{t-1}^{(N)})\} \overset{a}{\sim} p(\theta_{t-1}|D_{t-1})$, a natural Monte Carlo approximation (as $M \to \infty$) for the prior (1.5) is

$$\hat{p}(\theta_t|D_{t-1}) = \sum_{j=1}^{N} p\big(\theta_t|\theta_{t-1}^{(j)}\big) w_{t-1}^{(j)} \tag{3.3}$$

which, following Pitt and Shephard's (1999) terminology, is called the *empirical prediction density*. Combining this approximate prior with the observation equation produces, by Bayes' theorem, the following approximation for the state space vector posterior distribution at time t is

$$\begin{aligned}\hat{p}(\theta_t|D_t) &\propto p(y_t|\theta_t)\sum_{j=1}^{N} p\big(\theta_t|\theta_{t-1}^{(j)}\big) w_{t-1}^{(j)} \\ &\propto \sum_{j=1}^{N} p(y_t|\theta_t) p\big(\theta_t|\theta_{t-1}^{(j)}\big) w_{t-1}^{(j)}\end{aligned} \tag{3.4}$$

the *empirical filtering density* according to Pitt and Shephard (1999). A sampling scheme from the approximate posterior (3.4) is needed in order to complete the evolution/update cycle, very much like the SIR and the SIS filters. However, as it was discussed previously, the SIR method becomes ineffective either when the prior is relatively diffuse or the likelihood is highly informative.

ALGORITHM 6 (*Auxiliary particle filter*).

Posterior at $t-1$**:** $\{(\theta_{t-1}^{(1)}, \omega_{t-1}^{(1)}), \ldots, (\theta_{t-1}^{(N)}, \omega_{t-1}^{(N)})\} \overset{a}{\sim} p(\theta_{t-1}|D_{t-1})$
Sampling (k, θ_t)**:** *For* $i = 1, \ldots, N$

Indicator: *sample* k^i *such that* $Pr(k^i = k) \propto p(y_t|\boldsymbol{\mu}_t^{(k)})w_{t-1}^{(k)}$

Evolution: *sample* $\theta_t^{(i)}$ *from* $p(\theta_t|\theta_{t-1}^{(k^i)})$

Weights: *compute* $w_t^{(i)} \propto p(y_t|\theta_t^{(i)})/p(y_t|\boldsymbol{\mu}_t^{(k^j)})$

Posterior at t**:** $\{(\theta_t^{(1)}, \omega_t^{(1)}), \ldots, (\theta_t^{(N)}, \omega_t^{(N)})\} \overset{a}{\sim} p(\theta_t|D_t)$

Pitt and Shephard improve on particle filter methods by addressing to practical and important issues: (i) efficiently sampling from the approximate posterior distribution (Eq. (3.4)) and (ii) efficiently approximating tails' behavior of the approximate prior (Eq. (3.3)). They developed a generic filtering algorithm that is currently well known as *auxiliary particle filtering*. The basic feature of a auxiliary particle filter is to take advantage of the mixture of densities (3.4) to obtain draws from $p(\theta_t|D_t)$ by introducing latent indicator variables to identify the terms in the mixture (an idea commonly used in mixture modeling, Diebolt and Robert, 1994). In other words, if (θ_t, k) is sampled from $p(\theta_t, k) \propto p(y_t|\theta_t)p(\theta_t|\theta_{t-1}^{(k)})w_{t-1}^{(k)}$, the resulting θ_t is a sample from (3.4). If $\{(\theta_{t-1}^{(1)}, \omega_{t-1}^{(1)}), \ldots, (\theta_{t-1}^{(N)}, \omega_{t-1}^{(N)})\} \overset{a}{\sim} p(\theta_{t-1}|D_{t-1})$, then for $i = 1, \ldots, N$, sample $(\theta_t^{(i)}, k^i)$ from $g(\theta_t, k|D_t)$ and compute weights $w_t^{(i)} \propto p(y_t|\theta_t^{(i)})p(\theta_t^{(i)}|\theta_{t-1}^{(k^i)})g(\theta_t^{(i)}, k^i|D_t)$. By following these steps, which are essentially SIR steps, $\{(\theta_t^{(1)}, \omega_t^{(1)}), \ldots, (\theta_t^{(N)}, \omega_t^{(N)})\} \overset{a}{\sim} p(\theta_t|D_t)$.

Pitt and Shephard use $g(\theta_t, k|D_t) \propto p(y_t|\boldsymbol{\mu}_t^{(k)})p(\theta_t|\theta_{t-1}^{(k)})w_{t-1}^{(k)}$, as a generic importance function, where $\boldsymbol{\mu}_t^{(k)}$ is an estimate of θ_t given $\theta_{t-1}^{(k)}$, for instance the mean, the mode or any other likely value from $p(\theta_t|\theta_{t-1})$), such that $g(k|D_t) \propto p(y_t|\boldsymbol{\mu}_t^{(k)})w_{t-1}^{(k)}$.

Choosing $g(\cdot)$ is a nontrivial task and this is inherent to all Monte Carlo methods (SIR, adaptive, MCMC, etc.). Pitt and Shephard argue that the simulation algorithm will favor particles with larger predictive likelihoods. By doing so, the resampling step will have lower computational cost and will improve on statistical efficiency of the procedure.

3.3. *Parameter estimation and sequential Monte Carlo*

In this section the uncertainty about the hyperparameter ψ is incorporated in the filtering analysis. The main algorithm we present here is due to Liu and West (2001), who extend Pitt and Shephard's auxiliary particle filter. They combine kernel density estimation techniques with artificial parameter evolution and propose a novel algorithm to sequentially treat fixed parameters in general dynamic model settings.

Initially, let (1.7) be rewritten as

$$p(\theta_t, \psi|D_t) \propto p(y_t|\theta_t, \psi)p(\theta_t|\psi, D_{t-1})p(\psi|D_{t-1}), \tag{3.5}$$

where the uncertainty about ψ is assessed by adding the term $p(\psi|D_{t-1})$. As before, and conditional on ψ, the evolution density $p(\theta_t|\psi, D_{t-1})$ can be approximated by

$$\hat{p}(\theta_t|\psi, D_{t-1}) = \sum_{j=1}^{N} p\big(\theta_t|\psi, \theta_{t-1}^{(j)}\big)w_{t-1}^{(j)}, \tag{3.6}$$

where $\{(\theta_{t-1}^{(1)}, \psi_{t-1}^{(1)}, w_{t-1}^{(1)}), \ldots, (\theta_{t-1}^{(N)}, \psi_{t-1}^{(N)}, w_{t-1}^{(N)})\} \overset{a}{\sim} p(\theta_{t-1}, \psi|D_{t-1})$.

A natural solution, firstly explored by Gordon et al. (1993), is to pretend that the fixed parameters are states in the dynamic modeling, for instance, by adding small random disturbances to artificial evolutions, and proceed the analysis with the auxiliary particle filters presented in the previous section. Such artificial evolution reduces the sample degeneracy problems. However, it imputes unnecessary uncertainty into the model and also creates artificial loss of information resulting on overdispersion of the posterior distributions.

Liu and West (2001) reinterpret Gordon, Salmond, and Smith's artificial parameter evolution idea and combine it with West's kernel smoothing techniques. Approximations for $p(\psi|D_t)$ based on mixtures of multivariate normals were suggested by West (1993b),

$$\hat{p}(\psi|D_{t-1}) = \sum_{j=1}^{M} N\big(\psi|\boldsymbol{m}_{t-1}^{(j)}, h^2\boldsymbol{V}_{t-1}\big)w_{t-1}^{(j)}, \tag{3.7}$$

where h is a smoothing parameter, $\boldsymbol{V}_{t-1} = \mathrm{Var}(\psi|D_{t-1})$, and $\boldsymbol{\mu}_{t-1}^{(j)}$ are the locations of the components of the mixture. In standard kernel methods, $\boldsymbol{\mu}_{t-1}^{(j)} = \psi_{t-1}^{(j)}$. Also, for large M, it is also common practice to have h as a decreasing function of M. West (1993b) introduces a shrinkage rule for the locations,

$$\boldsymbol{m}_{t-1}^{(j)} = a\psi_{t-1}^{(j)} + (1-a)\bar{\boldsymbol{\psi}}_{t-1}, \tag{3.8}$$

where $\bar{\boldsymbol{\psi}}_{t-1} = E(\psi_{t-1}|D_{t-1})$. The variance of the resulting mixture of normals is $(a^2+h^2)\boldsymbol{V}_{t-1}$, which is always larger then $\boldsymbol{V}_{t-1}$ for $a^2+h^2 > 1$. West (1993b) suggests using $a^2 = 1-h^2$ to guarantee that the correct variance is used in the approximation, crucial in sequential schemes. Liu and West (2001) show that if δ is the discount factor used in Gordon, Salmond, and Smith's artificial evolution method, then defining $h^2 = 1 - [(3\delta-1)/2\delta]^2$ produces an algorithm that links the kernel estimation (3.6) with the shrinkage idea (3.8).

Liu and West (2001) apply their novel strategy in a couple of situations, including a simpler version of Aguilar and West's (2000) dynamic factor model for financial time series, which are similar to models previously described in this chapter. Among other empirical findings, they argue that MCMC methods should be combined with sequential algorithms in real applications. They show that, when performed for longer periods of time the filtering algorithm starts to deteriorate and diverges from the "gold standard" MCMC results.

ALGORITHM 7 (*Liu–West filter*).

At $t-1$**:** $\{(\theta_{t-1}^{(1)}, \psi_{t-1}^{(1)}, w_{t-1}^{(1)}), \ldots, (\theta_{t-1}^{(N)}, \psi_{t-1}^{(N)}, w_{t-1}^{(N)})\} \overset{a}{\sim} p(\theta_{t-1}, \psi | D_{t-1})$.

Kernel: $\boldsymbol{V}_{t-1} = \sum_{j=1}^{M} (\psi_{t-1}^{(j)} - \bar{\boldsymbol{\psi}}_{t-1})(\psi_{t-1}^{(j)} - \bar{\boldsymbol{\psi}}_{t-1})' w_{t-1}^{(j)}$ *and* $\bar{\boldsymbol{\psi}}_{t-1} = \sum_{j=1}^{M} \psi_{t-1}^{(j)} w_{t-1}^{(j)}$

Shrinkage: *For* $j = 1, \ldots, N$, $\boldsymbol{\mu}_t^{(j)} = E(\theta_t | \theta_{t-1}^{(j)}, \psi^{(j)})$ *and* $\boldsymbol{m}_{t-1}^{(j)} = a\psi_{t-1}^{(j)} + (1-a)\bar{\boldsymbol{\psi}}_{t-1}$

Sampling (k, θ_t, ψ)**:** *For* $i = 1, \ldots, N$

Indicator: *Sample* k^i *such that* $Pr(k^i = k) \propto w_{t-1}^{(k)} p(y_t | \boldsymbol{\mu}_t^{(k)}, \boldsymbol{m}_{t-1}^{(k)})$

Parameter: *Sample* $\psi_t^{(i)}$ *from* $N(\boldsymbol{m}_{t-1}^{(k^i)}; h^2 \boldsymbol{V}_{t-1})$

State: *Sample* $\theta_t^{(i)}$ *from the* $p(\theta_t | \theta_{t-1}^{(k^i)}, \psi_t^{(i)})$

Weights: $w_t^{(i)} \propto p(y_t | \theta_t^{(i)}, \psi_t^{(i)}) / p(y_t | \boldsymbol{\mu}_t^{(k^i)}, \boldsymbol{m}_{t-1}^{(k^i)})$

At t**:** $\{(\theta_t^{(1)}, \psi_t^{(1)}, w_t^{(1)}), \ldots, (\theta_t^{(N)}, \psi_t^{(N)}, w_t^{(N)})\} \overset{a}{\sim} p(\theta_t, \psi | D_t)$.

Computing predictive densities

Regardless of which filter is used, the predictive density

$$p(y_t | D_{t-1}) = \int p(y_t | \theta_t, \psi) p(\theta_t, \psi | D_{t-1}) \, d\theta_t \, d\psi$$

can be approximated by

$$\hat{p}(y_t | D_{t-1}) = \frac{1}{N} \sum_{j=1}^{N} p\big(y_t | \tilde{\theta}_t^{(l)}, \tilde{\psi}_t^{(l)}\big),$$

where $\{(\tilde{\theta}_t^{(1)}, \tilde{\psi}_t^{(1)}, \frac{1}{N}), \ldots, (\tilde{\theta}_t^{(N)}, \tilde{\psi}_t^{(N)}, \frac{1}{N})\} \overset{a}{\sim} p(\theta_t, \psi | D_{t-1})$, the joint prior distribution at time t of the states and hyperparameters. The pair $(\tilde{\theta}_t^{(i)}, \tilde{\psi}_t^{(i)})$ can be sampled from $p(\theta_t, \psi | D_{t-1})$ by following three simple steps: (i) sample k^i such that $\Pr(k^i = j) \propto w_t^{(j)}$, (ii) make $\tilde{\psi}_t^{(i)} = \psi_{t-1}^{(k^i)}$, and (iii) sample $\tilde{\theta}_t^{(i)}$ from the evolution equation $p(\theta_t | \theta_{t-1}^{(k^i)}, \tilde{\psi}_t^{(i)})$. If one is interested, for instance, in computing $E(g(y_t) | D_{t-1})$, an approximation based on $\{(\tilde{\theta}_t^{(1)}, \tilde{\psi}_t^{(1)}, \frac{1}{N}), \ldots, (\tilde{\theta}_t^{(N)}, \tilde{\psi}_t^{(N)}, \frac{1}{N})\}$ is $\widehat{E}(g(y_t) | D_{t-1}) = \frac{1}{N} \sum_{i=1}^{N} E(g(y_t) | \tilde{\theta}_t^{(i)}, \tilde{\psi}_t^{(i)})$. These approximations are extremely useful for sequential model comparison.

EXAMPLE 3.1. Lopes and Marinho (2002) studied the Markov switching stochastic volatility (MSSV) model, where $y_t = e^{\lambda_t/2} u_t$ and $\lambda_t = \alpha_{S_t} + \phi\lambda_{t-1} + \varpi_t$, for $u_t \sim N(0, 1)$ and $\varpi_t \sim N(0, \sigma^2)$ serially and temporally uncorrelated. S_t followed a homogeneous k-state first-order Markov process with transition matrix P where

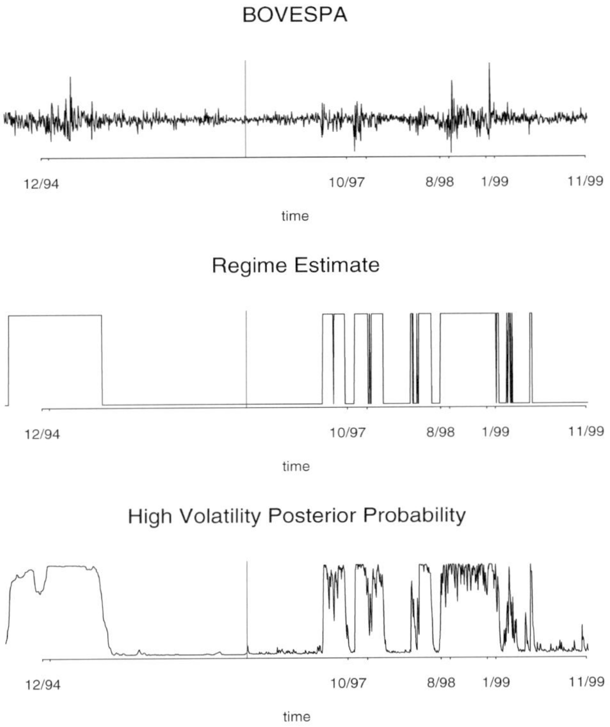

Fig. 4. Brazilian Bovespa: time series (top), regimes posterior mean (middle) and posterior mean of probability that the series in the "high volatility" regime at any given time t.

$p_{ij} = P[S_t = j|S_{t-1} = i]$, for $i, j = 1, \ldots, k$. The $k^2 + 2$-dimensional parameter vector is $\psi = (\alpha_1, \ldots, \alpha_k, \phi, \sigma^2, p_{11}, \ldots, p_{1,k-1}, \ldots, p_{k1}, \ldots, p_{k,k-1})$, while the state vector is $\theta_t = (S_t, \lambda_t)$. Figure 4 exhibits the sequential Monte Carlo estimates of $p(y_t|D_{t-1})$ for all t when y_t is the Bovespa stock index (São Paulo Stock Exchange). An extension of this model appears in Lopes (2002) who uses the MSSV structure to model common factor log-volatilities in a factor stochastic volatility model (Lopes, 2000).

3.4. *Recent developments*

There have been several theoretical developments that try to explain the behavior of sequential Monte Carlo filters as the number of particles, M in our notation, tends to infinity (see, for instance, Berzuini et al., 1997; Crisan and Doucet, 2002; Künsch, 2001). However, little is formally known about particle filter behavior when M is kept fixed and the sample size increases.

Alternative approaches to deal with parameter estimation in sequential particle filtering is already flourishing. Storvik (2002) proposed histogram-like filters that take advantage of sufficient statistics present in the state equations of several existing dynamic models. Polson et al. (2002) generalizes Storvik (2002) by using a fixed-lag approximation that simplifies the computation and improves the performance of the filter both in terms of estimation and computational speed.

Doucet and Tadić (2003) combines gradient algorithms with particle filters to estimate, by recursive maximum likelihood, the fixed parameters, ψ in our notation, in a fairly broad class of dynamic models.

4. Extensions

In the last decade, there have been new areas of statistical research derived from the extension of dynamic linear models to topologies other than time as, for example, space-time, and dyadic and quad trees. Moreover, there has been increasing understanding of the connections between DLMs and Gaussian Markov random fields. These developments have pointed out to exciting new applications to areas such as finance, epidemiology, climatology and image analysis.

In this final section of the chapter, the development of dynamic models for spatio-temporal processes is addressed in Section 4.1, the development of multi-scale models for processes living at different levels of resolution is addressed in Section 4.2, and some connections between Markov random fields and DLMs are addressed in Section 4.3.

4.1. Dynamic spatio-temporal models

There has been increasing interest in the statistics community on the development of models for spatio-temporal processes. Within the statistical community, initial efforts to analyze spatio-temporal data include the work of Haslett and Raftery (1989) on the assessment of wind power resource in Ireland and Handcock and Wallis (1994) on the analysis of meteorological fields. Cressie and Huang (1999) have worked on the development of nonseparable spatio-temporal variograms.

Ghil et al. (1981) were the first to consider the use of state space models to spatio-temporal data. Let $y(s_1, t_1), \dots, y(s_N, t_T)$ be data obtained from a spatially continuous process $y(s, t)$ where s belongs to some spatial domain R^2 or R^3 and t indexes discrete times. Let $y_t = (y(s_1, t), \dots, y(s_N, t))$ be the vectorized observed spatial data at time t. A dynamic linear spatio-temporal model can be written as (1.1)–(1.2), where the interpretation of the different elements of the model is analogous to the interpretation in traditional multivariate dynamic linear models. θ_t is the latent vectorized spatio-temporal process. F_t is the matrix that connects the latent process with the observations. ε_t is an idiosyncratic random error at time t. G_t describes the evolution of the latent spatio-temporal process through time. ω_t is a random error with effect not only at time t but also at subsequent times. The matrices V_t and W_t play the very important role of describing the spatial dependence of the process at the observation and state space levels. Eqs. (1.1) and (1.2) define a very flexible class of models for spatio-temporal processes, the key feature for the successful application of these models being the specification of the latent process θ_t and of the matrices F_t, G_t, V_t and W_t.

In the simplest dynamic linear spatio-temporal model, F_t and G_t are identity matrices, that is, θ_t is the level of the observed process and follows a random walk process through time. More interesting cases include some type of dimension reduction from the observed field y_t to the latent process θ_t. When there is dimension reduction, the latent process has much smaller dimension than the observations, and the matrix F_t makes the connection between them. This dimension reduction can be achieved through some type of factor modeling or through some type of spatial aggregation. Other interesting spatio-temporal dynamic models use the matrix G_t to incorporate expert knowledge of the physical characteristics of the process under study (e.g., Wikle et al., 2001).

The specification of V_t and W_t as covariance matrices of spatial processes is quite a challenge because spatio-temporal datasets are usually large and the computations necessary for the estimation of the parameters of the model become prohibitive. Thus, specifying V_t and W_t using geostatistical models such as the Matérn class is unfeasible because of the computational burden. An alternative approach that is being investigated by one of the authors is the use of Markov random fields in the specification of V_t and W_t (Vivar-Rojas and Ferreira, 2003).

Other alternative approaches that reduce the computational burden include some type of dimension reduction from the observations to the latent process through the matrix F_t. In this direction, the literature includes approaches based on principal components and factor analysis (Cressie, 1994; Goodall and Mardia, 1994; Mardia et al., 1998; Wikle and Cressie, 1999). In the other hand, dimension reduction based on multiscale ideas is being developed by Johannesson et al. (2003). As the approaches based on principal components and factor analysis are at this time better established, let us now describe them in more detail. In particular, let us review the quite general approach proposed by Wikle and Cressie (1999). Their model is the following:

$$y(s,t) = z(s,t) + \varepsilon(s,t), \tag{4.1}$$

$$z(s,t) = \theta(s,t) + \gamma(s,t), \tag{4.2}$$

$$\theta(s,t) = \int w_s(u)\theta(u,t-1)\,\mathrm{d}u + \eta(s,t), \tag{4.3}$$

where $y(s,t)$ are data from a continuous spatial process observed at discrete times, $z(s,t)$ is a nonobservable smooth process, $\varepsilon(s,t)$ is the measurement error, $\gamma(s,t)$ is a variance component, $\theta(s,t)$ is the latent spatio-temporal process that evolves through time, $w_s(u)$ is an interaction function between site s and its *neighbors* through time, and $\eta(s,t)$ are independent Gaussian errors. They assume that $z(s,t)$ can be decomposed into K dominant components $z(s,t) = \sum_{k=1}^{K} \phi_k(s)a_k(t)$ where $a_k(\cdot)$, $k = 1,\dots,K$, are zero mean time series and $\phi_1(\cdot), \phi_2(\cdot), \dots$ are complete and orthonormal basis functions. Moreover, taking advantage of the completeness of the ϕ's, they expand the interaction function as $w_s(u) = \sum_{l=1}^{\infty} b_l(s)\phi_l(u)$. In addition, using the orthonormality property and after some algebraic manipulation, they arrive at a DLM-like formulation:

$$y(s,t) = \phi(s)'a(t) + \gamma(s,t) + \varepsilon(s,t), \tag{4.4}$$

$$a(t) = Ha(t-1) + J\eta(t), \tag{4.5}$$

where H and J are matrices that depend on the ϕ's and the b's. The identifiability issue on the roles of $\varepsilon(s,t)$ and $\gamma(s,t)$ is resolved by assuming that $\varepsilon(s,t)$ represents a nugget effect and $\gamma(\cdot,t)$ follows a L_2-continuous random field process. In order to save computational time, they propose an empirical Bayes procedure to estimate the unknown matrices and then use the Kalman filter to estimate the state space parameters. See Wikle and Cressie (1999) for more details.

Wikle et al. (2001) propose a dynamic spatio-temporal model to tropical ocean surface winds. They incorporate scientific knowledge about tropical winds into the model through the evolution matrix G_t. Moreover, they take a hierarchical approach to incorporate two different types of data: (1) satellite data obtained with a scatterometer; (2) wind fields generated by global scale numerical weather prediction models from sparse in situ observations. Motivated also by the combination of data from different sources, Brown et al. (2001) develop a space-time model for the calibration of radar rainfall data. Another example of incorporation of scientific knowledge in a dynamic spatio-temporal model is given by Wikle (2003), where a diffusion process is included in the evolution equation to model the spread of a species population.

Wikle (2003) proposes a class of dynamic spatio-temporal models with evolution equation based on kernel convolutions.

Stroud et al. (2001) propose a locally weighted mixture of linear regressions, allowing the regression surfaces to change over time. Using a similar idea, Huerta et al. (2004) present a nice application of spatio-temporal models to the analysis of ozone levels, with F_t being regressors and θ_t temporally varying regressor coefficients.

These ideas involve either spatially or time varying regression coefficients. Banerjee et al. (2003) propose the use of spatially varying coefficients evolving in time according to a multivariate random walk process while making use of general forms for the evolution disturbance processes through coregionalization ideas. Related to that work, Gamerman (2002) considers spatio-temporally varying coefficients and also nonnormality, irregularities in the process of data collection and structural changes in the spatio-temporal process.

Dynamic spatio-temporal modeling is an area of active research, and we expect several important new developments on the subject in the following years.

4.2. Multi-scale modeling

Multi-scale modeling has mainly appeared in the engineering literature. Basseville et al. (1992) introduced isotropic multi-scale models defined on dyadic trees using a definition analogous to the one of autoregressive time series models. Chou et al. (1994) introduced the state-space representation of multi-scale models as an extension of the dynamic linear models, inference being efficiently carried out with a variant of the Kalman filter. Basseville et al. (1992) and Chou et al. (1994) define the model from coarser to finer levels and assume that the nodes of a given level are conditionally independent given the immediate coarser level. Consider a dyadic or quad tree with a corresponding latent process denoted by θ and observations y. Denote by $t = (m, n)$ the index of the nth node of the mth scale and by $\Delta(t-1)$ its parent node. The model proposed by Chou et al. (1994) can be written in a DLM formulation as follows:

$$y_t = F_t\theta_t + v_t, \tag{4.6}$$

$$\theta_t = G_t\theta_{\Delta(t-1)} + w_t, \tag{4.7}$$

where as usual w_t and v_t are independent zero mean processes with covariance matrices W_t and V_t respectively. The matrix F_t connects the state process θ_t to the observation y_t. The matrix G_t relates the tth node of the tree with its parent.

If the matrices F_t, G_t, W_t and V_t are known, then the estimation of the state parameters can be performed with a Kalman filter-like algorithm with some important modifications. First, while for processes on time the algorithm is composed by a filtering step forward in time and a smoothing step backward in time, the algorithm for processes on trees is composed by a filtering sweep from finer to coarser levels followed by a coarser to finer levels smoothing sweep. Second, the tree structure of the model and thus the algorithm have a pyramidal structure that lends itself to efficient parallel implementation.

While the assumption of conditionally independence of the nodes of a given level given the immediate coarser level leads to an efficient estimation algorithm, it also leads to blocky behavior, as pointed out by Irving et al. (1997). Recently, Ferreira (2002) has proposed a new class of multi-scale random fields also defined from coarser to finer levels, but assuming that conditional on its neighbors at the same resolution level and on its parent a node is independent of the remaining nodes at the same level. Full Bayesian analysis was developed with the help of Markov chain Monte Carlo techniques and the novel multi-scale framework was applied to the estimation of multi-scale permeability fields.

Other developments related to multi-scale models include: extension of the multi-scale models of Willsky by allowing arbitrary numbers of children and parents for each node (Huang and Cressie, 1997); multi-scale models for discrete valued latent processes considered on quadtrees (Laferté et al., 2000) and on pyramidal structures (Bouman and Shapiro, 1994; Kato et al., 1996a, 1996b), in the context of image segmentation. Johannesson and Cressie (2003) have considered the modeling and estimation of variances and covariances for multi-scale spatial models.

4.3. Connections between Gaussian Markov random fields and DLMs

Gaussian Markov random fields (GMRF) can be seen as generalizations of dynamic linear models to two or more dimensions. Even more interesting, Lavine (1999) pointed out that a Gaussian Markov random field prior on a lattice is the same as the posterior distribution of the state parameters of a particular dynamic linear model updated with special observations. Besides allowing fast computation of likelihoods, this idea leads to a fast exact simulation algorithm for Gaussian Markov random fields. For first-order neighborhood structures, this simulation algorithm is as efficient as the one proposed by Rue (2001) and is much more efficient than the Gibbs sampler simulation of GMRFs. For general discussion of Markov Chain Monte Carlo techniques and the Gibbs sampler in particular, see Gamerman (1997). This chapter is concluded with presentation of Lavine's idea.

Let us consider a first-order isotropic Gaussian Markov random field on a finite two-dimensional lattice (Besag, 1974). Let θ_{ij} be the latent process at site (i, j), $i = 1, \dots, I$, $j = 1, \dots, J$. Let $\theta_i = (\theta_{i1}, \dots, \theta_{iJ})'$ be the latent process at row i. Note that because of the Markovian property, θ_{i-1} and θ_{i+1} are conditionally independent given θ_i. This statement can be made more explicit by considering the system equation

$$\theta_i = \theta_{i-1} + w_i, \quad w_i \sim N(0, \psi^{-1} I), \tag{4.8}$$

where ψ is the partial autocovariance of neighbor sites. Together with a flat prior for θ_1, this system equation takes care of the interactions between neighbors in the same column.

In order to take care of the interactions between neighbors in the same row, Lavine (1999) considers pseudo-observations $Y = \{Y_{i,j}: i = 1, \ldots, I; j = 1, \ldots, J-1\}$, $Y_i = (Y_{i1}, \ldots, Y_{iJ})'$ and the observation equation:

$$Y_i = F\theta_i + v_i, \quad v_i \sim N(0, \psi^{-1} I), \tag{4.9}$$

where

$$F = \begin{pmatrix} 1 & -1 & 0 & \cdots & 0 \\ 0 & 1 & \ddots & \ddots & \vdots \\ \vdots & \ddots & \ddots & -1 & 0 \\ 0 & \cdots & 0 & 1 & -1 \end{pmatrix},$$

that is, $E(Y_{ij}|\theta) = \theta_{ij} - \theta_{i,j+1}$.

Taking the pseudo-observations to be equal to zero, then the posterior distribution of θ in the DLM defined by Eq. (4.8) and (4.9) is the same as the prior distribution of the considered GMRF.

Acknowledgements

The first, second and fourth authors wish to thank the financial support of research grants from CNPq. The authors thank Romy Rodriguez and Leonardo Bastos for the computations leading to Figure 1 and 2, respectively.

References

Aguilar, O., West, M. (2000). Bayesian dynamic factor models and variance matrix discounting for portfolio allocation. *J. Business Econom. Statist.* **18**, 338–357.

Ameen, J.R.M., Harisson, P. (1985). Normal discount Bayesian models. In: Bernardo, D.M.H.L.D.V., Smith, A.F.M. (Eds.), *Bayesian Statistics, vol. 2*. North-Holland and Valencia University Press.

Ameen, J.R.M., Harrison, P.J. (1983). Discount Bayesian models. In: Anderson, O. (Ed.), *Time Series Analysis: Theory and Practice, vol. 5*. North-Holland.

Anderson, B.O.O., Moore, V.B. (1979). *Optimal Filtering*. Prentice-Hall, Englewood Cliffs.

Banerjee, S., Gamerman, D., Gelfand, A.E. (2003). Spatial process modelling for univariate and multivariate dynamic spatial data. Technical Report, Department of Statistical Methods, Federal University of Rio de Janeiro.

Barbosa, E.P., Migon, H.S. (1996). Longitudinal data analysis of animal growth via multivariate dynamic models. *Comm. Statist. Simulation* **25**, 369–380.

Basseville, M., Benveniste, A., Willsky, A.S. (1992). Multiscale autoregressive processes, part i: Schur–Levinson parameterizations. *IEEE Trans. Signal Process.* **40**, 1915–1934.

Berzuini, C., Best, N., Gilks, W., Larizza, C. (1997). Dynamic conditional independence models and Markov chain Monte Carlo methods. *J. Amer. Statist. Assoc.* **92**, 1403–1412.

Besag, J. (1974). Spatial interaction and the statistical analysis of lattice systems (with discussion). *J. Roy. Statist. Soc., Ser. B* **36**, 192–236.

Bolviken, E., Storvik, G. (2001). Deterministic and stochastic particle filters in state-space models. In: Doucet, A., de Freitas, N., Gordon, N. (Eds.), *Sequential Monte Carlo Methods in Practice*. Springer-Verlag, New York.

Bouman, C.A., Shapiro, M. (1994). A multiscale random field model for Bayesian image segmentation. *IEEE Trans. Image Process.* **3** (2), 162–177.

Box, G., Jenkins, G. (1976). *Time Series Analysis: Forecasting and Control*, second ed. Holden-Day, San Francisco.

Brown, P.E., Diggle, P.J., Lord, M.E., Young, P.C. (2001). Space-time calibration of radar-rainfall data. *J. Roy. Statist. Soc., Ser. C (Applied Statistics)* **50**, 221–241.

Carlin, B.P., Polson, N.G., Stoffer, D.S. (1992). A Monte Carlo approach to nonnormal and nonlinear state-space modeling. *J. Amer. Statist. Assoc.* **87**, 493–500.

Carter, C., Kohn, R. (1994). On Gibbs sampling for state space models. *Biometrika* **81**, 541–553.

Chou, K.C., Willsky, A.S., Benveniste, A. (1994). Multiscale recursive estimation, data fusion, and regularization. *IEEE Trans. Automat. Control* **39**, 464–478.

Cressie, N. (1994). Comment on "An approach to statistical spatial-temporal modeling of meteorological fields" by M.S. Handcock and J.R. Wallis. *J. Amer. Statist. Assoc.* **89** (426), 379–382.

Cressie, N., Huang, H.C. (1999). Classes of nonseparable spatio-temporal stationary covariance functions. *J. Amer. Statist. Assoc.* **94** (448), 1330–1340.

Crisan, D., Doucet, A. (2002). A survey of convergence results on particle filtering for practitioners. *IEEE Trans. Signal Process.* **50** (3), 736–746.

Diebolt, J., Robert, C. (1994). Estimation of finite mixture distributions through Bayesian sampling. *J. Roy. Statist. Soc., Ser. B* **56**, 163–175.

Doucet, A. (2003). Sequential Monte Carlo methods for Bayesian analysis. *ISBA Bull.* **10** (1), 2–4.

Doucet, A., Tadić, V. (2003). Particle filters in state space models with the presence of unknown static parameters. *Ann. Inst. Statist. Math.* Submitted for publication.

Durbin, J., Koopman, S.J. (2001). *Time Series Analysis by State Space Methods*. Oxford University Press, London.

Durbin, J., Koopman, S. (2002). A simple and efficient simulation smoother for state space time series analysis. *Biometrika* **89**, 603–615.

Ferreira, M.A.R. (2002). Bayesian multi-scale modelling. Ph.D. thesis, Institute of Statistics and Decision Sciences, Duke University.

Ferreira, M.A.R., Gamerman, D. (2000). Dynamic generalized linear models. In: Ghosh, J.K., Dey, D., Mallick, B. (Eds.), *Generalized Linear Models: A Bayesian Perspective*. Marcel Dekker.

Frühwirth-Schnatter, S. (1994). Data augmentation and dynamic linear models. *J. Time Series Anal.* **15**, 183–202.

Gamerman, D. (1997). *Markov Chain Monte Carlo: Stochastic Simulation for Bayesian Inference*. Chapman & Hall, London.

Gamerman, D. (1998). Markov chain Monte Carlo for dynamic generalized linear models. *Biometrika* **85**, 215–227.

Gamerman, D. (2002). A latent approach to the statistical analysis of space-time data. In: *Proceedings of the XV International Workshop on Statistical Modeling*, pp. 1–15.

Gamerman, D., Migon, H.S. (1993). Dynamic hierarchical models. *J. Roy. Statist. Soc., Ser. B* **55**, 629–642.

Gamerman, D., Moreira, A.R.B. (2002). Bayesian analysis of econometric time series models using hybrid integration rules. *Comm. Statist.* **31**, 49–72.

Gamerman, D., Moreira, A.R.B., Rue, H. (2003). Space-varying regression models: Specifications and simulation. *Comput. Statist. Data Anal.* **42**, 513–533.

Gargallo, P., Salvador, M. (2002). Joint monitoring of several types of shocks in dynamic linear models: A Bayesian decision approach. Technical Report, Faculdad de Ciencias Económicas y Empresariales, Universidad de Zaragoza.

Geman, S., Geman, D. (1984). Stochastic relaxation, Gibbs distributions and the Bayesian restoration of images. *IEEE Trans. Pattern Anal. Machine Intelligence* **6**, 721–741.

Geweke, J. (1989). Bayesian inference in econometric models using Monte Carlo integration. *Econometrica* **57**, 1317–1339.

Ghil, M., Cohn, S., Tavantzis, J., Bube, K., Isaacson, E. (1981). Applications of estimation theory to numerical weather prediction. In: Bengtsson, L., Ghil, M., Källén, E. (Eds.), *Dynamic Meteorology: Data Assimilation Methods*. Springer-Verlag, New York, pp. 139–224.

Goodall, C., Mardia, K. (1994). Challenges in multivariate spatio-temporal modeling. In: *Proceedings of the XVIIth International Biometric Conference*. Hamilton, Ontario, Canada, p. 17.

Gordon, N., Salmond, D., Smith, A. (1993). Novel approach to nonlinear/non-Gaussian Bayesian state estimation. *IEE Proc. F* **140**, 107–113.

Handcock, M., Wallis, J. (1994). An approach to statistical spatial-temporal modeling of meteorological fields. *J. Amer. Statist. Assoc.* **89** (426), 368–390.

Harrison, P.J., Lai, I.C. (1999). Statistical process control and model monitoring. *J. Appl. Statist.* **26**, 273–292.

Harrison, P.J., Stevens, C. (1976). Bayesian forecasting (with discussion). *J. Roy. Statist. Soc., Ser. B* **38**, 205–247.

Harvey, A. (1989). *Forecasting Structural Time Series Models and the Kalman Filter*. Cambridge University Press, London.

Harvey, A., Ruiz, E., Shephard, N. (1994). Multivariate stochastic variance models. *Rev. Econom. Studies* **61**, 247–264.

Haslett, J., Raftery, A. (1989). Space-time modeling with long-memory dependence: Assessing ireland's wind power resource (with discussion). *J. Roy. Statist. Soc., Ser. C* **38**, 1–50.

Hastings, W.K. (1970). Monte Carlo sampling methods using Markov chains and their applications. *Biometrika* **57**, 97–109.

Hetherington, J. (1984). Observations on the statistical iteration of matrices. *Phys. Rev. A* **30**, 2713–2719.

Higuchi, T. (1997). Monte Carlo filter using the genetic algorithm operator. *J. Statist. Comput. Simulation* **59**, 1–23.

Higuchi, T. (2001). Self-organizing time series model. In: Doucet, A., de Freitas, N., Gordon, N. (Eds.), *Sequential Monte Carlo Methods in Practice*. Springer-Verlag, New York.

Huang, H.-C., Cressie, N. (1997). Multiscale spatial modeling. In: *ASA Proceedings of the Section on Statistics and the Environment*. Amer. Statist. Assoc., Alexandria, VA, pp. 49–54.

Huerta, G., Sansó, B., Stroud, J.R. (2004). A spatio-temporal model for Mexico city ozone levels. *Appl. Statist.* Submitted for publication.

Irving, W.W., Fieguth, P.W., Willsky, A.S. (1997). An overlapping tree approach to multiscale stochastic modeling and estimation. *IEEE Trans. Image Process.* **6**, 1517–1529.

Johannesson, G., Cressie, N. (2003). Variance–covariance modeling and estimation for multi-resolution spatial models. Technical Report, Department of Statistics, Ohio State University.

Johannesson, G., Cressie, N., Huang, H.-C. (2003) Dynamic multi-resolution spatial models. In: Higuchi, T., Iba, Y., Ishiguro, M. (Eds.), *Science of Modeling – The 30th Anniversary of the AIC*.

Jones, R. (1993). *Longitudinal Data with Serial Correlation – A State Space Approach*. Chapman-Hall, London.

Kato, Z., Berthod, M., Zerubia, J. (1996a). A hierarchical Markov random field model and multi-temperature annealing for parallel image classification. *Graph. Models Image Process.* **58** (1), 18–37.

Kato, Z., Zerubia, J., Berthod, M. (1996b). Unsupervised parallel image classification using Markovian models. *Graph. Models Image Process.* **58** (1), 18–37.

Kim, S., Shephard, N., Chib, S. (1998). Stochastic volatility: Likelihood inference and comparison with ARCH models. *Rev. Econom. Studies* **65**, 361–393.

Kitagawa, G. (1987). Non-Gaussian state-space modeling of nonstationary time series (with discussion). *J. Amer. Statist. Assoc.* **82**, 1032–1063.

Kitagawa, G. (1993). A Monte Carlo filtering and smoothing method for non-Gaussian nonlinear state space models. In: *Proceedings of the 2nd US–Japan Joint Seminar on Statistical Time Series Analysis*. Honolulu, Hawaii, pp. 110–131.

Kitagawa, G. (1996). Monte Carlo filter and smoother for non-Gaussian nonlinear state space models. *J. Comput. Graph. Statist.* **5**, 1–25.

Knorr-Held, L. (1999). Conditional prior proposal in dynamic models. *Scand. J. Statist.* **26**, 129–144.

Künsch, H. (2001). State space and hidden Markov models. In: Barndorff-Nielsen, D., Klüppelberg, C. (Eds.), *Complex Stochastic Systems*. Chapman and Hall, Boca Raton.

Laferté, J.-M., Pérez, P., Heitz, F. (2000). Discrete Markov image modeling and inference on the quadtree. *IEEE Trans. Image Process.* **9** (3), 390–404.

Landim, F., Gamerman, D. (2000). Dynamic hierarchical models – an extension to matricvariate observations. *Comput. Statist. Data Anal.* **35**, 11–42.

Lavine, M. (1999). Another look at conditionally Gaussian Markov random fields. In: Bernardo, J., Berger, J.O., Dawid, A.P., Smith, A.F.M. (Eds.), *Bayesian Statistics, vol. 6*. Oxford Universtity Press, pp. 577–585.

Lindley, D.V., Smith, A.F.M. (1972). Bayes estimates for the linear model (with discussion). *J. Roy. Statist. Soc., Ser. B*, 1–43.

Liu, J., West, M. (2001). Combined parameter and state estimation in simulation-based filtering. In: Doucet, A., de Freitas, N., Gordon, N. (Eds.), *Sequential Monte Carlo Methods in Practice*. Springer-Verlag, New York.

Lopes, H. (2000). Bayesian analysis in latent factor and longitudinal models. Ph.D. thesis, Institute of Statistics and Decision Sciences, Duke University.

Lopes, H. (2002). Sequential analysis of stochastic volatility models: Some econometric applications. Technical Report, Department of Statistical Methods, Federal University of Rio de Janeiro.

Lopes, H., Marinho, C. (2002). A particle filter algorithm for the Markov switching stochastic volatility model. Technical Report, Department of Statistical Methods, Federal University of Rio de Janeiro.

Lopes, H., Moreira, A., Schmidt, A. (1999). Hyperparameter estimation in forecast models. *Comput. Statist. Data Anal.* **29**, 387–410.

Mardia, K., Goodall, C., Redfern, E., Alonso, F. (1998). The Kriged Kalman filter (with discussion). *Test* **7**, 217–285.

Metropolis, N., Rosenbluth, A.W., Rosenbluth, M.N., Teller, A.H., Teller, E. (1953). Equation of state calculations by fast computing machine. *J. Chem. Phys.* **21**, 1087–1091.

Migon, H.S. (2000). The prediction of Brazilian exports using Bayesian forecasting. *Investigation Operativa* **9**, 95–106.

Migon, H.S., Gamerman, D. (1993). Generalised exponential modell – Δ Bayesian approach. *J. Forecasting* **12**, 573–584.

Migon, H.S., Gamerman, D. (1999). *Statistical Inference: An Integrated Approach*. Arnold, London.

Migon, H.S., Harrison, P.J. (1985). An application of nonlinear Bayesian forecasting to TV advertising. In: Bernardo, J.M., DeGroot, M.H., Lindley, D., Smith, A.F.M. (Eds.), *Bayesian Statistics, vol. 2*. North-Holland and Valencia University Press.

Migon, H.S., Mazuchelli, J. (1999). Bayesian garch models: Approximated methods and applications. *Rev. Brasil. Econom.* **1999**, 111–138.

Monteiro, A., Migon, H.S. (1997). Dynamic models: An application to rainfall-runoff modelling. *Stochastic Hydrology and Hydraulics* **11**, 115–127.

Müller, P. (1991). Monte Carlo integration in general dynamic models. *Contemp. Math.* **115**, 145–163.

Müller, P. (1992). Posterior integration in dynamic models. *Comput. Sci. Statist.* **24**, 318–324.

Nelder, J.A., Wedderburn, R.W.M. (1972). Generalized linear models. *J. Roy. Statist. Soc., Ser. A* **135**, 370–384.

Pitt, M., Shephard, N. (1999). Filtering via simulation: Auxiliary particle filters. *J. Amer. Statist. Assoc.* **94**, 590–599.

Pole, A., West, M. (1990). Efficient Bayesian learning in nonlinear dynamic models. *J. Forecasting* **9**, 119–136.

Pole, A., West, M., Harrison, P.J. (1994). *Applied Bayesian Forecasting and Time Series Analysis*. Chapman-Hall, London.

Polson, N., Stroud, J. Müller, P. (2002). Practical filtering with sequential parameter learning. Technical Report, Graduate School of Business, University of Chicago.

Reyna, F.R.Q., Migon, H.S., Duarte, A. (1999). Missing data in optimization models for financial planning. *Stochastic Hydrology and Hydraulics* **8**, 9–30.

Rubin, D. (1988). Using the SIR algorithm to simulate posterior distributions. In: Bernardo, J., DeGroot, M., Lindley, D., Smith, A. (Eds.), *Bayesian Statistics, vol. 3*.
Rue, H. (2001). Fast sampling of Gaussian Markov random fields. *J. Roy. Statist. Soc., Ser. B* **65**, 325–338.
Schmidt, A.M., Gamerman, D., Moreira, A.R.B. (1999). An adaptive resampling scheme for cycle estimation. *J. Appl. Statist.* **26**, 619–641.
Shephard, N., Pitt, M.K. (1997). Likelihood analysis of non-Gaussian measurement time series. *Biometrika* **84**, 653–667.
Singh, A.C., Roberts, G.R. (1992). State space modelling of cross-classified time series of counts. *Internat. Statist. Rev.* **60**, 321–336.
Smith, A., Gelfand, A. (1993). Bayesian statistics without tears: A sampling–resampling perspective. *Amer. Statist.* **46**, 84–88.
Storvik, G. (2002). Particle filters in state space models with the presence of unknown static parameters. *IEEE Trans. Signal Process.* **50** (2), 281–289.
Stroud, J.R., Müller, P., Sansó, B. (2001). Dynamic models for spatio-temporal data. *J. Roy. Statist. Soc., Ser. B* **63**, 673–689.
Triantafyllopoulos, K., Harrison, P. (2002). Stochastic volatility forecasting with dynamic models. Research Report sta02-06, School of Mathematics and Statistics, University of Newcastle.
Vivar-Rojas, J.C., Ferreira, M.A.R. (2003). A new class of spatio-temporal models. Technical Report, Universidade Federal do Rio de Janeiro.
West, M. (1986). Bayesian model monitoring. *J. Roy. Statist. Soc., Ser. B* **48**, 70–78.
West, M. (1993a). Approximating posterior distributions by mixtures. *J. Roy. Statist. Soc., Ser. B* **55**, 409–422.
West, M. (1993b). Mixture models, Monte Carlo, Bayesian updating and dynamic models. In: Newton, J. (Ed.), *Computing Science and Statistics: Proceedings of the 24th Symposium of the Interface*. Interface Foundation of North America, Fairfax Station, VA, pp. 325–333.
West, M., Harrison, P.J. (1986). Monitoring and adaptation in Bayesian forecasting models. *J. Amer. Statist. Assoc.* **81**, 741–750.
West, M., Harrison, P.J. (1997). *Bayesian Forecasting and Dynamic Models*, second ed. Springer-Verlag, London.
West, M., Harrison, P.J., Migon, H. (1985). Dynamic generalized linear model and Bayesian forecasting. *J. Amer. Statist. Assoc.* **80**, 73–97.
Wikle, C., Cressie, N. (1999). A dimension-reduced approach to space-time Kalman filtering. *Biometrika* **86**, 815–829.
Wikle, C.K. (2003). Hierarchical Bayesian models for predicting the spread of ecological processes. *Ecology* **84**, 1382–1394.
Wikle, C.K. (2003). A kernel-based spectral approach for spatiotemporal dynamic models. Technical Report, Department of Statistics, University of Missouri.
Wikle, C.K., Milliff, R.F., Nychka, D., Berliner, L.M. (2001). Spatio-temporal hierarchical Bayesian modeling: Tropical ocean surface winds. *J. Amer. Statist. Assoc.* **96**, 382–397.

Handbook of Statistics, Vol. 25
ISSN: 0169-7161
© 2005 Elsevier B.V. All rights reserved.
DOI 10.1016/S0169-7161(05)25020-4

20

Bayesian Thinking in Spatial Statistics

Lance A. Waller

Abstract
In the sections below we review basic motivations for spatial statistical analysis, review three general categories of data structure and associated inferential questions, and describe Bayesian methods for achieving inference. Our goal is to highlight similarities across spatial analytic methods, particularly with regards to how hierarchical probability structures often link approaches developed in one area of spatial analysis to components within other areas. By choosing to highlight similarities, we focus on general concepts in spatial inference, and often defer details of several interesting and current threads of development to the relevant literature. We conclude with a listing of some of these developing areas of interest and references for further reading.

1. Why spatial statistics?

Data are always collected at particular locations whether these be in a forest, at a particular street address, in a laboratory, or in a particular position on a gene expression array. In many cases, the location may provide additional insight into circumstances associated with the data item, in short "where" we collect a measurement may inform on "what" or "how much" we measure. The field of *spatial statistics* involves statistical methods utilizing location and distance in inference. Some methods are extensions of familiar techniques such as regression, generalized linear models, and time series, while others derive from particular models of stochastic processes in space.

Historically, spatial issues lurk in the background of many classical statistical settings, including the design and analysis of agricultural field trials. The recognition of spatial gradients in soil properties impacting yield led to various randomization schemes to effectively remove the spatial effect from consideration. Explicit models of spatial pattern often drew motivation from similar methods in time series, with the recognition that, in space, one no longer has a simple ordering of observations. This loss of ordering complicates the extension of popular temporal models to the spatial domain, for example, Markov processes and autoregressive schemes, while important, take on additional complexity in higher dimensions, as we will see below. Key early work in spatial analysis appears in the work of Moran (1948, 1950), Whittle (1954), and Bartlett (1964, 1975). These are followed by foundational work in spatial prediction (Matheron, 1963; Gandin, 1963), spatial autoregressive models (Besag, 1974), and spatial point processes (Ripley, 1977; Diggle, 1983). One measure of the increasing interest in and development of the field is a comparison of Ripley's (1981) 252-page and Cressie's (1993) 900-page texts addressing spatial statistics as a whole, in addition to the recent and rapid increase in texts addressing particular areas of application and/or theory (Stein, 1999; Chilès and Delfiner, 1999; Lawson, 2001; Lawson and Denison, 2002; Webster and Oliver, 2001; Waller and Gotway, 2004).

Much of the recent explosion in the use of spatial statistics can be tied to the relatively simultaneous increases in both spatial data availability, including accurate location measurements via global positioning system (GPS) technology, and rapid improvements in computational power allowing advances in both spatial data structures (particularly within geographic information systems) and in the fitting of extraordinarily complex models via sophisticated algorithms. Similar advances in computational speed and algorithm development drive the increases in the use of Bayesian methods in this setting, in particular the recognition and further development of Markov chain Monte Carlo (MCMC) algorithms allowing great flexibility in fitting spatially structured statistical models. Hierarchical models provide not only a means to fit spatial models, but the hierarchical structure itself allows the statistical modeler new ways of conceptualizing spatial patterns and processes.

In the sections below, we set the stage with a review of three basic types of spatial data, their associated inferential questions, and features of spatial data which drive many of the models described below. For each data type, we outline how Bayesian hierarchical structures address the associated inferential questions. While lengthy, this review addresses only basic questions and models and we end with a brief list of current research areas in spatial statistics where Bayesian ideas aid development.

2. Features of spatial data and building blocks for inference

A primary feature driving many methods of spatial analysis is described by Tobler's "First Law of Geography": "Everything is related to everything else, but near things are more related than far things" (Tobler, 1970). Statistically, Tobler's "law" refers to positive *spatial autocorrelation* in which pairs of observations taken nearby are more alike than those taken farther apart. Allowing correlated observations weakens the usual statistical assumption of independent observations and complicates analysis in several ways. First, if we assume independent observations, any observed spatial patterns can be modeled as a spatial *trend* in the expected values of the observations. If we allow correlation, observed spatial patterns may be due to a trend in expected values, correlation among observations with the same expectation, or some combination of the two. Second, we lose statistical precision in correlated data resulting in a reduced "effective sample size" (the number of independent observations containing the same amount of information as our set of correlated observations), see Cressie (1993, pp. 14–15) for an example. The presence of spatial correlation drives two major areas of spatial analysis outlined below: fitting models with residual spatial correlation and using spatial correlation to predict observations at locations where no measurement was taken.

In addition, spatial data complicate usual notions of statistical asymptotics. At least two sorts of asymptotics exist: *increasing domain* wherein we observe data in a study area with increasing area, or *infill* asymptotics wherein we observe more and more data within the same, fixed study area. Increasing domain asymptotics reflect traditional ideas more closely than infill asymptotics, but are not realistic if the study area under question already covers a large portion of the earth. As a result, there is considerably less reliance on asymptotic results in spatial statistics than in other branches of statistics.

Spatial statistical methods may be classified by the inferential questions of interest. These questions are motivated by application and often fall into categories based on the type of data available. Cressie (1993) provides three useful categories of spatial data that also serve to categorize both inferential questions and inferential approaches. We present these in order of data complexity.

First, consider *spatial point process* data consisting of a set of observed locations in a defined study area. We consider the locations themselves as the realization of some random process and seek inference regarding the properties of this process. Examples include the locations of trees in a forest, neurons in the brain, and galaxies in the universe. Questions of interest include:

- Are observations equally likely at all locations? If not, where are observations more or less likely?
- Are there (spatially-referenced) covariates that drive the probability of observations occurring at particular locations?
- Does the presence of an observation at a particular location either encourage or inhibit further observations nearby?
- If observations do impact the probability of nearby observations, what is the range of influence of observations?

Much of the literature on spatial point processes involves modeling of stochastic processes, and comparison of competing models describing observed spatial patterns. Statistical questions address estimation of model parameters and assessment of fit of various models.

Next, suppose we have *geostatistical* data consisting of a set of measurements taken at a fixed set of locations, e.g., ozone levels measured at each of a set of monitoring stations. In this case, the locations are set by design and not random. An inferential question of interest is prediction of the same outcome at locations where no measurement was taken. Examples of such predictions appear each day in weather maps of current temperatures interpolated from a set of official monitoring stations. The literature on spatial prediction builds from a fairly simple concept: spatial correlation suggests that one should give more weight to observations near the prediction location than to those far away. Spatial prediction theory explores how to optimally set these weights based on estimates of the underlying spatial autocorrelation structure.

Finally, we may observe data from a set of regions partitioning the study area. Such data are referred to as *lattice data* by Cressie (1993) and *regional data* by Waller and Gotway (2004). Lattices may be regularly or irregularly spaced, for example, pixels in an image or counties within a state, respectively, so we use the term "regional" to avoid confusion with literature that assumes the term "lattice" implies a regular lattice. Regional data generally involve summary measures for each region, e.g., number of residents in an enumeration district, average income for residents of the region, or number of items delivered within a postal delivery zone. Inferential questions often involve accurate estimation of summaries from regions with small sample sizes ("small area estimation"), or regression or generalized linear modeling linking outcomes and covariates measured on the same set of regions. In the first case, statistical methods involve how best to "borrow strength" from other regions in order to improve estimates within each region. In the second, methods involve accurate estimation of model parameters with adjustment for spatial correlation between nearby regions.

Since the inferential questions vary with data type, we describe models for each data type separately. We address methods for the data types in the reverse of the order of data complexity above since typically the analytic methods become more complex and less familiar as we move from regional regression/generalized linear model approaches, to geostatistical prediction, to point process models. While the approaches outlined derive from the different data types, we present associated models in a framework illustrating similarities in structure illustrating methodological commonalities in the statistical approaches underlying the three data structures.

3. Small area estimation and parameter estimation in regional data

We begin by addressing two primary analytic questions commonly encountered in the analysis of regional (lattice) data, namely, small area estimation and parameter estimation in generalized linear models with (residual) spatial correlation. We will illustrate how hierarchical models address both issues.

There is a large literature on small area estimation including the comprehensive review by Ghosh and Rao (1994) and the recent text by Rao (2003). We do not attempt a thorough review here, but borrow from several key approaches and link them to Bayesian hierarchical models for spatially correlated regional data. However, most of the attention in the small area estimation literature involves accurate estimation for subsets of larger samples, rather than on spatial patterns *per se*, although, as we will see below, adding spatial structure links the small area estimation literature and that addressing models for "disease mapping", i.e., small area estimation of local disease incidence or prevalence rates.

In a nutshell, the structure we describe involves a hierarchical model with random effects (intercepts) inducing "borrowed" strength. The widely-cited approach of Fay and Herriot (1979) involves a Gaussian linear model with a Gaussian random intercept and provides a base model from which many small area estimation approaches derive. Briefly, consider the setting where we observe Y_i for each region i, $i = 1, \ldots, I$, and we assume

$$Y_i \overset{\text{ind}}{\sim} N\left(\mu_i, \sigma_Y^2\right).$$

If $\mu_1 = \mu_2 = \cdots = \mu_I = \mu$, estimation proceeds either via maximum likelihood or some design-based estimate of the shared, global population mean.

However, suppose we aren't willing to assume equal means and instead wish to estimate each μ_i? Without further assumptions, we can only base our local estimates on the observation (or perhaps observations) from region i, resulting in considerable reduction in sample size and accompanying loss in statistical precision. This is a limitation of design-based estimation procedures and is the basis of the small area estimation problem (where "small" refers to the local sample size, not the geographic area).

If we are willing to add additional modeling assumptions, we can provide model-based or perhaps composite estimators of the local means that "borrow strength" from data in other areas to supplement each local sample size. The precise balance of input from the local and other areas depends on the particular model used. Suppose we consider the following random effects model:

$$Y_i | v_i \overset{\text{ind}}{\sim} N\left(\mu + v_i, \sigma_Y^2\right),$$
$$v_i \overset{\text{ind}}{\sim} N\left(0, \sigma_v^2\right),$$

where the zero-mean Gaussian random effects v_i provide random variation among the local means, centered around the global mean μ. Standard derivation of Bayesian estimators minimizing squared error loss result in a compromise "shrinkage" estimator based on a weighted average of the global sample mean and the local data where the weights depend on the relative sizes of the two variance components σ_Y^2 and σ_v^2.

The basic idea extends from the fairly straightforward Gaussian–Gaussian linear model to hierarchical generalized linear mixed model (GLMM). As an illustration, we consider a family of hierarchical Poisson regression models first proposed by Clayton and Kaldor (1987) to provide local (small area) estimates of disease risk, based on observed disease counts.

To begin, consider the standard development of Poisson regression as a generalized linear model (GLM) with a log link and an offset. Suppose we observe a number of events, Y_i, and a vector of covariates, $\boldsymbol{X}_i$, for each region i, $i = 1, \ldots, I$. In addition, let $\boldsymbol{\beta}$ denote the vector of parameters corresponding to covariate effects and let E_i denote the number of events "expected" in the absence of any covariate effects. The E_i are assumed to be fixed and known and serve as offsets in the GLM effectively removing known differences in expected counts due to, e.g., varying numbers at risk or varying age structures in each region. The model is then

$$Y_i \overset{\text{ind}}{\sim} \text{Poisson}\left(E_i \exp(\zeta_i)\right),$$
$$\zeta_i = \boldsymbol{X}_i'\boldsymbol{\beta}, \quad \text{for } i = 1, \ldots, I.$$

The expected number of events E_i is typically defined as the product of the number of units under study in region i, say n_i, and a constant "baseline" risk, and covariates provide multiplicative increases or decreases in this risk defined by the parameters in $\boldsymbol{\beta}$. The resulting Poisson GLM defines the natural logarithm of the mean count as

$$\log\left[E(Y_i)\right] = \log(E_i) + \boldsymbol{X}_i'\boldsymbol{\beta},$$

yielding a model with an offset E_i and a multiplicative impact of each covariate on the model-based expected observations.

The small area estimation question of interest is whether the baseline risk of disease for individuals within a region is the same between all regions. In addition, we seek estimates of the model parameters in $\boldsymbol{\beta}$.

Consider the small area estimation problem in the absence of covariates. In this case we are simply fitting an intercept-only GLM (with offset E_i), but with a separate intercept in each region so only the data from region i can contribute to the estimate of ζ_i. The addition of a mean-zero Gaussian random intercept v_i to each region results in the following model

$$Y_i | v_i \overset{\text{ind}}{\sim} \text{Poisson}\big(E_i \exp[\beta_0 + v_i]\big),$$
$$v_i \overset{\text{ind}}{\sim} N\big(0, \sigma_v^2\big), \quad \text{for } i = 1, \ldots, I.$$

Note that the observations are conditionally independent given the values of the random effects v_i, but will be marginally correlated. Similar models appear in the analysis of longitudinal data and the basic two-stage hierarchy allows one to consider the probability in two stages. A Bayesian interpretation of the model views the first stage as the likelihood (built from the product of the conditionally independent components) and views the distributions of the v_is as a set of exchangeable prior distributions. To complete the second stage, we assign vague (even flat) priors to the "fixed" effect β_0, which is well identified through the likelihood.

The addition of the random effect helps our estimation problem because, in effect, we induce some commonality between the local ζ_is and change our estimation problem from estimating I separate ζ_is to estimation of two parameters: β_0 and σ_v^2. The end result is a local estimate defined by a weighted average of the observed data in location i and the global overall mean. Clayton and Kaldor (1987), Marshall (1991), and Waller and Gotway (2004, Section 4.4.3) provide details of an empirical Bayes approach using data estimates of β_0 and σ_v^2. In a fully Bayesian approach we assign a hyperprior distribution to σ_v^2 (e.g., a conjugate inverse gamma distribution) to complete the model.

We note the approach outlined above does not incorporate any spatial similarity between observations or error terms. All local estimates are a compromise between the local data and a *global* estimate based on all of the data. Clayton and Kaldor (1987) modified the hierarchical structure defining a spatially structured prior distribution for the random effects to create local empirical Bayes estimates based on a compromise between the local data and the average of regions nearby, essentially borrowing strength *locally* rather than *globally*. Besag et al. (1991) extend the approach to a fully Bayesian setting using MCMC and clarify some technical points with the spatially structured prior. While the model remains the same, to avoid confusion with the exchangeable v_is above, we follow Besag et al. (1991) and denote the spatially correlated random effects by $u_i, i = 1, \ldots, I$.

We describe two classes of spatially structured prior formulations. First, consider a joint multivariate Gaussian prior distribution for $\boldsymbol{u}$ with spatial covariance matrix Σ_u, i.e.,

$$\boldsymbol{u} \sim MVN(\mathbf{0}, \Sigma_u). \tag{1}$$

Diggle et al. (1998) illustrate this approach defining the elements of Σ_u via an exponential correlation function with one parameter governing the rate of decay with distance and another defining overall variation (impacting the overall level of smoothness in the resulting estimates). The rate of decay and smoothness parameters are then assigned independent uniform priors. While conceptually more straightforward, the jointly-specified spatial prior is only beginning to see broad application for reasons that become apparent below.

As an alternative to the joint prior distribution defined in Eq. (1), Clayton and Kaldor (1987) propose a set of *conditionally autoregressive priors* (CAR priors) for the u_i, $i = 1, \ldots, I$. In addition, Breslow and Clayton (1993) apply CAR priors as random effects distributions within likelihood approximations for GLMMs. The CAR priors see wide use in the area of disease mapping, or regional smoothing of rates (proportions) of rare events. Clayton and Bernardinelli (1992), Mollié (1996) and Wakefield et al. (2000) review the application of such models to disease mapping issues.

The widespread application of CAR models merits a closer look at their structure and implementation, paraphrased from the development in Waller and Gotway (2004, pp. 409–423). First, consider a subset of Gaussian CAR priors defining the prior mean of each u_i as a weighted average of the other u_j, $j \neq i$,

$$u_i | u_{j \neq i} \sim N\left(\frac{\sum_{j \neq i} c_{ij} u_j}{\sum_{j \neq i} c_{ij}}, \frac{1}{\tau_{CAR} \sum_{j \neq i} c_{ij}}\right), \quad i = 1, \ldots, N. \tag{2}$$

Here, the c_{ij}s denote spatial dependence parameters defining which regions j are "neighbors" to region i, and τ_{CAR} denotes a hyperparameter related to the conditional variance of u_i given the other elements of $\boldsymbol{u}$. We set $c_{ii} = 0$ for all i, so regions are not neighbors to themselves. Many applications utilize adjacency-based weights where $c_{ij} = 1$ if region j is adjacent to region i, and $c_{ij} = 0$, otherwise. The c_{ij} are typically considered fixed and defined by the physical or political geography of the study area. Other weighting options appear in the literature (Best et al., 1999) but are much less widely applied.

While the hierarchical spatial GLMM defined above contains sensible pieces and MCMC algorithms offer a means to fit them, the end result is a rather complicated hierarchical model and several features merit further attention.

First, consider the relationship between the joint and the conditional specifications of spatially structured priors given in Eqs. (1) and (2), respectively. While both the joint and conditional specifications are possible, the connection between the user-defined c_{ij}s and the resulting (prior or posterior) spatial covariance structure of random effects in $\boldsymbol{u}$ rarely, if ever, provides a closed-form, functional relationship between the spatial dependence parameters $\{c_{ij}\}$, and the elements of the covariance matrix Σ_u, as we illustrate below.

Besag and Kooperberg (1995) clarify the connection between the autoregressive spatial dependence parameters and the spatial covariance matrix, and we highlight several key points from their development here. First, if $\boldsymbol{u}$ follows a multivariate Gaussian distribution with covariance Σ_u, then the density, $f(\boldsymbol{u})$, follows

$$f(\boldsymbol{u}) \propto \exp\left(-\frac{1}{2}\boldsymbol{u}' \Sigma_u^{-1} \boldsymbol{u}\right). \tag{3}$$

Multivariate Gaussian theory defines the associated conditional distributions via

$$u_i|u_{j\neq i} \sim N\left(\sum_{j\neq i}\left(\frac{-\Sigma^{-1}_{u,ij}}{\Sigma^{-1}_{u,ii}}\right)u_j, \frac{1}{(\Sigma^{-1}_{u,ii})}\right), \tag{4}$$

where $\Sigma^{-1}_{u,ij}$ denotes the (i, j)th element of the precision matrix Σ^{-1}_u. Notable features include: the conditional mean for u_i is a weighted sum of u_j, $j \neq i$, and the conditional variance is inversely proportional to the diagonal of the inverse of Σ_u. Going from a set of user-specified conditional Gaussian distributions to the associated joint distribution is less straightforward, requiring constraints on the weights defining the conditional mean and variance to ensure both a Gaussian joint distribution and a valid covariance matrix Σ_u (cf. Besag, 1974; Besag and Kooperberg, 1995; Arnold et al., 1999).

Both the conditional mean and the conditional variance in Eq. (4) depend on elements of the *inverse* of the covariance matrix Σ_u. Computationally, this implies an MCMC implementation of the joint specification based on straightforward updates from full conditional distributions will involve matrix inversion at each update of the covariance parameters. This suggests some computational advantage to the CAR prior formulation, since the CAR priors (effectively) limit modeling to the elements of Σ^{-1}_u, rendering the inversion unnecessary. In fact, the elegant theory in Besag (1974) provides the basis for both conditionally-specified spatial modeling and conditional updates in MCMC algorithms, although the latter component was not widely appreciated for some time. In essence, one could say that CAR priors are almost "custom-made" for MCMC implementation.

If the conditional formulation is more convenient for MCMC implementation, but the relationship between the c_{ij}s and the elements of Σ_u is complicated by the matrix inversion, how do we define sets of c_{ij}s that correspond to valid and reasonable spatial covariance structures?

To address this issue, we equate the spatial dependence parameters in Eq. (2) to their counterparts in Eq. (4) yielding

$$\frac{c_{ij}}{\sum_j c_{ij}} = \frac{-\Sigma^{-1}_{u,ij}}{\Sigma^{-1}_{u,ii}}, \tag{5}$$

and

$$\tau_{CAR}\sum_j c_{ij} = \Sigma^{-1}_{u,ii}.$$

Therefore, $c_{ij} = \Sigma^{-1}_{u,ij}/\tau_{CAR}$ and symmetry of Σ_u requires symmetry of the spatial dependence parameters, c_{ij}, in the collection of CAR priors defined by Eq. (2).

More generally, the family of CAR priors is broader than those defined by Eq. (2). The weights defining the conditional mean need not be symmetric themselves so long as the diagonal elements of Σ^{-1}_u compensate appropriately via

$$\left(\frac{-\Sigma^{-1}_{u,ij}}{\Sigma^{-1}_{u,ii}}\right)\Sigma^{-1}_{u,ii} = \left(\frac{-\Sigma^{-1}_{u,ji}}{\Sigma^{-1}_{u,jj}}\right)\Sigma^{-1}_{u,jj}. \tag{6}$$

(While Eq. (6) may seem trivial, recall that in practice we define the additive weights $(-\Sigma_{u,ij}^{-1}/\Sigma_{u,ii}^{-1})$, and conditional variances $1/(\Sigma_{u,ii}^{-1})$ without regard to the specific non-diagonal elements of Σ_u^{-1}, so verification is important for any proposed conditional structure.)

The class of CAR priors defined by Eq. (2) (with associated symmetric c_{ij}s) includes the widely-applied adjacency weights ($c_{ij} = 1$ when regions i and j share a boundary, $c_{ij} = 0$ otherwise), and (symmetric) distance-decay weights such as those considered by Best et al. (1999), and we continue to limit attention to this subset of CAR priors here, for simplicity and to raise some practical issues associated with these particular prior distributions.

Results in Besag (1974) indicate the set of CAR priors defined in Eq. (2) uniquely defines a corresponding multivariate normal joint distribution

$$\boldsymbol{u} \sim MVN(\boldsymbol{0}, \Sigma_u), \tag{7}$$

with $\Sigma_{u,ii}^{-1} = \sum_j c_{ij}$ and $\Sigma_{u,ij}^{-1} = -c_{ij}$. However, for symmetric c_{ij}s, the sum of any row of the matrix Σ_u^{-1} is zero, indicating Σ_u^{-1} is singular (noninvertible), and the corresponding covariance matrix Σ_u in Eq. (7) is not well-defined. This holds for any symmetric set of spatial dependence parameters c_{ij} (including the adjacency-based c_{ij}s appearing in many applications). Remarkably, the singular covariance does not preclude application of the model with such weight matrices (cf., Besag and Kooperberg, 1995), but it does prevent easy transition from the spatial dependence parameters to closed-form spatial covariances.

CAR priors with singular associated covariance matrices can still be applied since pairwise *contrasts* $u_i - u_j$ are well-identified. More explicitly, note the full class of CAR priors (including those defined by Eq. (2)) falls into the class of *pairwise difference* prior distributions defined by Besag et al. (1995). These distributions are improper priors since they only define contrasts (differences) between pairs of values $u_i - u_j$, $j \neq i$, but they do not identify an overall mean value for the elements of $\boldsymbol{u}$ (since such distributions define the value of each u_i relative to the values of the others). In this case, any likelihood function based on data allowing estimation of an overall mean also allows the class of improper pairwise difference priors to generate proper posterior distributions. In practice, we assure this by adding the constraint

$$\sum_{i=1}^{N} u_i = 0.$$

The term "intrinsic" is used when defining statistics on contrasts and autoregressions with singular covariance matrices are often referred to as "intrinsic autoregressions" (Künsch, 1987; Besag and Kooperberg, 1995).

The descriptions above reveal a set of conditional prior distributions defined by a sensible set of spatial dependence parameters that do not translate directly into a closed-form model of spatial correlation or covariance. In addition, we also defined a joint specification that allows direct parametric modeling of the covariance function but is computationally intensive to fit (due to the matrix inversion component). Both approaches induce spatial similarity between observations by "borrowing strength" from neighboring values, both provide posterior estimates of model parameters, and both allow for small area estimation via posterior predictive distributions. However, the conditionally specified model is much more attractive for practical MCMC implementation. At present, the literature contains many more examples of CAR priors than the multivariate normal formulation, but both are actively applied in many settings.

To this point, we have considered borrowing strength either globally or locally, but Besag et al. (1991) point out that we could include both effects in the same model via a convolution prior yielding the following model:

$$
\begin{aligned}
&Y_i|u_i \overset{\text{ind}}{\sim} \text{Poisson}\big(E_i \exp[\beta_0 + u_i + v_i]\big),\\
&u_i|u_{j\neq i} \sim N\left(\frac{\sum_{j\neq i} c_{ij}u_j}{\sum_{j\neq i} c_{ij}}, \frac{1}{\tau_{CAR}\sum_{j\neq i} c_{ij}}\right), \quad i = 1, \ldots, N,\\
&v_i \overset{\text{ind}}{\sim} N\big(0, \sigma_v^2\big), \quad \text{for } i = 1, \ldots, I.
\end{aligned}
$$

To complete the hierarchy, we assign hyperpriors to parameters τ_{CAR} and σ_v^2. As mentioned above, typical applications define conjugate inverse gamma hyperpriors, and Ghosh et al. (1999) and Sun et al. (1999) define conditions on the inverse gamma distributions necessary to ensure proper posterior distributions. When we include both $\boldsymbol{u}$ and $\boldsymbol{v}$ in the model some care is required since τ_{CAR} is related to the conditional variance of $u_i|u_{j\neq i}$ and σ_v^2 is related to the marginal variance of each v_i, and we want to take care not to assign "unfair" prior weight to either global or local smoothing. This issue is explored by Bernardinelli et al. (1995) who note that the prior marginal standard deviation of v_i is roughly equal to the conditional standard deviation of $u_i|u_{j\neq i}$ divided by 0.7, although this is only a rule of thumb and merits closer scrutiny (Banerjee et al., 2004).

While hierarchical models with CAR priors see broad application for parameter estimation and associated small area estimation for regional data, they certainly are not the only models for such data, nor necessarily optimal in any particular way. In addition to the features noted above regarding prior/posterior propriety and computational complexity, CAR-based hierarchical models are defined for the given set of regions and do not aggregate or disaggregate sensibly into larger or smaller regions, respectively, and should not be viewed as "discretized" versions of some latent, smooth random process.

In summary, the hierarchical modeling approach described above allows us to incorporate spatial correlation into generalized linear models, but to conveniently defer it to the second level of the model, thereby maintaining a relatively straightforward likelihood based on conditional independence given the random effects. In some cases it is possible to incorporate spatial correlation directly in the likelihood, but when we move away from Gaussian likelihoods, this exercise becomes much more complex (and, in fact, motivates some of the development and interesting examples/counterexamples in Besag, 1974).

We now turn from Bayesian spatial extensions of familiar regression and generalized linear models to the problem of statistical prediction based on spatially located data.

4. Geostatistical prediction

Statistical prediction has long had a Bayesian flavor since under most standard criteria the optimal predictor is the conditional expectation of the predictand given the observed data, i.e., the posterior predictive mean of the unknown predictand based on a fully specified model for observations (cf. Bernardo and Smith, 2000, pp. 482–483). In this section, we review the typical geostatistical data structure and briefly outline "classical" geostatistical prediction methods (discussed at an introductory level in Cressie, 1989; Webster and Oliver, 2001; Chapter 8 of Waller and Gotway, 2004, and considered more fully in Cressie, 1993; Chilès and Delfiner, 1999), then discuss aspects of this approach that may be addressed by Bayesian thinking. Our outline follows closely development in Le and Zidek (1992), Handcock and Stein (1993), and the recent text by Banerjee et al. (2004). One advantage of the Bayesian approach, as expanded by Banerjee et al. (2004), is a very general model-based view of prediction blurring distinctions between the regional models of the last section and models for geostatistical prediction, a point we return to at the conclusion of this section.

The basic structure of geostatistical data is as follows: we observe a vector of n measurements $\boldsymbol{Z} = (Z(\boldsymbol{s}_1), \ldots, Z(\boldsymbol{s}_n))$ at locations $\boldsymbol{s}_1, \ldots, \boldsymbol{s}_n$ in some domain D. We consider the locations to be fixed by design (with the term "design" including a convenience sample of locations), and for the remainder of this section assume the vector of observations follows a multivariate Gaussian distribution with mean $\boldsymbol{\mu}$ and covariance matrix Σ_Z, i.e.,

$$\boldsymbol{Z} \sim MVN(\boldsymbol{\mu}, \Sigma_Z).$$

It is typical to assume $\boldsymbol{\mu} = (\mu_1, \ldots, \mu_n)$ and $\mu_i = \boldsymbol{x}(\boldsymbol{s}_i)'\boldsymbol{\beta}$, where $\boldsymbol{x}(\boldsymbol{s}_i)$ represents a vector of p covariates observed at location $\boldsymbol{s}_i$, and $\boldsymbol{\beta}$ the vector of associated parameters. The elements of $\boldsymbol{x}(\boldsymbol{s}_i)$ may represent polynomial functions of location coordinates to accommodate polynomial trends in the data (a trend surface analysis), but can be viewed as any location-specific covariates linearly related to the expectation of $Z(\boldsymbol{s})$. In addition, following the descriptions in Handcock and Stein (1993) and Banerjee et al. (2004), suppose that the elements of Σ_Z can be described through a parametric *covariance function*

$$\operatorname{Cov}\big(Z(\boldsymbol{s}_i), Z(\boldsymbol{s}_j)\big) = \alpha K_\theta(\boldsymbol{s}_i, \boldsymbol{s}_j),$$

where α denotes a (positive) scale parameter and θ a vector of parameters (noting that elements of θ may also influence scale). As a result, we assume the observations $\boldsymbol{Z}$ represent a realization of a Gaussian random field and seek to use the structure of this stochastic process to provide predictive inference on elements of the field, say $Z(\boldsymbol{s}_0)$, at location $\boldsymbol{s}_0$ where we observe no measurement.

4.1. *Covariance functions and variograms*

Before moving to prediction, a few comments on the covariance function are in order. First, if we consider a spatially-referenced data set as a partial realization of a spatial stochastic process, e.g., a Gaussian random field, and if we have a single set of observations, we essentially have a single observation of the *process*, even though we have multiple observations *within* the process. In this case, spatial autocorrelation can provide some "repeatability" within the process allowing estimation and prediction. More formally, the process is said to be *second-order stationary* if the covariance function $K_\theta(\boldsymbol{s}_i, \boldsymbol{s}_j)$ depends only on the vector $\boldsymbol{s}_i - \boldsymbol{s}_j$ separating locations $\boldsymbol{s}_i$ and $\boldsymbol{s}_j$, and not the locations themselves. Furthermore, if $K_\theta(\boldsymbol{s}_i, \boldsymbol{s}_j)$ depends only on the distance $\|\boldsymbol{s}_i - \boldsymbol{s}_j\|$, we say the covariance function is *isotropic*. Neither stationarity nor isotropy are required for spatial prediction, but results are streamlined in these cases.

Second, to link results to many classical geostatistical techniques, we also consider the *variogram* defined as

$$2\gamma(\boldsymbol{h}) = \mathrm{Var}\big[Z(\boldsymbol{s}) - Z(\boldsymbol{s}+\boldsymbol{h})\big],$$

where $\boldsymbol{h}$ denotes a *spatial lag* or relocation vector. Also important is the *semivariogram*, $\gamma(\boldsymbol{h})$, defined as one-half of the variogram. Note that the variogram is defined as the variance of contrasts of observations, similar to the pairwise difference priors defined in the preceding section, and the term *intrinsic stationarity* implies the existence of a valid variogram for all locations, depending on the spatial lag $\boldsymbol{h}$ but not on particular locations $\boldsymbol{s}$ and $\boldsymbol{s}+\boldsymbol{h}$ themselves, as suggested by the notation $\gamma(\boldsymbol{h})$. Intrinsic stationarity and second-order stationarity are similar but not identical concepts where intrinsic stationarity defines a slightly broader class of spatial processes (see Cressie, 1993, p. 68; Banerjee et al., 2004, pp. 22–24). The notion of isotropy (rotation invariance) can apply to variograms in the same manner as it does for covariance functions. The semivariogram is related to the covariance function via the relationship

$$\gamma(\boldsymbol{h}) = \mathrm{Cov}\big(Z(\boldsymbol{s}), Z(\boldsymbol{s})\big) - \mathrm{Cov}\big(Z(\boldsymbol{s}), Z(\boldsymbol{s}+\boldsymbol{h})\big),$$

where $\mathrm{Cov}(Z(\boldsymbol{s}), Z(\boldsymbol{s}))$ naturally refers to $\mathrm{Var}(Z(\boldsymbol{s}))$ and represents

$$\lim_{\|\boldsymbol{h}\|\to\infty} \gamma(\boldsymbol{h}),$$

the limit of the semivariogram as the lag goes to infinity. This relationship holds for the set of second-order stationary processes, but does not automatically hold for intrinsically stationary processes unless they are also second-order stationary. We restrict attention to second-order stationary processes for the remainder of this review.

Finally, the covariance function $\mathrm{Cov}(\boldsymbol{s}, \boldsymbol{s}+\boldsymbol{h})$ (which we assume is defined by parametric family $\alpha K_\theta(\boldsymbol{s}, \boldsymbol{s}+\boldsymbol{h})$) must define a valid covariance function, i.e., must be a positive-definite function, i.e.,

$$\sum_{i=1}^{n}\sum_{j=1}^{n} a_i a_j \mathrm{Cov}(\boldsymbol{s}, \boldsymbol{s}+\boldsymbol{h}) \geqslant 0,$$

for any $a_1, \ldots, a_n$ and strict inequality if not all a_i are zero. Similarly, a valid semivariogram must be negative definite, i.e.,

$$\sum_{i=1}^{n}\sum_{j=1}^{n} a_i a_j \gamma(\boldsymbol{s}, \boldsymbol{s}+\boldsymbol{h}) \leqslant 0,$$

for $a_1, \ldots, a_n$ such that $\sum_{i=1}^{n} a_i = 0$.

Determining the validity of a proposed covariance function or semivariogram often is not a straightforward task, but there is a long list of (often isotropic) parametric families that generate valid forms for particular spatial dimensions. These are reviewed in Cressie (1993), Banerjee et al. (2004), and Waller and Gotway (2004) among other places. For illustration, we define three such families here and defer details on others to the references listed.

First, letting $h = \|\boldsymbol{h}\|$, consider the *exponential* covariance function defined by:

$$K_\theta(h) = \exp(-\theta h)$$

for $h > 0$. Here the covariance declines as a smooth exponential function of lag distance. Next, consider the *Gaussian* covariance function defined by:

$$K_\theta(h) = \exp\left(-\theta^2 h^2\right)$$

for $h > 0$. This covariance function levels off as $h \to 0$, implying an area of fairly strong and consistent spatial correlation for a small range of distances. In fact, the Gaussian covariance function guarantees a predicted surface that is infinitely differentiable, yielding an exceedingly smooth surface. As such, many question its applicability as a parametric covariance family (Wackernagel, 2003; Banerjee et al., 2004, pp. 36–37). Finally, Stein (1999) and Banerjee et al. (2004, p. 36) among others recommend the use of the *Matérn* covariance function which includes both the exponential and Gaussian covariance functions as special cases, moving between them via a parameter governing the smoothness of the underlying process. The Matérn covariance function is defined as

$$K_\theta(h) = \frac{1}{2^{\nu-1}\Gamma(\nu)}\left(2h\phi\sqrt{\nu}\right)^{\nu} B_\nu\left(2h\phi\sqrt{\nu}\right)$$

for $h > 0$ where θ denotes the vector (ν, ϕ), $\Gamma(\cdot)$ the gamma function, and $B_\nu(\cdot)$ the modified Bessel function of order ν (for details see Abramowitz and Stegun, 1965, Chapter 9). The parameter $\nu > 0$ represents a smoothness parameter where the Matérn model reduces to the exponential model for $\nu = 1/2$ and the Gaussian model as $\nu \to \infty$. Note that the functions $K_\theta(h)$ are multiplied by the scale parameter α to complete definition of $\mathrm{Cov}(h)$.

For each definition, the case of $h = 0$ merits special consideration. For a spatially continuous process, $\gamma(0) = 0$ since the value $[Z(\boldsymbol{s}) - Z(\boldsymbol{s})]$ is identically zero (and its variance zero) for any given realization of the process. However, in practice, measurements are often made with error, or there may be very small, local discontinuities such as a small "nugget" in a distribution of gold ore in a mining application. As a result $\lim_{h\to 0}\gamma(h) = \eta$ may not be zero, but rather some positive value, referred to in the geostatistical literature as a *nugget effect*. Practically speaking, a nonzero nugget effect allows local, nonspatially correlated noise in the process (somewhat in the spirit of the exchangeable noise of the regional data modeling section above) and we will note its role in prediction below.

Next, consider the semivariograms corresponding the parametric families of covariance functions defined above. For second-order stationary processes, the corresponding semivariograms are

$$\gamma(h;\eta,\alpha,\theta) = \eta + \alpha\bigl[1 - \exp(-\theta h)\bigr] \quad \text{(exponential)},$$
$$\gamma(h;\eta,\alpha,\theta) = \eta + \alpha\bigl[1 - \exp\bigl(-\theta^2 h^2\bigr)\bigr] \quad \text{(Gaussian)},$$

and

$$\gamma(h;\eta,\alpha,\theta) = \eta + \alpha\left[1 - \frac{(2h\theta\sqrt{\nu})^{\nu}}{2^{\nu-1}\Gamma(\nu)} B_{\nu}\bigl(2h\theta\sqrt{\nu}\bigr)\right] \quad \text{(Matérn)},$$

all for $h > 0$, and all equal η, the nugget effect, as $h \to 0$.

4.2. Kriging: Classical spatial prediction

Building from these ideas of covariance and semivariograms, we next review classical geostatistics. To begin, suppose all covariance/variogram parameters are known, the family $K_\theta(\boldsymbol{s}_i, \boldsymbol{s}_j)$ is correctly specified, all $Z(\boldsymbol{s})$ share the same mean μ (an assumption we later weaken), and we seek the best linear unbiased predictor (BLUP) of $Z(\boldsymbol{s}_0)$ for location $\boldsymbol{s}_0$,

$$\begin{aligned}\widetilde{Z}(\boldsymbol{s}_0) &= \sum_{i=1}^{n} \ell_i Z(\boldsymbol{s}_i)\\ &= \boldsymbol{\ell}'\boldsymbol{Z}.\end{aligned}$$

The goal is to find the vector of weights ℓ_i, $i = 1,\ldots,n$, minimizing the mean square prediction error

$$MSPE = E\bigl[\bigl(\widetilde{Z}(\boldsymbol{s}_0) - Z(\boldsymbol{s}_0)\bigr)^2\bigr].$$

Unbiasedness requires $\sum_{i=1}^{n} \ell_i = 1$ so we have a constrained optimization problem, traditionally solved via addition of a Lagrange multiplier m. Such predictions are referred to as *kriging* predictions based on the early work of South African mining engineer D.G. Krige (1951), and given a formal mathematical basis by Matheron (1963). Traditionally, we find the optimal predictions by solving the *kriging equations*:

$$\boldsymbol{\ell} = \boldsymbol{\Gamma}^{-1}\boldsymbol{\gamma} \tag{8}$$

where

$$\boldsymbol{\ell} = (\ell_1,\ldots,\ell_n,m)',$$
$$\boldsymbol{\gamma} = \bigl(\gamma(\boldsymbol{s}_0 - \boldsymbol{s}_1),\ldots,\gamma(\boldsymbol{s}_0 - \boldsymbol{s}_n),1\bigr)'$$

and the elements of $\boldsymbol{\Gamma}$ are

$$\boldsymbol{\Gamma}_{ij} = \begin{cases} \gamma(\boldsymbol{s}_i - \boldsymbol{s}_j), & i = 1,\ldots,n;\\ & j = 1,\ldots,n;\\ 1 & i = n+1;\ j = 1,\ldots,n;\\ & j = n+1;\ i = 1,\ldots,n;\\ 0 & i = j = n+1. \end{cases}$$

So (8) becomes

$$\begin{bmatrix} \ell_1 \\ \ell_2 \\ \vdots \\ \ell_n \\ m \end{bmatrix} = \begin{bmatrix} \gamma(\boldsymbol{s}_1 - \boldsymbol{s}_1) & \cdots & \gamma(\boldsymbol{s}_1 - \boldsymbol{s}_n) & 1 \\ \gamma(\boldsymbol{s}_2 - \boldsymbol{s}_1) & \cdots & \gamma(\boldsymbol{s}_2 - \boldsymbol{s}_n) & 1 \\ \vdots & \ddots & \vdots & \vdots \\ \gamma(\boldsymbol{s}_n - \boldsymbol{s}_1) & \cdots & \gamma(\boldsymbol{s}_n - \boldsymbol{s}_n) & 1 \\ 1 & \cdots & 1 & 0 \end{bmatrix}^{-1} \begin{bmatrix} \gamma(\boldsymbol{s}_0 - \boldsymbol{s}_1) \\ \gamma(\boldsymbol{s}_0 - \boldsymbol{s}_2) \\ \vdots \\ \gamma(\boldsymbol{s}_0 - \boldsymbol{s}_n) \\ 1 \end{bmatrix}.$$

Note that, in this formulation, the semivariogram defines the optimal weights. The matrix $\boldsymbol{\Gamma}$ utilizes the semivariogram to summarize spatial covariances between data values while the vector $\boldsymbol{\gamma}$ summarizes covariances between each data value and the prediction location $\boldsymbol{s}_0$. Also note that we require a separate set of weights $\boldsymbol{\ell}$ for each prediction location, $\boldsymbol{s}_0$. However, only the right-hand side of Eq. (8) changes with the prediction locations (through $\boldsymbol{\gamma}$). Since $\boldsymbol{\Gamma}$ depends only on the data locations and not on the prediction locations, we need only invert $\boldsymbol{\Gamma}$ once and then multiply by the associated $\boldsymbol{\gamma}$ vector to obtain a prediction for any $\boldsymbol{s}_0$ in D.

The minimized MSPE, also known as the *kriging variance*, also derives from the kriging weights $\boldsymbol{\ell}$ and the vector $\boldsymbol{\gamma}$, specifically,

$$\begin{aligned} \sigma^2_{\text{krige}}(\boldsymbol{s}_0) &= \boldsymbol{\ell}'\boldsymbol{\gamma} \\ &= \sum_{i=1}^{n} \ell_i \gamma(\boldsymbol{s}_0 - \boldsymbol{s}_i) + m \\ &= 2\sum_{i=1}^{n} \ell_i \gamma(\boldsymbol{s}_0 - \boldsymbol{s}_i) - \sum_{i=1}^{n}\sum_{j=1}^{n} \ell_i \ell_j \gamma(\boldsymbol{s}_i - \boldsymbol{s}_j). \end{aligned} \tag{9}$$

Taken together, Eqs. (8) and (9) define *ordinary kriging* prediction where we have a fixed and spatially constant (but unknown) mean. Next, let us consider the more general case considered above where the mean is a linear function of location-specific covariates, i.e., $\boldsymbol{\mu} = (\mu_1, \ldots, \mu_n)$, $\mu_i = \boldsymbol{x}(\boldsymbol{s})'\boldsymbol{\beta}$. In this case, suppose our observations may be modeled as

$$Z(\boldsymbol{s}) = \boldsymbol{x}(\boldsymbol{s})'\boldsymbol{\beta} + \delta(\boldsymbol{s}) + \varepsilon(\boldsymbol{s}), \tag{10}$$

the sum of a fixed (local) mean defined by observed covariates and unobserved parameters, a random intercept, $\delta(\boldsymbol{s})$, itself modeled by a mean-zero Gaussian random field, and spatially-independent noise, $\varepsilon(\boldsymbol{s})$. Note that the spatial correlation structure is captured through the covariance structure of the random intercept $\delta(\boldsymbol{s})$, and additional measurement error or small-scale variation (i.e., a nugget effect) is reflected in the variance of $\varepsilon(\boldsymbol{s})$. The basic structure mirrors that discussed for longitudinal models by Diggle et al. (1994, Section 5.2), and for spatial models by Diggle et al. (1998), Stein (1999), and Banerjee et al. (2004).

We now assume the parametric form of K_θ is known, but the parameter values in θ are not. In other words we seek the vector of prediction weights $\boldsymbol{\ell}(\theta)$ defining the best linear unbiased predictor

$$\widetilde{Z}(\boldsymbol{s}_0) = \boldsymbol{\ell}(\theta)'\boldsymbol{Z},$$

where the optimal weights depend on the unknown covariance parameters in θ. Stein (1999, pp. 7–9) reviews classical linear prediction theory noting that the best linear prediction of $Z(\boldsymbol{s}_0)$, assuming $\boldsymbol{\beta}$ known is

$$\widetilde{Z}(\boldsymbol{s}_0) = \boldsymbol{x}(\boldsymbol{s}_0)'\boldsymbol{\beta} + \boldsymbol{k}\boldsymbol{K}^{-1}(\boldsymbol{Z} - \boldsymbol{X}\boldsymbol{\beta}),$$

where $\boldsymbol{X}$ denotes the $n \times p$ design matrix with ith row $\boldsymbol{x}(\boldsymbol{s}_i)$, $\boldsymbol{k}$ the $n \times 1$ vector with ith element $K_\theta(\boldsymbol{s}_0, \boldsymbol{s}_i)$, and $\boldsymbol{K}$ the $n \times n$ matrix of covariances $\alpha K_\theta(\boldsymbol{s}_i, \boldsymbol{s}_j)$, $i, j = 1, \dots, n$. For $\boldsymbol{\beta}$ unknown, we can replace $\boldsymbol{\beta}$ by its generalized least squares estimate (the best linear unbiased estimate of $\boldsymbol{\beta}$), namely,

$$\hat{\boldsymbol{\beta}} = \left(\boldsymbol{X}'\boldsymbol{K}^{-1}\boldsymbol{X}\right)^{-1}\boldsymbol{X}'\boldsymbol{K}^{-1}\boldsymbol{Z},$$

yielding

$$\begin{aligned}\widetilde{Z}(\boldsymbol{s}_0) &= \boldsymbol{x}(\boldsymbol{s}_0)'\left(\boldsymbol{X}'\boldsymbol{K}^{-1}\boldsymbol{X}\right)^{-1}\boldsymbol{X}'\boldsymbol{K}^{-1}\boldsymbol{Z} \\ &\quad + \boldsymbol{k}\boldsymbol{K}^{-1}\left(\boldsymbol{Z} - \boldsymbol{X}\left(\boldsymbol{X}'\boldsymbol{K}^{-1}\boldsymbol{X}\right)^{-1}\boldsymbol{X}'\boldsymbol{K}^{-1}\boldsymbol{Z}\right).\end{aligned} \tag{11}$$

Rearranging terms yields the prediction weights

$$\begin{aligned}\boldsymbol{\ell}(\theta) &= \left[\boldsymbol{K}^{-1} - \boldsymbol{K}^{-1}\boldsymbol{X}\left(\boldsymbol{X}'\boldsymbol{K}^{-1}\boldsymbol{X}\right)^{-1}\boldsymbol{X}'\boldsymbol{K}^{-1}\right]\boldsymbol{k} \\ &\quad + \boldsymbol{K}^{-1}\boldsymbol{X}\left(\boldsymbol{X}'\boldsymbol{K}^{-1}\boldsymbol{X}\right)^{-1}\boldsymbol{x}(\boldsymbol{s}_0),\end{aligned} \tag{12}$$

which Stein (1999, pp. 7–8) shows are identical to those minimizing the MSPE under the unbiasedness constraint

$$\begin{aligned}&E\left(\boldsymbol{\ell}(\theta)'\boldsymbol{Z}\right) = E\left(Z(\boldsymbol{s}_0)\right), \\ &\Rightarrow \boldsymbol{\ell}(\theta)'\boldsymbol{X}\boldsymbol{\beta} = \boldsymbol{x}(\boldsymbol{s}_0)'\boldsymbol{\beta}.\end{aligned}$$

The associated (minimized) mean square error is given by

$$\begin{aligned}&K_\theta(\boldsymbol{s}_0, \boldsymbol{s}_0) - \boldsymbol{k}'\boldsymbol{K}^{-1}\boldsymbol{k} \\ &\quad + \left(\boldsymbol{x}(\boldsymbol{s}_0) - \boldsymbol{X}'\boldsymbol{K}^{-1}\boldsymbol{X}\right)'\left(\boldsymbol{X}'\boldsymbol{K}^{-1}\boldsymbol{X}\right)^{-1}\left(\boldsymbol{x}(\boldsymbol{s}_0) - \boldsymbol{X}'\boldsymbol{K}^{-1}\boldsymbol{X}\right).\end{aligned} \tag{13}$$

Eqs. (12) and (13) define the *universal kriging equations*. In the case of the constant mean used to present the kriging equations above, we have $\boldsymbol{x}(\boldsymbol{s}) = 1$ for any location $\boldsymbol{s}$. In this setting, referred to in the literature as *ordinary kriging*, the unbiasedness constraint implies $\boldsymbol{k}'\boldsymbol{K}^{-1}\boldsymbol{X} = 1$, and Eq. (11) reduces to $\boldsymbol{\ell}(\theta) = \boldsymbol{K}^{-1}\boldsymbol{k}$, the covariance function analogue to the semivariogram-based ordinary kriging equations above.

We note that the optimal predictors are only optimal if we consider the covariance function (equivalently the semivariogram for intrinsically stationary processes) fixed and known. In practice, this is not the case and most applications of classical geostatistical prediction "plug in" estimates of semivariogram parameters as the unknown values. As a result, such approaches often omit the variation associated with semivariogram estimation in the prediction process and the prediction errors do not reflect this uncertainty. Classical solutions involve exploration of "empirical" best linear unbiased prediction or EBLUPs, considered in some detail by Zimmerman and Cressie (1992) and Harville and Jeske (1992). The desire to account for uncertainty in covariance parameter uncertainty motivated both Le and Zidek (1992) and Handcock and Stein (1993) in their proposals of Bayesian kriging which we outline in the next section.

However, before leaving our (admittedly brief) review of classical kriging, we note that there is a very wide literature on *variography*, the modeling and estimation of (semi)variograms and their associated parameters. Such issues are of modeling interest in its own right as well as for their role in prediction and we refer interested readers to Cressie (1993) and Chilès and Delfiner (1999) for details and relevant references.

4.3. Bayesian kriging

A Bayesian approach brings additional flexibility to the classical prediction framework outlined above. First of all, the issue of incorporating uncertainty in covariance parameters follows directly from posterior inference. More specifically, Bayesian prediction derives from the posterior predictive distribution which integrates over the posterior distribution of all model parameters, i.e.,

$$\begin{aligned}
&p\big(Z(\boldsymbol{s}_0)|\boldsymbol{Z}, \boldsymbol{X}, \boldsymbol{x}(\boldsymbol{s}_0)\big)\\
&\quad = \int p\big(Z(\boldsymbol{s}_0), \boldsymbol{\beta}, \alpha, \eta, \theta|\boldsymbol{Z}, \boldsymbol{X}, \boldsymbol{x}(\boldsymbol{s}_0)\big)\, \mathrm{d}\boldsymbol{\beta}\, \mathrm{d}\alpha\, \mathrm{d}\eta\, \mathrm{d}\theta\\
&\quad = \int p\big(Z(\boldsymbol{s}_0)|\boldsymbol{Z}, \boldsymbol{X}, \boldsymbol{x}(\boldsymbol{s}_0), \boldsymbol{\beta}, \alpha, \eta, \theta\big) p(\boldsymbol{\beta}, \alpha, \eta, \theta|\boldsymbol{Z}, \boldsymbol{X})\, \mathrm{d}\boldsymbol{\beta}\, \mathrm{d}\alpha\, \mathrm{d}\eta\, \mathrm{d}\theta.
\end{aligned}$$

In many cases, this integration may be numerically or theoretically intractable leading to the use of MCMC methods for evaluation.

In addition to providing a mechanism for accounting for uncertainty in variance–covariance modeling, Bayesian thinking enters the spatial prediction process in additional ways as well. One simple advantage is the conceptual ease of accommodating functions of model parameters. For example, many researchers model environmental exposure measurements via a log-normal distribution where one treats the natural logarithm of observed values as observations from a Gaussian distribution. Classical geostatistical prediction as outlined above provides BLUP of the log-exposure values, but such are no longer linear or even unbiased predictions on the original (untransformed) scale. Adjustments exist (Cressie, 1993, pp. 135–137), but must be derived for any reparameterization. In the Bayesian context, transformations are less problematical, as predictions derive from the posterior predictive distribution rather than a necessarily linear combination of observations. This is particularly evident in MCMC implementation where a sample from the posterior distribution of model parameters may be transformed to a sample from the posterior distribution of the transformed parameters with little effort.

Bayesian thinking also aids in constructing spatial models, providing a means of inference for complex, multi-level, hierarchical models. Recalling the preceding section's discussion of how a hierarchical structure provides convenient spatial modeling opportunities for regional data (by deferring spatial structure to a second stage of the model), we see similar opportunities in Bayesian geostatistical prediction. For example, if we consider the model formulation in Eq. (10), the prior distribution of the spatial random effects $\delta(\boldsymbol{s})$ defines the spatial structure of the data (or, more accurately, the spatial structure of the expected value of the data). That is, thoughtful development of prior distributions for the random effects allows a variety of modeling options in addition to a means of "bookkeeping" components of uncertainty. In addition, hierarchical structures also allow options for efficient computation of MCMC schemes.

To see the impact of the choice of hierarchical structure consider the following formulations of the model in Eq. (10) outlined by Banerjee et al. (2004, pp. 131–132). First, define the initial stage of the model as

$$\mathbf{Z}|\boldsymbol{\beta}, \alpha, \theta, \eta \sim N\big(\boldsymbol{X}'\boldsymbol{\beta}, \alpha\boldsymbol{K}(\theta) + \eta I\big),$$

where I is the identity matrix and we specifically denote the dependence of the matrix $\boldsymbol{K}$ on the covariance parameters in θ. To complete the model one assigns prior distributions to model parameters $\boldsymbol{\beta}$, α, θ, and η. Next, consider a structure with first-stage conditional on model parameters and on the conditional random effects $\delta(\boldsymbol{s})$, i.e.,

$$\mathbf{Z}|\boldsymbol{\beta}, \alpha, \theta, \eta, \boldsymbol{\delta} \sim N\big(\boldsymbol{X}'\boldsymbol{\beta} + \boldsymbol{\delta}, \eta I\big),$$

where $\boldsymbol{\delta}$ denotes the vector with elements $\delta(\boldsymbol{s}_i)$, $i = 1, \ldots, n$. Now, we define a second stage of the model giving prior distributions to parameters $\boldsymbol{\beta}$ and η, and

$$\boldsymbol{\delta}|\alpha, \theta \sim N\big(\mathbf{0}, \alpha\boldsymbol{K}(\theta)\big),$$

then complete the model with prior distributions for α and θ. Briefly deferring a discussion of the prior distributions for model parameters for a moment, consider implications of these two model formulations. While both formulations ultimately base inference on the joint posterior of model parameters, $p(\boldsymbol{\beta}, \alpha, \theta, \eta|\mathbf{Z})$, this distribution is (assigning independent prior distributions for notational convenience)

$$p(\mathbf{Z}|\boldsymbol{\beta}, \alpha, \theta, \eta)p(\boldsymbol{\beta})p(\alpha)p(\theta)p(\eta)$$

in the first case and

$$p(\mathbf{Z}|\boldsymbol{\beta}, \eta, \boldsymbol{\delta})p(\boldsymbol{\delta}|\alpha, \theta)p(\boldsymbol{\beta})p(\eta)p(\alpha)p(\theta)$$

in the second case, where the former is marginalized over the random effects vector $\boldsymbol{\delta}$. The difference between the two formulations appears when considering MCMC implementation where Banerjee et al. (2004, p. 132) note that the marginalized formulation is often better behaved since the matrix $\alpha\boldsymbol{K}(\theta)$ is often nearly singular while $\alpha\boldsymbol{K}(\theta) + \eta I$ often is not.

While a Bayesian setting provides some conceptual and computational niceties, in application we must complete the hierarchical model by defining reasonable prior distributions that (a) do not overly influence predictions and (b) ensure proper posterior distributions. Models typically support vague or even uniform priors for the linear mean parameters $\boldsymbol{\beta}$ but prior specification for the nugget effect and other covariance parameters (η, α, and the elements of θ) requires more thought, and a variety of approaches appear in the literature. We next briefly review a few general strategies for prior specification for covariances.

Starting broadly, one could ignore parametric covariance functions and instead define a prior distribution on the entire dispersion matrix Σ_Z, perhaps using a conjugate inverse (or inverted) Wishart distribution (cf. Box and Tiao, 1973, pp. 459–470, for general results). While conjugacy offers some theoretical attraction, note that the inverse Wishart prior on Σ_Z no longer limits attention to specifically *spatial* covariance structures, and Daniels and Kass (1999) outline questionable performance of the inverse Wishart prior in some hierarchical modeling settings. There is increasing attention paid to nonconjugate Bayesian analysis of covariance/correlation matrices in general (Daniels and Kass, 1999, 2001; Wong et al., 2003; Liechty et al., 2004) and with respect to particular applications in time series (Daniels and Cressie, 2001) and longitudinal analysis (Daniels and Zhao, 2003). Such approaches often provide "shrinkage" between an independence structure and specified correlation structures.

Moving more specifically to spatial hierarchical models of the sort defined in Eq. (10) with an assumed parametric family for pairwise covariances, Banerjee et al. (2004, p. 131) recommend the general pragmatic solution of selecting proper, informative priors for each variance–covariance parameter, often starting with conjugate inverse Gamma distributions for variance/scale parameters such as α and η. The authors also caution that very vague priors on such parameters may lead to essentially improper posterior distributions hence MCMC convergence failure, so some care is still required to maintain posterior propriety without unduly limiting prior ranges of parameters.

Expanding on this issue, Berger et al. (2001, 2003) move toward objective Bayesian spatial analysis by considering *reference* and *Jeffreys' priors* for variance–covariance parameters in a Gaussian random field with no nugget effect. The authors illustrate that the Gaussian random field structure raises several interesting aspects not previously encountered in the reference prior literature. In particular, the model provides an example where Jeffreys' prior applied independently to each component yields an improper posterior. In addition, a popular approximation in the derivation of reference priors does not hold for the multivariate Gaussian data arising in a Gaussian random field. The results of Berger et al. (2001, 2003) provide tantalizing information regarding assignment of prior distributions for spatial covariance parameters, but also point to the need for additional development.

As a final note, the model formulation outlined above extends to non-Gaussian models by adding spatial random effects inside the link function of a generalized linear model, creating a generalized linear mixed model. More specifically, consider a generalized linear model where observations $Z(\boldsymbol{s})$ follow a distribution in the exponential family and

$$g\big[E\big(Z(\boldsymbol{s}_i)\big)\big] = \boldsymbol{x}(\boldsymbol{s}_i)'\boldsymbol{\beta} + \delta(\boldsymbol{s}_i)$$

for link function $g(\cdot)$ where $\delta(\boldsymbol{s})$ still denotes a set of random effects modeled by a zero-mean Gaussian random field (noting the observations are not necessarily Gaussian). In this case, the functional form of the likelihood changes to reflect the assumed distribution within the exponential family, but remains in an attractive product over the observations, conditional on the spatial random effects in $\delta(\boldsymbol{s})$ which are modeled at the second stage via a familiar spatial Gaussian form, very similar to the appealing structure of the Bayesian models for regional data presented in the preceding section. Such models fall into the class of *model-based geostatistics* introduced by Diggle et al. (1998) with relevant issues raised in the accompanying discussion of this paper, and further considered by Zhang (2004) and Banerjee et al. (2004).

The link in model structure between geostatistical and regional inference is appealing, but does not precisely equate the two categories of models. First, key inferential differences remain: in geostatistical data we assume that we could predict $Z(\boldsymbol{s})$ at any (point) location $\boldsymbol{s} \in D$, while regional data represent *summaries* over regions, and predicting a county-specific outcome for some hypothetical county falling between existing counties makes little sense. Furthermore, geostatistical models assume a field comprised of an infinite number of random variables, some of which are observed, while regional models often consist of a finite number of random variables, most or all of which are observed. Approaches layering latent geostatistical fields under observed regional data sets have been considered and offer areas of further development (Best et al., 2000). Finally, shifting between point and regional data support involves special consideration falling under the term "change of support", see Cressie (1993, pp. 284–289) for a description of classical approaches, Gotway and Young (2002) for a recent review, and Mugglin and Carlin (1998), Mugglin et al. (2000), and Banerjee et al. (2004, pp. 168–169, and Chapter 6) for a hierarchical Bayesian perspective.

We next consider the third area of inference, namely the analysis of spatial point processes.

5. Bayesian thinking in spatial point processes

A *spatial point process* is a stochastic process whose realizations consist of a set of n events occurring at locations $(\boldsymbol{x}_1, \dots, \boldsymbol{x}_n)$ in the study area of interest D, where the locations and total number of events (n) are random variables. We use the notation $\boldsymbol{x}$ for realized random locations to contrast with the fixed-by-design locations $\boldsymbol{s}$ in geostatistical data. Diggle (2003) and Waller and Gotway (2004, Chapter 5) provide general introductions to the statistical analysis of spatial point processes, and Stoyan et al. (1995), Lawson and Denison (2002), and Møller and Waagepetersen (2004) provide more detailed developments and applications. Here we briefly review definitions of key elements of a few particular classes of spatial point processes, namely spatial Poisson processes, Cox processes, and log Gaussian Cox processes, illustrating Bayesian concepts arising in their definition and application. Statistical issues in spatial point processes often wander farther from traditional statistical inference and stray more into probabilistic modeling than do methods for regional modeling or geostatistical prediction, but we highlight some familiar themes from preceding sections to link the corresponding literatures.

5.1. *Some spatial point processes of interest*

5.1.1. *Homogeneous Poisson processes*

To begin, consider a *homogeneous spatial Poisson process* defined by the following criteria (Diggle, 1983, p. 50; Stoyan et al., 1995, p. 33):

(i) The number of events occurring within a finite region D is a random variable following a Poisson distribution with mean $\lambda|D|$ for some positive constant λ and $|D|$ denoting the area of A, and
(ii) Given the total number of events, n, occurring within an area D, the locations of the n events represent an independent random sample of n locations, where each point (a location where an event could occur) is equally likely to be chosen as an event.

Criterion (ii) represents the general concept of *complete spatial randomness* (CSR) (events uniformly distributed across the study area), and criterion (i) introduces the idea of an *intensity* λ representing the expected number of events per unit area. The Poisson distribution allows the total number of events observed to vary from realization to realization while maintaining a fixed (but unknown) expected number of events per unit area. Dividing the total number of events observed by the total area provides a straightforward estimate of λ, i.e., $\hat{\lambda} = n/|D|$. This estimate serves well for many applications, however we note that other estimators can provide better performance in certain situations, particularly those involving estimates based on sparse samples of the set of observed events (Diggle and Cox, 1981).

For further insight into the properties of homogeneous Poisson processes, consider an equivalent definition listed in Cressie (1993, p. 634):

(a) the numbers of events in nonoverlapping regions are statistically independent,
(b) for any region $A \subseteq D$,

$$\lim_{|A|\to 0} \frac{\Pr[\text{exactly one event in } A]}{|A|} = \lambda > 0,$$

where $|A|$ = the area of region A, D is the domain of interest (study area), and

(c)

$$\lim_{|A|\to 0} \frac{\Pr[\text{two or more events in } A]}{|A|} = 0.$$

Component (a) is particularly important to the analysis of regional counts, e.g., the number of events observed in a partition of the study area into small areas. Diggle (1983, p. 50) formally establishes the link between criteria (i) and (ii) and component (a) above. Component (b) implies the probability of a single event in an increasingly small area A (adjusted for the area of A) is a constant (λ) independent of the location of region A within the study area of interest. Component (c) implies the probability of two or more events occurring in precisely the same location is zero. As above, the quantity λ is the Poisson parameter, or the intensity of the process, and is equal to the mean number of points per unit area.

Since the intensity of events is constant at all locations in the study area, we say the process is *homogeneous*. Mathematically, stationarity and homogeneity are related but separate concepts (e.g., a process defined within a finite study area cannot be stationary since translation outside of the study area is undefined).

While the homogeneous Poisson process provides a mathematical model for the lack of spatial pattern, various extensions to the Poisson process provide additional structure for modeling spatial pattern. We limit attention to a set of processes maintaining independence between event locations, i.e., the occurrence of an event at a particular location does not impact the probability of other events occurring nearby. Interested readers are directed to Stoyan et al. (1995), Diggle (2003), Geyer (1999), and Møller and Waagepetersen (2004) for models of processes with interevent dependence.

5.1.2. *Heterogeneous Poisson processes*

Our first generalization is to allow the intensity function to be a spatially varying function $\lambda(\boldsymbol{x})$, yielding a *heterogeneous Poisson process* defined by

(i*) The number of events occurring within a finite region A is a random variable following a Poisson distribution with mean $\int_A \lambda(\boldsymbol{x})\,\mathrm{d}\boldsymbol{x}$, and

(ii*) Given the total number of events, n, occurring within an area A, the n events represent an independent random sample of n locations, with the probability of sampling a particular point $\boldsymbol{s}$ proportional to $\lambda(\boldsymbol{x})$.

Note that the number of events observed in disjoint regions still follow independent Poisson distributions but now with expectation $\int_A \lambda(\boldsymbol{x})\,\mathrm{d}\boldsymbol{x}$. In addition, events are distributed according to a spatial density function proportional to $\lambda(\boldsymbol{x})$. In a heterogeneous Poisson process observed events may "cluster" together in areas with higher values of $\lambda(\boldsymbol{x})$, but, as noted above, the events remain mutually independent of one another.

Linking this definition back to the ideas of stationarity and isotropy defined for geostatistical data, note that heterogeneity (as defined by a spatially varying intensity function) necessarily implies nonstationarity as the point process is no longer translation invariant (stationarity). Since we define isotropy with respect to rotation invariance around a single point (typically referred to as the origin), heterogeneity in the intensity function results in an anisotropic process only if $\lambda(\boldsymbol{x})$ itself is anisotropic (i.e., not symmetric around the origin).

5.1.3. *Cox processes*

To further generalize our model, consider a *Cox process* (Cox, 1955) where the (heterogeneous) intensity function itself represents a realization of a random process, $\Lambda(\boldsymbol{x})$. That is, suppose we wish to generate a realization from a Cox process. First, we generate the number n of events from a Poisson distribution, and generate a particular intensity surface $\lambda(\boldsymbol{x})$ from the process $\Lambda(\boldsymbol{x})$ (ignoring for the moment the precise algorithm for doing so). Next, conditional on n and $\lambda(\boldsymbol{x})$, we generate n locations from the spatial probability density proportional to $\lambda(\boldsymbol{x})$.

To further crystallize the various types of processes considered, generating repeated realizations with n and $\lambda(\boldsymbol{x})$ fixed corresponds to a *binomial process* (Stoyan et al., 1995), generating repeated realizations with $\lambda(\boldsymbol{x})$ fixed and n drawn from a Poisson distribution corresponds to a *heterogeneous Poisson process*, and generating repeated realizations drawing $\lambda(\boldsymbol{x})$ from $\Lambda(\boldsymbol{x})$ and n from a Poisson distribution each time corresponds to a *Cox process*.

In a sense, a Cox process represents a hierarchical probability model with a spatially defined prior distribution on the intensity function. However, unless $\Lambda(\boldsymbol{x})$ is defined parametrically, the problem is a nonparametric one.

5.1.4. log Gaussian Cox processes

Next, suppose we limit attention to the set of *log Gaussian Cox processes*, i.e., Cox processes where the intensity surface process is defined by $\Lambda(\boldsymbol{X}) = \exp(Z(\boldsymbol{x}))$ where $Z(\boldsymbol{x})$ denotes a Gaussian random field with mean $E(Z(\boldsymbol{x})) = \mu(\boldsymbol{x})$ and covariance function $K(\boldsymbol{x}_1, \boldsymbol{x}_2)$ perhaps parameterized as in the geostatistical models above, and with some smoothness restrictions defined in Møller et al. (1998). $Z(\boldsymbol{x})$ need not be stationary (i.e., have a stationary covariance function), but stationarity is often convenient assumption. The family of log Gaussian Cox processes places a by-now familiar Gaussian random field model within the hierarchical structure of Cox processes, linking some ideas from preceding sections to our point process models.

5.2. *Inferential issues*

The models outlined above build increasing layers of complexity and variation onto the basic stochastic allocation of events to locations. Questions of interest include estimation of the intensity surface $\lambda(\boldsymbol{x})$ giving rise to the observed data in order to identify areas with higher or lower probabilities of event locations for the observed realization, and description of the intensity generating process $\Lambda(\boldsymbol{x})$. Approaches may be nonparametric or parametric depending on the sorts of models considered. We may be interested in inserting a parametric structure in $\lambda(\boldsymbol{x})$, for instance inserting a parametric distance decay component around a source of environmental pollution. An estimate of the rate of decay aids in assessments of putative increased incidence of adverse health events in residents living nearby (Diggle, 1990). The (latent) Gaussian random field driving intensity estimation in log Gaussian Cox processes suggests assessment of parameters in the mean function $\mu(\boldsymbol{x})$ and the covariance function $K(\boldsymbol{x}_1, \boldsymbol{x}_2)$.

We review three approaches for predicting $\lambda(\boldsymbol{x})$ based on the observed event locations $(\boldsymbol{x}_1, \ldots, \boldsymbol{x}_n)$, based on the structure of heterogeneous Poisson and Cox processes. In all cases considered here, we build inference (at least conceptually) from the hierarchical nature of a Cox process by treating the unknown intensity surface as a random quantity and defining associated prior distributions and likelihoods, yielding posterior inference for $\lambda(\boldsymbol{x})$.

To begin, we consider an approach due to Heikkinen and Arjas (1998) which merits attention not only due to the approach in general, but also due to interesting conceptual issues involving posterior simulation and inference. Heikkinen and Arjas (1998) build on earlier work by Arjas and Gasbarra (1994) regarding Bayesian estimation of the intensity (hazard rate) of temporal Poisson processes. The basic idea is to build a prior probability model on the space of random step functions defining distributions for the number, locations, and sizes of steps. Coupled with the likelihood based on the observed locations, we aim to generate such step functions from the associated posterior distribution of the intensity function. Arjas and Gasbarra (1994) and Arjas and Andreev (1996) note that each sample from the posterior represents a piecewise constant function so that any assumed "true" smooth intensity function is not a member of the class of intensities generated by the approach, rather the posterior reflects a family of approximations. More specifically, Arjas and Andreev (1996) note the pointwise posterior mean intensity need not belong to the class of sampled posterior intensity functions, but pointwise averages across sampled values typically yield smooth surfaces. We return to this issue after outlining the Heikkinen and Arjas (1998) formulation for spatial intensities.

Suppose the data represent a realization from a heterogeneous Poisson process with unknown intensity function $\lambda(\boldsymbol{x})$ defined in a study window $D \subset \boldsymbol{R}^2$. Our goal is posterior inference on $\lambda(\boldsymbol{x})$. Based on the properties of a heterogeneous Poisson process, the likelihood of an observed point pattern $(\boldsymbol{x}_1, \ldots, \boldsymbol{x}_n)$ is

$$p(\boldsymbol{x}|\lambda(\cdot)) \propto \exp\left\{-\int_D \lambda(\boldsymbol{x})\,\mathrm{d}\boldsymbol{x}\right\} \prod_{i=1}^{n} \lambda(\boldsymbol{x}_i),$$

where integration is with respect to the Lebesgue measure on $\boldsymbol{R}^2$.

We need to define a prior distribution for $\lambda(\boldsymbol{x})$. Heikkinen and Arjas (1998) define a prior distribution on the space of piecewise constant surfaces defined by creating a tessellation (tiling) of D and defining a constant intensity within each tile. This is accomplished in three steps. First, we define a set of K *generating points* $(\xi_1, \ldots, \xi_K)$ in an area $E \supseteq D$. We allow generating points outside of D to reduce the impact of *edge effects* in the estimation of $\lambda(\boldsymbol{x})$, and to allow posterior prediction of $\lambda(\boldsymbol{x})$ for points outside of D (noting that as one moves away from D, the data will have relatively less and the prior relatively more influence on inference). Second, we define the *Voronoi tessellation* (also known as the set of *Theissen polygons*), representing a partitioning of D into a tile (region) E_k for each generating point ξ_k representing the locus of all locations nearer to ξ_k than to any other generating point. Third, we define the constant intensity λ_k associated with each Voronoi tile. Taken together, the three steps yield a probability model defining a class of piecewise-constant intensity functions

$$\lambda(\boldsymbol{x}) = \sum_{k=1}^{K} \lambda_k \mathbf{1}\{\boldsymbol{x} \in E_k\},$$

where $\mathbf{1}\{\cdot\}$ represents the indicator function. The choice of the Voronoi tessellation provides a computational advantage, since updating the tessellation to reflect the addition or deletion of a single generating point can be accomplished very efficiently (Preparata and Shamos, 1985, pp. 204–225; Edelsbrunner, 1987, Chapter 13; Okabe et al., 2000, Chapter 4).

Defining the prior distribution for such piecewise-constant intensities requires defining the number and distribution of the set of generating points and the distribution of intensity values for each tile. This is accomplished in a hierarchical fashion by defining the set of generating points as the realization of a homogeneous Poisson process with constant intensity $\lambda_\xi \in (0, \infty)$ and zero probability assigned to the empty pattern. The properties of a Poisson process imply that the number K of generating points follows a Poisson distribution, and, given K, the generating points are uniformly distributed across E so $p(\boldsymbol{\xi}) \propto \lambda_\xi^K$ for $K > 0$, $p(\boldsymbol{\xi}) = 0$ otherwise.

Heikkinen and Arjas (1998) define the prior distribution of $\boldsymbol{\lambda} = (\lambda_1, \ldots, \lambda_K)$ conditional on the set of generating points $\boldsymbol{\xi}$, by defining a spatial Gaussian process for $\eta = \log(\boldsymbol{\lambda})$. The spatial Gaussian process (as illustrated above for both regional and geostatistical data) provides a mechanism for incorporating spatial correlations, in this case due to the assumed smoothness of the underlying intensity function. More specifically, Heikkinen and Arjas (1998) utilize a conditional autoregressive (CAR) prior for $\boldsymbol{\eta}$ with neighbors defined by the Delauney triangulation of generating points (the graph connecting any pair of generating points whose associated Voronoi tiles share a common point or boundary). As a Delauney triangulation, the set of neighbors represents the "dual graph" to the Voronoi tessellation and is again computationally straightforward to update following the addition/deletion of a single generating point. Heikkinen and Arjas (1998) define a more general set of weights than the adjacency weights defined in the discussion of regional models above, but nonetheless still within the class of weights meeting the symmetry requirements defined in Eq. (5).

This procedure yields a joint prior distribution

$$p(\boldsymbol{\xi}, \boldsymbol{\eta}) = p(\boldsymbol{\eta}|\boldsymbol{\xi})p(\boldsymbol{\xi})$$

with two hyperpriors to be defined, namely the constant intensity λ_ξ in $p(\boldsymbol{\xi})$ and an overall scale parameter in the conditional autoregressive prior $p(\boldsymbol{\eta}|\boldsymbol{\xi})$. Both hyperpriors influence the overall smoothness of the resulting posterior samples, with λ_ξ influencing the (expected) number of generating points (more points tending to give "lumpier" surfaces), and the scale parameter (analogous to τ_{CAR} in Eq. (2)) the conditional variance of η_k relative to its neighboring values.

Assembling the components, the joint posterior becomes

$$p(\boldsymbol{\xi}, \boldsymbol{\eta}|\boldsymbol{x}) \propto p(\boldsymbol{x}|\boldsymbol{\xi}, \boldsymbol{\eta})p(\boldsymbol{\eta}|\boldsymbol{\xi})p(\boldsymbol{\xi}),$$

where, for the piecewise-constant intensities under consideration, the likelihood becomes

$$p(\boldsymbol{x}|\boldsymbol{\xi}, \boldsymbol{\eta}) = \exp\left[\sum_{k=1}^{K} N(E_k)\eta_k - \text{area}(E_k \cap D)\exp(\eta_k)\right],$$

where $N(E_k)$ represents the number of observed events in $\boldsymbol{x}$ occurring within tile E_k and area$(E_k \cap D)$ represents the Lebesgue measure of the area of the intersection of tile E_k with the study area D.

To simulate samples from the posterior requires reversible jump Markov chain Monte Carlo iterations where each update requires either:

- a change in λ_k for a tile k,
- adding a new generating point and associated tile, or
- deleting a generating point and adjusting the neighboring tiles appropriately to maintain a Voronoi tessellation of the remaining generating points.

The probabilities of which action to take depends on the current number of tiles, and details in Heikkinen and Arjas (1998) reveal each update only requires local calculations due to the use of the Voronoi tessellation.

As mentioned above with respect to Arjas and Gasbarra's (1994) temporal intensity estimates, the set of posterior intensity functions is comprised (by definition) entirely of piecewise-constant surfaces, analogous to generating samples from a posterior distribution of histograms to approximate a density function (Arjas and Andreev, 1996).

While no single sampled value represents a smooth surface, the pointwise average of M postconvergence samples, $\lambda^{(1)}, \ldots, \lambda^{(M)}$,

$$\hat{\lambda}(\boldsymbol{x}) = \frac{1}{M} \sum_{m=1}^{M} \lambda^{(m)}(\boldsymbol{x})$$

does tend to yield a smooth surface approximating the posterior "mean" surface, even though such a surface is not a member of the class of surfaces generated by the posterior.

The end result of Heikkinen and Arjas's (1998) approach is an algorithm for posterior inference for intensity functions based on a series of clever approximations, each of which has strong computational advantages (e.g., the Voronoi tessellation, Delauney neighborhoods, and the use of the CAR prior for $\boldsymbol{\eta}$). Several examples involving both observed and simulated examples appear in Heikkinen and Arjas (1998, 1999) and illustrate the flexibility of the approach to accurately detect general modes, trends, and sharp transitions. More recently, Heikkinen (2003) provides updated suggestions for such models, discussing in particular alternatives to the CAR prior to streamline implementation.

Heikkinen and Arjas's (1998) use of a CAR prior for $\boldsymbol{\eta} = \log(\boldsymbol{\lambda})$ suggests a log-Gaussian Cox process structure. Independently, but almost simultaneously to Heikkinen and Arjas (1998), Møller et al. (1998) considered Bayesian prediction of $\lambda(\boldsymbol{x})$ in their definition of log Gaussian Cox processes, using a joint formulation of the Gaussian process defined for a discrete grid of locations within the study area D. The Møller et al. (1998) implementation utilizes a grid of locations spanning D and provides prediction of $\lambda(\boldsymbol{x})$ at each grid point via a Metropolis–adjusted Langevin algorithm (Besag, 1994; Grenander and Miller, 1994; Roberts and Tweedie, 1996).

The probabilistic hierarchy mirrors that of Heikkinen and Arjas (1998), beginning with the likelihood (defined more explicitly here in terms of the latent Gaussian process)

$$p(\boldsymbol{x}|Z) \propto \exp\biggl(-\int_D \exp\bigl(Z(\boldsymbol{x})\bigr)\,\mathrm{d}\boldsymbol{x}\biggr) \prod_{i=1}^{n} \exp\bigl(Z(\boldsymbol{x}_i)\bigr).$$

In general, the posterior distribution of $\lambda(\boldsymbol{x})$ is defined by transformation of the posterior distribution of $Z(\boldsymbol{s})$ given by

$$p(Z|\boldsymbol{x}) \propto p(\boldsymbol{x}|Z)p(Z),$$

where $p(Z)$ is defined as a prior over the family of Gaussian random fields. Heikkinen and Arjas (1998) operationalize this prior (hence posterior) via the definition of generating points and discretizing the study area by Voronoi tessellation.

In contrast, Møller et al. (1998) operationalize the prior (hence posterior) via the posterior distribution of $\lambda(\boldsymbol{s}_{0,1}), \ldots, \lambda(\boldsymbol{s}_{0,K})$ for a set of locations $\boldsymbol{s}_{0,k}$, $k = 1, \ldots, K$. We note that locations $s_{0,k}$, $k = 1, \ldots, K$, are fixed, user-defined prediction locations (typically a set of grid points) not random realizations from a point process so we revert to geostatistical notation for prediction locations $\boldsymbol{s}_0$. However, we also note the prediction locations also play a role similar to Heikkinen and Arjas's (1998) generating points $\{\xi_k\}$ in that they support predictions of $\lambda(\boldsymbol{x})$, so we again index by $k = 1, \ldots, K$, although here the total number of prediction locations K remains fixed throughout.

Waagepetersen (2004) summarizes the approach of Møller et al. (1998) by noting the posterior distribution of $(Z(s_{0,1}), \ldots, Z(s_{0,K}))$, given the observed locations $x_1, \ldots, x_n$ (and assuming $K \geqslant 1$) can be expressed as

$$p\big(Z(s_{0,1}), \ldots, Z(s_{0,K})|x\big) \\ \propto E\big[p\big(x|Z(s_{0,1}), \ldots, Z(s_{0,K})\big)\big]p\big(Z(s_{0,1}), \ldots, Z(s_{0,K})\big). \quad (14)$$

The posterior of the entire *process* $Z|x$ is given by the set of finite-dimensional posterior distributions in Eq. (14) for $K \geqslant 1$ and all sets of grid points $s_{0,1}, \ldots, s_{0,K}$ in D. The details of the MCMC approach are beyond the scope of this article, but are discussed in Møller et al. (1998), Møller and Waagepetersen (2004), and Waagepetersen (2004).

As in the approach of Heikkinen and Arjas (1998), the posterior defined in (14) is analytically intractable and Monte Carlo approaches are not possible without discretizing Z. Møller et al. (1998) approximate Z via a piecewise constant surface defined at each grid point. Waagepetersen (2004) shows that such an approximation converges to the true continuous process Z as the number of grid points increases and grid spacing decreases.

A third, independent and simultaneously developed approach to modeling spatial intensities is given by Wolpert and Ickstadt (1998), whose motivation included a desire to develop a scalable hierarchical model for regional counts, i.e., a model whose spatial correlation structure is preserved under aggregation and disaggregation of regions, unlike the CAR models defined for regional data above. As with Heikkinen and Arjas (1998) and Møller et al. (1998), Wolpert and Ickstadt (1998) consider a hierarchical Cox process structure. To begin, the number of events observed in the study area follows a Poisson distribution with mean defined by the intensity function $\lambda(x)$, $x \in D$. The intensity, in turn, represents the realization of a generating process $\Lambda(x)$.

Rather than a log Gaussian definition of $\Lambda(x)$, Wolpert and Ickstadt (1998) consider the intensity process to be defined by a weighted sum of kernels, i.e.,

$$\Lambda(x) = \sum_c \delta_c \omega(\kappa_c, x),$$

where δ_c denotes a set of weights, $\omega(\kappa_c, x)$ a kernel function centered at point κ_c and evaluated at x, and κ_c, $c = 1, \ldots, C$, denote points arising from a separate spatial point process. As a result, the observed points x represent the superposition of events generated by C independent heterogeneous Poisson processes with intensities defined by the kernels $\omega(\kappa_c, x)$ centered at points κ_c, $c = 1, \ldots, C$.

Conceptually, it is helpful to think of such a process as a *cluster process* (Neyman, 1939; Neyman and Scott, 1958) wherein *parent* events κ_c, $c = 1, \ldots, C$, are scattered across D, and the cth parent generates a random number of *offspring* events which are distributed according to an intensity $\omega(\kappa_c, x)$ around the location of the cth parent, κ_c. The observed process consists only of the offspring points, x. More recently, Møller (2003) considered the broader class of *shot noise Cox processes* (think of parent locations as the center of a shotgun blast and each offspring location as a point of impact of a particular pellet), which include a wide variety of cluster processes, including that proposed by Wolpert and Ickstadt (1998). While applications of such (cluster or shot noise) processes often involve assessment of spatial interactions between events from the same (unobserved) parent location, without additional information, the patterns of (correlated) locations from a cluster process are analytically indistinguishable from patterns of (independent) locations arising under a general Cox process with appropriately constructed intensity generating process, a relationship first formally derived by Bartlett (1964). In other words, we cannot determine from the observed pattern alone whether event locations are uniformly distributed in space but correlated with one another, are independent and heterogeneously distributed in space, or some combination of the two. However, even though the Wolpert and Ickstadt (1998) process can be viewed as a cluster process, a Bayesian analysis still allows posterior prediction of an underlying intensity surface, our underlying inferential goal.

Wolpert and Ickstadt's (1998) proposed prior for $\Lambda(\boldsymbol{x})$ represents a gamma random field in contrast to the Gaussian random field utilized in the other approaches. Some advantages of the gamma field prior include: conjugacy of the random field, the desired "scalability", and a convenient way to model uncertainty in location as a discrete sum of infinitely many point masses. For implementation, Wolpert and Ickstadt (1998) detail construction of a general MCMC approach with Metropolis steps based on an inverse Lévy measure.

To review, the primary difference between the three approaches outlined is in their respective definitions of a prior distribution for the unknown process $\Lambda(\boldsymbol{x})$ generating the intensity function $\lambda(\boldsymbol{x})$, and the resulting MCMC implementation. The family of log Gaussian Cox processes has many nice mathematical properties (Møller et al., 1998; Møller and Waagepetersen, 2004), and fits most seamlessly with the geostatistical models outlined above. Møller and Waagepetersen (2002), Richardson (2003), and Møller (2003) provide additional discussion comparing approaches, noting the amount and type of prior knowledge coupled with the user's analytic goals provide the primary motivations for choosing one approach over another. In particular, log Gaussian Cox processes offer an attractive structure for modeling mutually independent events whose locations result from a smoothly varying intensity function (e.g., due to smoothly varying environmental factors). The embedded Gaussian random field provides a direct link between log Gaussian Cox process models and the hierarchical prediction models described briefly above and outlined extensively in Banerjee et al. (2004). Shot noise Cox processes provide additional flexibility and seem most natural for modeling clustered phenomena. In addition, the weighted kernel mixture defining the intensity process for shot noise Cox processes bears similarity to Bayesian nonparametric models such as those incorporating Dirichlet process priors for density estimation (Escobar and West, 1995). Even though inference for density and intensity functions are (proportionally) equivalent, there has been relatively little overlap to date between the Bayesian point process literature and the Bayesian density estimation literature.

The discussion above focused on inference for the intensity function as the most straightforward connection to models considered in preceding sections. For example, a log Gaussian Cox process differs from hierarchical geostatistical models primarily through the addition of an additional (lower) level to the hierarchy reflecting the Poisson process generating event locations. As in other arenas, the advantage of a Bayesian formulation is fully model-based inference allowing assessment of additional features of the model, whether that be counts of events in regions or estimates of parameters driving spatial heterogeneities or correlations.

6. Recent developments and future directions

Our review of Bayesian thinking in spatial statistics only touches on a series of common themes across major inferential issues in spatial statistics. The central component to all approaches (indeed for all Bayesian analyses) is the development and application of a fully-defined probability model from which posterior inference flows. As noted by Box and Tiao (1973, pp. 9–10), the appropriateness of the model may (should) always be challenged, but fully Bayesian (posterior-based) inference focuses criticism on the appropriateness of the probabilistic structure of the model, rather than the mode of inference, accuracy of approximation, or appropriateness of the particular choice of estimator.

While lengthy, the above review is by no means comprehensive. There are many interesting branches from the main path and we conclude our tour by highlighting related areas and current developments relating to Bayesian analysis of spatially-referenced data.

Extensions of the CAR models for regional data for include spatio-temporal extensions (Waller et al., 1997; Knorr-Held and Besag, 1998; Xia and Carlin, 1998). Such models allow posterior inference regarding temporal evolution of spatial structure, spatial patterns in temporal trends, and combinations of the two. Additional spatio-temporal work includes expansion of the class of spatio-temporal covariance structures to allow "nonseparable" relationships involving interactions between spatial and temporal relationships (Knorr-Held, 2000; Sun et al., 2000).

Another class of extensions to the family of CAR priors addresses multivariate CAR (MCAR) models allowing spatial similarity between and across multiple random quantities measured over the same study area (e.g., allowing correlation between levels of particulate matter composition with neighboring values of ozone levels). Relevant work includes Kim et al. (2001) and Carlin and Banerjee (2003).

A variety of additional applications utilize CAR priors for parameters governing local deviations in survival rates and covariate effects. The former typically assigns spatial prior distributions to frailty models (Banerjee and Carlin, 2002, 2003). The latter seeks to provide model-based inference wherein covariate effects may vary from location to location. The growing popularity of *geographically weighted regression* (GWR, Fotheringham et al., 2002) in the social sciences illustrates the wide audience for such analyses but current implementations provide descriptive estimates and fall somewhat short of a comprehensive framework for inference. Bayesian approaches appear in Schootman and Sun (2004) who explore a specific regional data example, and in Banerjee et al. (2004, pp. 356–366) who provide a more general model-based structure.

Advances in the Bayesian analysis of geostatistical data include a thorough discussion of model-based geostatistics in Diggle et al. (1998) and the methods and models explored in detail in Banerjee et al. (2004). Particular examples include Ecker and Gelfand's (1997) Bayesian inference for semivariograms as a preface to their treatment of spatial prediction under anisotropy (Ecker and Gelfand, 1999, 2003); and inference for spatial gradients in the predicted surface (Banerjee et al., 2003).

The preceding section illustrated the use of hierarchical models for inference regarding the intensity of heterogeneous Poisson processes in the Cox process setting. Other applications include inference regarding pairwise interaction parameters for correlated point processes (Bognar and Cowles, 2004; Bognar, 2004). Unlike many MCMC-based Bayesian analyses, the models in Bognar and Cowles (2004) involve algorithms complicated by a variable proportionality constant, lessening one of the convenient advantages usually expected in MCMC approaches. Recent developments in the analysis of spatial point processes also include spatio-temporal formulations of both log Gaussian Cox processes (Brix and Diggle, 2001) and shot noise Cox processes (Brix and Chadoeuf, 2002). On a more theoretical note, Schoenberg (1999) provides a means to transform general spatial point processes into inhomogeneous Poisson processes, a result that may allow application of the Cox process approaches above to a much wider class of spatial point process models.

Finally, there is also a growing literature of Bayesian approaches to address spatial classification problems. Knorr-Held and Rasser (2000), Knorr-Held (2000), and Denison and Holmes (2001) both consider such problems with regional data, classifying regions into spatially contiguous categories to identify similar summary values. Byers and Raftery (1998, 2002) address a similar problem with point data using Voronoi tiles to estimate the underlying intensity function (in a manner very similar to the approach of Heikkinen and Arjas, 1998) in order to classify the study are into areas of high and low intensity.

In conclusion, we find great appeal for the application of Bayesian approaches to the analysis of spatial data, particularly through the use of hierarchical structures that play important roles in the analysis of regional, geostatistical, and point process data. We hope the discourse above provides a thorough (although by no means complete!) introduction to Bayesian ideas in spatial statistical analysis and gives readers multiple resources for further study.

References

Abramowitz, M., Stegun, I.A. (1965). *Handbook of Mathematical Functions*. Dover, New York.

Arjas, E., Andreev, A. (1996). A note on histogram approximation in Bayesian density estimation. In: Bernardo, J.M., Berger, J.O., Dawid, A.P., Smith, A.F.M. (Eds.), *Bayesian Statistics, vol. 5*. Oxford University Press, Oxford, pp. 487–490.

Arjas, E., Gasbarra, D. (1994). Nonparametric Bayesian inference from right censored survival data, using the Gibbs sampler. *Statistica Sinica* **4**, 505–524.

Arnold, B.C., Castillo, E., Sarabia, J.M. (1999). *Conditional Specification of Statistical Models*. Springer, New York.

Banerjee, S., Carlin, B.P. (2002). Spatial semiparametric proportional hazards models for analyzing infant mortality rates in Minnesota counties. In: Gatsonis, C., et al. (Eds.), *Case Studies in Bayesian Statistics, vol. VI*. Springer-Verlag, New York, pp. 127–151.

Banerjee, S., Carlin, B.P. (2003). Semiparametric spatio-temporal frailty modeling. *Environmetrics* **14**, 523–535.

Banerjee, S., Gelfand, A.E., Sirmans, C.F. (2003). Directional rates of change under spatial process models. *J. Amer. Statist. Assoc.* **98**, 946–954.

Banerjee, S., Carlin, B.P., Gelfand, A.E. (2004). *Hierarchical Modeling and Analysis for Spatial Data*. Chapman and Hall/CRC, Boca Raton.

Bartlett, M.S. (1964). The spectral analysis of two-dimensional point processes. *Biometrika* **51**, 299–311.

Bartlett, M.S. (1975). *The Statistical Analysis of Spatial Pattern*. Chapman and Hall, London.

Berger, J.O., De Oliveira, V., Sansò, B. (2001). Objective Bayesian analysis of spatially correlated data. *J. Amer. Statist. Assoc.* **96**, 1361–1374.

Berger, J.O., De Oliveira, V., Sansò, B., Ren, C., Wittig, T.A. (2003). Correction. *J. Amer. Statist. Assoc.* **98**, 779.

Bernardinelli, L., Clayton, D., Montomoli, C. (1995). Bayesian estimates of disease maps: How important are priors?. *Statistics in Medicine* **14**, 2411–2431.

Bernardo, J.M., Smith, A.F.M. (2000). *Bayesian Theory*. Wiley, Chichester.

Besag, J. (1974). Spatial interaction and the statistical analysis of lattice systems (with discussion). *J. Roy. Statist. Soc., Ser. B* **36**, 192–236.

Besag, J. (1994). Discussion of paper by Grenander and Miller. *J. Roy. Statist. Soc., Ser. B* **56**, 591–592.

Besag, J., Kooperberg, C. (1995). On conditional and intrinsic autoregressions. *Biometrika* **82**, 733–746.

Besag, J., York, J., Mollié, X. (1991). Bayesian image restoration, with two applications in spatial statistics (with discussion). *Ann. Inst. Statist. Math.* **43**, 1–59.

Besag, J., Green, P., Higdon, D., Mengersen, K. (1995). Bayesian computation and stochastic systems (with discussion). *Statist. Sci.* **10**, 3–66.

Best, N.G., Arnold, R.A., Thomas, A., Waller, L.A., Conlon, E.M. (1999). Bayesian models for spatially correlated disease and exposure data. In: Bernardo, J.M., Berger, J.O., Dawid, A.P., Smith, A.F.M. (Eds.), *Bayesian Statistics, vol. 6*. Oxford University Press, Oxford, pp. 131–156.

Best, N.G., Ickstadt, K., Wolpert, R.L. (2000). Spatial Poisson regression for health and exposure data measured at disparate resolutions. *J. Amer. Statist. Assoc.* **95**, 1076–1088.

Bognar, M.A. (2004). Bayesian inference for spatially inhomogeneous pairwise interacting point processes. *Comput. Statist. Data Anal.* Submitted for publication.

Bognar, M.A., Cowles, M.K. (2004). Bayesian inference for pairwise interacting point processes. *Statist. Comput.* **14**, 109–117.

Box, G.E.P., Tiao, G.C. (1973). *Bayesian Inference in Statistical Analysis*. Wiley, New York.

Breslow, N.E., Clayton, D.G. (1993). Approximate inference in generalized linear mixed models. *J. Amer. Statist. Assoc.* **88**, 9–25.

Brix, A., Chadoeuf, J. (2002). Spatio-temporal modeling of weeds by shot-noise Cox processes. *Biometrical J.* **44**, 83–99.

Brix, A., Diggle, P.J. (2001). Spatiotemporal prediction for log-Gaussian Cox processes. *J. Roy. Statist. Soc., Ser. B* **63**, 823–841.

Byers, S.D., Raftery, A.E. (1998). Nearest neighbor clutter removal for estimating features in spatial point processes. *J. Amer. Statist. Assoc.* **93**, 577–584.

Byers, S.D., Raftery, A.E. (2002). Bayesian estimation and segmentation of spatial point processes using Voronoi tilings. In: Lawson, A.B., Denison, D.G.T. (Eds.), *Spatial Cluster Modeling*. Chapman and Hall/CRC, Boca Raton, pp. 109–121.

Carlin, B.P., Banerjee, S. (2003). Hierarchical multivariate CAR models for spatio-temporally correlated survival data (with discussion). In: Bernardo, J.M., Bayarri, M.J., Berger, J.O., Dawid, A.P., Heckerman, D., Smith, A.F.M., West, M. (Eds.), *Bayesian Statistics, vol. 7*. Oxford University Press, Oxford, pp. 45–63.

Chilès, J.-P., Delfiner, P. (1999). *Geostatistics: Modeling Spatial Uncertainty*. Wiley, New York.

Clayton, D.G., Bernardinelli, L. (1992). Bayesian methods for mapping disease risk. In: Elliott, P., Cuzick, J., English, D., Stern, R. (Eds.), *Geographical and Environmental Epidemiology: Methods for Small-Area Studies*. Oxford University Press, London, pp. 205–220.

Clayton, D., Kaldor, J. (1987). Empirical Bayes estimates of age-standardized relative risk for use in disease mapping. *Biometrics* **43**, 671–681.

Cox, D.R. (1955). Some statistical methods related with series of events. *J. Roy. Statist. Soc., Ser. B* **17**, 129–157.

Cressie, N. (1989). Geostatistics. *Amer. Statist.* **43**, 197–202.

Cressie, N.A.C. (1993). *Statistics for Spatial Data*, revised ed. Wiley, New York.

Daniels, M., Kass, R. (1999). Nonconjugate Bayesian estimation of covariance matrices in hierarchical models. *J. Amer. Statist. Assoc.* **94**, 1254–1263.

Daniels, M.J., Kass, R.E. (2001). Shrinkage estimators for covariance matrices. *Biometrics* **57**, 1173–1184.

Daniels, M., Cressie, N. (2001). A hierarchical approach to covariance function estimation for time series. *J. Time Series Anal.* **22**, 253–266.

Daniels, M., Zhao, Y. (2003). Modeling the random effects covariance matrix in longitudinal data. *Statistics in Medicine* **22**, 1631–1647.

Denison, D.G.T., Holmes, C.C. (2001). Bayesian partitioning for estimating disease risk. *Biometrics* **57**, 143–149.

Diggle, P.J. (1983). *Statistical Analysis of Spatial Point Patterns*. Academic Press, London.

Diggle, P.J. (1990). A point process modeling approach to raised incidence of a rare phenomenon in the vicinity of a prespecified point. *J. Roy. Statist. Soc., Ser. A* **153**, 349–362.

Diggle, P.J. (2003). *Statistical Analysis of Spatial Point Patterns*, second ed. Arnold Publishers, London.

Diggle, P.J., Cox, T.F. (1981). On sparse sampling methods and tests of independence for multivariate spatial point patterns. *Bull. Internat. Statist. Inst.* **49**, 213–229.

Diggle, P.J., Liang, K.-Y., Zeger, S.L. (1994). *Analysis of Longitudinal Data*. Oxford University Press, New York.

Diggle, P.J., Tawn, J.A., Moyeed, R.A. (1998). Model-based geostatistics (with discussion). *J. Roy. Statist. Soc., Ser. C* **47**, 299–350.

Ecker, M.D., Gelfand, A.E. (1997). Bayesian variogram modeling for an isotropic spatial process. *J. Agricultural, Biological, and Environmental Statist.* **2**, 347–369.

Ecker, M.D., Gelfand, A.E. (1999). Bayesian modeling and inference for a geometrically anisotropic spatial process. *Mathematical Geology* **31**, 67–83.

Ecker, M.D., Gelfand, A.E. (2003). Spatial modeling and prediction under stationary non-geometric range anisotropy. *Environmental and Ecological Statist.* **10**, 165–178.

Edelsbrunner, H. (1987). *Algorithms in Computational Geometry*. Springer-Verlag, Berlin.

Escobar, M.D., West, M. (1995). Bayesian density estimation and inference using mixtures. *J. Amer. Statist. Assoc.* **90**, 577–588.

Fay, R.E., Herriot, R.A. (1979). Estimates of income for small places: An application of James–Stein procedures to census data. *J. Amer. Statist. Assoc.* **74**, 269–277.

Fotheringham, A.S., Brunsdon, C., Charlton, M. (2002). *Geographically Weighted Regression*. Wiley, Chichester.

Gandin, L.S. (1963). *Objective Analysis of Meteorological Fields*. Gidrometeorologicheskoe Izdatel'stvo (GIMIZ), Leningrad [Translated by Israel Program for Scientific Translations, Jerusalem].

Geyer, C.J. (1999). Likelihood inference for spatial point processes. In: Barndorff-Nielsen, O.E., Kendall, W.S., van Lieshout, M.N.M. (Eds.), *Stochastic Geometry: Likelihood and Computation*. Chapman and Hall/CRC, Boca Raton, pp. 79–140.

Ghosh, M., Rao, J.N.K. (1994). Small area estimation: An appraisal (with discussion). *Statist. Sci.* **9**, 55–93.

Ghosh, M., Natarajan, K., Waller, L.A., Kim, D. (1999). Hierarchical GLMs for the analysis of spatial data: An application to disease mapping. *J. Statist. Plann. Inference* **75**, 305–318.

Gotway, C.A., Young, L.J. (2002). Combining incompatible spatial data. *J. Amer. Statist. Assoc.* **97**, 632–648.

Grenander, U., Miller, M.I. (1994). Representation of knowledge in complex systems (with discussion). *J. Roy. Statist. Soc., Ser. B* **56**, 549–603.

Handcock, M.S., Stein, M.L. (1993). A Bayesian analysis of kriging. *Technometrics* **35**, 403–410.

Harville, D.A., Jeske, D.R. (1992). Mean squared error of estimation or prediction under a general linear model. *J. Amer. Statist. Assoc.* **87**, 724–731.

Heikkinen, J. (2003). Trans-dimensional Bayesian non-parametrics with spatial point processes. In: Green, P.J., Hjort, N.L., Richardson, S. (Eds.), *Highly Structured Stochastic Systems*. Oxford University Press, Oxford, pp. 203–206.

Heikkinen, J., Arjas, E. (1998). Nonparametric Bayesian estimation of a spatial Poisson intensity. *Scand. J. Statist.* **25**, 435–450.

Heikkinen, J., Arjas, E. (1999). Modeling a Poisson forest in variable elevations: A nonparametric Bayesian approach. *Biometrics* **55**, 738–745.

Kim, H., Sun, D., Tsutakawa, R.K. (2001). A bivariate Bayes method for improving the estimates of mortality rates with a twofold conditional autoregressive model. *J. Amer. Statist. Assoc.* **96**, 1506–1521.
Knorr-Held, L. (2000). Bayesian modeling of inseparable space-time variation in disease risk. *Statistics in Medicine* **19**, 2555–2567.
Knorr-Held, L., Besag, J. (1998). Modeling risk from a disease in time and space. *Statistics in Medicine* **17**, 2045–2060.
Knorr-Held, L., Rasser, G. (2000). Bayesian detection of clusters and discontinuities in disease maps. *Biometrics* **56**, 13–21.
Krige, D.G. (1951). A statistical approach to some basic mine evaluation problems on the Witwatersrand. *J. Chemical, Metallurgical and Mining Society of South Africa* **52**, 119–139.
Künsch, H.R. (1987). Intrinsic autoregressions and related models on the two-dimensional lattice. *Biometrika* **74**, 517–524.
Le, N.D., Zidek, J.V. (1992). Interpolation with uncertain spatial covariances. A Bayesian alternative to kriging. *J. Multivariate Anal.* **43**, 351–374.
Lawson, A.B. (2001). *Statistical Methods in Spatial Epidemiology*. Wiley, Chichester.
Lawson, A.B., Denison, D.G.T. (2002). *Spatial Cluster Modeling*. Chapman and Hall/CRC, Boca Raton.
Liechty, J.C., Liechty, M.W., Muller, P. (2004). Bayesian correlation estimation. *Biometrika* **91**, 1–14.
Marshall, R.J. (1991). Mapping disease and mortality rates using empirical Bayes estimators. *Appl. Statist.* **40**, 283–294.
Matheron, G. (1963). Principles of geostatistics. *Economic Geology* **58**, 1246–1266.
Møller, J. (2003). Shot noise Cox processes. *Adv. Appl. Probab.* **35**, 614–640.
Møller, J. (2003). A comparison of spatial point process models in epidemiological applications. In: Green, P.J., Hjort, N.L., Richardson, S. (Eds.), *Highly Structured Stochastic Systems*. Oxford University Press, Oxford, pp. 264–269.
Møller, J., Waagepetersen, R. (2002). Statistical inference for Cox processes. In: Lawson, A.B., Denison, D.G.T. (Eds.), *Spatial Cluster Modeling*. Chapman and Hall/CRC, Boca Raton, pp. 37–60.
Møller, J., Waagepetersen, R. (2004). *Statistical Inference and Simulation for Spatial Point Patterns*. Chapman and Hall/CRC, Boca Raton, FL.
Møller, J., Syversveen, A.R., Waagepetersen, R.P. (1998). Log Gaussian Cox processes. *Scand. J. Statist.* **25**, 451–482.
Mollié, A. (1996). Bayesian mapping of disease. In: Gilks, W.R., Richardson, S., Spiegelhalter, D.J. (Eds.), *Markov Chain Monte Carlo in Practice*. Chapman and Hall/CRC, Boca Raton, pp. 360–379.
Moran, P.A.P. (1948). The interpretation of statistical maps. *J. Roy. Statist. Soc., Ser. B* **10**, 243–251.
Moran, P.A.P. (1950). Notes on continuous stochastic phenomena. *Biometrika* **37**, 17–23.
Mugglin, A.S., Carlin, B.P. (1998). Hierarchical modeling in geographic information systems: Population interpolation over incompatible zones. *J. Agricultural, Biological, and Environmental Statist.* **3**, 111–130.
Mugglin, A.S., Carlin, B.P., Gelfand, A.E. (2000). Fully model-based approaches for spatially misaligned data. *J. Amer. Statist. Assoc.* **95**, 877–887.
Neyman, J. (1939). On a new class of contagious distributions, applicable in entomology and bacteriology. *Ann. Math. Statist.* **10**, 35–57.
Neyman, J., Scott, E.L. (1958). Statistical approach to problems in cosmology. *J. Roy. Statist. Soc., Ser. B* **20**, 1–43.
Okabe, A., Boots, B., Sugihara, K., Chiu, S.N. (2000). *Spatial Tessellations: Concepts and Applications of Voronoi Diagrams*, second ed. Wiley, Chichester.
Preparata, F.P., Shamos, M.I. (1985). *Computational Geometry: An Introduction*. Springer-Verlag, New York.
Rao, J.N.K. (2003). *Small Area Estimation*. Wiley, New York.
Richardson, S. (2003). Spatial models in epidemiological applications. In: Green, P.J., Hjort, N.L., Richardson, S. (Eds.), *Highly Structured Stochastic Systems*. Oxford University Press, Oxford, pp. 237–259.
Ripley, B.D. (1977). Modeling spatial patterns. *J. Roy. Statist. Soc., Ser. B* **39**, 172–212.
Ripley, B.D. (1981). *Spatial Statistics*. Wiley, New York.
Roberts, G.O., Tweedie, R.L. (1996). Geometric convergence and central limit theorems for multidimensional Hastings and Metropolis algorithms. *Biometrika* **83**, 95–110.
Schoenberg, F. (1999). Transforming spatial point processes into Poisson processes. *Stochastic Process. Appl.* **81**, 155–164.

Schootman, M., Sun, D. (2004). Small-area incidence trends in breast cancer. *Epidemiology* **15**, 300–307.

Stein, M.L. (1999). *Interpolation of Spatial Data: Some Theory for Kriging*. Springer-Verlag, New York.

Stoyan, D., Kendall, W.S., Mecke, J. (1995). *Stochastic Geometry and Its Applications*, second ed. Wiley, Chichester.

Sun, D., Tsutakawa, R.K., Speckman, P.L. (1999). Posterior distribution of hierarchical models using CAR(1) distributions. *Biometrika* **86**, 341–350.

Sun, D., Tsutakawa, R.K., Kim, H., He, Z. (2000). Spatio-temporal interaction with disease mapping. *Statistics in Medicine* **19**, 2015–2035.

Tobler, W. (1970). A computer movie simulating urban growth in the Detroit region. *Economic Geography* **46**, 234–240.

Waagepetersen, R. (2004). Convergence of posteriors for discretized log Gaussian Cox processes. *Statist. Probab. Lett.* **66**, 229–235.

Wackernagel, H. (2003). *Multivariate Geostatistics: An Introduction with Applications*, second ed. Springer-Verlag, New York.

Wakefield, J., Best, N.G., Waller, L.A. (2000). Bayesian approaches to disease mapping. In: Elliott, P., Wakefield, J.C., Best, N.G., Briggs, D.J. (Eds.), *Spatial Epidemiology: Methods and Applications*. Oxford University Press, Oxford, pp. 106–127.

Waller, L.A., Carlin, B.P., Xia, H., Gelfand, A. (1997). Hierarchical spatio-temporal mapping of disease rates. *J. Amer. Statist. Assoc.* **92**, 607–617.

Waller, L.A., Gotway, C.A. (2004). *Applied Spatial Statistics for Public Health Data*. Wiley, New York.

Webster, R., Oliver, M.A. (2001). *Geostatistics for Environmental Scientists*. Wiley, Chichester.

Whittle, P. (1954). On stationary processes in the plane. *Biometrika* **41**, 434–449.

Wolpert, R.L., Ickstadt, K. (1998). Poisson/gamma random field models for spatial statistics. *Biometrika* **85**, 251–267.

Wong, F., Carter, C.K., Kohn, R. (2003). Efficient estimation of covariance selection models. *Biometrika* **90**, 809–830.

Xia, H., Carlin, B.P. (1998). Spatio-temporal models with errors and covariates: Mapping Ohio lung cancer mortality. *Statistics in Medicine* **17**, 2025–2043.

Zhang, H. (2004). Inconsistent estimation and asymptotically equal interpolations in model-based geostatistics. *J. Amer. Statist. Assoc.* **99**, 250–261.

Zimmerman, D.L., Cressie, N. (1992). Mean squared prediction error in the spatial linear model with estimated covariance parameters. *Ann. Inst. Statist. Math.* **44**, 27–43.

Handbook of Statistics, Vol. 25
ISSN: 0169-7161
© 2005 Elsevier B.V. All rights reserved.
DOI 10.1016/S0169-7161(05)25021-6

Robust Bayesian Analysis

Fabrizio Ruggeri, David Ríos Insua and Jacinto Martín

Abstract
We provide an overview of robust Bayesian analysis with emphasis on foundational, decision oriented and computational approaches. Common types of robustness analyses are described, including global and local sensitivity analysis and loss and likelihood robustness.

Keywords: Bayesian robustness; sensitivity analysis; prior/likelihood/loss robustness; algorithms; applications

Subjectivists should feel obligated to recognize that any opinion (so much more the initial one) is only vaguely acceptable... So it is important not only to know the exact answer for an exactly specified initial problem, but what happens changing in a reasonable neighborhood the assumed initial opinion.
De Finetti, as quoted by Dempster (1975)

1. Introduction

As described in various chapters in this book, the Bayesian approach to inference and decision making, see French and Ríos Insua (2000), essentially implies:

- Modelling beliefs about a parameter θ of interest through a prior $\pi(\theta)$ which, in presence of further information $f(x|\theta)$, is updated to the posterior $\pi(\theta|x)$ (note that we use π to denote a density, whereas Π denotes the probability measure).
- Modelling preferences and risk attitudes about consequences (a, θ), when we take the alternative a and the parameter is θ, with the loss function $L(a, \theta)$.
- Associate with each alternative a its posterior expected loss

$$T(L, \pi, a) = \int L(a, \theta)\pi(\theta|x)\, \mathrm{d}\theta = \frac{\int_\Theta L(a, \theta) f(x|\theta)\pi(\theta)\, \mathrm{d}\theta}{\int_\Theta f(x|\theta)\pi(\theta)\, \mathrm{d}\theta}.$$

- Propose as optimal the alternative which minimises the posterior expected loss.

As in any quantitative approach, there are many reasons, see, e.g., Saltelli et al. (2000), to undertake a sensitivity analysis, that is of checking sensitivity of the output – the decision rule – with respect to the inputs: the model, the loss and the prior. In addition, since in this framework inputs to the analysis encode the DM's judgements, she should wish to explore their implications and possible inconsistencies. This is an especially important point, as her judgements will evolve through the analysis until they are *requisite*, see Philipps (1984). Sensitivity techniques, or robust Bayesian analysis, may guide this process.

The need for sensitivity analysis is further emphasised by the fact that the assessment of beliefs and preferences is a difficult task, particularly in the case of several DMs and/or experts. Some would even argue that priors, models and losses can never be quantified exactly, i.e., without error, especially in a finite amount of time. In most quantitative fields, there have been considerable efforts to develop sensitivity analysis techniques, catalysed by the increasing growth in computing power.

As suggested in some chapters in this volume, the usual practical motivation underlying robust Bayesian analysis is the difficulty in assessing the prior distribution. Consider the simplest case in which it is desired to elicit a prior over a finite set of states $\theta_i, i = 1, \ldots, I$. A common technique to assess a precise $\pi(\theta_i) = p_i$, with the aid of a reference experiment, proceeds as follows: one progressively bounds $\pi(\theta_i)$ above and below until no further discrimination is possible and then takes the midpoint of the resulting interval as the value of p_i. Instead, however, one could directly operate with the obtained constraints $\alpha_i \leqslant \pi(\theta_i) \leqslant \beta_i$, acknowledging the cognitive limitations. Constraints might also come in different forms, such as knowledge that $\pi(\theta_i) \geqslant \pi(\theta_j)$. The situation is similar in the continuous case; a common elicitation technique discretises the range Θ of the random variable and assesses the probabilities of the resulting sets. Again, however, one actually ends up only with some constraints on these probabilities. This is often phrased in terms of having bounds on the quantiles of the distribution. A related idea is having bounds on the distribution function corresponding to the random variable. Once again, qualitative constraints, such as unimodality, might also be added. Perhaps most common for a continuous random variable is to assess a parametrised prior, say a conjugate prior, and place constraints on the parameters of the prior.

The same situation holds in modelling preferences. One might, say, assess the loss of some consequences through the certainty equivalent method, and then fit a loss function. However, in reality, one only ends up with upper and lower constraints on such losses, possibly with qualitative features such as monotonicity and convexity, if preferences are increasing and risk averse. These constraints can often be approximated by an upper and a lower loss function, leading to the consideration of all loss functions that lie between these bounds (and that possibly satisfy the other qualitative constraints). If a parametrised loss function is assessed, say an exponential loss, the constraints are typically placed on the parameters of the loss, say the risk aversion coefficient. Of course, in developing the model for the data itself there is typically great uncertainty, and a need for careful study of loss robustness.

A final comment concerning the limits of elicitation concerns the situation in which there are several decision makers and/or experts involved in the elicitation. Then it is not even necessarily possible theoretically to obtain a single model, prior, or loss; one might be left with only classes of each, corresponding to differing expert opinions.

Our objective here is, therefore, to provide tools to check the impact of the loss, the prior and the model on the Bayes decision rule or the Bayes alternative, and their posterior expected loss. Note that these studies have concentrated mainly on inference and data analytic issues and, therefore, on sensitivity to the prior, thus acknowledging the standard criticism that the prior is the most difficult element to assess. Former robust Bayes analysis reviews may be seen in Berger (1984, 1985, 1990, 1994), Berger et al. (2000) or Ríos Insua and Ruggeri (2000). We emphasise basic concepts and computational issues.

The rest of this chapter is structured as follows. In Section 2, we introduce basic concepts, such as local versus global robustness, prior/likelihood/loss robustness vs. joint robustness, decision vs. expected loss robustness and others. These ideas might prove robust Bayes as a very heterogeneous paradigm, but they may be given some unity through foundations, as described in Section 3. Then, we describe issues concerning robust Bayesian computations in Section 4 and relate the robust Bayesian approach with other statistical paradigms in Section 5. We end up with some conclusions and pointers to future topics in Section 6.

2. Basic concepts

We review key issues in prior/model/loss sensitivity, looking at the major classes entertained in the literature and the most common measures of sensitivity. We illustrate relevant notions like global and local robustness.

2.1. *Different approaches*

A plethora of methods have been proposed to deal with the effects of changes in prior/likelihood/loss on decisions and inferences. It is difficult to classify the methods in homogeneous classes but, nonetheless, some criteria have been proposed.

One approach distinguishes priors (likelihoods and losses are treated similarly) if there is (or there is not) a baseline prior which is deemed to be a reasonable choice and is perturbed considering what happens in a neighbourhood around it.

Another, well-established criterion distinguishes three main approaches to Bayesian robustness. We illustrate it considering robustness with respect to changes in the prior, but similar issues are raised when considering likelihoods and losses. A "guided tour" through these three approaches is presented in Berger et al. (2000); see the references therein.

The first approach is the *informal* one, which considers a few priors and compare the quantity of interest (e.g., the posterior mean) under them. The approach has been (and is) very popular because of its simplicity. The rationale with this informal approach is that since we shall be dealing with messy computational problems, why not analyse sensitivity by trying only some alternative pairs of losses and priors? While this is a healthy practise and a good way to start a sensitivity analysis, in general this will not be enough and we should undertake more formal analyses such as later described: the limited number of priors chosen might not include some which are compatible with the prior knowledge and could lead to very different values of the quantity. It is worth mentioning that the consideration of a finite number of models and losses is typical, respectively, of model selection and multi-objective decision making problems.

EXAMPLE 1. Suppose we have a no data problem given by the table:

	θ_1	θ_2
a	c_1	c_2
b	c_2	c_3

Let $\pi(\theta_1) = p_1$. Suppose we have assessed that $0.4 \leqslant p_1 \leqslant 0.6$, $0.5 \leqslant L(c_1) \leqslant 1$, $0.25 \leqslant L(c_2) \leqslant 0.75$, and $0 \leqslant L(c_3) \leqslant 0.5$. We consider the following four loss-probability pairs associated with the bounds on losses and probabilities

p_1	Loss (L)	$T(L,\pi,a)$	$T(L,\pi,b)$
0.4	$L(c_1)=1,\ L(c_2)=0.75,\ L(c_3)=0.5$	0.85	0.6
0.4	$L(c_1)=0.5,\ L(c_2)=0.25,\ L(c_3)=0$	0.35	0.1
0.6	$L(c_1)=1,\ L(c_2)=0.75,\ L(c_3)=0.5$	0.9	0.65
0.6	$L(c_1)=0.5,\ L(c_2)=0.25,\ L(c_3)=0$	0.4	0.15

Since in all four cases, $T(L,\pi,a) > T(L,\pi,b)$, we could conclude that alternative b is preferred to alternative a. Note, though, that this is not true: if, e.g., we make $p_1 = 0.6$, $L(c_1) = 0.5$, $L(c_2) = 0.75$, $L(c_3) = 0.5$, then $T(L,\pi,a) = .6$ and $T(L,\pi,b) = .65$, raising doubts about the preference of b over a.

The most popular approach in Bayesian robustness is called *global sensitivity*. All probability measures compatible with the prior knowledge available are considered and robustness measures are computed as the prior varies in a class. Computations are not always easy since they require the evaluation of suprema and infima of quantities of interest. The last approach looks for *local sensitivity* and studies the rate of change in inferences and decisions, using differential techniques.

An important and occasionally controversial issue is the distinction between decision robustness and expected loss robustness: see Kadane and Srinivasan (1994). A variety of situations may hold. For instance, when performing sensitivity analysis, it may happen that the expected loss changes considerably with virtually no change in the optimal Bayes action, even if the loss is fixed.

2.2. *Prior robustness*

We first show various ways of modelling imprecision in the prior through classes of priors. This is only the first part in a sensitivity analysis; it must be followed by the identification of a tool capable of measuring the effects of the uncertainty on the quantities of interest. Different approaches can be taken, depending, e.g., on the existence or not of a baseline prior and a neighbourhood class around it. It is possible to consider a neighbourhood class around a prior Π_0 and check what happens when allowing for either infinitesimal or more relevant departures from Π_0. In the former case, tools from functional analysis, e.g., Fréchet derivatives, are used, whereas the latter leads to some less sophisticated tools. The former approach leads to *local sensitivity analysis*, since it is a local study around a prior Π_0, whereas the other is called *global sensitivity*, since it considers changes in the whole class.

2.2.1. *Classes of priors*

The choice of the class of priors should be driven by the following goals:

1. the class should be related with the elicitation method used;
2. the class should contain only "reasonable" priors, avoiding unreasonable priors which might erroneously lead to lack of robustness;
3. computation of sensitivity measures should be as easy as possible.

Consider, for example, the class of priors defined by some quantiles. As described in the introduction, the class is related with a prior assessment method but the trade-off between the two other goals is complicated. If no further feature is introduced in the definition of the class, then Ruggeri (1990) showed that computation of sensitivity measures is straightforward and achieved considering the subset of discrete measures, which are not always "reasonable" priors in the problem at hand. Berger and O'Hagan (1988) and O'Hagan and Berger (1988) added unimodality (or quasi-unimodality) to the features defining the class, removing the discrete measures but dramatically worsen the computation of the sensitivity measures.

We review some important classes:

Priors with given functional forms. Most sensitivity studies, specially in the early days, considered classes of priors with the same functional form (e.g., normal priors with mean and/or variance varying in an interval). Typical choices are conjugate classes, e.g., a class of Gamma priors for the parameter of an exponential distribution. In the pre-MCMC days, Bayesian computations were such a burden that conjugate classes were very fashionable, providing closed forms for the quantities of interest.

The choice of a given functional form might be mathematically convenient but, in general, it does not correspond to what the expert knows about the prior. In fact, it is possible that the expert knows only the median of the distribution and the choice of a given distribution, say normal, is an artifact. Nowadays, the use of powerful computers and sophisticated algorithm (e.g., MCMC) has dramatically reduced the importance of conjugate classes, although it is worth mentioning that there are no general-purpose algorithm in Bayesian robustness and the extension of MCMC techniques in this environment is not straightforward, see Section 4 for some ideas.

As an example of class of priors with given form, we mention the work by Cagno et al. (2000a, 2000b) who dealt with failure rates of iron cast pipelines in a gas distribution network. In the first paper they considered homogeneous Poisson processes (HPPs) to describe failures in different pipelines (varying for diameter, depth and position in the street) and used Gamma priors for the parameter of the HPP. In Cagno et al. (2000b), the propensity to failure of different pipelines was studied under classes of Gamma priors with fixed variance and mean bounded in an interval.

Priors with specified generalised moments. The generalised moments class is defined through

$$\Gamma = \left\{\pi \colon \int_{\Theta} H_i(\theta)\pi(\theta)\,\mathrm{d}\theta \leqslant \alpha_i,\ i = 1, \ldots, n\right\},$$

where H_i are given π-integrable functions and α_i, $i = 1, \ldots, n$, are fixed real numbers. This class, considered first in Betrò et al. (1994) and Goutis (1994), is very rich and contains well-known classes like the one with given moments and the quantile class. The former class has been used, e.g., by Hartigan (1969) and Goldstein (1980), who considered the priors with given first two moments. Although the elicitation of the first two moments, actually mean and variance, is often very reasonable, it is difficult to specify further moments and reasonable distributions, e.g., Cauchy, might be ruled out. The latter class, the quantile class, has been considered by, e.g., Ruggeri (1990) and Moreno and Cano (1991). The class is defined by specifying some quantiles, i.e., the functions H_i are set functions over connected intervals forming a partition of Θ. The class can be extended to any partition of Θ. The elicitation of quantiles is, probably, one of the more standard ways used by the experts in eliciting priors.

The generalised moments class allows for other choices of the functions H_i. Betrò et al. (1994) considered the class defined by bounds on the marginal probabilities of given sets K_i, taking

$$H_i(\theta) = \int_{K_i} f(x|\theta)\,\mathrm{d}x,$$

so that, by Fubini's theorem, it follows that

$$\int_{\Theta} H_i(\theta)\pi(\theta)\,\mathrm{d}\theta = \int_{K_i} m_\pi(x)\,\mathrm{d}x,$$

where $m_\pi(x) = \int_\theta f(x|\theta)\pi(\theta)\,\mathrm{d}\theta$ is the marginal density of X under π.

Computation of sensitivity measures, as discussed later in Sections 2.2.2 and 2.2.3, is possible by restricting the search among discrete distributions taking, at most, $n + 1$ values (where n is the number of generalised moments constraints). Although discrete distributions might be an unreasonable choice, the computations are very simple since the functional optimisation problem becomes a constrained nonlinear optimisation problem.

Neighbourhood classes. Sometimes a baseline prior Π_0 is elicited and sensitivity is studied when a "neighbourhood" around it is considered.

ε-contamination class. The popularity of the ε-contamination class arises, in part, from the ease of its specification and, in part, from the fact that it is typically easily handled mathematically. The ε-contamination class is defined as

$$\Gamma_\varepsilon = \{\pi \colon \pi = (1-\varepsilon)\pi_0 + \varepsilon q,\ Q \in \mathcal{Q}\},$$

where the measures Q, with densities q, are in the class of contaminations $\mathcal{Q}$. The class has been considered by Huber (1973) in classical robustness and sometimes it has been called gross error model to specify that data are coming from a different distribution with probability ε. In a Bayesian setting, the ε-contamination class is made of priors which resemble the baseline prior, as close as allowed by the size of the class of contaminations and ε. In fact, it is worth mentioning that, for any measurable subset $A \in \Theta$, it follows that

$$(1-\varepsilon)\Pi_0(A) + \inf_{Q \in \mathcal{Q}} Q(A) \leqslant \Pi(A) \leqslant (1-\varepsilon)\Pi_0(A) + \sup_{Q \in \mathcal{Q}} Q(A), \tag{1}$$

for any $\Pi \in \Gamma_\varepsilon$.

Researchers focused their interest in finding suitable classes of contaminations $\mathcal{Q}$. The first, obvious choice is the class of all distributions, considered in Berger and Berliner (1986). Unfortunately, this class is often too large and robustness is hardly achieved. In the same paper, smaller classes were considered with the prior π_0 being contaminated by some other classes of distributions: symmetric unimodal, unimodal and unimodality preserving contaminations. Arbitrary contaminations with known quantiles were considered in Moreno and Cano (1991) whereas unimodal contaminations with known quantiles were considered in Moreno and Gonzalez (1990) and Moreno and Pericchi (1991, 1993). Betrò et al. (1994) considered the contamination class defined by bounds on the marginal probabilities of given sets.

Sometimes, there is more uncertainty in the specification of the tails of the contaminated priors. More uncertainty should lead to greater variation in (1) for the subsets A in the tails, i.e., ε should be larger for θ in the tails of the distribution. ε-contaminations with $\varepsilon = \varepsilon(\theta)$ have been considered in Moreno et al. (1996).

Moreno and Cano (1992) addressed the problem of robustness in a multidimensional parameter space, considering the contamination class with partially known marginals, i.e., with some quantiles of the unidimensional marginals given, or with one completely specified marginal and the others known just via some quantiles. The problem has been addressed by Lavine et al. (1991), as well.

As a related class, Gelfand and Dey (1991) proposed the geometric ε-contamination class, defined as

$$\Gamma_g = \{\pi\colon \pi = c_g(\varepsilon)\pi_0^{1-\varepsilon} q^\varepsilon,\ Q \in \mathcal{Q}\},$$

where q is the density of a measure Q in the class of contaminations $\mathcal{Q}$.

Topological neighbourhoods. One drawback of ε-contamination classes is that they are not true neighbourhood classes in a topological sense. A variety of other classes have been considered that do have a formal interpretation in this sense. We illustrate some of them.

A first, very important example (considered, e.g., in Fortini and Ruggeri, 1994a) is the total variation neighbourhood of a prior probability measure Π_0 which contains all the probability measures Π on $(\Theta, \mathcal{F})$ that satisfy $\sup_{A \in \mathcal{F}} |\Pi(A) - \Pi_0(A)| \leqslant \varepsilon$, given a fixed $\varepsilon \in [0, 1]$.

Another example is given by the neighbourhoods based on the concentration function, a tool defined by Cifarelli and Regazzini (1987) as a generalisation of the well-known Lorenz curve. Whereas the Lorenz curve compares a discrete distribution with a uniform one, the concentration function allows for the comparison between two probability measures Π and Π_0, considering the range spanned by the probability, under Π, of *all* the subsets with a given probability under Π_0. Fortini and Ruggeri (1994b) proposed a method, based on the concentration function, to define neighbourhoods of probability measures and applied it in Bayesian robustness (see Fortini and Ruggeri, 1994a). Their approach allows the construction of probability measures Π with functional forms close to a nonatomic baseline measure Π_0. The mathematical illustration of the concentration function is beyond the scope of the current paper; we refer the interested reader to the review paper by Fortini and Ruggeri (2000) and references therein. Concentration function neighbourhoods are very rich since they include well-known classes like the ε-contamination class with arbitrary contaminations and the total variation neighbourhood. Unfortunately, the elicitation of a concentration function neighbourhood is not very simple, as it requires the specification of a monotone nondecreasing, continuous, convex function with nontrivial meaning, in general. Furthermore, often computations are not straightforward.

Whereas the previous neighbourhood classes were defined using set probabilities or densities with respect to a dominating measures, Basu and DasGupta (1990, 1995) and Basu (1995) considered the distribution band class, defined as

$$\Gamma_{BDG} = \{F\colon\ F \text{ is a cdf and } F_L(\theta) \leqslant F(\theta) \leqslant F_U(\theta), \forall\theta\},$$

where F_L and F_U are given cumulative density functions (cdf), with $F_L(\theta) \leqslant F_U(\theta)$. This class is important from a mathematical point of view since it includes, as special cases, Kolmogorov and Lèvy neighbourhoods. The elicitation of this class is relatively simple and other features, such as symmetry and unimodality of the distributions, can be added as in Basu (1992), although computations are even harder.

Functional derivatives, namely Fréchet derivatives, are sometimes considered in Bayesian robustness and they are defined on a normed, linear space. In this case, given a baseline prior Π_0, then the class $\{\Pi;\ \Pi = \Pi_0 + \delta\}$ is considered, where δ is a signed measure with $\delta(\Theta) = 0$. Note that the set Δ of such δ is a normed, linear space with norm given by $\|\delta\| = d(\delta, 0)$ where $d(P, Q) = \sup_{A\in\mathcal{B}(\Theta)} |P(A) - Q(A)|$ is the total variation metric. Choosing $\delta = (1-\varepsilon)(Q - \Pi_0)$, with Q a probability measure, then $\Pi_0 + \delta$ becomes an ε-contaminated prior.

Other classes. We conclude this nonexhaustive illustration of classes of priors with some others.

DeRobertis and Hartigan (1981) considered the density ratio class Γ_{DR} defined by

$$\Gamma_{DR} = \{\pi\colon\ M(\theta) \leqslant c\pi(\theta) \leqslant U(\theta) \text{ for some } c > 0\},$$

where M and U are given nonnegative functions and π is the density with respect to a dominating measure. Ruggeri and Wasserman (1995) considered the density ratio neighbourhood around Π_0 (with density $\pi_0(\theta)$), taking $M(\theta) = \pi_0(\theta)$ and $U(\theta) = k\pi_0(\theta)$ for $k > 0$ and almost all θ.

Lavine (1991) considered the density bounded class Γ_B defined as

$$\Gamma_B = \{\Pi\colon\ M(A) \leqslant \Pi(A) \leqslant U(A)\ \forall A \in \mathcal{F}\},$$

with $\mathcal{F}$ a σ-algebra on Θ and M and U finite measures. Ruggeri and Wasserman (1991) considered the density bounded neighbourhood around Π_0, taking $M = (1/k)\Pi_0$ and $U = k\Pi_0$ for $k > 0$.

Another interesting class, although quite neglected in the literature, was proposed by Madansky (1990): the class of the monotonically increasing density functions $\pi(\theta)$ (or, equivalently with convex cdf).

As seen for ε-contaminations, multidimensional parameter spaces might lead to contaminating classes with given marginals (i.e., in a Fréchet class). De la Horra and Fernández (1995) considered the Farlie–Gumbel–Morgenstern model as a class of priors. This model has fixed marginals and the joint density depends on a univariate parameter.

Finally, mixture classes of the form

$$\Gamma_M = \left\{\pi\colon\ \pi(\theta) = \int \pi(\theta|\omega)\, \mathrm{d}G(\omega)\right\}$$

were considered by Bose (1994). These classes stem from the idea that many classes of priors are mixtures of priors. Their computational treatment is simple.

2.2.2. *Global robustness*

We now present some measures which have been introduced to study different aspects of the changes induced by the class under investigation. These measures provide, in general, a number that, in principle, should be interpreted in the following way:

- if the measure is "small", then robustness is achieved and any prior in the class (possibly one computationally simple) can be chosen without relevant effects on the quantity of interest;
- if the measure is "large", then new should be acquired and/or further elicitation narrow the class, recomputing the robustness measure and stopping as in the previous item; o.w. ...
- ... if the measure is "large" and the class cannot be modified, then a prior can be chosen in the class but we should be wary of the relevant influence of our choice on the quantity of interest. Alternatively, we may use an ad hoc method such as the Γ-minimax, later described, to select an alternative.

Given a class Γ of prior measures, global sensitivity analysis will usually pay attention to the range of variation of a posterior (or predictive) functional of interest as the prior ranges over the class, i.e.,

$$\sup_{\pi \in \Gamma} T(h, \pi) - \inf_{\pi \in \Gamma} T(h, \pi),$$

where the posterior expectation of the function $h(\theta)$, i.e.,

$$T(h, \pi) = \frac{\int_\Theta h(\theta) f(x|\theta)\pi(\theta)\, \mathrm{d}\theta}{\int_\Theta f(x|\theta)\pi(\theta)\, \mathrm{d}\theta},$$

is the quantity of interest (typically, the posterior mean, a set probability or variance).

As an example, suppose we use a quadratic loss function in a problem. The Bayes rule is the posterior expectation. If we have imprecision about the prior, we would like to know the range of the posterior expectation as the prior ranges in the class, for the observed data.

The range is capable of measuring changes in the scale of the quantity of interest. Reporting only the range might be misleading, since a range of 1 has a different meaning, in terms of robustness, when the quantity of interest is of order 10^3 or 10. The calibration of these measures is still an unresolved problem since it is not clear when we can say that in a problem there is more robustness than in another just by comparing two numbers. It is not even possible to find an "objective" criterion which tells when robustness is achieved. These drawbacks of the range lead to the idea of scaling a sensitivity measure by a suitable posterior quantity, for the purpose of easy comparison or easy assessment. Gustafson (1994) and Sivaganesan (1991) suggested dividing the sensitivity measures of interest by the posterior mean with respect to the base prior, whereas Ruggeri and Wasserman (1995), and Gustafson (1994) suggested dividing the (local, in their case) sensitivity measures proposed, by the posterior standard deviation with respect to the *base* prior π_0. Supported by a decision theoretic interpretation, Ruggeri and Sivaganesan (2000) suggested dividing by the posterior s.d. (or variance) with respect to the "alternative" prior π. For instance, Ruggeri and Sivaganesan (2000) considered a scaled version of the range, for a target quantity of interest $h(\theta)$, called *relative sensitivity*, defined as

$$R_\pi = \frac{(T(h,\pi) - T(h,\pi_0))^2}{V^\pi},$$

where $T(h,\pi)$ and $T(h,\pi_0)$ equal $\mathbb{E}(h(\theta)|x)$ under π and the baseline prior π_0, respectively, and V^π is the posterior variance of $h(\theta)$ with respect to π. The idea is that, in estimation of $h(\theta)$, the posterior variance indicates the accuracy that is attainable from the experiment, so that one will often not worry if the (squared) range is small with respect to this posterior variance.

The decision theoretic justification derives from the following argument. Suppose we are interested in estimating $h(\theta)$ using the quadratic loss, i.e., considering the posterior expected loss of $h(\theta)$, and that π is the "true" prior. Let ρ_π and ρ_0 be the posterior expectations under π and π_0, respectively. The actual posterior risk of using ρ_0 would be

$$(\rho_\pi - \rho_0)^2 + V^\pi,$$

instead of V^π itself, if the ("true") prior π were used to obtain the estimate. Hence, the relative increase in risk when using π_0, relative to the risk, V^π, if the prior π were used, is

$$\frac{(\rho_\pi - \rho_0)^2}{V^\pi},$$

which is the measure R_π considered in Ruggeri and Sivaganesan (2000), along with its supremum over all priors π. An extension of this measure from quadratic loss to general ones is presented in Arias et al. (2004b).

Boratynska (1996) considered the "radius" (in the total variation metric) of the set of posterior distributions generated by the ε-contamination class of priors around Π_0, providing an upper bound on it and finding conditions under which it is achieved exactly.

The mentioned works consider single-valued measures where robustness is checked by analysing a summarising index. Fortini and Ruggeri (1995) compared the functional forms of the posterior measures to check if their concentration function with respect to a baseline posterior were above a given function, controlling the maximum variation in posterior probabilities for the sets with given probability content under the baseline posterior.

Further ideas on calibration of robustness measures may be seen in McCulloch (1989) and Ríos Insua (1990).

2.2.3. *Local robustness*

Stemming from work by Diaconis and Freedman (1986), Ruggeri and Wasserman (1993) measured the local sensitivity of a posterior expectation with respect to the prior by computing the norm of the Fréchet derivative of the posterior with respect to the prior over several different classes of measures. Furthermore, they used the Fréchet derivative to compute the derivative of the posterior upper expectation when the prior varied in a restricted ε-contamination class.

In their approach, Ruggeri and Wasserman (1993) consider a prior Π and contaminate it with any measure δ from the class Δ of all signed measures with zero mass spanning the set $\mathcal{M}$. Under the total variation metric, the space Δ is a normed space and $T_h = T(h, \Pi + \delta)$ can be considered as a nonlinear operator taking Δ into $\mathbb{R}$. $\dot{T}_h^{\Pi}$ will be the Fréchet derivative of the operator T_h, that is, a linear map on Δ satisfying $\dot{T}_h^{\Pi}(\delta) = T_h(\Pi + \delta) - T_h(\Pi) + o(\|\delta\|)$. $\dot{T}_h^{\Pi}$ measures how a small change in Π affects the posterior expectation. The norm of $\dot{T}_h^{\Pi}$ over $\mathcal{M}$ is defined by $\|\dot{T}_h^{\Pi}\|_{\mathcal{M}} = \sup_{\delta\in\Delta} |\dot{T}_h^{\Pi}|/\|\delta\|$. We consider it as a sensitivity measure of the posterior expectation as the prior ranges in $\mathcal{M}$. Further results can be found in Martín and Ríos Insua (1996). Different metrics (e.g., Prohorov, Levy, Kolmogorov) were considered in Basu (1996).

A similar notion, *total derivative*, has been considered by Basu et al. (1996) when a class of parametric priors is considered. They consider two different measures of robustness using this functional derivative, based on the its norm computed at a baseline prior Π_0 along all possible directions. The *maximum sensitivity* measures the supremum of the norm, whereas the *average sensitivity* is given by integrating the norms along all possible directions. The former measure is a conservative one, since it protects from the worst direction, whereas the latter reduces the influence of the least robust directions. The choice between these two measures depends on the problem at hand.

Another tool from functional analysis, the Gâteaux differential, has been considered by other authors, stemming from Ruggeri and Wasserman (1995). The Gâteaux differential of the operator T_h at Π_0 and in the direction of Δ (a signed measure with zero mass) is the limit (if it exists)

$$\lim_{\lambda\to 0^+} \frac{T_h(\Pi_0 + \lambda\Delta) - T_h(\Pi_0)}{\lambda}.$$

Ruggeri and Wasserman (1995) considered Gâteaux differentials for special cases of density bounded and density ratio classes. Gâteaux differential of the concentration function was computed in Fortini and Ruggeri (1997). A finding that is common when working with local sensitivity measures is that once their general formula is obtained using functional analytic methods, they are, in general, computed in practical cases more easily than global robustness measures. An example of their use in global robustness is provided by Ríos Insua et al. (1999) who considered a quantile class about Poisson parameters in a model about accidents in the Spanish building sector. Based on results in Martín and Ríos Insua (1996), they used Fréchet derivatives to find the best interval on which further elicitation could increase the (global) robustness.

Divergence measures have been considered to measure robustness in both global and local sensitivity, see, e.g., Ghosh and Dey (1994). Gelfand and Dey (1991) considered Kullback–Leibler divergence, whereas a more general class of ϕ-divergence measures were considered by Dey and Birmiwal (1994) and applied to both ε-contaminations and geometric ε-contaminations. The general ϕ-divergence between two posterior densities is defined as

$$D = \int \pi_0(\theta|x)\phi\left(\frac{\pi(\theta|x)}{\pi_0(\theta|x)}\right) \mathrm{d}\theta.$$

Appropriate choices of the function ϕ lead to well known measures, like Kullback–Leibler's distance, Hellinger's distance, χ^2, etc.

ϕ-divergence measures are intimately related to the concentration function as shown in Regazzini (1992). This fact has been considered by Carota and Ruggeri (1994) and by Fortini and Ruggeri (1995), who focused on Gini's and Pietra's indices.

2.3. Model robustness

We consider now issues in relation with model robustness. Again, we first review some key classes used in the literature and then mention specific ideas for model robustness analysis.

2.3.1. Classes of models

Classes of models are actually considered in all the parametric inferential problems when the parameter is estimated (via MLE's or Bayesian estimators) and posterior probabilities are given on the parameter, or, equivalently, on the parametric class of models.

Finite classes. A finite number of models can be considered. For example, Shyamalkumar (2000) considered two competing models, normal $\mathcal{N}(\theta, 1)$ and Cauchy $\mathcal{C}(\theta, 0.675)$, with classes of priors on the (well-defined) common parameter θ.

Parametric classes. There are many examples of parametric models. We just mention the very rich class of models due to Box and Tiao (1962), given by

$$\Lambda_{BT}=\left\{ f(y|\theta,\sigma,\beta) = \frac{\exp\left\{-\frac{1}{2}\left|\frac{y-\theta}{\sigma}\right|^{2/(1+\beta)}\right\}}{\sigma 2^{(1.5+0.5\beta)}\Gamma(1.5+0.5\beta)}; \text{ any } \theta, \sigma > 0, \beta \in (-1, 1]\right\}. \tag{2}$$

An application of this class is given in Shyamalkumar (2000).

Neighbourhood classes. ε-contaminations of models have been considered in, e.g., Sivaganesan (1993), although these classes work well only when it is possible to consider the *same* parameter for all the entertained models. A similar approach was taken by Ghosh and Dey (1994) and by Basu (1995) who considered normal scale mixtures as contaminating classes. Basu considered a density ratio class as well, i.e., all the densities f for which there exists $\alpha > 0$ such that $M(x - \theta_0) \leqslant \alpha f(x|\theta_0) \leqslant U(x - \theta_0)$, for all x, with M and U given nonnegative functions, with $M(\cdot) \leqslant U(\cdot)$.

Stemming from results in Cuevas and Sanz (1988), and similarly to the class of priors used by Ruggeri and Wasserman (1993) to compute Fréchet derivatives, Ruggeri (1991) considered a topological neighbourhood, in L_1-norm, of the likelihood $l(\theta)$. He considered the ε-contaminations of l given by $l_\varepsilon = (1 - \varepsilon)l + \varepsilon M$, with M in the class of all nonnegative, integrable functions or in its subset of functions with norm below a threshold. Furthermore, he considered the class of bounded likelihoods given by

$$\Gamma_{BL}(M, U) = \{l \in \mathcal{L}_P;\ M(\theta) \leqslant l(\theta) \leqslant U(\theta) \text{ a.e.}\},$$

where M and U are given measurable nonnegative function, not necessarily in $\mathcal{L}_P$, the space of all integrable functions. In particular, he considered the special case where $M = (1/k)\lambda$, $U = k\lambda$, $k \geqslant 1$, and $\lambda \in \mathcal{L}_P$ was given. Although this approach is mathematically nice, it is difficult to give a probabilistic interpretation of a perturbed likelihood.

Nonparametric classes. DeRobertis (1978) discussed one way of building nonparametric neighbourhoods of the likelihood, but the first thorough illustration of a nonparametric class is due to Lavine (1991). He considered a sample $X_1, X_2, \ldots, X_n$, where each independent random variable X_i takes values on the space $(\mathcal{X}, \mathcal{F}_\mathcal{X})$. The probability measure $\{P_\theta, \theta \in \Theta\}$ is defined on $(\mathcal{X}, \mathcal{F}_\mathcal{X})$. Lavine performed his robustness study considering a neighbourhood of P_θ. An initial prior Π_0 was defined on the space $(\Theta, \mathcal{F}_\Theta)$. He considered the product space $(\mathcal{X} \times \Theta, \mathcal{F}_\mathcal{X} \otimes \mathcal{F}_\Theta)$ with P_θ as the conditional probability given θ and Π_0 as the marginal of θ.

Lavine (1991) considered jointly a class of priors and a neighbourhood $N(\theta)$ of P_θ; for the former class, he suggested any of those considered in literature, whereas topological and other standard neighbourhoods, like ε-contaminations, could be used for the latter.

2.3.2. *Model robustness approaches*

Perturbations with respect to parametric models is more complex than changes in the prior and in the loss function, not only because of computational complexity but also because changes in the model should preserve the meaning of the parameter. The case of symmetric models is one of the most entertained since the median θ keeps the same meaning for all the models.

An informal approach can be followed in analysing likelihood robustness; one can simply try several models and see if the answer changes significantly. It is difficult to develop more formal approaches to likelihood robustness, for several reasons, mainly because parameters often change meanings across different models; hence changing models also often requires changing the prior distributions.

Sivaganesan (1993) considered a local influence measure for the sampling distribution for a class of ε-contaminations of models

$$f_\varepsilon(x;\theta) = (1-\varepsilon) f_0(x;\theta) + \varepsilon p(x;\theta),$$

with p in a class F_θ.

For a posterior quantity T_h, Sivaganesan defined the local influence measure as

$$IS(F_\theta) = \sup_{p \in F_\theta} \left(\frac{\partial T_h}{\partial \varepsilon}\right)^2_{\varepsilon=0}. \tag{3}$$

This measure follows from the appearance of the derivative in the Taylor expansion of T_h when considering the perturbation f_ε of f_0.

Ghosh and Dey (1994) considered, for the same class, a different notion of derivative and ϕ-divergence. Fréchet derivatives of the operator T_h were computed by Ruggeri (1991, manuscript) under contaminations from an $\mathcal{L}_1$ space, in the same fashion as in Ruggeri and Wasserman (1991).

2.4. *Loss robustness*

When addressing Decision Theoretic issues, we immediately think of considering classes of loss functions and extend sensitivity to the prior procedures to this new setting in a straightforward manner. For example, we could compute ranges of expected losses when the loss ranges in a class. These extensions demand some care, however. Suppose for example that the class of losses includes functions of the type $L(a,\theta) + k$, for $k \in [k_0, k_1]$. This happens for example in the universal class, see Hwang (1985). Then, the range of posterior expected losses is greater than $k_1 - k_0$, which, if "large", may suggest that we do not have robustness. Clearly, this is not true since all these loss functions are strategically equivalent, as they lead to the same optimal decision.

2.4.1. *Classes of losses*

Although some of the classes of losses are similar to those referring to priors, others arise from typical properties of the losses.

ε-contamination class. As shown earlier, *ε-contamination classes* are very popular in defining classes of priors. They can be used, as well, to define a neighbourhood around a loss function L_0 as follows:

$$\mathcal{L}^\varepsilon(L_0, \mathcal{W}) = \{L\colon L(c) = (1-\varepsilon)L_0(c) + \varepsilon M(c)\colon M \in \mathcal{W}\},$$

where ε represents imprecision over L_0, c is a consequence belonging to the state of consequences $\mathcal{C}$ and $\mathcal{W}$ is a class of loss functions, typically including L_0.

Partially known class. Stemming from the quantile class of priors, Martín et al. (1998) considered a finite partition $\{C_1, \ldots, C_n\}$ of the state of consequences $\mathcal{C}$ and gave upper and lower bounds on the losses over each element of the partition, i.e., they considered

$$\mathcal{L}_K = \{L\colon v_{i-1} \leqslant L(c) \leqslant v_i,\ \forall c \in C_i,\ i = 1, \ldots, n\},$$

where some of the sets C_i could be empty. Although the specification of the class is relatively simple as for the classes of priors, the computations become here more cumbersome.

Parametric class. The most famous parametric class of losses is LINEX (see Varian, 1974), defined by

$$L_\gamma(c) = \mathrm{e}^{\gamma c} - \gamma c - 1, \quad \gamma_L \leqslant \gamma \leqslant \gamma_U, \text{ with } c = a - \theta.$$

Some other examples can be seen in Bell (1995). In general, the parametric class is defined as

$$\mathcal{L}_\omega = \{L = L_\omega, \ \omega \in \Omega\}.$$

Mixtures of convex loss functions. Some classes of priors can be thought as mixtures of priors with respect to a probability measure (e.g., symmetric and unimodal priors as mixture of uniform distributions). Here it is possible to consider mixtures of convex loss functions, as follows.

Let $L_\lambda \in \Psi$ be a family of convex loss functions parametrised as $\lambda \in \Lambda$. Let $G \in \mathcal{P}$ be a class of all probability measures on Λ. Consider

$$\mathcal{L} = \left\{L\colon L(c) = \int_\Lambda L_\lambda(c)\,\mathrm{d}G(\lambda)\right\}.$$

Although the idea behind the class is simple (it "weighs" reasonable losses), the functional shape of the resulting loss function is quite difficult to recognise.

Bands of convex loss functions. Martín et al. (1998) considered losses which can be modelled as $L(\theta, a) = L(\theta - a)$. Therefore, it is possible to consider the loss as a function $L(t)$ and study its derivative, when it exists, to detect local increase/decrease in the behaviour of the loss. Let $L'(t) = l(t)$ and consider two functions $\lambda(t)$ and $u(t)$, both positive for positive t and negative for negative t. Then a band of convex loss functions is obtained by considering

$$\mathcal{L} = \{L\colon \lambda(t) \leqslant l(t) \leqslant u(t), \ \forall t\}.$$

2.5. *Loss robustness studies*

Loss robustness studies consider a class $\mathcal{L}$ of loss functions $L(\theta, a)$, defined for the action a in the action space $\mathcal{A}$ and θ in Θ. As L varies in $\mathcal{L}$, a measure of loss robustness is given by the range of the posterior expected loss of an action a, $\sup_{L\in\mathcal{L}} T(L, a) - \inf_{L\in\mathcal{L}} T(L, a)$.

Many studies in loss robustness concentrated on the search of the "best" loss, according to some criteria. A first criterion is based on Γ-minimax and it will be illustrated in Section 5.9. Some authors considered the influence of a loss function, defined as

$$\frac{\partial}{\partial x} \inf_a T(L, a) = \frac{\partial}{\partial x} T(L, a^*),$$

where a^* is the Bayes action.

Dey et al. (1998) proposed the following robustness measure:

$$D = \sup_x \left| \frac{\partial}{\partial x} T(L, a^*) \right| - \inf_x \left| \frac{\partial}{\partial x} T(L, a^*) \right|.$$

Makov (1994) used the influence to order loss functions in a class. In the first way, he said that the loss L_1 was preferred to L_2 if

$$\sup_x \left| \inf_a T(L_1, a) \right| < \sup_x \left| \inf_a T(L_2, a) \right|.$$

In the second method, L_1 was preferred to L_2 if

$$E^x\left[T(L_1, a)\right] < E^x\left[T(L_2, a)\right],$$

where the expectation was taken with respect to $m(x)$, the marginal distribution of x.

Martín et al. (1998) considered nondominated actions; their approach is thoroughly reviewed in Section 3.2.

2.6. *Joint robustness*

Joint robustness studies are essential, specially in decision theoretic contexts, as robustness studies with respect to only one of the elements varying, and the others fixed, may give a false sense of stability: a problem may be insensitive to changes only in the loss and changes only in the probability, but sensitive to simultaneous changes in both the loss and probability.

EXAMPLE 2 (*Continuation of Example* 1). Fix $p_1 = 0.5$. We have

$$\begin{aligned} T(L, \pi, a) - T(L, \pi, b) &= .5L(c_1) + .5L(c_2) - .5L(c_2) - .5L(c_3) \\ &= .5\big(L(c_1) - L(c_3)\big). \end{aligned}$$

Since $L(c_1) \geqslant L(c_3)$ for all values of L, we can conclude that the problem is robust against changes in L. Analogously, if we consider $L(c_1) = .75$, $L(c_2) = .5$ and $L(c_3) = .25$, we have

$$\begin{aligned} T(L, \pi, a) - T(L, \pi, b) &= .75p_1 + .5(1 - p_1) - .5p_1 - .25(1 - p_1) \\ &= .25. \end{aligned}$$

Then, we have robustness against changes in probabilities. However, we cannot conclude that the problem is robust, as shown before.

As before, the range of the posterior quantity is the most natural robustness measure when changes are allowed in prior/model/loss function.

Sivaganesan (1993) used the influence measure (3) not only for model robustness but also in prior/model robustness (with due changes) and he developed a robustness diagnostic to detect which features were more influential on the behaviour of the quantity T_h. He considered ε-contaminations for both priors and models. A similar approach was taken by Ghosh and Dey (1994) who considered a different notion of derivative and ϕ-divergence.

Clarke and Gustafson (1998) studied robustness with respect to changes in the prior, model and data. They proposed a comprehensive measure of sensitivity, and then partitioned it so that individual contributions of each input could be identified. All these measures are based on the notion of relative entropy. The main problem with these measures, as stated by the authors themselves, is that they cannot be used for formal diagnostics, since they are unable to calibrate the measures and interpret their actual values.

Martín et al. (2003) and Ruggeri et al. (2003) considered how joint perturbations of losses and priors could affect comparisons of alternatives a and b in a decision analytic framework. ε-contaminations and partially known classes of losses were considered in the former paper, whereas topological neighbourhoods, based upon norms, were considered in the latter.

3. A unified approach

We have overviewed various approaches to robust Bayesian analysis. In reading these lines, one may get an impression of a collection of ad hoc approaches. We may actually provide a unified setting by appealing to decision theoretic foundations which set up the stage for sensitivity analysis within Statistical Decision Theory. We then describe issues concerning the nondominated set and deal with qualitative aspects describing a theory of stability.

3.1. Foundations

To construct an appropriate framework for general sensitivity analysis, we should reconsider the standard Bayesian axiomatic foundations to account for imprecision in beliefs and preferences, see French and Ríos Insua (2000). Although this approach does not lead to such a well rounded development as in the precise case, in which various axiomatisations essentially lead to the subjective expected utility model, various partial results do exist, see, e.g., Girón and Ríos (1980), Ríos Insua (1990, 1991) or Walley (1991), leading to, essentially, the same conclusion: imprecise beliefs and preferences may be modelled by a class of priors and a class of loss functions, so that preferences among alternatives may be represented by inequalities of the corresponding posterior expected losses.

More technically, under certain conditions, we are able to model beliefs by a class Γ of priors $\pi(\cdot)$, which in the presence of additional information are updated one by one via Bayes theorem, and preferences by a class $\mathcal{L}$ of loss functions L, so that we find alternative a at most as preferred as b ($a \precsim b$) if and only if $(T(L,\pi,a) \geqslant T(L,\pi,b), \forall L \in \mathcal{L}, \pi \in \Gamma)$. The basic argument for such results assumes that the relation, rather than being a weak order, will be a quasi order. Then, assuming similar conditions to the Bayesian axioms, we are able to obtain the representation sketched above.

To give a flavour of these results, we summarise the representation in Girón and Ríos (1980). In their framework, they assume that we have been able to model preferences over consequences precisely with a loss function. Therefore we may associate with each action a the function $L(a,\theta)$, $\theta \in \Theta$. Let $\mathcal{D} = \{h \mid \exists a \in \mathcal{A}, h(\theta) = L(a,\theta), \forall\theta \in \theta\}$. Preferences $\precsim$ are established over these functions. The result under appropriate topological and measurability conditions is:

PROPOSITION 1. *Suppose that* $(\mathcal{D}, \precsim)$ *satisfies the following conditions*:

(GR1) $(\mathcal{D}, \precsim)$ *is a quasi order (reflexive and transitive).*
(GR2) *If* $L(a,\theta) < L(b,\theta), \forall\theta \in \Theta$, *then* $b \prec a$.
(GR3) *For* $a, b, c \in \mathcal{A}$, $\lambda \in (0,1)$, *then* $L(a,\theta) \leqslant L(b,\theta)$ *if and only if* $\lambda L(a,\theta) + (1-\lambda)L(c,\theta) \precsim \lambda L(b,\theta) + (1-\lambda)L(c,\theta)$.
(GR4) *For* $f_n, g, h \in \mathcal{D}$, *if* $f_n \to f$ *and* $f_n \precsim g$, $h \precsim f_n$, $\forall n$, *then* $f \precsim g$, $h \precsim f$.

Then, there exists a class Γ *of probability measures* $\pi(\cdot)$ *on* Θ *such that*

$$a \precsim b \iff \int L(a,\theta)\pi(\theta)\,\mathrm{d}\theta \geqslant \int L(b,\theta)\pi(\theta)\,\mathrm{d}\theta, \quad \forall\pi(\cdot) \in \Gamma.$$

Hence, we see that we model imprecision in beliefs by means of a class of probability distributions. Observe that, in this result, preferences are modelled with a single loss function. Interestingly enough, when $(\mathcal{D}, \precsim)$ is complete, i.e., a weak order, Γ is a singleton. Under additional conditions, we may see that beliefs are updated by updating each of the priors with Bayes formula.

These results have two basic implications. First, they provide a qualitative framework for sensitivity analysis in Statistical Decision Theory, describing under what conditions we may undertake the standard and natural sensitivity analysis approach of perturbing the initial probability-loss assessments, within some reasonable constraints. Second, they point to the basic solution concept of robust approaches, thus indicating the basic computational objective in sensitivity analysis, as long as we are interested in decision theoretic problems: that of *nondominated alternatives*. To wit, we say that an alternative a dominates another alternative b, denoted $b \prec a$, if $b \precsim a$ and $\neg(a \precsim b)$, or, in terms of posterior expected losses, $T(L,\pi,b) \geqslant T(L,\pi,a), \forall L \in \mathcal{L}, \pi \in \Gamma$, with strict inequality for one pair (L,π). We then say that an alternative a is *nondominated* if no other alternative in $\mathcal{A}$ dominates it.

Note the similarity with the admissibility concept. Both correspond to Pareto orderings of decision rules, see White (1982). Admissibility is based on inequalities on the risk function, therefore not making assumptions about the prior distribution and averaging over the data x (and assuming a precise loss function). Nondominance is based on inequalities on the posterior expected loss, therefore making assumptions over the prior distribution, using only the observed x (and, possibly, assuming imprecision in the loss function).

In summary, using a class Γ of priors over the set of states Θ, and a class $\mathcal{L}$ of loss functions, gives $a, b \in \mathcal{A}$, set of alternatives, we say that $b \preceq a$ if and only if

$$T(L,\pi,a) \leqslant T(L,\pi,b) \quad \forall\pi \in \Gamma, \forall L \in \mathcal{L},$$

where $\preceq$ is the preference relation between alternatives.

3.2. The nondominated set

As a consequence, the solution concept is the efficient set, i.e., the set of nondominated alternatives. When needed, only nondominated alternatives with respect to $\mathcal{L}$, with fixed $\pi \in \Gamma$, are considered; similarly, nondominated alternatives with respect to Γ, for fixed $L \in \mathcal{L}$, can be considered. They are defined in an obvious way.

3.2.1. Existence of efficient sets

Simple examples show that the nondominated set may be empty. Suppose, e.g., that $\mathcal{A} = (0, 1] \times (0, 1]$ with a generic alternative defined by $a = (a_1, a_2)$. Suppose also that $\mathcal{L} = \{L_1, L_2\}$, with $L_1(a, \theta) = a_1$ and $L_2(a, \theta) = a_2$ with a generic state space Θ. Then, whatever the prior and the model are, for any alternative a there is another alternative a' such that $a \prec a'$: the nondominated set is empty.

The following results show important statistical problems in which the nondominated set is nonempty. The first one refers to problems in which the set of alternatives is finite, a relevant example being that of multiple hypothesis testing.

PROPOSITION 2. *If the set $\mathcal{A}$ of alternatives is finite, the nondominated set is not empty.*

Similarly, we have:

PROPOSITION 3. *If for one loss function and one prior in the respective classes the set of Bayes alternatives is nonempty and finite, then there exists, at least, one nondominated alternative.*

By repeated application of the above result, we may find a representative set of nondominated alternatives. In some cases, the process ends in a finite number of steps. An important case is that in which the classes of losses and priors are finitely generated, that is, there is a set $\{L_1, L_2, \ldots, L_n\} \subset \mathcal{L}$ such that $\forall L \in \mathcal{L}$, $L = \sum_{i=1}^n w_i L_i$, $\sum w_i = 1$, $w_i \geqslant 0$, and, similarly, for the priors. An example is that of multi-objective designs, in which we have several loss functions and we find nondominated designs looking for designs which are optimal for convex combinations of the original loss functions: see Clyde and Chaloner (1996). Some mixture classes are examples of finitely generated classes of priors. We assume an appropriate topology for the next result, which refers only to losses, but with immediate extensions which refer, as well, to priors.

PROPOSITION 4. *If the set $\mathcal{A}$ of alternatives is compact and the class $\mathcal{L}$ is generated by a finite number of loss functions continuous in a, uniformly with respect to θ, then the nondominated set is not empty.*

Further results about existence of efficient sets may be seen in Arias (2002) and Arias and Martín (2002).

3.2.2. Bayes and nondominated alternatives

An important result relating the sets of Bayes and nondominated alternatives is:

PROPOSITION 5. *If for any loss and prior in the classes there is a unique Bayes alternative, then the set of Bayes alternatives is contained in the set of nondominated ones.*

An important consequence is given by

COROLLARY 1. *If the set of alternatives is compact and there exists one loss function L strictly convex, then the unique Bayes alternatives associated to L is a nondominated alternative.*

We will see later that both sets coincide in some important problems. Note, though, that both sets can be different even if there is a unique Bayes solution for each loss function and each prior in the class.

EXAMPLE 3. Let $\mathcal{A} = \{(a_1, a_2)\colon a_1^2 + a_2^2 \geqslant 1, 0 \leqslant a_1, a_2 \leqslant 1\}$. Suppose also that $\mathcal{L} = \{L_1, L_2\}$, with $L_1(a, \theta) = a_1$ and $L_2(a, \theta) = a_2$, with a generic state space Θ. Then, whatever the prior and the model are, the set of nondominated alternatives is $\mathcal{A} = \{(a_1, a_2)\colon a_1^2 + a_2^2 = 1, 0 \leqslant a_1, a_2 \leqslant 1\}$, whereas (1, 0) is Bayes for L_1 and (0, 1) is Bayes for L_2.

Under appropriate conditions both concepts are intimately related. For example, if the set of Bayes alternatives is finite for a loss function, one of those alternatives is nondominated.

The converse results are not true. For example, let

$$\mathcal{A} = \bigl\{(1, x),\ x \in (1/2, 1)\bigr\} \cup \bigl\{(x, 0),\ x \in (0, 1/2)\bigr\} \cup \bigl\{(3/4, 1/4)\bigr\}.$$

If $\mathcal{L} = \{L_1, L_2\}$ with $L_1(a, \theta) = a_1$ and $L_2(a, \theta) = -a_2$, then (3/4, 1/4) is the unique nondominated alternative and there is no Bayes alternative.

Under conditions similar to those of Proposition 4, and with basically the same proof, we have that if $\mathcal{L}$ is generated by a finite number of continuous, uniformly in θ, loss functions in a compact set $\mathcal{A}$, then, for any L_i there is one nondominated alternative that is Bayes for L_i.

The provision of more general results, concerning either the existence of nondominated alternatives or the relation between nondominated and Bayes alternatives, requires the adoption of abstract topological conditions similar to those commonly found in multi-objective optimisation, see, e.g., Yu (1986). However, it turns out to be more productive to consider specific statistical examples, such as the following from Martín et al. (1998).

Consider set estimation under the loss function

$$L_s(C, \theta) = s\bigl(\mathrm{vol}(C)\bigr) - I(\theta \in C),$$

where C is a subset, s is an increasing function and $I(\cdot)$ is the indicator function. Casella et al. (1994) studied admissibility of estimators under such loss functions.

For simplicity, we shall consider a problem with precision in the prior and posterior density nonconstant on any subset with positive Lebesgue measure. The latter condition implies that positive posterior probability will be given to any set with positive Lebesgue measure and allows for the existence of highest posterior density regions, $\{\theta: \pi(\theta|x) \geqslant k\}$, $k > 0$, with arbitrary volume. The posterior expected loss of an alternative C is given by

$$\int_{\Theta} L_s(C, \theta)\pi(\theta|x)\,\mathrm{d}\theta = s\bigl(\mathrm{vol}(C)\bigr) - \Pi(C|x).$$

We shall consider a class $\mathcal{L}$ of loss functions L_s depending on s, $\mathcal{L} = \{L_s\colon s_{Lo}(t) \leqslant s(t) \leqslant s_{Up}(t), \forall t\}$, where s_{Lo} and s_{Up} are given increasing functions. For this case, we have:

PROPOSITION 6. *The nondominated alternatives are necessarily* HPD *sets.*

It is not true, however, that any HPD set is a nondominated alternative, as shown in the following example.

EXAMPLE 4. Let $\pi(\theta|x)$ be a Beta distribution $\mathcal{B}(2, 1)$, corresponding, e.g., to a Bernoulli model under a uniform prior. The intervals $A_y = [y, 1]$ are HPD sets with $\Pi(A_y|x) = 1 - y^2$ and volume $1 - y$. Their corresponding expected loss is $s(1 - y) - 1 + y^2$. Consider the class $\mathcal{L}$ of all loss functions L_s, with $s_{Lo}(t) = t^2 \leqslant s(t) \leqslant s_{Up}(t) = t$. Then, the HPD sets A_y, with $y > (1 + \sqrt{1/2})/2$, are dominated by $A_{1/2}$. In fact, for every s such that $s_{Lo} \leqslant s \leqslant s_{Up}$, we have

$$\int_{\Theta} L_s(A_{1/2}, \theta)\pi(\theta|x)\,\mathrm{d}\theta - \int_{\Theta} L_s(A_x, \theta)\pi(\theta|x)\,\mathrm{d}\theta = -1/4 + 2y - 2y^2,$$

which is negative for such y.

3.3. Extracting additional information

In some cases, nondominance is a very powerful concept leading to a unique nondominated alternative. Here is a nontrivial case, taken from Martín et al. (1998).

EXAMPLE 5. Suppose that $\pi(\theta|x)$ is symmetric, unimodal at 0 and such that $\Pi(A|x) > 0$ for any measurable subset A with positive Lebesgue measure. Let $\mathcal{L}$ be the class of all convex, symmetric loss functions which are not constant on a subset with positive posterior probability. Then, we may see that the mode 0 is the Bayes alternative for any $L \in \mathcal{L}$ and the unique nondominated alternative.

However, in most cases, the nondominated set will be too large to imply a final decision. It may happen that there are several nondominated alternatives and differences in expected losses are nonnegligible. If such is the case, we should look for additional information that would help us to reduce the classes, and, perhaps, reduce the nondominated set. We outline tools based on functional derivatives to elicit additional information, see Martín and Ríos Insua (1996) for additional details. Tools based on distance analysis may be seen in Ríos Insua and French (1991).

First, we need to describe perturbations of a current assessment which preserve preferences. (L, π) is *ε-robust* for $b \precsim a$ within $\mathcal{L} \times \Gamma$, if $b \precsim a$ whenever we use $(tv + (1-t)L,\ tq_\theta + (1-t)\pi) \in \mathcal{L} \times \Gamma$ in place of (L, π), where $(v, q_\theta) \in \mathcal{L} \times \Gamma$ and $t < \varepsilon$. When $\varepsilon = 1$, the information $b \precsim a$ is robust. To determine the ε-robustness of an assessment, we shall use the Fréchet derivative of $N^{ab}(L, \pi) = \int_\Theta L(a, \theta) f(x|\theta)\pi(\theta)\,\mathrm{d}\theta - \int_\Theta L(b, \theta) f(x|\theta)\pi(\theta)\,\mathrm{d}\theta$.

PROPOSITION 7. *Let $\dot{N}^{ab}_{L,\pi}(\cdot, \cdot)$ be the Fréchet derivative of N^{ab} with respect to (L, π). Then*

$$\dot{N}^{ab}_{L,\pi}(m, \delta_\theta) = N^{ab}(L, \delta_\theta) + N^{ab}(m, \pi).$$

We may use Fréchet derivatives to bound the ε-robustness of a preference over a pair of alternatives, as shown by:

PROPOSITION 8. (a) *If*

$$-\inf_{\mathcal{L}\times\Gamma} \widehat{N}(v, \pi') > 2\sqrt{K N^{ab}(L, \pi)} \tag{4}$$

then (L, π) is ε-robust for $b \precsim a$ within $\mathcal{L} \times \Gamma$ with $\varepsilon \in [\varepsilon_1, \varepsilon_2]$ and

$$\varepsilon_1 = \min\left\{1, \frac{\inf_{\mathcal{L}\times\Gamma} \widehat{N}(v, \pi') + \sqrt{(\inf_{\mathcal{L}\times\Gamma} \widehat{N}(v, \pi'))^2 + 4K N^{ab}(L, \pi)}}{2K}\right\},$$

$$\varepsilon_2 = \min\left\{1, \sqrt{\frac{N^{ab}(L, \pi)}{K}}, \varepsilon_l, \varepsilon_\pi\right\}$$

with

$$K = 4 \sup\left|p_X(x|\theta)\right|,$$

$$\widehat{N}(v, \pi') = N^{ab}(v, \pi) + N^{ab}(L, \pi') - 2N^{ab}(L, \pi),$$

$$\varepsilon_L = \min\left\{1, \frac{T^{ab}(L)}{T^{ab}(L) - \inf_{w\in\mathcal{L}} T^{ab}(w)}\right\} \quad \textit{and}$$

$$\varepsilon_\pi = \min\left\{1, \frac{N^{ab}(\pi)}{N^{ab}(\pi) - \inf_{\pi'\in\Gamma} N^{ab}(\pi')}\right\}.$$

(b) *If does not hold, (L, π) is ε-robust for $b \precsim a$ within $\mathcal{L} \times \Gamma$, for $\varepsilon \in [\varepsilon_1, \varepsilon_3]$ and*

$$\varepsilon_3 = \min\{1, \varepsilon_L, \varepsilon_\pi\}.$$

Examples in Martín and Ríos Insua (1996) and Ríos Insua et al. (1999) show how these tools may help in guiding additional efforts.

The approach proposed so far may be summarised as follows: at a given stage of analysis, we elicit information on the DM's beliefs and preferences, and consider the class of all priors and losses compatible with such information. We approximate the set of nondominated solutions; if these alternatives do not differ too much in expected loss, we may stop the analysis; otherwise, we need to gather additional information, possibly with the tools outlined above. This would further constrain the class: in this case the set of nondominated alternatives will be smaller and we could hope that this iterative process would converge until the nondominated set is small enough to reach a final decision.

3.4. $\mathcal{L} \times \Gamma$-minimax alternatives

It is conceivable in these contexts that at some stage we might not be able to gather additional information yet there remain several nondominated alternatives with very different expected losses. In these situations, $\mathcal{L} \times \Gamma$*-minimax solutions* may be useful as a way of selecting a single robust solution, see, e.g., Makov (1994). We associate with each alternative its worst expected loss; we then suggest the alternative with minimum worst expected loss. Specifically, an alternative $a_M \in \mathcal{A}$ is $\mathcal{L} \times \Gamma$-minimax if $\sup_{L \in \mathcal{L}, \pi \in \Gamma} T(L, \pi, a_M) = \inf_{a \in \mathcal{A}} \sup_{l \in \mathcal{L}, \pi \in \Gamma} T(L, \pi, a)$. Note, again, the similarity with the classical concept of minimax alternatives. The latter is based on best worst performance on the risk function, ignores priors but averages over the sample space; the former is based on best worst performance of the posterior expected loss, but acknowledges prior beliefs and is conditional on the observed data.

Normally, minimax, and related concepts such as minimax regret are introduced as *ad hoc* concepts. In our context, we view them as heuristics that suggest a nondominated alternative in cases in which we are not able to gather additional information, as results such as the following show.

PROPOSITION 9. *If the set of $\mathcal{L} \times \Gamma$-minimax alternatives is finite, one of them is nondominated.*

As a corollary, if there is a unique $\mathcal{L} \times \Gamma$-minimax alternative, it is nondominated. Computation of minimax alternatives is not simple in general cases, but, again, we may use discrete approximation schemes. Drawing a discrete sample from the set of alternatives, we compute the corresponding nondominated set and the maximum posterior expected loss of each alternative – this is a conventional global sensitivity analysis problem – picking the one with the smallest maximum. Again, under mild conditions we may prove convergence of the approximate minimax solution to the actual minimax solution.

See Section 5.9, Vidakovic (2000), Ruggeri (2005) and references therein for results about Γ-minimax.

3.5. Stability theory

As mentioned at the beginning of the section, stability theory provides another unifying, general sensitivity framework. Stability theory formalises the idea that imprecision in elicitation of beliefs and preferences should not affect the optimal decision greatly. When *strong stability* holds, careful enough elicitation leads to decisions with expected loss close to the smallest achievable; when *weak stability* holds, at least one stabilised decision will have such property. However, when neither concept of stability applies, even small elicitation errors may lead to disastrous results in terms of large expected losses.

More technically, stability theory studies the convergence of alternatives a_n, 'nearly' optimal for the pair (π_n, L_n), to the optimal alternative a_0, for the loss L and the prior π, when the approximately specified beliefs and preferences (π_n, L_n) converge to (π, L): see, e.g., Kadane and Chuang (1978), Kadane and Srinivasan (1994, 1996, 1998), Chuang (1984), and Salinetti (1994). Convergence for priors is in the weak sense, denoted by $\pi_n \stackrel{w}{\to} \pi$. Convergence of the loss function relates to some topology. Hence, stability is strongly related to the continuity of the posterior expected loss functional.

These authors considered various notions of stability. The first is

DEFINITION 1 (*Strong stability I*). (L, p_X, π) is strongly stable I (SSI) if for any sequence $\pi_n \stackrel{w}{\to} \pi$ and $L_n \to L$

$$\lim_{\varepsilon\downarrow 0} \limsup_{n\to\infty} \left[\int L_n\big(\theta, a(\varepsilon)\big) p_X(x|\theta)\pi_n(\theta)\,\mathrm{d}\theta - \inf_{a\in\mathcal{A}} \int L_n(\theta, a) p_X(x|\theta)\pi_n(\theta)\,\mathrm{d}\theta \right] = 0 \tag{5}$$

for every $a(\varepsilon)$ such that:

$$\int L\big(\theta, a(\varepsilon)\big) p_X(x|\theta)\pi(\theta)\,\mathrm{d}\theta \leqslant \inf_{a\in\mathcal{A}} \int L(\theta, a) p_X(x|\theta)\pi(\theta)\,\mathrm{d}\theta + \varepsilon.$$

The decision problem (L, p_X, π) is *weakly stable I* (WSI) if (5) holds for a specific choice $a(\varepsilon)$.

DEFINITION 2 (*Strong stability II*). The decision problem (L, p_X, π) is *strongly stable II* (SSII) if for any sequence $\pi_n \stackrel{w}{\to} \pi$, $q_{\theta,n} \stackrel{w}{\to} \pi$, $L_n \to L$ and $w_n \to L$

$$\lim_{\varepsilon\downarrow 0} \limsup_{n\to\infty} \left[\int L_n\big(\theta, a_{q_n}(\varepsilon)\big) p_X(x|\theta)\pi_n(\theta)\,\mathrm{d}\theta - \inf_{a\in\mathcal{A}} \int L_n(\theta, a) p_X(x|\theta)\pi_n(\theta)\,\mathrm{d}\theta \right] = 0 \tag{6}$$

for every $a_{q_n}(\varepsilon)$ such that

$$\int w_n\big(\theta, a_{q_n}(\varepsilon)\big) p_X(x|\theta) q_{\theta,n}(\theta)\,\mathrm{d}\theta$$
$$\leqslant \inf_{a \in \mathcal{A}} \int w_n(\theta, a) p_X(x|\theta) q_{\theta,n}(\theta)\,\mathrm{d}\theta + \varepsilon.$$

The decision problem (L, p_X, π) is weakly stable II (WSII) if (6) holds for a specific choice $a_{q_n}(\varepsilon)$.

It can be easily seen that SSII implies SSI, while the converse is not true in general, as shown, e.g., in Kadane and Srinivasan (1998), who provided sufficient conditions ensuring the equivalence of SSI and SSII. Note that stability I is not always guaranteed, even in standard statistical problems, as shown, e.g., in Kadane et al. (2000). Such counterexamples show a need for conditions which ensure stability. Some of them were presented in Kadane and Chuang (1978). While these conditions simplify the task of verifying stability, it can still be hard to do so in practise. Some effort has been devoted in seeking more easily verified sufficient conditions ensuring strong (or weak) stability. Kadane and Srinivasan (1994) give several such sets of conditions drawing on some prior work of Salinetti (1994). Further convergence results may be seen in Kadane and Srinivasan (1998).

A different notion of stability of both posterior measures and posterior quantities was presented in Basu et al. (1996). Here we present only the latter, defined for priors Π in $\mathcal{M}$ and the posterior quantity T^{Π}, with distances $d(\cdot,\cdot)$ and $\nu(\cdot,\cdot)$ defined on the respective spaces.

DEFINITION 3. The posterior quantity T_h is called stable at $\Pi_0 \in \mathcal{M}$ if there exist $\delta > 0$ and $M > 0$ such that for any $\Pi \in \mathcal{M}$ satisfying $d(\Pi, \Pi_0) < \delta$, then $\nu(T^{\Pi}, T^{\Pi_0}) < M d(\Pi, \Pi_0)$.

This notion of stability comes from the notion of qualitative robustness (see Huber, 1981) developed in a classical framework. From a Bayesian viewpoint, it is given by

DEFINITION 4. The posterior quantity T_h is called qualitatively robust at $\Pi_0 \in \mathcal{M}$ if for any $\Pi \in \mathcal{M}$ and any $\varepsilon > 0$ there exists $\delta > 0$ such that $d(\Pi, \Pi_0) < \delta$ implies $\nu(T^{\Pi}, T^{\Pi_0}) < \varepsilon$.

4. Robust Bayesian computations

We illustrate now how to undertake various robust Bayesian computations described previously.

4.1. General computational issues

Given a robustness measure $R(\Pi)$, e.g., the range, which we assume for simplicity dependent only on the prior distribution Π in a class Γ, then most computations in Bayesian robustness rely on the following steps

- a class Ψ of extremal measures in Γ is identified, i.e., such that each measure in Γ is a mixture of measures in Ψ;
- it is shown that $\sup_{\Pi\in\Gamma} R(\Pi) = \sup_{\Pi\in\Psi} R(\Pi)$;
- $\sup_{\Pi\in\Psi} R(\Pi)$ is computed, in a simpler way than $\sup_{\Pi\in\Gamma} R(\Pi)$.

Here are some examples of classes Γ of probability measures with the subclass Ψ of their extremal measures.

- $\Gamma = \{$all priors$\}$ and $\Psi = \{$all Dirac priors$\}$
- $\Gamma = \{$all unimodal priors with mode at $\theta_0\}$ and $\Psi = \{$all uniform priors over intervals containing $\theta_0\}$
- $\Gamma = \{$all symmetric, unimodal priors with mode at $\theta_0\}$ and $\Psi = \{$all uniform priors over intervals with middle point at $\theta_0\}$
- $\Gamma = \{$all priors with n generalised moments constraints$\}$ and $\Psi = \{$all discrete priors on $n+1$ points at most$\}$
- $\Gamma = \{$all priors in a concentration function neighbourhood$\}$ and $\Psi = \{$all priors with concentration function coinciding with the monotone nondecreasing, continuous, convex function defining the neighbourhood$\}$

Once the classes Γ and Ψ have been identified, the computation of $R(\Pi)$ follows from a result in Sivaganesan and Berger (1989) which states, in a general form, that

$$\sup_{\Pi\in\Gamma} \frac{\int_\Theta f(\theta)\Pi(\mathrm{d}\theta)}{\int_\Theta g(\theta)\Pi(\mathrm{d}\theta)} = \sup_{\Pi\in\Psi} \frac{\int_\Theta f(\theta)\Pi(\mathrm{d}\theta)}{\int_\Theta g(\theta)\Pi(\mathrm{d}\theta)}.$$

Among the classes considered in the previous list, it is worth mentioning that the functional optimisation problem becomes a nonlinear one for all of them, except for the class based on concentration functions. For the class of all priors, the problem becomes

$$\sup_{\Pi\in\Gamma} \frac{\int_\Theta f(\theta)\Pi(\mathrm{d}\theta)}{\int_\Theta g(\theta)\Pi(\mathrm{d}\theta)} = \sup_{\theta\in\Theta} \frac{f(\theta)}{g(\theta)}.$$

For the generalised moments class, the search of the optima becomes the search for (at most) $n+1$ values of θ, as proved in Betrò et al. (1994). The results for the generalised moments constraints apply to the quantile class, where the specification of n quantiles is equivalent to assign probability to $n+1$ intervals and the discrete optimising measure takes one value on the closure of each interval. Ruggeri (1990) showed that most of the values were endpoints of interval and he thoroughly discussed strategies to detect them.

4.2. *Algorithms to compute ranges*

In Section 2, we defined some measures of robustness, with emphasis on the range of posterior quantities. The computation of this range is far from simple. So, there are techniques to approximate that value.

Lavine's algorithm

This algorithm is explained in articles by Lavine (1991) and Lavine et al. (1991, 2000). However, it was used before, among others, by DeRobertis and Hartigan (1981). We describe here the formulation from Lavine et al. (2000).

Consider a class $\mathcal{F}$ of likelihood functions $f(x|\theta_f)$, with θ_f denoting an unknown parameter and Γ_f a class of priors distributions π for θ_f. Then, we are interested in computing $\overline{\phi} - \underline{\phi}$, with

$$\overline{\phi} = \inf_{f \in \mathcal{F}} \inf_{\pi \in \Gamma_f} T(\pi, f) \quad \text{and} \quad \underline{\phi} = \sup_{f \in \mathcal{F}} \sup_{\pi \in \Gamma_f} T(\pi, f),$$

where $T(\pi, f) = N(\pi, f)/D(\pi, f)$ is the posterior expectation of some quantity $h(\theta_f)$.

We illustrate the method for $\overline{\phi}$. The idea is based on the following identity

$$\overline{\phi} > q \iff \sup_{\pi, f} \int \bigl(h(\theta_f) - q\bigr) f(x|\theta_f)\pi(\mathrm{d}\theta_f) > 0.$$

As a consequence, the initial nonlinear optimisation problem becomes a sequences of linear optimisation problems based on the following result:

PROPOSITION 10. *For every real* q, *we have*:

(1) $\sup_{\pi,f}(N(\pi, f) - qD(\pi, f)) > 0$ *if and only if* $\sup_{\pi,f} T(\pi, f) > q$;
(2) *if, in addition,* $\inf_{\pi,f} D(\pi, f) > 0$ *and* $\sup_{\pi,f} N(\pi, f) < +\infty$, *then* $\sup_{\pi,f}(N(\pi, f) - qD(\pi, f)) = 0$ *if and only if* $\sup_{\pi,f} T(\pi, f) = q$.

Lavine et al. (2000) summarise some examples about prior robustness and sampling distribution robustness, using this technique.

Betrò and Guglielmi's algorithm
Betrò and Guglielmi (1996, 2000) concentrated on the problem of global prior robustness under the generalised moment class. As we pointed out before, this is a rich class that includes several of the most common prior classes. Again, the idea is to obtain a linear operator. In this case, the authors consider a variable transformation approach.

Consider the class $\Gamma = \{\pi\colon \int_\Theta H_i(\theta)\pi(\theta)\,\mathrm{d}\theta \leqslant \alpha_i,\ i = 1, \ldots, n\}$ of specified generalised moments. Define the finite measure ν by:

$$\nu(A) = \frac{\int_A \pi(\mathrm{d}\theta)}{\int_\Theta f(x|\theta)\pi(\mathrm{d}\theta)}$$

for all measurable set A. Assuming that $0 < \int f(x|\theta)\pi(\mathrm{d}\theta) < +\infty$, we have

$$\int_\Theta f(x|\theta)\nu(\mathrm{d}\theta) = 1.$$

It is easy to show that

$$\pi(A) = \frac{\int_A \nu(\mathrm{d}\theta)}{\int_\Theta \nu(\mathrm{d}\theta)},$$

and

$$\sup_{\pi \in \Gamma} \frac{\int_\Theta h(\theta) f(x|\theta)\pi(\mathrm{d}\theta)}{\int_\Theta f(x|\theta)\pi(\mathrm{d}\theta)} = \sup_{\nu \in \Gamma'} \int_\Theta h(\theta) f(x|\theta)\nu(\mathrm{d}\theta),$$

where

$$\Gamma' = \left\{\nu\colon \int_{\Theta} l_i(\theta)\nu(\mathrm{d}\theta) \leqslant 0,\ i = 1, \ldots, n,\ \int_{\Theta} f(x|\theta)\nu(\mathrm{d}\theta) = 1\right\}$$

and $l_i(\theta) = H_i(\theta) - \alpha_i$.

4.3. Computations for loss robustness

We describe here some cases related with loss robustness computations.

ε-contaminated class. The computation of the supremum of $T(L)$ when $L \in \mathcal{L}^{\varepsilon}(L_0, \mathcal{W})$ leads to the computation of the supremum of $T(M)$ when $M \in \mathcal{W}$:

$$\sup_{L \in \mathcal{L}^{\varepsilon}(L_0, \mathcal{W})} T(L) = (1 - \varepsilon)T(L_0) + \varepsilon \sup_{M \in \mathcal{W}} T(M).$$

Parametric class. If we have an explicit expression for the posterior expected loss, the computation of the supremum is relatively simple, since we have a nonlinear programming problem. When we lack that expression, we propose a Monte Carlo method to approximate the supremum, see Shao (1989) for a related procedure to approximate the optimal alternative in a Bayesian decision analysis:

1. Generate n i.i.d. vectors $\theta_1, \ldots, \theta_n \sim \pi(\cdot|x)$.
2. Approximate $T(L_\omega)$ by

$$T_n(L_\omega) = \frac{1}{n}\sum_{i=1}^{n} L_\omega(a, \theta_i).$$

3. Approximate $\sup_{\omega \in \Omega} T(L_\omega)$ by

$$\widehat{T}_n = \sup_{\omega \in \Omega} T_n(L_\omega). \tag{7}$$

We have that if Ω is compact and $L_\omega(\cdot)$ is continuous in $\omega \in \Omega$, then $\widehat{T}_n$ converges almost surely (with respect to $\Pi(\cdot|x)$) to $\sup_{\omega \in \Omega} T(L_\omega)$.

We may use an interval estimate of $\sup_{\omega \in \Omega} T(L_\omega)$, to estimate an appropriate sample size n: For example, if L_{ω_n} designates the optimum of (7) and $SE(T_n(L_\omega))$ the standard error:

$$SE\bigl(T_n(\omega)\bigr) = \frac{1}{\sqrt{n}}\sqrt{\frac{\sum_{i=1}^{n}(L_\omega(a, \theta_i) - T_n(L_\omega))^2}{n - 1}}$$

we suggest a more efficient strategy:

1. Compute

$$L_n = T_n(L_{\omega_n}) = \sup_{\omega \in \Omega} T_n(L_\omega).$$

2. Estimate

$$SE\bigl(T_n(\omega)\bigr) = \widehat{SE}_n.$$

3. If $(L_n - 2\widehat{SE}_n, L_n + 2\widehat{SE}_n)$ is enough small, stop.
   In other case we sample more from $\pi(\cdot|x)$.

Partially known class. Let $\{C_1, \ldots, C_n\}$ be a finite partition of the state of consequences $\mathcal{C}$. We call partially known class of loss functions the class

$$\mathcal{L}_K = \big\{L\colon v_{i-1} \leqslant L(c) \leqslant v_i,\ \forall c \in C_i,\ i = 1, \ldots, n\big\},$$

where some of the sets C_i could be empty. We shall assume now that consequences are of type $f(a - \theta) \in \mathbb{R}$, with f nondecreasing, C_i will be intervals of $\mathbb{R}$ and we add to $\mathcal{L}_K$ the condition that $L \in \mathcal{L}_K$ must be nonincreasing in $(-\infty, 0)$ and nondecreasing in $(0, \infty)$.

If $\mathcal{C}$ is a finite set, we have a linear programming problem. Let $c_1 < c_2 < \cdots < c_m < 0 < c_{m+1} < \cdots < c_N$. Then, we have

$$\begin{aligned}
&\sup_{L \in \mathcal{L}_K} T(L) = \sum_{i=1}^{N} L(c_i)\pi\big(S_a(c)|x\big)\\
&\quad \text{s.t.}\\
&L(c_j) \geqslant L(0), \quad j \in \{m, m+1\},\\
&L(c_i) \geqslant L(c_{i+1}), \quad i = 1, \ldots, m-1,\\
&L(c_i) \leqslant L(c_{i+1}), \quad i = m+1, \ldots, N-1,\\
&\quad \forall c_i \in C_j \quad v_{j-1} \leqslant L(c_i) \leqslant v_j, \quad j = 1, \ldots, n,
\end{aligned}$$

where $S_a(c) = \{\theta \in \Theta\colon (a, \theta) = c\}$.

For the continuous case, we may prove that

$$\sup_{L \in \mathcal{L}} T(L) = \sum_{i=1}^{n} v_i \pi\big(S_a\big(\mathcal{C} \setminus C_i^*\big)|x\big),$$

where $S_a(C) = \{\theta \in \Theta\colon (a, \theta) \in C\}$, $\forall C \subset \mathcal{C}$.

4.4. Computing nondominated alternatives

Next we turn to the issue of computing the *nondominated set*. In most cases, it is not possible to compute the nondominated set exactly, and thus approximation schemes are necessary, such as the following:

1. Sample $A_n = \{a_1, \ldots, a_n\} \subset \mathcal{A}$.
2. Choose $(L, \pi) \in \mathcal{L} \times \Gamma$ and relabel the a's as $(e_1, \ldots, e_n)$ so that $T(L, \pi, e_{i+1}) \leqslant T(L, \pi, e_i), \forall i$
3. Let $d(i) = 0, \forall i$
4.

```
For i = 1 to n − 1
     If d(i) = 0
          For j = i + 1 to n
               If d(j) = 0
```

$\texttt{If}\ e_j \precsim e_i\ \texttt{then}\ d(j) = 1$
$\texttt{else,}\ \texttt{if}\ (T(L,\pi,e_i) = T(L,\pi,e_j)\ \texttt{and}\ e_i \precsim e_j)$
$\texttt{then}\ d(i) = 1\ \texttt{and next}\ i$

5. Let $N_n = \{e_i \colon d(i) = 0\}$.

N_n is the approximation of the nondominated set. Under mild conditions, we may prove that N_n converges to the nondominated set as n goes to infinity. Specifically, we have:

PROPOSITION 11. *Assuming that all alternatives are in the support of the sampling process from $\mathcal{A}$ and there are no isolated points, for each nondominated alternative a, almost surely, there is a convergent subsequence a_{n_j}, with $a_{n_j} \in N_{n_j}$, such that $a_{n_j} \to b$ and $a \sim b$.*

Note that in step 4 above we need to be able to check whether an alternative a dominates an alternative b. This leads us to study the operator $T^{ab} = T(L,\pi,a) - T(L,\pi,b)$ since

$$\text{if} \quad \inf_{(L,\pi)\in\mathcal{L}\times\Gamma} T^{ab}(L,\pi) \geqslant 0 \quad \text{then} \quad a \preceq b. \tag{8}$$

We are actually interested in knowing whether the infimum (8) is positive. This will be useful, since sometimes we are only able to bound that infimum. Hence, ignoring the common denominator, we could use the operator N^{ab}, defined in an obvious way, instead of T^{ab}.

There is little more to say in general about such problems, remaining only the issue of studying specific cases, especially those involving important classes of losses and priors, as we illustrate.

Convex loss functions. Let us suppose a problem of decision or inference where the preferences of the DM are modelled by a class $\mathcal{L}$ of loss functions L convex in $\mathcal{A}$. That is, for all θ, a, and b in the set of alternatives $\mathcal{A}$, and λ in the interval [0, 1], it holds that

$$L\big(\lambda a + (1-\lambda)b, \theta\big) \leqslant \lambda L(a,\theta) + (1-\lambda)L(b,\theta).$$

In the next we will suppose that the set of alternatives is $\mathcal{A} = \mathcal{R}$. Little modifications allow the use of the results for intervals of $\mathbb{R}$.

LEMMA 1. *Let (L,π) be a pair in $\mathcal{L} \times \Gamma$. Then the function $T(\cdot, L, \pi)\colon \mathcal{A} \to \mathbb{R}$ is convex.*

LEMMA 2. *Let (L,π) be a pair in $\mathcal{L}\times\Gamma$, such that the set of Bayes actions $B_{(L,\pi)}$ is not empty. Then, $T(\cdot, L, \pi)$ is strictly decreasing in $(-\infty, a_{(L,\pi)})$ and strictly increasing in $(a^{(L,\pi)}, +\infty)$, where*

$$a_{(L,\pi)} = \inf_{a\in B_{(L,\pi)}} a,$$

$$a^{(L,\pi)} = \sup_{a\in B_{(L,\pi)}} a.$$

Let us remember that if a function is convex in a set, then it is continuous in its interior, see Roberts and Varberg (1973). Therefore, alternatives $a_{(L,\pi)}$ and $a^{(L,\pi)}$ are Bayes alternatives. Then, for all $(L,\pi) \in \mathcal{L} \times \Gamma$, it holds that the function $T(\cdot, L, \pi)$ is strictly decreasing in $(-\infty, a_{(L,\pi)})$, constant in $[a_{(L,\pi)}, a^{(L,\pi)}]$ and strictly increasing in $(a^{(L,\pi)}, +\infty)$.

It is easy to note that the set of nondominated alternatives is included in the interval $[\mu_*, \mu^*]$, where μ_* and μ^* are the supremum and infimum of the set of Bayes alternatives respectively, that is to say,

$$\mu_* = \inf_{(L,\pi)\in\mathcal{L}\times\Gamma} a_{(L,\pi)},$$

$$\mu^* = \sup_{(L,\pi)\in\mathcal{L}\times\Gamma} a^{(L,\pi)}.$$

Using the range of the above interval as a measure of the robustness is a common practical in the literature. Moreover, using strictly convex loss functions this range coincides with the range of the nondominated set, but this is not true in general. The next result allows to obtain the set of nondominated alternatives in a rigorous way.

THEOREM 1. *Let $\mathcal{L}$ be a class of convex loss functions in $\mathcal{A}$ and Γ a class of probability distributions such that for all $(L,\pi) \in \mathcal{L} \times \Gamma$, the set of Bayes alternatives $B_{(L,\pi)}$ is no empty. Let $a_* = \inf_{(L,\pi)\in\mathcal{L}\times\Gamma} a^{(L,\pi)}$ and let $a^* = \sup_{(L,\pi)\in\mathcal{L}\times\Gamma} a_{(L,\pi)}$. Then, if a_* is smaller than a^*, it holds that*

$$\left(a_*, a^*\right) \subseteq \mathcal{ND}(\mathcal{A}) \subseteq \left[a_*, a^*\right].$$

Otherwise $\mathcal{ND}(\mathcal{A}) = [a^*, a_*]$.

This result characterises the set of nondominated alternatives using convex loss functions when the set of the Bayes actions exists and is bounded for each pair $(L,\pi) \in \mathcal{L} \times \Gamma$. Let us now see the other cases. If there exists (L_0, π_0) such that $B_{(L_0,\pi_0)}$ is empty, since T is convex in $\mathbb{R}$, the function $T(\cdot, L_0, \pi_0)$ is strictly decreasing or increasing:

- The function is strictly increasing in $\mathbb{R}$:
  - If $a_* < a^*$, then $(-\infty, a^*) \subseteq \mathcal{ND}(\mathcal{A}) \subseteq (-\infty, a^*]$.
  - If $a_* \geqslant a^*$, then $\mathcal{ND}(\mathcal{A}) = (-\infty, a^*]$.
- The function is strictly decreasing in $\mathbb{R}$:
  - If $a_* < a^*$, then $(a_*, +\infty) \subseteq \mathcal{ND}(\mathcal{A}) \subseteq [a_*, +\infty)$.
  - If $a_* \geqslant a^*$, then $\mathcal{ND}(\mathcal{A}) = [a_*, +\infty)$.

Sometimes the set of the Bayes actions can be unbounded, i.e., there exists a pair $(L,\pi) \in \mathcal{L} \times \Gamma$, such that $B_{(L,\pi)} = (-\infty, a^{(L,\pi)}]$ or $B_{(L,\pi)} = [a_{(L,\pi)}, +\infty)$. But this fact does not affect the result of the Theorem 1.

The result could be particularised for strictly convex loss functions:

COROLLARY 2. *Let $\mathcal{L}$ be a class of strictly convex loss functions in $\mathcal{A}$ and Γ a class of probability distributions π, such that for all $(L, \pi) \in \mathcal{L} \times \Gamma$, the set of Bayes alternatives $B_{(L,\pi)}$ is nonempty. Then*

$$\left(\mu_*, \mu^*\right) \subseteq ND(\mathcal{A}) \subseteq \left[\mu_*, \mu^*\right].$$

Note that if μ_* and μ^* are Bayes alternatives, then $ND(\mathcal{A}) = [\mu_*, \mu^*]$. So, in problems with strictly convex loss functions the range of $ND(\mathcal{A})$ is a measure of sensitivity usually used.

These results may be improved for particular loss functions. Arias et al. (2004a) study quadratic, absolute value, LINEX and L_p loss functions.

4.5. MCMC and robustness

Sensitivity analysis in MCMC-based models is very complex and there is very few work on it. Whereas the methodology is similar to other frameworks, the computations are very difficult. One possibility is to use sensitivity measures that can be approximated with the sample generated from MCMC algorithms. Typically, these measures are from local sensitivity. In this section, we use a example from Pérez et al. (2002) to illustrate this idea. Other approaches, also for local sensitivity, may be seen in Gustafson (1996) and Pérez et al. (2004).

Optimal screening schedules for breast cancer. Parmigiani (1993) considers the problem of optimal screening schedules for a chronic disease, for example, breast cancer. Design parameters are the age α at which to begin regular screening and the frequency δ of screenings, i.e., $a = (\alpha, \delta)$. He defines a four-state semi Markov process to describe the history of a chronic disease. The four states are "disease is absent or present but not detectable" (A), "detectable pre-clinical" (B), "clinical" (C) and "death" (D).

For breast cancer, Parmigiani (1993) uses the following specifications. He denotes with π the transition probability from A to B ($1 - \pi$ is the transition probability from A to D), with t_2 the transition time from A to B, t_3 the transition time from B to C, and t_4 the transition time from A to D. Also, $t_1 = 1$ or $t_1 = 0$ shows whether a patient's state changes from A to B or D, respectively. We denote the patient's history with $\theta = (t_1, t_2, t_3, t_4)$. For the sake of simplicity we assume no transitions are possible between B and D. If not detected, every preclinical case develops into a clinical case in t_3 years. Let $\hat{f}_2(t_2)$ denote the empirical distribution for t_2 based on historical data, and let y be a patient's age when entering state B. Parmigiani (1993) suggests using $\pi = 1/8$ and

$$\begin{aligned} t_2 &\sim \hat{f}_2(t_2), \\ t_4 &\sim \text{Weibull}(a_4, b_4), \\ \log(t_3) &\sim N\left(m_{t_2}, s^2\right), \\ s^{-2} &\sim \text{Gamma}(a_3, b_3) \end{aligned}$$

with $a_4 = 7.233$, $b_4 = 82.651$, $a_3 = 6.33$, $b_3 = 3.36$, $m_0 = 1.4$, $m_1 = 1.6$, $m_2 = -0.038$, and $m_{t_2} = m_0 - \exp(m_1 + m_2 t_2)$. Figure 1 shows the probability model.

Deciding on an optimal screening schedule requires specifying in the utility function a trade-off between the screening cost and the probability of early detection of breast

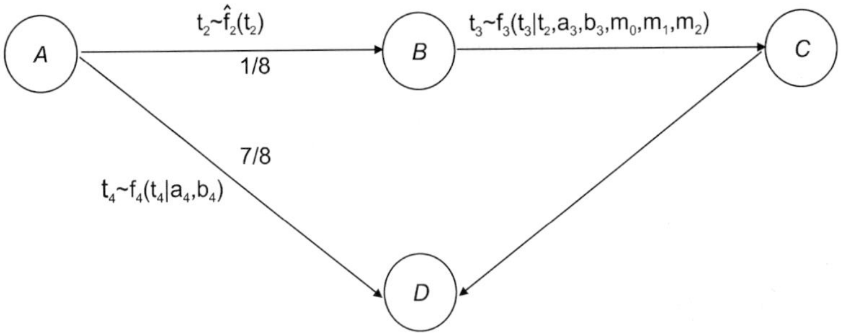

Fig. 1. Probability model for the transitions and transition times among the four states (A, B, C and D).

cancer. Let E denote the event of an early breast cancer detection and let $n_a(\theta)$ be the number of screenings under data θ and schedule a. The loss function is given by

$$L(a, \theta) = n_a(\theta) - r \cdot 1_E(a, \theta). \tag{9}$$

For a trade-off parameter $r = 0.001$, the maximum occurs around $\alpha^* = 42$, $\delta^* = 1.96$. We consider changes in expected loss when varying the prior probability model Π. Hence, in the following discussion the decision a^*, the loss L and the likelihood l are fixed.

Assume that the information about the probability distribution allows us to constrain it to a convex class Γ. Starting with a particular prior probability model Π we study the impact of changes in Π on the expected utility of a^*. We use neighbourhoods $\Gamma_\varepsilon \subseteq \Gamma$ of Π defined via ε-contaminations. These neighbourhoods are commonly used in sensitivity analysis, see Section 2.2.1.

For a given Π and a given neighbourhood Γ_ε of alternative prior models Π' around Π, Martín and Ríos Insua (1996) consider Fréchet derivatives, as in Section 2.2.3, and define as a sensitivity measure the supremum

$$m_\Pi^\varepsilon = \sup_{\Pi' \in \Gamma_\varepsilon} \left|\dot{T}_\Pi(\Pi' - \Pi)\right|. \tag{10}$$

The motivation is that this supremum corresponds to a bound on the possible changes in expected utility when varying the prior model over this class. When there is no data then $\dot{T}_\Pi(\cdot) = T(\cdot)$.

The remaining problem is to evaluate the supremum. For the quantile class, Γ_Q of probability measures and the ε-contamination neighbourhood $\Gamma_\varepsilon \subseteq \Gamma_Q$, Theorem 2 below gives an algorithm to compute the supremum via the solution of linear programming problems.

THEOREM 2. *Given $\Pi \in \Gamma_Q$ and $\Gamma_\varepsilon \subseteq \Gamma_Q$, then*

$$m_\Pi^\varepsilon = \sup_{\Pi' \in \Gamma_\varepsilon} \left|\dot{T}_\Pi(\Pi' - \Pi)\right| = \max\left\{H_1 - T(\Pi), T(\Pi) - H_2\right\}\varepsilon,$$

Table 1
Probabilities of the partition elements

p_i	$t_3 \in [0, 1.96]$	$t_3 \in (1.96, 80.18]$
$t_2 \in (0, 20]$	0.00770	0.00041
$t_2 \in (20, 30]$	0.02027	0.00122
$t_2 \in (30, 40]$	0.05310	0.01986
$t_2 \in (40, 60]$	0.20429	0.19538
$t_2 \in (60, 80]$	0.12850	0.27847
$t_2 \in (80, 105]$	0.02027	0.07053

where H_1, H_2 are, respectively, the optimal values of the linear programming problems:

$$\max \sum_{j=1}^{n} p_j \bar{h}_j$$

$$\text{s.t.} \sum_{j=1}^{n} p_j = 1 \quad \text{and} \quad \underline{p}_j \leqslant p_j \leqslant \bar{p}_j$$

$$\min \sum_{j=1}^{n} p_j \underline{h}_j$$

$$\text{s.t.} \sum_{j=1}^{n} p_j = 1 \quad \text{and} \quad \underline{p}_j \leqslant p_j \leqslant \bar{p}_j$$

with $\bar{h}_j = \sup_{\theta \in A_j} L(a, \theta)$ *and* $\underline{h}_j = \inf_{\theta \in A_j} L(a, \theta)$.

Pérez et al. (2002) consider uncertainty in the probability models for t_2 and t_3. Using the available information, they embed the prior distribution into a quantile class with two classes for t_3 and six classes for t_2. As the two variables are dependent, we obtain 12 classes, with quantiles as shown in Table 1.

They calculate $\bar{h}_j$ and $\underline{h}_j$ for every partition element. There are no bounds over the p_j values. They obtain $H_1 = \sum_{j=1}^{12} p_j \bar{h}_j = -50.249$ and $H_2 = \sum_{j=1}^{12} p_j \underline{h}_j = -97.978$. Moreover, $T(\Pi) = -52.009$, and applying Theorem 1

$$\begin{aligned} m_{\Pi}^{\varepsilon} &= \sup_{\Pi' \in \Gamma_\varepsilon} \left| \dot{T}_{\Pi}(\Pi' - \Pi) \right| \\ &= \max\{50.249 - 52.009, 52.009 - 97.978\}\varepsilon = 45.969\,\varepsilon. \end{aligned}$$

All the values are computed using the sample obtained through an MCMC method. We might conclude that there is no robustness. But note that the error is obtained when we consider the upper bound for the expected utility of the current optimal solution a^*. It is more important when the expected utility of a^* decreases, because this can lead to a change in the optimal decision. Therefore, we have to take a look at the lower bound. In this case, the measure is $(52.009 - 50.249)\varepsilon = 1.76\,\varepsilon$. In this sense, we can conclude that the solution is robust against changes in the prior distributions over t_2 and t_3. That is, we have decision robustness, but we do not have expected loss robustness; see Kadane and Srinivasan (1994) for a distinction.

5. Robust Bayesian analysis and other statistical approaches

We describe now the relation of robust Bayesian analyses with other statistical approaches, both within and outside the Bayesian arena.

5.1. *Robust priors*

As thoroughly discussed in Berger (1985, 1994), flat-tailed distributions tend to be more robust than standard conjugate choices (e.g., Cauchy vs. normal distributions). The construction of a class of priors should take into account tail behaviour in addition to symmetry, unimodality, etc. Tail modelling is straightforward for certain classes of priors, for instance the *density bounded* class, see Section 2.2.1 and Lavine (1991). With an appropriate choice of M and U, one can avoid priors with too-sharp tails. For instance, choosing M and U to be different multiples of a Cauchy density would preclude priors with normal-like tails. Additional constraints, such as unimodality, can be added to density bounded classes to further ensure that only reasonable prior densities are being considered.

5.2. *Robust models*

Many robust procedures in the literature are primarily concerned with obtaining resistance to outliers. For instance, the class (2), defined by Box and Tiao (1962), has been widely depicted as a robust class and its application for this purpose has been extensively discussed; see, e.g., Box and Tiao (1973). Another class of models, the extended power family, has been proposed by Albert et al. (1991). It can be viewed as a smooth alternative to the Box and Tiao power-series family, since its tails have similar functional forms but, unlike the power-series family, it is everywhere differentiable.

5.3. *Robust estimators*

As already discussed, different loss functions can lead to different Bayesian estimates. In particular, squared error and absolute error loss yield, respectively, the posterior mean and the posterior median as the optimal Bayesian estimator, see, e.g., French and Ríos Insua (2000). It is interesting to study which common estimators are more inherently robust. Sivaganesan (1991) investigated the behaviour of the posterior mean, median and mode when the prior is in the ε-contamination class, restricted to all bounded unimodal distributions with the same mode and median as the baseline π_0. He concludes that the posterior mean and posterior median are preferable to the posterior mode. Moreover, the posterior median seems to be worse, though only slightly, than the posterior mean.

Bayesian robustness has had interactions with other approaches, such as classical robustness, illustrated in Huber (1981), and upper and lower probabilities, presented in Walley (1991). We briefly review common points and differences in the next two subsections.

5.4. Frequentist robustness

Certain notions developed in classical robustness have been used in Bayesian robustness. We have already mentioned one, ε-contamination classes of distributions, that were introduced in Huber (1973) for classical robustness problems. In the classical context, they are primarily used to model outliers.

The local sensitivity approach, discussed in Gustafson (2000) and Sivaganesan (2000), borrows the notion of *influence function* from classical robustness theory (Hampel et al., 1986). Gustafson observes that the Gâteaux derivative can be written as

$$G_{\rho_0}(Q - \Pi_0) = \int I(z)\,\mathrm{d}[Q - \Pi_0](z),$$

where $I(z)$ is the influence function. Plots of the influence function can be very useful in visually assessing sensitivity. For most posterior quantities, the influence function can be easily determined.

Peña and Zamar (1996) present an asymptotic approach to Bayesian robustness that allows use of classical robustness tools, such as the influence function and the maximum bias function.

5.5. Imprecise probability

Imprecise probability is a generic term used to describe mathematical models that measure uncertainty without precise probabilities. This is certainly the case with robust Bayesian analysis, but there are many other imprecise probability theories, including upper and lower probabilities, belief functions, Choquet capacities, and upper and lower previsions; see Walley (1991) and the web page http://ippserv.rug.ac.be.

Some of these theories, such as fuzzy logic and belief functions, are only tangentially related to robust Bayesian analysis. Others are intimately related; for example, some classes of probability distributions that are considered in robust Bayesian analysis, such as distribution band classes, can also be interpreted in terms of upper and lower probabilities. Links between Choquet capacities and concentration functions were investigated by Fortini and Ruggeri (2000). Also, classes of probability distributions used in robust Bayesian analysis will typically generate upper and lower previsions as their upper and lower envelopes. Walley (1991) describes the connection between robust Bayesian analysis (in terms of sensitivity to the prior) and the theory of coherent lower previsions.

In a rough sense, the major difference between robust Bayesian analysis and these alternative theories is that robust Bayesian analysis stays with ordinary Bayesian intuition, considering, as ideal, classes (of priors, say) that consist only of those priors that are individually compatible with prior beliefs. In contrast, the alternative theories view the classes themselves (not the individual priors) as the basic elements of the theory.

5.6. Hierarchical approaches

Hierarchical priors are typically inherently robust in several senses, as discussed in Berger (1985). In part, this is because they are typically flat-tailed distributions. As an example, Moreno and Pericchi (1993) consider an ε-contamination neighbourhood of a $\mathcal{N}(\mu, \tau)$ prior and show that, when the prior information concerning τ is vague, the range of the desired posterior quantity is large. However, if one places a prior on τ, which is equivalent to considering a hierarchical model, then the posterior range is drastically reduced.

An approach to dealing with a lack of robustness with respect to, say, the prior is to place a hyperprior on the class of priors, which is a type of hierarchical analysis. Indeed, if there were no possibility of obtaining additional information to deal with the lack of robustness, we would recommend this technique, with the hyperprior being chosen in some default fashion. Note, however, that this effectively means one would be working with a single prior (that arising from marginalising over the hyperprior), so real robustness could not be claimed; in a sense, what has been accomplished is simply to use the inherent robustness of hierarchical modelling to develop a single prior that is hopefully reasonably robust. One could, of course, also embed the hyperprior in a class of priors and deal with robustness in that fashion; this is sensible if the hyperprior has clear intuitive meaning but otherwise can offer no real gain.

5.7. *Reference and objective Bayes approaches*

The primary applied form of Bayesian analysis, for over 200 years, has been the objective Bayes approach (often called the reference approach today), in which a single prior distribution is chosen in a default or noninformative fashion. A nice review of this approach is given in Kass and Wasserman (1996).

The objective Bayes approach is *not* related to the robust Bayesian approach through any type of formal attempt at being noninformative. (After all, if taken literally, *noninformative* would suggest that one consider the class of *all* possible prior distributions.) Rather, the objective Bayes approach is related in that it can be viewed as proposing a particular robust prior for default use of Bayesian methods.

There are two further relationships that should be mentioned. One has already been discussed: when robustness is not obtained, a good generic way to proceed is to determine a default hyperprior over the class of priors, and perform the (now single-prior) Bayesian analysis. Objective Bayesian techniques can be applied to determine a suitable default hyperprior. The second interesting relationship is the reverse: often there exists a number of possible default priors, and it is then natural to consider a Bayesian robustness analysis with respect to the class of default or noninformative priors.

5.8. *Asymptotics and robustness*

Asymptotic behaviour of robustness measures has received a lot of attention. The usual global sensitivity measure, i.e., the range, has a satisfactory asymptotic behaviour since even in very large classes, like the ε-contamination classes with arbitrary contaminations, it converges to 0 as the sample size grows to ∞. This behaviour is clearly what one would have intuitively anticipated, whereas this is not the case for the relative sensitivity defined by Ruggeri and Sivaganesan (2000). In fact, they showed that their sensitivity measure diverges at the rate of $\sqrt{n}$ for an ε-contamination class with arbitrary contaminations, whereas it converges to a positive finite constant for density bounded classes.

Convergence to 0 occurs only when further conditions are specified on the smoothness of the priors. It may be worth noting that this asymptotic behaviour closely resembles that of the local sensitivity measure, extensively studied in Gustafson and Wasserman (1995).

5.9. *Bayesian Gamma-minimax*

The Γ-minimax approach can be considered as an "automated" method which allows to choose actions or rules which guard against catastrophic consequences. A Γ-minimax approach could be suitable after a Bayesian sensitivity analysis has been unable to significantly narrow the range of variation, under changes in, say, the prior, of the quantity of interest. Faced with the need of having a unique estimate although the class of priors cannot be narrowed any further, the analyst can choose any possible action or, better, an optimal one, like the Γ-minimax which prevents from dramatic consequences.

The Γ-minimax approach has been questioned by some Bayesians, e.g., because it does not always guarantee the existence of a valid, data independent prior corresponding to the chosen action, or because there is no prior for which the conditional Γ-minimax action is a Bayes action. Under some mild regularity conditions, the Γ-minimax action is a Bayes action with respect to the least favourable prior. This fact has been criticised since the action corresponds to a prior which is deemed unreasonable, but this is nonsense: at the very same moment in which the prior belongs to the class elicited by the expert, then it is reasonable. The only possible way to get rid of it consists of further elicitation leading to a narrower class without it. A practical solution was provided by DasGupta and Studden (1989) who ruled out all the priors leading to a small marginal likelihood.

As mentioned in Vidakovic (2000), where a review of Γ-minimax in Bayesian robustness is illustrated, the Γ-minimax paradigm lies between Bayesian and minimax paradigms since it coincides with the former when a single prior gives all the available prior information, and with the latter when no prior information is available (and all priors are acceptable). Different Γ-minimax principles (e.g., conditional Γ-minimax expected loss or posterior Γ-minimax regret) have been considered.

Here we illustrate only the conditional Γ-minimax principle, under which a class Γ of priors is considered and the optimal action (*conditional Γ-minimax action*) a_0 is given by

$$a_0 = \arg \inf_{a \in \mathcal{A}} \sup_{\pi \in \Gamma} T(\pi, a).$$

Under convex losses, Betrò and Ruggeri (1992) were able to prove the conditional Γ-minimax action was Bayes. An earlier reference is the paper by DasGupta and Studden (1989).

The Γ-minimax posterior regret principle minimises the maximum loss when considering any action a instead of the Bayes one. Zen and DasGupta (1993) is the earliest paper in which the principle was thoroughly illustrated. The Γ-minimax posterior regret of an action a is

$$d(\pi, a) = T(\pi, a) - T(\pi, a_\pi),$$

where a_π is an action minimising $T(\pi, a)$.

The posterior regret Γ-minimax action a_M is given by

$$\sup_{\pi \in \Gamma} d(\pi, a_M) = \inf_{a \in \mathcal{A}} \sup_{\pi \in \Gamma} d(\pi, a).$$

Ríos Insua et al. (1995) proved that a_M is the middle point of the interval of Bayes actions obtained under a squared loss function and a parameter space represented by an interval.

While the previous papers were interested in the Γ-minimax principle under changes in the prior, Arias et al. (2004b) considered it under joint changes in prior and loss function. Further discussions, including links with the classical minimax, and an up-to-date literature can be found in Ruggeri (2005).

6. Conclusions

Robust Bayesian analysis is concerned with the sensitivity of the results of a Bayesian analysis to the inputs for that analysis. The early 90s was the golden age of robust Bayesian analysis, in that many statisticians were highly active in research in the area, and rapid progress was achieved. There was an explosion of publications focusing on studying sensitivity to the prior distribution, in part because non-Bayesians often view this sensitivity to be the major drawback of Bayesian analysis. Initially work focused on "global robustness", soon complemented by "local robustness" techniques. Interest naturally expanded into robustness with respect to the likelihood and loss function as well, with the aim of having a general approach to sensitivity towards all the ingredients of the Bayesian paradigm (model/prior/loss). Practical implementations of robust Bayesian ideas began to appear, including Godsill and Rayner (1996), Greenhouse and Wasserman (1996), Sargent and Carlin (1996) and Ríos Insua et al. (1999).

In the last half of the 90s, robust Bayesian analysis shifted from being a hot topic to being a mature field within Bayesian analysis, with continued gradual development, but with less of a sense of urgency. This change came about for several reasons. First, the initial flurry of fundamental theoretical advances naturally slowed. Second, few Bayesians continued to question the need to view robustness or sensitivity as a serious issue, so that the philosophical excitement with the idea waned. Indeed, the consensus was that the time was now ripe to develop user-friendly implementations of the existing robust Bayesian methodology. The timing was not great, however, in that, coincidentally, this period also marked the explosion of interest in MCMC computational methods for Bayesian statistics. As the MCMC methodology was not directly compatible with many of the robust Bayesian techniques that had been developed, it was unclear how formal robust Bayesian analysis could be incorporated into the future "Bayesian via MCMC" world.

Paradoxically, MCMC has dramatically increased the need for consideration of Bayesian robustness, in that the modelling that is now routinely used in Bayesian analysis is of such complexity that inputs (such as priors) can be elicited only in a very casual fashion. It is now time to focus again on Bayesian robustness and to attempt to bring its ideas into the Bayesian mainstream.

We have provided an overview of key results in robust Bayesian analysis. The usual approach in Bayesian inference emphasises prior robustness, arguing that the prior is the most difficult input to the analysis. Here we have been more general, considering loss and model robustness, and joint robustness. We finish this chapter with a brief discussion of potential topics for further research.

The most important challenge for the field of Bayesian robustness is to increase its impact on statistical practise; indeed, to make it a routine component of applied work. One avenue for doing so is simply to show, by example, the power of the methods. Many examples may be seen in Ríos Insua and Ruggeri (2000).

Perhaps the most important way to bring robust Bayesian methods to practitioners is to have these methods available in standard statistical software. There are several ongoing efforts to develop general Bayesian statistical software, and incorporation of robust Bayesian methodology into such software is clearly desirable. Indeed, Bayesian robustness can perhaps best be viewed as a component of general Bayesian software. Ideally, such software would proceed as follows:

- Elicit preferences and beliefs from the decision maker.
- Conduct a robustness study with classes compatible with the information elicited, and certain representative quantities of interest.
- If robustness appears to hold, conduct a standard Bayesian analysis with a representative prior/model/loss from the classes (or with respect to some hyperprior average); in part, performing the reported analysis with a single prior/model/loss would be for ease in communication.
- If robustness does not hold, elicit additional information to further constrain the classes (with the software perhaps suggesting which additional information would be best to elicit), or obtain additional data, and repeat.

The least understood aspect of this program is the possibility of interactive elicitation and is a topic that deserves much more research. Another aspect that requires clarification is the extent to which one can automate the process of determining whether or not robustness holds. Frequently, this will be a context-dependent decision, but generic calibration of robustness measures is often possible – some examples are given in McCulloch (1989) and Ríos Insua (1990) – and further efforts in this direction are clearly of interest.

We have referred to the enormous impact that MCMC methods have had on Bayesian analysis, and we mentioned that Bayesian robustness methodology will need to be compatible with MCMC methods to become widely used. There are some useful results in this direction. For instance, once one has obtained an MCMC sample from the posterior associated with a baseline prior, obtaining an MCMC sample from the posterior of a "close" prior can be done via importance sampling, see, for example, Smith and Gelfand (1992). But much additional work needs to be done to more closely relate robust Bayesian techniques with MCMC.

Another area of enduring importance to the field is foundations. There is still interest in axiomatic models that lead to robust Bayesian analysis. For instance, a direct implication of most axiomatic systems is that nondominated actions or estimators should be considered. Indeed, a topic of considerable importance is that of further developing methods for approximating the set of nondominated estimators or actions. On the other hand, axiomatics, such as in Nau (1995), lead to somewhat different models that would require new computational developments.

Acknowledgements

The work has been supported by grants from ESF, CICYT and the DMR Consulting Foundation at the Decision Engineering Lab.

References

Albert, J., Delampady, M., Polasek, W. (1991). A class of distributions for robustness studies. *J. Statist. Plann. Inference* **28**, 291–304.

Arias, J.P. (2002). The non-dominated set in robust Bayesian decision theory (in Spanish). Ph.D. Dissertation, Department of Mathematics, Universidad de Extremadura, Spain, 2002.

Arias, J.P., Martín, J. (2002). Uncertainty in beliefs and preferences: Conditions for optimal alternatives. *Ann. Math. Artificial Intelligence* **35**, 3–10.

Arias, J.P., Martín, J., Suárez, A. (2004a). Bayesian robustness with convex loss functions. Technical Report, Dept. of Mathematics, Universidad de Extremadura, Spain.

Arias, J.P., Martín, J., Ruggeri, F., Suárez, A. (2004b). Optimal actions in problems with convex loss functions. Technical Report 2-MI, CNR-IMATI Milano, Italy.

Basu, S. (1992). Variations of posterior expectations for symmetric unimodal priors in a distribution band. Technical Report 214, Department of Statistics and Applied Probability, University of California, Santa Barbara, USA.

Basu, S. (1995). Ranges of posterior probability over a distribution band. *J. Statist. Plann. Inference* **44**, 149–166.

Basu, S. (1996). Local sensitivity, functional derivatives and nonlinear posterior quantities. *Statist. Decisions* **14**, 405–418.

Basu, S., DasGupta, A. (1990). Bayesian analysis under distribution bands. Technical Report 90-48, Department of Statistics, Purdue University, USA.

Basu, S., DasGupta, A. (1995). Robust Bayesian analysis with distribution bands. *Statist. Decisions* **13**, 333–349.

Basu, S., Jammalamadaka, S.R., Liu, W. (1996). Local posterior robustness with parametric priors: Maximum and average sensitivity. In: Heidlbreder, G. (Ed.), *Maximum Entropy and Bayesian Methods*. Kluwer Academic, Dordrecht, The Netherlands.

Bell, D. (1995). Risk, return and utility. *Management Sci.* **41**, 23–30.

Berger, J.O. (1984). The robust Bayesian viewpoint (with discussion). In: Kadane, J. (Ed.), *Robustness of Bayesian Analysis*. North-Holland, Amsterdam, The Netherlands.

Berger, J.O. (1985). *Statistical Decision Theory and Bayesian Analysis*, second ed. Springer-Verlag, New York, USA.

Berger, J.O. (1990). Robust Bayesian analysis: Sensitivity to the prior. *J. Statist. Plann. Inference* **25**, 303–328.

Berger, J.O. (1994). An overview of robust Bayesian analysis. *Test* **3**, 5–58.

Berger, J.O., Berliner, L.M. (1986). Robust Bayes and empirical Bayes analysis with ε-contaminated priors. *Ann. Statist.* **14**, 461–486.

Berger, J.O., O'Hagan, A. (1988). Ranges of posterior probabilities for unimodal priors with specified quantiles. In: Bernardo, J.M., DeGroot, M.H., Lindley, D.V., Smith, A.F.M. (Eds.), *Bayesian Statistics, vol. 3*. Oxford University Press, Oxford, UK.

Berger, J.O., Ríos Insua, D., Ruggeri, F. (2000). Bayesian robustness. In: Ríos Insua, D., Ruggeri, F. (Eds.), *Robust Bayesian Analysis*. Springer-Verlag, New York, USA.

Betrò, B., Guglielmi, A. (1996). Numerical robust Bayesian analysis under generalized moments conditions. In: Berger, J.O., Betró, B., Moreno, E., Pericchi, L.R., Ruggeri, F., Salinetti, G., Wasserman, L. (Eds.), *Bayesian Robustness*. Institute of Mathematical Statistics, Hayward, CA, USA.

Betrò, B., Guglielmi, A. (2000). Methods for global prior robustness under generalized moments conditions. In: Ríos Insua, D., Ruggeri, F. (Eds.), *Robust Bayesian Analysis*. Springer-Verlag, New York, USA.

Betrò, B., Ruggeri, F. (1992). Conditional Gamma-minimax actions under convex losses. *Comm. Statist. Theory Methods* **21**, 1051–1066.

Betrò, B., Męczarski, M., Ruggeri, F. (1994). Robust Bayesian analysis under generalized moments conditions. *J. Statist. Plann. Inference* **41**, 257–266.

Boratynska, A. (1996). On Bayesian robustness with the ε-contamination class of priors. *Statist. Probab. Lett.* **26**, 323–328.

Bose, S. (1994). Bayesian robustness with mixture classes of priors. *Ann. Statist.* **22**, 652–667.

Box, G.E.P., Tiao, G.C. (1962). A further look at robustness via Bayes theorem. *Biometrika* **49**, 419–432 and 546.

Box, G.E.P., Tiao, G.C. (1973). *Bayesian Inference in Statistical Analysis*. Addison-Wesley, New York, USA.

Cagno, E., Caron, F., Mancini, M., Ruggeri, F. (2000a). Using AHP in determining prior distributions on gas pipeline failures in a robust Bayesian approach. *Reliab. Engin. System Safety* **67**, 275–284.

Cagno, E., Caron, F., Mancini, M., Ruggeri, F. (2000b). Sensitivity of replacement priorities for gas pipeline maintenance. In: Ríos Insua, D., Ruggeri, F. (Eds.), *Robust Bayesian Analysis*. Springer-Verlag, New York, USA.

Carota, C., Ruggeri, F. (1994). Robust Bayesian analysis given priors on partition sets. *Test* **3**, 73–86.

Casella, G., Hwang, G., Robert, C. (1994). Loss functions for set estimation. In: Gupta, S., Berger, J.O. (Eds.), *Statistical Decision Theory and Related Topics V*. Springer-Verlag, New York, USA.

Chuang, D.T. (1984). Further theory of stable decisions. In: Kadane, J. (Ed.), *Robustness of Bayesian Analyses*. North-Holland, Amsterdam, The Netherlands.

Cifarelli, D.M., Regazzini, E. (1987). On a general definition of concentration function. *Sankhyā, Ser. B* **49**, 307–319.

Clarke, B., Gustafson, P. (1998). On the overall sensitivity of the posterior distributions to its inputs. *J. Statist. Plann. Inference* **71**, 137–150.

Clyde, M., Chaloner, K. (1996). The equivalence of constrained and weighted designs in multiple objective design problems. *J. Amer. Statist. Assoc.* **91**, 1236–1244.

Cuevas, A., Sanz, P. (1988). On differentiability properties of Bayes operators. In: Bernardo, J.M., DeGroot, M.H., Lindley, D.V., Smith, A.F.M. (Eds.), *Bayesian Statistics, vol. 3*. Oxford University Press, Oxford, UK.

DasGupta, A., Studden, W.J. (1989). Frequentist behavior of robust Bayes estimates of normal means. *Statist. Decisions* **7**, 333–361.

De la Horra, J., Fernández, C. (1995). Sensitivity to prior independence via Farlie–Gumbel–Morgenstern model. *Comm. Statist. Theory Methods* **24**, 987–996.

Dempster, A.P. (1975). A subjectivist look at robustness. *Bull. Internat. Statist. Inst.* **46**, 349–374.

DeRobertis, L. (1978). The use of partial prior knowledge in Bayesian inference. Ph.D. Dissertation, Yale University, USA.

DeRobertis, L., Hartigan, J.A. (1981). Bayesian inference using intervals of measure. *Ann. Statist.* **1**, 235–244.

Dey, D.K., Birmiwal, L.R. (1994). Robust Bayesian analysis using entropy and divergences measures. *Statist. Probab. Lett.* **20**, 287–294.

Dey, D.K., Lou, K., Bose, S. (1998). A Bayesian approach to loss robustness. *Statist. Decisions* **16**, 65–87.

Diaconis, P., Freedman, D. (1986). On the consistency of Bayes estimates. *Ann. Statist.* **14**, 1–67.

Fortini, S., Ruggeri, F. (1994a). Concentration functions and Bayesian robustness. *J. Statist. Plann. Inference* **40**, 205–220.

Fortini, S., Ruggeri, F. (1994b). On defining neighbourhoods of measures through the concentration function. *Sankhyā, Ser. A* **56**, 444–457.

Fortini, S., Ruggeri, F. (1995). Concentration function and sensitivity to the prior. *J. Ital. Statist. Soc.* **4**, 283–297.

Fortini, S., Ruggeri, F. (1997). Differential properties of the concentration function. *Sankhyā, Ser. A* **59**, 345–364.

Fortini, S., Ruggeri, F. (2000). On the use of the concentration function in Bayesian robustness. In: Ríos Insua, D., Ruggeri, F. (Eds.), *Robust Bayesian Analysis*. Springer-Verlag, New York, USA.

French, S., Ríos Insua, D. (2000). *Statistical Decision Theory*. Arnold.

Gelfand, A., Dey, D.K. (1991). On Bayesian robustness of contaminated classes of priors. *Statist. Decisions* **9**, 63–80.

Ghosh, S.K., Dey, D.K. (1994). Sensitivity diagnostics and robustness issues in Bayesian inference. Technical Report 94-30, Department of Mathematics, University of Connecticut, USA.

Girón, J., Ríos, S. (1980). Quasi-Bayesian behaviour: A more realistic approach to decision making?. In: Bernardo, J.M., De Groot, M.H., Lindley, D.V., Smith, A.F.M. (Eds.), *Bayesian Statistics*. University Press, Valencia, Spain.

Godsill, S.J., Rayner, P.J.W. (1996). Robust treatment of impulsive noise in speech and audio signals. In: Berger, J.O., Betró, B., Moreno, E., Pericchi, L.R., Ruggeri, F., Salinetti, G., Wasserman, L. (Eds.), *Bayesian Robustness*. Institute of Mathematical Statistics, Hayward, CA, USA.

Goldstein, M. (1980). The linear Bayes regression estimator under weak assumptions. *Biometrika* **67**, 621–628.

Goutis, C. (1994). Ranges of posterior measures for some classes of priors with specified moments. *Internat. Statist. Rev.* **62**, 245–257.

Greenhouse, J., Wasserman, L. (1996). A practical robust method for Bayesian model selection: A case study in the analysis of clinical trials (with discussion). In: Berger, J.O., Betró, B., Moreno, E., Pericchi, L.R., Ruggeri, F., Salinetti, G., Wasserman, L. (Eds.), *Bayesian Robustness*. Institute of Mathematical Statistics, Hayward, CA, USA.

Gustafson, P. (1994). Local sensitivity of posterior expectations. Ph.D. Dissertation, Department of Statistics, Carnegie Mellon University, USA.

Gustafson, P. (1996). Local sensitivity of posterior expectations. *Ann. Statist.* **24**, 174–195.

Gustafson, P. (2000). Local robustness in Bayesian analysis. In: Ríos Insua, D., Ruggeri, F. (Eds.), *Robust Bayesian Analysis*. Springer-Verlag, New York, USA.

Gustafson, P., Wasserman, L. (1995). Local sensitivity diagnostics for Bayesian inference. *Ann. Statist.* **23**, 2153–2167.

Hampel, F.R., Ronchetti, E.M., Rousseuw, P.J., Stahel, W.A. (1986). *Robust Statistics: The Approach Based on Influence Functions*. Wiley, New York, USA.

Hartigan, J.A. (1969). Linear Bayesian methods. *J. Roy. Statist. Soc., Ser. B Stat. Methodol.* **31**, 446–454.

Huber, P.J. (1973). The use of Choquet capacities in statistics. *Bull. Internat. Statist. Inst.* **45**, 181–191.

Huber, P.J. (1981). *Robust Statistics*. Wiley, New York, USA.

Hwang, J.T. (1985). Universal domination and Stochastic domination: Estimation simultaneously under a broad class of loss functions. *Ann. Statist.* **13**, 295–314.

Kadane, J., Chuang, D.T. (1978). Stable decision problems. *Ann. Statist.* **6**, 1095–1110.

Kadane, J., Srinivasan, C. (1994). Discussion of Berger, J.O., An overview of robust Bayesian analysis (with discussion). *Test* **3**, 93–95.

Kadane, J., Srinivasan, C. (1996). Bayesian robustness and stability (with discussion). In: Berger, J.O., Betró, B., Moreno, E., Pericchi, L.R., Ruggeri, F., Salinetti, G., Wasserman, L. (Eds.), *Bayesian Robustness*. Institute of Mathematical Statistics, Hayward, CA, USA.

Kadane, J., Srinivasan, C. (1998). Bayes decision problems and stability. *Sankhyā, Ser. A* **60**, 383–404.

Kadane, J., Salinetti, G., Srinivasan, C. (2000). Stability of Bayes decision and applications. In: Ríos Insua, D., Ruggeri, F. (Eds.), *Robust Bayesian Analysis*. Springer-Verlag, New York, USA.

Kass, R., Wasserman, L. (1996). The selection of prior distributions by formal rules. *J. Amer. Statist. Assoc.* **91**, 1343–1370.

Lavine, M. (1991). Sensitivity in Bayesian statistics: The prior and the likelihood. *J. Amer. Statist. Assoc.* **86**, 396–399.

Lavine, M., Wasserman, L., Wolpert, R.L. (1991). Bayesian inference with specified marginals. *J. Amer. Statist. Assoc.* **86**, 400–403.

Lavine, M., Perone Pacifico, M., Salinetti, G., Tardella, L. (2000). Linearization techniques in Bayesian robustness. In: Ríos Insua, D., Ruggeri, F. (Eds.), *Robust Bayesian Analysis*. Springer-Verlag, New York, USA.

Madansky, A. (1990). Bayesian analysis with incompletely specified prior distributions. In: Geisser, S., et al. (Eds.), *Bayesian and Likelihood Methods in Statistics and Econometrics*. North-Holland, Amsterdam, The Netherlands.

Makov, U.E. (1994). Some aspects of Bayesian loss robustness. *J. Statist. Plann. Inference* **38**, 359–370.

Martín, J., Ríos Insua, D. (1996). Local sensitivity analysis in Bayesian decision theory (with discussion). In: Berger, J.O., Betró, B., Moreno, E., Pericchi, L.R., Ruggeri, F., Salinetti, G., Wasserman, L. (Eds.), *Bayesian Robustness*. Institute of Mathematical Statistics, Hayward, CA, USA.

Martín, J., Ríos Insua, D., Ruggeri, F. (1998). Issues in Bayesian loss robustness. *Sankhyā Ser. A* **60**, 405–417.

Martín, J., Ríos Insua, D., Ruggeri, F. (2003). Joint sensitivity in Bayesian decision theory. *Test* **12**, 1–22.

McCulloch, R. (1989). Local model influence. *J. Amer. Statist. Assoc.* **84**, 473–478.

Moreno, E., Cano, J.A. (1991). Robust Bayesian analysis for ε-contaminations partially known. *J. Roy. Statist. Soc., Ser. B Stat. Methodol.* **53**, 143–155.

Moreno, E., Cano, J.A. (1992). Classes of bidimensional priors specified on a collection of sets: Bayesian robustness. *J. Statist. Plann. Inference* **46**, 325–334.

Moreno, E., Gonzalez, A. (1990). Empirical Bayes analysis for ε-contaminated priors with shape and quantile constraints. *Rebrape* **4**, 177–200.

Moreno, E., Pericchi, L.R. (1991). Bayesian robustness for ε-contaminations with shape and quantile constrains. In: *Proc. Fifth Inter. Symp. on Applied Stochastic Models and Data Analysis*. World Scientific, Singapore.

Moreno, E., Pericchi, L.R. (1993). Bayesian robustness for hierarchical ε-contamination models. *J. Statist. Plann. Inference* **37**, 159–168.

Moreno, E., Martínez, C., Cano, J.A. (1996). Local robustness and influence for contamination classes of prior distributions (with discussion). In: Berger, J.O., Betró, B., Moreno, E., Pericchi, L.R., Ruggeri, F., Salinetti, G., Wasserman, L. (Eds.), *Bayesian Robustness*. Institute of Mathematical Statistics, Hayward, CA, USA.

Nau, R. (1995). The shape of incomplete preferences. Technical Report, Duke University, USA.

O'Hagan, A., Berger, J.O. (1988). Ranges of posterior probabilities for quasi-unimodal priors with specified quantiles. *J. Amer. Statist. Assoc.* **83**, 503–508.

Parmigiani, G. (1993). On optimal screening ages. *J. Amer. Statist. Assoc.* **88**, 622–628.

Peña, D., Zamar, R.H. (1996). On Bayesian robustness: An asymptotic approach. In: Rieder, H. (Ed.), *Robust Statistics, Data Analysis and Computer Intensive Methods*. Springer-Verlag, New York, USA.

Pérez, C., Martín, J., Müller, P. (2002). Sensitivity analysis in medical decision problems. In: Prado, Bolado (Eds.), *Proceedings SAMO 2001*, second ed. Editorial CIEMAT, Madrid, pp. 303–310.

Pérez, C., Martín, J., Rufo, M.J. (2004). MCMC-based local parametric sensitivity. Technical Report, Department of Mathematics, Universidad de Extremadura, Spain.

Philipps, L.D. (1984). A theory of requisite decision models. *Acta Psychologica* **56**, 29–48.

Regazzini, E. (1992). Concentration comparisons between probability measures. *Sankhyā, Ser. B* **54**, 129–149.

Ríos Insua, D. (1990). *Sensitivity Analysis in Multiobjective Decision Making*. Springer-Verlag, New York, USA.

Ríos Insua, D. (1991). On the foundations of decision making under partial information. *Theory Decisions* **33**, 83–100.

Ríos Insua, D., French, S. (1991). A framework for sensitivity analysis in discrete multi-objective decision making. *Eur. J. Oper. Res.* **54**, 176–190.

Ríos Insua, D., Ruggeri, F. (2000). *Robust Bayesian Analysis*. Springer-Verlag, New York, USA.

Ríos Insua, D., Ruggeri, F., Vidakovic, B. (1995). Some results on posterior regret Γ-minimax estimation. *Statist. Decisions* **13**, 313–331.

Ríos Insua, S., Martín, J., Ríos Insua, D., Ruggeri, F. (1999). Bayesian forecasting for accident proneness evaluation. *Scand. Actuar. J.* **99**, 134–156.

Roberts, A.W., Varberg, D.E. (1973). *Convex Functions*. Academic Press, New York, USA.

Ruggeri, F. (1990). Posterior ranges of functions of parameters under priors with specified quantiles. *Comm. Statist. Theory Methods* **19**, 127–144.

Ruggeri, F. (1991). Infinitesimal sensitivity of likelihood function. Unpublished manuscript.

Ruggeri, F. (2005). Gamma-minimax inference. Submitted for publication in: Balakrishnan, N., Read, C.B., Vidakovic, B. (Eds.), *Encyclopedia of Statistical Sciences*, second ed. Wiley, New York, USA.

Ruggeri, F., Sivaganesan, S. (2000). On a global sensitivity measure for Bayesian inference. *Sankhyā, Ser. A* **62**, 110–127.

Ruggeri, F., Wasserman, L. (1991). Density based classes of priors: Infinitesimal properties and approximations. Technical Report 528, Department of Statistics, Carnegie Mellon University, USA.

Ruggeri, F., Wasserman, L. (1993). Infinitesimal sensitivity of posterior distributions. *Canad. J. Statist.* **21**, 195–203.

Ruggeri, F., Wasserman, L. (1995). Density based classes of priors: infinitesimal properties and approximations. *J. Statist. Plann. Inference* **46**, 311–324.

Ruggeri, F., Martín, J., Ríos Insua, D. (2003). Optimal alternative robustness in Bayesian decision theory. *Rev. R. Acad. Cien. Serie A. Mat.* **97**, 407–412.

Salinetti, G. (1994). Stability of Bayesian decisions. *J. Statist. Plann. Inference* **40**, 313–320.

Saltelli, A., Chan, K., Scott, M. (2000). *Sensitivity Analysis*. Wiley, New York, USA.

Sargent, D.J., Carlin, B.P. (1996). Robust Bayesian design and analysis of clinical trials via prior partitioning (with discussion). In: Berger, J.O., Betró, B., Moreno, E., Pericchi, L.R., Ruggeri, F., Salinetti, G., Wasserman, L. (Eds.), *Bayesian Robustness*. Institute of Mathematical Statistics, Hayward, CA, USA.

Shao, J. (1989). Monte Carlo approximations in Bayesian decision theory. *J. Amer. Statist. Assoc.* **84**, 727–732.

Shyamalkumar, N.D. (2000). Likelihood robustness. In: Ríos Insua, D., Ruggeri, F. (Eds.), *Robust Bayesian Analysis*. Springer-Verlag, New York, USA.

Sivaganesan, S. (1991). Sensitivity of some standard Bayesian estimates to prior uncertainty – a comparison. *J. Statist. Plann. Inference* **27**, 85–103.

Sivaganesan, S. (1993). Robust Bayesian diagnostics. *J. Statist. Plann. Inference* **35**, 171–188.

Sivaganesan, S. (2000). Global and local robustness approaches: Uses and limitations. In: Ríos Insua, D., Ruggeri, F. (Eds.), *Robust Bayesian Analysis*. Springer-Verlag, New York, USA.

Sivaganesan, S., Berger, J.O. (1989). Ranges of posterior measures for priors with unimodal contaminations. *Ann. Statist.* **17**, 868–889.

Smith, A.F.M., Gelfand, A.E. (1992). Bayesian statistics without tears: A sampling–resampling perspective. *Amer. Statist.* **46**, 84–88.

Varian, H.R. (1974). A Bayesian approach to real state assessment. In: Fienberg, S., Zellner, A. (Eds.), *Studies in Bayesian Econometrics and Statistics in Honor of L.J. Savage*. North-Holland, Amsterdam, The Netherlands.

Vidakovic, B. (2000). Γ-minimax: A paradigm for conservative robust Bayesians. In: Ríos Insua, D., Ruggeri, F. (Eds.), *Robust Bayesian Analysis*. Springer-Verlag, New York, USA.

Walley, P. (1991). *Statistical Reasoning With Imprecise Probabilities*. Chapman and Hall, London, UK.

White, D.J. (1982). *Optimality and Efficiency*. Wiley, New York, USA.

Yu, P.L. (1986). *Multiple Criteria Decision Making*. Plenum Press, New York, USA.

Zen, M.M., DasGupta, A. (1993). Estimating a binomial parameter: Is robust Bayes real Bayes?. *Statist. Decisions* **11**, 37–60.

Handbook of Statistics, Vol. 25
ISSN: 0169-7161
© 2005 Elsevier B.V. All rights reserved.
DOI 10.1016/S0169-7161(05)25022-8

22

Elliptical Measurement Error Models – A Bayesian Approach

Heleno Bolfarine and R.B. Arellano-Valle

Abstract
The main object of this paper is to discuss Bayesian inference in elliptical measurement error models. We consider dependent and independent elliptical models. Some general results are obtained for the class of dependent elliptical models. Weak nondifferential and differential models are investigated. One special submodel is the Student-t family. Given the complexity of some of the independent models considered, we make use of Monte Carlo Markov chain techniques to make inference on the parameters of interest. An application to a real data set is considered to illustrate some of the main results of the chapter.

1. Introduction

Measurement error models (MEM), also called errors-in-variables models, are very important to develop inferences in most research areas, because they involve explanatory variables that are measured with errors or that can not be observed directly (latent variables). Many examples and applications of these models were considered in the books by Fuller (1987), Carroll et al. (1995) and Cheng and Van Ness (1999). The most popular MEM in the statistical literature is the simple linear model with additive measurement errors, which is defined by the equations

$$y_i = \beta_1 + \beta_2 \xi_i + e_i, \tag{1.1}$$

and

$$x_i = \xi_i + u_i, \tag{1.2}$$

$i = 1, \ldots, n$, where (y_i, x_i), $i = 1, \ldots, n$, are observed random quantities, (ξ_i, e_i, u_i), $i = 1, \ldots, n$, are unobserved random quantities and (β_1, β_2) are unknown parameters.

In the statistical literature we can find typically two interpretations of the MEM specified by (1.1)–(1.2). The first interpretation considers that (1.1) is a regression equation that relates a response variable y with a co-variate ξ, which is measured with an additive error, that is, assuming that we can obtain solely a measurement x of the true co-variate ξ with an (unobserved) additive error u as indicated in Eq. (1.2). The second interpretation considers that y and x are measurements of the same unknown quantity ξ, which are obtained by means of two different instruments, say, 1 and 2, respectively. In this case, the objective is the calibration of instrument 1 (a new instrument) by means of instrument 2 (a reference instrument). In both cases, it is assumed that the quantities (e, u) are random errors with a location vector $(0, 0)$ and a diagonal dispersion matrix $\text{diag}\{\sigma_e^2, \sigma_u^2\}$ of an (otherwise) unspecified distribution function. Moreover, since for a sample of size n, $\xi_1, \ldots, \xi_n$ are unobserved (unknown) quantities, we consider as model parameters in the MEM (1.1)–(1.2) the vectors $\boldsymbol{\xi} = (\xi_1, \ldots, \xi_n)^{\mathrm{T}}$ (incidental parameters) and $\boldsymbol{\theta} = (\beta_1, \beta_2, \sigma_e^2, \sigma_u^2)^{\mathrm{T}}$ (structural parameters).

The classical literature on the MEM inference problem is vast such as the books mentioned above. The classical analysis of MEM considers two approaches. One is the functional approach, where $\xi_1, \ldots, \xi_n$ are taken as unknown fixed quantities or incidental parameters, and the other is the structural approach, where these quantities are taken as random variables with mean μ_ξ and variance σ_ξ^2. In both cases, the estimation of the structural parameters is complex and requires additional assumptions to ensure consistency or identifiability (see, for example, Cheng and Van Ness, 1999). Hence, by considering these additional conditions (for example, that the variances ratio $\lambda = \sigma_e^2/\sigma_u^2$ is known exactly), corrections of naive approaches (replaces ξ by x directly) involving least squares, maximum likelihood, method of moments and estimating equations are taken by the classical approach and asymptotic properties are used for evaluating estimators' precision.

In this chapter, we consider Bayesian solutions for MEM. The pioneering Bayesian work in MEM was by Lindley and El Sayad (1968), where the main object was the estimation of the regression parameter β_2 in the MEM given by (1.1)–(1.2). Additional Bayesian results on MEM can be found in Zellner (1971), Villegas (1972), Florens et al. (1974), Reilly and Patino-Leal (1981), Bolfarine and Cordani (1993) and Bolfarine et al. (1999). In general the Bayesian approach to studying MEM has received less attention than the classical literature. This situation has been changed by the introduction of Markov chain Monte Carlo (MCMC) simulation methods. New Bayesian literature on MEM with more interesting applications, appeared in the last decade. See, for example, Dellaportas and Stephens (1995), Richardson and Gilks (1993). However, most of them deal with the simple linear additive MEM given by (1.1)–(1.2) and are restricted to the normality assumptions for the error terms (normal likelihood function). We consider the most general case of elliptical MEM, that is, elliptical distributions for the error terms in the MEM. The class of elliptical MEM has been studied by Arellano-Valle and Bolfarine (1996), Arellano-Valle et al. (1994), Bolfarine and Arellano-Valle (1994, 1998) and Vilca-Labra et al. (1998) from classical point of view. From a Bayesian perspective we are not aware of published results related to nonnormal elliptical models. Elliptical models are extensions of the normal model preserving most of its properties, but are more flexible than the normal model with respect to modeling the kurtosis. A comprehensive review of the properties and characterizations of elliptical distributions can be found in the book by Fang et al. (1990). See also Arellano-Valle (1994). Arellano et al. (2000, 2002) discussed aspects such invariance, influence measures and invariance of the Bayesian inference for ordinary linear models. Bayesian analysis for spherical linear models was developed by Arellano-Valle et al. (2005).

Dependent and independent elliptical MEM are considered. Following the definitions given by Bolfarine and Arellano-Valle (1998), we also consider elliptical MEM with nondifferential errors (NDE) and with weak nondifferential errors (WNDE). For the dependent elliptical MEM, some invariance results are established in both NDE and WNDE situations. Inference results are also established for independent elliptical MEM, with some emphasis on the class of representable elliptical models and, in particular, on the family of Student-t models.

The outline is as follows. In Section 2, dependent and independent elliptical MEM models are defined. In both situations, NDE and WNDE types of models are considered. Section 3 discusses consequences of considering a noninformative prior distribution for the vector $\boldsymbol{\xi} = (\xi_1, \ldots, \xi_n)^{\mathrm{T}}$. We discuss Bayesian inference for NDE and WNDE dependent elliptical MEM in Section 4. We show that under this type of elliptical MEM, the inference on the regression parameters $\boldsymbol{\beta} = (\beta_1, \beta_2)^{\mathrm{T}}$ is invariant in this class of models when a noninformative Jeffrey prior for the scale parameters $(\sigma_e^2, \sigma_u^2)^{\mathrm{T}}$ is taken. We also study the situation where $\sigma_e^2 = \lambda\sigma_u^2$, with λ known. Bayesian solutions for NDE and WNDE independent elliptical models are considered in Section 5. We discuss Bayesian solutions based on proper priors for the class of representable elliptical distributions, that is, scale mixture of the normal distribution. The Gibbs sampler algorithm is also presented to compute the necessary posterior distributions for the independent Student-t model. Finally, in Section 6 we present an application to a real data.

2. Elliptical measurement error models

In this section, we introduce different types of elliptical MEM, starting with some preliminary results on elliptical distributions.

2.1. *Elliptical distributions*

As in Fang et al. (1990), we use the notation $\boldsymbol{w} \sim El_p(\boldsymbol{\mu}, \Sigma; h)$ to indicate that the $p \times 1$ random vector $\boldsymbol{w}$ follows a p-dimensional elliptical distribution with $p \times 1$ location vector $\boldsymbol{\mu}$ and a $p \times p$, positive definite dispersion matrix Σ if its probability density function is

$$f(\boldsymbol{w}|\boldsymbol{\mu}, \Sigma; h) = |\Sigma|^{-1/2} h\big((\boldsymbol{w} - \boldsymbol{\mu})^{\mathrm{T}} \Sigma^{-1} (\boldsymbol{w} - \boldsymbol{\mu})\big), \tag{2.1}$$

where h is the generator density function (gdf), which is a nonnegative real function such that

$$\int_0^\infty u^{p/2-1} h(u)\, \mathrm{d}u = \frac{\Gamma(\frac{p}{2})}{\pi^{p/2}}. \tag{2.2}$$

Note that $h(\|\boldsymbol{t}\|^2)$, $\boldsymbol{t} \in R^p$, is a p-dimensional spherical density corresponding to the spherical random vector (distribution) $\boldsymbol{t} = \Sigma^{-1/2}(\boldsymbol{w} - \boldsymbol{\mu}) \sim El_p(\boldsymbol{0}, \boldsymbol{I}_p; h)$. Here $\|\boldsymbol{t}\|^2 =$

$\boldsymbol{t}^{\mathrm{T}}\boldsymbol{t}$, $\boldsymbol{t} \in R^p$, and $\boldsymbol{I}_p$ is the $p \times p$ identity matrix. From (2.2) it follows that the radial squared random variable $R^2 = \|\boldsymbol{t}\|^2$ has density function given by

$$f_{R^2}(u) = \frac{\pi^{p/2}}{\Gamma(\frac{p}{2})} u^{p/2-1} h(u).$$

This distribution will be called here radial-squared distribution with p degrees of freedom (the dimension of $\boldsymbol{t}$) and generator density function h and will be denoted by $\mathcal{R}_p^2(h)$. Note that $\boldsymbol{t} = R\boldsymbol{z}$, where $R = \|\boldsymbol{t}\|$ and $\boldsymbol{z} = \boldsymbol{t}/\|\boldsymbol{t}\|$ which (by symmetry) are independent. Since $R^2 \sim \mathcal{R}_p^2(h)$ and $\boldsymbol{z} = \boldsymbol{t}/\|\boldsymbol{t}\|$ has a uniform distribution on the p-dimensional unitary sphere $\{\boldsymbol{x} \in R^p\colon \|\boldsymbol{x}\| = 1\}$, which does not depends on h, it follows that any spherical (elliptical) distribution is determined by a radial-squared distribution only. Thus, since $\boldsymbol{w} = \boldsymbol{\mu} + \Sigma^{1/2}\boldsymbol{t}$, it follows that if $E(R) < \infty$, then $E(\boldsymbol{w}) = \boldsymbol{\mu}$ and if $E(R^2) < \infty$, then $V(\boldsymbol{w}) = \alpha_h \Sigma$, where $\alpha_h = E(\frac{R^2}{p})$. It is also well known that any marginal and conditional distributions of an elliptical distribution are elliptical distributions. Moreover, the p-dimensional density function generator $h(\cdot)$ is such that

$$\int_0^\infty \cdots \int_0^\infty \prod_{i=1}^k u_i^{p_i/2-1} h\left(\sum_{i=1}^k u_i\right) \mathrm{d}u_1 \cdots \mathrm{d}u_k = \frac{\prod_{i=1}^k \Gamma(\frac{p_i}{2})}{\pi^{p/2}}, \tag{2.3}$$

for all sequences of nonnegative integers $p_1, \ldots, p_k$ such that $\sum_{i=1}^k p_i = p$. As in the particular case of $k = 1$, this result follows from the fact that if $\boldsymbol{t} = (\boldsymbol{t}_1, \ldots, \boldsymbol{t}_k)' \sim El_p(\boldsymbol{0}, \boldsymbol{I}_p; h)$, then $R_i^2 = \|\boldsymbol{t}_i\|^2$, $i = 1, \ldots, k$, have joint density given by (see Fang et al., 1990; Arellano-Valle, 1994; Arellano-Valle et al., 2005)

$$f_{R_1^2,\ldots,R_k^2}(u_1, \ldots, u_k) = \frac{\pi^{p/2}}{\prod_{i=1}^k \Gamma(\frac{p_i}{2})} \prod_{i=1}^k u_i^{p_i/2-1} h\left(\sum_{i=1}^k u_i\right).$$

In this case, we say that $(R_1^2, \ldots, R_k^2)$ has a multivariate radial-squared distribution with $(p_1, \ldots, p_k)$ degrees of freedom and we denote this as $(R_1^2, \ldots, R_k^2) \sim \mathcal{M}R^2_{p_1,\ldots,p_k}(h)$. Note that $R^2 = \sum_{i=1}^k R_i^2 \sim \mathcal{R}_p^2(h)$, and $R_i^2 \sim \mathcal{R}_{p_i}^2(h)$, $i = 1, \ldots, k$, which are independent if and only if h is the normal generator density function, that is, $\mathcal{R}_p^2(h) = \chi_p^2$, the chi-square distribution with p degrees of freedom.

2.2. *Measurement error models*

From the class of the elliptical distributions, different types of elliptical MEM can be specified. For simplicity, the following notation will be used in the sequel. Let $\boldsymbol{y} = (y_1, \ldots, y_n)'$ and $\boldsymbol{x} = (x_1, \ldots, x_n)^{\mathrm{T}}$ be the observed vectors from a sample of size n. Likewise, $\boldsymbol{\xi} = (\xi_1, \ldots, \xi_n)^{\mathrm{T}}$ denotes the unobserved quantities of the true covariate vector, $\boldsymbol{e} = (e_1, \ldots, e_n)^{\mathrm{T}}$ and $\boldsymbol{u} = (u_1, \ldots, u_n)^{\mathrm{T}}$ the error vectors related with Eqs. (1.1) and (1.2), respectively. With this notation, Eqs. (1.1)–(1.2) can be rewritten as

$$\boldsymbol{y} = \beta_1 \boldsymbol{1}_n + \beta_2 \boldsymbol{\xi} + \boldsymbol{e} \tag{2.4}$$

and

$$x = \xi + u, \tag{2.5}$$

where $\mathbf{1}_n$ is an $n \times 1$ vector of ones. Moreover, the notation $[\boldsymbol{w}]$ will be used to denote the density or distribution of $\boldsymbol{w}$ and $[\boldsymbol{w}|\boldsymbol{v}]$ to denote the conditional density or distribution of $\boldsymbol{w}$ given $\boldsymbol{v}$. Thus, from this notation the posterior distribution (or density) of $(\boldsymbol{\xi}, \boldsymbol{\theta})$ can be written as

$$[\boldsymbol{\xi}, \boldsymbol{\theta}|\boldsymbol{y}, \boldsymbol{x}] \propto [\boldsymbol{y}, \boldsymbol{x}|\boldsymbol{\xi}, \boldsymbol{\theta}][\boldsymbol{\xi}, \boldsymbol{\theta}], \tag{2.6}$$

where $[\boldsymbol{y}, \boldsymbol{x}|\boldsymbol{\xi}, \boldsymbol{\theta}]$ is the (functional) likelihood function and $[\boldsymbol{\xi}, \boldsymbol{\theta}]$ is the prior distribution assigned to $(\boldsymbol{\xi}, \boldsymbol{\theta})$, where $\boldsymbol{\theta} = (\boldsymbol{\beta}^{\mathrm{T}}, \boldsymbol{\sigma}^{\mathrm{T}})^{\mathrm{T}}$, with $\boldsymbol{\beta} = (\beta_1, \beta_2)^{\mathrm{T}}$ being regression parameters and $\boldsymbol{\sigma} = (\sigma_e^2, \sigma_u^2)^{\mathrm{T}}$ being scale parameters. Here, the following types of (functional) elliptical likelihood functions will be considered:

(i) $[\boldsymbol{y}, \boldsymbol{x}|\boldsymbol{\xi}, \boldsymbol{\theta}]$ is an $(2n)$-dimensional elliptical density.
(ii) $[\boldsymbol{y}, \boldsymbol{x}|\boldsymbol{\xi}, \boldsymbol{\theta}] = [\boldsymbol{y}|\boldsymbol{\xi}, \boldsymbol{\theta}][\boldsymbol{x}|\boldsymbol{\xi}, \boldsymbol{\theta}]$, where $[\boldsymbol{y}|\boldsymbol{\xi}, \boldsymbol{\theta}]$ and $[\boldsymbol{x}|\boldsymbol{\xi}, \boldsymbol{\theta}]$ are n-dimensional elliptical densities;
(iii) $[\boldsymbol{y}, \boldsymbol{x}|\boldsymbol{\xi}, \boldsymbol{\theta}] = \prod_{i=1}^{n}[y_i, x_i|\xi_i, \boldsymbol{\theta}]$, where $[y_i, z_i|\xi_i, \boldsymbol{\theta}]$, $i = 1, \ldots, n$, are bivariate elliptical densities; and
(iv) $[\boldsymbol{y}, \boldsymbol{x}|\boldsymbol{\xi}, \boldsymbol{\theta}] = \prod_{i=1}^{n}[y_i|\xi_i, \boldsymbol{\theta}][x_i|\xi_i, \boldsymbol{\theta}]$, where $[y_i|\xi_i, \boldsymbol{\theta}]$ and $[x_i|\xi_i, \boldsymbol{\theta}]$, $i = 1, \ldots, n$, are univariate elliptical densities.

Note that models (i) and (iii) present dependent elliptical errors and models (ii) and (iv) present independent elliptical errors. Note also that models (ii) and (iv) assume that $\boldsymbol{y}$ and $\boldsymbol{x}$ are independent given $(\boldsymbol{\xi}, \boldsymbol{\theta})$. In this case it is said that the models have nondifferential errors (NDE) (see Carroll et al., 1995; Bolfarine and Arellano-Valle, 1998). This property does not hold in models (i) and (iii), which have weak nondifferential errors (WNDE) (see Bolfarine and Arellano-Valle, 1998), since it is assumed that the errors are uncorrelated. Moreover, under the normality assumption all these models are equivalent.

Further elliptical specifications may be obtained by modeling the joint distribution of $(\boldsymbol{y}, \boldsymbol{x}, \boldsymbol{\xi})$ given $\boldsymbol{\theta}$ within the elliptical family of distributions, and following the same ideas used in defining (i)–(iv). In the following, we assume in general that the incidental parameters $\boldsymbol{\xi}$ and the structural parameter $\boldsymbol{\theta}$ are independent.

3. Diffuse prior distribution for the incidental parameters

In this section, we consider (consequences of) the important special case where the prior distribution for $(\boldsymbol{\xi}, \boldsymbol{\theta})$ is given by

$$[\boldsymbol{\xi}, \boldsymbol{\theta}] = [\boldsymbol{\xi}|\boldsymbol{\theta}][\boldsymbol{\theta}] \propto [\boldsymbol{\theta}], \tag{3.1}$$

so that a diffuse prior distribution $[\boldsymbol{\xi}|\boldsymbol{\theta}] = 1$ is considered for the incidental parameter (or latent co-variables) $\boldsymbol{\xi} = (\xi_1, \ldots, \xi_n)^{\mathrm{T}}$ given the structural parameters

$\boldsymbol{\theta} = (\beta_1, \beta_2, \sigma_e^2, \sigma_u^2)^{\mathrm{T}}$. The prior specification in (3.1) may be also interpreted as follows:

$$\boldsymbol{\xi} \text{ and } \boldsymbol{\theta} \text{ are independent, and } [\boldsymbol{\xi}] = 1. \tag{3.2}$$

Similar assumptions are considered by Zellner (1971), Section 5.1, where properties of posterior distributions for the n-mean normal problem with improper prior distributions when a normal model is considered for the observed data.

We note by using (3.1) (or (3.2)) that

$$[\boldsymbol{x}|\boldsymbol{\theta}] = \int_{R^n} [\boldsymbol{x}|\boldsymbol{\xi}, \boldsymbol{\theta}][\boldsymbol{\xi}|\boldsymbol{\theta}]\, \mathrm{d}\boldsymbol{\xi} = \int_{R^n} [\boldsymbol{x}|\boldsymbol{\xi}, \boldsymbol{\theta}]\, \mathrm{d}\boldsymbol{\xi}. \tag{3.3}$$

Moreover, denoting by $[\boldsymbol{y}, \boldsymbol{x}|\boldsymbol{\theta}]$ the structural likelihood function of a MEM, we have also that

$$[\boldsymbol{y}, \boldsymbol{x}|\boldsymbol{\theta}] = [\boldsymbol{y}|\boldsymbol{x}, \boldsymbol{\theta}][\boldsymbol{x}|\boldsymbol{\theta}]. \tag{3.4}$$

Thus, the following results are direct consequences of (3.3) and (3.4).

LEMMA 3.1. *If* $[\boldsymbol{\xi}|\boldsymbol{\theta}] = 1$ *and* $[\boldsymbol{x}|\boldsymbol{\xi}, \boldsymbol{\theta}] = [\boldsymbol{\xi}|\boldsymbol{x}, \boldsymbol{\theta}]$ *is proper, then* $[\boldsymbol{x}|\boldsymbol{\theta}] = 1$.

LEMMA 3.2. *If* $[\boldsymbol{x}|\boldsymbol{\theta}] = 1$, *then* $[\boldsymbol{y}, \boldsymbol{x}|\boldsymbol{\theta}] = [\boldsymbol{y}|\boldsymbol{x}, \boldsymbol{\theta}]$.

LEMMA 3.3. *If* $[\boldsymbol{x}|\boldsymbol{\xi}, \boldsymbol{\theta}] = |\Sigma|^{-1/2} g(\Sigma^{-1/2}(\boldsymbol{x} - \boldsymbol{\xi}))$, *or* $[\boldsymbol{x}|\boldsymbol{\xi}, \boldsymbol{\theta}] = \prod_{i=1}^{n} [x_i|\xi_i, \boldsymbol{\theta}]$, *with* $[x|\xi, \boldsymbol{\theta}] = (1/\sigma) g((x - \xi)/\sigma)$, *where* $g(-u) = g(u)$, *then* $[\boldsymbol{x}|\boldsymbol{\xi}, \boldsymbol{\theta}] = [\boldsymbol{\xi}|\boldsymbol{x}, \boldsymbol{\theta}]$.

Clearly, for any elliptical MEM we have that $[\boldsymbol{\xi}|\boldsymbol{\theta}] = 1$ implies by Lemma 3.3 that $[\boldsymbol{x}|\boldsymbol{\xi}, \boldsymbol{\theta}] = [\boldsymbol{\xi}|\boldsymbol{x}, \boldsymbol{\theta}]$ and by Lemma 3.1 (or by (3.3)) that $[\boldsymbol{x}|\boldsymbol{\theta}] = 1$, so that by Lemma 3.2 (or by (3.4)) we can state the main result of the section.

THEOREM 3.1. *Let* $[\boldsymbol{y}, \boldsymbol{x}|\boldsymbol{\xi}, \boldsymbol{\theta}]$ *be the functional likelihood function of an (dependent or independent) elliptical MEM. Then, under prior distribution* (3.1) *we have that the structural likelihood function is such that* $[\boldsymbol{y}, \boldsymbol{x}|\boldsymbol{\theta}] = [\boldsymbol{y}|\boldsymbol{x}, \boldsymbol{\theta}]$, *so that the posterior distribution of* $\boldsymbol{\theta}$ *is given by*

$$[\boldsymbol{\theta}|\boldsymbol{y}, \boldsymbol{x}] \propto [\boldsymbol{y}|\boldsymbol{x}, \boldsymbol{\theta}][\boldsymbol{\theta}].$$

As consequence of Theorem 3.1, it is typically the case that a proper prior distribution (see also Zellner, 1971, Section 5.1) needs to be specified for the structural parameter $\boldsymbol{\theta}$ so that proper posterior distribution results are obtained.

EXAMPLE 3.1 (*The linear normal MEM*). Let us consider the linear MEM given by (1.1) and (1.2), with the assumption that, $e_i|\boldsymbol{\theta} \overset{\text{ind}}{\sim} N(0, \sigma_e^2)$ and $u_i|\boldsymbol{\theta} \overset{\text{ind}}{\sim} N(0, \sigma_u^2)$ are independent, $i = 1, \ldots, n$. In this case, for $i = 1, \ldots, n$, we have that $y_i|\xi_i, \boldsymbol{\theta} \overset{\text{ind}}{\sim} N(\beta_1 + \beta_2\xi_i, \sigma_e^2)$ and $x_i|\xi_i, \boldsymbol{\theta} \overset{\text{ind}}{\sim} N(\xi_i, \sigma_u^2)$ are independent. Thus, if $[\xi_i|\boldsymbol{\theta}] = 1$, $i = 1, \ldots, n$, then $\xi_i|x_i, \boldsymbol{\theta} \overset{\text{ind}}{\sim} N(x_i, \sigma_u^2)$, $i = 1, \ldots, n$, and so the structural likelihood function is given by the conditional normal regression model $y_i|x_i, \boldsymbol{\theta} \overset{\text{ind}}{\sim} N(\beta_1 + \beta_2 x_i, \sigma_e^2 + \beta_2^2\sigma_u^2)$, $i = 1, \ldots, n$. Clearly, to make inference on $\boldsymbol{\theta}$ (or functions of $\boldsymbol{\theta}$) its prior distribution in (3.1) needs to be proper.

4. Dependent elliptical MEM

In this section, we study some Bayesian solutions to inference problems for different classes of dependent elliptical MEM. Both the WNDE and NDE dependent elliptical MEM defined in (i) and (ii) in Section 2 are considered. The focus is on the linear regression MEM given by (2.3)–(2.4). Extensions to some multiple regression models are analogous. Under the general notation specified in Section 2, the class of MEM that we consider can be represented as:

(i) for WNDE elliptical MEM (case (i) in Section 2), we have that

$$[\boldsymbol{y}, \boldsymbol{x}|\boldsymbol{\xi}, \boldsymbol{\theta}] = \left(\frac{1}{\sigma_e^2\sigma_u^2}\right)^{n/2} h\left(\frac{\|\boldsymbol{y} - \beta_1\mathbf{1}_n - \beta_2\boldsymbol{\xi}\|^2}{\sigma_e^2} + \frac{\|\boldsymbol{x} - \boldsymbol{\xi}\|^2}{\sigma_u^2}\right), \tag{4.1}$$

for some $(2n)$-dimensional generator density function h; and

(ii) for NDE elliptical MEM (case (ii) in Section 2), we have that

$$[\boldsymbol{y}|\boldsymbol{\xi}, \boldsymbol{\theta}] = \left(\frac{1}{\sigma_e^2}\right)^{n/2} h_1\left(\frac{\|\boldsymbol{y} - \beta_1\mathbf{1}_n - \beta_2\boldsymbol{\xi}\|^2}{\sigma_e^2}\right)$$

and

$$[\boldsymbol{x}|\boldsymbol{\xi}, \boldsymbol{\theta}] = \left(\frac{1}{\sigma_u^2}\right)^{n/2} h_2\left(\frac{\|\boldsymbol{x} - \boldsymbol{\xi}\|^2}{\sigma_u^2}\right),$$

for some n-dimensional generator density functions h_1 and h_2, so that

$$[\boldsymbol{y}, \boldsymbol{x}|\boldsymbol{\xi}, \boldsymbol{\theta}] = \left(\frac{1}{\sigma_e^2\sigma_u^2}\right)^{n/2} h_1\left(\frac{\|\boldsymbol{y} - \beta_1\mathbf{1}_n - \beta_2\boldsymbol{\xi}\|^2}{\sigma_e^2}\right) h_2\left(\frac{\|\boldsymbol{x} - \boldsymbol{\xi}\|^2}{\sigma_u^2}\right). \tag{4.2}$$

Note that the model given by (4.1) follows by assuming in (2.4) and (2.5) that $[\boldsymbol{e}, \boldsymbol{u}|\boldsymbol{\theta}] \sim El_{2n}((\mathbf{0}, \mathbf{0}), \operatorname{diag}\{\sigma_e^2\boldsymbol{I}_n, \sigma_u^2\boldsymbol{I}_n\}; h)$ and by using properties of the elliptical distributions. Likewise, the model in (4.2) follows by considering relations (2.3) and (2.4) with the assumptions that $[\boldsymbol{e}|\boldsymbol{\theta}] \sim El_n(\mathbf{0}, \sigma_e^2\boldsymbol{I}_n; h_1)$ and $[\boldsymbol{u}|\boldsymbol{\theta}] \sim El_n(\mathbf{0}, \sigma_u^2\boldsymbol{I}_n; h_2)$ are independent.

EXAMPLE 4.1 (*The normal MEM*). Suppose that $[\boldsymbol{e}, \boldsymbol{u}|\boldsymbol{\theta}] \sim N_{2n}((\mathbf{0}, \mathbf{0}), \operatorname{diag}\{\sigma_e^2\boldsymbol{I}_n, \sigma_u^2\boldsymbol{I}_n\})$. Hence, we have that $[\boldsymbol{e}|\boldsymbol{\theta}] \sim N_n(\mathbf{0}, \sigma_e^2\boldsymbol{I}_n)$ and $[\boldsymbol{u}|\boldsymbol{\theta}] \sim N_n(\mathbf{0}, \sigma_u^2\boldsymbol{I}_n)$ are independent. That is, under the normal assumptions with uncorrelated errors it follows that the WNDE and NDE normal models are equivalent. This result follows from the fact that the normal distribution has p-dimensional generator density function given by $h(u) = (2\pi)^{-p/2} e^{-u/2}$, $u \geqslant 0$, which is such that $h(u_1 + u_2) = h_1(u_1)h_2(u_2)$ for each p_i-dimensional generator normal density h_i, $i = 1, 2$, with $p_1 + p_2 = p$. Indeed, as proved in Bolfarine and Arellano-Valle (1998), the normal linear MEM is the only MEM with the WNDE and NDE properties. Thus, we have that

$$
\begin{aligned}
[\boldsymbol{y}, \boldsymbol{x} | \boldsymbol{\xi}, \boldsymbol{\theta}] &= \left(\frac{1}{4\pi^2\sigma_e^2\sigma_u^2}\right)^{n/2} \exp\left\{-\frac{1}{2}\left(\frac{\|\boldsymbol{y} - \beta_1 \mathbf{1}_n - \beta_2\boldsymbol{\xi}\|^2}{\sigma_e^2} + \frac{\|\boldsymbol{x} - \boldsymbol{\xi}\|^2}{\sigma_u^2}\right)\right\} \\
&= \left(\frac{1}{2\pi\sigma_e^2}\right)^{n/2} \exp\left\{-\frac{1}{2}\left(\frac{\|\boldsymbol{y} - \beta_1 \mathbf{1}_n - \beta_2\boldsymbol{\xi}\|^2}{\sigma_e^2}\right)\right\} \\
&\quad \times \left(\frac{1}{2\pi\sigma_u^2}\right)^{n/2} \exp\left\{-\frac{1}{2}\left(\frac{\|\boldsymbol{x} - \boldsymbol{\xi}\|^2}{\sigma_u^2}\right)\right\} \\
&= [\boldsymbol{y} | \boldsymbol{\xi}, \boldsymbol{\theta}][\boldsymbol{x} | \boldsymbol{\xi}, \boldsymbol{\theta}].
\end{aligned}
$$

EXAMPLE 4.2 (*The Student-t MEM*). Consider now the density generator function given by $h(u) = c(\nu, p)\lambda^{nu/2}\{\lambda + u\}^{-(\nu+p)/2}$, $u \geqslant 0$, where $c(\nu, p) = \Gamma[(\nu + p)/2]/\Gamma[\nu/2]\pi^{p/2}$, providing the p-dimensional generalized Student-t distribution, denoted by $t_p(\boldsymbol{\mu}, \Sigma; \lambda, \nu)$, which is reduced to ordinary p-dimensional Student-t distribution when $\lambda = \nu$ (Arellano-Valle and Bolfarine, 1995). Thus:

(i) the WNDE Student-t MEM is given by

$$
\begin{aligned}
&[\boldsymbol{y}, \boldsymbol{x} | \boldsymbol{\xi}, \boldsymbol{\theta}] \\
&\quad = c(\nu, 2n)\lambda^{\nu/2}\left\{\lambda + \frac{\|\boldsymbol{y} - \beta_1 \mathbf{1}_n - \beta_2\boldsymbol{\xi}\|^2}{\sigma_e^2} + \frac{\|\boldsymbol{x} - \boldsymbol{\xi}\|^2}{\sigma_u^2}\right\}^{-(\nu+2n)/2},
\end{aligned}
$$

which follows by assuming in (1.1)–(1.2) that $[\boldsymbol{e}, \boldsymbol{u} | \boldsymbol{\theta}] \sim t_{2n}((\mathbf{0}, \mathbf{0}), \operatorname{diag}\{\sigma_e^2 \boldsymbol{I}_n, \sigma_u^2 \boldsymbol{I}_n\}; \lambda, \nu)$; and

(ii) the NDE Student-t MEM is given by

$$
\begin{aligned}
[\boldsymbol{y}, \boldsymbol{x} | \boldsymbol{\xi}, \boldsymbol{\theta}] &= c(\nu_1, n)\lambda_1^{\nu_1/2}\left\{\lambda_1 + \frac{\|\boldsymbol{y} - \beta_1 \mathbf{1}_n - \beta_2\boldsymbol{\xi}\|^2}{\sigma_e^2}\right\}^{-(\nu_1+n)/2} \\
&\quad \times c(\nu_2, n)\lambda_2^{\nu_2/2}\left\{\lambda_1 + \frac{\|\boldsymbol{x} - \boldsymbol{\xi}\|^2}{\sigma_u^2}\right\}^{-(\nu_2+n)/2},
\end{aligned}
$$

corresponding to the assumptions that $[\boldsymbol{e} | \boldsymbol{\theta}] \sim t_n(\mathbf{0}, \sigma_e^2 \boldsymbol{I}_n; \lambda_1, \nu_1)$ and $[\boldsymbol{u} | \boldsymbol{\theta}] \sim t_n(\mathbf{0}, \sigma_u^2 \boldsymbol{I}_n; \lambda_2, \nu_2)$ in (1.1) and (1.2), respectively, with $\boldsymbol{e}$ and $\boldsymbol{u}$ independent. In case (i) the WNDE property follows from the fact that $[\boldsymbol{y} | \boldsymbol{x}, \boldsymbol{\xi}, \boldsymbol{\theta}] \neq [\boldsymbol{y} | \boldsymbol{\xi}, \boldsymbol{\theta}]$.

Next, we consider the elliptical likelihood specifications (4.1) and (4.2) with the following noninformative prior distribution for $(\boldsymbol{\xi}, \boldsymbol{\theta})$,

$$
[\boldsymbol{\xi}, \boldsymbol{\theta}] \propto \frac{1}{\sigma_e^2\sigma_u^2}[\boldsymbol{\xi}, \boldsymbol{\beta}], \tag{4.3}
$$

where $[\boldsymbol{\xi}, \boldsymbol{\beta}]$ is the prior distribution for $(\boldsymbol{\xi}, \boldsymbol{\beta})$. Hence, the posterior distribution of $(\boldsymbol{\xi}, \boldsymbol{\beta})$ does not depend on the generator density functions that define the particular elliptical model considered in (4.1) or (4.2). The proof of this result follows from the fact that

$$[\boldsymbol{\xi}, \boldsymbol{\beta} | \boldsymbol{y}, \boldsymbol{x}] \propto [\boldsymbol{y}, \boldsymbol{x} | \boldsymbol{\xi}, \boldsymbol{\beta}][\boldsymbol{\xi}, \boldsymbol{\beta}], \tag{4.4}$$

where

$$[\boldsymbol{y}, \boldsymbol{x} | \boldsymbol{\xi}, \boldsymbol{\beta}] = \int_0^\infty\int_0^\infty \left(\frac{1}{\sigma_e^2\sigma_u^2}\right)^{n/2+1} h\left(\frac{\|\boldsymbol{y} - \beta_1\mathbf{1}_n - \beta_2\boldsymbol{\xi}\|^2}{\sigma_e^2} + \frac{\|\boldsymbol{x} - \boldsymbol{\xi}\|^2}{\sigma_u^2}\right) \mathrm{d}\sigma_e^2\, \mathrm{d}\sigma_u^2$$

for the model (4.1), and

$$[\boldsymbol{y}, \boldsymbol{x} | \boldsymbol{\xi}, \boldsymbol{\beta}] = \int_0^\infty \left(\frac{1}{\sigma_e^2}\right)^{n/2+1} h_1\left(\frac{\|\boldsymbol{y} - \beta_1\mathbf{1}_n - \beta_2\boldsymbol{\xi}\|^2}{\sigma_e^2}\right) \mathrm{d}\sigma_e^2 \times \int_0^\infty \left(\frac{1}{\sigma_u^2}\right)^{n/2+1} h_2\left(\frac{\|\boldsymbol{x} - \boldsymbol{\xi}\|^2}{\sigma_u^2}\right) \mathrm{d}\sigma_u^2,$$

for the model (4.2). In both cases, we have that (2.2) (or (2.3)) implies that these integrals are such that

$$[\boldsymbol{y}, \boldsymbol{x} | \boldsymbol{\xi}, \boldsymbol{\beta}] = [\boldsymbol{y} | \boldsymbol{\xi}, \boldsymbol{\beta}][\boldsymbol{x} | \boldsymbol{\xi}], \tag{4.5}$$

where

$$[\boldsymbol{y} | \boldsymbol{\xi}, \boldsymbol{\beta}] = \frac{\Gamma(\frac{n}{2})}{\pi^{\frac{n}{2}}}\left\{\|\boldsymbol{y} - \beta_1\mathbf{1}_n - \beta_2\boldsymbol{\xi}\|^2\right\}^{-n/2} \quad \text{and}$$

$$[\boldsymbol{x} | \boldsymbol{\xi}] = \frac{\Gamma(\frac{n}{2})}{\pi^{\frac{n}{2}}}\left\{\|\boldsymbol{x} - \boldsymbol{\xi}\|^2\right\}^{-n/2}, \tag{4.6}$$

so that Bayesian inference does not depend on the particular density generator function under consideration. Hence, by considering (4.4), (4.5) and (4.6) we have proved the following important result.

THEOREM 4.1. *Consider the WNDE (NDE) dependent elliptical MEM given by* (4.1) ((4.2)), *with the prior distribution given in* (4.3). *Then,*

$$[\boldsymbol{\xi}, \boldsymbol{\beta} | \boldsymbol{y}, \boldsymbol{x}] \propto \left\{\|\boldsymbol{y} - \beta_1\mathbf{1}_n - \beta_2\boldsymbol{\xi}\|^2\right\}^{-n/2}\left\{\|\boldsymbol{x} - \boldsymbol{\xi}\|^2\right\}^{-n/2}[\boldsymbol{\xi}, \boldsymbol{\beta}], \tag{4.7}$$

which is independent on the choice of the density generator function(s) h (h_1 and h_2) under consideration.

The above results show that under the prior distribution (4.3), Bayesian inference on $(\boldsymbol{\xi}, \boldsymbol{\beta})$ is invariant within the class of WNDE or NDE dependent elliptical MEM. One important consequence of this fact is that whatever the dependent elliptical model under consideration is, we can base our inference on the normal linear MEM, with the same prior specification. Moreover, by using (4.7) and the well known results that

$$\begin{aligned}\|\mathbf{y} - \beta_1 \mathbf{1}_n - \beta_2 \boldsymbol{\xi}\|^2 &= \|\mathbf{y} - \boldsymbol{\Upsilon}\boldsymbol{\beta}\|^2 \\ &= SSE(\boldsymbol{\xi}) + \big(\boldsymbol{\beta} - \hat{\boldsymbol{\beta}}(\boldsymbol{\xi})\big)^{\mathrm{T}} \boldsymbol{\Upsilon}^{\mathrm{T}} \boldsymbol{\Upsilon} \big(\boldsymbol{\beta} - \hat{\boldsymbol{\beta}}(\boldsymbol{\xi})\big),\end{aligned}$$

where $\boldsymbol{\Upsilon} = [\mathbf{1}_n, \boldsymbol{\xi}]$, $\hat{\boldsymbol{\beta}}(\boldsymbol{\xi}) = (\boldsymbol{\Upsilon}^{\mathrm{T}}\boldsymbol{\Upsilon})^{-1}\boldsymbol{\Upsilon}^{\mathrm{T}}\mathbf{y}$ and $SSE(\boldsymbol{\xi}) = \|\mathbf{y} - \boldsymbol{\Upsilon}\hat{\boldsymbol{\beta}}(\boldsymbol{\xi})\|^2$, we obtain the following special case.

COROLLARY 4.1. *Under the conditions in Theorem* 4.1, $[\boldsymbol{\xi}, \boldsymbol{\beta} | \mathbf{y}, \mathbf{x}] = [\boldsymbol{\beta} | \boldsymbol{\xi}, \mathbf{y}][\boldsymbol{\xi} | \mathbf{y}, \mathbf{x}]$, *where*

$$[\boldsymbol{\beta} | \mathbf{y}, \boldsymbol{\xi}] \propto \left\{1 + \frac{(\boldsymbol{\beta} - \hat{\boldsymbol{\beta}}(\boldsymbol{\xi}))^{\mathrm{T}} \boldsymbol{\Upsilon}^{\mathrm{T}} \boldsymbol{\Upsilon} (\boldsymbol{\beta} - \hat{\boldsymbol{\beta}}(\boldsymbol{\xi}))}{SSE(\boldsymbol{\xi})}\right\}^{-n/2} [\boldsymbol{\beta} | \boldsymbol{\xi}]$$

and

$$[\boldsymbol{\xi} | \mathbf{y}, \mathbf{x}] \propto \{S^2(\boldsymbol{\xi})\}^{-1/2} \{SSE(\boldsymbol{\xi})\}^{-(n-2)/2} \{\|\mathbf{x} - \boldsymbol{\xi}\|^2\}^{-n/2} [\boldsymbol{\xi}].$$

where $\bar{\xi} = \sum_{i=1}^n \xi_i / n$ *and* $S^2(\boldsymbol{\xi}) = |\boldsymbol{\Upsilon}^{\mathrm{T}}\boldsymbol{\Upsilon}|/n = \|\boldsymbol{\xi} - \bar{\xi}\mathbf{1}_n\|^2$. *Moreover, if* $[\mathbf{x}, \boldsymbol{\beta}] \propto [\boldsymbol{\xi}]$, *then* $[\boldsymbol{\beta} | \mathbf{y}, \boldsymbol{\xi}]$ *corresponds to the Student-t distribution given by*

$$\boldsymbol{\beta} | \mathbf{y}, \boldsymbol{\xi} \sim t_2\big(\hat{\boldsymbol{\beta}}(\boldsymbol{\xi}), \overline{SSE}(\boldsymbol{\xi})\big(\boldsymbol{\Upsilon}^{\mathrm{T}}\boldsymbol{\Upsilon}\big)^{-1}; n - 2\big),$$

where $\overline{SSE}(\boldsymbol{\xi}) = SSE(\boldsymbol{\xi})/(n - 2)$.

From Corollary 4.1 it follows that the conditional posterior distribution of $\boldsymbol{\beta}$ given $\boldsymbol{\xi}$ does not depend on $\mathbf{x}$. It also follows that we can generate a sample $\boldsymbol{\xi} = (\xi_1, \ldots, \xi_n)^{\mathrm{T}}$ from $[\boldsymbol{\xi} | \mathbf{y}, \mathbf{x}]$ and then generate samples from $[\boldsymbol{\beta} | \mathbf{y}, \boldsymbol{\xi}]$, using the generated $\boldsymbol{\xi}$ value. Moreover, if $[\boldsymbol{\beta} | \boldsymbol{\xi}] \propto c$ (constant), then this conditional posterior distribution corresponds to the Student-t distribution, say $t_2(\hat{\boldsymbol{\beta}}(\boldsymbol{\xi}), \overline{SSE}(\boldsymbol{\xi})(\boldsymbol{\Upsilon}^{\mathrm{T}}\boldsymbol{\Upsilon})^{-1}; n - 2)$. We note that propriety of $[\boldsymbol{\xi} | \mathbf{y}, \mathbf{x}]$ is guaranteed for proper prior distributions $[\boldsymbol{\xi}]$.

4.1. *Equal variances case*

One special and interesting case follows by considering that the ratio $\lambda = \sigma_e^2/\sigma_u^2$ is known, a situation that can also be treated generally in the whole dependent elliptical family. As is well known, considering that λ is known, is equivalent to consider that $\sigma_e^2 = \sigma_u^2 = \sigma^2$, with σ^2 unknown.

In this section, we only consider the dependent elliptical WNDE model in (4.1) with the assumption that $\sigma_e^2 = \sigma_u^2 = \sigma^2$, because, as shown in the following, we can get some more specific results in this special case. In fact, according to (4.1), we have

$$[\mathbf{y}, \mathbf{x} | \boldsymbol{\xi}, \boldsymbol{\theta}] = \left(\frac{1}{\sigma^2}\right)^{2n} h\left(\frac{\|\mathbf{y} - \beta_1 \mathbf{1}_n - \beta_2 \boldsymbol{\xi}\|^2 + \|\mathbf{x} - \boldsymbol{\xi}\|^2}{\sigma^2}\right), \tag{4.8}$$

for some $(2n)$-dimensional generator density function h. Hence, using (2.2) and considering the prior (4.3), with the noninformative prior on $(\boldsymbol{\beta}, \boldsymbol{\xi})$ given by

$$[\boldsymbol{\beta}, \boldsymbol{\xi}] \propto 1, \tag{4.9}$$

it follows from (4.4) that the posterior density of $(\boldsymbol{\xi}, \boldsymbol{\beta})$ is given by

$$[\boldsymbol{\beta}, \boldsymbol{\xi} | \boldsymbol{y}, \boldsymbol{x}] = \int_0^\infty \left(\frac{1}{\sigma^2}\right)^{(2n+2)/2+1} h\left(\frac{\|\boldsymbol{y} - \beta_1 \mathbf{1}_n - \beta_2 \boldsymbol{\xi}\|^2 + \|\boldsymbol{x} - \boldsymbol{\xi}\|^2}{\sigma^2}\right) \mathrm{d}\sigma^2$$

$$= \frac{\Gamma\left(\frac{2n+2}{2}\right)}{\pi^{\frac{2n+2}{2}}} \left\{\|\boldsymbol{y} - \beta_1 \mathbf{1}_n - \beta_2 \boldsymbol{\xi}\|^2 + \|\boldsymbol{x} - \boldsymbol{\xi}\|^2\right\}^{-(2n+2)/2}. \tag{4.10}$$

Letting

$$\hat{\boldsymbol{\xi}} = \frac{\boldsymbol{x} + \beta(\boldsymbol{y} - \mathbf{1}_n \alpha)}{1 + \beta_2^2},$$

which is the value of $\boldsymbol{\xi}$ that maximizes the last expression of (4.10) for any given $\boldsymbol{\beta}$, it is easy to see that (4.10) can be rewritten as

$$[\boldsymbol{\beta}, \boldsymbol{\xi} | \boldsymbol{y}, \boldsymbol{x}] = \frac{\Gamma\left(\frac{2n+2}{2}\right)}{\pi^{\frac{2n+2}{2}}} \left\{\frac{\|\boldsymbol{y} - \beta_1 \mathbf{1}_n - \beta_2 \boldsymbol{x}\|^2}{1 + \beta_2^2} + \left(1 + \beta_2^2\right) \|\boldsymbol{\xi} - \hat{\boldsymbol{\xi}}\|^2\right\}^{-(2n+2)/2}. \tag{4.11}$$

A similar fact is used in Zellner (1971), p. 165, under the normal assumption. After some standard algebraic manipulations, it can also be shown that the mode of the posterior distribution in (4.11) is given by

$$\hat{\beta}_1 = \bar{y} - \hat{\beta}_2 \bar{x} \quad \text{and} \quad \hat{\beta}_2 = \frac{S(\boldsymbol{y}) - S(\boldsymbol{x}) + \sqrt{(S(\boldsymbol{y}) - S(\boldsymbol{x}))^2 + 4S(\boldsymbol{y}, \boldsymbol{x})^2}}{2S(\boldsymbol{y}, \boldsymbol{x})}, \tag{4.12}$$

where $S(\boldsymbol{y}) = \|\boldsymbol{y} - \bar{y}\mathbf{1}_n\|^2/n$, $S(\boldsymbol{x}) = \|\boldsymbol{x} - \bar{x}\mathbf{1}_n\|^2/n$ and $S(\boldsymbol{y}, \boldsymbol{x}) = (\boldsymbol{x} - \bar{x}\mathbf{1}_n)^{\mathrm{T}}(\boldsymbol{y} - \bar{y}\mathbf{1}_n)/n$.

The estimators in (4.12) has been previously obtained in the literature (see, for example, Fuller, 1987) and are known as the generalized least squares estimators (GLSE). They correspond to the maximum likelihood estimators of β_1 and β_2. In the likelihood function given by (4.8) it is considered that $h(\cdot)$ is a nonincreasing and continuous function. It can also be shown that the posterior mode of σ^2 is given by

$$\hat{\sigma}^2 = \hat{\delta} \frac{\|\boldsymbol{y} - \hat{\beta}_1 \mathbf{1}_n - \hat{\beta}_2 \boldsymbol{x}\|^2}{2n(1 + \hat{\beta}_2^2)}, \tag{4.13}$$

with $\hat{\delta}$ maximizing the function

$$g(\delta) = \delta^{-n/2} h\left(\frac{n}{\delta}\right).$$

In the special case of the Student-t distribution, it follows that $\hat{\delta} = 1$.

Estimators (4.12)–(4.13) has been previously derived as a maximum likelihood estimators for functional and structural dependent elliptical models using an orthogonal parameterization in Arellano-Valle and Bolfarine (1996).

Collecting the above results, we have:

THEOREM 4.2. *Consider the WNDE dependent elliptical MEM in* (4.1), *with the assumption that* $\sigma_e^2 = \sigma_u^2 = \sigma^2$ *and the improper prior distribution specified by* $[\boldsymbol{\xi}, \boldsymbol{\beta}, \sigma^2] \propto \frac{1}{\sigma^2}$. *Then, it follows that the modal estimates of* β_1, β_2 *and* σ^2 *are as given in* (4.12) *and* (4.13).

Hence, we have proved that the GLSE of $\boldsymbol{\beta} = (\beta_1, \beta_2)$ given by (4.12) is the Bayes (modal) estimate in the class of dependent elliptical MEM under the above assumptions. Lindley and El Sayad (1968) concluded that the GLSE is the large sample Bayes estimator of $\boldsymbol{\beta}$ for the normal linear MEM. We have shown that this result indeed holds more generally, that is, within the class of WNDE dependent elliptical MEM.

5. Independent elliptical MEM

As considered in Section 2, two types of independent elliptical models for $[\boldsymbol{y}, \boldsymbol{x}|\boldsymbol{\xi}, \boldsymbol{\theta}]$ may be specified:

(iii) The WNDE independent elliptical MEM defined by

$$[\boldsymbol{y}, \boldsymbol{x}|\boldsymbol{\xi}, \boldsymbol{\theta}] = \prod_{i=1}^{n} \left(\frac{1}{\sigma_e^2 \sigma_u^2}\right)^{1/2} h\left(\frac{(y_i - \beta_1 - \beta_2 \xi_i)^2}{\sigma_e^2} + \frac{(x_i - \xi_i)^2}{\sigma_u^2}\right), \tag{5.1}$$

for some bivariate generator density h; and

(iv) The NDE independent elliptical MEM defined by

$$[\boldsymbol{y}, \boldsymbol{x}|\boldsymbol{\xi}, \boldsymbol{\theta}] = [\boldsymbol{y}|\boldsymbol{\xi}, \boldsymbol{\theta}][\boldsymbol{x}|\boldsymbol{\xi}, \boldsymbol{\theta}], \tag{5.2}$$

with

$$[\boldsymbol{y}|\boldsymbol{\xi}, \boldsymbol{\theta}] = \prod_{i=1}^{n} \left(\frac{1}{\sigma_e^2}\right)^{1/2} h_1\left(\frac{(y_i - \beta_1 - \beta_2 \xi_i)^2}{\sigma_e^2}\right) \quad \text{and}$$

$$[\boldsymbol{x}|\boldsymbol{\xi}, \boldsymbol{\theta}] = \prod_{i=1}^{n} \left(\frac{1}{\sigma_u^2}\right)^{1/2} h_2\left(\frac{(x_i - \xi_i)^2}{\sigma_u^2}\right),$$

for some univariate generator densities h_1 and h_2.

No general results are possible for such independent elliptical models. However, under the representable class of elliptical distributions we can introduce general procedures to obtain these conditional posterior distributions in order to apply large sample methods or MCMC methodology. In fact, it is well known that a representable elliptical random vector $\boldsymbol{w}|\boldsymbol{\mu}, \boldsymbol{\Sigma} \sim EL_p(\boldsymbol{\mu}, \Sigma; h)$ can be specified following two steps:

(a) $[\boldsymbol{w}|v, \boldsymbol{\mu}, \boldsymbol{\Sigma}] \sim N_p(\boldsymbol{\mu}, v\Sigma)$; and
(b) $[v] \sim H$,

where H is a cumulative distribution function that does not dependent on p and such that $H(0) = 0$.

In order to illustrate the MCMC approach, we concentrate the Bayesian inference for the case of the independent Student-t MEM, which certainly is one of the most important representable models.

5.1. Representable elliptical MEM

A general specification for the functional representable elliptical MEM is given by

$$[y_i|v_{1i}, \xi_i, \boldsymbol{\theta}] \overset{\text{ind}}{\sim} N(\alpha + \beta\xi_i, v_{1i}\sigma_e^2), \quad \text{with } [v_{i1}] \overset{\text{i.i.d.}}{\sim} H_1, \tag{5.3}$$

$$[x_i|v_{2i}, \xi_i, \boldsymbol{\theta}] \overset{\text{ind}}{\sim} N(\xi_i, v_{2i}\sigma_u^2), \quad \text{with } [v_{i2}] \overset{\text{i.i.d.}}{\sim} H_2, \tag{5.4}$$

where H_j is a cumulative distribution function such that $H_j(0) = 0$, $j = 1, 2$. In the special case of $v_{1i} = v_{2i} = v_i$, with $[v_i] \overset{\text{i.i.d.}}{\sim} H$, $i = 1, \ldots, n$, the WNDE *independent representable elliptical* MEM follows. On the other hand, if $v_{1i} \neq v_{2i}$, the NDE *independent representable elliptical* MEM follows. In addition, we can consider a representable elliptical prior distribution for ξ_i by specifying that

$$[\xi_i|v_{3i}, \mu_\xi, \sigma_\xi^2] \overset{\text{i.i.d.}}{\sim} N(\mu_\xi, v_{3i}\sigma_\xi^2), \quad \text{with } [v_{i3}] \overset{\text{i.i.d.}}{\sim} H_3, \tag{5.5}$$

where H_3 is a cumulative distribution function such that $H_3(0) = 0$. With this prior representation, the MEM obtained when in (5.3)–(5.5) it is assumed that $v_{ji} = v_i$, $j = 1, 2, 3$, with $[v_i] \overset{\text{i.i.d.}}{\sim} H$, $i = 1, \ldots, n$, is equivalent to the WNDE *independent structural elliptical* MEM given by

$$[y_i, x_i, \xi_i|\boldsymbol{\theta}, \mu_\xi, \sigma_\xi^2] \overset{\text{i.i.d.}}{\sim} El_3(\boldsymbol{\eta}, \boldsymbol{\Omega}; h^{(3)}), \tag{5.6}$$

$i = 1, \ldots, n$, where $h^{(k)}$ is just a k-dimensional generator density function associated with the cumulative distribution function H, and

$$\boldsymbol{\eta} = \begin{pmatrix} \alpha + \beta\mu_\xi \\ \mu_\xi \\ \mu_\xi \end{pmatrix} \quad \text{and} \quad \boldsymbol{\Omega} = \begin{pmatrix} \beta^2\sigma_\xi^2 + \sigma_e^2 & \beta\sigma_\xi^2 & \beta\sigma_\xi^2 \\ \beta\sigma_\xi^2 & \sigma_\xi^2 + \sigma_u^2 & \beta\sigma_\xi^2 \\ \beta\sigma_\xi^2 & \beta\sigma_\xi^2 & \sigma_\xi^2 \end{pmatrix}. \tag{5.7}$$

As a consequence, we have the marginal observed structural independent elliptical MEM given by

$$[y_i, x_i|\boldsymbol{\theta}, \mu_\xi, \sigma_\xi^2] \overset{\text{i.i.d.}}{\sim} El_2(\boldsymbol{\mu}, \Sigma; h^{(2)}), \tag{5.8}$$

$i = 1, \ldots, n$, where

$$\boldsymbol{\mu} = \begin{pmatrix} \alpha + \beta\mu_\xi \\ \mu_\xi \end{pmatrix} \quad \text{and} \quad \Sigma = \begin{pmatrix} \beta^2\sigma_\xi^2 + \sigma_e^2 & \beta\sigma_\xi^2 \\ \beta\sigma_\xi^2 & \sigma_\xi^2 + \sigma_u^2 \end{pmatrix}, \tag{5.9}$$

with likelihood function

$$[\boldsymbol{y}, \boldsymbol{x}|\boldsymbol{\theta}, \mu_\xi, \sigma_\xi^2] = \prod_{i=1}^{n} |\Sigma|^{-1/2} h^{(2)}\big((\boldsymbol{d}_i - \boldsymbol{\mu})^{\mathrm{T}} \Sigma^{-1} (\boldsymbol{d}_i - \boldsymbol{\mu})\big), \tag{5.10}$$

where $\boldsymbol{d}_i = (y_i, x_i)^{\mathrm{T}}$.

Note that the above model is different from the WNDE dependent elliptical model specified by

$$[\boldsymbol{y}, \boldsymbol{x}, \boldsymbol{\xi}|\boldsymbol{\theta}, \mu_\xi, \sigma_\xi^2] \sim El_{3n}(\mathbf{1}_n \otimes \boldsymbol{\eta}, \boldsymbol{I}_n \otimes \boldsymbol{\Omega}; h^{(3n)}),$$

which yields the structural marginal observed model

$$\left[y, x|\theta, \mu_\xi, \sigma_\xi^2\right] \sim El_{2n}\left(\mathbf{1}_n \otimes \boldsymbol{\mu}, \boldsymbol{I}_n \otimes \Sigma; h^{(2n)}\right),$$

with structural likelihood function

$$\left[y, x|\theta, \mu_\xi, \sigma_\xi^2\right] = |\Sigma|^{-n/2} h^{(2n)}\left(\sum_{i=1}^{n}(z_i - \boldsymbol{\mu})^{\mathrm{T}} \Sigma^{-1}(z_i - \boldsymbol{\mu})\right).$$

The most popular representable elliptical model is the Student-t model, which follows by considering an inverted gamma for the mixing distribution. Two types of independent Student-t MEM are discussed in the following sections. We illustrate a general methodology based on MCMC from these models.

5.2. A WNDE Student-t model

An important WNDE Student-t MEM follows by considering the mixture representation in (5.3)–(5.5), with $v_{ji} = v_i$, $j = 1, 2, 3$, and

$$[v_i] \overset{\text{i.i.d.}}{\sim} IG\left(\frac{\nu}{2}, \frac{\nu}{2}\right), \tag{5.11}$$

$i = 1, \ldots, n$, which yields the following 3-dimensional Student-t distribution in specification (5.6):

$$[y_i, x_i, \xi_i|\theta, \mu_\xi, \sigma_\xi^2] \overset{\text{i.i.d.}}{\sim} t_3(\boldsymbol{\eta}, \boldsymbol{I}_3 \otimes \boldsymbol{\Omega}; \nu), \tag{5.12}$$

$i = 1, \ldots, n$, with $\boldsymbol{\eta}$ and $\boldsymbol{\Omega}$ as in (5.7), implying in (5.8) the structural observed bivariate Student-t model

$$\left[y_i, x_i|\theta, \mu_\xi, \sigma_\xi^2\right] \overset{\text{i.i.d.}}{\sim} t_2(\boldsymbol{\eta}, \boldsymbol{I}_3 \otimes \boldsymbol{\Omega}; \nu), \tag{5.13}$$

$i = 1, \ldots, n$, so that the likelihood function that follows from (5.10) is given by

$$\begin{aligned}
&\left[y, x|\theta, \mu_\xi, \sigma_\xi^2\right] \\
&\quad = (2\pi)^{-n}|\boldsymbol{\Sigma}|^{-n/2}\nu^{n(\nu+2)/2}\prod_{i=1}^{n}\left\{\nu + (z_i - \boldsymbol{\mu})^{\mathrm{T}}\boldsymbol{\Sigma}^{-1}(z_i - \boldsymbol{\mu})\right\}^{-(\nu+2)/2},
\end{aligned} \tag{5.14}$$

with $z_i = (y_i, x_i)^{\mathrm{T}}$ and $\boldsymbol{\mu}$ and Σ are as in (5.9). Hence, considering μ_ξ and σ_ξ^2 known, the posterior distribution of $\boldsymbol{\theta}$ is given by

$$\left[\theta|y, x, \mu_\xi, \sigma_\xi^2\right] = \left[y, x|\theta, \mu_\xi, \sigma_\xi^2\right][\theta], \tag{5.15}$$

where $[y, x|\theta, \mu_\xi, \sigma_\xi^2]$ is given by (5.14) and $[\theta]$ is the prior distribution for θ. Notice that it is somewhat complicated to deal with the posterior distribution (5.15) in the sense of obtaining the marginal posterior distribution for β_2. Modal estimators, under noninformative priors, can be computed by directly maximizing the posterior distribution or by using the EM algorithm as done in Bolfarine and Arellano-Valle (1994).

To implement an alternative estimating approach for the above model, we consider the functional version given by (5.3)–(5.4) and the prior specification in (5.5) with $v_{ij} =$

v_i, $j = 1, 2, 3$, that is,

$$[y_i | v_i, \xi_i, \boldsymbol{\theta}] \overset{\text{ind}}{\sim} N\big(\beta_1 + \beta_2 \xi_i, v_i \sigma_e^2\big), \tag{5.16}$$

$$[x_i | v_i, \xi_i, \boldsymbol{\theta}] \overset{\text{ind}}{\sim} N\big(\xi_i, v_i \sigma_u^2\big), \tag{5.17}$$

$$\big[\xi_i | v_i, \mu_\xi, \sigma_\xi^2\big] \overset{\text{i.i.d.}}{\sim} N\big(\mu_\xi, v_i \sigma_\xi^2\big), \tag{5.18}$$

with v_i distributed as in (5.11), $i = 1, \ldots, n$. The parameters μ_ξ and σ_ξ^2 are considered known. If they are considered unknown, then prior distributions need to be assigned for them. We assume also that β_1, β_2, σ_e^2, σ_u^2 and (ξ_i, v_i), $i = 1, \ldots, n$ are a priori independent, and we consider the following prior distributions for the structural parameters:

$$[\beta_j] \sim N(a_j, b_j), \quad j = 1, 2, \tag{5.19}$$

$$\big[\sigma_k^2\big] \sim IG\left(\frac{c_k}{2}, \frac{d_k}{2}\right), \quad k = e, u. \tag{5.20}$$

The joint distribution that follows from (5.16)–(5.20) is then

$$\begin{aligned}
&\big[\boldsymbol{y}, \boldsymbol{x}, \boldsymbol{\xi}, \boldsymbol{v}, \beta_1, \beta_2, \sigma_e^2, \sigma_u^2 | \mu_\xi, \sigma_x\big] \\
&= \prod_{i=1}^{n} \big[y_i | v_i, \xi_i, \beta_1, \beta_2, \sigma_e^2\big]\big[x_i | v_i, \xi_i, \sigma_u^2\big]\big[\xi_i | v_i, \mu_\xi, \sigma_\xi^2\big][v_i] \\
&\quad \times [\beta_1][\beta_2]\big[\sigma_e^2\big]\big[\sigma_u^2\big] \\
&\propto \left(\frac{1}{\sigma_e^2}\right)^{n/2} \exp\left\{-\frac{1}{2\sigma_e^2} \sum_{i=1}^{n} \frac{(y_i - \beta_1 - \beta_2 \xi_i)^2}{v_i}\right\} \\
&\quad \times \left(\frac{1}{\sigma_u^2}\right)^{n/2} \exp\left\{-\frac{1}{2\sigma_u^2} \sum_{i=1}^{n} \frac{(x_i - \xi_i)^2}{v_i}\right\} \\
&\quad \times \left(\frac{1}{\sigma_\xi^2}\right)^{n/2} \exp\left\{-\frac{1}{2\sigma_\xi^2} \sum_{i=1}^{n} \frac{(\xi_i - \mu_\xi)^2}{v_i}\right\} \\
&\quad \times \prod_{i=1}^{n} \left(\frac{1}{v_i}\right)^{\nu/2+2} \exp\left\{-\sum_{i=1}^{n} \frac{\nu}{2v_i}\right\} \\
&\quad \times \exp\left\{-\frac{(\beta_1 - a_1)^2}{2b_1}\right\} \exp\left\{-\frac{(\beta_2 - a_2)^2}{2b_2}\right\} \\
&\quad \times \left(\frac{1}{\sigma_e^2}\right)^{c_e/2+1} \exp\left\{-\frac{d_e}{2\sigma_e^2}\right\} \left(\frac{1}{\sigma_u^2}\right)^{c_u/2+1} \exp\left\{-\frac{d_u}{2\sigma_u^2}\right\},
\end{aligned}$$

where $\boldsymbol{v} = (v_1, \ldots, v_n)^{\mathrm{T}}$. It can be shown that the Gibbs sampler algorithm is implemented by considering the following conditional distributions:

$$\big[\beta_1 | \beta_2, \boldsymbol{\sigma}^2, \boldsymbol{\xi}, \boldsymbol{v}, \boldsymbol{y}, \boldsymbol{x}\big] \sim N\left(\frac{\sum_{i=1}^{n} \frac{y_i - \beta_2 \xi_i}{\sigma_e^2 v_i} + \frac{a_1}{b_1}}{\sum_{i=1}^{n} \frac{1}{\sigma_e^2 v_i} + \frac{1}{b_1}}, \frac{1}{\sum_{i=1}^{n} \frac{1}{\sigma^2 v_i} + \frac{1}{b_1}}\right),$$

$$[\beta_2|\beta_1,\sigma^2,\boldsymbol{\xi},\boldsymbol{v},\boldsymbol{y},\boldsymbol{x}] \sim N\left(\frac{\sum_{i=1}^n \frac{(y_i-\beta_1)\xi_i}{\sigma_u^2 v_i} + \frac{a_2}{b_2}}{\sum_{i=1}^n \frac{\xi_i^2}{\sigma_u^2 v_i} + \frac{1}{b_2}}, \frac{1}{\sum_{i=1}^n \frac{\xi_i^2}{\sigma_u^2 v_i} + \frac{1}{b_2}}\right),$$

$$[\sigma_e^2|\beta_1,\beta_2,\sigma_u^2,\boldsymbol{\xi},\boldsymbol{v},\boldsymbol{y},\boldsymbol{x}] \sim IG\left(\frac{n+c_e+1}{2}, \sum_{i=1}^n \frac{(y_i-\beta_1-\beta_2\xi_i)^2}{2v_i} + \frac{d_e}{2}\right),$$

$$[\sigma_u^2|\beta_1,\beta_2,\sigma_e^2,\boldsymbol{\xi},\boldsymbol{v},\boldsymbol{y},\boldsymbol{x}] \sim IG\left(\frac{n+c_u+1}{2}, \sum_{i=1}^n \frac{(xi_i-\xi_i)^2}{2v_i} + \frac{d_u}{2}\right),$$

$$[\xi_i|\beta_1,\beta_2,\sigma^2,\boldsymbol{\xi}_{-i},\boldsymbol{v},\boldsymbol{y},\boldsymbol{x}] \overset{\text{ind}}{\sim} N\left(\frac{\frac{(y_i-\beta_1)\beta_2}{\sigma_e^2 v_i} + \frac{x_i}{\sigma_u^2 v_i} + \frac{\mu_\xi}{\sigma_\xi^2 v_i}}{\frac{\beta_2^2}{\sigma_e^2 v_i} + \frac{1}{\sigma_u^2 v_i} + \frac{1}{\sigma_\xi^2 v_i}}, \frac{1}{\frac{\beta_2^2}{\sigma_e^2 v_i} + \frac{1}{\sigma_u^2 v_i} + \frac{1}{\sigma_\xi^2 v_i}}\right),$$

$i = 1,\dots,n$, where $\boldsymbol{a}_{-i} = (a_1,\dots,a_{i-1},a_{i+1},\dots,a_n)^{\mathrm{T}}$, for any vector $\boldsymbol{a} = (a_1\dots,a_n)^{\mathrm{T}}$. If we consider that $v_i = 1$, $i = 1,\dots,n$, then the normal model follows. If

$$[v_i|\nu] \overset{\text{i.i.d.}}{\sim} IG\left(\frac{\nu}{2},\frac{\nu}{2}\right),$$

$i = 1,\dots,n$, then the WNDE Student-t model with ν degrees of freedom follows. In this case, the conditional posterior distribution of v_i is

$$[v_i|\beta_1,\beta_2,\sigma_e^2,\sigma_u^2,\boldsymbol{\xi},\boldsymbol{v}_{-i},\boldsymbol{y},\boldsymbol{x}] \overset{\text{ind}}{\sim} IG\left(\frac{\nu+1}{2}, \frac{\nu}{2} + \frac{(y_i-\beta_1-\beta_2\xi_i)^2}{2\sigma_e^2} + \frac{(x_i-\xi_i)^2}{2\sigma_u^2} + \frac{(x_i-\mu_\xi)^2}{2\sigma_\xi^2}\right),$$

$i = 1,\dots,n$. In the case where ν is unknown with $\nu \sim \exp(a)$, the conditional posterior distribution is given by

$$[\nu|\beta_1,\beta_2,\sigma_e^2,\sigma_u^2,\boldsymbol{v},\boldsymbol{\xi},\boldsymbol{y},\boldsymbol{x}] \propto \exp\left\{\sum_{i=1}^n\left[\frac{\nu}{v_i} + \left(\frac{\nu}{2}+1\right)\log\left(\frac{\nu}{v_i}\right)\right] - a\nu\right\}.$$

The algorithm starts with initial values for β_1 and $\boldsymbol{\xi}$ (all zeros, for example), β_2, σ_e^2, σ_u^2 and $\boldsymbol{v}$ (all ones, for example) and cycles, generating samples from the above conditional distributions, until convergence, which has to be checked by special procedures.

5.3. A NDE Student-t model

In this section, we consider the independent NDE Student-t model that follows by considering $H_j = IG(\nu_j/2,\nu_j/2)$, $j = 1,2,3$, in (5.3)–(5.5), which is equivalent to considering in (1.1)–(1.2) that e_i, u_i and ξ_i are independent with $[e_i|\sigma_e^2] \overset{\text{i.i.d.}}{\sim} t_1(0,\sigma_e^2,\nu_1)$,

$[u_i|\sigma_u^2] \overset{\text{i.i.d.}}{\sim} t_1(0, \sigma_u^2, \nu_2)$ and $[\xi_i|\mu_\xi, \sigma_\xi^2] \overset{\text{i.i.d.}}{\sim} t_1(\mu_\xi, \sigma_\xi^2, \nu_2)$, $i = 1, \ldots, n$. On combining above assumptions with (1.1)–(1.2), it then follows that

$$[y_i|\xi_i, \boldsymbol{\theta}] \overset{\text{ind}}{\sim} t_1(\alpha + \beta\xi_i, \sigma_e^2, \nu_1), \tag{5.21}$$

$$[x_i|\xi_i, \boldsymbol{\theta}] \overset{\text{ind}}{\sim} t_1(\xi_i, \sigma_u^2, \nu_2) \tag{5.22}$$

and

$$[\xi_i|\mu_\xi, \sigma_\xi^2] \overset{\text{i.i.d.}}{\sim} t_1(\mu_\xi, \sigma_\xi^2, \nu_2), \tag{5.23}$$

$i = 1, \ldots, n$.

To conduct the Bayesian analysis of (5.21)–(5.23) we have to implement MCMC type approaches since a full Bayesian approach is difficult to carry on for such models. Even asymptotic Bayesian inference as carried out in the previous section for the independent WNDE Student-t MEM is complicated. One special MCMC type approach which requires only the specification of the conditional posterior distribution for each parameter is the Gibbs sampler. In situations where those distributions are simple to sample from, as in the present case, the approach is easily implemented. In other situations, the more complex Metropolis–Hastings approach needs to be considered (see Gamerman, 1997). Thus, to implement the Gibbs sampler approach we need to obtain the conditional posterior distributions. This approach is facilitated by considering the normal mixture representation of the MEM used in the previous section. That is, we write the independent NDE Student-t MEM in (5.21)–(5.23) as

$$[y_i|v_{1i}, \xi_i, \boldsymbol{\theta}] \overset{\text{ind}}{\sim} N(\beta_1 + \beta_2\xi_i, v_{1i}\sigma_e^2), \tag{5.24}$$

$$[x_i|v_{2i}, \xi_i, \boldsymbol{\theta}] \overset{\text{ind}}{\sim} N(\xi_i, v_{2i}\sigma_u^2), \tag{5.25}$$

$$[\xi_i|v_{3i}, \mu_\xi, \sigma_\xi^2] \overset{\text{ind}}{\sim} N(\mu_\xi, v_{3i}\sigma_\xi^2), \tag{5.26}$$

with

$$[v_{ji}|\nu_j] \overset{\text{i.i.d.}}{\sim} IG\left(\frac{\nu_j}{2}, \frac{\nu_j}{2}\right), \tag{5.27}$$

$j = 1, 2, 3$ and $i = 1, \ldots, n$.

Thus, considering the same independence assumptions of the previous section and the prior specifications in (5.19)–(5.20) for the structural parameter, we obtain the following conditional posterior distribution for the case where μ_ξ and σ_ξ^2 are considered known:

$$[\beta_1|\beta_2, \boldsymbol{\sigma}^2, \boldsymbol{\xi}, \boldsymbol{v}, \boldsymbol{y}, \boldsymbol{x}] \sim N\left(\frac{\sum_{i=1}^n \frac{y_i - \beta_2\xi_i}{\sigma_e^2 v_{1i}} + \frac{a_1}{b_1}}{\sum_{i=1}^n \frac{1}{\sigma_e^2 v_{1i}} + \frac{1}{b_1}}, \frac{1}{\sum_{i=1}^n \frac{1}{\sigma^2 v_{1i}} + \frac{1}{b_1}}\right),$$

$$[\beta_2|\beta_1, \boldsymbol{\sigma}^2, \boldsymbol{\xi}, \boldsymbol{v}, \boldsymbol{y}, \boldsymbol{x}] \sim N\left(\frac{\sum_{i=1}^n \frac{(y_i - \beta_1)\xi_i}{\sigma_u^2 v_{1i}} + \frac{a_2}{b_2}}{\sum_{i=1}^n \frac{\xi_i^2}{\sigma_u^2 v_{1i}} + \frac{1}{b_2}}, \frac{1}{\sum_{i=1}^n \frac{\xi_i^2}{\sigma_u^2 v_{1i}} + \frac{1}{b_2}}\right),$$

$$
\begin{aligned}
&\left[\sigma_e^2 | \beta_1, \beta_2, \sigma_u^2, \boldsymbol{\xi}, \boldsymbol{v}, \boldsymbol{y}, \boldsymbol{x}\right] \\
&\quad \sim IG\left(\frac{n + c_e + 1}{2}, \sum_{i=1}^{n} \frac{(y_i - \beta_1 - \beta_2 \xi_i)^2}{2 v_{1i}} + \frac{d_e}{2}\right), \\
&\left[\sigma_u^2 | \beta_1, \beta_2, \sigma_e^2, \boldsymbol{\xi}, \boldsymbol{v}, \boldsymbol{y}, \boldsymbol{x}\right] \\
&\quad \sim IG\left(\frac{n + c_u + 1}{2}, \sum_{i=1}^{n} \frac{(x_i - \xi_i)^2}{2 v_{2i}} + \frac{d_u}{2}\right),
\end{aligned}
$$

and

$$
\begin{aligned}
&\left[\xi_i | \beta_1, \beta_2, \boldsymbol{\sigma}^2, \boldsymbol{\xi}_{-i}, \boldsymbol{v}, \boldsymbol{y}, \boldsymbol{x}\right] \\
&\quad \overset{\text{ind}}{\sim} N\left(\frac{\frac{(y_i - \beta_1)\beta_2}{\sigma_e^2 v_{1i}} + \frac{x_i}{\sigma_u^2 v_{2i}} + \frac{\mu_\xi}{\sigma_\xi^2 v_{3i}}}{\frac{\beta_2^2}{\sigma_e^2 v_{1i}} + \frac{1}{\sigma_u^2 v_{2i}} + \frac{1}{\sigma_\xi^2 v_{3i}}}, \frac{1}{\frac{\beta_2^2}{\sigma_e^2 v_{1i}} + \frac{1}{\sigma_u^2 v_{2i}} + \frac{1}{\sigma_\xi^2 v_{3i}}}\right),
\end{aligned}
$$

$i = 1, \ldots, n$, where $\boldsymbol{\sigma}^2 = (\sigma_e^2, \sigma_u^2)^{\mathrm{T}}$.

In this case, the conditional posterior distribution of v_{ji} is

$$
\begin{aligned}
v_{1i} | \beta_1, \beta_2, \boldsymbol{\sigma}^2, \boldsymbol{\xi}, \boldsymbol{v}_{(-j)}, \boldsymbol{y} &\overset{\text{ind}}{\sim} IG\left(\frac{\nu_1 + 1}{2}, \left(\frac{y_i - \beta_1 - \beta_2 \xi_i}{\sigma_e}\right)^2\right), \\
v_{2i} | \beta_1, \beta_2, \boldsymbol{\sigma}^2, \boldsymbol{\xi}, \boldsymbol{v}_{(-j)}, \boldsymbol{y} &\overset{\text{ind}}{\sim} IG\left(\frac{\nu_2 + 1}{2}, \left(\frac{x_i - \xi_i}{\sigma_u}\right)^2\right), \\
v_{3i} | \beta_1, \beta_2, \boldsymbol{\sigma}^2, \boldsymbol{\xi}, \boldsymbol{v}_{(-j)}, \boldsymbol{y} &\overset{\text{ind}}{\sim} IG\left(\frac{\nu_3 + 1}{2}, \left(\frac{\xi_i - \mu_\xi}{\sigma_\xi}\right)^2\right),
\end{aligned}
$$

$i = 1, \ldots, n$. In the case where ν is unknown with $\nu \sim \exp(a)$, the conditional posterior distribution is given by

$$
\nu | \boldsymbol{v}, \boldsymbol{y} \propto \exp\left\{\sum_{j=1}^{3} \sum_{i=1}^{n} \left[\frac{\nu}{v_{ji}} + \left(\frac{\nu}{2} + 1\right) \log\left(\frac{\nu}{v_{ji}}\right)\right] - a\nu\right\}.
$$

The case $\sigma_e^2 = \lambda \sigma_u^2$ with λ known can be dealt with similarly. Starting values for initializing the algorithm can be chosen as in Section 5.2.

6. Application

In this section we consider an application of the results considered in the previous sections to a real data set reported in Kelly (1984). The data set is presented in Miller (1980), and is related to simultaneous pairs of measurements of serum kanamycin level in blood samples drawn from twenty babies. One of the measurements was obtained by a heelstick method (X) and the other by using an umbilical catheter (Y). The basic question was whether the catheter values systematically differed from the heelstick values. Given the special nature of the data, it was reasoned that $\lambda = 1.0$ was correct. We present now a reanalysis of the data set by using a Student-t structural model described earlier. Considering first that the data follows a dependent elliptical MEM model with equal variances, modal estimates of β_2 and σ^2 that follow from (4.12) and (4.13), with noninformative prior (4.9) are $\hat{\beta}_2 = 1.071$ and $\hat{\sigma}^2 = 3.904$, which are also modal estimators under the normality assumption. Under the NDE Student-t model considered in Section 5.3, with $a_1 = 0.0$, $b_1 = 4.0$, $a_2 = 1.0$, $b_2 = 0.2$, and $\sigma = \sigma_e = \sigma_u \sim IG(c, d)$, with $c = 4.0$, $d = 3.0$, we obtain the following estimators:

Table 1
Posterior summaries for selected parameters

	Median	Mean	MC error	95% HPD
β_1	−0.121	−0.133	0.0026	(−0.534, 0.477)
β_2	1.112	1.113	0.0002	(0.977, 1.142)
σ^2	3.948	4.187	2.0055	(2.060, 6.820)
ν	10.030	10.033	0.975	(8.914, 10.813)

Convergence of the Gibbs algorithm was verified graphically and using the approach in Gelman and Rubin (1992). As seen from the table, a Student-t distribution with $\nu = 10$ degrees of freedom seems to present the best fit. Moreover, $\hat{\beta} = 1.07$ which is the estimate reported by Kelly (1984) corresponds to a very large value of ν (normality). On the other hand, at $\nu = 10$, $\hat{\beta} = 1.113$, indicating accommodation of extreme observations present in the data set. As in Kelly (1984) the hypothesis $\alpha = 0$ and $\beta = 1$ are not rejected, indicating that the two methods are equally efficient.

Acknowledgements

Authors acknowledge unknown referee for carefully reading and revising the paper. Partial financial support from CNPq (Brasil) and Fondecyt (Chile), Grant No. 1040865, is also acknowledged.

References

Arellano, R.B., Galea, M., Iglesias, P. (2000). Bayesian sensitivity analysis in elliptical linear regression models. *J. Statist. Plann. Inference* **86**, 175–199.

Arellano, R.B., Galea, M., Iglesias, P. (2002). Bayesian analysis in elliptical linear regression models. *Rev. Soc. Chilena Estadíst.* **16–17**, 59–100. (With discussion).

Arellano-Valle, R.B. (1994). Distribuições elípticas: propriedades, inferência e aplicações a modelos de regressão. Doctoral Thesis, IME–USP, Brasil.

Arellano-Valle, R.B., Bolfarine, H. (1995). On some characterizations of the t-distribution. *Statist. Probab. Lett.* **25**, 79–85.

Arellano-Valle, R.B., Bolfarine, H. (1996). Elliptical structural models. *Comm. Statist. Theory Methods* **25** (10), 2319–2342.

Arellano-Valle, R.B., Bolfarine, H., Vilca-Labra, F. (1994). Ultrastructural elliptical models. *Canad. J. Statist.* **24**, 207–216.

Arellano-Valle, R.B., del Pino, G., Iglesias, P.L. (2005). Bayesian inference for spherical linear models. *J. Multivariate Anal.* Submitted for publication.

Bolfarine, H., Arellano-Valle, R.B. (1994). Robust modeling in measurement error models using the t distribution. *Brazilian J. Probab. Statist.* **8** (1), 67–84.

Bolfarine, H., Arellano-Valle, R.B. (1998). Weak nondifferential measurement error models. *Statist. Probab. Lett.* **40**, 279–287.

Bolfarine, H., Cordani, L. (1993). Estimation of a structural regression model with known reliability ratio. *Ann. Inst. Statist. Math.* **45**, 531–540.

Bolfarine, H., Gasco, L., Iglesias, P. (1999). Pearson type II error-in-variables-models. In: Bernardo, J.M., Berger, J.O., David, A.P., Smith, A.F.M. (Eds.), *Bayesian Statistics, vol. VI*, pp. 713–722.

Carroll, R.J., Ruppert, D., Stefanski, L.A. (1995). *Measurement Error in Nonlinear Models*. Chapman & Hall, London.

Cheng, C., Van Ness, J. (1999). *Statistical Regression with Measurement Error*. Arnold, London.

Dellaportas, P., Stephens, D.A. (1995). Bayesian analysis of error in variable regression models. *Biometrics* **51**, 1085–1095.

Fang, K.T., Kotz, S., Ng, K.W. (1990). *Symmetric Multivariate and Related Distributions*. Chapman and Hall, London.

Florens, J.P., Mouchart, M., Richard, J.F. (1974). Bayesian inference in error-in-variable models. *J. Multivariate Anal.* **4**, 419–452.

Fuller, W. (1987). *Measurement Error Models*. Wiley, New York.

Gamerman, D. (1997). *Markov Chain Monte Carlo*. Chapman and Hall.

Gelman, A., Rubin, D. (1992). Inference from iterative simulation using multiple sequences (with discussion). *Statist. Sci.* **7**, 457–511.

Kelly, G. (1984). The influence function in the error in variables problem. *Ann. Statist.* **12**, 87–100.

Lindley, D.V., El Sayad, G.A. (1968). The Bayesian estimation of a linear functional relationship. *J. Roy. Statist. Soc., Ser. B* **30**, 190–202.

Miller, R.G.Jr. (1980). Kanamycin levels in premature babies. In: *Biostatistics Casebook*, vol. III, Technical Report No. 57, Division of Statistics, Stanford University, pp. 127–142.

Reilly, P., Patino-Leal, H. (1981). A Bayesian study of the error in variable models. *Technometrics* **23** (3), 221–231.

Richardson, S., Gilks, W.R. (1993). A Bayesian approach to measurement error problems in epidemiological studies with covariate measurement error. *Statistics and Medicine* **12**, 1703–1722.

Vilca-Labra, F., Bolfarine, H., Arellano-Valle, R.B. (1998). Elliptical functional models. *J. Multivariate Anal.* **65**, 36–57.

Villegas, C. (1972). Bayesian inference in linear relations. *Ann. Math. Statist.* **43**, 1767–1791.

Zellner, A. (1971). *An Introduction to Bayesian Inference in Econometrics*. Wiley, New York.

Handbook of Statistics, Vol. 25
ISSN: 0169-7161
© 2005 Elsevier B.V. All rights reserved.
DOI 10.1016/S0169-7161(05)25023-X

23

Bayesian Sensitivity Analysis in Skew-elliptical Models

I. Vidal, P. Iglesias and M.D. Branco

Abstract

The main objective of this chapter is to investigate the influence of introducing skewness parameter in elliptical models. First, we review definitions and properties of skew distributions considered in the literature with emphasis on the so-called skew elliptical distributions. For univariate skew-normal models we study the influence of the skew parameter on the posterior distributions of the location and scale parameters. The influence is quantified by evaluating the L_1 distance between the posterior distributions obtained under the skew-normal model and normal model respectively. We then examine the problem of computing Bayes factors to test skewness in linear regression models, evaluating the performance trough simulations.

Keywords: skew models; Bayesian inference; L_1 distance; Bayes factor

1. Introduction

This chapter deals with Bayesian sensitivity analysis of statistical models involving skew distributions. Different forms of defining skew distribution has been considered recently in the literature (see Arellano-Valle and Genton, 2005, for a review), which are in general constructed from departures of symmetric distributions or multivariate elliptical distributions.

Researches from different areas have paid attention to the study of definitions and probabilistic properties of distribution families including the normal distribution. The aim is to generate statistical models more flexible and suitable to describe practical situations, which traditionally have been treated using the normal theory. On the other hand the computational advances have facilitated the work with more complex models. Particularly, the Bayesian analysis of this models has been beneficiary with the development of the Markov Chain Monte Carlo (MCMC) methods.

Departures from normality can be generated in several ways. In the univariate case, the first extensions from normal distributions family is constructed through a mixture in the scale parameter of normal distributions. The heaviness of the distribution tails depends on the choice of the mixture measure. Contaminated normal distributions and Student-t distributions are well known special cases. Formally, a random variable X

has a distribution which is a mixture of normal distributions with scale parameter η and mixture measure defined by G if its probability density function (p.d.f.) can be written as

$$f(x|\eta, G) = \sqrt{\eta}\int_0^\infty (2\pi v)^{-1/2}\, \mathrm{e}^{-\eta x^2/(2v)}\, \mathrm{d}G(v), \tag{1}$$

where G is a cumulative density function (c.d.f) with $G(0) = 0$. Alternatively, we can express (1) as

$$f(x|\eta, G) = \sqrt{\eta}\phi\bigl(x\sqrt{\eta}|0, 1\bigr)\omega(x; \eta, G), \tag{2}$$

where $\phi(\cdot|\mu, \sigma^2)$ is the p.d.f. of the $N(\mu, \sigma^2)$ distribution and

$$\omega(x; \eta, G) = \int_0^\infty v^{-1/2}\, \mathrm{e}^{-\frac{\eta x^2}{2}(\frac{1}{v}-1)}\, \mathrm{d}G(v).$$

This representation is useful for making sensitivity analysis. In fact, computational implementation to assess divergence between models given by (1) and normal models is simplified if we use the representation (2) (see, for example, Weiss, 1996). Note that $\omega(x; \eta, G) = 1$ if, and only if, G is defined by $\delta_{\{1\}}$ (the Dirac measure in $\{1\}$). However, this representation is not so useful when we want to make inferences on model parameters. Random variables with p.d.f. given by (1) can also be characterized by the stochastic representation

$$X \stackrel{d}{=} RY, \tag{3}$$

where $R \geqslant 0$ is a random variable independent of Y with distribution defined by G, and Y has $N(0, \eta)$ distribution. We will employ the usual symbol $\stackrel{d}{=}$ to denote the equality in distribution. This representation has been used for making inferences in regression models with independent errors and distribution with p.d.f. (1) (see, for example, Branco et al., 1998, 2001).

A more general class than that given by (1) is that defined by the p.d.f.

$$f(x|\eta, h) = \sqrt{\eta}h\bigl(\eta x^2\bigr), \tag{4}$$

where $h \geqslant 0$ and it satisfies

$$\int_0^\infty \frac{\sqrt{\pi}}{\Gamma(\frac{1}{2})} u^{-1/2} h(u)\, \mathrm{d}u = 1,$$

that is, a distributions family defined by symmetric densities. Also, the p.d.f. (4) can be expressed as a perturbation from normal distribution,

$$f(x|\eta, \omega) = \sqrt{\eta}\phi\bigl(x\sqrt{\eta}\bigr)\omega\bigl(\eta x^2\bigr), \tag{5}$$

where $\omega \geqslant 0$ satisfies

$$\int_0^\infty \frac{1}{\Gamma(\frac{1}{2})\sqrt{2}} u^{-1/2} \mathrm{e}^{-u/2} \omega(u)\, du = 1.$$

In other words, ω is a function such that $\mathbb{E}[\omega(U)] = 1$ with $U \sim Ga(\frac{1}{2}, \frac{1}{2})$. If $\omega \equiv 1$ we obtain the normal model. The stochastic representation of a random variable X with p.d.f. defined by (5) is given by

$$X \stackrel{d}{=} RU,$$

where $R \geqslant 0$ is a random variable independent of U, and U has uniform distribution on the set $\{-1, 1\}$. Each definition or characterization has its own advantages. In general when a density class is defined, we must take into account that

1. The analytic treatment and/or the computational implementation.
2. The class contains the perturbed distributions family and it is simple to interpret.
3. The new class does not increase the mathematical complexity significantly.

Although the range of applications of (4) is wider than normal family, it allows variations from normality to nonnormality which preserve symmetry. Azzalini (1985) defined a variation of normality including skewness with density given by

$$f(x|\lambda) = 2\phi(x)\Phi(\lambda x), \tag{6}$$

where ϕ and Φ are the p.d.f. and c.d.f. of the $N(0, 1)$ distribution, respectively. By simplicity we will omit the location and scale parameters. Henze (1986) provides the stochastic representation

$$Z \stackrel{d}{=} \frac{\lambda}{\sqrt{1+\lambda^2}}|X| + \frac{1}{\sqrt{1+\lambda^2}}Y, \tag{7}$$

where X and Y are independents $N(0, 1)$ random variables. He applied this representation to get further results about $SN(\lambda)$, as well as to simplify some proofs. We denote that $Z \sim SN(\lambda)$ where λ is the skewness parameter. This class established a starting point for many univariate and multivariate extensions. In fact, as it is shown in Lemma 1 of Azzalini (1985), if f is a symmetric p.d.f. about 0, and G an absolutely continuous distribution function such that its first derivative G' is symmetric about 0, then

$$2f(x)G(\lambda x) \tag{8}$$

is a p.d.f. for any $\lambda \in \mathbb{R}$. The lemma introduces an extended class of skew distributions which includes the skew normal ones and the symmetric distributions. The notation $Z \sim S(f, G, \lambda)$ is typically used for denoting this class of distributions. Skew distributions has been also constructed starting from a stochastic representation following the ideas of Henze (1986). In fact, if Z is a random variable

$$Z \stackrel{d}{=} a|X| + bY,$$

where the distribution of (X, Y) satisfies some assumptions, different classes of distributions are obtained:

1. If (X, Y) is a random vector spherically distributed, then the distribution introduced by Branco and Dey (2001) is obtained.
2. If X and Y are independent random variables with symmetric distribution at zero, the Azzalini and Capitanio (1999) are obtained.
3. If (X, Y) is a $\mathcal{C}$-random vector, that is $(X, Y) \mid |X| = t_1, |Y| = t_2 \sim U\{O_t\}$, where $O_t = \{(x, y) \in \mathbb{R}^2\colon |x| = t_1, |y| = t_2\}$, then we obtain the skew distribution introduced by Arellano-Valle et al. (2002).

Arellano-Valle et al. (2002) present a detailed discussion of different ways to define skew distributions and their probabilistic properties, with emphasis in multivariate skew distributions (see also Liseo and Loperfido, 2003; Gupta et al., 2004). A advantage of definitions based on an stochastic representation is that the computational implementation to Bayesian inference is simplified.

The focus of this chapter is to study the effects of introducing skewness on the Bayesian inference. First, we consider i.i.d. data from distributions with density given by

$$\frac{2}{\sigma} f\left(\frac{x-\mu}{\sigma}\right) G\left(\lambda \frac{x-\mu}{\sigma}\right)$$

and we compare the inference obtained when $\lambda = 0$ versus $\lambda \neq 0$ in terms of the L_1-distance. Secondly, we approach the problem of testing asymmetry in linear regression models by computing Bayes factor and considering the skew distribution introduced by Branco and Dey (2001), hereafter referred to as skew elliptical distributions.

The chapter is organized as follows. In Section 2 we review the properties of skew elliptical distributions and univariate skew distributions given by (8). The influence of skewness parameters on the posterior distribution for the location and scale parameters is examined. Section 3 is devoted to the problem of examining the effects of asymmetry on inferences about the parameter in a linear models. In Section 4 we present simulation results. Some comments are presented in Section 5.

2. Definitions and properties of skew-elliptical distributions

As mentioned in the previous section, skew distributions can be defined in many different ways. Arellano-Valle et al. (2002) presented definitions and probabilistic properties of skew distributions, see also Arellano-Valle and Genton (2005). The objective of this sections is to review additional properties of skew distributions with focus on definition given by Branco and Dey (2001) which will be used on throughout of this chapter.

2.1. *The skew elliptical distribution*

The skew elliptical distribution is a very flexible class of multivariate probability distribution proposed by Azzalini and Capitanio (1999) and Branco and Dey (2001) using different approaches. Although we can not say both propose are equivalent in general, we can see that in many cases they generate the same class of distributions. For more information about the relationship between the two propose see Azzalini and Capitanio (2003).

In this chapter we follow the Branco and Dey (2001) definition and consider the conditioning method to obtain the skew elliptical distribution.

DEFINITION 2.1. Let U_0 be a scalar random variable and $\boldsymbol{U} = (U_1, \ldots, U_d)^t$ a d-dimensional variable such that $(U_0, \boldsymbol{U})^t \sim El_{d+1}(\boldsymbol{0}, \boldsymbol{\Omega}^*, h^{(d+1)})$, $\boldsymbol{\delta} = (\delta_1, \ldots, \delta_d)^t$,

$$\boldsymbol{\Omega}^* = \begin{pmatrix} 1 & \boldsymbol{\delta}^t \\ \boldsymbol{\delta} & \boldsymbol{I}_d \end{pmatrix}$$

and $\boldsymbol{\delta}^t\boldsymbol{\delta} < 1$. The d-dimensional random variable $\boldsymbol{Z} = (\boldsymbol{U} \mid U_0 > 0)$ is called standard skew elliptical distribution.

PROPOSITION 2.1. *The density function of* $\boldsymbol{Z}$ *is given by*

$$f_Z(z) = 2 f_{h^{(d)}}(z) F_{h_{q(z)}}(\boldsymbol{\lambda}^t z),$$

where

$$\boldsymbol{\lambda} = \frac{\boldsymbol{\delta}}{(1 - \boldsymbol{\delta}^t\boldsymbol{\delta})^{1/2}},$$

$f_{h^{(d)}}(\cdot)$ *is the p.d.f. of the elliptical distribution* $El_d(\boldsymbol{0}, \boldsymbol{I}_d, h^{(d)})$, $F_{h_{q(z)}}(\cdot)$ *is the c.d.f. of a* $El_1(0, 1, h_{q(z)})$ *for each fixed value of* z *and* $q(z) = z^t z$.

The density generator functions have the following relationship,

$$h_{q(z)}(u) = \frac{h^{(d+1)}[u + q(z)]}{h^{(d)}[q(z)]},$$

with

$$h^{(d)}(u) = \frac{2\pi^{d/2}}{\Gamma(d/2)} \int_0^\infty h^{(d+1)}(r^2 + u) r^{d-1}\, \mathrm{d}r.$$

We consider here the following notation $\boldsymbol{Z} \sim SE_d(\boldsymbol{0}, \boldsymbol{I}_d, \boldsymbol{\lambda}, h^{(d+1)})$, where $\boldsymbol{\lambda}$ is a skewness (or shape) parameter.

The skew elliptical distribution have interesting properties, like closeness under by linear transformations and marginalization. An important property is that the quadratic form $q(\boldsymbol{Z}) = \boldsymbol{Z}^t\boldsymbol{\Omega}^{-1}\boldsymbol{Z}$ has the same distribution than that derived under the corresponding symmetrical elliptical distribution. We present these properties below (for proofs, see Branco and Dey, 2001).

PROPOSITION 2.2. *Let* $\boldsymbol{Z} \sim SE_d(\boldsymbol{0}, \boldsymbol{I}_d, \boldsymbol{\lambda}, h^{(d+1)})$ *be a skew elliptical variable,* $\boldsymbol{\mu}$ *a* d*-dimensional vector and* $\boldsymbol{A}$ *a* $d \times d$ *matrix, then* $\boldsymbol{Y} = \boldsymbol{\mu} + \boldsymbol{A}\boldsymbol{Z}$ *has the following density function*

$$f_Y(\boldsymbol{y}) = 2A^{-1} f_{h^{(d)}}\big(\boldsymbol{A}^{-1}(\boldsymbol{y} - \boldsymbol{\mu})\big) F_{h_{q(A^{-1}(y-\mu))}}\big(\boldsymbol{\lambda}^t\boldsymbol{A}^{-1}(\boldsymbol{y} - \boldsymbol{\mu})\big).$$

We consider the notation $SE_d(\boldsymbol{\mu}, \boldsymbol{\Omega}, \boldsymbol{\lambda}, h^{(d+1)})$, *where* $\boldsymbol{\Omega} = \boldsymbol{A}^t\boldsymbol{A}$.

PROPOSITION 2.3. *Let $\mathbf{Z}_A[\boldsymbol{\lambda}_A]$ be the vector with the first m components of the original vector $\mathbf{Z}[\boldsymbol{\lambda}]$ and*

$$\boldsymbol{\Omega} = \begin{bmatrix} \boldsymbol{\Omega}_{11} & \boldsymbol{\Omega}_{12} \\ \boldsymbol{\Omega}_{21} & \boldsymbol{\Omega}_{22} \end{bmatrix},$$

then $\mathbf{Z}_A \sim SE_m(\mathbf{0}, \boldsymbol{\Omega}_{11}, \boldsymbol{\lambda}_A, h^{(m+1)})$.

PROPOSITION 2.4. *The quadratic form $q(\mathbf{Z})$ has the same distribution of R^2, where $R > 0$ is the radial variable from stochastic representation of the elliptical vector $\boldsymbol{U} \sim El_d(\mathbf{0}, \boldsymbol{I}_d, h^{(d+1)})$.*

Under the skew normal density, $q(\mathbf{Z})$ has a chi-square distribution with d degrees of freedom and under the skew Student-t density, $q(\mathbf{Z})$ has a F-Snedecor distribution with d and ν degrees of freedom.

A subclass of interest is the skew mixture of normals defined below.

DEFINITION 2.2. The d-dimensional variable $\boldsymbol{U}$ has a skew scale mixture of normal (or representable skew elliptical distribution) if its p.d.f. can be written as

$$f_U(\boldsymbol{u}) = \int_0^\infty \phi_d\big(\boldsymbol{u}|\mathbf{0}, K(\omega)\boldsymbol{\Omega}\big)\,\mathrm{d}H(\omega),$$

where ω is a random variable with c.d.f. H and $K(\omega)$ is a weight function. In this case the skewness density is given by

$$f_Z(z) = 2\int_0^\infty \boldsymbol{\phi}_d\big(z|\mathbf{0}, K(\omega)\boldsymbol{\Omega}\big)\Phi\big(\boldsymbol{\lambda}^t z K^{-1/2}(\omega)\big)\,\mathrm{d}H(\omega).$$

An important particular case is the skew normal which is obtained by making $K(\omega) = 1$, $f_Z(z) = 2\boldsymbol{\phi}_d(z|\mathbf{0}, \boldsymbol{\Omega})\Phi(\boldsymbol{\lambda}^t z)$.

Thus, the skew scale mixture of normal is also a scale mixture of skew normal densities. The same happens with the moment generating function (m.g.f.) as we will see. The m.g.f. of the skew normal distribution is given by

$$M_{SN}(\boldsymbol{t}) = 2\exp\left(\frac{1}{2}\boldsymbol{t}^t\boldsymbol{\Omega}\boldsymbol{t}\right)\Phi\big(\boldsymbol{\lambda}^t\boldsymbol{\Omega}\boldsymbol{t}(1+\boldsymbol{\lambda}^t\boldsymbol{\Omega}\boldsymbol{\lambda})^{1/2}\big)$$

and the m.g.f. of the scale mixture of normal is given by

$$M_{SMN}(\boldsymbol{t}) = 2\int_0^\infty \exp\left(\frac{1}{2}\boldsymbol{t}^t K^{1/2}(\omega)\boldsymbol{\Omega}\boldsymbol{t}\right)\Phi\left(\boldsymbol{\lambda}^t\boldsymbol{\Omega}K(\omega)\boldsymbol{t}\left(1+\frac{\boldsymbol{\lambda}^t\boldsymbol{\Omega}\boldsymbol{\lambda}}{K(\omega)}\right)^{1/2}\right)\mathrm{d}H(\omega).$$

From the m.g.f. the expectation vector and the covariance matrix are

$$\mathbb{E}_{SMN}(\boldsymbol{Z}) = \left(\frac{2}{\pi}\right)^{1/2} \mathbb{E}_{\omega}\big[K^{1/2}(\omega)\big]\boldsymbol{\delta}$$

and

$$\mathrm{Cov}_{SMN}(\boldsymbol{Z}) = \mathbb{E}_{\omega}\big[K(\omega)\big]\boldsymbol{\Omega} - \frac{2}{\pi}\mathbb{E}_{\omega}^{2}\big[K^{1/2}(\omega)\big]\boldsymbol{\delta}\boldsymbol{\delta}^{t}.$$

REMARK 2.1. Another advantage of representable skew elliptical distribution is that Bayesian computation rely upon the introduction of strategic latent variables as it has shown in Sahu et al. (2003). See also Arellano-Valle et al. (2004).

Other interesting cases are obtained when we let $K(\omega) = \omega^{-1}$, like the skew Student-t distribution discussed in the next example.

EXAMPLE 2.1. The multivariate skew Student-t distribution is obtained when ω follows a Gamma distribution with parameters $\nu/2$ and $\nu/2$. The corresponding density is given by

$$f_Z(z) = 2t_d(z|\boldsymbol{0}, \boldsymbol{\Omega}, \nu)T_1\big(\boldsymbol{\lambda}^t z\big(\nu + q(z)\big)^{-1/2}\big|0, 1, \nu + d\big),$$

where $t_d(\cdot|\boldsymbol{\mu}, \boldsymbol{\Omega}, \nu)$ represents the p.d.f. of the d-dimensional Student-t distribution with location and scale parameters $\boldsymbol{\mu}$ and $\boldsymbol{\Omega}$, respectively, and ν degree of freedom. Also, $T_1(z|\mu, \Omega, \nu)$ is the c.d.f. of the univariate Student-t distribution with parameters μ, Ω and ν.

Denote $\boldsymbol{\psi} = (\frac{\nu}{\pi})^{1/2}\frac{\Gamma[(\nu-1)/2]}{\Gamma(\nu/2)}\boldsymbol{\delta}$. The first two moments of skew Student-t distribution are given by

$$\mathbb{E}(\boldsymbol{Z}) = \boldsymbol{\psi}, \quad \nu > 1,$$

and

$$\mathbb{E}\big(\boldsymbol{Z}\boldsymbol{Z}^t\big) = \frac{\nu}{\nu - 2}\boldsymbol{\Omega}, \quad \nu > 2.$$

When $d = 1$ we also have expression for the skewness and the kurtosis coefficients, given respectively by

$$\gamma_1 = \psi\left[\frac{\nu(3-\delta^2)}{\nu-3} - \frac{3\nu}{\nu-2} + 2\psi^2\right]\left[\frac{\nu}{\nu-2} - \psi^2\right]^{-3/2}, \quad \nu > 3,$$

and

$$\gamma_2 = \left[\frac{3\nu^2}{(\nu-2)(\nu-4)} - \frac{4\psi^2\nu(3-\delta^2)}{\nu-3} + \frac{6\psi^2\nu}{\nu-2} - 3\psi^4\right]\left[\frac{\nu}{\nu-2} - \psi^2\right]^{-3}, \quad \nu > 4.$$

Another way to obtain the skew Student-t distribution is from the skew Pearson type VII (for details see Azzalini and Capitanio, 2003).

REMARK 2.2. When $d = 1$ and $K(\omega) = \omega^{-1}$ we obtain an univariate representable skew elliptical distribution under c.d.f. H whose density can be written by

$$f_{X|\lambda,\mu,\sigma}(x) = \int_0^\infty \frac{2}{\sigma\sqrt{\omega}}\phi\left(\frac{x-\mu}{\sigma\sqrt{\omega}}\right)\Phi\left(\lambda\frac{x-\mu}{\sigma\sqrt{\omega}}\right)\mathrm{d}H(\omega),$$

where H is the c.d.f. of a nonnegative random variable ω such that $\omega \perp\!\!\!\perp (\lambda, \mu, \sigma)$. The notation $x \perp\!\!\!\perp y$ means that x and y are independent random variables.

An equivalent definition, useful for Bayesian computations, is given by: $X|\lambda, \mu, \sigma$ is a representable skew elliptical distribution if and only if there is $\omega \sim H$ and $\omega \perp\!\!\!\perp (\lambda, \mu, \sigma)$ such that $X|\lambda, \mu, \sigma, \omega \sim SN(\lambda, \mu, \sigma\sqrt{\omega})$. The next section deals with a class of univariate skew distribution that includes the univariate representable skew distribution. In particular we will attempt to quantify the departure from the corresponding symmetric model.

2.2. *Univariate case*

Here we approach the problem of evaluating the distance between the probability measures corresponding to the densities

$$\frac{2}{\sigma}f\left(\frac{x-\mu}{\sigma}\right)G\left(\lambda\frac{x-\mu}{\sigma}\right), \tag{9}$$

and

$$\frac{1}{\sigma}f\left(\frac{x-\mu}{\sigma}\right). \tag{10}$$

In other words, we want to compute the distance between the symmetric and the perturbed model. The approach follows Vidal et al. (2005).

The L_1 distance has an easy and nice interpretation (see, for example, Peng and Dey, 1995; Weiss, 1996; Arellano-Valle et al., 2000) and it is defined by

$$L_1(f_1, f_2) = \frac{1}{2}\int \left|f_1(x) - f_2(x)\right| \mathrm{d}x = \sup_{A\in\mathcal{B}}\left|\mathbb{P}(A|f_1) - \mathbb{P}(A|f_2)\right|,$$

where $\mathcal{B}$ are the Borel's sets. The L_1 distance is bounded and takes values in [0, 1], where $L_1(f_1, f_2) = 0$ implies that $f_1(x) = f_2(x)$ a.s., and $L_1(f_1, f_2) = 1$ indicates that the support of the two densities is disjoint, indicating maximal discrepancy. Also, $L_1(f_1, f_2)$ is an upper bound from the differences $|\mathbb{P}(A|f_1) - \mathbb{P}(A|f_2)|$ for any set A, where $\mathbb{P}(\cdot|f)$ denote the probability measure defined by f. Generally, it is difficult to obtain explicit expressions for the L_1 distance, even in simple cases. The following proposition provides an useful expression to compute and understand this distance.

PROPOSITION 2.5. *For any μ and σ fixed, the L_1 distance between $\frac{1}{\sigma}f(\frac{x-\mu}{\sigma})$ and $\frac{2}{\sigma}f(\frac{x-\mu}{\sigma})G(\lambda\frac{x-\mu}{\sigma})$ is*

$$L_1(\lambda) = \mathbb{E}_{f^*}\left[G\left(|\lambda|Z\right)\right] - \frac{1}{2}, \tag{11}$$

where $f^*(z) = 2f(z)I_{[0,+\infty)}(z)$ is the p.d.f. of f truncated on zero.

Table 1
Examples of L_1-distance

Skew distribution	$L_1(\lambda)$
Skew normal	$\frac{1}{\pi}\arcsin\left(\frac{\lvert\lambda\rvert}{\sqrt{1+\lambda^2}}\right)$
Skew uniform	$\begin{cases} \frac{1}{2}-\frac{1}{4\lvert\lambda\rvert} & \text{if } \lvert\lambda\rvert > 1 \\ \frac{\lvert\lambda\rvert}{4} & \text{if } \lvert\lambda\rvert \leqslant 1 \end{cases}$
Skew double exponential	$\frac{\lvert\lambda\rvert}{2(1+\lvert\lambda\rvert)}$
Finite mixture of skew normal	$\frac{1}{\pi}\arcsin\left(\frac{\lvert\lambda\rvert}{\sqrt{1+\lambda^2}}\right)$

PROOF. See Vidal et al. (2005). □

REMARK 2.3. We note that the distance depends on λ and G, but it does not depend on μ and σ. Furthermore, the class of distributions introduced in Definition 2.1 letting $d = 1$, that is the univariate case, is a particular case of the p.d.f. (9). Therefore, the L_1 distance between any $f_{h^{(1)}}(z)$ and $2f_{h^{(1)}}(z)F_{h_{q(z)}}(\lambda z)$ can be computed using (11).

The following corollary shows that the supreme of $L_1(\lambda)$ as function of λ, f and G is invariant under different choices of f and G.

COROLLARY 2.6. *If L_1 is as in* (11), *then*

$$\sup_{\lambda} L_1(\lambda) = \frac{1}{2}.$$

PROOF. See Vidal et al. (2005). □

Table 1 presents some examples for the L_1 distance when f is the density function defined by G. See Vidal et al. (2005) for details.

We note that the L_1 distance for skew normal and finite mixture of skew normal is the same. This fact is a particular case of the following proposition which uses representable skew elliptical distributions.

Let

$$\begin{aligned} M_0&: \int_0^\infty \frac{1}{\sigma\sqrt{\omega}}\phi\left(\frac{x-\mu}{\sigma\sqrt{\omega}}\right)\mathrm{d}H(\omega), \\ M_1&: \int_0^\infty \frac{2}{\sigma\sqrt{\omega}}\phi\left(\frac{x-\mu}{\sigma\sqrt{\omega}}\right)\Phi\left(\lambda\frac{x-\mu}{\sigma\sqrt{\omega}}\right)\mathrm{d}H(\omega). \end{aligned} \tag{12}$$

PROPOSITION 2.7. *The L_1 distance between M_0 and M_1, specified in* (12), *is*

$$L_1(\lambda) = \frac{1}{\pi}\arcsin\left(\frac{|\lambda|}{\sqrt{1+\lambda^2}}\right).$$

PROOF. See Vidal et al. (2005). □

In general, it is not possible to evaluate the L_1 distance and we need numerical methods. The special case where f is the Student-t p.d.f. and G is its c.d.f. is considered in Vidal et al. (2005).

Another important question is how the introduction of skewness parameter affects the Bayesian inference on location and scale parameters.

2.3. *The L_1-distance for posterior distribution of (μ, σ) under skew-normal model*

An important way to measure the sensitivity of the λ parameter is consider its effect on the posterior distribution of (μ, σ). With this objective in mind we obtained the following results to compute the posterior distributions. Hereafter, $\mathbb{1}_n$ will denote a vector of n ones.

PROPOSITION 2.8. *If $X|\lambda, \mu, \sigma \sim SN(\lambda, \mu, \sigma)$, under the prior assumptions λ is independent of (μ, σ), $\mu|\sigma \sim N(m, \frac{\sigma^2}{v})$ and $\sigma^{-2} \sim Ga(a, b)$, then*

$$\pi(\mu, \sigma|\lambda, \boldsymbol{x}) = k(\mu, \sigma, \lambda, \boldsymbol{x}) \times \pi(\mu, \sigma|\lambda = 0, \boldsymbol{x}),$$

where

$$k(\mu, \sigma, \lambda, \boldsymbol{x}) = \frac{\prod_{i=1}^{n} \Phi(\lambda \frac{x_i - \mu}{\sigma})}{\boldsymbol{T}_n(\lambda\sqrt{n+2a}\frac{\mathbf{x} - \hat{\mu}\mathbb{1}_n}{r}|\mathbf{0}, \boldsymbol{\Sigma}, n+2a)},$$

with $\hat{\mu} = \frac{n\bar{x}+mv}{n+v}$, $r^2 = ns^2 + \frac{nv}{n+v}(m - \bar{x})^2 + 2b$, $s^2 = n^{-1}\sum_{i=1}^{n}(x_i - \bar{x})^2$ and $\boldsymbol{T}_n$ is the c.d.f. of the $t_n(\mathbf{0}, \boldsymbol{\Sigma}, n + 2a)$ with $\boldsymbol{\Sigma} = \boldsymbol{I}_n + \frac{\lambda^2}{v+n}\mathbb{1}_n\mathbb{1}_n^t$.

PROOF. See Vidal et al. (2005). □

COROLLARY 2.9. *Under conditions of Proposition 2.8 we have*

(1)

$$\pi(\mu|\lambda, \boldsymbol{x}) = k_1(\mu, \lambda, \boldsymbol{x}) \times \pi(\mu|\lambda = 0, \boldsymbol{x}),$$

where

$$k_1(\mu, \lambda, \boldsymbol{x}) = \frac{\boldsymbol{T}_n(\frac{\lambda\sqrt{n+2a+1}}{\sqrt{r^2+(n+v)(\mu-\hat{\mu})^2}}(\boldsymbol{x} - \mu\mathbb{1}_n)|\mathbf{0}, \boldsymbol{I}_n, n+2a+1)}{\boldsymbol{T}_n(\lambda\sqrt{n+2a}\frac{\boldsymbol{x} - \hat{\mu}\mathbb{1}_n}{r}|\mathbf{0}, \boldsymbol{\Sigma}, n+2a)}$$

and

(2)

$$\pi(\sigma|\lambda, \boldsymbol{x}) = k_2(\sigma, \lambda, \boldsymbol{x}) \times \pi(\sigma|\lambda = 0, \boldsymbol{x}),$$

where

$$k_2(\sigma, \lambda, \boldsymbol{x}) = \frac{\boldsymbol{\Phi}_n(\lambda\frac{\boldsymbol{x} - \hat{\mu}\mathbb{1}_n}{\sigma}|\mathbf{0}, \boldsymbol{\Sigma})}{\boldsymbol{T}_n(\lambda\sqrt{n+2a}\frac{\boldsymbol{x} - \hat{\mu}\mathbb{1}_n}{r}|\mathbf{0}, \boldsymbol{\Sigma}, n+2a)},$$

with $\hat{\mu}$ and r^2 are in Proposition 2.8.

$\boldsymbol{\Phi}_n(\cdot|\boldsymbol{\mu}, \boldsymbol{\Sigma})$ denotes the c.d.f. of the $N_n(\boldsymbol{\mu}, \boldsymbol{\Sigma})$ distribution. The terms $k(\mu, \sigma, \lambda, \boldsymbol{x})$, $k_1(\mu, \lambda, \boldsymbol{x})$ and $k_2(\sigma, \lambda, \boldsymbol{x})$ are the perturbation functions, introduced by Kass et al. (1989) and Weiss (1996). These terms can be interpreted as sensitivity factors because the sensitivity of the skewness parameter depends only on them.

Note that for $\lambda = 0$ (normal model), $\mu|\lambda = 0, \boldsymbol{x} \sim t_1(\hat{\mu}, \frac{r^2}{(n+v)(n+2a)}, n + 2a)$ and $\sigma^{-2}|\lambda = 0, \boldsymbol{x} \sim Ga(\frac{n}{2} + a, \frac{r^2}{2})$. The previous proposition allows us to calculate the two conditional posterior distributions, $\pi(\mu|\sigma, \lambda, \boldsymbol{x})$ and $\pi(\sigma|\mu, \lambda, \boldsymbol{x})$ which are nesessary, for example, in Gibbs sampler-type algorithms.

Also Corollary 2.9 enable us to calculate the L_1 for considering these posterior distributions. The L_1 distance between $\pi(\sigma|\lambda = 0, \boldsymbol{x})$ and $\pi(\sigma|\lambda, \boldsymbol{x})$ is given by

$$L_1(\lambda) = \frac{1}{2}\mathbb{E}\big[\big|1 - k_2\big(S^{-1}, \lambda, \boldsymbol{x}\big)\big|\big],$$

where the expected value is taken assuming $S^2 \sim Ga(\frac{n}{2} + a, \frac{r^2}{2})$. On the other hand, the L_1 distance between $\pi(\mu|\lambda = 0, \boldsymbol{x})$ and $\pi(\mu|\lambda, \boldsymbol{x})$ is given by

$$L_1(\lambda) = \frac{1}{2}\mathbb{E}\big[\big|1 - k_1(M, \lambda, \boldsymbol{x})\big|\big],$$

where the expectation is taken under $M \sim t_1(\hat{\mu}, \frac{r^2}{(n+v)(n+2a)}, n + 2a)$.

These last results depend on the sensitivity factors and can be used to study the influence of the skewness parameter over the posterior distribution of μ and σ. Figure 1 shows $\log[\bar{k}_2(\sigma, \lambda, \boldsymbol{x})]$, where $\bar{k}_2(\sigma, \lambda, \boldsymbol{x})$ is the average of $k_2(\sigma, \lambda, \boldsymbol{x}_1), \ldots, k_2(\sigma, \lambda, \boldsymbol{x}_{50})$, and $\boldsymbol{x}_1, \ldots, \boldsymbol{x}_{50}$ are data sets with sample size 100 drawn from the $N(0, 1)$ distribution. The $k_2(\sigma, \lambda, \boldsymbol{x}_j)$, $j = 1, \ldots, 50$, were computed for λ from 0 to 5 and σ from 0 to 10, both with step 0.25. From Figure 1(b) we can see a great influence of the λ parameter.

There are values of λ and σ for which the posterior distribution of σ equals to $\pi(\sigma|\lambda = 0, \boldsymbol{x})$. They arise for the $k_2(\sigma, \lambda, \boldsymbol{x}) = 1$ contour line, or equivalently, $\log[k_2(\sigma, \lambda, \boldsymbol{x})] = 0$. Figure 1(a) shows several contour lines from $\log[\bar{k}_2(\sigma, \lambda, \boldsymbol{x})]$.

3. Testing of asymmetry in linear regression model

In this section we present some results related with the Bayes factors, with the purpose of detecting asymmetry on the error distribution in a linear regression model.

3.1. Bayes factor

In this subsection we assume that the data set comes from the linear regression model,

$$\boldsymbol{Y} = \boldsymbol{X}\boldsymbol{\beta} + \boldsymbol{\varepsilon}, \tag{13}$$

where $\boldsymbol{\beta} \in \mathbb{R}^k$, $\boldsymbol{\varepsilon} \sim SE_n(\boldsymbol{0}, \phi^{-1}\boldsymbol{I}_n, \lambda\mathbb{1}_n, h^{(n)})$, $\phi > 0$ and $\lambda \in \mathbb{R}$. Our goal is to search for evidences in the data that allow us to choose between a symmetric or asymmetric model for the errors. In other words, we look for evidence in the data in favor of either $\lambda = 0$ or, $\lambda \neq 0$.

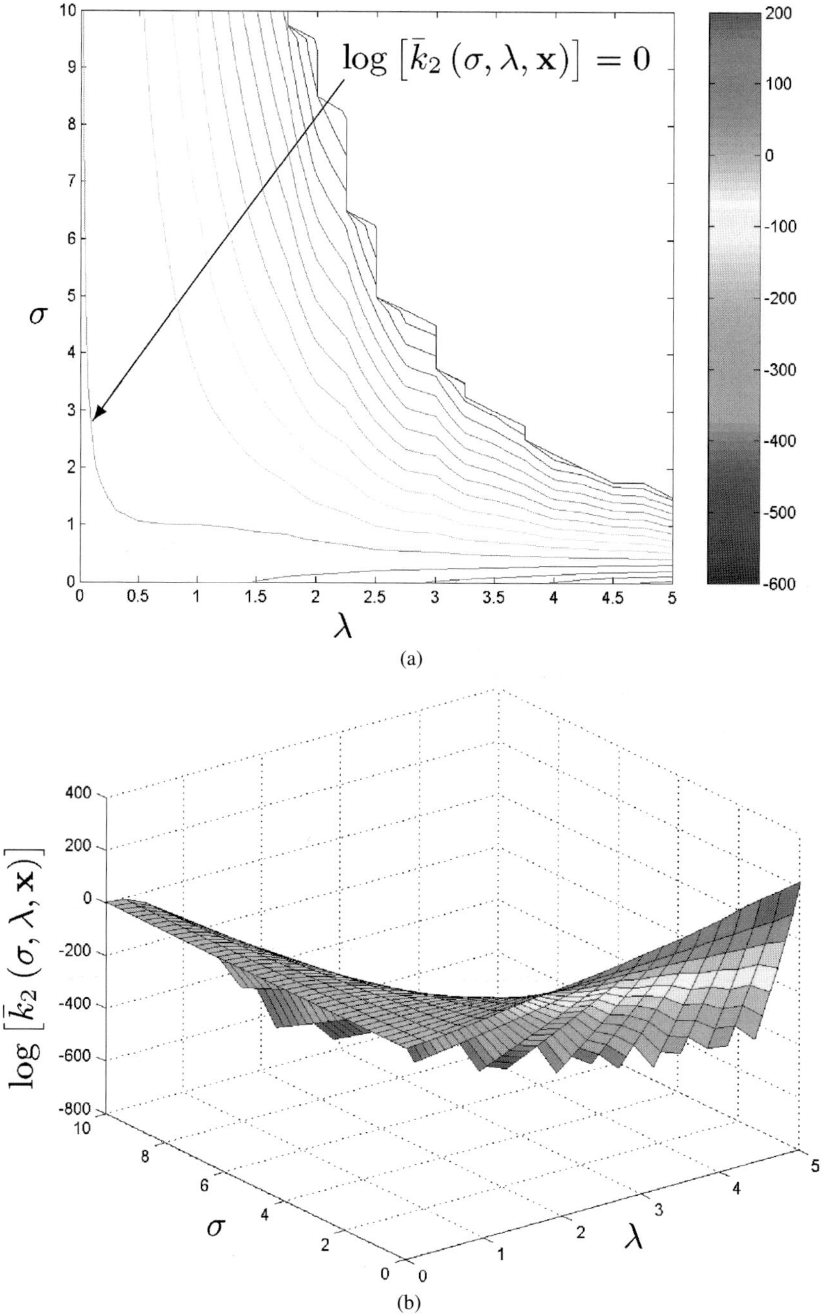

(b)

Fig. 1. $\log[\bar{k}_2(\sigma, \lambda, \boldsymbol{x})]$ for generated data: (a) contour lines; (b) surface. For a color reproduction of this figure see the color figures section, page 1001.

Note that if $\lambda = 0$, the data comes from a distribution with p.d.f. given by

$$\phi^{n/2} h^{(n)}\big(\phi \|\boldsymbol{y} - \boldsymbol{X}\boldsymbol{\beta}\|^2\big).$$

This case has been studied by Arellano-Valle et al. (2003) and if $\lambda \neq 0$, the data coming from

$$2\phi^{n/2} h^{(n)}\big[q(\boldsymbol{y})\big] \int_{-\infty}^{\lambda\sqrt{\phi}\mathbb{1}_n^t(\boldsymbol{y}-\boldsymbol{X}\boldsymbol{\beta})} h_{q(\boldsymbol{y})}\big(u^2\big)\,\mathrm{d}u,$$

where $q(\boldsymbol{y}) = \phi\|\boldsymbol{y} - \boldsymbol{X}\boldsymbol{\beta}\|^2$. In this way, for the data $\boldsymbol{y} = (y_1, \ldots, y_n)$, the Bayes factor in favor of $\lambda = 0$ is

$$BF = \frac{\int \phi^{n/2} h^{(n)}[q(\boldsymbol{y})]\pi(\boldsymbol{\beta}, \phi)\,\mathrm{d}\boldsymbol{\beta}\,\mathrm{d}\phi}{2\int\big\{\phi^{n/2} h^{(n)}[q(\boldsymbol{y})]\int_{-\infty}^{\lambda\sqrt{\phi}\mathbb{1}_n^t(\boldsymbol{y}-\boldsymbol{X}\boldsymbol{\beta})} h_{q(\boldsymbol{y})}(u^2)\,\mathrm{d}u\big\}\pi(\boldsymbol{\beta}, \phi, \lambda)\,\mathrm{d}\boldsymbol{\beta}\,\mathrm{d}\phi\,\mathrm{d}\lambda}, \tag{14}$$

where $\pi(\cdot)$ represents the prior distribution for the respective parameters. From (14), it is not generally feasible to obtain a closed form of the Bayes factor. Numerical computations are also complex. However, when $\boldsymbol{\beta}$ and ϕ are known, (14) takes the nice form

$$BF = \frac{1}{2\int\big[\int_{-\infty}^{\lambda\sqrt{\phi}\mathbb{1}_n^t(\boldsymbol{y}-\boldsymbol{X}\boldsymbol{\beta})} h_{q(\boldsymbol{y})}(u^2)\,\mathrm{d}u\big]\pi(\lambda)\,\mathrm{d}\lambda}.$$

If we consider a prior distribution to λ symmetric around zero, it is means we do not have any idea about the sign of λ. Then the marginal density for the data is symmetric, as we can see in the next lemma.

LEMMA 3.1. *If $f_X(x|\mu, \sigma, \lambda) = \frac{2}{\sigma} f(\frac{x-\mu}{\sigma}) G(\lambda\frac{x-\mu}{\sigma})$ and $\lambda \sim \pi(\lambda)$, where $\pi(\lambda)$ is any symmetric density with respect to zero, then*

$$f_X(x|\mu, \sigma) = \frac{1}{\sigma} f\bigg(\frac{x-\mu}{\sigma}\bigg).$$

PROOF.

$$\begin{aligned} f_X(x|\mu, \sigma) &= \int_{-\infty}^{\infty} \frac{2}{\sigma} f\bigg(\frac{x-\mu}{\sigma}\bigg) G\bigg(\lambda\frac{x-\mu}{\sigma}\bigg)\pi(\lambda)\,\mathrm{d}\lambda \\ &= \frac{1}{\sigma} f\bigg(\frac{x-\mu}{\sigma}\bigg) \int_{-\infty}^{\infty} 2\pi(\lambda) G\bigg(\lambda\frac{x-\mu}{\sigma}\bigg)\,\mathrm{d}\lambda = \frac{1}{\sigma} f\bigg(\frac{x-\mu}{\sigma}\bigg). \end{aligned}$$ □

An immediate consequence of this lemma is the following proposition which is a particular case where the Bayes factor cannot discriminate between a symmetric and asymmetric model.

PROPOSITION 3.2. *Let* $\mathbf{y} = (y_1, \ldots, y_n)$ *a random sample from the model* (13), *where* $\boldsymbol{\varepsilon} \sim SE_n(\mathbf{0}, \phi^{-1}\boldsymbol{I}_n, \lambda\mathbb{1}_n, h^{(n)})$. *If* $\lambda \perp\!\!\!\perp (\boldsymbol{\beta}, \phi)$ *and* $\lambda \sim \pi(\lambda)$, *where* $\pi(\lambda)$ *is a symmetric p.d.f. with respect to zero then, the Bayes factor* (14) *equals to* 1.

PROOF. From $\lambda \perp\!\!\!\perp (\boldsymbol{\beta}, \phi)$, then (14) can be written as

$$BF = \frac{\int \phi^{n/2} h^{(n)}[q(\mathbf{y})]\pi(\boldsymbol{\beta}, \phi)\, \mathrm{d}\boldsymbol{\beta}\, \mathrm{d}\phi}{\int\{\int 2\phi^{n/2} h^{(n)}[q(\mathbf{y})] F_{h_{q(\mathbf{y})}}[\lambda\sqrt{\phi}\mathbb{1}_n^t(\mathbf{y} - \boldsymbol{X}\boldsymbol{\beta})]\pi(\lambda)\, \mathrm{d}\lambda\}\pi(\boldsymbol{\beta}, \phi)\, \mathrm{d}\boldsymbol{\beta}\, \mathrm{d}\phi\, \mathrm{d}\lambda}.$$

Then, from Lemma 3.1, we obtain the result. □

Other particular cases of the Bayes factor to compare a symmetric distribution with an asymmetric one are given in the propositions below, where it is necessary to know the sign of λ. This assumption is a typically fulfilled in practice. The following lemma is well-known and it will be useful for the proof of the next propositions.

LEMMA 3.3. *If* $\boldsymbol{Y}|\boldsymbol{\mu}, \boldsymbol{\Sigma}, \tau \sim N_n(\boldsymbol{\mu}, \tau^{-1}\boldsymbol{\Sigma})$ *and* $\tau|a, b \sim Ga(a, b)$, *then* $\boldsymbol{Y}|\boldsymbol{\mu}, \boldsymbol{\Sigma}, a, b \sim t_n(\boldsymbol{\mu}, \frac{b}{a}\boldsymbol{\Sigma}, 2a)$.

PROPOSITION 3.4. *Let* $\mathbf{y} = (y_1, \ldots, y_n)$ *be a random sample from* (13), *where* $\boldsymbol{\varepsilon} \sim SN_n(\mathbf{0}, \phi^{-1}\boldsymbol{I}_n, \lambda\mathbb{1}_n)$ *with* $\lambda \perp\!\!\!\perp (\boldsymbol{\beta}, \phi)$, $\lambda \geqslant 0$ *and* $\lambda^2 \sim Ga(a, b)$ *when* $\lambda > 0$. *Then, the Bayes factor* (14) *for* $\lambda = 0$ *vs.* $\lambda > 0$ *is given by*

$$BF = \frac{\int \phi^{n/2} h^{(n)}[q(\mathbf{y})]\pi(\boldsymbol{\beta}, \phi)\, \mathrm{d}\boldsymbol{\beta}\, \mathrm{d}\phi}{2\int \phi^{n/2} h^{(n)}[q(\mathbf{y})] T_1\big[\mathbb{1}_n^t(\mathbf{y} - \boldsymbol{X}\boldsymbol{\beta})\sqrt{\frac{a\phi}{b}}|0, 1, 2a\big]\pi(\boldsymbol{\beta}, \phi)\, \mathrm{d}\boldsymbol{\beta}\, \mathrm{d}\phi},$$

where T_1 *is the c.d.f. of the* $t_1(0, 1, 2a)$ *and* $q(\mathbf{y}) = \phi\|\mathbf{y} - \boldsymbol{X}\boldsymbol{\beta}\|^2$.

PROOF. Notice that if $\lambda^2 \sim Ga(a, b)$ when $\lambda > 0$, then λ has probability density function $f(\lambda|a, b) = \frac{2b^a}{\Gamma(a)}(\lambda^2)^{a-1/2}\exp(-b\lambda^2) I_{(0,+\infty)}(\lambda)$. Therefore,

$$\int\left[\int_{-\infty}^{\lambda\sqrt{\phi}\mathbb{1}_n^t(\mathbf{y}-\boldsymbol{X}\boldsymbol{\beta})} h_{q(\mathbf{y})}(u^2)\, \mathrm{d}u\right]\pi(\lambda)\, \mathrm{d}\lambda$$

$$= \frac{2b^a}{\sqrt{2\pi}\Gamma(a)}\int_0^\infty \lambda^{2a-1}\, \mathrm{e}^{-b\lambda^2}\int_{-\infty}^{\lambda\sqrt{\phi}\mathbb{1}_n^t(\mathbf{y}-\boldsymbol{X}\boldsymbol{\beta})} \mathrm{e}^{-\frac{u^2}{2}}\, du\, \mathrm{d}\lambda$$

$$= \frac{2b^a}{\sqrt{2\pi}\Gamma(a)}\int_0^\infty \lambda^{2a}\, \mathrm{e}^{-b\lambda^2}\int_{-\infty}^{\sqrt{\phi}\mathbb{1}_n^t(\mathbf{y}-\boldsymbol{X}\boldsymbol{\beta})} \mathrm{e}^{-\frac{\lambda^2 r^2}{2}}\, dr\, \mathrm{d}\lambda.$$

Then, with the change of variable $l = \lambda^2$, we obtain

$$\int\left[\int_{-\infty}^{\lambda\sqrt{\phi}\mathbb{1}_n^t(\mathbf{y}-\boldsymbol{X}\boldsymbol{\beta})} h_{q(\mathbf{y})}(u^2)\, \mathrm{d}u\right]\pi(\lambda)\, \mathrm{d}\lambda$$

$$= \frac{b^a}{\sqrt{2\pi}\Gamma(a)} \int_0^\infty l^{a-1/2} e^{-bl} \int_{-\infty}^{\sqrt{\phi}\mathbb{1}_n^t(y-X\beta)} e^{-lr^2/2}\, dr\, dl,$$

and, from Lemma 3.3,

$$\int \left[\int_{-\infty}^{\lambda\sqrt{\phi}\mathbb{1}_n^t(y-X\beta)} h_{q(y)}(u^2)\, du \right] \pi(\lambda)\, d\lambda = T_1\left[\sqrt{\phi}\mathbb{1}_n^t(y - X\beta) | 0, \frac{b}{a}, 2a \right],$$

where T_1 is the c.d.f. of the $t_1(0, \frac{b}{a}, 2a)$. Then, the result of the proposition is immediate. □

If in the previous proposition we assume β and ϕ known, then

$$BF = \frac{1}{2T_1\left[\mathbb{1}_n^t(y - X\beta)\sqrt{\frac{a\phi}{b}} | 0, 1, 2a\right]}.$$

Numerical calculations of the previous Bayes factor can be done using the fact that the Student-t distribution is a mixture of normal distributions. In a similar way the following result is also obtained:

PROPOSITION 3.5. *Let $y = (y_1, \dots, y_n)$ be a random sample from* (13), *where $\varepsilon \sim SN_n(\mathbf{0}, \phi^{-1}I_n, \lambda\mathbb{1}_n)$ with $\lambda \perp\!\!\!\perp (\beta, \phi)$, $\lambda \leqslant 0$ and $\lambda^2 \sim Ga(a, b)$ when $\lambda < 0$. Then the Bayes factor* (14) *for $\lambda = 0$ vs. $\lambda < 0$ is given by*

$$BF = \frac{\int \phi^{n/2} h^{(n)}[q(y)]\pi(\beta, \phi)\, d\beta\, d\phi}{2\int \phi^{n/2} h^{(n)}[q(y)] T_1\left[\mathbb{1}_n^t(X\beta - y)\sqrt{\frac{a\phi}{b}} | 0, 1, 2a\right]\pi(\beta, \phi)\, d\beta\, d\phi},$$

where T_1 is the c.d.f. of the $t_1(0, 1, 2a)$ and $q(y) = \phi\|y - X\beta\|^2$.

PROOF. Notice that if $\lambda^2 \sim Ga(a, b)$ when $\lambda < 0$, then λ has probability density function $f(\lambda|a, b) = \frac{2b^a}{\Gamma(a)}(\lambda^2)^{a-1/2}\exp(-b\lambda^2) I_{(-\infty,0)}(\lambda)$. Therefore, making the change of variable $l = -\lambda$,

$$\int \left[\int_{-\infty}^{\lambda\sqrt{\phi}\mathbb{1}_n^t(y-X\beta)} h_{q(y)}(u^2)\, du \right] \pi(\lambda)\, d\lambda$$

$$= -\int_\infty^0 \left[\int_{-\infty}^{-l\sqrt{\phi}\mathbb{1}_n^t(y-X\beta)} h_{q(y)}(u^2)\, du \right] f(-l|a, b)\, dl$$

$$= \int_0^\infty \left[\int_{-\infty}^{l\sqrt{\phi}\mathbb{1}_n^t(X\beta-y)} h_{q(y)}(u^2)\, du \right] f(l|a, b)\, dl.$$

Then, it is enough to follow the proof of the previous proposition. □

When the parameters $\boldsymbol{\beta}$ and ϕ are unknown, the previous propositions do not solve the problem due to the great analytic and numeric complexity that the computation of the Bayes factors implies. A more tractable case is presented in the next subsection.

3.1.1. Bayes factor for representable skew elliptical linear model

The propositions of the previous section use skew-normal distributions for the error distribution of the model (13). In this section we will work with a wider class of skew-elliptical distributions than the skew-normal class.

The random vector $\boldsymbol{Y}|\boldsymbol{\mu}, \boldsymbol{\Omega}, \boldsymbol{\lambda}$ has representable skew elliptical distribution if, and only if, there is $\omega \sim H$ and $\omega \perp\!\!\!\perp (\boldsymbol{\mu}, \boldsymbol{\Omega}, \boldsymbol{\lambda})$ such that $\boldsymbol{Y}|\boldsymbol{\mu}, \boldsymbol{\Omega}, \boldsymbol{\lambda}, \omega \sim SN_n(\boldsymbol{\mu}, \omega\boldsymbol{\Omega}, \boldsymbol{\lambda})$. $\boldsymbol{Y}|\boldsymbol{\mu}, \boldsymbol{\Omega}, \boldsymbol{\lambda} \sim RSE_n(\boldsymbol{\mu}, \Omega, \boldsymbol{\lambda})$ will be used to denote that $\boldsymbol{Y}|\boldsymbol{\mu}, \boldsymbol{\Omega}, \boldsymbol{\lambda}$ has multivariate representable skew-elliptical distribution and its p.d.f. was given in Definition 2.2. Properties and examples of this class of distributions can be found in Branco and Dey (2001).

For the proof of the next proposition we need the following well-known result (see Azzalini and Dalla-Valle, 1996).

LEMMA 3.6. *If* $\boldsymbol{U} \sim N_k(\boldsymbol{0}, \boldsymbol{\Sigma})$, *then*

$$\mathbb{E}\left[\Phi\left(u + \boldsymbol{v}^t\boldsymbol{U}\right)\right] = \Phi\left(\frac{u}{\sqrt{1 + \boldsymbol{v}^t\boldsymbol{\Sigma}\boldsymbol{v}}}\right)$$

for any scalar u *and* $\boldsymbol{v} \in \mathbb{R}^k$.

PROPOSITION 3.7. *Let* $\boldsymbol{y} = (y_1, \ldots, y_n)$ *be a random sample from the model* (13), *where* $\boldsymbol{\varepsilon} \sim RSE_n(\boldsymbol{0}, \phi^{-1}\boldsymbol{I}_n, \lambda\mathbb{1}_n)$, $\lambda \perp\!\!\!\perp (\boldsymbol{\beta}, \phi)$, $\boldsymbol{\beta}|\phi \sim N_k(\boldsymbol{m}, \phi^{-1}\boldsymbol{B})$ *and* $\phi \sim Ga(a, b)$. *Then the Bayes factor* (14) *is given by*

$$BF(\boldsymbol{y}) = \frac{\int \omega^{(k-n)/2}|\boldsymbol{X}^t\boldsymbol{X} + \omega\boldsymbol{B}^{-1}|^{-1/2} r^{-n-2a}\, \mathrm{d}H(\omega)}{2\int \omega^{(k-n)/2}|\boldsymbol{X}^t\boldsymbol{X} + \omega\boldsymbol{B}^{-1}|^{-1/2} r^{-n-2a} g(\omega)\, \mathrm{d}H(\omega)},$$

where

$$g(\omega) = \int T_1\left[\frac{\lambda}{r}\sqrt{\frac{\omega(n+2a)}{1 + \lambda^2\mathbb{1}_n^t\boldsymbol{X}(\boldsymbol{X}^t\boldsymbol{X} + \omega\boldsymbol{B}^{-1})^{-1}\boldsymbol{X}^t\mathbb{1}_n}} \times \mathbb{1}_n^t\left(\boldsymbol{y} - \boldsymbol{X}\hat{\boldsymbol{\beta}}\right)\middle|\, 0, 1, n+2a\right]\pi(\lambda)\, \mathrm{d}\lambda,$$

T_1 *is the c.d.f. of the* $t_1(0, 1, n+2a)$, $\hat{\beta} = (\boldsymbol{X}^t\boldsymbol{X} + \omega\boldsymbol{B}^{-1})^{-1}(\boldsymbol{X}^t\boldsymbol{y} + \omega\boldsymbol{B}^{-1}\boldsymbol{m})$ *and* $r^2 = \frac{\boldsymbol{y}^t\boldsymbol{y}}{\omega} + \boldsymbol{m}^t\boldsymbol{B}^{-1}\boldsymbol{m} - \hat{\boldsymbol{\beta}}^t(\omega^{-1}\boldsymbol{X}^t\boldsymbol{X} + \boldsymbol{B}^{-1})\hat{\boldsymbol{\beta}} + 2b$.

PROOF. See Appendix A. □

The previous proposition presents an expression of the Bayes factor to detect evidence of the data with respect to the skewness in a representable skew-elliptical model. In the particular case where H is degenerated at $\omega = 1$, we obtain skew-normal distribution for the errors. In this case the Bayes factor is given by

$$BF(\boldsymbol{y}) = \frac{1}{2\int T_1\Big[\frac{\lambda}{r}\sqrt{\frac{n+2a}{1+\lambda^2\mathbb{1}_n^t X(X^tX+B^{-1})^{-1}X^t\mathbb{1}_n}}\mathbb{1}_n^t(\boldsymbol{y}-X\hat{\boldsymbol{\beta}})\Big|0,1,n+2a\Big]\pi(\lambda)\,\mathrm{d}\lambda}, \tag{15}$$

where $\hat{\boldsymbol{\beta}} = (\boldsymbol{X}^t\boldsymbol{X} + \boldsymbol{B}^{-1})^{-1}(\boldsymbol{X}^t\boldsymbol{y} + \boldsymbol{B}^{-1}\boldsymbol{m})$ and $r^2 = \boldsymbol{y}^t\boldsymbol{y} + \boldsymbol{m}^t\boldsymbol{B}^{-1}\boldsymbol{m} - \hat{\boldsymbol{\beta}}^t(\boldsymbol{X}^t\boldsymbol{X} + \boldsymbol{B}^{-1})\hat{\boldsymbol{\beta}} + 2b$. The expression (15) is easier to calculate through numerical methods.

4. Simulation results

In this section we describe simulation results in order to study the behavior of the Bayes factor given by (15). We used the *quad* integration subroutine of MATLAB package, which is based on the recursive adaptive Simpson quadrature method.

In this simulation study we generated data from the linear regression model (13), where $\boldsymbol{\beta} = (2, 1)^t$, $\boldsymbol{\varepsilon} \sim SN_n(\boldsymbol{0}, \phi^{-1}\boldsymbol{I}_n, \lambda\mathbb{1}_n)$ with the following values variety: $n = 50$ and 100; $\phi = 0.01$, 1 and 100; and $\lambda = 0, \ldots, 5$. Also, the design matrix was given by

$$\boldsymbol{X}^t = \mathbb{1}^t_{\frac{n}{10}} \otimes \begin{pmatrix} & & \mathbb{1}^t_{10} & & \\ -4 & -3 & \cdots & 4 & 5 \end{pmatrix},$$

where $\otimes$ denotes the Kronecker product of two matrices.

For each one of the 36 previously described models, we made 100 replicates and, for each one of these, we calculated the Bayes factor given by (15) under the following prior distributions, $\boldsymbol{\beta}|\phi \sim N_2[(2, 1)^t, \phi^{-1}v\boldsymbol{I}_2]$, $\phi \sim Ga(a, b)$ and $\lambda \sim Ga(a_\lambda, b_\lambda)$. Tables 2 to 4 display the first, second and third quartile (Q_1, Q_2 and Q_3, respectively) of the Bayes factor estimated from the 100 replicates in each one of the 36 models. Each table shows the results for different prior conditions where a_λ and b_λ are chosen so that they fulfill $\mathbb{E}(\lambda) = a_\lambda/b_\lambda$ and $\mathbb{V}(\lambda) = a_\lambda/b_\lambda^2$, where $\mathbb{E}(\lambda)$ and $\mathbb{V}(\lambda)$ are initially given.

Results for concentrated prior distributions are shown in Table 2. In this case, we are assuming $\boldsymbol{\beta} \sim t_2((2, 1)^t, \boldsymbol{I}_2, 2)$, $\phi \sim Ga(1, 100)$ and $\lambda \sim Ga(62.5, 0.025)$. In general, the results of this table are good, although the values of the Bayes factor are not very different from 1. In this sense, the results for the case $\phi = 100$ are the worst, but this was expected because the prior distribution taken for ϕ has mean equals to 0.01 and variance, 10^{-4}.

Prior conditions of Table 3 are similar to the previous table, except that $\lambda \sim Ga(0.625, 2.5)$. Due to this, nice results are obtained even when $\phi = 100$. We expected this improvement since this prior distribution for λ is more in agreement with the true values of λ which we used to generate the data.

In Table 4 we increased the prior variance of λ with the purpose of observing the behavior of the Bayes factor with respect to a vague informative prior distribution for λ. As we expected, the Bayes factor does not work when we use a vague informative prior distributions. This fact has been broadly reported, see, for example, Kass and Raftery (1995). Also, in order to obtain these figures, the computing time was large.

These results corroborate the well known result about the great influence that the prior distributions exert on the Bayes factor. Indeed, in these results the prior distributions almost determine the behavior of the Bayes factor given by (15). Nice results are

Table 2
Simulation results with $v = 0.01$, $a = 1$, $b = 100$, $\mathbb{E}(\lambda) = 2.5$ and $\mathbb{V}(\lambda) = 0.1$

		$\phi = 0.01$			$\phi = 1$			$\phi = 100$		
n	λ	Q_1	Median	Q_3	Q_1	Median	Q_3	Q_1	Median	Q_3
	0	0.6552	0.9870	2.0963	0.8498	1.0306	1.4500	0.9715	1.0034	1.0351
	1	0.5868	0.7353	1.0429	0.6795	0.7850	0.8893	0.9523	0.9680	0.9812
50	2	0.5883	0.7492	1.0438	0.6894	0.7806	0.8626	0.9429	0.9682	0.9832
	3	0.5799	0.7445	0.9679	0.6826	0.7830	0.8802	0.9585	0.9733	0.9869
	4	0.5868	0.7381	1.0663	0.7355	0.8237	0.9177	0.9524	0.9694	0.9842
	5	0.5544	0.7225	0.9950	0.6664	0.7786	0.8759	0.9558	0.9677	0.9843
	0	0.7211	1.0683	1.7372	0.8510	1.0087	1.3086	0.9662	0.9922	1.0253
	1	0.6365	0.7553	0.8552	0.7183	0.8058	0.8997	0.9568	0.9734	0.9906
100	2	0.6209	0.7392	0.8509	0.7157	0.8105	0.9029	0.9597	0.9797	0.9907
	3	0.6393	0.7240	0.8196	0.7281	0.8096	0.8928	0.9523	0.9766	0.9856
	4	0.6168	0.7031	0.8688	0.7444	0.8201	0.9035	0.9567	0.9750	0.9874
	5	0.6193	0.7119	0.8347	0.7217	0.8411	0.9125	0.9574	0.9734	0.9875

Table 3
Simulation results with $v = 0.01$, $a = 1$, $b = 100$, $\mathbb{E}(\lambda) = 2.5$ and $\mathbb{V}(\lambda) = 1$

		$\phi = 0.01$			$\phi = 1$			$\phi = 100$		
n	λ	Q_1	Median	Q_3	Q_1	Median	Q_3	Q_1	Median	Q_3
	0	0.7112	1.0213	2.8531	0.8502	1.0305	1.4481	0.9715	1.0005	1.0268
	1	0.5492	0.6399	0.7831	0.6801	0.7854	0.8896	0.9528	0.9683	0.9812
50	2	0.5502	0.6361	0.7390	0.6899	0.7811	0.8630	0.9430	0.9673	0.9832
	3	0.5551	0.6342	0.7696	0.6832	0.7835	0.8805	0.9586	0.9743	0.9873
	4	0.5893	0.6832	0.8349	0.7361	0.8241	0.9179	0.9532	0.9700	0.9842
	5	0.5387	0.6156	0.7588	0.6670	0.7791	0.8762	0.9556	0.9678	0.9844
	0	0.7549	1.0765	1.9039	0.7951	1.0213	1.2740	0.9735	1.0041	1.0318
	1	0.6087	0.6974	0.8090	0.7496	0.8332	0.9082	0.9560	0.9763	0.9882
100	2	0.6506	0.7441	0.8546	0.7240	0.7896	0.8909	0.9540	0.9728	0.9869
	3	0.6611	0.7486	0.8435	0.7116	0.7989	0.8917	0.9570	0.9718	0.9844
	4	0.6258	0.7617	0.8715	0.7579	0.8314	0.9249	0.9521	0.9696	0.9894
	5	0.6241	0.7215	0.8494	0.7209	0.8204	0.8876	0.9540	0.9711	0.9860

only obtained when, a priori, the idea is had about the true values of the parameters is nearly right.

5. Conclusions

In this chapter we evaluate the sensitivity of the skewness parameter using the L_1-distance between symmetric and asymmetric models. These models are given by the p.d.f. $\frac{2}{\sigma} f(\frac{x-\mu}{\sigma})G(\lambda\frac{x-\mu}{\sigma})$, where λ is the skewness parameter, f is a symmetric p.d.f. about 0, and G an absolutely continuous distribution function such that its first derivative G' is symmetric about 0. We present explicit expressions of the L_1 distance for

Table 4
Simulation results with $\upsilon = 0.01$, $a = 1$, $b = 100$, $\mathbb{E}(\lambda) = 2.5$ and $\mathbb{V}(\lambda) = 100$

		$\phi = 0.01$			$\phi = 1$			$\phi = 100$		
n	λ	Q_1	Median	Q_3	Q_1	Median	Q_3	Q_1	Median	Q_3
	0	0.9821	1.1157	1.3673	1.0198	1.0986	1.1782	1.0855	1.0967	1.1062
	1	0.8714	0.9344	1.0135	0.9746	1.0112	1.0447	1.0811	1.0894	1.0939
50	2	0.8750	0.9366	0.9924	0.9544	1.0095	1.0490	1.0803	1.0855	1.0924
	3	0.8790	0.9343	1.0049	0.9872	1.0256	1.0609	1.0812	1.0886	1.0925
	4	0.9046	0.9631	1.0362	0.9757	1.0144	1.0518	1.0790	1.0859	1.0920
	5	0.8653	0.9245	1.0029	0.9794	1.0097	1.0529	1.0808	1.0864	1.0923
	0	0.9735	1.0639	1.2652	1.0185	1.0861	1.1996	1.0885	1.0949	1.1050
	1	0.9154	0.9755	1.0450	0.9818	1.0140	1.0586	1.0787	1.0864	1.0909
100	2	0.9416	0.9988	1.0414	0.9980	1.0272	1.0616	1.0778	1.0864	1.0912
	3	0.9092	0.9747	1.0281	0.9921	1.0349	1.0648	1.0804	1.0862	1.0893
	4	0.9237	0.9874	1.0326	0.9881	1.0203	1.0567	1.0779	1.0851	1.0895
	5	0.9240	0.9626	1.0349	0.9801	1.0095	1.0630	1.0777	1.0855	1.0890

some skew models. Also, the L_1 distance between representable skew elliptical densities equals to the L_1 distance between normal and skew normal densities.

The influence of the skewness parameter on the posterior distribution of (μ, σ) only depends on a factor which is the perturbation function between the posterior distributions when $\lambda = 0$ and $\lambda \neq 0$.

Finally, we found closed expressions for the Bayes factor to detect asymmetry on error terms of linear regression models. The results are illustrated with simulations.

Acknowledgements

We thank Guido del Pino for his enlightening comments. The first two authors acknowledge the financial support of the DIAT, Universidad de Talca and the FONDECYT Projects 1050995 and 1030588, Fondo Nacional de Ciencia y Tecnología de Chile.

Appendix A: Proof of Proposition 3.7

Notice that

$$\begin{aligned} & f(\boldsymbol{y}|\phi, \lambda, \omega) \\ & \quad = \int_{\mathbb{R}^k} f(\boldsymbol{y}|\boldsymbol{\beta}, \phi, \lambda, \omega)\pi(\boldsymbol{\beta}|\phi)\,\mathrm{d}\boldsymbol{\beta} \\ & \quad = \int_{\mathbb{R}^k} 2N_n\left(\boldsymbol{y}|\boldsymbol{X\beta}, \frac{\omega}{\phi}\boldsymbol{I}_n\right)\Phi\left[\left(\frac{\phi}{\omega}\right)^{1/2}\lambda\mathbb{1}_n^t(\boldsymbol{y}-\boldsymbol{X\beta})\right]N_k(\boldsymbol{\beta}|\boldsymbol{m}, \phi^{-1}\boldsymbol{B})\,\mathrm{d}\boldsymbol{\beta} \end{aligned}$$

$$= \frac{2\phi^{(n+k)/2}|\boldsymbol{B}|^{-1/2}}{(2\pi)^{(n+k)/2}\omega^{n/2}} \int_{\mathbb{R}^k} \exp\left\{-\frac{\phi}{2\omega}\left[\|\boldsymbol{y} - \boldsymbol{X\beta}\|^2 + \omega(\boldsymbol{\beta} - \boldsymbol{m})^t \boldsymbol{B}^{-1}(\boldsymbol{\beta} - \boldsymbol{m})\right]\right\} \Phi\left[\left(\frac{\phi}{\omega}\right)^{1/2} \lambda \mathbb{1}_n^t(\boldsymbol{y} - \boldsymbol{X\beta})\right] \mathrm{d}\boldsymbol{\beta},$$

but since

$$\begin{aligned} \|\boldsymbol{y} - \boldsymbol{X\beta}\|^2 + \omega(\boldsymbol{\beta} - \boldsymbol{m})^t \boldsymbol{B}^{-1}(\boldsymbol{\beta} - \boldsymbol{m}) &= (\boldsymbol{\beta} - \hat{\boldsymbol{\beta}})^t(\boldsymbol{X}^t\boldsymbol{X} + \omega\boldsymbol{B}^{-1})(\boldsymbol{\beta} - \hat{\boldsymbol{\beta}}) \\ &\quad - \hat{\boldsymbol{\beta}}^t(\boldsymbol{X}^t\mathbf{X} + \omega\boldsymbol{B}^{-1})\hat{\boldsymbol{\beta}} + \boldsymbol{y}^t\boldsymbol{y} + \omega\boldsymbol{m}^t\boldsymbol{B}^{-1}\boldsymbol{m}, \end{aligned}$$

where

$$\hat{\boldsymbol{\beta}} = (\boldsymbol{X}^t\boldsymbol{X} + \omega\boldsymbol{B}^{-1})^{-1}(\boldsymbol{X}^t\boldsymbol{y} + \omega\boldsymbol{B}^{-1}\boldsymbol{m}),$$

then

$$\begin{aligned} f(\boldsymbol{y}|\phi, \lambda, \omega) &= \frac{2\phi^{n/2}|\boldsymbol{B}|^{-1/2}}{(2\pi)^{n/2}\omega^{(n-k)/2}|\boldsymbol{X}^t\boldsymbol{X} + \omega\boldsymbol{B}^{-1}|^{1/2}} h(\phi, \lambda, \omega) \\ &\quad \times \exp\left\{-\frac{\phi}{2\omega}\left[\boldsymbol{y}^t\boldsymbol{y} + \omega\boldsymbol{m}^t\boldsymbol{B}^{-1}\boldsymbol{m} - \boldsymbol{\beta}^t(\boldsymbol{X}^t\boldsymbol{X} + \omega\boldsymbol{B}^{-1})\hat{\boldsymbol{\beta}}\right]\right\}, \end{aligned} \tag{A.1}$$

where

$$\begin{aligned} h(\phi, \lambda, \omega) &= \int_{\mathbb{R}^k} \frac{(\phi/\omega)^{k/2}}{(2\pi)^{k/2}|\boldsymbol{X}^t\boldsymbol{X} + \omega\boldsymbol{B}^{-1}|^{-1/2}} \\ &\quad \times \exp\left\{-\frac{\phi}{2\omega}(\boldsymbol{\beta} - \hat{\boldsymbol{\beta}})^t(\boldsymbol{X}^t\boldsymbol{X} + \omega\boldsymbol{B}^{-1})(\boldsymbol{\beta} - \hat{\boldsymbol{\beta}})\right\} \\ &\quad \times \Phi\left[\left(\frac{\phi}{\omega}\right)^{1/2} \lambda \mathbb{1}_n^t(\boldsymbol{y} - \boldsymbol{X\beta})\right] \mathrm{d}\boldsymbol{\beta} \\ &= \int_{\mathbb{R}^k} \Phi\left\{\left(\frac{\phi}{\omega}\right)^{1/2} \lambda \mathbb{1}_n^t\left[\boldsymbol{y} - \boldsymbol{X}(\tilde{\boldsymbol{\beta}} + \hat{\boldsymbol{\beta}})\right]\right\} \\ &\quad \times \boldsymbol{\phi}_k\left[\tilde{\boldsymbol{\beta}}|\mathbf{0}, \frac{\omega}{\phi}(\boldsymbol{X}^t\boldsymbol{X} + \omega\boldsymbol{B}^{-1})^{-1}\right] \mathrm{d}\tilde{\boldsymbol{\beta}} \\ &= \mathbb{E}_{\tilde{\boldsymbol{\beta}}}\left\{\Phi\left[\left(\frac{\phi}{\omega}\right)^{1/2} \lambda \mathbb{1}_n^t(\boldsymbol{y} - \boldsymbol{X}\hat{\boldsymbol{\beta}}) - \left(\frac{\phi}{\omega}\right)^{1/2} \lambda \mathbb{1}_n^t \boldsymbol{X}\tilde{\boldsymbol{\beta}}\right]\right\}. \end{aligned}$$

From Lemma 3.6 we have

$$h(\phi, \lambda, \omega) = \Phi\left\{\frac{\left(\frac{\phi}{\omega}\right)^{1/2} \lambda \mathbb{1}_n^t(\boldsymbol{y} - \boldsymbol{X}\hat{\boldsymbol{\beta}})}{[1 + \lambda^2 \mathbb{1}_n^t \boldsymbol{X}(\boldsymbol{X}^t\boldsymbol{X} + \omega\boldsymbol{B}^{-1})^{-1}\boldsymbol{X}^t\mathbb{1}_n]^{1/2}}\right\}.$$

Now, if we substitute $h(\phi, \lambda, \omega)$ in (A.1), we get

$$f(\boldsymbol{y}|\phi, \lambda, \omega) = \frac{2\phi^{n/2}|\boldsymbol{B}|^{-1/2}}{(2\pi)^{n/2}\omega^{(n-k)/2}|\boldsymbol{X}^t\boldsymbol{X} + \omega\boldsymbol{B}^{-1}|^{1/2}}$$

$$\times \Phi\left\{\frac{\left(\frac{\phi}{\omega}\right)^{1/2}\lambda \mathbb{1}_n^t(\boldsymbol{y}-\boldsymbol{X}\hat{\boldsymbol{\beta}})}{[1+\lambda^2\mathbb{1}_n^t\boldsymbol{X}(\boldsymbol{X}^t\boldsymbol{X}+\omega\boldsymbol{B}^{-1})^{-1}\boldsymbol{X}^t\mathbb{1}_n]^{1/2}}\right\}$$
$$\times \exp\left\{-\frac{\phi}{2\omega}\left[\boldsymbol{y}^t\boldsymbol{y}+\omega\boldsymbol{m}^t\boldsymbol{B}^{-1}\boldsymbol{m}-\boldsymbol{\beta}^t\left(\boldsymbol{X}^t\boldsymbol{X}+\omega\boldsymbol{B}^{-1}\right)\hat{\boldsymbol{\beta}}\right]\right\}.$$

But since $\phi \sim Ga(a,b)$, then

$$f(\boldsymbol{y}|\lambda,\omega) = \int_0^\infty f(\boldsymbol{y}|\phi,\lambda,\omega)Ga(\phi|a,b)\,\mathrm{d}\phi$$
$$= \frac{2b^a|\boldsymbol{B}|^{-1/2}}{(2\pi)^{n/2}\omega^{(n-k)/2}|\boldsymbol{X}^t\boldsymbol{X}+\omega\boldsymbol{B}^{-1}|^{1/2}\Gamma(a)}$$
$$\times \int_0^\infty \Phi\left\{\frac{\left(\frac{\phi}{\omega}\right)^{1/2}\lambda \mathbb{1}_n^t(\boldsymbol{y}-\boldsymbol{X}\hat{\boldsymbol{\beta}})}{[1+\lambda^2\mathbb{1}_n^t\boldsymbol{X}(\boldsymbol{X}^t\boldsymbol{X}+\omega\boldsymbol{B}^{-1})^{-1}\boldsymbol{X}^t\mathbb{1}_n]^{1/2}}\right\}$$
$$\times \phi^{n/2+a-1}\exp\left\{-\frac{\phi r^2}{2}\right\}\mathrm{d}\phi,$$

where $r^2 = \frac{\boldsymbol{y}^t\boldsymbol{y}}{\omega} + \boldsymbol{m}^t\boldsymbol{B}^{-1}\boldsymbol{m} - \hat{\boldsymbol{\beta}}^t(\omega^{-1}\boldsymbol{X}^t\boldsymbol{X}+\boldsymbol{B}^{-1})\hat{\beta} + 2b$. Now, from Lemma 3.3, we obtain

$$f(\boldsymbol{y}|\lambda,\omega) = \frac{2^{n/2+a+1}b^a\Gamma\left(\frac{n}{2}+a\right)|\boldsymbol{B}|^{-1/2}}{(2\pi)^{n/2}\omega^{(n-k)/2}|\boldsymbol{X}^t\boldsymbol{X}+\omega\boldsymbol{B}^{-1}|^{1/2}\Gamma(a)r^{n+2a}}$$
$$\times T_1\left\{\frac{\lambda}{r}\sqrt{\frac{\omega(n+2a)}{1+\lambda^2\mathbb{1}_n^t\boldsymbol{X}(\boldsymbol{X}^t\boldsymbol{X}+\omega\boldsymbol{B}^{-1})^{-1}\boldsymbol{X}^t\mathbb{1}_n}}\right.$$
$$\left.\times \mathbb{1}_n^t\left(\boldsymbol{y}-\boldsymbol{X}\hat{\boldsymbol{\beta}}\right)|0,1,n+2a\right\},$$

where T_1 is the c.d.f. of the $t_1(0,1,n+2a)$.

On the other hand, when $\lambda = 0$:

$$f(\boldsymbol{y}|\phi,\omega) = \int_{\mathbb{R}^k} f(\boldsymbol{y}|\boldsymbol{\beta},\phi,\omega)\pi(\boldsymbol{\beta}|\phi)\,\mathrm{d}\boldsymbol{\beta}$$
$$= \int_{\mathbb{R}^k} N_n\left(\boldsymbol{y}|\boldsymbol{X}\boldsymbol{\beta},\frac{\omega}{\phi}\boldsymbol{I}_n\right)N_k(\boldsymbol{\beta}|\boldsymbol{m},\phi^{-1}\boldsymbol{B})\,\mathrm{d}\boldsymbol{\beta}$$
$$= \frac{\phi^{(n+k)/2}|\boldsymbol{B}|^{-1/2}}{(2\pi)^{(n+k)/2}\omega^{n/2}}$$
$$\times \int_{\mathbb{R}^k}\exp\left\{-\frac{\phi}{2\omega}\left[\|\boldsymbol{y}-\boldsymbol{X}\boldsymbol{\beta}\|^2+\omega(\boldsymbol{\beta}-\boldsymbol{m})^t\boldsymbol{B}^{-1}(\boldsymbol{\beta}-\boldsymbol{m})\right]\right\}\mathrm{d}\boldsymbol{\beta}$$
$$= \frac{\phi^{(n+k)/2}|\boldsymbol{B}|^{-1/2}}{(2\pi)^{(n+k)/2}\omega^{n/2}}$$

$$\times \exp\left\{-\frac{\phi}{2}\left[\frac{\mathbf{y}^t\mathbf{y}}{\omega} + \mathbf{m}^t\mathbf{B}^{-1}\mathbf{m} - \hat{\boldsymbol{\beta}}^t\left(\omega^{-1}\mathbf{X}^t\mathbf{X} + \mathbf{B}^{-1}\right)\hat{\boldsymbol{\beta}}\right]\right\}$$

$$\times \int_{\mathbb{R}^k} \exp\left\{-\frac{\phi}{2\omega}\left[(\boldsymbol{\beta} - \hat{\boldsymbol{\beta}})^t\left(\mathbf{X}^t\mathbf{X} + \omega\mathbf{B}^{-1}\right)(\boldsymbol{\beta} - \hat{\boldsymbol{\beta}})\right]\right\} \mathrm{d}\boldsymbol{\beta}$$

$$= \frac{\phi^{n/2}|\mathbf{B}|^{-1/2}}{(2\pi)^{n/2}\omega^{(n-k)/2}|\mathbf{X}^t\mathbf{X} + \omega\mathbf{B}^{-1}|^{1/2}}$$

$$\times \exp\left\{-\frac{\phi}{2}\left[\frac{\mathbf{y}^t\mathbf{y}}{\omega} + \mathbf{m}^t\mathbf{B}^{-1}\mathbf{m} - \boldsymbol{\beta}^t\left(\omega^{-1}\mathbf{X}^t\mathbf{X} + \mathbf{B}^{-1}\right)\hat{\boldsymbol{\beta}}\right]\right\}.$$

Considering that $\phi \sim Ga(a, b)$, we obtain

$$f(\mathbf{y}|\omega) = \frac{b^a|\mathbf{B}|^{-1/2}}{(2\pi)^{n/2}\omega^{(n-k)/2}|\mathbf{X}^t\mathbf{X} + \omega\mathbf{B}^{-1}|^{1/2}\Gamma(a)}$$

$$\times \int_0^\infty \phi^{n/2+a-1} \exp\left\{-\frac{\phi r^2}{2}\right\} \mathrm{d}\phi$$

$$= \frac{2^{n/2+a}b^a\Gamma\left(\frac{n}{2} + a\right)|\mathbf{B}|^{-1/2}}{(2\pi)^{n/2}\omega^{(n-k)/2}|\mathbf{X}^t\mathbf{X} + \omega\mathbf{B}^{-1}|^{1/2}\Gamma(a)r^{n+2a}}.$$

Since the Bayes factor is $BF(\mathbf{y}) = \frac{\int f(\mathbf{y}|\omega)\,\mathrm{d}H(\omega)}{\iint f(\mathbf{y}|\lambda,\omega)\pi(\lambda)\,\mathrm{d}\lambda\,\mathrm{d}H(\omega)}$, then we obtain the result.

References

Arellano-Valle, R.B., Galea-Rojas, M. (2005). On fundamental skew distributions. *J. Multivariate Anal.* **96**, 93–116.

Arellano-Valle, R.B., Galea-Rojas, M., Iglesias, P. (2000). Bayesian sensitivity analysis in elliptical linear regression models. *J. Statist. Plann. Inference* **86**, 175–199.

Arellano-Valle, R.B., del Pino, G., Martín, E.S. (2002). Definition and probabilistic properties of skew-distributions. *Statist. Probab. Lett.* **58**, 111–121.

Arellano-Valle, R.B., Iglesias, P., Vidal, I. (2003). Bayesian inference for elliptical linear models: Conjugate analysis and model comparison. In: Bernardo, J.M., Bayarri, M.J., Berger, J.O., Dawid, A.P., Heckerman, D., Smith, A.F.M., West, M. (Eds.), *Bayesian Statistics, vol. 7*. Oxford Univ. Press, Oxford.

Arellano-Valle, R.B., Gómez, H., Iglesias, P. (2004). Shape mixtures of skewed distributions: A Bayesian interpretation. Technical Report, Pontificia Universidad Católica de Chile, Chile.

Azzalini, A. (1985). A class of distributions which includes the normal ones. *Scand. J. Statist.* **12**, 171–178.

Azzalini, A., Capitanio, A. (1999). Statistical applications of the multivariate skew normal distribution. *J. Roy. Statist. Soc., Ser. B* **61**, 579–602.

Azzalini, A., Capitanio, A. (2003). Distributions generated by perturbation of symmetry with emphasis on a multivariate skew *t* distribution. *J. Roy. Statist. Soc., Ser. B* **65**, 367–389.

Azzalini, A., Dalla-Valle, A. (1996). The multivariate skew-normal distribution. *Biometrika* **83**, 715–726.

Branco, M.D., Dey, D.K. (2001). A general class of multivariate skew-elliptical distributions. *J. Multivariate Anal.* **79**, 99–113.

Branco, M.D., Bolfarine, H., Iglesias, P. (1998). Bayesian calibration under a Student-*t* model. *Comput. Statist.* **13**, 319–338.

Branco, M.D., Bolfarine, H., Iglesias, P., Arellano-Valle, R.B. (2001). Bayesian analysis of the calibration problem under elliptical distributions. *J. Statist. Plann. Inference* **90**, 69–95.

Gupta, A.K., González-Farías, G., Domínguez-Molina, J. (2004). A multivariate skew normal distribution. *J. Multivariate Anal.* **89** (1), 181–190.

Henze, N. (1986). A probabilistic representation of the 'skew-normal' distribution. *Scand. J. Statist.* **13**, 271–275.

Kass, R.E., Raftery, A.E. (1995). Bayes factors. *J. Amer. Statist. Assoc.* **90**, 773–795.

Kass, R.E., Tierney, L., Kadane, J.B. (1989). Approximate methods for assessing influence and sensitivity in Bayesian analysis. *Biometrika* **76**, 663–674.

Liseo, B., Loperfido, N. (2003). A Bayesian interpretation of the multivariate skew-normal distribution. *Statist. Probab. Lett.* **61**, 395–401.

Peng, F., Dey, D.K. (1995). Bayesian analysis of outlier problems using divergences measures. *Canad. J. Statist.* **23**, 199–213.

Sahu, S., Dey, D.K., Branco, M.D. (2003). A new class of multivariate skew distributions with applications to Bayesian regression models. *Canad. J. Statist.* **31**, 129–150.

Vidal, I., Iglesias, P., Branco, M.D., Arellano-Valle, R.B. (2005). Bayesian sensitivity analysis and model comparison for skew elliptical models. *J. Statist. Plann. Inference*. In press.

Weiss, R.E. (1996). An approach to Bayesian sensitivity analysis. *J. Roy. Statist. Soc., Ser. B* **58**, 739–750.

Handbook of Statistics, Vol. 25
ISSN: 0169-7161
© 2005 Elsevier B.V. All rights reserved.
DOI 10.1016/S0169-7161(05)25024-1

24

Bayesian Methods for DNA Microarray Data Analysis

Veerabhadran Baladandayuthapani, Shubhankar Ray and Bani K. Mallick

Abstract
DNA microarray technology enables us to monitor the expression levels of thousands of genes simultaneously and has emerged as a promising tool for disease diagnosis. We present a review of recent developments in Bayesian statistical methods for microarray data. In particular, we focus on Bayesian gene selection for and survival analysis. Owing to the large number of genes and the complexity of the data, we use Markov chain Monte Carlo (MCMC) based stochastic search algorithms for inference. Other recent technical developments in Bayesian modeling for microarray data are summarized. The methodology is illustrated using several well-analyzed cancer microarray datasets.

1. Introduction

Genomics study the complex interplay between a large number of genes and their products (i.e., RNA and proteins) and furthers our understanding of the biological processes in a living organism. Traditional methods in molecular biology work under "one gene per experiment"-al setups with a paucity of information, that fails to provide a larger picture of gene functions. The past few years have seen the development of the technology of DNA microarrays that increases the throughput of gene expression analysis to the level of the entire genome. A microarray is a convenient tool for analyzing gene expression that consists of a small membrane or glass slide containing samples of many genes (usually between 500–20,000) arranged in a regular pattern. DNA Microarrays allow simultaneous study of expressions for a large bunch of genes (Duggan et al., 1999; Schena et al., 1995). The mere prospect of analyzing the whole genome on a single chip is tempting for a researcher who is looking for gene interactions with possible biological implications and as a result the technology has emerged popularly into a diagnostic tool.

Microarrays may be used to assay gene expression within a single sample or to compare genes. This technology is still being developed and many studies as of now, using microarrays, have represented simple surveys of gene expression profiles in a variety of cell types. Nonetheless, these studies represent an important and necessary 'first step' in our understanding and cataloging of the human genome. With new advances, researchers will be able to infer probable functions of new genes based on similarities in expression patterns with those of known genes. Ultimately, these studies promise to expand the size of existing gene families, reveal new patterns of coordinated gene expression across gene families, and uncover entirely new categories of genes. Furthermore, because the product of any one gene usually interacts with those of many others, our understanding of how these genes coordinate will become clearer through such analyses, and precise knowledge of these inter-relationships will emerge. The use of microarrays allows faster differentiation of the genes responsible for certain biological traits or diseases by enabling scientists to examine a much larger number of genes. This technology will also aid the examination of the integration of gene expression and function at the cellular level, revealing how multiple gene products work together to produce physical and chemical responses to both static and changing cellular needs.

In this chapter on Bayesian statistical methodology for DNA microarray data analysis, we first start off by briefly reviewing the latest microarray technology in Section 2. Statistical analyses that follows from the availability of such rich data are then listed in Section 3. Bayesian gene selection, which will be the focus of this review, is undertaken in Section 4 for different instances of modeling such as selection for the multicategory case and in the context of survival analysis. The sections thereafter (5, 6 and 7), skim through other areas of Bayesian analysis for microarray data and popular methods in literature are compared. The concluding remarks are drawn in Section 8.

2. Review of microarray technology

2.1. *Biological principles*

As mentioned before microarrays measure gene expression, i.e., a gene is expressed if its DNA has been transcribed to RNA and gene expression is the level of transcription of the DNA of the gene. The primary biological processes can be viewed as information transfer process (Nguyen et al., 2002). All the necessary information for functioning of the cells are encoded in molecular units called genes. The information transfer processes are crucial processes mediating the characteristic features or phenotypes of the cells (e.g., normal and diseased cells). A schematic representation of the information transfer process is:

$$\text{DNA} \Longrightarrow \text{mRNA} \Longrightarrow \text{amino acid} \longrightarrow \text{cell phenotype}$$
$$\longrightarrow \text{organism phenotype}$$

This model shows how the genes (DNAs) are linked to the organism phenotype and the reason for measuring mRNA (transcript abundance), the direct product of DNA transcription. There are different levels of gene expression, one at transcription level where mRNA is made from DNA and another at protein level where protein is made from mRNA. Microarrays measure gene expression at the transcription level, although protein arrays have also been developed (Haab et al., 2001).

2.2. Experimental procedure

There are two primary microarray technologies: cDNA microarrays (Schena et al., 1995) and high density Oligonucleotide arrays or Oligoarrays (e.g., Affymetrix arrays). This categorization is based mainly on the way the arrays are manufactured and the way they are used. We provide here a short description of the former technology, while referring the reader to Lockhart et al. (1996) for Affymetrix arrays. The experimental procedure for cDNA microarray consists of four basic steps.

1. *Array fabrication.* This process involves preparing the glass slide, obtaining the DNA sequences (probes) and depositing the cDNA sequences onto the slide. A set of potentially relevant genes is first obtained, through the use of cDNA clones and a cDNA library. Each clone is then amplified to obtain multiple copies through a process called polymerase chain reaction (PCR) in order to have sufficient amount of each cDNA for each gene. The amplified clones are then spotted on the array using microspotting pins. Then the array goes through a series of postprocessing to enhance the binding of the cDNA to the glass surface and heated to denature the double-stranded cDNA probes. For more information see Kahn et al. (1999) and DeRisi et al. (1999).

2. *Sample preparation.* Biological samples of interest are prepared and processed, whereby the RNA (mRNA and other RNAs) from the samples is isolated.

3. *cDNA synthesis and labeling.* cDNA Arrays are designed to measure the gene expression in cells in a experimental sample relative to the same genes in a fixed reference sample. Thus, two pools of cDNA's are synthesized, one from mRNAs of the experimental cells and the other from mRNAs of the reference or control cells. cDNAs obtained from experimental and reference mRNAs are often labeled with red or green fluorescent dye called Cy5 or Cy3 respectively.

4. *Hybridization of experimental and reference cDNA targets.* The central dogma of microarrays is based on a process called hybridization, which refers to the binding of two complimentary DNA strands by base pairing. Each single stranded DNA fragment is made up of four different nucleotides, adenine (A), thymine (T), guanine (G), and cytosine (C), that are linked end to end. Adenine is the complement of, or will always pair with, thymine, and guanine is the complement of cytosine. Therefore, the complementary sequence to G–T–C–C–T–A will be C–A–G–G–A–T. When the mixed solution of the experimental and reference target cDNAs is applied to the array, the target cDNA probes should bind together by base pairing to the complementary to the probe cDNA on the array. Ideally in a given spot on the array, there will be bound experimental and reference target cDNAs of a gene of interest (say G) and the amount of the gene G's expression will show up at Cy5 and Cy3 dye intensities respectively.

2.3. Image analysis, data extraction and normalization

As mentioned before the spot intensities of the Cy5 (red) and Cy3 (green) dyes represent the expression levels of a gene in experimental or reference cells respectively. After hybridization, the array is scanned at two wavelengths or "channels", one for the Cy5 fluorescent tagged sample and another for the Cy3 tagged sample. The fluorescent tags are excited by a laser and the fluorescent intensities are obtained by scanning the array using a confocal laser microscope. For cDNA arrays the raw data consists of two gray-scale images, each through the red and green channels. Image processing tools are then applied to extract numerical data from these images. The primary purpose of the image analysis step is to extract numerical foreground and background intensities for the red and green channels for each spot on the microarray. The background intensities are used to correct the foreground intensities for local variation on the array surface, thus giving the corrected red and green intensities for each spot which becomes the primary data for subsequent analysis.

Let $\boldsymbol{F}^R = (f_{ij}^R)$ and $\boldsymbol{F}^G = (f_{ij}^G)$ be the $n \times p$ matrices containing the red and green foreground spot intensities of genes $j = 1, \ldots, p$ in samples (arrays) $i = 1, \ldots, n$. Similarly, $\boldsymbol{B}^R = (b_{ij}^R)$ and $\boldsymbol{B}^G = (b_{ij}^G)$ be the corresponding matrices for the red and green background spot intensities. It is almost universal practice to correct the foreground intensities by subtracting the background, $R = (r_{ij}) = (b_{ij}^R) - (b_{ij}^R)$ and $G = (g_{ij}) = (g_{ij}^R) - (g_{ij}^R)$, and the adjusted intensities then form the primary data for further analysis. The motivation for background adjustment is the belief that a spot's measured intensity includes a contribution not specifically due to the hybridization of the target to the probe, e.g., cross-hybridization and other chemicals on the glass. An undesirable side-effect of background correction is that negative intensities may be produced for some spots and hence missing values if log-intensities are computed, resulting in loss of information associated with low channel intensities.

Another critical step before analyzing the microarray data, is array normalization. The purpose of normalization is to adjust for any bias arising from variation in the experimental steps rather than from biological differences between the RNA samples. Such systematic bias can arise, among others, from red–green bias due to differences in target labeling methods, scanning properties of two channels perhaps by the use of different scanner settings or variation over the course of the print-run or nonuniformity in the hybridization. Thus it is necessary to normalize the intensities before any subsequent analysis is carried out. Normalization can be carried out within each array or between arrays. The simplest and most widely used within array normalization assumes that the red–green bias is constant with respect to log-scale across the array. The log-ratios are corrected by subtracting a constant c to get normalized values, i.e., $\log(r_j/g_j) \leftarrow \log(r_j/g_j) - c$. There are various methods of estimating c (see Ideker et al., 2000; Chen et al., 1997).

3. Statistical analysis of microarray data

With its inception a short while ago, early statistical literature for microarray data tended to focus on frequentist approaches that involved linear models, bootstrapping and cluster analysis (Kerr and Churchull, 2000; Black and Doerge, 2002; Van der Laan and Bryan, 2001; Eisen et al., 1998; Chen et al., 1997). Data structures for DNA microarrays are typically bulky and frequentist inference using parametric models are often not viable. Moreover, owing to the complexity, approximate inferences based on asymptotic theory are not reasonable, if not inconceivable.

The Bayesian paradigm is very well suited for analyzing DNA microarrays because the desired inferences for quantiles, credible sets, variability and predictions, directly follow from the computation of the posterior distribution. Bayesian methods have gradually gained popularity and the analysis has evolved into several well recognized areas. We only explore four broad areas of Bayesian analysis for microarray data, namely,

(i) *Gene selection and classification.* The aim is to narrow down the analysis of the enormous number of genes (from a microarray data of a diseased tissue) to a few that have significantly distinct expression profiles and are indicative of a causal relationship with certain biological phenomenon. In this review, gene selection is addressed in great detail with several instances of modeling and is well-supported by instructive examples. The various modeling cases considered are the binary variable selection case, the multicategory selection case and variable selection in survival analysis.

(ii) *Differential gene expression analysis.* Although the central theme is similar to gene selection, i.e., to identify altered gene expression profiles; the area of Bayesian models for studying differential gene expression over different conditions are notably diverse and are considered under a different heading.

(iii) *Gene clustering.* When there is no well-defined target (phenotypical) and it is of interest to break down the genes into separate classes with significantly different expression profiles. This is helpful in categorizing a variety of simultaneous (and possibly previously unknown) biological activities as they occur over time.

(iv) *Regression in grossly overparametrized models.* Regression models dealing with microarray data are typified by overparameterization, i.e., the set of interesting predictors is very large compared to the observations. Literature on this area is still sparse, therefore we review a few models which try to overcome these limitations.

4. Bayesian models for gene selection

One of the key goals of microarray data is to perform classification via different expression profiles. In principle, gene expression profiles might serve as molecular fingerprints that would allow for accurate classification of diseases. The underlying assumption is that samples from the same class share expression profile patterns unique to their class (Yeang et al., 2001). In addition, these molecular fingerprints might reveal newer taxonomies that previously have not been readily appreciated. Several studies have used microarrays to profile colon, breast and other tumors and have demonstrated the potential power of expression profiling for classification (Alon et al., 1999; Golub et al., 1999; Hedenfalk et al., 2001).

The problem we focus on here is on finding sets of genes that relate to different kinds of diseases, so that future samples can be classified correctly. Such models may allow identification of the genes that best discriminate between different types of tissue, i.e., gene (variable) selection. In common gene selection or classification analyses, one deals with a large collection of genes and one is looking for a small subset of genes that best discriminate between normal and diseased tissues. Bayesian statistical methods for modeling differences in gene expression profiles are predominantly parametric in nature and we shall contain ourselves with this area for the larger part of this review.

Some work on parametric models to characterize patterns of gene expression are by Newton et al. (2001), Chen et al. (1997) and West et al. (2003). Related work also includes that of Efron et al. (2001), who used a nonparametric empirical Bayes procedure for gene profiling. Bayesian variable selection methods for regression were proposed by Leamer (1978), Mitchell and Beauchamp (1988) and more recently by George and McCulloch (1997). Extension to a multivariate setting have been proposed by Brown et al. (1998). Albert (1996) proposed an approach to loglinear models, focusing on Poisson loglinear models with applications to multiway contingency tables, and to generalized linear models by Raftery (1996), who proposed methods to calculate approximate Bayes factors. Bayesian variable selection for categorical data have been proposed by Lee et al. (2003) for the binary case and extended to multicategory case by Sha et al. (2004).

In a number of published studies, the number of selected genes is large, for, e.g., 495 (Khan et al., 1998) and 2000 genes (Alon et al., 1999). Even in studies where smaller number of genes are obtained, the numbers of genes are excessive compared to the sample size (number of microarrays), for, e.g., 50 genes by Golub et al. (1999) and 96 genes (Khan et al., 2001). Such large number of genes combined with the small sample size creates a unreliable selection process (Dougherty, 2001). Various approaches have been proposed in literature for this gene selection problem (see Dudoit et al., 2002; Tusher et al., 2001; Kim et al., 2002 among others). In this section, we present model-based approaches for the gene selection problem. We start with the simplest case of gene selection when the response is binary and later extend the methodology for the multicategory case and in survival analysis models. We illustrate the methodology by analyzing the most recent microarray datasets in literature.

4.1. Gene selection for binary classification

We present here a Bayesian hierarchical model based approach to gene selection for binary classification problems following Lee et al. (2003). Rather than fixing the number of selected genes, we assign a prior distribution over it, which induces greater flexibility. Additionally, constraints can be placed on the number of genes to selected. For example, it might be feasible to restrict the model dimension to an acceptable number (say 10). As our model space is very large, i.e., with p genes we have 2^p possible models, exhaustive computation over this model space is not possible. Hence Markov chain Monte Carlo (MCMC; Gilks et al., 1996) based stochastic search algorithms are used. We are interested in identifying significant set(s) of genes over this vast model space. The microarray dataset from Alon et al. (1999) is used as an example to illustrate the methodology.

For a binary class problem the response is usually coded as $Y_i = 1$ for class 1 (diseased tissue) and $Y_i = 0$ (normal tissue) for the other class, where $i = 1, \ldots, n$ and where n is the number of samples (arrays). Gene expression data for p genes for n samples is summarized in an $n \times p$ matrix, X, where each element x_{ij} denotes the expression level (gene-expression value) of the jth gene in the ith sample where $j =$

$1, \ldots, p$ as,

$$\begin{bmatrix} & \text{Gene 1} & \text{Gene 2} & \cdots & \text{Gene } p \\ Y_1 & X_{11} & X_{12} & \cdots & X_{1p} \\ Y_2 & X_{21} & X_{22} & \cdots & X_{2p} \\ \vdots & \vdots & \vdots & \ddots & \vdots \\ Y_n & X_{n1} & X_{n2} & \cdots & X_{np} \end{bmatrix}. \tag{1}$$

Our objective is to use the training data $Y = (Y_1, \ldots, Y_n)'$ to estimate $p(X) = \Pr(Y = 1|X)$, the probability that a sample is diseased. Assuming Y_i has an independent binary distribution so that $Y_i = 1$ with probability p_i, independent of other $Y_j,\ j \neq i$, the expression levels of the genes can then be related to the response using a probit regression model as,

$$\Pr(Y_i = 1|\beta) = \Phi\big(X_i'\beta\big),$$

where X_i is the ith row of the matrix X (vector of gene expression levels of the ith sample), β is the vector of regression coefficients and Φ is the normal cumulative distribution function.

Following Albert and Chib (1993), n independent latent variables $Z_1, \ldots, Z_n$ are introduced with each $Z_i \sim \mathrm{N}(X_i'\beta, \sigma^2)$ and where

$$Y_i = \begin{cases} 1 & \text{if } Z_i > 0, \\ 0 & \text{if } Z_i < 0. \end{cases}$$

Hence, we assume the existence of a continuous unobserved or latent variable underlying the observed categorical variable. When the latent variable crosses a threshold, the observed categorical variable takes on a different value. The key to this data augmentation approach is to transform the model into a normal linear model on the latent responses, as now $Z_i = X_i'\beta + \varepsilon_i$ where $\varepsilon_i \sim \mathrm{N}(0, \sigma^2)$ where ε_i are the residual random effects. The residual random effects account for the unexplained sources of variation in the data, most probably due to explanatory variables (genes) not included in the study. We fix $\sigma^2 = 1$ as it is unidentified in the model.

A Gaussian mixture prior for β enables us to perform a variable selection procedure. Define $\gamma = (\gamma_1, \ldots, \gamma_p)$, such that $\gamma_i = 1$ or 0 indicates whether the ith gene is selected or not ($\beta_i = 1$ or 0). Given γ, let $\beta_\gamma = \{\beta_i \in \beta\colon\ \beta_i \neq 0\}$ and X_γ be the columns of X corresponding to β_γ. The prior distribution on β_γ is taken as $\mathrm{N}(0, c(X_\gamma' X_\gamma)^{-1})$ with c as a positive scalar (usually between 10 and 100. See Smith and Kohn, 1997). The indicators γ_i are assumed to be apriori independent with $\Pr(\gamma_i = 1) = \pi_i$, which are chosen to be small to restrict the number of genes. The Bayesian hierarchical model for gene selection can then be summarized as,

$$\begin{aligned} (Z|\beta_\gamma, \gamma) &\sim \mathrm{N}\big(X_\gamma\beta_\gamma, \sigma^2\big), \\ (\beta_\gamma|\gamma) &\sim \mathrm{N}\big(0, c\big(X_\gamma' X_\gamma\big)^{-1}\big), \\ \gamma_i &\sim \mathrm{Bernoulli}(\pi_i). \end{aligned}$$

With the above model and prior setup the conditional distributions of all the unknown random variables and parameters, (Z, β, γ) have closed form posterior conditional distributions. We use MCMC techniques (Gilks et al., 1996) for inference, specifically Gibbs sampling (Gelfand and Smith, 1990) to generate posterior samples. The latent variables Z, conditional on β, has simple form,

$$(Z_i|\beta, Y_i = 1) \sim \mathrm{N}(X_i'\beta, \sigma^2) \quad \text{truncated at the left by 0,}$$
$$(Z_i|\beta, Y_i = 0) \sim \mathrm{N}(X_i'\beta, \sigma^2) \quad \text{truncated at the right by 0.}$$

The conditional distribution of Z is a truncated normal distribution, and the optimal exponential accept–reject algorithm could be used (Robert, 1995). Using standard linear model results, with the conjugate normal prior on β, the conditional distribution of β is also normal, $p(\beta|\gamma, Z) \sim \mathrm{N}(V_\gamma X_\gamma' Z_\gamma, V_\gamma)$ where $V_\gamma = (c/1+c)(X_\gamma' X_\gamma)^{-1}$. Rather than drawing γ from its complete conditional, we draw γ from it marginal distribution, $p(\gamma|Z)$ (integrating β out) thereby gaining considerable computational advantage. Note here that the modified Gibbs sampler still leaves the target posterior distribution invariant. We draw γ componentwise from its marginal $\Pr(\gamma_i|Z, \gamma_{j\neq i})$ given by,

$$\Pr(\gamma_i|Z, \gamma_{j\neq i}) \\ \propto \exp\left[-1/2\left(Z'Z - \frac{c}{1+c} Z'X_\gamma\left(X_\gamma'X_\gamma\right)^{-1}X_\gamma'Z\right)\right]\pi_i^{\gamma_i}(1-\pi_i)^{1-\gamma_i}.$$

The relative importance of each gene can be assessed by the frequency of its appearance in the MCMC samples, i.e., the number of times the corresponding component of γ is 1. This gives us an estimate of the posterior probability of inclusion of a single gene as a measure of the relative importance of the gene for classification purpose. For a new sample with gene expression X_{new}, the marginal posterior distribution of the new disease state, Y_{new} is given by

$$\Pr(Y_{\text{new}} = 1|X_{\text{new}}) = \int_{\mathcal{M}} P(Y_{\text{new}} = 1|X_{\text{new}}, \mathcal{M}) P(\mathcal{M}|Y)\, \mathrm{d}\mathcal{M}, \tag{2}$$

where $\mathcal{M} = \{Z, \beta, \gamma\}$, the unknowns in the model, $\Pr(\mathcal{M}|Y)$ is the posterior probability of the unknown random variables and parameters in the model. The integral given in (2) is analytically intractable and needs approximate computational procedures. We approximate (2) by its Monte Carlo estimate by

$$\Pr(Y_{\text{new}} = 1|X_{\text{new}}) = \frac{1}{m}\sum_{j=1}^{m} P\left(Y_{\text{new}} = 1|X_{\text{new}}, \mathcal{M}^{(j)}\right), \tag{3}$$

where $\mathcal{M}^{(j)}$ for $j = 1, \ldots, m$ are the m MCMC posterior samples of the model unknowns $\mathcal{M}$ and can be easily evaluated using normal CDF. The approximation (3) converges to the true value (2) as $m \to \infty$.

In order to select from different models, we generally use misclassification error. When a test set is provided, we first obtain the posterior distribution of the parameters based on training data, Y_{trn} (train the model) and use them to classify the test samples. For a new observation from the test set, $Y_{i,\text{test}}$ we will obtain the probability $\Pr(Y_{i,\text{test}} = 1 | Y_{\text{trn}}, X_{i,\text{test}})$ by using the approximation to (2) given by (3). If there is no test set available, we will use a hold-one-out cross-validation approach. We follow the technique described in Gelfand (1996) to simplify the computation. For the cross validation predictive density, in general, let Y_{-i} be the vector of Y_j's without the ith observation Y_i,

$$\Pr(Y_i | Y_{-i}) = \frac{\Pr(Y)}{\Pr(Y_{-i})} = \left[\int \{\Pr(y_i | Y_{-i}, \mathcal{M})\}^{-1} \Pr(\mathcal{M} | Y) \, d\mathcal{M} \right]^{-1}.$$

The MCMC approximation to this is

$$\widehat{\Pr}(Y_i | Y_{-i,\text{trn}}) = m^{-1} \sum_{j=1}^{m} \{\Pr(y_i | Y_{-i,\text{trn}}, \mathcal{M}^{(j)})\}^{-1},$$

where $\mathcal{M}^{(j)}$ for $j = 1, \ldots, m$ are the m MCMC posterior samples of the model unknowns $\mathcal{M}$. This simple expression is due to the fact that the Y_i's are conditionally independent given the model parameters $\mathcal{M}$.

EXAMPLE 1. The data set used here from Alon et al. (1999), is composed of 40 colon tumor samples and 22 normal colon tissue samples, analyzed with an Affymetrix oligonucleotide array complementary to more than 6,500 human genes and expressed sequence tags (EST's). Of these genes, the 2,000 with highest minimal intensity across the tissue are selected for analysis. We randomly split the data into two parts, 42 observations are used as the training set and other 20 are used as the test set.

Table 1 lists the top 50 genes that best differentiate between the colon and normal cancer tissues. To visualize the data better, Figure 1 shows the "heat" map of 50 most significant genes. Heat maps have become popular in microarray literature, Eisen et al. (1998), as graphical representations of the primary data where each point is associated with a color that reflects its value. Increasingly positive values are represented with reds and increasing negative values with blues of increasing intensity. For this data set we obtained a misclassification error of 0.06, i.e., we misclassify 4 out of the 22 test samples.

4.2. *Gene selection for multicategory classification*

In this section, we setup the framework for analysis microarray data when the response is polychotomous, extending the approach for the binary case, following Sha et al. (2004). Again we introduce the model from point of view of data augmentation, i.e., using latent variables and use mixture priors on the regression coefficients for gene selection.

To set up the notation, let $Y_{n\times 1}$ be the categorical response vector coded as $0, \ldots, C-1$, where C denotes the number of classes (e.g., tissue types). Let $X_{n\times p}$ be the gene expression data matrix as in (1). Each polychotomous outcome Y_i is associated with a vector of probabilities $(p_{i,0}, \ldots, p_{i,C-1})$, where $p_{i,j}$ represents the probability

Table 1

Colon cancer data: list of top 50 genes that best differentiate between colon cancer and normal tissues. The genes have been sorted by the frequency of times they appear int the posterior MCMC samples

Gene ID	Gene description	Frequency
H08393	COLLAGEN ALPHA 2(XI) CHAIN (Homo sapiens)	0.932
M19311	Human calmodulin mRNA, complete cds	0.916
H17897	ADP, ATP CARRIER PROTEIN, FIBROBLAST ISOFORM	0.910
X12671	Human gene for heterogeneous nuclear ribonucleoprotein	0.892
R85558	INORGANIC PYROPHOSPHATASE (Bos taurus)	0.872
control	Control gene	0.848
H49870	MAD PROTEIN (Homo sapiens)	0.832
T68098	ALPHA-1-ANTICHYMOTRYPSIN PRECURSOR (HUMAN)	0.786
M16029	Human ret mRNA encoding a tyrosine kinase, partial cds	0.756
T57079	HIGH AFFINITY IMMUNOGLOBULIN GAMMA FC RECEPTOR I 'A FORM'	0.738
M35252	Human CO-029	0.712
H89477	G1/S-SPECIFIC CYCLIN D3 (Homo sapiens)	0.688
X74874	H. sapiens gene for RNA pol II largest subunit, exon 1	0.666
X68314	H. sapiens mRNA for glutathione peroxidase-GI	0.660
H89481	214 KD NUCLEOPORIN (Homo sapiens)	0.644
T41207	COME OPERON PROTEIN 3 (Bacillus subtilis)	0.632
R78934	ENDOTHELIAL ACTIN-BINDING PROTEIN (Homo sapiens)	0.586
R34701	TRANS-ACTING TRANSCRIPTIONAL PROTEIN ICP4	0.556
H11054	PROTEIN KINASE C, DELTA TYPE (HUMAN)	0.540
X90828	H. sapiens mRNA for transcription factor, Lbx1	0.532
U26401	Human galactokinase (galK) mRNA, complete cds	0.520
H65425	FIBRONECTIN RECEPTOR BETA SUBUNIT PRECURSOR	0.514
Z35093	H. sapiens mRNA for SURF-1	0.510
X57346	H. sapiens mRNA for HS1 protein	0.502
L06328	Human voltage-dependent anion channel isoform 2 (VDAC)	0.496
K03001	Human aldehyde dehydrogenase 2 mRNA	0.474
X79198	H. sapiens HCF-1 gene	0.446
control	Control gene	0.434
R44418	EBNA-2 NUCLEAR PROTEIN (Epstein-barr virus)	0.424
X59871	Human TCF-1 mRNA for T cell factor 1 (splice form C)	0.420
H69834	KININOGEN, HMW PRECURSOR (Homo sapiens)	0.400
H24754	FRUCTOSE-BISPHOSPHATE ALDOLASE A (HUMAN)	0.378
M74903	Human hematopoietic cell phosphatase mRNA	0.376
R16077	HOMEOTIC GENE REGULATOR (Drosophila melanogaster)	0.364
R94588	CELL DIVISION PROTEIN KINASE 2 (HUMAN)	0.330
T72582	GLUTAMATE RECEPTOR 5 PRECURSOR (Homo sapiens)	0.328
R42374	IMMEDIATE-EARLY PROTEIN IE180 (Pseudorabies virus)	0.308
T58756	POL POLYPROTEIN (Lelystad virus)	0.300
H46732	14 KD PROTEIN OF SIGNAL RECOGNITION PARTICLE (HUMAN)	0.276
H79349	VERPROLIN (Saccharomyces cerevisiae)	0.270
X53461	Human mRNA for upstream binding factor (hUBF)	0.264
X77548	H. sapiens cDNA for RFG	0.260
H69819	GENERAL NEGATIVE REGULATOR OF TRANSCRIPTION SUBUNIT 1	0.258
X81372	H. sapiens mRNA for biphenyl hydrolase-related protein	0.256
X55177	H. sapiens ET-2 mRNA for endothelin-2	0.254
H89092	H. sapiens mRNA for 17-beta-hydroxysteroid dehydrogenase	0.252
R39681	EUKARYOTIC INITIATION FACTOR 4 GAMMA (Homo sapiens)	0.252
M55422	Human Krueppel-related zinc finger protein (H-plk) mRNA, complete cds	0.234
L34840	Human transglutaminase mRNA, complete cds	0.224
X69550	H. sapiens mRNA for rho GDP-dissociation Inhibitor 1	0.220

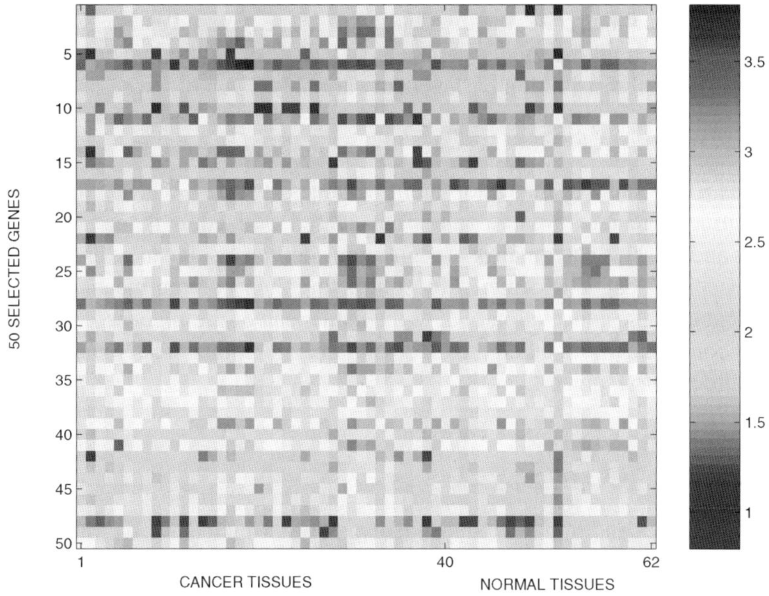

Fig. 1. Binary classification: Heat Map of the 50 most significant genes for classification between normal and colon cancer samples. For a color reproduction of this figure see the color figures section, page 1002.

that the ith sample falls in the jth category. We consider here only the case when the categories are unordered. As for the binary case, we will introduce latent variables into the model, transforming the model into a normal linear model in the latent responses. Let $Z_{n\times q}$, with $q = C - 1$ be the latent matrix for the $Y_{n\times 1}$ observed categorical vector. The element $z_{i,j}$ is the unobserved propensity of the ith sample to belong to the jth class. The latent variables Z are assumed to have multivariate normal distribution and consequently, the underlying relationship between the response Y_i and the latent variables Z becomes,

$$Y_i = \begin{cases} 0 & \text{if } z_{i,j} < 0\,\forall j, \\ j & \text{if } z_{i,j} = \max_{1\leqslant k\leqslant C-1}\{y_{i,k}\}. \end{cases} \tag{4}$$

Note that for the special case of $C = 2$, the model simplifies to binary case as explained in the previous section.

We use the notation for matrix variate distributions as introduced by Dawid (1981), that has the advantage of preserving the matrix structures without having to resort to Kronecker products and also simplifies the calculations through use of symbolic Bayesian manipulations. Let $Y_{p\times q}$ be a random matrix where every element has mean zero. Then the random matrix is said to have a matrix normal distribution, $\mathcal{N}(P, Q)$, if $p_{ii}Q$ and $q_{ii}P$ are the covariance matrices of the ith row and the jth column, respectively. Provided P and Q are positive definite, the matrix normal may be represented by the probability density function,

$$f_Y(y) = (2\pi)^{-pq}|P|^{-q/2}|Q|^{-p/2}\exp\bigl\{(-1/2)\mathrm{tr}\bigl(P^{-1}YQ^{-1}Y^{\mathrm{T}}\bigr)\bigr\}.$$

For a nonzero mean matrix M, the distribution of Y is expressed as $Y - M \sim \mathcal{N}(P, Q)$.

Conditionally on all the model parameters $\{\alpha, B, \Sigma\}$, using the above notation, the multivariate normal distribution of the latent variables Z can be expressed as,

$$Z - \mathbf{1}\alpha' - XB \sim \mathcal{N}(I_n, \Sigma), \tag{5}$$

where Z is $n \times q$ random matrix of latent variables, $\mathbf{1}$ is a $n \times 1$ vector of 1's, X is a $n \times p$ matrix of gene expression values considered as fixed, α is the $q \times 1$ vector of intercepts and B is the $p \times q$ matrix of regression coefficients. It can be assumed without loss of generality that X is centered by subtracting their column means, thus defining the intercept α as the expectation of Z at the data mean of the p genes.

We set conjugate priors on all the model parameters as this greatly eases the computational burden. First, given Σ,

$$\alpha^{\mathrm{T}} - \alpha_0 \sim \mathcal{N}(h, \Sigma), \tag{6}$$

$$B - B_0 \sim \mathcal{N}(H, \Sigma), \tag{7}$$

where α and B are conditionally independent given Σ, which has a marginal distribution as

$$\Sigma \sim \mathcal{IW}(\delta; Q). \tag{8}$$

Here $\mathcal{IW}(\delta; Q)$ refers to the matrix variate inverse Wishart distribution with shape parameter $\delta = n - p + 1$. If $V_{q\times q}$ follows a Wishart distribution with degrees of freedom ν and scale matrix $Q^*_{q\times q}$ with $Q^* > 0$, $\nu \geqslant q$ and V is positive definite almost surely, then the distribution of $U = V^{-1}$ is Inverse Wishart with shape parameter $\delta = \nu - q + 1$ and scale matrix $Q = (Q^*)^{-1}$ and is represented as $\mathcal{IW}(\delta; Q)$. For $\delta > 0$ the density function is defined as,

$$f(U) = c(q, \delta)|Q|^{(\delta+q-1)/2}|U|^{(\delta+2q)/2} \exp\{(-1/2)\mathrm{tr}(U^{-1}Q)\}, \quad U > 0,$$

with $c(q, \delta) = 2^{-q(\delta+q-1)/2}/\Gamma_q[(\delta = q - 1)/2]$.

As for the binary classification case, gene selection is done by via a Gaussian mixture prior on B by introducing a binary p-vector γ such that $\gamma_i = 1$ or 0 indicates whether the ith gene is selected or not. Such a mixture prior was employed for variable selection by Brown et al. (1998) in multivariate regression models, generalizing previous work of George and McCulloch (1997) for the multiple regression case. This selection prior will have as zeros those diagonal elements of H corresponding to $\gamma_i = 0$, so that the corresponding regression coefficients are zero with probability 1, provided that $B_0 = 0$. Thus for such a selection prior on B, each column has a singular p-variate distribution and given γ the prior of B is of the form,

$$B_\gamma - B_{0\gamma} \sim \mathcal{N}(H_\gamma, \Sigma),$$

where B_γ and H_γ selects the rows and columns of B and H respectively that have $\gamma_i = 1$. The complementary rows of B are fixed at their B_0-value with probability 1, in both the prior and posterior distribution. The prior mean B_0 of B will typically be take to be a $p \times q$ matrix of 0's.

A simple prior distribution on γ is of the form,

$$\pi(\gamma) = w^{p_\gamma}(1-w)^{p-p_\gamma}, \tag{9}$$

where p_γ indicates the number of genes chosen, i.e., the number of ones in γ. With this prior, the number of nonzero elements of γ has a binomial distribution with expectation pw.

Thus with the Bayesian model and prior setup (5)–(9), the unknown parameters of interest in the model are $\mathcal{M} = (\alpha, B, \Sigma, \gamma)$ along with the unobserved latent random variables Z. Inference on the joint posterior $f(Z, \alpha, B, \Sigma, \gamma | Y, X)$ can be carried out via a Gibbs sampling algorithm however with the mixture prior of B, the inference becomes hard owing to the variable nature of the parameter (model) space: different genes are included in different models. Dellaportas et al. (2000) review possible Gibbs sampling strategies, such as those involving the use of "pseudopriors" (Carlin and Chib, 1995). Another possible alternative is the variable dimension reversible jump algorithm outlined in Green (1995). Since the main focus here is on gene selection, a faster inference scheme can be obtained by carrying out marginal inference on γ given (X, Z). This marginal distribution gives the posterior probability of each γ vector (set of genes) that are most important for classification purposes. A hybrid MCMC sampler is proposed to update the latent matrix Z, that is a Gibbs sampler on the marginal for γ and Z with a Metropolis step (Metropolis et al., 1953; Hastings, 1970) on γ.

The latent variables Z are treated as missing and imputed from its marginal distribution. Conditional on Σ, γ, we have

$$Z - \mathbf{1}\alpha_0' - X_\gamma B_\gamma \sim \mathcal{N}(P_\gamma, \Sigma), \tag{10}$$

where $P_\gamma = I_n + h\mathbf{1}\mathbf{1}' + X_\gamma H_\gamma X_\gamma'$. Averaging over Σ, and setting the prior means $\alpha_0 = 0$ and $B_0 = 0$, with constraint (4) we have,

$$Z \sim \mathcal{T}(\delta; P_\gamma, Q), \tag{11}$$

where $\mathcal{T}$ represents the matrix Student T distribution. Given $\Sigma_{(q\times q)}$ be $\mathcal{IW}(\delta; Q)$ and given $\Sigma, T \sim \mathcal{N}(P, Q)$. Then the induced marginal distribution for $T_{(p\times q)}$ is a matrix-T distribution denoted by $\mathcal{T}(\delta; P, Q)$. The density function exists if $\delta > 0$, $P > 0$, $Q > 0$ and for $T \sim \mathcal{T}(\delta; P, Q)$ is given by,

$$f(T) = c(p, q, \delta)|P|^{-q/2}|Q|^{(\delta+q-1)/2}\left|Q + T'P^{-1}T\right|^{-(\delta+p+q-1)/2}$$

with $c(p, q, \delta) = \pi^{-pq/2}\Gamma_q[(\delta + p + q - 1)/2]/\Gamma_q[(\delta + q - 1)/2]$. Note that given constraint in (4), (11) is a truncated matrix Student t-distribution. Sampling schemes such as acceptance-rejection method employed by Albert and Chib (1993) in their latent variable approach could be employed. Other alternatives include drawing from a sub-chain Gibbs sampler as in McCulloch and Rossi (1994) and Geweke (1991).

The marginal distribution of γ, $\pi(\gamma | X, Y, Z)$ is obtained after integrating out α, B and Σ. Since we adopt conjugate priors on these parameters, using standard linear model theory (Lindley and Smith, 1972), we have

$$\pi(\gamma | X, Y, Z) \propto \pi(\gamma) f(Z | \gamma, X, Y)$$

$$\propto \pi(\gamma)|P_\gamma|^{-q/2}\left|Q + Z'P_\gamma^{-1}Z\right|^{-(\delta+n+q-1)/2},$$

where $P_\gamma^{-1} = I_n + r\mathbf{1}\mathbf{1}' + X_\gamma K_\gamma^{-1} X_\gamma'$, $r = 1/(n + h^{-1})$, $K_\gamma = X_\gamma' X_\gamma + H_\gamma^{-1}$ and $Q_\gamma = Q + Z' P_\gamma^{-1} Z$. This marginal distribution is obtained by setting $\alpha_0 = 0$, for large h and $B_0 = 0$. As this posterior distribution is of a nonstandard form, we sample γ using a Metropolis algorithm as suggested by Madigan and York (1995) and applied to variable selection by Brown et al. (1998), George and McCulloch (1997) and Raftery et al. (1997). Inference on relative importance of each gene and model selection is carried in a similar manner to the binary case.

EXAMPLE 2. We consider the dataset from Khan et al. (2001), who studied small, round blue cell tumors (SRBCT's) of childhood, which include neuroblastoma (NB), rhabdomyosarcoma (RMS), non-Hodgkin lymphoma (NHL) and Ewing family of tumors (EWS). Accurate diagnosis of SRBCT's is essential because the treatment options, response to therapy and prognosis vary widely depending on the type of tumor. The dataset contains expression levels from 6567 genes are measured for 88 samples, where 63 are labeled calibration (training) samples and 25 represent the test samples. The 63 training samples included both tumor biopsy material (13 EWS and 10 RMS) and cell lines (10 EWS, 10 RMS, 12 NB) and 8 Burkitt lymphomas (BL; a subset of NHL). Filtering for a minimal level of expression reduces the number of genes to 2308 and the natural logarithm of the relative red intensity is used for further analysis.

We use the methodology described above to use the expression levels of the 2308 genes to classify the samples into the four tumor types (EWS, BL, NB and RMS). A heat map of the top 50 genes that are most significant (occur most frequently in our posterior MCMC samples) is shown in Figure 2.

4.3. *Gene selection for survival methods*

Survival times of patients in clinical studies may shed light into the genetic variation. Sometimes it is possible to relate the expression levels of subset of genes with the survival time of the patients. Apart from this it is interesting to estimate the patient survival probabilities as a function of covariates such as levels of clinical risk. Frequentist approaches for variable selection in survival models include asymptotic procedures based on score tests and other approximate chi-square procedures, that do not work well when n is much smaller than p.

Bayesian analysis of survival data has received much attention recently due to advances in computational and modeling techniques. Ibrahim et al. (2001) provide a comprehensive review of Bayesian survival analysis. Bayesian variable selection for survival models is still in its infancy. In the survival data context, where the responses are time-to-event, Lee and Mallick (2004) perform gene selection by extending the work of George and McCulloch (1997) to a non-Gaussian mixture prior framework. A "small n large p" situation is considered for selection and survival analysis with the Weibull regression or Proportional Hazards (PH) model. The use of a random residual component in the model is notable, which is consistent with the belief that there may be

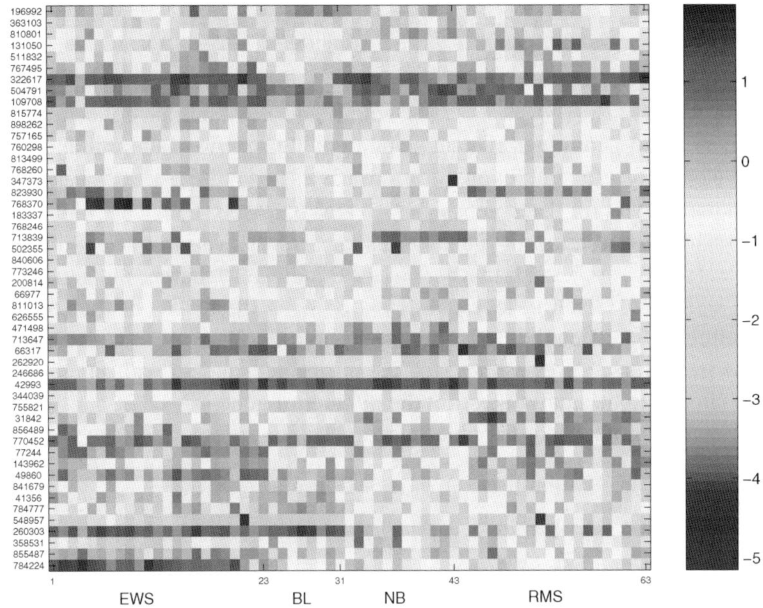

Fig. 2. Multicategory classification: Heat Map of the 50 most significant genes for classification between the four tumor types: EWS, BL, NB and RMS. For a color reproduction of this figure see the color figures section, page 1002.

unexplained sources of variation in the data perhaps due to explanatory variables that were not recorded in the original study.

Weibull regression model

First we consider the parametric model where the independently distributed survival times for n individuals, $t_1, t_2, \ldots, t_n$ follow a Weibull distribution with parameter α and γ_i^*. The data for the jth individual in the experiment, in addition to his survival time t_j consists of an observation vector $(Y_j, X_{j1}, \ldots, X_{jp})$ where Y_j indicates the binary or multicategory phenotype covariate followed by the p gene expressions $X_{j1}, \ldots, X_{jp}$. A reparameterized model is more convenient ($\lambda = \log \gamma^*$), whereby

$$f(t|\alpha, \lambda) = \alpha t^{\alpha-1} \exp\bigl(\lambda - \exp(\lambda) t^{\alpha}\bigr), \qquad t, \alpha > 0.$$

The hazard and survival function follow as $h(t|\alpha, \lambda) = \alpha t^{\alpha-1} \exp \lambda$ and $S(t|\alpha, \lambda) = \exp(-t^{\alpha} \exp \lambda)$. For the n observations at times $t = (t_1, \ldots, t_n)$, let the censoring variables be $\delta = (\delta_1, \ldots, \delta_n)'$, where $\delta_i = 1$ indicates right censoring and $\delta_i = 0$ indicates death of time in patient i. The joint likelihood function for all the n observations is

$$L(\alpha, \lambda|\theta) = \prod_{i=1}^{n} f(y_i|\alpha, \lambda_i)^{\delta_i} S(y_i|\alpha, \lambda_i)^{(1-\delta_i)}$$

$$= \alpha^d \exp\left\{\sum_{i=1}^{n} \delta_i \lambda_i + \delta_i(\alpha_i) \log y_i - y_i^{\alpha} \exp \lambda_i\right\}, \tag{12}$$

where $\lambda = (\lambda_1, \dots, \lambda_n)$, $\theta = (n, t, \delta)$ with t and δ as similar collections of the n parameters t_i and δ_i. In a hierarchical step, a conjugate prior can be induced as $\lambda_i = X_i'\beta + \varepsilon_i$ where X_i are gene expression covariates, $\beta = (\beta_1, \dots, \beta_n)$ is a vector of regression parameters for the p genes and $\varepsilon_i \sim \mathrm{N}(0, \sigma^2)$.

A Gaussian mixture prior for β enables us to perform a variable selection procedure. Define $\gamma = (\gamma_1, \dots, \gamma_p)$, such that $\gamma_i = 1$ or 0 indicates whether the gene is selected or not ($\beta_i = 1$ or 0). Given γ, let $\beta_\gamma = \{\beta_i \in \beta\colon \beta_i \neq 0\}$ and X_γ be the columns of X corresponding to β_γ. Then the β_γ is proposed as $\mathrm{N}(0, c(X_\gamma' X_\gamma)^{-1})$ with c as a positive scalar. The indicators γ_i are assumed to be a priori independent with $\Pr(\gamma_i = 1) = \pi_i$, which are chosen to be small to restrict the number of genes.

The hierarchical model for variable selection can be summarized as

$$\begin{aligned} &(T_i|\alpha, \lambda_i) \sim \mathrm{Weibull}(\alpha, \lambda_i), \quad \alpha \sim \mathrm{Gamma}(\alpha_0, \kappa_0), \\ &(\lambda_i|\beta, \sigma) \sim \mathrm{N}(X_i'\beta, \sigma^2), \quad (\beta_i \gamma_i) \sim \mathrm{N}(0, \gamma_i c\sigma^2), \quad \sigma^2 \sim \mathrm{IG}(a_0, b_0), \\ &\gamma_i \sim \mathrm{Bernoulli}(\pi_i). \end{aligned}$$

As the posterior distributions of the parameters are not of explicit form so we need to use MCMC based approaches, specifically Gibbs sampling alongside Metropolis algorithms to generate posterior samples. The unknowns are $(\lambda, \alpha, \beta, \gamma, \sigma^2)$ and separate M–H steps are used to sample λ_i, the indicators γ_i (one at a time), $\varphi = \log \alpha$ and finally (β, σ^2) is drawn as $(\beta_\gamma, \sigma^2|\gamma, \lambda) \sim \mathrm{NIG}(V_\gamma X_\gamma' \lambda, V_\gamma; a_1, b_1)$ where a_1, b_1 are defined as in O'Hagan (1994). For the details of the conditional distributions, the reader is referred to Lee and Mallick (2004).

Proportional hazards model

The Cox PH model (Cox, 1972) represents the hazard function $h(t)$ as

$$h(t|x) = h_0(t) \exp W,$$

where $h_0(t)$ is the baseline hazard function and $W = x'\beta$ where β is a vector of regression coefficients. The Weibull model in the previous section is a special case of Cox's proportional hazard model with $h_0(t) = \alpha t^{\alpha-1}$. Kalbfleisch (1978) suggests the nonparametric Bayesian method for the PH model. Lee and Mallick (2004) apply Bayesian variable selection approach to this model. As before, they assume $W_i = x_i'\beta + \varepsilon_i$ where $\varepsilon_i \sim \mathrm{N}(0, \sigma^2)$ where x_i forms the ith column of the design matrix X.

Let T_i be a independent random variable with conditional survival function

$$\Pr(T_i \geqslant t_i|W_i, \Lambda) = \exp\{-\Lambda(t_i) \exp W_i\}, \quad i = 1, \dots, n.$$

Kalbfleisch (1978) suggested a Gamma process prior for $h_0(t)$. The assumption is $\Lambda \sim \mathcal{GP}(a\Lambda^*, a)$ where Λ^* is the mean process and a is a weight parameter about the mean. If $a \approx 0$, the likelihood is approximately proportional to the partial likelihood and if $a \uparrow \infty$, the limit of likelihood is same to the likelihood when the gamma process is replaced by Λ^*. The joint unconditional marginal survival function is directly derivable as

$$P(T_1 \geqslant t_1, \ldots, T_n \geqslant t_n | W) = \exp\left\{-\sum_i \Lambda(t_i)\,\mathrm{e}^{W_i}\right\},$$

where $W = (W_1, \ldots, W_n)$. The likelihood with right censoring is then

$$L(W|\theta) = \exp\left\{-\sum_i a B_i \Lambda^*(t_i)\right\} \prod_{i=1}^{n} \left\{a\lambda^*(t_i) B_i\right\}^{\delta_i},$$

where $A_i = \sum_{l \in R(t_i)} \exp W_l$, $i = 1, \ldots, n$, $R(t_i)$ is the set of individuals at risk at time $t_i - 0$, $B_i = -\log\{1 - \exp W_i/(a + A_i)\}$ and $\theta = \{n, t, \delta\}$ as before. The hierarchical structure is as follows:

$$(W|\beta_\gamma) \sim \mathrm{N}\left(X_\gamma \beta_\gamma, \sigma^2 I\right), \quad \beta_\gamma \sim \mathrm{N}\left(0, c\sigma^2 \left(X_\gamma' X_\gamma\right)^{-1}\right),$$
$$\sigma^2 \sim \mathrm{IG}(a_0, b_0), \qquad \gamma_i \sim \mathrm{Bernoulli}(\pi_i).$$

The MCMC simulation that follows is similar to the Weibull regression model.

EXAMPLE 3. We consider the breast carcinomas data set of Sorlie et al. (2001), where novel subclasses of breast carcinomas based on gene expression patterns were discovered. A group of patients with the same therapy had different overall and relapse-free survival patterns. We use overall survival times of 76 samples each with 3097 gene expressions. There is additional subgroup information: Basal-like, ERBB2+, Normal Breast-like, Luminal subtype A, B, or C. We only consider binary covariate Y *as* $y_i = 1$ *if* ith sample is Luminal subtype A, B, or C and 0 otherwise. Later, this is extended to the multicategory case adding other covariates in the model.

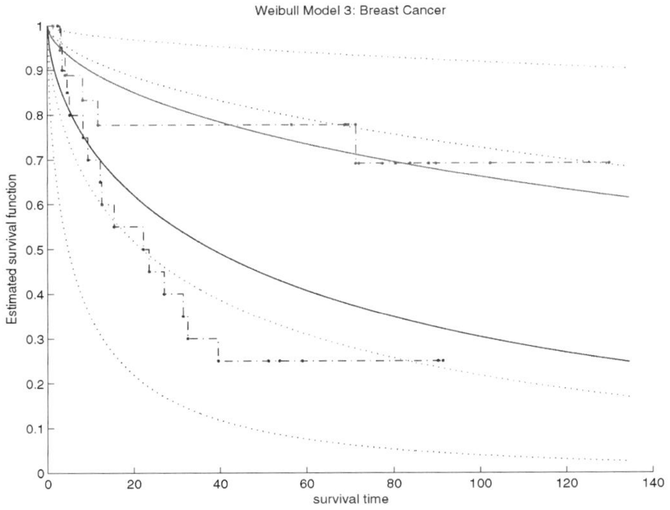

Fig. 3. Survival function for breast carcinomas data using Weibull model. (Reprinted with permission from Lee and Mallick, 2004.) For a color reproduction of this figure see the color figures section, page 1003.

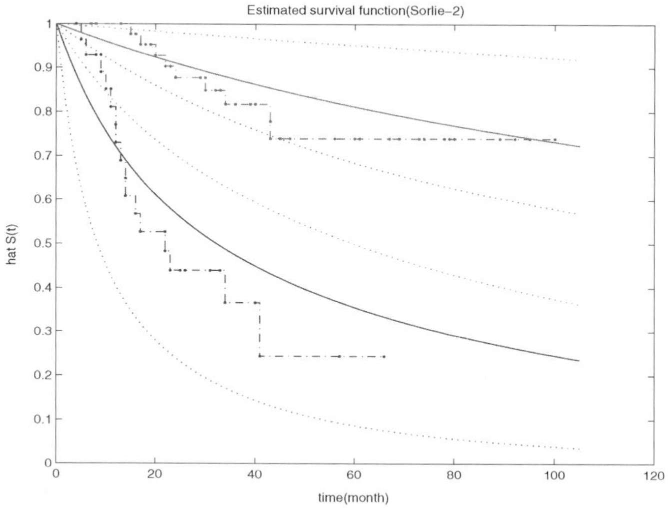

Fig. 4. Survival function for breast carcinomas data using semiparametric model. (Reprinted with permission from Lee and Mallick, 2004.) For a color reproduction of this figure see the color figures section, page 1003.

In the Weibull model the best subset found by the MCMC simulations consisted of 6 genes. The 2 log (Bayesfactor) of this model compared to the no-gene model was found to be 33.46. The posterior survival curve has been plotted in Figure 3, which again shows the need for improvement in the model-fit. In the semiparametric case, 5 genes comprise the best subset. The 2 log (Bayesfactor) value compared to the best Weibull model is 19.38. The posterior survival curves based on this model are given in Figure 4. The selected genes for this data has been tabulated in Table 2.

5. Differential gene expression analysis

In early studies, differential gene expression was assessed using the fold change of genes under different conditions (Chen et al., 1997). However, any procedure which uses the raw intensity ratios alone to infer differential expression may be inefficient and may lead to excessive errors. As a result, there has been space for more evolved models and over the years differential gene expressions have been accessed from various statistical methodologies. For a review of the frequentist methods the reader is referred to Dudoit et al. (2002).

In a possible analysis any genes whose transcripts are reset at a low total abundance can be ignored. This could possibly lead to throwing away valuable data and a threshold based on one channel may not conform with the thresholds in the other channel. If R and G correspond to the red and green intensities at a given spot on the microarray. In the initial stages of Bayesian analysis for differential gene expression analysis, Newton et al. (2001) use a hierarchical model, where they model R and G as independent gamma

Table 2
Important genes found for estimating the survival function for breast carcinomas data (Reprinted with permission from Lee and Mallick, 2004)

Freq.	Clone ID	Gene symbol	Gene description
3826	340826		"Homo sapiens cDNA FLJ12749 fis, clone NT2RP2001149"
1225	772890		Homo sapiens mRNA; cDNA DKFZp434N2412
1203	272018		ESTs, Weakly similar to AF126743 ...
1028	772304	SLC25A5	solute carrier family 25
593	813841	PLAT	plasminogen activator, tissue
573	745402	PRO2975	hypothetical protein PRO2975
396	40299	GDF10	growth differentiation factor 10
395	79045	CTL1	transporter-like protein
365	950578	NDUFA5	NADH dehydrogenase (ubiquinone) 1 alpha subcomplex
263	782760	CYP1B1	cytochrome P450, subfamily I polypeptide 1
229	768299	BRF1	butyrate response factor 1 (EGF-response factor 1)
198	951125	PECI	peroxisomal D3,D2-enoyl-CoA isomerase
185	131653	MRPS12	mitochondrial ribosomal protein S12
151	771323	PLOD	procollagen-lysine, 2-oxoglutarate 5-dioxygenase
143	755599	IFITM1	interferon induced transmembrane protein 1 (9–27)
132	214448	RNF10	ring finger protein 10
113	83610	FLJ22378	hypothetical protein FLJ22378
111	134476	SYBL1	synaptobrevin-like 1

r.v.s with constant shape parameters and a common scale parameter. Both R and G have the same coefficient of variation even though they may have different scales. They then derived the density of the their ratio T which is parameterized by a parameter $\rho = \theta_g/\theta_r$ where θ_r and θ_g are the shape parameters. The hierarchical model is completed by specifying a prior for the gamma scale parameters and carrying out the inference based on the posterior distribution of ρ. The argument is inspired from Chen et al. (1997) is that the sampling distribution of $T = R/G$ depends only on a constant coefficient of variation under the null hypothesis that R and G have the same expectation. If $R \sim \text{Gamma}(a, \theta_r)$ and $G \sim \text{Gamma}(a, \theta_g)$, the distribution of T is given by

$$(T|\rho, a) \sim \frac{\Gamma(2a)}{\Gamma^2(a)} \frac{1}{\rho} \frac{(t/\rho)^{a-1}}{(1+t/\rho)^{2a}}, \qquad t > 0, a > 1,$$

where $\rho = \theta_g/\theta_r$ and the distribution is the scale multiple of a Beta distribution of the second kind (Kendall and Stuart, 1969). When $\rho = 1$, there is no real differential expression (Chen et al., 1997) and the distribution of T_j depends on the coefficient of variation $a^{-1/2}$, only. Next, the model parameters θ_r and θ_g are specified by an identical gamma prior distribution $\text{Gamma}(a_0, \nu)$ and the posterior distribution of the differential gene expression is given by

$$(\rho|R, G, \eta) \sim \rho^{-(a+a_0+1)} \left(\frac{1}{\rho} + \frac{G+\nu}{R+\nu}\right)^{-2(a+a_0)},$$

where $\eta = (a, a_0, \nu)$. This leads to the posterior estimate of the differential gene expression $\hat{\rho} = (R+\nu)/(G+\nu)$. The hyperparameters in the hierarchical model are estimated from the data using the marginal likelihood of the data, which is also considered to assert significantly large differences in expression, and is given by

$$f_1(R, G|\eta) = \left[\frac{\Gamma(a+a_0)}{\Gamma(a)\Gamma(a_0)}\right]^2 \frac{\nu^{2a_0}(RG)^{a-1}}{[(R+G)(G+V)]^{a+a_0}}.$$

Under the null hypothesis that no differential expression, this distribution reduces to

$$f_0(R, G|\eta) = \frac{\Gamma(2a+a_0)}{\Gamma(2a)\Gamma(a_0)} \frac{\nu^{a_0}(RG)^{a-1}}{[R+G+V]^{2a+a_0}}$$

and the final decision is based on the factor

$$\frac{P(Z=1|R,G,p)}{P(Z=0|R,G,p)} = \frac{f_1(R,G|\eta)}{f_0(R,G|\eta)} \frac{p}{1-p},$$

where $Z \in \{0, 1\}$ indicates true differential expression and p is the prior confidence on one of the models.

5.1. *Censored models*

Ibrahim et al. (2002) introduces binary variable selection for the 'censored' data case, where genes are considered to be expressed only if they exceed a certain threshold. They examine samples for 3214 genes from two different tissues (4 normal and 10 diseased) from 14 individuals. The censored expression level from tissue i $(= 1, 2)$, gene j $(= 1, 2, \ldots, n)$ and individual k $(= 1, 2, \ldots, r)$, follows a mixture distribution

$$y_{ijk}^c \sim c^p (c + y_{ijk})^{1-p},$$

where y_{ijk} is the corresponding continuous covariate or expression level follows the log-normal distribution, $(y_{ijk}|\mu_{ij}, \sigma_{ij}^2) \sim \text{lognormal}(\mu_{ij}, \sigma_{ij}^2)$ and p expresses the chance that y_{ijk} is censored.

Using latent variables γ_{ijk} to indicate censoring the joint likelihood is written as

$$f(Y_1, Y_2|\mu, \sigma^2, p) = \prod_{i=1}^{2}\prod_{j=1}^{n}\prod_{k=1}^{r} p_{ij}^{\gamma_{ijk}} (1-p_{ij})^{1-\gamma_{ijk}} f\left(y_{ijk}|\mu_{ij}, \sigma_{ij}^2\right)^{1-\gamma_{ijk}},$$

where $Y_i = (y_{i11}, \ldots, y_{i1r}, \ldots, y_{in1}, \ldots, y_{inr})'$ for $i = 1, 2$, $\mu = (\mu_{11}, \mu_{21}, \ldots, \mu_{1n}, \mu_{2n})'$, $\sigma^2 = (\sigma_{11}^2, \sigma_{21}^2, \ldots, \sigma_{1n}^2, \sigma_{2n}^2)'$ and $p = (p_{11}, p_{21}, \ldots, p_{1n}, p_{2n})'$. The prior elicitation consists of specifying the $(\mu_{ij}, \sigma_{ij}^2)$ as

$$\left(\mu_{ij}, \sigma_{ij}^2|\mu_i\right) \sim \text{NIG}\left(\mu_i, \tau\sigma_{ij}^2/\bar{n}_i; a_i, b_i\right), \quad \forall i = 1, 2;\ j = 1, \ldots, n,$$

$$\text{where } \bar{n}_i = n^{-1}\sum_{j=1}^{n}\left(n - \sum_{k=1}^{r}\gamma_{ijk}\right) \text{ and } \tau > 0.$$

The model is further enriched by specifying $\mu_i \sim \text{N}(\mu_0, \sigma_0^2)$. The model indicator probabilities p_{ij} are drawn through a normal random variable δ_{ij}, which are related as $\delta_{ij} = \text{logit}(p_{ij})$.

A comparison is sought using the ratio of the predictive gene expression obtained as the posterior expectation $\mu_{ij}^* = E((c\gamma_{ijk} + (1-c)(c + y_{ijk}))|p_{ij}, \mu_{ij}, \sigma_{ij}^2)$ under the two tissue types, i.e., the jth gene is selected if

$$\Pr(\rho > 1 | p, \mu, \sigma^2),$$

where $\rho = \mu^*_{1j}/\mu^*_{2j}$.

5.2. Nonparametric empirical Bayes approaches

Efron et al. (2001) considered a nonparametric approach to gene profiling. Their problems concern a comparative experiment, involving characterization of gene expression for a group of patients receiving treatment versus those not receiving treatment. The approach starts with assuming that the observed expression scores are generated from a mixture of two distributions that can be interpreted as distributions for affected and unaffected genes. The desired inference about differential expression for a particular gene amounts to solving a deconvolution problem corresponding to this mixture. This idea has been extended for comparison of multiple groups using replicated gene expression profiles (Kendziorski et al., 2003) and a general review of Empirical Bayes methods for microarrays can be found in (Newton and Kendziorski, 2002).

A brief discussion of the design methodology in Efron et al. (2001) is in order. There are two protocols used for collecting the samples. Two matrices D_0 and D_1 are formed from the microarray data. D_1 contains all the possible differences between affected and normal tissues with the same protocol, while D_0 contains all possible differences between all pairs of affected columns and between all pairs of normal columns under the same protocol. Possible effects due to differential expression in affected versus normal tissues are included only in D_1, whereas the inter-patient variation as well as any other noise are included in both, D_0 and D_1. The goal then is to separate the differentially expressed genes that show an additional effect in D_1 from the nondifferentially expressed genes for which D_1 reports only noise as in D_0. If $\overline{D}_i$ denotes the row-wise average of D_i, then two sets of scores are defined as

$$Z_{ij} = \overline{D}_{ij}/(\alpha_i + S_i), \quad i = 0, 1, \ j = 1, \ldots, n,$$

where S_i are respective the standard deviations with offset α_i as the correction scores. The distribution of expression scores Z_{1i} is written as a mixture of two density functions f_0, and f_1 for the unaffected and affected case respectively. Thus

$$f(Z_{ij}) = p_0 f_0(Z_{ij}) + (1 - p_0) f_1(Z_{ij}), \tag{13}$$

where p_0 is the proportion of the genes that are not differentially expressed across the two experimental conditions. Given the densities f_0, f_1 and the mixing probability p_0, the posterior probability of differential expression is given by

$$P_1(Z_{ij} | f_0, f_1, p_0) = (1 - p_0) f_1(Z_{ij})/f(Z_{ij}) \tag{14}$$

and $P_0(Z_{ij} | f_0, f_1, p_0) = 1 - P_1(Z_{ij} | f_0, f_1, p_0)$ is the complementary probability of nondifferential gene expression.

Efron et al. (2001) follow an empirical Bayes approach to obtain a nonparametric estimate of the probability that a particular gene was affected by the treatment, i.e., P_0 by substituting point estimates of $q = f_0/f$ and p_0. To estimate q they construct a logistic regression experiment, set up such that the odds are $\pi(Z_{ij}) = f(Z_{ij})/(f(Z_{ij}) + f_0(Z_{ij}))$ and the resulting estimate of $\hat{\pi}$ implies an estimate for $\hat{q} = (1 - \hat{\pi})/\hat{\pi}$. The estimate of p_0 follows from the observation that $P_1 \geqslant 0$ implies $p_0 \leqslant \min_Z 1/q(Z)$ and thus $\hat{p}_0$ is proposed to be $\min_Z 1/q(Z)$. The overall estimates $\hat{P}_0$, and $\hat{P}_1$ is thus biased.

5.3. Nonparametric Bayesian approaches

The limitation of the empirical Bayes approach above, can be overcome by a full model-based Bayesian approach with prior probability models for p_0, f_0, f_1 that later provide posterior probabilities of differential gene expression. With ideas from the above treatment, a nonparametric Bayesian approach based on the Mixture of Dirichlet Processes (MDP) prior is taken in Do et al. (2005). They focus on 2000 genes from the Alon colon Oligoarray data set consisting samples from 40 patients, 22 patients supply both tumor and colon tissue samples. Thus the mixed samples Z_{1j}'s measure the noise + tumor effect whereas the null samples Z_{0j}'s are the measurement errors. In this model, the mixture (13) is realized at a hierarchical level as follows.

Let the score Z_{ij} is distributed as $Z_{ij} \sim \mathrm{N}(\mu_{ij}, \sigma^2)$, where μ_{ij} is proposed as a mixture of two Dirichlet process priors via an binary variable γ_i indicator the type of score. We write

$$\mu_{ij} \sim I(\gamma_{ij} = 0)P_0 + I(\gamma_{ij} = 1)P_1,$$

where $P(\gamma_{ij} = 0) = I(j = 0) + p_0 I(j = 1)$, $P_0 \sim \mathrm{DP}(\alpha_0, G_0)$ and $P_1 \sim \mathrm{DP}(\alpha_1, G_1)$. The probability p_0 is proposed as $p_0 \sim U(0.05, 1)$ and the two Dirichlet Process priors have base measures

$$G_0 = \Phi\left(\mu_0, \sigma_0^2\right), \quad \text{and}$$
$$G_1 = 0.5\Phi\left(-\mu_1, \sigma_1^2\right) + 0.5\Phi\left(\mu_1, \sigma_1^2\right)$$

respectively. The base measure for the null scores is unimodal and centered at zero, whereas for the differentially expressed genes it is symmetric bimodal considering the possibility that the expression could stray in both directions with equal chance. The model is further enriched by mixing distributions $\mu_i \sim \mathrm{N}(m_i, \tau_i^2)$, $\sigma^2 \sim \mathrm{IG}(a, b)$, and $\tau_i^2 \sim \mathrm{IG}(a_i, b_i)$, where i takes lies in $\{0, 1\}$ indicating the mixture component.

The MCMC simulations follow from the Gibbs Sampling from the posterior distribution of μ_{ij} conditional on $\mu_{-ij} = \{\mu_{kl}\colon k \neq i, l \neq j\}$. Thus

$$\mu_{ij}\sigma^2, \mu_{-ij} \sim c_{i0}\mathrm{N}_{Z_{ij}}\left(\mu_{ij}, \sigma^2\right)\mathrm{d}G_i(\mu_{ij}) + \sum_{i \neq k} c_{k1} \sum_{j \neq l} \delta(\mu_{ij}|\mu_{kl}),$$

where $c_{i0} = c\int \mathrm{N}_{Z_{ij}}(\mu_{ij}, \sigma^2)\,\mathrm{d}G_i(\mu_{ij})$ and $c_{i1} = c N_{Z_{ij}}(\mu_{ij}, \sigma^2)$ with c as a normalizing constant. From this the joint distribution of (μ_{ij}, γ_{ij}) can be derived in a simple form. The remaining hyperparameters are generated by the Gibbs sampler. Plug-in estimates are avoided by directly estimating $\overline{P}_1(z) = E\{P_1(z|f_0, f_1, p_0)|Y\}$ by collecting relevant MCMC samples in the representation (14). This computational step to estimate the posterior probability of differential expression is detailed in Do et al. (2005).

6. Bayesian clustering methods

With the advent of genome sequencing projects we have access to DNA sequences of the promoter regions that contain the binding sites of transcription factors that regulate gene expression. Development of microarrays have allowed researchers to measure very many mRNA targets simultaneously providing a genomic viewpoint of gene expression. As a consequence, this technology facilitates new experimental approaches for understanding gene expression and regulation.

The main biological hypothesis underlying cluster analysis is: genes with a common functional role have similar expression patterns across different experiments. This similarity of expression patters is due to co-regulation of genes in the same functional group by specific transcription factors. In the analysis following this assumption, genes are assorted by their expression levels (D'Haeseleer et al., 2000), even though there are gene functions for which these assumptions do not hold and there are co-expressed genes that are not co-regulated. Once potential clusters have been identified, then shorter DNA strings can be searched that appear in significant over-abundance in the promoter regions of these genes (Tavazoie et al., 1999). Such an approach can discover new binding sites in promoter regions. The biological significance of the results of such analysis has been demonstrated in numerous studies.

Bayesian clustering analysis succeeds the traditional clustering procedures based on a variety of strategies (often heuristical) from different scientific streams, notable among which is the use of self-organizing maps (Tamayo et al., 1999), simple agglomerative hierarchical methods (Eisen et al., 1998) and k-means procedures (Tavazoie et al., 1999). Tibshirani et al. (2001) use the gap statistic to estimate the number of clusters by comparing within-cluster dispersion to that of a reference null distribution.

In a model-based approach to clustering, the probability distribution of observed data is approximated by a statistical model. Parameters in such a model define clusters of similar observations. The cluster analysis is performed by estimating these parameters from the data and offers a principled alternative to heuristic-based algorithms. The issues of selecting a reasonable clustering method and determining the correct number of clusters are reduced to model selection problems in the probability framework.

6.1. *Finite mixture models*

In this context, a Gaussian finite mixture model (McLachlan and Basford, 1987) based approach has been used to cluster expression profiles (Yeung et al., 2001). The mixture model assumes that each component (of the mixture) of the data is generated by an underlying probability distribution. Suppose the data $Y = (Y_1, Y_2, \ldots, Y_n)$ represents n independent multivariate observations and C is the number of components in the data. The likelihood for the mixture model is

$$L(\theta_1, \ldots, \theta_C | Y) = \prod_{i=1}^{n} \sum_{j=1}^{C} p_j f(Y_i | \theta_j),$$

where $f(Y_i|\theta_j)$ is the density conditional on the parameters θ_j of the jth mixture component and p_j is the probability that an observation belongs to that component

($\sum_{j=1}^{C} p_j = 1$). In the Gaussian mixture model, we assume that $E(Y_i|\beta_j) = X\beta_j$ and the exact distributions are given by $f(Y_i|\beta_j, \Sigma_j) = \mathrm{N}(X\beta_j, \Sigma_k)$ where Σ_k is the model covariance.

To ease prior elicitation, Σ_j is represented as $\Gamma_j \Lambda_j \Gamma_j^t$ where Γ_j is orthogonal and Λ_j is diagonal. Different choices of Σ_j (such as $\Sigma_j = \lambda I$ or $\lambda_j I$) are used to model different geometric features of clusters (Banfield and Raftery, 1993). Each combination of a different specification of the covariance matrices and a different number of clusters corresponds to a separate probability model. Model selection is also performed with the Bayes Information Criterion (BIC; Schwartz, 1978). If M_r and M_s are two different models with parameters θ_r and θ_s, the integrated likelihood is defined as $f(Y|M.) = \int p(Y|\theta., M.)p(\theta.|M.)\,\mathrm{d}\theta.$ and $p(\theta.|M.)$ is the prior distribution of $\theta.$. The Bayes factor $p(Y|M_r)/p(Y|M_s)$, the posterior odds the data supports model M_r against model M_s (assuming equal a priori preference) is approximated by BIC (Fraley and Raftery, 2002) as

$$\begin{aligned} &2\log\big(p(Y|M_r)/p(Y|M_s)\big) \\ &\quad\approx 2\log\big(p(Y|\hat{\theta}_r, M_r)/p(Y|\hat{\theta}_s, M_s)\big) - (n_r - n_s)\log n, \end{aligned}$$

where $n.$ is the number of parameters in a model and $\hat{\theta}.$, is the corresponding maximum likelihood estimate. The BIC therefore adjusts the classical likelihood ratio criterion to favor more strongly the model with fewer parameters.

The model is attractive from the computational viewpoint, however, some questions as to the specification of the number of components in the mixture, remain unanswered. These problems are considered in the following generalization to infinite mixture models.

6.2. *Infinite mixture models*

Assuming that the number of mixture components is correctly specified, the finite mixture model offers reliable estimates of confidence in assigning individual profiles to particular clusters. In the situation where the number of clusters is not known, it relies on ones ability to identify the correct number of mixture components generating the data. All conclusions thereafter are then made assuming that the correct number of clusters is known. Consequently, estimates of model parameters do not take into account the uncertainty in choosing the correct number of clusters.

In this context, Medvedovic and Sivaganesan (2002) use a clustering procedure based on a Bayesian infinite mixture model for expression data. This model is asymptotically equivalent (Neal, 2000) to the mixture model of Ferguson (1973) and Antoniak (1974) using Dirichlet process priors. The conclusions about the probability that a set of gene expression profiles is generated by the same pattern are based on the posterior probability distribution of the clusterings given the data.

The normal mixture model can be briefly outlined as follows. Using notation as before, the expression profile Y_i for the ith gene, given that it comes from class j ($j = 1, \ldots, C$) is given by, $(Y_i|c_i = j) \sim \mathrm{N}(\mu_j, \sigma_j^2 I)$. Prior distributions for the parameters $\{(\mu_j, \sigma_j^2)\colon j = 1, \dots, C\}$ and the class indicators is given by

$$(c_i | p_1, \dots, p_C) \sim \prod_{j=1}^{C} p_j^{I(c_i = j)}, \qquad (\mu_j | \lambda, \tau^2) \sim \mathrm{N}(\lambda, \tau^2 I),$$
$$(\sigma_j^2 | \beta, w) \sim \mathrm{IG}(\beta, \beta w),$$

where the probabilities $(p_1, \dots, p_C)$ follow a Dirichlet distribution $\mathrm{D}(\alpha Q^{-1}, \dots, \alpha Q^{-1})$ and suitable priors are specified for the hyperparameters $(w, \beta, \lambda, \tau^2)$ for conjugacy. Cluster analysis is based on the approximated joint posterior distribution of the classification indicators, $f(c_1, \dots, c_n | Y)$.

The clusters of genes with similar expression patterns (and the number of clusters) are identified from the posterior distribution of clusterings implicit in the model. In addition to generating groups of similar expression profiles at a specified confidence, it also identifies outlying expression profiles that are not similar to any other profile in the data set. The infinite mixture model is reasonably 'precise' and in contrast to the finite mixture approach, this model does not require specifying the number of mixture components. Furthermore, Bayesian model is averaged over all possible clusters and number of clusters.

6.3. *Functional models*

Wakefield et al. (2003) use dynamic linear models (Harrison and Stevens, 1976) for situations where the microarray experiments are indexed by an ordered variable such as time. The formulation may lead to valuable insight into those genes that behave similarly over the course of the experiment.

If y_{it} is the expression level of gene i at time t, i.e., $Y_i = (y_{i1}, y_{i2}, \dots, y_{iT})$, then the observed data at some experimental time t is assumed to come from an underlying function f, such that

$$y_i(t) = f(\theta_j, t) + e_i(t),$$

where $e_i(t) \sim \mathrm{N}(0, \sigma^2)$ and $f(\theta_j, t)$ represents the unknown function of gene-specific parameters θ_j at time t, indicated as before by $c_i = j$ $(j = 1, \dots, C)$. These indicators $c_1, c_2, \dots, c_n$ are specified independently conditional on the class probabilities $(p_1, \dots, p_C)$ and the number of classes C. In addition, the probability vector $(p_1, \dots, p_C)$ is specified a Dirichlet distribution prior.

If the observation equation is simply

$$y_i(t_k) | c_i = j \sim \mathrm{N}(\mu_j(t_k), \sigma^2),$$

$i = 1, \dots, N$, $j = 1, \dots, C$ and t_k is the kth of the T time points, then the model relates the parameters through the system equation

$$\mu_j(t_k) = \mu_j(t_{k-1}) + u_j(t) \quad \text{and}$$
$$u_j(t_k) \sim \mathrm{N}(0, (t_k - t_{k-1})\tau^2)$$

so that the observations closer in time are more likely to be similar. For the first time point t_{k_1}, we propose $u_j(t_1) \sim \mathrm{N}(0, \tau_1^2)$. The system equation defines the evolution of μ_j as a Markov chain. We do not however observe the underlying system parameters, but the expression profiles $y_i(t)$ at time t is related to the them through the observation equation.

As a special application for gene cell cycle data, Wakefield et al. (2003) suggests representation of the function f in a periodic basis. This is accommodated readily with minor modifications in the above linear model. The model so defined is attractive from a computational viewpoint and it is well-suited to assort cyclic gene expression profiles based on observed periodic phenomena.

Luan and Li (2003) explore functional relationships in the B-spline mixed-effects model:

$$y_i(t_k) = \sum_{m=1}^{p} \zeta_m^{(j)} \overline{B}_m(t_k) + \sum_{m=1}^{q} \gamma_{im} B_m(t_k) + \varepsilon_{ij},$$

where the first term is used to model the average gene expression profile for the jth cluster with $\overline{B} = \{\overline{B}_m(\cdot), m = 1, \dots, p\}$ as a basis for the B-spline basis and $\zeta_m^{(j)}$ is the p-vector of coefficients corresponding to cluster j. The second term is a random effect curve for ith gene with $B = \{B_m(\cdot), m = 1, \dots, q\}$ as a spline basis for possibly different space of functions and γ_{im} are normal random coefficients with mean zero and covariance matrix Σ. This term models the gene-specific deviation from the average profile and is independent of the class indicators.

B-spline based clustering provides a flexible way to model possible nonlinear relationships between gene expression levels and is applicable to the nonequispaced data case, a feature that is lacking in the previous dynamic model. The model, however, is partially Bayesian and the EM algorithm is employed to obtain estimates to the parameters $(\zeta_m^{(j)}, \Sigma, \sigma^2)$. A particular gene is identified with the jth cluster if the estimated posterior probability $P(c_i = j|Y)$ is maximized for $c_i = j$. Posterior probabilities below a certain cutoff are declared as outlying values with no cluster membership. The number of clusters is estimated using the BIC criterion as detailed before.

7. Regression for grossly overparametrized models

West et al. (2003) considered an approach based on probit and linear regression for characterizing differential gene expression. They started out with a probit model for the data for which there is a binary clinical outcome response variable and the covariates consist of gene expression levels. The binary response variable is then transformed into a continuous variable via latent variables (Albert and Chib, 1993) and a linear regression model is used to develop the methodology. Because the linear regression is overparametrized, West et al. (2003) described techniques based on the singular value decomposition to make the model identifiable and considered a class of generalized g priors (Zellner, 1986) for the regression coefficients. We summarize the principal component regression model to accommodate the "Small n, Large p" situations while dealing with microarray data.

Assume the standard n-to-p linear model $Y = X\beta + \varepsilon$, where Y is the n-vector of responses, X is the $n \times p$ matrix of predictors, β is the p-vector regression parameter and $\varepsilon \sim \mathrm{N}(0, \sigma^2 I)$. When $n \gg p$, the design X is be singular-value decomposed (SVDed) as $X = ZR$ where Z is a $n \times k$ factor matrix satisfying $Z'Z = \Lambda^2$ where Λ is nonsingular positive diagonal matrix and R is the $k \times p$ matrix with $R'R = I$. The singularities or zero eigenvalues are ignored in this representation. The dimension reduction is embodied in a reparameterized model can be written as $Y = Z\gamma + \varepsilon$ with $\gamma = R\beta$ as a k-vector of regression parameters for the factor variables.

Bayesian prediction follows in a fairly standard manner, upon specifying the elements of γ independently as $\gamma_i \sim \mathrm{N}(0, c_i\sigma_i^2)$ where $\sigma^2 \sim \mathrm{IG}(a_0, b_0)$. With the least-norm inversion of $\gamma = R\beta$, these priors imply a prior on β with a familiar look of Zellner's g-priors (1986) is of interest. The implied prior on β is proportional to $\exp(-\beta' R' G R \beta/2)$, where $G = \mathrm{diag}(c_1\sigma_1^2, \ldots, c_k\sigma_k^2)$.

8. Concluding remarks

Several sources of variation and systematic bias are introduced at each stage of a microarray experiment. To ensure the reliability of the conclusions drawn from the statistical methods, careful experimentation is essential. In this chapter, we have reviewed Bayesian statistical methods to deal with the high throughput microarray data. Bayesian methods are particularly appealing with the availability of new MCMC methods allowing analysis of large data structures within complex hierarchical models. In particular, we explore the area of Bayesian gene selection and show its application through several microarray studies.

Acknowledgements

The authors work was supported by a grant from the National Cancer Institute (CA-57030) and the National Science Foundation (NSF DMS-0203215). The authors also wish to thank the editor to *Sankhya* for permission to reproduce some of the figures and tables from their journal.

References

Albert, J., Chib, S. (1993). Bayesian analysis of binary and polychotomous response data. *J. Amer. Statist. Assoc.* **88**, 669–679.

Albert, J.H. (1996). The Bayesian selection for log-linear models. *Canad. J. Statist.* **24**, 327–347.

Alon, U., Barkai, N., Notterman, D.A., Gish, K., Ybarra, S., Mack, D., Levine, A.J. (1999). Broad patterns of gene expression revealed by clustering analysis of tumor and normal colon tissues probed by oligonucleotide arrays. *Proc. Nat. Acad. Sci.* **96**, 6745–6750.

Antoniak, C.E. (1974). Mixtures of Dirichlet processes with applications to nonparametric problems. *Ann. Statist.* **2**, 1152–1174.

Banfield, J.D., Raftery, A.E. (1993). Model-based Gaussian and non-Gaussian clustering. *Biometrics* **49**, 803–821.

Black, M.A., Doerge, R. (2002). Calculation of the minimum number of replicate spots required for detection of significant gene expression fold change in microarray experiments. *Bioinformatics* **18**, 1609–1616.

Brown, P.J., Vannuci, M., Fearn, T. (1998). Multivariate Bayesian variable selection and prediction. *J. Roy. Statist. Soc., Ser. B* **60**, 627–641.

Carlin, B.P., Chib, S. (1995). Bayesian model choice via Markov chain Monte Carlo. *J. Roy. Statist. Soc., Ser. B* **77**, 473–484.

Chen, Y., Dougherty, E.R., Bittner, M.L. (1997). Ratio-based decisions and the quantitative analysis of cDNA microarray images. *J. Biomedical Optics* **4**, 364–374.

Cox, D.R. (1972). Regression models and life tables. *J. Roy. Statist. Soc., Ser. B* **34**, 187–220.

Dawid, A.P. (1981). Some matrix-variate distribution theory: Notational considerations and a Bayesian application. *Biometrika* **68**, 265–274.

Dellaportas, P., Forster, J.J., Ntzoufras, I. (2000). Bayesian variable selection using Gibbs sampler. In: Dey, D.K., Ghosh, S., Mallick, B.K. (Eds.), *Generalized Linear Models: A Bayesian Perspective*. Marcel Dekker, New York, pp. 273–286.

DeRisi, J., Iyer, V., Brown, P.O. (1999). *The Mguide: A Complete Guide to Building Your Own Microarrayer* version 2.0. Biochemistry Department, Stanford University, Stanford, CA. http://cmgm.stanford.edu/pbrown/mguide.

D'haeseleer, P., Liang, S., Somogyi, R. (2000). Genetic network inference: From co-expression clustering to reverse engineering. *Bioinformatics* **16**, 707–726.

Do, K., Müller, P., Tang, F. (2005). A nonparametric Bayesian mixture model for gene expression. *J. Roy. Statist. Soc., Ser. C* **54** (3), 1–18.

Dougherty, E.R. (2001). Small sample issues in microarray-based classification. *Comparative and Functional Genomics* **2**, 28–34.

Dudoit, S., Yang, Y.H., Speed, T.P., Callow, M.J. (2002). Statistical methods for identifying differentially expressed genes in replicated cDNA microarray experiments. *Statistica Sinica* **12**, 111–139.

Duggan, D.J., Bittner, M.L., Chen, Y., Meltzer, P.S., Trent, J.M. (1999). Expression profiling using cDNA microarrays. *Nature Genet.* **21**, 10–14.

Efron, B., Tibshirani, R., Storey, J.D., Tusher, V. (2001). Empirical Bayes analysis of a microarray experiment. *J. Amer. Statist. Assoc.* **96**, 1151–1160.

Eisen, M.B., Spellman, P.T., Brown, P.O., Bolstein, D. (1998). Cluster analysis and display of genome-wide expression patterns. *Proc. Nat. Acad. Sci.* **95**, 14863–14868.

Ferguson, T. (1973). A Bayesian analysis of some nonparametric problems. *Ann. Statist.* **1**, 209–230.

Fraley, C., Raftery, A.E. (2002). Model-based clustering, discriminant analysis, and density estimation. *J. Amer. Statist. Assoc.* **97**, 611–631.

Gelfand, A. (1996). Model determination using sampling-based methods. In: Gilks, W.R., Richardson, S., Spiegelhalter, D.J. (Eds.), *Markov Chain Monte Carlo in Practice*. Chapman and Hall, London.

Gelfand, A., Smith, A.F.M. (1990). Sampling-based approaches to calculating marginal densities. *J. Amer. Statist. Assoc.* **85**, 398–409.

George, E., McCulloch, R. (1997). Approaches for Bayesian variable selection. *Statistica Sinica* **7**, 339–373.

Geweke, J. (1991). Efficient simulation from the multivariate normal and Student-t distributions subject to linear constraints. In: *Comput. Sci. Statist.: Proc. of the* 23rd *Symposium on Interface*. Amer. Stat. Assoc., pp. 571–578.

Gilks, W.R., Richardson, S., Spiegelhalter, D.J. (1996). *Markov Chain Monte Carlo in Practice*. Chapman and Hall, London.

Green, P.J. (1995). Reversible jump Markov chain Monte Carlo computation and Bayesian model determination. *Biometrika* **82**, 711–732.

Golub, T.R., Slonim, D., Tamayo, P., Huard, C., Gaasenbeek, M., Mesirov, J., Coller, H., Loh, M., Downing, J., Caliguiri, M., Bloomfield, C., Lender, E. (1999). Molecular classification of cancer: Class discovery and class prediction by gene expression monitoring. *Science* **286**, 531–537.

Haab, B.B., Dunham, M.J., Brown, P.O. (2001). Protein microarrays for highly parallel detection and quantization of specific proteins and antibodies in complex solutions. *Genome Biology* **2**. 0004.1-0004.13.

Harrison, P.J., Stevens, C.F. (1976). Bayesian forecasting (with discussion). *J. Roy. Statist. Soc., Ser. B* **38**, 205–247.

Hastings, W.K. (1970). Monte Carlo sampling methods using Markov chains and their applications. *Biometrika* **57**, 87–109.

Hedenfalk, I., Duggan, D., Chen, Y., Radmacher, M., Bittner, M., Simon, R., Meltzer, P., Gusterson, B., Esteller, M., Kallioniemi, O.P., Wilfond, B., Borg, A., Trent, J. (2001). Gene expression profiles in hereditary breast cancer. *New England J. Medicine* **344**, 539–548.

Ibrahim, J.G., Chen, M.H., Sinha, D. (2001). *Bayesian Survival Analysis*. Springer, New York.

Ibrahim, J.G., Chen, M.H., Gray, R.J. (2002). Bayesian models for gene expression with DNA microarray data. *J. Amer. Statist. Assoc.* **97**, 88–99.

Ideker, T., Thorsson, V., Siegel, A.F., Hood, L.R. (2000). Testing for differentially-expressed genes by maximum-likelihood analysis of microarray data. *J. Comput. Biol.* **6**, 805–817.

Kahn, J., Saal, L.H., Bittner, M.L., Chen, Y., Trent, J.M., Meltzer, P.S. (1999). Expression profiling in cancer using cDNA microarrays. *Electrophoresis* **20**, 223–229.

Kalbfleisch, J.D. (1978). Nonparametric Bayesian analysis of survival time data. *J. Roy. Statist. Soc., Ser. B* **40**, 214–221.

Kendall, M.G., Stuart, A. (1969). *The Advanced Theory of Statistics, vol. 1*, third ed. Hafner, New York.

Kendziorski, C.M., Newton, M.A., Lan, H., Gould, M.N. (2003). On parametric empirical Bayes methods for comparing multiple groups using replicated gene expression profiles. *Statistics in Medicine* **22**, 3899–3914.

Kerr, K., Churchull, G. (2000). Bootstrapping cluster analysis: Assessing the reliability of conclusions from microarray experiments. *Proc. Nat. Acad. Sci.* **95**, 14863–14868.

Khan, J., Simon, R., Bittner, M., Chen, Y., Leighton, S.B., Pohida, P.D., Jiang, Y., Gooden, G.C., Trent, J.M., Meltzer, P.S. (1998). Gene expression profiling of alveolar rhabdomyosarcoma with cDNA microarrays. *Cancer Research* **58**, 5009–5013.

Khan, J., Wei, J., Ringner, M., Saal, L., Ladanyi, M., Westermann, F., Berthold, F., Schwab, M., Antinescu, C., Peterson, C., Meltzer, P. (2001). Classification and diagnostic prediction of cancers using gene expression profiling and artificial neural networks. *Nat. Med.* **7**, 673–679.

Kim, S., Dougherty, E.R., Barrera, J., Chen, Y., Bittner, M., Trent, J.M. (2002). Strong feature sets from small samples. *J. Comput. Biol.* **9**, 127–146.

Leamer, E.E. (1978). Regression selection strategies and revealed priors. *J. Amer. Statist. Assoc.* **73**, 580–587.

Lee, K.E., Mallick, B.K. (2004). Bayesian methods for gene selection in the survival model with application to DNA microarray data. *Sankhya, Ser. A, Indian Journal of Statistics* **66**, 756–778.

Lee, K.Y., Sha, N., Doughetry, E.R., Vanucci, M., Mallick, B.K. (2003). Gene selection: A Bayesian variable selection approach. *Bioinformatics* **19**, 90–97.

Lindley, D.V., Smith, A.F.M. (1972). Bayes estimates for the linear models (with discussion). *J. Roy. Statist. Soc., Ser. B* **34**, 1–41.

Lockhart, D.J., Dong, H., Byrne, M.C., Follettie, M.T., Gallo, M.V., Chee, M.S., Mittmann, M., Wang, C., Kobayashi, M., Horton, H., Brown, E.L. (1996). Expression of monitoring by hybridization to high-density oligonucleotide arrays. *Nature Biotechnology* **14**, 1675–1680.

Luan, Y., Li, H. (2003). Clustering of time-course gene expression data using a mixed-effects model with B-splines. *Bioinformatics* **19**, 474–482.

Madigan, D., York, J. (1995). Bayesian graphical models for discrete data. *Internat. Statist. Rev.* **63**, 215–232.

McCulloch, R., Rossi, P.E. (1994). An exact likelihood analysis of multinomial probit model. *J. Econometrics* **64**, 207–240.

McLachlan, J.G., Basford, E.K. (1987). *Mixture Models: Inference and Applications to Clustering*. Marcel Dekker, New York.

Medvedovic, M., Sivaganesan, S. (2002). Bayesian infinite mixture model based clustering of gene expression profiles. *Bioinformatics* **18**, 1194–1206.

Metropolis, N., Rosenbluth, A.W., Rosenbluth, M.N., Teller, A.H., Teller, E. (1953). Equations of state calculations by fast computing machines. *J. Chem. Phys.* **21**, 1087–1091.

Mitchell, T.J., Beauchamp, J.J. (1988). Bayesian variable selection in linear regression. *J. Amer. Statist. Assoc.* **83**, 1023–1036.

Neal, R.M. (2000). Markov chain sampling methods for Dirichlet process mixture models. *J. Comput. Graph. Statist.* **9**, 249–265.

Newton, M.A., Kendziorski, C.M. (2002). Parametric empirical Bayes methods for microarrays. In: Parmigiani, G., Garrett, E.S., Irizarry, R., Zeger, S.L. (Eds.), *The Analysis of Gene Expression Data: Methods and Software*. Springer-Verlag, New York.

Newton, M.A., Kendziorski, C.M., Richmond, C.S., Blattner, F.R., Tsui, K.W. (2001). On differential variability of expression ratios: Improving statistical inference about gene expression changes from microarray data. *J. Comput. Biology* **8**, 37–52.

Nguyen, D.V., Arpat, A.B., Wang, N., Carroll, R.J. (2002). DNA microarray experiments: Biological and technological aspects. *Biometrics* **58**, 701–717.

O'Hagan, A. (1994). *Bayesian Inference. Kendall's Advanced Theory of Statistics, vol. 2B*, second ed. Cambridge University Press.

Raftery, A.E. (1996). Approximate Bayes factors and accounting for model uncertainty in generalized linear models. *Biometrika* **83**, 251–266.

Raftery, A.E., Madigan, D., Hoeting, J.A. (1997). Bayesian model averaging for linear regression models. *J. Amer. Statist. Assoc.* **92**, 172–191.

Robert, C. (1995). Simulation of truncated normal variables. *Statist. Comput.* **5**, 121–125.

Schena, M., Shalon, D., Davis, R., Brown, P. (1995). Quantitative monitoring of gene expression patterns with a complementary DNA microarray. *Science* **270**, 467–470.

Schwartz, G. (1978). Estimating the dimension of a model. *Ann. Statist.* **6**, 497–511.

Sha, N., Vannucci, M., Tadesse, M.G., Brown, P.J., Dragoni, I., Davies, N., Roberts, T.C., Contestabile, A., Salmon, N., Buckley, C., Falciani, F. (2004). Bayesian variable selection in multinomial probit models to identify molecular signatures of disease stage. *Biometrics* **60** (3), 812–819.

Smith, M., Kohn, R. (1997). Nonparametric regression using Bayesian variable selection. *J. Econometrics* **75**, 317–344.

Sorlie, T., Perou, C.M., Tibshirani, R., Aas, T., Geisler, S., Johnsen, H., Hastie, T., Eisen, M.B., van de Rijn, M., Jeffrey, S.S., Thorsen, T., Quist, H., Matese, J.C., Brown, P.O., Botstein, D., Eystein Lonning, P., Borresen-Dale, A.L. (2001). Gene expression patterns of breast carcinomas distinguish tumor subclasses with clinical implications. *Proc. Nat. Acad. Sci.* **98**, 10869–10874.

Tamayo, P., Slonim, D., Mesirov, J., Zhu, Q., Kitareewen, S., Dmitrovsky, E., Lander, E.S., Golub, T.R. (1999). Interpreting patterns of gene expression with self-organizing maps: Methods and application to hematopoietic differentiation. *Proc. Nat. Acad. Sci.* **96**, 2907–2912.

Tavazoie, S., Hughes, J.D., Campbell, M.J., Cho, R.J., Church, G.M. (1999). Systematic determination of genetic network architecture. *Natural Genetics* **22**, 2815.

Tibshirani, R., Walther, G., Hastie, T. (2001). Estimating the number of clusters in a data set via the gap statistic. *J. Roy. Statist. Soc., Ser. B* **63**, 411–423.

Tusher, V.G., Tibshirani, R., Chu, G. (2001). Significance analysis of microarrays applied to the ionizing radiation response. *Proc. Nat. Acad. Sci.* **98**, 5116–5121.

Van der Laan, M., Bryan, J. (2001). Gene expression analysis with the parametric bootstrap. *Biostatistics* **2**, 445–461.

Wakefield, J., Zhou, C., Self, S. (2003). Modelling gene expression over time: Curve clustering with informative prior distributions. In: Bernardo, J.M., Bayarri, M.J., Berger, J.O., Dawid, A.P., Heckerman, D., Smith, A.F.M., West, M. (Eds.), *Bayesian Statistics, vol. 7*. Oxford University Press.

West, M. (2003). Bayesian factor regression models in the "Large p, Small n" paradigm. In: Bernardo, J.M., Bayarri, M.J., Berger, J.O., Dawid, A.P., Heckerman, D., Smith, A.F.M., West, M. (Eds.), *Bayesian Statistics, vol. 7*. Oxford University Press.

Yeang, C.H., Ramaswamy, S., Tamayo, P., Mukherjee, S., Rifkin, R.M., Angelo, M., Reich, M., Lander, E., Mesirov, J., Golub, T. (2001). Molecular classification of multiple tumor types. *Bioinformatics* **17**, 316–322.

Yeung, K.Y., Fraley, C., Murua, A., Raftery, A.E., Ruzzo, W.L. (2001). Model-based clustering and data transformations for gene expression data. *Bioinformatics* **17**, 977–987.

Zellner, A. (1986). On assessing prior distributions and Bayesian regression analysis with distributions. In: Goel, P.K., Zellner, A. (Eds.), *Bayesian Inference and Decision Techniques: Essays in Honor of Bruno de Finetti*. North-Holland, Amsterdam, pp. 233–243.

Handbook of Statistics, Vol. 25
ISSN: 0169-7161
© 2005 Elsevier B.V. All rights reserved.
DOI 10.1016/S0169-7161(05)25025-3

25

Bayesian Biostatistics

David B. Dunson

Abstract
With the rapid increase in biomedical technology and the accompanying generation of complex and high-dimensional data sets, Bayesian statistical methods have become much more widely used. One reason is that the Bayesian probability modeling machinery provides a natural framework for integration of data and information from multiple sources, while accounting for uncertainty in model specifications. This chapter briefly reviews some of the recent areas in which Bayesian biostatistical research has had the greatest impact. Particular areas of focus include correlated and longitudinal data analysis, event time data, nonlinear modeling, model averaging, and bioinformatics. The reader is referred elsewhere for recent Bayesian developments in other important areas, such as clinical trials and analysis of spatially correlated data. Certainly the many practical and conceptual advantages of the Bayesian paradigm will lead to an increasing impact in future biomedical research, particularly in areas such as genomics.

1. Introduction

In recent years, there have been enormous advances in biomedical research and technology, which have lead directly to fundamental shifts in the process of data collection and analysis. Historically, the focus of basic science and epidemiologic studies has been predominantly on investigating the relationship between a small set of predictors (e.g., dose of a chemical, age) and a single outcome (e.g., occurrence of a specific disease, age at onset of disease). In such simple settings, nonstatistical articles almost uniformly report p-values and results from frequentist analyses, and Bayesian approaches are seldom used. One reason for this is that, for simple models, vague priors, and sufficiently-large data sets, Bayesian and frequentist analyses typically yield nearly identical results. Hence, there has not been a compelling enough motivation for a large scale shift away from the frequentist paradigm towards the less familiar, but (arguably) more intuitive Bayesian approach.

However, this is all changing rapidly with the increasing movement in the scientific community away from the simple reductionist approach and towards an approach that embraces the new technology that have become available for generating and analyzing high-dimensional biologic data. Perhaps the most striking example of this trend has been the unprecedented focus of resources and creative energy on gene expression array data. Clearly, gene expression data are potentially highly-informative about biologic pathways involved in the development of cancer and other diseases, and in the metabolism of drugs used to treat these diseases. The difficulty is in elucidating these pathways and in identifying important genes and their interactions based on the extremely high-dimensional gene expression profiles that are collected, particularly given that the number of subjects is typically very small relative to the number of genes under consideration. Due to the dimensionality problem, one cannot hope to make progress without incorporating substantial prior information, either through a formal Bayesian analysis or through an ad hoc frequentist approach.

Gene expression arrays are one of the more visible examples of the increasing focus of biomedical studies on complex and high-dimensional data. Some other examples of emerging fields, which are also critically dependent on new statistical developments, include neurophysiology, proteomics, and metabolomics. However, the trend has not been confined to emerging fields. Traditional sciences, such as epidemiology and toxicology, also routinely collect complex and multivariate data. To provide an example of some of the challenges which are routinely encountered, consider the study of uterine leiomyoma (fibroids), a tumor that rarely becomes malignant but that leads to substantial morbidity, including bleeding, pain, pregnancy complications, infertility, and hysterectomy. Although fibroids are extremely common, they often go undetected for many years, while growing to a clinically significant size and spreading throughout the uterus. A recent epidemiologic study of fibroids attempted to obtain information on factors related to incidence and progression using a cross-sectional design in which a sample of women provided medical records and were given an ultrasound to determine current status of fibroids (Baird et al., 2003). The question is how do we do inferences on incidence and progression of fibroids based on the multivariate and informatively-censored information that we have available.

Bayesian methods are ideally suited for addressing challenges of this type, and this chapter will attempt to give a flavor of the Bayesian approach to complex biostatistical problems by focusing on some recent developments. This is not meant to be a comprehensive overview of the literature on Bayesian biostatistics, and the focus will be on a few areas meant to highlight the types of insights into difficult biomedical problems that can be obtained by using a Bayesian approach. For additional references on the Bayesian approach to biostatistics, refer to the book by Berry and Stangl (1996) and to commentaries by Dunson (2001), Etzioni and Kadane (1995), Freedman (1996), Gurrin et al. (2000), Gustafson et al. (2005), Lilford and Braunholtz (1996; 2000), Spiegelhalter et al. (1999).

2. Correlated and longitudinal data

2.1. Generalized linear mixed models

In longitudinal follow-up studies, multi-center clinical trials, animal experiments collecting data for pups within a litter, and in many other biomedical applications, response data are correlated. Conceptually, such correlation can be thought to arise from unmeasured covariates that are shared for observations within a cluster. For example, there may be unmeasured features of the individual in a longitudinal study of blood pressure that lead to residual dependency in blood pressure levels collected at different ages. A natural approach to account for such unmeasured covariates in the analysis is to use a random effects model (Laird and Ware, 1982). The Laird and Ware (1982) model allows the regression coefficients in a linear model to vary for different individuals according to a normal distribution with unknown mean and variance. This normal distribution is effectively a prior distribution for the individual-specific regression coefficients. However, Laird and Ware (1982) avoid a full Bayesian approach by not choosing prior distributions for the unknown mean and variance of the random effect distribution.

For a full Bayesian treatment of the linear random effects model and closely-related normal hierarchical models, refer to Gelfand et al. (1990) and Gilks et al. (1993). In addition, Zeger and Karim (1991) consider a Bayesian approach to the generalized linear mixed model (GLMM), which extends the linear random effects model to account for outcomes in the exponential family. Bayesian analyses of these models and much more complex extensions, involving multi-level data structures, censoring, and missing data, can be implemented routinely using Gibbs sampling in the WinBUGS software package (Lunn et al., 2000). Routine frequentist analyses of GLMMs typically relies on use of SAS Proc NLMIXED, which is notoriously unstable and sensitive to starting values, particularly in more complex models. This instability is reflective of general difficulties in developing algorithms for reliable optimization in complex nonlinear models. In addition, the estimated standard errors can be unreliable due in part to failure of standard large sample approximations.

The Bayesian Markov chain Monte Carlo (MCMC)-based approach has several important advantages from a practical perspective. First, the MCMC algorithm (assuming sufficient burn-in to allow convergence) produces not a point estimate and asymptotic standard error but instead samples from the exact posterior distribution of the unknowns. These samples can be used to obtain summaries of the exact posterior distribution of any functional of the model parameters. For example, in epidemiologic applications, one often wants to obtain point and interval estimates for exposure odds ratios, risk ratios, and the probability of disease for individuals in different categories. Frequentist interval estimates for such nonlinear functionals of the regression coefficients often rely on a sequence of asymptotic approximations, leading to potentially-unreliable confidence limits. Since epidemiologists often place heavy stock on whether the lower bound on the confidence limit for the exposure odds ratio is greater than one, it is important to have an accurate confidence bound.

However, the Bayesian MCMC-based approach to inferences in GLMMs is not without its difficulties. Firstly, it is well known that improper priors can lead to improper posteriors (Hobert and Casella, 1996) and that this problem is not solved by using "diffuse, but proper" priors as is often recommended (Natarajan and McCulloch, 1998). In fact, even for moderate variance priors on the variance components, there can be very poor computational efficiency due to slow mixing of the Gibbs sampler. This problem has led several authors to recommend strategies for speeding up computation, such as transformation of the random effects (Vines et al., 1996) and block updating (Chib and Carlin, 1999). However, even with the convergence and mixing issues inherent in MCMC computations, Bayesian hierarchical models are becoming more and more widely used due to their many advantages over frequentist methods. These advantages are particularly apparent in more complex settings, such as when the covariance structure and/or the parametric form of the random effect density are unknown.

2.2. *Covariance structure modeling*

Although random effects models are very widely used in longitudinal data analysis and other correlated data applications, standard implementations assume a known covariance structure. In many cases, there may be uncertainty in the covariance structure, which should be accounted for in the analysis. In other cases, a more flexible correlation structure than that typically assumed may be needed to adequately characterize the data, or there may be interest in inferences on the covariance structure. There have been numerous articles in recent years proposing Bayesian methods for addressing these problems.

In order to address the problem of selection of the subset of predictors having random effects in a linear mixed effects model, Chen and Dunson (2003) proposed a Bayesian variable selection approach. In particular, using a novel decomposition of the random effects covariance in the saturated model having random effects for every predictor, they proposed a mixture prior that allowed random effects to have zero variance and effectively drop out of the model. The conditionally-conjugate form of the prior leads to efficient posterior computation via a Gibbs sampling algorithm, and inferences on the fixed effects can be based on model averaging across different random effects structures.

Random effects models for longitudinal data typically assume that the random effects covariance matrix is constant across subjects, which may not be realistic in many biomedical applications. To address this problem, Daniels and Zhao (2003) proposed an approach for modeling of the random effects covariance matrix as a function of measured covariates. Following a Bayesian approach, they used a Gibbs sampler for posterior computation and applied the methods to data from depression studies.

In applications with a moderate to large number of response variables, parsimonious modeling of the covariance matrix is a very important problem, because the number of covariance parameters can be huge. Due to the dimensionality, traditional frequentist estimators can be unstable, motivating many authors to propose shrinkage estimators. In recent years, there has been a growing body of Bayesian work on this problem and the related problem of inference on the covariance structure using graphical models. Much of the recent work has been motivated by bioinformatics and gene expression array applications involving huge numbers of variables (refer to Jones et al., 2004, for a recent reference).

To give a flavor of what types of approaches have been proposed, we highlight papers by Smith and Kohn (2002), Daniels and Pourahmadi (2002), and Wong et al. (2003). Smith and Kohn (2002) proposed an approach for Bayesian estimation of the longitudinal data covariance matrix, parameterized through the Cholesky decomposition of its inverse. A variable selection-type prior is used to identify zero values for the off-diagonal elements, and a computationally efficient MCMC approach is developed that allows the dimension to be large relative to sample size. Daniels and Pourahmadi (2002) proposed a related approach, using the Cholesky decomposition, and developing a new class of conditionally-conjugate prior distributions for covariance matrices. These priors have a number of theoretical and practical advantages, including computation via Gibbs sampling and flexible shrinking of covariances towards a particular structure. Wong et al. (2003) proposed an approach to the somewhat different problem of estimating the precision matrix from Gaussian data. Using a prior that allows the off-diagonal elements of the precision matrix to be zero, they avoid assumptions about the structure of the corresponding graphical model. An efficient Metropolis–Hastings algorithm is used for posterior computation, and the parameters are estimated by model averaging.

2.3. *Flexible parametric and semiparametric methods*

Although Bayesian hierarchical models for correlated and longitudinal data have many appealing attributes, a common criticism is that the approach may be overly dependent on parametric assumptions about the prior distributions. A particular concern, which is also a potential problem in frequentist mixed models, is possible sensitivity to the assumed form for the random effects distribution. This concern has motivated many practitioners to use marginal models based on generalized estimating equations (Liang and Zeger, 1986; Zeger and Liang, 1986) instead of relying on Bayesian or non-Bayesian hierarchical models.

To address this concern, one can potentially utilize a flexible parametric mixture model (Richardson and Green, 1997), or a nonparametric Bayesian approach in which one accounts for uncertainty in the random effects or outcome distribution (refer to Walker et al., 1999, for a review of Bayesian nonparametrics). The majority of the literature on Bayesian nonparametrics has focused on Dirichlet process priors (Escobar, 1994) or related extensions, though there has also been some consideration of Polya trees (Lavine, 1992, 1994) and other nonparametrics processes. We focus here on articles proposing methods to account for uncertainty in the choice of distributions used to characterize heterogeneity in a hierarchical model.

One of the first articles to consider this type of approach was Bush and MacEachern (1996). They proposed a Bayesian semiparametric approach for analysis of randomized block experiments in which the block effects were assigned an unknown distribution G. This distribution was then assigned a Dirichlet process prior centered on the normal distribution, effectively shrinking the estimated block effect distribution towards the normal distribution, while allowing for unanticipated features of the data. Kleinman and Ibrahim (1998) later used a related strategy to allow the random effects distribution in a longitudinal random effects model to be unknown, again using a Dirichlet process prior. Posterior computation for this approach is straightforward using a Gibbs sampler.

Lopes et al. (2003) proposed an alternative approach for flexible modeling of the random effect distribution in a longitudinal data model. Instead of using a Dirichlet process prior, their approach relies on flexible mixtures of multivariate normals used to accommodate population heterogeneity, outliers, and nonlinearity in the regression function. In addition, they allow for joint inferences across related studies using a novel hierarchical structure. Posterior computation relies on a reversible jump MCMC algorithm, which allows the number of mixture components to be unknown.

One nice feature of the Bayesian probability modeling framework is that methods tend to be easy to generalize to more complex settings. This is very useful in biomedical studies, since complexity tends to be the norm. For example, it is often the case in prospective clinical trials and observational epidemiology studies, that data consist of both a repeatedly measured biomarker and an event time. In such cases, interest focuses on joint modeling of longitudinal and survival data. To address this problem using a Bayesian approach, while avoiding parametric assumptions on the distribution of the longitudinal biomarker, Brown and Ibrahim (2003) used Dirichlet process priors on the parameters defining the longitudinal model. This is just one example of a growing body of literature on flexible Bayesian approaches for time to event data, a topic discussed in the following section. For a book on Bayesian survival analysis, refer to Ibrahim et al. (2001).

3. Time to event data

3.1. Continuous right-censored time to event data

For simple right-censored survival data, Cox proportional hazards regression (Cox, 1972) is the default standard approach to analysis routinely reported in biomedical journals. One of the most appealing features of the Cox model partial likelihood-based approach is that (under the proportional hazards assumption) it is not necessary to specify the baseline hazard function. In contrast, fully Bayesian approaches to survival analysis require a complete specification of the likelihood function, and hence a model for the baseline hazard. However, one can still avoid a parametric specification of the baseline hazard by using a nonparametric prior process. For example, a widely used choice of nonparametric prior for the baseline cumulative hazard function is the gamma process (Kalbfleisch, 1978).

Clayton (1991) proposed a Bayesian approach for inference in the shared frailty model, which generalizes the Cox model to allow for dependency in multivariate failure time data by including a cluster-specific frailty multiplier, typically assumed to have a gamma distribution. Clayton used a gamma process prior for the baseline cumulative hazard function, and developed a Monte Carlo approach to posterior computation. The gamma process prior relies on the subjective choice of a best guess for the cumulative baseline hazard and for a precision parameter (c) controlling uncertainty in this guess. The estimated cumulative baseline hazard function is shrunk towards the initial guess to a degree determined by the precision parameter c. Since one often has little confidence in the initial guess, c is typically chosen to be close to 0, resulting in a low degree of shrinkage and an estimated hazard function that has large discontinuities.

Potentially, one can obtain a smoother estimate of the hazard function without obtaining results that are overly sensitive to subjectively-chosen hyperparameters by adding a level to the prior specification. This can be accomplished by specifying a parametric model for the initial guess, and then choosing hyperprior distributions for both the parameters in this model and the precision c. Dunson and Herring (2003) implemented this type of approach in developing an approach for Bayesian order-restricted inference in the Cox model. An alternative approach for smoothing of the baseline hazard function is to choose an autocorrelated prior process (Sinha and Dey, 1997; Aslanidou et al., 1998), such as the Markov gamma process proposed by Nieto-Barajas and Walker (2002).

Several recent articles have proposed alternative approaches for flexible Bayesian survival analysis, which do not require the proportional hazards assumption. For example, Gelfand and Kottas (2003) proposed a Bayesian semiparametric median residual life model, which used Dirichlet process mixing to avoid parametric assumptions on the residual survival time distribution. The reader can refer to their paper for additional references.

3.2. *Complications*

In biomedical studies, time to event data are often subject to a variety of complications, which can invalidate standard time-to-event analyses, such as the Cox (1972) model and Bayesian alternatives discussed in Section 3.1. Some of these complications include informative censoring, missing data, time-varying covariates and coefficients, and a proportion of long-term survivors who are not at risk of the event (as in cure rate models – refer to Tsodikov et al., 2003, for a recent review). In addition, many biomedical application areas that collect time to event data require specialized models consistent with the data structure and biology of the process. This is certainly true in modeling of tumorigenesis and in time to pregnancy studies.

To provide some intuition for the types of problems that routinely arise, consider an epidemiologic study relating one or more exposures to the hazard of onset of a disease or health condition. For example, the goal may be to relate alcohol consumption to the risk of cancer based on data from a detailed questionnaire on alcohol consumption at different ages along with follow-up (say, starting at age 40) until drop-out, occurrence of cancer, or death from other causes. There has been increasing evidence that early life exposures may have an important impact on the development of disease later in life. For example, in this case, binge drinking in college could potentially increase the risk of mutations in key genes involved in the regulation of cell division and death. This could in turn lead to an increased risk of cancer later in life after additional mutations have occurred, possibly also related to the level of alcohol consumption. Such mechanisms can lead to a complex dependency between an individual's history of exposure and the current risk of developing disease. Certainly, a simple Cox model analysis, which includes a single time-independent predictor summarizing an individual's lifetime exposure, can produce misleading results.

As a partial solution to this problem, one can use a generalized Cox model that allows for time-varying predictors. For example, in the alcohol and cancer example, one could use the cumulative amount of alcohol consumption over a person's lifetime as a predictor (though, of course, this would also have limitations). This can be done easily from either a Bayesian or frequentist perspective (refer to Ibrahim et al., 2001). The problem with this approach is that it assumes that the same coefficient applies to all ages. For example, a 40 year old heavy drinker who has the same cumulative lifetime exposure as a 70 year old moderate drinker will also have the same multiplicative increase in the risk of cancer attributable to drinking.

Potentially, one can relax this assumption by using a time-varying coefficient model, either with or without time-varying covariates. West (1992) proposed an innovative Bayesian approach for modeling time-varying hazards and covariate effects. Sinha et al. (1999) proposed a discretized hazard model for interval-censored data that allows the regression coefficients to vary between intervals. Mallick et al. (1999) proposed a multivariate adaptive regression spline approach for flexible modeling of the log hazard function, conditional on the covariates. McKeague and Tighiouart (2000) allowed the regression coefficients in a continuous-time Cox model to vary according to a step function, with jump times arising via a time-homogeneous Poisson process, and there have been many related approaches in the literature.

These time-varying coefficient approaches allow covariate effects on the age-specific incidence rate to vary according to an unknown function of the current age. However, in cases such as the alcohol and cancer example in which the entire history of exposure up to the current age can play a role in determining an individual's risk, it is not clear how to choose a time-varying covariate summarizing the exposure history. In such cases, both the age at exposure and the current age are potentially important in determining an individual's hazard rate. To address this problem, Dunson et al. (2003) proposed a Bayesian approach in which the effect of an exposure on an individual's hazard at a given age depends on an integral across the exposure history of an unknown age-specific weight function multiplied by a term that allows for possible waning. Hence, if early life exposures lead to a greater risk than later life exposures, the weight function would be declining with age.

In addition to difficulties in assessing the effects of time-varying exposures, there are other complications that routinely arise in biomedical studies collecting event time data. Missing values is one particularly vexing problem, and it is often the case that a substantial proportion of individuals fail to provide information on a least one of the predictors. In such cases, discarding subjects with missing values can lead to loss of information and, in some cases, biased inferences. Since accounting for missing data in the analysis typically requires assumptions about the distribution of the missing values, Bayesian approaches are natural in this setting. Since there is a large body of literature in this area, we focus on two recent papers which include references to previous work. In particular Scharfstein et al. (2003) proposed an approach for incorporating prior beliefs about nonidentifiable selection bias parameters into a Bayesian analysis of randomized trials with missing outcomes. Gagnon et al. (2003) instead focused on the problem of analyzing age at onset for diseases with a large genetic component, and hence a high proportion of non-susceptible individuals. Following a Bayesian parametric modeling approach, they allowed for missing information in a stochastic model of disease onset and survival after onset.

In many cases, noninformative censoring assumptions implicit in most survival models are violated in studies of diseases that can go undiagnosed for a substantial amount of time. For example, in studying factors related to the age at onset of cancer or benign tumors, such as uterine fibroids, it is typically not possible to obtain exact information on the age at onset. One instead relies on data on the age at diagnosis, with information on severity also available in many cases. In certain applications, as in animal studies of tumorigenesis, diagnosis does not occur until a pathological examination at the time of death. Clearly, if the risk of death is higher among animals with more tumors, then death does not provide a noninformative censoring mechanism, and some adjustment must be made. This is also true for studies that follow a cohort of individuals over time, providing periodic diagnostic tests at the clinic to obtain an interval-censored onset time. If the disease leads to symptoms causing the patient to go into the health clinic more often or earlier, then the examination times are informative about the disease onset times.

One approach for dealing with this problem is to use a multistate model in which the individual progresses through different states over time. For example, in a simple case, individuals all begin in the disease-free state (state 1), and then over time there is an age-specific rate of entering the preclinical disease state (state 2). Once in the preclinical disease state, there is an age-specific or latency-time specific risk of diagnosis or death (state 3). Such models have been widely used from a frequentist perspective, particularly in cancer and AIDS applications. A nice feature of multistate models is that they can be used to flexibly characterize the natural history of disease, including initial onset, progression, development of symptoms, diagnosis, and treatment. However, the available data from a single study may not be informative about all of the transition parameters needed to characterize the stochastic process, particularly when one allows for age- and latency-time specific transitions (as is a necessary complication in most cases).

For this reason and due to computational considerations, the Bayesian approach has important advantages. Instead of using an overly-simplified model, one can formulate a model consist with biological information and then choose informative prior distributions. Dunson and Dinse (2002) applied this strategy to multiple tumor site data from bioassay experiments in order to adjust inferences about chemical effects on multi-site tumor incidence for survival differences and tumor lethality. This same approach can be used for data from epidemiologic studies. Also using a discrete-time Markov model, Craig et al. (1999) developed a Bayesian approach for assessing the costs and benefits of screening and treatment strategies. Their model characterizes the natural progression of disease, including the possibility of both treatment intervention and death. Dunson and Baird (2002) developed alternative discrete-time stochastic models for onset, preclinical progression, and diagnosis of disease, using a latent growth model to characterize changes in multiple categorical and continuous measures of severity according to the time spent with disease.

3.3. Multiple event time data

In addition to the complications which arise due to missing and informatively-censored data and to the presence of nonsusceptible subpopulations, many biomedical studies collect information on the rate of event occurrence for an event that can occur repeatedly over time. For example, in animal studies of skin and breast tumors, each animal can potentially get many tumors with these tumors achieving a detectable size at different times during the study. A similar data structure arises in studies of bladder tumors and uterine fibroids, which can reoccur repeatedly after being removed. Data of this type are typically referred to as multiple event time data. Panel or interval count data, consisting of the number of events occurring between examinations, are an interval-censored version of multiple event time data. This is one area in which there has been relatively little Bayesian work, with a few important exceptions.

Sinha (1993) proposed a proportional intensity model and Bayesian approach to inference for multiple event time data. He chose a prior structure related to the gamma process prior used by Kalbfleisch (1978), and incorporated an unobserved random frailty to account for heterogeneity among subjects. A Gibbs sampler was then used for posterior computation. Much of the work in Bayesian methods for multiple event time data has been motivated by tumor studies. For example, in some types of cancer chemoprevention experiments and short-term carcinogenicity bioassays, the data consist of the number of observed tumors per animal and the times at which these tumors were first detected. In such studies, there is interest in distinguishing between treatment effects on the number of tumors induced by a known carcinogen and treatment effects on the tumor growth rate. Since animals may die before all induced tumors reach a detectable size, separation of these effects can be difficult.

Dunson and Dinse (2000) proposed a flexible parametric model for data of this type motivated by the biology of tumor initiation and promotion. The model accommodates distinct treatment and animal-specific effects on the number of induced tumors (multiplicity) and the time to tumor detection (growth rate). A Gibbs sampler is developed for estimation of the posterior distributions of the parameters. The methods are illustrated through application to data from a breast cancer chemoprevention experiment. A closely related model and Bayesian inferential approach was independently developed by Sinha et al. (2002).

4. Nonlinear modeling

Another area in which there have been many important advances in Bayesian methodology in recent years is nonlinear modeling. It is often the case in biomedical applications that linear regression models may provide a poor approximation, and more flexible regression methods are needed. For example, a common theme in toxicology and epidemiology applications is interest in the specific dose response relationship between the level of exposure and the risk of an adverse health outcome. It is clearly not enough to know that there is a positive association between a particular exposure and the occurrence of disease, one needs to know the level of risk at different doses. In addition, scientists are typically interested in making probability statements about whether there is an increased risk from low to moderate levels. For example, what is the probability of an increased hazard of heart attacks for individuals who are slightly overweight relative to thin individuals? Addressing such questions is critically important in making decisions about public health impact and the need for intervention, e.g., to regulate the level of chemical released into the environment or to strongly recommend dieting and exercise even for those patient who are only slightly overweight. In the following subsections we review some recent Bayesian methods that can be used to address questions of this type. For an excellent book on Bayesian methods for nonlinear classification and regression, refer to Denison et al. (2002).

4.1. *Splines and wavelets*

As in the frequentist literature on nonlinear regression, much of the Bayesian work has focused on regression splines. We focus here on a few recent papers containing many additional references to give the reader a flavor of the possibilities. Holmes and Mallick (2003) proposed an approach for nonparametric modeling of multivariate non-Gaussian response data using data-adaptive multivariate regression splines. They treated both the number and location of the knot points as unknown, and used a reversible-jump Markov chain Monte Carlo algorithm for posterior computation (a difficult problem in nonlinear regression). To improve computational efficiency they incorporated a latent residual term to model correlation. Kass et al. (2003) applied Bayesian adaptive regression splines (DiMatteo et al., 2001) to the problem of smoothing neuronal data to improve inferences on the instantaneous firing rate and related quantities. Wood et al. (2002) used a mixture of splines for Bayesian spatially adaptive nonparametric regression. By allowing each component spline to have its own smoothing parameter defined in a local region, they allow the regression function to be smoother in certain regions than in others. As in other work, efficient computation using Markov chain Monte Carlo is an important issue.

Much of the recent work on Bayesian methods for nonlinear regression has focused on models using wavelets, a flexible class of basis functions with many appealing properties. Clyde et al. (1998) proposed an approach for selecting the nonzero wavelet coefficients by assigning positive prior probability to wavelet coefficients being zero. This type of prior is closely related to priors used in subset selection in regression, but the dimensionality of the wavelet selection problem necessitates the use of innovative computational methods. Clyde et al. (1998) considered a variety of approaches for fast computation, including simple analytic approximations, importance sampling, and MCMC algorithms. Much of the appeal of wavelet-based models is their flexibility, and they have proven quite useful in a number of biomedical applications. We list only two recent applications here. Aykroyd and Mardia (2003) used wavelets to describe shape changes and shape difference in curves. Their approach was motivated by a longitudinal study of idiopathic scoliosis in U.K. children in which spinal curvature is an important response variable. Motivated by data from an experiment relating dietary fat to biomarkers of early colon carcinogenesis, Morris et al. (2003) proposed a wavelet-based approach to modeling of hierarchical functional data.

4.2. *Constrained regression*

Unfortunately, even though there have been substantial advances in both frequentist and Bayesian methodology for flexible regression, the default standard approach in biomedical journals is to assume linearity (e.g., on the logit scale) in assessing overall trends, and then to perform a separate categorical analysis to contrast effects at different exposure levels. Although results from frequentist generalized additive models (Hastie and Tibshirani, 1990) are commonly reported in biomedical journals, they are typically used primarily to produce smoothed estimates of the regression function and not as a basis for inferences on increases in risk at different exposure levels. In biomedical studies in which there is interest in inferences on a dose response curve, it is commonly the case that investigators have *a priori* knowledge of the possible direction of the effect. For example, in studying the relationship between body mass index (bmi) and heart disease, one does not expect heart disease incidence to decrease as bmi increases. However, it may be uncertain how high bmi needs to be before there is an increase in risk relative to thin individuals.

One could potentially put in the restriction that the risk is nondecreasing by choosing a prior distribution with support on the constrained space. Such constraints can easily be incorporated by proceeding with MCMC computation as if there was no constraint, and then discarding draws which violate the constraint (Gelfand et al., 1992). However, this approach can only be used directly when there are strict constraints on the regression parameters. Typically, in biomedical applications, interest focuses on assessing evidence of an association between a response variable Y and a predictor X, both over the range of X and in local regions. It is therefore important to allow *a priori* uncertainty in whether the regression function increases or is flat. Motivated by this problem, Dunson and Neelon (2003) proposed an approximate Bayesian approach in which samples from an unconstrained posterior distribution were projected to the constrained space via an isotonic regression transformation. Inference on local and global trends were then based on Bayes factors.

This approach still does not address the problem of inferences on an unknown continuous regression function $f(x)$, which is known *a priori* to be nondecreasing. Ramgopal et al. (1993) proposed a nonparametric approach for placing prior constraints on the form of a potency curve, restricting it to be convex, concave, or ogive. Lavine and Mockus (1995) developed an approach for Bayesian isotonic regression in which both the regression function and the residual distribution were modeled nonparametrically. These approaches are useful when the regression function $f(x)$ is known to be strictly increasing. To allow inferences on flat regions of the regression function, Holmes and Heard (2003) proposed an alternative approach. Starting with an unconstrained piecewise constant regression model, they allowed for unknown numbers and locations of knots, and developed a reversible jump MCMC approach for posterior computation. Then they proposed to use the proportions of samples for which the monotonicity constraint holds to estimate a Bayes factor for testing monotonicity vs. any other curve shape.

An important drawback of this approach is that it still does not provide a framework for inferences on flat regions of the regression function vs. increases. To address this problem, Neelon and Dunson (2004) proposed an alternative approach based on a constrained piecewise linear model. To smooth the curve and facilitate borrowing of information about the occurrence of flat regions, a Markov process prior was defined for latent slope parameters specific to the intervals between each of the data points. Flat regions of the curve were defined to correspond to intervals in which the latent slopes were less than a random threshold, and the actual slopes were equivalent to the latent slopes when the values were above this threshold. An efficient MCMC algorithm was then used for posterior computation. Point and interval estimates of the regression function, posterior distributions for thresholds, and posterior probabilities of an association for different regions of the predictor can be estimated from a single MCMC run.

5. Model averaging

One of the major advantages of the Bayesian approach is the ability to allow for model uncertainty in prediction through use of model averaging techniques (Raftery et al., 1997; Hoeting et al., 1999). In the presence of model uncertainty, it is clearly important to account for uncertainty in model selection to avoid underestimation of the errors in prediction and possible systematic biases. Although there has been some recent work on frequentist model averaging (Hjort and Claeskens, 2003), frequentist approaches to prediction typically rely on the final model selected by an ad hoc stepwise selection or criteria-based approach, without appropriately accommodating uncertainty and biases arising in the selection process. By placing a prior distribution on the model and averaging across different models, with weights dependent on the posterior model probabilities, Bayesian model averaging can be used to properly account for model uncertainty in performing predictions. To highlight some of the advantages, we consider a variety of recent applications in this section.

Clyde (2000) applied Bayesian model averaging to applications of health effect studies of particulate matter and other pollutants. In these types of studies, there is typically uncertainty in which pollutants and confounding variables should be included in the model, what type of time-dependency should be used for the covariates, whether interactions should be included, and how to incorporate nonlinear trends. Clyde noted that different model selection approaches can lead to completely different conclusions. She recommended a simple Bayesian model averaging approach, which combines the results of standard maximum likelihood estimation analyses for the separate models by using asymptotic approximations to the posterior model probabilities. Volinsky et al. (1997) applied a closely related approach to perform model averaging in proportional hazards regression, motivated by a study assessing the risk of stroke.

Bayesian model averaging is also very useful in forecasting. Stow et al. (2004) considered an interesting application in which there is interest in forecasting the levels of PCBs in Great Lake fish to assess progress towards the long-term goal that fish should be safe to eat. Applying Bayesian model averaging and dynamic linear models to historical data, they predicted a neglible probability that the short-term goal of a 25% reduction in PCB levels will be met. Gangnon and Clayton (2000) proposed an innovative approach, incorporating ideas from Bayesian model averaging, to address the very different problem of spatial disease clustering. Their approach can be used to obtain estimates for disease rates while allowing greater flexibility than traditional approaches in the type and number of clusters under consideration.

Bayesian model averaging provides an extremely useful framework for addressing complex biological problems and high-dimensional data sets in which a single model is unlikely to be flexible enough to fully characterize the data. For example, Conti et al. (2003) recently considered the problem of modeling of complex metabolic pathways, motivated by processes in which chronic diseases arise from a sequence of biochemical reactions involving exposures to environmental compounds metabolized by various genes. Clearly, standard epidemiologic analyses relying on contingency table or logistic regression models focusing on a few variables at a time can fail to capture the complex structure of the problem. Instead, they consider Bayesian model averaging and pharmacokinetic modeling approaches, structuring the analysis to take advantage of prior knowledge regarding the metabolism of compounds under study along with genes that regulate the metabolic pathway.

Seaman et al. (2002) presented an approach for the analysis of studies of highly polymorphic genes possibly involved in disease susceptibility. Since a large number of genotypes are possible, there will typically be few subjects having a given genotype, and a large amount of uncertainty in the genotype-specific risk. To address this problem, Seaman et al. (2002) proposed a clustering approach, imposing structure through an assumption on the joint effect of two alleles making up a genotype. Then, by using Bayesian model averaging across the partitions, they estimate genotype risks more accurately then under traditional approaches, while also ranking alleles by risk. Also addressing high-dimensional genetic data, Medvedovic and Sivagenesan (2002) proposed a Bayesian approach for clustering of profiles, a problem which has received abundant attention in the recent literature and will be considered further in Section 6. Medvedovic and Sivagenesan developed a clustering procedure for gene expression profiles by using a Bayesian infinite mixture model, and estimating the posterior distribution of clusterings using a Gibbs sampler.

6. Bioinformatics

As motivated in Section 1, one of the areas in which Bayesian approaches often have clear practical advantages in applications is the analysis of high-dimensional biological data sets. In this section, some of the recent work in this area is highlighted, though the literature is vast and a comprehensive overview is beyond the scope of this chapter. Some areas which have received abundant attention include analysis of gene expression arrays, protein structure modeling, and biological sequence alignment.

One important use of gene expression data is for predicting the clinical status of disease. West et al. (2001) developed Bayesian regression models that use gene expression profiles to predict the clinical status of human breast cancer. Their proposed approach can be used to estimate relative probabilities of clinical outcomes and to provide an assessment of the uncertainties present in such predictive distributions. Sha et al. (2003) proposed an approach for identification of multiple gene predictors of disease class using Bayesian variable selection applied to a binary probit classification model.

An important issue in analysis of gene expression array data is an inflated type I error rate due to the "large p, small n" problem. Motivated by this problem, Benjamini and Hochberg (1995) proposed controlling the false discovery rate (FDR) instead of the false positive rate. Storey (2003) later proposed a modification of the FDR, referred to as the pFDR, claiming that the pFDR is interpretable as a Bayesian posterior probability with a connection to classification theory. In recent work, Tsai et al. (2003) showed that different versions of the false discovery rate are equivalent under a Bayesian framework, in which the number of true hypotheses is modeled as a random variable. However, there is still no formal Bayesian decision theoretic justification for use of the FDR or its modifications.

Much of the recent work has focused on using Bayesian networks to infer relationships among genes, as highlighted by Friedman (2004) in his recent *Science* article. Friedman et al. (2000) proposed a framework for discovering interactions between genes based on multiple expression measurements by using Bayesian graphical models. Barash and Friedman (2002) proposed a class of models to relate transcription factors and functional classes of genes based on genomic data. They used an innovative search method for Bayesian model selection. Strimmer and Moulton (2000) used directed acyclic graphs expressed as Bayesian networks for analysis of phylogenetic networks.

The alignment of sequences problem is important in studying protein structure and function. Liu et al. (1999) developed an integrated Bayesian approach for the hidden Markov and block-based motif models commonly used in multiple sequence alignment. Boys et al. (2000) developed a Bayesian approach to the problem of detecting homogeneous segments within different DNA sequences. They developed a hidden Markov model, carefully considered the problem of prior elicitation, and used MCMC for posterior computation. Schmidler et al. (2000) developed a novel approach to the difficult problem of predicting the secondary structure of a protein from its amino acid sequence. Their approach is based on a probability model of protein sequence and structural relationships in terms of structural segments, with secondary structure prediction considered as a Bayesian inferential problem.

To address the problem of gene function prediction, Troyanskaya et al. (2003) proposed a framework that relies on a Bayesian approach to integrate heterogeneous types of high-throughput biological data for gene function prediction. The approach incorporates expert knowledge about the relative accuracies of different data sources.

Bayesian approaches have been also used to address a wide variety of difficult problems in genomics not covered in the above discussion, such as inferring haplotypes for a large number of linked single-nucleotide polymorphisms (SNPs) and estimation of the relative abundance of mRNA transcripts using serial analysis of gene expression data (Morris et al., 2003). Certainly, Bayesian methods will continue to play an important role in genomics and related areas of biomedical research involving extremely high-dimensional data sets.

7. Discussion

When faced with the challenges of modern biomedical research, in which data tend to be highly complex and multivariate and it is necessary to integrate information from multiple sources, the advantages of Bayesian methods become increasingly apparent. These advantages are both conceptual, in that the Bayesian approach provides a formal framework for accounting for multiple sources of information and uncertainty in model specifications, and practical, in that Bayesian models and computational algorithms are extremely flexible. MCMC and other algorithms are available for computation of exact posterior distributions and it is not necessary for Bayesian approaches to rely on asymptotic justifications, which typically form the backbone of frequentist inferences. Such justifications are often meaningless in biomedical studies involving small samples sizes relative to the number of unknowns in the model, an increasingly common scenario in practice.

This chapter has attempted to give the reader a sense of some of the possibilities by highlighting some of the recent developments in Bayesian biostatistics. The focus has been on clustered and longitudinal data analysis, event time data, nonlinear regression, model averaging, and bioinformatics. Certainly, there are many other areas of biostatistics in which Bayesian methods have had an important impact, such as measurement error modeling, clinical trial design and analysis, and analysis of spatio-temporal data,

References

Aslanidou, H., Dey, D.K., Sinha, D. (1998). Bayesian analysis of multivariate survival data using Monte Carlo methods. *Canad. J. Statist.* **26**, 33–48.

Aykroyd, R.G., Mardia, K.V. (2003). A wavelet approach to shape analysis for spinal curves. *J. Appl. Statist.* **30**, 605–623.

Baird, D.D., Dunson, D.B., Hill, M.C., Cousins, D., Schectman, J.M. (2003). High cumulative incidence of uterine leiomyoma in black and white women: Ultrasound evidence. *Amer. J. Obstetrics and Gynecology* **188**, 100–107.

Barash, Y., Friedman, N. (2002). Context-specific Bayesian clustering for gene expression data. *J. Comput. Biology* **9**, 169–191.

Benjamini, Y., Hochberg, Y. (1995). Controlling the false discovery rate – a practical and powerful approach to multiple testing. *J. Roy. Statist. Soc., Ser. B* **57**, 289–300.

Berry, D.A., Stangl, D.K. (1996). *Bayesian Biostatistics*. Marcel Dekker.

Boys, R.J., Henderson, D.A., Wilkinson, D.J. (2000). Detecting homogeneous segments in DNA sequences by using hidden Markov models. *Appl. Statist.* **49**, 269–285.

Brown, E.R., Ibrahim, J.G. (2003). A Bayesian semiparametric joint hierarchical model for longitudinal and survival data. *Biometrics* **59**, 221–228.

Bush, C.A., MacEachern, S.N. (1996). A semiparametric Bayesian model for randomised block designs. *Biometrika* **83**, 275–285.

Chen, Z., Dunson, D.B. (2003). Random effects selection in linear mixed models. *Biometrics* **59**, 762–769.

Chib, S., Carlin, B.P. (1999). On MCMC sampling in hierarchical longitudinal models. *Statist. Comput.* **9**, 17–26.

Clayton, D.G. (1991). A Monte-Carlo method for Bayesian-inference in frailty models. *Biometrics* **47**, 467–485.

Clyde, M. (2000). Model uncertainty and health effect studies for particulate matter. *Environmetrics* **11**, 745–763.

Clyde, M., Parmigiani, G., Vidakovic, B. (1998). Multiple shrinkage and subset selection in wavelets. *Biometrika* **85**, 391–401.

Conti, D.V., Cortessis, V., Molitor, J., Thomas, D.C. (2003). Bayesian modeling of complex metabolic pathways. *Human Heredity* **56**, 83–93.

Cox, D.R. (1972). Regression models and life tables (with Discussion). *J. Roy. Statist. Soc., Ser. B* **34**, 187–220.

Craig, B.A., Fryback, D.G., Klein, R., Klein, B.E.K. (1999). A Bayesian approach to modelling the natural history of a chronic conditional from observations with intervention. *Statistics in Medicine* **18**, 1355–1371.

Daniels, M.J., Pourahmadi, M. (2002). Bayesian analysis of covariance matrices and dynamic models for longitudinal data. *Biometrika* **89**, 553–566.

Daniels, M.J., Zhao, Y.D. (2003). Modelling the random effects covariance matrix in longitudinal data. *Statistics in Medicine* **22**, 1631–1647.

Denison, D.G.T., Holmes, C.C., Mallick, B.K., Smith, A.F.M. (2002). *Bayesian Methods for Nonlinear Classification and Regression*. Wiley, New York.

DiMatteo, I., Genovese, C.R., Kass, R.E. (2001). Bayesian curve-fitting with free-knot splines. *Biometrika* **88**, 1055–1071.

Dunson, D.B. (2001). Commentary: Practical advantages of Bayesian analyses of epidemiologic data. *Amer. J. Epidemiology* **153**, 1222–1226.

Dunson, D.B., Baird, D.D. (2002). Bayesian modeling of incidence and progression of disease from cross-sectional data. *Biometrics* **58**, 813–822.

Dunson, D.B., Dinse, G.E. (2000). Distinguishing effects on tumor multiplicity and growth rate in chemoprevention experiments. *Biometrics* **56**, 1068–1075.

Dunson, D.B., Dinse, G.E. (2002). Bayesian models for multivariate current status data with informative censoring. *Biometrics* **58**, 79–88.

Dunson, D.B., Herring, A.H. (2003). Bayesian inferences in the Cox model for order-restricted hypotheses. *Biometrics* **59**, 916–923.

Dunson, D.B., Neelon, B. (2003). Bayesian inference on order-constrained parameters in generalized linear models. *Biometrics* **59**, 286–295.

Dunson, D.B., Chulada, P., Arbes, S.J. (2003). Bayesian modeling of time varying and waning exposure effects. *Biometrics* **59**, 83–91.

Escobar, M.D. (1994). Estimating normal means with a Dirichlet process prior. *J. Amer. Statist. Assoc.* **89**, 268–277.

Etzioni, R.D., Kadane, J.B. (1995). Bayesian statistical methods in public health and medicine. *Ann. Rev. Public Health* **16**, 263–272.

Freedman, L. (1996). Bayesian statistical methods: A natural way to assess clinical evidence. *British Medical J.* **313**, 569–570.

Friedman, N. (2004). Inferring cellular networks using probabilistic graphical models. *Science* **303**, 799–805.

Friedman, N., Linial, M., Nachman, I., Pe'er, D. (2000). Using Bayesian networks to analyze gene expression data. *J. Comput. Biology* **7**, 601–620.

Gagnon, D.R., Glickman, M.E., Myers, R.H., Cupples, L.A. (2003). The analysis of survival data with a non-susceptible fraction and dual censoring mechanisms. *Statistics in Medicine* **22**, 3249–3262.

Gangnon, R.E., Clayton, M.K. (2000). Bayesian detection and modeling of spatial disease clustering. *Biometrics* **56**, 922–935.

Gelfand, A.E., Kottas, A. (2003). Bayesian semiparametric regression for median residual life. *Scand. J. Statist.* **30**, 651–665.

Gelfand, A.E., Hills, S.E., Racine-Poon, A., Smith, A.F.M. (1990). Illustration of Bayesian-inference in normal data models using Gibbs sampling. *J. Amer. Statist. Assoc.* **85**, 972–985.

Gelfand, A.E., Smith, A.F.M., Lee, T.M. (1992). Bayesian-analysis of constrained parameter and truncated data problems using Gibbs sampling. *J. Amer. Statist. Assoc.* **87**, 523–532.

Gilks, W.R., Wang, C.C., Yvonnet, B., Coursaget, P. (1993). Random-effects models for longitudinal data using Gibbs sampling. *Biometrics* **49**, 441–453.

Gurrin, L.C., Kurinczuk, J.J., Burton, P.R. (2000). Bayesian statistics in medical research: An intuitive alternative to conventional data analysis. *J. Evaluation of Clinical Practice* **6**, 193–204.

Gustafson, P., Hossain, S., McCandless, L. (2005). Innovative Bayesian methods for biostatistics and epidemiology. In: Dey, D.K., Rao, C.R. (Eds.), *Bayesian Thinking: Moldeling and Computation, Handbook of Statistics*, vol. 25. Elsevier, Amsterdam. This volume.

Hastie, T., Tibshirani, R. (1990). *Generalized Additive Models*. Chapman & Hall, New York.

Hjort, N.L., Claeskens, G. (2003). Frequentist model average estimators. *J. Amer. Statist. Assoc.* **98**, 879–899.

Hobert, J.P., Casella, G. (1996). The effect of improper priors on Gibbs sampling in hierarchical linear mixed models. *J. Amer. Statist. Assoc.* **91**, 1461–1473.

Hoeting, J.A., Madigan, D., Raftery, A.E., Volinsky, C.T. (1999). Bayesian model averaging: A tutorial. *Statist. Sci.* **14**, 382–401.

Holmes, C.C., Heard, N.A. (2003). Generalized monotonic regression using random change points. *Statistics in Medicine* **22**, 623–638.

Holmes, C.C., Mallick, B.K. (2003). Generalized nonlinear modeling with multivariate free-knot regression splines. *J. Amer. Statist. Assoc.* **98**, 352–368.

Ibrahim, J.G., Chen, M.-H., Sinha, D. (2001). *Bayesian Survival Analysis*. Springer, New York.

Jones, B., Carvalho, C., Dobra, A., Hans, C., Carter, C., West, M. (2004). Experiments in stochastic computation for high-dimensional graphical models. ISDS Working Paper Series 04-01, Duke University.

Kalbfleisch, J.D. (1978). Nonparametric Bayesian analysis of survival data. *J. Roy. Statist. Soc., Ser. B* **40**, 214–221.

Kass, R.E., Ventura, V., Cai, C. (2003). Statistical smoothing of neuronal data. *Network-Computation in Neural Systems* **14**, 5–15.

Kleinman, K.P., Ibrahim, J.G. (1998). A semiparametric Bayesian approach to the random effects model. *Biometrics* **54**, 921–938.

Laird, N.M., Ware, J.H. (1982). Random-effects models for longitudinal data. *Biometrics* **38**, 963–974.

Lavine, M. (1992). Some aspects of Polya tree distributions for statistical modeling. *Ann. Statist.* **20**, 1222–1235.

Lavine, M. (1994). More aspects of Polya tree distributions for statistical modeling. *Ann. Statist.* **22**, 1161–1176.

Lavine, M., Mockus, A. (1995). A nonparametric Bayes method for isotonic regression. *J. Statist. Plann. Inference* **46**, 235–248.

Liang, K.Y., Zeger, S.L. (1986). Longitudinal data-analysis using generalized linear-models. *Biometrika* **73**, 13–22.

Lilford, R.J., Braunholtz, D. (1996). The statistical basis of public policy: A paradigm shift is overdue. *British Medical J.* **313**, 603–607.

Lilford, R.J., Braunholtz, D. (2000). Who's afraid of Thomas Bayes? *J. Epidemiology and Community Health* **54**, 731–739.

Liu, J.S., Neuwald, A.F., Lawrence, C.E. (1999). Markovian structures in biological sequence alignments. *J. Amer. Statist. Assoc.* **94**, 1–15.

Lopes, H.F., Müller, P., Rosner, G.L. (2003). Bayesian meta-analysis for longitudinal data models using multivariate mixture priors. *Biometrics* **59**, 66–75.

Lunn, D.J., Thomas, A., Best, N., Spiegelhalter, D. (2000). WinBUGS – a Bayesian modeling framework: Concepts, structure, and extensibility. *Statist. Comput.* **10**, 325–337.

Mallick, B.K., Denison, D.G.T., Smith, A.F.M. (1999). Bayesian survival analysis using a MARS model. *Biometrics* **55**, 1071–1077.

McKeague, I.W., Tighiouart, M. (2000). Bayesian estimators for conditional hazard functions. *Biometrics* **56**, 1007–1015.

Medvedovic, M., Sivaganesan, S. (2002). Bayesian infinite mixture model based clustering of gene expression profiles. *Bioinformatics* **18**, 1194–1206.

Morris, J.S., Vannucci, M., Brown, P.J., Carroll, R.J. (2003). Wavelet-based nonparametric modeling of hierarchical functions in colon carcinogenesis. *J. Amer. Statist. Assoc.* **98**, 573–583.

Natarajan, R., McCulloch, C.E. (1998). Gibbs sampling with diffuse but proper priors: A valid approach to data-driven inference? *J. Comput. Graph. Statist.* **7**, 267–277.

Neelon, B., Dunson, D.B. (2004). Bayesian isotonic regression and trend analysis. *Biometrics* **60**, 398–406.

Nieto-Barajas, L.E., Walker, S.G. (2002). Markov beta and gamma process for modelling hazard rates. *Scand. J. Statist.* **29**, 413–424.

Raftery, A.E., Madigan, D., Hoeting, J.A. (1997). Bayesian model averaging for linear regression models. *J. Amer. Statist. Assoc.* **92**, 179–191.

Ramgopal, P., Laud, P.W., Smith, A.F.M. (1993). Nonparametric Bayesian bioassay with prior constraints on the shape of the potency curve. *Biometrika* **80**, 489–498.

Richardson, S., Green, P.J. (1997). On Bayesian analysis of mixtures with an unknown number of components. *J. Roy. Statist. Soc., Ser. B* **59**, 731–758.

Scharfstein, D.O., Daniels, M.J., Robins, J.M. (2003). Incorporating prior beliefs about selection bias into the analysis of randomized trials with missing outcomes. *Biostatistics* **4**, 495–512.

Seaman, S.R., Richardson, S., Stucker, I., Benhamou, S. (2002). A Bayesian partition model for case-control studies on highly polymorphic candidate genes. *Genetic Epidemiology* **22**, 356–368.

Schmidler, S.C., Liu, J.S., Brutlag, D.L. (2000). Bayesian segmentation of protein secondary structure. *J. Comput. Biology* **7**, 233–248.

Sha, N., Vannucci, M., Brown, P.J., Trower, M.K., Amiphlett, G., Falcianni, F. (2003). Gene selection in arthritis classification with large-scale microarray expression profiles. *Comparative and Functional Genomics* **4**, 171–181.

Sinha, D. (1993). Semiparametric Bayesian-analysis of multiple event time data. *J. Amer. Statist. Assoc.* **88**, 979–983.

Sinha, D., Dey, D.K. (1997). Semiparametric Bayesian analysis of survival data. *J. Amer. Statist. Assoc.* **92**, 1195–1212.

Sinha, D., Chen, M.-H., Ghosh, S.K. (1999). Bayesian analysis and model selection for interval-censored survival data. *Biometrics* **55**, 585–590.

Sinha, D., Ibrahim, J.G., Chen, M.-H. (2002). Models for survival data from cancer prevention studies. *J. Roy. Statist. Soc., Ser. B* **64**, 467–477.

Smith, M., Kohn, R. (2002). Parsimonious covariance matrix estimation for longitudinal data. *J. Amer. Statist. Assoc.* **97**, 1141–1153.

Spiegelhalter, D.J., Myles, J.P., Jones, D.R., et al. (1999). An introduction to Bayesian methods in health technology assessment. *British Medical J.* **319**, 508–512.

Stow, C.A., Lamon, E.C., Qian, S.S., Schrank, C.S. (2004). Will Lake Michigan lake trout meet the Great Lakes strategy 2002 PCB reduction goal? *Environmental Sci. Technol.* **38**, 359–363.

Storey, J.D. (2003). The positive false discovery rate: A Bayesian interpretation and the q-value. *Ann. Statist.* **31**, 2013–2035.

Strimmer, K., Moulton, V. (2000). Likelihood analysis of phylogenetic networks using directed graphical models. *Molecular Biology and Evolution* **17**, 875–881.

Troyanskaya, O.G., Dolinski, K., Owen, A.B., Altman, R.B., Botstein, D. (2003). A Bayesian framework for combining heterogeneous data sources for gene function prediction (in *Saccharomyces cerevisiae*). *Proc. Nat. Acad. Sci.* **100**, 8348–8353.

Tsai, C.-A., Hsueh, H.-M., Chen, J.J. (2003). Estimation of false discovery rates in multiple testing: Application to gene microarray data. *Biometrics* **59**, 1071–1081.

Tsodikov, A.D., Ibrahim, J.G., Yakovlev, A.Y. (2003). Estimating cure rates from survival data: An alternative to two-component mixture models. *J. Amer. Statist. Assoc.* **98**, 1063–1078.

Vines, S.K., Gilks, W.R., Wild, P. (1996). Fitting Bayesian multiple random effects models. *Statist. Comput.* **6**, 337–346.

Volinsky, C.T., Madigan, D., Raftery, A.E., Kronmal, R.A. (1997). Bayesian model averaging in proportional hazard models. Assessing the risk of a stroke. *Appl. Statist.* **46**, 433–448.

Walker, S.G., Damien, P., Laud, P.W., Smith, A.F.M. (1999). Bayesian nonparametric inference for random distributions and related functions. *J. Roy. Statist. Soc., Ser. B* **61**, 485–527.

West, M. (1992). Modelling time-varying hazards and covariate effects. In: Klein, J.P., Goel, P.K. (Eds.), *Survival Analysis: State of the Art*. Kluwer Academic, Boston, pp. 47–62.

West, M., Blanchette, C., Dressman, H., Huang, E., Ishida, S., Spang, R., Zuzan, H., Olson, J.A., Marks, J.R., Nevins, J.R. (2001). Predicting the clinical status of human breast cancer by using gene expression profiles. *Proc. Nat. Acad. Sci.* **98**, 11462–11467.

Wong, F., Carter, C.K., Kohn, R. (2003). Efficient estimation of covariance selection models. *Biometrika* **90**, 809–830.

Wood, S.A., Jiang, W.X., Tanner, M. (2002). Bayesian mixture of splines for spatially adaptive nonparametric regression. *Biometrika* **89**, 513–528.

Zeger, S.L., Karim, M.R. (1991). Generalized linear-models with random effects – a Gibbs sampling approach. *J. Amer. Statist. Assoc.* **86**, 79–86.

Zeger, S.L., Liang, K.Y. (1986). Longitudinal data-analysis for discrete and continuous outcomes. *Biometrics* **42**, 121–130.

Handbook of Statistics, Vol. 25
ISSN: 0169-7161
© 2005 Elsevier B.V. All rights reserved.
DOI 10.1016/S0169-7161(05)25026-5

26

Innovative Bayesian Methods for Biostatistics and Epidemiology

Paul Gustafson, Shahadut Hossain and Lawrence McCandless

Abstract
Complex data and models now pervade biostatistics and epidemiology. Increasingly, Bayesian methods are seen as desirable tools to tame this complexity and make principled inferences from the data at hand. In this chapter we try to convey the flavor of what Bayesian methods have to offer, by describing a number of applications of Bayesian methods to health research. We emphasize the strengths of these approaches, and points of departure from non-Bayesian techniques.

1. Introduction

Statistical science has seen a revolution in which Bayesian methods have moved from the realm of conceptual interest to the realm of practical interest, largely on the basis of advances in computational techniques. The impact of this revolution is being felt in biostatistics and epidemiology, and in health research generally. Simple examples of Bayesian thinking have long been taught to medical students, and form an integral part of clinical medicine. Conceptually, clinicians combine *prior* information on the prevalence of a specific condition with *data* – signs and symptoms elicited in a clinical exam, to determine the *posterior* evidence that a patient has the condition. But nowadays this is merely the tip of the iceberg for applications of Bayesian analysis in the health sciences.

The rise of Bayesian methods in biostatistics and epidemiology has been pronounced. Over the last few years, journals which emphasize new statistical methods for medical applications, such as *Biometrics*, *Biostatistics*, *Lifetime Data Analysis*, and *Statistics in Medicine*, have devoted substantial proportions of their space to articles developing Bayesian techniques for specific problems. An edited volume devoted to Bayesian methods in Biostatistics has appeared (Berry and Stangl, 1996), and the medical literature contains several recent editorials and articles on the advantages of Bayesian techniques (Dunson, 2001; Freedman, 1996; Goodman, 1999a, 1999b, 2001; Lilford and Braunholtz, 1996). Moreover, Bayes methods have penetrated subject-area journals in the health sciences, with examples of recent articles including Berry et al. (1997), Brophy and Joseph (1995, 1997), Brophy et al. (2001), Carlin and Xia (1999), Johnson et al. (1999), Joseph et al. (1999), Parmigiani et al. (1997), Pinsky and Lorber (1998), and Tweedie et al. (1996).

One can advance a number of arguments as to why the Bayesian approach is particularly helpful in health research. Some that we would emphasize as important are as follows. First, much of the data arising in modern health research is of complex structure. In particular, *nested* or clustered data structures are very common, as multiple measurements or observations may be nested within subjects, or subjects may be nested within treatment sites. The Bayesian paradigm is naturally suited to modeling such data, particularly via *hierarchical models* which are conditionally specified, i.e., the first stage describes multiple measurements per subject, the second stage describes subjects within sites, and so on. Even many non-Bayesian techniques for handling such data structures have a Bayesian flavor, as there is an inherent desire to invoke prior information stating that response mechanisms are neither identically the same nor entirely separate across clusters.

Second, many of the problems in health research involve multiple sources of uncertainty. For instance, in Section 4 we discuss situations where assessment of an exposure–disease relationship is complicated because only rough exposure measurements can be made on individuals. Thus there is uncertainty about the relationship between the health-based outcome and the actual exposure, linked with uncertainty about the relationship between the actual exposure and the measured exposure. A very nice feature of Bayesian analysis is the uncertainty is propagated seamlessly and appropriately across linked models, so that final inferential summaries properly account for all the uncertainties at play. There are numerous examples where Bayesian interval estimates are wider than those arising from non-Bayesian procedures, and this extra width is appropriate as an honest assessment of post-data knowledge. This advantage of Bayesian inference is particularly important in the health sciences where it is often vital to be realistic about how much is known about the relationship of interest, particularly given the obvious need to be convinced before approving a new treatment or issuing a health-related warning. We have noticed that in other areas where Bayesian methods are brought to bear, such reflection of uncertainty is less important. Applications of Bayesian inference in computer science, for instance, often involve making predictions based on the best point estimates at hand, and calibration of the quality of prediction is often less important.

Third, health research is full of situations where there is partial knowledge about important quantities. For instance, the experience of physicians with their patient populations often yields rough knowledge about the efficacy of a treatment or the impact of a covariate, without any formal data collection. Such knowledge, perhaps suitably discounted, can be encapsulated in prior distributions and used in various analyses. Other paradigms for statistical inference are not as flexible; quantities are either completely known or completely unknown. Of course if there is any doubt one can assume the latter for the sake of apparent objectivity. However, there are a number of situations in biostatistics and epidemiology where various limitations of measurement and data collection imply that not all the relevant parameters can be estimated consistently from the data. In such situations, prior information has a particularly important role to play.

Fourth, with modern Bayesian computation techniques, principally *Markov chain Monte Carlo* methods, there are very few constraints on the nature of models that can be considered. That is, modelers are free to specify models as realistically as possible, without undue concern over which models might be easier or harder to deal with computationally. Usually one can find reasonable MCMC implementation schemes for whatever model and prior distributions are deemed appropriate. That is, one can find a way to simulate samples of parameter values from the posterior distribution of unknown quantities given observed quantities. Of course this is a very general feature of modern Bayesian analysis, but one that has been quite liberating for the modeling of data in health research contexts. In fact, we make relatively few comments on computation in this chapter. We prefer to focus on conceptual issues, with the idea that by and large MCMC can be brought to bear for model fitting as needed.

Fifth, health research is rife with situations where decisions must be taken after observing data, and often these decisions must weigh very disparate quantities, such as efficacy and cost for a treatment. Decision–theoretic analysis can be brought to bear in such situations, and such analysis is fundamentally linked with Bayesian modeling and inference. That is, decisions which are optimal with respect to average loss (here the average is a weighted average over the values of unknown parameters, with the weighting corresponding to the prior distribution) can be obtained as outputs of a Bayesian analysis.

The goal of this chapter is to elucidate some scenarios where these advantages prevail, to give the reader a broad sense of what Bayesian methods can accomplish in biostatistical and epidemiological settings. In particular, in Sections 2 through 11 we survey ten methodological areas where Bayesian techniques are useful for health research. By design our coverage is broad rather than deep, though we have tried to supply sufficient and appropriate references for interested readers to investigate the topics further.

2. Meta-analysis and multicentre studies

It is often desirable to study a disease–exposure relationship or a disease–treatment relationship using data on patients at multiple clinical sites. This may be by design, as with a multi-centre clinical trial, or as a post-hoc *meta-analysis* which draws together uncoordinated single-site studies. We do not necessarily expect the relationship of interest to be identically the same at each site, due to across-site variation in patient populations and clinical practices. But nor do we think the relationship at one site is unrelated to the relationship at other sites. Thus we seek a middle ground between (i), an analysis which pools all the patients together and 'forgets' which site each came from, and (ii), separate analyses for each site. Bayesian hierarchical models provide an extremely natural way to find such a middle-ground. We frame our discussion of this in the meta-analysis context, but then make some remarks about how the ideas translate to the closely related scenario of a multi-centre study.

Meta-analysis has become an accepted and valued practice in medical research. It includes methodologies for combining information from a number of related but independent studies with the aim of being able to resolve issues that can not be concluded from a single study. The primary objective is usually increasing the power to detect an overall treatment effect or effect measure in order to aid clinical decision making, but estimating the extent of heterogeneity among studies is often of interest as well. Meta-analyses are sometimes motivated by disagreement in the results of different clinical trials aimed to address the same research question, with the idea that some of the discrepancy may arise from limited sample sizes in the individual studies. Meta-analysis is also helpful in assessing generalizability of conclusions reached in the individual studies. Besides combining results across studies, meta-analysis allows the incorporation of study-level covariates (DuMouchel and Harris, 1983). The dose of a pharmaceutical intervention would be an important example of such a covariate, as it will likely vary much more across studies than within studies.

Statistical models for meta-analysis involve individual parameters or *effects* for the separate studies. A first choice in modeling is whether to view these as fixed effects or random effects. The prevailing view is that fixed effects models are not sufficient to reflect the fact that the same disease and (roughly) the same treatment is being studied at each site. By itself, a preference for random effect models does not completely resolve the strategy for analysis, as such models can be approached from a classical perspective, an empirical Bayes perspective, or a fully Bayesian perspective. There are, however, several potential pitfalls with a classical approach. First, the requisite large-sample approximations required for inference may be dubious if the number of studies involved is moderate. Second, the chi-square test for homogeneity across studies may have low power (Fleiss, 1993), and third, some classical approaches have trouble ruling out a negative estimate of the variance component describing between-study variability. Arguably an empirical Bayes approach is better, but this is often criticized for not propagating statistical uncertainty about the values of variance components when making inferences about the study effects. A fully Bayesian approach is able to overcome this drawback.

Bayesian meta-analysis is discussed at length in several chapters in the volume edited by Stangl and Berry (2000). A prototypical full Bayesian random effects model for meta-analysis includes a three level hierarchy and can be given as follows.

$$Y_i|\theta, \mu, \tau^2 \sim N(\theta_i, \sigma_i^2), \tag{1}$$

$$\theta_i|\mu, \tau^2 \sim N(\mu, \tau^2), \tag{2}$$

$$(\mu, \tau^2) \sim f(\mu, \tau^2). \tag{3}$$

Here, Y_i is the observed effect measure of interest from the ith of m studies, while θ_i is the true effect measure. Conditional independence across studies is assumed in both (1) and (2). The variance σ_i^2 in (1) describes the precision of Y_i as an estimate of θ_i; as such, it will depend chiefly on the sample size for the ith study. Often σ_i^2 is artificially assumed to be known exactly. That is, σ_i is taken to be the standard error reported for the estimated effect measure in the ith study. One could take a more principled approach involving a prior on each σ_i with the study-specific standard errors then being data which inform about these parameters. The second level of the hierarchy (2) describes the extent to which the studies are homogeneous in terms of the effect of interest. While μ is the average effect across studies, τ describes the extent of heterogeneity. In assigning a prior distribution to (μ, τ^2), we will let the data inform us about the extent of heterogeneity through the posterior distribution of τ^2.

The hierarchical model presented in (1) through (3) can *borrow strength* across studies in estimating both the study specific effects, $\theta = (\theta_1, \ldots, \theta_m)$, and population effect μ. If warranted by the data (i.e., the posterior distribution on τ^2 concentrates near zero), the posterior estimates of the θ_i's will have little variation, and the analysis will be close to that achieved by pooling all the data together. On the other hand, a posterior distribution for τ^2 which favors large values will be similar to doing a separate analysis for each study. In most situations a middle-ground situation between these two extremes will prevail.

One important issue in this and other hierarchical models is the choice of the prior distribution in (3). Typically μ and τ^2 are treated as independent *a priori*; that is, $f(\mu, \tau^2) = f(\mu)f(\tau^2)$. The common choice for μ is a flat prior, either literally in the sense of the improper prior $\pi(\mu) \propto 1$, or approximately in the sense of a (proper) normal distribution with very large variance. Invariably the specific choice has little impact on the inference. The choice of prior for τ^2, however, is much more important, since this parameter governs how much borrowing of strength occurs. DuMouchel et al. (1995) and DuMouchel and Normand (2000) address some technical issues on choosing the prior for τ^2. Daniels (1999) proposes a 'uniform shrinkage' for τ^2, which has some desirable frequentist properties.

In practice it is very common to specify an inverse gamma prior for τ^2, as this is a 'conditionally conjugate' choice which simplifies MCMC fitting of the model. Often the parameters of the inverse gamma are chosen to centre the distribution near one, but yield a very large variance. In our view this is not a good specification, as any inverse gamma density is zero at $\tau^2 = 0$. In models such as (1) through (3), however, the data never definitively rule out the case of homogeneity ($\tau^2 = 0$). In this sense, an inverse gamma prior necessarily has a large impact of the shape of the posterior distribution for τ^2. Gustafson and Wasserman (1996) discuss this point, and argue for a prior density which is positive (but finite) at zero, and decreasing. Others have also emphasized such priors. See Spiegelhalter et al. (2003) for some recent discussion.

As alluded to above, hierarchical models such as (1) through (3) which are used for meta-analysis are also appropriate for multicentre studies, perhaps after some tweaking. Of course a normal model for the response variable of interest may not be appropriate. See, for instance, Skene and Wakefield (1990) for en example with a binary response variable, or Gray (1994) or Gustafson (1995, 1997) for examples with survival time responses. Of course a multicentre trial is coordinated with the intent that treatment be administered as similarly as possible across sites. Thus we anticipate generally smaller values of τ than with meta-analysis. Moreover, we might now tend to interpret τ as reflecting variation due to socio-economic and environmental factors across sites. But by and large this will not affect the plan of attack for the statistical analysis.

From a technical viewpoint, often the primary difference between meta-analysis and multicentre studies is that the latter analysis is based on individual-level data across sites. Roughly speaking, the first level of the hierarchy (1) might be changed to model Y_{ij}, the response variable for the jth subject observed at site i. In contrast, usually individual-level data from specific studies are not available to the meta-analyst, who must content himself with published estimates of treatment effects (and their standard errors) from the various studies.

3. Spatial analysis for environmental epidemiology

Studying geographic patterns of disease incidence or prevalence, and assessing the relationships between regional-level covariates and disease rates are typical goals in environmental epidemiology. It can be important to consider the spatial pattern of a disease to identify the areas in need of essential health services provisions and to allocate health service resources. Also, allowing for regional effects due to unobserved factors can be crucial for accurate assessment of exposure–disease relationships. In essence, one wants to reflect the reality that disease processes at nearby geographic regions are likely more similar than those at distant regions.

Let Y_i be the observed summary measure of a disease outcome in the ith of m regions. If Y_i is continuous, then a simple spatial model without covariates might take the following hierarchical form:

$$(Y_i|\theta, \mu, \sigma^2, \tau^2) \sim N(\mu + \theta_i, \sigma^2), \tag{4}$$

$$(\theta|\mu, \sigma^2, \tau^2) \sim N_m(0, \tau^2 V), \tag{5}$$

$$(\mu, \sigma^2, \tau^2) \sim f(\mu) f(\sigma^2) f(\tau^2), \tag{6}$$

with conditional independence across regions at the first stage (4). That is, conditioned on region-specific random effects $\theta = (\theta_1, \dots, \theta_m)$ the responses are modeled as independent. At the second stage (5), V is a known $m \times m$ correlation matrix chosen to reflect higher positive correlations for the random effects of closer regions. The use of a spatially structured covariance matrix is the main difference between this hierarchical model and that discussed in Section 2 for meta-analysis and multi-center studies. There the random effects where treated exchangeably, i.e., conditionally independent at the second stage. The unknown variance parameter τ^2 in (5), which governs the overall magnitude of the region-to-region variation, is assigned a prior distribution at the third stage (6). Issues surrounding the selection of this prior are very similar to those for the corresponding variance component of the model in Section 2. We also note in passing that the first stage (4) is easily modified to incorporate explanatory variables at the level of regions. That is, $\mu + \theta_i$ can be replaced with $\mu + \beta' x_i + \theta_i$, where x_i is a covariate vector for the ith region and β is a vector or regression coefficients.

Often in spatial epidemiology the response variable for region i is a count – the number of cases of a disease in the region. An obvious adaptation of (4) is then $Y_i \sim Bin(n_i, p_i)$, where n_i is the number at risk in the region, while $\text{logit}(p_i) = \mu + \theta_i$. Or, if the disease is rare, a Poisson specification may be appropriate. That is, $Y_i \sim Poisson\{E_i \exp(\theta_i)\}$, where E_i is the expected number of cases in region i under the assumption of a spatially homogeneous disease process.

A common specification for the second stage distribution (5) is the *conditional autoregressive* (CAR) prior distribution (Besag, 1974). This distribution is characterized by the conditional density of θ_i given $\theta_{-i} = \{\theta_k\colon k \neq i\}$, for $i = 1, \ldots, m$. One particular CAR prior assumes that the conditional density $f(\theta_i|\theta_{-i})$ depends on a relatively small set of neighbors δ_i for each region i, i.e., $f(\theta_i|\theta_{-i}) = f(\theta_i|\{\theta_k\colon k \in \delta_i\})$. Neighborhoods can be defined in different ways, with the literal notion of physical adjacency being the most common choice.

To be more specific, one common version of the CAR prior for θ vector assumes that the conditional distribution of $(\theta_i|\theta_{-i})$ follows a normal distribution with mean $n_i^{-1}\sum_{k\in\delta_i}\theta_k$ and variance $n_i^{-1}\tau^2$, where $n_i = |\delta_i|$ is simply the number of neighbors for the ith region. That is, the conditional expected value of θ_i is an average of the random effects at neighboring regions. Consideration of quadratic forms shows that formally this corresponds to setting $V^{-1} = B$, where $B_{ii} = n_i$, and, for $i \neq j$, B_{ij} is either -1 or 0 according to whether regions i and j are neighbors or not. In fact this B is singular (as the columns and rows sum to zero). While this specification can be used, it corresponds to an improper prior distribution, which for some purposes may be undesirable. Other variants of the CAR prior are discussed by Carlin and Louis (1998) and Cressie and Chan (1989).

To obtain a proper prior, and to allow for both spatial and unstructured variation across regions, Besag et al. (1991) suggest a so-called *nonintrinsic* CAR prior. The formulation of this offered by Leroux et al. (1999) takes V in (5) based on

$$V^{-1} = \lambda B + (1-\lambda)I_m, \tag{7}$$

where B is as above, while $\lambda \in [0, 1]$, which is usually regarded as an unknown parameter, governs the balance between spatially structured and unstructured variation across regions. Now the right-hand side of (7) is invertible, as long as $\lambda < 1$. This CAR prior is flexible in the sense that it can range from an exchangeable prior ($\lambda = 0$) to an intrinsic CAR model ($\lambda = 1$).

The full Bayesian version of the spatial hierarchical model is completed by specifying priors for μ, σ^2, τ^2, and also for λ if the nonintrinsic prior is used. Often it is hard to elucidate much prior knowledge about these parameters, so that noninformative priors are sought. Kelsall and Wakefield (1998) and MacNab (2003) provide some discussion on choosing these priors. With specifications in place, the common MCMC strategy of updating the parameters in a hierarchical model 'one level at a time' is usually quite easy to implement. Posterior inferences about $\theta = (\theta_1, \ldots, \theta_m)$ are usually of primary interest, and these might typically be displayed in graphical form. In particular, grey-scale maps of posterior means across space are commonly produced.

4. Adjusting for mismeasured variables

Many medical studies involve some kind of regression analysis to study the association between a response variable and explanatory variables, where the response variable indicates some aspect of patient health, while the explanatory variables might be comprised of putative *exposure* variables along with other patient characteristics. In epidemiological investigations particularly, it is common to encounter exposure variables which cannot be measured very well. Obvious examples of exposure variables which cannot be measured precisely at the level of an individual include smoking history, exposure to an airborne pollutant, and dietary intake of a particular foodstuff.

To be more specific, say one wishes to model the distribution of a response variable Y given explanatory variables X and Z, where both X and Z may have multiple components. In particular, say a parametric model indexed by parameter vector θ_1 is used to model $(Y|X, Z)$. This might be referred to as the *disease model*. But due to mismeasurement, X itself is not observed. A noisy surrogate X^* is recorded in its place. A *naive* analysis might then simply pretend that X^* is actually X, and fit the intended model to (Y, X^*, Z) data. It is well known, however, that even if this mismeasurement is free of bias, i.e., X^* is obtained from X by adding mean zero noise, then the naive modeling strategy involves a bias in parameter estimation. In most situations the estimated association between Y and X^* given Z understates the actual association between Y and X given Z. This is referred to as bias 'toward the null,' in reference to the null hypothesis that Y and X are not associated given Z.

If some knowledge about the distribution of the surrogate X^* given the true X is available, then we might attempt an analysis which accounts for the mismeasurement and avoids the bias described above. The distribution of $(X^*|Y, X, Z)$ might be termed the *measurement model*. If it is legitimate to view the difference between X^* and X as arising solely because of a noisy measurement process, then it may be reasonable to assume the mismeasurement is *nondifferential*, in that the distribution of $(X^*|Y, X, Z)$ in fact depends only on X. In the case of a continuous X, a further specialization might be $X^*|Y, X, Z \sim N(X, \tau^2)$, which is appropriate is X^* is thought to be unbiased and homoscedastic as a 'surrogate' for X. For the sake of discussion assume for the moment that this measurement model is adopted, and that in fact the parameter τ is known.

The other distribution at play here is the conditional distribution of X given Z, which for obvious reasons Gustafson (2003) refers to as a *conditional exposure model*. Say this distribution is modeled with a parametric family indexed by θ_2. Then schematically the posterior distribution of unknown quantities given observed quantities follows

$$
\begin{aligned}
f(\theta_1, \theta_2, x|x^*, y, z) \propto & \prod_{i=1}^{n} f(x_i^*|y_i, x_i, z_i) \\
& \times \prod_{i=1}^{n} f(y_i|x_i, z_i, \theta_1) \\
& \times \prod_{i=1}^{n} f(x_i|z_i, \theta_2) \\
& \times f(\theta_1, \theta_2),
\end{aligned}
$$

where the last term is a joint prior density for the parameters. Often it is quite straightforward to implement an MCMC algorithm for this posterior distribution. Typically the resulting posterior distribution for disease model parameters θ_1 will differ in two ways from the naive posterior distribution that arises from treating X^* as if it were X. First, the distribution will be shifted, typically away from zero, to correctly mitigate the bias described above. Second, the distribution will be more diffuse, to correctly reflect the increase in uncertainty associated with observing X^* rather than X.

One contentious point is that the approach above requires the specification of a conditional exposure model which is not usually of direct inferential interest. More generally, such an approach is often referred to as *structural*, as opposed to *functional* approaches which attempt to correct for measurement error without making assumptions about the distribution of X, or more specifically the distribution of X given Z. Some general discussion along these lines is given by Carroll et al. (1995). In general the evidence on how badly misspecification of the exposure model affects inference about disease model parameters is mixed. Richardson and Leblond (1997) and Richardson et al. (2002) give some discussion of this in the context of logistic regression disease models. Interestingly, when each of the measurement model, disease model, and exposure model involve normal and linear structure, misspecification of the exposure model does not yield a bias, or even an increase in variance, for estimators of disease model coefficients (Fuller, 1987; Gustafson, 2003).

Another important issue in adjusting for mismeasurement is the identifiability of parameters. It stands to reason that one cannot undo the effects of mismeasurement without some idea of the extent of mismeasurement. In the developments above we took the distribution of X^* given (Y, X, Z) to be known. This would almost never be the case in practice, although one could use it as a basis for sensitivity analysis. That is, one could compare inferences about θ_1 under a variety of assumed measurement model distributions which investigators deem to be realistic.

Much of the literature on mismeasurement demands some mechanism to make all the parameters identifiable. Beyond the artificial assumption of a known mismeasurement model, two situations commonly discussed involve a *validation subsample* and *replication*. The former scenario involves measuring both X and X^* for a small subset of the study sample, while only X^* is measured for the remainder of the sample. Such a design may arise if it is possible but very expensive to measure X, while measurement of X^* is inexpensive. Intuitively the validation subsample permits inference on unknown parameters in the mismeasurement model, which in turn permits adjustment for the mismeasurement. In the Bayesian framework this 'information flow' is quite seamless, with proper accounting for uncertainty *en route*.

Of course a validation subsample can only be used if it is feasible to measure X exactly, and for many exposures of interest this cannot be achieved at any price. Sometimes replication can help in this regard. Say, for instance, that the surrogate measurement X^* is based on a laboratory assay. If two assays can be completed per subject, yielding two surrogates X_1^* and X_2^* for X, then an identifiable model may result. Often whether this works rests on the assumption that X_1^* and X_2^* are conditionally independent given X, which permits estimation of the variance for the surrogate given the true exposure. Unfortunately this may be a delicate and untestable assumption in practice, since X cannot be observed.

An interesting feature of the Bayesian paradigm is its lack of distinction between identified and nonidentified models. That is, a prior distribution and a likelihood function lead to a posterior distribution, whether the likelihood function arises from a model which is identified or not. Of course with a nonidentified model one cannot expect the posterior distribution to have consistency properties, i.e., converge to a point mass at the true value of the parameter as the sample size grows. But if one believes in the prior distribution as a pre-data summary of knowledge about the parameter values, then the posterior distribution as a post-data summary is equally credible, even if the model is nonidentified. This has implications for mismeasured variable scenarios when investigators have some rough knowledge about the mismeasurement process, but not enough knowledge to yield a formally identifiable model.

To explore this notion further we consider a concrete setting from Gustafson et al. (2001). Say that a case-control study is focused on a binary exposure which cannot be measured precisely, with Y_0 out of n_0 controls, and Y_1 out of n_1 cases, being *apparently* exposed. Moreover, the *sensitivity* p and *specificity* q of the exposure assessment are unknown. To be more specific, p and q are simply the probabilities of correct assessment for a truly exposed and a truly unexposed subject respectively. The other two parameters are r_0 and r_1, the prevalences of exposure in the control and case populations respectively.

To shed light on the issue we note that the prevalences of apparent exposure in the two populations will be θ_0 and θ_1, where

$$\theta_i = r_i p + (1 - r_i)(1 - q). \tag{8}$$

In fact, reparameterizing from (r_0, r_1, p, q) to $(\theta_0, \theta_1, p, q)$ clearly reveals the nonidentified nature of the model, since the likelihood function depends on only two of the four parameters. That is, $Y_i \sim Bin(n_i, \theta_i)$, $i = 0, 1$. Thus the posterior marginal distribution of (θ_0, θ_1) is nicely behaved in the sense of converging to the true parameter values at rate $n^{-1/2}$. On the other hand, the posterior conditional distribution of (p, q) given (θ_1, θ_2) is unaffected by the data. Particularly, an immediate consequence of (p, q) being absent from the likelihood function is that $f(p, q|\theta_0, \theta_1, \text{data}) = f(p, q|\theta_0, \theta_1)$ – the posterior conditional distribution is the same as the prior conditional distribution. This gives one sense in which nothing can be learned about these parameters. It does *not* follow, however, that the posterior *marginal* distribution of (p, q) is the same as the prior marginal distribution. Particularly, no matter what prior distribution is originally specified for (r_0, r_1, p, q), upon reparameterization it must necessarily yield prior dependence between (θ_0, θ_1) and (p, q), since the parameter space is constrained via (8); p and $1 - q$ must straddle both θ_0 and θ_1. Thus

$$f(p, q|\text{data}) = \iint f(p, q|\theta_0, \theta_1) f(\theta_0, \theta_1|\text{data})\, d\theta_0\, d\theta_1$$

will depend on the data. Gustafson (2003, 2005) calls this *indirect learning* about the nonidentified parameters, as it is learning about (p, q) that arises only through learning about the identified parameters (θ_0, θ_1).

Gustafson et al. (2001) demonstrate that moderately informative prior distributions on (p, q) may in fact lead to useful inferences about the odds-ratio $\{r_1/(1 - r_1)\}/\{r_0/(1 - r_0)\}$ which summarizes the disease–exposure association. This comes about in large part because of the extra learning about the nonidentified parameters that is 'squeezed out' via indirect learning. Gustafson (2003, 2005) gives a more general characterization of indirect learning in nonidentified models, characterizes the large-sample performance of estimators arising in such models, and considers the impact in other mismeasured variable scenarios. Overall the findings in this work tend to support a principled Bayes approach of putting prior distributions on the parameters describing the mismeasurement process which are as precise as possible subject to being honest, and that often reasonable inferences will result despite the nonidentifiability. This work also demonstrates that making assumptions and model elaborations to obtain identifiable models at all costs can, in fact, lead to worse inferences.

5. Adjusting for missing data

Missing data due to survey nonresponse, censoring, or other factors occurs frequently in epidemiological research. Typically, investigators may choose to disregard the missing data by dropping incomplete records from the analysis. But this approach can be wasteful, or downright hazardous if the missing data mechanism is *nonignorable* in the sense given by Rubin (1976) (and discussed below). Models for missing data are often highly complex and inference about treatment effects which account for missing data can present major challenges to researchers. Traditionally adjustments for missing data involve specifying a parametric missing data mechanism and imputing the missing data using inference techniques such as the expectation–maximization (EM) algorithm (Dempster et al., 1977). Bayesian methods can be useful in missing data models because sampling from the posterior distribution of a parameter of interest will automatically reflect information that is lost due to missingness.

Bayesian inference in missing data problems involves the joint posterior distribution of model parameters and the missing data. To formalize this idea, let y denote all the data, whether missing or observed, and let I be a collection of indicator variables denoting which elements of y are observed. Further, let $(y_{\text{obs}}, y_{\text{mis}})$ be a partition of y into those elements which are observed (corresponding elements of I are one) and unobserved (corresponding elements of I are zero). Using this notation, the joint probability distribution of y and I can be written as

$$f(y, I|\theta, \phi) = f(y|\theta) f(I|y, \phi),$$

where $f(y|\theta)$ is the complete data likelihood indexed by parameters of interest θ, and $f(I|y, \phi)$ denotes the missing data mechanism which depends on the data y and a parameter ϕ. The missing data mechanism is said to be *ignorable* if (i), $f(I|y, \phi)$ depends on y only through y_{obs}, and (ii), θ and ϕ are judged to be independent in the prior distribution. It is easy to verify that under these conditions the marginal posterior distribution for θ is identically that arising from the likelihood of y_{obs} alone, so that one need not be concerned with modeling the missingness process (Rubin, 1976; Gelman et al., 2003).

In fact, talking about whether $f(I|y, \phi)$ depends only on y_{obs} seems somewhat loosely stated, since y_{obs} and y_{mis} only attain meaning via I. But, as a typical scenario, say that a study involves n subjects, with regression of a response variable Z on an explanatory variable W being of interest. And say that W is observed for all subjects, whereas Z is missing for some subjects. For the ith of n subjects then, I_i indicates whether Z_i is observed or not. If $Pr(I_i = 1|Z_i, W_i)$ depends on W_i alone, then the ignorable situation obtains. That is, missingness is only influenced by observed quantities. If, on the other hand, this probability also depends on Z_i, then we have the nonignorable situation.

In the nonignorable case, a standard Bayesian approach would involve drawing samples from the joint probability distribution

$$f(\theta, \phi, y_{\text{mis}}|y_{\text{obs}}, I) \propto f(I|y, \phi) f(y|\theta) f(\theta, \phi), \tag{9}$$

as a route to sampling from $f(\theta|y_{\text{obs}}, I)$, since the latter distribution is often not amenable to direct analysis. The usual approach to MCMC sampling from (9) is to sample from the full conditionals of y_{mis}, θ, and ϕ in turn. For some model specifications these full conditionals will all be standard distributions, so that Gibbs sampling is readily implemented. In particular, the full conditional for θ is simply the posterior distribution arising in a 'full data' problem, so it will often have a convenient form.

Bayesian adjustments for missing data bear some similarities to EM algorithm approaches. The EM algorithm can be viewed as computing the mode of the marginal posterior density for θ when flat priors are specified. Rather than sampling from $(y_{\text{mis}}|y_{\text{obs}}, \theta, \phi)$ at each iteration, the E step of the EM algorithm computes $E\{\log f(Y|\theta)|y_{\text{obs}}, \theta^*, \phi^*\}$, where θ^* and ϕ^* are parameter values at the current iteration. Thus the log-likelihood is being averaged with respect to plausible values of y_{mis}, in contrast to the repeated sampling of y_{mis} in the Bayes-MCMC analysis. A full discussion of the EM algorithm in missing data contexts is described by Little and Rubin (2002).

To give a more focused example of how Bayesian methods might adjust for missing data, we consider a regression model with incomplete data discussed in Little and Rubin (2002). Say (Y_i, X_i) are response and covariate vectors for subject i, with

$$Y_i \sim N(X_i\beta, \Sigma)$$

independently for $i = 1$ through $i = n$. Also, let $y_i = (y_{\text{mis},i}, y_{\text{obs},i})$ partition the observed and unobserved responses for subject i. Depending on the specification of the covariance matrix Σ, this model can be used to describe a variety of phenomena including correlated subject responses in a repeated measures design (Jenrich and Schluchter, 1986). For simplicity, we assume an ignorable missing data mechanism, but extensions to the nonignorable case are relatively straightforward.

Let $(y_{\text{mis}}^{(t)}, \beta^{(t)}, \Sigma^{(t)})$ be the values of the unobservable quantities at the t-th iteration of the Gibbs sampler. As an 'imputation' step we update the missing data by sampling $y_{\text{mis},i}^{(t+1)}$ from the multivariate normal distribution of $(y_{\text{mis},i}|y_{\text{obs},i}, \beta^{(t)}, \Sigma^{(t)})$, independently for each subject. In particular, this conditional distribution corresponds to a linear regression of $y_{\text{mis},i}$ on $(y_{\text{obs},i}, x_i)$, which makes intuitive sense for 'imputing' the missing data. Then as a 'parameter' step we can sample $(\beta^{(t+1)}, \Sigma^{(t+1)})$ from the posterior distribution of $(\beta, \Sigma|y_{\text{obs}}, y_{\text{mis}}^{(t+1)})$. Under appropriate conjugate priors, this distribution can be expressed as an inverse Wishart marginal distribution for Σ, along with a multivariate normal distribution for β given Σ. Thus both components of the MCMC scheme are readily implemented.

In situations such as these, the 'seamless information flow' of Bayesian methods that we alluded to in Section 1 is very useful. While the computations break down into manageable components, overall we obtain inferences which are properly and transparently adjusted for the missing data mechanism. While there are similarities with EM algorithm approaches, by itself the EM algorithm cannot correctly reflect the uncertainty associated with estimates, and further numerical work is required to capture this (see, for instance, Meng and Rubin, 1991).

6. Sensitivity analysis for unobserved confounding

The threat of unobserved confounding is frequently used to discredit the findings of observational studies investigating the effects of exposures on disease in human populations. Because study subjects are assigned to exposure levels in a nonrandom way, hidden differences may be induced between exposure groups which cannot be adjusted for in the analysis. If these differences affect the outcome under study, then effect estimates will be biased in a manner which is difficult to predict. A popular solution is to conduct a sensitivity analysis (SA) wherein the model of the relationship between exposure and disease is expanded to include sensitivity parameters which reflect the investigators assumptions about unobserved confounding. If the exposure effect estimates are insensitive to a broad range of sensitivity parameters, then concerns about unobserved confounding are ameliorated.

SA was first proposed in the 1950's during the debate over the possible role of hidden confounders in the observed effect of tobacco smoking on the risk of lung cancer (see Cornfield et al., 1959). The method has seen later development by a numerous authors, including, Bross (1966), Schlesselman (1978), Rosenbaum and Rubin (1983), Yanagawa (1984), and Lin et al. (1998). The books of Rothman and Greenland (1998) and Rosenbaum (2002) give in-depth coverage of the topic.

Most sensitivity analysis is characterized by the use of models for unobserved confounding which are not identifiable, in the sense that the sensitivity parameters cannot be estimated consistently from available data. Schlesselman (1978), for example, proposes a simple model relating the apparent relative risk of disease due to an exposure with the true relative risk acting under the influence of a confounding effect of an unmeasured variable U. His model incorporates three sensitivity parameters: the relative risk of disease due to U and the two exposure specific prevalences of U. But since U is never observed to begin with, estimating any of these three quantities is not possible, and the resultant model is not identifiable.

Early pioneers in SA sought to overcome the challenge of nonidentifiability in the simplest way possible – using information external to the study to either specify the sensitivity parameters exactly or to specify ranges of plausible values for the parameters. The researcher is then free to 'plug in' the chosen parameter values, thereby rendering the model identifiable. The remaining model parameters, including the exposure effect, can then be estimated using standard inference techniques. But such 'externally adjusted' estimates suffered from the obvious limitation that different choices of sensitivity parameters can result in markedly different effect estimates, depending on where information on sensitivity parameters is obtained. Rosenbaum and Rubin (1983) suggest a more conservative approach, involving the presentation of effect estimates in tabular form for different combinations of sensitivity parameter values. This method is advantageous because it requires less information to specify the sensitivity parameters and reflects a broad range of assumptions about unobserved confounding.

But specifying ranges of sensitivity parameters can also be problematic. One can argue that the results of such an analysis may be unnecessarily pessimistic, as investigators tend to focus on only the most extreme effect estimates, which correspond to the most extreme sensitivity parameters considered. What is lost is the fact that typically less extreme values of the sensitivity parameters will be much more plausible than the extreme values. Intuitively, a Bayesian analysis derived from a prior distribution over the sensitivity parameters can reflect such nuances.

A Bayesian approach to unobserved confounding follows the general strategy of including the unobserved confounding variable as an unknown described by the posterior distribution. This is very much in line with the previous two sections where an unobserved variable is treated in this manner. As alluded to in Section 4, there is no technical prohibition on bringing the Bayes-MCMC machinery to bear on problems which lack formal parameter identifiability.

Bayesian approaches to SA has been the focus of a number of recent papers (Greenland 2001, 2003; Lash and Fink, 2003). The approach of Greenland (2003) involves a dichotomous exposure (X), outcome (Y) and unobserved confounder (U), and starts with the classic epidemiological paradigm of a 2×2 table describing observed subject counts classified on X and Y. Greenland considers the unobserved $2 \times 2 \times 2$ table which is further stratified over levels of U. He then models the expected number of study subjects in each of the eight possible cells using a simple log-linear model with parameters modeling the dependence and interaction between each of the three variables X, Y and U. Letting E_{uxy} denote the expected cell count in the (u, x, y) cell he writes:

$$E_{u,x,y} = \exp(\theta_0 + \theta_u u + \theta_x x + \theta_y y + \theta_{ux} ux + \theta_{uy} uy + \theta_{xy} xy).$$

Here, $(\mathrm{e}^{\theta_u}, \mathrm{e}^{\theta_x}, \mathrm{e}^{\theta_y})$ and $(\mathrm{e}^{\theta_{ux}}, \mathrm{e}^{\theta_{uy}}, \mathrm{e}^{\theta_{xy}})$ model the prevalences and the odds ratios respectively of U, X and Y. Note that the model assumes that risk ratios are homogeneous across strata of U, and that the outcome Y is rare, implying that risk ratios and odds ratios can be considered roughly equal. Since U is unmeasured, it is immediately apparent that this model is nonidentifiable and has three sensitivity parameters $(\theta_u, \theta_{ux}, \theta_{uy})$ which dictate the bias in the observed exposure effect that is induced by U.

Having defined the model, prior distributions for the seven model parameters must be specified. The usual approach is to obtain external information on the model parameters, including possible dependencies, and specify an appropriate prior distribution reflecting this information. Once this is accomplished, the likelihood of the given data is readily obtained and standard MCMC techniques can be used to sample from the joint posterior distribution of all the model parameters and (unobserved) cell counts in the $2 \times 2 \times 2$ table. But Greenland instead considers an alternative approach. He asks, given the observed marginal (i.e., unadjusted for U) odds-ratio between X and Y, can one identify a reasonable prior distribution for the seven model parameters, such that the resulting posterior distribution on the adjusted (for U) odds-ratio e^{θ}_{xy} has appreciable mass on the 'other side' of one. If such a prior exists, it prevents one from refuting the possibility that X and Y are independent given U and that the observed effect is the result of confounding by U. Such an approach runs contrary to the usual method of specifying prior information *a priori*, but it is potentially convenient because it only requires ruling out a class of priors rather than specifying a single prior precisely.

In general such Bayesian approaches to sensitivity analysis seems appealing relative to simply stating combinations or ranges of sensitivity parameters and reporting the corresponding point estimates of parameters of interest. In particular, the posterior distribution of an interest parameter derived from a prior distribution over sensitivity parameters conveys more information than say simply a range of point estimates arising from a range of sensitivity parameter values, in the sense of giving more weight to some interest parameter values than others. Another important point is that the posterior distribution synthesizes uncertainty arising because the sensitivity parameters are not estimable along with the 'regular' statistical uncertainty arising from limited sample size. This is hard to do in a simple way otherwise. For instance, it might be necessary to present tables of point estimates and tables of confidence intervals under different combinations of sensitivity parameters, and it may be hard to synthesize such an unwieldy mass of information.

7. Ecological inference

In many studies concerned with human health, data are not available at as fine a level as would be ideal, and arguably Bayesian methods are good for managing the uncertainty resulting from this limitation. A prime example of this is so-called *ecological inference*. Say that interest is focused on the association between variables X and Y, where, *in principal*, observations of (X, Y) could be made on individual subjects. Moreover, say that the study population is dispersed across n well-defined geographic regions, with individuals being sampled within each region. In particular, say m_i subjects are sampled in the ith region, and conceptually think of (X_{ij}, Y_{ij}) being the observations for the jth subject in this region. Sometimes, however, such individual-level measurements are not made or are not disclosed. Rather, only *aggregated* data in the form of region-specific means are available. That is, the observed data are $(\widetilde{X}_i, \widetilde{Y}_i)$ for $i = 1, \ldots, n$, where $\widetilde{X}_i = m_i^{-1} \sum_{j=1}^{m_i} X_{ij}$ and $\widetilde{Y}_i = m_i^{-1} \sum_{j=1}^{m_i} Y_{ij}$.

In practice there are several possible reasons why one might have only aggregated data. In some instances the individual cell counts are withheld by the agency responsible for the data on the grounds of maintaining confidentiality, particularly if the number of subjects per region (m_i's) is small. More commonly, in fact, the cell counts are never really measured at all. One agency might sample X in different regions for one purpose, without any effort to measure Y. And correspondingly another agency might sample Y in different regions for another purpose, without any interest in X. In some rough sense, even if two agencies use disjoint samples, a variant of $(\widetilde{X}, \widetilde{Y})$ measurements are obtained.

The primary question of ecological inference is the extent to which an estimated association between $(\widetilde{X}, \widetilde{Y})$ at the level of regions can be used to infer an association between (X, Y) at the level of individuals. Such questions abound in epidemiology. For instance, say $\widetilde{X}_i$ is the per-capita consumption of a specific foodstuff in country i, whereas $\widetilde{Y}_i$ is the incidence of a specific disease in that country. We note in passing that this is an example where likely joint (X, Y) measurement are never made for subjects, and presumably different agencies contribute the $\widetilde{X}$ and $\widetilde{Y}$ measurements. If the data from n countries indicate a strong association between $\widetilde{X}$ and $\widetilde{Y}$, can we infer that a similar association exists between X and Y for individuals within each country? In some strict sense the answer is no, as one can construct worst-case scenarios where the $(\widetilde{X}, \widetilde{Y})$ association is totally misleading (i.e., perhaps even in the wrong direction) as a surrogate for the (X, Y) association. Such examples are closely tied to Simpson's paradox regarding the collapsing of stratum-specific 2×2 tables into an overall table. However, appropriate Bayesian modeling with realistic prior distributions should indicate what, if anything, can be inferred about associations for individuals from associations for regions.

To be more specific we consider recent work of Wakefield (2004). He considers the case where both X and Y are binary. Thus the data we would like to see are n 2×2 tables cross-classifying X and Y for each region. The data we actually see are only the marginal totals for each table. Of course for a single 2×2 table it is a relatively hopeless task to try to guess the cell values given only the marginal totals. The information is very weak in that typically only wide bounds for the cell values are obtained. The hope is that 'borrowing strength' across n similar tables will yield some concentration of information. Intuitively, Bayesian hierarchical models are the right sort of tool to accomplish this task.

One hierarchical model considered by Wakefield (2004) is as follows. Following the common approach to regression, the distribution of $Y|X$ is viewed as being of interest, while the distribution of X is a nuisance quantity. Specifically, let $p_{0i} = Pr(Y_{ij} = 1|X_{ij} = 0)$ and $p_{1i} = Pr(Y_{ij} = 1|X_{ij} = 1)$ describe the region-specific distribution of Y given X. The model for the observable data is in fact more easily expressed in terms of totals rather than means. Towards this end, let $X_i^* = \sum_{j=1}^{m_i} X_{ij}$ and $Y_i^* = \sum_{j=1}^{m_i} Y_{ij}$ be the total number of $X = 1$ and $Y = 1$ subjects sampled in region i. The first-stage of the hierarchical model is then expressed as

$$Y_i^* = Y_{0i}^* + Y_{1i}^*$$

for each region i, where the unobservable cell counts (Y_{0i}^*, Y_{1i}^*) are distributed as

$$\begin{aligned} Y_{0i}^* &\sim \text{Binomial}\big(m_i - X_i^*, p_{0i}\big), \\ Y_{1i}^* &\sim \text{Binomial}\big(X_i^*, p_{1i}\big). \end{aligned}$$

That is, Y_{0i}^* is the unobserved $(Y = 1, X = 0)$ count in the ith table, while Y_{1i}^* is the unobserved $(Y = 1, X = 1)$ count.

To engender borrowing of strength, the next stage of the hierarchical model postulates that the variation across regions in (p_{0i}, p_{1i}) is somewhat structured. Wakefield suggests modeling these pairs as conditionally independent and identically distributed across regions, with

$$\begin{pmatrix} g(p_{0i}) \\ g(p_{1i}) \end{pmatrix} \sim N\left(\begin{pmatrix} \mu_0 \\ \mu_1 \end{pmatrix}, \begin{pmatrix} \sigma_0^2 & \rho\sigma_0\sigma_1 \\ \rho\sigma_0\sigma_1 & \sigma_1^2 \end{pmatrix} \right) \tag{10}$$

for an appropriate transform $g()$, such as the logit transform. Note that the dependence allowed by a nonzero value of ρ in (10) is very important. In most situations a positive correlation would be expected, due to unmeasured factors that vary across regions and are correlated with Y.

The hierarchical model is completed with the specification of a prior distribution over the means (μ_0, μ_1) and the variance components $(\sigma_0, \sigma_1, \rho)$. Wakefield emphasizes that typically inferences will be sensitive to this choice of prior, as the information contained in the data about these parameters will not be terribly strong. Obvious choices are normal priors for μ_0 and μ_1, along with an inverted Wishart prior for the variance matrix parameterized by $(\sigma_0, \sigma_1, \rho)$, with the choice of hyperparameters for these distributions then being important. In many applications it would likely be reasonable to do some crude elicitation of prior distributions. For instance, investigators could likely speculate on values of σ_0 and σ_1 which constitute implausibly large region-to-region variation in the distribution of $Y|X$, and choose prior distributions accordingly.

As is typically the case, MCMC techniques are effective at exploiting hierarchical model structure. That is, separate updates to the first-level unknowns (Y^*_{0i}, Y^*_{1i}), the second level unknowns (p_{0i}, p_{1i}), and the third level unknowns $(\mu_0, \mu_1, \sigma_0, \sigma_1, \rho)$, are readily carried out. One limitation noted by Wakefield is that the requisite distributions for Y^*_{0i} and Y^*_{1i} given all other quantities are extended hypergeometric distributions, which are costly to simulated from when the m_i's are large.

Fitting such a hierarchical model can lead to inferences about various quantities that may be of interest in a given application. In some cases the inferential focus may be on the multivariate normal parameters in (10). For instance, $\psi = \mu_1 - \mu_0$ describes the centre of the across-region distribution of log-odds ratios, while $\sigma_0^2 + \sigma_1^2 - 2\rho\sigma_0\sigma_1$ describes the region-to-region variability in these quantities. On the other hand, the bivariate normal distribution gives equal weight to all regions regardless of population size. Thus Wakefield emphasizes quantities such as $\bar{p}_j = \sum_i w_i p_{ji}$ for $j = 0, 1$ as target quantities of interest, where the weights $w_1, \ldots, w_n$, which sum to one, are proportional to population size. In this manner one is studying the overall distribution of $Y|X$. Of course the hierarchical Bayesian formulation is very flexible in this regard, as one can obtain the posterior distribution of any function of one or more parameters at any level of the hierarchy.

8. Bayesian model averaging

One theme espoused in Section 1 is that often Bayesian analysis facilitates more honest uncertainty assessments than other techniques. A very common biostatistical situation where substantial uncertainty is often ignored lies in the data-driven selection of a small number of apparently relevant predictor variables from a large pool of measured variables. Particularly, say that study data includes a health-related outcome variable Y and a large number p of possible predictor variables $(X_1, \ldots, X_p)$. In actuality, however, investigators suspect that many of these variables are not actually associated with the outcome variable, and therefore they use some sort of *model selection* scheme. That is, they use the data to select q of the p predictors, and commonly q turns out to be substantially less than p. We denote the selected predictors as $(\widetilde{X}_1, \ldots, \widetilde{X}_q)$. Various *stepwise regression* schemes are available in standard software packages to accomplish such variable selection.

The problem arises in that typically inferences are reported as if the selected model were known to be the correct model. For instance, standard errors associated with regression coefficients reflect variation in the sampling distribution of Y given $(\widetilde{X}_1, \ldots, \widetilde{X}_q)$, but ignore the uncertain selection, i.e., if a new sample were drawn from the population, a different subset of predictors might well be selected. While there have been sporadic attempts to deal with the problem for non-Bayesian analyses (Zhang, 1992; Buckland et al., 1997), an overarching approach is not evident. In contrast, *Bayesian model averaging* provides a general and principled way to reflect this uncertainty. Hoeting et al. (1999) give a detailed overview of this approach.

In general, say that k different models are under consideration, where the ith model is parameterized by $\theta^{(i)}$, with likelihood $f_i(\text{data}|\theta^{(i)})$ and prior density $f_i(\theta^{(i)})$. With multiple models the Bayesian paradigm also demands a prior distribution across the models in addition to prior distributions for parameters within models. Conceptually, let $M \in \{1, \ldots, k\}$ indicate the 'true' model, with $\pi_i = Pr(M = i)$ being the prior probability ascribed to the truth of the ith model. Of course, $\sum_{i=1}^{k} \pi_i = 1$. In principal there is no difficulty in applying Bayes theorem to the 'overall' parameter (M, θ_M), which leads to

$$Pr(M = i|\text{data}) = \frac{Pr(M = i) \int f_i(\text{data}|\theta^{(i)}) f_i(\theta^{(i)})\, d\theta^{(i)}}{\sum_{j=1}^{k} Pr(M = j) \int f_j(\text{data}|\theta^{(j)}) f_j(\theta^{(j)})\, d\theta^{(j)}}, \tag{11}$$

while the parameter given model component of the posterior distribution is, as usual,

$$f_i\big(\theta^{(i)}|M = i, \text{data}\big) = \frac{f_i(\text{data}|\theta^{(i)}) f_i(\theta^{(i)})}{\int f_i(\text{data}|\theta^{(i)}) f_i(\theta^{(i)})\, d\theta^{(i)}}.$$

With respect to this paradigm, say we are interested in estimating a quantity γ which has a common meaning across models. In particular, say $\gamma = g_i(\theta^{(i)})$ for each $i = 1, \ldots, k$. As an example, γ might be a predicted value of Y for a fixed value of $(X_1, \ldots, X_p)$, with each function $g_i()$ chosen accordingly. Or γ might be the regression coefficient associated with a particular predictor, with the understanding that $\gamma = 0$ under models which exclude this predictor from consideration. If we wish to summarize inference about γ in terms of its posterior mean and variance then clearly

$$E(\gamma|\text{data}) = \sum_{i=1}^{k} E\big\{g_i\big(\theta^{(i)}\big)|M = i, \text{data}\big\} Pr(M = i|\text{data}) \tag{12}$$

and

$$
\begin{aligned}
&\mathrm{Var}(\gamma|\text{data}) \\
&\quad = \sum_{i=1}^{k} \mathrm{Var}\big(g(\theta^{(i)})|M = i, \text{data}\big) Pr(M = i|\text{data}) \\
&\qquad + \sum_{i=1}^{k} \big\{E\big(g(\theta^{(i)})|M = i, \text{data}\big) - E(\gamma|\text{data})\big\}^2 Pr(M = i|\text{data}). \qquad (13)
\end{aligned}
$$

The expressions above reveal that both our point estimate of γ and our assessment of the estimate's precision differ fundamentally from the corresponding quantities obtained by selecting a single model and making inferences conditioned on the truth of this model. In situations where estimating γ is a predictive endeavor, there is both theoretical and empirical evidence to suggest that model averaging leads to smaller predictive errors than model selection (Raftery et al., 1995). Moreover, the second term in (13) can be directly interpreted as reflecting the uncertainty that is ignored by the single model approach. Overall, it seems hard to argue against a procedure which simultaneously enhances predictive performance and gives a more honest assessment of uncertainty in predictions.

A nice demonstration of the advantages of model averaging for logistic regression in epidemiological studies is provided by Viallefont et al. (2001). They carry out an extensive simulation study under conditions that typify many epidemiological investigations. Particularly, p, the number of potential predictors, is large, but only a small number q of these are actually associated with the outcome. Moreover, these associations are set to be quite weak, as reflected by modest adjusted odds-ratios. These authors simulate repeated datasets under these conditions, and compare stepwise logistic regression procedures to Bayesian model averaging. In the former case they look at predictors which are declared to be significant on the basis of small P-values in the finally selected model, and find that typically many of these declarations are for predictors which are *not* actually associated with the outcome. That is, the combination of calibrating evidence via P-values plus ignoring the uncertainty in the selection process leads to a high 'false discovery' rate. Alternately, they consider declaring a predictor to be significant if it has a small posterior probability of being zero. That is, the sum of (11) over models excluding this predictor is small. This procedure tends to have a much lower false discovery rate. For instance, in one simulation these authors isolate all the instances where the posterior probability of predictor exclusion is between 0.01 and 0.05, and find that 92% of these instances correspond to an actual association between the predictor in question and the outcome. In contrast they apply stepwise procedures and P-values to the same simulated datasets and isolate all the instances where the P-value lies between 0.01 and 0.05. Depending on which of two stepwise selection schemes is used, only 68% or 55% of these instances correspond to real associations. Thus the evidence arising from the Bayesian model averaging is much better *calibrated* than that arising from P-values following model selection.

In situations where closed-form expressions do not exist, a general theme in Bayesian computation is that across-model computations are an order-of-magnitude more difficult than within-model computations. See Han and Carlin (2001) for some general discussion of across model calculations. Moreover, in general any schemes for model selection or model averaging that involve computing quantities for all possible models suffer from a curse of dimensionality. With p potential predictors there are $k = 2^p$ possible models based on including some subset of predictors. In a strict sense then it will rapidly become impossible to evaluate (12) or (13) as p becomes moderately large. However, these expressions can be well-approximated if the models with appreciable posterior probability can be identified. Indeed, much of the literature on model averaging uses the notion of *Occam's window*, which is defined to include only those models. Also, there have been numerous computational schemes suggested to identify the high posterior probability models without exhaustively considering all possible models. The software described in Raftery (1995) appears to be particularly simple and thus easy to apply. A 'leaps-and-bounds' procedure is used to find high posterior probability models, while Laplace approximations to the integrals in (11) are used when necessary.

9. Survival analysis

Survival analysis deals with time-to-event data, which is ubiquitous in biostatistics. As one example, it is often of interest to compare times to disease progression or times to death for patients on different treatment arms in a clinical trial. Statistical methods for survival analysis have evolved in some isolation from other statistical methods, largely because survival data have several characterizing features. First, the response variables (i.e., times) are inherently positive and often have skewed distributions. Second, most survival data is subject to some kind of *censoring*, the typical case being that for some subjects we only lower bounds on the response times are observed. A patient's response time might be censored because the patient drops out of the study or the study ends before the target event occurs.

Survival analysis departs from other techniques with its focus on the *hazard function* or *hazard rate* as characterizing the distribution of a time to event. The hazard function $h(t)$ is the density at time t of the conditional distribution of $(T \mid T > t)$. So the probability of failure in the interval $(t, t+\delta)$ given survival to time t is approximately $\delta h(t)$. The hazard function is obtained from the density function $f()$ via $h(t) = f(t)/\int_t^\infty f(s)\,\mathrm{d}s$. The pioneering idea of Cox (1972) is to model the impact of covariates X on response time T via a multiplicative effect on the hazard rate. That is,

$$h(t \mid x) = h_0(t)\exp\bigl(x'\beta\bigr), \tag{14}$$

where $h_0(t)$ is the so-called *baseline hazard*. This is referred to as the Cox proportional hazards model, as two individuals with different covariate values will have hazard functions proportional to one another. This model has proved to be immensely popular, and is the standard approach in most subject-area research involving time-to-event data.

One approach to proportional hazards models is to assume a parametric form for the baseline hazard, with a small number of unknown parameters. Baseline hazards corresponding to exponential, Weibull, and gamma distributions are the most common choices. An alternative strategy, which is widely viewed as appealing, avoids parametric assumptions about the baseline hazard by using *semiparametric* techniques. On the non-Bayesian side, the Cox partial likelihood (Cox, 1975) is a function of β alone which plays the role of a likelihood, and the maximum partial likelihood estimator of β can be shown to have good frequentist properties. On the Bayesian side, flexible models and priors for the baseline hazard have received considerable attention in the literature. As a starting point, a baseline hazard which is piecewise linear with respect to a partition of the time axis into small intervals can be considered. Here an appropriate prior distribution would reflect prior belief that hazard values on adjacent sub-intervals are likely to be similar. This possibility, and many more refined approaches based on stochastic process priors for the baseline hazard, are surveyed by Sinha and Dey (1997) and Ibrahim et al. (2001, Chapter 3). Many of these methods strike a nice balance, because (i) they do not make rigid parametric assumptions, about the baseline hazard, but (ii) they still permit inference about the baseline hazard, in contrast to the partial likelihood approach, and (iii) a suitable prior distribution can engender plausible smoothness in the estimated baseline hazard. In turns out there are interesting connections between flexible Bayesian treatments of the baseline hazard and the partial likelihood approach, as first delineated by Kalbfleisch (1978). This line of inquiry has recently been rekindled with the interesting work of Sinha et al. (2003).

Another advantage of Bayesian approaches to survival analysis is that usually it is very easy to handle complicated patterns of censored data. In particular, one can pursue the same tack as discussed for mismeasured data in Section 4, missing data in Section 5, and completely unobserved data in Section 6. That is, the 'pure' but unobserved data are treated as unknown quantities 'inside' the posterior distribution, which is conditioned on whatever rough information about these data that is available. Thus in the survival analysis context one component of the MCMC scheme will be to update the unobserved response times given current parameter values and censoring information. As a simple example, say that T_i, the response time for subject i, is right-censored at time c_i. That is, we know only that $T_i > c_i$. Then we simply update T_i by sampling from the conditional distribution of $(T_i|T_i > c_i)$ under the current parameter values. While such a scheme is not actually needed for simple censoring mechanisms, it is still simple to implement when much more complex patterns of censoring are at play. In particular, it compares very favorably to some of the difficult and highly-specialized suggestions for non-Bayesian approaches to dealing with complicated censoring.

Another common situation in medical settings is to find response times which are clustered, in the sense of multiple times per person or family, or multiple times per clinical site. As alluded to in Section 2, a hierarchical modeling approach can account for such clustering in a natural way. Often such models are referred to as *frailty models*, with the name derived from situations where there are multiple response times clustered within individuals, some of whom are more frail than others. This variation in frailty is represented by the variation in random effects at the second stage of a hierarchical model such as that described in Section 2. Ibrahim et al. (2001, Chapter 4) survey the sizable body of work on Bayesian frailty models.

As a final thought, we speculate that the gulf between survival analysis methods and other statistical methods may, in fact, be wider than necessary. Particularly under the

Bayesian paradigm, a more unified view is possible, and arguably desirable. We have already commented that the Bayesian approach via MCMC affords an easy connection between dealing with censored data and dealing with complete data. On another front, Gustafson et al. (2003) make a small step towards bridging the gap. Specifically, these authors focus on the central role of the gamma distribution as a prior and posterior distribution for a hazard rate, contrasted with the central role of the normal distribution as a prior and posterior distribution for parameters in linear models. It is well known that for a moderately large shape parameter the log-gamma distribution is well approximated by a normal distribution, a connection that they exploit as a computational engine for Bayesian inference in survival models. That is, learning about log-hazard rates in survival models is not actually so dissimilar from learning about parameters in linear models.

10. Case-control analysis

Many diseases are sufficiently rare that following an initially healthy cohort *prospectively* (i.e., forward in time) will lead to very few cases of the disease, and hence very limited ability to infer associations between explanatory (or 'exposure') variables and disease development. Thus many epidemiological studies employ a *retrospective* (or case-control) study design. Samples of n_0 disease-free subjects (the controls) and n_1 diseased subjects (the cases) are assembled, and then explanatory variables $X = (X_1, \ldots, X_p)$ are measured for subjects. Often the epidemiological question is whether an exposure might be associated with increased risk of getting the disease at a future point in time. Flipping this around, in a case-control study the measurement of explanatory variables may be very retrospective in nature, as some elements of X might pertain to a subject's behavior and environment in the quite distant past.

If we let Y be a $0/1$ indicator for absence/presence of disease, then a prospective study might be regarded as sampling subjects from the joint distribution of (Y, X), with the aim of learning about the conditional distribution of $(Y|X)$. In a retrospective study we still wish to learn about $(Y|X)$, but we sample subjects from $(X|Y)$. Commonly a logistic regression model is postulated for $Y|X$. Say this model is parameterized by θ. Then the contribution of the ith case to the likelihood function is

$$f_{X|Y}(x_i|y_i) = \frac{f_{Y|X}(y_i|x_i, \theta) f_X(x_i)}{\int f_{Y|X}(y_i|\tilde{x}, \theta) f_X(\tilde{x})\, \mathrm{d}\tilde{x}}. \tag{15}$$

Two things are immediately apparent. First, this depends on the marginal distribution of X as well as the conditional distribution of $(Y|X)$. Second, even if we viewed the marginal distribution of X as known, the likelihood contribution is not simply proportional to $f_{Y|X}(y_i|x_i, \theta)$, since the denominator of (15) also depends on θ.

In fact it is very common practice to analyze data from a retrospective study as if it were collected prospectively, i.e., take $f(y_i|x_i, \theta)$ rather than (15) as the contribution of the ith case to the likelihood function. In the case of maximum likelihood inference using large-sample theory, the well known work of Prentice and Pyke (1979) justifies this false pretense, which obviates the need to specify a marginal distribution for X, and avoids the integration in the denominator of (15). However, the Prentice and Pyke justification does not extend to small samples, nor to Bayesian analysis. As well, it does not extend easily to more complex modeling scenarios, the situation where X is subject to mismeasurement being a prime example. Thus there has been a good deal of recent attention on (i) Bayesian modeling of retrospective data without the convenience of pretending the data were collected prospectively, and (ii) examining the interplay between retrospective and prospective models more carefully in the Bayesian context.

In regard to implementing Bayesian inference with a likelihood based on (15) or a close variant, there have been a number of recent suggestions in the literature. Gustafson et al. (2000) consider a simple parametric approach, while Gustafson et al. (2002) implement a grid-based model for the distribution of X (see also the comments in Gustafson, 2003, about extending to more covariates using 'reverse exposure' models). Muller and Roeder (1997) use Dirichlet-process mixture models for the distribution of covariates, while Seaman and Richardson (2001) consider categorical covariates and multinomial models. We also note the recent work of Rice (2003) and Sinha et al. (2005) on Bayesian analysis of matched case-control studies.

As alluded to above, the Prentice and Pyke argument does not extend easily to situations where X is subject to mismeasurement. Moreover, this is a common situation in retrospective studies, as the attempts to ascertain X are made after-the-fact, often on the basis of subject recall on questionnaires. Generally speaking if X in unobserved then modeling it's distribution is unavoidable, so it behooves us to work with the retrospective likelihood. This same point was emphasized in Section 4 where it was necessary to posit a conditional exposure model. Following Gustafson et al. (2002) and Gustafson (2003), a convenient approach is to use the retrospective model defined by a parametric model for the distribution of $(Y|X)$, and a parametric model for the mixture distribution which gives weight n_0/n to the distribution of $(X|Y = 0)$ and weight n_1/n to the distribution of $(X|Y = 1)$. This mixture distribution can be regarded as the marginal distribution of X under a retrospective scheme where each Y is sampled with probabilities matching the case/control fractions. Then, with appropriate treatment of the intercept term, the posterior distribution assuming prospective sampling emerges as an approximation to the posterior which assumes retrospective sampling. In fact, one can draw posterior samples from the prospective posterior but then use importance sampling (see, for instance, Evans and Swartz, 2000) to convert this to a sample from the retrospective posterior. In the example of Gustafson et al. (2002), this conversion has negligible effect; that is, the prospective approximation is effectively indistinguishable from the retrospective posterior distribution. However, Gustafson (2003) shows that at modest sample sizes the difference can be nonnegligible.

In essence, the utility of Bayesian methods for case–control analysis stems from the seamless integration of probability to describe both the sampling scheme and the knowledge of unobservables given observables. In complex situations where the Prentice and Pyke justification does not apply, the Bayesian approach gives a principled strategy for inference.

11. Bayesian applications in health economics

Recent advances in Bayesian statistical methods and software has also motivated interest in Bayesian applications in the field of health economics. Health care providers currently face increasing financial restrictions, and resource limitations dictate that new spending for costly medical interventions must be balanced by appropriate improvements in clinical outcomes as compared to traditional interventions. Consequently, health care providers are turning to cost-effectiveness data in addition to efficacy data in the evaluation of new medical technologies. Typically, such information is obtained from randomized controlled trials where data on the costs of treatments, as well as efficacy, can be collected.

A comprehensive framework for Bayesian modeling of cost-effectiveness data in randomized clinical trials is given by O'Hagan and Stevens (2001, 2002). Approaching the problem in a general setting, the authors consider a standard parametric model used in the analysis of data from a randomized controlled trial. Specifically, the data $\{y_{ij} = (c_{ij}, e_{ij}) \mid i \in 1, 2, j \in 1, 2, \ldots, n_i\}$ are collected. Here e_{ij} and c_{ij} denote individual measurements of the efficacy and cost respectively on subject j receiving treatment i. Traditional models have assumed that the e_{ij} and c_{ij} measurements are normally distributed, or at least that large-sample normal approximations to sample means can be used (see, for instance, Van Hout et al., 1994). O'Hagan and Stevens assume only that the e_{ij} and c_{ij} are independent and identically distributed with probability density $f(y|\theta_i)$.

The quantities of interest in the analysis of the cost-effectiveness of the treatment are the parameters that govern the observed costs and efficacies on subjects receiving each of the two treatments. Since the distribution of the (e_{ij}, c_{ij}) is parameterized by θ_i, one can naturally model the mean cost of treatment i as $\mu_i = \mu(\theta_i)$ and mean efficacy of treatment i as $\tau_i = \tau(\theta_i)$. The mean incremental cost of treatment 2 relative to treatment 1 is then $\Delta C = \mu_2 - \mu_1$, and analogously the mean incremental efficacy of treatment 2 relative to 1 is $\Delta E = \tau_2 - \tau_1$. Hence, in a simple example where the observed costs and efficacies on study subjects are assumed to be normally distributed, ΔC and ΔE would represent the difference in mean costs and mean efficacies of each of the two treatments.

Comparing the cost-effectiveness of two treatments is often accomplished by making inferences about the incremental cost-effectiveness ratio (ICER) of one treatment relative to the other. The ICER statistic, defined specifically as $\rho = \Delta C/\Delta E$, provides an estimate of the dollar cost required for an improvement of one unit efficacy for treatment two as compared to treatment one (O'Hagan and Stevens, 2001, 2002). This ICER estimate must then be related to a decision making process on behalf of the health care provider, usually reflected in the *willingness to pay* coefficient K. This quantity reflects the minimum that the health care provider will pay for a medical intervention which provides a one unit increase in efficacy. The role of K then is to prevent the adoption of new technologies that are only modestly more cost-effective.

The use of ICER estimates and K forms the basis of what has been referred to as the 'Net Benefits Approach' towards cost-effectiveness evaluation (Stinnett and Mullahy, 1998). Specifically, this approach involves estimating ΔE and ΔC from the data and then attempting to determine if the function $\beta(k) = k\Delta E - \Delta C$ is strictly greater than zero. Here $\beta(k)$, referred to as the Net Monetary Benefit, converts ΔE units of efficacy into $K\Delta E$ units of money and hence $\beta(k) > 0$ corresponds to the instance in which treatment two is more cost-effective than treatment one (O'Hagan and Stevens, 2001, 2002).

Health care providers are sometimes reluctant to specify one value of k rather than another, and consequently, the cost effectiveness acceptability curve (CEAC) has become a popular tool for measuring cost-effectiveness. The CEAC, discussed extensively by O'Hagan et al. (2000), is based on consideration of $Pr\{\beta(K) > 0\}$. Typically it is presented as a plot of the probability that the net benefit is positive versus k. But, as these authors point out, the definition of CEAC is meaningless unless the parameters that index $\beta(k)$ are treated as random variables.

Armed with appropriate prior distributions, Bayesian estimation of (θ_1, θ_2) given the data is typically straightforward. Of course this induces a posterior distribution on $(\Delta E, \Delta C)$, and, for a given k, a posterior distribution on $\beta(k)$. The CEAC is then given as a plot of the posterior probability that $\beta(k)$ exceeds zero versus k. Thus we have a situation where there is little technical innovation in implementing Bayesian analysis, but doing so gives subject-area researchers quite interpretable inferential summaries. Moreover, there are clear routes to extending this kind of analysis to situations which require much more complex modeling of the available efficacy and cost data.

12. Discussion

Having touched on a variety of scenarios where Bayesian analysis can be fruitfully brought to bear, we hope we have conveyed the spirit in which Bayesian methods are valuable in biostatistics and epidemiology. To various extents, we feel these examples illustrate the five benefits outlined in Section 1.

Of course our choice of scenarios and methods is idiosyncratic, and to considerable extent reflects the interests and biases of the authors. Our discussion has largely been oriented towards observational rather than experimental studies, though Bayesian methods are useful in the latter case as well. As one example, Bayesian methods have been touted as particularly useful for consideration of early stopping for randomized clinical trials. In fact this is a point of some controversy, as traditional and Bayesian methods can lead to quite discrepant results in such settings. Another large area on which we have been silent is statistical genetics and genomics. Of course there is a tremendous amount of statistical research in this area currently, and for all the reasons outlined in Section 1, Bayesian methods are amongst those showing the most promise.

As mentioned in Section 1, we have also been quite skimpy with our discussions of computational techniques. This stems from our belief that by and large computation is no longer a limiting factor for applications of Bayesian methods to health research. Thus we have concentrated our discussion on conceptual issues, and examples of Bayesian model construction. We take for granted that with a modest amount of effort one can implement a MCMC algorithm to simulate draws from the posterior distribution of the unobserved quantities given what is observed, though invariably there are interesting challenges in trying to do so in as efficient manner as possible.

To say more about computation, the basic idea of MCMC is to simulate a Markov chain which has the desired posterior distribution as its stationary distribution. Since a Markov chain tends to converge quickly to its stationary distribution, after some burn-in iterations we can treat the chain output as a dependent sample from the desired distribution. Choosing an appropriate transition kernel for the chain is made quite easy by the Metropolis–Hastings algorithm, which starts with literally any transition kernel, and by introducing an accept/reject step converts this to a kernel possessing the desired stationary distribution. In essence the algorithm is almost too rich, as literally any initial transition scheme can be employed. But, of course, a 'reasonable' scheme must be chosen in order to yield a chain that mixes tolerably well. Another theme is that typically one attempts to convert a high-dimensional problem into a iterated series of low-dimensional problems, by applying MCMC to only a portion of the unknown quantities at a time. In the extreme case that each 'full conditional' of the target distribution, i.e., the distribution of a univariate component given all other components, has a nice form, one can sample from each full conditional in turn. This rejection-free MCMC algorithm is the Gibbs sampler. If such nice structure is not present, then other transitions can be brought to bear. Some, like the so-called 'random-walk' proposal, require very little information about the target distribution, but also require 'tuning' to get reasonable MCMC output. Other updating schemes attempt to use more information about the target distribution as a route to gaining computational efficiency.

In modern Bayesian practice, the primary question surrounding MCMC implementation for a given problem is whether to use special-purpose MCMC software, or whether to code a particular MCMC implementation 'from scratch' in a language such as C, MatLab, or R. The primary special-purpose software in WinBUGS (http://www.mrc-bsu.cam.ac.uk/bugs/). Essentially this software allows the user to specify models and priors as he sees fit, and then attempts to implement a particular MCMC algorithm. Initially there was skepticism about whether Bayesian inference could be automated to such an extent, but there does seem to be a growing list of success stories, i.e., instances of complex Bayesian modeling where WinBUGS is able to deal with the computations. While we ourselves have very little experience with WinBUGS, we have noticed from the WinBUGS e-mail list that this software has attracted a lot of interest amongst subject-area researchers, including health researchers, who want to use Bayesian methods, but do not have the background to implement MCMC from scratch.

Our own experience is more with scenarios where we feel a need to implement MCMC from scratch, for the sake of computational speed, or because we perceive a need for very specialized MCMC algorithms to handle a problem that is hard because of high-dimension and/or unusual posterior forms. There is no denying that it can be time-consuming to carry out such an implementation, and that in some cases the actual simulation of chains can take a considerable amount of computer time. Having said this, we feel criticisms of the Bayesian approach on the grounds of increased implementation time or computing time relative to non-Bayesian techniques are not particularly warranted. In any real medical study, the time and cost to implement and run either MCMC or a competing non-Bayesian technique will be a negligible fraction of the time and cost required to collect and pre-process the data. Direct comparisons of the times associated with various statistical methods need to be tempered by this fact. If the data took a year to collect, are we going to quibble over whether the analysis takes a few minutes or a few hours to compute?

As a final thought, an overarching advantage of the Bayesian approach to statistics in all application areas is the provision of first principles. The analyst thinks hard about the most appropriate models and priors, and then Bayes theorem provides a principled and fully specified route to inference about any quantity of interest. This provides a refreshing counterpoint to some views of statistics which are more of a 'cookbook' variety, i.e., a list of data types mapped to a list of specialized techniques. Such first principles ought to be a welcome in the difficult challenge of providing rigorous statistical analysis in the complex spheres of biostatistics and epidemiology.

References

Berry, D.A., Stangl, D.K. (Eds.) (1996). *Bayesian Biostatistics*. Marcel Dekker, New York.

Berry, D.A., Parmigiani, G., Sanchez, J., Schildkraut, J., Winer, E. (1997). Probability of carrying a mutation of breast-ovarian cancer gene BRCA1 based on family history. *J. Nat. Cancer Inst.* **88**, 227–238.

Besag, J. (1974). Spatial interaction and the statistical analysis of lattice systems (with discussion). *J. Roy. Statist. Soc., Ser. B* **36**, 192–236.

Besag, J., York, J., Mollie, A. (1991). Bayesian image restoration, with two application in spatial statistics. *Ann. Inst. Statist. Math.* **43**, 1–59.

Brophy, J.M., Joseph, L. (1995). Placing trials in context using Bayesian analysis: GUSTO revisited by reverend Bayes. *J. Amer. Med. Assoc.* **273**, 871–875.

Brophy, J.M., Joseph, L. (1997). Bayesian interim statistical analysis of randomised trials. *Lancet* **349**, 1166–1168.

Brophy, J.M., Joseph, L., Rouleau, J. (2001). Beta blockers in congestive heart failure – a Bayesian meta-analysis. *Ann. Int. Med.* **134**, 550–560.

Bross, I.D.J. (1966). Spurious effects from an extraneous variable. *J. Chronic Dis.* **19**, 637–647.

Buckland, S.T., Burnham, K.P., Augustin, N.H. (1997). Model selection: An integral part of inference. *Biometrics* **53**, 603–618.

Carlin, B.P., Louis, T.A. (1998). *Bayes and Empirical Bayes Methods for Data Analysis*, second ed. Chapman and Hall/CRC, Boca Raton.

Carlin, B.P., Xia, H. (1999). Assessing environmental justice using Bayesian hierarchical models: Two case studies. *J. Expos. Anal. Environ. Epi.* **9**, 66–78.

Carroll, R.J., Ruppert, D., Stefanski, L.A. (1995). *Measurement Error in Nonlinear Models*. Chapman and Hall/CRC, Boca Raton.

Cornfield, J., Haenszel, W., Hammond, E., Lilienfeld, A., Shimkin, M., Wynder, E. (1959). Smoking and lung cancer: Recent evidence and a discussion of some questions. *J. Nat. Cancer Inst.* **22**, 173–203.

Cox, D.R. (1972). Regression models and life tables. *J. Roy. Statist. Soc., Ser. B* **34**, 187–220.

Cox, D.R. (1975). Partial likelihood. *Biometrika* **62**, 269–276.

Cressie, N.A., Chan, N.H. (1989). Spatial modeling of regional variables. *J. Amer. Statist. Assoc.* **84**, 393–401.

Daniels, M.J. (1999). A prior for variance in hierarchical models. *Canad. J. Statist.* **27**, 567–578.

Dempster, A.P., Laird, N.M., Rubin, D.B. (1977). Maximum likelihood from incomplete data via the EM algorithm (with discussion). *J. Roy. Statist. Soc., Ser. B* **39**, 1–38.

DuMouchel, W., Harris, J. (1983). Bayes methods for combining the results of a cancer studies in humans and other species. *J. Amer. Statist. Assoc.* **78**, 293–308.

DuMouchel, W., Normand, S. (2000). Computer modelling and graphical strategies for meta-analysis. In: Stangl, D.K., Berry, D.A. (Eds.), *Meta-Analysis in Medicine and Health Policy*. Marcel Dekker, New York.

DuMouchel, W., Christine, W., Kinney, D. (1995). Hierarchical Bayesian linear models for assessing the effect of extreme cold weather on schizophrenic births. In: Berry, D.A., Stangl, D.K. (Eds.), *Bayesian Biostatistics*. Marcel Dekker, New York.

Dunson, D.B. (2001). Commentary: Practical advantages of Bayesian analysis of epidemiologic data. *Amer. J. Epi.* **153**, 1222–1226.

Evans, M., Swartz, T.B. (2000). *Approximating Integrals via Monte Carlo and Deterministic Methods*. Oxford University Press, Oxford.

Fleiss, J.L. (1993). The statistical basis for meta-analysis. *Statist. Meth. Med. Res.* **2**, 147–160.

Freedman, L. (1996). Bayesian statistical methods. *British Med. J.* **313**, 569–570.

Fuller, W.A. (1987). *Measurement Error Models*. Wiley, New York.

Gelman, A., Carlin, J.B., Stern, H., Rubin, D.B. (2003). *Bayesian Data Analysis*, second ed. Chapman and Hall/CRC, Boca Raton.

Goodman, S.N. (1999a). Toward evidence-based medical statistics. 1: The P value fallacy. *Ann. Int. Med.* **130**, 1019–1021.

Goodman, S.N. (1999b). Toward evidence-based medical statistics. 2: The Bayes factor. *Ann. Int. Med.* **130**, 1005–1013.

Goodman, S.N. (2001). Of P-values and Bayes: A modest proposal. *Epidem.* **12**, 295–297.

Gray, R. (1994). A Bayesian analysis of institutional effects in a multi-center cancer clinical trial. *Biometrics* **50**, 244–253.

Greenland, S. (2001). Sensitivity analysis, Monte Carlo risk analysis, and Bayesian uncertainty assessment. *Risk Analysis* **21**, 579–583.

Greenland, S. (2003). The impact of prior distributions for uncontrolled confounding and response bias: A case study of the relation of wire codes and magnetic fields to childhood leukemia. *J. Amer. Statist. Assoc.* **98**, 47–54.

Gustafson, P. (1995). A Bayesian analysis of bivariate survival data from a multi-center cancer clinical trial. *Statist. Med.* **14**, 2523–2535.

Gustafson, P. (1997). Large hierarchical Bayesian analysis of multivariate survival data. *Biometrics* **53**, 230–242.

Gustafson, P. (2003). *Measurement Error and Misclassification in Statistics and Epidemiology: Impacts and Bayesian Adjustments*. Chapman and Hall/CRC, Boca Raton.

Gustafson, P. (2005). On model expansion, model contraction, identifiability, and prior information: Two illustrative scenarios involving mismeasured variables (with discussion). *Statist. Sci.* Submitted for publication.

Gustafson, P., Wasserman, L. (1996). Comment on "Statistical inference and Monte Carlo algorithms," by G. Casella. *Test* **5**, 300–302.

Gustafson, P., Le, N.D., Vallée, M. (2000). Parametric Bayesian analysis of case-control data with imprecise exposure measurements. *Statist. Probab. Lett.* **47**, 357–363.

Gustafson, P., Le, N.D., Saskin, R. (2001). Case-control analysis with partial knowledge of exposure misclassification probabilities. *Biometrics* **57**, 598–609.

Gustafson, P., Le, N.D., Vallée, M. (2002). A Bayesian approach to case-control studies with errors in covariables. *Biostatistics* **3**, 229–243.

Gustafson, P., Aeschliman, D., Levy, A.R. (2003). A simple approach to fitting Bayesian survival models. *Lifetime Data Anal.* **9**, 5–19.

Han, C., Carlin, B.P. (2001). MCMC methods for computing Bayes factors: A comparative review. *J. Amer. Statist. Assoc.* **96**, 1122–1132.

Hoeting, J.A., Madigan, D., Raftery, A.E., Volinsky, C.T. (1999). Bayesian model averaging: A tutorial. *Stat. Sci.* **14**, 382–417. Corrected version available at http://www.stat.washington.edu/www/research/online/hoeting1999.pdf.

Ibrahim, J.G., Chen, M.-H., Sinha, D. (2001). *Bayesian Survival Analysis*. Springer, New York.

Jenrich, R.I., Schluchter, M.D. (1986). Unbalanced repeated measures models with structured covariance matrices. *Biometrics* **42**, 805–820.

Johnson, B., Carlin, B.P., Hodges, J.S. (1999). Cross-study hierarchical modeling of stratified clinical trial data. *J. Biopharmaceutical Statist.* **9**, 617–640.

Joseph, L., Wolfson, D.B., Belisle, P., Brooks, J.O., Mortimer, J.A., Tinklenberg, J.R., Yesavage, J.A. (1999). Taking account of between-patient variability when modeling decline in Alzheimer's disease. *Amer. J. Epi.* **149**, 963–973.

Kalbfleisch, J.D. (1978). Nonparametric Bayesian analysis of survival time data. *J. Roy. Statist. Soc., Ser. B* **40**, 214–221.

Kelsall, J.E., Wakefield, J.C. (1998). Modeling spatial variation in disease risk. Technical Report, Imperial College, London.

Lash, T.L., Fink, A.K. (2003). Semi-automated sensitivity analysis to assess systematic errors in observational data. *Epidem.* **14**, 451–458.

Leroux, B.G., Lei, X., Breslow, N. (1999). Estimation of disease rates in small areas: A new mixed model for spatial dependence. In: Halloran, M.E., Berry, D. (Eds.), *Statistical Models in Epidemiology, the Environment and Clinical Trials*. Springer-Verlag, New York.

Lilford, R.J., Braunholtz, D. (1996). The statistical basis of public policy: A paradigm shift is overdue. *British Med. J.* **313**, 603–607.

Lin, D.Y., Psaty, B.M., Kronmal, R.A. (1998). Assessing the sensitivity of regression results to unmeasured confounders in observational studies. *Biometrics* **54**, 948–963.

Little, R.J., Rubin, D.B. (2002). *Statistical Analysis with Missing Data*, second ed. Wiley, New York.

MacNab, Y.C. (2003). Hierarchical Bayesian modeling of spatially correlated health service outcome and utilization rates. *Biometrics* **59**, 305–316.

Meng, X.L., Rubin, D.B. (1991). Using EM to obtain asymptotic variance–covariance matrices: The SEM algorithm. *J. Amer. Statist. Assoc.* **86**, 899–909.

Muller, P., Roeder, K. (1997). A Bayesian semiparametric model for case-control studies with errors in variables. *Biometrika* **84**, 523–537.

O'Hagan, A., Stevens, J.W. (2001). A framework for cost-effectiveness analysis from clinical trial data. *Health Econ.* **10**, 303–315.

O'Hagan, A., Stevens, J.W. (2002). Bayesian methods for design and analysis of cost-effectiveness trials in the evaluation of health care technologies. *Statist. Meth. Med. Res.* **11**, 469–490.

O'Hagan, A., Stevens, J.W., Montmartin, J. (2000). Inference for the C/E acceptability curve and C/E ratio. *Pharmacoeconomics* **17**, 339–349.

Parmigiani, G., Samsa, G.P., Ancukiewicz, M., Lipscomb, J., Hasselblad, V., Matchar, D.B. (1997). Assessing uncertainty in cost-effectiveness analyses: Application to a complex decision model. *Med. Dec. Mak.* **17**, 390–401.

Pinsky, P.F., Lorber, M.N. (1998). A model to evaluate past exposure to 2,3,7,8-TCDD. *J. Expos. Anal. Environ. Epi.* **8**, 187–206.

Prentice, R.L., Pyke, R. (1979). Logistic disease incidence models and case-control studies. *Biometrika* **66**, 403–411.

Raftery, A.E. (1995). Bayesian model selection in social research (with discussion). In: Marsden, P.V. (Ed.), *Sociological Methodology 1995*. Blackwell, Cambridge, MA.

Raftery, A.E., Madigan, D., Volinsky, C.T. (1995). Accounting for model uncertainty in survival analysis improves predictive performance. In: Bernardo, J.M., Berger, J.O., Dawid, A.P., Smith, A.F.M. (Eds.), *Bayesian Statistics, vol. 5*. Oxford University Press, Oxford.

Rice, K. (2003). Full-likelihood approaches to misclassification of a binary exposure in matched case-control studies. *Statist. Med.* **22**, 3177–3194.

Richardson, S., Leblond, L. (1997). Some comments on misspecification of priors in Bayesian modelling of measurement error problems. *Statist. Med.* **16**, 203–213.

Richardson, S., Leblond, L., Jaussent, I., Green, P.J. (2002). Mixture models in measurement error problems, with reference to epidemiological studies. *J. Roy. Statist. Soc., Ser. A* **165**, 549–566.

Rosenbaum, P.R. (2002). *Observational Studies*, second ed. Springer, New York.

Rosenbaum, P.R., Rubin, D.B. (1983). Assessing sensitivity to an unobserved binary covariate in an observational study with binary outcome. *J. Roy. Statist. Soc., Ser. B* **45**, 212–218.

Rothman, K.J., Greenland, S. (1998). *Modern Epidemiology*, second ed. Lippincott, Philadelphia.

Rubin, D.B. (1976). Inference and missing data. *Biometrika* **63**, 581–592.

Schlesselman, J.J. (1978). Assessing effects of confounding variables. *Amer. J. Epi.* **108**, 3–8.

Seaman, S.R., Richardson, S. (2001). Bayesian analysis of case–control studies with categorical covariates. *Biometrika* **88**, 1073–1088.

Sinha, D., Dey, D.K. (1997). Semiparametric Bayesian analysis of survival data. *J. Amer. Statist. Assoc.* **92**, 1195–1212.

Sinha, D., Ibrahim, J.G., Chen, M.-H. (2003). A Bayesian justification of Cox's partial likelihood. *Biometrika* **90**, 629–642.

Sinha, S., Mukherjee, B., Ghosh, M., Mallick, B.K., Carroll, R. (2005). Bayesian semiparametric analysis of matched case-control studies with missing exposure. *J. Amer. Statist. Assoc.* Submitted for publication.

Skene, A.M., Wakefield, J.C. (1990). Hierarchical models for multicenter binary response studies. *Statist. Med.* **9**, 919–929.

Spiegelhalter, D.J., Abrams, K.R., Myles, J.P. (2003). *Bayesian Approaches to Clinical Trials and Health-Care Evaluation.* Wiley, Chichester.

Stangl, D.K., Berry, D.A. (Eds.) (2000). *Meta-Analysis in Medicine and Health Policy.* Marcel Dekker, New York.

Stinnett, A.A., Mullahy, J. (1998). Net health benefits: A new framework for the analysis of uncertainty in cost-effectiveness analysis. *Med. Dec. Mak.* **18S**, 467–471.

Tweedie, R.L., Scott, D.J., Biggerstaff, B.J., Mengersen, K.L. (1996). Bayesian meta-analysis, with application to studies of ETS and lung cancer. *Lung Cancer* **14** (Suppl 1), S171–S194.

Van Hout, B.A., Al, M.J., Gordon, G.S., Rutten, F. (1994). Costs, effects and C/E ratios alongside a clinical trial. *Health Econ.* **3**, 309–319.

Viallefont, V., Raftery, A.E., Richardson, S. (2001). Variable selection and Bayesian model averaging in case-control studies. *Statist. Med.* **20**, 3215–3220.

Wakefield, J. (2004). Ecological inference for 2 $\times$ 2 tables. *J. Roy. Statist. Soc., Ser. A* **167**, 385–445.

Yanagawa, T. (1984). Case-control studies: Assessing the effect of a confounding factor. *Biometrika* **71**, 191–194.

Zhang, P. (1992). Inference after variable selection in linear regression models. *Biometrika* **79**, 741–746.

Handbook of Statistics, Vol. 25
ISSN: 0169-7161
© 2005 Elsevier B.V. All rights reserved.
DOI 10.1016/S0169-7161(05)25027-7

27

Bayesian Analysis of Case-Control Studies

Bhramar Mukherjee, Samiran Sinha and Malay Ghosh

Abstract

Case-control studies are dominant tools in epidemiologic research and consequently, there is a substantial volume of frequentist literature on different aspects of case-control studies. The Bayesian framework for case-control studies offer possibilities for flexible, hierarchical modeling that may be necessary in many contexts but as of yet, the Bayesian pathways remain relatively less explored. The present article reviews existing Bayesian work for analyzing case-control data, some recent advancements and possibilities for future research.

Keywords: conditional likelihood; Dirichlet process; hierarchical model; likelihood inference; logistic regression; Mantel–Haenszel; Markov chain Monte Carlo; matching; missing data; measurement error; odds ratio; prospective study; retrospective study; semiparametric efficiency

1. Introduction: The frequentist development

The contribution of statisticians to the development of case-control methodology is perhaps the most important contribution that they have made to public health and biomedicine. The central theme of a case-control study is to compare a group of subjects (cases) having the outcome (typically a disease) to a control group (not having the outcome or disease) with regard to one or more of the disease's potential risk factors. The method gained popularity in the 1920's for studying rare diseases, especially cancer, where following a healthy population or a cohort over time is impractical. It was believed in the early era of case-control studies that such a study did not provide relevant quantitative information about the disease rates. Cornfield (1951) demonstrated that the exposure odds ratio for cases versus controls equals the disease odds ratio for exposed versus unexposed, and the latter in turn approximates the ratio of the disease rates if the disease is rare. The following discussion will clarify the issue. Consider two dichotomous factors, D, denoting the disease status and E, representing the exposure status that characterize individuals in a population. The collected data is represented by a 2×2 table of the following form:

	D	$\overline{D}$
E	a	b
$\overline{E}$	c	d

One measure of association between D and E is the odds ratio (OR) defined as

$$OR = \frac{Pr(D|E)}{Pr(\overline{D}|E)} \cdot \frac{Pr(\overline{D}|\overline{E})}{Pr(D|\overline{E})}. \tag{1}$$

The disease odds ratio may be estimated from any of the three popular study designs namely, cohort, cross-sectional and case-control. It can be easily seen from (1) by using the Bayes theorem, namely, by using the fact that $Pr(E|D) = Pr(D|E)Pr(E)/Pr(D)$, that the exposure odds ratio and the disease odds ratio are the same. Another measure of association is the relative risk of the disease for different exposure values, defined as $RR = Pr(D|E)/Pr(D|\overline{E})$. When the disease is rare, both $Pr(\overline{D}|\overline{E})$ and $Pr(\overline{D}|E)$ are close to one and the exposure odds ratio approximates the relative risk of the disease (Cornfield, 1951).

The landmark paper by Mantel and Haenszel (1959) further clarified the relationship between a retrospective case-control study and a prospective cohort study. The paper considers a series of I 2×2 tables of the following pattern:

Disease status	Exposed	Not exposed	Total
Case	n_{11i}	n_{10i}	n_{1i}
Control	n_{01i}	n_{00i}	n_{0i}
Total	e_{1i}	e_{0i}	N_i

The Mantel–Haenszel (MH) estimator of the common odds ratio across the tables is

$$\hat{\theta}_{mh} = \frac{\sum_{i=1}^{I} n_{11i}n_{00i}/N_i}{\sum_{i=1}^{I} n_{01i}n_{10i}/N_i} \tag{2}$$

and the test statistic to test H_0: $\theta_1 = \cdots = \theta_I$ is $\sum_{i=1}^{I}\{n_{11i} - E(n_{11i}|\hat{\theta}_{mh})\}^2/\mathrm{Var}(n_{11i}|\hat{\theta}_{mh})$ which follows an approximate χ^2 distribution with $I - 1$ degrees of freedom under the null hypothesis. The derivation of the variance of the MH estimator posed a challenge, and was addressed in several subsequent papers (see Breslow, 1996, for details).

The next era saw development of likelihood based inference methods for the odds ratio (Breslow and Day, 1980). Methods to evaluate the simultaneous effects of multiple quantitative risk factors on disease rates started appearing in 1960's.

In a case-control study the appropriate likelihood is the 'retrospective' likelihood of exposure given the disease status. Cornfield et al. (1961) noted that if the exposure distributions in case and control populations are normal with different means but a common covariance matrix, then the prospective probability of disease (D) given the exposure ($\mathbf{Z}$) turns out to be the logistic response curve,

$$P(D = 1|\mathbf{Z} = z) = H(\alpha + \boldsymbol{\beta}^{\mathrm{T}} z), \tag{3}$$

where $H(u) = 1/\{1 + \exp(-u)\}$. Cox (1966), and Day and Kerridge (1967) confirmed that the maximum likelihood estimate of $\boldsymbol{\beta}$ obtained by the logistic regression model in (3) was efficient in a semiparametric sense. The only conceptual complication was in using a likelihood based on $P(D|\boldsymbol{Z})$ whereas the case-control sampling design naturally leads to a likelihood containing terms of the form $P(\boldsymbol{Z}|D)$. Anderson (1972) and Prentice and Pyke (1979) investigated this problem and showed that the ordinary logistic regression coefficients actually yielded the desired estimates of the relative risk parameters. They also established that the estimating equations based on the prospective logistic likelihood is unbiased and the asymptotic standard error of the resulting estimators of the relative risk parameters are identical with those obtained from the retrospective model. Carroll et al. (1995) extended this prospective formulation to the situation of missing data and measurement error in the exposure variable. They showed that the estimator of $\boldsymbol{\beta}$ obtained through prospective formulation and ignoring the retrospective nature of the case control data is consistent and asymptotically normally distributed with prospective standard errors that are at worst conservative.

In a case-control set-up, matching is often used for selecting "comparable" controls to eliminate bias due to confounding factors. Statistical techniques for analyzing matched case-control data were first developed in Breslow et al. (1978). In the simplest setting, the data consist of s matched sets and there are M_i controls matched with a case in each matched set or stratum. We denote the ith matched set by S_i, $i = 1, \ldots, s$. As before, one assumes a prospective stratified logistic disease incidence model, namely,

$$P(D = 1|z, S_i) = H\{\alpha_i + \boldsymbol{\beta}^{\mathrm{T}} z\}, \tag{4}$$

where α_i's are stratum-specific intercept terms. The stratum parameters α_i are eliminated by conditioning on the unordered set of exposures for the cases and controls in each stratum. This is equivalent to conditioning on the number of cases in each stratum which is a complete sufficient statistic for the nuisance parameters α_i. The generated conditional likelihood, is free of the nuisance parameters and yields the optimum estimating function (Godambe, 1976) for estimating $\boldsymbol{\beta}$. Assuming, without loss of generality, the first subject in each stratum is a case and rest of the subjects are controls, the derived conditional likelihood is

$$L_c(\boldsymbol{\beta}) = \prod_{i=1}^{s} \frac{\exp(\boldsymbol{\beta}^{\mathrm{T}} z_{i1})}{\sum_{j=1}^{M_i+1} \exp(\boldsymbol{\beta}^{\mathrm{T}} z_{ij})}. \tag{5}$$

The above method is known as conditional logistic regression (CLR).

The usual prospective likelihood for unmatched data and conditional logistic regression for matched studies produce biased estimates of the relative risk parameters in the presence of measurement error or misclassification in the exposure variable. Armstrong et al. (1989) proposed a correction to the estimates obtained through the classical methods using normal discriminant analysis with normal differential measurement error (to be made precise in Section 2). Carroll et al. (1993) proposed a pseudo-likelihood method to handle this situation, where they assume a differential measurement error density with a finite set of parameters, and empirical distribution functions are used in place of needed distributions. Recently, McShane et al. (2001) proposed a conditional score method for estimating bias-corrected estimates of log odds ratio parameters in matched case control studies.

The classical methods for analyzing unmatched and matched studies suffer from loss of efficiency when the exposure variable is partially missing. Lipsitz et al. (1998) proposed a pseudo-likelihood method to handle missing exposure variable. Rathouz et al. (2002) developed a more efficient semiparametric method of estimation in presence of missing exposure in matched case control studies. Satten and Kupper (1993), Paik and Sacco (2000) and Satten and Carroll (2000) addressed the missingness problem from a full likelihood approach assuming a distribution of the exposure variable in the control population.

The above account of the frequentist development of case-control analysis is nowhere near complete and is beyond the scope of the current article. Here we only attempt to review the Bayesian contributions in this field. In spite of the enormously rich literature in the frequentist domain, Bayesian modeling for case-control studies did not start until the late 1980's. Bayesian analysis of case-control data seem particularly appealing with the rapid development of Markov chain Monte Carlo techniques. The possibilities include random effects, measurement error, missingness, flexibility to incorporate hierarchical structure and prior information in modeling the relative risk. In fact, epidemiology is indeed a science where accumulated knowledge from similar prior studies should be utilized in the analysis of a new one. In the subsequent three sections we attempt to review the available Bayesian literature on modeling data arising from case-control studies. Sections 2 and 3 focus on models for unmatched case-control data. In Section 2, we consider models for a single binary exposure. Section 3 contains a review of more involved modeling of continuous exposure with measurement error and categorical covariates. Section 4 is devoted to methods available for the analysis of matched case-control studies. Section 5 focuses on some recent work on equivalence of prospective and retrospective studies in a Bayesian framework and discusses a new Bayesian interpretation to the method of conditional likelihood. Section 6 presents concluding remarks and possible directions for future research.

2. Early Bayesian work on a single binary exposure

Altham (1971) is possibly the first Bayesian work which considered several 2×2 contingency tables with a common odds ratio, and performed a Bayesian test of association based on this common odds ratio. This paper is not specifically targeted towards a case-control design. Later, Zelen and Parker (1986), Nurminen and Mutanen (1987), Marshall (1988) and Ashby et al. (1993) all considered identical Bayesian formulations of a case-control model with a single binary exposure X. The model can be described as follows:

Let ϕ and γ be the probabilities of exposure in control and case populations respectively. The retrospective likelihood is

$$l(\phi, \gamma) \propto \phi^{n_{01}}(1-\phi)^{n_{00}}\gamma^{n_{11}}(1-\gamma)^{n_{10}}. \tag{6}$$

Where n_{01} and n_{00} are the number of exposed and unexposed observations in control population, whereas n_{11} and n_{10} denotes the same for case population.

Independent conjugate prior distributions for ϕ and γ are assumed to be $Beta(u_1, u_2)$ and $Beta(v_1, v_2)$ respectively. After reparametrization one obtains the posterior distribution of the log odds ratio parameter, namely, $\beta = \log\{\gamma(1-\phi)/\phi(1-\gamma)\}$ as

$$p(\beta|n_{11}, n_{10}, n_{01}, n_{00}) \propto \exp\{(n_{11}+v_1)\beta\} \int_0^1 \frac{\phi^{n_{11}+n_{01}+v_1+u_2-1}(1-\phi)^{n_{10}+n_{00}+v_2+u_1-1}}{\{1-\phi+\phi\exp(\beta)\}^{n_{11}+n_{10}+v_1+v_2}} \,\mathrm{d}\phi. \quad (7)$$

The above posterior density of β does not exist in closed form but may be evaluated by numerical integration.

Since interest often lies in the hypothesis $\beta = 0$, Zelen and Parker (1986) recommended calculating the ratio of the two posterior probabilities $p(\beta)/p(0)$ at selected deviates β. When β is set at the posterior mode, a large value of this ratio will indicate concentration of the posterior away from 0 and one would infer disease-exposure association. However the "critical value" suggested for this ratio to detect presence of association is completely arbitrary. Zelen and Parker also provided a normal approximation to the posterior distribution of β to avoid numerical computation, and discussed the problem of choosing a prior distribution based on prior data available on exposure values. Using their prior elicitation method and normal approximation to the posterior, they analyzed data showing association between exposure in utero to Diesthylstibestrol (a drug which came in 1940's to prevent pregnancy complications) and adenocarcinoma of vagina in young women. The ratio of the posterior density at the posterior mode and at 0 turn out to be overwhelmingly large at 302, suggesting strong evidence of association between disease and exposure. This paper also raised the possibility of performing a case-control analysis without the presence of control sample if adequate exposure information is available from prior studies.

Nurminen and Mutanen (1987) considered a more general parameterization based on a general comparative parameter such as the risk ratio or the risk difference, and not just the odds ratio. They provided a complicated exact formula for the cumulative density function of this general comparative parameter. The formula can be related to Fisher's exact test for comparing two proportions in sampling theory. The Bayesian point estimates are considered as posterior median and mode, whereas inference is based on the highest posterior density interval for the comparative parameter of interest.

Marshall (1988) provided a closed form expression for the moments of the posterior distribution of the odds ratio. The paper mentions that an approximation to the exact posterior density of the odds ratio parameter can be obtained by power series expansion of the hypergeometric functions involved in the expression for the density, but acknowledge the problem of slow convergence in adopting this method. Marshall then used Lindley (1964) result on the approximate normality of $\log(OR)$ which works very well over a wide range of situations. In the absence of any exposure information, he recommended using uniform independent priors for the parameters. The paper suggests that a perception about the value of the odds ratio should guide the choice of prior parameters rather than attempting to exploit the exposure proportions as suggested in Zelen and Parker. Inference again is based on posterior credible intervals. Marshall (1988) reanalyzed the Zelen–Parker data on association of vaginal cancer and Diesthylstibestrol (DES). With the same prior as Zelen and Parker, the posterior credible interval for the odds ratio turned out to be (7.8, 388). Since this was a dataset with very few observations, the author chose a more conservative set of priors which also suggested association between vaginal cancer and DES. The paper concludes that the exact magnitude of the risk may not be quantified realistically based on such a small dataset.

Ashby et al. (1993) analyzed a case-control to assess risk of leukemia following chemotherapy for Hodgkin's disease from a Bayesian perspective and used it as a source of prior information for a second study. They also proposed modeling the odds ratio directly through a normal model besides the usual beta-binomial modeling. The paper emphasized practical relevance of Bayesian perspective from a epidemiological standpoint. Epidemiology progresses by an accumulation of evidence about the existence and magnitude of risk factors; so a Bayesian approach offers a natural framework for integrating and updating the knowledge available at each stage.

In cancers for which survival is good, there is a growing interest in the long-term effects of treatment. Hodgkin's disease, which is cancer of the lymph nodes, has a 5 year survival of about 80% and survivors have excess risks for leukemias and lymphomas. Although these risks are very small, compared to the benefits of the treatment, there is interest in studying these effects. The paper considers three such studies: an initial cohort study (Kaldor et al., 1987), an international case-control study conducted in 1990 and a case-control study finished in 1993 as a follow-up to the 1990 study with the same protocol as the earlier study. The data for the 1990 case-control study is analyzed with no prior information, slightly informative prior and then using the 1987 cohort study as prior evidence. Then a fraction of data collected in a follow-up case-control study is analyzed with the initial 1990 study as prior information. With small amount of data, prior information plays a vital role. Interesting numerical differences may be noted for different choices of the prior. The paper makes a very clear point that before embarking on a study of a possible etiological risk for any disease, the investigators may consider existing evidence across a range of sources for arriving at a more efficient strategy for design and analysis of the data following Bayesian principles.

3. Models with continuous and categorical exposure

Müller and Roeder (1997) introduced a novel aspect to Bayesian treatment of case-control studies by considering continuous exposure with measurement error. This paper is a significant advance in Bayesian analysis of case-control data as it is the first attempt to address nontrivial, unorthodox modeling scenarios beyond the earlier work on binary exposure as discussed in Section 2.

Müller and Roeder used the validation group, for which error-free high quality measurements are available to form a model that indirectly links the error-prone measurement to the disease outcome through its relation to the error-free measurement. The set-up of their paper is as follows:

Let D be a binary response observed together with covariates X and Z, where Z may be vector-valued. The retrospective case-control sample is chosen by selecting n_1 cases with $D = 1$ and n_0 controls with $D = 0$. Errors in variables occur when for a subset of the data (referred as reduced data or $\mathcal{R}$ from now on) a proxy W is measured instead of the true covariate X. For a typical validation study, for the remaining subjects (referred to as complete data or $\mathcal{C}$ from now on) one has observations on both the gold standard X and the proxy W.

The approach proposed in their paper is based on a nonparametric model for the retrospective likelihood of the covariates and the imprecisely measured exposure. For reduced data, this nonparametric distribution models the joint distribution of (Z, W) and the missing covariate X_R. For complete data, the same distribution models the joint distribution of the observed covariates (X_C, Z, W). The nonparametric distribution is a class of flexible mixture distributions, obtained by using mixture of normal models with a Dirichlet Process prior on the mixing measure (Escober and West, 1995). The prospective disease model relating disease to exposure is taken as a logistic distribution characterized by a vector of log odds-ratio parameters $\boldsymbol{\beta}$. Also, if $\boldsymbol{\theta}$ denotes the parameters describing the marginal distributions of the exposure and covariates, under certain conditional independence type assumptions and nondifferential measurement error, i.e. $p(D|X, Z, W, \boldsymbol{\beta}) = p(D|X, Z, \boldsymbol{\beta})$, the retrospective likelihood is shown to be compatible with the prospective model.

With a prior $p(\boldsymbol{\beta}, \boldsymbol{\theta})$, one obtains the joint posterior as

$$p(\boldsymbol{\beta}, \boldsymbol{\theta}|X_C, W, D, Z) \propto p(\boldsymbol{\beta}, \boldsymbol{\theta}) \prod_{i \in \mathcal{C}} p(X_i, Z_i, W_i|D_i, \boldsymbol{\beta}, \boldsymbol{\theta}) \times \prod_{i \in \mathcal{R}} \left\{ \int p(X_i, Z_i, W_i|D_i, \boldsymbol{\beta}, \boldsymbol{\theta})\, \mathrm{d}X_i \right\}.$$

By augmenting the parameter with the latent vector one can write the joint posterior of parameters and missing exposure X_R as,

$$p(\boldsymbol{\beta}, \boldsymbol{\theta}, X_R|X_C, W, D, Z) \propto p(\boldsymbol{\beta}, \boldsymbol{\theta}) \prod_{i=1}^{n} p(X_i, Z_i, W_i|D_i, \boldsymbol{\beta}, \boldsymbol{\theta}).$$

Under the additional assumption of nondifferential measurement error,

$$p(\boldsymbol{\beta}, \boldsymbol{\theta}, X_R|X_C, W, D, Z) \propto p(\boldsymbol{\beta}, \boldsymbol{\theta}) \prod_{i=1}^{n} p(X_i, Z_i, W_i|\boldsymbol{\theta}) p(D_i|X_i, Z_i, \boldsymbol{\beta}) / p(D_i|\boldsymbol{\beta}, \boldsymbol{\theta}),$$

where $p(D_i|\boldsymbol{\beta}, \boldsymbol{\theta}) = \int p(D_i, X_i, Z_i, W_i, \boldsymbol{\beta})\, \mathrm{d}P(X_i, Z_i, W_i|\boldsymbol{\theta})$. Let $\boldsymbol{\theta} = (\theta_1, \theta_2, \ldots, \theta_n)$. Then one assumes the following mixture model with a multivariate normal kernel ϕ_{θ_i}. Specifically.

$$p(X_i, Z_i, W_i|\theta_i) = \phi_{\theta_i}(X_i, Z_i, W_i), \tag{8}$$

$$\theta_i \sim G,$$

$$G \sim DP(\alpha G_0).$$

Where $\theta_i = (\mu_i, \Sigma_i)$, the parameters of the trivariate normal kernel and $DP(\alpha G_0)$ denotes a Dirichlet process with a base measure G_0 and concentration parameter α. The hierarchy is completed by assuming a normal/Wishart hyperprior on the parameters of the base measure $G_0 = N(\mu_0, \Sigma_0)$ and a Gamma prior on the concentration parameter α. A noninformative prior is assumed on $\boldsymbol{\beta}$. A Markov chain Monte Carlo numerical integration scheme is designed for estimating the parameters. The computation scheme becomes either very expensive or inaccurate when the dimension of the (X, Z, W)-space increase as one has to evaluate $p(D_i|\boldsymbol{\beta}, \boldsymbol{\theta})$ by numerical integration over the (X, Z, W)-space. This paper breaks many new grounds in Bayesian treatment of case-control studies including consideration of continuous covariates, measurement error and flexible nonparametric modeling for the exposure distribution leading to robust, interpretable results quantifying the effect of exposure. This paper brings out the tremendous possibility of using modern Bayesian computational techniques to offer solutions to complex data scenarios in case-control studies.

Müller and Roeder analyzed a dataset extracted from Lipids Research Clinic's prospective study to estimate the risk of coronary heart disease given levels of LDL cholesterol and triglycerides (TRG). An individual is classified as a case if he has experienced a previous heart attack, an abnormal exercise electrocardiogram or a history of angina pectoris. Since direct measurement of LDL is expensive, total cholesterol (TC) is often used as a proxy. In their set-up, onset of coronary disease, log(TC), log(LDL) and log(TRG) play the roles of D, W, X and Z respectively. Based on their analysis with and without the reduced observations, it is clear that the reduced observations are highly informative and TC is an excellent proxy for LDL. The posterior distribution for the log-odds ratio parameter β_1 corresponding to X (LDL) has a much smaller variance when the reduced data is used.

Müller and Roeder pointed out that categorical covariates required a different treatment as assuming a multivariate normal kernel for the Dirichlet Process implicitly assumes continuous exposures. Seaman and Richardson (2001) extended the binary exposure model of Zelen and Parker to any number of categorical exposures simply by replacing the binomial likelihood in (6) by a multinomial likelihood, and then adopting a MCMC strategy to estimate the log odds ratio for the disease in each category with respect to a baseline category. The set of multinomial probabilities corresponding to exposure categories in case and control populations are assumed to have a discrete Dirichlet prior. The prior on the log odds ratios could be chosen to be any distribution.

Seaman and Richardson also adapted the Müller–Roeder approach to categorical covariates by simply modeling the multinomial probabilities corresponding to exposure categories in the entire source population as opposed to looking at case and control populations separately as in the generalized Zelen–Parker binary model described above. A noninformative Dirichlet $(0, 0, \ldots, 0)$ prior is assumed for the exposure probabilities. Seaman and Richardson establish the equivalence of the generalized binary and adapted Müller–Roeder in some limiting cases with noninformative Dirichlet $(0, 0, \ldots, 0)$ prior on ϕ.

Continuous covariates may be treated in the Seaman and Richardson framework by discretizing them into groups and little information is lost if discretization is sufficiently fine. This opens up the possibility of treating continuous and categorical covariates simultaneously which is not obvious in the Müller–Roeder approach.

Seaman and Richardson as well as Müller and Roeder assumed a prospective logistic likelihood, assumed a flexible prior for the exposure distribution, and derived the implied retrospective likelihood. In contrast, Müller et al. (1999) specified a retrospective likelihood, and then derived the implied prospective likelihood. Müller et al. (1999) specified both $p(\boldsymbol{X}|D = 0, \boldsymbol{\theta})$ and $p(\boldsymbol{X}|D = 1, \boldsymbol{\theta})$ as mixtures of multivariate normal distributions with M components. First we consider the simple normal case and begin with the notation, $p(\boldsymbol{X}|D, \boldsymbol{\theta}_D) = N(\boldsymbol{\mu}_D, \boldsymbol{\Sigma}_D)$, where $\boldsymbol{\theta}_D = (\boldsymbol{\mu}_D, \boldsymbol{\Sigma}_D)$ for $D = 0$ and 1; so $\boldsymbol{\theta}_0$ and $\boldsymbol{\theta}_1$ are parameters of the exposure distributions in the control and case populations respectively. The two mixtures of multivariate distributions are denoted as $p(\boldsymbol{X}|D, \boldsymbol{\theta}_D) = \sum_{m=0}^{M} \pi_{Dm} N(\boldsymbol{\mu}_{Dm}, \boldsymbol{\Sigma}_{Dm})$ where $\boldsymbol{\theta}_D = (\pi_{D0}, \boldsymbol{\mu}_{D0}, \boldsymbol{\Sigma}_{D0}, \ldots, \pi_{DM}, \boldsymbol{\mu}_{DM}, \boldsymbol{\Sigma}_{DM})$ for $D = 0$ and 1. The authors assumed $\gamma = P(D = 1)$, a constant prevalence of the disease. With this retrospective formulation, the following can be shown:

RESULT. (i) In the simple normal case, i.e., when $\boldsymbol{\mu}_{Dm} = \boldsymbol{\mu}_D$ and $\boldsymbol{\Sigma}_{Dm} = \boldsymbol{\Sigma}_D$ for $D = 0, 1$, the prospective log odds have a quadratic form $\eta_p(\boldsymbol{X}) = \boldsymbol{X}'\boldsymbol{A}\boldsymbol{X} + \boldsymbol{X}'b + c$, with $2A = \boldsymbol{\Sigma}_0^{-1} - \boldsymbol{\Sigma}_1^{-1}$, $b = \boldsymbol{\Sigma}_1^{-1}\boldsymbol{\mu}_1 - \boldsymbol{\Sigma}_0^{-1}\boldsymbol{\mu}_0$ and $2c = \boldsymbol{\mu}_0'\Sigma_0^{-1}\boldsymbol{\mu}_0 - \boldsymbol{\mu}_1'\Sigma_1^{-1}\boldsymbol{\mu}_1 + 2\log\{\gamma/(1-\gamma)\} + \log(|\Sigma_0|/|\Sigma_1|)$.

(ii) In the simple normal model, if in addition $\boldsymbol{\Sigma}_1 = \boldsymbol{\Sigma}_0$, then the prospective model is the logistic with linear predictor $\eta_p(\boldsymbol{X}) = \boldsymbol{X}'b + c$.

(iii) In the mixture of normals model, the prospective log odds are of the form $\eta_p(\boldsymbol{X}) = \log\{\sum_{m=0}^{M} \pi_{1m}^* \exp(\boldsymbol{X}'\boldsymbol{A}_{1m}\boldsymbol{X} + \boldsymbol{X}'\boldsymbol{b}_{1m} + c_{1m})\} - \log\{\sum_{m=0}^{M} \pi_{0m}^* \times \exp(\boldsymbol{X}'\boldsymbol{A}_{0m}\boldsymbol{X} + \boldsymbol{X}'\boldsymbol{b}_{0m} + c_{0m})\} + \log\{\gamma/(1-\gamma)\}$, where $\pi_{Dm}^* = \pi_{Dm}|\boldsymbol{\Sigma}_{Dm}|^{-1/2}$, $\boldsymbol{A}_{Dm} = -\frac{1}{2}\boldsymbol{\Sigma}_{Dm}^{-1}$, $\boldsymbol{b}_{Dm} = \boldsymbol{\Sigma}_{Dm}^{-1}\boldsymbol{\mu}_{Dm}$ and $c_{Dm} = -\frac{1}{2}\boldsymbol{\mu}_{Dm}^t\boldsymbol{\Sigma}_{Dm}^{-1}\boldsymbol{\mu}_{Dm}$ for $D = 0, 1$.

Result (ii) reiterates the classical result of Cornfield et al. (1961) about discriminating the exposure distributions in case and control populations using the prospective logistic model when the exposures have a multivariate normal distribution with different means but common covariance matrix in the case and control populations. Result (iii) is a generalization which states that with the mixture of M normals retrospective model, the prospective odds ratio turn out to be a weighted average of M quadratic regressions in the log odds which becomes a weighted average of M linear regressions when the case and control distributions have the same dispersion parameter. Their example shows that nonlinear modeling of the odds ratio as described in (iii) may be needed in some real situations. The choice of M, the number of components in the mixture model is guided by two competing principles, easier estimation and flexibility of modeling. In their example, Müller et al. use $M = 3$.

Müller et al. (1999) also addressed the problem of handling categorical and quantitative exposure simultaneously. Conditional on the quantitative covariates, a multivariate probit model is assumed for the binary covariates. They mentioned that this model could be generalized for general categorical covariates as well, but did not specify exactly how this might be done. They developed MCMC scheme for implementation of this model and for inference and prediction. The computational difficulty of evaluating the normalizing constant in using the prospective model as seen in Müller–Roeder is circumvented here through direct modeling of the retrospective likelihood. Two other attractive features of this paper are (1) to incorporate inter-center variability while combining data from eight tumor registries and (2) provide a natural framework of combining data coming from case-control and cohort studies. For the latter, the likelihood is based on both prospectively and retrospectively collected individuals. The risk factors for a retrospectively ascertained individual enter the likelihood conditionally on whether the person is a case or a control. For a prospectively ascertained individual, the disease outcome and individual risk factors are modeled jointly in terms of $p(\boldsymbol{X}, D|\boldsymbol{\theta}, \gamma)$ where $\gamma = p(D = 1)$ is the disease prevalence. Let P be the set of indices corresponding to all prospectively ascertained patients. Then the combined likelihood can be written as:

$$L(\boldsymbol{\theta}, \gamma) = \prod_{D_i=1} p(\boldsymbol{X}_i|D = 1, \boldsymbol{\theta}) \prod_{D_i=0} p(\boldsymbol{X}_i|D = 0, \boldsymbol{\theta}) \prod_{i \in P} \gamma^{D_i}(1-\gamma)^{1-D_i}.$$

The example using the Cancer and Steroid Hormone (CASH) study indicates that the two commonly adopted assumptions regarding the linearity of the log odds and homogeneity of log odds ratio across studies need not always be true. Models based on nonlinear and nonhomogeneous odds ratios may lead to a more complete account of uncertainties than the usual logistic regression model.

Recently, Gustafson et al. (2002) considered another Bayesian approach to handle exposure variables that are measured imprecisely. They considered approximation of the imprecise exposure distribution by a discrete distribution supported on a suitably chosen grid. In the absence of measurement error, the support is chosen as the set of observed values of the exposure, a device that resembles the Bayesian Bootstrap (Rubin, 1981). In the presence of measurement error, these authors proposed strategies for selection of these grid points depending on the validation design. A Dirichlet$(1, \dots, 1)$ prior was assigned to the probability vector corresponding to the set of grid points. The posterior distributions of the parameters of interest were derived directly from the retrospective likelihood of data and design, and a MCMC computing scheme was used for carrying out posterior inference. Interestingly, in the process of devising the MCMC algorithm, they showed that the posterior distribution as derived from a prospective likelihood can be used as an approximation to the posterior distribution directly derived from the retrospective likelihood. They also proposed an importance sampling scheme based directly on the posterior derived from retrospective data. It is observed that if the importance weights exhibit little variation, then the inference based on the prospective model suffices. In the subsequent data analysis, very small differences are noted between prospective and retrospective models for moderate sample sizes and the authors continue to work with the prospective model which is more amenable to the MCMC analysis. This paper is the first one to shed some light on the interplay between the equivalence of prospective and retrospective studies in a Bayesian perspective which we continue to discuss in Section 5. The equivalence established in Gustafson et al. (2002) is an approximate equivalence whereas later Seaman and Richardson (2004) established exact equivalence results (see Section 5).

4. Analysis of matched case-control studies

All the articles discussed so far in the Bayesian domain considered unmatched case-control studies. One requirement for a case-control study is that the selection of cases and controls is independent of their exposure status. In the simplest form of study designs, cases and controls are identified and information on exposure and relevant confounder variables (age, sex, race) are determined on the sampled subjects. The subsequent analysis tries to adjust for the confounding effects by including the additional terms in the model or by doing a stratified Mantel–Haenszel analysis. The design could potentially be somewhat inefficient if large number of controls fall within confounder-defined strata which contain only a few cases. In such a situation the control information is effectively wasted. Matched case-control designs try to circumvent this problem by keeping the ratio of cases to controls constant within each sub-group defined by the relevant confounders, called the *matching variables*. Frequency matching corresponds to maintaining constant case-control ratios across broad strata whereas individual matching selects a set of matched controls for each sampled case. As discussed in Section 1, frequentist inference in the matched set-up proceeds via conditional likelihood. However, Bayesian attempts to analyze a matched case-control study have started appearing only very recently.

Diggle et al. (2000) present the first Bayesian analysis for situations when cases are individually matched to the controls and as a result, nuisance parameters are introduced to represent the separate effect of matching in each matched set. Diggle et al. (2000) consider matched data when exposure of primary interest is defined by the spatial location of an individual relative to a point or line source of pollution.

The basic model of Diggle et al. (2000) for an unmatched set-up is as follows: Let the locations of individuals in the population at risk follow an nonhomogeneous Poisson process with intensity function $g(x)$ and the probability that an individual at location x becomes a case is $p^*(x)$. Then the odds of disease amongst individuals sampled at location x are:

$$r(x) = (a/b)p^*(x)/\big(1 - p^*(x)\big),$$

where a and b are the sampled proportions of cases and controls. Typically, $r(x) \equiv r(x, \theta)$ involves some unknown parameter θ, and is modeled as

$$r(x, \theta) = \rho h(x, \theta),$$

where $h(x, \theta)$ involves interpretable parameters of interest θ and ρ is a nuisance parameter reflecting background odds of disease. The extension of this model to J individually matched case-control pairs in a study region is made in the following way. Suppose, the J locations for cases and controls are denoted by x_{0j} and x_{1j} respectively, $j = 1, \ldots, J$. The probability of disease for an individual at location x in stratum j is modeled as:

$$p_j(x, \theta) = \frac{r_j(x, \theta)}{1 + r_j(x, \theta)} = \frac{\rho_j h(x, \theta)}{1 + \rho_j h(x, \theta)},$$

where ρ_j are varying baseline odds between matched pairs and are considered as nuisance parameters. Conditioning on the unordered set of exposures within each matched set, one has the conditional probability of disease of an individual at distance x in stratum j as,

$$p_c(x_{j0}, \theta) = \frac{h(x_{j0}, \theta)}{h(x_{j0}, \theta) + h(x_{j1}, \theta)}.$$

In case of $1:M$ matching in each stratum and in presence of q additional covariates $z_k(x_{ji})$ measured for the ith individual in jth stratum at location x_{ji}, $i = 1, \ldots, M$; $j = 1, \ldots, J$; $k = 1, \ldots, q$, the full fledged conditional probability can be modeled in the form:

$$p_c(x_{j0}, \theta, \phi) = \frac{h(x_{j0}, \theta) \exp\bigl(\sum_{k=1}^{q} z_k(x_{j0})\phi_k\bigr)}{\sum_{i=0}^{M} h(x_{ji}, \theta) \exp\bigl(\sum_{k=1}^{q} z_k(x_{ji})\phi_k\bigr)},$$

with the conditional log likelihood being of the form:

$$L(\theta, \phi) = \sum_{j=1}^{J} \log p_c(x_{j0}, \theta, \phi).$$

Diggle et al. (2000) considered the likelihood as well as Bayesian approach for estimation of parameters and inference in this model with specified form of $h(x, \theta)$. Often in studying environmental pollution, there is a putative point source at location x^* and interest is on how the risk surface changes in relation to x^*. Let $d = \|x - x^*\|$ be the distance of location x from the source. In such instances they propose h of the form

$$h(x) = 1 + \alpha \exp\bigl(-(d/\lambda)^2\bigr), \quad \text{for fixed } x^*.$$

This model has a natural interpretation with α being the proportional increase in disease odds at the source, while λ measures the rate of decay with increasing distance from the source in units of distance. Standard likelihood inference for such models based on the likelihood ratio statistic may not be reliable as the likelihood involved is highly nonregular. The Bayesian paradigm provides an alternative when likelihood surface is highly irregular. In the paper independent normal priors are assumed on the covariate related regression parameters ϕ and independent uniform priors are chosen for α and λ. Estimation is carried out by adopting a MCMC technique and inference is conducted by computing the Bayes factor for testing the null hypothesis $h(x) = 1$ which amounts to saying that the risk of the disease does not depend on the distance from the putative source.

Diggle et al. analyzed two sets of data, collected to study the relationship between proximity of residence to main roads and increased risk of respiratory disease. The first dataset consisted of 125 asthma cases, and the second 124 chronic obstructive airways disease (COAD). Cases were clear-cut diagnoses attending the casualty department at the Royal London Hospitals in 1991–92. Controls were selected from non-chest related admissions. Each case was individually matched to a single control using sex, race and consultant team. The analysis was based on the distance odds model as described above. For the Asthma data, no association could be detected between asthma and distance of residence to main roads. The COAD data exhibited a reasonable evidence of elevated risk close to main roads, but the functional form of the variation in risk with distance could not be estimated with precision. The paper indicates introducing a stochastic spatial component in the model which could be an interesting direction for further research. This paper indicates using conditional likelihood as a basis for Bayesian inference and what assumptions are implicitly made regarding the choice of priors when such an analysis is carried out. This issue is investigated in detail by Rice (2004) as discussed in Section 5.

Ghosh and Chen (2002) developed general Bayesian inferential techniques for matched case-control problems in the presence of one or more binary exposure variables. The model considered was more general than that of Zelen and Parker (1986). Also, unlike Diggle et al. (2000), the analysis was based on an unconditional likelihood rather than a conditional likelihood after elimination of nuisance parameters. Any methodology based on the conditional likelihood is applicable only for the logit link. The general Bayesian methodology based on the full likelihood as proposed by Ghosh and Chen worked beyond the logit link. Their procedure included not only the probit and the complementary log links but also some new symmetric as well as skewed links. The propriety of posteriors was proved under a very general class of priors which need not always be proper. Also, some robust priors such as multivariate t-priors were used. The Bayesian procedure was implemented via Markov chain Monte Carlo.

To illustrate their general methodology, Ghosh and Chen considered the L.A. Cancer data as given in Breslow and Day (1980), and studied the effect of gall-bladder disease on the risk of endometrial cancer as a single exposure variable, and both gall-bladder disease and hypertension as multiple exposure variables. In the case of one case and one control, it was demonstrated that the Bayesian procedure could yield conclusions different from the conditional likelihood regarding association between the exposures and the disease. This is because, while the conditional likelihood ignores all concordant pairs, the Bayesian procedure based on the full likelihood utilizes these pairs as well.

Rice (2003) considered a matched case-control analysis where a binary exposure is potentially misclassified. His method can be interpreted in a Bayesian framework with prior information incorporated on odds ratios, misclassification rates and other model parameters. Prescott and Garthwaite (2005) also consider the problem of analyzing matched case-control data with misclassified binary exposure.

Recently Sinha et al. (2004, 2005a, 2005b) have proposed a unified Bayesian framework for matched case-control studies with missing exposure and a semiparametric alternative for modeling varying stratum effects on the exposure distribution. They considered s, $1:M$ matched strata with a disease indicator D, a vector of completely observed covariate $\mathbf{Z}$ and an exposure variable X with possible missing values. The missingness is assumed to be at random (MAR in the sense of Little and Rubin, 1987). They assume a prospective model for the disease status as,

$$P(D_{ij} = 1 | X_{ij}, \mathbf{Z}_{ij}, S_i) = H\big(\beta_0(S_i) + \boldsymbol{\beta}_1^{\mathrm{T}} \mathbf{Z}_{ij} + \beta_2 X_{ij}\big),$$

where the index ij refers to the jth individual in the ith stratum S_i, $j = 1, \ldots, M+1$ and $i = 1, \ldots, s$. The distribution of the exposure variable X in a control population is assumed to be a member of general exponential family, namely

$$p(X_{ij}|\mathbf{Z}_{ij}, S_i, D_{ij} = 0) = \exp\big[\xi_{ij}\big\{\theta_{ij}X_{ij} - b(\theta_{ij})\big\} + c(\xi_{ij}, X_{ij})\big], \tag{9}$$

where the canonical parameter θ_{ij} is modeled in terms of the completely observed covariate $\mathbf{Z}_{ij}$ and stratum specific varying intercept terms, namely $\theta_{ij} = \gamma_{0i} + \boldsymbol{\gamma}^{\mathrm{T}}\mathbf{Z}_{ij}$. They propose two important lemmas which are useful to write down the likelihood.

LEMMA 1. $\frac{\mathrm{pr}(D_{ij}=1|\mathbf{Z}_{ij},S_i)}{\mathrm{pr}(D_{ij}=0|\mathbf{Z}_{ij},S_i)} = \exp[\beta_0(S_i) + \boldsymbol{\beta}_1^{\mathrm{T}}\mathbf{Z}_{ij} + \xi_{ij}\{b(\theta_{ij}^*) - b(\theta_{ij})\}]$, *where* $\theta_{ij}^* = \theta_{ij} + \xi_{ij}^{-1}\beta_2$.

LEMMA 2. *Based on the above two model assumptions, one can derive the distribution of the exposure variable in the case population as*:

$$p(X_{ij}|D_{ij} = 1, \mathbf{Z}_{ij}, S_i) = \exp\big[\xi_{ij}\big\{\theta_{ij}^*X_{ij} - b\big(\theta_{ij}^*\big)\big\} + c(X_{ij}, \xi_{ij})\big]. \tag{10}$$

Using the above two lemmas and assuming data missing at random, the prospective joint conditional likelihood of the disease status and the exposure variable given the completely observed covariate, the stratum effect and the conditional event that there is exactly one case in each stratum can be shown to be proportional to,

$$\begin{aligned}
&\prod_{i=1}^{S} P\Bigg(D_i = 1, D_{i2} = \cdots D_{iM+1} = 0, \{X_{ij}, \Delta_{ij}\}_{j=1}^{M+1}|S_i, \{\mathbf{Z}_{ij}\}_{j=1}^{M+1}, \\
&\qquad \sum_{j=1}^{M+1} D_{ij} = 1\Bigg) \\
&\quad\propto \prod_{i=1}^{S}\Bigg\{P\Bigg(D_{i1} = 1, D_{i2} = \cdots = D_{iM+1} = 0|\{\mathbf{Z}_{ij}\}_{j=1}^{M+1}\sum_{j=1}^{M+1} D_{ij} = 1, S_i\Bigg) \\
&\qquad \times p^{\Delta_{i1}}(X_{i1}|\mathbf{Z}_{i1}, D_{i1} = 1, S_i) \times \prod_{j=2}^{M+1} p^{\Delta_{ij}}(X_{ij}|\mathbf{Z}_{ij}, D_{ij} = 0, S_i)\Bigg\} \\
&\quad= \prod_{i=1}^{S}\Bigg\{\frac{P(D_{i1} = 1|\mathbf{Z}_{i1}, S_i)/P(D_{i1} = 0|\mathbf{Z}_{i1}, S_i)}{\sum_{j=1}^{M+1} P(D_{ij} = 1|\mathbf{Z}_{ij}, S_i)/P(D_{ij} = 0|\mathbf{Z}_{ij}, S_i)} \\
&\qquad \times p^{\Delta_{i1}}(X_{i1}|\mathbf{Z}_{i1}, D_{i1} = 1, S_i) \times \prod_{j=2}^{M+1} p^{\Delta_{ij}}(X_{ij}|\mathbf{Z}_{ij}, D_{ij} = 0, S_i)\Bigg\}.
\end{aligned}$$

In the above expression it is assumed that the first observation in each stratum is the one coming from case population, Δ_{ij} is the missing value indicator and takes value 0 if X_{ij} is missing and 1 otherwise.

The parameters were estimated in a Bayesian framework by using a nonparametric Dirichlet Process prior on the stratum specific effects γ_{0i} in the distribution of the exposure variable and parametric priors for all other parameters. Since the number of these stratum specific parameters grow with the sample size, estimating these effects from a conditional MLE perspective may lead to the Neyman–Scott phenomenon. The novel feature of the Bayesian semiparametric method is that it can capture unmeasured stratum heterogeneity in the distribution of the exposure variable in a robust manner. The proposed method has been extended to the situation when one has multiple disease states (see Sinha et al., 2004). The method is appealing as a unified framework as one can not only handle the missingness in the exposure variable but also possible measurement error and misclassification in the exposure variable. The method may also be extended to take into account the possible association schemes that may exist in a mixed set of continuous and categorical multiple exposures with partial missingness (Sinha et al., 2005b).

In the following subsections we present some of the examples presented in Sinha et al. (2004, 2005a) of Bayesian analysis of matched case control study.

4.1. A continuous exposure: The equine epidemiology example

The first example is equine epidemiology example earlier analyzed by Kim et al. (2002). The data consist of 498 strata with $1:1$ matching in each stratum. Participating veterinarians were asked to provide data monthly for one horse treated for colic and one horse that received emergency treatment for any condition other than colic, between March 1, 1997, and February 28, 1998. A case of colic was defined as the first horse treated during a given month for signs of intra-abdominal pain. A control horse was defined as the next horse that received emergency treatment for any condition other than colic. Age was considered as a single exposure variable (X) measured on a continuous scale with one binary covariate (Z) indicating whether the horse experienced recent diet changes or not. For scaling purposes the authors linearly transformed age so that X was on the interval [0, 1]. The proposed method for the general exponential family, as described in the previous section, applies here with continuous exposure X. Sinha et al. (2005a) assumed that conditional on $D_{ij} = 0$, i.e., for the control population the exposure variable age has a normal distribution of the following form:

$$[X_{ij} \mid D_{ij} = 0, Z_{ij}, S_i] \sim \text{Normal}\big(\gamma_{0i} + \gamma_1 Z_{ij}, \sigma^2\big), \tag{11}$$

where γ_{0i} are stratum effects on the exposure distribution and γ_1 is the regression parameter for regressing X_{ij} on the measured covariates Z_{ij}. They also assumed the logistic regression model for disease status. From Lemma 2 it follows the exposure distribution among the cases is Normal with mean $\gamma_{0i} + \gamma_1 Z_{ij} + \sigma^2 \beta_2$ and variance σ^2. The conditional likelihood can now be derived using these facts. Suitable priors are chosen on the parameters. For estimation of the parameters, a Markov chain Monte Carlo numerical integration scheme was used. For simulating observations from the posterior distribution of the γ_{0i}, the methods proposed in West et al. (1994), Escober and West (1995) and Neal (2000) was used. For generating random numbers from the conditional distribution of other parameters, the authors followed a componentwise Metropolis–Hastings scheme.

For each of the examples as presented in Sections 4.1–4.3 and in Sinha et al. (2005a), three analyses were conducted. One is the proposed Bayesian semiparametric analysis

(BSP). The other is a parametric Bayesian (PB) version of the work of Satten and Carroll (2000) by considering a constant stratum effect $\gamma_{0i} \equiv \gamma_0$ and then assuming a normal prior on this common stratum effect parameter γ_0. They emphasize here and in the tables that the parametric Bayesian method is simply the Bayesian version of the method of Satten and Carroll (2000). The authors also present the frequentist alternative for analyzing this matched data through a standard conditional logistic regression (CLR) analysis (Breslow and Day, 1980).

They ran the three analyses mentioned above on the equine dataset with all 498 strata and reran the analyses where they randomly deleted 40% of the age observations. The results are given in Table 1.

Briefly, with no missing data the results of the methods are roughly in accordance with one another. The slightly small standard error estimates for the Bayesian methods may be a reflection of the fact that unlike CLR, they are full likelihood based semiparametric and parametric methods. With missing data, of course, one can see that the Bayesian methods yield much smaller standard errors than CLR, reflecting the fact that their methods use all the data, and not just the pairs with no missing data. There is little difference between the semiparametric analysis and the parametric Bayes version of the method of Satten and Carroll (2000), a phenomenon that can be explained as follows. The average variance among the γ_{0i}'s is very small (about 0.0054), suggesting that the stratum effects are almost constant across strata. Thus the parametric model with constant stratum effect essentially holds for this example, and hence it is not surprising that the two methods give similar results.

4.2. A binary exposure: Endometrial cancer study

In the Los Angeles 1 : 4 matched case-control study on endometrial cancer as discussed in Breslow and Day (1980), the variable obesity may be considered as a binary exposure variable (X) for contracting endometrial cancer. This variable had about 16% missing observations. Sinha et al. (2005a) considered the presence of gall bladder disease in the subject as our completely observed dichotomous covariate Z.

In such a case with dichotomous exposure variable, the exposure distribution for control population is naturally assumed to have a logistic form: $\mathrm{pr}(X_{ij} = 1|Z_{ij}) = H(\gamma_{0i} + \gamma_1 Z_{ij})$, where $H(\cdot)$ is the logistic distribution function. From Lemma 2 the exposure distribution in the case population is again of the logistic form with: $\mathrm{pr}(X_{ij} = 1|Z_{ij}) = H(\gamma_{0i} + \gamma_1 Z_{ij} + \beta_2)$.

In the LA study there were 63 strata. The original data contained missing observations on the exposure variable obesity. They modeled the missingness in the exposure variable in both the parametric Bayes (the Bayesian version of the Satten and Carroll method with fixed stratum effects) and the semiparametric Bayes analysis. They also conducted classical conditional logistic analysis. Table 2 presents the results. Although there are certainly some interesting numerical differences, in the main the results are reasonably comparable.

The average variance among the γ_{0i}'s over the last 5000 MCMC runs is approximately 4.5. This is fairly large, and if there were major covariate effects one would expect the BSP and the parametric Bayes version of the method of Satten and Carroll

Table 1
Results of the equine data example. Here "Mean" is the posterior mean, "s.e." is the posterior standard deviation, "Lower HPD" and "Upper HPD" are the lower and upper end of the HPD region respectively,"Lower CL" and "Upper CL" are the lower and upper end of the confidence limit respectively, The parametric Bayesian method is the Bayesian version of the method of Satten and Carroll (2000)

Full equine data					
Method		β_1	β_2	$10 \times \gamma_1$	$10 \times \sigma^2$
Bayes semiparametric					
	Mean	2.16	2.18	0.18	0.23
	s.e.	0.33	0.39	0.14	0.01
	Lower HPD	1.57	1.36	−0.10	0.21
	Upper HPD	2.80	2.88	0.40	0.25
Bayes parametric					
	Mean	2.14	2.10	0.22	0.26
	s.e.	0.32	0.41	0.17	0.01
	Lower HPD	1.60	1.32	−0.20	0.24
	Upper HPD	2.87	2.88	0.60	0.28
CLR					
	Mean	2.13	2.05		
	s.e.	0.32	0.47		
	Lower CL	1.50	1.13		
	Upper CL	2.76	2.97		
Equine data with 40% missing data					
Method		β_1	β_2	$10 \times \gamma_1$	$10 \times \sigma^2$
Bayes semiparametric					
	Mean	2.16	2.30	0.14	0.23
	s.e.	0.32	0.48	0.24	0.01
	Lower HPD	1.70	1.30	−0.25	0.19
	Upper HPD	2.80	3.42	0.60	0.25
Bayes parametric					
	Mean	2.13	1.90	0.17	0.28
	s.e.	0.32	0.47	0.20	0.02
	Lower HPD	1.56	0.92	−0.20	0.24
	Upper HPD	2.85	2.85	0.67	0.31
CLR					
	Mean	2.81	2.79		
	s.e.	0.59	0.77		
	Lower CL	1.65	1.47		
	Upper CL	3.96	4.49		

(2000) to be different. However, in this example the analyses suggest that $\beta_2 = 0$ is plausible, so that D and X are essentially independent and whether the γ_{0i}'s are constant or not does not cause a model bias. This is an example where there is stratum variability and natural missingness in the data.

4.3. *Another binary exposure: Low birthweight study*

In Example 4.1 on equine data, there was no appreciable stratum effect variability though the exposure had a significant effect on disease probability. Example 4.2 on

Table 2
Results of the endometrial cancer data example. Here "Mean" is the posterior mean, "s.e." is the posterior standard deviation, "Lower HPD" and "Upper HPD" are the lower and upper end of the HPD region respectively,"Lower CL" and "Upper CL" are the lower and upper end of the confidence limit respectively, The parametric Bayesian method is the Bayesian version of the method of Satten and Carroll (2000)

		Gall bladder	Obesity	
Method		β_1	β_2	γ_1
Bayes semiparametric				
	Mean	1.29	0.70	−0.15
	s.e.	0.39	0.41	0.49
	Lower HPD	0.54	−0.10	−0.99
	Upper HPD	2.18	1.49	0.99
Bayes parametric				
	Mean	1.19	0.49	0.42
	s.e.	0.39	0.36	0.62
	Lower HPD	0.58	−0.15	−0.70
	Upper HPD	2.12	1.25	1.78
CLR				
	Mean	1.28	0.44	
	s.e.	0.39	0.38	
	Lower CL	0.52	−0.30	
	Upper CL	2.04	1.19	

the other hand has natural missingness, appreciable stratum variability but the exposure variable obesity did not have significant effect on disease probability. The example considered in this section illustrates a case when there is significant stratum variability and modeling it causes some interesting differences in inference about the exposure effect when compared to a constant stratum effect model.

This example involves a matched case-control dataset coming from a low birth weight study conducted by the Baystate Medical Center in Springfield, Massachusetts. The dataset is discussed in Hosmer and Lemeshow (2000, Section 1.6.2) and is used as an illustrative example of analyzing a matched case-control study in Chapter 7 of their book. Low birth weight, defined as birth weight less than 2500 grams, is a cause of concern for a newborn as infant mortality and birth defect rates are very high for low birth weight babies. The data was matched according to the age of the mother. A woman's behavior during pregnancy (smoking habits, diet, prenatal care) can greatly alter the chances of carrying the baby to terms. The goal of the study was to determine whether these variables were "risk factors" in the clinical population served by Baystate Medical Center. The matched data contain 29 strata and each stratum has one case (low birthweight baby) and 3 controls (normal birthweight baby). One can possibly think of many different models for explaining the disease in terms of the possible covariates recorded in the data set. The authors considered smoking status of mother as a single exposure variable. Two other covariates, a binary variable denoting presence of uterine irritability (UI) in mother and weight of the mother at last menstrual period (LWT) are also included in the model.

Table 3
Full data analysis of low birth weight data. BSP stands for Bayesian semiparametric method whereas PBC stands for parametric Bayes method assuming constant stratum effects. CLR stands for usual conditional logistic regression for analyzing matched data

	BSP			PBC			CLR	
Parameter	Mean	Sd	HPD region	Mean	Sd	HPD region	Estimate	S.E
SMOKE	1.21	0.56	(0.15,2.31)	0.96	0.46	(−0.01, 1.85)	0.86	0.45
LWT	−1.17	1.26	(−3.39, 1.78)	−1.34	1.25	(−3.75, 1.15)	−1.13	1.36
UI	0.91	0.49	(0.03,1.98)	0.79	0.52	(−0.31, 1.86)	0.85	0.51

Table 4
Analysis of low birth weight data with 30% missingness in the SMOKE variable. BSP stands for Bayesian semiparametric method whereas PBC stands for parametric Bayes method assuming constant stratum effects. CLR stands for usual conditional logistic regression for analyzing matched data

	BSP			PBC			CLR	
Parameter	Mean	Sd	HPD region	Mean	Sd	HPD region	Estimate	S.E
SMOKE	1.19	0.82	(−0.45, 2.85)	0.65	0.59	(−0.47, 1.98)	0.65	0.53
LWT	−1.12	1.27	(−3.51, 1.46)	−1.22	1.24	(−3.70, 1.21)	−1.06	1.62
UI	0.96	0.55	(−0.01, 2.20)	0.86	0.53	(−0.30, 1.91)	0.88	0.60

In this example the average stratum variability is about 3.82 justifying the need for a varying stratum effect model even in the absence of any missingness. Consequently, one can note in Table 3 that in the full data case, the significance of smoking is more pronounced with a varying stratum effect model when compared to the CLR and the parametric Bayes version of Satten and Carroll which assumes a constant stratum effect. The latter two models only exhibit marginal significance of the exposure. The association between smoking and low birth weight has been clinically well-established (Walsh, 1994) though the exact mechanism of decreased birth-weight is not yet known. With introduction of missingness, the BSP estimate for the log odds ratio in Table 4 for smoking remains very close to the full data counterpart whereas the estimates from the constant stratum effect models change appreciably.

4.4. Example of a matched case-control study with multiple disease states

For this example Sinha et al. (2004) again considered the low birth weight data discussed in Section 4.3. They divided the cases, namely, the low birth-weight babies into two categories, *very* low (weighing less than 2000 gms) and low (weighing between 2000 to 2500 gms) and tried to assess the impact of smoking habits of mother on the chance of falling in the two low birth-weight categories relative to the baseline category (normal birthweight, weighing more than 2500 grams). Presence of uterine irritability in mother and mother's weight at last menstruation period were considered as relevant covariates. It was noted that smoking mothers had a higher relative risk of having a low birth weight child when compared to a nonsmoking mother. However, the risk of having a *very* low birth weight child did not depend on smoking significantly.

For the analysis the conditional probabilities of the disease variable given the covariate, exposure and the stratum are given by,

$$P(D_{ij} = k|S_i, \mathbf{Z}_{ij}, X_{ij}) = \frac{\exp\{\beta_{0k}(S_i) + \boldsymbol{\beta}_{1k}^{\mathrm{T}}\mathbf{Z}_{ij} + \beta_{2k}X_{ij}\}}{1 + \sum_{r=1}^{K}\exp\{\beta_{0r}(S_i) + \boldsymbol{\beta}_{1r}^{\mathrm{T}}\mathbf{Z}_{ij} + \beta_{2r}X_{ij}\}}$$
$$\text{for } k = 1, \dots, K, \tag{12}$$

and binary variable smoking status is assumed to follow a Bernoulli distribution. In their example $K = 2$. Three analyses were performed, namely Bayesian Semiparametric (BSP), Parametric Bayes (PB) and the third one is i.i.d. parametric (PBV) where they assumed all the stratum effects are different and each of the n stratum effect parameters γ_{0i}, $i = 1, \dots, n$, are coming from n independent normal distributions.

Table 5 contains the posterior means, posterior standard deviations and 95% HPD credible intervals for the parameters of interest under the proposed Bayesian Semiparametric method (BSP) and the parametric Bayes (PB the PBV) methods as discussed before. To illustrate the methods in presence of missingness, the authors deleted 40% of exposure values completely at random and reran the three analyses. The results are presented in Table 6. The full data analysis indicates that smoking of mother is a significant

Table 5
Analysis of low birth weight data with two disease states, using full dataset. BSP stands for Bayesian semiparametric method whereas PB and PBV stand for parametric Bayes methods assuming constant and varying stratum effects respectively

		BSP			PB			PBV		
Logit	Parameter	Mean	Sd	HPD region	Mean	Sd	HPD region	Mean	Sd	HPD
	SMOKE	1.42	0.60	(0.33, 2.72)	1.26	0.56	(0.25, 2.50)	1.48	0.65	(0.26, 2.08)
1	LWT	−0.86	1.39	(−3.78, 1.81)	−1.03	1.35	(−3.58, 1.86)	−0.73	1.36	(−3.40, 2.01)
	UI	0.15	0.67	(−1.27, 1.46)	0.10	0.67	(−1.19, 1.52)	0.18	0.67	(−1.14,1.52)
Logit	Parameter	Mean	Sd	HPD region	Mean	Sd	HPD region	Mean	Sd	HPD region
	SMOKE	0.37	0.83	(−1.35, 2.05)	0.23	0.66	(−1.10, 1.54)	0.38	0.85	(−1.30, 2.17)
2	LWT	−0.52	1.61	(−3.76, 2.52)	−0.55	1.59	(−3.73, 2.41)	−0.55	1.62	(−3.65, 2.79)
	UI	1.81	0.83	(0.18, 3.51)	1.78	0.83	(0.30, 3.59)	1.81	0.87	(0.27, 3.72)

Table 6
Analysis of low birth weight data with two disease states after deleting 40% observations on smoking completely at random. BSP stands for Bayesian semiparametric method whereas PB and PBV stand for parametric Bayes methods assuming constant and varying stratum effects respectively

		BSP			PB			PBV		
Logit	Parameter	Mean	Sd	HPD region	Mean	Sd	HPD region	Mean	Sd	HPD
	SMOKE	0.86	0.88	(−0.88, 2.55)	0.55	0.78	(−0.84, 2.18)	0.56	0.86	(−0.71, 2.49)
1	LWT	−0.92	1.31	(−3.63, 1.51)	−1.03	1.34	(−3.50, 1.83)	−1.01	1.33	(−3.57, 1.75)
	UI	0.19	0.69	(−1.11, 1.49)	0.21	0.67	(−1.10, 1.55)	0.20	0.69	(−1.31, 1.48)
Logit	Parameter	Mean	Sd	HPD region	Mean	Sd	HPD region	Mean	Sd	HPD region
	SMOKE	0.54	1.04	(−1.46, 2.07)	0.13	0.93	(−1.85, 1.88)	0.12	0.93	(−1.10, 1.54)
2	LWT	−0.43	1.62	(−3.98, 2.38)	−0.59	1.63	(−3.75, 2.59)	−0.54	1.54	(−3.62, 2.45)
	UI	1.82	0.86	(0.34, 3.65)	1.80	0.83	(0.24, 3.46)	1.87	0.82	(0.43, 3.63)

risk factor for low birth-weight category (category 1) and is not very significant in the *very* low birth-weight category (category 2). UI on the other hand shows an opposite association, showing significance in category 2 and almost no significance in category 1. LWT does not seem to be a significant covariate in any of the categories. The BSP and the PBV methods are in closer agreement whereas the PB estimates show some numerical differences. Obviously, without the finer classification into two weight categories, the fact that smoking is not so significant for category 2 and UI is appreciably significant for category 2 cannot be concluded from looking at the overall analysis.

For the analysis with 40% missing observations on smoking one notices that the estimates corresponding to smoking in the BSP method comes closer to their full data counterparts even though the inferences are same in all three methods. As one might expect, with 40% missingness, the parameter estimates for smoking lose precision and the effect of smoking now appears to be insignificant in both categories 1 and 2. Inferences on the other two covariates remain essentially unchanged when compared to full data inferences.

5. Some equivalence results in case-control studies

5.1. Equivalence of retrospective and prospective analysis

Prentice and Pyke (1979) showed that if the disease risk is modeled by logistic regression function (3) and the subjects are selected into the study irrespective of their exposure value, prospective and retrospective analysis of the case-control data yield the same estimate of the association parameter $\boldsymbol{\beta}$. Moreover, the asymptotic standard error of the estimate are the same in both the methods. However, the intercept parameter of the prospective model of the disease risk is not identifiable in a retrospective likelihood unless we know the disease prevalence in the population. Generally, the prospective model involves fewer parameters, and hence is easy to implement. Prentice and Pyke (1979) provided the theoretical validity for prospective analysis of data collected retrospectively.

Roeder et al. (1996) proved a similar result when exposure variables are measured with error. They showed that the profile likelihood function of the association parameter obtained from a retrospective likelihood is the same as the one obtained through the joint distribution of the disease variable, true exposure variable, and its error prone surrogate variable. Carroll et al. (1995) extended Prentice–Pyke type results to situations with missingness and measurement error in the exposure variable.

Recently, Seaman and Richardson (2004) proved a Bayesian analogue of the Prentice–Pyke equivalence result. They considered a discrete exposure vector $\boldsymbol{X}$ with J support points $\{z_1, \dots, z_J\}$. Let n_{0j} and n_{1j} be the number of cases and controls having the exposure value $\boldsymbol{X} = z_j$, $j = 1, \dots, J$. Now if $\text{pr}(\boldsymbol{X} = z_j | D = 0) = \theta_j / \sum_{j=1}^{J} \theta_j$, and odds of disease associated with $\boldsymbol{X} = \boldsymbol{x}$ is $\exp(\boldsymbol{\beta}^{\mathrm{T}} \boldsymbol{x})$, then the natural retrospective likelihood for the case-control data is

$$L_{MR} = \prod_{d=0}^{1} \prod_{j=1}^{J} \left\{ \frac{\theta_j \exp(d\boldsymbol{\beta}^{\mathrm{T}} z_j)}{\sum_{k=1}^{J} \theta_k \exp(d\boldsymbol{\beta}^{\mathrm{T}} z_k)} \right\}^{n_{dj}}. \tag{13}$$

If one assumes that the data came from a cohort study, then the natural prospective likelihood is

$$L_{MP} = \prod_{j=1}^{J} \prod_{d=0}^{1} \left\{ \frac{\alpha^d \exp(d\boldsymbol{\beta}^{\mathrm{T}} z_j)}{\sum_{k=0}^{1} \alpha^k \exp(d\boldsymbol{\beta}^{\mathrm{T}} z_j)} \right\}^{n_{dj}}, \tag{14}$$

where α is the baseline odds of disease when exposure $\boldsymbol{X} = \mathbf{0}$. The authors proved the following two results.

THEOREM 1. *The profile likelihood of $\boldsymbol{\beta}$ obtained by maximizing L_{MR} with respect to $\boldsymbol{\theta} = (\theta_1, \ldots, \theta_J)$ is the same as the profile likelihood of $\boldsymbol{\beta}$ obtained by maximizing L_{MR} with respect to α.*

The next theorem proves the Prentice–Pyke type equivalence result in Bayesian context for a specific class of priors.

THEOREM 2. *Suppose that random variables n_{dj} $(d = 0, 1;\ j = 1, \ldots, J)$ are independently distributed as $n_{dj} \sim$ Poisson(λ_{dj}), where $\log \lambda_{dj} = d \log \alpha + \log \theta_j + d\boldsymbol{\beta}^{\mathrm{T}} z_j$. Assume independent improper priors, $p(\alpha) \propto \alpha^{-1}$ and $p(\theta_j) \propto \theta_j^{a_j - 1}$ for α and $\boldsymbol{\theta}$, and a prior, $p(\boldsymbol{\beta})$ for $\boldsymbol{\beta}$ that is independent of α and $\boldsymbol{\theta}$. Let $n_{+j} = n_{0j} + n_{1j}$, and $n_{d+} = \sum_{j=1}^{J} n_{dj}$. Also, let $\boldsymbol{n}^{\mathrm{T}} = (n_{01}, \ldots, n_{0J}, \ldots, n_{11}, \ldots, n_{1J})$.*

(i) *The joint posterior density of $(\omega, \boldsymbol{\beta})$, where $\omega = \log \alpha$, is*

$$p(\omega, \boldsymbol{\beta} | \boldsymbol{n}) \propto p(\boldsymbol{\beta}) \prod_{j=1}^{J} \frac{\{\exp(\omega + \boldsymbol{\beta}^{\mathrm{T}} z_j)\}^{n_{1j}}}{\{1 + \exp(\omega + \boldsymbol{\beta}^{\mathrm{T}} z_j)\}^{n_{+j} + a_j}}. \tag{15}$$

(ii) *The posterior density of $(\boldsymbol{\pi}, \boldsymbol{\beta})$, where $\boldsymbol{\pi} = (\pi_1, \ldots, \pi_J)$ and $\pi_j = \theta_j / \sum_{k=1}^{J} \theta_k$ is*

$$p(\boldsymbol{\pi}, \boldsymbol{\beta} | \boldsymbol{n}) \propto p(\boldsymbol{\beta}) \prod_{j=1}^{J} \pi_j^{a_j - 1} \prod_{d=0}^{1} \left[\frac{\prod_{j=1}^{J} \{\pi_j \exp(d\boldsymbol{\beta}^{\mathrm{T}} z_j)\}^{n_{dj}}}{\left\{\sum_{j=1}^{J} \pi_j \exp(d\boldsymbol{\beta}^{\mathrm{T}} z_j)\right\}^{n_{d+}}} \right]. \tag{16}$$

The proportionality is up to a constant.

COROLLARY. *The marginal posterior densities of $\boldsymbol{\beta}$ obtained from the joint densities in* (15) *and* (16) *are the same.*

The above corollary states that one may fit either the prospective model or the retrospective model and obtain the same posterior marginal distribution for the parameter of interest $\boldsymbol{\beta}$. One can note that the prospective model contains only one nuisance parameter α, whereas the retrospective model contains J nuisance parameters. Consequently, the prospective model is easier to fit.

The equivalence result of Seaman and Richardson (2004) is a significant contribution to the Bayesian literature on case-control studies. However, the results are limited in the sense that they only apply to the Dirichlet prior on the exposure probabilities in the control population. The equivalence results for any other type of prior still remains an open question. Another limitation is that the exposure variable is assumed to be discrete. For continuous exposure, the authors recommend categorization which is not an ideal solution. The extension of the results to individually matched case-control studies under missingness and/or measurement error in exposure values may provide deeper insight into the Bayesian equivalence of prospective and retrospective model formulation.

5.2. Equivalence between conditional and marginal likelihood for analyzing matched case-control data

The disease risk model (4) for matched case-control study involves a set of nuisance parameters α_i's which capture the stratification effect on the disease risk. A natural way of analyzing matched case-control data is the conditional likelihood approach, where one considers conditional likelihood of the data conditioned on an approximately ancillary statistic for the parameter of interest (see, e.g., Godambe, 1976). In a matched case-control study we condition on the complete sufficient statistic for the nuisance parameters α_i, namely, the number of cases in each matched set. Alternatively, one can work with a marginal likelihood obtained by integrating the nuisance parameters over a mixing distribution, say G. Rice (2004) showed that the full conditional likelihood can exactly be recovered via marginal likelihood approach by integrating the nuisance parameter with respect to a particular mixing distribution G. The author derived the necessary and sufficient conditions on the class of distributions G for the two approaches to agree. The conditions invoke certain invariance properties of the distribution G and such invariant distributions are shown to exist under certain mild natural condition on the odds ratios. In the light of this agreement, in a Bayesian framework, the posterior distribution of the disease-exposure association parameter, $\boldsymbol{\beta}$, of a matched case-control study using an invariant mixture distribution as a prior on the nuisance parameter and a flat prior on the association parameter, is proportional to the conditional likelihood. This provides a validation of using informative priors together with conditional likelihood as done in Diggle et al. (2000).

6. Conclusion

In this review article we have focused on Bayesian modeling and analysis of case-control studies. The Bayesian paradigm offers a great deal of flexibility to accommodate unusual, unorthodox data situations and incorporating prior information on risk related parameters but comes with certain computational challenges. The popularity and use of these methods is highly dependent on developing user friendly softwares for implementing the analysis.

There are other issues besides analysis, like that of Bayesian variable selection (Raftery and Richardson, 1996), sample size determination (De Santis et al., 2001) in case-control studies which are not discussed in this article but are interesting in their own right.

Case-control studies bear enough promise for further research, both methodological and applied. One potential area of research is to extend the methodology when longitudinal data are available for both cases and controls. Hierarchical Bayesian modeling, because of its great flexibility, should prove to be a very useful tool for the analysis of such data. A second important area of research is to adapt the method for the analysis of survival data. Some work in this regard has been initiated in a frequentist framework by Prentice and Breslow (1978), but Bayesian inference for such problems seems to be largely unexplored. For example, the methods could potentially be adapted to bivariate survival models in family based study design (see Oakes, 1986, 1989; Shih and Chatterjee, 2002). Finally, genetic case-control study is a new emerging area of research where one of the objectives is to study the association between a candidate gene and a disease. In many such instances, the population under study is assumed to be homogeneous with respect to allele frequencies, but comprises subpopulations that have different allele frequencies for the candidate gene (see Satten et al., 2001). If these subpopulations also have different disease risks, then an association between the candidate gene and disease may be incorrectly estimated without properly accounting for population structures. Accounting for population stratification may pose some interesting statistical challenges in case-control studies of disease-gene or disease-gene-environment association.

Bayesian approach could also be useful to analyze data coming from other designs used in an epidemiologic context (Khoury et al., 1993). For example, another frequently used design is a nested case-control design (Wacholder, 1996) where a case-control study is nested within a cohort study. This type of study is useful to reduce the cost and labor involved in collecting the data on all individuals in the cohort as well as to reduce computational burden associated with time-dependent explanatory variable. Unlike the case-control design this design allows us to estimate the absolute risk of the disease by knowing the crude disease rate from the cohort study. Some other study designs along this line are the case cohort design (Prentice, 1986), proband design (Gail et al., 2001) and case-only design (Armstrong, 2003). In a recent article, Cheng and Chen (2005) propose Bayesian analysis of case-control studies of genetic association, specifically for assessing gene-environment interaction via case-only design under independence of genetic and environmental factors. They use informative power prior (Ibrahim and Chen, 2000) which in their analysis is taken to be the retrospective likelihood based on historic data raised to a suitable power.

It is important to note that case-control designs are choice-based sampling designs in which the population is stratified on the values of the response variable itself. Among others, this was noticed in Scott and Wild (1986) who compared in such cases a maximum likelihood based approach with some other ad hoc methods of estimation. Breslow and Cain (1988) considered in such cases a two-phase sampling design. In later years, there is a long series of publications on inference for such designs, but any Bayesian approach to these problems is still lacking, and will be a worthwhile future undertaking.

References

Altham, P.M.E. (1971). The analysis of matched proportions. *Biometrika* **58**, 561–576.

Anderson, J.A. (1972). Separate sample logistic discrimination. *Biometrika* **59**, 19–35.

Armstrong, B.G. (2003). Fixed factors that modify the effects of time-varying factors: Applying the case-only approach. *Epidemiology* **14**, 467–472.

Armstrong, B.G., Whittemore, A.S., Howe, G.R. (1989). Analysis of case-control data with covariate measurement error: Application to diet and colon cancer. *Statistics in Medicine* **8**, 1151–1163.

Ashby, D., Hutton, J.L., McGee, M.A. (1993). Simple Bayesian analyses for case-controlled studies in cancer epidemiology. *Statistician* **42**, 385–389.

Breslow, N.E. (1996). Statistics in epidemiology: The case-control study. *J. Amer. Statist. Assoc.* **91**, 14–28.

Breslow, N.E., Cain (1988). Logistic regression for two-stage case-control data. *Biometrika* **75**, 11–20.

Breslow, N.E., Day, N.E. (1980). *Statistical Methods in Cancer Research, vol. 1*. International Agency for Research on Cancer, Lyon.

Breslow, N.E., Day, N.E., Halvorsen, K.T., Prentice, R.L., Sabai, C. (1978). Estimation of multiple relative risk functions in matched case-control studies. *Amer. J. Epidemiology* **108**, 299–307.

Carroll, R.J., Gail, M.H., Lubin, J.H. (1993). Case-control studies with errors in covariates. *J. Amer. Statist. Assoc.* **88**, 185–199.

Carroll, R.J., Wang, S., Wang, C.Y. (1995). Prospective analysis of logistic case-control studies. *J. Amer. Statist. Assoc.* **90**, 157–169.

Cheng, K.F., Chen, J.H. (2005). Bayesian models for population based case-control studies when the population is in Hardy–Weinberg equilibrium. *Genetic Epidemiology* **28**, 183–192.

Cornfield, J. (1951). A method of estimating comparative rates from clinical data: Applications to cancer of the lung, breast, and cervix. *J. Nat. Cancer Inst.* **11**, 1269–1275.

Cornfield, J., Gordon, T., Smith, W.W. (1961). Quantal response curves for experimentally uncontrolled variables. *Bull. Internat. Statist. Inst.* **38**, 97–115.

Cox, D.R. (1966). A simple example of a comparison involving quantal data. *Biometrika* **53**, 215–220.

Day, N.E., Kerridge, D.F. (1967). A general maximum likelihood discriminant. *Biometrics* **23**, 313–323.

De Santis, F., Perone Pacifico, M., Sambcini, V. (2001). Optimal predictive sample size for case-control studies. Rapporto Tecnico n. 17, Dipartmento di Statistica, Probabilita e Statistiche Applicate, Universita di Roma.

Diggle, P.J., Morris, S.E., Wakefield, J.C. (2000). Point-source modeling using matched case-control data. *Biostatistics* **1**, 89–105.

Escober, M.D., West, M. (1995). Bayesian density estimation and inference using mixtures. *J. Amer. Statist. Assoc.* **90**, 577–588.

Gail, M.H., Pee, D., Carroll, R.J. (2001). Effects of violations of assumptions on likelihood methods for estimating the penetrance of an autosomal dominant mutation from kin-cohort studies. *J. Statist. Plann. Inference* **96**, 167–177.

Ghosh, M., Chen, M.-H. (2002). Bayesian inference for matched case-control studies. *Sankhyā, Ser. B* **64**, 107–127.

Godambe, V.P. (1976). Conditional likelihood and unconditional optimum estimating equations. *Biometrika* **63**, 277–284.

Gustafson, P., Le, N.D., Vallee, M. (2002). A Bayesian approach to case-control studies with errors in covariables. *Biostatistics* **3**, 229–243.

Hosmer, D.A., Lemeshow, S. (2000). *Applied Logistic Regression*, second ed. Wiley, New York.

Ibrahim, J.H., Chen, M.H. (2000). Power prior distributions for regression models. *Statist. Sci.* **15**, 46–60.

Kaldor, J., Day, N., Band, P., Choi, N., Clarke, E., Coleman, M., Hakama, M., Koch, M., Langmark, F., Neal, F., Petterson, F., Pompe-Kirin, V., Prior, P., Storm, H. (1987). Second malignancies following testicular cancer, ovarian cancer and Hodgkin's disease: An international collaborative study among cancer registries. *Internat. J. Cancer* **39**, 571–585.

Khoury, M.J., Beaty, T.H., Cohen, B.H. (1993). *Fundamentals of Genetic Epidemiology*. Oxford University Press, Oxford.

Kim, I., Cohen, N.D., Carroll, R.J. (2002). A method for graphical representation of effect heterogeneity by a matched covariate in matched case-control studies exemplified using data from a study of colic in horses. *Amer. J. Epidemiology* **156**, 463–470.

Lindley, D.V. (1964). The Bayesian analysis of contingency tables. *Ann. Math. Statist.* **35**, 1622–1643.

Lipsitz, S.R., Parzen, M., Ewell, M. (1998). Inference using conditional logistic regression with missing covariates. *Biometrics* **54**, 295–303.

Little, R.J.A., Rubin, D.B. (1987). *Statistical Analysis with Missing Data*. Wiley, New York.

Mantel, N., Haenszel, W. (1959). Statistical aspects of the analysis of data from retrospective studies of disease. *J. Nat. Cancer Inst.* **22**, 719–748.

Marshall, R.J. (1988). Bayesian analysis of case-control studies. *Statistics in Medicine* **7**, 1223–1230.

McShane, L.M., Midthune, D.N., Dorgan, J.F., Freedman, L.S., Carroll, R.J. (2001). Covariate measurement error adjustment for matched case-control studies. *Biometrics* **57**, 62–73.

Müller, P., Roeder, K. (1997). A Bayesian semiparametric model for case-control studies with errors in variables. *Biometrika* **84**, 523–537.

Müller, P., Parmigiani, G., Schildkraut, J., Tardella, L. (1999). A Bayesian hierarchical approach for combining case-control and prospective studies. *Biometrics* **55**, 858–866.

Neal, R.M. (2000). Markov chain sampling methods for Dirichlet process mixture models. *J. Comput. Graph. Statist.* **9**, 249–265.

Nurminen, M., Mutanen, P. (1987). Exact Bayesian analysis of two proportions. *Scand. J. Statist.* **14**, 67–77.

Oakes, D. (1986). Semiparametric inference in a model for association in bivariate survival data. *Biometrika* **73**, 353–361.

Oakes, D. (1989). Bivariate survival models induced by frailties. *J. Amer. Statist. Assoc.* **84**, 487–493.

Paik, M.C., Sacco, R. (2000). Matched case-control data analyses with missing covariates. *Appl. Statist.* **49**, 145–156.

Prentice, R.L. (1986). A case-cohort design for epidemiologic cohort studies and disease prevention trials. *Biometrika* **73**, 1–11.

Prentice, R.L., Breslow, N.E. (1978). Retrospective studies and failure time model. *Biometrika* **65**, 153–158.

Prentice, R.L., Pyke, R. (1979). Logistic disease incidence models and case-control studies. *Biometrika* **66**, 403–411.

Prescott, G.J., Garthwaite, P.H (2005). Bayesian analysis of misclassified binary data from a matched case-control study with a validation sub-study. *Statistics in Medicine* **24**, 379–401.

Raftery, A.E., Richardson, S. (1996). Model selection for generalized linear models via GLIB: Application to nutrition and breast cancer. *Bayesian Biostatistics*, 321–353.

Rathouz, P.J., Satten, G.A., Carroll, R.J. (2002). Semiparametric inference in matched case-control studies with missing covariate data. *Biometrika* **89**, 905–916.

Rice, K.M. (2003). Full-likelihood approaches to misclassification of a binary exposure in matched case-control studies. *Statistics in Medicine* **22**, 3177–3194.

Rice, K.M. (2004). Equivalence between conditional and mixture approaches to the rash model and matched case-control studies, with application. *J. Amer. Statist. Assoc.* **99**, 510–522.

Roeder, K., Carroll, R.J., Lindsay, B.G. (1996). A semiparametric mixture approach to case-control studies with errors in covariables. *J. Amer. Statist. Assoc.* **91**, 722–732.

Rubin, D.B. (1981). The Bayesian bootstrap. *Ann. Statist.* **9**, 130–134.

Satten, G.A., Carroll, R.J. (2000). Conditional and unconditional categorical regression models with missing covariates. *Biometrics* **56**, 384–388.

Satten, G.A., Kupper, L. (1993). Inferences about exposure-disease association using probability-of-exposure information. *J. Amer. Statist. Assoc.* **88**, 200–208.

Satten, G.A., Flanders, W.D., Yang, Q. (2001). Accounting for unmeasured population substructure in case-control studies of genetic association using a novel latent-class model. *Amer. J. Human Genetics* **68**, 466–477.

Scott, A.J., Wild, C.J. (1986). Fitting logistic models under case-control or choice-based sampling. *J. Roy. Statist. Soc., Ser. B* **48**, 170–182.

Seaman, S.R., Richardson, S. (2001). Bayesian analysis of case-control studies with categorical covariates. *Biometrika* **88**, 1073–1088.

Seaman, S.R., Richardson, S. (2004). Equivalence of prospective and retrospective models in the Bayesian analysis of case-control studies. *Biometrika* **91**, 15–25.

Shih, J.H., Chatterjee, N. (2002). Analysis of survival data from case-control family studies. *Biometrics* **58**, 502–509.

Sinha, S., Mukherjee, B., Ghosh, M. (2004). Bayesian semiparametric modeling for matched case-control studies with multiple disease states. *Biometrics* **60**, 41–49.

Sinha, S., Mukherjee, B., Ghosh, M., Mallick, B.K., Carroll, R. (2005a). Semiparametric Bayesian modeling of matched case-control studies with missing exposure. *J. Amer. Statist. Assoc.* Submitted for publication.

Sinha, S., Mukherjee, B., Ghosh, M. (2005b). Modeling association among multivariate exposures in matched case-control study. Preprint.

Wacholder, S. (1996). The case-control study as data missing by design: Estimating risk difference. *Epidemiology* **7**, 144–150.

Walsh, R.D. (1994). Effects of maternal smoking on adverse pregnancy outcomes: Examination of the criteria of causation. *Human Biology* **66**, 1059–1092.

West, M., Müller, P., Escobar, M.D. (1994). Hierarchical priors and mixture models, with application in regression and density estimation. In: *Aspects of Uncertainty. A Tribute to D.V. Lindley*, pp. 363–386.

Zelen, M., Parker, R.A. (1986). Case-control studies and Bayesian inference. *Statistics in Medicine* **5**, 261–269.

Handbook of Statistics, Vol. 25
ISSN: 0169-7161
© 2005 Elsevier B.V. All rights reserved.
DOI 10.1016/S0169-7161(05)25028-9

Bayesian Analysis of ROC Data

Valen E. Johnson and Timothy D. Johnson

1. Introduction

When measured on a dichotomous scale, the performance of a diagnostic test can be summarized by two quantities: the false positive fraction (FPF) and the true positive fraction (TPF). The FPF is defined as the fraction of "healthy" subjects who test positive, but who do not have the condition that the test is designed to detect. The TPF is the fraction of subjects who are correctly diagnosed as having the condition (i.e., "diseased" subjects). In Table 1, FPF $= n_{+|-}/n_-$ and TPF $= n_{+|+}/n_+$. The sensitivity of a test equals the TPF, while the specificity of a test equals $1 - \text{FPF}$.

In many cases, however, test results cannot unambiguously be categorized as being either positive or negative, and so instead are reported either on a continuous or ordinal scale. Based on such a report, decision makers are faced with the choice of either taking a corrective action or not. Implicitly, this choice is made on the basis of whether or not the test result exceeded a threshold set by the decision maker. This choice of a threshold determines the relative tradeoff between false positive rates and true positive rates. Receiver operating characteristics (ROC) methodology provides a framework for analyzing this tradeoff as the threshold applied to the diagnostic test is varied.

An ROC curve describes the tradeoff between FPF and TPF as the disease threshold is changed. More specifically, an ROC curve can be defined as the curve defined by the set of points

$$\text{ROC}(\gamma) = \big(\text{FPF}(\gamma), \text{TPF}(\gamma)\big) \quad \text{for } \gamma \in \mathbb{R},$$

Table 1
Predicted status of patients based on a hypothetical diagnostic test versus true patient status

		Test +	Test −	
Truth	+	$n_{+\|+}$	$n_{-\|+}$	n_+
	−	$n_{+\|-}$	$n_{-\|-}$	n_-

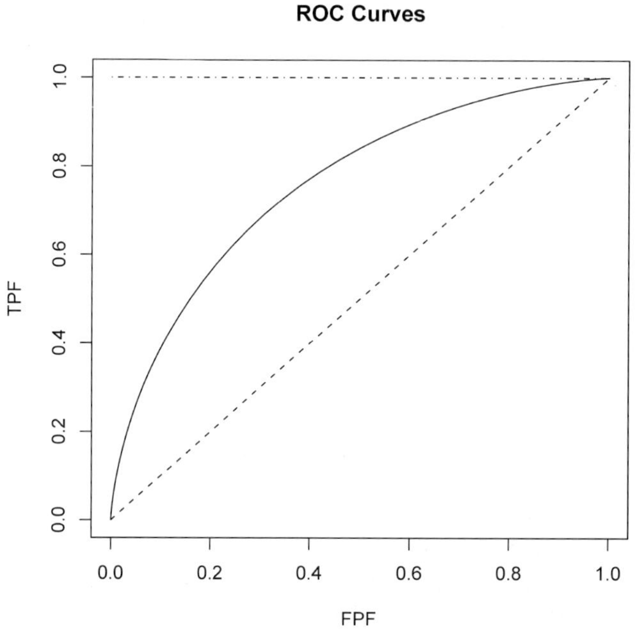

Fig. 1. The solid line shows an example of a typical ROC curve. The dashed line represents an ROC curve for a test with no diagnostic utility (i.e., $\mathrm{TPF}(\gamma) = \mathrm{FPF}(\gamma)$), while the dot-dashed line represents the ROC curve from a test with perfect diagnostic capabilities (i.e., $\mathrm{TPF}(\gamma) = 1$ for all γ).

where $\mathrm{FPF}(\gamma)$ and $\mathrm{TPF}(\gamma)$ are the values of the FPF and TPF achieved for a given value of the threshold γ. By convention, larger values of the diagnostic measure are assumed to be more indicative of disease.

A test has perfect discriminatory capability if $\mathrm{TPF}(\gamma) = 1$ for every $\mathrm{FPF}(\gamma) \in [0, 1]$, and has no diagnostic value if $\mathrm{TPF}(\gamma) = \mathrm{FPF}(\gamma)$ for all $\gamma \in \mathbb{R}$. Example ROC curves are displayed in Figure 1. Additional properties of ROC curves are summarized in, for example, Pepe (2003, Chapter 4).

A commonly used summary of the ROC curve is the area under the curve, which is often denoted by A_Z and defined formally as

$$A_Z = \int_0^1 \mathrm{ROC}(\gamma)\,\mathrm{d}\gamma.$$

The most common application of ROC methodology arises in the comparison of new diagnostic technologies. Generally speaking, when ROC analyses are applied for this purpose, such comparisons are based on the values of the area under the ROC curves for each of the tests, with tests associated with higher A_Z values being preferred. ROC methodology is especially prevalent in the evaluation of radiological devices, where it has been used extensively to determine whether one imaging modality is superior to another.

Several types of data can be used for estimating ROC curves. In the biomedical sciences, the "rating method" is the collection method of choice, and requires that each of several raters score subjects on an ordinal scale. Usually, four or five ordered categories are used. For example, a five-point scale might include the ordered categories "disease definitely absent," "disease possibly absent," "indeterminate," "disease possibly present," and "disease definitely present." From data collected using these categories, ROC analyses attempt to estimate TPF and FPF for all possible thresholds.

In addition to the scale upon which data are collected, researchers must also determine the ROC study design. Which design is selected depends on a number of factors, including the number of readers (i.e., raters) available to provide ratings, the number of diagnostic tests to be evaluated, the number of available subjects, and the extent to which data from the same subject can be used to generate an outcome under each test.

For purposes of illustration, we focus attention here on a design frequently used in biomedical studies, the multirater correlated ROC study. In this design, patients with known disease status are subjected to two or more diagnostic tests. Examples of diagnostic tests that might be compared under this study design include the comparison of two blood serum levels for the diagnosis of a bacterial infection, or two imaging techniques (e.g., CT and MRI) intended to detect lung nodules. Based on the measure provided by the test, several experts are asked to rate the measure on an ordinal scale similar to the one described above. Each diagnostic test is applied to every subject, and all of the experts score the tests from each patient. To minimize recall effects, readers are typically asked to rate test results from the same patient at different times (i.e., on different days) so that their ratings of test results on the same patient can be considered to be approximately independent. By comparing results from multiple experts, this study design facilitates analyses that extend to broader populations of readers. Also, the use of paired data (that is, the ratings generated by two diagnostic tests for the same subject) can reduce the uncertainty associated with differences in areas under the ROC curves.

Given data collected from this study design, ROC curves can be constructed in several ways. The most common method in the biomedical literature is to assume that individual ratings are generated from an unobserved Gaussian random variable. That is, an assumption is made that on an underlying scale, a continuous "extent of disease" variable exists, and experts base their ratings of the subject on their observation of this variable. For convenience, this variable is assumed to be normally distributed. Furthermore, when two diagnostic tests are compared, the latent variables upon which ratings from each test are based are assumed to be correlated. Under the normality assumption, this means that the latent traits for individuals drawn from the disease population is assumed to be drawn from a bivariate normal distribution with mean, say, $(\mu_1, \mu_2)^{\mathrm{T}}$ and covariance matrix, say, Σ. Similarly, the latent traits for individuals from the healthy population are assumed to be drawn from a bivariate normal distribution with mean $(0, 0)^{\mathrm{T}}$ and covariance matrix

$$\begin{pmatrix} 1 & \rho \\ \rho & 1 \end{pmatrix}.$$

Here, ρ denotes the correlation between the two modalities induced by the fact that the same subject was tested twice. The assumption that the healthy population has mean 0

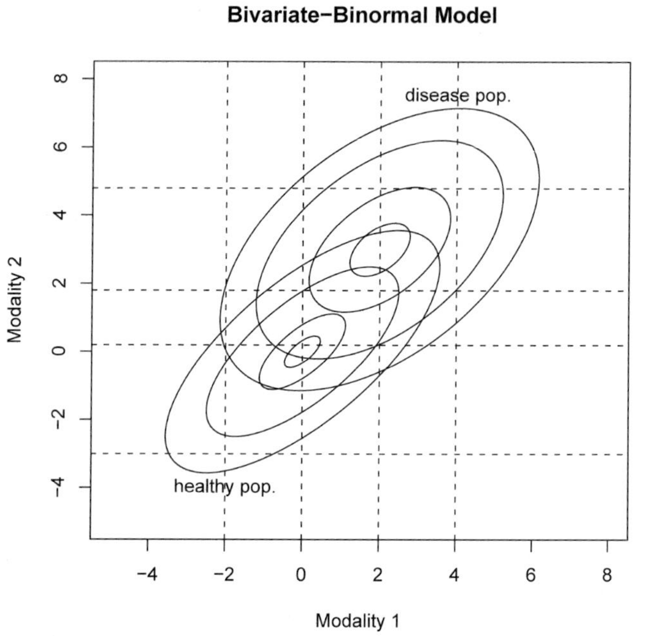

Fig. 2. Graphical depiction of the bivariate binormal ROC assumptions. Deviates from healthy and disease populations are assumed to be generated from the bivariate normal densities depicted. Vertical and horizontal dashed lines represent hypothetical rater thresholds for assignments of disease status to five categories.

is made only to establish a center for the scale of measurement; the assumption that the marginal variances are 1 of the latent traits observed under each modality serves to establish the scale of measurement.

This set of assumptions is called the bivariate binormal ROC model and is illustrated in Figure 2. In current practice, the bivariate binormal ROC model is fitted one rater at a time. The normality assumption on the latent traits is used to estimate the TPF and FPF for each threshold in the following way. From their definitions, it follows that at each possible value of the threshold γ, $\gamma \in (-\infty, \infty)$, FPF($\gamma$) represents the area under the marginal distribution for healthy individuals to the right of γ and TPF(γ) represents the area under the marginal distribution for diseased individuals to the right of γ. Figure 3 depicts an example of the marginal normal distributions and the values of FPF(γ) and TPF(γ) for three values of γ. Also displayed are the corresponding points on the associated ROC curve.

The values of A_Z, the area under the ROC curve, can be easily computed under this model and have a simple interpretation. If X denotes a latent variable drawn from the diseased population and Y denotes a latent random variable drawn from the healthy population, then A_Z is the probability that X is greater than Y (Bamber, 1975). That is $A_Z = \Pr(X > Y)$. From the bivariate–binormal assumptions, it follows that A_Z can be calculated as the probability that a Gaussian random variable with expected value μ and variance $1 + \sigma^2$ is greater than 0.

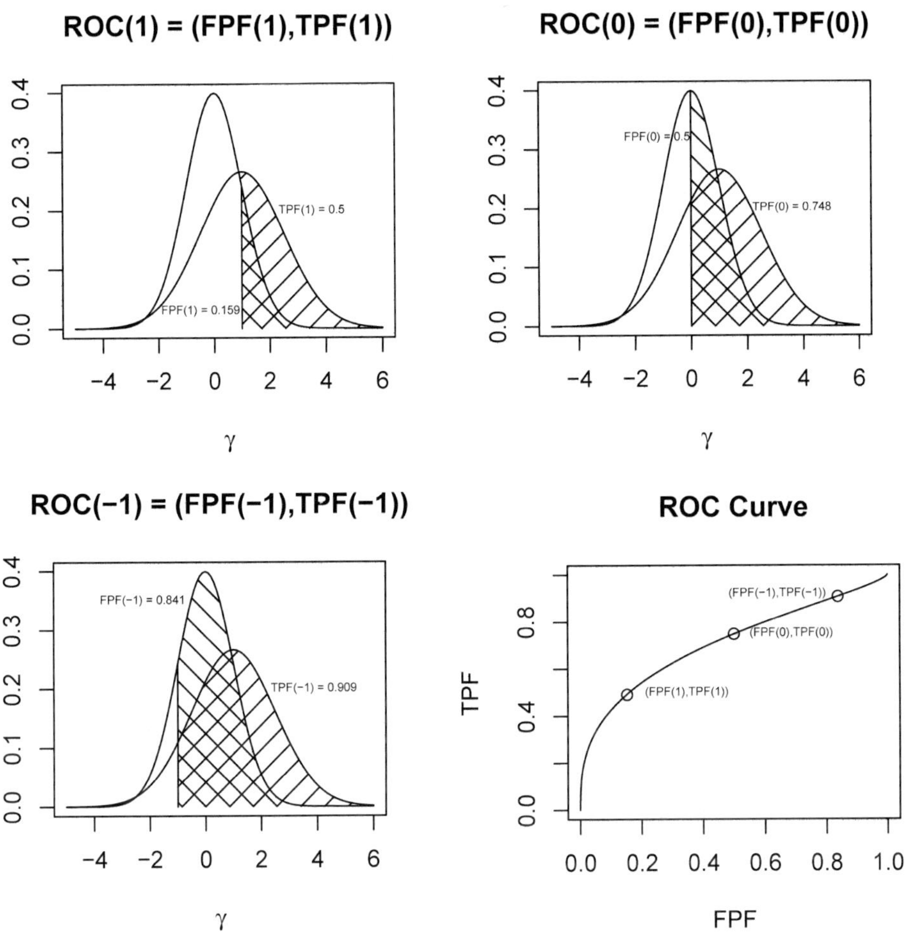

Fig. 3. Graphical depiction of the definition of the ROC curve for one modality under the bivariate–binormal assumption. Latent traits for healthy subjects are assumed drawn from a $N(0, 1)$ distribution. In this illustration, the latent traits from diseased subjects are drawn from a marginal $N(1, 1.5)$ distribution. The lower right panel is the ROC curve implied by these distributions of latent traits. As the threshold, γ runs from ∞ to $-\infty$ the parametrized ROC curve moves from $(0, 0)$, to $(1, 1)$. The upper left panel depicts the FPF(1) as the shaded area under the distribution of healthy individuals; the TPF(1) is depicted as the shaded area under the distribution of diseased individuals. The upper right and lower left panels have similar interpretations.

From a classical perspective, it is difficult to combine ROC data collected from several raters. This difficulty is caused by the absence of a model component for rater variability, although several methods have now been proposed to account for this source of variability (see, for example, Dorfman et al., 1992; Obuchowski and Rockette, 1995; Beiden et al., 2000). Each of these methods requires two stages of modeling. In the first stage, estimates of A_Z (and jackknife pseudo-values of A_Z) for each rater are obtained. In the second stage, these estimates are used as observations in a mixed-effects analysis of variance model. Further discussion of these approaches may be found in Zhou et al. (2002, Chapter 9).

The problem of combining information across raters is more transparent from the Bayesian perspective, and several Bayesian approaches have now also been explored. Among the earlier efforts in this direction are those detailed in Ishwaran and Gatsonis (2000) and Johnson and Albert (1999). In this article, we describe an hierarchical latent variable model for analyzing multirater correlated ordinal ROC data that combines modeling aspects from each of these approaches and others currently being developed. The primary innovation of this model over more commonly used ROC models is the manner in which it accounts for three sources of variation inherent in this study design; namely, variation in ratings attributable to differences in patient/subject characteristics, variation in ratings introduced by inaccuracies in the procedures used to define the diagnostic measure (modality effects), and variation attributable to readers of the diagnostic test. By explicitly modeling these three sources of variation, Bayesian models for ROC analysis are able to achieve substantial increases in power for detecting modality effects, which are the primary variables of interest in most ROC studies. This partitioning of error variances also facilitates the study of individual reader characteristics and provides a natural mechanism for predicting the diagnostic performance of the test when interpreted by a reader drawn randomly from the larger population of potential readers.

2. A Bayesian hierarchical model

For concreteness, in the remainder of this article we consider the multirater correlated ROC study design in which a panel of readers have assessed the disease status of a population of controls and cases to produce ordinal ratings based on two or more diagnostic tests (modalities). In performing our analysis, we assume that the true disease status of all subjects is known.

Let N_h denote the number of healthy subjects and let N_d denote the number of diseased subjects and define $N = N_h + N_d$. The set of diseased subjects is denoted by $\mathcal{D}$ and the set of healthy subjects is denoted by $\mathcal{H}$. Assume that each of $J > 1$ readers rate all N subjects using measurements derived from each of $K > 1$ diagnostic tests. Further assume that each subject is placed into one of $C > 1$ ordered categories by each reader under each test. Let Y_{ijk} denote the rating of case i from reader j, evaluated using measurements derived from test k. We adopt the convention that larger values of Y_{ijk} indicate a higher degree of confidence that the subject is "diseased." In accordance with the bivariate–binormal assumptions described above, assume also that the latent variables underlying the generation of the reader data result from the (possibly distorted) observations of a continuous, scalar-valued random variable representing the presence of a disease attribute.

In the first level of this model, the latent (disease) trait, for the ith subject, Z_i, is assumed to follow a normal distribution. For healthy cases, the latent trait is assumed to be marginally distributed as a $N(0, 1)$ random variable, while the latent value for a diseased individual is assumed to be marginally distributed as $N(\mu, \psi^2)$. The parameters μ and ψ^2 are not known a priori but are estimated from data (see Figure 4, panel A).

In the second level of the hierarchy, $N(0, \phi_k^2)$ errors are added to each latent disease trait. This error term accounts for inaccuracies and distortions introduced by the diagnostic modality. The parameter ϕ_k^2 denotes the variance of this error for modality k (Figure 4, panel B). Let Z_{ik} denote the value of the latent trait of case i that would be observed by an ideal rater using modality k.

At the final stage of the model hierarchy, the values of Z_{ik} are assumed to be further distorted by the addition of a $N(0, \theta_j^2)$ random variable that represents error attributable to the jth reader's observation of Z_{ik}. The parameter θ_j^2 denotes the error variance particular to the jth reader. Z_{ijk} denotes the sum of Z_{ik} and the error induced by reader j (Figure 4, panel C).

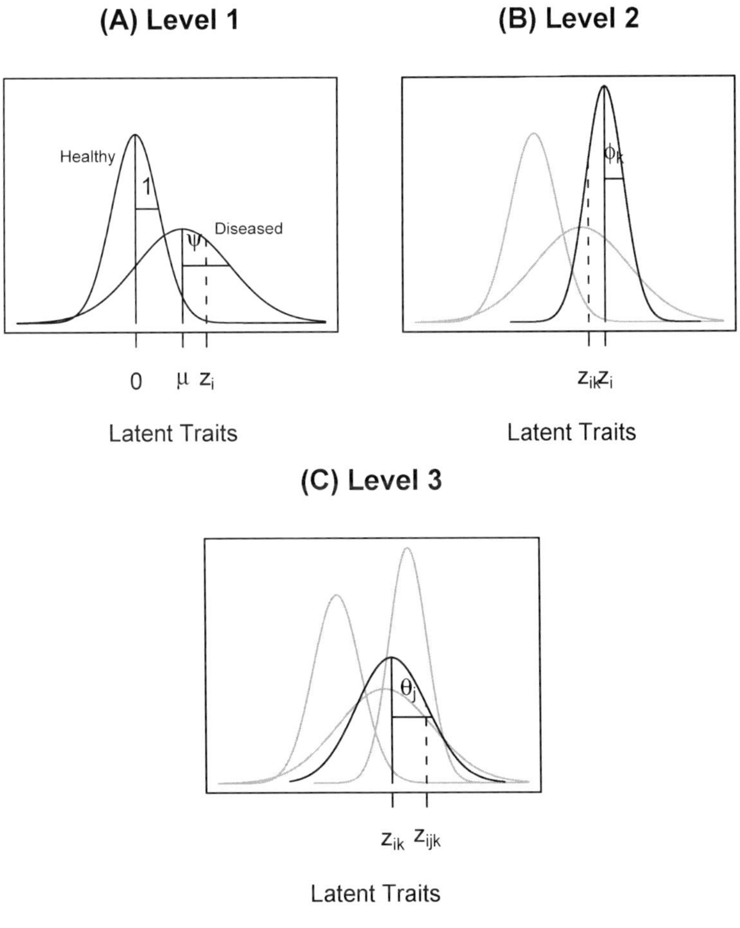

Fig. 4. Graphical representation of the hierarchical model. Panel A: Distributions of the latent traits. Panel B: Modality k introduces random error to the true latent trait Z_i. Panel C: The latent observation Z_{ijk} is the result of further error added by the imperfect rater j to Z_{ik}. The dashed vertical line represents a rater's perceived latent trait at the final stage in the hierarchy.

Each reader is assumed to use a unique set of thresholds in assigning cases to categories. Let the thresholds for reader j be denoted by $\boldsymbol{\gamma}_j$. The components of $\boldsymbol{\gamma}_j$ satisfy $-\infty = \gamma_{j0} < \gamma_{j1} < \cdots < \gamma_{jC-1} < \gamma_{jC} = \infty$. Reader j assigns case i, assessed with modality k, to category c if Z_{ijk} falls between the $(j-1)$st and jth threshold. That is, $Y_{ijk} = c$ if and only if $\gamma_{jc-1} < Z_{ijk} \leqslant \gamma_{jc}$.

To summarize the hierarchical model as specified thus far, we have

$$\text{Level 1:} \quad \begin{cases} [Z_i \mid i \in \mathcal{H}] \overset{\text{i.i.d.}}{\sim} N(0,1), \\ [Z_i|\mu, \psi^2, i \in \mathcal{D}] \overset{\text{i.i.d.}}{\sim} N(\mu, \psi^2), \end{cases} \tag{1}$$

$$\text{Level 2:} \quad [Z_{ik}|Z_i, \phi_k^2] \overset{\text{i.i.d.}}{\sim} N(Z_i, \phi_k^2), \quad \forall k, \tag{2}$$

$$\text{Level 3:} \quad \begin{aligned} p(Z_{ijk}|Z_{ik}, \theta_j^2) \propto \exp[-0.5(Z_{ijk} - Z_{ik})/\theta_j)^2]/\theta_j \\ \times \mathrm{I}(\gamma_{jy_{ijk}-1} < Z_{ijk} \leqslant \gamma_{jy_{ijk}}). \end{aligned} \tag{3}$$

The right-hand side of level 3 is an un-normalized, truncated normal density.

Heuristically, this model may be interpreted as follows. First, latent disease traits for individuals in the population have inherent variability, the magnitude of which is different among the diseased and nondiseased populations. Without loss of generality, we assume the mean of the latent traits among the nondiseased population is zero and has a variance of one. Among the diseased population, the mean of the latent traits is μ with a variance of ψ^2. Second, measurements of the latent disease traits obtained by a diagnostic test or modality are imperfect. The error introduced by diagnostic test k follows a normal distribution with zero mean and variance ϕ_k^2. Lastly, interpretations of diagnostic tests by readers are subject to further error. We allow for the possibility that different readers may have different expertise and, thus, different error variances. The error introduced by reader j has mean zero and variance θ_j^2.

To specify prior constraints on parameters appearing in (1)–(3), the following prior factorization can be assumed:

$$\pi(\mu)\pi(\psi^2)\left(\prod_{k=1}^{K} \pi(\phi_k^2|\theta_1^2, \ldots, \theta_J^2)\right)\left(\prod_{j=1}^{J} \pi(\theta_j^2)\pi(\boldsymbol{\gamma}_j)\right).$$

Accepting this factorization, in the absence of subjective prior information it makes sense to impose an improper uniform prior on μ, the disease population mean. A reasonably noninformative prior for the disease population variance, ψ^2, can be specified as an inverse gamma prior distribution with parameters 3 and 3 (i.e., $IG(3,3)$). This prior distribution has a mean of 1.5 and a mode of 0.75, reflecting a prior belief that variability in the disease population is usually larger than in the healthy population, but not too different from 1. Ninety percent of the mass of this prior lies between 0.48 and 3.7.

Some caution must be exercised in specifying priors for modality and reader variances. Because the scale of the "observed" latent variables Z_{ijk} is not well defined from the priors specified on the distribution of the latent disease traits Z_i, the use of improper priors on both the modality variances $\{\phi_k^2\}$ and reader variances $\{\theta_j^2\}$ can result in either (1) a rater or modality variance collapsing to zero – resulting in a marginal posterior density that becomes highly peaked around zero – or (2) variances that becomes arbitrarily large as category thresholds and all "observed" latent trait errors become large simultaneously.

These problems can be avoided by using uniform shrinkage priors for the modality variances. The uniform shrinkage prior – first proposed by Strawderman (1971) – represents a relatively vague, but proper, prior specification. This prior derives its name from the fact that it places a uniform distribution on the shrinkage of a posterior mean toward a prior mean in simple linear hierarchical models. More recently, this prior was investigated by Daniels (1999) and Natarajan and Kass (2000). Daniels showed the uniform shrinkage prior possesses favorable frequentist properties when compared to other commonly used noninformative priors for variance components. To define a version of the uniform shrinkage prior here, let H_θ denote the harmonic mean of the rater variances, $H_\theta = J / \sum_{j=1}^{J} \theta_j^{-2}$, and consider the conditional posterior expectation of Z_{ik}:

$$E\big(Z_{ik} | Z_i, \phi_k^2, Z_{ijk}, \theta_j^2, j = 1, \dots, J\big)$$
$$= \frac{H_\theta \phi_k^2}{J\phi_k^2 + H_\theta} \sum_{j=1}^{J} Z_{ijk}/\theta_j^2 + \frac{H_\theta}{J\phi_k^2 + H_\theta} Z_i.$$

The uniform shrinkage prior for ϕ_k^2, $k = 1, \dots, K$, can be induced by placing a uniform prior on the shrinkage parameter $H_\theta/(J\phi_k^2 + H_\theta)$ and applying the appropriate transformation of variables. Thus

$$\pi\big(\phi_k^2 | \theta_1^2, \dots, \theta_J^2\big) = H_\theta / \big(J\phi_k^2 + H_\theta\big)^2, \quad k = 1, \dots, K. \tag{4}$$

In contrast to other commonly used noninformative priors for variance components that place infinite mass at zero, the uniform shrinkage prior puts arbitrarily small mass in neighborhoods of zero, thus avoiding a collapse of one or more of the marginal posterior distributions on the variance parameters to zero. In conjunction with this uniform shrinkage prior on the components of $\{\phi_j^2\}$, we take improper uniform priors on the rater variances $\{\theta_j^2\}$.

With these choices of priors for the reader and modality variances, propriety of the posterior distribution depends on the choice of prior density for the reader thresholds. To avoid impropriety of the posterior distribution, a prior density for the category thresholds can be defined by assigning probabilities to the event that a randomly selected subject would be categorized into category c by an ideal rater (zero variance) under an ideal modality (zero variance). The prior density induced in this way has as its support the real line. To define the specific form of this prior, let

$$F(x) = \frac{N_h}{N} \Phi(x; 0, 1) + \frac{N_d}{N} \Phi(x; 1.07, 1.5).$$

Here, the disease group variance has been set to its prior mean and the disease group mean has been set to 1.07, which lead to an ideal A_Z of about 0.75. This value is midway between the minimum informative value of 0.5 and the maximum of 1.0. Define the probability that a subject is placed in category c by rater j under this idealized rating scheme by

$$p_{jc} = F(\gamma_{jc}) - F(\gamma_{jc-1}), \quad c = 1, \ldots, C.$$

Then the prior density for the category thresholds is obtained by placing a Dirichlet prior on $\boldsymbol{p}_j = \{p_{j1}, \ldots, p_{jC}\}$ and transforming to the category threshold scale to obtain

$$\pi(\boldsymbol{\gamma}_j) \propto \mathrm{J} \prod_{c=1}^{C} \{(N_h/N)[\Phi(\gamma_{jc}; 0, 1) - \Phi(\gamma_{jc-1}; 0, 1)] + (N_d/N)[\Phi(\gamma_{jc}; 1.07, 1.5) - \Phi(\gamma_{jc-1}; 1.07, 1.5)]\}^{(2-1)}. \quad (5)$$

Here, J is the Jacobian of the transformation. The parameters of the Dirichlet prior effectively place 2 observations, a priori, in each category for each reader. This prior prevents the escape of the reader thresholds to $\pm\infty$, but imposes only weak information about the values of the reader thresholds. This completes the specification of the model.

2.1. Connection to the bivariate–binormal model

It is interesting to compare the distributional assumptions implicit in this hierarchical Bayesian model for ROC data with the assumptions implicit to the classical bivariate–binormal model. To make this comparison, we adopt the following simplified notation and assume that interest focuses on the evaluation of only two diagnostic tests. Let A_j be a two-by-two diagonal matrix with diagonal elements $(1 + \phi_1^2 + \theta_j^2)^{-1/2}$ and $(1 + \phi_2^2 + \theta_j^2)^{-1/2}$. Marginalizing over $\{Z_i\}$ and $\{Z_{ik}\}$ in (1), (2), and (3) and applying the transformation of variables $(X_{ij1}, X_{ij2})^{\mathrm{T}} = A_j(Z_{ij1}, Z_{ij2})^{\mathrm{T}}$, it follows that the marginal distribution for the latent traits observed under each modality for a healthy case under the Bayesian hierarchical model can be expressed as

$$(X_{ij1}, X_{ij2})^{\mathrm{T}} \sim \Phi_2\big((0, 0)^{\mathrm{T}}, \Sigma_h\big), \quad (6)$$

where Φ_2 denotes a bivariate normal distribution. Similarly, the marginal distribution for the latent traits observed under each modality of a diseased case under the Bayesian model can be expressed

$$(X_{ij1}, X_{ij2})^{\mathrm{T}} \sim \Phi_2\big((\mu_1, \mu_2)^{\mathrm{T}}, \Sigma_d\big). \quad (7)$$

Conditionally on the observed values of y_{ij}, the latent trait distributions are truncated to the interval $(\gamma_{jy_{ij1}-1}, \gamma_{jy_{ij1}}]A_j \times (\gamma_{jy_{ij2}-1}, \gamma_{jy_{ij2}}]A_j$, where

$$\mu_1 = \frac{\mu}{\sqrt{1 + \phi_1^2 + \theta_j^2}} \quad \text{and} \quad \mu_2 = \frac{\mu}{\sqrt{1 + \phi_2^2 + \theta_j^2}},$$

and

$$\Sigma_h = \begin{pmatrix} 1 & \frac{1+\theta_j^2}{\sqrt{(1+\phi_1^2+\theta_j^2)(1+\phi_2^2+\theta_j^2)}} \\ \frac{1+\theta_j^2}{\sqrt{(1+\phi_1^2+\theta_j^2)(1+\phi_2^2+\theta_j^2)}} & 1 \end{pmatrix},$$

$$\Sigma_d = \begin{pmatrix} \frac{\psi^2+\phi_1^2+\theta_j^2}{1+\phi_1^2+\theta_j^2} & \frac{\psi^2+\theta_j^2}{\sqrt{(1+\phi_1^2+\theta_j^2)(1+\phi_2^2+\theta_j^2)}} \\ \frac{\psi^2+\theta_j^2}{\sqrt{(1+\phi_1^2+\theta_j^2)(1+\phi_2^2+\theta_j^2)}} & \frac{\psi^2+\phi_2^2+\theta_j^2}{1+\phi_2^2+\theta_j^2} \end{pmatrix}.$$

Eqs. (6) and (7) reflect the distributional assumptions made for the latent variables in the standard bivariate–binormal model. However, in the classical bivariate–binormal model, the covariance matrix of the latent traits for the disease population between the two diagnostic tests is completely arbitrary, as is the correlation between the two tests in the healthy population. In the hierarchical Bayesian model, the equations for the marginal covariance matrices given above imply that the correlations between traits is forced to be positive. This is a reasonable constraint to impose on the covariance between latent traits observed for the same subject, making this a feature, rather than a drawback, of the model. A further difference is that the standard bivariate–binormal model is fitted to one reader at a time. In order to combine information from several readers, estimates of the A_Z for each modality for each reader are used as the observed outcomes in mixed-effects ANOVA type model. Within this hierarchical Bayesian framework, all parameters are estimated jointly.

2.2. *MCMC details*

The complexity of the joint posterior distribution on model parameters precludes the analytical study of posterior expectations. For that reason, Markov chain Monte Carlo (MCMC) methodology must be used to generate samples from the posterior distribution. Model-based inferences can then be based on these sampled values. Because the hierarchical model specified in the previous sections shares many features with other Bayesian models for ROC analysis, we summarize the basic steps of an MCMC algorithm to sample from its posterior distribution below.

After initializing model parameters, the particular steps in such a MCMC scheme may be described as follows.

1. Sample $\psi^2 \sim IG(3 + 0.5N_d, 3 + 0.5\sum_{i\in\mathcal{D}}(Z_i - \mu)^2)$.
2. Sample $\mu \sim N(\sum_{i\in\mathcal{D}} Z_i/N_d, \psi^2/N_d)$.
3. For $i \in \mathcal{H}$, sample $Z_i \sim N(m_h, v_h^2)$ where $m_h = v_h^2 \sum_k Z_{ik}\phi_k^{-2}$ and $v_h^2 = (1 + \sum \phi_k^{-2})^{-1}$.
4. For $i \in \mathcal{D}$ sample $Z_i \sim N(m_d, v_d^2)$ where $m_d = v_d^2(\mu\psi^{-2} + \sum_k Z_{ik}\phi^{-2})$ and $v_d^2 = (\psi^{-2} + \sum_k \phi_k^{-2})^{-1}$.
5. For $k = 1, \ldots, K$ draw a candidate value $\phi_{k*}^2 \sim IG(-1 + N/2, 0.5\sum_{i=1}^N(Z_{ik} - Z_i)^2)$. Accept ϕ_{k*}^2 with probability $\min[1, (H_\theta + J\phi_k^2)^2/(H_\theta + J\phi_{k*}^2)^2]$.
6. For $i = 1, \ldots, N$ and $k = 1, \ldots, K$, sample $Z_{ik} \sim N(m_{ik}, v_{ik}^2)$ where $m_{ik} = v_{ik}^2(Z_i\phi_k^{-2} + \sum_j Z_{ijk}\theta_j^{-2})$ and $v_{ik}^2 = (\phi_k^{-2} + \sum_j \theta_j^{-2})^{-1}$.
7. Draw a candidate value $\theta_{j*}^2 \sim IG(-1 + NK/2, 0.5\sum_{i=1}^N \sum_{k=1}^K(Z_{ijk} - Z_{ik})^2)$ for $j = 1, \ldots, J$. Accept θ_{j*}^2 with probability $\min[1, (H_\theta + J\phi_k^2)^2/(H_{\theta*} + J\phi_k^2)^2]$.
8. For $j = 1, \ldots, J$ and for $c = 1, \ldots, C - 1$ draw γ_{jc*} from a truncated normal distribution with mean γ_{jc} and variance V, truncated to $(\gamma_{jc-1*}, \gamma_{jc+1})$ where

$\gamma_{j0*} \equiv -\infty$ and $\gamma_{jC*} \equiv \infty$. Let $\boldsymbol{\gamma}_{j*} = (\gamma_{j1*}, \ldots, \gamma_{jC*})$. Accept $\boldsymbol{\gamma}_{j*}$ as the new value of $\boldsymbol{\gamma}_j$ with probability

$$\min\left(1, \frac{\pi(\boldsymbol{\gamma}_{j*}|\mu, \psi^2, \theta_j^2, \phi_1^2, \ldots, \phi_K^2)}{\pi(\boldsymbol{\gamma}_j|\mu, \psi^2, \theta_j^2, \phi_1^2, \ldots, \phi_K^2)} \times \prod_{i=1}^{N}\prod_{k=1}^{K} \frac{\Phi(\gamma_{jy_{ijk}*}; Z_{ik}, V) - \Phi(\gamma_{jy_{ijk}-1*}; Z_{ik}, V)}{\Phi(\gamma_{jy_{ijk}}; Z_{ik}, V) - \Phi(\gamma_{jy_{ijk}-1}; Z_{ik}, V)} \times \prod_{c=1}^{C-1} \frac{\Phi(\gamma_{jc+1}; \gamma_{jc}, V) - \Phi(\gamma_{jc-1*}; \gamma_{jc}, V)}{\Phi(\gamma_{jc+1*}; \gamma_{jc*}, V) - \Phi(\gamma_{jc-1}; \gamma_{jc*}, V)}\right).$$

V was chosen so that the acceptance rate was approximately 35%. This Metropolis–Hastings strategy for updating the category thresholds was proposed by Cowles (1996).

9. For all i, j and k, draw $Z_{ijk} \sim N(Z_{ik}, \theta_j^2)$ truncated to the interval $(\gamma_{jy_{ijk}-1}, \gamma_{jy_{ijk}}]$.

Samples from the posterior distribution on the model parameters can be used to obtain posterior samples of A_Z values as follows. Let W_{jk} denote the latent variable of a randomly chosen healthy individual from rater j under condition k and let U_{jk} denote the latent variable of a randomly chosen diseased individual. Then, $A_{Z_{jk}} = \Pr(W_{jk} < U_{jk})$ (Bamber, 1975) yields a sample from the posterior distribution on A_Z for rater j under condition k. From the model assumptions, it follows that $A_{Z_{jk}} = 1 - \Phi(0; \mu, 1 + \psi^2 + \phi_k^2 + \theta_j^2)$. Similarly, we define A_Z for an ideal rater as A_Z for a rater with zero variance. The value of A_Z for such a rater is denoted by A_{Z_k} and is equal to $1 - \Phi(0; \mu, 1 + \psi^2 + \phi_k^2)$.

3. An example

In 1993, a ROC study was conducted in the UCLA Department of Radiology. The purpose of this study was to compare the diagnostic performance of chest film radiographs to digitized film displayed on a 1K × 1K video display. Radiologists' ability to detect lung nodules in the radiographs was the basis of comparison. Presence or absence of nodules (truth) on 772 archived chest radiographs (59 cases with nodules and 713 disease free cases) was determined by consensus of a panel of expert radiologists. Each of the 772 radiographs were then digitized for video display.

Two radiology residents and three experienced radiologists read each case under both systems (film vs. video display) and ranked the presence of nodules on a scale of 1 to 5. Large ratings represented high confidence that nodules were present. The outcome of interest was the difference in A_Z values. For these data, the estimated posterior mean difference in A_Z values (film – video display) for an ideal rater was -0.0002 with a 95% posterior probability interval of $(-0.003, 0.002)$. From this result, we conclude that A_Z values for the two methods were not substantively different. Figure 5 shows the estimated posterior distribution of the difference in A_Z values for an ideal rater.

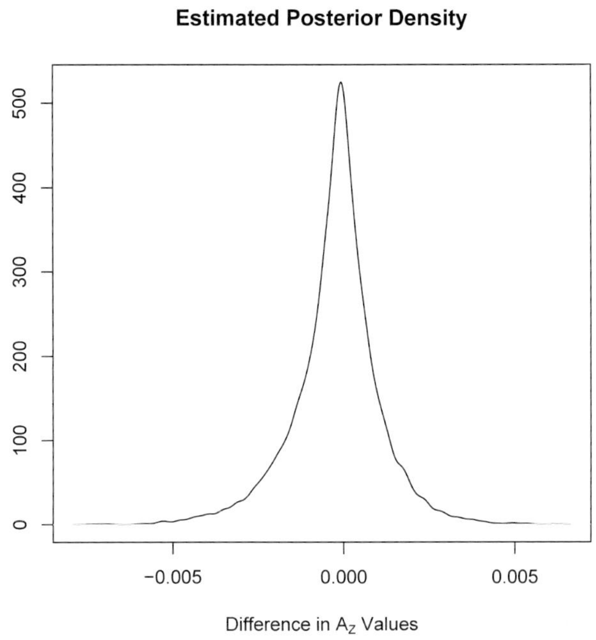

Fig. 5. Estimated posterior density of the difference in ideal rater A_Z values (film – video display).

References

Bamber, D. (1975). The area above the ordinal dominance graph and the area below the receiver operating characteristic graph. *J. Math. Psychol.* **12**, 387–415.

Beiden, S.V., Wagner, R.F., Campbell, G. (2000). Components-of-variance models and multiple bootstrap experiments: An alternative method for random-effects, receiver operating characteristic analysis. *Academic Radiology* **7**, 341–349.

Cowles, M.K. (1996). Accelerating Monte Carlo Markov chain convergence for cumulative-link generalized linear models. *Statist. Comput.* **6**, 101–111.

Daniels, M.J. (1999). A prior for the variance in hierarchical models. *Canad. J. Statist.* **27**, 567–578.

Dorfman, D.D., Berbaum, K.S., Metz, C.E. (1992). Receive operating characteristic rating analysis: Generalization to the population of readers and patients with the jackknife method. *Investigative Radiology* **27**, 723–731.

Ishwaran, H., Gatsonis, C.A. (2000). A general class of hierarchical ordinal regression models with applications to correlated ROC analysis. *Canad. J. Statist.* **28**, 731–750.

Johnson, V.E., Albert, J.H. (1999). *Ordinal Data Modeling*. Springer-Verlag, New York.

Natarajan, R., Kass, R.E. (2000). Reference Bayesian methods for generalized linear mixed models. *J. Amer. Statist. Assoc.* **95**, 227–237.

Obuchowski, N.A., Rockette, H.E. (1995). Hypothesis testing of diagnostic accuracy for multiple readers and multiple tests: An ANOVA approach with dependent observations. *Comm. Statist. Simulation* **24**, 285–308.

Pepe, M.S. (2003). *The Statistical Evaluation of Medical Tests for Classification and Prediction*. Oxford University Press, Oxford.

Strawderman, W.E. (1971). Proper Bayes minimax estimators of the multivariate normal mean. *Ann. Math. Statist.* **42**, 385–388.

Zhou, X., Obuchowski, N.A., McClish, D.K. (2002). *Statistical Methods in Diagnostic Medicine*. Wiley, New York.

Handbook of Statistics, Vol. 25
ISSN: 0169-7161
© 2005 Elsevier B.V. All rights reserved.
DOI 10.1016/S0169-7161(05)25029-0

29

Modeling and Analysis for Categorical Response Data

Siddhartha Chib

1. Introduction

In this chapter we discuss how Bayesian methods are used to model and analyze categorical response data. As in other areas of statistics, the growth of Bayesian ideas in the categorical data setting has been rapid, aided by developments in *Markov chain Monte Carlo* (MCMC) methods (Gelfand and Smith, 1990; Smith and Roberts, 1993; Tanner and Wong, 1987; Tierney, 1994; Chib and Greenberg, 1995), and by the framework for fitting and criticizing categorical data models developed in Albert and Chib (1993). In this largely self-contained chapter we summarize the various possibilities for the Bayesian modeling of categorical response data and provide the associated inferential techniques for conducting the prior–posterior analyses.

The rest of the chapter is organized as follows. We begin by including a brief overview of MCMC methods. We then discuss how the output of the MCMC simulations can be used to calculate the marginal likelihood of the model for the purpose of comparing models via Bayes factors. After these preliminaries, the chapter turns to the analysis of categorical response models in the cross-section setting, followed by extensions to multivariate and longitudinal responses.

1.1. Elements of Markov chain Monte Carlo

Suppose that $\boldsymbol{\psi} \in \Re^d$ denotes a vector of random variables of interest (typically consisting of a set of parameters $\boldsymbol{\theta}$ and other variables) in a particular Bayesian model and let $\pi(\boldsymbol{\psi}|\boldsymbol{y}) \propto \pi(\boldsymbol{\psi})p(\boldsymbol{y}|\boldsymbol{\psi})$ denote the posterior density, where $\pi(\boldsymbol{\psi})$ is the prior density and $p(\boldsymbol{y}|\boldsymbol{\psi})$ is the sampling density. Suppose that we are interested in calculating the posterior mean $\boldsymbol{\eta} = \int_{\Re^d} \boldsymbol{\psi}\pi(\boldsymbol{\psi}|\boldsymbol{y})\,\mathrm{d}\boldsymbol{\psi}$ and that this integral cannot be computed analytically, or numerically, because the dimension of the integration precludes the use of quadrature-based methods. To tackle this problem we can rely on Monte Carlo sampling methods. Instead of concentrating on the computation of the above integral we can proceed in a more general way. We construct a method to produce a sequence of draws from the posterior density, namely

$$\boldsymbol{\psi}^{(1)}, \ldots, \boldsymbol{\psi}^{(M)} \sim \pi(\boldsymbol{\psi}|\boldsymbol{y}).$$

Provided M is large enough, this sample from the posterior density can be used to summarize the posterior density. We can estimate the mean by taking the average of these simulated draws. We can estimate the quantiles of the posterior density from the quantiles of the sampled draws. Other summaries of the posterior density can be obtained in a similar manner. Under suitable laws of large numbers these estimates converge to the posterior quantities as the simulation-size becomes large.

The sampling of the posterior distribution is, therefore, the central concern in Bayesian computation. One important breakthrough in the use of simulation methods was the realization that the sampled draws need not be independent, that simulation-consistency can be achieved with correlated draws. The fact that the sampled variates can be correlated is of immense practical and theoretical importance and is the defining characteristic of Markov chain Monte Carlo methods, popularly referred to by the acronym MCMC, where the sampled draws form a Markov chain. The idea behind these methods is simple and extremely general. In order to sample a given probability distribution, referred to as the target distribution, a suitable Markov chain is constructed with the property that its limiting, invariant distribution, is the target distribution. Once the Markov chain has been constructed, a sample of draws from the target distribution is obtained by simulating the Markov chain a large number of times and recording its values. Within the Bayesian framework, where both parameters and data are treated as random variables, and inferences about the parameters are conducted conditioned on the data, the posterior distribution of the parameters provides a natural target for MCMC methods.

Markov chain sampling methods originate with the work of Metropolis et al. (1953) in statistical physics. A vital extension of the method was made by Hastings (1970) leading to a method that is now called the Metropolis–Hastings algorithm (see Tierney, 1994 and Chib and Greenberg, 1995 for detailed summaries). This algorithm was first applied to problems in spatial statistics and image analysis (Besag, 1974). A resurgence of interest in MCMC methods started with the papers of Geman and Geman (1984) who developed an algorithm, a special case of the Metropolis method that later came to be called the Gibbs sampler, to sample a discrete distribution, Tanner and Wong (1987) who proposed a MCMC scheme involving data augmentation to sample posterior distributions in missing data problems, and Gelfand and Smith (1990) where the value of the Gibbs sampler was demonstrated for general Bayesian problems with continuous parameter spaces.

The Gibbs sampling algorithm is one of the simplest Markov chain Monte Carlo algorithms and is easy to describe. Suppose that $\boldsymbol{\psi}_1$ and $\boldsymbol{\psi}_2$ denote some grouping (blocking) of $\boldsymbol{\psi}$ and let

$$\pi_1(\boldsymbol{\psi}_1|\boldsymbol{y}, \boldsymbol{\psi}_2) \propto p(\boldsymbol{y}|\boldsymbol{\psi}_1, \boldsymbol{\psi}_2)\pi(\boldsymbol{\psi}_1, \boldsymbol{\psi}_2), \tag{1.1}$$

$$\pi_2(\boldsymbol{\psi}_2|\boldsymbol{y}, \boldsymbol{\psi}_1) \propto p(\boldsymbol{y}|\boldsymbol{\psi}_1, \boldsymbol{\psi}_2)\pi(\boldsymbol{\psi}_1, \boldsymbol{\psi}_2) \tag{1.2}$$

denote the associated conditional densities, often called the full conditional densities. Then, one cycle of the Gibbs sampling algorithm is completed by sampling each of the full conditional densities, using the most current values of the conditioning block. The Gibbs sampler in which each block is revised in fixed order is defined as follows.

ALGORITHM (*Gibbs Sampling*).

1 `Specify an initial value` $\boldsymbol{\psi}^{(0)} = (\boldsymbol{\psi}_1^{(0)}, \boldsymbol{\psi}_2^{(0)})$:

2 `Repeat for` $j = 1, 2, \ldots, n_0 + G$.

  `Generate` $\boldsymbol{\psi}_1^{(j)}$ `from` $\pi_1(\boldsymbol{\psi}_1 | \boldsymbol{y}, \boldsymbol{\psi}_2^{(j-1)})$

  `Generate` $\boldsymbol{\psi}_2^{(j)}$ `from` $\pi_2(\boldsymbol{\psi}_2 | \boldsymbol{y}, \boldsymbol{\psi}_1^{(j)})$

3 `Return the values` $\{\boldsymbol{\psi}^{(n_0+1)}, \boldsymbol{\psi}^{(n_0+2)}, \ldots, \boldsymbol{\psi}^{(n_0+G)}\}$.

In some problems it turns out that the full conditional density cannot be sampled directly. In such cases, the intractable full conditional density is sampled via the Metropolis–Hastings (M–H) algorithm. For specificity, suppose that the full conditional density $\pi(\boldsymbol{\psi}_1 | \boldsymbol{y}, \boldsymbol{\psi}_2)$ is intractable. Let

$$q_1(\boldsymbol{\psi}_1, \boldsymbol{\psi}_1' | \boldsymbol{y}, \boldsymbol{\psi}_2)$$

denote a proposal density that generates a candidate $\boldsymbol{\psi}_1'$, given the data and the values of the remaining blocks. Then, in the first step of the jth iteration of the MCMC algorithm, given the values $(\boldsymbol{\psi}_1^{(j-1)}, \boldsymbol{\psi}_2^{(j-1)})$ from the previous iteration, the transition to the next iterate of $\boldsymbol{\psi}_1$ is achieved as follows.

ALGORITHM (*Metropolis–Hastings step for sampling an intractable* $\pi_1(\boldsymbol{\psi}_1 | \boldsymbol{y}, \boldsymbol{\psi}_2)$).

1 `Propose a value for` $\boldsymbol{\psi}_1$ `by drawing:`

$$\boldsymbol{\psi}_1' \sim q_1(\boldsymbol{\psi}_1^{(j-1)}, \cdot | \boldsymbol{y}, \boldsymbol{\psi}_2^{(j-1)})$$

2 `Calculate the probability of move` $\alpha(\boldsymbol{\psi}_1^{(j-1)}, \boldsymbol{\psi}_1' | \boldsymbol{y}, \boldsymbol{\psi}_2^{(j-1)})$ `given by`

$$\min\left\{1, \frac{\pi(\boldsymbol{\psi}_1' | \boldsymbol{y}, \boldsymbol{\psi}_2^{(j-1)}) q_1(\boldsymbol{\psi}_1', \boldsymbol{\psi}_1^{(j-1)} | \boldsymbol{y}, \boldsymbol{\psi}_2^{(j-1)})}{\pi(\boldsymbol{\psi}_1^{(j-1)} | \boldsymbol{y}, \boldsymbol{\psi}_2^{(j-1)}) q_1(\boldsymbol{\psi}_1^{(j-1)}, \boldsymbol{\psi}_1' | \boldsymbol{y}, \boldsymbol{\psi}_2^{(j-1)})}\right\}.$$

3 `Set`

$$\boldsymbol{\psi}_1^{(j)} = \begin{cases} \boldsymbol{\psi}_1' & \text{with prob } \alpha(\boldsymbol{\psi}_1^{(j-1)}, \boldsymbol{\psi}_1' | \boldsymbol{y}, \boldsymbol{\psi}_2^{(j-1)}), \\ \boldsymbol{\psi}_1^{(j-1)} & \text{with prob } 1 - \alpha(\boldsymbol{\psi}_1^{(j-1)}, \boldsymbol{\psi}_1' | \boldsymbol{y}, \boldsymbol{\psi}_2^{(j-1)}). \end{cases}$$

A similar approach is used to sample $\boldsymbol{\psi}_2$ if the full conditional density of $\boldsymbol{\psi}_2$ is intractable. These algorithms extend straightforwardly to problems with more than two blocks.

1.2. Computation of the marginal likelihood

Posterior simulation by MCMC methods does not require knowledge of the normalizing constant of the posterior density. Nonetheless, if we are interested in comparing alternative models, then knowledge of the normalizing constant is essential. This is because the formal Bayesian approach for comparing models is via *Bayes factors*, or ratios of *marginal likelihoods*. The marginal likelihood of a particular model is the normalizing constant of the posterior density and is defined as

$$m(\boldsymbol{y}) = \int p(\boldsymbol{y}|\boldsymbol{\theta})\pi(\boldsymbol{\theta})\,\mathrm{d}\boldsymbol{\theta}, \tag{1.3}$$

the integral of the likelihood function with respect to the prior density. If we have two models $\mathcal{M}_k$ and $\mathcal{M}_l$, then the Bayes factor is the ratio

$$\mathrm{B}_{kl} = \frac{m(\boldsymbol{y}|\mathcal{M}_k)}{m(\boldsymbol{y}|\mathcal{M}_l)}. \tag{1.4}$$

Computation of the marginal likelihood is, therefore, of some importance in Bayesian statistics (DiCiccio et al., 1997, Chen and Shao, 1998, Roberts, 2001). It is possible to avoid the direct computation of the marginal likelihood by the methods of Green (1995) and Carlin and Chib (1995), but this requires formulating a more general MCMC scheme in which parameters and models are sampled jointly. We do not consider these approaches in this chapter. Unfortunately, because MCMC methods deliver draws from the posterior density, and the marginal likelihood is the integral with respect to the prior, the MCMC output cannot be used directly to average the likelihood. To deal with this problem, a number of methods have appeared in the literature. One simple and widely applicable method is due to Chib (1995) which we briefly explain as follows.

Begin by noting that $m(\boldsymbol{y})$ by virtue of being the normalizing constant of the posterior density can be expressed as

$$m(\boldsymbol{y}) = \frac{p(\boldsymbol{y}|\boldsymbol{\theta}^*)\pi(\boldsymbol{\theta}^*)}{\pi(\boldsymbol{\theta}^*|\boldsymbol{y})}, \tag{1.5}$$

for any given point $\boldsymbol{\theta}^*$ (generally taken to be a high density point such as the posterior mode or mean). Thus, provided we have an estimate $\hat{\pi}(\boldsymbol{\theta}^*|\boldsymbol{y})$ of the posterior ordinate, the marginal likelihood can be estimated on the log scale as

$$\log m(\boldsymbol{y}) = \log p(\boldsymbol{y}|\boldsymbol{\theta}^*) + \log \pi(\boldsymbol{\theta}^*) - \log \hat{\pi}(\boldsymbol{\theta}^*|\boldsymbol{y}). \tag{1.6}$$

In the context of both single and multiple block MCMC chains, good estimates of the posterior ordinate are available. For example, when the MCMC simulation is run with B blocks, Chib (1995) employs the marginal–conditional decomposition

$$\pi(\boldsymbol{\theta}^*|\boldsymbol{y}) = \pi(\boldsymbol{\theta}_1^*|\boldsymbol{y}) \times \cdots \times \pi(\boldsymbol{\theta}_i^*|\boldsymbol{y}, \boldsymbol{\Theta}_{i-1}^*) \times \cdots \times \pi(\boldsymbol{\theta}_B^*|\boldsymbol{y}, \boldsymbol{\Theta}_{B-1}^*), \tag{1.7}$$

where the typical term is of the form

$$\begin{aligned}
&\pi(\boldsymbol{\theta}_i^*|\boldsymbol{y}, \boldsymbol{\Theta}_{i-1}^*) \\
&\quad = \int \pi(\boldsymbol{\theta}_i^*|\boldsymbol{y}, \boldsymbol{\Theta}_{i-1}^*, \boldsymbol{\Theta}^{i+1}, z)\pi(\boldsymbol{\Theta}^{i+1}, z|\,\boldsymbol{y}, \boldsymbol{\Theta}_{i-1}^*)\,\mathrm{d}\boldsymbol{\Theta}^{i+1}\,\mathrm{d}z.
\end{aligned}$$

In the latter expression $\boldsymbol{\Theta}_i = (\boldsymbol{\theta}_1, \ldots, \boldsymbol{\theta}_i)$ and $\boldsymbol{\Theta}^i = (\boldsymbol{\theta}_i, \ldots, \boldsymbol{\theta}_B)$ denote the list of blocks up to i and the set of blocks from i to B, respectively, and z denotes any latent data that is included in the sampling. This is the *reduced conditional ordinate*. It is important to bear in mind that in finding the reduced conditional ordinate one must integrate only over $(\boldsymbol{\Theta}^{i+1}, z)$ and that the integrating measure is conditioned on $\boldsymbol{\Theta}^*_{i-1}$.

Consider first the case where the normalizing constant of each full conditional density is known. Then, the first term of (1.7) can be estimated by the Rao–Blackwell method. To estimate the typical reduced conditional ordinate, one conducts a MCMC run consisting of the full conditional distributions

$$\begin{aligned}&\{\pi(\boldsymbol{\theta}_i|\boldsymbol{y}, \boldsymbol{\Theta}^*_{i-1}, \boldsymbol{\Theta}^{i+1}, z); \ldots; \pi(\boldsymbol{\theta}_B|\boldsymbol{y}, \boldsymbol{\Theta}^*_{i-1}, \boldsymbol{\theta}_i, \ldots, \boldsymbol{\theta}_{B-1}, z);\\ &\pi(z|\boldsymbol{y}, \boldsymbol{\Theta}^*_{i-1}, \boldsymbol{\Theta}^i)\},\end{aligned} \tag{1.8}$$

where the blocks in $\boldsymbol{\Theta}_{i-1}$ are set equal to $\boldsymbol{\Theta}^*_{i-1}$. By MCMC theory, the draws on $(\boldsymbol{\Theta}^{i+1}, z)$ from this run are from the distribution $\pi(\boldsymbol{\Theta}^{i+1}, z|\boldsymbol{y}, \boldsymbol{\Theta}^*_{i-1})$ and so the reduced conditional ordinate can be estimated as the average

$$\hat{\pi}(\boldsymbol{\theta}^*_i|\boldsymbol{y}, \mathcal{M}, \boldsymbol{\Theta}^*_{i-1}) = M^{-1}\sum_{j=1}^{M}\pi(\boldsymbol{\theta}^*_i|\boldsymbol{y}, \boldsymbol{\Theta}^*_{i-1}, \boldsymbol{\Theta}^{i+1,(j)}, z^{(j)})$$

over the simulated values of $\boldsymbol{\Theta}^{i+1}$ and z from the reduced run. Each subsequent reduced conditional ordinate that appears in the decomposition (1.7) is estimated in the same way though, conveniently, with fewer and fewer distributions appearing in the reduced runs. Given the marginal and reduced conditional ordinates, the marginal likelihood on the log scale is available as

$$\begin{aligned}&\log\hat{m}(\boldsymbol{y}|\mathcal{M})\\ &\quad= \log p(\boldsymbol{y}|\boldsymbol{\theta}^*, \mathcal{M}) + \log\pi(\boldsymbol{\theta}^*) - \sum_{i=1}^{B}\log\hat{\pi}(\boldsymbol{\theta}^*_i|\boldsymbol{y}, \mathcal{M}, \boldsymbol{\Theta}^*_{i-1}),\end{aligned} \tag{1.9}$$

where $p(\boldsymbol{y}|\boldsymbol{\theta}^*)$ is the density of the data marginalized over the latent data z.

Consider next the case where the normalizing constant of one or more of the full conditional densities is not known. In that case, the posterior ordinate is estimated by a modified method developed by Chib and Jeliazkov (2001). If sampling is conducted in one block by the M–H algorithm with proposal density $q(\boldsymbol{\theta}, \boldsymbol{\theta}'|\boldsymbol{y})$ and probability of move

$$\alpha(\boldsymbol{\theta}^{(j-1)}, \boldsymbol{\theta}'|\boldsymbol{y}) = \min\left\{1, \frac{\pi(\boldsymbol{\theta}'|\boldsymbol{y})q(\boldsymbol{\theta}', \boldsymbol{\theta}^{(j-1)}|\boldsymbol{y})}{\pi(\boldsymbol{\theta}^{(j-1)}|\boldsymbol{y})q(\boldsymbol{\theta}^{(j-1)}, \boldsymbol{\theta}'|\boldsymbol{y})}\right\}$$

then it can be shown that the posterior ordinate is given by

$$\pi(\boldsymbol{\theta}^*|\boldsymbol{y}) = \frac{E_1\{\alpha(\boldsymbol{\theta}, \boldsymbol{\theta}^*|\boldsymbol{y})q(\boldsymbol{\theta}, \boldsymbol{\theta}^*|\boldsymbol{y})\}}{E_2\{\alpha(\boldsymbol{\theta}^*, \boldsymbol{\theta}|\boldsymbol{y})\}},$$

where the numerator expectation E_1 is with respect to the distribution $\pi(\boldsymbol{\theta}|\boldsymbol{y})$ and the denominator expectation E_2 is with respect to the proposal density $q(\boldsymbol{\theta}^*, \boldsymbol{\theta}|\boldsymbol{y})$. This leads to the simulation consistent estimate

$$\hat{\pi}(\boldsymbol{\theta}^*|\mathbf{y}) = \frac{M^{-1}\sum_{g=1}^{M}\alpha(\boldsymbol{\theta}^{(g)}, \boldsymbol{\theta}^*|\mathbf{y})q(\boldsymbol{\theta}^{(g)}, \boldsymbol{\theta}^*|\mathbf{y})}{J^{-1}\sum_{j=1}^{M}\alpha(\boldsymbol{\theta}^*, \boldsymbol{\theta}^{(j)}|\mathbf{y})}, \tag{1.10}$$

where $\boldsymbol{\theta}^{(g)}$ are the given draws from the posterior distribution while the draws $\boldsymbol{\theta}^{(j)}$ in the denominator are from $q(\boldsymbol{\theta}^*, \boldsymbol{\theta}|\mathbf{y})$, given the fixed value $\boldsymbol{\theta}^*$.

In general, when sampling is done with B blocks, the typical reduced conditional ordinate is given by

$$\pi(\boldsymbol{\theta}_i^*|\mathbf{y}, \boldsymbol{\Theta}_{i-1}^*) = \frac{E_1\{\alpha(\boldsymbol{\theta}_i, \boldsymbol{\theta}_i^*|\mathbf{y}, \boldsymbol{\Theta}_{i-1}^*, \boldsymbol{\Theta}^{i+1}, z)q_i(\boldsymbol{\theta}_i, \boldsymbol{\theta}_i^*|\mathbf{y}, \boldsymbol{\Theta}_{i-1}^*, \boldsymbol{\Theta}^{i+1}, z)\}}{E_2\{\alpha(\boldsymbol{\theta}_i^*, \boldsymbol{\theta}_i|\mathbf{y}, \boldsymbol{\Theta}_{i-1}^*, \boldsymbol{\Theta}^{i+1}, z)\}}, \tag{1.11}$$

where E_1 is the expectation with respect to $\pi(\boldsymbol{\Theta}^i, z|\mathbf{y}, \boldsymbol{\Theta}_{i-1}^*)$ and E_2 that with respect to the product measure $\pi(\boldsymbol{\Theta}^{i+1}, z|\mathbf{y}, \boldsymbol{\theta}_i^*)q_i(\boldsymbol{\theta}_i^*, \boldsymbol{\theta}_i|\mathbf{y}, \boldsymbol{\Theta}_{i-1}^*, \boldsymbol{\Theta}^{i+1})$. The quantity $\alpha(\boldsymbol{\theta}_i, \boldsymbol{\theta}_i^*|\mathbf{y}, \boldsymbol{\Theta}_{i-1}^*, \boldsymbol{\Theta}^{i+1}, z)$ is the M–H probability of move. The two expectations are estimated from the output of the reduced runs in an obvious way.

2. Binary responses

Suppose that y_i is a binary {0, 1} response variable and $\boldsymbol{x}_i$ is a k vector of covariates. Suppose that we have a random sample of n observations and the Bayesian model of interest is

$$\begin{aligned} y_i|\boldsymbol{\beta} &\sim \Phi(\boldsymbol{x}_i'\boldsymbol{\beta}), \quad i \leqslant n, \\ \boldsymbol{\beta} &\sim \mathcal{N}_k(\boldsymbol{\beta}_0, \boldsymbol{B}_0), \end{aligned}$$

where Φ is the cdf of the $N(0, 1)$ distribution. We consider the case of the logit link below. On letting $p_i = \Phi(\boldsymbol{x}_i'\boldsymbol{\beta})$, the posterior distribution of $\boldsymbol{\beta}$ is proportional to

$$\pi(\boldsymbol{\beta}|\mathbf{y}) \propto \pi(\boldsymbol{\beta})\prod_{i=1}^{n} p_i^{y_i}(1 - p_i)^{(1-y_i)},$$

which does not belong to any named family of distributions. To deal with this model, and others involving categorical data, Albert and Chib (1993) introduce a technique that has led to many applications. The Albert–Chib algorithm capitalizes on the simplifications afforded by introducing latent or auxiliary data into the sampling.

Instead of the specification above, the model is re-specified as

$$\begin{aligned} z_i|\boldsymbol{\beta} &\sim \mathcal{N}(\boldsymbol{x}_i'\boldsymbol{\beta}, 1), \\ y_i &= I[z_i > 0], \quad i \leqslant n, \\ \boldsymbol{\beta} &\sim \mathcal{N}_k(\boldsymbol{\beta}_0, \boldsymbol{B}_0). \end{aligned} \tag{2.1}$$

This specification is equivalent to the binary probit regression model since

$$\Pr(y_i = 1|\boldsymbol{x}_i, \boldsymbol{\beta}) = \Pr(z_i > 0|\boldsymbol{x}_i, \boldsymbol{\beta}) = \Phi(\boldsymbol{x}_i'\boldsymbol{\beta}),$$

as required.

Albert and Chib (1993) exploit this equivalence and propose that the latent variables $z = \{z_1, \dots, z_n\}$, one for each observation, be included in the MCMC algorithm along with the regression parameter $\boldsymbol{\beta}$. In other words, they suggest using MCMC methods to sample the joint posterior distribution

$$\pi(\boldsymbol{\beta}, z|\boldsymbol{y}) \propto \pi(\boldsymbol{\beta}) \prod_{i=1}^{N} \mathcal{N}(z_i|\boldsymbol{x}_i'\boldsymbol{\beta}, 1) \Pr(y_i|z_i, \boldsymbol{\beta}).$$

The term $\Pr(y_i|z_i, \boldsymbol{\beta})$ is obtained by reasoning as follows: $\Pr(y_i = 0|z_i, \beta)$ is one if $z_i < 0$ (regardless of the value taken by $\boldsymbol{\beta}$) and zero otherwise – but this is the definition of $I(z_i < 0)$. Similarly, $\Pr(y_i = 1|z_i, \boldsymbol{\beta})$ is one if $z_i > 0$, and zero otherwise – but this is the definition of $I(z_i > 0)$. Collecting these two cases we have therefore that

$$\pi(z, \boldsymbol{\beta}|\boldsymbol{y}) \propto \pi(\boldsymbol{\beta}) \prod_{i=1}^{N} \mathcal{N}(z_i|\boldsymbol{x}_i'\boldsymbol{\beta}, 1)\{I(z_i < 0)^{1-y_i} + I(z_i > 0)^{y_i}\}.$$

The latter posterior density is now sampled by a two-block Gibbs sampler composed of the full conditional distributions

$$z|\boldsymbol{y}, \boldsymbol{\beta};\ \boldsymbol{\beta}|\boldsymbol{y}, z.$$

Even though the parameter space has been enlarged, the introduction of the latent variables simplifies the problem considerably. The second conditional distribution, i.e., $\boldsymbol{\beta}|\boldsymbol{y}, z$, is the same as the distribution $\boldsymbol{\beta}|z$ since z implies $\boldsymbol{y}$, and hence $\boldsymbol{y}$ has no additional information for $\boldsymbol{\beta}$. The distribution $\boldsymbol{\beta}|z$ is easy to derive by standard Bayesian results for a continuous response. The first conditional distribution, i.e., $z|\boldsymbol{y}, \boldsymbol{\beta}$, factors into n distributions $z_i|y_i, \boldsymbol{\beta}$ and from the above is seen to be truncated normal given the value of y_i. Specifically,

$$p(z_i|y_i, \boldsymbol{\beta}) \propto \mathcal{N}(z_i|\boldsymbol{x}_i'\boldsymbol{\beta}, 1)\{I(z_i < 0)^{1-y_i} + I(z_i > 0)^{y_i}\}$$

which implies that if $y_i = 0$, the posterior density of z_i is proportional to $\mathcal{N}(z_i|\boldsymbol{x}_i'\boldsymbol{\beta}, 1)I[z < 0]$, a truncated normal distribution with support $(-\infty, 0)$, whereas if $y_i = 1$, the density is proportional to $\mathcal{N}(z_i|\boldsymbol{x}_i'\boldsymbol{\beta}, 1)I[z > 0]$, a truncated normal distribution with support $(0, \infty)$. These truncated normal distributions are simulated by applying the *inverse cdf* method (Ripley, 1987). Specifically, it can be shown that the draw

$$\mu + \sigma\Phi^{-1}\left[\Phi\left(\frac{a-\mu}{\sigma}\right) + U\left(\Phi\left(\frac{b-\mu}{\sigma}\right) - \Phi\left(\frac{a-\mu}{\sigma}\right)\right)\right], \tag{2.2}$$

where Φ^{-1} is the inverse cdf of the $\mathcal{N}(0, 1)$ distribution and $U \sim \text{Uniform}(0, 1)$, is from a $\mathcal{N}(\mu, \sigma^2)$ distribution truncated to the interval (a, b).

The Albert–Chib algorithm for the probit binary response model may now be summarized as follows.

ALGORITHM 1 (*Binary probit link model*).

```
1 Sample
```

$$z_i|y_i, \boldsymbol{\beta} \propto \mathcal{N}(z_i|\boldsymbol{x}_i'\boldsymbol{\beta}, 1)\{I(z_i < 0)^{1-y_i} + I(z_i > 0)^{y_i}\}, \quad i \leqslant n$$

```
2 Sample
```

$$\boldsymbol{\beta}|z \sim \mathcal{N}_k\left(\boldsymbol{B}_n\left(\boldsymbol{B}_0^{-1}\boldsymbol{\beta}_0 + \sum_{i=1}^{n}\boldsymbol{x}_i z_i\right), \boldsymbol{B}_n = \left(\boldsymbol{B}_0^{-1} + \sum_{i=1}^{n}\boldsymbol{x}_i\boldsymbol{x}_i'\right)^{-1}\right)$$

```
3 Go to 1
```

EXAMPLE. Consider the data in Table 1, taken from Fahrmeir and Tutz (1994), which is concerned with the occurrence or nonoccurence of infection following birth by cesarean section. The response variable y is one if the cesarean birth resulted in an infection, and zero if not. The available covariates are three indicator variables: x_1 is an indicator for whether the cesarean was nonplanned; x_2 is an indicator for whether risk factors were present at the time of birth and x_3 is an indicator for whether antibiotics were given as a prophylaxis. The data in the table contains information from 251 births. Under the column of the response, an entry such as 11/87 means that there were 98 deliveries with covariates (1, 1, 1) of whom 11 developed an infection and 87 did not.

Let us model the binary response by a *probit* model, letting the probability of infection for the ith birth be given as

$$\Pr(y_i = 1|\boldsymbol{x}_i, \boldsymbol{\beta}) = \Phi(\boldsymbol{x}_i'\boldsymbol{\beta}),$$

where $\boldsymbol{x}_i = (1, x_{i1}, x_{i2}, x_{i3})'$ is the covariate vector, $\boldsymbol{\beta} = (\beta_0, \beta_1, \beta_2, \beta_3)$ is the vector of unknown coefficients and Φ is the cdf of the standard normal random variable. Let us assume that our prior information about $\boldsymbol{\beta}$ can be represented by a multivariate normal density with mean centered at zero for each parameter, and variance given by $5\boldsymbol{I}_4$, where $\boldsymbol{I}_4$ is the four-dimensional identity matrix.

Algorithm 1 is run for 5000 cycles beyond a burn-in of 100 iterations. In Table 2 we provide the prior and posterior first two moments, and the 2.5th (lower) and 97.5th (upper) percentiles, of the marginal densities of $\boldsymbol{\beta}$. Each of the quantities is computed from the posterior draws in an obvious way, the posterior standard deviation in the table is

Table 1
Cesarean infection data

y (1/0)	x_1	x_2	x_3
11/87	1	1	1
1/17	0	1	1
0/2	0	0	1
23/3	1	1	0
28/30	0	1	0
0/9	1	0	0
8/32	0	0	0

Table 2
Cesarean data: Prior–posterior summary based on 5000 draws (beyond a burn-in of 100 cycles) from the Albert–Chib algorithm

	Prior		Posterior			
	Mean	Std dev	Mean	Std dev	Lower	Upper
β_0	0.000	3.162	−1.100	0.210	−1.523	−0.698
β_1	0.000	3.162	0.609	0.249	0.126	1.096
β_2	0.000	3.162	1.202	0.250	0.712	1.703
β_3	0.000	3.162	−1.903	0.266	−2.427	−1.393

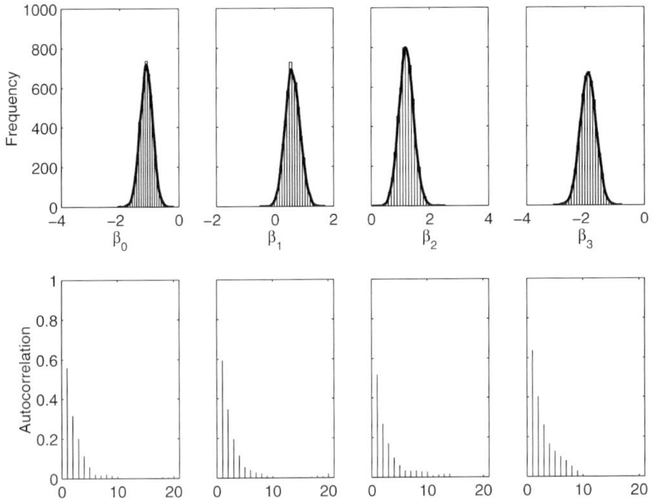

Fig. 1. Cesarean data with Albert–Chib algorithm: Marginal posterior densities (top panel) and autocorrelation plot (bottom panel).

the standard deviation of the sampled variates and the posterior percentiles are just the percentiles of the sampled draws. As expected, both the first and second covariates increase the probability of infection while the third covariate (the antibiotics prophylaxis) reduces the probability of infection.

To get an idea of the form of the posterior density we plot in Figure 1 the four marginal posterior densities. The density plots are obtained by smoothing the histogram of the simulated values with a Gaussian kernel. In the same plot we also report the autocorrelation functions (correlation against lag) for each of the sampled parameter values. The autocorrelation plots provide information of the extent of serial dependence in the sampled values. Here we see that the serial correlations decline quickly to zero indicating that the sampler is mixing well.

2.1. Marginal likelihood of the binary probit

The marginal likelihood of the binary probit model is easily calculated by the method of Chib (1995). Let $\boldsymbol{\beta}^* = E(\boldsymbol{\beta}|\boldsymbol{y})$, then from the expression of the marginal likelihood in (1.6), we have that

$$\ln m(\boldsymbol{y}) = \ln \pi(\boldsymbol{\beta}^*) + \ln p(\boldsymbol{y}|\boldsymbol{\beta}^*) - \ln \pi(\boldsymbol{\beta}^*|\boldsymbol{y}),$$

where the first term is $\ln \mathcal{N}_k(\boldsymbol{\beta}^*|\boldsymbol{\beta}_0, \boldsymbol{B}_0)$, the second is

$$\sum_{i=1}^{n} y_i \ln \Phi(\boldsymbol{x}_i'\boldsymbol{\beta}^*) + (1 - y_i)(1 - \ln \Phi(\boldsymbol{x}_i'\boldsymbol{\beta}^*))$$

and the third, which is

$$\ln \int \pi(\boldsymbol{\beta}|z)\pi(z|\boldsymbol{y})\,\mathrm{d}z$$

is estimated by averaging the conditional normal density of $\boldsymbol{\beta}$ in Step 2 of Algorithm 1 at the point $\boldsymbol{\beta}^*$ over the simulated values $\{z^{(g)}\}$ from the MCMC run. Specifically, letting $\hat{\pi}(\boldsymbol{\beta}^*|\boldsymbol{y})$ denote our estimate of the ordinate at $\boldsymbol{\beta}^*$ we have

$$\hat{\pi}(\boldsymbol{\beta}^*|\boldsymbol{y}) = M^{-1} \sum_{i=1}^{M} \mathcal{N}_k\left(\boldsymbol{\beta}^*|\boldsymbol{B}_n\left(\boldsymbol{B}_0^{-1}\boldsymbol{\beta}_0 + \sum_{i=1}^{n} \boldsymbol{x}_i z_i^{(g)}\right), \boldsymbol{B}_n\right).$$

2.2. Other link functions

The preceding algorithm is for the probit link function. Albert and Chib (1993) showed how it can be readily extended to other link functions that are generated from the scale mixture of normals family of distributions. For example, we could let

$$\begin{aligned} z_i|\boldsymbol{\beta}, \lambda_i &\sim \mathcal{N}(\boldsymbol{x}_i'\boldsymbol{\beta}, \lambda_i^{-1}), \\ y_i &= I[z_i > 0], \quad i \leqslant n, \\ \boldsymbol{\beta} &\sim \mathcal{N}_k(\boldsymbol{\beta}_0, \boldsymbol{B}_0), \\ \lambda_i &\sim \mathcal{G}\left(\frac{\upsilon}{2}, \frac{\nu}{2}\right) \end{aligned} \tag{2.3}$$

in which case $\Pr(y_i = 1|\boldsymbol{\beta}) = F_{t,\nu}(\boldsymbol{x}_i'\beta)$, where $F_{t,\nu}$ is the cdf of the standard-t distribution with ν degrees of freedom. Albert and Chib (1993) utilized this setup with $\nu = 8$ to approximate the logit link function. In this case, the posterior distribution of interest is given by

$$\pi(z, \boldsymbol{\lambda}, \boldsymbol{\beta}|\boldsymbol{y}) \propto \pi(\boldsymbol{\beta}) \prod_{i=1}^{N} \mathcal{N}(z_i|\boldsymbol{x}_i'\boldsymbol{\beta}, \lambda_i^{-1})\mathcal{G}\left(\lambda_i\left|\frac{\upsilon}{2}, \frac{\nu}{2}\right.\right) \\ \times \left\{I(z_i < 0)^{1-y_i} + I(z_i > 0)^{y_i}\right\}$$

which can be sampled in two-blocks as

$$(z, \boldsymbol{\lambda})|\boldsymbol{y}, \boldsymbol{\beta};\ \boldsymbol{\beta}|\boldsymbol{y}, z, \boldsymbol{\lambda},$$

where $\lambda = (\lambda_1, \ldots, \lambda_n)$. The first block is sampled by the method of composition by sampling $z_i|y_i, \boldsymbol{\beta}$ marginalized over λ_i from the truncated student-t density

$\mathcal{T}_\nu(z_i|\boldsymbol{x}_i'\boldsymbol{\beta}, 1)\{I(z_i < 0)^{1-y_i} + I(z_i > 0)^{y_i}\}$, followed by sampling $\lambda_i|z_i, \boldsymbol{\beta}$ from an updated Gamma distribution. The second distribution is normal with parameters obtained by standard Bayesian results.

ALGORITHM 2 (*Student-t binary model*).

```
1 Sample
```

(a)

$$z_i|y_i, \boldsymbol{\beta} \propto \mathcal{T}_\nu\left(z_i|\boldsymbol{x}_i'\boldsymbol{\beta}, 1\right)\left\{I(z_i < 0)^{1-y_i} + I(z_i > 0)^{y_i}\right\}$$

(b)

$$\lambda_i|z_i, \boldsymbol{\beta} \sim \mathcal{G}\left(\frac{\nu+1}{2}, \frac{\nu + (z_i - \boldsymbol{x}_i'\boldsymbol{\beta})^2}{2}\right), \quad i \leqslant n$$

```
2 Sample
```

$$\boldsymbol{\beta}|z, \boldsymbol{\lambda} \sim \mathcal{N}_k\left(\boldsymbol{B}_n\left(\boldsymbol{B}_0^{-1}\boldsymbol{\beta}_0 + \sum_{i=1}^n \lambda_i \boldsymbol{x}_i z_i\right), \boldsymbol{B}_n = \left(\boldsymbol{B}_0^{-1} + \sum_{i=1}^n \lambda_i \boldsymbol{x}_i \boldsymbol{x}_i'\right)^{-1}\right)$$

```
3 Goto 1
```

Chen et al. (1999) present a further extension of this idea to model a skewed link function while Basu and Mukhopadhyay (2000) and Basu and Chib (2003) model the distribution of λ_i by a Dirichlet process prior leading to a semiparametric Bayesian binary data model. The logistic normal link function is considered by Allenby and Lenk (1994) with special reference to data arising in marketing.

2.3. *Marginal likelihood of the student-t binary model*

The marginal likelihood of the student-t binary model is also calculated by the method of Chib (1995). It is easily seen from the expression of the marginal likelihood in (1.6) that the only quantity that needs to be estimated is $\pi(\boldsymbol{\beta}^*|\boldsymbol{y})$, where $\boldsymbol{\beta}^* = E(\boldsymbol{\beta}^*|\boldsymbol{y})$. An estimate of this ordinate is obtained by averaging the conditional normal density of $\boldsymbol{\beta}$ in Step 2 of Algorithm 2 at the point $\boldsymbol{\beta}^*$ over the simulated values $\{z^{(g)}\}$ and $\{\lambda_i^{(g)}\}$ from the MCMC run. Specifically, letting $\hat{\pi}(\boldsymbol{\beta}^*|\boldsymbol{y})$ denote our estimate of the ordinate at $\boldsymbol{\beta}^*$ we have

$$\hat{\pi}(\boldsymbol{\beta}^*|\boldsymbol{y}) = M^{-1}\sum_{i=1}^M \mathcal{N}_k\left(\boldsymbol{B}_n^{(g)}\left(\boldsymbol{B}_0^{-1}\boldsymbol{\beta}_0 + \sum_{i=1}^n \lambda_i^{(g)}\boldsymbol{x}_i z_i^{(g)}\right), \boldsymbol{B}_n^{(g)}\right),$$

where

$$\boldsymbol{B}_n^{(g)} = \left(\boldsymbol{B}_0^{-1} + \sum_{i=1}^n \lambda_i^{(g)}\boldsymbol{x}_i\boldsymbol{x}_i'\right)^{-1}.$$

3. Ordinal response data

Albert and Chib (1993) extend their algorithm to the ordinal categorical data case where y_i can take one of the values $\{0, 1, \ldots, J\}$ according to the probabilities

$$\Pr(y_i \leqslant j|\boldsymbol{\beta}, \boldsymbol{c}) = F(c_j - \boldsymbol{x}_i'\boldsymbol{\beta}), \quad j = 0, 1, \ldots, J, \tag{3.1}$$

where F is some link function, say either Φ or $F_{t,\nu}$. In this model the $\{c_j\}$ are category specific cut-points such that $c_0 = 0$ and $c_J = \infty$. The remaining cut-points $\boldsymbol{c} = (c_1, \ldots, c_{J-1})$ are assumed to satisfy the order restriction $c_1 \leqslant \cdots \leqslant c_{J-1}$ which ensures that the cumulative probabilities are nondecreasing. Given data $y_1, \ldots, y_n$ from this model, the likelihood function is

$$p(\boldsymbol{y}|\boldsymbol{\beta}, \boldsymbol{c}) = \prod_{j=0}^{J} \prod_{i:y_i=j} \left[F(c_j - \boldsymbol{x}_i'\boldsymbol{\beta}) - F(c_{j-1} - \boldsymbol{x}_i'\boldsymbol{\beta})\right] \tag{3.2}$$

and the posterior density is proportional to $\pi(\boldsymbol{\beta}, \boldsymbol{c})p(\boldsymbol{y}|\boldsymbol{\beta}, \boldsymbol{c})$, where $\pi(\boldsymbol{\beta}, \boldsymbol{c})$ is the prior distribution.

Simulation of the posterior distribution is again feasible by the tactical introduction of latent variables. For the link function $F_{t,\upsilon}$, we introduce the latent variables $z = (z_1, \ldots, z_n)$, where $z_i|\boldsymbol{\beta}, \lambda_i \sim \mathcal{N}(\boldsymbol{x}_i\boldsymbol{\beta}, \lambda_i^{-1})$. A priori, we observe $y_i = j$ if the latent variable z_i falls in the interval $[c_{j-1}, c_j)$. Define $\delta_{ij} = 1$ if $y_i = j$, and 0 otherwise, then the posterior density of interest is

$$\pi(z, \boldsymbol{\lambda}, \boldsymbol{\beta}, \boldsymbol{c}|\boldsymbol{y}) \propto \pi(\boldsymbol{\beta}, \boldsymbol{c}) \prod_{i=1}^{N} \mathcal{N}(z_i|\boldsymbol{x}_i'\boldsymbol{\beta}, \lambda_i^{-1}) \mathcal{G}\left(\lambda_i \middle| \frac{\upsilon}{2}, \frac{\upsilon}{2}\right) \times \left\{\sum_{l=0}^{J} I(c_{l-1} < z_i < c_l)^{\delta_{il}}\right\},$$

where $c_{-1} = -\infty$. Now the basic Albert and Chib MCMC scheme draws cut-points, the latent data and the regression parameters in sequence. Albert and Chib (2001) simplified the latter step by transforming the cut-points so as to remove the ordering constraint. The transformation is defined by the one-to-one map

$$a_1 = \log c_1; \qquad a_j = \log(c_j - c_{j-1}), \quad 2 \leqslant j \leqslant J - 1. \tag{3.3}$$

The advantage of working with $\boldsymbol{a}$ instead of $\boldsymbol{c}$ is that the parameters of the tailored proposal density in the M–H step for $\boldsymbol{a}$ can be obtained by an unconstrained optimization and the prior $\pi(\boldsymbol{a})$ on $\boldsymbol{a}$ can be an unrestricted multivariate normal. Next, given $y_i = j$ and the cut-points the sampling of the latent data z_i is from $\mathcal{T}_\nu(\boldsymbol{x}_i'\boldsymbol{\beta}, 1)I(c_{j-1} < z_i < c_j)$, marginalized over $\{\lambda_i\}$. The sampling of the parameters $\boldsymbol{\beta}$ is as in Algorithm 2.

ALGORITHM 3 (*Ordinal student link model*).

```
1 M-H
   (a) Sample
```

$$\pi(\boldsymbol{a}|\boldsymbol{y},\boldsymbol{\beta}) \propto \pi(\boldsymbol{a}) \underbrace{\prod_{j=0}^{J} \prod_{i:y_i=j} \left[F_{t,\nu}\left(c_j - \boldsymbol{x}_i'\boldsymbol{\beta}\right) - F_{t,\nu}\left(c_{j-1} - \boldsymbol{x}_i'\boldsymbol{\beta}\right)\right]}_{p(\boldsymbol{y}|\boldsymbol{\beta},\boldsymbol{a})}$$

with the M-H algorithm by calculating

$$\boldsymbol{m} = \arg\max_{\boldsymbol{a}} \log p(\boldsymbol{y}|\boldsymbol{\beta},\boldsymbol{a})$$

and $\boldsymbol{V} = \{-\partial \log f(\boldsymbol{y}|\boldsymbol{\beta},\boldsymbol{a})/\partial\boldsymbol{a}\partial\boldsymbol{a}'\}^{-1}$ the negative inverse of the hessian at $\boldsymbol{m}$, proposing

$$\boldsymbol{a}' \sim f_T(\boldsymbol{a}|\boldsymbol{m},\boldsymbol{V},\xi),$$

for some choice of ξ, calculating

$$\alpha(\boldsymbol{a},\boldsymbol{a}'|\boldsymbol{y},\boldsymbol{\beta}) = \min\left\{\frac{\pi(\boldsymbol{a}')p(\boldsymbol{y}|\boldsymbol{\beta},\boldsymbol{a}')}{\pi(\boldsymbol{a})p(\boldsymbol{y}|\boldsymbol{\beta},\boldsymbol{a})} \frac{\mathcal{T}_\nu(\boldsymbol{a}|\boldsymbol{m},\boldsymbol{V},\xi)}{\mathcal{T}_\nu(\boldsymbol{a}'|\boldsymbol{m},\boldsymbol{V},\xi)}, 1\right\}$$

and moving to $\boldsymbol{a}'$ with probability $\alpha(\boldsymbol{a},\boldsymbol{a}'|\boldsymbol{y},\boldsymbol{\beta})$, then transforming the new $\boldsymbol{a}$ to $\boldsymbol{c}$ via the inverse map $c_j = \sum_{i=1}^{j} \exp(a_i)$, $1 \leqslant j \leqslant J-1$.

(b) Sample

$$z_i|y_i,\boldsymbol{\beta},\boldsymbol{c} \propto \mathcal{T}_\nu\left(z_i|\boldsymbol{x}_i'\boldsymbol{\beta},1\right)\left\{\sum_{l=0}^{J} I(c_{l-1} < z_i < c_l)^{\delta_{il}}\right\}, \quad i \leqslant n$$

(c) Sample

$$\lambda_i|y_i,z_i,\boldsymbol{\beta},\boldsymbol{c} \sim \mathcal{G}\left(\frac{\nu+1}{2}, \frac{\nu+(z_i-\boldsymbol{x}_i'\boldsymbol{\beta})^2}{2}\right), \quad i \leqslant n$$

2 Sample

$$\boldsymbol{\beta}|z,\boldsymbol{\lambda} \sim \mathcal{N}_k\left(\boldsymbol{B}_n\left(\boldsymbol{B}_0^{-1}\boldsymbol{\beta}_0 + \sum_{i=1}^{n}\lambda_i\boldsymbol{x}_i z_i\right), \boldsymbol{B}_n = \left(\boldsymbol{B}_0^{-1} + \sum_{i=1}^{n}\lambda_i\boldsymbol{x}_i\boldsymbol{x}_i'\right)^{-1}\right)$$

3 Goto 1

3.1. Marginal likelihood of the student-t ordinal model

The marginal likelihood of the student-t ordinal model can be calculated by the method of Chib (1995) and Chib and Jeliazkov (2001). Let $\boldsymbol{\beta}^* = E(\boldsymbol{\beta}|\boldsymbol{y})$ and $\boldsymbol{c}^* = E(\boldsymbol{c}^*|\boldsymbol{y})$ and let $\boldsymbol{a}^*$ be the transformed $\boldsymbol{c}^*$. Then from the expression of the marginal likelihood in (1.6) we can write

$$\ln m(\boldsymbol{y}) = \ln\pi\left(\boldsymbol{\beta}^*,\boldsymbol{a}^*\right) + \ln p\left(\boldsymbol{y}|\boldsymbol{\beta}^*,\boldsymbol{c}^*\right) - \ln\pi\left(\boldsymbol{a}^*|\boldsymbol{y}\right) - \ln\pi\left(\boldsymbol{\beta}^*|\boldsymbol{y},\boldsymbol{a}^*\right).$$

The marginal posterior ordinate $\pi(\boldsymbol{a}^*|\boldsymbol{y})$ from (1.11) can be written as

$$\pi(\boldsymbol{a}^*|\boldsymbol{y}) = \frac{E_1\{\alpha(\boldsymbol{a}, \boldsymbol{a}^*|\boldsymbol{y}, \boldsymbol{\beta})\mathcal{T}_\nu(\boldsymbol{a}'|\boldsymbol{m}, \boldsymbol{V}, \xi)\}}{E_2\{\alpha(\boldsymbol{a}^*, \boldsymbol{a}|\boldsymbol{y}, \boldsymbol{\beta})\}},$$

where E_1 is the expectation with respect to $\pi(\boldsymbol{\beta}|\boldsymbol{y})$ and E_2 is the expectation with respect to $\pi(\boldsymbol{\beta}|\boldsymbol{y}, \boldsymbol{a}^*)\mathcal{T}_\nu(\boldsymbol{a}|\boldsymbol{m}, \boldsymbol{V}, \xi)$. The E_1 expectation can be estimated by draws from the main run and the E_2 expectation from a reduced run in which $\boldsymbol{a}$ is set to $\boldsymbol{a}^*$. For each value of $\boldsymbol{\beta}$ in this reduced run, $\boldsymbol{a}$ is also sampled from $\mathcal{T}_\nu(\boldsymbol{a}|\boldsymbol{m}, \boldsymbol{V}, \xi)$; these two draws (namely the draws of $\boldsymbol{\beta}$ and $\boldsymbol{a}$) are used to average $\alpha(\boldsymbol{a}^*, \boldsymbol{a}|\boldsymbol{y}, \boldsymbol{\beta})$. The final ordinate $\pi(\boldsymbol{\beta}^*|\boldsymbol{y}, \boldsymbol{a}^*)$ is estimated by averaging the normal conditional density in Step 2 of Algorithm 3 over the values of $\{z_i\}$ and $\{\lambda_i\}$ from this same reduced run.

4. Sequential ordinal model

The ordinal model presented above is appropriate when one can assume that a single unobservable continuous variable underlies the ordinal response. For example, this representation can be used for modeling ordinal letter grades in a mathematics class by assuming that the grade is an indicator of the student's unobservable continuous-valued intelligence. In other circumstances, however, a different latent variable structure may be required to model the ordinal response. For example, McCullagh (1980), and Farhmeir and Tutz (1994) consider a data set where the relative tonsil size of children is classified into the three states: "present but not enlarged", "enlarged", and "greatly enlarged". The objective is to explain the ordinal response as a function of whether the child is a carrier of the Streptococcus pyogenes. In this setting, it may not be appropriate to model the ordinal tonsil size in terms of a single latent variable. Rather, it may be preferable to imagine that each response is determined by two continuous latent variables where the first latent variable measures the propensity of the tonsil to grow abnormally and pass from the state "present but not enlarged" to the state "enlarged". The second latent variable measures the propensity of the tonsil to grow from the "enlarged" to the "greatly enlarged" states. Thus, the final "greatly enlarged" state is realized only when the tonsil has passed through the earlier levels. This latent variable representation leads to a *sequential* model because the levels of the response are achieved in a sequential manner.

Suppose that one observes independent observations $y_1, \ldots, y_n$, where each y_i is an ordinal categorical response variable with J possible values $\{1, \ldots, J\}$. Associated with the ith response y_i, let $\boldsymbol{x}_i = (x_{i1}, \ldots, x_{ik})$ denote a set of k covariates. In the sequential ordinal model, the variable y_i can take the value j only after the levels $1, \ldots, j-1$ are reached. In other words, to get to the outcome j, one must pass through levels $1, 2, \ldots, j-1$ and stop (or fail) in level j. The probability of stopping in level j $(1 \leqslant j \leqslant J-1)$, conditional on the event that the jth level is reached, is given by

$$\Pr(y_i = j|y_i \geqslant j, \boldsymbol{\delta}, \boldsymbol{c}) = F(c_j - \boldsymbol{x}_i'\boldsymbol{\delta}), \tag{4.1}$$

where $\boldsymbol{c} = (c_1, \ldots, c_{J-1})$ are unordered cutpoints and $\boldsymbol{x}_i'\boldsymbol{\delta}$ represents the effect of the covariates. This probability function is referred to as the discrete time hazard function

(Tutz, 1990; Farhmeir and Tutz, 1994). It follows that the probability of stopping at level j is given by

$$\begin{aligned}\Pr(y_i = j|\boldsymbol{\delta}, \boldsymbol{c}) &= \Pr(y_i = j|y_i \geqslant j, \boldsymbol{\delta}, \boldsymbol{c}) \Pr(y_i \geqslant j) \\ &= F\left(c_j - \boldsymbol{x}_i'\boldsymbol{\delta}\right) \prod_{k=1}^{j-1} \left\{1 - F\left(c_k - \boldsymbol{x}_i'\boldsymbol{\delta}\right)\right\}, \quad j \leqslant J-1, \end{aligned} \tag{4.2}$$

whereas the probability that the final level J is reached is

$$\Pr(y_i = J|\boldsymbol{\delta}, \boldsymbol{c}) = \prod_{k=1}^{J-1} \left\{1 - F\left(c_k - \boldsymbol{x}_i'\boldsymbol{\delta}\right)\right\}, \tag{4.3}$$

since the event $y_i = J$ occurs only if all previous $J - 1$ levels are passed.

The sequential model is formally equivalent to the continuation-ratio ordinal models, discussed by Agresti (1990, Chapter 9), Cox (1972) and Ten Have and Uttal (2000). The sequential model is useful in the analysis of discrete-time survival data (Farhmeir and Tutz (1994), Chapter 9; Kalbfleish and Prentice, 1980). More generally, the sequential model can be used to model nonproportional and nonmonotone hazard functions and to incorporate the effect of time dependent covariates. Suppose that one observes the time to failure for subjects in the sample, and let the time interval be subdivided (or grouped) into the J intervals $[a_0, a_1), [a_1, a_2), \ldots, [a_{J-1}, \infty)$. Then one defines the event $y_i = j$ if a failure is observed in the interval $[a_{j-1}, a_j)$. The discrete hazard function (4.1) now represents the probability of failure in the time interval $[a_{j-1}, a_j)$ given survival until time a_{j-1}. The vector of cutpoint parameters $\boldsymbol{c}$ represents the baseline hazard of the process.

Albert and Chib (2001) develop a full Bayesian analysis of this model based on the framework of Albert and Chib (1993). Suppose that F is the distribution function of the standard normal distribution, leading to the sequential ordinal model. Corresponding to the ith observation, define latent variables $\{w_{ij}\}$, where $w_{ij} = \boldsymbol{x}_i'\boldsymbol{\delta} + e_{ij}$ and the e_{ij} are independently distributed from $N(0, 1)$. We observe $y_i = 1$ if $w_{i1} \leqslant c_1$, and observe $y_i = 2$ if the first latent variable $w_{i1} > c_1$ and the second variable $w_{i2} \leqslant c_2$. In general, we observe $y_i = j$ $(1 \leqslant j \leqslant J-1)$ if the first $j - 1$ latent variables exceed their corresponding cutoffs, and the jth variable does not: $y_i = j$ if $w_{i1} > c_1, \ldots, w_{ij-1} > c_{j-1}, w_{ij} \leqslant c_j$. In this model, the latent variable w_{ij} represents one's propensity to continue to the $(j+1)$st level in the sequence, given that the individual has already attained level j.

This latent variable representation can be simplified by incorporating the cutpoints $\{\gamma_j\}$ into the mean function. Define the new latent variable $z_{ij} = w_{ij} - c_j$. Then it follows that $z_{ij}|\boldsymbol{\beta} \sim \mathcal{N}(\boldsymbol{x}_{ij}'\boldsymbol{\beta}, 1)$, where $\boldsymbol{x}_{ij}' = (0, 0, \ldots, -1, 0, 0, \boldsymbol{x}_i')$ with -1 in the jth column, and $\boldsymbol{\beta} = (c_1, \ldots, c_{J-1}, \boldsymbol{\delta}')'$. The observed data are then generated according to

$$y_i = \begin{cases} 1 & \text{if } z_{i1} \leqslant 0, \\ 2 & \text{if } z_{i1} > 0, z_{i2} \leqslant 0, \\ \vdots & \vdots \\ J-1 & \text{if } z_{i1} > 0, \dots, z_{iJ-2} > 0, z_{iJ-1} \leqslant 0, \\ J & \text{if } z_{i1} > 0, \dots, z_{iJ-1} > 0. \end{cases} \tag{4.4}$$

Following the Albert and Chib (1993) approach, this model can be fit by MCMC methods by simulating the joint posterior distribution of $(\{z_{ij}\}, \boldsymbol{\beta})$. The latent data $\{z_{ij}\}$, conditional on $(\boldsymbol{y}, \boldsymbol{\beta})$, are independently distributed as truncated normal. Specifically, if $y_i = j$, then the latent data corresponding to this observation are represented as $z_i = (z_{i1}, \dots, z_{ij_i})$ where $j_i = \min\{j, J-1\}$ and the simulation of z_i is from a sequence of truncated normal distributions. The posterior distribution of the parameter vector $\boldsymbol{\beta}$, conditional on the latent data $z = (z_1, \dots, z_n)$, has a simple form. Let $\boldsymbol{X}_i$ denote the covariate matrix corresponding to the ith subject consisting of the j_i rows $\boldsymbol{x}'_{i1}, \dots, \boldsymbol{x}'_{ij_i}$. If $\boldsymbol{\beta}$ is assigned a multivariate normal prior with mean vector $\boldsymbol{\beta}_0$ and covariance matrix $\boldsymbol{B}_0$, then the posterior distribution of $\boldsymbol{\beta}$, conditional on z, is multivariate normal.

ALGORITHM 4 (*Sequential ordinal probit model*).

1 Sample $z_i | y_i = j, \boldsymbol{\beta}$

(a) If $y_i = 1$, sample

$$z_{i1} | y_i, \boldsymbol{\beta} \propto \mathcal{N}(\boldsymbol{x}'_{i1}\boldsymbol{\beta}, 1) I(z_{i1} < 0)$$

(b) If $y_i = j$ $(2 \leqslant j \leqslant J-1)$, sample

$$z_{ik} | y_i, \boldsymbol{\beta} \propto \mathcal{N}(\boldsymbol{x}'_{ik}\boldsymbol{\beta}, 1) I(z_{ik} > 0), \quad k = 1, \dots, j-1$$
$$z_{ij} | y_i, \boldsymbol{\beta} \propto \mathcal{N}(\boldsymbol{x}'_{ij}\boldsymbol{\beta}, 1) I(z_{ij} < 0)$$

(c) If $y_i = J$, sample

$$z_{ik} | y_i, \boldsymbol{\beta} \propto \mathcal{N}(\boldsymbol{x}'_{ik}\boldsymbol{\beta}, 1) I(z_{ik} > 0), \quad k = 1, \dots, J-1$$

2 Sample

$$\boldsymbol{\beta} | z \sim \mathcal{N}_k\left(\boldsymbol{B}_n\left(\boldsymbol{B}_0^{-1}\boldsymbol{\beta}_0 + \sum_{i=1}^{n} \boldsymbol{X}'_i z_i\right), \boldsymbol{B}_n = \left(\boldsymbol{B}_0^{-1} + \sum_{i=1}^{n} \boldsymbol{X}'_i \boldsymbol{X}_i\right)^{-1}\right)$$

3 Goto 1

Albert and Chib (2001) present generalizations of the sequential ordinal model and also discuss the computation of the marginal likelihood by the method of Chib (1995).

5. Multivariate responses

One of the considerable achievements of the recent Bayesian literature is in the development of methods for multivariate categorical data. A leading case concerns that of multivariate binary data. From the frequentist perspective, this problem is tackled by Carey et al. (1993) and Glonek and McCullagh (1995) in the context of the multivariate logit model, and from a Bayesian perspective by Chib and Greenberg (1998) in the context of the multivariate probit (MVP) model (an analysis of the multivariate logit model from a Bayesian perspective is supplied by O'Brien and Dunson, 2004). In many respects, the multivariate probit model, along with its variants, is the canonical modeling approach for correlated binary data. This model was introduced more than twenty five years ago by Ashford and Sowden (1970) in the context of bivariate binary responses and was analyzed under simplifying assumptions on the correlation structure by Amemiya (1972), Ochi and Prentice (1984) and Lesaffre and Kaufmann (1992). The general version of the model was considered to be intractable. Chib and Greenberg (1998), building on the framework of Albert and Chib (1993), resolve the difficulties.

The MVP model provides a relatively straightforward way of modeling multivariate binary data. In this model, the marginal probability of each response is given by a probit function that depends on covariates and response specific parameters. Associations between the binary variables are incorporated by assuming that the vector of binary outcomes are a function of correlated Gaussian variables, taking the value one if the corresponding Gaussian component is positive and the value zero otherwise. This connection with a latent Gaussian random vector means that the regression coefficients can be interpreted independently of the correlation parameters (unlike the case of log-linear models). The link with Gaussian data is also helpful in estimation. By contrast, models based on marginal odds ratios Connolly and Liang (1988) tend to proliferate nuisance parameters as the number of variables increase and they become difficult to interpret and estimate. Finally, the Gaussian connection enables generalizations of the model to ordinal outcomes (see Chen and Dey, 2000), mixed outcomes (Kuhnert and Do, 2003), and spatial data (Fahrmeir and Lang, 2001; Banerjee et al., 2004).

5.1. Multivariate probit model

Let y_{ij} denote a binary response on the ith observation unit and jth variable, and let $\boldsymbol{y}_i = (y_{i1}, \ldots, y_{iJ})'$, $1 \leqslant i \leqslant n$, denote the collection of responses on all J variables. Also let $\boldsymbol{x}_{ij}$ denote the set of covariates for the jth response and $\boldsymbol{\beta}_j \in R^{k_j}$ the conformable vector of covariate coefficients. Let $\boldsymbol{\beta}' = (\boldsymbol{\beta}'_1, \ldots, \boldsymbol{\beta}'_J) \in R^k$, $k = \sum k_j$ denote the complete set of coefficients and let $\boldsymbol{\Sigma} = \{\sigma_{jk}\}$ denote a $J \times J$ correlation matrix. Finally let

$$\boldsymbol{X}_i = \begin{pmatrix} \boldsymbol{x}'_{i1} & \mathbf{0}' & \cdots & \mathbf{0}' \\ \mathbf{0}' & \boldsymbol{x}'_{i2} & \mathbf{0}' & \mathbf{0}' \\ \vdots & \vdots & \vdots & \vdots \\ \mathbf{0}' & \mathbf{0}' & \cdots & \boldsymbol{x}'_{iJ} \end{pmatrix}$$

denote the $J \times k$ covariate matrix on the ith subject. Then, according to the multivariate probit model, the marginal probability that $y_{ij} = 1$ is given by the probit form

$$\Pr(y_{ij} = 1 | \boldsymbol{\beta}) = \Phi\left(\boldsymbol{x}'_{ij}\boldsymbol{\beta}_j\right)$$

and the joint probability of a given outcome y_i, conditioned on parameters $\boldsymbol{\beta}$, $\boldsymbol{\Sigma}$, and covariates $\boldsymbol{x}_{ij}$, is

$$\Pr(y_i|\boldsymbol{\beta}, \boldsymbol{\Sigma}) = \int_{A_{iJ}} \ldots \int_{A_{i1}} \mathcal{N}_J(t|0, \boldsymbol{\Sigma})\, \mathrm{d}t, \tag{5.1}$$

where $\mathcal{N}_J(t|0, \boldsymbol{\Sigma})$ is the density of a J-variate normal distribution with mean vector 0 and correlation matrix $\boldsymbol{\Sigma}$ and A_{ij} is the interval

$$A_{ij} = \begin{cases} (-\infty, \boldsymbol{x}'_{ij}\boldsymbol{\beta}_j) & \text{if } y_{ij} = 1, \\ [\boldsymbol{x}'_{ij}\boldsymbol{\beta}_j, \infty) & \text{if } y_{ij} = 0. \end{cases} \tag{5.2}$$

Note that each outcome is determined by its own set of k_j covariates $\boldsymbol{x}_{ij}$ and covariate effects $\boldsymbol{\beta}_j$.

The multivariate discrete mass function presented in (5.1) can be specified in terms of latent Gaussian random variables. This alternative formulation also forms the basis of the computational scheme that is described below. Let $z_i = (z_{i1}, \ldots, z_{iJ})$ denote a J-variate normal vector and let

$$z_i \sim \mathcal{N}_J(\boldsymbol{X}_i\boldsymbol{\beta}, \boldsymbol{\Sigma}). \tag{5.3}$$

Now let y_{ij} be 1 or 0 according to the sign of z_{ij}:

$$y_{ij} = I(z_{ij} > 0), \quad j = 1, \ldots, J. \tag{5.4}$$

Then, the probability in (5.1) may be expressed as

$$\int_{B_{iJ}} \cdots \int_{B_{i1}} \mathcal{N}_J(z_i|\boldsymbol{X}_i\boldsymbol{\beta}, \boldsymbol{\Sigma})\, \mathrm{d}z_i, \tag{5.5}$$

where B_{ij} is the interval $(0, \infty)$ if $y_{ij} = 1$ and the interval $(-\infty, 0]$ if $y_{ij} = 0$. It is easy to confirm that this integral reduces to the form given above. It should also be noted that due to the threshold specification in (5.4), the scale of z_{ij} cannot be identified. As a consequence, the matrix $\boldsymbol{\Sigma}$ must be in correlation form (with units on the main diagonal).

5.2. *Dependence structures*

One basic question in the analysis of correlated binary data is the following: How should correlation between binary outcomes be defined and measured? The point of view behind the MVP model is that the correlation is modeled at the level of the latent data which then induces correlation amongst the binary outcomes. This modeling perspective is both flexible and general. In contrast, attempts to model correlation directly (as in the classical literature using marginal odds ratios) invariably lead to difficulties, partly because it is difficult to specify pair-wise correlations in general, and partly because the binary data scale is not natural for thinking about dependence.

Within the context of the MVP model, alternative dependence structures are easily specified and conceived, due to the connection with Gaussian latent data. Some of the possibilities are enumerated below.

- Unrestricted form. Here $\boldsymbol{\Sigma}$ is fully unrestricted except for the unit constraints on the diagonal. The unrestricted $\boldsymbol{\Sigma}$ matrix has $J^* = J(J-1)/2$ unknown correlation parameters that must be estimated.
- Equicorrelated form. In this case, the correlations are all equal and described by a single parameter ρ. This form can be a starting point for the analysis when one is dealing with outcomes where are all the pair-wise correlations are believed to have the same sign. The equicorrelated model may be also be seen as arising from a random effect formulation.
- Toeplitz form. Under this case, the correlations depend on a single parameter ρ but under the restriction that $\mathrm{Corr}(z_{ik}, z_{il}) = \rho^{|k-l|}$. This version can be useful when the binary outcomes are collected from a longitudinal study where it is plausible that the correlation between outcomes at different dates will diminish with the lag. In fact, in the context of longitudinal data, the $\boldsymbol{\Sigma}$ matrix can be specified in many other forms with analogy with the correlation structures that arise in standard time series ARMA modeling.

5.3. Student-t specification

Now suppose that the distribution on the latent z_i is multivariate-t with specified degrees of freedom ν. This gives rise to a model that may be called the multivariate-t link model. Under the multivariate-t assumption, $z_i|\boldsymbol{\beta}, \boldsymbol{\Sigma} \sim \mathcal{T}_\nu(\boldsymbol{X}_i\boldsymbol{\beta}, \boldsymbol{\Sigma})$ with density

$$\mathcal{T}_\nu(z_i|\boldsymbol{\beta}, \boldsymbol{\Sigma}) \propto |\boldsymbol{\Sigma}|^{-1/2}\left\{1 + \frac{1}{\nu}(z_i - \boldsymbol{X}_i\boldsymbol{\beta})'\boldsymbol{\Sigma}^{-1}(z_i - \boldsymbol{X}_i\boldsymbol{\beta})\right\}^{-(\nu+J)/2}. \quad (5.6)$$

As before, the matrix $\boldsymbol{\Sigma}$ is in correlation form and the observed outcomes are defined by (5.4). The model for the latent z_i may be expressed as a scale mixture of normals by introducing a random variable $\lambda_i \sim \mathcal{G}(\frac{\nu}{2}, \frac{\nu}{2})$ and letting

$$z_i|\boldsymbol{\beta}, \boldsymbol{\Sigma}, \lambda_i \sim \mathcal{N}_J\left(\boldsymbol{X}_i\boldsymbol{\beta}, \lambda_i^{-1}\boldsymbol{\Sigma}\right). \quad (5.7)$$

Conditionally on λ_i, this model is equivalent to the MVP model.

5.4. Estimation of the MVP model

Consider now the question of inference in the MVP model. We are given a set of data on n subjects with outcomes $\boldsymbol{y} = \{\boldsymbol{y}_i\}_{i=1}^n$ and interest centers on the parameters of the model $\boldsymbol{\theta} = (\boldsymbol{\beta}, \boldsymbol{\Sigma})$ and the posterior distribution $\pi(\boldsymbol{\beta}, \boldsymbol{\Sigma}|\boldsymbol{y})$, given some prior distribution $\pi(\boldsymbol{\beta}, \boldsymbol{\Sigma})$ on the parameters.

To simplify the MCMC implementation for this model Chib and Greenberg (1998) follow the general approach of Albert and Chib (1993) and employ the latent variables $z_i \sim \mathcal{N}_J(\boldsymbol{X}_i\boldsymbol{\beta}, \boldsymbol{\Sigma})$, with the observed data given by $y_{ij} = I(z_{ij} > 0)$, $j = 1, \ldots, J$. Let $\boldsymbol{\sigma} = (\sigma_{12}, \sigma_{31}, \sigma_{32}, \ldots, \sigma_{JJ})$ denote the $J(J-1)/2$ distinct elements of $\boldsymbol{\Sigma}$. It can be shown that the admissible values of $\boldsymbol{\sigma}$ (that lead to a positive definite $\boldsymbol{\Sigma}$ matrix) form a convex solid body in the hypercube $[-1, 1]^p$. Denote this set by C. Now let $z = (z_1, \ldots, z_n)$ denote the latent values corresponding to the observed data $\boldsymbol{y} = \{\boldsymbol{y}_i\}_{i=1}^n$.

Then, the algorithm proceeds with the sampling of the posterior density

$$\pi(\boldsymbol{\beta}, \boldsymbol{\sigma}, z|\boldsymbol{y}) \propto \pi(\boldsymbol{\beta}, \boldsymbol{\sigma}) f(z|\boldsymbol{\beta}, \boldsymbol{\Sigma}) \Pr(\boldsymbol{y}|z, \boldsymbol{\beta}, \boldsymbol{\Sigma})$$
$$\propto \pi(\boldsymbol{\beta}, \boldsymbol{\sigma}) \prod_{i=1}^{n} \big(\phi_J(z_i|\boldsymbol{\beta}, \boldsymbol{\Sigma}) \Pr(\boldsymbol{y}_i|z_i, \boldsymbol{\beta}, \boldsymbol{\Sigma})\big), \quad \boldsymbol{\beta} \in \Re^k, \boldsymbol{\sigma} \in C,$$

where, akin to the result in the univariate binary case,

$$\Pr(\boldsymbol{y}_i|z_i, \boldsymbol{\beta}, \boldsymbol{\Sigma}) = \prod_{j=1}^{J} \big\{I(z_{ij} \leqslant 0)^{1-y_{ij}} + I(z_{ij} > 0)^{y_{ij}}\big\}.$$

Conditioned on $\{z_i\}$ and $\boldsymbol{\Sigma}$, the update for $\boldsymbol{\beta}$ is straightforward, while conditioned on $(\boldsymbol{\beta}, \boldsymbol{\Sigma})$, z_{ij} can be sampled one at a time conditioned on the other latent values from truncated normal distributions, where the region of truncation is either $(0, \infty)$ or $(-\infty, 0)$ depending on whether the corresponding y_{ij} is one or zero. The key step in the algorithm is the sampling of $\boldsymbol{\sigma}$, the unrestricted elements of $\boldsymbol{\Sigma}$, from the full conditional density $\pi(\boldsymbol{\sigma}|z, \boldsymbol{\beta}) \propto p(\boldsymbol{\sigma}) \prod_{i=1}^{n} \mathcal{N}_J(z_i|\boldsymbol{X}_i\boldsymbol{\beta}, \boldsymbol{\Sigma})$. This density, which is truncated to the complicated region C, is sampled by a M–H step with tailored proposal density $q(\boldsymbol{\sigma}|z, \boldsymbol{\beta}) = \mathcal{T}_\nu(\boldsymbol{\sigma}|\boldsymbol{m}, \boldsymbol{V}, \xi)$ where

$$\boldsymbol{m} = \arg\max_{\sigma \in C} \sum_{i=1}^{n} \ln \mathcal{N}_J(z_i|\boldsymbol{X}_i\boldsymbol{\beta}, \boldsymbol{\Sigma}),$$
$$\boldsymbol{V} = -\left\{\frac{\partial^2 \sum_{i=1}^{n} \ln \mathcal{N}_J(z_i|\boldsymbol{X}_i\boldsymbol{\beta}, \boldsymbol{\Sigma})}{\partial \boldsymbol{\sigma}\, \partial \boldsymbol{\sigma}'}\right\}^{-1}_{\sigma=\boldsymbol{m}}$$

are the mode and curvature of the target distribution, given the current values of the conditioning variables.

ALGORITHM 5 (*Multivariate probit*).
1 Sample for $i \leqslant n$, $j \leqslant J$

$$z_{ij}|z_{i(-j)}, \boldsymbol{\beta}, \boldsymbol{\Sigma} \propto \mathcal{N}(\mu_{ij}, v_{ij})\big\{I(z_{ij} \leqslant 0)^{1-y_{ij}} + I(z_{ij} > 0)^{y_{ij}}\big\}$$
$$\mu_{ij} = \mathrm{E}(z_{ij}|z_{i(-j)}, \boldsymbol{\beta}, \boldsymbol{\Sigma})$$
$$v_{ij} = \mathrm{Var}(z_{ij}|z_{i(-j)}, \boldsymbol{\beta}, \boldsymbol{\Sigma})$$

2 Sample

$$\boldsymbol{\beta}|z, \boldsymbol{\beta}, \boldsymbol{\Sigma} \sim \mathcal{N}_k\left(\boldsymbol{B}_n\left(\boldsymbol{B}_0^{-1}\boldsymbol{\beta}_0 + \sum_{i=1}^{n} \boldsymbol{X}_i' \boldsymbol{\Sigma}^{-1} z_i\right), \boldsymbol{B}_n\right),$$

where

$$\boldsymbol{B}_n = \left(\boldsymbol{B}_0^{-1} + \sum_{i=1}^{n} \boldsymbol{X}_i' \boldsymbol{\Sigma}^{-1} \boldsymbol{X}_i^{-1}\right)^{-1}$$

3 Sample

$$\pi(\boldsymbol{\sigma}|z, \boldsymbol{\beta}) \propto p(\boldsymbol{\sigma}) \prod_{i=1}^{n} \mathcal{N}_J(z_i|\boldsymbol{X}_i\boldsymbol{\beta}, \boldsymbol{\Sigma})$$

by the M-H algorithm, calculating the parameters $(\boldsymbol{m}, \boldsymbol{V})$, proposing $\boldsymbol{\sigma}' \sim \mathcal{T}_\nu(\boldsymbol{\sigma}|\boldsymbol{m}, \boldsymbol{V}, \xi)$, calculating

$$\alpha(\boldsymbol{\sigma}, \boldsymbol{\sigma}'|\boldsymbol{y}, \boldsymbol{\beta}, \{z_i\}) = \min\left\{\frac{\pi(\boldsymbol{\sigma}') \prod_{i=1}^{n} \mathcal{N}_J(z_i|\boldsymbol{X}_i\boldsymbol{\beta}, \boldsymbol{\Sigma}')I[\boldsymbol{\sigma}' \in C]}{\pi(\boldsymbol{\sigma}) \prod_{i=1}^{n} \mathcal{N}_J(z_i|\boldsymbol{X}_i\boldsymbol{\beta}, \boldsymbol{\Sigma})} \frac{\mathcal{T}_\nu(\boldsymbol{\sigma}|\boldsymbol{m}, \boldsymbol{V}, \xi)}{\mathcal{T}_\nu(\boldsymbol{\sigma}'|\boldsymbol{m}, \boldsymbol{V}, \xi)}, 1\right\}$$

and moving to $\boldsymbol{\sigma}'$ with probability $\alpha(\boldsymbol{\sigma}, \boldsymbol{\sigma}'|\boldsymbol{y}, \boldsymbol{\beta}, \{z_i\})$

4 Goto 1

EXAMPLE. As an application of this algorithm consider a data set in which the multivariate binary responses are generated by a panel structure. The data is concerned with the health effects of pollution on 537 children in Stuebenville, Ohio, each observed at ages 7, 8, 9 and 10 years, and the response variable is an indicator of wheezing status. Suppose that the marginal probability of wheeze status of the ith child at the jth time point is specified as

$$\Pr(y_{ij} = 1|\boldsymbol{\beta}) = \Phi(\beta_0 + \beta_1 x_{1ij} + \beta_2 x_{2ij} + \beta_3 x_{3ij}), \quad i \leqslant 537,\ j \leqslant 4,$$

where $\boldsymbol{\beta}$ is constant across categories, x_1 is the age of the child centered at nine years, x_2 is a binary indicator variable representing the mother's smoking habit during the first year of the study, and $x_3 = x_1 x_2$. Suppose that the Gaussian prior on $\boldsymbol{\beta} = (\beta_1, \beta_2, \beta_3, \beta_4)$ is centered at zero with a variance of $10\boldsymbol{I}_k$ and let $\pi(\boldsymbol{\sigma})$ be the density of a normal distribution, with mean zero and variance $\boldsymbol{I}_6$, *restricted* to region that leads to a positive-definite correlation matrix, where $(\sigma_{21}, \sigma_{31}, \sigma_{32}, \sigma_{41}, \sigma_{42}, \sigma_{43})$. From 10,000 cycles of Algorithm 5 one obtains the following covariate effects and posterior distributions of the correlations.

Table 3
Covariate effects in the Ohio wheeze data: MVP model with unrestricted correlations. In the table, NSE denotes the numerical standard error, lower is the 2.5th percentile and upper is the 97.5th percentile of the simulated draws. The results are based on 10000 draws from Algorithm 5

	Prior		Posterior				
	Mean	Std dev	Mean	NSE	Std dev	Lower	Upper
β_1	0.000	3.162	−1.108	0.001	0.062	−1.231	−0.985
β_2	0.000	3.162	−0.077	0.001	0.030	−0.136	−0.017
β_3	0.000	3.162	0.155	0.002	0.101	−0.043	0.352
β_4	0.000	3.162	0.036	0.001	0.049	−0.058	0.131

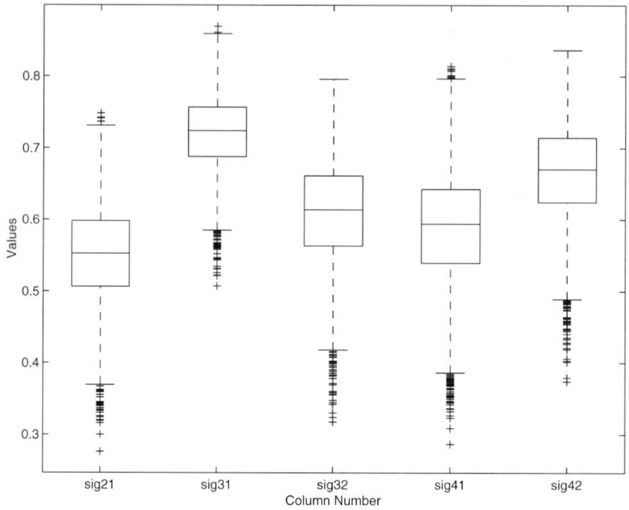

Fig. 2. Posterior boxplots of the correlations in the Ohio wheeze data: MVP model.

Notice that the summary tabular output contains not only the posterior means and standard deviations of the parameters but also the 95% credibility intervals, all computed from the sampled draws. It may be seen from Figure 2 that the posterior distributions of the correlations are similar suggesting that an equicorrelated correlation structure might be appropriate for these data.

5.5. Marginal likelihood of the MVP model

The calculation of the marginal likelihood of MVP model is considered by both Chib and Greenberg (1998) and Chib and Jeliazkov (2001). The main issue is the calculation of the posterior ordinate which may be estimated from

$$\pi\left(\boldsymbol{\sigma}^*, \boldsymbol{\beta}^* | \boldsymbol{y}\right) = \pi\left(\boldsymbol{\sigma}^* | \boldsymbol{y}\right) \pi\left(\boldsymbol{\beta}^* | \boldsymbol{y}, \boldsymbol{\Sigma}^*\right),$$

where the first ordinate (that of the correlations) is not in tractable form. To estimate the marginal ordinate one can apply (1.11) leading to

$$\pi\left(\boldsymbol{\sigma}^* | \boldsymbol{y}\right) = \frac{E_1\{\alpha(\boldsymbol{\sigma}, \boldsymbol{\sigma}' | \boldsymbol{y}, \boldsymbol{\beta}, \{z_i\}) \mathcal{T}_\nu(\boldsymbol{\sigma}' | \boldsymbol{m}, \boldsymbol{V}, \xi)\}}{E_2\{\alpha(\boldsymbol{\sigma}^*, \boldsymbol{\sigma} | \boldsymbol{y}, \boldsymbol{\beta}, \{z_i\})\}},$$

where $\alpha(\boldsymbol{\sigma}, \boldsymbol{\sigma}' | \boldsymbol{y}, \boldsymbol{\beta}, \{z_i\})$ is the probability of move defined in Algorithm 5. The numerator expectation can be calculated from the draws $\{\boldsymbol{\beta}^{(g)}, \{z_i^{(g)}\}, \boldsymbol{\sigma}^{(g)}\}$ from the output of Algorithm 5. The denominator expectation is calculated from a reduced run consisting of the densities

$$\pi\left(\boldsymbol{\beta} | \boldsymbol{y}, \{z_i\}, \boldsymbol{\Sigma}^*\right); \quad \pi\left(\{z_i\} | \boldsymbol{y}, \boldsymbol{\beta}, \boldsymbol{\Sigma}^*\right),$$

after $\boldsymbol{\Sigma}$ is fixed at $\boldsymbol{\Sigma}^*$. Then for each draw

$$\boldsymbol{\beta}^{(j)}, z^{(j)} \sim \pi\left(\boldsymbol{\beta}, z | \boldsymbol{y}, \boldsymbol{\Sigma}^*\right)$$

in this reduced run, we also draw $\boldsymbol{\sigma}^{(j)} \sim \mathcal{T}_\nu(\boldsymbol{\sigma}|\boldsymbol{m}, \boldsymbol{V}, \xi)$. These draws are used to average $\alpha(\boldsymbol{\sigma}^*, \boldsymbol{\sigma}|\boldsymbol{y}, \boldsymbol{\beta}, \{z_i\})$. In addition, the sampled variates $\{\boldsymbol{\beta}^{(j)}, z^{(j)}\}$ from this same reduced run are also used to estimate $\pi(\boldsymbol{\beta}^*|\boldsymbol{y}, \boldsymbol{\Sigma}^*)$ by averaging the normal density in Step 2 of Algorithm 5.

5.6. *Fitting of the multivariate t-link model*

Algorithm 5 is easily modified for the fitting of the multivariate-t link version of the MVP model. One simple possibility is to include the $\{\lambda_i\}$ into the sampling since conditioned on the value of λ_i, the t-link binary model reduces to the MVP model. With this augmentation, one implements Steps 1–3 conditioned on the value of $\{\lambda_i\}$. The sampling is completed with an additional step involving the simulation of $\{\lambda_i\}$. A straightforward calculation shows that the updated full conditional distribution of λ_i is

$$\lambda_i|z_i, \boldsymbol{\beta}, \boldsymbol{\Sigma} \sim \mathcal{G}\left(\frac{\nu + J}{2}, \frac{\nu + (z_i - \boldsymbol{X}_i\boldsymbol{\beta})'\boldsymbol{\Sigma}^{-1}(z_i - \boldsymbol{X}_i\boldsymbol{\beta})}{2}\right), \quad i \leqslant n,$$

which is easily sampled. The modified sampler thus requires little extra coding.

5.7. *Binary outcome with a confounded binary treatment*

In many problems, one central objective of model building is to isolate the effect of a binary covariate on a binary (or categorical) outcome. Isolating this effect becomes difficult if the effect of the binary covariate on the outcome is confounded with those of missing or unobserved covariates. Such complications are the norm when dealing with observational data. As an example, suppose that y_i is a zero-one variable that represents post-surgical outcomes for hip-fracture patients (one if the surgery is successful, and zero if not) and x_i is an indicator representing the extent of delay before surgery (one if the delay exceeds say two days, and zero otherwise). In this case, it is quite likely that delay and post-surgical outcome are both affected by patient factors, e.g., patient frailty, that are usually either imperfectly measured and recorded or absent from the data. In this case, we can model the outcome and the treatment assignment jointly to control for the unobservable confounders. The resulting model is similar to the MVP model and is discussed in detail by Chib (2003).

Briefly, suppose that the outcome model is given by

$$y_i = I\left(z'_{1i}\boldsymbol{\gamma}_1 + x_i\beta + \varepsilon_i\right), \quad i \leqslant n, \tag{5.8}$$

and suppose that the effect of the binary treatment x_i is confounded with that of unobservables that are correlated with ε_i, even after conditioning on z_{1i}. To model this confounding, let the treatment assignment mechanism be given by

$$x_i = I\left(z'_i\boldsymbol{\gamma} + u_i\right),$$

where the confounder u_i is correlated with ε_i and the covariate vector $z_i = (z_{1i}, z_{2i})$ contains the variables z_{2i} that are part of the intake model but not present in the outcome model. These covariates are called instrumental variables. They must be independent of (ε_i, u_i) given z_{1i} but correlated with the intake. Now suppose for simplicity that the joint distribution of the unobservables (ε_i, u_i) is Gaussian with mean zero and covariance matrix $\boldsymbol{\Omega}$ that is in correlation form. It is easy to see, starting with the joint distribution of (ε_i, u_i) followed by a change of variable, that this set-up is a special case of the MVP model. It can be estimated along the lines of Algorithm 5.

Chib (2003) utilizes the posterior draws to calculate the average treatment effect (ATE) which is defined as $E(y_{i1}|z_{1i}, \boldsymbol{\gamma}_1, \beta) - E(y_{i0}|z_{1i}, \boldsymbol{\gamma}_1, \beta)$, where y_{ij} is the outcome when $x_i = j$. These are the so-called potential outcomes. From (5.8) it can be seen that the two potential outcomes are $y_{i0} = I(z_{1i}'\boldsymbol{\gamma}_1 + \varepsilon_i)$ and $y_{i1} = I(z_{1i}'\boldsymbol{\gamma}_1 + \beta + \varepsilon_i)$ and, therefore, under the normality assumption, the difference in expected values of the potential outcomes conditioned on z_{1i} and $\boldsymbol{\psi} = (\boldsymbol{\gamma}_1, \beta)$ is

$$E(y_{i1}|z_{1i}, \boldsymbol{\psi}) - E(y_{i0}|z_{1i}, \boldsymbol{\psi}) = \Phi\big(z_{1i}'\boldsymbol{\gamma}_1 + \beta\big) - \Phi\big(z_{1i}'\boldsymbol{\gamma}_1\big).$$

Thus, the ATE is

$$\text{ATE} = \int \big\{\Phi\big(z_1'\boldsymbol{\gamma}_1 + \beta\big) - \Phi\big(z_1'\boldsymbol{\gamma}_1\big)\big\}\pi(z_1)\,\mathrm{d}z_1, \tag{5.9}$$

where $\pi(z_1)$ is the density of z_1 (independent of $\boldsymbol{\psi}$ by assumption). Calculation of the ATE, therefore, entails an integration with respect to z_1. In practice, a simple idea is to perform the marginalization using the empirical distribution of z_1 from the observed sample data. If we are interested in more than the difference in expected value of the potential outcomes, then the posterior distribution of the ATE can be obtained by evaluating the integral as a Monte Carlo average, for each sampled value of $(\boldsymbol{\gamma}_1, \beta)$ from the MCMC algorithm.

6. Longitudinal binary responses

Consider now the situation in which one is interested in modeling longitudinal (univariate) binary data. Suppose that for the ith individual at time t, the probability of observing the outcome $y_{it} = 1$, conditioned on parameters $\boldsymbol{\beta}_1$ and random effects $\boldsymbol{\beta}_{2i}$, is given by

$$\Pr(y_{it} = 1|\boldsymbol{\beta}_i) = \Phi(\boldsymbol{x}_{1it}\boldsymbol{\beta}_1 + \boldsymbol{w}_{it}\boldsymbol{\beta}_{2i}),$$

where Φ is the cdf of the standard normal distribution, x_{1it} is a k_1 vector of covariates whose effect is assumed to be constant across subjects (clusters) and $\boldsymbol{w}_{it}$ is an additional and *distinct* q vector of covariates whose effects are cluster-specific. The objective is to learn about the parameters and the random effects given n subjects each with a set of measurements $\boldsymbol{y}_i = (y_{i1}, \ldots, y_{in_i})'$ observed at n_i points in time. The presence of the random effects ensures that the binary responses are correlated. Modeling of the data and the estimation of the model is aided once again by the latent variable framework of Albert and Chib (1993) and the hierarchical prior modeling of Lindley and Smith (1972).

For the ith cluster, we define the vector of latent variable

$$z_i = \boldsymbol{X}_{1i}\boldsymbol{\beta}_1 + \boldsymbol{W}_i\boldsymbol{\beta}_{2i} + \boldsymbol{\varepsilon}_i, \quad \boldsymbol{\varepsilon}_i \sim \mathcal{N}_{n_i}(\mathbf{0}, \boldsymbol{I}_{n_i}),$$

and let

$$y_{it} = I[z_{it} > 0],$$

where $z_i = (z_{i1}, \ldots, z_{in_i})'$, $\boldsymbol{X}_{1i}$ is a $n_i \times k_1$, and $\boldsymbol{W}_i$ is $n_i \times q$.

In this setting, the effects $\boldsymbol{\beta}_1$ and $\boldsymbol{\beta}_{2i}$ are estimable (without any further assumptions) under the *sequential exogeneity* assumption wherein ε_{it} is uncorrelated with $(\boldsymbol{x}_{1it}, \boldsymbol{w}_{it})$ given past values of $(\boldsymbol{x}_{1it}, \boldsymbol{w}_{it})$ and $\boldsymbol{\beta}_i$. In practice, even when the assumption of sequential exogeneity of the covariates $(\boldsymbol{x}_{1it}, \boldsymbol{w}_{it})$ holds, it is quite possible that there exist covariates $\boldsymbol{a}_i$: $r \times 1$ (with an intercept included) that are correlated with the random-coefficients $\boldsymbol{\beta}_i$. These subject-specific covariates may be measurements on the subject at baseline (time $t = 0$) or other time-invariant covariates. In the Bayesian hierarchical approach this dependence on subject-specific covariates is modeled by a hierarchical prior. One quite general way to proceed is to assume that

$$\underbrace{\begin{pmatrix} \beta_{21i} \\ \beta_{22i} \\ \vdots \\ \beta_{2qi} \end{pmatrix}}_{\beta_{2i}} = \underbrace{\begin{pmatrix} \boldsymbol{a}_i' & \boldsymbol{0}' & \cdots & \cdots & \boldsymbol{0}' \\ \boldsymbol{0}' & \boldsymbol{a}_i' & \cdots & \cdots & \boldsymbol{0}' \\ \vdots & \vdots & \vdots & \vdots & \vdots \\ \boldsymbol{0}' & \boldsymbol{0}' & \cdots & \cdots & \boldsymbol{a}_i' \end{pmatrix}}_{A_i = \boldsymbol{I} \otimes \boldsymbol{a}_i'} \underbrace{\begin{pmatrix} \boldsymbol{\beta}_{21} \\ \boldsymbol{\beta}_{22} \\ \vdots \\ \vdots \\ \boldsymbol{\beta}_{2q} \end{pmatrix}}_{\beta_2} + \underbrace{\begin{pmatrix} b_{i1} \\ b_{i2} \\ \vdots \\ b_{iq} \end{pmatrix}}_{\boldsymbol{b}_i}$$

or in vector–matrix form

$$\boldsymbol{\beta}_{2i} = \boldsymbol{A}_i \boldsymbol{\beta}_2 + \boldsymbol{b}_i,$$

where $\boldsymbol{A}_i$ is a $q \times k_2$ matrix, $k = r \times q$, $\boldsymbol{\beta}_2 = (\boldsymbol{\beta}_{21}, \boldsymbol{\beta}_{22}, \ldots, \boldsymbol{\beta}_{2q})$ is a $(k_2 \times 1)$-dimensional vector, and $\boldsymbol{b}_i$ is the mean zero random effects vector (uncorrelated with $\boldsymbol{A}_i$ and ε_i) that is distributed according to the distribution (say)

$$\boldsymbol{b}_i | \boldsymbol{D} \sim \mathcal{N}_q(\boldsymbol{0}, \boldsymbol{D}).$$

This is the second stage of the model. It may be noted that the matrix $\boldsymbol{A}_i$ can be the identity matrix of order q or the zero matrix of order q. In this hierarchical model, if $\boldsymbol{A}_i$ is not the zero matrix then identifiability requires that the matrices $\boldsymbol{X}_{1i}$ and $\boldsymbol{W}_i$ have no covariates in common. For example, if the first column of $\boldsymbol{W}_i$ is a vector of ones, then $\boldsymbol{X}_{1i}$ cannot include an intercept. If $\boldsymbol{A}_i$ is the zero matrix, however, $\boldsymbol{W}_i$ is typically a subset of $\boldsymbol{X}_{1i}$. Thus, the effect of $\boldsymbol{a}_i$ on β_{21i} (the intercept) is measured by $\boldsymbol{\beta}_{21}$, that on β_{21i} is measured by $\boldsymbol{\beta}_{22}$ and that on β_{2qi} by $\boldsymbol{\beta}_{2q}$.

Inserting the model of the cluster-specific random coefficients into the first stage yields

$$\boldsymbol{y}_i = \boldsymbol{X}_i \boldsymbol{\beta} + \boldsymbol{W}_i \boldsymbol{b}_i + \boldsymbol{\varepsilon}_i,$$

where

$$\boldsymbol{X}_i = (\boldsymbol{X}_{1i}\ \boldsymbol{W}_i \boldsymbol{A}_i) \quad \text{with } \boldsymbol{\beta} = (\boldsymbol{\beta}_1 \boldsymbol{\beta}_2),$$

as is readily checked. We can complete the model by making suitable assumptions about the distribution of $\boldsymbol{\varepsilon}_i$. One possibility is to assume that

$$\boldsymbol{\varepsilon}_i \sim \mathcal{N}_{n_i}(\mathbf{0}, \boldsymbol{I}_{n_i})$$

which leads to the probit link model. The prior distribution on the parameters can be specified as

$$\boldsymbol{\beta} \sim \mathcal{N}_k(\boldsymbol{\beta}_0, B_0), \qquad \boldsymbol{D}^{-1} \sim \mathcal{W}_q(\rho_0, \boldsymbol{R}),$$

where the latter denotes a Wishart density with degrees of freedom ρ_0 and scale matrix $\boldsymbol{R}$.

The MCMC implementation in this set-up proceeds by including the $z = (z_1, \ldots, z_n)$ in the sampling (Chib and Carlin, 1999). Given z the sampling resembles the steps of the continuous response longitudinal model. To improve the efficiency of the sampling procedure, the sampling of the latent data is done marginalized over $\{\boldsymbol{b}_i\}$ from the conditional distribution of $z_{it}|z_{i(-t)}, y_{it}, \boldsymbol{\beta}, \boldsymbol{D}$, where $z_{i(-t)}$ is the vector z_i excluding z_{it}.

ALGORITHM 6 (*Gaussian–Gaussian panel probit*).

```
1 Sample
```

(a)

$$\begin{aligned} z_{it}|z_{i(-t)}, y_{it}, \boldsymbol{\beta}, \boldsymbol{D} &\propto \mathcal{N}(\mu_{it}, v_{it})\left\{I(z_{it} \leqslant 0)^{1-y_{it}} + I(z_{it} > 0)^{y_{it}}\right\} \\ \mu_{it} &= \mathrm{E}(z_{it}|z_{i(-t)}, \boldsymbol{\beta}, \boldsymbol{D}) \\ v_{it} &= \mathrm{Var}(z_{it}|z_{i(-t)}, \boldsymbol{\beta}, \boldsymbol{D}) \end{aligned}$$

(b)

$$\boldsymbol{\beta}|z, \boldsymbol{D} \sim \mathcal{N}_k\left(\boldsymbol{B}_n\left(\boldsymbol{B}_0^{-1}\boldsymbol{\beta}_0 + \sum_{i=1}^n \boldsymbol{X}_i'\boldsymbol{V}_i^{-1}z_i\right), \boldsymbol{B}\right)$$

$$\boldsymbol{B}_n = \left(\boldsymbol{B}_0^{-1} + \sum_{i=1}^n \boldsymbol{X}_i'\boldsymbol{V}_i^{-1}\boldsymbol{X}_i\right)^{-1}; \quad \boldsymbol{V}_i = \boldsymbol{I}_{n_i} + \boldsymbol{W}_i\boldsymbol{D}\boldsymbol{W}_i'$$

(c)

$$\begin{aligned} \boldsymbol{b}_i|\boldsymbol{y}, \boldsymbol{\beta}, \boldsymbol{D} &\sim \mathcal{N}_q\left(\boldsymbol{D}_i\boldsymbol{W}_i'(z_i - \boldsymbol{X}_i\boldsymbol{\beta}), \boldsymbol{D}_i\right), \quad i \leqslant N \\ \boldsymbol{D}_i &= \left(\boldsymbol{D}^{-1} + \boldsymbol{W}_i'\boldsymbol{W}_i\right)^{-1} \end{aligned}$$

```
2 Sample
```

$$\boldsymbol{D}^{-1}|\boldsymbol{y}, \boldsymbol{\beta}, \{\boldsymbol{b}_i\} \sim \mathcal{W}_q\{\rho_0 + n, \boldsymbol{R}_n\}$$

```
where
```

$$\boldsymbol{R}_n = \left(\boldsymbol{R}_0^{-1} + \sum_{i=1}^n \boldsymbol{b}_i\boldsymbol{b}_i'\right)^{-1}$$

```
3 Goto 1
```

The model with errors distributed as student-t

$$\begin{aligned}\boldsymbol{\varepsilon}_i|\lambda_i &\sim \mathcal{N}_{n_i}(\mathbf{0}, \lambda_i \boldsymbol{I}_{n_i}),\\ \eta_i &\sim G\left(\frac{\nu}{2}, \frac{\nu}{2}\right)\end{aligned}$$

and random effects distributed as student-t

$$\begin{aligned}\boldsymbol{b}_i|\boldsymbol{D}, \boldsymbol{\eta}_i &\sim \mathcal{N}_q\big(\mathbf{0}, \eta_i^{-1}\boldsymbol{D}\big),\\ \eta_i &\sim G\left(\frac{\nu_b}{2}, \frac{\nu_b}{2}\right)\end{aligned}$$

can also be tackled in ways that parallel the developments in the previous section. We present the algorithm for the student–student binary response panel model without comment.

ALGORITHM 7 (*Student–Student binary panel*).

1 Sample

(a)

$$\begin{aligned}z_{it}|\boldsymbol{z}_{i(-t)}, y_{it}, \boldsymbol{\beta}, \boldsymbol{D} &\propto \mathcal{N}(\mu_{it}, v_{it})\big\{I(z_{it} \leqslant 0)^{1-y_{it}} + I(z_{it} > 0)^{y_{it}}\big\}\\ \mu_{it} &= \mathrm{E}(z_{it}|\boldsymbol{z}_{i(-t)}, \boldsymbol{\beta}, \boldsymbol{D}, \lambda_i, \eta_i)\\ v_{it} &= \mathrm{Var}(z_{it}|\boldsymbol{z}_{i(-t)}, \boldsymbol{\beta}, \boldsymbol{D}, \lambda_i, \eta_i)\end{aligned}$$

(b)

$$\boldsymbol{\beta}|\boldsymbol{z}, \boldsymbol{D}\{\lambda_i\}, \{\eta_i\} \sim \mathcal{N}_k\left(\boldsymbol{B}_n\left(\boldsymbol{B}_0^{-1}\boldsymbol{\beta}_0 + \sum_{i=1}^{n} \boldsymbol{X}_i'\boldsymbol{V}_i^{-1}z_i\right), \boldsymbol{B}_n\right)$$

where

$$\boldsymbol{B}_n = \left(\boldsymbol{B}_0^{-1} + \sum_{i=1}^{n} \boldsymbol{X}_i'\boldsymbol{V}_i^{-1}\boldsymbol{X}_i\right)^{-1}; \quad \boldsymbol{V}_i = \lambda_i^{-1}\boldsymbol{I}_{n_i} + \eta_i^{-1}\boldsymbol{W}_i\boldsymbol{D}\boldsymbol{W}_i'$$

(c)

$$\boldsymbol{b}_i|z_i, \boldsymbol{\beta}, \boldsymbol{D}, \lambda_i, \eta_i \sim \mathcal{N}_q\big(\boldsymbol{D}_i\boldsymbol{W}_i'\lambda_i(z_i - \boldsymbol{X}_i\boldsymbol{\beta}), \boldsymbol{D}_i\big)$$

where

$$\boldsymbol{D}_i = \big(\eta_i\boldsymbol{D}^{-1} + \lambda_i\boldsymbol{W}_i'\boldsymbol{W}_i\big)^{-1}$$

2 Sample

(a)

$$\lambda_i|\boldsymbol{y}, \boldsymbol{\beta}, \{\boldsymbol{b}_i\} \sim \mathcal{G}\left(\frac{\nu + n_i}{2}, \frac{\nu + \boldsymbol{e}_i'\boldsymbol{e}_i}{2}\right), \quad i \leqslant n$$

where $\boldsymbol{e}_i = (z_i - \boldsymbol{X}_i\boldsymbol{\beta} - \boldsymbol{W}_i\boldsymbol{\beta}_i)$

(b)

$$\eta_i | \boldsymbol{b}_i, \boldsymbol{D} \sim \mathcal{G}\left(\frac{\nu_b + q}{2}, \frac{\nu_b + \boldsymbol{b}_i' \boldsymbol{D}^{-1} \boldsymbol{b}_i}{2}\right), \quad i \leqslant n$$

3 Sample

$$\boldsymbol{D}^{-1} | z, \boldsymbol{\beta}, \{\boldsymbol{b}_i\}, \{\lambda_i\}, \{\eta_i\} \sim \mathcal{W}_q\{\rho_0 + n, \boldsymbol{R}_n\}$$

where

$$\boldsymbol{R}_n = \left(\boldsymbol{R}_0^{-1} + \sum_{i=1}^{n} \eta_i \boldsymbol{b}_i \boldsymbol{b}_i'\right)^{-1}$$

4 Goto 1

Related modeling and analysis techniques can be developed for ordinal longitudinal data. In the estimation of such models the one change is a M–H step for the sampling of the cut-points, similar to the corresponding step in Algorithm 3. It is also possible in this context to let the cut-points be subject-specific; an example is provided by Johnson (2003).

6.1. *Marginal likelihood of the panel binary models*

The calculation of the marginal likelihood is again straightforward. For example, in the student–student binary panel model, the posterior ordinate is expressed as

$$\pi\left(\boldsymbol{D}^{-1*}, \boldsymbol{\beta}^* | \boldsymbol{y}\right) = \pi\left(\boldsymbol{D}^{-1*} | \boldsymbol{y}\right) \pi\left(\boldsymbol{\beta}^* | \boldsymbol{y}, \boldsymbol{D}^*\right),$$

where the first term is obtained by averaging the Wishart density in Algorithm 7 over draws on $\{\boldsymbol{b}_i\}$, $\{\lambda_i\}$ and z from the full run. To estimate the second ordinate, which is conditioned on $\boldsymbol{D}^*$, we run a reduced MCMC simulation with the full conditional densities

$$\pi\left(z | \boldsymbol{y}, \boldsymbol{\beta}, \boldsymbol{D}^*, \{\lambda_i\}\right); \pi\left(\boldsymbol{\beta} | z, \boldsymbol{D}^*, \{\lambda_i\}, \{\eta_i\}\right);$$
$$\pi\left(\{\boldsymbol{b}_i\} | z, \boldsymbol{\beta}, \boldsymbol{D}^*, \{\lambda_i\}, \{\eta_i\}\right); \pi\left(\{\lambda_i\}, \{\eta_i\} | z, \boldsymbol{\beta}, \boldsymbol{D}^*\right),$$

where each conditional utilizes the fixed value of $\boldsymbol{D}$. The second ordinate is now estimated by averaging the Gaussian density of $\boldsymbol{\beta}$ in Algorithm 7 at $\boldsymbol{\beta}^*$ over the draws on $(z, \{\boldsymbol{b}_i\}, \{\lambda_i\}, \{\eta_i\})$ from this reduced run.

7. Longitudinal multivariate responses

In some circumstances it is necessary to model a set of multivariate categorical outcomes in a longitudinal setting. For example, one may be interested in the factors that influence purchase into a set of product categories (for example, egg, milk, cola, etc.) when a typical consumer purchases goods at the grocery store. We may have a longitudinal sample of such consumers each observed over many different shopping occasions.

On each occasion, the consumer has a multivariate outcome vector representing the (binary) incidence into each of the product categories. The available data also includes information on various category specific characteristics, for example, the average price, display and feature values of the major brands in each category. The goal is to model the category incidence vector as a function of available covariates, controlling for the heterogeneity in covariates effects across subjects in the panel. Another example of multidimensional longitudinal data appears in Dunson (2003) motivated by studies of an item responses battery that is used to measure traits of an individual repeatedly over time. The model discussed in Dunson (2003) allows for mixtures of count, categorical, and continuous response variables.

To illustrate a canonical situation of multivariate longitudinal categorical responses, we combine the MVP model with the longitudinal models in the previous section. Let $\boldsymbol{y}_{it}$ represent a J vector of binary responses and suppose that we have for each subject i at each time t, a covariate matrix X_{1it} in the form

$$\boldsymbol{X}_{1it} = \begin{pmatrix} \boldsymbol{x}'_{11it} & \boldsymbol{0}' & \cdots & \boldsymbol{0}' \\ \boldsymbol{0}' & \boldsymbol{x}'_{12it} & \boldsymbol{0}' & \boldsymbol{0}' \\ \vdots & \vdots & \vdots & \vdots \\ \boldsymbol{0}' & \boldsymbol{0}' & \cdots & \boldsymbol{x}'_{1Jit} \end{pmatrix},$$

where each $\boldsymbol{x}_{1jit}$ has k_{1j} elements so that the dimension of $\boldsymbol{X}_{1it}$ is $J \times k_1$, where $k_1 = \sum_{j=1}^{J} k_{1j}$. Now suppose that there is an additional (distinct) set of covariates $\boldsymbol{W}_{it}$ whose effect is assumed to be both response and subject specific and arranged in the form

$$\boldsymbol{W}_{it} = \begin{pmatrix} \boldsymbol{w}'_{1it} & \boldsymbol{0}' & \cdots & \boldsymbol{0}' \\ \boldsymbol{0}' & \boldsymbol{w}'_{2it} & \boldsymbol{0}' & \boldsymbol{0}' \\ \vdots & \vdots & \vdots & \vdots \\ \boldsymbol{0}' & \boldsymbol{0}' & \cdots & \boldsymbol{w}'_{Jit} \end{pmatrix},$$

where each w_{jit} has q_j elements so that the dimension of $\boldsymbol{W}_{it}$ is $J \times q$, where $q = \sum_{j=1}^{J} q_j$. Also suppose that in terms of the latent variables z_{it}: $J \times 1$, the model generating the data is given by

$$z_{it} = \boldsymbol{X}_{1it}\boldsymbol{\beta}_1 + \boldsymbol{W}_{it}\boldsymbol{\beta}_{2i} + \boldsymbol{\varepsilon}_{it}, \quad \boldsymbol{\varepsilon}_{it} \sim \mathcal{N}_J(\boldsymbol{0}, \Sigma),$$

such that

$$y_{jit} = I[z_{jit} > 0], \quad j \leqslant J,$$

and where

$$\boldsymbol{\beta}_1 = \begin{pmatrix} \boldsymbol{\beta}_{11} \\ \boldsymbol{\beta}_{12} \\ \vdots \\ \boldsymbol{\beta}_{1J} \end{pmatrix} : k_1 \times 1 \quad \text{and} \quad \boldsymbol{\beta}_{2i} = \begin{pmatrix} \boldsymbol{\beta}_{21i} \\ \boldsymbol{\beta}_{22i} \\ \vdots \\ \boldsymbol{\beta}_{2Ji} \end{pmatrix} : q \times 1.$$

To model the heterogeneity in effects across subjects we can assume that

$$\underbrace{\begin{pmatrix} \boldsymbol{\beta}_{21i} \\ \boldsymbol{\beta}_{22i} \\ \vdots \\ \boldsymbol{\beta}_{2Ji} \end{pmatrix}}_{\beta_{2i}:\, q\times 1} = \underbrace{\begin{pmatrix} \boldsymbol{A}_{1i} & \mathbf{0} & \cdots & \cdots & \mathbf{0} \\ \mathbf{0} & \boldsymbol{A}_{2i} & \cdots & \cdots & \mathbf{0} \\ \vdots & \vdots & \vdots & \vdots & \vdots \\ \mathbf{0} & \mathbf{0} & \cdots & \cdots & \boldsymbol{A}_{Ji} \end{pmatrix}}_{A_i:\, q\times k_2} \underbrace{\begin{pmatrix} \boldsymbol{\beta}_{21} \\ \boldsymbol{\beta}_{22} \\ \vdots \\ \vdots \\ \boldsymbol{\beta}_{2J} \end{pmatrix}}_{\beta_2:\, k_2\times 1} + \underbrace{\begin{pmatrix} \boldsymbol{b}_{i1} \\ \boldsymbol{b}_{i2} \\ \vdots \\ \boldsymbol{b}_{iJ} \end{pmatrix}}_{b_i:\, q\times 1},$$

where each matrix $\boldsymbol{A}_{ji}$ is of the form $\boldsymbol{I}_{q_j} \otimes \boldsymbol{a}'_{ji}$ for some r_j category-specific covariates $\boldsymbol{a}_{ji}$. The size of the vector $\boldsymbol{\beta}_2$ is then $k_2 = \sum_{j=1}^{J} q_j \times r_j$. Substituting this model of the random-coefficients into the first stage yields

$$z_{it} = \boldsymbol{X}_{it}\boldsymbol{\beta} + \boldsymbol{W}_{it}\boldsymbol{b}_i + \boldsymbol{\varepsilon}_{it}, \quad \boldsymbol{\varepsilon}_{it} \sim \mathcal{N}_J(\mathbf{0}, \Sigma),$$

where

$$\boldsymbol{X}_{it} = (\boldsymbol{X}_{1it}, \boldsymbol{W}_{it}\boldsymbol{A}_i)$$

and $\boldsymbol{\beta} = (\boldsymbol{\beta}_1, \boldsymbol{\beta}_2)\colon\ k \times 1$. Further assembling all n_i observations on the ith subject for a total of $n_i^* = J \times n_i$ observations, we have

$$z_i = \boldsymbol{X}_i\boldsymbol{\beta} + \boldsymbol{W}_i\boldsymbol{b}_i + \boldsymbol{\varepsilon}_i, \quad \boldsymbol{\varepsilon}_i \sim \mathcal{N}_{n_i^*}(\mathbf{0}, \boldsymbol{\Omega}_i = \boldsymbol{I}_{n_i} \otimes \boldsymbol{\Sigma}),$$

where $z_i = (z_{i1}, \ldots, z_{in_i})'\colon\ n_i^* \times 1$, $\boldsymbol{X}_i = (\boldsymbol{X}'_{i1}, \ldots, \boldsymbol{X}'_{in_i})'\colon\ n_i^* \times k$, and $\boldsymbol{W}_i = (\boldsymbol{W}'_{i1}, \ldots, \boldsymbol{W}'_{in_i})'\colon\ n_i^* \times q$. Apart from the increase in the dimension of the problem, the model is now much like the longitudinal model with univariate outcomes. The fitting, consequently, parallels that in Algorithms 6 and 7.

ALGORITHM 8 (*Longitudinal MVP*).

```
1 Sample
```

(a)

$$\begin{aligned} z_{it}|z_{i(-t)}, y_{it}, \boldsymbol{\beta}, \boldsymbol{\Sigma}, \boldsymbol{D} &\propto \mathcal{N}(\mu_{it}, v_{it})\{I(z_{it} \leqslant 0)^{1-y_{it}} + I(z_{it} > 0)^{y_{it}}\} \\ \mu_{it} &= \mathrm{E}(z_{it}|z_{i(-t)}, \boldsymbol{\beta}, \boldsymbol{\Sigma}, \boldsymbol{D}) \\ v_{it} &= \mathrm{Var}(z_{it}|z_{i(-t)}, \boldsymbol{\beta}, \boldsymbol{\Sigma}, \boldsymbol{D}) \end{aligned}$$

(b)

$$\boldsymbol{\beta}|z, \boldsymbol{\Sigma}, \boldsymbol{D} \sim \mathcal{N}_k\left(\boldsymbol{B}_n\left(\boldsymbol{B}_0^{-1}\boldsymbol{\beta}_0 + \sum_{i=1}^{n} \boldsymbol{X}'_i\boldsymbol{V}_i^{-1}z_i\right), \boldsymbol{B}_n\right)$$

```
where
```

$$\boldsymbol{B}_n = \left(\boldsymbol{B}_0^{-1} + \sum_{i=1}^{n} \boldsymbol{X}'_i\boldsymbol{V}_i^{-1}\boldsymbol{X}_i\right)^{-1}; \quad \boldsymbol{V}_i = \boldsymbol{\Omega}_i + \boldsymbol{W}_i\boldsymbol{D}\boldsymbol{W}'_i$$

(c)

$$\boldsymbol{b}_i | z_i, \boldsymbol{\beta}, \boldsymbol{\Sigma}, \boldsymbol{D} \sim \mathcal{N}_q\big(\boldsymbol{D}_i \boldsymbol{W}_i' \boldsymbol{\Omega}_i^{-1}(z_i - \boldsymbol{X}_i\boldsymbol{\beta}), \boldsymbol{D}_i\big)$$

where

$$\boldsymbol{D}_i = \big(\boldsymbol{D}^{-1} + \boldsymbol{W}_i' \boldsymbol{\Omega}_i^{-1} \boldsymbol{W}_i\big)^{-1}$$

2 Sample

$$\boldsymbol{D}^{-1} | z, \boldsymbol{\beta}, \{\boldsymbol{b}_i\} \sim \mathcal{W}_q\{\rho_0 + n, \boldsymbol{R}_n\}$$

where

$$\boldsymbol{R}_n = \left(\boldsymbol{R}_0^{-1} + \sum_{i=1}^{n} \boldsymbol{b}_i \boldsymbol{b}_i'\right)^{-1}$$

3 Goto 1

8. Conclusion

This chapter has summarized Bayesian modeling and fitting techniques for categorical responses. The discussion has dealt with cross-section models for binary and ordinal data and presented extensions of those models to multivariate and longitudinal responses. In each case, the Bayesian MCMC fitting technique is simple and easy to implement.

The discussion in this chapter did not explore residual diagnostics and model fit issues. Relevant ideas are contained in Albert and Chib (1995, 1997) and Chen and Dey (2000). We also did not take up the multinomial probit model as, for example, considered in McCulloch et al. (2001), or the class of semiparametric binary response models in which the covariate effects are modeled by regression splines or other related means and discussed by, for example, Wood et al. (2002), Holmes and Mallick (2003) and Chib and Greenberg (2004). Fitting of all these models relies on the framework of Albert and Chib (1993).

In the Bayesian context it is relatively straightforward to compare alternative models. In this chapter we restricted our attention to the approach based on marginal likelihoods and Bayes factors and showed how the marginal likelihood can be computed for the various models from the output of the MCMC simulations. Taken together, the models and methods presented in this chapter are a testament to the versatility of Bayesian ideas for dealing with categorical response data.

References

Agresti, A. (1990). *Categorical Data Analysis*. Wiley, New York.

Albert, J., Chib, S. (1993). Bayesian analysis of binary and polychotomous response data. *J. Amer. Statist. Assoc.* **88**, 669–679.

Albert, J., Chib, S. (1995). Bayesian residual analysis for binary response models. *Biometrika* **82**, 747–759.

Albert, J., Chib, S. (1997). Bayesian tests and model diagnostics in conditionally independent hierarchical models. *J. Amer. Statist. Assoc.* **92**, 916–925.

Albert, J., Chib, S. (2001). Sequential ordinal modeling with applications to survival data. *Biometrics* **57**, 829–836.

Allenby, G.M., Lenk, P.J. (1994). Modeling household purchase behavior with logistic normal regression. *J. Amer. Statist. Assoc.* **89**, 1218–1231.

Amemiya, T. (1972). Bivariate probit analysis: Minimum chi-square methods. *J. Amer. Statist. Assoc.* **69**, 940–944.

Ashford, J.R., Sowden, R.R. (1970). Multivariate probit analysis. *Biometrics* **26**, 535–546.

Banerjee, S., Carlin, B., Gelfand, A.E. (2004). *Hierarchical Modeling and Analysis for Spatial Data*. Chapman and Hall/CRC, Boca Raton, FL.

Basu, S., Chib, S. (2003). Marginal likelihood and Bayes factors for Dirichlet process mixture models. *J. Amer. Statist. Assoc.* **98**, 224–235.

Basu, S., Mukhopadhyay, S. (2000). Binary response regression with normal scale mixture links. In: Dey, D.K., Ghosh, S.K., Mallick, B.K. (Eds.), *Generalized Linear Models: A Bayesian Perspective*. Marcel Dekker, New York, pp. 231–242.

Besag, J. (1974). Spatial interaction and the statistical analysis of lattice systems (with discussion). *J. Roy. Statist. Soc., Ser. B* **36**, 192–236.

Carey, V., Zeger, S.L., Diggle, P. (1993). Modelling multivariate binary data with alternating logistic regressions. *Biometrika* **80**, 517–526.

Carlin, B.P., Chib, S. (1995). Bayesian model choice via Markov chain Monte Carlo methods. *J. Roy. Statist. Soc., Ser. B* **57**, 473–484.

Chen, M.-H., Dey, D. (2000). A unified Bayesian analysis for correlated ordinal data models. *Brazilian J. Probab. Statist.* **14**, 87–111.

Chen, M.-H., Shao, Q.-M. (1998). On Monte Carlo methods for estimating ratios of normalizing constants. *Ann. Statist.* **25**, 1563–1594.

Chen, M.-H., Dey, D., Shao, Q.-M. (1999). A new skewed link model for dichotomous quantal response data. *J. Amer. Statist. Assoc.* **94**, 1172–1186.

Chib, S. (1995). Marginal likelihood from the Gibbs output. *J. Amer. Statist. Assoc.* **90**, 1313–1321.

Chib, S. (2003). On inferring effects of binary treatments with unobserved confounders (with discussion). In: Bernardo, J.M., Bayarri, M.J., Berger, J.O., Dawid, A.P., Heckerman, D., Smith, A.F.M., West, M. (Eds.), *Bayesian Statistics, vol. 7*. Oxford University Press, London.

Chib, S., Carlin, B.P. (1999). On MCMC sampling in hierarchical longitudinal models. *Statist. Comput.* **9**, 17–26.

Chib, S., Greenberg, E. (1995). Understanding the Metropolis–Hastings algorithm. *Amer. Statist.* **49**, 327–335.

Chib, S., Greenberg, E. (1998). Analysis of multivariate probit models. *Biometrika* **85**, 347–361.

Chib, S., Greenberg, E. (2004). Analysis of additive instrumental variable models. Technical Report, Washington University in Saint Louis.

Chib, S., Jeliazkov, I. (2001). Marginal likelihood from the Metropolis–Hastings output. *J. Amer. Statist. Assoc.* **96**, 270–281.

Connolly, M.A., Liang, K.-Y. (1988). Conditional logistic regression models for correlated binary data. *Biometrika* **75**, 501–506.

Cox, D.R. (1972). Regression models and life tables. *J. Roy. Statist. Soc., Ser. B* **34**, 187–220.

DiCiccio, T.J., Kass, R.E., Raftery, A.E., Wasserman, L. (1997). Computing Bayes factors by combining simulation and asymptotic approximations. *J. Amer. Statist. Assoc.* **92**, 903–915.

Dunson, D.B. (2003). Dynamic latent trait models for multidimensional longitudinal data. *J. Amer. Statist. Assoc.* **98**, 555–563.

Farhmeir, L., Tutz, G. (1994). *Multivariate Statistical Modelling Based on Generalized Linear Models*. Springer-Verlag, New York.

Fahrmeir, L., Lang, S. (2001). Bayesian semiparametric regression analysis of multicategorical time-space data. *Ann. Inst. Statist. Math.* **53**, 11–30.

Gelfand, A.E., Smith, A.F.M. (1990). Sampling-based approaches to calculating marginal densities. *J. Amer. Statist. Assoc.* **85**, 398–409.

Geman, S., Geman, D. (1984). Stochastic relaxation, Gibbs distributions and the Bayesian restoration of images. *IEEE Trans. Pattern Analysis and Machine Intelligence* **12**, 609–628.

Glonek, G.F.V., McCullagh, P. (1995). Multivariate logistic models. *J. Roy. Statist. Soc., Ser. B* **57**, 533–546.

Green, P.E. (1995). Reversible jump Markov chain Monte Carlo computation and Bayesian model determination. *Biometrika* **82**, 711–732.

Hastings, W.K. (1970). Monte Carlo sampling methods using Markov chains and their applications. *Biometrika* **57**, 97–109.

Holmes, C.C., Mallick, B.K. (2003). Generalized nonlinear modeling with multivariate free-knot regression splines. *J. Amer. Statist. Assoc.* **98**, 352–368.

Johnson, T.R. (2003). On the use of heterogeneous thresholds ordinal regression models to account for individual differences in response style. *Psychometrica* **68**, 563–583.

Kalbfleish, J., Prentice, R. (1980). *The Statistical Analysis of Failure Time Data*. Wiley, New York.

Kuhnert, P.M., Do, K.A. (2003). Fitting genetic models to twin data with binary and ordered categorical responses: A comparison of structural equation modelling and Bayesian hierarchical models. *Behavior Genetics* **33**, 441–454.

Lesaffre, E., Kaufmann, H.K. (1992). Existence and uniqueness of the maximum likelihood estimator for a multivariate probit model. *J. Amer. Statist. Assoc.* **87**, 805–811.

Lindley, D.V., Smith, A.F.M. (1972). Bayes estimates for the linear model. *J. Roy. Statist. Soc., Ser. B* **34**, 1–41.

McCullagh, P. (1980). Regression models for ordinal data. *J. Roy. Statist. Soc., Ser. B* **42**, 109–127.

McCulloch, R.E., Polson, N.G., Rossi, P.E. (2001). A Bayesian analysis of the multinomial probit model with fully identified parameters. *J. Econometrics* **99**, 173–193.

Metropolis, N., Rosenbluth, A.W., Rosenbluth, M.N., Teller, A.H., Teller, E. (1953). Equations of state calculations by fast computing machines. *J. Chem. Phys.* **21**, 1087–1092.

O'Brien, S.M., Dunson, D.B. (2004). Bayesian multivariate logistic regression. *Biometrics* **60**, 739–746.

Ochi, Y., Prentice, R.L. (1984). Likelihood inference in a correlated probit regression model. *Biometrika* **71**, 531–543.

Ripley, B. (1987). *Stochastic Simulation*. Wiley, New York.

Roberts, C.P. (2001). *The Bayesian Choice*. Springer-Verlag, New York.

Smith, A.F.M., Roberts, G.O. (1993). Bayesian computation via the Gibbs sampler and related Markov chain Monte Carlo methods. *J. Roy. Statist. Soc., Ser. B* **55**, 3–24.

Tanner, M.A., Wong, W.H. (1987). The calculation of posterior distributions by data augmentation. *J. Amer. Statist. Assoc.* **82**, 528–549.

Ten Have, T.R., Uttal, D.H. (2000). Subject-specific and population-averaged continuation ratio logit models for multiple discrete time survival profiles. *Appl. Statist.* **43**, 371–384.

Tierney, L. (1994). Markov chains for exploring posterior distributions (with discussion). *Ann. Statist.* **22**, 1701–1762.

Tutz, G. (1990). Sequential item response models with an ordered response. *British J. Math. Statist. Psychol.* **43**, 39–55.

Wood, S., Kohn, R., Shively, T., Jiang, W.X. (2002). Model selection in spline nonparametric regression. *J. Roy. Statist. Soc., Ser. B* **64**, 119–139.

Handbook of Statistics, Vol. 25
ISSN: 0169-7161
© 2005 Elsevier B.V. All rights reserved.
DOI 10.1016/S0169-7161(05)25030-7

30

Bayesian Methods and Simulation-Based Computation for Contingency Tables

James H. Albert

1. Motivation for Bayesian methods

There is an extensive literature on the classical analysis of contingency tables. Bishop et al. (1975) and Agresti (1990) illustrate the use of log-linear models to describe the association structure in a multidimensional contingency table. Chi-square statistics are used to examine independence and to compare nested log-linear models. P-values are used to assess statistical significance. Estimation and hypothesis testing procedures rest on asymptotic results for multinomial random variables.

Several problems with classical categorical data analyses motivate consideration of Bayesian methods. First, there is the problem of estimating cell probabilities and corresponding expected cell counts from a sparse multi-way contingency table with some empty cells. It is desirable to obtain positive smoothed estimates of the expected cell counts, reflecting the knowledge that the cell probabilities all exceed zero. Second, in comparing models, there is difficulty in interpreting the evidence communicated by a test statistic's p-value, as noted for instance by Diaconis and Efron (1985) in the simple case of a p-value for detecting association in a two-way table with large counts. This problem motivates a Bayesian approach to measuring evidence. Lastly, there is the potential bias in estimating association measures from models arrived at by classical model selection strategies, e.g., for choosing the best log-linear model. One typically uses a fitted model to estimate an association parameter and the variability of that estimate, while ignoring uncertainty in the process of arriving at the model on which estimation is based. Bayesian methods allow a user to explicitly model the uncertainty among a class of possible models by means of a prior distribution on the class of models, so that the posterior estimates of association parameters explicitly account for uncertainty about the "true" model on which estimates should ideally be based.

2. Advances in simulation-based Bayesian calculation

Although the Bayesian paradigm has been viewed as an attractive approach for model building and model checking, Bayesian methodology for categorical data for many years, say before 1990, was limited to relatively simple models. Much of the early work in Bayesian modeling for categorical data focused on the development of tractable approximations for marginal posterior distributions of parameters of interest.

In recent years, there have been great advances in the area of Bayesian computation ever since the introduction of Gibbs sampling in Gelfand and Smith (1990). Presently there exist general-purpose algorithms for constructing a Markov chain to simulate from a large class of posterior distributions. The program WinBUGS provides a convenient language for specifying a Bayesian model and simulating from the model using a Metropolis-within-Gibbs algorithm. The WinBUGS examples manual (Spiegelhalter et al., 2003) and Congdon (2001, 2003) illustrate the use of WinBUGS for fitting a wide variety of Bayesian models.

These developments in Bayesian computation will likely lead to the further development of Bayesian models for categorical data, especially for fitting multidimensional tables with many parameters and for selecting log-linear models. To illustrate the usefulness of simulation-based calculation, this chapter will use WinBUGS to perform some of the Bayesian calculations for the models that have been used to fit categorical data.

3. Early Bayesian analyses of categorical data

Early Bayesian analyses for categorical data focused on tractable approximations for posterior distributions and measures of evidence for two-way tables. For a multinomial random variable $\{y_{ij}\}$ with cell probabilities $\{\theta_{ij}\}$, Lindley (1964) considered the posterior distribution of the log contrast $\lambda = \sum\sum a_{ij} \log \theta_{ij}$ where $\sum\sum a_{ij} = 0$. In a 2×2 table with cell probabilities $(\theta_{11}, \theta_{12}, \theta_{21}, \theta_{22})$, one example of a log contrast is the log odds-ratio

$$\lambda = \log \theta_{11} - \log \theta_{12} - \log \theta_{21} + \log \theta_{22}.$$

If $\{\theta_{ij}\}$ is assumed to have a Dirichlet $(\{\alpha_{ij}\})$ distribution of the form $p(\{\theta_{ij}\}) \propto \prod \theta_{ij}^{\alpha_{ij}-1}$, Lindley showed that the posterior distribution for λ is approximately normal with mean and variance given respectively by

$$\lambda^* = \sum_i \sum_j a_{ij} \log(\alpha_{ij} + y_{ij}), \qquad v^* = \sum_i \sum_j a_{ij} (\alpha_{ij} + y_{ij})^{-1}.$$

Lindley used this approximation to obtain the posterior density of the log odds-ratio and to develop a Bayesian statistic for testing independence in a 2×2 table.

The formal way of comparing models from a Bayesian perspective is by the use of Bayes factors. If y denotes the data and θ denotes a vector of parameters, a Bayesian model is described by a sampling density for the data, $f(y|\theta)$, and a prior density for the parameter, $g(\theta)$. If one has two Bayesian models M_1: $\{f_1(y|\theta_1), g_1(\theta_1)\}$ and M_2: $\{f_2(y|\theta_2), g_2(\theta_2)\}$, then the Bayes factor in support of model 1 over model 2 is given by

$$BF_{12} = \frac{m_1(y)}{m_2(y)},$$

where $m_i(y)$ is the marginal or predictive density of the data for model M_i:

$$m_i(y) = \int f_i(y|\theta_i) g_i(\theta_i)\,\mathrm{d}\theta_i.$$

Jeffreys (1961) was one of the first to develop Bayes factors in testing for independence in a 2×2 table. Under the independence model H, the cell probabilities can be expressed as $\{\alpha\beta, \alpha(1-\beta), (1-\alpha)\beta, (1-\alpha)(1-\beta)\}$, where α and β are marginal probabilities of the table. Under the dependence hypothesis, Jeffreys expressed the probabilities as $\{\alpha\beta + \gamma, \alpha(1-\beta) - \gamma, (1-\alpha)\beta - \gamma, (1-\alpha)(1-\beta) + \gamma\}$. By assuming independent uniform $(0, 1)$ prior distributions on α, β, and γ, Jeffreys developed an approximate Bayes factor in support of the dependence hypothesis.

Suppose one observes multinomial $\{y_1, \ldots, y_t\}$ with cell probabilities $\{\theta_1, \ldots, \theta_t\}$ and total count $n = \sum_{i=1}^t y_i$. Good (1965) noted that simple relative frequency estimates of the probabilities θ_i can be poor when data are sparse, and studied the alternative estimates $(y_i + k)/(n + kt)$, where n is the fixed total count and k is a "flattening constant." This estimate is the mean of the posterior density assuming that the $\{\theta_i\}$ have a symmetric Dirichlet distribution of the form $p(\theta_i) \propto \prod_{i=1}^t \theta_i^{k-1}$, where $\theta_i > 0, i = 1, \ldots, t$ and $\sum_{i=1}^t \theta_i = 1$. In practice, it may be difficult for a user to specify the Dirichlet parameter k. Good then advocated use of a prior distribution $g(k)$ for k, resulting in a prior for the $\{\theta_i\}$ of the hierarchical form

$$g(\{\theta_i\}) = \int_0^\infty \frac{\Gamma(tk)}{(\Gamma(k))^t} \prod_{i=1}^t \theta_i^{k-1} g(k)\,\mathrm{d}k.$$

As will be seen later, Good also used this form of a mixture of symmetric Dirichlet distributions to develop Bayes factors for testing independence in contingency tables.

EXAMPLE 1 (*Smoothing towards an equiprobability model*). Jarrett (1979) considered the number of mine explosions in British mines between 1851 and 1962. He recorded the multinomial frequencies (14, 20, 20, 13, 14, 10, 18, 15, 11, 16, 16, 24) that represent the number of explosions during the days Sunday, Monday, ..., Saturday. Using a GLIM formulation for these data, suppose that the counts $\{y_i\}$ are independent Poisson with means $\{\mu_i\}$. Then the proportion parameters are defined as $\theta_i = \mu_i / \sum_j \mu_j$.

To model the belief that the $\{\theta_i\}$ are similar in size, we assume that $\{\theta_i\}$ has a symmetric Dirichlet distribution with parameter k. We assume that k has the vague, but proper density proportional to $(1+k)^{-2}$ that is equivalent to assuming a standard logistic density for $\log k$.

Below we show the WinBUGS code for fitting this exchangeable model. Currently in WinBUGS it is not possible to directly assign $\{\theta_i\}$ a Dirichlet prior density with unknown parameters. But we can equivalently specify this Dirichlet density by specifying independent gamma densities for $\{\mu_i\}$ with shape parameter k and then defining $\theta_i = \mu_i / \sum_j \mu_j$.

Table 1
Posterior means and standard deviations of the cell means under the exchangeable model for mine explosion data from Jarrett (1979)

y_i	$E(\mu_i)$	$SD(\mu_i)$
14	15.0	2.85
20	18.0	3.12
20	18.0	3.10
13	14.5	2.81
14	15.0	2.86
10	13.0	2.71
18	17.0	3.04
15	15.5	2.90
11	13.5	2.74
16	16.0	2.94
16	16.0	2.96
24	20.0	3.27

```
model
{
logk~dlogis(0,1)
k<-exp(logk)

for (i in 1:I) { mu[i] ~ dgamma(k,1)
                 x[i] ~ dpois(mu[i])
                 theta[i] <- mu[i]/mu.sum }
mu.sum <- sum(mu[]);
}

data
list(x=c(14,20,20,13,14,10,18,15,11,16,16,24), I=12)
```

Table 1 displays the posterior means and standard deviations of the cell means $\{\mu_i\}$. Under an equiprobability model, the expected cell counts are all 15.9. Also, from the WinBUGS output, we compute the posterior mean of k to be 16.0. The posterior mean of μ_i is approximately $(y_i + 16)/2$. Essentially, the posterior mean estimates of $\{\mu_i\}$ shrink the observed counts 50% towards the equiprobability estimate of 15.9.

4. Bayesian smoothing of contingency tables

One difficulty with large contingency tables is that observed sampling zeros in some cells may lead to poor estimates of the underlying cell probabilities. One *ad hoc* adjustment is to add 1/2 to each observed count, as in Good's (1965) approach with prespecified flattening constant $k = 1/2$. Fienberg and Holland (1973) were interested in developing better estimates for cell probabilities in these tables with sparse counts. They first considered the conjugate Dirichlet model $g(\{\theta_i\}) \propto \prod \theta_i^{K\lambda_i - 1}$ as a prior for the cell probabilities, where λ_i is the prior mean of θ_i and K is a precision parameter. The use of this conjugate prior results in the posterior mean estimate

$$\hat{\theta}_i = \left(\frac{n}{n+K}\right)\frac{y_i}{n} + \left(\frac{K}{K+n}\right)\lambda_i.$$

Since the hyperparameter K is unknown, Fienberg and Holland (1973) developed an empirical Bayes estimator. For fixed $\{\lambda_i\}$, they showed that the risk of $\hat{\theta}_i$, under squared error loss, is equal to

$$R(\hat{\theta}, \theta) = \big(n/(n+K)\big)^2\big(1 - \|\theta\|^2\big) + \big(K/(n+K)\big)^2 n\|\theta - \lambda\|^2,$$

where $\theta = (\theta_1, \ldots, \theta_t)$, $\lambda = (\lambda_1, \ldots, \lambda_t)$, and $\|\theta\|^2 = \sum_{i=1}^t \theta_i^2$. The value of K that minimizes this risk is $\widehat{K} = (1 - \|\theta\|^2)/(\|\theta - \lambda\|^2)$. If one replaces K with the estimate $\widehat{K}$ in the expression for $\hat{\theta}_i$, one obtains the empirical Bayes estimate

$$\theta_i^* = \left(\frac{n}{n+\widehat{K}}\right)\frac{y_i}{n} + \left(\frac{\widehat{K}}{\widehat{K}+n}\right)\lambda_i.$$

Fienberg and Holland (1973) showed that the estimates $\{\theta_i^*\}$ had good risk properties relative to the maximum likelihood estimates $\{y_i/n\}$. In practice, one can choose the prior means $\{\lambda_i\}$ to reflect one's prior beliefs, or choose data-dependent values for $\{\lambda_i\}$ based on the estimated expected counts from a log-linear model. For example, one might in this way shrink estimates towards the independence model for two-way contingency tables, or towards conditional independence or the no three-way interaction model for three-way tables.

A number of alternative fully Bayesian methods have been proposed for smoothing contingency table counts. One approach is based on normal prior distributions placed on components of a logit representation of a cell probability. For a two-way table with counts $\{y_{ij}\}$ and cell probabilities $\{\theta_{ij}\}$, Leonard (1975) defines the multivariate logit $\gamma_{ij} = \text{logit}\,\theta_{ij} = \log\theta_{ij} + D(\theta)$, where $D(\theta)$ is chosen to ensure that the probabilities sum to one. The logit is decomposed as

$$\gamma_{ij} = \alpha_i + \beta_j + \lambda_{ij},$$

where the terms correspond respectively to a row effect, a column effect, and an interaction effect in the two-way table. To model the belief that the set of row effects $\{\alpha_i\}$ is exchangeable, Leonard (1975) uses the two-stage prior

1. $\alpha_1, \ldots, \alpha_I$ are independent $\text{N}(\mu_\alpha, \sigma_\alpha^2)$;
2. $\mu_\alpha, \sigma_\alpha^2$ independent, with μ_α having a vague flat prior and $\nu_\alpha\tau_\alpha\sigma_\alpha^{-2}$ distributed chi-squared with ν_α degrees of freedom. The hyperparameter τ_α represents a prior estimate at σ_α^2 and ν_α measures the sureness of this prior guess.

Similar exchangeable prior distributions are placed on the sets of column effects $\{\beta_j\}$ and interaction effects $\{\lambda_{ij}\}$. Leonard (1975) used this model to find posterior modal estimates of the probabilities. When the interaction effects are set equal to zero, these Bayesian estimates smooth the table towards an independence structure. Nazaret (1987) extended this multivariate logit representation to three-way tables.

Table 2
Data from Cornfield (1962) classifying males with respect to blood pressure and absence or presence of heart disease

Blood pressure	Disease	No disease
<116	3	153
116–125	17	235
126–135	12	272
136–145	16	255
146–155	12	127
156–165	8	77
166–185	16	83
>185	8	35

Albert and Gupta (1982) and Epstein and Fienberg (1992) perform similar smoothing using mixtures of conjugate priors. To model the belief that the cell probabilities satisfy an independence structure, Albert and Gupta (1982) assign the $\{\theta_{ij}\}$ a Dirichlet distribution with precision parameter K and prior cell means $\{\lambda_{ij}\}$ satisfying the independence structure $\lambda_{ij} = \lambda_{i+}\lambda_{+j}$, with λ_{i+}, λ_{+j} respectively the prior row and column marginal means: $\lambda_{i+} = \sum_j \lambda_{ij}$, $\lambda_{+j} = \sum_i \lambda_{ij}$. At the second-stage of the prior, these marginal prior means $\{\lambda_{i+}\}$ and $\{\lambda_{+j}\}$ are assigned vague uniform distributions. The posterior mean of the cell probability θ_{ij} can be expressed as

$$\left(\frac{n}{n+K}\right)\frac{y_{ij}}{n} + \left(\frac{K}{K+n}\right)E(\lambda_{i+}\lambda_{+j}),$$

where $E(\cdot)$ denotes the expectation over the posterior distribution of $\{\lambda_{i+}\,\lambda_{+j}\}$. This estimate is a compromise between the observed relative frequencies y_{ij}/n, which are the usual unconditional maximum likelihood estimates, and the estimates under an independence model. Epstein and Fienberg (1992) generalized this conjugate approach by modeling the prior mean of θ_{ij} by a logit model. Laird (1978) and Knuiman and Speed (1988) perform Bayesian smoothing by applying a multivariate normal prior to the vector of logarithms of expected cell counts. (King and Brooks, 2001 show that a multivariate normal prior on the log-linear model parameters induces a multivariate normal prior on the expected cell counts of the contingency table.)

EXAMPLE 2 (*Smoothing a two-way table*). To illustrate smoothing a two-way table, we consider data from Cornfield (1962) displayed in Table 2 classifying males with respect to blood pressure and absence or presence of coronary heart disease. Due to the presence of some small cell counts in the table, it is desirable to shrink the counts towards expected counts from an independence model.

In WinBUGS it is convenient to assume that the cell counts $\{y_{ij}\}$ are independent Poisson $\{m_{ij}\}$, where the Poisson means satisfy the log-linear model

$$\log m_{ij} = u_0 + u_{1(i)} + u_{2(j)} + u_{12(ij)}.$$

We place vague normal priors on the constant and main effect terms. To perform the desired shrinkage, we assume that the interaction parameters $u_{12(ij)}$ are independent

Table 3
Posterior means and standard deviations using hierarchical log-linear model on the cell means

Blood pressure	Disease	No disease
<116	5.0 (2.1)	154.0 (12.4)
116–125	17.2 (3.9)	234.8 (15.2)
126–135	13.2 (3.5)	270.9 (16.4)
136–145	16.5 (3.9)	254 (15.9)
146–155	11.7 (3.2)	127.3 (11.2)
156–165	7.6 (2.5)	77.4 (8.7)
166–185	14.3 (3.7)	84.7 (9.2)
>185	6.4 (2.4)	36.5 (6.0)

from a normal distribution with mean 0 and precision τ. To complete the prior, $\log \tau$ is assigned a standard logistic distribution. The WinBUGS program for this model follows.

```
model {#  PRIORS
  for (i in 1:I) { u1[i] ~ dnorm(0,0.01)}
  for (i in 1:J) { u2[i] ~ dnorm(0,0.01)}
  u  ~ dnorm(0,0.01);
   for (i in 1:I) {
   for (j in 1:J) {
   u12[i,j] ~ dnorm(0,tau)
   }}

   logtau ~ dlogis(0,1)
   tau <- exp(logtau)

   for (i in 1:I) {
   for (j in 1:J) {
   m[i,j] ~ dpois(mu[i,j])
   log(mu[i,j]) <-  u + u1[i]+u2[j]+u12[i,j]
    }}
  }

Data
list(m=structure(.Data=c(3,156,17,235,12,272,16,255,12,127,8,
                 77,16,83,8,35),.Dim=c(8,2)),I=8,J=2)

Inits
list(u1=c(1,1,1,1,1,1,1,1),u2=c(1,1),  u=1,
u12=structure(.Data=c(1,1,1,1,1,1,1,1,1,1,1,1,1,1,1,1),
              .Dim=c(8,2)))
```

Table 3 displays posterior means and standard deviations for the cell means $\{m_{ij}\}$ using this hierarchical model. For this data, the independence model is a poor fit (chi-square statistic of 33.38 with 7 degrees of freedom) and the Bayesian estimates shrink the observed cell counts only a modest amount towards an independence structure.

5. Bayesian interaction analysis

Leonard and Novick (1986) describe the use of a Bayesian hierarchical model to explore the interaction structure of a two-way contingency table. Given cell counts $\{y_{ij}\}$ from independent Poisson distributions with respective means $\{\theta_{ij}\}$ they assume, at the first stage, that the θ_{ij} have independent Gamma distributions with respective means ξ_{ij} and precision parameter α. The means ξ_{ij} are presumed to satisfy the independence structure $\log \xi_{ij} = \mu + \lambda_i^A + \lambda_j^B$. At the second stage of the prior, all unknown parameters are given vague distributions. The posterior distribution of the precision parameter α is informative about the goodness of fit of the independence model. In addition, Leonard and Novick (1986) consider the posterior distributions of the "parametric residuals" $\{\log \theta_{ij} - \log \xi_{ij}\}$ to explore the dependence pattern in the table. Leonard et al. (1989) consider an alternative interaction analysis based on a nonhierarchical prior. They obtain approximations to the joint posterior distribution of $\{\theta_{ij}\}$, and dependence in the table is studied by considering the posterior distribution of the interaction parameters $\{\theta_{ij} - \theta_{i+} - \theta_{+j} + \theta_{++}\}$, with $\theta_{i+} = \sum_j \theta_{ij}$, $\theta_{+j} = \sum_i \theta_{ij}$, and $\theta_{++} = \sum_i \sum_j \theta_{ij}$.

EXAMPLE 3 (*Interaction analysis*). We illustrate Bayesian interaction analysis on a dataset derived from the March 1989 Current Population Survey. A number of husband-wife families were identified in a random sample from the survey by Freedman et al. (1991), and each husband was classified with respect to the region of the state of residence and his education.

One approach for studying interaction in this table is to fit a saturated log-linear model for the cell counts and study the posterior distributions of the interaction effects. The cell counts $\{y_{ij}\}$ are assumed independent Poisson $\{m_{ij}\}$ where the cell means satisfy the log-linear model $\log m_{ij} = u_0 + u_{1(i)} + u_{2(j)} + u_{12(ij)}$. To make the model identifiable, we apply the corner-point constraints $u_{1(1)} = u_{2(1)} = 0$ and $u_{12(1j)} = 0, u_{12(i1)} = 0$ for all i, j. All of the log-linear model terms are assumed a priori independent normal with zero means and small precision parameters. The WinBUGS code for this model is shown below.

```
model {
  for (i in 2:I) { u1[i] ~ dnorm(0,0.01)}
  for (i in 2:J) { u2[i] ~ dnorm(0,0.01)}
  u  ~ dnorm(0,0.01);
   for (i in 2:I) { for (j in 2:J) {
   u12[i,j] ~ dnorm(0,0.01) }}

   u1[1]<-0
   u2[1]<-0
   for (i in 1:I) {u12[i,1] <-0 }
   for (j in 2:J) {u12[1,j] <-0 }

   for (i in 1:I) {for (j in 1:J) {
   m[i,j] ~ dpois(mu[i,j])
   log(mu[i,j]) <-  u + u1[i]+u2[j]+u12[i,j]}}
  }
```

```
Data
list(m=structure(.Data=c(47,63,30,31,30,111,225,84,77,71,147,270,
106,72,63,46,122,53,42,29,173,235,110,81,76,102,85,32,18,16,98,
123,88,68,36,27,64,48,26,24,104,151,117,76,73),.Dim=c(9,5)),
I=9,J=5)

Inits
list(u1=c(NA,1,1,1,1,1,1,1,1),u2=c(NA,1,1,1,1),
u=1,u12=structure(.Data=c(NA,NA,NA,NA,NA,NA,1,1,1,1,NA,1,1,1,1,NA,
1,1,1,1,NA,1,1,1,1,NA,1,1,1,1,NA,1,1,1,1,NA,1,1,1,1,NA,1,1,1,1),
.Dim=c(9,5)))
```

Table 4 displays the posterior means of the interaction parameters $\{u_{12(ij)}\}$. Generally the posterior standard deviations of these parameters are in the .25 to .35 range, so one can informally judge the significance of various interaction terms by judging the size relative to .3.

Another way to study the interaction structure of this table is to fit an independence model to the table and view the posterior distributions of the Bayesian residuals. In the model specified by the WinBUGS code below, we assume that

$$\log m_{ij} = u_0 + u_{1(i)} + u_{2(j)},$$

where the constant and main effects are assigned vague normal prior distributions. We then focus on the posterior distributions of the Pearson residuals

$$r_{ij} = \frac{y_{ij} - m_{ij}}{\sqrt{m_{ij}}}.$$

```
model {
  for (i in 1:I) { u1[i] ~ dnorm(0,0.01)}
  for (i in 1:J) { u2[i] ~ dnorm(0,0.01)}
  u  ~ dnorm(0,0.01);

   for (i in 1:I)  {
   for (j in 1:J) {
   m[i,j] ~ dpois(mu[i,j])
   log(mu[i,j]) <-  u + u1[i]+u2[j]
   resid[i,j]<- (m[i,j]-mu[i,j])/sqrt(mu[i,j])
    }}
  }

Data
list(m=structure(.Data=c(47,63,30,31,30,111,225,84,77,71,147,270,
106,72,63,46,122,53,42,29,173,235,110,81,76,102,85,32,18,16,98,
123,88,68,36,27,64,48,26,24,104,151,117,76,73),.Dim=c(9,5)),
I=9,J=5)

Inits
list(u1=c(1,1,1,1,1,1,1,1,1),u2=c(1,1,1,1,1),  u=1)
```

Table 4
Posterior means of the interaction parameters from a fit of the saturated model to a table the March 1989 Current Population Survey

Region	Husband's education (years)				
	0–11	12	13–15	16	17–18
Northeast	0	0	0	0	0
Middle Atlantic	0	.39	.14	.02	−.02
East North Central	0	.29	.09	−.34	−.42
West North Central	0	.65	.55	.28	−.05
South Atlantic	0	.00	−.03	−.38	−.39
East South Central	0	−.50	−.75	−1.37	−1.45
West South Central	0	−.09	.31	.01	−.58
Mountain	0	.57	1.02	.36	.32
Pacific	0	.05	.53	.06	.07

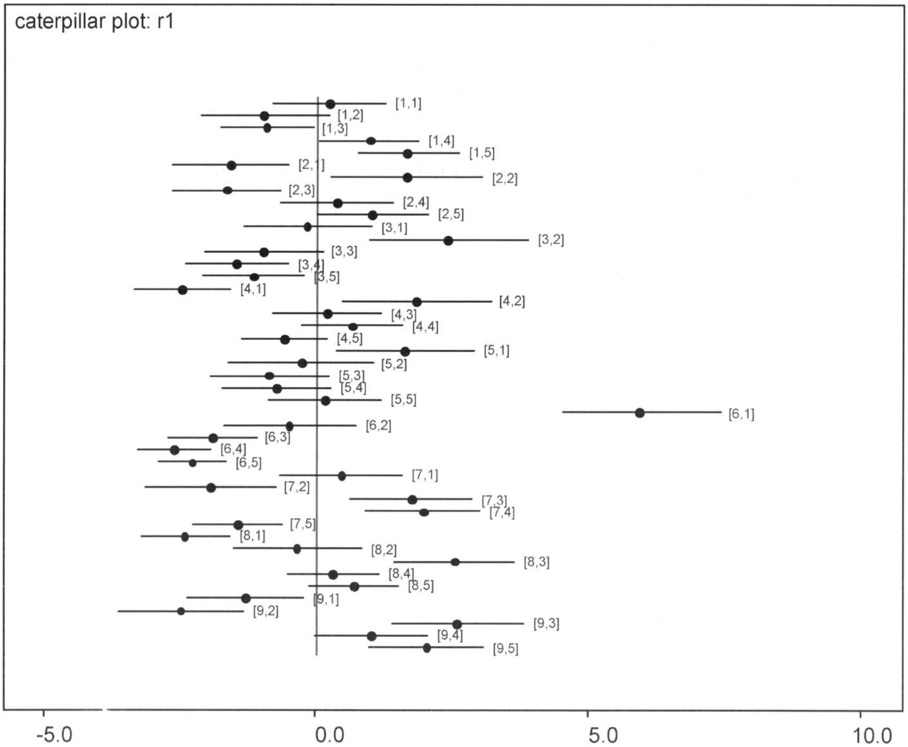

Fig. 1. Posterior densities of set of parametric Pearson residuals from a fit of an independence model for the Current Population Survey dataset from Freedman et al. (1991). The limits for each bar correspond to the 2.5% and 9% percentiles of the posterior distribution of the residual. For a color reproduction of this figure see the color figures section, page 1004.

Figure 1 illustrates the use of a caterpillar plot to simultaneously display the posterior densities of the Bayesian Pearson residuals for all cells of the table. The limits of each bar in the plot correspond to the endpoints of a 95 percent interval estimate for the residual. Bars that lie completely on one side of the origin correspond to significantly large residuals. Note that the residual [6, 1] stands out – this corresponds to the count in the East South Central region with a husband's education between 0 to 11 years. There are quite a few significant residuals in this table, indicating that an independence model is a poor fit to this data.

6. Bayesian tests of equiprobability and independence

I.J. Good, in a large number of papers, developed Bayes tests for contingency tables under a variety of sampling models. We illustrate the general approach by considering Good's (1967) construction of a significance test for equiprobability of a multinomial probability vector. As usual, one observes multinomial $\{y_1, \ldots, y_t\}$ with cell probabilities $\{\theta_1, \ldots, \theta_t\}$ and total count $n = \sum_{i=1}^{t} y_i$. The hypothesis of interest is H: $\theta_i = 1/t, i = 1, \ldots, t$, and the usual classical test statistic is Pearson's $X^2 = \sum_{i=1}^{t} \frac{(y_i - E(y_i))^2}{E(y_i)} = \frac{t}{n} \sum (y_i - \frac{n}{t})^2$, which is asymptotically chi-square with $t - 1$ degrees of freedom.

To develop a Bayes factor to test H against the alternative hypothesis $\overline{H}$: $\theta_i \neq 1/t$ for some $i = 1, \ldots, t$, note that the density of the data $\{y_1, \ldots, y_t\}$ is fully specified under the hypothesis H. Let H_k denote the hypothesis that the $\{\theta_i\}$ have a symmetric Dirichlet (k) distribution with density $g(\theta) \propto \prod \theta_i^{k-1}$. The hyperparameter k, since it is difficult to specify, is given a log Cauchy prior

$$g(k) = \left(\frac{1}{\pi k}\right)\left(\frac{\lambda}{\lambda^2 + \log(k/\mu)^2}\right).$$

Then the Bayes factor BF of $\overline{H}$ against H is given by $BF = E(BF(k)|\lambda, \mu) = \int BF(k) g(k)\, dk$, where $BF(k)$ is the Bayes factor of H_k against equiprobability,

$$BF(k) = \frac{t^n \Gamma(tk) \prod_{i=1}^{t} \Gamma(y_i + k)}{\Gamma(k)^t \Gamma(n + tk)}.$$

To illustrate and compare Bayesian and classical measures of evidence, we consider 150 voters of whom 61, 53, and 36 expressed preferences for candidates 1, 2, and 3, respectively. The chi-square statistic for equiprobability is $X^2 = 6.52$ with an associated p-value of .0384. Table 5 gives values of the Bayes factor of H_k against H for a range of values of the Dirichlet parameter k. A Dirichlet prior with a large value of k, say $k = 1000$, concentrates most of its probability on the equiprobability hypothesis and the Bayes factor is close to one. In contrast, the hypothesis H is more likely than the use a Dirichlet prior with a small value of k. In fact, the Bayes factor is not defined by the use of an improper prior with $k = 0$. Note that the maximum value of $BF(k)$ in the table is 2.8; for a wide range of log Cauchy priors that might be used in practice, BF will be

Table 5
Values of the Bayes factor of H_k against H for values of the Dirichlet parameter k

k	0.1	0.5	1	2	10	100	1000
$BF(k)$	0	0.2	0.5	0.9	2.8	2.0	1.1

smaller than 1. In contrast with the classical result, the Bayesian measures indicate that there is little evidence against equiprobability in these data. Generally, when testing a point null hypothesis, a classical p-value overstates the evidence against the hypothesis compared to a Bayes factor test statistic (Berger and Delampady, 1987).

EXAMPLE 4 (*Computing a Bayes factor*). The WinBUGS program can be used to compute the Bayes factor against equiprobability when a prior distribution is placed on the Dirichlet hyperparameter k. Suppose $\log k$ is assigned a standard logistic distribution that is equivalent to assigning k a prior density $g(k) = 1/(1+k)^2$. (This prior has a similar form to the log Cauchy form discussed above. We use it here since the logistic function is available in WinBUGS.) In the WinBUGS code below, the expression for $BF(k)$ is programmed and the integral $BF = \int BF(k)g(k)\,\mathrm{d}k$ is computed by simulating over the prior density and estimating the integral by the mean of the simulated values of $BF(k)$. (There are more efficient ways of computing the Bayes factor by simulation – see Kass and Raftery, 1995.) For our example, we find that the Bayes factor against equiprobability is $BF = 0.846$ that means that the two models are approximately equally supported by the data.

```
model
{
logk~dlogis(0,1)   # logistic prior on log precision
                     hyperparameter
k<-exp(logk)

for (i in 1:I) {
     logxk[i] <- loggam(x[i]+k) }
     sx <- sum(x[]);

lbf <- loggam(I*k)-I*loggam(k)+sum(logxk[])
       -loggam(I*k+sx)+sx*log(I)
bf<- exp(lbf)
}

data
list(x=c(61,53,36), I=3,k=10)
```

Good (1976), Crook and Good (1980) and Good and Crook (1987) extended the above methodology in developing Bayes tests for two-way contingency tables. Tests were constructed based on mixtures of conjugate priors for the three sampling models (multinomial, product-multinomial, multivariate hypergeometric) corresponding respectively to fixed overall table total n or to fixed totals along one or both marginal dimensions. An objective of this analysis was to assess whether the marginal totals convey any evidence for or against independence of rows and columns. In the multinomial sampling situation (model 1) where only the total count is fixed, the Bayes factor in support of the dependence hypothesis $\overline{H}$ over the independence hypothesis H is given by

$$BF_1 = \frac{P(\{y_{ij}\}|\overline{H})}{P(\{y_{ij}\}|H)},$$

where $P(\{y_{ij}\}|\overline{H})$ and $P(\{y_{ij}\}|H)$ are the marginal probabilities of the data under the hypotheses $\overline{H}$ and H, respectively. The marginal probability under $\overline{H}$ is computed using a prior on the vector of cell probabilities $\{\theta_{ij}\}$ that is a mixture of symmetric Dirichlet distributions, and the probability under H is computed by use of a similar mixture of Dirichlet distributions placed on the vectors of marginal cell probabilities $\{\theta_i\}$ and $\{\theta_j\}$. Good and Crook also developed Bayes factors against independence in the situations where either the row or column totals were fixed (model 2), and in the situation where both row and column totals were fixed (model 3). The evidence provided by the row and column totals alone is defined to be the ratio $FRACT = BF_1/BF_3$ of the Bayes factors under model 1, in which information about rows and columns is observed, and model 3, in which row and column margins are arbitrarily fixed. One conclusion from their studies was that *FRACT* is usually between .5 and 2.5, indicating that the row and column totals typically contain a modest amount of evidence against independence. These Bayesian measures have the advantage that they explicitly allow for the sampling model and do not depend on asymptotic theory. In addition, the Bayesian factor measures can be used as classical test statistics against independence and they can be shown to possess good power against alternative hypotheses. Gunel and Dickey (1974) and Albert (1989) also develop Bayes factors for two-way tables based on conjugate priors and mixtures of conjugate priors, respectively.

7. Bayes factors for GLM's with application to log-linear models

Raftery (1996) presents a general approach for testing within generalized linear models, with direct applications to comparisons of log-linear models for multiway contingency tables. Recall from Section 2 that if one observes data D and has two possible models M_1 and M_2, then the evidence in support of the model M_1 is given by the Bayes factor $B_{12} = \frac{P(D|M_1)}{P(D|M_2)}$, where $P(D|M_k)$ is the marginal probability

$$P(D|M_k) = \int P(D|\theta_k, M_k)p(\theta_k|M_k)\,\mathrm{d}\theta_k.$$

Raftery (1996) presents several methods for approximating $P(D|M_k)$. By the Laplace method for integrals, one has the approximation

$$P(D|M_k) \approx (2\pi)^{p_k/2}|\Psi_k|^{1/2}P\bigl(D|\tilde{\theta}_k, M_k\bigr)p\bigl(\tilde{\theta}_k|M_k\bigr),$$

Table 6
Data from Shapiro et al. (1979) from a case-control study

	Age group (A)									
	25–29		30–34		35–39		40–44		45–49	
	Myocardial infarction (M)									
Oral contraceptives (C)	No	Yes	No	Yes	No	Yes	No	Yes	No	Yes
Not used	224	2	390	12	330	33	362	65	301	93
Used	62	4	33	9	26	4	9	6	5	6

where $\tilde{\theta}_k$ is the posterior mode, Ψ_k is the inverse of the negative of the Hessian matrix of $\log P(D|\theta_k, M_k)p(\theta_k|M_k)$ evaluated at the mode, p_k is the number of parameters of the model, and $|A|$ indicates the determinant of the matrix A.

The following approximation was developed as an alternative that capitalizes on quantities available from standard generalized linear model software:

$$2\log BF_{12} \approx \chi^2 + \left(E_1^* - E_2^*\right),$$

$$E_k^* = -\log|I_k| + 2\log P\left(\hat{\theta}_k|M_k\right) + p_k\log(2\pi),$$

where $\hat{\theta}_k$ and I_k are respectively the maximum likelihood estimate and observed information matrix from fitting the model M_k, and X^2 is the classical χ^2 "drop in deviance" test statistic for comparing the two models M_1 and M_2. One important issue is the choice of prior on the regression coefficients. Raftery (1996) discusses suitable "vague" choices of hyperparameters to use in a testing situation.

To illustrate the use of Bayes factors in model choice, Raftery (1996) considers the data shown in Table 6 from a case-control study in which oral contraceptive histories were compared between groups of women having suffered myocardial infarction and control women who had not (Shapiro et al., 1979). The table shows a cross-classification of case (**M**yocardial infarction) or control status (M), **A**ge category (A), and history of oral **C**ontraceptive use (C).

Suppose we wish to compare the two loglinear models M_1 and M_2, where M_1 denotes the no three-way interaction model indicating that the relative risk relating disease and oral contraceptive history (estimated in this context by the odds ratio) is constant across age groups, and the more complicated model M_2 indicates that the relative risk is constant from ages 25–34 but may shift to a different constant during ages 35–49. Using a classical loglinear analysis, the difference in deviances is 4.7 on one degree of freedom and the p-value is .03, indicating different relative risks for the age groups 25–34 and 35–49. Computation of a Bayes factor, in contrast, slightly favors the simpler model M_1.

One advantage of this Bayesian approach is that it can explicitly allow for model uncertainty in the estimation of parameters of interest. In the case where there are many possible models $\{M_1, \ldots, M_K\}$, the posterior distribution of the parameter θ may be expressed as the mixture of posteriors:

$$p(\theta|D) = \sum_{i=1}^{K} p(\theta|D, M_k)p(M_k|D),$$

where $p(\theta|D, M_k)$ is the posterior of θ under model M_k and $p(M_k|D) \propto P(M_k)P(D|M_k)$ is the posterior probability of model M_k. Using the above example, Raftery (1996) illustrates this "model averaging" approach in estimating a relative risk parameter when there are two plausible models. (See Raftery and Richardson, 1996 for additional applications of this method.)

EXAMPLE 5 (*Testing for independence in a two-way table*). We first illustrate model selection in the simple setting where one is testing for independence in a two-way table. As in Section 4, we assume that the cell counts $\{y_{ij}\}$ are independent Poisson $\{m_{ij}\}$, where the Poisson means satisfy the log-linear model

$$\log m_{ij} = u_0 + u_{1(i)} + u_{2(j)} + u_{12(ij)}.$$

As before, we place vague normal priors on the constant and main effect terms. Assume that the interaction parameters $u_{12(ij)}$ are independent from a normal distribution with mean 0 and precision τ. We can define independence and dependence models by means of specifying values for the precision parameter τ. By setting a large value for τ, the interaction terms have priors concentrated near 0, and we have (approximately) an independence model. A dependence model corresponds to a "small" value of τ, where this hyperparameter value is chosen based on one's knowledge about the association in the table.

In the WinBUGS program, we compare the models $\tau = 10$ (dependence) and $\tau = 100$ (independence). We construct a categorical variate k that corresponds to the model – $k = 1$ is dependence and $k = 2$ is independence and place a uniform prior on these two models. After the MCMC run, we can compute the posterior probabilities of the two models that can be used to compute a Bayes factor in support of the dependence hypothesis. (With equal prior probabilities on the two models, the Bayes factor in support of dependence is $BF = p/(1 - p)$, where p is the posterior probability of dependence.)

```
model
{  for (i in 1:N) {for (j in 1:N) {
       y[i,j] ~ dpois(m[i,j]);

       log(m[i,j]) <- u0+u1[i]+u2[j]+u12[i,j];}}
# priors on main effects
   P0 <- 0.001
   u0 ~ dnorm(0,P0)

   for (i in 1:N) {u1[i] ~ dnorm(0,P0)
                   u2[i] ~ dnorm(0,P0)}
```

```
# discrete prior on models for interactions
  prior[1] <- 0.5; prior[2] <- 0.5
  k ~ dcat(prior[1:2]);
  P1 <- equals(k,1)*10+equals(k,2)*100
  for (i in 1:N){for (j in 1:N){  u12[i,j] ~ dnorm(0,P1)}} }

Data
list(y=structure(.Data=c(14,3,5,7,20,10,4,8,12),.Dim=c(3,3)),N=3)


Inits
list( k=1,u1 = c(0.1, 0.1,0.1),
    u0 = 2, u2 = c(0.1, 0.1,0.1),
    u12=structure(.Data=c(0.1,0.1,0.1,
                          0.1,0.1,0.1,
                          0.1,0.1,0.1),.Dim=c(3,3))          )
```

This program was used to compare the two models $\tau = 10$ (dependence) and $\tau = 100$ (independence) for the 3 by 3 table of counts [14, 3, 2; 7, 20, 10; 4, 8, 12]. From the WinBUGS output, the posterior probability of the model $\tau = 10$ was 0.94 and the Bayes factor in support of the dependence hypothesis is 15.76. This simulation-based computation of the Bayes factor agrees with the computation based on the Laplace method (Albert, 1989).

8. Use of BIC in sociological applications

If one ignores terms of order O(1) or smaller, one gets a further approximation to the log marginal density of the data under model M_k (Raftery, 1994):

$$\log P(D|M_k) \approx \log P\big(D|\hat{\theta}\big) - \frac{p_k}{2} \log n,$$

where, as above, p_k is the number of parameters in model M_k. The difference between values of this approximation for a reduced vs. a saturated model is the BIC (Bayesian Information Criterion) measure for assessing the overall fit of a model M_k:

$$BIC_k = L_k^2 - df_k \log n,$$

where L_k^2 is the usual deviance statistic and df_k is the associated degrees of freedom of this statistic. Two models M_j and M_k can be compared by the difference $BIC_k - BIC_j$. Raftery (1994) gives tables helpful for interpreting the significance of a computed BIC value and comparing it with traditional p-values. Specifically, a BIC measure gives precise guidelines on how one should adjust a significance level pertaining to model comparisons as the sample size increases, in order to avoid including trivial complexity in a final model.

Table 7
Data from Fienberg (1980) classifying rates with respect to dose of carcinogen, time of death, and presence or absence of cancer

	Time of death (T)			
	Premature		At sacrifice	
	Dose (D)			
Cancer (C)	Low	High	Low	High
Present	4	7	0	7
Absent	26	16	14	14

9. Bayesian model search for loglinear models

Recent advances in Bayesian computing have increased interest in using Bayesian models to search for the "best" loglinear model for a multidimensional contingency table. Madigan and Raftery (1994) define some general principles that should be expressed in the behavior of any model selection strategy. One principle is that models that predict the data far less well than the best model should be discarded; that is, models M_k such that $(\max_l P(M_l|D))/P(M_k|D) \geqslant C$ should be removed from consideration. A second principle, "Occam's Razor," states that if two models predict the data equally well, the simpler should be preferred. Madigan and Raftery (1994) describe a procedure where one can search through the space of models by use of Bayesian posterior model probabilities. In this approach, a model is represented by a directed graph with a node for each variable, and the dependence structure is represented by use of edges connecting pairs of nodes. Hyper-Dirichlet priors are used to represent prior opinion about the model parameters (Dawid and Lauritzen, 1993).

Albert (1996, 1997) describes Bayesian model selection based on priors placed directly on terms of the loglinear model. For a three-way table with a saturated log-linear model represented by

$$\log m_{ijk} = u + u_{1(i)} + u_{2(j)} + u_{3(k)}$$
$$+u_{12(ij)} + u_{13(ik)} + u_{23(jk)} + u_{123(ijk)},$$

Albert (1996) places a multivariate normal prior directly on the vector formed by stringing out the sets of u-terms $(u, u_1, u_2, u_3, u_{12}, u_{13}, u_{23}, u_{123})$, and models are defined by means of priors that constrain sets of u-terms to zero. The Laplace method (Tierney and Kadane, 1986) is used to compute the model probabilities.

To illustrate the use of a Bayesian model selection strategy, consider Table 7 (Fienberg, 1980) which classifies test rats in an experiment with respect to dose of a possible carcinogen (D), time of death (sacrificed at 132 weeks or age, or prematurely) (T), and presence or absence of cancer at necropsy (C).

A classical stepwise model search leads to the choice of the model [T], [CD], which indicates that cancer and dose are independent of survival to the end of the experiment. In Albert (1996), a Bayesian model search is performed over $2^4 = 16$ models consisting of combinations of the presence or absence of each of the four sets of interaction terms $u_{DT}, u_{DC}, u_{CT}, u_{DTC}$. Table 8 gives the posterior model probabilities for the six models

Table 8
Posterior model probability of the most likely models for a Bayesian model search for the Fienberg (1980) data set

Interactions included	Posterior probability
None	.05
u_{CD}	.39
u_{CA}, u_{CD}	.07
u_{CD}, u_{ACD}	.09
u_{CD}, u_{AD}	.18
u_{CA}, u_{CD}, u_{AD}	.04

with the largest values. This Bayesian analysis is similar to that of the classical analysis in the sense that the model with the highest posterior probability includes the interaction u_{CD} and no other interactions, but the probability of this model is only .39 and the remaining models have relatively small posterior probabilities. Bayesian estimates of the association between dose and cancer in the table will account for the uncertainty of the "best" log-linear model.

EXAMPLE 6 (*Model selection in WinBUGS*). We illustrate the model selection computations in WinBUGS for the Fienberg (1980) dataset. If we apply the corner-point constraints on the parameters of the log-linear model, then there are four nonzero interaction terms which are denoted in the WinBUGS program by $u_{12}, u_{13}, u_{23}, u_{123}$. The interaction term are assigned normal priors with means 0 and precision parameters $\tau_{12}, \tau_{13}, \tau_{23}, \tau_{123}$. Each precision is assumed apriori to be equal to 1 or 100 with equal probability. Letting $\tau = 100$ effectively removes the interaction term from the model, and $\tau = 1$ keeps the term in the model. There are 16 possible models corresponding to the inclusion or exclusion of each interaction term, and we use the MCMC run in WinBUGS to compute the posterior probabilities of the 16 models. The computed simulated posterior probabilities agree with the computations in Albert (1996) that were done using a Laplace approximation.

```
model     {for (i in 1:N1) {for (j in 1:N2) {for (k in 1:N3)
           {y[i,j,k] ~ dpois(m[i,j,k]);

log(m[i,j,k]) <- u0+u1[i]+u2[j]+u3[k]+ u12[i,j]+u23[j,k]+u13[i,k]
                 +u123[i,j,k]}}}

logodds1<-log(m[1,1,1]*m[2,1,2]/m[1,1,2]/m[2,1,1])
logodds2<-log(m[1,2,1]*m[2,2,2]/m[1,2,2]/m[2,2,1])

# Priors for included terms u0,u1,u2,u3, u12, u23
                                  (standard fixed effects priors)
   T <- 0.1
   u0 ~ dnorm(0,T); u1[1] <- 0; u2[1] <- 0; u3[1] <- 0
```

```
    for (i in 2:N1) {u1[i] ~ dnorm(0,T)}
    for (i in 2:N2) {u2[i] ~ dnorm(0,T)}
    for (i in 2:N3) {u3[i] ~ dnorm(0,T)}

    u12[1,1]<-0;  u12[1,2]<-0;  u12[2,1]<-0;  u13[1,1]<-0;
    u13[1,2]<-0;  u13[2,1]<-0
    u23[1,1]<-0 ; u23[1,2]<-0;  u23[2,1]<-0;  u123[1,1,1]<-0;
    u123[1,1,2]<-0;  u123[1,2,1]<-0;  u123[1,2,2]<-0;
    u123[2,1,1]<-0;  u123[2,1,2]<-0;  u123[2,2,1]<-0

prior[1] <- 0.5; prior[2] <- 0.5
k12 ~ dcat(prior[1:2]); k13 ~ dcat(prior[1:2]);
k23 ~ dcat(prior[1:2]); k123 ~ dcat(prior[1:2]);
P12<-equals(k12,1)*1+equals(k12,2)*100;
P13<-equals(k13,1)*1+equals(k13,2)*100;
P23<-equals(k23,1)*1+equals(k23,2)*100;
P123<-equals(k123,1)*1+equals(k123,2)*100;

u12[2,2]~dnorm(0,P12)
u13[2,2]~dnorm(0,P13)
u23[2,2]~dnorm(0,P23)
u123[2,2,2]~dnorm(0,P123)

 }

Data
list(N1=2,N2=2,N3=2,y=structure(.Data=c(4,7,0,7,26,16,14,14),
.Dim=c(2,2,2)))

Inits
list(u0=0, u1 = c(NA, 0),u2 = c(NA, 0), u3 = c(NA, 0),
     u12 = structure(.Data = c(NA, NA, NA, 0 ), .Dim = c(2, 2)),
     u23= structure(.Data = c(NA, NA, NA, 0 ), .Dim = c(2, 2)),
     u13 = structure(.Data = c(NA, NA, NA, 0 ), .Dim = c(2, 2))
     u123 = structure(.Data = c(NA, NA, NA, NA, NA, NA, NA, 0),
     .Dim=c(2,2,2) )
```

Dellaportas and Forster (1999) propose a simulation-based approach for finding the best log-linear model. In this approach, they assume that the vector of logarithms of the expected cell counts has a multivariate normal prior distribution. They work with a parameterization under which all parameters are identifiable and linearly independent, and choose priors for these parameters that reflect vague prior beliefs. For searching through the model space, Dellaportas and Forster (1999) propose a MCMC strategy based on the reversible jump Markov chain Monte Carlo (MCMC) algorithm of Green (1995). An attractive feature of this algorithm is that one can move between models of different dimension. These authors apply their model selection strategy, and those of Raftery (1996) and Madigan and Raftery (1994), to finding the best model for a three-way contingency table. Although there are differences in the computed posterior model probabilities, all of these approaches select the same loglinear model. In the normal regression context, George and McCulloch (1993) propose an alternative Bayesian algorithm, Stochastic Search Variable Selection (SSVS), for searching through the space of all possible models. Ntzoufras et al. (2000) extend the SSVS approach to loglinear modeling.

10. The future

Much of the early Bayesian methodology for contingency tables was devoted to issues regarding computation due to the difficulties in computing integrals of several variables. However, by virtue of great advances in computing posterior distributions by simulation, it is now possible to fit sophisticated Bayesian models for high-dimensional contingency tables. We have illustrated the relative ease of using WinBUGS of simulating from posterior distributions of categorical data models and comparing models by Bayes factors. Further Bayesian advances may be expected, especially with respect to criticism of single loglinear models, and model selection among large classes of hierarchical and graphical models.

References

Agresti, A. (1990). *Categorical Data Analysis*. Wiley, New York.

Albert, J.H. (1989). A Bayesian test for a two-way contingency table using independence priors. *Canad. J. Statist.* **18**, 347–363.

Albert, J.H. (1996). Bayesian selection of log-linear models. *Canad. J. Statist.* **24**, 327–347.

Albert, J.H. (1997). Bayesian testing and estimation of association in a two-way contingency table. *J. Amer. Statist. Assoc.* **92**, 685–693.

Albert, J.H., Gupta, A.K. (1982). Mixtures of Dirichlet distributions and estimation in contingency tables. *Ann. Statist.* **10**, 1261–1268.

Berger, J.O., Delampady, M. (1987). Testing precise hypotheses. *Statist. Sci.* **2**, 317–335.

Bishop, Y.M.M., Fienberg, S.E., Holland, P.W. (1975). *Discrete Multivariate Analysis*. MIT Press, Cambridge, MA.

Congdon, P. (2001). *Bayesian Statistical Modeling*. Wiley, Chichester.

Congdon, P. (2003). *Applied Bayesian Modelling*. Wiley, Chichester.

Cornfield, J. (1962). Joint dependence of risk of coronary heart disease on serum cholesterol and systolic blood pressure: A discriminant function analysis. *Federal Proceedings* **21**, 58–61.

Crook, J.F., Good, I.J. (1980). On the application of symmetric Dirichlet distributions and their mixtures to contingency tables: Part II. *Ann. Statist.* **8**, 1198–1218.

Dawid, A.P., Lauritzen, S.L. (1993). Hyper Markov laws in the statistical analysis of decomposable graphical models. *Ann. Statist.* **21**, 1272–1317.

Dellaportas, P., Forster, J.J. (1999). Markov chain Monte Carlo model determination for hierarchical and graphical log-linear models. *Biometrika* **86**, 615–633.

Diaconis, P., Efron, B. (1985). Testing for independence in a two-way table: New interpretations of the chi-square statistic. *Ann. Statist.* **13**, 845–874.

Epstein, L.D., Fienberg, S.E. (1992). Bayesian estimation in multidimensional contingency tables. In: Goel, P.K., Sreenivas Iyengar, N. (Eds.), *Bayesian Analysis in Statistics and Economics*. Springer-Verlag, New York.

Fienberg, S.E. (1980). Using loglinear models to analyze cross-classified categorical data. *Math. Sci.* **5**, 13–30.

Fienberg, S.E., Holland, P.W. (1973). Simultaneous estimation of multinomial cell probabilities. *J. Amer. Statist. Assoc.* **68**, 683–689.

Freedman, D.R., Pisani, R., Purves, R., Adhikari, A. (1991). *Statistics*. W.W. Norton, New York.

Gelfand, A., Smith, A. (1990). Sampling-based approaches to calculating marginal densities. *J. Amer. Statist. Assoc.* **85**, 398–409.

George, E.I., McCulloch, R.E. (1993). Variable selection via Gibbs sampling. *J. Amer. Statist. Assoc.* **88**, 881–889.

Good, I.J. (1965). *The Estimation of Probabilities*. MIT Press, Cambridge, MA.

Good, I.J. (1967). A Bayesian significance test for multinomial distributions. *J. Roy. Statist. Soc., Ser. B* **29**, 399–431.

Good, I.J. (1976). On the application of symmetric Dirichlet distributions and their mixtures to contingency tables. *Ann. Statist.* **4**, 1159–1189.

Good, I.J., Crook, J.F. (1987). The robustness and sensitivity of the mixed-Dirichlet Bayesian test for "independence" in contingency tables. *Ann. Statist.* **15**, 670–693.

Green, P.J. (1995). Reversible jump Markov chain Monte Carlo computation and Bayesian model determination. *J. Roy. Statist. Soc., Ser. B* **82**, 711–732.

Gunel, E., Dickey, J.M. (1974). Bayes factors for independence in contingency tables. *Biometrika* **61**, 545–557.

Jarrett, R.G. (1979). A note on the intervals between coal-mining disasters. *Biometrika* **66**, 191–193.

Jeffreys, H. (1961). *Theory of Probability*, third ed. Oxford University Press, Oxford.

Kass, R.E., Raftery, A.E. (1995). Bayes factors. *J. Amer. Statist. Assoc.* **90**, 773–795.

King, R., Brooks, S.P. (2001). Prior induction in log-linear models for general contingency table analysis. *Ann. Statist.* **29**, 715–747.

Knuiman, Speed (1988). Incorporating prior information into the analysis of contingency tables. *Biometrics* **44**, 1061–1071.

Laird, N.M. (1978). Empirical Bayes methods for two-way contingency tables. *Biometrika* **65**, 581–590.

Leonard, T. (1975). Bayesian estimation methods for two-way contingency tables. *J. Roy. Statist. Soc., Ser. B* **37**, 23–37.

Leonard, T., Novick, M.R. (1986). Bayesian full rank marginalization for two-way contingency tables. *J. Educational Statist.* **11**, 33–56.

Leonard, T., Hsu, J.S.J., Tsui, K.W. (1989). Bayesian marginal inference. *J. Amer. Statist. Assoc.* **84**, 1051–1058.

Lindley, D.V. (1964). The Bayesian analysis of contingency tables. *Ann. Math. Statist.* **35**, 1622–1643.

Madigan, D., Raftery, A.E. (1994). Model selection and accounting for model uncertainty in graphical models using Occam's window. *J. Amer. Statist. Assoc.* **89**, 1535–1546.

Nazaret, W.A. (1987). Bayesian log-linear estimates for three-way contingency tables. *Biometrika* **74**, 401–410.

Ntzoufras, I., Forster, J.J., Dellaportas, P. (2000). Stochastic search variable selection for log-linear models. *J. Statist. Comput. Simulation* **68**, 23–38.

Raftery, A.E. (1994). Bayesian model selection in social research. In: Marsden, P.V. (Ed.), *Sociological Methodology 1995*. Blackwell Publishing, Cambridge, MA, pp. 111–163.

Raftery, A.E. (1996). Approximate Bayes factors and accounting for model uncertainty in generalised linear models. *Biometrika* **83**, 251–266.

Raftery, A.E., Richardson, S. (1996). Model selection for generalized linear models via GLIB, with application to nutrition and breast cancer. In: Berry, D.A., Stangl, D.K. (Eds.), *Bayesian Biostatistics*. Marcel Dekker, New York, pp. 321–353.

Shapiro, S., Slone, D., Rosenberg, L., Kaufman, D.W., Stolley, P.D., Miettinen, O.S. (1979). Oral-contraceptive use in relation to myocardial infarction. *Lancet* **313**, 743–747.

Spiegelhalter, D.J., Thomas, A., Best, N., Lunn, D. (2003). *WinBUGS User Manual*.

Tierney, L., Kadane, J.B. (1986). Accurate approximations for posterior moments and marginal densities. *J. Amer. Statist. Assoc.* **81**, 82–86.

Handbook of Statistics, Vol. 25
ISSN: 0169-7161
© 2005 Elsevier B.V. All rights reserved.
DOI 10.1016/S0169-7161(05)25031-9

31

Multiple Events Time Data: A Bayesian Recourse

Debajyoti Sinha and Sujit K. Ghosh

Abstract
In many long-term clinical trials subjects often experience a number of events all of which might serve as important endpoints for medical studies. The analysis of such multiple events can be beneficial from both medical and statistical perspectives. We review and discuss some recent developments in the area of Bayesian methodologies for the analysis of multiple event times data, focusing more on semiparametric models based on the intensity functions. The nonparametric part of such models is modeled as a realization of a stochastic process; while the parametric part, which may include a regression parameter and a parameter quantifying the heterogeneity (i.e., frailty) of a population, is assumed to have a parametric distribution with possibly unknown hyperparameters. Numerical computations required to obtain posterior estimates of the parameters are generally based on a class of sampling based methods, broadly known as the Markov Chain Monte Carlo (MCMC) algorithms. The methodologies discussed in this chapter are motivated by and aimed at analyzing some common types of multiple event times data that are often observed in medical studies. Bayesian Exploratory Data Analysis (EDA) methods and diagnostics for model selection and model assessment are considered for each of these cases.

Keywords: censored data; Markov Chain Monte Carlo; mixture models; prior processes; semiparametric models; survival analysis

1. Introduction

In medical studies, interest often centers on the relation between time to an event or 'failure' of interest (such as occurrence of a tumor) and the explanatory variables or covariates (such as treatments, design variables, etc.). During a given follow-up time, a subject may experience more than one events; which may consist of repeated occurrences (recurrent events) or events from different sources. In such cases, we may be interested in the process $N(t)$, counting the number of these important but possibly nonfatal events up to time t. This counting process $N(t)$ for each subject may serve as a measure of disease morbidity. There are several kinds of multiple event times data encountered by statisticians in medical studies. We discuss some major categories of multiple events data primarily based on informative and noninformative termination. Several statistical methods are available to analyze multiple events data from both frequentist and Bayesian perspectives. Wei and Glidden (1997) and Hougaard (2000) have provided some recent overviews of the frequentist methods for multiple events data. In this review, we focus on the semiparametric Bayesian models primarily based on the so-called intensity functions which is the risk of a new event given the history of events in the past. In semiparametric Bayes methods, the nonparametric part of the model is assumed to be a realization of a stochastic process summarizing the available prior information about the unknown intensity function. The parametric part is assumed to have a prior distribution with possibly unknown hyperparameters. The semiparametric nature of the model allows considerable generality and applicability but enough structure for useful physical interpretation and understanding for particular applications in the medical research. The advantages of models based on the intensity functions are explained briefly by Wang et al. (2001).

In Section 2, we provide several motivating examples to illustrate various types of multiple events data encountered in practice. These examples serve as a basis to consider complex multiple events models and indicate the challenges to the statisticians posed by typical modern day multiple event times data problems. In addition, such complex models though suitable for complex data structure are not easy to handle and usually requires computational tools sophisticated enough to deal with complex data structures. The semiparametric Bayesian methodologies, despite being comparatively new concepts, have enormous potential for success in handling such complex models.

Section 3 reviews some existing and newly developed semiparametric models to address various features of complex data structures and designs as illustrated by the examples in Section 2. In Section 4 we briefly review some of the recently developed frequentist statistical methods to analyze these type of complex data structures, along with their limitations in the scope of applications and related computational difficulties.

Any Bayesian inference requires prior specification and the inference based on such Bayesian methods depends crucially on such prior specifications. In Section 5, we briefly discuss the commonly used prior processes to model the nonparametric part (e.g., the intensity function). Section 6 presents some existing and newly developed semiparametric Bayesian models suitable for the examples introduced in Section 2.

In Section 7, we present a case study to illustrate the implementation of a Bayesian semiparametric model. Section 8 summarizes and concludes with a discussion about future research directions and potential new problems.

2. Practical examples

This section presents examples of the major categories of multiple event times data that are commonly encountered in practice. These examples are not intended to cover all possible types of such data encountered in medical studies, but at least these examples illustrate the three major classifications among these data problems: (1) multiple events data with single event type and noninformative termination; (2) multiple events data with multiple event types and noninformative termination; and (3) multiple events data with dependent termination. These examples also illustrate one important and challenging aspect of multiple events data, viz., termination and censoring of different kinds depending on the particular design and method of data collection used in medical studies.

(1) An example of *multiple event times data with noninformative censoring* is given by Gail et al. (1980). The observations are the times to occurrence of mammary tumors in 48 rats injected with a carcinogen and subsequently randomized to receive either the treatment or the control. In addition, this study had regular follow-ups (only twice in each week) and hence for each rat only the number of new tumors between two consecutive follow up times was recorded. These highly grouped data contain many ties among event times within the same rat and across different rats due to imprecise measurements (specifically, twice a week scheduled observations) of tumor times. All rats were terminated from observation (right censored) after the 182nd day.

(2) Many methods for analyzing multiple events data are based on the assumption that any subject's risk of being under observation at any time is independent of the history of recurrent events experienced by the subject. However, this assumption of independent censoring may not be valid when the termination is caused by informative dropouts or failure events. These terminating events may not be of direct interest, however, the informative censoring induced by such terminating events, if not accounted for, may result into biased estimates. An example of such data is the cohort study of AIDS Link to Intravenous Experiences (ALIVE) discussed by Wang et al. (2001), where the history of recurrent hospitalizations for each patient is recorded. The interest is to study the effects of race and drug user status on the rate of hospitalization, however, for each patient, the time of dropout from the study is believed to be associated with the history of hospitalization.

    These type of data structures are not limited to medical studies. For instance in industrial settings, for a reliability study of repairable systems, each system may experience recurrent repairable failures until it is censored from observation due to some reason related to its failure pattern. In product testing, the recurrent adverse events may be detection of repairable faults and the censoring may be the final release of the product.

(3) Finally, we consider a case where the recurrent nonfatal events as well as the informative termination both are of interest. For example, in some bladder–cancer trials, after the surgical removal of a patients' primary tumors, each cancer patient is monitored through regularly scheduled clinic visits for occurrence of recurrent but benign superficial bladder tumors till their termination from observation due to cure from the disease. For such a study, the recurrent adverse events are nonfatal, but their intensity may signify the underlying quality-of-life experience and disease state of the patient, whereas the termination due to cure is a positive outcome. Such a study often includes regular follow-ups, say, monthly, and for each patient, the number of new recurrent events between each pair of consecutive clinic visits is recorded. Bladder cancer data from the VACURG (Sun and Wei, 2000) is a good example of such a multiple-events data with dependent termination. The data set is available at the web-site http://www.blackwellpulishers.co.uk/rss. The

time-independent covariates available for each patient consist of a binary variable of treatment (e.g., 1 for treatment and 0 for placebo), the number of tumors at the entry and the size of the largest tumor at the entry. One of our main focuses is to investigate the effects of treatment on the risk of superficial tumor as well as on the risk of discharge from monitoring. In addition, we are also interested in the possible association between the history of nonfatal tumor occurrences and the risk of termination from regular monitoring of tumors.

In this review article, we also consider methodologies applicable for time-varying covariates, but, unless mentioned otherwise, the covariates below are assumed to be constant over time within the follow up time. After developing different competing methodologies for time-independent covariates, we extend the methodologies to models with time-dependent covariate. We also explore methodologies that deal with interval censoring and panel count data based on regularly scheduled follow-up times.

3. Semiparametric models based on intensity functions

One of the early scholarly reviews of several models and methodologies proposed for multiple event time data in the context of frequentist inference is available in Oakes (1992). Most of the early multiple events models, as described and proposed in Andersen and Gill (1982), often make the unrealistic assumption that the risk of an event for a given subject is unaffected by all of the earlier events that a given subject has experienced. For a more recent comprehensive review of frequentist models for multiple events, see Hougaard (2000). Some of these models cannot assess heterogeneity between individuals nor can they handle ties and grouping of event times. We review some fully specified stochastic models which avoid such limitations.

A fully specified semiparametric model of multiple event times data, the so-called *proportional intensity model*, is presented in Sinha (1993). In this model, number of recurrent events experienced by subject i, within the time interval $(0, t]$, denoted by $N_i(t)$ is modeled as a mixed nonhomogeneous Poisson process (Ross, 1983) with the conditional intensity function,

$$\lim_{\delta \downarrow 0} \frac{\Pr[N_i(t, t+\delta) = 1 | w_i, x_i]}{\delta} = b_i(t|x_i, w_i) = b_0(t)\exp(\beta x_i) w_i, \tag{1}$$

given the unobserved subject-specific random 'frailty' w_i and the observed time-independent covariate x_i, where $N_i(t, t+\delta) = N_i(t+\delta) - N_i(t)$. The frailty random variables w_i's are assumed to be independent and identically distributed (i.i.d.) as Gamma variates in Sinha (1993). However, any random variable with unit mean and support on $[0, \infty)$ can be used to model the frailty w_i. Alternatively, we can characterize the frailty distribution using the moment generating function (mgf) of the w_i,

$$\Psi(t) = \mathrm{E}\big[\exp(tW)\big],$$

assuming that the mgf exists. To assure identifiability, we assume that $\mathrm{E}[W] = \Psi^{(1)}(0) = 1$, where $\Psi^{(k)}(t) = \frac{d^k}{dt^k}\Psi(t)$ denotes the kth derivative of the mgf $\Psi(t)$.

One advantage of characterizing the frailty distribution by its mgf is that it helps to express the conditional risk of a new event given history of events $\mathcal{N}(t) = \{N(s)\colon s < t\}$ at any time t as,

$$\lim_{\delta \downarrow 0} \frac{1}{\delta} \Pr\big[N(t, t+\delta) = 1 | \mathcal{N}(t)\big] = h\big(t|\mathcal{N}(t)\big) = \frac{\Psi^{(N(t)+1)}(-B(t))}{\Psi^{(N(t))}(-B(t))}, \tag{2}$$

where $N(t, t+\delta)$ is the number of recurrent events within the time interval $(t, t+\delta]$, $N(t) = N(0, t)$ and $B(t) = \int_0^t b(u)\,du$ is the cumulative intensity of the nonhomogeneous Poisson process. In (2), we have suppressed the role of covariate x in the expression to simplify the notations. It is to be noted that in absence of covariate x we have $B(t) = \int_0^t b_0(u)\,du$, otherwise we have $b(u) = b_0(u)\exp(\beta x)$. This result expresses the Markovian structure of the recurrent events process in the sense that the risk of future event at a given time depends on the past history $\mathcal{N}(t)$ only through $N(t)$, current count of the past events.

We denote the unknown variance of W by κ, quantifying the dependencies among inter-event times within each subject and the heterogeneity of the intensity processes in different subjects. Under the Gamma frailty model of Sinha (1993), $\Psi(t) = (1-t\kappa)^{-1/\kappa}$ and it easily follows that,

$$h\big(t|\mathcal{N}(t), x\big) = \big\{1 + \kappa^{-1}N(t)\big\}h_1(t|x), \tag{3}$$

where $h_1(t|x) = \{1+B(t|x)\}^{-1}$ is the hazard for the first event. One can clearly see how the variance, κ of frailty establishes the relationship between the risk of a future event and the number of past events. Motivated by the result in (3), Oakes (1992) proposed a marginal model for multiple events data with conditional hazard,

$$h\big(t|\mathcal{N}(t), x\big) = \big\{1 + \kappa^{-1}N(t)\big\}h_0(t)\exp(\beta x).$$

It is clear that in the presence of any covariate x, this model is not equivalent to the mixed Poisson process model of Sinha (1993) because $h_1(t|x)$ is not in a proportional hazards structure when the conditional intensity is given by (1).

The model in (1) is useful when the focus is on the influence of the covariates (or treatments) and on the heterogeneity among subjects related to the rate of occurrence of events in each subject. The model also assumes that the heterogeneity among subjects remains constant over time, and hence it works best when each subject experiences several events, but a moderate number of subjects are observed in a relatively short span of time.

The M-site model of Gail et al. (1980) assumes that each subject has fixed number, M, of potential i.i.d. events at risk. Assuming that the underlying hazard for each potential event at risk is $h_0(t)\exp(\beta x)$, the conditional hazard for a new event at time t, given the history $\mathcal{N}(t)$, is given by,

$$\begin{aligned}\lim_{\delta \downarrow 0} \frac{P[N(t, t+\delta) = 1|\mathcal{N}(t)]}{\delta} &= h\big(t|\mathcal{N}(t), x\big) \\ &= \big\{M - N(t)\big\}h_0(t)\exp(\beta x).\end{aligned} \tag{4}$$

This model may not be applicable for most of the clinical studies involving multiple events as we can see from the above expression, that the marginal risk of the future event to any subject decreases with the number of previous events $N(t)$. However, it may be an ideal feature for some other situations, for example, when we are dealing with the number of crashes to a software with fixed but unknown M number of undetected bugs at the onset of testing.

An extension of M-site model is when M is assumed to vary among subjects with probability distribution $P[M = m|\phi] = g(m|\phi)$ for $m = 0, 1, 2, \ldots$, where ϕ is the parameter of this discrete parametric distribution. If we look at only the time to first event T_1 under this model, we can show that $S_1(t) = P[T_1 > t] = \Psi_M[\log S(t)]$ where $\Psi_M(t)$ is the mgf of M and $S(t)$ is the common survival function corresponding to each one of M i.i.d. latent events at risk. Note that when $g(0|\phi) > 0$ there is a probability that a subject may not be at risk of any recurrent event at any time point. When M follows a Poisson distribution, this $S_1(t)$ is same as the cure-rate model of Chen et al. (2002). To avoid this positive probability of a subject being not at risk of any event, we consider only the distributions with $g_M(0) = 0$ (for example, a Poisson distribution truncated at 0).

For this extended M-site model, the conditional hazard for a new event at time t given the history $\mathcal{N}(t)$, can be obtained as,

$$h\big(t|\mathcal{N}(t), x\big) = h(t|x)\frac{\mathrm{E}[M \times \cdots \times (M - N)\{S(t)\}^M 1_{[M>N]}]}{\mathrm{E}[M \times \cdots \times (M - N + 1)\{S(t)\}^M 1_{[M \geqslant N]}]}, \tag{5}$$

where $h(t|x) = h_0(t)\exp(\beta x)$ and 1_A is the indicator function. We can further simplify this as,

$$h\big(t|\mathcal{N}(t), x\big) = h(t|x)\mathrm{E}\big[M - N|M \geqslant N, N(t) = N\big], \tag{6}$$

where the expectation in (6) is taken with respect to the conditional probability distribution $P[M - N = m|M \geqslant N, N(t) = N] = g^*(m|N)$ given by

$$g^*(k|N) = \frac{g(N + k)\{S(t)\}^{N+k}\binom{N+k}{k}}{\sum_{k'=0}^{\infty} g(N + k')\{S(t)\}^{N+k'}\binom{N+k'}{k'}}. \tag{7}$$

The expression in (6) shows that the relationship between risk of a future event and events history is complicated for this extended M-site model and it depends to a great extend on the distribution $g_M(\cdot)$ of M.

Markov and semi-Markov models of Lagakos et al. (1978) often have many parameters and can be hard to interpret. Another class of models – 'modulated renewal processes' (Cox, 1972b) – has not been widely used, perhaps because of the difficulty in estimating the asymptotic covariance matrix of the parameter estimates (Oakes and Cui, 1994).

Besides the assumption of (1), the methodology described in Sinha (1993) further assumes that the hazard of the termination time T_i at time t

$$\lim_{\delta \downarrow 0} \frac{P[T_i \in (t, t + \delta)|\mathcal{N}_i(t), x_i, T_i > t]}{\delta} = h_T\big(t|\mathcal{N}_i(t), x_i\big) \tag{8}$$

is independent of the history of point process $\mathcal{N}_i(t) = \{N_i(s)\colon\ s < t\}$. This underlying assumption, not often precisely mentioned, can be expressed as

$$h_T\big(t|\mathcal{N}_i(t), x_i, T_i > t\big) = h_T(t|T_i > t, x_i). \tag{9}$$

For various practical situations, when the termination T_i is induced by a terminal or fatal event such as death, this assumption in (9) may not be valid at all. Because, in such a case, the pattern of recurrent events in $N_i(\cdot)$ may indicate the progression of the morbidity of the subject. This is the situation when there is an informative termination (e.g., Wang et al., 2001).

A generalization of (1) in Sinha and Maiti (2004) assumes that the hazard of T_i at time t depends on the previous nonfatal events via the frailty w_i. But, given the random effect w_i, the hazard of termination at time t does not depend on the history ($\mathcal{N}_i(t)$) of recurrent events. For the case with only a discrete distribution for the termination time, the discretized conditional hazard is given by,

$$P\big[T_i = a_j | x_i, w_i, \mathcal{N}_i(a_j), T_i > a_{j-1}\big] = 1 - \phi_j^{\exp(\beta_2 x_i) w_i^{\alpha}}, \tag{10}$$

where $\phi_j = P[T = a_j | x = 0, w = 1, \mathcal{N}(a_j), T > a_{j-1}]$ is the discretized baseline conditional probability of not termination at time point a_j for a patient at risk of termination at that time point. This is assuming that the termination can occur only at one of the scheduled inspection times $a_1 < \cdots < a_g$. Also, β_2 is the regression parameter of the covariate effect on the conditional hazard of termination. We are using a discretized version of the proportional hazards (Cox, 1972a) structure in (10). Also, $\alpha \in \mathcal{R}$ quantifies the nature of dependence between $\mathcal{N}_i(t)$ and the hazard of T_i.

For $\alpha = 0$, we have $P(T_i = a_j | x_i, T_i > a_{j-1}, \mathcal{N}_i(t))$ free of $\mathcal{N}_i(t)$ and we are in the situation similar to (9) when the risk of termination is independent of the events history. For $\alpha > 0$, we have positive association between the N_i and T_i and hence, given the covariate, subjects with higher than usual rates of nonfatal events are more likely to be terminated earlier than others. Similarly, $\alpha < 0$ indicates a negative relationship between termination time and rate of nonfatal events. The special case $\alpha = 0$ further indicates that the heterogeneity among subjects in terms of rates of nonfatal events is same as the heterogeneity in terms of their termination times. Hence, $\alpha > 1$ implies more variability in their termination times compared to the variability in their rates of nonfatal events. Similarly, $0 < \alpha < 1$ implies more variability in the later. For the special case of $\alpha = 1$ and $\phi_j = \exp[-\int_{a_{j-1}}^{a_j} b_0(u)\,du]$, we get the model of Lancaster and Intrator (1998). Their model has the restrictive assumptions that the hazard function of termination is proportional to the conditional baseline intensity $B_0(t)$ and more importantly, the variability of the subjects in terms of their proneness to nonfatal events is same as their variability in termination times. In Sinha and Maiti (2004), independent $Ga(\eta, \eta)$ density is used for the frailty w_i.

4. Frequentist methods for analyzing multiple event data

Statistical methodologies for multiple events data can be classified into two groups: *conditional* and *marginal* approaches. Conditional models and approaches are used mainly in semiparametric Bayesian and likelihood based methods of analysis. Marginal methods are more popular with *generalized estimating equations* (GEE) methodologies. This section discusses the present status of the frequentist methods to analyze the examples in Section 2 using the above mentioned models. We focus on the main difficulties faced by the frequentist methods and later describe how these difficulties are overcome by Bayesian analyses.

First, we discuss existing frequentist methods for analysis of multiple events data under noninformative or independent terminating event. Oakes (1992) gives a comprehensive review of the several frequentist methods for analyzing multiple event time data and discusses limitations of the methodologies of Andersen and Gill (1982), Lagakos et al. (1978) and the modulated renewal process (Cox, 1972b). His proposed marginal model for multiple event time data applies only to nongrouped data. Thall's (1988) Poisson regression model for analyzing grouped multiple event-time data does not use the hierarchical structure for the random frailty and he estimates the frailty for each subject instead of a single hyperparameter quantifying the overall variability of the random frailties. Also, Thall and Vail's (1990) quasi-likelihood method to analyze grouped data cannot be used to estimate efficiently the parameter of the frailty distribution and the standard error of the treatment effect. One common problem present in all these methods for multiple event time data is the lack of a suitable method to verify the modeling assumptions. Most recent review of the conventional frequentist methods for analysis of multiple data can be found in Hougaard (2000). These methods may allow a time-dependent covariate, but usually assume the covariate effect is constant over time. However, there is no specific frequentist methodology to deal with interval-censored data when the scheduled inspection times are not same for each subject. The frequentist methodologies to verify the modeling assumptions, such as the assumption of proportional hazards, constancy of the covariate effect over time, etc., are also scarce for multiple events data.

For many currently popular statistical procedures for multiple events data, the final estimates are often highly dependent on starting values, and in practice, it is often difficult to agree on a starting value. In the presence of many nuisance parameters, the asymptotic properties of the maximum likelihood estimate (MLE) and other frequentist estimate of the parameter of interest, say, the regression parameter β, await more extensive results. The small sample properties of these frequentist estimates have yet to be determined.

When the risk of termination depends on the history of recurrent events, existing methods in literature are not so common. When the terminating event is of no interest and it is treated only as a sort of informative censoring of recurrent event times, there are some existing frequentist methods to deal with such cases, most notably of them is Wang et al. (2001). However, these approaches are not based of fully specified models. Depending on their goals of analysis, these procedures can be categorized in three groups – (1) based on the inference about the rate function, which is $P[N(t, t + \delta) = 1|x]$; (2) based on the mean function $E[N(t)|x]$; (3) based on the intensity function $P[N(t, t + \delta) = 1|x, \mathcal{N}(t)]$. However, these methods do not precisely mention the model for $h_T(t|\mathcal{N}(t), x)$. One can not automatically estimate this hazard function using these approaches. It is difficult to know the loss in efficiency for not specifying the model for termination and it is also not possible to check the sensitivity of the assumptions about termination, though the expectation is that the approach used is somewhat 'robust' as it is free of any precise modeling assumptions on termination.

When we have multiple events with dependent important terminating events (e.g., Sinha and Maiti, 2004) and we are interested about both types of events, there is even bigger need to use a fully specified stochastic model for $h_T(t|\mathcal{N}(t), x)$. One possible strategy is to focus on either the mean function or the rate function of $1_{[T>t]}N(t)$ (called the termination modulated recurrent events process, e.g., Ghosh and Lin, 2004). However, this can be misleading if we are interested about understanding the effects of covariates on recurrent events as well as on the termination events. We instead will focus on fully specified stochastic models which can be used for complete understanding of the covariate-effects, prediction, model selection and for understanding the process generating the data.

5. Prior processes in semiparametric model

The most popular way to model a prior process on the baseline cumulative intensity function (as well as for the baseline cumulative hazard function) is to use the Levy process introduced by Kalbfleisch (1978), Ferguson and Phadia (1979) and Burridge (1981). They advocated the use of a process with positive independent increments in disjoint intervals (Levy process) to model the prior process of the cumulative hazard and intensity function $B_0(t) = \int_0^t b_0(u)\,\mathrm{d}u$. Following Sinha (1993) we can assume that $B_0(t)$ is the realization of a positive nondecreasing independent increment process (Levy process). That is, the increments of $B_0(t)$ in disjoint intervals are positive and mutually independent. Similarly, the cumulative hazard $\int_0^t h_0(u)\,\mathrm{d}u = H_0(t)$ can be modeled by a Levy process when the cdf is absolutely continuous.

5.1. Gamma processes

The most commonly used Levy process for modeling $B_0(t)$ is the Gamma process (Sinha, 1993), denoted by $B_0(t) \sim GP(\Lambda^*, c)$, when $B_0(t_2) - B_0(t_1)$ has a gamma distribution with mean $\Lambda^*(t_2) - \Lambda^*(t_1)$ and variance $\frac{\Lambda^*(t_2)-\Lambda^*(t_1)}{c}$ for $t_2 > t_1$ and $c > 0$. Here the nondecreasing positive function, Λ^*, is the mean function of the random prior process and the hyper-parameter c can be interpreted as a measure of confidence attached to the initial guess $\Lambda^*(\cdot)$ of $B_0(\cdot)$, the expected number of events by time t at baseline covariate value and in absence of any frailty effect. It is assumed that Λ^* is completely specified except possibly for a few unknown parameters. For example, one can consider a Gamma process model with $\Lambda^*(t) = \lambda t$ for some hyperparameter λ, possibly unknown.

For ordinary univariate survival data, Burridge (1981) explains how a Gamma process model may give wrong inferences if applied to continuous survival data. Similar caution should be used for application of the Gamma process while dealing with continuously observed multiple events data with no ties. Two unsatisfactory features of a Gamma process are: (1) independent increments in adjacent intervals, and (2) realization of the sample path is discrete with probability one. However, this process has found a vast number of applications in literature for modeling different types of multiple events data.

5.2. *Correlated prior processes*

This model was introduced by Leonard (1978) for modeling density functions with prior information on smoothness. A discretized approximation of this process assumes a piecewise constant function $b_0(t)$. Over a set of prespecified grouping intervals $I_j = (a_{j-1}, a_j]$ for $j = 1, 2, \ldots, g$ with $0 = t_0 < a_1 < a_2 < \cdots < a_{g-1} < a_g$, the baseline intensity function is given as $b_0(t) = \lambda_j$ for $t \in I_j$. We introduce some smoothness for the values of the λ_j's in adjacent intervals using a first order random walk process for $\alpha_j = \log \lambda_j$ with

$$\alpha_{j+1} = \alpha_j + e_{j+1} \quad \text{for } j = 1, \ldots, g-1. \tag{11}$$

The e_j's are i.i.d. normal random variables with 0 mean and variance σ^2. Higher values of σ imply less prior information on the discretized linear spline smoothing of the λ_j's. The prior information can even be incomplete, because without the specification of the marginal prior of α_1, the joint prior distribution of the α's remains undefined. Original model of Gamerman (1991), introduced in the context of modeling hazard of an univariate survival time, used the so-called *linear Bayesian method* which requires only a partial specification (mean 0 and finite variance) of the distribution of e_{j+1} in (11). It is also important to note that, Gray (1994) again in the context of hazard function, used the prior process structure similar to (11) on λ_j's itself and used a 'flat' prior for λ_1. See also Leonard (1978) for a more formal justification for using this kind of incomplete prior information and prior process in theoretical nonparametric Bayesian inference.

Various other types of correlated prior processes involving linear and higher-order discretized spline smoothing of the α_j's can be used to summarize different types of prior information about the underlying baseline function. One interesting class of prior process is where a local linear trend is modeled as

$$\alpha_{j+1} = \alpha_j + \delta_j + e_j \quad \text{and} \quad \delta_{j+1} = \delta_j + e'_j \tag{12}$$

for $j = 1, \ldots, g$, where $e_j \sim N(0, \sigma_1^2)$, $e'_j \sim N(0, \sigma_2^2)$, and e_j's and e'_j's are mutually independent. The model in (12) will reduce to a smoothed linear trend model by taking $\sigma_2 = 0$. This process allows the intensity to be somewhat linear and smooth without forcing the form to be completely parametric. The completely parametric form with $\sigma_1 = \sigma_2 = 0$ will be insensitive to any surprising trend of the baseline intensity function suggested by the data. Despite some of their nice physical properties, correlated prior processes have so far enjoyed only limited applications for modeling multiple events data.

The correlated prior process is also related to the so-called *martingale process* introduced by Arjas and Gasbarra (1994), where $b_0(t)$ is modeled as a piecewise constant function with jumps. The jump times are realizations of a time homogeneous Poisson process, and the sequence of $b_0(t)$ values in different jump intervals form a discrete martingale. This martingale process is very similar to the model of (11) if we relax the condition of the jump times being random, and instead assume that the hazard rates change in each grouping interval. The condition of the hazard rates being martingale can be relaxed also to get a sub-martingale process similar to that in (12). A slight modification of the prior process in (11), with $\alpha_{j+1} - \alpha_{0j} = \gamma(\alpha_j - \alpha_{0j}) + e_{j+1}$, where α_{0j}

is the prior mean of α_j and γ is another hyperparameter of deviation, is a discretized integrated Ornstein–Uhlenbeck (DIOU) process discussed by Cumberland and Sykes (1982). However, to the best of our knowledge in the current literature, these martingale processes have not been used in multiple events data.

6. Bayesian solution

As an illustration of the implementation and applicability of the Bayesian solution for the analysis of multiple events data, we concentrate on the analysis of the bladder–cancer data (Sun and Wei, 2000). Bladder cancer data from the VACURG is a good example of multiple events data with dependent termination process. After their entry at the trial, the patients in this study were monitored every month during their clinic visit for number of new (since the previous clinic visit) superficial bladder tumors till their termination time due to significant improvement in health status. The time-independent covariates used for each patient are binary variable of treatment (1 for treatment and 0 for placebo), number of tumors at the entry and the size of the largest tumor at the entry.

To model the recurrent events process $N(t)$, we use the mixed Poisson process model of Sinha (1993) with conditional intensity function of (1). We use the Gamma frailty for w_i. For the discrete termination time T, we use the model (10). The baseline intensity function $b_0(t)$ is assumed to have the Gamma process prior. Because the observed recurrent events history is based only on observations recorded at regular time intervals (based on regularly scheduled clinic visits), the likelihood based on the sampling distributions of the observed history of recurrent bladder-tumors given the frailty variables $\boldsymbol{w} = (w_1, \ldots, w_n)$ and termination times is given by

$$L_1(\beta_1, \mu, \boldsymbol{w}|D) \propto \prod_{i=1}^{n} \prod_{j \in M_i} \left[e^{x_i \beta_1} w_i \mu_j\right]^{N_{ij}} \exp\left[-e^{x_i \beta_1} w_i \mu_j\right], \tag{13}$$

where M_i is the set of inspection intervals when the subject i is under observation and $N_{ij} = N_i(a_j) - N_i(a_{j-1})$ is the number of new bladder-tumors to subject i observed during the interval between a_{j-1} and a_j. It is obvious that the inspection intervals in M_i are all below the observed termination time T_i. For this case, the baseline cumulative intensity $B_0(t)$ enters the likelihood L_1 via the increments in these pre-specified time points, say, $0 = a_0 < a_1 < \cdots < a_g$. One consequence of this fact is that the property of the Gamma process to have discrete sample path does not turn out to be any worrying factor in this application. When $B_0(t)$ is a priori $GP(\Lambda^*, c)$, the priors of $\mu_j = B_0(a_j) - B_0(a_{j-1})$ for $j = 1, \ldots, g$ are independent Gamma densities with mean $\gamma_j = \Lambda^*(a_j) - \Lambda^*(a_{j-1})$ and variance γ_j/c. Instead of the Gamma priors, one can also use alternative prior processes such as the correlated prior processes defined via the joint prior distribution of $(\mu_1, \ldots, \mu_g)$.

Similarly, the likelihood $L_2(\beta_2, \pi, \boldsymbol{w}|D)$ based on the sampling distributions of the T_i is given as

$$L_2(\beta_2, \phi, \boldsymbol{w}|D) \propto \prod_{i=1}^{n} \left(1 - \phi_{(i)}^{w_i^{\alpha} \exp(\beta_2 x_i)}\right) S_{(i)}^{\exp(\beta_2 x_i) w_i^{\alpha}}, \tag{14}$$

where $t_i = a_{(i)}$ is the discrete termination time of the subject i. The joint posterior for this model is given by

$$\begin{aligned} p(\beta, \boldsymbol{w}, \eta, \mu, \phi, \alpha|D) \\ \propto L_1(\beta_1, \mu, \boldsymbol{w}|D) \times L_2(\beta_2, \phi, \boldsymbol{w}|D) \\ \times f(\boldsymbol{w}|\eta) \times \pi(\beta, \phi, \eta, \alpha) \times \prod_j [\pi(\mu_j|\gamma_j, c)], \end{aligned} \tag{15}$$

where $f(\boldsymbol{w}|\eta) = \prod_i [\eta^\eta w_i^{\eta-1} \exp(-w_i \eta)/\Gamma(\eta)]$, the product of Gamma densities of the $w_1, \ldots, w_n$ given $\eta = \kappa^{-1}$, and $\pi(\beta, \phi, \eta, \alpha)$ is obtained via the joint prior distribution for $\phi = (\phi_1, \ldots, \phi_g)$ and specification of the conditional priors of other parameters given ϕ. Note that this model here is identical to the one used by Sinha and Maiti (2004). We further refer this article for details about how to specify the priors and about the method of eliciting such priors. The details about numerical integration of (15) to obtain the marginal posterior using MCMC tools are also given in Sinha and Maiti (2004). At each iterative stage of the MCMC, we only need to sample from the conditional posteriors. For further convenience we use $N(\beta_{10}, \sigma_1)$ for $\pi(\beta_1)$, $N(\beta_{20}, \sigma_2)$ for $\pi(\beta_2)$, and independent $Beta(\phi_j|A_j, B_j)$ densities for $\pi(\phi_j)$. The $\pi(\alpha)$ can be assumed as a $N(\alpha_0, \sigma_3)$ and we need to take $\alpha_0 = 0$ as we are not sure whether the association between the two classes of events should be either positive or negative. We have also assumed that these parameters are apriori independent of each other. To use the MCMC method, one prefers the conditional posteriors that are quite easy to sample from. These forms of the prior are flexible, yet they facilitate easy to sample conditional posteriors.

The expressions and the derivations of the conditional posteriors of the parameters used in the MCMC are given in Sinha and Maiti (2004). Except μ_j all other full conditional distributions are not in standard form. The full conditional distributions for ϕ and α are log-concave and an adaptive rejection sampling method by Gilks and Wild (1992) was used to sample from these log-concave densities. Metropolis–Hastings algorithm was used to sample w_i's.

7. Analysis of the data-example

Sinha and Maiti (2004) illustrates the kind of approximate Bayesian method with ten parallel Gibbs chain and 20,000 iterations in each chain. The previous analyses of this data ignored the possible association between the history of nonfatal tumor occurrences and the risk of termination. We would like to emphasize that to deal with missing observations in some of the months, Sinha and Maiti (2004) had to make some appropriate changes in our likelihood based on the assumption of noninformative missingness. Priors with zero mean and high variance (1000) for the regression parameters β_1 and β_2 have been used to reflect prior opinions of no covariate effects on the nonfatal events intensity as well as on the risk of termination. For the gamma priors of λ, ν and η, selecting small scales and shapes make the prior distributions very diffuse. One needs to evaluate the sensitivity of the posterior analysis over these choices by varying the values of the hyperparameters over a range of possible values.

Table 1
Summary of posterior analysis

Parameter	Mean	S.E.	95% credible interval
β_1	1.491	0.577	(0.508, 2.77)
β_2	−0.0227	0.249	(−0.520, 0.468)
α	−0.0137	0.0295	(−0.081, 0.037)

In this analysis, another crucial elicitation step is to choose the hyperparameters associated with the gamma process, we set, $\tau = 1/c$ and $\gamma_1, \ldots, \gamma_g$, where $\gamma_j = \gamma_0(a_j) - \gamma_0(a_{j-1})$. One possible rational behind eliciting these hyperparameters is given in the original paper.

Table 1, reproduced from Sinha and Maiti (2004) presents the results of the data analysis. There is not much evidence for the treatment effect on time of termination, because the credible interval for β_2 contains zero. But, there is good evidence for a treatment effect on the rate of nonfatal events. The posterior mean of α lies below zero indicating some evidence of negative association between nonfatal events intensity and the termination time. Negative value of α implies that patients who are more prone to nonfatal events are likely to be terminated later. But, the evidence for negative α is not strong at the 95% posterior interval for α does contain zero. There is also strong evidence that the baseline hazard rate for the termination is increasing over time. Please note that one of the major advantages of a Bayesian analysis is that one can compute the prediction distributions of the future nonfatal events and termination time given the values of the covariate.

Model validation diagnostics are very important in any data analysis. Many model diagnostics have been developed in the literature. For detailed reviews of Bayesian model validation methods in multiple events data analysis, we may refer to Sinha and Dey (1997), Ibrahim et al. (2001) and Sinha and Maiti (2004). Most popular is a method for Bayesian model validation via conditional predictive ordinate (CPO) (see Gelfand et al., 1992). For the subject i, the cross-validated posterior predictive probability evaluated at the observed data point is given by

$$\begin{aligned} CPO_{(i)} &= P(\boldsymbol{N}_i = \boldsymbol{N}_{i,\mathrm{obs}}, T_i = t_i \mid D_{(-i)}) \\ &= \big\{E\big[P(\boldsymbol{N}_i = \boldsymbol{N}_{i,\mathrm{obs}}, T_i = t_i \mid \theta)^{-1} \mid D\big]\big\}^{-1}, \end{aligned} \tag{16}$$

where $D_{(-i)}$ is the observed data with the patient i removed, $\theta = (\mu, \beta_1, \beta_2, \eta, \alpha, \lambda, \nu)$ and $\boldsymbol{N}_i = (N_{i1}, \ldots, N_{iT_i})$. The second expression of the CPO in (16) follows from a result of Gelfand et al. (1992). This expression actually enables us to use the samples from the posterior $p(\theta|D)$ to compute the CPOs. Sinha and Maiti (2004) again illustrates the algorithm we used for calculating CPO's for the bladder–cancer data example. CPOs are treated just as the same way residuals are treated in frequentist survival analysis and CPOs are plotted against either their covariate values or to the termination times to check whether the CPO values have any relationship with the covariates and the termination times.

8. Discussions and future research

We presented several Bayesian approaches to model the nonparametric part of a survival model based on intensity functions. There is no clear choice for the best model among these different models. The selection of the model depends mostly on the goal of the study, type of the data set and the amount of the computations required. Some of the merits and demerits of different processes are discussed below.

Two unsatisfactory features of a Gamma process are (i) independent increments, and (ii) the realizations are discrete with probability one, so that the resulting marginal distributions imply a positive probability of tied failure times. As pointed out by Burridge (1981), this disadvantage mentioned in (ii) above and some other deficiencies of the Levy process model disappear when applied to grouped data.

The specification of any one of the Levy processes and random mixture process requires the knowledge of the mean function of the prior process and the prior uncertainty around the mean. The correlated prior process does not necessarily require the direct knowledge of the mean of the prior process nor any confidence around the mean. But, the implementations of the Bayesian methodologies using the correlated prior processes are often computationally more intensive due to the apparent loss of the conjugacy of the conditional posteriors of the process parameters in most of the practical situations that we usually come across. But, our personal experience suggests that in general the methodologies based on correlated process are less computationally intensive compared to the methodologies based on random mixture processes. Random mixture processes have the advantages that the posterior estimate of the hazard (or intensity) function is continuous and one can compute the posterior estimates of the hazard curve and also any transformation thereof.

Another type of semiparametric approach of Bayesian methods of survival data are being explored in the literature of competing risks models. See Cox and Oakes (1984) for the background of a typical competing risks problem. Recent work of McIntosh (1996) and Ghosh and Gelfand (1998b) consider few problems of this type, that is when the failure of the subject is due to one of many competing causes (also see Gelfand et al., 2000). The subject may be observed to fail from one of these causes (failure types), otherwise the subject will be observed to be censored at some point of time. The interest is in evaluating the effect of the treatment which is intended to act in only one of the possible failure types. Such failure types are often dependent (Moeschberger and Klein, 1995), and in fact, the order of the types might determine the appropriate model. In such contexts, Ghosh and Gelfand (1998a) propose a number of complex shock models incorporating covariates and time-dependent intensities.

Semiparametric Bayesian analysis of multiple events is an expanding field of research. Lot of the works mentioned in this review are still in the process of development and getting more refined. Some of the subtopics, where there are enormous scopes of future research, can be very easily identified. We are yet to see a comprehensive Bayesian analysis with model diagnostics for some of the important multiple event survival models, such as stable frailty models and copula models. Another interesting future research problem is the Bayesian analysis of multivariate survival data when there is multiple level of clustering (Peng and Oakes, 1996).

The modeling techniques, computational tools, model diagnostics and model selection methods discussed can also be extended to more complicated problems in survival analysis, and there lies a wide cope of the future researches on adaptive and widely applicable methodologies.

References

Andersen, P.K., Gill, R.D. (1982). Cox's regression models for counting processes: A large sample study. *Ann. Statist.* **10**, 1100–1120.

Arjas, E., Gasbarra, D. (1994). Nonparametric Bayesian inference from right censored survival data, using the Gibbs sampler. *Statistica Sinica* **4**, 505–524.

Burridge, J. (1981). Empirical Bayes analysis of survival time data. *J. Roy. Statist. Soc., Ser. B* **43**, 65–75.

Chen, M.H., Ibrahim, J.G., Sinha, D. (2002). Bayesian inference for multivariate survival data with a surviving fraction. *J. Multivariate Anal.* **80**, 101–126.

Cox, D.R. (1972a). Regression models and life tables. *J. Roy. Statist. Soc., Ser. B* **34**, 187–220.

Cox, D.R. (1972b). The statistical analysis of dependencies in point processes. In: Lewis, P.A.W. (Ed.), *Stochastic Point Processes*. Wiley, New York, pp. 55–66.

Cox, D.R., Oakes, D.O. (1984). *Analysis of Survival Data*. Chapman & Hall, London.

Cumberland, W.G., Sykes, Z.M. (1982). Weak convergence of an autoregressive process used in modeling population growth. *J. Appl. Probab.* **19**, 450–455.

Ferguson, T.S., Phadia, E.G. (1979). Bayesian nonparametric estimation based on censored data. *Ann. Statist.* **7**, 163–186.

Gail, M.H., Santner, T.J., Brown, C.C. (1980). An analysis of comparative carcinogenesis experiments with multiple times to tumor. *Biometrics* **36**, 255–266.

Gamerman, D. (1991). Dynamic Bayesian models for survival data. *Appl. Statist.* **40**, 63–79.

Gelfand, A.E., Dey, D.K., Chang, H. (1992). Model determination using predictive distributions with implementation via sampling-based methods. In: Bernardo, J.M., Berger, J.O., Dawid, A.P., Smith, A.F.M. (Eds.), *Bayesian Statistics, vol. 4*. Oxford University Press, pp. 147–167.

Gelfand, A.E., Ghosh, S.K., Christiansen, C., Soumerai, S.B., McLaughlin, T.J. (2000). Proportional hazards models: A latent competing risk approach. *J. Roy. Statist. Soc., Ser. C (Appl. Statist.)* **49**, 385–397.

Ghosh, D., Lin, D.Y. (2004). Semiparametric analysis of recurrent events data in the presence of dependent censoring. *Biometrics* **59**, 877–885.

Ghosh, S.K., Gelfand, A.E. (1998a). A latent risk approach for modeling individual level data consisting of multiple event times. *J. Statist. Res.* **32**, 23–39.

Ghosh, S.K., Gelfand, A.E. (1998b). Latent waiting time models for bivariate event times with censoring. *Sankhya Ser. B* **60**, 31–47.

Gilks, W.R., Wild, P. (1992). Adaptive rejection sampling for Gibbs sampling. *Appl. Statist.* **41**, 337–348.

Gray, R.J. (1994). A kernel method for incorporating information on disease progression in the analysis of survival. *Biometrika* **81**, 527–539.

Hougaard, P. (2000). *Analysis of Multivariate Survival Data*. Springer-Verlag, New York.

Ibrahim, J.G., Chen, D., Sinha, M.-H. (2001). *Bayesian Survival Analysis*. Springer-Verlag.

Kalbfleisch, J.D. (1978). Nonparametric Bayesian analysis of survival time data. *J. Roy. Statist. Soc., Ser. B* **40**, 214–221.

Lagakos, S.W., Sommer, C.J., Zelen, M. (1978). Semi-Markov models for partially censored data. *Biometrika* **65**, 311–317.

Lancaster, T., Intrator, O. (1998). Panel data with survival: Hospitalization of HIV-positive patients. *J. Amer. Statist. Assoc.* **93**, 46–51.

Leonard, T. (1978). Density estimation, stochastic processes and prior information. *J. Roy. Statist. Soc., Ser. B* **40**, 113–146.

McIntosh, M.W. (1996). The population risk as an explanatory variable in research synthesis of clinical trials. *Statist. Medicine* **15**, 1713–1728.

Moeschberger, M.L., Klein, J.P. (1995). Statistical methods for dependent competing risk. *Lifetime Data Anal.* **1**, 195–204.

Oakes, D. (1992). Frailty models for multiple event times. In: Klein, J.P., Goel, P.K. (Eds.), *Survival Analysis State of the Art*. Kluwer Academic, Dordrecht, pp. 371–379.

Oakes, D., Cui, L. (1994). On semiparametric inference for modulated renewal processes. *Biometrika* **81**, 83–90.

Peng, F.P., Oakes, D. (1996). Bayesian analysis of multiple event time data under a nested frailty model. Technical Report, Department of Biostatistics, University of Rochester.

Ross, S.M. (1983). *Stochastic Process*. Wiley, New York.

Sinha, D. (1993). Semiparametric Bayesian analysis of multiple event time data. *J. Amer. Statist. Assoc.* **88**, 979–983.

Sinha, D., Dey, D.K. (1997). Semiparametric Bayesian analysis of survival data. *J. Amer. Statist. Assoc.* **92**, 1195–1212.

Sinha, D., Maiti, T. (2004). Analyzing panel-count data with dependent termination: A Bayesian approach. *Biometrics* **60**, 34–40.

Sun, J., Wei, L.J. (2000). Regression analysis of panel count data with covariate-dependent observation and censoring times. *J. Roy. Statist. Soc., Ser. B* **62**, 293–302.

Thall, P.F. (1988). Mixed Poisson likelihood regression models for longitudinal count data with overdispersion. *Biometrics* **44**, 197–209.

Thall, P.F., Vail, S.C. (1990). Some covariance models for longitudinal count data with overdispersion. *Biometrics* **46**, 657–671.

Wang, M.-C., Qin, J., Chang, C.-T. (2001). Analyzing recurrent event data with informative censoring. *J. Amer. Statist. Assoc.* **96**, 1057–1065.

Wei, L.J., Glidden, D.V. (1997). An overview of statistical methods for multiple failure time data in clinical trials (with discussion). *Statistics in Medicine* **16**, 833–839.

Handbook of Statistics, Vol. 25
ISSN: 0169-7161
© 2005 Elsevier B.V. All rights reserved.
DOI 10.1016/S0169-7161(05)25032-0

32

Bayesian Survival Analysis for Discrete Data with Left-Truncation and Interval Censoring

Chong Z. He[1] *and Dongchu Sun*[2]

Abstract

In this chapter, a Bayesian model for discrete left-truncation and right censoring data is presented with applications to the age-specific nest survival rates under an irregular visit. Independent beta priors are placed on the age-specific hazard rates and conditional encounter rates. A two-stage hierarchical prior is placed on the number of encountered nests. Four types of latent variables are introduced to simplify the likelihood function. A class of informative and noninformative priors is given for the unknown number of nests encountered. We find some mild condition so that the posterior is proper. All the conditional posterior densities are standard densities such as beta, gamma, Poisson, negative binomial, and multinormal. Therefore, the Gibbs sampling computation is efficient. The reference prior under this model is derived, and the hyperparameter specification problem is solved to some extent. The simulation result shows that the proposed Bayesian model provides accurate and stable estimates of survival rates. The model also is applied to a real-nest data set.

Keywords: Bayesian estimation; beta priors; Poisson distribution; Gibbs sampling; latent variable; double-interval censoring; truncation; age-specific survival; nest study

1. Introduction

In the last several decades, wildlife conservation continues to attract more and more attention because of public awareness of conservation issues. However, wildlife conservation studies bring many unique statistical problems and pose new challenges to statisticians. It is well known that the populations of many bird species are declining, and wildlife biologists have been examining reproductive potential. One goal of bird–nest survival studies is to obtain accurate estimates of discrete nest-survival rates. Bayesian analysis has shown to be successful in such study.

[1]Supported by Federal Aid in Wildlife Restoration Project W-13-R.
[2]Supported by the NSF grants SES-0095919 and SES-0351523 and a grant from Missouri Department of Conservation.

In a nest-survival study, biologists choose a study area and search for active nests. Once a nest is found, it will be revisited at least once more. At each visit, biologists note whether nests have failed (get destroyed or abandoned), are still active, or have succeeded (at least one egg hatches or at least one young bird fledges, as appropriate to the study in question). Typically nests are not found until after incubation has begun, and nests that failed are of unknown age (although in some studies eggs are candled to establish age).

Nesting study data are often left-truncated because nests that failed or hatched before they would otherwise be discovered never come under observation. The consequence is that the observed nests are a nonrandomly screened subset of all nests. It is also common that nesting data are double-interval censored. General double-interval censoring arises when neither the origin (date incubation started) nor failure date are known exactly. Uncertainty about the origin date arises if nests cannot be aged exactly. Uncertainty about the failure date arises if the nests are not revisited daily; the failure date may be known only to an interval if the failure occurred between visits.

The Mayfield (1961, 1975) method and it modifications (Miller and Johnson, 1978; Johnson, 1979; Hensler and Nichols, 1981; Nichols et al., 1984; Bart and Robson, 1982) have been the standard and popular methods for estimating nest success rates. The Mayfield method was developed to accommodate truncation bias, but it assumes that the daily failure hazard rate is constant regardless of the nest's age and that the nests are revisited daily. If nests are observed less frequently than daily, which is often the case, Mayfield (1961, 1975) proposed a method assuming that an unsuccessful nest failed at the midpoint between the last observation when it was still alive and the first observation when it was found to have failed. Miller and Johnson (1978) and Johnson (1979) suggested some modifications to improve the Mayfield method. They proposed that when days are skipped between nest visits and failure is subsequently observed, the 40% point rather than the mid-point is assigned as the day of failure. The assumption of constant daily failure hazard rates is obviously unrealistic in many situations. Several stratified Mayfield methods have been developed to overcome the problem (Green, 1977; Johnson, 1979; Klett and Johnson, 1982). The stratified Mayfield estimator requires knowing the nest age at least to age interval, but in many situations such age information is not available.

As a solution to this problem, Pollock and Cornelius (1988) and Cornelius (1993) proposed a discrete-survival model that allows survival to be age specific but does not require knowing the nest ages. The model, called the $M_{\delta q}$ model, involves unknown age-specific failure probabilities and age-specific encounter probabilities. Encounter and failure probabilities are assumed independent. The $M_{\delta q}$ model assumes that the nests are revisited daily. Let J be the number of days that a nest is required to survive to be considered successful. The $M_{\delta q}$ model requires estimation of $J-1$ parameters δ_j and J parameters q_j, which may lead to unstable MLE for large J. The Pollock and Cornelius (1988) approach to this problem is to group the nest life span into long intervals so that J is small enough to avoid instability problems. Heisey and Nordheim (1990) pointed out that the weakness of the Pollock and Cornelius' approach is that the nesting period is divided into too few intervals, resulting in unacceptable bias in the daily success probability estimates. Bromaghin (1991) and Bromaghin and McDonald

(1993) proposed a weighted-nest success model formulation, which is a generalization of the Pollock and Cornelius model. Their model allows the encounter probabilities to be determined by specification of the sampling design, permitting estimation of the number of nests not seen.

Heisey and Nordheim (1995) presented a bivariate contingency table approach for modeling age-specific survival with left-truncated and double-interval censored data (i.e., nests are revisited less frequently than daily). They used the bivariate approach to model both the age of the nest at recruitment and the age at failure. The likelihood is maximized using the EM algorithm. This algorithm is often slow to converge and may produce unstable estimates when J is large. To avoid a large number of unknown parameters, Heisey and Nordheim (1995) take a different approach to this problem by assuming piecewise constant encounter probabilities and piecewise constant hazard rates.

He et al. (2001) presented a Bayesian analysis for age-specific nest survival studies when the nests are revisited daily. The Bayesian approach needs neither the grouping of data nor the making of structural assumptions about unknown parameters. Bayesian analysis uses a researcher's knowledge about unknown parameters. This knowledge is expressed as a prior distribution. When data are observed, the prior information is updated in the posterior distribution. The inference is then made based on this posterior distribution. Once the posterior distribution is available, the critical issue of instability mentioned by Heisey and Nordheim (1995) is avoided.

He et al. (2001) placed the Dirichlet priors on the encounter probabilities and the failure probabilities. This approach has several short comings. First, the conditional posterior densities of unknown parameters are complex. Therefore, the method involves an intensive Gibbs sampling computation. After some suitable transformations, the conditional posterior densities of transformed unknown parameters are log-concave functions. Then the Gibbs sampling is done based on these log-concave functions using the adaptive rejection sampling method by Gilks and Wild (1992). Second, the choice of the hyperparameters is very difficult because it is hard to get the prior information on age-specific failure probabilities. Furthermore, He et al. (2001) addressed the daily visit case only. In practice, it is common that nests are revisited less frequently than daily and irregularly. Bad weathers or schedule conflicts (such as weekends) make the regular revisit impossible. Sometimes researchers revisit nests less frequently than daily to reduce the effect of disturbances.

In this chapter, we present a new Bayesian model for the age-specific nest survival rates to handle general irregular visit cases. First, we introduce some additional latent variables so that the likelihood function based on argumented data is very simple. Then we place independent beta priors on the age-specific hazard rates and conditional-encounter rates. A two-stage hierarchical prior is placed on the number of encountered nests. All the full conditional posterior distributions are standard ones such as beta, gamma, Poisson, negative binomial, and multinormal distributions. Therefore, the Gibbs sampling computation is more efficient. To avoid the hyperparameter specification problem, we consider Berger and Bernardo's (1989, 1992a, 1992b) reference prior and an improper vague prior. Under a mild condition, the posterior is proved to be proper. The simulation result shows that the proposed Bayesian models provide quite accurate and stable estimates of age-specific survival rates.

In Section 2, the model and likelihood function is presented. Latent variables are used so that the conditional distributions are standard distributions. The full conditional distributions are given in Appendix C. In Section 3, both informative and noninformative priors are discussed. The choice of the hyperparameters is considered, and the reference prior is derived. In Section 4, the property of the posterior is investigated. In Section 5, the method is applied to real bobwhite data. A simulation result also is presented to compare the Bayesian method with the Mayfield method. Finally, some comments are given in Section 6.

2. Likelihood functions

2.1. *Models and latent variables*

Suppose that there are J number of days that a nest is required to survive to be considered successful. Let u be the age of a nest at first encounter, z be the number of observed days before a failure or a success occurs, and t be the age of a nest at the time the outcome (either a failure or a success) is determined. Note that $z = t - u + 1$. Let y be the indicator that a nest is a success. If indeed $y = 1$, we have $t = J$; if $y = 0$, u and t are both unobservable.

Assume that there are N_{pop} nests in the study area. The nest population can be divided into two sub-populations: nests are encountered between age 1 and age J and nests are not encountered. The nests encountered between age 1 and age J can be further divided into observed nests and truncated nests. The observed nests are those nests encountered before failure or success, i.e., $u \leqslant t$ and $u \leqslant J$. The truncated nests are those nests failed before otherwise encountered, i.e., $u > t$ and $u \leqslant J$. Let N be the number of nests encountered (include both observed nests and truncated nests) and n be the number of observed nests. Then N is an unknown parameter, and n is an observed data.

Clearly a nest is encountered if $u \leqslant J$. Let δ_i denote the conditional encounter probability given the nest is encountered and q_i be the failure probability of the nest at age i. That is to say,

$$\delta_i = P(u = i | u \leqslant J) \quad \text{and} \quad q_i = P(t = i,\ y = 0). \tag{1}$$

Note that $\delta_1 + \cdots + \delta_J = 1$ and $0 \leqslant q_1 + \cdots + q_J \leqslant 1$. Then $q_{J+1} = 1 - (q_1 + \cdots + q_J)$ is the probability of success. We will define $\boldsymbol{\delta} = (\delta_1, \ldots, \delta_{J-1})$ and $\boldsymbol{q} = (q_1, \ldots, q_J)$.

Let k index the kth observed nest. If the nest is revisited daily, we know z_k, the number of observed days before a failure or a success occurs. Then $u_k = J - z_k + 1$ for succeeded nests and $1 \leqslant u_k \leqslant J - z_k + 1$ for failed nests.

Case I: Regular visit. Consider n as a fixed number. Let $\boldsymbol{y} = (y_1, \ldots, y_n)$ and $\boldsymbol{z} = (z_1, \ldots, z_n)$. The data observed is of the form $(\boldsymbol{y}, \boldsymbol{z})$. The likelihood function of $(\boldsymbol{\delta}, \boldsymbol{q})$ based on $(\boldsymbol{y}, \boldsymbol{z})$ is

$$L_1(\boldsymbol{\delta}, \boldsymbol{q}) = \frac{1}{A^n} \prod_{k=1}^{n} (\delta_{J-z_k+1} q_{J+1})^{y_k} \left(\sum_{l=z_k}^{J} \delta_{l-z_k+1} q_l \right)^{1-y_k}, \tag{2}$$

where A is the probability that a nest is observed, given by

$$A = P(u \leqslant J) = \sum_{i=1}^{J} \delta_i \left(\sum_{j=i}^{J+1} q_j \right). \tag{3}$$

Case II: Irregular visit. If the nest is revisited irregularly, we know z_k only to an interval. Assume that the visit schedule is prespecified. Let $[Z_{L_k}, Z_{R_k}]$ be the interval that z_k falls in, and $[U_{L_k}, U_{R_k}]$ the interval that u falls in. Z_{L_k} is the smallest number of days that the nest could have possible been under observation and Z_{R_k} is the greatest number of observed days. U_{L_k} is the youngest that the nest could have possible been discovered, and U_{R_k} is the oldest. Define $\boldsymbol{Z}_L = (Z_{L_1}, \ldots, Z_{L_n}), \boldsymbol{Z}_R = (Z_{R_1}, \ldots, Z_{R_n}), \boldsymbol{U}_L = (U_{L_1}, \ldots, U_{L_n}), \boldsymbol{U}_R = (U_{R_1}, \ldots, U_{R_n})$. Then the data observed is of the form $(\boldsymbol{y}, \boldsymbol{Z}_R, \boldsymbol{Z}_L, \boldsymbol{U}_R, \boldsymbol{U}_L)$. Again, n is considered as a fixed number.

The likelihood function of $(\boldsymbol{\delta}, \boldsymbol{q})$ based on observed data $(\boldsymbol{y}, \boldsymbol{Z}_L, \boldsymbol{Z}_R, \boldsymbol{U}_L, \boldsymbol{U}_R)$ is

$$L_2(\boldsymbol{\delta}, \boldsymbol{q}) \propto \frac{1}{A^n} \prod_{k=1}^{n} \left(\sum_{i=U_{L_k}}^{U_{R_k}} \delta_i q_{J+1} \right)^{y_k} \left(\sum_{i=U_{L_k}}^{U_{R_k}} \sum_{j=i+Z_{L_k}}^{i+Z_{R_k}} \delta_i q_j \right)^{1-y_k}. \tag{4}$$

Here A also is given by (3).

Clearly L_1 in (2) is a special case of L_2 by taking $(U_{L_k}, U_{R_k}; Z_{L_k}, Z_{R_k}) = (J - z_k + 1, J - z_k + 1; z_k, z_k)$ for a succeeded nest and $(U_{L_k}, U_{R_k}; Z_{L_k}, Z_{R_k}) = (1, J - z_k + 1; z_k, z_k)$ for a failed nest.

The likelihood function L_2 is complex with A in the denominator and a double summation in the numerator. Therefore, the posterior distribution of $(\boldsymbol{\delta}, \boldsymbol{q})$ based on this likelihood will be hard to simulate. We need to add some latent variables to simplify the likelihood function.

Case III: Irregular visit with latent variables $(\boldsymbol{u}, \boldsymbol{t})$. Recall that u_k is unknown and t_k is known for succeeded nests only. Let $\boldsymbol{u} = (u_1, \ldots, u_n)$ and $\boldsymbol{t} = (t_1, \ldots, t_n)$.

For a regular visit case, He et al. (2001) used the latent variables $\boldsymbol{u}$ to simplify the computation so the summation in the numerator of the likelihood function of $(\boldsymbol{q}, \boldsymbol{\delta})$ based on the augmented data $(\boldsymbol{y}, \boldsymbol{u})$ disappeared. In fact, for a succeeded nest k, t_k is determined.

In irregular cases, adding latent variables $(\boldsymbol{u}, \boldsymbol{t})$ makes the double summation in the numerator disappeared. In fact, the likelihood function of $(\boldsymbol{\delta}, \boldsymbol{q})$ based on the latent data $(\boldsymbol{u}, \boldsymbol{t})$ and the observed data $(n, \boldsymbol{y}, \boldsymbol{Z}_L, \boldsymbol{Z}_R, \boldsymbol{U}_L, \boldsymbol{U}_R)$ is

$$L_3(N, \boldsymbol{\delta}, \boldsymbol{q}) \propto \frac{1}{A^n} \prod_{k=1}^{n} (\delta_{u_k} q_{J+1})^{y_k} (\delta_{u_k} q_{t_k})^{1-y_k}. \tag{5}$$

Case IV: Irregular visit with latent variables $(\boldsymbol{u}, \boldsymbol{t}, \boldsymbol{g})$. Let g_{ij} be the number of nests failed at age j before otherwise encountered at age i $(1 \leqslant j < i \leqslant J)$. Note that g_{ij} is unknown because of the truncation. Define $\boldsymbol{g} = (g_{ij},\ 1 \leqslant j < i \leqslant J)$. To get rid of A in the denominator, latent variable $\boldsymbol{g}$ and a nuisance unknown parameter N (the number of nests encountered) are added in the likelihood function. Consider n as an observed-random variable. The likelihood function of $(N, \boldsymbol{\delta}, \boldsymbol{q})$ given the latent data $(\boldsymbol{u}, \boldsymbol{t}, \boldsymbol{g})$ and the observed data $(n, \boldsymbol{y}, \boldsymbol{Z}_L, \boldsymbol{Z}_R, \boldsymbol{U}_L, \boldsymbol{U}_R)$ is then

$$L_4(N, \boldsymbol{\delta}, \boldsymbol{q}) \propto \frac{N!}{\prod_{1 \leqslant j<i \leqslant J} g_{ij}!} \prod_{1 \leqslant j<i \leqslant J} (\delta_i q_j)^{g_{ij}} \times \prod_{k=1}^{n} (\delta_{u_k} q_{J+1})^{y_k} (\delta_{u_k} q_{t_k})^{1-y_k}, \tag{6}$$

where g_{ij} are nonnegative integers satisfying $\sum_{1 \leqslant j<i \leqslant J} g_{ij} = N - n$, and

$$U_{L_k} \leqslant u_k \leqslant U_{R_k} \quad \text{and} \quad u_k + Z_{L_k} - 1 \leqslant t_k \leqslant u_k + Z_{R_k} - 1.$$

2.2. *Reparameterization*

Note that the ranges of $\boldsymbol{\delta}$ and $\boldsymbol{q}$ is irregular and computation would have to be done componentwise. Alternatively, we reparameterize δ_i and q_j into discrete–hazard rates so that the ranges of new parameters are unit intervals. For $i = 1, \ldots, J-1$ and $j = 1, \ldots, J$, define the hazard rates corresponding to $\boldsymbol{\delta}$ and $\boldsymbol{q}$ by

$$\alpha_i = \frac{\delta_i}{\delta_i + \cdots + \delta_J}, \qquad \xi_j = \frac{q_j}{q_j + \cdots + q_{J+1}}, \tag{7}$$

and $\alpha_J = \xi_{J+1} = 1$. Note that $0 \leqslant \alpha_i \leqslant 1$ and $0 \leqslant \xi_j \leqslant 1$. We also have closed-form expressions between δ_i and α_i, q_j and ξ_j, respectively,

$$\delta_i = \alpha_i \prod_{l=1}^{i-1} (1 - \alpha_l) \quad \text{for } i = 1, 2, \ldots, J-1, \qquad \delta_J = \prod_{l=1}^{J-1} (1 - \alpha_l),$$

$$q_j = \xi_j \prod_{l=1}^{j-1} (1 - \xi_l) \quad \text{for } i = 1, 2, \ldots, J, \qquad q_{J+1} = \prod_{l=1}^{J} (1 - \xi_l).$$

We now give the likelihood function of $(N, \boldsymbol{\alpha}, \boldsymbol{\xi})$ based on augmented data. The following notations are needed. Let $n^S = \#\{y_k = 0\colon k = 1, \ldots, n\}$ (the total number of successful nests), $n_i^S = \#\{(y_k, u_k) = (1, i)\colon k = 1, \ldots, n\}$ (the number of successful nests encountered at age i), and $n_{ij}^F = \#\{(y_k, z_k, u_k) = (0, j, i)\colon k = 1, \ldots, n\}$ (the number of failed nests encountered at age i and observed j days). Define

$$m_i = \begin{cases} n_1^S + \sum_{j=1}^{J} n_{1j}^F, & \text{if } i = 1; \\ n_i^S + \sum_{j=1}^{J-i+1} n_{ij}^F + \sum_{j=1}^{i-1} g_{ij}, & \text{if } i = 2, \ldots, J-1; \end{cases} \tag{8}$$

$$l_i = \sum_{l=i+1}^{J} \left(n_l^S + \sum_{j=1}^{J-l+1} n_{lj}^F + \sum_{j=1}^{l-1} g_{lj} \right), \quad \text{if } i = 1, \ldots, J-1; \tag{9}$$

$$m_j^* = \begin{cases} \sum_{l=1}^{j} n_{l,j-l+1}^F + \sum_{i=j+1}^{J} g_{ij}, & \text{if } j = 1, \ldots, J-1; \\ \sum_{l=1}^{J} n_{l,J-l+1}^F, & \text{if } j = J. \end{cases} \tag{10}$$

$$l_j^* = \begin{cases} n^S + \sum_{l=j+1}^{J} \left(\sum_{i=l+1}^{J} g_{il} \right) + \sum_{i+l-2 \geqslant j} n_{il}^F, \\ \quad \text{if } j = 1, \ldots, J-1; \\ n^S, \quad \text{if } j = J. \end{cases} \tag{11}$$

Then the likelihood function of $(N, \boldsymbol{\alpha}, \boldsymbol{\xi})$ based on the latent data $(\boldsymbol{u}, \boldsymbol{t}, \boldsymbol{g})$ and observed data $(n, \boldsymbol{y}; \boldsymbol{Z}_L, \boldsymbol{Z}_R, \boldsymbol{U}_L, \boldsymbol{U}_R)$ is

$$L_5(N, \boldsymbol{\alpha}, \boldsymbol{\xi}) \propto \frac{N!}{\prod_{1 \leqslant j < i \leqslant J} g_{ij}!} \left\{ \prod_{i=1}^{J-1} \alpha_i^{m_i} (1-\alpha_i)^{l_i} \right\} \left\{ \prod_{j=1}^{J} \xi_j^{m_j^*} (1-\xi_j)^{l_j^*} \right\}, \tag{12}$$

where $g_{ij} \geqslant 0$ and $\sum_{1 \leqslant j < i \leqslant J} g_{ij} = N - n$.

3. Bayesian analysis

We will discuss both informative and noninformative priors in this section.

3.1. Informative priors

Priors for $(\boldsymbol{\delta}, \boldsymbol{q})$. Because the probability $(\boldsymbol{\delta}, \boldsymbol{q})$ and the hazard rate $(\boldsymbol{\alpha}, \boldsymbol{\xi})$ have a one-to-one relationship, one may choose to place priors on either $(\boldsymbol{\delta}, \boldsymbol{q})$ or $(\boldsymbol{\alpha}, \boldsymbol{\xi})$. It is often assumed that $\boldsymbol{\delta}$ and $\boldsymbol{q}$ are independent. It is also common to assume that multinomial probabilities follow Dirichlet priors. For example, He et al. (2001) placed the Dirichlet $(a_1, \ldots, a_J)$ and Dirichlet $D(b_1, \ldots, b_{J+1})$ on $\boldsymbol{\delta}$ and $\boldsymbol{q}$, respectively. So the prior density of $(\boldsymbol{\delta}, \boldsymbol{q})$ is

$$\pi(\boldsymbol{\delta}, \boldsymbol{q}) \propto \left\{ \prod_{i=1}^{J} \delta_i^{a_i - 1} \right\} \left\{ \prod_{j=1}^{J+1} q_j^{b_j - 1} \right\}. \tag{13}$$

Here a_j and b_j are positive constants. As we mentioned in Section 1, the full conditional densities in MCMC simulations under this setting is quite complex due to the irregular range of $\boldsymbol{\delta}$ and $\boldsymbol{q}$.

Priors for $(\boldsymbol{\alpha}, \boldsymbol{\xi})$. Unlike the ranges of $\boldsymbol{\delta}$ and $\boldsymbol{q}$, the ranges of the components of $\boldsymbol{\alpha}$ and $\boldsymbol{\xi}$ are all unit intervals. It is natural to assume independent beta priors for α_i and ξ_j. So the prior density of $\boldsymbol{\alpha}$ and $\boldsymbol{\xi}$ is

$$\pi(\boldsymbol{\alpha}, \boldsymbol{\xi}) \propto \left\{ \prod_{i=1}^{J-1} \alpha_i^{c_i - 1} (1-\alpha_i)^{c_i^* - 1} \right\} \left\{ \prod_{j=1}^{J} \xi_j^{d_j - 1} (1-\xi_j)^{d_j^* - 1} \right\}. \tag{14}$$

Here c_i, c_i^*, d_j and d_j^* are positive constants.

We can prove that the Dirichlet priors on $\boldsymbol{\delta}$ and $\boldsymbol{q}$ are just a special case of the independent beta priors on $\boldsymbol{\alpha}$ and $\boldsymbol{\beta}$. Specifically, we have the following proposition. The proof of this proposition is straightforward and omitted.

PROPOSITION 1. (a) *If $\boldsymbol{\delta}$ has a Dirichlet $(a_1, \ldots, a_{J-1})$ distribution, the corresponding hazard rates $\alpha_1, \ldots, \alpha_J$ are independent, and α_j has a Beta$(a_j, \sum_{i=j+1}^{J} a_i)$ distribution.*

(b) *If $\boldsymbol{q}$ has a Dirichlet $(b_1, \ldots, b_{J+1})$ distribution, the corresponding hazard rates $\xi_1, \ldots, \xi_J$ are independent, and ξ_j has a Beta$(a_j, \sum_{i=j+1}^{J+1} b_i)$ distribution.*

Priors for N. For the unknown parameter N, we propose a class of hierarchical priors. For given $\lambda > 0$, N has a Poisson density with mean λ. Then λ follows a gamma distribution with parameters (e_1, e_2). So the joint prior of (N, λ) is

$$\pi(N, \lambda) \propto \frac{\lambda^N}{N!} \lambda^{e_1 - 1} \mathrm{e}^{-(1+e_2)\lambda}. \tag{15}$$

It turns out that the marginal prior of N is

$$\pi(N) = \int_0^\infty \pi(N, \lambda)\, \mathrm{d}\lambda \propto \frac{\Gamma(N + e_1)}{N!(e_2 + 1)^{N+e_1}}. \tag{16}$$

Here e_1 and e_2 are positive constants. We will see that the hierarchical structure (15) has nice computational advantages.

3.2. Noninformative priors

There are $2J - 1$ hyperparameters to be specified in the Dirichle prior and $4J - 2$ in the Independent beta prior. If researchers have detailed beliefs about the hyperparameters in the prior distributions, it is wise to use subjective priors reflecting their beliefs. But, in practice, this is not always the case. One danger is that if prior information developed from previous experience is unwittingly inappropriate for a new data set, it might produce a flawed analysis. It is then interesting to use a variety of so-called noninformative priors that allow us much of the benefit of a Bayesian analysis without the difficulty of obtaining an informative prior. A common noninformative prior is the Jeffreys prior (Jeffreys, 1961), which is proportional to the square root of the determinant of the Fisher information matrix. The Jeffreys prior is quite useful for a single parameter but can be seriously deficient in multiparameter problems (cf. Berger and Bernardo, 1992a, 1992b). As alternatives, Berger and Bernardo's (1989, 1992a, 1992b) reference priors have been shown to be successful in various statistics models, especially for the i.i.d. cases. A recent review of various approaches to noninformative priors can be found in Kass and Wasserman (1996).

It is almost intractable to find reference priors for all unknown parameters $(N, \boldsymbol{\alpha}, \boldsymbol{\xi})$ under the proposed model. Instead of searching for the joint reference prior for $(N, \boldsymbol{\alpha}, \boldsymbol{\xi})$, one possibility is to take a vague prior for N, the number of nests encountered within J days, and a conditional reference prior for $\boldsymbol{\alpha}$ and $\boldsymbol{\xi}$ given N. The following result can be found from He (2003).

PROPOSITION 2. *The reference prior for* $(\boldsymbol{\alpha}, \boldsymbol{\xi})$ *given* N *is*

$$\pi_R(\boldsymbol{\alpha}, \boldsymbol{\xi}) \propto \prod_{i=1}^{J-1} \frac{1}{\sqrt{\alpha_i(1-\alpha_i)}} \prod_{j=1}^{J} \frac{1}{\sqrt{\xi_j(1-\xi_j)}}, \tag{17}$$

i.e., α_i *and* ξ_j *are independent beta distributions with parameters* $(1/2,\ 1/2)$.

He et al. (2001) used constant priors on the encounter probabilities $\boldsymbol{\delta}$ and failure probabilities $\boldsymbol{q}$ by choosing $a_1 = \cdots = a_J = 1$ and $b_1 = \cdots = b_{J+1} = 1$ for the Dirichlet priors (13). Using Proposition 1, we see that these constant priors are equivalent to that α_j has independent $Beta(1, J-j)$ and ξ_j has independent $Beta(1, J-j+1)$. Alternatively, we may choose the constant priors on the encounter hazard rates α_i and failure hazard rates ξ_j. That is, specify $c_1 = \cdots = c_{J-1} = 1$, $c_1^* = \cdots = c_{J-1}^* = 1$, $d_1 = \cdots = d_j = 1$, and $d_1^* = \cdots = d_j^* = 1$, i.e., α_j has independent $Beta(1,1)$ and ξ_j has independent $Beta(1,1)$.

The reference prior, the constant prior on probabilities and the constant prior on hazard rates are all noninformative priors or vague priors. From Berger (1985), Bayesian estimators are often quite robust in terms of changes of hyperparameters, especially when noninformative priors or vague priors are used. Furthermore, Bayesian estimators are consistent when the sample sizes are large enough, regardless of the choice of hyperparameters. Our simulation study shows that all these three priors work quite well and produce similar estimates.

If one does not have enough prior information on N, a noninformative prior can be chosen by assuming $e_2 = 0$ in (16), for which,

$$\pi_{e_1}(N) \propto \frac{\Gamma(N+e_1)}{N!}. \tag{18}$$

Special cases are as follows,

$$\pi_{e_1}(N) \propto \begin{cases} \frac{1}{N}, & \text{if } e_1 = 0, \\ \frac{\Gamma(N+0.5)}{N!}, & \text{if } e_1 = 0.5, \\ 1, & \text{if } e_1 = 1. \end{cases} \tag{19}$$

Note that $\Gamma(N+0.5)/N!$ is quite close to the prior $1/\sqrt{N}$. In Figure 1, we plot the ratio of two priors, i.e., $\Gamma(N+0.5) * \sqrt{N}/N!$. It is seen even if $N = 1$, it is about .886; when $N = 10$, the ratio is .9876. Considering the normalizing constants of posteriors using these two priors, the actual posteriors based on the two priors could be closer.

3.3. Propriety of posterior under the noninformative priors

Note that the noninformative prior $\pi(N)$ defined by (18) is improper. It is essential in a Bayesian analysis that the posterior distribution be proper. The following theorem gives a condition for a proper posterior.

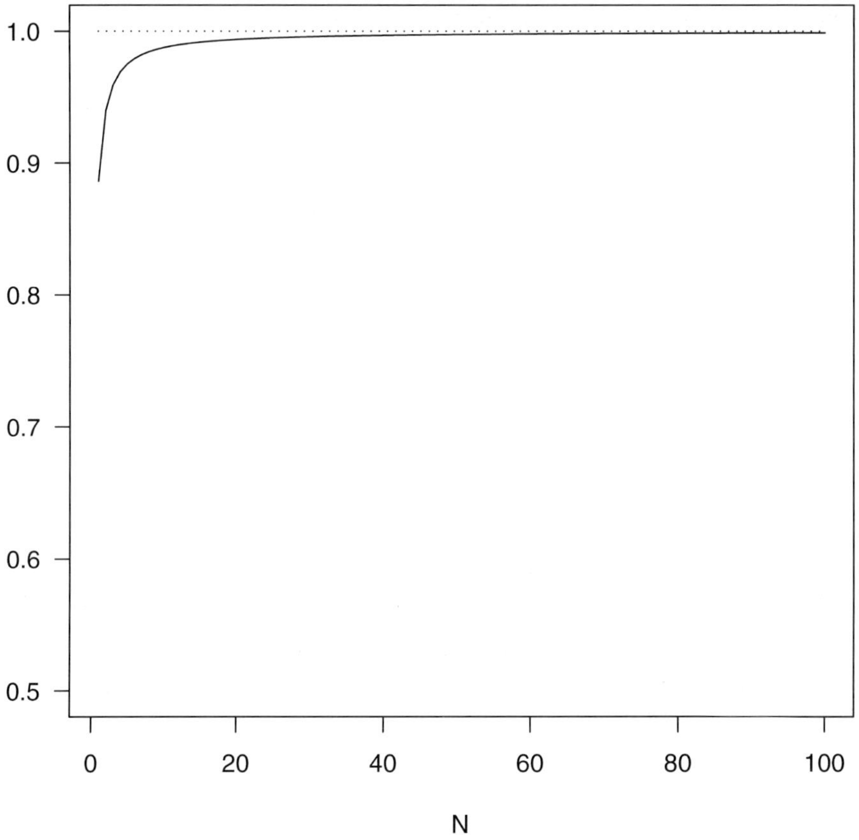

Fig. 1. Plot of $\pi_{0.5}(N) * \sqrt{N}$, the ratio of the priors $\pi_{0.5}(N)$ and $\pi(N) = 1/\sqrt{N}$.

THEOREM 1. *Under the prior* (14) *and* (18), *the joint posterior distribution of* $(\boldsymbol{\alpha}, \boldsymbol{\xi}; \boldsymbol{u}, \boldsymbol{t}, \boldsymbol{g}, N, \lambda)$ *given the observed data* $(n, \mathbf{y}; \mathbf{Z}_L, \mathbf{Z}_R, \boldsymbol{U}_L, \boldsymbol{U}_R)$; *is proper if*

$$\sum_{i=1}^{J-1} n_i^S - \frac{J-3}{2} - e_1 > 0. \tag{20}$$

PROOF. Write

$$\begin{aligned} &W(N, \boldsymbol{\alpha}, \boldsymbol{\xi}, \boldsymbol{u}, \boldsymbol{t}, \boldsymbol{g}) \\ &\quad = \frac{\Gamma(N + e_1)}{\prod_{1 \leqslant j < i \leqslant J} g_{ij}!} \prod_{i=1}^{J-1} \alpha_i^{m_i + c_i - 1} (1 - \alpha_i)^{l_i + c_i^* - 1} \\ &\quad \times \prod_{j=1}^{J} \xi_j^{m_j^* + d_j - 1} (1 - \xi_j)^{l_j^* + d_j^* - 1}. \end{aligned}$$

We need to show that for given data $(n, \boldsymbol{y}, \boldsymbol{Z}_L, \boldsymbol{Z}_R, \boldsymbol{U}_L, \boldsymbol{U}_R)$,

$$W_1 \equiv \sum_{N=n}^{\infty} \int_0^1 \cdots \int_0^1 \sum_{\boldsymbol{u}} \sum_{\boldsymbol{t}} \sum_{\boldsymbol{g}} W(N, \boldsymbol{\alpha}, \boldsymbol{\xi}, \boldsymbol{u}, \boldsymbol{t}, \boldsymbol{g}) \prod_{l=1}^{J-1} \mathrm{d}\alpha_l \prod_{l=1}^{J} \mathrm{d}\xi_l < \infty. \tag{21}$$

Without loss of generality, assume that $n > 0$. It is easy to see that

$$\sum_{\boldsymbol{g}} W(N, \boldsymbol{\alpha}, \boldsymbol{\xi}, \boldsymbol{u}, \boldsymbol{t}, \boldsymbol{g}) = \left\{ \prod_{i=1}^{J-1} \alpha_i^{m_{0i}+c_i-1} (1-\alpha_i)^{l_{0i}+c_i^*-1} \right\} \left\{ \prod_{j=1}^{J} \xi_j^{m_{0j}^*+d_j-1} (1-\xi_j)^{l_{0j}^*+d_j^*-1} \right\} \times \frac{\Gamma(N+e_1)(1-A)^{N-n}}{n!(N-n)!},$$

where

$$m_{0i} = \begin{cases} n_1^S + \sum_{j=1}^{J} n_{1j}^F, & \text{if } i = 1; \\ n_i^S + \sum_{j=1}^{J-i+1} n_{ij}^F, & \text{if } i = 2, \ldots, J-1; \end{cases}$$

$$l_{0i} = \sum_{l=i+1}^{J} \left(n_l^S + \sum_{j=1}^{J-l+1} n_{lj}^F \right), \quad \text{if } i = 1, \ldots, J-1;$$

$$m_{0j}^* = \begin{cases} \sum_{l=1}^{j} n_{l,j-l+1}^F, & \text{if } j = 1, \ldots, J-1; \\ \sum_{l=1}^{J} n_{l,J-l+1}^F, & \text{if } j = J; \end{cases}$$

$$l_{0j}^* = \begin{cases} n^S + \sum_{i+l-2 \geqslant j} n_{il}^F, & \text{if } j = 1, \ldots, J-1; \\ n^S, & \text{if } j = J. \end{cases}$$

Define

$$W_2(\boldsymbol{u}, \boldsymbol{t}) = \sum_{N=n}^{\infty} \int_0^1 \cdots \int_0^1 \sum_{\boldsymbol{g}} W(N, \boldsymbol{\alpha}, \boldsymbol{\xi}, \boldsymbol{u}, \boldsymbol{t}, \boldsymbol{g}) \prod_{l=1}^{J-1} \mathrm{d}\alpha_l \prod_{l=1}^{J} \mathrm{d}\xi_l. \tag{22}$$

From Fubini theorem,

$$W_1 = \sum_{\boldsymbol{u}} \sum_{\boldsymbol{t}} W_2(\boldsymbol{u}, \boldsymbol{t}).$$

Note that W_1 is finite if W_2 is finite for each fixed $(\boldsymbol{t}, \boldsymbol{u})$ because both $\boldsymbol{u}$ and $\boldsymbol{t}$ are bounded within 1 and J. Now we will prove that W_2 is finite. Note that

$(m_{0i}, l_{0i}, m^*_{0j}, l^*_{0j})$ are bounded for given data and do not depend on N. Clearly,

$$(1-A)^{N-n} = \sum_{\sum_{j=1}^{J-1} r_j = N-n} \frac{(N-n)!}{r_1! \cdots r_{J-1}!} w_1^{r_1} \cdots w_{J-1}^{r_{J-1}}$$
$$= \sum_{\sum_{j=1}^{J-1} r_j = N-n} \frac{(N-n)!}{r_1! \cdots r_{J-1}!}$$
$$\times \left\{ \prod_{i=1}^{J-1} (1-\alpha_i)^{\sum_{l=i}^{J-1} r_l} \prod_{j=1}^{J-2} \xi_j^{r_j} (1-\xi_j)^{\sum_{l=j+1}^{J-1} r_l} \xi_{J-1}^{r_{J-1}} \right\}.$$

We have that

$$W_2(\boldsymbol{u}, \boldsymbol{t}) = \sum_{N=n}^{\infty} \frac{\Gamma(N+e_1)}{n!}$$
$$\times \left\{ \sum_{\sum_{j=1}^{J-1} r_j = N-n} \frac{1}{r_1! \cdots r_{J-1}!} \left(\prod_{i=1}^{J-1} \frac{\Gamma(m_{0i}+c_i)\Gamma\big(\sum_{l=i}^{J-1} r_l + l_{0i} + c_i^*\big)}{\Gamma\big(m_{0i}+c_i+\sum_{l=i}^{J-1} r_l + l_{0i} + c_i^*\big)} \right) \right.$$
$$\times \left(\prod_{j=1}^{J-2} \frac{\Gamma(r_j + m^*_{0j} + d_j)\Gamma\big(\sum_{l=j+1}^{J-1} r_l + l^*_{0j} + d^*_j\big)}{\Gamma\big(r_j + m^*_{0j} + d_j + \sum_{l=j+1}^{J-1} r_l + l^*_{0j} + d^*_j\big)} \right)$$
$$\times \frac{\Gamma(r_{J-1} + m^*_{0,J-1} + d_{J-1})\Gamma(l^*_{0,J-1} + d^*_{J-1})}{\Gamma(r_{J-1} + m^*_{0,J-1} + d_{J-1} + l^*_{0,J-1} + d^*_{J-1})}$$
$$\left. \times \frac{\Gamma(m^*_{0J} + d_J)\Gamma(l^*_{0J} + d^*_J)}{\Gamma(m^*_{0J} + d_J + l^*_{0J} + d^*_J)} \right\}.$$

Note that $x! = \sqrt{2\pi} x^{x+1/2} \mathrm{e}^{-x+\theta/(12x)}$ for $x > 0$, where $0 < \theta < 1$ and $1 < \mathrm{e}^{\theta/(12x)} < e$. Also, note that $0! = 1! = 1$. So,

$$W_2 \leqslant M \sum_{N=n}^{\infty} \frac{1}{N^{1-e_1}(N-n)^{\sum_{i=1}^{J-1}(m_{0i}+c_i) + l^*_{0,J-1} + d^*_{J-1} - n - (J-2)/2 - 1/2}}.$$

Here M is a positive constant for given data. Therefore, W_2 is finite if

$$\sum_{i=1}^{J-1} (m_{0i} + c_i) + l^*_{0,J-1} + d^*_{J-1} - n - \frac{J-2}{2} - \frac{1}{2} + 1 - e_1 > 1.$$

Because

$$\sum_{i=1}^{J-1} m_{0i} + l^*_{0,J-1} \geqslant n + \sum_{i=1}^{J-1} n_i^S,$$

W_2 is finite if (20) holds. □

If any of the three improper priors $\pi_{e_1}(N)$ in (19) is used, the posterior of $(N, \boldsymbol{\alpha}, \boldsymbol{\xi})$ given $(n, \boldsymbol{y}; \boldsymbol{Z}_L, \boldsymbol{Z}_R, \boldsymbol{U}_L, \boldsymbol{U}_R)$ is proper.

4. Posterior distributions and Bayesian computation

In Bayesian analysis, an unknown parameter is often estimated by its posterior mean. The posterior mean is the best estimator of a parameter under a squared-error loss (Berger, 1985). The evaluation of the posterior means requires high-dimensional integrations. Such a computation is quite intractable. Techniques have been developed, such as the Markov chain Monte Carlo (MCMC), including Gibbs sampler (Geman and Geman, 1984; Gelfand and Smith, 1990; Vounatsou and Smith, 1995), to overcome these difficulties. We use the Gibbs sampler to get samples from the full conditional posterior distributions. Posterior quantities can be computed based on the samples from the MCMC.

Because (19) and (18) are special cases of (16), we will consider the posterior under the prior (16) using the latent variable λ in (15).

PROPOSITION 3. *Under the prior* (14) *and* (15), *the posterior density* $(\boldsymbol{\alpha}, \boldsymbol{\xi}; \boldsymbol{u}, \boldsymbol{t}, \boldsymbol{g}, N, \lambda)$ *given the observed data* $(n, \boldsymbol{y}; \boldsymbol{Z}_L, \boldsymbol{Z}_R, \boldsymbol{U}_L, \boldsymbol{U}_R)$ *is*

$$[N, \lambda, \boldsymbol{\alpha}, \boldsymbol{\xi}; \boldsymbol{u}, \boldsymbol{t}, \boldsymbol{g} | \text{data}] \\ \propto \frac{\lambda^N \lambda^{e_1 - 1} \mathrm{e}^{-(1+e_2)\lambda}}{\prod_{J \geqslant i > j \geqslant 1} g_{ij}!} \\ \times \left\{ \prod_{i=1}^{J-1} \alpha_i^{m_i + c_i - 1} (1 - \alpha_i)^{l_i + c_i^* - 1} \right\} \left\{ \prod_{j=1}^{J} \xi_j^{m_j^* + d_j - 1} (1 - \xi_j)^{l_j^* + d_j^* - 1} \right\}. \tag{23}$$

We would see that the Gibbs sampling procedure based on augmented data and latent variable λ allows us to have much simpler full-conditional posterior distributions.

PROPOSITION 4. *Assume the proper prior* (14) *and improper priors* (18) *and condition* (20). *Let* m_i, l_i, m_j^* *and* l_j^* *as defined in* (8) *to* (11). *The full conditional distributions of* $(\boldsymbol{\alpha}, \boldsymbol{\xi}, \boldsymbol{u}, \boldsymbol{t}, \boldsymbol{g}, N)$ *given data* $(n, \boldsymbol{y}, \boldsymbol{Z}_L, \boldsymbol{Z}_R, \boldsymbol{U}_L, \boldsymbol{U}_R)$ *are as follows.*

(a) *Given* $(\boldsymbol{\xi}, \boldsymbol{u}, \boldsymbol{t}, \boldsymbol{g}, N, \lambda;$ data$)$, $\alpha_1, \ldots, \alpha_{J-1}$ *are independent and*

$$(\alpha_i | \boldsymbol{\xi}, \boldsymbol{u}, \boldsymbol{t}, \boldsymbol{g}, N; \text{data}) \sim Beta(m_i + c_i, l_i + c_i).$$

(b) *Given* $(\boldsymbol{\alpha}, \boldsymbol{u}, \boldsymbol{t}, \boldsymbol{g}, N, \lambda;$ data$)$, $\xi_1, \ldots, \xi_J$ *are independent and*

$$(\xi_j | \boldsymbol{\alpha}, \boldsymbol{u}, \boldsymbol{t}, \boldsymbol{g}, N, \lambda; \text{data}) \sim Beta(m_i + c_i, l_i + c_i).$$

(c) *Given* $(\boldsymbol{\alpha}, \boldsymbol{\xi}, \boldsymbol{g}, N, \lambda;$ data$)$, $(u_1, t_1), \ldots, (u_n, t_n)$ *are independent and*

$$[u_k, t_k | \boldsymbol{\alpha}, \boldsymbol{\xi}, \boldsymbol{g}, N, \lambda; \text{data}] \\ = [t_k | u_k, \boldsymbol{\alpha}, \boldsymbol{\xi}, \boldsymbol{g}, N, \lambda; \text{data}][u_k | \boldsymbol{\alpha}, \boldsymbol{\xi}, \boldsymbol{g}, N, \lambda; \text{data}].$$

Define

$$Q_i = \frac{\alpha_i \prod_{l=1}^{i-1}(1-\alpha_l) \sum_{j=\min(J,\, i+Z_{L_k}-1)}^{\min(J,\, i+Z_{R_k}-1)} \xi_j \prod_{l=1}^{j-1}(1-\xi_l)}{\sum_{h=U_{L_k}}^{U_{R_k}} \{\alpha_h \prod_{l=1}^{h-1}(1-\alpha_l)\} \sum_{m=\min(J,\, i_1+Z_{L_k}-1)}^{\min(J,\, i_1+Z_{R_k}-1)} \{\xi_m \prod_{l=1}^{m-1}(1-\xi_l)\}}.$$

Then

$$P(u_k = i|\boldsymbol{\alpha}, \boldsymbol{\xi}, \boldsymbol{g}, N; \text{data}) = \begin{cases} Q_i, & \text{if } y_k = 0,\ U_{L_k} \leqslant i \leqslant U_{R_k}, \\ \frac{\alpha_i \prod_{l=1}^{i-1}(1-\alpha_l)}{\sum_{h=U_{L_k}}^{U_{R_k}} \alpha_h \prod_{l=1}^{h-1}(1-\alpha_l)}, & \text{if } y_k = 1,\ U_{L_k} \leqslant i \leqslant U_{R_k} \end{cases}$$

and

$$P(t_k = j|u_k, \boldsymbol{\alpha}, \boldsymbol{\xi}, \boldsymbol{g}, N, \lambda; \text{data}) = \begin{cases} \frac{\xi_j \prod_{l=1}^{j-1}(1-\xi_l)}{\sum_{j_1=\min(J,\ u_k+Z_{L_k}-1)}^{\min(J,\ u_k+Z_{R_k}-1)} (\xi_{j_1} \prod_{l=1}^{j_1-1}(1-\xi_l))}, \\ \quad \text{if } y_k = 0,\ \min(J,\ u_k + Z_{L_k} - 1) \\ \quad \leqslant j \leqslant \min(J,\ u_k + Z_{R_k} - 1), \\ 1, \quad \text{if } y_k = 1,\ j = J. \end{cases}$$

(d) *The conditional distribution of* $(\boldsymbol{g}, N, \lambda)$ *given* $(\boldsymbol{\alpha}, \boldsymbol{\xi}, \boldsymbol{u}, \boldsymbol{t}, \lambda; \text{data})$ *is*

$$[\boldsymbol{g}, N|\boldsymbol{\alpha}, \boldsymbol{\xi}, \boldsymbol{u}, \boldsymbol{t}, \lambda; \text{data}] = [\boldsymbol{g}|\boldsymbol{\alpha}, \boldsymbol{\xi}, \boldsymbol{u}, \boldsymbol{t}, N; \text{data}][N|\boldsymbol{\alpha}, \boldsymbol{\xi}, \boldsymbol{u}, \boldsymbol{t}, \lambda; \text{data}],$$

where

$$[\boldsymbol{g}|N, \boldsymbol{\alpha}, \boldsymbol{\xi}, \boldsymbol{u}, \boldsymbol{t}; \text{data}] \propto \prod_{1\leqslant j<i\leqslant J} p_{ij}^{g_{ij}} \Big/ \prod_{1\leqslant j<i\leqslant J} g_{ij}!,$$

for $g_{ij} \geqslant 0$, $\sum_{1\leqslant j<i\leqslant J} g_{ij} = N - n$ *and*

$$[N - n|\boldsymbol{\alpha}, \boldsymbol{\xi}, \boldsymbol{u}, \boldsymbol{t}, \lambda; \text{data}] \sim \textit{Poisson}((1 - A)\lambda).$$

It is equivalent to

$$[N|\boldsymbol{\alpha}, \boldsymbol{\xi}, \boldsymbol{u}, \boldsymbol{t}; \text{data}] \sim \textit{negative binomial } (n, A)$$
$$[\boldsymbol{g}|N, \boldsymbol{\alpha}, \boldsymbol{\xi}, \boldsymbol{u}, \boldsymbol{t}; \text{data}] \sim \textit{multinomial } (N - n,\ \boldsymbol{p}),$$

where

$$A = 1 - \sum_{j=1}^{J-1} \xi_j(1 - \alpha_j)\left\{\prod_{l=1}^{j-1}(1 - \alpha_l)(1 - \xi_l)\right\},$$

$$p_{ij} = \begin{cases} \frac{\alpha_i\xi_j}{1-A} \prod_{l=1}^{i-1}(1 - \alpha_l) \prod_{h=1}^{j-1}(1 - \xi_h), & \text{if } 1 \leqslant j < i \leqslant J - 1, \\ \frac{\xi_j}{1-A} \prod_{l=1}^{J-1}(1 - \alpha_l) \prod_{h=1}^{j-1}(1 - \xi_h), & \text{if } 1 \leqslant j < J - 1. \end{cases}$$

(e) *The conditional distribution of* λ *given* $(\boldsymbol{\alpha}, \boldsymbol{\xi}, \boldsymbol{u}, \boldsymbol{t}, \boldsymbol{g}, N; \text{data})$ *depends only on* N, *and is* $\textit{Gamma}(e_1 + N, e_2 + 1)$.

In contrast to He et al. (2001) model, the full-conditional posterior distributions are all standard ones such as beta, Poisson, multinormal, and discrete distributions. The Gibbs sampling is done easily without using any other special sampling methods such as the adaptive rejection sampling or Metropolis–Hastings sampling.

REMARK 1. In Proposition 4(a)–(d), the conditional distributions of $(\boldsymbol{\alpha}, \boldsymbol{\xi}, \boldsymbol{u}, \boldsymbol{t}, \boldsymbol{g})$ do not depend on λ. If we choose the prior $\pi(N) \propto 1$, there is no need for the latent variable λ. In fact, under the prior $\pi(N) \propto 1$, the posterior density of N given others is proportional to

$$[N|\boldsymbol{\alpha}, \boldsymbol{\xi}, \boldsymbol{u}, \boldsymbol{t}; \text{data}] \propto \frac{N!}{(N-n)!}(1-A)^{N-n}$$
$$\propto \frac{N!}{(N-n)!n!}A^{n+1}(1-A)^{N-n},$$

which is negative binomial $(n+1, A)$. Similarly, if we choose the prior

$$\pi(N) \propto \frac{1}{N},$$

then the posterior $(N|\boldsymbol{\alpha}, \boldsymbol{\xi}, \boldsymbol{u}, \boldsymbol{t}; \text{data})$ follows negative binomial (n, A).

5. Applications

5.1. Bobwhite example

The Missouri Department of Conservation conducted a nest-success study of bobwhite during 1990, 1991, and 1992 in Knox and Macon counties, Missouri. The goal of this study was to estimate the survival rate of bobwhite nests from the time of nest initiation until hatching. Nests in this data set were not aged. Nests were revisited daily in weekdays. There were no visits during weekends. The average number of days of the incubation period is about 24 days. We assume that $J = 24$. There are 157 observed nests with 68 successes and 89 failures. 65 or 41.4% of observed nests are double-interval censored.

We apply the likelihood L_4 in (6) for the example data. The conditional reference prior (14) on $\boldsymbol{\alpha}$ and $\boldsymbol{\xi}$ and the vague prior (18) on N are used. That is, we specify that all hyperparameters in (14) to be 1/2. The computation is done by Gibbs sampling with 10,000 cycles after the 1,000 burn-in cycles. The computation program is written in FORTRAN. It takes about 4 seconds to run 11000 Gibbs cycles on an Intel 3.06 GHz Pedium 4 processor. Define $s_i = 1 - \xi_i$ and $s = \prod_{i=1}^{J} s_i$. Our primary interest is to estimate s_i (age-specific daily survival rate) and s (the total survival rate). The model also produces the estimates of $\boldsymbol{\delta}$ (age-specific conditional encounter probabilities). A list of the estimates of the parameters and posterior standard deviations, along with results from the Mayfield method, is given in Table 1.

The Bayesian estimate of the total survival rate is 35.88% with standard deviation 3.75%. The Mayfield estimate is 39.50%. Note that the Bayesian estimate of s_{24}, the daily survival rate at age 24, is much smaller than the rest of s_i. This indicates that the nest-survival rate drops at the end of the incubation period. The histograms of marginal posterior densities of s and some selected s_i are given in Figure 2. The marginal posterior distributions of s and s_{24} are quite symmetric about the posterior means. The rest of them are not symmetric because the posterior means are large and bounded by 1.

Table 1
Bayesian estimates and Mayfield estimates for bobwhite example

Parameter	Bayes est. (std)	Parameter	Mayfield est.	Bayes est. (std)
δ_1	.41037 (.05757)	s_1	.96204	.98660 (.01677)
δ_2	.12960 (.04073)	s_2	.96204	.97014 (.02823)
δ_3	.09775 (.03591)	s_3	.96204	.97265 (.02772)
δ_4	.04311 (.03105)	s_4	.96204	.97888 (.02418)
δ_5	.05657 (.03083)	s_5	.96204	.94192 (.04074)
δ_6	.04409 (.03126)	s_6	.96204	.97831 (.02572)
δ_7	.06146 (.03187)	s_7	.96204	.98102 (.02265)
δ_8	.03827 (.02729)	s_8	.96204	.95238 (.03837)
δ_9	.00832 (.01095)	s_9	.96204	.93427 (.04978)
δ_{10}	.01858 (.01520)	s_{10}	.96204	.95506 (.04422)
δ_{11}	.03982 (.02408)	s_{11}	.96204	.94980 (.04763)
δ_{12}	.01293 (.01658)	s_{12}	.96204	.93184 (.05798)
δ_{13}	.00806 (.01105)	s_{13}	.96204	.92806 (.05620)
δ_{14}	.00402 (.00603)	s_{14}	.96204	.97732 (.02890)
δ_{15}	.01003 (.00940)	s_{15}	.96204	.96694 (.03412)
δ_{16}	.00271 (.00429)	s_{16}	.96204	.90775 (.05402)
δ_{17}	.00241 (.00383)	s_{17}	.96204	.95016 (.04764)
δ_{18}	.00310 (.00477)	s_{18}	.96204	.97659 (.02916)
δ_{19}	.00679 (.00715)	s_{19}	.96204	.98630 (.01853)
δ_{20}	.00117 (.00264)	s_{20}	.96204	.98894 (.01558)
δ_{21}	.00042 (.00114)	s_{21}	.96204	.95302 (.03680)
δ_{22}	.00022 (.00080)	s_{22}	.96204	.98503 (.01985)
δ_{23}	.00011 (.00046)	s_{23}	.96204	.98064 (.02367)
δ_{24}	.00011 (.00044)	s_{24}	.96204	.88493 (.04219)
$\sum_{i=1}^{J} \delta_i$	1.00000	$\prod_{i=1}^{J} s_i$	.39502	.35883 (.03752)

5.2. *Simulation studies*

Assume that the nest lifetime J is 21 days. The nests are visited daily in weekdays and no visits are made on weekends. The same number of nests are laid on each day of a week.

For the given conditional encounter probabilities $(\delta_1, \ldots, \delta_{21})$ and the daily survival rate $(s_1, \ldots, s_{21})$, we generate u_k and t_k independently until we get n nests with $u_k \leqslant t_k$. If $u_k > t_k$, the nest is discarded.

We choose the hyperparameters in (14) to be $1/2$ for the simulation data. The results are presented in Tables 2 and 3. In these tables, δ_i and s_i are the true conditional encounter probabilities and daily survival rates, respectively. $\hat{\delta}_i$ and $\hat{s}_i$ are the Bayesian estimates of δ_i and s_i, respectively. The *std* in the parentheses are the corresponding standard deviations of the Bayesian estimates. The Mayfield estimates also are presented in these tables for the comparison.

The true conditional encounter probabilities are chosen to be similar to the estimated conditional encounter probabilities in the bobwhite example. Two sets of true daily survival rates are used. These two sets of true daily survival rates have the same pattern that the survival rate increases a little after the first two days and decreases on the last three days. Set II has a greater decrease than Set I on the last three days. Tables 2 and 3 present results for Set I and Set II, respectively. Both sets are based on a sample size of 300.

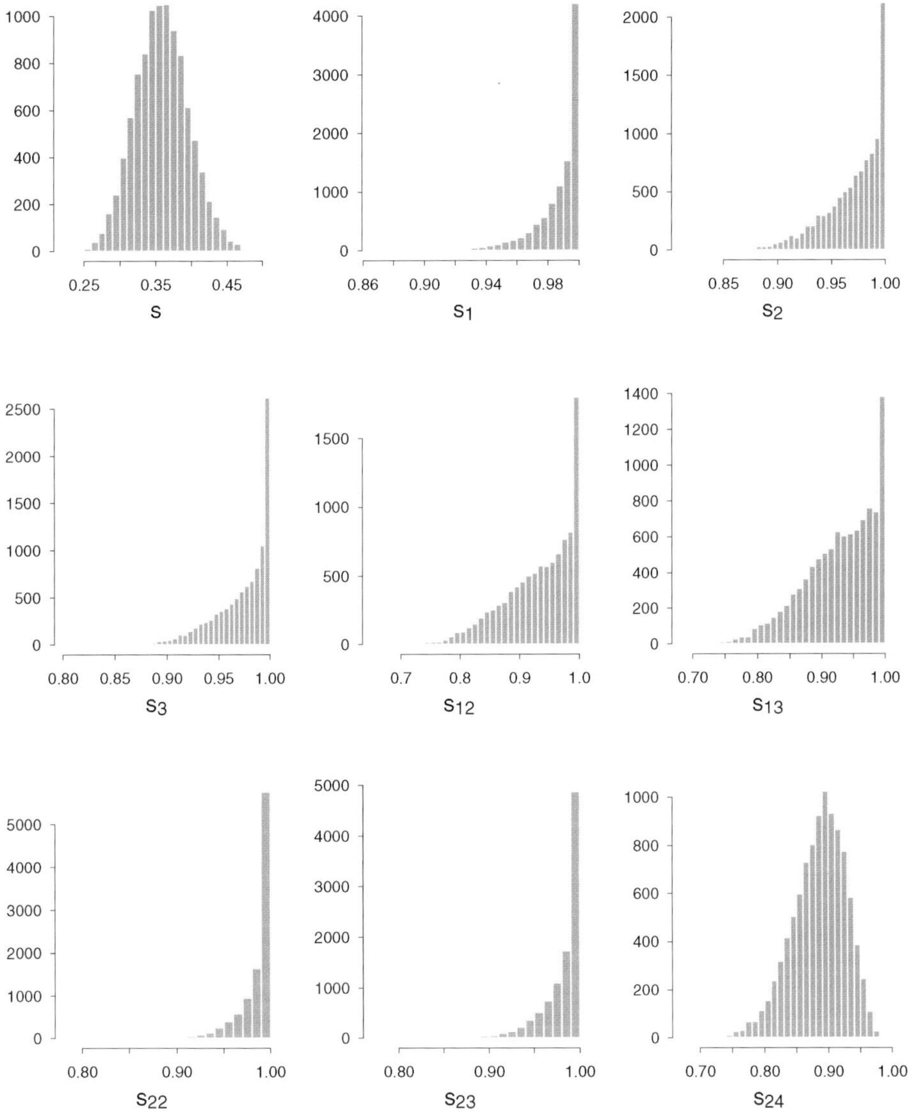

Fig. 2. Marginal posterior densities of the total survival rate s and some daily survival rate s_i for the bobwhite example.

For the given parameter Set I (Table 2), the Bayesian estimate of the total survival rate 49.44% is very close to the true value 47.19% with a small standard deviation of 3.10%. The Mayfield estimate of the total survival rate is 56.08%. The Bayesian estimates of age-specific survival rates also are close to the true values with small standard deviations. The Bayesian estimates capture the increase and decrease changes in the true daily survival rates although the corresponding standard deviations are relatively larger at the change times compared to the others. For the given parameter Set II (Ta-

Table 2
Bayesian and Mayfield estimates for simulation data set I

Age	δ_i	$\hat{\delta}_i$ (std)	s_i	Mayfield	$\hat{s}_i$ (std)
1	.40001	.30339 (.04389)	.95	.97283	.95047 (.03874)
2	.24001	.31127 (.05545)	.95	.97283	.97427 (.02555)
3	.14400	.16525 (.04776)	.99	.97283	.96334 (.02783)
4	.08640	.15163 (.04165)	.99	.97283	.98081 (.01929)
5	.05184	.02590 (.01863)	.99	.97283	.99212 (.00976)
6	.03110	.01764 (.01069)	.99	.97283	.98895 (.01180)
7	.01866	.00255 (.00369)	.99	.97283	.98758 (.01187)
8	.01120	.00364 (.00493)	.99	.97283	.98988 (.01074)
9	.00672	.00476 (.00582)	.99	.97283	.99178 (.00933)
10	.00403	.00729 (.00669)	.99	.97283	.98584 (.01371)
11	.00242	.00262 (.00363)	.99	.97283	.98970 (.01125)
12	.00145	.00108 (.00188)	.99	.97283	.99229 (.00953)
13	.00087	.00074 (.00138)	.99	.97283	.98544 (.01462)
14	.00052	.00055 (.00106)	.99	.97283	.98386 (.01548)
15	.00031	.00128 (.00185)	.99	.97283	.98710 (.01388)
16	.00019	.00021 (.00058)	.99	.97283	.98997 (.01244)
17	.00011	.00011 (.00034)	.99	.97283	.97593 (.02476)
18	.00007	.00005 (.00023)	.99	.97283	.96442 (.03531)
19	.00004	.00003 (.00015)	.85	.97283	.94044 (.05110)
20	.00002	.00001 (.00008)	.85	.97283	.91056 (.06416)
21	.00001	.00001 (.00007)	.85	.97283	.80960 (.04702)
Total	.99998	1.00000	.47191	.56080	.49442 (.03105)

Table 3
Bayesian and Mayfield estimates for the simulation data set II

Age	δ_i	$\hat{\delta}_i$ (std)	s_i	Mayfield	$\hat{s}_i$ (std)
1	.40001	.31515 (.06537)	.95	.95360	.94903 (.04060)
2	.24001	.34380 (.07264)	.95	.95360	.97418 (.02410)
3	.14400	.15564 (.06124)	.99	.95360	.96612 (.02556)
4	.08640	.14719 (.04531)	.99	.95360	.98137 (.01873)
5	.05184	.01063 (.01422)	.99	.95360	.99218 (.00960)
6	.03110	.00588 (.00845)	.99	.95360	.98851 (.01188)
7	.01866	.00349 (.00515)	.99	.95360	.98852 (.01110)
8	.01120	.00299 (.00460)	.99	.95360	.99059 (.01005)
9	.00672	.00449 (.00601)	.99	.95360	.99241 (.00886)
10	.00403	.00299 (.00456)	.99	.95360	.98666 (.01338)
11	.00242	.00297 (.00482)	.99	.95360	.98648 (.01381)
12	.00145	.00116 (.00217)	.99	.95360	.99000 (.01140)
13	.00087	.00082 (.00156)	.99	.95360	.98400 (.01581)
14	.00052	.00067 (.00146)	.99	.95360	.98859 (.01267)
15	.00031	.00164 (.00269)	.99	.95360	.98275 (.01740)
16	.00019	.00024 (.00072)	.99	.95360	.97733 (.02280)
17	.00011	.00012 (.00045)	.99	.95360	.96633 (.03237)
18	.00007	.00006 (.00028)	.99	.95360	.94725 (.04921)
19	.00004	.00003 (.00017)	.65	.95360	.90921 (.07999)
20	.00002	.00002 (.00010)	.65	.95360	.54727 (.08838)
21	.00001	.00002 (.00010)	.65	.95360	.63994 (.09120)
Total	.99998	1.0000	.21104	.36872	.21355 (.02349)

ble 3), the Bayesian estimate of the total survival rate 21.36% also is very close to the true value 21.10% with a small standard deviation of 2.35%. The Mayfield estimate of the total survival rate is 36.87%. The Bayesian estimates of age-specific survival rates have the similar pattern as for Set I.

The Bayesian estimates of the conditional encounter probabilities δ_i (nuisance parameters) are reasonable but not as accurate as those of the daily survival rates. All the true

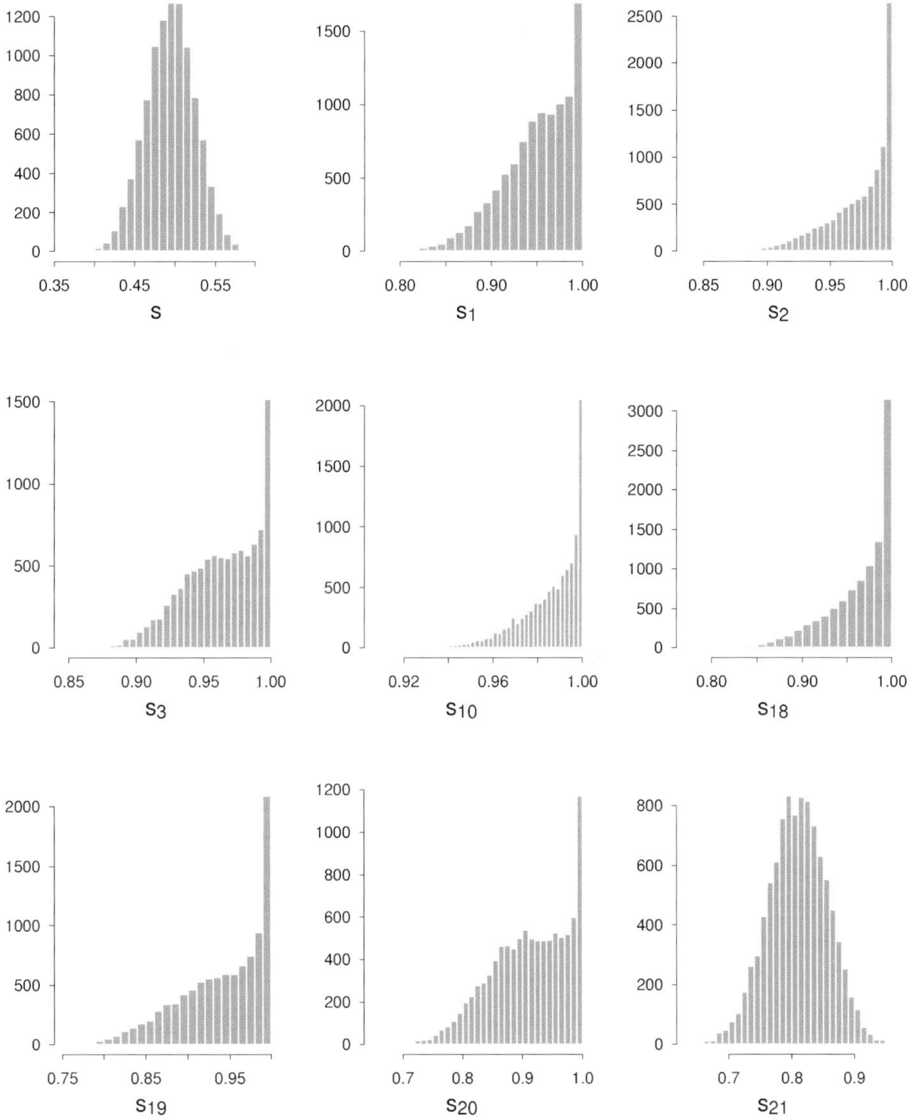

Fig. 3. Marginal posterior densities of the total survival rate s and some daily survival rate s_i for the simulation data set I.

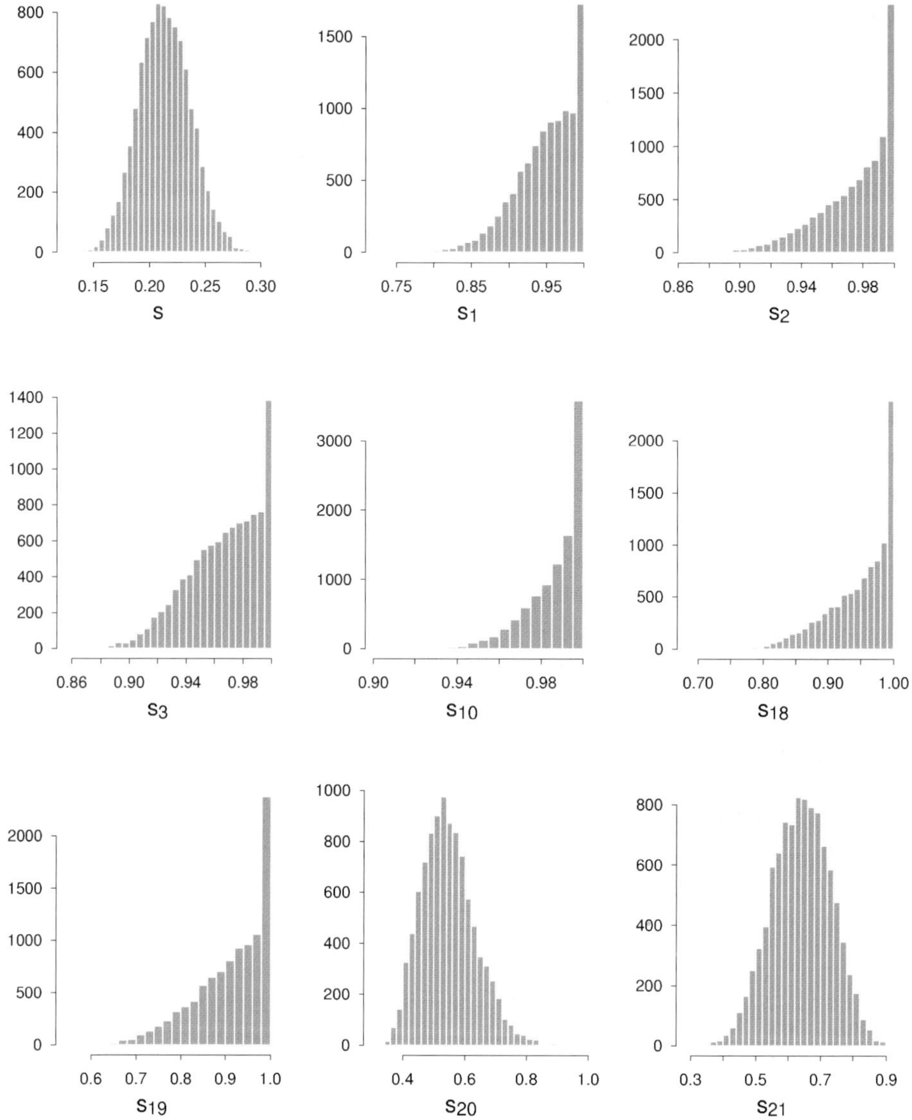

Fig. 4. Marginal posterior densities of the total survival rate s and some daily survival rate s_i for the simulation data set II.

conditional encounter probabilities are within two standard deviations of the Bayesian estimates.

The graphs of marginal posterior histograms of s and some selected s_i are given in Figures 3 and 4. The marginal posterior distributions of s are symmetric about the posterior means, but some posterior densities are quite skewed because they are bounded by 1.

6. Comments

In this chapter, we have presented a Bayesian method for estimating survival rates for discrete data, especially for age-specific nest survival probabilities. A large class of priors, including noninformative priors, is used for the model. If an investigator has some prior information about unknown parameters $\boldsymbol{\alpha}$ and $\boldsymbol{\xi}$, he or she may choose hyperparameters to match the prior information. If the investigator knows nothing about those unknown parameters, he or she may choose noninformative priors such as reference priors or uniform priors. Unlike Pollock and Cornelius (1988) and Heisey and Nordheim (1995), the Bayesian approach needs neither the grouping of data nor the making of structural assumptions about unknown parameters when J is large. The Bayesian method provides a much more accurate estimate of the total survival rate than the standard Mayfield method. This Bayesian method also enables the biologist to look for high and low survival during the whole nesting period.

In practice, it is common for data of several types to be gathered in a single study. That is, some nests may be aged, others are not. Some nests are visited regularly, others are visited irregularly. The Bayesian method accommodates any mix of these samplings.

In this chapter, we utilize the nest-age information in estimating the survival rate. The nest-survival rates might vary across the surrounding habitat as well. Many biologists are collecting habitat data in the nest study. Our future work is to develop Bayesian models including habitat variables as covariates.

Acknowledgements

The authors would like to thank Tom Dailey and Wes Burger for sharing their bobwhite data, and Steven L. Sheriff for helpful comments and discussions.

References

Bart, J., Robson, D.S. (1982). Estimating survivorship when the subjects are visited periodically. *Ecology* **63**, 1078–1090.

Berger, J.O. (1985). *Statistical Decision Theory and Bayesian Analysis*, second ed. Springer, New York.

Berger, J.O., Bernardo, J.M. (1989). Estimating a product of means: Bayesian analysis with reference priors. *J. Amer. Statist. Assoc.* **84**, 200–207.

Berger, J.O., Bernardo, J.M. (1992a). On the development of reference priors (with discussion). In: Bernardo, J.M., Berger, J.O., Dawid, A.P., Smith, A.F.M. (Eds.), *Bayesian Statistics, vol. 4*. Oxford University Press, pp. 35–60.

Berger, J.O., Bernardo, J.M. (1992b). Ordered group reference priors with application to the multinomial problem. *Biometrika* **79**, 25–37.

Bromaghin, J.F. (1991). Sampling and parameter estimation in avian nesting studies. Ph.D. Dissertation, University of Wyoming, Laramie, Wyoming.

Bromaghin, J.F., McDonald, L.L. (1993). Weighted nest survival models. *Biometrics* **49**, 1164–1172.

Cornelius, W.L. (1993). An avian nest survival modeling scheme and comparison of nest survival probability estimation methods. Ph.D. Dissertation, North Carolina State University, Raleigh.

Gelfand, A., Smith, A.F.M. (1990). Sampling based approaches to calculating marginal densities. *J. Amer. Statist. Assoc.* **85**, 398–409.

Geman, S., Geman, A. (1984). Stochastic relaxation, Gibbs distributions and the Bayesian restoration of images. *IEEE Trans. Pattern Analysis and Machine Intelligence* **6**, 721–740.
Gilks, W.R., Wild, P. (1992). Adaptive rejection sampling for Gibbs sampling. *Appl. Statist.* **41** (2), 337–348.
Green, R.H. (1977). Do more birds produce fewer young? A comment on Mayfield's measure of nest success. *Wilson Bulletin* **89**, 173–175.
He, Z. (2003). Bayesian modeling age-specific survival in bird nesting studies under Irregular visits. *Biometrics* **59**, 952–973.
He, Z., Sun, D., Tra, Y. (2001). Bayesian modeling age-specific survival in nesting studies under Dirichlet priors. *Biometrics* **57**, 281–288.
Heisey, D.M., Nordheim, E.V. (1990). Biases in the Pollock and Cornelius method of estimating nest survival. *Biometrics* **46**, 855–862.
Heisey, D.M., Nordheim, E.V. (1995). Modeling age-specific survival in nesting studies, using a general approach for doubly-censored and truncated data. *Biometrics* **51**, 51–60.
Hensler, G.L., Nichols, J.S. (1981). The Mayfield method of estimating nesting success: A model, estimators and simulation results. *Wilson Bulletin* **93**, 42–53.
Jeffreys, H. (1961). *Probability Theory*. Oxford University Press, New York.
Johnson, D.H. (1979). Estimating nest success: The Mayfield method and an alternative. *The Auk* **96**, 651–661.
Kass, R.E., Wasserman, L. (1996). The selection of prior distributions by formal rules. *J. Amer. Statist. Assoc.* **91**, 1343–1370.
Klett, A.T., Johnson, D.H. (1982). Variability in nest survival rates and implications to nesting studies. *The Auk* **99**, 77–87.
Mayfield, H. (1961). Nesting success calculated from exposure. *Wilson Bulletin* **73**, 255–261.
Mayfield, H. (1975). Suggestions for calculating nest successes. *Wilson Bulletin* **87**, 456–466.
Miller, H.W., Johnson, D.H. (1978). Interpreting the results of nesting studies. *J. Wildlife Management* **42**, 471–476.
Nichols, J.D., Percival, H.F., Coon, R.A., Conroy, M.J., Hensler, G.L., Hines, J.E. (1984). Observer visitation frequency and success of mourning dove nests: A field experiment. *The Auk* **101**, 398–402.
Pollock, K.H., Cornelius, W.L. (1988). A distribution-free nest survival model. *Biometrics* **44**, 397–404.
Vounatsou, P., Smith, A.F.M. (1995). Bayesian analysis of ring-recovery data via Markov chain Monte Carlo simulation. *Biometrics* **51**, 687–708.

Handbook of Statistics, Vol. 25
ISSN: 0169-7161
© 2005 Elsevier B.V. All rights reserved.
DOI 10.1016/S0169-7161(05)25033-2

Software Reliability

Lynn Kuo

Abstract
Bayesian inference for several models in software reliability is discussed. These models include the Jelinski and Moranda, Littlewood and Verall, and their related nonhomogeneous Poisson point processes (the Goel and Okumoto, the Musa and Okumoto, the Duane, the Cox and Lewis, and their superposed processes). Model selection based on predictive likelihood is also discussed.

1. Introduction

Software reliability models are used to describe the evolution of the software debugging process and to measure the quality of the software. In software testing, we execute the software with a test suite that resembles the anticipated usage profile. We observe failures (unacceptable performance) caused by bugs (faults) in the program. The bugs are largely due to errors in requirement specification, design, or coding.

Given the growing economic importance of software development, software reliability research has continued to be active and expanding. This chapter concentrates on Bayesian inference for software reliability models.

Various techniques are employed to detect and to correct the bugs in software. The quality of the software is therefore expected to improve. We continue testing until the software reliability reaches a level according to specifications, subject to time or cost constraints. Various stochastic models have been used to monitor software quality changes due to debugging. The objectives of using these models include estimating the number of residual faults, predicting the mean time between failures, assessing the software reliability, and determining the optimal release policy. We roughly classify these models into two types: the *dynamic models* and the *static models*.

The *dynamic models*, also called the *software reliability growth models*, follow the changes of the software throughout the whole testing period. Most of the models employed in software reliability are the dynamic models. Software reliability is defined to be the probability of failure-free operation of a computer program in a specified environment for a specified period of time. Most models assume that when a bug is detected in the software, it is immediately fixed and the time for fixing it is negligible. The dynamic models can be further divided into *time domain models* and *counting process models*. The time domain models are probability models for the sequence of interfailure times that are caused by faults. Many of the time domain models can be related to their dual, the counting process that counts the number of failures found in testing. Nonhomogeneous Poisson point processes (NHPP) have been used extensively for the counting process.

The *static models* usually measure the quality of the software by taking one snapshot. It can be taken at the beginning of the debugging phase to assist managers in planning or at the validation phase that follows the testing and debugging phase. The static models can be further divided into *data domain models* and *structure models*. In the data domain model, the reliability is defined to be the probability that the software will perform according to specifications for a given number of runs. Survey sampling (cf. Bastani and Ramamoorthy, 1988), fault seeding, and capture and recapture methods (cf. Eick et al., 1993) are usually used. In structure models, both regression methods and fault classification membership method (Pasquini et al., 1996) are employed. The regression method uses software metrics (size, complexity, running environment, etc.) as explanatory variables for the software reliability. The studies include Khoshgoftaar and Munson (1990), Akiyama (1971), and a nonlinear regression study by Lipow (1982), a pattern recognition approach by Briand et al. (1992), and tree-based methods by Tian (1995).

Software reliability has been developed by both software engineers and statisticians. It is a natural interface subject between the two groups. The books by Musa et al. (1987), Xie (1991), and Singpurwalla and Wilson (1999) provide a comprehensive account of software reliability. The review articles by Singpurwalla and Wilson (1994), Singpurwalla and Soyer (1996), Chen and Singpurwalla (1997), and Kuo (1999) give an overview and insight into the area.

Section 2 discusses several commonly used models in more details. Section 3 focuses on Bayesian inference aspects. Section 4 discusses the model determination issues. Section 5 briefly discusses the optimal release policies.

2. Dynamic models

In this chapter, we will focus on the dynamic models because they are the core models for software reliability. We classify the dynamic models into two groups: the time domain model and the counting process model. We first discuss them separately. Then we discuss their connections in the model unification subsection.

2.1. Time domain models

Let $\boldsymbol{t} = (t_1, t_2, \ldots)$ denote the set of interfailure times, where t_i is the time executed by the software between the $(i-1)$th failure and the ith failure. Let N denote the initial number of bugs in the software. One type of model, the bug-counting-time-domain model, defines the probability distribution of t_i as a function of $N-i+1$, the number of residual bugs. This type, explicitly depending on N, includes the Jelinski and Moranda

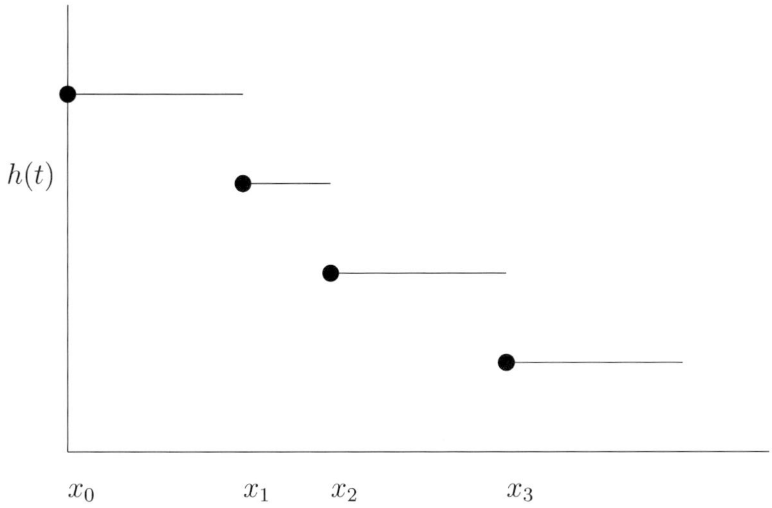

Fig. 1. The concatenated failure rate function of the Jelinski and Moranda model.

(1972), Schick and Wolverton (1973, 1978), and Xie and Bergman (1988) models. The objective of these models is to estimate N, the number of initial faults. However, it can be argued that the number of faults may not be a good measure of reliability because a program with many faults, each with a very low rate of causing failures, may be more reliable than the one with fewer faults but with high occurrence rates of failures. Therefore, we consider a second type of model where we measure the failure intensity directly from the interfailure times or their equivalent waiting times. These models treat N as a nuisance parameter or do not consider N at all. They include the Littlewood and Verall (1973), record value statistics models in Kuo and Yang (1996), and some time series models in Singpurwalla and Soyer (1992). These models are more concerned with predicting the next interfailure time and determining whether or not each debugging action is beneficial. Next we elaborate on these two types.

2.1.1. Models depending on the initial number of bugs

A natural way of defining the joint density of the interfailure times $t_1, t_2, \ldots$ would be defining the following conditional densities $f(t_1)$, $f(t_2|t_1)$, $f(t_3|t_2, t_1)$, etc. However, this is rarely done directly in software reliability contexts. Instead, it is done implicitly by defining the change points of the concatenated constant failure rate. Suppose we are testing until we observe and correct the nth bug. The Jelinski–Moranda (JM) model (Jelinski and Moranda, 1972) assumes the interfailure times $t_1, \ldots, t_n$ are conditionally independent given N and ϕ (a fault detection rate) with t_i exponentially distributed with mean $1/[\phi(N - i + 1)]$. To represent the changes of the constant failure rates, we can use the concatenated hazard function plot as in Figure 1. That is, we plot these hazard rates in serial in the same picture with the hazard rate for T_i only plotted in between the $(i-1)$st and ith epochs of failures for all i. So, the concatenated hazard function for the JM model can be described by $h(t)=\phi(N-i+1)$ for $x_{i-1}\leqslant t<x_i$, where $x_i=\sum_{j=1}^{i}t_j$, the ith waiting time or the epoch of the ith failure, and $x_0=0$. It is a decreasing step function that changes only at the observed epochs of failures.

Goel and Okumoto (1978) propose an imperfect debugging model that is like the JM model except there is a probability of p, $0\leqslant p\leqslant 1$, for fixing a bug when it is encountered. So, the hazard for the ith interfailure time in the model is $h_{T_i}(t)=\phi(N-p(i-1))$ for $x_{i-1}\leqslant t<x_i$.

The JM model assumes that all faults are of equal size. This assumption has been criticized because, in practice, larger faults will likely be detected earlier. To remedy this, Xie and Bergman (1988) generalize the JM model to the following two models using size-biased sampling. The first model, the *power type decreasing failure rate model*, assumes the constant hazard function for T_i is $h_{T_i}(t)=\phi(N-i+1)^{\alpha}$, $x_{i-1}\leqslant t<x_i$, where α is usually assumed to be a constant greater than one to reflect that the decrease in hazard rate, as a function of i. The second model, the *exponential decreasing failure rate model*, assumes $h_{T_i}(t)=\phi[\exp\{-\beta(N-i+1)\}-1]$ for $x_{i-1}\leqslant t<x_i$.

Schick and Wolverton (1973) propose a sawtooth concatenated hazard function that modifies the JM model by assuming $h(t)=\phi(N-i+1)(t-x_{i-1})$, for $x_{i-1}\leqslant t<x_i$. That is, they assume the interfailure times $t_1,\ldots,t_n$ are conditionally independent given N and ϕ with t_i Raleigh distributed with hazard function $h_{T_i}(t)=\phi(N-i+1)t$. Schick and Wolverton (1978) generalize the sawtooth function to a quadratic-component function where $h(t)=\phi(N-i+1)[-a(t-x_{i-1})^2+b(t-x_{i-1})+c]$ for $x_{i-1}\leqslant t<x_i$. Note these concatenated hazard functions are constructed from hazards determined by different random variables $T_1,T_2,\ldots$, as opposed to one random variable. They are useful in describing the evolution of the hazard rates in software testing. Given the software has no aging, the step functions of JM and others reflect this phenomenon. The sawtooth or quadratic function of Schick and Wolverton perhaps can be justified by the subjective opinion that the hazard gets worse as a function of time to reflect anticipated failure; then it immediately drops to 0 after a correction action.

Raftery (1987) considers a general order statistics model (GOS) where the observed epochs of failures $x_1,\ldots,x_n$ are assumed to be the first n order statistics taken from N independent and identically distributed (i.i.d.) observations with density f supported in $\mathbb{R}^+$. Then the JM model is a special case of the GOS with f being an exponential density with mean $1/\phi$. Nayak (1986) uses a multivariate Lomax distribution to model the joint density of the epochs of failures.

All the above models can be considered as death processes where one bug or a fraction of it is removed at each epoch of failure. There are other more sophisticated models that incorporate imperfect repairs. For example, Kremer (1983) considers birth–death processes. Ross (1985a) proposes superposed Poisson processes where the bugs may have different failure detection rates. Fakhre-Zakeri and Slud (1995) extend the Ross model with a mixture distribution with emphasis on imperfect repairs. In addition to imperfect repair, van Pul (1991) models the changing size of the software. He proposes Poisson growth and imperfect repair models where addition to the size of the new software at each repair instance is observed.

2.1.2. Other concatenated failure rate models

Miller (1986) considers order statistics models of independent nonidentically distributed exponential random variables. Littlewood and Verall (1973) study a hierarchical Bayesian model with reliability growth. They assume the interfailure times $t_1, \ldots, t_n$ are conditionally independent with t_i exponentially distributed with mean $1/\lambda_i$, and λ_i are independent with gamma density $\Gamma(\alpha, \beta_0 + \beta_1 i)$ (with mean $\alpha/(\beta_0 + \beta_1 i)$). Therefore, marginally, the t_i are conditionally independent given α and β's, and t_i has the Pareto density $f(t_i|\alpha, \beta_0, \beta_1) = \alpha(\beta_0+\beta_1 i)^\alpha/(t_i+\beta_0+\beta_1 i)^{\alpha+1}$. The concatenated hazard rate function is $h(t) = \alpha/(\beta_0 + \beta_1 i + t - x_{i-1})$ for $x_{i-1} \leqslant t < x_i$. It is a strictly piecewise decreasing function of t with jumps of decrements at x_i. This function reflects that the correction of each bug significantly reduces the failure rate. In between, when no bugs are found, the decreasing hazard rate reflects the belief that the software reliability also gets better. Moranda (1975) proposes the concatenated hazard function to be a step function, decreasing geometrically in i. That is, $h(t) = \alpha\beta^{i-1}$ for $x_{i-1} \leqslant t < x_i$, $\alpha > 0$ and $0 < \beta < 1$. Al-Mutairi et al. (1998) propose $h(t) = 1/\{[(t-x_{i-1})/\alpha]+x_{i-1}/[(i-1)\beta]\}$ for $x_{i-1} \leqslant t < x_i$. The hazard is a piecewise decreasing function and takes a sharp upward jump upon the occurrence of failure. Moreover, it has desirable features for reliability growth: the longer the interfailure time, the smaller the failure rate; and the more frequent the failures, the higher the jumps, and vice versa.

2.1.3. Record value statistics models

In addition to GOS models, Kuo and Yang (1996) argue the need to add record value statistics (RVS) models to describe the failure epochs in software testing, because the RVS are the epochs of failures of the NHPP that is often considered in software reliability. These models treat $x_1, x_2, \ldots$ as the record-breaking statistics of unobserved i.i.d. outcomes from a density f. The density can be exponential, Pareto, Weibull, Gompertz, etc. Let $S_1, S_2, \ldots$ denote the unobserved i.i.d. random variables distributed according to f. We first define the index of record values statistics by $R_1 = 1$, and $R_{k+1} = \min\{i\colon S_i > S_{R_k}\}$, where $k = 1, 2, 3, \ldots$. Then we obtain the record value statistics by setting $X_n = S_{R_n}$ for all n to model the epochs of software failures.

2.1.4. Time series models

Crow and Singpurwalla (1984) consider a Fourier series model for clustered failure data with cyclic behavior. Singpurwalla and Soyer (1992) propose an AR(1) model for the log of the interfailure times. They consider $t_i = \delta_i t_{i-1}^{\theta_i}$, for $i = 1, 2, 3, \ldots$. The coefficients θ_i are unknown and the error terms δ_i are added to account for the slight deviation from the relationship $t_i = t_{i-1}^{\theta_i}$. The coefficient $\theta_i > (<)1$ represents the reliability growth (decay) for the ith period. Chen and Singpurwalla (1994) study a non-Gaussian Kalman filter model that is better suited for the skewed data often encountered in software reliability. They assume $t_i|\theta_i$ has a gamma distribution $\Gamma(\omega_i, \theta_i)$ with ω_i known and θ_i evolving according to beta distributions.

2.2. Counting process models

In addition to modeling the failure time, we can also model its dual, the number of failures $M(t)$ discovered in the interval $(0, t]$. Many authors assume $M(t)$ is a nonhomogeneous Poisson point process (NHPP) with a particular mean function $m(t)$, where $m(t) = EM(t)$, a nondecreasing function of t. NHPP has attractive features that it is closed under superposition and time-transformation. Therefore, it can be useful for software testing with different kinds of failures, with changing software, and for the combination of software and hardware failures. There are two types of NHPP, one type called NHPP-I, where $\lim_{t\to\infty} m(t) < \infty$; the other type called NHPP-II, where $\lim_{t\to\infty} m(t) \to \infty$. NHPP-I includes the Goel–Okumoto process (Goel and Okumoto, 1979) where $m(t) = \theta(1 - \mathrm{e}^{-\beta t})$ and the Yamada et al. (1983) model where $m(t) = \theta[1 - (1 + \beta t)\,\mathrm{e}^{-\beta t}]$. In this kind of models, there is a positive probability that no failures will ever occur even when $t \to \infty$. NHPP-II includes the Musa–Okumoto process (Musa and Okumoto, 1984) with the intensity function $\lambda(t) = \mathrm{d}m(t)/\mathrm{d}t = \alpha/(t + \beta)$, the Duane process (Duane, 1964) with $\lambda(t) = \alpha\beta t^{\alpha-1}$, and the Cox and Lewis process (Cox and Lewis, 1966) with $\lambda(t) = \exp(\alpha + \beta t)$. The NHPP-II process, unlike NHPP-I, allows the possibility of introducing additional errors during debugging. To relax the assumption of monotone intensity functions in all the above processes, Yang and Kuo (1999) extend them to the superposed NHPP where the intensity function can be bathtub shaped, unimodal, polynomial shaped, etc. In addition, Kuo et al. (1997) consider s-shaped software reliability models that were earlier studied by Yamada et al. (1983) and Khoshgoftaar and Woodcock (1991). Lin and Chen (1993) consider multiple versions of the NHPP with linear dependence in a fault-tolerant situation.

2.3. *Model unification*

Due to the prolific number of models proposed in software reliability, some efforts have been made in model unification. Al-Mutairi et al. (1998) unify the concatenated hazard function processes using the shot-noise process. Kuo and Yang (1996) unify the GOS, RVS, and NHPP by considering GOS and RVS as the time domain model for the NHPP. Chen and Singpurwalla (1997) unify essentially all the dynamic models using the self-exciting process. We will now elaborate on these approaches.

All the concatenated hazard rate processes described earlier can be considered as the shot-noise process (Cox and Isham, 1980, p. 135). The shot-noise process consists of two processes, the “shot process” and the “stress process”. The former produces a sequences of shots (failures) possibly according to NHPP. Each shot generates a stress of random magnitude that can remain at a constant, decay over time, or increase over time until the next shot.

For both the GOS and the RVS models, we can relate their corresponding point process $M(t)$ to the NHPP. For the GOS model, one needs the assumption that N has a Poisson distribution with mean θ. Then one can show that $M(t)$ is NHPP-I with $m(t) = \theta F(t)$ where F is the cumulative distribution function with density f that is used to derive a GOS model. In particular, when $F(t) = 1 - \mathrm{e}^{-\beta t}$, which is the JM model, then $M(t)$ is NHPP with the mean function $m(t) = \theta(1 - \mathrm{e}^{-\beta t})$. This is the Goel–Okumoto (Goel and Okumoto, 1979) process as first pointed out in Langberg and Singpurwalla (1985). If $F(t)$ is the cumulative gamma distribution with shape parameter 2 and scale parameter β used in the GOS model, then $M(t)$ is the NHPP Ohba–Yamada model with $m(t) = \theta[1 - (1 + \beta t)\,\mathrm{e}^{-\beta t}]$ studied in Kuo et al. (1997).

Given a RVS model derived from unobserved i.i.d. outcomes from f, one can show the corresponding $M(t)$ is a NHPP-II with mean function $m(t) = -\ln[1 - F(t)]$. Conversely, one can show that failure epochs of a NHPP-II with intensity function $\lambda(t)$ obey the RVS model derived from the i.i.d. unobserved outcomes with the common density $f(t) = \lambda(t)\exp(-\int_0^t \lambda(u)\,du)$. In particular, the set of observed failure epochs of the Musa–Okumoto process is the RVS from a Pareto distribution. The Duane process is the counting process of the RVS from a Weibull distribution. The Cox and Lewis process is the counting process of the RVS from a truncated extreme value (Gompertz) distribution. Moreover, the joint density of the record value statistics can be derived via the following Markov property: $f_{T_{i+1}}(t_{i+1}|x_1, \ldots, x_i) = f_{T_{i+1}}(t_{i+1}|x_i) = \lambda(x_i + t_{i+1})\exp\{m(x_i) - m(x_i + t_{i+1})\}$.

Chen and Singpurwalla (1997) consider the self-exciting point process (SEPP) as in Synder and Miller (1991) for software reliability. They show that self-exciting point processes not only include NHPP but also include the concatenated failure processes and time series processes discussed earlier. A unique feature of the self-exciting process is that the probability of having exactly one event in an infinitesimal interval is proportional to its length and a constant of proportionality that depends on the history (self-exciting). Singpurwalla and Wilson (1999, Chapter 2) discuss the various models and point out the relationships among the processes. That is HPP $\subset$ NHPP $\subset$ DSPP $\subset$ SEPP, where HPP stands for the homogeneous Poisson process and DSPP stands for the doubly stochastic Poisson process.

3. Bayesian inference

3.1. The Jelinski and Moranda model

The Jelinski and Moranda (JM) model can be considered as the earliest and the most fundamental one in software reliability. Jelinski and Moranda (1972) apply the maximum likelihood method. The maximum likelihood estimates for JM may be unstable and unreasonable due to a flat likelihood. Consequently, Bayesian methods with suitably chosen priors have been advocated by many authors. The Bayesian method has the added advantage of allowing software engineers to incorporate information from prior testing. The work includes Bayesian study by Langberg and Singpurwalla (1985), Meinhold and Singpurwalla (1983), and Littlewood and Sofer (1987); hierarchical Bayesian study by Jewell (1985) and Kuo and Yang (1995) and Bayes empirical Bayes by Raftery (1987). Most of these models assume for tractability that N (the initial number of bugs) has a Poisson distribution and ϕ (a fault detection rate) has a gamma distribution as the first stage prior. With the advance of the Markov chain Monte Carlo (MCMC) algorithm, other realistic priors, e.g., priors constructed from expert opinions, can be explored.

We consider two situations: The test is monitored (a) until the nth failure, or (b) until time t. Let $D_{t_n} = \{t_1, t_2, \ldots, t_n\}$ denote the observed successive interfailure times. Observe $t_i = x_i - x_{i-1}$. Define $T = \sum_{i=1}^n x_i + (N - n)x_n$, i.e. the total time-on-test for N items tested in parallel. Observe that $T = \sum_{i=1}^n (N - i + 1)t_i$. Then the likelihood function for the JM model monitored until the nth failure is

$$L(N, \phi; D_{t_n}) = \prod_{i=1}^{n}(N - i + 1)\phi\, \mathrm{e}^{-(N-i+1)\phi t_i}$$
$$= \left\{\prod_{i=1}^{n}(N - i + 1)\right\}\phi^n\, \mathrm{e}^{-\phi T}. \tag{3.1}$$

Observe this is also the likelihood of the first n order statistics of the survival times for N items tested in parallel. This general order statistics model (GOS) allows us to extend the exponential distribution to other distributions, such as the Pareto distribution (the Littlewood model), and the Weibull distribution. Other extensions include the Bayes empirical Bayes approach studied by Raftery (1987).

For the second situation, testing until time t with the observed $D_t = \{t_1, t_2, \ldots, t_n, t\}$, the likelihood is

$$L(N, \phi; D_t) = \left\{\prod_{i=1}^{n}(N - i + 1)\right\}\phi^n\, \mathrm{e}^{-\phi T'}, \tag{3.2}$$

where $T' = \sum_{i=1}^{n} x_i + (N - n)t$. Because both likelihood functions are similar, we will only use the latter one for illustration.

We consider both the simple Bayesian analysis and the hierarchical Bayesian analysis. We use the notation $X \perp Y$ to mean X and Y are independent, and $X \perp Y \perp Z$ to mean X, Y, and Z are jointly independent. In the following description of the Gibbs sampling procedures, we only specify the conditional distributions in the Markov chain Monte Carlo technique. Please refer to Geman and Geman (1984), Tanner and Wong (1987), Gelfand and Smith (1990), Tierney (1994), Casella and George (1992), Chen and Singpurwalla (1994), Kuo and Yang (1995, 1996), and Kuo et al. (1997) for detailed discussions of Gibbs sampling.

3.1.1. *Simple Bayesian approach*

For the simple Bayesian approach, we consider the following priors assuming independent N and ϕ. (1) N has a Poisson distribution $P(\theta)$ and ϕ has a gamma distribution $\Gamma(\gamma, \delta)$ with mean γ/δ. (2) N has a negative binomial distribution $NB(a, p)$ with mean $a(1 - p)/p$ and ϕ has a gamma distribution $\Gamma(\gamma, \delta)$ with mean γ/δ. (3) The prior is similar to (1), except that N has a logarithmic distribution with parameter θ. For the simple Bayesian analysis, all the hyperparameters $\theta, \gamma, \delta, a, p$ are assumed known.

Prior: $N \sim P(\theta)$; $\phi \sim \Gamma(\gamma, \delta)$; $N \perp \phi$.

Let $N' = N - n$. This definition of N' will be used throughout this chapter.

- Gibbs sampling:

$$N'|\phi, D_t \sim P\left(\theta\, \mathrm{e}^{-\phi t}\right) \tag{3.3}$$

and

$$\phi|N, D_t \sim \Gamma\left(\gamma + n, \delta + N't + \sum_{i=1}^{n} x_i\right).$$

The derivation of (3.3) is given in Langberg and Singpurwalla (1985, p. 784). Meinhold and Singpurwalla (1983) provide an exact expression for the posterior joint density of N and ϕ. Their expression includes an infinite series that requires numerical computation. We find it is easier to program the above Gibbs procedure than the one given in Meinhold and Singpurwalla.

Instead of $N \sim P(\theta)$, we can consider $N \sim NB(a, p)$ in the above setup. Then the Gibbs procedures follow similar steps except (3.3) is replaced by

$$N' | \phi, D_t \sim NB\big(a + n, 1 - (1 - p)\,\mathrm{e}^{-\phi t}\big).$$

Similarly, we can consider the logarithmic prior for N, i.e., $f(N|\theta) = \frac{\theta^N}{N(-\ln(1-\theta))}$ for $0 < \theta < 1$. Then the same Gibbs procedures follow except (3.3) is replaced by

$$N' | \phi, D_t \sim NB\big(n, 1 - \theta\,\mathrm{e}^{-\phi t}\big).$$

We consider, in addition, a hierarchical Bayesian approach, where the hyperparameters are unknown with a known distribution (second stage distribution). The hierarchical prior is often used when it is convenient to model priors in stages. It often has advantages in calculations and in specifying a more robust prior than a single stage prior.

3.1.2. Hierarchical Bayesian approach

Now we treat the previous prior models as the first stage priors, where all of the hyperparameters are randomly distributed from a second stage prior distribution.

Model, Prior, and Hyperprior: $N|\theta \sim P(\theta)$; $\theta \sim \Gamma(\alpha, \beta)$; $\phi|\gamma, \delta \sim \Gamma(\gamma, \delta)$; $\delta \sim \Gamma(\zeta, \eta)$; $\gamma \sim \pi$ and $\theta \perp \delta \perp \gamma$. The hyperparameters α, β, ζ, and η are known. The prior π on the shape parameter γ can be quite arbitrary, for example $U(a, b)$ or $\Gamma(a, b)$, since the Metropolis algorithm will be employed to generate γ.

- Gibbs sampling:

$$\begin{aligned}
&N' | \theta, \phi, \delta, \gamma, D_t \sim P\big(\theta\,\mathrm{e}^{-\phi t}\big),\\
&\theta | N, \phi, \delta, \gamma, D_t \sim \Gamma(\alpha + N, \beta + 1),\\
&\phi | N, \theta, \delta, \gamma, D_t \sim \Gamma\left(\gamma + n, \delta + N't + \sum_{i=1}^{n} x_i\right),\\
&\delta | N, \theta, \phi, \gamma, D_t \sim \Gamma(\zeta + \gamma, \eta + \phi),
\end{aligned}$$

and $\gamma | N, \theta, \phi, \delta, D_t$ is generated by the Metropolis algorithm, where

$$f(\gamma) \propto \frac{\pi(\gamma)\delta^{\gamma}\phi^{\gamma-1}}{\Gamma(\gamma)}.$$

Instead of the above model, we can also consider reduced models as in Kuo and Yang (1995), where some (not all) of the hyperparameters are assumed to have a second stage prior distribution.

We can also derive the Gibbs algorithm for the hierarchical Bayesian case with the negative binomial distribution $N|a, p \sim NB(a, p)$ by placing an arbitrary prior on a and a beta prior on p. The details are given in Kuo and Yang (1995).

3.1.3. Bayesian inference and prediction for the JM model

In this section, we describe the Bayes estimators. In addition to making inference on the unknown parameters, we are also interested in predicting the future interfailure time, its conditional predictive density, and the future reliability function.

For inference on the parameters, such as N, ϕ, θ, δ, and γ, we can use the empirical measure of the Gibbs samplers or the mixture of conditional densities of the parameter of interest given the rest of the parameters (the Rao–Blackwell version). The highest posterior density credible sets can also be obtained using the empirical measure of the Gibbs samplers.

For prediction, it would be easier to consider the situation of testing until the nth failure. All the posterior distributions described in earlier subsections need to be modified by replacing t with x_n.

For the JM model, $E(T_{n+1}|D_{t_n})$ is undefined, since there is a nonzero probability that the last fix removes the last error. Inference for the future survival function evaluated at x distance away from x_n is

$$\begin{aligned} E\big(S(x)|D_{t_n}\big) &= E\big[E\big(P(T_{n+1} > x)|N, \phi, D_{t_n}\big)|D_{t_n}\big] \\ &= E\big[\mathrm{e}^{-(N-n)\phi x}|D_{t_n}\big] \\ &= \iint \mathrm{e}^{-N'\phi x} p\big(N', \phi|D_{t_n}\big)\, \mathrm{d}N'\, \mathrm{d}\phi. \end{aligned} \tag{3.4}$$

Recall $N' = N - n$. The measure $\mathrm{d}N'$ refers to the counting measure on the nonnegative integers.

The prequential conditional predictive ordinate (PCPO) for the future interfailure time t_{n+1} is defined to be $c_{n+1} = p(t_{n+1}|D_{t_n})$, the conditional density of T_{n+1} evaluated at the future observed time given $(t_1, \ldots, t_n)$. The PCPO is particularly relevant for model selection, because it assesses the predictability of T_{n+1} given the past data. The PCPO c_{n+1} can be computed by

$$\begin{aligned} p(t_{n+1}|D_{t_n}) &= \iint p(t_{n+1}|N, \phi, D_{t_n}) p(N, \phi|D_{t_n})\, \mathrm{d}N\, \mathrm{d}\phi \\ &= \iint (N-n)\phi\, \mathrm{e}^{-(N-n)\phi t_{n+1}} p(N, \phi|D_{t_n})\, \mathrm{d}N\, \mathrm{d}\phi. \end{aligned} \tag{3.5}$$

Both (3.4) and (3.5) can be evaluated from the Gibbs samplers or the Rao–Blackwell versions. For example, we can estimate c_{n+1} by direct substitution from the Gibbs sampler:

$$\hat{p}(t_{n+1}|D_{t_n}) = \frac{1}{r} \sum_{s=1}^{r} N'^{(k,s)} \phi^{(k,s)}\, \mathrm{e}^{-N'^{(k,s)} \phi^{(k,s)} t_{n+1}}, \tag{3.6}$$

or applying the Gelman and Rubin method in a similar fashion, where $N'^{(k,s)}$ and $\phi^{(k,s)}$ denote the values of the parameters N' and ϕ in the kth iteration and the sth replication of the Gibbs sampler. Two other versions of the estimates of c_{n+1} can be obtained by means of the conditioning distribution of N'. They are given in Kuo and Yang (1995).

Now switch back to the second testing situation, testing until time t. Suppose we are interested in the predictive density $p(t_{n+1}|D_t)$. Then we can easily modify (3.5) to obtain

$$p(t_{n+1}|D_t) = \iint p(t_{n+1}|N, \phi, D_t)p(N, \phi|D_t)\,\mathrm{d}N\,\mathrm{d}\phi$$
$$= \iint (N-n)\phi\,\mathrm{e}^{-(N-n)\phi(t_{n+1}\ \ t\,|\,x_n)}p(N, \phi|D_t)\,\mathrm{d}N\,\mathrm{d}\phi,$$

which can be estimated similarly from the Gibbs sampler.

3.2. *The Littlewood and Verall model*

This model was proposed by Littlewood and Verall (1973) to relax the assumption of perfect repair in the JM model. It assumes that the interfailure times $t_1, t_2, \ldots, t_n$ are conditional independent given the parameter and t_i is exponentially distributed with mean $1/\lambda_i$. Let $\boldsymbol{\lambda}$ denote $(\lambda_1, \ldots, \lambda_n)$. Then the likelihood function is

$$L(\boldsymbol{\lambda}, D_{t_n}) = \prod_{i=1}^{n} \lambda_i\,\mathrm{e}^{-\lambda_i t_i}. \tag{3.7}$$

Littlewood and Verrall describe a hierarchical Bayesian approach to this problem. They consider the first stage prior on $\boldsymbol{\lambda}$ with $\lambda_i \sim \Gamma(a, \psi(i))$, for $i = 1, \ldots, n$, and given the hyperparameters a and $\{\psi(i)\}_{i=1}^{n}$, the λ_i's are independent. Furthermore, they assume that $\psi(i)$ is an increasing function of i to reflect that repairs tend to make the software more reliable. We consider $\psi(i)$ to be the polynomial $\psi(i) = \beta_0 + \beta_1 i + \cdots + \beta_k i^k$, where $\beta_j \geqslant 0$ for all j, $j = 0, \ldots, k$. For the second stage prior, we assume that a has a prior distribution π and β_j has a distribution $\Gamma(\alpha_j, \gamma_j)$, where a, β_j, $j = 1, \ldots, k$, are independent. The assumption that all the coefficients β_j are nonnegative is more restrictive than the increasing function constraint on $\psi(i)$. Nevertheless, it still allows a wide class of priors. Moreover, it simplifies the specification of the second stage prior, for example, permitting gamma distributions. It is worthwhile noting that Littlewood and Verall consider (3.7) together with the above first stage prior as the LV model and consider the above second stage prior as the prior.

Figure 2 is a graphical representation of the model for t_i. Similar representation can be obtained for other $t's$. In this figure, we use directed graphs as discussed in Whittaker (1990) and Gilks et al. (1994) to describe the model. The notation $X \to Y$ means that the distribution of Y is generated from the X variable, i.e., it specifies the conditional distribution of Y given X. The directed graphs clarify the dependence as well as the conditional independence relationships in the model and assist us in specifying the Gibbs sampler. The square or rectangular nodes indicate constants. The circles indicate the unknown parameters or the data.

3.2.1. *Gibbs sampler for the LV model*

Let $\boldsymbol{\beta} = (\beta_0, \ldots, \beta_k)$. The construction of the Gibbs sampler consists of specifying $p(\boldsymbol{\lambda}|\boldsymbol{\beta}, a, D_{t_n})$ and $p(\boldsymbol{\beta}, a|\boldsymbol{\lambda}, D_{t_n})$. It is straightforward to specify the former. It is more complicated to specify the latter. First, it is difficult to specify the posterior distribution of a given $\boldsymbol{\beta}, \boldsymbol{\lambda}, D_{t_n}$. We use the Metropolis law to sample this variate a. Moreover, $p(\boldsymbol{\beta}|a, \boldsymbol{\lambda}, D_{t_n})$ is a very complicated mixture of products of gamma densities. However,

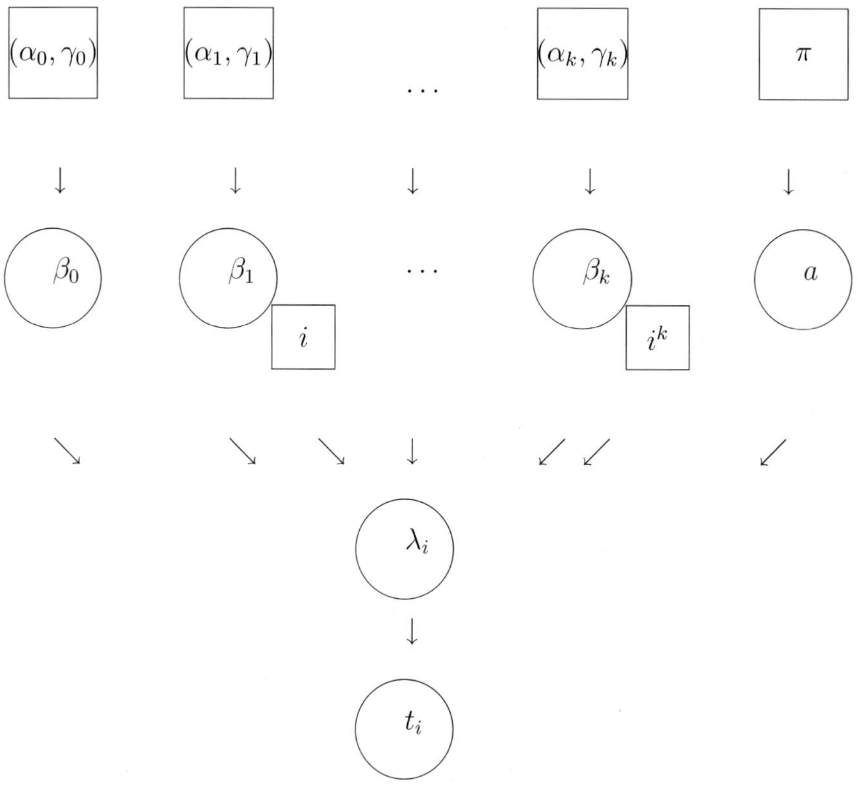

Fig. 2. Directed graph of the LV model for t_i.

this can be circumvented by introducing the latent variables $z_i = (z_{i0}, \ldots, z_{ik})$. Let $r_{ij} = \beta_j i^j / (\beta_0 + \beta_1 i + \cdots + \beta_k i^k)$. Consider z_i to have a multinomial distribution *MN* with parameters a and cell probabilities $(r_{i0}, \ldots, r_{ik})$. Let z denote $z_1, \ldots, z_n$. Then, the posterior distribution of $\boldsymbol{\beta}$ given $a, \boldsymbol{\lambda}, D_{t_n}$ can be constructed by sampling $\boldsymbol{\beta}$ from $p(\boldsymbol{\beta}|a, z, \boldsymbol{\lambda}, D_{t_n})$ and sampling z from $p(z|\boldsymbol{\beta}, a, \boldsymbol{\lambda}, D_{t_n})$ iteratively. Both conditional distributions can be specified as in the following. We use " " in the following to surround the variable if the conditional density does not depend on that variable. Note that we need to restrict our prior on a to be a discrete measure with support on the positive integers, because a is also used as the sample size in our multinomial variates.

- Gibbs sampling
  generate $\lambda_i,\ i = 1, \ldots, n$, independently, where

$$\lambda_i | \boldsymbol{\beta}, a, D_{t_n} \sim \Gamma\left(a + 1, t_i + \sum_{j=0}^{k} \beta_j i^j\right);$$

  generate $z_i,\ i = 1, \ldots, n$, independently, where

$$z_i | \boldsymbol{\beta}, a, \text{“}\boldsymbol{\lambda}\text{”}, \text{“}D_{t_n}\text{”} \sim MN(a, r_{i0}, \ldots, r_{ik});$$

generate $\beta_j,\ j = 0, \ldots, k$, independently, where

$$\beta_j | a, \boldsymbol{\lambda}, z, \text{“}D_{t_n}\text{”} \sim \Gamma\left(\alpha_j + \sum_{i=1}^{n} z_{ij}, \gamma_j + \sum_{i=1}^{n} i^j \lambda_i\right); \tag{3.8}$$

and generate $a|\boldsymbol{\beta}, z, \boldsymbol{\lambda}, \text{“}D_{t_n}\text{”}$ from the Metropolis' algorithm, where

$$f(a) \propto \pi(a) \prod_{i=1}^{n} \frac{a\lambda_i^{a-1}(\beta_k i^k)^{a-\sum_{j=0}^{k-1} z_{ij}}}{\left(a - \sum_{j=0}^{k-1} z_{ij}\right)!}. \tag{3.9}$$

Alternatively, the last variate a can also be generated simply from $p(a|\boldsymbol{\beta}, \boldsymbol{\lambda}, \text{“}D_{t_n}\text{”})$, where

$$f(a) \propto \frac{\pi(a)}{\Gamma^n(a)} \prod_{i=1}^{n} \lambda_i^a \left(\beta_0 + \cdots + \beta_k i^k\right)^a.$$

A derivation of these conditional densities is given in Kuo and Yang (1995).

3.2.2. Bayesian inference and prediction for the LV models

For the LV model, the inference for the future interfailure time T_{n+1} is computed by

$$E(T_{n+1}|D_{t_n}) = E\left(\frac{\beta_0 + \cdots + \beta_k (n+1)^k}{a-1}\middle| D_{t_n}\right), \tag{3.10}$$

which is obtained in two steps, first conditioning on λ_{n+1}, then conditioning on a and $\psi(n+1)$. Similarly, the conditional variance of the future interfailure time can be computed by

$$\begin{aligned} &V(T_{n+1}|D_{t_n}) \\ &\quad = E\left(T_{n+1}^2|D_{t_n}\right) - E^2(T_{n+1}|D_{t_n}) \\ &\quad = E\left(\frac{2(\beta_0 + \cdots + \beta_k (n+1)^k)^2}{(a-1)(a-2)}\middle| D_{t_n}\right) - E^2(T_{n+1}|D_{t_n}). \end{aligned} \tag{3.11}$$

Applying the same conditioning arguments as in (3.10) to the future survival function, we can show

$$E\left(S(x)|D_{t_n}\right) = E\left(\frac{1}{\left(1 + \frac{x}{\beta_0 + \cdots + \beta_k (n+1)^k}\right)^a}\middle| D_{t_n}\right), \tag{3.12}$$

where x is measured from x_n.

The PCPO c_{n+1} for the LV model can be computed by

$$\begin{aligned} &p(t_{n+1}|D_{t_n}) \\ &\quad = \int p(t_{n+1}|a, \boldsymbol{\beta}, \boldsymbol{\lambda}, \lambda_{n+1}) p(a, \boldsymbol{\beta}, \boldsymbol{\lambda}, \lambda_{n+1}|D_{t_n})\, \mathrm{d}a\, \mathrm{d}\boldsymbol{\beta}\, \mathrm{d}\boldsymbol{\lambda}\, \mathrm{d}\lambda_{n+1} \\ &\quad = \int \lambda_{n+1} \mathrm{e}^{-\lambda_{n+1} t_{n+1}} p(a, \boldsymbol{\beta}, \boldsymbol{\lambda}, \lambda_{n+1}|D_{t_n})\, \mathrm{d}a\, \mathrm{d}\boldsymbol{\beta}\, \mathrm{d}\boldsymbol{\lambda}\, \mathrm{d}\lambda_{n+1} \end{aligned}$$

$$= \int \frac{a(\beta_0 + \cdots + \beta_k (n+1)^k)^a}{(t_{n+1} + \beta_0 + \cdots + \beta_k (n+1)^k)^{a+1}} p(a, \boldsymbol{\beta}, \boldsymbol{\lambda} | D_{t_n}) \, da \, d\boldsymbol{\beta} \, d\boldsymbol{\lambda}. \tag{3.13}$$

Recall $\boldsymbol{\lambda} = (\lambda_1, \ldots, \lambda_n)$. The last equality of (3.13) is obtained by using

$$p(a, \boldsymbol{\beta}, \boldsymbol{\lambda}, \lambda_{n+1} | D_{t_n}) = p(\lambda_{n+1} | a, \boldsymbol{\beta}, \boldsymbol{\lambda}, D_{t_n}) p(a, \boldsymbol{\beta}, \boldsymbol{\lambda} | D_{t_n}), \tag{3.14}$$

$\lambda_{n+1} | a, \boldsymbol{\beta}, \boldsymbol{\lambda}, D_{t_n} \sim \Gamma(a, \psi(n+1))$, and by integrating out λ_{n+1}.

Eqs. (3.10)–(3.13) can be computed from the Gibbs sampler by methods similar to (3.6).

3.3. NHPP models

There are numerous papers on the inference for NHPP. Goel and Okumoto (1979), Musa and Okumoto (1984), Cox and Lewis (1966), and Xie and Hong (2001) derive the maximum likelihood estimators. Kyparisis and Singpurwalla (1985), Guida et al. (1989), and Bar-Lev et al. (1992) study Bayesian inference for the Weibull process. Campodónico and Singpurwalla (1995) study Bayesian inference for the Musa–Okumoto process using expert opinions. MCMC methods for inference and model selection for NHPP and superposed NHPP are given in Kuo and Yang (1996), Achcar et al. (1998), Kuo et al. (1997), Yang and Kuo (1999), and Achcar (2001).

The likelihood function of a NHPP is given in this section. Moreover, we clarify the relationships between the GOS and the NHPP-I and between the RVS and the NHPP-II.

Both the mean function $m(t)$ and the intensity function $\lambda(t)$ are indexed by the unknown parameters θ and $\boldsymbol{\beta}$, where θ is a nonnegative real number and $\boldsymbol{\beta}$ can be a vector of unknown parameters. For simplicity, we write $m(t)$ for $m(t|\theta, \boldsymbol{\beta})$ and $\lambda(t)$ for $\lambda(t|\theta, \boldsymbol{\beta})$.

Recall $D_t = \{n, t_1, t_2, \ldots, t_n; t\}$. It is also equivalent to $\{n, x_1, x_2, \ldots, x_n; t\}$. The likelihood for the time truncated model is

$$L_{\text{NHPP}}(\theta, \boldsymbol{\beta} | D_t) = \left(\prod_{i=1}^{n} \lambda(x_i) \right) \exp\bigl(-m(t)\bigr). \tag{3.15}$$

It is developed in many textbooks, for example, Cox and Lewis (1966), Basawa and Prakasa Rao (1980), and Crowder et al. (1991).

For the failure truncated model, a similar expression to (3.15) can be applied with D_t replaced by D_{X_n} and t replaced by x_n. Note $D_{X_n} = \{t_1, \ldots, t_n\}$. It is also equivalent to $\{x_1, x_2, \ldots, x_n\}$.

In the following, we consider two types of NHPP processes according to the limiting behavior of $m(t)$: (1) NHPP-I, if $\lim_{t\to\infty} m(t) < \infty$; (2) NHPP-II, if $m(t) \to \infty$ as $t \to \infty$.

3.3.1. General order statistics and NHPP-I

We assume $\lim_{t\to\infty} m(t) = \theta < \infty$ for the NHPP-I. We will show the likelihood used in the GOS model given in Raftery (1987) and Joe (1989) is the conditional likelihood of θ, $\boldsymbol{\beta}$ given N in the NHPP-I. Moreover, the likelihood in (3.15) is the marginal likelihood with N 'integrated out' by assuming N has a Poisson distribution.

Let us first consider the GOS model. The GOS model assumes that the epochs of failures are the first n order statistics taken from N i.i.d. observations distributed according to a density of $f(\cdot|\boldsymbol{\beta})$ with distribution function $F(\cdot|\boldsymbol{\beta})$. Therefore, the likelihood function for the failure truncated model is

$$L_{\text{GOS}}(N, \boldsymbol{\beta}|D_{x_n}) = \frac{N!}{(N-n)!}\left(\prod_{i=1}^{n} f(x_i|\boldsymbol{\beta})\right)\left(1 - F(x_n|\boldsymbol{\beta})\right)^{N-n}. \tag{3.16}$$

The likelihood $L_{\text{GOS}}(N, \boldsymbol{\beta}|D_t)$ for the time truncated model is similar to (3.16) except the x_n argument in F is replaced by t. Suppose the failure epochs follow the GOS model and N has a Poisson distribution with mean θ; then one can show by a straightforward argument that $M(t)$ is a NHPP-I with the mean value function $m(t) = \theta F(t|\boldsymbol{\beta})$ (cf. Musa et al., 1987, p. 269 or Kuo and Yang, 1996).

Now, let us consider the conditional likelihood of $\theta, \boldsymbol{\beta}$ given N, D_t for the time truncated model in NHPP-I with $\lim_{t\to\infty} m(t) = \theta < \infty$:

$$L_{\text{NHPP-I}}(\theta, \boldsymbol{\beta}|N, D_t) = \frac{N!}{(N-n)!}\left(\prod_{i=1}^{n} \frac{\lambda(x_i)}{\theta}\right)\left(1 - \frac{m(t)}{\theta}\right)^{N-n}. \tag{3.17}$$

Let $p(N|\theta)$ denote the Poisson probability mass function with mean θ. It is straightforward to verify

$$\sum_{N=n}^{\infty} L_{\text{NHPP-I}}(\theta, \boldsymbol{\beta}|N, D_t) \times p(N|\theta) = L_{\text{NHPP}}(\theta, \boldsymbol{\beta}|D_t). \tag{3.18}$$

Hence, the likelihood in (3.15) is the marginal likelihood when N has a Poisson distribution.

Letting $m(t) = \theta F(t)$ (consequently, $\lambda(t) = \theta f(t)$), we see that $L_{\text{GOS}}(N, \boldsymbol{\beta}|D_t)$ and the conditional likelihood in (3.17) are identical. Similar results apply to the failure truncated testing.

To summarize, suppose the failure epochs follow the GOS model in (3.16) and N has a Poisson distribution with mean θ; then the number of failures $M(t)$ in $(0, t]$ is a NHPP-I with the mean value function $m(t) = \theta F(t|\boldsymbol{\beta})$. Conversely, we can derive the GOS model from NHPP-I by letting $m(t) = \theta F(t)$ and assuming a total number N of faults at the beginning of testing (cf. Parzen, 1962, p. 143).

The Jelinski and Moranda model is a special case of the GOS model with $f(t) = \beta\exp(-\beta t)$, and the corresponding point process which counts the number of failures in $(0, t]$ is a NHPP-I with mean $\theta[1 - \exp(-\beta t)]$. This NHPP is proposed by Goel and Okumoto (1979) to model software reliability. Similarly, we can obtain the Pareto GOS/NHPP-I, the Weibull GOS/NHPP-I, and the truncated extreme value GOS/NHPP-I. We list the densities in Table 1 and the mean functions for the corresponding NHPP-I, where $t > 0$, $\beta > 0$, and $\alpha > 0$ except in the extreme value distribution, where $-\infty < \alpha < \infty$.

We can also consider many other distributions, such as the gamma, Rayleigh, truncated exponential, and lognormal distributions. Note that the Yamada et al. (1983) S-shaped software reliability models are contained in the NHPP-I with f as a gamma density with shape parameter equal to 2.

Table 1
The mean functions of the NHPP-I and NHPP-II processes

Density	GOS/NHPP-I	RVS/NHPP-II
Exponential: $\beta\exp(-\beta t)$	$\theta[1-\exp(-\beta t)]$	βt
Pareto: $\alpha\beta^\alpha/(\beta+t)^{\alpha+1}$	$\theta[1-\beta^\alpha/(t+\beta)^\alpha]$	$\alpha\ln(1+t/\beta)$
Weibull: $\beta\alpha t^{\alpha-1}\exp(-\beta t^\alpha)$	$\theta[1-\exp(-\beta t^\alpha)]$	βt^α
Extreme-value: $\exp(\alpha+\beta t-\frac{e^{\alpha+\beta t}-e^\alpha}{\beta})$	$\theta[1-\exp(-\frac{e^{\alpha+\beta t}-e^\alpha}{\beta})]$	$(e^{\alpha+\beta t}-e^\alpha)/\beta$

The second column lists the mean functions of the exponential NHPP-I (Goel–Okumoto), the Pareto NHPP-I, the Weibull NHPP-I, and the extreme value NHPP-I. The third column lists the mean functions of the homogeneous Poisson process, the Pareto NHPP-II (Musa–Okumoto), the Weibull NHPP-II (Duane) and the extreme value NHPP-II (Cox–Lewis).
Prepared for *Handbook of Statistics*, vol. 25.

Observe we have that $m(t) < \infty$ as $t \to \infty$ in this section. Therefore, GOS models cannot be applied to situations where new faults might be introduced in the process of debugging. Let X_1 denote the time of first failure. Then $p(X_1 > t) = p(M(t) = 0) = \exp[-\theta F(t)]$. Observe this probability does not approach 0 as $t \to \infty$ unless $\theta \to \infty$. Therefore, in NHPP-I, the distribution of X_1 is defective and there is a positive probability that we may never observe any failures in the debugging process. To extend the NHPP-I to the NHPP-II, where $m(t) \to \infty$ as $t \to \infty$, we need to replace the GOS model with a record value statistics model (RVS).

3.3.2. Record value statistics and NHPP-II

In the RVS model, we assume the epochs of failures are the record values of i.i.d. random variables distributed according to a density $f(S|\boldsymbol{\beta})$ with distribution function $F(S|\boldsymbol{\beta})$. Let $S_1, S_2, \ldots$ denote the i.i.d. random variables distributed according to $f(S|\boldsymbol{\beta})$. We define the sequence of record values, $\{X_n\}_{n\geqslant 1}$ as given in Subsection 2.1.3. We model the observed epochs of failures $x_1 < x_2 < \cdots$ as the record values $X_1, X_2, \ldots$.

Glick (1978) shows that this sequence of record values can be infinite. Therefore, we can model failure times when the number of failures is unbounded as $t \to \infty$.

To relate the RVS model to the NHPP-II, we quote a result from Dwass (1964) or Resnick (1987, p. 165): "Suppose F is continuous with full support on R^+. Then the record values in $(0, t]$ constructed above from F are the points of a NHPP-II in $(0, t]$ with the mean measure $m(t) = -\ln[1 - F(t)]$." Consequently, the intensity function of the NHPP-II is $\lambda(t) = m'(t) = f(t)/[1 - F(t)]$, which is also the hazard function of F.

Now let us consider the various f listed in Table 1. It is straightforward to verify that $f(t)/[1-F(t)]$ is β, $\alpha/(\beta+t)$, $\beta\alpha t^{\alpha-1}$, and $\exp(\alpha+\beta t)$, respectively. Therefore, the point processes associated with the RVS models constructed from the exponential, Pareto, Weibull, and truncated extreme value densities are exactly the homogeneous Poisson process, the Musa–Okumoto, the Weibull (Duane), and the Cox and Lewis processes, where the latter three are the nonhomogeneous Poisson processes. Table 1 summarizes these results where the mean value functions for the corresponding Poisson point processes are listed.

The intensity function of a Musa–Okumoto process is a decreasing function of t. Therefore, the process can be used to model a "happy" system. For the Weibull process, we can choose $0 < \alpha < 1$ to model the "happy" system and $\alpha > 1$ to model the "sad" system. Note that the scale parameter β in the Cox and Lewis model must be positive to have a valid density. This model is used for "sad" processes. "Sad" processes are appropriate for a repairable system with extremely rapid deterioration. For software reliability, these processes may be appropriate when it is difficult to find the faults at the beginning; then after a burn-in period, the faults are found rapidly. Bain and Engelhardt (1986) also consider NHPP with log-linear intensity function, which is a special case of the Cox and Lewis process.

Now let us consider the likelihood function. Plugging in $\lambda(t) = f(t)/[1 - F(t)]$ in (3.15), the likelihood of NHPP-II is

$$L_{\text{NHPP-II}}(\boldsymbol{\beta} \mid D_t) = \left(\prod_{i=1}^{n} \frac{f(x_i \mid \boldsymbol{\beta})}{1 - F(x_i \mid \boldsymbol{\beta})} \right) \big(1 - F(t \mid \boldsymbol{\beta})\big). \tag{3.19}$$

For the failure truncated model, we obtain the likelihood $L_{\text{NHPP-II}}(\boldsymbol{\beta} \mid D_{x_n})$, similar to (3.19) with t replaced by x_n.

Next, we divide our discussion on Bayesian inference into two subsections, one for the NHPP-I, the other for the NHPP-II. For the NHPP-I, data augmentation with the unknown remaining number of faults helps us specify the conditional distributions in the Gibbs sampler. For the NHPP-II, no data augmentation is proposed. The Metropolis algorithm is often used in the Gibbs algorithm.

3.3.3. Gibbs sampling for the general order statistics model

The expression $\exp\{-\theta F(t \mid \boldsymbol{\beta})\}$ for the NHPP-I in the likelihood (3.15) usually prevents us from specifying a convenient form for the conditional density of $\boldsymbol{\beta}$ given θ and D_t needed in the Gibbs sampling. Therefore, we introduce the latent variable $N' = N - n$ which has a Poisson distribution with parameter $\theta[1 - F(t \mid \boldsymbol{\beta})]$. Then, the posterior distribution $p(\theta, \boldsymbol{\beta} \mid D_t)$ can be obtained from the joint density $p(\theta, \boldsymbol{\beta}, N' \mid D_t)$ by marginalization. The latter density is approximated from the Gibbs sample drawn from the following conditional densities iteratively: $p(N' \mid \theta, \boldsymbol{\beta}, D_t)$, $p(\theta \mid N', \boldsymbol{\beta}, D_t)$, and $p(\boldsymbol{\beta} \mid N', \theta, D_t)$, where all can be specified straightforwardly.

We can also develop a Gibbs sampler for the GOS model in (3.16) as in Kuo and Yang (1996). The algorithm is the same as given below, where N has the first stage Poisson prior distribution with mean θ and the second stage prior identical with the prior for θ in NHPP. Therefore, we won't repeat the algorithms for the GOS model. This is because in the NHPP-I with data augmentation N', we consider the following product:

$$\begin{aligned} &L_{\text{NHPP-I}} \times p\big(N' \mid \theta\big) \\ &= \left(\prod_{i=1}^{n} \theta f(x_i \mid \boldsymbol{\beta}) \right) \mathrm{e}^{-\theta F(t \mid \boldsymbol{\beta})}\, \mathrm{e}^{-\theta[1 - F(t \mid \boldsymbol{\beta})]} \big\{\theta\big[1 - F(t \mid \boldsymbol{\beta})\big]\big\}^{N'} / N'! \\ &= \frac{N!}{(N-n)!} \left(\prod_{i=1}^{n} f(x_i \mid \boldsymbol{\beta}) \right) \big[1 - F(t \mid \boldsymbol{\beta})\big]^{N'} \times \mathrm{e}^{-\theta} \theta^{N} / N!. \end{aligned} \tag{3.20}$$

This is exactly the product of the likelihood of the GOS model and the Poisson prior for θ.

We use the vector $\boldsymbol{\beta}$ to denote either β or (α, β). The class of priors often includes the family of gamma densities. The priors are chosen so that we can easily specify the conditional densities in the Gibbs algorithm. The family of gamma densities is versatile enough to incorporate both increasing and decreasing failure rates. Priors based on expert opinion and engineering information on software testing for the Musa–Okumoto process can be seen in Campodónico and Singpurwalla (1994, 1995). In addition to the following Bayesian inference, hierarchical Bayesian methods can be developed similarly as in the JM and LV models.

The exponential order statistics model. The likelihood function follows from (3.15),

$$L_{\text{NHPP-I}}(\theta, \boldsymbol{\beta} | D_t) = \left(\prod_1^n \theta\beta\, \mathrm{e}^{-\beta x_i} \right) \exp\{-\theta(1 - \mathrm{e}^{-\beta t})\}.$$

Prior: $\theta \sim \Gamma(a, b)$; $\beta \sim \Gamma(c, d)$; $\theta \perp \beta$.

Gibbs algorithm:

$$\begin{aligned}
&N' | \theta, \beta, D_t \sim P(\theta\, \mathrm{e}^{-\beta t}); \\
&\theta | N', \beta, D_t \sim \Gamma(a + n + N', b + 1); \\
&\beta | N', \theta, D_t \sim \Gamma\left(c + n, d + tN' + \sum_{i=1}^n x_i\right).
\end{aligned} \tag{3.21}$$

This is exactly the same algorithm given in Kuo and Yang (1995), where the Jelinski–Moranda model is assumed. This is expected because of the argument given after (3.20).

The Pareto order statistics model. The likelihood function is

$$L_{\text{NHPP-I}}(\theta, \boldsymbol{\beta} | D_t) = \left(\prod_1^n \theta \frac{\alpha\beta^\alpha}{(\beta + x_i)^{\alpha+1}} \right) \exp\left\{-\theta\left(1 - \frac{\beta^\alpha}{(t + \beta)^\alpha}\right)\right\}.$$

Prior: $\theta \sim \Gamma(a, b)$; $\alpha \sim \Gamma(c, d)$; $\beta \sim \pi_1$; where π_1 is an arbitrary prior on $\beta > 0$, and $\theta \perp \alpha \perp \beta$.

Gibbs algorithm:

$$\begin{aligned}
&N' | \theta, \alpha, \beta, D_t \sim P\left(\theta \frac{\beta^\alpha}{(t + \beta)^\alpha}\right); \\
&\theta | N', \alpha, \beta, D_t \sim \Gamma(a + n + N', b + 1); \\
&\alpha | N', \theta, \beta, D_t \sim \Gamma\left(n + c, d - n \ln \beta + N' \ln\left(1 + \frac{t}{\beta}\right) + \sum_{i=1}^n \ln(\beta + x_i)\right); \\
&\beta | \theta, \alpha, N', D_t \propto \frac{\beta^{n\alpha}}{\prod_{i=1}^n (\beta + x_i)^{\alpha+1}} \left(1 + \frac{t}{\beta}\right)^{-\alpha N'} \times \pi_1(\beta).
\end{aligned}$$

The last variate β is generated by the Metropolis algorithm using the above equation as the target distribution.

The Weibull order statistics model. The likelihood function is

$$L_{\text{NHPP-I}}(\theta, \boldsymbol{\beta}|D_t) = \left(\prod_{1}^{n} \theta\alpha\beta x_i^{\alpha-1} \exp(-\beta x_i^{\alpha})\right) \exp\{-\theta(1 - e^{-\beta t^{\alpha}})\}.$$

Prior: $\theta \sim \Gamma(a, b)$; $\beta \sim \Gamma(c, d)$; $\alpha \sim \pi_2$; where π_2 is a prior on $\alpha > 0$, and $\theta \perp \alpha \perp \beta$.

Gibbs algorithm:

$$N'|\theta, \alpha, \beta, D_t \sim P(\theta\, e^{-\beta t^{\alpha}});$$
$$\theta|N', \alpha, \beta, D_t \sim \Gamma(a + n + N', b + 1);$$
$$\beta|N', \alpha, \theta, D_t \sim \Gamma\left(n + c, N' t^{\alpha} + d + \sum_{i=1}^{n} x_i^{\alpha}\right);$$
$$\alpha|N', \theta, \beta, D_t \propto \alpha^n \left(\prod_{i=1}^{n} x_i^{\alpha}\right) e^{-\beta(N' t^{\alpha} + \sum_{i=1}^{n} x_i^{\alpha})} \times \pi_2(\alpha).$$

The last variate α is generated by the Metropolis algorithm using the above equation as the target distribution.

The extreme value order statistics model. The likelihood function is

$$L_{\text{NHPP-I}}(\theta, \boldsymbol{\beta}|D_t) = \left(\prod_{1}^{n} \theta \exp\left\{\alpha + \beta x_i - \frac{e^{\alpha+\beta x_i} - e^{\alpha}}{\beta}\right\}\right)$$
$$\times \exp\left\{-\theta\left[1 - \exp\left(-\frac{e^{\alpha+\beta t} - e^{\alpha}}{\beta}\right)\right]\right\}.$$

Prior: $\theta \sim \Gamma(a, b)$; $\alpha \sim \pi_3$; $\beta \sim \pi_4$; where π_3 and π_4 are priors on $-\infty < \alpha < \infty$ and $\beta > 0$, respectively, and $\theta \perp \alpha \perp \beta$.

Gibbs algorithm:

$$N'|\theta, \alpha, \beta, D_t \sim P\left(\theta \exp\left(-\frac{e^{\alpha+\beta t} - e^{\alpha}}{\beta}\right)\right);$$
$$\theta|N', \alpha, \beta, D_t \sim \Gamma(a + n + N', b + 1);$$
$$\alpha|N', \theta, \beta, D_t \propto \exp\left\{n\alpha + \frac{e^{\alpha}}{\beta}\left(N - \sum_{i=1}^{n} e^{\beta x_i} - N' e^{\beta t}\right)\right\} \times \pi_3(\alpha);$$
$$\beta|N', \theta, \alpha, D_t \propto \exp\left\{\beta \sum_{i=1}^{n} x_i + \frac{e^{\alpha}}{\beta}\left(N - \sum_{i=1}^{n} e^{\beta x_i} - N' e^{\beta t}\right)\right\} \times \pi_4(\beta),$$

where α and β can be generated by the Metropolis algorithm.

3.3.4. Gibbs sampling for record value statistics models

The exponential process. The likelihood function is

$$L_{\text{NHPP-II}}(\boldsymbol{\beta}|D_t) = \beta^n \exp(-\beta t).$$

Prior: $\beta \sim \Gamma(c, d)$.

Posterior distribution:

$$\beta | D_t \sim \Gamma(n + c, t + d).$$

The Pareto process. The likelihood function is

$$L_{\text{NHPP-II}}(\boldsymbol{\beta} | D_t) = \left(\prod_1^n \frac{\alpha}{x_i + \beta} \right) \exp\left\{ -\alpha \ln\left(1 + \frac{t}{\beta} \right) \right\}.$$

Prior: $\alpha \sim \Gamma(a, b)$; $\beta \sim \pi_1$; where π_1 is a prior on $\beta > 0$, and $\alpha \perp \beta$.

Gibbs algorithm:

$$\alpha | \beta, D_t \sim \Gamma\left(n + a, \ln\left(1 + \frac{t}{\beta} \right) + b \right);$$

$$\beta | \alpha, D_t \propto \frac{1}{(1 + t/\beta)^\alpha \prod_{i=1}^n (x_i + \beta)} \times \pi_1(\beta),$$

where β can be generated by the Metropolis algorithm.

The Weibull process. The likelihood function is

$$L_{\text{NHPP-II}}(\boldsymbol{\beta} | D_t) = \left(\prod_1^n \beta \alpha x_i^{\alpha - 1} \right) \exp(-\beta t^\alpha).$$

Prior: $\beta \sim \Gamma(c, d)$; $\alpha \sim \pi_2$, where π_2 is a prior on $\alpha > 0$, and $\alpha \perp \beta$.

Gibbs algorithm:

$$\beta | \alpha, D_t \sim \Gamma(n + c, t^\alpha + d);$$

$$\alpha | \beta, D_t \propto \alpha^n \left(\prod_{i=1}^n x_i \right)^\alpha \exp(-\beta t^\alpha) \times \pi_2(\alpha),$$

where α can be generated by the Metropolis algorithm.

The extreme value process. The likelihood function is

$$L_{\text{NHPP-II}}(\boldsymbol{\beta} | D_t) = \left(\prod_{i=1}^n \exp(\alpha + \beta x_i) \right) \exp\left\{ -\left(e^{\alpha + \beta t} - e^\alpha \right) / \beta \right\}.$$

Prior: $\alpha \sim \pi_3$; $\beta \sim \pi_4$, where π_3 and π_4 are priors for $-\infty < \alpha < \infty$ and $\beta > 0$, respectively, and $\alpha \perp \beta$.

Gibbs algorithm:

$$\alpha | \beta, D_t \propto \exp(n\alpha) \exp\left\{ -\left(e^{\alpha + \beta t} - e^\alpha \right) / \beta \right\} \times \pi_3(\alpha);$$

$$\beta | \alpha, D_t \propto \exp\left(\beta \sum_{i=1}^n x_i \right) \exp\left\{ -\left(e^{\alpha + \beta t} - e^\alpha \right) / \beta \right\} \times \pi_4(\beta),$$

where α (β) can be generated by the Metropolis algorithm.

3.3.5. Bayesian inference for NHPP

In this subsection, we use the Weibull process for illustration. Similar methods can be applied to other models.

For inferences on the parameters, such as N, θ, and $\boldsymbol{\beta}$, we can use the empirical measure of the sample generated by the Gibbs sampler (Gelman and Rubin, 1992). Now we study prediction. We need to consider the failure truncated situation, i.e., testing until the nth failure. All the posterior distributions described in previous subsection should be modified by replacing t with x_n. Moreover, instead of the full data D_{x_n}, we can consider the sequential setup where Bayesian inference is applied for each $D_{x_i} = (x_1, \dots, x_i)$, $i = 1, \dots, n$.

Inference for the future survival function evaluated at x distance away from x_i can be obtained by

$$\begin{aligned} E\big(S(x)|D_{x_i}\big) &= E\big[E\big(P(X_{i+1} > x_i + x)|\boldsymbol{\beta}, D_{x_i}\big)|D_{x_i}\big] \\ &= E\big[\exp\big\{-m(x_i + x) + m(x_i)\big\}|D_{x_i}\big] \\ &= \iint \exp\big\{-\beta(x_i + x)^\alpha + \beta x_i^\alpha\big\} p(\alpha, \beta|D_{x_i})\,\mathrm{d}\alpha\,\mathrm{d}\beta. \end{aligned} \tag{3.22}$$

The second equality of (3.22) is due to the Markovian property of the waiting times in NHPP (Çinlar, 1975, p. 97).

Inference for the future mean time between failures can be computed by means of

$$\begin{aligned} E(T_{i+1}|D_{x_i}) &= E\big[E(T_{i+1}|\boldsymbol{\beta}, D_{x_i})|D_{x_i}\big] \\ &= \int E(T_{i+1}|\boldsymbol{\beta}, D_{x_i}) p(\boldsymbol{\beta}|D_{x_i})\,\mathrm{d}\boldsymbol{\beta}. \end{aligned} \tag{3.23}$$

The expectation in (3.23) can be computed from the tail distribution of T_{i+1} given D_{x_i}, (Parzen, 1962, p. 138),

$$E(T_{i+1}|\boldsymbol{\beta}, D_{x_i}) = \int_0^\infty \exp\big\{-m(x_i + t) + m(x_i)\big\}\,\mathrm{d}t, \tag{3.24}$$

which can be evaluated in closed form for many processes. For example,

$$E(T_{i+1}|\boldsymbol{\beta}, D_{x_i}) = \begin{cases} \frac{1}{\beta} & \text{for the exponential NHPP-II,} \\ \frac{\beta + x_i}{\alpha - 1} & \text{for the Pareto NHPP-II with } \alpha > 1, \\ \frac{\exp(\beta x_i^\alpha)}{\alpha \beta^{\frac{1}{\alpha}}}\big(\Gamma\big(\frac{1}{\alpha}\big) - \mathit{ING}\big(\beta x_i^\alpha; \frac{1}{\alpha}\big)\big) & \\ \quad \text{for the Weibull NHPP-II,} & \end{cases}$$

where *ING* denotes the incomplete gamma function.

Both (3.22) and (3.23) can be evaluated using the Gibbs sample or the Rao–Blackwell version. We will illustrate the Monte Carlo integration method using (3.22). We apply the Gelman and Rubin (1992) approach to the Monte Carlo method. Let $\alpha^{(r,l)}$ and $\beta^{(r,l)}$ denote the variates for α and β generated in the rth replication and the lth iteration of the MCMC algorithm based on the data set $\{x_1, \dots, x_i\}$. We are running a separate Gibbs sampler for each partial data set. We can derive an estimator of (3.22) using direct substitution from the Gibbs sample with L a sufficiently large even number for the number of iterations:

$$\widehat{S}(x|D_{x_i}) = \frac{2}{RL}\sum_{r=1}^{R}\sum_{l=\frac{L}{2}+1}^{L} \exp\{-\beta^{(r,l)}(x+x_i)^{\alpha^{(r,l)}} + \beta^{(r,l)}x_i{}^{\alpha^{(r,l)}}\}. \quad (3.25)$$

Bayesian credible regions for the unknown parameters and Bayesian confidence intervals for the functionals such as the future survival function and the mean time between failures can be computed using the quantiles of the unknown parameters and the functionals evaluated from the second half of the Gibbs sample. Presumably nonasymptotic Bayesian credible regions could have closer-to-nominal coverage probabilities than asymptotic likelihood regions, for data sets of realistic size.

3.4. Superposed NHPP processes

The intensity function of the nonhomogeneous Poisson process is often assumed to be a monotonic function of t. Such is the case, *inter alia* for the homogeneous Poisson, the Musa–Okumoto (Musa and Okumoto, 1984), the Weibull (Power Law), and the Cox–Lewis (Cox and Lewis, 1966) processes, whose intensity function is respectively constant, a fractional function, a power function, and a log-linear function of time. To relax the monotonic intensity function assumption, we can consider a model that superimposes several independent nonhomogeneous Poisson processes with simple intensity functions.

Let $M_j(t)$ denote the nonhomogeneous Poisson process that counts the number of failures from the jth component in $(0, t]$, with intensity function $\lambda_j(t|\boldsymbol{\theta}_j) = \mathrm{d}E\{M_j(t)\}/\mathrm{d}t$, where the functional form of $\lambda_j(t|\boldsymbol{\theta}_j)$ is assumed to be specified but with unknown parameter $\boldsymbol{\theta}_j$ that may be a vector. Moreover, the numbers of failures from the jth component $M_j(t)$, $j = 1, \dots, J$, are assumed to be independent. Then a process $M(t) = \sum_{j=1}^{J} M_j(t)$, which counts the number of failures in the interval $(0, t]$ for the superposition model, is also a nonhomogeneous Poisson process with intensity function $\lambda(t|\boldsymbol{\theta}) = \mathrm{d}E\{M(t)\}/\mathrm{d}t = \lambda_1(t|\boldsymbol{\theta}_1) + \cdots + \lambda_J(t|\boldsymbol{\theta}_J)$, where $\boldsymbol{\theta} = \{\boldsymbol{\theta}_1, \dots, \boldsymbol{\theta}_J\}$ (cf. Çinlar, 1975, pp. 87–88). In our notation, we often omit $\boldsymbol{\theta}$ and $\boldsymbol{\theta}_j$ for simplicity.

The family of intensity functions for the superposition model is richer than that of its components. It includes bathtub-shaped functions, polynomials with peaks and valleys, and simple monotonic functions. For example, the intensity function for a superposition model with two nonhomogeneous Poisson processes with Weibull intensities is $\lambda(t) = \alpha_1\beta_1 t^{\alpha_1-1} + \alpha_2\beta_2 t^{\alpha_2-1}$. If $\alpha_1 < 1$ and $\alpha_2 > 1$, then the intensity function has a bathtub shape. When $\alpha_1 = \alpha_2$, the intensity function of the superposition model reduces to that of a simple Weibull intensity, and thus β_1 and β_2 are not separately identifiable. However, Bayesian analysis with continuous prior distribution assigns zero probability to the case $\alpha_1 = \alpha_2$. Therefore, the identifiability issue in β_1 and β_2 is not a serious problem in common Bayesian analysis. On the other hand, if one assumes a superposition model and the data actually come from a nonhomogeneous Poisson process with a simple intensity function, then the posterior distributions of α_1 and α_2 would heavily overlap and the separation in β_1 and β_2 is only inherent in the prior distribution. To avoid this problem, we assume that the number of components J is known. A similar identifiability issue has been addressed in Berger and Sun (1993). They study inference for poly-Weibull densities. In fact, we could also reproduce poly-Weibull densities in our framework. Let $\lambda(t) = \sum_{j=1}^{J} \alpha_j \beta_j t^{\alpha_j - 1}$ for the intensity of a superimposed model of J Weibull nonhomogeneous Poisson processes. Applying the unified approach of Kuo and Yang (1996), the points of failures for the superposition model are the record value statistics derived from an i.i.d. unobserved sample distributed with $f(t) = \lambda(t) \exp\{-\int_0^t \lambda(u)\, du\}$. Then $f(t) = (\sum_{j=1}^{J} \alpha_j \beta_j t^{\alpha_j - 1}) \exp(-\sum_{j=1}^{J} \beta_j t^{\alpha_j})$. This is the poly-Weibull density considered by Berger and Sun except we consider different likelihoods. See Yang and Kuo (1999) for more discussion of the different likelihoods.

Bayesian inference for the superposition model is accessible now by means of a Markov chain Monte Carlo technique with data augmentation. The technique consists of iteratively generating latent variables using multinomial distributions and applying the technique in Kuo and Yang (1996) to update the parameters given the latent variables.

In the full model, we assume none of the λ_j's can be zero. Therefore, continuous priors on the unknown parameters suffice. In order to incorporate nested submodels with positive prior probability, we consider a mixture of discrete and continuous (discrete/continuous) distributions for λ_j. That is, λ_j equals zero with prior probability p_{0j} and $\lambda_j > 0$ is distributed according to a continuous density $\pi_j(\lambda_j)$ on $(0, \infty)$ with probability $1 - p_{0j}$. When λ_j is indexed by $\boldsymbol{\theta}_j$ (vector), it is often sufficient to assign a discrete/continuous prior to just one of the parameters in $\boldsymbol{\theta}_j$. Consequently, we develop a Gibbs sampler for the Bayesian test of the full model against a nested submodel.

3.4.1. Gibbs sampling for the full model

In this section, we develop the conditional distributions used in the Gibbs sampler for the full model.

Suppose that the software is to be tested until time t. We assume the prior has independent components for the $\boldsymbol{\theta}_j$s. The posterior density of $\boldsymbol{\theta}$ is

$$f(\boldsymbol{\theta}|D_t) \propto \left[\prod_{i=1}^{n} \{\lambda_1(x_i) + \cdots + \lambda_J(x_i)\}\right] \times \exp\bigl[-\{m_1(t) + \cdots + m_J(t)\}\bigr] \times \prod_{j=1}^{J} \pi(\boldsymbol{\theta}_j), \tag{3.26}$$

where $\pi(\boldsymbol{\theta}_j)$ denotes the prior marginal density of $\boldsymbol{\theta}_j$.

The expression $\prod_{i=1}^{n}\{\lambda_1(x_i) + \cdots + \lambda_J(x_i)\}$ in (3.26) usually prevents us from specifying a convenient form for the conditional density of $\boldsymbol{\theta}$ given D_t. We, therefore, introduce latent variables, denoted by $\boldsymbol{I}$, so we can obtain the posterior distribution of $\boldsymbol{\theta}$ given D_t by iteratively sampling $\boldsymbol{I}$ from $f(\boldsymbol{I}|\boldsymbol{\theta}, D_t)$ and sampling $\boldsymbol{\theta}$ from $f(\boldsymbol{\theta}|\boldsymbol{I}, D_t)$. We use the notation f to denote a generic name for the densities.

For each i, $i = 1, \ldots, n$, let us consider the latent variable $\boldsymbol{I}_i = (I_{i1}, \ldots, I_{iJ})^{\mathrm{T}}$, where $I_{ij} = 1$ if the ith failure is caused by the jth type and $I_{ij} = 0$ otherwise.

Note $\sum_{j=1}^{J} I_{ij} = 1$ for $i = 1, \ldots, n$. Let the conditional density of $\boldsymbol{I}_i$ given $\boldsymbol{\theta}$ and D_t be a multinomial distribution MN with parameters 1 and the cell probabilities $(p_{i1}, \ldots, p_{iJ})$, where $p_{ij} = \lambda_j(x_i)/\{\lambda_1(x_i) + \cdots + \lambda_J(x_i)\}$. Let $\boldsymbol{I} = (\boldsymbol{I}_1, \ldots, \boldsymbol{I}_n)$, an $J \times n$ matrix with the ith column indicating which component causes the ith failure. Then we simulate $\boldsymbol{I}_1, \ldots, \boldsymbol{I}_n$ independently, where $\boldsymbol{I}_i \sim MN\{1, (p_{i1}, \ldots, p_{iJ})\}$ for $i = 1, \ldots, n$. The posterior density of $\boldsymbol{\theta}$ given $\boldsymbol{I}$ and D_t is

$$\begin{aligned} f(\boldsymbol{\theta}|\boldsymbol{I}, D_t) &\propto L(\boldsymbol{\theta}|D_t) \times \prod_{i=1}^{n} p(\boldsymbol{I}_i|\boldsymbol{\theta}, D_t) \times \prod_{j=1}^{J} \pi(\boldsymbol{\theta}_j) \\ &\propto \prod_{j=1}^{J} \prod_{i:I_{ij}=1} \lambda_j(x_i) \times \prod_{j=1}^{J} \exp\{-m_j(t)\} \times \prod_{j=1}^{J} \pi(\boldsymbol{\theta}_j). \end{aligned} \tag{3.27}$$

This conditional density is a joint density composed of independent posterior densities of the $\boldsymbol{\theta}_j$'s given $\boldsymbol{I}$ and D_t. We can therefore update the $\boldsymbol{\theta}_j$'s independently for $j = 1, \ldots, J$ as in Kuo and Yang (1996), thus providing a practical way of computing the Bayes estimates for superposition models.

We can also use the above algorithm to obtain the maximum likelihood estimator (MLE) of $\boldsymbol{\theta}$. This can be done by choosing $\pi(\boldsymbol{\theta}_j) = 1$, for all j, with $\boldsymbol{\theta}_j$ in the suitable domain, and using the posterior mode to estimate $\boldsymbol{\theta}$. Because this flat prior may be improper, the condition $\int L(\boldsymbol{\theta}|D_t)\,\mathrm{d}\boldsymbol{\theta} < \infty$ is required to make this method valid.

Next we provide more detailed algorithms for two examples.

Superposition of Musa–Okumoto and Weibull processes. Let the intensities for the two component processes be $\lambda_1(t) = \beta_1/(\alpha_1+t)$ and $\lambda_2(t) = \alpha_2\beta_2 t^{\alpha_2-1}$, where $\alpha_i > 0$ and $\beta_i > 0$ for $i = 1, 2$. Hjorth (1980) points out that this intensity function $\lambda_1(t) + \lambda_2(t)$ has three possible types of shapes: increasing, decreasing, and bathtub. We use $\boldsymbol{\theta}_1$ to denote (α_1, β_1) and $\boldsymbol{\theta}_2$ to denote (α_2, β_2), and $\boldsymbol{\theta}$ to denote $(\boldsymbol{\theta}_1, \boldsymbol{\theta}_2)$.

The likelihood function of the superposition model follows from (3.15),

$$\begin{aligned} L(\boldsymbol{\theta}|D_t) = &\left\{\prod_{i=1}^{n}\left(\frac{\beta_1}{x_i + \alpha_1} + \alpha_2\beta_2 x_i^{\alpha_2-1}\right)\right\} \\ &\times \exp\left\{-\beta_1 \ln\left(1 + \frac{t}{\alpha_1}\right) - \beta_2 t^{\alpha_2}\right\}. \end{aligned} \tag{3.28}$$

We consider the prior that has four independent components: $\alpha_1 \sim \pi(\alpha_1)$; $\beta_1 \sim \Gamma(a_1, b_1)$; $\alpha_2 \sim \pi(\alpha_2)$; $\beta_2 \sim \Gamma(a_2, b_2)$, where $\pi(\alpha_1)$ and $\pi(\alpha_2)$ are arbitrary priors on $\alpha_1 > 0$ and $\alpha_2 > 0$, respectively. We use $\Gamma(a, b)$ to denote the gamma density $f(\beta) = b^a\beta^{a-1}\exp(-b\beta)/\Gamma(a)$. Although the independent marginals and the gamma distribution for the β's are chosen for convenience, this prior is quite flexible. The prior on $\boldsymbol{\theta}_2 = (\alpha_2, \beta_2)$ includes the class of customary priors for the Weibull family as discussed in Sinha (1986, p. 160). In the following we list the conditional densities used in the Gibbs algorithm.

- Construct $\boldsymbol{I}$ given $\boldsymbol{\theta}$ and D_t by generating independent I_{i1} from the Bernoulli distribution with parameter p_{i1}, for $i = 1, \ldots, n$, where $p_{i1} = \{\beta_1/(x_i + \alpha_1)\}/\{\beta_1/(x_i + \alpha_1) + \alpha_2\beta_2 x_i^{\alpha_2-1}\}$. Let $I_{i2} = 1 - I_{i1}$;

- $\alpha_1|\beta_1, \alpha_2, \beta_2, \boldsymbol{I}, D_t \propto \frac{1}{\prod_{i:I_{i1}=1}(x_i+\alpha_1)} \exp\{-\beta_1 \ln(1 + t/\alpha_1)\} \times \pi(\alpha_1)$;
- $\beta_1|\alpha_1, \alpha_2, \beta_2, \boldsymbol{I}, D_t \sim \Gamma\{a_1 + \sum_{i=1}^n I_{i1}, b_1 + \ln(1 + t/\alpha_1)\}$;
- $\alpha_2|\alpha_1, \beta_1, \beta_2, \boldsymbol{I}, D_t \propto \alpha_2^{\sum_{i=1}^n I_{i2}} (\prod_{i:I_{i2}=1} x_i)^{\alpha_2} \exp(-\beta_2 t^{\alpha_2}) \times \pi(\alpha_2)$;
- $\beta_2|\alpha_1, \beta_1, \alpha_2, \boldsymbol{I}, D_t \sim \Gamma(a_2 + \sum_{i=1}^n I_{i2}, b_2 + t^{\alpha_2})$.

The variates α_1 and α_2 can be generated by the Metropolis algorithm with the above expressions in the conditional densities for α_1 and α_2 as the target densities.

Polynomial intensity function. Let us consider a nonhomogeneous Poisson process with intensity function $\lambda(t) = \beta_0 + \beta_1 t + \cdots + \beta_{J-1} t^{J-1}$. It can be considered as a superposition of a homogeneous Poisson and $J - 1$ Weibull processes with known shape parameters $2, \ldots, J$. We assume $\beta_j > 0$ for $j = 0, \ldots, J - 1$. For better notation, we are changing the index space on j from 1 to J to 0 to $J - 1$. We can apply the Gibbs sampler developed earlier in this section with $p_{ij} = \beta_j x_i^j / (\sum_{j=0}^{J-1} \beta_j x_i^j)$, for $j = 0, \ldots, J - 1$. Then the conditional posterior density of $\boldsymbol{\theta}$ in (3.27) is

$$f(\boldsymbol{\theta}|\boldsymbol{I}, D_t) \propto \prod_{j=0}^{J-1} \left\{ \beta_j^{\sum_{i=1}^n I_{ij}} \exp\left(-\frac{t^{j+1}}{j+1} \beta_j \right) \pi(\beta_j) \right\}.$$

Therefore, we can update β_j, $j = 0, \ldots, J - 1$, independently given $\boldsymbol{I}$ and D_t. If $\pi(\beta_j)$ is a gamma density or a flat prior density for β_j, we can easily generate β_j using an updated gamma density.

3.4.2. Bayesian inference for the superposed models

Bayesian inference for any functionals can be developed easily from the posterior distribution. The predictive reliability function evaluated at $x_n + x$ derived similar to (3.22) is

$$E\big(S(x)|D_{x_n}\big) = \int \exp\{-m(x_n + x|\boldsymbol{\theta}) + m(x_n|\boldsymbol{\theta})\} f(\boldsymbol{\theta}|D_{x_n})\, \mathrm{d}\boldsymbol{\theta}. \tag{3.29}$$

Let $\boldsymbol{\theta}^{(r,\ell)}$ denote the Gibbs variate for $\boldsymbol{\theta}$ generated in the rth replication and the ℓth iteration of the MCMC algorithm based on the whole data D_{x_n}. Then, we can obtain a Monte Carlo estimate of the predictive reliability function in (3.29) for the superimposed Musa–Okumoto and Weibull process with L being a sufficiently large even number:

$$\begin{aligned}
\widehat{S}(x|D_{x_n}) &= \frac{2}{RL} \sum_{r=1}^{R} \sum_{\ell=\frac{L}{2}+1}^{L} \exp\{-m(x_n + x|\boldsymbol{\theta}^{(r,\ell)}) + m(x_n|\boldsymbol{\theta}^{(r,\ell)})\} \\
&= \frac{2}{RL} \sum_{r=1}^{R} \sum_{\ell=\frac{L}{2}+1}^{L} \exp\bigg[-\beta_1^{(r,\ell)} \ln\bigg(1 + \frac{x_n + x}{\alpha_1^{(r,\ell)}}\bigg) \\
&\quad - \beta_2^{(r,\ell)} (x_n + x)^{\alpha_2^{(r,\ell)}} + \beta_1^{(r,\ell)} \ln\bigg(1 + \frac{x_n}{\alpha_1^{(r,\ell)}}\bigg) + \beta_2^{(r,\ell)} x_n^{\alpha_2^{(r,\ell)}}\bigg].
\end{aligned}$$

3.4.3. *Gibbs sampling for nested models*

The priors $\pi(\beta_j)$ given previously are assumed to be continuous. They have the drawback that they assign zero probability *a priori* to any reduced model. To incorporate the possibility that a reduced model can be a reasonable contender in model fitting, we need to assign reduced models *a priori* positive probabilities. Therefore, we consider discrete/continuous priors for some parameters in the model.

Yang and Kuo (1999) consider the superposition model of Musa and Okumoto and Weibull with intensities $\beta_1/(t+\alpha_1)$ and $\alpha_2\beta_2 t^{\alpha_2-1}$, respectively. Suppose we are interested in comparing the full model and the reduced model with the Musa–Okumoto model only. We can consider the prior for β_2 to be a mixture distribution, where β_2 is equal to 0 with probability p_0 and is distributed according to a continuous density $\pi(\beta_2)$ with probability $1-p_0$. We assume independent continuous priors for the other parameters. So this model assigns *a priori* the reduced model ($\beta_2=0$) with probability p_0. Therefore, we would prefer the reduced model ($\beta_2=0$) to the full model ($\beta_2>0$) provided $p^*=\Pr(\beta_2=0|D_t)>1/2$, where p^* can be approximated from the Gibbs sample.

To describe the Gibbs sampler, let us first consider the conditional density of β_2 given $\beta_1,\alpha_1,\alpha_2$ and D_t. It is a mixture of a point mass at 0 and a continuous density $\pi_c(\beta_2|\alpha_1,\beta_1,\alpha_2,D_t)$ with mass p^{**} and $1-p^{**}$, respectively. The continuous part of the posterior density of β_2, $\pi_c(\beta_2|\alpha_1,\beta_1,\alpha_2,D_t)$, can be shown to be proportional to $L(\boldsymbol{\theta}|D_t)\pi(\beta_2)$, where $L(\boldsymbol{\theta}|D_t)$ is given in (3.28).

Moreover, let

$$m_1(\alpha_1,\beta_1,\alpha_2,D_t)=\int_0^\infty L(\boldsymbol{\theta}|D_t)\pi(\beta_2)\,\mathrm{d}\beta_2,$$

and

$$\begin{aligned}L(\boldsymbol{\theta}_1|D_t)&=L(\boldsymbol{\theta}_1,\alpha_2,\beta_2=0|D_t)\\&=\left(\prod_{i=1}^n\frac{\beta_1}{x_i+\alpha_1}\right)\exp\left\{-\beta_1\ln\left(1+\frac{t}{\alpha_1}\right)\right\},\end{aligned}$$

where $\boldsymbol{\theta}_1=(\alpha_1,\beta_1)$. The Bayes factor (*BF*) in favor of H_0: $\beta_2=0$ (the Musa–Okumoto model only), against the alternative H_1: $\beta_2>0$ (the full model), given the other parameters $\alpha_1,\alpha_2,\beta_1$, and D_t is given by

$$BF=\frac{L(\boldsymbol{\theta}_1|D_t)}{m_1(\alpha_1,\beta_1,\alpha_2,D_t)}.$$

Then, the posterior probability p^{**} for $\beta_2=0$ is given by $p^{**}=p_0/\{p_0+(1-p_0)/BF\}$. The probability p^{**} is a function of $\alpha_1,\beta_1,\alpha_2$, and D_t. We omit its arguments for simplicity of notation. Most mathematical software provides routines to compute numerically one-dimensional integrals such as needed to evaluate m_1. In the absence of such software, we can approximate this integral using numerical integration techniques, for example, the Simpson's rule or the trapezoidal rule.

The MCMC algorithm which consists of simulating each parameter from the full conditional density can be described as follows:

1. Generate β_2 given $\alpha_2, \boldsymbol{\theta}_1$, and D_t:
   Simulate $\beta_2 = 0$ or $\beta_2 > 0$ with probabilities p^{**} and $1 - p^{**}$, respectively. If $\beta_2 > 0$, then generate β_2 from the conditional density $\pi_c(\beta_2|\alpha_1, \beta_1, \alpha_2, D_t)$ which is proportional to $L(\boldsymbol{\theta}|D_t)\pi(\beta_2)$. This can be done by a Metropolis algorithm or by following the procedure in Subsection 3.4.1 with data augmentation.
2. Generate α_2 given $\beta_2, \alpha_1, \beta_1$, and D_t:
   (a) If $\beta_2 = 0$, generate α_2 conditional on α_1, β_1 and D_t from the conditional density $\pi(\alpha_2|\alpha_1, \beta_1, D_t)$ (that has β_2 integrated out). The conditional density $\pi(\alpha_2|\alpha_1, \beta_1, D_t)$ is proportional to $p_0\pi(\alpha_2)L(\boldsymbol{\theta}_1|D_t) + (1 - p_0)\pi(\alpha_2) \times m_1(\alpha_1, \beta_1, \alpha_2, D_t)$. So α_2 can be sampled by the Metropolis algorithm.
   (b) If $\beta_2 > 0$, simulate α_2 from the conditional density $\pi_c(\alpha_2|\beta_2, \alpha_1, \beta_1, D_t)$ that is proportional to $L(\boldsymbol{\theta}|D_t)\pi(\alpha_2)$. This can be done by a Metropolis algorithm, or by following the procedure in Subsection 3.4.1 with data augmentation.
3. Simulate α_1 and β_1 given $\boldsymbol{\theta}_2$ and D_t as follows:
   If $\beta_2 > 0$, then generate $\boldsymbol{\theta}_1$ from the full model as in Subsection 3.4.1; if $\beta_2 = 0$, then generate $\boldsymbol{\theta}_1$ from the reduced model $L(\boldsymbol{\theta}_1|D_t)$ as in Kuo and Yang (1996).
4. Iterate and replicate steps 1–3 to obtain R multiple chains of length L. Then estimate p^* by using the relative frequency of $\beta_2 = 0$ in the second half of the Gibbs sample.

Bayesian inference for the nested model. We can develop a Bayesian procedure for testing the nested Weibull model versus the full superposition model. We assume that β_1 is assigned a discrete/continuous prior with a continuous density $\pi(\beta_1)$ of prior probability $1 - q_0$ and with a discrete point mass at zero with prior probability q_0. Then we can derive a new test similarly as before, by interchanging β_1 and β_2 and interchanging α_1 and α_2 in steps 1–3 with the appropriate changes on prior and the Bayes factor: $BF = L(\boldsymbol{\theta}_2|D_t)/\{m_2(\alpha_1, \alpha_2, \beta_2, D_t)\}$, where

$$L(\boldsymbol{\theta}_2|D_t) = \left(\prod_{i=1}^{n} \alpha_2\beta_2 x_i^{\alpha_2 - 1}\right) \exp\left(-\beta_2 t^{\alpha_2}\right) \quad \text{and}$$

$$m_2(\alpha_1, \alpha_2, \beta_2, D_t) = \int_0^\infty L(\boldsymbol{\theta}|D_t)\pi(\beta_1)\,\mathrm{d}\beta_1.$$

Then we use q^* to denote the posterior probability assigned to the Weibull submodel ($\beta_1 = 0$).

If we consider discrete/continuous priors for both β_1 and β_2 simultaneously, this will allow us to select among the Musa–Okumoto, the Weibull, and the superposition models. However, the independence assumption for the parameters in the prior implies that we choose an intensity to be identically equal to zero *a priori* with probability p_0q_0. It is not a very attractive prior unless we believe the intensity can be identically zero. Dependent priors can be constructed to eliminate this drawback.

4. Model selection

We have presented several models of software reliability. Which model or models are appropriate? Many Bayesian predictive model selection criteria can be explored. Three criteria are singled out here for their ease in implementation and interpretation: (a) the standardized deviation, (b) the prequential conditional predictive ordinate (PCPO), and (c) the predictive likelihood for the partial block.

The standardized deviation d_i computes the one stage forecasting error for the ith interfailure time. It is defined by

$$d_i = \frac{t_i - E(T_i|D_{t_{i-1}})}{\sqrt{V(T_i|D_{t_{i-1}})}},$$

provided $E(T_i|D_{t_{i-1}}) < \infty$ and $V(T_i|D_{t_{i-1}}) < \infty$. Note D_{t_0} denotes the prior measure. For the LV model, the mean and variance are computed from (3.10) and (3.11) respectively. The deviation d_i depends on the model. Let $d_i(l)$ denote the ith deviation of the lth model. Then we choose the model that minimizes $D_0(l) = \sum_{i=1}^n d_i^2(l)$. Alternatively, we can choose the model that minimizes $D(l) = \sum_{i=2}^n d_i^2(l)$. The latter is less sensitive to a poor prior choice, because it excludes d_1 which is highly influenced by the prior choice.

The PCPO defines the conditional predictive ordinate (CPO) of t_i by conditioning on the set of past failure times $t_1, \ldots, t_{i-1}$. It is slightly different from the CPO for cross-validated data where the conditioning is on all the data past and future except t_i. The prequential approach is adopted because software testing is usually conducted in real-time. At each epoch of failure, we are interested in predicting the time of next failure. The CPO rules provide an assessment of the entertained model in light of current data. The desirability of using the CPO rules from cross-validated data (usually used in exchangeable models) for model determination has been studied by Stone (1974), Geisser and Eddy (1979), Box (1980), Pettit and Young (1990), Gelfand et al. (1992), and Kuo and Cohen (1999). Dawid (1984) has argued for the prequential likelihood criterion in model selection. In addition to using the Bayesian conditional predictive ordinates described here, Dawid also proposes a plug-in rule, where forecasting future from present data is based on treating the estimate of the parameter as the true parameter. We prefer the Bayesian PCPO approach because it may take a large data set to justify the efficiency of the plug-in rule.

The PCPO for the ith observation is defined by $c_i = p(t_i|t_1, \ldots, t_{i-1})$ for $i \geqslant 2$. For $i = 1$, we define $c_1 = p(t_1|\text{prior})$. Usually, we select diffuse priors to let data do the talking. We can plot c_i versus i, $i = 1, \ldots, n$, for each model. The bigger the c are on the average, the better the model. Moreover, the plot will reveal outliers, if any, where the c values are unusually small. The predictive Bayesian likelihood criterion is to select the model that maximizes the likelihood $C_0(l) = \prod_{i=1}^n c_i(l) = p(t_1, \ldots, t_n)(l)$ where l is the index on the models. Alternatively, we can choose the model that maximizes $C(l) = \prod_{i=2}^n c_i(l)$. The computation of c_i is straightforward from (3.5) and (3.13) for the JM and LV model. For the NHPP, we generally use the PCPO for x_{i+1} given the past. That is, the conditional predictive density for the future failure epoch x_{i+1} given

$D_{x_i} = (x_1, \ldots, x_i)$ can be computed by

$$p(x_{i+1}|D_{x_i}) = \int p(X_{i+1}|\boldsymbol{\theta}, D_{x_i}) p(\boldsymbol{\theta}|D_{x_i})\,\mathrm{d}\boldsymbol{\theta}$$
$$= \int \lambda(X_{i+1}) \exp\{-m(X_{i+1}) + m(x_i)\} p(\boldsymbol{\theta}|D_{x_i})\,\mathrm{d}\boldsymbol{\theta}.$$

The third approach is the predictive likelihood approach that modifies the posterior Bayes factor (Aitkin, 1991). Yang and Kuo (1999) discussed the advantage of using the predictive likelihood for model selection. We divide our data into two parts for a fixed i: the first part, or the "past", consists of the first i failure times; the second part, or the "future", consists of the remaining data. It may be a reasonable decision to define the "future" as the last half portion of the whole data and the "past" as the first half portion of the data. We use $\overline{D}_{x_i}$ to denote the second part, that is, $(x_{i+1}, \ldots, x_n)$. Then we define the predictive likelihood of the "future" $\overline{D}_{x_i}$ given the "past" D_{x_i} for a model denoted by M:

$$p_M(\overline{D}_{x_i}|D_{x_i})$$
$$= \int p(\overline{D}_{x_i}|\boldsymbol{\theta}_M, D_{x_i}) \pi_M(\boldsymbol{\theta}_M|D_{x_i})\,\mathrm{d}\boldsymbol{\theta}_M$$
$$= \int \left\{ \prod_{\ell=i+1}^{n} \lambda(x_\ell|\boldsymbol{\theta}_M) \right\} \mathrm{e}^{m(x_i|\boldsymbol{\theta}_M) - m(x_n|\boldsymbol{\theta}_M)} \pi_M(\boldsymbol{\theta}_M|D_{x_i})\,\mathrm{d}\boldsymbol{\theta}_M. \tag{4.1}$$

The last equality follows from the Markovian property of the inter-arrival times for the nonhomogeneous Poisson process. Here $\boldsymbol{\theta}_M$ is the set of all unknown parameters in the model M. The posterior density of $\boldsymbol{\theta}_M$ given D_{x_i} is π_M. The modified posterior Bayes factor criterion prefers model M_1 to M_2 if the ratio $p_{M_2}(\overline{D}_{x_i}|D_{x_i})/p_{M_1}(\overline{D}_{x_i}|D_{x_i}) < 1$. This ratio can easily be approximated by applying the Monte Carlo integration technique to $p_{M_1}(\overline{D}_{x_i}|D_{x_i})$ and $p_{M_2}(\overline{D}_{x_i}|D_{x_i})$ separately; that is, for $j = 1, 2$, the predictive likelihood $p_{M_j}(\overline{D}_{x_i}|D_{x_i})$ is estimated by averaging $p_{M_j}(\overline{D}_{x_i}|\boldsymbol{\theta}_{M_j}, D_{x_i})$ over the $\boldsymbol{\theta}_{M_j}$s drawn in the second half of the iteration and all replications of the Gibbs sampler based on the data D_{x_i}. For model selection, the modified posterior Bayes factor (predictive likelihood) criterion would choose the model with the largest $\hat{p}_M(\overline{D}_{x_i}|D_{x_i})$.

Recently, deviance information criterion (DIC) (Spiegelhalter et al., 2002) has been used by Ray et al. (2003) for model selection.

We are also interested in the model adequacy issue. For each i, $i = 1, \ldots, n$, the conditional predictive density for the future failure epoch X_{i+1} given $D_{x_i} = (x_1, \ldots, x_i)$ can be computed by

$$p(X_{i+1}|D_{x_i}) = \int p(X_{i+1}|\boldsymbol{\theta}, D_{x_i}) p(\boldsymbol{\theta}|D_{x_i})\,\mathrm{d}\boldsymbol{\theta}$$
$$= \int \lambda(X_{i+1}) \exp\{-m(X_{i+1}) + m(x_i)\} p(\boldsymbol{\theta}|D_{x_i})\,\mathrm{d}\boldsymbol{\theta}. \tag{4.2}$$

Therefore, we can construct a $1 - \alpha$ level prediction interval for X_{i+1} from the Gibbs sample of X_{i+1} using (4.2). For example, the 95% prediction interval for X_{i+1} can be constructed from the 2.5% and 97.5% quantiles of the sample $\{x_{i+1}^{(r,\ell)}\}_{r=1,\ell=L/2+1}^{R\quad L}$ generated from (4.2). We construct intervals for each X_i based on past data $D_{x_{i-1}}$ for $i = 2, \ldots, n$. We judge the model adequate if 95% of the intervals contain the actual observed data x_i.

Now we elaborate on how to generate X_{i+1} from the predictive density $p(X_{i+1}|D_{x_i})$ in (4.2). Observe the expression $\lambda(X_{i+1})\exp\{-m(X_{i+1}) + m(x_i)\}$ is a density for a truncated variable with value greater than x_i. Therefore, we can simulate X_{i+1} from the truncated density with parameters $\boldsymbol{\theta}^{(r,\ell)}$ that are Gibbs variates from the posterior density $p(\boldsymbol{\theta}|D_{x_i})$. Let G denote the untruncated distribution, where $G(x) = 1 - \exp\{-m(x)\}$ that is a function of $\boldsymbol{\theta}^{(r,\ell)}$. Then x_{i+1} can be generated by setting $x_{i+1}^{(r,\ell)} = G^{-1}[G(x_i) + U^{(r,\ell)}\{1 - G(x_i)\}]$, with $U^{(r,\ell)}$ being the independent uniform variates generated in the ℓth iteration and rth replication of the Markov chain. When this inverse G^{-1} is complicated to solve explicitly, we can obtain x_{i+1} by solving for it as the root of the equation $G(x) = c$, where $c = G(x_i) + U^{(r,\ell)}\{1 - G(x_i)\}$, a known constant. The root can be solved easily using standard techniques or software. This method is more efficient and therefore to be preferred to the rejection method for generating a truncated random variate.

5. Optimal release policy

Statistical inference for software reliability can be considered as a means to a goal, determining the optimal policy for releasing the software. Chapter 8 of Xie (1991) and Chapter 6 of Singpurwalla and Wilson (1999) review several criteria and approaches to this problem. We can roughly classify the work into the following three categories.

(1) *Determining the minimal testing time to achieve the reliability requirement.* The reliability can be either measured through the number of remaining faults, the failure intensity, or the time to next failure. The work can be found in Caspi and Kouka (1984), Forman and Singpurwalla (1977, 1979), Okumoto and Goel (1980), Ross (1985b) Nayak (1988), and Yamada and Osaki (1986).

(2) *Optimal testing policy that depends on a cost criterion.* Ideally, we would like to test the software as long as possible, so we not only gain information, but also correct bugs. However, if we test it for too long, the software may become obsolete. Several loss functions have been proposed to incorporate the trade-off between these two aspects. Okumoto and Goel (1980) write the expected cost as the sum of the expected cost of removing faults during testing, the expected cost of removing faults during the operation phase (after testing), and the expected cost of testing (that include the cost of delay in releasing the software). Levin and Yadid (1990) extend the formulation to optimal release policies for several (improved) versions of the software. Chao et al. (1993) consider a stopping rule for recapture debugging. In terms of solutions to these problems, most authors derive simple deterministic rules by assuming the mean function of the NHPP or the characteristic of the failure process is completely known. These simple rules give insight on how different costs in fixing the bugs before and after releasing the software affect the optimal stopping. However, they fail to address the uncertainty associated with the assessment of software reliability. To remedy this drawback, many

authors develop Bayesian sequential decision methodology. It is worthwhile noting the Bayesian sequential decision solutions are different from the usual Bayesian sequential decisions where the observations are assumed to be i.i.d. In software reliability, the time required until observing an additional fault usually increases stochastically, so the observations are not i.i.d. Therefore, different treatments are required. Dalal and Mallows (1988, 1992) present the first solution to the problem. They use a dynamic programming approach to derive an asymptotic solution. Dalal and Mallows (1990) and Dalal and McIntosh (1994) further provide some graphical aids to the problem. Singpurwalla (1991) gives a two stage procedure. Ferguson (1997) provides a general framework and conditions for the one-stage look-ahead rules to be optimal. Zacks (1995) derives sequential stopping rules for both time-domain and data-domain models that include the Littlewood–Verall model and a capture-recapture model. Fakhre-Zakeri and Slud (1995) consider an optimal stopping rule for size-dependent successive sampling. Özekici and Soyer (2001) develop adaptive optimal testing strategies based on operational profile. McDaid and Wilson (2001) develop a single stage, a two stage and a next fixed time look ahead procedure. Morali and Soyer (2003) determine a one-stage look ahead optimal rule for a non-Gaussian Kalman filter type of model.

(3) *Minimizing the expected cost function subject to a reliability goal.* Okumoto and Goel (1980), Yamada and Osaki (1987), and Kapur and Garg (1990) pursue a constrained minimization problem where the expected cost is minimized subject to the reliability meeting the specified requirement.

6. Remarks

The models we have described consist of the dynamic models and the static models. The dynamic models have also been termed the *black-box models* because they only use the interfailure times, not the structure of the software. Some of the static models are called the *white-box models* because of their use of metrics. Integration of these two types, modeling the intensities of the dynamic models using the metric information has received some attention. Proportional hazard models are one approach to integration where the intensity functions of the failure process depends on the metric information. A few references exploring Cox's proportional hazard models for software reliability can be found in Xie (1991). Slud (1997) recently considers a proportional hazard model that can be considered as Poisson regression with a random effect model. He also considers a semiparametric model with time dependent covariates. The semiparametric models incorporating renewals and time trends considered by Lawless and Thiagarajah (1996) would also be useful. Other recent advances in the area include the martingale processes studied by Basu and Ebrahimi (2003) and dynamic reliability models using time-dependent covariates developed by Ray et al. (2003).

References

Achcar, J.A. (2001). Bayesian analysis of software reliability data. In: Balakrishnan, N., Rao, C.R. (Eds.), *Advances in Reliability*, *Handbook of Statistics*, vol. 20. Elsevier Science B.V., Amsterdam, pp. 733–748.

Achcar, J.A., Dey, D.K., Niverthi, M. (1998). A Bayesian approach using nonhomogeneous Poisson Process for software reliability models. In: Basu, A.P., et al. (Eds.), *Frontiers of Reliability Analysis*. In: *World Scientific Series on Quality, Reliability and Engineering Statistics*, pp. 1–18.

Aitkin, M. (1991). Posterior Bayes factors (with discussion). *J. Roy. Statist. Soc., Ser. B* **53**, 111–142.

Al-Mutairi, D., Chen, Y., Singpurwalla, N.D. (1998). An adaptive concatenated failure rate model for software reliability. *J. Amer. Statist. Assoc.* **93**, 1150–1163.

Akiyama, F. (1971). An example of software system debugging. In: *Proceedings of IFIP Congress, vol. 1*. North-Holland, Amsterdam, pp. 353–359.

Bain, L.J., Engelhardt, M. (1986). On the asymptotic behavior of the mean time between failures for repairable systems. In: Basu, A.P. (Ed.), *Reliability and Quality Control*. Elsevier Applied Science, London, pp. 1–7.

Bar-Lev, S.K., Lavi, I., Reiser, B. (1992). Bayesian inference for the power law process. *Ann. Inst. Statist. Math.* **44**, 623–639.

Basawa, I.V., Prakasa Rao, B.L.S. (1980). *Statistical Inference for Stochastic Processes*. Academic Press, London.

Bastani, F.B., Ramamoorthy, C.V. (1988). Software reliability. In: Krishnaiah, P.R., Rao, C.R. (Eds.), *Quality Control and Reliability, Handbook of Statistics*, vol. 7. North-Holland, Amsterdam, pp. 7–26.

Basu, S., Ebrahimi, N. (2003). Bayesian software reliability models based on martingale processes. *Technometrics* **45**, 150–158.

Berger, J.O., Sun, D. (1993). Bayesian analysis for poly-Weibull distribution. *J. Amer. Statist. Assoc.* **88**, 1412–1418.

Box, G. (1980). Sampling and Bayes inference in scientific modeling and robustness (with discussion). *J. Roy. Statist. Soc., Ser. A* **143**, 382–430.

Briand, L., Basili, V.R., Thomas, W.M. (1992). A pattern recognition approach for software engineering data analysis. *IEEE Trans. Software Engineering* **18**, 931–942.

Campodónico, S., Singpurwalla, N.D. (1994). A Bayesian analysis of the logarithmic-Poisson execution time model based on expert opinion and failure data. *IEEE Trans. Software Engineering* **20**, 677–683.

Campodónico, S., Singpurwalla, N.D. (1995). Inference and predictions from Poisson point processes incorporating expert knowledge. *J. Amer. Statist. Assoc.* **90**, 220–226.

Casella, G., George, E.I. (1992). Explaining the Gibbs sampler. *Amer. Statist.* **46**, 167–174.

Caspi, P.A., Kouka, E.F. (1984). Stopping rules for a debugging process based on different software reliability models. In: *Proceedings of International Conference on Fault-Tolerant Computing*, pp. 114–119.

Chao, A., Ma, M.C., Yang, M.C.K. (1993). Stopping rules and estimation for recapture debugging with unequal failure rates. *Biometrika* **80**, 193–201.

Chen, Y., Singpurwalla, N.D. (1994). A non-Gaussian Kalman filter model for tracking software reliability. *Statistica Sinica* **4**, 535–548.

Chen, Y., Singpurwalla, N.D. (1997). Unification of software reliability models by self-exciting point processes. *Adv. Appl. Probab.* **29**, 337–352.

Çinlar, E. (1975). *Introduction to Stochastic Process*. Prentice-Hall, New Jersey.

Cox, D.R., Isham, V. (1980). *Point Processes*. Chapman and Hall, London.

Cox, D.R., Lewis, P.A. (1966). *Statistical Analysis of Series of Events*. Methuen, London.

Crow, L.H., Singpurwalla, N.D. (1984). An empirically developed Fourier series model for describing software failures. *IEEE Trans. on Reliability* **R-33**, 176–183.

Crowder, M.J., Kimber, A.C., Smith, R.L., Sweeting, T.J. (1991). *Statistical Analysis of Reliability*. Chapman and Hall, London.

Dalal, S.R., Mallows, C.L. (1988). When should one stop testing software. *J. Amer. Statist. Assoc.* **83**, 872–879.

Dalal, S.R., Mallows, C.L. (1990). Some graphical aids for deciding when to stop testing software. *IEEE J. Selected Area in Communications* **8** (2), 169–175.

Dalal, S.R., Mallows, C.L. (1992). When to stop testing software? some exact results. In: Goel, P.K., Iyengar, N.S. (Eds.), *Bayesian Analysis in Statistics and Econometrics*. Springer-Verlag, New York, pp. 267–276.

Dalal, S.R., McIntosh, A.A. (1994). When to stop testing for large software systems with changing code. *IEEE Trans. Software Engineering* **20**, 318–323.

Dawid, A.P. (1984). Statistical theory: The prequential approach. *J. Roy. Statist. Soc., Ser. A* **147**, 278–292.

Duane, J.T. (1964). Learning curve approach to reliability monitoring. *IEEE Trans. on Aerospace* **AS-2** (2), 563–566.

Dwass, M. (1964). Extremal processes. *Ann. Math. Statist.* **35**, 1718–1725.

Eick, S.G., Loader, C.R., Vander Wiel, S.A., Votta, L.G. (1993). How many errors remain in a software design document after inspection?. In: *Computing Science and Statistics: Proceedings of the 25th Symposium on the Interface*, pp. 195–202.

Fakhre-Zakeri, I., Slud, E. (1995). Mixture models for reliability of software with imperfect debugging: Identifiability of parameters. *IEEE Trans. on Reliability* **44**, 104–113.

Ferguson, T. (1997). A Poisson fishing model. In: *Festschrift for Lucien Le Cam*. Springer, New York, pp. 234–244.

Forman, E.H., Singpurwalla, N.D. (1977). An empirical stopping rule for debugging and testing computer software. *J. Amer. Statist. Assoc.* **72**, 750–757.

Forman, E.H., Singpurwalla, N.D. (1979). Optimal time intervals for testing hypotheses on computer software errors. *IEEE Trans. on Reliability* **28**, 250–253.

Geisser, S., Eddy, W. (1979). A predictive approach to model selection. *J. Amer. Statist. Assoc.* **74**, 153–160.

Gelfand, A.E., Smith, A.F.M. (1990). Sampling-based approaches to calculating marginal densities. *J. Amer. Statist. Assoc.* **85**, 398–409.

Gelfand, A.E., Dey, D.K., Chang, H. (1992). Model determination using predictive distributions with implementation via sampling-based methods (with discussion). In: Bernardo, J.M., Berger, J.O., Dawid, A.P., Smith, A.F.M. (Eds.), *Bayesian Statistics, vol. 4*. Oxford University Press, Oxford, pp. 147–169.

Gelman, A.E., Rubin, D. (1992). Inference from iterative simulation using multiple sequences. *Statist. Sci.* **7**, 457–472.

Geman, S., Geman, D. (1984). Stochastic relaxation, Gibbs distributions, and the Bayesian restoration of images. *IEEE Trans. Pattern Analysis and Machine Intelligence* **6**, 721–741.

Gilks, W.R., Thomas, A., Spiegelhalter, D.J. (1994). A language and program for complex Bayesian modelling. *The Statistician* **43**, 169–177.

Glick, N. (1978). Breaking records and breaking boards. *Amer. Math. Monthly* **85**, 2–26.

Goel, A.L., Okumoto, K. (1978). An analysis of recurrent software failures on a real-time control systems. In: *Proceedings of the ACM Annual Technical Conference*. ACM, Washington, DC, pp. 496–500.

Goel, A.L., Okumoto, K. (1979). Time dependent error detection rate model for software reliability and other performance measures. *IEEE Trans. on Reliability* **28**, 206–211.

Guida, M., Calabria, R., Pulcini, G. (1989). Bayes inference for a non-homogeneous Poisson process with power intensity law. *IEEE Trans. on Reliability* **38**, 603–609.

Hjorth, U. (1980). A reliability distribution with increasing, decreasing, constant and bathtub-shaped failure rates. *Technometrics* **22**, 99–107.

Jelinski, Z., Moranda, P.B. (1972). Software reliability research. In: Freiberger, W. (Ed.), *Statistical Computer Performance Evaluation*. Academic Press, New York, pp. 465–497.

Jewell, W.S. (1985). Bayesian extension to a basic model of software reliability. *IEEE Trans. Software Engineering* **11**, 1465–1471.

Joe, H. (1989). Statistical inference for general-order-statistics and nonhomogeneous-Poisson-process software reliability models. *IEEE Trans. Software Engineering* **15**, 1485–1490.

Kapur, P.K., Garg, R.B. (1990). Cost-reliability optimum release policies for a software system with testing effort. *Opsearch* **27**, 109–114.

Khoshgoftaar, T.M., Munson, J.C. (1990). Predicting software development errors using software complexity metrics. *IEEE J. Selected Areas in Communications* **8** (2), 253–261.

Khoshgoftaar, T.M., Woodcock, T.G. (1991). Software reliability model selection: A case study. In: *Proc. Int'l Symp. Software Reliability Engineering*, pp. 183–191.

Kremer, W. (1983). Birth–death and bug counting. *IEEE Trans. on Reliability* **R-32** (1), 37–47.

Kuo, L. (1999). Software reliability (update). In: *Encyclopedia of Statistical Sciences, Update Volume 3*, pp. 671–680.

Kuo, L., Cohen, M. (1999). Bayesian analysis for multi-stage linearized models in quantal bioassay. *Biometrical J.* **41**, 53–69.

Kuo, L., Yang, T.Y. (1995). Bayesian computation of software reliability. *J. Comput. Graph. Statist.* **4**, 65–82.

Kuo, L., Yang, T.Y. (1996). Bayesian computation for nonhomogeneous Poisson processes in software reliability. *J. Amer. Statist. Assoc.* **91**, 763–773.

Kuo, L., Lee, J., Choi, K., Yang, T. (1997). Bayes inference for s-shaped software reliability growth models. *IEEE Trans. on Reliability* **46**, 76–80.

Kyparisis, J., Singpurwalla, N.D. (1985). Bayesian inference for the Weibull process with applications to assessing software reliability growth and predicting software failures. In: Billard, L. (Ed.), *Computer Science and Statistics: Proceedings of the Sixteenth Symposium on the Interface*. Elsevier Applied Science, pp. 57–64.

Langberg, N., Singpurwalla, N.D. (1985). A unification of some software reliability models. *SIAM J. Sci. Statist. Comput.* **6**, 781–790.

Lawless, J.F., Thiagarajah, K. (1996). A point-process model incorporating renewals and time trends, with application to repairable systems. *Technometrics* **38**, 131–138.

Levin, K.D., Yadid, O. (1990). Optimal release time of improved versions of software packages. *Inform. Software Techn.* **32**, 65–70.

Lin, H.H., Chen, K.H. (1993). Nonhomogeneous Poisson process software-debugging models with linear dependence. *IEEE Trans. on Reliability* **42**, 613–617.

Lipow, M. (1982). Number of faults per line of code. *IEEE Trans. Software Engineering* **SE-8**, 437–440.

Littlewood, B., Sofer, A. (1987). A Bayesian modification to the Jelinski–Moranda software reliability growth model. *IEE/BCS Software Engineering J.* **3**, 30–41.

Littlewood, B., Verall, J.L. (1973). A Bayesian reliability growth model for computer software. *Appl. Statist.* **22**, 332–346.

McDaid, K., Wilson, S.P. (2001). Deciding how long to test software. *The Statistician* **50**, 117–134.

Meinhold, R.J., Singpurwalla, N.D. (1983). Bayesian analysis of a commonly used model for describing software failures. *The Statistician* **32**, 168–173.

Miller, D.R. (1986). Exponential order statistics models of software reliability. *IEEE Trans. Software Engineering* **SE-12**, 12–24.

Moranda, P.B. (1975). Prediction of software reliability and its applications. In: *Proceedings of the Annual Reliability and Maintainability Symposium*. Washington, DC, pp. 327–332.

Morali, N., Soyer, R. (2003). Optimal stopping in software testing. *Naval Res. Logistics* **50**, 88–104.

Musa, J.D., Okumoto, K. (1984). A logarithmic Poisson execution time model for software reliability measurement. In: *Proceedings Seventh International Conference on Software Engineering*. Orlando, pp. 230–238.

Musa, J.D., Iannino, A., Okumoto, K. (1987). *Software Reliability: Measurement, Prediction, Application*. McGraw-Hill, New York.

Nayak, T.K. (1986). Software reliability: Statistical modeling & estimation. *IEEE Trans. on Reliability* **35**, 566–570.

Nayak, T.K. (1988). Estimating population size by recapture sampling. *Biometrika* **75**, 113–120.

Okumoto, K., Goel, A.L. (1980). Optimum release time for software systems based on reliability and cost criteria. *J. Systems and Software* **1**, 315–318.

Özekici, S., Soyer, R. (2001). Bayesian testing strategies for software with an operational profile. *Naval Res. Logistics* **48**, 747–763.

Parzen, E. (1962). *Stochastic Process*. Holden-Day, San Francisco.

Pasquini, A., De Agostino, E., Di Marco, G. (1996). An input-domain based method to estimate software reliability. *IEEE Trans. on Reliability* **45**, 95–105.

Pettit, L.I., Young, K.D.S. (1990). Measuring the effect of observations on Bayes factors. *Biometrika* **77**, 455–466.

Raftery, A.E. (1987). Inference and prediction for a general order statistic model with unknown population size. *J. Amer. Statist. Assoc.* **82**, 1163–1168.

Ray, B.K., Liu, Z., Ravishanker, N. (2003). Dynamic reliability models for software. Technical Report 03-24, Statistics Department, University of Connecticut.

Resnick, S.I. (1987). *Extreme Values, Regular Variation, and Point Process*. Springer-Verlag, Berlin.

Ross, S.M. (1985a). Statistical estimation of software engineering. *IEEE Trans. Software Engineering* **SE-11**, 479–483.

Ross, S.M. (1985b). Software reliability: The stopping rule problem. *IEEE Trans. Software Engineering* **SE-11**, 1472–1476.
Schick, G.J., Wolverton, R.W. (1973). Assessment of software reliability. In: *Proceedings of Operations Research*. Physica-Verlag, Wiezberg–Wien, pp. 395–422.
Schick, G.J., Wolverton, R.W. (1978). An analysis of computing software reliability models. *IEEE Trans. Software Engineering* **4**, 104–120.
Singpurwalla, N.D. (1991). Determining an optimal time interval for testing and debugging software. *IEEE Trans. Software Engineering* **17**, 313–319.
Singpurwalla, N.D., Soyer, R. (1992). Non-homogeneous autoregressive processes for tracking (software) reliability growth, and their Bayesian analysis. *J. Roy. Statist. Soc., Ser. B* **54**, 145–156.
Singpurwalla, N.D., Soyer, R. (1996). Assessing the reliability of software: An overview. In: Özekici, S. (Ed.), *Reliability and Maintenance of Complex Systems*. Springer, Berlin.
Singpurwalla, N.D., Wilson, S.P. (1994). Software reliability modeling. *Internat. Statist. Rev.* **62**, 289–317.
Singpurwalla, N.D., Wilson, S.P. (1999). *Statistical Methods in Software Engineering: Reliability and Risk*. Spinger-Verlag, New York.
Sinha, S.K. (1986). *Reliability and Life Testing*. Wiley, New York.
Slud, E. (1997). Some applications of counting process models with partially observed covariates. *Telecommunications Systems* **7**, 95–104.
Spiegelhalter, D.J., Best, N.G., Carlin, B.P., van der Linde, A. (2002). Bayesian measures of models complexity and fit (with discussion). *J. Roy. Statist. Soc., Ser. B* **64**, 583–639.
Stone, M. (1974). Cross-validatory choice and assessment of statistical predictions. *J. Roy. Statist. Soc., Ser. B* **36**, 111–147.
Synder, D.L., Miller, M.I. (1991). *Random Point Processes in Time and Space*. Springer-Verlag, New York.
Tanner, M., Wong, W. (1987). The calculation of posterior distributions by data augmentation. *J. Amer. Statist. Assoc.* **82**, 528–550.
Tian, J. (1995). Integrating time domain and input domain analysis of software reliability using tree-based models. *IEEE Trans. Software Engineering* **21**, 945–958.
Tierney, L. (1994). Markov chains for exploring posterior distributions (with discussion). *Ann. Statist.* **22**, 1701–1758.
van Pul, M.C. (1991). Statistical models in software reliability. In: *Proceedings of the 23rd Symposium on the Interface*, pp. 106–109.
Whittaker, J. (1990). *Graphical Models in Applied Multivariate Statistics*. Wiley, Chichester.
Xie, M. (1991). *Software Reliability Modeling*. World Scientific, Singapore.
Xie, M., Bergman, B. (1988). On modeling reliability growth for software. In: *Proceedings of the 8th IFAC Symposium on Identification and System Parameter Estimation*.
Xie, M., Hong, G.Y. (2001). Software reliability modeling, estimation and analysis. In: Balakrishnan, N., Rao, C.R. (Eds.), *Advances in Reliability, Handbook of Statistics*, vol. 20. Elsevier Science B.V., Amsterdam, pp. 707–731.
Yamada, S., Osaki, S. (1986). Optimal software release policies for a nonhomogeneous software error detection rate model. *Microelectronics and Reliability* **26**, 691–702.
Yamada, S., Osaki, S. (1987). Optimal software release policies with simultaneous cost and reliability requirements. *European J. Oper. Res.* **31**, 46–51.
Yamada, S., Ohba, M., Osaki, S. (1983). S-shaped reliability growth modeling for software error detection. *IEEE Trans. on Reliability* **32**, 475–478.
Yang, T., Kuo, L. (1999). Bayesian computation of the superposition of the nonhomogeneous Poisson process. *Canad. J. Statist.* **27**, 547–556.
Zacks, S. (1995). Sequential procedures in software reliability testing. In: Balakrishnan, N. (Ed.), *Recent Advances in Life-Testing and Reliability*. CRC Press, Boca Raton, pp. 107–126.

Handbook of Statistics, Vol. 25
ISSN: 0169-7161
© 2005 Elsevier B.V. All rights reserved.
DOI 10.1016/S0169-7161(05)25034-4

34

Bayesian Aspects of Small Area Estimation

Tapabrata Maiti

1. Introduction

Small area estimation has become an indispensable tool for policy making in recent years. The need for reliable small area statistics is felt by many agencies, both public and private, for making useful policy making decisions. The term "small area", or "small domain" usually refers to small geographical area such as state, municipality, or county etc., or a small age–sex–race group of people. A state may be a small area with respect to a nation while a county may be a small area with respect to a state. It is now widely recognized that the direct survey estimates for small areas are usually unreliable, being accompanied with large standard errors and coefficient of variation. This makes it necessary to use models, either explicit or implicit, to connect the small areas, and obtain estimates of improved precision by "borrowing strength". Models play a major role in modern small area estimation. There are several advantages of model based inference over the traditional indirect estimators. In this article we will focus mainly on empirical Bayes and hierarchical Bayes estimators related to small area models. The Bayes estimators are usually obtained under squared error loss. Empirical Bayes estimators are obtained as estimated posterior mean while for hierarchical Bayes usually subjective prior distributions are assumed for the model parameters.

The organization of this chapter is as follows. Section 2 discusses some applications where small area statistics are heavily used. The small area models are described in Section 3. Empirical Bayes estimators are described in Section 4 and hierarchical Bayes estimators are given in Section 5. Section 5 draws the conclusion with direction of future research.

2. Some areas of application

By now small area statistics are being used almost in all aspects of real life activities, such as health, agriculture, social science, business, economics and environmental science.

Socio-economic application. Fay and Herriot (1979) first used model based small area methods to estimate per capita income (PCI) for several small places. US Treasury Department use these statistics to determine apportionment funds to local governments within a state by dividing the corresponding state fundings.

The US Census Bureau has initiated the "Small Area Income and Poverty Estimates" (SAIPE) project to meet growing demands for regularly updated estimates of key income and poverty measures for sub national areas, such as states, counties, and school districts. These estimates are being used by the US Department of Education to allocate over $7 billion of funds for poor and educationally backward school children under Title I of Elementary and Secondary Education Act. These estimates are produced every two years by the Census Bureau and being used under Title I allocations in place of estimates from the recent most census, unless the Secretary of Commerce and Education found them 'inappropriate' or 'unreliable' on the basis of a review by a panel of committee on National Statistics at the National Research Council (2000).

The European Community funded a project called "Enhancing Small Area Estimation Techniques to meet European Needs" (EURAREA) to improve the small area estimation methods currently used in a consortium of six European National Statistical Institutes and five universities, covering seven European countries. The methods are being tested on data relating to unemployment status, income, and household composition.

Health application. Small area statistics has become very important in health system of developed countries like U.S. and Canada. The U.S. National Center for Health Statistics is the pioneer of using small area statistics to estimate disability and other health characteristics for different groups of people from the National Health Interview Survey (NHIS). Malec et al. (1999) studied overweight prevalence for adults by states, using data from NHANES III. Folsom et al. (1999) produced prevalence rates for states and age groups for several variables including drug use using data from National Household surveys on Drug Abuse.

Small area methodology is popular in disease mapping. Disease mapping is useful for formulating and assessing aetiological hypotheses, resource allocation and identification of areas of unusually high risk lockets. Standardized mortality ratios are unstable and unreliable due to variation of different population sizes. Thus model based small area estimates seems to be more appropriate. Currently U.S. National Cancer Institute is using sophisticated model for mapping cancer and prediction of number of cancer deaths. More details are available in the special issue of *Statistics in Medicine*, October, 2000.

There are several other applications which are not mentioned in this chapter for brevity.

3. Small area models

Traditionally small area methods use implicit models. However, models are well established in modern small area literature. We will discuss here only a few of the basic models, and there may be large number of variations given a particular application. The general form of a small area model is as follows:

The direct survey estimator is connected with the true small area parameter through a link function and then in second stage, the true parameter is linked with a set of auxiliary variables and a random effect, usually known as small area effect.

3.1. Basic area level model

Often microdata is not available for many reason, such as confidentiality. Let y_i be the direct survey estimate of a small area parameter θ_i for area i, $i = 1, \dots, m$. The small area parameter θ_i, may be area total or area mean or some other function of the response variable. Then assume

$$y_i = \theta_i + e_i, \quad i = 1, \dots, m, \tag{3.1}$$

where e_i's are sampling error, independently distributed with $E_\pi(e_i|\theta_i) = 0$ and $V_\pi(e_i|\theta_i) = \psi_i$ and ψ_i's are assumed to be known sampling variances. Here π denotes the probability distribution from randomization. Let $x_i = (x_{i1}, \dots, x_{ip})^{\mathrm{T}}$ be a set of p-auxiliary variables that relates to θ_i as

$$\theta_i = \boldsymbol{x}_i^{\mathrm{T}}\boldsymbol{\beta} + u_i, \quad i = 1, \dots, m, \tag{3.2}$$

where $\boldsymbol{\beta}$ is the regression coefficient and u_i's are random effect, distributed independently with $E(u_i) = 0$ and $V(u_i) = \sigma_u^2$. e_i's and u_i's are mutually independent and normality is assumed in general.

Combining (3.1) and (3.2) one obtain the mixed linear model

$$y_i = \boldsymbol{x}_i^{\mathrm{T}}\boldsymbol{\beta} + u_i + e_i, \quad i = 1, \dots, m. \tag{3.3}$$

The model (3.3) often called as Fay–Herriot model since Fay and Herriot (1979) first used this model to estimate income of small places.

3.2. Basic unit level model

Suppose y_{ij} is the response from jth unit of ith area $j = 1, \dots, N_i$; $i = 1, \dots, m$ and $x_{ij} = (x_{ij1}, \dots, x_{ijp})^{\mathrm{T}}$ is the set of auxiliary variables. Here N_i is the population size of ith area then the basic unit level model is given by

$$y_{ij} = \boldsymbol{x}_{ij}^{\mathrm{T}}\boldsymbol{\beta} + u_i + e_{ij}, \tag{3.4}$$

where the error e_{ij}'s are independent and identical random variables with $E(e_{ij}) = 0$, $V(e_{ij}) = \sigma_e^2$ and u_i's are area specific random effects distributed independently with $E(u_i) = 0$, $V(u_i) = \sigma_u^2$. e_{ij}'s and u_i's are mutually independent. Normal distribution is assumed in general for both u_i and e_{ij}.

The small area parameter, say population mean $\overline{Y}_i = \frac{1}{N_i}\sum_{j=1}^{N_i} y_{ij}$ is then given by

$$\overline{Y}_i = f_i\bar{y}_i + (1 - f_i)\bar{y}_i^*,$$

where $f_i = n_i/N_i$ = sampling fraction in ith area and $\bar{y}_i$ and $\bar{y}_i^*$ denoting the means of the sampled and nonsampled units respectively. If the area size N_i is large, then one can define the small area mean as

$$\overline{Y}_i = \overline{\boldsymbol{X}}_i^{\mathrm{T}} \boldsymbol{\beta} + u_i$$

by ignoring $\bar{e}_i = \frac{1}{N_i} \sum_{j=1}^{N_i} e_{ij}$. Here $\overline{\boldsymbol{X}}_i$ is the mean of x for ith area. Then the problem is equivalent to estimate the regression coefficient $\boldsymbol{\beta}$ and predict random effect u_i.

Battese et al. (1988) first used this model to estimate Iowa county areas under corn and soybeans.

There are number of extensions of above two basic models available in the literature which are not mentioned here. However, some of them will be discussed to some extent in the context of particular examples.

3.3. Generalized linear mixed models

The basic two models mentioned above are particularly suitable for continuous response variable. Often the variable of interest is discrete, such as binary or count data. In such cases, generalized linear mixed models are more suitable for small area estimation than the linear mixed model. For area level data, a generalized linear mixed model is given by

$$f(y_i|\theta_i) = \exp\{(y_i\theta_i + b(\theta_i))/\phi_i + c(y_i, \phi_i)\}, \tag{3.5}$$

where θ_i is the canonical parameter. The exponential family (3.5) cover many known distributions like normal, binomial, Poisson, etc. Then θ_i is modeled as

$$\eta(\theta_i) = \boldsymbol{x}_i^{\mathrm{T}} \boldsymbol{\beta} + u_i, \tag{3.6}$$

where u_i's are i.i.d. random variable with $E(u_i) = 0$ and $V(u_i) = \sigma_u^2$. Parameter of interest is θ_i or some function of θ_i.

Ghosh et al. (1998) and Maiti (1998) used this type of model for disease mapping. Malec et al. (1997) use the methodology for estimating proportion of persons who have visited a physician in the past year in subnational level.

4. Inference from small area models

Estimation of small area models was first rigorously discussed by Prasad and Rao (1990) in a frequentist framework. They first obtained the best linear unbiased predictor of small area parameter which involves unknown model parameters. Then by replacing the unknown parameters by corresponding estimates, the empirical best linear unbiased predictor (EBLUP) was obtained. To obtain EBLUP, normal distribution of random terms are not required. However, normality is required to obtain mean squared error (MSE) estimate up to a desire level of accuracy. Under normality empirical Bayes (EB) and EBLUP results are equivalent and produce second order correct MSE estimate. On the other hand hierarchical Bayes procedure produce accurate MSE estimate and usually no approximation is needed. In this article EBLUP approach will not be discussed. For detail of the procedures, we refer to Rao (2003).

4.1. *Empirical Bayes small area estimation*

In EB approach, first the posterior distribution of small area parameters are obtained in a given data. Then the model parameters are estimated from the marginal distribution of data and plug-in into the posterior distribution. Inference is made from the estimated posterior distribution. Morris (1983) gives an excellent account of the EB approach and significant application.

The basic two level hierarchical model can be written as

(i) $y_i|\theta_i \overset{\text{ind}}{\sim} N(\theta_i, \psi_i)$ and

(ii) $\theta_i \overset{\text{ind}}{\sim} N(\boldsymbol{x}_i^{\mathrm{T}}\boldsymbol{\beta}, \sigma_u^2)$, $i = 1, \ldots, m$.

Then the posterior distribution of θ_i given y_i, $\boldsymbol{\beta}$ and σ_u^2 is normal with mean $\theta_i^B = \gamma_i y_i + (1-\gamma_i)\boldsymbol{x}_i^{\mathrm{T}}\boldsymbol{\beta}$ and variance $g_{1i}(\sigma_u^2) = \gamma_i\psi_i, i = 1, \ldots, m$, where $\gamma_i = \sigma_u^2/(\sigma_u^2 + \psi_i)$. Under squared error loss θ_i^B is the 'Bayes' estimator of small area parameter θ_i. θ_i^B is also known as the 'best predictor' of (BP) θ_i. Note that the Bayes estimator θ_i^B is not useable since it depends on the model parameters $\boldsymbol{\beta}$ and σ_u^2. The model parameters are estimated from the marginal distribution of y_i using maximum likelihood (ML), restricted maximum likelihood (REML) or the method of moments. It is worth mentioning that marginally y_i's are independent $N(\boldsymbol{x}_i^{\mathrm{T}}\boldsymbol{\beta}, \sigma_u^2 + \psi_i)$. The estimated posterior distribution is $N(\hat{\theta}_i^{EB}, g_{1i}(\hat{\sigma}_u^2))$, where $\hat{\theta}_i^{EB}$ is $\theta_i^B(\hat{\boldsymbol{\beta}}, \hat{\sigma}_u^2)$ and $\hat{\boldsymbol{\beta}}$ and $\hat{\sigma}_u^2$ are the estimate of $\boldsymbol{\beta}$ and σ_u^2. A naive measure of uncertainty of $\hat{\theta}_i^{EB}$ is the estimated posterior variance

$$V\big(\theta_i|y_i, \hat{\boldsymbol{\beta}}, \hat{\sigma}_u^2\big) = g_{1i}\big(\hat{\sigma}_u^2\big). \tag{4.1}$$

It is now well understood that (4.1) leads to underestimation of the true posterior variance $V(\theta_i|\boldsymbol{y})$, although $\hat{\theta}_i^{EB} = E(\theta_i|y_i, \hat{\boldsymbol{\beta}}, \hat{\sigma}_u^2)$ is approximately equal to the true posterior mean $E(\theta_i|\boldsymbol{y})$, where $\boldsymbol{y} = (y_1, \ldots, y_m)^{\mathrm{T}}$. This point will be well understood from the fact that

$$E(\theta_i|\boldsymbol{y}) = E_{\boldsymbol{\beta},\sigma_u^2}\big\{E\big(\theta_i|y_i, \boldsymbol{\beta}, \sigma_u^2\big)\big\} \quad \text{and} \tag{4.2}$$

$$V(\theta_i|\boldsymbol{y}) = E_{\boldsymbol{\beta},\sigma_u^2}\big\{V\big(\theta_i|y_i, \boldsymbol{\beta}, \sigma_u^2\big)\big\} + V_{\boldsymbol{\beta},\sigma_u^2}\big\{E\big(\theta_i|y_i, \boldsymbol{\beta}, \sigma_u^2\big)\big\}, \tag{4.3}$$

where $E_{\boldsymbol{\beta},\sigma_u^2}$ and $V_{\boldsymbol{\beta},\sigma_u^2}$ respectively denote the expectation and variance with respect to posterior distribution of $\boldsymbol{\beta}$ and σ_u^2 given the data $\boldsymbol{y} = (y_1, \ldots, y_m)^{\mathrm{T}}$.

Note that $E(\theta_i|y_i) = \hat{\theta}_i^{EB}(1 + \mathrm{o}(m^{-1}))$ regardless of the prior distributions of $\boldsymbol{\beta}$ and σ_u^2. Hence $\hat{\theta}_i^{EB}$ is a good approximation of $E(\theta_i|\mathrm{y})$. On the other hand, $g_{1i}(\hat{\sigma}_u^2)$ is a good approximation of first term of the right-hand side of (4.3), but the second term is ignored in the naive EB approach, and, therefore, ignore the uncertainty about the parameters $\boldsymbol{\beta}$ and σ_u^2.

There are some procedures available in literature to account for this under estimation. We will discuss only two methods – one based on bootstrap (Laird and Louis, 1987) and the other is based on asymptotic approximation to the posterior variance irrespective of the form of the prior distribution of $\boldsymbol{\beta}$ and σ_u^2 (Kass and Steffey, 1989).

Bootstrap method:

(i) Generate B independent bootstrap sample $\{y_i^{*(b)}\}_{b=1}^B$ from the marginal distribution of y_i: $N(\boldsymbol{x}_i^{\mathrm{T}}\hat{\boldsymbol{\beta}}, \hat{\sigma}_u^2 + \psi_i)$, $i = 1, \ldots, m$.

(ii) Estimate $\{\hat{\boldsymbol{\beta}}^{*(b)}\}_{b=1}^B$ and $\{\hat{\sigma}_u^{*2(b)}\}_{b=1}^B$ from the bootstrap sample and construct empirical Bayes estimates $\{\hat{\theta}_i^{*EB(b)}\}_{b=1}^B$.

(iii) Construct the EB estimate as

$$\hat{\theta}_i^* = \frac{1}{B}\sum_{b=1}^{B}\hat{\theta}_i^{*EB(b)}.$$

(iv) Construct the posterior variance as

$$V_{LL}(\theta_i|\boldsymbol{y}) = \frac{1}{B}\sum_{b=1}^{B} g_{1i}\big(\hat{\sigma}_u^{*2(b)}\big) + \frac{1}{B-1}\sum_{b=1}^{B}\big(\hat{\theta}_i^{*EB(b)} - \hat{\theta}_i^*\big)^2. \tag{4.4}$$

The second term on the right-hand side of (4.4) tracks the second term on the right-hand side of (4.3).

Butar and Lahiri (2001) studied the performance of V_{LL} as an estimator of $\mathrm{MSE}(\hat{\theta}_i^{EB})$ and found that V_{LL} still under estimate $\mathrm{MSE}(\hat{\theta}_i^{EB})$, of order $o(m^{-1})$. They proposed a bias corrected formula which is approximately

$$V_{BL}\big(\hat{\theta}_i^{EB}\big) = g_{1i}\big(\hat{\sigma}_u^2\big) + g_{2i}\big(\hat{\sigma}_u^2\big) + g_{3i}^*\big(\hat{\sigma}_u^2\big) + g_{3i}\big(\hat{\sigma}_u^2\big), \tag{4.5}$$

where

$$g_{2i}\big(\hat{\sigma}_u^2\big) = (1-\gamma_i)^2\boldsymbol{x}_i^{\mathrm{T}}\bigg(\sum_i \frac{\boldsymbol{x}_i\boldsymbol{x}_i^{\mathrm{T}}}{\psi_i + \sigma_u^2}\bigg)^{-1}\boldsymbol{x}_i,$$

$$g_{3i}\big(\hat{\sigma}_u^2\big) = \psi_i^2\big(\psi_i + \sigma_u^2\big)^{-3}V\big(\hat{\sigma}_u^2\big),$$

and

$$g_{3i}^*\big(\hat{\sigma}_u^2\big) = \psi_i^2\big(\psi_i + \hat{\sigma}_u^2\big)^{-4}\big(y_i - \boldsymbol{x}_i^{\mathrm{T}}\hat{\boldsymbol{\beta}}\big)^2\widehat{V}\big(\hat{\sigma}_u^2\big),$$

$$\widehat{V}\big(\hat{\sigma}_u^2\big) = \big(I\big(\hat{\sigma}_u^2\big)\big)^{-1} = \text{the asymptotic variance of } \hat{\sigma}_u^2.$$

The Kass–Steffey method adds a positive term to the estimated positive variance $V(\theta_i|y_i, \hat{\boldsymbol{\beta}}, \hat{\sigma}_u^2)$ to account for under estimation. In particular their formula for $\mathrm{MSE}(\hat{\theta}_i^{EB})$ is

$$\begin{aligned} V_{KS}(\theta_i|\boldsymbol{y}) = {} & g_{1i}\big(\hat{\sigma}_u^2\big) + \bigg(\frac{\partial\theta_i^B}{\partial\boldsymbol{\beta}}\bigg)^{\mathrm{T}} V(\hat{\boldsymbol{\beta}})\bigg(\frac{\partial\theta_i^B}{\partial\boldsymbol{\beta}}\bigg)\bigg|_{\boldsymbol{\beta}=\hat{\boldsymbol{\beta}},\sigma_u^2=\hat{\sigma}_u} \\ & + \bigg(\frac{\partial\theta_i^B}{\partial\sigma_u^2}\bigg)^2 V\big(\hat{\sigma}_u^2\big)\bigg|_{\boldsymbol{\beta}=\hat{\boldsymbol{\beta}},\sigma_u^2=\hat{\sigma}_u^2}, \end{aligned} \tag{4.6}$$

where $V(\hat{\boldsymbol{\beta}}) = (\sum \frac{\boldsymbol{x}_i\boldsymbol{x}_i^{\mathrm{T}}}{\psi_i+\sigma_u^2})^{-1}$ is the asymptotic covariance matrix of $\boldsymbol{\beta}$. After some simplifications, the Kass–Steffey formula becomes

$$V_{KS}(\theta_i\,|\,\boldsymbol{y}) = g_{1i}(\hat{\sigma}_u^2) + g_{2i}(\hat{\sigma}_u^2) + g_{3i}^K(\hat{\sigma}_u^2, y_i) \tag{4.7}$$

comparing (4.7) with (4.5), V_{KS} still under estimate the true MSE by the quantity $g_{3i}(\sigma_u^2)$. Kass and Steffey (1989) have given a more accurate approximation to $V(\theta_i\,|\,y)$ and ensures that the neglected terms are of lower order than m^{-1}. But this approximation depends on the prior density of $\boldsymbol{\beta}$ and σ_u^2. For more details see Singh et al. (1998).

Turning to the basic unit level model (3.4), an EB estimate of small area mean $\overline{Y}_i$ is same as EBLUP

$$\hat{\theta}_i^{EB} = \hat{\gamma}_i\big(\bar{y}_i + (\overline{\boldsymbol{X}}_i - \bar{\boldsymbol{x}}_i)^{\mathrm{T}}\boldsymbol{\beta}\big) + (1-\hat{\gamma}_i)\overline{\boldsymbol{X}}_i^{\mathrm{T}}\hat{\boldsymbol{\beta}}, \tag{4.8}$$

where $\bar{y}_i$ and $\bar{x}_i$ is the sample mean of y and x for ith area. $\hat{\gamma}_i = \hat{\sigma}_u^2/(\hat{\sigma}_u^2 + \frac{\hat{\sigma}_e^2}{n_i})$. The estimated posterior variance is given by $(1-f_i)^2 g_{1i}(\hat{\sigma}_u^2, \hat{\sigma}_e^2)$ which again underestimate the true variance $V(\theta_i\,|\,\boldsymbol{y})$ by the amount of uncertainty associated with $\boldsymbol{\beta}, \sigma_u^2$ and σ_e^2. Again the bootstrap and Kass and Steffey methods can be applied to account for the under estimation.

Now we will discuss the EB approach when the response variable is binary or count data. In particular we consider natural exponential family-quadratic variance function (NEF-QVF) populations. Beginning with Dempster and Tomberlin (1980), there are several attempts towards EB analysis of binomial data (see, for example, MacGibbon and Tomberlin, 1989; Farrell et al., 1997; Jiang and Lahiri, 2001), but a general systematic approach based on exponential family models which handles both discrete and continuous data is mostly lacking.

Sarkar and Ghosh (1998) initiated such an EB analysis for the natural exponential family (NEF) with quadratic variance functions (QVF) models. Morris (1983) characterized the NEF-QVF distributions. Its members are the binomial, Poisson, normal with known variance, gamma, negative binomial and the generalized hyperbolic secant distributions.

Let y_i denote the direct estimator of the ith area mean ($i = 1, \ldots, k$). We assume that y_i to be independent with pdf

$$f(y_i\,|\,\theta_i) = \exp\big\{\xi_i\big(y_i\theta_i - \psi(\theta_i)\big) + C(y_i, \xi_i)\big\}, \tag{4.9}$$

where ξ_i (> 0) are known constants. From McCullagh and Nelder (1989, p. 28) we have

$$\begin{aligned} E(y_i\,|\,\theta_i) &= \psi'(\theta_i) = \mu_i \quad \text{(say)},\\ V(y_i\,|\,\theta_i) &= \psi''(\theta_i)/\xi_i = V(\mu_i) \quad \text{say}. \end{aligned} \tag{4.10}$$

For the NEF-QVF family of distributions, $V(\mu_i) = v_0 + v_1\mu_i + v_2\mu_i^2$ where v_0, v_1, and v_2 are not simultaneously zeroes. In particular, for the binomial model, $v_0 = 0$, $v_1 = 1$, and $v_2 = -1$. For the Poisson model, $v_0 = v_2 = 0$ and $v_1 = 1$.

For θ_i, we consider the conjugate prior

$$\pi(\theta_i) = \exp\big\{\lambda\big[m_i\theta_i - \psi(\theta_i)\big]\big\}K(\lambda, m_i). \tag{4.11}$$

The prior mean and variance of μ_i defined in (4.10) are then given by (cf. Morris, 1983) $E(\mu_i) = m_i$, $\text{var}(\mu_i) = \frac{V(m_i)}{\lambda - v_2}$ provided $\lambda > v_2$. It is easy to check from (4.9) and (4.11) that the posterior $\pi(\theta_i|y_i)$ also belongs to the NEF-QVF family with

$$\pi(\theta_i|y_i) \propto \exp\bigl[(\xi_i y_i + \lambda m_i)\theta_i - (\xi_i + \lambda)\psi(\theta_i)\bigr] \tag{4.12}$$

which leads to

$$E(\mu_i|y_i) = (1 - B_i)y_i + B_i m_i, \quad B_i = \frac{\lambda}{\lambda + \xi_i},$$

$$\text{Var}(\mu_i|y_i) = \frac{V[E(u_i|y_i)]}{\xi_i + \lambda - v_2}.$$

Also the prior mean m_i are modeled as $g(m_i) = x_i^{\mathrm{T}}\boldsymbol{\beta}$, where g is a strictly increasing function. The above formulation has the advantage over standard generalized linear mixed model is that the posterior expectation has a closed form expression. This is particularly useful in obtaining the EB estimators, and then finding closed form approximations to their MSE's for very large datasets.

For simplicity we take $g \equiv (\psi')^{-1}$ and $\xi_i = \eta_i$ which reflects the tacit assumption that y_i is the average on n_i i.i.d. random variables each having a distribution of the form (4.9) with $\xi_i = 1$. To obtain the EB estimates of θ_i's, we need to estimate $\boldsymbol{\beta}$ and λ from the marginal distribution of y_i's. We use the theory of optimal estimating functions (Godambe and Thompson, 1989) for simultaneous estimation of $\boldsymbol{\beta}$ and λ.

Noting that $E(y_i) = m_i$ and $\text{Var}(y_i) = E(y_i - m_i)^2 = \phi_i V(m_i)$, where $\phi_i = \frac{\lambda + n_i}{n_i(\lambda - v_2)}$, following Godambe and Thompson (1989), we begin with the elementary estimating functions $g_i = (g_{1i}, g_{2i})^{\mathrm{T}}$, where $g_{1i} = y_i - m_i$ and $g_{2i} = (y_i - m_i)^2 - \phi_i V(m_i)$, $i = 1, \ldots, k$. Let

$$D_i = \begin{bmatrix} -E(\frac{\partial g_{1i}}{\partial \boldsymbol{\beta}}) & -E(\frac{\partial g_{2i}}{\partial \boldsymbol{\beta}}) \\ -E(\frac{\partial g_{1i}}{\partial \lambda}) & -E(\frac{\partial g_{2i}}{\partial \boldsymbol{\beta}}) \end{bmatrix} = \begin{bmatrix} V(m_i)x_i & V(m_i)V'(m_i)\phi_i x_i \\ 0 & V(m_i)\frac{n_i + v_2}{n_i(\lambda - v_2)} \end{bmatrix}.$$

Next, writing $\mu_{ri} = E(y_i - m_i)^r$ $(r = 1, 2, \ldots)$, let

$$V_i = \begin{pmatrix} \mu_{2i} & \mu_{3i} \\ \mu_{3i} & \mu_{4i} - \mu_{2i}^2 \end{pmatrix}.$$

Let

$$\eta = (\boldsymbol{\beta}, \lambda)^{\mathrm{T}} \quad \text{and} \quad S_k(\eta) = \sum_{i=1}^{k} D_i^{\mathrm{T}} V_i^{-1} g_i. \tag{4.13}$$

From Godambe and Thompson (1989), the optimal estimating equation are given by $S_k(\eta) = 0$ and its solution, say $\hat{\eta}$ is used to estimate η. Solving (4.13), one obtain $\hat{\eta} = (\boldsymbol{\beta}, \lambda)^{\mathrm{T}}$ and the EB estimator as

$$\hat{\mu}_i^{EB} = \bigl(1 - \widehat{B}_i\bigr)y_i + \widehat{B}_i\psi'\bigl(x_i^{\mathrm{T}}\hat{\boldsymbol{\beta}}\bigr), \tag{4.14}$$

where $\widehat{B}_i = \frac{\hat{\lambda}}{\hat{\lambda} + n_i}$.

Crescenzi et al. (2002) gave a valid estimate of MSE $(\hat{\mu}_i^{EB})$ which is correct up to $o(k^{-1})$. The formula is complicated and involves large expression and hence omitted from this chapter.

An example

Efron (1978, 1986) considered an example showing the proportion of subjects testing positive for the disease toxoplasmosis in 34 cities of El Salvador. Efron's prime objective was to assess the effect of rainfall on the proportions positive. Efron (1978) met the objective by fitting a logistic regression to the data, and subsequently (Efron, 1986), by fitting an overdispersed generalized linear model.

Our objective, however, is more modest. Since the sample sizes in the local areas are mostly small, it is anticipated that the EB estimates will have greater precision than the usual estimates, namely, the sample proportions. This is substantiated in the following table which provides the sample sizes (n_i), the sample proportions ($\hat{p}_i$), the EB estimates ($\hat{p}_i^{EB}$), the estimated standard errors of the sample proportions, namely, $SE(\hat{p}_i) = [\hat{p}_i(1-\hat{p}_i)/n_i]^{-1/2}$, the naive standard errors of the EB estimators, namely, $NSE(\hat{p}_i^{EB}) = [V((1-\widehat{B}_i)y_i + \widehat{B}_i\hat{m}_i)/(n_i+\hat{\lambda}+1)]^{1/2}$, and the approximate root mean squared errors of the EB estimators as derived in Section 3, denoted by $RMSE(\hat{p}_i^{EB})$.

Following Efron (1978), we consider the model $m_i = b_0 + b_1 x_i + b_2 x_i^2 + b_3 x_i^3$, $i = 1, 2, \ldots, 34$. Thus, based on (4.13), we need to solve five nonlinear equations to derive the estimates.

Table 1 exhibits several interesting features. First, the naive standard errors grossly underestimate the precisions of the estimates, especially for the strata where the sample sizes are very small. Naturally, these are the strata where the $RMSE(\hat{p}_i^{EB})$ correct these naive standard errors the most. For example, in city 3, the correction is of the order of 16% over the naive estimate, whereas in cities 13, 14, 26, 29 and 32, there is no correction at all. Also, there are several strata where the sample sizes are 1, leading necessarily to zero standard errors of the classical MLE's. The EB RMSE's provide more credible measures of uncertainty in these cases. For other strata, $RMSE(\hat{p}_i^{EB})$ are always smaller than $SE(\hat{p}_i)$'s and sometimes are substantially smaller when the strata sizes are very small (but not 1 or $\hat{p}_i \neq 1$). This is evidenced, for example, in city 1, where the improvement is of the order of 33%. Also, the MLE's for the population proportions are shrunk most for strata with small sample sizes rather than those with moderate or large sample sizes.

4.2. Hierarchical Bayes small area estimation

Hierarchical Bayes approach, sometime referred to as fully Bayesian approach assume a prior distribution on the model parameters and then inference is made based on the posterior distribution. In particular, under squared error loss, the small area parameter is estimated by posterior means and its uncertainty is measured by posterior variance. The hierarchical Bayes method is simple to understand and easy to apply. However the application is computation intensive and required evaluation of high-dimensional integrations. Recent advances in computational aspects, particularly Gibbs sampling (cf. Gelfand and Smith, 1990) the Bayesian approach becomes quite popular in every field of statistics. The basic principle is very simple and application of Bayes theorem. Let $\boldsymbol{\theta}$ is the parameter of interest and $\boldsymbol{\lambda}$ is the set of all other parameters. Then the joint posterior distribution of $\boldsymbol{\theta}$ and $\boldsymbol{\lambda}$ is obtained by applying Bayes theorem as

$$f(\theta, \boldsymbol{\lambda} | \boldsymbol{y}) = \frac{f(y, \theta | \lambda) f(\lambda)}{f(y)},$$

Table 1
The estimates and the standard errors for the toxoplasmosis data

City(i)	n_i	$\hat{p}_i$	$\hat{p}_i^{EB}$	$SE(\hat{p}_i)$	$NSE(\hat{p}_i^{EB})$	$RMSE(\hat{p}_i^{EB})$
1	4	.500	.394	.250	.143	.165
2	5	.200	.389	.179	.144	.167
3	2	1.000	.513	.000	.162	.189
4	8	.250	.256	.153	.114	.127
5	6	.500	.469	.204	.140	.160
6	24	.292	.353	.093	.075	.078
7	30	.500	.511	.091	.075	.082
8	1	.000	.501	.000	.193	.231
9	1	.000	.338	.000	.197	.233
10	9	.444	.501	.166	.120	.137
11	12	.167	.191	.134	.114	.130
12	11	.727	.690	.134	.114	.130
13	51	.471	.462	.070	.058	.059
14	82	.561	.569	.055	.042	.042
15	43	.535	.525	.076	.064	.066
16	13	.615	.530	.135	.094	.101
17	6	.167	.327	.152	.140	.161
18	10	.300	.346	.145	.119	.133
19	10	.300	.374	.145	.118	.132
20	5	.600	.445	.219	.129	.147
21	19	.368	.436	.111	.080	.083
22	10	.800	.643	.126	.104	.113
23	1	.000	.215	.000	.177	.207
24	22	.182	.268	.082	.077	.080
25	11	.545	.468	.150	.101	.109
26	54	.611	.586	.066	.051	.051
27	8	.625	.672	.171	.109	.131
28	1	.000	.474	.000	.196	.260
29	77	.532	.518	.057	.043	.043
30	16	.437	.414	.124	.093	.100
31	13	.692	.713	.128	.090	.096
32	75	.707	.685	.053	.045	.045
33	10	.300	.322	.145	.115	.128
34	37	.622	.595	.080	.067	.073

where $f(y)$ is the marginal density of y and $f(\lambda)$ is the prior distribution of λ. Then the posterior distribution of θ is given by

$$\begin{aligned} f(\theta|y) &= \int f(\theta, \lambda|y)\,\mathrm{d}\lambda \\ &= \int f(\theta|y, \lambda) f(\lambda|y)\,\mathrm{d}\lambda. \end{aligned} \tag{4.15}$$

It is clear from (4.15) that inference for the posterior distribution of θ may require multi-dimensional integration. For details we refer to Gilks et al. (1996), Gelman et al. (1996) and Chen et al. (2000).

Consider the basic area level model (3.3). Here we need to define prior distribution on the model parameters $\boldsymbol{\beta}$ and σ_u^2. In presence of prior knowledge one would work with sensitive prior distributions. In lack of prior information, noninformative prior is often used. Using noninformative improper prior, one has to be careful about the propriety of posterior distribution (cf. Hobert and Casella, 1996). Assuming $\boldsymbol{\beta}$ and σ_u^2 are independent *apriori*, we take 'flat' prior of $\boldsymbol{\beta}$ and inverse gamma distribution on σ_u^2. In particular, we take $f(\boldsymbol{\beta})\alpha 1$ and $f(\sigma_u^2) \sim IG(\frac{a}{2}, \frac{b}{2}), a > 0, b > 0$. The positive constants a and b are usually set to be very small to make the prior distribution diffuse. To apply Gibbs sampler, one is to find the full conditional distributions, which are easy to find in this case.

Bell (1999) used hierarchical Bayes method in basic area level model to estimate median income and number of poor school-age children in all the states of the U.S. Datta et al. (1996) used multivariate version of basic area level models to develop HB estimators of median income of four-person families for U.S. states. You et al. (2001) applied above HB method in even more complex model to estimate monthly unemployment rate for small places in Canada. Datta et al. (1999) applied HB small area estimation method to estimate unemployment rates for all the U.S. states by combing cross-sectional and time series data.

Now we will present a small data analysis using both EB and HB approach of basic area level model. We take the data from Battese et al. (1988), where survey data and satellite data are combined to estimate area under corn and soybeans for twelve counties of north central Iowa. Let $\bar{y}_i = \frac{1}{n_i}\sum_{j=1}^{n_i} y_{ij}$ be the sample average corn acres for county i $(i = 1, \ldots, m)$ and $x_i = (1, \overline{X}_i)^{\mathrm{T}}$, where $\overline{X}_i$ be the average number of pixels classified as corn in the ith county. We then model

$$\text{(I) } \bar{y}|\theta_i \sim N(\theta_i, \psi_i), \quad \psi_i = \frac{\sigma_e^2}{n_i}, \sigma_e^2 \text{ known,}$$

$$\text{(II) } \theta_i|\boldsymbol{\beta}, \sigma_u^2 \sim N\left(\boldsymbol{x}_i^{\mathrm{T}}\boldsymbol{\beta}, \sigma_u^2\right),$$

$$\text{(III) } \boldsymbol{\beta}\alpha 1,$$

$$\text{(IV) } \sigma_u^2 \sim IG\left(\frac{a}{2}, \frac{b}{2}\right).$$

The Gibbs conditional distributions are given by

$$\text{(i) } [\boldsymbol{\beta}|\text{rest}] \sim N_p\left(\boldsymbol{\beta}^*, \sigma_u^2\left(\sum_i \boldsymbol{x}_i\boldsymbol{x}_i^{\mathrm{T}}\right)^{-1}\right),$$

$$\text{(ii) } \left[\sigma_u^2|\text{rest}\right] \sim IG\left(\frac{m+a}{2}, \frac{1}{2}\left\{b + \sum_i\left(\theta_i - \boldsymbol{x}_i^{\mathrm{T}}\boldsymbol{\beta}\right)^2\right\}\right),$$

$$\text{(iii) } [\theta_i|\text{rest}] \sim N\left(\theta_i^B, \frac{\sigma_u^2\psi_i}{\sigma_u^2 + \psi_i}\right), \quad i = 1, \ldots, m,$$

where $\boldsymbol{\beta}^* = (\sum_i x_i x_i^{\mathrm{T}})^{-1}(\sum_i x_i\theta_i)$ and $\theta_i^B = \gamma_i y_i + (1-\gamma_i)x_i^{\mathrm{T}}\boldsymbol{\beta}$, $\gamma_i = \frac{\sigma_u^2}{\sigma_u^2+\psi_i}$.

Table 2 describes the result for EB and HB analysis of the corn data.

Table 2
EB and HB estimates and their standard error for basic area level model

County	n_i	Estimate		Standard error	
		EB	HB	EB	HB
1	1	143.31	143.31	27.74	27.85
2	1	107.51	107.51	27.74	27.85
3	1	94.82	94.82	27.74	27.86
4	2	142.75	142.80	29.39	29.39
5	3	152.15	152.15	29.38	29.39
6	3	104.60	104.65	37.86	38.10
7	3	112.50	112.55	25.03	26.50
8	3	141.23	141.23	29.81	29.90
9	4	117.30	117.30	31.62	32.60
10	5	112.36	112.36	25.23	25.25
11	5	112.02	112.02	22.84	22.84
12	6	116.64	116.64	25.60	25.60

Next we turn to basic unit level mode. Note that here we need prior distribution of $(\boldsymbol{\beta}, \sigma_u^2, \sigma_e^2)$. Again in absence of prior information and apriori independence, we take 'flat' prior for $\boldsymbol{\beta}$ and diffuse prior for σ_u^2 and σ_e^2. In particular $f(\boldsymbol{\beta})\alpha 1$ and $f(\sigma_e^2) = IG(\frac{a}{2}, \frac{b}{2})$, $f(\sigma_e^2) = IG(\frac{c}{2}, \frac{d}{2})$. The full conditional distributions are

$$\text{(I) } [\boldsymbol{\beta}|\text{rest}] \sim N_r\left[\left(\sum_i\sum_j \boldsymbol{x}_{ij}\boldsymbol{x}_{ij}^{\mathrm{T}}\right)^{-1}\sum_i\sum_j x_{ij}(y_{ik}-u_i),\right.$$
$$\left.\sigma_e^2\left(\sum_i\sum_j \boldsymbol{x}_{ij}\boldsymbol{x}_{ij}^{\mathrm{T}}\right)^{-1}\right],$$

$$\text{(II) } [u_i|\text{rest}] \sim N\left[r_i(\bar{y}_i - \bar{\boldsymbol{x}}_i^{\mathrm{T}}\boldsymbol{\beta}), \gamma_i\frac{\sigma_e^2}{n_i}\right], \quad i = 1, \ldots, m,$$

$$\text{(III) } \left[\sigma_e^2|\text{rest}\right] \sim IG\left[\frac{n+a}{2}, \frac{1}{2}\left\{\sum_i\sum_j \left(y_{ij} - x_{ij}^{\mathrm{T}}\boldsymbol{\beta} - u_i\right)^2 + b\right\}\right],$$

$$\text{(IV) } \left[\sigma_u^2|\text{rest}\right] \sim IG\left[\frac{m+e}{2}, \frac{1}{2}\left\{\sum_i u_i^2 + d\right\}\right],$$

where $n = \sum_{i=1}^m n_i$.

Rao (2003) applied this method to the original data of Battese et al. (1988) and the result is shown in Table 3.

For generalized linear mixed model (4.5), Ghosh et al. (1998) gave a general treatment. For the regression parameter $\boldsymbol{\beta}$, they take 'flat' prior $f(\boldsymbol{\beta})\alpha 1$ and for variance components inverse gamma prior. Maiti (1998) gave a sufficient condition for the propriety of posterior distribution under log-normal model and Ghosh et al. (1998) made this result general valid for more complex model. For selection of noninformative prior in generalized mixed model we refer to Kass and Natarajan (2000). There are numerous example where HB small area estimation techniques have been applied in some form of generalized linear mixed models. Nandram et al. (1999) applied the HB method

Table 3
EB and HB estimates and their standard error for unit level model

County	n_i	Estimate		Standard error	
		EB	HB	EB	HB
1	1	122.41	122.41	8.17	8.19
2	1	123.29	123.10	8.10	8.50
3	1	113.96	114.90	8.13	8.20
4	2	115.55	115.10	8.37	8.50
5	3	136.12	135.50	7.94	8.10
6	3	108.46	108.46	8.00	8.10
7	3	116.92	116.92	7.97	7.99
8	3	122.69	124.10	8.05	8.15
9	4	111.01	111.01	7.76	7.76
10	5	124.49	125.50	7.35	7.35
11	5	113.49	113.70	7.35	7.35
12	6	131.54	131.54	7.15	7.16

to estimate age-specific and age-adjusted mortality rates for US Health Service Areas. Malec et al. (1999) applied small area HB method to estimate health-related proportions from the National Health Interview Survey (NHIS) data for the U.S. states. Folsom et al. (1999) used logistic linear mixed HB model to produce HB estimates of small area prevalence rates for states and small group of peoples. Maiti (1998) applied HB method to lip cancer counts in 56 counties of Scotland. In particular his model is as follows: Let θ_i be the true incidence rate of a disease for area $i, i = 1, \ldots, m$. Let y_i be the observed number of incidence and E_i be the number of persons at risk in area $i, i = 1, \ldots, m$. The hierarchical model MI is given by

$$\text{(I) } y_i|\theta_i \overset{\text{ind}}{\sim} Poisson(\theta_i, E_i), \quad i = 1, \ldots, m,$$

$$\text{(II) } \log\theta_i|(\mu, r) \overset{\text{ind}}{\sim} N(\mu, r^{-1}), \quad i = 1, \ldots, m,$$

$$\text{(III) } \mu \sim uniform(-\infty, +\infty),$$

$$\text{(IV) } r \sim Gamma\left(\frac{a}{2}, \frac{b}{2}\right), \quad a > 0, b > 0.$$

The full conditional distributions are given by

$$\text{(I) } [\theta_i|\text{rest}]\alpha\theta_i^{y_i^{-1}} \exp\left[-\theta_i E_i - \frac{r}{2}(\log\theta_i - \mu)^2\right], \quad i = 1, \ldots, m,$$

$$\text{(II) } [\mu|\text{rest}] \sim N\left(\frac{1}{m}\sum_{i=1}^{m}\log\theta_i, (mr)^{-1}\right), \quad i = 1, \ldots, m,$$

$$\text{(III) } [r|\text{rest}] \sim Gamma\left[\frac{1}{2}(m+a), \frac{1}{2}\left\{\sum_1^m (y\theta_i - \mu)^2 + b\right\}\right].$$

The conditional distribution (i) is known only up to a multiplicative constant. The sample is generated using Metropolis–Hastings algorithm. The result from a 10 Gibbs chain each of length 200 (with proper burning period) is given in Table 4.

Table 4
Hierarchical Bayes estimates of lip cancer incidence in Scotland by county

County	y	x	HB estimate		Standard error		Adjacent counties
			MI	MII	MI	MII	
1	9	652.2	458.3	413.3	163.0	137.9	5, 9, 11, 19
2	39	450.3	424.6	404.7	70.1	66.2	7, 10
3	11	361.8	303.0	292.0	92.6	81.9	6, 12
4	9	355.7	285.9	256.1	96.8	75.6	18, 20, 28
5	15	352.1	304.6	314.5	80.9	83.7	1, 11, 12, 13, 19
6	8	333.3	270.4	269.2	96.1	81.2	3, 8
7	26	320.6	302.5	301.5	56.7	58.0	2, 10, 13, 16, 17
8	7	304.3	249.3	238.2	92.0	72.9	6
9	6	303.0	239.1	254.1	92.1	86.9	1, 11, 17, 19, 23, 29
10	20	301.7	275.6	277.8	61.8	59.3	2, 7, 16, 22
11	13	295.5	259.4	278.8	70.3	70.3	1, 5, 9, 12
12	5	279.3	225.4	259.4	93.3	86.7	3, 5, 11
13	3	277.8	201.3	245.0	105.9	91.2	5, 7, 17, 19
14	8	241.7	216.3	189.2	75.1	54.9	31, 32, 35
15	17	216.8	202.6	189.9	45.9	44.2	25, 29, 50
16	9	197.8	180.5	194.5	58.8	50.0	7, 10, 17, 21, 22, 29
17	2	186.9	161.6	209.7	84.5	76.6	7, 9, 13, 16, 19, 29
18	7	167.5	156.5	164.5	55.0	47.5	4, 20, 28, 33, 55, 56
19	9	162.7	152.8	190.7	47.1	48.0	1, 5, 9, 13, 17
20	7	157.7	149.2	166.6	52.0	45.0	4, 18, 55
21	16	153.0	147.0	148.9	34.6	32.4	16, 29, 50
22	31	136.7	136.0	142.3	22.8	22.7	10, 16
23	11	125.4	125.7	118.2	34.8	31.9	9, 29, 34, 36, 37, 39
24	7	124.6	126.9	100.2	42.8	31.2	27, 30, 31, 44, 47, 48, 55, 56
25	19	122.8	122.7	127.7	28.2	24.4	15, 26, 29
26	15	120.1	119.7	118.5	29.4	26.6	25, 29, 42, 43
27	7	115.9	118.6	123.8	39.5	35.1	24, 31, 32, 55
28	10	111.6	115.4	121.4	33.9	31.9	4, 18, 33, 45
29	16	111.3	111.5	109.0	27.3	24.9	9, 15, 16, 17, 21, 23, 25, 26, 34, 43, 50
30	11	107.8	110.2	95.1	29.8	25.1	24, 38, 42, 44, 45, 56
31	5	105.3	111.4	96.8	43.1	33.6	14, 24, 27, 32, 35, 46, 47
32	3	104.2	110.7	123.7	49.6	43.9	14, 27, 31, 35
33	7	99.6	101.7	112.2	31.9	31.2	18, 28, 45, 56
34	8	93.8	99.3	71.2	31.8	22.4	23, 29, 39, 40, 42, 43, 51, 52, 54
35	11	89.3	92.8	94.8	25.3	22.7	14, 31, 32, 37, 46
36	9	89.1	94.3	93.4	27.8	24.3	23, 37, 39, 41
37	11	86.8	89.4	89.0	23.5	22.3	23, 35, 36, 41, 46
38	8	85.6	91.1	80.6	28.9	23.1	30, 42, 44, 49, 51, 54
39	6	83.3	91.3	86.2	30.4	26.7	23, 34, 36, 40, 41
40	4	75.9	87.0	73.6	32.8	25.1	34, 39, 41, 49, 52
41	10	53.3	59.6	55.0	15.1	14.2	36, 37, 39, 40, 46, 49, 53
42	8	50.7	59.7	60.9	17.0	15.9	26, 30, 34, 38, 43, 51
43	2	46.3	74.7	84.2	34.6	30.0	26, 29, 34, 42
44	6	41.0	51.3	54.6	16.9	14.1	24, 30, 38, 48, 49
45	19	37.5	42.3	47.9	9.1	9.5	28, 30, 33, 56
46	3	36.6	52.1	55.8	19.6	18.0	31, 35, 37, 41, 47, 53
47	2	35.8	62.2	52.0	27.0	18.6	24, 31, 46, 48, 49, 53
48	3	32.1	50.1	52.8	19.0	17.0	24, 44, 47, 49

Table 4
(*Continued*)

County	y	x	HB estimate		Standard error		Adjacent counties
			MI	MII	MI	MII	
49	28	31.6	34.2	32.1	6.4	5.8	38, 40, 41, 44, 47, 48, 52, 53, 54
50	6	30.6	42.5	58.2	13.8	15.7	15, 21, 29
51	1	29.1	65.7	67.6	37.6	24.0	34, 38, 42, 54
52	1	27.6	64.5	60.9	37.1	25.5	34, 40, 49, 54
53	1	17.4	51.5	53.9	28.3	20.5	41, 46, 47, 49
54	1	14.2	50.8	48.3	22.1	17.9	34, 38, 49, 51, 52

Note: The reported standard errors are scale transformed.

He has also considered conditional autoregressive (CAR) model for the log relative risk.

Model MII

(I) $y_i|\theta_i \sim Poisson(\theta_i E_i)$ independently, $\quad i = 1, \ldots, m$,

(II) $\boldsymbol{\beta}_i|\boldsymbol{\beta}_j, j \in S_i, \rho, v \sim N\left[\mu + \rho \sum_{j \in S_i} (\boldsymbol{\beta}_j - \mu), v^{-1}\right]$,

$i = 1, \ldots, m$,

(III) ρ, μ and v are mutually independently distributed, respectively, with $\rho \sim U(0, 1/\lambda_{\max})$, $\quad \mu \sim Uniform(-\infty, +\infty)$ and $v \sim Gamma(g, h)$, $\quad g > 0, h > 0$.

We assume both g and h are known. Here also we have used noninformative prior. As in Section 2 we use Gibbs sampler to obtain the posterior distribution of θ_i, $i = 1, \ldots, m$. The full conditional distributions are as follows:

(i) $\left[\theta_i|(\underset{\sim}{y}, v, \mu, \rho, \theta_j, j \in S_i)\right]$

$$\propto \theta_i^{y_i - 1} \exp\left[-\theta_i E_i - \frac{r}{2}(\log\theta_i - A_i\mu - B_i)^2\right], \quad i = 1, \ldots, m,$$

(ii) $\mu|(\underset{\sim}{\theta}, \underset{\sim}{y}, v, \rho) \sim N\left(\frac{\sum_{i=1}^m A_i(\log\theta_i - B_1)}{\sum_{i=1}^m A_i}, \left(v \sum_{i=1}^m A_i^2\right)^{-1}\right)$,

(iii) $v|(\boldsymbol{\theta}, \boldsymbol{y}, \mu, \rho)$

$$\sim Gamma\left[\frac{1}{2}(m + a), \frac{1}{2}\left\{\sum_{i=1}^m (\log\theta_i - A_i\mu - B_i)^2 + b\right\}\right],$$

(iv) $\rho|(\boldsymbol{\theta}, \boldsymbol{y}, \mu, v) \sim TN\left(\frac{\sum_{i=1}^m k_i d_i}{\sum_{i=1}^m k_i^2}, \left(r \sum_{i=1}^m k_i^2\right)^{-1}\right)$,

where *TN* denotes the truncated normal distribution. The posterior distribution (iv) is truncated outside the range $(0, 1/\lambda_{\max})$. $A_i = 1 - \rho \sum_{j \in S_i} W_{ij}$, $B_i =$

$\rho \sum_{j \in S_i} W_{ij} \log \theta_j$, $d_i = \log \theta_i - \mu$, $k_i = \sum_{j \in S_i} W_{ij} d_j$. In this case also the posterior distribution (i) is known only up to a multiplicative constant. The conditional distribution (i) is not log-concave. So, we use the Metropolis–Hastings algorithm to generate samples from (i). The posterior distributions are proper, assuming at least one y_i is strictly positive.

An illustration with lip cancer data

In this section a HB analysis is carried out in the example of Clayton and Kaldor (1987). The observed (y) and the expected (E) number of cases of lip cancer registered during the six years period (1975–1980) for each of 56 counties of Scotland are given. We consider the first 54 counties as we obtain the "expected" number of incidence E_i for ith county using the relation $E_i = y_i/x_i$, $i = 1, \ldots, m$, where x_i denotes the SMR for county i, $i = 1, \ldots, m$. Here x_i's are multiplied by 100. The y_i's are zero for the last two counties. We exclude those counties as E_i's become zero. The adjacency matrix W is obtained from the last column of Table 1 of Clayton and Kaldor (1987). Our data analysis is presented in Table 4. Table 4, column 4, corresponds to the estimates from Model MI and column 5 corresponds to the estimates from Model MII. The standard errors (s.e.) of the estimates are given in columns 6 and 7, respectively, for Model MI and Model MII.

To implement the Gibbs sampler of Gelman and Rubin (1992), we consider 10 independent sequences each of length $2t = 200$. The convergence of Gibbs sampler is assessed by using the potential scale reduction of Gelman and Rubin (1992) which is near unity for all the scalar estimands. The procedure is not sensitive to the initial choices of the parameters.

The estimates of θ_i's ($i = 1, \ldots, m$) are very similar for both the models. The dispersions among the estimates are slightly larger than those for the EB estimates. We use $(\lambda_{\max})^{-1} = 0.175$ which is the maximum value of ρ in a CAR process (see Clayton and Kaldor, 1987). In our case the estimate of ρ comes out as 0.164. The spatial effects are reflected from the estimates of some of the counties, e.g., the counties 24, 30, 31 are affected by several low-risk areas while the counties 17, 32 and 33 are affected by several high-risk areas.

The standard errors of the estimates for Model II are smaller than those for Model I. This is because Model II exploits the spatial structure of the data. For the counties 1, 13, 17, 32, 43, 46, 47, 48, 51, 52, 53, and 54 the standard errors are high due to insufficient representation of observed number of incidences. However, the counties 2, 7, 22, 45, and 49 have small standard errors due to availability of more information.

The values of the potential scale reduction of Gelman and Rubin (1992) are close to unity for all the scalar estimands.

5. Conclusion

In this chapter, we have discussed only some aspects of small area estimation, particularly the model based EB and HB approach. EB produce approximate estimates whereas HB provide accurate estimates. We have discussed only basic models, and mentioned some extensions of those models. The detail of other available models, like, constrained Bayes, spatial model, time series models are available in Rao (2003). The basic idea remain same and measure of uncertainty is always an issue.

Since small area estimation problem often arises from sample survey, sampling design is an issue and must be taken into account if necessary. Work has begun in this area. However, small area estimation problem may arise in other areas of statistical application, and design may not be an issue. Furthermore, for complex model EB estimation may be problematic and large data set HB estimation may not be suitable.

Availability of good covariates and model selection is important. The research on model selection and model diagnostic in the context of small area estimation is not enough. For HB analysis, the standard Bayesian model diagnostics tools are applied so far. Research in this area should be encouraged. As noted by Schaible (1996) "indirect estimators should be considered when better alternatives are not available, but only with appropriate caution and in conjunction with substantial research and evaluation efforts. Both producers and users must not forget that, even after such efforts, indirect estimates may not be adequate for the intended purpose".

Acknowledgements

The author is thankful to the editors for their invitation to write this chapter. The research is partially funded by a NSF grant No. SES-0318184.

References

Battese, G.E., Harter, R.M., Fuller, W.A. (1988). An error component model for prediction of county crop areas using survey and satelite data. *J. Amer. Statist. Assoc.* **83**, 28–36.

Bell, W.R. (1999). Accounting for uncertainty about variances in small area estimation. *Bull. Internat. Statist. Inst.*

Butar, F.B., Lahiri, P. (2001). On measures of uncertainty of empirical Bayes small area estimators. Technical Report, Department of Mathematics and Statistics, University of Nebraska-Lincoln.

Chen, M.-H., Shao, Q.-M., Ibrahim, J.G. (2000). *Monte Carlo Methods in Bayesian Computation*. Springer-Verlag, New York.

Clayton, D., Kaldor, J. (1987). Empirical Bayes estimates of age standardized relative risks for use in disease mapping. *Biometrics* **43**, 671–681.

Crescenzi, M., Ghosh, M., Maiti, T. (2002). Empirical Bayes estimators with uncertainty measures for NEF-QVF populations. Technical Report, University of Florida.

Datta, G.S., Ghosh, M., Nangia, N., Natarajan, K. (1996). Estimation of median income of four person families: A Bayesian approach. In: Berry, W.A., Chaloner, K.M., Geweke, J.K. (Eds.), *Bayesian Analysis in Statistics and Econometrics*. Wiley, New York, pp. 129–140.

Datta, G.S., Lahiri, P., Maiti, T., Lu, K.L. (1999). Hierarchical Bayes estimation of unemployment rates for the U.S. states. *J. Amer. Statist. Assoc.* **94**, 1074–1082.

Dempster, A.P., Tomberlin, T.J. (1980). The analysis of census undercount from a post-enumeration survey. In: *Proceedings of the Conference on Census Undercount*, p. 94.

Efron, B. (1978). Regression and ANOVA with zero-one data: Measures of residual variation. *J. Amer. Statist. Assoc.* **73**, 113–121.

Efron, B. (1986). Double exponential families and their use in generalized linear regression. *J. Amer. Statist. Assoc.* **81**, 709–721.

Farrell, P.J., MacGibbon, B., Tomberlin, T.J. (1997). Empirical Bayes estimators of small area proportions in multistage designs. *Statistica Sinica* **7**, 1065–1083.

Fay, R.E., Herriot, R.A. (1979). Estimation of income from small places: An application of James–Stein procedures to census data. *J. Amer. Statist. Assoc.* **74**, 269–277.

Folsom, R., Shah, B.V., Vaish, A. (1999). Substance abuse in states: A methodological report on model based estimates from the 1994–1996 National Household Surveys on Drug Abuse. In: *Proceedings of the Section on Survey Research Methods*. Amer. Statist. Assoc., Washington, DC, pp. 371–375.

Gelman, A., Rubin, D.B. (1992). Inference from interactive simulation using multiple sequences. *Statist. Sci.* **7**, 457–472.

Gelfand, A.E., Smith, A.F.M. (1990). Sample-based approaches to calculating marginal densities. *J. Amer. Statist. Assoc.* **85**, 972–985.

Gelman, A., Carlin, J.B., Stern, H.S., Rubin, D.B. (1996). *Bayesian Data Analysis*. Chapman & Hall.

Ghosh, M., Natarajan, K., Stroud, T.W.F., Carlin, B.P. (1998). Generalized linear models for small area estimation. *J. Amer. Statist. Assoc.* **93**, 273–282.

Gilks, W.R., Richardson, S., Spiegelhalter, D.J. (Eds.) (1996). *Markov Chain Monte Carlo in Practice*. Chapman & Hall, London.

Godambe, V.P., Thompson, M.E. (1989). An extension of quasi-likelihood estimation (with discussion). *J. Statist. Plann. Inference* **22**, 137–152.

Hobert, J.P., Casella, G. (1996). The effect of improper priors on Gibbs sampling in hierarchical linear mixed models. *J. Amer. Statist. Assoc.* **91**, 1461–1473.

Jiang, J., Lahiri, P. (2001). Empirical best prediction for small area inference with binary data. *Ann. Inst. Statist. Math.* **53**, 217–243.

Kass, R.E., Natarajan, R. (2000). Reference Bayesian methods for generalized linear mixed models. *J. Amer. Statist. Assoc.* **95**, 227–237.

Kass, R.E., Steffey, D. (1989). Approximate Bayesian inference in conditionally independent hierarchical models. *J. Amer. Statist. Assoc.* **84**, 717–726.

Laird, N.M., Louis, T.A. (1987). Empirical Bayes confidence intervals based on bootstrap samples. *J. Amer. Statist. Assoc.* **82**, 739–750.

MacGibbon, B., Tomberlin, T.J. (1989). Small area estimation of proportions via empirical Bayes technique. *Survey Methodology* **15**, 237–252.

Maiti, T. (1998). Hierarchical Bayes estimation of mortality rates for disease mapping. *J. Statist. Plann. Inference* **69**, 339–348.

Malec, D., Davis, W.W., Cao, X. (1999). Model based small area estimates of overweight prevalence using sample selection adjustment. *Statistics in Medicine* **18**, 3189–3200.

Malec, D., Sedrannsk, J., Moriarity, C.L., LeClere, F.B. (1997). Small area inference for binary variables in National Health Interview Survey. *J. Amer. Statist. Assoc.* **92**, 815–826.

McCullagh, P., Nelder, J.A. (1989). *Generalized Linear Models*, second ed. Chapman and Hall, New York.

Morris, C.A. (1983). Parametric empirical Bayes inference: Theory and applications. *J. Amer. Statist. Assoc.* **78**, 47–54.

Nandram, B., Sedransk, J., Pickle, L. (1999). Bayesian analysis of mortality rates for U.S. health service areas. *Sankhya B* **61**, 145–165.

National Research Council (2000). *Small Area Estimates of School Age Children in Poverty: Evaluation of Current Methodology*. Citro, C.F., Kalton, G. (Eds.), Committee on National Statistics, Washington, DC. National Academy Press.

Prasad, N.G.N., Rao, J.N.K. (1990). The estimation of the mean squared error of small area estimators. *J. Amer. Statist. Assoc.* **85**, 163–171.

Rao, J.N.K. (2003). *Small Area Estimation*. Wiley.

Sarkar, S., Ghosh, M. (1998). Empirical Bayes estimation of the local area means for NEF-QVF superpopulations. *Sankhya, Ser. B* **60**, 464–487.

Schaible, W.A. (Ed.) (1996). *Indirect Estimation in U.S. Federal Programs*. Wiley, New York.

Singh, A.C., Stukel, D.M., Pfeffermann, D. (1998). Bayesian versus frequentist measures of error in small area estimation. *J. Roy. Statist. Soc., Ser. B* **60**, 377–396.

You, Y., Rao, J.N.K., Gambino, J. (2001). Model-based unemployment rate estimation for the Canadian labour force survey: A hierarchical Bayes approach. Technical Report, Household Survey Methods Division, Statistics Canada.

Handbook of Statistics, Vol. 25
ISSN: 0169-7161
© 2005 Elsevier B.V. All rights reserved.
DOI 10.1016/S0169-7161(05)25035-6

35

Teaching Bayesian Thought to Nonstatisticians

Dalene K. Stangl

Abstract
This paper discusses goals and strategy for teaching Bayesian thought without the use of calculus in undergraduate courses and in workshops for academics in non-mathematical fields. The primary goal is making these audiences better consumers of the statistical output presented in both the popular press as well as professional journals. The strategy makes use of case studies as a primary motivational and conceptual tool, course mapping, active learning, and spaced practice with repetition. Examples of case studies, assignments, and assessment methods are provided.

Keywords: mapping; case study; active learning

1. Introduction

Nonstatisticians are now eager to understand Bayesian thinking. From the design of clinical trials to the development of spam filters, Bayes' theorem and Bayesian thinking is permeating the daily activities of everyone. Academics in fields as diverse as environmental science, economics, engineering and political science as well as lawyers, health-care professionals and policymakers in government and industry are taking advantage of what Bayesian methods have to offer. What has enabled such progress in the last decade? The answer: Bayesian thought is now being taught in courses beyond graduate-level statistics courses. Tools for teaching Bayesian methods to undergraduates and to professionals in nonstatistical fields are developing.

This paper will review some of the literature on teaching Bayesian methods to nonstatisticians and make suggestions based on the author's own experiences teaching undergraduates in majors such as literature, biology, public policy, psychology, biology, sociology, history, economics, computer science, engineering, and political science as well as teaching nonstatisticians in academics, government, and industry. Despite their different backgrounds, interests, and goals, these groups share many commonalities in learning about Bayesian methods. This paper brings together many ideas (Stangl, 1998, 2000, 2001, 2002, 2003) and will discuss the commonalities and present guidelines that apply to teaching all these groups about Bayesian thinking.

2. A brief literature review

Jim Albert and Don Berry have been two of the most prolific writers and proponents of teaching Bayesian methods both to statisticians and nonstatisticians alike. They were among the earliest to enter the discussion of didactical issues in teaching Bayesian inference to undergraduates (Albert, 1997; Berry, 1997; Moore, 1997). Moore argued that it is, at best, premature to teach the ideas and methods of Bayesian inference in a first statistics course for general students. He argued that: (1) Bayesian techniques are little used, (2) Bayesians have not yet agreed on standard approaches to standard problem settings, (3) Bayesian reasoning requires a grasp of conditional probability, a concept confusing to beginners, and (4) an emphasis on Bayesian inference might impede the trend toward experience with real data and a better balance among data analysis, data production, and inference in first statistics courses.

Berry counter argues that students in a Bayesian course are served well, because the Bayesian approach fits neatly with a scientific focus and takes a larger view, not limited to data analysis. His elementary statistics book, *Statistics: A Bayesian Perspective* (Berry, 1996), is well suited for undergraduates or professionals who use statistics, but who have not been trained as statisticians. His book relies only on high school algebra. I echo Berry's counter argument, and my reasons will become clear later in this chapter.

Albert (1993, 1995, 1996, 1997, 2000) has described what he calls a "Statistics for Poets" class. He argues that the inferential concepts that should be taught in such a course include the distinction between statistics and parameters, the inherent variability in data, that sample data provide an incomplete description of the population, the dependence of statistical procedures on the underlying assumptions of the model, the distinction between inference procedures including estimation, testing, prediction, and decision making, and the interpretation of statistical "confidence." He argues that the primary advantage of teaching from a Bayesian viewpoint is that Bayes' thinking is more intuitive than the frequentist viewpoint and better reflects the commonsense thinking about uncertainty that students have before taking a statistics class.

Albert also discusses the finding of Gigerenzer and Hoffrage (1995) who found that people generally are more successful in Bayes' rule calculations when presented with tables of counts rather than tables of probabilities. He illustrates a graphical method of doing Bayes' rule calculations in the well-known diagnostic testing context.

Albert and Rossman (2001) have published a collection of activities that assist in teaching statistics from a Bayesian perspective. Like the Berry textbook, it covers estimating proportions and means using discrete and continuous models. Web-based javascript software to illustrate probability concepts and perform Bayesian calculations is also available.

3. Commonalities across groups in teaching Bayesian methods

When teaching Bayesian methods to nonstatisticians I follow five guidelines that build upon commonalities among all students. The guidelines include: motivating students, providing conceptual explanations, linking ideas through conceptual mapping, making learning active, and encouraging spaced practice and repetition to improve long-term retention.

Not all faculty believe that motivating students to learn should be part of a teacher's responsibilities. Ideally, this is true. Ideally students would come to our classes eager to learn with a full understanding of the importance of mastering statistics. However realistically, only a few students arrive to our classrooms with such enthusiasm or understanding. For many reasons, most students view the learning of statistics as a necessary evil, a hurdle to meet curriculum requirements or a literacy required to read the modern research protocols and journal articles of others. Add to this perceived general burden of taking statistics, the fact that what will be taught, the Bayesian paradigm, is in addition to the more mainstream frequentist topics. At Duke, introductory courses cover all the topics in *Statistics*, by Freedman, Pisani, Purves first, and then the Bayesian paradigm is added on as supplemental material. It takes more motivation and energy to learn something that is viewed as 'extra' or 'different.' This is especially true when what is new is at odds with what a student already knows and when it requires new 'adjacent' skills, for example, different computing packages. It is also much easier to teach Bayesian thought if the students were not initially indoctrinated that frequentist methods are objective and Bayesian methods are subjective. This polarizing and inaccurate teaching makes it difficult for students to be open to learning new ways of thinking. Assumptions in the frequentist paradigm that students took as "fact" are now called into question by the Bayesian paradigm. For some students, this juxtaposition is exciting, but for others it can be quite unnerving. Taking these hurdles in combination, it may seem that mustering student motivation to learn Bayesian thinking is an insurmountable task. I'm here to say it is not. If students see the big picture and if you keep the classroom intellectually engaging, you will motivate them to learn.

It is impossible to be intellectually engaging if you do not teach to the level of the group, understanding what they know and how they learn. Students at all levels learn best when they receive conceptual explanations, an overview of the big picture. Where many instructors lose their audiences is by teaching via mathematical formula. While this may work for the mathematically inclined, it is not the way to go for those whose first love is not mathematics. For this group it is important to teach conceptual intuitive explanations before teaching formula. My gold standard for this sort of teaching is *Statistics* by Freedman et al. (1998). While the book does not teach statistics from a Bayesian perspective, the book is still one the best conceptual books on the market for teaching introductory frequentist statistics. I use it as a precursor for teaching Bayesian methods in my undergraduate courses (covering it in the first 10 weeks of the semester) as well as a model for the Bayesian component of my course (the last 4 weeks of the semester). Concepts are introduced intuitively with a minimal reliance on formula. Exercises carefully take the student through a conceptual process first, and then later exercises require the student to pull pieces together in less transparent ways.

While much of the experience described here, comes from Duke undergraduates majoring in social science and humanities who may be viewed as unrepresentative because of the school's selectivity, Duke students are not that different when it comes to motivation and need for conceptual explanations. While more intelligent than average, Duke students are not well prepared to think mathematically or abstractly, and surprisingly, their study skills are often in need of serious overhaul. Many students struggle with the concepts presented in Freedman, Pisani, and Purves, not because of intellectual deficit, but because of poor study habits and weak math and analytic skills. Many cannot add simple fractions or take the square root of 36 without a calculator. While extremely ambitious and engaged in extracurricular activities, many are not intellectually engaged. Many seek out the courses and instructors with low demand and generous grading (Johnson, 2003). Courses requiring 3 hours per week outside of class are rated as having an average demand level. Many students are masters at procrastination and cramming. Because cramming for exams worked well for them in high school, it is difficult for them to understand why this tact doesn't work in learning statistics in college. So while students at Duke may not be representative of students at other universities, they have characteristics that make motivating them and their need for conceptual explanations remarkably similar to students at other universities.

In addition to needing external motivation and conceptual explanations, students at all levels need conceptual mapping. Ideas need to be linked to each other and to what students already know, and ideas need to be examined critically in order for higher-level learning to occur. We should not teach Bayesian methods or any statistical methods in a cookbook or template fashion. Indeed it is this sort of teaching that is likely responsible for the sorry state of statistical analysis in many substantive journals. While eight years ago Moore argued that teaching Bayesian methods to undergraduates was premature, because there were not standard approaches to standard problems, I wonder whether most fields might not be better off if they did not rely on what they have been taught are standard approaches. Had they been taught to think rather than copy templates of standard approaches, perhaps there would not be so many egregious statistical errors in their journals.

Lastly, students at all levels need active learning and repetition. Active learning helps focus students and helps them to understand how what they are learning is applicable to daily life. Repetition helps students generalize their learning inductively to many contexts, and repetition increases retention.

4. Motivation and conceptual explanations: One solution

I will discuss motivation and conceptual explanations together, because there is one solution to both of these needs. That solution is a great case study. A case study needs to clearly convince the audience about the importance of knowing the material. The case study needs to be presented first in general without technical details.

One of the best examples I have found and the one that I continue to use after many years of teaching is the GUSTO clinical trial. The beauty of this case study includes (1) it is a very simple controlled clinical trial, (2) it is controversial, (3) the embedded decision problem is transparent, (4) it is published in top medical journals, and (5) everyone makes choices about medications they take, so it is easy for students of all types to see how it is relevant to their personal lives.

The GUSTO trial compares tissue plasminogen activator (t-PA) and streptokinase (SK) for the treatment of myocardial infarction. The results of the trial were first presented in the *New England Journal of Medicine* (The GUSTO Investigators, 1993) and were subsequently reanalyzed by Brophy and Joseph (1995) in the *Journal of the American Medical Association*. The statistical argument in the *NEJM* paper uses confidence intervals and tests of significance. Finding an increased survival of 1% and rejecting the null hypothesis of no difference between treatments, the GUSTO investigators conclude that t-PA provides a survival benefit. In the *JAMA* paper, Brophy and Joseph use Bayesian statistical arguments to argue that the jury is still out. They find that the posterior probability that survival on t-PA is greater than survival on streptokinase by at least 1% ranges from 0% to 36% depending on how much weight is placed on previous trials. A third source for the case is an article, "The Mathematics of Making up Your Mind," by Hively (1996). The article appeared in the popular science magazine *Discover* in May 1996. It covers the differences between inferential paradigms and highlights the controversies that can arise between them. The *Discover* article uses the GUSTO trial as their primary example.

The primary focus of the case study discussion is on the following questions: (1) What is the research question we are trying to answer? (2) To answer the question of interest, what is more useful, the probability of the data given the null hypothesis, or the probability of the hypothesis given the data? (3) What is magical about a 1% survival difference? (4) Should data from previous studies be synthesized into the results of the GUSTO trial? (5) Are tests of significance being used as decision rules, and if so, why? (6) Should medication costs affect treatment decisions? (7) What are the biases in the GUSTO study and previous studies?

After this case study my audience is usually hooked. It is easy for them to understand the design of this simple controlled trial. They understand why the trial is controversial. They understand the impact of the prior data and how it affects conclusions. They have seen the importance of the t-PA versus streptokinase debate validated by publication in top medical journals.

A second more hands-on case study that I use asks students to decide whether the proportion of yellow M&Ms in a jar is 20% or 30%. The question is set up as a decision problem for which the cost of a wrong answer is that the student buys me lunch, while a correct answer yields the student a lunch on me. Students are walked through the problem first using a hypothesis testing framework. Students purchase a sample of M&Ms to provide information about the proportion of yellows. The student determines the sample size, knowing that each additional M&M improves their information and increases the chance of a correct answer, but at same time the cost of each M&M decreases the net benefit of a correct answer. Students must also choose their significance level by balancing the costs of errors. Rarely do students choose to buy enough M&Ms that they are able to reject their null hypothesis. From this perspective, I nearly always win a lunch. While they have been taught not to use significance levels as decision rules, when they approach the problem from a frequentist perspective it is difficult for them to get their thinking around the notion of balancing type I and type II errors. They revert to using a significance level of .05 as their decision rule.

After approaching the M&M problem from a frequentist perspective, we start over approaching it from a Bayesian perspective. Now instead of calculating the probability of the data or more extreme given the null hypothesis is true, they must calculate the probability of each hypothesis given the data they have observed. In this approach students are allowed to analyze information and increase their sample of M&Ms sequentially. Again they are required to pay for their samples. Now they are calculating posterior probabilities, and it is easy for them to see how this probability guides their decision making.

5. Conceptual mapping

In presenting the GUSTO case study, I have avoided any mathematical detail except for some weights and basic probability statements. What students need next is linkage between what this case study teaches and what they already know. I provide a conceptual map (Schau and Mattern, 1997) linking where we are going with where we have been. At this point I re-teach Bayes' theorem in a context that most people are familiar – diagnostic testing. I add a wee bit of algebra to crank through Bayes' theorem. I explain why we need Bayes' theorem and introduce some terminology including prior and posterior probabilities and discuss probability in terms of beliefs. Next I argue that Bayes' theorem isn't just for diagnostics. Bayesian methods are useful in more general contexts too.

I try to link Bayes' theorem to what they already know about conditional probabilities and frequentist methods and hypothesis testing. This is where I pull in case studies of estimating the proportion of yellows in a bag of M&Ms, estimating the proportion of Duke basketball players that will shoot better than 35% in the next game, or more seriously, estimating survival in a clinical trial. I walk students through Bayes' theorem for finding the probability of the hypothesis or parameters given the data. And then, we do use a little math. We walk through a simple example with binomial data and a discrete parameter space. Then I show them (with graphics, without formula) a simple example with binomial data and a continuous parameter space with a conjugate beta prior distribution. I can relate this simple example back to the GUSTO trial. Finally, we work through the algebra needed to get from prior to posterior in a simple conjugate beta/binomial analysis of the GUSTO data.

Depending on the audience's mathematical sophistication, I next show simple normal/normal models with either simple graphical visualizations or with simple graphical visualizations and formula. Again I relate this model back to the GUSTO analysis presented by Brophy and Joseph.

After showing these discrete/binomial, beta/binomial, and normal/normal models, we return to comparing and contrasting Bayesian and Frequentist methods of inference more abstractly.

6. Active learning and repetition

To actively engage students in learning, I include activities. I believe it is George Cobb who coined a term that describes why these activities are so useful. The term is "authentic play," meaning you engage students in activities that simulate real life work. Children like to engage in activities that simulate what they see adults doing. Similarly, students whether adolescents or adults, enjoy applying what they are learning to activities that simulate what professionals do in real life.

One activity is role-playing. After introducing the GUSTO study, students role-play a mock medical malpractice lawsuit trial. The background of the suit is that a doctor prescribes streptokinase and the patient dies. The patient's wife is a rather educated woman, and she gets her hands on the NEJM paper and decides the doctor was negligent for not prescribing t-PA, so she launches a medical malpractice suit. Students play roles of doctor, wife, lawyer, and expert statistical witness, one for each side. Often the deceased also becomes a character. Students may bring in other experts if they want such as HMO representatives, hospital management, etc., but the basis of their arguments must be primarily statistical. However, I have never been completely successful holding students to this. They enjoy dramatizing and bringing in emotional arguments on how special 'Frank' was and value-laden ethical arguments on the value of human life. Despite this, every audience has provided feedback that this is an extremely useful exercise for getting them to think about Bayesian versus frequentist methods and how these paradigms differ in bringing evidence to bear upon decisions.

Another fun useful activity combines writing and debate. One-half the students write a policy statement to Congress on why government health plans should pay for t-PA, the more expensive drug, and the other one-half do the opposite: they write on why government health plans should not pay for the more expensive drug. Again arguments are expected to be statistical not emotional. After writing the policy statements, representatives from both sides debate the issue.

One last activity brings repetition and brings the issues home. Students are asked to suppose that a parent has had a heart attack and was treated with t-PA and survives. However, the insurance company reimburses for the cost of streptokinase and the hospital expects the family to come up with the remainder. Students are asked to write a letter to the insurance company arguing for full payment. Students must hand in their letters. I read some of the best to the class. On the next day, students are told they work for the insurance company that refused to pay for t-PA. Now they must write a reply letter to themselves arguing against full payment of t-PA. Arguments must be statistical. Arguments such as "This is company policy," get negative points unless they have a statistical argument for why this is company policy.

I've had guest faculty come into my undergraduate class. They are impressed by the level of questions my students ask and by the level of student engagement. These activities get the credit.

After introducing students to the basics of Bayesian thinking, I encourage students to buy books and read more. I encourage them to start off easy and build. Depending on the level of the student, the books I recommend include *Statistics: A Bayesian Perspective* by Berry (1996), *Scientific Reasoning* by Howson and Urbach (1993), *Bayesian Data Analysis* by Gelman et al. (2003), *Bayesian Biostatistics*, by Berry and Stangl (1996), *Bayes and Empirical Bayes Methods for Data Analysis* by Carlin and Louis (2000), *Modeling and Medical Decision Making: A Bayesian Approach* by Parmigiani (2002), and the BUGS website hosted by the MRC Biostatistics Unit, Cambridge, UK. Depending on the background and interest of the students I supplement with papers that use Bayesian methods in particular areas of interest, whether it be epidemiology, law, environment, political science, or another field. Some of these are included in the reference list.

7. Assessment

The good news is that the question is no longer should we teach Bayesian methods, but rather how should we teach Bayesian methods. I've argued that across diverse audiences we should pay attention to motivation and conceptual explanations, use course mapping to link Bayesian ideas to what students already know, and to use active learning and repetition. But how does all this really impact learning? How does teaching Bayesian methods change how my students think?

I base my beliefs about this question on exams I have administered over the past few years, and one imparticular. I gave the same exam to two classes – one class was exposed only to frequentist methods and the other was exposed to both Bayesian and frequentist methods. The exam covered only frequentist ideas. The exam consisted of reading a simple journal article on the impact of lead exposure on the neurological functioning of children. The paper presented averages, percents, standard deviations, t-tests for differences between means, simple regression, and contingency tables. I asked students to replicate some tests of significance, to answer questions about the authors' interpretation of the results, and to decide whether the results supported a particular policy decision.

I categorize the difference between the groups' performances into perception of question, use of prior information, how results were used in decision-making, and critique of the study design. Overall, students that were exposed to both frequentist and Bayesian methods were more likely to answer questions in ways that reflected an understanding that questions are multifaceted versus a question of statistical significance. They viewed the whole research process as subjective and wanted to discuss the implications of this subjectivity rather than give only a p-value summation. They were also more likely to question whether any decision should be based on a single study, were less likely to make decisions based on tests-of-significance, and were much better at explaining what a p-value does tell us. Finally they were much more likely to challenge the research at every step. The bottom line: the students exposed to Bayesian methods were better consumers of the statistical output given in a research paper.

Since this semester, I have been assessing students understanding of both frequentist and Bayesian ideas. My exams include both computational and interpretive questions. For the frequentist paradigm students must be able to set up simple hypotheses, calculate simple test statistics and confidence intervals, and determine the associated p-values. They must be able to interpret each of these concepts within the context of research question pertaining to a real-world data set.

For the Bayesian paradigm students must be able to set up a prior and calculate likelihoods and posterior and predictive distributions for simple discrete parameter spaces and binomial data. They must be able to interpret priors, likelihoods, posterior and predictive distributions for both discrete and continuous parameter spaces for binomial and normal data; although they do not need to be able to calculate them except in a couple of simple conjugate cases. Berry (1996) is full of fun examples. One that Duke students enjoy is: "Three Duke students were interested in whether basketball players are more effective when under pressure. They considered the three-point shots attempted by Dukes's 1992–1993 basketball team in the first half versus second half of games, thinking there would be more pressure in the second half. Regard the following as random samples from larger populations: Of the 211 three-point shots attempted in the first half, 71 or 33.6%, were successful; of the 255 attempted in the second half, 90 or 35.5% were successful. Assuming beta(1, 1) prior densities for both success proportions, find the 99% posterior probability interval for $d = p_f - p_s$."

Finally, exams always include a question that asks students to compare and contrast frequentist and Bayesian ideas. Often I use a frequently cited paper by Landrigan et al. (1975) titled, "Neuropsychological Dysfunction in Children with Chronic Low-Level Lead Absorption." The paper uses frequentist statistics of most every type covered in Freedman et al. (1998). Students are asked to critique the paper and comment on its value for making decisions about living close to a lead smelter. Their thinking proceeds down very different tracks when they think as a frequentist versus when they think as a Bayesian.

8. Conclusions

When I discuss my experiences about teaching Bayesian statistics to nonstatisticians, I frequently get skepticism that one can teach these ideas to those without much mathematical training. I hear, "but Duke students aren't like my students." My response is your students are very much like Duke students. They procrastinate, they cram, they are dependent upon calculators. Duke students may be smarter in some respects, but this just means they can get by with procrastinating more, cramming more, and having more sophistication with their calculators. They still need the same things to learn. They need motivation to begin engagement with abstract ideas. They need conceptual explanations and mapping to link ideas. They need active learning to stay engaged and work through processes. And, they need spaced practice and repetition to improve long-term retention. Using these techniques all students can understand the basics of Bayesian statistics.

References

Albert, J. (1993). Teaching Bayesian statistics using sampling methods and MINITAB. *Amer. Statist.* **47**, 182–191.

Albert, J. (1995). Teaching inference about proportions using Bayes and discrete models. *J. Statist. Education* **3** (3).

Albert, J. (1996). *Bayesian Computation Using Minitab*. Duxbury Press.

Albert, J. (1997). Teaching Bayes' rule: A data-oriented approach. *Amer. Statist.* **51** (3), 247–253.

Albert, J. (2000). Using a sample survey project to compare classical and Bayesian approaches for teaching statistical inference. *J. Statist. Education* **8** (1).

Albert, J., Rossman, A. (2001). *Workshop Statistics: Discover with Data, A Bayesian Approach*. Key College.

Berry, D.A. (1996). *Statistics: A Bayesian Perspective*. Duxbury Press.

Berry, D.A. (1997). Teaching elementary Bayesian statistics with real applications in science. *Amer. Statist.* **51** (3), 241–246.

Berry, D.A., Stangl, D. (Eds.) (1996). *Bayesian Biostatistics.* Marcel Dekker.

Brophy, J.M., Joseph, L. (1995). Placing trials in context using Bayesian analysis: Gusto revisited by reverend Bayes. *J. Amer. Medical Assoc.* **273** (11), 871–875.

Carlin, G., Louis, T. (2000). *Bayes and Empirical Bayes Methods for Data Analysis*, second ed. Chapman & Hall.

Freedman, D., Pisani, R., Purves, R. (1998). *Statistics.* Norton.

Gelman, A., Carlin, J., Stern, H., Rubin, D. (2003). *Bayesian Data Analysis*, second ed. Chapman & Hall.

Gigerenzer, G., Hoffrage, U. (1995). How to improve Bayesian reasoning without instruction: Frequency formats. *Psychological Review* **102**, 684–704.

Hively, W. (1996). The mathematics of making up your mind. *Discover*, May, 90–97.

Howson, C., Urbach, P. (1993). *Scientific Reasoning the Bayesian Approach.* Open Court.

Johnson, V. (2003). *Grade Inflation: A Crisis in College Education.* Springer-Verlag.

Landrigan (1975). Neuropsychological dysfunction in children with chronic low-level lead absorption. *Lancet*, 708–712.

Moore, D. (1997). Bayes for beginners? Some reasons to hesitate. *Amer. Statist.* **51** (3), 254–261.

Parmigiani, G. (2002). *Modeling in Medical Decision Making A Bayesian Approach.* Wiley.

Schau, C., Mattern, N. (1997). Use of map techniques in teaching statistics courses. *Amer. Statist.* **51** (2), 171–175.

Stangl, D. (1998). Classical and Bayesian paradigms: Can we teach both?. In: Pereira-Mendoza, L., Kea, L.S., Kee, T.W., Wong, W. (Eds.), *Proceedings of the Fifth International Conference on Teaching Statistics, vol. 1.* International Statistics Institute, pp. 251–258.

Stangl, D. (2000). Design of an Internet course for training medical researchers in Bayesian statistical methods. In: Batenero, C. (Ed.), *Training Researchers in the Use of Statistics.* International Association for Statistical Education.

Stangl, D. (2001). A case study for teaching Bayesian methods. In: *Proceedings of JSM 2001.* ASA Section on Education.

Stangl, D. (2002). From testing to decision making: Changing how we teach statistics to health care professionals. In: *Proceedings of the Sixth International Conference on Teaching Statistics.* International Statistics Institute.

Stangl, D. (2003). Do's and don'ts of teaching Bayesian methods to healthcare professionals. In: *Proceedings of JSM 2003.* ASA Section on Teaching Statistics in Health Sciences.

The GUSTO Investigators (1993). An international randomized trial comparing four thrombolytic strategies for acute myocardial infarction. *New England J. Medicine* **329**, 673–682.

Colour figures

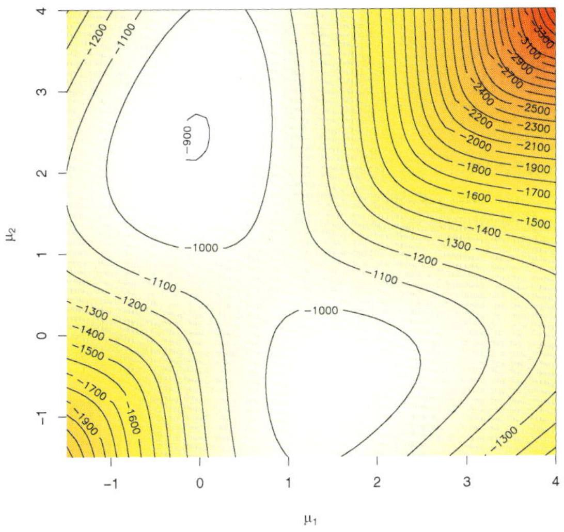

Chapter 16, Fig. 4, p. 467. R image representation of the log-likelihood of the mixture (7) for a simulated dataset of 500 observations and true value $(\mu_1, \mu_2, p) = (0, 2.5, 0.7)$.

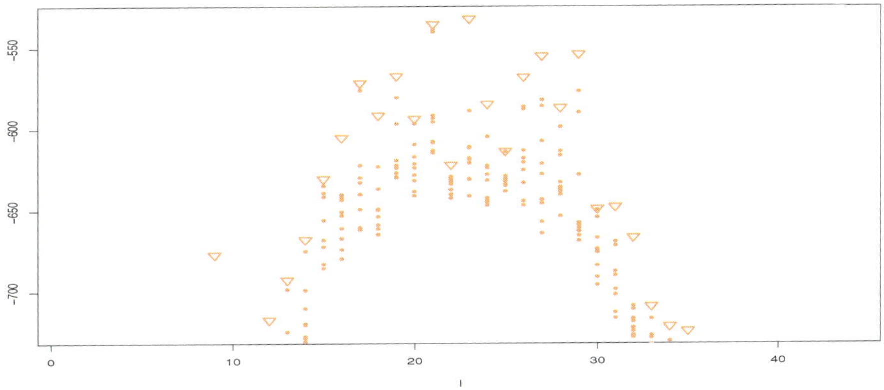

Chapter 16, Fig. 6, p. 470. Ten highest log-weights $\omega(\underline{z})$ (up to an additive constant) found in the simulation of random allocations for each partition size l for the same simulated dataset as in Figure 5. (Triangles represent the highest weights.)

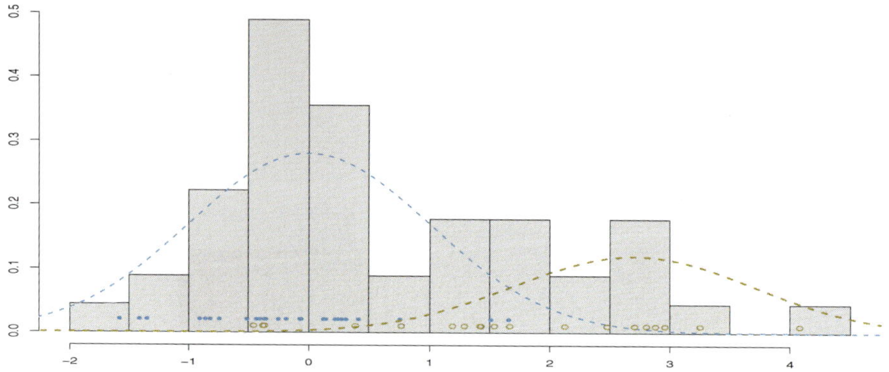

Chapter 16, Fig. 7, p. 471. Histogram, true components, and most likely allocation found over 440,000 simulations of $\underline{z}$'s for a simulated dataset of 45 observations and true values as in Figure 5. Full dots are associated with observations allocated to the first component and empty dots with observations allocated to the second component.

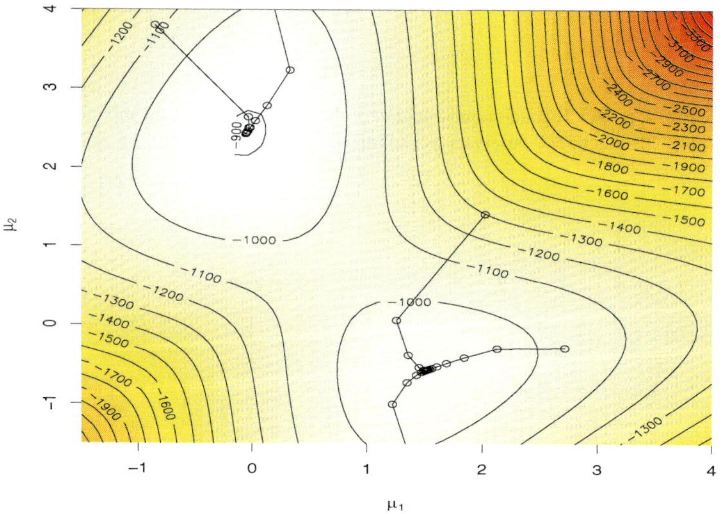

Chapter 16, Fig. 8, p. 473. Trajectories of five runs of the EM algorithm on the log-likelihood surface, along with R contour representation.

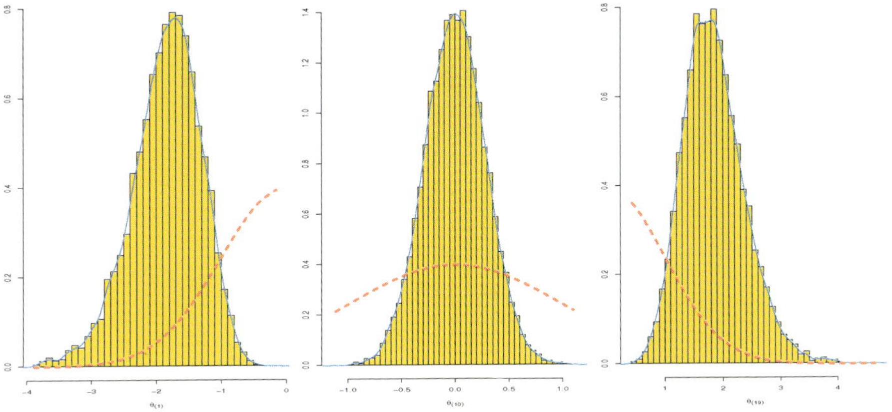

Chapter 16, Fig. 9, p. 474. Distributions of $\theta_{(1)}$, $\theta_{(10)}$, and $\theta_{(19)}$, compared with the $\mathcal{N}(0, 1)$ prior.

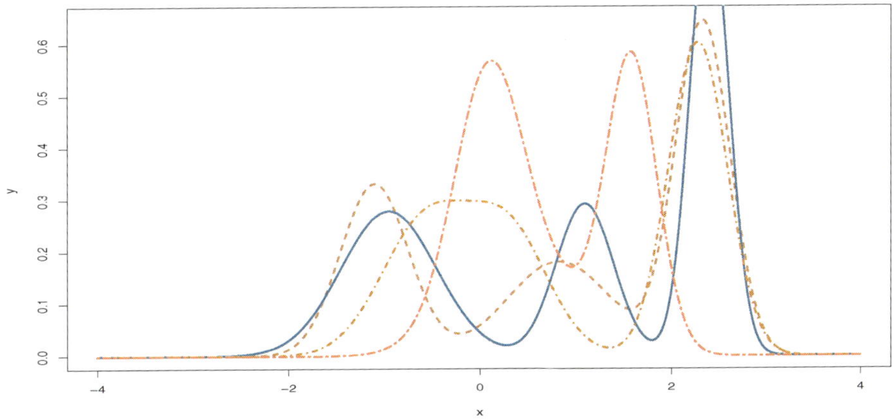

Chapter 16, Fig. 11, p. 479. Comparison of the plug-in densities for the estimations of Table 1 and of the true density (full line).

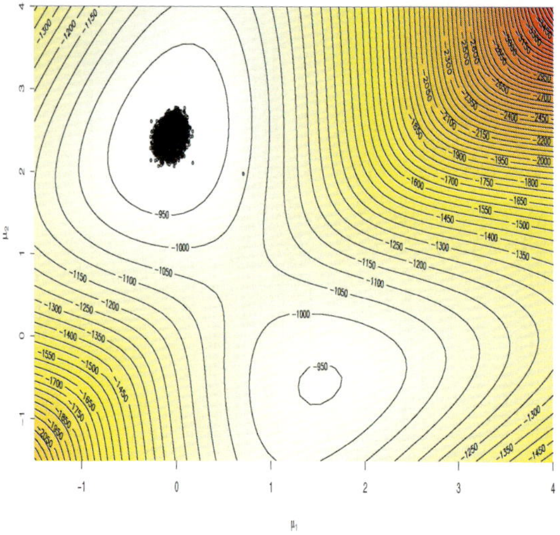

Chapter 16, Fig. 12, p. 484. Log-posterior surface and the corresponding Gibbs sample for the model (7), based on 10,000 iterations.

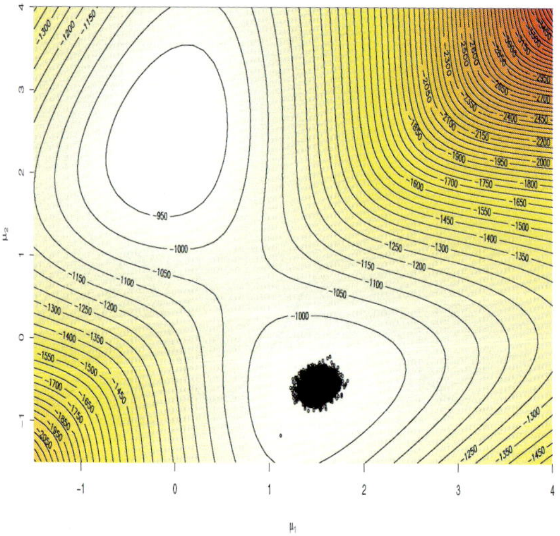

Chapter 16, Fig. 13, p. 486. Same graph, when initialised close to the second and lower mode, based on 10,000 iterations.

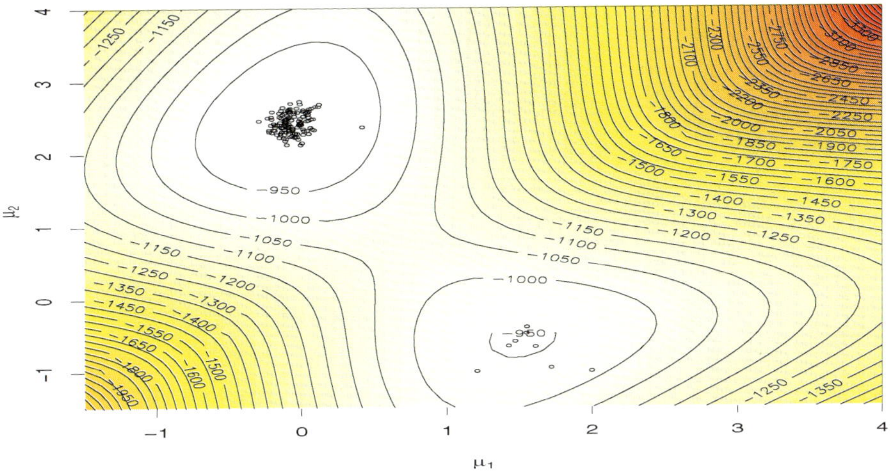

Chapter 16, Fig. 17, p. 490. Track of a 10,000 iterations random walk Metropolis–Hastings sample on the posterior surface, the starting point is equal to $(2, -1)$. The scale of the random walk ζ^2 is equal to 1.

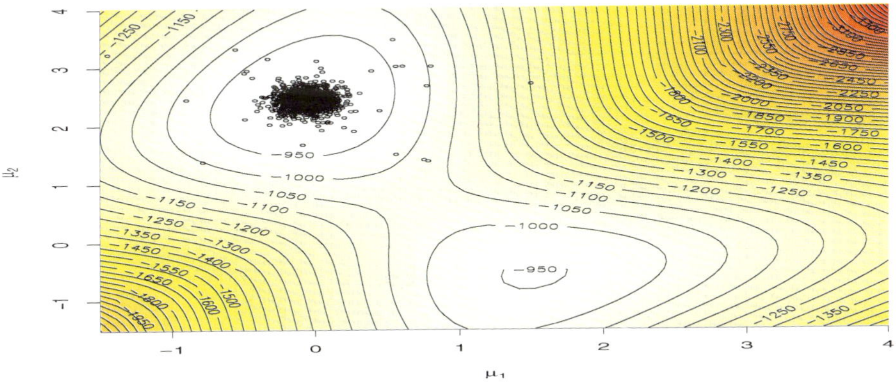

Chapter 16, Fig. 20, p. 496. Representation of the log-posterior distribution with the PMC weighted sample after 10 iterations (the weights are proportional to the circles at each point).

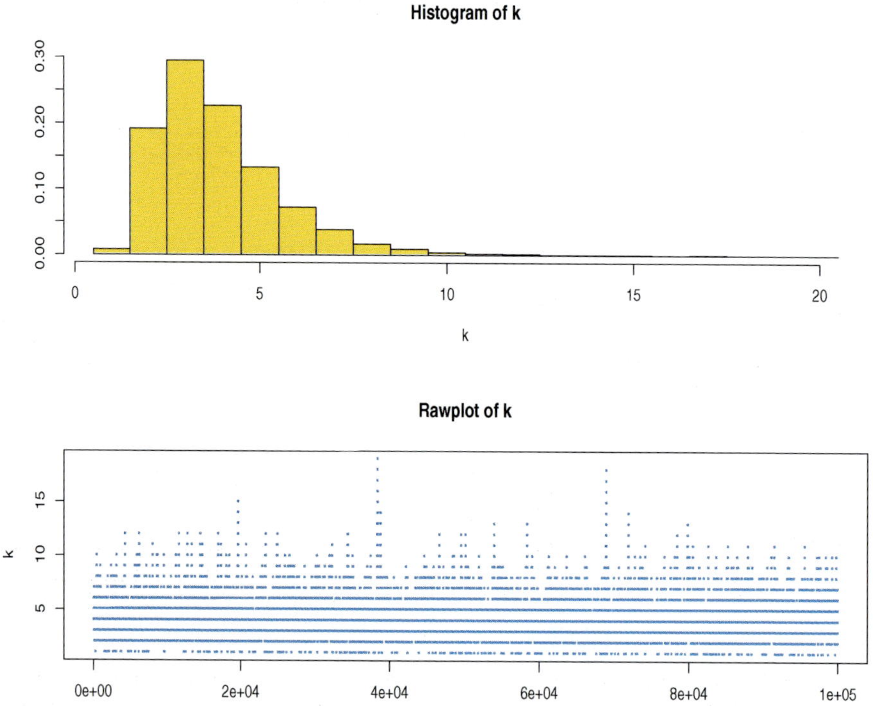

Chapter 16, Fig. 21, p. 499. Histogram and raw plot of 100,000 k's produced by a reversible jump MCMC algorithm for the Galaxy dataset (*Source:* Robert and Casella, 2004).

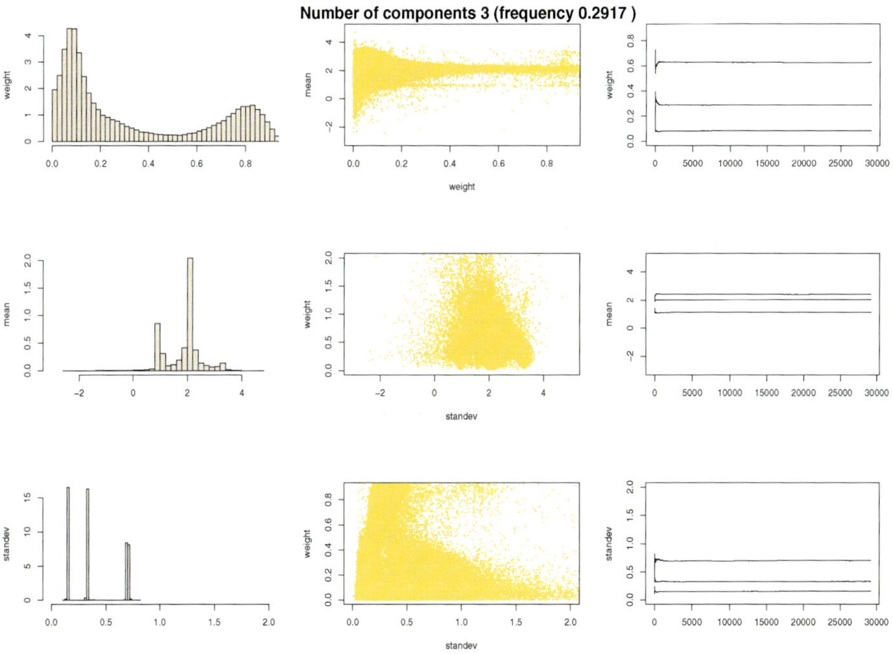

Chapter 16, Fig. 22, p. 500. Reversible jump MCMC output on the parameters of the model $\mathcal{M}_3$ for the Galaxy dataset, obtained by conditioning on $k = 3$. *The left column gives the histogram of the weights, means, and variances; the middle column the scatterplot of the pairs weights–means, means–variances, and variances–weights; the right column plots the cumulated averages (over iterations) for the weights, means, and variances (Source:* Robert and Casella, 2004).

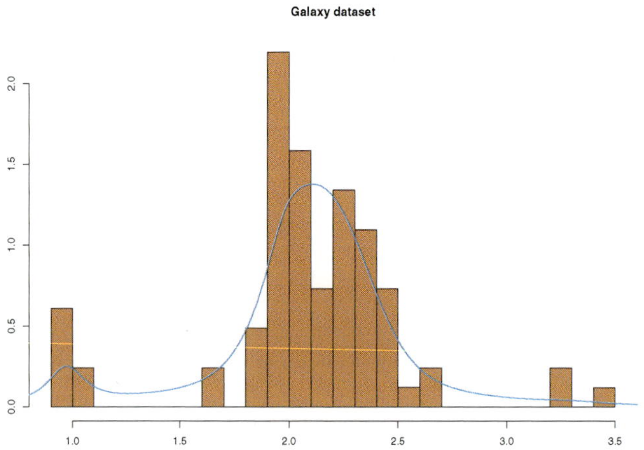

Chapter 16, Fig. 23, p. 500. Fit of the dataset by the averaged density (*Source:* Robert and Casella, 2004).

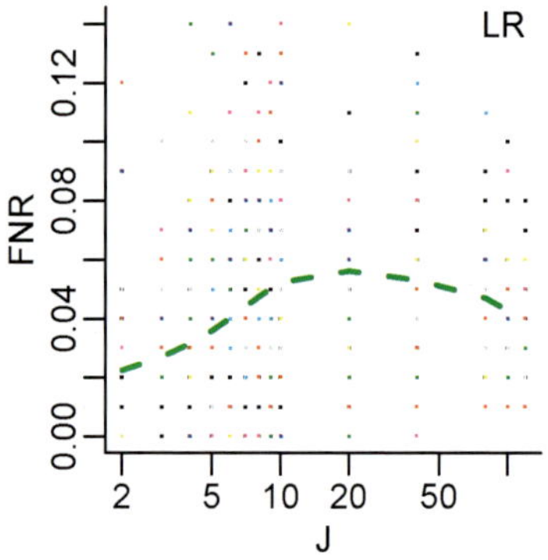

Chapter 17, Fig. 1, p. 512. A typical expected utility function $U(d)$. The figure shows estimated expected utility $\widehat{U}(d)$ (labeled FNR) against decisions d (labeled J). The dots show realized utilities $u(d, \theta_j, y_j)$. The dashed line shows a fitted smooth curve $\widehat{U}(d)$. The example is taken from Müller et al. (2004a). The decision is the number of arrays in a microarray experiment. The utility function is a combination of false disovery and false negative rate.

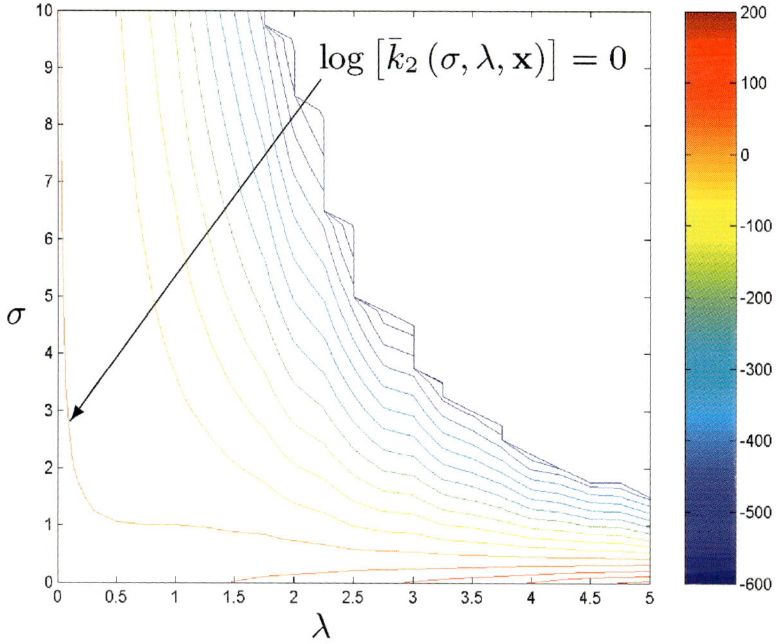

(a)

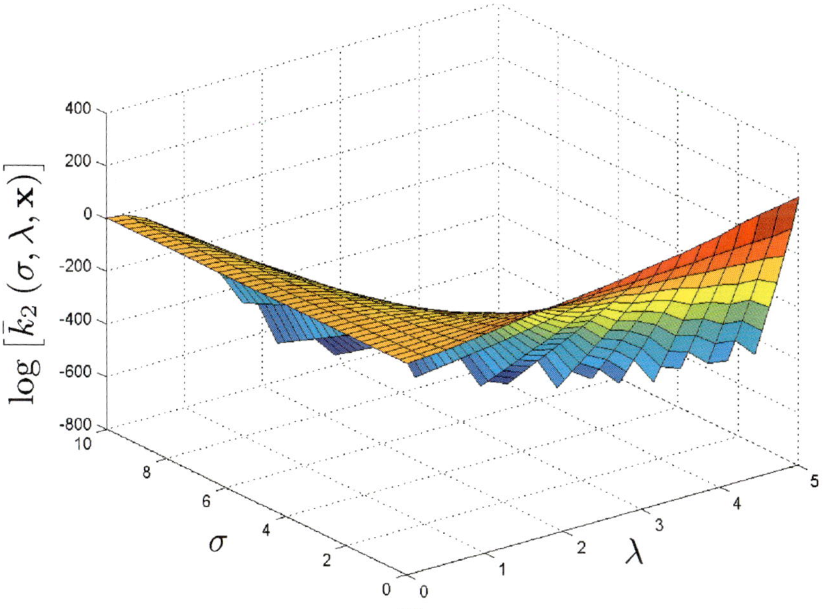

(b)

Chapter 23, Fig. 1, p. 700. $\log[\bar{k}_2(\sigma, \lambda, \boldsymbol{x})]$ for generated data: (a) contour lines; (b) surface.

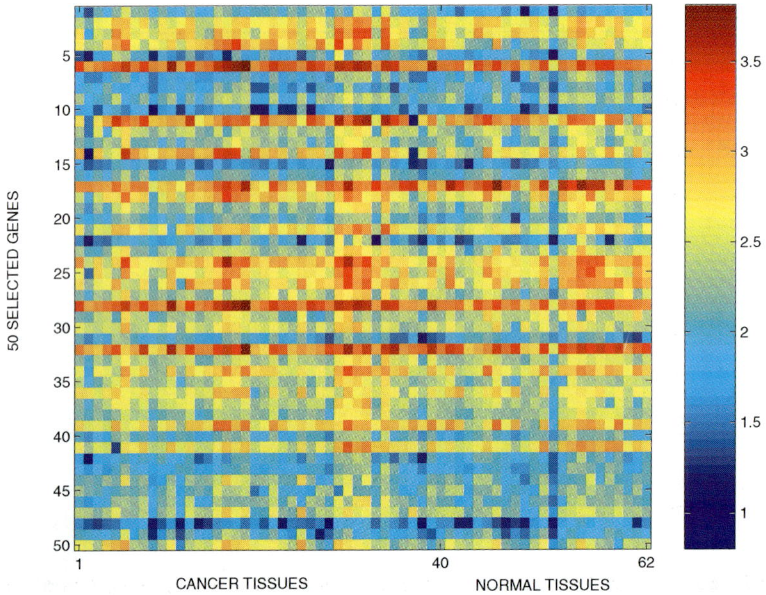

Chapter 24, Fig. 1, p. 723. Binary classification: Heat Map of the 50 most significant genes for classification between normal and colon cancer samples.

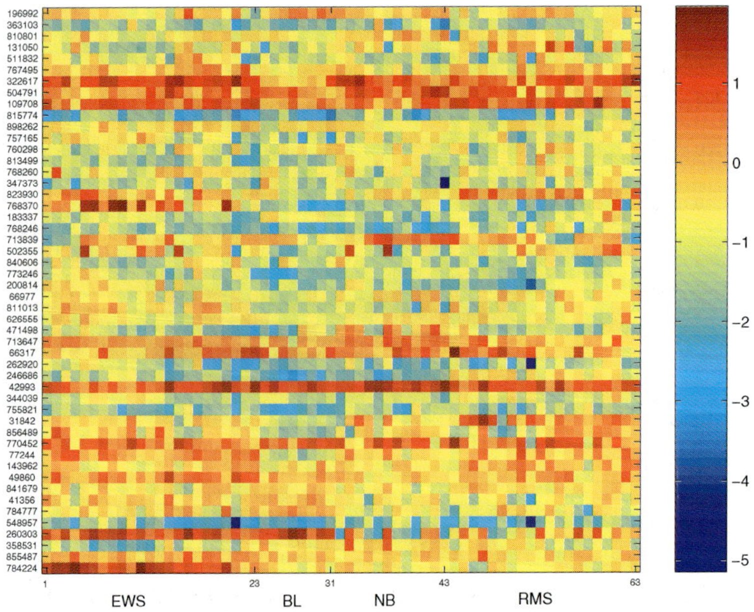

Chapter 24, Fig. 2, p. 727. Multicategory classification: Heat Map of the 50 most significant genes for classification between the four tumor types: EWS, BL, NB and RMS.

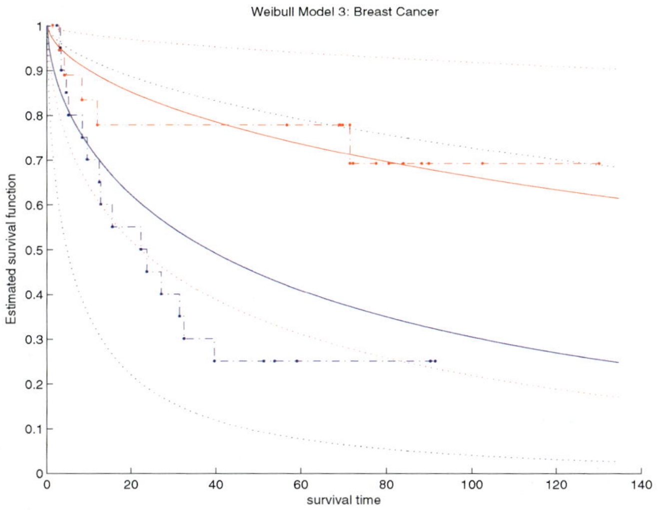

Chapter 24, Fig. 3, p. 729. Survival function for Breast carcinomas data using Weibull model. (Reprinted with permission from Lee and Mallick, 2004.)

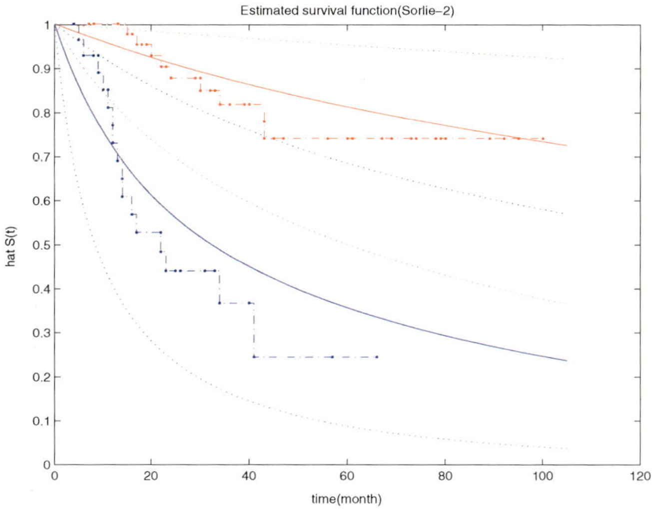

Chapter 24, Fig. 4, p. 730. Survival function for Breast carcinomas data using semiparametric model. (Reprinted with permission from Lee and Mallick, 2004.)

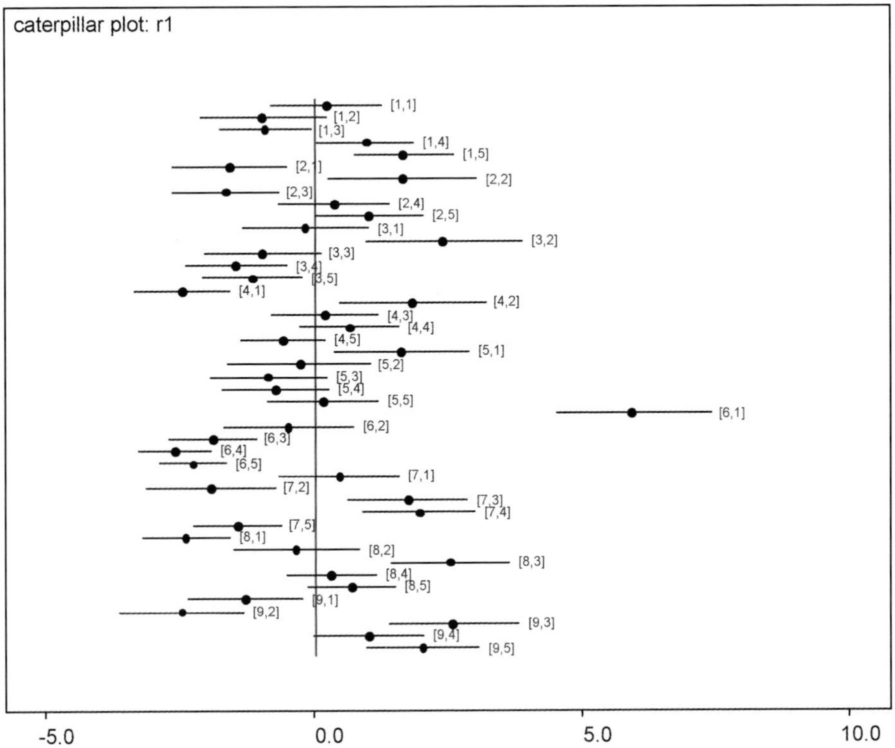

Chapter 30, Fig. 1, p. 878. Posterior densities of set of parametric Pearson residuals from a fit of an independence model for the Current Population Survey dataset from Freedman et al. (1991). The limits for each error bar correspons to the 25% and 97% percentiles of the posterior distribution of the residual.

Subject Index

Handbook of Statistics
Contents of Previous Volumes

Volume 1. Analysis of Variance
Edited by P.R. Krishnaiah
1980 xviii + 1002 pp.

Volume 2. Classification, Pattern Recognition and Reduction of Dimensionality
Edited by P.R. Krishnaiah and L.N. Kanal
1982 xxii + 903 pp.

Volume 3. Time Series in the Frequency Domain
Edited by D.R. Brillinger and P.R. Krishnaiah
1983 xiv + 485 pp.

Volume 4. Nonparametric Methods
Edited by P.R. Krishnaiah and P.K. Sen
1984 xx + 968 pp.

Volume 5. Time Series in the Time Domain
Edited by E.J. Hannan, P.R. Krishnaiah and M.M. Rao
1985 xiv + 490 pp.

Volume 6. Sampling
Edited by P.R. Krishnaiah and C.R. Rao
1988 xvi + 594 pp.

Volume 7. Quality Control and Reliability
Edited by P.R. Krishnaiah and C.R. Rao
1988 xiv + 503 pp.

Volume 8. Statistical Methods in Biological and Medical Sciences
Edited by C.R. Rao and R. Chakraborty
1991 xvi + 554 pp.

1. Methods for the Inheritance of Qualitative Traits by J. Rice, R. Neuman and S.O. Moldin
2. Ascertainment Biases and their Resolution in Biological Surveys by W.J. Ewens
3. Statistical Considerations in Applications of Path Analytical in Genetic Epidemiology by D.C. Rao
4. Statistical Methods for Linkage Analysis by G.M. Lathrop and J.M. Lalouel
5. Statistical Design and Analysis of Epidemiologic Studies: Some Directions of Current Research by N. Breslow
6. Robust Classification Procedures and their Applications to Anthropometry by N. Balakrishnan and R.S. Ambagaspitiya
7. Analysis of Population Structure: A Comparative Analysis of Different Estimators of Wright's Fixation Indices by R. Chakraborty and H. Danker-Hopfe
8. Estimation of Relationships from Genetic Data by E.A. Thompson
9. Measurement of Genetic Variation for Evolutionary Studies by R. Chakraborty and C.R. Rao
10. Statistical Methods for Phylogenetic Tree Reconstruction by N. Saitou
11. Statistical Models for Sex-Ratio Evolution by S. Lessard
12. Stochastic Models of Carcinogenesis by S.H. Moolgavkar
13. An Application of Score Methodology: Confidence Intervals and Tests of Fit for One-Hit-Curves by J.J. Gart
14. Kidney-Survival Analysis of IgA Nephropathy Patients: A Case Study by O.J.W.F. Kardaun
15. Confidence Bands and the Relation with Decision Analysis: Theory by O.J.W.F. Kardaun
16. Sample Size Determination in Clinical Research by J. Bock and H. Toutenburg

Volume 9. Computational Statistics
Edited by C.R. Rao
1993 xix + 1045 pp.

1. Algorithms by B. Kalyanasundaram
2. Steady State Analysis of Stochastic Systems by K. Kant
3. Parallel Computer Architectures by R. Krishnamurti and B. Narahari
4. Database Systems by S. Lanka and S. Pal
5. Programming Languages and Systems by S. Purushothaman and J. Seaman
6. Algorithms and Complexity for Markov Processes by R. Varadarajan
7. Mathematical Programming: A Computational Perspective by W.W. Hager, R. Horst and P.M. Pardalos

Volume 10. Signal Processing and its Applications
Edited by N.K. Bose and C.R. Rao
1993 xvii + 992 pp.

Volume 11. Econometrics
Edited by G.S. Maddala, C.R. Rao and H.D. Vinod
1993 xx + 783 pp.

Volume 12. Environmental Statistics
Edited by G.P. Patil and C.R. Rao
1994 xix + 927 pp.

Volume 13. Design and Analysis of Experiments
Edited by S. Ghosh and C.R. Rao
1996 xviii + 1230 pp.

Volume 14. Statistical Methods in Finance
Edited by G.S. Maddala and C.R. Rao
1996 xvi + 733 pp.

Volume 15. Robust Inference
Edited by G.S. Maddala and C.R. Rao
1997 xviii + 698 pp.

Volume 16. Order Statistics – Theory and Methods
Edited by N. Balakrishnan and C.R. Rao
1997 xix + 688 pp.

Volume 17. Order Statistics: Applications
Edited by N. Balakrishnan and C.R. Rao
1998 xviii + 712 pp.

Volume 18. Bioenvironmental and Public Health Statistics
Edited by P.K. Sen and C.R. Rao
2000 xxiv + 1105 pp.

Volume 19. Stochastic Processes: Theory and Methods
Edited by D.N. Shanbhag and C.R. Rao
2001 xiv + 967 pp.

Volume 20. Advances in Reliability
Edited by N. Balakrishnan and C.R. Rao
2001 xxii + 860 pp.

Volume 21. Stochastic Processes: Modelling and Simulation
Edited by D.N. Shanbhag and C.R. Rao
2003 xxviii + 1002 pp.

Volume 22. Statistics in Industry
Edited by R. Khattree and C.R. Rao
2003 xxi + 1150 pp.

Volume 23. Advances in Survival Analysis
Edited by N. Balakrishnan and C.R. Rao
2003 xxv + 795 pp.

Volume 24. Data Mining and Data Visualization
Edited by C.R. Rao, E.J. Wegman and J.L. Solka
2005 xiv + 643 pp.